9th International Scientific Symposium on Electric Power Engineering (ELEKTROENERGETIKA 2017)

Stara Lesna, Slovakia
12 – 14 September 2017

ISBN: 978-1-5108-4653-1

Printed from e-media with permission by:

Curran Associates, Inc.
57 Morehouse Lane
Red Hook, NY 12571

Some format issues inherent in the e-media version may also appear in this print version.

Printed by Curran Associates, Inc. (2017)

For permission requests, please contact
Technical University of Kosice – Department of Electrical Power Engineering
at the address below.

Technical University of Kosice – Department of Electrical Power Engineering
Letna 9
042 00 Kosice
Slovak Republic

Phone: 421 55 602 1111

http://www.tuke.sk

Additional copies of this publication are available from:

Curran Associates, Inc.
57 Morehouse Lane
Red Hook, NY 12571 USA
Phone: 845-758-0400
Fax: 845-758-2633
Email: curran@proceedings.com
Web: www.proceedings.com

TECHNICAL UNIVERSITY OF KOŠICE
Faculty of Electrical Engineering and Informatics
Department of Electrical Power Engineering

Proceedings of the 9th International Scientific Symposium on Electrical Power Engineering

ELEKTROENERGETIKA 2017

September 12–14, 2017, Stará Lesná, Slovak Republic

GENERAL SPONSOR

Člen innogy

Východoslovenská energetika a.s.

SPONSORS AND SUPPORTERS

SLOVENSKÉ ELEKTRÁRNE

Slovenské elektrárne, a.s.

ZSE

Západoslovenská energetika, a.s.

Slovenská elektrizačná prenosová sústava, a. s.

SCHRACK TECHNIK

Schrack Technik s.r.o.

MEDIA AND OTHER PARTNERS

AUTOMA

ENERGETIKA

elektrotechnika v praxi

Polish Journal of Management Studies

World Energy Council
Slovenský výbor svetovej energetickej rady

TECHNICAL UNIVERSITY OF KOŠICE
Faculty of Electrical Engineering and Informatics
Department of Electrical Power Engineering

Proceedings of the 9th International Scientific Symposium on Electrical Power Engineering

ELEKTROENERGETIKA 2017

September 12–14, 2017, Stará Lesná, Slovak Republic

Name: Proceedings of the 9th International Scientific Symposium on Electrical Power Engineering ELEKTROENERGETIKA 2017

Publisher: Technical University of Košice
Faculty of Electrical Engineering and Informatics
Department of Electrical Power Engineering

Published: September 2017

Editors: Michal Kolcun, Iraida Kolcunová, Juraj Kurimský

ELEKTROENERGETIKA 2017 – General Informations

Symposium Guarantee

Dr. h.c. prof. Ing. Michal Kolcun, Ph.D., Technical University of Košice, SK

International Scientific Committee

- Dauren S. Akhmetbaev, L.N.EA. National University, Astana, KAZ
- Juraj Altus, University of Žilina, SK
- Anton Beláň, Slovak University of Technology in Bratislava, SK
- Július Bems, Czech Technical University in Prague, CZ
- Ľubomír Beňa, Technical University of Košice, SK
- István Berta, BME Budapest, HU
- Peter Birkner, Mainova AG, DE
- Roman Cimbala, Technical University of Košice, SK
- Ivo Doležel, AV ER Praha, CZ
- Nikolay Djagarov, N. Vaptsarov Naval Academy, BG
- Emil Dvorský, University of West Bohemia, CZ
- Anna Gawlak, Czestochowa University of Technology, PL
- Sergej Gerasimov, St. Peterburg, RUS
- Georgi Georgiev, Plethora Ltd, Varna, BG
- Janusz K. Grabara, Czestochowa University of Technology, PL
- Valerij G. Goldstein, TU Samara, RUS
- Radomír Goňo, VSB - Technical University of Ostrava, CZ
- Rainer Haller, Regensburg, DE
- Zdeněk Hradílek, VSB - Technical University of Ostrava, CZ
- Iurii Chukreyev, Ural Branch of the RAS, Syktyvkar, Russia, RUS
- František Janíček, Slovak University of Technology in Bratislava, SK
- Zdislaw Kabza, TU Opolska, PL
- Péter Kádár, Óbuda University, Budapest, HU
- Rumen Kirov, Technical University of Varna, BG
- Jaroslav Knápek, Czech Technical University in Prague, CZ
- Iraida Kolcunová, Technical University of Košice, SK
- Sebastian Kot, Czestochowa University of Technology, PL
- Hristo Krachunov, Technical University of Varna, BG
- Dušan Krokavec, Technical University of Košice, SK
- Juraj Kurimský, Technical University of Košice, SK
- Jaroslav Lelák, Slovak University of Technology in Bratislava, SK
- Tatiana Lomane, Riga Technical University, LV
- Anatolij Mahnitko, Riga Technical University, LV
- Tatjana Malinina, St. Petersburg State Polytech.University, RUS
- Lubomir Marciniak, Czestochowa University of Technology, PL

- Zbyněk Martinek, University of West Bohemia, CZ
- Alexander Mészáros, Technical University of Košice, SK
- Ján Michalík, University of Žilina, SK
- György Morva, University of Budapest, HU
- Hans Michael Muhr, Technical University, Graz, AT
- Zdeněk Müller, Czech Technical University in Prague, CZ
- Vladimir Nagay, NPI, RUS
- Alexander Nazaricev, St. Peterburg, RUS
- Vladimir Nepomnyashchy, St. Peterburg, RUS
- Karel Noháč, University of West Bohemia, CZ
- Maria Nowicka-Skowron, Czestochowa University of Technology, PL
- Irina Oleinikova, Smart Grid Research Centre, Riga, LV
- Daniela Perduková, Technical University of Košice, SK
- Andrzej Rusek, Czestochowa University of Technology, PL
- Stanislav Rusek, VSB - Technical University of Ostrava, CZ
- Pavel Santarius, VSB - Technical University of Ostrava, CZ
- Antanas Sauhatas, Riga Technical University, LV
- Rasid Sitdikov, TSTU Tashkent, UZB
- Karel Sokanský, VSB - Technical University of Ostrava, CZ
- Viktor Starovojtenkov, Saint Petersburg, RUS
- Tynys B. Suleimenov, L.N.EA. National University, Astana, KAZ
- Jerzy Szkutnik, Czestochowa University of Technology, PL
- Alexey Tadzhibaev, St. Peterburg, RUS
- Kolio Tasev, Technical University of Varna, BG
- Vasily V. Titkov, St. Petersburg State Polytech. University, RUS
- Josef Tlustý, Czech Technical University in Prague, CZ
- Petr Toman, VUT Brno, CZ
- Toomas Vaimann, Technical University of Tallin, EE
- Margreta Vasileva, Technical University of Varna, BG
- Zdeněk Vostracký, University of West Bohemia, CZ
- Irena Wasiak, Polytechnika Lodz, PL
- Oldřich Záviška, Siemens KWU Erlangen, DE
- Markus Zdrallek, Bergische Universität Wuppertal, DE

Scientific Review Committee

Chairman

Dr. h.c. prof. Ing. Michal Kolcun, Ph.D., Technical University of Košice, SK

Members

- Mikołaj Bartłomiejczyk, Gdańsk University of Technology, PL
- Anton Beláň, Slovak University of Technology in Bratislava, SK
- Július Bems, Czech Technical University in Prague, CZ
- Ľubomír Beňa, Technical University of Košice, SK
- Kristina Berzina, Riga Technical University, LV
- Roman Cimbala, Technical University of Košice, SK
- Zsolt Čonka, Technical University of Košice, SK
- Jiří Drápela, Brno University of Technology, CZ
- Emil Dvorský, University of West Bohemia, CZ
- Jaroslav Džmura, Technical University of Košice, SK
- Žaneta Eleschová, Slovak University of Technology in Bratislava, SK
- Anna Gawlak, Czestochowa University of Technology, PL
- Georgiev Georrgi, Plethora Ltd, Varna, BG
- Radomir Goňo, VSB - Technical University of Ostrava, CZ
- Janusz K. Grabara, Czestochowa University of Technology, PL
- Miroslav Gutten, University of Žilina, SK
- Zoltán Hogya, ABB, SLOVAKIA, SK
- Stanislav Ilenin, Technical University of Košice, SK
- František Janíček, Slovak University of Technology in Bratislava, SK
- Martin Kanálik, Technical University of Košice, SK
- Michal Kolcun, Technical University of Košice, SK
- Iraida Kolcunova, Technical University of Košice, SK
- Mirosław Kornatka, Czestochowa University of Technology, PL
- Michal Kosterec, Technical University of Košice, SK
- Sebastian Kot, Czestochowa University of Technology, PL
- Jozef Kováčik, VUJE, a.s., SK
- Slavomír Kučera, SAT Automation Ltd, SK
- Juraj Kurimsky, Technical University of Košice, SK
- Jaroslav Lelák, Slovak University of Technology in Bratislava, SK
- Tatjana Lomane, Riga Technical University, LV
- Anatolijs Mahnitko, Riga Technical University, LV
- Veleslav Mach, VSB - Technical University of Ostrava, CZ
- Martin Marci, Východoslovenská distribučná, a.s. , SK
- Zbyněk Martinek, University of West Bohemia, CZ
- Karel Máslo, ČEPS, a.s., Prague, CZ

- Petr Mastný, Brno University of Technology, CZ
- Dušan Medveď, Technical University of Košice, SK
- Marián Mešter, Východoslovenská distribučná, a.s. , SK
- Alexander Mészáros, Technical University of Košice, SK
- Daniel Minařík, VSB - Technical University of Ostrava, CZ
- Stanislav Mišák, VSB - Technical University of Ostrava, CZ
- Gyorgy Morva, Óbuda Unovesrity, Budapest, HU
- Karel Noháč, University of West Bohemia, CZ
- Lucie Noháčová, University of West Bohemia, CZ
- Irina Oleinikova, Smart Grid Research Centre, Riga, LV
- Jaroslava Orságová, Brno University of Technology, CZ
- Alena Otčenášová, University of Žilina, SK
- Marek Pavlík, Technical University of Košice, SK
- Daniela Perduková, Technical University of Košice, SK
- Jaroslav Petras, Technical University of Košice, SK
- Marek Roch, University of Žilina, SK
- Vladimír Šály, Slovak University of Technology in Bratislava, SK
- Peter Semančík, Východoslovenská distribučná, a.s. , SK
- Ľuboš Skurčák, VUJE, a.s., SK
- Marek Šmilňák, ABB, SLOVAKIA, SK
- Jindřich Stuchlý, VSB - Technical University of Ostrava, CZ
- Jerzy Szkutnik, Czestochowa University of Technology, PL
- Vasilii Titkov, Saint Petersburg State Polytechnical University, RUS
- Petr Toman, Brno University of Technology, CZ
- Marián Uher, VSB - Technical University of Ostrava, CZ
- Petr Vaculík, VSB - Technical University of Ostrava, CZ
- Georgii Vagapov, Kazan State Power Engineering University, RUS
- Toomas Vaimann, Technical University of Tallin, EE
- Michal Váry, Slovak University of Technology in Bratislava, SK
- Václav Veselý, Czech Academy of Sciences, CZ
- Zdeněk Vostracký, University of West Bohemia, CZ
- František Vranay, Technical University of Košice, SK
- Ján Zbojovský, Technical University of Košice, SK
- Inga Zicmane, Riga Technical University, LV
- Iveta Zolotová, Technical University of Košice, SK

Symposium Steering Committee

- Iraida Kolcunová, Technical University of Košice, SK
- Juraj Kurimský, Technical University of Košice, SK
- Jaroslav Džmura, Technical University of Košice, SK
- Jana Varnavčinová, Technical University of Košice, SK
- Dagmar Kramolišová, Technical University of Košice, SK

Table of Contents

Index of Authors

Invited Lectures

Trends of Development of Power Systems in the Czech and Slovak Republic Past, Present and Future

Jiří Ptáček, Petr Modlitba
EGÚ Brno, a.s.
Česká republika
jiri.ptacek@egubrno.cz, petr.modlitba@egubrno.cz

Milan Straka, Ľuboš Samsely
Slovenská elektrizačná prenosová sústava, a.s.
Slovenská republika
milan.straka@sepsas.sk lubos.samsely@sepsas.sk

Abstract: **The common long-term development of the Czech (CR) and Slovak Republic's (SR) power systems (PS) grids has created similar conceptual bases. Since the division into two separate states, the development of both PS has begun to differ. Ambitious EU targets require high demands on both PS in the future. Both systems are already preparing for these challenges.**

Keywords: *development of power systems, consumption, international exchange, diagrams of system loads, development of renewables*

I. INTRODUCTION

The power systems of the Czech and Slovak Republic have been developed hand in hand for a long time. After the splitting of Czechoslovakia into two independent states, originally common power grids split too. We can detect the common basis of both systems that determined common starting points and conditions. Independent development of the two grids led to some differences which can be visible in the development strategies as well as in the operation of the power grids. Today we can see common operational rules but also different attitudes and solutions. The Political goals derived from the EU climate and environment protection regulations have a significant influence on further development of the power systems. Interesting is the comparison of the features of both state energy concepts, which determine the long term development of power systems. The common conceptual basis, the strong interconnection and similarity of both systems create conditions for further closer cooperation in the future.

II. THE COMMON DEVELOPMENT OF THE BOTH PARTS OF THE POWER SYSTEMS TILL 1992

The main objectives for the power grid development after the establishment of Czechoslovakia (1918) was the "development of the large centralized power plants, utilizing the natural resources (coal from the mines and water from rivers), connected to the common power grid and efficient transmission of the electricity in the whole State". These objectives were specified in the Act of the "Systematic state electrification" (1919).

At first the 100 kV voltage level "transmission system" was built, first line from Třinec to Žilina was built in 1930. In the plans since the 1929 the Morava region and Slovakia were connected by the three 100 kV lines. Unified dispatch operation of the grid was established for the Bohemia in 1945 in Prague, for the Moravia region in 1945 in Ostrava and for Slovakia in 1948 in Žilina. Since 1962 the power grids in Central and Eastern Europe were synchronously connected with the USSR. The whole "Eastern block" inter-connected power grid was operated from Moscow, since 1962 by means of the CDO (Central Dispatch Organization) situated in Prague. Cooperation of the central Europe power grids with the "Western" power grids (through the Iron curtain) were done through the isolated operation of the power plants (e.g. Opatovice power plant) and separated 220 kV line to Austria and later through the embedded back to back DC inter-connectors Dürnrohr (1983), Wien SO (1993) and Etzenricht (1993). In 1992 the CENTREL association was established (power systems of Poland, the Czech Republic, the Slovak Republic and Hungary) mainly in order to cooperate with UCPTE. In 1992 also synchronous operation of the CENTREL and the USSR was interrupted.

Transmission systems with 220 kV voltage level and later 400 kV in Czechoslovakia were developed jointly and conceptually. At first 220 kV voltage level grid was built from the fifties to the seventies, later – at the beginning of the sixties – the construction of the 400 kV voltage level grid began, which has been further developed significantly. The transmission system of Czechoslovakia was developed in the East-West direction. This concept was based on the natural conditions (huge mines - power plants in western part in the Podkrušnohoří region and close distance to reserves of fuels, resources and fuel import in the eastern part of Slovakia) and on the idea of common development of the power grids (with the USSR), based on the huge amount of electricity transmission to Central and Southern Europe (by 750 kV lines). The conceptual basis of the transmission grid has been developed in former Czechoslovakia and provides a robust foundation of nowadays transmission systems in both states (CR, SR). Proof of that is the fact that the Moravian – Slovakian profile has not been topologically modified and corresponds to the contemporary requirements for cross-border power transmissions.

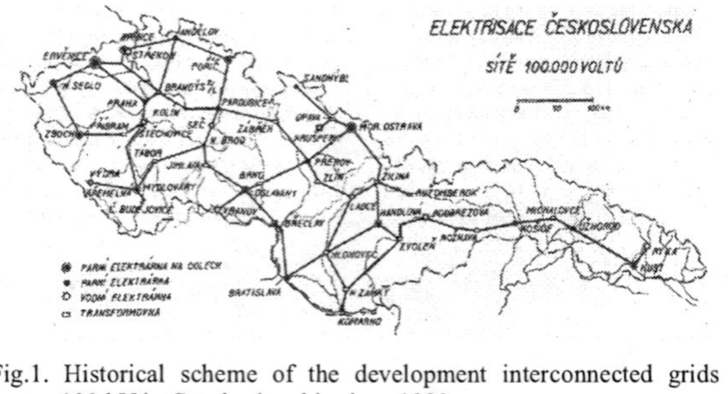

Fig.1. Historical scheme of the development interconnected grids 100 kV in Czechoslovakia since 1929

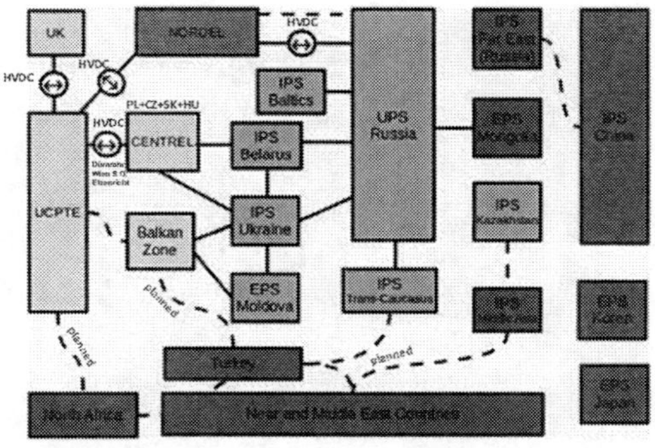

Fig. 2. Scheme of interconnected power systems in Europe in 1992

III. Individual development of the Slovak and Czech power systems till now

Cooperation of the both power systems:

The main goal of the power system operation after 1993 was the synchronous interconnection between CENTREL and UCPTE, reached in August 1995 after the number of the preparatory works (connection through closure of the 400 kV line Hradec – Röhrsdorf). After that the Czech and Slovak Republics became full members of UCPTE (1999). The needs for higher market cooperation with neighbours led to further development of the cross-border interconnection of the Czech and Slovak power systems (e.g. to Germany – Etzenricht, to Poland – Krosno). Not all development plans for further cross-border interconnections are realized, but both transmission systems are moving forward also in the empowerment of the cross-border profiles (new cross-border 400 kV transmission lines on the SK-HU profile and reinforcement of the CZ-AT profile). The new objective for the cross-border interconnections development both in the Czech and Slovak Republic are the transits of the electricity through the power systems and unexpected transits of the power, which can jeopardize security and reliability of both transmission systems operation. In order to prevent the negative effects caused by the huge and fluctuating import power flows from Germany on the CZ-DE cross-border profile (high power generation of wind power plants in the Baltic and North Sea) the 4 Phase Shifting transformers (PST) were installed on the CZ-DE cross-border profile in the Hradec substation in period 2016-2017. The major task of these PSTs is to limit the transmission flows on the cross-border lines on CZ-DE profile, due to a risk of internal CZ transmission grid overloading. Limiting of the unexpected power flows is increasing the security and reliability of the Czech and also neighbouring transmission systems operation. Since 2012 the common market area CZ – SK has been established (interconnection of the electricity markets) [5].

Development of the internal Czech and Slovak power systems:

The electricity markets in both countries have been fully liberalised (from 2002 to 2006, 2007). The energy sectors were unbundled (separation of transmission and distribution from generation) in 2006. The predominant electricity producer in the Slovak Republic - Slovenské elektrárne (SE) were privatized by ENEL in 2006 (66 % ownership) and after that in 2015, 33 % share was sold to Czech company EPH. The state still owns 34 % share of stocks of SE. The predominant electricity producer in the Czech Republic – ČEZ, a.s. –has remained with the majority stake in state ownership (70 % of stocks). The major part of the distribution companies have been privatized in both countries: In the Czech Republic - E.ON Distribuce, PREdistribuce, only ČEZ Distribuce has been state owned as a daughter company of ČEZ, a.s. In the Slovak Republic – Západoslovenská distribučná, a.s. ZSD (owned E.ON), Stredoslovenská energetika - Distribúcia, a.s. SSE-D (formerly owned by EDF, nowadays EPH), Východoslovenská distribučná, a.s. VSD (owned by RWE).

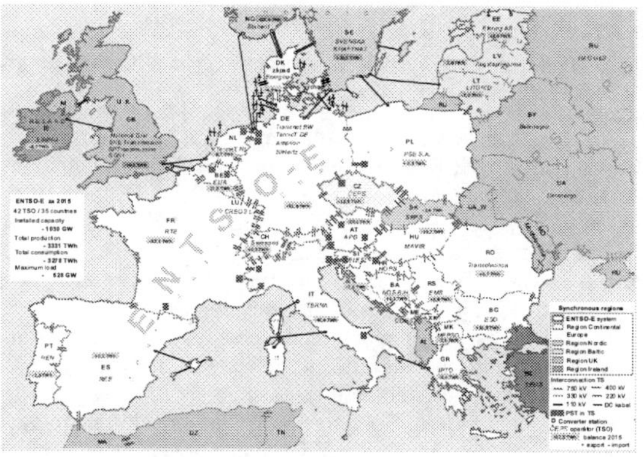

Fig. 3. Cooperation of interconnected power systems in Europe - year 2016

Significant changes have also taken place in generation mix:

In both power systems there is an ongoing decommissioning of the large centralized power plants (e.g. in the Slovak Republic: nuclear power plant (NPP) Bohunice V1, unit 1 -440 MW was decommissioned in 2006, unit 2 - 440 MW in 2008, Power plants Vojany 2 (6 x 110 MW), Vojany 1 (4 x 110 MW) (only 2x 110 MW has remained), Nováky (2 x 110 MW), CCGT Malženice (430 MW) have stopped generation and are conserved, CCGT Bratislava (210 MW) is in operation but only for ancillary services provision, in the past in base load covering operation 180 MW.

In the Czech Republic – some units of the power plants Prunéřov II, Mělník I were decommissioned and the next group of power plants are prepared for decommissioning (Mělník II a III, Prunéřov I, Ledvice – 110 MW units).

The Czech Republic has maintained its export balance by the commissioning of the NPP Temelín (in 2000 and 2001). The Slovak Republic, for the first time in its history, became an exporter of electricity after the commissioning of the first two units of the NPP Mochovce in 1998 and 2000. After the decommissioning of NPP Bohunice V1, it has become a rather balanced country from the electricity point of view. Moderate import of electricity was economically more beneficial in comparison with generation of "expensive" electricity in thermal power plants internally.

In the Czech Republic there was a spontaneous development („boom") of renewable energy sources (RES) from 2009 to 2010. About 2000 MW new installed capacity of photovoltaic power plans (FVE) were commissioned (the plans were significantly higher). In the Slovak Republic RES development was partly regulated and slowed. The installed capacity of the FVE power plants in the Slovak Republic is 530 MW.

Different natural conditions in Czech Republic and the Slovak Republic leads to different requirements of RES share in generation mix. During the accession negotiation in the process of entrance for the Czech Republic to the EU, the share of the RES generation on the gross domestic consumption was set to 8 % in 2010 (and was fulfilled) and

13 % in 2020 (will be fulfilled). The Slovak Republic has a commitment to have a 14 % share in 2020. Commissioning of the CCGT power plants happened in both power systems, but they have not been significantly used due to the high price of fuel. While CCGT Malženice is now out of operation (conserved – not able to run in a short time), a comparable CCGT power plant Počerady has been operating for a 4 years (even though with lower usage than it was planned).

Development of the transmission systems of the Czech and Slovak Republic since 1993:

During the last 24 years the transmission systems of the Czech and Slovak Republic have been developed qualitatively and also quantitatively, what is proven by the next figure 4 (the topological changes since 1993 are highlighted). Cross-border interconnectors have been strengthened (to Germany, Poland and Austria). The internal transmission system has been reinforced in new routes. The development of the large centralized new power plants connected into the transmission system has had the significant impact on the transmission system, for instance, the connection of new (e.g. Temelín) or decommissioning (e.g. Bohunice V1). The Slovak transmission system transition from 220 kV voltage level to 400 kV is significantly advanced in comparison with the Czech system.

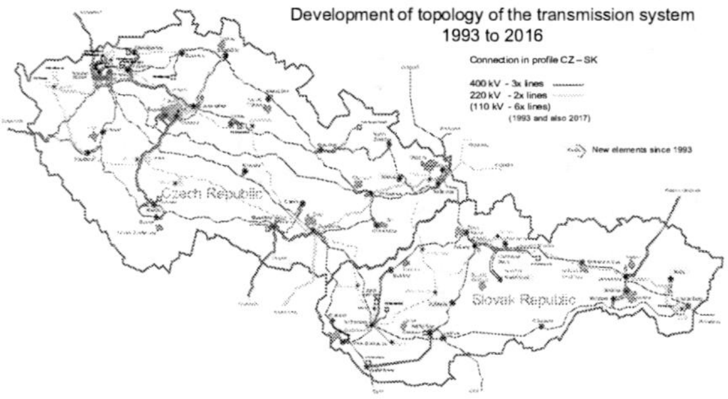

Fig. 4. Development of transmission network in the Czech and Slovak Republic since 1993 to present

Total net domestic consumption in the Czech Republic has now and in a future will have slightly ascending trend. The demand will grow with the wholesale consumers, retail consumers, with the exception of households. Along with the reduction of the consumption caused by the decreasing of the energetic efficiency of the apparent the usage of electricity for heat pumps is expanding as a replacement of the solid fuels and moreover usage of the electricity for transport (electro mobility and further electrification of the railway transport). In the Slovak Republic the consumption after the 1989 stagnated. The demand reaches the values from 1989 in 2006-2008. After the economic crisis demand decreased again and in 2016 the magical value of gross demand 30 TWh was reached. The comparison of the gross demand of Czech and Slovak Republic till 2016 is depicted in the following figure 5 [3], [4]. Character of the consumption development in Czech Republic and Slovakia is very similar from the shape point of view. The possibilities of the generation mix in Czech Republic are higher, what corresponds to the continuously higher export balance of the power system.

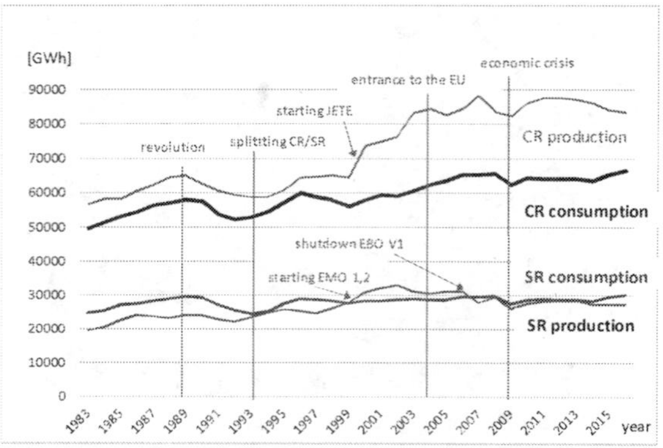

Fig. 5. Gross consumption (brutto) and production of the electricity in CR and SR in period 1983 – 2016

By the increased consumption during the last three years and by the gradual decommissioning of the thermal power plants in the Slovak Republic, the country is becoming an importer of electricity. The change will occur after the commissioning of the third and fourth unit of the NPP Mochovce. Generation of the NPP Mochovce units 3 and 4 with installed capacity 2 x 471 MW is planned in December 2018, resp. December 2019. The Czech and Slovak Republic are important transit countries. In Slovak Republic transit is equal to half of the generation/demand of the electricity. The following figure 6 is proving the above mentioned facts [3], [4]. Energy transits are mainly in the North-South direction and respond to distribution of generation and consumption in the region of Central Europe.

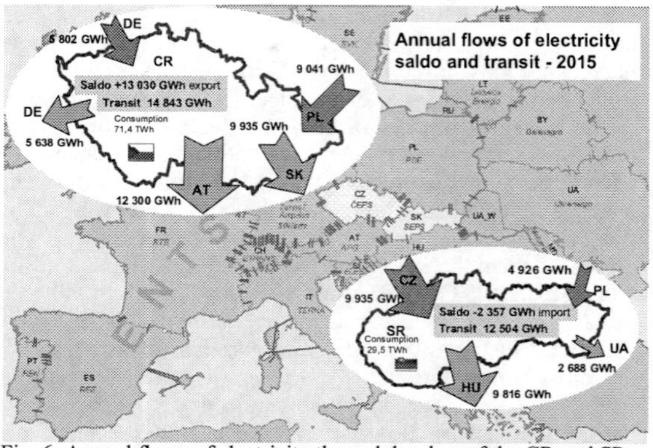

Fig. 6. Annual flows of electricity through borders of the CR and SR, achieved balance and transit – year 2015

IV. Comparison of some operational indicators of Czech and Slovak power systems

Operation of both power systems looks similar from many points of view, but there are also some differences. Below you can see the comparison of some operational indicators [3], [4].

It is very interesting to compare the load curve of both power systems, Fig. 7. The next picture presents a comparison of the daily load curve and weekly load curve in the year 2015. Shapes of daily load curve for maximum and minimum load are very similar. This fact shows the likeness of both countries. There is some time shift of loads. This time shift of several hours could be used by the electricity market. Earlier morning peak load in the Slovak republic results from the higher participation of heavy industry. Weekly maximum and minimum difference during the entire year is higher in the

Czech Republic. Minimal load is only 48% of maximal load. In the Slovak Republic minimal load reaches 56% of maximal load.

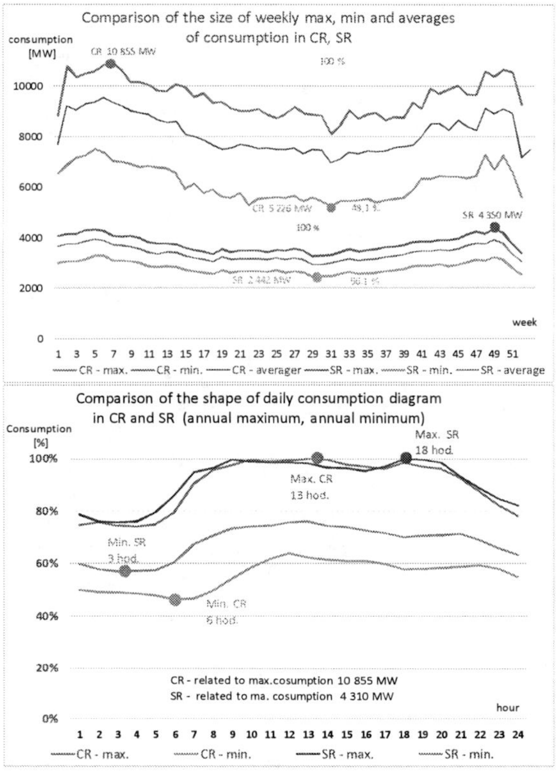

Fig. 7. Comparison of the size and shape of electricity consumption diagrams in the Czech and Slovak Republic - year 2015

The next way to compare both power systems could be by comparison of operation of 110 kV areas. There are big differences between power supply in the Czech Republic compared to the Slovak Republic as the next picture presents. In the Slovak Republic 110 kV areas are very often supplied by parallel operated transformers from one substation or also from more different substations. In the Czech Republic this style is in minority or under preparation.

- areas in ZSD are supplied from two different substations by two transformers. Areas in VSD are supplied from three different substations by three transformers. The only difference is T401 Levice and T401 Spišská Nová Ves. In SSE-D are utilized different kinds of supplying of areas and actual status is comparable to the Czech Republic.

- In order to increase the reliability of supply of consumers SEPS prefers parallel transformers operation in whole Slovak Republic.

In the Czech Republic parallel operation of transformers is being verified and supply reliability benefits are being compared to possible risks of 110 kV network overloading during outages and a risk of black-out in bigger area.

The next big difference is supplying wholesale consumers direct from the power grid. In the Slovak Republic there was a 220 kV and later also 400 kV network build due to reliable supply of electricity to the wholesale consumers for instance USSK, OFZ Široká, Duslo Šaľa, Slovalco.

Fig. 8. Coupling of transformers transmission network/110 kV through 110 kV networks in CR, SR

V. Main Trends for the Further Power Systems Development – Common and Different

Long term characteristics of the further development of both of the power systems are possible to see from the state approved energy policies for the Czech Republic and Slovak Republic (State energy policy for the Czech Republic approved in 12/2014 and for the Slovak Republic in 11/2014). Both state energy policies main characteristics are very similar with high importance on security, competitiveness and sustainability. The basis are the natural conditions (availability of the resources for „classic" energy, natural conditions for RES). An important driver is also the state of the economies in both countries, which are very similar (good position in the EU, expected development of the demand). From the changes in energy sectors point of view the most important are the commitments and the EU requirements in reaching common environmental goals.

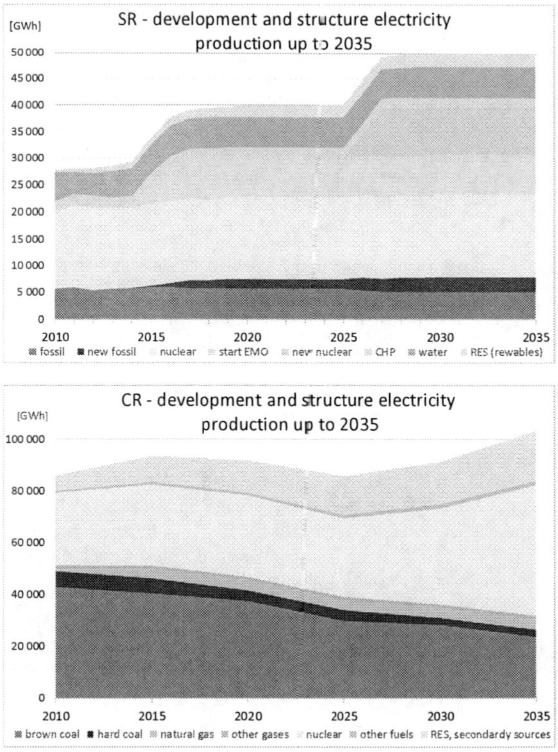

Fig. 9. Expected long-term development of generation mix in the SR and CR up to 2035

The figure 9 shows the comparison of the long term development generation mix in the Czech Republic and Slovak Republic till the 2035 based on the approved energy policies [1], [2]. The significant share of the nuclear power plants, which on the long term basis covers over 50 % of the consumption in Slovakia will be extended not only by the NPP Mochovce 3 and 4, but also by the next new NPP. The deadlines of the finishing of the new NPP Mochovce units are still being postponed which corresponds also with technical problems with finishing of other NPPs in Europe (e.g. Flamanville, Olkiluoto).

The share of the nuclear power plants in the Czech Republic will be increased by the new units extension with the installed capacity 1 200 MW and 1 700 MW in each of the present localities. The tender release and the decision about the construction of the new units is still being postponed, therefore their realisation cannot be expected earlier than 2032 or 2035.

In the Czech Republic the significant long term reduction of the coal burning power plants is expected, on the contrary Slovak energy policy assume the preservation and further moderate development of the fossil (coal burning) power plants.

The natural conditions in the Slovak Republic for the hydro power plants are better in comparison with those in the Czech Republic. Development of the hydro power plants in the Czech Republic is restricted and for the flexibility of the power system assurance the development of the natural gas burning power plants is expected.

The type of the generation which in the future will most significantly influence the operation of the power systems are the RES. The comparison of the development and prediction of the RES generation till 2035 based on the approved energy policies is shown in the following figure 10 [1], [2]. In the Slovak Republic there is higher generation from the hydro power plants due to the better natural conditions (River Váh cascade and HPP Gabčíkovo). The wind generation in the Slovak Republic is negligible. The most important RES generation type in the Czech Republic is FVE, where 4000 MW of the new FVE (mainly on the roofs of the houses) are assumed. The lower share of the hydro power plants in the Czech Republic is replaced by the generation of the biomass and biogas burning power plants.

It is supposed only 400 kV network development in both countries. The gradual process of replacing the 200 kV network by 400 kV network has already started in the Slovak Republic. In the Czech Republic this process is now in the preparatory phase, as the 220 kV grid is used nowadays as the backup reserve system. Replacement of the 220 kV grid in the Slovak Republic has begun in 2006, where the main impulse was decommissioning of the NPP EBO V1 and compulsory operation of the EVO 1. Network 220 kV was step by step replaced in Križovany, Lemešany, USSK, Medzibrod and Voľa. 220 kV substation in Bystričany and in Senica should be replaced by 400 kV system till the 2021. After the replacing of 220 kV substation in Považská Bystrica, the Czech Republic and the Slovak Republic will lose the 220 kV interconnection.

The result of the ongoing Ukrainian power system preparation for synchronous interconnection to ENTSO-E will significantly affect the further cooperation of the power systems in the region.

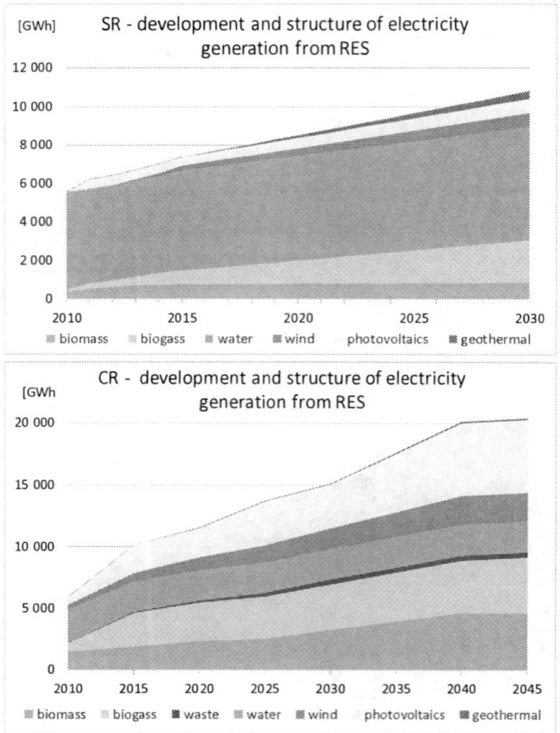

Fig.10. Comparison of the development and forecast of electricity generation from RES in the SR and CR up to 2030, 2040

VI. WHAT IS THE FUTURE FOR DEVELOPMENT OF ENERGY SECTOR?

The main challenges facing the Czech and Slovak electricity grid in the future can be characterized:

- significant environmental objectives (EU commitments), up to the limit of technical feasibility
- a significant downturn in classical large (and carbon) energy and the development of decentralized energy based on renewable sources
- developing decentralized energy sources in the regions, mainly at lower voltage levels
- big uncertainties in the organization of the European energy sector (mainly in the organization of the European electricity market)
- expected deficiencies in regulatory power and drive for new technologies and the management organization (battery, system flexibility)
- the need to cope with the high power transfers in Europe from the intermittent RES (for example technical measures e.g. redispatch, reconfiguration or PST transformers)
- fast development of new technologies, mainly at the lower voltage levels in networks

These influences already affect both the Czech Republic and the Slovak Republic in the Central European region.

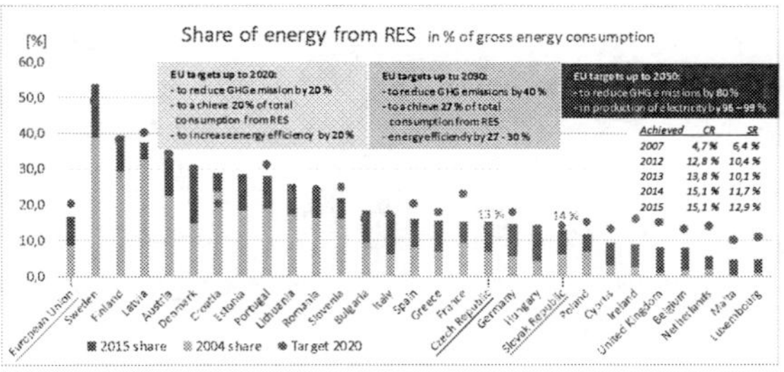

Fig. 11. Achieved RES shares in gross final energy consumption (2013) and EU environmental targets up to 2020, 2030 and 2050

As shown by the figure 11 of the achieved RES shares for gross final energy consumption by 2013 [6], the EU environmental targets by 2020 will be met in most countries. But next goals (up to 2050) are really very ambitious, and their fulfilment is both on the edge of today's technologies, and will fundamentally change the face of today's "big energy sector". Despite the large development of decentralized sources, the rundown of today's fossil power sources and the progressive development of new technologies, the transmission networks in the Czech Republic and the Slovak Republic will continue to play their irreplaceable role and the corresponding concept of their development by 2050.

The bases for the further development of both SR and CR systems are very good, but both power systems need a stable, long-term and transparent (for all states the same) legislative and regulatory framework for the energy sector, especially the European one.

In future there will need to be close cooperation between the both countries in the development of the transmission infrastructure. This cooperation will be need as on the bilateral and also on the regional level. Both transmission system operators are members of the Regional group Continental Central East (under the „System Development Committee" in ENTSO-E) where they cooperate on the coordinated development of the transmission systems in mid and long term time horizons, where the above mentioned challenges of the grid development are taken into consideration. The results of the common activities are summarized in the documents "Ten Year Network Development Plan ENTSO-E 2016" [7], which part is also the Regional Investment Plan Continental Central East.

REFERENCES

[1] State energy policy of the Czech Republic (Ministry of industry and trade of the Czech Republic, Decemeber 2014)

[2] Energy policy of the Slovak Republic (Ministry of economy of the Slovak Republic, November 2014)

[3] "ČEPS, a.s.". [Online]. Available: http://www.ceps.cz/ENG/Pages/default.aspx. [Accessed: 30-Jun-2017].

[4] "Slovak electricity transmission system, Plc." [Online]. Available: http://www.sepsas.sk/en_index.asp. [Accessed: 30-Jun-2017].

[5] The Anticipated Long-term Balance between Electricity Supply and Demand (OTE CR, December 2016, www.ote-cr.cz)

[6] Czech energy industry – Reality and future in the context of transition of the European energy industry (WEC, CME – Czech Member Committee of WEC, June 2016)

[7] "Ten-Year Network Development Plan 2016 - How did we improve". [Online]. Available: https://www.entsoe.eu/major-projects/ten-year-network-development-plan/ten%20year%20network%20development%20plan%202016/Pages/default.aspx. [Accessed: 30-Jun-2017].
'ENTSO-E — European Network of Transmission System Operators', ENTSO-E — European Network of Transmission System Operators. [Online]. Available: https:// https://www.entsoe.eu/. [Accessed: 30-Jun-2017].

Smog – Are Power Plants Guilty ?

Janusz Grabara
The Management Faculty, Czestochowa UT
Czestochowa, Poland
janusz@grabara.eu

Sebastian Kot
The Management Faculty, Czestochowa UT
Czestochowa, Poland,
Faculty of Economic Sciences and IT, North-West University,, South Africa.
sebacat@zim.pcz.czest.pl

Abstract — **This article is devoted to smog, which is increasingly annoying and dangerous phenomenon occurring both in Poland and other countries in the world. The article presents the phenomenon of smog in Poland defining its concept and presenting the state of air pollution analysis, also sources of the pollution resulting in smog are studies. The next part of the article presents possible "classic" solutions such as increasing the availability of system heat for individual consumers as well as long-term solutions by increasing the number of pumping plants as the energy reservoirs for use in individual households. A significant part of the article is devoted to the methods of smog decreasing in the concept of "innovation" and "future", which are implemented in China, already concerns the ecological production of ethanol from hard coal as well as American solutions on ethanol production directly from carbon dioxide those can be a clean energy sources. In the article's conclusions, the authors suggest that significant negative pressures and expenditures should be put on the greater use of traditional heat sources in the form of system heat, electricity production and the use of state-of-the-art technology by the creation of complex sources of energy and heat their storage in energy warehouses.**

Keywords: smog; particulate matters; power plants; domestic heating.

I. Introduction

In recent years the phenomenon of smog dominated the problem of the cleanness of air in Poland. Special concern is being devoted to the influence of particulate matters on human organisms and other living creatures. The article shows sources generating this state of affairs and alternative options of avoiding or reducing the scale of influence of smog in Poland. The definition of smog is as follows, smog [Eng. 'smoke', 'fog'], fog containing pollution of the atmospheric air. Smog divides into: photochemical smog, oxidizing smog, smog, which compounds are organic compounds chemically active (peroxides, aldehydes) and ozone, carbon monoxide, nitrogen oxides; sour smog, industrial fog, smog containing mainly sulphur dioxide and carbon dioxide (forming droplets of acids with the fog) and dusts and soot; atmospheric aerosols, aerosols, in which air is a continuous phase, and microscopic particulates (solid and liquid) constituting pollutants of the earth's atmosphere are a dispersed phase [1, 2]. Therefore the paper aim is determination of smog sources especially in relation to power plants and proposals of solutions for the smog problem using less known methods as well.

II. Problem of Smog in Poland

According to data of the Department of the Environment in Poland main factor generating these threats is so-called "low emission", which comes from furnaces used in households most often using fuels, which are cheap, low calorific and not-fulfilling applicable standards in the area of the exhaust emission. It is being estimated that in Poland there are about 3 million of such furnaces.

Low emission causes the appearance of many harmful substances in the air, amongst which it is possible to specify:

• particulate matters (depending on the cut of molecules they are PM10, PM5 or PM2,5 –the lower the value is the lower is the cut and simultaneously the harmfulness is greater) with grains of soot etc.,

• sulphur dioxide (SO2) – exceptionally harmful both for the health of the man, as well as the entire environment (one of main causes of acid rains),

• nitrogen oxides (NOx) – including nitrogen dioxide - are one of reasons of ozone hole or smog,

• heavy metals (Hg – mercury, Cd – cadmium, Pb – lead, Mn – manganese, Cr – chromium) – harmful for people, animals and plants,

• polycyclic aromatic hydrocarbons (WWA – i.e. α-benzopyrene) - carcinogens and causing strong poisonings,

• dioxins – poisoning substances, often responsible for the appearance of cancers or infertility.

Particulate matter PM 10 - consists of the mixture of particles hung in air, being a blend of various substances. Particulate matter can contain intoxicants so as polycyclic aromatic hydrocarbons (e.g. carcinogenic benzo[a]pyrene), heavy metals and dioxins and furans. PM 10 dust are particles with a diameter smaller than 10 micrometers (about one fifth fatness of the human hair), which can reach the upper respiratory tract and lungs.

- Maximum level for the daily mean value of the concentration is 50 µg/m3 and can be crossed no more than 35 days within one year.
- Maximum level for annual concentration is 40 µg/m3,
- Informing level - 200 µg/m3
- Alert level - 300 µg/m3

Particulate matter PM 2,5 - these are superfine particles: with a diameter smaller than 2,5 micrometers, which can reach the upper respiratory tract, lungs and penetrate into blood. Maximum level for annual concentration for PM 2,5 particular matter is 25 µg/m3

Benzo(a)pyrene – it is a carcinogenic and mutagenic substance being an example of polycyclic aromatic hydrocarbons (WWA). Benzo(a)pyrene demonstrates small acute toxicity, whereas great chronic toxicity what is connected with its ability of the accumulation in the body. Maximum annual level of B(a)P is 1 ng/m3

Households are responsible for 52% of emission of dusts while the participation of energetics is on the level of 9% and emission from industrial plants and the motor transport is more responsible for smog (Fig. 1).

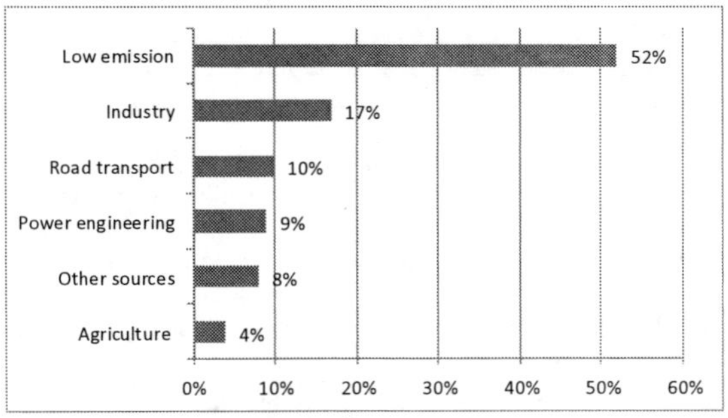

Figure 1. Sources of PM10 dust emission

How it was mentioned, the most important "producers" of pollutants are: households, in which apart from fuel pieces of rubbish of the organic origin [3, 4] and what is worse inorganic e.g. plastics of different form are also most often burnt (Fig.2).

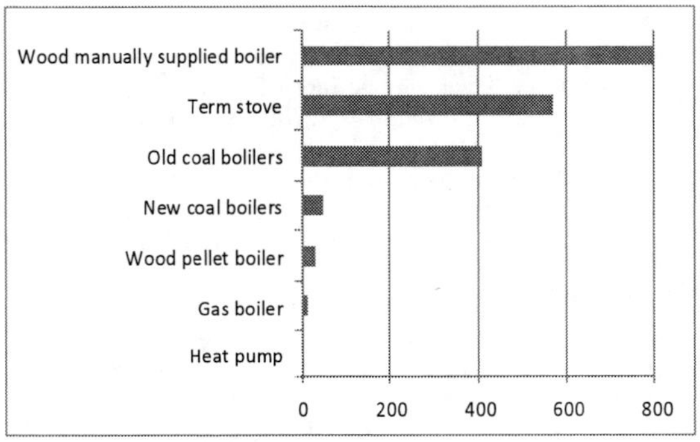

Figure 2 Emission of dusts in mg/m3.

Polish air quality index is being calculated exclusively on the basis of one-hour not-verified data from automatic stations functioning as part of the National Environmental Monitoring (PMŚ). Polish air quality index is being counted directly in the Chief Inspectorate of Environmental Protection JPOAT2,0 database of air quality on the base of data obtained from measuring PMŚ stations via databases of provincial inspectorates of environmental protection. Polish air quality index is being counted based on one-hour results from measurements of concentration in air: sulphur dioxide (SO_2), nitrogen dioxide (NO_2), PM10 dust, PM2,5 dust, carbon monoxide (CO), benzene (C_6H_6), ozone (O_3). Air quality indexes for individual pollutants are being counted on the basis of one-hour concentrations of these pollutants (only from automatic stations). In case of the individual index at the lack of sufficient measuring information needed for calculating the index the given point is not being shown. The value of Polish air quality index is being counted from table of scopes (one-hour averages) shown above, where individual indexes are being counted for individual pollutants by comparing the measured value of pollutant with set trough value. Next the general index assumes the value of the worst index for given pollutant. Before the comparison with limits of ranges for individual classes of the index, values are rounded off to integral numbers, according to mathematical principles. On the map of current data Polish air quality index and indexes of individual pollutants from the last hour are visible (if there is no data from the las hour, it is possible to see index from the second and maximum from the third hour backwards).

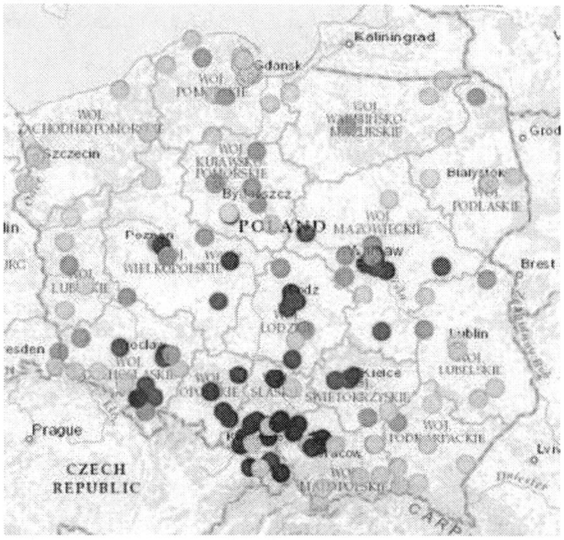

Figure 3 Map of threats in the air in Poland

The map shows the degree of threats from the lighter tone meaning very good state of air through good, moderate, sufficient, bad and the darker tone means very bad condition. PM 10 and PM 2,5 dusts and the following substances: SO_2, CO, O_3, NO_2 and benzene are subject to an evaluation.

An acceptable average twenty-four hour emission standard PM10 particulate matter is 50 micrograms per cubic meter. Crossing of trough value of 200 micrograms per cubic meter must be made public, however crossing 300 micrograms per cubic meter means the danger point, when one should try to limit staying in the open air.

Air quality index specifies six levels:

-Very good – when the concentration is below 20 micrograms,
-Good – when the concentration is from 21 to 60 micrograms,
-Moderate - 61-100 micrograms,
-Sufficient - 101-140 micrograms,
-Bad – when the concentration is 141 to 200 micrograms,
-Very bad - above 200 micrograms, crossing 300 micrograms of dust per cubic meter means smog alert.

Concentration of particulate matter in Poland for 17.01.2017r. in - Warsaw. (Measuring station: 68 Marszałkowska Street)
Air quality: bad
PM10: 115,7 µg/m3
CO: 1,6 mg/m3
PM2,5: 99,2 µg/m3
NO2: 50,6 µg/m3

Two Polish cities are a flagship example. Żywiec delightfully situated next to Silesian Beskid, is inhabited by 32 thousand people. In the newest report of the World Health Organization (WHO), which fixes global standards of air quality, this town was number one in Poland, next to Pszczyna which is being called the Pearl of Upper Silesia. What is more, residents of both towns are breathing the most polluted air in the entire European Union. A Bulgarian town Dimitrowgrad was only on the third place.

Presented data demonstrate in the distinct and explicit way directions of action, which should be taken or are already being carried out in this area. There are a lot of tasks to be carried out, and the biggest "poisoners" which are domestic furnaces and local boiler rooms should be subjects to start with.

There are already tested solutions, both from traditional as well as innovative ones, which are possible to be exploited.

Eliminating black coal and other solid fuels from individual households would be the simplest solution.

III. CHOSEN SOLUTION FOR SMOG PROBLEM

Smog poisons both large cities and small towns, even the spa ones. The Ministry of the Energy wants to fight with smog, encouraging electric heating. Electric heating is rightly regarded as the most expensive way of ensuring the warmth in buildings, more expensive than gas heating, which next is more expensive than coal and wood heating. But the Ministry of Energetics wants to promote electric heating in detached houses, seeing it as the clean source of the warmth, without emission of any gasses (not including these ones, which are emitted earlier by the power station). The department of the power industry counted that at production of 1 kWh in the power station (even fossil-fuel one) 10-times fewer dusts is being made than in the domestic coal stove, even the most modern one.

Of course it is not about warming itself up with driers and heating units of this type with the spiral of resistance wire and the whiff of the heated air activated by the fan. It is rather about modern stoves fed with the electric energy, but having an extended accumulation of the warmth. Similar solutions are applied in the old municipal construction, where implementing the new heating installation constitutes the problem, however it is possible to put electric cells in old tiled stoves accumulating the warmth. Night rate for heating purposes. How "Pulse of the Business" writes, the ministry targets the offer especially to people warming houses with coal and rubbish, that is to about The Government is going to create enough attractively-priced solutions so that even people of modest means can afford it. It is about energy night rate intended only for heating purposes - for all households warming up with electricity. – The price will be much lower than in other rates, we want to eliminate all overheads and payments imposed on the electric energy – says Andrzej Piotrowski, deputy minister of the energy. Poland, next to Italy and Germany, is the biggest issuer of dust in Europe. According to the Ministry of the Energy the reduction in the release of pollutants by popularizing the electric heating will considerably improve the medical condition of the sizeable crowd of Poles. In this particular case it is about developing the energy generated mainly by fossil-fuel power stations from the twenty-four hour period, in which the demand for the energy drops under 40% of the nominal power. It is being known that keeping the efficient effect of the power station requires the minimum of keeping production on the level of recalled 40%. There are periods when such a demand does not exist and the produced energy goes "into air". Some European countries use such a state of affairs for filling water bodies of the pumped storage power stations. In Poland we have only 6 power stations of this type. Polish pumped storage power stations can produce 10 GWh of the energy.

Table 1. Pumped storage power stations in Poland

Power Plant Name	Power (GW)	Water drop height (m)	Capacity (million m³)	Stored Energy (GWh)
Żarnowiec	0,68	100	13,6	3,6
Porąbka-Żar	0,50	430	2,3	2,6
Solina-Myczkowce	0,20	60	472	1,3 (50)
Czorsztyn–Niedzica	0,09	50	200	1,0 (15)
Żydowo	0,16	83	3,3	0,7
Dychów	0,09	18	3,6	0,2

Upper tank of the Solina-Myczkowce power station (Lake Solińskie) is being used mainly as the storage reservoir. The capacity of the bottom tank in Myczkowce lets the Solina power station to operate with the full validity within 5-6 hours, what means storing up of 1.3 GWh. If the bottom tank was equally big as upper, the pumped storage power station in Solina would be able to store the energy of about 50 GWh.

Czorsztyn-Niedzica power station due to minor differences in daily and nightly energy prices currently works in turbine mode.

Also in this case a restriction is the size of the bottom tank in Sromowce. If the bottom tank was equally big as upper, the pumped storage power station could store the energy of about 15 GWh. Simultaneous activation of all pumped storage power stations allows to generate 1.7 GW of the electricity. Current possibilities of the storage of energy in Poland are also completely insufficient.

Considering the possibilities of the execution of tasks associated with the liquidation of threats posed by smog as well as sources of "smog causing" factors it is possible to accept, reduction the negative influence of smog can be done by popularizing networks delivering system heat and using electric energy generated in periods of the minimal demand for the energy in the form of warming with using storage heaters for heating purposes.

Unfortunately these means are possible to be described as temporary, rather than system solutions. Therefore it is necessary to implement the politics of closer and more distant perspective from "innovation" and "future" areas [5, 6].

Today we already have a lot of good information which let for daring looking to the future. Here a technology of acquiring the ethanol from the black coal is being talked about. At the end of 2016 such installation on an industrial scale was started in China.

Recently China's coal chemical industry (especially in the year 2007) has seen faster progress due to high oil prices in the world market. The Chinese energy industry has been centered on speeding up the development of a new coal chemical industry. The national energy progress strategy, states that new coal chemical industry mainly includes the manufacturing of clean energy and alternatives such as diesel oil, gasoline, etc.

In the next twenty years the coal chemical industry will be very important in the sustainable progress of China's energy resources which will relieve environmental contamination and lower China's dependence on oil imports. In essence, there will be great requirement and chances for China's coal chemical industry.

A short while ago, newest coal - based chemical technologies, in which is coalliquefaction, coal – to - methanol including its derived products and co-production based on the coal chemicals industry were being adopted speedily in China. Some national coal - mining companies and regions use plentiful resources to achieve great economic returns. The top priority for development of the local economy is being given to newcoal chemical engineering. Great efforts have been made during the adaptation of the industrial structure based on the coal chemical industry to advance towards new paths of industrialization.

Entry and progress of C1 chemical technology has made significant advancement. The growth has been achieved in the size and production capability for carbonylation of methanol to acetic acid (including anhydride, acetyl cellulose), as well as the progress has been made in the technology for the synthesis of lower carbon alcohols from gasified coal gas [7].

Since the 1980s, when China continued its exploration of coal-derived liquid fuels, scientists in Shanxi Institute of Coal Chemistry, Chinese Academy of Sciences (CAS) have taken the leadership in the research field involving the synthesis of oil. Chinese researchers using the analysis of the Fischer-

Tropsch (F-T) and Methanol-to-Gasoline (MTG) processes from abroad, discovered the fixed-bed two-stage (MFT) method and the slurry-bed and fixed-bed two-stage methods. These methods unify conventional F-T synthesis with the unique shape choice functions of molecular sieves. Small-scale tests and pilotscale experiments of manufacturing hundred tonnes of the MFT process (SMFT) annually ended with success. Both its oil yield and action proved to be good. At the same time, a commercial test of 2000 tonnes annually and a simulation test of the SMFT process were made. They also gave good outcomes in a long-term run (3000 h) of two sorts of catalysts of the scientists own conception.

At present the pilot-scale experiment of a thousand tonnes annual output for the slurry-bed synthesis process is being conducted. The mature technology of coal direct liquefaction has now been mastered in China, as well the path for the coal liquefaction has been improved.

In the initial engineering stage 9.7 million tonnes of coal will be consumed and 3.2 million tonnes of oil can be produced, including 0.5 million tonnes of gasoline, 2.15 million tonnes of diesel oil, 0.31 million tonnes of liquefied gas and 0.24 million tonnes of benzene, mixed xylene etc. once it is finished and commissioned.

The Sasol Company of South Africa possesses a mature technology on coal indirect liquefaction. Now this technology is being commercially exploited. The Shenhua Group and Ningxia Coal Industry Group are making negotiations with the Sasol Company, which are crucial for implementing this mature technology in China. In China indirect coal liquefaction technology has made spectacular breakthroughs. Shandong Yanzhou Mining Group and Shanxi Institute of Coal Chemistry, CAS, have made real advance in the indirect liquefaction F-T process. Two pilot-scale facilities, which have the capacity of 1000 tonnes/year and 10,000 tonnes/year, respectively, have been constructed on an individual basis and are being run by them in the right way.

The second very interesting problem solving is being suggested by employees from Department of Energy's Oak Ridge National Laboratory in the USA. Accidentally this laboratory managed to get the ethanol directly of the carbon dioxide.

We put into an electrocatalyst, which runs at room temperature and in water for the electroreduction of dissolved CO2 with great selectivity for ethanol. The overpotential (which might be brought down with the right electrolyte, and by dividing the hydrogen manufacturing to another catalyst) likely prevents economic capacity for this catalyst. However the great selectivity for a 12-electron reaction means that nanostructured surfaces with multiple reactive sites in close proximity can give fresh reaction mechanisms [8]. This suggests that the synergistic effect from interactions between Cu and CNS demonstrates a novel strategy, which can be used for developing highly selective electrocatalysts. While the whole reaction mechanism has not yet been explicated, future work will discover further details from changeover of possible intermediates (e. g. CO, formic acid and acetaldehyde).

IV. Conclusion

How it results from material presented above means of the liquidation of smog and threats made by it can be different, nevertheless comprehensive action using all recalled methods is recommended.

In fact one of the most important solution is decreasing of low emission from households but this proposal is hard to achieve because of lack of financial means for refunding the old technology boilers exchange for new more environmental friendly heating systems in households. There is a need of hard coal supply with reasonable price, now because of too many intermediaries in this supply chain, the price is more than twice higher between exploitation and the final customer. There still a lack of hard coal quality norms in the market.

Of course the energetics role in smog problem solution cannot be omitted. Lower price of power for heating can be a good choice for some of households. Now it is rather more expensive than other possibilities.

The fact of the change of coal or of carbon dioxide into ethanol is a crucial factor, because we are able to store without special investments because we have seized storing alcohol in centuries-old tradition of our nations.

This solution is not only against to smog problem but it is also sustainable nature [9]. Such a form of fuel is perfectly suitable for a road transport. Brazil, where practically all cars are being driven with ethanol what as a result of the burn gives water and carbon dioxide, is the best example.

Comprehensive using of the solar power for generating the electricity like in the Australian project carried out by Elon Musk applied for producing the ethanol, what allows for the accumulation, storing very efficient and eco-friendly fuel on time, when solar power stations do not work for example at night.

The issue of smog solution is not closed because everyday news can bring us new ideas for that problem. So future researches must be focused on innovation as well as organizational changes. Concluding power plants are not very guilty for the smog problem however they can do much more to solve it.

References

[1] M. Shuangchen, C. Jin, J. Kunling, M. Lan, Z. Sijie, W. Kai, "Environmental influence and countermeasures for high humidity flue gas discharging from power plants". Renewable and Sustainable Energy Reviews, 73, 2017, pp. 225-235.

[2] J. Janekova, J. Fabianova, A. Rosova, "Environmental and economic aspects in decision making of the investment project "wind park" Polish Journal of Management Studies, 13 (1), 2016, pp. 90-100.

[3] B. Ślusarczyk, M. Baryń, S. Kot, „Tire industry products as an alternative fuel" Polish Journal of Environmental Studies, 25 (3), 2016, pp. 1263-1270

[4] B. Atilgan, A. Azapagic, "Renewable electricity in Turkey: Life cycle environmental impacts". Renewable Energy, 89, 2016, pp. 649-657

[5] N. Vagapova, A. Fedotov, A. Sidorov, G. Vagapov, "Mechanisms of effective management in electric power companies" Polish Journal of Management Studies, 13 (2), 2016, pp. 183-191.

[6] S. Kot, B. Slusarczyk, "The role of biomass in the energy production in the European union countries" Proceedings of the 7th International Scientific Symposium on Electrical Power Engineering, ELEKTROENERGETIKA 2013, pp. 559-561.

[7] Y. Kameya, T. Hayashi, M. Motosuke, "Oxidation-resistant graphitic surface nanostructure of carbon black developed by ethanol thermal decomposition", Diamond and Related Materials, 2016, 65, pp. 26-31.

[8] Y.El. Fouih, C. Bouallou, "Recycling of carbon dioxide to produce ethanol", Energy Procedia, 2013, 37, pp. 6679-6686.

[9] J.K., Grabara, "The another point of view on sustainable management" Quality - Access to Success, 18, 2017 pp. 344-349.

Preventing Large-Scale Emergencies in Modern Power Systems: AI approach

Michael Negnevitsky
School of Engineering and ICT
University of Tasmania
Hobart, Australia
michael.negnevitsky@utas.edu.au

Abstract — **In recent years, due to liberalization, power systems are being operated closer and closer to their limits. At the same time, they have increased in size and complexity. Both factors increase the risk of major power outages and blackouts. In emergency and abnormal conditions, a power system operator has to deal with large amounts of data. However, due to emotional and psychological stress, an operator may not be able to respond to critical conditions adequately and make correct decisions promptly. Mistakes can damage very expensive power system equipment or worse lead to major emergencies and catastrophic situations. Intelligent systems can play an important role by alarming the operator and suggesting the necessary actions to be taken to deal with a given emergency. This paper outlines some experience obtained at the University of Tasmania, Australia, in developing intelligent systems for preventing large-scale emergencies and blackouts in modern power systems.**

Keywords: power system, emergency condition, computational intelligence, disaster management.

I. Introduction

Modern power systems are vulnerable to large-scale blackouts. During the past ten years events in Europe, North America and Asia have clearly demonstrated an increasing likelihood of large-scale blackouts, which cause huge economic losses, and disruptions in communication and transport, heating and cooling, water supply, emergency services, and financial trading. Blackouts now typically spread across borders, escalating their impacts, and the trend towards national grid integration is likely to increase the risk of even larger power outages in the future. There are many examples. The 2003 blackout in the US and Canada affected some 50 million people and 61,800 MW of electric load in eight states and the province of Ontario. The US Department of Energy estimated total losses of about $6 billion [1]. In November 2006, more than 15 million households in Germany, France, Italy, Belgium, Spain, and Portugal were left without power, and there were local outages in the Netherlands, Poland, Switzerland, the Czech Republic, Greece and Morocco. The root cause was an overload when the German electricity company E.ON switched off an electricity line over the river Ems to allow a cruise ship to pass through. The impact of this disconnection resulted in the European transmission grid splitting into three independent parts for two hours; the blackouts were caused by the imbalance between generation and demand in each section. The July 2012 India blackout, the biggest ever power blackout in the world, left half of India without electricity, twice in two days, affecting millions of households and businesses. In October 2012, Hurricane Sandy brought high winds and coastal flooding to a large part of the eastern United States, leaving an estimated 8 million customers without power. The blackout directly affected nine states. While these are spectacular examples, short blackouts occur daily: in the US alone, they cause annual economic losses of $104–164 billion [1].

Australia is far from immune from large-scale blackouts, and examples abound over the past decade. In January 2007, power was cut to 200,000 people in Victoria when bushfires caused the state's electricity connection to the national grid to shut down. In April 2008, around 420,000 households were left without power in Melbourne and in other parts of Victoria after the state was hit by winds of up to 130 km/h. In January 2009, hundreds of thousands of homes in Victoria, including Melbourne, suffered power failures as a result of a record heat wave, in which more than 500,000 residents in central Melbourne were without power for a whole evening, with train and tram services cancelled, evacuations, traffic disruption, and people trapped in lifts.

Over the past decade, power industries worldwide have experienced two major changes: liberalization of the electricity market and the expansion of renewable energy.

As wholesale markets have developed, congestion in the transmission network has increased significantly. In many countries, states and regions that have liberalized their electricity sectors, investments in transmission capacity, have not kept pace with the expansion in demand and generation. In Europe, the United States and Australia there has been almost no investment in interregional transmission capacity since the early 1990s. As a result, modern power systems have to operate close to their operational limits; this intensifies their vulnerability to blackouts.

The increasing penetration of renewable energy has led to increasing dependency on the volatile nature of renewable energy sources. It is not only a shortage of power supply that can result in a blackout; oversupply can lead to frequency and voltage instability problems. For example, wind farms in Eastern Germany during strong wind conditions can supply up to 12 GW, which is more than all of the coal- and gas-fired power plants in that part of the country combined. This situation presents significant risks to system stability even when demand is low.

Because load centers are often remote from the sources of renewable generation large amounts of energy often have to be transported over long distances or even through distribution grids that were not designed for this purpose. This leads to higher loading of the grid, and the necessity for costly upgrades. However, due to several factors – including public resistance – planning and extending the electricity network can take many years. In the meantime, power systems have to operate closer to their operational limits, and the probability of large-scale emergencies and blackouts is inevitably increasing. Power systems need to become much "smarter" to handle these challenges.

This paper outlines some experience obtained at the University of Tasmania, Australia.

II. Basics of Power system Operation and Control

Power system security includes measures which intend to keep the system in an operating condition when one or even a few components fail to operate. For example, a generating unit may fail but the remaining units of the system can increase power output to make up the deficit, and thus prevent the load shedding. Similarly, a power transmission line may be damaged and switched-off by the power system protection but the remaining transmission lines can take the increased load.

In order to prevent a power system from collapse, its elements are operated within certain constraints and protected by automatic devices that cause equipment to be taken off if those constraints are violated. However, due to some failures, equipment may still be left in the operating state with constraints violated. Such conditions may lead in turn to switching-off other equipment out of service and if this process continues, large parts or even the entire system can collapse completely.

A major emergency in a power system might start with opening-off a single line as a result of a short-circuit. The remaining transmission system takes up the power that was flowing on that line. However, if even one of the remaining lines is now overloaded, it also may open due to an operation of the protection system, and thus an additional load is imposed on the remaining network. Moreover, a certain transmission line outage can cause some serious voltage problems as well. Therefore, power system operators are required to take immediate actions to avoid a further deterioration of the situation.

In emergency and abnormal conditions, a power system operator has to deal with a large amount of data and apply the most appropriate remedial actions. However, due to emotional and psychological stress, an operator may not be able to adequately respond to critical conditions and make correct decisions. Moreover, in emergency conditions, a vital decision must often be taken in a matter of minutes, sometimes even seconds. Mistakes can damage very expensive power system equipment or worse lead to the major emergencies and even catastrophic events. There is no time to examine advantages and disadvantages of different approaches to the restoration process and look into long columns of results. Nevertheless, an operator must respond to an emergency and only his or her experience can help to find a right decision. Clearly, there is a strong need for intelligent decision support systems for preventing major emergencies and blackouts in modern power systems.

III. Intelligent System for Clearing Overloads

An overload of transmission plant (power transmission lines and transformers) is a common problem in modern power systems. In fact, it is impossible to avoid overloads which may lead to a failure or damage of the power system equipment as well as to a collapse of the entire system. Therefore, power system operators must take safety measures to clear overloads so that their duration and magnitude do not exceed the allowable values. The problem of an overload clearing requires the performance of two tasks.

The first task is to ascertain an overloading condition. Supervisory Control and Data Acquisition (SCADA) systems provide a power system operator with information if any equipment is overloaded. Then the permissible duration of the overload must be determined based on the existing loading guides and practical experience.

The second task in the clearing of overloads is the decision making. The overload clearing process is amenable to formalization with difficulty and cannot be expressed mathematically. Its successful performance largely depends on the operator's skills and his or her ability to respond reasonably to unpredictable situations. Human behavior under such conditions is essentially determined by the experience obtained in similar situations and the operator's capability to use heuristic logic in order to compare a given complex situation with already known emergencies and thereby to make a correct decision. However, under emergency conditions a power system operator may act either wrongly by violating the guides and instructions or may not act at all due to emotional stress. As experience has shown, there is a strong need for on-line support to assist operators in making prompt and correct decisions under emergency and abnormal conditions. This support can be obtained by employing an expert system based aid described in this section.

In order to clear an overload, a power system operator has to perform the following tasks:

- identify overloaded elements and determine their permissible loading time;
- select necessary remedial actions for clearing overloads;
- execute the selected actions;
- monitor conditions of the overloaded equipment and, if required, revise and adjust the remedial actions to be taken.

Decision making under emergency and abnormal power system conditions is intimately bound up with an increasing sophistication of information and alarms to be examined. However, the main difficulty is the shortage of the time available for decision making. An appropriate solution must be found and implemented in a matter of minutes, sometimes even seconds.

An accurate determination of loading capabilities of power system equipment allows operators a greater flexibility in decision making and in choosing suitable remedial actions. From an operator point of view, the most hazardous overloads are these which occur on tie power transmission lines connecting different parts of power systems. Therefore, it is vital to examine factors setting limits on the overloading time of power transmission lines.

The conductor operating temperature is limited by allowable mid-span sag or conductor ground clearance and also by creep in the conductor and long-term mechanical effects [2]. Usually the conductor ground clearance consideration is critical because it gives the lower limits. The loading capacity or thermal rating can be described as a function of conductor temperature, ambient temperature, wind velocity, elevation, ground reflection, solar radiation, etc. [2]. In transmission power systems, load currents during and following system disturbances may reach values well above the steady-state thermal rating of the line. Such conditions may be safe for a period of up to 30 minutes. This time can be used to perform the generation rescheduling or load shedding in order to clear or at least reduce overloading. Therefore, it is necessary to determine how long a conductor can carry an overload without exceeding its maximum permissible operating temperature. This problem can be solved by conventional methods [4] or by employing intelligent knowledge-based systems [5]. As a result, an operator can obtain the permissible time for clearing overloads on transmission lines.

Once the permissible overloading duration is determined, an operator has to choose an appropriate plan for clearing the overload.

Actions directed to relieve overloads are usually applied in the following order [6]:

- switch-on or reclose the reserve or disconnected generators, transmission lines and transformers;
- provide full loading of the generators in the receiving part of the power system;
- provide unloading of the generators in the sending part of the power system;
- adjust phase-shift transformers and switch capacitor banks and reactors;
- transfer the load from one part of the system to another;
- increase output of the active and reactive power of generators and synchronous compensators via their permissible short-term overloading;
- adjust bus voltages in order to decrease the power demand;
- curtail loads of the lowest priority;
- switch off radial transmission lines;
- perform emergency load shedding at selected locations.

To determine a remedial action to be taken first, it is necessary to know its execution time and the overload relief caused by the application of this action (sensitivity factor).

Equations of load flow are solved by the application of the Newton-Raphson algorithm which can be expressed in the matrix form as a set of bus active and reactive power mismatches,

$$\begin{bmatrix}\Delta P\\ \Delta Q\end{bmatrix}=\begin{bmatrix}H & N\\ J & L\end{bmatrix}\begin{bmatrix}\Delta\theta\\ \Delta V/V\end{bmatrix} \quad (1)$$

where the solution variables are the unknown voltage angles and magnitudes.

If the coupling sub-matrixes [*N*] and [*J*] are neglected, two separated equations may be obtained and solved [7]. The first equation expresses the relationship between active power and bus voltage angles and can be used for sensitivity calculations.

$$[\Delta P] = [H]\,[\Delta\theta] \quad (2)$$

In the determination of the control actions for clearing overloads, elements which affect VAR flows (reactive power flows), i.e. shunt reactances and off-nominal in-phase transformer taps can be omitted.

Based on the decoupling properties, overloads can be maintained within permissible limits by controlling the MW generation in the receiving and sending parts of the system, reconfiguration of the transmission network and load shedding.

The generator shift factors can be expressed as:

$$\alpha_{ij} = \frac{\Delta S_i}{\Delta P_{Gj}} \quad (3)$$

where ΔS_i is the change in MVA flow on element *i* when a change in generation, ΔP_{Gj}, occurs at bus *j*.

If it is assumed that all other generators remain fixed, then the α_{ij} factor represents the sensitivity of the flow on element *i* to a change in generation at bus *j*.

The network reconfiguration factors are used in a similar manner. The line outage distribution factor has the following meaning:

$$\beta_{i,k} = \frac{\Delta S_i}{\Delta S_i^0} \quad (4)$$

where $\beta_{i,k}$ is the line outage distribution factor when monitoring element *i* after an outage on line *k*; ΔS_i is the change in MVA flow on element *i*; ΔS_k^0 is the original flow on line *k* before it was opened.

The load shedding factors can be obtained as:

$$\gamma_{ij} = \frac{\Delta S_i}{\Delta P_{Ln}} \quad (5)$$

where ΔS_i is the change in MVA flow on element *i* when a change in load, ΔP_{Ln}, occurs at bus *n*.

The overload clearing is subject to the following security constraints:

$$P_{Gj}^{\min} \le P_{Gj} + \Delta P_{Gj} \le P_{Gj}^{\max} \quad (6)$$

$$-S_{k+1}^{\max} \le S_{k+1}^0 + \Delta S_{k+1} \le S_{k+1}^{\max} \quad (7)$$

$$P_{Ln} - \Delta P_{Ln} \ge P_{Ln}^{\min} \quad (8)$$

$$V_j^{\min} \le V_j \le V_j^{\max} \quad (9)$$

where $P_{Gj}^{\max}$, $P_{Gj}^{\min}$ and P_{Gj} are respectively the maximum and minimum limits, and operating value of the generator MW output at bus *j*; S_{k+1}^0, ΔS_{k+1} and $S_{k+1}^{\max}$ are respectively the original MVA flow, change in MVA flow on line (*k*+1) after an outage on line *k* and long-term rating (or emergency rating) of line (*k*+1) (with regard to that a line flow can be positive or negative); $P_{Ln}^{\min}$ and P_{Ln} are respectively the guaranteed minimum load supply and original load at bus *n*; $V_j^{\max}$, $V_j^{\min}$ and V_j are respectively the maximum and minimum voltage limits, and actual voltage at bus *j*.

The corrective actions are not equally effective and, therefore, required adjustments are different for each of them. In order to find the most effective corrective actions for the given situation, the sensitivity tree method can be used. It provides the relationship between element loads and the controllers.

A sensitivity matrix relating the changes in the controlled variables, ΔP, to the changes in the controllers, $\Delta\theta$, is defined by *Eq.*(2). For a power system that includes n controlled elements (power transmission lines and transformers) and m controllers, the relationship between overloaded elements and available controllers can be represented as shown in Fig. 1, where $S_{i,j}$ is sensitivity factor between the element load and controller.

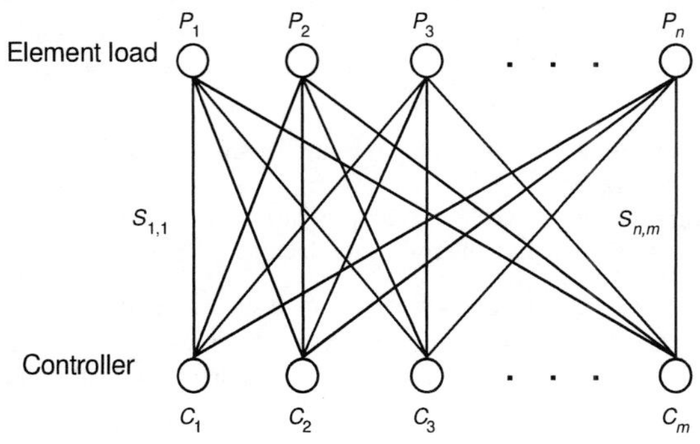

Figure 1. The sensitivity tree.

The sensitivity tree shows that a load correction on any overloaded element can be achieved via the application of several different controllers. In the same time, an adjustment of any controller may result in load changes of several elements.

The expert system was developed and implemented on an IBM personal computer using VP_EXPERT system shell. It provided interactions with power flow analysis software to calculate network sensitivity factors and line flows, and other external programs and a database.

Any overload can be classified using answers to the following questions [6]:

1. What is the type of plant?
2. What is the overload duration?
3. What is the overload permissible duration?
4. What are the means available for clearing the overload?
5. What is the action to be applied first?

These provide the basis on which the database and the knowledge base are organized. The expert system for clearing overloads uses the following data:

The upper load limits (short-term rating) for each supervised element.

- The permissible overload duration for each supervised element as a function of the load. The permissible time for clearing an overload can be obtained using the data provided in loading guides, calculated, e.g. [3] or employing intelligent systems as it was discussed above.
- The upper and lower limits for each power station (or, if required, for each generator).
- The long-term ratings (emergency ratings) for each power transmission line.
- The guaranteed minimum load supply (or the highest priority load level) at each bus.
- The sensitivity factors for each supervised element and controller, and also the execution time for each controller.

The expert system applies the following heuristic rules specified in [6]:

***Rule 1*:**
IF the real power flow on element i is more than its short-term rating, THEN element i is overloaded.

***Rule 2*:**
IF element i is overloaded, THEN include element i in the list of overloaded elements.

***Rule 3*:**
IF element i is in the list of overloaded elements, THEN determine permissible overload duration t_i for element i.

***Rule 4*:**
IF permissible overload duration t_i is the lowest over all elements in the list of overloaded elements, THEN calculate the sensitivity factors (generator shift, network reconfiguration and load shedding).

***Rule 5*:**
IF the generator shift list is not empty, THEN select the most effective controller AND check its limits AND calculate the overload relief on element i.

***Rule 6*:**
IF the overload relief caused by the selected controller is less than the overload magnitude of element i AND the controller execution time is less than permissible overload duration t_i, THEN include the controller in the list of feasible actions.

***Rule 7*:**
IF element i is still overloaded, THEN select next available controller on the generator shift list until all controllers are taken.

***Rule 8*:**
IF element i is still overloaded AND the network reconfiguration list is not empty, THEN select the most effective controller AND check long-term rating violations on the other elements after the controller is applied AND calculate the overload relief on element i

***Rule 9*:**
IF the selected controller does not cause long-term rating violations on the other elements AND the overload relief is less than the overload magnitude of element i AND the controller execution time is less than permissible overload duration t_i, THEN include the controller in the list of feasible actions.

***Rule 10*:**
IF element i is still overloaded, THEN select next available controller on the network reconfiguration list until all controllers are taken.

***Rule 11*:**
IF element i is still overloaded AND the load shedding list is not empty, THEN select the most effective bus (a bus where the load shedding causes the most significant effect on the overload relief of element i) AND calculate the load shedding required to clear the overload on element i.

***Rule 12*:**
IF the required load shedding is less than the load at the selected bus minus the guaranteed minimum load supply AND

the shedding execution time is less than permissible overload duration t_i, THEN include the required load shedding at the selected bus in the list of feasible actions.

***Rule 13*:**
IF the required load shedding is greater than the load at the selected bus minus the guaranteed minimum load supply AND the shedding execution time is less than permissible overload duration t_i, THEN include the allowable load shedding at the selected bus in the list of feasible actions.

***Rule 14*:**
IF element *i* is still overloaded, THEN select next bus on the shedding list until the overload on element *i* is cleared.

***Rule 15*:**
IF the overload on element *i* is cleared, THEN take next element on the list of the overloaded elements until the list becomes empty.

The proposed approach allows improving computational performance of the intelligent system by reducing the size of a power system and eliminating the less effective controllers. This should enable intelligent systems to meet requirements for real-time applications using data provided by the SCADA system.

IV. EXPERT SYSTEM FOR VOLTAGE CONTROL

Several studies identified overloading of transmission line and voltage instability as two major reasons of blackouts [8], [9]. A typical blackout scenario develops as follows: high system loading (due to heavy transfers across the grid) is followed by events that initiate protection system actions. As a result, some lines are disconnected, the grid becomes even more overloaded, and consumption of reactive power is increased, causing a cascading effect in which voltages drop even further.

Therefore, control actions are required to deal with voltage problems. This section describes an expert system application for voltage control for preventing voltage collapse.

The main objective of voltage control is to maintain power system voltages within specified limits under various loading conditions and changes in the transmission network. Classical analytical methods based on different optimization techniques intend to minimize system losses and improve voltage profiles. However, most of these methods are too complex, provide analytical solutions only, and cannot recommend appropriate control actions required to overcome voltage violations in a given emergency condition.

In order to improve the computational performance, a reduced system can be built using the subsystem reduction and equivalencing techniques [10]. The reduced system consists of an internal "three-tier" subsystem that introduces the voltage problem area, and the external subsystem that represents the rest of the power system. This approach allows to minimize the system size and eliminate the less effective voltage controllers. Therefore, voltage magnitudes can be predominantly controlled by the reactive power injections into the internal subsystem. An expert system technique can be used to provide a set of practically effective control actions to solve a given voltage problem in real-time.

Voltage controllers can change their values either continuously or in steps, they are not equally effective and adjustments required from an individual controller are different under different power system conditions. The sensitivity tree method is used to find the effective order of control actions to be applied. It provides the relationship between bus voltages and the controller settings. A sensitivity matrix which relates the changes in the controlled variables, ΔX, to the changes in the controller settings, ΔU, may be described as suggested in [11]:

$$[\Delta X] = [S]\,[\Delta U] \tag{10}$$

where matrix [S] provides sensitivity factors between settings of the controllers and bus voltages. Voltage at any bus in the system can be adjusted by several different controllers. On the other hand, a change in any controller setting results in voltage changes on several buses.

The developed voltage control expert system interacts with the power flow analysis package, network sensitivity analysis and other external programs and databases. The expert system uses the following data:

- Upper and lower limits of voltage at each bus;
- Upper and lower limits of each controller;
- Sensitivity factors for each load bus and each controller.

When a bus voltage exceeds its specified limits, either high or low, a power system operator can adjust transformer taps, change generator bus voltages and switch capacitor banks and reactors on or off.

The voltage control expert system performs the following tasks:

1. Determine the problem area and build an internal "three-tier" subsystem.
2. Calculate the network sensitivity factors and build the sensitivity tree.
3. Identify all buses with abnormal voltages.
4. Select the bus with the maximum level of voltage violation.
5. Select the most effective controller, estimate controller effect and check security constraints.
6. If security constraints are not violated then implement the controller selected. If security constraints are violated then proceed to the next task.
7. If the voltage problem still exists then select the next available controller until all controllers are taken.
8. If the voltage on the maximum violated bus has become normal then take the next bus with the maximum abnormal voltage and repeat the procedures.

A complete set of rules can be found in [12]. Results are demonstrated on the modified IEEE 30-bus test system shown in Fig. 2. The most severe case is an outage of transformer T1 connected between buses 27 and 28. The "three-tier" internal subsystem incorporates the following buses:

First tiers:	27, 28.
Second tiers:	6, 8, 25, 29, 30.
Third tiers:	2, 4, 7, 9, 10, 24, 26.

Table 1 provides results (voltage levels and optimal settings of the controllers) given by the expert system for the full 34-bus system and the equivalent system, respectively, in the case of the outage of transformer T1.

The results obtained for the equivalent system are very close to those obtained for the entire 34-bus system. It indicates that the proposed approach provides a satisfactory solution. Recommendations given by the expert system on voltage control actions to be taken under the outage of transformer T1 are shown in Fig. 3. The expert system performs well and provides correct decisions.

V. CONCLUSIONS

With rapidly increasing complexity of power grids, liberalization of electricity markets and increasing penetration of renewable energy, the risk of large-scale emergencies and blackouts increases. This paper presented some experience obtained in the development of intelligent systems for preventing large-scale emergencies and blackouts in modern power systems. Prototype intelligent systems for clearing overloads, voltage control and detecting voltage instability problems on a large-scale have been developed and successfully evaluated. The expert system for clearing overloads applies the network sensitivity factors to determine appropriate actions, which include generation rescheduling, network reconfiguration and load shedding. The expert system for voltage control is used for detecting voltage violations and providing recommendations for solving voltage problems in real-time.

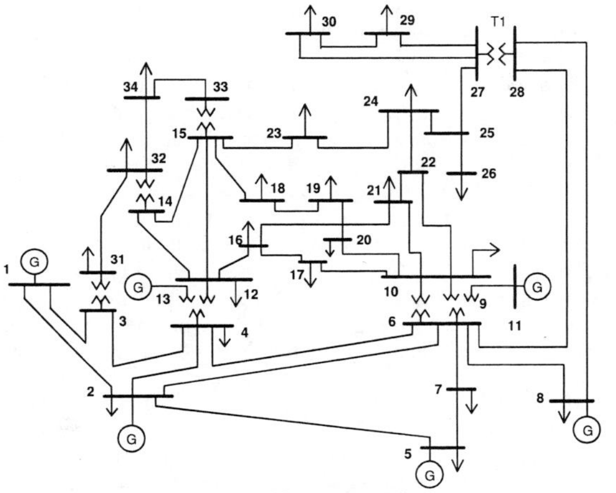

Figure 2. The modified IEEE 30-bus system.

Table 1. Bus voltages and optimal control settings.

Remarks	Full system		Equivalent		Remarks	Full system		Equivalent	
	Initial	Final	Initial	Final		Initial	Final	Initial	Final
Load voltages					Load voltages				
V3	1.026	1.014	1.026	1.014	V29	*0.763	0.955	*0.762	0.954
V4	1.020	1.003	1.020	1.003	V30	*0.747	0.950	*0.746	0.950
V6	1.016	0.999	1.016	0.999	V31	0.975	1.027	0.975	1.027
V7	1.005	0.995	1.005	0.995	V32	0.975	1.027	0.975	1.027
V9	0.999	1.059	0.999	1.059	V33	0.982	0.960	0.982	0.960
V10	0.955	1.041	0.955	1.041	V34	0.962	0.978		
V12	0.972	1.051	0.972	1.051					
V14	0.950	1.023	*0.950	1.020	Gen. setting				
V15	*0.943	1.032	*0.943	1.031	V1	1.050	1.060	1.050	1.060
V16	0.957	1.040			V2	1.045	1.045	1.045	1.045
V17	0.950	1.035			V5	1.010	1.010	1.010	1.010
V18	*0.934	1.023			V8	1.025	1.025	1.025	1.025
V19	*0.932	1.021			V11	1.050	1.050	1.050	1.050
V20	*0.937	1.025			V13	1.050	1.050	1.050	1.050
V21	*0.934	1.026			Trans. setting				
V22	*0.933	1.025			N:3 - 31	1.069	1.000	1.069	1.000
V23	*0.915	1.013			N:4 - 12	1.032	0.901	1.032	0.901
V24	*0.888	0.998	*0.887	0.998	N:6 - 9	0.978	0.903	0.978	0.903
V25	*0.818	0.974	*0.818	0.973	N:6 - 10	1.069	0.900	1.069	0.900
V26	*0.796	0.955	*0.795	0.955	N:14 - 32	0.950	0.950	0.950	0.950
V27	*0.789	0.971	*0.788	0.970	N:15 - 33	0.950	1.094	0.950	1.094
V28	1.018	1.001	1.018	1.001	N:27 - 28	--- outage ---		--- outage ---	

Note: (*) bus with voltage constraints violated.

Task 1: Calculate the sensitivity factors

Task 2: Detect voltage violations

* number of voltage violations: 13
* the lowest voltage occurs at bus: 30 { V(30) = 0.747 p.u. }

Task 3: Correct voltage violations

Action 1: select controller: transformer 6 - 9
adjust its setting: 0.978 ⟶ 0.903 { V(30) = 0.769 p.u. }
Action 2: select controller: transformer 6 - 10
adjust its setting: 1.069 ⟶ 0.900 { V(30) = 0.810 p.u. }
Action 3: select controller: transformer 4 - 12
adjust its setting: 1.032 ⟶ 0.901 { V(30) = 0.839 p.u. }
Action 4: select controller: generator 1
adjust its setting: 1.050 ⟶ 1.060 { V(30) = 0.842 p.u. }
Action 5: select controller: transformer 3 - 31
adjust its setting: 1.069 ⟶ 1.000 { V(30) = 0.846 p.u. }
Action 6: select controller: transformer 15 - 33
adjust its setting: 0.950 ⟶ 1.094 { V(30) = 0.850 p.u. }

** This voltage violation cannot be corrected by the available controllers
** VAR compensation is required

Figure 3. Voltage control action screen.

REFERENCES

[1] M. Bruch, V. Münch, M. Aichinger, M. Kuhn, M.Weymann, G. Schmid. *Power Blackout Risks: Risk Management Options – Emerging Risk Initiative.* Position Paper. CRO Forum, November 2011.

[2] B. M. Weedy, B. J. Cory, N. Jenkins, J. B. Ekanayake, G. Strbac. *Electric Power Systems*, 5th edition, Wiley, 2012.

[3] V. T. Morgan, (1991) *Thermal Behavior of Electrical Conductors*, Research Studies Press, John Wiley, 1991.

[4] F.Vakili and M.R. Viles, "Dynamic thermal line loading monitor", *IEEE Trans. Power Systems*, Vol. PWRS-1, No.2, May 1986.

[5] T. L. Le, M. Negnevitsky and M. Piekutowski, "Expert System Application for the Loading Capability Assessment of Transmission Lines", *IEEE Transactions on Power Systems*, Vol. 10, No. 4, November 1995, pp. 1805 - 1812.

[6] M. Negnevitsky, "An Expert System Application for Clearing Overloads", *International Journal of Power and Energy Systems*, Vol. 15, No. 1, 1995, pp. 9-13.

[7] B. Stott, "Decoupled Newton Load Flows", *IEEE Trans. on Power Apparatus and Systems*, Vol. PAS-91, September/October 1972, pp. 1955-1957.

[8] IEEE PES PSDP Task Force on Blackout experience, mitigation, and role of new technologies, blackout experiences and lessons, Best practices for system dynamic performance, and the role of new technologies, *IEEE Special Publication* 07TP190, July 2007.

[9] IEEE PES CAMS Task Force on Understanding, Prediction, Mitigation and Restoration of Cascading Failures "Initial Review Of Methods For Cascading Failure Analysis In Electric Power Transmission Systems," *In Proc. IEEE PES General Meeting*, Pittsburgh, PA USA July 2008 1.

[10] E. C. Housos, et. al. "Steady state network equivalents for power system planning applications", *IEEE Trans. on Power Systems*, 1980, PAS-99, (6).

Russian Case Study: the Influence of Severe Weather Conditions on Reliability of Overhead Line Infrastructure

Aleksandr Nazarychev, Aleksei Tadzhibaev, Yury Solovyev
Petersburg power engineering institute of professional development (PEIPK)
Saint-Petersburg, Russia
nazarythev@mail.ru, rector@peipk.spb.ru, yury.solovyev@yahoo.com

Abstract – **Due to the opening of new fields and construction of responsible interstate long-distance systems for transportation of oil and gas in difficult climatic conditions it is required to provide reliable power supply of field equipment. Therefore, it is necessary to provide the reliable operability of overhead lines (OHL) as key installations in electricity transmission and distribution systems. The challenge is significant due to severe climatic conditions which are very different and territory dependent in Russia. Therefore, the geographical factor has importance and influence on OHL reliability under severe ice and wind loads. Generally the geographical factor has an impact on the value of ice deposits depth and includes a number of parameters: absolute and relative heights of OHL route passing through territory of a region, land relief, arrangement of highland slopes related to preferred wind direction, etc.**

Considering the simultaneous influence of each parameter on OHL infrastructure it is difficult to assess its service condition. Moreover, the task is not solved up to now. In the paper the assessment of ice and wind load intensity depending on the mentioned-above geographical factor is proposed. Obtained results showed the significant influence of relative height of OHL route on the type and repeatability of ice deposits on OHL conductors and wires. It was found that territories with the most frequent icing repeatability are located at relative heights of more than 200 m above the ground level. Frequent repeatability of icing along OHL routes passing mountain tops relates to high density of icing flux, changing density of precipitation nearby bottom of highlands, high value of moisture around highland tops, movement of cloud, low air temperature on height levels.

Icing repeatability depends not only on absolute and relative height, but type of land relief changing along OHL route. Land relief is defined by its form, land surface slope and degree of shielding from icing flux. The highest icing repeatability is observed on well-blown territory, in particular, on high plateaus and highland tops, windward highland and mountain slopes, etc.

Ice deposits on OHL conductors and wires can be characterized as solid ice layer formed at a temperature from 0 to minus 3°C. The main reason of ice deposits is a freezing of cooled water drops on conductor surface. Ice deposits become more difficult, but less dense with reduction of drop volume and temperature. Contrariwise, glassy and transparent ice deposits observed at ice rain are formed at zero temperature.

Ice deposits on OHL conductors and wires increase due to wind loads. Vibration of OHL conductors and wires is due to unilateral ice deposits formed at wind speed of 10 m/s. Icing can arise both in homogeneous air mass, and in result of interaction of cold and warm air masses along frontal surfaces. According to the nature icing is classified on air-mass and frontal. Icing seldom arises in homogeneous air mass. Frontal icing formation is due to warm fronts following by strong winds. It has the average intensity of 100 g/h per 1 m of conductor length. Its maximum value is about 300 g/hour per 1 m of conductor length. The average intensity of air-mass icing does not exceed 5 g/h per 1 m of conductor length; its maximum intensity is 22 g/h per 1 m of conductor length. Therefore, the frontal icing determines the maximum depth of ice deposits on OHL conductors and wires. Cold air fronts favor to freezing of wet snow deposits with the subsequent icing on conductor surface. The most intensive formation of ice deposits is due to the air frontal sections arising at the exit of the southern cyclones to the European part of Russia after initiation of anti-cyclone with low air temperature. Strong anti-cyclone creates suitable conditions for increasing of wind strength and icing growth intensity. In Russia nominal rating values of wind and ice loads on OHL conductors and wires have been classified on the basis of special classification maps forming climatic zonation inside Russia's boarders (fig. 1, fig. 2).

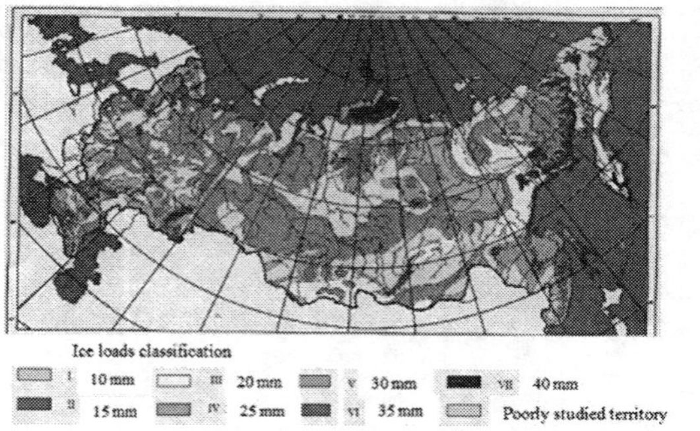

Figure 1. Zone classification according to the value of ice deposits depth, mm

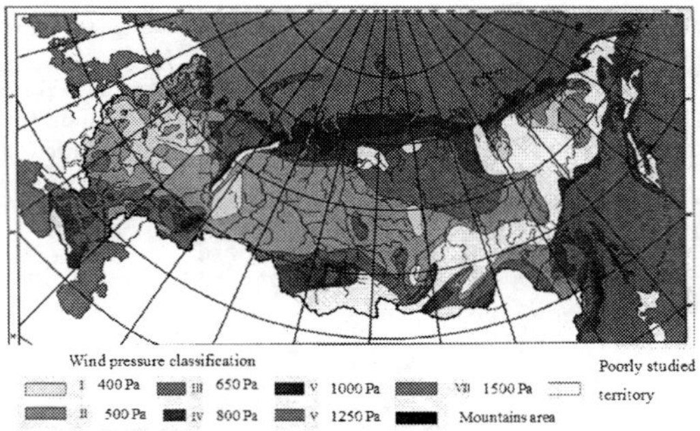

Figure 2. Zone classification according to the value of wind pressure, Pa

Every year power supply outages take place due to severe wind and ice loads observed in Rostov, Krasnodar and Stavropol, Sochi, North Caucasus regions. Heavy ice deposits on OHL conductors and wires are formed due to southern cyclones passing over the Black and Azov seas. These cyclones interact with anti-cyclones passing over east and southeast of the European part of Russia. Icing can be a result of cyclones going from Scandinavian Peninsula: air with high pressure above the east and southeast territories of the European part of Russia helps to passing the cyclones from Scandinavia. Warm air flux interacting with coming cold anti-cyclones leads to high humidity values with consequent precipitation. It is caused by the existence of highlands and land relief features in the mentioned above regions. Besides the existence of rivers and lakes closed to OHL routes will provides heavy ice deposits on OHL conductors.

In particular, Rostov region represents the territory where absolute height changes from 150 to 260 m due to existence of mountains belonging to Donetsk chain of hills. One of the reasons leading to heavy icing on OHL conductors is increasing wind speed with upward absolute heights. In fig. 3 recorded data on ice deposits depth versus height of OHL passing route is given.

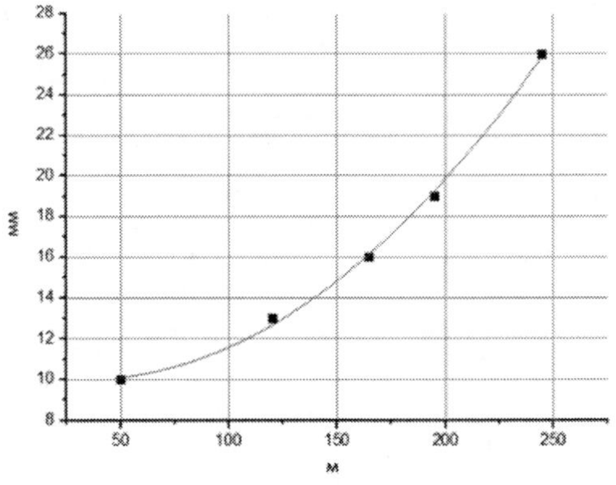

Figure 3. Depth of ice deposits on OHL conductors versus height of passing route

It is necessary to highlight the complex of weather conditions affecting the frequency of ice deposits on OHL conductors: changing air temperature; increasing repeatability and intensity of precipitation events; lack of OHL wind screening due to the absence of trees and buildings; land relief features.

According to fig. 2 high wind loads usually exist in Yamalo-Nenets Autonomous Area where a number of distribution overhead lines are under construction now. This region is characterized by subarctic climate with inherent weak heat supply infrastructure, long and heavy winter, cold summer, permafrost. Close proximity to Barents and Karsky seas, existence of numerous rivers, lakes and bogs lead to high humidity of climate mainly forming under influence of arctic air masses. Geographical feature is also important factor: northern part of the region is strongly extended from the west to the east. Increased frequency of air masses changing is the reason of variability of weather conditions. During winter and autumn seasons prevailing winds are coming from the south; during summer season – from north and north-east due to outbreak of cold arctic air masses on preheated inland with decreased air pressure. In summer air temperature depends on solar radiation.

Near Naryan-Mar Region the average temperature of July is +11°C. During the cold period the main factor which responsible for temperature condition is a heat transfer from Atlantic. Therefore, temperature decreasing from the west to the east is distinctly observed. The average temperature of January is –18°C. Duration of the winter period is 230 days. The most part of the territory is located in the zone with high humidity. The annual amount of precipitation is about 500 mm. Besides, the minimum of precipitation is observed in February, its maximum is in August and September. Not less than 50% of precipitation has a form of snow. Weather condition in the region can be accepted on the basis of data taken from measurement in weather station "Varandey" (tab. 1).

Table 1. General weather conditions in Naryan-Mar region

Climatic parameters	**Value**
Climatic parameters during cold seasons	
Air temperature of the coldest day, °C, – reliability index 0.98 – reliability index 0.92	 –41 –40
Air temperature of the coldest five-day stretch, °C, – reliability index 0.98 – reliability index 0.92	 –38 –37
Air temperature, °C, with reliability index 0.94	–24
Ambient air absolute minimum temperature, °C	–45
Daily average amplitude of air temperature of the coldest month during a year, °C	8.9
Average monthly relative air humidity of the coldest month during a year, %	87
Precipitation amount during November-March period, mm	128
Prevailing wind direction during December – February period	south-west
Average wind speed during a period with average daily temperature less than 8°C, m/s	6.5
Climatic parameters during warm seasons	
Air temperature with reliability index 0.95, °C	11.3
Air temperature with reliability index 0.98, °C	15.7
Average maximum air temperature of the warmest month during a year, °C	13.2
Absolute maximum air temperature, °C	31
Average monthly relative air humidity of the warmest month during a year, %	87
Precipitation amount during April – October, mm	278
Daily maximum of rainfall amount, mm	47
Prevailing wind direction during June-August	north-east
Average monthly and annual average air temperature	
Annual average air temperature, °C	–5.7
Average monthly air temperature in July, °C	8.7
Average monthly air temperature in February, °C	–19

Intensity of solar radiation within a year is non-uniform due to the existence of polar day and night. Specific high cloud amount above the territory decreases influence of direct radiation, but increases scattered radiation. Total annual solar radiation equals to 69–80 kcal/cm^2. Air temperature changing from zero values to the positive values is observed in the middle of May. During the autumn period air temperature changing from zero to negative values is observed in the end of September – in the beginning of October. Geographical factor has the significant influence on air temperature distribution within a year: in winter air temperature on coasts of seas is higher in comparison with long-distance territories. Inversely, in summer its value is lower. Average monthly and annual average air temperature distribution according to data taken from weather station "Varandey" is presented in tab. 2.

Table 2. Average monthly and annual average air temperature in Naryan-Mar region

Month	Average monthly air temperature, °C	Annual average air temperature, °C
January	–17.7	–5.7
February	–19.3	
March	–15.9	
April	–10.2	
May	–3.4	
June	3.3	
July	9.2	
August	8.7	
September	4.7	
October	–2.3	
November	–10.1	
December	–14.5	

Intensity of precipitation amount is defined by vigorous cyclonic activity. Maximum precipitation amount is observed when cyclones move from the northwest (from Scandinavia and Barents Sea) and the north (from Arctic, Kara Sea). Northwest cyclones are specified by high intensity and depth. Arctic air masses are colder and drier. Average annual amount of precipitation equals to 430 mm. During the cold period from November to March the average annual amount of precipitation is about 35%, during the warm period it is about 65% of annual values. Minimum of annual amount of precipitation is observed in April. Minimum monthly amount of precipitation (7.1 mm) was observed in February, 1998. Maximum amount of precipitation is observed in September. Maximum of monthly amount of precipitation was observed in 1998, and it was equaled to 81.3 mm. About 45% of precipitation value has the form of snow, 40% – the form of rain, other 15% – mixed precipitation the form of wet snow, frozen rain, etc. Maximum duration of precipitation is observed on the coast in summer. Overall duration of precipitation during a year equals to 1600 hours. Average monthly and annual average precipitation distribution according to data taken from weather station "Varandey" is presented in tab. 3.

Table 3. Average monthly and annual average amount of precipitation in Naryan-Mar region

Month	Average monthly amount of precipitation, mm	Annual average amount of precipitation, mm
January	35	422
February	26	
March	23	
April	20	
May	27	
June	38	
July	40	
August	52	
September	59	
October	45	
November	29	
December	28	

Maximum value of relative air humidity in warm seasons equals to 85–94%. In cold seasons relative air humidity equals to 85–88% due to low temperatures. During a year deep cloud dominates on the territory of the region. Blizzards in winter season (November – March) are observed during 23–28 days per month.

On coast high intensity of precipitation in the form of fogs takes place in a warm season particularly. Maximum of daily fog variation is observed during winter period mainly in the afternoon, during summer period – in the morning. Average fog duration is about 7 hours. Repeatability of fogs is higher during winter period. The average annual number of days with fog on the coast equals to 80, but inland it equals to 50. Thunderstorms are rather rare for the region. During the period from May to September the average number of days with thunderstorm activity is 5. Average number of days and maximum number of days per year with different forms of precipitation in the region is shown in tab. 4.

Table 4. Average and maximum number of days per year with different forms of precipitation

Number of days	Fog	Thunderstorm	Blizzard	Hail
Average	68	5	95	1
Maximum	101	15	130	2

Snow is observed mainly during a period from the end of September to the beginning of October. At this time temperature is below zero. Snow cover appears within 2 months. Steady snow cover is formed in the second half of October and in the beginning of November. Maximum snow cover is observed in the second half of March and in the beginning of April. Depth of snow cover in April equals to 65 –70 cm. Due to wind changing the average height of snow cover reaches is 1.5 m at topographic low because of land relief. On heights of land relief snow cover depth equals to 0.2 –0.3 m. Winter thaws with strong winds lead to snow cover compaction, formation of a crust, snow and ice crusts on OHL conductors and poles. In the region icing is mostly happened in November and December. Experience showed that ice crust can remain till spring due to fast decrease of temperature (from 120 to 140 days). Snow descends only by third decade of May and in beginning of June. Snow melting is slow due to lowered land relief. Average number of days with snow cover is 238. Average density of snow cover is 0.25 g/cm^3. Wind intensity in Nenets Autonomous Area is defined by intensity of cyclonic activity in different seasons. Particularly, during winter period wind intensity is defined by interaction of Icelandic cyclone and Siberian anti-cyclone. There is a deep pressure trough over the territory which forms a wind direction. Mainly winds have southern and southwest directions. On coast near the western part of the region the average monthly wind speed is about 10 m/s, in the eastern part – 6–7 m/s. In winter the average duration of strong winds equals to 8–10 h. The maximum duration of wind blowing is about 40–60 h. Preferred wind directions are south-west and western. Average wind speed is about 15 m/s. Its repeatability equals to 40 times. Wind repeatability decreases with distance from the coast of Barents Sea. In the region during winter winds are active till May. In spring and summer wind force weakens. On the most part of the territory repeatability of winds with a speed of 5 m/s is reduced to 2.5 %. During summer wind direction changes and takes east, southeast (in the west) and northeast (in the east of the region) directions. Wind speed decreases and equals to 5–6 m/s. Intensity of winds having speed more than 15 m/s increases in autumn. There is a frequent repeatability of hazard weather conditions (low temperatures and strong winds). Average duration of windy weather with low temperatures is 146 days per year. Values of maximum wind speeds within a year taken from weather station "Varandey" are given in tab. 5.

Table 5. Maximum wind speed values within a year

Month	**Maximum wind speed, m/s**
January	31
February	35
March	29
April	25
May	25
June	25
July	26
August	35
September	25
October	25
November	29
December	35

From October to March preferred wind directions are southern and south-west. In January repeatability of preferred wind directions is about 26–33% (fig. 4). Accounting of wind direction helps to calculate ice depth on a conductor in case of icing flux directed at the angle to OHL conductor. Particularly, it is suggested to use a coefficient K_φ equals to ratio of the maximum depth of ice deposits on OHL conductor to icing weight:

$$\mu_{90} = \mu_\varphi \cdot K_\varphi \quad (1)$$

where m_φ is the weight of ice deposits on OHL conductor which is located at the angle φ to wind direction; m_{90} is the weight of ice deposits on OHL conductor, located perpendicularly to wind direction. The values of K_φ are given in tab. 6.

Table 6. Value of coefficient K_φ for different wind directions

Wind direction when icing	K_φ
North-East, South-East, North-West, South-West	1.82
East-North-East, North-North-East, East-South-East, South-South-East, South-South-West, West-South-West, West-North-West, North-North-West	1.33

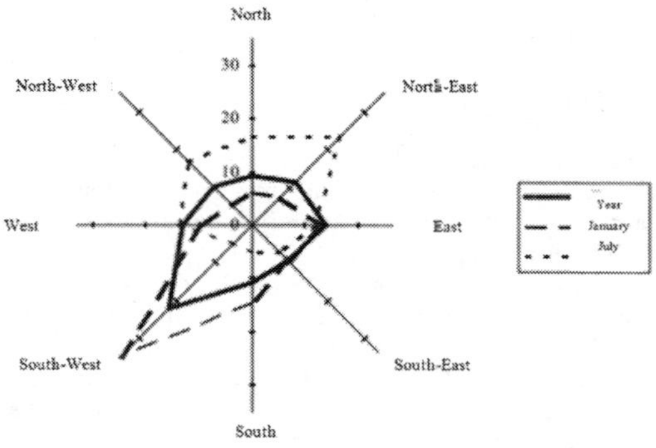

Figure 4. Distribution of wind directions within a year

Summarize, geographical and meteorological factors with land relief features significantly affect local wind and ice loads. At the same time there is a feedback when land relief features define local climatic conditions along OHL route due to changing of air circulation (air turbulence). It is offered to divide OHL route into elementary parts with a land relief of various horizontal and vertical length (macro-, meso-, microrelief) that is shown in fig. 5. For definition of morphometric characteristics of various types of land relief the following parameters are proposed:

h – absolute height above sea level, m;

Δh – relative height of territory is the exceeding of the highest point of relief headland over the average level of the next flat relief part, m;

l – the shortest distance from the highest point to a flat relief part, km;

α – the steepness of slopes defined by a ratio:

$$si^n(\alpha) = \frac{\Delta h}{l} \quad (2)$$

For every type of land relief the parameters are given in tab. 7.

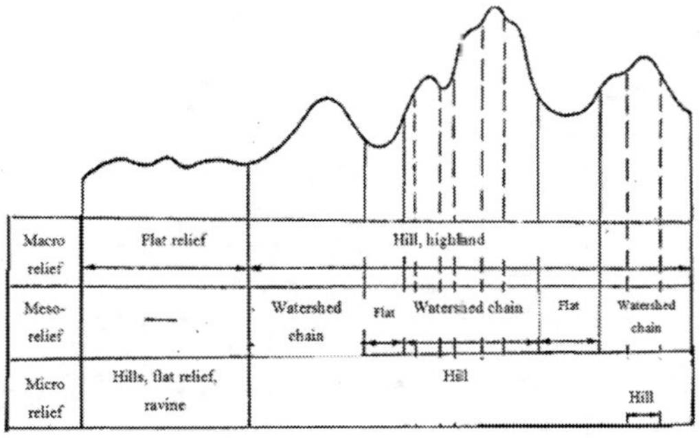

Figure 5. Special features of macro - meso - and microrelief

Table 7. General morphometric characteristics of various types of land relief

Type of land relief	***h*, m**	**Δ*h*, m**	***l*, m**	**α, degree**
Macrorelief	0 – 755	0 – 755	> 55	< 1
Mesorelief	0 – 755	0 – 755	4 – 55	2 – 15
Microrelief	0 – 755	< 82	< 4	> 4

For possibility to detail a land relief every type can be divided into subtype with definition of specific relief elements (tops, slopes, valleys, etc.) position in relation to preferred wind direction (tab. 8). Definition of land relief types together with subtypes is made on the hypsometric form with scale of 1:500 000. Macrorelief types are determined by the size of relative heights (Δh): flats and highlands. Classification of macrorelief subtypes on "leeward, windward and parallel to wind slopes" is prepared on the basis of making the schemes of slopes position against preferred wind directions. For definition of slope position related to direction of climatic factor action there are eight expositions: northern, north-west, western, south-west, southern, south-east, east, north-east (fig. 6). On the basis of given types of land relief the schematic map of types and subtypes of macrorelief and microrelief for each climatic factor is constructed: wind speed, depth of ice deposits, wind load.

During map construction it is required to consider possible shielding of separate tops and slopes by other nearby highlands which are settling down from the preferred wind direction. For assessment of shielding level the corresponding parameter can be offered. It is equal to the ratio of excess of slopes of one highland (mountain) in relation to another (Δh) to distance between the considered tops (S) can be offered.

Table 8. Classification of macro- and mesorelief considering flat and low mountain territories

Type of relief	Characteristics	Relief subtype	Character of displacement against preferred wind direction
Macrorelief	A. Flat, $h = 0 - 750$, Δh less than 100 m	–	Open
	B. Highland, h up to 750 m, Δh not less than 100 m	Top	Open
		Leeward slope	Open
		Slope parallel to wind	Open
		Windward slope	Shielded
		Opened flat	Open
		Closed flat	Shielded
Mesorelief	Watershed chain, h up to 750 m, Δh more than 100 m	Top	Open
		Windward slope	Open
		Slope parallel to wind	Open
		Opened flat	Open
		Leeward slope	Shielded
		Closed flat	Shielded

A slope can be considered as shielded, if Δh ≥ S/20. Microrelief of the territory which is crossed by OHL route is determined according to geographical position of the OHL route and topographic maps.

It will define values of climatic (weather) factors affecting OHL.

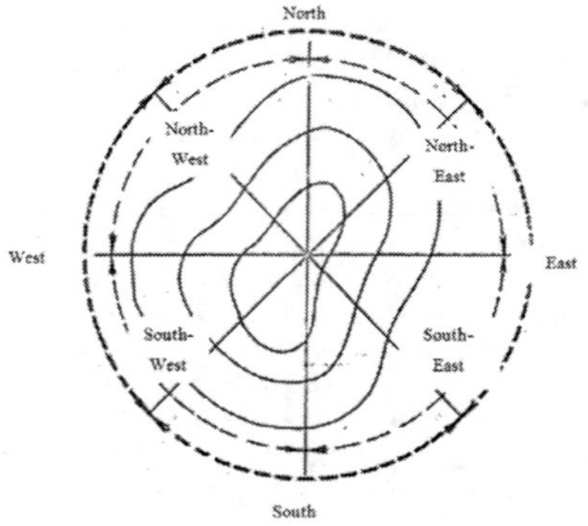

Figure 6. Scheme of slope positions

Therefore, it is offered to use the specific coefficient to convert data taken from measurements on weather stations installed at open sites to real conditions with aspect to considered features of land relief:

$$\Pi = \Pi_0 \cdot K, \quad (2)$$

where Π is a value of climatic factor with aspect to land relief; Π_0 is a value of climatic factor for open sites; K is a coefficient taking into account the influence of land relief on a value of climatic factor in comparison with open site. The values of coefficient K is defined for every climatic factor separately (tab. 9).

Table 9. Coefficient K for wind speed calculation with aspect to types of land relief

Types of land reliefs with its characteristic parameters		K
Tops	$\Delta h \geq 50$ m	1.35
	$\Delta h < 50$ m	1.25
Windward slopes	Upper side	1.3
	Middle side	1.1
	Bottom side	0.95
Slopes parallel to wind	Upper side	1.15
	Middle side	0.9
	Bottom side	0.8
Leeward slopes	Upper side	0.75
	Middle side	0.9
	Bottom side	0.8
Valley bottom	With wind blow	1.35
	Without wind blow	0.7
Highland with flat tops and mild slopes	Tops, upper side of leeward and windward slopes	1.3
	Middle and bottom sides of leeward and windward slopes	0.9–1.2

The values of coefficient K taking into account the influence of land relief on a value of ice deposits depth are given in tab. 10.

Table 10. Coefficient *K* for ice deposits depth calculation with aspect to types of land relief

Type of land relief	Ice deposits zone classification (open sites)	K							
		Δh, m							
		0	10	20	30	40	50	60	70
Flat		1,0	–	–	–	–	–	–	–
Windward or leeward slope	I – III	1,0	1,05	1,10	1,12	1,15	1,2	1,25	1,30
	IV – special	1,0	1,05	1,10	1,15	1,20	1,25	1,30	1,35
Shielded flat	I – II	0,9–1,0	–	–	–	–	–	–	–
	III – special	1,1	–	–	–	–	–	–	–

The values of coefficient K taking into account the influence of land relief on a value of wind loads are given in tab. 11.

Table 11. Coefficient *K* for wind loads calculation with aspect to types of land relief

Type of land relief	Δh, m							
	0	10	20	30	40	50	60	70
Windward slope	1,0	1,16	1,32	1,46	1,62	1,74	1,86	1,98

CONCLUSION

Summarize, in case of severe climatic conditions it is necessary to divide OHL route into local elementary lengths where increasing of loads is possible due to influence of land relief. These parts of OHL require special attention and supervision during service operations on OHL. It needs to eliminate the defects which are fast growing in OHL conductors and wires under severe ice and wind loads.

[1] N.I. Voropay, G.F. Kovalyov, Y.N. – M. Kucherov. *The concept of reliability in the electricity industry.* OOO ID "ENERGY", 2013, p. 304.

[2] A.N. Nazarychev. *Methods and repair optimization model of electrical energy, taking into account the technical condition.* Ivan. gos. energ. un-t. Ivanovo, 2002, p. 168.

[3] A.N. Nazarychev, D.A. Andreev. *Methods and mathematical models of complex evaluation of technical condition of electrical equipment.* Ivan. gos. energ. un-t. Ivanovo, 2005, p. 224.

[4] A.N. Nazarychev, D.A. Andreev. *Methodical bases of determining deadlines for operation and sequence of technical re-equipment of power.* Ivan. gos. un-t. Ivanovo, 2005, p. 168

[5] A.N. Nazarychev, A. Tadzhibaev, Y. Solovyev. "The Inspection of the Technical State of Critical Infrastructure Facilities," *International Conference on Problems of Critical Infrastructures, Proceedings*. Edited by Z.A. Styczynski and N.I.Voropai. Irkutsk, Energy Systems Institute, pp.144 – 151, 2015.

[6] A. Nazarychev, M. Huhtinen. *Learning from Differences in Energy Sector.* Proc. International Seminar, Saint-Petersburg, Kotka: PEIPK, KYAMK, 2016. p. 108.

[7] Y. Solovyev, A. Nazarythev, A. Tadzhibaev. "Details on technical condition assessment of overhead covered conductors under extreme climatic impacts," Proc. Electric Power Quality and Supply Reliability Conference (PQ 2014), Rakvere, Estonia, 11-13 June 2014, IEEE Catalog Number CFP1452E-POD, ISBN 978-1-4799-5023-2, pp. 379 – 383.

[8] I.N. Sulynenkov, A.N. Nazarychev, A.I. Tadzhibayev. "Reliability Model of High-Voltage Circuit Breakers With Taking Into Account Breaker Failure Protection," *Scientific – Industrial Journal Electroenergetics, Electrotechnics, Electromechanics + Control*, Vol. 4, No. 4, pp. 46 – 55, Baki 2013.

[9] Iu.V. Soloviev, A.I Tadzhibaev, A.N. Nazarythev. *Analyz technicheskogo sostoyaniya zaschischernych provodovм liniyм electroperedachiъм.* Saint Petersburg. Izd-voм PEIPK, 2015, p. 193.

[10] A.N. Nazarychev, A.I. Tadjibaev, V.V. Titkov, F.KH. Khalilov. *Basic technical condition of electrical equipment: Stud. Manual.* Spb. : Izd-vo Polytechnic. un-t., 2015, p. 204.

A: Generation of Electricity, Transmission, Distribution and Consumption of Electricity

Power Quality in Industrial Isolated Generation Power Systems with Powerful Nonlinear Consumers

Ivan I. Artyukhov, Irina I. Bochkareva, Svetlana V. Molot, Svetlana G. Kalganova, Sergey S. Stepanov, Sergey V. Trigorly
Department of Electric Power Engeneering
Yuri Gagarin State Technical University of Saratov
Saratov, Russian Federation
ivart54@mail.ru; irina.bochkareva1986@icloud.com; sve_1992@mail.ru; sgkalganova@sstu.ru; stepanov460915@yandex.ru; trigorly55@mail.ru

Abstract — **Massive retrofitting of all sorts of industrial units with power electronics currently progresses. However, nonlinear pulsing characteristics of electronic switches result in power quality deterioration in the supply mains. Retrofitting often occurs without ensuring electromagnetic compatibility (EMC).**

The problem of EMC ensuring becomes crucial when a facility or a separate technological unit is powered by an isolated generation power plant with the wattage capacity commensurable with the load. A typical example is the use of variable-frequency driven gas cooling units in the oil and gas industries. Field experience revealed necessity of EMC ensuring for these units and the power supply. The reason for this is that a variable-frequency drive conserves electricity concomitantly with degrading the power quality. Simultaneous operation of a certain number of variable-frequency driven gas cooling units leads to a significant voltage waveform distortion in the supply mains.

Similar problems occur in the industrial power systems, consisting of such powerful nonlinear loads as arc furnaces, welding units, electric heaters with thyristor power regulation. Alongside with these consumers, auxiliary variable-frequency drives for transportation and ventilation may be powered from the same system as well.

The power quality indices analysis for an industrial power system with a variable-frequency driven gas cooler powered from a gas-turbine power plant was performed with a simulation model built in the MATLAB+Simulink environment.

Power feeders of the actual industrial units such as set temperature electric furnaces and variable-frequency drives were examined for conformity of the power quality indices and the current standard corresponding values.

Engineering solutions on maintaining power quality in industrial isolated generation power systems with powerful nonlinear consumers were recommended.

Keywords: power supply system; power quality; electromagnetic compatibility; variable-frequency drive; frequency converter; gas cooling plant; gas-turbine generator

I. Introduction

Intensive implementation of power electronics devices in different technological processes leads to necessity of some measures to ensure the required power quality in a power supply system (PSS). The consumption of non-sinusoidal current from the power source leads to the fact that the voltage drops of non-sinusoidal shape occur on the consistent elements of the PSS. As a result, the voltage at the PSS also becomes non-sinusoidal [1, 2].

Non-sinusoidal voltage and current cause additional losses and heating, as well as accelerated aging of the electrical equipment. This fact proved by the results of diagnostic tests on variety of power equipment: transformers, cable lines, motors, compensating capacitors, converters, and generators of PSS [3]. In addition, voltage and current distortion causes operational disturbances and nuisance tripping of relay protection and automation, leading to failures in electronic control systems and computing machinery, create noise in the telemetry and communication, distort the readings of electric energy meters.

Under certain combinations of PSS elements parameters and nonlinear loads, the increasing of voltage curves distortion, caused by resonance phenomena, is possible. It leads to malfunction of the electrical equipment.

II. Problems with Power Quality in Power Supply Systems of Gas Cooling Units Variable Frequency Drives

At the enterprises of gas industry works are ongoing to equip the gas cooling units (GCU) by variable-frequency drive (VFD) of the fan [4, 5]. The experience of conducting these works showed the need to take measures of providing EMC of such technological complexes with power sources. Along with the energy savings, usage of VFD leads to deviations of the power quality indices characterizing the voltage shape in the network. Significant waveform distortion of the supply voltage registered when a certain number of GCU with VFD of fans are simultaneously working. The problem gets worse if a power supply source for a gas transport plant is replaced with an autonomous one, for example, gas-turbine power station [6, 7].

The composition of the one GCU includes from 24 to 32 motors of 37 kW each, therefore the total power consumed by the unit can exceed 1 MW. In accordance with the requirements of regulatory documents related to power supply category of GCU scheme of the motor control needs to be released on two symmetrical sections, each is connected to the secondary winding of a corresponding step-down transformer 10/0,4 kV of a complete transformer substation (CTS). Thus, both sections must be able to obtain electricity from one of the transformers of CTS.

Fig. 1 shows a single line diagram of power supply system for several sections of gas cooling units. Each of the shown transformers supplies power to a group of cooling fan VFDs.

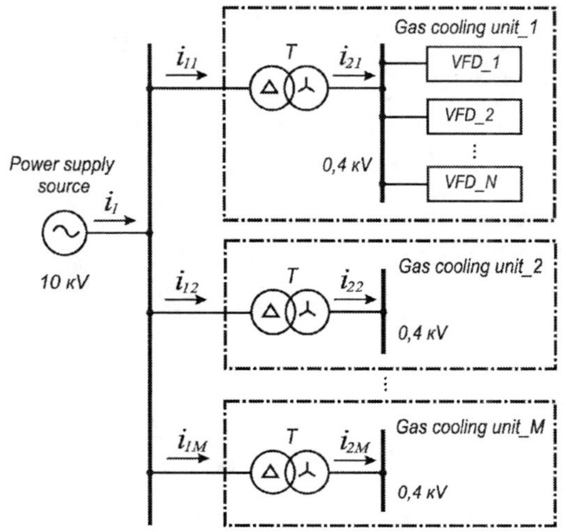

Figure 1. Single line diagram of gas cooling units power supply system

Modern frequency converters for VFD built mainly according to the scheme: uncontrolled rectifier – smoothing filter - autonomous voltage inverter with IGBT components (Fig. 2). Therefore, the harmonic current composition induced by inverter in supply mains with the particular parameter set depends on the configuration and parameters of the smoothing filter at the output of the rectifier, phase of rectifier and control algorithm [8, 9]. Since rectifiers are mainly made in three phase bridge, the input current of frequency converter is representing a sum of odd harmonics except multiples of three. In this case the most significant are 5th and 7th harmonics.

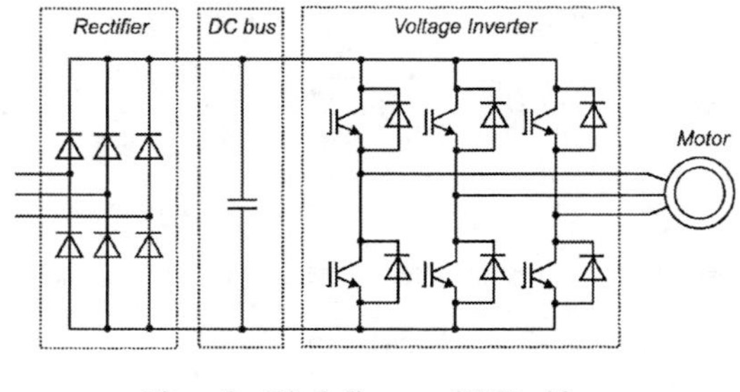

Figure 2. Block diagram of VFD of fans

Converters are mainly produced with a capacitor filter within the DC-link. Thus, value of harmonic current in the overall converter input current is considerable in comparison with the fundamental current. This leads to voltage distortion at all levels of a gas compressor station PSS.

Fig. 3 shows voltage waveform at the 0,4 kV bus and current waveform of a CTS feeding VFDs shown at fig.1. The voltage and current diagrams were experimentally taken with digital memory oscilloscope Fluke 196C with two galvanic isolated channels. Transformer with the nominal power of 630 MVA was supplying power to 12 fully operational converters Altivar 58. The distortion of the input current feeding these converters was 40,13 %. Meanwhile the 5th harmonic exceeded 37,4 % of the fundamental current. Voltage total harmonic distortion (voltage THD) was 6,36 %.

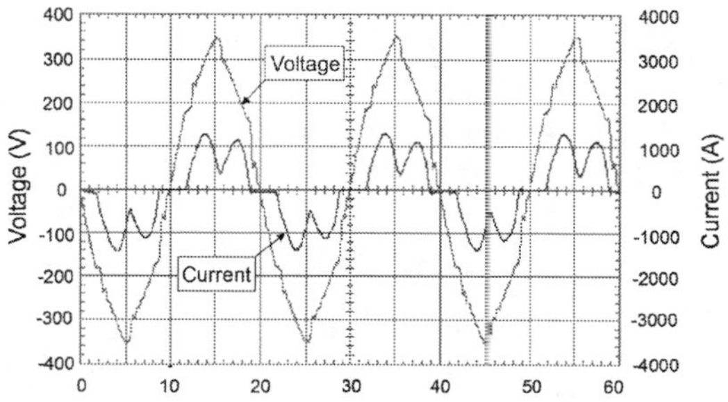

Figure 3. Voltage and current oscillograms of the secondary winding of the transformer when 12 VFD operate

A model of a GCU with VFD powered by power source was developed in MATLAB/Simulink environment to investigate EMC (Fig. 4). The model performs the case where three GCU connecting to the PSS. Power elements of the technological complex are represented by the Three-Phase Source 10 kV, Three-Phase Transformer and Variable Frequency Drive. Current&Voltage Analizer estimates the power quality indices [10].

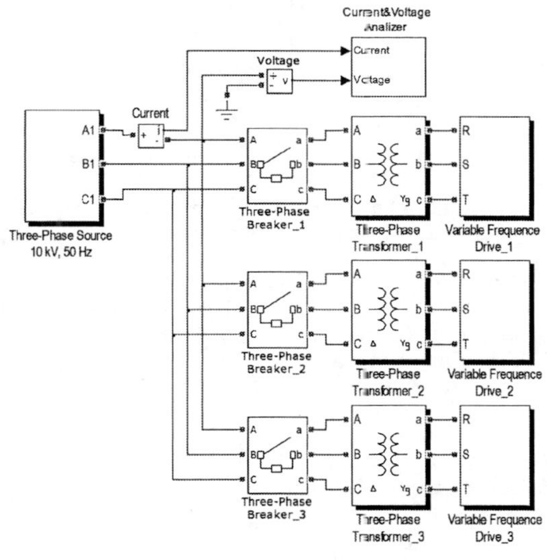

Figure 4. Simulation model of GCU with VFD power supply system

The model contains such units as Universal Bridge – three-phase rectifier; RS – line reactors; LD, RD, CF – parameters of inductive capacitive filter; Rin – inverter equivalent resistance (Fig. 5). Method [10] was used to develop the model. The method affords to supersede a voltage inverter and attached motors with the equivalent resistance. It could be done because in the case under consideration the purpose of modeling is to analyze the influence of canonical harmonics in the spectrum of the frequency converter input current to the power quality in the mains. According to the relevant standard for power quality in analysis of the power quality indices, harmonics with a frequency of up to 2 kHz should be taken into account. Mainly IGBT modules are used in modern voltage inverters as power semiconductor switches. Clock rate of the module is at least 8 kHz. Therefore, the harmonic spectrum of the frequency converter input current, caused by switching of the IGBT modules, is in the high-frequency range, which is considerably higher than 2 kHz.

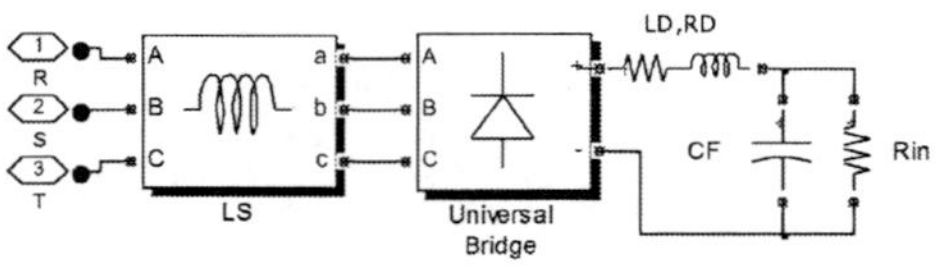

Figure 5. Simulink model of VFD

Two models of power source (Three-Phase Source 10 kV) were used to investigate power quality in PSS of VFD group. If the electric drives were powered by a 110/10 kV power transformer, then the ordinary block from the SimPowerSystems library could be used as a power supply model. When the drivers were powered by autonomous power source, a model of a gas turbine generator was used (Fig. 6).

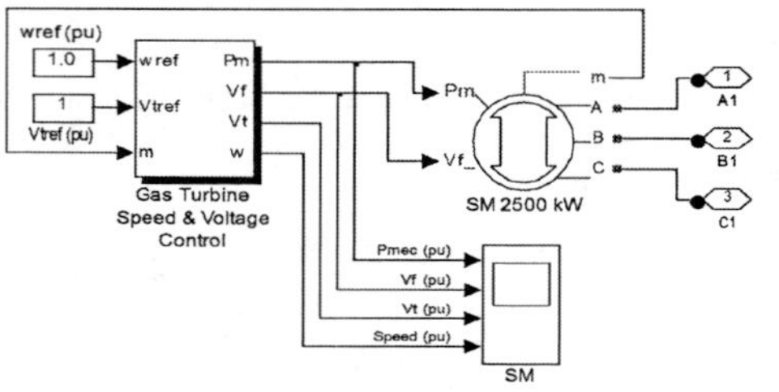

Figure 6. Simulink model of gas turbine generator

An analysis of power quality in dependence on the number of gas cooling units with VFD connected to a power supply was performed for the PSS shown in Fig. 1 using the simulation model. It was presumed, that the 10/0,4 kV transformer on each CTS was supplying power to 12 VFDs in nominal mode. The nominal power of the transformers was 630 kVA.The simulation covers the cases where as power supply sources were considered whether 110/10 kV transformers with nominal power of 10 000 and 16 000 kVA or a 2 500 kW gas-turbine power station. The analysis was performed for both cases to stress aggravation of the power quality maintaining problem in the case of stand-alone power supply systems.

Fig. 7, 8, and 9 show the values of the voltage THD, 5th and 7th voltage harmonics respectively for a different number of gas cooling units with a VFD connected to the 10 kV bus. The red lines represent permissible values of the shown parameters.

In accordance with the simulation results, only one gas cooling unit can be retrofitted with a VFD, if the gas compressor station with an isolated PSS is powered by a 2500 kW gas-turbine power station. If two such units are powered by the PSS, THD and 5th harmonic current level exceed their permissible values.

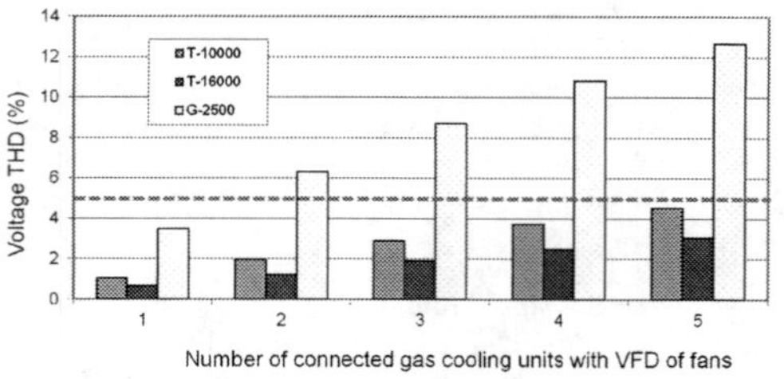

Figure 7. The value of the total hanmonic distortion of voltage

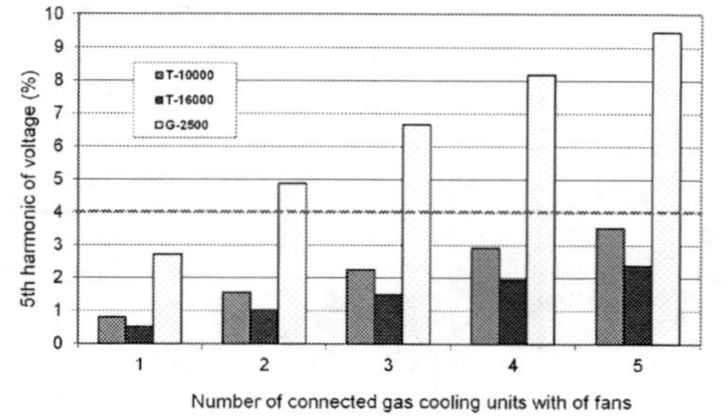

Figure 8. The value of the 5th hanmonic of voltage

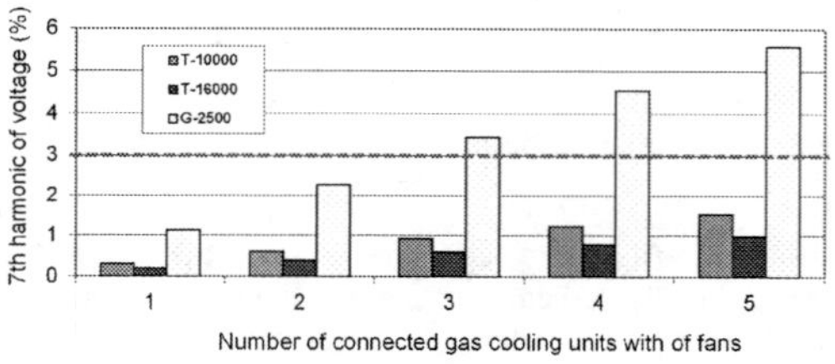

Figure 9. The value of the 7th hanmonic of voltage

This problem does not appear in case of the power supply through the 110/10 kV transformers. Up to 5 units could be powered without derating of the power quality.

III. Electrical power quality management in power supply systems for gas cooling units with VFD

Power quality of industrial mains electricity is regulated in Russia by GOST 32144-2013 "Electric energy. Electromagnetic compatibility of technical equipment. Power quality limits in the public power supply systems". In accordance with the norm, total harmonic distortion (THD) of 6-25 kV electrical power supply system should not exceed 5% during 95% of a week interval. Meanwhile the 5th and 7th harmonic current should not exceed 4% and 3% of the fundamental current respectively.

If harmonic distortion, according to the analysis, surpasses the desired value, there are a number of ways to mitigate it. Harmonics are caused by the first stage of a VFD (the rectifier). The distortion of input current feeding this rectifier on a standard (6-pulse) VFD ranges between 70 - 100%.

AC line reactors and DC link chokes (reactors) will make the current drawn from the power line more sinusoidal. With an

AC line reactor or DC link choke, the distortion of the input current will typically be 30-40% compared to the 70-100% of a drive with no reactor. Reactors will increase the initial cost of the drive 10-20% and are included as standard.

AC line reactors have the dual benefit of convenience and providing additional attenuation of AC power supply disturbances such as surges or poor phase balance. They do however cause a reduction in DC link voltage so they can create voltage drop issues. An oversized AC line reactor with a low line voltage can create enough of a voltage drop that not enough output voltage can be created to provide full motor torque and power.

As a general rule, the total AC reactance should not exceed 5%. This is sufficient to meet the harmonic levels required by international standard IEC-61000-3-12 for a balanced 3 phase rectifier with Rsce (the ratio of the drive-rated current-to-supply fault level at the point of common coupling with other supply users) greater than or equal to 120 (i.e. primarily a 5th harmonic not exceeding 40% and a current THD not exceeding 48%).

DC link chokes can be used individually (typically on the positive DC bus) or in pairs with one each on the positive and negative bus. When two DC reactors are used on the bus, the inductance is additive. Equivalent impedances can be achieved using either one larger reactor on the positive or negative bus or two smaller reactors on both the positive and negative bus.

A key advantage of DC link inductors is the minimal drop in DC voltage with increasing load. For reactance values on the order of 4%, the benefit for harmonics is similar to that of the AC reactor but with slightly greater benefit for the lower order harmonics such as the 5th. However, the benefit is subject to diminishing returns since the harmonics can never be reduced below the levels for an infinite inductance (e.g. 20% for the 5th harmonic) and in practice, the 5th harmonic is unlikely to be reduced cost-effectively to below 30%.

The disadvantage of DC inductors is that the rectifier becomes more susceptible to the effects of supply surges and poor balance and must therefore be equipped with adequate surge suppression. To avoid undue sensitivity to supply unbalance, it is important to choose inductor values such that the inherent resonant frequencies with the DC capacitors do not coincide with the characteristic frequencies of an unbalanced supply.

Passive filters comprised of a tuned configuration of capacitors, inductors and in some cases resistors, are sometimes used to eliminate specific harmonics (typically the 5th, 7th, etc.). Depending on the level of mitigation, use of passive filters increases the total cost of the installed drives. These filters are slightly more tolerant of line imbalances but also have losses associated with using them [13].

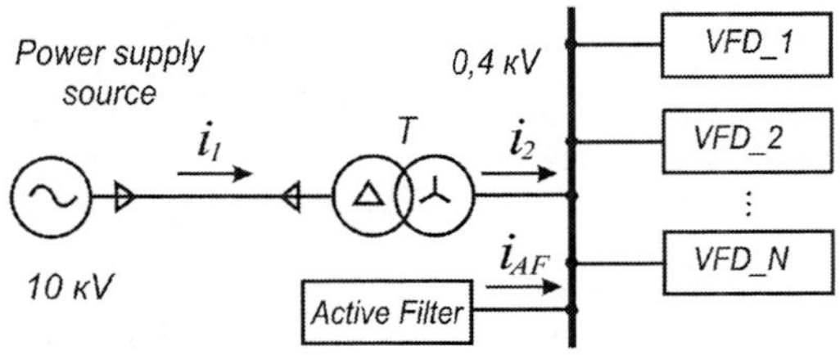

Figure 10. Diagram of connection of VFD with active harmonic filter

Newer technologies are the active harmonic filter (AHF) VFD. A single AHF can filter the harmonics of several VFDs or an entire facility. The AHF is in parallel with the other loads and only handles the corrective harmonics required to offset those generated by the nonlinear loads. The current distortion at the input of the filter will be less than 4% and these filters are relatively unaffected by line imbalances (Fig. 10) [14].

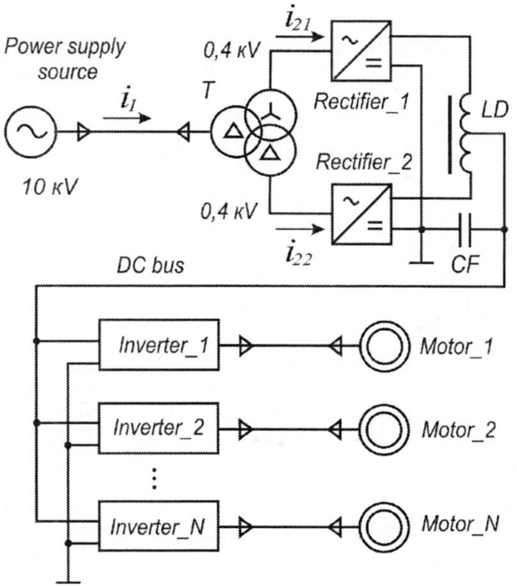

Figure 11. Single line diagram of connection VFD with integrated bus of direct current

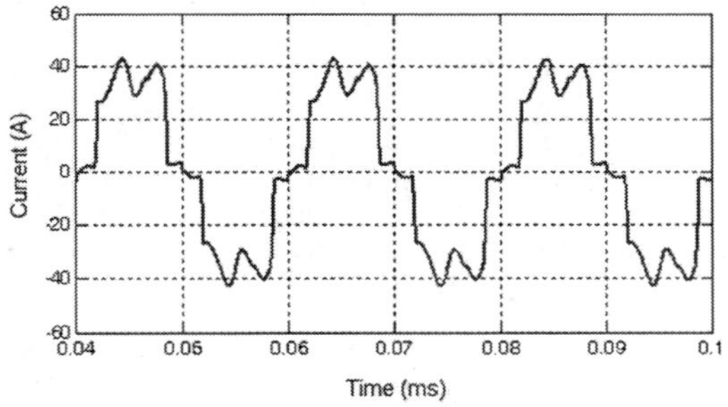

A

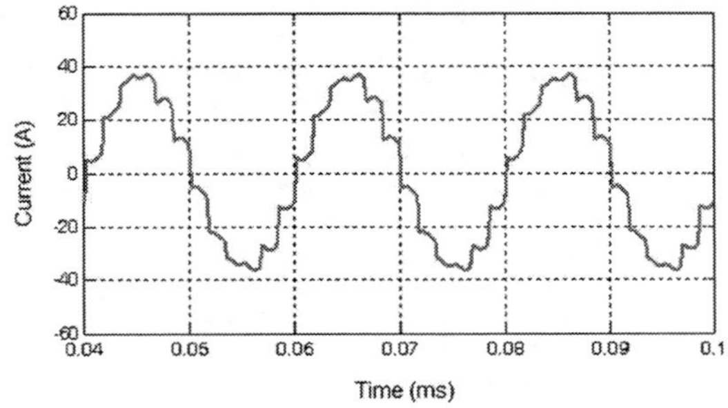

B

Figure 12. Simulated input current of VFD: six-pulse circuit (A) and twelve-pulse circuit (B)

Filtering of the 5th and 7th voltage harmonics in CTS feeder can be achieved via 12-pulse rectification (Fig. 11). VFD converters are supplied with the rectifier voltage output through a compensation coil and a snubber. In this case each of the transformers at CTS should have two secondary windings with the y- and d- connection. Since such transformer produces a 30 degrees phase shift between voltages of its secondary, an equivalent electrical load on the windings cancels the 5th and 7th harmonics in the magnetic flux. Therefore, the indicated harmonics would be filtered out of the primary winding voltage as well.

Simulated current diagrams of different CTS circuit structures are shown in Fig. 12. The first simulated current-time diagram (Fig. 12, a) was obtained for the case when twelve converters are connected to the same bus section in accordance with the circuit structure in Fig. 10. THD of the current in this diagram exceeds 26%. Simulation of the two-section circuit structure (Fig. 11) shows a power quality improvement (Fig. 12, B). THD of this curve is 11%.

Use of the circuit structure in Fig. 11 allows significantly improve electrical power quality in the 10 kV power supply system. If CTS is supplied via 2,5 MW gas-turbine power station, THD exceeds the permissible value of 5% in the 10 kV PSS, even with 4 VFD connected to the bus sections

IV. Experimental Investigation of Electirical Supply Power Systems with Nonlinear and Unbalanced Consumers

At some industrial enterprises power quality is experimental investigated to compare it with the permissible values GOST 32144-2013. The measurements are carried out on 0,4 kV buses, supplying variable frequency drives and electrothermal units, in which voltage converters are used to maintain a technological regime. These electrothermal units are used in different technological complexes. Power quality indices were measured using analyzer "Resurs-UF2C".

Fig. 13 shows fragment of a single line diagram for isolated power supply system. Power supply source presented by 1700 kW gas-turbine power station. 10/0,4 kV transformer with power of 630 kVA feeds an industrial electro technical unit consists of two 5 kW VFDs and a 100 kW resistance heated furnace with the thyristor voltage converter.

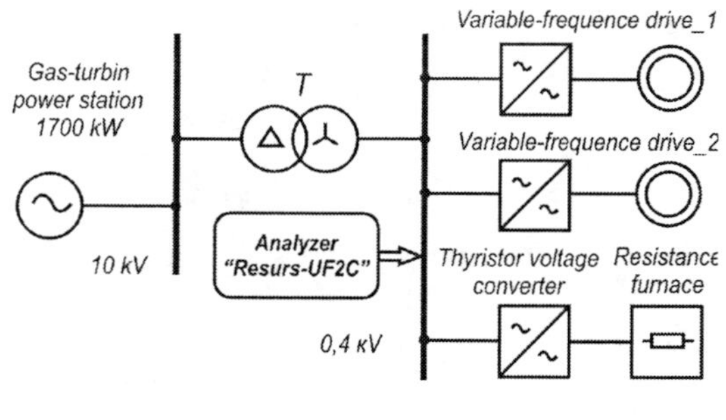

Figure 13. Isolated PSS single line diagram fragment

The diagram at fig. 14 demonstrates the influence of the resistance furnace with thyristor converter on the 0,4 kV bus power quality. The diagram shows 3-hours measurement segment. Measurements fixed with 1-minute interval.

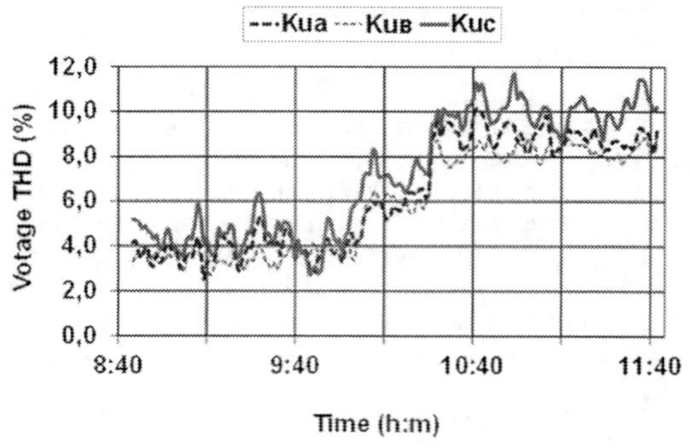

Figure 14. Measured value of voltage THD

During this 3 hour interval the resistance furnace launches and rumps up to nominal mode. This leads to voltage THD magnification to 11,6%. The value exceeds the permissible value of 8% for power supply system 0,4 kV.

A large number of single-phase electrical sets connected to the 380/220 V mains are used in technological complexes for various purposes. Such electrical sets include single-phase electrothermal units use for automated heating of process facilities, and also for maintaining the required air temperature in the premises. As a result of experimental investigation, it was found that on 380/220 V buses, to which single-phase electrothermal units are connected, the zero-sequence voltage unbalance factor is 10.1-10.7% with a maximum permissible value of 4%. As can be seen from the above, according to the results of experimental investigation, in 380/220 V mains that supplies electrothermal units and VFD, the electric power quality indices, such as THD of the phase voltages and the zero-sequence voltage unbalance factor, exceeded the permissible value.

Non-sinusoidal and unbalanced mains voltage increases power losses in the PSS elements such as transformers, asynchronous and synchronous motors, power lines. The losses were estimated using the software [15] for the technological complex consists of 58% of motors, 37% of electrothermal units, and 5% of illuminating and other equipment.

A structure of the additional power losses of the technological complex equipment caused by harmonic current and voltage unbalance was built in accordance with the estimation results (Fig. 15).

When voltage THD increases from 3 to 6%, additional power losses increase 4-fold. The main part of the additional losses caused by harmonic current (75%) emerge in the 0.4 - 10 kV power lines (Fig. 15, A).

When zero-sequence voltage unbalance factor increases from 3 to 10%, additional power losses increase 11-fold. The largest additional losses caused by voltage unbalance in technological complex equipment (81%) emerge in asynchronous motors (Fig. 15, B).

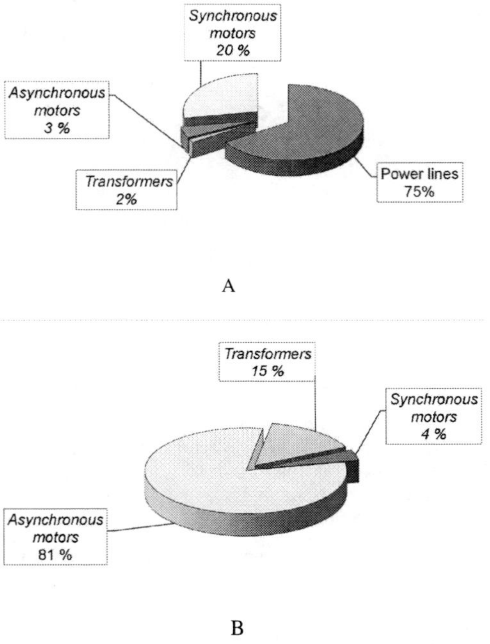

Figure 15. The structure of additional power losses caused by harmonic voltages (A) and voltage unbalance (B) for the technological complex equipment.

Conclusion

Retrofitting of the existing equipment with a variable-frequency drive should be done alongside with ensuring electromagnetic compatibility in the newly established PSS. Otherwise, the probability of electrical power quality deterioration in an industrial power supply system will be substantially high.

Changeover towards the circuit structures allowing suppression of the 5th and 7th harmonic voltages through the multipulse rectification and one sectional DC bus or active harmonic filtering can be a solution to this problem for unitized transformer substations powering gas cooling units with variable-frequency drives. Therefore, carrying out a technical and economic comparative analysis of these engineering solutions – is the initial task.

Acknowledgment

The work was supported by the Ministry of Education and Science of the Russian Federation (grant number 10.1434.2017).

References

[1] E. F. Fuchs, and M. A. S. Masoum, "Power quality in electrical machines and power systems". USA: Elsevier/Academic Press, 2008.

[2] A. Y. Abdelaziz, S. F. Mekhamer, and S. M. Ismael, "Sources and mitigation of harmonics in industrial electrical power systems: state of the art," The Online Journal on Power and Energy Engineering (OJPEE) , vol. 3, No. 4, 2012, pp. 320-332.

[3] Horia Andrei, Costin Cepisca, and Sorin Grigorescu, "Power quality and electrical arc furnaces", Power Quality, A. Ebrehard, InTech, 2011. [Online], Available at: https://www.intechopen.com/books/power-quality/power-uality-and-electrical-arc-furnaces, last see (2017, May)

[4] T. Bjazic, F. Kolonic, and P. Crnosija, "Experimental identification of natural gas cooling process," IECON2006, The 32nd Annual Conference of the IEEE Industrial Electronics Society, 2006. – pp. 1 – 6.

[5] I. I. Artyukhov, I. I. Arshakyan, and A. A. Trimbach, "Analysis of circuits of frequency control of gas cooling device's fans," 2008 International Conference on Actual Problems of Electron Devices Engineering, 2008, pp. 381-389, doi: 10.1109/APEDE.2014.6958207.

[6] S. F. Stepanov, I. I. Artyukhov, P. V. Kovalenko, and V. V. Kovalenko, "Construction concept of cluster honeycomb systems decentralized electric power supply under the technology platform "small energy"," 2016 International Conference on Actual Problems of Electron Devices Engineering, 2016, pp. 1-7, doi: 10.1109/APEDE.2016.7879005

[7] Gas-turbine generator sets. [Online], Available at: http://www.pmz.ru/eng/products/gtu-for-stations, last see (2017, May)

[8] J. Novák, J. Šimánek, O. Černý, and R. Doleček, "EMC of frequency controlled electric drives," Radioengineering, vol. 17, No. 4, 2008, pp. 101-105.

[9] J. Novák, R. Doleček, and O. Černý, "EMC Specific problems of frequendy converters used at interlocking device," Radioengineering, vol. 18, No. 4, 2009, pp. 606-610.

[10] I. I. Artyukhov, I. I. Bochkareva, and S. V. Molot, "Assurance of electromagnetic compatibility of group frequency controlled electro drives with power supply sources," 2014 International conference on Actual Problems of Electron Devices Engineering, APEDE 2014, Institute of Electrical and Electronics Engineers Inc., 2014, pp. 11-17, doi: 10.1109/APEDE.2014.6958207.

[11] Wu Zhu, Wei-Ya Ma, Yuan Gui, and Hua-Fu Zhang, "Modelling and simulation of PQ disturbance based on Matlab," International Journal of Smart Grid and Clean Energy, vol. 2, no. 1, January 2013, pp. 18-24.

[12] Mehdi Taleshian, Hasan Rastegar, and Hossein Askarian Abyaneh, "Modeling and power quality improvement of grid connected induction generators driven by turbo-expanders," International Journal of Energy Engineering, 2(4), 2012, pp. 131-137, doi: 10.5923/j.ijee.20120204.04.

[13] T. Devaraju, V.C. Veera Reddy, and M. Vijay Kumar, "Modeling and simulation of custom power devices to mitigate power quality problems," International Journal of Engineering Science and Technology, Vol. 2(6), 2010, pp. 1880-1885.

[14] M. D. Kharad, and N. Kumar, "Modeling and simulation of unified power quality conditioner (UPQC)," International Journal of Engineering and Advanced Technology (IJEAT) ISSN: 2249 – 8958, Volume-3 Issue-6, August 2014, pp. 133-137.

[15] S. V. Trigorly, I. I. Artyukhov, and A.V. Getmanskya, "The software complex for calculating additional losses of electrical energy in the elements of the power supply system due to non-sinusoidal voltage and current, unbalance of voltages", 2016, August 26, Certificate of state registration of computer program, RU No. 2016619716.

New Trends in MV Grids and Their Impacts

Daniel Kouba
Grid Development West
E.ON Distribution
Ceske Budejovice, Czech Republic
daniel.kouba@eon.cz

Jiri Celeda
Grid Development West
E.ON Distribution
Ceske Budejovice, Czech Republic
jiri.celeda@eon.cz

Abstract — **The paper is focused on the new trends in development of current medium voltage (MV) distribution systems. We considered two approaches of the future development. Firstly, we evaluated building-up of new distribution cable lines and their impact and secondly installation of new automatic switching devices.**

I. General Objectives

Current medium voltage distribution systems are more demanding on area of grid reliability, connection of renewable energy sources or high industry parks, and targets of The Smart Grid Concept.

Smart Grid objects are usually focused on the low voltage grid. In our opinion, initial creation of Smart Grid at higher voltage level is more efficient, because there more customers can be influenced by smart technologies. Considering that the high voltage (HV) grid has already reached advanced requirements such as security, reliability, protection, remote control etc., we tend to improve the current medium voltage (MV) grid. It shows us the measurable impact of the new smart devices for the customers with the reasonable costs.

The MV grid has own specific deficiencies which are needed to be eliminated. We should achieve the HV grid properties as an ideal MV grid. Certainly, we do not propose to use HV towers in MV grid. There is a cheaper and more effective way how to install cables. We consider these main targets as the MV Smart Grid:

- Increase of reliability
- Operation cost reduction
- Ability to connect renewable energy sources
- Reduction of losses

In connection with the Smart Grid idea, the advanced distribution automation is to represent intelligent network control that will, in real time (or quasi-real time), respond to changes in load and production, as well as to occurrence of fault state, usually without the controller's intervention. In this contribution, pole-mounted reclosers are considered as automatic switching devices for overhead MV lines and smart MV switchgears are presumed for cable grid.

The primary aim of our work is to propose a gradual development of existing MV equipment with expectation of partially automated grid in the future and to show the impacts of new approaches.

II. Replacement Of Overhead Lines By Cables

The first step to achieve more reliable MV grid is usage of underground cables. Average year fault frequency of overhead MV lines (OL) is about 100 faults per 1000 kilometres excluding calamites and in cable grid, there are approximately only 5 faults per 1000 km (considering insolating faults only). Therefore, we can dramatically reduce outages by installing cables.

Next CEER report diagrams prove that the countries with more than 50 % of LV and MV grid in cables reach the lowest values of fault interruption indexes: SAIDI (System Average Interruption Duration Index), SAIFI (System Average Interruption Frequency Index).

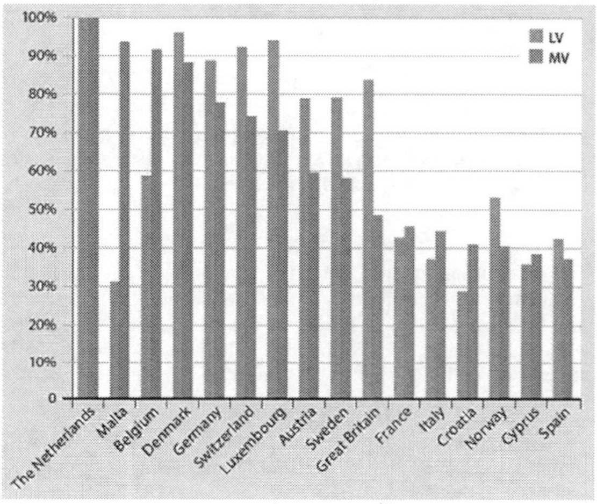

Figure 1. Rate of LV and MV Underground Cables [3]

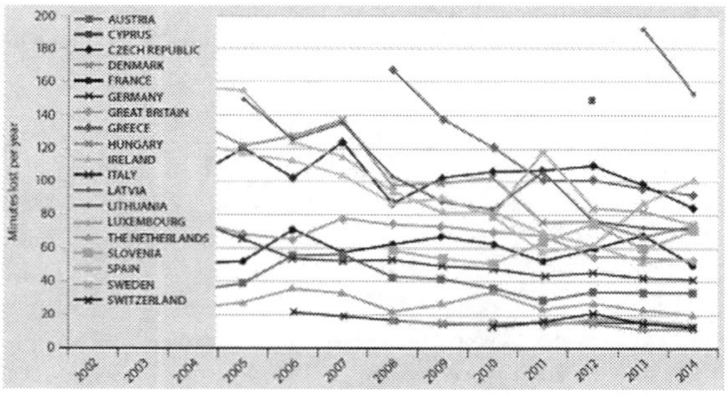

Figure 2. Unplanned long interruptions [3]

On the other hand, the disadvantage of the MV cable grids might be high capacitive current in resonant neutral earthing system. Safe operating of MV grid depends on a residual current during a single-phase earth fault. Reducing of this current requires reactive power of existing Petersen's coils and their more accurate tuning. Furthermore, we must build the OL with a high quality earthing system.

The reactive power of cables influences negatively a reactive power flows between the distribution and the transmission system. All mentioned disadvantages lead to additional cost.

A. Cities and countryside

There is no question about installing cables in a built-up area but in order to protect people, who are moving in vicinity of electrical equipment, it is required to come up with an underground solution of grid development. Hence, replacement of existing OL in built-up area by cables is demanded by law in several countries.

We must consider all advantages and disadvantages of cables installation in countryside. Investment's costs of cables are a bit higher. It is mainly influenced by the length of cable lines and pavement reconstruction (in case of cities). If we replace OL by cables we can assume the length would be $\sqrt{2}$ times longer because OL lines usually cross the field and cables are located along the routes.

Next figure shows the capacitive current development in grid E.ON Czech. There is nearly linear increase from year 1970 up to this day. We have counted two possible scenarios of its further development. The green curve respects continuous replacement of OL by cables in built-up areas and surroundings. The replacement was progressed from the oldest OL to the newer ones, thus it starts with the equipment from year 1960 and finishes with lines from 1997 which will have been replaced by the year 2040. The purple one represents the replacement of OL in existing built-up areas only and shows similar growth as before.

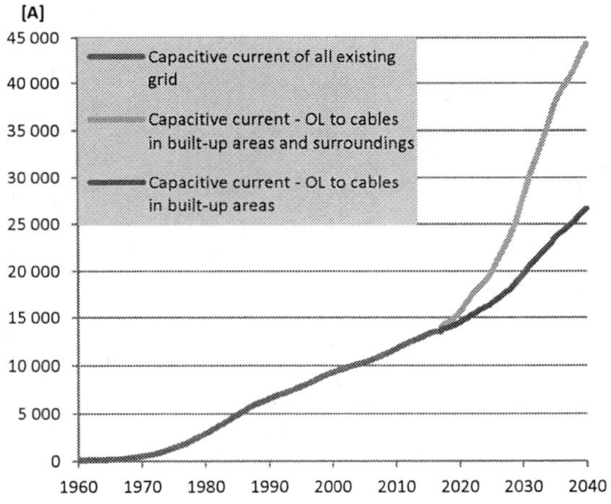

Figure 3. Capacitive current development – scenario I.

A little bit different approach will follow up the capacitive current development if we replace old OL by cables. We contained the same equipment from years 1960 to 1997 to the reconstruction and counted capacitive current increase caused by replacement 25 % - 100% of OL.

It is important to note that replacement of old cables with paper-oil insulation by new cables with insulation based on polyethylene materials cause a significant decrease of capacitive current because new cables have its capacitive current approximately 1 A/km less than the old cables. The diagrams respect this fact. There is also computed with decrease of capacitive current by removal of existing lines.

The results show us that in the highest scenario we will need to add reactive power 720 MVAr to compensate capacitive current during a single-phase earth fault if we want to operate resonant earthed neutral system everywhere. In fact, the neutral point of supply transformer is usually earthed through resistor but in cable grid only.

The problem could come into existence in OL grid with a lot of cables which has to be operated with Petersen's coils. This scenario is described better in results 50 % of OL replaced by cables in Fig. 4 or green curve in Fig. 3. The reactive power of Petersen's coils increase is about 350 MVAr in these cases.

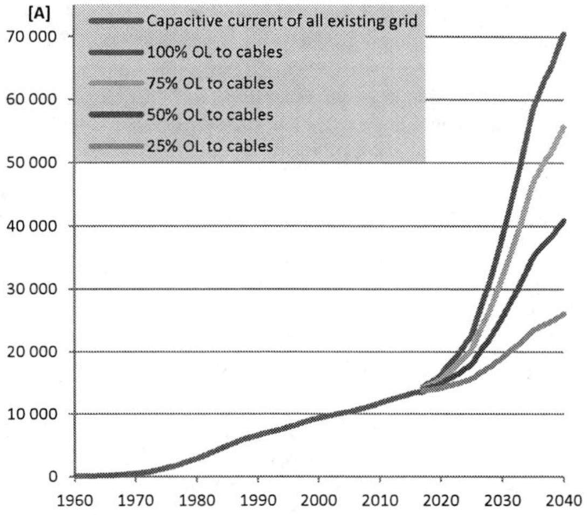

Figure 4. Capacitive current development – scenario II.

The installed power of Petersen's coils must be much higher than the additional reactive power in year 2040. We can consider two times more power because 50 % reserve is needed while one of two supply transformers is in active preventive maintenance time in case of central compensation of capacitive current during a single-phase earth fault. That means additional 700 MVAr installed power of Petersen's coils. If we built 20 new substations with two HV/MV transformers until 2040 we will add only 200 MVAr. Therefore we have to change our approach of central compensation to the decentralized.

Table 1. Parametres of standard MV cables

22 kV		**Catalogue**		**Standard**
Polyethylen	*mm²*	*C[uF/km]*	*Ic [A/km]*	*Ic[A/km]*
AXEKCY AXEKCEY AXEKVCEY	70	0,2	2,39	2,4
	120	0,24	2,87	2,9
	150	0,26	3,11	3,15
	240	0,3	3,59	3,7
NA2XS(F)2Y	150	0,251	3	3,15
	240	0,301	3,6	3,7
HATPS 120	120	0,28	3,35	2,9
Paper-oil				
ANKOPV, ANKOY	120	0,28	3,35	3,65
ANKOYPV, ANKTOPV	150	0,3	3,63	3,95
ANKTOYPV	185	0,33	3,92	4,3
ANKTOYPVs	240	0,36	4,32	4,7
AMKTOYPV	120	0,24	2,84	3,65
AMKTOYPVs	240	0,3	3,61	4,7

B. Forest corridors

The important information to determine the potential for installation of cables in forest corridors is based on the comparison of maintenance costs for overhead and cable lines in forest corridors, including their lifetime investment costs. The outcome of technical and economic comparison is that cables should be installed in a forest corridor with minimum length of 800m. [1]

III. Interconnections of Long Radial Feeders

Building new interconnections of long radial feeders is also included in the concept of Smart Grids and automation of DSO (Distribution System Operators) on MV level. These new interconnections will be primarily used while the alternative configuration of network is being operated.

Realization of new interconnections is described in following methodology which has three main points of view.

First point of view is a fault rate on both feeders. The final value consists of two parameters: total length and average age of lines. The total length of feeders means higher probability of fault. The average age of feeders shows us the future inevitability of supply interruption, based on technical service life.

Second aspect of view solves the reliability on both feeders. This parameter involves total sum of customers and total power consumption. First parameter describes the number of customers which are connected to both branches. This parameter is also used to determine SAIDI and SAIFI. Second parameter represents type of customers and therefore the importance of the branch.

Last point of view is usability of interconnections. This view is focused on requirements from control centre and department of grid development reflecting the requirements of the new connection of customers or switching abilities.

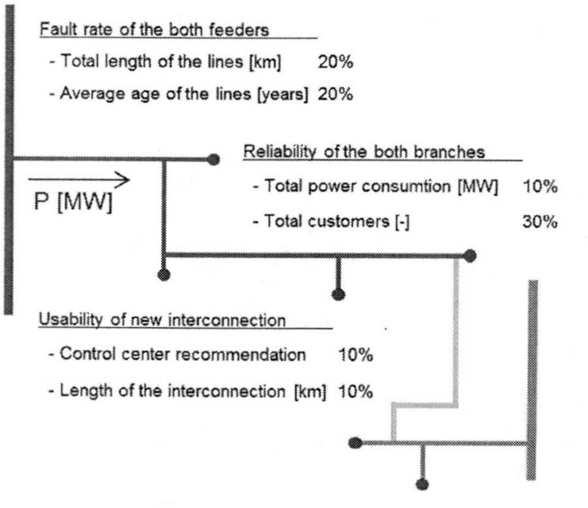

Figure 5. Aspects of building new interconnections

Final result is defined by percentage of the individual points of view. Failure rate is represented by 40 % of total. Reliability consists of 40 % of total, as well. And last point of view, usability, is represented by 20 % of final result.

Interconnections with the greatest potential (final result) will be chosen for implementation. This network development brings us more possibilities for installation of reclosers according to described methodology because the new circuits are created. Additional reclosers in grid will offer a challenge to the protection settings.

It is worth mentioning that it is often necessary to use cables for the interconnections.

IV. Automation of Overhead Lines

Recloser is a circuit-breaker with reclosing ability, located on the poles of overhead line. It is equipped with remote terminal unit as well as the elements with protective, monitoring and automatic functions.

Recloser complies with the function of feeder protection and thus it responds to faults before action of protection in the substation.

Due to the fact that the recloser can be mounted almost anywhere in the line (limitation is caused by breaking capacity of short-circuit current), it also clears the faults that would hardly be recorded by feeder protection in the substation due to a large distance of the point of fault (e.g. interrupted conductor).

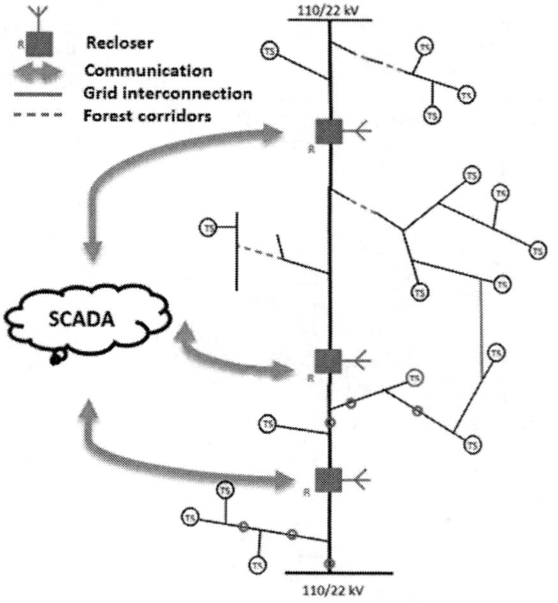

Figure 6. Installation of reclosers in overhead lines

However, the main benefit of the reclosers lies in their ability to execute a certain autonomous operations that in co-operation with other elements ensure automation of a part of distribution network line.

We propose three devices as an ideal number of the reclosers between two supply substations or busbars (in case of the same substation). The number is limited by the required selectivity of protection settings.

Firstly, we install one recloser in one line and then we will add recloser at the end of the line where the interconnection to other line is possible.

Finally, two feeders supplied by same busbar will be operated in parallel. This operation enables the highest contribution of new automatic devices in grid.

In case of two supply substations, feeders have to be operated in radial with a bit less contribution to system reliability caused by additional switching. However, the communication settings from SCADA to all three reclosers provide the similar function as parallel operation of the feeders. The advanced behaviour may be denoted as quasi-parallel operation.

Parallel network operation will assure that in case of fault, only the smallest possible number of customers will be disconnected from electric power – this is ensured by power supply from both sides and automatic functions of the recloser and its protections, enabling handling of only necessary number of customers. In case of radial network operation, the whole MV line affected by fault is switched off.

The recloser is also convenient to install to the radial branch, where the faults occur very often.

A. *Methodology of installation*

In this chapter, the described methodology is based on the discussed installation of reclosers but it is possible to use it for general placement of automatic devices.

The first step was to select feeders which have been most affected by faults over the last six years.

We also checked if the occurrence of the faults is repeated. Then we found the faulted part of each line and analysed the fault type, while focusing on short-circuit faults and earth faults (*F* coefficient).

Next criterion for each fault describes how many customers were without electricity (*C* coefficient). From this analysis, we got the list of the lines where we want to install reclosers.

The placement of automatic devices to each line is determined by using the two parameters: *FC* selects the line and *HO* finds the optimum position of device.

$$FC_i = \sum_{j=1}^{n} F_j \cdot C_j \quad (1)$$

Where index "*i*" selects the line and "*j*" means the number of faults. Feeders with the highest *FC* parameter are the most convenient for installation of an automatic device.

The *HO* parameter is formed by the product of two coefficients. Health coefficient *H* takes into account the number of faults which would be reduced in case of installation of automatic device.

Second coefficient *O* represents the sum of supply points which would be turned on during the fault owing to reaction of automatic device.

$$HO_m = \sum_{k=0} H_k \cdot \sum_{l=0} O_l \quad (2)$$

Where index "*m*" selects the position and "*k*" means the number of fault between the counted position and the end of line. The index "*l*" reflects the summarization of supply points in the whole feeder excluding the branch that begins with counted position.

The following figure shows a graphical representation of selecting the optimum position for installation. The red curve is given by the value of fault reduction in each node along the line. The blue curve involves the protected customers in each node of the line. We get the optimal location for installing of automatic device by finding the maximum of *HO* parameter which represents the green curve.

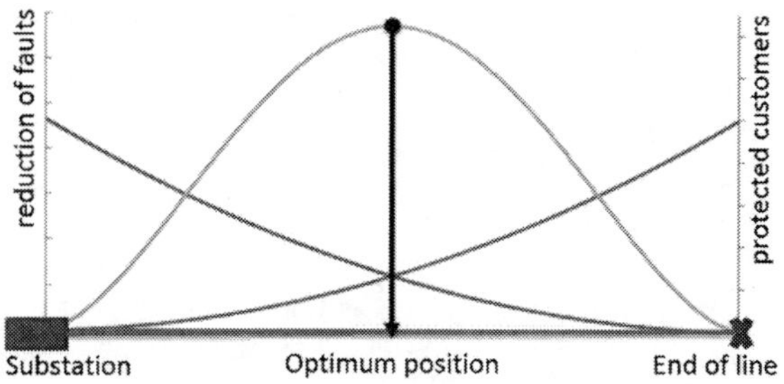

Figure 7. Settings of optimum position

It is important to note that, in fact the functions are not continuous but they are comprised of discreet nodes, referring to the poles where is convenient to install recloser, as well as poles with current switch-disconnectors.

B. *Rollout and expectation*

After reviewing state of all MV overhead lines in the distribution grid, we assessed the total need for reclosers and calculated their respective contribution towards SAIDI and SAIFI mitigation.

The results can be seen on the chart in Fig. 8, which presents a significant reliability boost after automated parallel (or quasi-parallel) operation comes into practice. This is expected by the end of 2020, when the final reclosers in the line interconnection will have been installed.

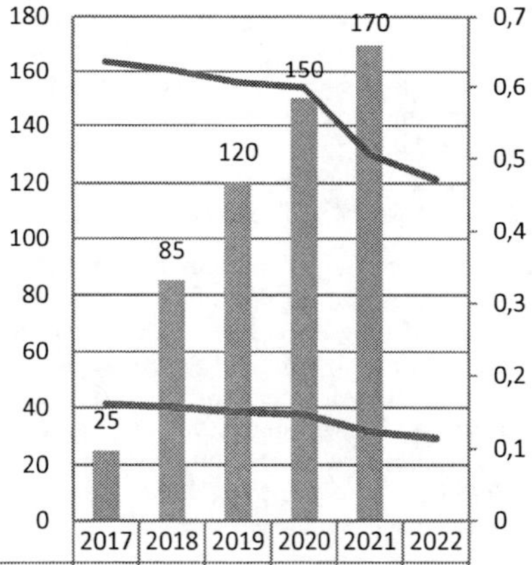

	2017	2018	2019	2020	2021	2022
REC [pc]	25	85	120	150	170	
SAIDI MV [min. /cust./year]	41,15	40,18	38,49	37,65	31,57	29,32
SAIFI MV [- /cust./year]	0,636	0,624	0,606	0,599	0,505	0,472

Figure 8. Assumed decreasing of SAIDI, SAIFI

V. Automation of Existing Cable Lines

The aforementioned process of installation of automated grid elements can be also utilized in MV cable networks. Optimal installation points lie in MV switchgears in MV/LV substations, especially in those with larger number of MV cable feeders.

Smart MV Switchgear has to comply with all relevant requirements for compact switchgear construction (such as for the SF6 switchgears already in operation), communication with SCADA system, remote control of the switching element, protection or fault location device, and potential for automatic functions.

We identify two basic types of MV/LV substation with such Smart MV switchgear: A separation point and a manipulation point. The main difference is that MV switchgears in a separation point have remote-controlled circuit breaker with protection, whereas MV switchgears in a manipulation point have remotely controlled switch and a fault current detector.

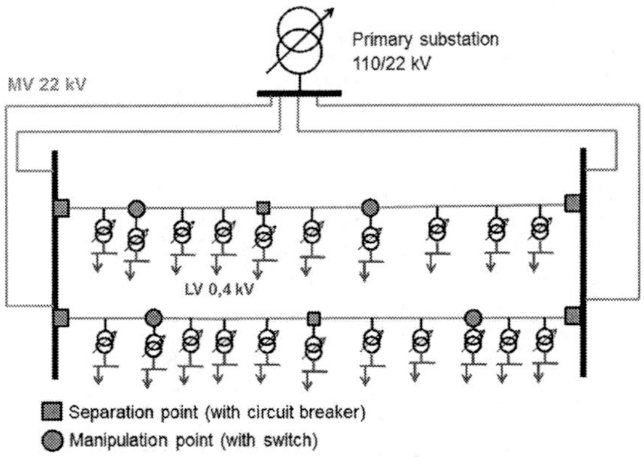

Figure 9. Installation of Smart MV Switchgear

Fig. 9 illustrates an example of a cable network with partially automated MV/LV substations.

Separation points can be seen as equivalents to reclosers in that they also automatically decrease number of affected consumers in case of a fault. The lowest possible number of such devices in a line between two feeders is two, but we assume the same principle as with reclosers will be used and there will be three such installations per line. This is also highest acceptable number for which a satisfactory selectivity of protection can be achieved.

Degree of grid automation in MV cable lines can be increased further by installing manipulation points; especially in case of long lines, lines with higher number of consumers, or lines with high reliability requirements. Dispatcher can use these points for faster grid reconfiguration, according to the priority of power supply for segments in large urban grids. The SAIDI and SAIFI mitigation is not a crucial priority in cable networks due to relatively low number of reoccurring faults, and convenience of remote grid reconfiguration is the main driver here.

In the last type of MV line – one with both cable and overhead part, Smart MV switchgear has to communicate reliably with a recloser via SCADA system.

VI. Conclusion

The trend in MV grid development expects using cables, therefore reconstruction of current overhead lines to underground and interconnections of long radial feeders are included in the proposal for a future grid.

However, the construction of new distribution lines in new routes is very expensive, thus the benefits of these interconnections have to be carefully evaluated according to the described methodology.

The impact of using cables is measurable in capacitive current size in resonant earthed MV grid. Increase of those currents will cause problems in a future distribution system operation and higher costs. The results show that we have to change our approach of central compensation of capacitive currents to the decentralized.

Pole-mounted reclosers are considered as automatic switching devices for overhead MV lines and smart MV switchgears are presumed for cable grid. The benefits of our work are, among other things, a proposal of a placement methodology of the new automatic switching devices and calculation of their respective contribution towards SAIDI and SAIFI mitigation.

References

[1] D. Kouba, D. Mezera, M. Kopt, F. Brož, 2015, "Smart Improvement of Distribution Grid Reliability", Proceedings CIRED conference, ISSN 2032-9644, paper 1240.

[2] D. Kouba, L. Kolar, J. Celeda, M. Jurik, "MV Grids Development And Automation", Proceedings CIRED conference 2017, Glasgow, paper 0750.

[3] 6th CEER Benchmarking Report on the Quality of Electricity and Gas Supply, 2016, [Online], Available: http://www.ceer.eu/, last see (2017, May)

[4] M. Kašpírek, D. Šimáček, "Active and reactive power flows between the distribution and the transmission system", Proceedings of the 2015 16th International Scientific Conference on ELECTRIC POWER ENGINEERING, ISBN 978-1-4673-6787-5, IEEE Catalog Number: CFP1573X – USB

[5] L. Kolář, D. Kouba, J. Čeleda, P. Múdry, 2014, "Automatizace VN sítí - Recloser", Proceedings 18th Conference of the Czech Committee of CIRED, ISBN 978-80-905014-3-0.

[6] M. Tesařová, O. Král, V. Majer, L. Frank, 2011, "Automatizace VN sítí v konceptu Smart Grid", Study of the University of West Bohemia in Pilsen.

Advanced Calculation of the Catenary Length in the Spans of Overhead Lines

Alen Hatibović
Department of Electrical Engineering
Optimum Solar Kft.
Baja, Hungary
hatibovic.alen@gmail.com

Abstract — **Taking into consideration the sag of overhead lines (OHL), the conductor within the span is always longer than the span itself. This way the conductor length calculation has an importance when constructing overhead lines. The current paper shows the mentioned calculation in the case when the conductor curve in a span is considered as a catenary. Correspondingly the universal formula for the conductor length calculation has been derived, which is applicable in inclined and level spans as well, in both full span and arbitrarily selected span–part. The use of the new formula is presented in a practical example.**

Keywords: Overhead Lines; Catenary; Conductor Curve; Inclined Span; Level Span

I. Introduction

In the case of high voltage overhead lines (Fig. 1) the conductor curve in a span is generally considered as a catenary, but not as a parabola. Independently of this fact, literatures for OHL design generally give a solution for the conductor length calculation in level spans, but very rarely in inclined ones [1]. Furthermore, the available length formulas are defined for determining the conductor length in a full span, i.e. for frequent conventional tasks, but not for rare unconventional tasks; for instance, the conductor length calculation in an arbitrary part of the span, either a level or inclined one. These are the reasons for deriving the algorithm for calculation of conductor length, which ensures adequate solution in each case, i.e. in level and inclined spans as well, and also in a full span and in its part. Such a complex task can be effectively solved by the application of the integral calculus. The used symbols are listed below.

S	span length, (m)
h_1	height of the left–hand side support point, (m)
h_2	height of the right–hand side support point, (m)
c	parameter of the catenary, (m)
$[x_1,x_2]$	part of a span (i.e. span–part), $[x_1,x_2] \subseteq [0, S]$
x_1	start point of a part of a span, (m)
x_2	end point of a part of a span, (m)
$y(x)$	conductor curve (catenary)
ψ	angle of the span inclination, (°)
L	catenary length in a full span, (m)
$L_{[x1,x2]}$	catenary length in a part of a span, (m)
D_{max}	maximum sag, (m)
x_{MIN}	x–coordinate of the low point, (m)
y_{MIN}	y–coordinate of the low point, (m).

Figure 1. High voltage overhead line

Considering the fact that the conductor length changes with temperature, thus the length calculation in each concrete case considers to that conductor's temperature, for which the used catenary's parameter (c) is regarded. The latter depends on temperature. Datum c has to be taken from the mechanical design of OHL. It is not the subject of this paper, as it is widely available and well explained in literatures, as in [2] [3].

The following sections of the paper show the derivations of the four different formulas for computing of the catenary length in a span, covering both very frequent and very rare tasks in practice:

- Formula for the catenary length in part of an inclined span
- Formula for the catenary length in a full inclined span
- Formula for the catenary length in part of a level span
- Formula for the catenary length in a full level span.

The first of these four formulas is a universal one, while the other three are its simplifications for some special cases.

Computing in a part of a span generally appears as a part–task within the calculations connected to the electromagnetic compatibility (EMC) or corona effects [4].

II. Calculations in Inclined Spans

When the support points of the conductor are on different elevations ($h_1 \neq h_2$) then we talk about an inclined span [5] [6]. Such a case is shown in Fig. 2, where $y(x)$ is a conductor curve (catenary). Supporting a strictly mathematical approach, the coordinate system is applied in a way that the origin is set on the line of the left–hand side support, on the elevation of the bottom of the lower–positioned support of the span [7]. Points $A(0;h_1)$ and $B(S;h_2)$ are the support points, while $E(x_1;y(x_1))$ and $F(x_2;y(x_2))$ are the start and end points of the conductor in a freely selected span–part. The low point is remarked as *MIN*.

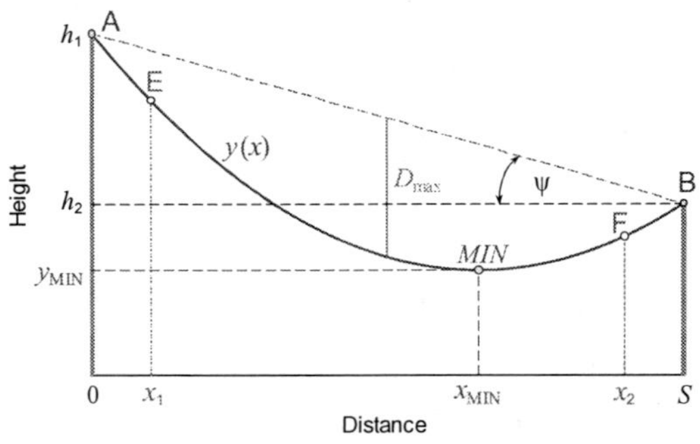

Figure 2. Inclined span with $h_1 > h_2$

A. Conductor (Catenary) Length in Part of an Inclined Span

According to [8] [9] the equation of the conductor curve in Fig. 2 is given by (1), considering (2) [10] and (3):

$$y(x) = 2c \cdot \mathrm{sh}^2 \frac{x - x_{\mathrm{MIN}}}{2c} + y_{\mathrm{MIN}}, \quad x \in [0, S] \tag{1}$$

$$x_{\mathrm{MIN}} = \frac{S}{2} - c \cdot \mathrm{arsh} \frac{h_2 - h_1}{2c \cdot \mathrm{sh}(S/2c)} \tag{2}$$

$$y_{\mathrm{MIN}} = h_1 - 2c \cdot \mathrm{sh}^2 \frac{x_{\mathrm{MIN}}}{2c} \tag{3}$$

The length of the curve $y(x)$ on the interval $[x_1, x_2]$, shown in Fig. 2 can be determined by the following mathematical formula for the *arc length* [11] [12] [13]:

$$L_{[x_1,x_2]} = \int_{x_1}^{x_2} \sqrt{1 + \left(\frac{d}{dx} y(x)\right)^2} \cdot dx \tag{4}$$

Squaring the first derivative of (1) results in (5). Above it is substituted in (4) in aim to obtain (9) using (7):

$$\left(\frac{d}{dx} y(x)\right)^2 = \left(\mathrm{sh} \frac{x - x_{\mathrm{MIN}}}{c}\right)^2 \tag{5}$$

$$L_{[x_1,x_2]} = \int_{x_1}^{x_2} \sqrt{1 + \left(\mathrm{sh} \frac{x - x_{\mathrm{MIN}}}{c}\right)^2} \cdot dx \tag{6}$$

$$\mathrm{ch} u = \sqrt{1 + \mathrm{sh}^2 u} \tag{7}$$

$$L_{[x_1,x_2]} = \int_{x_1}^{x_2} \mathrm{ch} \frac{x - x_{\mathrm{MIN}}}{c} \cdot dx = c \cdot \mathrm{sh} \frac{x - x_{\mathrm{MIN}}}{c} \Bigg|_{x_1}^{x_2} \tag{8}$$

$$L_{[x_1,x_2]} = c \cdot \left(\mathrm{sh} \frac{x_2 - x_{\mathrm{MIN}}}{c} - \mathrm{sh} \frac{x_1 - x_{\mathrm{MIN}}}{c}\right) \tag{9}$$

Applying identity (10) [14] [15], the previous expression can be transformed into (11), then into (12) after substitution of (2).

$$\mathrm{sh} x - \mathrm{sh} y = 2 \cdot \mathrm{ch} \frac{x + y}{2} \cdot \mathrm{sh} \frac{x - y}{2} \tag{10}$$

$$L_{[x_1,x_2]} = 2c \cdot \mathrm{sh} \frac{x_2 - x_1}{2c} \cdot \mathrm{ch}\left(\frac{x_1 + x_2}{2c} - \frac{x_{\mathrm{MIN}}}{c}\right) \tag{11}$$

$$L_{[x_1,x_2]} = 2c \cdot \mathrm{sh} \frac{x_2 - x_1}{2c} \cdot \mathrm{ch}\left(\frac{x_1 + x_2 - S}{2c} + \mathrm{arsh} \frac{h_2 - h_1}{2c \cdot \mathrm{sh}(S/2c)}\right) \tag{12}$$

Thus (12) is a formula for the length of the catenary in arbitrarily selected part of an inclined span, $[x_1, x_2]$. Actually (12) is a universal formula for the conductor length calculation based on the catenary model, since it can be directly used for deriving the final formulas for calculations in a full inclined span, but also in a level span (a full or its part). It is shown in other subchapters and chapters of the paper.

The applicability of (12) is properly presented in the following practical numerical example with two curves drawn on the common figure:

- $y_1(x)$ catenary in an inclined span with $h_1 < h_2$
- $y_2(x)$ catenary in an inclined span with $h_1 > h_2$.

The necessary input data are given in Table 1. The lengths of the two catenaries on the intervals [0,100], [100,300], [300,400] and [0,400] are calculated by the use of (12) and then presented in Table 2. The mathematical equations of the two catenaries can be seen under the Table 1.

Example 1:

Table 1. Input data for the catenary conductor curves

Catenary	S (m)	h_1 (m)	h_2 (m)	c (m)
$y_1(x)$	400	20	60	10^3
$y_2(x)$	400	60	20	10^3

$$y_1(x) = 2 \cdot 10^3 \cdot \mathrm{sh}^2 \frac{x - 100{,}826}{2 \cdot 10^3} + 14{,}913 \quad x \in [0{,}400] \tag{13}$$

$$y_2(x) = 2 \cdot 10^3 \cdot \mathrm{sh}^2 \frac{x - 299{,}174}{2 \cdot 10^3} + 14{,}913 \quad x \in [0{,}400] \tag{14}$$

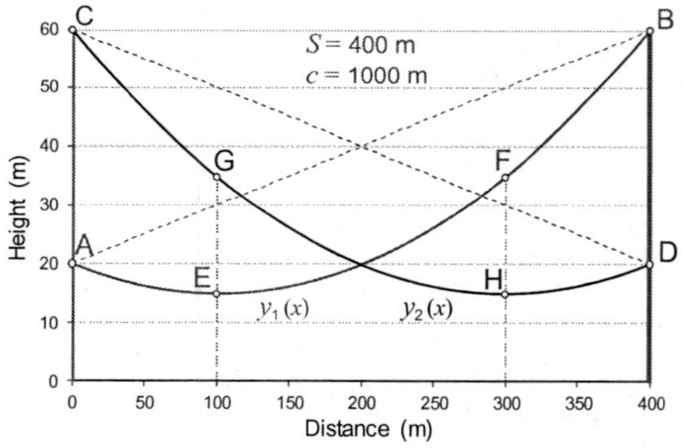

Figure 3. Catenaries $y_1(x)$ and $y_2(x)$ on a common figure

Table 2. Data x_1, x_2 and results obtained by using (12) in Example 1

x_1 (m)	x_2 (m)	Catenary $y_1(x)$		Catenary $y_2(x)$	
0	100	L_{AE} (m)	100,17	L_{CG} (m)	103,16
100	300	L_{EF} (m)	201,32	L_{GH} (m)	201,32
300	400	L_{FB} (m)	103,16	L_{HD} (m)	100,17
0	400	L_{AB} (m)	404,65	L_{CD} (m)	404,65

Let us mention that the sum of the lengths L_{AE}, L_{EF} and L_{FB} is equal to the length L_{AB}:

$$L_{\mathrm{AE}} + L_{\mathrm{EF}} + L_{\mathrm{FB}} = 100{,}17\,\mathrm{m} + 201{,}32\,\mathrm{m} + 103{,}16\,\mathrm{m} = \\ = 404{,}65\,\mathrm{m} = L_{\mathrm{AB}} \tag{15}$$

This result improves the mathematical correctness of (12). Notice that the equalities in (16) also refer to the previous example.

$$L_{AE} = L_{HD}, \quad L_{EF} = L_{GH}, \quad L_{FB} = L_{CG}, \quad L_{AB} = L_{CD} \tag{16}$$

Considering the results in Table 2, (12) is usable in both types of an inclined span, i.e. when $h_1 < h_2$ and $h_1 > h_2$.

B. Conductor (Catenary) Length in a Full Inclined Span

Having obtained the formula for the length of the catenary in part of an inclined span, given by (12), it can be simply transformed into an adequate formula for the catenary length in a full inclined span, setting $x_1=0$ and $x_2=S$. (The reverse order is not possible.) Deriving the final formula is detailed below, where at first (17) is obtained.

$$L = 2c \cdot \mathrm{sh}\frac{S}{2c} \cdot \mathrm{ch}\left(\mathrm{arsh}\frac{h_2 - h_1}{2c \cdot \mathrm{sh}(S/2c)} \right) \tag{17}$$

Taking into consideration (7) and (18), the previous length formula can be transformed into (20), as follows:

$$\mathrm{sh}^2(\mathrm{arsh}\, u) = u^2 \tag{18}$$

$$L = 2c \cdot \mathrm{sh}\frac{S}{2c} \cdot \sqrt{1 + \frac{(h_2 - h_1)^2}{4c^2 \mathrm{sh}^2(S/2c)}} \tag{19}$$

$$L = \sqrt{4c^2 \mathrm{sh}^2(S/2c) + (h_2 - h_1)^2} \tag{20}$$

This formula is corresponding to the one which is presented in [16].

III. Calculations in Level Spans

The special case of an inclined span is a level one. The support points of the conductor in a span are then on the same elevation and this way $\psi=0$. It is shown in Fig. 4. As it can be seen, the x–coordinate of the low point is located in the middle of the span.

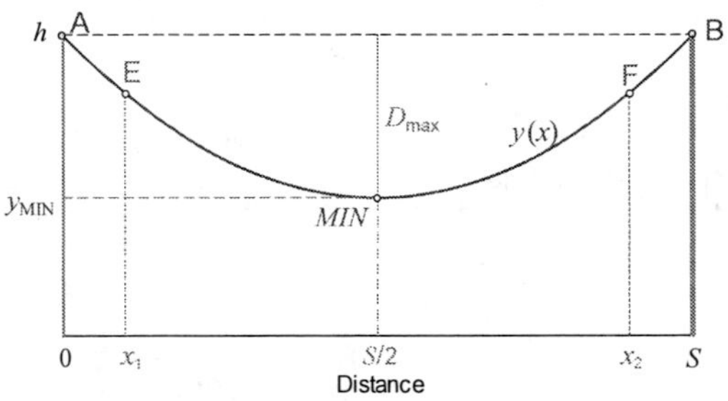

Figure 4. Catenary in a level span

Similarly as it is done in the previous chapter, in the case of level span it is necessary to define the formulas for the length of the catenary both in part of the span and in a full span. The latter formula is a simplification of the first one.

A. Conductor (Catenary) Length in Part of a Level Span

When $h_1 = h_2$, then $x_{MIN}=S/2$ and hence (9) changes into (21). The use of identity (10) yields (22).

$$L_{[x_1,x_2]} = c \cdot \left(\mathrm{sh}\frac{x_2 - S/2}{c} - \mathrm{sh}\frac{x_1 - S/2}{c} \right) \tag{21}$$

$$L_{[x_1,x_2]} = 2c \cdot \mathrm{sh}\frac{x_2 - x_1}{2c} \cdot \mathrm{ch}\frac{x_1 + x_2 - S}{2c} \tag{22}$$

The previous formula is also obtainable by considering that $h_1=h_2$ in (12). The use of (22) is presented below.

Example 2:

Table 3. Input data for the catenary conductor curve

Catenary	S (m)	h_1 (m)	h_2 (m)	c (m)
$y(x)$	400	30	30	10^3

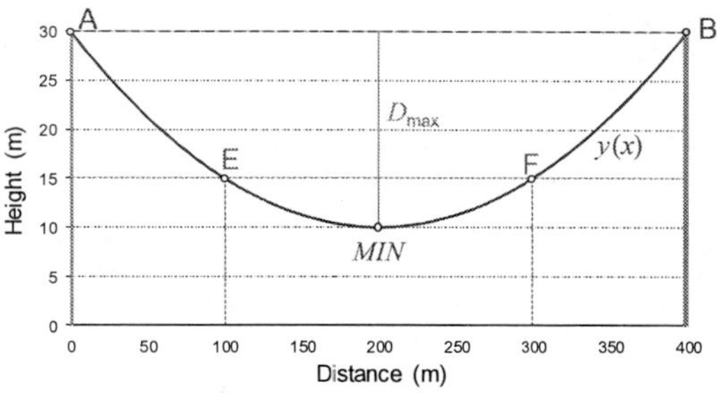

Figure 5. Catenary $y(x)$ in a level span (Example 2)

The mathematical equation of the catenary shown in Fig. 5 is given by (23):

$$y(x) = 2 \cdot 10^3 \cdot \mathrm{sh}^2 \frac{x - 200}{2 \cdot 10^3} + 9{,}933 \quad x \in [0{,}400] \tag{23}$$

Table 4. Data x_1, x_2 and results obtained by using (22) in Example 2

x_1 (m)	x_2 (m)	Catenary $y(x)$	
0	100	L_{AE} (m)	101,17
100	300	L_{EF} (m)	200,33
300	400	L_{FB} (m)	101,17
0	400	L_{AB} (m)	402,67

Let us mention that the sum of the lengths L_{AE}, L_{EF} and L_{FB} is equal to the length L_{AB}:

$$L_{AE} + L_{EF} + L_{FB} = 101{,}17\,\mathrm{m} + 200{,}33\,\mathrm{m} + 101{,}17\,\mathrm{m} = 402{,}67\,\mathrm{m} = L_{AB} \tag{24}$$

This result improves the mathematical correctness of (22). Notice that the following equality also refers to the previous example:

$$L_{AE} = L_{FB} \tag{25}$$

B. Conductor (Catenary) Length in a Full Level Span

Returning to (4) and knowing that $\mathrm{sh}^2(x)$ is an *even* function, i.e. $\mathrm{sh}^2(-x) = \mathrm{sh}^2(x)$, the length of the catenary curve in a full level span can be determined as a double length of the curve in one half of the span, according to (26):

$$L = \int_0^S \sqrt{1 + \left(\frac{d}{dx} y(x) \right)^2} \cdot dx = 2 \int_0^{S/2} \sqrt{1 + \left(\frac{d}{dx} y(x) \right)^2} \cdot dx \tag{26}$$

However, using (12), (20) or (22), the length formula in a full level span can be easily obtained in the following three ways:

- substituting $h_1=h_2$, $x_1=0$ and $x_2=S$ into (12)
- substituting $h_1=h_2$ into (20)
- substituting $x_1=0$ and $x_2=S$ into (22).

Each of these methods yields the same result, (27).

$$L = 2c \cdot \mathrm{sh}\frac{S}{2c} \tag{27}$$

Expression (27) is a formula for the conductor length calculation in a full level span. Differently to (12), (20) and (22), formula (27) is widely available in literatures regarded to OHL design. Here it is important to mention that (27) can also be derived from the new universal formula for the length calculation given by (12).

IV. Mathematical Relation Between the Catenary Lengths in Full Inclined and Level Spans

All derivations and calculations in this chapter are done under an assumption that the parameter of the catenary is equal in both spans (inclined and level) and the span length is as well:

$$c_{\mathrm{inc}} = c_{\mathrm{lev}} = c \tag{28}$$

$$S_{\mathrm{inc}} = S_{\mathrm{lev}} = S \tag{29}$$

This means that two conductor curves are in fact different parts of the same catenary in the spans of the same length. In such cases it makes sense to define the mathematical relation between the catenary lengths in inclined and level spans. The following two new symbols are introduced here:

- L_{inc} – catenary length in a full inclined span
- L_{lev} – catenary length in a full level span.

Returning to (20), it can also be expressed by (30) considering (27):

$$L_{\mathrm{inc}} = \sqrt{L_{\mathrm{lev}}^2 + (h_2 - h_1)^2} \tag{30}$$

Squaring (30) results (31), which makes possible computing L_{inc} from L_{lev} and vice versa, when $\Delta h=h_2-h_1$ is a given datum. Data S and c are not needed here.

$$L_{\mathrm{inc}}^2 = L_{\mathrm{lev}}^2 + (h_2 - h_1)^2 \tag{31}$$

It has to be mentioned that in the case of the parabola it is not possible to derive a similar mathematically correct relation. In literatures an approximate one is available in the case of the parabola. The use of (30) is shown in the following example.

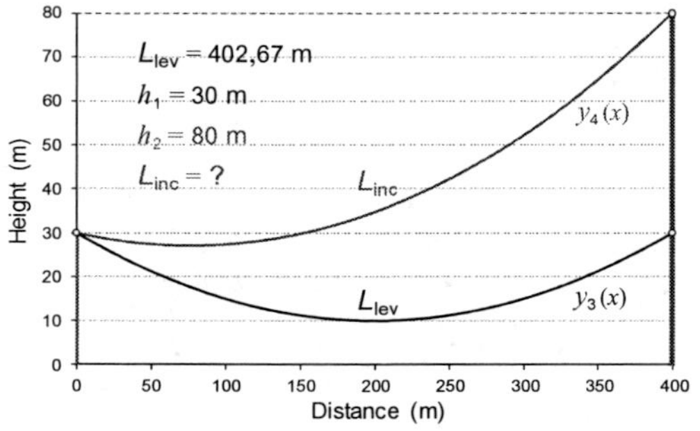

Figure 6. Catenaries $y_3(x)$ and $y_4(x)$ in level and inclined spans

Example 3:

Considering assumptions (28) and (29), two catenaries $y_3(x)$ and $y_4(x)$, in level and inclined spans, are shown in Fig. 6. Data L_{lev}=402,67 m and h_2-h_1=50 m are given ones. Applying them is necessary to determine L_{inc} by (30).

$$L_{\mathrm{inc}} = \sqrt{402{,}67^2 + (80-30)^2} = 405{,}76\ \mathrm{m} \tag{32}$$

Knowing the fact that data S and c are taken from Example 1, the previous result can be checked by (20), as follows.

$$L_{\mathrm{inc}} = \sqrt{4\cdot(10^3)^2 \cdot \mathrm{sh}^2\left(\frac{400}{2\cdot 10^3}\right) + 50^2} = 405{,}76\ \mathrm{m} \tag{33}$$

Since (30) yields the same result as (20) does, it means that (30) is applicable in all those cases when $\Delta h=h_2-h_1$ and L_{lev} are given data.

V. Conclusion

Taking into consideration that currently there are not any publications for OHL design which deals widely with the calculation of the conductor length, and also there is no publication which gives the length formulas for all cases or, if there is, it gives only approximate ones; this paper shows the derivation of the universal formula based on the catenary model, which is usable in all possible cases (inclined or level span, full span or span–part). The following facts have been taken into account:

- Full span is a special case of the span–part, when the start and end points of the latter are the x–coordinates of the two support points in a given span.
- Level span is a special case of an inclined span, when the support points are on the same elevation.

Besides the new universal formula for the catenary length calculation, its three different simplifications have also been shown for some special cases, which can occur in practice.

References

[1] C. Bayliss, B. Hardy, "Transmission and Distribution Electrical Engineering 4th edition", Elsevier, Great Britain, 2012.

[2] CIGRÉ Technical Brochure No. 324, "Sag–Tension Calculation Methods for Overhead Lines", CIGRÉ Working Group B2.12, 2007.

[3] CIGRE Green Book, "Overhead Lines", CIGRE Study Committee B2, 2014.

[4] F. Kiessling, P. Nefzger, J. F. Nolasco, U. Kaintzyk, "Overhead Power Lines, Planning, Design, Construction", Springer, 2003.

[5] F. Novothny, "Electricity Supply Part 1", ["Villamosenergia–ellátás I."], BMF KVK, Budapest, 2009.

[6] F. Novothny, "Electricity Supply Part 2", ["Villamosenergia–ellátás II."], BMF KVK, Budapest, 2012.

[7] A. Hatibovic, "Derivation and Analysis of the Relation between Conductor Sags in Inclined and Levelled Spans Based on Known Data of the Latter", CIGRÉ 2014, 45th International Conference on Large High Voltage Electric Systems, Paris, 2014.08.24–29, Paper B2–202–2014, pp. 1–8

[8] A. Hatibovic, "Derivation of Equations for Conductor and Sag Curves of an Overhead Line Based on a Given Catenary Constant", Periodica Polytechnica Electrical Engineering, Vol. 58, No. 1, 2014, pp. 23–27, doi: 10331/PPee.6993.

[9] A. Hatibovic, "Determination of the Lowest Point of the Conductor in Inclined Spans Based on a Known Maximal Sag of the Parabola", CIRED 2013, 22nd International Conference on Electricity Distribution, Stockholm, 2013.06.10–13, pp. 1–4, Paper No. 0150, ISBN: 978–1–84919–732–8, doi: 10.1049/cp.2013.0583.

[10] CTC Global Brochure, "Engineering Transmission Lines with High Capacity Low Sag ACCC Conductors", USA, 2011.

[11] [Online], Available: www.mathwords.com/a/arc_length_of_a_curve.htm, last see (2017 , May)

[12] A. Hatibovic, "Integral Calculus Usage for Conductor Length Determination on the Basis of Known Maximal Sag of a Parabola", Periodica Polytechnica Electrical Engineering, Vol. 56, No. 2, 2012, pp. 35–41, doi: 10331/PPee.7076.

[13] E. Costello, "Length of a Hanging Cable", Undergraduate Journal of Mathematical Modelling: One + Two, Vol. 4, No. 1, Article 4, 2011, pp. 1–13, doi: http://dx.doi.org/10.5038/2326-3652.4.1.4

[14] J. Stewart, "Single Variable Calculus", CENGAGE Learning, 2011.

[15] A. D. Polyanin, A. V. Manzhirov, "Handbook of Mathematics for Engineers and Scientists", Taylor & Francis Group, 2007.

[16] V. Jones, J. Howells, "Suspension Bridges", ICE Manual of Bridge Engineering © 2008 Institution of Civil Engineers, pp. 383–419, doi: 10.1680/mobe.34525.03

Balancing of Stochastic Generation of a Renewable Park by Means of Thermal, Hydro Power Plants and Bi-Directional Flows from Neighboring Energy Networks

Georgi Georgiev
Plethora Ltd
Varna, Bulgaria
Email: georgidg@gmail.com

Dmitriy Kim
Kazakh National Research Technical University
EcoRisk LLP,
Almaty, Kazakhstan
E-mail: kdk26@mail.ru

Abstract —**We study specific problem of integrating large renewable solar park to electrical power system based on coal fuel and thermal electric power plants (TPP), which have low power ramp rates. We suppose that capacity of hydro power plants (HPP) is insufficient to cover the daily schedule load peaks and the balance in the power system is achieved by expensive and undesired bi-directional flows from neighboring energy networks. The perfect control of the power system is obtained when the flows are close to zero or at least below a certain threshold. We study the use of TPP with its slow and monotonic increase or decrease power and HPP with its restricted volume of energy to balance the random generation of solar park.**

Keywords: renewables, optimal balancing, simulating model, probabilistic forecast, demand response.

I. Introduction

The construction of power stations producing renewable energy leads to the problem of their integration into the existing electricity networks, taking into account the random fluctuations in their power generation. It is required the development and analysis of mathematical models for the control of an energy system with renewables (RES), and the development of new approaches for balancing in the system. There are thus significant control problems related to the stochastic nature of renewable generation; this is in contrast to the management of traditional energy networks, whose behaviour is in general highly predictable and therefore much more easily manageable. Known methods of management of traditional power system do not reflect the random nature of the RES and are insufficient for use in the new environment. It is required to develop new approaches for the analysis of the essence of the problem related to the random nature of the RES. The problems of integrating renewables have been discussed in [1]-[3].

It is almost impossible to achieve accurate prediction of renewable generation. All forecasts will have a certain degree of accuracy, for example, the accuracy of the forecast of the generation of a solar park will strongly depend on whether a day is clear or cloudy. For example, for solar power park and for unshaded day the most likely generation during a day will represent a strip of small width around the maximum daily power (see Fig. 1). In a variably cloudy day the situation is opposite: solar power can take both large and small values (see Fig.2-3). Quite often, the System Operator (SO), which responsible for power balance and voltage maintenance, wants to be ready for any situations of renewable power, i.e. regardless the day is clear or cloudy the parameters of the power system are at the levels complying with the state standards for any RES power.

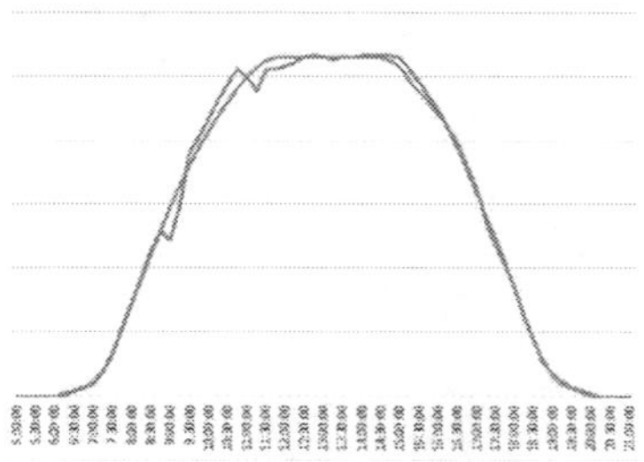

Figure 1. Real output and forecast for solar power plant "Burnoye", 2 May 2017. Blue line is forecast. Brown line is real output.

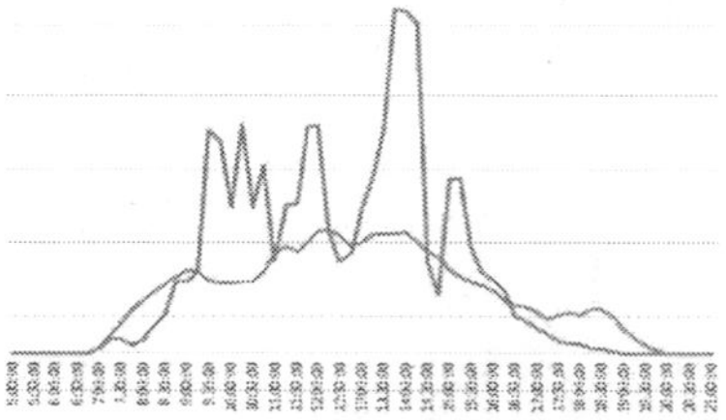

Figure 2. Real output and forecast for solar power plant "Burnoye", 16 April 2017. Blue line is forecast. Brown line is real output.

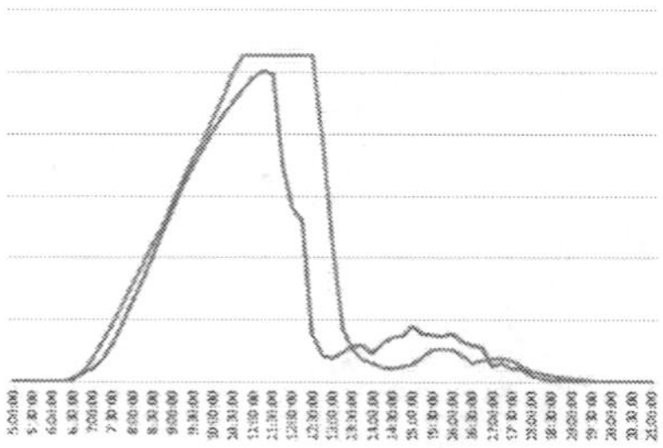

Figure 3. Real output and forecast for solar power plant "Burnoye", 27 April 2017. Blue line is forecast. Brown line is real output.

At the same time the larger the range of RES values required to be balanced, the greater the amount of power is used from the neighboring grid. In most cases the imported/exported power from or to neighboring grid is very expensive and SO aims to reduce its value. We think that probabilistic description of renewable generation instead of exact prediction of RES generation can be applied. This approach allows us to take into account only possible RES values and the optimal power system management based on the most likely scenarios of renewable generation.

Problems of efficient balancing depends on the properties of each power system and are of particular relevance to the energy networks with lack of balancing capacity, lack of modern compensators of reactive power and single entity which responsible for voltage and frequency maintenance. We consider such an example of high-voltage electric power system (based on simplified model of Kazakhstan's power grid) with large solar power park (SPP), thermal power plant with slow ramp rates, hydro power plant with restricted capacity and neighboring energy network with restrictions on bi-directional flows which must be close to zero or at least below a certain threshold.

The problem of integrating renewables in electrical power system with thermal electric power plants and some solutions described in [4]. The response of thermal power plants to the change of the networks parameters was studied in [5].

The simulation method will be applied for our research. We develop a simulating stochastic model of studied grid as add-on structure on the known equivalent power flow model for the existing power system. This approach allows to analyze the properties of the system in time and its parameters as voltage, active and reactive power by means of simulation techniques. We apply dynamic programming method based on probabilistic description of renewable generation of solar park. As a result we obtain estimation of the behavior of the power system under various optimal control algorithms and probabilistic forecasts of renewable generation. It helps us to understand how the system parameters can change their values during a day, how they depend on power of solar station and how to provide balance in the system in optimal way.

II. Model Description

Time is slotted and we study power system during N moments of time. The power system is managed by SO, which responsible for power balance and voltage maintenance. The SO controls thermal power plant (TPP), hydro power plant (HPP) and two non-regulated capacitor batteries (CB). It means that their schedules are defined by SO. Assume also that the studied power system is connected with neighboring grid (NG). We suppose that except TPP, HPP, two CB, NG and SPP the other power loads or generations of power system might be predicted in N times in advance. We will call them scheduled nodes.

We put that renewable node produce only active power R_n, $0 \leq R_n \leq \bar{R}_n$, where $\bar{R}_n$ is maximal power of SPP at time n. We suppose that SO has to manage power system in such a way that during N times for any $n = 1,2,...,N$ and for any $R_n \in [0, \bar{R}_n]$ power system is in a steady state and voltages lie within the permissible limits, i.e. for all nodes of the power system the absolute value of the ratio of the bus voltage to its rated value is less 1.05 and more 0.95.

Definition 1. Power system is in normal steady state if it is in steady state and voltages of all nodes lie within the permissible limits.

Definition 2. if we consider $\{R_n\}_{n=1}^{N}$ as a sequence of discreet random variables with known distribution

$$\mathbf{P}(R_n = r) = p_n(r), \quad r = 0,1,2,..., \bar{R}_n,$$

then sequences of distributions $p_n(r)$, $r = 1,2,...,R_n$, $n = 1,2,...,N$, we will call as probabilistic forecast.

Let TP_n be active power of TPP, HP_n – active power of HPP, $CQ_n^{(j)}$, $j = 1,2$, – reactive power of two capacitor banks and BP_n – active power of the NG at time n.

Suppose that in accordance with agreement with neighboring grid bi-directional flows of active power are limited by certain constants:

$$\underline{BP} \leq BP_n \leq \overline{BP}, \tag{1}$$

where $\underline{BP} \leq 0 \leq \overline{BP}$. Assume that the price of one unit of NG's power is C_b, i.e. if SO is used BP_n units of neighboring grid's power, it costs for SO as $C_b \times |BP_n|$.

Due to technical characteristics of TPP there are constraints on minimal and maximal output:

$$0 \leq \underline{TP} \leq TP_n \leq \overline{TP} \tag{2}$$

and so-called "ramp constraints":

$$|TP_n - TP_{n-1}| \leq K, \tag{3}$$

where $K > 0$. Assume that price of one unit of TPP's power is C_a.

For a HPP there are constraints on minimal and maximal output at each time instant and during all N times:

$$0 \leq HP_n \leq H, \tag{4}$$

$$\sum_{n=1}^{N} HP_n \leq \overline{H}$$

where H is a maximal power of HPP and H relates to the water level in the reservoir and the water regime. We put that the price of one unit of HPP's power is C_h.

There are also some constraints on capacitor banks:

$$0 \leq CQ_n^{(j)} \leq CQ^{(j)}, \quad j = 1,2, \tag{5}$$

which consists of a certain set of capacitors with various parameters. We put that reactive power of capacitor banks is free for SO. We omit many other technical restrictions on various parameters of power system, e.g. on reactive power or transmission lines, etc. as not essential for detailed consideration.

Suppose that SO manages the power system as follows: at initial time instant, all scheduled nodes send SO their data about load or generating power for the next N times and SPP send its probabilistic forecast. We put that all sending data describe the behavior of nodes in a correct way.

We can formulate the problem as follows: for some initial power state at initial time instant SO has to find such sequences TP_n, TP_n, $CQ_n^{(1)}, CQ_n^{(2)}$, $n = 1,2,...,N$, for each constraints (1) –
(5) are satisfied and for any n and $0 \leq R_n \leq R_n$ the power system is in normal steady state and cost function

$$F = \sum_{n=1}^{N} (C_a \times TP_n + C_h \times HP_n + C_b \times |BP_n|).$$

is minimal.

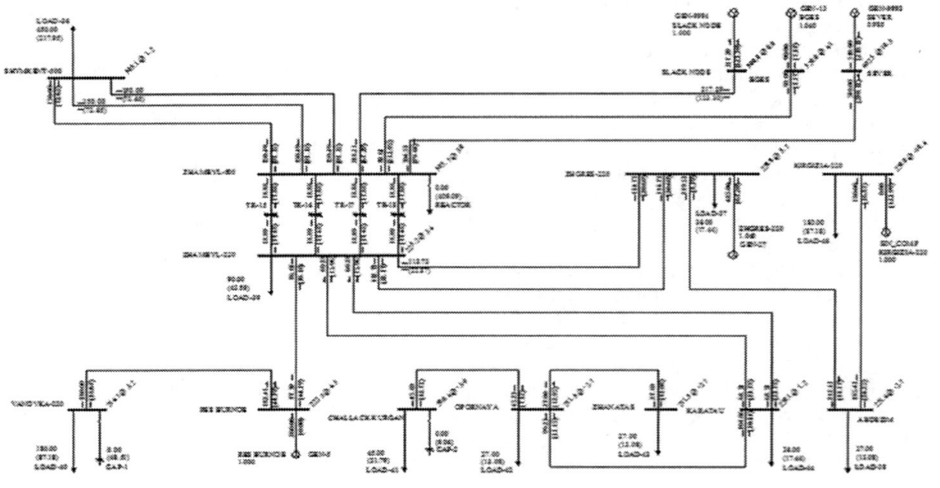

Figure 4. Equivalent circuit of the electric network in Zhambyl region, Kazakhstan, with EasyPower software.

With this problem's formulation, we apply dynamic programming method: for $n = N$ and for any TP_N, TP_N, $CQ_N^{(1)}, CQ_N^{(2)}$ find $F_N = C_a \times TP_N + C_h \times HP_N + C_b \times |BP_N|$. In the next step for $n = N-1$ and for any TP_n, TP_n, $CQ_n^{(1)}, CQ_n^{(2)}$ we find such TP_N, TP_N, $CQ_N^{(1)}, CQ_N^{(2)}$ that constraints (1) – (5) are satisfied and $F_n = C_a \times TP_n + C_h \times HP_n + C_b \times |BP_n| + F_N$ is minimal. If we repeat this procedure N times in a similar way we obtain optimal F for some for initial state of the power system.

III. EXAMPLE

We consider a simplified model of Zhambyl region with SPP of Kazakhstans power system (Fig. 4). Our model consists of 15 nodes (buses), 21 transmission lines, 4 transformators, 1 TPP, 1 HPP, 1 SPP. Power system is connected with Russian grid. Suppose that maximal power of SPP is 180 MW. The SO can control TPP "ZHGRES", HPP, capacitor bank at Vannovka (3 capacitors – 25 MW, 25 MW and 50 MW) and capacitor bank at Chulak-Kurgan (2 capacitors – 25 MW and 25 MW).

We study the behavior of the power system during the day with an hourly time step, i.e. $N = 24$.

Assume that $C_a = C_h = 0$, $C_b = 1$ and $\underline{BP} = -200\,MW$, $\overline{BP} = 200\,MW$. Suppose that SPP does not have any information about weather forecast the next day or the weather forecast is partly cloudy. In this case SPP thinks that $\{R_n\}_{n=1}^{N}$ is a sequence of independent random variables with uniform distribution, i.e.

$$p_n(r) = r/R_n,\ n = 1,2,\dots,N.$$

The following figure shows active and reactive power of neighboring grid if the day was clear and power output was maximal.

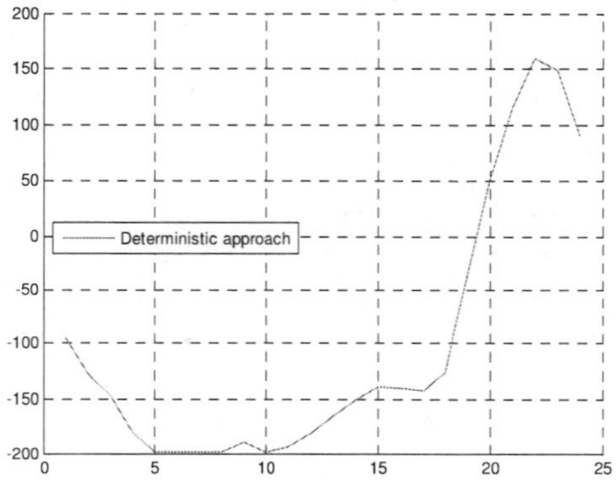

Figure 5. Active power of NG for clear day and under uniform forecast (deterministic approach).

Suppose now that the weather forecast says that the next day's sky is clear. SO assumes that $\{R_n\}_{n=1}^{N}$ is also a sequence of independent random variables, but their values belong to strip between red line and purple one (Fig. 6)

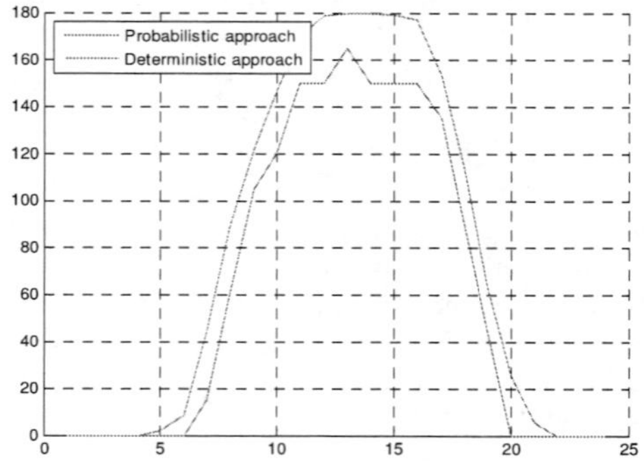

Figure 6. The strip of possible values of SPP for clear day.

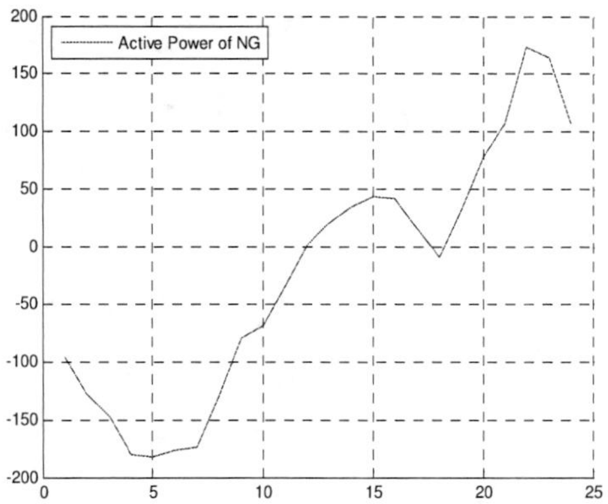

Figure 7. Active power of NG for clear day and under probabilistic approach (deterministic approach).

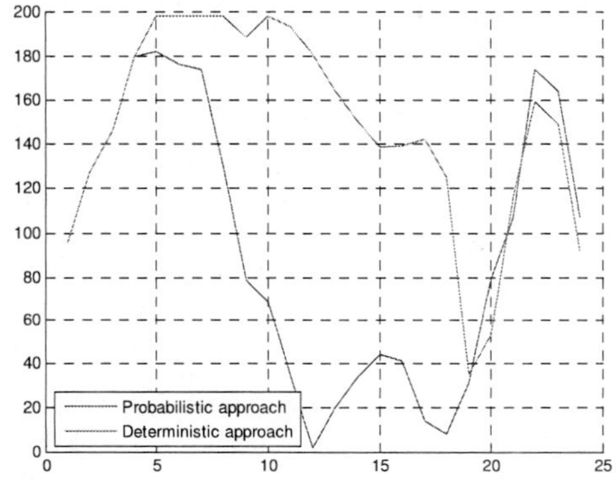

Figure 8. Absolute values of active power of NG for clear day and under uniform forecast (deterministic approach) and more detailed forecast (probabilistic approach).

The result of calculation of active power of neighboring grid if the forecast was that the day would be clear and power output was really maximal is shown in the figure 7. Comparison of absolute values of neighboring power is on the figure 8.

IV. Conclusion

We study the management's problem of power system with renewable generation. The power system is managed by SO which is able to provide normal steady state for any power of RES. In the work we showed that the using of probabilistic description of renewable's power, so called probabilistic forecast, can reduce cost function of optimal balancing of power system during finite period of time.

Acknowledgment

The authors like to thank the grant 1030/GF4 of Kazakhstan Ministry of Education and Science.

References

[1] G. Bhatt, S. Affijulla, R. Roy, "Analysis of Large Scale PV Penetration Impact on IEEE 39-Bus Power System," 2017 IEEE International Conference on Environment and Electrical Engineering and 2017 Industrial and Commercial Power System Europe (EEEC/ I&CPS Europe) Milan, Italy, 6-9 June 2017, pp.1243-1248.

[2] L.Bird, M. Milligan and D. Lew, "Integrating variable renewable energy: challenges and solutions," Technical Report NREL/TP-6A2060451, 2013.

[3] A. von Meier, "Challenges to the Integration of Renewable Resources at High System Penetration," Project Report, California Institute for Energy and Environmen, [Online] Available: https://uc-ciee.org/downloads/CEC-500-2014-042.pdf. Last see (2014, May)

[4] C. Morris, M. Pehnt, "How is Germany integrating and balancing renewable energy today?," (The German energiewende book), [Online] Available: https://energytransition.org/2015/02/how-germany-integrates-renewable-energy/. Last see (2015, Feb.)

[5] V. Vijay, "The Impact of Renewable Resources on the Performance and Reliability of the Electricity Grid," National Academy of Engineering Publications, Vol. 40 No. 1[Online] Available: http://www.nae.edu/Publications/TheBridge/Archives/TheElectricityGrid/18587.aspx. last see (2010, March)

Management of the Kazakh High-Voltage Power System in the Presence of Solar Power Park

Dmitriy Kim
Kazakh National Research Technical University,
EcoRisk LLP,
Almaty, Kazakhstan
E-mail: kdk26@mail.ru

Georgi Georgiev
Plethora Ltd
Varna, Bulgaria
Email: georgidg@gmail.com

Abstract —**We study the problem of balancing of renewable generation of solar power park in Kazakh high-voltage electric power system, which characterized with lack of balancing capacity, large spatial diversity of energy production and consumption facilities, the extension of high-voltage transmission lines and lack of modern compensators of reactive power. We develop a simulating power flow model of Kazakhstan grid as add-on structure on the known equivalent power flow model for the existing power system. This approach allows to analyze the properties of the real system in time and its parameters as voltage, active and reactive power by means of simulation techniques. We apply dynamic programming method and obtain estimation of the maximum power of renewables, which can be integrated in the power system.**

Keywords: renewables, balancing of power system, simulating model, maximal renewable output.

I. Introduction

Entering the renewable energy sources (RES) in the traditional power system, which is characterized by a high degree of predictability (the accuracy of the forecast), will bring new elements in its management related to the random and unpredictable nature of wind speed and solar radiation. Known methods of management of traditional power system do not reflect the random nature of the RES and are insufficient for use in the new environment. It is required to develop new approaches for the analysis of the essence of the problem related to the random nature of the RES, and the search for solutions and answers to many questions, such as recommendations to modernize the power system for more flexible management in the new conditions, methods of forecasting operation of the RES, etc. For example, Germany, which already has 30% renewable sources, is preparing its coal power plants to operate in a more flexible mode, smoothly reducing and increasing their power on an hourly basis [1]. This approach could extremely important in integrating RES into Kazakhstan, where there are mainly thermal plants.

First of all, we need to understand and evaluate how the power system will behave when increasing generation of the RES, which problems can arise and what are the possible solutions for them. The development of simulation model of the power system and carrying out of its multiple automated modelling will allow answering a lot of questions. It is proposed to use a well-known method of simulation modelling to study the problem appeared of the power system operation with the RES, i.e., experimental method for studying the real system in its simulation model that combines features of the experimental approach and specific conditions for the use of computing machinery. Using the method of simulation modelling involves the study of the power system, the development of its logical-mathematical model and the modelling itself based on the Monte Carlo simulations.

We consider a simplified equivalent model of high-voltage power network of Kazakhstan, which consists of conventional power plants (including that of hydropower station), solar power park (SPP), aggregated power loads and slack generator (NG), represented the balancing role played by the neighboring country's electrical power system. Various types of power plants have different characteristics and restrictions (ramp constraints, limited volume of water, limits on imported energy, etc.). We suppose that the power system is managed by System Operator (SO), which can control some elements of the grid and the aim of the research is to define maximal power of renewables, which might be integrated in the system.

II. General Problem Formulation

One may represent a power electrical system shortly as a graph with nodes (buses) and links (power transmission lines, transformers). The graph is assumed to be connected (any two nodes are connected by a link or a chain of links). For simplicity we identify each node with the bus and the generators or loads connected to it. It allows us to say that the node produces or consumes power.

We operate in discrete time and study a power system during a finite time period of length N, where $N > 0$ is a given integer.

We put that the power system consists of s node. Every node i at any time n is characterized by the vector

$$V_n^{(i)} = \left(P_n^{(i)}, Q_n^{(i)}, U_n^{(i)}\right),$$

where P is active power, Q – reactive power and U – absolute value of the ratio of the bus voltage to its rated value. We suppose that $P > 0$ ($Q > 0$) if the node is producing active (reactive) power and $P < 0$ ($Q < 0$) if the node is consuming power.

Every node and link has own characteristics as voltage angle, angle difference, length of the line, resistance, reactance, transformation ratio, etc. For brevity we omit many details of the power system.

At present the load-flow model is used to calculate the voltage and power levels of the steady state of energy system at any given time, i.e. $V_n^{(i)}$, $i = 1,2,\dots,s$, at time n. In this work we develop a simulating model as add-on structure to this load flow model, which will allow modelling the power system not only for fixed moment of time but over a certain period of time such as day, week, month, etc. with a sufficiently small time step. The new model, in addition to existing traditional calculations, will carry out of automatic calculation states of energy system in time. As a result of modelling various scenarios of the RES performance together with the daily schedules of loads and generations, it will be possible to study the properties of functioning of the energy system taking into account the RES, to check the requirements for the balance of active power and voltage, to study overloads of the lines and transformers, features to control the regulation power, etc. After statistical processing of the received modelling results array, it is possible to draw conclusion and make recommendations on the possibilities of the RES integration and its quantitative values (for example, estimate the maximum power of the RES, which can be integrated in the system) as well as the control methods of the active power balance. The modelling results will make it possible to select the power system control method under various scenarios of the situation development (production of RES) and verification of findings obtained. The proposed approach will require the development of a mathematical model of the power system functioning, its analytical study and repeated modelling of its work for different situations, taking into account the different meteorological, historical and predicted input data, time periods and time steps. The task of constructing a mathematical model of the power system, describing its behavior within a certain period of time (day, week, month, quarter, etc.) with or without the stochastic nature of the RES will depend on many factors. In addition to describing the power system it is required to determine the types and the volume of regulatory capacity, to know the rules of interaction with the neighboring power systems, the structure of determining the value of the use of different tools. Moreover, it is necessary to consider operation rules of the RES in the power system, the possibilities of forecasting of the RES production and more.

Remark that simulating model of the real system should take into account the specificity and peculiarities of the studied energy system. The next section is devoted to the application of this approach for studying the maximum power of a solar power plant that can be integrated into energy grid using the example of a simplified model of Zhambyl region of Kazakhstan's power system.

III. MODEL

We study specific problem of integrating large solar power plant (SPP) to electrical power system of Kazakhstan based on coal fuel and thermal electric power plants (TPP). TPPs generate more 85% of electricity and have low power ramp rates. Capacity of hydro power plants (HPP) is insufficient to cover the daily schedule load peaks and the balance in the power system is achieved by expensive and undesired bi-directional flows from neighboring energy networks. The perfect control of the power system is obtained when the flows are close to zero or at least below a certain threshold.

The problem of integrating and efficient balancing of renewables on the example of Kazakhstan's power grid can be considered like a special case of more general topic of efficient balancing of energy system with lack of balancing capacity, large spatial diversity of energy production and consumption facilities, the extension of high-voltage transmission lines and lack of modern compensators of reactive power. The problems of integrating renewables have been discussed in [2]-[4]. The authors of [5] propose to make a preliminary examination of the variation of nodal voltages under the impact of loads and to determine the "weak" nodes, which are the best place to attach the PV to improve the voltage profile. The analysis performed on the New England 39-bus test system. Our preliminary study has shown that the integration of PV can also lead to a worsening of the voltage profile. The reason is in the higher PV power transported on longer lines results in large losses of reactive power. Covering these losses by additional reactive flows from other nodes could aggravate the voltage profile. Therefore, the outcome of the integration of the PV is much more multivalued and requires a comprehensive and careful study.

We consider high-voltage electric power system of Kazakhstan with large solar power plant (SPP) and study the use of TPP, HPP and neighboring grid to balance the random generation of renewable energy. We develop a simulating power flow model of Kazakhstan grid and apply dynamic programming method based on deterministic schedules of traditional generation and demand and real output of solar power plant "Burnoye". As a result, we obtain estimation of the maximal power of SPP and behavior of the power system under optimal control algorithms which provide integration of SPP. It helps us to understand how the system parameters can change their values during a day and how they depend on power of solar plant.

We suppose that in the power system, managed by System Operator (SO), there are only four nodes (thermal (TPP) and hydro power plants (HPP), and two non-regulated capacitor batteries (CB)), which can be used by SO for balancing, i.e. SO defines their schedules. The remaining nodes produce or consume power according to some schedule independently on SO, defined for any $n = 1,2,...,N$ at initial moment of time, and SO is not able to manage their power directly. If the power parameters are changed by voltage, then this is an indirect control.

We study the behavior of the power system during the day with an hourly time step, i.e. $N = 24$. We put that renewable node produce only active power R_n which satisfy the following inequalities:

$$0 \leq R_n \leq \overline{R} \times \delta_n,$$

where $\overline{R}$ is unknown maximal power of SPS and $0 \leq \delta_n \leq 1$ is known ratio of maximal power of SPS at time n to $\overline{R}$ which is due to time of the day (see Fig. 1).

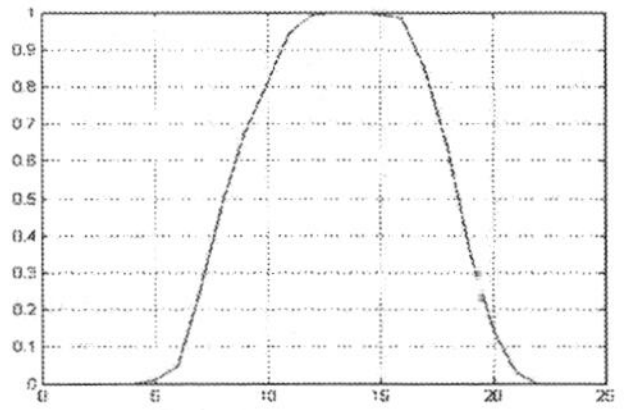

Figure 1. Values of δ_n during a day for SPP "Burnoye", Kazakhstan.

Let TP_n be active power of TPP, HP_n – active power of HPP, $CQ_n^{(j)}$, $j = 1,2,$ – reactive power of two capacitor batteries and BP_n – active power of the NG at time n. Suppose that in accordance with agreement with neighboring grid bi-directional flows of active power are limited by certain constants:

$$\underline{BP} \leq BP_n \leq \overline{BP}, \quad (1)$$

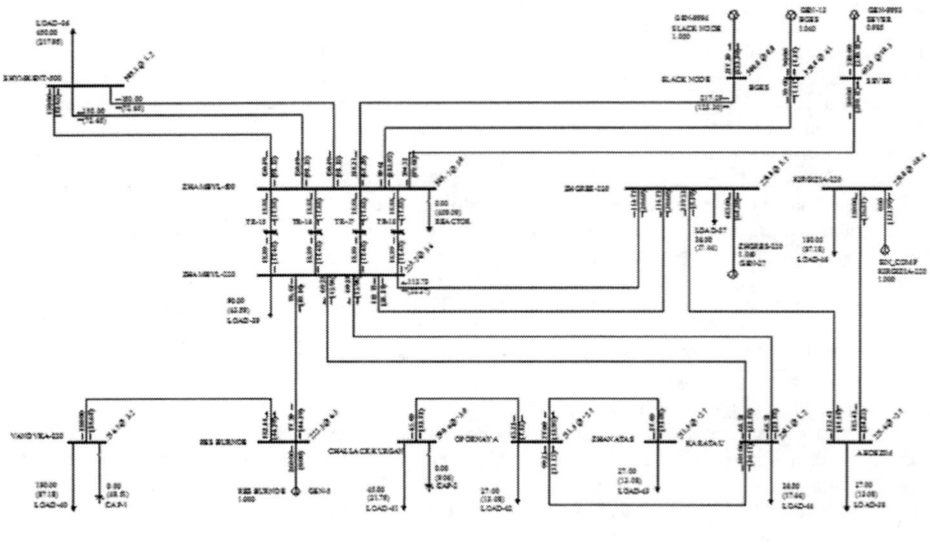

Figure 2. Equivalent circuit of the electric network in Zhambyl region, Kazakhstan, with EasyPower software.

where $\underline{BP} \leq 0 \leq \overline{BP}$. Due to technical characteristics of TPP there are constraints on minimal and maximal output:

$$0 \leq \underline{TP} \leq TP_n \leq \overline{TP} \tag{2}$$

and so-called "ramp constraints":

$$|TP_n - TP_{n-1}| \leq K, \tag{3}$$

where $K > 0$.

For a HPP there are constraints on minimal and maximal output at each time instant and during all N times:

$$0 \leq HP_n \leq H, \tag{4}$$

$$\sum_{n=1}^{N} HP_n \leq \overline{H}$$

where H is a maximal power of HPP and H relates to the water level in the reservoir and the water regime.

There are also some constraints on capacitor batteries

$$0 \leq CQ_n^{(j)} \leq CQ^{(j)}, \qquad j = 1,2, \tag{5}$$

which consists of a certain set of capacitors with various parameters.

SO is responsible to provide a reliable electricity supply and the voltage and frequency maintenance at the levels complying with the state standards. Remark that frequency of Kazakh power system is provided by more powerful neighboring Russian grid. So, the problem of integrating SPP reduces to the correct management of nodes' voltage.

Definition 1. Power system at time n is in normal steady state if it is in steady state and voltages of all nodes and

$$min\left\{U_n^{(1)}, U_n^{(2)}, \ldots, U_n^{(s)}\right\} \geq 0.95,$$
$$max\left\{U_n^{(1)}, U_n^{(2)}, \ldots, U_n^{(s)}\right\} \leq 1.05.$$

In our view various mathematical statements of the problem can take place. We consider the following deterministic formulation: to find maximal R that for some initial state of the power system there exists such control of the system, i.e. sequences $\left(TP_n, HP_n, CQ_n^{(1)}, CQ_n^{(2)}\right)$, $n = 1,2,\ldots,N$, that for all n power system is in a normal steady state for any $0 \leq R_n \leq \overline{R} \times \delta_n$.

Suppose that at least for $R = 0$ power system is in a normal steady mode for all n. Kazakh power system had been operated in a normal steady mode before as SPP was launched, therefore there exists a normal steady mode for all n and $\overline{R} = 0$.

For some state of power system, we can define interval $[0, S_n]$ such that for any $R_n \in [0, S_n]$ the power system is in a normal steady mode. Further we apply dynamic programming method: for $n = N$ and for any $\left(TP_N, HP_N, CQ_N^{(1)}, CQ_N^{(2)}\right)$, find

$$S_n = S_n\left(TP_n, HP_n, CQ_n^{(1)}, CQ_n^{(2)}\right).$$

Then for $\left(TP_n, HP_n, CQ_n^{(1)}, CQ_n^{(2)}\right)$ and other parameters $\overline{S}_n$ is maximal power of SPP for time instant $n = N$. Let $\overline{R}_n$ be maximal power of SPP for one last unit of time, it is clear that $\overline{R}_n = \overline{S}_n$. In the next step for $n = N - 1$ and for any $\left(TP_n, HP_n, CQ_n^{(1)}, CQ_n^{(2)}\right)$ we find S_n and define maximal power of SPP $\overline{S}_n$ for time instant $n = N - 1$. Then we can find a steady normal state of the system that satisfied restrictions (1) – (5) for $n = N$ and where $\overline{R}_n$ that is maximal power of SPP for two units of time $N - 1$ and N, i.e.

$$\overline{R}_{N-1} = min\{\overline{S}_{N-1}, \overline{R}_N\}$$

is maximal. If we repeat this procedure N times in a similar way we obtain

$$\overline{R} = \overline{R}_1$$

for some initial state of the power system $\left(V_s^{(s)}, V_0^{(s)}, \ldots, V_0^{(s)}\right)$. After simulating this approach for various initial states and other parameters like power of scheduled nodes, H, periods of time, time steps, etc., which repeat the parameters' fluctuations of the real power system we can assess the sought maximal power of SPS.

IV. Example

We consider a simplified model shown on Fig. 2. Our model consists of 15 nodes (buses), 21 transmission lines, 4 transformers, 1 TPP, 1 HPP, 1 SPP. Power system is connected with Russian grid. The SO can control TPP, HPP, capacitor battery at Vannovka (2 capacitors – 50 MVAR and 50 MVAR) and capacitor battery at Chulak-Kurgan (2 capacitors – 25 MVAR and 25 MVAR). Assume that $\underline{BP} = -200\,MW, \overline{BP} = 200\,MW$. The following figure shows the results of one of the calculations.

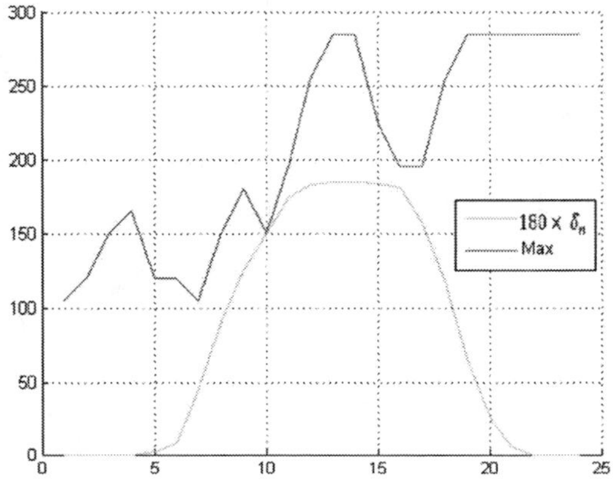

Figure 3. The results of calculations: green line maximal generation of SPP during a day with $\overline{R} = 180$, red line - $\overline{S}_n$.

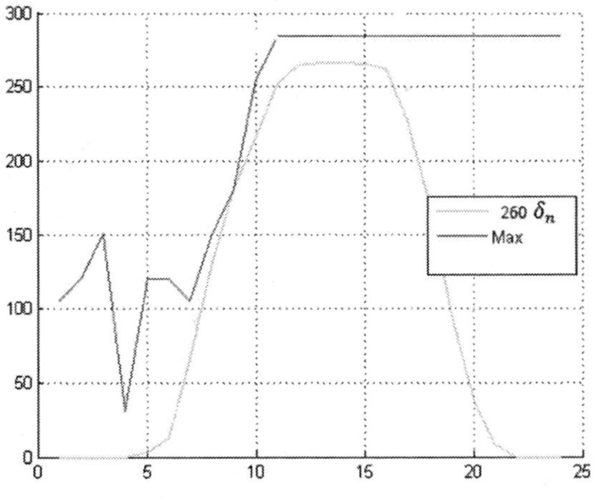

Figure 4. The results of calculations: green line maximal generation of SPP during a day with $\overline{R} = 260$, red line - $\overline{S}_n$.

The results of calculations show that maximal power of SPP, which can be integrated in the power system is 180 MW. It is clear that the value of maximal power of SPS is restricted by the problem at 10 hours. The problem is that the capacitor battery at Vannovka node has only two capacitors which restrict flexibility of their using: if one capacitor is turned on, the voltage is very low, if both are turned on, the voltage is very high. If we 'split' one capacitor on two parts and consider capacitor battery from 3 capacitors (25 MVAR, 25 MVAR and 50 MVAR) but with the same total capacity the maximal power of SPP can be significantly increased till 260 MW (Fig. 4).

V. Conclusion

We study the problem of integrating renewables in energy grid and applied simulating method. This method based on the study of the real system, the development of its logical-mathematical model and the modelling itself. We develop a simulating power flow model of power system as add-on structure on the known equivalent power flow. This approach allows to analyze the properties of the real system in time and its parameters as voltage, active and reactive power by means of simulation techniques.

As an example, we apply such approach for estimation of maximal power of solar park that can be integrated in the simplified model of Kazakh power system. It allows to obtain not only maximal power of solar park but also to analyze its various parameters under unpredictable renewable generation.

Acknowledgment

The authors like to thank the grant 1030/GF4 of Kazakhstan Ministry of Education and Science.

References

[1] C. Morris, M. Pehnt, "How is Germany integrating and balancing renewable energy today?," (The German energiewende book), I [Online] Available: https://energytransition.org/2015/02/how-germany-integrates-renewable-energy/. Last see (2015, Feb.)

[2] L.Bird, M. Milligan and D. Lew, "Integrating variable renewable energy: challenges and solutions", Technical Report NREL/TP-6A2060451, 2013.

[3] A. von Meier, "Challenges to the Integration of Renewable Resources at High System Penetration", Project Report, California Institute for Energy and Environmen [Online] Available: https://uc-ciee.org/downloads/CEC-500-2014-042.pdf. last see (2014, May)

[4] V. Vijay, "The Impact of Renewable Resources on the Performance and Reliability of the Electricity Grid." National Academy of Engineering Publications, Vol. 40 No. 1, [Online] Available: http://www.nae.edu/Publications/TheBridge/Archives/TheElectricityGrid/18587.aspx. Last see (2010, March)

[5] G. Bhatt, S. Affijulla, S. Debbarma. "Voltage Stability Improvement of Transmission Network Integrated with PV Plants," 2017 IEEE International Conference on Environment and Electrical Engineering and 2017 Industrial and Commercial Power System Europe (EEEC/ I&CPS Europe) Milan, Italy, 6-9 June 2017, pp. 2224-2228.

Fault identification at microgrid-connected converter and its ride-through capability

Viktor Valouch, Martin Čerňan
Faculty of Electrical Engineering
Czech Technical University in Prague
160 00 Prague 6, Czech Republic
viktor.valouch@fel.cvut.cz, cernama6@fel.cvut.cz

Petr Šimek, Jiří Škramlík
Institute of Thermomechanics
Academy of Sciences of the Czech Republic
182 00 Prague 8, Czech Republic
simek@it.cas.cz, skramlik@it.cas.cz

Abstract — **A model of the three-phase voltage converter for simulation of its modes of operation in standard as well as fault states is presented The presented model can be used to simulate steady-states and also transients at *dc* and *ac* side of the grid-connected converter. Based on simulation, an algorithm was designed to identify and localize a fault in some phase of the three-phase converter. The algorithm was experimentally verified.**

Keywords: grid-connected converter; fault identification and localization; reconfiguration; right-thought; simulation; experimental verification

I. Introduction

Fault tolerant converters have been intensively studied in several last years. Among main problems analyzed and tested belong fault diagnosis and reconfiguration of converter power circuitry and of its control strategy after a fault [1]-[6].

In the contribution, a model of the three-phase voltage converter for simulation of its modes of operation in standard as well as fault states is presented, which makes their analysis possible.

The contribution presents voltage and current responses at dc as well ac converter side for typical cases of faults. Based on analysis of these responses an algorithm of fault localization and its place has been developed. The strategy is based on evaluation of a converter ac current trajectory.

Further, rules for reconfiguration of the converter topology are presented. Transient responses in the interval between an instant of a fault and the HW and SW reconfiguration are discussed as well.

II. Converter model

The converter scheme in the full B6 configuration is presented in Fig. 1. The phase voltages v_{ga}, v_{gb}, v_{gc} are grid voltages.

The converter generates a PWM (Pulse Width Modulated) voltage, where switching functions S_{j+}, S_{j-}, $j = a, b, c$ of individual elements determine whether a firing pulse is applied to the respective switching element.

$$S_{j+-} = \begin{cases} 1 & \text{firing pulse is applied} \\ 0 & \text{firing pulse is not applied} \end{cases} \qquad (1)$$

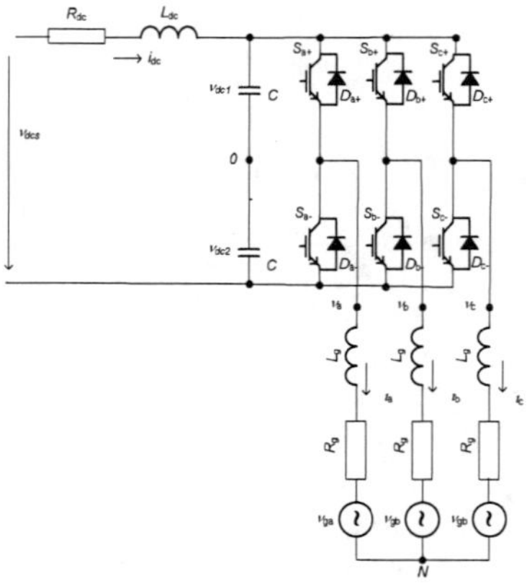

Figure 1. Circuit scheme of B6 converter. In case of fault, respective load phase is connected to midpoint 0 of two condensers

We can introduced generalized switching functions that determine to which dc bar a load phase j ($j = a, b, c$) is connected. The generalized switching functions consider also directions of phase load currents and the fact whether the switching element in respective converter phase is faulted or not.

$$\begin{aligned} \hat{S}_{j+} &= [h_{j+}S_{j+} \wedge (i_j > 0)] \vee [h_{j-}\overline{S}_{j-} \wedge (i_j < 0)] \\ \hat{S}_{j-} &= [h_{j-}S_{j-} \wedge (i_j < 0)] \vee [h_{j+}\overline{S}_{j+} \wedge (i_j > 0)] \end{aligned} \qquad (2)$$

$$\overline{S}_{j+-} = \text{NOT}(S_{j+-}) = S_{j-+}$$

where

$$h_{j+-} = \begin{cases} 1 & \text{switching element is not in fault} \\ 0 & \text{switching element is in fault} \end{cases} \qquad (3)$$

The model can be extended by including the cases of combined fault of a switching element and its free-wheeling diode or even of the whole phase damaged.

Now, we can describe the inverter circuit in Fig.1 by the following differential equations

$$\frac{di_{dc}}{dt} = \frac{1}{L_{dc}}(v_{dcs} - v_{dc1} - v_{dc2} - R_{dc}i_{dc}) \quad (4)$$

$$\frac{dv_{dc1}}{dt} = \frac{1}{C}(i_{dc} - \hat{\boldsymbol{S}}_{h+}\boldsymbol{i}) \quad (5)$$

$$\frac{dv_{dc2}}{dt} = \frac{1}{C}(i_{dc} - \hat{\boldsymbol{S}}_{h-}\boldsymbol{i}) \quad (6)$$

where

$$\hat{\boldsymbol{S}}_{h+} = (\hat{S}_{a+}h_a, \hat{S}_{b+}h_b, \hat{S}_{c+}h_c)^T \quad (7)$$
$$\hat{\boldsymbol{S}}_{h-} = (\hat{S}_{a-}h_a, \hat{S}_{b-}h_b, \hat{S}_{c-}h_c)^T$$

$$\boldsymbol{i} = (i_a, i_b, i_c)^T \quad (8)$$

and it holds for the phase currents $\boldsymbol{i}$ of the inverter load

$$\frac{d\boldsymbol{i}}{dt} = \frac{1}{L_g}(\boldsymbol{v} - \boldsymbol{v}_g - R_g\boldsymbol{i}) \quad (9)$$

where the phase voltages (between *ac* inverter terminals and the load neutral point N) are described by

$$\boldsymbol{v} = (v_a, v_b, v_c)^T = \boldsymbol{T}_h(\hat{\boldsymbol{S}}_{h+}v_{dc1} - \hat{\boldsymbol{S}}_{h-}v_{dc2}) \quad (10)$$

The matrix $\boldsymbol{T}_h$ shows the relation among the voltages $\boldsymbol{v}$ and those measured between the ac inverter terminals and the midpoint 0, where the identifiers h_j, $j = a, b, c$ are also included

$$\boldsymbol{T}_h = \frac{1}{3}\begin{pmatrix} 2h_a & -h_b & -h_c \\ -h_a & 2h_b & -h_c \\ -h_a & -h_b & 2h_c \end{pmatrix} \quad (11)$$

III. Converter voltage and curent responses

The following results in the paragraph have been obtained for the parameters of the scheme in Fig. 1

$v_{dcs} = 200$ V, $R_{dc} = 0.1\ \Omega$, $L_{dc} = 0.005$ H
$v_{ga} = v_{gb} = v_{gc} = 0$, $R_g = 10\ \Omega$, $L_g = 0.005$ H

Thus, it is the load RL without an inner voltage (generated by an ac machine or represented a grid voltage).

Fig. 2 shows responses of the load phase currents i_a, i_b, i_c for all three possible faults: a) fault of the switching element S_{a+} only; b) fault of the upper leg S_{a+}, D_{a+}; c) the whole phase a is out of the order. The times of successive changes are 0.175 s a 0.24 s.

Fig. 3 presents the respective phase voltage v_a, v_b, v_c responses. It is evident that in case of the faults a), b), the voltage responses have very similar waveforms, which is not the case c).

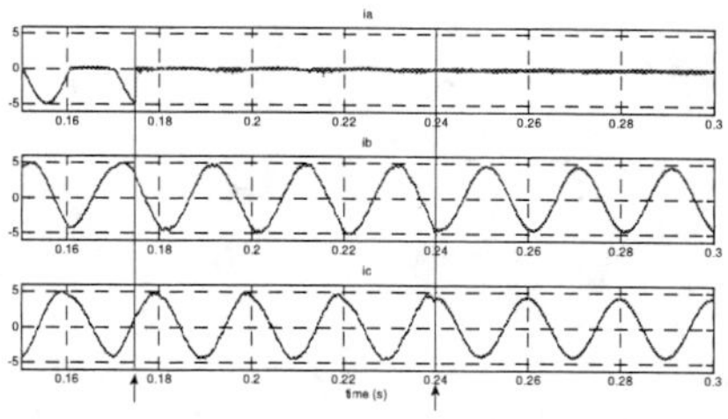

Figure 2. Load phase currents response: a) fault of switching element S_{a+} only (till 0.175s); b) fault of upper leg S_{a+}, D_{a+} (between 0.175 and 0.24s); c) whole phase *a* is out of the order (after 0.24s)

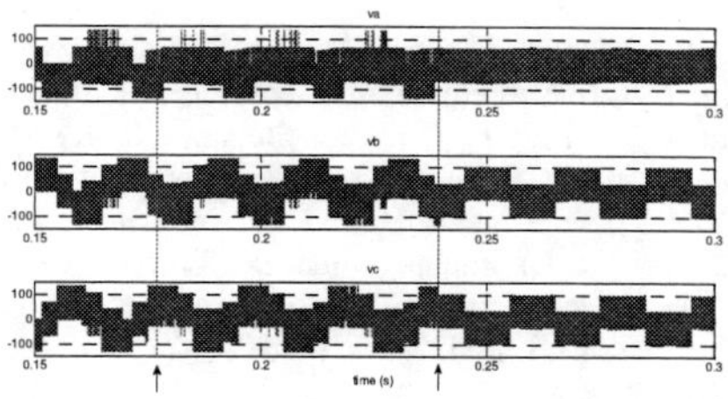

Figure 3. Phase voltage v_a, v_b, v_c responses for the same fault sequence as that shown in Fig. 2 where load current waveforms were presented

IV. Fault identification

Fig. 4 shows an example how a load current i trajectory changes in case of a fault in the phase a. If we want to evaluate a fault, it is recommended to observe the vector of filtrated phase currents (here a simple LP filter with the time constant 0.5 ms was applied).

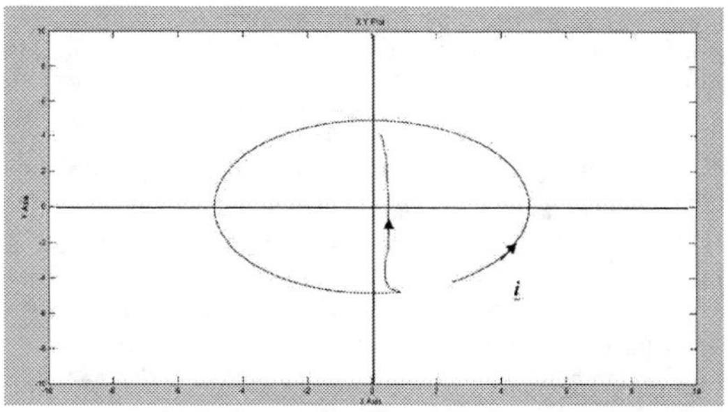

Figure 4. Load current i trajectory in case of fault in phase *a*

If only the switching element S_{a+} is damaged, the current vector draws only the left side part of the circle ($i_a < 0$) in a steady state, because $i_a > 0$ can not flow through the element S_{a+} (Fig. 1). But, if also the diode D_{a+} or even the second switching element Sa- is in fault, the current vector will move along a line perpendicular to the axis of the phase *a*. In case of a fault in the phase *b* or *c*, the current vector trajectory will look similarly, but will be rotated by the angle ±120°.

A difference between a fault of the elements S_{a+}, D_{a+} and that of both the legs in the phase a manifests itself by a different phase voltage trajectory in this phase.

We can formulate the conditions for occurrence faults in all three phase

$$i_{\text{atest}} = \left|\bar{i}_{\alpha\text{filt}}\right| \leq \varepsilon \quad (12)$$

$$i_{\text{btest}} = \left|\sqrt{3}\bar{i}_{\beta\text{filt}} - \bar{i}_{\alpha\text{filt}}\right| \leq \varepsilon$$

$$i_{\text{ctest}} = \left|\sqrt{3}\bar{i}_{\beta\text{filt}} + \bar{i}_{\alpha\text{filt}}\right| \leq \varepsilon$$

where $i_{\alpha\text{filt}}$, $i_{\beta\text{filt}}$ are normalized orthogonal components of the load current vector $\boldsymbol{i}$

$$\bar{i}_{\alpha\text{filt}} = i_{\alpha\text{filt}} / \sqrt{i_{\alpha\text{filt}}^{2} + i_{\beta\text{filt}}^{2}} \quad (13)$$

$$\bar{i}_{\beta\text{filt}} = i_{\beta\text{filt}} / \sqrt{i_{\alpha\text{filt}}^{2} + i_{\beta\text{filt}}^{2}}$$

Fig. 5 shows the responses of the calculated current indicators i_{atest}, i_{btest}, i_{ctest} (13) for the same fault sequence as that shown in Fig. 3 where the load current waveforms were presented. Two times a period the indicator i_{jtest}, $j=a,b,c$ touches zero at moments when the current vector i is perpendicular to the axis of the respective phase. It is clear that the responses in healthy phases b, c and that in the damaged phase a are dramatically different. It may be used to locate the place of the fault.

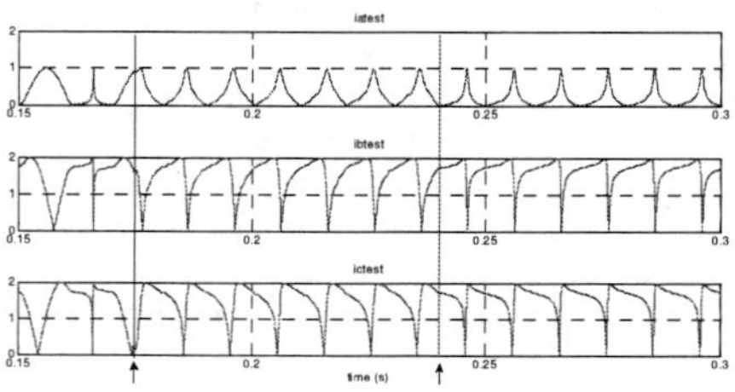

Figure 5. Responses of calculated current indicators i_{atest}, i_{btest}, i_{ctest} (16) for the same fault sequence as that shown in Fig. 3 where load current waveforms were presented

Let us introduce a quantity „*fault_j_ident*", *j=a,b,c*. It is the output of a counter that counts how often the condition (12) is fulfilled within a time interval T_s. At the end of the interval T_s the counter is reset. It is obvious that the sum „*fault_j_ident*" is high if the variable i_{jtest} is located around zero during this interval. Thus the value of „*fault_j_ident*", *j=a,b,c* can be used for occurrence of a fault in the phase *j*.

Fig. 6 shows the captured experimental responses of the quantities „*fault_j_ident*", *j=a,b,c* in case of the fault in the phase *a*. Now, the converter was connected to the grid and the parameters were as follows

$v_{\text{dcs}} = 200$ V, $R_{\text{dc}} = 0.1\ \Omega$, $L_{\text{dc}} = 0.005$ H

$v_{\text{ga}} = v_{\text{gb}} = v_{\text{gc}} = 45$ V, $R_{\text{g}} = 0.1\ \Omega$, $L_{\text{g}} = 0.005$ H

To synchronize the converter to the positive sequence of the grid voltage the developed method CDSC+DSOGI-FLL (Cascaded Delayed Signal Cancellation + Dual Second Order Generalized Integrator-Frequency Locked Loop) has been applied [7].

It is evident that the introduced variable „*fault_a_ident*" can without doubts identify this phase.

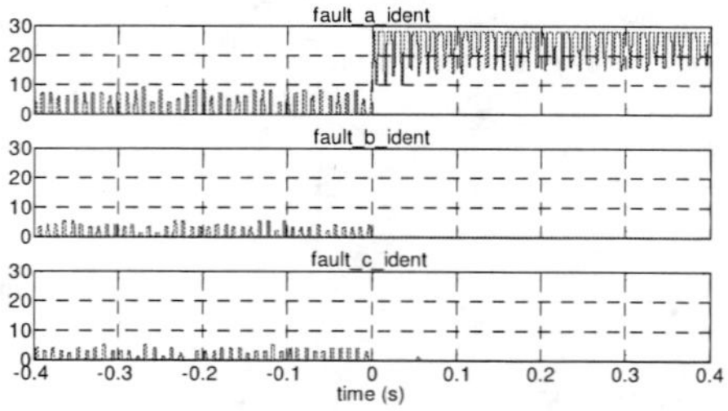

Figure 6. Captured experimental responses of quantities „*fault_j_ident*", *j=a,b,c* in case of fault in phase *a*

V. Converter Reconfiguration and Operation in Regime B4

If a fault occurs, it is necessary to change the converter topology to the B4 structure where only two healthy phases are active. The phase under the fault should be connected to the midpoint 0 of the pair of two dc condensers. In a time interval between the occurrence of the fault and its identification a transient process appears. Only after that a new steady state in the structure B4 is reached. The behaviour of the converter B4 structure has been extensively studied in past years, e.g. [8].

Fig. 7 shows the responses of the phase grid currents, the reference and real active and non-active powers (calculated from non-filtered voltages and currents), and the phase grid voltages. The grid-connected converter was equipped with the power controllers working in the stationary reference frame. The arrows mark the instants of the changes in the reference powers, in the grid voltages, and mark also the interval 0.2-0.25 s when the identification of the fault in the phase *a* and the converter reconfiguration in the B4 structure was performed. Before the reconfiguration performed the power references were nullified. We can see that the current transients in the B6 and B4 structure are almost identical for all combinations of power references and grid voltages, but the power fluctuation of the frequency of 100 Hz is more pronounced in the B4 configuration than that in the full B6 structure.

VI. Conclusions

The presented model of the three-phase converter can be used to simulate steady-states and also transients at dc and ac side of the grid-connected converter. The model allows simulate also failure states. Based on simulation, an algorithm was designed to identify and localize a fault in some phase of the three-phase converter. The algorithm was experimentally verified.

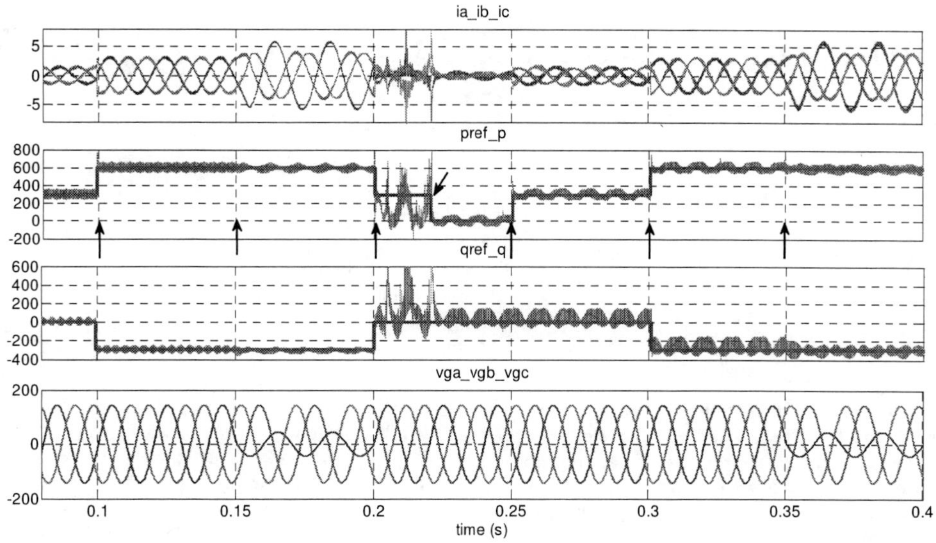

Figure 7. Responses of phase grid currents, reference and real active and non-active powers, and phase grid voltages before and after fault in phase *a*, which was identified in time interval 0.2-0.25 s, when reconfiguration from B6 to B4 converter structure was also performed

ACKNOWLEDGMENT

The financial support of the Institute of Thermomechanics (project RVO//:61388998) is highly acknowledged.

REFERENCES

[1] R. Peuget, S. Courtine, and J.-P. Rognon, "Fault detection and isolation on a PWM inverter by knowledge-based model," *IEEE Transactions on Industry Applications*, 1998, pp. 1318 –1326.

[2] A. Tanwani, A.D. Dominguez-Garcia, and D. Liberzon, "An Inversion-Based Approach to Fault Detection and Isolation in Switching Electrical Networks," *IEEE Transactions on Control Systems Technology*, 2011, pp. 1059 –1074.

[3] A. Yazdani, H. Sepahvand, M.L. Crow, and M. Ferdowsi, "Fault Detection and Mitigation in Multilevel Converter STATCOMs," *IEEE Transactions on Industrial Electronics*, 2011, pp. 1307–1315.

[4] Z. Zeng, W. Zheng, R. Zhao, C. Zhu, and Q. Juan," The Comprehensive Design and Optimization of the Post-Fault Grid-Connected Three-Phase PWM Rectifier," *IEEE Transactions on Industrial Electronics,* 2016, pp. 1629–1642.

[5] P. S. Das, K.-H. Kim, "Open-Switch Fault-Tolerant Control of a Grid-Side Converter in a Wind Power Generation System," International Journal of Power Electronics and Drive System (IJPEDS), Vol. 6, No. 2, June 2015, pp. 293-304.

[6] C. Cecati, A. O. DiTommaso, F. Genduso, R. Miceli, and G. R. Galluzzo,"Comprehensive Modeling and Experimental Testing of Fault Detection and Management of a Nonredundant Fault-Tolerant VSI," *IEEE Transactions on Industrial Electronics,* 2015, pp. 3945–3953.

[7] Bejvl, M., Šimek, P., Fandi, G., Čerňan, M., Muller, M., Muller, Z., Tlustý, J., Valouch, V.: Contribution to Control Strategies for Converter Connected to Unbalanced Grid. SETIT 2016 – The 7th Inter. Conference Sciences of Electronics, Technologies of Information and Telecommunications, Hammamet, Tunisia, December 18-20, 2016, pp. 1–6, ISBN: 978-1-5090-4712-3

[8] Klíma, J., Škramlík, J., Valouch, V.: Four-Switch Three-Phase Space-Vector PWM AC-DC Converter. International Journal of Electronics, Vol. 95, No. 4, April 2008, pp. 383–397

Simulation Method for Determining the SAIDI and SAIFI Indicators

Mirosław Kornatka
Faculty of Electrical Engineering,
Częstochowa University of Technology
Częstochowa, Poland
kornatka@el.pcz.czest.pl

Abstract — **One of the commonly recognized targets of Smart Grid aspect is reliability of electricity supply. Nowadays new approaches are required for evaluating the reliability of electricity distribution as well as planning the reliability of supply new grid. This paper presents methodology developed to calculate in these networks the most commonly used reliability indices which are the SAIDI (System Average Interruption Duration Index) and SAIFI (System Average Interruption Frequency Index). The developed analysis method is based on kernel density estimation (KDE) and Monte Carlo (MC) simulation algorithm, which provides an opportunity to assess statistically the reliability performance of a network. This methodology has been applied on a real MV distribution network.**

Keywords: simulation; reliability indices; Monte Carlo; SAIDI; SAIFI

I. Introduction

It is possible to analyse power distribution networks in a number of respects, including the optimization of exploitation cost, loss minimization, or impact of connecting new power generation sources and energy inventories [1-5]. Recently, special attention is paid to the issue of reliability of power supply [1, 6-8]. Especially, in the case of production processes that require very high supply continuity, the importance of a reliable supply increases. Issues concerning the power system reliability can be analyzed in many aspects, for example: assessment of the existing levels of reliability, security and power quality.

In distribution network planning, the supply reliability is traditionally approached through deterministic reliability analyses and depicted with average indicators, such as customer-based reliability indices SAIFI and SAIDI [6-9].

The proposed stochastic reliability analysis of SAIFI and SAIDI is based on a MC simulation algorithm. The algorithm is used for simulating the annual customer interruptions on the basis of the probability density function (PDF) of fault frequencies, fault duration and power lines length. As a result, the simulations produce the PDF of the numbers and the durations of customer supply interruptions.

Evaluation of reliability of the MV overhead lines is one of the main points considered during the distribution power system analysis. In the case of failure, both the energy producers and its customers can suffer very important economic losses. Especially, in the case of production processes that require very high supply continuity, the importance of a reliable supply increases.

Research shows that 84% of the total values of the distribution network index SAIDI are caused by the failure rate of the MV network, about 15% are caused by the LV network and only 1 % by the HV network. Special attention should be paid to the analysis of the condition of the MV network in order to locate its weak points. The scope of this paper is thus restricted to fault events occurring in the MV overhead line, and faults in the high and low voltage networks are not dealt with.

II. Reliability Indices

The reliability of electricity supply can be defined based on the numbers and durations of interruptions in a customer's voltage supply. The concept of supply reliability has to be separated from the concept of network reliability. The reliability of the electricity distribution depends on the reliability of the network and the security of operating the network. From the viewpoint of customers, the reliability of power supply is the most essential.

Reliability can be, for instance, defined on annual basis or as the mean value over a set of years. When considering the supply reliability experienced by a customer, a very often statistical distribution having a mean value and a standard variation can be defined based on annual variation.

The fundamental reliability indices calculated for the electrical distribution system are the following:

- average failure frequency;
- average duration of outage;
- annual unavailability of the receiving node;
- expected value of undelivered energy;
- indices SAIDI and SAIFI.

SAIDI and SAIFI are the most important ones in the distribution network of various voltage levels. SAIDI gives the average amount of time per year while the supply of a customer is interrupted. It is expressed in minutes per customers per year and calculated by means of the following expression:

$$\mathrm{SAIDI} = \left(\sum r_i \, N_i\right) / N_T \,, \qquad (1)$$

where the summation is taken over all incidents, either at all voltage levels or only at selected voltage levels; r_i gives the restoration time for each incident; N_i gives the number of customers interrupted by each incident; N_T gives the total number of customers in the system for which the index is calculated [6].

SAIFI gives the average number of times per year while the supply of a customer is interrupted. It is expressed in interruptions per customer per year and calculated by means of the following expression [6]:

$$\mathrm{SAIFI} = (\sum N_i) / N_T . \quad (2)$$

In Poland, according to the Ordinance of the Minister of Economy of 2007 on the specific conditions of the functioning of the power system, by 31 March of every year, the Distribution Network Operations (DNOs) are obliged to announce on their website the indicators concerning the time of interruptions in electricity supply calculated for the preceding calendar year. The role of DNOs is to guarantee the distribution network reliability and ensuring the security of supply energy to each consumer within its own area of operation [9].

SAIDI and SAIFI are calculating including or excluding specific cause of interruption such as force majeure, exceptional events etc. In Poland breaks in electricity supply are divided into: short – between 1 second and 3 minutes; long - between 3 minutes and 12 hours; very long - between 12 and 24 hours and disastrous – longer than 24 hours [9].

III. Description Of The Saidi And Saifi Estimation Method

Analytical methods may be unsuitable for solving reliability problems, possibly for the reason of time constraints. In such situations, the Monte Carlo method (MC) can be applied. MC consists in constructing probabilistic models, algorithms or computer programmes simulating random processes.

The exploitation process of the power system provides important data on the operation of component objects of the system performing their functions. Therefore, it is possible to use an MC simulation to estimate the reliability indices SAIDI and SAIFI of objects under scrutiny in the power distribution system.

In this method of SAIDI and SAIFI estimation it is assumed that:

- The total time of fault localization and repair is subject to a statistical distribution obtained on the basis of data provided by an object. The time necessary for reaching the place and carrying out a repair is assumed as uniform.
- The probability of a fault occurring in an object is equally distributed over the object area and is proportional only to that area/length/number. A PDF of an object fault index is obtained on the basis of data on faults that have previously occurred in this object.
- Consumers are distributed equally on the area of the object under scrutiny.
- The number of consumers fed from the part of the distribution system under scrutiny is proportional to the share the area/length/number of the object(s) takes in the total area/length/number of objects in the distribution system.

Software development tools such as R or MatLab are equipped with pseudorandom number generators, such numbers being subject to classical probability distribution. It is however necessary to obtain pseudorandom numbers with probability distribution corresponding to empirical data.

The starting point is an estimation of the density function distribution and at the same time its cumulative distribution function (CDF) on the basis of the empirical data available. To this end, this study uses a commonly applied method of non-parametric kernel density estimation, described e.g. in [10]. Of great importance is the accuracy of representing the CDF. The division of the CDF values into equal intervals affects the magnitude of the matrix and the numerical values that can be generated in the MC simulation. To overcome this limitation, it is possible to approximate the CDF values by means of a continuous function. Then it is necessary to verify the accuracy of the approximation of the real CDF values by means of the function. The method of obtaining a random sample from any distribution by means of a MC simulation is presented in Fig. 1.

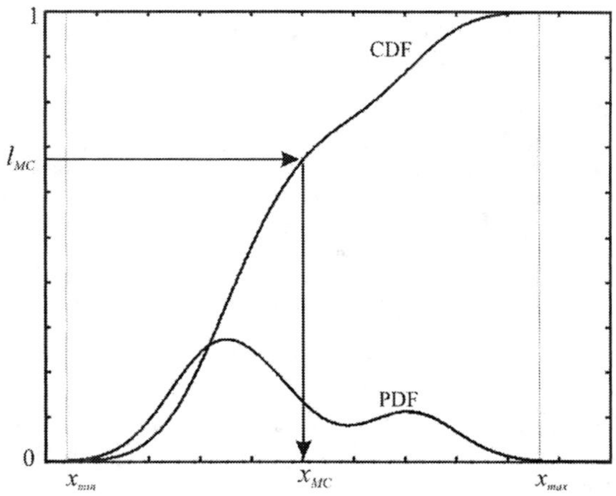

Figure 1. A method for obtaining a random sample by means of the Monte Carlo simulation with the probability distribution determined by an empirical CDF, as described in the text

A value of the CDF for the variable x_{MC} corresponds to the pseudorandom number l_{MC} drawn from the interval 0 ÷ 1 with uniform distribution. After n MC simulations have been performed, n values are obtained of numbers being subject to probability distribution determined by the empirical CDF. All the values of the random variable are in the interval x_{min} to x_{max}.

Below a method of estimating the SAIDI and SAIFI indices for medium-voltage overhead lines will be presented. The indices have been calculated by means of the following procedure:

- Obtain the PDF for the estimated variables in overhead MV lines, i.e. for the line length, failure rate and interruption duration.
- Perform an MC simulation n times, by drawing n pseudorandom numbers from the interval <0,1>, separately for each random variable. Then, determine line lengths, failure rate and interruption durations for the drawn numbers on the basis of the corresponding CDFs.
- Determine the values of SAIDI and SAIFI on the basis of their definitions for the drawn numbers. After n MC simulations there will be n values of SAIDI and SAIFI, which will be further used for determining the PDF.
- The number of consumers supplied from the part of the power distribution network under scrutiny is proportional to the share of the overhead MV lines in the total length of lines. In the distribution system analysed in the study this share is 0.23.

IV. Stochastic Reliability Analysis Of Mv Overhead Line – A Case Study

The MV overhead lines are representing a very important part of the distribution network. This network generally works as a radial network, divided into so-called sections of the network.

A real distribution system is used for the simulations. The analyzed network supplies 10 658 thousand customers (N_T) using the 128 655 km of MV overhead lines (the total length of the MV lines is 173 299 km). In the years 2012-2014, 38 869 disconnections of lines (on the average, yearly) occurred in the fields of MV overhead lines. Essentially, the MV power lines belong to the 19 independent DNOs.

A. *Assessment of the MV Lines Duration of Interruptions*

Users commonly interpret the reliability of the distribution power network through the prism of the duration of interruptions in supply. The duration of failure is defined as the time interval since its beginning till the return of voltage, to which the transmission capability, meaning the user's electrical power requirement, simultaneously corresponds. On the basis of the author's research, concerning the average duration of permanent interruptions of the MV overhead lines in the selected DNO in the years 2012÷2014, the empirical PDF has been determined with use of the KDE (Fig. 2).

In this paper, all PDFs have been estimated with use of the R functions using normal kernel and distribution limited to positive values [11].

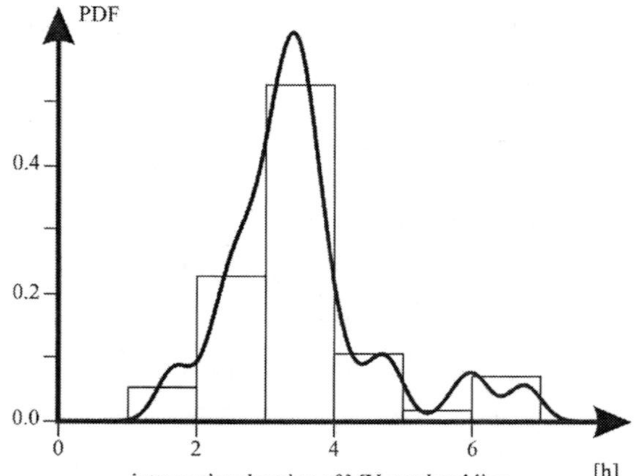

Figure 2. The PDF of the average duration of interruptions of the MV overhead lines (the smoothing parameter in KDE was equal to 0.24)

Analyzing the above graph, one should notice a great variability of the average outage time duration in the tested division. The modal value of the outage time duration (during the analyzed period) is equal to 3.4 h. Certainly, the time durations of interruptions are different for different groups of customers involved in the same incident. Therefore, the sum must be extended to each group of customers suffering the same interruptions time.

B. *Assessment of the MV Lines Failure Rate*

Analysis of failure rate is one of the methods of the MV power lines reliability evaluation. Disregarding its imperfection that neglects the differences for particular line tracks, many distribution companies commonly apply this indicator. The empirical PDF of damage index has been determined on the basis of the author's research, concerning the failure rate of the MV overhead lines in the selected DNO in the years 2012÷2014. The calculation results are presented in Fig. 3.

The KDE has been applied for the normal kernel with the smoothing parameter equal to 1.11. The variability of the MV overhead lines failure rate is significant. Such great differences may be caused by the variable conditions of networks. The modal value of the index is equal to 9.5 outages annually per 100 km of the MV overhead line.

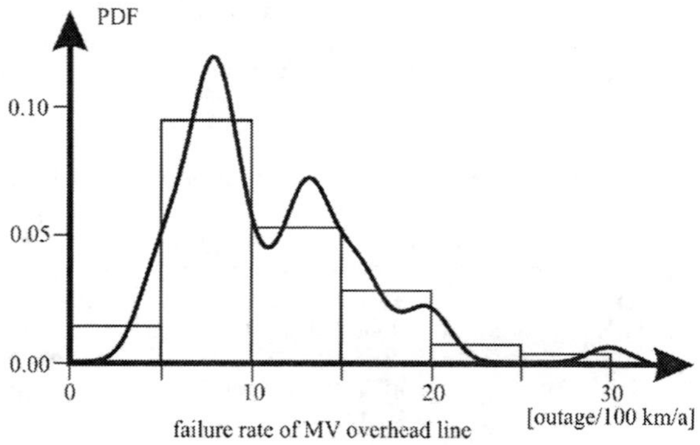

Figure 3. The PDF of the average failure rate of the MV overhead lines

C. *Units Length of the MV Overhead Lines*

The average length of the MV overhead lines has been determined on the basis of the author's research of the same DNOs data. Fig. 4 shows the PDF of the MV power lines length. The smoothing, being a calculated parameter, has reached the value 7.05.

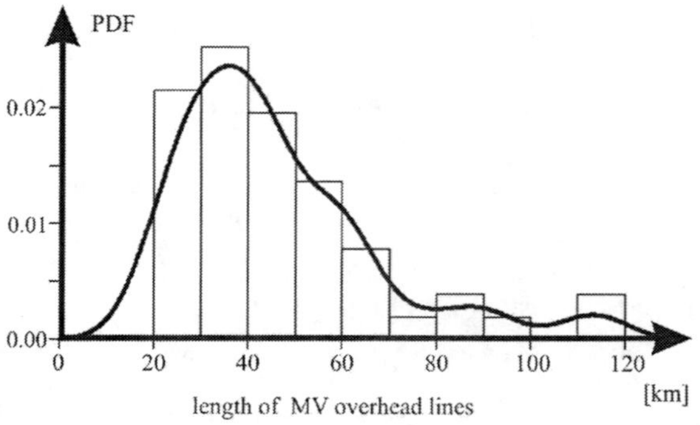

Figure 4. The PDF of the MV overhead lines length

The overhead tracks of a length about 40 km dominate but there are also tracks of the length up to 120 km. Such significant differences in the length of the lines may result from the unevenness of the DNO areas covered by the MV/LV transformer stations.

D. *Assessment of the SAIDI and SAIFI of MV Overhead Lines*

The values of SAIDI and SAIFI of the network under study were used as the basis of model verification. The number of simulation sequences was chosen to be 20 thousand. The network used in testing of the model is a typical MV overhead network in Poland. The PDF of SAIDI for the network is presented in Fig. 5. The smoothing parameter was equal to 15.7.

The deviation in the values of SAIDI (average interruption duration per customer) are large due to varying operating conditions of networks. Fig. 5 shows that customers supplied by the MV overhead lines have on average suffered about 166 minutes of non-planned interruption per year. In these calculations, neither generation plant breakdown nor the LV and MV lines interruptions were taken into account.

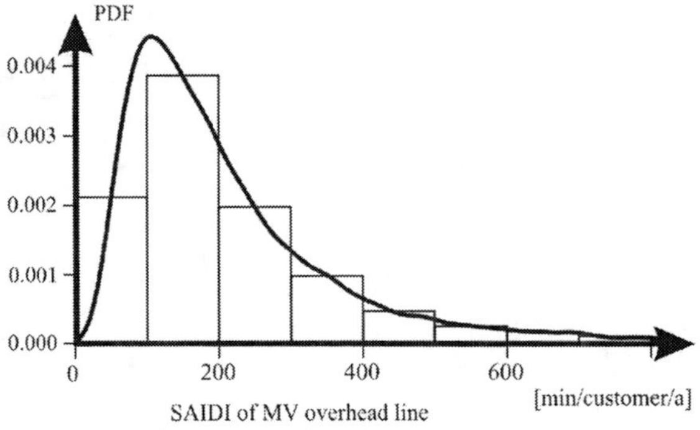

Figure 5. The PDF of the SAIDI of MV overhead lines

The results of SAIFI indices calculations for 20 thousand Monte Carlo simulations are presented in Fig. 6. The smoothing parameter was equal to 0.12.

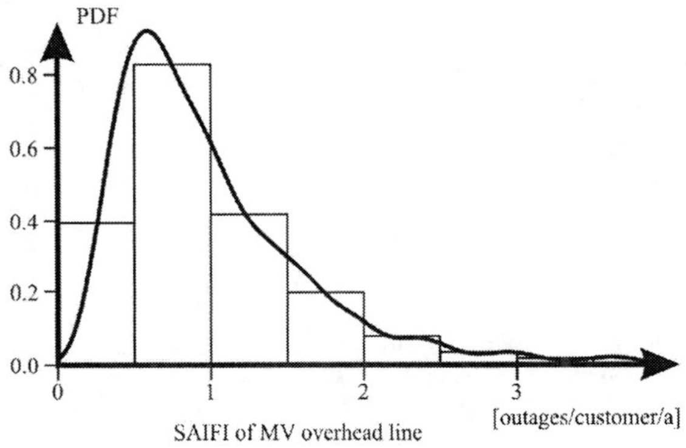

Figure 6. The PDF of the SAIFI of MV overhead lines

During one year, these customers can expect less than one long and very long interruptions in overhead lines (the modal values equal to 0.84 outages annually per customer).

Unfortunately, it is impossible to compare the presented simulations with the real data of SAIDI and SAIFI . In Poland, the DNOs are obliged to announce on their website only the total values of SAIDI and SAIFI for the preceding calendar year, without separating them into the values for HV, MV and LV networks.

V. Conclusion

The need to reduce the value of the SAIDI and SAIFI indicators requires from the Distribution Network Operators to perform ongoing analysis of the network condition in order to improve its reliability.

The method presented in the study enables estimation of the SAIDI and SAIFI indices for any network on the basis of easily available empirical data on failure frequency and duration in this network.

Monte Carlo simulations are a suitable tool for performing statistical analyses of the distribution of SAIDI and SAIFI for any distribution of faults in a distribution network, including theoretical ones.

From the viewpoint of the network operator making a decision about future investments, it is of crucial importance to obtain PDFs of SAIDI and SAIFI for the particular types of network.

The analysis proves that the PDFs of the particular random quantities being estimated are not generally submitted to the classical parametric distributions.

Knowledge about the reliability of MV overhead lines is very important and vital for taking decisions concerning investment.

References

[1] M. Kolcun, and P. Frak, "Grid reliability analysis as a tool for effective asset management", Proc. 5th International Scientific Symposium Elektroenergetika 2009 Stará Lesná, 2009 pp. 60-62.

[2] J.Sowiński and R. Tomaszewski, "Modelling of power engineering processes under risk", Proc. 8th International Scientific Symposium on Electric Power Engineering "Elektroenergetika2015", Stara Lesna, Slovak Republic, 16-18 September 2015, pp. 396-399.

[3] J. Szkutnik and A. Gawlak, „Dynamiczna efektywność rozdziału energii elektrycznej w sieciach rozdzielczych". Rynek energii No. 2 (81) ISSN 1425-5960 April 2009 pp. 36-41.

[4] J. Sowiński, R. Tomaszewski and A. Wachtarczyk, "Electricity generation from RES under uncertainty," Proc. 17th International Scientific Conference on Electric Power Engineering, Prague, 2016, pp. 1-4. doi: 10.1109/EPE.2016.7521733.

[5] A. Gawlak, R. Gono, S. Rusek and J. Szkutnik, "Method of calculating energy losses in distribution networks". Rynek energii No. 5 (98) ISSN 1425-5960 September 2012, pp.126-131.

[6] R. Billinton, and R. N. Allan, Reliability Evaluation of Power Systems, (2nd ed.), New York, Plenum Press, 1996.

[7] R. Briš, P. Byczanski, R. Goňo and S. Rusek, "Discrete maintenance optimization of complex multi-component systems", Reliability Engineering and System Safety (2017), http://dx.doi.org/10.1016/j.ress.2017.04.008, in press.

[8] 6th CEER Benchmarking Report on the Quality of Electricity and Gas Supply Ref: C16-EQS-72-03, CEER, Brussels, 2016, (also previous editions).

[9] Minister of Economy order in the matter of detailed operation of the power system, 4th may 2007, § 41.

[10] B. W. Silverman, "Density estimation for statistics and data analysis", London: Chapman & Hall/CRC. 1986.

[11] R Development Core Team "R: A language and environment for statistical computing".

Influence of Load on Harmonic Emission of Single Phase Rectifiers from Modern Lighting Lamps

Róbert Istók, István Szén
Óbuda University, Kálmán Kandó Faculty of Electrical Engineering
Power System Department Budapest, Hungary
istok.robert@kvk.uni-obuda.hu, szen.istván@kvk.uni-obuda.hu

Abstract — **Modern lighting lamps have an important harmonic emission. Compact fluorescent lamps typically have a higher harmonic emission than LED- lighting emission diodes. In case of LED could be find important differences in case of harmonic emission. Emission of some LEDs type are as high as emissions of compact fluorescent lamps. The most important source of harmonic emission is the rectifier stage of electronic ballast. The compact fluorescent lamp could be divided in two parts. One part is the electronic ballast and the second part is fluorescent tube. The fluorescent tube is a nonlinear load. The harmonic emission of electronic ballast loading by fluorescent tube is higher. If the electronic ballast has a linear resistive load, the harmonic emission is lower than in case when the load is the fluorescent tube.**

Keywords: harmonic emission; total harmonic distortion; compact fluorescent lamp, lighting emission diode, electronic ballast, single phase rectifiers, modern lighting, electromagnetic compatibility, power factor correction

I. Introduction

Replacing of the incandescent bulbs are doing continuously. The main reason of replacing is the low efficiency of the incandescent bulbs. Using of modern lamps conduct to a decrease of energy consumption in lighting field. The energy saving is an important thing, which the modern lighting try to fulfil, but the apparition of modern lamps determines new questions. One of them is electromagnetic compatibility – EMC [1]. All the modern lamps: compact fluorescent lamps, induction lamps, LED, are driven by electronic ballasts. It is well known that the functionality of the electronic ballast is accompanied by noise emissions. This noise emission are low and high frequencies. The spreading path of electromagnetic noise are next: conducted through power and signal cable and radiated. The conducted emission is divided in two groups: low frequency conducted emission as harmonic and flicker emission. In case of modern lighting the low frequency, noise is mean typically the harmonic emission. The flicker emission does not represent a problem in case of lighting lamps.

II. Harmonic and total harmonic distortion

A linear load connected to an electric power system is defined as a load which draws current from the supply which is proportional to the applied voltage. The incandescent lamp is a linear load and does not produce harmonics. In Fig. 1 is presented the voltage and the current wave form of an incandescent lamp. The current and voltage wave form are booth sinusoidal. A non-linear load impedance changes with the applied voltage. Due to this changing impedance, the current drawn by the non-linear load is also non-linear, even when it is connected to a sinusoidal voltage source. The compact fluorescent lamp is a non-linear load, and when is connected to power system, it draws a current that is not sinusoidal. A dominant part of the electronic ballast which drive the fluorescent lamp is a diode rectifier. To reduce the ripple in the dc output voltage, a suitable filter capacitor and/or an inductor is used at the rectifier output [2] [3]. But due to these reactive components, the current drawn by this converter is a series of relatively narrow positive and negative pulses rather than the more extended waveform of a pure sinewave. This waveform is rich in harmonics, particularly at low frequencies. The most important harmonics are: 3rd, 5th, 7th and 9th harmonics. The current waveform can become quite complex, depending on the type of load and its interaction with other components of the system. In Fig. 2 is presented the voltage and current wave form of a compact fluorescent lamp.

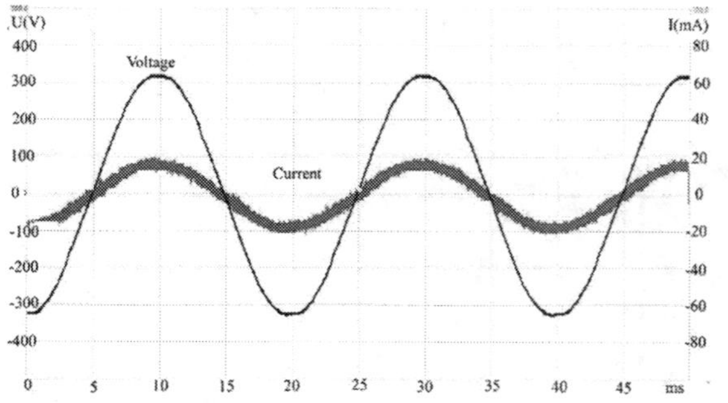

Figure 1. Voltage and current wave of an incandescent lamp

The current waveform is described with help of Fourier series analysis [4]. It is divided into a series of simple sinusoids, which start at the power system fundamental frequency and occur at integer multiples of the fundamental frequency.

The voltage from power source is distorted by current harmonics due to source impedance. If the voltage source has small source impedance, the current harmonics will cause only small voltage harmonics. This is the case when voltage harmonics are small compared to current harmonics. It is the casse for lighting electronic ballast.

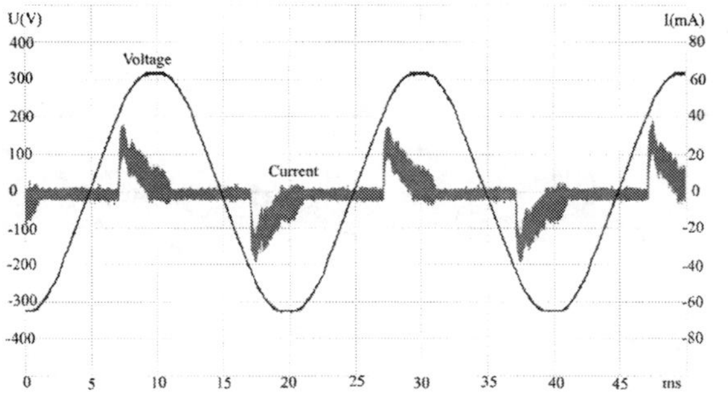

Figure 2. Voltage and current wave of a compact fluorescent lamp

Total Harmonic Distortion (THD) is a measurement of the level of harmonic distortion present in power systems. Total harmonic distortion can be related to either current harmonics or voltage harmonics, and it is defined as the ratio of total harmonics to the value at fundamental frequency times 100% [5].

$$THD_V = \frac{\sqrt{V_2^2 + V_3^2 + \lambda + V_n^2}}{V_1} \cdot 100\% = \frac{\sqrt{\sum_{k=2}^{n} V_k^2}}{V_1} \cdot 100\% \text{, (1)}$$

$$THD_I = \frac{\sqrt{I_2^2 + I_3^2 + \lambda + I_n^2}}{I_1} \cdot 100\% = \frac{\sqrt{\sum_{k=2}^{n} I_k^2}}{I_1} \cdot 100\% \text{ ,(2)}$$

where V_n is the RMS voltage of nth voltage harmonic; I_n is the RMS current of the nth current harmonic; n=1 fundamental frequency.

III. Effects of harmonics

A. Motors and generators

The voltage and current harmonics in motors and generators increase heating due to iron and cooper losses at the harmonic frequencies, and affects the efficiency and developed torque too. The harmonic currents in a motor can rise the audible noise emission, and produce a resultant flux distribution in the air gap. The fifth and seventh harmonics can create mechanical oscillation in a turbine-generator combination or in a motor load system.

B. Transformers

Current harmonics cause an increase in cooper loses and stray flux losses. Voltage harmonics cause an increase in iron losses. The result is transformer heating increases. The losses increase with frequency. The higher frequency harmonic components are more important that lower frequency component. Another effect of harmonics is increasing audible noises.

C. Power cables

If cables take part in a resonance system, could be affected by voltage stress and corona, which can produce dielectric failure.

D. Capacitors

The reactance of a capacitor bank decrease with frequency, and the bank acts as a sink for higher harmonic currents. This effect increases the heating and dielectric stresses.

E. Electronic equipment

Electronic devices, which are sensitive to the zero-crossing point of the voltage waveform, can be affected by harmonic distortion. Harmonics can shift the voltage zero crossing or the point at which one phase-to-phase voltage becomes greater than another phase-to-phase voltage.

Other types of electronic devices can be affected by transmission of ac supply harmonics through the equipment power supply or by magnetic coupling of harmonics, which can produce malfunctions of device. The harmonic voltage distortion factor has to be no more than 5% and the largest single harmonic no more than 3% [6].

F. Lighting

Lighting is highly sensitive to rms voltage changes; even a slight deviation (of the order of 0.25%) is perceptible to the human eye in some types of lamps. Superimposed interharmonic voltages in the supply voltage are a significant cause of light flicker in both incandescent and fluorescent lamps [7].

IV. Measurement system

IEC 61000-3-2 is the standard, which describes the harmonic measurements method for lighting equipment. The measurement system is presented in Fig. 3.

The power supply of the measurement system has to be power supply with a stable voltage and frequency. The harmonic emission has to be very low.

The output voltage of power supply has to be the same with the rated voltage of the device under test. In the case of a voltage range, the test voltage has to be 230 V. The variation of test voltage has to be ±2.0% and the frequency ±0.5% of the nominal value.

The harmonic emissions have not exceed the next values.

- 0.9 % for harmonic of order 3;
- 0.4 % for harmonic of order 5;
- 0.3 % for harmonic of order 7;
- 0.2 % for harmonic of order 9;
- 0.2 % for even harmonics of order from 2 to 10;
- 0.1 % for harmonics of order from 11 to 40

The peak value of the test voltage shall be within 1.40 and 1.42 times its RMS value and shall be reached within 87° to 93° after the zero crossing [8].

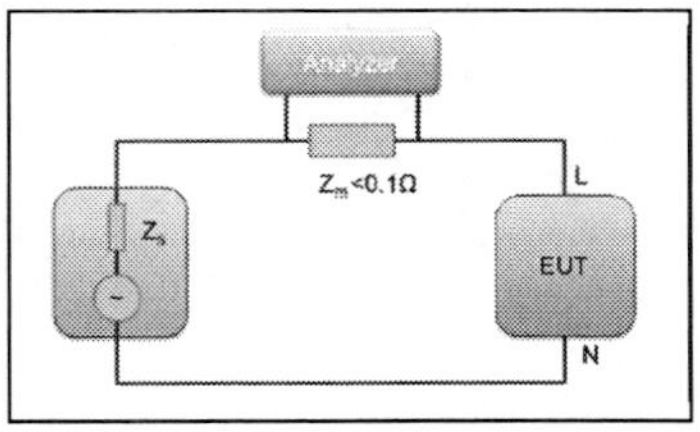

Figure. 3. Harmonic emission measurement system

V. Harmonic emission of modern lamps

Power Factor PF and Total Harmonic Distortion THD provide a quantitative measure of the power quality in an electrical system. The amount of useful power being consumed by an electrical system is predominantly decided by the PF of the system. Typically, in case of LED lamps the harmonic emission in lower than in case of CFL lamps [9][10]. The reason is that the electronic ballast of LED contains active power factor correction circuits. In Fig. 6 is presented the wave forms of current and voltage of a 7W compact fluorescent lamp and a 7W LED lamp. It could be seen that the current wave form of LED is close to sinusoidal shape what is mean the harmonics emissions are on a low level. The current wave form of compact fluorescent lamp has a spike form, what is mean this lamp type has a high harmonic emission. Unfortunately, not all LEDs have a good power factor correction. In case of cheap LEDs and compact fluorescent lamps the power factor correction is done with inductors which are used on DC side or AC side, Fig. 4 and Fig. 5.

C_i - capacitor could improve the power factor.

For higher quality LED lamps a better power factor is achieved with active power factor correction circuits[11]. In Fig. 7 is presented the measurement results for a 15W LED with good power factor correction and a 15W LED with poor factor correction. The voltage wave forms are the same. In case of modern lightings, the voltage total harmonic distortion is very low. The current wave forms are different. The LED with low power factor correction has a spike current wave form which is similar with current wave form of compact fluorescent lamps. In table 1 is presented the total current harmonic distortion and power factors of different compact fluorescent and LED lamps. The total harmonic distortion emission of LED lamp with a good power factor correction is more than 3 times lower than in case of LED with poor power factor correction.

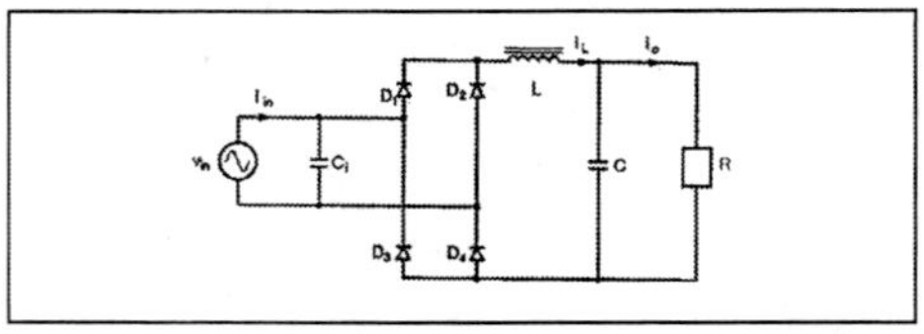

Figure 4. Full-wave rectifier with LC filter dc-side inductor

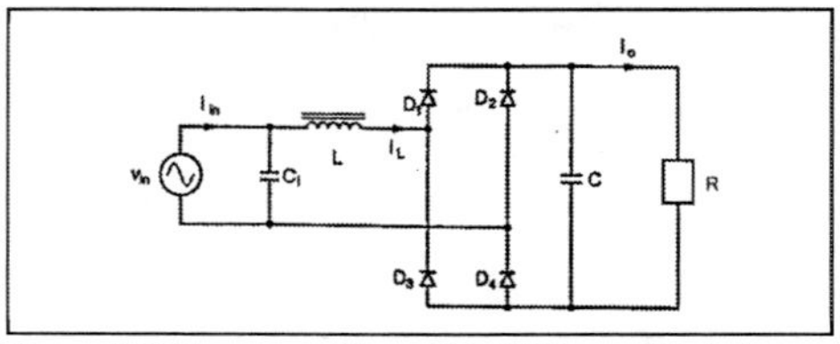

Figure 5. Full-wave rectifier with LC filter ac-side inductor

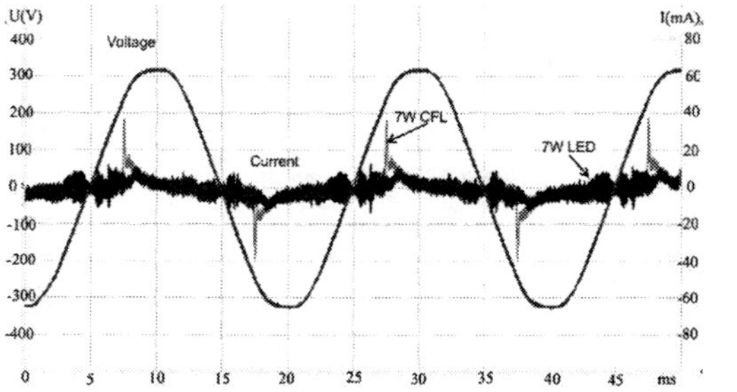

Figure 6. Current and voltage wave form of a 7W CFL and 7W LED

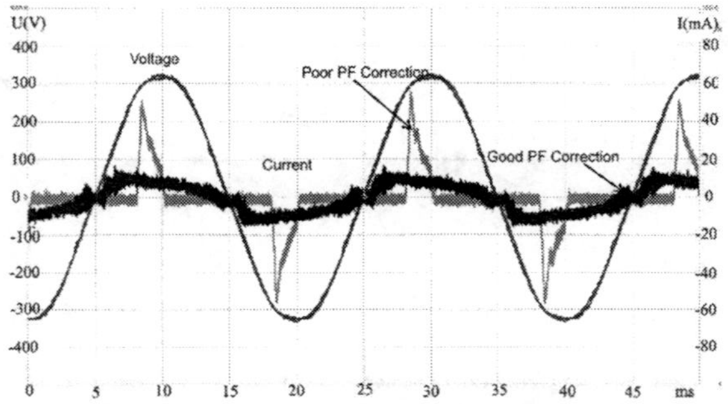

Figure 7. Current and voltage wave form of a 15W LED with good PF correction and a 15W LED with poor PF correction

VI. Influence of fluorescent tube on harmonic emission

The compact fluorescent lamp could be divided in two parts:

- Fluorescent tube
- Electronic ballast

From harmonic emission point of view the electronic ballast is the main source. The load of electronic ballast is the fluorescent tube. An important question is: Has the fluorescent tube any role in harmonic emission?

To know the influence of fluorescent tube a compact fluorescent lamp was measured. In Fig. 8 is presented the current and the voltage wave shape of a 30W compact fluorescent lamp. The fluorescent tube of the same lamp was replaced with a variable resistor. The resistor load which is a linear load assure a constant load for electronic ballast. With the help of variable resistor the consumed power of electronic ballast was adjust to be the same as in situation when was loaded with fluorescent tube [12]. In Fig. 9 is presented the voltage and current wave form for 30W electronic ballast with a resistor load. In the case of linear load, the current wave form is a bit different from situation when the load is the fluorescent tube. In table 1 can be seen that the total harmonic distortion emission is quite the same in both cases. That is mean the load of electronic ballast do not has a high influence on harmonic emission.

Table 1. THD and PF of different lamps

Lamp Type	THD %	PF
30W normal CFL	104,1	0,61
30W resistor load CFL	101,8	0,613
7W LED	40,7	0,84
7W CFL	113,7	0,58
15W LED Good PFC	16,4	0,95
15W LED Poor PFC	140	0,552

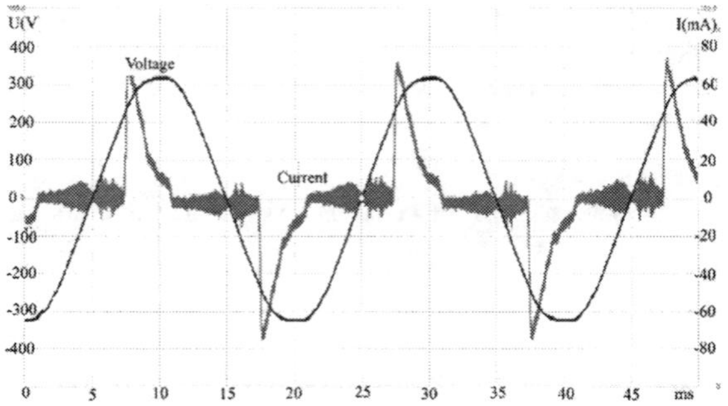

Figure 8. Current and the voltage wave shape of a 20W compact fluorescent lamp.

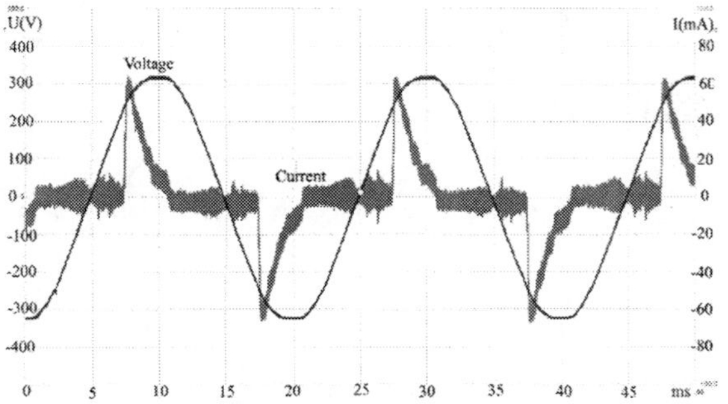

Figure 9. Current and the voltage wave shape of a 20W electronic ballast with linear load.

The current wave forms from LED and compact fluorescent lamps measurements shows noise components not only in low range frequencies, but in higher frequency ranges too. The next frequency range where the noise emission is measured is 9kHz – 30MHz. For this range the measurement systems are different from measurement system used for harmonic emission.

VII. CONCLUSION

Some of modern lamps type have an important harmonic emission. Typically the harmonic emission of compact fluorescent lamps is high. Harmonic emission of some cheaper LED lamps could be higher than emission of compact fluorescent lamps. The electronic ballast is the main source of harmonic emission. In case of compact fluorescent lamps the fluorescent tube do not influence significantly the harmonic emission.

REFERENCES

[1] I. Azcarate, J.J. Gutierrez, A. Lazkano, P. Saiz, K. Redondo, L.A. Leturiondo, “Experimental study of the response of efficient lighting technologies to complex voltage fluctuations”, *International Journal of Electrical Power and Energy Systems*, Vol. 63, 2014, pp. 499- 506.

[2] Gerry Moschopoulos and Praveen Jain, “Single-Phase Single-Stage Power-Factor-Corrected Converter Topologies”, *IEEE Transactions On Industrial Electronics*, Vol. 52, No.1, February 2005, pp 23-35.

[3] B. Singh, G. D. Chaturvedi, “Analysis, Design, Modeling, Simulation and Development of Single-Switch AC-DC Converters for Power Factor and Efficiency Improvement”, *Journal of Power Electronics*, Vol. 8, No.1, January 2008, pp 51-59.

[4] S. Abagiu, I. Lepadat, E. Helerea, C Mihai, “Assesment of voltage quality indicators in unbalanced and harmonic distorted regime”, *Pollack Periodica*, Vol. 11, No. 1, 2016, pp. 67–79.

[5] M. Domagk, O. Zyabkina, J. Meyer, P. Schegner, “Trend Identification” in *Power Quality Measurements 2015 Australasian Universities Power Engineering Conference* (AUPEC) Wollongong, Australia 27-30 Sept. 2015, pp.1-6

[6] *IEEE recommended practices and requirements for harmonic control in electrical power systems*, IEEE Std 519: 2014.

[7] N. Shah, “Harmonics in power systems Causes, effects and control”, Siemens Industry, Inc. 2013

[8] *Electromagnetic compatibility (EMC) - Part 3-2: Limis - Limits for harmonic current emissions (equipment input current ≤ 16 A per phase)*, IEC 61000-3-2: 2014,

[9] A. Gil-de-Castro ; S. Rönnberg ; M. H. J. Bollen ; A. Moreno-Muñoz ; V. Pallres-Lopez, “Harmonics from a domestic customer with different lamp technologies”, *2012 IEEE 15th International Conference on Harmonics and Quality of Power* , Hong Kong 17-20 June 2012, pp. 585 – 590

[10] S. K. Rönnberg, M. Wahlberg and M. H. J. Bollen, “Harmonic emission befor and after changing to LED lamps — Field measurements for an urban area”, *2012 IEEE 15th International Conference on Harmonics and Quality of Power*, Hong Kong, 17-20 June 2012, pp. 552-557.

[11] “Power Factor Correction (PFC) basics, Fairchild Semiconductor”, Application note 42047, 2004.

[12] A.Varga, E. Ráczand, P. Kádár, “New Experimental Method for Measuring Power Characteristics of Photovoltaic Cells at Given Light Irradiation”, *IEEE 8th International Symposium on Applied Computational Intelligence and Informatics*, SACI 2013, Timisoara, Romani, Print ISBN: 978-1-4673-6397-6., DOI: 10.1109/SACI.2013.6609008., pp. 405-409., May 2-25, 2013.

Impact of Distributed Energy Resources on the Low Voltage Distribution Network

Martin Kašpírek
Grid management
E.ON Distribution
České Budějovice, Czech Republic
martin.kaspirek@eon.cz

David Mezera
Grid operation
E.ON Czech Republic
České Budějovice, Czech Republic
david.mezera@eon.cz

Abstract — **This paper describes the impact of operation of distributed energy resources on voltage quality parameters in low voltage distribution grids. The survey is based on the evaluation of 60 voltage quality measurements which were carried out in 30 different low voltage distribution grids with connected power sources. These 60 measurements were made in 2016 and were evaluated according to the standard EN 50160. The impact on voltage variations and other voltage quality parameters was analysed.**

Keywords: standard EN 50160; voltage quality; distribution grid; distributed energy resources

I. Introduction

The supply territory of the E.ON Distribution in the Czech Republic, operated by the company E.ON Czech Republic, accounts for approximately 1.5 million customers. More than 1000 MW of distributed energy resources (DER) were connected to this network mainly in photovoltaic (PV) sources. The mass operation of generating plants also results in poorer voltage quality parameters in the distribution network. In case of accumulation of sources in one part of distribution system, it is possible that overvoltage will be detected or else voltage variations will not comply with the requirements of the standard EN 50160 [1]. The theoretical impact of DER on voltage quality (VQ) has already been described [2]. Practical experience of the operation of DER from the distribution network operator´s (DNO) side is not available or the measured data are not from the area E.ON Distribution. This paper brings a current study of VQ measurements which were carried out in 2016.

II. Description of Measured Grids

Thirty other representative LV distribution grids with connected PV plants were chosen and these grids were measured. Each measurement lasted at least one week according to the EN 50 160 standard. Two VQ measurements were made at the same time in each grid, first in the substation (on the LV level, on the output of the transformer 22kV/0.4kV – node U2 in the Fig. 1) and second in the point of connection of PV plant to LV distribution grid (supply terminal of PV plant – node U3 in the Fig. 1). The nominal active power (AP) of each PV plant, rated power of supply transformer (T) 22kV/0.4kV and measurement period are available – see the Tab. 1. The measurements were made in summer with the try to discover impact of DER on the upper voltage limit according to the EN 50160 standard.

Table 1. Rated power of the transformer MV/LV, rated power of PV plant, measurement period

Grid No.	Measurement period	power of T [kVA]	AP of PV [kW]
1	30.6.-12.7.2016	400	7
2	4.5.-18.5.2016	400	13
3	7.6.-22.6.2016	160	5
4	19.6.-29.6.2016	400	3
5	26.5.-7.6.2016	400	24
6	2.5.-10.5.2016	250	30
7	11.5.-19.5.2016	160	19
8	23.5.-31.5.2016	400	24
9	24.5.-1.6.2016	400	19
10	22.6.-30.6.2016	250	4,8
11	7.6.-22.6.2016	400	65
12	11.7.-21.7.2016	400	21
13	13.6.-22.6.2016	400	7
14	14.6.-22.6.2016	630	6
15	22.7.-1.8.2016	630	9
16	18.7.-9.8.2016	400	4,5
17	15.7.-22.7.2016	250	4,6
18	30.5.-10.6.2016	400	12
19	13.5.-30.5.2016	400	30
20	12.7.-28.7.2016	400	6
21	1.7.-11.7.2016	400	9
22	30.6.-11.7.2016	250	10,8
23	11.5.-18.5.2016	400	11
24	22.7.-1.8.2016	250	5
25	4.5.-16.5.2016	160	5
26	19.5.-26.5.2016	400	16
27	28.6.-11.7.2016	400	5,7
28	10.5.-19.5.2016	250	8
29	1.7.-14.7.2016	160	25
30	10.6.-1.7.2016	250	70

Measured PV plants are connected in the node U3 when T1 is distribution MV/LV transformer and C1, C2 and C3 are customers with the load profile – see general block diagram in the Fig. 1. It was not measured at the end of the LV feeder (node U4 in the Fig. 1) because of a lot of data. We wanted to discover the problem with overvoltage, if this problem occurs. For that reason we suggested to measure in the point of connection of PV plant to gird. For decision about responsibility for potential overvoltage we suggested to measure in the substation too (on the output of transformer MV/LV). Sometimes overvoltage occurs in the substation due to incorrect set tap changer on the supply transformer.

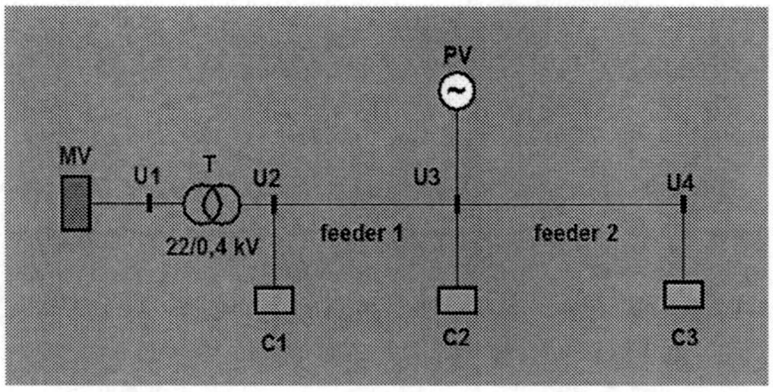

Figure 1. General bloc diagram of measured grids

III. Requirements of the EN 50160 Standard

A. Voltage Variations

All the LV measurements were evaluated according to the standard EN 50160 [1]. Under normal operating conditions excluding the periods with interruptions, supply voltage variations should not exceed ± 10% of the nominal voltage Un. A test method for low voltage level (under normal operating conditions) follows these rules/requirements [1]:

- During each period of one week 95% of the 10 min mean r.m.s. values of the supply voltage shall be within the range of Un ± 10%; and
- All 10 min r.m.s. values of the supply voltage shall be within the range of Un + 10%/-15%

Also when one 10 min mean r.m.s. value of the supply voltage exceeds during the week the limit Un + 10% (110% Un), it results in overvoltage and voltage quality or voltage variations will not comply with the requirements of the standard EN 50160.

B. Flicker

Under normal operating conditions, during each period of one week the long term flicker severity P_{lt} caused by voltage fluctuation should be less than or equal to 1 for 95 % of the time.

C. Harmonic Voltage

Under normal operating conditions, during each period of one week, 95 % of the 10 min mean r.m.s. values of each individual harmonic voltage shall be less than or equal to the values given in [1] when the limit value for the 3rd voltage harmonic is 5%, for the 5th voltage harmonic is 6%, for the 7th voltage harmonic is 5% and for the 15th voltage harmonic is 0.5% of nominal voltage (only the most important harmonic voltages were selected). Moreover, the THD of the supply voltage (including all harmonics up to the order 40) shall be less than or equal to 8 %.

D. Unbalance

Under normal operating conditions, during each period of one week, 95 % of the 10 min mean r.m.s. values of the negative phase sequence component (fundamental) of the supply voltage shall be within the range 0 % to 2 % of the positive phase sequence component (fundamental).

E. Frequency

The nominal frequency of the supply voltage shall be 50 Hz but frequency is set from the transmission system so it is not evaluated in this paper because of its small role for LV systems with synchronous connection. It is important to evaluate frequency in LV systems with no synchronous connection when the power source operates in an island mode [3].

IV. Reference Impedance

The standard IEC 60725 [4] sets the reference impedance value, or more precisely, a reference (relative) short circuit impedance (short circuit power) value for the LV distribution network and electric equipment of nominal current less than 16/75 A. It is assumed that if the required impedance value is met at the point of network connection, an electric appliance will not produce any adverse retroactive impacts on the network therefore no interference in the network will occur.

Table 2. Reference impedance for equipment with current ratings ≤16A/75A

Electric equipment with nominal current	Reference impedance $Z_{phase-earth}$	Reference short circuit power Sk”
up to 16A	0.47 Ω	570 kVA
up to 75 A	0.35 Ω	760 kVA

The question is when the DNO and when the customer is responsible for poor voltage quality (VQ). It can be used the reference impedance as the guideline for decision about responsibility for poor VQ. The experience of DNO shows that keeping of the reference impedance will not guarantee full voltage quality for all LV grids. In this paper the short circuit impedance was evaluated and compared with the reference value, see the Fig. 2.

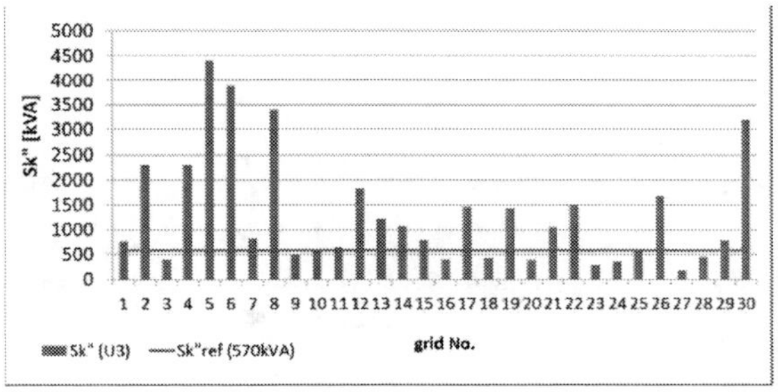

Figure 2. Evaluation of short circuit power in the supply terminal of PV plant (node U3 in the Fig. 1), comparison with the reference value 570 kVA (reference impedance 0.47 Ω)

V. Evaluation of VQ Measurements

A. Voltage Variations

In the Fig. 3 and Fig. 4 you can see the evaluation of maximal/minimal 10 min mean r.m.s. values of supply voltage (V100%max/V100%min) in 30 LV grids during week measurements in nodes U2 (substation) and U3 (PV). Fig. 3 shows conforming voltage variations in all grids. Voltage approaches the upper limit 110% Un in grids No. 20, 25, 28 but the upper limit is not exceeded. This operation state is not optimal and can be caused by bad set tap changer. Fig. 4 shows exceeding of the upper voltage limit in the grid No. 28. There was detected overvoltage only in the supply terminal of PV plant, voltage variations in the substations comply with requirements of the standard EN 50160. The overvoltage in this case is caused by operation of PV plant and solution of this problem has to be found.

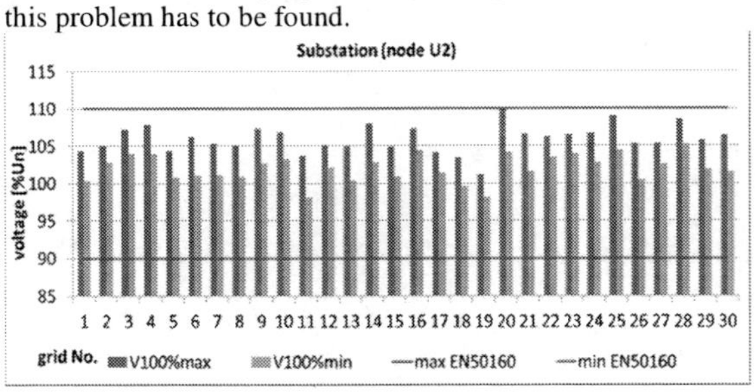

Figure 3. Evaluation of voltage variations (substation)

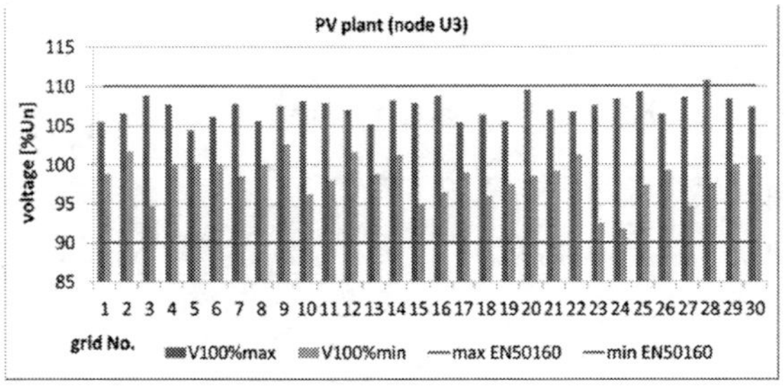

Figure 4. Evaluation of voltage variations (PV plant)

It was evaluated the voltage difference dU between maximal and minimal voltage values during the measurement too – see the Fig. 5. The voltage difference reaches the value 5% Un for substation and 16% Un for PV plant.

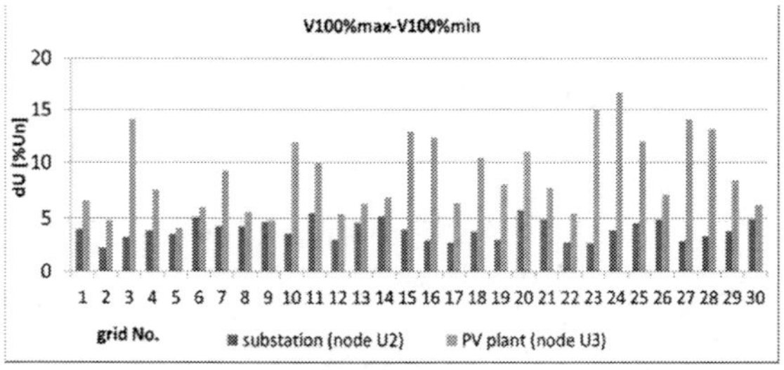

Figure 5. Voltage difference between maximal and minimal voltage values

B. Flicker

In the Fig. 6 and Fig. 7 you can see the evaluation of long-term flicker (Plt) and short-term flicker (Pst) in 30 LV grids during week measurements in nodes U2 (substation) and U3 (PV).

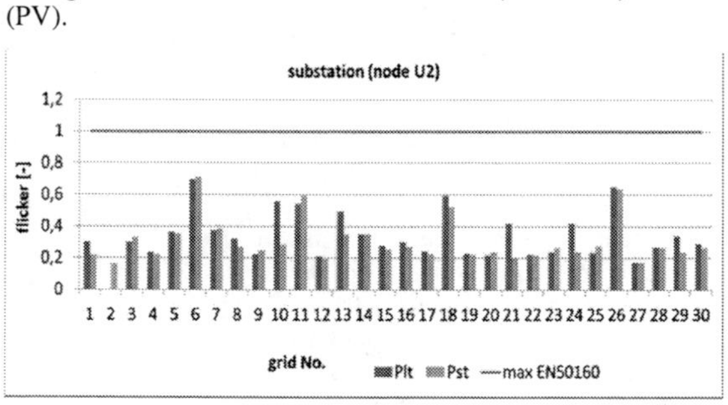

Figure 6. Evaluation of flicker (substation)

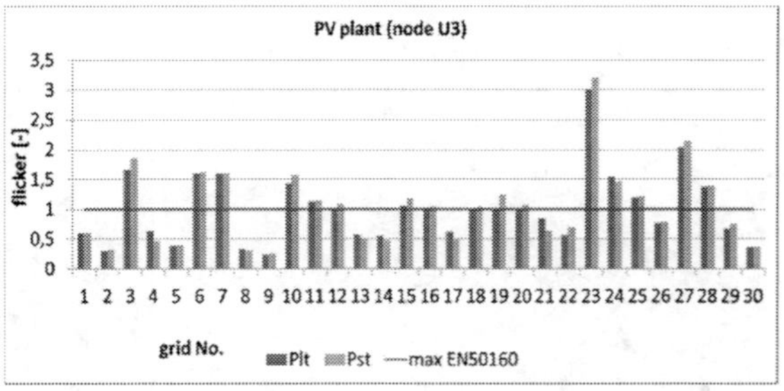

Figure 7. Evaluation of flicker (substation)

The flicker level was exceeded in sixteen cases (grids No. 3, 6, 7, 10, 11, 12, 15, 16, 18, 19, 20, 23, 24, 25, 27, 28) only in the supply terminal of PV plant (node U3), flicker level in substations (node U2) complies with requirements of the standard EN 50160. It is the question if the flicker problem is caused by operation of PV plants or not. It means that in 53% (16/30*100%) of LV grids with PV plants, flicker does not comply with the requirements of the standard EN 50160 and solutions of this problem have to be found.

There is lower value of short circuit power than reference value in grids No. 3, 10, 16, 18, 20, 23, 24, 25, 27, 28 with exceeded flicker level.

C. Harmonic Voltages and THD

The Fig. 8 shows no problem with the 3rd voltage harmonic, the Fig. 9 shows no problem with the 5th voltage harmonic and the Fig. 10 shows no problem with the 7th voltage harmonic. The Fig. 11 shows the problem with the 15th harmonic voltage in two cases (grids No. 23, 24). This could be caused by high penetration of appliances producing 15th harmonics current which isn't our case, the measurement was done in standard rural and small city area. The second reason could be a low limit value for the 15th harmonic voltage when measurement uncertainty for generally used power quality analyzers (class S according to the IEC 61000-4-30:2008 standard [5]) is +/- 0.5% of nominal voltage. The standard IEC 62749 [6] recommends limit values up to the 50th voltage harmonic and the limits for some harmonics (with the multiple of three) are only 0.2% of the nominal voltage. Precise VQ analyzers with low system noise should be used for measuring of low harmonic voltage values. For some measuring equipment the internal noise of the instrument could disqualify VQ analyzer from measuring of such low harmonic levels [7]. The experience of the DNO shows that no complaints regarding harmonic voltage were obtained up till now. Exceeding of the limit value for the 15th harmonic voltage is hardly perceived by the customer and has no influence on operation of commonly used electrical devices. The Fig. 12 shows no problem with total harmonic distortion (THD).

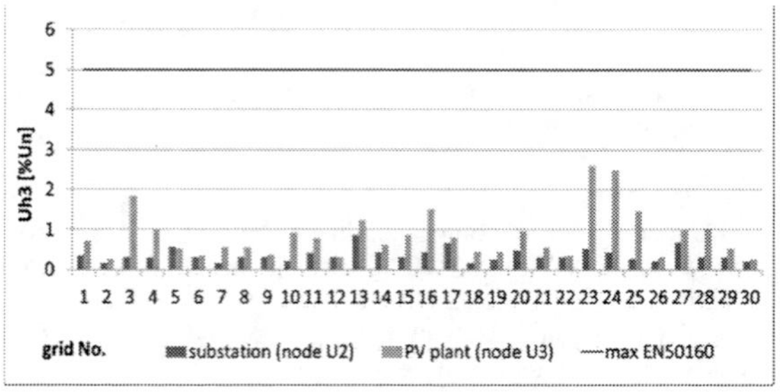

Figure 8. Evaluation of the 3rd voltage harmonic

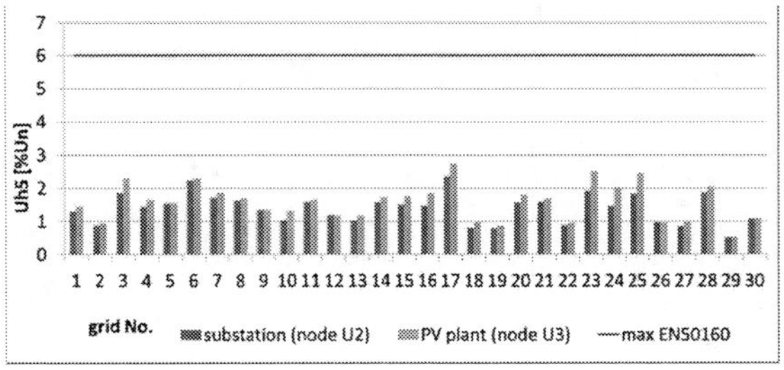

Figure 9. Evaluation of the 5th voltage harmonic

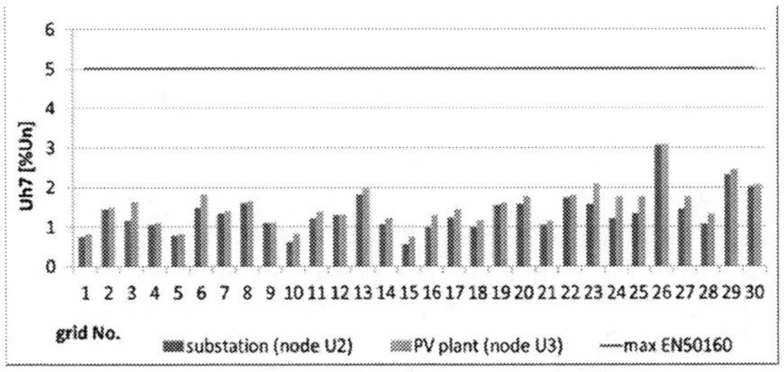

Figure 10. Evaluation of the 7th voltage harmonic

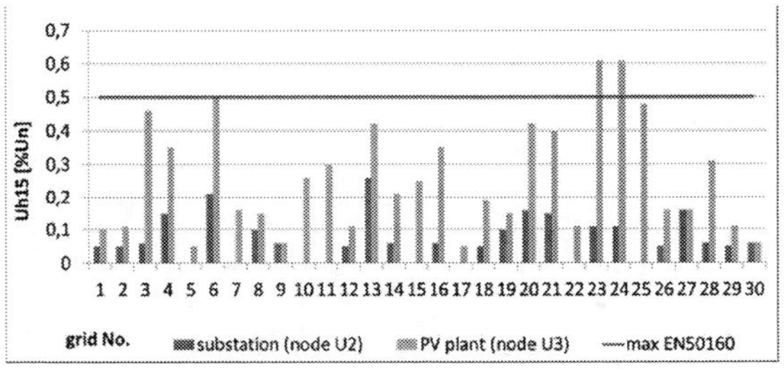

Figure 11. Evaluation of the 15th voltage harmonic

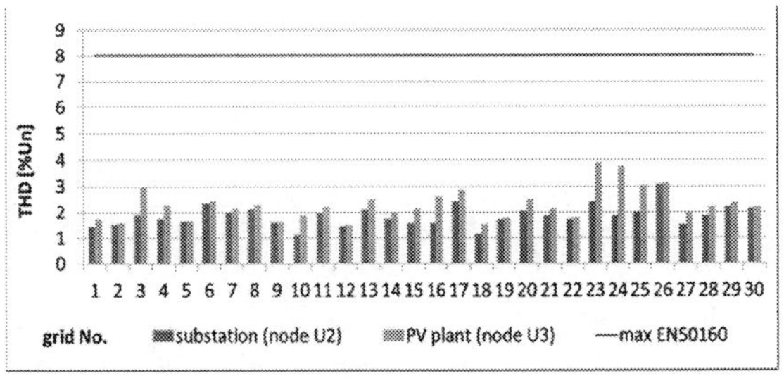

Figure 12. Evaluation of total harmonic distortion THD

There is lower value of short circuit power than reference value in grids No. 23, 24 with exceeded level of 15th harmonic voltage.

D. Voltage Unbalance

The Fig. 13 shows the evaluation of voltage unbalance (95% percentile) in 30 LV grids during week measurements in nodes U2 (substation) and U3 (PV). There was evaluated no problem with voltage unbalance in LV grids. The maximal level of voltage unbalance reaches the value 1.5% in case of grid No. 24.

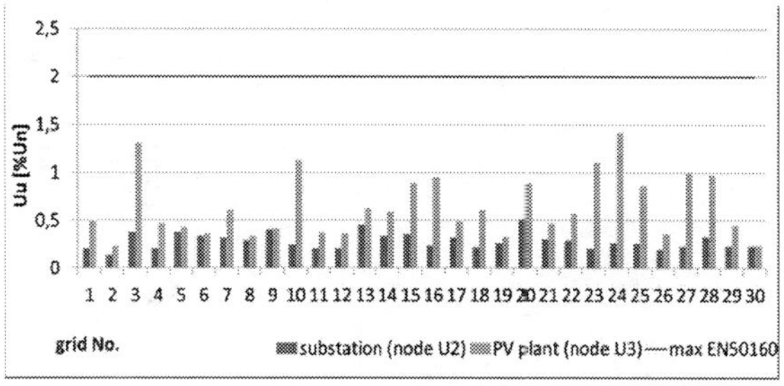

Figure 13. Evaluation of voltage unbalance

VI. Conclusion

It was demonstrated that operation of DER in the LV distribution grids can cause poor voltage quality and the problem with overvoltage has to be solved. The problem with flicker was detected approximately in 53% of LV grids and the problem with the 15th harmonic voltage approximately in 7% of grids (2/30*100%) but it is the question how the operation of DER influences these voltage parameters. It is also a question how voltage quality will develop in the LV distribution network due to the operation of a growing number of DER. The experience of DNO shows that the number of LV grids with nonconforming VQ will grow due to the operation of DER so that new and stricter conditions for the connection of DER to the distribution network have to be introduced. The operation of battery storages is the current topic in the Czech Republic. This storage can influence the voltage so it is necessary to calculate if it is connectable. The rules for operation of battery storages are not described yet but Czech DNO´s prepare new grid code which should include it.

Acknowledgment

The authors would like to thank colleagues from the department power quality for cooperation regarding voltage quality measurements.

References

[1] EN 50160 Ed.3 Voltage characteristics of electricity supplied by public distribution systems. Brussels: European Committee for Electrotechnical Standardization, 2010. 20 p.

[2] M. Kaspirek, "Impact of RES in distribution networks", Proceedings of the 12th International Scientific Conference ELECTRIC POWER ENGINEERING 2011, VSB – Technical University of Ostrava, Faculty of Electrical Engineering and Computer Science, 17.-19.5.2011, Kouty nad Desnou, ISBN 978-80-248-2393-5

[3] M. Tesarova and K. Nohac, "Voltage and frequency stability analysis of an island mode LV distribution network" In Proceedings of the 15th International Scientific Conference Electric Power Engineering 2014. Brno: Brno University of Technology, 2014. ISBN: 978-1-4799-3806-3

[4] IEC/TR 60725:2012 Consideration of reference impedances and public supply network impedances for use in determining the distrurbance characteristics of electrical equipment having a rated current ≤75A per phase, IEC, ISBN 978-2-83220-161-9

[5] IEC 61000-4-30 Ed.3.0 Electromagnetic compatibility (EMC): Part 4-30: Testing and measurement techniques – Power quality measurement methods. Geneva: IEC

[6] IEC/TS 62749 Ed.1 Assessment of power quality – Characteristics of electricity supplied by public networks

[7] P. Bilik and J. Zidek, "Precise Harmonic Component Analyzer based on Virtual Instrumentation" In Conference proceeding of the International Conference of Applied Electronics APPEL 2010, Pilsen, 8.-9.9.2010, FEE, ZCU Plzen, 2010, p. 39 - 42. ISBN 978-80-7043-865-7

Impact of Asymmetry on Energy Loss and Voltage Drops in Low-Voltage Lines

Anna Gawlak, Romuald Larisch
Institute of Electrical Power Engineering
Technical University of Częstochowa
Częstochowa, Poland
gawlak@el.pcz.czest.pl, romuald@larisch.pl

Abstract — **The paper presents the impact of asymmetry in the load of rural consumers on energy loss and voltage drops in low-voltage lines. The results of the study were obtained by analysing the operation of a low-voltage line of a known length, location, shape, conductors' cross-section and power consumed in the particular phases.**

Keywords: energy losses; asymmetry; low-voltage network

I. Introduction

Load asymmetry occurs frequently in rural power low-voltage networks and is the main cause of current flow in the neutral wire. This condition is notably hard to eliminate due to the character of appliances used in farms: three-phase appliances of large power are used only sporadically and switched on for short periods of time, whereas single-phase appliances are used for extended periods of time and are often connected to randomly chosen phases. Such single-phase appliances take the greatest share in total power consumption. The phenomenon of voltage and current asymmetry affects the power system negatively in a number of ways: it increases active power loss in transmission lines and contributes to power loss in transformers as well [1-3]. When an inverse component appears, devices heat beyond the regular level and their transmission capacity decreases; when an inverse component appears in the windings of rotating machines, thermal protection systems may switch a machine off to prevent excessive heating [4-7]. Voltage and current asymmetry can also exert negative impact on the operation of power protection control systems, possibly increasing errors of short circuit localisation.

II. Object under study

The object under study is an asymmetrically loaded power low-voltage line feeding rural consumers. The diagram of the line is presented in Fig. 1.

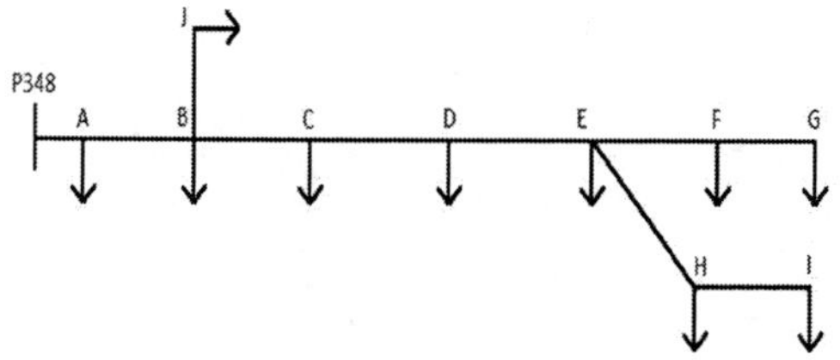

Figure 1. Low-voltage line diagram

Fig. 2. presents an example of a 24-hour phase current curves as well as a neutral wire current curve at an initial section of a low-voltage line fed from a medium-to-low-voltage transformer station. As can be observed, the phase load is highly imbalanced mostly due to consumer load asymmetry. In the system under scrutiny, at 7.00, 18.00 and 20.00 hours, the current in two phases was 0 [A].

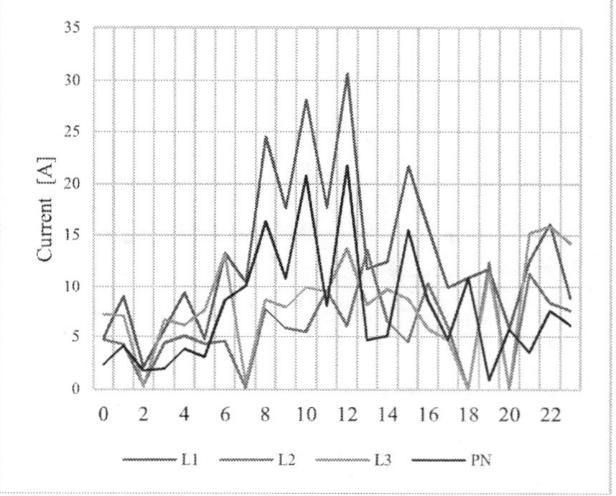

Figure 2. 24-hour phase current and neutral wire current (PN) curves at the initial section of the low-voltage line

Fig.3. presents energy loss at the consecutive line sections for the real (asymmetrical) circuit and for the symmetrical circuit. The difference in loss with respect to the symmetrical circuit is indicated as percentage and varies from 48% to 196%. The lower the total loss in the circuit, the greater the difference. At the final sections of the line there are fewer consumers, or there may even be only one consumer with a single-phase meter. Asymmetry in such a section is very high.

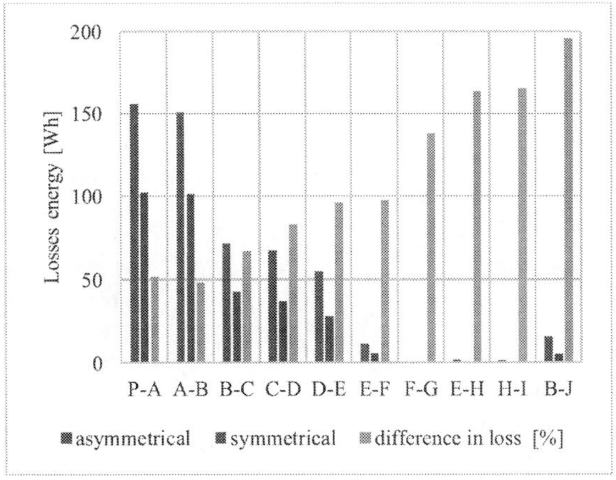

Figure 3. Energy loss at line sections and difference in loss between the asymmetrical and symmetrical circuit

Fig.4. presents power loss at the particular times of the day for the symmetrical and asymmetrical systems. At some hours (e.g. 10.00 or 12.00), the loss in the asymmetrical circuit is almost twice as high as that in the symmetrical system. This is caused by the existence of a single-phase consumer at the point E, with significantly high load at these hours. Another consumer contributing to asymmetry is the three-phase consumer connected at the point J of the system. At some time of the day, the load of this consumer is higher than the total load of all the other consumers.

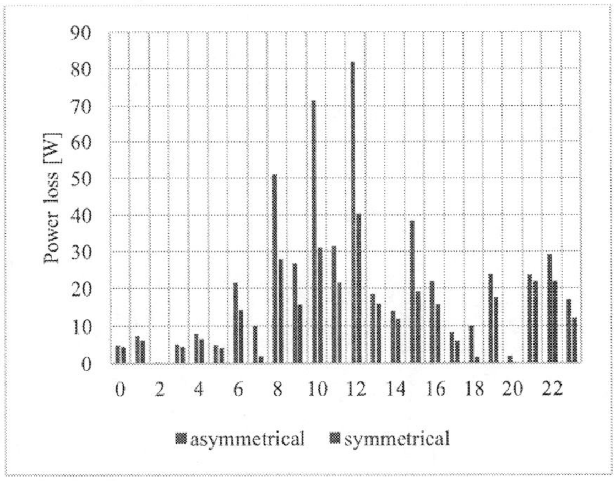

Figure 4. 24-hours power loss

Fig.5. presents voltage drops in the phases of the main track of the line. As evident, there are great differences in voltage drops between the phases – in the most heavily loaded phase the drop is up to five times higher than in the least loaded phase (e.g. at 10.00 hours).

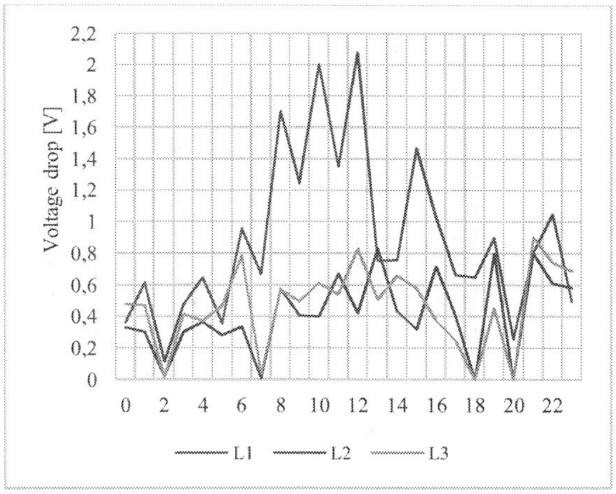

Figure 5. 24-hours voltage drops in the phases of the main track of the line

A. *Quantities characterising asymmetry and methods of their determination*

The phenomenon of asymmetry can be qualitatively described by means of a number of coefficients [6, 8]. So far, a few methods for determining asymmetry coefficients have been developed. These methods can be divided into two groups: direct and indirect. The former include analytical, graphic-analytical, nomographic and measurement methods, whereas the latter include the method based on calculations in the reference frame and the method based on calculations using the symmetrical component system [7, 9]. The most popular method of investigating asymmetry is the direct analytical method involving decomposition of the three-phase system of voltage and current vectors into a sum of three symmetrical systems: direct, inverse and zero. The description of the transformation of an asymmetrical system into the set of symmetrical components can be found in the literature [4, 6, 10]. When the values of the symmetrical components of the currents are known, it is possible to define the required coefficients.

- the inverse phase sequence asymmetry current coefficient:

$$k_{i2} = \frac{I_2}{I_1} 100\% \quad (1)$$

where: I_1, I_2 – the symmetrical components of current of direct (positive) and inverse sequence

- the zero asymmetry current coefficient (current imbalance):

$$k_{i0} = \frac{I_0}{I_1} 100\% \quad (2)$$

where: I_0, I_1 – the symmetrical components of current of zero and direct sequence

- the coefficient of total current asymmetry:

$$k_i = \frac{|I_0 + I_2|}{I_1} 100\% \quad (3)$$

where: I_0, I_1, I_2 – the symmetrical components of current of direct and inverse sequence

Figures 6 and 7 present 24-hours variation in current asymmetry coefficients

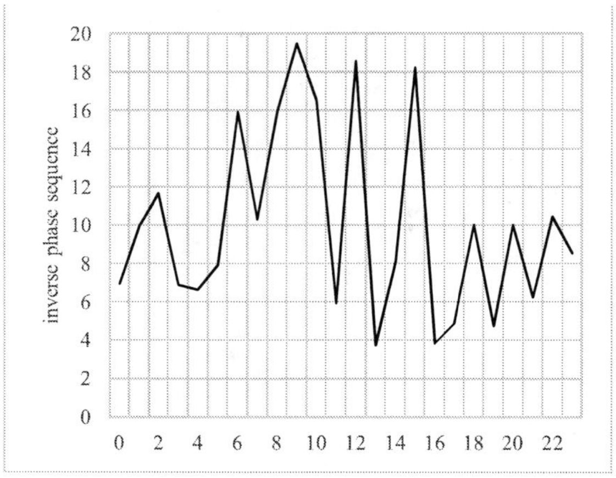

Figure 6. Inverse sequence phase current asymmetry coefficient

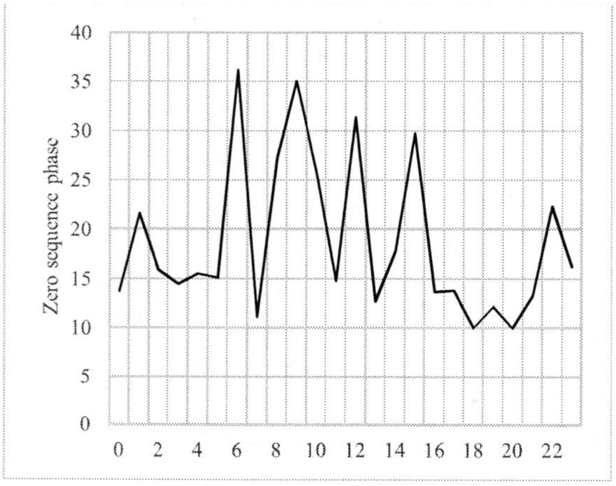

Figure 7. Zero sequence phase current asymmetry coefficient

B. *Current magnitude in the neutral wire*

The most important factor affecting the value of current flowing in the neutral wire is load asymmetry. The current can be estimated by means of the following coefficient denoting the percentage ratio of this current to the mean value of phase currents:

$$I_{N\%} = \frac{3\, I_{ms0}}{I_{msL1}+I_{msL2}+I_{msL3}}\, 100\% \qquad (4)$$

where: $I_{msL1}, I_{msL2}, I_{msL3}$ - rms values of the phase currents, I_{ms0} - rms value of the current in the neutral wire.

Fig.8. presents a 24-hours curve of the coefficient of the neutral-wire current to phase current ratio. The current in the neutral wire ranges from 7% to 300% of the mean value of the phase currents. The highest values of the coefficients occur at 18.00 hours and at 20.00 hours, at which times the load in two phases was equal to zero.

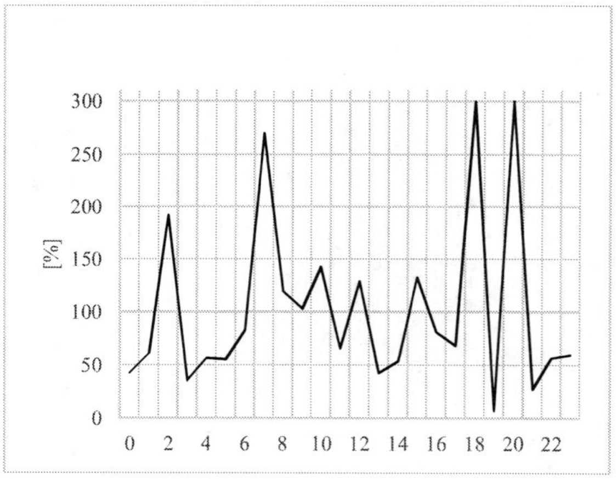

Figure 8. 24-hours variation of the neutral wire current with respect to the mean value of phase currents

C. *Voltage drop asymmetry*

The degree of voltage drops in distribution networks can be obtained by measurements. The voltage drop asymmetry coefficient is calculated from the formula:

$$k_{a\Delta u} = \frac{\Delta u_{max} - \Delta u_{min}}{\Delta u_1 + \Delta u_2 + \Delta u_3} \qquad (5)$$

where: $\Delta u_{max}, \Delta u_{min}$ - the greatest and the smallest voltage drop, respectively, $\Delta u_1, \Delta u_2, \Delta u_3$ – the voltage drops in the phases

Fig.9. presents a 24-hours variation in the voltage drops asymmetry coefficient. It varies from 0.04 at 21.00 hours, when the line load can be considered symmetrical, to 1 at 18.00 and 20.00 hours, when the asymmetry in the system is very high.

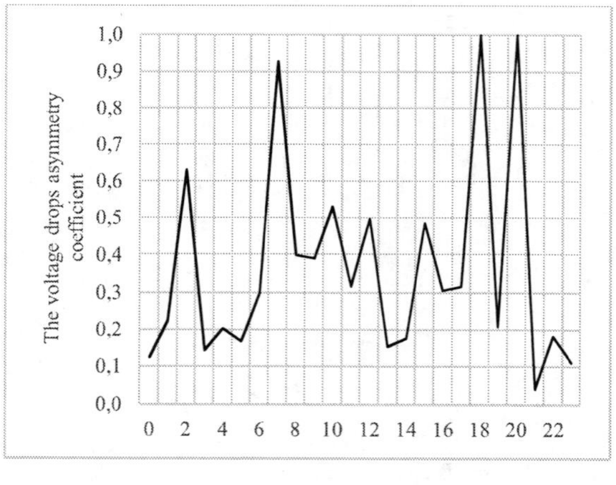

Figure 9. 24-hours variation of the voltage drops asymmetry coefficient

III. NUMERICAL SIMULATIONS

The line under scrutiny feeds 13 three-phase consumer circuits and 4 single-phase circuits. The single-phase circuits can then be fed from any phase. Taking this into account, simulations were performed with single-phase consumers being randomly fed from various phases. The results of energy loss calculations are shown in Fig.6. Depending on load asymmetry, i.e. on which consumers are fed by randomly chosen phases, energy losses in asymmetrical circuits are higher by 34% to 73% with respect to the loss in the symmetrical circuit. This implies that thanks to this analysis, it is possible to connect single phase consumer circuits in such a way as to minimise losses.

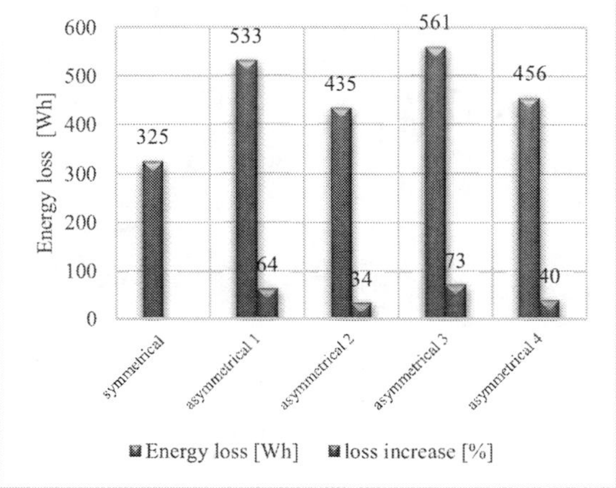

Figure 10. Energy loss [Wh] in a symmetrical and asymmetrical system and loss increase [%] with respect to the symmetrical system

Fig.11 shows the dependence between the asymmetry coefficient and the difference in power loss for all the systems. As has been established, the correlation between the two is very high. It can be therefore said that by calculating the voltage asymmetry coefficient the magnitude of energy loss can be obtained.

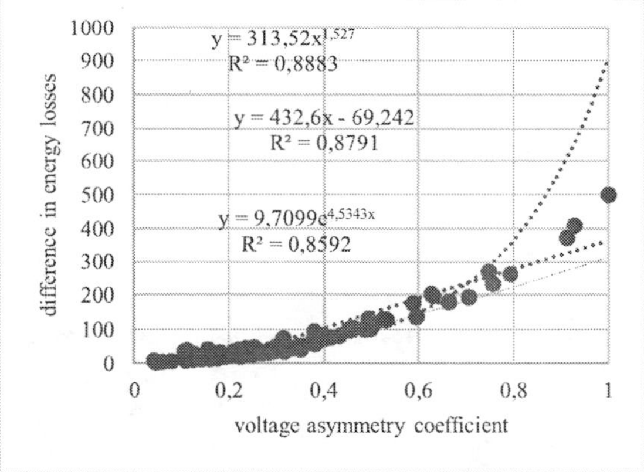

Figure 11. Dependence between the difference in energy losses and the voltage asymmetry coefficient

Conclusion

- Due to high load asymmetry occurring in rural low-voltage networks, the magnitude of current flowing in the neutral wire may reach up to 300 % of phase currents. This happens when two phases of the circuit are non-loaded.
- High root mean square and deformation of the current flowing in the neutral wire contributes to power loss, constituting a significant share of the total power loss in the three-phase power line.
- The level of asymmetry is affected by a number of factors, the most important of which are line configuration, its length and the value of load current.
- Maintaining full symmetry in low-voltage networks is not possible, but excessing a certain level of asymmetry has a markedly negative effect on the network operation by causing high increase in voltage drops and in power loss. To keep the effect of asymmetry on the network operation fairly harmless, the coefficient of voltage drops asymmetry should not exceed 0.25.

References

[1] A. Gawlak, "Technological aspects of electrical energy distribution". 14th International Scientific Conferenc Electric Power Engineering May 2014 Brno Czech Republic ISBN 978-80-214-4514-7 pp. 45-48

[2] M. Kornatka, „Some methods of improving reliability medium voltage overhead lines" Rynek Energii, Issue 1 pp. 50-56, 2010.

[3] R. Gono, S. Rusek, M. Kratky, and M. Slivka, M "Component reliability parameters of distribution network", (2015) Proceedings of the 8th International Scientific Symposium on Electrical Power Engineering, ELEKTROENERGETIKA 2015, pp. 364-367.

[4] Z. Kowalski, "Jakość energii elektrycznej". Wydawnictwo Politechniki Łódzkiej, Łódź, 2007 .

[5] S. Bolkowski, "Teoria Obwodów Elektrycznych". Wydanie piąte, WNT, Warszawa, 1995.

[6] A. Gawlak, and L. Poniatowski, "Main directions of investment in the sector of electrical energy distribution". 14th International Scientific Conference Electrical Power Engineering - EPE 2014, Brno Czech Republic, ISBN 978-80-214-4514-7 pp. 39-43

[7] M. Sobierajski, and W. Rojewski, "Obliczanie asymetrii napięć w sieci niskiego napięcia wprowadzanej przez jednofazową mikrogenerację z wykorzystaniem równań rozpływu mocy". XVI Międzynarodowa Konferencja Naukowa Aktualne Problemy w Elektroenergetyce APE'13 Jurata, 12-14 czerwca 2013.

[8] G. Dudek, "Extreme Learning Machine as a Function Approximator: Initialization of Input Weights and Biases" 9th International Conference on Computer Recognition Systems (CORES 2015), Wrocław 2016, pp 59-69.

[9] M. Kornatka, „Analysis of unreliability of low voltage power networks some branches of Distribution System Operators" Przegląd Elektrotechniczny Volume 88 Issue 10B pp. 303-306, 2012.

[10] J. Sowiński and R. Tomaszewski, "Modelling of power engineering processes under risk", Proceedings of The 8th International Scientific Symposium on Electric Power Engineering "Elektroenergetika2015", Stara Lesna, Slovak Republic, 16-18 September 2015, pp. 396-399.

Development of Low Voltage Distribution Grids With the Focus on Maximal Value of Short Circuit Impedance

Martin Kašpírek
Grid management
E.ON Distribuce
České Budějovice, Czech Republic martin.kaspirek@eon.cz

Abstract — **The supply territory of the E.ON Distribution in the Czech Republic accounts for approximately 1.5 million customers. The most of them are supplied from the low voltage (LV) distribution network when E.ON Distribution runs 40 thousand km of LV power lines. Distribution network operator (DNO) should guarantee voltage quality according to the EN 50160 standard in delivery points of all the LV customers. The level of individual voltage quality (VQ) parameters depends on the real value of short circuit power and the value of short circuit power depends on the length of the feeder. The paper describes a proposal of minimal values of short circuit power in LV grid with the focus on keeping with the EN 50160 standard. Proposed minimal value of short circuit power is tested in practice when E.ON Distribution has in database more than 2 thousand VQ measurements from LV distribution grid with known value of short circuit impedance which is measured by each VQ measurement in LV grid. The minimal value of short circuit power (maximal value of short circuit impedance) is recalculated to the maximal length of the commonly used types of conductors. Keeping of the maximal length of the feeder is necessary condition (but not sufficient condition) for meeting of the standard EN 50160 and gives the technician a smart tool by grid projecting.**

Keywords: distribution grid; low voltage; short circuit impedance; EN 50160 standard

I. Reference Impedance

The standard IEC 60725 [1] sets the reference impedance value, or more precisely, a reference (relative) short circuit impedance (short circuit power) value for the LV distribution network and electric equipment of nominal current less than 16/75 A. It is assumed that if the required impedance value is met at the point of network connection, an electric appliance will not produce any adverse retroactive impacts on the network therefore no interference in the network will occur.

Table 1. Reference impedance for equipment with current ratings ≤16A/75A [1]

Electric equipment with nominal current	Reference (loop) impedance $Z_{phase\text{-}earth}$
up to 16A	0.47 Ω
up to 75 A	0.35 Ω

The question is when the DNO and when the customer is responsible for poor voltage quality (VQ). It can be used the reference impedance as the guideline for decision about responsibility for poor VQ. The condition for using of the reference impedance from side of DNO is that keeping of the reference impedance will guarantee full voltage quality for all LV grids (or for majority of grids) according to the EN 50160 standard [2]. The proposal of DNO is to keep reference impedance 0.47 Ω (value of loop impedance) in rural and reference impedance 0.35 Ω (value of loop impedance) in urban area. Using of these values recommends the standard PNE 333430-0 [3]. It is necessary to verify these proposed values because of danger of problems with voltage quality in a part of low voltage distribution grids.

II. Evaluated Data

A. *Selected VQ Measurements*

More than 2 thousand of VQ measurements were evaluated. These measurements were made from 2011 to 2016 in low voltage distribution grid.

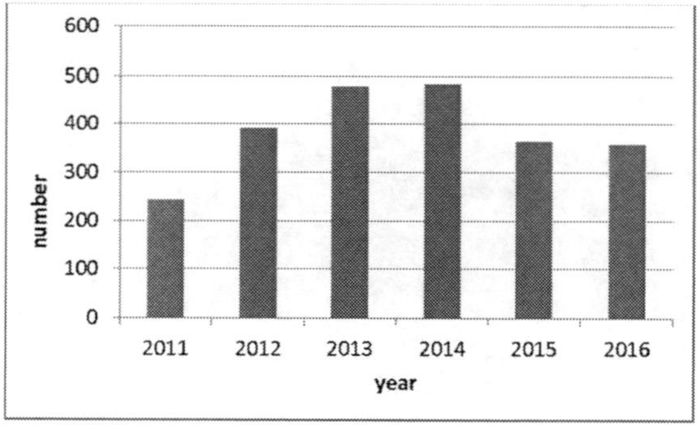

Figure 1. Number of evaluated VQ measurements from LV level

B. *Measuring of Loop Impedance*

All measurements are systematically archived and it is possible to filter them for example according to the voltage level or exceeded VQ parameters. If possible, the system impedance (real value of loop impedance) phase-earth is measured in the point of the VQ measurement. A special measuring instrument Zerotest 46N is used – see Fig. 2. The system impedance in all three phases is measured and the average value is calculated and used in following graphs. Values of the measured system impedance are included in the archival system (see the Fig. 3) so that the link between the VQ measurement and the system impedance value is available for each VQ measurement. The Zerotest shows a relatively big measurement error but it is not possible to draw 2,000 different circuit diagrams and calculate the short circuit impedance in these grids. Moreover, an impedance module is measured. Let consider the evaluation as approximate. However, the Zerotest measurements are commonly used in practice and give an approximate idea of the grid quality.

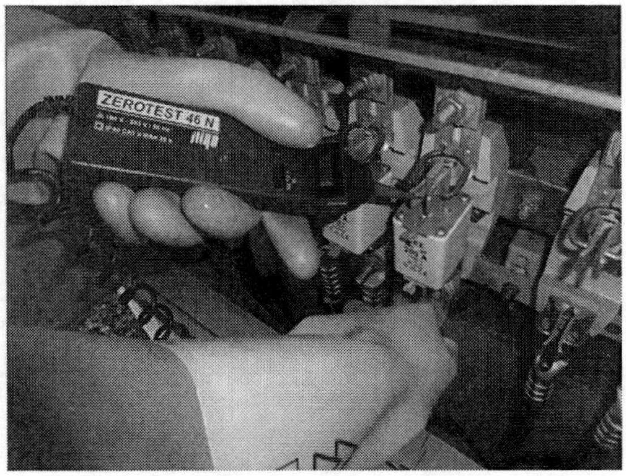

Figure 2. Measuring instrument Zerotest 46N for the system impedance (loop impedance) measurement

III. Standard EN 50160

Before the comparing of VQ measurements with the reference impedance it is necessary to describe requirements and limit values of the standard EN 50160.

A. *Voltage Variations*

Under normal operating conditions excluding the periods with interruptions, supply voltage variations should not exceed ± 10% of the nominal voltage Un. A test method for low voltage level (under normal operating conditions) follows these rules/requirements [2]:

- During each period of one week 95% of the 10 min mean r.m.s. values of the supply voltage shall be within the range of Un ± 10%; and
- All 10 min r.m.s. values of the supply voltage shall be within the range of Un + 10%/-15%

Also when one 10 min mean r.m.s. value of the supply voltage exceeds during the week the limit Un + 10% (110% Un), it results in overvoltage and voltage quality or voltage variations will not comply with the requirements of the standard EN 50160.

B. *Flicker*

Under normal operating conditions, during each period of one week the long term flicker severity P_{lt} caused by voltage fluctuation should be less than or equal to 1 for 95 % of the time.

C. *Harmonic Voltages*

Under normal operating conditions, during each period of one week, 95 % of the 10 min mean r.m.s. values of each individual harmonic voltage shall be less than or equal to the values given in [1] when the limit value for the 3rd voltage harmonic is 5%, for the 5th voltage harmonic is 6%, for the 7th voltage harmonic is 5%, for the 9th voltage harmonic is 1.5% and for the 15th voltage harmonic is 0.5% of nominal voltage (only the most important harmonic voltages were selected). Moreover, the THD of the supply voltage (including all harmonics up to the order 40) shall be less than or equal to 8 %.

D. *Unbalance*

Under normal operating conditions, during each period of one week, 95 % of the 10 min mean r.m.s. values of the negative phase sequence component (fundamental) of the supply voltage shall be within the range 0 % to 2 % of the positive phase sequence component (fundamental).

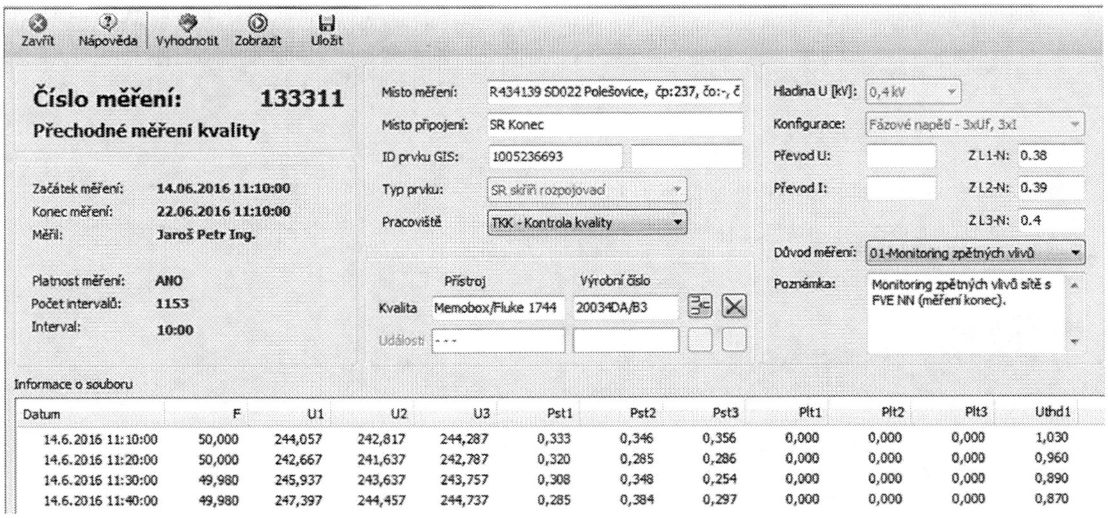

Datum	F	U1	U2	U3	Pst1	Pst2	Pst3	Plt1	Plt2	Plt3	Uthd1
14.6.2016 11:10:00	50,000	244,057	242,817	244,287	0,333	0,346	0,356	0,000	0,000	0,000	1,030
14.6.2016 11:20:00	50,000	242,667	241,637	242,787	0,320	0,285	0,286	0,000	0,000	0,000	0,960
14.6.2016 11:30:00	49,980	245,937	243,637	243,757	0,308	0,348	0,254	0,000	0,000	0,000	0,890
14.6.2016 11:40:00	49,980	247,397	244,457	244,737	0,285	0,384	0,297	0,000	0,000	0,000	0,870

Figure 3. Example of a heading of one selected VQ measurement in database with the illustration of loop impedance values

E. Frequency

The nominal frequency of the supply voltage shall be 50 Hz but frequency is set from the transmission system so it is not evaluated in this paper because of its small role for LV systems with synchronous connection. It is important to evaluate frequency in LV systems with no synchronous connection when the power source operates in an island mode [4].

IV. Dependence of VQ Parametrs on System Impedance

The following graphs show the dependence of selected VQ parameters on system impedance (real value of loop impedance). The red horizontal line shows the limit value according to the EN 50160 standard.

A. Voltage Variations

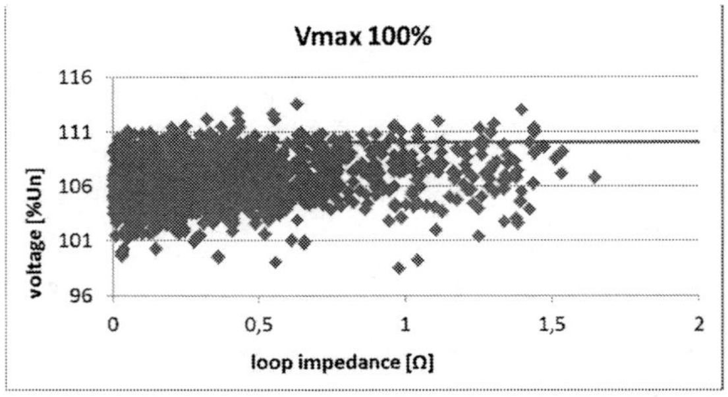

Figure 4. Evaluation of maximal voltage variations (100% percentile)

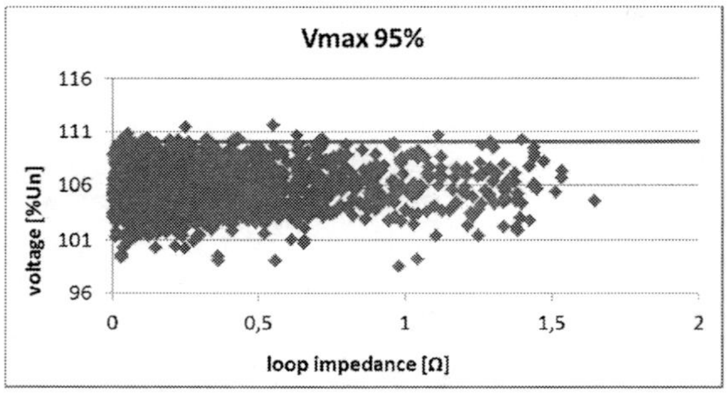

Figure 5. Evaluation of maximal voltage variations (95% percentile)

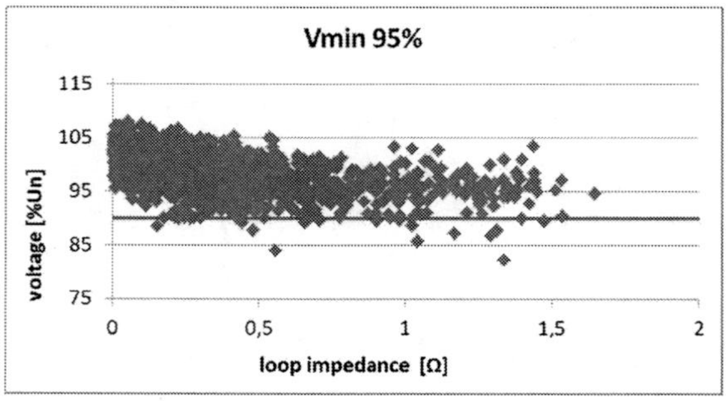

Figure 6. Evaluation of minimal voltage variations (95% percentile)

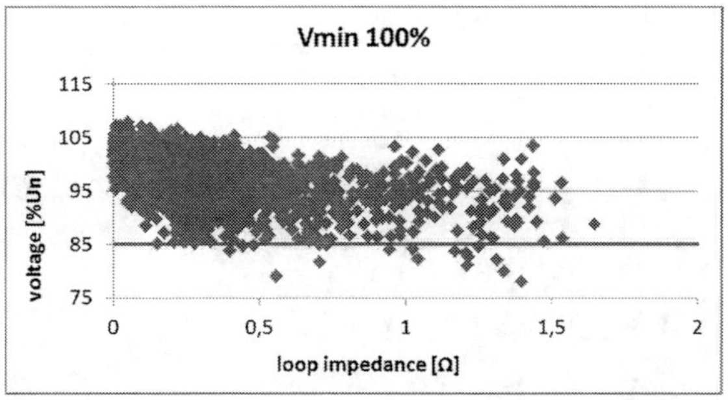

Figure 7. Evaluation of minimal voltage variations (100% percentile)

Additionally it was evaluated the voltage difference dU between maximal and minimal voltage values during the measurement too – see the Fig. 8 and Fig. 9.

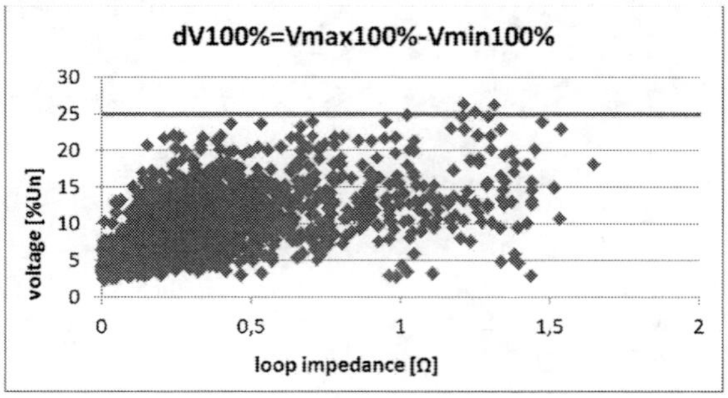

Figure 8. Voltage difference between maximal and minimal voltage values (100% percentile)

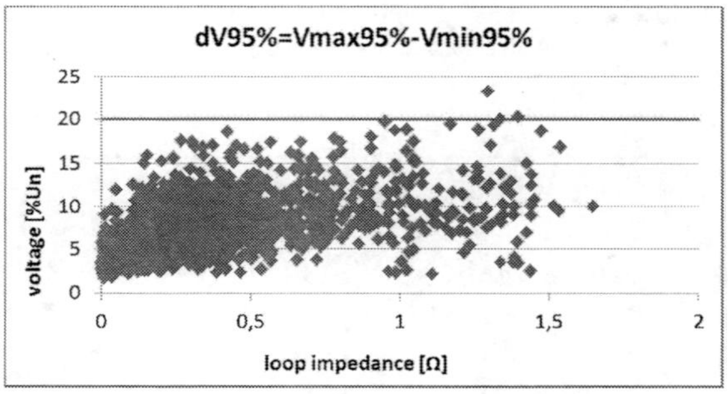

Figure 9. Voltage difference between maximal and minimal voltage values (95% percentile)

B. Flicker

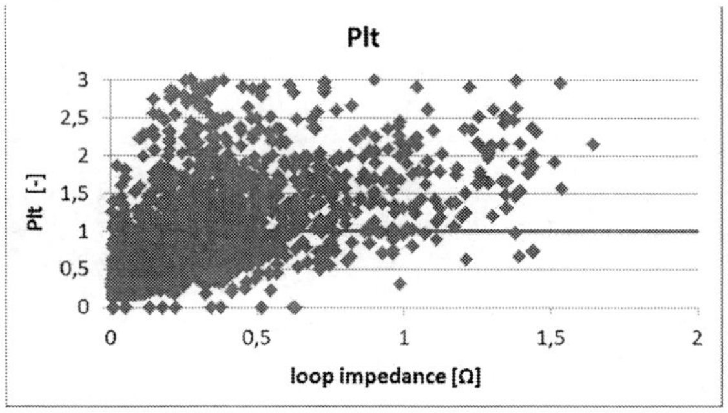

Figure 10. Evaluation of flicker Plt (95% percentile)

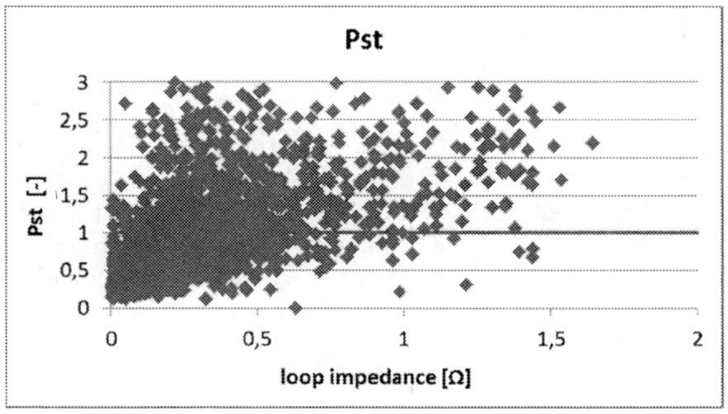

Figure 11. Evaluation of flicker Pst (95% percentile)

C. Voltage Unbalance

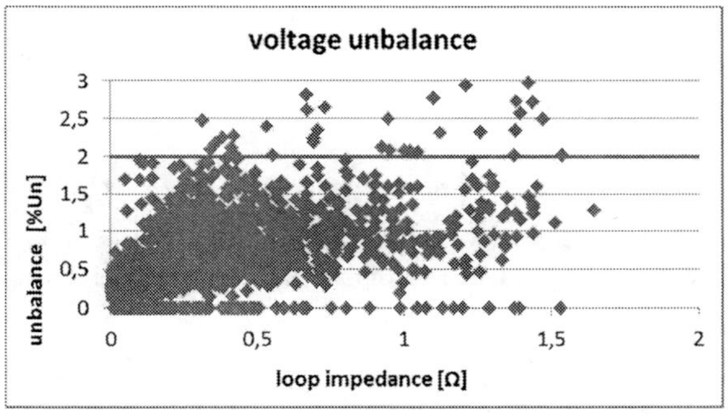

Figure 12. Evaluation of voltage unbalance (95% percentile)

D. Voltage Harmonics

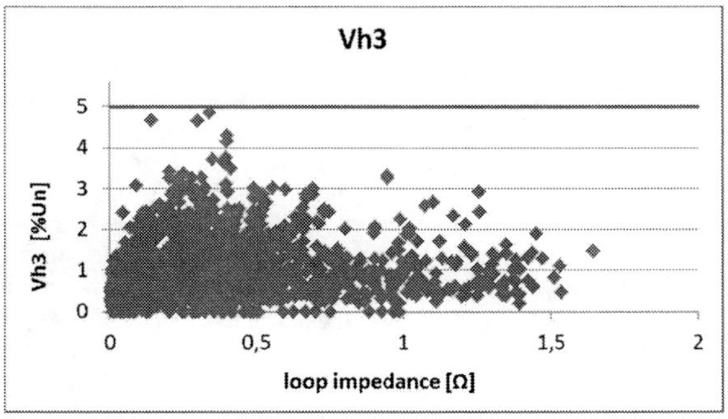

Figure 13. Evaluation of the 3rd voltage harmonic (95% percentile)

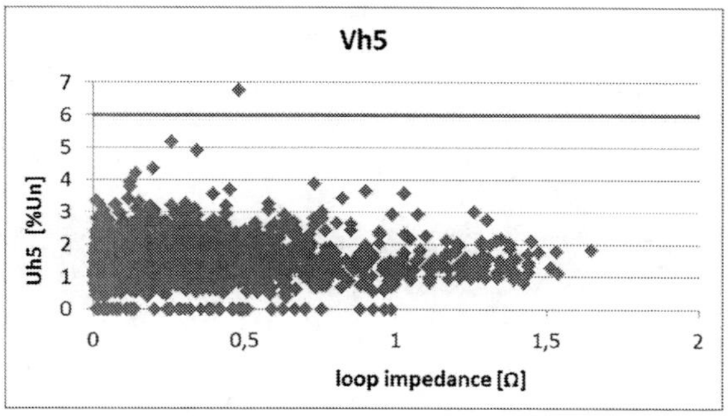

Figure 14. Evaluation of the 5th voltage harmonic (95% percentile)

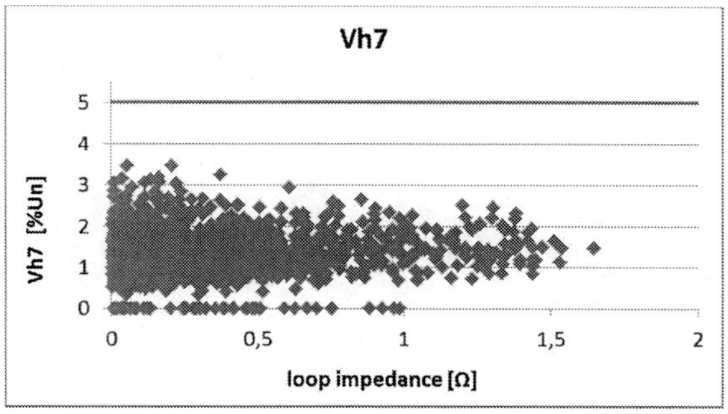

Figure 15. Evaluation of the 7th voltage harmonic (95% percentile)

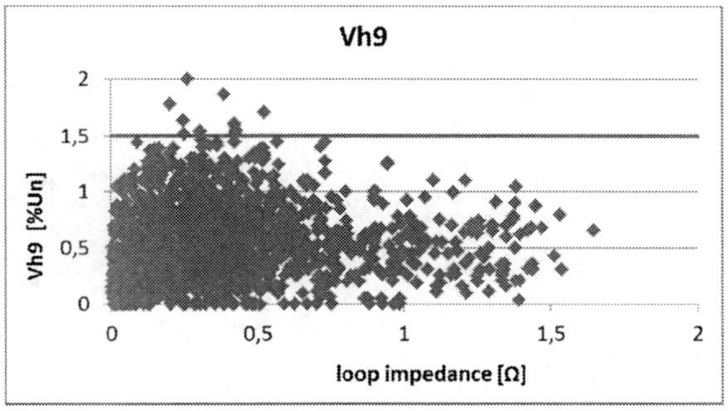

Figure 16. Evaluation of the 9th voltage harmonic (95% percentile)

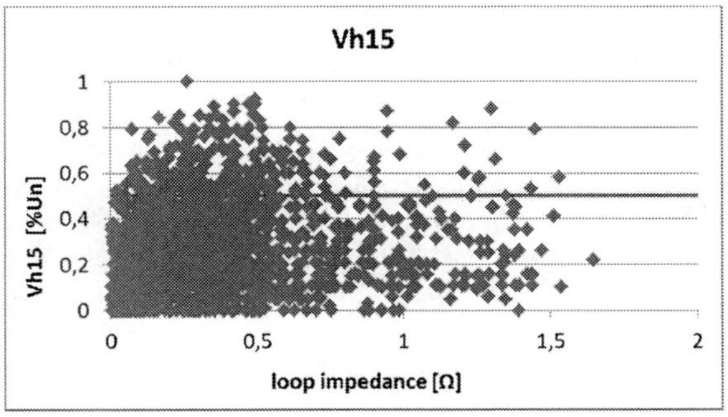

Figure 17. Evaluation of the 15th voltage harmonic (95% percentile)

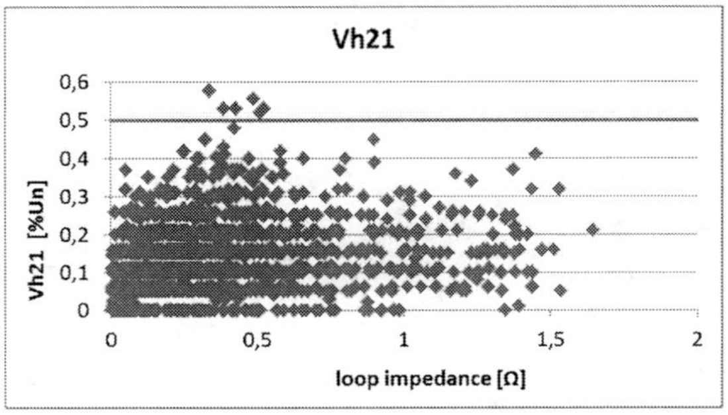

Figure 18. Evaluation of the 21st voltage harmonic (95% percentile)

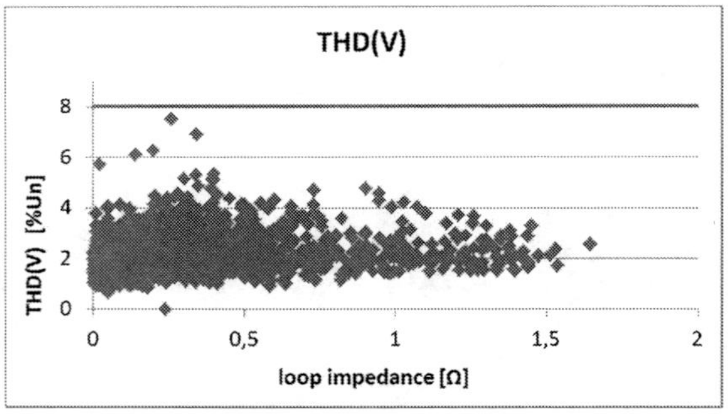

Figure 19. Evaluation of total harmonic distortion THD (95% percentile)

V. SUMMARY

A. Voltage Variations

The Fig. 4 and Fig. 5 show overvoltage in some grids where the values of system impedance are lower (better) than the reference value. It is a problem because it was expected that keeping of the reference impedance will guarantee full voltage quality for all LV grids (or for majority of grids). The problem of overvoltage in these grids (with the value of system impedance lower than reference value) does not depend on impedance value because it is caused by incorrectly set tap changer (ratio) on the MV/LV distribution transformer. The problem can be successfully solved in practise by resetting of the transformer ratio as the experience of DNO shows.

The Fig. 6 and Fig. 7 show low voltage in some grids where the values of system impedance are lower (better) than the reference value. The problem can be partially caused by incorrectly set tap changer (ratio) on the MV/LV distribution transformer or the LV grid is loaded a lot of when too much customers are connected in a part of grid. The problem of low voltage is not so important (so common) as the problem with overvoltage. The problem can be solved by uniform distribution of customer connections.

Voltage difference exceeds the limit value only in grids where the values of system impedance are higher (worse) than the reference value.

B. Flicker

The Fig. 10 and Fig. 11 show the problem with flicker in approximately 37% of grids – see the Tab. 2.

Table 2. Number of grids (in %) with the exceeded flicker limit Plt(Pst)=1

Flicker > 1	All grids	$Z_{loop} \leq 0.47$	$Z_{loop} \leq 0.35$
Plt	37 %	21 %	15 %
Pst	37 %	22 %	15 %

Tab. 2 shows exceeding of the flicker level Plt=1 together in 37% of grids (without the focus on the impedance value). Flicker level Plt=1 is exceeded in 21% of grids where the values of system impedance are lower (better) than the reference value 0.47 Ω and is exceeded in 15% of grids where the values of system impedance are lower (better) than the reference value 0.35 Ω. Because of a high number of inconvenient grids it was theoretically proposed the flicker limit Plt=1.5 and it was made the analysis – see the Tab. 3.

Table 3. Number of grids with the exceeded flicker limit Plt(Pst)=1

Flicker > 1.5	All grids	$Z_{loop} \leq 0.47$	$Z_{loop} \leq 0.35$
Plt	18 %	9 %	6 %
Pst	18 %	9 %	6 %

Tab. 3 shows exceeding of the flicker level Plt=1.5 together in 18% of grid (without the focus on the impedance value). Flicker level Plt=1.5 is exceeded in 9% of grids where the values of system impedance are lower (better) than the reference value 0.47 Ω and is exceeded in 6% of grids where the values of system impedance are lower (better) than the reference value 0.35 Ω.

Analysis according to the Tab. 1 and Tab. 2 shows that flicker limit Plt=1 is set incorrect because in case of increasing of the limit value to 1.5 a lot of grids do not comply with the requirements according to the EN 50160 standard. It is necessary to make an extensive analysis and to revise limit value for flicker.

C. Voltage Unbalance

The Fig. 12 shows problems with unbalance in some grids where the values of system impedance are lower (better) than the reference value. The problem occurs in grids where too much customers are connected in a part of grid. The problem can be solved by uniform distribution of customer connections.

D. Voltage Harmonics

It was detected the problem with the 15th voltage harmonic in approximately 12% of grids, in grids with good system impedance too. This could be caused by high penetration of appliances producing 15th harmonics current which isn't our case, the measurement was done in standard rural and small city area. The second reason could be a low limit value for the 15th harmonic voltage when measurement uncertainty for generally used power quality analyzers (class S according to the IEC 61000-4-30:2008 standard [5]) is +/- 0.5% of nominal voltage. The standard IEC 62749 [6] recommends limit values up to the 50th voltage harmonic and the limits for some harmonics (with the multiple of three) are only 0.2% of the nominal voltage. The experience of the DNO shows that no complaints regarding harmonic voltage were obtained up till now. Exceeding of the limit value for the 15th harmonic voltage is hardly perceived by the customer and has no influence on operation of commonly used electrical devices. The Fig. 19 shows no problem with total harmonic distortion (THD).

VI. MAXIMAL LENGTH OF THE FEEDER

The maximal values of short circuit impedance are recalculated to the maximal length of the commonly used types of conductors – see the Fig. 20. Keeping of the maximal length of the feeder gives the technician a smart tool by grid projecting. The reference values according to the Tab. 1 were used as maximal values.

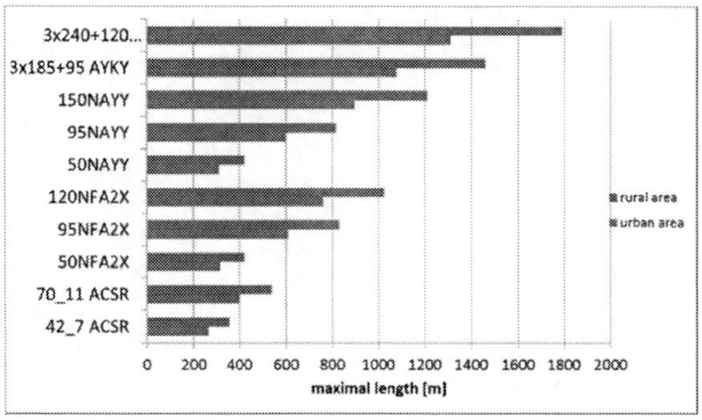

Figure 20. Maximal length of conducotrs with the focus on keeping of refernce impedance

VII. Conclusion

Keeping of the maximal length of the feeder (keeping of the reference impedance value) is necessary condition, but not sufficient condition, for meeting of the standard EN 50160. Additionally it is necessary to correctly set tap changer (ratio) on the MV/LV distribution transformer and to distribute uniform customer connections. Proposed values of reference impedances were verified expecting the flicker parameter. It was demonstrated that limit value for flicker is set incorrect and the topic of limit values for flicker and some voltage harmonics (mainly for 15th voltage harmonic) should be opened again.

Acknowledgment

The author would like to thank colleagues from the department power quality for cooperation regarding voltage quality measurements.

References

[1] IEC/TR 60725:2012 Consideration of reference impedances and public supply network impedances for use in determining the distrurbance characteristics of electrical equipment having a rated current ≤75A per phase, IEC, ISBN 978-2-83220-161-9

[2] EN 50160 Ed.3 Voltage characteristics of electricity supplied by public distribution systems. Brussels: European Committee for Electrotechnical Standardization, 2010. 20 p.

[3] PNE 333430-0 ed.5 Výpočetní hodnocení zpětných vlivů odběratelů a zdrojů distribučních soustav, E.ON Česká republika, s.r.o., 2015

[4] M. Tesarova and K. Nohac, "Voltage and frequency stability analysis of an island mode LV distribution network" In Proceedings of the 15th International Scientific Conference Electric Power Engineering 2014. Brno: Brno University of Technology, 2014. ISBN: 978-1-4799-3806-3

[5] IEC 61000-4-30 Ed.3.0 Electromagnetic compatibility (EMC): Part 4-30: Testing and measurement techniques – Power quality measurement methods. Geneva: IEC

[6] IEC/TS 62749 Ed.1 Assessment of power quality – Characteristics of electricity supplied by public networks

Human Interfaces Used for Smart Electric Installation Control

Jaroslav Petráš
Department of Electric Power Engineering
FEI TU in Košice
Košice, Slovakia
jaroslav.petras@tuke.sk

Jaroslav Džmura
Department of Electric Power Engineering
FEI TU in Košice
Košice, Slovakia
jaroslav.dzmura@tuke.sk

Abstract — **"Smart House" means a building with structuralized installation systems, especially electric installation systems with connection to heating and air-conditioning systems, or to systems with multiple energy sources and, in addition, this system should react to outer environment changes independently and it should learn, i.e. adapt its reactions according to the way, how it reacted. Smart electric installation requires control interface that is easy enough to use even by unqualified personnel, users with disabilities or senior users. An application for KNX EIB standard electric installation system that can be connected to any proposed human interface is presented. This capability means freedom in design of control system according to user requirements. Thus complex visualization system for these installations is designed and it is prepared to connect to any innovative human interface appearing even in the future.**

Keywords: Smart electric installation; KNX EIB; human interface

I. Introduction

Latest available world statistic data show a trend of population ageing and of quick increasing in number of people over 65 years number and people with physical disabilities, or people requiring external aid and assistance in their everyday life. This can create serious problems in near future. Increased need for older people care and care for people with physical disabilities will become more serious in future and as with this trend current social care for seniors and immobile people model will become unsustainable. Therefore it is necessary to search for innovative solutions [1].

A possible future problem of care shortage or unavailability for people that need it and require it cannot be solved by assistance service provider number increase in long terms. Currently available innovative solutions include building architecture modification and usage of modern technologies in buildings, which take a "smart houses" status by application of modern smart technologies.

II. Smart Systems and Smart builidings

The term "Smart Building" means a building with structuralized installation systems, especially electric installation systems with connection to heating and air-conditioning systems, or to systems with multiple energy sources with some additional features. A smart electric installation enables to changes installation even after it has been installed by simple re-programming, simpler systems for difficult functions in comparison with classical electric installation systems. It enables to combine different power sources depending on the current consumption and current power source availability.

Moreover remote control feature and the smart house visualization in different target devices are additional extra features.

But what a smart building makes smart, it is a capability to react to outer stimulus independently without any user interaction and this reaction should work as a standalone system. Moreover, the reactions should be changeable or adaptable according to former reactions and situations in such a system. Thus the system should have a learning feature, which helps to improve the reactions of such a system without any external reprogramming or correction.

External stimulus can be understood as e.g. weather conditions such as air temperature, humidity, wind strength and direction, light amount, sunshine amount and direction, energy consumption, and energy source availability at any moment, building parameters, daytime or nighttime etc.

Even user interaction can be considered as external stimulus as far the user does not control the system directly.

Main benefits of such smart building systems are:

- Positive trend in life quality improvement in general, and especially for seniors, immobile users and users with physical disabilities.
- Economic cost decrease for hospital care for these people thanks to remote monitoring of selected life functions.
- Economic cost decrease in general because of smart energy source combination usage.
- Environmental protection increase by energy savings in house energy requirements and effective usage of available.
- Visualization and remote smart home control and monitoring in general.

III. New Trends And Innovative interfaces

The above mentioned features of smart systems require the application of new methods along with existing solutions. New types of interfaces which get much closer to the human being are currently in development.

Thus the control by EEG (electro – encephalography, brain activity monitoring) signal, i.e. control based on emotional state of an individual as well as on targeted brain activity and thoughts, based on head orientation, mimics and muscle activity.

Another method of control that can be determined as human control is e.g. the control by EMG activity that means the control by muscle activity measured by muscle activity measuring armbands or other devices.

Voice control implemented in our system is another way of installation control, it does not require pre-learning of user voice or the sentence the installation should react. It is completely language and voice independent, the programmed sentence to react is simply interchangeable by rewriting the sentence in user interface (the user can do it without any reprogramming of the installation).

These control methods are more natural for people in general and are especially comfortable and advantageous for people with physical disabilities and immobile people which would have more facilitated access to smart house device control.

Currently available commercial systems of smart electric installation offer classical method of control, voice control (although very limited, the voice must be first memorized and the recognition rate is relatively low, in addition it has to be trained to the voice of individual, reprogramming requires relatively high effort), remote control via internet (often a part of visualization system, this solution is often very expensive and tied to specific piece of hardware, thus it is often outperformed by newer hardware as it is used 1 or 2 years and these systems are closed technologies and have small personalization and setting possibilities and can be expensive) or via SMS messages or other telecommunication channels.

Authors do not know any commercially available products for abovementioned innovative ways of smart electric installation control and any interfaces enabling the connection of these products to the installation with comparable parameters with designed system.

Currently we research the usability of EEG activity measurement interface for our visualization system, i.e. control by targeted brain activity, by emotional state of an individual, by head orientation change, mimics, muscle activity.

IV. The KNX/EIB Standard

EIB (European Installation Bus) system serves as a unified system for information transmission between devices. It is an industrial communication standard for information connection of devices (sensors, actors, regulation and control devices, service and measurement devices). KNX/EIB as European installation bus is a world standard for building devices. It was included in EN 500090 as the only public standard for electronic systems for buildings. The association includes manufacturers and commercial organizations and companies in the field of electric engineering, heating and air-conditioning technology. Existence of standard, world-wide usage of the KNX/EIB system, component availability from various manufacturers (manufacturer diversification), connection simplicity, programming and control simplicity are main benefits of KNX/EIB making it an ideal platform for a smart building system with the extension of conventional and unconventional human interfaces.

V. Visualization of Smart Building System

A visualization of automated electric installation system was developed for KNX/EIB systems on our department which include touchscreen gestures, mobile device specific controls using accelerometer such as space orientation change of the device, voice control with high reliability and reprogramming simplicity, movement gesture control. The system is ready to integrate any other even future type of human interface developed in the future and this makes our system flexible, extensible and open.

The system enables to visualize smart electric installation system components on a remote device. This target remote device requires features that are now widely commercially available – internet connection and web browser installed camera and microphone. For more sophisticated control interfaces such as voice, movement gestures, brain activity, muscle activity, face mimics, head movement and orientation change require more demanding target displaying device equipment (microphone, camera, special sensors) which are currently available or will be available in near future (as they are currently in development).

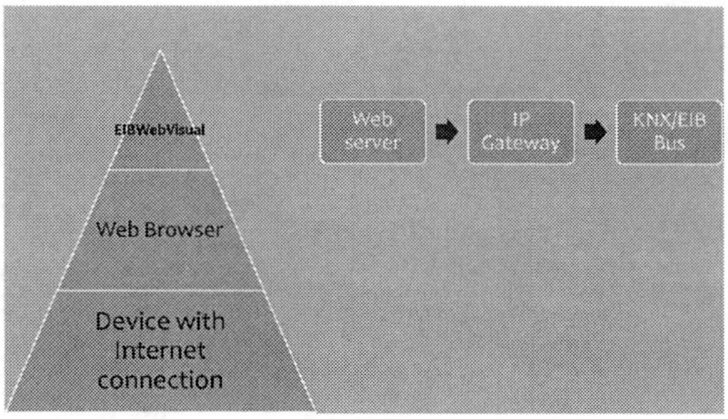

Figure 1. Principal diagram of EIBWebVisual

Our visualization system has the possibility of continual or manual (i.e. on demand) parameter monitoring and it is able to store and archive measured data for later analysis. In the context of smart electric installation usage with complex visualization system by seniors and immobile users or patients requiring continuous physiological parameter monitoring has the benefit of remote monitoring.

EIBWebVisual, as we named our visualization system, is a web application for KNX/EIB smart installation system. It runs on a web server, stores data in SQLite type database.

There are two main parts: core of the web application and EIB functionality made as a JQuery plugin.

It is possible to rewrite or redesign or change the core par independently of EIB functionality and vice versa.

Currently involved features and control user interface elements and methods are:

- Simple writing widget which sends one value to one function (simulating buttons).

- Simple reading widget reading one value from one function or sensor or actuator.

- Feedback widget on/off displaying feedback, reading on/off value from a function, actuator or sensor state (e.g. used to visualize if the light is on or off).

- Feedback value displaying feedback with real number reading value (displaying e.g. temperature or any other physical parameter measured by a sensor in the installation). This parameter can be displayed as a simple text or gauge.

- Cyclical reading widget reading one value from one group address cyclically, it is an extension of basic reading widget (displaying e.g. temperature or any other physical parameter measured by a sensor in the installation). This parameter can be displayed as a simple text or gauge.

- Multi writing widget sending multiple values to multiple group addresses, functions or actuators made for scene programming, which is independent from hardcoded scenes in installation software.

- Toggle on/off widget which is toggling on/off value, write 1, if status is 0 and vice versa.

- Slider widget with slider for sending and writing value from a certain value range made especially for light dimming.

- Text widget for simple text description widget with no other functionality.

- All reading widgets can be set as data recording widgets for read data archiving to text or any other file type or any other target archive device type.

- Widgets with archived data can be set to display archived data in a graph generated on the fly.

- Logical widget for combination of several input parameters in a logical expression involving logical AND, NOT and OR operators. This widget enables to combine these input values and conditions.

- Relative dimming widget writing value relative to current status, suitable especially for light dimming functions.

- Push button widget sending values continuously to group address until button is pushed or kept mouse button clicked over it.

- Speech recognition widget detecting speech and trigger action on corresponding widget exists as data-speech attribute, not as a separate widget. Speech recognition is language independent and voice independent.

- Touch screen widgets include orientation and swipe gestures on touch screen. These actions trigger corresponding action.

- Camera gesture widgets for camera gestures to trigger widget.

- Live monitoring camera feature with iframe live camera for monitoring of controlled room if available.

Selected widgets can have custom icons and all of them can have a tooltip text set by a room designer to make clear what the function of the widget is.

Visualization is divided into rooms representing one control entity, e.g. room, floor or building. The layout of room widget is stored in Room files. All rooms are defined in room directory; each room has its own directory, e.g. myRoom.room. Room is easily portable thanks its compact structure in one directory.

VI. Innovative Human Interfaces

Among innovative human interfaces used in our visualization system there are:

- touchscreen gestures, mobile device specific controls using accelerometer such as space orientation change of the device,
- voice control with high reliability and reprogramming simplicity,
- movement gesture control,
- interface for muscle activity sensing,
- interface for brain activity sensing.

The gestures made on touchscreen of mobile devices depend on application interface (API) that these devices provide. As of now, there are gestures e.g. swipe, touch, and screen orientation change.

Voice control in our system is completely independent on person's voice, gender or language used. Any word or sentence can be set to trigger appropriate action in the installation.

Muscle activity is sensed by Myo device which is a wearable device that captures electro myo-graphical signals

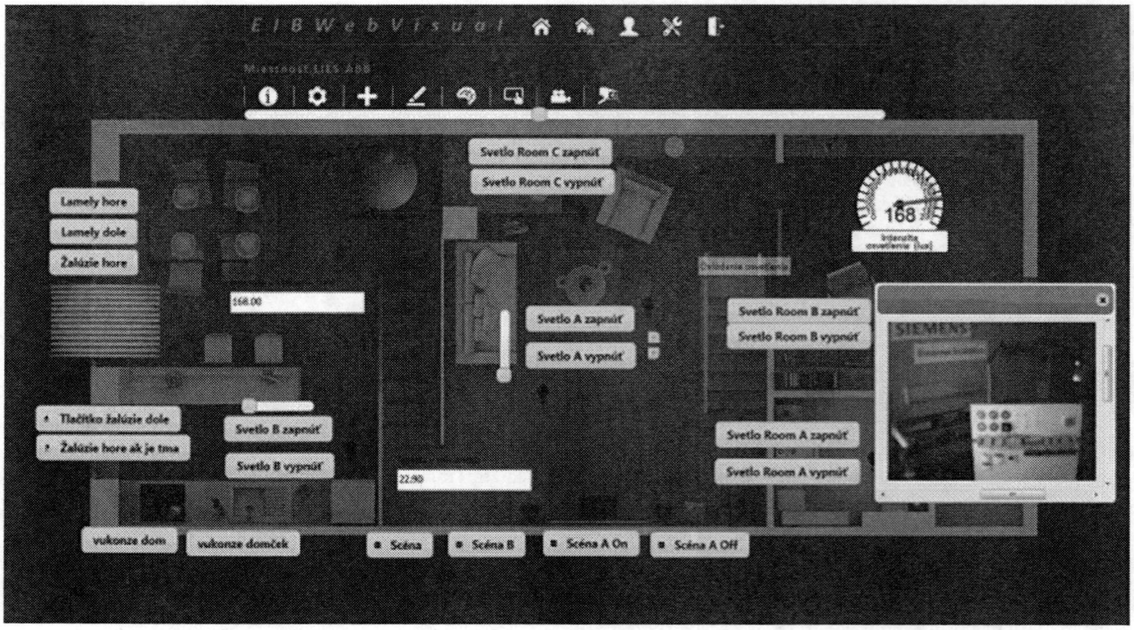

Figure 2. Visualization of one room on room footprint background

generated by muscles of hand. There are 5 gestures built in and in addition we can measure the position, acceleration and orientation of the hand and map these data to a set of parameters that trigger a specific action in electric installation.

The brain activity or electro encephalographic signals are captured by a special headset with 16 sensors. Thus, interactions of non-conscious cues are translated into human-computer dialog to emulate human to human interaction. These can be again used for triggering an action in electric installation. This module of our visualization system is still under development because of long term experiment and the need of calibration for every person that wants to use the headset [2, 3, 4, 5, 6].

VII. Innovative Human Interfaces

One of the innovative control types is the movement gesture control used in combination with camera built in the device. However it depends on many external conditions, such as the person's personality, movement speed, skin color, ambient lighting, and background color. We have designed, tested and chosen a set of optimized parameters for gesture recognition method and algorithm. In result, we can achieve between 72% and 94% of correct gesture recognition by algorithm used in our visualization system.

The optimization process was based on theoretically stated parameters, such as:

- Skin color determined in HSV color coordinate system (most typical theoretical skin colors can be seen in Table 1).

Table 1.

Hue	Saturation	Value	Skin color
0.075	0.231	1.000	
0.078	0.208	0.941	
0.075	0.248	0.933	
0.072	0.320	0.882	
0.092	0.336	0.898	
0.050	0.566	0.776	
0.053	0.608	0.729	
0.058	0.473	0.647	

- Ambient lighting intensity.
- Camera position in relation to the controlling person
- Background color (best result achieved with black or dark background).
- The type of the device and camera.

In Table 2 the default, theoretical and found optimal parameters can be seen for gesture detection algorithm.

Table 2.

Name of parameter	Description	Default plugin parameters	Theoretical values	Optimal values	Success in percent
huemin	Minimal skin color hue value	0.0	0.050	0.0	94 %
huemax	Maximal skin color hue value	0.1	0.092	0.05	76 %
satmin	Minimal skin color saturation value	0.0	0.208	0.25	92 %
satmax	Maximal skin color saturation value	1.0	0.608	0.6	84 %
valmin	Minimal skin color value	0.4	0.647	0.1, 0.7	90 %
valmax	Minimal skin color value	1.0	1.0	0.85	90 %
thresh	Color change needed to trigger movement	150	-	150	88 %
movethresh	Number of pixels changed needed to trigger movement	2.0	-	2.0	94 %
brightthresh	Minimal level of brigthness to trigger movement	300	-	400	78 %
overthresh	Parameter for direction detection	1000	-	1000	72 %
avg	Average number of changed pixels in time dimmension	0	-	100	85 %

VIII. Conclusion

Our visualization system is a contribution to the smart home comfort especially for users with movement disabilities. Default interfaces available on the market can still be enhanced and improved as we have shown. In addition human interface are currently under development and it is expectable that these interfaces will be soon commercially available and they can be implemented in our system, as it is open and ready for extension.

Acknowledgment

The authors would like to acknowledge Scientific Grant Agency of The Ministry of Slovak Republic and Slovak Academy of Science for funding experimental works in the frame of VEGA No. 1/0311/15 and VEGA No. 2/0141/16.

This work was supported by the Slovak Research and Development Agency under the contract No. APVV-15-0438.

References

[1] D. H. Stefanov, Z. Bien, W. Bang, "The Smart House for Older Persons and Persons With Physical Disabilities: Structure, Technology Arrangements, and Perspectives", IEEE TRANSACTIONS ON NEURAL SYSTEMS AND REHABILITATION ENGINEERING, VOL. 12, NO. 2, JUNE 2004, ISSN 1534-4320

[2] G. Edlinger, C. Holzner, Ch. Guger, C. Groenegress, M. Slater, "Brain-Computer Interfaces for Goal orientated Control of a Virtual Smart Home Environment", Proceedings of the 4th International IEEE EMBS Conference on Neural Engineering Antalya, Turkey, April 29 - May 2, 2009, ISBN 978-1-4244-2073-5

[3] C. Holzner, C. Guger, G. Edlinger, C. Grönegress, M. Slater, "Virtual Smart Home Controlled By Thoughts", 2009 18th IEEE International Workshops on Enabling Technologies: Infrastructures for Collaborative Enterprises, ISBN 1524-4547

[4] A. Kanemura, Y. Morales, M. Kawanabe, H. Morioka, N. Kallakuri, T. Ikeda, T. Miyashita, N. Hagita, S. Ishii, "A Waypoint-Based Framework in Brain-Controlled Smart Home Environments: Brain Interfaces, Domotics, and Robotics Integration", 2013 IEEE/RSJ International Conference on

[5] "Intelligent Robots and Systems (IROS)" November 3-7, 2013. Tokyo, Japan, ISBN 978-1-4673-6358-7

[6] S.P.Pande, P. Sen, "Review On: Home Automation System For Disabled People Using BCI", IOSR Journal of Computer Science (IOSR-JCE) e-ISSN: 2278-0661, p-ISSN: 2278-8727

The Assessment of the Implementation of the Concept of Sustainable Development on the European Energy Market

Katarzyna Grondys
Czestochowa University of Technology
Czestochowa, Poland
katarzyna_grondys@vp.pl

Abstract – **The energy market is one of the determinants of striving for the balance between the economic and natural development. In order to assess the implementation of the concept of sustainable development on the European energy market, the selected data associated with the energy market, having stimulative or non-stimulative impact on the natural and social environment, have been subjected to the analysis. The use of factor analysis and cluster analysis in relation to the identified variables has allowed to isolate the groups of the EU countries whose energy markets are more or less favorable for the implementation of the concept of sustainable development. The research results have allowed to observe that the vast majority of the analyzed EU countries have a great potential of renewable energy, which they frequently use, while simultaneously reducing the use of the sources of traditional fossil fuels in generating energy. The obtained results also extend knowledge on energy markets which must be subjected to further analysis and may serve as an example for other countries in the process of undertaking pro-environmental activities.**

Keywords: sustainable development, renewable and non-renewable energy, European energy market

I. The Energy Market in the Concept of Sustainable Development

Both the predicted significant increase in the demand for energy in the EU countries (up to half of the current demand) and the structure of energy carriers, still based on a large share of solid and liquid fuels [10,12], are not favorable for maintaining the major priorities of the pro-environmental economy of the European Union. Striving for sustainable development is also hindered by high energy dependence of Europe on the countries of the Middle East or Russia as well as more than 90% share of the sector in carbon dioxide emissions [13,14]. On the other hand, energy is essential for economic development of each country. The tasks formulated in the framework of the Community concerning the future of the energy market, among others, include an improvement in energy efficiency and environmental protection and development of renewable energy sources [10,15]. Sustainable development on the energy market consists in ensuring the proper balance between the natural progress and the use of more energy-efficient or alternative types of energy [10,11], which will result in more effective use of resources, being more environmentally friendly.

II. Research Methodology

The assessment of the implementation of the concept of sustainable development on the European energy markets has been made using the assumed statistical analyses. In the paper, there are presented only the final results essential for the process of the final conclusions. In the first stage, there has been made a selection of the EU countries and there have been identified the primary data associated with the European energy market (section III). In the subsequent stage, there has been conducted the statistical analysis of the data (section IV). In order to identify the key factors characterizing the energy market, there has been applied exploratory factor analysis (section IV A). The isolated factors have been the basis for making a division of the analyzed EU countries with cluster analysis in the area of stimulation and destimulation of sustainable development of the energy market (section IV B). The summary of the obtained results and the summary of the observations as well as the future scope of the research have been presented in the final part of the paper (section V).

III. Selection of the Research Sample

The first stage consisted in the selection of the European countries based on two key sources of the data on the energy market, i.e. Eurostat statistical database and the report "BP Statistical Review of World Energy June 2015 [16,17]". All the data referring to both renewable and non-renewable as well as conventional and innovative energy have been taken into account. For the purposes of the research, there have been selected 21 EU countries, which have been characterized by all the figures in the field of the energy market. Full range of data has been one of the conditions for conducting statistical analysis. In the initial stage, there have been specified 31 variables in the area of production and consumption with a breakdown by renewable and conventional energy, where the former ones have been defined as those stimulating sustainable development whereas the latter as the ones threatening ecological balance (Table 1). The division of the variables into stimulants and destimulants has allowed to carry out an assessment of the potential of the EU countries in the category of the implementation of the concept of sustainable development on the energy market.

Table 1. The division of the identified variables into stimulants and destimulants [16,17]

No.	Stimulants	Destimulants
1	Share of renewable energy in gross final energy consumption %	Market share of the largest generator in the electricity market (% of the total generation)
2	Electricity generated from renewable sources (% of gross electricity consumption)	Energy intensity of the economy (Gross inland consumption of energy divided by GDP (kg of oil equivalent per 1000 euro)
3	Renewable energies	Greenhouse gas emissions intensity of energy consumption (Index (2000 = 100))
4	Share of renewable energy in fuel consumption of transport %	Solid fuels
5	Natural Gas: Production (Million tonnes oil equivalent)	Crude oil (without NGL) production
6	Natural Gas: Consumption (Million tonnes oil equivalent)	Nuclear heat
7	Hydroelectricity: Consumption (Terawatt-hours)	Total gross electricity generation (GWh)
8	Renewables: Consumption - Solar (Terawatt-hours)	Final Energy Consumption - Industry
9	Renewables: Consumption - Wind (Terawatt-hours)	Final Energy Consumption - Transport
10	Renewables: Consumption - Geothermal, Biomass and Other (Terawatt-hours)	Final Energy Consumption - Other Sectors
11	Renewable energy - Biofuels production (housand b/doe)	Energy dependence %
12		Oil: Consumption (Million tonnes)
13		Coal: Production (Million tonnes)
14		Coal: Consumption (Million tonnes oil equivalent)
15		Nuclear: Consumption (Terawatt-hours)
16		Electricity Generation (Terawatt-hours)
17		Carbon Dioxide Emissions (Million tonnes carbon dioxide)

Finally, there have been identified 11 variables as stimulants and 18 as destimulants. Numerous distinguished variables would hinder cluster analysis at once, therefore an important stage in the assessment of the implementation of sustainable development of the European energy market has been to carry out factor analysis whose task is to reduce the number of variables to the most significant ones. For the purposes of the whole statistical analysis the data have been subjected to the process of standardization.

IV. Statistical Analysis of the Data

A. *Factor analysis of the variables*

Due to a large number of the variables of a similar nature, there has been used data reduction with the method of exploratory factor analysis which allows to specify a much smaller number of mutually independent (uncorrelated) factors. Firstly, in order to determine the number of factors, there has been used the criterion of scree plot (Figure1a), on the basis of which it has been established that there may be extracted three main components.

It can be seen in the graph that the steepness of the graph ends with the third factor but it includes only one variable, therefore, eventually only two key components that explain more than 70% of variability of all the stimulants can be extracted (Table 2).

After carrying out the analysis with factor non-orthogonal rotation with Promax method, it has been concluded that the first factor significantly "loads" 6 variables and refers to the quantitative level of consumption of energy coming from renewable sources, therefore a new component adopts the name "Consumption of renewable energy".

Table 2. Matrix of stimulants structure using Promax rotation method with Kaiser normalization.

Zestandaryzowane zmienne	Czynnik		
	1	2	3
Consumption of Geothermal, Biomass and Other	,917	-,031	,080
Consumption of Solar	,915	-,032	-,094
Using of renewable energies	,911	,319	,220
Consumption of Natural Gas	,900	-,098	-,241
Consumption of Wind	,855	,004	-,115
Consumption of Hydroelectricity	,821	-,075	,015
Share of renewable energy in gross final energy consumption	-,197	,916	,446
Electricity generated from renewable sources	-,077	,914	,296
Production of Biofuels	,215	,892	,178
Production of Natural Gas	,019	,721	-,305
Share of renewable energy in fuel consumption of transport	-,010	,227	,937

The second component mainly relates to the indices associated with generation and consumption of energy in a more general context, therefore the name of the second component is "Potential of renewable energy". The values of the indices have been prepared with regression analysis taking

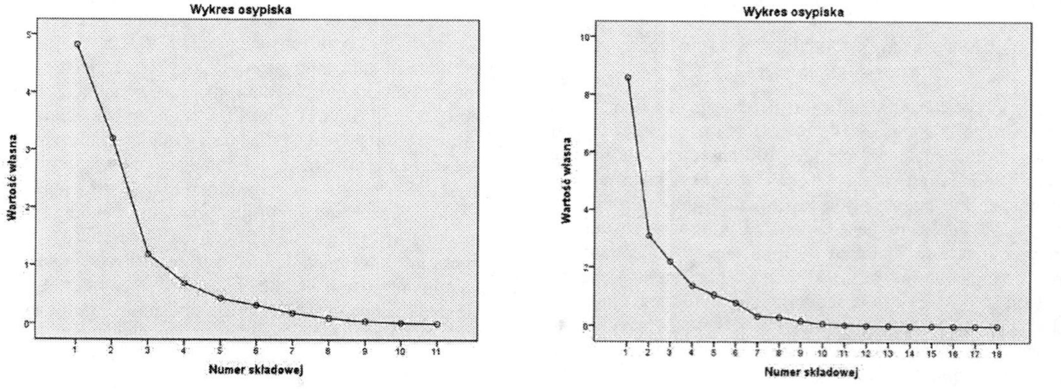

(a) (b)

Figure1. Scree plot for stimulants (a) and destimulants (b)

into account factor loadings of individual variables within each factor.

An identical process of extracting the principal components has been carried out for the variables which are destimulants in nature. Scree plot for these variables is presented in Figure 1b.

The steepness of the graph ends with the fifth factor, therefore it can be acknowledged that there have been successfully extracted four key components which, in total, explain more than 90% of variability of all the destimulants.

Table 3. Matrix of destimulants structure using Promax rotation method with Kaiser normalization.

Zestandaryzowane zmienne	Factor			
	1	2	3	4
Total gross electricity generation	,963	,390	,495	-,049
Electricity Generation	,962	,386	,495	-,052
Final Energy Consumption by Industry	,962	,487	,231	,003
Final Energy Consumption by Other Sectors	,959	,287	,500	-,072
Carbon Dioxide Emissions	,938	,594	,171	,102
Final Energy Consumption by Transport	,923	,338	,381	-,091
Consumption of Oil	,576	,977	,304	-,004
Production of Coal	,441	,974	,015	,277
Solid fuels	,362	,971	-,013	,254
Consumption of Coal	,688	,897	,010	,220
Nuclear heat	,552	-,002	,903	-,247
Consumption of Nuclear Energy	,557	-,001	,899	-,243
Greenhouse gas emissions intensity of energy consumption	-,002	,251	-,022	,893
Market share of the largest generator in the electricity market	,012	-,236	,400	-,805
Energy intensity of the economy	-,452	,339	,255	,631

As a result of the use of factor orthogonal rotation with Promax method, it has been concluded that the first factor significantly "loads" 7 variables which include energy consumption both in a general sense and in breakdown by sectors. These variables strongly correlate with each other, therefore they create a common set. The first component has been generally named as "Energy consumption". The second component includes 3 variables referring to the type of non-renewable fuel which is used for energy production, therefore the second component has adopted the name "Types of non-renewable fuel". The third component includes the subsequent two variables under the name of "Nuclear energy". The fourth component "loads" three variables which can be given a common name "Energy intensity of the country". The values of the indices, as previously, have been prepared with regression analysis taking into account factor loadings of individual variables within each factor.

Finally, as a result of the conducted factor analysis, there have been extracted 7 main components which are stimulants and destimulants in their nature (Table 4).

Table 4. Names of new indices created as a result of factor analysis.

Numer wskaźnika	Nazwa wskaźnika	
	Czynniki składowe - stymulanty	Czynniki składowe - destymulanty
1	Konsumpcja energii odnawialnej	Zużycie energii elektrycznej
2	Potencjał energii odnawialnej	Rodzaje paliwa nieodnawialnego
3		Energia nuklearna
4		Chłonność energetyczna kraju

B. *K-means clustering*

The task of the next stage is to determine the potential of sustainable development of the selected EU countries using the indices mentioned above. The method of k-means clustering has been used for the classification. There has been assumed the division into two clusters for each of the previously obtained factors. Figure 2 illustrates the breakdown of the countries by the factors stimulating sustainable development.

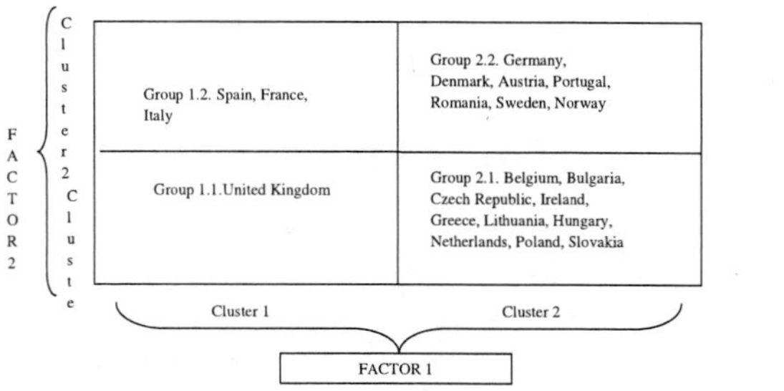

Figure 2. The share of the EU countries in the clusters by the stimulants of sustainable development obtained with the method of k-means clustering

When analyzing the share of the analyzed countries in individual groups by the factors stimulating their sustainable development associated with the use of renewable energy, there are observed four main clusters of the countries. In the first group, there is only the United Kingdom, which differs from the other countries with both the possessed potential of renewable energy and its use in the economy. The most numerous group is 11 countries, also including Poland. They are similar to each other both with reference to the possessed potential and its use. More detailed conclusions concerning the characteristics of individual groups can be drawn using the input data which, in the form of structural indices, are included in Table 5. The table shows the percentage share of the factors, estimated by averaging the numerical values of the analyzed variables for individual groups of the countries, broken down by two components.

Table 5. Indices of the structure of primary data stimulating sustainable development on the energy market

Numer grupy	Konsumpcja energii odnawialnej (Factor 1)	Potencjał energii odnawialnej (Factor 2)
Grupa 1.1.	21%	22%
Grupa 1.2.	45%	24%
Grupa 2.2.	28%	41%
Grupa 2.1.	6%	13%
Razem	100%	100%

The analysis of the tabular data allows for the conclusion that the groups 1.2. and 2.2. are characterized by the highest potential of renewable sources of energy as well as their use. They include Spain, France, Italy and Germany, Denmark, Austria, Portugal, Romania, Sweden and Norway. The first three countries lead in the level of consumption (45%), the contribution of the others is primarily visible in involvement in the production of energy from renewable sources (41%). In total, these countries produce and consume more than half of total energy coming from renewable energy sources in Europe. In this group, most of energy comes from hydro power plants and wind power plants and natural gas. Great Britain also has a significant contribution to generation and consumption of renewable energy, both components are at the level of 21%. These countries are distinguished by the greatest potential in generating energy from renewable sources while, at the same time, consciously minimizing the effects of the negative impact of the activity of this sector. The countries from group 2.1, including Hungary, Poland or Slovakia, require the greatest changes in the approach to sustainable development. They share both a low level of consumption and the limited facilities for the production of renewable energy.

Table 6 shows the division of the countries into groups by the factors destimulating sustainable development on the energy market.

Table 6. The share of the EU countries in the clusters by the destimulants

Numer skupienia i czynnika	Factor 1	Factor 2	Factor 3	Factor 4
Cluster 1	Germany, Spain, France, Italy, United Kingdom	Germany, Poland	France	Bulgaria, Lithuania
Cluster 2	Belgium, Bulgaria, Czech Republic, Denmark, Ireland, Greece, Lithuania, Hungary, Netherlands, Austria, Sweden, Slovakia, Poland, Portugal, Romania, Norway	Belgium, Bulgaria, Czech Republic, Denmark, Ireland, Greece, Lithuania, Hungary, Netherlands, Austria, Sweden, Slovakia, Portugal, Romania, Norway, Spain, France, Italy, United Kingdom	Belgium, Bulgaria, Czech Republic, Denmark, Ireland, Greece, Lithuania, Hungary, Netherlands, Austria, Sweden, Slovakia, Portugal, Romania, Norway, Spain, Italy, United Kingdom, Germany, Poland	Belgium, Czech Republic, Denmark, Ireland, Greece, Hungary, Netherlands, Austria, Sweden, Slovakia, Portugal, Romania, Spain, Italy, United Kingdom, Germany, Poland

Eventually, as a result of the division of the countries into two clusters, according to the values of five extracted factors, there have been obtained 8 groups diversified by the conditions destimulating sustainable development. The identified clusters indicate high homogeneity of the analyzed countries by the analyzed variables. This means that the vast majority of them is characterized by similar conditions in the field of searching for the balance on the energy market. In order to assess individual groups there have been determined structural indices for individual groups (Table 7).

Table 7. Indices of the structure of primary data destimulating sustainable development on the energy market

Numer skupienia	Zużycie energii elektrycznej (Factor 1)	Rodzaje paliwa nieodnawialnego (Factor 2)	Energia nuklearna (Factor 3)	Chłonność energetyczna kraju (Factor 4)
Cluster 1	86%	95%	96%	62%
Cluster 2	14%	5%	4%	38%
Razem	100%	100%	100%	100%

When analyzing the indices of the structure of input data referring to the activities destimulating sustainable development on the analyzed energy markets, it is observed that the vast majority of the EU countries use less energy coming from non-renewable sources and nuclear energy. At the same time, these countries are also less energy-intensive, which results from the implementation of innovation and energy-efficient solutions in industry, transport and services [18,19]. Generally, Poland and Germany reach for energy coming from non-renewable sources. In turn, France dominates in using nuclear energy (96% of the total use). The vast majority of the EU countries reduce the consumption of energy, among others, by aiming at reduction in consumption of renewable fuels (due to EU requirements). At the same time, the greatest demand for energy is in Germany, Spain, France, Italy, United Kingdom and therefore they are the ones that emit most of carbon dioxide. The majority of the analyzed countries do not have access to nuclear energy.

V. CONCLUSIONS

The division of the EU countries by the energy market factors stimulating and not stimulating their sustainable development, allowed to differentiate and assess the efforts in the field of conducting the eco-friendly and pro-social energy policy. The energy markets, identified as the ones which are the most favorable for the natural environment, consume a substantial portion of energy coming from conventional sources of its generation (Spain, Germany, France, Italy). The optimal group of the countries which both maximize the opportunities of renewable sources of energy and minimize the negative impact of its conventional sources are Denmark, Austria, Portugal, Romania, Sweden and Norway. These countries are in the group which implements sustainable development through generating environmentally-friendly energy, rarely uses non-renewable fuels and shows low energy consumption in the sector of industry, transport and services, thereby reducing emissions of greenhouse gases. The economic development of these countries has been directed to such consumption of energy which is favorable for sustainable development. Proper observation and analysis of the optimal solutions on these energy markets will allow increasing energy independence and more effectively adapting to the existing EU requirements in the field of the energy policy. Therefore, in the further research, it is planned to conduct the analysis of specific solutions on the energy market of the countries conducting optimal management of energy resources.

REFERENCES

[1] M. Burchard-Dziubińska, A. Rzeńca, D. Drzazga, Zrównoważony rozwój – naturalny wybór, Wydaw. Uniwersytetu Łódzkiego, Łódź 2014, p. 9.

[2] I. Kott, Oddziaływanie transportu na działalność gospodarczą i środowisko naturalne w Unii Europejskiej, Zarządzanie łańcuchem dostaw w teorii i praktyce. Monogrph. B. Skowron-Grabowska (ed.), Wydaw. WZPCz, Częstochowa 2010, pp. 102.

[3] V. Šoltés, B. Gavurová, Modification of performance measurement system in the intentions of globalization trends, Polish Journal of Management Studies, Vol. 11, No 2/2015,, pp. 163.

[4] Wskaźniki zrównoważonego rozwoju Polski, GUS, Katowice, Katowice 2011, p. 7.

[5] United Nations General Assembly, Report of the world commission on environment and development: Our common future. Oslo, Norway 1987, p. 43.

[6] R. Emas, Brief for GSDR 2015 The Concept of Sustainable Development: Definition and Defining Principles, Florida International University, p. 3 [Online], Available: https://sustainabledevelopment.un.org/content/documents/5839GSDR%202015_SD_concept_definiton_rev.pdf last see (2017, May)

[7] [Online], Available:https://www.rand.org/content/dam/rand/pubs/rgs_dissertations/RGSD151/RGSD151.chap2.pdf, p.2 last see (2017, May)

[8] T. Strange, A. Bayley, Sustainable Development, Linking economy, society, environment, OECD Insights, 2008, p.25.

[9] R.W. Flint, Practice of Sustainable Community Development, Springer Science+Business Media New York 2013, p. 43

[10] OECD Contribution to the United Nations Commission on Sustainable Development 15, Energy for sustainable development, Organization for economic co-operation and development, [Online], Available https://www.oecd.org/greengrowth/38509686.pdf, p. 9 last see (2017, May)

[11] Z. Bohátová, P. Schwarcz, L. Schwarczová, A. Bandlerová, V. Tĺčik, L. Mura: "Multifunctionality and regional economy management in Kopaniciarsky Region of the Slovak Republic," Polish Journal of Management Studies, Vol. 12, No 2/2015,, pp. 27.

[12] B. Ślusarczyk, M. Baryń, S. Kot, Tire industry products as an alternative fuel (2016) Polish Journal of Environmental Studies, 25 (3), pp. 1263.

[13] S.Kot, B.Ślusarczyk, The role of biomass in the energy production in the European union countries, Proceedings of the 7th International Scientific Symposium on Electrical Power Engineering, ELEKTROENERGETIKA 2013, (559-561), pp. 560;

[14] [Online], Available: http://www.energetyka-w-ue.cire.pl/st,9,185,tr,78,0,0,0,0,0,charakterystyka-energetyki-w-unii-europejskiej.html, last see (2017, May)

[15] J. Malko, A. Wilczyński, H. Wojciechowski, Bezpieczeństwo energetyczne, dostępność energii i zrównoważony rozwój a strategia Unii Energetycznej, Rynek Energii, 2015/No. 2 , p.10.

[16] BP Statistical Review of World Energy June 2015, http://www.bp.com/statisticalreview

[17] Eurostat, [Online], Available: http://ec.europa.eu/eurostat/web/energy/data/main-tables last see (2017, May)

[18] A. Hildebrandt, Europejska energetyka – skąd poszczególne kraje czerpią energię i jak ją wykorzystują? Pomorski Przegląd Gospodarczy 2013; [Online], Available: http://ppg.ibngr.pl/pomorski-przeglad-gospodarczy/europejska-energetyka-skad-poszczegolne-kraje-czerpia-energie-i-jak-ja-wykorzystuja last see (2017, May)

[19] P. Tomski, Partnering in Modern Economy. An Investigation on Walmart's Sustainable Value Networks, Contemporary and Future Trends in Management. Edited by Jerzy Lewandowski, Irena Jałmużna, Anna Walaszczyk, Wydawnictwo Politechniki Łódzkiej, Łódź 2011, pp.158.

Paris Agreement and the Development of Ecological Energy in Poland

Iva Pavlova-Marciniak
Faculty of Electrical Engineering
Czestochowa University of Technology
Czestochowa, Poland
E-mail: iva@el.pcz.czest.pl

Abstract — **The paper will present the latest documents adopted on a global scale, such as the agreement adopted at COP 21 in Paris on December 2015, that replaces the expired in 2012 Kyoto Protocol, directives adopted by the European Commission and approved national normative documents, that apply to climate – energy policy. Implementation of the objectives of these documents is associated with further efforts to improve the climate and the pursuit of competitive low-carbon economy in the EU.**

Keywords: Paris Agreement, Climate Packet, renewable sources of energy (RSE), Green Climate Fund, climate summit.

I. Introduction

The rapid industrial development of highly developed countries, in the second half of the 20th century, significantly changed the carbon dioxide (CO_2) and other greenhouse gases concentration and climatic conditions on Earth. Particularly rapid increase of greenhouse gases concentration has been observed since the mid-twentieth century. It has led to climate change, and consequently also to the extinction of whole species of flora and fauna and genetic and disease changes of human population. The greenhouse gas concentration index in the atmosphere is the number of particles of a specific gas per million particles (ppm - parts per million) or per billion particles (ppb – parts per billion) of clean air. Today, the methane concentration increased by almost a thousand particles - from 715 to 1774 ppb, the average sea temperature increased from 13,65°C to 14,55°C, and the amount of CO_2 particles in the air is one hundred more than in the middle of the 19th century (380 ppm), as shown in Fig. 1 [1].

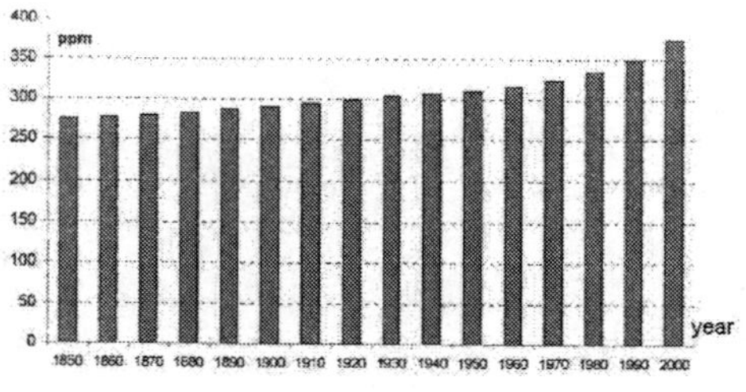

Figure 1. Changes in carbon dioxide concentration in individual years from 1850 to 2000 [1]

The necessary step for most industrialized countries was to meet in the mid-nineties of the last century, join international climate organizations and sign a number of emission reduction agreements, like Kyoto Protocol in December 1997. However, it began to apply on February 16, 2005, after its ratification by 55 countries, including Poland (without major polluters - China and USA).

The Kyoto Protocol was in force until 2012. However, neither on COP 14 (Poznan 2008) nor on COP 17 (Durban, South Africa, 2011) no document to apply after its expiration was signed. In Durban (South Africa) not only the United States and China, but also developing countries, were against signing of a new document. After long negotiations at this conference, the deadline for adoption of this important document was postponed to 2015, and the Kyoto Protocol extended to 2020.

II. New Climate Agreement

The greatest success achieved at the Poznan summit is an approved document called "Climate Package". The purpose of this document was to reduce CO_2 emissions by 20% until 2020 and to increase electric power production from renewable sources to 20% (3x20 Program). In the Package it was stated, that from 2013 all CO_2 emission permissions would be sold on auctions and could be traded freely on stock markets. Such decision would have a great impact on Poland, because more than 80% of its electric power is produced from conventional coal-fired power plants. As the result of the Polish government negotiations with the EC in December 2008, immediately after the end of COP 14, a decision was made to gradually increase the number of CO_2 permissions purchased on free market over the transitional period until 2020 and thus to avoid a rapid jump in electric energy prices. By 2020, power production from renewable sources in Poland must reach at least 15%.

There were no significant agreements at the Copenhagen summit (COP-15).

Another UN climate summit took place in 2010 at the Mexican resort of Cancun (COP 16). Deputies from 194 countries around the world took part in it. The biggest success was the foundation of the Green Climate Fund (GCF), mainly by rich countries with greater financial capacity. The fund's goal is to distribute $100 billion by 2020 for poor equatorial regions (New Guinea, Brazil, and Indonesia) in the form of compensation for losses resulting from the limitation of the tropical jungle logging.

The Kyoto Protocol was in force until 2012, but a new document to supersede it was not signed neither at COP 14 (Poznan 2008) nor at COP 17 (Durban, South Africa, 2011). In Durban (South Africa), not only the United States and

China, but also developing countries, were against signing a new document. After long negotiations at this conference, the deadline for adoption of a new document was postponed to 2015, and the Kyoto Protocol extended to 2020.

In March 2011, the European Commission adopted a document called: "Road Map for a Competitive Low Carbon Economy", covering a long-term CO_2 emission reduction plan of 80-95% by 2050 (compared to 1990).

At the UN COP 18 summit in Doha, Qatar in 2012, with 17 000 participants (from 194 countries), representatives of the IAEA's International Atomic Energy Agency (IAEA) were invited. An expert on nuclear energy Alan McDonald presented a report titled: "Climate Change and Nuclear Power 2012", in which nuclear power was stated as a clean energy [2].

In November 2013, the next UN climate summit, COP 19, took place. It was attended by representatives from about 200 countries. It was proposed to reduce the CO_2 emission to 40% in 2030. Each obliged to develop the way to meet these requirements and present it at the key COP 21 summit in Paris 2015, when the deadline was to come for the new agreement replacing the expiring Kyoto Protocol. During summits organized beforehand, COP 19 in Warsaw and COP 20 on December 2014 in Lima (Peru), the necessary documents should have been preliminarily prepared [3].

On 12 December 2015, at the UN climate conference COP 21 in Paris, with the collaboration of representatives from almost 200 countries the global climate agreement was made. Its goal is to stop the global warming significantly below $2^{\circ}C$ ($1,5^{\circ}C$). The participating countries would aim to balance CO_2 and other greenhouse gases emissions with their removal, including absorption, as shown in Fig. 2 [4].

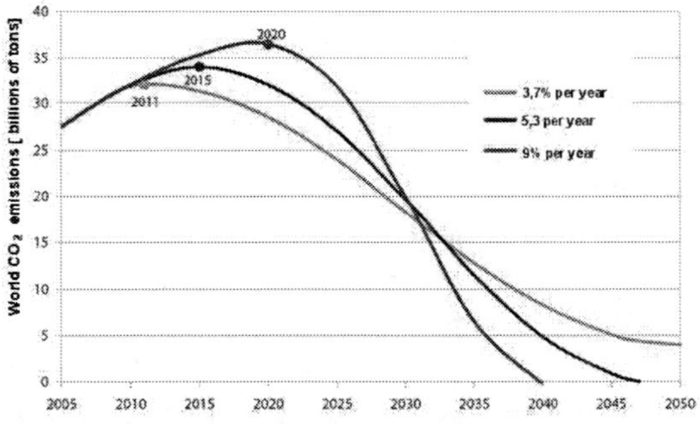

Figure 2. Maximum rate reduction of CO_2 emission reduction to keep the temperature rise below 2^0C

The agreement adopted in Paris creates a chance for sustainable development around the world and, unlike the Kyoto Protocol, seeks to reduce CO_2 and greenhouse gas concentration in two ways - by reducing emissions with new technologies and by sequestering greenhouse gases by forests. The Paris Agreement from 12 December 2015 is the first ever international agreement that obliges all countries in the world to work towards climate protection.

In October 7, 2016 in New York, Polish Prime Minister Beata Szydlo submitted a complete documentation according to the schedule, which meant acceptance and ratification of the Paris Agreement. Until the next UN COP 22 conference in Marrakesh, Morocco the agreement was also ratified by: China, United States, Brazil, Canada, Mexico and the following EU Member States: Austria, France, Germany, Hungary, Malta, Portugal and Slovakia. Since Poland was one of the few EU Member States that ratified the agreement, it was possible for it to have an individual representation at COP 22.

The UN Conference COP 22 was held in Marrakech, Morocco from 7 to 18 November 2016. During its duration the meeting of the Paris Agreement (CMA) participants took place for the first time and a discussion on the implementation of the climate agreement was launched. Poland actively participated in the negotiations because it wanted to spread the postulate to use forests to reduce carbon dioxide concentration in the atmosphere. The Polish delegation, which included, inter alia, representatives of industry, science, State Forests and trade unions, showcased the forest carbon farms and waste-free coal energy. It has been particularly emphasized that a reduction of CO_2 concentration through its absorption leads to improved water, air and soil quality, also protects biodiversity, prevents hunger and migration.

III. THE ANTI-SMOG ACT

Photochemical smog, also known as *oxidizing smog*, is formed by photochemical processes in the environment including high concentrations of nitrogen oxides, hydrocarbons, especially unsaturated ones (alkenes) and other ingredients of the exhaust gases (mainly automotive) during strong sunlight exposure. *Acid smog*, also called *industrial mist*, creates in humid air when highly contaminated with so-called acid gases from burning coal and other solid fuels, mainly SO_2 and carbon dioxide (CO_2) and PM10 and PM2,5 (particulate matter with a grain diameter of less than 10 µm and 2,5 µm). The greatest pollution in Europe is in Poland, in Silesia and Małopolskie Voivodeships. There is also an increased pollution in other regions, and among the most polluted cities in Europe, as many as 9 are Polish, according to a report by the World Health Organization (WHO), as shown in Fig. 3 [5].

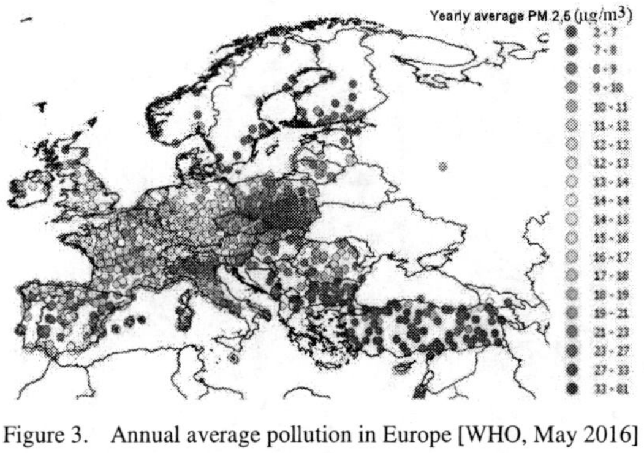

Figure 3. Annual average pollution in Europe [WHO, May 2016]

According to the data presented in May 2016 the air in Żywiec has the highest pollution level. Even every fifth lung cancer in Poland may be caused by smog. The high concentration of PM 2,5, that gets easily inside the human body and its blood system is far more dangerous and carcinogenic than PM10 coarse dust. The analysis showed that the top places on the WHO list are large cities in the Silesian and Malopolskie Provinces, such as Żywiec, Kraków, Bielsko-Biała, Sosnowiec and Zakopane as shown in Tab. 1 [5].

Table 1. Most polluted cities in Poland [5]

City/province	Amount of dust PM 2,5, µg/m³
Żywiec, province Silesian	43
Pszczyna, province Silesian	43
Rybnik, province Silesian	40
Wodzisław Śląski, province Silesian	39
Opoczno, province Łódzkie	39
Sucha Beskidzka, province Małopolskie	39
Godów, province Silesian	38
Kraków, province Małopolskie	37
Skawina, province Małopolskie	36
Katowice, province Silesian	35
Zakopane, province Małopolskie	32

The main cause of that high pollution levels is the traditional, coal burning heating method of Polish houses during the heating period, not traffic, like in other European cities. Lack of strong actions of local governments has led the southern part of Poland to become the most polluted region in Europe. During the winter months in Zakopane the PM10 concentration value in the air exceeds the limit several times, which endangers the health and life of people, and is the reason for protests of tourists against the climate tax included in accommodation price. According to the standards the permissible PM10 level per day is 50 µg/m³, above 200 µg/m³ air quality is considered very bad. Just within the first days of 2016 in Zakopane six times the PM10 exceeded 400 µg/m³. The law requires officials to inform citizens if the average daily PM10 level in the air exceeds 300 µg/m³. In other EU countries these norms are even more restrictive.

The anti-smog law enacted by the parliament on August 5, 2015 assumes inter alia that local governments will determine what fuel can be used in furnaces and set technical standards for furnaces . That law makes it possible to apply at local level legal solutions that will contribute to improving air quality and noise protection. At the beginning of September, the Minister for the Environment announced the National Air Protection Program (NAPP), the main objective of which is to achieve safe concentrations of certain hazardous substances in the air, especially the PM 2,5 and benzo(a)pyrene (BAP) – Fig. 4 [5].

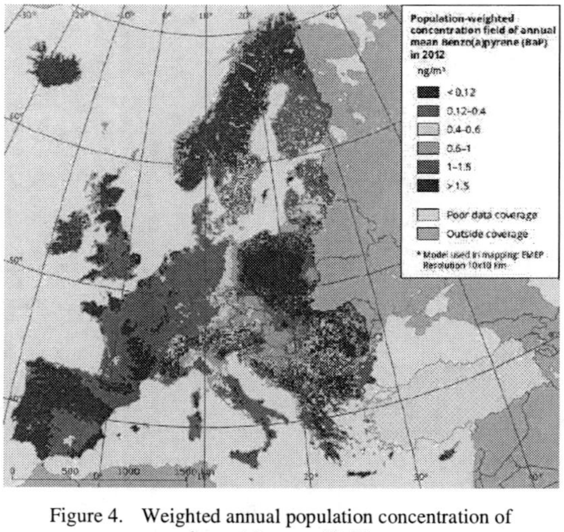

Figure 4. Weighted annual population concentration of benzo(a)pyrene (BaP)

According to the amendment of the Environmental Protection Law of October 15, 2015, covering the provisions of the Act, voivodship senators have the right to approve resolutions specifying the type and quality of conventional fuels approved for use. They should set the technical parameters and allowable emission of equipment. Also approve resolutions prohibiting the use of specific installations, emitting hazardous substances. They should also determine the boundaries of the restricted area.

IV. New Coal Technologies

Poland has large resources of hard coal and brown coal. The provisions of the Paris Agreement guarantee the continued use of those. The emissions reduction can be achieved using the latest technologies in building power units and by introducing modern furnaces using ecological fuel – Fig. 5 [6].

Figure 5. Defro BIOMAX type 150 kW furnace for eco-pea coal [6]

The fuel of 5 to 25 mm is made of high-calorific brown or hard coal with heating value of $Q_d = 24$ MJ/kg, low sulfur content $S \leq 1\%$ and low humidity, which results in low SO_2 and ash combustion. It is fed automatically into retort burners, combustion in which reduces CO_2, aromatic hydrocarbons, dioxins, furans and dusts emissions. The accomplishment of low emission for this fuel is certified by Institute for Chemical Processing of Coal in Zabrze (IChPC).

The latest achievement of scientists from the IChPC is the so-called Blue Coal or Low Carbon Coal, obtained by subjecting ordinary hard coal types 31 and 32 of heating value $Q_d = 24 \div 29$ MJ/kg value of the almost complete degassing process – Fig. 6 [7].

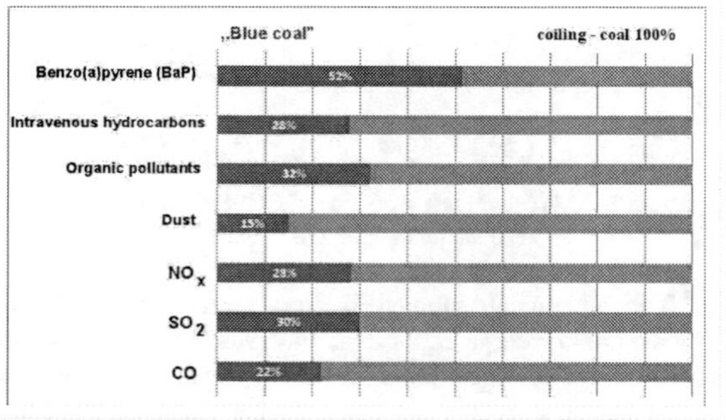

Figure 6. Emission factor of pollutants

For solid fuels, the emission of sulfur dioxide, carbon dioxide, nitrogen dioxide, carbon dioxide, BaP is calculated by the formula:

$$E = B \cdot w \cdot S \cdot A \quad (1)$$

where: B – amount of fuel burned in Mg (tones), w – unit emissions for particular pollutants, S – sulfur content, %, A – ash content, %.

Thanks to the previous degassing for combustion in classic simple furnaces (charcoal furnaces, coal cookers, conventional ovens, fireplaces) dust emissions are reduced by 40%, tarry substances by 75% and toxic benzo(a)pyrene by up to 90%. Those results were obtained during burning tests in one of tenement houses in Zabrze. The amount of other substances as percentage of the amount from burning an ordinary coal: NO_X = 28%, SO_2 = 30% and CO = 22%. The Blue Coal is tested among others in Cracow (in the Swoszowice spa resort) after the allowance of local authorities in Roszkowo (in the commune of Krzyżanowice near Raciborz) and in Jedlina-Zdrój near Wałbrzych. According to preliminary estimates from IChPC researchers, Blue Coal can be just 10 by 20% more expensive than eco-pea coal, which in turn requires much more expensive, automated and more difficult to use, especially for older people, furnaces.

Technological solutions for automotive are associated with gas-electric hybrid and electric motors to cut down the pollution as much as possible. Most cities have already introduced such solutions in public transport.

V. Energetic Efficiency and RES Act

The directive 2012/27/EU of the European Parliament and of the Council of 25 October 2012 on energy efficiency calls for a thorough rethinking of the concept of treating energetic efficiency as an outright energy source, representing the value of energy saved. By focusing on energetic efficiency as a way to reduce the energy demands, we strive to achieve security of supply, competitiveness and sustainability goals, and enable savings for consumers and industry [8]. The energy union strategy reaffirmed the 20% energy efficiency target by 2020, which means that final energy consumption will be less than 1086 Mtoe or primary energy consumption will drop below 1483 Mtoe. That is the cause for the target of reducing consumption by at least 27% by 2030 to be verified by 2020 so a level of even 30% could be reached.

The European Council session on 20 – 21 March 2014 outlined the EU's main climate and energy priorities so the 3x20 Climate Package targets (estimated at 24%, 21% and 17%, respectively) and energy efficiency guidelines could be met [9].

On 20 February 2015, Polish parliament passed the Renewable Energy Sources Act (RES), which came to force on 4 May of the same year [10]. The law determines renewable energy development in the country, and also regulates and simplifies the procedure of selling the energy excess produced by home installations by individual investors. The government has the ability to decide how much renewable energy it needs to meet the EU climate policy requirements. The law includes a proposal to introduce an auction system for "green" energy that would be won lowest price offeror, who in return would receive a support guarantee for 15 years. Auctions will be carried out separately for large and small installations. The Minister of Economy will set the so-called "reference price" (maximum) at which you will be able to purchase renewable energy in a given year. Support for renewable energy will be transferred to consumers who will have additional "renewable energy" charge in their energy bill, which in 2015 was 2, 27 PLN/MWh.

The amendment of the renewable energy law, which entered into force on December 31, 2015, prolonged the life of green certificates by half a year, thus postponed the start of the RES support auction system until July 1, 2016. The support system for micro-installations up to 10 kW was postponed as well by guaranteed tariffs. In addition, according to the amendment, the micro-installations up to 40 kW owners who started off their installations before 1 January 2016 were also covered by the so-called net-metering balancing system.

The parliament adopted the next amendment to the Energy Act on June 10, 2016, which entered into force on July 1 of the same year. Completely new support rules for prosumers were introduced and the auction system was changed to support more biomass and co-firing energy.

In Poland, primarily RES technologies are based on water, biomass, wind, solar and photovoltaic panels and heat pumps.

Poland has the greatest potential among all EU countries for the production of biomass, which is supplied by agriculture (eg. straw), animal breeding, forest waste and the woodworking industry. The reduction of carbon dioxide emissions in agriculture has already been done to a large extent. Further reduction can also be achieved through lower fertilizer use, construction of biogas plants for waste treatment, improvement of natural fertilizer management, improved feed quality. The firing value of 1,5-2 tones of biomass, depending on its the type, allows to obtain the amount of heat equivalent to 1 tone of hard coal. Renewable energy is also produced by burning biomass or co-firing it with coal. Several large biomass power plants have already been established in Poland, such as the 190 MW biomass power plant near Połaniec. Forecasts show that in 2020 the demand for biomass will be 8 million tons. Also the greenhouse gas emissions reduction is a result of municipal waste segregation, which can then fuel the thermal waste utilization furnaces, producing electricity and heat. The construction of those has already started in larger cities as Konin, Bialystok, Bydgoszcz, Cracow, Poznan, Szczecin and Rzeszow. All these incineration plants are expected to be put into service in 2017-2018.

The National Fund for Environmental Protection and Water Management (NFEPiWM) strongly co-finances the renewable energy sector investments such as small hydro and wind power projects up to 10 MW, biogas, photovoltaics, energy saving and passive houses –Tab. 2 [11].

Table 2. The installed power in RES [11]

Type of RES Power Plants	31.12.2014r *MW*	30.06.2015r *MW*	Change *MW*	Change %
- for biogas	188,549	191,381	2,832	1,5
- for biomass	1 008,245	1 008,245	0,000	0,0
- photovoltaic	21,004	35,586	14,582	69,4
- wind	3 833,832	4 117,421	283,589	7,4
- watery	977,007	980,323	3,316	0,3
Total	**6 028,637**	**6 332,956**	**304,319**	**5,0**

The output of a wind turbine within the working wind range of 4 to approx. 20 m/s can be estimated according to the approximate formula:

$$E = A \cdot K_{el} \cdot \eta, \qquad \mathrm{kW \cdot h/a} \qquad (2)$$

where: A – wheel surface, m^2, K_{el} – energy potential $kWh/(m^2a)$, η – turbine efficiency (η = 0,25).

In the winter holiday season, on December 26, 2016, the Orkan Barbara, which was rampaging over Poland, was able to produce a record amount of green energy - wind turbines delivered almost 5 GW of power to the National Power System (NPS). That means that wind turbines operated at 86% of rated power, delivering one third of the energy needed in Poland. This result was achieved despite the fact that some of the power lines had been damaged by the wind and some turbines were turned off because of excessive wind speed. With such a high share of green energy in the NPS, the power from conventional coal power plants dropped to 37%. The largest installed windmill at present in Poland is the type V112 with a capacity of 3,075 MW (Power Plant Pągów commune of Wolves, Wroclaw, 17 x 3,075 = 52,275 MW). The life span of such turbines is 20÷25 years. It was also planned that the first offshore wind farm would be built until 2021–22. The planned power is 600 MW [12].

Figure 7. The first Polish wind farm on the Baltic Sea [12]

The energy density of solar radiation in Poland on a horizontal plane is within 950-1250 kWh/m^2, and the period of highest sunshine exposure, from April to September, is about 1600 hours per year. The best sunshine exposure is in the seaside and eastern part of Poland. The International Energy Agency (IEA) forecast assumes that by 2050 solar energy, solar collectors and photovoltaic (PV) cells will be able to deliver about 21% of energy needed. By the end of 2016, 17 545 photovoltaic systems of 192,82 MW total power have been installed in Poland, including 473 ones with the Office of Energy Regulation (OER) license of 99,1 MW total power– Fig.8 [13].

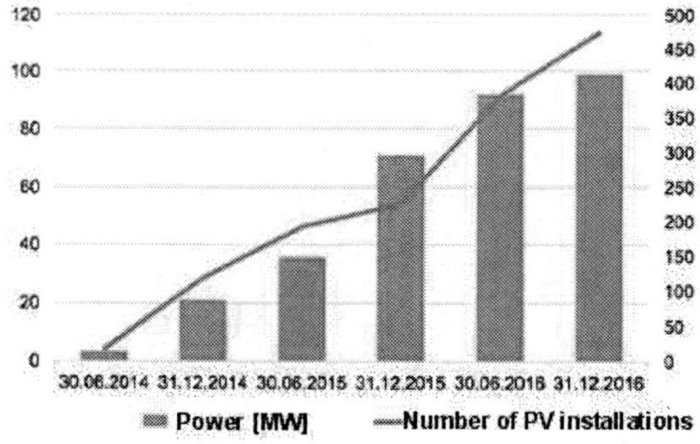

Figure 8. Photovoltaic power stations in Poland (OER – Office of Energy Regulation))

Directive 2007/60/EC of the European Parliament and of the Council of Europe of October 23, 2007 (so-called *Flood Directive)* and EU subsidies support the development of hydro-power. However, administrative barriers discourage investors, even though projects can be funded by up to 75%. According to official data, 727 *hydroelectric power plants* (including Small Hydro Power Plants) currently operate with a combined capacity of 937 MW, which means that hydroelectricity usage is about 12%.

High salinity of *geothermal waters*, requiring special treatment, low to medium temperatures, 65°C tops, frequent lack of spontaneous discharge and deep sedimentation, cause major technical difficulties during their operation. Despite those difficulties in Poland, several large *geothermal* heating plants have already been built.

In recent years, very high interest and application have also been found in *heat pumps*, which are one of the forms of use of *low temperature geothermal.*

According to energy development scenarios until 2050 the share of RES will gradually increase [14]. On the other hand, we will use CO_2 sequestration by forests. This will allow us to lower the carbon dioxide concentration in the air and use it to recover soils and forests and thus improve water and air quality and protect biodiversity. These actions address sustainable development policy and can be a model for other countries.

Actions to reduce CO_2 emissions will be undertaken by all participants on the Paris Agreement, respecting their specificity and socio-economic opportunities. These will include, inter alia, the introduction of new and development of existing low-carbon technologies or the increase of renewable energy share in the energy balance. In turn, the balancing of emissions will take place, inter alia, through increased afforestation.

During its meeting on February 21, 2017, the European Parliament decided to reform the ETS (European Trading Scheme), which aims to increase the price of CO_2 emission allowances to 15 and even up to 30 €/t, which would force member states to seek coal combustion restrictions. Currently the emission rights for 1 tone of CO_2, are about 5 ÷ 7 €/t on the stock market [15]. This way the regulations can be met for the Roadmap for a Low Carbon Competitiveness from March 2011, which covers a long-term reduction (CO_2) reduction plan, which is expected to reach 80-95% by 2050 compared to 1990. Poland up to this point gets a lot of free CO_2 emission allowances because the EC is aware that our power is coal-based up to 90%. An increase in prices to 20 ÷ 35 €/t in 2030 will cause huge losses for Polish power plants and a sharp increase in electricity prices for consumers. According to the project, Poland has to invest in RES and gas power plants, which can make it dependent on fuel supplies from Russia, whereas for now, thanks to our own reserves, we are one of the least dependent on energetic resources import EU countries.

The revised Atomic Law became entered into force on July 1, 2011, and in September 2014, its last amendments were announced in the Journal of Laws. There have also been discussions with Japanese representatives about the project to build a nuclear power plant in Poland, because Japanese companies such as Hitachi, Toshiba and Mitsubishi have the best and safest technologies in nuclear power production. As the cost of building a 3GW nuclear power plant (2x1500 MW) is estimated at 50-60 billion PLN, the government estimates that for this money three conventional coal blocks can be built. However, it has to be stressed that the CO_2 emission fees may, however, reach PLN 40-50 billion. The final decision on nuclear plant construction in Poland may be postponed until the first half of 2017, as the amount of energy required to ensure the country's energetic security by 2030 is estimated at 10 GW. From economic efficiency point of view the traditional coal power plants win clearly. However, taking climatic requirements into consideration, the nuclear ones are the priority.

Nuclear Power can solve the problem of increasing electricity demand. Due to prolonged talks and preparatory work, the date of launching of the first nuclear power plant in Poland was set for 2028 or later, but for now, no location and technology was chosen - Fig. 9 [16].

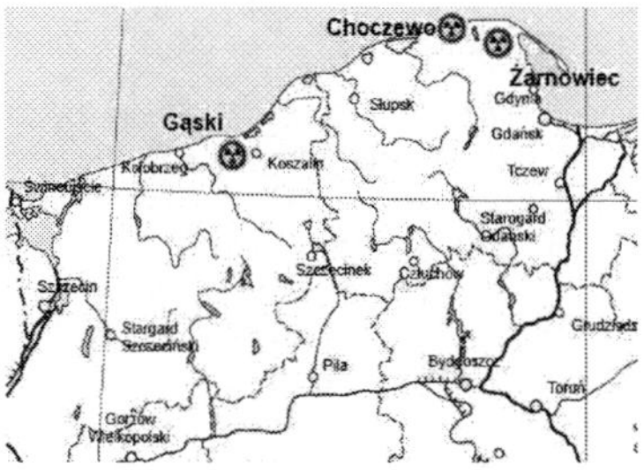

Figure 9. Proposed locations of a nuclear power plant in Poland

Final Conclusions

1. The Kyoto Protocol was in force until 2012, but due to a number of unsuccessful efforts at the UN COP 14 ÷ COP20, the signing of the new document did not take place. On December 2015, at the UN climate conference COP 21 in Paris, with representatives from nearly 200 countries, a global climate agreement was reached, with the aim of stopping global warming below 2ºC. Countries present at COP 21 in the second half of the 21st century will seek for solutions to reduce the greenhouse gas emissions and to remove them, including their absorption through increased afforestation.

2. Smog, because of the high concentration of aggressive chemicals, states a threat to human and animal health, causes plant diseases and destroys materials. The greatest pollution in Europe is in Poland, specifically in Silesia, according to a report by the World Health Organization (WHO).

3. According to the amendment of the Environmental Protection Law from October 15, 2015, covering the provisions of the Act, the regional councils are entitled to state laws specifying the type and quality of conventional fuels approved for use, technical parameters and emission parameters of devices carrying out the combustion process.

4. The adopted law gives the possibility of further development of modern technologies not only in power industry and mining, as well as in other related branches of the economy. New technological solutions, such as electric vehicles, new technologies and fuels such as eco-pea and low carbon coal, are aimed at supporting the implementation of not only the anti-smog law, but also the global climate agreement adopted at COP 21.

5. Directive 2012/27 / EU of the European Parliament and of the Council from October 25, 2012 on energy efficiency calls for a thorough consideration of the concept of treating it as a fully-fledged energy source representing the value of energy saved. By focusing on energy efficiency as a way to reduce energy demand, we strive to achieve security of supply, competitiveness and sustainability goals, and enable savings for consumers and industry. The Energy Union Strategy reaffirms the 20% energy efficiency target by 2020. This is the basis for the goal of reducing consumption by at least 27% by 2030, to be verified by 2020 strive to reach even 30% level.6. At its meeting on February 21, 2017, the European Parliament decided to reform the ETS (European Trading Scheme), which aims to increase the price of CO_2 emission rights to 15 and even up to 30 €/t, which will force member states to reduce the burning of coal. The current emission rights for CO_2, cost about 5 ÷ 7 €/ton the stock market.7. Until now, Poland has been given a large pool of free CO_2 emission allowances because the EC is aware that our power is as 90% coal-based. An increase in prices to 20 ÷ 35 €/t in 2030 will cause huge losses for Polish power plants and a sharp increase in electricity prices for consumers. According to the project, Poland has to invest in RES and gas power plants, which can make it dependent on fuel supplies from Russia. For now thanks to our own reserves we are one of the least import dependent EU countries when it comes to energetic resources. Nuclear Power can solve the problem of increasing power demand.

References

[1] I. Pavlova-Marciniak, "Implementation of climate policy assumptions in Poland", 8th International Scientific Symposium on Electrical Power Engineering, 16 – 18 September 2015, Stara Lesna, Slovak Republic, pp. 455-458

[2] [Online], Available: https://www.iaea.org/newscenter/multimedia/videos/climate-change-and-nuclear-power-interview-alan-mcdonald last see (2017, May)

[3] [Online], Available: http://www.cop20lima.org/about/cop20 (4.06.2017)

[4] [Online], Available: http://naukaoklimacie.pl/aktualnosci/porozumienie-klimatyczne-w-paryzu-124 last see (2017, May)

[5] [Online], Available: http://tvn24bis.pl/ze-swiata,75/raport-who-najbardziej-zanieczyszczone-miasta,643576.html last see (2017, May)

[6] [Online], Available: http://piece-kotly.eu/kolty-o-duzej-mocy last see (2017, May)).

[7] [Online], Available: http://www.agh.edu.pl/blog-naukowy/info/article/blekitny-wegiel-blog-naukowy/ last see (2017, May)

[8] A. Gawlak, "Noninvestment Forms of Reducing Energy Losses in Distribution Networks" - 8th International Scientific Symposium on Electrical Power Engineering, 16-18 September 2015, Stara Lesna, Slovak Republic, pp. 57-60

[9] [Online], Available: http://docplayer.pl/6027131-Rada-europejska-bruksela-21-marca-2014-r-or-en-delegacje-otrzymuja-w-zalaczniku-konkluzje-rady-europejskiej-20-21-marca-2014-r.html last see (2017, May)

[10] [Online], Available: http://isap.sejm.gov.pl/DetailsServlet?id=WDU20150000478 last see (2017, May)

[11] [Online], Available: http://www.cire.pl/item,110502,1,0,0,0,0,0,ponad-615-gw-mocy-w-oze-na-koniec-pierwszego-kwartalu.html last see (2017, May)

[12] [Online], Available: http://gramwzielone.pl/energia-wiatrowa/22794/pierwsze-farmy-wiatrowe-na-polskim-morzu-maja-powstac-za-5-lat last see (2017, May)

[13] [Online], Available: http://gramwzielone.pl/energia-sloneczna/25391/moc-elektrowni-fotowoltaicznych-w-polsce-przekroczyla-190-mw last see (2017, May)

[14] J. Sowiński, R. Tomaszewski and A. Wachtarczyk, "Electricity generation from RES under uncertainty", 17th International Scientific Conference on Electric Power Engineering (EPE), Prague, 2016, pp. 1-4. doi: 10.

[15] I. Pavlova-Marciniak, "Światowe i unijne normatywne dokumenty a rozwój OZE w Polsce", Przegląd Elektrotechniczny, nr 7, 2014, pp. 115-118.

[16] [Online], Available: http://www.nuclear.pl/ last see (2017, May)

GIS-based Visualization of Power Flows and Voltages of Electrical Network

Kairat Bostanbekov, Dmitriy Kim, Roman Lyssenko
National Research Laboratory of Information and Space Technologies
Kazakh National Research Technical University,
EcoRisk LLP, Almaty, Kazakhstan
kairat.boss@gmail.com, kdk26@mail.ru, lyssenkori@gmail.com

***Abstract* — To ensure a stable operation of the modern power system it is necessary to analyze an increasing number of parameters in time. In this work we propose to use the Geoinformation System (GIS) and its tools for displaying and studying the information concerning the system. On the example of a simplified model of the electric network in Zhambyl region we show how it is possible to visualize power flows and voltage values and obtain the animation of these for a 24-hour period using the GIS.**

Keywords: Visualizing the Electrical Structure of Power Systems; Geoinformation system

I. Introduction

For the purposes of ensuring a stable functioning of the electric power system it is necessary to take into account a large amount of technical and operational information on the facilities and equipment of the networks. In particular, in the course of computation of the operational mode of the energy system on the basis of the model diagram with units, generators, power lines, etc., it is necessary to determine a variety of parameters of the network, e.g. power flows, voltage values, energy losses, etc. Based on the available and calculated data various activities on management and assurance of power system sustainability are developed.

Let us note that the parameters of the power supply network in visualization do not have any georeferencing to the real landscape, and this fact makes the analysis of the obtained results unclear. Besides, computation of operational modes is performed for a certain fixed moment using unenlightening single line diagrams that does not allow studying the behavior of the system in time. As a rule, the behavior of a traditional power supply network without renewables is highly predictable and the number of possible modes of operation and scenarios is not large, therefore for the analysis of the operation of the system it is sufficient to consider finite number of modes.

At present various software complexes (RastrWin, Dakar Eleks, PSSE/E, EasyPower, Etap, PowerFactory, etc.) that include a graphical editor are used for the analysis and computation of power supply networks. The main task of such editor is a graphical representation of a power supply network as a one-line diagram ensuring visual demonstration and ease of perception.

At the same time the process of renewables (RES) integration into energy system will lead to an increase in the number and frequency of fluctuations between electricity consumption and generation that requires more advanced techniques to analyze various scenarios of the system's behavior as a whole: in space and in time.

To study the operation of RES objects, it is necessary to automatically recalculate the steady-state operating mode taking into account various generations and load values.

For this problem, the simulating stochastic power flow model of grid as add-on structure on the known equivalent power flow model for the existing power system might be applied. This approach allows analyzing the properties of the real system in time and its parameters, such as voltage, active and reactive power by means of simulation techniques. Another problem arises: the results of simulating form datasets of a big volume. How to deal with these sets of output data, to analyze them and to find out what is the information hidden behind lots of numbers calculated by software? In our mind, good visualization and animation tools will help grid managers, power traders and others to work with data sets produced by the modern computers and to interpret the data more rapidly and more accurately in an intuitive and natural manner.

Visualization of the power supply network parameters was studied in the works of many authors, see, for example, references in [1]. The problem of visualization of power system data was emphasized in [2-9], three-dimensional visualization and animation was considered in [10-11], the issue of showing real-time data about the power market was discussed in [12].

In this work we suggest using the Geoinformation systems (GIS) and wide range of tools of these systems for geographical mapping of power systems behavior in the form of separate layers for each time point. The use of the GIS reveals a number of advantages compared to the usual procedure. First, all the units and lines of the diagram are distributed similar to their real location on the territory under study allowing visual estimation of the distances. Goereferencing can be used for superposition of meteo fields on the map of energy system and studying of the demand and supply parameters under weather conditions. At the same time georeferencing is optional and the diagram may be considered in a modified configuration, more convenient for studying. Second, we may freely navigate the whole diagram choosing the most convenient scale or consider the required part of the diagram to thoroughly study it. Third, we may store and display additional information (attributes) on one or another element of the map, e.g. the length of a power line, the capacity of an electric substation or generator, etc. Besides, the GIS tools allow animation where the values and directions of power flows dynamically change in time. This approach allows studying the whole diagram of the network in time and space choosing the required scale at any moment.

Let us note, that the most advanced software complexes for computation of the network operational modes include additional modules that allow superposing line diagrams with geographic information systems (the GIS maps), but, as a rule, software tools of these complexes are rather limited.

II. STRUCTURE OF VISUALIZATION DESIGN

In this work we consider an example of visualization and animation of computations based on the GIS for the purposes of simulating a model of a power system allowing automatic computation of the network parameters for various time periods and scenarios. We suggest that the given simulating model upon completing each computation saves the obtained results, i.e. the parameters of capacity, voltage, directions of a power flow and other data in separate files, e.g. text files. Further, to visualize the results in the GIS it is required to perform the following sequential steps shown in Fig.1. The results of computations are automatically downloaded from the text file into the geodatabase using a developed script. At the last stage the loaded data are read from the geodatabase and displayed in the GIS as separate layers. The GIS tools enable the user to visualize the results and perform the animation for a better perception of the simulation results.

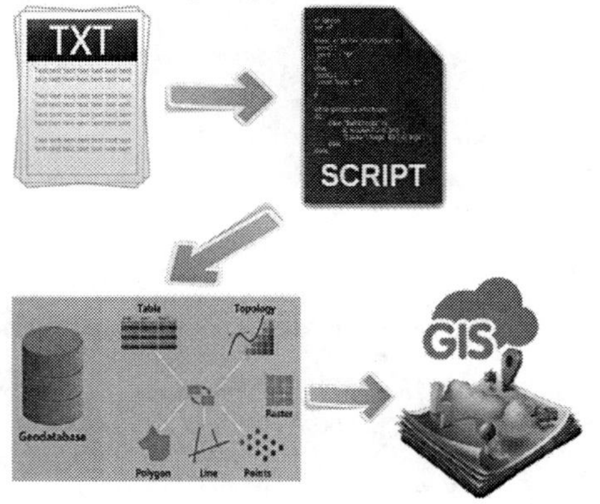

Fig. 1 Logic of visualization in GIS

III. EXAMPLE

As an example we have considered the diagram of the equivalent circuit of a power supply network in the Zhambyl region (Fig. 2), while as the GIS we have used a well-known ESRI ArcGIS software.

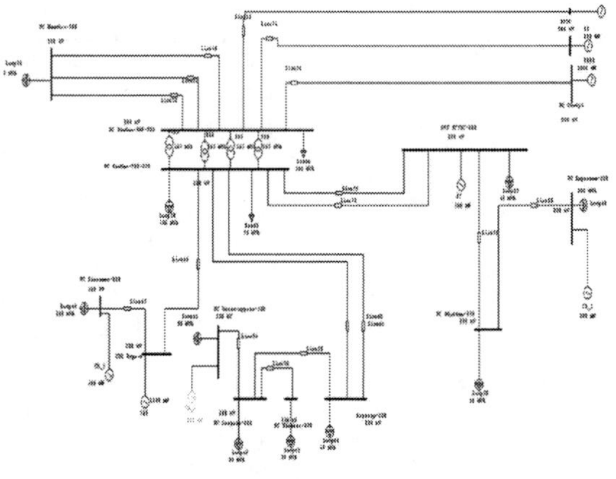

Fig. 2 Equivalent circuit of the electric network in Zhambyl region

At the first step we have to prepare a map with the diagram of power supply of the Zhambyl region in ArcGIS. The diagram consists of three elements – substations, power plants and power lines with georeferencing to the area. In Fig.3 a part of the diagram with the substations of Shymkent, ZhGRES, Karatau and Vannovka with nominal voltages of 500 kV and 220 kV is shown.

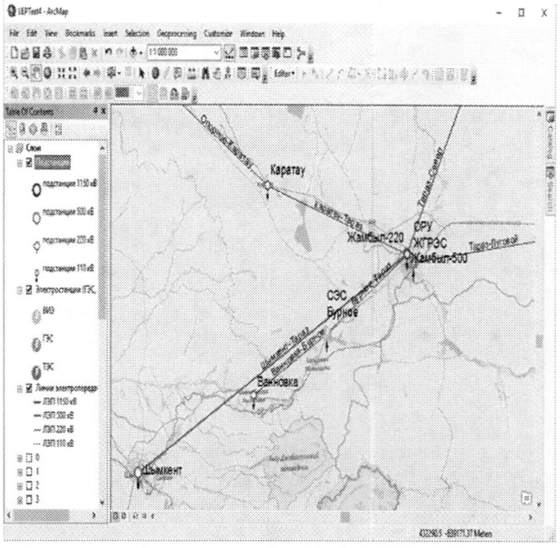

Fig. 3 The diagram of the power supply system of Zhambyl region on ArcGIS. Power stations are shown by the circles with a lightning, the green circle is solar power plant (SPP) Burnoye, the blue circle – ZhGRES, 500 kV power transmission lines are indicated with the red line and 220 kV lines - with the blue line

Further it is necessary to create vector layers to reflect power flows, voltage values, loads and generations. For this purpose, three layers are developed: “Voltage”, “Flows” and “Loads and Generations” (Fig. 4). The layer “Voltage” reflects current voltage values in each unit, and in case the permissible values are exceeded, the color of numerical values may change. The layer “Flows” shows in the form of an arrow the direction of the active power flow and the active and reactive power values for each line. In case the permissible values of current in the lines are exceeded, the color of the lines also may change. Besides, the size of the arrow indicating the direction of the power flow may also change depending on the power value. In addition to the layer “Flows” the layers reflecting only the flows of active and reactive power may be considered separately. The values of energy consumption and generation in the units are displayed in the form of vertical arrows, the direction of which indicates consumption or generation. To see the dynamics of hourly changes of the values, the layers are grouped in a certain order from 0 to 23 that corresponds to each hour of the day. Thus, we may see the state of the electrical system at any moment with an hourly interval.

The next step is to convert the number results of computation of the mode of the electrical network at any fixed moment. These results include the voltages in the units, active and reactive power flows in the lines, the values of loads and generations in the units. The results are presented in the form of text files. At this stage a software (script) is created in the Python programming language to introduce the received data into the map layers in ArcGIS. The Python programming language has been chosen for the reason of the availability of the arcpy library, which allows gaining access, through the Python script, to the tables of the layers in the ArcGIS geodatabase and changing the values in these tables.

This script is launched in the cycle with time control and allows obtaining visualization and animation of power flows and voltages of the power system diagram (Fig. 5).

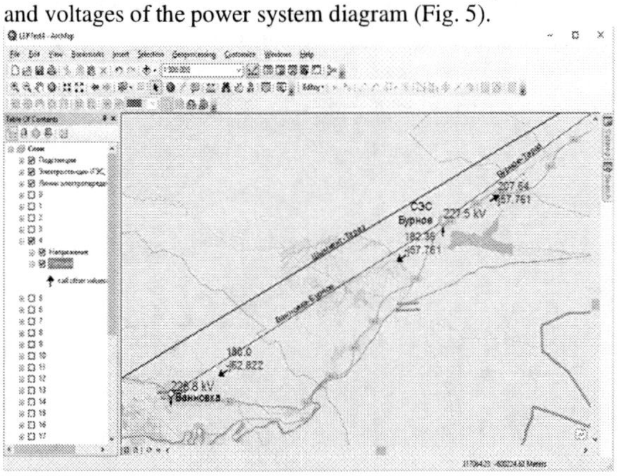

Fig. 4 The diagram of the electrical system reflecting power flows and voltages

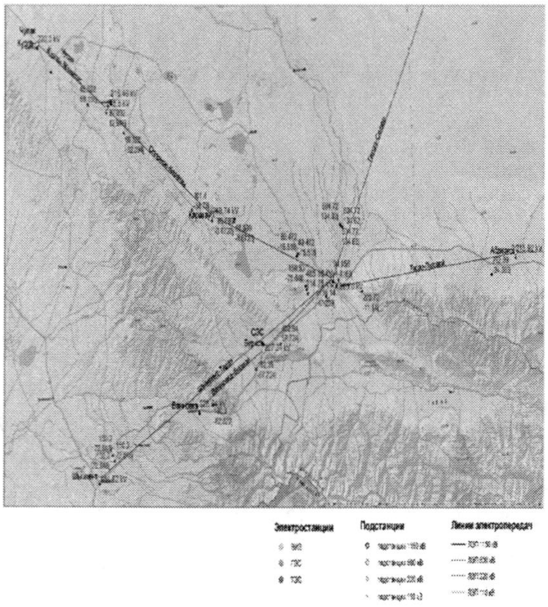

Fig. 5 The diagram of the power system in Zhambyl region with power flows and voltages

IV. Conclusion

The inevitable development of alternative and renewable energy sources and their integration into power system will require novel principles for the construction of the associated systems and the development of intelligent management systems using new information and communication software. The increasing share of RES into power system leads to changes of its management because of the spatially distribution of RES in the territory and high variability over time (see Fig. 6). Furthermore, the forecast of wind and solar radiation very often has a low accuracy leading to high levels of uncertainty (see Fig. 6-7 for real output and forecast for solar power plant "Burnoye").

Existing method of load flow calculations for fixed moment of time using single line diagrams does not provide convenient tool for operational dispatching of power system. There are many scenarios, which can occur, and fluctuations of RES can be so unpredicted in time. The development of a new methodology which allows the monitoring and control of power system is required.

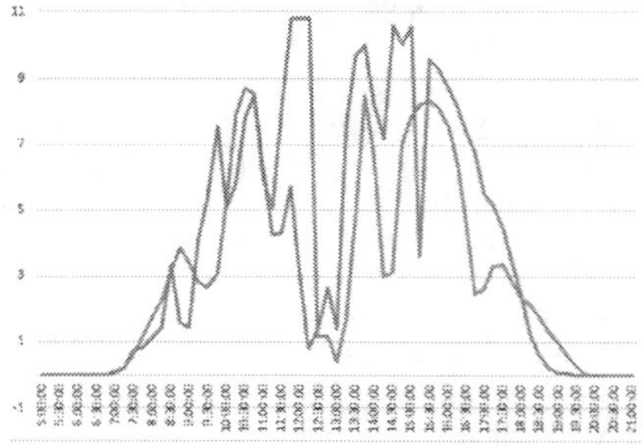

Fig. 6 Real output and forecast for solar power plant "Burnoye", 11 April 2017. Blue line is forecast. Brown line is real output.

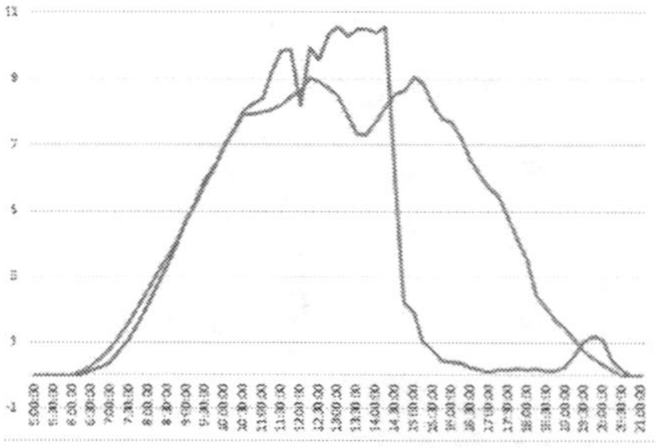

Fig. 7 Real output and forecast for solar power plant "Burnoye", 16 May 2017. Blue line is forecast. Brown line is real output.

This will include the ability to respond quickly to changes in various parameters in the power system and provide a reliable energy supply for distribution networks in the presence of large proportion of wind and solar generation.

The simulation method can be applied to develop a simulating stochastic model of studied grid as add-on structure on the known equivalent power flow model for the existing energy system. This approach allows to analyze the properties of the system in time and its parameters as voltage, active and reactive power by means of simulation techniques. As a result the estimation of the behavior of the power system under various optimal control algorithms and scenarios of renewable generation can be obtained. It helps to understand how the system parameters can change their values during a day, how they depend on power of solar station and how to provide balance in the system in optimal way. In this case, it is necessary to monitor and analyze a large number of parameters in time in a simple and intuitive way.

In this work we have shown how the application of the GIS technologies for the analysis of the operations of the power systems using the tools for visualization and animation of the network parameters may facilitate its analysis and studying. This approach may be used when it is necessary to investigate a significant number of various modes and scenarios for the behavior of energy system in time.

The use of the GIS reveals a number of advantages compared to the usual procedure: all the units and lines of the diagram could be located in a user-friendly form, for example according to their real location on the territory or as any other single diagram (see Fig. 8). It is possible to store and display additional information (attributes) on any element of the map, e.g. the length of a power line, the capacity of an electric substation or generator, etc. GIS allows freely navigating the whole diagram choosing the most convenient scale or considering the required part of the diagram for thorough studying.

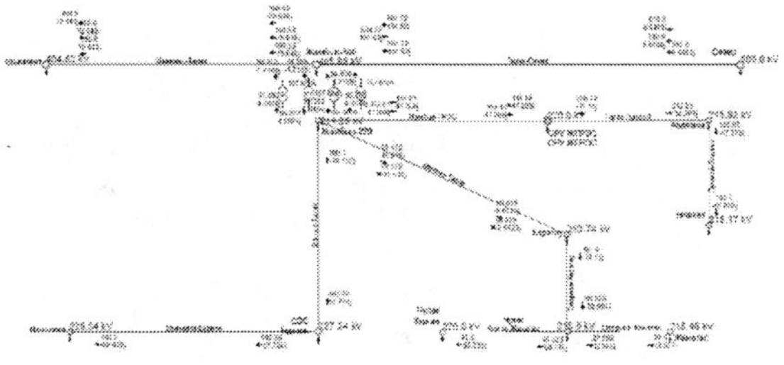

Fig. 8 An example of another single diagram of the power system in Zhambyl region with power flows and voltages.

Moreover, the GIS tools allow animation where the values and directions of power flows dynamically change in time. This approach allows studying the whole diagram of the network in time and space choosing the required scale at any moment.

Acknowledgment

This work was supported by the grant 1030/GF4 of Kazakhstan Ministry of Education and Science.

References

[1] P. Cuffe, A.Keane, "Visualizing the Electrical Structure of Power Systems", IEEE Systems Journal, Issue 99, pp. 1-12, 2015, doi: 10.1109/JSYST.2015.2427994.

[2] P. M. Mahadev and R. D. Christie, "Envisioning power system data: concepts and a prototype system state representation," Power Systems, IEEE Transactions on, vol. 8, pp. 1084-1090, 1993, doi: 10.1109/59.260890.

[3] E. R. Tufte, Visual Explanations: Images and Quantities, Evidence and Narrative, Graphics Press, 1997.

[4] E. R. Tufte, The visual display of quantitative information, Graphics Press, 1983.

[5] E. R. Tufte, Envisioning Information, Graphics Press, 2001.

[6] C. Mikkelsen, J. Johansson, and M. Cooper, "Visualization of Power System Data on Situation Overview Displays," in Information Visualisation (IV), 2012 16th International Conference on, pp. 188-197, 2012, doi: 10.1109/IV.2012.41.

[7] T. J. Overbye and J. D. Weber, "Visualizing the electric grid," Spectrum, IEEE, vol. 38, pp. 52-58, 2001, doi: 10.1109/6.898800.

[8] T. J. Overbye, D. A. Wiegmann, A. M. Rich, and S. Yan, "Human factors aspects of power system voltage contour visualizations," Power Systems, IEEE Transactions on, vol. 18, pp. 76-82, 2003, doi: 10.1109/PES.2003.1270376.

[9] J. D. Weber and T. J. Overbye, "Voltage contours for power system visualization," Power Systems, IEEE Transactions on, vol. 15, pp. 404-409, 2000, doi 10.1109/59.852151.

[10] S. Yan and T. J. Overbye, "Visualizations for power system contingency analysis data," Power Systems, IEEE Transactions on, vol. 19, pp. 1859-1866, 2004, doi: 10.1109/TPWRS.2004.836193.

[11] F. Milano, "Three-dimensional visualization and animation for power systems analysis," Electric Power Systems Research, vol. 79, pp. 1638-1647, 2009.

[12] M. J. Laufenberg, "Visualization approaches integrating real-time market data," in Power Systems Conference and Exposition, IEEE PES, vol. 3, pp. 1550-1555, 2004, doi: 10.1109/PSCE.2004.1397546.

The Real Test of Black-start Capacity and Island Operation in East Part of Slovakia

Michal Grega, Stanislav Priežný,
Martin Jedinák, Peter Gamboš
Slovenská elektrizačná prenosová sústava, a.s.
Bratislava, Slovak Republic
michal.grega@sepsas.sk, stanislav.prielozny@sepsas.sk,
martin.jedinak@sepsas.sk, peter.gambos@sepsas.sk

Michal Kolcun, Martin Kanálik
Dept. of Electrical Power Engineering, Faculty of
Electrical Engineering and Informatics
Košice, Slovak Republic
michal.kolcun@tuke.sk, martin.kanalik@tuke.sk

Abstract — **The defense and restoration plans development is a part of a transmission system operator's responsibilities and requires thorough calculations, simulations and publication of instructions. The plans focus on extensive failures that have a heavy impact on the system operation and could lead to a black-out. To restore the system after such fatal failure generation units that can start-up without outside help are essential. This ability is fairly easy to check. The units are certified regularly and represent black-start ancillary services for transmission system operator (TSO). However, the only way to check the practical applicability of the restoration plan as a whole is to test it in parallel with normal operation of the rest of the system. This process is called the real black start test and this paper deals with a one such test in Slovak power system. Specifically a thermal power plant EVO1 started from a pump-storage power plant (PVE) Ružín and island operation with available loads followed by synchronization of the island to the rest of the power system that was done on October 27th, 2016.**

Keywords: black-out, restoration, black-start, test, voltage supply, self-consumption, frequency

I. INTRODUCTION

Despite the endeavor of transmission system operators (TSO) to design and operate the power system securely, i.e. at least by fulfillment of N-1 criterion, and a low probability of an occurrence of such serious failure as is a partial or a complete black-out, experience from real operation shows that this threat is real and quite possible. [1]

Every year numerous disturbances within the contemporary power system are detected and recorded. The vast majority of them are handled without serious consequences by automatic designated systems or by human operators. Nevertheless, some of them can lead to a complete failure of a part or even a whole system. Changes in conditions related to energy market liberalization create high demands on technical ability of formerly stable system and consequently affect reliability of power generation, transmission and distribution and finally increase the likelihood of a critical event occurrence. [2]

All black-out cases that happened so far were caused by an accumulation of several factors in combination of staff failure. [1]

The objective of power system restauration is to bring the system to the point at which as much as possible within the capacity of the remaining generation and transmission plant is being supplied at normal frequency, voltage and security levels. This will be a moving feast; it will depend upon the initial plan margins, but is often taken as all demand being supplied. In the event, restauration will be a combination of operator decisions and automatic control actions. [4]

II. BLACK-START TESTS IN SLOVAKIA

SEPS as a TSO of the control area of Slovakia is according to the Dispatch rules for control of power system of Slovak republic [1] obligated to ensure a control of Slovakia power system during emergency situations which a necessity to secure a coordination and direct governance of defense and restoration plans of Slovak power system relates to. A verification of an ability to successfully restore the system is also done in accordance with SEPS's obligations resulting from RGCE ENTSO-E Operation Handbook.

A readiness for black-start of all entities involved in restoration plan is absolutely necessary for a successful recovery of the Slovakia's power system. One of the most important resulting tasks is a preparation of dispatchers and other operational staff for system recovery procedures. For this reason a regular black-start tests are essential because they provide a great basis for dispatchers' preparation for real events and consequently a preparation of trainings, adjustment of training simulators, adaptation of regulations and a verification of all kind of communication e.g. satellite communication and simultaneously recorded data can be used to verify a theoretical calculations of modeled situations. The restoration plan's philosophy in case of a black-out state is to, in the first step, restore the Slovak electricity system from neighboring systems (top-down strategy), if it is not possible for whatever reason, the second step is to recover it from internal sources (bottom-up strategy) according to standards in Policy 5 RGCE ENTSO-E [1].

The principle of the second step is to supply voltage to a system power plant from an electricity producer providing a black-start ancillary service. In case of a black-out state black-start ancillary provider has to be able to start-up his devices without any outside help and provide voltage for the system power plant for its self-consumption devices and sequential start-up, using transmission or distribution system operator's network elements. The supplied power is then used to initiate various pumps, fans, drives etc. Because they belong to a self-consumption of a system power plant, which is a large source (installed power greater than 100 MW), their input power is accordingly significant therefore, the source providing voltage has to be sufficient.

Simulations and calculations showed a need to test in reality another steps needed to be done in a situation when an extensive part of the system is without voltage. That means to practically try out to supply voltage from black-start ancillary service providing generation unit to the system power plant and consequently an island operation with a connection of a consumption. Besides two mentioned entities, smaller black-start source and system power plant, partaking in the test is needed a consumer, who needs to disconnect from normal operation and stepwise connect his devices, which represents the island's load (as a substitution of commercial consumers). The goal is to test a frequency, power and voltage control inside the island. Transmission and distribution system operators are not less important because they need to create a route in a state without voltage using their devices (lines and transformers) among all involved parties.

After evaluation of the test results from May 10th, 2016 [7] and other tests [8] [9], involved entities agreed that repeated complex test, i.e. try out of pump-storage power plant (PVE) Ružín start from dark, connection of thermal power plant Vojany (EVO1) unit to the island, island operation with consumption and finally connection of the island to the electricity system of Slovakia (working in normal operation in interconnected system of continental Europe) would be done on October 27th, 2016. After power regulation of EVO1 unit to its self-consumption it would be phased via the test route with PVE Ružín and then the PVE Ružín – EVO island would be loaded by consumptions in pump-storage power plant (PVE) Dobšiná in pump operation and steel company (U. S. Steel) Košice. If EVO1 unit would be incapable to regulate to its self-consumption the test would continue with a procedure verified on May 10th, 2016. In the end would the created island be phased via bus bar breaker in 400 kV Lemešany substation to the rest of the Slovak power system in normal operation. This way not only a capability to start-up EVO1 from PVE Ružín would be tested but also a complex black-start including island loading and phasing to the rest of the system in the area of east Slovakia. Slovenská elektrizačná prenosová sústava, a.s. (SEPS), Východoslovenská distribučná, a.s. (VSD), Slovenské elektrárne, a.s. (SE), U. S. Steel Košice (USSK) and Technická univerzita Košice (TUKE) were involved in this test.

III. The Course of the Test

The test was performed on October 27th, 2016 using the route in Fig. 1. Rms voltage, current, active and reactive power and frequency from fixed measuring devices installed in the individual SEPS or VSD substations were automatically recorded by their SCADA systems with 1 s interval. Afterwards the data were joined with measurements from portable A class power quality analysers installed in chosen nodes. For periodic data recording interval of 200 ms the registration module "Disturbance" within the analysers was used. Above mentioned data were as follows:

- rms value and the phase angle of the first harmonic voltage at each phase,
- rms value and phase angle of the first harmonic of actual current at each phase,
- total rms value of the phase voltage and current at each phase,
- total rms value of an associated voltage,
- power factor and frequency at each phase.

Having finished preparatory works for the test a conference call among test supervisors of individual entities was created. At 8:31 starting a loss of power supply to PVE Ružín self-consumption on 0.4 kV level and then started its restoration from own diesel generators. This was finished at 8:32 and five minutes later at 8:37 the PVE Ružín generator (TG1) was started to nominal speeds and voltage. After switching on of 110 kV V6301 line voltage was supplied to 110 kV Lemešany busbar and then transformed to 220 kV Lemešany substation by 220/110 kV T201 transformer in Lemešany substation at 8:40.

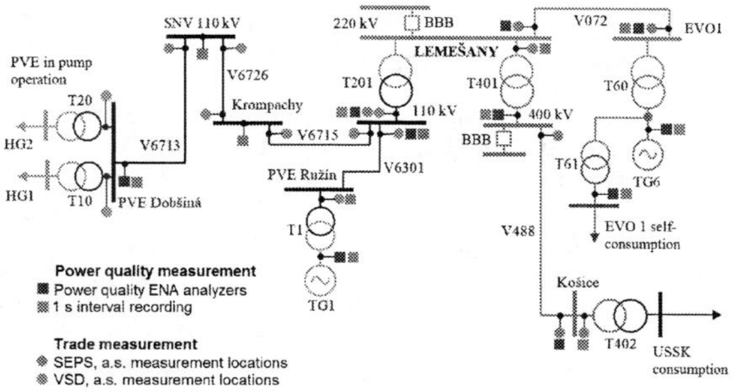

Figure 1. Single-line diagram of the island operation during complete black-start test with individual measurement points

At 8:56 was EVO1 unit cut off by switching the N60 circuit breaker and successfully regulated to its self-consumption which is about 6 MW. At 9:01 was EVO 1 unit phased to PVE Ružín following creation of island operation PVE Ružín – EVO with load of unit 6 of EVO1 self-consumption. Because the loading of PVE Ružín united was about 0 MW (even going into minus values) there was a risk of activation of reverse power watt-protection. Therefore, EVO1 decreased the production in order a part of load would be covered by PVE Ružín TG1. The next step was recover of the route to PVE Dobšiná by switching on 110 kV lines V6715 (Lemešany – Krompachy), V6726 (Krompachy – Spišská Nová Ves) and V6713 (Spišská Nová Ves – PVE Dobšiná). At 9:13 first generator (TG1) PVE Dobšiná in pump operation with a consumption of app. 9 MW was connected to the island. At 9:28 the second generator (TG2) PVE Dobšiná was connected which increased the load of the island by another 9 MW. After TG1 and TG2 were connected frequency in the island was improved (increased) by an increase of EVO1 unit output power (a takeover of part of the PVE Ružín power). Next step consisted of a creation of the route to U. S. Steel Košice i.e. connection of 400/220 kV T401 Lemešany transformer, 400 kV Line V488 Lemešany – Košice and 400/110 kV T402 Košice transformer to the PVE Ružín – EVO1 – PVE Dobšiná island. Connection of loads in U. S. Steel Košice started at 10:13 and continued until 11:23. By that time a load of app. 24 MW was connected to the island. A critical point was a connection of a high-pressure part of an EVO 6 unit turbine (HP), which connects automatically after meeting a set of conditions and means a discrete boost of unit's output. Therefore, a power reserve had to be created (on PVE Ružín) before the estimated connection of HP part, so that the power boost could be regulated by a decrease of a PVE Ružín generation. HP part was connected at 10:41, when also the highest frequency 50.25 Hz occurred (a short increase of the EVO1 unit output power and a consequent power surplus in the island until regulation by PVE Ružín). After the connection of HP part on EVO1 unit a frequency containment reserve (FCR) on it was activated in order to help regulation of the frequency to 50 Hz, which was necessary for the island connection to the rest of the Slovak power system. After the connection of the whole load of more than 50 MW (2 x 9 MW on PVE Dobšiná, EVO1 self-consumption and U. S. Steel load of 24 MW) to the island, conditions for synchronization with the power system via 400 kV Lemešany bus-bar coupler (which meant interconnection of W1 bus – U1: Slovak power system voltage and W2 bus – U2: island voltage) were being created. The synchronization conditions settings for phasing via 400 kV Lemešany bus-bar coupler were: delta U = 10 %, delta f = 100 mHz, angle difference = 2°. After actions necessary for meeting the conditions were done, the island was successfully synchronized to Slovak power system at 11:28. After shutting down of the PVE Dobšiná and U. S. Steel loads the test was finished at 12:02.

IV. Test findings

In Fig. 2 is a frequency characteristics during island operation and in Fig. 3 are characteristics of active power on PVE Ružín TG1 terminal, EVO1 TG6, 110 kV bus-bar in PVE Dobšiná and 400 kV bus-bar in Košice substation and a frequency characteristics during the whole black-start test. Frequency was during the island operation within an interval from 49.48 Hz to 50.26 Hz, while the lowest values was reached during the highest load of PVE Ružín TG1 (app. 8 MW), which occurred after the connection of loads PVE Dobšiná TG2 and an induction furnace in USSK. In Fig. 4 are detailed characteristics of active power and frequency during synchronization of EVO1 TG6 to the island. They show that after the synchronization of the EVO1 TG6 an oscillation of active power with a magnitude of app. 1 MW occurred between the working generators. After the synchronization at 09:02 the output power of EVO1 TG6, which had covered the whole EVO1 self-consumption, was decreased by app. 2 MW. In Fig. 5 is a detail of an active power and a frequency characteristics during a connection and in increasing of PVE Dobšiná TG1 and TG2 load between 9:17 and 9:18. In Fig. 6 is a detail of active power characteristics and frequency during a connection of USSK load. Before that connection, EVO1 TG6 worked only with low-pressure part while HP part was bypassed. In this mode TG6 turbine can generate only up app. 30 MW. After the start of induction furnace and electro furnace (app. 10 MW combined) was the 30 MW limit crossed and the use of HP part was necessary, after which the output of TG6 turbine increased by app. 8 MW almost stepwise to which the PVE Ružín TG1 reacted with a decrease of its power output. Because PVE Ružín TG1 was before the connection of EVO1 TG6 HP part loaded with app. 8 MW, after the connection its load decreased to 0 MW. Since the active power slightly oscillated between generators TG1 PVE Ružín and TG6 EVO1, TG1 PVE Ružín got in some moments into a motor mode at which point a disconnection of the generator by reverse watt protection was an imminent threat. Given the speed of the change of an active power generation on EVO1 TG6 by app. 8 MW and a speed of the PVE Ružín TG1 revolution controller response, island frequency increased to the 50.26 Hz after the connection of HP part of EVO1 TG6. Consequently there were several (three or four) disconnections and reconnections of HP part because of a small hysteresis between values of opening of control valves for HP part connection and disconnection. After subsequent increase of the load in the USSK at 9:43 was the generation of active power in TG1 PVE Ružín increased to more than 2 MW. At this time, EVO1 TG6 didn't contribute to the island frequency regulation (it was around 49.8 Hz) so it started to fall to the value of 49.8 Hz until a proportional component of EVO1 TG6 speed control started to have an influence. An active power and frequency characteristics detail of a moment of the HP part connection is in Fig. 7. At 11:17was in EVO activated an frequency corrector (primary control, frequency containment control) after which EVO TG6 took over the load of PVE Ružín TG1 and so increased the island frequency to the value nearly of 50 Hz. At 11:28 the island was synchronized with the rest of the interconnected power system via 400 kV bus bar coupler in Lemešany substation – Fig. 8. In Fig. 9 is a reactive power characteristics recorded by power quality analyzers in individual measurement points.

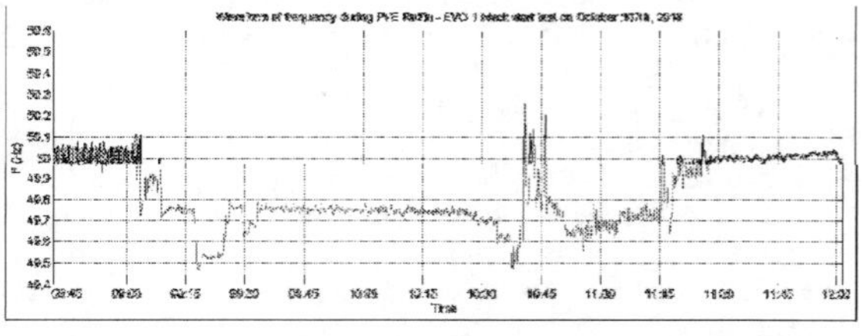

Figure 2. Frequency characteristics of island operation during black-start test.

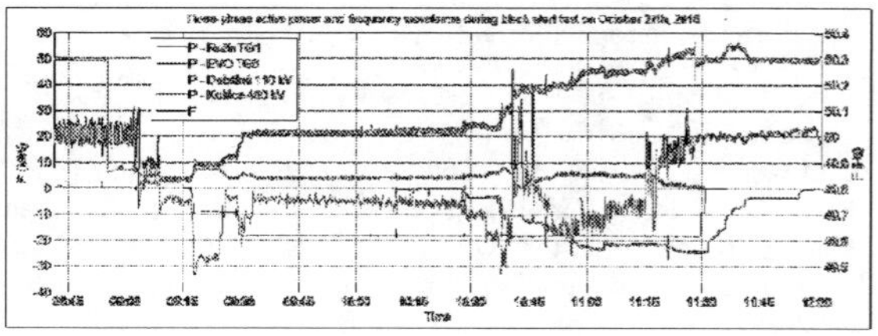

Figure 3. Three-phase active power characteristics during black-start test from October 27th, 2016.

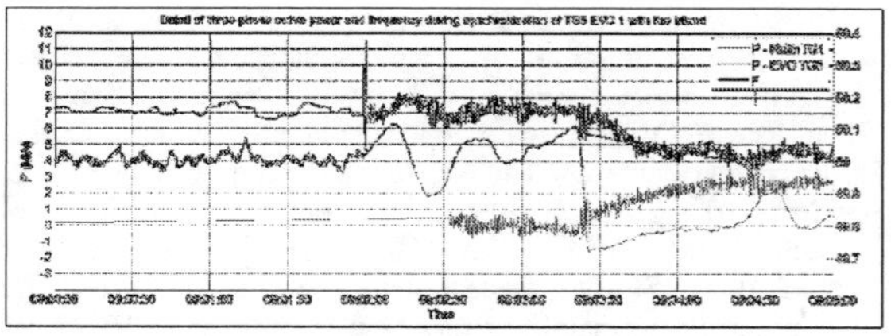

Figure 4. Detail of active power and frequency characteristics during TG6 EVO 1 synchronization with the island.

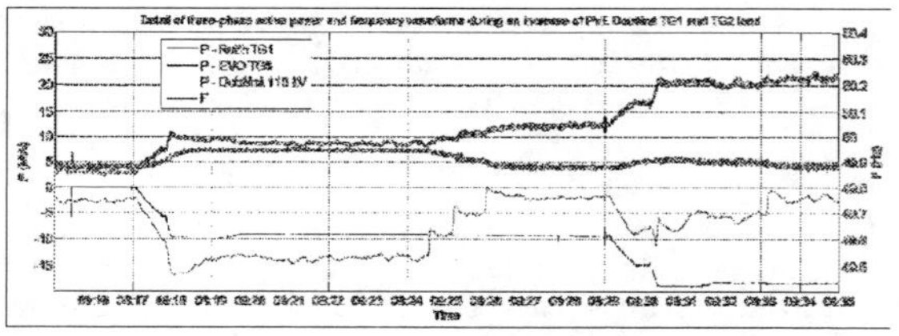

Figure 5. Detail of active power and frequency characteristics during an increase of PVE Dobšiná TG1 and TG2 load.

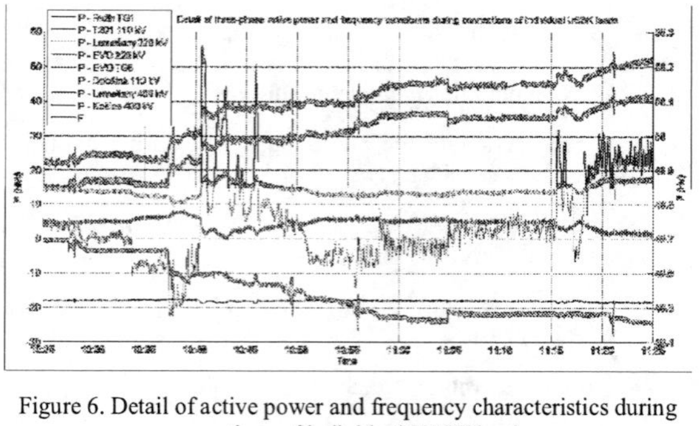

Figure 6. Detail of active power and frequency characteristics during connections of individual USSK loads.

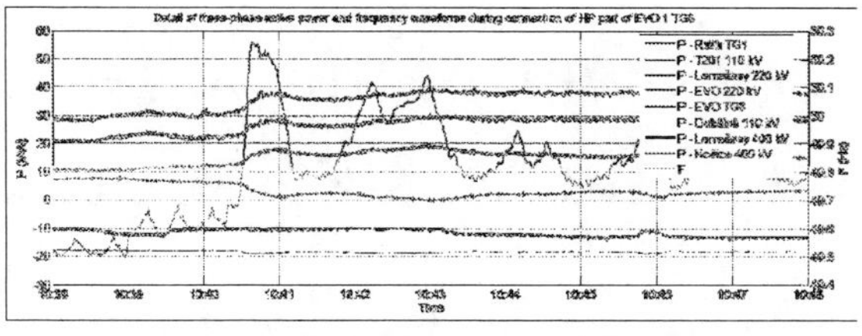

Figure 7. Detail of active power and frequency characteristics during a connection of HP part of EVO 1 TG6.

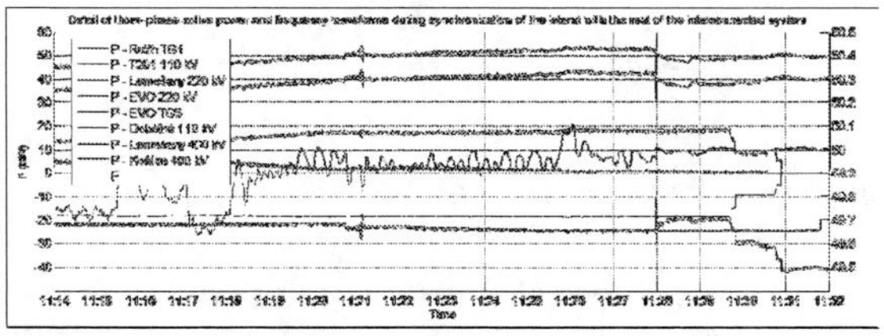

Figure 8. Detail of active power and frequency characteristics during synchronization of the island with the rest of the interconnected system.

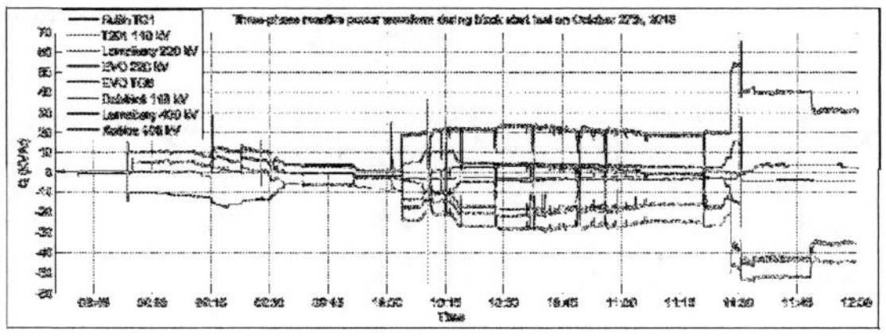

Figure 9. Three-phase reactive power characteristics during black-start test on October 27th, 2016.

V. Conclusion

Black-start test was done on October 27th, 2016 according to the schedule without significant problems and with a successful synchronization of the island with the rest of the Slovak system working in synchronously connected continental Europe, which verified the readiness and feasibility of restoration of Slovak power system within Slovakia in case of occurrence of black-out failure. PVE Ružín – EVO1 black-start test provided important findings regarding EVO1 technology behavior mainly in terms of options of condition settings to minimalize the boost of produced output after a connection of HP part (Russian type turbine setting) and in terms of turbine controller operation, which when the frequency is outside of an interval 49.8 Hz to 50.2 Hz automatically changes its mode to proportional speed control with 4 % statics even in spite of both manual mode being set and blocked setting of defence plan on frequency relays. Nevertheless, after an evaluation of recorded measurement data, the automatic transition to proportional speed control from power control of EVO1 TG6 turbine controller proved to be beneficial. However, the unawareness of this TG6 turbine control operation could in case of an island cooperation of TG6 with another generator with faster response can cause active power oscillation between individual power sources which could lead to failure and even to break-up of the island.

One of the important things for the black-start test of PVE Ružín - EVO1 was a checking of all kind of communication channels between individual involved entities. The communication was a basis for a smooth run of the test. In case of this black-start test on October 27th, 2016 a teleconference via commercial phone line as well as via satellite phones was made amongst all of the test participants. During the whole test the connection was in a very good quality without any interruptions.

References

[1] K. Máslo: Rízení a stabilita elektrizační soustavy, 2003, AEM, ISBN 978-80-260-44671-1, 272 pages.

[2] J. Tuma, S. Rusek, Z. Martínek, I. Chemišinec, R. Goňo – Spolehlivosť v elektroenergetice, ISBN 80-239-6483-6, 291 strán, CONTE spol s.r.o., ČVUT Praha

[3] M. Eremia, M. Shahidehpour, "Handbook of Electrical Power System Dynamics", IEEE Press, 2013, p. 976, ISBN 978-1-118-49717-3.

[4] U.G.Knight, Power Systems in Emergencies, From Contingency Planning to Crisis Management, John Wiley&sons LTD, 2001, ISBN 0-471-49016-4, p.393.

[5] Dispach rules for control of Slovak power system, SEPS, Available at http://www.sepsas.sk.

[6] RG CE Operation Handbook, Policy 5: Emergency Operations Security, ENTSO-E, 2009. Available at: http://www.entsoe.eu.

[7] M. Grega, S. Prieložný, M. Jedinák, P. Gamboš, M. Kanálik, M. Kolcun.: The real test of Vojany power plant self-consumption restoration after a black-out, paper, International symposium, Stará Lesná 2017.

[8] S. Prieložný, M. Jedinák, M. Kolcun, F. Janíček: The real test of „Black Start" units in Eastern Slovakia, In: Elektroenergetika 2015: 8th International Scientific Symposium on Electrical Power Engineering. Stará Lesná, Slovakia, September 16-18, 2015. - Košice: Technical University of Košice.

[9] P. Gamboš, M. Jedinák, S. Prieložný: The real test of „Black Start" units in Eastern Slovakia. In: Blackout a Krajowy System Elektroenergetycny. Poznań 2016, Poland, June 1-3, 2016. Osrodek Wydawnictw Naukowych ICHB PAN, ISBN 978-83-7712-152-8. pp. 37-46.

The Real Test of Vojany Power Plant Self-Consumption Restoration After a Black-out

Michal Grega, Stanislav Prieložný,
Martin Jedinák, Peter Gamboš
Slovenská elektrizačná prenosová sústava, a.s.
Bratislava, Slovak Republic
michal.grega@sepsas.sk, stanislav.prielozny@sepsas.sk,
martin.jedinak@sepsas.sk, peter.gambos@sepsas.sk

Michal Kolcun, Martin Kanálik
Dept. of Electrical Power Engineering, Faculty of Electrical Engineering and Informatics
Košice, Slovak Republic
michal.kolcun@tuke.sk, martin.kanalik@tuke.sk

Abstract — **The defense and restoration plans development is a part of a transmission system operator's responsibilities and requires thorough calculations, simulations and publication of instructions. The plans focus on extensive failures that have a heavy impact on a system operation and could lead to a black-out. To restore the system after such a fatal failure generation units that can start-up without outside help are essential. This ability is fairly easy to check. The units, which are able to start up independently after black-out are certified regularly and represent black-start ancillary services for TSO. However, the only way to check the practical applicability of the restoration plan as a whole is to test it in parallel with normal operation of the rest of the system. This process is called the real black start test and this paper deals with one such a test in Slovak power system. Specifically power plant Vojany1 self-consumption start test.**

Keywords: black-out, restoration, black start, self-consumption

I. INTRODUCTION

Despite the endeavor of transmission system operators (TSO) to design and operate the power system securely, i.e. at least by fulfillment of N-1 criterion, and a low probability of an occurrence of such serious failure as is a partial or a complete black-out, experience from realtime operation shows that this threat is real and quite possible. [1]

Interrupted power supply means significant economic losses with society-wide aspects. Short power loss (a few hours) has an impact on above standard services and in case of industry with continual production its technology can malfunction. If the power loss is longer (1 – 2 days) supplies of heat and water are cutoff, mobile network, cash machines and whole banks are unavailable, medical care is limited, transport, delivery of food and also fuel are stopped, signs of aggressive behavior starts to show up (violent crime, looting), which are the result of a stressful situation. If the power loss lasts few days or even weeks, violence and looting is on daily basis and society is threatened by chaos and panic. Therefore, to ensure the shortest possible time of a system restoration is unquestionably a necessity. [1]

Every year numerous disturbances within the contemporary power system are detected and recorded. The vast majority of them are handled without serious consequences by automatic designated systems or by human operators. Nevertheless, some of them can lead to a complete failure of a part or even a whole system. Changes in conditions related to energy market liberalization create high demands on technical ability of formerly stable system and consequently affect reliability of power generation, transmission and distribution and finally increase the likelihood of critical event occurrence. [2]

However, all black-out cases that happened so far were caused by an accumulation of several factors in combination of staff failure. [1] They usually start with an initiatory event during escalated operation of the system caused mainly by:

- natural factors – storm, lightning, contact between line and tree and so on,
- technical factors – short circuit, component failure, overloading, break-down with age and so on,
- human factors – switching/communication mistakes, lack of training (especially for emergency situations), sabotage etc.,
- trading factors – export/import over limits of profiles/ elements or unsuitably defined bidding zones.

II. SLOVAK POWER SYSTEM

Deteriorating situation within European interconnected power system affects also Slovak system operation which has recently been manifested by frequent N-1 security criterion violation and in some cases even by lines overload. Without accessible measures designated for dealing with such critical situations the real-time operation can end up in the total collapse of the system.

If this happens a restoration of the operation is, in the first step, based on a recovery from the neighboring transmission systems (TS), which is possible only if at least one neighboring TS is still operational. If this option is unavailable the restoration is done using the domestic sources. Several "black start" units, i.e. power plants that are able to initiate their generators without outside help and jump start system power plants that need power for their self-consumption devices and finally sequential start-up, are connected to the Slovak power system and are certified for providing "black start" ancillary service.

Various procedures of defense and restoration of Slovak system that make use of individual "black start" units are described by restoration plans. These have been in Slovakia updated regularly since 2008 based on calculations and simulations that consider current system development and modernization as well as update of relevant operational instructions and legislation based on which internal operational instructions of transmission system operator that contain defense plans have been consequently updated. Nevertheless, those studies cover only theoretical level and it is reasonable to prove them in reality.

Slovak TSO did this in 2016 by the real test of Vojany power plant self-consumption restoration. The test required two attempts to be successful and proved to be a huge contribution to an understanding of conditions during the recovery of the power plant production. Slovenská elektrizačná prenosová sústava, a.s. (SEPS), Východoslovenská distribučná, a.s. (VSD), Slovenské elektrárne, a.s. (SE) and Technická univerzita Košice (TUKE) were involved in this test.

III. TEST PREPARATION PHASE

Two units providing a "black start" ancillary service, i.e. DG Moldava and pumped-storage hydro power plant Ružín (PVE Ružín) and two system power plants – thermal power station Vojany (EVO1) and a combined heat and power plant Košice (TEKO) are located in the area of east Slovakia. System power plants are power plants with an installed output power higher than 100 MW and are crucial for system restoration, because they can maintain large islands and therefore, restore a significant part of the power system. After a successful black start test in 2014 for system power station TEKO Košice [4][5] SEPS decided to carry out a similar test in 2015 with the involvement of the biggest system power station in the east part of Slovakia i.e. Vojany power plant in which two power generating blocks are operational – units 5 and 6.

Before the real test a confirmative study was done (2015) by TUKE within which four different variants of plant restoration were analyzed. Routes of power supply for EVO1 self-consumption were variably analyzed for the supply from PVE Ružín or diesel generators (DG) in Moldava nad Bodvou to the devices of EVO1 self-consumption (VS) via 220 kV power lines or via 110 kV lines only. The analyses showed a feasibility of the test if the power for EVO1 would be supplied only from PVE Ružín via 220 kV and also 110 kV power lines. In the end, the supply of the power to the EVO1 from PVE Ružín via 220 kV lines was chosen as the most suitable variant.

The original date of the test in 2015 had to be canceled because of the emergency situation in the transmission system. It was shifted to March 2016, when the black start test was attempted but was evaluated as unsuccessful because of the inability to restore the self-consumption of EVO1 unit 6. Consequently a detailed analyses of conditions was done and measures in the area of optimization of voltage conditions on route, adjustment of settings of protections in the Vojany power plant self-consumption and a shutdown of frequency converters, were proposed. After their implementation the test of Vojany plant self-consumption restoration was executed again on May 10th, 2016.

In the Figure 1 is a single-line diagram of the power supply route for the test of VS EVO1 restoration with marked locations of an A class power quality analyzers. For the test TG2 of pumped storage plant Ružín and EVO1 unit 6 was reserved. The route of the test lead via 110 kV line V6301 Ružín – Lemešany, transformer 220/110 kV T201 Lemešany,

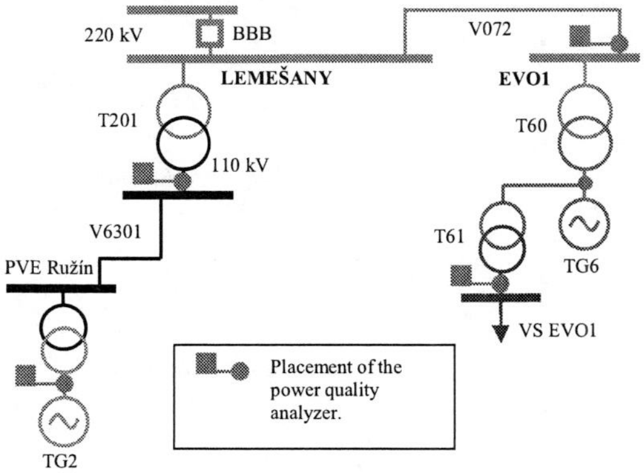

Figure 1. Route of test Ružín - Vojany

line 220 kV V072 Lemešany – Vojany, block transformer 220/13.8 kV T60 Vojany and transformer 13.8 kV/6.3 kV T61 Vojany to the Vojany self-consumption.

A measurement in Vojany plant was done by two power quality analyzers BK-Elcom type ENA330 in 6.3 kV substation of VS EVO1. One device was installed on the feed to the substation from transformer T61 and the second was installed on the output terminal of the feed pump n. 1 (NČ1). For all analyzers was used the same module for periodic data recording with 1 s interval, module "disturbance" with 0.02 s interval and module "transient" with an interval of 192 samples per period. Module "disturbance" recorded rms values of voltage and current and also rms values and phase angles of voltage and current of 1st harmonic. Module "transient" recorded instantaneous values of voltage and current in the individual phases (oscilloscopic record).

Rms voltage, current, active and reactive power and frequency from fixed measuring devices installed in the individual SEPS or VSD substations were automatically recorded by their SCADA systems with 1 s interval. Afterwards the data were joined with measurements from portable A class power quality analysers installed in chosen nodes. For periodic data recording interval of 200 ms the registration module "Disturbance" within the analysers was used. Above mentioned data were as follows:

- rms value and the phase angle of the first harmonic voltage at each phase,
- rms value and phase angle of the first harmonic of actual current at each phase,
- total rms value of the phase voltage and current at each phase,
- total rms value of an associated voltage,
- power factor and frequency at each phase.

IV. HYDRO POWER PLANT PVE RUŽÍN – VOJANY UNIT 6 SELF-CONSUMPTION TEST

The test of the VS EVO1 start was performed on May 10th, 2016 from 8:00 to 12:07 and after a successful start of all VS EVO1 devices was at 10:54 evaluated as successful. The connection of the loads of self-consumption was done between 9:35 and 10:54.

The total power connected in EVO1 during test was approximately 6 MW. For summary of connected load in each step see the Table 1.

V. TEST RESULTS

In Figure 2 is the frequency characteristics of island operation during whole test of VS EVO1. The biggest frequency decline was recorded after connecting the largest loads in self-consumption of power station Vojany - feeding pump NČ1 at 10:04 h., exhaust fan at 10:23 h., secondary fan at 10:36h. and primary fan at 10:39 h. After connection of feeding pump NČ1, secondary fan and primary fan the frequency decreased under 49 Hz.

In Figure 3 is a three phase active power characteristics in the individual measurement points during whole island operation during test and in the Fig. 4 is a comparison of active power and frequency on PVE Ružín TG2 terminal. Reactive power characteristics in the individual measurement points during island operation during the whole test with 20 ms record interval is shown in Fig. 5.

Waveforms of the current rms value of all phases on PVE Ružín TG2 terminals, comparison of waveforms of rms values of voltage positive sequence in p.u. and comparison of reactive power MVAr in individual measurement points during connection of the largest loads of Vojany self-consumption (feeding pump, exhaust fan, secondary and primary fan) are in the Fig. 6, 7, 8, 9.

Table 1. Connected Vojany self-consumption devices during test

Connection time	Connected EVO1 VS load
9:53	Oil vapors suction fan – greasing oil
9:53	Oil start-up pump
9:54	Elevating oil pump
9:55	Turning device
9:56	Cooling water pump
9:59	Raw cooling water pump – inner cooling circuit
10:01	Cooling water pump – outside circuit
10:02	Condensation pump, water supply to NNV etc.
10:04	Electro feeder for feeding pump n. 1
10:10	Cooling water pump – generator cooling
10:12	Regulation oil pump
10:12	Oil vapors suction fan - regulation oil
10:13	350kPa compressor
10:18	700kPa compressor
10:21	3 x electrostatic separator
10:23	Exhaust fan
10:35	Blower 31
10:35	Blower 32
10:36	Secondary fan and oil pump
10:39	Primary fan and oil pump
10:44	Blower 51
10:45	Blower 41
10:46	Blower 42
10:46	Blower 52
10:48	Coal feeder 51 (short inclined)
10:49	Coal feeder 51 (long inclined)
10:50	Coal crusher 51
10:51	Raw coal feeder 51 (straight)
10:52	Coal feeder 52 (short inclined)
10:53	Coal feeder 52 (long inclined)
10:54	Coal crusher 52
10:54	Raw coal feeder 52 (straight)
10:54	**Successful start of whole EVO1 TG6 unit self-consumption**

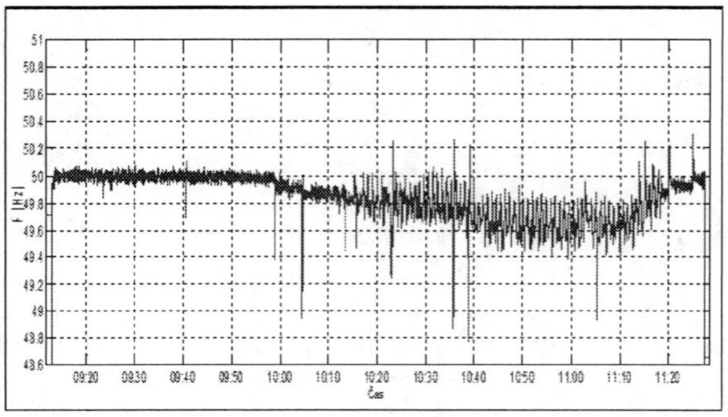

Figure 2. Frequency characteristics in the island during VS EVO1 start test on May 10th, 2016

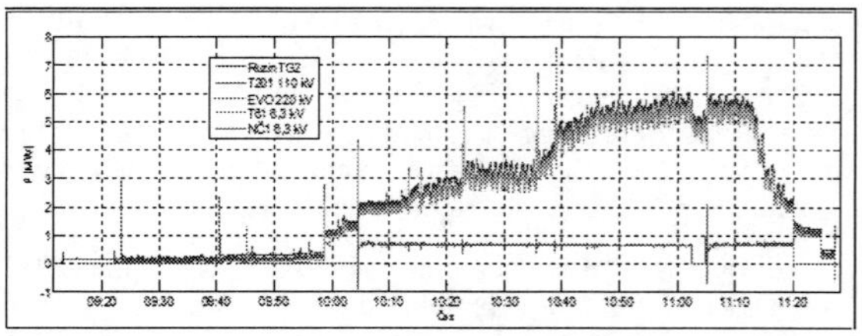

Figure 3. Course of active power in individual nodes during test PVE Ružín – EVO1 10th of May 2016

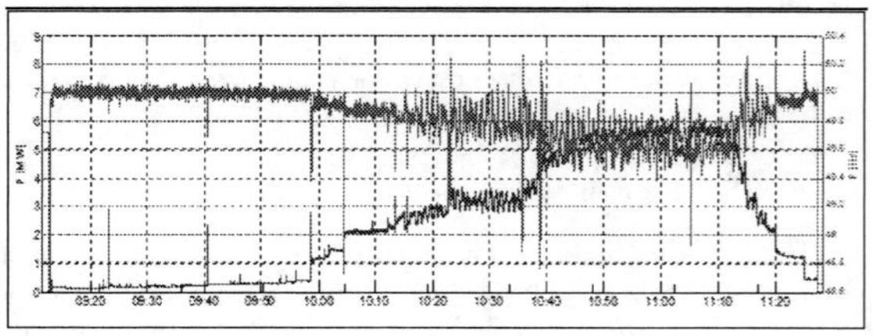

Figure 4. Active and reactive power and frequency waveforms on PVE Ružín TG2 terminal within the island during VS EVO1 start test on May 10th, 2016

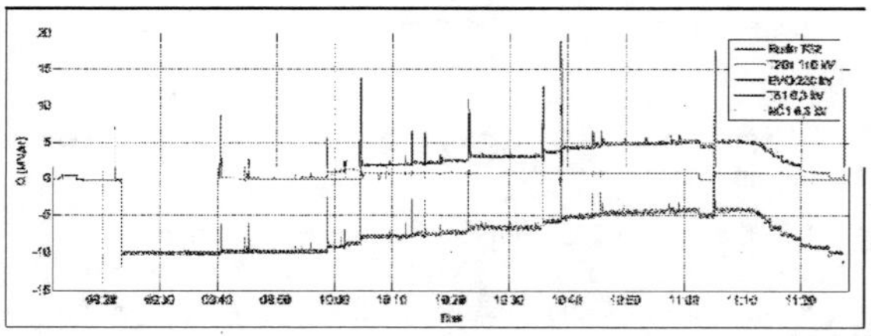

Figure 5. Course of reactive power in individual nodes during test PVE Ružín – EVO1

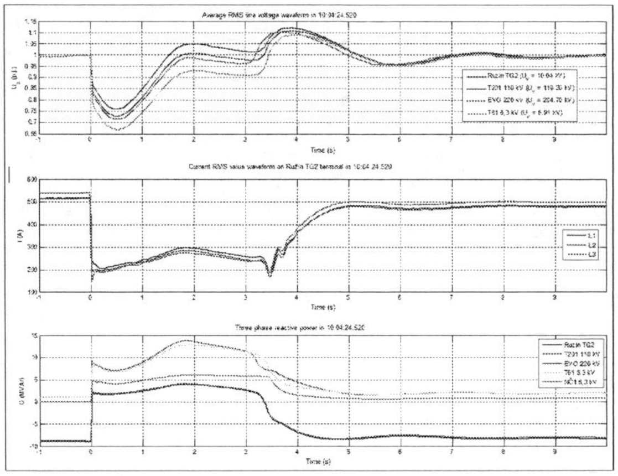

Figure 6. Current RMS waveforms of individual phases on PVE Ružín TG2 terminal and a comparison of Line voltage RMS values in p.u. and reactive power in MVAr in the individual measurement points during start of the feeding pump NČ1 of EVO1 self-consumption.

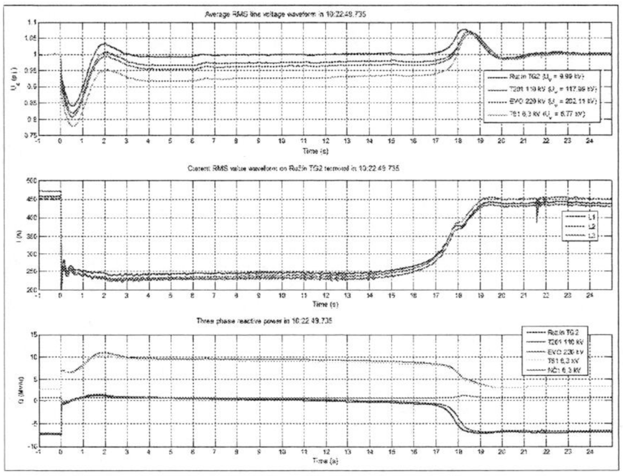

Figure 7. Current RMS characteristics in individual phases on PVE Ružín TG2 terminal and a comparison of line voltage RMS characteristics in p.u. and reactive power in MVAr in the individual measurement points during start of the exhaust fan of EVO1 self-consumption.

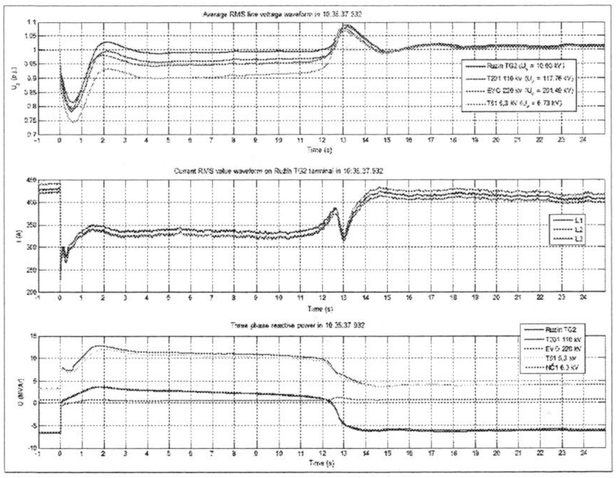

Figure 8. Current RMS characteristics in individual phases on PVE Ružín TG2 terminal and a comparison of line voltage RMS characteristics in p.u. and reactive power in MVAr in the individual measurement points during start of the secondary fan of EVO1 self-consumption.

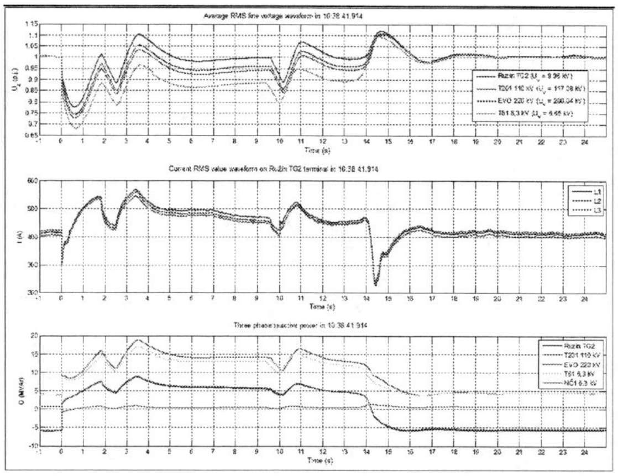

Figure 9. Current RMS characteristics in individual phases on PVE Ružín TG2 terminal and a comparison of line voltage RMS characteristics in p.u. and reactive power in MVAr in the individual measurement points during start of the primary fan of EVO1 self-consumption.

VI. Test Findings

After unsuccessful test from March 24th, 2016 some measures were implemented. First of all a tap settings of transformers in island were changed so the RMS voltage on 220 kV would be cca 200 kV during the test and on the 6.3 kV level of EVO1 self-consumption the RMS voltage would be cca 6.9 kV. This measures lead to a significant decrease of a reactive power generated by the 220 kV line Lemešany – EVO1 V072 which caused a smaller voltage difference between PVE Ružín TG 2 and EVO1 220 kV substation and a decrease of under-excitation of the generator. Secondly, settings of an under voltage protection in EVO1 6.3 kV substation was changed to 0.5 x U_n with a 6 s time delay. The third measure was switching off of the exhaust fan frequency converter and its control by a control rim. Thanks to implementation of these measures the second repeated test was conducted successfully.

Results of the test showed that the change of transformers tap settings worked as expected and Ružín TG 2 was under-excited as much as possible. In the voltage drop during start of individual drives the tap settings manifested itself only slightly.

One of the important things for the “black start” test of EVO1 was a creation of a communication channels between individual involved entities. The communication was a basis for a smooth run of the test. In case of VS EVO1 black start test on May 10th, 2016 a teleconference via commercial phone line as well as via satellite phones was made amongst all of the test participants. During the whole test the connection was in a very good quality without any interruptions.

VII. Conclusion

The test from May 10th, 2016 showed a capability of the Vojany self-consumption to be restored with the use of power provided from pumped storage hydro power plant Ružín in case of black-out failure. Alongside, conclusions of the analyses done before the test, mainly in the field of voltage optimization in the island were confirmed which is a significant contribution to the increase dispatch centres in Slovak electricity system readiness to deal with system restoration. Also, the necessity of optimal protection settings on individual elements was showed (normal operation versus restoration after black-out).

REFERENCES

[1] K. Máslo: Rízení a stabilita elektrizační soustavy, 2003, AEM, ISBN 978-80-260-44671-1, p. 272.

[2] J. Tuma, S. Rusek, Z. Martínek, I. Chemišinec, R. Goňo: Spolehlivosť v elektroenergetice, ISBN 80-239-6483-6, p. 291. CONTE spol s.r.o., ČVUT Praha.

[3] M. Eremia, M. Shahidehpour: "Handbook of Electrical Power System Dynamics", IEEE Press, 2013, p. 976, ISBN 978-1-118-49717-3.

[4] S. Prieložný, M. Jedinák, M. Kolcun, F. Janíček,: The real test of „Black Start" units in Eastern Slovakia, In: Elektroenergetika 2015: 8th International Scientific Symposium on Electrical Power Engineering. Stará Lesná, Slovakia, September 16-18, 2015. - Košice: Technical University of Košice.

[5] P. Gamboš, M. Jedinák, S. Prieložný: The real test of „Black Start" units in Eastern Slovakia. In: Blackout a Krajowy System Elektroenergetycny. Poznań 2016, Poland, June 1-3, 2016. Osrodek Wydawnictw Naukowych ICHB PAN, ISBN 978-83-7712-152-8. pp. 37-46.

Analysis of VQ Parameters in Delivery Points of Selected Customers Supplied from MV Distribution Grid

Martin Kašpírek
E.ON Distribuce, a.s.
Grid management
České Budějovice, Czech Republic
martin.kaspirek@eon.cz

Helena Drdáková
E.ON Česká republika, s.r.o.
Management of Distribution Grid
České Budějovice, Czech Republic
helena.drdakova@eon.cz

Abstract — **This paper describes the impact of operation of big customers on voltage quality parameters in medium voltage distribution grids. The survey is based on the evaluation of 45 voltage quality measurements which were carried out in 15 different delivery points of other customers. These 45 measurements were made in years 2014, 2015 and 2016 (15 measurements a year) and were evaluated according to standard EN 50160. The impact on voltage variations and other voltage quality parameters was analysed.**

I. Introduction

The supply territory of the company E.ON Distribution in the Czech Republic (run by the company E.ON Czech Republic) accounts for approximately 1.5 million customers when approximately 6000 customers are supplied from medium voltage (MV) level. Problems with flicker, voltage variations and harmonic voltage were detected in low voltage (LV) distribution grids but the analysis from the MV level is not available. The electricity market liberalisation brings considerable pressure to introduce penalties for insufficient voltage quality (VQ) parameters. In case of poor voltage quality, these penalties should be of the electricity rebate payment nature. The analysis of VQ parameters in the distribution grid is necessary for definition of interface between the customer and distribution network operator (DNO) from view of responsibility for poor VQ. The 22 kV voltage level is the level for connection of bigger customers to the distribution grid and it is necessary to know the conditions in it. This paper brings a detailed analysis of VQ parameters in this MV grid.

II. Description of Measured Grids

Fifteen customers connected to the MV distribution grid are measured for many years when VQ measurement is temporary with the duration of one week according to the standard EN 50160 [1]. The measured period during the year is chosen accidentally with respect to possibility of the DNO. The MV customers have own substation with the own transformer 22/0.4 kV. Voltage quality measurement has to be done in the delivery point of the customer so that voltage quality is measured on the output of the voltage instrument transformer (VT) 22/0.1 kV. This instrument transformer is a part of MV switchboard and it is in the possession of customer. The distance between the customer and the supply HV/MV substation can be different – see the Fig. 1.

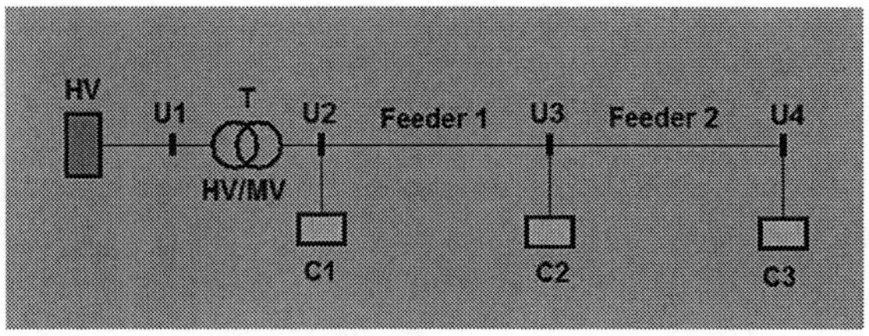

Figure 1. General block diagram of grid (C1, C2, C3 – measured customers connected to the MV grid)

For comparison of the distance from HV/MV substation we used calculation of short circuit power in delivery point of customers – see the Fig. 2.

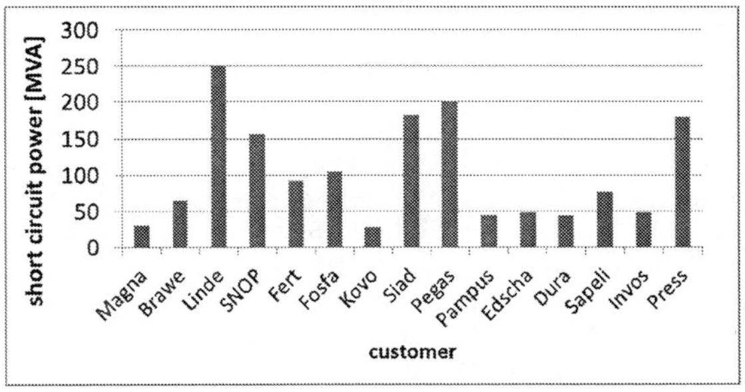

Figure 2. Short circuit power in delivery points of measured customers

Reference value of short circuit power (short circuit impedance) does not exist for MV distribution grids in contrast to LV distribution grids [2]. The experience of DNO shows that the distance about 30 km between the customer and supply HV/MV substation is too long when this distance is equivalent to approximately 30 MVA of short circuit power (in case of ACSR conductor). According to author's more than 10 year experience in the field of voltage quality, the reference short circuit power of 50 MVA is proposed as the reference value. This reference value of short circuit power should be used as the guideline for decision about responsibility for poor VQ. It is the question, when the DNO and when the customer is responsible for poor voltage quality. This topic is permanently actual.

III. Requirements of the EN 50160 Standard

A. Voltage variations

All the MV measurements were evaluated according to the standard EN 50160 [1]. Under normal operating conditions excluding the periods with interruptions, supply voltage variations should not exceed ± 10% of the nominal voltage Un. A test method for medium voltage level (under normal operating conditions) follows these rules/requirements [1]:

- During each period of one week at least 99% of the 10 min mean r.m.s. values of the supply voltage shall be below the upper limit of Un + 10%; and
- During each period of one week at least 99% of the 10 min mean r.m.s. values of the supply voltage shall be above the lower limit of Un - 10%; and
- All 10 min r.m.s. values of the supply voltage shall be within the range of Un + 15%/-15%

Also when one 10 min mean r.m.s. value of the supply voltage exceeds during the week the limit Un + 15% (115% Un) or when more than 1% of 10 min mean r.m.s. values xceed during the week the limit Un + 10% (110% Un), it results in overvoltage and voltage quality (voltage variations) will not comply with the requirements of the standard EN 50160.

B. Flicker

Under normal operating conditions, during each period of one week the long term flicker severity P_{lt} caused by voltage fluctuation should be less than or equal to 1 for 95 % of the time.

C. Harmonic voltage

Under normal operating conditions, during each period of one week, 95 % of the 10 min mean r.m.s. values of each individual harmonic voltage shall be less than or equal to the values given in [1] when the limit value for the 3rd voltage harmonic is 5%, for the 5th voltage harmonic is 6%, for the 7th voltage harmonic is 5% and for the 15th voltage harmonic is 0.5% of nominal voltage (only the most important harmonic voltages were selected). Moreover, the THD of the supply voltage (including all harmonics up to the order 40) shall be less than or equal to 8 %.

D. Unbalance

Under normal operating conditions, during each period of one week, 95 % of the 10 min mean r.m.s. values of the negative phase sequence component (fundamental) of the supply voltage shall be within the range 0 % to 2 % of the positive phase sequence component (fundamental).

E. Frequency

The nominal frequency of the supply voltage shall be 50 Hz but frequency is set from the transmission system so it is not evaluated in this paper because of its small role for LV systems with synchronous connection. It is important to evaluate frequency in LV systems with no synchronous connection when the power source operates in an island mode [3].

IV. Evaluation of VQ Measurements

A. Voltage variations

In the Fig. 3 and Fig. 4 you can see the evaluation of maximal/minimal 10 min mean r.m.s. values of supply voltage (V99%max and V99%min) in 15 delivery points of MV customers.

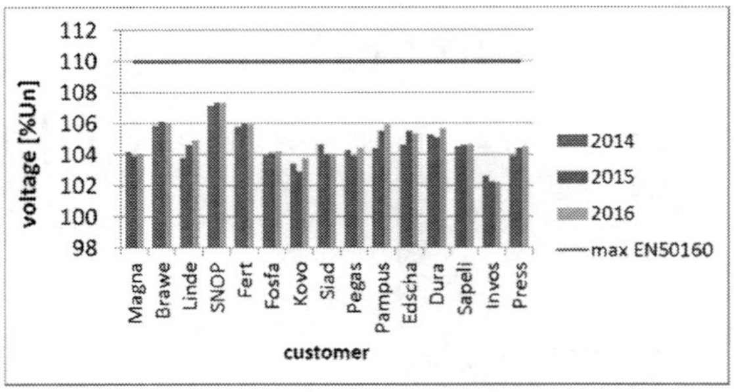

Figure 3. Evaluation of maximal voltage variations (99% percentile), comparison to the upper voltage limit

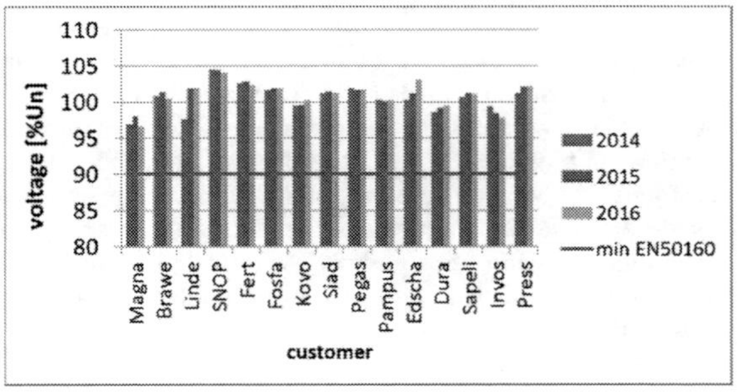

Figure 4. Evaluation of minimal voltage variations (99% percentile), comparison to the lower voltage limit

The Fig. 3 and Fig. 4 show conforming voltage variations in all grids. Maximal voltage reaches the value 107.3 % Un in case of the customer Snop (see the Fig. 3) and minimal voltage reaches the value 96.6 % Un in case of the customer Magna (see the Fig. 4).

Voltage difference dU between maximal and minimal voltage values during the measurement was evaluated too – see the Fig. 5. The voltage difference reaches the value 7.4% Un in case of the customer Magna which is supplied by means of ACSR conductor with the length of approximately 30 km (distance from HV/MV supply substation). Voltage difference depends not only on short circuit power (which is very low in case of the customer Magna) in the delivery point but also on load of grid. Load was not evaluated but we recommend taking account of load in future survey.

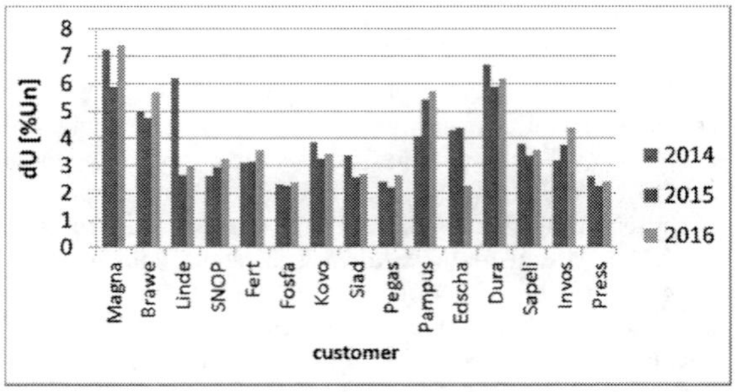

Figure 5. Voltage difference between maximal (99% percentile) and minimal (99% percentile) voltage values

B. Flicker

In the Fig. 6 and Fig. 7 you can see the evaluation of long-term flicker (Plt) and short-term flicker (Pst) in delivery points of MV customers. The flicker limit is exceeded in case of the customer Magna (both Plt and Pst) and Kovo (both Plt and Pst in 2014, only Plt in 2015). The algorithm for evaluation of long-term flicker can be negatively influenced by voltage dips caused by faults occurring in the public network. If no faults during the measurement period occur, then long-term flicker values do not differ from short-term flicker values (for example the customer Magna). If faults during the measurement period occur, then long-term flicker values are higher than short-term flicker values (for example the customer Kovo in 2015). Thus we recommend evaluating both long-term flicker and short-term flicker together. Short-term flicker has higher information value than long-term flicker although the EN 50160 defines limit value only for long-term flicker.

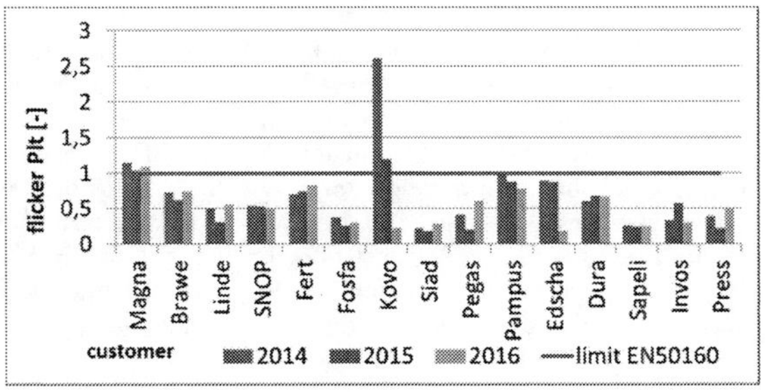

Figure 6. Evaluation of flicker Plt

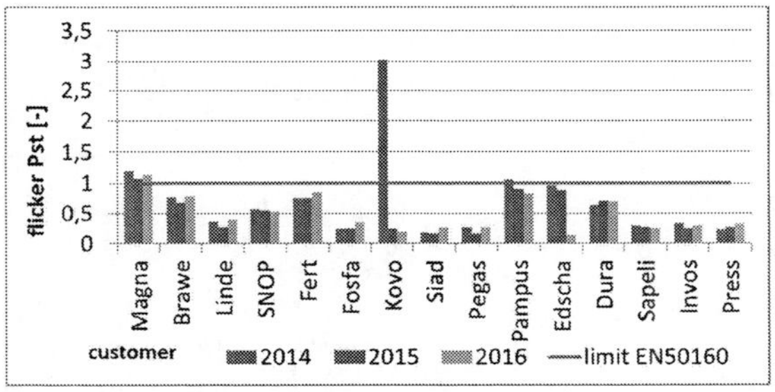

Figure 7. Evaluation of flicker Pst

C. *Harmonic voltage and THD*

The Fig. 8 shows no problem with the 3rd voltage harmonic, the Fig. 9 shows no problem with the 5th voltage harmonic (with the exception of the customer Pegas in 2016), the Fig. 10 shows no problem with the 7th voltage harmonic and the Fig. 11 shows no problem with the 15th harmonic voltage.

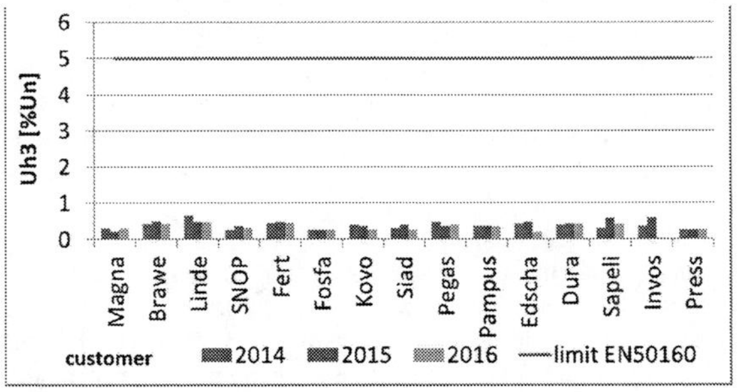

Figure 8. Evaluation of the 3rd voltage harmonic

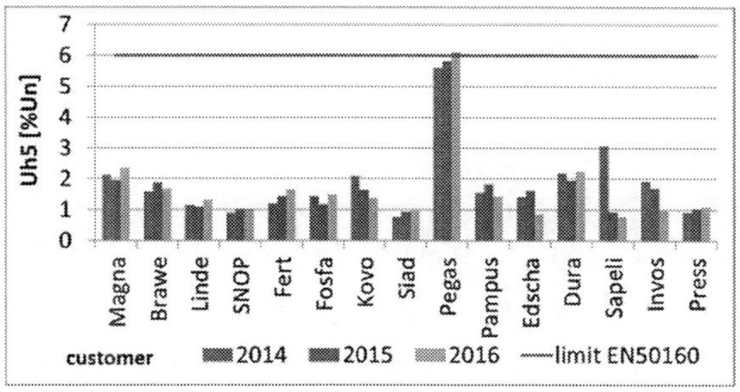

Figure 9. Evaluation of the 5th voltage harmonic

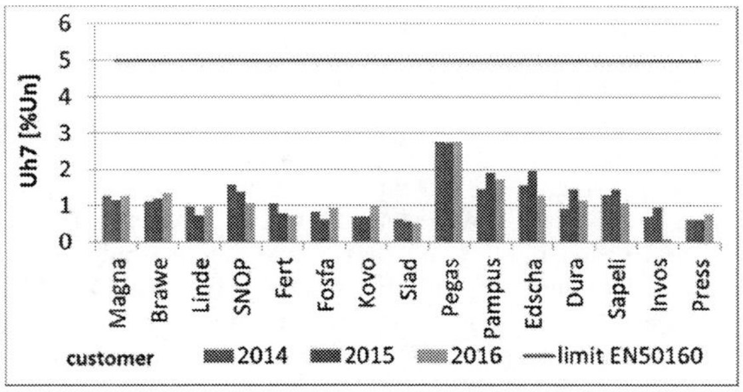

Figure 10. Evaluation of the 7th voltage harmonic

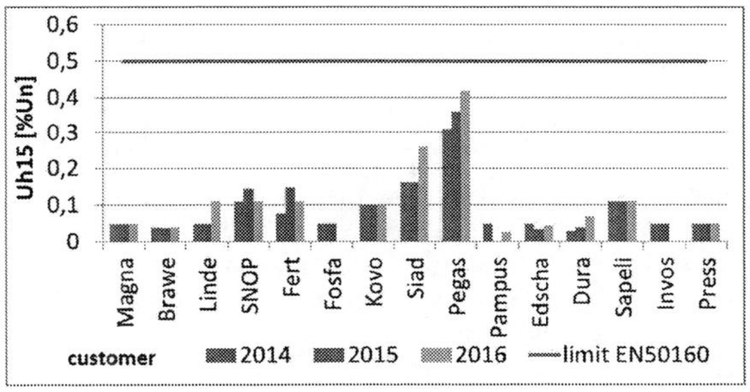

Figure 11. Evaluation of the 15th voltage harmonic

The Fig. 12 shows no problem with total harmonic distortion (THD) in medium voltage grid.

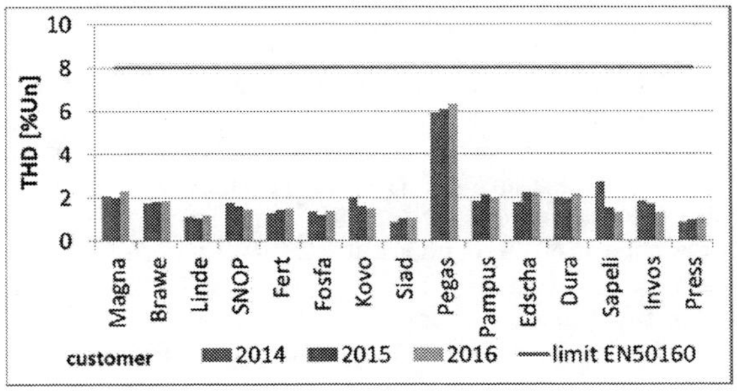

Figure 12. Evaluation of total harmonic distortion THD

D. *Voltage unbalance*

The Fig. 13 shows the evaluation of voltage unbalance (95% percentile). There was evaluated no problem with voltage unbalance in MV grids.

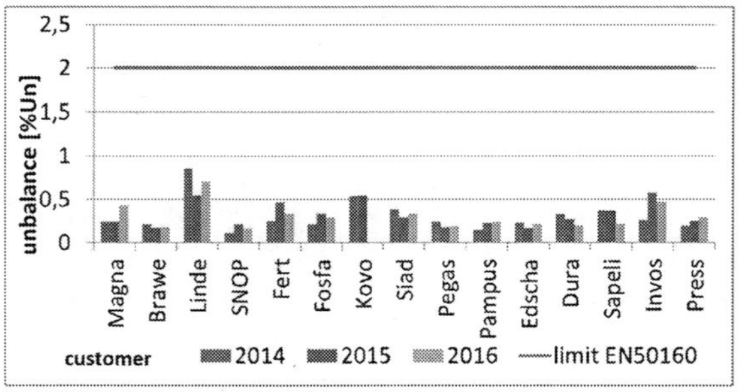

Figure 13. Evaluation of voltage unbalance

V. Customer Magna – Detailed Analysis

Flicker level in delivery point of the customer Magna exceeds the limit. Flicker is produced by operation of new forging presses which were put into operation in 2012. The customer Magna was informed about the problem of flicker because it was proved that flicker was caused by the customer's operation. The flicker intensity depends on the current intensity. The customer did not believe the DNO that the flicker was produced by its operations and so it was decided that the university should perform an independent assessment of the cause of the flicker.

A. Assessment of the cause of flicker

University experts made independent VQ measurements and they discovered the cause of the flicker [4]. The customer operates two forging presses and the university experts compared the flicker level in two states (see Tab. 1.). In the first state both the forging presses were operated and the flicker level reached the value of 1.16 on the MV side. In the second state both the forging presses were out of operation and the flicker level reached the value of 0.63 on the MV side and so the dependence of the flicker on the operation of the forging presses was proved.

Table 1. Flicker on the MV level in different states of the operation

State of the operation	Flicker Plt [-]
Forging presses are in operation	1.16
Forging presses out of operation	0.63

After this independent study the customer accepted that he was responsible for the flicker in the MV grid and he proposed a remedy. This remedy is based on adjusting the forging presses. It is possible to remedy the situation from the side of the DNO too. The DNO suggested building a new HV/MV substation near the customer so that the short circuit power should increase and the flicker level should decrease. This solution would take many years due to the justification of construction of a new high-voltage distribution feeder.

B. Remedy from the side of the customer

The forging presses operate with a definite reserve for performing the pressing. This reserve was 100% because of defective work. With the reduction of this reserve the current value should decrease. A decrease in the current value should reduce the flicker level. The customer proposed reducing the reserve R from R=100% to R=75% and R=60%. The defective work has to be evaluated. Analysis of measurement show that the RMS peak of the current is approximately 700 A for R=100%, approximately 540 A for R=75% and approximately 400 A for R=60%. The amplitude of the voltage changes depends on the amplitude of the current peak and the flicker level depends on the amplitude of voltage changes. Thus the flicker level decreased because of smaller voltage changes, but not enough. After the adjustment of both the forging presses to R=60%, the flicker level reached the value approximately of 1.1 and the number of defective products did not increase. Voltage quality in the MV distribution grid does not comply with the requirements of the EN 50160 standard because of flicker but Magna has no other way to solve the problem apart from halting production. Fortunately, no complaints regarding the flicker were received.

C. Remedy from the side of DNO

It is possible to remedy the situation from the side of the DNO too. The DNO is planning to build a new HV/MV substation near the customer so that the short circuit power should grow and the flicker level should drop. The expected value of the short circuit power after the building of the new HV/MV substation is approximately S_k"=130MVA. So the flicker level should drop below the limit according to the EN 50160 standard – see the Tab. 2. The investment costs for building the new HV/MV substation, including a supply HV feeder, are approximately one million EUR. This solution will take many years due to the hearing of construction of a new high-voltage distribution feeder.

Table 2. Short circuit power in the delivery point of Magna

Way of supply	Current MV feeder	New HV/MV substation
S_k" [MVA]	30	130
Flicker P_{lt} [-]	1.1	0.25

VI. Conclusion

The paper brings a view on voltage quality in MV distribution grids and deals in detail with the problem of flicker in case of the customer Magna. It is a problem when the VQ in the MV distribution grid does not comply with the requirements of the EN 50160 standard because VQ is influenced by many other customers. The question is who is responsible for poor VQ. A simple method based on the correlation of the flicker and current was described in this paper. The remedy does not always have to be expensive from the point of view of the customer but sometimes it might be not sufficiently efficient. In this study the DNO is going to build a new HV/MV substation but it takes about three years. No other affordable solution is possible apart from halting the production but the DNO has limited tools to realise it. It can therefore be stated that in the case of customers who consume high power with a considerable dynamic shape of loading that causes flicker outside the permitted tolerance (and where production optimisation has already taken place, for instance, by the distribution of manufacturing over time in order to prevent the simultaneous running of those appliances that contribute to the flicker the most), the most suitable way to reduce the flicker is by increasing the short circuit power, for instance, by building a new transformer station, etc. For this type of power take-off, the authors recommend performing a connectivity study before connecting the industrial plant to the network, in which the elaborator should take into account a future increase in the reserved power input over the period of approximately the following 10 years.

References

[1] EN 50160 Ed.3 Voltage characteristics of electricity supplied by public distribution systems. Brussels: European Committee for Electrotechnical Standardization, 2010. 20 p.

[2] IEC/TR 60725:2012 Consideration of reference impedances and public supply network impedances for use in determining the distrurbance characteristics of electrical equipment having a rated current ≤75A per phase, IEC, ISBN 978-2-83220-161-9

[3] M. Tesarova and K. Nohac, "Voltage and frequency stability analysis of an island mode LV distribution network". In Proceedings of the 15th International Scientific Conference Electric Power Engineering 2014. Brno: Brno University of Technology, 2014. ISBN: 978-1-4799-3806-3

[4] M. Kaspirek, P. Krejci, P. Santarius and K. Prochazka, "Flicker caused by operation of industrial technology" Advances in Electrical and Electronic Engineering, Volume 14, Issue 5, 2016, Pages 475-480, ISSN:1336-1376, E-ISSN:1804-3119

Medium-Term Forecast of Monthly Supply of Electricity and Maximum-Demand-Day Power Using Adaptive Neuro-Fuzzy Inference System

Janusz Sowiński
Faculty of Electric Engineering
Czestochowa University of Technology
Czestochowa, Poland
jansow@el.pcz.czest.pl

Abstract — **The statistical data used in the research were prepared by Agency of Energy Market, Inc. (ARE SA). The data constitute a database on the monthly balances of electricity and maximum-demand-day power in each month of the year in the period 2004-2015. Database offers a possibility of analysing monthly consumption of electrical energy, at the same time being able to obtain a yearly consumption. Modern forecasting methods, such as artificial neural networks, evolutionary algorithms and fuzzy logic require a large number of sets for the process of learning and then for verifying the model obtained. The presented in the paper method can yield forecasts at monthly intervals, with the forecasting process using adaptive neuro-fuzzy inference system (ANFIS). Some results of medium-term forecast of electricity supply and power demand have been presented.**

Keywords: electricity balance; forecast; fuzzy logic

I. Introduction

Statistics on the monthly balances of power generation in the period 2004 – 2015 in Poland are included in publications issued by Agency of Energy Market, Inc. (ARE S.A.) [1]. Using this database, it is possible to forecast the power generation in conventional power plants and renewable sources of energy, and on this basis to analyse the energy mix. At present, the creation of energy-environment policy in the European Union until 2030 generates a discussion on the main challenges for the EU and Polish power system. In Poland we would like to construct a long-term vision for power system development. The development of power plants should be correlated with the investment in transmission and distribution networks [2, 3] taking into account reliability of electricity supply [4]. A compromise between conventional power generation and renewable energy should be achieved considering environmental requirements [5].

There are a lot of modern forecasting methods, such as artificial neural networks, evolutionary algorithms and fuzzy logic, which are used in power system analysis. The models require a large number of sets for the process of learning and then for verifying the model obtained [6, 7]. The paper proposes to use an adaptive neuro-fuzzy inference system (ANFIS) in forecasting process. The method requires preparing statistical sample of input variables. A yearly load variation in the power system [8] is used in analysis of statistical data. Only the information of the process is applied in this kind of approach. The forecasting method presented in section IV can be used for obtaining medium-term forecast of electricity supply as well as generation in various types of power plants. These forecast shed light on phenomena occurring at the power generation on the electricity market.

II. Statistical Data

The statistical data on electricity supply and power generation presented in [1] were prepared by Agency of Energy Market, Inc. (ARE S.A.). The data constitute a database on the power generation in thermal power plants, wind power plants, hydro power plants, thermal power plants of independent power producers, renewable sources of energy, industrial power plants, and import of electricity in the period 2004 - 2015. Selected statistical data for one year are presented in Table 1.

The important information in investment decisions of the power system development is the knowledge of maximal power demand (peak load). The maximal power demands in the period 2005-2015 are presented in Table 2.

III. Yearly Variation of the Electricity Supply and Power Demand

The main causes of load variation are customer habits, work schedule, and astronomical causes (sunrise and sunset time). The load is influenced both by non-random and random factors. The former include geographical location and climate, variation in the sunrise and sunset times, variation in the angle of incidence of sunrays, as well as the features of the power system, such as legal working time and the system of working shifts, the structure of consumers, and the dynamics of economic development [9, 10]. Random factors include temperature changes, wind speed, precipitation, cloud cover, changes in the structure of energy consumption, breakdowns of devices, etc. The volatility of the load and electricity supply is seasonal in character. This is confirmed by the curves presented in the figures 1-3.

The basic indicator of the yearly load variation could be defined as the mean yearly load ratio m_r described by formula:

$$m_r = \frac{P_{r\,sr}}{P_r} = \frac{A_r}{T_r \cdot P_r} = \frac{T_{rs}}{T_r}, \quad (1)$$

where: A_r - yearly electricity consumption, $P_{r\,sr}$ - yearly mean load, P_r - yearly peak load, T_r - duration of a year in hours (8760 h), T_{rs} - duration of peak power consumption.

Table 1. Electricity in (GWh) in selected year 2015 [1]

Months	1	2	3	4	5	6	7	8	9	10	11	12
Global supply of electricity	15973	14855	15412	13818	13829	13634	14811	14542	14819	15855	15697	16158
Generation	14827	13817	14307	12905	12767	12520	13504	13238	13475	14444	14319	14821
Power plants	12786	12163	12532	11209	11382	11336	12133	11932	12071	12857	12289	12351
Thermal PP	12184	11753	12027	10694	10991	11055	11799	11659	11772	12549	11768	11631
Wind PP	340	205	260	269	181	132	203	154	192	183	360	494
Hydro PP	262	205	245	246	210	149	131	119	107	125	161	226
Independent power producers	1235	872	1033	1021	732	594	768	657	798	880	1245	1630
Thermal IPP	253	233	242	208	160	151	144	106	154	217	233	252
RES	982	639	791	813	572	443	624	551	644	663	1012	1378
Industrial power plants	806	782	742	675	653	590	603	649	606	707	785	840
Import	1146	1038	1105	913	1062	1114	1307	1304	1344	1411	1378	1337

As was demonstrated in publications [9, 10], a number of quantities characterising the yearly load variation can be approximated by means of cosine curves (see Fig.1).

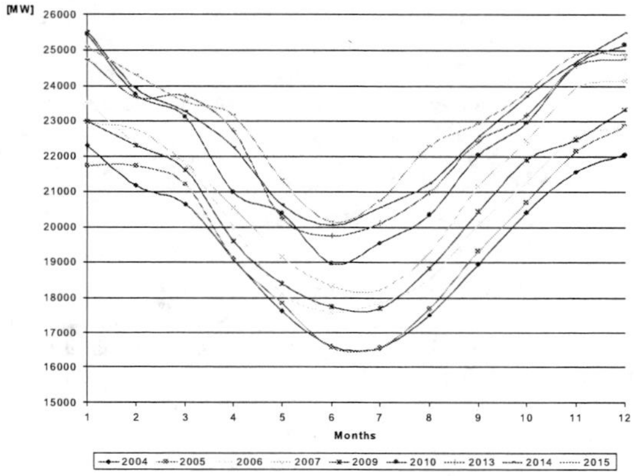

Figure 1. Yearly variation of peak loads

The past values of the mean yearly load ratio m_r are the most crucial statistical data in the power system (Table 2). It is important to emphasize the high stability of the mean yearly load ratio course in the years 2004-2015.

The chronological courses of the monthly global supply of electricity and peak loads from the statistical data are presented in Fig. 2 and 3. The years 2008, 2011 and 2012 are omitted due to lack of data.

Table 2. Maximal power demand and the mean yearly load ratio in National Power System (KSE). Prepared on the basis of [1]

Year	Maximum Power Demand	Mean Yearly Load Ratio
-	MW	-
2004	22 319	0,755
2005	22 849	0,752
2006	23 523	0,753
2007	24 157	0,722
2008	25 121	0,675
2009	23 343	0,707
2010	25 449	0,670
2011	24 780	0,715
2012	25 845	0,679
2013	24 761	0,716
2014	25 535	0,670
2015	25 101	0,707

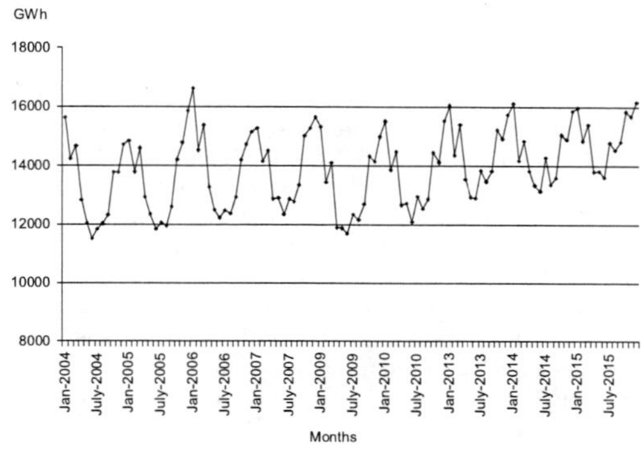

Figure 2. Monthly values of global supply of electricity in the selected years of the period 2004-2015

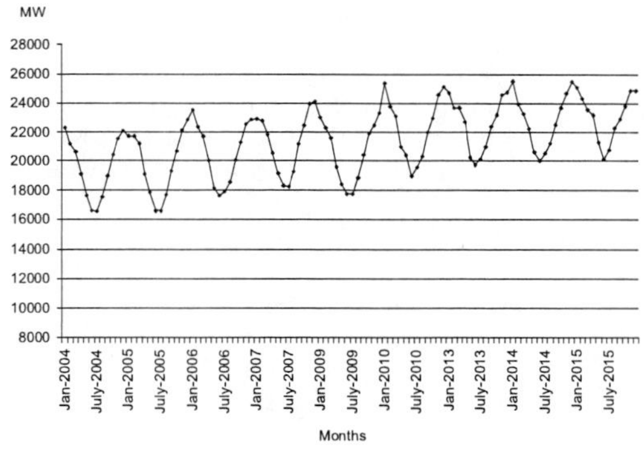

Figure 3. Monthly values of peak loads in the selected years of the period 2004-2015

IV. ANFIS AS FORECASTING METHOD

The global supply of electricity and peak loads can be forecast with the use of an adaptive neuro-fuzzy inference system (ANFIS) [6, 7]. A fuzzy inference system (FIS) was constructed on the basis of the input and output data sets of the model using the function from Fuzzy Logic Toolbox of Matlab®, MathWorks, Inc. A backpropagation algorithm, combined with the least squares method is used to estimate the parameters of the membership function. Estimation is performed on a fuzzy set, constructed on the basis of the modelled data. The modelling in ANFIS is similar to many other identification techniques. The ANFIS model is trained on a part of data. The main goal of the training process is to select the parameters of a membership function, in accordance with the prescribed criterion of minimising an error. The other part of data (i.e. those which were not employed for model estimation) is used in the process of verifying the model. In the next step of the research the prediction is made and on this basis the fit of the model can be assessed. The output membership functions can only be linear or constant because the ANFIS model uses the Sugeno method of fuzzy inferencing. The method of weighted arithmetic mean defuzzification is used to obtain output variable. The form of the defuzzification function and of the membership function is defined by ANFIS.

V. Forecasting Process

In the process of time-series prediction using ANFIS it was assumed that the values of monthly global supply of electricity and peak loads $X(t)$ are known for $t=1, ..., T$. A 12-month shift in data for the input variable was chosen because of the annual load variation. Due to a limited number of statistical data ($T=108$), a two-dimensional input vector of training data $w(t)=[X(t-12)\ X(t)]$ was selected in a number of trials, with the input data of the training set corresponding to the prediction trajectory $s(t)=X(t+12)$. In this way 84 values were obtained of the input/output data set using the data from the period January 2004 to December 2015. The first 48 values were used in the process of training and the other 36 in the process of verification of the prediction model. In the model there are only two input variables, that's why the number of fuzzy rules is 4. The estimation of the ANFIS model and forecasting process were performed. Fig. 4 and 5 present a membership functions after estimation, and Fig. 6 and 7 the root mean squared error RMSE for the sample under verification. Comparison of the real data on global supply of electricity per month and peak loads with the prediction obtained by means of ANFIS (Fig. 8 and 9) indicates that differences are relatively small. Absolute errors of the prediction are presented in Fig. 10 and 11, and *MAPE* obtained on their basis is 2.47 % for global supply of electricity and 1.49 % for monthly peak loads.

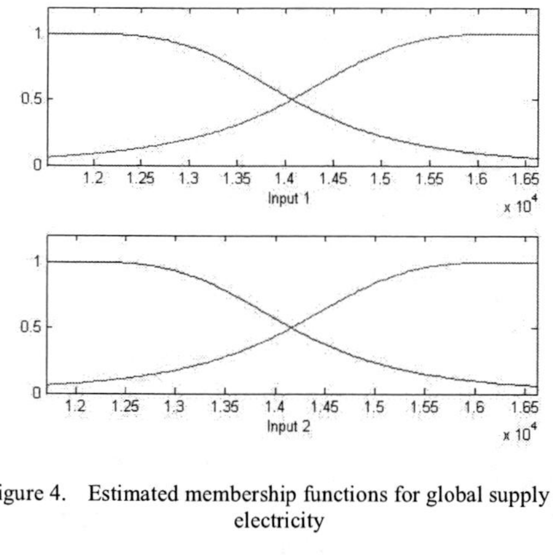

Figure 4. Estimated membership functions for global supply of electricity

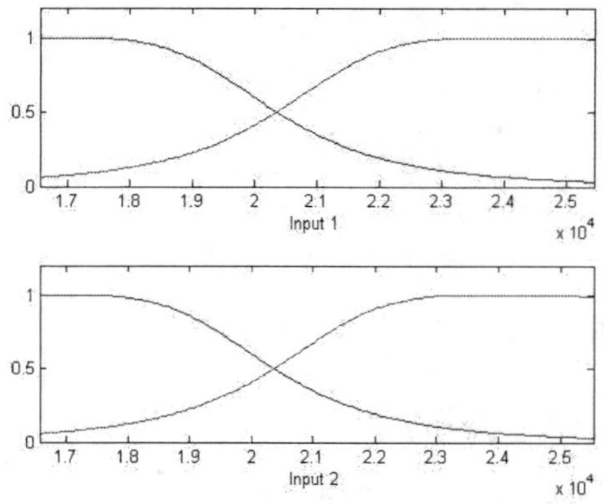

Figure 5. Estimated membership functions for peak loads

After constructing the ANFIS models, a forecast of global supply of electricity and peak loads was carried out in the horizon until 2020, as presented in Fig. 12 and 13. The statistical data covered the selected years of the period from 2004 till 2015, and the values of monthly global supply of electricity and peaks loads were indexed from $t=1$ to $t=108$, hence the indices of the monthly supply of electricity and peak loads for the 2016 - 2020 forecast are from $t=109$ to $t=169$.

Similar research were conducted for the remaining components of the electricity balance (see Table I). The positive results of the ex-post forecasts were obtained for variables characterized by seasonal (annual) variability. On the other hand, for unstable generation sources, eg RES, large errors of expired forecasts have been occured in the process of models verification.

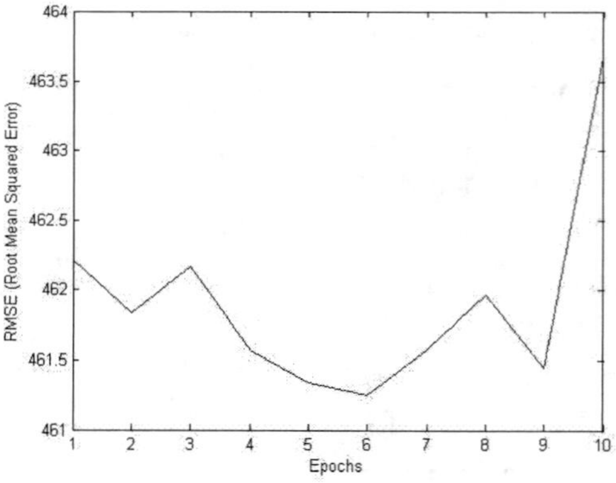

Figure 6. Root mean squared error RMSE of global supply of electricity

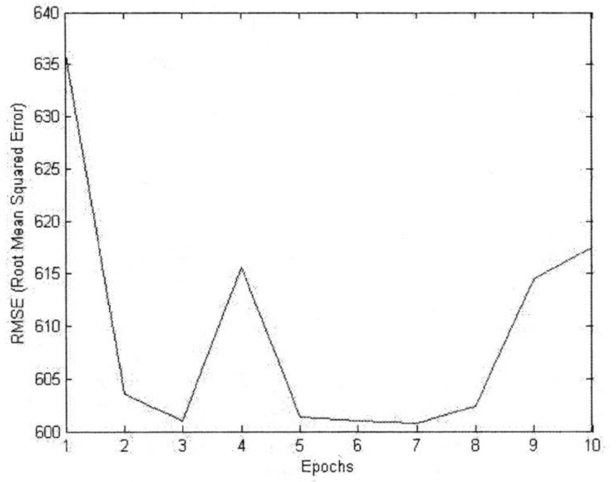

Figure 7. Root mean squared error RMSE of peak loads

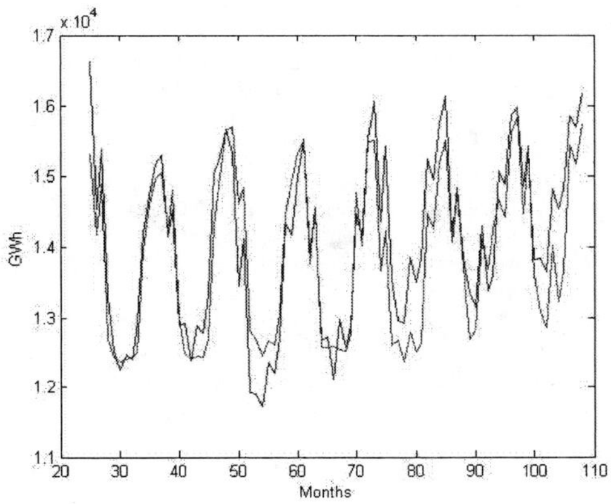

Figure 8. Comparison of global electricity supply data (blue line) with prediction curve (green line)

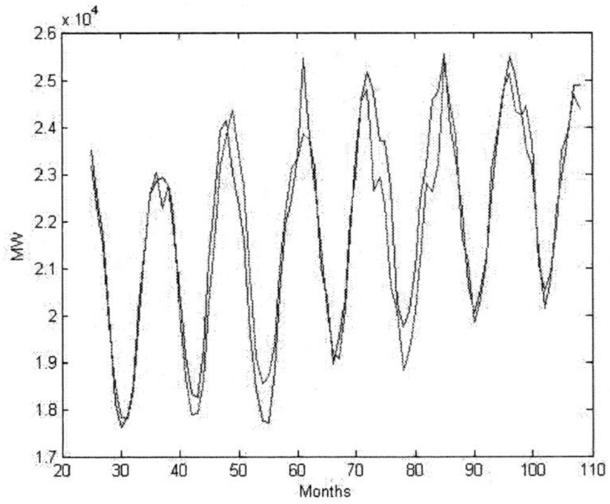

Figure 9. Comparison of peak loads data (blue line) with prediction curve (green line)

VI. Concluding Remarks

A lot of publications present the problem of load variation in the power system. The thorough investigation resulting in detailed characteristics and analyses for National Power System in Poland was described in [9, 10]. The global supply of electricity and peak loads are characterized by seasonal variability as it follows from an analysis of Fig. 2 and 3. This type of curves is suitable for forecasting using the ANFIS model. The seasonal character of electricity supply and peak loads is crucial in estimating these variables in a monthly period.

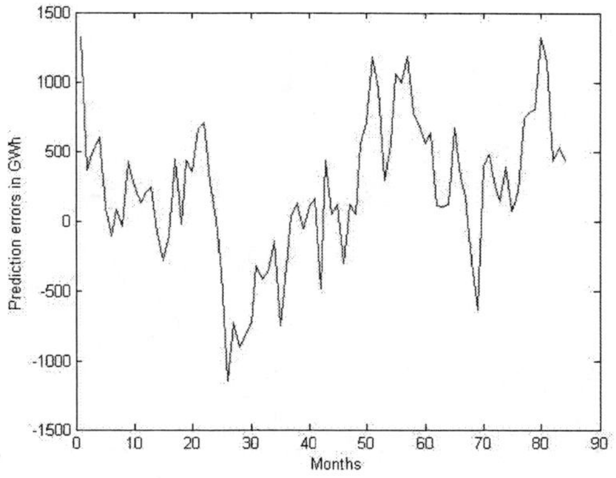

Figure 10. ANFIS errors of global electricity supply prediction

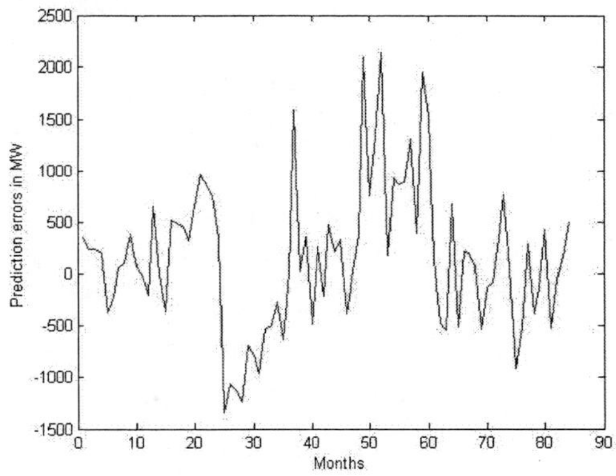

Figure 11. ANFIS errors of peak loads prediction

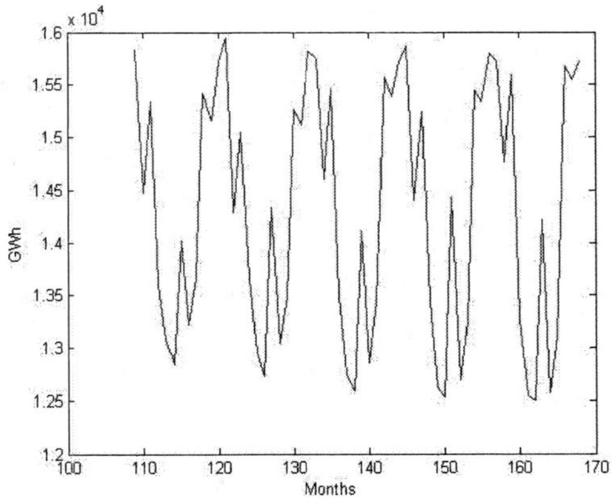

Figure 12. Forecast of monthly global supply of electricity from 2016 to 2020

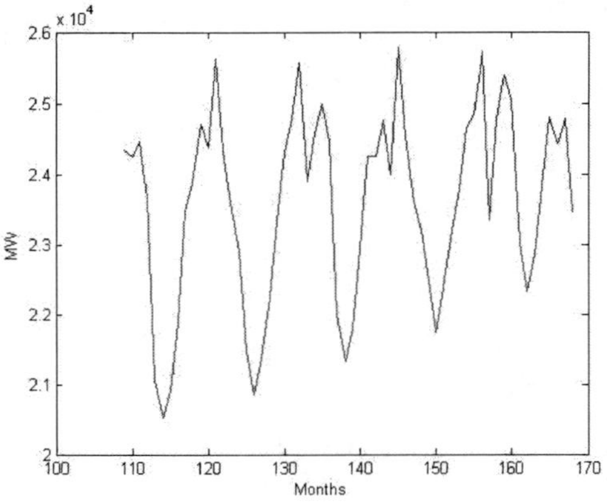

Figure 13. Forecast of monthly load peaks from 2016 to 2020

ANFIS appears to be a highly promising forecasting tool. The forecast of global supply of electricity and peak loads until 2020 presented in the paper (Fig. 12 and 13) reflects the behaviour of the variable in the period 2004-2015 (Fig. 2 and 3). The small values of *MAPE* equal to 2.67 % in an ex post forecast of electricity supply and 1.49 % for peak loads indicate that the forecast prepared with the use of ANFIS is of good quality.

The ANFIS model managed satisfactorily in the context of some gaps in the statistics on electricity supply and peak loads. This is an undoubted advantage of the ANFIS model.

References

[1] ARE SA, Statystyka elektroenergetyki polskiej (Statistical data of Polish power sector. *(in Polish)*). Warsaw, 2004-2016

[2] A. Gawlak, „Noninvestment forms of reducing energy losses in distribution networks" - *Proceedings of the The 8th International Scientific Symposium ELEKTROENERGETIKA* 2015, 16-18.9.2015, Stara Lesna, Slovak Republic, pp. 61-64.

[3] M. Kornatka, and A. Gawlak, „Comparative analysis of operating conditions in Polish medium-voltage and 110 kV networks," *Proceedings of the The 8th International Scientific Symposium ELEKTROENERGETIKA* 2015, 16-18.9.2015, Stara Lesna, Slovak Republic, pp. 57-60

[4] M. Kornatka, "The weighted kernel density estimation methods for analysing reliability of electricity supply," *17th International Scientific Conference on Electric Power Engineering (EPE)*, Prague, 2016, pp. 1-4. doi: 10.1109/EPE.2016.7521729.

[5] I. Pavlova-Marciniak, "Implementation of climate policy assumptions in Poland," *Proceedings of the The 8th International Scientific Symposium ELEKTROENERGETIKA* 2015, 16-18.9.2015, Stara Lesna, Slovak Republic, pp.455-458

[6] J.-S. R. Jang, "ANFIS: Adaptive-Network-based Fuzzy Inference Systems," *IEEE Transactions on Systems, Man, and Cybernetics*, Vol. 23 (3), pp. 665-685, 1993.

[7] L.-Ch. Ying, and M.-Ch. Pan, "Using adaptive network based fuzzy inference system to forecast regional electricity loads," Energy Conversion and Management, Vol 49 (2), pp. 205-211, 2008.

[8] A. Hamadi. and S. A Soliman., "Long-term/mid-term electric load forecasting based on short-term correlation and annual growth," Electric Power Systems Research, Vol. 74 (3), pp. 353-361, 2005.

[9] I. Dobrzańska, K. Dąsal, J. Łyp, T. Popławski, and J. Sowiński, Prognozowanie w elektroenergetyce. Zagadnienia wybrane. (Forecasting in power engineering. Selected issues. *(in Polish)*), Wydawnictwo Politechniki Częstochowskiej, Częstochowa, 2002.

[10] T. Popławski (ed), K. Dąsal, J. Łyp, and J. Sowiński, Wybrane zagadnienia prognozowania długoterminowego w systemach elektroenergetycznych. (Selected problems of long-term forecasting in power systems. *(in Polish)*). Wydawnictwo PCz, Częstochowa, 2011.

Smart Grid in Practise - Implementation of the Billaterall Supply in Gdynia Trolleybus Network

Mikołaj Bartłomiejczyk
Gdańsk University of Technology,
Faculty of Electrical and Control Engineering
G. Narutowicza 11/12, 80-233 Gdańsk, Poland
mikolaj.bartlomiejczyk@pg.gda.pl

Abstract—**Since 2001, trolleybus system in Gdynia has been involved in many activities related to the reduction of power consumption, both in terms of implementation and research and development. In 2011 was introduced the first supercapacitor storage energy system. In PKT started applications of Smart Grid technologies in supply network: the bilateral supply. The paper presents results of these this novel investitions.**

Keywords: Trolleybuses, supercapacitors, energy storage, energy recuperation, smart grid, energy savings, emobility

I. Introduction

The possibility to work in reverse mode, as generators of electric energy, is a characteristic feature of electric motors [1-4]. This feature is applied in electric vehicles and used during braking, when kinetic energy of a vehicle is changed into electric energy, which can then be returned to the traction supply system and consumed by other vehicles. However, when there is no possibility to receive recuperation energy, it is dissipated in braking resistors. In this situation we lose the energy which could potentially be recovered.

The city of Gdynia, Poland, exploits an extensive, 50-km-long trolleybus network, which is serviced by 85 vehicles. This network is supplied from 10 traction substations with 600 V DC. The trolleybus system is operated by Przedsiębiorstwo Komunikacji Trolejbusowej Sp. z o.o. (PKT) which, for a number of years, has been working towards limiting energy consumption [5]. Vehicles with power electronic drive system are gradually introduced and, consequently, energy consumption has decreased by more than 20% over the past 10 years. In the years 2009-2012 PKT conducted comprehensive modernisation of the supply system, which resulted in the increase of its reliability and in the decrease of transmission losses [6]. PKT also carries out scientific and research activities connected with reducing energy consumption. In the years 2010-2012 a super-capacitor storage system was being tested by PKT in cooperation with the Electrotechnical Institute in Warsaw [7]. In 2014 was installed the second storage energy system in Wielkopolska substation

II. Bilateral supply of contact line

In a classic supply system each of the contact line sections is supplied solely from one substation, so the supply areas of individual substations are galvanically separated from each other. In the case of low traffic intensity or small substation areas, there may occur a situation where there is no vehicle which could absorb the recuperation energy from a braking vehicle. Then this energy will be dissipated in braking resistors. However, it is possible to increase the possibility of finding a recipient of the generated energy by joining the supply area with the neighbouring substation, thus allowing for the flow of recuperation energy to the area of that neighbouring substation. Such a situation has been presented schematically in Fig. 1.

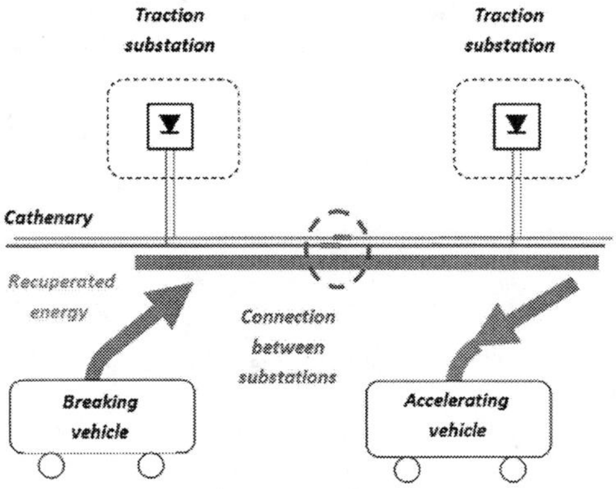

Figure. 1. Diagram of recuperation energy flow between braking vehicle and accelerating vehicle in bilateral contact line supply system.

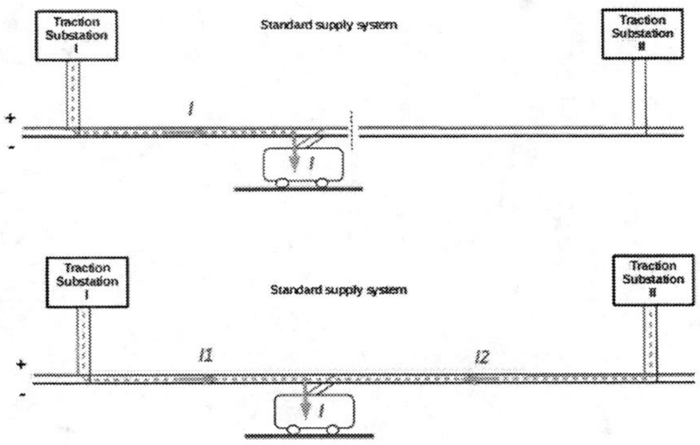

Figure. 2. Comparison of vehicle current distribution between supply traction substations.

Billaterall supply also allows to reduce the voltage losses in overhead catenary, what is presented on fig. 2. In billaterall supply the vehicle can be supplied in two ways, so the vehicle current I can be distributed between two smaller substation load currents I_1 and $I\,I_2$, which reduces the load of catenary and voltage losses.

An important issue relating to the implementation of bilateral power supply in traction supply system is to ensure effective short circuit breaking capacity. In the case of long power sections supplied from two substations, there may be a

Complicated protection systems cause low popularity of bilateral power supply in municipal transport power systems. However, technology of modern protection devices can solve this problem. It is now possible to equip traction substations with time-current protections and differential protections, which help to locate even the distant short-circuits in the overhead line.

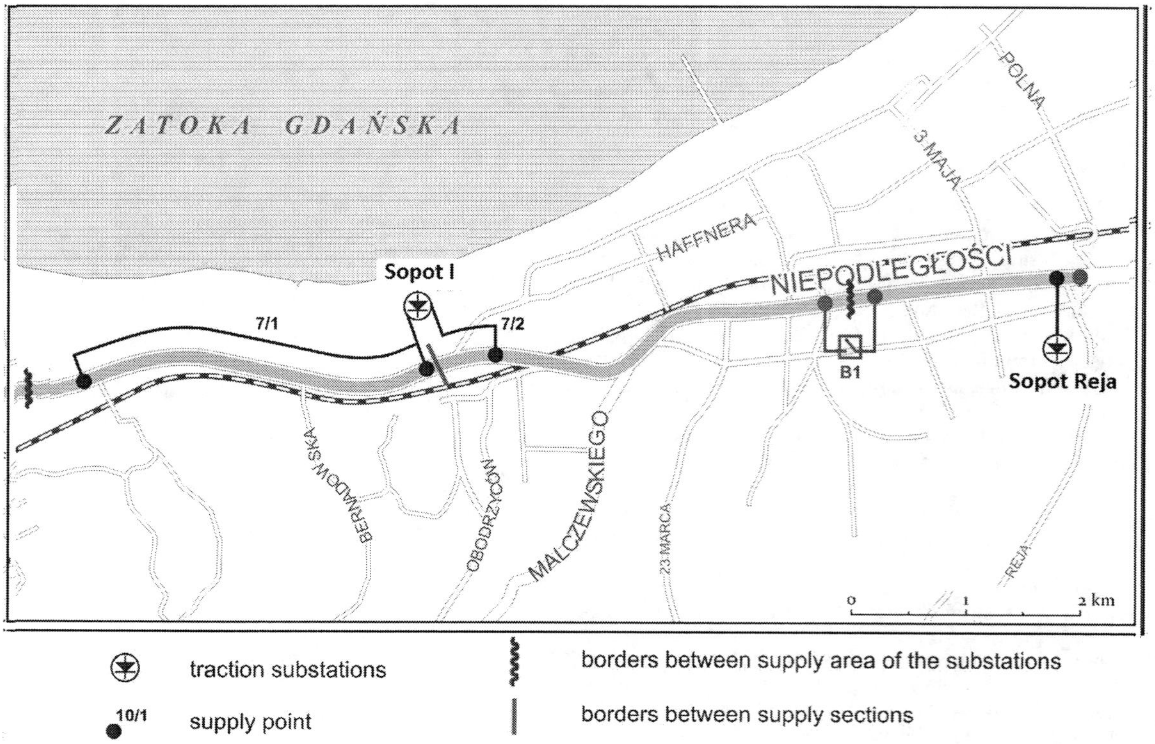

Figure. 3. The supply area scheme of substation Sopot I and Sopot Reja, B1 - connection between substations

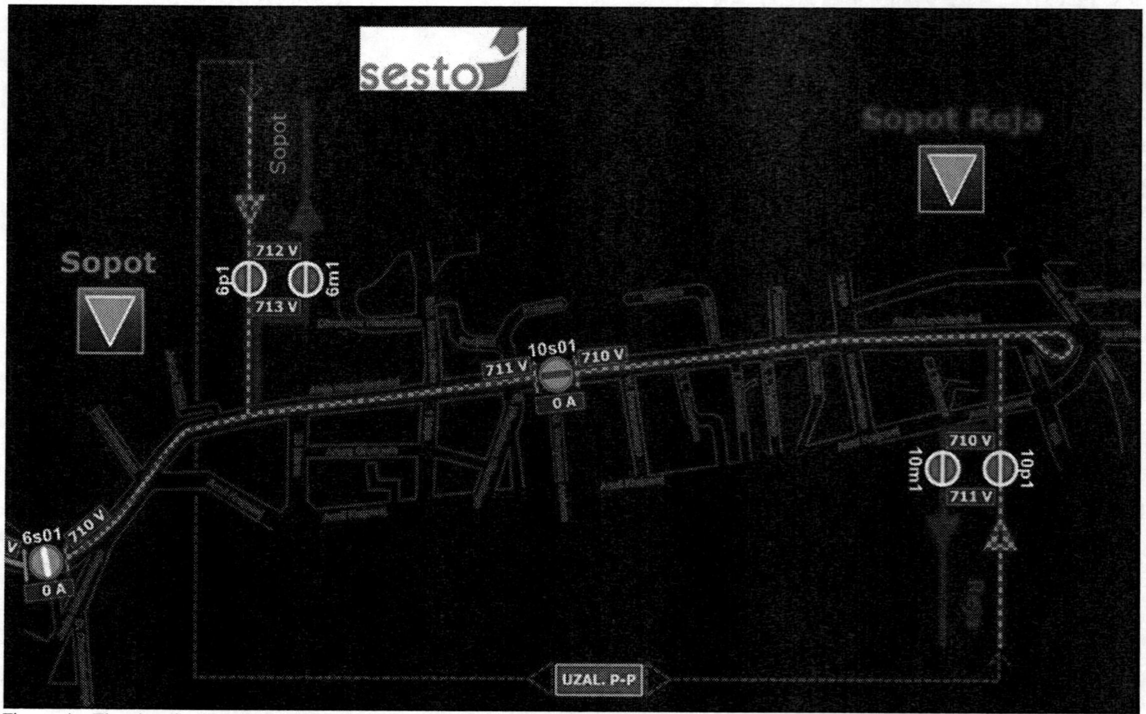

Figure. 4. The visualisation of the bilaterall suppy system on the controll panell of dispath controll centre of trolleybus sunstation, button "UZAL. P-P" and dashed line symbolize the interconnection of control system, substations Sopot – Sopot Reja

Those protections may be connected by means of GSM network with a remotely-controlled overhead line disconnectors which, in the case of detecting a short-circuit and its interruption, automatically reconfigure the supply system. This allows for reduction of the area without power in the event of damage to the overhead line.

In June 2016 PKT Gdynia began practical implementation of Smart Grid technology within the exploited trolleybus system. This task was realised in the year 2016, within the Eliptic "Electrification of public transport in cities" H2020 project. The first stage of the task realisation was to introduce bilateral supply of the trolleybus network in the city of Sopot. The neighbouring city of Sopot is a well-known Polish sea resort. A suburban trolleybus route Gdynia – Sopot, was built in 1949. The section located within the area of Sopot is supplied from two traction substations Sopot I and Sopot Reja. Due to the low frequency of trolleybus service on the suburban route Gdynia – Sopot, recuperation of braking energy is utilised to a small degree [7].

The billaterall supply system was realised as a software integrated with the remote control system of trolleybus substation in Gdynia. The supply areas of substations Sopot I and Sopot Reja were connected (connector B1, fig. 3). The software was developed by SESTO company, which was delivers control system equipment and software of all trolleybus substation in Gdynia. The billaterall supply system and substation protection devices were based only on the internet connection between substation, which was the main challenge of this system. The final reliability of it were the results were surprisingly good. The screenshot of display remote control system SESTO is shown on fig. 4, where the interconnection between substation is visible. The SESTO bilateral supply system software is symbolised by button "UZAL. P-P" and dashed line, which connects symbols of two coupled supply outputs. In August 2016 the similar system was implemented between substations Dworzec and Grabówek (fig. 5, 6), in one of the most exploited trolleybus line in Gdynia, due to high voltage drops in catenary.

The introduction of bilateral supply between substation Sopot I and Sopot Reja resulted visible in the increase of recuperation effectiveness and slightly in the reduction of energy transmission losses (Table 1). The drops and fluctuations of voltage in the contact line were also significantly reduced (Fig. 7, 8). In case of substations Dworzec – Grabówek, the implementation of billaterall supply significantly reduced the voltage drops (table 1, fig. 9 and 10). Voltage stability was improved, what is visible on graph of the voltage dispersion (fig. 10).

It should also be underlined the possibility of use of the voltage dispersion graph in the traction network (fig. 8, 10). These graphs were made on the basis of vehicle data loggers registrations, which include values of electrical parameters of the vehicle and GPS localisation in every second. It allows to conduct the spatial analysis of electrical parameters of supply system [8-10]. In case of supply area of substations Sopot and Sopot Reja there are visible two horizontal groups of points (fig. 8). Upper group, in the shape of "line" means the upper voltage limit during recuperation breaking. In another way, these are situations when there is no receivers for breaking energy. Lower group of the points corresponds to standard situation of supply system with active load.

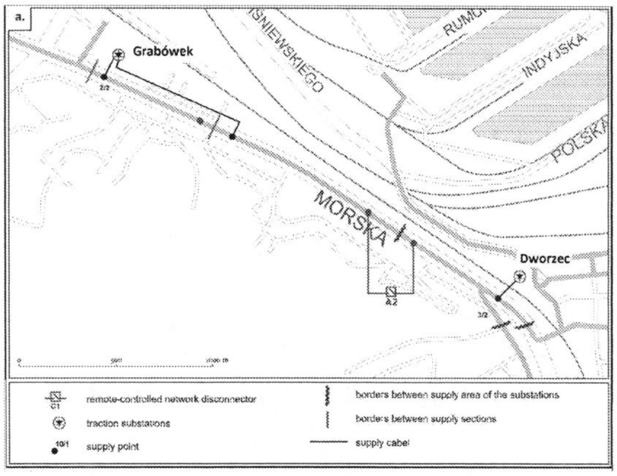

Figure 5. The supply area scheme of substation Grabówek and Dworzec, A2 - connection between substations

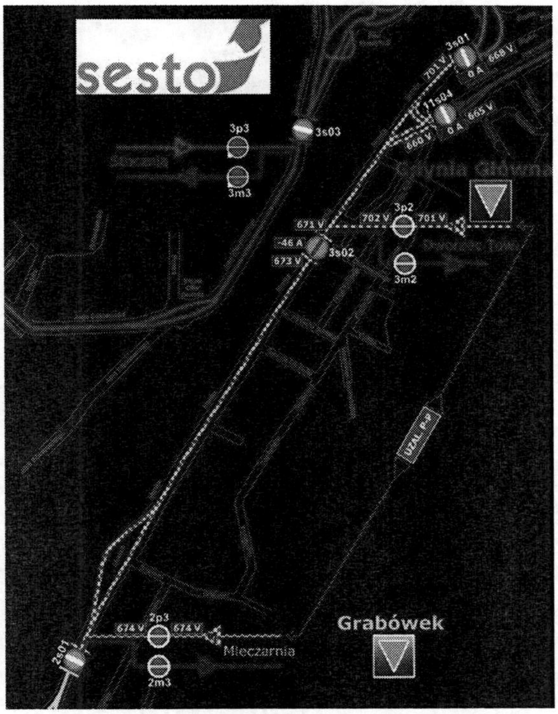

Figure. 6. The visualisation of the bilaterall suppy system between substations Dworzec (Gdynia Główna) and Grabówek

The horizontal upper limit of this group means the no-load substation voltage. In case of substations Dworzec and Grabówek after implementation of billaterall supply system the spread of voltage points was narrowed from both side - lower and upper part. Reduction of the upper part means the reduction of voltage drops caused by voltage loses in catenary. Limitation of upper values is caused by improving of energy recuperation after implementation of billaterall supply - the number of situation, when there is no receiver of recuperated energy, which reflects by high value of voltage, was reduced.

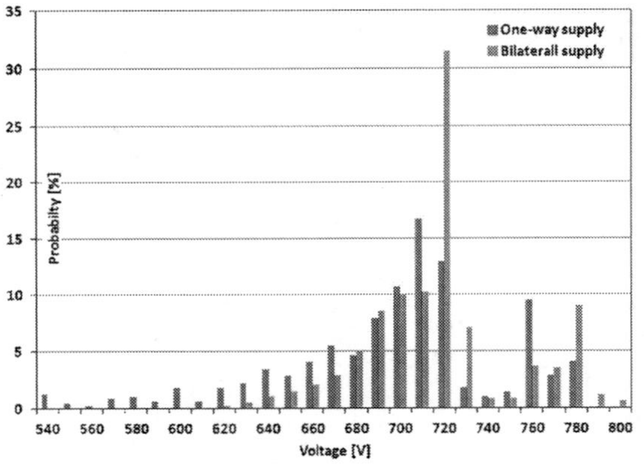

Figure. 7. Histogram of the voltage of catenary in supply area of substations Sopot and Sopot II before and after the introduction of bilateral supply system

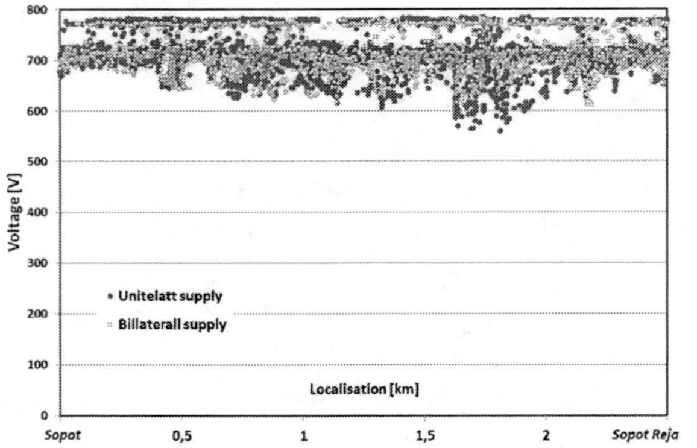

Figure. 8. Voltage dispersion in area of substations Sopot – Sopot Reja in catenary before and after of implementation of bilateral supply

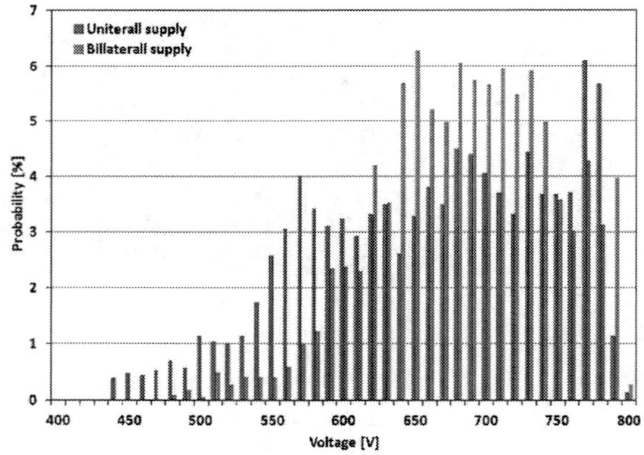

Figure. 9. Histogram of the voltage of catenary in supply area of substations Dworzec (Gdynia Główna) - Grabówek before and after the introduction of bilateral supply system

Table. 1. Effects of introducing bilateral supply between Sopot and Sopot Reja substations

	Unilateral supply	Bilateral supply
	Sopot – Sopot I	
Recovery effectiveness	2,5%	10%
Transmission Losses	4,6%	3,2%
Voltage drops <550 V	0,23%	0%
Voltage drops <500 V	0,01%	0%
	Dworzec - Grabówek	
Recovery effectiveness	19%	22%
Transmission losses	11,2%	7,1%
Voltage drops <550 V	0,93%	0,01%
Voltage drops <500 V	0,38%	0,09%

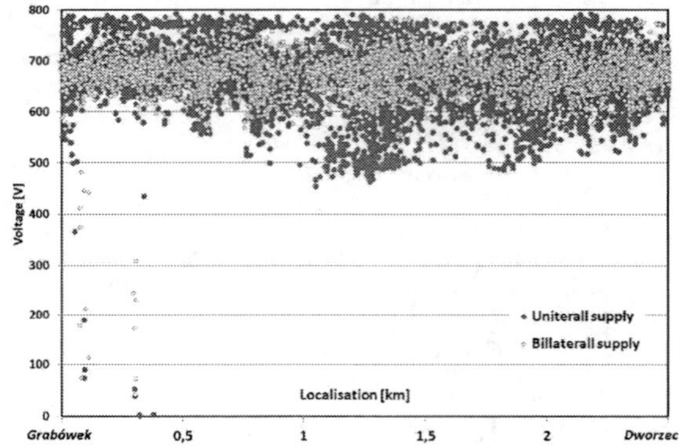

Figure. 10. Voltage dispersion in area of substations Dworzec (Gdynia Główna) – Grabówek in catenary before and after of implementation of bilateral supply

III. Conclusions

Billaterall supply system allows to reach the energy savings and improvement of supply parameters with small investments. It can be recommended especially for weak supply systems or decentralized systems with many small substations, which makes the flow of recuperated energy difficult. Moreover, the billaterall supply allows to reduce voltage drops in supply network.

To increase the use of energy recovery, at first it is necessary to consider the possibility of reconfiguring the supply system, which will facilitate the flow of braking energy. In many situations, very good results can be achieved at a low cost. Small substation power supply areas galvanically isolated from the rest of the network should be avoided. Supply areas of such substations should be connected to create the largest area of recuperation energy flow. Therefore, it is recommended to introduce bilateral supply of the overhead line to allows for an increase in the utilization rate of recuperation and reduced transmission losses. The power supply system designed taking into account the flow of recuperation energy is able to accept the vast majority of the recovered energy [11].

ACKNOWLEDGMENT

This paper has been elaborated in the framework of the project ELIPTIC "Electrification of public transport in cities" co-financed by the European Union's Horizon 2020 research and innovation programme under the grant agreement No 636012

REFERENCES

[1] P. Kacor, M. Marek, J. Palecek, „Analysis Of Magnetic Field Distribution Inside Tram-Car", Proceedings Of The IV International Scientific Symposium On Electric Power Engineering – Elektroenergetika, 2007, pp. 128-129

[2] L. Jarzebowicz, K. Karwowski, W.J. Kulesza" Sensorless algorithm for sustaining controllability of IPMSM drive in electric vehicle after resolver fault", Control Engineering Practice, vol. 58, 2017, pp. 117-126. doi: 10.1016/j.conengprac.2016.10.004.

[3] L. Jarzebowicz and S. Judek, "3D Machine Vision System for Inspection of Contact Strips in Railway Vehicle Current Collectors," in 2014 International Conference on Applied Electronics (AE), Pilsen, Czech Republic, 2014, pp. 139–144. http://dx.doi.org/10.1109/AE.2014.7011686.

[4] S. Hamacek, M. Bartlomiejczyk, R. Hrbac, S. Misak, V. Styskala, "Energy recovery effectiveness in trolleybus transport", Electric Power Systems Research, vol. 112, 2014, pp. 1–11, doi:10.1016/j.epsr.2014.03.001

[5] K. Hebel, M. Wołek, "Perception of modes of public transport compared to travel behaviour of urban inhabitants in light of marketing research", Scientific Journal of Silesian University of Technology. Series Transport, vol. 92, 2016, pp. 65-75, doi: 10.20858/sjsutst.2016.92.7.

[6] M. Bartłomiejczyk, M. Połom, "The impact of the overhead line's power supply system spatial differentiation on the energy consumption of trolleybus transport: planning and economic aspects", Transport, vol. 32(1), 2017, pp. 1–12, doi:10.3846/16484142.2015.1101611, ISSN 1648-4142

[7] M. Bartłomiejczyk, M. Połom,"Multiaspect measurement analysis of breaking energy recovery", Energy Conversion and Management, vol. 127, 2016, pp. 35-42, doi:10.1016/j.enconman.2016.08.089

[8] R. Hrbáč, V. Kolář, T. Mlčák, "Distributed measurement system with GPS synchronisation and its use in electric traction". Elektronika ir elektrotechnika, Kaunas university of technology, Kaunas, Lithuania, vol. 21 (6), 2015, pp. 8-13, ISSN: 1392-1215.

[9] J. Skibicki, M. Bartłomiejczyk, "Analysis of measurement uncertainty for contact-less method used to measure the position of catenary contact wire, performed with the use of Monte Carlo method", Measurement, Volume: 97 Pages: 203-217 Published: FEB 2017, 10.1016/j.measurement.2016.11.008, ISSN: 0263-2241

[10] R. Hrbáč, T. Mlčák, V. Kolář, "Improving Power Quality with the Use of a New Method of Serial Active Power Filter (SAPF) Control", Elektronika ir Elektrotechnika, Vol. 23, NO. 1, pp: 15-20. ISSN: 1392-1215, 2017, DOI: 10.5755/j01.eie.23.1.17578

[11] B. Destraz, P. Barrade, and A. Rufer, Klohr, "Study and simulation of the energy balance of an urban transportation network," in Proc. European Conf. Power Electronics and Applications 2007 (EPE 2007), Aalborg, 2007.

Transmission Losses Spatial Analysis of the Supply System of Electrified Urban Transport Network

Mikołaj Bartłomiejczyk
Gdańsk University of Technology,
Faculty of Electrical and Control Engineering
G. Narutowicza 11/12, 80-233 Gdańsk, Poland
mikolaj.bartlomiejczyk@pg.gda.pl

Abstract—**The paper presents mathematical model TOPSIS which was applied for MCDA benchmark of trolleybus supply system. Moreover, paper presents the novel method of transmission losses analysis in electrified urban transport system. Research work was based on measurements realized in Gdynia trolleybus network.**

Keywords: Assigning indicators, transmission losses, trolleybuses, MCDA, TOPSIS, traction substation, energy saving

I. Introduction

Trolleybuses are trackless vehicles, thus what it entails is that a trolleybus traction network serves at the same time as the power supplying and the return one. It is composed of two parallel wires, as a standard hung at the height of 5,5 m over the street level. A trolleybus traction network is powered from traction substations placed in 2 – 5 km distances. In substations this voltage is decreased and straightened. Rated voltage of traction network power supply is 600 or 750 V. Analogically to the rail or tram traction the trolleybus traction network is divided into power supply sections. Each section is powered by a dual wire feeder from a traction substation. The tracks of the traction network for both travel directions are constantly connected with each other by means of diagonal equalizing links (so called equalizing bridges), located at every 300 – 500 metres.

A characteristic feature of a trolleybus power supply system is a considerable changeability of specific feeders loading. It is a consequence of an influence of traffic congestion on the running of trolleybuses, the effect of which loading of the trolleybus power supply system has a random and difficult to forecast character. In reference to the above statistical analysis methods constitute a perfect tool for the analysis of city traction power supply systems and allow for finding dependencies between numerous factors having impact on the power supply system operation [1-5].

In many situations, there is a need to perform the analysis of the supply system. In the case of urban traction power systems there are no clearly defined limits for the power supply system. Supply quality can be describe by several factors: average value of voltage, voltage drops or transmission losses. Each of these parameters expresses another aspect of the power supply. Moreover, due to random nature of traction load the state of energy system changes dynamically. For this reason, the power quality should be described by a combination of several parameters. The article presents the method of traction power supply system benchmark Multi Criteria Method Analysis. The several indicators of power supply quality were combined in one Composed Indicator by Topsis method. In addition, the article presents an innovative method of measuring the transmission losses in the urban traction power supply system, which is an important element of MCDA supply system analysis.

II. Measurement of the Transmission Losses

The spatial structure of the overhead line power supply system affects the energy transmission losses in the overhead line. Transmission losses are dependent on resistance of the conductor, or power cables and the overhead line. This in turn depends on their length. A significant source of energy dissipation in the supply systems are losses in long power cables, whose length can be up to several kilometers [6].

The most of scientific works publish the values of transmission losses based only on theoretical simulations . The measurement of transmission losses in tram or trolleybus supply system is difficult due to complicated structure of city transport network. This loses can be estimated basing on the energy balance of the supply system as a difference between energy taken by traction substation and the energy consumed by vehicles. Unfortunately, in many cities not every vehicles are equipment with energy measuring system and there is no precisely information about the energy consumption of the vehicles. Moreover, this method doesn't allow to eliminate energy of own needs if substations and energy consumed by other non-vehicles receivers which connected to traction cathenary (fg. ticket machines, steering and heating of rail switches).

The author of present paper propose the method based on measuring of the voltage and current of the vehicles, which doesn't requires knowledge of the supply system resistance. Also, there is no need to proceeded measurements in every vehicles.

Trolleybus current consumption of value I_{veh} increases the substation load current by the same amount I_{veh}. Likewise, the energy consumption of the trolleybus defined as:

$$E_{veh} = U_{veh} \cdot I_{veh} \tag{1}$$

where U_{veh} - supply voltage of vehicle on cathenary, increases the loading of the substation by value

$$\Delta E_{TS} = U_{TS} \cdot I_{veh} \tag{2}$$

where U_{TS} - output voltage of substation.

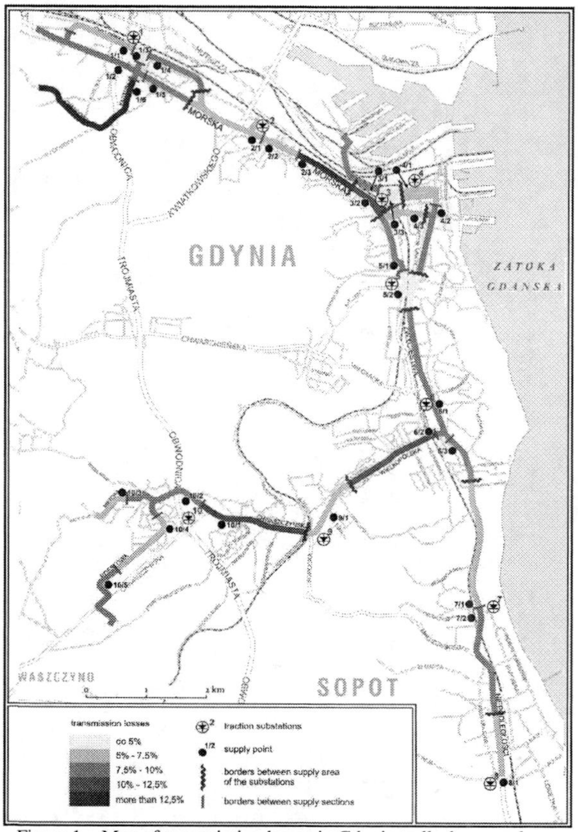

Figure 1. Map of transmission losses in Gdynia trolleybus supply system

Transmission losses E_{losses} in the supply system during provision of energy to the particular traction vehicle can be define as:

$$E_{loss} = \Delta E_{TS} - E_{veh} \tag{3}$$

This formula can be modify into the following form:

$$E_{loss} = I_{veh} \cdot (U_{TS} - U_{veh}) \tag{4}$$

The relative value e of the transmission losses in time period from t_1 to t_2 may be defined as:

$$e = U_{TS} \cdot \frac{\int_{t_1}^{t_2} I_{veh}(t)dt}{\int_{t_1}^{t_2} I_{veh}(t) \cdot U_{veh}(t)dt} - 1 \tag{5}$$

What is important, all relations are linear, so will be valid for any number of vehicles. For the purposes of the analysis, the values I_{veh} and U_{veh} are based on measurements of the energy consumption in trolleybuses, while the value U_{TS} on the basis of measurements conducted in traction substations. The expected value of the voltage of busbar in the DC switchgear was assumed as the value of substation voltage U_{TS}. The graphical representation of transmission losses calculation is presented on fig. 1.

III. Measurement of Voltage Drops

Voltage drops are the another, very important, indicator of state of supply system. They are caused by the current flow through the resistance of the power supply. The traction network load varies randomly, therefore the value of voltage drops in the traction network has a random nature and for quantitative evaluation it is necessary to use statistical indicators. For this purpose, as the main indicator of voltage drops was assumed the relative duration $t_{\text{u\_min}}$ of the minimum voltage level $U_{\min}$ in the traction network, which is expressed as the ratio of the time $T_{\text{u\_min}}$ when the voltage in the network was lower than the value $U_{\min}$ to the total time T, which can be written as:

$$t_{u\_\min} = \frac{T_{u\_\min}}{T} \tag{6}$$

In order to eliminate the effect of short voltage drops caused by passing through section insulators, the minimum voltage values of at least 3 seconds have been exceeded for calculations. Calculations were made on the basis of voltage registrations by on-board data acquisition systems in trolleybuses [7-8], the analysis was carried out in the separation into individual power sections. Two thresholds have been adopted $U_{\min}$:

- 550 V,
- 500 V.

Calculations were made for the month of November 2016. The result is shown in Table 1, where the relative voltage drops time in the promiles is given. The average values of voltage in the traction network in the given power section are also given there.

IV. TOPSIS Method

Four sub-indicators of have been identified: two relative times of voltage drops, average supply voltage and transmission losses, which provide information on the power status of the supply sections. In order to express this value in a more precise way, a composed indicator containing information on four sub-indices will be used. The composed indicator will be drawn up using the TOPSIS method.

A composite indicator (CI) is an mathematical aggregation of a set of individual indicators that measure multi-dimensional concept. There are m comparised alternatives, each alternative consist n sub-indicators x_{ij}. For each alternative is evaluated CI. CI is used for performance measurement, benchmarking, via providing an aggregated performance index in various fields such as Human Development Index, Road Safety Index [9-12].

The TOPSIS method is used to analyze a multi-criteria decision making problem with m alternatives with n criteria. In TOPSIS method, the best alternative should have the shortest Euclidean distance from the positive ideal solution (PIS) and the longest distance from the negative ideal solution (NIS). The PIS is a hypothetical solution which maximum values from database of all alternatives, the NIS is a hypothetical solution which minimum values from database of all alternatives. TOPSIS defines an index called relative closeness to the PIS and remoteness from the NIS [2, 13]. This index can be used as a CI of alternatives. The main procedure of the TOPSIS method is described in the following steps:

Step 1: Define a decision matrix

The decision matrix $\boldsymbol{D}$ of m x n dimension consists values of n sub-indicators for m alternatives.

Step 2: Normalize the decision matrix

The values of sub-indicators are normalized to scale 0-1. In case of "benefit type" indicators, what means a higher value is better, is used formula:

$$x'_{ij} = \frac{x_{ij} - \min_i\{x_{ij}\}}{\max_i\{x_{ij}\} - \min_i\{x_{ij}\}} \tag{7}$$

"Cost type" sub-indicators, what means the lower value is better, are normalized in the following way:

$$x'_{ij} = \frac{\max_i\{x_{ij}\} - x_{ij}}{\max_i\{x_{ij}\} - \min_i\{x_{ij}\}} \tag{8}$$

As a result is obtained the normalized decision matrix ***D'***.

Step 3: Compute the weighted normalized decision matrix

Elements of the normalized decision matrix ***D'*** are multiplied by weight vector ***W***, which consist n weight factors w. These factors express the relatively importance of criteria. The elements of weighted normalized decision matrix ***V*** are expressed as:

$$v_{ij} = w_j x'_{ij} \tag{9}$$

Step 4: Identify the PIS and NIS

The positive ideal solution ***A+*** and the negative ideal solution ***A-*** can be expressed as:

$$A^+ = \left(\max_i\{v_{i,1}\},\ldots,\max_i\{v_{i,n}\}\right) = \left(v_1^+,\ldots,v_n^+\right) \tag{10}$$

$$A^- = \left(\min_i\{v_{i,1}\},\ldots,\min_i\{v_{i,n}\}\right) = \left(v_1^-,\ldots,v_n^-\right) \tag{11}$$

Step 5: Calculate the distance to PIS and NIS

For each alternative *i* the Euclidean distance d_i^+ to the positive ideal solution and distance d_i^- to the negative ideal solution is define.

Step 6: Compute the relative closeness data to CI

Values d_j^+ and d_j^- are combined to relative closeness index C_i:

$$C_i = \frac{d_i^-}{d_i^+ - d_i^-} \tag{12}$$

The C_i is a composed indicator CI of alternative *i*.

The purpose of the composed index is to assess the power quality of a given section, so the partial loss and voltage drops are assumed to be destimulants, since the greater average value of voltage is the stimulant. The values of input parameters and composed indicator are presented in table 1. Relative time values of voltage drops are shown in promiles. Graphic representation of CI values is shown on fig. 2. Excepts values of voltage and transmission losses, there was added the average value of load current I_{av} of each section, which gives information about traffic density.

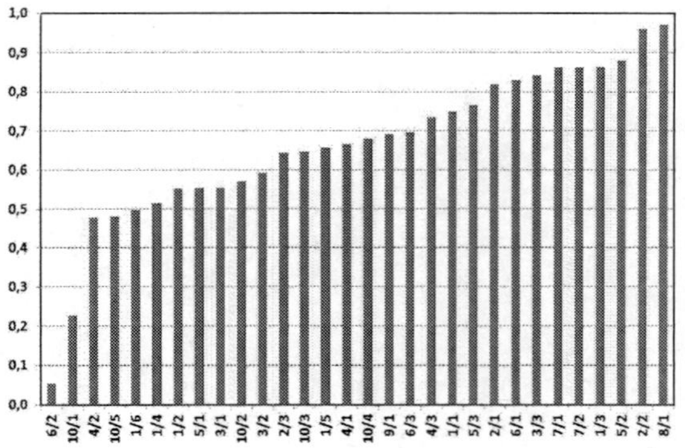

Figure 2. Values of CI for supply sections, the numbering corresponds to fig. 1.

Table 1. Sections parameters

Section No	Losses	V_{av}	t_{550V}	t_{500V}	I_{av}	CI
1/1	3,9	668	0,00	0,00	127	0,75
1/2	7,1	655	2,84	0,11	196	0,55
1/3	2,0	676	0,00	0,00	54	0,86
1/4	7,0	657	4,04	0,53	115	0,52
1/5	6,0	663	0,92	0,00	120	0,66
1/6	9,0	655	4,24	0,12	85	0,50
2/3	5,9	660	0,41	0,00	158	0,64
3/1	9,1	660	1,27	0,04	100	0,56
3/2	7,7	659	0,53	0,00	136	0,59
3/3	3,5	679	0,00	0,00	116	0,84
4/1	4,8	659	0,08	0,00	180	0,67
4/2	8,5	645	2,90	0,21	281	0,48
4/3	3,7	666	0,09	0,00	113	0,74
5/1	7,0	659	2,15	0,47	188	0,55
5/2	2,0	678	0,00	0,00	65	0,88
5/3	3,7	670	0,51	0,02	153	0,77
6/1	2,3	673	0,19	0,03	165	0,83
6/2	9,2	647	13,73	1,95	272	0,05
6/3	5,1	665	0,12	0,00	150	0,70
7/1	2,1	676	0,00	0,00	50	0,86
7/2	2,3	676	0,00	0,00	34	0,86
8/1	1,5	687	0,00	0,00	62	0,97
9/1	5,4	666	0,55	0,00	82	0,69
10/1	10,2	647	9,44	1,22	208	0,23
10/2	7,0	658	2,47	0,18	159	0,57
10/3	6,9	665	0,33	0,00	54	0,65
10/4	5,7	666	0,58	0,00	83	0,68
10/5	8,9	657	5,31	0,24	125	0,48

The correlation between input parameters is shown on table 2. There is need to underline the correlation between value of transmission losses and average voltage in cathenary. This relationship is also presented of fig. 3 and it allows to easily estimate the transmission losses by simply measurement of voltage of vehicle collectors. Also noticeable is strong negative correlation between average voltage and average current, which flows from Ohm's low.

Table 2. Correlation between input parameters

	Losses	V_{av}	t_{550V}	t_{500V}	I_{av}
Losses	1,00				
V_{av}	**-0,91**	1,00			
t_{550V}	0,68	-0,69	1,00		
t_{500V}	0,53	-0,57	**0,95**	1,00	
I_{av}	0,54	**-0,74**	0,59	0,59	1,00

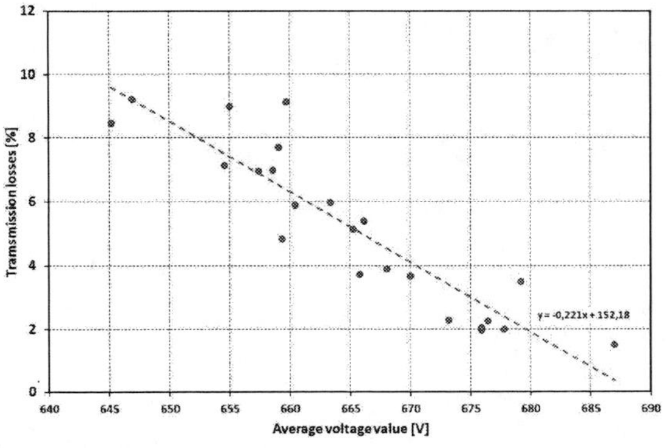

Figure 3. Dependence between average voltage in cathenary and transmission losses.

Modernization of power supply system of the trolleybus network in Gdynia, conducted in the years 2009–2012, significantly improved conditions of supplying the trolleybus transport. This is reflected in the energy efficiency of the power system. Average transmission losses in feeders and the overhead line amount to 8.5%, which, in the case of trolleybus system, can be considered a very good result. Losses in electrical overhead line are generally assumed at the level of 10%. This value is exceeded in Gdynia only in the case of several supply sections.

Supply sectors which requires improvement are sectors which CI lower than 0,5: 6/2, 10/1, 10/5, 4/2 and 1/6. These results confirm exploit experience: sectors 6/2 and 10/1 supply trolleybus line in hilly terrains and powered by long cable feeders. Sections 1/6 and 10/5 are long and often occurs voltage drops in terminus stations. 4/2 is the section with the biggest trolleybus traffic in the city.

V. CONCLUSIONS

Paper is based on author's analysis realized for PKT Gdynia, which was concentrated on defining the "road map" for develop of Gdynia's trolleybus system. It was realised by application of the presented novel method of benchmark of urban transport supply system. It can be applied for purposes of another energy audits or feasibility studies. Presented method can be used for defining of the weakness elements point of energy system, which requires the modernization.

An important element of paper is presented novel method allows measurement of transmission losses in traction supply system, which requires the registration of the vehicle current, supply voltage of vehicle, voltage of substation and GPS localisation of vehicle. These values can be easily measurable. The energy efficient of supply system in particular place can be set. The method can be used for diagnostic purposes of supply systems and energy efficiency assessment as a part of energetic audit purposes of public transport system.

The third, important element of the paper if simplified method of estimation of transmission losses in supply system, based on current voltage registrations (fig. 2). In combination with voltage drops analysis, it can be useful tool for simply way of supply system control.

ACKNOWLEDGMENT

This paper has been elaborated in the framework of the project ELIPTIC "Electrification of public transport in cities" co-financed by the European Union's Horizon 2020 research and innovation programme under the grant agreement No 636012

REFERENCES

[1] K. Hebel, M. Wołek, "Perception of modes of public transport compared to travel behaviour of urban inhabitants in light of marketing research", Scientific Journal of Silesian University of Technology. Series Transport, 2016, 92, 65-75. ISSN: 0209-3324. DOI: 10.20858/sjsutst.2016.92.7.

[2] S. Hamacek, M. Bartlomiejczyk, R. Hrbac, S. Misak, V. Styskala, "Energy recovery effectiveness in trolleybus transport", Electric Power Systems Research 112 (2014) 1–11.

[3] M. Bartłomiejczyk, M. Połom,"Multiaspect measurement analysis of breaking energy recovery", Energy Conversion and Management, Volume 127, 1 November 2016, Pages 35-42

[4] L. Jarzebowicz, K. Karwowski, W.J. Kulesza" Sensorless algorithm for sustaining controllability of IPMSM drive in electric vehicle after resolver fault", Control Engineering Practice 58 (2017), pp. 117-126. DOI: 10.1016/j.conengprac.2016.10.004.

[5] L. Jarzebowicz and S. Judek, "3D Machine Vision System for Inspection of Contact Strips in Railway Vehicle Current Collectors," in 2014 International Conference on Applied Electronics (AE), Pilsen, Czech Republic, pp. 139–144. http://dx.doi.org/10.1109/AE.2014.7011686.

[6] M. Bartłomiejczyk, M. Połom, "The impact of the overhead line's power supply system spatial differentiation on the energy consumption of trolleybus transport: planning and economic aspects", Transport, 2017 Volume 32(1): 1–12, doi:10.3846/16484142.2015.1101611, ISSN 1648-4142

[7] R. Hrbáč, V. Kolář, T. Mlčák: "Distributed measurement system with GPS synchronisation and its use in electric traction". Elektronika ir elektrotechnika, Kaunas university of technology, Kaunas, Lithuania. Volume: 21/2015, Issue: 6, pp. 8-13. ISSN: 1392-1215.

[8] J. Skibicki, M. Bartłomiejczyk, "Analysis of measurement uncertainty for contact-less method used to measure the position of catenary contact wire, performed with the use of Monte Carlo method", MEASUREMENT Volume: 97 Pages: 203-217 Published: FEB 2017, 10.1016/j.measurement.2016.11.008, ISSN: 0263-2241

[9] M. Gutten," Diagnostic of Distribution Transformers by SFRA Method", Przegląd Elektrotechniczny= Electrical Review, Vol. 83, 2007, No. 4, p. 144-146, ISSN 0033-2097

[10] T. Figlus, J. Gnap, T. Skrucany, "The Use of Denoising and Analysis of the Acoustic Signal Entropy in Diagnosing Engine Valve Clearance", Entropy, Volume: 18 Issue: 7 Article Number: 253 Published: JUL 2016

[11] R. Hrbáč, T. Mlčák, V. Kolář, "Improving Power Quality with the Use of a New Method of Serial Active Power Filter (SAPF) Control", Elektronika ir Elektrotechnika, Vol. 23, NO. 1, pp: 15-20. ISSN: 1392-1215, 2017, DOI: 10.5755/j01.eie.23.1.17578

[12] P. Kacor, M. Marek, J. Palecek, „Analysis Of Magnetic Field Distribution Inside Tram-Car", Proceedings Of The IV International Scientific Symposium On Electric Power Engineering - Elektroenergetika 2007, Pages: 128-129, Published: 2007

Mitigation of Interferences Between Electrical Regulated Drainage and Track Circuits

Vaclav Kolar, Roman Hrbac, Tomas Mlcak
Department of electrical engineering
VSB – Technical University of Ostrava
17. listopadu 15, 708 33, Czech Republic
vaclav.kolar@vsb.cz, roman.hrbac@vsb.cz, tomas.mlcak@vsb.cz

Abstract — **Electrical drainage is a device used to protect metal objects buried underground (usually pipelines or water supply) from electrochemical corrosion caused by DC stray currents. DC stray currents come from DC electric traction, such as electric railway or tramway. Electrical drainage is a conductive connection between the rails and underground device, so current does not flow from the device into the ground (causing corrosion), but it flows through this connection instead. The most advanced type of electrical drainage is regulated electrical drainage, which usually uses pulse width modulation (PWM). This modulation creates current harmonic components with different frequencies. In railway tracks there are often track circuits, which use a certain frequency. For secure operation, it is necessary to make sure that frequencies generated by the drainage do not interfere with track circuits. This paper deals with the mitigation of interferences between regulated electrical drainage and track circuits. It contains simulation results and also some measurement results acquired on a prototype installed on a tram track.**

Keywords: electric traction, track circuit, electromagnetic compatibility, harmonic mitigation, regulated electrical drainage, simulation, measurement

I. Introduction

Electric traction, which uses rails as conductors conducing electricity, e.g. electrified railways or trams, cause "stray currents" – currents flowing through the ground. This happens because rails are never perfectly isolated from the ground. From this point of view trolleybuses are more convenient than trains or trams, as they create no stray currents [1]. If there is some metal object in the ground (for example pipeline) stray current flows through it and contributes to its corrosion. The most critical area in terms of corrosion is the so-called "anode area" or "anodic area", i.e. the place where current leaves the device [2].

This area can be protected by electrical drainage. The oldest type of drainage is direct drainage - a simple conductive connection between the underground device and rails in the place where the anode area is found. In this case, current flows through the drainage and does not enter the soil and cause corrosion. However, since trains are moving and their current consumption is variable, the position of anode area is not stable, due to which direct drainage sometimes makes the situation even worse. Nowadays, direct drainage is not used anymore.

"Polarized electrical drainage" is a more modern solution. Here the connection between the rail and the underground device is ensured by a one-way element, e.g. a diode or combination of MOSFET transistors.

The latest type of electrical drainage is "regulated electrical drainage". It combines polarized drainage and current regulation. When current flowing through drainage is too high or the potential of the underground device is too negative. (When the potential of underground iron object is lower than -2V or -3V, hydrogen is released - this process, called electrolysis, leads to metal embitterment). Regulation is usually controlled by a proportional or proportional-integral controller. There are also other solutions, based on adaptive regulation and self learning [3].

Current regulation is usually done with the use of pulse width modulation (PWM), or by means of switching different resistors (that is an older solution). When PMW is used, the flowing current contains harmonic components of different frequencies.

If drainage is installed on a tramway track, harmonic components do not cause any problem, but when the drainage is installed on a railway track equipped with track circuits, it is necessary to make sure that frequencies generated by the drainage do not interfere with the frequency of the track circuit. Nowadays, most of the main railway tracks are equipped by track circuits, so regulated electrical drainage designed for the railway must be compatible with the track circuit.

The team of the authors of this paper is working on the development of a new prototype of electric regulated drainage. This drainage is designed primarily for tramways, but because of a future possibility of adapting it to the railway, the drainage is designed to meet the conditions of electromagnetic compatibility with track circuits.

So far, drainage has been tested in laboratory conditions and on a tramway track. Until the problem of electromagnetic compatibility between drainage and track circuits has been solved in a reliable way, drainage must not be tested on railway tracks. There is also a theoretical possibility of testing drainage on an electric railway track where there are no track circuits, but unfortunately our team do not have access to any such location. In addition, a large number of computer simulations of this problem have been done. Some results of previous measurements and simulations have been published in [2], [4].

The main difference between the connection of drainage to a tramway track and a railway track is that in the case of railway track the drainage is connected via an inductor with an inductance of 10 - 40 mH (see Fig. 1). The most adverse situation is when the inductance is low, therefore the smallest value 10 mH must be considered for simulations. These coils have usually a rated current of 100 A, so drainage control must ensure that current will not be higher. In simulations and testing measurements we usually work with currents approximately 100 A.

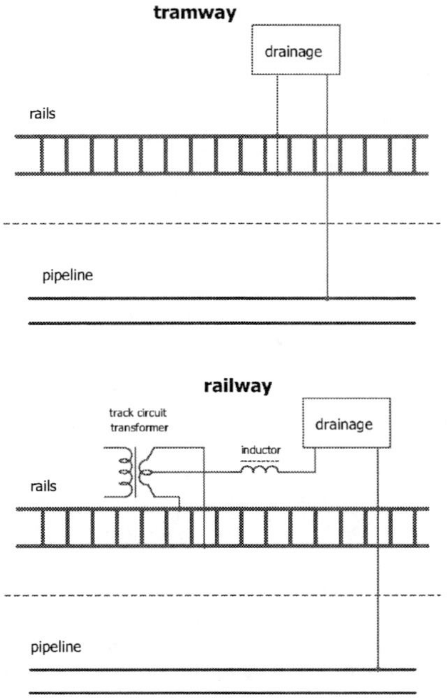

Figure 1. Connection of drainage on tramway and railway track

II. Track circuits – reserved frequencies

In many countries, including the Czech Republic and Slovakia, some of the following frequencies are used in most track circuits: 50 Hz (can be used only for DC tracks, oldest track circuits), 75 Hz and 275 Hz.

Requirements regarding track circuits are stated in the Czech standard ČSN 342613. Among others, there are defined protective frequency bands for the specific frequency of the track circuit. Protective frequency bands and security limits of current are shown in Table I. This standard states security current limits for analogue track circuit receivers (which are lower) and for digital track circuit receivers (which are higher). This means that if drainage meets the requirements for analogue receivers, it will also meet the requirements for digital receivers.

Table 1. Protective frequency bands and current limits for analogue track circuit receivers acording to Standard ČSN 342613

Protective frequency band (Hz)	Duration	
	permanent	< 0.52 s
44 – 54	0.26 A	0.50 A
68 - 80	0.11 A	0.50 A
262 - 280	0.13 A	0.50 A
For duration < 0.12 s, current values are not restricted.		

III. Regulated electrical drainage PWM switching diagram and disturbing frequencies

Different PWM switching frequencies were tested, in computer simulation and in real measurement. Several possible switching diagram sequences were also tested, as described in [4]. For the sake of convenience, 100 Hz PWM frequency was chosen, and also that 12 PWM periods form one control cycle (120 ms), after which PWM duty cycle may change. PWM duty cycle can wary in 17 steps, 0/16 (permanently opened circuit), 1/16, 2/16 ... 15/16, 16/16 (permanently closed circuit). Duty cycle change between two consecutive control cycles will always be one step at the most. This is given by the control algorithm in the microprocessor.

The result was that in steady state, current harmonic components are within limits according to Table 1. However, problems start when the PWM duty cycle is changing periodically, for example 16/16-15/16-16/16-15/16 or 15/16-14/16-15/16-14/16. In this case, current limits in Table 1. can be broken.

Fig. 2 shows a current waveform as a measurement result for PWM 15/16 and Fig. 3 for PWM varying between 15/16 and 16/16. These waveforms were measured on a tramway track.

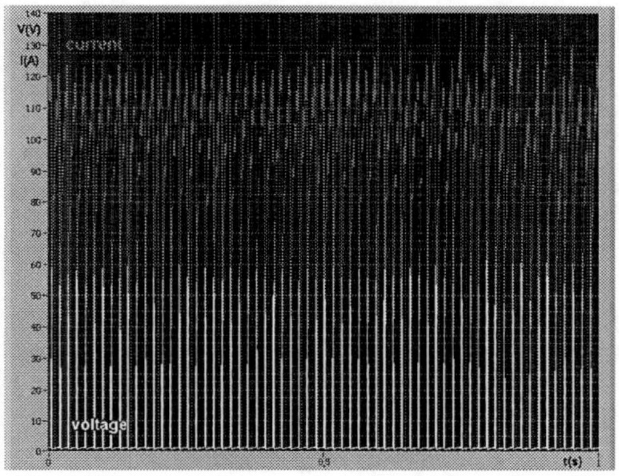

Figure 2. Current and voltage waveforms, PWM duty cycle 15/16

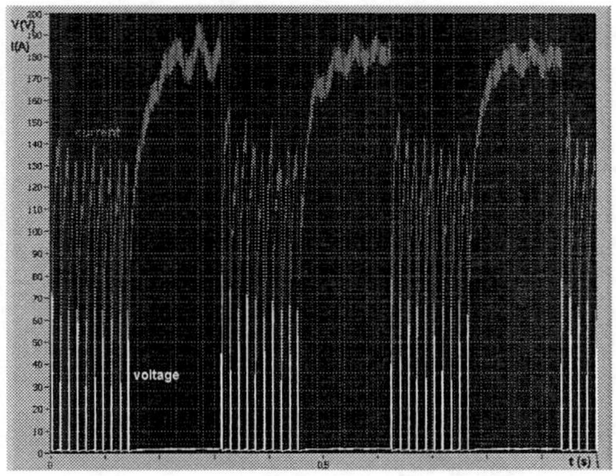

Figure 3. Current and voltage waveforms when PWM duty cycle is oscillating between 15/16 and 16/16 (permanently closed)

Because waveforms shown in Figs. 2 and 3. were measured on a tramway track (where there is no inductor), the current harmonic spectrum is different than in a railway circuit and it would be no use presenting its harmonic spectrum here. The figures are here as an example.

IV. Computer simulation of current harmonics

Computer simulations were done in the Microcap 9 software. The rail - crosstie - soil - piping circuit was created as a cascade of series-parallel combinations of resistances and inductances. Inductance and resistance values were estimated based on standards and experience from previous measurements and simulations [5]. The circuit diagram created in Microcap 9 can be found in Fig. 4.

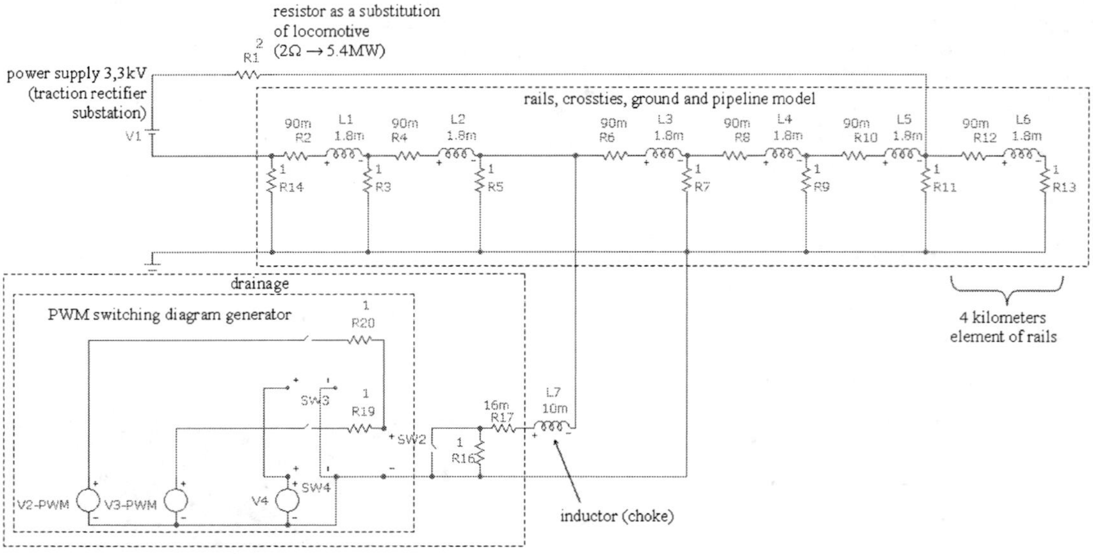

Figure 4. Model created in the Microcap simulation software

Many PWM switching diagrams were simulated. It was found out that the most problematic situation occurs when PWM duty cycle is periodically changing. For example 14/16-15/16-14/16-15/16-... . The current waveform and corresponding harmonic spectrum are presented in Figs. 5 and 6.

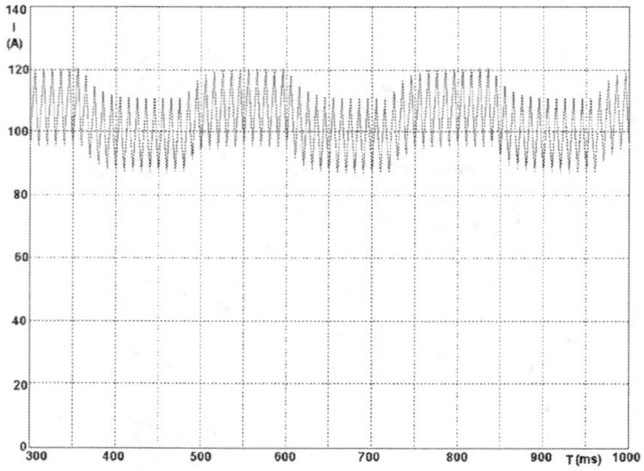

Figure 5. Current waveform for PWM duty cycle alternating betwen 14/16-15/16, f=100 Hz, I_{av} = 101 A

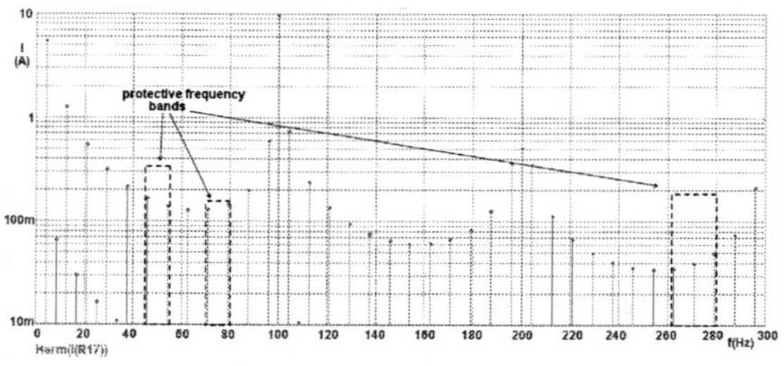

Figure 6. Harmonic spectrum for PWM duty cycle alternating between 14/16-15/16, f = 100 Hz, I_{av} = 102 A

Because each protective band contains 2 or 3 current harmonic components, it is necessary to calculate an aggregate RMS values. Then the Microcap software calculates the harmonic spectrum as amplitudes, so the obtained values must be converted to RMS values for comparison with limits from Tab 1. Calculation of aggregate RMS value is presented in (1).

$$I_{RMS} = \frac{1}{\sqrt{2}} \sqrt{\sum_{k=1}^{n} I_k^2} \quad (1)$$

Aggregate RMS values for the waveform in Fig. 5 are presented in Table 2.

Table 2. Simulated current harmonics values from Fig. 6.

Frequency band (Hz)	Simulation (A)	Standard limit (A)
44-54	0.153	0.26
68-80	0.138	0.11
262-280	0.051	0.13

From Table 2 it can be seen that the simulated value for the 68-80Hz band is higher than what is allowed by the standard. This problem needs to be solved.

A simple and effective method how to mitigate undesirable harmonic components is to use a passive filter. For the circuit in Fig. 4, it can be an RLC filter, where we connect RC combination in parallel to inductor L7. Capacitor value should be determined so that the resonant frequency is 75 Hz. Resistor value should be several Ohms, to suppress sharp current peaks. A capacity of 400 μF and resistance of 3.9 Ω were chosen. Filter schematic diagram is shown in Fig. 7.

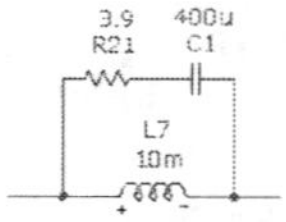

Figure 7. Passive filter for current harmonics mitigation

All possible PWM duty cycle changes were simulated with and without passive filter. Current flowing through drainage was always 101 A, which was achieved by adjusting the value of resistor R1.

Current waveform for the same situation as in Fig. 5, but with filter is shown in Fig. 8.

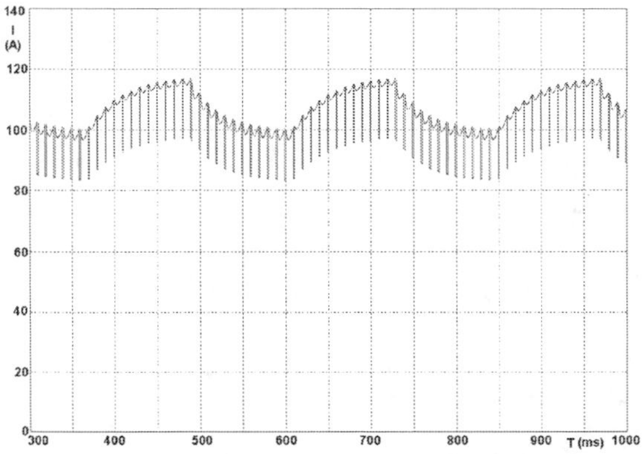

Figure 8. Current waveform for PWM duty cycle alternating between 14/16-15/16, f = 100 Hz, I_{av} = 101 A, with passive filter

All important simulation results are summarised in Table 3.

Table 3. Simulation results for circuit with and without passive filter.

Duty cycle change	**Aggregate RMS current (A)**					
	without filter			**with filter**		
	44-54 Hz	**68-80 Hz**	**262-280 Hz**	**44-54 Hz**	**68-80 Hz**	**262-280 Hz**
15/16-16/16	0.133	0.138	0.042	0.084	0.066	0.122
14/16-15/16	0.158	0.149	0.047	0.105	0.089	0.144
13/16-14/16	0.157	0.144	0.048	0.107	0.089	0.149
12/16-13/16	0.154	0.14	0.049	0.11	0.076	0.149
11/16-12/16	0.152	0.135	0.052	0.11	0.87	0.154
10/16-11/16	0.15	0.135	0.052	0.108	0.085	0.159
9/16-10/16	0.15	0.133	0.051	0.112	0.071	0.152
8/16-9/16	0.15	0.136	0.05	0.11	0.083	0.149
7/16-8/16	0.152	0.137	0.051	0.108	0.086	0.153
6/16-7/16	0.15	0.142	0.051	0.11	0.073	0.154
5/16-6/16	0.152	0.145	0.053	0.112	0.088	0.157
4/16-5/16	0.151	0.149	0.051	0.108	0.092	0.158
3/16-4/16	0.154	0.152	0.05	0.116	0.084	0.152
2/16-3/16	0.151	0.152	0.049	0.116	0.098	0.15
1/16-2/16	0.152	0.153	0.048	0.116	0.103	0.156
0/16-1/16	0.148	0.146	0.047	0.122	0.098	0.159
Limit from ČSN342613	0.26	0.11	0.13	0.26	0.11	0.13

As can be seen from Table 3, current RMS values in the 68-80 Hz frequency band without filter exceed the limit set by standard ČSN342616. Values in other frequency bands are in accordance with the standard. When designed passive filter is applied, current RMS values in frequency band 68-80 Hz decrease, so they are under the standard limit. This filter also mitigates current in the 44-54 Hz frequency band, but this is not necessary. On the contrary, current in the 262-280 Hz frequency band is increased by this filter.

Therefore, the designed filter can be used for harmonic mitigation, when there is a track circuits with a frequency of 75 Hz. If track circuit frequency is 50 Hz or 275 Hz, no filter is needed.

Filter design and mentioned simulation results were done for inductor value 10mH. If the inductance is different, values of the filter, especially capacitance, must be recalculated.

V. Conclusion

The authors have developed a prototype of regulated electrical drainage for tramway traction, in which current regulation is done by PWM. As there might be a possibility to use this drainage on railways, the authors have also investigated the compatibility of this prototype with track circuits. Current harmonics were simulated and compared with limits stated by Standard ČSN 342613. It was found out that for the 68-80 Hz frequency band, this type of drainage generates higher current harmonics than what is allowed by this standard. A simple passive filter was developed to solve this problem, and its efficiency was verified by simulation. This filter would be a good solution for track circuits with a frequency of 75 Hz. For other track circuit frequencies, it is not necessary to use this filter. Filter component values must always be designed based on the values of the drainage inductor used on the particular location. If the device is used on a railway track, this filter must first be tested in laboratory conditions and an evaluation must be made as to whether connecting other components to the drainage inductor (an RC circuit which forms the filter, together with the inductor) is acceptable in terms of the corresponding standards and regulations.

Acknowledgment

This paper is supported by the following projects: LO1404: Sustainable development of ENET Centre; CZ.1.05/2.1.00/19.0389 Development of the ENET centre research infrastructure; SP2017/159 Students Grant Competition and TACR TH01020426, Czech Republic.

References

[1] M. Bartłomiejczyk, V. Styskala, R. Hrbac, M. Połom. "Trolleybus with traction batteries for autonomous running," in *Proceedings of the 7th International Scientific Symposium on Electrical Power Engineering*, Elektroenergetika 2013, Technical University of Kosice, Slovakia; pp. 204-206, ISBN: 978-805531442-6

[2] V. Kolar, R. Hrbac, A. Neumann, T. Mlcak. "Simulation and measurement of selected electrical parameters for electric drainage," in *11th International Conference ELEKTRO 2016*, Strbske Pleso, Slovakia; 16-18 May 2016, pp. 320-324, ISBN: 978-146738698-2.

[3] M. Repka, R. Danel, Z. Neustupa. „Intelligent control of treatment technological processes in preparation plants using neural networks," in *SGEM2013 Conference Proceedings*, pp. 441-447, June 16-22, 2013, ISBN 978-954-91818-9-0 / ISSN 1314-2704.

[4] V. Kolar, R. Hrbac, T. Mlcak, J. Dudek. "Selecting a suitable controlling algorithm for a regulated electric drainage," in *18th International conference EPE 2017*, Dlouhe Strane, Czech Republic, 17-19 May, 2017.

[5] S. Zajaczek, P. Orsag, J. Ciganek, R. Hrbac. „Localization of current fields in the vicinity of a railway platform," in *Proceedings of the 8th International Scientific Symposium on Electrical Power Engineering*, Elektroenergetika 2015, Technical University of Kosice, pp. 101-104, ISBN: 978-805532187-5.

Proposal of WAMS Controlled TCSC for Improvement of Transfer Capacity in Power System

Zsolt Čonka, Michal Kolcun, Martin Vojtek, Michal Kosterec, Azamat Suleymenov
Department of electric power engineering
Technical University of Košice
Košice, Slovakia
zsolt.conka@tuke.sk, michal.kolcun@tuke.sk, martin.vojtek@tuke.sk, michal.kosterec@tuke.sk, suleymenovaz@gmail.com

Abstract — **Implementation of new technologies into existing power system can solve the problems associated with increasing power consumption. Increasing demand of electricity in combination with unplanned power flows from north part of interconnected power system to south creating serious problems for TSO. This paper investigates the use of WAMS systems for management of TCSC equipment to increase net transmission capacity in Central East Europe power system. Optimal control and allocation of TCSC is very important for TSO. Optimal allocation, effectiveness, and utilization of phasor measurement units (PMUs) for TCSC device which was designed for power flow control and increasing of net transfer capacity were investigated**

Keywords: TCSC, WAMS, transfer capacity, FACTS)

I. INTRODUCTION

An unresolved issue today is the ever-increasing number of unpredictable renewable energy sources in the northern part of the European power system and the lack of electricity production in the south part. We have gained new opportunities for electricity market participants by liberalization of electricity market. On the other hand, liberalization also causes significant problems for TSO operators.

As a result of an unplanned power flow the N-1 criterion is unfulfilled cause of overloading of transmission lines. These facts bring the operation of existing transmission systems closer to their physical limits.

One solution to this problem was to build new transmission power lines or efficiently use existing transmission capacity through dedicated systems such as flexible AC systems. For increasing of efficiency of AC transmission lines we can use FACTS devices. These special devices can control the flow of energy through transmission lines and help operators maintain a stable operation of power systems. Combined with sophisticated WAMS, this can be solved.

Any TSO transmission system refers to the need to improve the reliability and security of the electricity system. Transmission system operators must sometimes be guided by the operational limitations of power systems. In addition, in some cases, there may be large black out areas that have a negative impact on society as a whole. For the sake of fact, monitoring of energy systems is carried out through SCADA systems and EMS (Energy Management Systems) that do not work in real time and therefore the results are not usable for fast transient processes in the system. Additionally, both of these control schemes are not synchronized over time. We now have the opportunity to create a modern WAMS-based control system for transmission system operators.

The first GPS-based PMU has been found 1988 Synchrophasor technology has been developed for several years. During this period, many new and promising concepts, such as extensive measurement and monitoring (WAMS), have been designed and implemented. This is a great potential for upgrading power control management, surveillance and protection of power systems.

In addition to real-time surface surveillance, visualization and subsequent off-line analysis of these subsystems, they are also used for network management (for example, using the entire transmission line capacity), system stability monitoring, swing and early warning. The use of WAMS systems can be found in distribution networks, such as the management of distributed energy sources. Errors measuring the synchronous tendency to reduce future calculations and control systems, resulting in improved control of the network dispatcher.

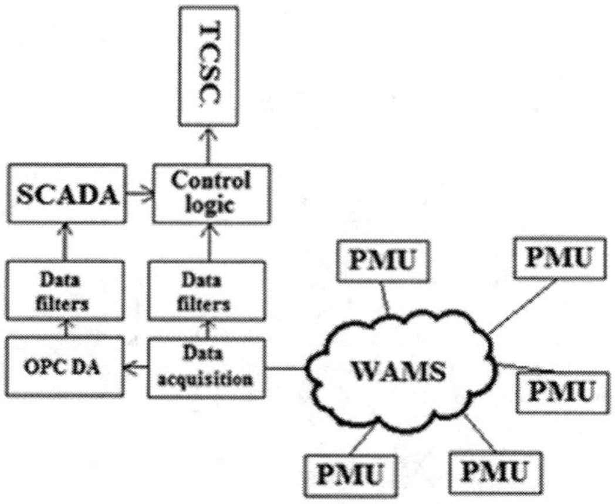

Figure 1. TCSC control scheme with WAMS

The use of the PMU to obtain data to regulate FACTS can bring new opportunities for further development of network management. Measurement of phasors in different parts of the energy system can be used to optimize energy flow, minimize energy loss, or maximize transfer capacity between different TSOs. An example of how to use WAMS to control TCSC is shown in Figure 2.

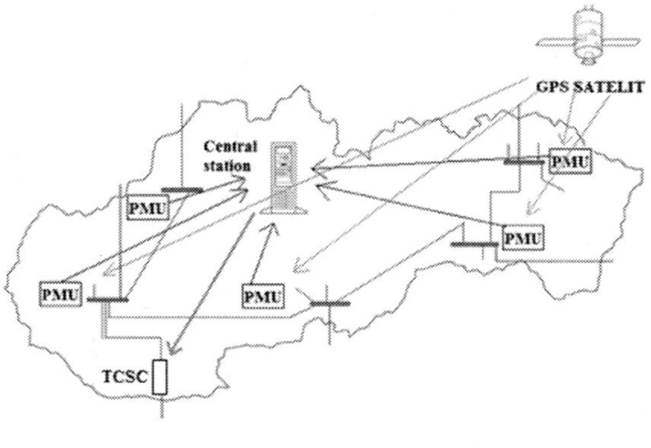

Figure 2. Slovak power system with PMUs and TCSC

II. THYRISTOR CONTROLLED SERIES CONDENSER (TCSC)

TCSC is one of the best FACTS devices for controlling power flow in the power system. TCSC consists of a capacitor which is parallel connected to a thyristor regulated reactor. This layout allows you to constantly change the impedance of the power lines and continuously monitor the frequency of the capacity wide. For that proper operation control of TCSC is very important.

TCSC reactance Control is a very complex and dynamic process. The efficacy and effect of TCSC depend on the accuracy of the TCSC reactance calculation currently. In the event of an incorrect calculation of the reactance, there may also be an opposite effect to what is provided by the TCSC.

It may also have an adverse impact on TCSC's capabilities. TCSC consists of 3 parts and these are The Fig.1 shows the main diagram of TCSC. It consists of 3 parts: capacitor C, parallel connected reactor L and the thyristor valves VT1 and VT2 [5].

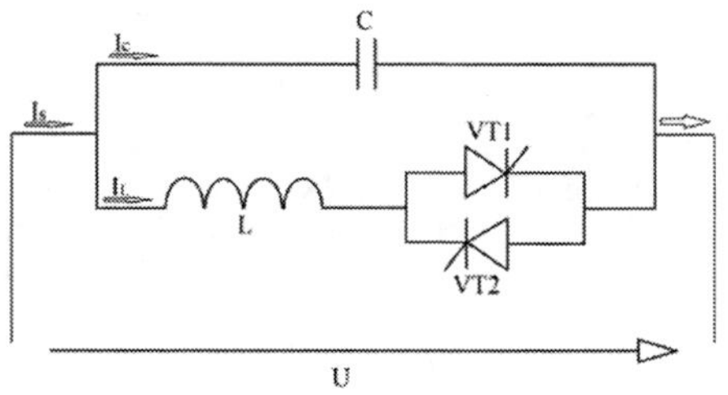

Figure 3. TCSC connection scheme

$$i_c = C \cdot \frac{dv}{dt} \tag{1}$$

$$i_L = L \cdot \frac{di_L}{dt} \tag{2}$$

$$i_S = i_C + i_L \tag{3}$$

where

I_C and I_L - is the instantaneous capacitor and reactor currents

i_S - is the instantaneous current of controlled line

U - is the voltage of TCSC.

TCSC can be controlled by adjusting the opening angle of thiristors (α) thyristors, which alters the frequency of the capacitor. Connection shows the relationship between the thyristor opening angle (α) and a reagent throughout the TCSC - XTCSC (α)

$$X_{TCSC}(\alpha) = X_C - \frac{X_C^2}{(X_C - X_P)} \frac{\sigma + \sin\sigma}{\pi} + \frac{4 \cdot X_C^2}{(X_C - X_P)} \frac{\cos^2(\frac{\sigma}{2})}{(k^2-1)} \frac{\left(k \cdot tg\left(\frac{k \cdot \sigma}{2}\right) - \tan\left(\frac{\sigma}{2}\right)\right)}{\pi} \tag{4}$$

where

X_C - is the capacitor reactance

X_P - is the inductive reactance of the reactor connected in parallel to the condenser.

$\sigma = 2(\pi - \alpha)$ - opening angle of TCSC controller

$k = \sqrt{\frac{XC}{XP}}$ - compensation ratio.

The unique property of XTCSC (α) is that we can continuously change the TCSC reactance. The TCSC can operate in capacitive or inductive mode, but switching from one mode to another should be avoided by resonance. The TCSC variable is modeled as a capacitor, limited by the scope of the α-angle control.

The Control range of TCSC is:

$$XTCSC\ (min) \le XTCSC\ (\alpha) \le XTCSC\ (max), \tag{5}$$

where

XTCSC (min) = XTCSC (180°) – The thyristor is in a permeable state

XTCSC (max) = XTCSC (αmin), – thyristor in trailing state [5].

III. CALCULATION OF TRANSFER CAPACITY WITH WAMS CONTROLLED TCSC

Prior to determining the total transmission capacity, it is necessary to model the effects of the current transmission between the two interconnected power systems and it simulation. TTC is the maximum available for power exchange, which can be transmitted reliably and it is consistent with the system stability criteria between the two interconnected power grids. This is usually done by simulating load flows within and between two systems. Starting with the expected network configurations, cross-border exchanges, consumption scenarios and electricity generation, the power generated between these two systems is shifting to creating further cross-border flows. This is achieved by increasing the output in the X system and reducing production in the Y system by the same steps. Consumption loads in both systems remain unchanged. Generation shifts stop when security issues occur in X, Y, or interconnection lines [3], [4], [7], [9].

The total transmission capacity obtained in this case must be evident that it expects that the maximum power flow that can be transmitted by connection of the two systems that do not lead to power constraints in both systems if the future network conditions.

TRM cover flow forecast uncertainty of proposed power flow due to imperfect information on market participants and unexpected events in real time. Information of market players in the lack of transmission capacity there is also uncertainty about some parameters of the electricity system as well as uncertainty discharge curves due to unexpected events in real time, which is always possible. In this study, the value of TRM is expected to be 200 MW.

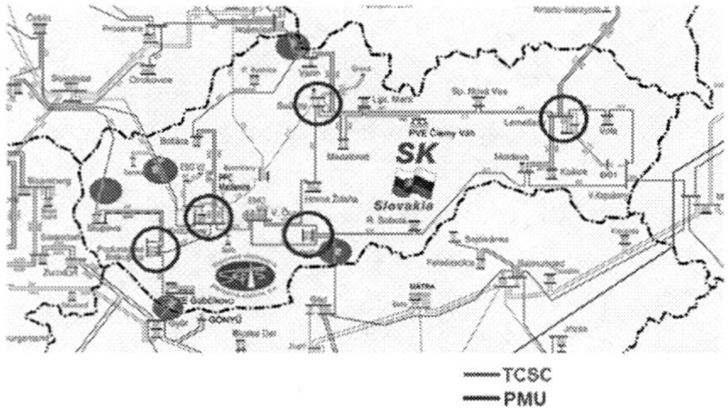

Figure 4. TCSC and PMU placements in Slovak power system

The calculation of the maximum transmission capacity between Slovakia - Hungary and Slovakia - Czech Republic was secured by a simulation program. In the simulation program, the Central Eastern European energy system was built with all branches and generators. The transmission capacity calculation was performed by maintaining the N-1 safety criterion in the entire energy system. To maintain the maximum energy flow in the cross-section between Slovakia and Hungary, electricity production in the southern part of the entire energy system was reduced and increased in the northern part.

There were 4 variants of cross-border transaction between Slovakia - Czech Republic and 3 variants between Slovakia - Hungary. All variant were defined below.

4.1 Variant SK-HU I.

The transmission capacities of the SK - HU cross - border connection for the default grid without TCSC installed on the power system are shown in Table 1. The limit element to maintain the N - 1 criterion was the line V 448 (V Gabčíkovo - Györ) when the line V449 (Levice - Göd) was switched off.

Table 1. Transfer capacity for variant SK-HU I.

Branch	Line loads [MW]	TTC [MW]	NTC [MW]	Limit element (N-1)
448 (Gabčíkovo-Györ)	1097,2	1948,4	1748,4	448 (449)
449 (Levice-Göd)	851,2			

4.2 Variant SK-HU II.

Table 2 shows the SK - HU cross - border transmission capacity for the pre - selected grid with the TCSC installed in line series with V 449 (Levice - Göd). To maintain the N-1 criterion, the limit element was the line V 490 (EMO - Levice), when there was disconnected line V491 (EMO - Levice).

Table 2. Transfer capacity for variant SK-HU II.

Branch	Line loads [MW]	TTC [MW]	NTC [MW	Limit element (N-1)
448 (Gabčíkovo-Györ)	1075,8	2372,8	2172,8	490 (491)
449 (Levice-Göd)	1296,9			

4.3 Variant SK-HU III.

Table 3 shows the SK - HU cross - border transmission capacity for the pre - selected grid with the TCSC installed in line series with V 448 (VE Gabčíkovo - Györ) To maintain the N-1 criterion, the limit element was the line V 440 (Lemešany - Mukačevo), when there was disconnected line V449 (Levice - Göd).

Table 3. Transfer capacity for variant SK-HU III.

Branch	Line loads [MW]	TTC [MW]	NTC [MW]	Limit element (N-1)
448 (Gabčíkovo-Györ)	1312,8	2509	2309	440 (449)
449 (Levice-Göd)	1194,2			

4.4 Variant SK-CZ I.

The transmission capacities of the SK – CZ cross - border connection for the default grid without TCSC installed on the power system are shown in Table 4. The limit element to maintain the N - 1 criterion was the line V 270 (P. Bystrica - Lískovec) when the line V404 (Varín - Nošovice) was switched off.

Table 4. Transfer capacity for variant SK-CZ I.

Branch	Line loads [MW]	TTC [MW]	NTC [MW]	Limit element (N-1)
404 (Nošovice – Varín)	-889	2556,7	2356,7	270 (404)
424 (Sokolnice – Križovany)	-532,2			
497 (Sokolnice – Stupava)	-921,1			
280 (Sokolnice – Senica)	-101,2			
270 (Liskovec – P. Bystrica)	-113,4			

4.5 Variant SK-CZ II.

Table 5 shows the SK - CZ cross - border transmission capacity for the pre - selected grid with the TCSC installed in line series with V 404 (Varín - Nošovice) To maintain the N-1 criterion, the limit element was the line V 280 (Senica - Sokolnice), when there was disconnected line V497 (Stupava - Sokolnice).

Table 5. Transfer capacity for variant SK-CZ II.

Branch	**Line loads [MW]**	**TTC [MW]**	**NTC [MW]**	**Limit element (N-1)**
404 (Nošovice – Varín)	-1345,2	2682,5	2482,5	280 (497)
424 (Sokolnice – Križovany)	-336,9			
497 (Sokolnice – Stupava)	-661,3			
280 (Sokolnice – Senica)	-242			
270 (Liskovec – P. Bystrica)	-97,1			

4.6 Variant SK-CZ III.

Table 6 shows the SK - CZ cross - border transmission capacity for the pre - selected grid with the TCSC installed in line series with V 424 (Križovany - Sokolnice). To maintain the N-1 criterion, the limit element was the line V 270 (P. Bystrica - Lískovec), when there was disconnected line V404 (Varín - Nošovice).

Table 6. Transfer capacity for variant SK-CZ III.

Branch	**Line loads [MW]**	**TTC [MW]**	**NTC [MW]**	**Limit element (N-1)**
404 (Nošovice – Varín)	-586,3	2740	2540	270 (404)
424 (Sokolnice – Križovany)	-1338,3			
497 (Sokolnice – Stupava)	-492,1			
280 (Sokolnice – Senica)	-172,2			
270 (Liskovec – P. Bystrica)	-151,1			

4.7 Variant SK-CZ IV.

Table 7 shows the SK - CZ cross - border transmission capacity for the pre - selected grid with the TCSC installed in line series with V 497 (Sokolnice – Stupava). To maintain the N-1 criterion, the limit element was the line V 270 (P. Bystrica - Lískovec), when there was disconnected line V404 (Varín - Nošovice).

Table 7. Transfer capacity for variant SK-CZ IV.

Branch	**Line loads [MW]**	**TTC [MW]**	**NTC [MW**	**Limit element (N-1)**
404 (Nošovice – Varín)	-682	2741,5	2541,5	270 (404)
424 (Sokolnice – Križovany)	-352,2			
497 (Sokolnice – Stupava)	-1336			
280 (Sokolnice – Senica)	-217			
270 (Liskovec – P. Bystrica)	-155,2			

IV. Economic aspects

On the TCSC final price has the greatest impact its nominal power [MVA]. These TCSC characteristics, especially for large transmission capabilities, are important for an economic solution. Investing in TCSCs with the same experience, reaching approximately 400% of the Phase Shift Trasformers price, may be considered with a value of 20,000 for 1 MVA installed PST capacity. For the assumption, the maximum current limit of 2000A corresponds to approx. 1400 MW of active power transmission. Installing TCSC requires the construction of specialized fields in switchboards. The TCSC field must be equipped with a comparison with standard sockets (eg For a line or transformer) - a measure that is protected and especially when the full field device approaches (Unlike conventional fields, the TCSC array is equipped with 2 or 3 A- insulators). The cost of one average field in the 400 kV distribution system is approximately EUR 3.0 million. The cost per TCSC field is approximately € 7.5 million.

Conclusion

The results of the calculation show that the best place to install TCSC on the border line between Slovakia and Hungary on line V 448 (VE Gabčíkovo - Győr). The TTC value with TCSC installed in series with line V 448 (VE Gabčíkovo - Gyðr) is 2509 MW. It was 136.2 MW larger than with TCSC, installed on line 449 (Levice - G□d). The best place to implement TCSC at the border between Slovakia and the Czech Republic was on line V 497 (Stupava - Sokolnice). The TTC value with TCSC installed on line V 497 was 2741, 9 MW. TTC with TCSC installed on line 424 (Krožovany - Sokolnice) was 2741,5, which was smaller by about 1,5 MW and the total transport capacity with TCSC on line 404 (Varín - Nošovice) was 2682,5 MW. It was smaller by 59,4 MW against TTC with TCSC on line 497.

ACKNOWLEDGMENT

This work was supported by grant award FEI no. FEI-2015-28 Proposal of the possibility of implementation of Wide Area Monitoring (WAM) systems into protection system of the electric power system

REFERENCES

[1] I. Dobson, S. Green, R. Rajaraman, Ch. L. DeMarco, F. L. Alvarado, R. Zimmerman.: Electric Power Transfer Capability: Concepts, Applications, Sensitivity, Uncertainty, Power Systems Engineering Research Center November 2001 online [12.05.2015] <http://www.pserc.cornell.edu/tcc/tutorial/TCC_Tutorial.pdf>

[2] Calculation methods online [12.05.2015] <http://www.elia.be/en/products-and-services/cross-border-mechanisms/transmission-capacity-at-borders/calculation-methods>

[3] Ministry of Economy of Slovak Republic: Návrh Energetickej politiky Slovenskej republiky, May 2014 Online: cit.[19. feb. 2015] <http://www.rokovania.sk/File.aspx/ViewDocumentHtml/Mater-Dokum-165253?prefixFile=m_>

[4] FAQ CEPS, a.s. online [19. Feb. 2015] <http://www.ceps.cz/ENG/Media/Pages/FAQ.aspx>

[5] Definitions of Transfer capacities in liberalised Electricity Markets online [19. Feb. 2015] <https://www.entsoe.eu/fileadmin/user_upload/_library/ntc/entsoe_transferCapacityDefinitions.pdf>

[6] C. Grande-Moran: Phase-shifting Transformer Modeling in PSS®E online [12.05.2015] <http://ewh.ieee.org/soc/pes/newyork/NewSite/PDFs/Protection%20Related/Phase_Shifting_Transformer_Modeling.pdf>

[7] ENTSO-E, Network Code on Capacity Allocation and Congestion Management, September 2012.

[8] ENTSO-E, Network Code on Forward Capacity Allocation, October 2013.

[9] R. Jakubčák, Ľ. Beňa, M. Kmec: Possibilities of Using Facts Devices In Power System. In: Acta Electrotechnica et Informatica. Roč. 13, č. 3 (2013), s. 8-11. - ISSN 1335-8243

[10] S. Ilenin, L. Varga, "*Swing Up of Conductor Depending on Icing Area, Weight of Ice and Span Size*" Elektroenergetika 2015. - Košice : TU, 2015 S. 121-124. - ISBN 978-80-553-2187-5

[11] M. Pavlík, I. Kolcunová, B. Dolník, J. Kurimský, A. Mészáros, D. Medveď, M. Kolcun, J. Zbojovský, "*Measuring of Dependence of Shielding Effectiveness of Wet Materials on The Frequency of Electromagnetic Field in the High Frequency Range*", Acta Electrotechnica et Informatica. Vol. 13, No. 3 (2013), pp. 12-16. ISSN 1335-8243.

[12] A.G. Kasembe, Z. Muller, J. Svec, J. Tlusty, "*Synchronous phasors monitoring system application possibilities*", 2012 IEEE 27th Convention of Electrical & Electronics Engineers in Israel (IEEEI), p. 1 - 3, DOI: 10.1109/EEEI.2012.6376936, ISBN: 978-1-4673-4682-5.

[13] M. Kornatka "The weighted kernel density estimation methods for analysing reliability of electricity supply", 17th International Scientific Conference on Electric Power Engineering (EPE), Prague, pp. 2-5, doi: 10.1109/EPE.2016.7521729 2016.

[14] M. Kornatka, „Some methods of improving reliability medium voltage overhead lines" Rynek Energii, Issue 1 pp. 50-56, 2010.

[15] Program rozvoja SEPS, a.s., na roky 2011 – 2020. [online]. [cit. 08.08.2011]. Available on internet: http://www.sepsas.sk/seps/Dokumenty/PR2020.pdf

[16] J. Zbojovský, A. Meszáros, D. Medveď, *"Simulation of electromagnetic fields propagation through the various materials"* - 2014. In: EPE 2014 : proccedings of the 2014 15th International Scientific Conference on Electric Power Engineering : Brno, 12. - 14.5.2014. - Brno : University of Technology, 2014 P. 257-260. - ISBN 978-1-4799-3806-3

[17] FAQ CEPS, a.s. online [19. Feb. 2015] http://www.ceps.cz/ENG/Media/Pages/FAQ.aspx

[18] Z. Čonka, M. Kolcun: Impact of TCSC on the Transient Stability In: Acta Electrotechnica et Informatica. Roč. 13, č. 2 (2013), s. 50-54. - ISSN 1335-8243 Available on internet :www.versita.com/aei. (2013)

Estimation of Overhead Conductor Temperature Using Artificial Neural Networks

Klemen Deželak, Jože Pihler
Faculty of Electrical Engineering and Computer Science, Power Laboratory
University of Maribor
2000 Maribor, Slovenia
klemen.dezelak@um.si, joze.pihler@um.si

Abstract — **In paper the estimation of overhead conductor temperature using artificial neural networks is dealt. Artificial neural networks, seen as systems of mutually connected neurons that exchange information among each other, are mainly used for evaluation of functions that are based on a large number of input data. In that manner they could be applied inside the dynamic thermal ratings of overhead lines. Aforementioned could represent a promising approach to increase the transmission capacity by considering the weather dependent thermal ratings of overhead power lines.**

Keywords: overhead line; conductor; temperature; artificial neural network;

I. Introduction

The primary function of overhead lines is transmission of electricity. The voltage is usually constant; therefore, the amount of transmitted energy depends on the type of load, i.e. on current, as well as on the environmental impacts [1] - [5]. When talking about environmental impact on a conductor, we have to bear in mind ambient temperature, wind and solar radiation [6], [7]. Heating of a conductor is also influenced significantly by the current amplitude and duration. Heating of the conductor causes its expansion, which results in an increased sag that my lead to the threatening of safety distance and safety height of the conductor. Increased temperatures also cause a reduction of a conductor's lifespan, which may lead to the destruction, or even tearing of the conductor. In the Slovenian electrical power system the Al/Je 490/65 conductors are used for electricity transmission on a 400 kV grid, therefore, our focus will be on this type of conductors [8].

This paper aimed to use the artificial neural network (ANN) to estimate the conductor temperature of an overhead power line. With using both, the measured and calculated data, the accuracy of the artificial neural network is evaluated. In that manner, to train the artificial neural network, ambient temperature, solar radiation and wind speed values are inputs, and the conductor temperature is an output. So, the artificial neural network is used to estimate the conductor temperature, while the feed forward artificial neural network is trained by using a Levenberg-Marquardt algorithm. The input learning signal (pattern) of the artificial neural network in the learning process is the value of the aforementioned variables. According to the conductor temperature, the output learning signal of the presented solution (target) is set to the value zero or one [9], [10].

II. Steady and Transient States Heat Balance

The heat balance equation in the stationary condition is described by the equation (1), where P_j is the power of heating due to joule losses in the conductor, P_s is the power received by solar radiation, P_c is the power emitted by convection and P_r power emitted by radiation [3], [6].

The stationary state of the conductor's temperature is, in practice, impossible to be achieved, since the currents are not constant and are changing constantly, as well as other influential factors. The stationary condition can be understood as the state where transient phenomena ended. The basic equation (2) defines the current conductor's temperature that is denoted in a differential form, where m is the conductor's mass, c_p the specific heat of the conductor and T_s the conductor's temperature [2] - [4].

The equation (1) differs from equation (2) only in the added differential term that represents the power that can be compensated by the conductor due to its thermal inertia.

$$P_j + P_s = P_c + P_r \quad (1)$$

$$P_j + P_s = P_c + P_r + mc_p \frac{dT_s}{dt} \quad (2)$$

A. Joule Heating Calculation

The power of heating due to joule losses in the conductor can be determined using the equation (3), where I is current, R_{dc} ohmic resistance per unit of length, k so called skin effect factor, α_{20} temperature coefficient at 20°C and T_s the conductor temperature.

$$P_j = I^2 k R_{dc} \left(1+\alpha_{20}\left(T_s - 20\right)\right) \quad (3)$$

B. Solar Heating Calculation

Heating of the conductor due to the solar radiation P_s can be described using (4), where α is the absorption factor of the conductor, H is the power of solar radiation at the Earth's surface and d is the conductor diameter.

$$P_s = \alpha H d \quad (4)$$

C. Convective Cooling Calculation

It is well-known that the conductor also dissipates heat by means of movement of the surrounding air, which improves conductor cooling in the case of strong wind – forced convection. On the other hand, the conductor can also be cooled by natural convection. The wind direction and speed v_w determine whether the conductor is cooled by forced or natural convection. The forced convection cooling P_c is calculated using (5) - (6), where λ_f is the thermal conductivity of air, T_s the conductor temperature, T_{am} the ambient temperature and Nu the Nusselt number [6].

$$P_c = f\left(v_w\right) \tag{5}$$

$$P_c = \pi\lambda_f\left(T_s - T_{am}\right)Nu \tag{6}$$

D. Radiative Cooling Calculation

The heat dissipation due to radiation P_r is defined by the Stefan-Boltzmann law, which is described by (7), where d is the conductor diameter, σ_{sb} the Stefan-Boltzmann constant amounting to 5,67·10-8 [W/(m^{2}K^4)], T_s the absolute conductor temperature, T_{am} absolute ambient temperature and ε emission factor of the conductor.

$$P_r = \pi d\varepsilon\sigma_{sb}\left[\left(T_s + 273\right)^4 - \left(T_{am} + 273\right)^4\right] \tag{7}$$

III. Artificial Neural Networks

The artificial neural networks are mathematical models based on the imitation of biological neural networks. A biological neuron is composed of three basic elements: the body of the cell, axon and dendrites. The role of the axon is to transfer impulses from the body of the cell, while the dendrites receive signals from other cells. Mathematical models of an artificial neuron were presented for the first time already in 1943. They are based on the assumption that the inputs are weighted and that the composed signal depends on the product of the input values of the weights. The composed signal is then led to a special switching (threshold) function, the result of which is at the same time the output of the artificial neuron [9], [10].

ANN are used for evaluation of functions that are based on a large number of input data. Artificial neuron networks can be seen as systems of mutually connected neurons that exchange information among each other. Connections have numerical weights that can be changed based on experiences. Due to this fact, the neural networks are adaptable and able to learn. If we, for example, have a system with three layers, the first layer contains input neurons that send data through synapses to the hidden layer of neurons, which send data to the third layer of output neurons. In this process, the synapses store the parameters of weights that operate the computation of data. Artificial neural networks are usually defined by three types of parameters: the pattern of connections between layers, the learning process that refreshes weights between the connections, and the activation function that transforms the input weights of a neuron to the output activation. As the other methods of machine learning, the methods of neural networks are used for solving of various problems that are difficult to be solved by the usual rules of logical programming [10] - [12].

The present paper will be focused on oriented ANN, while the following properties are considered:

- Connections between the neurons at the same level are not allowed;

- Feed-back connections to neurons are not allowed; and

- Jumping over layers is not allowed.

Additionally, the neural networks are distinguished according to the following characteristics:

- Topology type,

- Orientation of connections,

- Learning method,

- Type of input variables, and

- Type of activation functions.

The basic of the artificial neural networks' operation is learning. The learning process can be divided into two groups, the learning set and the testing set. The learning is performed on the learning set. It is being learned as long as it yields the desired output values (results) with regard to the input data. In the next step, the learned network is tested on the input data of the testing set.

IV. Results

In real conditions the conductor temperature varies all the time, which means that the differential term can be added to the heat equation. This enables determination of the time course of heating. In the computations, the following assumptions were used: an average current loading of 0-850 A, a wind speed 0-4 m/s, and the solar radiation 0-1100 W/m^2. The emission factor and the absorption factor had the same value of 0,92.

As already mentioned, the ANN accumulates knowledge during the learning process. Figs. 1 - 7 show signals, which are involved in the learning process. The input learning signal (pattern) of the ANN in the learning process consists of the temperature values for 35 hours rate (Fig. 1), dependent on solar radiation (Fig. 2), wind speed (Fig. 3), heating due to joule losses (Fig. 4) and heat dissipation due to radiation (Fig. 5), as well. All of the aforementioned values were normalized during the learning process.

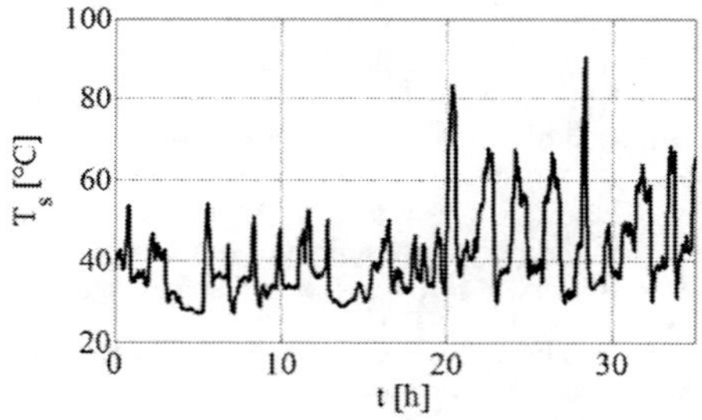

Figure 1. The temperature values for 35 hours rate.

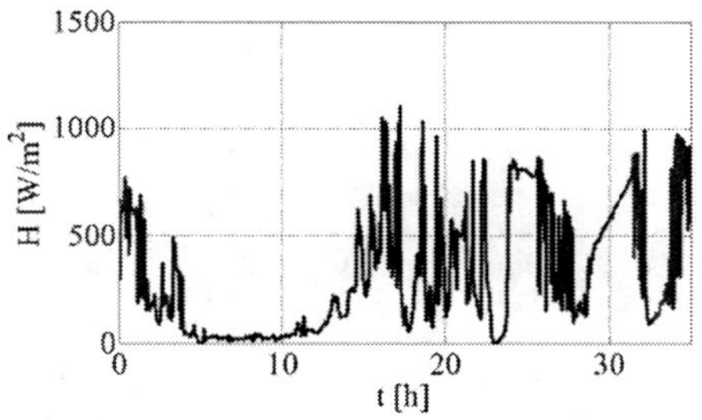

Figure 2. The solar radiation values for 35 hours rate.

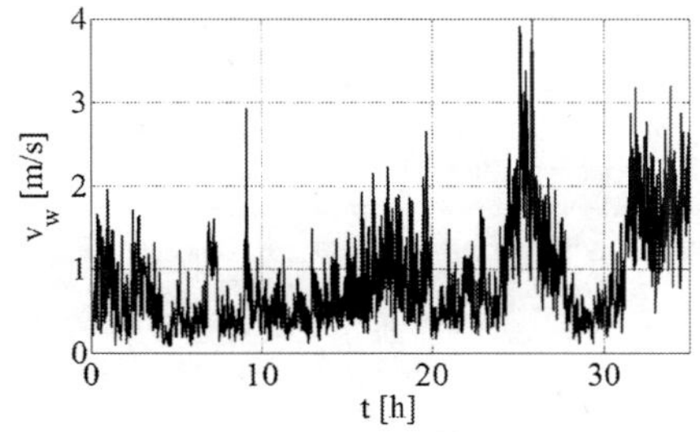

Figure 3. The wind velocity values for 35 hours rate.

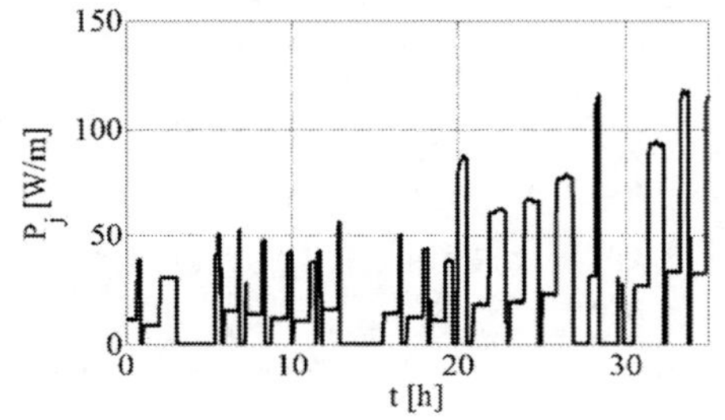

Figure 4. The power of heating due to joule losses.

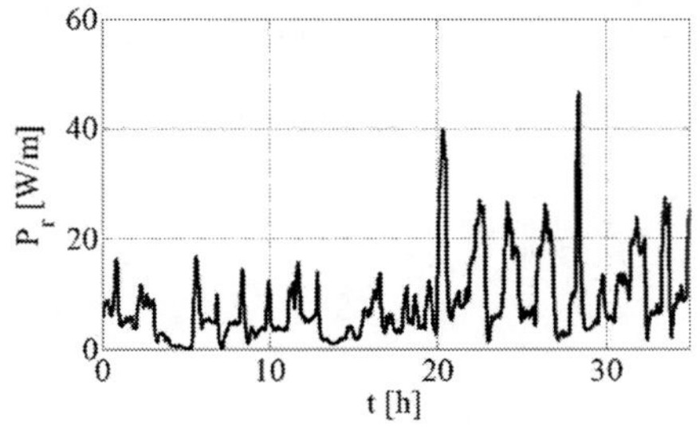

Figure 5. The heat dissipation due to the radiation.

According to the temperature level the output learning signal (target) was set to zero or one, dependent on the six temperature rates ($T_{70,80}$, $T_{60,70}$, $T_{50,60}$, $T_{40,50}$, $T_{30,40}$ and $T_{20,30}$). For example $T_{20,30}$ (Fig. 6) stands for the period between 20 °C and 30 °C, while $T_{70,80}$ (Fig. 7) stands for the period between 70 °C and 80 °C. It should be mentioned that all values higher than 80 °C are added into $T_{70,80}$ temperature rate, while all values lower than 20 °C are added into the $T_{20,30}$ temperature rate.

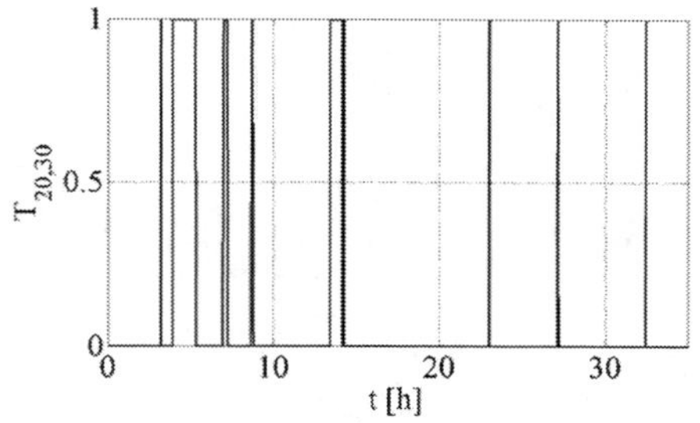

Figure 6. ANN target signal; period between 20 °C and 30 °C.

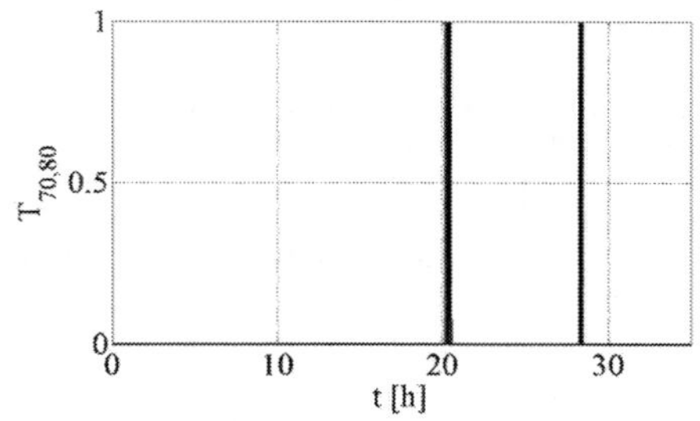

Figure 7. ANN target signal; period between 70 °C and 80 °C.

Results (Figs. 8 and 9) of the ANN are very dependent on the ANN net configuration, therefore an extensive testing of different net configurations was performed by the Matlab software tool. The proper net structure can be defined with the proper model of a whole system. Based on results of the analysis, the three-layer ANN structure with 20 neurons in hidden layer was chosen. In addition to determining the ANN structure, the model was applied also for determining the most appropriate learning rule, so the Levenberg-Marquardt backpropagation algorithm was chosen in that manner.

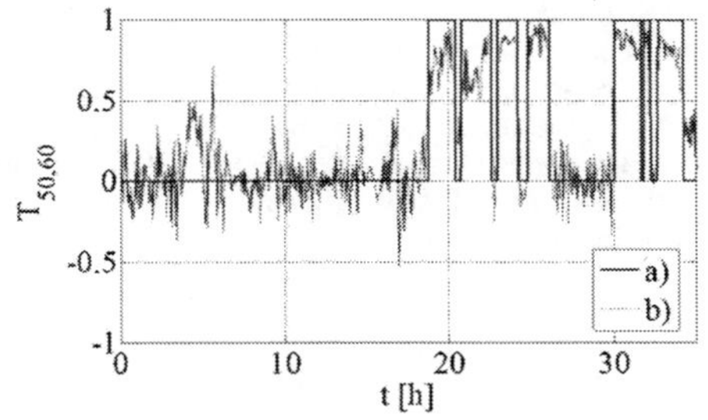

Figure 8. Target a) and ANN signal b) for period between 50 °C and 60 °C.

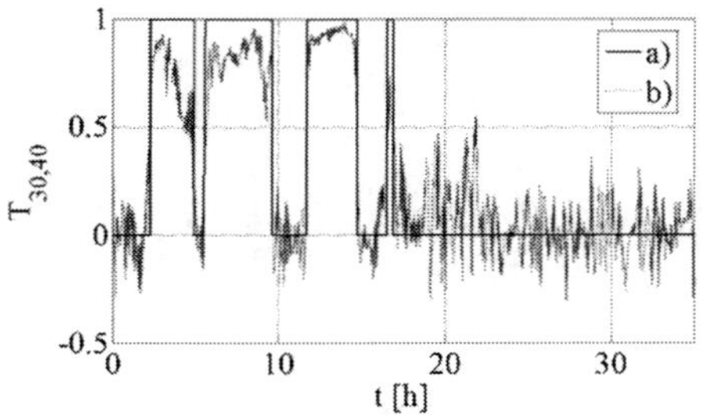

Figure 9. Target a) and ANN signal b) for period between 30 °C and 40 °C.

The results in form of ANN signals for periods between 50 °C and 60 °C and 30 °C and 40 °C presented in Figs. 8 and 9 show that the ANN could be reliable enough for using inside the conductor temperature of an overhead power line estimation. However, it could be very useful tool as a supplement to the existing algorithms for estimation of overhead conductor temperature.

V. Conclusion

In general, there is need for maximizing the use of electrical infrastructures, allowing them to operate at higher temperatures to increase the maximal power flow of power lines. In thermally-limited overhead power lines, the maximal transmitted power is limited by the maximal allowable temperature of the conductor. Dynamic thermal ratings of overhead lines could represent a promising approach to increase the transmission capacity by considering the weather dependent thermal ratings of overhead power lines in real-time instead of using constant ratings. Knowledge about the aforementioned is not only needed in real-time observation, but also on a day-ahead basis within the network operational planning and in order to assess network security. The conductor heating and, consequently, its temperature should not exceed the permitted values, both in terms of deterioration of the conductor's mechanical properties and the permitted sag. The stationary conductor temperature is reached when the equilibrium between the supplied and dissipated heat is reached, as is evident from the heat balance equation. It is shown that the application of an artificial neural network, as part of the aforementioned problem, could represents a promising solution.

References

[1] R. Tilman, P. Schäfer, and A. Moser, "Probabilistic ampacity forecasting for overhead lines using weather forecast ensembles," Electrical Engineering, vol. 95 (2), 2013.

[2] S. D. Kim, and M. M. Morcos, "An application of dynamic thermal line rating control system to up-rate the ampacity of overhead transmission lines," IEEE Transactions on power delivery, vol. 28(2), pp. 1231-1232, 2013.

[3] L. Staszewski, and W. Rebizant, "The differences between IEEE and CIGRE heat balance concepts for line ampacity considerations," Modern Electric Power Systems (MEPS), Proceedings of the International Symposium, IEEE, 2010.

[4] T. Krontiris, A. Wasserrab, and Gerd Balzer, "Weather-based loading of overhead lines—Consideration of conductor's heat capacity," Modern Electric Power Systems (MEPS), Proceedings of the International Symposium, IEEE, 2010.

[5] J. Zhan, C. Y. Chung, and E. Demeter, "Time Series Modelling for Dynamic Thermal Rating of Overhead Lines," IEEE Transactions on Power Systems, 2016.

[6] CIGRE, "Guide for thermal rating calculations of overhead lines," Working Group B2.43, 2014.

[7] R. Alberdi, E. Fernandez, I. Albizu, V. Valverde, M. T. Bedialauneta, and K. J. Sagastabeitia, "Statistical methods and weather prediction for ampacity forecasting in smart grids," In PowerAfrica, IEEE PES, pp. 21-25, 2016.

[8] F. Jakl, and A. Jakl, "Effect of elevated temperatures on mechanical properties of overhead conductors under steady state and short-circuit conditions," IEEE Transactions on Power Delivery, vol. 15 (1), pp. 242-246, 2000.

[9] K. Deželak, and J. Pihler, "Artificial Neural Network as Part of a Saturation-Level Detector Within the Transformer's Magnetic Core," IEEE Transactions on Magnetics, vol. 52 (5), pp. 1-4, 2016.

[10] S. E. Noh, Y. G. Kim, S. H. Lim, and I. D. Kim, "Dynamic Line Rating Prediction in Overhead Transmission Lines Using Artificial Neural Network," Journal of the Korean Institute of Illuminating and Electrical Installation Engineers, vol. 28 (1), pp. 79-87, 2014.

[11] A. Nag, and A. Yadav, "Fault classification using Artificial Neural Network in combined underground cable and overhead line," In Power Electronics, Intelligent Control and Energy Systems (ICPEICES), IEEE International Conference, pp. 1-4, 2016.

[12] I. Sadeghkhani, A. Ketabi, and R. Feuillet, "An intelligent switching overvoltages estimator for power system restoration using artificial neural network," International Journal of Innovative Computing, Information and Control, vol. 10 (5), pp. 1791-1808, 2014.

Photovoltaic Power Generation Forecasting in Slovak Transmission System

Jaroslav Beňo, Miroslav Kret, Rastislav Šmidovič
SED Žilina
Slovak Transmission System Operator
Žilina, Slovakia
rastislav.smidovic@sepsas.sk

Peter Braciník
Department of Power Electrical Systems
Faculty of Electrical Engineering, University of Žilina
Žilina, Slovakia
peter.bracinik@fel.uniza.sk

Abstract — **Photovoltaic power generation forecasting is an important task for a renewable energy power system's output planning and operation. This paper introduces an application of a forecasting software that is implemented by the Slovak Transmission System Operator (SEPS, a. s.) into its operation planning. A total capacity of 530 MW in photovoltaics has been installed in the Slovak power system. A cooperation with the Department of Power Electrical System of the University of Zilina, aimed on a software development for the photovoltaic output forecast, started in 2015. The developed program runs as an online application and provides the data for the day ahead forecast, that is then used as a one of inputs for the operation planning procedure of the Slovak transmission system.**

Keywords: photovoltaic generation; forecasting model; operation planning

I. Introduction

Photovoltaic (PV) power forecasting involves the knowledge of atmosphere's condition and the characteristics of used PV systems. PV power plants use a solar energy to generate electricity. The generated output depends on available solar irradiance as well as on solar panels' characteristics. Because a photovoltaic power production has been increased in recent years, a correct PV power forecast information is essential for the management of the electricity grid and the allocation of auxiliary services for a power regulation.

II. Current Status in Slovakia

Average annual values of solar irradiance in Slovakia are shown in Fig. 1. As it can be seen from Fig. 1, different regions of Slovakia have different conditions for PV systems operation, thus resulting naturally in difference spatial density of their installation.

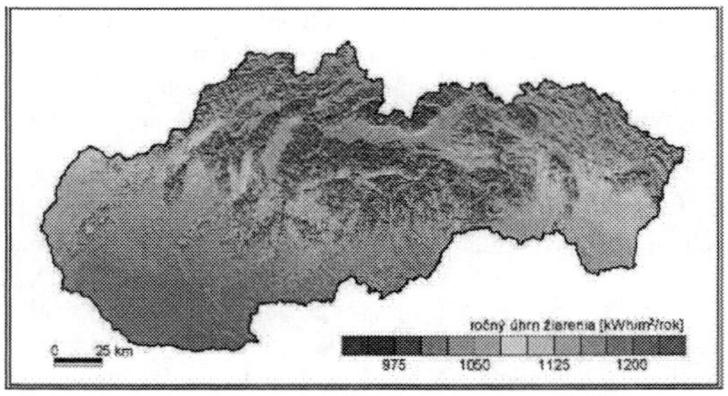

Figure 1. Average annual values of solar irradiance in Slovakia.

An overview of PV systems' installed capacity for different distribution areas in Slovakia is presented in Table 1.

Table 1. PV installed capacity of distribution areas (2013)

Distribution area	Installed capacity (MW)
Senica	16.86
P. Biskupice - Stupava	6.43
P. Biskupice - Gabčíkovo	34.18
Križovany	33.73
Levice	18.90
Bošáca - Bystričany	13.36
Bystričany - H. Ždaňa	41.77
P. Bystrica - Varín	4.95
Sučany	11.26
L. Mara	12.93
Medzibrod	18.06
R. Sobota SSE-D	170.66
R. Sobota VSD	13.83
Sp. N. Ves	12.12
Lemešany - Sp. N. Ves	53.87
Lemešany – Voľa	27.79
Moldava	25.02
Spolu	**515.72**

The production of a PV system depends on solar conditions. Table 2 presents the values of maximum electricity generation in PV systems in Slovakia during recent years. Although the output did not reach the installed capacity presented in Table 1, the achieved value of 380 MW is big enough to be taken seriously for a transmission system operation.

Table 2. Maximum electricity generation in PV systems

Year	***2011***	***2012***	***2013***	***2014***	***2015***	***2016***
Maximum generation (MW)	311	368	359	348	359	381
Day	20.8.	9.4.	16.4.	6.5.	23.4.	23.6.
Hour	13:00	13:00	13:00	13:00	13:00	13:00

The Fig. 2 shows a share of different type of sources on the annual electricity generation in 2016 in Slovakia. The largest share is of course covered by nuclear power plants. The electricity generation in PV power plants is small, only 2.1 % of the total production, but it represents 2 430 GWh in 2016.

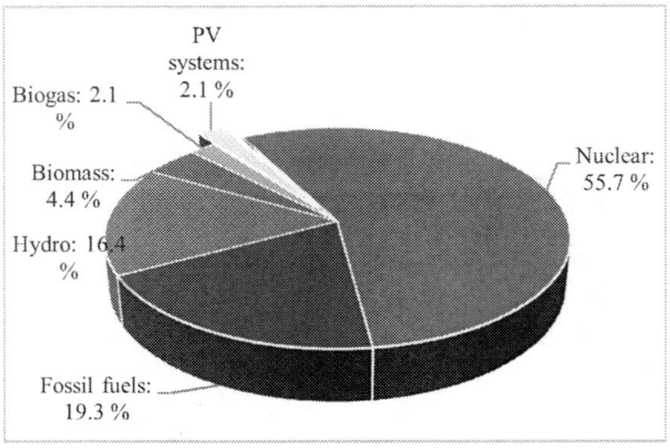

Figure 2. Share of sources on the electicity production in year 2016.

III. Forecasting Method

Intra□day forecasts are an important component of the integration of intermittent renewable energy sources to the electric power system.

Forecasts may be applied to a single PV power plant, or refer to the aggregation of large numbers of power plants spread over an extended geographic area. Forecasts may focus on the output power or on its rate of change (also known as the ramp rate). Accordingly, different forecasting methods are used. Forecasting methods also depend on the tools and information available to forecasters, such as data from weather stations and satellites, PV system data and outputs from numerical weather prediction (NWP) models [1].

Forecasting methods can be broadly characterized as physical or statistical. The physical approach uses solar and PV models to generate PV forecasts, whereas the statistical approach relies primarily on past data. Any output from any model described above must then be converted to the electricity generation that a particular solar PV plant will produce.

Meanwhile, purely statistical approaches do not use the solar and PV models. Their starting point is a training dataset that contains a really generated PV output, as well as various inputs or potential inputs, such NWP outputs, ground station or satellite data, PV system data and so on. This dataset is used to train models – such as autoregressive or artificial intelligence models – that output a forecast of PV power at a given time based on past inputs available at the time when the model is run.

IV. Input Data

There was not any experience with the operation and output prediction of PV power plants in Slovakia a few years ago. A first attempt to make a software program for the PV output forecasting was based on data provided by the Slovak Hydrometeorological Institute (SHMU)[3].

The SHMU uses a numerical model to calculate weather conditions in the specific areas of the Slovakia. Numerical calculations are based on the results of a European numerical model. Obtained results are then checked by a specialist, who defines the final weather forecast. The forecast is burden with a specific error. Its value is given both by the error of numerical prediction as well as by an experience and skills of the specialist. The error can be neglected for shorter time periods, due to the comparison with selected measured weather parameter in the field.

The Slovak transmission system operator (SED) started to use the software program for the PV power plants' output forecast in 2015. Instead of using data from SHMU, the program is based on data provided by Foreca.

Foreca Company is a leading provider of global weather and has its headquarters in Finland. It uses data provided by the European Center for Medium-Range Weather Forecast (ECMWF) for its predictions. The main product of the ECMWF are medium-term probabilistic forecasts that are derived from a 51-member file, calculated for 10 days ahead with a horizontal resolution of 50 km and 60 vertical levels [4]. An example of the Foreca prediction system for one locality is shown in Fig. 3.

Figure 3. Example of FORECA prediction.

The Foreca web page (www.foreca.sk/Slovakia) is used to provide data for 18 carefully selected places, each of them representing one distribution area from Table 1. The forecast is updated every hour and provide information about the type of weather, temperature, wind, humidity and rain probability for each selected place. This data can be downloaded for free. Together with a good forecast accuracy, this availability was a very important reason why the Foreca system was chosen as a one input for the forecasting software program used by SED.

The other input is data representing the electricity generation in installed PV power plants. It is stored in a SCADA system used by SED to control power transmission system. A SED's archiving system stores data about the real PV output in specific power system nodes as well as the total output of all PV power plants in Slovakia.

Data from both sources, the Foreca web page and SED's archive, is used in the crated program to calculate the expected output of all PV power plants for the current day D. The program is also capable of downloading meteorological data for a day ahead (D+1) and two days ahead (D+2) forecast and use them to calculate expected D+1 and D+2 PV power plants output. Such information can be than used to improve the power transmission system operation planning.

V. A PV Output Forecast Algorithm

A developed algorithm for the PV power plants output forecast works with values of weather forecast and the real PV output (weighted average per hour) of n previous consecutive days. The calculation of the PV output prediction for the current day starts with a comparison of predicted types of weather conditions of the current day and n previous consecutive days for each hour of the day. Compared types of weather conditions are listed in Table 3.

Table 3. Different types of weather conditions

Type of weather conditions
Partly cloudy, possible thunderstorms with rain
Partly cloudy
Mostly clear
Partly cloudy and light rain
Overcast, thunderstorms with rain
Clear
Overcast and light rain
Cloudy and light rain
Cloudy
Partly cloudy and showers
Overcast
Overcast and rain
Overcast and showers
Cloudy and showers
Overcast and light snow
Cloudy and light snow
Cloudy and snow showers
Partly cloudy and light snow
Overcast and snow showers
Overcast and snow
Cloudy and light wet snow
Overcast and light wet snow
Overcast and wet snow showers
Partly cloudy and light wet snow
Partly cloudy and snow showers
Overcast and wet snow
Partly cloudy and wet snow showers
Cloudy and wet snow showers

If a match is found, the value of the real PV output for that hour is taken to the calculation, but it is modified with other weather parameters with following weights – the humidity 20 %, the temperature 10 % and the wind 5 %.

When the comparison is over, there are none, one, two or more pairs consisting of the values of rain probability and modified PV output for each hour of the day. If there is none or only one pair for a specific hour, the additional pairs are created for the maximum and minimum PV power plants output. Obtained pairs are then used as inputs for a linear regression.

The used linear regression of two-dimensional statistical data is based on a least-square method. Statistical data are in the form of pairs, each pair on a separate line. The first number in each pair (the odd one) is taken as x, the other number (the even one) is taken as y. The output of the linear regression are coefficients A and B of the linear function:

$$\mathrm{A} \cdot x + \mathrm{B} = y, \tag{1}$$

which approximates the given dependence by a linear function (line) such that at the given nodes the minimum power (square) of deviations is reached (Fig. 4). The coefficients are calculated by following equations:

$$\mathrm{A} = \frac{\sum_{i=1}^{n} y_i - \mathrm{B} \cdot \sum_{i=1}^{n} x_i}{n}, \tag{2}$$

$$\mathrm{B} = \frac{n \cdot \sum_{i=1}^{n} x_i \cdot y_i - \sum_{i=1}^{n} x_i \cdot \sum_{i=1}^{n} y_i}{n \cdot \sum_{i=1}^{n} x_i^2 - \left(\sum_{i=1}^{n} x_i\right)^2}, \tag{3}$$

where n is the number of pairs,

x is the value of rain probability,

y is the value of modified real PV output.

The calculated coefficients of linear function (1) and the value of rain probability are used to calculate the forecasted PV power plants output (Fig. 4) in the specific distribution area (Table 1) for the specific hour of the day. The total forecasted value of PV power plants output in whole Slovakia is calculated as a sum of values calculated for all distribution areas [5].

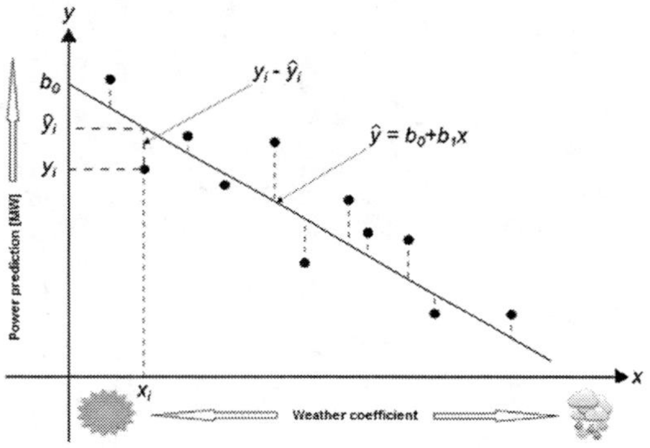

Figure 4. An example of the forecast calculation.

The accuracy of the PV output forecast depends on the number of days (n) used for the linear regression. Both longer as well as shorter history can be considered. An intensive testing has proved that an interval of 14 days provides sufficient forecast accuracy and does not slow down a calculation speed. Therefore, it has been also implemented to the developed forecasting software program. This value can be of course changed if needed.

VI. The Forecast of PV Power Plants Output

The developed forecasting software program has been integrated into the MES (Manufacturing Execution System) application supporting the dispatching control of Slovak power transmission system. Thus, providing the highest possible comfort for the users, because all necessary calculations are done automatically.

The first step of the forecast process is an automatic download of weather forecast data for each of 18 defined places. SED does not need to pay for it, as all data is obtained from freely available Internet resources. Although, the data needed for the forecast calculation is downloaded with a required quality. The example of a downloaded weather forecast sheet and its presentation in the forecasting program is shown in Fig. 5.

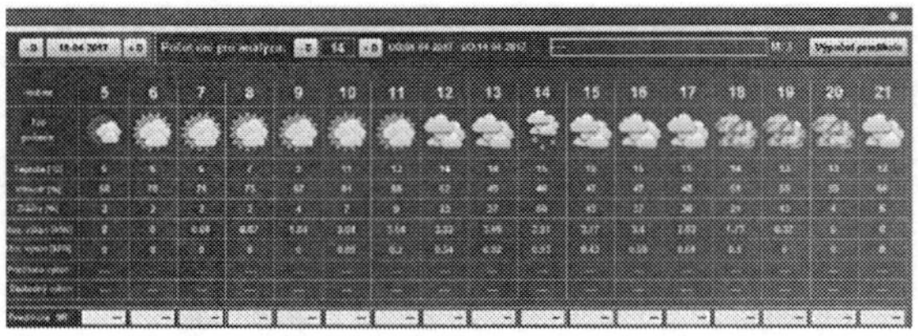

Figure 5. A presentation of downloaded weather forecast data.

Calculated results of PV power plants output forecast are presented to users either by graphs or spreadsheets. The user can choose, if he wants to see PV forecast for a special distribution area or the whole Slovakia. The time resolution is one hour. The example of a three-day PV output forecast is shown in Fig. 6.

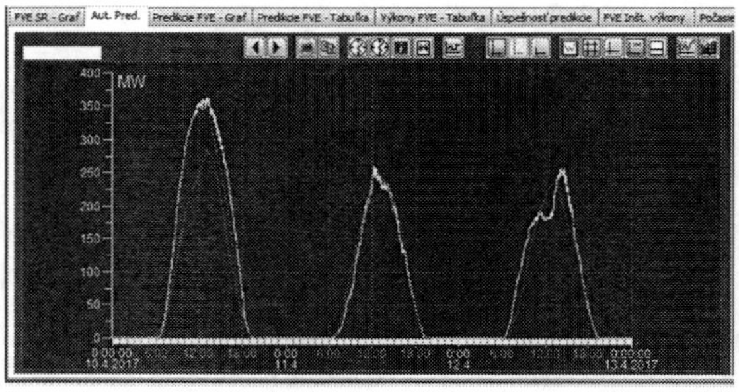

Figure 6. The example of the three-day forecast presentation.

The forecast value of PV power plants output is entered to a trading system (Damas Energy System) before 2 p.m. After this time, all preparations of the operation of all available electricity sources as well as sources providing ancillary services are completed. Consequently, cross-border balances are also closed at 2:30 p.m., and the overall picture of the production and consumption of electricity in the Slovakia is completed. After this time, the expected difference between production and consumption is seen.

VII. Conclusions

The short-term PV power plants output forecast is a very important for distribution utilities, as they responsible for an elimination of production deviations caused by renewable energy sources.

But the good and accurate PV power plants output forecast is also very important for the dispatching center of transmission system operator to make correct decisions supporting a power system security management. The operation planning of power transmission system operation (an allocation of electricity sources and sources providing ancillary services) gives a good overview of the situation expected in the balance between the electricity generation and demand. Especially a comparison of demand forecast and contracted diagrams of generation facilities for the D+1 day provide dispatchers with a comprehensive view of prerequisites for the activation of individual types of ancillary services and the amount of balancing electricity.

The demand forecast is done with a long-term precision better than 2 %, which has represented a deviation of 0 – 70 MW for the last 2 years. To fully determine the total electricity generation in Slovakia, it is not necessary to know only the generation diagrams of facilities providing this information mandatory one day ahead, but also of facilities that do not need to provide such information. The second group of facilities consists of small power plants and renewable energy sources with constant operation. Based on measured data and historical data, this part of the generation is estimated with a high degree of precision.

The interesting as well as important task is the determination of electricity generation in renewable energy sources with intermittent and highly weather dependent operation. In the case of Slovakia, the intermittent generation is provided solely with photovoltaic power plants. Their total output varies from 40 MW in bad to 380 MW in ideal weather conditions.

The transmission system operator is responsible for the frequency control within its regulation area. The frequency control is covered by automatically operated ancillary services (FCR and FRR). Because an active power allocated for the FCR and FRR is relatively small, with an average value only +/- 170 MW, what is a smaller value than possible PV power plants output, it is important to make the precise forecast of PV power plants generation. Moreover, the dynamic changes of PV power plants operation caused by their spatial distribution can reach the values from -80 MW up to +100 MW. Such fluctuations imply increased requirements on a dispatcher's attention, and may lead to the deployment of manual ancillary services (mFRR), in Slovakia called as tertiary frequency regulation.

The developed forecasting software program has been used for PV power plants output forecast by Slovak TSO since the year 2015. It is used every day and has proven its merit and reliability. Of course, as any other software, it can be improved in the future. Main changes can relate to the calculation method's modifications. For this purpose, the quality testing of the PV system prediction is currently being performed.

References

[1] S. Pelland, B. Noble, at all. "Photovoltaic and Solar Forecasting," IEA, Paris, October 2013.

[2] J. Turček, J. Oravec, R. Šmidovič. "Operation of Power System with Photovoltaic Power Plans," in *Proc. EEE*, June 2011, High Tatras, Slovakia.

[3] P. Bracinik, M. Roch, R. Šmidovič. "Možnosti predpovede výkonu z fotovoltických elektrární na Slovensku (The possibilities of the photovoltaic power plants output forecast in Slovakia)", *Journal EE*, vol. 17, pp. 8-11, april 2011.

[4] [Online] Available: http://www.foreca.com/, last see (Apr. 30, 2017)

[5] J. Altus. "Štruktúra systému predikcie výroby z fotovoltických elektrární v elektrizačnej sústave SR," University of Zilina, KVES, 2015

[6] J. Hofierka, T. Cebecauer. "Priestorová a časová distribúcia slnečného žiarenia na georeliéfe Slovenska", *International Scientific Conference*, Poľana nad Detvou, Slovakia, September 17 - 20, 2007, ISBN 978-80-228-17-60-8

[7] Z. Chen, Z. Yang, Q. Peng. "Study of photovoltaic short-term output predicting based on optimal algorithm", *2016 IEEE Advanced Information Management, Communicates, Electronic and Automation Control Conference (IMCEC)*, 2016, pp. 1442 - 1446, DOI: 10.1109/IMCEC.2016.7867453

[8] Ch. Li, Q. Hui, Y.Teng, Ch. Xiong, Y. Zhu. „Nonlinear time series forecast model for grid-connected photovoltaicstation" *IEEE International Conference on Progress in Informatics and Computing (PIC)*, 2015, pp. 443 - 448

Grid Frequency Regulation by Recycling Electrical Energy in Flywheels

Emil Dvorsky, Lenka Rakova, Pavla Hejtmankova
Department of Electrical Power Engineering and Environmental Engineering
University of West Bohemia
Pilsen, Czech Republic
dvorsky@kee.zcu.cz, lencar@kee.zcu.cz, hejtman@kee.zcu.cz

Abstract — **Grid frequency regulation function is based on the fact that the amount of power generated and the amount of power consumed must be in exact balance at all times. The system operator usually generates a signal which is called, Area Control Error (ACE), based on the difference between these two parameters. The expected goal is to keep the system near nominal frequency value. Traditionally, frequency regulation is managed by varying the output of power stations connected to the electric grid. The article presents a method in which electric energy, which is necessary for power balance keeping, is recycled, in the system. First, the system absorbs energy when it is in abundance, then discharges the same for the desired frequency regulation effect. The method is based on kinetic energy storage technology in high-speed flywheels (FES). The modeling results, based on actual ACE signals, should, indicate that a modular flywheel system can be effective in providing frequency regulation in power grids.**

Keywords: grid frequency regulation; primary frequency control; secondary frequency control; droop characteristics; flywheels

I. Introduction

Today's transmission and distribution grid has been described as the greatest technological achievement of the century. The objective function of the grid is to keep the amount of power generated and the amount of power consumed in exact balance at all times. When imbalances, in AC network occur, the frequency that the users of electricity expect (60 Hz or 50Hz depending on the continent) will not be maintained. When generation exceeds consumption, the frequency increases, and when generation is less than the aggregate load, the frequency decreases. This constant balancing of load and generation to maintain frequency is called Frequency Regulation. It is necessary to make remark, that described method is valid only for transmission system where we can assume only reactive part of the line impedance between generators and loads [1], [2]. Fig. 1 [3] illustrates the difference example between Frequency Regulation and daily load of active power changes. The balance deviation is called the System Deviation (δ) and Regulation Power (RP) must correct it – Fig. 2 [4].

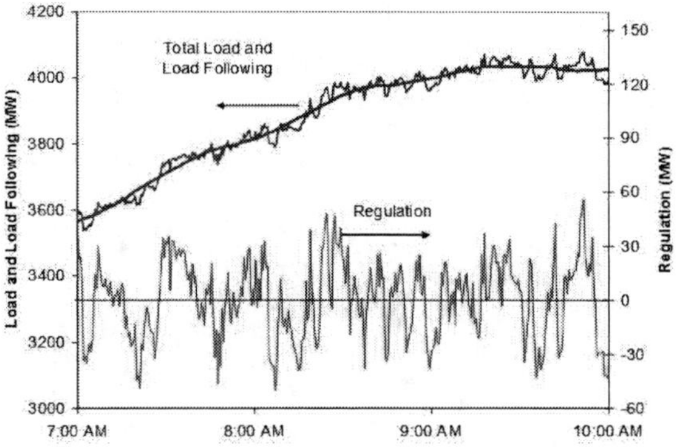

Figure 1. Regulation is a zero-energy service that compensates for minute-to-minute fluctuations in total system load and uncontrolled generation [3]

Generators that supply power to the grid have a large sum moment of inertia. The impact of this inertia is that a large amount of surplus power will only slightly increase their speed. The generators within a specific control area are also all synchronized to the same frequency.

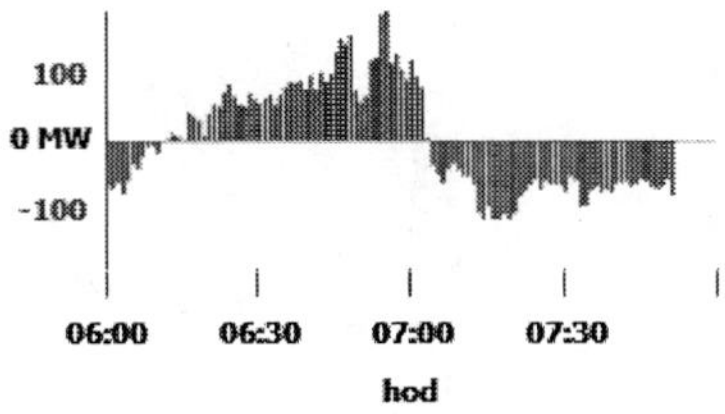

Figure 2. Regulation Power in CZ power transmit network, 08.05.2017 - [4]

This combination of facts provides an effective buffer against the inevitable imbalances that occur when loads are quickly added or subtracted, or generators drop off line. Imbalances that occur more slowly increase or decrease the frequency of the grid and must be corrected for by making constant adjustments to generators to keep the frequency within required limits. This constant cycling increases the wear on the generators and they operate run off the design parameters at a lower efficiency. The cost impact of this reduced efficiency, lower generation, and increased maintenance has a direct impact on the profitability of the generators.

Usually, regulation uses on-line generation, but there is possible to use a storage system (it is discussed in the paper), or load, that is equipped with automatic generation control (AGC) that can change output quickly (MW/min) to track the moment-to-moment fluctuations in customer loads and to correct for the unintended fluctuations in generation. Regulation helps to maintain interconnection frequency, manage differences between actual and scheduled power flows between control areas, and match generation to load within the control area.

Generators differ dramatically in how well they follow the system operator's commands to supply regulation. This is one reason that the amount of regulation required by two control areas of similar size can differ substantially. The quality of the available regulation resources can differ, and consequently, the amount of regulating resources (MW) that must be deployed to obtain similar satisfactory reliability performance differs. Fig. 3 compares the requested and actual regulation movements for two large coal-fired generators. Generator A is considered to be fairly good (for a coal burner) at regulating. Generator B is clearly not very good. But there are no metrics for regulation provision. All regulation providers are paid on the basis of the capacity they supply, not on the performance of that capacity. At most, system operators reserve the right to disqualify a generator from providing regulation if performance degrades too badly.

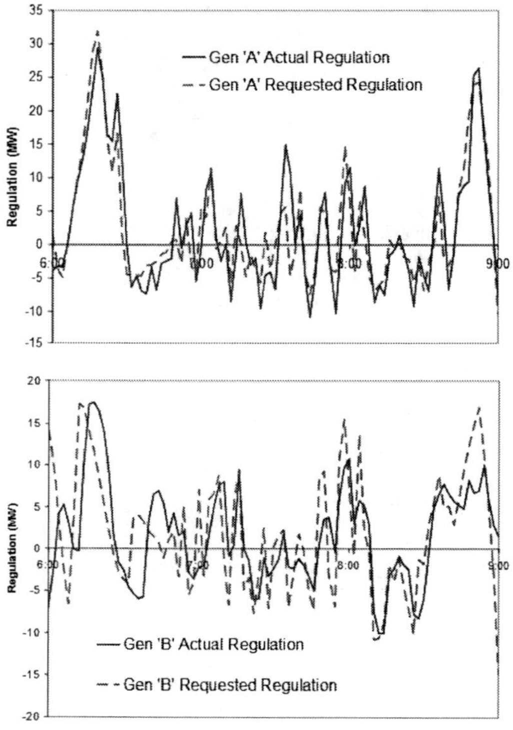

Figure 3. Compering of regulation abilities of two coal-fired generators [3]

The concept proposed would recycle energy (store energy when generation exceeds loads; discharge energy when load exceeds generation) instead of trying to constantly adjust generator output. This cyclic characteristic of regulation services makes a flywheel energy storage system uniquely suited to the application. An innovative means, which are able to provide frequency regulation with the use of flywheel energy storage rather than by cycling the output of a generator should be better to ensure ancillary services for electric power networks.

The aim of the article is to assess the ability of flywheels to meet the ancillary services requirements of power system operator.

II. Ancillary Services for Power System Reliability

Maintenance of the system function (the energy balance between the appliances and the sources at a given time with the relevant quality parameters) are used ancillary services which can be divided:

Ancillary services required during normal conditions

- ancillary services required during normal conditions,
  - continuous regulation - primary regulation (PR),
  - energy imbalance management - secondary regulation (SR),
- ancillary services used during system contingencies,
  - instantaneous contingency reserve,
  - replacement reserve ,
- other ancillary services,
  - voltage support
  - black start

A. *Primary frequency regulation*

The primary power unit frequency regulation is a local automatic function provided by primary circuit's regulation, consisting of a precisely defined change in power of the power block depending on the frequency deviation from the nominal value [5]. Power output changes of power unit required by the primary regulation circuits, depending on the frequency deviation, which are given by the equation (1).

$$\Delta P = -\frac{100}{\delta}\frac{P_n}{f_n}\Delta f \ [MW] \qquad (1)$$

where:

P_n – nominal unit power output [MW],
f_n – nominal grid frequency [Hz],
δ – slope of the static (drop) characteristic [%],
Δf – frequency deviation [Hz].

The PR provider must ensure that the required RE (regulation back up) is released within 30 seconds of the occurrence of the frequency deviation. The maximum reserve size of PR (RRPR) per block is released when the frequency is changed by 200 mHz from the nominal value (applies to blocks up to 300 MW). For blocks over 300 MW, the maximum reserve size RRPR is released when the frequency is changed by 100 mHz nominal values.

To limit the effect of block failures, that provide PR, the maximum RRPR from a single block is 10 MW. The minimum size RRPR provided by one block is 3 MW. Assuming the value of static characteristic is 5 %, the min. nominal value of block power output must be:

- 25 MW - maximum value RRPR,
- 0.75 MW – minimal value RRPR.

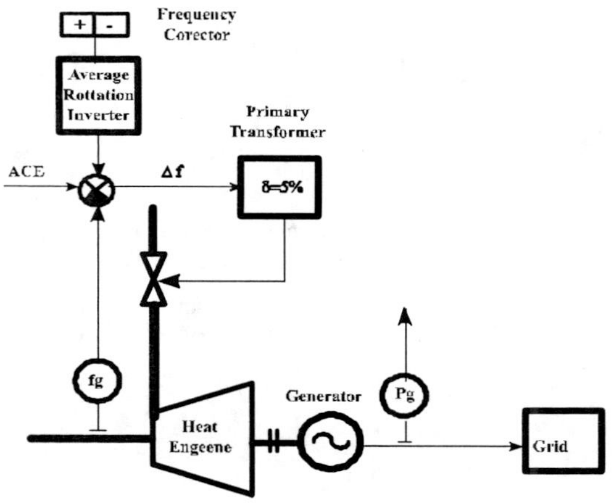

Figure 4. Connection of thermal power unit to f regulation [6]

The principal technology scheme of frequency ancillary services during normal condition for typical thermal unit is presented on Fig. 4. Frequency corrector is the powerful element of PR – Fig. 5. It corrects the specified value of the power generating unit when the frequency deviations in the grid occurs and it is connected to the average rotate inverter. The corrector insensitivity zone of the f corrector determines the minimum magnitude of the change f at which the block participates in the PR. The limitation of the output signal from the corrector determines the maximum size of the block power change.

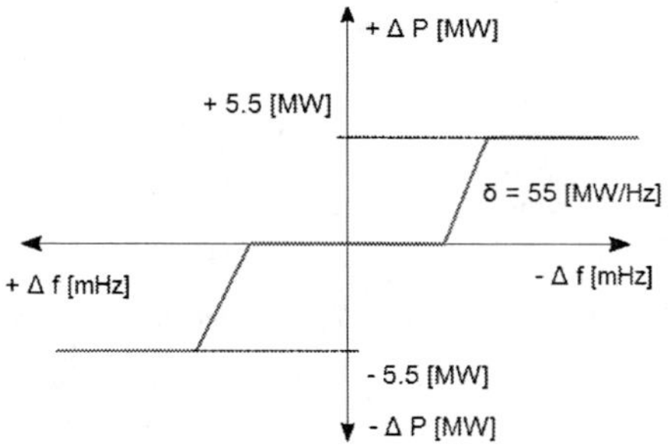

Figure 5. Static (drop) characterists of frequency corector [6]

The limitation of the output signal from the corrector determines the maximum size of the block power change. After adjusting the power balance in PR, the grid operator (transmission system operator - TSO) generates the frequency ACE deviation, which increases the power of the block delivered to the network, so that the frequency can be on the nominal value. It is principle of secondary frequency regulation.

B. *Secondary frequency regulation*

The secondary f control and P is at the level of the grid control automatically controlled by the secondary frequency controller and transmitted power. The controller is located at the control room of the TSO and its terminals are in the regulated power stations and at the border switch stations. The controller itself works according to the drop characteristics method. Method of the network static characteristics is equal to the control deviation of the secondary AC controller is composed by the sum of two components:

- frequency deviations from the specified value multiplied by the constant,
- frequency deviations from the specified value multiplied by the constant,
- deviation from the planned value of cross border value among the interconnected grids.

If the set constant of the regulator corresponds exactly to the power of the regulated system, its regulator does not respond to the power imbalance outside the regulated system, thus ensuring the principle of non-intervention. The process of the secondary control f and P is realized by sending the set-point power value from the secondary controller to the control blocks. The operation of the secondary control f and P should reset the nominal frequency and transmit power (ACE = 0) to 15 min. from the start point of the imbalance.

The substantial conclusion is that every power unit, which would like to provide ancillary services during normal operation conditions, must be equipped by PI power output regulator.

C. *Voltage support*

Control of voltage levels, on the power supplying part of the grid, is accomplished by controlling the production, absorption, and flow of reactive power at all levels in the system. As was motioned in introduction, there are other possibilities of voltage controlling [1], [2]. Usually generating units provide the basic means of voltage control – Fig. 6. The power unit trays to keep voltage level, in defined point of the grid (generators, pilot node) by static voltaic regulation. For this purpose usually primary and secondary voltaic regulation is used.

Additional means are usually required to control voltage throughout the system:

- sources or sinks of reactive power, such as shunt capacitors, shunt reactors, synchronous condensers, and static var compensators (SVCs),
- line reactance compensators, such as series capacitors,
- regulating transformers, such as tap-changing transformers and boosters.

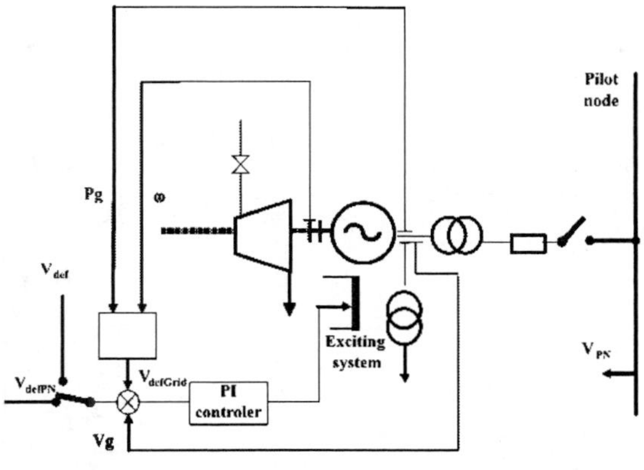

Figure 6. Schema of frequency and voltage control [6]

III. ANCILLARY SERVICES BY FLYWHEELS

The main objective of the paper is evaluating the possibilities of FES units using them to ensure the system requirements of the power system for ancillary services required during normal conditions. Therefore, there is necessary to evaluate especially performance FES characteristic.

A. *FES design*

FES consists of a rotary flywheel and an induction machine that works both as a motor and a generator. During normal operation, electricity is transferred from the grid to the kinetic energy through the engine. If necessary, the stored kinetic energy is converted back to the electric generator – Fig. 7.

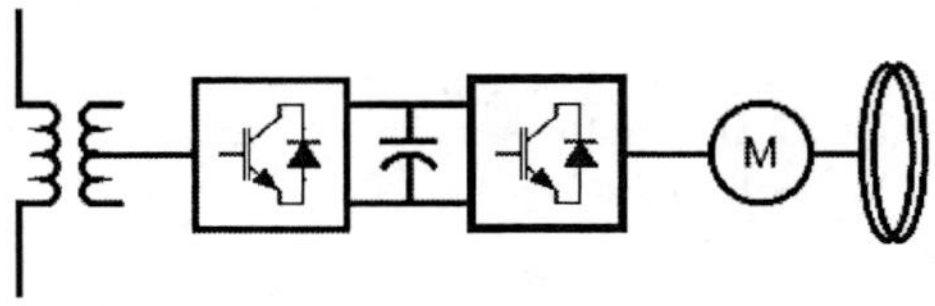

Figure 7. Principle of FES [7]

The flywheel stores kinetic energy of rotation, where the stored energy E_n depends on the moment:

$$E = \frac{1}{2} m.\omega^2 \tag{2}$$

The flywheel can be used as a power handling device, in result from advances in key enabling technologies. Power is the rate at which energy is converted, given by:

$$P = \frac{\Delta E_n}{\Delta t} = \frac{1}{2} J \frac{\omega_1^2 - \omega_2^2}{\Delta t} \tag{3}$$

The torque T for a given (instantaneous) power output in electrical machines can be calculated as:

$$T = \frac{P}{\omega} \tag{4}$$

The instantaneous power injected by the torque depends only on the instantaneous angular speed and not on whether the angular speed increases, decreases, or remains constant while the torque is being applied.

The flywheel can be charged or discharged during the operating period T at a constant speed, or at a rate increase or a decrease in speed. The rate of speed change [7], [8], [9] is exponential and can be expressed for discharging in the given period by the equation:

$$\omega(t) = \begin{cases} \omega_{max} & t \leq T \\ \omega_{max}.e^{-\frac{K_0}{J}.t} & t \geq T \end{cases} \tag{5}$$

where:

K_0 – torque factor from mechanical to electrical,
T – period for which the flywheel works at ω_{max},

If this time is not available, it can be roughly determined using the rule of dependence between moment of inertia and flywheel power:

$$T_{max} = \frac{J.\omega_{max}^2}{2.P_{max}} \tag{6}$$

The flywheel performance at exponential speed change can be deduced from equation (7) while respecting individual times:

$$P = \begin{cases} P_{max} & t \leq T \\ \eta_{setr}\omega_{max}^2.e^{-\frac{K_0}{J}.t} & t \geq T \end{cases} \tag{7}$$

For modeling performance characteristic FES, we can use already created FES models that are available on the network. - http://www.calculatoredge.com/mech/flywheel.htm. It is also possible to use published models see [8].

The presented model (Fig. 9) is made up of two basic parts. The first is a flywheel model (Fig. 8), that processes the voltage and current of the busbar to which the load is connected and controls the power output from the flywheel. In the flywheel model it is possible to specify its parameters represented by maximum and minimum angular speeds (rad/s), power (kW), power capacity (kW.s), efficiency, charging time (s), discharge time from maximum power. The output from the flywheel model is its output power (kW), and angular velocity (rad/s). Typical flywheel discharging times are between 10-30 seconds according to manufacturer's data, depending on the type of flywheel and the technology used to charge the flywheel for 1-10 minutes.

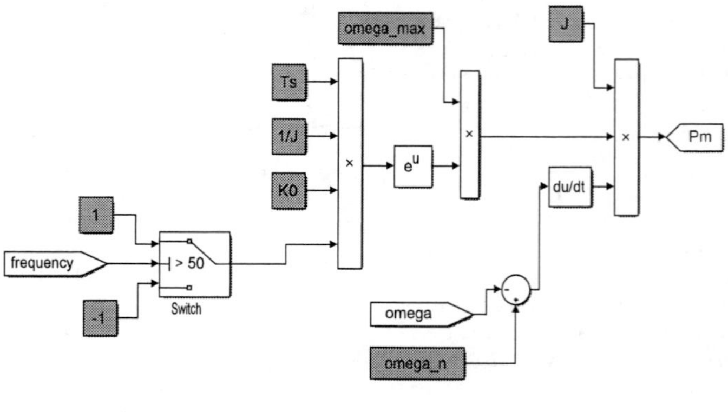

Figure 8. Control circuit for calculating the mechanical power of the FES.

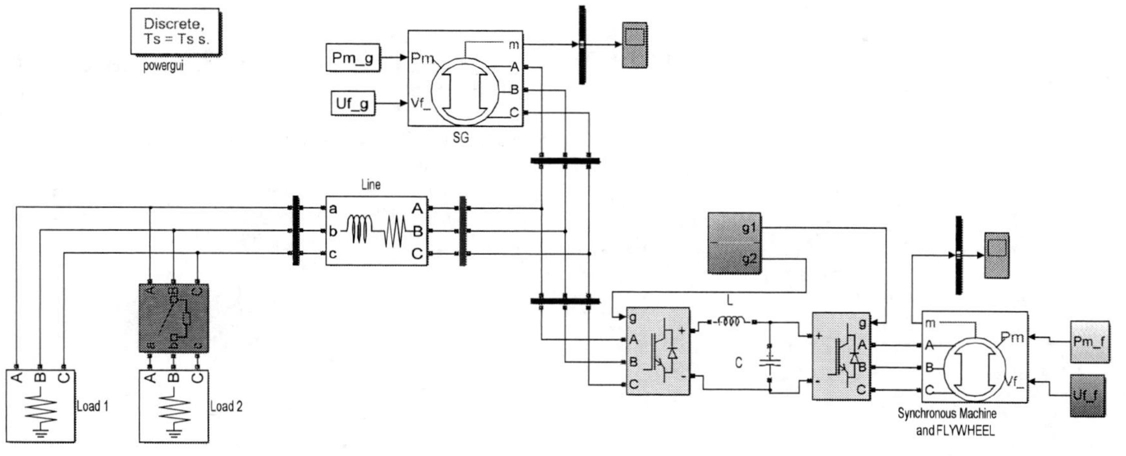

Figure 9. Basic FES model for grid supplying and anxilary service use

IV. Simulation Results

For the purpose of verifying FES functionality, the model of performance characteristic was applied to four types of flywheels of varying outputs (120, 150, 160 and 250 kW). The modeled discharge characteristics were compared with the data presented by the manufacturers based on the test measurements. The presented results in [7] are shown in Fig. 10 and in Fig. 11.

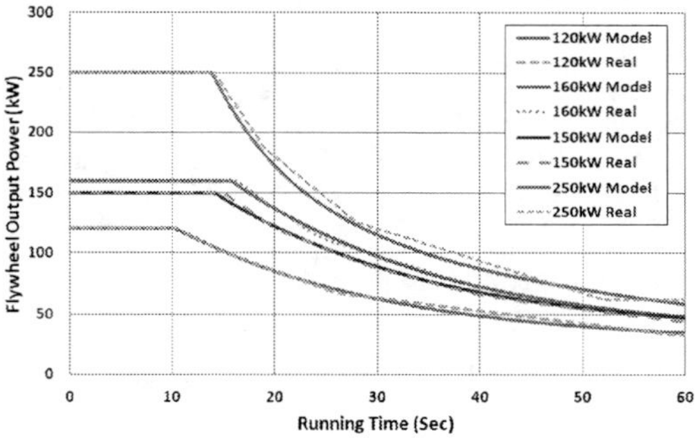

Figure 10. FES discharging characteristics [7]

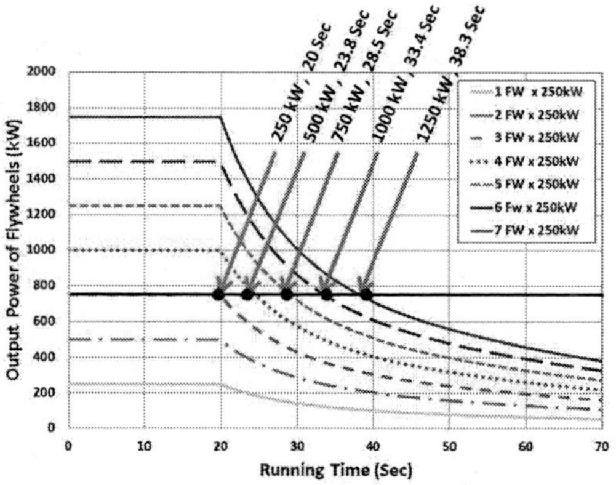

Figure 11. FES paralel combination [7]

In this paper was simulated a load change in microgrid and the response of the synchronous generator and the flywheel. The results from Basic model of the energy system using FES, was applied on the grid with two loads (5 kW, 3 kW) and synchronous generator with nominal power output 15 kW, FES system has nominal power output 10 kW. Time of connection of 3 kW load is 4 s. From Fig. 12 it is seen that, the active power from synchronous generator is constant even after the load change, because this source does not have a primary regulator connected. If this controller was connected, the power from the synchronous controller would increase to offset the power balance.

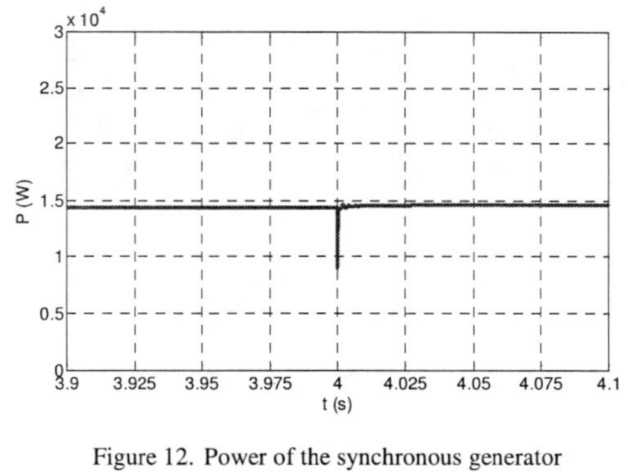

Figure 12. Power of the synchronous generator

The power output and rotation of the FES system are presented on Fig. 13 and on Fig. 14.

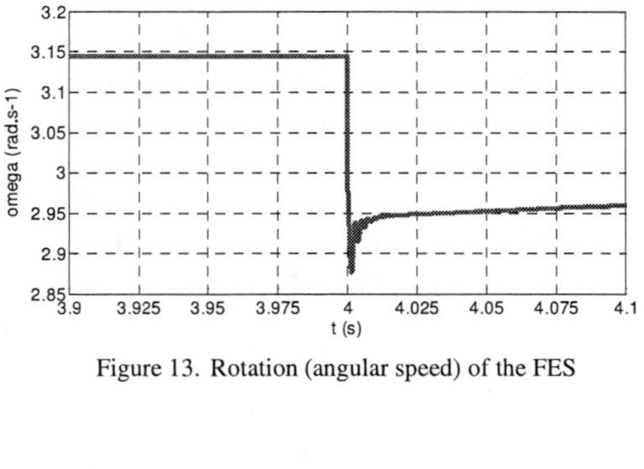

Figure 13. Rotation (angular speed) of the FES

As shown in Fig. 13, with the increasing power deficit of the synchronous generator, the frequency in the network is reduced and therefrom the angular velocity of the flywheel decreases.

From the curve in Fig 14, it is seen, that in the time 4 s the active power value of the flywheel decreased, due to a drop frequency value in the microgrid. In this case, the flywheel does not have a control circuit and behaves like a synchronous motor. According to static frequency characteristics, the resulting waveforms are correct. When the network frequency drops, the flywheel is braked thereby consumes less power from the network.

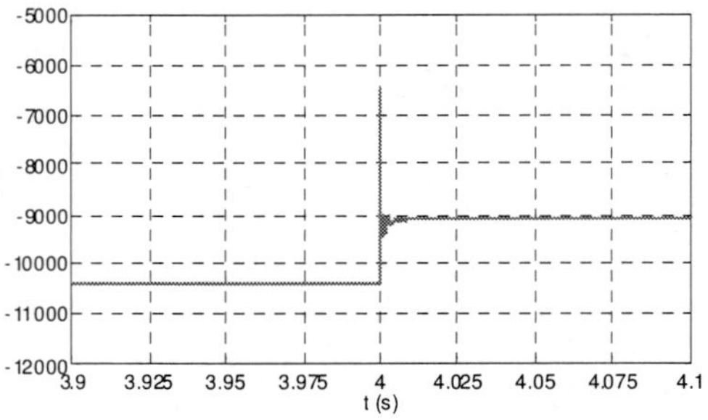

Figure 14. Power of the FES

Voltage on the SG terminals is presented in Fig. 15. This voltage waveforms show that the load change has an effect on both frequency and voltage.

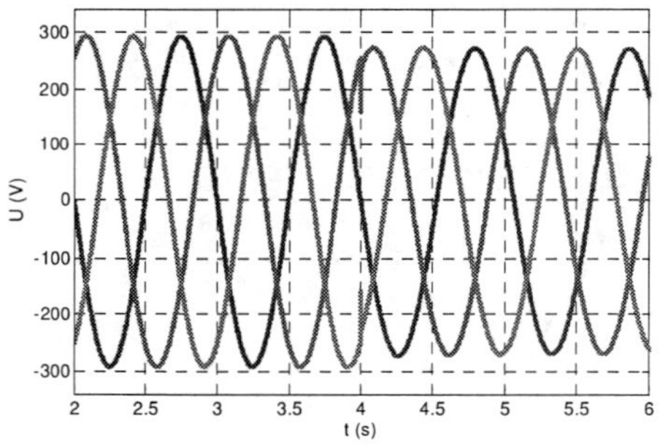

Figure 15. Voltage of SG

Conclusion

The analysis of the grid network regulatory power requirements and the model that has been developed have shown that the use of FES in conjunction with other sources used to power the network is possible. The possibility of applicability will of course depend on the calculation of the whole network with using FES and the solution of the economic model.

In this article, the cooperation of the synchronous generator with the flywheel has been verified, provided that primary frequency and voltage regulation is not implemented. The results show the necessity of this primary frequency and voltage control and the possibility of the using flywheels to participate in the primary control, i.e. to the compensation of the power balance. Further work will focus on verification of the proposed FES control circuit (Fig. 8) for the primary control.

Acknowledgment

This paper was supported by the student research project SGS - 2015 – 031.

References

[1] L. Rakova, E. Dvorsky, P. Hejtmankova, "Utilization of modified virtual method for frequency and voltage primary control in microgrid with photovoltaic power plants," In Power Engineering 2016, Bratislava: Slovak University of Technology, 2016, pp. 1-5, ISBN: 978-80-89402-81-6.

[2] E. Dvorsky, P. Hejtmankova, L. Rakova, "Frequency and voltage restoration in microgrid with renewable energy sources," In Power Engineering 2016, Bratislava: Slovak University of Technology, 2016, pp. 1-5, ISBN: 978-80-89402-81-6.

[3] B. J. Kirby, "Frequency Regulation Basics and Trends," Department of Energy (DOE) Information Bridge, ORNL/TM-2004/291.

[4] Regulation energy. [online]. Prag: ČEPS. [cit.: 2017-04-30]. Avalaible on: ‹https://www.ceps.cz/ENG/Data/Vsechna-data/Pages/Regulacni-energie.aspx?lang=CZE›

[5] Code of Transmission System, Part II. - Ancilary Services (AS), Czech Transmission System Operator, Revised edition, March 16, 2016.

[6] E. Dvorsky, "Electric Power Measurement and Control – lectures". [online]. Pilsen: WBU Pilsen. 2016 [cit.: 2017-04-20]. Avalaible on: ‹http://home.zcu.cz/~dvorsky/›

[7] R. Arghandeh, M. Pipattanasomporn and S. Rahman, "Flywheel Energy Storage Systems for Ride-through Applications in a Facility Microgrid," in IEEE Transactions on Smart Grid, vol. 3, no. 4, pp. 1955-1962, Dec. 2012. doi: 10.1109/TSG.2012.2212468

[8] M. Ahrens, L. Kucera, R. Larsonneur, "Performance of a magnetically suspended flywheel energy storage device", IEEE Transactions on Control Systems Technology, Vol. 4, Issue 5, pp: 494 – 502, Sept. 1996.

[9] T. T. Leung, "Concept of a modified flywheel for megajoule storage and pulse conditioning ", IEEE Transactions on Magnetics, Vol. 27, Issue 1, pp. 403 – 408, Jan 1991

[10] P. Kanakasabapathy, I. V. V. Rao, "Control strategy for inverter based micro-grid," In Power and Energy Systems Conference: Towards Sustainable Energy, 2014, Bangalore, 2014, pp. 1-6. ISBN: 978-1-4799-3421-8.

Voltage Control and Load Redistribution Between Decentralized Power Sources in Microgrids in Island Mode

Lenka Rakova
Department of Electrical Power Engineering and Environmental Engineering
University of West Bohemia
Pilsen, Czech Republic
lencar@kee.zcu.cz

Emil Dvorsky
Department of Electrical Power Engineering and Environmental Engineering
University of West Bohemia
Pilsen, Czech Republic
dvorsky@kee.zcu.cz

Pavla Hejtmankova
Department of Electrical Power Engineering and Environmental Engineering
University of West Bohemia
Pilsen, Czech Republic
hejtman@kee.zcu.cz

Abstract — **Decentralized electricity generation from photovoltaic power plants (PVPPs) has an impact on the operation and regulation of the electric power systems. For the maintaining of the grid quality parameters and for the correct voltage and frequency primary control is important the appropriate load redistribution between individual generators. The aim of this paper is to evaluate whether it is possible load redistribution between two parallel cooperating generators by using the modified virtual method. The basis of the innovation of the virtual frequency and voltage method is the virtual active and reactive power method. The proposed control circuit of the inverter is verified by the simulations of the islanded microgrid (MG) operation in MATLAB Simulink. The results of the simulations show that the proposed control circuit operates properly because the load change is redistributed between the decentralized sources with help the coefficients of the static characteristics uniform. The modified virtual frequency and voltage method can be used to keep the voltage stability in the islanded microgrid, as well.**

Keywords: virtual frequency and voltage method; primary control; photovoltaic power plants; droop characteristics; load redistribution

I. Introduction

As described for example in [1], [2], [3], [4] and [5] in low voltage (LV) and in high voltage (HV) microgrids, where the line reactance is small in comparison to the line resistance ($X << R$) or both components of the line impedance cannot be neglected, voltage U and frequency f values are regulated by means of the virtual methods:

- virtual active and reactive power,
- virtual frequency and voltage,
- virtual impedance.

The aim of this article is to evaluate whether it is possible in islanded MG the redistributed load change between two parallel cooperating generators uniformly in a similar way as it is carried out in the power transmission system. For the primary voltage and frequency control it is used the modified virtual frequency and voltage method. Its simplification was proposed in [6] and in [7]. In these articles this method is verified by use simulation of the primary and secondary controls in islanded MG with one energy distributed source.

II. Virtual Frequency and Voltage Method

Principle of the virtual frequency and voltage method is same as a principle of the virtual active and reactive method. The main idea is the rotation of the frequency and voltage axes by angle φ around the origin one so that the new positions correspond to their location in the situation where $X >> R$. Then the active power P axis is parallel to the virtual frequency f′ axis and the reactive power Q axis is parallel to the virtual voltage U′ axis as shown in Fig. 1.

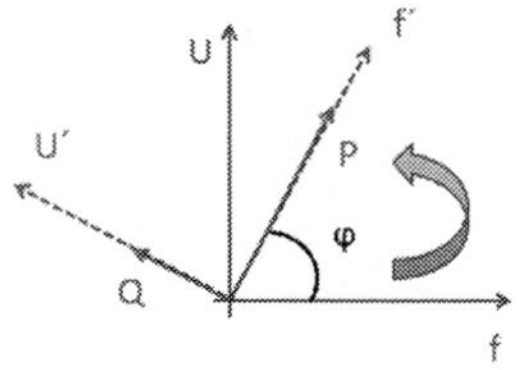

Figure 1. Virtual frequency and voltage transformation

Virtual frequency f′ and voltage U′ values results from transformation matrix T, it is seen in (1).

$$\begin{vmatrix} f' \\ U' \end{vmatrix} = T \begin{vmatrix} f \\ U \end{vmatrix} = \begin{vmatrix} \cos\varphi & \sin\varphi \\ -\sin\varphi & \cos\varphi \end{vmatrix} \cdot \begin{vmatrix} f \\ U \end{vmatrix} \quad (1)$$

The fundamental idea of the proposed simplification this method is described in [6], [7] and presented in Fig. 2

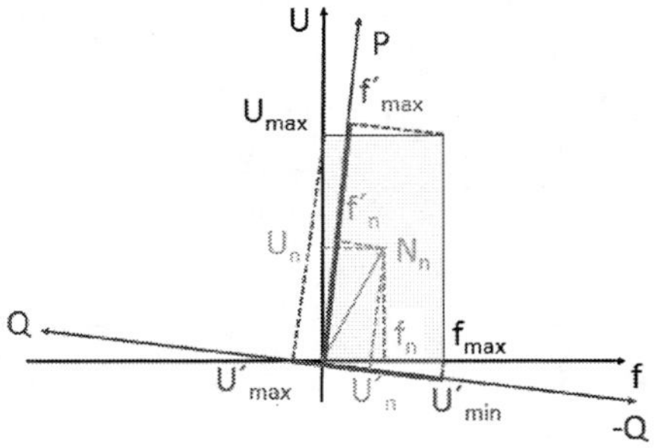

Figure 2. Principle of proposed virtual frequency and voltage method

Calculation of the static characteristic coefficients k_p and k_q, U′ and f′ values uses the fictitious vector Nn expressing nominal voltage and frequency values in MG. Function of this vector is possible partly compared to apparent power in the virtual power method. The resulting virtual frequency P = f (f′) and voltage Q = f (U′) droop characteristics are defined by the following relationships (4) and (5) that are obtained from equations (2) and (3).

$$f'-f'_n = -k_p(P - P_n) \quad (2)$$

$$U'-U'_n = -k_q(Q - Q_n) \quad (3)$$

$$f - f_n = k_p \frac{Z}{X}(P_n - P) + \frac{R}{X}(U_n - U) \quad (4)$$

$$U - U_n = k_q \frac{Z}{X}(Q_n - Q) + \frac{R}{X}(f - f_n) \quad (5)$$

The complied primary control circuit is in Fig. 3.

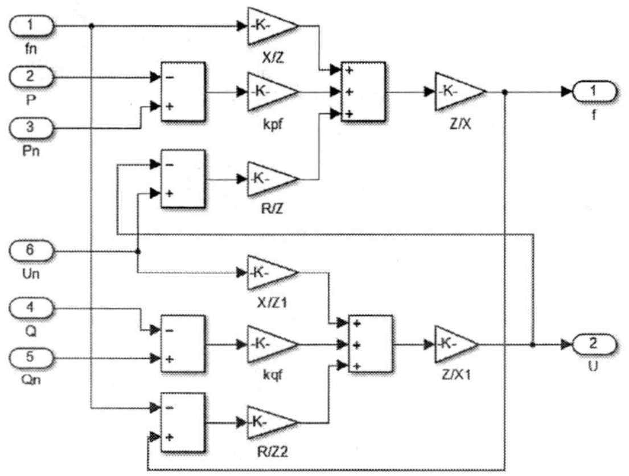

Figure 3. Primary control circuit

III. Verification of Load Redistribution Between Two Decentralized Sources

The load distribution between two decentralized sources and the proposed primary control method are verified by means simulation model of the island MG created in program MATLAB Simulink. The created model is presented in Fig. 4 and it is composed of two PV power plants with the storage devices, power lines and variable loads. These photovoltaic systems are labeled as PVPP 1 and PVPP 2 and there are connected to the point of common coupling (PCC) through 10 m long lines. In the given MG all lines have the impedance ratio R/X = 3.739 and the rated load is 32 kW. Parameters of all model parts are listed in the Tab. 1 or in Fig. 4.

Table 1. Parameters of microgrid elements

Parameter	Dimension	Value
Connected filter inductance (L)	[mH]	17
Connected filter capacity (C)	[μF]	15
Power line reactance (X)	[Ω /km]	0.083
Power line resistance (R)	[Ω /km]	0.642
Nominal voltage value	[V]	230

Both PV systems with storage devices are replaced by a DC voltage sources to which are connected the three-phase IGBT inverters and LC filters. The schema of one photovoltaic power plant is in Fig. 5. The generated voltage from the IGBT inverter is calculated by the control circuit using the voltage loop through the LC filter and the measured current and voltage in the PCC1 or in the PCC2. After the Park and Clark transformations into the two-phase system d, q the generated voltage $u_{invd}(t)$ and $u_{invq}(t)$ can be calculated according to equations (6) and (7):

$$u_{invd}(t) = u_{PCCd}(t) + L\frac{di_d(t)}{dt} + Ri_d(t) - L\omega i_q(t), \quad (6)$$

$$u_{invq}(t) = u_{PCCq}(t) + L\frac{di_q(t)}{dt} + Ri_q(t) + L\omega i_d(t). \quad (7)$$

where $L\omega i_d(t)$ a $L\omega i_q(t)$ represent so-called movement or rotational voltage. The output current from the inverter $i_{inv}(t)$ flowing through the inductance of the LC filter is expressed by components $i_q(t)$ and $i_d(t)$. L and R are components of the LC filter impedance and inverter and $i_g(t)$ in Fig. 5 is the current on the line PVPP 1 or on the line PVPP 2.

Voltage signals are adjusted by use of the pulse width modulation (PWM) and then sent to the inverters.

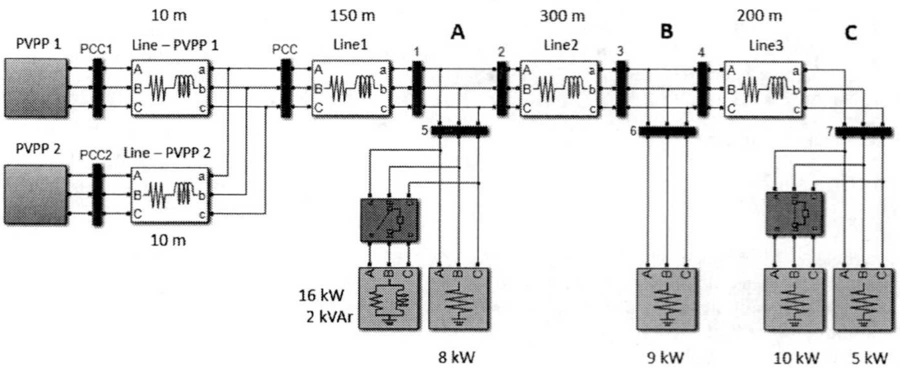

Figure 4. Simulation model of the microgrid

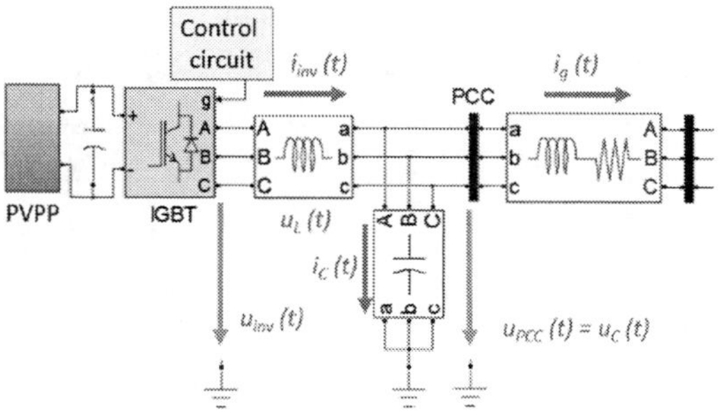

Figure 5. Simulation model of one PPVP

The main objective of the simulations is to verify in program MATLAB Simulink the load redistribution between two cooperating parallel PVPPs using of the primary control that is based on the virtual frequency and voltage method. The solved MG operates in the island mode and therefore the frequency control is not necessary. The frequency value is maintained at 50 Hz due to inverters of PVPPs. Permitted range of voltage is ± 10% U_n. Functionality of the virtual control method and a control circuit are evaluated by the active powers measurement and by the voltage waveforms on the line that depend on the load change.

There are analyzed two variants:

- PV systems with the same coefficients of static voltage characteristics,
- PV systems with different coefficients of static voltage characteristics.

In the first variant, the static voltage characteristics of the PVPPs have the same coefficients, i.e. both rated active and reactive powers of the decentralized sources are the same. In the second case the PVPPs have the different maximum and rated active and reactive powers and therefore their coefficients of the static voltage characteristics are not equal.

IV. Simulation Results

During simulation of both solved variants there were measured voltage, current and active power in the point PCC1, PCC2, PCC and points labeled number 1 to point number 7. The waveforms names of these parameters are called the same in the resulting graphs so as in Fig. 4.

A. *PV systems with the same coefficients of static voltage characteristics*

In this simulated variant of the MG the both photovoltaic power sources have the nominal active power P_n = 16.5 kW and nominal reactive power Q_n = 66.9 VAr. At the beginning of the simulation, the power supply was equal to the demand and during the first load change from t_1 = 1 s to t_2 = 2.3 s, the load is increased by 16 kW at the point "A". In the second case, in the time from t_3 = 3.5 s to t_4 = 4.5 s, there was the load with the power input of 10 kW disconnected at the end of the power line. From the results shown in the graph in Fig. 6, it can be seen that the proposed control circuit works correctly because the change in load power has been redistributed equally between the decentralized power sources. The curve of the active power P_{PCC} in the PCC1 is overlapped with the waveform of the active power in the PCC2. Exactly, as it can be assumed in two parallel cooperating generators that have the same coefficients of static characteristics. In these MG model their values are k_p = 1.4 mHz/VA a k_q = 0.15 mV/VA.

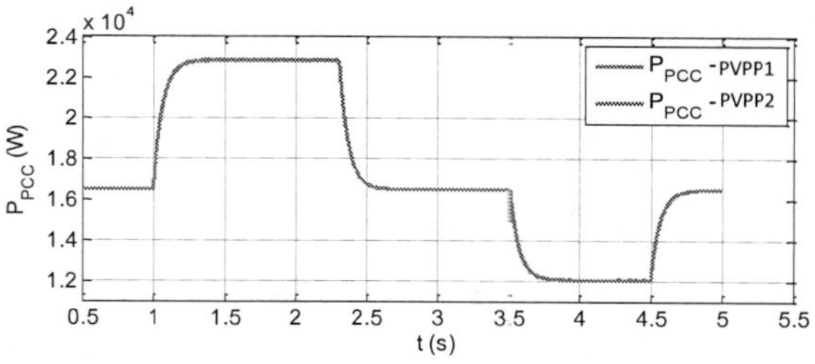

Figure 6. Waveforms of active power sources with the same coefficients of static characteristics

As can be seen from the blue curve in Fig. 7, the value of the active power in the PCC is twice higher than by photovoltaic power plant one. When the power consumption is increasing, the powers output of both PVPPs are about 22.9 kW, and therefore as well the active power in the PCC is raising to 46 kW.

From purple and green curve in Fig 7, it is seen, that in the time from t_1 to t_2 the active power values in the point "B" and "C" decreased, because the loads to that they are connected at the point "A" consume the more active power. At the time t_3 = 3.5 s, the load is disconnected at the end of the line then by using of the primary control the active powers of both sources are decreased to 12 kW and thereby the power transmitted along all long of the line, see Fig. 6 and Fig. 7.

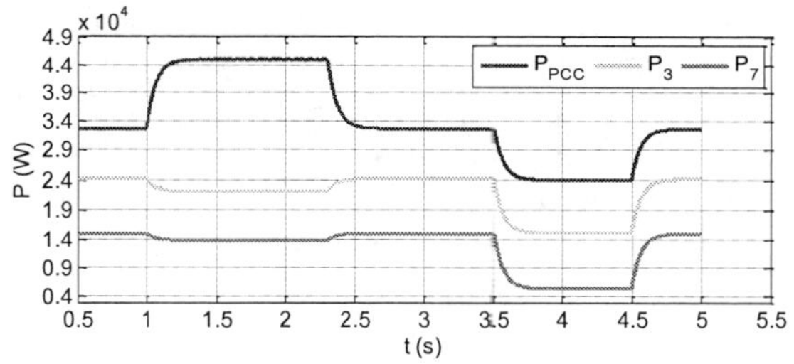

Figure 7. Active power waveforms along the line - PC sources with the same static characteristics

The active powers correspond to the voltage waveforms are plotted in Fig. 8, where U_{PCC1} and U_{PCC2} are voltages at the point of connection of each PV system to the MG. These voltages are measured on the bus called as PCC1 and PCC2. Since the power balance is restored with the help of the primary control circuit, the voltage is maintained within the allowable limits too at the end of the line. The primary control circuit stabilizes the voltage at a new constant value but does not provide the required power thereby the permanent voltage deviation occurs in the network until load is not equal to nominal value or until the secondary control circuit is not activate. But in this inverter circuit control solved model the secondary control it is not apply.

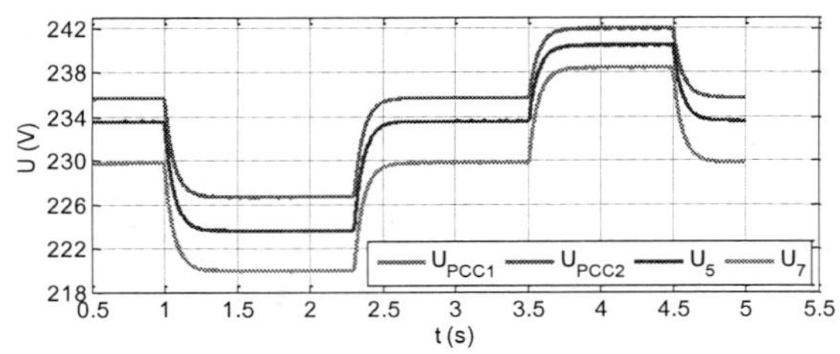

Figure 8. Voltage waveforms along the line - PC sources with the same static characteristics

From Fig. 8 it is clear that the voltage drop along the line rises with increasing the amount of the transmitted power.

B. PV systems with different coefficients of static voltage characteristics

In the second variation, the model parameters of the microgrid remain the - the active power range of the PV systems. Photovoltaic power plant PVPP1 has nominal powers P_{nPVPP1} = 13 kW and Q_{nPVPP1} = 62 VAr and the second source PVPP2 has rated powers P_{nPVPP2} = 19.6 kW and Q_{nPVPP2} = 93.5 VAr. Since the power range of the sources is different, the coefficients of static characteristics are different, for PVPP1 are k_{pPVPP1} = 1.8 mHz/VA and k_{qPVPP1} = 0.18 mV/VA and for PVPP2 are k_{pPVPP2} = 1.2 mHz/VA and k_{qPVPP2} = 0.12 mV/VA.

In the period from t_1 = 1 s to t_2 = 2.3 s the load is increased by 13 kW in the point "A" and at the time t_3 = 3.5 s the load with input power 10 kW is disconnected for one second at the end of the line. Fig. 9 shows how the PVPPs redistribute the load change exactly according to their coefficient of the static characteristics. Since the proportion of the rated P is:

$$\frac{P_{nPVPP1}}{P_{nPVPP2}} = \frac{13}{19.6} \cong 0.7 \quad (8)$$

a load change should be redistributed between these sources in this ratio.

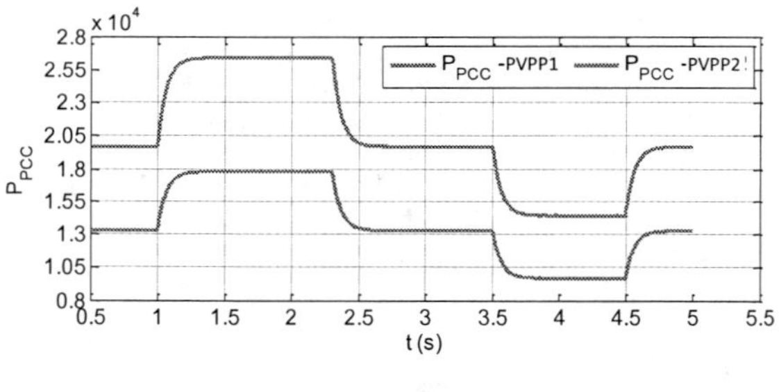

Figure 9. Waveforms of active power sources with different coefficients of static characteristics

The simulation results presented in Fig. 10 show that during the first load change the output power of the PVPP1 is increased by 4.8 kW and the active powers value of the PVPP2 is higher by 6.3 kW and when power load 10 kW was disconnected, the output power of the PVPP1 is decreased approximately to 9.7 kW and the power of the PVPP2 is about 14.4 kW. It is clear that PV systems redistributed load change between themselves in the correct ratio and designed primary control circuit according to the virtual frequency and voltage method works properly too for parallel cooperation of the several PV systems with the different slopes of the static characteristics.

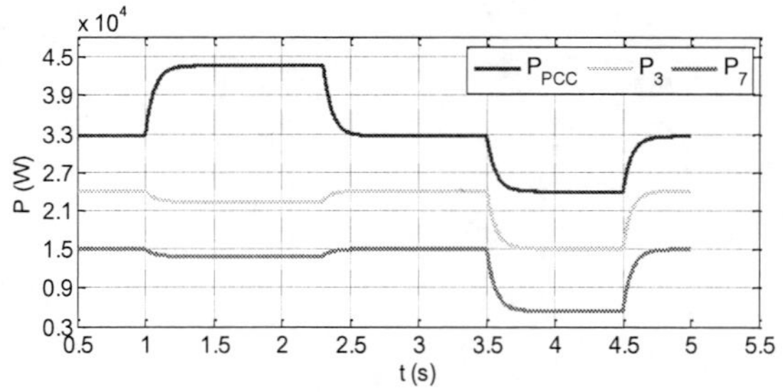

Figure 10. Active power waveforms along the line - PC sources with different static characteristics

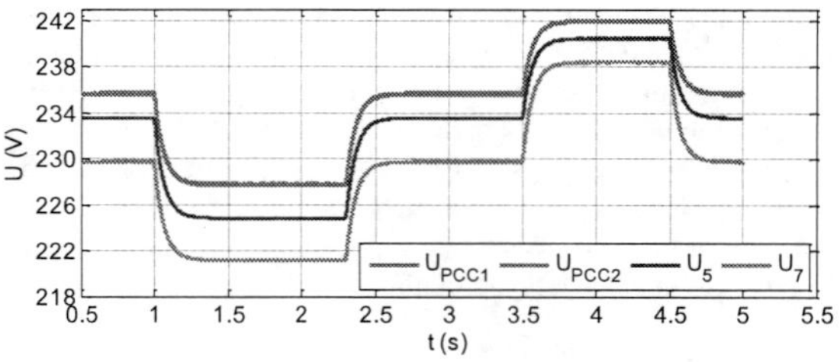

Figure 11. Voltage waveforms along the line - PC sources with different static characteristics

V. Conclusion

From the comparison of Fig. 7 with Fig. 10 and Fig. 8 with Fig. 11, it is clear, that when the 10 kW load is disconnected, the waveforms of an active power and a voltage along the lines are the same for both cases simulation. It follows that the proposed circuit control of the inverter with the primary control redistributes the load between the sources correctly according to their power range. In islanded MG there are measured same voltage and active power values irrespective of whether the load is powered by two identical or two differ power decentralized sources. This is confirmed that in the islanded MG in LV and in HV the required load can be redistributed between the individual sources in an analogous way which is ensured within the transmission system.

Acknowledgment

This paper was supported by the student research project SGS 2015 031.

References

[1] Y. LI, Y. W. LI, "Virtual frequency-voltage frame control of inverter based low voltage microgrid," In Electrical Power & Energy Conference (EPEC), 2009 IEEE, pp. 1-6, 22-23 Oct. 2009.

[2] S. M. Ashabani; Y. A. –R.-I. Mohamed, "General Interface for Power Management of Micro-Grids Using Nonlinear Cooperative Droop Control," In Power Systems, IEEE Transactions on, vol.28, no.3, pp. 2929-2941, Aug. 2013.

[3] Y. Li, Yun Wei Li,"Power Management of Inverter Interfaced Autonomous Microgrid Based on Virtual Frequency-Voltage Frame," In 2011 IEEE Trans. on Smart Grid, vol. 2, no. 1, pp. 30 – 40, March 2011, ISSN: 1949-3053

[4] J.M. Guerrero; M. Chandorkar, T. Lee; P. C. Loh, "Advanced Control Architectures for Intelligent Microgrids—Part I: Decentralized and Hierarchical Control," In Industrial

Electronics, IEEE Transactions on, vol.60, no.4, pp. 1254-1262, April 2013.

[5] P. Kanakasabapathy, I. V. V. Rao, "Control strategy for inverter based micro-grid," In Power and Energy Systems Conference: Towards Sustainable Energy, 2014, Bangalore, 2014, pp. 1-6. ISBN: 978-1-4799-3421-8.

[6] L. Rakova, E. Dvorsky, P. Hejtmankova, "Utilization of modified virtual method for frequency and voltage primary control in microgrid with photovoltaic power plants," In Power Engineering 2016, Bratislava: Slovak University of Technology, 2016, pp. 1-5, ISBN: 978-80-89402-81-6.

[7] E. Dvorsky, P. Hejtmankova, L. Rakova, "Frequency and voltage restoration in microgrid with renewable energy sources," In Power Engineering 2016, Bratislava: Slovak University of Technology, 2016, pp. 1-5, ISBN: 978-80-89402-81-6.

Behavior of Anchor-Mooring Research Operated by Induction and Permanent Magnet Synchronous Motor

Nikolay Djagarov
Nikola Vaptsarov Naval Academy
Varna, Bulgaria
jagarov@ieee.bg

Milen Bonev
Nikola Vaptsarov Naval Academy
Varna, Bulgaria
m.bonev@nvna.eu

Zhivko Grozdev
Nikola Vaptsarov Naval Academy
Varna, Bulgaria
grozdew@yahoo.com

Hristo Milushev
Nikola Vaptsarov Naval Academy
Varna, Bulgaria
h.milushev@nvna.eu

Julia Djagarova
Technical University
Varna, Bulgaria
julija_d@abv.bg

***Abstract* — In the paper is studied the work of ship's electrical power system in dynamic mode caused by the work of the anchor-mooring device for hoisting the anchor. Permanent magnet synchronous motor is used in the studied power electrical drive. Using a mathematical model, the oscillograms of the mode parameters are obtained. They are compared with the results obtained using an induction motor. The comparison shows the greater effectiveness of PMSM.**

***Keywords* — anchor mooring drive, permanent magnet synchronous motor, transient process, mathematical model.**

I. Introduction

The marine power supply systems are characterized by commensurability of loads and synchronous generators. Similar loads are propulsion for bow thrusters, electric propulsion motors, anchor-mooring devices and etc. These devices most affect the stability of the ship's power plant operation and quality of electricity [1,2]. Moreover the electricity is generated and consumed without using of transformers which means commensuration to electromagnetic and mechanical time constants of generators and electrical drives. These specific features of ship's power plant determined more stringent requirements for regulators and control systems.

One of operation mode which is studied at designing of ship's power plant is process of starting of such commensurate loads at supply from ships power station. The aim of such studies is to examine the stability of marine synchronous generators, stability at connection of powerful electric loads and the quality of transient processes [3,4].

The ship electrical drives use different types of electric drives and electronic converters. Anchor-mooring electric drive is one of those that ensures the reliability of ship operation. In recent years, modern management methods have also been widely used - vector control and direct torque control.

Synchronous, induction motors and permanent magnet motors are used. The synchronous motor is highly technological and suitable for large ships, most used in recent years. It is more efficient than the induction motor, but it can lose synchronism during large loads and fast maneuvering. Induction motor is the most used engine not only on ships but also on shore due to its qualities: simple and robust construction, low price, easy maintenance. It is used throughout the range of various electric-powered vessels. Motors with permanent magnets (usually in the rotor) made from high-energy alloys such as samarium-cobalt SmCo or neodymium-iron-boron NdFeB. Permanent magnet motors provide high power density over the entire speed and load range. They are about 30% lighter than conventional synchronous motors. They provide a maximum torque at very low speed, which is not always achieved by induction motors.

The above-mentioned advantages of PMSM explain their rapid entering into modern electric drives. The current task is also their study in dynamic transition modes and the comparison of the characteristics of induction motors and PMSM.

The present article has developed a mathematical model of a ship power station, which includes the electric drive of the anchor-mooring device using an induction motor and PMSM. The model was implemented in Matlab space and with its help its operating modes were studied.

II. Lifting of Ships Anchor Mode

Basic mode of operation is lifting an anchor from a normal depth of the gate. The loading diagram of electric drive is separated into four stages as in each of them of the operation of electric motor is characterized from a special law of torque of the shaft variation. In the beginning of lifting the anchor is in point O of bottom (Fig.1) and hawsehole of a ship standing on an anchor holds in position A_1 [4,5]. In this part of the anchor chain equal to the straight line OB_1 lies on bottom and other part A_1B_1 hangs freely in the water, preserving the form

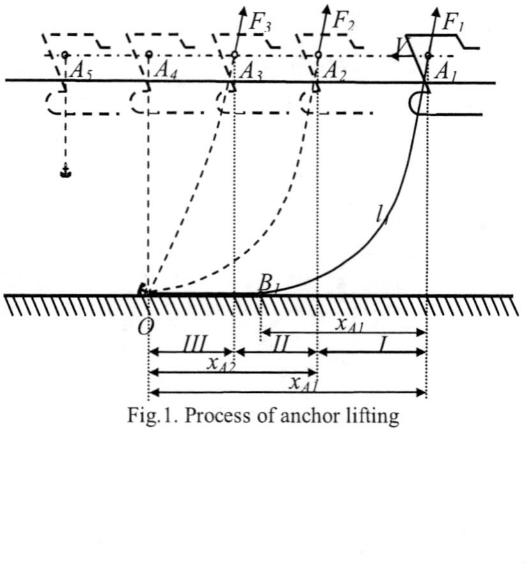

Fig.1. Process of anchor lifting

of a chain line.

After connection of electrical drive the ship becomes to move in forward with permanent speed V=const because of the effect of traction effort F_1. Up to moment of lifting of last link of the anchor chain from bottom the form of lengths of anchor chain, freely suspended in the water is not altered, because the number of lifted links is equal to the number of entered links into hawse hole. Also the tight of chain into hawse hole remain constant F_1=const, therefore the torque of electrical drive also remain constant T_L=const. This situation is kept to holding point A_2 of the hawse hole, where the last link of the chain is raised from the bottom, which ends the first stage. It is characterized by tightening the ship at constant tractive effort to the anchor star.

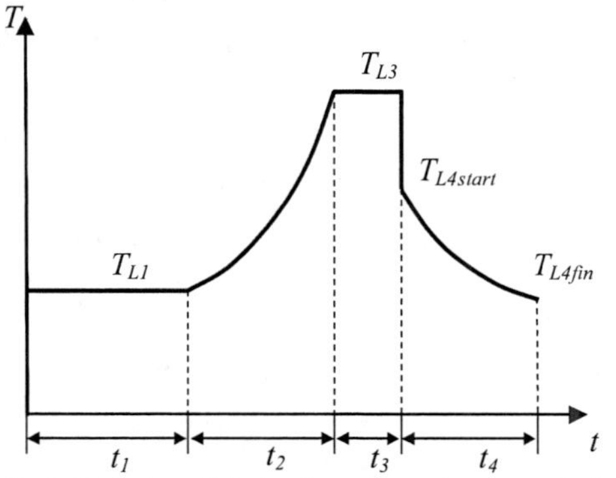

Fig.2. Variation of the load torque on the motor shaft

In second stage the vessel is moving under the influence of electric drives as the form of hanging part from the anchor chain is changing, chain is pulling, efforts in hawsehole in the anchor point are increasing. In point A_3 the tension in hawsehole reaches a value equal to the maximum traction effort created by electric drive. The time since the end of the first stage to the moment of maximum traction force is the duration of the second stage. At the end of this stage the engine speed decreases significantly due to increased torque resistance. In this way the movement of the ship will be mainly at the expense of its stocks from kinetic energy. The second stage is characterized by retraction of the anchor in its place at variable traction effort.

In part A_3A_4 the torque of electric motor is maximal – this corresponds to third stage. On the end of that stage the anchor is divided from the sea bottom. After that the length of anchor chain decrease which allows decreasing of torque of electric drive - T_L.

The fourth stage is the time from the moment of separate from the bottom of the anchor until the anchor entering into hawsehole. Loading diagram is shown on Fig.2.

III. Mathematical Model of Ships Electrical Power System

The mathematical model of the studied ships power supply station (SPSS) was constructed using a non-iterative method [6]. The model of element SPSS are recorded in the form of Cauchy in terms of currents. The bus voltage of the main switchboard is calculated by the Kirchoff first law in differential form. Below is described the created model.

Synchronous generator

$$\frac{d}{dt}\begin{bmatrix} \boldsymbol{I}_s \\ \boldsymbol{I}_r \end{bmatrix} = \begin{bmatrix} \boldsymbol{A}_{ss} & \boldsymbol{A}_{sr} \\ \boldsymbol{A}_{rs} & \boldsymbol{A}_{rr} \end{bmatrix} \cdot \begin{bmatrix} \boldsymbol{I}_s \\ \boldsymbol{I}_r \end{bmatrix} + \begin{bmatrix} \boldsymbol{B}_{ss} & \boldsymbol{B}_{sr} \\ \boldsymbol{B}_{rs} & \boldsymbol{B}_{rr} \end{bmatrix} \cdot \begin{bmatrix} \boldsymbol{U}_s \\ u_f \end{bmatrix} =$$
$$= \begin{bmatrix} \boldsymbol{H}_s \\ \boldsymbol{H}_r \end{bmatrix} + \begin{bmatrix} \boldsymbol{B}_{ss} & \boldsymbol{B}_{sr} \\ \boldsymbol{B}_{rs} & \boldsymbol{B}_{rr} \end{bmatrix} \cdot \begin{bmatrix} \boldsymbol{U}_s \\ u_f \end{bmatrix}; \qquad (1)$$
$$\frac{d}{dt}\omega_k = \frac{1}{\tau_m}\left(T_{pm} + T_e\right)$$

where: $\boldsymbol{A}$ and $\boldsymbol{B}$ are parameter-depended matrices; through subscribes s and r note stator and rotor variables and parameters respectively; $\boldsymbol{I}_s = [i_d, i_q]$; $\boldsymbol{U}_s = [u_d, u_q]$; T_{pm} is the prime mover torque; τ_m is the rotor mechanical time constant; $T_e = x_{ad} \cdot (i_d + i_f + i_g) i_q - x_{aq} \cdot (i_q + i_h) i_d$ is the electromagnetic torque; $\boldsymbol{I}_r = [i_f, i_g, i_h]$.

Induction motor

$$\frac{d}{dt}\begin{bmatrix} \boldsymbol{I}_s \\ \boldsymbol{I}_r \end{bmatrix} = \begin{bmatrix} \boldsymbol{A}_{ss} & \boldsymbol{A}_{sr} \\ \boldsymbol{A}_{rs} & \boldsymbol{A}_{rr} \end{bmatrix} \cdot \begin{bmatrix} \boldsymbol{I}_s \\ \boldsymbol{I}_r \end{bmatrix} + \begin{bmatrix} \boldsymbol{B}_s \\ \boldsymbol{B}_r \end{bmatrix} \cdot \boldsymbol{U}_s =$$
$$= \begin{bmatrix} \boldsymbol{H}_s \\ \boldsymbol{H}_r \end{bmatrix} + \begin{bmatrix} \boldsymbol{B}_s \\ \boldsymbol{B}_r \end{bmatrix} \cdot \boldsymbol{U}_s; \qquad (2)$$
$$\frac{d}{dt}\omega_r = \frac{1}{\tau_m}\left(T_e - T_l\right)$$

where: $\boldsymbol{A}$ and $\boldsymbol{B}$ are parameter-depended matrices; through subscribes s and r note stator and rotor variables and parameters respectively; $T_e = x_{ad} \cdot (i_{rd} \cdot i_q - i_{rq} \cdot i_d)$ is the electromagnetic torque; T_l is the load torque; τ_m is the mechanical time constant. $\boldsymbol{I}_s = [i_d, i_q]$; $\boldsymbol{I}_r = [i_{rd}, i_{rq}]$;

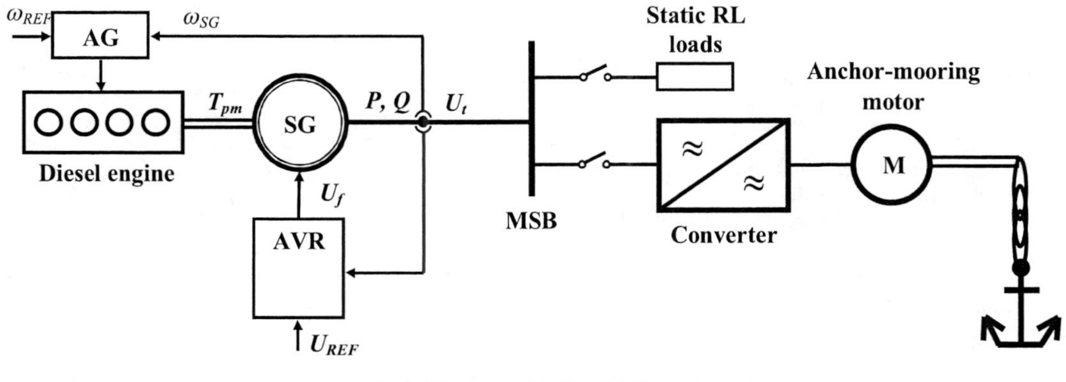

Fig.3. Diagram of studied ship's power system

$U_s = [u_d, u_q]$.

Permanent magnet synchronous motor [7]

$$\frac{d}{dt}\begin{bmatrix} i_{sd} \\ i_{sq} \end{bmatrix} = \begin{bmatrix} a_{11} & a_{12} \\ a_{21} & a_{22} \end{bmatrix} \cdot \begin{bmatrix} i_{sd} \\ i_{sq} \end{bmatrix} + \begin{bmatrix} 0 \\ a_{23} \end{bmatrix} \psi_{PM} + \begin{bmatrix} b_{11} & 0 \\ 0 & b_{22} \end{bmatrix} \cdot \begin{bmatrix} u_{sd} \\ u_{sq} \end{bmatrix} =$$
$$= \frac{d}{dt} \boldsymbol{I_s} = \boldsymbol{A_1}.\boldsymbol{I_s} + \boldsymbol{A_2}\psi_{PM} + \boldsymbol{B}.\boldsymbol{U_s};$$
$$\frac{d}{dt}\omega_r = \frac{1}{\tau_m}(T_e - T_l) \tag{3}$$

where: $a_{11} = -r_s / L_d$; $a_{22} = -r_s / L_q$; $a_{12} = (L_q / L_d)\omega_r$; $a_{21} = (L_d / L_q)\omega_r$; $b_{22} = 1/L_q$; $a_{23} = \omega_r / L_q$; $b_{11} = 1/L_d$; T_e is the electromagnetic torque; T_l is the load torque; τ_m is the mechanical time constant.

Static R-L load

$$\frac{d}{dt}\boldsymbol{I} = \boldsymbol{A}.\boldsymbol{I} + \boldsymbol{B}.\boldsymbol{U} = \boldsymbol{H} + \boldsymbol{B}.\boldsymbol{U}; \tag{4}$$

where: $\boldsymbol{A}$ and $\boldsymbol{B}$ are parameter-depended matrices: $a_{11} = a_{22} = -\frac{r_l}{l_{l1}}$; $a_{12} = -a_{21} = \omega_k$; $b_{11} = b_{22} = \frac{1}{l_{l1}}$.

The bus voltage of the main switchboard is calculated by the Kirchoff first law in differential form:

$$\frac{d}{dt}\boldsymbol{I_s^G} + m_M \frac{d}{dt}\boldsymbol{I_s^M} + m_L \frac{d}{dt}\boldsymbol{I^L} = 0; \tag{5}$$

where: m_M, m_L - scale coefficients adjusting the motor and static load parameters to the base power of the generator; $m_M = \frac{S_M}{S_G}$; $m_L = \frac{S_L}{S_G}$.

after substitution the derivatives of the currents in (5) through the corresponding right parts of (1), (2), (3) and (4) and obtain the following algebraic system of equations:

$$\begin{aligned} &\boldsymbol{H_s^G} + \boldsymbol{B_s^G}.\boldsymbol{U_{MSB}} + m_M.\left(\boldsymbol{H_s^M} + \boldsymbol{B_s^M}.\boldsymbol{U_{MSB}}\right) + \\ &+ m_L.\left(\boldsymbol{H^L} + \boldsymbol{B^L}.\boldsymbol{U_{MSB}}\right) = 0 \end{aligned} \tag{6}$$

Then convert (6) to the unknown voltage vector on the main switchboard:

$$\boldsymbol{U_{MSB}} = -\frac{\boldsymbol{H_s^G} + m_M.\boldsymbol{H_s^M} + m_L.\boldsymbol{H^L}}{\boldsymbol{B_s^G} + m_M.\boldsymbol{B_s^M} + m_L.\boldsymbol{B^L}}. \tag{7}$$

Frequency converter

The Frequency converter consists of an uncontrolled rectifier, a DC converter used to utilize the energy returned via back rectifier, and invertor realized by IGBT transistors.

The rectifier is a system with variable structure (topology), since switching on and off of the valves changes it. For the modeling of the valves are used so-called switching functions:

$$\frac{d}{dt}\begin{bmatrix} i_a \\ i_b \\ i_c \end{bmatrix} = \frac{1}{L}.\begin{bmatrix} u_a \\ u_b \\ u_c \end{bmatrix} - \frac{R}{L}.\begin{bmatrix} i_a \\ i_b \\ i_c \end{bmatrix} -$$
$$-\frac{U_{dc}}{L}.\begin{bmatrix} (2S_a - S_b - S_c)/3 \\ (2S_b - S_c - S_a)/3 \\ (2S_c - S_a - S_b)/3 \end{bmatrix} = \tag{8}$$
$$= \frac{d}{dt}\boldsymbol{I_{abc}} = \frac{1}{L}.\boldsymbol{U_{abc}} - \frac{R}{L}.\boldsymbol{I_{abc}} - \frac{U_{dc}}{L}.\boldsymbol{S}$$

where: $\boldsymbol{U_{abc}}$, $\boldsymbol{I_{abc}}$ - vectors of voltage and current; $\boldsymbol{S}$ - vector of switching functions, elements of its accept values 0 or 1 corresponding to the open or close condition of valves; U_{dc} - output rectifier voltage; R, L - active and reactive resistance of rectifier.

After transformation to the axes $d, q, 0$ the system of equations will be:

$$\frac{d}{dt}\boldsymbol{I_{dq}} = \frac{1}{L}.\boldsymbol{U_{dq}} - \frac{R}{L}.\boldsymbol{I_{dq}} - \frac{U_R}{L}.\boldsymbol{P}.\boldsymbol{S} - \boldsymbol{W}.\boldsymbol{I_{dq}} \tag{9}$$

where: $\boldsymbol{P}$ - matrix of direct transformation of Park.

$$\boldsymbol{W} = \begin{bmatrix} 0 & -\omega_k \\ \omega_k & 0 \end{bmatrix}.$$

The voltage of condenser is calculated by following expression:

$$\frac{d}{dt}U_{dc} = \frac{S_{dc}.I_C}{C}; \tag{10}$$

where: S_{dc} is switching function of converter, which have two levels and control the state condition of converter transistors – *1* for conducting state and *0* for no conducting state; $I_C = I_R - I_I$.

Equations for the output converter voltage:

$$\begin{bmatrix} u_a \\ u_b \\ u_c \end{bmatrix} = \frac{1}{3}.\begin{bmatrix} 2 & -1 & -1 \\ -1 & 2 & -1 \\ -1 & -1 & 2 \end{bmatrix}.\begin{bmatrix} S_{1a}.S_{2a} \\ S_{1b}.S_{2b} \\ S_{1c}.S_{2c} \end{bmatrix}.U_{dc}; \tag{11}$$

where: S_{kj} - switching function accepting values 1 or 0 according state conditions of valves (conducting state or non-conducting state); k - name of frame(a, b, c); j - number of valve in the frame$(j = 1,2)$.

Current of direct current side:

$$I_I = S_{1a}.S_{2a}.i_a + S_{1b}.S_{2b}.i_b + S_{1c}.S_{2c}.i_c. \tag{12}$$

Output voltage of inverter in axes $d, q, 0$ is calculated through next expression:

$$\begin{bmatrix} u_d \\ u_q \\ u_0 \end{bmatrix} = \frac{2}{3} \cdot \begin{bmatrix} \cos\theta_k & \cos(\theta_k - 120^\circ) & \cos(\theta_k + 120^\circ) \\ -\sin\theta_k & -\sin(\theta_k - 120^\circ) & -\sin(\theta_k + 120^\circ) \\ 1/2 & 1/2 & 1/2 \end{bmatrix} \bullet$$

$$\bullet \begin{bmatrix} u_a \\ u_b \\ u_c \end{bmatrix} = \boldsymbol{P}.\boldsymbol{U}_{abc} = \boldsymbol{U}_{dq0}. \qquad (13)$$

IV. Control of Lifting of Ships Anchor Mode

For control of synchronous generator is used standard models of IEEE automatic voltage regulator (AVR) [8] и automatic governor (AG) [9].

Vector control provides tighter speed control, higher starting torque, and higher low-speed torque [10,11,12]. The most significant benefit of this control is that it allows up to 200% of the motor's rated torque to be produced at zero speed. This is especially useful in the anchor hoisting.

The vector control have ability to control both the amplitude and phase of the AC excitation in the motor. The vector control of currents and voltages results in control of spatial orientation of rotor and stator magnetic fields. Hence, it is sometime also referred as field orientation control. The requirements necessary to control the torque production in a motor are: an independently controlled stator current to overcome the effects of stator winding resistances, leakage induct-

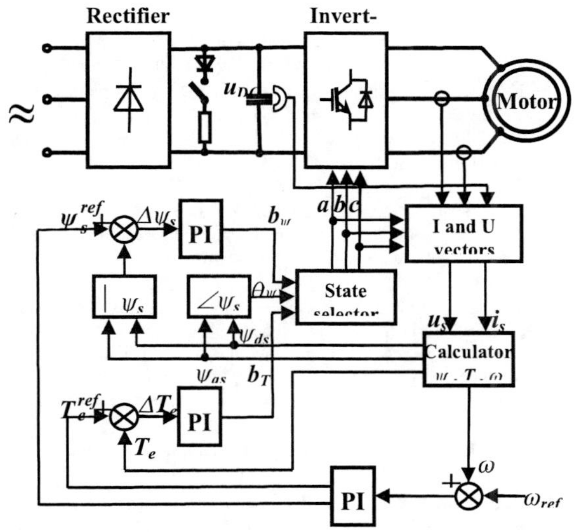

Fig.4. Block diagram of vector control of anchor-

ances and induced voltage. an independently controllable spatial angle.

On the fig.4 is shown block diagram of sensorless vector control of anchor-mooring drive. As can be seen from the figure, using the measured two stator currents and the motor model is calculated its flux ψ, torque T and speed ω. The control is carried out on two mutually perpendicular and independent variables: the flux and torque with the help of PI regulators. Then the control signals for inverter transistors are generated.

III. Experiments Results

The created model was implemented in Matlab space [13]. The lifting mode of the anchor was followed, at the engine shaft resistance moment, according to the time diagram of fig.2. The perspective figures show part of the results obtained in the form of time diagrams of the basic parameters for the transition process. The curves for changing the basic parameters of the regime are shown.

The comparison of the obtained results using an induction motor and PMSM show that synchronous driving consumes 30% less power than the induction motor. This provides less disturbance in the ship's power station - smaller deviations in the frequency and operating voltage of the anchor-mooring device.

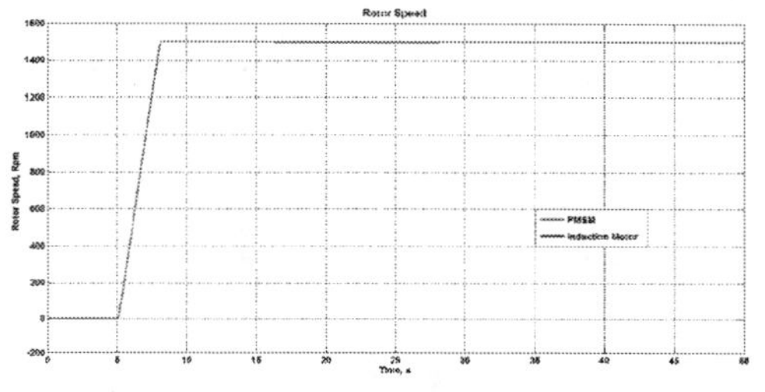

Fig.5. Rotor speed

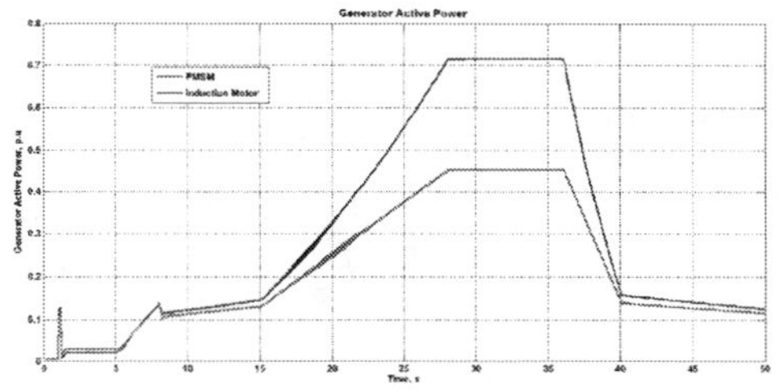

Fig.6. Active power of generator

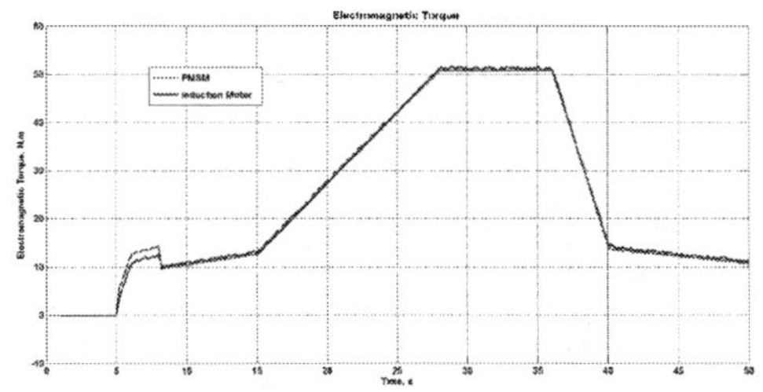

Fig.7. Motor electromagnetic

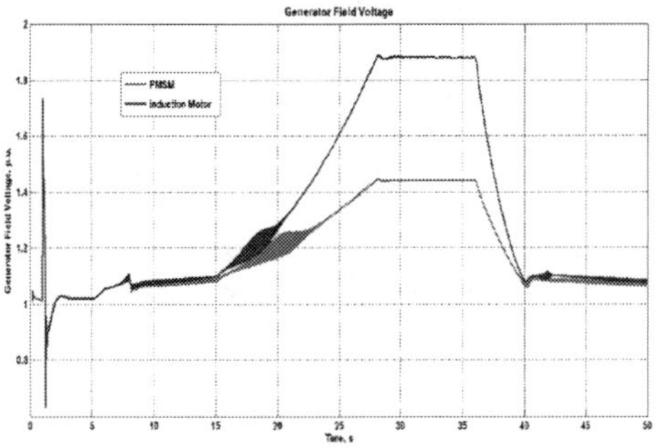

Fig.8. Generator field voltage

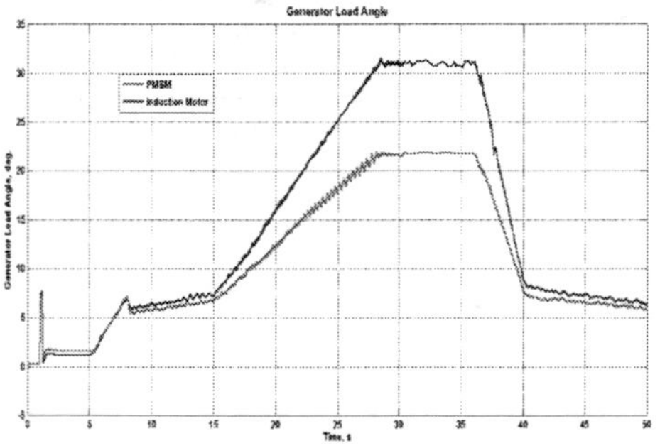

Fig.9. Generator load angle

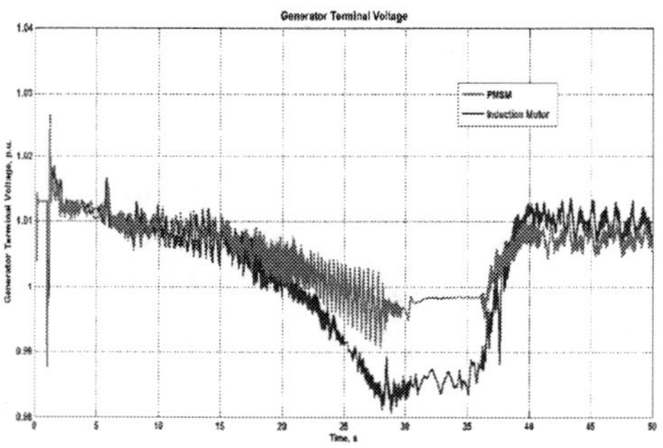

Fig.10. Generator terminal voltage

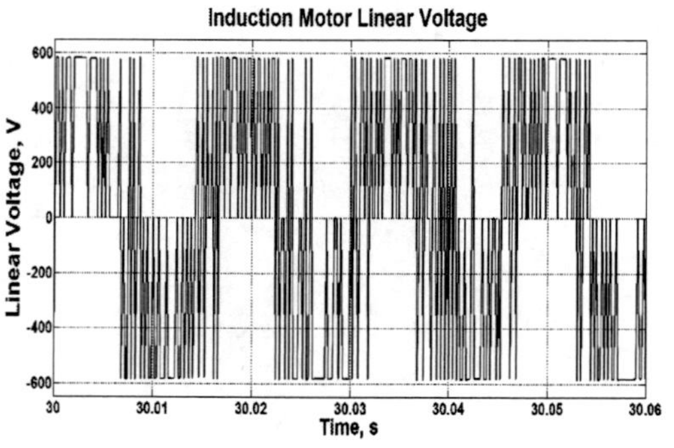

Fig.11. Induction motor linear voltage

Conclusion

It is proposed to use PMSM to drive an anchor-mooring device using vector control. Vector control provides tighter speed control, higher starting torque, and higher low-speed torque. The most significant benefit of this control is that it allows up to 200% of the rated torque to be produced at zero speed. This is especially useful in the anchor hoisting.

A mathematical model of the studied anchor-mooring device was developed, with the help of which the anchor lifting modes were simulated The modeling results show the high efficiency of the used PMSM drive as compared to an asynchronous drive.

The power consumption of PMSM is 1/3 less than that of an asynchronous drive, which has a beneficial effect on the operation of the synchronous generator: it does not consume reactive power; Smaller voltage and frequency deviations, better transient processes.

References

[1] Hebner R. E., "Electric Ship Power System – Research at the University of Texas at Austin", IEEE Electric Ship Technologies Symposium IEEE ESTS 2005, 25-27 July 2005, Philadelphia, PA USA.

[2] J. Clerk Maxwell, A Treatise on Electricity and Magnetism, 3rd ed., vol. 2. Oxford: Clarendon, 1892, pp.68-73.

[3] Nord T., "Voltage Stability in an Electric Propulsion System for Ships", Thesis of Royal Institute of Technology, Stockholm, Sweden 2006.

[4] N.Djagarov, Ship's Electrical Power Systems, Technical University of Varna, Bulgaria, 1997, 375 p. (in Bulgarian).

[5] N.Djagarov, M.Bonev, Zh.Grozdev, "Study of the work of ship electrical power station at anchor-mooring device operation," 10th IEEE International Conference on Environment and Electrical Engineering (EEEIC), Rome, 8-11 May 2011, pp.1221-1224.

[6] N.F.Djagarov, "A method of transient electromechanical processes modeling in power systems", Proceedings of 2009 IEEE Bucharest PowerTech Conference, June 28th - July 2nd, Bucharest, Romania, 6 p.

[7] N.Djagarov, M.Panayotov, S.Ficlchev, J.Djagarova, M.Bonev, Zh.Grozdev, D.Hristov, D.Cvetanov, "Mathematical models of wind permanent magnet synchronous generator", IIIth International Conference Power Engineering by Eyes of Youth, 22-26 October 2012, Yekaterinburg, Russia, pp.51-62 (in Russian).

[8] IEEE Recommended Practice for Excitation System Models for Power System Stability Studies, IEEE Standard 421.5-2016

[9] IEEE Guide for Synchronous Generator Modeling Practices and Applications in Power System Stability Analyses, IEEE Standard -1110-2002.

[10] P. Vas, Sensorless Vector and Direct Torque Control, Oxford University Press, 1998, 729 p.

[11] B.K. Bose, Modern Power Electronics and AC Drives, Prentice-Hall, 2002, 711 p.

[12] P.C. Krause, O. Wasynczuk, S.D.Sudhoff. Analysis of electric machinery and drive systems, IEEE Press, 2013, 659 p.

[13] Using MATLAB, Version 6, Revised for MATLAB 2013b, MATLAB® The Language of Technical Computing.

Investigation of Hoisting Electrical Drive with Series Electromechanic Brake Mode Operation

Nikolay Djagarov
Nikola Vaptsarov Naval Academy
Electrical Department,
Varna, Bulgaria
jagarov@ieee.bg

Julia Djagarova
Technical University,
Electrical Department
Varna, Bulgaria
julija_d@abv.bg

***Abstract* — Using a mathematical model of an electric hoist asynchronous electric drive with direct torque control using an electromechanical brake, the winding of which is series with one of the phases, are examined the operating modes. The modes of starting, operating and stopping the electric drive were studied. A part of the experimental results has been added. The obtained results give the possibility to design electric hoists and their control using frequency converters.**

Keywords — hoisting electrical drive; induction motor; series electromechanical brake; direct torque control; mathematical model; transient modes investigation.

I. INTRODUCTION

Electric hoists are an important part of lifting mechanisms and cranes. Frequent starts and stops are typical for them. Modern electric hoists usually use asynchronous motors with vector control or direct torque control. An important part of the electric hoist is the electromagnetic brake, which greatly determines the efficiency of its operation. Safety and control is crucial in hoisting electrical drives. They require high accuracy and do not allow any errors. The critical starting and stopping can lead to harmful trusts and incorrect tripping, and accidental dropping of the load.

The modeling of the processes in the electric hoist allows to examine the interaction processes between the different parts of the electric hoist, to choose the control method, to design the optimal lifting device.

Modern electric hoists perform complex tasks. They must control the speed of lifting and lowering the load, limit the starting current, hold the load stationary on command or when there is no power supply. To fulfill these tasks, a brake is essential to provide reliable and accurate braking to decelerate, stop or hold the load. Braking systems available for use on AC induction motors can be divided into two basic categories [1]: external braking; internal braking. External braking uses electrical or mechanical devices mounted on the exterior of the motor and includes such methods as: friction braking-spring-set, electrically released; magnetic braking-electrically set, spring-released; eddy current braking.

It is not possible to ensure normal operation of electric drives of electric hoists without reliably acting brakes. The overview of the braking devices used shows that there are no universal braking devices. Therefore, the creation and testing of electromechanical brakes is topical, ensuring effective rapid and accurate braking.

The electromechanical transient processes in the electric hoists at start, power off, and subsequent braking have not been sufficiently investigated. The specificity of the transient processes is that they consist of different stages, accompanied by the commutation of stator circuits and braking devices. Transient modes are: power on and load lifting; Shutting down the stator power or brake in series respectively; mechanical braking.

The main distinguishing feature of the investigated electric hoist is that the brake electromagnet is in series between the frequency converter and the asynchronous motor stator. That is why it is important to investigate the mutual operation and influence of the drive and the brake.

Electromechanical brakes of asynchronous electric hoists are simple and inexpensive, providing:

- Fastest braking compared to other methods;
- Reliable braking when power is lost. The other methods stop braking process;
- There are no additional motor losses, which allows frequent stops.

This study uses the direct torque control method [2÷8]. The difference between vector control and direct torque control is that direct torque control does not have fixed patterns to switch the inverter's switches. The switching of the inverter's switches is in accordance with the load requirements. Due to the dropping of the inverter switching patterns (characterizing scalar and vector control), it`s performance is much higher when the load is sharply changed. Since high accuracy of speed control is ensured, the need for feedback devices is eliminated.

II. DIRECT TORQUE CONTROL OF INDUCTION MOTOR

The characteristic feature of hoisting applications is that the torque that puts a motor under load is always an active one. The electromagnetic brake holds the motor shaft when its electromagnetic when the supply is turned off. Such logic of control is necessary as during any break-down the full stop of the motor shaft must occur. The brake mustn't release the motor so long as the motor is not capable to keep the load. The brake control takes place at the start and braking of a load. When starting the brake control is possible in two modes: in speed mode, where a moment when the brake is released is determined by the motor current value - mechanical brake control mode; in torque mode, the value of the torque that keeps the motor load is developed in the torque mode control – hoist mechanical brake mode.

In fig.1 is shown block diagram of direct torque control of hoisting induction drive with series electromechanic brake. To control the asynchronous drive, direct sequence selection of the inverter states, called direct torque control, is used. Therefore, the most commonly used is pulse-width modulation, controlled by a state variables vector. The hysteresis modulation is used as a feedback current control method where the motor current tracks the reference current within a hysteresis band.

The DC voltage u_{DC} and the two stator currents i_a and i_b are measured, the stator voltage $\boldsymbol{u}_s^s$ and current $\boldsymbol{i}_s^s$ vectors are defined by calculation. The voltage vector is calculated via $\boldsymbol{U}_i$ and inverter switching variables a,b,c . The produced moment T_e and stator flux vector ψ_s are calculated by $\boldsymbol{u}_s$ and $\boldsymbol{i}_s$. The stator flux amplitude ψ_s is compared to the reference ψ_s^{ref} and T_e - to the reference T_e^{ref} in the control circuit.

Deviations $\Delta\psi_s$ and ΔT_e respectively are advanced to relay controllers. The flux b_ψ controller output signal can accept values 0 and 1, and for the moment b_M -1 , 0 , 1 . The choice of the inverter state is based on the b_ψ and b_T values. It also depends on the current position of the stator flux ψ_s vector, i.e. form the angle θ_ψ and the direction of the motor rotation.

Only the stator resistance R_s is necessary for the direct torque control algorithm, whose actual value is critical to ensure a good characteristic at low rotation speeds of the drive. At low speeds the stator voltage is low and commensurable with the voltage drop on R_s . Therefore, modern adjustable direct torque control drives are equipped with evaluators or observers of this resistance.

III. Investigatsion of Hoisting Electrical Drive

In [9] is proposed complex mathematical model of induction motor hoisting electric drive systems with series brake for determining the main quality parameters of the transient processes.

The modeling of the start and stop modes of the asynchronous drive provides characteristics acquisition of the transition conditions, which helps to select the method, control algorithm and the technical implementation of the controller performing the startup, operation and stopping of the electric hoist.

By using mathematical modeling, can be investigated the braking modes of an asynchronous drive by means of an electro-magnetic brake, the power of which is switched on consecutively in one of the phases.

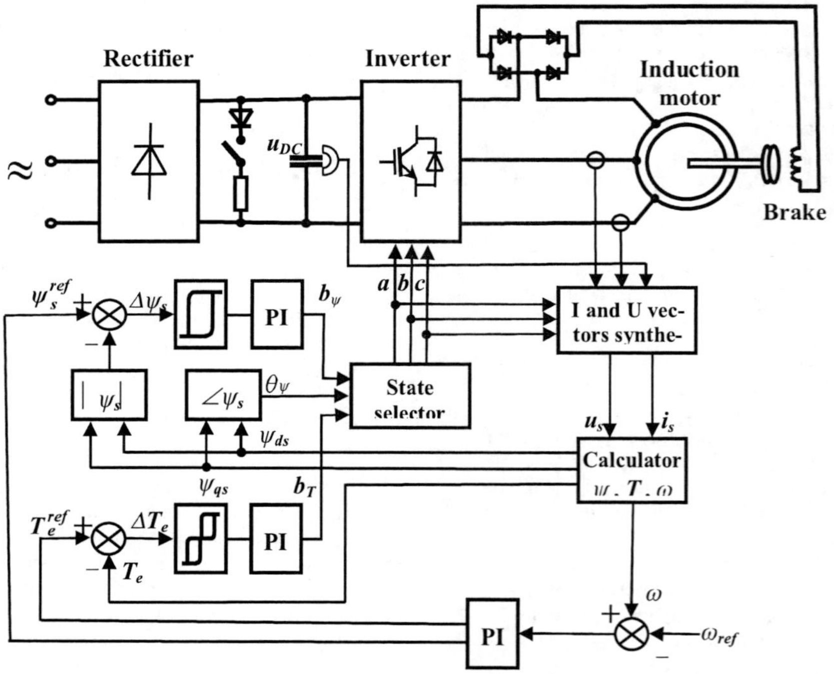

Fig.1. Block diagram of direct torque control of hoisting induction drive with series electromechanic brake

The designed in [9] mathematical model is simulited in Matlab.

At first, the pulling force caused by the brake coil increases, but cannot overcome the force of the spring. With the further increase of the force, it becomes larger than the spring force and the brake disc separates from the stationary disk, supporting an air gap, the motor rotates.

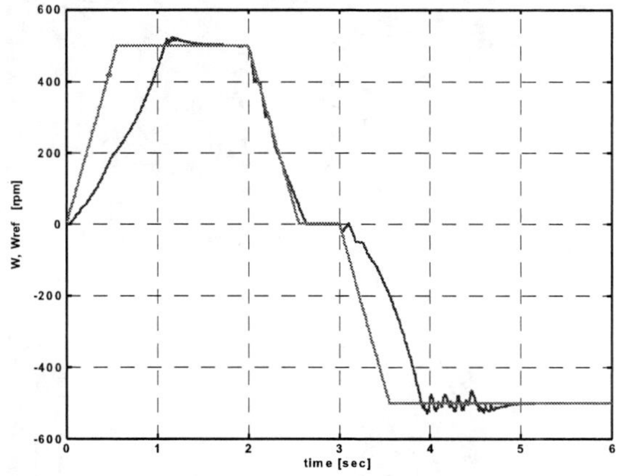

Fig.2. Speed of rotation ω and speed reference ω_{ref} (W, Wref)

To stop the motor, it is sufficient to turn off the power supply. At the first moment the force of restraint decreases but is still greater than the force of the spring. When the breaking force is reduced under the force of the spring, the brake disc is pressed against the stationary disc and the mechanical braking starts.

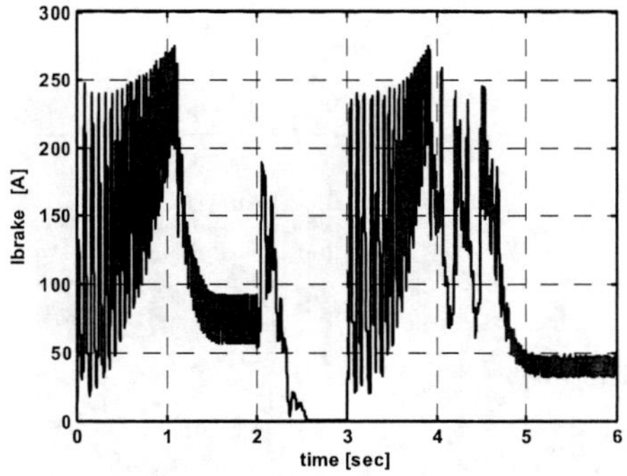

Fig.3. Brake current i_{br} (Ibrake)

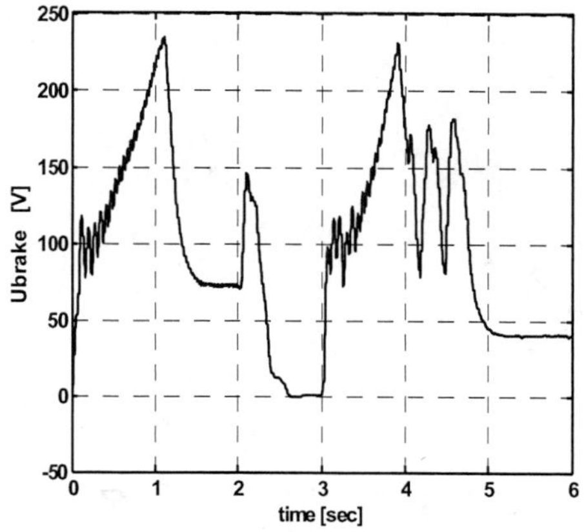

Fig.4. Brake voltage u_{br} (Ubrake)

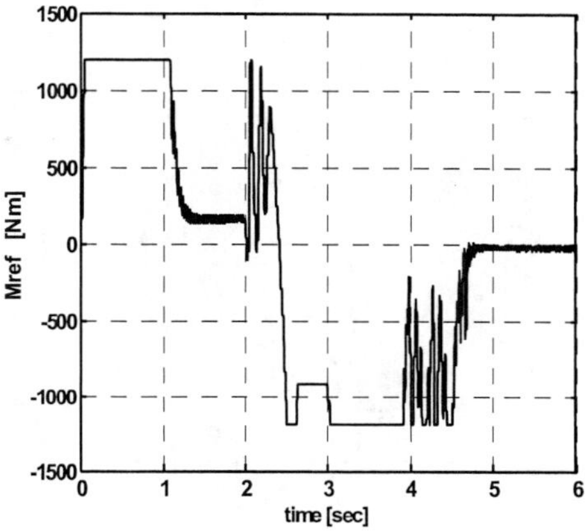

Fig.5. Torque reference of motor T_{ref} (Tref)

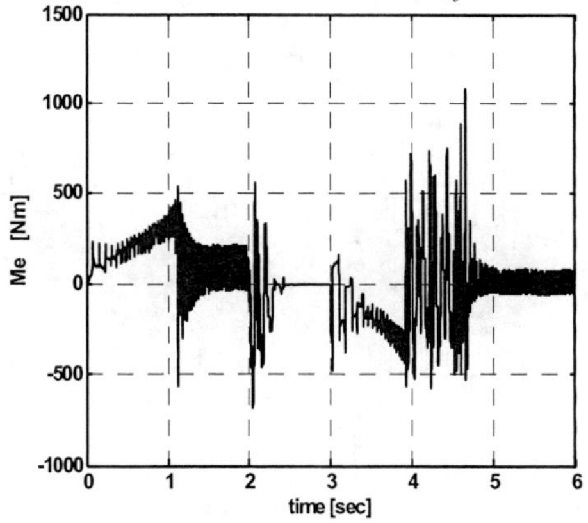

Fig.6. Electromagnetic torque of motor T_{em}

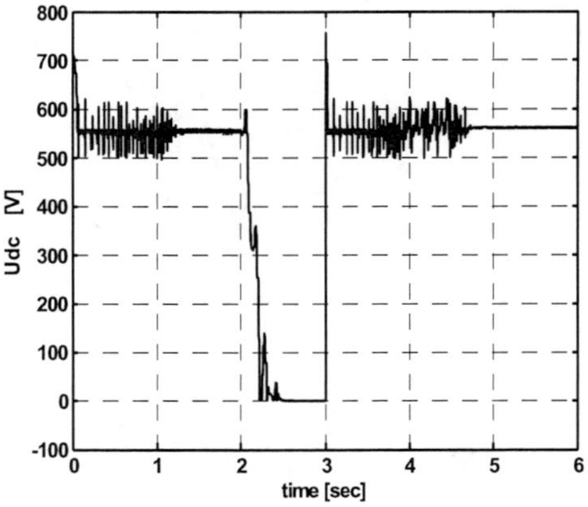

Fig.7. DC voltage of converter U_{DC} (Udc)

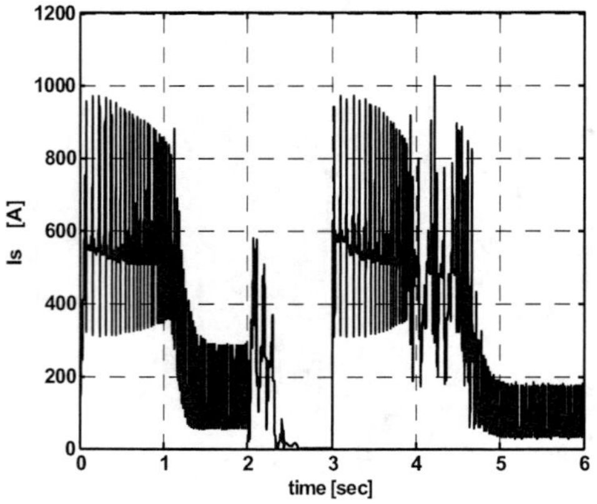

Fig.8. Amplitude of stator current $I_s = \sqrt{i_d^2 + i_q^2}$

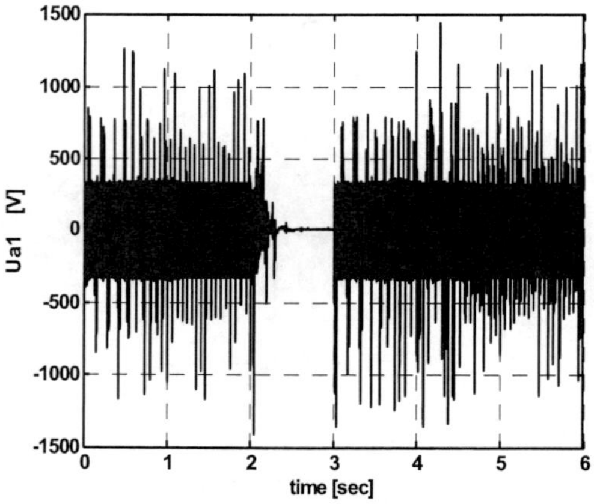

Fig.9. Phase A voltage in which the brake is included u_a

The electromechanical brake releases the motor shaft when the supply of motor is turned on and when the motor current creates a torque that exceeds the braking torque. The hoisting winch start-up in the speed mode, during which the current reaches the value necessary to take over and to keep the load.

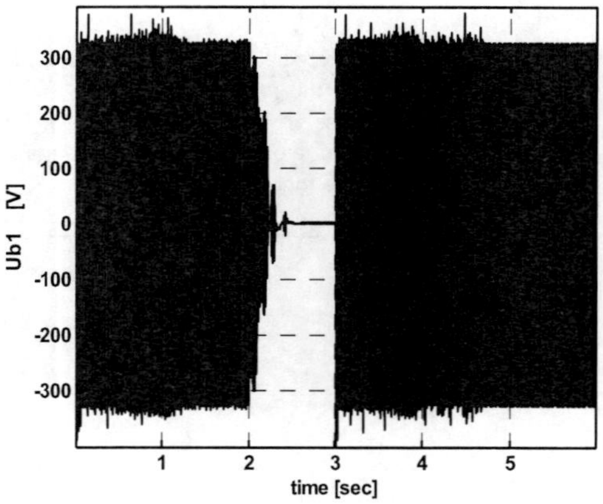

Fig.10. Phase B voltage u_b (Ub1)

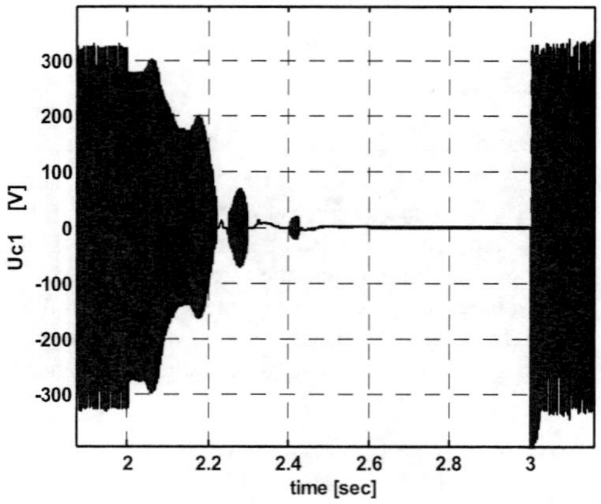

Fig.11. Phase C voltage u_c (Uc1)

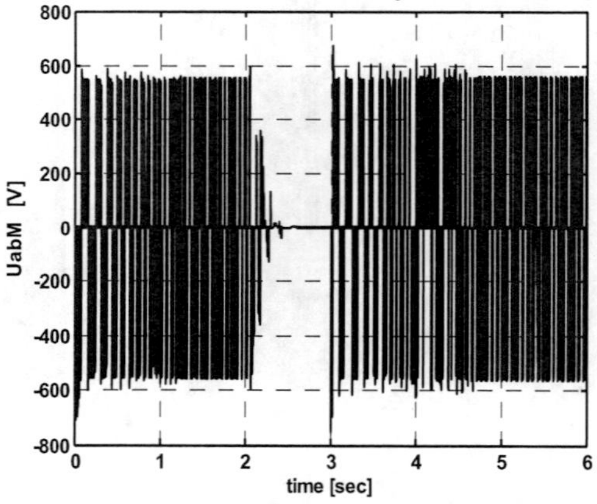

Fig.12. Phase to phase voltage of motor u_{abM} (UabM)

The electromechanical brake is on (it 'catches' the motor), when the motor is in the ramp-down mode (after activating the stop mode), and when the output speed is lower than the reference value.

The basic difference between controlling the brake in the speed and in the torque mode relies in the fact that the moment of releasing the brake is not decided by the current value anymore.

The results of the investigation of hoisting electric drive with a series electromechanical brake using mathematical modeling show the adequacy of the model used. With its help, the parameters and the characteristics of the transient modes of the electric hoist can be obtained.

The obtained results can be used to study the processes in the electric hoist in normal and emergency modes of operation. The results obtained from the investigation can be used to select optimal operating modes and law and control system, and also to design an electric hoist with the converter, the brake and the control system.

Conclusions

Using a mathematical model are investigated the operating modes of a modern electric hoist using a DC frequency converter, direct torque control and a series electromechanical brake provide accurate and reliable braking.

A complex mode of operation has been investigated: start, load lifting, braking, lowering the load. The obtained results can be used to design the electric hoist and its control. By the quality parameters of the system transients is the control system is suggested.

References

[1] R. Shemanske, "Electronic motor braking", IEEE Transactions on Industry Applications, Vol.IA-19, Apr 1983, pp.824-831.

[2] J.N. Nash, "Direct torque control, induction motor vector control without an encoder", IEEE Transactions on Industry Applications, Vol.33, No.2, Mar/Apr 1997, pp.333–341.

[3] A.M. Trzynadlowski, Control of Induction Motors, Academic Press, 2001, 228 p.

[4] Torque Control, Editor M.T. Lamchich, InTech, 2011 InTech, 304 p.

[5] M.P. Kazmierkowski, L.G. Franquelo, J. Rodriguez, M.A. Perez, J.I. Leon, "High-performance motor drives", IEEE Industrial Electronics Magazine, Vol.5, No.3, Sept 2011, pp.6–26.

[6] F. Giri, AC Electric Motors Control: Advanced Design Techniques and Applications, 2013, John Wiley & Sons, 555 p.

[7] P.R. Mankad, A.R. Chudasama, "Review of improved direct torque control methodologies for induction motor drives", International Journal of Innovative Research in Advanced Engineering (IJIRAE), Vol.1, No.6, July 2014, pp.393-399.

[8] N.V.S.K. Avinash, M. Avinash, B. Bharath, Y. Teja, B.Sai, V. Vidya, "Direct torque control of induction motor - a comparison", International Journal of Control Theory and Applications, Vol.9, 2016, pp.7547-7556.

[9] N. Djagarov, J. Djagarova, "Hoisting electrical drive mathematical model with series electromechanic brake", Proceedings of International Conference "Elektroenergetika", 2017, Stara Lesnia, Slovakia.

Wind Load of Overhead Electrical Lines Exceeding AC 1 kV

Stanislav Ilenin
Department of Electric Power Engineering
Faculty of Electrical Engineering and Informatics
Technical University of Košice
Košice
stanislav.ilenin@tuke.sk

Ladislav Varga
Department of Electric Power Engineering
Faculty of Electrical Engineering and Informatics
Technical University of Košice
Košice
ladislav.varga@tuke.sk

Abstract — **The article deals with the influence of wind on wires and poles of overhead power lines. Wind causes additional load; thereby, it increases the mechanical stress in the wires. The calculation of wind force on wires depends on its velocity, span size, direction, and various factors affecting it. Wind may be also exerting flagpoles; therefore, they must be dimensioned to the total force converted to the top of the mast.**

Keywords: overhead power lines, wires, wind speed, wind force

I. Introduction

Load impacting poles of the overhead electrical lines may be constant and random. Random load can be short-term and extraordinary. Load-bearing parts of the flagpole structure, consoles and brackets of earthing ropes, insulating hinges, wires, earthing and anchoring ropes, and pulling from anchorages of the anchored flagpoles are considered to be a constant load. Heavy icing on wires and earthing ropes, furthermore wind pressure, insulting hinges, conductors and anchoring ropes are important from a short-term perspective. The article deals with the effects of wind overloading external power lines. It focuses mainly on overloading the wires and the flagpoles. The calculation is made for specific 22 kV overhead electrical line based on different standards, whereas the results are compared. The dimensions of the flagpoles, consoles and wire mount are shown in Figure 1.

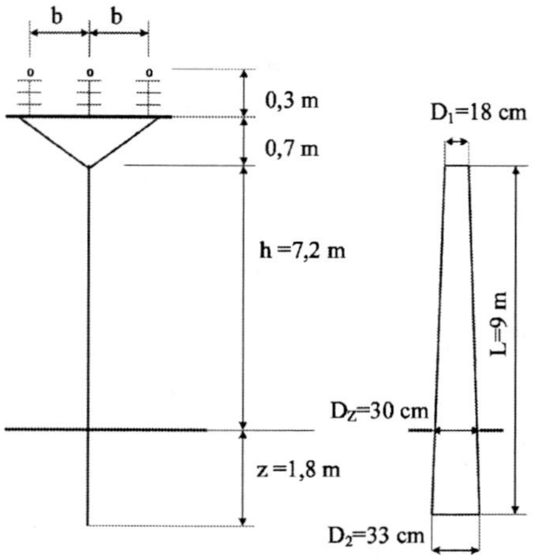

Figure 1. Dimensions of the pole

For the conductors of 22 kV overhead electrical line, the AlFe 42/7 rope is used [1]:

- diameter of rope d = 9,0 mm,
- mathematical cross - section S = 49,48 mm^2,
- cross - section ratio S_{Al}:S_{Fe} = 6,
- nominal weight q_1 = 0,17159kg.m^{-1},
- weight density γ = 0,03401 MN.m^{-3},
- maximum allowable stress σ_{dov} = 115,866 MPa,
- temperature coefficient of expansion α = 18,714.10^{-6} °C^{-1},
- length of spans L = 80 m.

II. Calculation according to STN 33 3300 (to 2006)

Under wind load, it is always assumed that wind direction is horizontal. The static component of wind load W in N applied perpendicular to the wires and utility poles [2].

$$W = w^n \cdot A \tag{1}$$

where

w^n standard wind pressure on line element in N.m^{-2},

A area of the line element exposed to wind in m^2.

A. Wind Forces on Conductors

$$W_v = w_v^n \cdot A_v \tag{2}$$

where

W_v static component of wind load conductors in N,

w_v^n standard wind pressure for wires in N.m^{-2},

A_v surface of conductor exposed to the wind in m^2.

According to STN 33 3300, table 18, standard wind pressure for conductors with diameter less than 16 mm, if the hinging point of the conductors of the terrain is up to 20 m w_v^n = 484 N.m^{-2}.

The area of the conductors exposed to the wind is calculated as the product of the driver diameter and the length of the span $A_v = d \cdot a$ = 0,72 m^2.

The wind force per one conductor is W_v = 348,48 N.

Wind force acting on three conductors is W_{v3} = 1045,44 N.

B. *Wind Force on Flagpoles*

The area of the flagpole A_s in the wind in m^2 shall be calculated:

$$A_s = \frac{(D_1 + D_z)}{2} \cdot h \qquad (3)$$

where

D_1 the diameter of the flagpole at the top in m,

D_z the diameter of the flagpole in the ground in m,

h height of the pole above the ground in m.

The diameter of the pole at the ground level is calculated:

$$D_z = \frac{(D_2 - D_1)}{L} \cdot h + D_1 \qquad (4)$$

where D_2 refers to the diameter of the pole in the heel.

Area of the pole exposed to wind A_s = 1,728 m^2.

The force of the wind W_s acting on the flagpole in N is calculated:

$$W_s = w^n \cdot A_s \qquad (5)$$

where

w^n standard wind pressure on the pole in N.m^{-2},

A_s area exposed to wind in m^2.

According to STN 33 3300, table 19, standard wind pressure on the flagpole is w^n = 385 N.m^{-2}.

The force acting on the flagpole is W_s = 665,28 N

C. *Wind Force at the Top of the Pole*

Wind forces acting on the overhead line conductors W_v and on the pole W_s should be recalculated for the top of the pole F_V.

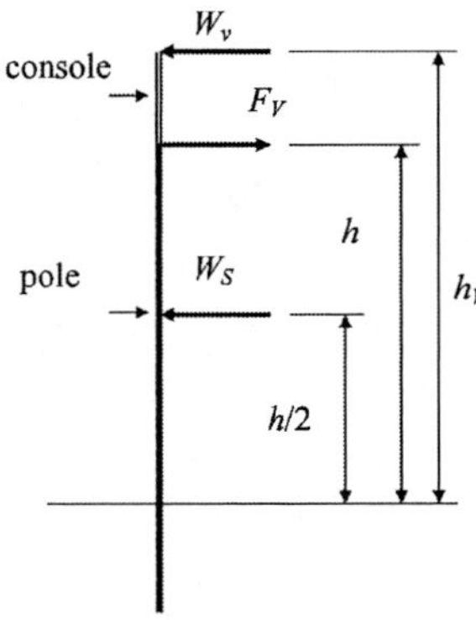

Figure 2. Recalculation of forces by the moment sentence

The force at the top of the pole F_V is determined by the moment balance of forces, according to Figure 2.

$$F_V \cdot h = W_{v3} \cdot h_v + W_s \cdot \frac{h}{2} \qquad (6)$$

$$F_V = \frac{W_{v3} \cdot h_v + W_s \cdot \frac{h}{2}}{h} \qquad (7)$$

The force at the top of the pole F_V = 1523 N.

III. Calculation of Forces According to Standards STN EN 50 341-1 and STN EN 50 423-1 (to 2016)

Standard STN EN 50 431 defines rules necessary to determine the designed load of the wind acting on the part of the overhead electric line on the basis of meteorological data. These rules apply to the height of the support points up to the values specified in the national normative aspect (NNA). If no requirements are specified in this aspect, the generally acceptable value of 60 m is accepted [3].

A. *Wind Speed*

When designing external power lines for load capacities, it is determined either by the mean wind speed V_{mea} or by its velocity V_g. Based on the standard, it is possible to use the average wind speed or the wind speed as the basis for determining the extreme wind speed in accordance with the practice in each country [3].

Average wind velocity V_{mea} is defined as the mean wind speed in m.s^{-1} measured over a 10 meter interval over 10 minutes above the ground in a relatively open terrain. Calculated according to (NNA) the STN EN 50341-1, article 4.2.2.1.2:

$$V_{mea} = c_{AL} \cdot V_{mea,0} \qquad (8)$$

where c_{AT} refers to the coefficient of altitude and $V_{mea,0}$ is the base reference speed of the wind. According to the table 4.2.2/SK1 in (NNA) for wind zone 2, c_{AT} = 1, $V_{mea,0}$ = 26 m.s^{-1} (for altitude H ≤ 700 m).

The reference wind speed V_R is the wind speed at 10 m above the ground at the appropriate location. In countries calculating with the average wind speeds, the reference wind speed at the appropriate site of the given terrain category can be calculated from the wind speed reference at the nearest metric station $V_R(II)$ in the terrain of category II based on the following formula:

$$V_R = k_T \cdot ln \frac{10}{z_0} \cdot V_R(II) \qquad (9)$$

where k_T refers to the factor of the terrain and z_0 is rub of height. Value of k_T and z_0 are given in STN EN 50341-1, table 4.2.1.

When constructing overhead electrical lines, terrain categories I and II are usually applied. For terrain category II k_T = 0,19, z_0 = 0,05

The reference wind speed is reported in NNA. In cases where the wind speed reference data are not sufficient, other data sources should be used to determine the extreme wind speed at the appropriate location. In most European countries, wind speed maps are being processed for different regions. For regions not illustrated in these maps, the services of meteorologists should be used.

The wind speed at any altitude above ground h shall be determined in accordance with STN EN 50341-1, article 4.2.2.1.6.

$$V_h = k_T \cdot ln \frac{h}{z_0} \cdot V_R(II) \qquad (10)$$

For overhead electrical line elements up to 10 m above the ground, the wind speed V_h is directly equal to the reference wind speed ($V_h = V_R$). In case of elements of the overhead electrical line at a height of more than 10 m above the ground, logarithmic dependence is used to increase the wind speed.

According to STN EN 50423-1 art. 4.2.2.3, lines up to 45 kV for elements with a maximum height of up to 20 m, the fixed dynamic wind pressure given for height of 10 m is given.

According to STN EN 50341-1, article 4.2.2.1.5 (in NNA): $V_R(II) = V_{mean}$

Then:

$$V_h = V_R = k_T \cdot ln\frac{10}{z_0} \cdot V_{mean} = 26{,}17 \text{ m·s}^{-1} \quad (11)$$

B. Wind Pressure

According to STN EN 50341-1, article 4.2.2.2 the dynamic wind pressure q_h in N.m^{-2} at height h over the ground is determined from the formula:

$$q_h = \frac{1}{2} \cdot \rho \cdot V_h^2 = \frac{1}{2} \cdot 1{,}225 \cdot 26{,}17^2 = 419{,}48 \text{ Pa} \quad (12)$$

where ρ refer to air density equal to 1,225 kg.m^{-3} at 15 °C and at 1013 hPa. At other temperature and atmospheric pressure values, the corresponding air density is either calculated or proportional values can be used.

C. Wind Forces on Conductors

Wind pressure on conductors causes the forces to act transversely to the direction of the line and at the same time it increases the pull in conductors. The force of the wind at a support point in the direct route caused by each separate conductor from both neighboring spans is:

$$Q_{wc} = q_h \cdot G_q \cdot G_c \cdot C_c \cdot d \cdot \frac{L_1 + L_2}{2} \cdot cos^2\varphi \quad (13)$$

where φ refers to the angle of rise of the decisive wind direction, $\varphi = 0$ ($\cos\varphi = 1$), G_q is the impact coefficient of the wind, according to STN EN 50341-1 table 4.2.4 $G_q = 2{,}05$, G_c is conductor response factor, according to STN EN 50341-1 table 4.2.5 (comment 3) $G_c = 0{,}936$, C_c is a drag factor for conductor, according STN EN 50341-1, article 4.2.2.4.1/SK, $C_c = 1{,}2$, L_1 and L_2 are span length, $(L_1 + L_2)/2$ is wind span, d is conductor diameter.

Then

$$Q_{wc} = 419{,}48 \cdot 2{,}05 \cdot 0{,}936 \cdot 1{,}2 \cdot 0{,}009 \cdot 80 \cdot 1 = 695{,}43 \text{ N}$$

The wind force per one conductor is $Q_{wc} = 695{,}43$ N

Wind force acting on three conductors is $Q_{wc3} = 2086{,}30$ N

D. Wind Force on Pole

The power of the wind acting on the pillars (steel, concrete, wood) comes from the wind force on the conductors and insulators and also from the wind pressure on the flagpole itself. The wind force directly on the pole is equal:

$$Q_{wpol} = q_h \cdot G_q \cdot G_{pol} \cdot C_{pol} \cdot A_{pol} \quad (14)$$

where G_q is the impact coefficient of the wind, according to STN EN 50341-1 table 4.2.4 $G_q = 2{,}05$, G_{pol} is the dynamic pole coefficient, according to STN EN 50341-1 (comment - typical value) $G_{pol} = 1{,}15$, C_{pol} is drag factor of pole, according to STN EN 50423-1, article 4.2.2.4.4 $C_{pol} = 0{,}7$, A_{pol} is the area of the pole projection, $A_{pol} = 1{,}728$ m^2 [4].

Then

$$Q_{wpol} = 419{,}48 \cdot 2{,}05 \cdot 1{,}15 \cdot 0{,}7 \cdot 1{,}728 = 1196{,}03\ N$$

The force acting on the pole is $Q_{wpol} = 1196{,}03$ N

The force at the top of the pole $F_V = 2974$ N

IV. Calculation of Forces According to STN EN 50341-1 and STN 50341-2-23 (from 2017)

A. Wind Speed

The basic wind speed is defined as a 10-minute average wind speed independent of wind direction and seasons, at a height of 10 m above the terrain level in an open low-vegetation landscape such as grass and isolated obstacles spaced at least 20 times their height. The basic wind speed is a characteristic value with an annual probability of 0.02 being exceeded, corresponding to a mean return time of 50 years. According to geographic location, the territory of Slovakia is divided into two basic wind areas. Furthermore, it is divided by altitude. This division defines four basic wind areas for which different basic wind speeds according to STN EN 50341-2-23 table 4/SK.1 are defined [5].

Table 1. Basic wind speeds

Wind area	I	II	III	IV
Basic wind speed	24	26	30	33
Areas I and II are basic geographic locations. Area III corresponds to locations with an altitude of 700 to 1300 m. Area IV corresponds to locations with an altitude above 1300 m.				

Average wind speed V_h depends on:

- basic wind speed $V_{b,0}$,
- wind direction coefficient c_{dir},
- the reference height of the considered component over the terrain h,
- length of roughness z_0,
- orography coefficient c_0.

The wind direction coefficient may have different values for different wind directions. The average wind speed is independent of wind direction in Slovakia. The value of the wind direction coefficient is used $c_{dir} = 1{,}0$. The reference height above the terrain h to be considered depends on the winding component of the line. The smallest reference height above the terrain for each lead component to be considered is 10 m. The length of roughness z_0 and the terrain coefficient, k_t that are directly linked to the length of roughness, characterize the roughness of the terrain, that modifies the mean velocity and intensity of the turbulence $l_v(h)$. Values z_0 and k_t are stated in STN 50341-1 table 4.1 ($z_0 = 0{,}05$, $k_t = 0{,}189$) [6].

Assuming that the average slope of the terrain side of the terrain is small, the effect of the topography may be neglected. In this case, the recommended value of the orography coefficient c_0 is 1. The mean wind speed $V_h(h)$ in m.s^{-1} at the reference height above terrain h can be determined from the relationship:

$$V_h(h) = V_{b,0} \cdot c_{dir} \cdot c_0 \cdot k_r \cdot \ln\left(\frac{h}{z_0}\right) \quad (15)$$

Then

$$V_h(h) = 25 \cdot 1 \cdot 1 \cdot 0.189 \cdot ln\left(\frac{10}{0{,}05}\right) = 25{,}03 \text{ m·s}^{-1}$$

For elements connected to structures with a rated voltage above 1 kV and up to and including 45 kV and with a maximum height of 20 m, it is permissible to use a constant average wind speed calculated at a height of 10 m above the ground.

B. Wind Pressure

Wind pressure $q_h(h)$ in Pa at the height h is designated as:

$$q_h(h) = \frac{1}{2} \cdot \rho \cdot V_h^2(h) = \frac{1}{2} \cdot 1{,}225 \cdot 25{,}04^2 = 383{,}73 \text{ Pa} \quad (16)$$

where ρ refers to the density of air in kg.m^{-3}, which depends on the altitude, temperature and barometric pressure that can be expected in the region during wind storm. ρ = 1,225 kg.m^{-3} at temperature 15°C and atmospheric pressure 1013 hPa. For other temperature and atmospheric pressure values, the air density can be calculated or the relative values from the table can be used.

C. *Turbulence Intensity and Peak Wind Pressure*

The turbulence intensity $l_v(h)$ at the reference height above terrain h is defined as the standard deviation of the turbulence divided by the mean wind velocity. The intensity of the turbulence can be calculated based on the following formula:

$$l_v(h) = \frac{1}{[c_0 \cdot ln(h/z_0)]} = \frac{1}{[1 \cdot ln(10/0{,}05)]} = 0{,}189 \quad (17)$$

The peak wind pressure $q_p(h)$ at the reference height above terrain h, that takes into account the intensity of the turbulence:

$$q_p(h) = [1 + 7 \cdot l_v(h)] \cdot q_h(h) \quad (18)$$

Then

$$q_p(h) = (1 + 7 \cdot 0{,}189) \cdot 383{,}73 = 890{,}70 \text{ Pa}$$

D. *Wind Forces on Conductors*

The wind pressure on the wires produces forces acting transversely on the direction of the line and simultaneously increasing the pull in the wires. The total wind force for the phase conductor bundle is defined as the sum of the forces on the individual bundle conductors, without considering the possible shielding effect on the leeward conductors.

In general, the force of the wind at the support point of each individual driver from both adjacent spans is given as:

$$Q_{Wc} = q_p(h) \cdot G_c \cdot C_c \cdot A_c \quad (19)$$

where G_c stands for the conductor response factor, according to STN EN 50341-1, table 4.4c G_c = 0,7, C_c is the drag factor for conductors, according to STN EN 50341, article 4.4.1.3 C_c =1, A_c = 0,72 m^2.

Then

$$Q_{Wc} = 890{,}70 \cdot 0{,}7 \cdot 1 \cdot 0{,}72 = 448{,}91 \text{ N}$$

The wind force per one conductor is Q_{Wc} = 448,91 N.

Wind force acting on three conductors is Q_{Wc3} = 1346,74 N.

E. *Wind Force on Poles*

The strength of the wind on the pole (steel, concrete, wood, etc.) results from the wind force on the conductors and insulators and from the wind pressure on the pole itself. The reference height to be considered when calculating the wind force on the pillars can be determined by one of the methods [6].

The force of the wind directly on the pole in a direction perpendicular to the wind direction is determined by the formula:

$$Q_{Wpol} = q_p(h) \cdot G_{pol} \cdot C_{pol} \cdot A_{pol} \quad (20)$$

where G_{pol} is coefficient of pole design, according to STN EN 50341, article 4.4.4 G_{pol} = 1,. C_{pol} is drag factor of pole, according to STN EN 50341, article 4.4.4 C_{pol} = 0,7, A_{pol} = 1,728 m^2.

Then

$$Q_{Wpol} = 890{,}70 \cdot 1 \cdot 0{,}7 \cdot 1{,}728 = 1077{,}39 \text{ N} \quad (21)$$

The force acting on the pole is Q_{Wpol} = 1077,39 N.

The force at the top of the pole F_V = 2072 N.

V. Conclusion

In conclusion, it is possible to compare the results obtained by the calculation. Calculations of wind forces on wires and poles have been made using three different methodologies listed in the standards. Calculated values are shown in Table 2.

Table 2. Wind Force Value

Calculation according to:	Wind force on conductors [N]	Wind force on pole [N]	The force at the top of the pole [N]
STN 33 3300	1045	665	1523
STN EN 50341-1 STN EN 50423-1	2086	1196	2974
STN EN 50341-1 STN EN 50341-2-23	1347	1077	2072

On the basis of these forces, it is possible to properly dimension the poles of the overhead electrical line, thus limiting the disturbances in the electricity system [7].

Acknowledgment

This work was supported by the Ministry of Education, Science, Research and Sport of the Slovak Republic and the Slovak Academy of Science under contract No. VEGA 1/0132/15.

References

[1] Fecko, Š., Reváková, D., Varga, L., Lago, J., Ilenin, S.: Vonkajšie elektrické vedenia. Vydavateľstvo Renesans Bratislava 2010. 599 pp. ISBN 80-89402-35-9.

[2] STN 33 3300 Stavba vonkajších silových vedení, Slovenský ústav technickej normalizácie, 1983

[3] STN EN 50431-1: Elektrické vonkajšie vedenia so striedavým napätím nad 45 kV. Časť 1: Všeobecné požiadavky. Spoločné špecifikácie., Slovenský ústav technickej normalizácie, 2006

[4] STN EN 50423-1: Vonkajšie elektrické vedenia so striedavým napätím nad 1 kV do 45 kV vrátane. Časť 1: Všeobecné požiadavky. Spoločné špecifikácie., Slovenský ústav technickej normalizácie, 2006

[5] STN EN 50341-2-23: Elektrické vonkajšie vedenia so striedavým napätím nad 1 kV. Časť 2-23: Národné normatívne hľadiská (NNA) pre Slovensko (založené na EN 50341-1:2012), Úrad pre normalizáciu, metrológiu a skúšobníctvo, 2017

[6] Z. Čonka, M. Kolcun,"Impact of the regulators on the size of the short-circuit currents" In: Elektroenergetika 2013 : proceedings of the 7th International Scientific Symposium on Electrical Power Engineering, Stará Lesná, Slovak Republic. - Košice : TU, FEI, 2013 S. 86-89. - ISBN 978-80-553-1441-9

[7] STN EN 50341-1: Elektrické vonkajšie vedenia so striedavým napätím nad 1 kV. Časť 1: Všeobecné požiadavky. Spoločné špecifikácie., Slovenský ústav technickej normalizácie, 2013

[8] D. Medveď, M. Kolcun, Z. Čonka, P. Hocko, M. Novaák. "Simulation of Faults in the Power System and Their Elimination in the EMTP-ATP" In: Elektroenergetika 2013: Proceedings from the Seventh International Scientific Symposium on Electrical Power Engineering : September 18-20, 2013, Stará Lesná, Slovak Republic. - Košice : TU, FEI, 2013 S. 484-487. - ISBN 978-80-553-1441-9.

Conditions for ECV Charging Infrastructure Development in Poland

Beata Ślusarczyk
Czestochowa University of Technology
Czestochowa, Poland
Faculty of Economic Sciences and IT North-West University
South Africa
jagoda@zim.pcz.pl

Abstract — **The issue of using alternative fuels in the functioning of the trans-European transport network (TEN-T) is a topic that is frequently discussed both at the level of the EU and Member States. The major matter that forces the search for effective solutions is the broadly defined protection of the environment. The use of alternative fuels is especially important in the context of the reduction of smog, among others, which in Poland exceeds several hundred times the norm. One of the types of fuel, that is currently gaining popularity, as a substitute for traditional propulsion products of petroleum derivatives is electricity. The number of electric cars, both in the world and in Europe is growing gradually. However, the growth dynamic of this type of vehicles' users, depending on a country, is quite diverse. One of the key factors that determines the purchase of an electric car, apart from its price, is the access to the charging infrastructure. The aim of this article is to indicate the determinant of the development of the charging network for electric vehicles in Poland. On the basis of comparison with other European countries and the verification of legal sources, at the EU and state level, it will be possible to evaluate planned solutions for the development of charging infrastructure for electric cars, and to propose additional improvements.**

Keywords: ECV, electric vehicle, charging infrastructure, political support

Introduction

"Resource-efficient Europe" is one of the projects listed in the document: "The Europe 2020 strategy for smart, sustainable and inclusive growth". Its main idea is to take efforts for the benefit of social and economic growth of the EU countries, based on the key priorities, including [1]: development of economy based on knowledge and innovation; supporting the economy with a high level of employment, ensuring social and territorial cohesion and the above mentioned need to support the economy which more efficiently uses the resources, which is more environmentally friendly and more competitive [2]. One of the solutions enabling the implementation of the assumptions made is electromobility. This issue is currently the area of great interest of both EU institutions, national ones but also local authorities or even individual users. And this is not only the consequence of the arrangements established at the legislative level on the share of alternative fuels in transport [3] but also or perhaps, most of all, the concern for the natural environment and excessive exploitation of energy resources in the world.

However, road transport is still largely based on the use of petroleum-derived fuels. Currently, 98% of the transportation sector in the EU depends on fossil fuels [4]. The sector is responsible for approx. 21% of the greenhouse gas (GHG) emissions, with more than half of the emissions produced by passenger cars. The EU Directive (2009/33/EC) on the promotion of clean and energy efficient road transport vehicles has been released to foster a broad market penetration of environmentally-friendly vehicles in order to decarbonize the transportation sector and to reduce oil dependency [5].

Electricity has the potential to increase the energy efficiency of road vehicles and contribute to the reduction in CO2 emissions from transport. It is a source of energy necessary for the spread of electric vehicles, which may contribute to the improvement in air quality and reduction in noise level in urban, suburban and other densely populated areas.

In the White Paper of the European Commission "Roadmap to a Single European Transport Area — Towards a Competitive and Resource Efficient Transport System" there are indicated the ways to reduce the dependence of transport on petroleum. The achievement of the goals set will be possible by taking many different policy initiatives, including the development of the strategy concerning sustainable alternative fuels as well as the development of appropriate infrastructure [6].

I. The Development of Electromobility in the Countries of the European Union

Electric driven vehicles offer a solution to reduce greenhouse gas emissions and local air pollution, but their market penetration is still marginal. The widespread deployment of electric vehicles in the EU depends on a large variety of factors. To be a successful mobility alternative several obstacles and challenges have to be overcome first. Various studies identified the following main obstacles and challenges for a broad market penetration [7]:

- Battery technology (costs, energy density, recycling etc.)
- Availability and preparedness of relevant industrial capacity (business models, new vehicle models etc.)
- Charging infrastructure (minimum density of charging points, identification of early adopter hotspots etc.)
- Customers' requirements (awareness, willingness-to-pay, range anxiety, safety concerns etc.)
- Electric grid (capacity and connectivity issues etc.)

- Impacts on energy efficiency and green house gas emissions (electricity mix, load management etc.)

- Standardization issues (charging plug, billing systems, data protocol etc.).

Despite the fact that the interlinkages between the aforementioned factors can play a major role for the market penetration of electric cars, the customers' requirements and expectations are the most decisive aspects for the market deployment of alternative propulsion systems.

In most EU countries there is observed growth in sales of ECV vehicles (Electric Chargeable Vehicles = battery electric vehicles (BEV) + extended-range electric vehicles (EREV) + fuel cell electric vehicles (FCEV) + plug-in hybrid electric). Table 1 includes the data for all the Member States as well as for Norway and Switzerland.

Table 1. New passenger ECV registrations by market in the EU+EFTA*

	2013	2014	2015	2016
Austria	654	1 718	2 787	5 068
Belgium	500	2 047	3 837	8 958
Bulgaria	1	2	21	13
Czech Republic	37	197	298	200
Denmark	650	1 616	5 298	2 063
Estonia	149	340	34	35
Finland	50	445	658	1 431
France	8 779	12 497	22 867	29 189
Germany	6 051	13 118	23 557	25 214
Greece	0	59	67	32
Hungary	10	39	130	343
Ireland	49	256	583	690
Italy	864	1 420	2 343	2 827
Latvia	4	194	35	43
Lithuania	n.a.	9	37	64
The Netherlands	2 619	14 805	44 448	22 801
Poland	31	141	337	556
Portugal	166	289	1 166	1 821
Romania	4	7	24	74
Slovakia	6	117	52	59
Spain	811	1 405	2 246	3 654
Sweden	432	4 667	8 663	13 221
United Kingdom	2 719	14 608	28 715	36 917
European Union	**24 586**	**69 996**	**148 203**	**155 273**
Norway	7 882	19 771	33 770	44 908
Switzerland	1 156	2 688	6 289	6 403
EFTA	9 038	22 459	40 059	51 311
EU + EFTA	**33 624**	**92 455**	**188 262**	**206 584**

* Only countries for which sourced data is available are listed
Authors' elaboration based on [8]

When assuming 2013 as the base year, the largest increase in ECV vehicles was recorded in the following countries: Sweden, Belgium, Ireland, Great Britain, Portugal and also Poland. In Poland, every year there are registered more and more electric cars, although their share in the group of new car registrations amounted only to: in 2012 – 0.29%, 2013 – 0.65%, 2014 – 1.29%. In spite of the apparent trend of sales growth the rate of market development of electric vehicles is still lower than in the European Union [9]. There can also be noticed the countries in which the popularity of ECV vehicles is slightly lower and their purchase is falling. Among others, this is the case of: Estonia, Denmark, Greece, Lithuania or Slovakia. The leader among the European countries in the number of registered electric cars is Norway (in the analyzed period, their number amounted to 106 331 vehicles). Norway, where currently 25% of cars are electric ones and which built its wealth on the extraction of oil and gas, is taking next steps towards the intensification of electromobility [10,11]. According to "The Independent", in the Norwegian policy, there has been reached the agreement on the gradual shift from the use of fuels in road transport and ultimately only electric cars are to be driven on Norwegian roads. Another example is the Netherlands where there is planned the introduction of a ban on sales of fuel vehicles from 2025 [12].

Many European countries have introduced national targets for the EV driving stock, the expansion of charging infrastructure, or production targets of electric vehicles [13,14]. Most EU member states overcome the cost disadvantage of alternative vehicles by introducing policy instruments such as an up-front price support in order to increase the affordability of electric vehicles by reducing the marginal capital cost, which is considered as one of the key barriers for consumers. Gass, Schmidt and Schmid presented the incentives used by selected European countries (Tab.2).

Table 2. Policy instruments to supporting EV in the selected countries passenger

Country	Economic instruments for the support of EV
Austria	Exemption from fuel consumption tax Exemption from monthly vehicle
Belgium	Purchasers of electric cars receive a personal income tax reduction of 30% of the purchase price (with a maximum of EUR 9,000)
Finland	Exemption of fuel tax
Italy	A tax incentive of EUR800 and a two year exemption from annual circulation tax is granted for the purchase of a new passenger
Denmark	Exemption from registration tax and annual circulation tax. Further EV qualify for free parking
Germany	EV exempt from the annual road tax for a period of five years from the date of the first registration
Spain	Various regional governments grant tax incentives for the purchase of alternative fuel vehicles including EV – approx. EUR 6,000
France	Bonus-Malus System; New Cars with CO2 emissions below 125 g/km receive a premium. EV receive currently EUR 5,000
Greece	EV exempt from registration tax. If engine capacity below 1929 cc, exemption from road tax. Further EV are even allowed to drive in Athens when parts of the city are restricted to ICE to reduce traffic congestion
Ireland	EV exempt from registration tax – approx. EUR 2,500
The Netherlands	approx. EUR 6,400 reduction from the registration tax
Portugal	Exemption from registration tax
Sweden	Exemption from annual road tax for a period of 5 years upon first registration
United Kingdom	Exemption from annual road tax
Luxemburg	Annual circulation taxes based on CO2 emissions

[5]

Clearly, the involvement in the use of environmentally friendly forms of road transport in the European countries is significant. Also in Poland, the problem has received the enthusiastic response of the government, the result of which is "Electromobility Development Plan for Poland" developed by the Ministry of Energy [15].

According to the government promises, the plan is to be implemented in three stages (Tab.3).

Table 3. Stages of Electromobility Development in Poland by 2025

STAGE	PERIOD	ACTIONS TAKEN
I	by 2018	- implementing pilot programs promoting electromobility, - determining legal conditions for development of electromobility, - developing the first prototypes of the vehicle adapted to the needs of the Polish and European market, - appointing the Measurement Information Operator, which will integrate information on the behavior of all the users of power grid.
II	2019 - 2020	- building charging infrastructure for electric vehicles in selected metropolitan areas and along TEN-T (Trans-European Transport Network). In 32 selected agglomerations, there are to be 50 thousand electric vehicles and 6 thousand normal power charging points and 400 high power ones, - intensifying the incentives for the purchase of electric vehicles (in announcements by the government these are to be, among others, VAT-zero rate for electric vehicles, elimination of the excise tax, subsidies for the first 100 thousand emission-free cars, dependence of registration fee on emissivity and price of a vehicle, - producing the catalogue of good practices of social communication in the field of electromobility, - beginning the production of short series of electric vehicles based on prototypes developed in the first stage.
III	2020 - 2025	- increasing the demand for electric vehicles, - the network fully prepared to provide energy for 1 million electric vehicles and adapted to use vehicles as power system stabilizers, - manufacturing high-quality components for electric vehicles by the Polish industry, production of vehicles and necessary instrumentation and infrastructure.

Authors' elaboration based on [15]

These ambitious plans of the government, predicting that in 2025 there will be 1 million of electric vehicles being driven on Polish roads, often faces polemics, and even criticism. Doubt in the success of this project, among others, is formulated by P. Maciążek, referring to the report by the consulting company Facts Global Energy (FGE), which has conducted an interesting analysis of the future of electromobility in the world. According to it, every year in the coming decade there will come into operation 2 million cars, of which the overwhelming majority will be gasoline vehicles, which will gradually increase the consumption of oil. At the same time, in a short term, FGE does not provide for a significant increase in the price of this raw material, which may additionally increase the demand for conventional vehicles. As a result, by 2040 there will have been a significant increase in the number of new vehicles in the world, in total, there will be 1.8 billion vehicles, however only 10% of them will be electric cars and 20% - hybrids [16].

One may also come across critical voices concerning individual ideas formulated in the government document, for example, the idea of VAT-zero rate for electric vehicles has no chance for implementation since this would require the agreement of the European Union, which is impossible to achieve as it would give the cars assembled in Poland unjustified competitive advantage. Norway, where the sale of electric cars has been VAT exempt, is not a member of the European Union. On the other hand, the elimination of the excise taxi is possible as it does not require the acceptance of the other Member States. There are also no obstacles so as to eliminate the peculiarities of the Polish tax system in relation to electric cars by means of [17]:

- 100% VAT deduction on the purchase of a vehicle,
- full deduction of VAT included in the operating costs with no need to keep records of vehicle mileage,
- elimination of the obligation to pay income tax on the use of cars for private purposes.

II. SUPPORT FOR THE DEVELOPMENT OF CHARGING INFRASTRUCTURE FOR ELECTRIC CARS

Define Despite many uncertainties, the development of electromobility is prioritized and both the European Union and individual countries of our continent make every effort to allow consumers to purchase and use ECV vehicles. The plans to support electromobility listed in Table 2 include a wide range of incentives but now the most important issues include the development of charging infrastructure for electric cars [18]. In Poland, the problem of poor distribution of stations and charging points is raised as the most significant one, right next to high prices of electric vehicles (the price of an electric car is significantly higher than a conventional car – it ranges from PLN 80 to 220 thousand).

The market for electric vehicle charging services in Europe is still in its initial stage of development, and the activity in this field is conducted by different actors in different business models, often in the form of free service aimed at building the image of a specific company or property (among others, shopping centers, hotels).

At the EU level, the issue of the development of infrastructure for electric vehicles was regulated in the document *"The Directive of the European Parliament and the Council 2014/94/UE of 22 October 2014 on the development of infrastructure for alternative fuels"* [19]. In Article 4 (paragraph 1) of this document, the Member States shall ensure that, by means of their national policy frameworks, by 31.12.2020 there will have been developed an adequate number of publically accessible charging points. The Directive indicates that there should be one publically accessible charging point per every 10 registered electric cars (according to the Directive (article 4 paragraph 9) a publically accessible charging point is the one which allows the users of electric vehicles for ad hoc charging without a contract with the electricity supplier or operator.).

Table 4 shows the number of charging stations (normal ones of the power not exceeding 22 kW and high-speed and high-power ones, of more than 22 kW, which allow for charging a vehicle in 30 minutes) in the European countries covered by the research.

Table 4. The number of electric vehicle chargers (normal and high-speed) and the number of cars per charging station in each country (2Q2017)*

	Charging Power			PEV per position (vehicle/total position)
	Normal Power <= 22	High Power > 22	Total	
Austria	2356	437	2793	4
Belgium	1335	480	1815	10
Bulgaria	18	4	22	1
Czech Republic	294	156	450	3
Denmark	2114	422	2 536	3
Estonia	191	187	378	3
Finland	706	265	971	3
France	14290	1593	15883	7
Germany	22857	1 810	24667	3
Greece	31	2	33	5
Hungary	163	42	205	3
Ireland	837	156	993	2
Italy	1796	211	2007	5
Latvia	60	12	72	3
Lithuania	15	11	26	5
The Netherlands	26651	614	27265	4
Poland	290	34	324	2
Portugal	1192	58	1 250	3
Romania	95	9	104	3
Slovakia	255	80	335	1
Spain	1378	370	1748	7
Sweden	1654	1198	2852	11
United Kingdom	9983	2178	12161	7
European Union	**88 561**	**10 329**	**98 890**	**4**
Norway	7367	1367	8734	14
Switzerland	3399	484	3883	4
EFTA	10 766	1 851	12 617	9
EU + EFTA	99 327	12 180	111 507	**6,5**

*Only countries for which sourced data is available are listed
Authors' elaboration based on [20]

According to the index of the last column, which indicates the number of ECV vehicles per one charging point, most countries may boast with excellent results, for example, in Poland, there are 2 cars per one charging point. However, these data should be treated indicatively due to the lack of uniform database where there is information concerning ECV vehicles and infrastructure for charging them. Paradoxically, Norway, whose inhabitants purchase the largest number of electric cars in Europe, has this index at the level of 14. It is probably the result of lower rate of infrastructure development compared to the sales dynamic of ECV vehicles. On the other hand, unfortunately, at least in Poland, the problem is the geographical distribution of charging points. Nowadays, the largest number of them is in Warsaw, Cracow, Poznan and Gdansk.

As indicated in the project "The National Policy Framework for Alternative Fuels Infrastructure Development" [21], there are no legal regulations specifically designed for infrastructure for charging electric cars. There are applied universally binding laws, particularly there should be indicated the provisions of the Act – Construction Law, the Act – Energy Law and the Law on public roads. There is also no sales model of electric car charging service, which would determine the principles for sales of electricity, accounting rules between the car owner and the charging service provider or, if the seller and the provider are the same entity, then between this entity and the power distributor. Most charging points are provided to users free of charge, they do not pay for the consumed electricity. Some entities have such points only and exclusively for their own use. There is a small number of charging points used commercially, however there are no accurate data concerning this issue. Sharing the charging service free of charge is aimed at the popularization of electric cars in Poland.

Particularly important for the development of charging infrastructure for electric vehicles are the provisions in the EU Directive 2014/94/UE concerning the activity of Distribution System Operators (OSD) as well as the operators of publically accessible charging points. This document indicates that the Member States, through legislation at a national level, should oblige electricity distribution system operators to cooperate with each person who establishes a publically accessible charging point. The creation and operation of charging points for electric vehicles should be developed by the entities operating on the free market, and the activity in this field should be allowed for all interested parties. The development of charging infrastructure in the conditions of market activity appears to be an effective solution since it provides appropriate competitive environment.

On the other hand, while discussing the use of charging points by users, in the Directive, it is underlined that the operators of publically accessible charging points are free to buy electricity from any electricity supplier in the European Union, including the agreements with the supplier. The operators of charging points can provide consumers with services in terms of charging electric vehicles on the basis of agreements, including on behalf of and for the benefit of other service providers. All the publically accessible charging points should also allow the users of electric vehicles for ad hoc charging without concluding the contract with the specific electricity provider or operator. On the other hand, the prices given by the operators of publically accessible charging points should be reasonable, easily and clearly comparable, transparent and non-discriminatory [22].

III. Conclusions

Number As underlined in "National policy frameworks for the development of alternative fuels and their infrastructure", the e-mobility market in Poland (infrastructure and electric cars) is very poorly developed. In fact, one can talk more about the promotion activities of individual companies than the organized operating market. Therefore, it is necessary for both government institutions as well as economic entities to take efforts in order to change this situation.

Among experts there is held the discussion on the role and tasks of Distribution System Operator (DSO), particularly with reference to how far these entities are to get involved in providing services associated with charging, how much they are to concentrate only on connecting charging infrastructure to the power grid. Clearly, according to the EU Directive, the latter concept prevails, i.e. limiting the role of DSO to connecting charging points to a network and ensuring such a development of power grid that allows the development of charging infrastructure in the future. However, in some European countries (Austria, Luxemburg and Slovenia) there has been implemented the model where the tasks of DSO include the creation and management of charging infrastructure (it does not include the sales of energy from this infrastructure). In these countries, DSO is the owner of charging infrastructure and provides measurement data whereas the technical and trade operation of infrastructure is provided by third parties under contracts concluded with DSO. A slightly different example is Ireland where DSO independently manages charging infrastructure and only trade activity is undertaken by the third party. In other countries, the activities associated with the development of infrastructure and provision of charging service have been entrusted to entities operating on the free market [23].

For the development of electromobility in Poland an effective solution seems to be to allocate DSO a wide range of functions as is the case of Austria or Luxemburg. Providing the service of charging electric vehicles from publically accessible infrastructure requires the commitment of significant funds for the purchase and installation of necessary infrastructure and the operators of distribution systems have the financial and organizational capabilities to accomplish these tasks. As the ones managing the infrastructure for charging ECV vehicles, DSOs would have an opportunity to implement appropriate procedures for its continuous monitoring and providing customers with information on the status and availability of chargers. The applications and accounting models that allow drivers to easily locate the nearest charging points, verification of the ability to use the service, the possibility of reservation of a stand and then smooth settlement of the purchased service. This requires the implementation of appropriate IT systems, applications for communication with customers, continuous surveillance of the operation of charging infrastructure and ensuring the highest level of its performance. Certainly, the accumulation of the largest number of the listed activities within the competences of one entity would be advisable for the development of charging infrastructure of electric vehicles, standardization of the system of providing services and bearing responsibility for its effective operation.

Additionally, a very important issue is proportional distribution of chargers across the country. Publically accessible charging points should be located in a way that allows free use of electric vehicles. An adequate number of publically accessible chargers should be installed particularly along highways and freeways, at points of public transport, such as passenger port terminals, airports or railway stations. The owners of private electric vehicles are largely dependent on the access to charging points at parking lots, e.g. at blocks of flats, office buildings and business premises. Public authorities should take measures supporting the users of such vehicles by ensuring that investors and managers of premises create appropriate infrastructure with sufficient number of suitable charging points for electric vehicles.

REFERENCES

[1] "The Europe 2020: Strategy for smart, sustainable and inclusive growth", Communication from the European Commission, Brussels 2010, [Online] Available: http://ec.europa.eu/eu2020/pdf/1_PL_ACT_part1_v1.pdf last see (2017, May)

[2] J. Janekova, J. Fabianova and A., Rosova, "Environmental and economic aspects in decision making of the investment project "wind park", Polish Journal of Management Studies, 13 (1), 2016, pp. 90-100, DOI 10.17512/pjms.2016.13.1.09.

[3] G. Kovacs and S. Kot, "New Logistics and Production Trends as the Effect of Global Economy Changes", Polish Journal of Management Studies 14 (2), 2016, pp. 115-126, DOI 10.17512/pjms.2016.14.2.11.

[4] S. Kot, "Cost Structure in Relation to the Size of Road Transport Enterprises", PROMET-Traffic&Transportation 27 (5), 2015, pp. 387-394, DOI 10.7307/ptt.v27i5.1687.

[5] V. Gass, J. Schmidt, and E. Schmid, "Analysis of alternative policy instruments to promote electric vehicles in Austria" Renewable Energy, Vol. 61, 2014, pp. 96-101.

[6] WHITE PAPER of the European Commission "Roadmap to a Single European Transport Area — Towards a Competitive and Resource Efficient Transport System", Brussels 2011, [Online] Available: http://eur-lex.europa.eu/LexUriServ/LexUriServ.do?uri=COM:2011:0144:FIN:pl:PDF last see (2017, May)

[7] J. A. Bühne, D. Gruschwitz, J. Hölscher, M. Klötzke, U. Kugler, and C. Schimeczek, "How to promote electromobility for European car drivers? Obstacles to overcome for a broad market penetration" European Transport Research Review, 7(3), 30, 2015. DOI 10.1007/s12544-015-0178-0.

[8] National Automobile Manufacturers' Associations http://www.acea.be (access: 13.03.2017).

[9] "The National Policy Framework for the Development of Infrastructure for Alternative Fuels", September 2016, p. 13, http://www.me.gov.pl/files/upload/26450/ME_KRAJOWE_RAMY_projekt.pdf

[10] E. Figenbaum, T. Assum, and M. Kolbenstvedt, "Electromobility in Norway: experiences and opportunities" Research in Transportation Economics, Vol. 50, 2015, pp. 29-38.

[11] K.Y. Bjerkan, T.E. Nørbech, and M.E. Nordtømme, "Incentives for promoting battery electric vehicle (BEV) adoption in Norway" Transportation Research Part D: Transport and Environment 43, 2016, pp. 169-180.

[12] „Rekordowa sprzedaż samochodów elektrycznych", Akademia Solsum; [Online] Available: http://gramwzielone.pl/auto-ekologiczne/22330/rekordowa-sprzedaz-samochodow-elektrycznych (access: 06.04.2017).

[13] J. Begley, N. Berkeley, T. Donnelly and D. Jarvis, "National policy-making and the promotion of electric vehicles", International Journal of Automotive Technology and Management, 16 (3), 2016, pp. 319-340.

[14] B. Holtsmark and A. Skonhoft, "The Norwegian support and subsidy policy of electric cars. Should it be adopted by other countries?" Environmental science & policy, 42, 2014, pp. 160-168.

[15] "Electromobility Development Plan for Poland", September 2016, [Online] Available: http://bip.me.gov.pl/node/26453 (access: 10.03.2017).

[16] P. Maciążek, „Milion polskich „Tesli"? Kasandryczna wizja globalnego rozwoju elektromobilności" [Online] Available: http://www.energetyka24.com/566458,milion-polskich-tesli-kasandryczna-wizja-globalnego-rozwoju-elektromobilnosci-komentarz, march 2017.

[17] R. Ostrowski, „Ładowanie samochodu elektrycznego w świetle polskiego prawa" 24 października 2016, [Online] Available: http://polskiemedia.org/ladowanie-samochodu-elektrycznego-w-swietle-polskiego-prawa/ (access: 03.04.2017).

[18] K. van Montfort, M. Kooi, G. van der Poel and R. van den Hoed, "Which factors influence the success of public charging stations of electric vehicles?", 6th Hybrid and Electric Vehicles Conference (HEVC 2016), November 2016, pp. 1-5.

[19] The Directive of the European Parliament and the Council 2014/94/UE of 22 October 2014 on the development of

infrastructure for alternative fuels, Official Journal of the European Union No 307, 28.10.2014.

[20] European Alternative Fuels Observatory [Online] Available: http://www.eafo.eu/electric-vehicle-charging-infrastructure (access: 10.04.2017).

[21] „The National Policy Framework for Alternative Fuels Infrastructure Development" [Online] Available: http://www.me.gov.pl/files/upload/26450/ME_KRAJOWE_RAMY_projekt.pdf.

[22] The Directive of the European Parliament and the Council 2014/94/UE of 22 October 2014 on the development of infrastructure for alternative fuels, Official Journal of the European Union No 307, 28.10.2014.

[23] I. Chojnacki, „Rafał Czyżewski, GIP: OSD powinni zapewniać szybkie przyłączanie punktów ładowania", 28.02.2017, [Online] Available: www.wnp.pl (access: 29.03.2017).

Economic Analysis of Power Transformer Selection in Electrical Power Systems

Alexander Mészáros
Department of Electric Power engineering
Technical University of Košice
Košice, Slovakia
alexander.meszaros@tuke.sk

Vladimír Gáll
Department of Electric Power engineering
Technical University of Košice
Košice, Slovakia
vladimir.gall.@student.tuke.sk

Abstract — **the proper selection and operation of the transformer is important in the power industry. Also this article shows the whole energy system influence on the operation of the transformer. At the same time, here is shown, how significant are the relations between technical and economical parameters of transformer. The aim of this research work is represent new comprehensive approach to the transformers as one the most important part of whole power system.**

Keywords: annual costs; economy; electric energy; power industry; specific costs; transformers;

I. Introduction

Power transformers belong to the most used devices in power industry in electricity transmission and distribution on the economically most appropriate voltages. The proper selection and operation of transformer can greatly contribute to the economy of entire power system. Continuous increase of transmitted power and increase the voltage levels makes construction of transformers more complex and more expensive. To solve the problem economy of transformers has a major significance the method of pricing of the losses electric energy, which in transformers occur during their operation. In this case as solution, we can use the equation for annual production costs and equation for relative costs for transformation [1]. At the same, it is necessary to know the power losses, relative power losses and power efficiency of transformer. When we calculate the power losses of transformer in different times, we can consider the knowledge about his energy losses, relative energy losses and energy efficiency. All of these issues create extremely complex tissue of relation to achieving the proper selection and operation of the transformer in the power industry. In the final result it is a need for a new comprehensive approach to the transformers as well as the economy of the whole power system [2] [3] [4].

II. Derivation of Equations for Transformer

Relative cost for c_1 and c_2 for power losses are defined with bicomponent expression, where c_{st} is fixed value and c_{po} is variable value of price of the electric energy.

$$c_1 = \frac{c_{st}}{T_{pr}} + c_{po} \tag{1}$$

$$c_2 = \frac{c_{st}}{\tau_\Delta} + c_{po} \tag{2}$$

Where T_{pr} is an operation time and τ_Δ is a time of full losses. The apparent no-load losses ${}_\Delta S_{0n}$ and the apparent short-circuit losses ${}_\Delta S_{kn}$ are defined by equations:

$${}_\Delta S_{0n} = i_0 \; S_n \tag{3}$$

$${}_\Delta S_{kn} = u_k \; S_n \tag{4}$$

Where i_o is the relative no-load current, u_k is the relative short-circuit voltage, and S_n is the nominal apparent power of transformer. The reactive no-load losses ${}_\Delta Q_{0n}$ and the reactive short-circuit losses ${}_\Delta Q_{kn}$ are defined by equations:

$${}_\Delta Q_{0n} = \sqrt{{}_\Delta S_{0n}^{\;2} - {}_\Delta P_{0n}^{\;2}} \tag{5}$$

$${}_\Delta Q_{kn} = \sqrt{{}_\Delta S_{kn}^{\;2} - {}_\Delta P_{kn}^{\;2}} \tag{6}$$

Where ${}_\Delta P_{0n}$ means active no-load losses and ${}_\Delta P_{kn}$ active short-circuit losses. Also we can define tg ξ_{0n} and tg ξ_{kn}:

$$tg\,\xi_{0n} = \frac{{}_\Delta Q_{0n}\; n}{\left({}_\Delta P_{0n} + k_{\Delta 0}\; {}_\Delta Q_{0n}\right) n} = \frac{{}_\Delta Q_{0n}}{{}_\Delta P_{0n} + k_{\Delta 0}\; {}_\Delta Q_{0n}} \tag{7}$$

$$tg\,\xi_{kn} = \frac{{}_\Delta Q_{kn}\; n}{\left({}_\Delta P_{kn} + k_{\Delta k}\; {}_\Delta Q_{kn}\right) n} = \frac{{}_\Delta Q_{kn}}{{}_\Delta P_{kn} + k_{\Delta k}\; {}_\Delta Q_{kn}} \tag{8}$$

Where n is number of identical parallel transformers, which have the same parameters, $k_{\Delta 0}$ is a specific loss factor for no-load run, and $k_{\Delta k}$ is a specific loss factor for short-circuit run of transformer. Transmitted energy A is defined by equation:

$$A = P_m \; \tau \tag{9}$$

Where P_m is transferred power and is τ is a time of using the maximum. Annual costs N_{tr} for n identical parallel transformers we can calculate like this

$$N_{tr} = \left(N_i\; k + \left({}_\Delta P_{0n} + k_{\Delta 0}\; {}_\Delta Q_{0n}\right) T_{pr}\; c_1\right) n + \frac{\left({}_\Delta P_{kn} + k_{\Delta k}\; {}_\Delta Q_{kn}\right) P_m^{\;2}\; \tau_\Delta\; c_2}{S_n^{\;2} \cos^2 \varphi} \frac{1}{n} \tag{10}$$

Where N_i is initial costs, k is annual quota and cos φ is power factor [1] [5]. For specific costs n_{tr} we can defined equation for n identical parallel transformers like this

$$n_{tr} = \frac{N_{Pi}\; k + \left({}_\Delta P_{0n} + k_{\Delta 0}\; {}_\Delta Q_{0n}\right) T_{pr}\; c_1}{P_m\; \tau} n + \frac{\left({}_\Delta P_{kn} + k_{\Delta k}\; {}_\Delta Q_{kn}\right) P_m\; \tau_\Delta\; c_2}{S_n^{\;2}\; \tau \cos^2 \varphi} \frac{1}{n} \tag{11}$$

According transferred power P_m we can calculate losses ${}_\Delta P$ and relative losses ${}_\Delta p$ for n identical parallel transformers

$${}_\Delta P = \left({}_\Delta P_{0n} + k_{\Delta 0}\; {}_\Delta Q_{0n}\right) n + \frac{\left({}_\Delta P_{kn} + k_{\Delta k}\; {}_\Delta Q_{kn}\right) P_m^{\;2}}{S_n^{\;2} \cos^2 \varphi} \frac{1}{n} \tag{12}$$

$$ {}_\Delta p = \frac{{}_\Delta P}{P_m} = \frac{({}_\Delta P_{0n} + k_{\Delta 0}\, {}_\Delta Q_{0n})\, n}{P_m} + \frac{({}_\Delta P_{kn} + k_{\Delta k}\, {}_\Delta Q_{kn})\, P_m}{S_n^2 \cos^2 \varphi} \frac{1}{n} \qquad (13) $$

According transferred power P_m we can calculate reactive losses ${}_\Delta Q$ for n identical parallel transformers

$$ {}_\Delta Q = ({}_\Delta P_{0n} + k_{\Delta 0}\, {}_\Delta Q_{0n})\, tg\, \xi_{0n}\, n + \frac{({}_\Delta P_{kn} + k_{\Delta k}\, {}_\Delta Q_{kn})\, P_m^2\, tg\, \xi_{kn}}{S_n^2 \cos^2 \varphi} \frac{1}{n} \qquad (14) $$

Power efficiency η_{Ptr} of transformer is defined [6] as the ratio of the transferred power P_m and the input power P_{vst}

$$ \eta_{Ptr} = \frac{P_m}{P_{vst}} = \frac{P_m}{P_m + {}_\Delta P} \qquad (15) $$

After substitution of equation (12) into equation (15) we get equation for power efficiency η_{Ptr}

$$ \eta_{Ptr} = \frac{P_m}{P_m + ({}_\Delta P_{0n} + k_{\Delta 0}\, {}_\Delta Q_{0n})\, n + \frac{({}_\Delta P_{kn} + k_{\Delta k}\, {}_\Delta Q_{kn})\, P_m^2}{S_n^2 \cos^2 \varphi} \frac{1}{n}} \qquad (16) $$

According transferred power P_m we can calculate losses of energy ${}_\Delta A$ and relative losses of energy ${}_\Delta a$ for n identical parallel transformers

$$ {}_\Delta A = ({}_\Delta P_{0n} + k_{\Delta 0}\, {}_\Delta Q_{0n})\, T_{pr}\, n + \frac{({}_\Delta P_{kn} + k_{\Delta k}\, {}_\Delta Q_{kn})\, \tau_\Delta\, P_m^2}{S_n^2 \cos^2 \varphi} \frac{1}{n} \qquad (17) $$

$$ {}_\Delta a = \frac{{}_\Delta A}{A} = \frac{{}_\Delta A}{P_m \tau} = \frac{({}_\Delta P_{0n} + k_{\Delta 0}\, {}_\Delta Q_{0n})\, T_{pr}\, n}{P_m \tau} + \frac{({}_\Delta P_{kn} + k_{\Delta k}\, {}_\Delta Q_{kn})\, \tau_\Delta\, P_m}{\tau\, S_n^2 \cos^2 \varphi} \frac{1}{n} \qquad (18) $$

According transferred power P_m we can calculate reactive losses of energy ${}_\Delta A_j$ for n identical parallel transformers

$$ {}_\Delta A_j = ({}_\Delta P_{0n} + k_{\Delta 0}\, {}_\Delta Q_{0n})\, T_{pr}\, tg\, \xi_{0n}\, n + \frac{({}_\Delta P_{kn} + k_{\Delta k}\, {}_\Delta Q_{kn})\, \tau_\Delta\, P_m^2\, tg\, \xi_{kn}}{S_n^2 \cos^2 \varphi} \frac{1}{n} \qquad (19) $$

Energy efficiency η_{Atr} of transformer is defined [6] as the ratio of the transferred energy A and the input energy A_{vst}

$$ \eta_{Atr} = \frac{A}{A_{vst}} = \frac{A}{A + {}_\Delta A} = \frac{P_m \tau}{P_m \tau + {}_\Delta A} \qquad (20) $$

After substitution of equation (17) into equation (20) we get equation for energy efficiency η_{Atr}

$$ \eta_{Atr} = \frac{P_m \tau}{P_m \tau + ({}_\Delta P_{0n} + k_{\Delta 0}\, {}_\Delta Q_{0n})\, T_{pr}\, n + \frac{({}_\Delta P_{kn} + k_{\Delta k}\, {}_\Delta Q_{kn})\, \tau_\Delta\, P_m^2}{S_n^2 \cos^2 \varphi} \frac{1}{n}} \qquad (21) $$

III. Numerical Calculation for the Transformer

For the numerical calculation were considered four cases of $k_{\Delta 0}$ as specific loss factor for no-load run, and $k_{\Delta k}$ as specific loss factor for short-circuit run of transformer, which are shown in the table 1. For transformation we are used two or three transformers [7] [8].

Table 1. Variants the values of specific loss factors

Number of case	$k_{\Delta 0}$ (kW.kVAr^{-1})	$k_{\Delta k}$ (kW.kVAr^{-1})
Case 1	0,01433	0,01433
Case 2	0,01507	0,01507
Case 3	0,00183	0,01182
Case 4	0,00166	0,01274

For next numerical calculation is necessary to ascertain properties of actual transformer [9] [10] [11]. All necessary input values for this transformer we can see in the table 2.

Table 2. Input parameters of transformer

Name of value	Mark	Physical quantity
Operation time	T_{pr}	8760 h.year^{-1}
Time of using maximum	τ	5410 h.year^{-1}
Time of full losses	τ_Δ	3579 h.year^{-1}
Fixed component of costs	c_{st}	13,774 €.kW^{-1}.year^{-1}
Variable component of costs	c_{po}	0,1448 €.kWh^{-1}
Investment costs of transformer	N_i	20 000 €
Annual quota	k	0,2 year^{-1}
Primary voltage	U_1	110 kV
Secondary voltage	U_2	22 kV
Nominal apparent power	S_n	40000 kVA
Power factor	cos φ	0,95
Relative no-load current	i_0	0,00109
Relative short-circuit voltage	u_k	0,1124
Active no-load losses	${}_\Delta P_{0n}$	31 kW
Active short-circuit losses	${}_\Delta P_{kn}$	148 kW

We can see calculated annual costs N_{tr} on Figure 1 and specific costs n_{tr} on Figure 2. These graphs are made for different loads P_m of transformer.

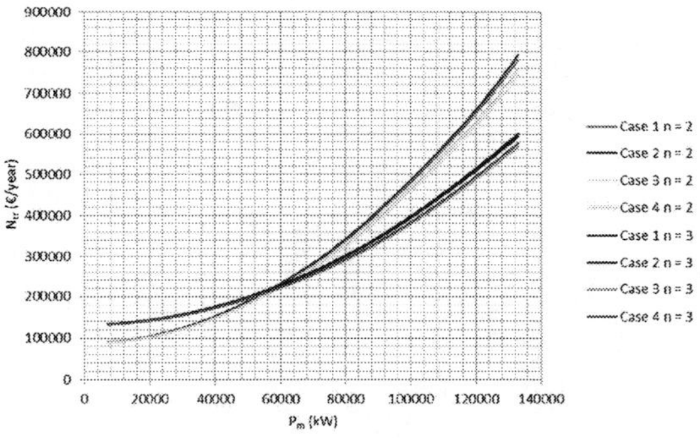

Figure 1. The annual costs on transformers

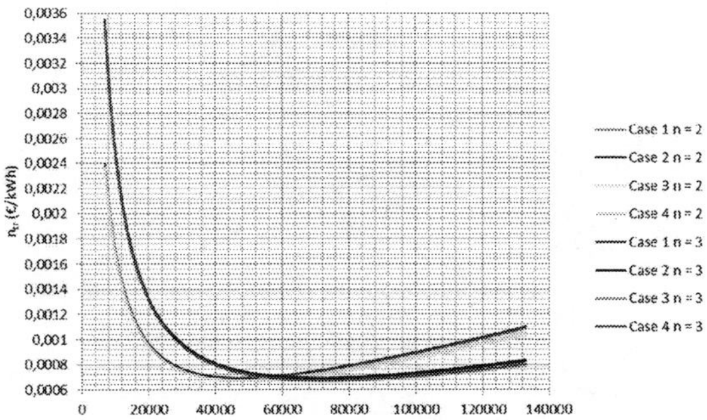

Figure 2. The specific costs on transformers

We can see calculated losses ${}_\Delta P$ on Figure 3, relative losses ${}_\Delta p$ on Figure 4, reactive losses ${}_\Delta Q$ on Figure 5 and power efficiency η_{Ptr} Figure 6. The reactive losses ${}_\Delta Q$ are independent of the specific loss factor for no-load run $k_{\Delta 0}$ and the specific loss factor for short-circuit run $k_{\Delta k}$. These graphs are made for different loads P_m of transformer.

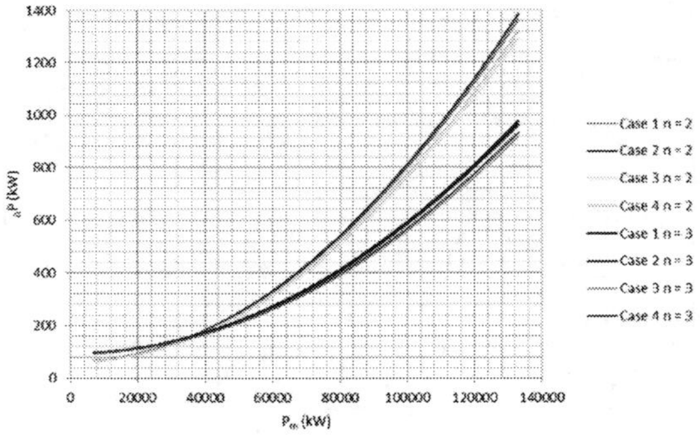

Figure 3. The losses on transformers

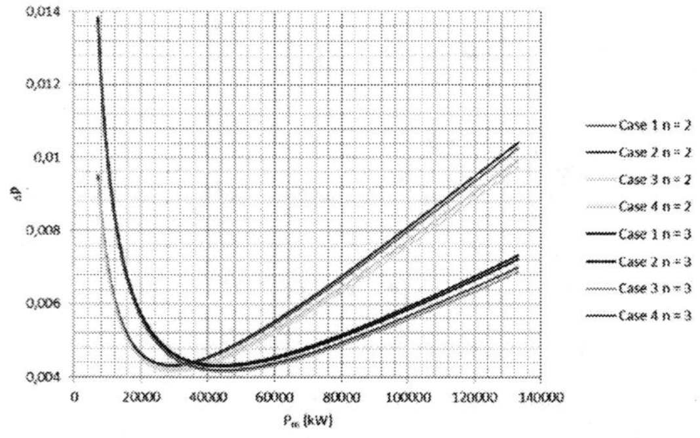

Figure 4. The relative losses on transformers

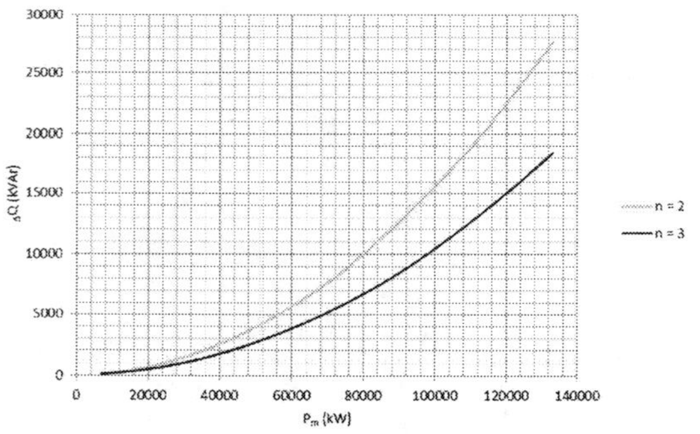

Figure 5. The reactive losses on transformers

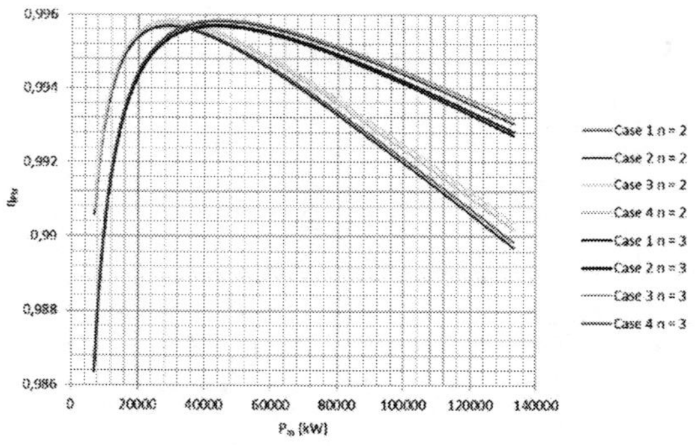

Figure 6. The power effeciency on transformers

We can see calculated losses of energy $_{\Delta}A$ on Figure 7, relative losses of energy $_{\Delta}a$ on Figure 8, reactive losses of energy $_{\Delta}A_j$ on Figure 9 and energy efficiency η_{Atr} on Figure 10. The reactive losses of energy $_{\Delta}A_j$ are independent of the specific loss factor for no-load run $k_{\Delta 0}$ and the specific loss factor for short-circuit run $k_{\Delta k}$. These graphs are made for different loads P_m of transformer.

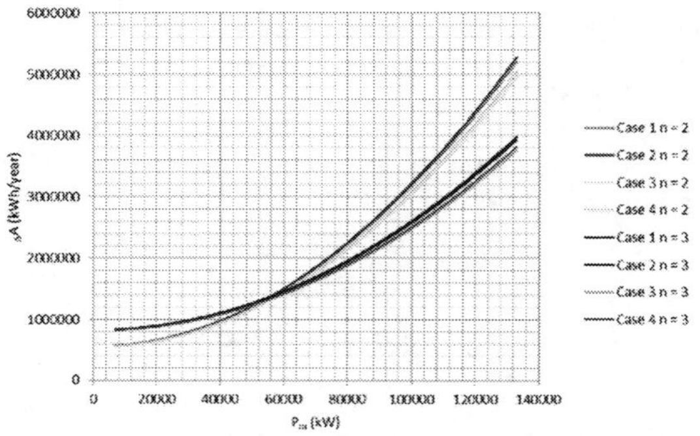

Figure 7. The losses of energy on transformers

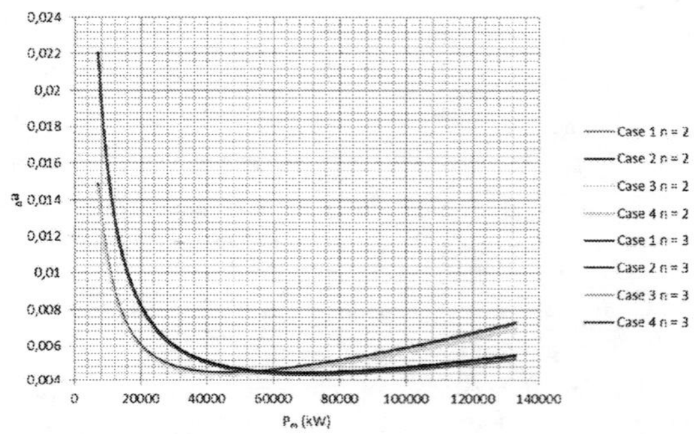

Figure 8. The relative losses of energy on transformers

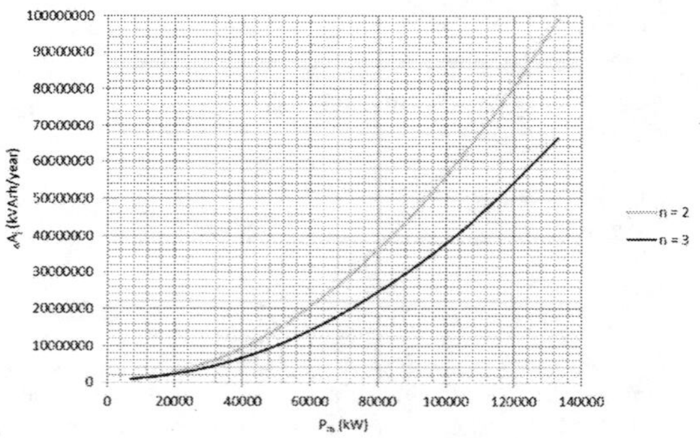

Figure 9. The reactive losses of energy on transformers

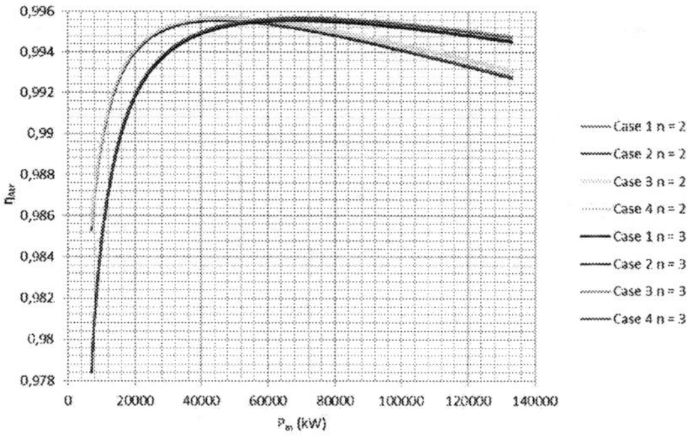

Figure 10. The energy effeciency on transformers

Minimum of specific costs n_{tr} on the transformer is when transformer is loaded by economic power P_e. Economic power P_e for minimum of specific costs n_{tr} is defined by equation

$$P_e = n\, S_n \cos\varphi \sqrt{\frac{N_i\, k + \left({}_{\Delta}P_{0n} + k_{\Delta 0}\ {}_{\Delta}Q_{0n}\right) T_{pr}\ c_1}{\left({}_{\Delta}P_{kn} + k_{\Delta k}\ {}_{\Delta}Q_{kn}\right) \tau_{\Delta}\ c_2}} \tag{22}$$

Maximum of power efficiency η_{Ptr} on the transformer is when transformer is loaded by power $P_{\eta Pmax}$. Power $P_{\eta Pmax}$ for maximum of power efficiency η_{Ptr} is defined by equation

$$P_{\eta P\max} = n\, S_n \cos\varphi \sqrt{\frac{\left({}_{\Delta}P_{0n} + k_{\Delta 0}\ {}_{\Delta}Q_{0n}\right)}{\left({}_{\Delta}P_{kn} + k_{\Delta k}\ {}_{\Delta}Q_{kn}\right)}} \tag{23}$$

Maximum of energy efficiency η_{Atr} on the transformer is when transformer is loaded by power $P_{\eta Amax}$. Power $P_{\eta Amax}$ for maximum of energy efficiency η_{Atr} is defined by equation

$$P_{\eta A\max} = n\, S_n \cos\varphi \sqrt{\frac{\left({}_{\Delta}P_{0n} + k_{\Delta 0}\ {}_{\Delta}Q_{0n}\right) T_{pr}}{\left({}_{\Delta}P_{kn} + k_{\Delta k}\ {}_{\Delta}Q_{kn}\right) \tau_{\Delta}}} \tag{24}$$

On the table 3 we can see calculated values of transferred power P_e, $P_{\eta Pmax}$ and $P_{\eta Amax}$ for number n of identical transformers is two.

Table 3. The calculated values of transferred power

Number of case	P_e (kW)	$P_{\eta Pmax}$ (kW)	$P_{\eta Amax}$ (kW)
Case 1	47590,6338393976	29240,2071878076	45743,3830092344
Case 2	47237,9085943090	29024,4335935006	45405,8267770136
Case 3	48634,9885573672	29865,2410516208	46721,1860474450
Case 4	48139,0260815650	29560,4593885548	46244,3855836768

On the table 4 we can see calculated values of transferred power P_e, $P_{\eta Pmax}$ and $P_{\eta Amax}$ for number n of identical transformers is three.

Table 4. The calculated values of transferred power

Number of case	P_e (kW)	$P_{\eta Pmax}$ (kW)	$P_{\eta Amax}$ (kW)
Case 1	71385,9507590964	43860,3107817114	68615,0745138516
Case 2	70856,8628914635	43536,6503902509	68108,7401655204
Case 3	72952,4828360508	44797,8615774312	70081,7790711675
Case 4	72208,5391223475	44340,6890828322	69366,5783755152

In following case, when is equality of specific costs n_{tr} between different number of identical transformers n_1 and n_2, we can ascertain boundary power P_h from the equality of specific costs.

$$P_h = \sqrt{n_1\, n_2}\ S_n \cos\varphi \sqrt{\frac{N_i\, k + \left({}_{\Delta}P_{0n} + k_{\Delta 0}\ {}_{\Delta}Q_{0n}\right) T_{pr}\ c_1}{\tau_{\Delta}\ c_2 \left({}_{\Delta}P_{kn} + k_{\Delta k}\ {}_{\Delta}Q_{kn}\right)}} \tag{25}$$

In following case, when is equality of power efficiency η_{Ptr} between different number of identical transformers n_1 and n_2, we can ascertain boundary power $P_{\eta Ph}$ from the equality of power efficiency.

$$P_{\eta Ph} = \sqrt{n_1\, n_2}\ S_n \cos\varphi \sqrt{\frac{\left({}_{\Delta}P_{0n} + k_{\Delta 0}\ {}_{\Delta}Q_{0n}\right)}{\left({}_{\Delta}P_{kn} + k_{\Delta k}\ {}_{\Delta}Q_{kn}\right)}} \tag{26}$$

In following case, when is equality of energy efficiency η_{Atr} between different number of identical transformers n_1 and n_2, we can ascertain boundary power $P_{\eta Ah}$ from the equality of energy efficiency.

$$P_{\eta Ah} = \sqrt{n_1\, n_2}\ S_n \cos\varphi \sqrt{\frac{\left({}_{\Delta}P_{0n} + k_{\Delta 0}\ {}_{\Delta}Q_{0n}\right) T_{pr}}{\left({}_{\Delta}P_{kn} + k_{\Delta k}\ {}_{\Delta}Q_{kn}\right) \tau_{\Delta}}} \tag{27}$$

On the table 5 we can see calculated values of transferred power P_h, $P_{\eta Ph}$ and $P_{\eta Ah}$ between two and three identical transformers [1] [5].

Table 5. The calculated values of transferred power

Number of case	P_h (kW)	$P_{\eta Ph}$ (kW)	$P_{\eta Ah}$ (kW)
Case 1	58286,3847210772	35811,7937916948	56023,9737406610
Case 2	57854,3862861446	35547,5261886856	55610,5534764423
Case 3	59565,4528058241	36577,3008108461	57221,5329969405
Case 4	58958,0253071827	36204,0210321118	56637,5740742633

IV. Calculation for the Transformer with Different Load

The calculations can be proved by the forecast transferred power P_m. On the Figure 11 we can see the example of transferred power P_m in time t for the next five days [6] [12].

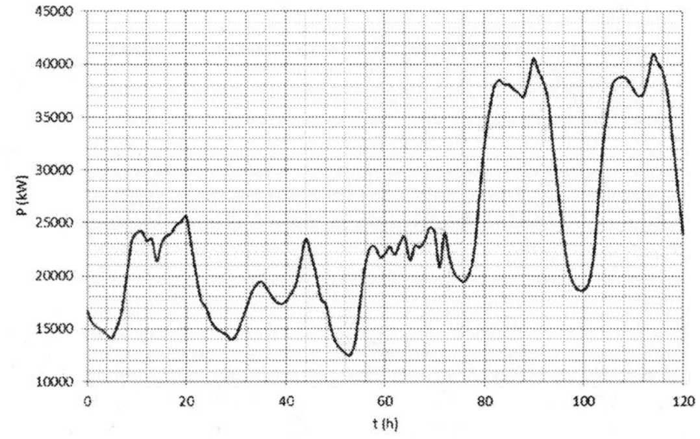

Figure 11. The forecast of transferred power

On the Figure 12 we can see calculated specific costs n_{tr}. This graph is made according the example of transferred power P_m which flow through transformers.

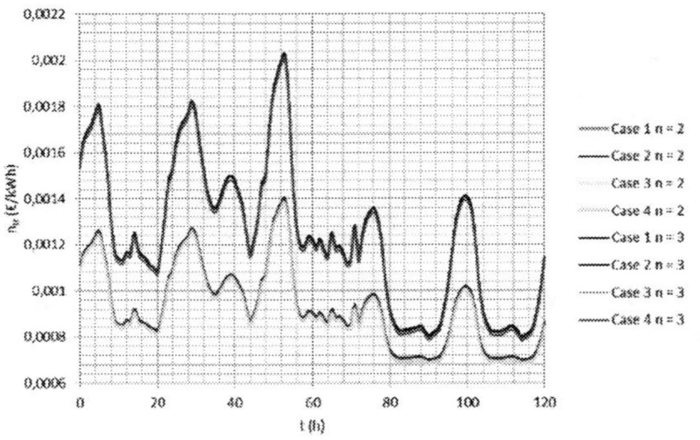

Figure 12. The specific costs on transformers

On the Figure 13 we can see calculated relative losses $_{\Delta}p$ and on the Figure 14 we can see calculated power efficiency η_{Ptr}. These graphs are made according the example of transferred power P_m which flow through transformers.

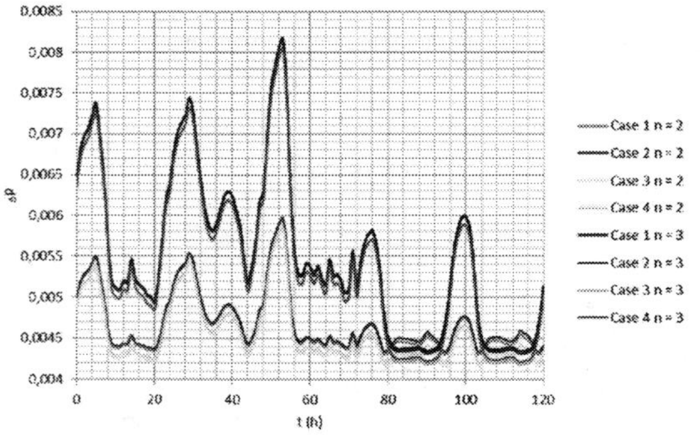

Figure 13. The relative losses on transformers

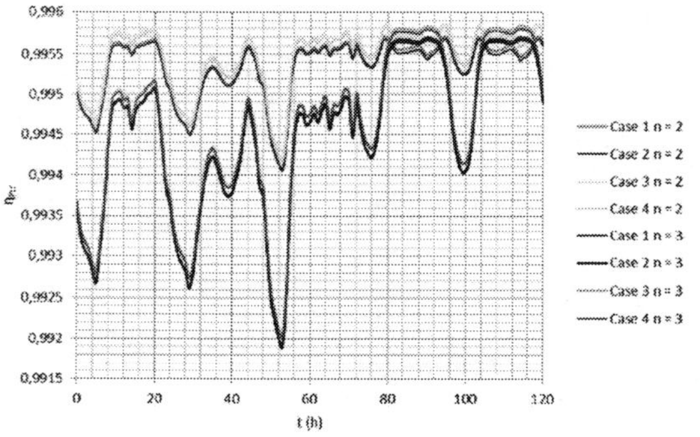

Figure 14. The power effeciency on transformers

On the Figure 15 we can see we can see calculated relative losses of energy $_{\Delta}a$ and on the Figure 16 we can see we can see calculated energy efficiency η_{Atr}. These graphs are made according the example of transferred power P_m which flow through transformers.

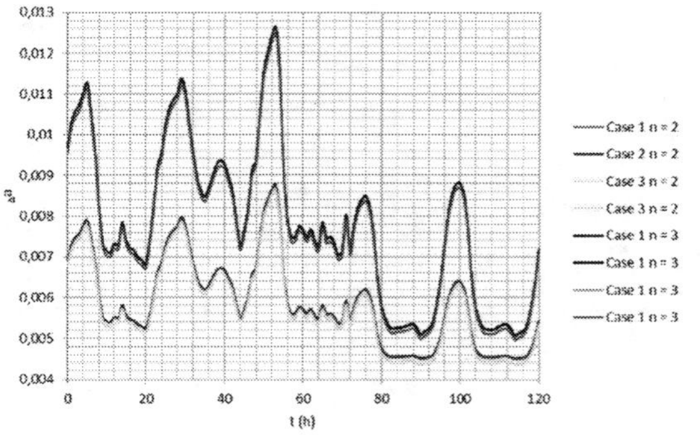

Figure 15. The relative losses of energy on transformers

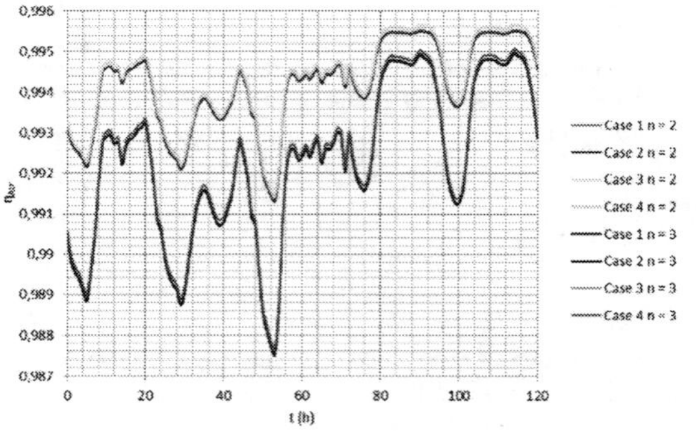

Figure 16. The energy effeciency on transformer

When we consider specific costs n_{tr} on the Figure 12 for five days, it is appropriate to consider using two transformers. In this case, if here is forecasted the rise of transferred power P_m in the future, it is appropriate to consider using of three transformers. When we consider relative losses $_{\Delta}p$ on the Figure 13 or power efficiency η_{Ptr} on the Figure 14 for five days, we can see approximate equality of these values, so we can use two or three transformers. In this case, if better technical parameters are necessary, it is appropriate to consider using of three transformers. When we consider relative losses of energy $_{\Delta}a$ on the Figure 15 or energy efficiency η_{Atr} on the Figure 16 for five days, it is appropriate to consider using two transformers. In this case, if here is forecasted the rise of transferred power P_m in the future, it is appropriate to consider using of three transformers. The proposed analysis can be easily expanded to take into account different scenarios of pricing of the losses electric energy in transformers and variation of transferred power P_m according to the conditions, that occur during winter or summer season and the type of the demand for electric power P_m [13].

V. Conclusion

A key problem of the economics of the power industry is the variation in the demand for electric power during each day, each week and throughout year [6]. On the other side, electric energy is versatile, universal, clean to use and easy to distribute and supreme to control. That is important, it is now established that electric energy has better productivity in many applications in industries than most other energy forms [14]. Understanding these calculations is decisive if they are to be used in making decisions how to solve proper selection and operation of the transformer in the power industry. At the same, it is equally important to recognize that the maximum of power in the power system is limited [15]. These practical examples illustrate that the selection of number identical parallel transformers are based on comprehensive approach and calculation of different loads of transformer. At present, this article also discusses the problem economy of transformer and technical parameters of transformer. Proper selection and operation of the transformer scientifically is becoming an effective way to reduce power losses and losses of energy in the power system [16]. Different designs of transformer can be developed to meet the requirements of a particular specification by the power system [13]. In the future, the power systems will cooperate with renewable resources to a larger extent. Currently, there is a change to natural conditions and is expected to further increase in the temperature [2]. This means, there will be less cold winters, but hotter summers. It also changes the consumption of electric energy, which in winter will be the less heat, but in the summer the electric energy will be used to drive air conditioners for cooling buildings. Experience from countries where a large part of consumption of electric energy is used to drive air conditioners show that not all the problems associated resolved through example Smart Grid in power systems [3]. On the other side, the variations in the demand for electric power have contributed to the need to develop a new technology strategy called the smart grid systems [2]. At the same, power systems are vulnerable to extreme atmospheric phenomena, which may destabilize in the case disturbance of the power system and cause power failure for huge areas [4].

Acknowledgment

This work was supported by the Scientific Grant Agency of the Ministry of Education of Slovak Republic and the Slovak Academy of Sciences under the contract No. VEGA 1/0132/15.

References

[1] M. Kolcun, Ľ. Beňa, A. Mészáros, "Optimalizácia prevádzky elektrizačnej sústavy", Technical University of Košice, 2009, ISBN 978-80-553-0323-9

[2] P. Staněk, P. Ivanová, "Súčasné tendencie ekonomickej globalizacie", Bratislava 2015, ISBN 978-80-970135-7-8

[3] P. Staněk, P. Ivanová, "Štvrtá priemyselná revolúcia a piaty civilizačný zlom", Bratislava 2016, ISBN 987-80-970135-8-5

[4] P. Staněk, P. Ivanová, "Európska únia na križovatke postrehy a inšpirujúce riešenia", Bratislava 2016, ISBN 978-80-8168-482-1

[5] L. Varga, S. Ilenin, P. Leščinský, "Prenos a rozvod elektrickej energie", Technical University of Košice, 2003

[6] C. R. Dorf, "Energy, resources and policy", Addinson Wesley Publishing Company Inc., 1978, ISBN 0-201-01673-7

[7] J. Chlebný, P. Beneš, J. Langer, J. Král, M. Martinásková, "Automatizace a automatizační technika", Computer Press Brno, 2014

[8] J. Desmet, G. Delaere, L. Lemecko, "Volba a dimenzování transformátorů", Technical University of Ostrava, 2006

[9] M. Kolcun, V. Chladný, M. Mešter, R. Cimbala, J. Tkáč, M. Hvizdoš, J. Rusnák, "Elektrárne", Technical University of Košice, 2006, ISBN 80-8073-704-5

[10] P. Novák, "Elektrické teplo", Technical University of Košice, 2010, ISBN 978-80-89284-48-1

[11] A. Pietriková, J. Banský, "Základy inžinierstva materiálov", Technical University of Košice, 2007, ISBN 80-8086-001-7

[12] M. Kolcun, V. Griger, Ľ. Beňa, J. Rusnák, "Prevádzka elektrizačnej sústavy", Technical University of Košice, 2007

[13] E. I. Amoiralis, M. A. Tsili, A. Kladas, "Economic evaluation of transformer selection in electrical power systems", International Conference on Electrical Machines, IEEE, 2010

[14] H. Khatib, "Economic evaluation of projects in the electricity supply industry", London 2008,

[15] B. C. Lesieutre, J. H. Eto, "Electricity transmission congestion costs a review of recent reports", University of California Berkeley, 2003

[16] W. Jinli, S. Wanxing, W. Li, Y. Honglei, "Study on technical and economical efficiency of amorphous alloy transformer and on-load capacity regulating transformer in distribution network application", China International Conference on Electricity Distribution, 2014

Investigation of Influence of the Additional Electromagnetic Shielding on Domestic Appliances

Dušan Medveď
Department of Electrical Power Engineering
Mäsiarska 74
041 20 Košice, Slovak Republic
Dusan.Medved@tuke.sk

Abstract — **This paper deals with the modeling and analysis of additional shielding influence against the harmful electromagnetic field on domestic appliances. There were used two different materials especially for shielding against the magnetic field and against electric field that were applied on induction cooker. The simulations were realized in ANSYS and evaluated which one of the shielding materials is more suitable.**

Keywords: electromagnetic field, shielding, simulation, ANSYS

I. Introduction

At present, the people are exposed to permanent influence of electromagnetic radiation. It consists of both natural radiation and artificial electromagnetic fields. Each field is characterized by physical quantities, e.g. frequency and wavelength. We recognize these fields in the form as light, heat, and so on. Earth, in general, has an electromagnetic field around that does not adversely affect the human being. Therefore, the natural electromagnetic field exists for longer. Artificially created electromagnetic field arises around the electrical devices as a side-effect. The sources of this radiation are, for example, Wi-Fi devices, electrical equipment, and many others.

The sources of the electromagnetic field are located both in the exterior and in the interior, so they can be detected also inside the residential buildings. These sources include electrical appliances such as hair dryer, induction cooker, electric floor heating, and others. For humans, fields created from these devices could be harmful. At present, the International Commission on Non-Ionizing Radiation Protection (ICNIRP) deals with the impact of electromagnetic fields on human.

From this need to determine the magnitude of the electromagnetic field and possible options for shielding against the influence of the electromagnetic field will be solved and presented in this paper.

II. Proposal and Selection of Shielding materials

In order to be able to shield the electromagnetic field, it is necessary to choose suitable material to shield the assumed harmful effects. Since the proposed shielding of a particular electric and especially magnetic field was proposed in this paper, it was necessary to select two types of materials that were used to shield a specific field. Two materials were chosen for each type of field, and a comparison was realized which one of these shielding materials is more appropriate.

A. *Shielding materials against the influence of magnetic fields*

In order to select a suitable shielding material for a magnetic field, it was necessary to find the physical parameters such as permeability, resistivity, and possibly its cost. The next part of the paper presents a brief description of the selected shielding materials from Magnetic Shield Corporation.

A1. Co-Netic® AA Perfection Annealed

This material was selected for simulation in the ANSYS program. Co-Netic® AA Perfection Annealed is used most often to attenuate the magnetic field. It provides the high attenuating without further processing. The thickness of the material has variations from 0.36 mm to 2 mm. There is an annealed plate that is used worldwide for planar shading such as shading of doors, windows or walls. This material is also in the form of foils. In this configuration, it is available with or without a PST (Pressure Sensitive Tape). PST is a pressure-sensitive tape added [7]. Both products are heat-treated in challenging conditions, and they form a grain structure important for a final shielding. The necessary parameters for the correct simulation were read from the manufacturer's catalog [1].

TABLE I. Table of Magnetic Data of Material Co-Netic AA Perfection Annealed

Electrical resistivity $\times 10^{-6}$ [$\Omega \cdot m^2 \cdot m^{-1}$]	55
Initial permeability [$H \cdot m^{-1}$]	30 000
Maximal permeability [$H \cdot m^{-1}$]	450 000

A2. Co-Netic® B Stress Annealed

The second material was chosen similar to the former one but with the different material properties. Stress Annealed is a sheet material for further processing, stamping, heavy melting or welding. In production, the granularity particle size is weakened. After final modifications, the grain structure changes due to air annealing. This is done to mitigate the loss of material grain. This leads to a thicker grain structure, softer tempering and final shading [1].

TABLE II. TABLE OF MAGNETIC DATA OF MATERIAL CO-NETIC B STRESS ANNEALED

Electrical resistivity $\times 10^{-6}$ [$\Omega \cdot m^2 \cdot m^{-1}$]	48
Initial permeability [$H \cdot m^{-1}$]	8 000
Maximal permeability [$H \cdot m^{-1}$]	150 000

For the correct field simulation, electrical resistivity and initial permeability were used from the table data. All the necessary parameters were available from the manufacturer catalogue.

B. *Shielding materials against influence of electrical field*

Similarly, two materials with different electrical parameters were chosen for the choice of shielding materials against the influence of the magnetic field as well as the electric field. When selecting the shielding material, it is especially important to take in account the permitivity and resistivity of the material for the proper simulation in the ANSYS program. All materials were selected from the Yshield Company catalogue.

B1. YSHIELD® HNG 80/66

A material of this type is a woven, metallized, polyester meshwork. It is used for shielding both on high-frequency radiation and on low-frequency radiation of electric fields. It is commonly used on walls, ceilings and floors. Its advantage is that it is easy to stick. Its application was mainly in banks and laboratories and currently is used also for private purposes [2].

TABLE III. TECHNICAL PARAMETERS OF SHIELDING MATERIAL HNG 80/66

Width [cm]	66
Attenuation [dB]	80 (with 2 layers 106 dB)
Weight [g/m^2]	80
Material thickness [mm]	0,07
Composition	PE, Cu, Ni, protective layer
Surface conductivity [Ω]	0,02

Based on the composition of the material (according to the percentage representation of the relevant elements in the material), average (mean) relative permitivity was calculated, and its value is 4,515. The surface conductivity was read from the catalog [5], [6].

B2. YHSIELD® HNG 100/66

This material from company Yshield® was selected as the second one. HNG 100/66 is a compact, woven, heavily metallized, polyester, crosslinked system for the shielding high-frequency radiation as well as a low-frequency electric field. It is usually used for defense ministries, banks and laboratories. It is used to bond internal walls, floors, ceilings, construction and free placing [2].

TABLE IV. TECHNICAL PARAMETERS OF SHIELDING MATERIAL HNG 100/66

Width [cm]	66
Attenuation [dB]	100 (with 2 layers is 134 dB)
Weight [g/m^2]	140
Thickness of material [mm]	0,08
Composition	PE, Cu, Ni, protective layer
Surface conductivity [Ω]	0,003

As with the first material, the manufacturer also did not presented the value the relative permitivity, so the average permitivity was calculated in a similar way (mean value).

C. *Creation of a model geometry for simulation*

Electrolux EHH6240ISK induction cooker was selected to simulate the influence of electromagnetic field. This cooker was mounted on an electrical oven of 600 mm depth and 590 mm height (Fig. 1). The considered induction cooker consists of 4 induction plates. Two induction plates with a diameter of 210 mm and 180 mm were considered. The rear plate with a 210 mm diameter has an installed power of 2800 W. A front plate with a diameter of 180 mm has an installed power of 1800 W. The whole model is located in a finite space, which is 1500 × 1500 mm in dimension (Fig. 2). This model was considered in 2-dimensional space to simplify the calculation.

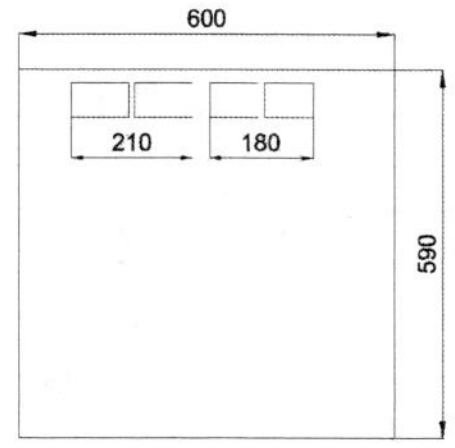

Figure 1. Geometry of the model

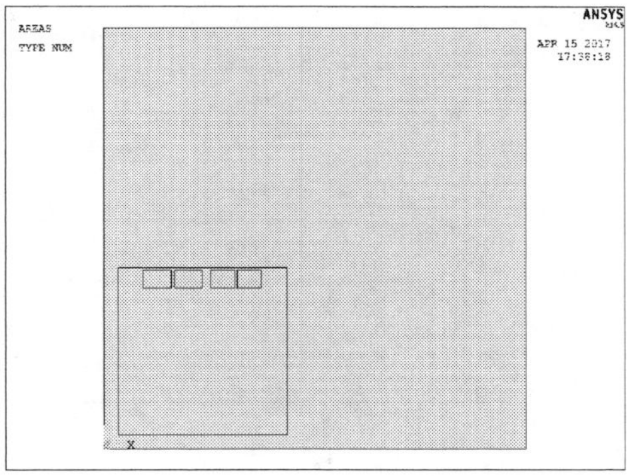

Figure 2. Considered model of induction cooker in space

After creating the basic geometry, the shielding was proposed for a given model of induction cooker. Because the model is placed close to the wall, the proposed shielding was applied from the top and the right side. Each thickness of the shielding material was different. When shielding the magnetic field, the shielding material was considered of thickness 1 mm, and for the electric field it was considered thickness of material 0.07 mm. When enlarging the overall geometry, one can see the application of shielding on the model.

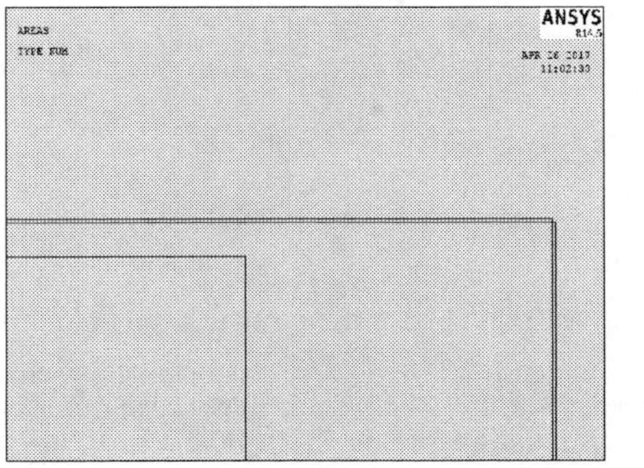

Figure 3. Detail at supposed shielding (thin layer around cooker)

In order to correctly proceed the whole calculation, it was necessary to define the material properties of the particular objects. The table (Table V) shows the material properties that represent the relative permeability, relative permitivity and resistivity of the materials. Magnetic field determines mostly the relative permeability and relative permittivity for the electric field. The resistivity is the same for both fields. These values were obtained from sources [3].

TABLE V. MATERIAL PROPERTIES

			magnetic field	***electric field***
area no.	material	resistivity ($\Omega \cdot m$)	relative permeability ($H \cdot m^{-1}$)	relative permitivity ($F \cdot m^{-1}$)
A1	air	$2 \cdot 10^{14}$	1,00000037	1,00058986
A2	el. oven	1	1	1
A3	copper	$1{,}68 \cdot 10^{-8}$	0,999994	5,6

Similarly, it was necessary to define the material properties of shielding materials. In the following table there are presented as well as magnetic as electric properties of shielding materials [3].

TABLE VI. MATERIAL PROPERTIES OF SHIELDING MATERIALS

	area no.	material	resistivity ($\Omega \cdot m$)	relative permeability ($H \cdot m^{-1}$)	relative permitivity ($F \cdot m^{-1}$)
electric field	A4	YSHIELD HNG 80/66	50	–	4,515
		YSHIELD HNG 100/66	333,333	–	4,515
magnetic field		Co-NETIC AA PERFECTION ANNEALED	$5{,}5 \cdot 10^{-7}$	30000	–
		Co-NETIC B STRESS ANNEALED	$4{,}8 \cdot 10^{-7}$	8000	–

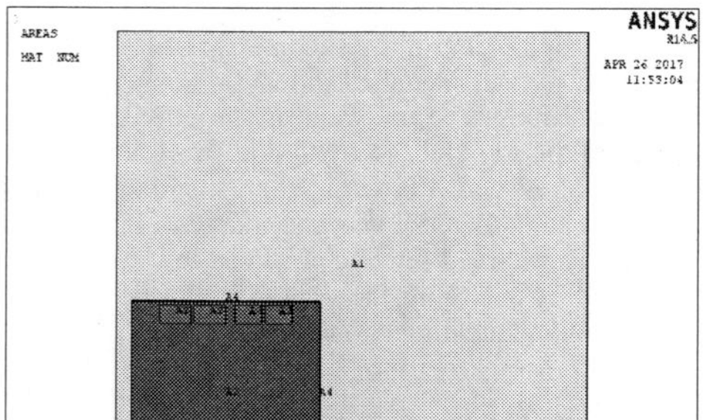

Figure 4. Geometry and labeling of materials

III. RESULTS OF MODELING THE ELECTROMAGNETIC FIELD

In this part of the paper, distribution of electric and magnetic field, the graphical dependence and direction of lines of forces are displayed. In each distribution of magnetic and electric fields, the appropriate detail was chosen to better observe the shielding efficiency, as well as a clearer illustration of the magnetic and electric fields effects. All calculations were performed at 25 kHz (current and voltage frequency). The magnitude of the current densities of both plates was calculated according to the expressions mentioned below.

Calculation of the current flowing through the first induction plate with a diameter of 210 mm and a power of 2800 W.

$$I = \frac{P}{U} = \frac{2800}{230} = 12{,}17 \text{ A} \quad (1)$$

and then the calculation of current density

$$J = \frac{I}{\frac{\pi \cdot d^2}{4}} = \frac{12{,}17}{\frac{\pi \cdot 0{,}21^2}{4}} = 620013 \text{ A} \cdot \text{m}^{-2} \quad (2)$$

Similarly, for the calculation of the current flowed through the second induction plate of the diameter 180 mm and of power 1800 W it was $I = 7{,}83$ A, $J = 398\,579$ $A \cdot m^{-2}$.

A. *Magnetic field modeling results on induction cooker without magnetic shielding*

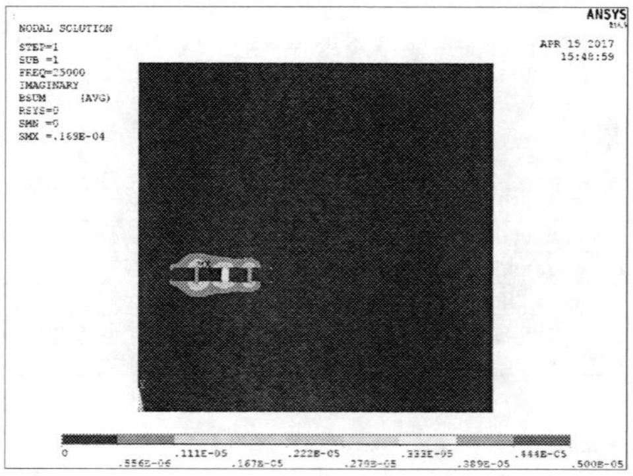

Figure 5. Magnetic induction distribution around the induction cooker without shield

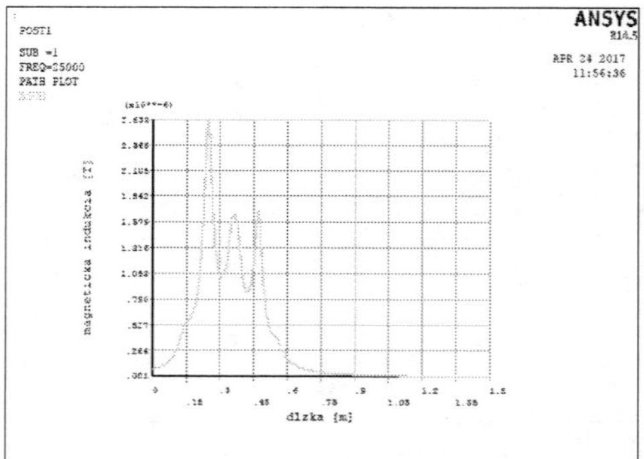

Figure 6. Graphical dependence of magnetic fields distribution

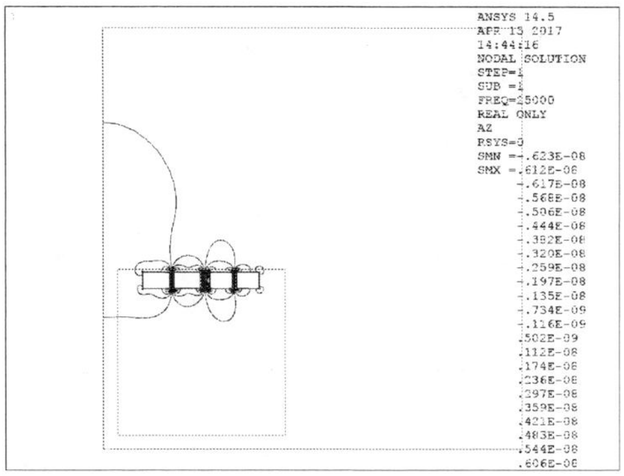

Figure 7. Lines of force of magnetic field distribution

Graphical dependence (Fig. 6) was realized at a height of 645 mm from the floor, which is 5 mm from the surface of the induction cooker. It can be seen that the first inductor plate had a stronger effect than the second plate. This is because the first plate is more efficient and thus has a higher current density (2). From Fig. 5 and Fig. 6 it can be seen that the magnetic field is also induced between the first and second induction plates.

*B. **Magnetic field** modeling results on induction cooker with shielding no. 2*

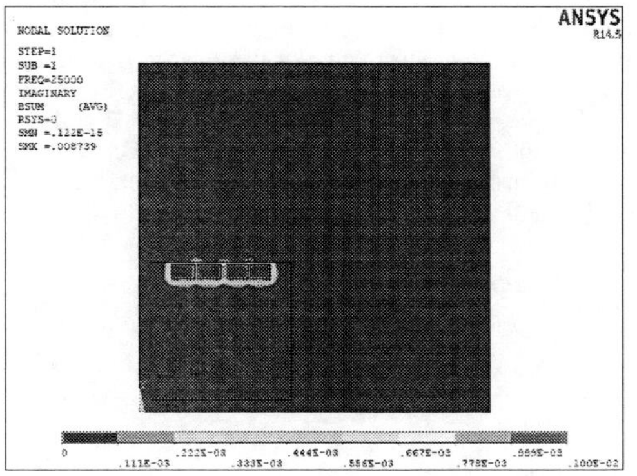

Figure 8. Magnetic induction distribution around the induction cooker with shield no. 2

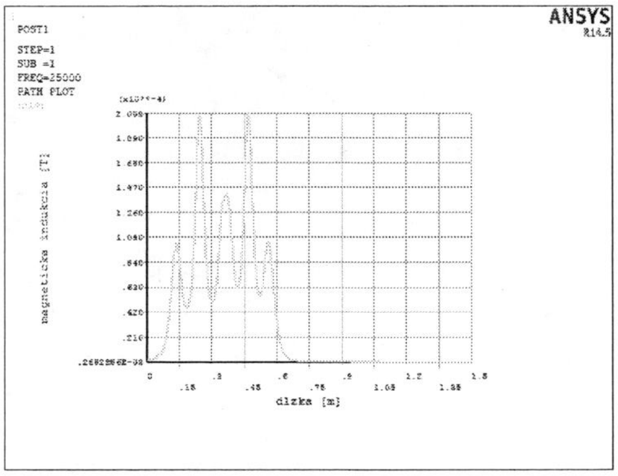

Figure 9. Graphical dependence of magnetic fields distribution

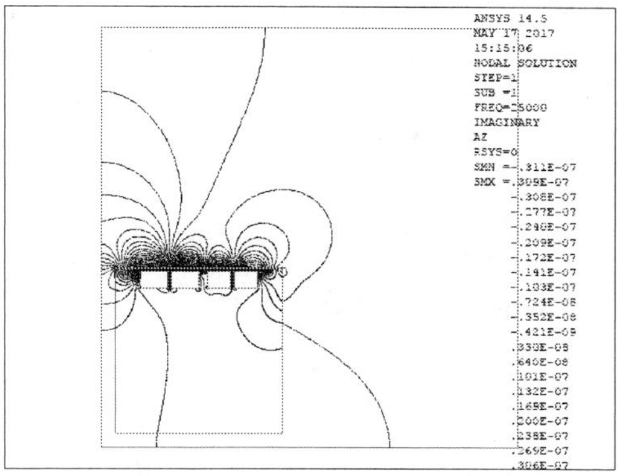

Figure 10. Lines of force of magnetic field distribution

With shield no. 2, it can be seen that this type of shielding material almost attenuated the entire magnetic field, which tends to spread to the environment. In this case, the type of shielding material "Co-Netic B Stress Annealed" was used. The properties of the material are expressed in the table Tab. VI. The graphical dependence was made at a height of 645 mm from the ground (Fig. 9). It can be seen that the magnetic induction does not extend over the cooking zone of the induction cooker. This is due to the addition of shielding material that attenuated the unwanted magnetic field. The maximum measured value of the magnetic induction in the cooking zone is lower than in the case of without shielding. Adding shielding material to a magnetic field source reduced the magnetic field propagation and reduced the magnitude of induction ***B***. This can be seen in Fig. 9 that its maximum magnetic induction value is $2{,}099 \cdot 10^{-4}$ T. This value exceeds the value set in decree 534/2007 Coll. of laws.

*C. The results of the **electric field** modeling around the induction cooker without shielding*

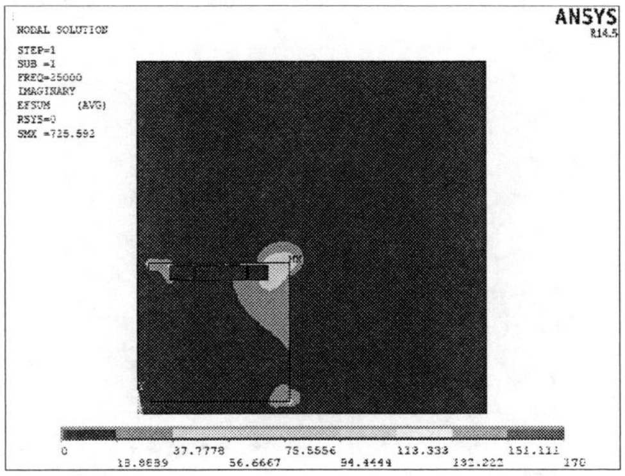

Figure 11. Distribution of electric intensity around the induction cooker without shielding

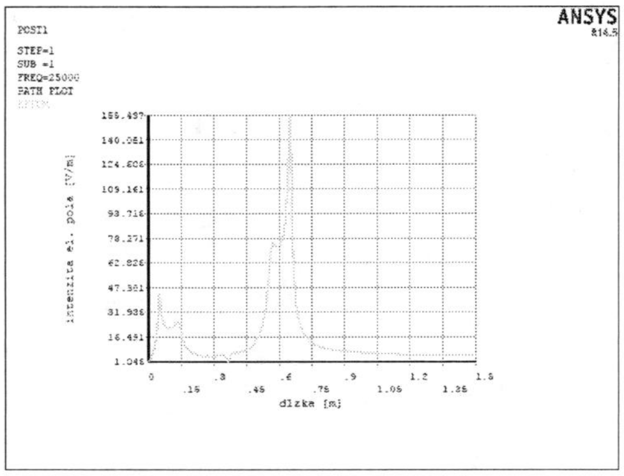

Figure 12. Graphical dependence of the action of the electric field

From the picture Fig. 11, it can be seen that the action of the electric field mainly extends along the edges of the induction cooker plates, respectively from the wound of coils. The greatest effect is seen in the upper right corner that is the area where a man obviously stands. This greatest effect is indicated by green to red color. The highest field intensity was noticed in the top right corner of the model. Its value stopped at 714,592 $V \cdot m^{-1}$. This field intensity value exceeds more than 800 % of the value that is in the Decree 534/2007 of the determined value without harmful effects. Graphical dependence on Fig. 12 was realized at a height of 645 mm from the floor. From this graphical dependence, it can be seen that at a height of 645 mm from the floor the largest effect was at a distance of 620 mm from the wall. Its measured value of 155,497 $V \cdot m^{-1}$ exceeds 1.7-times the maximum allowable value of 87 $V \cdot m^{-1}$ set by the Decree. It can be said that at such a distance of 620 mm from the wall usually stay a human who operate the induction cooker.

D. The results of the ***electric field*** *modeling around the induction cooker with shielding material no. 1*

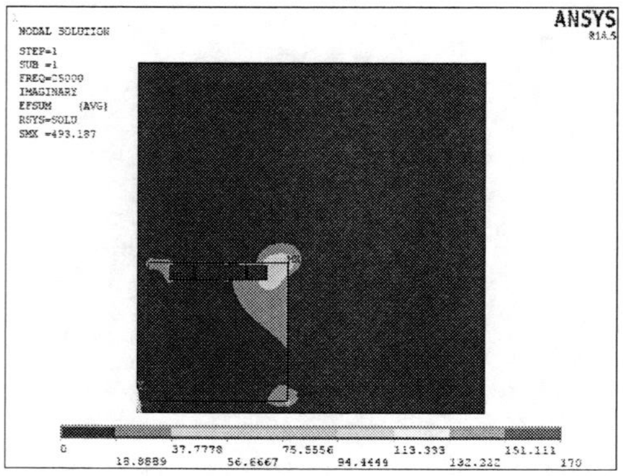

Figure 13. Distribution of electric intensity around the induction cooker with shield no. 1

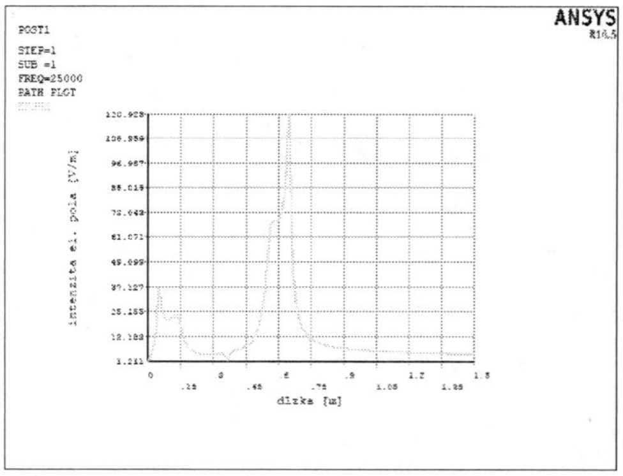

Figure 14. Graphical dependence of the action of the electric field

Shielding material Yshield HNG 80/66 was used for shielding no. 1. Its material properties are presented in Tab. VI. It can be seen that after application of shielding material no. 1, the electric field propagates as well as without shielding. Only the maximum field intensity is changed. This value is 493,187 $V \cdot m^{-1}$. It is located in the upper right corner and represents only a small area where such a large value exists. Permitivity and voltage values were similar as for non-shielded models.

The graphic dependence (Fig. 14) was realized at a height of 645 mm from the floor. From this dependence, it can be observed that the highest value of the electric field intensity is at the same spot as compared to no shielding (Fig. 12), but its value dropped to 120,928 $V \cdot m^{-1}$. It is a value that still does not comply with the value registered in the decree 534/2007. Smaller influence is also seen under the inducer, but this value is not harmful to the human organism.

E. The results of the ***electric field*** *modeling around the induction cooker with shielding no. 2*

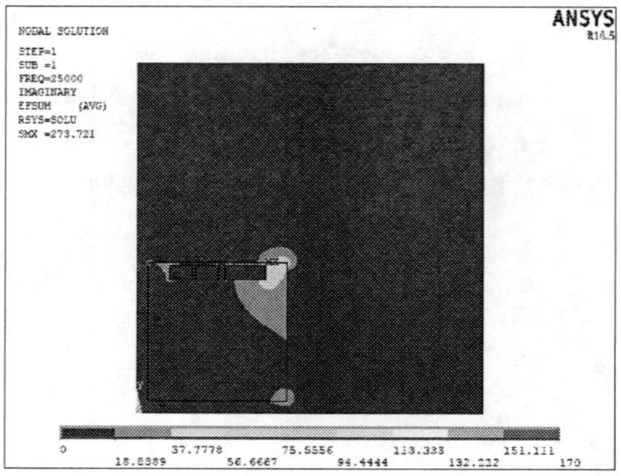

Figure 15. Distribution of electric intensity around the induction cooker with shield no. 2

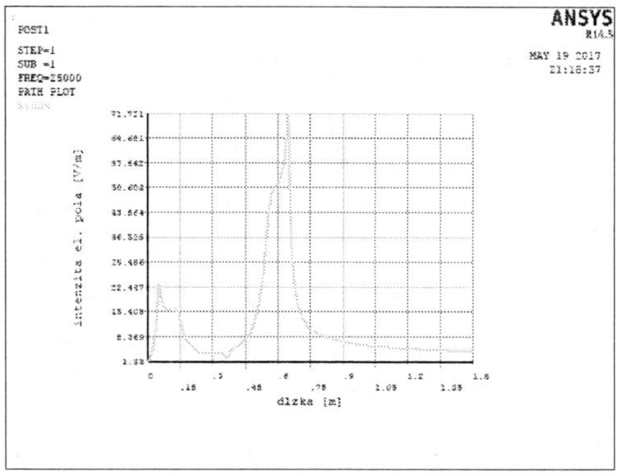

Figure 16. Graphical dependence of the action of the electric field

In this case the shielding material no. 2 Yshield HNG 100/66 was used. The electric field distribution (Fig. 15) is the similar as in the previous cases. The maximum value of the electric intensity ***E*** changed its influence with the winding coil. The value stopped at 273.721 V·m^{-1}. As it can be seen, this value does not interfere with the space outside the induced plate, and therefore it cannot cause harm to humans. Graphical dependence (Fig. 16) was realized at a height of 645 mm from the floor. This height was determined as the height at which a person is operating the induction cooker. It was observed that the used shield attenuated the electrical intensity to 71.721 V·m^{-1}. This value is less than the maximum allowed value of 87 V·m^{-1} specified by the decree 534/2007.

Electric field simulation evaluation

Comparison of both shielding materials tested under the same conditions can be claimed that only the shielding material HNG 88/66 exceeded permissible value of ***E*** in the cooking zone (according to 534/2007 Coll. of laws) [8]. Therefore, if this material is used it is suitable to use two or more layers for effective shielding. The HNG 100/66 shielding material has not exceeded the permitted field intensity and therefore it appears to be a reliable shielding material.

IV. Conclusion

By measuring on a non-shielding induction cooker, it was discovered that at height of 645 mm from the floor, the permissible values of ***E*** and ***B*** were exceeded according to 534/2007 Coll. of laws. In this decree, the maximum allowed value of the magnetic induction is 6.25 μT and the maximum electric intensity is 87 V·m^{-1}. In the case of magnetic field shielding, the magnetic induction ***B*** 94-times exceeded the permissible value and in case of electric fields the electric field intensity was at height of at 645 mm 1.7-times higher than permissible value. Magnetic induction values were highest perpendicular to the induction plates. The proposed shielding for the magnetic field has not been sufficient but can be avoided by applying multiple layers. However, in electric fields, the electric field intensity was greatest at the sides of the induction cooker. At those sides, where the shielding was proposed, one mentioned shielding material was suitable to protect against the harmful effects of the electric field [9], [10].

Therefore, it is necessary to recommend that manufacturers of electrical appliances, such as an induction cooker, have to pay more attention to shielding of electromagnetic fields, and those vendors of these induction cookers have to familiarize buyers with their health hazards and possible shielding against the electromagnetic fields. It is also recommended that producer will carefully analyze the electromagnetic field for the appliance in order to avoid insufficient shielding.

Acknowledgement

This work was supported by the Ministry of Education, Science, Research and Sport of the Slovak Republic and the Slovak Academy of Sciences under the contract No. VEGA 1/0132/15.

References

[1] Decree of the Ministry of Health of the Slovak Republic on details of requirements for sources of electromagnetic radiation and limits of exposure of the population to electromagnetic radiation in the environment 534/2007, Collection of laws.

[2] B. Dolník, *"Electromagnetic compatibility"*, Technical University of Košiciach, December 2013, 240 pages, elfa, s.r.o., ISBN 978-80-8086-221-3.

[3] *Yshield Shielding Products Miscellaneous*, [online], Available at: < http://www.safesolution.ro/Technical_Details/YSHIELD_ShieldingProductsMiscellaneous_EN.pdf >.

[4] M. Pavlík, et. al., *"Measuring of dependence of shielding effectiveness of wet materials on the frequency of electromagnetic field in the high frequency range"*, Acta Electrotechnica et Informatica, Vol. 13, No. 3, pp. 12-16 (2013).

[5] M. Pavlík, I. Kolcunová, L. Lisoň, *"Measuring the shielding effectiveness and reflection of electromagnetic field of building material"*, In: Electric Power Engineering (EPE). Ostrava: VŠB-TU, 2015 p. 56-59. ISBN 978-1-4673-6787.

[6] J. Zbojovský, P. Liptai, M. Moravec, *"Modelling and calculating the shielding effectiveness of building materials"*, In: Technical sciences and technologies. Vol. 6, no. 4 (2016), p. 205-210. ISSN 2411-5363.

[7] M. Kanálik, M. Pavlík, M. Kolcun, *"The impact of multi-system overhead lines operation with different voltage levels to voltage unbalance"*, In: Elektroenergetika 2015. Košice: TU, 2015, p. 73-76. ISBN 978-80-553-2187-5.

[8] J. Zbojovský, M. Pavlík, *"The propagation of electromagnetic fields through the various materials"*, In: Scientific Letters of Academic Society of Michal Baludansky. Vol. 4, No. 6A (2016), p. 107-109. ISSN 1338-9432.

[9] D. Medveď, O. Hirka, *"Impact of electromagnetic fields in residential areas"*, In: Elektroenergetika. Vol. 9, No. 2 (2016), p. 13-16. ISSN 1337-6756.

[10] D. Medveď, et. al., *"Additional Modification of Thermomagnetic Properties of Objects of Low Relative Permeability in Electromagnetic Field"*, Acta Physica Polonica A, Vol. 131 (2017), No. 4, DOI: 10.12693/APhysPolA.131.1138.

[11] Z. Čonka, et. al., *"Impact of renewable energy sources on stability of EWIS transmision system"*, In: 14th Int. Conf. on Env. and El. En.: conf. proc.: Krakow, Poland, 10-12 May, 2014. Danvers: IEEE, 2014 P. 75-79. ISBN 978-1-4799-4661-7.

Teaching Generation and Transmission of Electricity With an Interactive Whiteboard

Ján Pavlovkin
Faculty of Natural Sciences
UMB Banská Bystrica, Slovak republic
Jan.Pavlovkin@umb.sk

Ľubomír Žáčok
Faculty of Natural Sciences
UMB Banská Bystrica, Slovak republic
Lubomir.Zacok@umb.sk

Milan Bernát
FHPV Prešov, University of Prešov
in Prešov, Slovak republic,
milan.bernat@unipo.sk

Renáta Bernátová
PgF Prešov University of Prešov
in Prešov, Slovak republic
Renata.Bernatova@unipo.sk

Jaroslav Džmura
FEI TU Kosice.
Mäsiarska 74, Košice, Slovakia,
Jaroslav.Dzmura@tuke.sk

Milan Ďuriš
Faculty of Natural Sciences
UMB Banská Bystrica, Slovak republic
Lubomir.Zacok@umb.sk

Abstract
The paper describes the possibilities of innovation in teaching students in the field of electricity generation and transmission at faculties of teaching. The article contains a description of the interactive whiteboard, interactive whiteboard software, and ways to use the interactive whiteboard in the classroom. The authors find out the effectiveness of the teaching of these topics traditionally (using real electro physical models) and the innovative computer simulation model presented by the interactive chart as a support of a real electro-physical model. The individual teaching systems were tested in the student sample. Based on the results of the tests, it was shown that the instruction supported by the interactive chart showed better quantitative and qualitative results in the knowledge tests than in the traditional teaching system.

***Key words**: interactive whiteboard, teaching techniques, electricity, generation of electricity*

I. Introduction

The penetration of information and communication technologies into all areas of life increases the demands for education. However, their presence does not automatically change the teaching process, as it depends on creative and innovative teachers [1]. Information and communication technologies make a significant contribution to school change, changing how we learn:

- We learn visually and interactively, With visuals or interactive boards,
- Learning through a variety of media and tools, for example, using various multimedia resources and the internet.

The prerequisite for the development of digital literacy is the modernization of education using information and communication technologies, it is the mutual integration of traditional and digital educational materials using modern didactic techniques that will enable teachers to use digital educational content in the learning process, interactive whiteboard. Both the teacher and the student can actively engage in teaching while influencing and adapting to current needs.

It is natural that the changes that are currently taking place in the learning and cognitive process are significant and extensive and also relate to textbooks, teaching aids, methods and forms of teaching. The form and content of textbooks and other learning materials is greatly changed [1]. According to Brestenská et al [1], we have seen the following changes in the context of traditional textbooks in recent years:

- Creative, innovative teachers use a wide range of different learning materials, textbooks use only as one source of their preparation,
- The material they use often fails to provide their pupils with work at home,
- Several new textbooks already contain a digital media with tasks and additional data, spreadsheets, text, etc., and links to web portals,
- In addition to textbooks, digital resources are also distributed, multimedia CDs or learning materials on the web,
- Students' access to web resources from home,
- Students do not want to memorize textbooks and require textbooks to be lightweight, not to contain a lot of text to be boring, tentative, and socialized,
- Increased student demand for unrestricted access to learning resources (trends leading to mobile learning),
- Changing the overall approach to learning and learning.

II. Interactive whiteboard in the education

The Interactive whiteboard is a touch-sensitive area where interactive communication between the user and the computer takes place in order to ensure the maximum possible level of clarity of the displayed content. Using an interactive whiteboard, we can:

- Share all necessary information and educational materials with all participants in education,
- Tap an electronic pen or index finger to control the computer program directly from the board,

- Add notes to all applications by directing on desktop, in the case of bad additions you can easily delete and overwrite,
- Interact with the Internet and use all the possibilities of computer programs.

Working on an interactive whiteboard moves learning forward as it engages both the student and the learner. From this, we can conclude that: "The Interactive whiteboard is a modern learning tool that helps us streamline the learning process through presentations with a thorough use of ICT. It's an electronic device that lets you interact live with your PC or Notebook directly from the board by clicking on the projected image using an interactive pen. Using it, it is possible to control learning programs, office applications as well as teaching aids and teaching texts about subjects that the educator himself prepares in external graphic or text environments, teaching pictures or videos, all directly from the board. "[2]

Didactic benefits of using an interactive whiteboard [3]:

- Use of IT leads to motivation to learn,
- Using the principle of clarity - visualization, interactive moving,
- Better attention,
- Reuse of the already prepared subject,
- Active involvement of students and pupils in teaching,
- Networking and Internet sharing,
- Development of computer and information literacy,
- Direct use of resources from the Internet.

Didactic Disadvantages of Using an Interactive Board [3]:

- Threatens to slip into the encyclopaedias and thus suppress interactivity,
- Suppression of abstract thinking,
- The table is only a technical supplement, with frequent use falling,
- The table is not used interactively, but only as a "screenplay"
- High demands on time when teaching,
- The necessary knowledge in the field of computing,
- Absence of finished multimedia programs,
- Absence of e-textbooks and i-textbooks, their high price,
- The problem of low pupils, low teachers,
- High cost of elements, energy intensity, vulnerability to damage,
- Refusal to write and work by fingers younger pupils,
- Suppress the meaning of classical textbooks and lessons,
- Written and oral representation,
- Suppress the use of real aids and experiments,
- Simulation only of reality,
- Poor readability from a greater distance,
- Low light area problem, shaded shadow.

III. POSSIBILITIES OF USING THE INTERACTIVE WHITEBOARD SOFTWARE

The differences between the individual types of whiteboard consist mainly of the way of transferring the changes made on the board to the digitized dimension. Active Inspire software is a program specially designed for the school environment. The main workspace is the presentation of a workbook in which we can create presentation content using interactive tools. The number of pages is unlimited, there are several open ones, and we can export the content to other formats or print it out. One main toolbar allows you to control the entire software and choose the features we need and use; all other tools are clearly arranged and described in the appropriate panels.

Writing and highlighting on the interactive whiteboard allows the user to write on their desktop, Change colours, use the highlight feature to create notes for text and images, charting, and write simple notes on a blank page. The pen allows you to change the size, colour, and line properties, e. g. add the arrow to the beginning and end of the line. Most interactive whiteboard software includes digital rubber that is used to remove text from the screen.

Moving objects and joining them is the easiest operation on an interactive whiteboard, on the screen by dragging the mouse or directly onto the desktop that has a wide use.

Text tools allow you to add blocks of text, words, or longer text to the screen. They include all common word processor tools that can change size, fonts, and highlight parts of text by colouring or highlighting. Hide and uncover text, e.g. write the word the same colour as the background of the screen and then change the colour to make the word visible.

Spotlight/Spotlight allows the user to explore individual parts or object properties on the screen. The user can change the size of the surface or change its shape - to a square or rectangle, e.g. when viewing pictures or text.

Hidden and uncovered tools/shades, blinds are a simple screen cover that a teacher can use to completely or partially cover it, respectively. The gradual discovery of individual points or activities during teaching, e.g. as a roller shutter that can also be used horizontally.

Handwriting recognition enables you to write a pen on an interactive plaque and automatically convert the handwriting to text. Typically, the software offers multiple transcript alternatives that take into account different handwriting styles, converting hand-written text into a specific target language.

Creating Object Copies/Cloning and Duplication is a simple feature that allows you to create multiple copies of one object. Cloning and duplication allows the user to quickly create an unlimited number of copies of the same word or image within an hour. The user selects the object and sets its properties so that it automatically copes. Duplicates can also be used by individual parties and advanced users can use this feature to prepare learning materials to save time.

Timer devices/clocks and clocks allow the teacher to control the length of the individual parts of the clock and set the playback of the sound or melodies that mark the end of the activity. Hours can be set to display time in 12 or 24-hour format and are used instead of traditional clocks. The interactive whiteboard software, in addition to the timer, also includes a calculator and a thermometer.

Uploading a screen or page is a tool that teachers use to record activities on an interactive whiteboard during a lesson. You can record a whole screen or just a certain area. The resulting video can be saved in various formats and viewed with most players. Teachers can use it to upload, for example, A lecture they delivered during a lesson.

Layering/Rubbing and Uncovering allows you to place objects on the screen in the desired order in the so-called Layers, e.g. to create different tasks in which students reveal hidden images, covered texts, etc. The user hides an object, e.g. Text or image, under the colour layer. Colour can be the same colour as the background of the page or create colour contrast. The user simply uses the rubber to remove the colour to expose the word or image below it.

Animation allows animation of objects on the screen in different ways, Create a page and the amount of a copy of it, move each object slightly, similar to the animator block, if it scrolls, the object moves or changes. This feature can be used well in explaining processes or cycles. The second animation technique used is to program objects on the screen to move, disappear, or change [4].

Advantages of an interactive whiteboard against the classic blackboard

The Interactive whiteboard unifies the classroom instruction with a computer-aided instruction, enabling the creation of dynamic content. The interactive whiteboard area provides plenty of space for work while teaching, with tools on the main toolbar being used when the teacher can use, for example, the width of the pen and the colour it needs. Students do not have to distribute their attention to the computer monitor and the teacher's interpretation. Mimics and teacher gestures are an important part of teaching, and important knowledge escapes with distant attention. An interactive table offers the possibility to work with an inexhaustible amount of images that can be accessed, edited, saved with the changes captured during the classroom. Interactive whiteboard means more in one, e. g. includes a recorder, player, video recorder, and the like. We can animate the lesson with video, sound effects, and so on.

IV. INTERACTIVE WHITEBOARD FOR THE LEARNING OF ELECTROTECHNICS TO TEACHERS ORIENTED FACULTY

A. Teaching of the Thematic Unit Production and transmission of electricity by application of an interactive board

The subject of Electrical Engineering for Teaching-oriented Faculties provides students with knowledge from the field of electrical engineering and technology that deals with the transformation of other types of energy into electricity, transmission of electricity (both wire and non-conductive) and transformation of electricity into other types of energy, In the acquisition of heat, light and motion, as well as its use for electrochemical, information and communication purposes.

The aim of the subject is to teach students how to characterize the technological equipment of each type of power plant, to identify what it consists of, to explain the principle of electricity transmission, to explain the function of the transformer in electricity transmission, to explain the principle of electricity consumption measurement and to define basic principles, Which must be observed in electrical installations.

Students will learn about basic forms of energy conversion to electrical energy and its use in practice. They learn how to solve electrical circuits. The methods, forms and means of teaching the subject should stimulate the development of cognitive abilities of students; promote their aim, autonomy and creativity. The laboratory in which practical training takes place is equipped with technologies that provide students with the appropriate environment in the process of learning. The learning material contains all the principles of transforming different forms of energy into electricity. In Figures 1 to 3, there is a demonstration of the energy conversion to electricity, its transmission from the power plant to the point of consumption and utilization. Figure 4 is a repetition task. In repeating tasks, if a student chooses the correct answer, he or she will get another job, and if he/she chooses the wrong answer, he/she automatically gets into the text to find the correct answer. After pointing to numbers 1 through 10, the student will see the explanatory text in the bubble, e. g. by placing the cursor on the number 1 in Fig. 1 shows the information of generator. Fig. 1 to 3 illustrate the graphical interfaces of the key sequences in animation on a system of interactive animation simulation models presented by the interactive chart, which the authors of the article use in the subject of the thematic unit Electricity Distribution and Use of Electricity in the subject Electrical Engineering for Teaching Study. The interactive whiteboard presents in particular the complexity of the view of the production, distribution and use of electric energy, but also the so Functional dynamics of the individual processes of the mentioned topic.

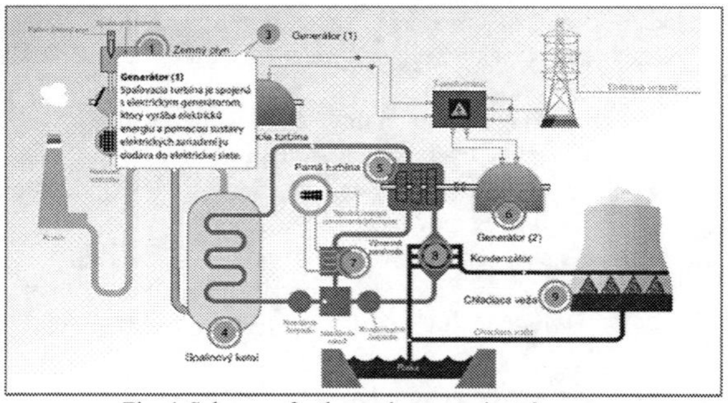

Fig. 1 Scheme of a thermal power plant for gas

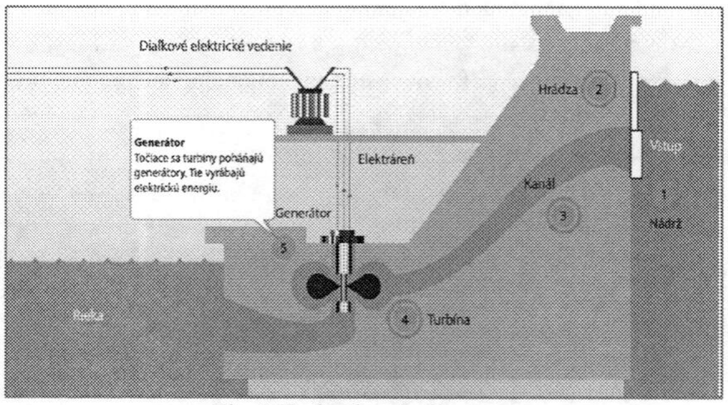

Fig. 2 Hydroelectric diagram

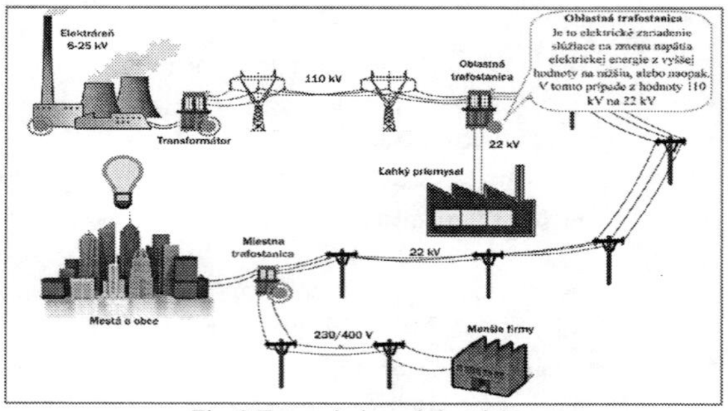

Fig. 3 Transmission of electricity

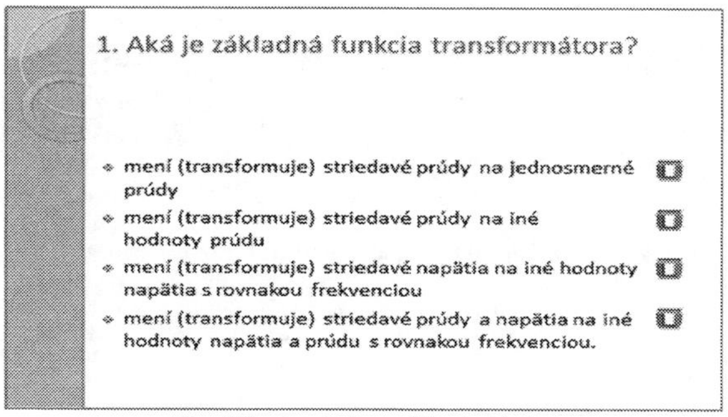

Fig. 4 Demonstration of the test task

A particularly didactically efficient use (based on the conclusions of our didactic experiments in order to determine the didactic efficiency of teaching) has the use of an interactive chart in the subject of home electrical homework. For this purpose, the authors of the contribution suggested and use the innovative experimental teaching system (hereinafter referred to as NIESVE). The innovative innovation part of NIESVE is the application of a virtual way of visualizing processes and phenomena based on an interactive whiteboard.

B. Teaching of the thematic unit the basics of home electrical installation by application of an interactive whiteboard

The teaching of the thematic group takes place in 4 consecutive phases:

Phase 1, on the platform of traditional visualization methods and techniques for visualizing processes going on in the electro physical systems (characteristic of the traditional teaching system). The visualization of the course of electromagnetic processes at this stage in the realized lessons is carried out on a real model of the electro physical system (based on the use of the kits Electro-assembly kit for primary and secondary schools): see Fig. 5.

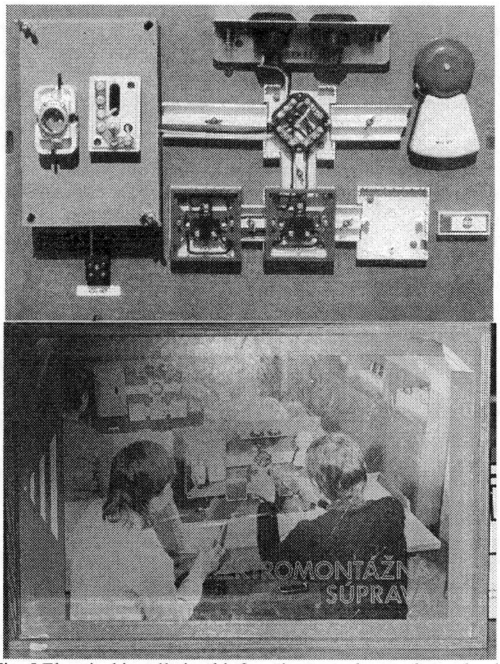

Fig. 5 Electrical installation kit for primary and secondary schools

Phase 2 is the dominant application of an interactive whiteboard on the platform of innovative methods and visualization techniques (characteristic of NIESVE). For NIESVE, it is a typical symbiosis of visualization on the real model (traditional visualization techniques and techniques) and on a computer model of the same electro-physical system (based on interactive graphic animation simulation computer models).

It should be reminded that the essence of the strategy of access to our design of the innovation of traditional systems of visualization of processes and phenomena in the curriculum is not to "duplicate" in practice didactically proven ways of visualization (e. g demonstration by measuring on an electro technical kit or a measuring instrument) and to replace them with new, But on the contrary - to keep them and only to enrich them organically with new innovative elements in an innovative base. Thus, in the concrete, it remains to be a priority that what is possible and appropriate to demonstrate by measuring on a traditional real-world measuring device or an electro technical kit can do so and what extends the visualization possibilities of the preparation or the kit to be realized virtually. In other words, the use of so- Intelligent computer graphics allows the visualization of the subject to a degree of clarity that is otherwise unreachable (traditional visualization, e. g on a circuit-based demonstration).

The aim of this experimental lesson is to deepen and expand the knowledge inter thematically causal and application. The computer models described in this part of the work were mainly synthesized on a schematic basis, which contains instructions for use with a pupil electrical kit – a wiring kit designed for primary and secondary schools.

In the didactically oriented simulation of the "basic" types of electrical circuits of residential electrical installation, certain didactic priorities, principles and didactic demonstration sequence must be respected.

First of all, it is necessary to highlight in particular the differences in the working patterns of individual electrical contact systems forming the serial, alternating and crossover switches and, secondly, the connection of these "differences" to the way of controlling the electrical lighting elements. A high degree of clarity in explaining the essence of this problem is achieved (based on the conclusions of our didactic experiments) by simultaneous simulation on the functional submodule (on the principle diagram) and on the submodule assembly of the individual basic circuits of the residential electrical installation (in the sense of emphasizing the essence of the causal relationship between phenomena). At first, however, students' attention should be focused on the processes running in the functional submodule, and then, when these processes are readily understood, it is possible to demonstrate the design and assembly features of the individual basic types of switches and switches in the electrical distribution system. The basis of this didactic activity with high degree of clarity is captured in Fig. 6).

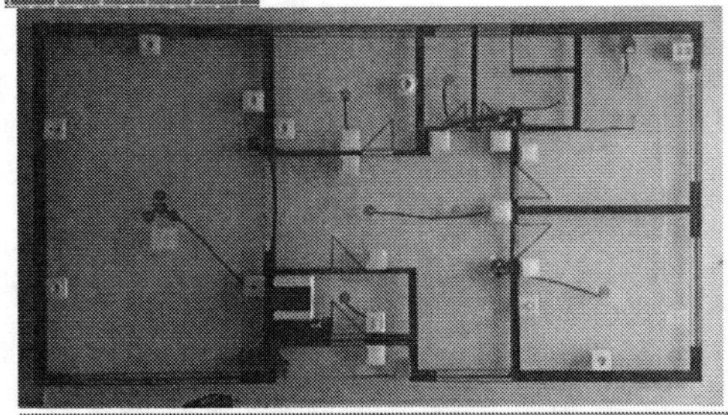

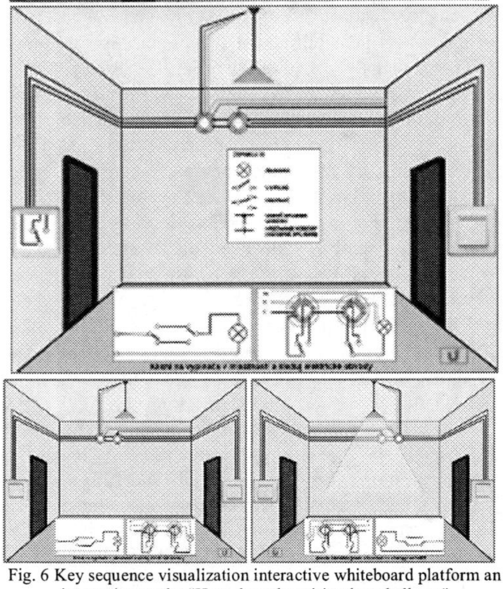

Fig. 6 Key sequence visualization interactive whiteboard platform an interactive applet "How does the wiring long hallway"

Fig. 7 Demonstration of symbiosis of virtual visualization and visualization on real models of housing distribution of cottage (small house)

Phase 3 is project-applied teaching. The NIESVA support pillar is also the use of elements of the project-application teaching in comorbidity with the inquiry-best instruction, whose purpose is to give the teaching pragmatic-practical and, in our case, mainly application sense. The above mentioned concept represents educational projects "Electrical installation of cottage (small house)" (see Fig. 7).

The specifics of didactic strategy and learning tactics

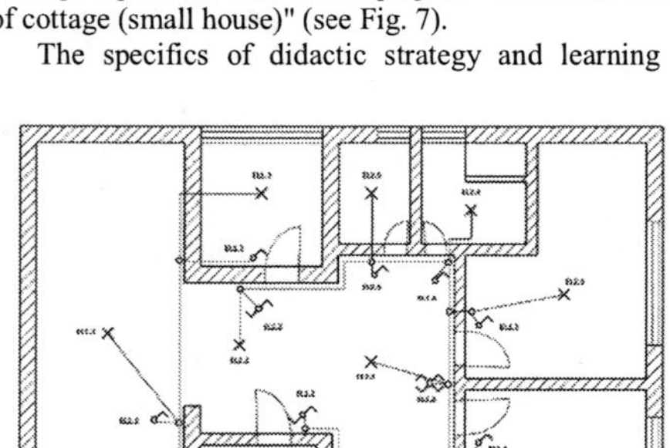

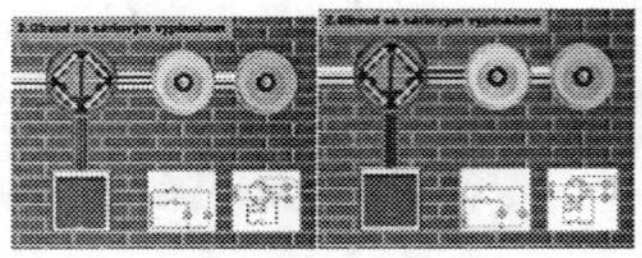

using computer models of residential electrical installation are that the teacher conducts a demonstration on computer simulation models to first deepen and extend the familiar knowledge of the students, ie knowledge related to the basic circuits of this installation, and then demonstrated The operation of complex systems of these basic circuits (e. g on the home electrical wiring model or cottage).

The teacher should emphasize before demonstrating the computer model of home electrical wiring that although this installation looks like a very complicated system at first glance, it is nothing else, it only knows how the system of basic circuits (already demonstrated, which students should already have mastered) To a larger unit.

Teacher of his previous claims (after initial acquaintance of the students with the scheme of the housing distribution of the cottage) subsequently demonstrates the demonstration of his statement by sharing with the students in detail the whole residential electrical installation of the cottage on the above mentioned individual elementary circuits. The presentation of the models from the previous section may reinforce the presentation of the deduction from general to specific whole.

The didactic importance of the strategy and tactics of the above didactic procedure is also that in the broader context, students are systematically involved in the analysis of residential electrical wiring circuits and also that in the above-mentioned context, the computer model of the cabinet's electrical installation is a sensible application unit composed of Known basic circuits.

Computer simulation as a method of didactic visualization gives an illustration of the teaching of home electrical wiring circuits, compared to the simplicity realized by traditional methods, an unachievable dimension. Its benefits, which are already beyond the scope of traditional visualization, can be summarized in the following paragraphs.

The foregoing relates to another didactic aspect of said computer simulation of electrical circuitry of residential electrical installation, namely, that the electrical processes running in the above mentioned models can be slowed or even divided into steps. Furthermore, on the models there are preserved the colour differentiation of the drivers according to the valid standards, but at the same time it is also possible to "highlight" the colour of the driver if the live conductor is under tension. Such indication of the driver, if under voltage, is also important in didactic training of diagnostics, while such "extensive" visualization cannot be achieved in traditional ways.

Phase 4.of the test, by means of feedback whiteboard.

According to our opinion-based research, the role of designing and creating home electrical distribution is a strong motivational charge, which obviously stems from the undeniable pragmatism of a task, in addition to the needs of ordinary life, but it also has an integrative and inters thematic nature.

V. Empirical Research Conducted Interactive Whiteboard Application in Teaching Process

The arrival of computer technology has offered unprecedented opportunities for the application of computer simulation and animation in the teaching process. It has raised our awareness of the necessity of a new quality platform creation for visualisation of objects, processes and phenomena in teaching technical subjects

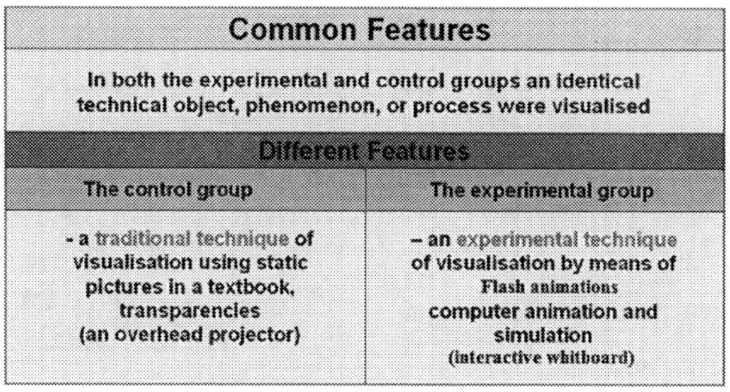

Fig. 8 The principle of the pedagogical experiment

We made a database of Flash animations that served as a platform for the creation of the experimental innovative teaching system called NIESV. It was designed for visualisation of teaching processes and phenomena through applets. In the process of our research the NIESV system (in the form of concrete models designed for teaching selected thematic sections in teaching was also experimentally verified. The method of pedagogical experiment was used to compare the two teaching systems in the experimental group (the NIESV system) and the control group (traditional teaching system). The principle of the pedagogical experiment is demonstrated in Figure 7. The concrete teaching system (the lift operation control) is demonstrated in Figure 8.

The main aim of the experimental research was to investigate the possibilities of the NIESV system application in order to increase the effectiveness of the teaching process.

VI. Research Sample

The research sample consisted of 36 research samples were in educational subject Electrical engineering for teacher education. The basis of results achieved in the input didactic tests divided into experimental and control group. 21 students were placed in the experimental group and 15 students in the control group. Pedagogical experiment was carried out from January to June (2015-16). In the experimental group, there was applied the computer supported (using the computer visualization of sing an interactive whiteboard in teaching the theme of production, distribution and use of electricity, there was education carried out without computer support (using traditional way of teaching). The comparison of results showed that pupils of experimental group achieved in the final didactic test (total) from the Electrical engineering for teacher education of 86,8 % success in solving, pupils of control group of 71,6 % (see Graph 1). In the Niemerka subtest system, the following difference between the experimental and the control group was reached - 10.4 % in the memorandum, 17.9 % in the understanding and 16.1 % in the knowledge application. Thus, in our experiment, we have achieved not only a quantitative increase in knowledge through NIESV but what is essential and qualitative for us (knowledge is better understood and have better application ability).

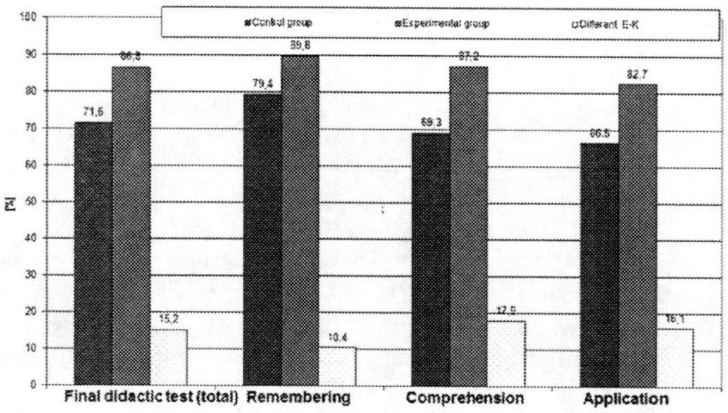

Graph 1. Percentage of successful solution of the final didactic test by experimental and control group

Conclusion

In conclusion, we would like to say that we present only partial results of the continuous pedagogical experiment in the article, which we implement in educational subject Electrical engineering for teacher education. The current results described in our article show that computer-supported learning of technical science subjects in has a positive impact on the acquisition of the educational contents of the technical sciences by school students.

The interactive tab, operating on the principle of electromagnetic sensing, which is controlled by an electronic pen, also offers the possibility of dual control. This interactive table can also be used as a classic tab, we can write it with erasable fixes. We can say that a teacher of a subject working on an interactive whiteboard should be creative and creative to be able to capture the most important feature of interactive boards, and that is the simplicity. In particular, this didactic principle can help pupils to motivate new curricula and help them understand the curriculum. In addition to working on an interactive whiteboard, both teachers and students are expanding their knowledge of computer literacy and IT skills. Another important factor is that teachers themselves try to educate themselves in the field of information and communication technologies. This article describes some of the options offered by interactive whiteboard software. The slideshow contains all of the illustrative elements such as, for example, Pictures, videos and animations, just need an Internet connection and speaker connections to play audio from videos and sound effects. In its proper use, teachers should have more effective educational and educational goals in the learning process. Finally, we can summarize that the interactive table is a modern didactic tool that cannot be used in all the topics that are taught in schools, but it will certainly contribute to the professional subjects to improve and make more effective the acquisition of pupils' knowledge, skills and habits in the teaching process, And is a very good means of motivating pupils.

Acknowledgment

This work was supported by scientific Grant Agency of the ministry of Education of the Slovak Republic projects: KEGA No. 017UMB-4/2017 Formal evaluation of pupils in subject Technology in lower secondary education focusing on the cognitive field

This work was supported by scientific Grant Agency of the ministry of Education of the Slovak Republic projects: KEGA No. 02UMB-4/2015 The creation modern textbooks and workbooks for technical subjects in lower secondary education

References

[1] B. Brestenská, M. Kabátová, I. Kalaš, K. Mikolajová, J. Pekárová, K. Szarka, J. Vaníček, "Premena školy s využitím informačných a komunikačných technológií," 1. vydanie. Košice: ELFA, s.r.o., 2009. ISBN 978-80-8086-143-8.

[2] M. Bitterová, J. Mních, "Didakticko-motivačné využitie interaktívnych tabuľových systémov z aspektu manažmentu škôl," In Moderní vzdělávání: Technika a informační technologie. Olomouc: PU, 2011. ISBN 978-80-244-2912-0.

[3] J. Pavlovkin, D. Novák, "Interaktívna tabuľa vo výučbe elektrotechniky," In: Acta Universitas Matthiae Belii Ser.: Technická výchova No 13 (2013). Banská Bystrica: FPV UMB, 2013 s. 65-75, ISBN 978-80-557-0624-5.

[4] D. Bannister, "Ako čo najlepšie využívať interaktívnu tabuľu." [online]. Bratislava: ÚIPŠ [cit. 2015-11-22].

[5] M. Bernát, "Visualization of some electro-physical processes through computer for didactic purposes and its application in teaching electro technical subjects." PhD. thesis, PdF UKF Nitra 2005 (in Slovak).

[6] R. Adamek, R. et al. "Moderná didaktická technika v práci učiteľa." 1. vydanie. Košice: ELFA, s.r.o., 2010. ISBN 978-80-8086-135-3.

[7] I. Baran, "Interaktívne tabule vo výučbe." [online]. Košice: TU. [cit. 2015-11-10].

[8] L. Cuban, "Teachers and Machines: The Classroom Use of Technology Since 1920." Amsterdam: Teacher College Press, 1986. ISBN 0-8077-2792-X.

[9] J. Dostál, "Interaktívna tabuľa - významný prínos pre vzdelávanie." In: Česká škola [online].

[10] G. Novacká, J. Hnatová, E. Fryková, "Interaktívna tabuľa a softvér Activstudio na hodinách prírodovedných predmetov." Bratislava: MPC, 2011. 78 s. ISBN 987-80-8052-366-4.

The Deviation of Optimization Calculations of the HPP in Cascade while is not Considered the Hydraulic Connections

Anatolijs Mahnitko
Tatjana Loman
Institute of Power Engineering
Riga Technical University
Riga, Latvia
mahno@eef.rtu.lv; loman@eef.rtu.lv

Janis Gerhards
Renata Varfolomejeva
Institute of Power Engineering
Riga Technical University
Riga, Latvia
gerhards@eef.rtu.lv; renata.varfolomejeva@rtu.lv

Abstract — **Latvia, among the Baltic countries, is the leader in the production of "green" energy, since the basis of its energy generation is formed from the cascade of three HPPs on the Daugava river. This fact lends topicality to the task of effective control of HPPs in cascade, because those are directly participating in electricity trading on the Nord Pool. The present paper considers issues related to the influence of the initial characteristics of HPP reservoirs within this HPP cascade on the determination of the amount of electricity generated by the cascade ensuring maximum profit in the case of participation in the electricity market.**

Keywords: hydraulic power plants; optimization; cascade.

I. Introduction

The efficiency of use of energy resources by humanity is an important prerequisite for its existence, development, improvement of living standards and welfare. The last-mentioned things are largely measured by the level (kWh/person) of electricity consumption per inhabitant of the country. This rate depends on the level of development of the energy sector, which is a basic parameter for the economies of all the developed countries. In recent years, significant progress has been made towards creating a so-called "green" economy, which is operating on the basis of closed cycles of use of various kinds of resources, including energy resources.

Besides, a trend can be observed towards the limitation of the use of energy resources accumulated by nature that are mostly non-renewable (oil, coal, shales, peat, natural gases). This is due to the negative impact of burning these, which ultimately leads to environment pollution and climate change in general. As a counterbalance to the above trend, we increasingly see the strategy of increasing the share of resources that are not accumulated in nature but are constantly renewed, within the power industry (solar radiation, wind, water of rivers and lakes, sea waves and tides, the Earth's internal heat, biomass), and exist continuously or emerge periodically in the environment. Renewable energy sources (RES) lie at the basis of the development of "green" energy, which does not have a negative environmental impact on the external environment.

At present, the European Union (EU), within its fight against climate change, is implementing a number of strategies, such as energy use efficiency and decarbonisation. These strategies have been laid down in corresponding EU directives. The most widespread support mechanism for the decarbonisation policy is the mandatory procurement (MP) of "green" electricity. In practice, this means that the system operators (SOs) sign agreements regarding the purchase of electricity for a fixed amount of money with the operators of power plants that mainly use RES. Support to climate conservation policies for this kind of electricity producers is implemented in the form of various tax relief schemes. On the other hand, the electricity generation that pollutes the environment is imposed with additional taxes.

Today's hydropower industry is the most economical and environment-friendly way of obtaining electricity as compared with other fields of the electric power industry [1, 2, 18-21]. One doubtless advantage of hydropower plants (HPPs) lies in their stable operation within the power system regardless of the time of the day.

Small hydropower moves even further in this direction [19]. Small hydropower plants help to preserve the natural landscape and do not have an environmental impact, be it at the operation stage or in the construction process. Unlike other renewable energy sources (RES), for example, solar or wind energy, the small-scale hydropower industry is practically independent of the weather conditions and is capable of ensuring stable deliveries of cheap electricity to the consumer. The operation of small HPPs has no harmful impact on water quality, fully preserving the initial natural properties of the water. The fish in the rivers survive and the water can be used for water supply.

Large-scale HPPs impact the environment, changing the thermal, hydraulic and climatic conditions of the area. Such changes are not typical of small HPPs since, in most cases, these use natural water head and do not require construction of large hydrotechnical structures (dams, sluices).

Latvia, among the Baltic countries, is the leader in the production of "green" energy, since the basis of its energy generation is formed from the cascade of three HPPs on the Daugava river. This fact lends topicality to the task of effective control of HPPs in cascade, because those are directly participating in electricity trading on the Nord Pool.

The present paper considers issues related to the influence of the initial characteristics of HPP reservoirs within this HPP cascade on the determination of the amount of electricity generated by the cascade ensuring maximum profit in the case of participation in the electricity market. Results of experimental calculations are shown, considering the variation of the above characteristics when determining the modes of the cascade by the statistical tests method [22 - 26].

II. General Information of the HPPs in Cascade Mode Calculation

The method for calculating the mode of a cascade of hydropower plants located on the same river if each of them has a water reservoir (Fig. 1) that ensures at least 24-hour regulation, basically does not differ from the method of calculating the operating mode of a power system with hydropower plants located on different rivers. The calculation processes differ in that apart from the electric link between the HPPs of the cascade, it is also necessary to consider their hydraulic link via the watercourse, which considerably complicates the task of determining the operating mode of the cascade. This is due to the fact that changes in the load and water flow of the upper HPP do not influence the mode of the next downstream HPP at once but after some period of time, during which the flow wave travels to that HPP. For the sake of objectivity, it should be pointed out that there is also a different point of view, the essence of which is that "consideration of the wave travel time, while making the calculation more complicated, seldom leads to a noticeable improvement in the accuracy of the solution" [1]. Usually, the consideration of the travel time is done in the form of an average value (mathematical expectation), which results in a certain error in the determination of capacities.

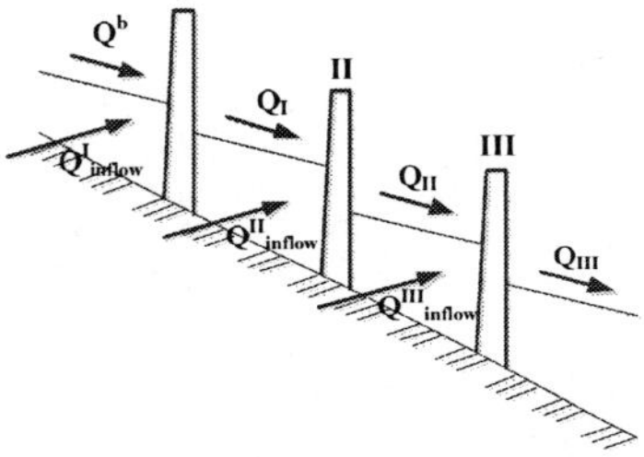

Figure 1. The basic diagram of a cascade consisting of three HPPs.

In addition, it should be mentioned that the steps of calculation of the most economically efficient mode in the case of a HPP cascade does not differ from the case of one HPP. Appropriate management of the reservoirs could provide an additional hydroelectric power generation amount of up to 5%, which ensures significant fuel savings for the power plants operating in a parallel mode in the power system [1].

Various mathematical modeling methods are known, which are used to calculate the modes of the reservoirs of HPPs for the whole period of regulation (day, week, season), in which the optimum condition of the power system is provided [1-17]. In the classical formulation of the problem, it is required to solve a non-linear programming task with linear constraints in the form of equalities and inequalities. For the solution of such a task, mathematical programming methods are used, such as: dynamic, gradient, stochastic, etc. [1]. Recently, the tendency to reduce the nonlinear programming problem to a linear one has increased in order to make the decision process more transparent and to provide a guaranteed global solution [1]. It can be surmised that such an approach to the solution of the considered task will make the solving process more demonstrable, considering that there are a great number of industrially developed solvers that implement the simplex method of solution, which ensures a global solution, which is not always guaranteed by the majority of the above-mentioned mathematical programming methods.

III. The Task Statement

A. *The parameters characterizing HPP operation*

All the HPPs in the cascade are numbered from the top down along the flow of the river. All in all, n calculation intervals and m HPPs are considered (the indices i and j designate the number of the calculation interval and the number of the HPP).

The procedure of the mathematical formulation and solution of HPP optimum mode calculation problems is the same for different regulation cycles of water reservoirs. For all of these problems, a calculation period [0, T] is set, which is divided into n discrete intervals (which are of different length in the common case). The calculation is made at set discharge records of inflow to the uppermost HPP of the cascade and side inflow to each of the following HPPs. The difference here is only the need to account for the hydraulic link via the watercourse between the HPPs of the cascade, which considerably complicates the task of determining the operating mode of the cascade. This is to be explained by the fact that changes in the loading and water flow of the upper HPP do not influence the mode of the next downstream HPP at once; rather, it happens after some time, which is required for the flow wave to arrive (time flow lag of water).

B. *The task formulation*

The goal of this study is to choose the generated capacities P_{ij} of the HPP cascade over a 24-hour period that would provide the maximum income I_{ij}. The expenses E_{ij} of the HPPs are regarded to be constant and are not involved in the optimization process and the computation has to do with the maximization of only the revenue from the sales of the generated electricity. It can be reflected by the following function:

$$I_{sum} = \sum_{i=1}^{n} \sum_{j=1}^{m} c_i \cdot P_{ij} \cdot \Delta t_i = \sum_{i=1}^{n} \sum_{j=1}^{m} c_i \cdot 9{,}81 \cdot \eta_{jt} \cdot Q_{jt} \cdot H_{jt} \cdot \Delta t_i , \qquad (1)$$

where P_{ij} — the electric capacity (kW) of the j-th HPP in the i-th interval; Δt_i — the duration of the calculation interval in hours; i — the number of intervals; j — the number of HPPs in the cascade; c_i — the price of electricity at the i-th time interval; g = 9.81 m/s^2 — the gravity acceleration; Q_{jt} — the flow of water through the turbines of the HPP, determined by the expression

$$Q_{jt} = Q_{\inf low\, jt} \pm Q_{r\, jt} , \qquad (2)$$

where $Q_{\inf low\, jt}$ — the everyday inflow in the water reservoir, $Q_{r\, jt}$ — the regulation flow of the water reservoir. In case of water reservoir drawdown $Q_{r\, jt} > 0$; in case of filling $Q_{r\, jt} < 0$; H_{jt} — the hydraulic head of the HPP, determined as the difference between the levels of the head race and the tail race, m; η_{jt} — the efficiency of the HPP, in relative units.

The objective function (1) includes limitations regarding the operating mode of the cascade in the form of equations and inequalities. The inequalities represent the following limitations: the levels of each water reservoir $Z^{min}_{top\ jt} \le Z_{top\ jt} \le Z^{max}_{top\ jt}$; the volumes of each water reservoir $V^{min}_{jt} \le V_{jt} \le V^{max}_{jt}$; the water flows into the tail race of each HPP $Q^{\min}_{bottom\ jt} \le Q_{bottom\ jt} \le Q^{\max}_{bottom\ jt}$; regarding the water flows from each water reservoir $Q^{min}_{r\ jt} \le Q_{r\ jt} \le Q^{max}_{r\ jt}$; regarding the summary capacity of all the HPPs of the cascade (the electric capacity generated by the cascade over every i–th time interval $\sum_{j=1} P^{min}_{jt} \le \sum_{j=1} P_{jt} \le \sum_{j=1} P^{max}_{jt}$, which link the technological limitations of the generated capacities of each HPP of the form $P^{min}_{jt} \le P_{jt} \le P^{max}_{r\ jt}$.

The equation (2) includes the water-balance equations of every j-th HPP for any i-th time interval Δt of the form [7]:

$$Q_{top\,jt} = Q_{bottom\,jk}\left(1 - \frac{t_{g,ik}}{\Delta t_i}\right) + \\ + Q_{buttom(i-1)k} \cdot \frac{t_{g(i-1)k}}{\Delta t_i} + Q_{\inf low+,ij} \tag{3}$$

where $Q_{top\ jt}$ — the water flow in the i-th interval for the j-th HPP; $t_{g,ik}$ — the travel time of the water flow in the i-th interval from the η -th (upstream) HPP located directly adjacent to the considered j-th HPP; $Q_{\inf\ low+,ij}$ — the inflow of the river to the j-th HPP in the i-th interval: if the j-th HPP is the uppermost one then the value of this inflow is a sum of the everyday inflow and the side inflow (see Figure 1). Otherwise, it is equal to side inflow; $Q_{bottomij}$ — the average value of water flow over the i-th interval into the tail race for the j-th HPP, determined by the expression [7]:

$$Q_{bottom,jt} = Q_{\inf low+ij} + Q_{rij} + \\ + \left(1 - \frac{t_{g,ik}}{\Delta t_i}\right) \cdot Q_{buttom\ ik} + \\ + \frac{t_{g(i-1)k}}{\Delta t_i} Q_{buttom(i-1)k} \tag{4}$$

where Q_{rij} — the average water flow from the water reservoir of the j-th HPP over time interval i.

The limitations regarding the set filling volumes of the HPP water reservoirs at the end of the regulation period:

$$\sum_{i=1}^{n} Q_{ij} = \frac{V_{0j} - V_{end\ j}}{\Delta t_i}, \quad j = \overline{1,m} \tag{5}$$

where V_{0j} — the volume of the water reservoir of the j-th HPP at the beginning of the regulation cycle; $V_{end\ j}$ — the same, at the end of the regulation cycle.

C. *Water reservoir volume change*

Given the surface area of the j-th HPP, the water volume change ΔV_j in the water reservoir of the j-th HPP enables calculation of the change in its level ΔZ_j, based on the correlation

$$\Delta Z_j = \frac{Q_{inf\ low} - Q_{r\ ij}}{S(Z_j)} \cdot \Delta t_i, \tag{6}$$

where ΔZ_j — the water level change in the water reservoir of the j-th HPP, m; $S(Z_j)$ — the surface area of the water reservoir, km^2; Q_{inflow} — the water inflow in the water reservoir, m^3/s; $Q_{r\ ij}$ — the water flow from the water reservoir over time interval i.

By adding ΔV to the value of water volume in the water reservoir at the end of the previous interval, it is easy to find the water volume in the water reservoir at the end of the calculation interval and then, by the curve linking the volumes and levels of the water reservoir, to find the water level in the water reservoir in the next interval.

IV. The Case Study

Modelling of the operating mode of a HPP cascade containing three HPPs on the Daugava river in Latvia was done on the basis of initial data regarding the characteristics and parameters of HPP reservoirs, known from official sources of information. The following limitations and conditions were considered:

- Regarding the minimum permissible water level in the water reservoirs — the power units of the HPPs are started only if in the case of a sufficient water level in the top reservoir, i.e. at levels that are lower than the minimum permissible ones, considering the reserve coefficient, the plant does not operate;
- Regarding the maximum permissible water level in the water reservoirs — all the power units of the HPPs operate at full capacity if the water level in the top reservoir has reached the maximum permissible level (to prevent idle water discharge over the dam);
- Regarding the rated capacity of the power units — the power units are started only with their rated capacities. The power units do not operate if the requested capacity is lower than the rated capacity of the power unit;
- Regarding the minimum electricity selling prices — at times of low electricity prices during the night hours, the power units do not operate.

As a result of the optimization of the operation of the HPPs in cascade, we obtained the optimal operating modes of the three HPPs of the cascade at different values of water reserve in the reservoirs at the moment of plant start-up.

Figures 2, 3, and 4 show graphs of the optimal generation of active power of each HPP in the cascade as well as drawdown of the reservoirs for every hour of the day and night provided maximum water levels in the reservoirs of the HPPs at the moment of plant start-up, which correspond to the levels of mode 4 as shown in Table 1.

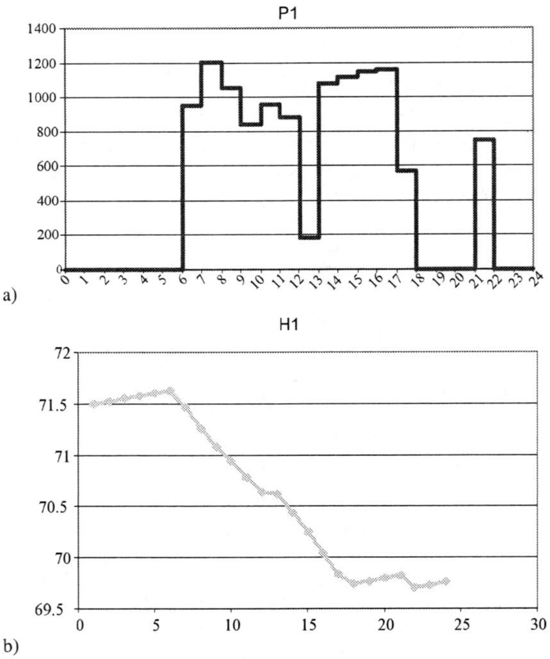

Figure 2. The time dependences of generated active power (a)) and reservoir level drawdown (b)) of the first HPP in the cascade.

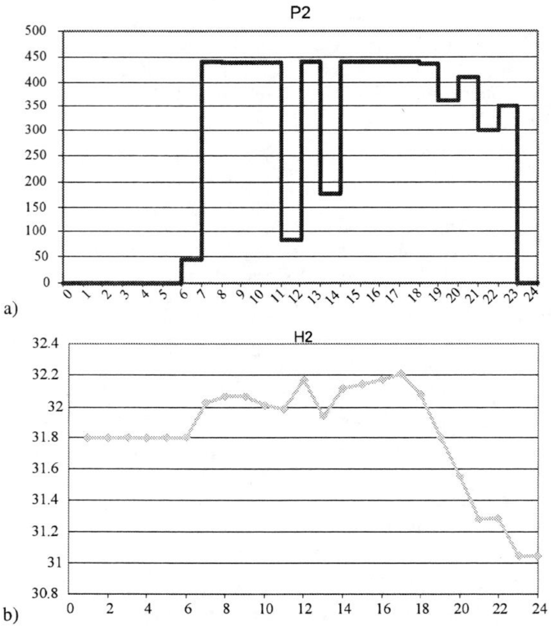

Figure 3. The time dependences of generated active power (a)) and reservoir level drawdown (b)) of the second HPP in the cascade.

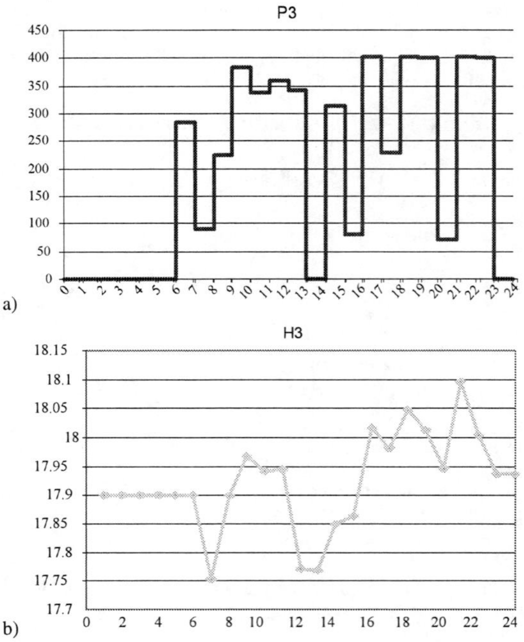

Figure 4. The time dependences of generated active power (a)) and reservoir level drawdown (b)) of the third HPP in the cascade.

Table 1 shows the results of the examination of various cascade operating modes, which correspond to four alternatives of initial water levels in the reservoirs (H0I, H0II, H0III).

For each case, the optimal modes were obtained and the summary capacity generations for the 24-hour period were shown (P1+P2+P3), along with the summary income from energy sales (SumF) in EUR, using a mode calculation programme that did not consider the wave travel time and considered the moment of wave arrival. The generated capacity and income values of the cascade corresponding to these calculations depending on the initial water levels in the reservoirs and the hydraulic link between the HPPs are illustrated in Fig. 5.

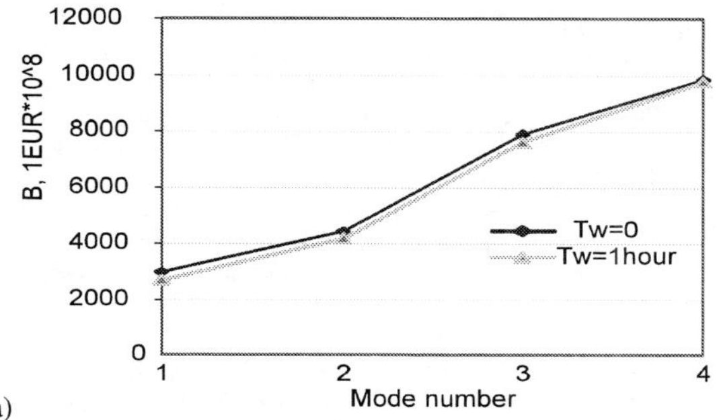

a)

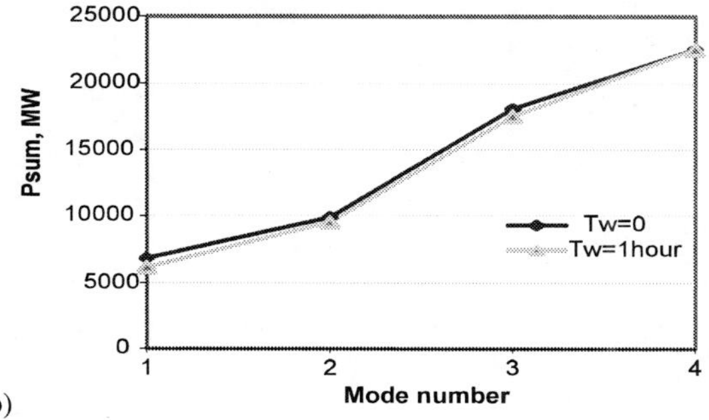

b)

Figure 5. The dependences of income (a) active power generation (b) on various modes differing by water reserve

Table 1. Case study

Mode	Start Water Level			Tw=0	
	H0_I	H0_II	H0_III	B, 1EUR*10^8	P sum,MW
1	69.8	31	17.6	2980	6903
2	70	31.3	17.9	4432	9940
3	70.8	31.8	17.9	7906	18136
4	71.6	31.8	17.9	9865	22557

Table 1. Case study

Mode	Start Water Level			Tw=1 hour	
	H0_I	H0_II	H0_III	B, 1EUR*10^8	P sum,MW
1	69.8	31	17.6	2677	6186
2	70	31.3	17.9	4154	9521
3	70.8	31.8	17.9	7626	17515
4	71.6	31.8	17.9	9777	22509

Conclusions

Consideration of the flow wave travel time from an upstream HPP to a downstream HPP makes the cascade mode calculation more complicated yet does not have a significant impact on the final result. This is due to the fact that the wave travel time is considered in the form of an average value, which depends on a multitude of reasons (reservoir levels, reservoir inflows, wind speed, etc.). Since this value is not the same even for corresponding time intervals at various HPP cascade control periods, also the mode calculated for it will contain a certain error, which will be reflected in the values of the generated capacity of the cascade HPPs. For the discussed example, the maximum error does not exceed 10%.

References

[1] Г.Г.Горбунов, Недельное и суточное регулирование на гидроэлектростанциях. Госэнергоиздат. p.116 , 1951.

[2] Н.А.Картвелишвили, Регулирование речного стока. Л.: Гидрометеоиздат. p. 218, 1970.

[3] А.Ш. Резниковский, М.И. Рубинштейн, Управление режимами водохранилищ гидроэлектростанций. М.: Энергия. p..176, 1974.

[4] V.M. Gornstein, "Optimal Operation Hydro Power Plants in Power Systems" Moscow: Gosenergoizdat, 1959. - 248 p. (In Russian).

[5] V.M.Sinkov, A.V. Bogoslovski, V.G. Grigorenko, "Optimal Operation of Power Systems" Kiev, Visca skola, 1976. - 308 p. (In Russian).

[6] V.A. Venikov, V.G. Zhuravljov, T.A. Filippova, "Optimal Operation of Power Plants and Electric Systems" M.: Energoizdat, 1981. – 464 p. (In Russian).

[7] Цветков Е.В., Алябышева Т.М., Парфёнов Л.Г. Оптимальные режимы гидроэлектростанций в энергетических системах. –М.: Энергоатомиздат, 1984. – 304 с.

[8] D.I.Legalov, S.I.Palamarchuk. "Hydro generation scheduling with electricity price calculation", IEEE Trans. Power Tech, St.-Peterburg, 2005

[9] E.Faria, L.A.Barroso, R.Kelman, S.Granville, M.V.Pereira, " Allogation of Firm-Energy Rights Among Hydro Plants: An Aumann-Shapley Approach", IEEE Trans.on Power Systems, Vol. 24, No. 2, May 2009.

[10] Летун В.М., Обоскалов В.П. Оптимизация режимов работы гидротепловых энергетических систем при краткосрочном планировании графиков нагрузки электростанций. – Электричество, 2015, № 9, с. 12 – 19.

[11] G. Shenglian, L. Xiang, L. Pan, G. Fuqiang "Optimal Operation of Cascade Hydropower Plants" Available: MAY http://www.researchgate,net/publication/251886120 last see (2017, May)

[12] J.P.S.Catalao, H.M.I.Pousinho, V.M.F. Mendes.Short-Term Hydro Scheduling: Mixed-Integer Nonlinear Optimisation Methodology. Available: http://www.aedie.org/11chlie-papers/RESUMEN/133-Catalao-summary.pdf

[13] P. Radu, P. Florica, P. Bogdan, M. Nicolescu "Optimal Operation of Cascade Hydropower Plants"U.P.B. Sci.Bull., Series D, Vol. 72, Iss. 1, 2010, pp. 93-100. ISSN 1454-2358

[14] A. Mahnitko, J. Gerhards, O. Linkevics, R. Varfolomejeva, I. Umbrasko, "Small Hydropower in Latvia and Intellectualization of its Operating Systems" Latvian Journal of Physics and Technical Sciences, 2013, vol.50 (6), pp. 3 – 15.

[15] M. Pereira, N. Campodonico, R. Kelman, "Long-Term Hydro Scheduling Based on Stochastic Models", Internat.Conf. On Electrical Power Systems Operation and Management (EPSOM ' 98), September, 1998, Zurich, Switzerland.

[16] R. Kelman, L.A.Barroso, M.V.Pereira, "Market Power Assessment and Mitigation in Hydrothermal Systems", IEEE Trans. on Power Systems, Vol. 16, No.3, August 2001.

[17] L.T. Biegler, A. Wachter "On the implementation of a primal-dual interior point filter line search algorithm for large-scale nonlinear programming". Mathematical Programming, 106, 2006, c.25-57.

[18] J.W. Labadie, M. Asce. "Optimal Operation of Multireservoir Systems: State-of-the-Art Review", Journal of water resources planning and management © ASCE / March/April, 2004 / 93.

[19] J.R. Buch, B.N. Parlett, "Direct methods for solving symmetric indefinite systems of linear equations" SIAM J. Numer. Anal., Vol. 8, 1971, pp.639 – 655.

[20] A. Mahnitko,J. Gerhards, T. Lomane, R. Varfolomejeva,V. Oboskalov, K. Koljasnikov, "Maximizing the Profit of a HPP Cascade Considering Hydraulic Link via Reservoirs.",

Proceedings of PPE-2016 Conference, Ukraine, Kiev, June 6-10, 2016, pp. 246-249, (SCOPUS).

[21] I. Zicmane, A. Mahnitko, S. Kovalenko, G. Georgiev, D.Kim. " Algoritmization in the Task of Optimization of Cascade Hydro Power Plants", 2017, pp. 1-6.

[22] K. Baltputnis, A. Sauhats, O. Linkevičs, R. Petričenko,R. Varfolomejeva, Z. Broka," Modeling of Water Utilization in Hydroelectric Power Plants on the Daugava River" No: *Proceedings of the 56th International Scientific Conference on Power and Electrical Engineering of Riga Technical University (RTUCON) : Proceedings*, Latvia, Riga, 14.-14. oktobris, 2015. Riga: RTU, pp.47.-52, 2015.

[23] A. Sauhats, R. Varfolomejeva, R. Petričenko, J. Kucajevs, "A Stochastic Approach to Hydroelectric Power Generation Planning in an Electricity Market" No: *2015 IEEE 15th International Conference on Environment and Electrical Engineering (EEEIC 2015): Conference Proceedings*, Italy, Rome, 10.-13. June, 2015.

[24] A. Mahnitko, J. Gerhards, T. Lomane, R. Varfolomejeva, V. Oboskalov, "A. Model for Maximizing the Profit of a HPP Cascade Considering Hydraulic Link via Reservoirs" Proceedings of IEEE International Conference on Environment and Electrical Engineering (EEEI 2016), Italy, Florence, June 7.-10, 2016, pp. 1071-1075.

[25] A. Mahnitko, J. Gerhards, T. Lomane, R. Varfolomejeva, V. Oboskalov, K. Koljasnikov, "Maximizing the Profit of a HPP Cascade Considering Hydraulic Link via Reservoirs" Proceedings of PPE-2016 Conference, Ukraine, Kiev, June 6-10, 2016, pp. 246-249.

[26] A.Sauhats, R. Petričenko, K. Baltputnis, Z. Broka, R. Varfolomejeva, "A Multi-Objective Stochastic Approach to Hydroelectric Power Generation Scheduling," Proc. 2016 Power Systems Computation Conference (PSCC 2016), Italy, Genoa, June, 2016, pp. 20.-24.

Research of Electrical Power Consumers Reaction in Power Consumption Management System

Anatolijs Mahnitko, Kristina Berzina
Riga Technical University
Riga, Latvia
mahno@eef.rtu.lv, Kristina.Berzina@rtu.lv

Anatoly Zamulko, Yurii Veremiichuk,
Volodymyr Nakhodov
National Technical University of Ukraine
«Igor Sikorsky Kyiv Polytechnic Institute»,
Kyiv, Ukraine,
zai_71@mail.ru, y.veremiichuk@kpi.ua, nakhodov@ukr.net

Abstract — **In the United Energy System (UES) of Ukraine retained the negative trend of significant irregularity of daily electricity consumption schedule, especially in autumn and winter period. UES Ukraine daily load schedule irregularity in 2016 amounted from the 4500 MW in summer up to 6500 MW in the winter period. This situation requires finding effectiveness evaluation methods of management decisions in the energy sector implementation, particularly at power consumption management. One available power consumption managing option is tariff system, differentiated by time of use periods, which effectiveness is ensured by usage of consumers potential in power consumption reduction.**

An important step in research of such consumers opportunities is to determine its response to proposed changes in the tariff system. This approach significantly reduces the risk of management methods implementation. Allows to make adjustments at initial stages. Basic for such assessment implementation are communication methods. A further step is processing of expert surveys results and usage of fuzzy logic elements. Article reviewed methodical approaches for defining existing differentiated tariffs improving priorities. Conducted research of dependence between or tariffs coefficients level established for allocate day zones and consumers electrical power demand changes that occurred or could occur through the use of differentiated tariffs in terms of their multivariate characteristics.

Expert data and fuzzy logic methods usage allows to recive fuzzy inference models, that are used in analysis of differentiated tariffs impact at energy system by changing the price rates. Obtained dependencies allow to change electricity consumption and receive predicted values with a given probability.

Keywords: power supply, мethods of controlling, group of electric energy consumers, electricity tariff.

I. Introduction

In the United Energy System (UES) of Ukraine retained the negative trend of significant electricity consumption profile irregularity, especially in autumn and winter. Configuration of daily power consumption profile in Ukraine is characterized by peak load at evening peak hours of UES Ukraine, load profile falling in mid-day and night zones and almost steady electricity consumption during morning peak hours and in half-peak area (Fig.1.). This unevenness daily schedule load UES Ukraine ranges from 4,500 MW in summer to 6500 MW in winter. In terms of market, daily electricity consumption profile regulation the most effective is application of economic management, one of which is a tariffs system, differentiated by time periods. Consumption of electricity by consumers that are calculated by differentiated tariffs for the period shown on Fig.2.

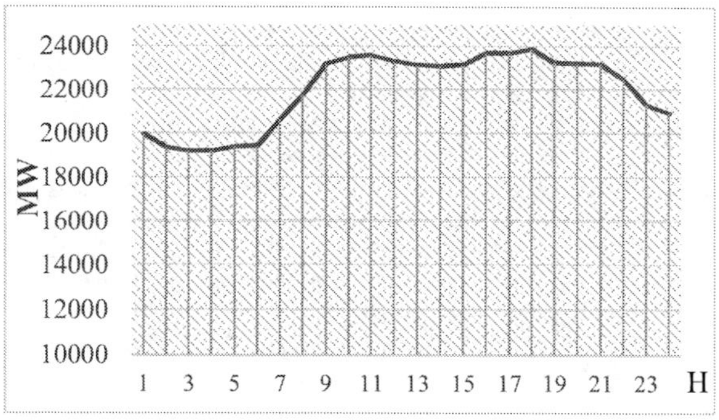

Figure 1. Ectricity consumption profile in winter regime day 21.12.2016

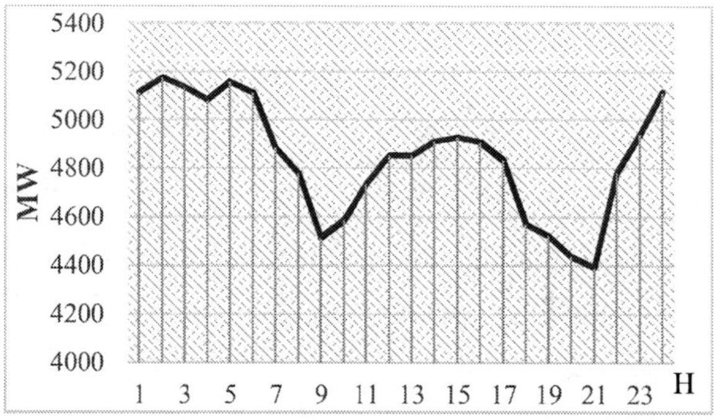

Figure 2. Hourly consumers electricity consumption, calculated by time of use periods in winter regime day 21.12.2016.

Application of such electricity demand managing method can be provided by usage of customers potential to reduce power consumption [1,2].

An important step in efficiency of power consumption management assessment is to determine consumers reaction to proposed changes. Usage of this approach significantly reduces the risks of management implementation and enables corrections at the initial stages. The base for the implementation of this assessment is further communicative methods of results processing using elements of fuzzy logic.

II. Problem Definitions

Effective use of power consumption management needs to take into account various factors [3-5]. These factors, in most cases are probabilistic in nature and depend on the characteristics of the consumer and its attitude to proposed implement measures.

Considering the electricity consumers quantitative indicators, their economic opportunities and social background, it is necessary to investigate the mechanisms of influence on them. The approaches to ensure the balance between generation, electricity market and demand [6].

In recent years, in Ukraine, the use of differentiated tariffs for electric energy, especial time of use rates, changed slightly a few times, available statistical data is not enough for the construction and analysis of relevant stochastic dependencies as proposed by the author [7].

Results of using such system in Ukraine for nearly 20 years requires in-depth analysis of possibility to finding further customers attraction to use it, and the tariff system improvements.

Number of consumers who are using differentiated rates is big. In most cases, they are powerful industry and a significant number of residential customers. In order to prevent negative economic effects for the state and social dissatisfaction at the electricity market. It is necessary to implement rates changes which should be based on experimental research proposals with regard to the consumer and the elasticity of electricity demand. [8] Efficiency of tariff system for UES of Ukraine heavily depends on the reaction of consumers and the possibility of rapid changes in its components.

Therefore, to determine ways to improve existing differentiated tariffs there is scientific-practical research problem of electricity consumers reactions including the use of the existing tariff system in Ukraine.

The basic approach for this task realization defined communication methods with subsequent processing results using elements of fuzzy logic.

III. Proposed Solution

Requirements for the formation of modern power consumption management is introducing an extensive system of electricity tariffs. One of such systems is the tariff system differentiated by time periods used in Ukraine since 1995. Ground formation rules for this tariff regulatory system approved by the national regulator commission of Ukraine #1241 from 20.12.2001 (as amended). The resolution outlines:

- the level of tariff ratios applying to average market retail tariff;
- the length of tariff periods for the day;
- requirements for the tariff periods setting procedure in furthering seasons.

However, it should be noted that the effectiveness of the tariff system essentially depends on the consumers reaction and the possibility of rapid changes in its components. For the research, questionnaires were developed and experts chosen. The general scheme of possible changes assessment in the tariffs system conducted according to the Delphi method.

Experts (energy sector employees, production managers, engineers, economical units employees) were invited based on their expertise and professional experience to assess the possible consumption of electricity at an appropriate site in each day zone in the case of each differentiated tariffs option usage. The poll was conducted for two types of enterprises:

- industrial enterprises that use differentiated by day periods electricity tariffs;
- industrial enterprises which use usual tariff system and perspective in terms of regulation of power demand and energy throughout the day.

Experts received a questionnaire in which they set various options for differentiated tariffs in accordance to theoretically possible levels of electricity consumption in each tariff period of the day. The experts provided separately for winter and summer characteristic (or operational) business days. For the base case experts selected characteristic structure of power and energy consumption for their enterprise under the existing circumstances.

Analyzing received indicators of consumers load profile configurations during the day, taking in to account that the change in electricity consumption for different day zones is a random value with normal distribution. In this case, the procedure for assessing the survey consisted of several stages.

Step 1. Study questionnaire survey, analysis and evaluation of experts levels. The general opinion of experts calculated using the expression:

$$\Delta W_v = \frac{1}{m} * \sum_{i=1}^{m} \Delta W_i \quad (1)$$

where m – the number of experts in group;

ΔW_i – electrical energy consumption changes in a certain zone of the day.

Determine the dispersion and standard deviation of expert opinion using the following expression:

$$\overline{D}(\Delta W) = \frac{1}{m-1} * \sum_{i=1}^{m} (\Delta \overline{W}_v - \Delta W_j)^2 \quad (2)$$

$$\overline{\sigma} = \sqrt{\overline{D}(\Delta W)} \quad (3)$$

Selects a group of experts closest to the estimate point of consumption in a particular day zone, using the expression:

$$\Delta \overline{W}_v - t * \frac{\overline{\sigma}}{\sqrt{m}} \le \Delta W \le \Delta \overline{W}_v + t * \frac{\overline{\sigma}}{\sqrt{m}} \quad (4)$$

where t – Student test at a probability level of $p = 0.95$ with expected value μ and variance.

$$t = \frac{\bar{\mu} - \mu}{\frac{\bar{D}(\Delta W)}{\sqrt{m}}} \quad (5)$$

Information received from experts is subjective about daily load profile configuration . Since the amount and configuration of daily profile, except differentiated tariffs, significantly influenced by other factors. Among them: seasonality, the ambient temperature, the volume of orders on finished products and others. Therefore, none of the experts cannot with 100% certainty predict changes in load demand of the enterprise.

So the answers received in the questionnaires, should be regarded as approximate value around which can be found real value of power consumption in different day zones. In addition, the information is inaccurate, since interviewed a significant number of consumers with different technological processes and production conditions. Therefore, subjective opinions of experts can considerably vary between them.

Further for the analysis of the questionnaires used methods of fuzzy mathematics and constructing expert knowledge bases. Based on works about use of fuzzy logic in various areas of research [9-11]. Formed a number of methodological principles that will be used while processing expert information and modeling expert knowledge base.

Step 2. Processing expert information and modeling expert knowledge base.

The following designations are necessary for the mathematical formulation of the problem:

n – the number of partial state parameters;

x_i – particular state parameter with a serial number i ($i = \overline{1,n}$);

$d_1, d_2, \ldots d_m$ – possible changes in electricity consumption at certain day zone ;

num_t - the total number of fuzzy terms that are used to assess the linguistic parameters of state;

num_r – the total number of rules in hierarchically related fuzzy knowledge bases;

b_l та c_l – function parameters - coordinate of maximum and concentration factor of 1st fuzzy term;

$B = (b_1, b_2, \ldots, b_{num\_t})$ and $C = (c_1, c_2, \ldots, c_{num\_t})$ – vector parameters of functions;

W_v – v-th rules weight coefficient ($v = \overline{1, num\_r}$);

$W = (w_1, w_2, \ldots, w_{num\_t})$ – vector of fuzzy rules weighting coefficients.

(X^r, d^r) – teaching sample, which is a set of experimental parameters of condition parameters pairs.

$X^r = (x_1^r, x_2^r, \ldots, x_n^r)$ verified change in volume of electricity consumption in a particular day zone $d^r \in \{d_1, d_2, \ldots d_m\}, r = \overline{1, M}$, M – size of teaching sample.

Using teaching methods requires state parameters in a quantitative form. The task of training of fuzzy models with discrete output is reduced to solving this optimization problem: find a vector *(B,C,W)*:

$$\sqrt{\frac{1}{M} \sum_{r=\overline{1,M}} (q(X^r, d^r, B, C, W))^2} \to \min, \quad (6)$$

where $q(X^r, d^r, B, C, W)$ – the discrepancy between the experimental release d_r and the discrepancy between the experimental release *(B,C, W)* for object X_r.

There are several ways to calculate discrepancies $q(X^r, d^r, B, C, W)$, among which commonly used [12]:

$$q(X^r, d^r, B, C, W) = \sum_{j=\overline{1,m}} (\mu^{d_j}(d^r) - \mu^{d_j}(X^r, B, C, W))^2 \quad (7)$$

where $\mu^{d_j}(X^r, B, C, W)$ – degrees of fuzzy inference results for the model, given vector *(B,C,W),* for object X_r to causes d_j, $(j = \overline{1,m})$; These degrees of membership are calculated as follows:

$$\mu^{d_j}(x_1, x_2, \ldots, x_n) = \bigvee_{p=1}^{k_j} (w_{pj} \cdot \bigwedge_{i=1}^{n} \mu^{a_i^{jp}}(x_i)), j = \overline{1,m}, \quad (8)$$

$\mu^{d_j}(d^r)$ – degrees of membership in the volume of electricity consumption at particular day zone d_r to fuzzy terms $d_j, j = \overline{1,m}$; according to [12] defined as:

$$\mu^{d_j}(d^r) = \begin{cases} 1, & d_j = d^r \\ 0, & d_j \neq d^r \end{cases} \quad (9)$$

To ensure transparency of fuzzy model after training on controlled variables in the optimization problem (9) impose such limits:

- coordinates of the vector W should be found in the unit interval: $W_v \in [0,1]$, $v = \overline{1, num\_r}$;
- coordinates of the vector B should be in range of possible values of the relevant Particle condition parameter;
- coordinates of the vector B shall not violate the terms of term-sets linear order;
- coordinates of the vector C should be nonnegative: $C_l > 0$, $l = \overline{1, num\_t}$.

Realized approach using Matlab software (Fig.3.). Received models of fuzzy inference, which analyzes impact of differentiated tariffs on energy system according to rates changes. They allow to change the electricity consumption and get the forecast values with a certain probability Table 1.

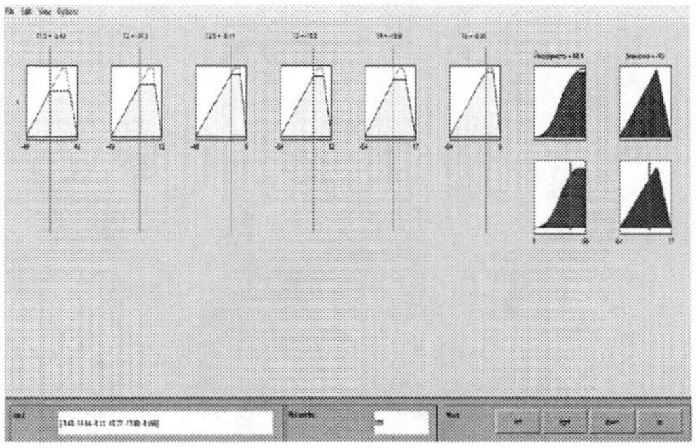

Figure 3. Simulation results definition

Table 1. Results of the analysis reduce electric power consumption during peak hours of the operational winter day.

	The value of membership function of fuzzy evaluation	**Value of load reduction at peak hours**
ΔT=1,5 □/ kW-h	$\mu\ (\Delta W\ ) = 0{,}96$	3,43%
ΔT=2,0 □/ kW-h	$\mu\ (\Delta W\ ) = 0{,}95$	3,45%
ΔT=2,5 □/ kW-h	$\mu\ (\Delta W\ ) = 0{,}5$	5,34%
ΔT=3,0 □/ kW-h	$\mu\ (\Delta W\ ) = 0{,}5$	5,34%
ΔT=4,0 □/ kW-h	$\mu\ (\Delta W\ ) = 0{,}5$	5,47%
ΔT=5,0 □/ kW-h	$\mu\ (\Delta W\ ) = 0{,}5$	5,52%

As a result of the influence we can simulate the configuration of consumers power load profile by changing price rates ΔT in different periods of the day. Model allows to determine the potential in terms of future consumer power and energy demand management. For effective work, model need continuously improvements with defining new day zones and formation new economic proposals for consumers.

IV. Conclutions

Proposed approach allows to create unified model based on experts data and methods of fuzzy logic. This model allows to evaluate and make analysis of opportunities and efficiency of tariff changes, differentiated by time zones.

The improvement directions of existing electricity tariffs differentiated over time, which must meet the current needs of power demand management should be considered as:

- introduction of monitoring system and in-depth analysis of the differentiated tariffs usage results;
- periodic adjustments to length and limits of tariff zones;
- reasonable, periodic review of tariff coefficients in order to enhance the impact of differentiated tariffs at consumers demand for electric power during the day.

Further methodical approach can be implemented for efficiency analysis of solving a number of problems associated with the use of tariffs, differentiated by time periods and different tariff menu.

References

[1] Y. Zhou, F. Gao, S. Ci, Y. Yang, Y. Xu. "Time-of-use pricing in retail electricity market: Step tariff vs. usage-based schemes", 2016 International Conference on Probabilistic Methods Applied to Power Systems (PMAPS) 16-20 Oct. 2016.

[2] L. Herre, L. Soder, "On the Flexibility of Electricity Consumers: Introducing Notice Time", 2016 13th International Conference on the European Energy Market (EEM). 6-9 June 2016.

[3] A. Zamulko, Y. Veremiichuk, "Methods of controlling power consumption in terms of reforming market conditions", Scientific Journal of Riga Technical University, series: Power and Electrical Engineering. – 2014. – № 32. – pp. 41–46.

[4] A. Zamulko, Y. Veremiichuk, "Evaluation of efficient usage of methods of the electric power consumption control under the condi-tions of uncertainty", Visnyk of Vinnytsia Politechnical Institute, [S.l.], pp. 16-21, july 2013. ISSN 1997-9274.

[5] A. Zamulko, Y. Veremiichuk "The research of the charts of electrical loads of groups of electric energy users". Visnyk of Vinnytsia Politechnical Institute, [S.l.], pp. 82-85, june 2014. ISSN 1997-9274.

[6] J. A. Qureshi, M. Gul, M. A. Faruqi, W. A. Qureshi. Intelligent demand management for end user benefits. 2011 IEEE PES Conference on Innovative Smart Grid Technologies - Middle East, 17-20 Dec. 2011

[7] D. Fischer, B. Stephen, A. Flunk, N. Kreifels, K. Byskov Lindberg, B. Wille-Haussmann, E. Owens, "Modelling the Effects of Variable Tariffs on Domestic Electric Load Profiles by Use of Occupant Behavior Submodels", IEEE Transactions on Smart Grid (Volume: PP, Issue: 99)

[8] A. Asadinejad, K. Tomsovic, C. Chen, "Sensitivity of incentive based demand response program to residential customer elasticity", 2016 North American Power Symposium (NAPS). 18-20 Sept. 2016

[9] M. Hanss, "Applied fuzzy arithmetic. An introduction with engineering applications", Berlin : Springer-Verlag, 2005. – P. 270.

[10] K. Asai, M. Sugeno, T. Terano, "Applied Fuzzy Systems", New York : Academic Press, 1994. – P. 302

[11] L. Smith, "A tutorial on Principal Components Analysis" [Online] Available: http://www.cs.otago.ac.nz/ cosc453/ student_tutorials/ principal_ components.pdf last see (2017, May)

[12] S. Haykin "Neural Networks a comprehensive foundation second edition", Prentice Hall, Inc., 1999. P. 823, ISBN 81-7808-300-0

Autonomous Power Supply Model for Passive Building

Kristina Berzina, Inga Zicmane, Tatjana Lomane
Faculty of Power and Electrical Engineering,
Riga Technical University,
Riga, Latvia
Kristina.Berzina@rtu.lv, Inga.Zicmane@rtu.lv,
loman@eef.rtu.lv

Georgi Georgiev
Energy Solutions Centre
Varna, Bulgaria
georgidg@hotmail.com

***Abstract* — The objective of the paper is to analyze theoretically technological solutions of power supply for modular passive building based on the PV energy generation and looking the optimal storage. It is planned to develop a model of power supply system for mobile modular commercial building up to 20m2 with smart system elements using solar, power with no connections to external engineering and verify the micro grid stability. Design of power supply system's model will be developed and load analysis will be carried out for the modular building. Following tasks will be decided in the paper: development of model of power supply system for modular building for the mobile and stationary consumers, analysis of optimal location and technical parameters of PV energy sources, development of model of power supply system for technologically highly efficient mobile building.**

Keywords: modular passive house, autonomous power supply, load analysis

I. INTRODUCTION

Environmental problems of a global scale and the foreseeable prospect of depleting the fuel and energy resources of the Earth are causing worldwide growth in new energy based on the large-scale use of renewable energy resources of the planet: solar radiation, wind, water flows, geothermal energy and biomass energy.

As strategic goals for the use of renewable energy sources (RES) and local types of fuel, in this document are identified:

- reduction of the growth rates of anthropogenic impact on the environment and counteraction to climate changes, if it is necessary to meet the growing energy consumption;

- rational use and reduction of growth rates of consumption of available resources of fossil fuel under conditions of inevitable depletion of its reserves;

- maintaining public health and quality of life by reducing of growth rates of environmental pollution by using fossil fuel, reducing national-level expenditures on health care;

- reduction of the growth rates of costs for the distribution and transportation of electricity and fuel and the resulting losses;

- involvement in the fuel and energy balance of additional fuel and energy resources;

- increasing the level of energy security and reliability of energy supply by increasing the level of its decentralization.

According to existing estimates, the technical resource of renewable energy sources (the dominant share in which has the potential of using solar and wind energy) is at least 4.5 billion tons of standard fuel per year, which is more than four times higher than the consumption of all fuel and energy resources.

Compared to an autonomous system, connection to a centralized power supply network is financially costly. Laying new power lines requires significant costs, but if installation of a substation is required as well, the amount of connection will increase at times. So, the money will go to purchase equipment, which will belong to local power networks. Thus, the autonomous system can cost less (when accounting for electricity payments in long-term perspective) than connection to power lines.

Such a method of generating electricity is absolutely safe for the environment, so it is always environment-friendly. Ecology care can and should be shown in all available ways. [5]

I. THE LOAD ANALYSIS FOR MODULAR PASSIVE BUILDING

The paper considers the process of forming the load schedule of a group of electric receivers of autonomous consumers. The switching on and off of individual electrical receivers can be considered as random events. As it is known, a random process, a load schedule, is characterized by a function of distribution, expectation value and a correlation function that reflects the relationship between process parameters. If we assume that the electric receivers are switched on independently of each other, then the number of electric receivers of their total number working simultaneously and making up the group load can be determined by binomial law. In this case, the probability of operation of m electric receivers from their total number n is determined using the following expression [12]

Currently variety of solutions for modular buildings (small commercial buildings, workshops on the children's playgrounds, etc.) is available that requires providing connections to central networks, that makes installation of such building complicated.

The research main aim and concept is: this research application is to develop a set of technological solutions for modular mobile commercial building up to 20 m2 considering climatic conditions of Baltic region, without connections to central power networks.

For the research was chosen modular house with residential consumers as it shown in the tabula 1. The modular house working cycle is chosen 6-8 hours in the summer months and 4-6 in the winter months.

Fig. 1. The location of modular house

Table 1. The modular house consumers

Nr.	Type of consumer	Current type	Voltage, V	Power, W
1	Basic lighting	DC	24	200
2	Background lighting	DC	24	140
3	Power network with mobile consumer Nr.1.	AC	~230	300
4	Power network with mobile consumer Nr.2.	AC	~230	300
5	Power network with mobile consumer Nr.3.	AC	~230	350
6	Cash machine with terminal	AC	~230	250
7	Refrigerator	AC	~230	0
8	Electric kettle	AC	~230	1000
9	TV advertising modules	DC	24	500

A load analysis was performed, using mathematical model [12]. The result data are represented in Figure 2. The power supply system's consumption in our case is constant throw the season in summer and winter months.

To model the electrical load of local power supply system, we use the probability and statistical model:

$$P_{pi} = \overline{P_l} + \beta\sigma \cdot (P_i)\,,$$

(1)

where P_i - wattage rating on the i-th hour of the daily schedule; $\overline{P_l}$ - mathematical expectation on the i-th hour of the daily schedule; β – reliability factor of the calculation, which determines the probability with which the random load values will remain lower than the assumed estimated value P_{pi}; $\sigma(P_i)$ – average square deviation for the i-th stage of the daily schedule.

Under the normal probability distribution law of the load values under $\beta = 2$, $\sigma(P_i) = 0.025$.

$$P_{pi} = \overline{P_i} P_{\max} \cdot (1 \pm \beta\sigma(P_i))K_s\,,$$

(2)

where K_S – seasonal factor, the value of which in our case is equal to1 [12].

As a basic model of the load, the load schedule of consumers of modular house space was adopted. The consumption sees on the fig.4.

II. Calculation of Power Balance and Optimal Battery Storage for Autonomous Power System

Assuming the transfer of the power of modular house to alternative sources of generation, potential balance of capacities is shown on Fig. 2.

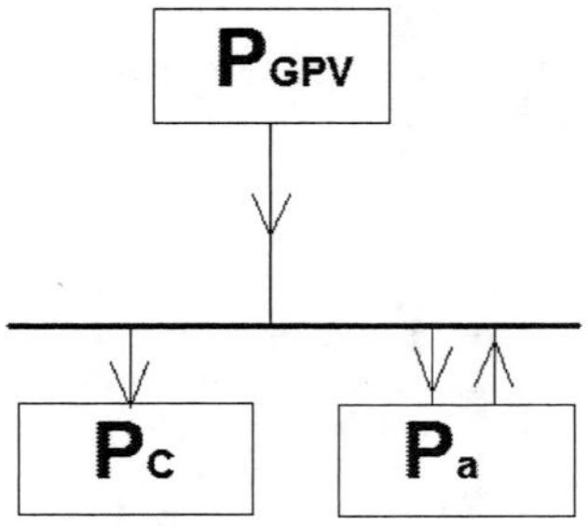

Fig. 2. Power Balance

The main point for the working capacity of the planned closed-loop system is the ability to provide the consumer with sufficient power.

$$P_{GPV} + P_a \geq P_c K_s$$

(3)

where, P_{GPV} – generated PV power; P_a – accumulated power, P_c – consumed power; K_s – safety factor.

The amount of energy possible to save at the current time (t) is determined by two parameters of "free" capacity of the accumulation system (D_a) and the maximal charge current value (I_c).

The main condition for the working capacity of the planned closed-loop system is the ability to provide the consumer with sufficient power.

$$P_{GPV} + P_a \geq P_c K_s$$
(4)

where P_{GPV} is generated PV power; P_a – accumulated power, P_c – consumed power; K_s – safety factor.

The amount of energy being able to save at the current time (t) is determined by two parameters of "free" capacity of the accumulation system (D_a) and the maximal charge current value (I_c).

If the battery capacity is equal to Ca, then

$$0.8 \cdot C_a \geq D_a$$
(5)

Coefficient 0.8 is taken from equation: (1-K_P), where K_p is coefficient of allowable battery charge (K_p~0.8) [12].

Thus, the mathematical expression for the required accumulation in terms of calculating the power balances of the system of communal lighting in the charging mode of accumulating devices will be equal to:

$$P_a = P_{GPV} - \frac{P_c}{E_i} \geq 0$$
(6)

where P_{GPV} is generated PV power, P_C – power consumed by the system, E_i – efficiency of the inverter.

Consequently, within the battery discharge mode:

$$P_a = P_{GPV} - \frac{P_c}{E_i} < 0$$
(7)

The mathematical expression for calculating the power balance in the charging regime of the accumulating unit will be equal to:

$$\begin{cases} P_a t \leq D_a U_a K_a \\ I_a = \dfrac{P_a}{n_a U_a} \leq I_C \end{cases},$$
(8)

$$\begin{cases} P_a t > D_a U_a K_a \\ I_a = \dfrac{P_a}{n_a U_a} \leq I_C \end{cases},$$
(9)

where n_a is the number of battery units, U_a and I_a – the voltage and the charge current of one battery, K_a <1– the efficiency of the battery, [12]

Optimizing the expression will look like:

$$\begin{cases} \overline{D}_a = D_a - I_a t K_a \\ P_C = P_a - I_a U_a \end{cases}$$
(10)

$$\begin{cases} \overline{D}_a = D_a - I_a t K_a \\ P_C = [(P)]_a - I_a U_a \end{cases},$$
(11)

Mathematical expressions for calculating the power balance in the discharge mode of the accumulating unit will be equal to: [12]

$$\begin{cases} |P_a| t \leq [(1-K_p) C_a - D_a] U_a \\ I_a = \dfrac{|P_a|}{n_a U_a} \leq I_p \end{cases}$$
(13)

The C_a and I_a values determine the working capacity of an autonomous electrical power supply system applying renewable energy sources.

Adding of battery storage, it is possible to use the solar power more efficiently, but there is the unknown variable, how much exactly has to be stored.

By looking back to Fig. 3 and adding the energy produced by photovoltaic elements (Figure 4), it can be seen, that in summer months there is potentially a lot of wasted energy, which can be stored. And for the time of the year, when there is not enough solar energy to cover the whole daily consumption, it can be used to shave off the demand peaks.

The minimum storage capacity can be calculated, using procedure:

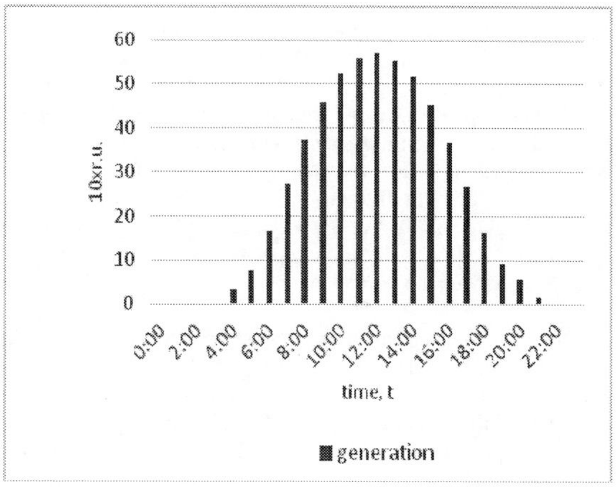

Fig. 3. Generated power of PV at the building's roof in Latvia, Riga

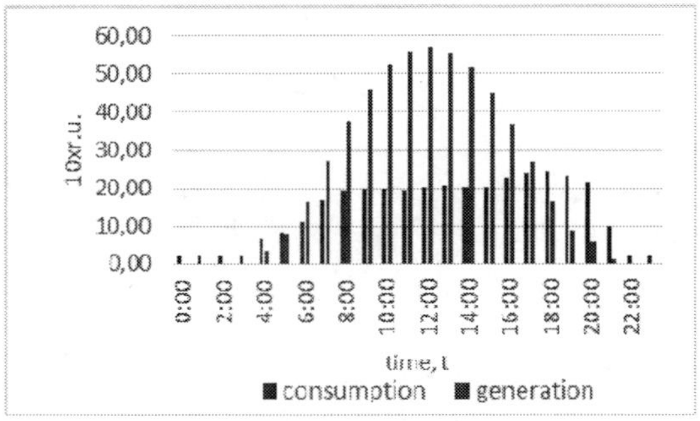

Fig. 4. Consumption and generation of energy: summer months

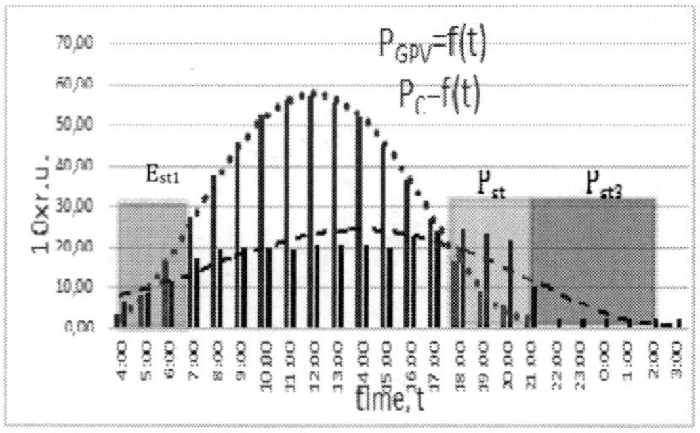

Fig. 5. Representation of areas to be calculated

IV. Conclusion

To ensure power supply of solar electricity in Northern European countries, it is necessary to use the optimal storage system and optimisation between consumption and generation as it was shown in the paper.

During the summer period, the residual energy appears more than the required energy in winter, for which it is necessary to look for the summer load. This can be, for example local air conditioners.

The right size for energy storage in photovoltaic system can prove itself to be difficult, especially in northern European countries, where the initial investment can be expensive and should be considered.

In this paper, the authors have looked into a way to calculate the needed storage capacity on the basis of potential PV system and existing power consumption of modular house.

The minimal storage system for planned modular house power supply system should be 1 AB with U=24 V and 100 Ah.

The usage of direct current (DC) and wireless dimming sensors can bring great benefits in the local low voltage system.

The next optimization step is the possibility to combine of local DC system distribution.

References

[1] A. Polman, M. Knight, E. C. Garnett, B. Ehrler, and W. C. Sinke, "Phototoltaic materials: Present efficiencies and future challenges," Science (80-.)., vol. 352, no. 6283, 2016.

[2] R. Fares, "The Price of Solar Is Declining to Unprecedented Lows," Scientific America, 2016. [Online]. Available: https://blogs.scientificamerican.com/plugged-in/the-price-of-solar-is-declining-to-unprecedented-lows/. [Accessed: 30-Jan-2017].

[3] T. Huld and E. D. Dunlop, "PVGIS." European Commission, Joint Research Centre Institute for Energy and Transport, Renewable Energy Unit, 2012.

[4] Y. K. Tan, T. P. Huynh, and Z. Wang, "Smart Personal Sensor Network Control for Energy Saving in DC Grid Powered LED Lighting System," vol. 4, no. 2, pp. 669–676, 2013.

[5] D. Salomonsson, S. Member, and A. Sannino, "Low-Voltage DC Distribution System for Commercial Power Systems With Sensitive Electronic Loads," vol. 22, no. 3, pp. 1620–1627, 2007.

[6] G. Notton, V. Lazarov, L. Stoyanov, Optimal sizing of a grid-connected PV system for various PV module technologies and inclinations, inverter efficiency characteristics and locations,Renewable Energy, Volume 35, Issue 2, February 2010, Pages 541-554.

[7] M. Mandalaki, , K. Zervas, , T. Tsoutsos, , A. Vazakas, Assessment of fixed shading devices with integrated PV for efficient energy use Original Research Article, Solar Energy, Volume 86, Issue 9, September 2012, Pages 2561-2575.

[8] V.Voynov, G.Pudovkina, A.Telegin. Modeling of Economic Systems and Processes. Experience in Constructing of ARIS Models: Monograph, - Chelyabinsk: Publishinh house ЮУрГУ, 2002 - elibrary.ru.

[9] A. Hoke, R. Butler, J. Hambrick, B. Kroposki, "Steady-state analysis of maximum photovoltaic penetration levels on typical distribution feeders" IEEE Transactions on Sustainable Energy 4 350-357.

[10] S G Obukhovetal, "Method for Prediction of the Power Output from Photovoltaic Power Plant under Actual Operating Conditions" 2017 IOP Conference Series: Materials Science and Engineering 189 012008 IOPscience

[11] I O Muravlev, Fuse Selection for the Two-Stage Explosive Type Switches 2017 IOP Conference Series: Materials Science and Engineering 189 012004 IOP science.

[12] SG. Lukutin, BV. Obuhov, EB. Papdarova "Autonomous power supply from the micro hydro power station", Tomsk – 2001 - 104.p.

[13] [Online] Available: http://re.jrc.ec.europa.eu/pvgis/apps4/pvest php?lang=en&map=Europe, last see (2017, May)

PV Electrical Energy Storage of Residential Buildings Communal Space Lighting

Kristina Berzina, Inga Zicmane, Tatjana Lomane
Faculty of Power and Electrical Engineering,
Riga Technical University,
Riga, Latvia
Kristina.Berzina@rtu.lv, Inga.Zicmane@rtu.lv,
loman@eef.rtu.lv

Georgi Georgiev
Energy Solutions Centre
Varna, Bulgaria
georgidg@hotmail.com

***Abstract* — In terms of the advances in photo-voltaic material research, solar (PV) panel price reduction over last years, the usage of this alternative energy source and the angle of slope of PV panels towards the horizon in Northern countries seems more attractive. Namely, the usage of the angle of slope of PV panels control or optimization devices can help use the solar power more efficiently and smarter. The paper deals with the optimization of suggested solar panel array of renovated residential building's communal lighting in Riga, applying storage devices and demand side management of the produced power.**

Keywords: PV panels, power balance, solar energy

I. Introduction

The conversion factor of the incident solar energy of modern solar photovoltaic power plants is not great. In clear sunny weather, about 1 kW of solar energy per square meter of area falls in perpendicular to the solar vector, but from the output of autonomous photovoltaic energy systems and plants, a much smaller amount of energy is supplied to the consumer.

Factors that significantly reduce the amount of energy generated are the low real average efficiency coefficient of silicon photocells of mass production (12-14%) [1] and underutilization of the generating capabilities of the selected solar panel. At the same time, the underutilization of solar panels in terms of energy (at least by 30-50% is explained by the fact that majority of power systems do not have systems for forced cooling and continuous automatic panels monitoring the Sun, as well as systems for controlling the maximum power in terms of the volt-ampere characteristic of the solar panel. The absence of systems for the continuous automatic panels monitoring the Sun and forced cooling is in turn explained by the fact that in many cases solar panels are located on the roofs of houses and the implementation of such systems is technically difficult and not justified economically, but in the case solar panels are located in open areas – the indecision of consumers in the usefulness and economic benefit of their use, because the technical implementation is complex and areas of rational use of the continuous automatic Sun monitoring systems are not defined, depending on the level of output power or the area of the solar panel [13].

As a result, the total energy efficiency of most photovoltaic power plants and power supply systems does not exceed 5-10%.

Energy efficiency (Ke) is analytically expressed by the ratio of the difference between the energy produced by the solar panel (WSB) with the solar battery voltage regulation (SB) adopted in the system, as well as the energy losses in the converter units (ΔWP) and AB (ΔWAB) to the energy that can be generated by solar panel in the continuous regulation of the solar panel voltage at the peak power point (W_{SBmax}) [13].

$$K_e = \frac{(W_{SB} - \Delta W_P - \Delta W_{AB})}{W_{SB\max}} = \frac{W_P}{W_{SB\max}} = \frac{\int_0^T P_P dt}{\int_0^T P_{SB\max} dt}, \quad (1)$$

where W_P is energy delivered to the load; $P_{SB\ max}$ – the extreme current power value of the solar panel; P_L – current value of load power; T – period of functioning.

Thus, the maximum selection of the solar panel is possible only if the solar panel voltage is continuously regulated at the optimal operating point (at the peak power point).

II. The Problem Definition

1. The power of a solar panel in winter is less than in summer; in its turn, the load, on the contrary, is greater than in the summer. Consequently, if you provide autonomy in the winter, then it certainly will "work" in the summer. However, in the winter the solution is simple, it does not have degrees of freedom for optimization. This suggests a different approach:

- The choice of the optimal slope angle of the module planes relative to the horizon in such a way as to maximize winter consumption, but not to lose a good balance in the summer with the same number of batteries (the own specificity of the northern countries).

- As one of the possibilities, it may be advisable to add a simple mechanism for seasonally changing the angle of the panels, although such a mechanism will require some maintenance, as well as the need to find additional load for the summer season (for example, in the form of capacitors and fans, etc.).
- Calculation of solar power for the winter period should be carried out in two options: a) without changing the winter slope of the modules; and c) with winter and summer slope (taking into account the search for additional consumption), after which both options are compared. The drive for changing the slope 2 times per year is not required, everything can be done simply and manually, having 2 holes on the mounting bus; the bolt is unscrewed and screwed in another hole. Everything can be done in a half a day, but it will be more difficult if it's on the roof. However, you can combine with the prevention (cleaning the surface). With the drive, it will be a little more expensive and difficult.
- In all cases, but especially in winter, it is necessary to use DC lamps (24 volts) in order not to have a loss during inverting. In summer, you can use the inverter if the power is more (in this case you can enter the operation mode compatible with the distribution network and sell the surplus). In southern countries, such a regime can be in winter).
- Also, it should be taken into account that the winter slope (more than 57 degrees to the horizon) is such that if the panels will be installed on the horizontal plane (in the garden, etc.), the first row will close the second row from the sun, i.e., the second row must be higher than the first one, etc. If the panels will be on the roof, then you need to have a steep roof - with a slope of about 60 degrees and higher.

2. The number of batteries depends on their type, i.e., acid ones will be needed about 1.6 times more than lithium-ion batteries.

III. THE SUN INSULATION AND PV GENERATION ANALYSIS

The intensity of solar radiation depends on a variety of factors, such as climate, air pollution, sea level elevation, season, geographical latitude and the angle of inclination of the receiving surface towards the Sun. The study assumes the installation of PV on a roof of the existing building in Riga. In this regard, all the above-mentioned parameters were taken into account.

The direct solar radiation Q_1 is found on the basis of Kostrov's formula:

$$Q_1 = \frac{Q_0 \sin\alpha}{\sin\alpha + c}, \qquad (2)$$

where Q_0 is the solar constant of 1370 W/m^2; c - the value characterized by the degree of transparency of the atmosphere (in our case c = 0.81).

Calculation of the Sun's altitude α, geographical latitude, inclination towards the horizon is calculated employing the Kupar approximation formula.

Diffuse solar radiation is calculated based on the formula:

$$Q_2 = Q_{2h}\left[0.55 + 0.434\cos\theta + 0.313(\cos\theta)^2\right] \qquad (3)$$

where Q_{2h} is the flow of diffuse solar energy falling on the horizontal plane (according to Berlage's expression) [11].

Thus, the orientation angles of the site were set using the above-mentioned expressions in order to determine the given energy falling on the inclined platform (at an angle of 45°). After the data analysis was carried out, it was decided to place 72 photovoltaic panels on the building's roof; the generation is represented at the Fig. 1, the location of building is represented on Fig.2.[3].

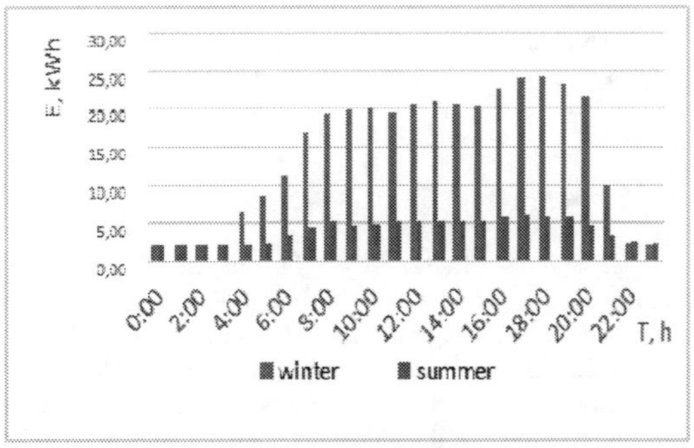

Fig. 1. Consumed energy comparison between summer and winter

Fig.2.The location of PV panel. [15]

IV. COMPARISON OF TWO POWER SUPPLY OPTIONS

Undoubtedly, in the general case, energy consumption for lighting of residential areas in summer is less than in winter. Only at the equator, winter and summer consumption are approximately the same. The north is the localization the summer day is longer than the winter one, which leads to a difference in energy consumption for household lighting during these two seasons.

The installation site has latitude of 56.9°, resulting in peak hourly electricity power for lighting in winter is 4 times higher than in summer, but the ratio between daily winter and summer energy consumption is 3.4 times (Fig.1). At the same time, the energy received from the Sun has the opposite relation: in winter, the energy received by solar panels is several times less than in summer in this northern latitude. Thus, the design calculation of an alternative solar source should be based on the winter season, with 2 options possible:

A. The angle of inclination of the panels to the horizon is constant throughout the year.

This is a classic option for the Central and Southern European countries, since it turns out that the initial and operating financial costs for the automatic drive, which changes the tilt of the modules towards the horizon do not pay off. Then it is possible to determine the optimal angle of inclination of the fixed panels, which ensures the maximum annual production of electricity. At first glance, this angle should be equal to the geographical latitude. Then the solar panel will be perpendicular to the sun's rays at midday in the spring and autumn equinox. The winter and summer solar trajectories across the sky will deflect symmetrically with respect to the spring trajectory, which will determine this slope as the optimal one. For example, for the geographical latitude of the city of Riga with a mounting angle of 57° with 1m2 of the solar panel, you can get:

$$E = A \times r \times H \times P_r \qquad (4)$$

where A is total solar panel area (m^2), r - solar panel yield (%, usually r =15%), H - annual average irradiation on tilted panels (kWh/m^2), Pr - performance ratio (between 0.9 and 0.5, average value 0.75).

Obviously, in the case when more energy is consumed from an autonomous system in winter, the optimal angle should be greater than 57°. Using (4), we found the optimal angle of slope of the panels at 64.35° and the following characteristics:

- Daily possible consumption in winter (from November to March) is 22.15 kWh, maximum power is 1.5 kW; number of lamps of 20 W each - 75 pieces. Accumulator capacity is 2784 Ah for 3 autonomous days at 60% discharge. Number of charges - 8.
- Daily consumption for lighting in summer (from May to September) is 6.5 kWh; number of lamps – 75. Accumulator capacity is 1653 Ah for 3 autonomous days at 60% discharge. For this energy, only 11 modules are sufficient.
- Daily consumption in April and October is 14.4 kWh; the number of lamps is 75. Their energy is provided only by 24 modules. Required accumulator capacity is 1810 Ah for 3 autonomous days with 60% of the allowable discharge.
- Power system – 48 V, Pr = 82.88%, real energy daily – 26.73 kWh; number of lamps – 75. An accumulator group 24 V with 2784 Ah for 3 autonomous days, number of charge controllers 8, efficiency – 72%

Winter consumption is basic one. The number of lamps used is determined on its basis. Their consumption will decrease in spring, autumn and summer due to shortening of the dark part of the day. This leaves unused energy. Its amount:

- April, October – 51.35 kWh per day
- May-September – 49.8 kWh per day

B. The angle of slope of the panels to the horizon varies depending on the season 2 times a year - in the autumn and in the spring.

1) Winter period: October-March.

The optimal winter angle is 73°. The daily possible consumption from 83 modules is 26.3 kWh, the maximum power is 1.77 kW. Required battery capacity is 2784 Ah for 3 autonomous days at 60% discharge. Possible number of lamps of 20 W - 89 pieces.

2) Summer period: April-September.

The optimal summer angle is 48.1°. The daily possible consumption is 49.0 kWh. Necessary number of modules for 89 lamps - 11 pieces; their maximum power is 3.46 kW. Required battery capacity is 1653 Ah for 3 autonomous days at 60% discharge. Number of lamps of 20 W - 89 pieces.In the summer period, the unused daily energy is 42.5 kWh.

It should be noted that the angle change can be made 2 times per year manually, which will require no more than 2 resource days for 83 modules.

C) Comparison of two options.

The results obtained for the two options are p in Table 1.

Table 1. The resuts of research

Option	Parameter	Winter	Summer
1	Angle of slope	64.35°	64.35°
	Number of lamps	75	75
	Unused energy (kWh)	0	49.8
2	Angle of slope	73°	48.1°
	Number of lamps	89	89
	Unused energy (kWh)	0	42.5

The second option, as expected, is more efficient in terms of using primary energy. Within it, the possible number of lamps is more by 19%, but the unused daily energy in summer is less by 14.6%. As a more detailed analysis shows, the advantage of the second option disappears, taking into account the additional operational costs for changing the angle of slope 2 times per year.

IV. Conclusion

In the paper was shown the comparison of two power supply options for the North European countries. For the power

supply with PV panels calculation in North countries should be increased attention to the PV panels installation angle.
To ensure communal lighting of solar electricity in Northern European countries, it is necessary to use the optimal unchanged angle of slope of the modules, determined by the annual consumption characteristic of lamp consumption.
During the summer period, the residual energy appears, the volume of which is almost two times more than the required energy in winter, for which it is necessary to look for the summer load. This can be, for example: air conditioners in nearby municipal organizations, water pumps for watering in parks and fountains, electric cultivators and cutting tools in parks, etc.

References

[1]. A. Polman, M. Knight, E. C. Garnett, B. Ehrler, and W. C. Sinke, "Photololtaic materials: Present efficiencies and future challenges," Science (80-.)., vol. 352, no. 6283, 2016.

[2] R. Fares, "The Price of Solar Is Declining to Unprecedented Lows," Scientific America, 2016. [Online]. Available: https://blogs.scientificamerican.com/plugged-in/the-price-of-solar-is-declining-to-unprecedented-lows/. [Accessed: 30-Jan-2017].

[3] T. Huld and E. D. Dunlop, "PVGIS." European Commission, Joint Research Centre Institute for Energy and Transport, Renewable Energy Unit, 2012.

[4] Y. K. Tan, T. P. Huynh, and Z. Wang, "Smart Personal Sensor Network Control for Energy Saving in DC Grid Powered LED Lighting System," vol. 4, no. 2, pp. 669–676, 2013.

[5] D. Salomonsson, S. Member, and A. Sannino, "Low-Voltage DC Distribution System for Commercial Power Systems With Sensitive Electronic Loads," vol. 22, no. 3, pp. 1620–1627, 2007.

[6] G. Notton, V. Lazarov, L. Stoyanov, Optimal sizing of a grid-connected PV system for various PV module technologies and inclinations, inverter efficiency characteristics and locations,Renewable Energy, Volume 35, Issue 2, February 2010, Pages 541-554.

[7] M. Mandalaki, , K. Zervas, , T. Tsoutsos, , A. Vazakas, Assessment of fixed shading devices with integrated PV for efficient energy use Original Research Article, Solar Energy, Volume 86, Issue 9, September 2012, Pages 2561-2575.

[8] V.Voynov, G.Pudovkina, A.Telegin. Modeling of Economic Systems and Process es. Experience in Constructing of ARIS Models: Monograph, - Chelyabinsk: Publishinh house ЮУрГУ, 2002 - elibrary.ru.

[9] A. Hoke, R. Butler, J. Hambrick, B. Kroposki "Steady-state analysis of maximum photovoltaic penetration levels on typical distribution feeders" IEEE Transactions on Sustainable Energy 4 350-357.

[10] Method for Prediction of the Power Output from Photovoltaic Power Plant under Actual Operating Conditions S G Obukhovetal 2017 IOP Conference Series: Materials Science and Engineering 189 012008 IOPscience

[11] Fuse Selection for the Two-Stage Explosive Type Switches I O Muravlev et al 2017 IOP Conference Series: Materials Science and Engineering 189 012004 IOPscience.

[12] SG. Lukutin, BV. Obuhov, EB. Papdarova "Autonomous power supply from the micro hydro power station",Tomsk – 2001 -104.p.

[13] http://calculationsolar.com/calculate.php

[14] Yu.A. Shinyakov, Yu.A. Shurygin, O.E. Arcatova Increase of energy efficiency of autonomous photovoltaic power plants, УДК 621.31

[15] [Online] Available:http://re.jrc.ec.europa.eu/pvgis/apps4/pvest.php?lang=en&map=europe, last see (2017, May)

The Influence of Power Balancer on the Electricity Meter Results in Residential Houses with Photovoltaic Power Plant

Petr Bilik
FEECS
VSB – Technical University of Ostrava
Ostrava, Czech Republic
petr.bilik@vsb.cz

Radek Martinek
FEECS
VSB – Technical University of Ostrava
Ostrava, Czech Republic
radek.martinek@vsb.cz

Vaclav Kus
FEE
University of West Bohemia
Plzen, Czech Republic
kus@kev.zcu.cz

Martin Kaspirek
E.ON Ceska Republika s.r.o.
Ceske Budejovice, Czech Republic
martin.kaspirek@eon.cz

Abstract - **Many small photovoltaic power plants (PVPP) exists on the roofs of family houses in the Czech Republic. The Czech Republic state legislation for such a photovoltaic power plants emphasis on-site consumption of generated energy. To force the power plant owner for such behavior the Distribution System Operator (DSO) charges for energy supplied to house 10 times more in comparison with energy supplied by photovoltaic power plants to the distribution grid. Described situation leads to high effort to consume all generated energy in-house and this is reached by using device called or Power Balancer (PB) which ensures fast switching of appropriate house appliances. This fast loads control brings fast changes in energy flow direction and if such changes are timed properly this could lead to situation when electricity meter (EM) shows no energy flow in any direction. The paper describes functionality of Power Balancer, internal functionality of electronic electricity meters and describes the results of on-site measurement.**

***Keywords:** electricity meter; power balancer; power quality*

I. Introduction

Inexpensive residential electricity meters design reflects strong pressure to the instrument price. The compensated resistor divider is used as the voltage sensor. As the current sensor either shunt or current transformer is used. Shunt is quite linear for low frequencies. The central processing unit (CPU) calculation power of such an instrument does calculations in time domain with samples taken 500-2500 times per second. The calculation contains correction of delayed samples in case if multiplexer is used in front of AD converter and calculation of non-calibrated values of U, I, P, Q. Then calibration for fundamental power grid frequency (50Hz) is applied. The calibration is done according the line equation (offset and constant). The precision is sufficient from fundamental frequency, harmonic components are filtered in signal conditioning block by low pass to prevent aliasing, [13], [14], and [15].

To achieve better accuracy class of industrial electricity meter the internal architecture has to be changed and the sampling frequency has to be increased to 10kS/s or even more. For corrections of amplitude frequency and phase frequency characteristics of voltage and current sensors and input circuits the calibration has to be done in frequency domain. In global this means to use faster AD converters, more powerful microprocessor. The result of such measures is the electricity meter which can measure correctly power and energy for distorted voltage and current signals. The appropriate antialiasing filter is also important. For such enhanced procedure see block diagram in Fig. 1.

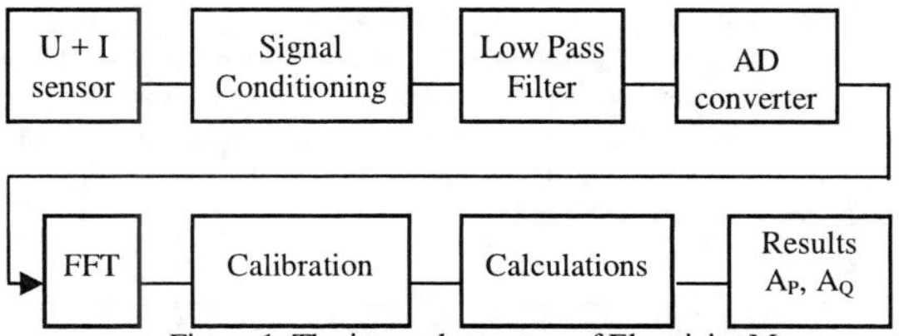

Figure 1. The internal structure of Electricity Meter

The electronic meter does not process the signal samples continuously, but cyclically, after a period of time defined by the manufacturer, which is called a time window T_W. ADC provides samples of the voltage and current signal to the microprocessor. The microprocessor collects samples in the buffer and performs all calculations periodically with a set of samples, provided by an AD converter within the time window T_W. The basis for calculations in the microprocessor are instantaneous voltage and current values, from which the value of the instantaneous active power is calculated.

$$p(n) = u(n) \cdot i(n) \qquad (1)$$

The microprocessor of the static electricity meter collects samples of instantaneous voltage and current values and once within the time window T_W, it will calculate the mean value of active power (2). On principle, the shortest time window T_W, for which the active power mean value can be calculated, is equal to half the period of the fundamental harmonic. The length of the time window T_W for electricity meters is not determined by any international standard and it has been experimentally found on a number of different major manufacturers of electricity meters that T_W is 1 second [1], [2].

$$P = \frac{1}{N}\sum_{0}^{N-1} p(n) \quad (2)$$

From the mean active power value, the active energy A_P value is calculated per time window T_W (3). This value is added to the existing value of energy in storage in the registry and displayed on the electric meter.

$$Ap = P \cdot Tw \quad (3)$$

Active energy consumed A_{P+} and generated A_{P-} depends on time integration of consumed power and generated power. Due to discrete window integration, time variations between consumption and generation can result in lower registered energies than the actual ones. This phenomenon can occur at the customer with simultaneous production and consumption when power sign changes rapidly [3].

II. Functionality Of The Power Balancer

The Power Balancer is able to regulate the surplus of energy produced in photovoltaic panels in two ways. The first way (slow regulation) consists in switching on the accumulation appliance (e.g. a boiler) at the moment when the performance of the photovoltaic power plant is larger than the power of the accumulation appliances. For switching, contactors are used. Subsequently, while the appliance is switched on, it is powered exclusively from the photovoltaic power plant. At the moment when the performance of the photovoltaic power plant falls below the level of the appliance power, the Power Balancer switches the appliance off. The appliances are switched on within minutes and the electric meter in this case correctly evaluates the consumed or produced energy [4]. The problem is the rapid regulation using the imperfections of the time window T_W of the electric meter of nonzero length. The fast regulation takes place in the case that the instantaneous power of the photovoltaic power plant is smaller than the power of the switched on accumulation appliances, and the semiconductor Solid State Relay is used for switching on; its shortest period of switching on or switching off is period 1 of the fundamental harmonic. A wiring diagram of a house with a photovoltaic power plant and the Power Balancer is shown in Fig. 2

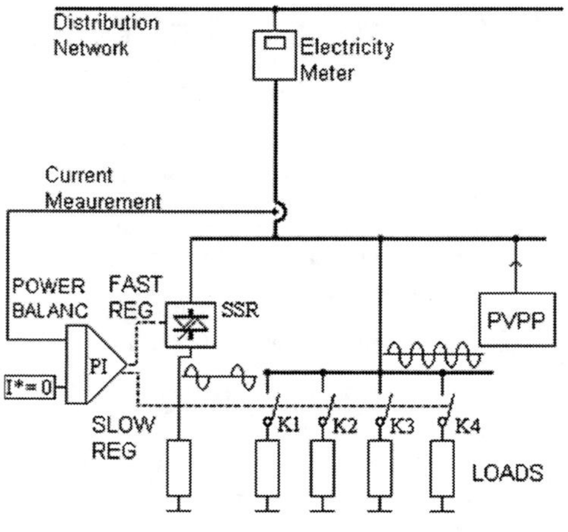

Figure 2. Schematic o Power Balancer in Building with Photovoltaic Power Plant

Theoretically, the Power Balancer should not switch the appliance on to prevent power consumption from the grid. However, the Power Balancer provides zero mean power value during the time window T_W as described below. The Power Balancer switches the appliance on for the time period of tens of milliseconds and then the appliance consumes all the power from the solar power plant, and at the same time, the difference between the actual power output of the power plant and the power demand of the appliance is taken from the grid. Consequently, the appliance is switched off by the Power Balancer and the output of the photovoltaic power plant is supplied to the grid so that the power taken by the appliance from the grid was returned. Switching the appliance on and off takes place several times per second, and within the time window of the electric meter T_W, the balance of power flow and balance of energy consumed A_{P+} and energy supplied A_{P-} is zero. Consequently, the meter records zero or only small values of consumption and supply of energy into its memory registers [5], [16], and [17].

$$A_{P+} = A_{P-} = 0 \quad (4)$$

Power Balancer activity leads to rapid changes in sign of instantaneous active power p(t), see Fig. 3. If the loads are controlled selectively in individual phases while the voltage passes zero, the curve of the active power p(t) has different shape, see Fig. 4.

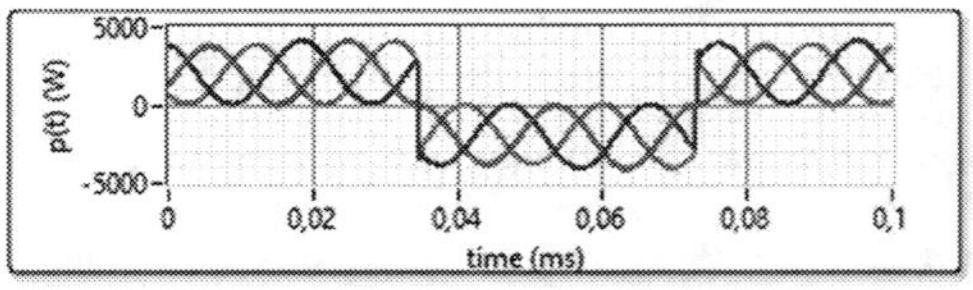

Figure 3. Simulation of instantaneous active power p(t), the sign of power is changed at the same time on all phases.

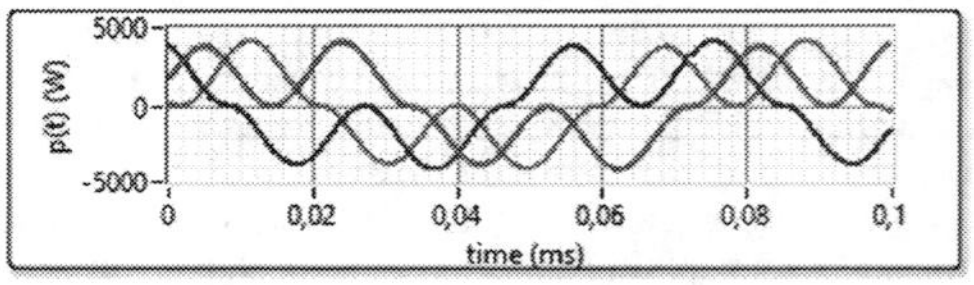

Figure 4. Simulation of instantaneous active power p(t), the sign of power is changed in sequence selectively phase per phase.

For illustration, the following example is described: The Power Balancer switches the accumulation appliance with a power of 2.5 kW. The photovoltaic power plant generates power of 0.5 kW. The Power Balancer appliance switches the appliance for the time of 100 ms, during which the appliance takes full power of the photovoltaic power plant 0.5 kW and the remaining power 2 kW is taken from the grid, see Fig. 5. During the time of 100 ms, the house takes energy of 200 Ws from the grid. After that, the appliance is switched off by the Power Balancer and the power of 0.5 kW of the photovoltaic power plant is supplied to the grid, when for a period of 400 ms, the power plant will supply energy of -200 Ws to the grid. Then the whole action is repeated periodically, see Fig. 6. The energy balance between consumption and supply, as it is evaluated by the electricity meter, is equal to zero within the time window T_W 1s. In fact, during 1 s time window T_W the residential home consumed 400 Ws and supplied to grid - 400 Ws, see Fig. 7 and Fig. 8

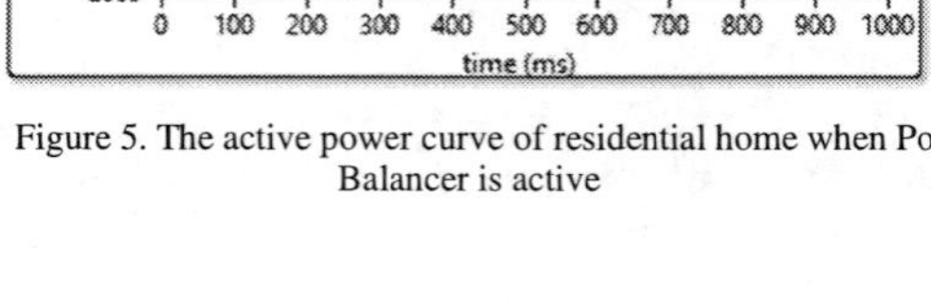

Figure 5. The active power curve of residential home when Power Balancer is active

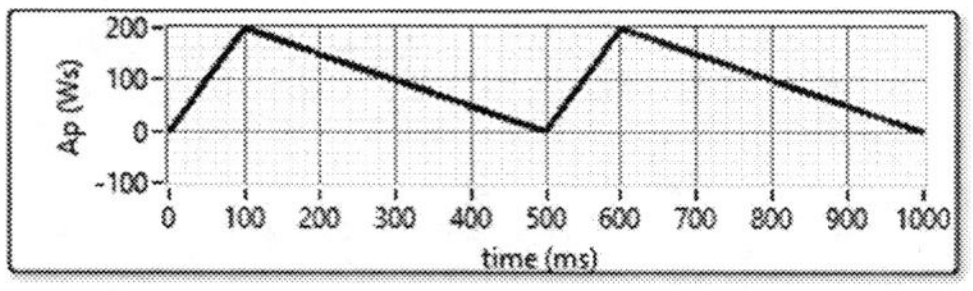

Figure 6. The active energy curve of residential home when Power Balancer is active

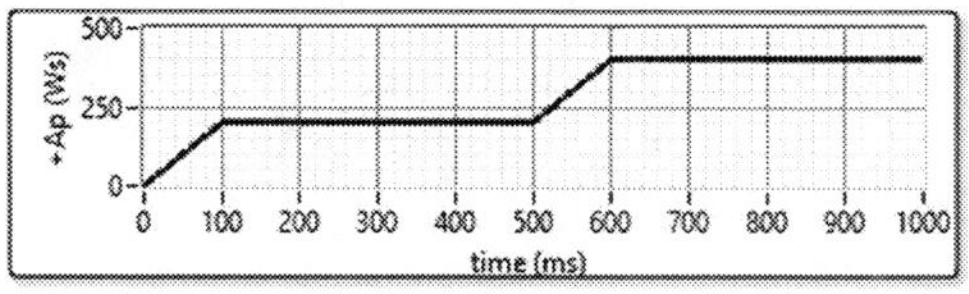

Figure 7. Integrated consumed active energy of residential home when Power Balancer is active

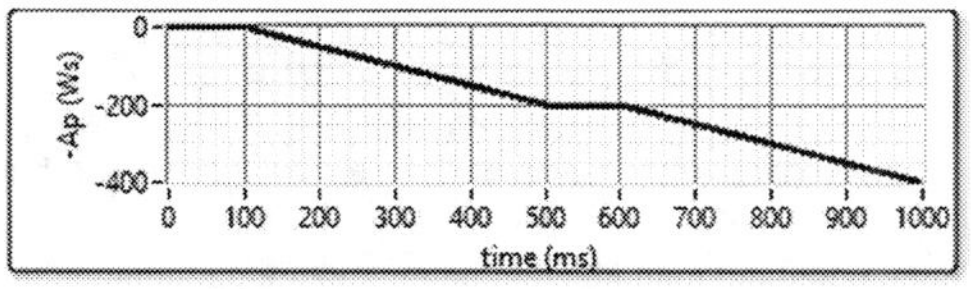

Figure 8. Integrated supplied active energy of residential home when Power Balancer is active

The ratio of time over which the appliance is switched on t_{ON} to the time during which it is switched off t_{OFF} corresponds to the ratio of power produced by the power plant P_{PVPP} to the power consumed from the distribution system P_{DS} during the time when the appliance with the performance P_{LOAD} is switched on. This fact is expressed by (5), (6) and (7).

$$A_{P+} = A_{P-} \tag{5}$$

$$P_{DS} \cdot t_{ON} = P_{PVPP} \cdot t_{OFF} \tag{6}$$

$$\frac{t_{ON}}{t_{OFF}} = \frac{P_{PVPP}}{P_{LOAD} - P_{PVPP}} = \frac{P_{PVPP}}{P_{DS}} \tag{7}$$

III. Results Of The Energy Meter Depending on Time Window Length

To verify information from previous chapters special measurement instrument was designed and implemented. Such instrument is capable to analyze electrical powers and energies from the time window T_W long of half period of fundamental frequency. Special power meter was built on special compact PC equipped by DAQ board with 16-bit AD converters and signal conditioning module for voltage and current signals. As a current sensor the Chauvin Arnoux MN39 current clamps has been used. The firmware was implemented in graphical programming environment National Instruments LabVIEW. Since LabVIEW version 2013 there exists the Electrical Power Module which helps to accelerate prototyping of application dealing with signal processing for Electrical Power engineering. After the firmware was finished the developed power meter was calibrated using OMICRON CMC 256+ calibration source, see Fig. 9.

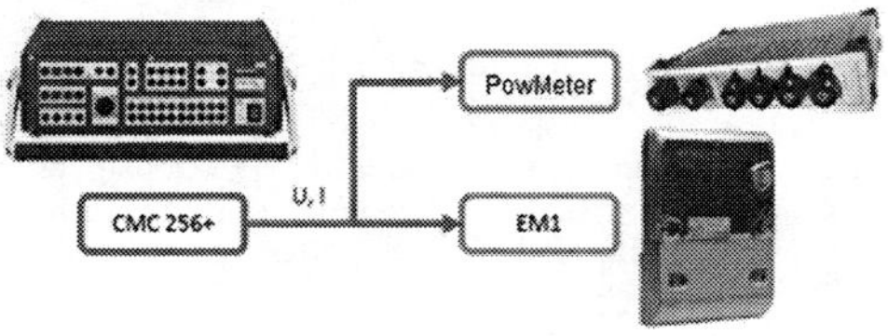

Figure 9. Schematic of testing and calibration stand

After many tests verifying correct power meter functionality and precision the on-site measurement was done at the residential house equipped with roof photovoltaic power plant and power balancer. Developed power meter stores voltage, current, active power and apparent power each half cycle (10 ms) of 50 Hz fundamental frequency. The instrument stores data on local solid state disc and this allows various off-line analysis. On the Fig. 10 can be seen almost 3 hours long curve of active power analyzed by the time window T_W = 10 ms. To make graphs clear the phase L1 is displayed only. On Fig. 11 can be seen detail of active power P curve. From the graph it is clear that power changes directions very fast.

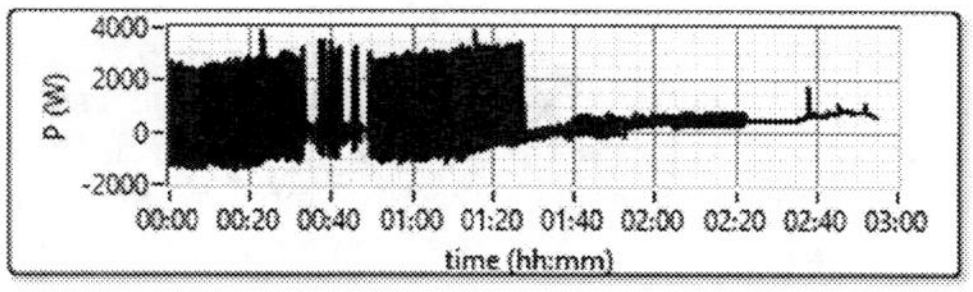

Figure 10. Recorded active power P, T_W = 10 ms, (real data)

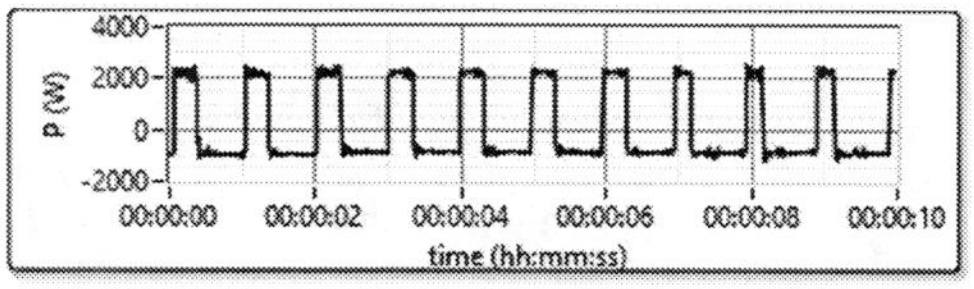

Figure 11. Short time detail of active power P, T_W = 10 ms

When active power P is analyzed by time window T_W = 1 s as the standard electricity meter does see Fig. 12, the curve of active power seems very different in comparison with the one on Fig. 10. From the time 01:30 there is no produced power because of heavy clouds on the sky, so there is consumption only.

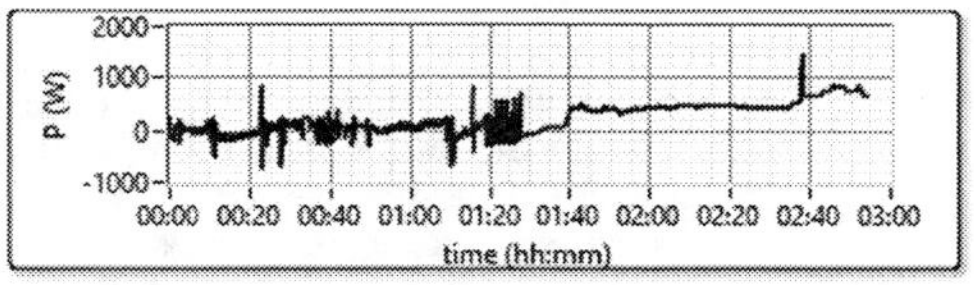

Figure 12. Active power recorded by power analyzer, T_W = 1s

Then the energies A_{P+} and A_{P-} were calculated from active power P. The curve of energy taken from distribution network (black curve) and energy measured by electricity meter (red curve) can be seen on Fig. 13 and Fig. 14.

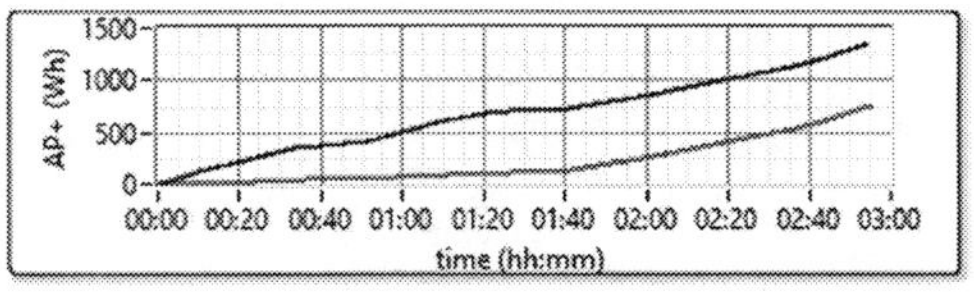

Figure 13. Active energy taken from distribution network A_{P+}, T_W = 10ms (black curve) a T_W = 1s (red curve)

From the Fig. 13 it is apparent a different shape of integrated consumed power in time, but most importantly the calculated final values of consumed active energy is also different: for the time window T_W = 10 ms is the final value of active energy A_{P+} = 1351 Wh and for the time window T_W = 1 s is the final value of active energy A_{P+} = 744 Wh

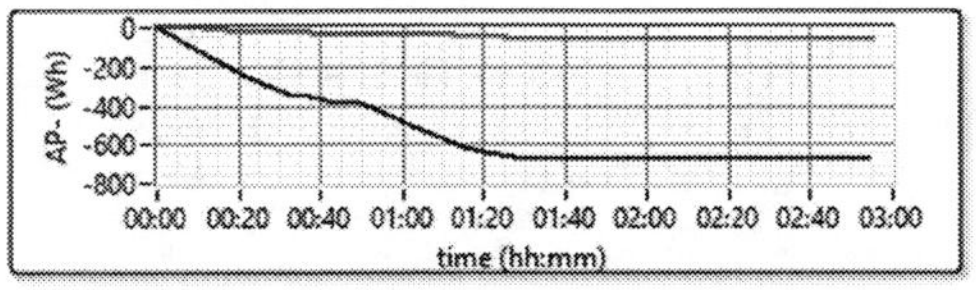

Figure 14. Active energy supplied to distribution network A_{P-}, T_W = 10 ms (black curve) a T_W = 1 s (red curve)

The same data analysis was done for produced (supplied) active energy, see Fig. 14. For the time window T_W = 10 ms is the final value of active energy A_{P-} = -670 Wh and for the time window T_W = 1 s is the final value of active energy A_{P-} = -63 Wh. If both energy A_{P+} and A_{P-} are summed for T_W = 1 s or T_W = 10 ms the same result is obtained, what was expected. The average value of energy remains the same (8) and (9).

$$A_{P+1s} + A_{P-1s} = 1351 + (-670) = 681 \text{ Wh} \tag{8}$$

$$A_{P+10ms} + A_{P-10ms} = 744 + (-63) = 681 \text{ Wh} \tag{9}$$

From presented results it is clear that the electronic electricity meter suffers from the integration character of 1 s long time window T_w. The power balancer uses this weakness of electricity meter to balance the consumption and production of electrical energy in residential home in such a way to minimize the values of electricity meter registers even the both directional energy flow is much higher.

IV. Conclusion

The facts described above clearly show a paradox caused by two factors: a) the legislative effort to achieve a situation where there will be only a minimal flow of energy produced in houses to the distribution network, b) the impact of the deployment of the power balancer simultaneously with the imperfections of the static electricity meter. In fact, in this case, there is only a minimal measured produced energy, because through appropriate management, the power balancer reaches a stage where from a short-time perspective (t <1 s), the distribution system is used as an accumulator. The residential house briefly borrows energy in order to supply a large appliance, even if the output of the photovoltaic power plant is not sufficient at that time, but within the evaluation time window of the electric meter T_W = 1 s, it returns energy back into the grid. In terms of the distribution system, there is no loss due to incorrect electric meter data on the measured energy, the difference in the mean value of the supplied and consumed energy is correct. However, from the short-time perspective (t <1 s), there is a significantly higher power flow in both directions with a number of negative effects, such as losses in the distribution system and flicker [6], [7]. Slow regulation of power balancer leads to slow power changes which cause voltage dips and rapid voltage changes. This brings high attention to power quality monitoring and evaluation [8], [9], [10], [11]. Depending on used method of fast power changes could be also produced current harmonics which make voltage drop on network impedance originating voltage harmonics [12] and again the power quality gets worse and need to be evaluated.

Acknowledgment

This work was supported by the Student Grant System of VSB-TU Ostrava [grant number SP2017/128].

References

[1] IEC 62053-22:2003 Electricity metering equipment (a.c.) – Particular requirements – Part 22: Static meters for active energy (classes 0,2 S and 0,5 S). Geneva : International Electrotechnical Commission, 2003.

[2] EN 50470-3:2007 Electricity metering equipment (a.c.). Part 3: Particular requirements. Static meters for active energy (class indexes A, B and C). Brussel: CENELEC, 2007.

[3] A. Bernieri, G. Betta, L. Ferrigno and M. Laracca: "Electrical energy metering in compliance with recent european standards," Proc. Instrumentation and Measurement Technology Conference (I2MTC), 2012 IEEE International. IEEE, 2012. p. 1541-1545.

[4] J. Jiricka and M. Tesarova, "Determination of Coincidence Factors for PV Plant Generation Depending on Constancy of Sun Exposure," Proc. of the 13th International Scientific Conference Electric Power Engineering 2012. Brno: Brno University of Technology, May 2012, pp. 469-474. ISBN: 978-80-214-4514-7

[5] J. Novotny, J. Drapela and D. Topolanek, "Frequency response of revenue meters in measured active energy," Proc. Harmonics and Quality of Power (ICHQP), 2016 17th International Conference on. IEEE, 2016. p. 524-529, doi: 10.1109/ICHQP.2016.7783309

[6] M. Kaspirek: "The issue of striking demonstration of reversing the effects of contributors," Proc. of the 14th International Scientific Conference Electric Power Engineering 2013, VSB – Technical University of Ostrava, Faculty of Electrical Engineering and Computer Science, Department of Electrical Power Engineering, May 28-30, 2013, Kouty nad Desnou, pp. 531-534, ISBN 978-80-248-2988-3

[7] A. Otcenasova, J. Altus, P. Hecko and M. Roch: "Measurement characteristics of voltage in practice and possibilities for improvement of voltage," Przeglad elektrotechnniczny (Electrical Review), Warsaw, Poland, 2012, R. 88 NR 9a/2012, pp.: 103-106, ISSN 0033-2097 WOS:000308318000021

[8] J. Jiricka and M. Tesarova: "Assessment of Voltage Changes of Single-phase and Double-phase Sources Connected into LV Network," In Proceedings of the 13th International Scientific Conference Electric Power Engineering 2012. Brno: Brno University of Technology, May 2012, pp. 671-676. ISBN: 978-80-214-4514-7

[9] M. Tesarova and M. Kaspirek, "Evaluation of long-term voltage dip monitoring in the distribution system," Proc. of 7th International Scientific Symposium on Electrical Power Engineering Elektroenergetika 2013 Košice: Technical University of Kosice, 2013. pp. 268-271. ISBN: 978-80-553-1441-9

[10] M. Tesarova: "Methods for the prediction of short voltage dips and interruptions and the possibility of their practical use," In Proceedings of the 12 th International Scientific Conference Electric Power Engineering 2011. Ostrava: VSB - Technical University of Ostrava, May 2011, pp. 415-418. ISBN: 978-80-248-2393-5

[11] A. Otcenasova, R. Bodnar, M. Regula, M. Hoger and M. Repak, "Methodology for Determination of the Number of Equipment Malfunctions Due to Voltage Sags," in Energies, 2017, ISSN 1996-1073, doi:10.3390/en10030401

[12] A. Belan, Z. Eleschova: Harmonic Distortion Produced by Synchronous Generators. In Proceedings of the 4th International Scientific Symposium on Electrical Power Engineering (Elektroenergetika 2007). Košice: Technical University of Košice, September 2007. pp. 432-439. ISBN: 978-80-8073-844-0

[13] R. Martinek, J. Vanus, M. Kelnar and P. Bilik, "Control methods of active power filters using soft computing techniques," Proceedings of the 8th International Scientific Symposium on Electrical Power Engineering, ELEKTROENERGETIKA 2015, pp. 351–354, ISBN: 978-80-5532-187-5

[14] R. Martinek, J. Vanus and P. Bilik, "New strategies for application of recursive least square algorithm in active power filters," Proceedings of the 8th International Scientific Symposium on Electrical Power Engineering, ELEKTROENERGETIKA, pp. 332–335, ISBN: 978-80-5532-187-5

[15] R. Martinek, et al. (2016). An efficient control method of shunt active power filter using ADALINE. IFAC-PapersOnLine, 49(25), 352-357. doi:10.1016/j.ifacol.2016.12.064

[16] P. Bilik, P., et al. (2015). Results of multiple phasor measurement units testing procedures. Proceedings of the 8th International Scientific Symposium on Electrical Power Engineering, ELEKTROENERGETIKA, pp. 132-135, ISBN: 978-80-5532-187-5.

[17] P. Valicek, P., et al. (2016). Measurement of illuminance of interior lighting system automatically dimmed to the constant level depending on daylight. Paper presented at the EEEIC 2016, International Conference on Environment and Electrical Engineering, doi:10.1109/EEEIC.2016.755560

Experimental Investigation Based on Analyzation of Electric Current-Voltage Characteristics on Pasan-Flasher Flashlight Irradiated Photovoltaic Cells Aged by Various Processes

Ervin Racz
Power Systems Department
Kando Kalman Faculty of Electrical Engineering
at the Obuda University
Budapest, Hungary
racz.ervin@kvk.uni-obuda.hu

Andrea Varga
Power Systems Department
Kando Kalman Faculty of Electrical Engineering
at the Obuda University
Budapest, Hungary
varga.andrea@kvk.uni-obuda.hu

Balazs Horompoli
Power Systems Department
Kando Kalman Faculty of Electrical Engineering
at the Obuda University
Budapest, Hungary
balazs.horompoli@gmail.com

Erik Rene Neuchel
Institut für Energie- und Antriebstechnik
Hochschule Ulm
Ulm, Germany
neuchel@hs-ulm.de

Abstract — **Electric current-voltage characteristics and generated electric power-voltage characteristics of artificially aged solar cells were recorded experimentally. Aging of the photovoltaic cells were made different ways using Kesternich and Climate chamber tests using more identical solar cells. The characteristics were generated using Pasan-Flasher professional irradiation source. Analyzing the measurement plots some information for the Silicon material structure modification after different aging processes can be taken.**

Keywords: photovoltaic panels; irradiation measurements; electric current-voltage characteristics; electric power-voltage plots; solid state of Silicone.

I. Introduction

Renewable energy sources play more and more important role at energy production in Europe. One of the best developing sector between the renewables is photovoltaic technology. Nowadays, looking around more and more solar modules can be observed on the roof tops. These are mainly monocrystalline and polycrystalline Silicone based solid state photovoltaic panels. These panels on the roofs installed are exposed to extreme weather conditions. For example, cold and hot temperatures modify, rain or probably in some cases acid rain can wash the surface of the solar modules, etc. These natural weather effects can affect the operation of the solar panels. In some cases, the extreme temperatures, or the temperature difference, or SO_2 in the acid rain can damage the material of the photovoltaic structure, causing modifications in their operations. In the most cases, the modifications mean efficiency of the solar modules decrease. For these reasons, it is important to investigate how the weather effects can influence the operation of the solar panels. This paper deals with this area using experimental method for the investigation. [1]

II. Aging of Our Photovoltaic Cells

A. Preparing the aging experiments

To prepare and design our experiments we purchased four pieces of identical small size monocrystalline photovoltaic cells. Open circuit voltage and short circuit current of every purchased cell were 0.5 V and 400 mA, respectively. Using these data, the nominal power of every single cell was 0.2 W. Three cells from the four ones were artificially aged, one panel was kept on its original status as reference cell. The three cells were aged by the following aging processes:

B. Description of aging effect applied on solar cells

To age our brand new solar cells, the following devices was used:

a.) CON 300-FL AIR/SO_2 test chamber:

Nowadays, acid rain can cause serious damages on terrain features. For example: metals can be destroyed, polish layers or metals can be faded, or glasses can be found as opaline and not transparent material. In other words, materials in the nature aged by acid rains. Fortunately, there is a way how to simulate acid rain medium with laboratory experiments: Kesternich chamber [2] can be used for.

Using this kind of test chamber vaporized air tests, condensation tests and SO_2 tests or in other words Kesternich tests [2] can be ran. During a well-defined time interval, SO_2 vapor is interacting with samples. The experimental time interval can be multiplied, so this way longer time periods or many-cycle experiments also can be running. Amount and concentration of the SO_2 and the temperature of the chamber can be varied during the whole experiment. Some parameters of the Kesternich test chamber are summarized in Table 1. [4]

Table 1. Main parameters of the Kesternich test chamber [2], [4]

Important parameters of Kesternich test chamber		
Parameter	***Value***	***Dimension***
Inner temperature of the chamber (T)	outer temperature < T < 50	°C
Constant Humidity	100	%
Temperature Stability	±0.5	%
Salt-free water supply	<20	$\mu S/cm$
Water	through ion exchange unit	
Voltage	230/50	V/Hz
Electric power	3000	W
Compressed air	6-8	bars
Chamber material	stainless steel	

b.) ACS Discovery DY110C climate test chamber [3]:

Materials in the nature can be damaged not only from the acid rains but from other extreme weather conditions, such as extreme temperatures as well. Nowadays, extreme weather conditions appear more and more times. For example, in Hungary December in the year 2016 and January of the year 2017 were the coldest months in the last 25 years in Hungary temperature between -25 °C and -15 °C could be measured for 2-3 weeks long time. From other side, thanking to the global warming dry and hot summer times could be lived through. +40 °C or higher temperature could be measured from summer in the year 2016 in Hungary. These types of extreme temperatures can affect the operation of the photovoltaic panels as well. [3], [4]

In laboratory conditions, there is a way to produce extreme temperatures in test chamber. For this applications climate chambers can be used. In our laboratory, an ACS Discovery DY110C climate chamber [3] can be found. Control of the test chamber works with microprocessors, so during the test with a touchpad works with KeyRatos control system can we do the operation. Programming the KeyRatos touchpad monitor test parameters can be uploaded to the system memory. Running the test works with running the programmed test parameters, everything is automated during the run. Table 2. summarizes the most important parameters of our ACS Discovery DY110C climate chamber. [3], [4]

Table 2. Main parameters of our climate chamber [3], [4]

Important parameters of ACS Discovery DY110C Climate chamber		
Parameter	***Value***	***Dimension***
Inner temperature range of the chamber (T)	-70 ... +180	°C
Temperature stabiliy	±0.1 ... ± 0.3	°C
Speed of the temperature change	-2.8, +3.2	$°C/min$
Relative humidity range	10...98	%
Relative humidity accuracy	±1 ... ± 3	%
Voltage	230/50	V/Hz

C. Aging and numbering of our photovoltaic cells

a.) CON 300-FL AIR/SO_2 test chamber:

During the Kesternich test 2 deciliters of gas was insufflated into the test chamber. The insufflated gas has been interacting with the sample during 8 hours in the chamber. After the 8 hours of experimental time the ventilation of the chamber has been started. The ventilation time has been kept for 16 hours long time. This is 1 experimental cycle. During the whole aging period 6 cycles have been made to age the selected sample. Test period was designed by the standard DIN 50018 and ISO 6988. [4]

b.) ACS Discovery DY110C climate test chamber:

This has been kept running for 8 hours in 1 cycle. Temperature has been modified between -20 °C and +80 °C. Speed of the temperature change was $\pm 0.5\ °C/min$. Beginning of the test started at 20 °C. Duration of segments in 1 cycle was 24 hours (for segments 4-5-6-7). Number of cycles was set to 6, or in other words 6×24 hours. Time duration of the whole aging cycle was ~146 hours. Cycle program of the climate chamber can be seen in Figure 1. [4]

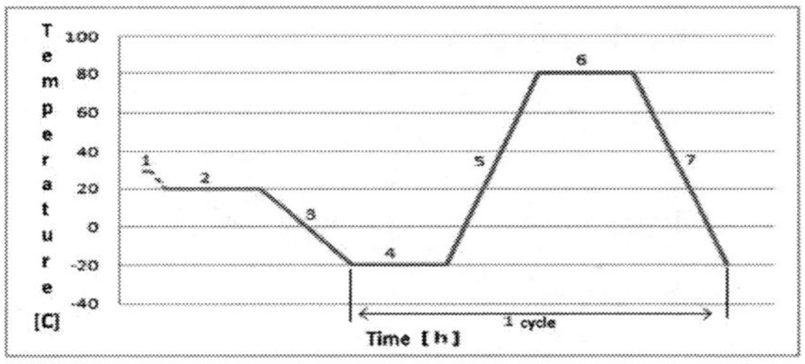

Figure 1. KeyRatos program cycle of the climate chamber. [4]

Finally, our solar cells have been numbered and aged according to the following Table content. See it in Table 3.

Table 3. Numbering the photovoltaic cells and types of the aging tests have been made on the solar cells.

Numbering of the used solar cells		
Name or number of the cells	***Type of test***	***Number of aging runs***
Cell #1.	Climate chamber test	2
Cell #2.	Kesternich chamber test	2
Cell #3.	Climate + Kesternich chamber tests	1 + 2.5
Cell #4.	no test = reference cell	0

III. Experimental

Aim of running our main experiment was to see and record electric current-voltage (I-U) plots and generated electric power-voltage (P-U) plots of our aged photovoltaic cells. *Using the plots, we wanted to analyze how the mentioned weather effects can affect the operation of the solar cells.* To see and record the plots mentioned before, our solar cells had to be irradiated by light sources. Using natural light sources such as Sun light was not possible, because parameters of the radiated Sun light are not stable in time and space as well. To insure the stable light source in time and space, artificial light source had to be used. For artificial light source so many lights, lamps can be considered. In cooperation with Institut für Energie- und Antriebstechnik at the Hochschule Ulm, Ulm, Germany, a professional and calibrated powerful flash light source had been chosen. The powerful flash light was a Pasan-Flasher flashing radiation light source, as Sun simulator device. In the next paragraph the Pasan-Flasher system will be introduced shortly.

In one of the laboratory of Institut für Energie- und Antriebstechnik at the Hochschule Ulm, in Ulm, in Germany a Photovoltaic Flasher, or in other words a Pasan-Flasher system with the accuracy class AAA+ is available. Marking AAA considers: homogeneity of the solar irradiation on the test field; homogeneity of the solar irradiation in the measuring time; conformity of the spectrum of the irradiated light with the AM 1.5 (air mass of 1.5) spectrum of the sunlight. The flasher consists of a flashlight with its power supply and a solar module characteristic measurement unit. [5]

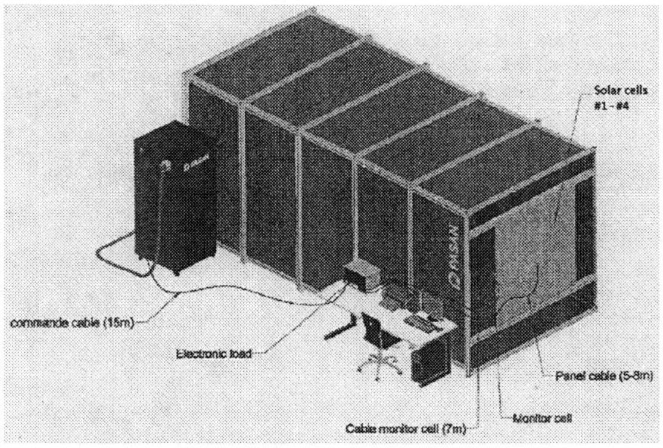

Figure 2. Schematics of the Pasan-Flasher system with its most important parts. Our solar cells have been set up to the position "Solar cells #1-#4. as it is shown in the Figure." [5]

The flash light part of the system creates a homogeneous light pulse of 1000 W/m^2 with a duration of 10 ms onto a test field of 2 m × 2 m. Taking a test object – in our case the solar cells – to the test field of the device, nice and precise irradiation measurements can be made. Using the flasher device, intensity of the irradiating flash light can be modified by using filter series and by varying power of the light source. With pushing in/out different filters in front of the flash light source intensities of 200 W/m^2, 400 W/m^2, 700 W/m^2, 800 W/m^2, are available at the test filed. With the variation of the light power or intensity of the built in light source intensity of 900 W/m^2 is also available at the test field. Naturally, irradiation intensity of 1000 W/m^2 is also available as standard irradiation intensity of the device. Choosing the necessary filter, necessary irradiation could be set. The flasher and testing field parts of the system are installed in a huge, room-size black box. The black box is used as shielding box, shielding the measurements from any natural or scattered lights originating from outside. [5]

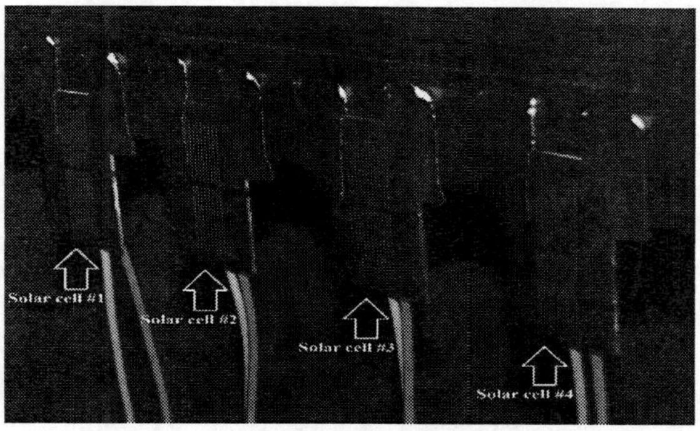

Figure 3. Photograph of the experimental setup of our solar cells #1-#4. The cells were fixed by black color duct tapes at the test field of Pasan-Flasher system in the black shadowing box.

Another part of the Photovoltaic Flasher is the photovoltaic module characteristic measurement unit, which holds some fast electronics as measurement chain and a personal computer (PC) with a special software. With the help of the electronics chain ohmic loads can set to the test cell. The fast electronics can modify the ohmic loads on the tested solar cell very quickly. During 10 ms hundreds of ohmic loads on the solar cell can be applied for the measurement. The experimenter can set the load chain with programming the PC software and run the measurement from the PC. The cells as the targets are set as Figure 2. and Figure 3. shows it. The tested solar cell is connected to the system by electric wires. During the flasher light is flashing, the ohmic loads are uploading to the output of testing solar cell and this way electric current-voltage (I-U) and electric power-voltage (P-U) plots can be measured. One or two seconds after the flash and the measurement, the mentioned plots are plotted to the monitor screen of the software (see in Figure 4.). The plots and the measuring data can also be recorded in a defined directory on the hard disk of the PC. Later, data files can be analyzed by any data analyzer target software. Figure 3. introduces how the photovoltaic cells have been fixed at the test field during the experiment. Figure 4. shows a monitor snapshot taken after one experimental run.

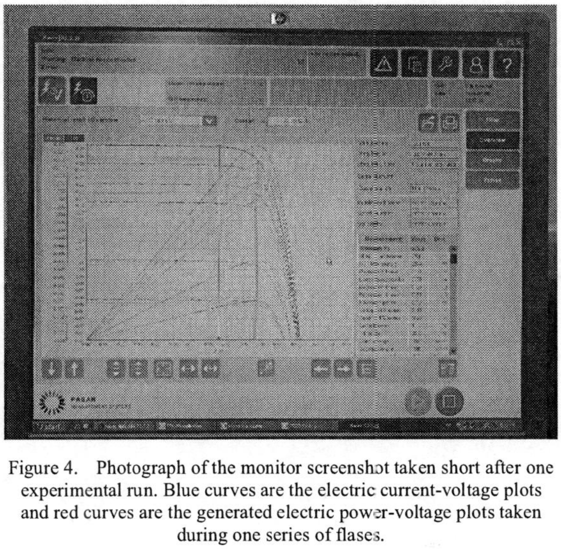

Figure 4. Photograph of the monitor screenshot taken short after one experimental run. Blue curves are the electric current-voltage plots and red curves are the generated electric power-voltage plots taken during one series of flases.

In the next chapter, the plots after analyzing data will be shown.

IV. Measured Electric current-Voltage (I-U) and Electric power-Voltage (P-U) plots

A. Electric current-voltage characteristics of solar cells #1-4.

After finishing the experiments, we recorded all the measured data to the hard disk and started to analyze them. In order to investigate the measured values, the raw data have been read in the Microsoft Excel software. After defining the appropriate columns in the software, the data could be plotted out easily. In the coming part our plotted curves will be introduced. Data belongs to the experiments were the test photovoltaic cells have been irradiated by the flash light intensities of three times 1000 W/m^2, one time 900 W/m^2, 800 W/m^2, 700 W/m^2, 400 W/m^2, 200 W/m^2, respectively. In this paragraph, the plots are separated according to the intensity of the irradiated light. Starting from Figure 5. to Figure 10. show the characteristics for the dedicated case.

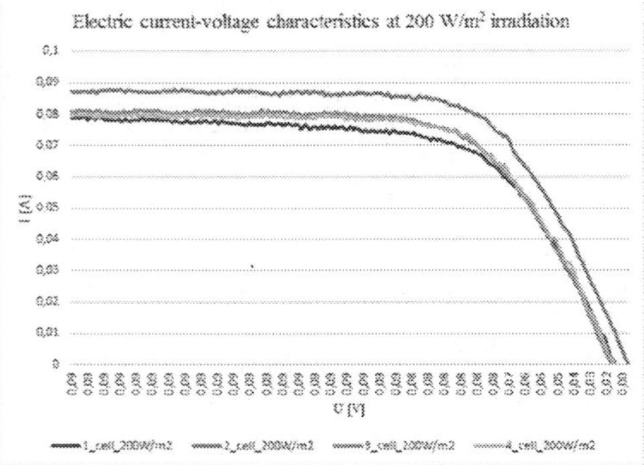

Figure 5. Electric current-voltage plots for cells #1- #4 at 200 W/m^2 flasher irradiation.

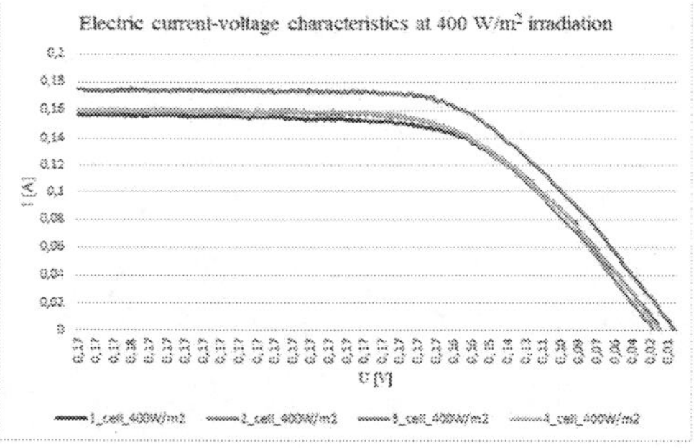

Figure 6. Electric current-voltage plots for cells #1- #4 at 400 W/m^2 flasher irradiation.

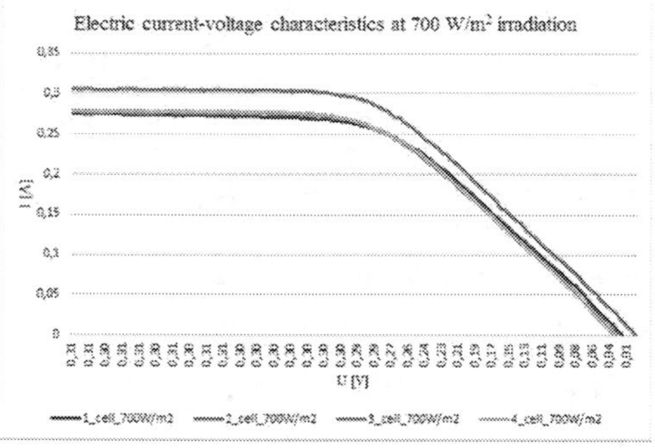

Figure 7. Electric current-voltage plots for cells #1- #4 at 700 W/m^2 flasher irradiation.

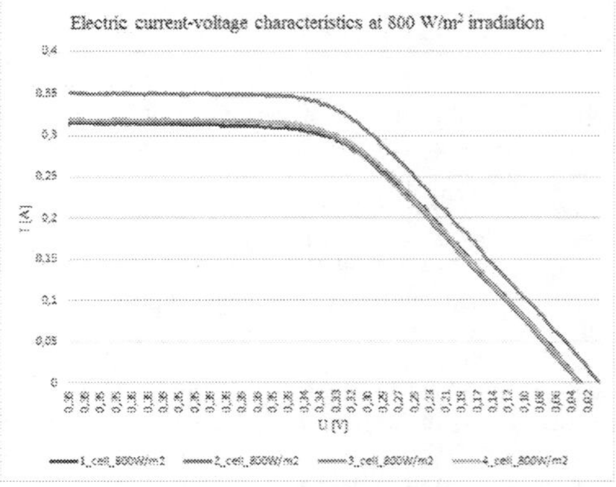

Figure 8. Electric current-voltage plots for cells #1- #4 at 800 W/m^2 flasher irradiation.

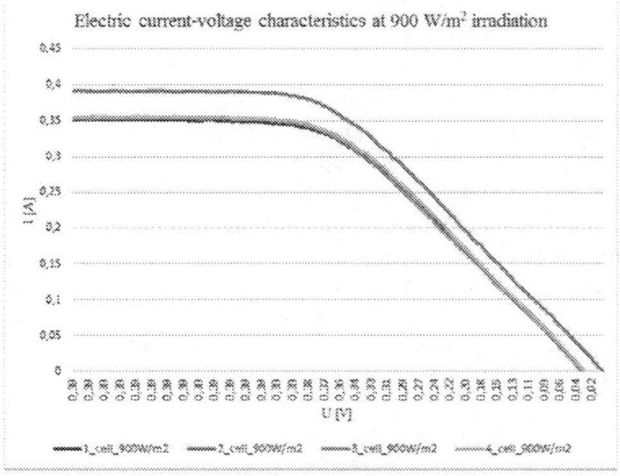

Figure 9. Electric current-voltage plots for cells #1- #4 at 900 W/m^2 flasher irradiation.

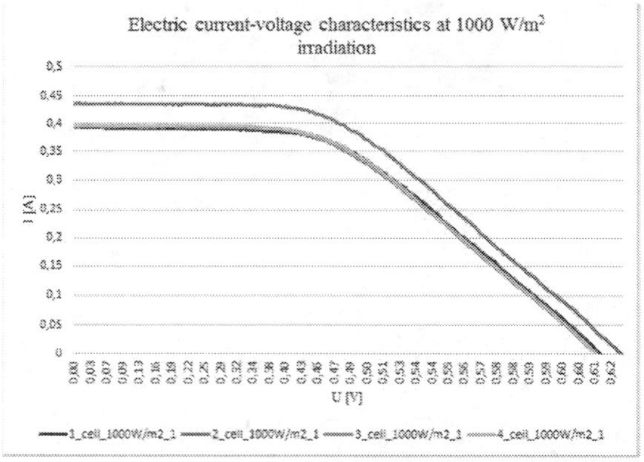

Figure 10. Electric current-voltage plots for cells #1- #4 at 1000 W/m^2 flasher irradiation.

B. Electric power-voltage characteristics of solar cells #1-4.

In this part, the generated electric power-voltage characteristics for photovoltaic cells #1 to #4 recorded at the irradiation intensities of three times 1000 W/m^2, one time 900 W/m^2, 800 W/m^2, 700 W/m^2, 400 W/m^2, 200 W/m^2 (written here according to the series of the experimental runs) will be introduced. In this paragraph, the plots are separated according to the intensity of the irradiated light, again. Starting from

Figure 11. to Figure 16. show the characteristics for the dedicated case.

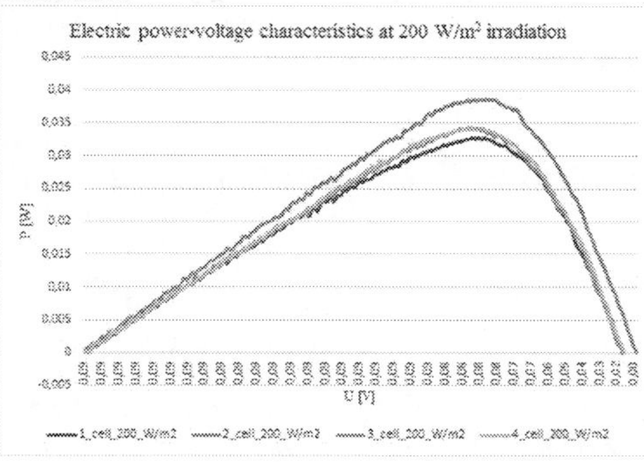

Figure 11. Generated electric power-voltage plots for cells #1- #4 at 200 W/m^2 flasher irradiation.

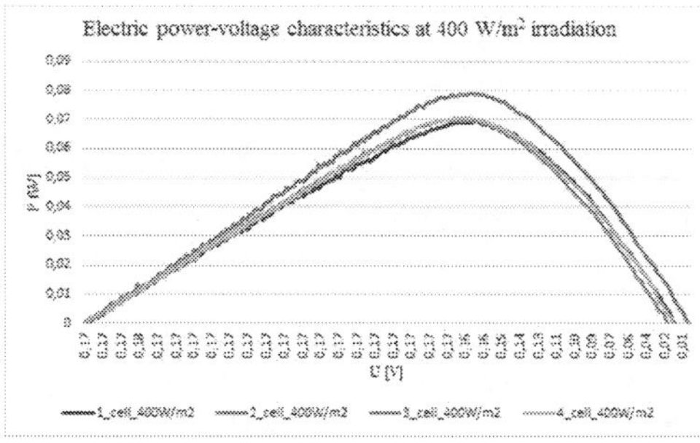

Figure 12. Generated electric power-voltage plots for cells #1- #4 at 400 W/m^2 flasher irradiation.

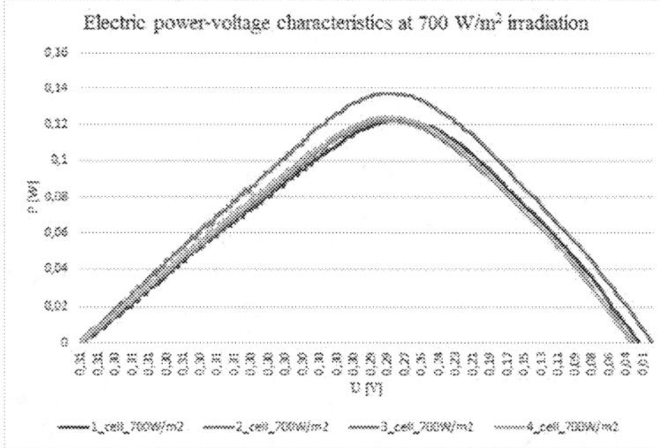

Figure 13. Generated electric power-voltage plots for cells #1- #4 at 700 W/m^2 flasher irradiation.

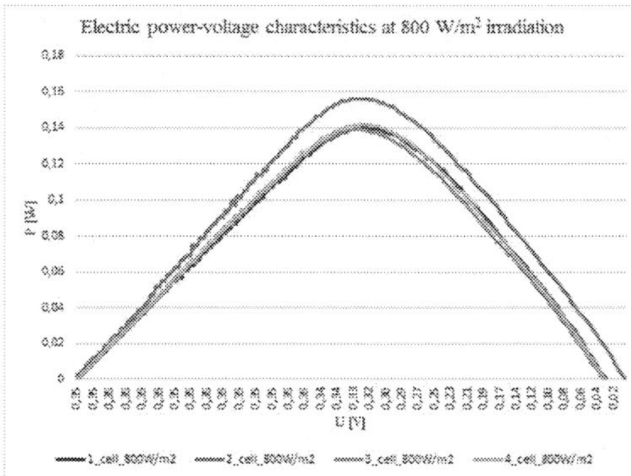

Figure 14. Generated electric power-voltage plots for cells #1- #4 at 800 W/m^2 flasher irradiation.

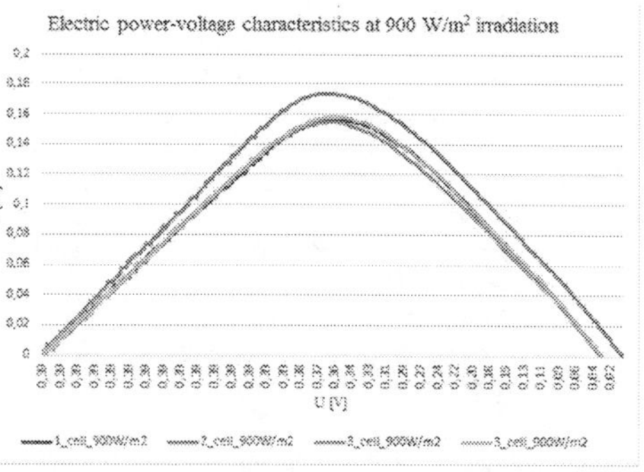

Figure 15. Generated electric power-voltage plots for cells #1- #4 at 900 W/m^2 flasher irradiation.

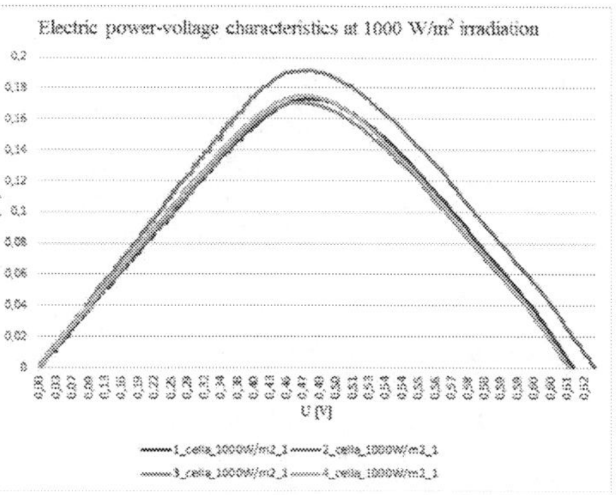

Figure 16. Generated electric power-voltage plots for cells #1- #4 at 1000 W/m^2 flasher irradiation.

V. Results, Analyzation of the Results

A. Electric current-voltage characteristics of solar cells #1-4.

Analyzing the plotted curves the following conclusions can be determined. For better clarity and reading the concluded results will be itemized here.

- Seeing in Figures 5-10. we can see that curves belong to solar cell #1 run at the bottom in every Figures. It means that, the lowest electric current could be generated at the valued electric voltage. It might mean that, the electric resistance for solar cell #1, might be the highest from all the investigated solar cell #1-#4. For solar cells #1 two runs with climate chamber aging periods were made. Comparing the results with other solar cells we might say that, **the climate chamber test (periodic temperature change) might cause the increase of the electric resistance of the photovoltaic cell.**
- Seeing in Figures 5-10. we can see that curves belong to solar cell #3 and solar cell #4 run very close to each other, or in other words, these curves are almost overlapping each other. It means that, electric current-voltage (I-U) plots are almost the same in the investigated voltage range. Because photovoltaic cell #4 was the reference cell, the result may mean that, **one run with climate chamber test and 2.5 runs with Kesternich chamber tests were not enough**

impact or influence to modify the electric resistance of the solar cell.

- Seeing in Figures 5-10. we can see that curves belong to solar cell #2 run at the top of the graph. This result came out from the experiments at all irradiances, and the result seems to be significant. The result shows that, the electric currents have the highest values at the given voltages in the investigated voltage range. In other words, in this case the electric resistance of the solar panel may have the lowest value. Summarizing the observations here, the result might mean that **two runs with Kesternich chamber tests did not affect, or it is better to say that the Kesternich chamber test could not decrease the electric resistance of the solar panel.**

B. Electric power-voltage characteristics of solar cells #1-4.

Analyzing the plotted P-U curves the following conclusions can be determined. For better clarity and reading the concluded results will also be itemized here.

- At lowest, 200 W/m^2 irradiations, the P-U plot belongs to cell #1 runs at the bottom, but in case of higher irradiances all plots belong to cell #1, cell #3, and cell #4 run very close to each other, or the curves are almost overlapping each other on the Figures (see blue, grey and light yellow curves in the Figures).
- Significant observation is that, the **generated electric powers (P) show the highest values at the function of measured voltages (V) at the whole range in case of solar cell #2.**
- Also, significant observation is that the maximum electric power could be generated in case of the use of photovoltaic cell #2. (See orange plots in the Figures). Cell #2 could produce the highest electric power during the experiments. Highest generated power is consistent with the highest currents could be generated in the same solar cell. $P=U \cdot I$, so this way power plots and electric current-voltage plots are consistent with each other.

  The results observed here are correlated with the results taken at point *A* in this chapter.

VI. Summary

Summarizing our results the followings can be written:

Two runs with Kesternich chamber tests may could decrease the electric resistance of the material of the photovoltaic cell. So, medium of SO_2 vaporized air could decrease the resistance of the cell. It may go back to reasons taken from the field of material science. One of the next goals of our next experiments could be that to investigate reasons of material reason of the measured data.

Two runs with the climate chamber tests may could increase the electric resistance of the photovoltaic cell. Reason of this observation may go back to the material sciences as well, so the structure of the solid-state Silicon – this is the base material of our photovoltaic cells – may change with the significant and fluctuated change of the temperature of the cell material. -20 °C - +80 °C can cause modification in the solid crystalline state in the monocrystalline material. +80 °C might be to high temperature for the material. This temperature value might affect the crystalline structure and may cause modification in electric resistance. In order to give correct answer the suspicion written above, new experiments will be designed and theoretical models have to be ran in the future. [6], [7]

Acknowledgment

Authors for Hungary say great thank you for Hochschule Ulm, Institut für Energie und Antriebstechnik, Ulm, Germany for using their experimental apparatuses, for the great help and assistance provided during the experiments and for the long-term friendly cooperation.

Authors gratefully say thanks for the support of the Óbuda University, Doctoral School of Applied Informatics and the Alternative Energy Sources Knowledge Center.

References

[1] M.A.. Green, K. Emery, Y. Hishikawa, W. Warta, E.D. Dunlop, "Solar cell efficiency tables (Version 45)". Progress in Photovoltaics: Research and Applications 2014; 20: 12–20.

[2] CON 300-FL AIR/SO_2 Kesternich test chamber users manual

[3] ACS Discovery DY110C climate test chamber users manual

[4] Zs. Csegezy, "Investigation of power change of photovoltaic cells using climate and Kesternich chamber test" Óbuda University, Kandó Kálmán Faculty of Electrical Engineering, name of supervisor: Andrea Varga, pp. 14-18, June 2016. written in Hungarian

[5] E. R. Neuchel, "Measurement of the characteristics of PV modules", Dezentrale Energiesysteme, Laborversuch, Institut für Energie- und Antriebstechnik, Hochschule Ulm, Ulm, Germany, July 2016, pp.I.-1.

[6] A. Skoczek,T. Sample, E. D. Dunlop, "The results of performance measurements of field-aged crystalline silicon photovoltaic modules", Progress in Photovoltaics, Volume 17, Issue 4, June 2009, pp. 227–240., 24 December 2008., DOI: 10.1002/pip.874.

[7] E. Skoplaki, J.A. Palyvos, "On the temperature dependence of photovoltaic module electrical performance: A review of efficiency/power correlations", Solar Energy (Elsevier), Volume 83, Issue 5, May 2009, Pages 614–624, doi:10.1016/j.solener.2008.10.008

Electricity Markets Integrations – What is the Current Status and Future Outlook of Bidding Zones Reconfiguration?

Tomáš Králík, Július Bemš, Oldřich Starý
Department of Economics, Management and Humanities,
Faulty of Electrical Engineering, Czech Technical University in Prague
Prague, Czech Republic
tomas.kralik@fel.cvut.cz, bemsjuli@fel.cvut.cz, staryo@fel.cvut.cz

Abstract — **The aim of this paper is to summarize the current status of the EU Electricity Market Integration, especially address the highly discuss issue of Bidding Zones Reconfiguration, analyse proposed criterions for the potential reconfiguration and also discuss possible future development.**

Speaking of Bidding Zones Reconfiguration the mainly discussed issue is the German-Austria-Luxemburg trading zone (DE-AT-LU) where the problems with internal congestion occur and correspondingly affect surrounding bidding zones, especially Czech Republic, Slovakia and Poland. Transmission capacities inside this zone are insufficient and unscheduled electricity flows through neighbouring grids and constrains transmission capacities for local market participants. This has resulted, among others, into installation of phase shifting transformer at Czech-German and Polish-German borders. There is therefore an urgent pressure to solve this situation and most promising and investigated option is to split the DE-AT-LU zone in to two and even the Germany's power grid regular address all 4 Germany TSO to prepare the introduction of capacity allocation on the border with Austria. On the other hand, there are arguments that go against this and they support the idea of even larger bidding zones (further integration) rather than division of current bidding zones (that is especially the case of DE-AT-LU zone). These arguments are mainly based on the liquidity of the market. If divided there is also a problem of current traders´ open positions on derivate contracts (within DE-AT-LU zone these contracts are estimated at over €20 bn.).

This article is therefore investigating the whole process of proposed splitting process with the focus on presented criterions that should be used to find the optimal configuration. The special focus is paid to the applicability of social welfare as one of the "unclear" criterion that has the potential to strongly influence the whole decision making process.

Based on the analyses we can conclude that the main approach to the process of reconfiguration is solid however, the results will strongly depend on the final settings (limiting values) of individual criterions and final weights of each criterion. Without this information, the whole process can be rightfully opposed by any involved participant. (Abstract)

Keywords: bidding; zones; energy; markets; social; welfare

I. Introduction

A. Problem Description-current situation

The integration of European Electricity Market is on ongoing process. It has been effectively started by the First Energy Package (1996) and the main aim was to develop common effective market platform. Unfortunately the market integration has also shown some drawbacks. One of the biggest problems has been identified within the Central East Europe (CEE). Businesses operation on the electricity market in the CEE Europe Region, mostly in geographical location of Czech Republic and Poland, meet the problem of insufficient electricity transmission capacities on the border with Germany. [1], [2] Available transfer capacities are severely limited especially due to the influence of so-called loop-flows (flows within the Germany-Austrian (DE-AT) bidding zone, which flow through the transmission system of neighbouring countries). This situation is caused mainly by following reasons:

- very high generation capacities based mainly on intermittent RES located in the northern part of Germany,
- lack of transmission capacities from northern part to southern part of Germany (where are located significant consumption capacities),
- Energiewende program that will result into full phase out of nuclear power plants in Germany [3], which will lead to intensification of distribution of electricity generation and consumption.

The affected business entities (electricity traders) are currently losing business opportunities and their losses are not compensated [4], [5]. Loop-flows are unscheduled power flows and currently are almost not a subject to payments for usage of transmission system. This situation is caused mainly by insufficient transmission capacities in Germany. However construction of new power transmission lines has long delay (especially thanks to the fact that the major parts of new transmission lines must be put underground) and the forecasted cost will be significantly higher (up to four times more).[6]. This means that current situation will not be solved (from the technical point of view) in the short term period.

The most promising and effective solution of this problem can be seen in splitting of Germany-Austrian bidding zone into several smaller bidding zones. Current situation is profitable for participants from Germany-Austrian bidding zones because cheap electricity from wind power plants from northern Germany can be transferred to southern Germany and Austria without payment for transmission fees. Whole Germany-Austrian bidding zone benefits from lower electricity prices. If this zone were divided, there would become a difference in electricity price between new zones (e.g. northern and southern Germany) [7]–[9]. This price difference would be caused by requirement of transmission capacity allocation and related costs. Czech and Polish transmission system operators would obtain revenues from transmission capacity utilization. Moreover, companies (mostly electricity traders) from Czech Republic and Poland could take place in capacity allocation mechanism and they could have opportunities of trading abroad.

On the other hand there are also efforts from the electricity traders to maintain the current configuration of DE-AT bidding zone. The main concern is the negative effect of zonal splitting to forward market liquidity, wholesales market power and retail competition. The European Federation of Energy Traders (EFET) issued, among others, an investigation [10] describing what happened in Sweden after splitting into smaller bidding zones in 2011 from the electricity traders point of view. Even though Swedish Energy Market Inspectorate stated that the reconfiguration had not any major negative effects, EFET argues that based on the hard data evidence from Sweden there, in fact, were negative effects in terms of the amount of future contracts traded on Nasdaq as well as the drop in the liquidity. Based on this evidence EFET strongly suggest that economic consequences have to be taken into account.

This situation can serve as a great example of contradictory views on current discussion about DE-AT bidding zone reconfiguration. Therefore it is obvious that even though the technical criterion will play significant role in the potential splitting process, economic and social criterion must be taken also in to account - the right question is how and to what extent.

B. Historical Background

Bidding zones theory was developed in 90s in US by [11]–[16] as the consequence of power grid congestion (insufficient electricity transmission capacities). Market liberalization and integration process in Europe begun significantly later.

Bidding zone can be defined as an area without internal business congestion. It means that transaction can be completed between any two points inside this area and electricity can be transferred without requirement of transmission capacity allocation. Bidding zones borders in many countries are the same as the political borders because countries were more isolated in the past than in present time. This situation is changing with the increase of international cooperation. Countries like Czech Republic, Slovakia, Poland, Hungary and France are examples of bidding zones that are identical with political borders. On the other hand, there is a single bidding zone that includes Germany, Austria and Luxembourg. [17]

Germany-Austrian (DE-AT) bidding zone was established in 2005, but the process of its legal origin is not almost described. We assume that DE-AT bidding zone does not fulfil condition for being full independent bidding zone. The main reason is that DE-AT zone cannot be understood as "copper plate" due to the congestion of transmission lines. ACER (Agency for the Cooperation of Energy Regulators) has adopted a legally non-binding opinion to split the German-Austrian single bidding zone. However this opinion could not start the splitting process (especially when Austria is strongly against) therefore the comprehensive bidding zone review is being done by European Network of Transmission System Operators (ENTSO-E). It should be finished by the end of 2017 and if it supports the break-up ACER, according to declaration of head of wholesale markets at the European Commission, should be responsible *"to define a new split model against the wishes of Austria"*.[18]

Once the splitting process is started there is also a question how to settle current open positions of electricity traders. It is estimated that these positions on DE-AT border are worth up to €20 - 25bn. According to [18] there are three potential options:

- *"The introduction of new futures products specifically for Germany or Austria, with existing positions rolled into the new ones. "*
- *"The introduction of a system price – as is used in the Nordic and Italian markets – which would be used to settle against, and would be generated by trade in the two separate price zones."*
- *""An "adjustment" of the existing Phelix futures to reflect the price zone split."*

The whole process is therefore lying on the results of above mentioned bidding zone review. It is obvious that the main focus is will be on defining critical grid elements that violate the "copper plate" principle. To do that the whole European electricity grid has to be modelled and therefore the following section is devoted to the main principles of electricity flows calculation as a primary manner for the identification of congested grid elements.

II. Core Idea

The main reason for bidding zones reconfiguration is congestion of power lines resulting in unscheduled power flows in neighbouring grids. Central European power grid was recently modelled in [2] and results present congested power lines. Comprehensive review of bidding zones definitions, history, development in Europe and criteria for reconfiguration, were introduced in [17] and its brief summary was presented in introductory section.

There are several issues related with calculations of power flows, identification of unscheduled flows and identification of critical branches.

1. Power flows in grid are usually decomposed using Kirchhoff laws, however other method "Natural Flows" (NF) was used in material [19]. It is necessary to compare results of power flow decomposition (PFD) to NF.
2. The other issue is finding relevant and critical power grid elements required for power flows calculation. European grid is large and using all elements would lead to very complex calculation. Rational aggregation and simplification will be needed.
3. To be able to find relevant grid elements, their load and congestion has to be reported in meaningful way. Load of the grid elements is changing in time. Critical load can last short or long time, it can occur rarely or frequently.

Based on above specified issues, part of this article will be devoted to description of grid element load and selection of relevant elements for further analysis.

III. Grid Element Load Description

The aim is to find a set of statistics and tools that accurately describe the load of the elements while being simple to understand, compute, and interpret.

In the statistical representation of the elements load, we assume that we will have a time series of physical flows on grid elements. To describe the situation on an element, we propose a combination of the following statistics.

1. Arithmetic mean or other measure of central tendency (modus, median). Describes the predominant direction of electricity flow. In case of zero average, we can use median or modus. If there were a situation when the average and the median were zero, it can be concluded that there is no predominant flow direction.
2. Standard deviation. It is possible to estimate the changing intensity of the flow of the element.
3. Maximum and minimum flow value.
4. Variation coefficient. It is defined as the ratio of standard deviation and arithmetic mean. This is a standardized standard deviation, making statistics comparable between calculations on lines with different capacities. The problem may occur in situations where the average is close to zero, because the value of the coefficient may be high. In this case, we can use absolute flow values.
5. Quantiles (10%, 90%, possibly others) and the quantile in which the sign of the flow changes (if it changes). This characteristic provides information on the statistical distribution of flows. It can serve to determine the likelihood of changing flows.
6. The number of changes in the sign of the flow over a period.
7. The number of exceedances of the predetermined flow threshold value on the element. Circular flows cause flow fluctuations on the network element, and we can identify how often this fluctuation occurs when the threshold value is set correctly.

IV. Relevant Grid Elements Selection

In this section, we focused on appropriate selection of relevant elements for loop flow analysis. In our opinion, it makes sense to monitor the frequency and intensity of flows occurring on a network element. An undesirable state can be understood as exceeding a set flow limit on an element.

1. The limit can be defined as:
   a. The percentage of the nominal load.
   b. The percentage of the load in a base scenario.
   c. The percentage of the difference between nominal load and load in a base scenario. This approach is appropriate if we assume that loop flows are not harmful if there is free capacity for them on specific grid element.
   d. The standard deviation (or its multiple) from the average value of the load in a base scenario.
2. Evaluation of limit exceeding:
   a. Simple (binary) cross-border information.
   b. The extent of the limit exceedance (absolute, relative).
   c. Duration of the limit exceedance.
   d. Intensity of the limit exceedance as the combination of exceedance extent and duration. We may also consider the product of the power and duration, i.e. energy, which flew through grid elements if limit was exceeded.
3. Determine the occurrence of undesirable states, e.g. period of year (number or percentage of 8760 hours per year) in order to consider grid element as relevant in further calculations.

Similarly, N-1 rule violation could be monitored. The number of hours (or percentage) of 8760 hours per year in which the N-1 rule was violated can give us also good information about relevancy of specific grid elements. The only problem is the data availability, because we would need large amount of data besides power flows. We think that the appropriate combination of suggested approaches can be used to select relevant elements for loop flow analysis. Specific combination of above-mentioned approaches (rules) can be found after deeper statistical analyses of reasonable amount of data.

Identification of relevant grid elements is only first (but very important) step towards the setting of criterion for bidding zone reconfiguration. However it is already known that the final decision will be made upon set of criterion with different weights. The most of the criterion result from a guideline on Capacity Allocation and Congestion Management (CACM) [20] and are:

- Congestion Rent
- Remedial Actions
- Difference in Marginal Prices
- Price Volatility
- Transition Costs
- Social Welfare

Detailed description of above mentioned criterion is discussed in [17]. From our point of view the most "problematic" criteria is the social welfare.

V. Social Welfare Aplication and Conclusion

One of the most robust economic criteria is the social welfare. As defined in [17] the social welfare "*is the concept based on evaluation of all possible effects resulting either from some action (e.g. creation of new interconnecting capacity between two regions) or consumption of given commodity or service. According to the economic theory, social welfare is a measure of well-being of the entire society. In relation to given action or commodity (or service) consumption social welfare can be understood as the sum of effects resulting from the consumption. It can include direct and indirect economic impact.*"

Unfortunately, there is no unequivocal definition of social welfare and calculation of social welfare is very often (particularly in bidding zones reconfiguration) simplified to calculation of economic welfare. Economic welfare is defined as sum of consumers' and producers' surpluses and in the context of bidding zones was described in [21]. Prices of electricity have strong impact on whole society because electricity is used widely and industrial production is very sensitive on electricity price. Economic welfare calculation does not include secondary (indirect) impacts on society and it is assuming homogenous region (e.g. Central Europe or Central and Eastern Europe). The other problem is purchase power in countries with different economic development [22]. Increase or decrease of electricity prices by one euro has different impact on German residents and Czech residents. This discrepancy should be reflected in social welfare calculation [17].

Even though the social welfare definition is still unclear, there are efforts to include social welfare criterion into current decision processes. We do not cast doubt on the main ideas of social welfare as a solid criterion, but we do believe it is necessary to continue discussions and research regarding achievement of clear definition including methodology for identification and calculation of all necessary inputs. Moreover, according to correlation between Social Welfare, Congestion Rents and Remedial Action criteria, it is possible to conclude that any optimization using Congestion Rents and Remedial Action criteria should lead also to increasing of social welfare.

Based on the materials that have been already published regarding the proposed criterion for bidding zones evaluation, we do believe that the best way is to keep all criteria as simple as possible with very unequivocal definition. Only in such case, results can be taken as a solid foundation for future reconfiguration of bidding zones in CEE region.

REFERENCES

[1] ČEPS, "Critical Situation in the Transmission System of the Czech Republic as a Result of Enormous Production in Wind Parks in Germany at the Turn of 2014/2015," 2015.

[2] K. Janda, J. Malek, and L. Recka, "Influence of Renewable Energy Sources on Transmission Networks in Central Europe." 2016.

[3] REUTERS, "Germany to phase out nuclear power: deputy minister | Reuters," 2011. .

[4] J. Morecroft, "GERMAN POWER LOOP FLOWS THREATEN EU'S INTERNAL MARKET: REGULATORS," 2012. [Online]. Available: http://www.platts.com/latest-news/electric-power/brussels/german-power-loop-flows-threaten-eus-internal-8878410. [Accessed: 12-Dec-2015].

[5] V. Szabo, "The German–Austrian bidding zone and its compatibility with the free movement of goods in the EU," *J. World Energy Law Bus.*, vol. 10, no. 1, p. jww034, Dec. 2016.

[6] "RPT-INTERVIEW-German transmission line delays to curb wind power use | Reuters." .

[7] J. Egerer, J. Weibezahn, and H. Hermann, "Two price zones for the German electricity market — Market implications and distributional effects," *Energy Econ.*, vol. 59, pp. 365–381, 2016.

[8] Consentec, "Economic efficiency analysis of introducing smaller bidding zones," 2015.

[9] K. Trepper, M. Bucksteeg, and C. Weber, "An integrated approach to model redispatch and to assess potential benefits from market splitting in Germany," *EWL Work. Pap.*, 2014.

[10] E. F. of E. Traders, "A reality check on the market impact of splitting bidding zones," 2016.

[11] W. W. Hogan, "Contract networks for electric power transmission," *J. Regul. Econ.*, vol. 4, no. 3, pp. 211–242, 1992.

[12] R. D. Tabors, "A market-based proposal for transmission pricing," *Electr. J.*, vol. 9, no. 9, pp. 61–67, Nov. 1996.

[13] S. Walton and R. D. Tabors, "Zonal transmission pricing: methodology and preliminary results from the WSCC," *Electr. J.*, vol. 9, no. 9, pp. 34–41, Nov. 1996.

[14] S. Stoft, "Transmission pricing zones: simple or complex?," *Electr. J.*, vol. 10, no. 1, pp. 24–31, Jan. 1997.

[15] W. W. Hogan, "Competitive Electricity Market Design: A Wholesale Primer," 1998.

[16] S. Stoft, "Congestion Pricing with Fewer Prices Than Zones," *Electr. J.*, vol. 11, no. 4, pp. 23–31, May 1998.

[17] J. Bemš, T. Králík, J. Knápek, and A. Kradeckaia, "Bidding zones reconfiguration - Current issues: Literature review, criterions and social welfare," in *Proceedings of the 2nd International Conference on Intelligent Green Building and Smart Grid, IGBSG 2016*, 2016.

[18] "Review could end German-Austrian power bidding zone split issue." [Online]. Available: https://www.icis.com/resources/news/2017/02/08/10077228/review-could-end-german-austrian-power-bidding-zone-split-issue/. [Accessed: 05-Apr-2017].

[19] Vukasovic Milan, "Simple nodal-PTDF Deviation Method: Solution for the Assessment of Unallocated and Unexpected Flows and Redispatch Cost-Sharing among TSOs in the European Zonal Market Design?," *Int. J. Power Renew. Energy Syst.*, vol. 2, no. 2, p. 8, 2015.

[20] European Commission, "Commission Regulation (EU) 2015/1222 of 24 July 2015 establishing a guideline on capacity allocation and congestion management," 2015. [Online]. Available: http://eur-lex.europa.eu/legal-content/EN/TXT/?uri=uriserv:OJ.L_.2015.197.01.0024.01.ENG. [Accessed: 07-Jan-2016].

[21] I. Andročec and S. Krajcar, "MECHANISMS FOR EFFICIENT INVESTMENTS AND OPTIMAL ZONES IN REGIONAL POWER MARKET," *Therm. Sci.*, vol. 18, no. 3, pp. 755–770, 2014.

[22] M. Czajkowski and M. Ščasný, "Study on benefit transfer in an international setting. How to improve welfare estimates in the case of the countries' income heterogeneity?," *Ecol. Econ.*, vol. 69, no. 12, pp. 2409–2416, 2010.

An Alternative Method of Evaluating Voltage Deviations with Regard to the Quality of Electricity in Distribution Networks

Richard Velicka, Pavel Santarius
Department of cybernetics and biomedical engineering
VSB – Technical University of Ostrava
Czech Republic
richard.velicka@vsb.cz, pavel.santarius@vsb.cz

Petr Krejci, Tadeusz Sikora
Department of electrical power engineering
VSB – Technical University of Ostrava
Czech Republic
petr.krejci@vsb.cz, tadeusz.sikora@vsb.

Abstract—**In the revision of the standard ČSN EN 50160, tightening the evaluation of supply voltage deviations in low-voltage networks is considered. As part of the verification of possible methods of evaluating supply voltage deviations in low-voltage networks, yearly monitoring of voltage deviations was performed at selected locations of low-voltage distribution networks. From the measured values, 95% and 99% deviation values were evaluated and compared in time intervals of 1 minute and 10 minutes.**

Keywords—distribution network, voltage quality, supply voltage, voltage deviation, voltage quality monitor

I. Introduction

Power quality problems are closely connected with gradual development technologies for future smart grids. Power quality issues are very topical in national grids range and of course in local microgrids and off-grids systems [1-4]. The recent period is characterized by efforts to tighten the conditions for the supply of electricity, especially the limits on some parameters of the supply voltage. The most discussed issue relates to stricter limits on supply voltage deviations, which is obviously both technical and economic problem. The research department of the Faculty of Electrical Engineering and Computer Science at the VSB - Technical University of Ostrava, in cooperation with ČEZ Distribuce, a.s., therefore verified the possible ways of evaluating voltage deviations in real distribution networks [5].

II. Parameters of Electric Power Quality

The basic standard that defines the quality parameters is ČSN EN 50160 (ed. 3) "Voltage characteristics of electricity supplied from public distribution networks". This standard also represents a basis for "Rules for the operation of distribution networks" (Annex 3 - Quality of electricity in the distribution system, methods of its detection and evaluation).

Regarding low-voltage networks, the quality parameters of the supplied electricity are defined in standard ČSN EN 61000-2-2 Compatibility levels for low-frequency conducted disturbances and signaling in public low-voltage power supply systems.

In terms of measuring the quality parameters of supplied electricity, the decisive standard is ČSN EN 61000-4-30 - Power quality measurement methods [6]. This technical standard deals with the properties of devices of Class A and B in terms of the measurement methodology used to measure electricity quality parameters.

According to standard ČSN EN 50160, characteristic voltage parameters for low-voltage networks are as follows:

A. *Magnitude of supply voltage*

Standardized nominal voltage U_n for the public low-voltage network is $U_n = 230$ V, either between the phase and node or between phases.

(For HV and VHV, the voltage magnitude is determined by the agreed supply voltage U_c).

B. *Supply frequency*

Nominal supply voltage frequency is 50 Hz. Under normal operating conditions, the mean value of fundamental harmonic frequency measured at intervals of 10 sec must be in the following ranges – in systems with synchronous connection to the interconnected system:

- 50 Hz ± 1 % during 99.5 % of the year,
- 50 Hz +4 %/-6 % at 100 % of the time.

C. *Supply voltage deviations*

Under normal operating conditions, except the interruption of supply, 95 % of the average effective values (rms values) of supply voltage at measuring intervals of 10 minutes must be in the range of $U_n \pm 10$ % during each week.

Except voltage drops, interruptions and overvoltage, all the average effective values of supply voltage at measuring intervals of 10 minutes must be within

U_n +10 %/ - 15 %. However, in cases of power supply in areas with long low-voltage lines, they can sometimes be in the range of U_n +10 %/-20 %. Consumers should be informed about this fact.

D. Rapid voltage changes

Under normal operating conditions, rapid voltage changes do not generally exceed 5 % of U_n; under certain circumstances, however, rapid voltage changes up to10 % of U_n can occur several times a day.

Under normal operating conditions, in any weekly period, the long-term rate of flicker perception must be $P_{lt} \leq 1$ at 95 % of the time.

E. Supply voltage imbalance

Under normal operating conditions, in any weekly period, 95 % of ten-minute mean effective values of negative component (basic) of supply voltage must be in the range of 0 - 2 % of positive component.

F. Harmonic voltages

Under normal operating conditions, in any weekly period, 95 % of ten-minute mean effective values of the voltage of each harmonic must be less than or equal to the value listed in Table 1. For each harmonic, resonances may cause higher voltages.

TABLE I. THE LEVELS OF HARMONIC VOLTAGE IN PERCENTAGE OF U_1

Odd harmonics				Even harmonics	
Not multiples of 3		Multiples of 3			
Harm. order h	*Harm. voltage U_h*	*Harm. order h*	*Harm. voltage U_h*	*Harm. order h*	*Harm. voltage U_h*
5	6 %	3	5 %	2	2 %
7	5 %	9	1.5 %	4	1 %
11	3.5 %	15	0.5 %	6...24	0.5 %
13	3 %	21	0.5 %		
17	2 %				
19	1.5 %				
23	1.5 %				
25	1.5 %				
Note: Levels for harmonics of orders higher than 25 are not shown because they are usually small but difficult to predict due to resonance phenomena.					

Furthermore, the total harmonic distortion (THD) [7-10], of the supply voltage (including all harmonics up to order of 40) must be less than or equal to 8 %.

G. Interharmonic voltages

The level of interharmonics is growing with the development of the use of frequency converters and similar equipment. The values are currently studied and additional experience is gained.

H. Levels of signal voltage in the supply voltage

In some countries, distribution network operators use the public distribution network to transfer information. For 99 % of the day, the mean value of the signal voltage measured for three seconds must be less than or equal to the values given in Figure 1.

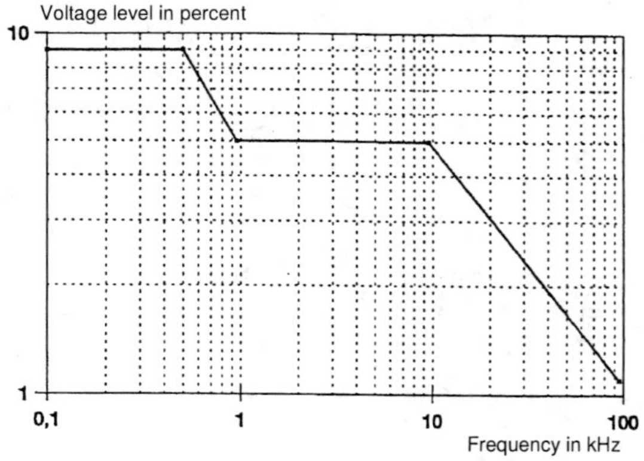

Figure 1. Voltage levels for signal frequencies

I. Interruption of the supply voltage

Interruption of the supply voltage is the condition in which

the voltage at the delivery point is less than 5 % of the agreed voltage.

a) short-term (within 3 minutes),

b) long-term (more than 3 minutes).

(Edition 2 defines interruption below 1 %.)

J. Supply voltage drops

A temporary drop in the supply voltage at the feed point of the distribution system below the given threshold initial value. Application: For the purposes of this standard, the threshold initial value equals to 90 % of the agreed voltage.

The voltage drop duration is the time between the moment at which the voltage at a certain point of the power supply system declines below the threshold value of the drop beginning and the moment when it rises to the threshold value of the drop end. For cases of this standard, the time of voltage drop duration is from 10 ms up to 1 min inclusive.

Voltage changes not reducing the supply voltage to less than 90 % of the agreed voltage are not considered as voltage drops.

K. Temporary increases in voltage with power frequency

It is a temporarily increased voltage at a point of the power supply system, over the given initial threshold. For the purposes of this standard, the threshold of temporary increased voltage equals to 110 % of the reference voltage.

Temporary increases in voltage are usually caused by switching operations or short-circuit faults. According to the node layout, earth faults can also increase overvoltage between healthy phases and the neutral.

L. *Transient overvoltage*

Transient overvoltage is an oscillation or non-oscillation short-term overvoltage, usually heavily damped, with duration of a few milliseconds or less. Transient overvoltage is usually caused by atmospheric surge, switching operations or the activity of fuses.

III. Measurement of Parameters of Electric Power Quality

In terms of measuring the quality parameters of the supplied electricity, the decisive standard is ČSN EN 61000-4-30 Power quality measurement methods.

Three classes of measurement performance are defined for each measured parameter. The measuring devices may measure some or all of the parameters identified in this standard and have different classes of performance for different parameters. The device manufacturer must declare which parameters are measured and which classes of performance are used for various parameters, the range of U_{din} (agreed input voltage) to which each class conforms, and all necessary requirements and accessories (synchronization, testing, calibration period, temperature range, etc.) for each class.

A. *Function class A*

This function class is used if accurate measurements are needed, e.g. for contract applications, verification of conformity to standards, removing doubts, etc. If the same signals are measured, any parameter measurements carried out by two different devices complying with class A requirements will generate comparable results within the specified uncertainty.

To ensure comparable results, the devices of function class A require bandwidth characteristics and a sampling rate sufficient for the specified uncertainty of each parameter.

B. *Function class B*

This function class can be used for statistical surveys, troubleshooting applications and other applications where low uncertainty is not required.

C. *Function class S*

This function class can be used for statistical surveys and contracts where litigations are not anticipated. It is a version of class B, specified in more detail.

IV. Normalization of Supply Voltage Deviations in Low-Voltage Networks

- Supply voltage deviations in low-voltage networks are defined by standard ČSN EN 50160 (Voltage characteristics of electricity supplied from public distribution networks). In Edition 3 of this standard from 2015, the voltage deviations are defined as follows:
  - Under normal operating conditions, except during the interruption of supply, the supply voltage deviation should not exceed ± 10 % of nominal voltage U_n.
  - Where the power supply in networks is not connected to transmission networks or in the case of special remotely controlled users, the supply voltage deviations should not exceed +10 %/-15 % of U_n. Network users should be informed about these conditions.

Under normal operating conditions:

- **95 %** of the average effective values of supply voltage at measuring intervals of 10 minutes must be in the range of U_n ± 10 % during each week; and
- all the average effective values of supply voltage at measuring intervals of 10 minutes must be in the range of U_n +10 %/-15 %.

The latest version of standard ČSN EN 50 160 (Edition 3)

- defines supply voltage deviations for **high voltage** networks – under conditions in 5.2.2.1, the following limits are used:

99 % of the average effective values of supply voltage at measuring intervals of 10 minutes must be in the range less than the threshold limit of +10 % specified in 5.2.2.1 during each week and;

- **99 %** of the average effective values of supply voltage at measuring intervals of 10 minutes must be in the range greater than the threshold limit of -10 % specified in 5.2.2.1 during each week and;
- none of the average effective values of supply voltage at measuring intervals of 10 minutes can be outside the ranges of ±15 % of U_c.

- does not define supply voltage deviations for **very high voltage** (VHV) networks. As the number of users supplied directly from the VHV network is limited and as they usually have individual contracts, this standard does not provide any limits for voltage deviations.

V. Possibilities to Evaluate Supply Voltage Deviations in Low-Voltage Networks

According to ČSN EN 50160, Ed. 3, the voltage deviations are currently evaluated during each week using 95% average effective values of supply voltage at measuring intervals of 10 minutes. The following possibilities of tightening the evaluation are discussed [11-15]:

- evaluation of voltage using 99% average effective values of supply voltage at measuring intervals of 10 minutes,
- evaluation of voltage using 95% or 99% average effective values of supply voltage at measuring intervals of 1 minute.

TABLE II. MINIMUM VOLTAGE DEVIATIONS – 4TH QUARTER OF 2014

Week	40	41	42	43	44	45	46
PR_1.2 - 10'	-3.00	-3.27	-3.93	-3.15	-4.47	-7.59	-7.32
PR_1.2- 1'	-4.09	-5.29	-4.87	-4.94	-5.57	-84.40	-8.91
PR_1.3 - 10'	-10.43	-12.02	-9.11	-7.94	-10.35	-12.45	-11.83
PR_1.3 - 1'	-13.85	-17.63	-13.07	-12.53	-13.23	-64.94	-13.73
PR_2 - 10'	-8.99	-9.07	-8.29	-6.46	-7.74	-6.50	-8.72
PR_2 - 1'	-11.01	-10.04	-13.27	-9.18	-11.63	-7.16	-11.13
PR_3 - 10'	-5.33	-6.42	-6.73	-5.68	-5.45	-7.98	-7.16
PR_3 - 1'	-7.47	-7.59	-8.64	-7.63	-7.47	-9.92	-8.13
VS_1 - 10'	-9.26	-9.69	-10.27	-10.66	-11.32	-9.38	-10.39
VS_1 - 1'	-11.52	-45.17	-35.13	-13.31	-14.59	-12.14	-12.96
VS_2 - 10'	-3.15	-2.26	-2.61	-5.56	-4.94	-3.97	-4.01
VS_2 - 1'	-6.23	-5.17	-4.40	-7.31	-6.89	-8.60	-6.65
VS_3 - 10'	-9.38	-11.25	-9.14	-8.99	-9.77	-8.95	-7.98
VS_3 - 1'	-11.98	-42.14	-11.75	-14.17	-11.24	-10.47	-11.24
Week	**47**	**48**	**49**	**50**	**51**	**52**	
PR_1.2 - 10'	-4.16	-5.37	-5.06	-4.16	-3.46	-4.63	
PR_1.2- 1'	-5.99	-6.46	-54.13	-5.64	-4.75	-5.53	
PR_1.3 - 10'	-10.00	-11.25	-9.49	-8.83	-7.86	-8.83	
PR_1.3 - 1'	-13.35	-13.66	-52.69	-11.09	-11.83	-12.41	
PR_2 - 10'	-6.23	-5.45	-12.80	-6.11	-5.91	-11.95	
PR_2 - 1'	-7.31	-8.25	-69.34	-10.43	-9.61	-12.72	
PR_3 - 10'	-6.65	-8.02	-7.98	-8.25	-9.34	-8.52	
PR_3 - 1'	-42.57	-9.46	-90.16	-8.83	-10.86	-11.48	
VS_1 - 10'	-10.12	-9.69	-9.42	-13.70	-11.79	-8.37	
VS_1 - 1'	-12.53	-11.32	-25.60	-29.18	-13.11	-11.32	
VS_2 - 10'	-73.70	-3.19	-2.72	-3.35	-3.15	-2.92	
VS_2 - 1'	-77.31	-5.25	-4.55	-6.03	-6.50	-5.41	
VS_3 - 10'	-11.60	-10.51	-9.22	-8.02	-8.52	-8.21	
VS_3 - 1'	-13.66	-13.19	-11.94	-10.90	-10.35	-9.73	

VI. EVALUATION SUPPLY VOLTAGE DEVIATIONS IN REAL LOW-VOLTAGE NETWORKS

A. Methodology of measuring the supply voltage deviations

Measurements of the voltage deviations are implemented in two stages. The first stage of measuring the voltage deviations was started in the spring of 2014 and completed in the spring of 2015. In the first stage, measurements of the voltage deviations were implemented at six selected localities of Central Moravia. One locality contained a connected photovoltaic power plant (PPP); at this locality, the measurements were performed in three places (DTS, PPE and line end). Measurements were carried out continuously (except a few-minutes-interruption of measurement during the readout of data with a reading cycle of about every 3 weeks); the voltage was measured with storage intervals of 1 minute and 10 minutes. The second stage of measuring has taken place at selected locations in North Moravia since the spring of 2015. The values are measured and recorded using MEG40 and PQ monitors. Currently processed data are those from the first stage of measurement.

TABLE III. MAXIMUM VOLTAGE DEVIATIONS – 4TH QUARTER OF 2014

Week	40	41	42	43	44	45	46
PR_1.2 - 10'	7.12	7.39	6.11	5.56	5.76	6.15	5.95
PR_1.2- 1'	7.43	7.78	6.97	6.11	6.30	6.65	6.42
PR_1.3 - 10'	8.21	7.47	6.15	5.45	5.84	6.42	5.95
PR_1.3 - 1'	8.52	8.33	7.04	6.46	7.24	6.73	7.28
PR_2 - 10'	6.96	7.90	6.61	6.58	6.58	7.51	6.03
PR_2 - 1'	8.95	9.07	7.78	9.03	7.43	7.94	6.93
PR_3 - 10'	5.56	5.72	5.18	5.60	4.94	5.06	4.82
PR_3 - 1'	5.99	6.23	5.99	6.42	5.64	5.80	5.10
VS_1 - 10'	8.87	8.99	8.48	8.52	9.69	7.86	7.59
VS_1 - 1'	9.61	9.84	9.69	8.68	10.04	8.64	8.87
VS_2 - 10'	5.64	5.76	5.80	5.72	6.65	5.72	5.41
VS_2 - 1'	6.65	6.07	7.78	6.23	7.67	6.34	6.34
VS_3 - 10'	8.72	8.87	7.47	5.72	6.34	6.85	7.00
VS_3 - 1'	9.14	9.42	8.09	7.00	7.82	7.20	7.24
Week	**47**	**48**	**49**	**50**	**51**	**52**	
PR_1.2 - 10'	5.99	5.53	6.42	6.03	5.29	5.84	
PR_1.2- 1'	6.30	5.83	6.93	6.61	6.65	6.89	
PR_1.3 - 10'	5.88	5.53	6.07	5.88	5.18	7.00	
PR_1.3 - 1'	7.98	5.91	6.73	7.63	6.50	8.83	
PR_2 - 10'	5.68	5.80	5.91	5.53	5.41	5.95	
PR_2 - 1'	6.77	6.73	6.85	7.28	6.38	7.12	
PR_3 - 10'	5.72	5.45	5.02	4.75	5.25	4.98	
PR_3 - 1'	6.03	6.19	6.34	5.25	6.30	5.53	
VS_1 - 10'	7.78	6.46	6.07	5.91	6.93	7.08	
VS_1 - 1'	8.33	7.94	7.63	7.08	7.94	7.78	
VS_2 - 10'	5.64	5.80	5.25	5.60	5.68	5.21	
VS_2 - 1'	6.81	6.19	6.19	7.00	6.65	6.30	
VS_3 - 10'	6.46	6.07	5.72	6.46	6.65	6.65	
VS_3 - 1'	7.24	6.42	6.07	7.16	8.02	7.47	

B. Methodology of evaluating the supply voltage deviations

For the first case of evaluation, we only try to identify whether 95 % or 99 % of the values of voltage deviations exceed or do not exceed the limit of ±10 %. Therefore, the deviations were determined as the absolute value of the relative voltage deviation, and we assessed whether this value exceeded or did not exceed 10 %.

From the measured ten-minute and one-minute values, we determined percentage voltage deviations and evaluated the following data for each weekly period:

- 95% values from ten-minute values,
- 99% values from ten-minute values,
- 95% values from one-minute values.

As these values are only positive, we can calculate the average value of the three phases to simplify the presentation of results. A sample evaluation for the 4th quarter of 2014 is shown in Figure 2. (The figures do not present PR_1.1 place where only 10-minute voltage values are available).

Throughout the monitored period, we identified only one place (VS_1) with a 95% voltage deviation of about 75% in one phase in three consecutive weeks (week 16 – week 18 of 2014); it is therefore possible to assume that it was an error in the measurement.

Furthermore, these data were used to determine differences between individual values – see Figure 3:

- (95%, 1‘) – (95%, 10‘),
- (99%, 10‘) – (95%, 10‘).

These data suggest that a transition from evaluating 95% values to evaluating 99% values would provide poorer outcomes.

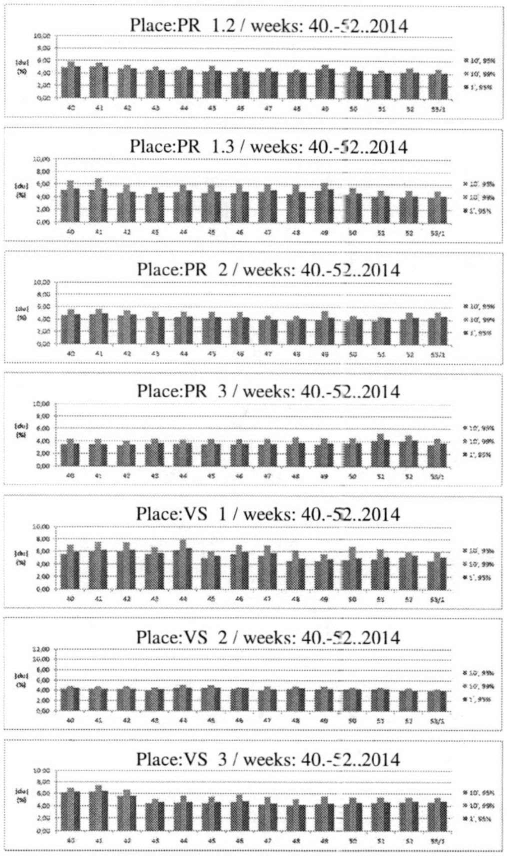

Fig. 2 Absolute values of relative voltage deviations – 4th quarter of 2014

From the measured ten-minute and one-minute values, we also determined minimum and maximum percentage voltage deviations for every weekly period. Given the fact that the standard ČSN EN 50160 requires that all voltage values are in the range of -15 %/+10 %, the minimum and maximum value for each week was always searched for from all three phases. A sample evaluation of the minimum and maximum value for the 4th quarter of 2014 is shown graphically in Figure 4 whereas the respective values are given in Table 2 and Table 3 with a colour indication of values exceeding the minimum value of -15 % and maximum value of +10 %.

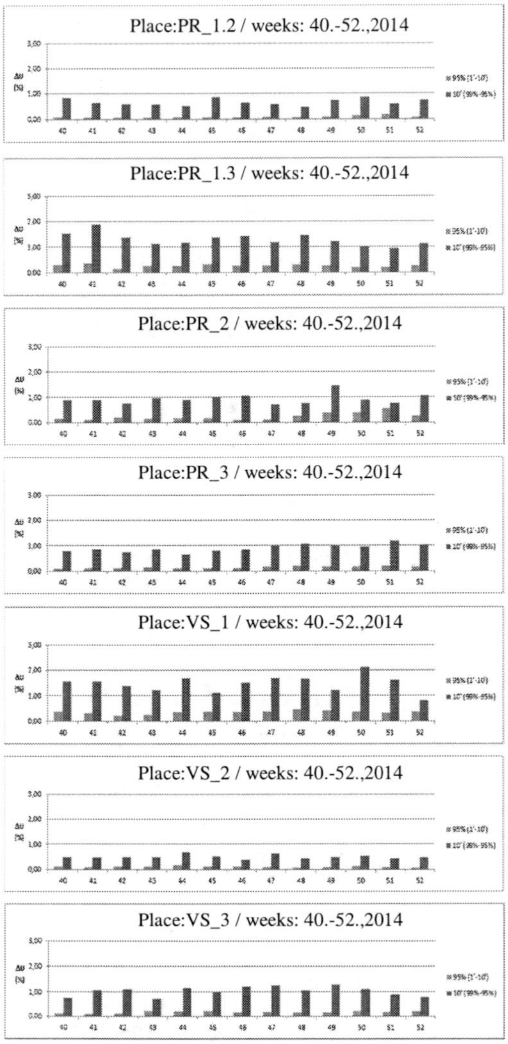

Fig. 3 Difference in absolute values of relative voltage deviations – 4th quarter of 2014

VII. Conclusions

Figures 2 and 3 demonstrate that that the transition from evaluating 95% values to evaluating 99% values would provide worse results; the absolute value of the relative voltage deviation is higher by about 1 %, as shown in Figure 3. The deterioration in the transition from 10-minute values to 1-minute values is in the order of tenths of a percent. At the monitored localities, during the entire period observed (1 year), the voltage excess of +10 % of U_n occurred very rarely; the excess of voltage was usually in tenths of a percent. The situation differs for voltage drops below -15 %; the incidence of such drops was higher and they were much deeper, even to -90 %.

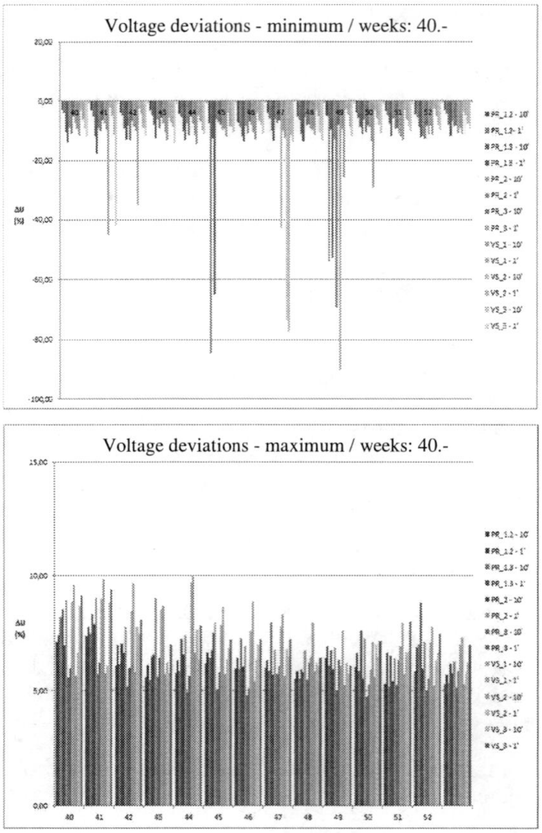

Fig. 4 Voltage deviations – minimum/maximum – 4th quarter of 2014

Acknowledgment

This research was partially supported by the SGS grant from VSB – Technical University of Ostrava [Grant No. SP2016/95, SP2017/54 and SP2017/127].

References

[1] J. Kosmak, S. Misak, L. Prokop, "Power quality dependence on connected appliances in an Off-Grid system," Proceedings of the 2015 16th International Scientific Conference on Electric Power Engineering, EPE 2015, pp. 423-428.

[2] S. Misak, L. Prokop, P. Bilik, P. Krejci, "Evaluation of power quality in off-grid power system," Komunikacie, 17 (1A), pp. 71-76.

[3] S. Misak, J. Stuchly, J. Vramba, L. Prokop, M. Uher, "Power quality analysis in off-grid power platform," Advances in Electrical and Electronic Engineering, 12 (3), pp. 177-184.

[4] S. Misak, L. Prokop, P. Bilik, "Power quality analysis in off-grid power system," 10th International Conference, ELEKTRO 2014 - Proceedings

[5] J. Vanus, R. Martinek, P. Bilik, J. Zidek, J. Machac, "Energy management strategies in intelligent office building using PI system," In: Proceedings of the 8th International Scientific Symposium on Electrical Power Engineering (Elektroenergetika). 2015. p. 416-419.

[6] P. Bilik, M. Horinek, R. Martinek, J. Vanus, "Results of multiple phasor measurement units testing procedures," In Conference Proceedings: 8th International Scientific Symposium on Electrical Power Engineering (Elektroenergetika). Stara Lesna, Slovakia, 16-18 September 2015, pp. 136-139, ISBN 978-80-553-2187-5.

[7] R. Martinek; J. Vanus, M. Kelnar, P. Bilik, „Control methods of active power filters using soft computing techniques," In

Conference Proceedings: 8th International Scientific Symposium on Electrical Power Engineering (Elektroenergetika). Stara Lesna, Slovakia, 16-18 September 2015, pp. 363-366, ISBN 978-80-553-2187-5.

[8] R. Martinek, J. Vanus, P. Bilik, "New strategies for application of recursive least square algorithm in active power filters," In Conference Proceedings: 8th International Scientific Symposium on Electrical Power Engineering (Elektroenergetika). Stara Lesna, Slovakia, 16-18 September 2015, pp. 344-347, ISBN 978-80-553-2187-5.

[9] R. Martinek, J. Manas, J. Zidek, P. Bilik, „Power quality improvement by shunt active performance filters emulated by artificial intelligence techniques," In Conference Proceedings: 2nd International Conference on Advances in Computer Science and Engineering (CSE 2013). Los Angeles, CA, USA, 1-2 July 2013, pp. 157-161, ISSN 1951-6851, ISBN 978-90786-77-70-3, DOI:10.2991/cse.2013.3.

[10] R. Martinek, J. Vanus, P. Bilik, T. Stratil, J. Zidek, "An efficient control method of shunt active power filter using ADALINE," IFAC-PapersOnLine, 2016, 49.25: 352-357, DOI: 10.1016/j.ifacol.2016.12.064.

[11] V. Ajodhia, R. Hakvoort, "Economic regulation of quality in electricity distribution networks," Utilities Policy, 2005, 13.3: 211-221.

[12] P. Richardson, D. Flynn, A. Keane, "Optimal charging of electric vehicles in low-voltage distribution systems," IEEE Transactions on Power Systems, 2012, 27.1: 268-279.

[13] C.-G. Duan, S. Ouyang, Z.-X. Song, D.-G. Chen, J.-H. Wang, „Design of a new online and real-time power quality monitor," Power System Technology, 2004, 2: 014.

[14] H. Akagi, H. Fujita, K. Wada, "A shunt active filter based on voltage detection for harmonic termination of a radial power distribution line," IEEE Transactions on Industry Applications, 1999, 35.3: 638-645.

[15] H. Fujita, H. Akagi, "The unified power quality conditioner: the integration of series and shunt-active filters," IEEE transactions on power electronics, 1998, 13.2: 315-322.

Modification of Pull Characteristics of Lever Magnet with Variable Reluctance

Petr Bernat, Petr Kacor, Zdenek Hytka, Tomas Pavelek
FEI, VSB-TU Ostrava, 17. listopadu 15
Ostrava, Czech Republic
petr.bernat@vsb.cz, petr.kacor@vsb.cz, zdenek.hytka@vsb.cz, tomas.pavelek@vsb.cz

Abstract — **Lever magnets, often defined like magnets with variable reluctance belong to significant components of automation system. The pull force is generated by convenient curvature of flux lines within the magnetic field gained through specific geometrical layout of the magnetic circuit. Their main advantages include a relatively high value of force at the initial position of armature, especially with the noiseless operation that determines such magnets for utilization in automation systems, electric-pneumatic circuits as well as in electric-pneumatic controls with noise-less condition. Our paper presents one of the possibilities how to change output acting force of lever magnets by using of permanent magnet in magnetic circuit. Comparison of output results are made by the real measurement and numerical simulation of usual design of lever magnet.**

Keywords: open C frame magnet; reluctance; magnetomotive force; cut-away armature; lever magnet; action magnet; electromagnetic valve; pull down lever magnet

I. Introduction

Electromagnets belong to the essential action components of power appliances and other action elements used in automation systems [1]. The effect of magnetic field generates force effects on the movable armature the motion of which is further transmitted to the controlled element [2, 9]. Electromagnets with variable reluctance create a separate part of industrial electromagnets with movable armatures [3]. Their most common design comprises a single solenoid coil on a steel core; applications requiring installation of the smallest design magnet's width use then the distribution of the winding on two coils with the same polarity [4], see Fig. 1. The magnetic circuit of magnet is made of solid steel; the coil is powered using a direct-current power supply unit. The coil is further provided with components to ensure contact arc extinguishing properties and excess voltage reduction during shutdown, which would be mostly a parallel resistor, a varistor, a diode or a transient voltage-suppression diode [5]. The movable part of magnet - the armature, is pivoted inside a bearing that enables its swinging motion within a small angle range, usually within approx. 10-20 degrees.

The movable part of armature is attached to an action element with almost linear motion ability; the link may be direct – the action element being a component of the armature assembly, or indirect, whereas the action element is controlled with pull rod, lever or combination of multiple components similar to those. The motion itself may exit the magnet from either it front - active part - or even the rear; with the armature forming a lever of the second or third kind, see Fig. 2. The lever principle then helps to adjust the resultant path by suitable means, shifting the motion output point on the armature while maintaining the same torque of force.

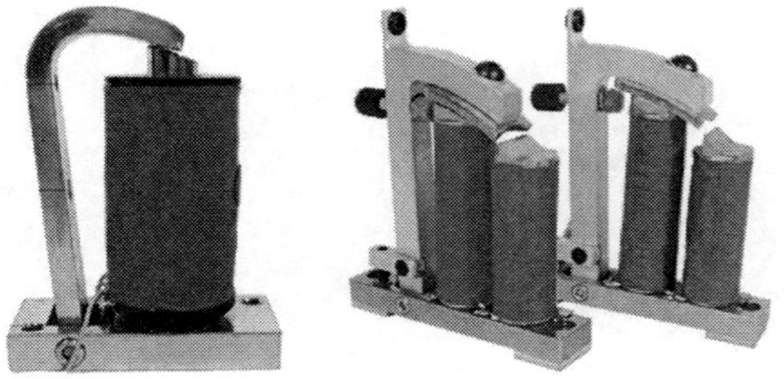

Figure 1. Basic design of lever electromagnet with one and two coils

II. Design And Operation Principle

There are actually several ways to design a magnet with variable reluctance. The basic principle is shown in Fig. 1 and Fig. 2. The movable part of magnetic circuit (the armature) is mechanically connected to keep an air gap between pole and armature ends regardless of its position [4]. Unlike other types of electromagnets with movable armature, the terminal position does not involve any mechanical contact of metal parts of the magnetic circuit that is followed by a characteristic noise, while also bringing other issues – wear of the contact point, the risk of magnetic "sticking" etc. When in open position the flux lines pass through the space with higher magnetic conductivity, whereas they vault at the air gap adjacent to the armature to reach out of the magnetic circuit and curvature. The curvature of field at the point of magnetic leakage path would then act as "the source" of force effect subsequently demonstrated by generation of the pull force – the torque and movement of armature. When in the terminal position – the balance conducting state, the force acting on armature within the electromagnetic field will be weak, the magnetic leakage is minimal and most of the flow will be defined by direct flux lines [5].

Figure 2. Output of lever move on the side of coil and behind of bearing

When the power supplied to the excitation solenoid is turned off, the magnetic field will vanish as well as force effects on the armature, until it reaches the front position – the open armature condition reached by the deactivated magnetic circuit only due to an external force, mostly exerted by a spring or action of any controlled mechanism – e.g. pressure of a fluid against a controlled valve attached to the armature [6].

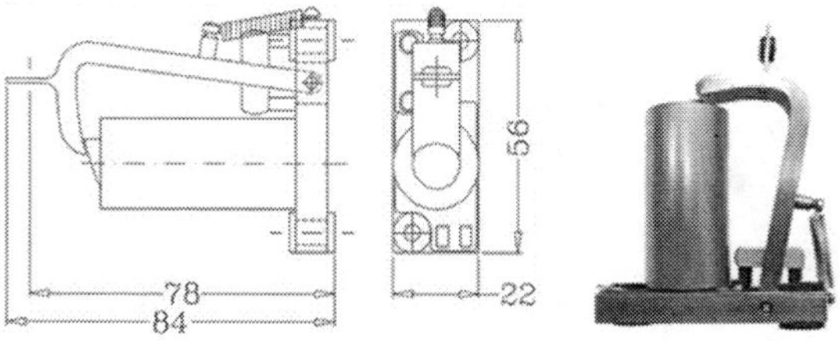

Figure 3. Basic dimension of analyzed lever electromagnet

The force effects of magnetic field generated by passing of electric current through coils of a solenoid electromagnet also depend on the material properties of the magnetic circuit, its geometrical layout – mechanical dimensions and the air gap size. The resultant force at the output point is neither constant nor monotonous along the armature travel path; its maximum is located at a point in the vicinity of open armature, while decreasing along the magnetic circuit closing path. The shape of these characteristics can be affected by the armature geometry, especially the shape of edges and adaptors defining the air gap. Common modifications would include alterations of the armature end by rounding, recessing or even pointing.

Then verification of the actual course and methods for potential alteration of the pull characteristics performed with an injection of additional magnetic field into the armature circuit has been scheduled with the commercially available pull down magnet manufactured by Aug. Laukhuff GmbH & Co. Its main dimensions are shown in the Fig. 3. The default specifications of magnet are as follows – the power of U= 24V, I= 0.5A enables the magnet generate the pull force of F= 10N long the path of L= 10mm with respect to the output point on the end of armature arm. The magnetic circuit is made of nickel plated steel, the solenoid coils are made of d= 0.18mm thick copper wire. The solid steel foundation is integrated with a mandrel to form the solenoid core. The mandrel is inserted with the plastic core of the cylindrical coil, covered with a tubular shielding layer made of nickel plated copper sheet. The loose end of this mandrel is rounded off to a radius with its center inside the armature bearing, matching the shape of armature on this loose end, forming the constant-sized air gap between its body and the core mandrel. The bearing is formed by a bronze peg encapsulated in a steel housing. The armature is made of a rolled and milled intermediate featuring a single unit with the output loop. The assembly of these components onto the controlled mechanism is accommodated by two openings, the need to reduce transmission of vibrations and keeping the system operation noiseless, the openings are cushioned with rubber grommets inserted within to form a silent block in combination with the fastening screws. When in its neutral position, the armature is held by means of a spring and the terminal positions of the path are limited using adjustable silicone rubber stops.

Magnets for this particular application would usually have a similar layout, differing by certain engineering details and the precision of individual components only. The limiting stops are mostly made of rubber or felt, the armature is usually made of bent metal sheet intermediate, some manufacturers would not produce these from single piece of material due to technological and especially economic reasons; it would be therefore assembled using parts fixed to each other by riveting, bolting or even spot welding. The bearing designs could also differ, cheap low-end magnets would often have a sole steel mandrel right in the armature lever opening, precision magnets would have their casings or mandrels made of bronze. As far as the service life and the number of operation cycles considered for magnets are concerned, the bearing is a vital part, the magnet operation depends on keeping the alignment between the fixed and mobile segments; any wear of the bearing resulting in a change to the air gap or armature path, such condition would also bring changes to the pull characteristics and even potential seizure of the magnet.

These types of magnets are very reliable from the operation prospective, the most common failures would include interruption - blow out - of the solenoid coil wiring, as well as armature bearing failures (seizure, wear and large play), magnets operated in corrosive environments are also prone to effects of corrosive products or magnetically conductive particles (e.g. chips, cuts or dust) inside the air gap.

III. Experiment and Mathematical Background

Analytical determination of the magnet pull force and plotting of the pull characteristics are only possible with the simplest alignment of the magnetic circuit – a cylindrical armature moving inside the solenoid hollow in a linear direction. The concept of analytical interpretation to express the magnetic conductivity within the variable air gap is actually causing certain issues. There are some formulas describing other layouts, e.g. conical armatures, however, these formulas are based on a whole range of simplifications of these problems and they had been generated empirically in the past. One of the options would be employing methods of statistical modeling [7]. Starting from the basic analytical formula (1), the force exerted against the armature in the vicinity of the magnetic point is defined for homogeneous fields and constant air gap only. The force effects cannot be described by analytical means for a variable air gap, movement of the armature along the general path and the change in magnetic conductivity. With respect to all the variables, the analytical expression would have different shapes for each point along the armature path, the pull characteristics scheme can be hence demonstrated using certain numerical methods only.

$$F = \frac{B_{AG}^2 S_{AG}}{2\mu_0} \tag{1}$$

Where: B_{AG} is the magnetic flux density in air gap, S_{AG} is the surface of body under force F. For basic solution of electromagnetic field we used general description of Maxwell´s equation in numerical type of computing.

$$\nabla \times \mathrm{H} = J + \frac{\partial D}{\partial t} \tag{2}$$

$$\nabla \times \mathrm{E} = -\frac{\partial B}{\partial t} \tag{3}$$

$$\nabla \cdot \mathrm{B} = 0 \tag{4}$$

$$\nabla \cdot \mathrm{D} = \rho \tag{5}$$

Force acting on arm is computed from Maxwell stress tensor. The Maxwell stress approach computes local stress at all points of a bounding surface then sums the local stresses by means of a surface integral to find the global magnetic force [8]. Thus if the total field surrounding a body is known, then the force exerted on the body in terms of the field alone can be expressed by (5).

$$F_{MST} = \oint \left[(B_n H_t)\boldsymbol{t} + \frac{1}{2}\left(\frac{1}{\mu_0} B_n^2 - \mu_0 H_t^2\right)\boldsymbol{n} \right] dS \tag{6}$$

Where S is the surface enveloping the body under force, B the flux density, n and t are unit vectors in the normal and tangential direction on the surface S. The torque T may be obtained by the cross product T= r×F, where the force F is given by (6) and r is vector from the point of application of F to the axis of rotation.

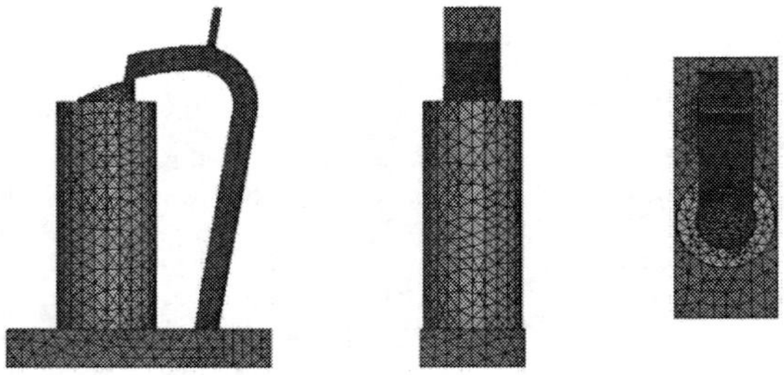

Figure 4. FEM model of lever electromagnet

One of the potential methods is based on utilization of finite elements method - FEM to describe the magnetic circuit characteristics [10]. The model of magnet subject to analysis is shown in the Fig. 4, whereas Fig. 5 shows the paths of flux lines of the magnetic field with the cut-away magnet and reference to three different positions of armature.

Fig. 6 shows the behavior of force at the magnet output point, depending on its path, generated by FEM calculation of the armature angle of yaw. The course shows an evident change in magnitude of the magnetic flux along the entire armature movement path. The flow closes at the path beginning over the edge of the armature lever only, which is strongly overloaded, resulting in vaulting of the magnetic field and increase of force with the armature in open position. The force starts by increasing and the reduction of load would be followed by is slight decrease during further movement. Due to the overload, the maximum magnet pull is not achieved at the path beginning yet only at the point, where the edges of magnetic circuit overlap at the direction perpendicular to the direction of movement. The dimensions and geometrical layout of the model match the above-mentioned commercially available electromagnet put into the experimental testing for measurement and verification of pull characteristics under conditions applied for modeling.

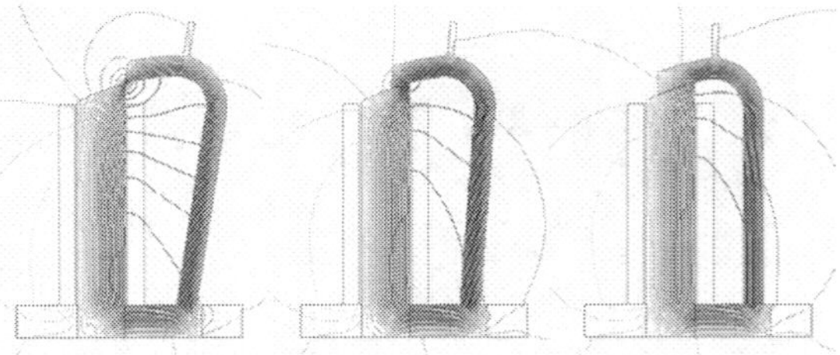

Figure 5. Flux lines of magnetic field at three position of lever

Following measurements of individual points for pull characteristics, the electromagnet was fitted into the jig made of magnetically non-conductive material in order to rule out potential effects of magnetic fields in the vicinity. The jig also held the electronic digital strain-gauge for force measurements. The gauge was in a sliding fit to allow the adjustment screw for measurement of the force at the magnet output point within the entire range of armature movement.

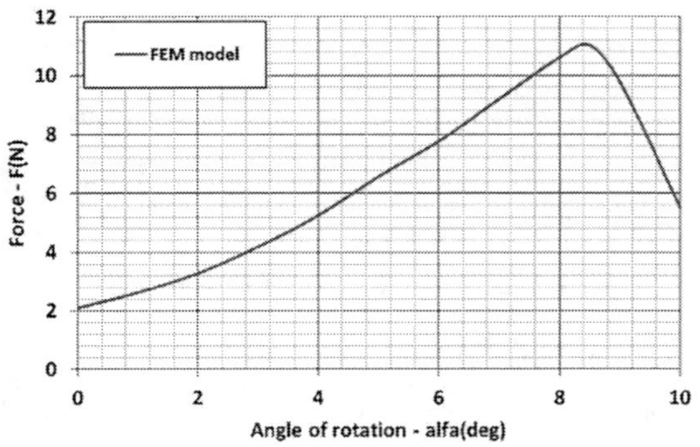

Figure 6. Dependence of acting force at angle of rotation from FEM analysis

The electromagnet was powered from a direct-current power supply unit with the current characteristic, using constant current I= 0.5A. Instead of the predefined power supply unit delivering U= 24V, the current feed was selected to rule out potential impact on the measurement result due to changes in solenoid coil resistance caused by heat-up during the experiment. The resultant pull characteristic is shown in Fig. 7, compared to the characteristic obtained through the numerical model. Both the measured and simulated characteristics show a certain similarity in shape, yet they differ in magnitude of the characteristic forces, the initial force of a real magnet equals to F_{ST}= 6.4N, whereas this magnet achieves its maximum force F_{MAX}= 11.6N at the stage of geometrical overlap of the armature cut-out edges with the edge of magnet pole (the stage of maximum edge overload).

IV. Layout Configuration

Hydraulic or pneumatic applications of the magnet require that the maximum force exerted is situated at the beginning of the magnet path. Such magnified initial force is required to detach the valve subject to the passive force from the spring, in closed position, as well as the static pressure of the controlled fluid (liquid or gas) to vanish after opening of the valve and balancing of the pressure before and behind it. Other applications also feature a force peak at the path beginning, caused by dry friction, for example.

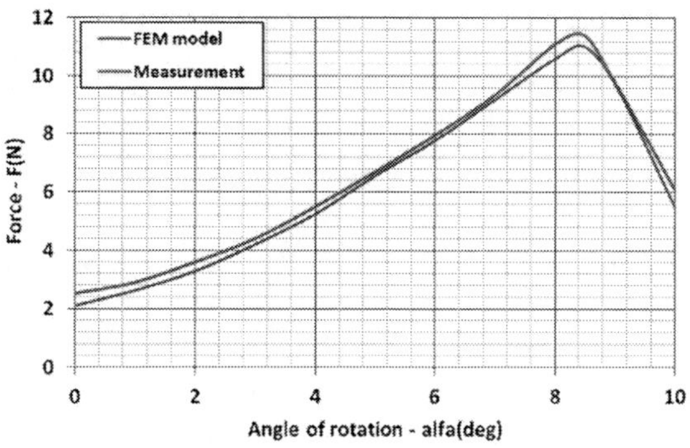

Figure 7. Comparison of real measurement and FEM analysis of magnet

There are certain manufacturers - e.g. Peterson EMP - seeking the increase of initial force through the principle described by the U.S. Patent No. 4,851,80, fitting the end of lever arm with an extra permanent magnet.

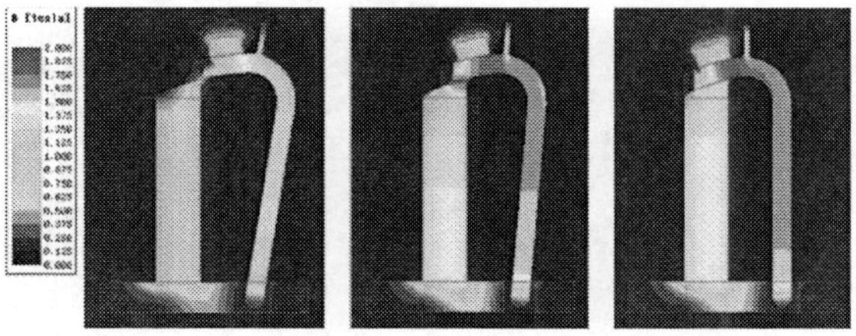

Figure 8. Flux lines of magnetic field at three position of lever with PM

The shape of resultant field within the magnetic circuit following a similar adjustment in the open position is the showed by cross-sections in the Fig. 8, the Fig. 9 then shows the final shape of the pull characteristics. The pull characteristics present after the completed adjustment shows an evident difference, when compared to the initial design of the electromagnet, especially due to the change in initial conditions within the magnetic circuit of the magnet used.

The pull characteristics of such adjusted electromagnet were then verified through an experiment. The armature end was provided with an adhesive neodymium magnet with subsequent performance of the same measurement as in the first case. The verification process also focused on the impact of magnet polarization. The Fig. 9 shows pull characteristics for the identical PM+ orientation and opposite PM- orientation. The intensity of magnetic field generated by the extra magnet must be comparable to the magnitude of induction created by the field around the control coil of electromagnet to obtain a significant effect on the resultant mechanical properties, i.e. mainly the absolute magnitude of force and the total pull characteristics. The magnet itself must be sized accordingly to avoid any serious interference with the magnetic circuit geometry and the course of flux lines in the magnetic field, especially outside the magnetic circuit here, along leakage paths. That is why the most suitable materials suggested for these applications are magnetically hard substances based on rare earth minerals that allow obtaining significantly higher intensity and smaller sizes when comped to ferrites. Since the magnet is also a part of the armature, which are a moving part, its dimensions and weight are limiting factors even for the moment of inertia and the relevant dynamic properties of the resultant electromagnet. Both the measured and experimental characteristics then show the impact of extra magnet on the initial and the maximum magnitude of resultant force as well as the position of curve peak.

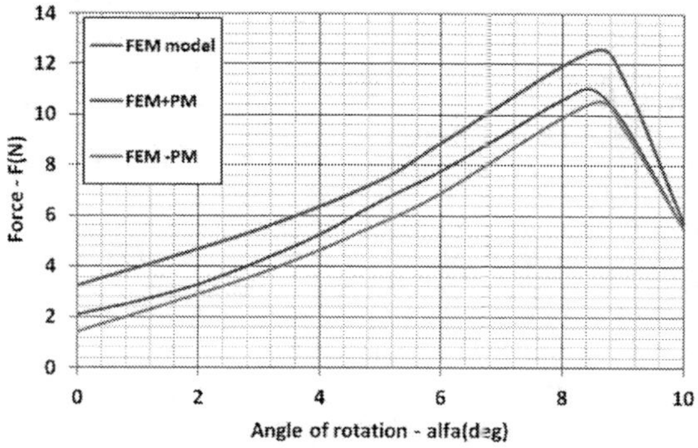

Figure 9. Force characteristic with the same and opposite orientation of PM at real measurement by force-meter

The additional magnetic field helps extend the area, where the armature edge is overloaded and the initial pull force of magnet rises, whereas the impact of extra magnet will be minimal towards the end of armature path. Thanks to the shape of resultant field and the course of flux lines, the extra permanent magnet will have almost no influence on forces acting on the mobile part of the armature when stationary.

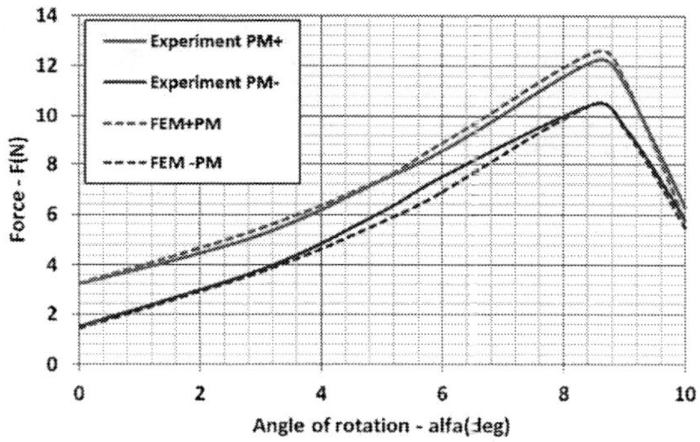

Figure 10. Comparison of real measurment and FEM model with the same and opposite orientation of PM

While other magnet types would be prone to "armature sticking", for example, following addition of an extra permanent magnet, the magnets with variable reluctance can never get into such state with the air gap size kept constant. The Fig. 10 shows comparison of resultant characteristics of the initial and modified magnets.

V. Conclusion

Lever electromagnets represent a fairly common type among electromagnets with their actual characteristics often determined using alignment obtained through empirical methods only, since there had been no statistical solution in place due to the reasons described above. Those were only the results generated by the mathematical model and confirmed by practical verification in form of actual electromagnet measurement providing pull characteristics so that the resultant product can be used as desired and also in designing of other equipment, mainly electrical appliances, enabling them noiseless swinging or sliding motion of particular components using this type of magnet.

The configuration with magnetic circuit and the extra permanent magnet provides new opportunities for their application in design of action elements, especially in terms of potential miniaturization of components with preservation of identical force effects, higher potential control dynamics, savings on material and energy. The gradual progression in utilization of new materials, mainly for permanent magnets therefore induces a potential change in design and utility characteristics of action elements that had been considered finished in development time ago. The said procedure then enriches the concept with the necessary element of certainty with the possibility to replace or supplement the formerly regular or intuitive or empirical modifications in design and to improve effectiveness of engineering, manufacturing and application of the products fitted with the said electromagnet.

VI. References

[1] G. L. Hedges, "Solenoid and electromagnet windings," in Proceedings of the American Institute of Electrical Engineers, vol. 34, no. 11, pp. 2595-2614, Nov. 1915, doi: 10.1109/PAIEE.1915.6590492

[2] D. D. Miller, "Design characteristics of electromagnets for telephone relays," in The Bell System Technical Journal, vol. 3, no. 2, pp. 206-231, April 1924, doi: 10.1002/j.1538-7305.1924.tb01355.x

[3] M. York and J. M. Stephenson, "An electronically-operated, normal-force, variable-reluctance actuator," IEE Colloquium on Limited Motion Electrical Actuation Systems (DIgest No. 1998/494), London, 1998, pp. 5/1-5/4, doi: 10.1049/ic:19980930

[4] N. C. Cheung, "A compact and robust variable reluctance actuator for grasping applications," Industrial Electronics Society, 1998. IECON '98. Proceedings of the 24th Annual Conference of the IEEE, Aachen, 1998, pp. 1072-1076 vol.2., doi: 10.1109/IECON.1998.724244

[5] S. Gibson, G. W. Jewel and R. E. Clark, "Variable-airgap, cylindrical, linear variable reluctance actuators for high-force, medium-stroke applications," in IET Electric Power Applications, vol. 3, no. 4, pp. 352-362, July 2009, doi: 10.1049/iet-epa.2006.0304

[6] J. L. G. Janssen, J. J. H. Paulides, E. A. Lomonova and A. J. A. Vandenput, "Analysis of a Variable Reluctance Permanent Magnet Actuator," 2007 IEEE Industry Applications Annual Meeting, New Orleans, LA, 2007, pp. 502-509, doi: 10.1109/07IAS.2007.45

[7] J. Skibicki and M. Bartłomiejczyk, "Analysis of measurement uncertainty for contact-less method used to measure the position of catenary contact wire, performed with the use of Monte Carlo method," Measurement, Vol. 97, February 2017, pp. 203-217, ISSN 0263-2241, doi.org/10.1016/j.measurement.2016.11.008

[8] A. Benhama, A. C. Williamson and A. B. J. Reece, "SRM torque computation from 3D finite element field solutions," 1997 Eighth International Conference on Electrical Machines and Drives (Conf. Publ. No. 444), Cambridge, 1997, pp. 59-63, doi: 10.1049/cp:19971038

[9] J. Skibicki and M. Bartlomiejczyk, "Analysis of measurement uncertainty for contact-less method used to measure the position of catenary contact wire, performed with the use of Monte Carlo method," MEASUREMENT, vol 97, pp. 203-217, 2017, ISSN: 0263-2241, doi.org/10.1016/j.measurement.2016.11.008

[10] P. Kacor and P. Bernat, "Analysis of force characteristic of short-circuit release in low voltage circuit breaker," Proceedings of the 2014 15th International Scientific Conference on Electric Power Engineering (EPE), Brno, 2014, pp. 515-519, September 2015, doi: 10.1109/EPE.2014.6839414

[11] K. Leubner, R. Laga, and I. Dolezel, "Advanced Model of Electromagnetic Launcher," Advances In Electrical And Electronic Engineering, vol 13, no. 3, pp. 223-229, doi:10.15598/aeee.v13i3.1419

Locally Produced Solid Biofuels for Individual Space Heating – Competitiveness With Brown Coal and Heat Pumps

Jiří Beranovský
EkoWatt CZ, s.r.o.
Švábky 52, Praha 8, Czech Rep.
jiri.beranovsky@ekowatt.cz

Jaroslav Knápek
CTU in Prague, FEE
Technická 2, 16627 Praha 6, Czech Rep.
knapek@fel.cvut.cz

Kamila Vávrová
VÚKOZ, v.v.i.
Květnové nám. 391, Průhonice, Czech Rep.

Abstract — **Brown coal still plays an important role in space heating in single-family houses, particularly in smaller municipalities with under 2000 inhabitants. Coal combustion in obsolete boilers with low efficiency is an important cause of local air pollution. Natural gas is not available in many smaller municipalities; pellets made from locally available biomass are one of the possible substitutes for brown coal combustion. Other options include heat pumps and modern automatic coal boilers. Analysis results shows that replacement of obsolete boilers with pellet boilers is potentially competitive with modern automatic coal boilers. In comparison with pellet or coal boilers, heat pumps are only competitive in buildings with a high heat demand for space heating. Simultaneously, analysis results show that pellet boilers or automatic coal boilers significantly increase the price of the heat produced compared to operation of obsolete coal boilers, namely by CZK 110 to 180/GJ depending on building type.**

Keywords: energy crop; heating costs; brown coal; pellets

I. Introduction

There are 4.75 million dwellings in the Czech Republic at present. Out of that, approx. 47% are in single-family houses. The dominant method of space heating in single-family houses is local in-house heating systems: approx. 89% in 2011 [1]. Approximately 1/3 of the households in single-family houses are heated using solid fuels; this is often domestic brown coal. Brown coal plays an important role particularly in space heating in single-family houses in smaller municipalities. The proportion of brown coal in space heating decreases with the growing municipality size and vice versa. For example, in municipalities with under 1000 inhabitants, brown coal is used for heating in approx. 30% of households; conversely, in municipalities with over 20,000 inhabitants, the proportion of brown coal is only max. 5%. Biomass also makes up a considerable share of space heating in smaller municipalities with under 2000 inhabitants, typically in the form of firewood, and a small proportion of pellets and briquettes. Solid fuels have a relatively significant share also in heating of newly built single-family houses; it is approx. 16% (largely firewood, with approx. 40% of the solid fuels being brown coal; the share of pellets is almost negligible) [2].

Combustion of domestic brown coal, often in obsolete boilers with low efficiency, causes problems with local air pollution in municipalities. This is often even multiplied by combustion of inappropriate fuels, including various types of waste. For example, the specific SO2 emissions from brown coal are approx. 60x higher compared to combustion of wood or pellets [3]. Moreover, reserves of brown coal suitable for combustion in local fireplaces are limited and their gradual depletion is going to occur [4].

Approximately 1/4 of all inhabited dwellings are in smaller municipalities with under 2000 municipalities. Connection to the natural gas system is not available in many of these smaller municipalities. Substitution for coal (combusted in often highly obsolete and inefficient boilers) is thus possible essentially with the following options: (1) heat pump, (2) pellet boiler, (3) possibly modern brown coal boiler with automated feeder. A decrease in the interest in space heating using firewood is expected in future due to the considerably lower comfort compared to the above options.

The key point of view in households' decision-making concerning the heating method will undoubtedly be the heating costs. Here, we need to consider the fact that in many cases, heating substitution concerns old houses, which have a high heat demand for space heating. Only a part of the dwellings in these smaller municipalities have undergone more or less extensive renovation connected with improving thermal and technical parameters of the buildings.

In smaller municipalities, brown coal can be replaced with locally available biomass, processed in the form of pellets. The following can primarily serve as local biomass sources: (1) biomass from forest residuals; (2) residual biomass from conventional agriculture (cereal straw, etc.); (3) biomass from purposely grown energy crops (short rotation coppice plantations, woody perennial energy crops, such as reed canary grass, Chinese silver-grass, etc.). An advantage of local pellet production is the ability of using free workforce capacities in winter, and the ability to use the pelletising plant flexibly for production of fodder and other types of pellets. At present, there are techniques for pelletising biomass with annual production capacity of approx. 800-1200 t/year, which can cover the demand for heating fuel in these smaller municipalities (see, e.g., technology offered by ProPelety – www.propelety.cz).

Utilisation of locally available biomass is a frequently discussed alternative for future. It is an interesting question, however, what the competitiveness of single-family house heating based on locally produced pellets is compared to heat pumps as well as modern automatic brown coal boilers, which are currently supported in the form of "boiler subsidy". Another important question is the comparison of these three options with the still common heating method using an obsolete brown coal boiler with manual stoking.

The paper describes the results of an economic analysis of local pellet production based on locally available biomass and compares the economic effectiveness of space heating with pellets with heating using a heat pump and a modern coal boiler while considering the status quo.

II. METHODOLOGY OF ECONOMIC EVALUATION

A. *Minimum price*

The economic assessment is made using the minimum price methodology, based on the economic assessment concept using net present value (NPV) of future cash flows caused by implementation of a project. The general rule is that a project is implemented if the NPV is positive, and rejected if the NPV is negative. If the NPV is zero, the assessed project brings the same economic effectiveness as other alternative options of investment by the investor. The minimum price is then a price for which (at least) production has to be sold for the investor to realise the required amount per capital invested. The formula for determining the minimum price can be expressed as follows:

$$NPV = \sum_{t=0}^{Tn} (c_{\min,t} \cdot Q_t - E_t) \cdot (1 + r_n)^{-t} = 0 \quad (1)$$

where

Tn	...	is the assessment period [years]
t	...	serial year of the assessment period
cmin,t	...	minimum production price in year t [CZK]
Qt	...	quantity of production in the year t [CZK]
Et	...	expenditures in the year t [CZK]
rn	...	nominal discount rate [-]

The minimum price methodology makes it possible, e.g., to derive the price of dedicated energy crops on agricultural land, price of pellets produced, etc. Its logic consists in the fact that it is the lower limit price from the perspective of the producer (investor in the project). A rationally reasoning investor will not accept prices for the production lower than the minimum price.

The minimum price methodology can also be used to derive an imaginary price for which one produces heat for space heating of one's own single-family house, considering the costs of installation of the equipment (e.g., pellet boiler) and costs of its operation for the expected lifetime (fuel, repair, maintenance, inspection and other costs).

B. *Minimum price of pellets*

Derivation of the minimum price of pellets produced based on locally available biomass involves the entire cycle from biomass acquisition (e.g., by growing it) to its transport, storage and reprocessing into pellets. We also need to consider energy loss incurred due to biomass storage (biodegradation) and biomass reprocessing.

Dedicated energy crops on agricultural land in the Czech Republic is usually represented by the following four energy crops: reed canary grass, miscanthus (Chinese silver-grass), dock (schavnat) and SRC (short rotation coppice) plantations. The derivation of the price of dedicated energy crops uses economic models covering the entire life cycle of stands/plantations of energy crops; see, e.g. [5] for more detail.

The analysis presented in this paper makes use of results obtained from solving of research projects focused on identification of biomass potential [6] and analysis of options for using locally available biomass [7].

Considering the typical yield range of biomass of 3.8-7.5 t(FM)/ha for reed canary grass, 3.5-9.4 t(FM) /ha for schavnat, 5-12.5 t(FM)/ha for miscanthus, and 5-12.5 t(FM)/ha for SRC plantations the minimum price range for grown biomass is 60 - 202 CZK/GJ (depending on the soil and climate conditions of the site), and the typical range is 108-135 CZK/GJ. Calculations of minimum price of grown biomass assume the use of standard cultivation procedures; they apply a discount rate of 9% (see [6], [7] for more detail).

Reed canary grass and miscanthus are harvested in winter, when the biomass moisture content is low (approx. 20%); schavnat is harvested in summer (July) at a moisture content of 15%. Additional drying before the pelletisation is not assumed. SRC plantations are also harvested in winter; however, the wood chips have a moisture content of approx. 52%. The moisture content of the chips can be reduced to approx. 30% by approx. one year storage. Here, additional drying before the pelletisation is necessary.

The price of residual straw from conventional agriculture is substantially locally variable (depending on local demand, livestock numbers, structure of conventional agriculture in the area, etc.), and is typically between 800 and 900 CZK/t.

The price of wood chips from forest residuals is again strongly locally variable, and can be estimated at approx. 100-130 CZK/GJ.

The energy content loss in the course of storage can be estimated at approx. 7% pro perennials and 10% for wood chips from SRC plantations [8]. The loss during pelletisation can be estimated at approx. 5%. The energy content loss increases the minimum price of the final product – pellets.

The minimum price of pellet production can be derived, e.g., from the parameters of the pelletising plant offered by ProPelety (see www.propelety.cz). The production plant capacity is approx. 1000-1200 t/year. The plant acquisition costs are approx. 5.4 million CZK (including installation), and the annual operating costs are approx. 1.7 million CZK (excluding costs of input raw material). The lifetime of the equipment is considered to be 10 years; this time is used as the project assessment period Tn (see Equation 1). If we consider the concept of local pellet production as a municipal project, a significantly lower discount rate, say 3%, can be applied. The project is then primarily not focused on business profit but on production of affordable pellets for local consumption. With the discount rate of 3%, the minimum price of pelletisation (excluding distribution to end consumers) is approx. 120 CZK/GJ (excl. VAT). Considering the same discount rate as for dedicated energy crops, the minimum price of pelletisation increases to approx. 130 CZK/GJ. The reason for the relatively low effect of increased discount rate on the minimum price of pelletisation is the high share of annual operating costs in the total annual costs (including depreciation).

The minimum price of pellets from dedicated energy crops (including the effect of energy content loss and costs of biomass storage totalling approx. 8-10 CZK/GJ) will then be approx. 280-310 CZK/GJ (including VAT). The minimum price of pellets produced from forest residuals can be estimated in a similar range. As for pellets produced from residual straw from conventional crops, the minimum price (considering the price of straw of 800 CZK/t) is approx. 50-80 CZK/GJ lower. However, the question is availability of such residual biomass in the area, and how the price of straw would increase if demand for it increased. Generally, we can assume that the price of straw would converge to the price of dedicated energy crops.

C. Heat price calculation methodology

Unlike dedicated energy crops and subsequent pelletisation, which are business activities and thus have to respect principles of business environment (depreciation, income tax, business entity as a VAT payer), heat production in a single-family house is a non-entrepreneurial activity. Thus, we have to respect specific characteristics of households in the economic assessment – calculation of the minimum price of heat produced. Above all, there is no depreciation of the input investment, low discount rate (a household has a lower time value of money due to the real possibility of investing it by a typical household) and VAT payment. Moreover, the heat production is not burdened by income tax. The minimum price of heat thus assumes the form of total costs of production of 1 GJ of heat used for heating the building (respecting the given equipment lifetime and time value of money).

The minimum price of heat was assessed for 3 basic options of renovation of an existing system based on brown coal – a pellet boiler, a modern automatic brown coal boiler, and a heat pump. These three renovation options were compared with the status quo, comprising a brown coal boiler with manual stoking and low efficiency (see below). Besides the costs of fuel, equipment, etc., the minimum price of heat is significantly affected by the absolute heat consumption. That is why 4 alternatives of single-family houses were defined for the assessment with a wide range of specific heat demand for space heating. These 4 alternatives reflect the typical house options in terms of energy demand depending on age, degree of renovation if any, etc.; see below for more details.

The calculation of the minimum price of heat uses the brown coal and electricity prices at the 2017 price level. Brown coal is not a generally traded energy commodity, and is also not a homogenous category (see [9] for more detail). Its parameters, such as calorific capacity, ash content, etc., may differ greatly from supplier to supplier. Generally speaking, households can only use coal with better qualitative parameters than coal used for power and/or heat generation in large power plants and heating plants. The price of brown coal for households reflects, to a great extent, the demand for this coal and the price of its potential substitutes (pellets, etc.). The calculations worked with two types of coal: ordinary lower-grade brown coal (for standard boilers with manual stoking), which is significantly cheaper; and high-grade sorted brown coal for boilers with automated fuel replenishment.

III. HEATING OF SINGLE-FAMILY HOUSES

A. Analysis of present and future situation in heating

Statistical surveys feature the following types of heating: central, in-house boiler, flat boiler, and stoves. Furthermore, statistics different the following solid fuels: coal, coke, coal briquettes, wood, wood briquettes, gas, electricity.

The analysis shows that pellets are economic on a par with natural gas; that said, heating using pellets provides comparable operator comfort. Pellets can reliably economically replace (displace) costlier energies, meaning that they easily substitute for electricity. Their level of operator comfort and higher combustion efficiency help them compete with brown coal too. Nevertheless, competition with heat pumps can be considered in both renovated and new buildings where operator comfort is a preference. In economic terms, heat pumps typically compete with natural gas; again, heating using pellets provides comparable operator comfort.

Based on empirical practical findings, installation of heat pumps can be expected in single-family houses, smaller apartment buildings, hotels and guesthouses. According to Marek Bláha, for example, new buildings typically use heat pumps in 15-30% of sites where electricity the available high-comfort energy media. With growing interest in low-energy and passive construction, we can expect an expansion in the market segment of small heat pumps suitable for domestic hot water and auxiliary heating. The reasons include the fact that air-to-air heat pumps do not require a heating system. In older developments, we can historically expect a shift to heat pump in areas that have used electricity for heating. The high original operating costs of these buildings make the substitution of electric heating with heat pumps economically attractive while retaining the same operator comfort. Due to the higher operator comfort, we can even expected a higher use of heat pumps in the wood heating segment, to a certain degree, replacing fireplaces and stove fireplaces. In this market segment, the practice shows an approx. 20% substitution. A similar situation is observed in the gas heating segment, where up to 20% of installations are being replaced. The reason is that replacement of an old boiler with a new one implies, in some cases, increased costs of chimney renovation, heating system renovation or additions, higher price of natural gas condensing boilers, etc. In future, we may even expect gradual reverse replacement of the pellet segment. The reason is the relatively high price of pellets, which may still increase over time due to the gradual decrease of available biomass. The case of Sweden is a proof of this assertion: there, replacement of boilers with pellet boilers was supported on a massive scale. The experiment failed as a consequence of increasing prices of pellets, large spatial demand and lower user comfort compared to heat pumps. Even as soon as 3 years afterwards, households shifted from the boilers to heat pumps [10], [11].

When replacing electric heating (electric boilers, storage heaters, radiant heating panels), the user comfort is slightly reduced because one needs to consider fuel and ash handling. In this market segment, heat pumps will have a competitive advantage due to their high user comfort. In the brown coal segment, pellet combustion offers greater handling comfort, better availability of pellets, cleaner operation, less storage space demand, lower ash content (0.5 – 2%, [12], [13]), higher combustion efficiency of approx. 87% [14] compared to approx. 65% in original old brown coal boilers. In modern automatic boilers, which offer comparable seasonal efficiency of up to 84% [15] and often enable alternative pellet combustion, the competitive is more between brown coal vs. pellets as the fuel. Compared to pellets, brown coal has a higher ash content of 10 – 12% [16], lower combustion efficiency and generally lower handling comfort. Brown coal can be expected to carry higher environmental taxation burden in future.

B. Model single-family houses

The definition of the 4 model single-family houses (the same dwelling size of 120 m2 is assumed in all for the sake of comparability) used data from the 2011 Census [1] and data from studies dealing with analysis of the housing capacity in the CR [17], [18]. These data were then confronted with the reality of the housing capacity of the municipality of Zbilidy, situated near the regional capital Jihlava. The municipality currently has 180 inhabitants and 92 houses. A total of 65 houses are permanently inhabited, the rest are used for recreational purposes. Out of the 92 houses, 52 is heated by brown coal. The analysis of the municipality's housing capacity estimated the total current energy demand for heating at approx. 6500 GJ/year; a large part of the houses are not renovated, or only with minor renovations. Considering a fundamental renovation of all the single-family houses, the total energy demand for heating could be reduced to approx. 2800 GJ/year. As it is, the energy demand for heating (considering the efficiency of pellet boilers) would be covered by less than 500 tonnes of pellets a year.

In order to compare the economic effectiveness of the different heating methods, we defined four model houses with a dwelling size of 120 m2 with the following heat demands for heating: D1 330 kWh/m2.year, D2 110 kWh/m2.year, D3 70 kWh/m2.year, D4 50 kWh/m2.year.

The first building, D1, is a representative of old development (before 1930), not renovated and no major energy-aware renovation is expected in future. The second house, D2, is a representative of the building category implemented in the 1990s; its heat demand for heating corresponds to partly renovated buildings of the first type. Both house types currently use brown coal for heating (central heating plus a coal boiler with manual stoking).

The houses D3 and D4 represent contemporary construction; D4 corresponds approximately to a low-energy house, or houses built after 2020. At the same time, D3 and D4 may also represent older buildings that have undergone fundamental renovation. The comparison excludes single-family houses in the passive standard, which typically use heating methods different than those listed here.

Besides the demand for heating, the energy demands of D1 to D4 also include energy demands for domestic hot water, at 2.5 kWh/person/day (the average number of persons per dwelling is 2.45 [1] in all the cases). For the sake of simplicity, the energy demand for DHW is added to the energy demand for heating.

C. *Heating equipment parameters*

The model calculations use typical parameters of the heating equipment, adopted from the literature [19] and the authors' own experience of implementation of energy optimisations, audits and assessments. The electricity prices are adopted from the calculator in the TZB-info website [20]. The brown coal prices are adopted from coal suppliers' offers – coal is considered in two categories, namely ordinary brown coal at 173 CZK/GJ and high-grade sorted coal for automatic boilers at 273 CZK/GJ [19]. The price of pellets is assumed at 283 CZK/GJ, which approximately corresponds to the lower bound of the minimum price of pellets produced from locally available biomass.

The prices of equipment are assumed in the calculations inclusive of VAT as follows: Brown coal boiler with manual stoking 28,000 CZK (seasonal efficiency approx. 65%), brown coal (efficiency 80%) and/or pellet (efficiency 85%) boiler with automated stoking 112,000 CZK, and air-to-water heat pump (seasonal heating factor 2.9) 265,000 CZK for the larger version (buildings D1 and D2) and 200,000 CZK for the smaller one (D3 and D4). The above costs include costs of installation, inspection, etc. Change of the whole heating system is not assumed when the heating method changes.

As a permissible simplification, we assume that a typical brown coal boiler made of "sheet steel" has a lifetime of 15 years, the same as a heat pump. Heat pumps typically require a reinvestment of approx. 40-60% for water-to-water designs and 60-80% for ait-to-air (water) designs after 15 years. Besides fuel/electricity costs, the operating costs include the costs of repairs, maintenance and periodic inspections. For example, the operation of a pellet boiler is cheaper compared to a heat pump, as it involves only a flue inspection amounting to approx. 700 CZK/year.

IV. Results of Comparison of Different Heating Methods

The results of the model calculations are summarised in Tables 1-4.

Table 1. Prices of heat for heating single-family house D1 with different methods

	D1, 330 kWh/m2/year, energy need 151 GJ			
	coal, old boiler	***coal, new boiler***	***pellet boiler***	***heat pump***
Energy consumption [GJ]	232	188	177	52
Inv. costs [th. CZK]	28	112	112	265
Fuel costs [th. CZK]	41.2	51.4	50.2	39.2
Heat min. price [CZK/GJ]	280	390	380	380

Table 2. Prices of heat for heating single-family house D2 with different methods

	D2, 110 kWh/m2/year, energy need 56 GJ			
	coal, old boiler	***coal, new boiler***	***pellet boiler***	***heat pump***
Energy consumption [GJ]	86	70	65	19
Inv. costs [th. CZK]	28	112	112	265
Fuel costs [th. CZK]	15.2	19.0	18.5	17.6
Heat min. price [CZK/GJ]	300	480	470	640

Table 3. Prices of heat for heating single-family house D3 with different methods

	D3, 70 kWh/m2/year, energy need 38 GJ		
	coal, new boiler	***pellet boiler***	***heat pump***
Energy consumption [GJ]	48	45	13
Inv. costs [th. CZK]	112	112	200
Fuel costs [th. CZK]	13.0	12.8	12.1
Heat min. price [CZK/GJ]	540	540	680

Table 4. Prices of heat for heating single-family house D4 with different methods

	D4, 50 kWh/m2/year, energy need 30 GJ		
	coal, new boiler	*pellet boiler*	*heat pump*
Energy consumption [GJ]	37	35	10
Inv. costs [th. CZK]	112	112	200
Fuel costs [th. CZK]	10.1	9.9	9.4
Heat min. price [CZK/GJ]	590	590	770

It follows from the analysis results that it is much cheaper for buildings D1 and D2 to operate a brown coal boiler with manual stoking. The heat price (calculating using the minimum price method) is 280 and 300 CZK/GJ for these options, respectively. The annual operating costs are approx. 41,000 and 15,000 CZK, respectively; the difference is caused by the triple energy consumption of D1 compared to D2. If the heat source is replaced and a modern automatic coal or pellet boiler is installed, the price of heat increases significantly to 390 and 480 CZK/GJ, respectively, for brown coal, and 380 and 470 CZK/GJ, respectively, for pellets. Even though a modern coal/pellet boiler has a substantially higher efficiency than a manually operated boiler, there is no operating cost saving. The reason is the significantly higher price of coal suitable for the automatic boiler, or higher price of pellets. The comparison is somewhat distorted by the fact that we are comparing options with fundamentally different levels of comfort; see the discussion above.

A heat pump has a heat price similar to that of a modern coal/pellet boiler only for the building D1 (380 CZK/GJ). The high investment costs are felt adversely in the case of D2, which has a substantially lower heat demand. If the comparison includes the buildings D3 and D4, then the option with a heat pump leads to significantly higher heat prices (680 and 770 CZK/GJ, respectively), but the annual operating expenditures are similar to those of coal or pellets.

V. CONCLUSIONS

Pellets produced from locally available biomass are an attractive alternative to domestic brown coal. The minimum price analysis shows that, under certain conditions, they can be competitive with sorted brown coal combusted in modern boilers with automated stoking. In the case of single-family houses with a high energy demand for heating, however, the comparison shows that the heat pump is by far the best – it provides much better comfort compared to both a modern coal boiler and a pellet boiler with approximately the same minimum heat price. Conversely, a heat pump becomes economically non-competitive with coal/pellet boilers with decreasing heat demand.

Simultaneously, the calculation results show that a shift to more efficient heating methods (replacement of old coal boiler) leads to significantly higher minimum heat prices. That greatly reduces house owners' motivation to renovate the heating method, even where it would result in much better comfort. The house owners' motivation to replace obsolete brown coal boilers with modern technologies can be radically increased with suitable economic incentives, which typically assumes the form of investment subsidy at present.

The competitiveness of pellets produced from locally available biomass is affected by a number of factors, including primarily availability of residual biomass from forest logging and conventional agriculture, ability to grow energy crops, and last but not least, soil and climate conditions of the site, which fundamentally affect the minimum price of the biomass grown. The minimum price of dedicated energy crops shows a considerable variance and, in many areas with low biomass yields, the higher price might cause non-competitiveness of pellets with coal or heat pumps.

ACKNOWLEDGMENT

This work was supported by the Technology Agency of the Czech Republic [grant number TD03000039, TA04020970].

REFERENCES

[1] Czech Republic - Census 2011: Houses and flats – doc. 170216-14. Czech Statistical Office. Available at https://www.czso.cz/documents/10180/20551777/17021614.pdf/6bf03ae5-3196-464e-9200-611c97ba8484?version=1.0, (Accessed May 2, 2017), in Czech

[2] Statistics on small heating devices using solid fuels for decentralized heating. Ministry of Industry and Trade 2015, available at http://www.mpo.cz/dokument169398.html, October 11, 2016, in Czech

[3] Emission factors for evaluation of environmental feasibility of energy audits. Ministry of the Environment, 2017. In Czech. Available at http://www.mzp.cz/C1257458002F0DC7/cz/emisni_faktory_pomery_castic/$FILE/OOO-SdeleniEmisniFaktory-20170214.pdf (Accessed May 2, 2017)

[4] Update of State Energy Policy of the Czech Republic.2013. Available from: http://portal.cenia.cz/eiasea/detail/SEA_MZP148K (Accessed May 2, 2017)

[5] K. Havlíčková, J. Weger, J.Knápek, "Modelling of biomass prices for bio-energy market in the Czech Republic". Simulation Modelling Practice and Theory. 2011, vol. 19, no. 9, p. 1946-1956. ISSN 1569-190X

[6] K. Vávrová et al. Report of the research project TA04020970 (Technology Agency of the Czech Republic). 2016, VUKOZ, v.v.i.

[7] K. Vávrová et al. Report of the research project TD03000039 (Technology Agency of the Czech Republic). 2016, VUKOZ, v.v.i.

[8] J. Knápek et al. "Effectiveness of biomass for energy purposes: a fuel cycle approach". Wiley Interdisciplinary Reviews: Energy and Environment. 2015, 4(6), pp. 575-586. ISSN 2041-8396

[9] J. Bejbl, J. Bemš, T. Králík, O. Starý, J. Vastl, "New approach to brown coal pricing using internal rate of return methodology". Applied Energy, 2014, vol. 133, p. 289-297

[10] Kolik uspoříme tepelným čerpadlem. Typické příklady instalací. [online] Opočno: PZP Komplet website. Available from: http://www.tepelna-cerpadla-pzp.cz/cs/tepelna-cerpadla-kolik-usporime-90.html (Accessed May 2, 2017)

[11] M. Bláha: Face-to-face interview with a representative of Tepelná čerpadla IVT s.r.o. (2012)

[12] M. Řebíček, "Výroba a distribuce dřevěných pelet v ČR." [online] Company article of CDP IVORY, s.r.o. TZB-info, Available from: http://oze.tzb-info.cz/peletky/8965-vyroba-a-distribuce-drevenych-pelet-v-cr (Accessed May 2, 2017)

[13] Company materials of PELLETIA s.r.o. (2001) Test certificate of Fuel Research and Utilisation Institute in Praha-Běchovice. [online] TZB-info, professional Internet portal. Available from:

http://www.tzb-info.cz/485-peletky-od-firmy-pelletia (Accessed May 2, 2017)

[14] Z. Lyčka, "Náklady na vytápění dřevními peletami". [online] TZB-info, professional Internet portal. Available from: http://oze.tzb-info.cz/biomasa/6263-naklady-na-vytapeni-drevnimi-peletami (Accessed May 2, 2017)

[15] Company article of KOVARSON s.r.o. "Automatický kotel GEKON na hnědé uhlí, pelety v 5. emisní třídě a Ekodesignu s účinností 95,5 %". [online] TZB-info, professional Internet portal. Available from: http://vytapeni.tzb-info.cz/vytapime-tuhymi-palivy/13423-automaticky-kotel-gekon-na-hnede-uhli-pelety-v-5-emisni-tride-a-ekodesignu-s-ucinnosti-95-5 (Accessed May 2, 2017)

[16] Company article of BENEKOVterm s.r.o. "Automatické kotle Benekov – Úspora není jen v nákladech na palivo (I)." [online] TZB-info, professional Internet portal. Available from: http://vytapeni.tzb-info.cz/vytapime-tuhymi-palivy/11138-automaticke-kotle-benekov-uspora-neni-jen-v-nakladech-na-palivo-i. (Accessed May 2, 2017)

[17] J. Antonín, "Průzkum fondu budov a možností úspor energie". Desk research of existing studies and calculation verification for residential buildings. With contribution of Šance pro budovy working group. Šance pro budovy, Praha. 2014

[18] J. Antonín and P. Holub, "Analýza fondu nerezidenčních budov v České republice a možností úspor v nich". Šance pro budovy, 2016 Praha.

[19] [19] Comparison of cost for space heating in family houses. TZB-info website – 2016 update. Available from http://vytapeni.tzb-info.cz/tabulky-a-vypocty/138-porovnani-nakladu-na-vytapeni-tzb-info (Accessed May 2, 2017)

[20] Calculator of electricity prices. TZB-info website – 2016 update. Available from http://kalkulator.tzb-info.cz/cz/dodavka-elektricke-energie-vyber-kraje, (Accessed May 2, 2017)

Conducted and Radiated Emissions of LED Lamps and Intelligent Lighting Systems

Róbert Istók
Óbuda University
Kálmán Kandó Faculty of Electrical Engineering
Power System Department
Budapest, Hungary
istok.robert@kvk.uni-obuda.hu

Abstract — **In case of RF outdoor lighting control the EMC electromagnetic compatibility question become more important, because there we have two different devices with different EMC requirements which have to work together. For EMC characterization, we have to follow the next steps. The first step is to determine which standard requirements have to be fulfilled by the intelligent lighting system besides the EN55015 [1] and EN61547 [2] lighting standard. The next step is how to do the tests, which kind of measurement system can be used. If we are doing radiated emission, do we use an anechoic chamber or an open area test site? The third steps are: how to decrease the noise emission under the limits if these does not passed, the tests.**

Keywords: controller; EMC; LED; RF outdoor lighting controls; antennas, CDN, radiated emission, conducted emission

I. INTRODUCTION

Today the outdoor lighting is in a total transformation. The most important scope of that transformation is the energy saving. The first step in this direction was the replacement of high energy consumption lamps with modern low energy consumption lamps like HID high-intensity discharge, HPS high pressure sodium and LED light emitting diodes lamps. These lamps could have up to 50% energy saving [3]. The next step is ongoing and that is the utilization of lighting control[4]. A simple definition of lighting control could be the next: Optimized switch on/off the lighting in conformity with actually outdoor status: weather, level of darkness, and light with full brightness in case of motion detection, otherwise light with low brightness.

Only the optimized switch on/off control produced an 8-10% energy savings [5]. Control system which can dim the lamp save up to 35% power. A combination of control system and the energy efficient lamps could reach a 67% energy saving . To see the importance of the lighting control let is see an example about USA energy consumption for outdoor lighting in 2010. A report of Department of Energy estimates that in United State are 52,6 million roadway fixtures installed from which 26,5 million streetlights and 26.1 million highway fixtures. Conventional lamps use 0,25 kWh every hour. Assuming that lights are on an average of 12 hours/day, each lamp uses 3 kWh/day, or around of 1100 kWh/year. The result is around of 57 billion kWh per year for the US. With a charge average of 0,11$/kWh. That is mean around of 120$ for each lamp per year:

The total annual energy consumption for USA is bigger than 6 billion $/yr[5]. If the combination of intelligent system and modern lamp is used there could be save around of 3 billion $/yr.

II. RADIO FREQUENCY LIGHTING CONTROL SYSTEM

Typically, a lighting control system contains the next parts:

- Lamp + dimmable electronic ballast
- controller. The role of controller is to switch on/off and dim the lamp. For lamp dimming, is used DALI protocol or 0-10V analog signal [6].
- sensors: movement detector, daylight sensor,
- gateway drive the controllers. In case of RF lighting control systems, the communication between gateway and controllers could be done through SRD and ISM radio or mobile phone network frequencies [7].
- Energy management system. The communication between gateway and energy management system typically is done through Ethernet [8].

III. EELECTROMAGNETIC COMPATIBILITY MEASUREMENT OF INTELLIGENT LIGHTING CONTROL SYSTEM

Because we have a lighting equipment, the lamps and a radio equipment the controller, the intelligent lamp has to be tested not only with lighting EMC standards but with radio equipment EMC standards too [9].

For radio equipment (controllers) the next standard has to be used for EMC emission and immunity tests: ETSI EN 301 489-1 Electromagnetic compatibility and Radio Spectrum Matters (ERM); Electro Magnetic Compatibility (EMC) standard for radio equipment and services.

For lighting equipment (lamps) the next standards has to be used for emission: CISPR 15, Limits and methods of measurement of radio disturbance characteristics of electrical lighting and similar equipment, for immunity: EN 61547 Equipment for general lighting purposes. EMC immunity requirements. In Fig. 1 it can be seen the measurement set up for conducted noise emission. The measurement receiver has to fulfil the next requirements from bandwidth point of view Table 1. The measurement frequency range has to start from 9kHz. The amplitude accuracy has to be ±2 dB. For this measurement, the quasi-peak and the average detectors are used. Using the peak detector, the measurement time could be reduced [10]. The measurement time typically is chosen at half of used bandwidth.

Table 1. Bandwidth requirements CISPR 16-1-1

Bandwidth	Frequency range
200 Hz	9 kHz to 150 kHz
9 kHz	150 kHz to 30 MHz
120 kHz	30 MHz to 1000 MHz
1 MHz impulse	1 GHz to 18 GHz

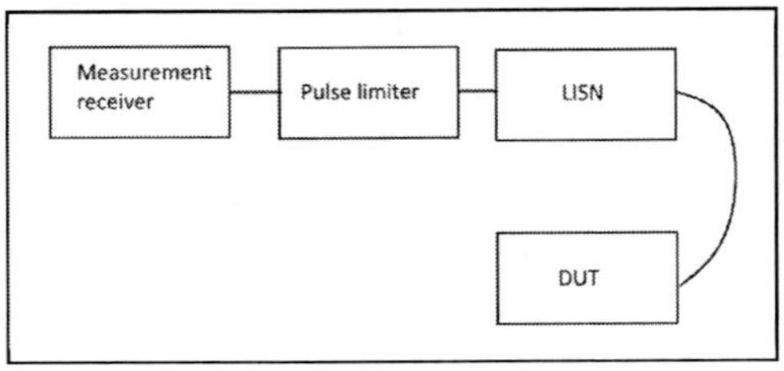

Figure 1. - Conducted emission measurement netwok

The Line Impedance Stabilization Networks (LISN) Fig. 2 have to fulfil the next requirements:

- Ensure a stable impedance for the DUT through the measured frequency range 50Ω.
- Filter the noise signals from power supply network

The transient limiter protect the EMI receiver against of pulses. In Fig. 3 is presented the conducted emission measurement results of controller and LED lamps. The upper limit is the QP quasi peak limit, the lower limit is AV average limit [11]. In spite of fact that the lamp and the controller passed the emission test separately as can be seen in Fig. 4 and Fig. 5. The combination of these two devices failed the test Fig. 3.

It is mean that the measurement result of a device which provide from a combination of two conducted EMC test passed devices is not necessarily a device which will pass automatically the conducted EMC test.

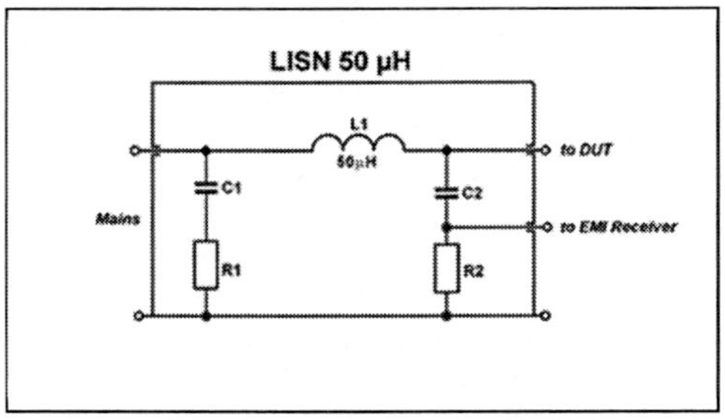

Figure 2. - Line impedance stabilization netwok

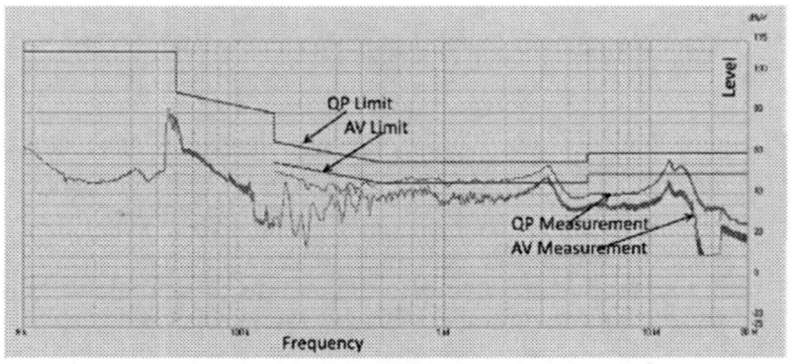

Figure 3. - Controller +LED lamp conducted emission measurement result

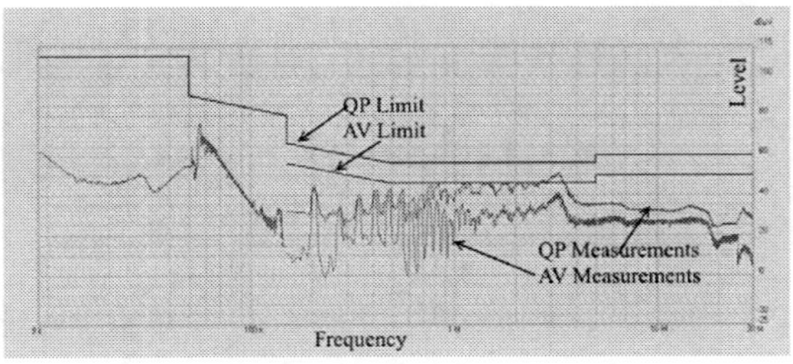

Figure 4. - Controller conducted emission measurement result

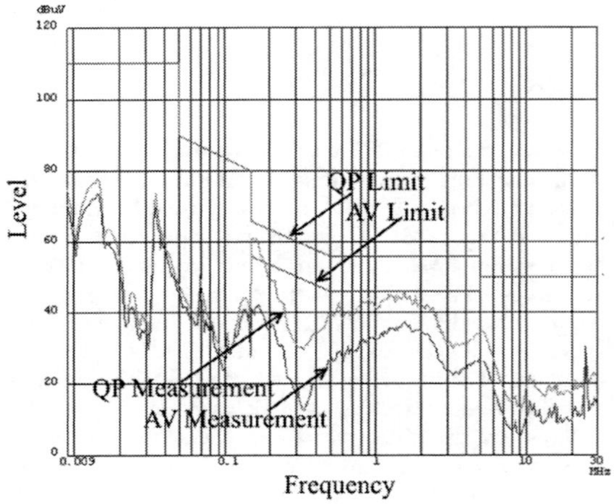

Figure 5. - LED lamps conducted emission measurement result

To solve this problem the first step is to check the frequencies where the device failed the test. In this cases, the frequency is around of 3MHz. For this frequency we can not use ferrite rings for noise suppression because the frequency of the signal is low.

Typically the ferrite rings are used in 5MHz-500MHz frequency band. A common-differential mode filter Fig. 6 is a better choice [12]. If we use a filter we do not know if the noise is a differential or a common mode noise. The ferrite ring suppresses only common mode noise.

As we seen in Fig. 7 the filter reduce the conducted emission under a limit which is acceptable for a positive conducted measurement result. In case of gateway the situation is simple because it is used alone. The conducted noise emission is not influenced by another device.

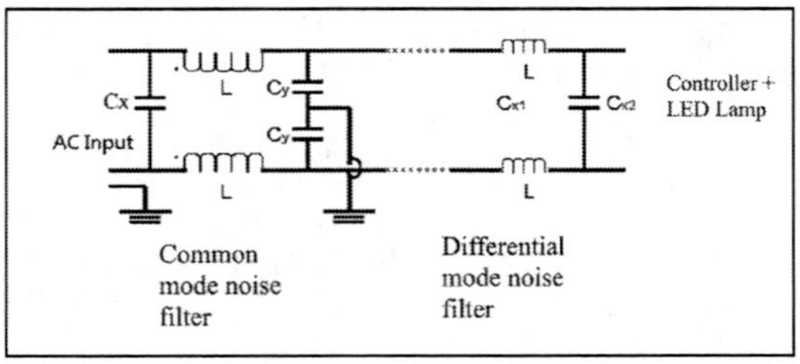

Figure 6. - Common – differential mode filter

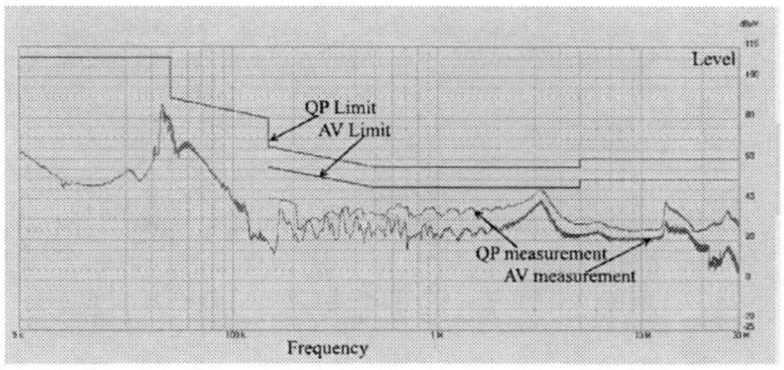

Figure 7. - Common – differential mode filter Controller +LED lamp+filter conducted emission measurement result

A measurement method for noise emission in 30MHz-300MHz frequency range could be CDN – coupled decoupled network measurement system. This measurement system Fig. 8 is used in Lighting industry. The CDN working priciple is the next:

Decoupling the power netwok and filter the high frequency noises from the LED lamp and controller and protect the measurement system from high pulses wich can appear from power supply network. The coupling network is a high pass band filter wich drive the high frequency signal to the receiver. The construction of the CDN is presented in Fig. 9. The CDN measurement method use quasi-peak detector which measure the common mode current on the power supply line or on the controller lines. In Fig. 10 it can be see a measurement result with CDN method. It is better if this measurement is done in a anechoic chamber. In this case the ambient noise cannot influence the measurement results.

The 30MHz-300MHz frequency range could be measured with antenna measurement method too. For the radiated measurement noise emission are used an OATS – open area test site or an anechoic chamber and different antenna types.

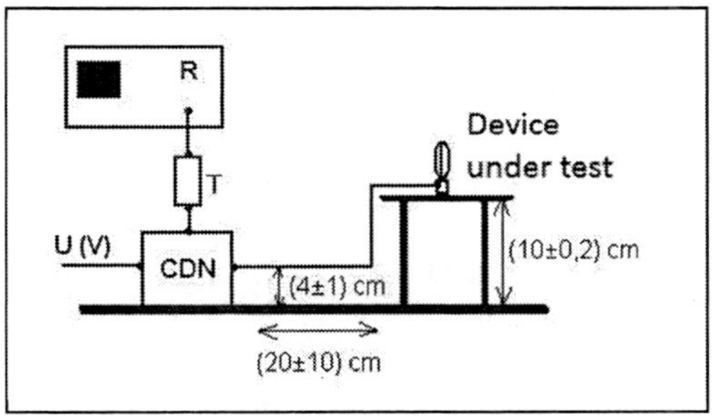

Figure 8. - CDN measurement method

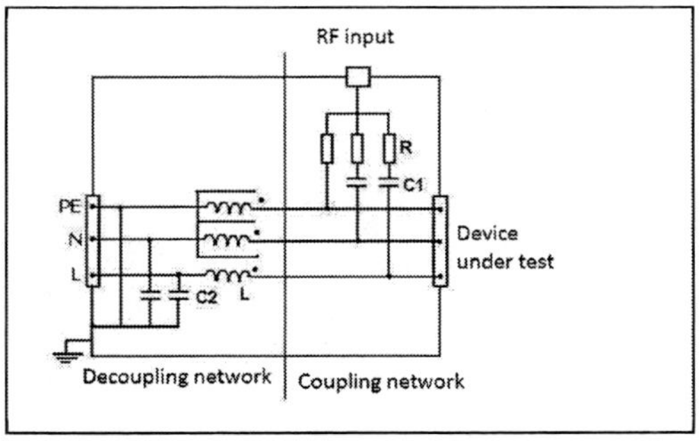

Figure 9. - Coupling decoupling network

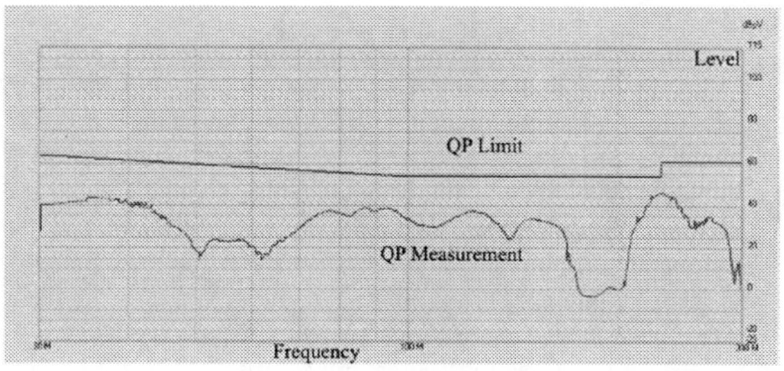

Figure 10. - Controller+ LED lamp CDN method noise emission

During the measurements, the anechoic chamber was used. In Fig. 11 is presented a measurement results of controller and LED lamp. The main signal is the communication signal. Except the communication signal frequency there are only some harmonics of this signal which are much lower than it. That is mean with an adequate design the spurious emissions could be keep on low level.

As could be seen in Fig. 12 the radiated power of signal has different frequencies, that is because the communication used the FHSS mode. Frequency Hopping Spread Spectrum (FHSS): technique in which the transmitter signal occupies a number of frequencies in time, each for some period of time, referred to as the dwell time. The difference amplitude between the two signals from Fig. 12 are coming from the antenna of controller which radiated in horizontal polarization.

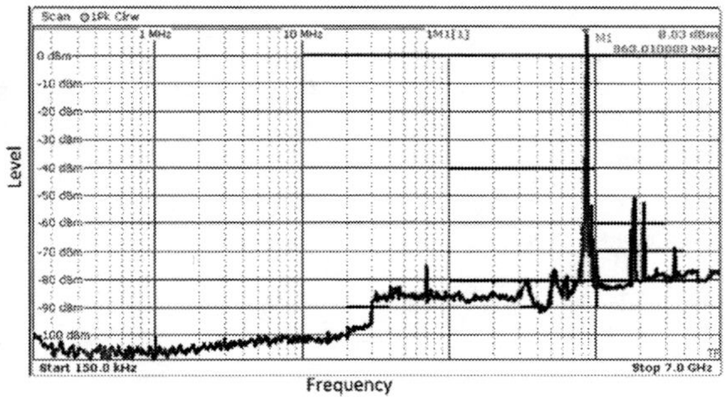

Figure 11. - Controller and LED Lamp spectrum of frequency emission

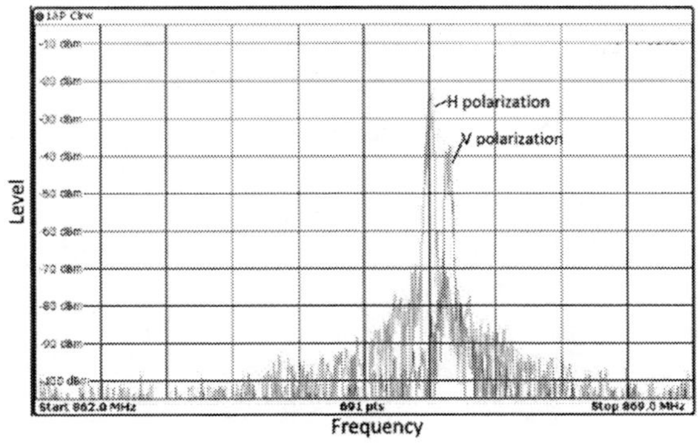

Figure 12. – Power of signal radiated in vertical and horizontal polarization

IV. Conclusion

In case of RF intelligent lighting network for EMC noise emission test besides the EN55015 lighting standard the ETSI EN 301 489-1 standard have to be used for testing of radio equipment. As we seen the lamp and the controller separately pass the test requirements, but together there failed the test. The lamp and the controller are a system, and have to be tested together. In this case the system fail the test. For reducing of noise emission under the limits a proper filter have to be used. Typically a common-differential mode filter with cutting band in 2 - 4MHz frequency band are used in this case. If the frequency of the noise is over 20MHz a ferrite coil could be used. The ferrite are placed on power supply cable. For 30-300MHz frequency band the radiated emission noise could be measured with two measurement systems: the CDN and the antenna method. The CDN method are used for simple electronic circuits. Over 300MHz are used only the antenna method. Antenna method give more information about noise emission, as polarization of radiated noise. The communication signal of the intelligent lighting system could be measured only with antenna measurement system.

References

[1] Limits and methods of measurement of radio disturbance characteristics of electrical lighting and similar equipment EN 55015:2006

[2] Equipment for general lighting purposes - EMC immunity requirements EN 61547:2009

[3] A. N. Shaikh, N. Shah, A. Tripathy, M. H. Naikwadi "Intelligent Monitoring and Control Rendered to Street" Lighting Advances in Technology and Engineering (ICATE), Mumbai, India 23-25 Jan. 2013 pp. 1–5

[4] Y. Liang, ; Z. Jia ; H. Yao ; J. Chen, "Housing intelligent lighting control strategy research" 2014 IEEE 3rd International Conference on Cloud Computing and Intelligence Systems, Shenzhen, China 27-29 Nov. 2014 DOI: 10.1109/CCIS.2014.7175829 Page(s):728 – 731

[5] System Design Considerations When Planning a Network Controlled Street Light Project, Cimcon Lighting, www.cimconlighting.com

[6] R. Lenk, C. Lenk, "Practical Lighting Design with LEDs", Second edition, John Wiley & Sons, 2017. March

[7] A. Martian, "Evaluation of spectrum occupancy in urban and rural environments of Romania", Revue roumaine des sciences techniques Série Électrotechnique et Énergétique 59, 4, pp. 87–96, 2014.

[8] B. Baranyai, I. Kistelegdi, "Energy management monitoring and control of public buildings", Pollack Periodica, Akadémiai Kiadó, Budapest, Hungary, 2014 Vol. 9, No. 2, pp. 77–88

[9] D. S. Lee, Chang-Zhe Sung "RFID based intelligent lighting system development for the convenience store", Systems Man and Cybernetics (SMC), Istanbul, Turkey, 10-13 Oct. 2010 pp. 4012 – 4017

[10] T. Williams, "EMC for Product Designers", Fourth Edition, Elsevier, 2007

[11] IEEE 802.15.4-2011 IEEE Standard for Local and metropolitan area networks--Part 15.4: Low-Rate Wireless Personal Area Networks (LR-WPANs)

[12] M. Miloudi, A. Bendaoud ; H. Miloudi ; S. Nemmich ; H. Slimani, "Analysis and reduction of common-mode and differential-mode EMI noise in a Flyback switch-mode power supply (SMPS)", Telecommunications Forum (TELFOR), 2012 20th Belgrad, Serbia 20-22 Nov. 2012 pp. 1080 – 1083

Study on Possibilities for Improvement of Energy Efficiency of Power Transformers in Electric Distribution Networks

Valentin Gyurov
Department of Electric Power Supply and Equipment
Technical University of Varna
Varna, Republic of Bulgaria
e-mail: valentin.giurov@tu-varna.bg

Ilian Iliev
Department of Electrification of Mining Industry
University of mining and geology "St. Ivan Rilski"
Sofia, Republic of Bulgaria
e-mail: ilian.iliev@mgu.bg

Rumen Kirov
Department of Electric Power Supply and Equipment
Technical University of Varna
Varna, Republic of Bulgaria
e-mail: kvc_electroinvest@yahoo.com

Georgi Georgiev
Plethora Ltd.
Sofia, Republic of Bulgaria
e-mail: georgidg@gmail.com

Abstract—**Transformer power in the electricity supply systems (ESS) of the Republic of Bulgaria exceeds the consumed power of industrial sites several times. The average load of the power transformers (PT) ranges from 0.4-0.65 and each 1 MW maximum load corresponds to 1.5-2 MW installed transformer power. The reasons for such low-load of transformers are usually the incorrect methodological approach in choosing the number and power of the power transformers and the erroneous forecasts of the dynamics in the development of the electric consumption. The low load of the power transformers leads to additional power and energy losses and inefficient use of the financial investment. The report presents methods for optimal load of power transformers in order to minimize power losses in power supply systems.**

Keywords: power and energy losses; electric power supply systems, energy efficiency

I. Introduction

The optimum load of the power transformers can be determined by three different criteria [1]:

- Minimum power losses ΔP achieved through the implementation of the so-called "Economically viable mode of operation" (EVMO) of power transformers.

$$\Delta P = \Delta P_0 + \beta^2 \cdot \Delta P_{SC(R)} \to \min \qquad (1)$$

- Minimum energy loss ΔW

$$\Delta W = T_0 \cdot \Delta P_0 + \tau \cdot \beta^2 \cdot \Delta P_{SC} \to \min \qquad (2)$$

- Minimum adjusted discounted annual costs

$$AC = r \cdot C + C_E \cdot \Delta W \to \min \qquad (2)$$

Where: ΔP_0 – no-load loss [kW]; $\Delta P_{SC(R)}$ – rated short circuit loss [kW]; β – load factor; T_0 – hours of work per year [h]; τ -working time at maximum loss [h]; C – capital investments [euro]; C_E – cost of electricity [euro/ kWh]; r – discount rate.

$$r = (F_n - b)/(1 + b) \qquad (3)$$

Where: F_n – nominal increase rate [%]; b-annual inflation rate [%]

In this study, the first criterion of economically viable operation of power transformers is applied.

II. Theoretical Study

In the case of working of one transformer, a maximum efficiency index η corresponding to the minimum loss ΔP is obtained at $\Delta P_0 = \Delta P_{SC}$ and β=0.71 (Figure 1). At low load levels, the loss curve ΔP is determined predominantly by ΔP_0, therefore in the low-load mode, typical of the present moment in Bulgaria, it is necessary to use transformers with lower no-load loss.

In the parallel operation of more than one transformer, so-called "limit powers" S_{Li} are defined. Figure 2 and Figure 3 shows parallel operation of 2 and 4 identical power transformers. The limit powers are defined by formula 4.

$$S_{Ln} = S_R \cdot \sqrt{n \cdot (n-1) \cdot K} \qquad K = \Delta P_0 / \Delta P_{SC} \qquad (4)$$

The limit powers S_{Ln} show the borders of optimal operation of the transformers. In case of lower consumption than the limit power value, it is necessary to switch to a smaller number of working transformers (n-1) and, respectively, at higher value to a larger number.

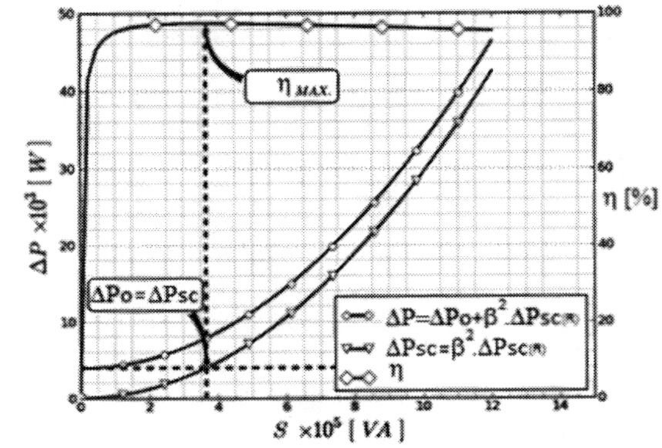

Figure 1. Efficiency index for one power transformer

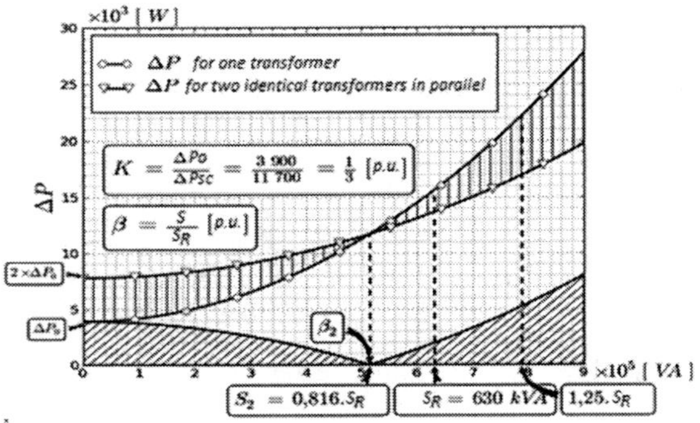

Figure 2. Parallel work of two power transformers (S_R=630kVA)

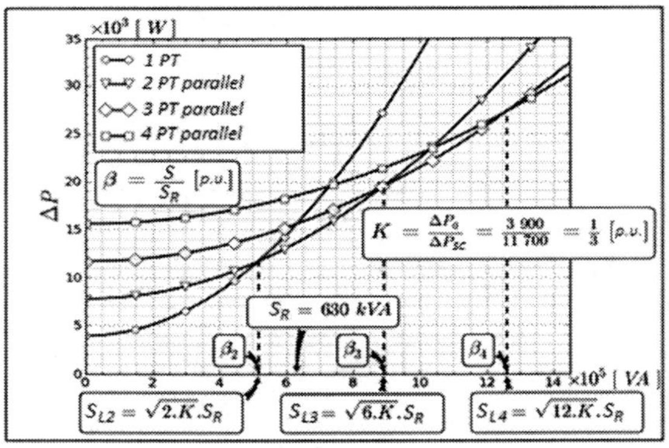

Figure 3. Parallel work of four identical power transformers (S_R=630kVA)

In case of n power transformers with different rated power, the limit power is determined by Formula 5 [1].

$$S_{Ln} = \left(\sum_{i=1}^{n} S_{Rn}\right) \cdot \sqrt{\frac{\Delta P_{0n}}{\left(\frac{\sum_{i=1}^{n} S_{Rn}}{\sum_{i=1}^{n-1} S_{Rn}}\right)^{2} \cdot \left(\sum_{i=1}^{n-1} \Delta P_{SCn}\right) - \left(\sum_{i=1}^{n-1} \Delta P_{SCn}\right)}} \tag{5}$$

III. Experimental Study

The study was conducted in the power supply system of the Technical University of Varna. The power supply system is constructed as a closed-loop scheme and consist of 5 transformer substations 10/0.4 kV with different rated power (Table 1). The survey was conducted for summer and winter period with the use of Iskra PNA 760 portable analyzer. The study involves determining the different relative power losses $\delta \equiv \Delta P^*$ [%] at a different load factor β.

$$\delta = \Delta P^* = 1 - \eta \tag{5}$$

Power transformers have more than 35 years of operation, resulting in deterioration of values for ΔP_0 and ΔP_{SC}. Table 1 shows the data for the nominal and the observed values. Figure 4 illustrates a graphical interpretation of Delta P, presenting the possibility of replacing some power transformers operating at low load. This opportunity is most pronounced for transformers model TM1000. At an average load of this type of transformer at 20% or S=200kVA, the losses determined in Figure 4 are ΔP=2%. It is proposed to replace TM1000 with a transformer TM400, which will work at a load of 50%. In this case, the power loss will be 1.8%, i.e. such reconfiguration is acceptable. In accordance with Formula 1, ΔP* can be defined with an expression (6).

Table 1. Values for power losses in distribution transformers

Type and units	Rated and experimental values for power transformers in Electric power supply system of TU-Varna		
	Rated $\Delta P_0/\Delta P_{SC}$ ***[kW]***	***Measured*** $\Delta P_0/\Delta P_{SC}$ ***[kW]***	***Deviation*** $\Delta P_0/\Delta P_{SC}$ ***[%]***
TM1000 1 unit	2.1	3.28	35.97
	12.75	14.5	12.06
TM630 5 units	1.44	16.9	14.79
	8.5	12.85	33.85
TM560 1 unit	2.1	3.45	39.13
	9.5	11.3	15.92
TM400 1 unit	0.98	1.68	41.66
	5.92	8.13	27.18
TM320 1 unit	1.44	2.75	49.09
	5.6	6.78	17.04

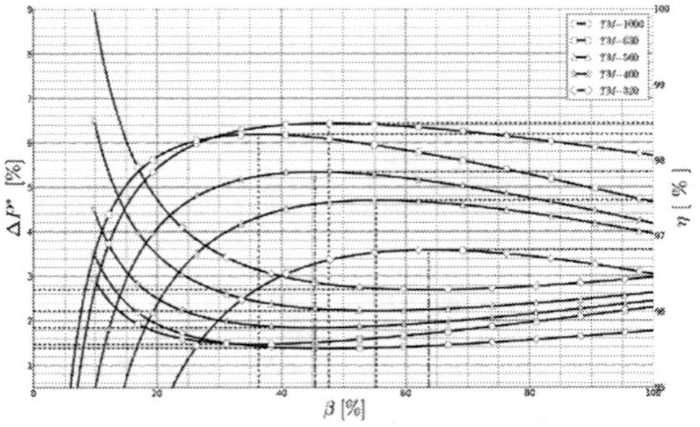

Figure 4. ΔP*=f(β) for different power transformers

$$\delta = \Delta P^* = \frac{\Delta P_0 + \beta^2 \cdot \Delta P_{SC}}{S_2} = \frac{\Delta P_0 + \beta^2 \cdot \Delta P_{SC}}{\eta \cdot S_1} \quad (6)$$

Where: $S_1 \approx S_R$ and S_2 [kVA] – apparent power on primary and secondary side of transformer; η – efficiency index.

$$\eta = 1 - \frac{\Delta P_0 + \beta^2 \cdot \Delta P_{SC}}{\beta \cdot S_R \cdot \cos\varphi_R + \Delta P_0 + \beta^2 \Delta P_{SC}} \quad (7)$$

If in Formula 6 $\eta = \eta_{MAX}$, then $\Delta P^* = \Delta P^*_{MIN}$

$$\delta_{MIN} = \Delta P_{MIN}{}^* = \frac{\Delta P_0 + \beta^2 \cdot \Delta P_{SC}}{\eta_{MAX} \cdot S_R} \quad (8)$$

According to studies in reference [2,3,4]for optimal load of the power transformers, a value of β obtained at $\eta = \eta_{MAX}$ corresponds to the fulfillment of condition $\Delta P_0 = \beta^2_{OPTIMAL} \cdot \Delta P_{SC}$

$$\beta_{OPTIMAL} = \sqrt{\frac{\Delta P_0}{\Delta P_{SC}}} \quad (9)$$

For transformers within the range 250÷1600kVA, β is assumed to $\beta_{OPTIMAL} = 0.71$, because of $\Delta P_0 = 0.5 \Delta P_{SC}$. Thus, according to Formula 9 is obtained δ_{MIN}.

$$\delta_{MIN} = \Delta P_{MIN}{}^* = \frac{\Delta P_0 + \beta_{OPTIMAL}{}^2 \cdot \Delta P_{SC}}{\eta_{MAX} \cdot S_R} = \frac{2\Delta P_0}{\eta_{MAX} \cdot S_R} \quad (10)$$

Table 2 shows calculated by (6) values of theoretical minimal loss ΔP^*_{Tmin} [%], actual minimal loss ΔP^*_{Amin} [%] defined by Figure 4 and deviation in [%], and also theoretical and actual values of $\beta_{OPTIMAL}$, calculated by (9)

Table 2. Calculated values of parameters for different transformers

Type and units	Calculated values of parameters for different transformers				
	ΔP^*_{Tmin} [%]	ΔP^*_{Amin} [%]	Deviation [%]	$\beta_{Toptimal}$	$\beta_{Aoptimal}$
TM1000	0.66	1.4	52	0.4	0.47
TM630	0.54	1.5	64	0.41	0.36
TM560	1.26	2.22	43	0.47	0.55
TM400	0.85	1.82	53	0.4	0.45
TM320	1.8	2.7	33	0.5	0.63

In most cases, actual losses are more than twice higher than theoretical. The analysis of $\beta_{OPTIMAL}$ shows, that it varies within the range of $\beta_{OPTIMAL} = 0.36 \div 0.63$, and also $\beta_{Aoptimal} > \beta_{Toptimal}$. Consequently, the concept of optimal load of power transformers in theory should not be related to the experimentally obtained values for $\beta_{Aoptimal}$. There is no universal solution to achieve optimal load of the power transformers. Such consideration of the issue is only appropriate in a specific schematic situation and by comparing different schematic solutions under different loads. Therefore, defining the optimal working area of the active load P of a power transformer only by the load factor β is not sufficiently reliable. Estimation of the influence of the reactive load is carried out with Formula 11, which is obtained by formula 1 [5,6].

$$\delta(P, \cos\varphi) = \left[\frac{\Delta P_0 + \left(\frac{\Delta P_{SC}}{S_R^2}\right) \cdot \left(\frac{P^2}{\cos^2\varphi}\right)}{P + \Delta P_0 + \left(\frac{\Delta P_{SC}}{S_R^2}\right) \cdot \left(\frac{P^2}{\cos^2\varphi}\right)} \right] \quad (11)$$

From Formula 11, a function of the change of power factor and active power for different loss levels is obtained. This expression is called the φ-P plan (12).

$$cos\,\varphi(P) = \sqrt{\frac{k \cdot P^2 \cdot \left(1 - \frac{\delta}{100}\right)}{\frac{\delta}{100} \cdot (P + \Delta P_0) - \Delta P_0}} \quad (12)$$

The surface of $\delta = f(\cos\varphi, P)$ and the $\cos\varphi$-P plan of transformer TM630 are shown in Figure 5 and Figure 6.

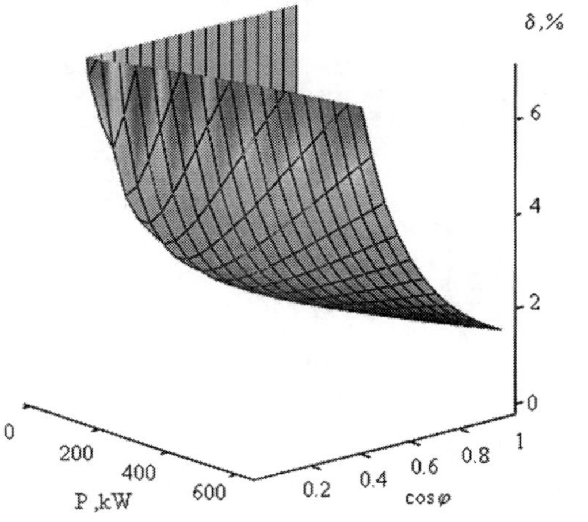

Figure 5. $\delta = f(\cos\varphi, P)$ for TM630kVA 20/0.4kV

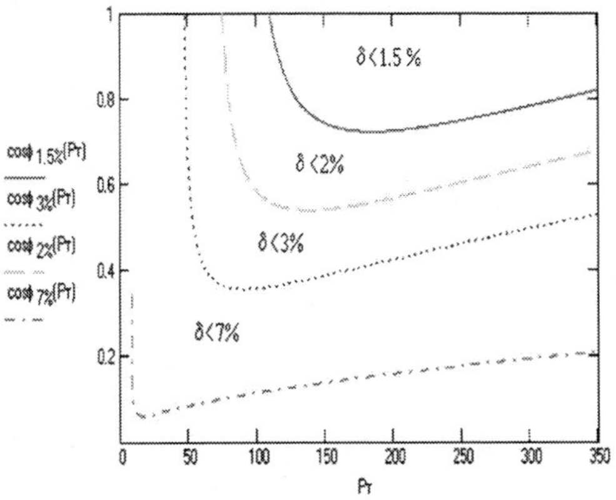

Figure 6. $\cos\varphi$-P plan for TM630kVA 20/0.4kV

Where: k – constant

$$k = \Delta P_{SC} / S_R^2 \qquad (13)$$

The connection of different single-phase loads or non-symmetry loads to different phases, in most cases determines the presence of pulsating power (unbalance power - N) with significant value. The connection of non-linear loads determine the presence of distortion power (harmonic power - D). The inclusion of unbalanced power N and the harmonics power D into cosφ-P plan leads to a change in the loss levels curves and a narrowing of the effective working areas of the power transformers. The method shown is applicable for the calculation of practical tasks relating to the determination of losses in power transformers as well as the efficiency of reactive power compensation systems. There are no such practices in the urban electricity supply systems in the Republic of Bulgaria and the method can be used for preliminary assessment of the effectiveness of their application. The definition of the cosφ-P plan, taking into account the unbalanced power and harmonics power, is shown in Formula 14.

$$\cos\varphi_{ND}(P) = \sqrt{\frac{k \cdot P^2 \cdot \left(1 - \frac{\delta_{ND}}{100}\right)}{\frac{\delta_{ND}}{100} \cdot P + \left(\frac{\delta_{ND}}{100} - 1\right) \cdot \left(\Delta P_0 + k \cdot N^2 + k \cdot D^2\right)}} \qquad (14)$$

Experimental data from the application of the presented method for TM630 transformer from an urban power supply system are shown in Figure 7 and Figure 8

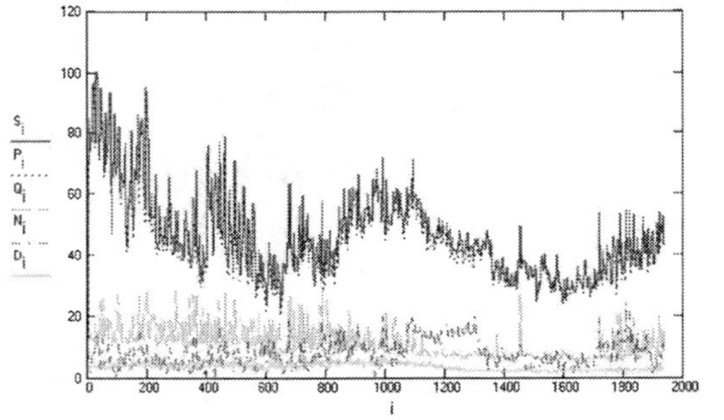

Figure 7. Deviation of active power (P), apparent power (S), reactive power (Q), unbalanced power (N) and harmonics power (D) for TM630 for 24 hours in 2 minutes interval (i=2min)

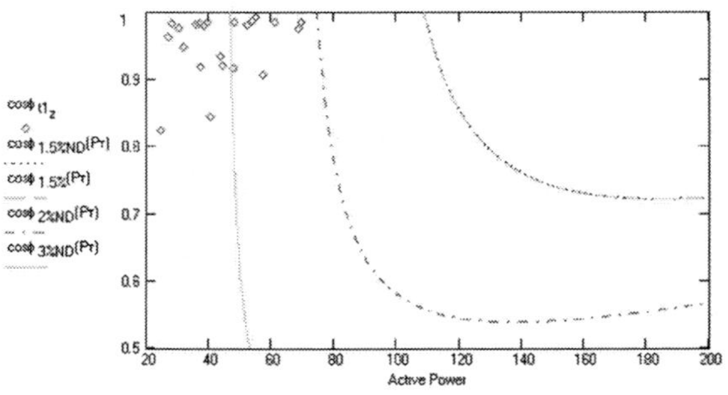

Figure 8. Operation areas and operation points of TM630 (for N=0.1S; D=0.05S)

In this case, the presence of unbalance and harmonics power does not affect the variation of the loss levels curves. This is explained by the low values of N and D (7.8 kVA and 3.5kVA, which although significant compared to apparent power (10% for N and 5% for D), are not commensurate with the rated power of the transformer. Figure 8 shows the location of the transformer operating points, averaged over each hour for a 24 hour period. In this case, all work points are in a very inefficient operation area. This is due to the fact that, at loads lower than 10% of the rated power, the curves are very steep and close to each other. This means that even the smallest power reduction will cause an increase in losses by several percents.

From the point of view of the compensation of reactive loads, the effectiveness of compensating systems can be estimated by δ(P) at certain fixed levels of cosφ. For a TM630 transformer at cosφ=0.5 and cosφ=0.9, the function δ(P) is shown in Figure 9.

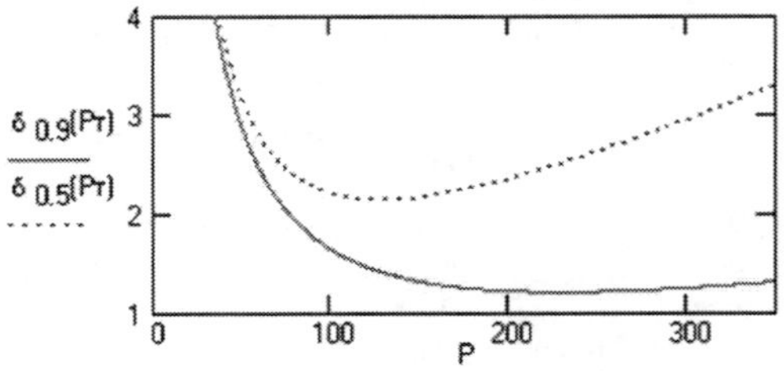

Figure 9. δ(P) of TM630 for cosφ=0.5 and cosφ=0.9

It can be seen that under different regime situations, maintaining a high cosφ results in lower values of δ at the same load.

IV. Conclusion

This paper explores the possibilities of achieving energy efficiency in the operation of power transformers. An economically feasible mode of operation is defined, and it is associated with the optimal load of the power transformers. The subject of the study is the power supply system of TU-Varna in operation condition. The research presents an approach to achieve an economically viable regime that reduces power consumption and energy losses. A concept of the cosφ-P plan has been developed with the inclusion of unbalanced and harmonics power, which allows to define optimal working areas of power transformers.

References

[1] R. Kirov, "Electric Power Supply (Fundamentals of the Design)", Technical University of Varna, 2013

[2] R. Kirov, D. Matev, V. Gyurov, V. Chikov and N. Makedonski, "Summary analysis of the electricity consumption regimes in TU-Varna", Scientific Congress "50 years TU-Varna", Proceedings, 2012

[3] A. Nazaritchev, A. Tadzhibaev, "Justification of the lifetime of the electrical equipment", Industrial Energetic, vol.4., 2005

[4] V. Zaugol'nikov, A. Balabin and A. Savinkov, "Some aspects of economical operation of power transformers", Industrial Energetic, vol.4, 2006

[5] V. Gyurov, "Rationalization of reactive power compensation in unbalanced and non-linear contirions of electric power supply systems", PhD Thesys, Technical University of Varna, 2010

[6] M. Yang, Y. Shi, Z. Jing, "Efficent Operation Regions of Power Distribution Transformers", IEEE Trans. on Power Delivery, vol.19, No4, 2004

Effects of Small Non-Linear Loads on the Power Grid

Vaclav Kus
Faculty of Electrical Engineering
University of West Bohemia
Pilsen, Czech Republic
kus@kev.zcu.cz

Petr Bilik
Faculty of Electrical Engineering and Computer Science
VSB – Technical University of Ostrava
Ostrava, Czech Republic
petr.bilik@vsb.cz

***Abstract* — The paper deals with the results and analysis of measurement of non-linear appliances connected to the low voltage distribution system. It compares the results of block of flats measurements and measurements in a satellite town of family houses. For comparison, the results of measurements of one family house in terms of power and harmonic analysis are presented. On the basis of statistical calculations, we analyze the average consumption of the housing unit with respect to harmonic components.**

Keywords: voltage harmonics; current harmonics; non-linear loads

I. Introduction

Semiconductor convertors are not only typical of industry, but are increasingly used in households. Even though these are usually small loads, the sum of the installed power of non-linear household loads can reach values that require attention, [1-3].

Loads with the phase current to 16 A must comply with the standard [4], which stipulates limits for harmonic currents that a particular load can take from the grid. Experience shows that most of these loads comply with the standard. The standard, however, does not solve simultaneity of work of these loads, which can be very high in particular daily cycles or seasons of the year.

With respect to the little experience with a great number of small loads working simultaneously in low voltage distribution network and a relatively small number of publications in the world literature, e.g. [5-7]), a series of measurements in the grid was started with the purpose of verifying the influence of an increasing number of semiconductor household loads on the quality of electric power in the grid.

Firstly, the measurement in a typical housing estate of blocks of flats was made. It lasted more than a week. The essential information on this measurement is included in [8,9]. For the second measurement, a newly built satellite town with family houses was chosen. This kind of housing development is becoming more wide-spread. Moreover, the connecting conditions may be insufficient with regard to the distances of some locations, given that we are concerned with the harmonic voltage of higher orders. Then limits defined in the standard [10] are exceeded and so the conditions for the quality of the distributed power according to the standard [11] are not fulfilled.

The aim of the article is to summarize the findings concerning the issues mentioned above. First conclusions drawn from the experience from the massive use of these loads are presented. In contrast to conventional measurements, statistical methods are used at the same time.

II. Measured Areas

A. *Measurement in a housing estate*

Parameters of the examined system are stated in [8], including the basic diagram. There are 15 blocks in the system, which means 361 flats.

B. *Measurement in a satellite town*

The value for the short circuit power with regard to the housing development is very high and amounts to S_k = 1644 MVA. Further the high voltage line of the total length of 8 870 metres and the 400 kVA distribution transformer are connected. From this spot individual houses are powered. There are 35 family houses in the powered part of the measured housing development.

C. *Locations and the measurement method*

The measurement was always carried out by 2 analysers. One analyser was placed at the output of the distribution transformer and the other analyser was placed at a block of flats or a family house. The measurement always lasted more than one week and the selection was made for one week measurement. The measurement took place simultaneously in all 3-phases with the recording period of 30 seconds.

III. Basic Measurement

During the measurement voltages, currents and powers including their complete analysis with respect to the harmonics were observed. As the article aims at the results of the harmonic analysis, this chapter focuses especially on harmonic voltages.

A. *Harmonic analysis of voltage at distant nodes*

Regarding harmonic voltages, the measurement results should be observed especially at distant nodes, where the highest distortion values are expected. Figures 1 and 2 show the results of the harmonic analysis of voltage, detected at the house entrance, with the maximum measured value and the average value from all the week's measurement. For comparison purposes, the value permitted in the standard is included in the chart as well, [12].

It is evident in the charts that in case of the housing estate, the permitted value of distortion for the 15th harmonic is exceeded in both the maximum and the average values. With the 21st harmonic (but also the 9th harmonic), the permitted value is exceeded at certain times, but the average value remains below the allowed limit.

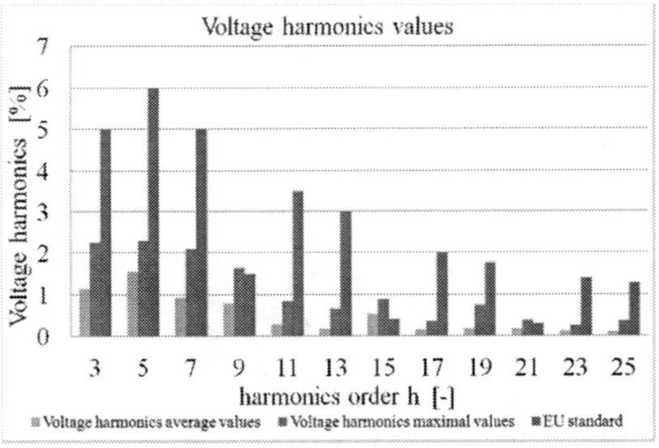

Figure 1. Voltage harmonics at the block of flats

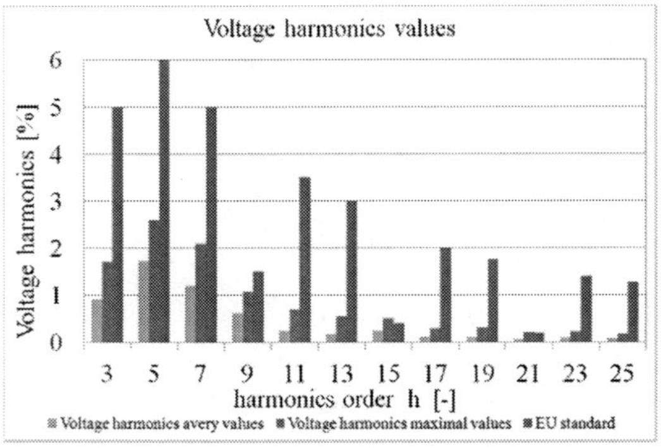

Figure 2. Voltage harmonics at the family house

In case of the satellite town, the permitted value for the 15th harmonic is only exceeded in several intervals, namely to the value of 0,51%. In all the other cases, the values of harmonics are below the values given in the standard.

B. *Magnitudes of the 15th and 17th voltage harmonics*

Considering the results of the harmonic analysis (Fig. 1,2), the time curves of these values were analysed. The permissible level for voltage was exceeded by more than half of the measured samples. In the family houses, on the contrary, the limit levels for the 15th harmonic were exceeded sporadically. As emerged from the analysis, it was caused by a greater number of 3-phase appliances and appliances with resistive character in family houses. In order to compare the "negative results" for the 15th harmonic, the levels for the 17th harmonic are shown in the figures as well. This harmonic is a characteristic harmonic and the permissible levels are high. The detected values are, however, lower than for the 15th harmonic.

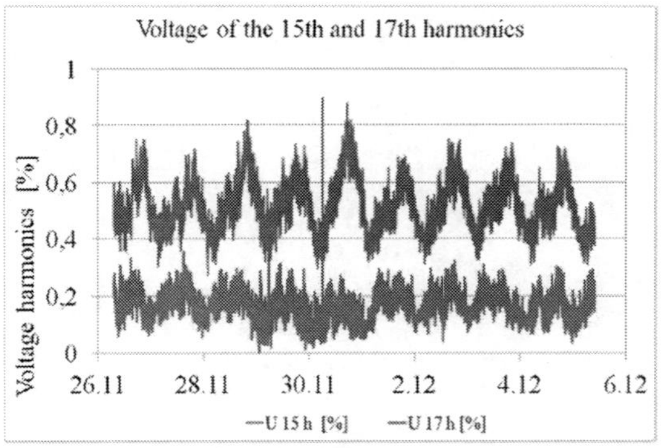

Figure 3. The 15th and 17th harmonics, measured at the input of the block of flats

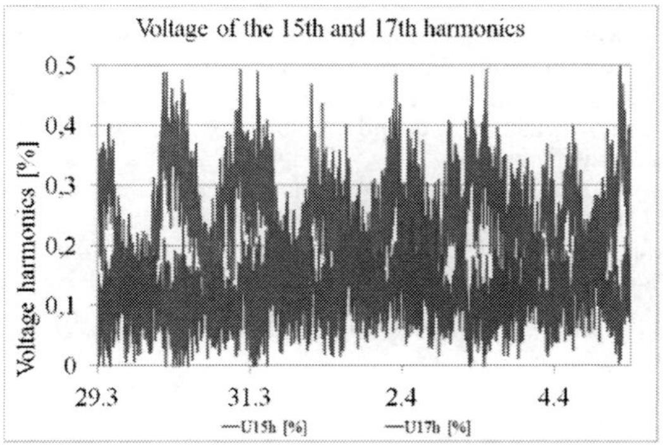

Figure 4. The 15th and 17th harmonics, measured at the input of the family house

IV. Statistical Analysis of the Measured Data

This part includes results of the selected values histograms, illustrating the simultaneous character of connection of non-linear appliances. During the week's measurement, the data was recorded as average half-minute values. It means that for every measured or calculated parameter, 20 160 values were available.

A. *Active power*

The histogram of active power P is used in order to consider the character of the installed appliances. Figures 5 and 6 show a principal difference between the consumption in the blocks of flats and the family houses. During the measurement in the blocks of flats, the consumption is relatively low. The histogram of active power shows a considerably big number of households with a very low power consumption. On the other hand, the load of family houses is uniform. (Note: negative values are determined by the operation of FV cells.)

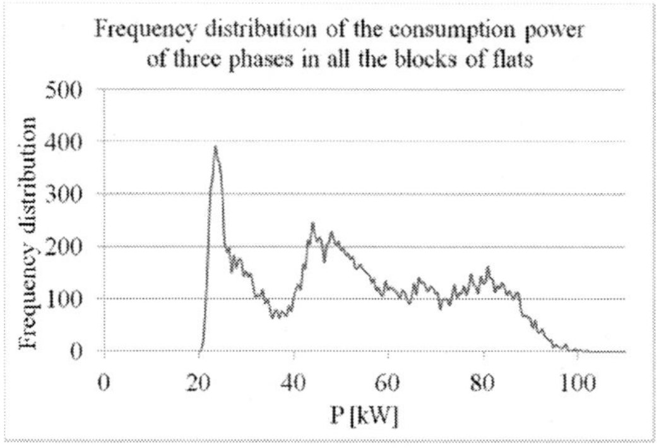

Figure 5. Histogram of the sum of three phases active power in all the blocks of flats

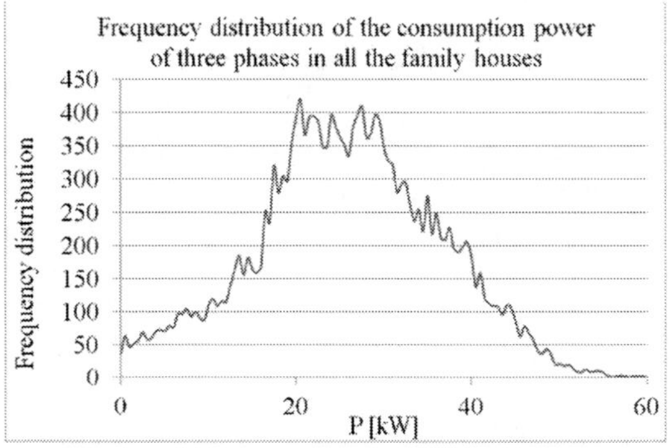

Figure 6. Histogram of the sum of three phases active power in all the family houses

B. *Currents in individual phases*

This part deals with the histogram of currents in individual phases. Both figures indicate a relatively uniform load in individual phases. The measurement in a housing estate of block of flats also shows the same distribution of the power consumption. The measurement in a housing estate of family houses is effected by the connected photovoltaic panels. If the sunshine is sufficient, the value of the current from the grid is low or negative. This is evident in the value deviation in the single phase consumption. In the other parts, there is an evident similarity with the trend of active power, and at the same, a uniform load in individual phases.

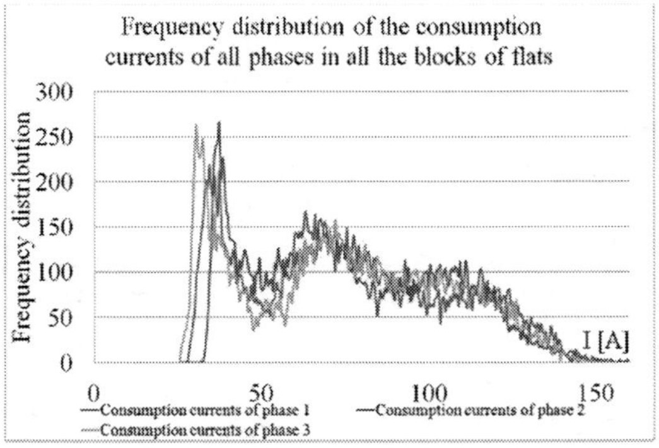

Figure 7. Histogram of currents on all phases in all the blocks of flats

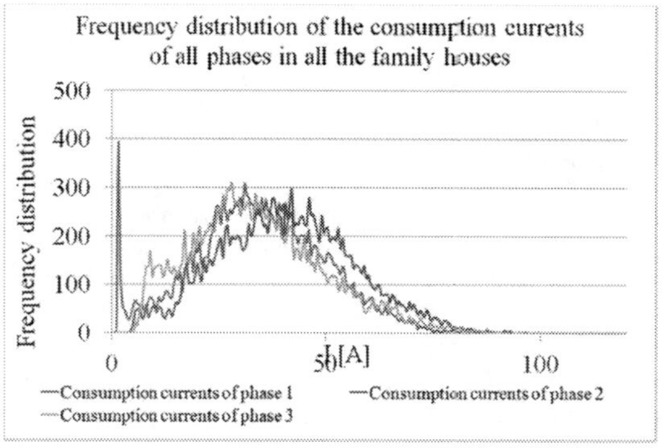

Figure 8. Histogram of currents on all phases in all the family houses

C. *Currents of the 5 th harmonics*

The currents of the 5th harmonic are observed in order to find the share of the three-phase non-linear appliances. Thus it is not possible to follow a direct connection either to the drawn RMS current or the power. With family houses, the distribution is steadier, and it copies the total three-phase consumption more. This indicates a higher usage of three-phase non-linear appliances. With blocks of flats, the portion of three- phase appliances is lower.

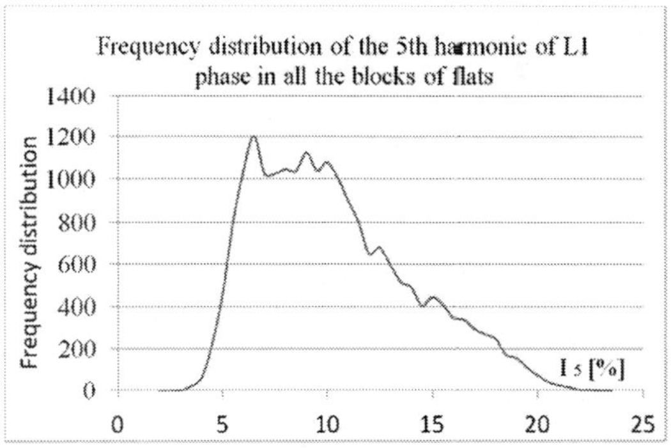

Figure 9. Histogram of the 5th harmonic of L1 phase in all the blocks of flats

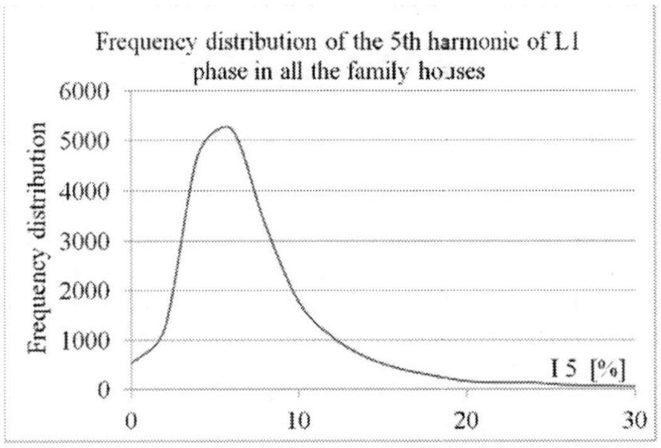

Figure 10. Histogram of the 5th harmonic of L1 phase in all the family houses

V. Discussion on the Measurements Results

A. *Power analysis*

The findings are very interesting. On one hand, the consumption in the blocks of flats is very low. The average consumption of 144 W per one flat is one half smaller than the average household consumption in the Czech Republic. On the other hand, the average consumption in the family houses is very high. The power factor equals 0.95 with the blocks of flats, and 0.98 with the family houses. The values for distortion power are insignificant.

Table 1 Power analysis of measurements at the blocks of flats transformers, converted to units

Power analysis (average per unit)	P_{max} [W]	P_{avg} [W]	Q_{max} [var]	Q_{avg} [var]	D_{max} [VA]	D_{avg} [VA]
1 flat	286	144	50	20	41	23
1 house	1699	714	435	128	241	31

B. *Power analysis and the total voltage distortion*

With regard to the amount and quality of voltage, it is necessary to analyse voltage at individual houses, both the blocks of flats and the family houses. Tab. 2 shows a relatively low amount of the total voltage distortion for the THDu [9-11]. An insignificant increase in voltage (the recorded maximum of 254 V) occurs sporadically due to the effect of the photovoltaic power panels.

Table 2 Comparison of the voltage values and the total voltage distortion coefficient at the house measurement

Voltage analysis	U_{max} [V]	U_{avg} [V]	THD_{Umax} [%]	THD_{Uavg} [%]
Blocks of flats	241	236	3.411	2.54
Satellite town	254	242	3.299	2.35

C. *Analysis of lower order harmonics*

It is evident from Figure 13 that at the places which are distant from the distribution transformer the harmonics of the 3rd, 5th, 7th, 11th, and 13th orders are comparatively low and within the limits set in the standards.

Table 3 Analysis of lower order harmonics in the housing estate measurements

Voltage analysis	*U3 max* [%]	*U3 avg* [%]	*U5 max* [%]	*U5 avg* [%]
Blocks of flats	2.26	1.14	2.31	1.56
Satellite town	1.71	0.9	2.59	1.72

D. *Harmonic voltages of higher order harmonics*

In the case of the measurements at the transformers, the measured values do not usually exceed the standards. The situation is, however, different with individual houses. In the housing estate of blocks of flats, the permitted values for the 15th harmonic are exceeded both for the maximum and average values. The average values for the 21st harmonic are closely below the permitted value, but the maxima exceed the permitted value twice. In the case of family houses the permitted values are only exceptionally exceeded.

Table 4 Average and maximum values for the 15th – 17th harmonics at individual houses

Voltage analysis	*U15 max* [%]	*U15 avg* [%]	*U17 max* [%]	*U17 avg* [%]
Blocks of flats	0.897	0.516	0.36	0.163
Satellite town	0.51	0.241	0.302	0.12

Table 5 Average and maximum values for the 19th – 21th harmonics at individual houses

Voltage analysis	*U19 max* [%]	*U19 avg* [%]	*U21 max* [%]	*U21 avg* [%]
Blocks of flats	0.731	0.192	0.391	0.174
Satellite town	0.32	0.11	0.211	0.064

VI. Conclusion

The article focuses on the harmonic analysis of currents and voltages in public grids. Based on two one-week measurements, the article aims at determining, whether the increasing number of non-linear household loads causes any voltage distortion, and whether the limits defined in the standard are exceeded.

In both the cases a large short-circuit power was available in the system. The measurements showed a great difference in power consumption. In the old blocks of flats, the average power consumption was about 150 W, whereas in the satellite town of family houses it was over 700 W. With respect to the big number of non-linear loads, the current distortion can be regarded as low. In relative quantities, the biggest distortion is in night hours, when most appliances of active load character are switched off. This mainly applies to the blocks of flats.

The distortion of lower order harmonics is small. The values for the distortion of the 9th, 15th, and 21st harmonics, i. e. the odd multiples of 3, are arguable. Despite the relatively low measured values, the limits defined in the standard are exceeded. The total current distortion coefficient is low, too

The article indicates that non-linear appliances are not problematic if the quality of the grid is sufficient. In the case of long cable lines, resonances in the area of the 15th harmonic can occur, and the increase in the neutral conductor load needs to be observed. The "limit" for the permitted values for the 9th, 15th, and 21st harmonics is the issue to be discussed.

Acknowledgment

This work was supported by the project of University of West Bohemia [grant number SGS-2015-038] and Student Grant System of VSB-TU Ostrava [grant number SP2017/128].

References

[1] R. Martinek, J. Vanus and P. Bilik, "New strategies for application of recursive least square algorithm in active power filters," Proceedings of the 8th International Scientific Symposium on Electrical Power Engineering, ELEKTROENERGETIKA, pp. 332–335, ISBN 978-80-5532-187-5

[2] R. Martinek, J. Vanus, M. Kelnar and P. Bilik, "Control methods of active power filters using soft computing techniques," Proceedings of the 8th International Scientific Symposium on Electrical Power Engineering, ELEKTROENERGETIKA 2015, pp. 351–354, ISBN 978-80-5532-187-5.

[3] R. Martinek, J. Vanus, P. Bilik, T. Stratil and J. Zidek, "An efficient control method of shunt active power filter using ADALINE," IFAC-PapersOnLine," vol. 49, pp. 352–357, 2016.

[4] IEC 61000-3-2:2014 Electromagnetic compatibility (EMC) - Part 3-2: Limits - Limits for harmonic current emissions (equipment input current ≤ 16 A per phase). Geneva : International Electrotechnical Commission, 2014.

[5] S. Hardie and N. Watson, "The effect of new residential appliances on Power Quality,". Proc. Of Universities Power Engineering Conference (AUPEC), 2010 20th Australasian. IEEE, 2010. p. 1-6.

[6] H. Sharma, W. Sunderman and A. Gaikwad, "Harmonic impacts of widespread use of CFL lamps on distribution systems," Proc.

Of Power and Energy Society General Meeting, 2011 IEEE. IEEE, 2011. p. 1-5. doi: 10.1109/pes.2011.6039496.

[7] A. Gil De Castro, S. Ronnberg, M. Bollen and A. Moreno-Munoz, "Harmonics from household equipment and different lamp technologies," Proc. Of Compatibility and Power Electronics (CPE), 2013 8th International Conference on. IEEE, 2013. p. 1-6. ISBN 978-1-4673-4913-0. doi: 10.1109/CPE.2013.6601119.

[8] V. Kus, J. Pikous, "The Effects of Connecting High Amounts of Nonlinear Appliances to the Supply Network," Proc. Power Engineering, Energy and Electrical Drives (POWERENG), 2015 IEEE 5th International Conference on. IEEE, 2015. p. 201-206. doi: 978-1-4799-9978-1/151$31.00.

[9] V. Kus, "Current and Voltage Harmonics in High-Rise Prefabricated Buildings," Proc. Of Electric Power Engineering (EPE), 2016 17th International Scientific Conference on. IEEE, 2016. p. 1-6. doi: 10.1109/EPE.2016.7521730

[10] IEC 61000-2-2:2002 Electromagnetic compatibility (EMC) - Part 2-2: Environment - Compatibility levels for low-frequency conducted disturbances and signalling in public low-voltage power supply systems. Geneva : International Electrotechnical Commission, 2002.

[11] EN 50160 Ed.3 *Voltage characteristics of electricity supplied by public distribution systems.* Brussels: European Committee for Electrotechnical Standardization, 2010. 20 p.

[12] P. Bilik, M. Horinek, R. Martinek and J. Vanus, "Results of multiple phasor measurement units testing procedures," Proceedings of the 8th International Scientific Symposium on Electrical Power Engineering, ELEKTROENERGETIKA 2015, pp. 132-135, ISBN 978-80-5532-187-5

B: High Voltage Technique and Diagnostics in Power Engineering

Theoretical and Experimental Problems of the PD Parameters Estimation

Nikolay Kinsht
Institute of Automation & Control Processes of Far Eastern Branch of RAS, Vladivostok, Russia
kin@dvo.ru

Natalia Petrunko
Institute of Automation & Control Processes of Far Eastern Branch of RAS, Vladivostok, Russia
pnn@dvo.ru

***Abstract*—The subject of discussion are the methodological aspects of the use of the equivalent circuit parameters and assign them as the models used in the diagnosis of high-voltage equipment based on a study of partial discharge (PD). The problem of "apparent charge" as applied regulatory setting, and the true charge of the PD. Analyzed equivalent calibration circuit and measuring the apparent charge and the corresponding matrix model based measurement circuitry and calibration. It is shown that the "apparent" and true charges are reflected in the system of the equivalent electrical circuit equations independent of each other parameters, the connection between them is merely a heuristic. We analyzed the traditional model of a spherical bubble in a uniform external electric field and assessed the limits of its correctness. Analytically calculated value of the maximum charge air bubble in oil, initiated by an external electric field. It is shown that the available analytical and experimental evaluation of the apparent charge significantly different from the theoretical possibility of a true charge. It was concluded that from a scientific point of view, the model of the PD on the basis of the concept of "apparent charge" has exhausted itself.**

Keywords*—*HV equipment; partial discharges; electric field; equivalent circuit

I. Statement of the Problem

Technical state of the elements of high voltage power transformer equipment to the greatest extent determined by the state of its insulation. It associated with the presence, intensity and distribution of electrical partial discharge (PD) arising as during both normal operation and the degradation of insulation and other equipment structural elements. There is currently regulated estimation procedure of the technical state of the high voltage equipment based on the experimental determination of the integral characteristics of the PD - "apparent charge" and a number of functions based on this value. Detailed analysis shows that the uncertainties inherent in the traditional model of PD together constitute at least several orders of magnitude, when the basis for assessing technical state of equipment is the concept of the apparent charge [1]. A number of features of the processes, connected with the phenomena of the PD, were analyzed with the electrophysical parameters of media materials [2, 3]. But they do not constitute a comprehensive analysis of the phenomena, being rather an illustration of the diversity of factors between the parameters.

The aims of this study are:

- analyze of formal connections between the true and apparent charges on the basis on matrix description of the equivalent circuits.
- analytically calculating value of the maximum charge air bubble in oil, initiated by an external electric field.

II. Traditional Calibration and Measurement of Apparent Charges

Let's describe in general terms the technology of the PD diagnosing. the PD. Let us turn to Fig. 1. The problem is that PDs originate in the depth of the transformer construction and are not accessible to direct monitoring. The IEC Standard [1] explicitly states that the value of the true charge, localized in the discharge circuit, "... can not be measured directly". Electrical measurements are possible only at external terminals (usually at the inputs). In this regard, historically, the following scheme was adopted for estimating the true charge accumulated in the inclusion under the action of the ground electric field $E(x,y,z)$ and causing the PD current.

Let consider Fig.2. There are two independent nodes, point 1 represents the input, point 2 is the pole of the potential defect.

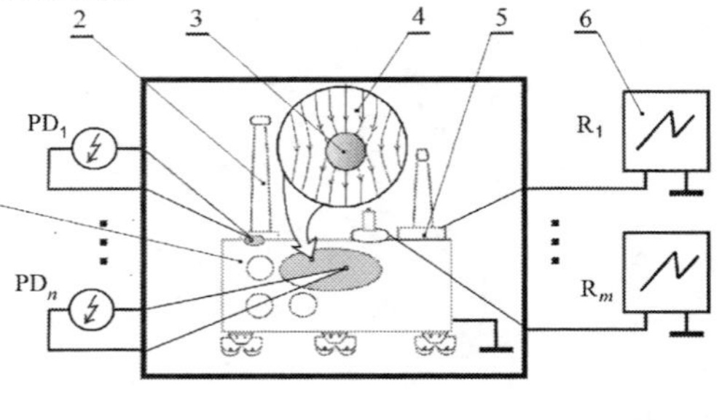

Fig. 1. Schematic representation of the PD diagnosis process. 1 - transformer casing (tank), 2 - high-voltage input, 3 - potential defect (gas inclusion in oil insulation), 4 - electric field in the vicinity of the inclusion, 5 - PD sensor, 6 – PD registrar

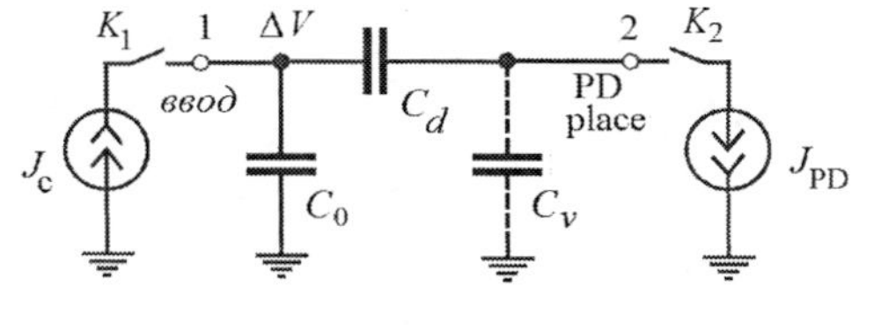

Fig. 2. The simplest equivalent calibration circuit.

Capacitor C_0 represents the total capacitance between input and ground, capacitor C_d capacitance between input and defect. The equivalent of the potential defect is the capacitor C_{PD}, which accumulates the charge Q. When the breakdown voltage (i.e. closure of the key K_2) comes, a partial discharge occurs and, on the equivalent circuit it is depicted as a current source J_{PD}. Due to the PD current at the input, the voltage ΔV can be recorded.

As soon as the PD starts, the discharge of the charged bubble can be simulated by the current source J_{PD}, and in the operator form of the record we have the expression for ΔV_{PD}:

$$\Delta V_{PD} = J_{PD} / pC_0.$$

Similarly, when key K_2 is open, applying a calibration current J_c (carrying the total calibration charge) to the circuit and taking into account that the capacitance C_0 is much larger than the capacity of the chain (C_d, C_v), we obtain the calibration voltage which initiated by calibration current J_c (or calibration charge q_c :

$$\Delta V_c = J_c / pC_0.$$

Thus, in the simplest case, when measuring in operating mode the voltage jump ΔV, if $\Delta V_c = \Delta V_{PD}$, it is assumed that the apparent charge q_c is numerically equal to the true charge Q. In general, use the proportion:

$$Q \approx q_c \, \Delta V / \Delta V_c.$$

Note that the estimate of the apparent charge could be tried to clarify, knowing the parameters C_b and C_d. Unfortunately, such objective data are usually not available. Let's take a step to complicate the task. Let us imagine the possibility of the existence of at least 2 independent PDs (which from the methodological point of view is equivalent to the possibility of the existence of a single PD in one of the two different points of the active space). To increase the information content of the diagnostic process, we allow the registration of the PD at 2 points (LV and HV), as shown in Fig. 3.

It is clear that the equivalent circuit in Fig. 3 already for 2 PD and 2 registration points is quite cumbersome, but at the same time it is absolutely correct. It is appropriate to recall here that researchers introduce a number of substitution schemes of varying complexity (for example, to using a three-dimensional RC grid for formalizing the partial discharge process in [5]). All these substitution schemes appear due to the inability to correctly apply the simplest circuit of Fig. 2.

In the other words because of uncertainty of the notion of "apparent" charge, all efforts have the only

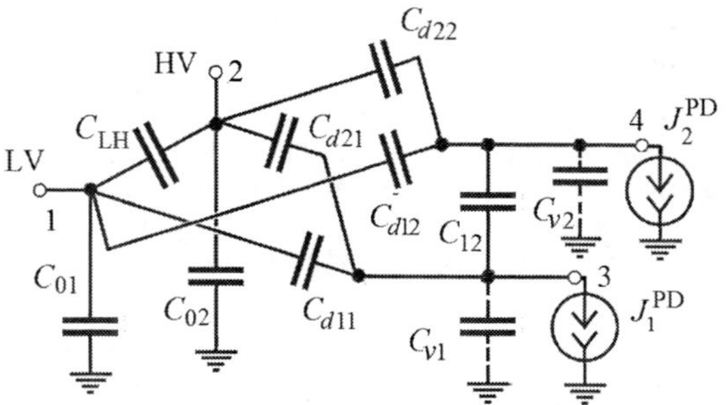

Fig. 3. Equivalent circuit for 2 independent PDs and 2 registration points.

purpose - to simulate the complexity of the real electric field, with the aim of changing from the parameters of the electric field to the parameters of some electrical circuit.

Unfortunately, none of them described the formalized process of calibration and interpretation of measurement results. It is not clear how to determine the "apparent" charge and what is its relation to the true.

III. Matrix Description of Calibration and Measurements

Let us try to understand the real situation in a fairly general setting, when several (n) potential PD sources exist in the transformer and there is a principal possibility to register their signals with the help of several (m) recorders R_1, ... R_m (Fig. 1) and formalize it using standard methods for diagnosing electrical circuits. The diagnostic methods could be applied by presenting the structure of the system of basic equations and having available a sufficient number of known parameters.

As a natural mathematical connection between the sources of PD and the results of measurements is the model of a multipolar network.

The calibration process is described by the matrix equation:

$$\mathbf{V}^{C} = \mathbf{Z}^{C}\,\mathbf{J}^{C}, \quad \mathbf{Z}^{C}=[z_{kj}], \ (k, j = 1,\ldots m), \qquad (1)$$

Here are:

$\mathbf{J}^C$ is the column matrix of calibration currents;

$\mathbf{V}^C$ is the column matrix of stresses arising during calibration, and recorded at the same points as UP;

$\mathbf{Z}^C$ is a square calibration impedance matrix.

And matrix equation reflecting PD registration process can be described in the following form:

$$\mathbf{V}^{R} = \mathbf{Z}_{nm}\,\mathbf{J}^{PD}, \ \mathbf{Z}_{nm}=[z_{ik}], \ (i=1,\ldots n; \ k=1,\ldots m), \qquad (2)$$

where are:

$\mathbf{V}^R = \mathrm{col}[V_1^R,\ldots,V_m^R]$ is the column matrix of the voltages (or their increments) associated with the PD, which arise at the points of their registration,

$\mathbf{J}^{PD} = \mathrm{col}[J_1^{PD}, \ldots, J_n^{PD}]$ is the column matrix of the PD currents; (without loss of generality, for simplicity we assume that the currents of the PD are closed to ground)

$\mathbf{Z}_{nm}$ $n\times m$ is a rectangular matrix of transfer (mutual) impedances between n potential defects. In general, the numerical values of the matrix elements are unknown to us, they depend on the shape of the electric field and the mutual geometrical location of the places of origin of the PD and the points of registration.

(Here, for the compactness of the notation, we use operator notations that represent both currents and voltages, which in reality are arbitrary functions of time.

The relationship between the discussed matrices is clarified by introducing inaccessible voltages on potential defects $\mathbf{V}^{\mathrm{PD}}$ that are not available to immediate measurement. Assuming that for the simplicity, in the absence of the PDs, the potential defects itself are breakdown, and in the PD process they are current sources, we get a complete system of equations that reflects both the calibration mode and the multiple PD mode:

$$\begin{bmatrix} \mathbf{V}^C \\ \mathbf{V}^{PD} \end{bmatrix} = \begin{bmatrix} \mathbf{Z}^C & \mathbf{Z}_{nm} \\ \mathbf{Z}_{mn} & \mathbf{Z}^{PD} \end{bmatrix} \begin{bmatrix} \mathbf{J}^C \\ \mathbf{J}^{PD} \end{bmatrix} = \mathbf{ZJ} \qquad (3)$$

where are:

$\mathbf{V}^{\mathrm{PD}}=$ col$[V_1^{\mathrm{PD}},\ldots,V_m^{\mathrm{PD}}]$ is the column matrix of the potential defect voltages,

$\mathbf{V}^{\mathrm{C}}$ is the column matrix of the calibration voltages

$\mathbf{J}^{\mathrm{C}}$ is the column matrix of the calibration sources,

$\mathbf{J}^{\mathrm{PD}}$ is is the column matrix of the PD currents.

Take the attention that in the calibration mode we set $\mathbf{J}^{\mathrm{PD}} = \mathbf{0}$, and in the diagnostic mode we set $\mathbf{J}^{\mathrm{C}} = \mathbf{0}$.

All of above analytical expression can be illustrate for equivalent circuit is shown in Fig. 3. In accordance to nodal circuit analysis we have main equation:

$$\mathbf{YV} = \mathbf{J}, \qquad (4)$$

and in detail:

$$j\omega \begin{bmatrix} \begin{pmatrix} C_{01}+C_{LH}+ \\ C_{d11}+C_{d12} \end{pmatrix} & -C_{LH} & -C_{d11} & -C_{d12} \\ -C_{LH} & \begin{pmatrix} C_{02}+C_{LH}+ \\ C_{d21}+C_{d22} \end{pmatrix} & -C_{d21} & -C_{d21} \\ -C_{d11} & -C_{d21} & \begin{pmatrix} C_{12}+C_{d11}+ \\ C_{d21}+(C_{11}) \end{pmatrix} & -C_{d22} \\ -C_{d12} & -C_{d12} & -C_{d22} & \begin{pmatrix} C_{12}+C_{d12}+ \\ C_{d22}+(C_{22}) \end{pmatrix} \end{bmatrix} \begin{bmatrix} V_1 \\ V_2 \\ V_1^{PD} \\ V_1^{PD} \end{bmatrix} = \begin{bmatrix} J_1^C \\ J_2^C \\ J_1^{PD} \\ J_2^{PD} \end{bmatrix}.$$

Here the large matrix associates with admittance matrix $\mathbf{Y}$,

$\mathbf{V} =$ col$[V_1,\ V_2\ V_1^{\mathrm{PD}},\ V_2^{\mathrm{PD}}]$ is generalized voltage vector,

$\mathbf{J} =$ col$[J_1^{\mathrm{C}},\ J_2^{C}\ J_1^{\mathrm{PD}},\ J_2^{\mathrm{PD}}]$ is generalized current sources vector.

After solving this equation we have:

$$\begin{bmatrix} V_1 \\ V_2 \\ V_1^{PD} \\ V_1^{PD} \end{bmatrix} = \mathbf{Y}^{-1} \begin{bmatrix} J_1^C \\ J_2^C \\ J_1^{PD} \\ J_2^{PD} \end{bmatrix} = \begin{bmatrix} Z_{11} & Z_{12} & Z_{13} & Z_{14} \\ Z_{21} & Z_{22} & Z_{23} & Z_{24} \\ Z_{31} & Z_{32} & Z_{33} & Z_{34} \\ Z_{41} & Z_{42} & Z_{43} & Z_{44} \end{bmatrix} \begin{bmatrix} J_1^C \\ J_2^C \\ J_1^{PD} \\ J_2^{PD} \end{bmatrix} = \mathbf{ZJ}. \qquad (5)$$

It's easy to see that $\mathbf{Z}^{\mathrm{C}}$, $\mathbf{Z}^{\mathrm{PD}}$, and $\mathbf{Z}_{mn}$ are correspond 2×2 fragments of the 4×4 $\mathbf{Z}$ – matrix:

$$\mathbf{Z}^C = \begin{bmatrix} Z_{11} & Z_{12} \\ Z_{21} & Z_{22} \end{bmatrix},\ \mathbf{Z}^{PD} = \begin{bmatrix} Z_{33} & Z_{34} \\ Z_{43} & Z_{44} \end{bmatrix},\ \mathbf{Z}_{nm} = \begin{bmatrix} Z_{13} & Z_{14} \\ Z_{23} & Z_{24} \end{bmatrix} = \mathbf{Z}'_{mn},$$

As can be seen, both $\mathbf{Z}^{\mathrm{C}}$ and $\mathbf{Z}^{\mathrm{PD}}$ are block elements of the generalized matrix of input and transmission impedances $\mathbf{Z}$. It should be noted that the $\mathbf{Z}^{\mathrm{C}}$ and $\mathbf{Z}^{\mathrm{PD}}$ matrices can not only in principle have different sizes, but in general, none of the $\mathbf{Z}^{\mathrm{PD}}$ matrix elements has an equivalent among the matrix $\mathbf{Z}^{\mathrm{C}}$ elements.

Referring to the example in Fig. 3, for the formalization of the PD processes and calibration, it would be necessary to use the relations (1, 2). $\mathbf{Z}^{\mathrm{C}}$ and $\mathbf{Z}_{mn}$ matrices have size 2×2, but they reflect different realities: the $\mathbf{Z}^{\mathrm{C}}$ matrix is the matrix of the input and transfer impedances of the pair of clamps LV and HV, and the matrix $\mathbf{Z}_{mn}$ is the matrix of the transfer impedances between the pair of clamps (LV, HV) and pair (PD_1, PD_2), respectively.

And it is necessary to pay attention to term in the smallest brackets in the equation (4). The terms (C_{11}) and (C_{22}) must be erased if corresponds are non zero. I.e. in fact it is need to write specify equation (5) for calibration and each phenomena PD.

In other words, the estimate of the apparent charge (based on the elements of the gauge matrix $\mathbf{Z}^{\mathrm{C}}$) formally has nothing to do with the estimate of the true charge (which could be obtained by knowing the values of the $\mathbf{Z}_{mn}$ matrix). From this point of view, without formulating any sufficiently strong a priori constraints, it is generally incorrect to raise the question of the relationship between apparent and true charges. Concluding the section, we conclude that there are no effective tools in the form of reliable and simple equivalent circuits or some integral parameters for the diagnosis of true PD charges.

IV. True Charge Accumulation in the Spherical Inclusion

Let prepare the analytical expressions describing the parameters of the electric field inside the spherical inclusion (potential defect) in a homogeneous field [3 , 4]. It is schematically shown on the Fig. 4.

There are here:

a – radius of sphere,

R, z, θ – spherical coordinates

E_0 – primary electric intensity,

ε_i – sphere interior permittivity

The analytical expression required for further analysis for electric intensity in the sphere interior is known (3):

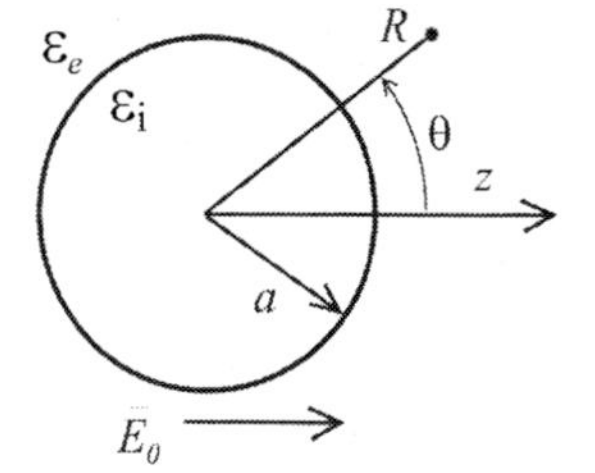

Fig. 4. The spherical inclusion in a homogeneous field.

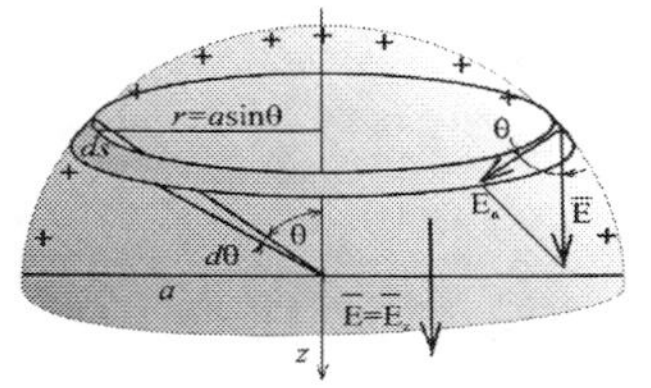

Fig. 5. The inner of upper inclusion hemisphere for charge calculation.

$$E_i(R,\theta) = E_z = E_0 \frac{3\varepsilon_e}{\varepsilon_i + 2\varepsilon_e} \qquad (6)$$

Let us estimate value of the charge accumulated by the time of the PD beginning. Let consider the upper inclusion hemisphere for charge calculation (Fig. 8).

Elementary connected charge dQ on the inner surface of hemisphere may be written:

$$dQ = \sigma ds = \varepsilon_0 E_n ds,$$

where are:

σ – area charge dencity,

E_n– normal electric field dencity on the inner surface.

Let us highlight elementary thin ring of upper inclusion hemisphere $ds=2\pi a^2 \sin\theta d\theta$ (fig. 5).

According to (6) E_i = const,

$$E_n=E_i\cos\theta, \quad dQ = \varepsilon_0 E_i ds\ 2\pi a^2 \sin\theta \cos\theta\ d\theta,$$

and integrating, we obtain an induced charge of the upper hemisphere, which can participate in a single act of the PD:

$$Q = \varepsilon_i E_i 2\pi a^2 \int_0^{\pi/2} \sin\theta\cos\theta d\theta = \varepsilon_i E_i \pi a^2 \qquad (7)$$

Let take for an example the air bubble with a diameter 1mm (a = 0.5 mm). Let also take inner field intensity E_i equal breakdown field intensity E_b by the moment previous breakdown:

E_b = 3 MV/m = 3·10^{6}V/m and as ε_0 = 8,86 10^{-12} F/m.

And we have estimation value for maximal accumulated charge in the 1-mm bubble:

$$Q|_{d=1} = \pi \times 3\ 10^6 \times 8.86\ 10^{12} \times (0.5\ 10^{-3})^2 = 20.88\ pC \qquad (8)$$

The expression (7) for a charge can be accumulated in inclusion in the electric field can be useful.

V. Comparison of the Results with Published

In [5] it is said: "... do not difficult to determine that the normalized values of apparent charge in 300 pC corresponds to the PD in the bubbles with a radius of 1 mm" (2 mm in diameter); according to the formula (6) the true charge of 300pC corresponds to the bubble diameter of 3.9 mm. Let recalculate this value by 1 mm in diameter:

$$q_4 = 75\ pC.$$

Let us refer to other papers, which sends us a report [6]. The experimental data (V. Swinka, F. R. Robert), a bubble diameter of 1 mm has apparent charge 130 pC,

$$q_{Sw} = 130\ pC.$$

We repeat the report graph (Fig. 5, in the original data points marked by circles). Let's try to add the estimated quadratic parabola, based on the higher experimental point (d = 1mm). The quadratic function $q = 130d^2$ passes almost exactly on 10 of the 13 data points. It seems that the experimental data are in satisfactory agreement with the quadratic increase in the charge in diameter function but it's something wrong in the measurement calibration.

Next, in the same [5, 6]. According to a study [7] (G. Niasar, and other), if $d \approx 2.1$ mm, the apparent charge is equal to $q \approx 200 \ldots 2000$ pC, ie if attributed to the data d ≈ 1mm, we obtain:

$$q_{Nia} \approx (200 \ldots 2000)/2.1^2 = 45.3 \ldots 453\ pC..$$

Note that in [5] the true charge must be even more than apparent. On the other hand, on the contrary, all of the above regulatory or experimental data on the apparent charge, q_4, q_{Sw}, q_{Nia}, related to d = 1mm greater than the magnitude of $Q|_{(d=1)}$. In short, all these figures are radically different from the simple and correct theoretical result (7, 8). It adds that, in accordance with [2], the uncertainty in the assessment of the apparent charge may even be several orders of magnitude.

At the end of this section we state that the known regulatory and experimental values of the "apparent" charges are a significant spread, thus significantly exceeding the possible values of the true charge of the potential defect.

The phenomenon of the PD as a whole is a multivariable, essentially depends on such parameters as the pattern of the electric field in the core of the transformer, the location of the defect, the electrical properties of the defect, and the properties of the insulation. [3]

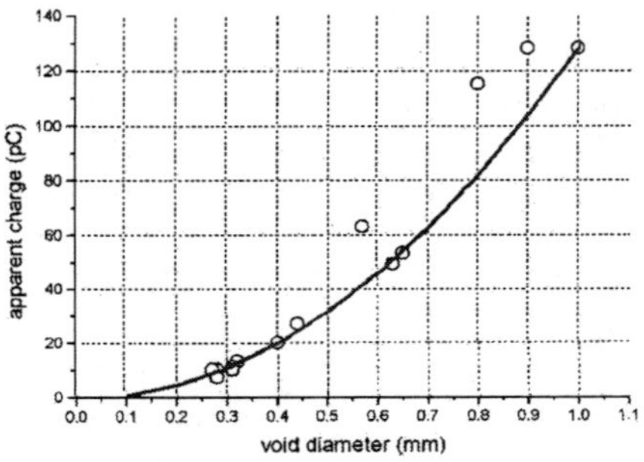

Fig. 6. Data from V. Swinka, F. R. Robert (circles) and our quadratic approximation.

VI. Conclusions

1. The conception of apparent charge can be used only in the simplest cases. In general this conception is no good for real application
2. The experimental estimations of the true PD are very unreliable. Strict theoretical assessment can be very useful.

Acknowledgment

The investigations are partly sponsored by Russian Foundation Basic Researches (grant № 17-08-00693).

References

[1] International IEC Standard 60270 – 2000.

[2] N. V. Kinsht and M. A. Katz, "Some Problems of the Partial Discharge Burning Time," TEEM, Vol. 7, No. 6, pp. 319-323, 2006.

[3] N.V. Kinsht and N.N. Petrunko, "Some Equivalent Circuits and Analytical Estimation of the PD Parameters" Proc. of Int. Conf. on Condition Monitoring and Diagnosis 2016 (CMD2016), pp. 327-330, September 25-29, 2016, Xi'an, China

[4] G.P.Harnwell, Principles of Electricity and Electromagnetism, McGraw-Hill Book Co., New York, 1949.

[5] S. Korobeynikov, A. G. Ovsyannikov "Apparent and true charges of partial discharges" Proc. of the 2016 IEEE int. conf. on dielectrics, ICD 2016, Montpellier, France, 3–7 July 2016. – [France] : IEEE, 2016. – Vol. 2. – pp. 485-489 .

[6] V. Swinka, F. R. Robert. "Imaging of voids in dielectric materials by x-ray induced partial discharge" Review of scientific instruments, Vol. 73 (No 3) (2002) pp. 1288 – 1292.

[7] M. G. Niasar, H. Edin, X. Wang and R. Clemence. "Partial discharge characteristics due to air and water vapor bubbles in oil" Proc. of XVII International Symposium on High Voltage Engineering, Hannover, Germany, August 22-26, 2011. – Paper D-067.

Detection of Transformer Windings Failures With Frequency Response Analysis Method Obtained With Different Signals

Maciej Kuniewski
Department of Electrical Engineering
and Electrical Power
AGH University of Science and Technology
Kraków, Poland
maciej.kuniewski@agh.edu.pl

Jakub Furgał
Department of Electrical Engineering
and Electrical Power
AGH University of Science and Technology
Kraków, Poland
furgal@agh.edu.pl

***Abstract*—Power transformers are subjected to detailed diagnostic investigations. Investigations include among others mechanical condition of windings. Detection of windings failures is made with the frequency response analysis method (FRA). As a base of this method are the frequency characteristics of transfer function of transformer winding obtained with sinusoidal signals, (Sweep Frequency Response Analysis) SFRA method. SFRA method is off-line method and thus has some limitations in application. Usage of transient voltage signals occurring in power systems during normal work can extend application of FRA method. This paper presents comparison of laboratory measurement results of transformer transfer function, used for analysis of windings technical condition, obtained with SFRA and Low voltage impulse (LVI) method.**

Keywords: power transformers, FRA, winding failures, diagnostics

I. INTRODUCTION

Proper and efficient diagnostic procedures of electrical power equipment have strong importance in increasing of electrical power delivery reliability. Actual trends in exploitation process are subjected to predicting probability of device failure. Prediction is based on results from constant monitoring of technical condition of strategic power equipment in power grids, such strategy is called Condition Based Monitoring (CBM). CBM strategy uses on-line diagnostics methods for determination of technical condition of power equipment continuously [1-3]. Main issues related with usage of on-line methods are improvement of Data Mining procedures [4] used for analysis of obtained results for concluding on technical state of object and analysis of impact of field measurements on obtained results [5]. One of the most important objects in power delivery infrastructure are power transformers. Mechanical and electrical condition of power transformers have high influence on reliability of electrical energy delivery. Power transformers used in electrical power systems are subjected to detailed diagnostics procedures [6-12]. Generally used and standardized on-line transformer diagnostics methods are: Estimation of oil ageing degree, Dissolved gas analysis (DGA) in oil, Partial Discharge (PD) measurements [11].

Determination of winding technical condition with FRA as an on-line method is still under consideration, and is scope of work in many scientific centers [13-15]. Proper determination of kind and range of transformer winding fault has high practical importance [16]. Key feature on mechanical condition of windings have electrodynamic forces, which cause buckling and displacement of winding [6]. During fault current flow through transformer winding electromagnetic forces act on winding radially. Radial component of leakage magnetic field around winding cause forces acting along axis of the winding, this forces cause flattering of winding due to changes of spaces between turns or group of turns. Windings and magnetic core are exposed to mechanical forces also during transportation of power transformer. Impact of winding geometry deformation has important influence on changes of oil ducts in winding which result in overheating of some parts of winding. Operating temperature of paper-oil insulation has strong influence on life time of insulation system, on account to acceleration of chemical reaction in dielectrics. Weakening of electrical properties of insulation can cause short circuits between turns, which influence on winding ratio and can create high inter turn currents which overheat winding and its insulation.

Analysis of mechanical condition of windings are mainly based on frequency characteristics measurements which are basis for Frequency Response Analysis Method [7-10]. Frequency characteristics can be defined as a ratio of different signals [17,18]. Transfer function of transformer in frequency domain can be obtained with different methods SFRA or Low Voltage Impulse LVI [19-21]. Evaluation of technical condition of windings is made during transformer exploitation (Fig. 1) in constant time periods.

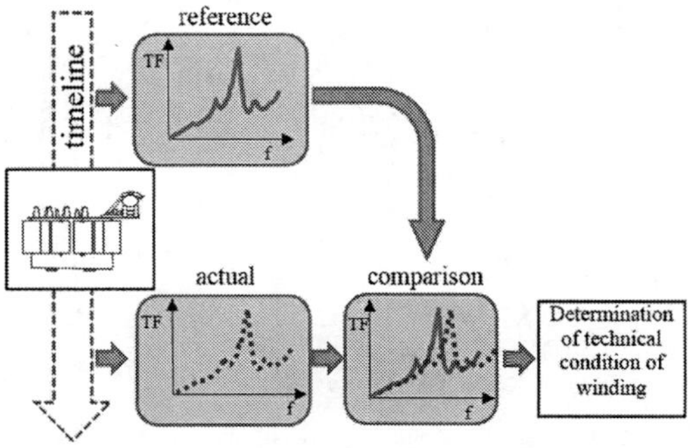

Fig 1. Block scheme of transformer winding fault analysis with FRA method.

Aim of this article is to compare frequency characteristics of transformer windings obtained with sinusoidal signal SFRA and with steep voltage surge LVI. Measurements of characteristics are in configuration of healthy winding and in winding with simulated failures like short circuit between turns, differences in frequency characteristics are caused by winding failure. Comparison between results is made on the background of usage of transient voltages appearing in power systems as a diagnostics signals for transformer windings fault detection.

II. Diagnostic of Mechanical Condition of Transformer Winding

A. Influence of Winding Failure on Frequency Characteristics

Physical principles for winding technical condition monitoring with FRA method are electro-magnetical couplings between elements inside transformer tank. Strength of coupling is determined by parameters of used electrotechnical materials, actual winding geometry and placement of elements inside transformer tank. Diagnostics of winding fault with frequency response analysis uses correlation between impact of geometry changes on changes in capacitive and inductive couplings in winding. Displacement, deformation of winding shape, short circuit between turns causes locally changes of leakage inductance and capacitances [7]. Equivalent RLC circuit of transformer (Fig. 2a) used for wide band frequency analysis include electrical and magnetic couplings between elements, such RLC circuit have unique frequency characteristics for specific geometry. While mechanical forces are acting on elements inside transformer tank, causing deformation of windings and displacements, electro-magnetic couplings are changed what influence on values of RLC parameters in equivalent circuit and shifts frequency characteristic [7].

Transfer functions (TF) of transformer windings are defined as ratios of equivalent voltages and currents in frequency domain, measured in standardized configurations [18]. Shapes of frequency characteristics of transformers are different for various transformers types and depends on design of power transformers and its equipment. Peaks appearing in characteristics are related to series and parallel resonances in equivalent RLC circuit.

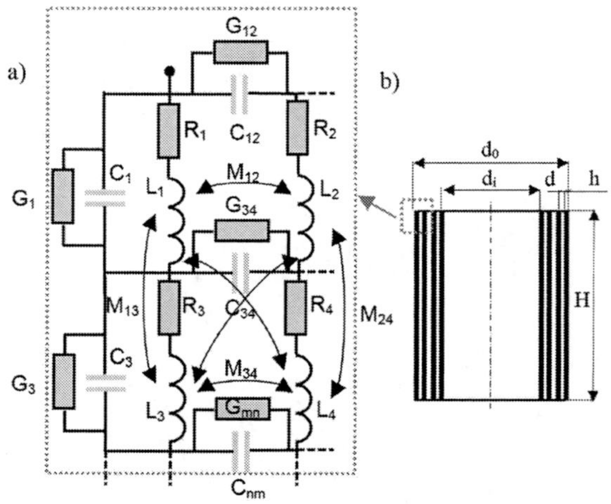

Fig. 2 a) RLC equivalent model of transformer winding used for wideband frequency analysis, b) simplified cross-section of transformer layer winding

Technical condition of power transformer is usually based on comparative analysis of TF obtained in different time periods (Fig. 1) [7-9]. In case of lack of measurements performed for healthy transformer it is possible to compare actual TF with TF from twin transformer. Basic importance in evaluation of winding condition it to proper interpretation of differences in TF changes. IEC [6] standard provides division of frequency bands in to four ranges which are correlated to particular components, low frequency band (up to 2 kHz) is correlated to changes in magnetic core, middle frequency band (2-20 kHz) is related to interaction between windings, changes of shape in high frequency band (20 kHz – 2 MHz) are related to changes in winding structure, shifts in range above 2 MHz are correlated with measurement stand [6]. Detailed influence of changes inside transformer tank on frequency characteristics differences are subject of research in many institutes [17-19].

Frequency characteristics of transformer windings can be obtained mainly with two different methods. First method uses sinusoidal signal with variable frequency for transformer excitation (Sweep Frequency Response Analysis) SFRA method [7,22]. Second kind of method uses steep voltage signals as a excitation like Low Voltage Impulse (LVI) method. For diagnostic purposes LVI method uses voltage impulses with maximal value up to several hundred of volts.

B. Diagnostics Methodology of SFRA and LVI Method

Power transformers are subjected to detailed diagnostic investigations. Investigations include among others mechanical condition of windings. Detection of windings failures is made with the frequency response analysis method. As a base of this method are the frequency characteristics of transfer function of transformer winding obtained with sinusoidal signals (SFRA method (Fig. 3a)) or with impulse voltage signals (LVI (Fig. 3b)). SFRA method is off-line method thus test transformer must be switched off, this procedure is inconvenient while providing constantly electrical energy for customers should be ensured in different way. Frequency of this kind of measurement is rare and is defined in distribution company exploitation procedures. Usage of random signals occurring in power systems during exploitation can extend usage of FRA method to on-line method and provide constantly information's about mechanical condition of winding [13-14]. This signals should be wide band signals like lightning surges or steep transients which can stimulate higher frequencies [22,23]. Energy in frequency spectrum of wide band signal depends on its maximal steepness (Fig. 4c), higher frequencies are energized with lower energy, therefore frequency characteristics obtained with different measurement signals can differ in some frequency bands, limitations of usage depends on voltage shapes. Sinusoidal signals provide constant energy for specific frequency, measurements of frequency characteristics with SFRA method have good quality, and every frequency is stimulated with the same energy, limitations of this method are due to used signal generator and measurement equipment.

In both methods signals of excitation and response are measured and basic signal information are computed. SFRA method provides transformation to frequency domain by used hardware (Fig. 3a) (sinusoidal signal generator), in LVI method frequency domain is achieved by application of Fast Fourier Transform (Fig. 3b). While LVI method needs only one measurement for determination of frequency characteristic implementation of this method can be used as an on-line diagnostic method [19].

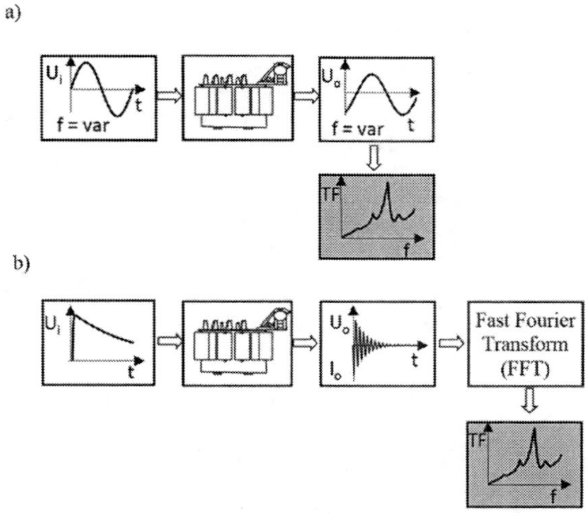

Fig. 3 Block scheme for transfer function measurement with a) sinusoidal voltage stimulus, SFRA method, b) Impulse voltage stimulus, LVI Method

Main assumptions for LVI method as a method which can provide information during transformer work, are usage of transients occurring in power systems as a diagnostic signals. Voltage transients reaching transformer terminals are caused mainly by failures and elimination of failures, side effects of working principles of some electrical power apparatus (i.e. strikes between circuit breaker contacts). Overvoltages propagating in power systems have wide band spectrum of frequencies which are stimulated. Energy in this spectrum is determined by maximal steepness in voltage transient i.e. for normalized lightning voltage impulse maximal frequencies in energy spectrum are stimulated up to 200 kHz (Fig. 4c).

C. *Test Object and Measurement Stands*

Laboratory measurements covered analysis of mechanical condition of transformer windings with two methods of transfer function measurement: SFRA and LVI method. Main objective was to compare obtained results and analyze possibilities of usage LVI, which can use transient voltages appearing in power network as a diagnostic signal. In analysis short circuit between turns at High Voltage (HV) winding were simulated. Test object was 250 kVA distribution transformer (see parameters in Table 1), high voltage winding has form of layers connected in series. (Fig. 2b).

The experimental setup of SFRA method consisted of a programmable function generator Tektronix type AFG310, a digital oscilloscope Tektronix type TDS1000/2000 sampling at 100 MHz, 1 Gs/s connected to the host computer by a GPIB-USB interface [22]. The software based control, processing and analysis has been implemented in LabView™ from National Instruments. The sinusoidal supply voltage with amplitude of 20 V and stepwise variable frequency was applied to the terminals of the transformer windings. The frequency characteristics of the voltage ratio were recorded at discrete frequency steps for 200 measuring points in the frequency range from 1 kHz up to 1 MHz. The software prepared in LabView™ was used to store the measurement results and to determine the values of transformers phase-amplitude TF for each frequency step.

Impulse generator used in LVI stand (Fig. 4a) is constructed with passive RC elements, it provides voltage impulses with shape 50/180 ns (Fig 4b) and with maximal value up to 500 V, this type of impulse can stimulate frequencies up to 10 MHz (Fig 4c), but for further analysis 1 MHz limit was chosen (according to [18]). Measurements of Voltage Impulse and response of the system are made with Tektronix 784D with 250k samples per channel. Post processing of measured signal is made in MATLAB® software, FFT algorithm for transformation from time domain to frequency domain is used.

Transfer function of voltages at LV winding to HV winding ratio was chosen as a diagnostic frequency response. Measurements were performed in case with no failure, and with simulated failures of 1%, 2% and 4% of turns in one layer shorted.

Table 1. Basic parameters of test transformer 250 kVA, 15/0,4 kV

Parameter	Value
rated power, kVA	250
rated voltage, kVA	15/0,4 kV
type of winding, (HV)	layer type
winding height, H, mm	732
number of layers, -	14
inner diameter, d_i, m	0,201/0,233
outer diameter d_0, m	0,255/0,303
number of turns in layer, -	191
distance between layers g, mm	0,3

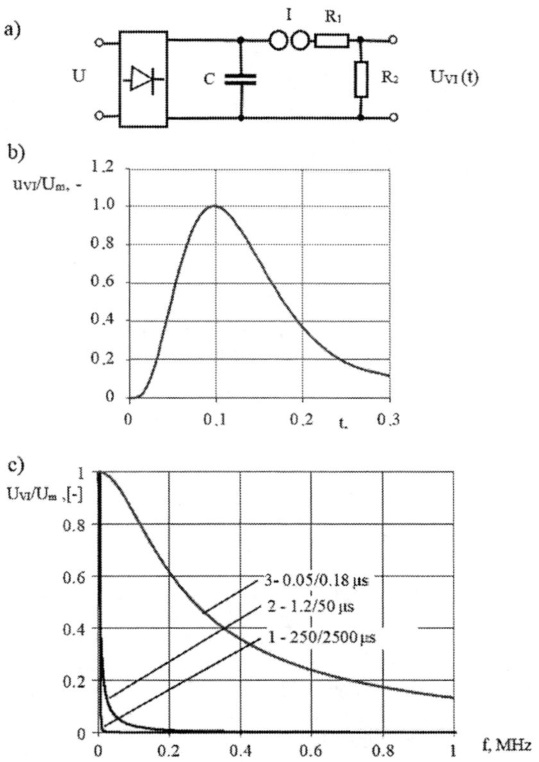

Fig. 4. Generation of low voltage impulses (U_{VI}) used in LVI method a) simplified LVI generator, where C=10 nF, R_1= 0,1 Ω, R_2 = 5 Ω, b) measured 50/180 ns voltage impulse normalized to maximal value (U_m), c) comparison of frequency spectrum calculated for: 1 - normalized switching voltage impulse, 2 - normalized lightning voltage impulse, 3 - 50/180 ns LVI

III. Measurement Results of Transfer Functions

A. *Frequency Characteristics of Winding*

Measurement results of transfer functions obtained for different simulated short circuits between turns are presented in Fig. 5 and Fig. 6. In Figure 5. Are presented frequency characteristics of LV to HV site voltage ratio measured with SFRA method. Differences in characteristics are related to changes of couplings in winding caused by winding turn short circuit, highest modifications are correlated with resonant frequency area (400 kHz to 750 kHz), changes in amplitude are observed, shifts of resonance peak at 400 kHz and 750 kHz are seen.

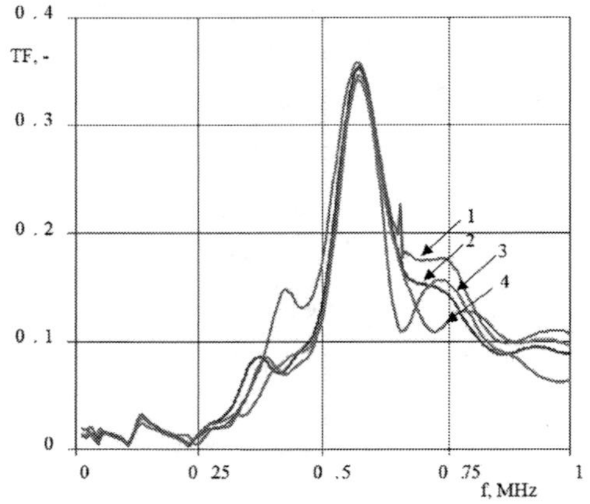

Fig. 5. Measured transfer functions of 250kVA transformer winding, during simulated fault between turns, phase A, obtained with SFRA method, 1 – healthy winding, 2 – 1% turns shorted , 3 – 2% turns shorted, 4 – 4% turns shorted

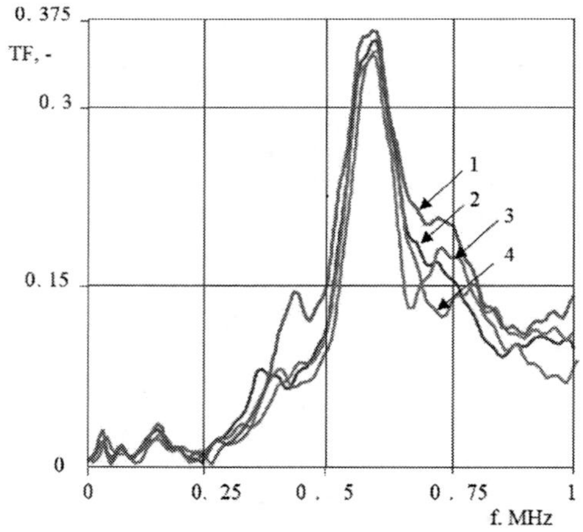

Fig. 6. Measured transfer functions of 250kVA transformer winding, during simulated fault between turns, phase A, obtained with LVI method, 1 – healthy winding, 2 – 1% turns shorted , 3 – 2% turns shorted, 4 – 4% turns shorted

Figure 6 presents frequency characteristics obtained for LVI Method with stimulus 200 V, 50/180 ns Voltage Impulse. Measured response are similar to SFRA method results. Characteristics differences are related with frequency range 400-750 kHz. Comparison of measurements form Fig. 5 to measurements from Fig. 6 shows that both methods are suitable for winding failure detection.

B. *Coherence Function for LVI Method*

Last case in analyzed LVI method was to check frequency range for which external noise and influence of used measurement equipment provides deformations in achieved frequency characteristics. Estimation of this frequency band is possible with calculation of coherence function for particular case. Coherence (1) is calculated for n-number of measurements in unchanged conditions. Frequencies for which value of coherence function differs from unity are subjected to influence of noise and have high uncertainly of measurements [21].

$$\gamma^2 = \frac{G_{xy}^2}{G_{xx}G_{yy}} \tag{1}$$

where: γ – coherence function of measurement stand, G_{xx} – self correlation of source voltage, G_{yy} - self correlation of response voltage, G_{xy} - cross correlation between analyzed signals.

Calculated coherence function for measurement setup with transformer 250 kVA is shown in Fig. 7. Calculations were made for 5 measurements with healthy winding configuration. Results shows that frequency band at which noise affect measurement results is below frequency 300 kHz, thus changes in analyzed frequency area (400-750kHz) are caused only by simulated winding faults.

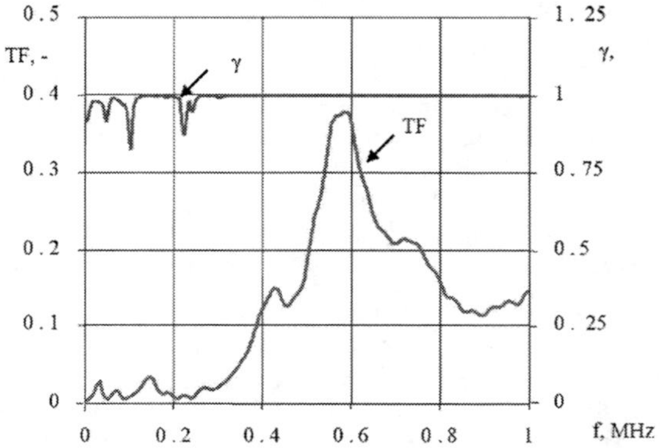

Fig. 7. Frequency characteristics (TF) measured for 250 kVA transformer with LVI method, and coherence function (γ) calculated for LVI method measurement stand,

IV. Conclusions

Results of laboratory measurement presented in this article shows that usage of voltage transients appearing at transformer terminals is possible for measurements of frequency characteristics which can be used as diagnostics information for determination of transformer windings mechanical condition. Steeper signal appeared at transformer terminals provides better quality of measured frequency characteristic. Limitations of LVI method are related with measurement stand which can provide uncertainly in frequency regions for which noise ratio to diagnostic signal is significant. Comparison of results obtained with SFRA and LVI shows similarities in characteristics, usage of LVI for diagnostic procedure of mechanical condition of winding is possible.

V. Bibliography

[1] X. Zhang and E. Gockenbach, "Asset-Management of Transformers Based on Condition Monitoring and Standard Diagnosis [Feature Article]," in IEEE Electrical Insulation Magazine, vol. 24, no. 4, pp. 26-40, July-Aug. 2008.

[2] H. Borsi and E. Gockenbach, "Monitoring, diagnosis and life management of power transformers", Conf. Maintenance

Operation High-Voltage Electrical Equipment, (TechCon Asia-Pacific), Sydney, Australia, May 2007.

[3] M. H. Samimi and S. Tenbohlen, "The Numerical Indices Proposed for the Interpretation of the FRA Results: a Review," VDE High Voltage Technology 2016; ETG-Symposium, Berlin, Deutschland, 2016, pp. 1-7.

[4] Z. Xue, X. Sun and Y. Liang, "Application of Data Mining Technology Based on FRS and SVM for Fault Identification of Power Transformer," 2009 International Conference on Artificial Intelligence and Computational Intelligence, Shanghai, 2009, pp. 452-455.

[5] L. Satish and A. Saravanakumar, "Identification of Terminal Connection and System Function for Sensitive Frequency Response Measurement on Transformers," in IEEE Transactions on Power Delivery, vol. 23, no. 2, pp. 742-750, April 2008.

[6] CIGRE Working Group 12.05, "An international survey on failures in large power transformers in service," Electra, no. 88, pp. 21–47, 1983.

[7] M. Florkowski, J. Furgał, "Detection of Transformer Windings Deformations Based on the Transfer Function - Measurements and Simulations," in Meas. Scien. and Techn., No. 14, 2003, pp. 1986 - 1992

[8] E. Rahimpour, J. Christian, K. Feser and H. Mohseni, "Transfer function method to diagnose axial displacement and radial deformation of transformer windings," in IEEE Transactions on Power Delivery, vol. 18, no. 2, pp. 493-505, April 2003.

[9] J. Christian, K. Feser, "Procedures for Detecting Winding Displacements in Power Transformers by the Transfer Function Method," in IEEE Trans. on Pow. Deliv., No. 19, 2004, pp. 214 - 220

[10] M. Wang, A. J. Vandermaar, K. D. Srivastava, "Improved Detection of Power Transfomer Winding Movement by Extending the FRA High Frequency Range," in IEEE Trans. on Pow. Deliv., No. 20, 2005, pp. 1930 - 1938

[11] S. Tenbohlen, S. Coenen, M. Djamali, A. Müller, "Samimi, M.H.; Siegel, M. Diagnostic Measurements for Power Transformers," in *Energies* 2016, *9*, 347 pp. 1-25

[12] E. Kornatowski and S. Banaszak, "Diagnostics of a Transformer's Active Part With Complementary FRA and VM Measurements," in *IEEE Transactions on Power Delivery*, vol. 29, no. 3, pp. 1398-1406, June 2014.

[13] C. Gonzalez, J. Pleite, V. Valdivia and J. Sanz, "An overview of the On Line Application of Frequency Response Analysis (FRA)," 2007 IEEE International Symposium on Industrial Electronics, Vigo, 2007, pp. 1294-1299.

[14] S. Birlasekaran and F. Fetherston, "Off/on-line FRA condition monitoring technique for power transformer," in IEEE Power Engineering Review, vol. 19, no. 8, pp. 54-56, Aug 1999.

[15] E. Gómez-Luna, G. Aponte Mayor, J. Pleite Guerra, D. F. Silva Salcedo and D. Hinestroza Gutiérrez, "Application of Wavelet Transform to Obtain the Frequency Response of a Transformer From Transient Signals—Part 1: Theoretical Analysis," in IEEE Transactions on Power Delivery, vol. 28, no. 3, pp. 1709-1714, July 2013.

[16] H. El-Hajjar, Z. D. Wang, Z. D. Sofian, P. N. Jarman, S. A. Ryder, A. W. Darwin: , "Analysis and Inretpretation of FRA Measurement Results using Transfer Function Application on Large Power Transformers," 14th Int. Symp. on High Voltage Engin. ISH'2005 (Beijng, China), 2005, paper G – 098

[17] J. Christian, K. Feser, U. Sundermann and T. Leibfried, "Diagnostics of power transformers by using the transfer function method," 1999 Eleventh International Symposium on High Voltage Engineering, London, 1999, pp. 37-40 vol.1.

[18] IEC 60076-18 – Power transformer – Part 18: Measurement of frequency response

[19] Malewski R., Poulin B.: Impulse Testing of Power Transformers using the Transfer Function Method, IEEE Trans. on Power Delivery, No. 2, 1988, pp. 476 – 489

[20] Wang M., Vandermaar A. J., Srivastava K. D.: Condition Monitoring of Transformers in Service by the Low Voltage Impulse Test Method. Eleventh Int. Symp. on High-Volt. Eng., London, Aug. 23 - 27, 1999, Vol. I., paper 1.45.S3

[21] Malewski R., Claudi A., Josephy C., Jud S. : "Checking Electromagnetic Compatibility of a HV Impulse Measuring Circuit with Coherence Functions", ERA Technology Conference High Voltage Measurements and Calibration Arnhem 19 and 20 October 1994, E1-82, pp 1-11

[22] M. Florkowski, J. Furgał and M. Kuniewski, "Propagation of overvoltages in distribution transformers with silicon steel and amorphous cores," IET Gen. Trans. & Distr., vol. 9, no. 16, pp. 2736-2742, 2015

[23] M. Florkowski, J. Furgał and M. Kuniewski, "Propagation of overvoltages transferred through distribution transformers in electric networks," IET Gen. Trans. & Distr., vol. 10, no. 10, pp. 2531-2537, 2016.

Experimental Measurement and Analysis of Frequency Responses SFRA for Rotating Electrical Machines

Martin Brandt
Department of Measurement and Applied Electrical Engineering
Faculty of Electrical Engineering, University of Žilina
Žilina, Slovakia
martin.brandt@fel.uniza.sk

Abstract — **The paper deals with the possibilities of measuring frequency responses using SFRA (Sweep Frequency Response Analysis) for rotating electrical machines. There are defined conditions and procedures for measuring the frequency responses, methods of measurement and analysis of measured frequency responses using cross correlation coefficients. In the paper are compared the results of frequency responses using SFRA between several types of rotating electrical machines (both AC and DC). On one of the investigated machines is analyzed internal winding fault. The fault is diagnosed and confirmed by measuring using SFRA method, winding resistance, capacitance, insulation resistance and the polarization index.**

Keywords: SFRA; rotating machine; measurement;analysis

I. INTRODUCTION

A rotating electrical machine – motors and generators are important machines in the industry. In production, they serve as a drive mechanism for the other machines and devices and generate electricity that is subsequently transformed and transmitted to the consumers. In households, they are used to drive hand tools and kitchen appliances, etc. In the case of both power and large-scale rotary machines it is important to have information about their current state because they can serve in strategic locations such as gas and oil refineries, power plants and production factories. In the event of an accidental outage caused by a faulty device the economic and technical damage can be very extensive. Currently, there are several online and offline diagnostic and monitoring methods with which we can measure and detect possible malfunctions arising on the rotating machines during their operation. In this case, the high ability of the methods used, the time consuming of the measurements, the difficulty of identification and the detection of the type of failure, and last but not least the expertise and practical experience of the technicians, play a big role. The most often methods use measurements of winding resistance, insulation resistance, winding capacities are also measured in some cases. If it is possible in terms of machine dimension partial discharges at the leads are measured or by means of acoustic sensors implemented directly on the machine. The thermal camera can capture the distribution of the temperature fields of machines in their surface under load, etc.

II. SWEEP FREQUENCY RESPONSE ANALYSIS METHOD

In practice, the SFRA (Sweep Frequency Response Analysis) method is used for frequency response analysis of power transformers. Both the measurement methodology and the basic evaluation of measured frequency responses are based on the applicable standard EN 60076-18:2013 [1]. The SFRA may indicate faults, changes in windings and the core of the electric rotating and non-rotating machine. The principle of measurement is that the varying low voltage frequency pulses are supplied to the input winding of the electric machine. At the end of the winding the response is measured and shown as a frequency curve (frequency dependent waveform). The responses depend on the overall impedance of the machine being measured which is dependent on the winding resistance, its own, mutual and leakage inductance as well as the individual capacities given by the machine construction.

A fault detection or damage by SFRA is most effective when reference data of the machine are available for evaluating measured frequency responses. By comparing the frequency responses (reference and after the time of operation) of the SFRA characteristics of the same electric machines it is possible to determine whether there have been mechanical changes in its interior. If different response forms are found, it is necessary to investigate and identify what change was recorded. Subsequently, the type of failure is determined.

The procedure for frequency response measurement is done by feeding a low voltage reference signal onto a terminal (generally the input terminal of measuring winding) of the measured machine through a coaxial cable (measuring cable) whose earthing is grounded by a transformer tank (connection to the bushing). On the other terminal (output of the measured winding) the output voltage signal (response signal) is measured via a shielded coaxial cable (measuring cable). The relative value of the frequency response amplitude (RHA) is the scalar ratio between the response signal (U_{out}) and the reference voltage (U_{in}) as the frequency function expressed in dB, eq. (1). The phase of the frequency response is the phase difference between the reference voltage and the response voltage expressed in degrees. The voltage response measurement is made through a 50 Ω impedance. [2][3]

$$RHA[dB] = 20\log_{10}\left(\frac{U_{\text{out}}}{U_{\text{in}}}\right) \quad (1)$$

The output voltage is grounded across a 50 Ω coaxial cable, with respect to:

$$\frac{U_{\text{out}}}{U_{\text{in}}} = \frac{50}{50+Z}, \quad (2)$$

Where Z is the frequency-dependent impedance by the inductance or capacitance function, or a combination of both.

III. Methods of measuring rotating machines by SFRA

Rotating electric machines are very different from the construction point of view as compared to the transformers. However, it is important to realize that they have a winding which is located in the grooves, or on the pole posts of the stator or rotor. [4] These are made up of mutually insulated silicon steel sheets (with properties similar to transformer sheets forming the core of the transformer). There is an air gap between the stator and the rotor, the size of which is different for each type of rotating machine. Its size will thus affect the value of the mutual induction between stator and rotor windings (depending on the type of the rotating machine). The rotating electric machine has then all the basic elements from the point of view of the frequency response measurements [5] [6]. These are resistors, inductances and capacities. In rotating electric machines, the shape and magnitude of the measured frequency responses will influence:

- the total impedance of the measured circuit (formed by the coil with resistance, inductance and capacity between thread),
- mutual inductance between coils,
- the capacities between the windings and the capacities between the winding and the ground,
- leakage inductances
- length of supply leads from the terminal to the winding,
- the laying of the supply leads,
- way of making the stator and rotor (sheet type, thickness, etc.).

Since it is a rotating machine whose rotor rotates about its axis in the stator space, we must define the concept of the rotor's basic position relative to the stator - initialization of the position. This is because we need to achieve repeatability of measurements on individual machines. Depending on the type and wiring of the winding on the terminal block of the machine, the following combinations of SFRA frequency response measurements for Asynchronous Motors can be made according to the following tables (similar combinations of measurements also apply to synchronous and DC motors / generators):

Tab. 1 Standard measurement of the winding of Asynchronous (Induction) Machine with squirrel cage rotor in delta connection

Number	Input and reference cable (U_{in}) connected to	Response cable (U_{out}) connected to	Measured winding
1	A, U, R, X or 1	B, V, S, Y or 2	stator
2	B, V, S, Y or 2	C, W, T, Z or 3	stator
3	C, W, T, Z or 3	A, U, R, X or 1	stator

Tab. 2 Standard measurement of the winding of Asynchronous (Induction) Machine with open circuited wound rotor in delta connection

Number	Input and reference cable (U_{in}) connected to	Response cable (U_{out}) connected to	Measured winding
1	A, U, R, X or 1	B, V, S, Y or 2	stator
2	B, V, S, Y or 2	C, W, T, Z or 3	stator
3	C, W, T, Z or 3	A, U, R, X or1	stator
4	A, U, R, X, x or 1	B, V, S, Y, y or 2	rotor
5	B, V, S, Y, y or 2	C, W, T, Z, z or 3	rotor
6	C, W, T, Z, z or 3	A, U, R, X, x or 1	rotor

Tab. 3 Standard measurement of the winding of Asynchronous (Induction) Machine with short circuited wound rotor in delta connection

Number	Input and reference cable (U_{in}) connected to	Response cable (U_{out}) connected to	Measured winding
1	A, U, R, X or 1	B, V, S, Y or 2	stator
2	B, V, S, Y or 2	C, W, T, Z or 3	stator
3	C, W, T, Z or 3	A, U, R, X or 1	stator

Tab. 4 Standard measurement of the winding of Asynchronous (Induction) Machine with open circuited wound rotor in star connection

Number	Input and reference cable (U_{in}) connected to	Response cable (U_{out}) connected to	Measured winding
1	A, U, R, X or 1	Neutral, X0, 0	stator
2	B, V, S, Y or 2	Neutral, X0, 0	stator
3	C, W, T, Z or 3	Neutral, X0, 0	stator
4	A, U, R, X, x or 1	B, V, S, Y, y or 2	rotor
5	B, V, S, Y, y or 2	C, W, T, Z, z or 3	rotor
6	C, W, T, Z, z or 3	A, U, R, X, x or 1	rotor

Note: The input and reference cable in the case of the manufacturer DOBLE is processed as one cable with red marking of the terminal, the response cable is marked with a black terminal (also applies to other commercial manufacturers - TETTEX, MEGGER, etc.)

A. *Rotor position initialization*

Each rotating machine has a shaft mechanically mounted on the rotor which is seated in bearings located in the front and rear in the machine frame. Each shaft has a rotor groove at its end. The mechanical load is connected to the rotor by mechanical coupling. The basic position of the rotor relative to the stator is determined by the position of the groove like it is depicted in Fig. 1.

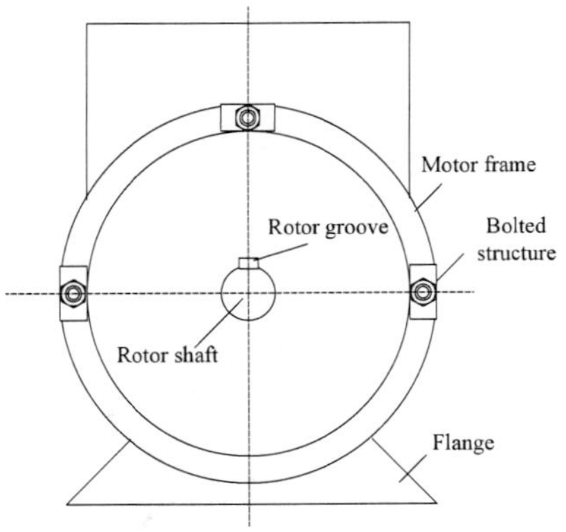

Figure 1. Illustrative view to the intialization of the rotor position for the SFRA measurement

B. Connecting the measuring cables

Before measuring, it is necessary to determine the method of connecting the measuring cables. Since the rotating machines from the constructional point of view have the ends of the winding connected to the terminal it is not possible to ensure the method of connecting the measuring cables according to the standard [1].

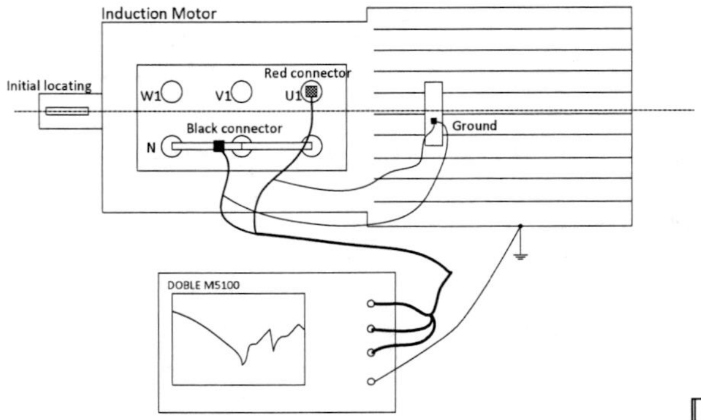

Figure 2. Example of test cable connection to the measured motor (the shielding wires are connected together to the motor frame)

End of the measuring cable supplying the reference signal (red terminal) is connected to input terminal of the measured winding. For measuring the output signal from a measured winding the black cable must be connected to other end of the winding. The shielding cables are connected to the lifting frame of the machine which are usually located at the top of the frame. The lifting frame must be free of impurities so that a perfect conductive connection is created. The earth cable of the measuring instrument is connected to the main earthing terminal of the rotating machine.

C. Measured machines and their frequency responses

Tab. 5 Nameplate data of measured Induction motor (Motor 1)

Manufacturer	MEZ Vsetín
Model	Sn50-4
Power	110 kW
Rated Voltage	380 V
Phases, windings connection	3, Y
Serial number	159672

Tab. 6 Table of SFRA measurements for the Motor 1 s.n. 159672:

Stator winding connection Y		
Test 1	Test 2	Test 3
X – Y	Y – Z	Z – Y

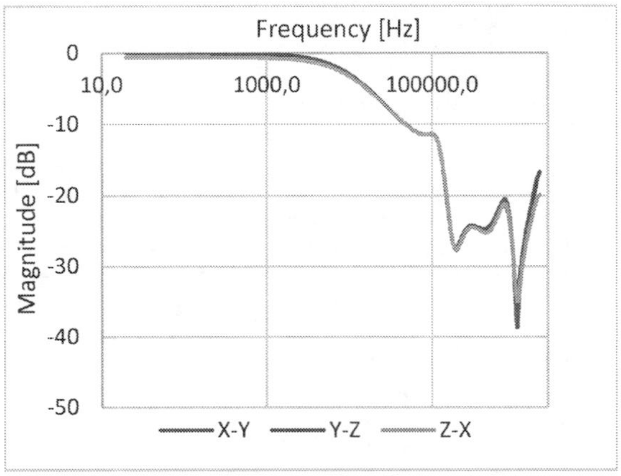

Figure 3. Frequency responses of IM s.n. 159672

Tab. 7 Nameplate data of measured Synchronous generator (Generator 1)

Manufacturer	MEZ Frenštát
Model	A12A4 0G0
Power	100 kW
Rated Voltage	380 V
Phases, windings connection	3, Y + excitation
Serial number	1524953

Tab. 8 Table of SFRA measurements for the Generator 1 s.n. 1524953:

Stator winding connection Y			Excitation winding
Test 1	Test 2	Test 3	Test 4
X – X0	Y – X0	Z – X0	B1 – B2

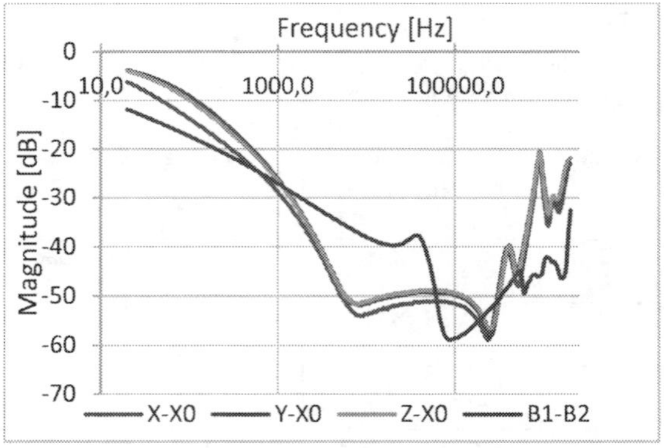

Figure 4. Frequency responses of SG s.n. 1524953

Tab. 9 Nameplate data of measured DC motor with separate excitation (Motor 2)

Manufacturer	MEZ Vsetín
Model	MS 2821-4

Power	100 kW
Rated Voltage	300 V
Serial number	213657

Tab. 10 Table of SFRA measurements for the Motor 2 s.n. 213657:

Main winding	Excitation winding
Test 1	Test 2
K – 0	B1 – B2

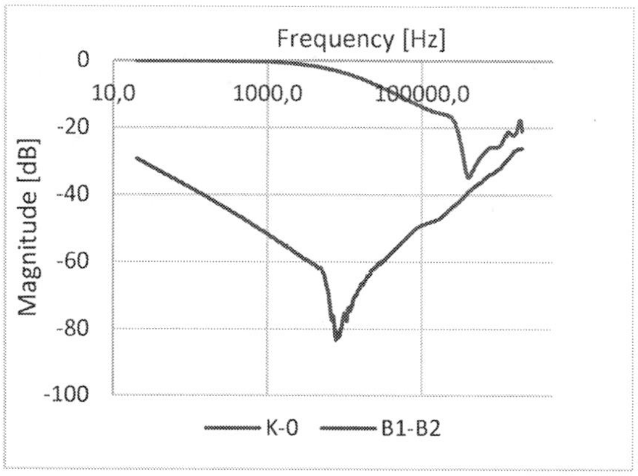

Figure 5. Frequency responses of DC motor s.n. 213657

Tab. 11 Nameplate data of measured Induction motor (Motor 3)

Manufacturer	MEZ Mohelnice
Model	4AP 80-2
Power	1,1 kW
Rated Voltage	230 V
Phases, windings connection	3, Y
Serial number	320122

Tab. 12 Table of SFRA measurements for the Motor 3 s.n. 320122:

Stator winding connection Y		
Test 1	Test 2	Test 3
U – N	V – N	W – N

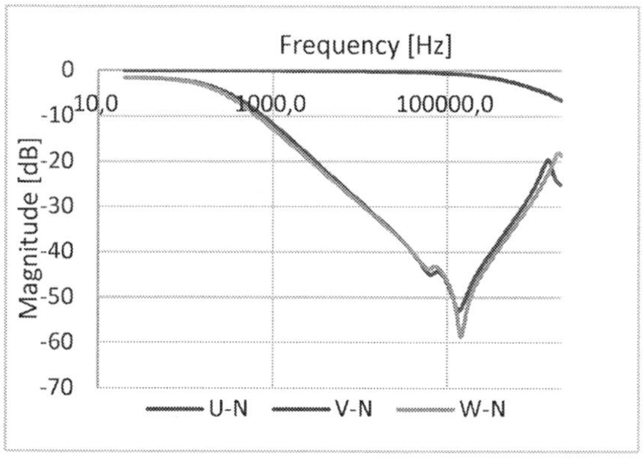

Figure 6. Frequency responses of IM s.n. 320122

D. Frequency Response Analysis of Motor 3 by Cross Correlation Coefficients

A fault acknowledgement of Motor 3 was done by means of Cross Correlation Coefficients (CCF) analysis, which calculate a conformity coefficients of two frequency responses for the same measurement. CCF is often used in industry, telecommunications and where the precise analysis of signals is needed. Its use in the SFRA is in place for an analysis of the two waveforms. If the result of CCF is equal to 1.0, it means the absolute correlation. The zero value means the absolute discorrelation. The negative correlation coefficients are meaningless in this case [7] [9].

Tab. 13 . An explanation of CCF examples

	CCF
good agreement	0,95 – 1.0
cuttoff agreement	0,90 – 0 94
bad agreement	< 0,89
no agreement	<= 0,0

CCFs are defined [7] according to:

$$CCF = \frac{\sum_{i=1}^{n}\left(X_i - \overline{X}\right)\left(Y_i - \overline{Y}\right)}{\sqrt{\sum\left(X_i - \overline{X}\right)^2} \cdot \sqrt{\sum\left(Y_i - \overline{Y}\right)^2}} \tag{3}$$

where Xi and Yi are two real rows (or graphs – in case of SFRA) compared to each individual frequency „i" X and Y are the value axes. In a case of mathematical signal processing, which is more complicated, the coefficient values are in a range of 1 and -1 still exact for needed conclusions.

Calculated CCFs for an Motor 3 between all phases of motor according to tab. 12 are in tab. 14.

Table 14. Calculating CCF's for Motor 3

Bands	10 Hz-100 Hz	100Hz-1kHz	1kHz-10kHz	10kHz-100kHz	100kHz-1MHz
CCF U-V	0,916	0,965	0,917	-0,96	-0,99
CCF U-W	0,999	1,00	0,998	0,976	0,999
CCF V-W	0,919	0,964	0,917	-0,889	-0,999

From Fig. 6 is clearly seen the difference in measured responses of the machine 3. The middle phase measurement (V-N) indicates fault as a short circuit. The short circuit fault is also confirmed by the calculation of the CCF (Tab. 14) in the phase V. The other two phases (U and W) have identical CCF values according Tab. 13.

It follows that a different shape of the measured response of the second phase of the stator winding (Motor 3) shows the large change in impedance values Z. Its shape and values corresponds with the curve which is measured for the real resistor [2], [8]. It is possible to determine that the measured part of the stator winding (second phase V) is only resistive. In this case, we can definitely confirm that the measured Motor 3 has inter-winding short circuit. From the impedance values, which is very low, it can be obtained via the evaluation software that it is possible to determine a locality of a short circuit. Short circuit must connect the input and output of the second phase winding coils.

This is confirmed by the measured resistance value of the individual coils (Tab. 15) and insulating resistor (Tab. 16). For measuring the winding resistance was used MULTIAMP Ohmeter , the test conditions was set to 1 Amp and voltage of 30V DC. All the results are converted to 20°C (Tab. 16) by the following equation:

$$R_{20} = R_{\upsilon} \frac{230+20}{230+\upsilon}, \text{ where } \upsilon = 24°C \quad (4)$$

For measuring the insulating resistor was used measuring instrument Megger MIT 520. [10]

Tab. 15 The Values of Winding Resitance of Motor 3

U1 – U2	8,2 Ω
V1 – V2	1,2 Ω
W1 – W2	7,8 Ω

Tab. 16 The Values of Insulation Resitance of Motor 3*

U – G	264 GΩ
V – G	170 GΩ
W – G	135 GΩ

*Value for 1000 V DC test voltage.

IV. CONCLUSION

The use of the SFRA method on rotating electrical machine has importance. As can be seen from the results obtained in the article, this method can detect an internal fault. What type of malfunction and where it is located is still hard to confirm at this time. If a large change in frequency response is measured, as it was in this case, it is not difficult to determine the type and location of the fault. For SFRA to be relevant in rotary machine measurements, it is necessary to make new measurements on multiple types of rotary machines to form a comprehensive database of SFRA frequency response patterns. Their results should be confirmed by simulation models for rotating machines and computational methods of substitution schemes to facilitate interpretation of measured results. Further results will be gradually published by authors participating in the internal research project at the Department of Measurement and Applied Electrical Engineering at the Faculty of Electrical Engineering University of Zilina in the near future.

ACKNOWLEDGMENT

This work was supported by the Grant Agency VEGA from the Ministry of Education of Slovak Republic under contract 1/0602/17

REFERENCES

[1] EN 60076-18 – Power transformer – Part 18: Measurement of frequency response

[2] M. Brandt. *Bezdemontážna diagnostika distribučných transformátorov*, Disertation thesis, Žilina 2009

[3] M. Gutten, J. Jurcik, D. Korenciak, R. Janura. "Analysis of Transformer Winding considering Transient Effects", *Journal of Electrical and Control Engineering*, Vol. 4, Issue 3, Jun 2014

[4] S. Kascak, B. Dobrucky, M. Prazenica. "A new approach for estimation of speed/position of two-phase induction machine using virtual high frequency injection method", In: *International review of electrical engineering (IREE)*, Vol. 8, no. 4 (2013), s. 1156-1160.

[5] R. Cimbala, J. Kurimský, I. Kolcunová. "Determination of Thermal Ageing Influence on Rotating Machine Insulation System Using Dielectric Spectroscopy", In: *Przegląd Elektrotechniczny*. Vol. 87, no. 8 (2011), p. 176-179. - ISSN 0033-2097

[6] Y. Yagami, C. Araki, Y. Mizuno, H. Nakamura. „Turn-to-Turn Insulation Failure Diagnosis of Stator Winding of Low Voltage Induction Motor with the Aid of Support Vector Machine", *IEEE TRANSACTIONS ON DIELECTRICS AND ELECTRICAL INSULATION* Volume: 22 Issue: 6 Pages: 3099-3106 Published: DEC 2015

[7] G. M. Kennedy, A. J. McGrail, J. A. Lapworth. "Using cross-correlation coefficients to analyze transformer Sweep Frequency Response Analysis (SFRA) traces", *IEEE-Power-Engineering-Society Conference and Exposition in Africa Location*: Johannesburg, SOUTH AFRICA Date: JUL 16-20, 2007

[8] B. G. Xu. "Intelligent fault inference for rotating flexible rotors using Bayesian belief network", *EXPERT SYSTEMS WITH APPLICATIONS* Volume: 39 Issue: 1 Pages: 816-822 Published: JAN 2012

[9] M. Gutten, R. Janura, D. Korenciak. "Experimental Analysis of Short-Circuit Effects in Transformer Winding by Impact Test and SFRA", *International Conference on Diagnostic of Electrical Machines and Insulating Systems in Electrical Engineering (DEMISEE)* Location: Papradno, SLOVAKIA Date: JUN 20-22, 2016

[10] M. Brandt, S. Kascak. "Failure Identification of Induction Motor using SFRA Method", *ELEKTRO 11th International Conference* Location: SLOVAKIA Date: MAY 16-18, 2016

Examination of Insulation in Fire Retardant Cables by Time Domain Methods

Michal Váry, Jaroslav Lelák, Vladimír Ďurman, Juraj Packa
Institute of Power and Applied Electrical Engineering
Faculty of Electrical Engineering and Information Technology, Slovak University of Technology
Bratislava, Slovak Republic
michal.vary@stuba.sk

Abstract — **Dielectric properties of halogen free fire retardant cables were investigated with aim of choosing the most suitable measuring and analyzing method which can characterize the insulation state of these cables. The work was directed to the utilization of time domain methods. Namely, it was the absorption current method and recovery voltage method. During experiments a very slow polarization process was found with relaxation time greater than 1000 s, which is probably caused by the interfacial polarization on the transition between the individual cores insulation. The detected insulation resistance was in general relative high in consequence of an intermittent contact at insulation interfaces. For the future work, the time domain methods were appreciated as more reliable with less demand on the measuring equipments comparing with the frequency domain methods.**

Keywords: cable diagnostics; absorption current; insulation resistance; recovery voltage

I. Introduction

In agreement with the IEEE standard No. 400, the cable tests can be categorized into two major types: destructive withstand tests and non-destructive diagnostic [1]. The first type includes DC, power frequency AC, and very low frequency (VLF) AC withstands tests. The second type includes general condition assessment, such as dissipation factor diagnostic test, and defect specific diagnostics, such as a partial discharge diagnostic test. The purpose of a non-destructive diagnostic test is to indicate the condition of a cable system and establish whether or not it is reliable in service. Generally, in the field of non-destructive tests the dielectric spectroscopy methods are widely used [2]. The dielectric spectroscopy is based on current or voltage measurements in the time domain and also impedance measurements in the frequency domain. By investigating the system properties, we are dealing with phenomena originating at the layer interfaces. The interface phenomena are often manifested as slow polarization processes i.e. the processes with high relaxation time. Monitoring of the slow processes is useful in praxis also for investigation of materials ageing.

II. Time Domain Dielectric Spectroscopy

The time domain spectroscopy comprises the absorption and resorption current measurement and the recovery voltage measurements. The output of this type of spectroscopy provides obviously the information on the insulation resistance and also the parameters of relaxation process (if it exists). The insulation resistance is the basic indicator of insulation quality and stability. The relaxation process parameters show us the type of relaxation (dipolar, interfacial) and its influence on the insulating system properties at working frequency. Knowledge of the relaxation process is of great importance at dc applications [3] and also during the long-term insulation ageing tests.

A. *Absorption current in dielectrics*

In the theory of dielectrics the measured object is characterized by the dielectric response *h(t)* which is response of the system resulting from unit (Dirac) impulse $\delta(t.)$. The formulation of dielectric response is better understood from definition of absorption current density i.e. the current density *i(t)* through dielectric after applying a step electric field E_0:

$$i(t)=\gamma_{dc}(t)E_0+\varepsilon_0\frac{dE(t)}{dt}+\varepsilon_0 E_0\frac{d}{dt}\int_0^t h(\lambda)d\lambda, \qquad (1)$$

where $\gamma_{dc}(t)$ denote the dielectric conductivity response to a step electric field and ε_0 is permittivity of free space. Dividing (1) by E_0 we get a formal expression for the time dependence of conductivity γ after application of a step electric field:

$$\gamma(t)=\gamma_{dc}(t)+\varepsilon_0\delta(t)+\gamma_a(t). \qquad (2)$$

The absorption part of conductivity $\gamma_a(t)$ is simply a product of ε_0 and *h(t)*. When switching from the time domain to the frequency domain we must calculate the transfer function, which is the Fourier transform of impulse response. The necessary impulse response can be calculated as the first derivative of the step response. Finally we get the formula for the specimen complex permittivity $\varepsilon^*(\omega)$ as

$$\varepsilon^*(\omega)=\varepsilon'-j\varepsilon''=\varepsilon_\infty+\frac{1}{\varepsilon_0}\int_0^\infty \gamma(t)\exp(-j\omega t)dt, \qquad (3)$$

where ε_∞ is the optical permittivity, ω is angular frequency and j is the imaginary unit.

In practice the measured data are available in a form of discrete values of time and conductivity. Taking this fact into account, the former continuous-time Fourier transform must be replaced by the discrete Fourier transform (DFT). Instead of the straightforward application of DFT, many approximation methods are used in this field. Here we can mention the famous Hamon transform [4]. It is based on assumption that the time dependence of absorption conductivity equals to

$$\gamma_a(t)=\gamma_1 t^{-n}, \tag{4}$$

where n is a parameter and γ_1 is value of conductivity for time $t=1$. For this special case the Fourier transform can be calculated by means of Γ function. For the imaginary part of complex permittivity $\varepsilon''(\omega)$ the next equation is valid:

$$\varepsilon''(\omega)=\frac{\gamma_1}{\omega^{1-n}\varepsilon_0}\Gamma(1-n)\cos\frac{n\pi}{2} \quad 0<n<2. \tag{5}$$

Hamon choose a special linkage between the time and frequency f:

$$f=\frac{y(n)}{t}, \tag{6}$$

where the function $y(n)$ is:

$$y(n)=\frac{1}{2\pi}\left[\Gamma(1-n)\cos\frac{n\pi}{2}\right]^{-\frac{1}{n}} \tag{7}$$

It was found that if unequality $0.3 < n < 1.2$ is satisfied then $y(n)$ equals to 0.1 within 3 % precision. Finally, combining equations (5) and (7) we get:

$$\varepsilon''(\omega)=1.8\ 10^{11}\gamma_a(t)t. \tag{8}$$

The Hamon transform provides a very useful tool for the absorption and resorption currents analysis.

B. *Recovery voltage measurement*

The recovery voltage is defined as the voltage generated on the capacitance-like specimen after long charging with the voltage U_0 and an instant short-circuiting. This kind of voltage is sometime called as return voltage. In the past, the examination of recovery voltage was connected with the safety issues of electrical devices. Later, it was established as an alternative to the methods of time-domain dielectric measurements. Unlike at very small current measurements, here we are dealing with measurements of relative high voltages, which are less disturbed by the surroundings influence.

The recovery voltage analysis is generally directed at calculation of the complex permittivity parameters and also at calculation of insulating resistance. Transformation of recovery voltage data from the time domain into the frequency domain can be performed in two ways [5]:

a) by modeling the process using an equivalent circuit,
b) by using the Maxwell equations for solving the transient of recovery voltage .

The equivalent circuit comprises parallel connection of a single capacitor (capacitance C_0), single resistor (resistance R_0) and a set of n series R_i C_i elements connected in parallel with C_0 and R_0. In this circuit C_0 is the capacitance corresponding to the fast polarizations and R_0 is the resistance corresponding to the steady state conductivity (insulation resistance). The set of R_i C_i elements represents a slow relaxation process with spread of relaxation times.

The transient of recovery voltage $u_r(t)$ in this circuit can be solved by using the Laplace transform. Denoting the Laplace transform of $u_r(t)$ by $u_r(p)$ and the time constants R_iC_i by τ_i, we have

$$u_r(p)=\frac{\dfrac{U_0}{C_0}\sum_{i=1}^{n}\dfrac{C_i}{1+p\tau_i}}{p+\dfrac{1}{C_0R_0}+\dfrac{p}{C_0}\sum_{i=1}^{n}\dfrac{C_i}{1+p\tau_i}}. \tag{9}$$

The inverse Laplace transform of $u_r(p)$ can be computed from the rational function of (9) by standard procedure of partial fractions expansion.

The calculation procedure in the case of equivalent circuit includes an initial estimation of all parameters (except C_0). This is followed by the inverse Laplace transform of the expression at the right side of equation (9). The calculated values of recovery voltage are compared with the measured values and the initial estimation is modified until the best fit of calculated and measured values is achieved. The number of parameters and also the precision of calculated values in this procedure are questionable.

Similar result as we get from equation (9) can be reached also in another way. The transient of recovery voltage is in this case calculated by solving equation (1) for all three segments of the recovery voltage test (charging, short circuit and open circuit). The result is in the form of Laplace transform $u_r(p)$:

$$u_r(p)=\frac{U_0C_g\chi(p)}{pC_g\chi(p)+\dfrac{1}{R_0}+p\,C_0}, \tag{10}$$

where C_g is geometric or vacuum capacitance of the measured object and $\chi(p)$ is the dielectric susceptibility (i.e. permittivity without the term ε_∞). The calculation procedure in this case includes an initial estimation of the susceptibility parameters and R_0. This is followed by the inverse Laplace transform of the expression on the right side of equation (10) and comparison with the measured recovery voltage values. The estimation is modified until achieving the best fit of the compared expressions. Before performing the calculation we must explicitly define the formula of dielectric susceptibility. Often we choose the well-known susceptibility function of Havriliak-Negami [6]:

$$\chi(p)=\frac{\Delta\varepsilon}{\left[1+(p\tau)^{(1-\alpha)}\right]^{\beta}}, \tag{11}$$

where $\Delta\varepsilon$ is the permittivity increment, τ is the relaxation time of polarization process, α, β are the distribution parameters.

Here we must emphasize that the susceptibility function can be also obtained straight from the Fourier transform $u_r^*(\omega)$ without any fitting procedure if the insulation resistance is already known (e.g. from absorption current measurement). But the frequency range of the susceptibility function is limited by the Nyquist frequency of $u_r^*(\omega)$ [7].

III. Experiment

Three specimens of the length of 1 m were used in our experiment. They were cut from three types of flame retardant non-corrosive cables. The construction details are summarised in Tab. I. The cables A and B were compatible in construction, but the specimen B was declared as providing enhanced circuit integrity at fire.

TABLE I. CONSTRUCTION ELEMENTS OF THE FLAME RETARDANT CABLES

Construction element	Cable A, B	Cable C
Core	Cu, round shape, 2x1.5 mm^2	Cu, round shape, 3x2.5 mm^2
Insulation	glass-mica tape+EPR	cross-linked halogen free ceramic forming, 2 layer insulation HXI 2
Filling parts	halogen free, flame retarding	not specified
Sheath	halogen free, flame retarding	halogen free polymer compound

The measurements in the time domain were performed in the temperature range from 40 °C to 70 °C. The absorption and resorption currents were measured with Keithley 6517A electrometer. The charging voltage 100 V was applied to each specimen during the period of 1000 s. After this period the resorption current was measured during the next 1000 s. The recovery voltage was then measured separately with the same instrument after 1000 s charging period.

IV. RESULTS AND DISCUSSION

The measured values of absorption and resorption currents for one selected cable are plotted in Fig. 1. The difference of the charging and discharging current is a constant at longer times. In the short time region (approximately to 100 s), a small memory effect was observed which deforms the linear behavior of the mentioned difference

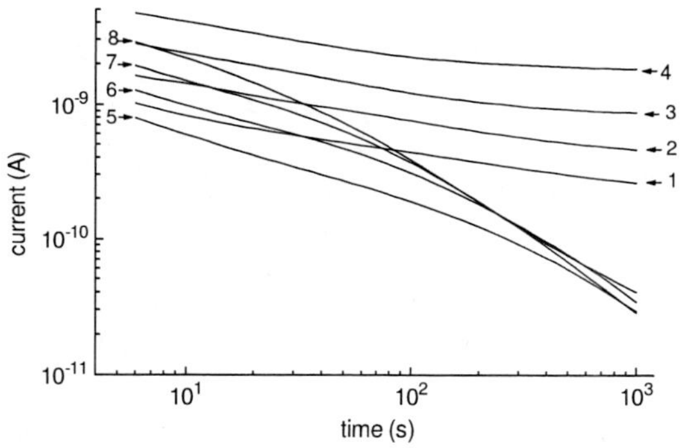

Figure 1. Cable C - absorption currents (1 - 4) and resorption currents (5 - 8) at temperature 40 °C (1, 5), 50 °C (2, 6), 60 °C (3, 7), 70 °C (4, 8)

The insulation resistances calculated from the differences of absorption and resorption currents are in Tab. II.

TABLE II. TEMPERATURE DEPENDENCE OF INSULATION RESISTANCE

temperature (°C)	R_0 (10^{11} Ω)		
	Cable A	Cable B	Cable C
40	7.77	7.27	4.03
50	5.77	5.10	2.19
60	9.55	7.42	1.14
70	5.70	3.87	0.53

Next investigation was concerned to the polarization processes in cable specimens. They were examined by means of the Hamon transform of resorption currents with help of equation (8). As the precise dimension of the cable insulation was not known, we used the values of imaginary part of complex capacitance (C”) instead of loss factor $\varepsilon''(\omega)$. The calculated values of C” for one selected cable are plotted in Fig. 2.

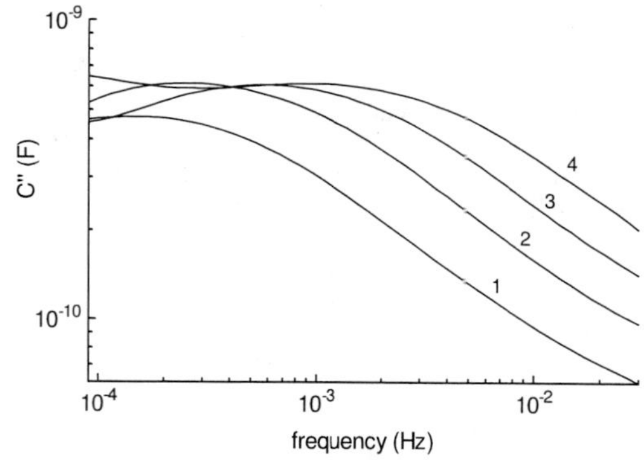

Figure 2. Cable C – imaginary part of complex capacitance calculated by Hamon transform at temperature 40 °C (1), 50 °C (2), 60 °C (3), 70 °C (4)

The parameters of polarization processes were also examined by means of recovery voltage measurements. As it was remarked above, these measurements were performed separately (after the absorption and resorption currents measurement). The charging voltage was 100 V. The measured values of recovery voltages for one selected cable are plotted in Fig. 3.

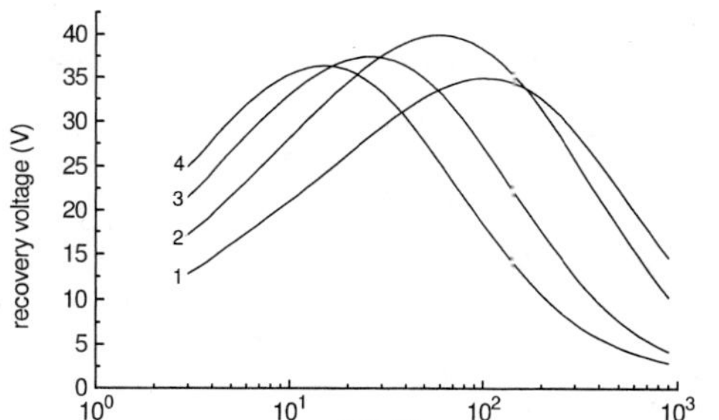

Figure 3. Cable C – recovery voltages at temperature 40 °C (1), 50 °C (2), 60 °C (3), 70 °C (4)

For calculation of the susceptibility parameters we used the procedure described in the previous part (equation (10) and (11)). The optimization procedure included estimation of five parameters: $C_g\Delta\varepsilon$, τ, α, β and R_0. A very important step at optimization is the initial estimation of parameters. While the parameters α and β are limited by their definition to the interval from 0 to 1, the other parameters are mostly unknown. As we are dealing with slow relaxation processes, the parameter τ can be search in interval from 10 s to 1000 s. For initial estimation of $C_g\Delta\varepsilon$ and R_0 we proposed a procedure which is based on the equivalent circuit described above. Our equivalent circuit comprises parallel connection of a single capacitor (capacitance C_0), single resistor (resistance R_0) and one series $R\,C$ element (assigned R_1 C_1). Let assign the charge on capacitor C_1 as Q_1 and the charge on capacitor C_0 as Q_0. Further we assigned the charging voltage as U_0, the maximum of recovery voltage as U_{max} and the time matching to the voltage maximum as t_{max}. After the instant short-circuiting at the beginning of voltage build-up, the whole initial charge is placed on C_1. In every instant of the test the sum of charges on capacitors equals to the difference of initial charge and the charge effluent through the resistance R_0:

$$Q_1 + Q_0 = C_1 U_0 - \frac{1}{R_0} \int_0^t u_r(\lambda)\, d\lambda. \tag{12}$$

The formula for charge equivalence can be applied for the time, when the voltage reaches its maximum U_{max}. At that time the charges Q_1 and Q_0 are equal. On the other hand, we can apply the formula at the end of the test when both charges equals to zero. In this way, we have two equations with two unknowns: C_1 and R_0. We suppose that C_0 is known from the measurements at higher frequency. The final formulae for estimation of C_1 and R_0 are as follows:

$$C_1 = \frac{C_0 U_{max}}{U_0 - U_{max} - U_0 \dfrac{\int_0^{t_{max}} u_r(t)dt}{\int_0^{\infty} u_r(t)dt}}, \tag{13}$$

$$R_0 = \frac{\int_0^{\infty} u_r(t)dt}{C_1 U_0}. \tag{14}$$

The integral of recovery voltage at infinity can be solved by approximation of the voltage at longer times in lin – log coordinates. The calculated C_1 is a good initial approximation of $C_g\Delta\varepsilon$.

The susceptibility parameters were calculated from recovery voltages by the optimization procedure for all cables. The procedure was not successful at some temperatures because of lack of complex capacitance maximum mainly at the lower temperatures. Another reason for the procedure failure was the presence of two maxima as it is seen in Fig. 2. In the cases when the procedure was successful we compared the magnitude of imaginary part of complex capacitance (C''_{max}) and the frequency matching to the maximum (f_{max}) with the same parameters acquired from the Hamon transform. They are listed in Table III.

TABLE III. COMPARISON OF THE RELAXATION PARAMETERS CALCULATED FROM THE HAMON TRANSFORM AND RECOVERY VOLTAGE MEASUREMENTS

Cable / temperature (°C)	Hamon transform		Recovery voltage	
	f_{max} (10^{-4} Hz)	C''_{max} (10^{-10} F)	f_{max} (10^{-4} Hz)	C''_{max} (10^{-10} F)
Cable A/ 50	1.8	5.3	2.9	5.1
Cable B/ 50	2.1	6.5	2.6	6.8
Cable C/ 40	1.4	4.7	2.3	4.8
Cable C/ 50	2.6	6.1	3.3	6.1

V. CONCLUSION

The work was aimed at research of dielectric properties of flame retardant cables. Research was directed into two stages. At first we tried to find a stable parameter which can characterize the insulation state of the cables. On the second stage we compare the applicability of two methods from the category of time domain measurements.

Generally the basic dielectric parameter is insulating resistance. The insulation resistances were calculated from the differences of absorption and resorption currents. Here we observed an expected behavior of insulation resistance only for the specimen of Cable C. As for the temperature dependence it follows the well-known Arrhenius law. The other two specimens indicate an irregular dependence with a small maximum at 60 °C. The most probable explanation of this phenomenon resides in the change of cable cores distance due to enhanced temperature. The stable behavior of the Cable C is probably a result of its three cores structure. From the measurement of resorption current we obtained also information on the relaxation process in the measured cables. The process is very slow with relaxation time about 800 s. The capacitance increment from the process is about ten times greater than the basic capacitance of the cable. From the values of relaxation time and capacitance increment we can conclude that the process is of the interface nature, which is given by the layered structure of insulation. The problem is that the process is hardly identified at temperatures below 50 °C in Cable A and Cable B as the relaxation maxima are not present in the frequency course of the complex capacitance imaginary part. For the future work at observing the insulation state of the cables we recommend, if possible, to choose the temperature of measurement for each cable individually. Other solution is extension of the measuring time over 1000 s to catch the required relaxation maximum properly also at low temperature.

As for the measuring method, we suppose, that the measurements of absorption and resorption currents are sufficient for acquiring the essential parameters. If more details are required on the relaxation process then we can confirm the previous results by recovery voltage measurements. Here we must take into consideration also the differences between results of these methods which can achieve values of 40 %.

ACKNOWLEDGMENT

This work was supported by the Slovak Research and Development Agency (APVV) [Contract No. APVV-15-0110, APVV-15-0108 and APVV-15-0326].

REFERENCES

[1] G. Hartshorn, B. Lanz, and B. Broussard, "Medium voltage cable predictive diagnostics technique," IEEE Paper No. PCIC-2007-30, 2007.

[2] W. S. Zaengl, "Dielectric spectroscopy in time and frequency domain for HV power equipment," 12th Internat. Symposium on High Voltage Engineering (ISH), Bangalore, India, 20 - 24 August 2001, pp. 1-10.

[3] V. Ďurman, M. Perný, V. Šály, and M. Pípa, "Space charge distribution in photovoltaic cell insulation," Journal of El. Eng., vol. 65, pp. 34-37, 2014.

[4] B. V. Hamon, "An approximate method for deducing dielectric loss factor from direct-current measurements," Proc. IEE., vol. 99, pp. 151-155, 1952.

[5] V. Der Houhanessian, "Measurement and analysis of dielectric response in oil-paper insulation systems," Dissertation No. 12832, ETH Zurich, 1998.

[6] S. Havriliak and S. Negami, "A complex plane analysis of α-dispersions in some polymer systems," Journal of Polymer Sci.: C No.14, pp. 99-117, 1966.

[7] V. Ďurman and M. Váry, "Assessment of insulation resistance from recovery voltage data," 17th International Scientific Conference on Electric Power Engineering, Prague, Czech Republic, 16-18 May 2016, pp. 351-354.

Identification of Power Cable Conditions Based on Broadband Powerline Communication Characteristics in Distribution Grids

Nikolai Hopfer, Markus Zdrallek
University of Wuppertal
Wuppertal, Germany
nikolai.hopfer@uni-wuppertal.de,
zdrallek@uni-wuppertal.de

Ulrik Dietzler, Markus Krampf
Energieversorgung Leverkusen GmbH & Co. KG
Leverkusen, Germany
ulrik.dietzler@evl-gmbh.de,
markus.krampf@evl-gmbh.de

Christoph Raquet, Marilen Ronczka
Power Plus Communications AG
Mannheim, Germany
c.raquet@ppc-ag.de
m.ronczka@ppc-ag.de

Hamed Rezaei
Nexans Power Accessories Germany GmbH
Hof, Germany
hamed.rezaei@nexans.com

Abstract — **The current paper deals with characteristics of broadband powerline communication to support condition assessment of power cables. This technology meets the growing demand for communications in electrical distribution grids. Since it uses the existing electrical grid infrastructure as transmission media, in particular power cables fulfill two functions: first of all the reliable distribution of electrical energy with mains frequency, essentially and as originally planned, and additionally the data transmission based on high frequencies. To satisfy the demand for high quality electrical energy supply, maintenance of power cables according to the actual condition is fundamental. Nevertheless, the actual condition of cables is widely unknown, because diagnostic measurements, which are conventionally applied offline, are performed neither extensively nor continuously in the course of condition assessments. This paper illustrates, how broadband powerline communication generates new insights into power cable conditions using data transmission characteristics. Therefore, theoretical background and the results of laboratory tests are presented. Thus, broadband powerline communication characteristics will be analyzed in order to determine the cable condition as shown in this paper. This procedure will benefit condition assessments of power cables in distribution grids, since it will be applied during service and generates valuable information with low effort.**

Keywords: broadband powerline communication; power cable; condition assessment; cable diagnosis

I. Introduction

The broadband powerline communications technology (BPL) is used to perform communication tasks in electrical distribution grids, e.g. as needed for the growing demand of smart grid applications. For this purpose, the extant electrical infrastructure of distribution grids and power cables in particular become a transmission medium also for communications in this manner. Besides communications another benefit of this technology is investigated regarding condition assessment of power cables and identification of power cable conditions respectively [1-3]. Although the technical condition of power cables affects the reliability of electrical energy supply, cable conditions remain widely unknown since power cables are inaccessible to visual inspections. Furthermore, diagnostic measurements are not used extensively in distribution grids to determine technical conditions accurately.

To provide a remedy, the characteristics of broadband powerline communication can be analyzed. Since BPL transmission is based upon OFDM multi-carrier modulation scheme, signals are transmitted using 917 independent carriers in the frequency range from 1.8 to 30 MHz. The most significant parameter that is related to any link quality and achievable data rates is the signal to noise ratio (SNR) at any receiving BPL modem. BPL modems determine such characteristics of all carriers automatically and on a regular basis for every link as a part of standard operations. The general functioning of the BPL technology is described in the IEEE Std 1901-2010 [4].

The SNR can be used as input data for further analyses regarding cable conditions. It is assumed that critical power cable conditions influence their high frequency characteristics, i.e. attenuation or noise, and thus allow detection via BPL after analyses of its transmission characteristics [1, 3]. Hence, the signal attenuation as an attribute of transmission lines like power cables and the noise power at the receiver affect SNR and should be taken into account for condition assessment.

II. Theoretical Background

A. Attenuation as a function of ageing and temperature

High frequency characteristics of power cables can be derived from transmission line theory. Each power cable has a characteristic impedance and propagation constant that determine behaviour of electromagnetic waves, e.g. BPL signals. The propagation constant contains attenuation constant per unit length (p.u.l.). There are individual transmission line models that fit to different properties of power cables and are used to determine these parameters [5].

The attenuation of signals in power cables has a low pass characteristic. This is caused by losses due to the skin effect and additionally by dielectric losses. Dielectric losses are caused by polarization and conductivity of the dielectric materials and increase approximately linear with frequency, while losses in the conductors are proportional to the square root of frequency due to the skin effect [5]. Since dielectric losses are dominant at BPL frequencies, correlations between dielectric condition, i.e. ageing and temperature, and attenuation in general are expected [1, 3].

The attenuation constant α (in Np/m) describes the attenuation p.u.l. of signals in transmission lines like power cables. It is defined, e.g. in [5], through the p.u.l. parameters inductivity (L), capacity (C), resistance (R) and conductance (G) as shown in (1):

$$\alpha = \frac{\sqrt{LC}}{2}\left(\frac{R}{L}+\frac{G}{C}\right) \tag{1}$$

Because these parameters are frequency dependent, the attenuation constant becomes a function of frequency (f), as originally intended, and additionally, it depends on temperature (T) and ageing (A) as further influencing factors as defined in (2) and also shown in [1].

$$\alpha = \alpha(f,T,A) \tag{2}$$

Additional signal losses due to signal reflections at discontinuities of characteristic impedance are possible and described in [1, 3] in this context. There, it is shown that cable conditions as a kind of local weak points yield further losses and can be determined with corresponding BPL characteristics.

B. *Noise in conjunction with partial discharges*

Partial discharges (PD) are a hint for critical asset conditions, if they are emitted from local weak points in power cables or other assets. They can be detected with different electrical and non-electrical diagnostic measurement techniques.

Each PD pulse has a high frequency content typically ranging into the BPL frequency spectrum. Hence, PD appear as impulsive noise at the BPL receiver. Equation (3) is extracted from [6] and shows the relation between total noise power (P_n), background noise power (P_g) and impulsive noise power (P_i).

$$P_n = P_g + P_i = P_g\left(1+\frac{P_i}{P_g}\right) \tag{3}$$

As a result from (3), the recognition of PD using BPL communications becomes possible, if the corresponding noise of PDs (P_i) increases the total noise (P_n) at receiver. Thus, impulsive noise of PDs must be at least of the same magnitude as the background noise to increase total noise power and in order decrease SNR by at least 3 dB, i.e. SNR is halved.

III. Laboratory Results

Laboratory testing was performed to investigate the influence of several conditions of XLPE and PILC cables in conjunction with (2) and (3). Therefore, attenuation of differently aged medium voltage cables was measured using time-domain-reflectometry as described in [7] and the SNR of a BPL modem influenced by PDs was measured.

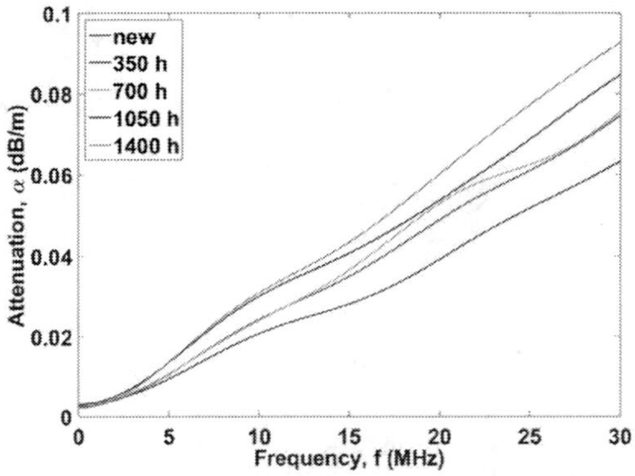

Figure 1. Attenuation constants α of thermally aged XLPE medium voltage cable as a function of frequency and ageing [3].

A. *Ageing*

The ageing dependency of the attenuation of XLPE medium voltage cables was investigated for both influence of thermal ageing in a climate air oven and ageing in a tempered water bath. Investigations about ageing of PILC cables are still in progress.

Fig. 1 shows the attenuation p.u.l. of a 10 kV XLPE cable for several cycles of thermal ageing. Each cycle consists of 350 h heating at 90 °C and cool down to 20 °C for measurements.

The results show that low pass characteristic of attenuation in XLPE cables is intensified due to increasing duration of thermal ageing. The attenuation especially at high BPL frequencies is sensitive to ageing, so investigated ageing conditions can be determined using BPL. Thermal ageing of 1400 h leads to an increase of attenuation by 0.03 dB/m at 30 MHz, which is an increase by 50 %. Hence, the effect is very significant especially for long cables.

Another 20 kV XLPE cable was aged for two years in a water bath with 40 °C temperature under a test voltage of 36 kV/50 Hz. The outer sheath was removed before conditioning, so that the screen got strongly corroded. The results of attenuation measurements at this cable are depicted in Fig. 2. The attenuation of the aged cable is in this manner much higher, though its graph is less flat than results of other thermally aged cables in Fig. 1. A further influencing factor occurred due to heavy corrosion of screen wires. Corrosion led

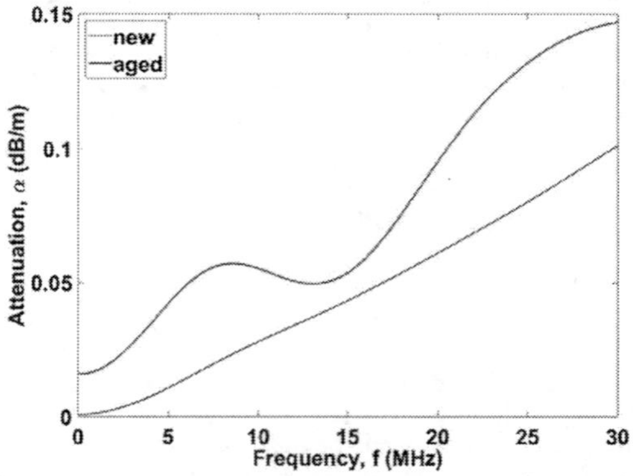

Figure 2. Attenuation constants α of XLPE medium voltage cable new and aged in tempered water bath as a function of frequency [3].

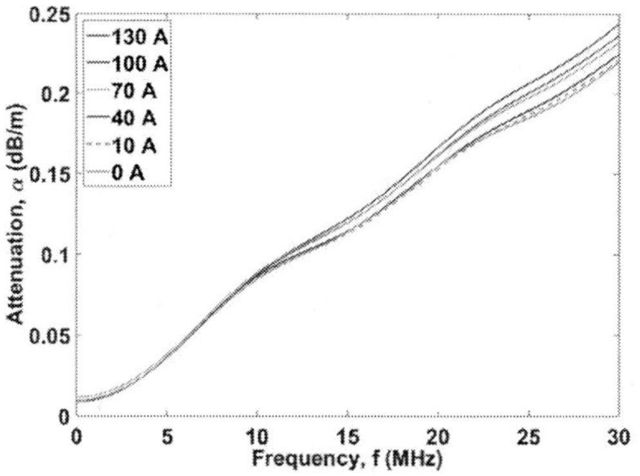

Figure 3. Attenuation constants α of PILC medium voltage cable as a function of frequency and electrical current [3].

to interruptions of wires and caused little reflections at these points. Because of this effect, exact attenuation constant is influenced at its extraction from measurements.

The results of laboratory tests prove that the attenuation, in particular of upper BPL frequencies, increases due to ageing of XLPE cables. Since attenuation constants are p.u.l. parameters, increased attenuation due to ageing becomes more significant the longer the aged cables are. Thus, the attenuation is an important parameter to support condition assessment of power cables.

B. Temperature

The effect of temperature on the attenuation was examined for PILC and XLPE medium voltage cables. PILC cables were heated using high currents. As shown in Fig. 3 their default low pass characteristic is intensified. It was determined that PILC cables return to their default (no current) attenuation values after every heat cycle. The difference between attenuation without current and risen temperature respectively and nominal current (130 A) is 0.03 dB/m at 30 MHz. After transmission along 100 m cable length, signal power would be attenuated by further 3 dB due to increased temperature, i.e. signal power is additionally halved.

XLPE cables were heated in a climate air oven to examine the influence of different temperatures on the attenuation measurement results. In contrast to PILC cables, XLPE cables show a thermal balancing effect, as default attenuation is not reached after cooling down. This balancing effect results from the contribution of semiconducting layers to attenuation in XLPE cables. Their behavior under thermal stress was

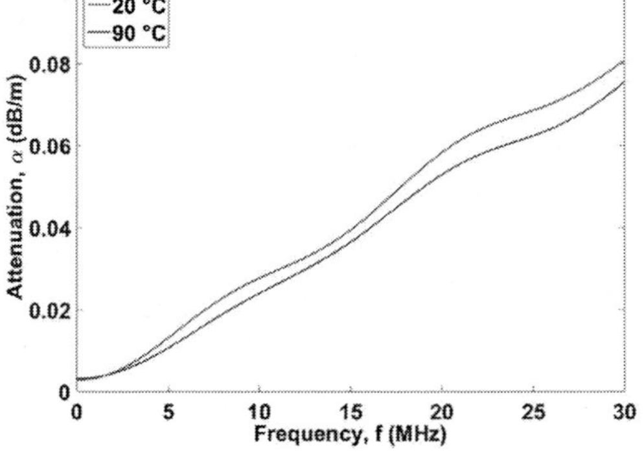

Figure 4. Attenuation constants α of a 700 hours thermally aged XLPE medium voltage cable as a function of frequency and temperature [1].

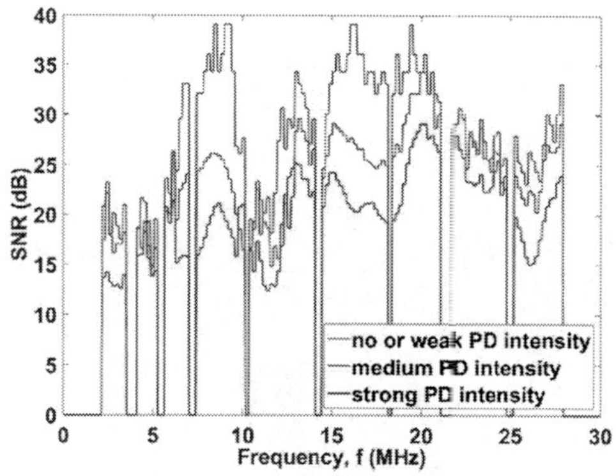

Figure 5. SNR at a BPL modem as a function of frequency and PD intensity [3].

examined in [8] and temperature dependency of XLPE cables in this context has also been published in [1] and shown in Fig. 4. Nevertheless, temperature behavior of attenuation is similar to PILC cables as shown in Fig. 4, where attenuation of a 700 h thermally aged 10 kV XLPE is shown, according to Fig. 1. Heating up from 20 °C to 90 °C cable temperature is associated with a further attenuation of only less than 0.01 dB/m at 30 MHz.

It can be concluded that especially the highest BPL frequencies are sensitive to temperature influence. The influence of temperature on p.u.l. attenuation is very similar to ageing. An analysis of the transmission characteristics in the upper frequency range of BPL communication can be used to get valuable information about ageing and temperature.

C. Partial Discharges

Partial discharges were measured for different cables under test. The results of the measurements using BPL modems are shown in Fig. 5 in principle. There, the SNR has specific values dependent on the actual transmission characteristics for each carrier frequency, i.e. attenuation and background noise. If a PD resultant impulsive noise power exceeds background noise in a specific BPL frequency spectrum the PDs become recognizable.

Laboratory tests show the influence of PDs on SNR measurements using BPL modems especially in the lower frequency range up to 10 MHz. Test results, i.e. the relation between PD intensity and BPL communications are published in more detail in [1, 3]. Beside the topic of cable diagnosis, PDs in other assets are also detectable. Although the origin of PD cannot be detected with BPL, since total noise and SNR do not contain any information about it, some valuable information about the presence of PD is gathered. A BPL communication might act as an indicator for PD, which occur in proximity to the BPL modem, and benefit condition assessment.

IV. Field Testing

After laboratory testing investigated effects will be analyzed in a field test environment in a part of the distribution grid of Leverkusen, Germany. For this purpose, probably 36 BPL modems will be installed there, that perform SNR measurements every 15 minutes for any carrier frequency resulting in a kind of monitoring of cables in distribution grids. Fig. 6 shows the results of SNR measurements as a time series exemplary for a first trial. At this, the information resulting from Fig. 5 is illustrated in another way for each instant of time, where high SNR is marked by blue and low SNR by red zones. It can be seen that SNR measurements are time

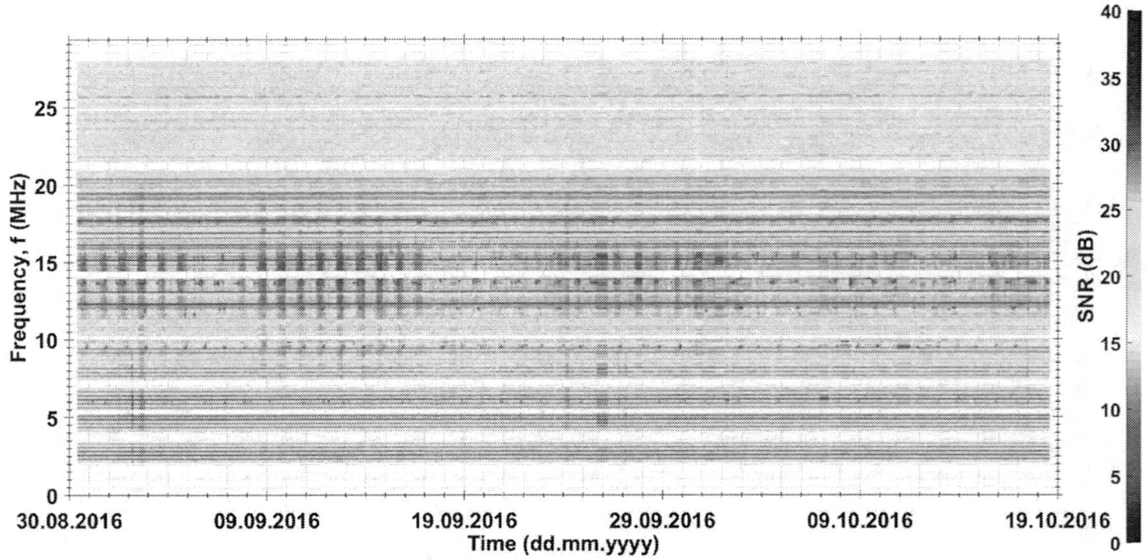

Figure 6. Time series of SNR measurements as a function of time and frequency. All time axis marks are at 0:00 a.m.

dependent and show cyclic behavior in this case. This visualization is the first step of analyzing BPL communication quality. Afterwards, further calculations are necessary to determine the examined effects, e.g. time series analysis of SNR to separate similar effects of temperature and ageing.

V. Conclusions

In this paper the influence of several technical conditions and increasing temperature of medium voltage power cables on BPL communication is examined. The intention is to establish a method to identify those conditions and assess their technical condition using SNR measurements of BPL communication as a basis. The results show that PDs and different kinds of ageing which are critical cable conditions, will be identified.

Power cables have a natural low pass characteristic. This is intensified through thermal ageing and increasing temperature of medium voltage cables, i.e. the cable's p.u.l. attenuation is increased especially at upper BPL frequencies. In contrast to this, PDs act as additional impulsive noise dependent on PD intensity especially at the lower BPL frequencies. They intensify total noise and thus lead to a decreased SNR.

Ageing, partial discharges and temperature influence can be detected in the corresponding frequency ranges. Hence, it is valid that BPL communication has the potential to determine different kinds of cable conditions using SNR measurements and analyses. Electrical distribution grid operators will be supported as condition assessment of power cables becomes possible with low effort and critical cable conditions will be identified before an outage occurs.

The first results of an extensive field testing confirm the practicability of the new method. All results of the field trial, in which SNR measurements of several BPL modems are performed every 15 minutes, will be available at 2018. They will be the basis of a best practice in order to use BPL characteristics for condition assessment.

Acknowledgment

This work was supported by the German Federal Ministry of Education and Research [grant number 03EK3540B].

References

[1] N. Hopfer, D. Beerboom, M. Zdrallek, C. Raquet, M. Ronczka, U. Dietzler and M. Krampf, "Influence of Cable Conditions on Broadband Powerline Communications in Medium Voltage Grids", VDE-Fachtagung Hochspannungs-technik, Berlin, Germany, 2016.

[2] J. Auer, C. Freudenmann, C. Raquet, M. Ronczka, N. Hopfer, D. Beerboom and M. Zdrallek, "Identifikation von Netz- und Anlagenzuständen mittels Analyse der Kanalcharakteristik der Breitband-Powerline-Datenübertragung auf dem Stromnetz", Konferenz "Zukunftsfähige Stromnetze für Erneuerbare Energien (4. OTTI-Konferenz)", Berlin, Germany, 2017.

[3] N. Hopfer, M. Zdrallek, U. Dietzler, M. Krampf, C. Raquet and M. Ronczka, "New Approach for medium voltage power cable assessment using broadband powerline communications", The 20th International Symposium on High Voltage Engineering, Buenos Aires, Argentina, 2017, unpublished.

[4] H. Latchman et al.: Homeplug AV and IEEE 1901. Hoboken, New Jersey, USA, John Wiley & Sons, 2013.

[5] G. Mugala, "High Frequency Characteristics of Medium Voltage XLPE Power Cables", Doctoral Thesis, KTH Royal Institute of Technology, Stockholm, Sweden, 2005.

[6] Y. Ma, P. So and E. Gunawan, "Performance Analysis of OFDM Systems for Broadband Power Line Communications Under Impulsive Noise and Multipath Effects", IEEE Transactions on Power Delivery. Vol. 20, No. 2, pp. 674-682, April 2005.

[7] R. Papazyan and R. Eriksson, „Calibration for Time Domain Propagation Constant Measurements on Power Cables", IEEE Transactions on Instrumentation and Measurement, Vol. 52, No. 2, pp. 415-418, April 2003.

[8] R. Heinrich, S. Bonisch, D. Pommerenke, R. Jobava, and W. Kalkner. "Broadband Measurement of the Conductivity and the Permittivity of Semiconducting Materials in High Voltage XLPE Cables", *IEEE Dielectric Materials, Measurements and Applications*, Conference Publication No. 473, pp. 212-217, September 2000.

A Concept for an Ice Detection System on Overhead Power Lines, Theory and Practical Results

Georgii Vagapov
Department "Methods of increase of power supply reliability and power quality in distributive electrical grids".
Kazan State Power Engineering University
Kazan, Russia
vagapov@list.ru

Alexsander Fedotov
Department "Electrical power systems and grids"
Kazan State Power Engineering University
Kazan, Russia
fed.ai@mail.ru

Steffen Kurth
Department manager "Multi device Integration"
Fraunhofer-Institut für Elektronische Nanosysteme
ENAS
Chemnitz, Germany
steffen.kurth@enas.fraunhofer.de

Sven Voigt
Group "Metrology".
Fraunhofer-Institut für Elektronische Nanosysteme
ENAS
Chemnitz, Germany
sven.voigt@enas.fraunhofer.de

Abstract — **The detection of the mechanical load of high voltage overhead power lines caused by ice is a crucial point for decisions to activate de-icing and for monitoring of high current based de-icing procedures. This paper reports about solutions for ice load and high current load monitoring by wireless sensors at overhead power lines of distribution grids of 110 kV level voltage and more. A theoretical methodic of calculating of the sag of the conductors depending on the ice load and on the elevated temperature due to Joules loss is implemented into the algorithms of the sensor system. The contribution shows the practical realization of such sensors and of theoretical methods by practical results. It reports the results of a field tests at a real distribution grid of 110 kV voltage level in different weather conditions.**

Keywords: power line monitoring, icing of power line, sag, model of overhead power line.

I. Introduction

External mechanical and climatic factors have a significant influence at the efficiency, reliability and safety of overhead power lines. Some of the climatic factors, such as ice-load or high current load can lead to damage of the wires due following effects:

- The distance to the ground becomes lower than the minimum distance to maintain safe operation because of the sag of the line at high temperature expansion or under elastic extension under ice load.
- The tension stress of the conductor exceeds the maximum allowable tension stress and the conductor mechanically breaks.
- High temperature induced weakness and plastic deformation of the steel core permanently deform the conductor.

Therefore, it is important to detect critical values of sag, temperature and of the tension stress in critical situations of ice load [1]. One of the ways of practical solution is to use different kinds of sensors as discussed in [2, 3, 4].

Mathematical modeling of sag is starting point of theoretical analyzing of mode of overhead power line. There are some quantity of papers [5, 6], standards and technical brochures [7] of calculating and measurement of sag. However, most of them show the calculation of critical modes of wires. This papers shows a method of calculation of the sag and of the tension in real-time, which is based on data from sensors located at the conductor and the practical realization in a real 110 kV power line.

II. CONCEPT

The principle of the ice load detection is based on the measurement of the increased inclination of the conductor due to mechanical weight of the ice, which is accumulated on the wires. This load leads to elastic extension of the wires and to higher sag and to an increased inclination angle of the conductors at the location of isolators (Fig. 1).

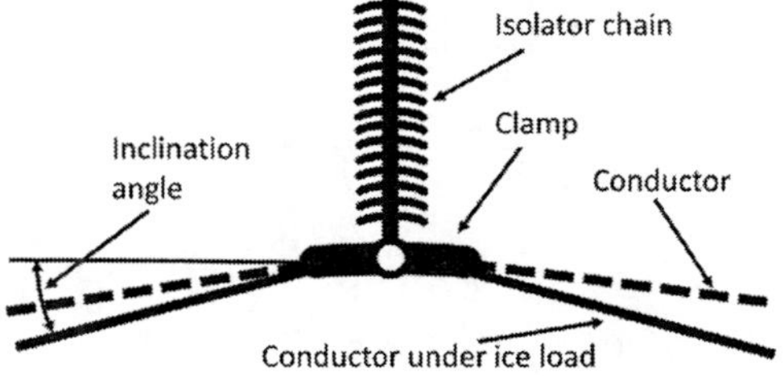

Figure 1: Schematic drawing of the clamping point of the conductor and the inclination angle in regular situation and under ice load at the same conductor temperature.

A sensor, which is mounted on the conductor, measures the inclination angle and the temperature and transmits the data to a receiver in a substation. In regular situations without additional load at the conductor, the inclination correlates to the temperature of the conductor due to the thermal expansion. The inclination increases with increasing temperature and visa versa. When ice is loading the conductor, the inclination angle becomes larger. Increased inclination angle may also occur with increased temperature. These situations need to be discriminated by observing of the temperature and considering ice load only if the temperature is less than the freezing point. Inclination and temperature do not correlate during the time span of ice accumulation. Moreover, modeling of the mechanical system including temperature expansion and elasticity of the conductor leads to an algorithm for determination of the ice weight when the ice is equally distributed over the full length within the span field. The measurement and calculation are done in real time and ensure an instantaneous detection of the ice.

III. Mathematical Modeling

A. *Power line equation*

The value of angle between the wire of an overhead power line and the horizontal line (inclination angle) is one of the measuring criteria for dynamic analyzing of influence of temperature, ice load, wind, and long-term creep. All these loads influence the length of the wire due to thermal expansion and elastic extension [8].

The following catenary equation in hyperbolic function describes the shape of the wire under the condition of equally distributed wire parameters, equal temperature and load.

$$y = a(cosh\frac{x}{a} - 1). \tag{1}$$

The parameter of the canonical equation of the catenary

$$a = \frac{T}{q}, \tag{2}$$

determines the shape of the wire sag where T is the effective value of wire tension and q is the weight of unit length of the wire. The parameters T and a in (2) are primarily unknown.

B. *Mathematical determine of the sag of the wire*

Fig.2 shows the typical case of overhead power line wire in a span field. The wire is clamped at the pylons at the fix-points A and B that may have different height.

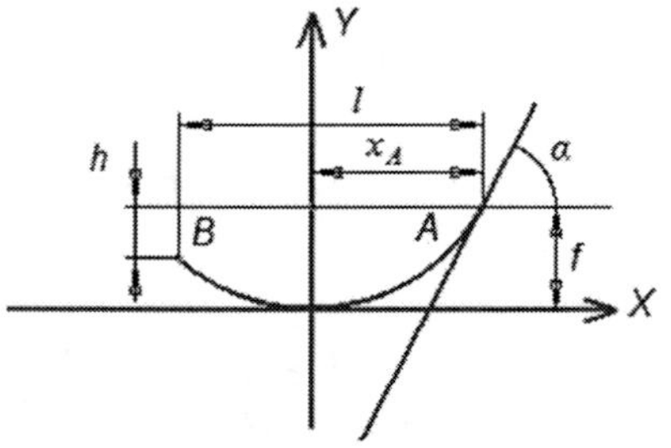

Figure 2: Overhead power line wire in span

The straight length of the span field (distance between the pylons) is l, and the distance between point A and the axis is X_A. The sag of the wire is f with the lowest point of the wire corresponding to the axes origin in Fig. 2.

The angle α characterizes the angle of inclination of the tangent at the fix-point of the wire. Obviously, if $h \neq 0$, $x_b=-(l-x_a)$, than

$$h = a\left(cosh\frac{x_A}{a} - cosh\frac{l-x_A}{a}\right), \tag{3}$$

where h is the difference of height between the fix-points of the wire

$$h = y_A - y_B. \tag{4}$$

The canonical equation of the catenary becomes

$$y = a(cosh\frac{x_A}{a} - 1). \tag{5}$$

The tangent of the angle α is

$$\tan\alpha = \sinh\frac{x_A}{a}. \tag{6}$$

Combining the equations (5) and (6) results in the sag

$$f = a \cdot \left(\cosh\frac{x_A}{a} - 1\right) \tag{7}$$

And in the length

$$L = a \cdot \left(\sinh\frac{x_A}{a} + \sinh\frac{l-x_A}{a}\right) \tag{8}$$

of the span field.

The tension at fix-points is

$$T = a \cdot q \cdot \left(\cosh\frac{x_A}{a}\right). \tag{9}$$

The horizontal tension H_0, at the lowest points Y_0 is

$$T_0 = H_0 = a \cdot q, \tag{10}$$

and the stress becomes

$$\sigma = \frac{T}{F} \tag{11}$$

with the cross-section of the wire F.

IV. Practical Realization and Test

The principle of ice load detection was evaluated by a field test at a 110 kV power line 270 m above sea level in a geographic region where the hills intercept the moisture-laden air coming in and ice rain occurs occasionally as a result in winter. A power line monitoring system that measures the temperature of the conductor, the inclination of the wires close to the isolator chain, and the electric current in the wires was applied for collecting of the raw data.

In this field test, 59 sensors are mounted at the conductors and are spread over approximately 11 km line length in a distance to each other, that basically corresponds to the distance of the pylons. The measured data is wirelessly transferred from sensor to sensor up to the final destination at a data receiving base station, which is located in the substation. The sensors have an implemented receiver so that the can be arranged in a sensor grid or they are cascaded. The full length of the power line can be monitored even when only a few sensors have direct radio link to the substation.

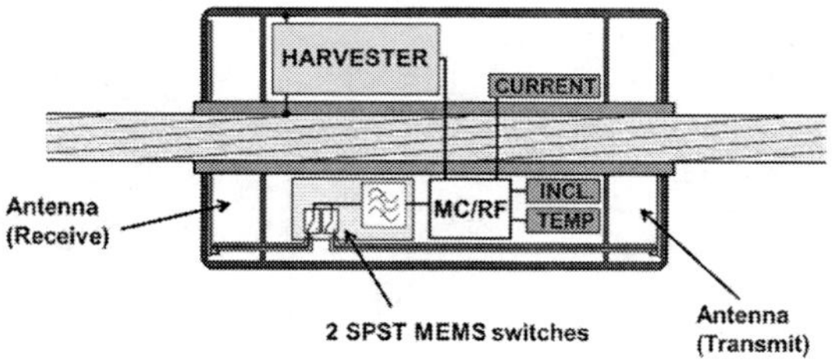

Figure 3: Set-up of the sensors for power line monitoring

Each sensor consists of a microcontroller, a 2.4 GHz transceiver, antennas, sensor modules, an energy harvesting system and a shielding and voltage overload protection system (Fig. 3). It is powered by the electric fringing field of the high voltage conductor. An internal maintenance-free power collector provides enough power for data transmission, and it is holding the power for more than 20 minutes if the high voltage power line is switched off. In case of longer off-periods and after a first installation at a power line, a self-initializing procedure is started when the power line is switched on.

A key component of the sensor is a high precision inclinometer with capacitive read-out and a resolution of 0.01° inclination angle. It is specially designed for very low temperature influence on the measurement and fabricated by a MEMS fabrication process. A particular ASIC is applied for reading out the inclinometer and for error correction.

A summary of the general specification of the sensors is given by Table 1.

Table 1: Specifications of the applied sensors

Description	Value	Unit
Diameter of the sensor	170	mm
Length of the sensor	280	mm
Length of the sensor with mounting helix	1140	mm
Resolution of inclination sensor	0.01	°
Measurement range of inclination sensor	±30	°
Measurement range of current sensor	4	kA
Resolution of the current sensor	10	A
Measurement range of temperature sensor	-25 … 100	°C
Resolution of the temperature sensor	0.1	°C

Fig. 4 shows an example of installation of the sensor node on a wire close to the insulator in a power grid. The sensor is attached to the conductor by spiral spring clamping systems which holds the sensors in their designated position on the conductor.

Figure 4: Example of sensor node of measure inclination of wire

V. Realization of the Methodology with Matlab/Simulink

A. Model in Matlab/Simulink

The software Matlab/Simulink is used for the implementation of the algorithms for the detection of the ice load and sag. It calculates automatically the geometrical parameters of the wire based on the measured inclination angle α and on the parameters of the span field given by the project documents.

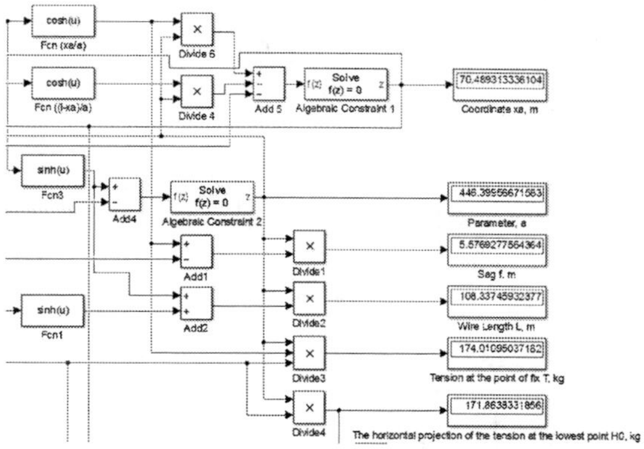

Figure 5: Part of model in Matlab/Simulink

The following blocks from the library *Math Operations* are used with their functionality for implementation of the methodology: Constant, Sum, Trigonometric and Hyperbolic functions, Display and Gain.

B. Numerical example in Matlab/Simulink

The numerical example shows practical realization of the methodology. Initial data of the numerical example are given in the following.

Table 2: Parameters of the power line span field investigated as an example

Parameter (from project documentation of the power line)	Value
Length of span field	189.1 m
Height of first fixed point above sea level	221.16 m
Height of second fixed point above sea level	215.65 m
Conductor type	AlSt 150/25
Wire cross-section	173.1 mm²
Wire diameter	17.1 mm
Steel core cross section	24.2 mm²
Wire mass	0.601 kg/m
E-module	77 kN/mm²
Coefficient of thermal expansion	18.5e-6 /K
Maximum allowable tension (breaking load)	72.28 kN

The mean value of the inclination according to the sensor readings is α = 7.05 °. Denoting X_A. by x, and the parameter a by y with concurrently substituted numeric values *h*, *l* и *tan*(α).

[x, y] = solve ('(sinh(x/y)) = tan(a)', 'y*(cosh(x/y)-cosh((l-x)/y)) =h'), (12)

where *h*=4.5 m.

The system of equations (5) and (6) has single-valued solutions x and y; X_A and a.

x = 120.24, y = 974.76.

Thus, it is possible to determine the uniquely coordinate x_A and parameter a, by the value of the inclination angle at the fix-point, received from the sensor and by the values of the heights of the fix-points of the wire and the span field length of the overhead line, obtained from the project documentation.

The next step is to determine the exact values of the sagging arrow f, the length of the wire L, and the wire tension T at the fix-point of the wire and at the low point H_0 by the before defined parameters X_A and a.

$$f = a \cdot \left[\cosh\left(\frac{x_A}{a}\right) - 1\right] = 7.43\ m. \quad (13)$$

$$L = a \cdot \left[\sinh\left(\frac{x_A}{a}\right) + \sinh\left(\frac{l - x_A}{a}\right)\right] = 189.36\ \text{m}. \quad (14)$$

$$T = a \cdot q \cdot (\cosh\frac{x_A}{a}) = 5903\ \text{N}, \quad (15)$$

where q = 6,01 N/m.

The horizontal projection of the thread tension H_0 at the lowest point of the wire

$$H_0 = a \cdot q = 5858.3\ \text{N}, \quad (16)$$

The stress in the wire σ in the fix-point of wires with value of the cross section of wire F = 173,1 mm^2 is

$$\sigma = \frac{T}{F} = 33.9\ N/mm^2. \quad (17)$$

The maximum allowable stress with low temperature and ice load is 156 N/mm^2 for this type of wire, (for calculating breaking load 264.8 kN).

VI. Discussion of the Results

The above given example is taken from a real situation with ice load obeservation by the field test during winter 2014/2015. The temperature and inclination were recorded and are given by the diagrams within Fig. 6.

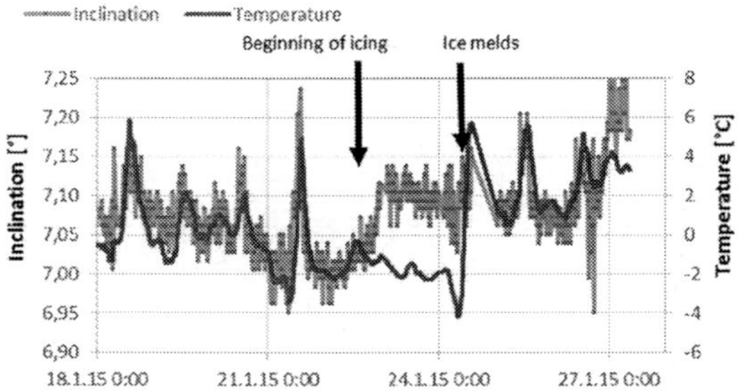

Fig. 6: Recorded data from observation of ice during the field test

Before the icing occured, temperature and inclination correlates well. In case of icing, we observed increasing inclination and decreasing temperature. Some hours later, the ice melds at higher temperature, and the inclination again follows the temperature. In order to evaluate the sag and the tension of the conductor, these values have been calculated for different states during this obervation period of time.

Table 3: Summary of the measured and calculated data during ice observation during the field test

State	Inclination angle	Tempe-rature	Stress [N/mm²]	Tension [N]
Before icing	6.9°	-2 °C	34.8	6022
During icing	7.1°	-2 °C	33.5	5850
After melding	7,1°	2 °C	33.9	5903

VII. Conclusion.

The proposed mathematical model allows quantitative and qualitative analysis of mechanical processes in the wire of overhead power line during mechanical load changes due to ice. The considered model can furthermore be used for the analysis of dynamic loads on the wire and in the design of the overhead lines for determining critical loads under normal conditions.

Acknowledgement

This work was supported by the German Federal Ministry for Economic Affairs and Energy [contract number E3ET5720E]. The authors acknowledge the cooperation with MITNETZ Strom GmbH and Fraunhofer IZM during the field tests at the 110 kV power line.

References

[1] L. Yongfu, Z. Liguo, W. Jun. "Computation of standard sag of overhead lines for power grid based on mathematical model of iterative technique," IEEE Transactions on Power Delivery, Dec. 2010, pp. V1-445 - V1-448, doi: 10.1109/ICCET.2010.5486049.

[2] A.H. Khawaja, Q. Huang, J. Li, Z. Zhang. "Estimation of Current and Sag in Overhead Power Transmission Lines With Optimized Magnetic Field Sensor Array Placement," IEEE Transactions on Magnetics, IEEE Press, Jan. 2017, pp. 897 - 909, doi: 10.1109/TIM.2017.2676140.

[3] W. de Villiers, J.H. Cloete, L.M. Wedepohl, A. Burger. "Real-Time Sag Monitoring System for High-Voltage Overhead Transmission Lines Based on Power-Line Carrier Signal Behavior", IEEE Transactions on Power Delivery, January 2008, pp. 389 - 395, doi: 10.1109/TPWRD.2007.905550.

[4] E. Golinelli, U. Perini, G. Ogliari. "A new IR laser scanning system for power lines SAG measurements", 18th Italian National Conference on Photonic Technologies, June 2016,. doi: 10.1049/cp.2016.0919.

[5] L. Yongfu, Z. Liguo, W. Jun. "Computation of standard sag of overhead lines for power grid based on mathematical model of iterative technique," IEEE Transactions on Power Delivery, December 2010, pp. V1-445 - V1-448, doi: 10.1109/ICCET.2010.5486049.

[6] J. Quintana, V. Garza, C. Zamudio. "Sag-tension calculation program for power substations", Transmission & Distribution Conference and Exposition-Latin America (PES T&D-LA), 2016 IEEE PES, Sept. 2016, doi: 10.1109/TDC-LA.2016.7805629.

[7] CIGRE B2-12 Brochure (Ref. No. 324) Sag-tension calculation methods for overhead lines, 2007.

[8] R. Fischer. Freileitungen: Planung, Berechnung, Ausfiihrung; mit 86 Tabellen / R. Fischer; F. Kiessling.- 4. Aufl. - Berlin; Heidelberg; New York; London; Paris; Tokyo; Hong Kong; Barcelona; Budapest: Springer, 1993, doi: 10.1007/978-3-642-97924-8.

[9] Y. Motlis, J. S. Barrett, G. A. Davidson, D. A. Douglass, P. A. Hall, J. L. Reding, T. O. Seppa, F. R. Thrash, H. B. White. "Limitations of the Ruling Span Method for Overhead Line Conductors at High Operating Temperatures", IEEE Transactions on Power Delivery, Vol. 14, No. 2, pp.: 549 – 560, April 1999, doi: 10.1109/61.754102.

Localization of E-M Interference Sources in Air-Insulated Substation using VHF/UHF Antenna Arrays

Jozef Roehrich, Pawel Zydron, Bartlomiej Szafraniak
Department of Electrical and Power Engineering
AGH University of Science and Technology
Krakow, Poland
roehrich@agh.edu.pl, pzydron@agh.edu.pl, szafrani@agh.edu.pl

Abstract —**The demand for high reliability of electricity supply, which is commonly required nowadays, forces the need for diagnostic tests and continuous monitoring of the proper functioning of electrical power equipment. Various test and measurement methods are used in the diagnosis of high-voltage insulating systems. Their selection depends on the construction materials and the design of tested objects. An important group of methods used in diagnostics are measurements using antenna detection of electromagnetic interferences. The presented work deals with the detection and localization of E-M pulse sources caused by partial discharges generated in the air-insulated high voltage transmission systems. Defects occurring in overhead transmission lines and in equipment installed in power substations, which are characterized by local increasing of the electric field strength over the critical value, result in generation of partial discharges in various forms (corona, surface discharges, and internal discharges). Formed as a result E-M pulses of a very short rise times and a narrow widths can be used to detection and location of insulating systems defects by broadband detection of disturbances in VHF/UHF bands. The paper presents an analysis of the E-M pulse sources location in air-insulated substations using basic antenna array systems. This is illustrated by the results of laboratory measurements collected with selected directional antenna detection systems used for E-M disturbance detection.**

Keywords: substations, transmission lines, E-M pulses, antenna detection, E-M sources localization

I. Introduction

Modern requirements for the quality and high reliability of electricity supply in electrical power systems cause the need for diagnostic tests and monitoring of the correct working of the electrical equipment in-service [1]. The defects in the overhead lines and the air-insulated substation equipment, which are characterized by an increase of the local electric field strength over the critical value, cause the formation of partial discharges (PD) in different forms (corona, surface discharges, internal discharges). Various diagnostic methods, both electrical and non-electrical, can be used to detect and locate such defects [1-14]. In non-electrical group of methods often detection of acoustic emission signals is used (for diagnostics of transformers, gas insulated switchgears, transmission lines, and other high voltage objects), as well as methods of selective detection of visible and ultraviolet radiation generated by corona discharges in the air, as one of the symptoms of electric discharges on defects of transmission line wires, fittings, accessories or other equipment. Among the electrical methods of detection, recognition and localization of the PD sources introduced into diagnostic practice, the methods based on the detection of electromagnetic disturbances, using various types of antenna sensors, should be mentioned. The range of frequencies used for this aim is very wide, ranging from High Frequency *HF* (3 MHz÷30 MHz), through the Very High Frequency *VHF* (30 MHz÷300 MHz) to the Ultra High Frequency *UHF* (0.3 GHz÷3 GHz) [1, 4, 5]. The frequency of operation of the detection and measurement systems depends on the type of object and the time-frequency parameters of the impulse generated in insulating defects.

For localization of sources of electromagnetic disturbances in overhead transmission lines and air-insulated substations (Fig. 1) various types of antennas can be used [2-9, 15-18], which can be divided into two main groups:

- directional antennas;
- arrays of omnidirectional antennas.

Directional antenna with sensitive wideband or tuned narrow-band receiver works as a *radio direction finder* (RDF), useful for detecting interferences sources that are significantly higher than the surrounding electromagnetic environment. Omni-directional antennas can be used for defect localization only as *distributed* or *centralized antenna arrays* using, for example, *the time differences of arrival* (TDOA) algorithms for signal source finding [17-20].

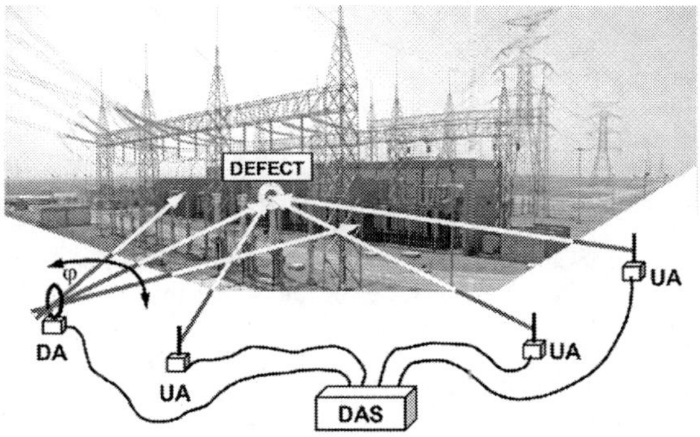

Figure 1. Electromagnetic interference source localization in the air-insulated substation; DA – directional antenna, φ – source directon angle, UA – omni-directional antenna, DAS – data acquisition system

II. PD Pulses Parameters

Partial discharge pulses are specific signals, usually of very short rise-time, e.g. for PDs in gaseous inclusions or in GIS installations, the shortest registered rise-times of pulses are in $10^{-11} \div 10^{-9}$ [s] range. Their *FWHM* (*Full Width at Half Maximum*) parameter is in range $10^{-9} \div 10^{-10}$ [s]. As a direct consequence of such extremely short time parameters PD pulses represent very high equivalent bandwidth, enabling its detection in a broad frequency range. This is important for selection of the detection methods, conventional or *VHF*/*UHF*, and for proper selection of *E-M* sensor parameters (Fig. 2). "Wideband" electrical methods of PD detection defined in IEC 60270 standard refer to measurements performed in the frequency range not higher than 1 MHz. However, measurements in the VHF/UHF ranges are better for on-line PD detection as they allow for optimizing the signal-to-noise ratio.

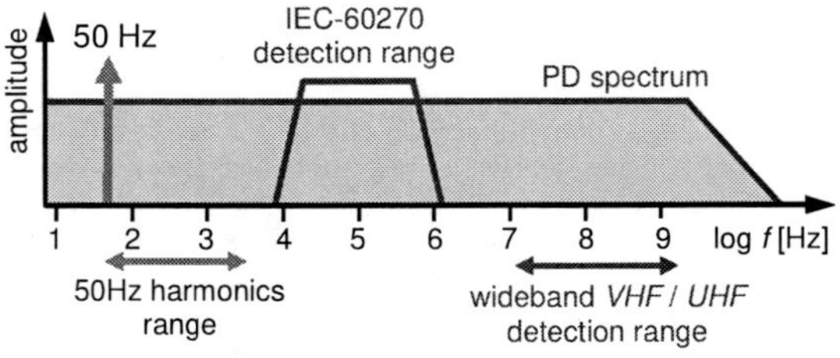

Figure. 2. Frequency spectrum of working voltage (50 Hz with harmonics) and PD pulses together with PD detection ranges of conventional (IEC 60270) and *VHF*/*UHF* methods

III. Directional Antennas

Directional antenna is a type of antenna that exhibits grater gain in one specific direction, that gain can be observed in transmit and receive mode. Well known type of directional antenna is *Yagi antenna* (*Yagi-Uda*) which is widely used in TV reception techniques. Yagi's can be built for variety of bands beginning from HF bands like 20 m band until very short waves like couple of centimeters which stands for frequencies of GHz. Other widely used antennas are *log-periodic antennas* which are similar to Yagi's but are wideband directional antennas. Antennas that exhibit the best directional properties are dish antennas which are used for weak signal reception in terrestrial communication or satellite communication as well. Those type of antennas are also used in radar techniques to radiate signals and receive bounced signals as well. *Dish antennas* are built to work on frequencies above 10 GHz. In modern radar systems *antenna arrays* are used to obtain directive properties of antenna.

IV. Antenna Arrays

An antenna array is a set of simple antennas that are arranged as a matrix, in the simplest application it can be omni-directional, monopole antennas that are energized by coherent signal. The phase and amplitude of energizing signal to form a gain of the antenna in a specific direction [21-24]. The gain of an array is obtained because in some directions radiated waves adds up and some other directions cancels. Moreover those mechanism works in both way when wave is transmitted and when is received as well. Antenna arrays may be formed as linear, planar or even 3D arrangements. If the distances between consisting antennas are equal a regular array is formed, if distances are different an array is not regular. The main advantage of antenna array is that radiated beam may be controlled in electronic way without moving the antenna array in space [1]. The beam of array antenna can be changed very quickly in contrast to beam of dish antenna which have to be moved in space (e.g. radar antenna which rotates around or moves periodically). Mechanisms of interference of waves radiated by two isotropic antennas arranged in array when distance between antennas is set to $\lambda/2$ is described by (1), where θ is a direction of radiation:

$$f(\theta) = cos\left(\frac{\pi}{2} cos\theta\right) \tag{1}$$

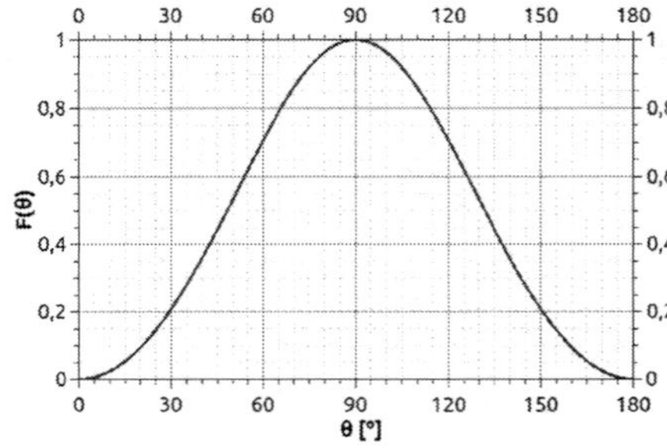

Figure 2. Normalized characteristic of radiation for two isotropic antennas situated in distance of $\lambda/2$.

The propagation velocity of *E-M* wave in the air, which is almost $c \approx 300$ m/μs, determines the time of propagation of PD pulse signals in free space. Using modern measurement instruments (high bandwidth, high sampling frequency DSOs) it is possible to measure the time of propagation even on short distances like a couple of meters. This technique makes possible to determine direction of incoming wave using antenna array build of two or more basic antennas (Fig. 3).

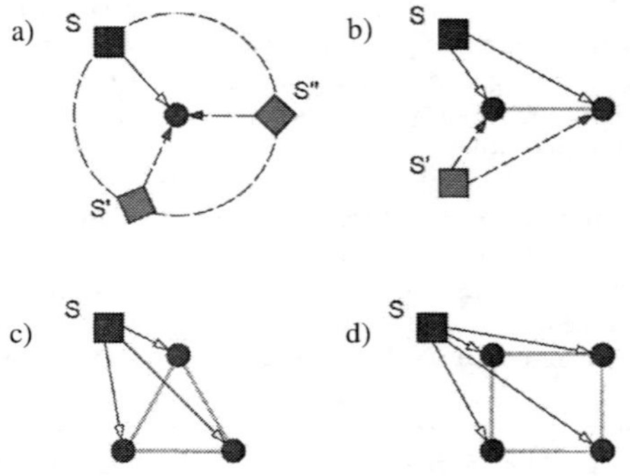

Figure 3. Signal source localization properties of different antenna configurations: a) single omni-directional antena; b) two monopole array; c) three monopole array; d) four monopole array. S, S', S" (black rectangle) – signal sources; black circle – monopole antenna

In Figure 4 the simplest linear antenna array with two monopoles is presented. If it is assumed that incoming wave is a plane wave which means that vectors E and H are perpendicular to each other and both are perpendicular to direction of propagation of the wave. It may be assumed this way if a distance that wave travels is greater than $\lambda/2\pi$ [24]. Accordingly when $\theta = 0$ wave reaches both antennas in the same moment. If the wave is a plane wave it is possible to calculate the distance *b* and also a time lag of incoming wave between A1 and A2 antennas which is described by (2), where *a* is a distance between antennas in array and *c* is speed of light.

$$t(\theta) = \left(a \cdot sin(\theta)\right)/c \tag{2}$$

The above formula looks simple but its real application is more complicated because of:

- the time resolution of waveform measurements,
- the bandwidth of signals generated by PD impulses,

- the influence of relative permittivity of insulating materials to speed of propagation of E-M waves,
- bounce effects and boundary effects

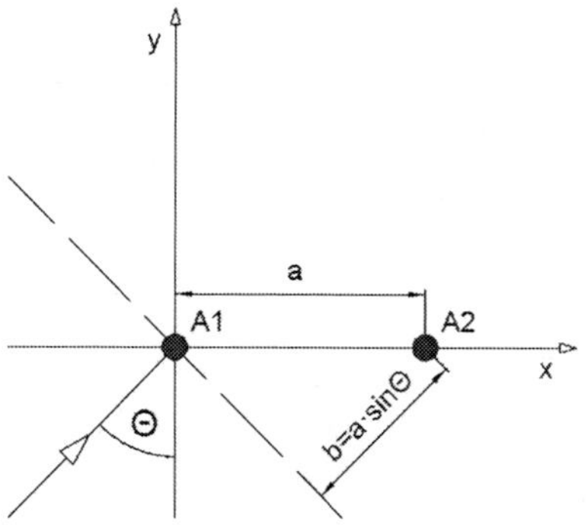

Figure 4. Linear antenna array consisting of two isotropic antennas situated in distance of *a* to each other (double-monopole antenna array). Θ - angle of arrival of E-M wave

V. Measurements Results

In laboratory experiments, E-M impulse signals generated by commercial apparent charge calibrators with the very short rise time, in the order of 10^{-10} - 10^{-8} s (Power Diagnostix model CAL2B with rise time ≤ 200 ps and CAL1D with rise time of about 20 ns) were used. Calibrators worked as pulse wave transmitters equipped with a monopole antennas, sources of repetitive disturbances (Fig. 5). Detected signals from basic double-monopole antenna array were acquired by digital storage oscilloscope (max. analogue bandwidth 1 GHz; max. sampling frequency 4 GHz), controlled via IEEE488 interface.

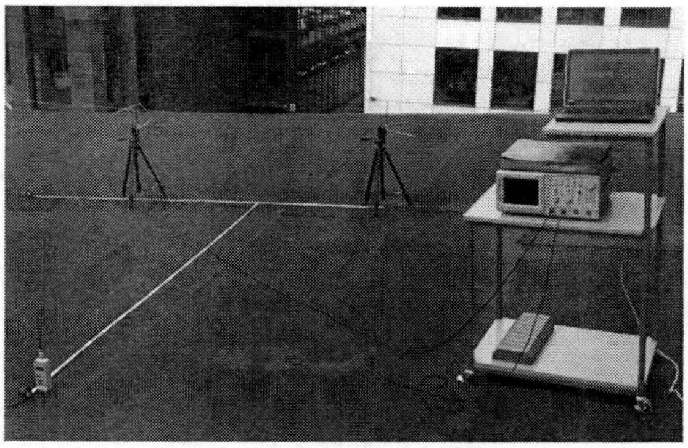

Figure 5. Laboratory setup of experimental stand with E-M pulse source and double-monopole antenna array

The basic linear antenna array with 2 isotropic antennas (double-monopole array), Fig. 4, is the simplest antenna configuration that allows to estimate *the angle of arrival* θ (AOA) of electromagnetic wave. The algorithm for calculation of this angle in the case of impulse signals is based on the measurement of *the time difference of arrival* (TDOA). For plane wave, the angle of arrival θ is a function of the measured TDOA and antenna separation distance *a.* In general, the problem of precision of the AOA evaluation using antenna arrays is the matter of resolution of time differences measurement. The speed of E-M wave, which determines time lag between of two signals, causes that in tested antenna arrays measured time differences are in a range of nanoseconds. Time lags calculated for linear antenna array (Fig. 4) built of two isotropic antennas with normalized 1 m separation from each other are presented in table 1.

The results of laboratory measurements of E-M pulse waveforms, generated by CAL2B calibrator, made for an antenna array with two monopole antennas with 2 m separation are shown in Figure 6. The determined time lags for different AOA values correlate well with the data obtained from the calculations. Waveform were acquired with full 1 GHz bandwidth.

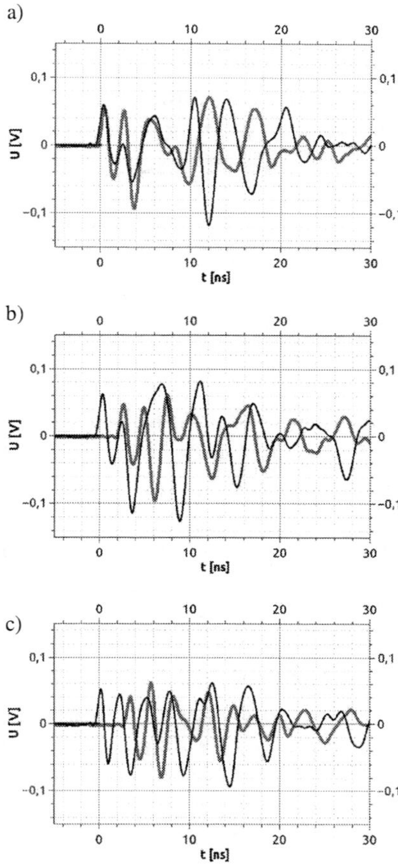

Figure 6. Time lags measured in linear antenna array where separation distance $a = 2$ m : a) θ = 0°, Δt ≈ 0 ns; b) θ = 15°, Δt ≈ 2 ns c) θ = 30°, Δt ≈ 3.1 ns

Table 1. Time lags calculated for linear antenna array consisted of 2 isotropic antennas, with separation distance a =1 m (c ≈ 3·10^8 m/s)

θ [°]	Distance and time lag calculated	
	b [m]	Δ*t* [ns]
0	0	0
10	0.17	0.58
30	0.5	1.67
45	0.71	2.36
60	0.87	2.89
90	1.00	3.34

In next step, the receiver system with two monopole antennas located at a distance of d = 1 m from each other was examined, placing the E-M impulse source (CAL1D) at points

X_n on line L_1 and X_n 'on line L_2 (Fig. 7). In this case, two 350 MHz preamplifiers with calibrated gain were used. Due to the proximity of the antennas from the signal source they locate in the near field (induction zone). During the measurements, the influence of the position of the signal source on the peak voltage of the recorded pulses was investigated. Results of measurement in Fig. 8 and Fig. 9 show that this parameter also gives information about the position of the defect.

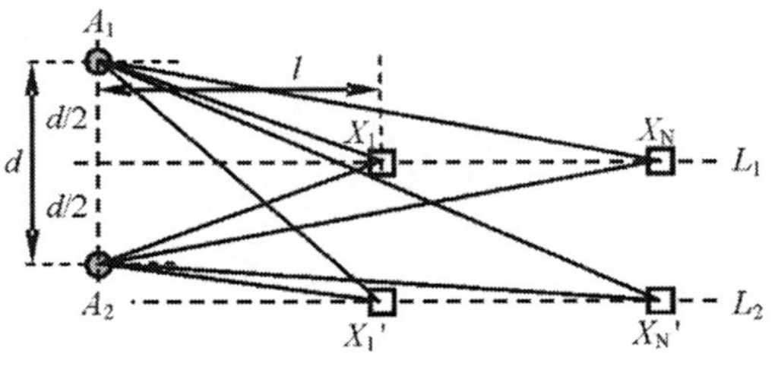

Figure 7. Geometry of the mutual position of the antenna array and signal source for the study of the influence of the source position on the peak value of the recorded pulse signals

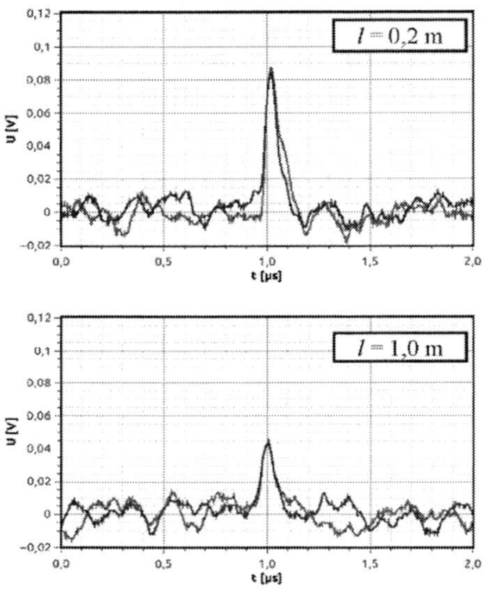

Figure 8. The waveforms acquired from the antennas A1 and A2 for the pulse signal source on the L_1 line; the increase of the distance l causes the lowering of the pulses (the amplitudes are equal due to the equal distance of the antennas from the source)

VI. Summary

The presented report discusses the possibility and properties of antenna array detecting of electromagnetic pulse signals generated by defects in overhead transmission lines and air-insulated substations, which results in the generation of partial discharges. Two methods of analysis of digitally recorded pulse signals were used for the estimation of the direction / location of the defect:

- method based on the time difference of arrival (TDOA) of pulse signals measurement, used for the angle of arrival (AOA) estimation;
- a method based on comparing the peak values (signal level) of registered E-M pulses, processed in two calibrated analogue measurement paths.

The TDOA method requires the use of digital signal acquisition systems with very wide bandwidths, of the order of gigahertz (UHF range). For a simple, centralized 2-element antenna array, with a spacing of 1-2 m between antennas, the measured time differences for determining the angle of arrival (AOA) are even below 1 ns.

Measurements in the VHF range, near the source of partial discharges, may be used to estimate the defect position based on the analysis of the level of the acquired impulses, but at present it does not appear to be very accurate approximation for defect localization Further studies in the field are planned, also using impulse energy analysis.

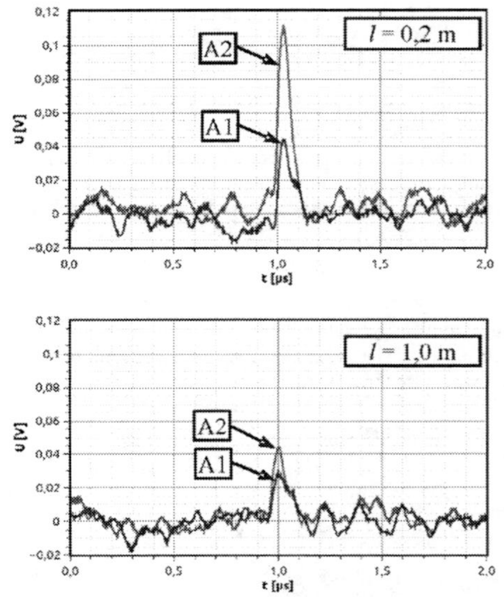

Figure 9. The waveforms acquired from the antennas A1 and A2 for the pulse signal source on the L_1 line; the increase of the distance l causes the lowering of the pulses (the amplitudes differ due to the different antenna distances from the source)

Acknowledgment

The researches described in the paper were financed by the Polish Ministry of Science and Higher Education - grants for Faculty of Electrical Engineering, Automatics, Computer Science and Biomedical Engineering of AGH University of Science and Technology, Krakow, Poland.

References

[1] Q. Su and R. E. James, Condition Assessment of High Voltage Insulation in Power System Equipment, IET, London, 2008.

[2] B. Florkowska, P. Zydron, and M. Florkowski, "Localization and identification of corona forms based on phase-resolved images", Measurement Science and Technology, vol. 12, pp. 1304-1310, 2001.

[3] P.J. Moore, I.E. Portugues, and I.A. Glover, "Remote diagnosis of overhead line insulation defects", IEEE Power Eng. Society General Meeting, 2004, vol. 2, pp. 1831-1835, doi: 10.1109/PES.2004.1373196.

[4] M. D. Judd, "Radiometric partial discharge detection", Int. Conf. on Condition Monitoring and Diagnosis, pp. 1025-1030, 2008, doi: 10.1109/CMD.2008.4580457.

[5] B.G. Stewart, A. Nesbitt, and L. Hall, "Triangulation and 3D location estimation of RFI and partial discharge sources within a 400kV substation", 2009 IEEE Electr. Insul. Conf., pp. 164-168, 2009, doi: 10.1109/EIC.2009.5166337.

[6] I.E. Portugues, P.J. Moore, I.A. Glover, C. Johnstone, R.H. McKosky, M.B. Goff, L. van der Zel. "RF-based partial

discharge early warning system for air-insulated substations", IEEE Trans. on Power Delivery, vol. 24, no. 1, pp. 20-29, 2009, doi: 10.1109/TPWRD. 2008.2005464.

[7] Q. Shan, I.A. Glover, R.C. Atkinson, S.A. Bhatti, I.E. Portugues, P.J. Moore, R. Rutherford, M. De Fátima Queiroz Vieira, A. Marcus N. Lima, B.A. De Souza. "Estimation of impulsive noise in an electricity substation", IEEE Trans. on EMC, vol. 53, no. 3, pp. 653-663, 2011, doi: 10.1109/TEMC.2010.2092782.

[8] K. Walczak, "Defect localization in post insulator using capacitive sensors", Przegląd Elektrotechniczny, vol. 88, nr 5a, pp. 55-58, 2012 (in Polish).

[9] H. Hou, G. Sheng, X. Jiang, "Robust time delay estimation method for locating UHF signals of partial discharge in substation", IEEE Trans. Power Delivery, vol. 28, no. 3, 2013, pp. 1960-1968, doi: 10.1109/TPWRD.2013.2262315.

[10] J. Zhao, C. D. Smith, B. R. Varlow, "Substation monitoring by acoustic emission techniques", IEE Proc. Science, Measurement and Technology, vol. 148, no. 1, pp. 28-34, 2001, doi: 10.1049/ip-smt:20010134.

[11] T. Boczar, A. Cichon, S. Borucki, "Diagnostic expert system of transformer insulation systems using the acoustic emission method", IEEE Trans. Dielectr. Electr. Insul., vol. 21, no. 2, pp. 854-865, 2014, doi: 10.1109/TDEI.2013.004126.

[12] M. Lindner, S. Elstein, P. Lindner, J. Topaz, A. Phillips, "Daylight corona discharge imager", 11th ISH, London, vol. 4, pp. 349-352, 1999, doi: 10.1049/cp:19990864.

[13] C. Zang, Z. Xinjie, H. Shuang, H. Lei, J. Zhenglong, Y. Huisheng, J. Zhiwen. "Research on mechanism and ultraviolet imaging of corona discharges of electric device faults", Conf. Proc. 2008 IEEE ISEI, pp. 690-693, doi: 10.1109/ELINSL.2008.4570424.

[14] P. Fracz, "Measurement of optical signals emitted by surface discharges on bushing and post insulator", IEEE Trans. Dielectr. Electr. Insul., vol. 20, no. 5, pp. 1909-1914, 2013 doi: 10.1109/ TDEI.2013.6633724.

[15] Q. Wu, G. Liu, Z. Xia, L. Lu. "The study of Archimedean spiral antenna for partial discharge measurement", 4th Int. Conf. Intellig. Control and Inform. Proc., 2013, pp. 694-698, doi: 10.1109/ ICICIP.2013.6568162.

[16] G. Robles, M. Sanchez-Fernandez, R. Albarracin Sanchez, M.V. Rojas-Moreno, E. Rajo-Iglesias, J.M. Martinez-Tarifa. "Antennas parameterization for the detection of partial discharges", IEEE Trans. Instrum. Meas., vol. 62, no. 5, pp. 932-941, 2013, doi: 10.1109/TIM.2012.2223332.

[17] H. Hou, G. Sheng, X. Jiang, "Localization algorithm for the PD source in substation based on L-shaped antenna array signal processing", IEEE Trans. Power Delivery, vol. 30, no. 1, 472-479, 2015, doi: 10.1109/TPWRD.2014.2344014.

[18] M.-X. Zhu, Y.-B. Wang, L. Yuan, H.-B. Mu, J.-B. Deng, X.J. Shao, G.-J. Zhang. "Detection and localization of partial discharge in air-insulated substations using UHF antenna array", 3rd ICPERE, pp. 221-224, 2016, doi:10.1109/ICPERE.2016. 7904875.

[19] H.-C. Chen, T.-H. Lin, H. T. Kung; C.-K. Lin, Y. Gwon, "Determining RF angle of arrival using COTS antenna arrays: A field evaluation", MILCOM 2012 IEEE Military Communications Conf., pp. 1-6, doi: 10.1109/MILCOM.2012.6415851.

[20] M.-X. Zhu, Y.-B. Wang, Q. Liu, J.-N. Zhang, J.-B. Deng, G.-J. Zhang, X.-J. Shao, W.-L. He. "Localization of multiple partial discharge sources in air-insulated substation using probability-based algorithm", IEEE Trans. Dielectr. Electr. Insul., vol. 24, no. 1, pp. 157-166, 2017, doi: 10.1109/TDEI.2016.005964.

[21] R. C. Hansen, Phased Array Antennas, 2nd ed., Wiley, 2009.

[22] R. J. Mailloux, Phased Array Antenna Handbook, Artech House, 2005.

[23] C. A. Balanis (ed.), Modern Antenna Handbook, Wiley, 2008.

[24] P. Dębicki, Introduction to Theory of Antennas and Antenna Arrays, Akademia Morska w Gdyni. Gdynia, 2012 (in Polish).

Influence of Temperature and High-Energy Stresses on Selected Wideband Characteristics of Metal-Oxide Varistor Parameters

Michał Bonk, Łukasz Fuśnik, Bartlomiej Szafraniak, Pawel Zydron
Department of Electrical and Power Engineering
AGH University of Science and Technology
Krakow, Poland
bonkm@agh.edu.pl, lfusnik@agh.edu.pl, szafrani@agh.edu.pl, pzydron@agh.edu.pl

Abstract—**Gapless Metal-Oxide Varistors (MOV) are used for many years in systems limiting the surges arising in electrical power networks, at all voltage levels. Surge arresters used for protection of electrical installations and devices are currently produced mainly on the basis of ZnO varistors. Each overvoltage causes high energy stress, which degrades the internal structure of the varistor. As a result, it changes its properties, and by cumulative effects of aging processes reduces the lifetime. Diagnosis of surge arresters is performed to determine the degree of aging of their varistors structures. For this purpose, different testing methods are used. The paper describes issues related to the application of wideband impedance spectroscopy (IS) for the experimental determination of some electrical and material parameters of low voltage surge arresters, subjected to the influence of aging factors. They are illustrated with results of tests performed to determine the effect of temperature and high energy stresses (current pulses 8 µs/20 µs or 0.5 s AC burst wave-forms) on some parameters describing the condition of the varistors. Modeling of varistor properties using electrical schemes is used for analysis of its structural changes.**

Keywords: metal-oxide varistor; energetic stresses; degradation processes; diagnostics; impedance spectroscopy

I. Introduction

The application of *metal-oxide varistors* (MOVs) for the protection of electrical equipment in electrical networks and installations significantly changed equipment operating conditions because of the significant reduction of the value of overvoltages affecting their insulating systems. The application of surge protective devices SPD (surge arresters) with technical parameters properly selected for the location and real work conditions for many years seems to be commonly used overvoltage suppression method [1, 2]. Modern varistors, which are the basic working element of SPDs, are produced mainly on the basis of polycrystalline ZnO ceramics together with suitable additives (e.g. bismuth, magnesium, cobalt, chromium oxides), which allow for proper parameters of the production process, optimizing the operating properties, in particular shaping the voltage and current characteristics [3, 4].

Metal-oxide varistors most often made as cylindrical disks of various diameters and thicknesses, are enclosed in hermetic enclosures, which protect them against the effects of external environmental factors, especially against penetration of moisture (Fig. 1). The varistor disk geometry is related to the assumed discharge current (disk diameter / disk area), continuous operating and residual voltages (disk thickness), and its energy absorption capability (volume of the disk). Low voltage SPDs usually contain a single varistor disk, but surge arresters for medium and high voltage applications always consist of a large number of stacked varistors disks.

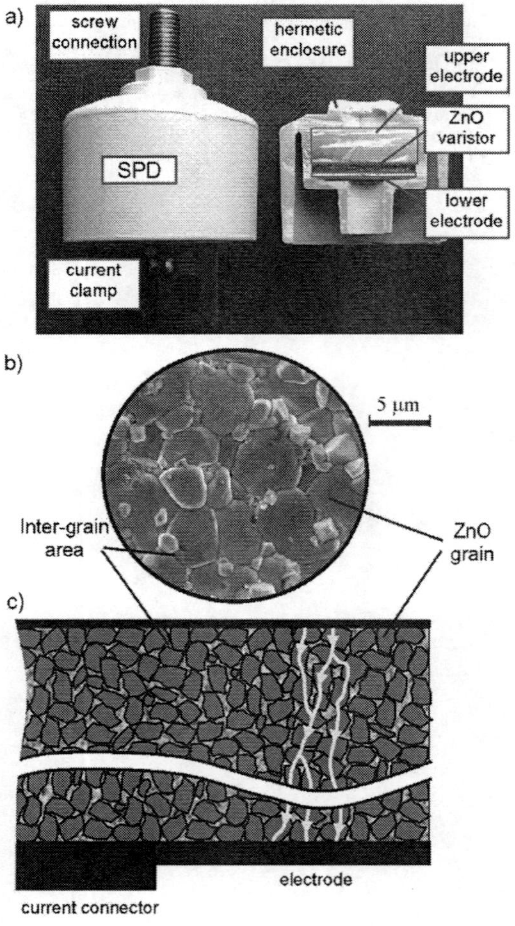

Figure 1. Construction of low voltage surge arrester with metal-oxide varistor: a) cross-section of the 280V/5kA SPD, b) microscope view of the polycrystalline MOV structure, c) MOV multi-grain structure model with examples of current flow paths

The basic components of the varistor structure are crystalline ZnO grains, which should be of adequate size, homogeneity and resistivity. The current flowing through the varistor is the sum of the partial currents flowing through its volume through different paths with the numerous double-Schottky barriers that exist between randomly distributed grains in varistor polycrystalline structure (Fig. 1c). The mechanism of current conduction in varistors is complex, due to the influence of material properties and nonlinear phenomena occurring on the boundary of polycrystalline grains [5-7]. These results in a strongly nonlinear relationship between the current flowing through the varistor and the voltage applied to its electrodes causes, that it can be used as a voltage stabilizer.

Most commonly, the shape of the varistor static *I-U* characteristic is approximated by the power law formula:

$$I = k\,U^{\alpha} \tag{1}$$

where: I – current through the varistor, U – voltage on varistor, k, α – constants, dependent on the material and parameters of the technological process of varistor production.

The *I-U* characteristics are usually normalized with respect to the geometric dimensions of the varistor and represented in *J-E* (current density – electric field strength) coordinates. In this varistor characteristic three specific current ranges are present (Fig. 2):

- *Pre-breakdown range*, which is a normal operating range, with a high varistor resistance; the low leakage current is less than the capacitive current of the varistor. The temperature has a strong effect on the leakage current value.
- *Breakdown range*, which is the range of voltage stabilization; the voltage parameter describing SPD in this state is the residual voltage.
- *Saturation range* in which the current flowing through the varistor is limited by external circuit parameters. A further increase of the voltage depends on the conductivity of the ZnO grains.

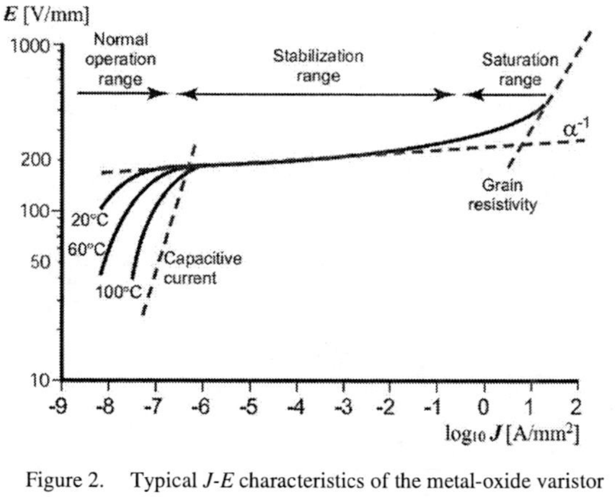

Figure 2. Typical *J-E* characteristics of the metal-oxide varistor

Internal aging changes of the polycrystalline varistor structure caused by the energy interactions of the current pulses absorbed during the overvoltages and their results for SPDs parameters can be detected/observed by various measurement methods. Researches on the influence of aging processes on: the value of the leakage current at the operating voltage, the presence of higher harmonics in the varistor current, the value of the residual voltage, and even heating of SPDs at working conditions were conducted [8-12]. One of the varistor condition assessment methods is to analyze the wideband characteristics of selected dielectric parameters. It is based on the premise that for the linear range of *I-U* characteristics changes in varistor structure result in noticeable changes of: electrical conductivity, permittivity, and, in general, relaxation processes parameters [13-15].

II. Measurement Equipment and Setup

All presented measurements were made in High Voltage Laboratory of Electrical & Power Engineering Department of AGH University of Science and Technology Krakow, Poland. For realization of the test program, a current pulse generator (Fig. 3) and a laboratory stand equipped with an alternating current (AC) source with an exposure time programmer (Fig. 4) were used for stressing of tested SPDs.

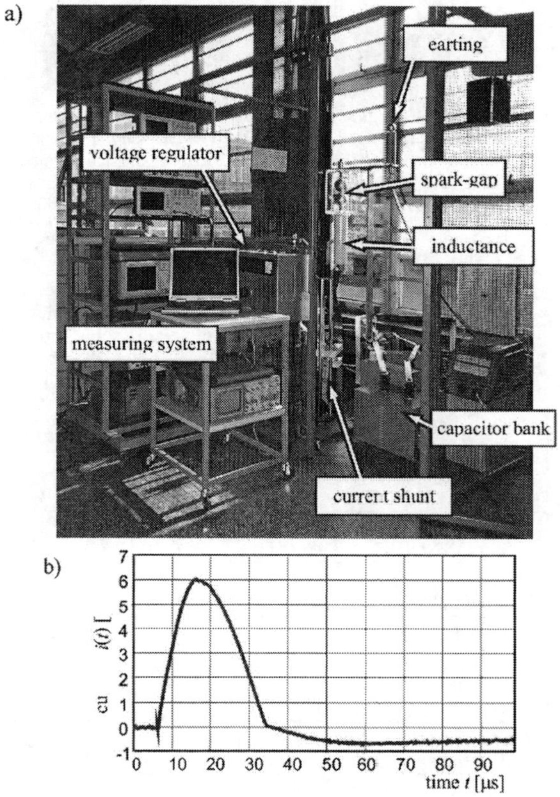

Figure 3. The laboratory stand of high-current pulse generator: a) overwiev, b) registered oscillogram of 6 kA 8/20 μs current pulse

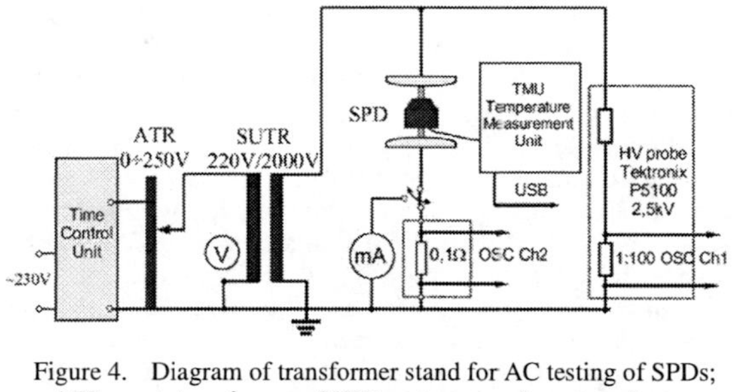

Figure 4. Diagram of transformer stand for AC testing of SPDs; ATR – autotransformer, SUTR – step-up voltage transformer

For the determination of dielectric parameters of varistors at different temperatures, a special thermal chamber with Teflon® bushings and a Solartron 1260+1296 measuring system were used to record the impedance characteristics of the varistors over a wide frequency range (measurements ranging from 10^{-3} Hz to 10^{6} Hz), Fig. 5.

Figure 5. Configuration of a laboratory stand for measuring dielectric parameters of tested SPDs: 1) wideband impedance measurement system (Solartron 1260 + 1296); 2) thermal chamber; 3) host computer with a program controlling the measurements; 4) SPD located inside the thermal chamber with Teflon® bushings

III. Tested Objects, Procedures and Results

In described studies the effect of energetic stresses and temperature on changing the wideband characteristics of dielectric parameters of varistors of low voltage surge arresters (SPDs) have been investigated. Table 1 shows the selected technical parameters of tested SPDs.

Table 1. Selected parameters of tested SPD

Parameter	SPD type		
	280	***440***	***660***
Continuous operating voltage U_c [V]	280	440	660
Nominal discharge current (for 8 μs /20 μs pulse) I_n [kA]	5		
Voltage protection level U_{pn} at I_n [V]	1100	1800	2500
Maximum discharge current (for 8μs /20μs pulse) I_{max} [kA]	25		
Max. voltage protection level U_{pmax} at I_{max} [V]	1500	2500	3200
Energy absorption capability (1* 8μs /20μs; I_{max}) W_{max} [J]	1800	3000	4000

During laboratory experiments varistors of tested SPDs were exposed to energetic stresses performed by: nominal 5 kA 8 μs /20 μs current pulses or 50 Hz alternating current pulses with a duration of about 500 ms (25/26 AC periods). Waveforms of voltages and currents were recorded for determining the value of energy absorbed by individual varistors. Figure 6 shows the voltage and current waveforms recorded during initial measurements on one of the varistors (SPD type 280) that was damaged during the AC impulse test, at absorbed energy about 2000 J. In the main test cycle, when tested with 0.5 s AC pulse (Fig. 7), energies equal to about 50% of the energy absorption capability of specific SPD were applied, at initial varistors temperatures (measured on bottom SPD's connector) equal to ambient temperature. Figures 8 and 9 show varistor temperature profiles for SPDs energized by one or two AC energy pulses, respectively. In the latter case, the cumulative thermal effect of both pulses is observed. Miniature contact thermocouple of K type (connected to AD 595 integrated circuit and precision voltmeter with serial interface), mounted in the bottom metal electrode of the varistor, were used for temperature measurements.

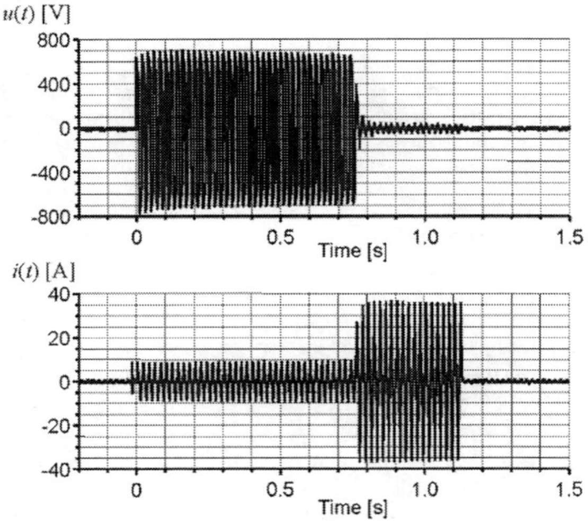

Figure 6. Registered 1.25 s AC burst, acquired for varistor damaged during AC voltage action

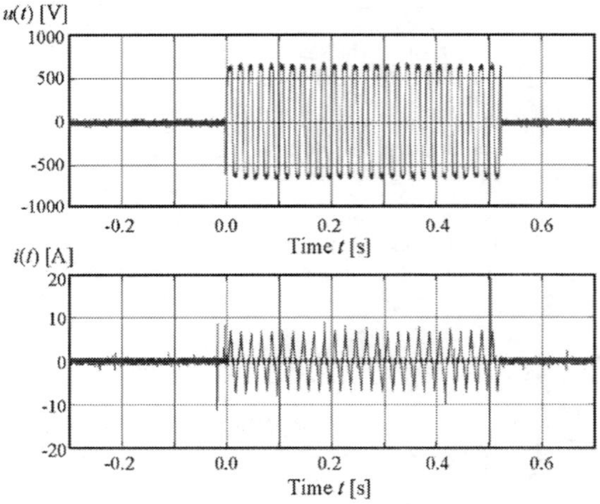

Figure 7. Registered ~0.5 s AC burst, applied during SPDs tests

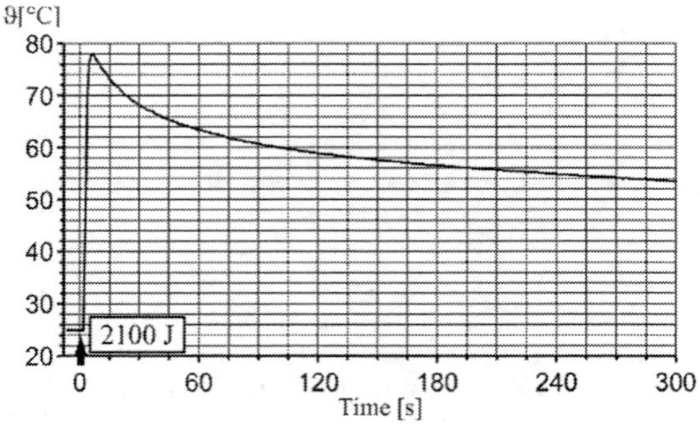

Figure 8. An example of the temperature profile on the lower connection of SPD type 280, for a single AC energy pulse

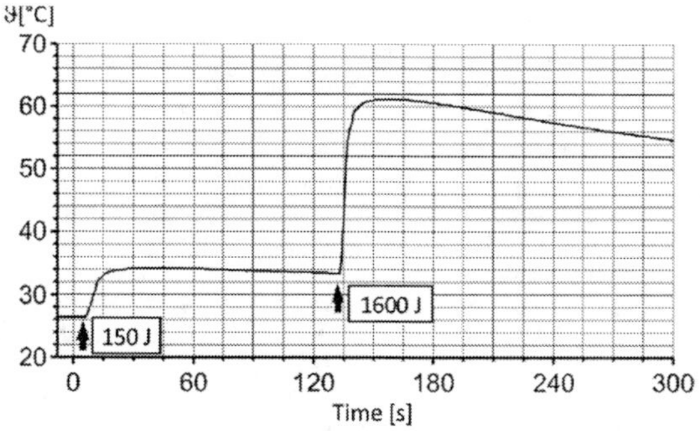

Figure 9. An example of the temperature profile on the lower connection of SPD type 440, for two AC energy pulses

Four varistors of each type were measured in the main part of the study, three of which were exposed and the last one served as a reference object. Each stressed varistor was subjected to energetic exposure with 10 standardized 5 kA current pulses 8 µs / 20 µs or 5 AC pulses, in successive series. After each series, the capacitance and dielectric loss characteristics vs. frequency (from 10^{-3} to 10^{6} Hz) were recorded. Changes in recorded characteristics for different SPDs were similar to those in Fig. 10. Under the influence of energy stresses, there are a slight increases in capacitances and dielectric loss factors in the lower frequency range. However, these changes are significantly less than those caused by the temperature increase used in the tests. It has been also noted that increases in capacitance and tgδ due to AC stresses, with higher energies than for current pulses 8/20, were lower than those caused by high-current pulses.

The varistors measurement results were analyzed by estimating the obtained wideband characteristics using the modeling of the varistor with its equivalent circuit shown in diagram at Fig. 11. When modeling, in branch parallel to resistor R2, capacitor C2 or CPE element (Constant Phase Element) was placed [16]. An example of matching the tgδ(*f*) characteristics for both models to the varistors' measurement results is shown in Fig. 12. For each of the analyzed cases, a better fit for the model with the CPE was obtained. Wideband characteristics were fitted and analyzed using equivalent circuit modelling in ZView program (SAI Scribner Associates Inc.).

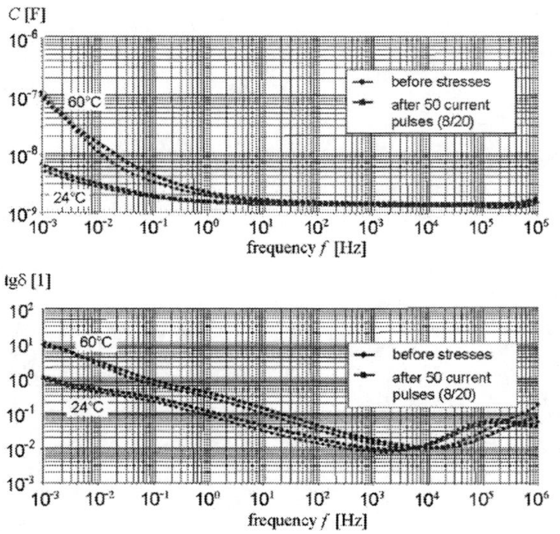

Figure 10. Exemplary wideband characteristics of capacitance *C*(*f*) and dissipation factor tgδ(*f*) for one of stressed SPDs type 280

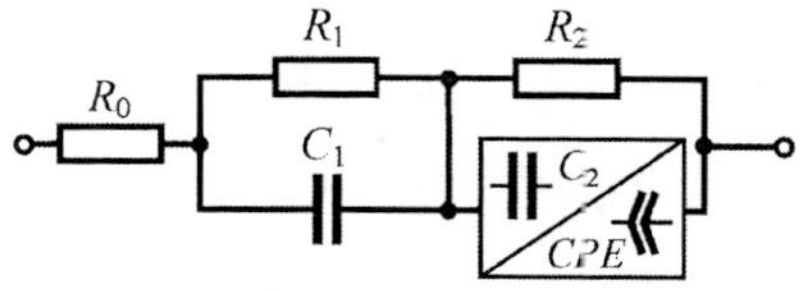

Figure 11. Proposed equivalent circuit of SPD's varistor

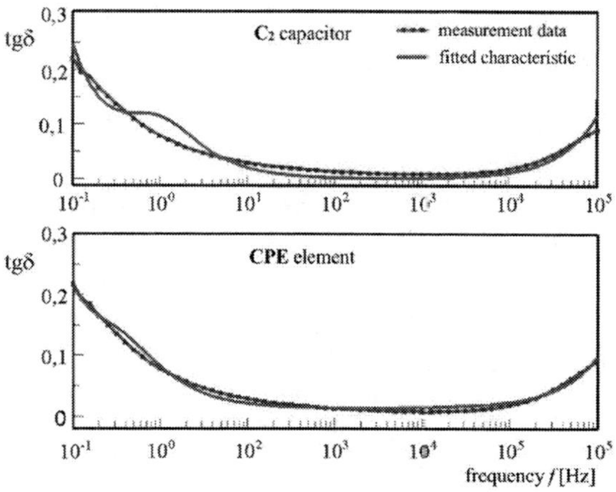

Figure 12. Example of tgδ(*f*) wideband characteristic fitting to measurement data for model circuit presented in Fig. 11, with C_2 (top graph) or CPE (bottom graph) elements

IV. SUMMARY

The result of the high energy stresses is a permanent change in the internal structure of the varistors, modifying their electrical characteristics, described by the varistors response, both for low and high signal excitation. During varistor studies using impedance spectroscopy, they show changes in the shape of wideband characteristics of dielectric parameters. This is one of the non-destructive testing methods which, along with other methods, can be used for diagnostic evaluation of this type of device. Based on the results of the measurements, the electrical characteristics of the varistor structures were modelled using equivalent schemes.

In the layered topology model of varistor, taking into account the phenomena occurring in the ZnO grains and at their boundaries, a CPE element was used in one of the layers, which reduced the mean square error of the characteristics fitting. This demonstrates a better representation of the physical phenomena occurring in varistors and the distribution of the time constants of the relaxation process taking place in their polycrystalline structure. Results of measurements showed also a greater effect of high current pulses on the internal varistor structure, which determines its electrical parameters than 50 Hz AC pulses (with significantly lower current density).

Temperature and structural changes cause, for example, an increase in conductive losses, which is evident in the very low frequency range of the tgδ(*f*) characteristics. Varistors studies have shown that the temperature has significant influence on their results, so it can be regarded as a disturbance that reduces the diagnostic efficacy of this test method. Temperature and aging structural changes in the same way modify the analysed wideband characteristics. Even small temperature changes (order of single degrees) produce an effect similar to the effects of structural changes caused by multiple current pulses.

Varistors' low-signal analysis shows that the use of CPE elements allows for a more accurate representation of real dielectric loss characteristics along with the observed effects of their wideband dispersion. This direction of study will be continued to describe changes in varistor parameters caused by the synergic effects of different stresses.

Acknowledgment

The researches described in the paper were financed by the Polish Ministry of Science and Higher Education - grants for Faculty of Electrical Engineering, Automatics, Computer Science and Biomedical Engineering of AGH University of Science and Technology, Krakow, Poland.

References

[1] W. D. Niebuhr, "MOV surge arresters: Improved substation equipment protection", IEEE Trans. Industry Applic., vol. 21, no. 4, pp. 1081-1086, 1985, doi: 10.1109/TIA.1985.349582.

[2] D. Paul and S.I. Venugopalan, "Power distribution system equipment overvoltage protection", IEEE Trans. Industry Applic., vol. 30, no. 5, pp. 1290-1297, 1994, doi: 10.1109/28.315241.

[3] M.R. Meshkatoddini, "Investigation of the influence of the ZnO varistor preparation method on its characteristics", Conf. Rec. IEEE Int. Symp. on Electr. Insul. ISEI, 2008, pp. 320-323, doi: 10.1109/ELINSL.2008.4570338.

[4] A. Gubański, W. Mielcarek, K. Prociów, J. Warycha, and J. Wróbel, "The effect of aluminium additive on the electrical properties of ZnO varistors", Materials Science – Poland, vol. 27, no. 4/2, pp. 1207-1218, 2009.

[5] L. M. Levinson and H. R. Philipp, "The physics of metal oxide varistors", J. Appl. Phys., vol. 46, pp. 1332-1341, 1975, doi: 10.1063/1.321701.

[6] G. D. Mahan, L. M. Levinson, and H. R. Philipp, "Theory of conduction in ZnO varistors", J. Appl. Phys., vol. 50, pp. 2799-2812, 1979, doi: 10.1063/1.326191.

[7] E. Canessa and V.L. Nguyen, "Non-linear I-V characteristics of double Schottky barriers and polycrystalline semiconductors", Physica B: Condensed Matter, vol. 179, no. 4, pp. 335-341, 199, doi: 10.1016/0921-4526(92)90634-5.

[8] X. Yan, Y. Wen, and X. Yi, "Study on the resistive leakage current characteristic of MOV surge arresters", IEEE/PES Trans. & Distrib. Conf. and Exhib., vol. 2, pp. 683-687, 2002, doi: 10.1109/TDC.2002.1177555.

[9] Z.N. Stojanovic and Z.M. Stojkovic, "Evaluation of MOSA condition using leakage current method", Int. J. Electr. Power Energy Syst., vol. 52, no. 2, pp. 87-95, 2013, doi: 10.1016/j.ijepes.2013.03.027.

[10] Z. Xu, L. Zhao, A. Ding, and F. Lu, F., "A current orthogonality method to extract resistive leakage current of MOSA", IEEE Trans. Power Deliv., vol. 28, no. 1. pp. 93–101, 2013, doi: 10.1109/TPWRD.2012.2221145.

[11] J. Wu, J. Hu, C. Shuiming, and J. He, "Harmonic characteristics of leakage currents of ZnO varistors under impulse ageing", IEEE Trans. Power Deliv., vol. 32, pp. 1758 – 1765, 2017, doi: 10.1109/TPWRD.2016. 2585501.

[12] C.A.L. Almeida, at al., "Intelligent thermographic diagnostic applied to surge arresters. A new approach", IEEE Trans. Power Delivery, vol. 24, no. 2, pp. 751–757, 2009, doi: 10.1109/TPWRD.2009.2013375.

[13] M. Jaroszewski and, J. Pospieszna, "An assessment of ageing of oxide varistors exposed to pulse hazards using dielectric spectroscopy", 2004 IEEE ICSD, vol. 2, pp. 727-730, Toulouse, France, 2004, doi: 10.1109/ICSD.2004.1350534.

[14] P. Zydron, M. Bonk, Ł. Fuśnik, and B. Szafraniak, "Influence of temperature on parameters of low-voltage surge arresters' metal-oxide varistors investigated by impedance spectroscopy methods" (in Polish), Przegląd Elektrotechniczny, vol. 92, no. 10, pp. 158-162, 2016, doi: 10.15199/48.2016.10.38.

[15] W. Bassi and H. Tatizawa, "Early prediction of surge arrester failures by dielectric characterization", IEEE Electr. Insul. Mag., vol. 32, no. 2, pp. 35-42, 2016, doi: 10.1109/MEI.2016.7414229.

[16] E. Borsukov and J. R. Macdonald, Impedance Spectroscopy. Theory, Experiment, and Applications, John Wiley & Sons, Inc., Hoboken, New Jersey, 2005.

Increasing the Efficiency of Transmission of Electricity in the Transmission System

Ľuboš Pavlov, Ľuboš Skurčák
VUJE a.s.
Trnava, Slovakia
lubos.pavlov@vuje.sk, lubos.skurcak@vuje.sk

Juraj Altus, Marek Roch
KVES
Žilina, Slovakia
juraj.altus@fel.uniza.sk, marek.roch@fel.uniza.sk

Abstract — The present article discusses the possibility of increasing the efficiency of transmission of electricity in SR transmission system by determining the losses caused by impedance asymmetry of selected electrical elements (transformers, external lines and compensating chokes) of the transmission system as a generally unsymmetrical system.

Keywords: efficiency of power transmission system; power transmission system; unbalance; losses reduction

I. Introduction

Reducing the loss of electricity is a constant effort of TSOs (220 kV, 400 kV), as well as DSOs (22 kV, 110 kV). The reason is to conserve primary energy sources (coal, natural gas, oil, uranium) which are currently partly replaced by renewable sources, which ultimately contributes to emission reductions. The issue of reducing losses is closely linked to increasing energy efficiency in all developed economies that have joined the Paris Convention on Climate Change. Increasing energy efficiency in the transmission and distribution of electricity is not new topic even in the Slovak Republic. Reducing the losses of electricity is a permanent endeavour. This issue is solved in the framework of Act no. 321/2014 Coll. and in the Energy Efficiency Action Plan of the Slovak Republic, which brings together the requirements of Directive 2002/36/ EC on energy services as well as the new requirements of Directive 2012/27 / EU on energy efficiency. In line with the new Directive 2012/27 / EU, the third action plan already includes the area of energy conversion, transmission and distribution (reducing the energy difficulty of transmission and distribution of electricity), while the savings achieved on the transmission and distribution side can be partly offset by the set savings targets energy.

Reducing the loss of electricity and its quantification is a very complicated task, which has been solved in practice for many years. However, some irregularities, the cause of which is not known, are frequently found in their evaluation. If we omit operating states, measured energy transmission, calculation methodology and similar causes, there is still a reality of hidden losses. For the operator, there is a need to find out how they occur, how much they are, and how the presence of these losses can be reduced or eliminated. Progress in measuring or computing technology brings new possibilities for more accurate analysis of total electricity losses and individual components. One of the components of total loss of electricity is the loss caused by asymmetry of phasor star of voltages and currents [1].

II. Asymmetry in Transmission System

Determining the exact values of the asymmetry of voltage and current star phasor requires a new approach to calculations and measurements of network operating quantities, e.g. based on synchronic phasor monitoring. The balance of transmission losses has so far been evaluated in the calculations on the basis of a simplified assumption of TS symmetry, using a single-line substitution scheme for the transmission system. By this assumption was automatically (and negligently) negated the zero and back components of voltages and currents. For example, the existence of the zero component of the current of the asymmetric phasor star as well as the voltage can be easily verified by measuring the currents flowing to the ground through the grounded transformer nodes and from there through the ground and the ground cables of the OHL between the TS substations. The resultant losses are untenable in present methodology, respecting the simplistic assumption of the symmetrical star of the consistent components of operating voltages and currents. The results of such transmission loss calculations do not make it possible to objectively explore all possibilities for increasing energy efficiency [1]

Losses in HV and EHV grids are determined by the use of physical formulas using measure of active power and voltage on individual network elements and actual network parameters. These values are in practice provided by various measuring devices. Mostly, they are electrometers. These devices allow recording of the effective values of the above-mentioned variables. At present, however, it is also possible to realize a vector record of these electrical quantities. The vector record thus provides a more complex picture of the recorded electrical quantities in the phasor form. The phasor values of the electrical quantities allow more detailed quantification, but also an analysis of the individual components of the loss of electricity on any element in the transmission system. There are several possible ways to determine overall losses in the transmission system. The losses on any element of the transmission system are based on the measured phasor values calculated from the difference of the column vectors with the apparent power phasors at its starting and ending nodes according to equation (1) [2,3]:

$$\left\lfloor \Delta^{S}\bar{S}^{C}_{A,B,C} \right\rfloor = \left\lfloor {}^{S}\bar{S}^{Z}_{A,B,C} \right\rfloor - \left\lfloor {}^{S}\bar{S}^{K}_{A,B,C} \right\rfloor = = \begin{bmatrix} \Delta^{S}\bar{S}^{C}_{A} \\ \Delta^{S}\bar{S}^{C}_{B} \\ \Delta^{S}\bar{S}^{C}_{C} \end{bmatrix} = \begin{bmatrix} {}^{S}\bar{S}^{Z}_{A} \\ {}^{S}\bar{S}^{Z}_{B} \\ {}^{S}\bar{S}^{Z}_{C} \end{bmatrix} - \begin{bmatrix} {}^{S}\bar{S}^{K}_{A} \\ {}^{S}\bar{S}^{K}_{B} \\ {}^{S}\bar{S}^{K}_{C} \end{bmatrix} \quad (1)$$

As mentioned above, total losses in the transmission system are expressed by the sum of several loss components. One of the components of total losses is loss due to asymmetry in the transmission system. Asymmetry losses can then be subdivided into losses due to impedance and operational asymmetry. The transmission system consists of a set of three-phase, impedance-symmetric and impedance-asymmetrical electrical machines (power transformer), apparatuses (instrument transformer of current and voltage, disconnectors, lightning arresters), regulating (protection and automation) and compensating (choke) devices in electric stations, which are interconnected by asymmetrical or symmetrical OHL and cable lines.

The total loss of asymmetry of the apparent power on the transmission system branch is given by the difference of the generally asymmetrical apparent powers at the start and end nodes into which the branch of the network is drawn in accordance with equation (2) [4]:

$$\Delta^N\bar{S}^C_{A,B,C} = {}^N\bar{S}^Z_{A,B,C} - {}^N\bar{S}^K_{A,B,C} = 3\left(\bar{S}_0^Z - \bar{S}_0^K\right) + 3\left(\bar{S}_1^Z - \bar{S}_1^K\right) + 3\left(\bar{S}_2^Z - \bar{S}_2^K\right) \quad (2)$$

Losses caused by the impedance asymmetry of the transmission system examined element are expressed by the difference of the asymmetrical losses obtained by measurement and symmetrical losses expressed by computation (decomposition on symmetrical star, components 012) at its start and end according to equation (3):

$$\begin{aligned}
\Delta\bar{S}_{nesymetriou\ A,B,C} &= \Delta^N\bar{S}^C_{A,B,C} - \Delta^S\bar{S}^C_{A,B,C} = \\
&= 3\left(\bar{S}_0^Z - \bar{S}_0^K\right) + 3\left(\bar{S}_1^Z - \bar{S}_1^K\right) + \\
&+ 3\left(\bar{S}_2^Z - \bar{S}_2^K\right) - 3\left(\bar{S}_1^Z - \bar{S}_1^K\right) = \\
&= 3\left(\bar{S}_0^Z - \bar{S}_0^K\right) + 3\left(\bar{S}_2^Z - \bar{S}_2^K\right) = \\
&= \mathrm{Re}\left\{3\left(\bar{S}_0^Z - \bar{S}_0^K\right) + 3\left(\bar{S}_2^Z - \bar{S}_2^K\right)\right\} + \\
&+ \mathrm{jm}\left\{3\left(\bar{S}_0^Z - \bar{S}_0^K\right) + 3\left(\bar{S}_2^Z - \bar{S}_2^K\right)\right\} = \\
&= \Delta P_{nesymetriou\ A,B,C} + j.\Delta Q_{nesymetriou\ A,B,C}
\end{aligned} \quad (3)$$

III. METHODOLOGY FOR CALCULATION THE LOSSES CAUSED BY IMPEDANCE ASYMMETRY

The following is a brief description of the methodology for the calculation of losses due to impedance asymmetry of selected elements of the transmission system (OHL, transformers and compensating chokes) by means of synchronous measurements.

By means of a synchronous monitoring system, the voltage and current time courses are synchronously recorded in chosen time interval, at the start (Z) and at the end (K), at the input and output of any TS element voltages values ($u_A^Z, u_B^Z, u_C^Z, u_A^K, u_B^K, u_C^K$) and current values ($i_A^Z, i_B^Z, i_C^Z, i_A^K, i_B^K, i_C^K$) . From these are determined by calculating the current flowing through the grounded node into the ground and the potential difference between the ground and the ground of the respective TS node. The monitoring of these nodes $U_Z=3.U_0$ and $I_Z=3.I_0$ gives an image of the size of the grid asymmetry (zero voltage and current components) in its particular node.

From the voltage and current time courses at the start and end generally at inputs and outputs of any TS element, harmonic analyses will be made, resulting in time courses with phase shifts of all harmonics (n = 1, 2, 3 till 40) that are contained in them

The reference time course is first harmonic course of voltage on first phase-A, at the start of the examined element, as well as for all other measured time courses of the first and higher harmonics of the other phases

The phase shifts of the other time courses $u^Z_{A,B,C}, u^K_{A,B,C}, i^Z_{A,B,C}, i^K_{A,B,C}$ of the voltages, currents and their higher harmonics (and thus the rotation of their phasors in the complex plane) are read off relative to the intersection with the time axis of the first harmonic time course of phase voltage u_A^Z (at the nearest zero transition point with positive first derivation).

The magnitude of the phasor of each harmonic voltage and current is considered to be the effective value $X_{ef} = X_{max}/\sqrt{2}$ of its time course.

The result is synchronous phasors of three voltages and three currents in the versor shape at both the start and the end of the examined TS element.

The calculation of losses due to impedance asymmetry is performed in a cycle for each harmonic (n = 1,2,3 till 40), which is not indexed in the following steps in the following formulas.

From the measured time courses of the following variables:

- voltage at start $u_A^Z(t)$, $u_B^Z(t)$, $u_C^Z(t)$, of any TS element,
- current at start $i_A^Z(t)$, $i_B^Z(t)$, $i_C^Z(t)$ of any TS element,
- voltage at the end $u_A^K(t)$, $u_B^K(t)$, $u_C^K(t)$, of any TS element,
- current at the end $i_A^K(t)$, $i_B^K(t)$, $i_C^K(t)$, of any TS element.

Their phasors are found in a complex plane in a versor shape:

- phasors of (phase) voltages at the start of any TS element,

$$\bar{U}_A^Z = U^Z_{ef\_A}.e^{j.\varphi_{uZ\_A}} \quad , \quad \bar{U}_B^Z = U^Z_{ef\_B}.e^{j.\varphi_{uZ\_B}} \quad ,$$

$$\bar{U}_C^Z = U^Z_{ef\_C}.e^{j.\varphi_{uZ\_C}}$$

- phasors of (phase) currents and their conjugate values (*) at the start of any TS element,

$$\bar{I}_A^Z = I^Z_{ef\_A}.e^{j\varphi_{iZ\_A}} \quad ; \quad \bar{I}_A^{*Z} = I^Z_{ef\_A}.e^{-j\varphi_{iZ\_A}} \quad ;$$

$$\bar{I}_B^Z = I^Z_{ef\_B}.e^{j\varphi_{iZ\_B}} \quad ; \quad \bar{I}_B^{*Z} = I^Z_{ef\_B}.e^{-j\varphi_{iZ\_B}} \quad ;$$

$$\bar{I}_C^Z = I^Z_{ef\_C}.e^{j\varphi_{iZ\_C}} \quad ; \quad \bar{I}_C^{*Z} = I^Z_{ef\_C}.e^{-j\varphi_{iZ\_C}}$$

- phasors of (phase) voltages at the end of any TS element,

$$\bar{U}_A^K = U^K_{ef\_A}.e^{j.\varphi_{uK\_A}} \quad , \quad \bar{U}_B^K = U^K_{ef\_B}.e^{j.\varphi_{uK\_B}} \quad ,$$

$$\bar{U}_C^K = U^K_{ef\_C}.e^{j.\varphi_{uK\_C}}$$

- phasors of (phase) currents and their conjugate values (*) at the end of any TS element,

$$\bar{I}_A^Z = I_{ef\_A}^Z . e^{j\varphi_{iZ\_A}} \quad ; \quad \bar{I}_A^{*Z} = I_{ef\_A}^Z . e^{-j\varphi_{iZ\_A}} \quad ;$$

$$\bar{I}_B^Z = I_{ef\_B}^Z . e^{j\varphi_{iZ\_B}} \quad ; \quad \bar{I}_B^{*Z} = I_{ef\_B}^Z . e^{-j\varphi_{iZ\_B}} \quad ;$$

$$\bar{I}_C^Z = I_{ef\_C}^Z . e^{j\varphi_{iZ\_C}} \quad ; \quad \bar{I}_C^{*Z} = I_{ef\_C}^Z . e^{-j\varphi_{iZ\_C}}$$

In the next step, are compiled row vector of voltage phasors and column vector of current phasors and their conjugate values at the start and end of the investigated TS element in matrix form:

- row vectors of phase voltage phasors at the start of the investigated element,

$$\left[\bar{U}^Z_{A,B,C}\right] = \left[\bar{U}^Z_A \bar{U}^Z_B \bar{U}^Z_C\right] = \left[\ \left|\bar{U}^Z_A\right| . e^{j\varphi uZ\_A} \ \left|\bar{U}^Z_B\right| . e^{j\varphi uZ\_B} \ \left|\bar{U}^Z_C\right| . e^{j\varphi uZ\_C}\ \right]$$

- row vectors of phase voltage phasors at the end of the investigated element,

$$\left[\bar{U}^K_{A,B,C}\right] = \left[\bar{U}^K_A \bar{U}^K_B \bar{U}^K_C\right] = \left[\ \left|\bar{U}^K_A\right| . e^{j\varphi uK\_A} \ \left|\bar{U}^K_B\right| . e^{j\varphi uK\_B} \ \left|\bar{U}^K_C\right| . e^{j\varphi uK\_C}\ \right]$$

- Column vectors of phase current phasors and their conjugate values at the start of the investigated element

$$\left[\bar{I}^Z_{A,B,C}\right] = \begin{bmatrix} \bar{I}^Z_A \\ \bar{I}^Z_B \\ \bar{I}^Z_C \end{bmatrix} = \begin{bmatrix} \left|\bar{I}^Z_A\right| . e^{j\varphi iZ\_A} \\ \left|\bar{I}^Z_B\right| . e^{j\varphi iZ\_B} \\ \left|\bar{I}^Z_C\right| . e^{j\varphi iZ\_C} \end{bmatrix}$$

$$\left[\bar{I}^{*Z}_{A,B,C}\right] = \begin{bmatrix} \bar{I}^{*Z}_A \\ \bar{I}^{*Z}_B \\ \bar{I}^{*Z}_C \end{bmatrix} = \begin{bmatrix} \left|\bar{I}^Z_A\right| . e^{-j\varphi iZ\_A} \\ \left|\bar{I}^Z_B\right| . e^{-j\varphi iZ\_B} \\ \left|\bar{I}^Z_C\right| . e^{-j\varphi iZ\_C} \end{bmatrix}$$

- Column vectors of phase current phasors and their conjugate values at the end of the investigated element

$$\left[\bar{I}^K_{A,B,C}\right] = \begin{bmatrix} \bar{I}^K_A \\ \bar{I}^K_B \\ \bar{I}^K_C \end{bmatrix} = \begin{bmatrix} \left|\bar{I}^K_A\right| . e^{j\varphi iK\_A} \\ \left|\bar{I}^K_B\right| . e^{j\varphi iK\_B} \\ \left|\bar{I}^K_C\right| . e^{j\varphi iK\_C} \end{bmatrix}$$

$$\left[\bar{I}^{K*}_{A,B,C}\right] = \begin{bmatrix} \bar{I}^{K*}_A \\ \bar{I}^{K*}_B \\ \bar{I}^{K*}_C \end{bmatrix} = \begin{bmatrix} \left|\bar{I}^K_A\right| . e^{-j\varphi iK\_A} \\ \left|\bar{I}^K_B\right| . e^{-j\varphi iK\_B} \\ \left|\bar{I}^K_C\right| . e^{-j\varphi iK\_C} \end{bmatrix}$$

From complex phasors of (phase) voltages and currents, a column vector of phasor of (phase) apparent power is calculated:

- at the start of the examined element according to equation (4):

$$\begin{bmatrix} {}^N\bar{S}^Z_A \\ {}^N\bar{S}^Z_B \\ {}^N\bar{S}^Z_C \end{bmatrix} = \begin{bmatrix} {}^N\bar{U}^Z_A . {}^N\bar{I}^{Z*}_A \\ {}^N\bar{U}^Z_B . {}^N\bar{I}^{Z*}_B \\ {}^N\bar{U}^Z_C . {}^N\bar{I}^{Z*}_C \end{bmatrix} \tag{4}$$

- at the end of examined element according to equation (5):

$$\begin{bmatrix} {}^N\bar{S}^K_A \\ {}^N\bar{S}^K_B \\ {}^N\bar{S}^K_C \end{bmatrix} = \begin{bmatrix} {}^N\bar{U}^K_A . {}^N\bar{I}^{K*}_A \\ {}^N\bar{U}^K_B . {}^N\bar{I}^{K*}_B \\ {}^N\bar{U}^K_C . {}^N\bar{I}^{K*}_C \end{bmatrix} \tag{5}$$

The apparent power of the generally asymmetrical examined element at the start and end is calculated according to equations (6 and 7):

- at the start of examined element

$$\begin{aligned}
\left[{}^N\bar{S}^Z_{A,BC}\right] &= \left[\bar{U}^Z_{A,BC}\right]^T . \left[\bar{I}^{Z*}_{A,BC}\right] = \\
&= \left(\left[T\right]\left[\bar{U}^Z_{012}\right]\right)^T . \left(\left[T\right]\left[\bar{I}^Z_{012}\right]\right)^* = \\
&= \left[\bar{U}^Z_{012}\right]^T . \left[T\right]^T . \left[T\right]^* . \left[\bar{I}^Z_{012}\right]^* = \\
&= \left[\bar{U}^Z_{012}\right]^T . \left[J\right] . \left[\bar{I}^Z_{012}\right]^* = \\
&= \left[\bar{U}^Z_0 + \bar{U}^Z_1 + \bar{U}^Z_2\right] \begin{bmatrix} 3 & 0 & 0 \\ 0 & 3 & 0 \\ 0 & 0 & 3 \end{bmatrix} \begin{bmatrix} \bar{I}^{Z*}_0 \\ \bar{I}^{Z*}_1 \\ \bar{I}^{Z*}_2 \end{bmatrix} = \\
&= 3\bar{U}^Z_0\bar{I}^{Z*}_0 + 3\bar{U}^Z_1\bar{I}^{Z*}_1 + 3\bar{U}^Z_2\bar{I}^{Z*}_2 = \\
&= 3\bar{S}^Z_0 + 3\bar{S}^Z_1 + 3\bar{S}^Z_2 = \\
&= 3\left(P^Z_0 + jQ^Z_0\right) + 3\left(P^Z_1 + jQ^Z_1\right) + 3\left(P^Z_2 + jQ^Z_2\right)
\end{aligned} \tag{6}$$

- at the end of examined element:

$$\begin{aligned}
\left[{}^N\bar{S}^K_{A,BC}\right] &= \left[\bar{U}^K_{A,BC}\right]^T . \left[\bar{I}^{K*}_{A,BC}\right] = \\
&= \left(\left[T\right]\left[\bar{U}^K_{012}\right]\right)^T . \left(\left[T\right]\left[\bar{I}^K_{012}\right]\right)^* = \\
&= \left[\bar{U}^K_{012}\right]^T . \left[T\right]^T . \left[T\right]^* . \left[\bar{I}^K_{012}\right]^* = \\
&= \left[\bar{U}^K_{012}\right]^T . \left[J\right] . \left[\bar{I}^K_{012}\right]^* = \\
&= \left[\bar{U}^K_0 + \bar{U}^K_1 + \bar{U}^K_2\right] \begin{bmatrix} 3 & 0 & 0 \\ 0 & 3 & 0 \\ 0 & 0 & 3 \end{bmatrix} \begin{bmatrix} \bar{I}^{K*}_0 \\ \bar{I}^{K*}_1 \\ \bar{I}^{K*}_2 \end{bmatrix} = \\
&= 3\bar{U}^K_0\bar{I}^{K*}_0 + 3\bar{U}^K_1\bar{I}^{K*}_1 + 3\bar{U}^K_2\bar{I}^{K*}_2 = \\
&= 3\bar{S}^K_0 + 3\bar{S}^K_1 + 3\bar{S}^K_2 = \\
&= 3\left(P^K_0 + jQ^K_0\right) + 3\left(P^K_1 + jQ^K_1\right) + 3\left(P^K_2 + jQ^K_2\right)
\end{aligned} \tag{7}$$

By the decomposition of the phasor star of the apparent phase powers at the node at the start $\bar{S}^Z_A$, $\bar{S}^Z_B$, $\bar{S}^Z_C$ and at the end $\bar{S}^K_A$, $\bar{S}^K_B$, $\bar{S}^K_C$ of examined element, the symmetrical components are found, zero (0), the consecutive (1) and the reverse (2) of apparent power at the start $\bar{S}^Z_0$, $\bar{S}^Z_1$, $\bar{S}^Z_2$ and the end $\bar{S}^K_0$, $\bar{S}^K_1$, $\bar{S}^K_2$ of the examined TS element.

Losses caused by impedance asymmetry in the start and end nodes of the generally asymmetrical examined element examined are obtained by the sum of the zero and the reverse symmetric component at the start and end of the generally asymmetrical examined element according to (8) and (9):

$$\bar{S}^Z_N = 3(\bar{S}^Z_0 + \bar{S}^Z_2) \tag{8}$$

$$\bar{S}^K_N = 3(\bar{S}^K_0 + \bar{S}^K_2) \tag{9}$$

The resulting losses of impedance asymmetry on the generally asymmetrical element of TS are obtained by the relationship (10) by difference in losses due to impedance asymmetry at the start and end nodes of the generally asymmetrical examined element.

$$\Delta\overline{S}_N = \overline{S}_N^Z - \overline{S}_N^K \tag{10}$$

The ratio of reverse and consecutive component (voltage and current) is the coefficient of (voltage, current) asymmetry and the ratio of zero and consecutive component (voltage, current) is the coefficient of (voltage, current) imbalance.

IV. Impedance Asymmetry of Chosen Transmission System Elements

The calculations of losses by own impedance asymmetry is dependent on input parameters of TS type elements. Each type element (power line, transformer and compensator choke) influences the size of the losses by impedance asymmetry by other design arrangement and location. For this reason, it is necessary to determine their input parameters for the calculation of losses by their own impedance asymmetry [4].

Non-transposed HV and EHV OHL are a typical asymmetrical element of the transmission system. The elements of their generally asymmetrical impedance [Z_{ABC}] and admittance [Y_{ABC}] matrices are dependent on the geometry of the masts and the sequence of phase conductors and earth ropes in the mast crown along the line of the conductor, while the longitudinal impedance in homogeneities caused by the change in configuration of masts types (PORTAL, F , MAČKA, DONAU, SÚDOK, H, STROMČEK) contribute to asymmetry, by changing the sequence of phases on the mast brackets, changing the type of phase conductors and ground ropes in particular sections of the OHL, presence of concurrent with other lines.

The primary construct, impedance and admittance asymmetry of line is highlighted (especially in the 110 kV distribution network) by its asymmetrical operating modes, which are forced by asymmetry of the star phasors of nodal voltages and branch currents of individual phases A, B, C, single or parallel 3-phase systems.

The calculation of losses due to impedance asymmetry of the OHL results from the 3-phase impedance [Z_{ABC}] and admittance [Y_{ABC}] matrix of the individual phases in general of the 3 * n-conductor system with ground wires in matrix form, which have their own (again ground) and mutual interphase inductive and capacitance bonds between all conductors. These bonds can be obtained by calculating the impedance and admittance matrices based on the geometric distribution of the individual phases in the crown of the electric mast. For the calculation of impedance and admittance matrices, input data such as line type, voltage level, cline length, phase and ground conductor type, geometric distribution of phase conductors to masts, type and number of masts, size of the transmitted power by the respective electric line [5] are required.

Two-and three-wind power HV and EHV transformers are commonly considered in calculations of steady-state as symmetrical elements in ES. However, the transformer impedance asymmetry exists and depends on the design of the magnetic circuit of its core (core or shell construction), further on from the asymmetry of the design of the 3-f primary and secondary windings, the clock angle and the corresponding internal current connections of the transformer windings. This fact is highlighted by the number of measurements and published publications on the issue (V. Mentlík, P. Prost, R. Polanský, J. Michalík, M. Brandt: "Instruments for On-line Monitoring of Transformers", Z. Paukner: "On-Line Transformer Monitoring, Experience, Models").

The size of current asymmetry also depends on the transformer load. For this reason, it is necessary, inter alia, to carry out the monitoring of the time courses, respectively. phase phasors values $\left(\overline{U}_A, \overline{U}_B, \overline{U}_C, \overline{I}_A, \overline{I}_B, \overline{I}_C\right)$ of the primary, secondary and tertiary voltages and currents of the transformers in real operation. The symmetric components of the phasors of primary, secondary and tertiary voltages ($\overline{U}_0, \overline{U}_1, \overline{U}_2$) and currents ($\overline{I}_0, \overline{I}_1, \overline{I}_2$) are then determined and calculation determines the voltage and current asymmetry and the imbalance of the 3-phase system on primary, secondary and tertiary at different transformer loads. Impedance matrices of own and mutual impedances [Z_{ABC}] between primary / secondary, primary / tertiary and secondary / tertiary will be determined from it. For this reason, it is necessary to first analyse their replacement schemes and calculate the size of their R, L elements based on label data. For the calculation of impedance and admittance matrices, the input data on the transformer label (clock angel, transformer ratio, and short circuit current and no-load current) type of the transformer and core geometry are therefore required [5].

The size of the impedance asymmetry of the compensation chokes depends on the design of the geometric arrangement, and the electrical connection in the transmission system.

If the compensating chokes are connected separately for each phase individually as three separate units (air coils) and would be arranged in a row (asymmetrical arrangement) and not in a triangle (symmetrical arrangement) it is possible to observe the presence of impedance asymmetry. The existence of this asymmetry is due to another magnitude of the electromagnetic field, which is generated by the passage of electric current through the extreme coils, as in the coil located in the centre of the array of three separate coils of compensating choke.

From a design point of view, the compensating choke coils can be installed in a common metal container, or each coil of the compensating choke can be installed separately in the metal box itself. In the first case, impedance asymmetry can be detected. If, in this case, the three coils of the compensation choke were arranged in row in a common metal container, the impedance asymmetry would be higher than that of the triangle arrangement by a higher magnetic bond than the common metal container. Compensation chokes in a common metal container are mostly in a 110 kV distribution system. Three separate air compensation chokes are mostly in a 400 kV transmission system.

The asymmetry of compensating chokes is one of the partial asymmetries across the ES. Therefore, it is necessary to consider it. In the impedance-symmetric compensation choke, voltage and current asymmetry will not occur unless the phasor star of the voltage and current is symmetrical. For asymmetrical load (asymmetrical consumption currents at ES nodes), asymmetry is present even in the case of impedance symmetric compensation choke [5].

For calculating impedance and admittance matrices, input data such as geometric layout, design layout, and connection of the compensation choke in the transmission system are required.

Acknowledgment

This work was supported by the Slovak Research and Development Agency under the contract No. APVV-15-0464.

REFERENCES

[1] W.U. Anguan, N. Baoshan. *Line Loss Analysis and Calculation of Electric Power Systems*, October 2015, ISBN: 9781118867235

[2] A.G. Phadke, J.S. Thorp. *Synchronized Phasor Measurements and Their Applications*. Springer. USA.

[3] S.Di. *Utilizing synchrophasor technology to determine transmission line impedance parameters. Arizona State University*. USA.

[4] J. C. Das. *Understanding Symetrical Components for Power System Modeling*, February 2016, Wiley-IEEE Press, ISBN: 9781119226857

[5] B. Singh, K.N. Sharma, A. Tiwari, S.K. Verma, N.S. Singh. "Applications of phasor measurement units (PMUs) in electric power system networks incorporated with FACTS controllers", *International Journal of Engineering, Science and Technology* Vol. 3, No. 3, 2011

[6] IEEE Standard for Synchrophasors for Power Systems. IEEE Std C37.118™-2005. USA.

Natural Ionizing Radiation Absorbed by Ceramic Insulators

Rafał Sobota
Institute of Optoelectronics and Measurement Systems
Czestochowa University of Technology
Czestochowa, Poland
rsobota@el.pcz.czest.pl

Abstract—**The paper presents dosimetric properties and results of measurements of electrical porcelain C120 and C130, of which power insulators are produced. The properties of the protective enamel of such insulators are also discussed. Some insulators under scrutiny have been used in medium-voltage networks for over fifty years, which resulted in a significantly long period of exposure to natural radiation.**

Keywords: thermoluminescence; natural radiation; electrical porcelain; electrical insulator

I. Introduction

The most recent area of application of thermoluminescence (TL) is dosimetry of ionizing radiation. By means of TL it is possible to establish a dose of absorbed radiation within a large range: from 10^{-6} to 10^{6} Gy. To perform measurements based on thermoluminescence two elements are needed: dosimeters and equipment for reading them. [7]

The amount of ionizing energy absorbed by living or inanimate matter, i.e. doses of such energy, are monitored at places where ionizing radiation is emitted. Typically, the consequences of absorbing ionizing radiation depend on a specific dose absorbed. No wonder that together with the development in applications of ionising radiation, research on precise methods of measuring radiation doses was developed in parallel. Based on the knowledge of various properties of radiation, a number of measurement methods, also known as dosimetric methods, have been developed. [3, 7]

Taking this into account, an attempt was made to examine dosimetric properties of electrical porcelains C120 and C130, of which power insulators used in medium-voltage networks are produced, as well as of their protective enamels. Throughout the exploitation period the insulators are exposed to natural ionizing radiation present in the environment.

II. Natural Ionizing Radiation

Both people and materials are constantly exposed to radiation originating in the outer space and inside the Earth, which are natural sources of ionizing radiation. Apart from that, there may occur events being sources of temporary radiation, such as intensive decay of radioactive elements, breakdowns in nuclear power stations and in companies producing nuclear fuel or nuclear power experiments.

One of the first studies on absorption of ionizing radiation by porcelain insulators was conducted by the Institute of Radiation Hygiene in Saint Petersburg (Russia). In this study a TL signal was measured at the location where a controlled underground nuclear explosion "Kraton-3" took place. The main objective of the study was to determine a cumulative event dose for porcelain insulators using the TL method [2]. The results of these measurements were used for obtaining additional retrospective data, such as the gamma radiation dose to which the personnel participating in the experiment was exposed. Estimating the radiation dose accumulated in ceramic material is also useful for assessing the overall environmental impact of radiation, including the effect on animals and plants. [1]

In the light of the above-mentioned research, the dose D_M absorbed by a human being or by material, obtained on the basis of the TL method applied to porcelain insulators is equal to:

$$D_M = D_T - D_N \quad (1)$$

where:

D_T – total TL dose, Gy,

D_N – dose coming from natural sources, Gy.

The dose D_N coming from natural radiation sources is calculated as:

$$D_N = A(f_\alpha \cdot e_\alpha \cdot D_\alpha + f_\beta \cdot e_\beta \cdot D_\beta + f_\gamma \cdot e_\gamma \cdot D_\gamma + f_{Ext} \cdot e_{Ext} \cdot D_{Ext} + f_c \cdot e_c \cdot D_c) \quad (2)$$

where:

A – age of a sample, years,

f_i – fraction of the total dose obtained for a sample,

e_i – TL radiation efficiency,

D_i – total dose, Gy/year,

i – index related to radiation α, β, γ, Ext – external β+γ, c – of space.

Depending on the actual TL measurement technique, the minimal detectable dose of ionizing radiation absorbed by various materials is 0,01 ÷ 0,5 Gy, whereas the maximal measurable dose is 12,9 C kg^{-1} (50,000 R). [1]

III. TL as a Dating Tool

In some areas of science, the TL phenomenon is used for dating materials and also for examining crystalline and defect structures. In these applications, TL is commonly referred to as a dating tool.

In archaeology TL is used for determining the time when excavated ceramic finds were produced. In geology, in turn, TL is used for determining the age of rocks. In both cases, the method relies on the presence of minerals capable of TL, on the ubiquity of ionizing radiation in the environment and also on separating the dose absorbed by the material tested due to a specific event (such as high temperature or sun radiation exposure) and measuring the dose accumulated due to natural radiation. [4]

In the case of ceramic material, all electron traps that had been filled due to natural radiation were then emptied when the pottery was fired [8]. Since then, the radionuclides present in the material together with earth radiation gamma and space radiation (the latter contributing up to a few per cent of the dose) start the process of filling the traps again. The TL results measuring these factors depend on the properties of the material *S*, mean power *D* of the dose at the location of the material and exposure time *T* [5]:

$$TL_N = S \cdot T \cdot D \quad (3)$$

where:

S – thermoluminescence sensitivity of the material (mainly SiO_2) per unit of a dose absorbed.

Hence, the time can be obtained as:

$$T = \frac{TL_N}{S \cdot D} \quad (4)$$

The dose absorbed by the ceramic objects originated from three types of radiation: α, β and γ, so considering (4), calculating the time appears to be very simple under one condition, that the doses coming from individual sources are known. This can be achieved by analysing the content of particular radionuclides in the ceramic material and in the surrounding soil. [5]

The above-presented description of the TL as a dating tool offers only its brief characterization, as the procedure for dating materials is much more detailed. Nowadays it is used for dating products from the Palaeolithic era to the present and can be a way of confirming or disconfirming results obtained by means of other methods.

In geology TL is used for dating magmatic rocks and calcium or loess sediments. In magmatic rocks the process of trap filling starts when lava solidifies and in sedimentary rocks when they crystallize. In loess sediments the process of trap filling starts when after many years of exposure to solar radiation these rocks become bleached and covered by others layers of sediment. [6]

To contribute new findings to the applications of TL, an attempt was made to measure the dose of natural radiation absorbed by porcelain insulators during their exploitation in power networks, in the period spanning 50 years. The results will be presented in the next section.

IV. TL RESULTS

The material under test were samples of electrical porcelain and protective enamels of insulators used in medium-voltage networks since 1960, 1963, 1999, 2003, 2005, 2010 and 2012. Initially, insulators were made of the ceramic mass C120. Nowadays, insulators are standardized and the mass used for their manufacturing is C130.

Figs. 1 ÷ 4 present the TL measurements results of natural samples taken from the location of the core, cap and enamel of porcelain insulators produced in 1960, 1999, 2003 and in 2010. The curve representing the enamel sample from 1960 shows one flat peak in the temperature range 305 ÷ 420 K. Also in the insulator core sample from 1999 there is a single peak at the temperature about 340 K. The curve representing the C130 porcelain from 2003 shows two peaks, corresponding to the temperatures about 325 K and about 385 K. It has to be noted that all the signals of natural TL in the samples under test are very weak and strongly noised, which makes their examination difficult.

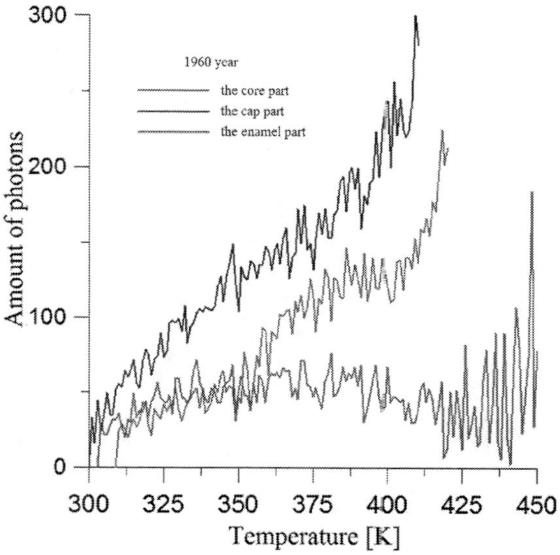

Figure 1. Natural TL intensity of porcelain insulator samples manufactured from C120 in 1960. [9]

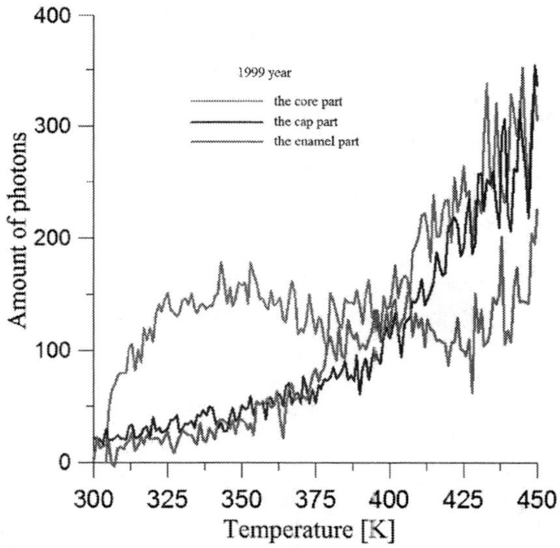

Figure 2. Natural TL intensity of porcelain insulator samples manufactured from C120 in 1999. [9]

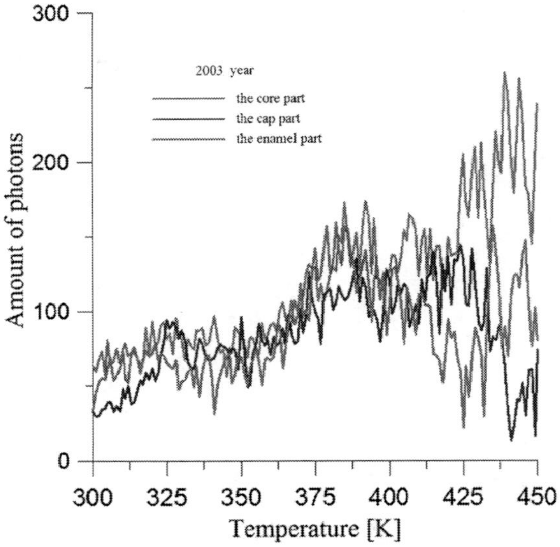

Figure 3. Natural TL intensity of porcelain insulator samples manufactured from C130 in 2003. [9]

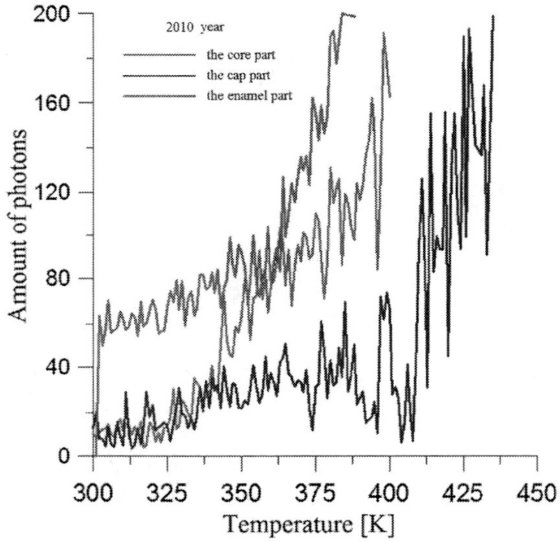

Figure 4. Natural TL intensity of porcelain insulator samples manufactured from C130 in 2010. [9]

Since the TL glowing curves in their natural form displayed low intensity of TL signal, to reduce background thermal emission in the high temperature, an optical filter was used in further measurements. The optical band-pass filter used was BG-39 manufactured by Schott. Taking into account the characteristics of the photomultiplier, the effective range of registration was about 330 ÷ 650 nm [8].

Figs. 5 ÷ 7 present the results of TL measurements of samples taken from cores, caps and enamel of porcelain insulators manufactured in 1963, 2005 and 2012. The result obtained from the 1963 samples is a very weak and noised TL signal. For a 2005 insulator core one peak can be seen at the temperature about 425 K, whereas for a cap two peaks appear at about 390 K and about 475 K. The 2012 cap sample displays one peak at the temperature about 450 K. All the natural TL signals obtained using the optical filter BG-39 are very weak and noised and for this reason they will not be analysed any further.

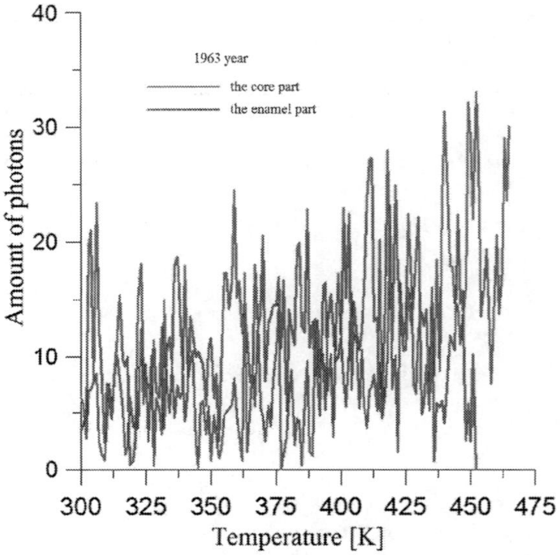

Figure 5. Natural TL intensity of porcelain insulator samples manufactured from C120 in 1963, measurement with optical filter BG-39. [9]

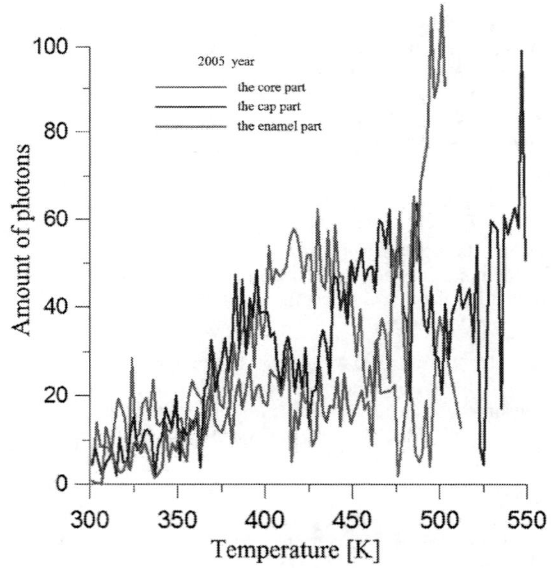

Figure 6. Natural TL intensity of porcelain insulator samples manufactured from C130 in 2005, measurement with optical filter BG-39. [9]

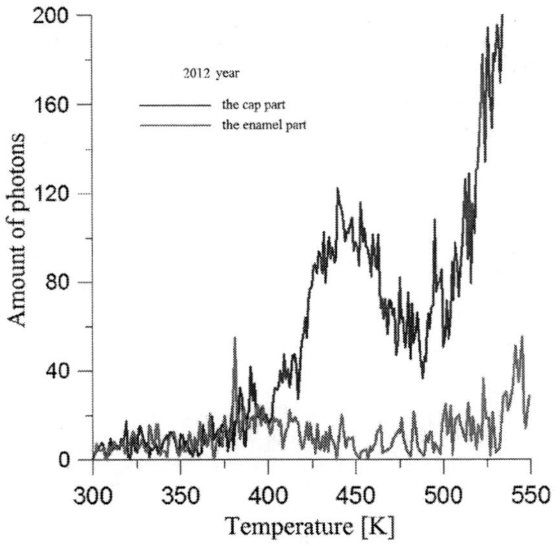

Figure 7. Natural TL intensity of porcelain insulator samples manufactured from C130 in 2012, measurement with optical filter BG-39. [9]

Even with the use of the optical filter BG-39 then, the results for both the enamel and the porcelain are similar to those obtained without the filter – the natural TL signal is still weak and noised.

At this point, it is worth mentioning that in the extended version of the TL tests on porcelain insulators in which they were additionally radiated by a beta minus source (the dose being about 260 Gy), the aim was successfully achieved.

V. SUMMARY

Luminescence phenomena and methods based upon them, including thermoluminescence are widely applied in a number of fields, such as dosimetry, dating, examination of defect structure and of the number of electron traps [10] in insulation materials and in semiconductors.

TL measurements conducted in the present study were aimed to determine the amount of natural radiation absorbed by porcelain insulators during their exploitation in power networks as well as to derive a formula for determining the production year for an insulator.

Due to the fact that the TL signal obtained in the measurements was very weak and significantly noised, no further mathematical analysis was applied to it. The TL emission intensity at the level of 0 ÷ 400 photons is insufficient – it has to be noted that results obtained from quality detectors are in the range 10K ÷ 10M photons.

It can be definitely stated that electrical porcelain used for manufacturing power insulators may be considered as a very week detector. The material is not suitable for determining low doses of ionizing radiation (in this case natural ionizing radiation during the exploitation time).

Even the period of 50 years is too short for the thermoluminescence method to be successfully used in dating these materials. The use of an optical filter for eliminating the background thermal radiation did not improve the signal in any significant measure.

It is worth noting that in the alternative TL research an additional radiation source was used to increase the radiation dose absorbed by electrical porcelain. In this case, the desired effect was achieved and the results have already been published.

ACKNOWLEDGMENT

THE PROJECT WAS FOUNDED BY THE NATIONAL SCIENCE CENTRE UNDER DECISION NUMBER DEC-2012/07/B/ST8/03890.

REFERENCES

[1] V. Ramzaev, A. Mishin, V. Golikov, T. Argunova, V. Ushnitski, A. Zhuravskaya, P. Sobakin, J. Brown, P. Strand, "Radioecological studies at the Kraton-3 underground nuclear explosion site in 1978 - 2007," Journal of Environmental Radioactivity - Vol. 100, Issue 12, 2009, pp. 1092-1099.

[2] V. Ramzaev, Y. Goksu, "Cumulative dose assessment using thermoluminescence properties of porcelain isolators as evidence of a severe radiation accident in the Republic of Sakha (Yakutia), Russia, 1978," Health Physics - Vol. 91, Nr 3, Lippincott Williams and Wilkins, Baltimore, 2006, pp. 263-269.

[3] R. Chen, S. W. S. McKeever, "Theory of Thermoluminescence and Related Phenomena," World Scientific, Singapore, 1997.

[4] V. Pagonis, R. Chen, G. Kits, "On the intrinsic accuracy and precision of luminescence dating techniques for fired ceramics," Journal of archaeological science, Vol. 38, Issue 7, 2011, pp. 1591-1602.

[5] W. D. Wang, "Study and progress of the thermoluminescence dating of the ancient pottery and porcelain," Science in China series e-technological sciences, Vol. 52, Issue 6, Beijing, 2009, pp. 1613-1640.

[6] I. Liritzis, "Surface dating by luminescence: An overview," Geochronometria, Vol. 38, Issue 3, Warsaw, 2011, pp. 292-302.

[7] T. Niewiadomski, "Dozymetria termoluminescencyjna w praktyce," Instytut Fizyki Jądrowej, Raport nr 1550/D, Wyd. 1, Kraków, 1991.

[8] R. Sobota, A. Mandowski, E. Mandowska, "Luminescencja izolatorów ceramicznych sieci energetycznych średnich napięć," - Elektronika - konstrukcje, technologie, zastosowania - 11/2012, Sigma - Not, Warszawa, 2012, pp. 109-111.

[9] R. Sobota, "Wykorzystanie zjawiska thermoluminescenceji w badaniach porcelany elektrotechnicznej izolatorów energetycznych do 24 kV," Rozprawa Doktorska, Politechnika Częstochowska, Instytut Elektroenergetyki, Częstochowa, 2014.

[10] R. Sobota, A. Mandowski, E. Mandowska, M. Kornatka, "Struktura pułapkowa izolatorów ceramicznych sieci energetycznych średnich napięć," Przegląd Elektrotechniczny, nr 9, Warszawa, 2015, pp. 121-124.

The Deconvolution Analysis of TL Curves for Current Ceramic Insulators

Rafał Sobota
Institute of Optoelectronics and Measurement Systems
Czestochowa University of Technology
Czestochowa, Poland
rsobota@el.pcz.czest.pl

Abstract—**The paper presents the deconvolution analysis method of thermoluminescence (TL) curves for porcelain insulators made of currently used ceramic mass called C130. Additionally, detailed characteristics of the electrical porcelain is given which is used for producing power line standing insulators in nowadays.**

Keywords: thermoluminescence; deconvolution; electrical porcelain; electrical insulator

I. Introduction

The ceramic material known as "C130" is currently widely used for manufacturing electrotechnical products of high mechanical strength and high durability and reliability, such as power line HV and LV insulators. support HV and MV insulators of high mechanical strength requirements, railways insulators and apparatus insulation of high mechanical and pressure strength requirements [1]. The manufacturing technology for C130 is the same as for other types of electric porcelain but the most important problem is ensuring optimum rheological properties, which can affect the crystalline structure of an insulator.

II. Current electrical porcelain

Since Poland joined the European Union, manufacturers of electrical porcelain began to produce electrical insulators only mass C130. High-aluminum electric porcelain C130 was developed in the 1960's to meet the demand for high mechanical strength material which could be used for producing shield insulators for high-voltage pneumatic switches requiring high bending resistance. Among all the types of electric porcelain, C130 has one of the best strength parameters and high aging resistance. This is mostly due to corundum content and also to the Al_2O_3:SiO_2 ratio in the chemical composition [2]. The contents of the particular components is as follows: 55% - Al_2O_3, 40% - SiO_2, 5% - other components (i.e., TiO_2, CaO, MgO, K_2O, Na_2O, 0,5% - Fe_2O_3) [3].

The presence of loam guarantees elasticity and high mechanical strength before firing the material. Many of the C130 components are non-elastic, such as ceramic Al_2O_3 and feldspars. There are not many fluxes used in the production of the mass but they are strong. Apart from potassium feldspar orthoclase, up to a few % of other fluxes are used, such as nepheline syenite (nepheline and alkali feldspar) and sodium feldspar (albite). These components reduce the amount of quartz relics in the insulator head. Those with smaller size particles get dissolved in the liquid feldspar during the firing process and other fluxes melt. The greatest problem is caused by SiO_2 grains, present in feldspars, with plate morphology. The problem is due to the fact that the grains interact with the mass only partially, mostly on the edges. After the firing, the grains are weakly integrated with the mass and can be a source of defects and cracks occurring in exploitation. Microcracks may appear both inside and in the outside layer from which they can expand to the matrix. Corundum should melt only to a limited extend during the firing and should remain located in the insulator head. This concerns mainly the α-corundum variant (> 95 %) and sodium oxide (< 0,5 %). [6]

Al_2O_3 occurs in the form of small, longitudinal grains of the size of a few µm, constituting 10 ÷ 25 % of the C130 mass. This improves mechanical and thermomechanical parameters of electric porcelain and provides dispersive and fibre strengthening by means of the crystalline phase. It is essential for the mechanical strength and durability of the mass that the crystalline phases are evenly distributed in the material. Their integration with the matrix as well as the size and separation of grains is also of great importance. In comparison to the electric porcelain C120, the strength margin with respect to the rated value is much greater. The mechanical strength of C130 is higher than that of C120 by about a half, and the life of the former is longer by several years. [7]

A large number of structural defects in electrical porcelain may indicate low quality insulator [8]. Since defects give rise to localized energy states within the band gap, it seems possible to develop a method that identifies these defects on the basis of registration of the light emitted during the radiative recombination of defects on a role of recombination centers. The purpose of this research is to gain knowledge about the structural defects of ceramic insulators grids medium voltages using the method of thermoluminescence (TL).

III. Additional information

The measurements of electrical porcelain C130 used for manufacturing power insulators were conducted using thermoluminescence, being a dosimetric method. A full characteristics of this phenomenon, as well as research methods in which it is used have been thoroughly described in the literature [1, 3, 8].

Since the insulator glaze is of highly complex chemical composition and it is not the subject proper of the present analysis, the TL measurement result of the glaze of the insulator manufactured in 2003 and 2010 are presented only in its raw form (Fig. 1, 2). The TL results obtained for electric porcelain are presented together with deconvolution analysis in the next part of the paper.

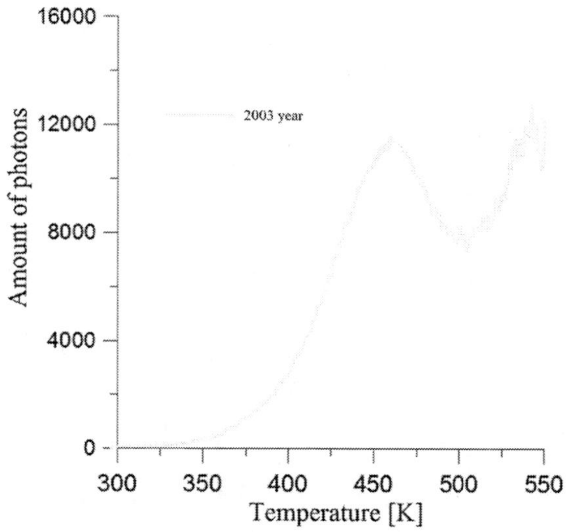

Figure 1. TL spectrum of radiated protective insulator glaze made in 2003. [9]

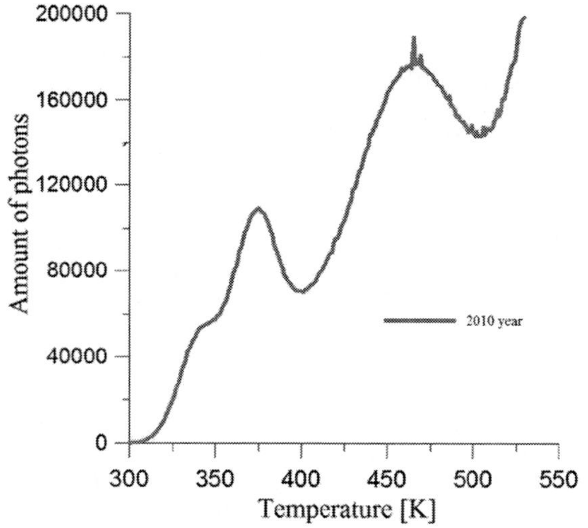

Figure 2. TL spectrum of radiated protective insulator glaze made in 2010. [9]

IV. Deconvolution analysis of C130 TL curves

TL curves, also referred to as real thermoluminescence spectra, were obtained by means of the linear interpolation method, which in this case involved subtracting the value of the thermal background of a sample from its glow intensity curve at a given temperature.

Received thermoluminescence curves for the electric porcelain C130 were subjected to numerical analysis based on the deconvolution method. This method involves numerical matching and decomposition of a complex curve into first-order elementary peaks (b = 1) TL. [4] It is typically used in dosimetry on the basis of the Randall - Wilkins equations, that is, first-order kinetics [5].

The analysis made it possible to model previously computed TL curves and to deconvolute them into elementary peaks, representing first-order kinetics solutions. The curves are deconvoluted by means of a nonlinear regression method, in which the sum of deviation squares is minimised [4]:

$$S = \frac{1}{N_p}\sum_{n=1}^{N_p}\left[y_n - J(x_n)\right]^2 \quad (1)$$

N_p represents the number of measuring points (x_n, y_n) taken into consideration when the function is being matched. First, the user introduces the initial values, which are subsequently modified by establishing a vector in the parameter space with the dimension 4Np and minimising the sum S of squares along this vector. The length of the vector is determined as the variable *Krok* (Polish for "Step" - the names are given as in the original application). If a number of consecutive iterations are successful, an attempt is made to increase the step length by multiplying the variable *Krok* by *Czynnik Zmiany Kroku* (Step Variation Factor). If *Krok* is too long and the sum of squared deviations for the newly established values is larger than previously obtained parameters, the attempt is aborted. The variable *Krok* is divided by *Czynnik Zmiany Kroku* and the computations are repeated. [9]

The computations and the TL curve modelling are performed until the quality of the match characterised by the parameter *FOM* (*Figure Of Merit*) (2) reaches a desired value. [5]

$$FOM = \frac{\sum_{i=1}^{N_p}\left[y(T_i) - y_{fit}(T_i)\right]}{\sum_{i=1}^{N_p} y(T_i)} \cdot 100\% \quad (2)$$

In equation (2) N_p is the number of points on the curve and y and y_{fit} are the TL curve and the matched curve, respectively. The TL curve decomposition can be considered valid when the match coefficient *FOM* is lesser than 2 % (the stringent case) or up to 3 % (the less stringent case). [5] The deconvolution analysis below has been carried out for the spectra of radiated electric porcelain C130.

Figs. 3 ÷ 4 present the numerical matching of the TL curve to the results obtained for C130 manufactured in 2003. It was possible to decompose the TL spectra obtained both for the insulator core and cap samples into five elementary peaks. The peaks occur in different temperature ranges for the two samples.

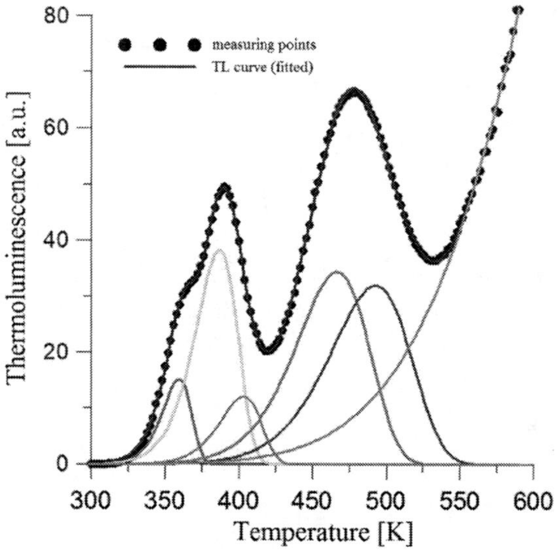

Figure 3. Numerical matching and deconvolution of the TL curve for a sample collected from a porcelain insulator core made of the C130 ceramic mass in 2003. [9]

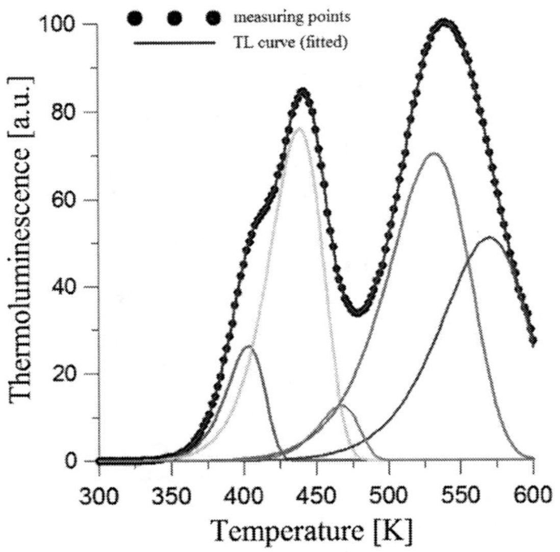

Figure 4. Numerical matching and deconvolution of the TL curve for a sample collected from a porcelain insulator cap made of the C130 ceramic mass in 2003. [9]

Figs. 5 ÷ 6 present deconvolution results for the electric porcelain C130 manufactured in 2005. The TL spectrum obtained for the insulator core samples were decomposed into four elementary peaks. The TL spectrum for the cap samples was deconvoluted into five peaks. The peaks obtained for the two porcelain samples occur in different temperature ranges.

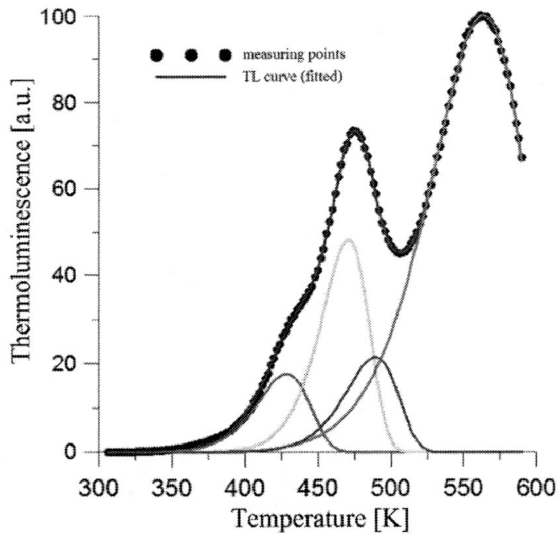

Figure 5. Numerical matching and deconvolution of the TL curve for a sample collected from a porcelain insulator core made of the C130 ceramic mass in 2005. [9]

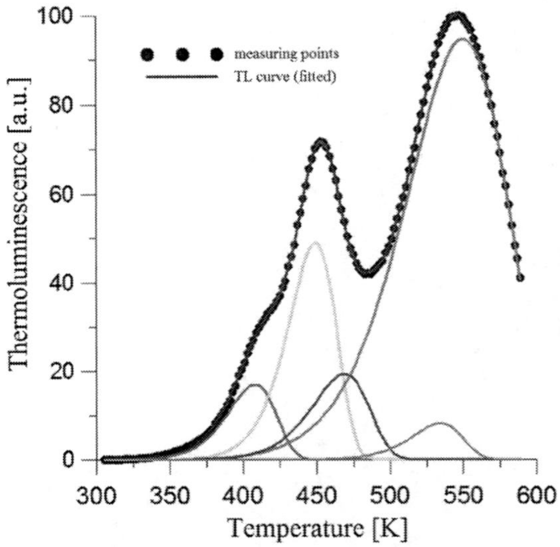

Figure 6. Numerical matching and deconvolution of the TL curve for a sample collected from a porcelain insulator cap made of the C130 ceramic mass in 2005. [9]

Figs. 7 ÷ 8 present numerical matching of the TL curve to the results obtained for the samples of electric porcelain C130 produced in 2010. If was possible to deconvolute the TL spectrum into five elementary peaks both for the core and cap samples. For both samples of porcelain tested the elementary peaks occur in the similar temperature ranges: 350 K, about 375 K, about 400 K, about 450 K and about 475 K.

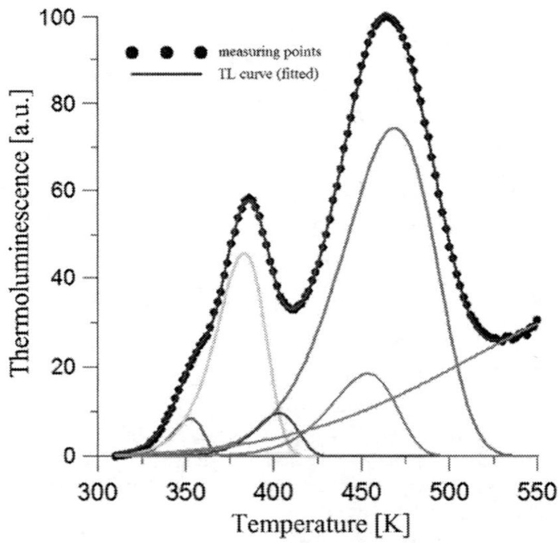

Figure 7. Numerical matching and deconvolution of the TL curve for a sample collected from a porcelain insulator core made of the C130 ceramic mass in 2010. [9]

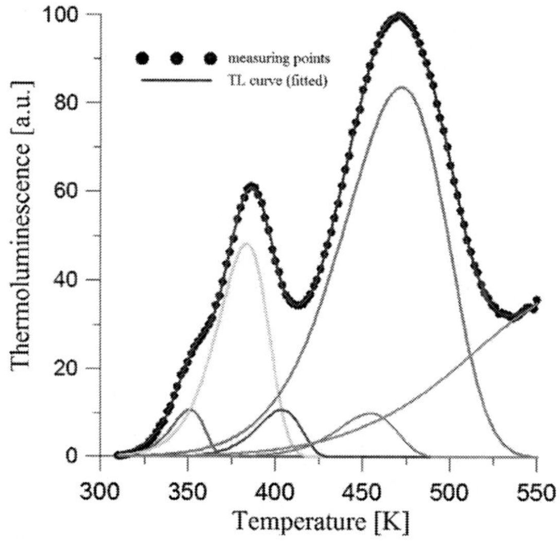

Figure 8. Numerical matching and deconvolution of the TL curve for a sample collected from a porcelain insulator cap made of the C130 ceramic mass in 2010. [9]

Fig. 9 presents the deconvolution results for a TL curve obtained for the samples of electric porcelain C130 produced in 2012. The spectrum was decomposed into five elementary peaks.

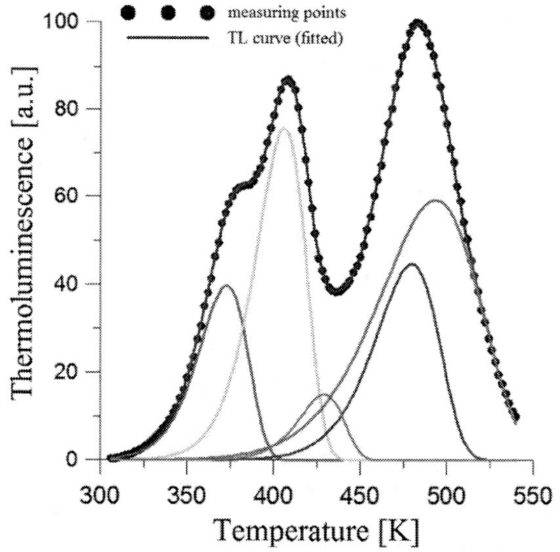

Figure 9. Numerical matching and deconvolution of the TL curve for a sample collected from a porcelain insulator cap made of the C130 ceramic mass in 2012. [9]

All the deconvoluted TL spectra obtained in the investigation of electric porcelain C130 samples displayed five elementary peaks.

In Figs. 3, 7 and 8 one more peak can be seen (marked in orange), representing background thermal radiation. The peak should be disregarded, it is however necessary for carrying out the numerical deconvolution analysis.

V. SUMMARY

The thermoluminescence method with deconvolution analysis of TL curves offers a new diagnostic possibilities for testing ceramic insulators structure and number of defects in a electrical porcelain. Due to the novelty of the measurement method, it is necessary to develop in-depth analyses of the results obtained.

Using the deconvolution method, it was possible to decompose the TL curves obtained for each sample of the C130 porcelain into 5 elementary peaks. It is a very important information in dosimetry measurements.

The second important information is that in the majority of cases there were two dominant peaks (green and pink), which can indicate the presence of two dominant components of the material under scrutiny: Al_2O_3 and SiO_2. [3]

The obtained results demonstrate that the quality and structure of currently produced insulators are stable, since the TL measurement results are repeated across a wide range of samples collected from one production batch. This observation has to be confirmed on the basis of a larger sample of the material and such investigation will be conducted in the nearest future.

ACKNOWLEDGMENT

THE PROJECT WAS FOUNDED BY THE NATIONAL SCIENCE CENTRE UNDER DECISION NUMBER DEC-2012/07/B/ST8/03890.

REFERENCES

[1] R. Sobota, "Thermoluminescence phenomena of electrical porcelain," Proceedings of the 8th International Scientific Symposium on Electrical Power Engineering - ELEKTROENERGETIKA 2015, September 16-18, Stará Lesná, Slovak Republic, 2015, pp. 311-314.

[2] J.A.C. Broekaert, T. Graule, H. Jenett, G. Tölg, P. Tschöpel, "Analysis of advanced ceramics and their basic products," Analytische Chemie, Vol. 332, Issue 8, Springer -Verlag, 1989, pp. 825-838.

[3] A. Mandowski, R. Sobota, "Possible relation between thermoluminescence and chemical composition of electrical porcelain," Proceedings of the 8th International Scientific Symposium on Electrical Power Engineering - ELEKTROENERGETIKA 2015, September 16-18, Stará Lesná, Slovak Republic, 2015, pp. 315-318.

[4] A. Mandowski, E. Mandowska, J. Świątek, "Some simple methods for the analysis of TL glow curves," Solid state crystals in optoelectronics and semiconductor technology, Vol. 3179, Bellingham, 1997, pp. 294-297.

[5] A. Mandowski, J. Świątek, "On the numerical deconvolution of thermoluminescence spectra," Single crystal growth, characterization and application, Vol. 3724, Bellingham 1999, pp. 203-208.

[6] Z. Pohl, "Izolatory Elektroenergetyczne,' Oficyna Wydawnicza Politechniki Wrocławskiej, Wrocław, 1995.

[7] Z. Celiński, "Materiałoznawstwo elektrotechniczne," Oficyna Wydawnicza Politechniki Warszawskiej, Warszawa, 2011.

[8] R. Sobota, A. Mandowski, E. Mandowska, M. Kornatka, "Struktura pułapkowa izolatorów ceramicznych sieci energetycznych średnich napięć," Przegląd Elektrotechniczny, nr 9, Warszawa, 2015, pp. 121-124.

[9] R. Sobota, "Wykorzystanie zjawiska thermoluminescenceji w badaniach porcelany elektrotechnicznej izolatorów energetycznych do 24 kV," Rozprawa Doktorska, Politechnika Częstochowska, Instytut Elektroenergetyki, Częstochowa, 2014.

The Influence of Voltage on Surface Resistivity of Outdoor Insulator

Roman Cimbala
Department of Electric Power Engineering, Faculty of Electrical Engineering and Informatics
Technical University in Košice
Košice, Slovak Republic
roman.cimbala@tuke.sk

Abstract — **This article identifies the calculation of the parameters for an expanded equivalent model of insulation system based on Maxwell - Wagner theory for ceramic outdoor insulator. Debye independent polarization processes were under consideration. The values of a steady component of elementary polarization currents and time of responses were estimated. The main focus was aimed on voltage varied from 1 to 1000 V influence on of a steady element of charging current represented insulation resistivity in infinity time. Comparison between results from the time and frequency domain was carried out.**

Keywords: polarization; charging current; dielectric model; test voltage; steady element

I. Introduction

The most commonly used material in ceramic group is porcelain, which is used as insulating material for insulators and power equipment, respectively in high frequency technology for the production of capacitors. These materials are resistant to high temperatures. [1]. The impedance relaxation spectroscopy IRS also includes dielectric relaxation spectroscopy (DRS), which occupies a special place among many modern non-destructive methods used for physical and chemical analyses of materials in many fields (electrotechnics, physics, chemistry, material engineering, and biomedicine) because it allows the exploration of dielectric relaxation Processors in an extremely wide range of 10^4 to 10^{-12} sec. The principle of the IRS method is to investigate the reflections of the microphysical mechanisms of the test substance on external stimuli. In our case, it is the indirect measurement of the polarization of polar or polarizable materials by means of a response generated by changing the external electric field. In general, we can characterize the dielectric by means of unidirectional conductivity σ_0, relative ε_∞ and dielectric response f_r. [2] [3].

These polarizing mechanisms can be measured by alternating and DC fields, dividing impedance spectroscopy into frequency domain measurement methods where polarization mechanisms are tracked in the alternating electric field and methods measured in the time domain where the polarizing mechanisms are monitored in a DC electric field. [2].

In the time domain we can monitor the polarization and depolarization current of the dielectric, which was inserted into the DC electric field, respectively. The time course of the dielectric current is recorded for a step change of voltage. If the dielectric can be described as a linear system, the measured values can be converted from time to frequency domain. This is advantageous because the measured values of the dielectric studied are easier and easier to compare in the frequency domain. The IRS's interest is to obtain a complex permeability of the substance on the frequency (the so-called dielectric spectrum) and on other variables whose properties are significantly influenced. Here we can radiate the ambient temperature with the possible combination of the temperature aging of the sample and the relative humidity of the surroundings that influences the sample [1] [2] especially liquids [4].

II. Worksurface

The degree of protection afforded by a worksurface is strongly related to the time needed to discharge an object. In practice, some form of resistance value is commonly given as an indication of the effectiveness of the worksurface. Strictly speaking, this description is incomplete since discharge time depends on several other factors such as the effective capacitance of the worksurface, contact resistance and the actual discharge path. It can be one cause of earth fault when breakdown appears according to surface resistance decreasing [5].

However, the capacitance does not usually vary as much as resistance. The other effects are very dependent on the individual situation. As a result, resistance seems the best single predictor of performance of ESD-protective worksurfaces. The standard [6] relies on resistive measurements, utilizing standard instruments, to provide a means of evaluation materials or installed worksurfaces.

III. Surface Resistivity and Surface Resistance Measurements

According to [6] concepts of surface resistance and surface resistivity can be sometimes confusing. Definitions of both terms can be found in many books and standards [7–11].

$$R_s = \frac{U}{I_s} \tag{1}$$

$$\rho_s = \frac{U/L}{I_s/D} \tag{2}$$

Surface resistance, $\boldsymbol{R_s}$, is defined in all of the aforementioned literature sources as the ratio of a DC voltage $\boldsymbol{U}$ to the current, $\boldsymbol{I_s}$ flowing between two electrodes of specified configuration that are in contact with the same side of a material under test (Figure 1)

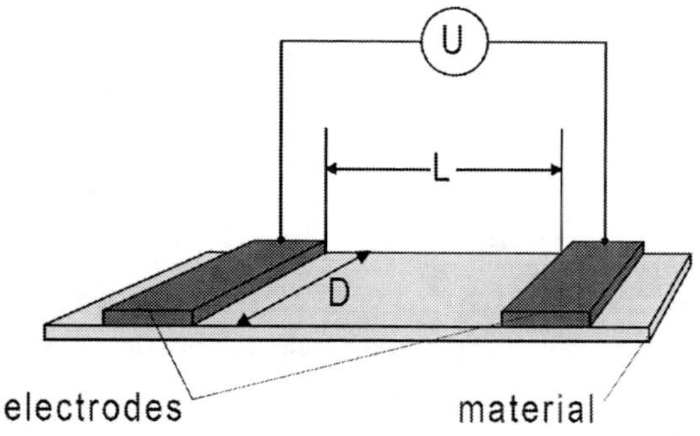

Figure 1. Basic setup for surface resistance and surface resistivity measurement [6]

Surface resistivity is a property of a material. Theoretically, it should remain constant regardless of the method and configuration of the electrodes used for the surface resistivity measurement. A result of the surface resistance measurement depends on both the material and the geometry of the electrodes used in the measurement. The physical unit of surface resistivity is Ohm (Ω). The legitimate unit of the surface resistance is also Ohm. Because of that surface resistivity and the surface resistance are often mixed up. In order to differentiate between the two, surface resistivity is often expressed also in Ωm^{-2} which is not a valid unit from the dimensional analysis point of view.

It is possible to establish a relationship between the surface resistance and surface resistivity for any electrode configuration.

The flow density, be it electric current, is the amount of flow passing through a unit area of the sample of the material. The surface area is perpendicular to the direction of the flowing current.

When the voltage U is kept constant, the current density for the thin and the thick bar is the same. The electric current density is often expressed by:

$$J = \frac{I}{S} \tag{3}$$

where I is the current and S is the surface area, J measured in [A/m2].

Surface current density is the next concept helpful in understanding the relationship between the surface resistance and surface resistivity. Consider Figure 1, where both electrodes are on the same side of the material. It is assumed that electric current flows on the surface of the material only. In reality, this is not exactly true. There is always a portion of that current flowing through the bulk of the material [12]. However, in order to make it possible to compare surface properties of various materials, it had been presumed that the surface current flows through infinitesimally thin surface layer. This layer is so thin, that the thickness of it can be neglected. Surface current density J_s is therefore defined as:

$$J_s = \frac{I}{D} \tag{4}$$

where D is a width of the electrode.

The relationship between surface resistivity and the surface resistance for a concentric ring probe geometry can be found by defining a surface current density in the area between the rings. Knowing the surface current density, it is possible to find an electric field intensity between the electrode rings (Figure 2).

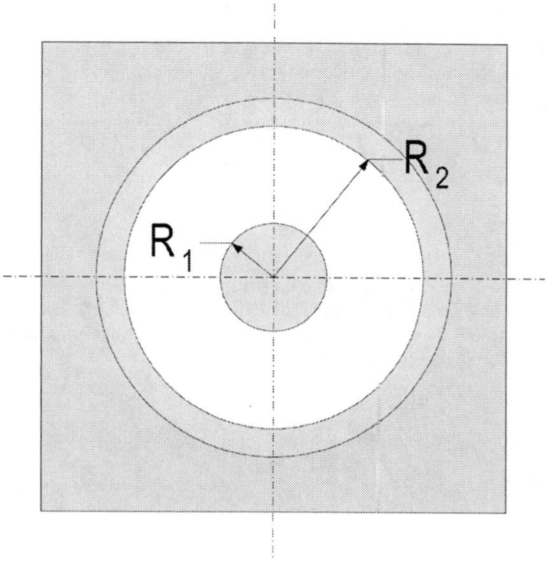

Figure 2. Surface resistance and surface resistivity measurement configuration of concentric ring electrodes [6]

where R_1 outer radius of the centre electrode, R_2 inner radius of the outer ring electrode,

The surface current density J_s for a concentric ring configuration is determined as:

$$J_s = \frac{I_s}{2.\pi.r} \tag{5}$$

where the radius r varies from R_1 to R_2. It is important to remember that when testing the surface resistivity (or resistance) of any material, it is assumed that all the currents flow between electrodes along the surface and do not penetrate into the bulk of the material. In order to ensure that the surface currents are measured properly, some more advanced techniques for surface resistivity measurements have been developed [9][10][13]. The Ohm's law describes the relationship between a current density J and an electric field intensity E. It is also valid for the surface currents:

$$J_s = \frac{E}{\rho_s} \tag{6}$$

Therefore, it is possible to find electric field between the concentric rings by solving the following dependency (using equation 5 and 6):

$$E = \frac{\rho_s I_s}{2.\pi.r} \tag{7}$$

The voltage between the electrodes can be found by integrating the electric field E from R_1 to R_2:

$$U_{R1,R2} = \int_{R2}^{R1} E dr = \int_{R2}^{R1} \frac{\rho_s I_s}{2.\pi.r} dr = \frac{\rho_s I_s}{2.\pi} \int_{R2}^{R1} \frac{1}{r} dr$$

$$U_{R1,R2} = \frac{\rho_s I_s}{2.\pi} \ln \frac{R_2}{R_1} \quad (8)$$

After rearrangements, the surface resistivity is related to the surface resistance by a constant that depends on the geometry of the electrodes only [6]:

$$\rho_s = R_s \frac{2\pi}{\ln \frac{R_2}{R_1}} \quad (9)$$

IV. DC CURRENT MEASUREMENT ON DIELECTRICS

Not so far ago the polarization spectrum of insulation material was scanned only by polarization indices. They are described very easy as a ratio of observing absorptive or resorptive currents at the determined time.

$$p_1 = \frac{I_{15}}{I_{60}} \quad (10)$$

$$p_{10} = \frac{I_{60}}{I_{600}} \quad (11)$$

where I_{15} is magnitude of observing current at 15s, I_{60} is magnitude of observing current at 60 s and I_{600} is magnitude of observing current at 600 s

According to Debye's relaxation theory the total current is a macroscopic event of independent polarization processes. There are represented as exponential decreasing current curves. Then it is possible to write

$$i(t) = i_v + \sum_{i=1}^{n} I_{mi} \exp\left(\frac{-t}{\tau_i}\right)$$

(12)

$$i_v = \frac{U_0}{R_0} \quad (13)$$

$$I_{mi} = \frac{U_i}{R_i} \quad (14)$$

$$\tau_i = R_{i.} C_i \quad (15)$$

where U is applied DC voltage, R_i is DC insulation resistance,

I_{mi} is maximal amplitude of i-th elementary Debye's current, τ_I is time relaxation constant of i-th elementary Debye's current.

The response time constant of the geometrical capacity C_0 current is about 10^{-12} s and it is possible to neglect it practically. In the case of dry insulation system the steady current is reduced in value than absorption current. In the case of material with moisture the steady current component appears in calculation data. On this base it is possible to distinguish low insulation resistance caused by humidity or by ageing.

The equivalent model of the insulating material at Fig.3 is based on *n* independent Debye's polarization processes.

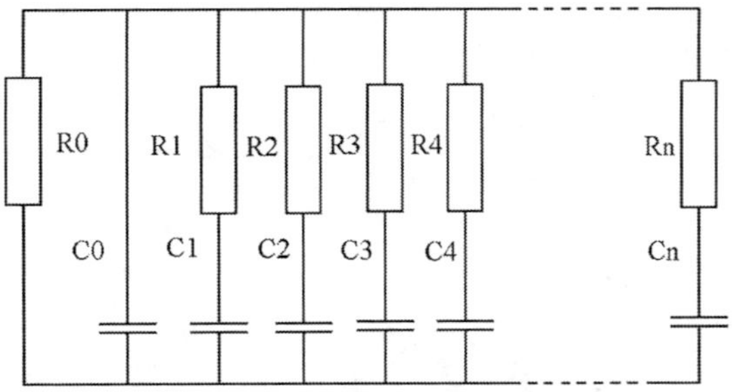

Figure 3. The equivalent model of dielectric material

V. EXPERIMENT

Calculating the stabilization times of Debye independent polarization processes is one way to test the state of the insulation system. Another way to observe the polarization spectrum is transforming the current response to the frequency domain.

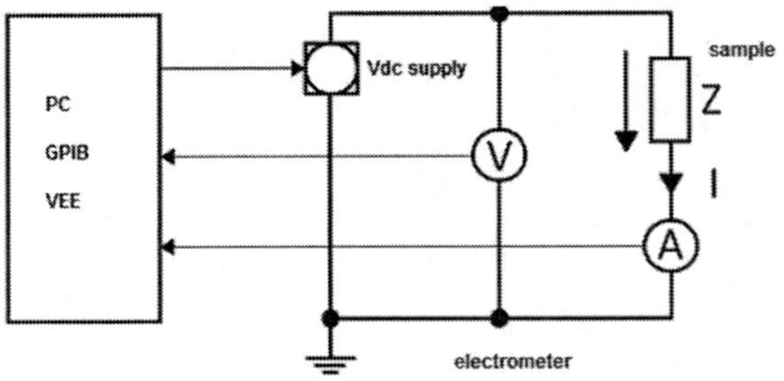

Figure 4. The measuring system with electrometer Keithley 6517B

The final decision was measurement of real object. The new outdoor insulator was examined at a temperature of 20 °C. The sample was made from glaze porcelain 23 kV voltage operating level. It was charged with direct voltage at 1, 10, 100 and 1000 V and the current response was observed for a period of 1000 s for charging. In the level of 1 V the disturbances were so high that it was impossible to carry out any analysis of data on it.

The main idea was observing the influence of voltage on a steady element in the polarization current. Data were analysed with the presence of the DC component that represents surface resistance.

Measurement set of instrument is shown in Fig 5.

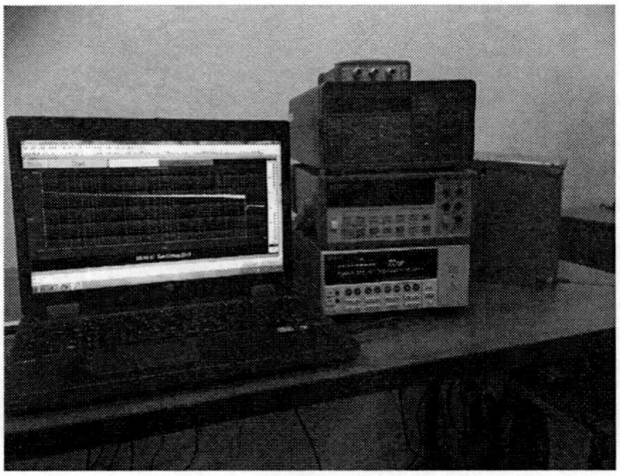

Figure 5. The measurement set of instruments

Two developments in elementary currents were performed with Matlab. Number of elements was upto 7. The first calculation was made for exponential currents only – see Ie. The second calculation was performed for exponential and steady element together – see I_v. There are Table 1 for 10 V, Table 2 for 100 V and Table 3 for 1000 V of test voltage.

There is measured outdoor insulator in Figure 6. Radius R_1 was 0.103 m and radius R_2 was 0.13 m. There are measured currents during 1 000 s charging in Figure 7. Test voltages were 10, 100 and 1000 V. There are measured current during 1 000 s discharging in Figure 8. Test voltages were 1, 10, 100 and 1000 V.

Figure 6. The outdoor insulator

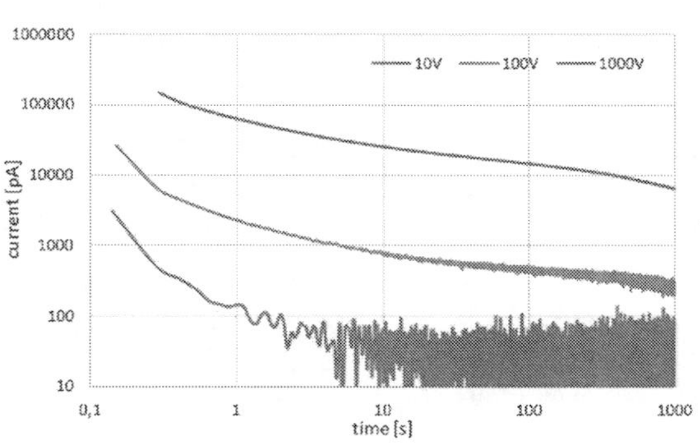

Figure 7. The measured currents during charging

Table 1

Degree element	10 V: Ie_i [pA]	τe_i [s]	Iv_i [pA]	τv_i [s]
1	14845	0,09	15914	0,08
			31	
2	15920	0,08	29011	0,06
	31	2,8E+15	286	1,1
			31	
3	15789	0,09	61553	0,04
	2,1E+11	-3E+13	697	0,36
	-2E+11	2,5E+16	53	5,6
			31	
4	15789	0,09	34813	0,06
	2,1E+11	-3E+13	379	0,73
	-2E+11	2,5E+16	-27289	98,21
	0	1,2E+17	27298	98,23
			34	
5	15789	0,09	34963	0,06
	2,1E+11	-3E+13	382	0,73
	-2E+11	2,5E+16	-8E+10	1,3E+07
	0	1,2E+17	1E+11	1,6E+07
	0	6,2E+17	32	53,77
			-2E+10	
6	15789	0,09	55784	0,05
	2,1E+11	-3E+13	641	0,39
	-2E+11	2,5E+16	-9E+09	7282777
	0	1,2E+17	4,7E+10	3,7E+07
	0	6,2E+17	46	6,66
	0	3,1E+18	28	118,37
			-4E+10	
7	15789	0,09	55827	0,05
	2,1E+11	-3E+13	643	0,39
	-2E+11	2,5E+16	-7E+09	7600865
	0	1,2E+17	5,1E+10	5,2E+07
	0	6,2E+17	46	6,7
	0	3,1E+18	1318	190,35
	0	1,6E+19	-1307	194,24
			-4E+10	

Table 2

Degree element	100 V: Ie_i [pA]	τe_i [s]	Iv_i [pA]	τv_i [s]
1	84623	0,13	91837	0,12
			364	
2	4,5E+17	0	109566	0,1
	6602	1,22	813	32,16
			338	
3	7E+10	0,01	172058	0,07
	4897	1,41	2651	2,56

	513	1298,8	308	254,51
			281	
4	7,2E+07	0,02	280714	0,06
	9183	0,42	4625	0,74
	929	10,86	774	10,45
	478	1543,62	273	414,55
			251	
5	1551204	0,03	388403	0,05
	10760	0,28	6173	0,44
	1557	3,25	1230	3,78
	278	46,1	264	36,22
	448	1836,86	279	727,22
			197	
6	433187	0,05	625952	0,04
	6736	0,34	9105	0,24
	1469	1,88	2002	1,25
	517	9,89	686	7,15
	167	83,09	186	58,24
	435	1991,92	305	1007,3
			153	
7	826967	0,04	672671	0,04
	11419	0,2	9456	0,23
	2438	0,93	2123	1,04
	798	4,98	746	5,21
	201	23,8	192	22,05
	120	113,59	126	102,02
	429	2065,77	389	1743,17
			45	

Table 3

Degree element	1000 V			
	Ie_i [pA]	τe_i [s]	Iv_i [pA]	τv_i [s]
1	18446	745,58	18277	122,95
			8071	
2	78512	2,87	97523	1,71
	16885	864,64	13745	222,63
			7118	
3	138290	0,75	166980	0,5
	17951	24,7	24696	10,26
	15018	1066,86	11839	335,1
			6213	
4	227508	0,3	297036	0,21
	37884	3,76	48971	2,11
	9729	62,26	12425	24,7
	13872	1234,16	11163	427,82
			5547	
5	354812	0,18	519495	0,13
	55588	1,56	67486	0,91
	14471	14,4	20530	5,54
	6590	141,39	8335	39,19

	12378	1510,1	10897	485,38
			5170	
6	579434	0,12	867181	0,1
	70427	0,8	81214	0,51
	22727	4,41	31456	2,23
	9022	29,06	11456	10,65
	6614	248,25	6299	51,35
	10504	1992,44	10763	516,27
			4985	
7	844808	0,1	955994	0,09
	80173	0,52	83106	0,43
	31029	2,25	34856	1,58
	11376	10,71	14423	6,05
	6281	51,33	7332	27,46
	10692	512,64	3059	97,6
	5063	104808	10418	575,62
			4681	

Resistances for 15, 60 and 600 sec were calculated also. They were used to determine of 1 and 10 minute polarization indices. Results are in Table 4.

Table 4

	1000 V	100 V	10 V
R15[GΩ]	43,73	143,47202	21277
R60[GΩ]	60,44	186,56716	55556
R600[GΩ]	120,08	320,51282	83333
P1[-]	1,38	1,30	2,61
P10[-]	1,99	1,72	1,50

The steady elements of charging current representing DC resistance were calculated for voltage levels 10, 100 and 1000 V and they are changing according to the number of degree element is shown in Table 5.

Table 5

Degree element	1000 V	100 V	10 V
	R_{1000}[GΩ]	R_{100}[GΩ]	R_{10}[GΩ]
1	124	275	323
2	140	296	323
3	161	356	323
4	180	398	294
5	193	508	
6	201	654	
7	214	2222	

According to equation 9 the surface resistivity for differential voltage levels was calculated. The results are in Table 6 and they are shown in Figure 8.

Table 6

Degree element	1000 V	100 V	10 V
	ρ_{1000} [TΩm^{-2}]	ρ_{100} [TΩm^{-2}]	ρ_{10} [TΩm^{-2}]
1	3,34	7,41	8,71
2	3,79	7,98	8,71
3	4,34	9,60	8,71
4	4,87	10,75	7,94
5	5,22	13,70	
6	5,41	17,64	
7	5,77	59,98	

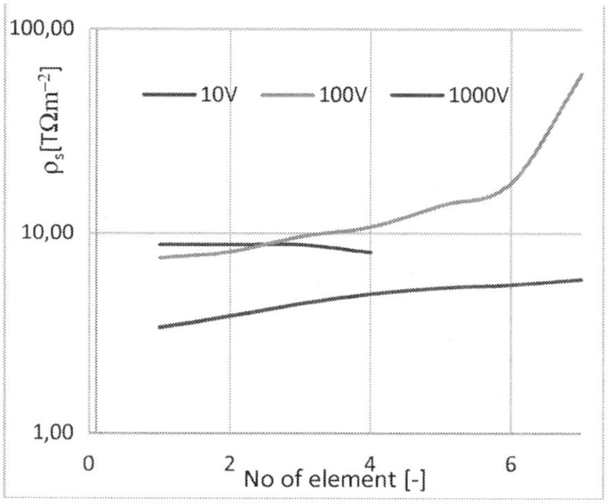

Figure 8. Surface resistivity

VI. Conclusion

This paper deals with calculating steady elements of dielectric material equivalent models for surface resistivity determination. The basic principles were described.

The real object has a steady element of current which is DC resistance in macroscopic point of view. As it was mentioned, it is impossible to differentiate between surface and volume current during measurement. This is evident the element current of Maxwell-Wagner equivalent model is not linearly dependent on the applied voltage. Magnitude of 1000 V has to be used to excite all polarization processes and is the least dependent on the stage of development. There are voltage nonlinear dielectric phenomena on them. Additionally, it is necessary to calculate steady element calculation for time as long as the elements have a real physical explanation. It is the case of 10 V measurement especially. Negative value of elementary current or time higher than 10^{+6} sec, if measuring time was 1000 sec, is out of sense.

As we compare all results, the higher test voltage, the less dependence of resistivity on it. The higher stage of element development has to be considered. It is the case of 1000 V test voltage and 5.77 surface resistivity TΩm^{-2}.

Knowledge of the component reliability parameters such as the outdoor isolator undoubtedly in power networks is necessary for the reliability computation and also for reliability-oriented maintenance system [14] [15]. Measurement of quality parameters as a steady current element which is part of current – voltage characteristics one of them [16].

Acknowledgment

This paper was developed with support of operating program Research and development for the project: University Science Park Technicom for innovative applications with support of knowledge technologies, ITMS: 26220220182, co-financed from European funds and Scientific Grant Agency of Slovak Republic and Slovak Academy of Science for funding of experimental works in the frame of VEGA No. 1/0311/15 and 2/0141/16. This work was supported by the Slovak Research and Development Agency under the contract No. APVV-15-0438

References

[1] I. Kolcunová. *Diagnostika elektrických strojov*, TU Košice, 2006, ISBN 80-8073-550-6.

[2] M. Farahani, H. Borsi and E. Gockenbach. "Dielectricspectroscopy in Time and Frequency Domain on Insulation System of High Voltage Rotating Machines." In *8th IEEE International Conference on Solid Dielektrics (ICSD)*. Toulose, France, 2004.

[3] S. W. Zaengl. *Dielectric spectroscopy in time and frequency domain for HV power equipment, Part I: Theoretical Considerations*. Zurich: Swiss federal institute of technology, 2001

[4] R., Cimbala, I. Kolcunova, L. Dedinska and V.Čačková. "Thermal Dependency of Dielectric Properties of Liquid Insulating Materials",*11th International Scientific Conference on Electric Power Engineering*. 2010 Brno, Pages: 743-747

[5] D. Topolánek, M. Lehtonen, M. Adzman, P. Toman. "Earth fault location based on evaluation of voltage sag at secondary side of medium voltage/ low voltage transformers". IET GENER TRANSM DIS, 2015, roč.9, č.14, s.2069-2077. ISSN: 1751-8687.

[6] W.A. Maryniak, T. Uehara, M.A. Noras. "Surface Resistivity and Surface Resistance Measurements Using a Concentric Ring Probe". Technique Trek Number 1005

[7] An American National Standard Approved July 17, 2006 Electrostatic Discharge Association 7900 Turin Road, Bldg. 3 Rome, NY 13440

[8] I. Kolcunová, M. Hrinko, J. Kurimský, "Short-term thermal stress of high voltage coil", Przegląd Elektrotechniczny. Vol. 89, no. 11 (2013), p. 16-19.

[9] ASTM Standard D 257-99. Standard test methods for D-C resistance or conductance of insulating materials, 1999.

[10] ESD STM 11.11-2001 Standard. Surface resistance measurement of static dissipative planar materials, 2001.

[11] M.B. Heaney. *The Measurement, Instrumentation and Sensors Handbook*, chapter "Electrical Conductivity and Resistivity". CRC Press, 1999.

[12] J. Balogh, J. Petráš, J. Džmura. "Measurement of polypropylene rope electrical strength". In: EPE 2014 : proccedings of the 2014 15th International Scientific Conference on Electric Power Engineering : May 12-14, 2014, Brno. - Brno : University of Technology, 2014 P. 427-430. - ISBN 978-1-4799-3806-3

[13] IEC 61340-5-1 Standard. Electrostatics - part 5-1: Protection of electronic devices from electrostatic phenomena - general requirements, 1998.

[14] Z. Martínek, T. Klor, J. Holý: "Reliability of the electrical power system", 15-th International Scientific Conference Electric Power Engineering 2014, May 12-14, 2014 Hotel SANTON, Brno, Czech Republic, ISBN978-1-4799-3807-0 2014 IEEE

[15] R. Goňo, M. Krátky, S. Rusek, "Reliability of Distribution Network Components Based on Failure Databases", AASRI Conference on Power nad Energy Systems, Book Series: AASRI Procedia Volume: 2 Pages: 75-80 DOI: 10.1016/j.aasri.2012.09.017 Published: 2012

[16] J. Kurimský, B. Dolnik, M. Kolcun. "Unipolar characteristics of ZnO ceramics". J. Electrost. 71, 418–421.

Influence of contamination on the integral surface conductivity of insulators

Iraida Kolcunová
Department of Electric Power Engineering, Technical University of Kosice, Slovakia
iraida.kolcunova@tuke.sk

Jaroslav Briančin, Martin Fabián
Institute of Geotechnics, Slovak Academy of Sciences, Kosice, Slovakia
briancin@saske.sk, fabian@saske.sk

Pavol Csáji
Kosice, Slovakia
palbarca@azet.sk

Abstract -**Outdoor insulation of a high voltage equipment is exposed to a contamination and a sporadic atmospheric moistening. It is well known that dry contamination of the surface of the high voltage insulation has a very limited influence on its electrical breakdown strength. However, even a moderate moistening of the contaminated surface layer considerably decreases the insulating level of the overhead line. In a critical case the insulating level of the overhead line may be lower than the operating voltage which may lead to a flashover. From the perspective of the electricity supplier, determining the precise level of insulator contamination is important in order to achieve continuous supply of electricity and to determine the optimal insulation of the overhead line. In this paper we describe the methodology for measuring the leakage current of contaminated insulators. We also describe an appropriate method for moistening of the insulators. These methods are based on analysis of the natural build-up of the insulator contamination, including its moistening in the critical climatic conditions.**

Keywords: insulator, conductivity, leakage current, pollution, wetting

I. Introduction

The isolator is one of the important elements of the transmission system. It electrically isolates and mechanically joins two points of the different potential. During operation the insulator is permanently exposed to the weather conditions. In recent years, along with the rapid development of the industry and the growth of the economy, the air quality deteriorates. Due to the surface contamination of the high-voltage insulators, the surface flashover occurs, which can cause extensive power outages [1, 2]. Recently, many studies have been conducted on the topic of the occurrence of flashovers on the surface of polluted insulators [3]. Experiments have shown that the layer of the accumulated pollution on the surface of the operating insulators is always uneven. The value of the flashover voltage is influenced by the uneven wetness of the top and the bottom of the insulator surface. [4, 5].

When studying the contamination of the external insulation, special attention should be paid to the pollution of the troposphere. The main contributor to the pollution of the troposphere is the industry. Lately, we can also see an increase in the chemical pollution. This has been mainly contributed to the increase in the agriculture production, which heavily rely on chemical spraying on fields. Additionally, intensified spraying of meadows and forests using aircrafts and helicopters can be observed. When the spraying is being applied, the wind can carry small particles which are then condensated on the external surface of the insulators. These chemical compounds are very well soluble in water, while they contain large amount of ion-forming substances. Pollutants in the atmosphere can be found in all three states: solid, liquid (aerosols), and gas.

II. Contamination of Outdoor Insulators

A. *Properties and formation of the surface layer of pollution*

Creation of the layer of the contamination on the surface of the insulator is influenced by many processes. The intensity of the growth of this layer is determined mainly by the concentration of the industrial air pollutants, however, it is also influenced by the local climatic conditions, the shape and size of the insulator, etc. The effect of the wind and rain usually either slows down or reverse the growth of the contamination layer, even though, wind can in some circumstances increase this growth. The distribution of the electric field along the insulators as well as the processes of charging the polluting particles in the field of corona discharges near the high voltage electrodes greatly affects the sedimentation of contamination on the surface of the external insulation under voltage stress. Last but not least, triboelectric phenomena and the transport of already charged particles in the electric field also affects sedimentation of pollutants on the surface of insulators.

The following findings are known from the operation of the external lines, especially at the level of 200kV and 400kV [6]:

- thicker layers of pollutants are found near the parts under the high voltage stress
- the pollution increases with increasing of the applied voltage,
- the degree of unevenness of the polluting layer along the insulator chain increases with increasing of the nominal voltage of transmission line.

The wetting-drying cycles caused by the leakage current play an important role in the formation of the surface polluting layer of insulators under voltage stress during various forms of atmospheric precipitation. The geometric dimensions and the shape of the insulator affects the leakage current value.

Additionally, the formation of the surface polluting layer depends on its composition. Adhesion of the pollution layer on the insulator surface depends in addition to the properties of the air pollutants and the meteorological conditions also on the type of the material and the surface quality of the insulators themselves (smooth glass glazing of porcelain, non-homogenous glass surface of glass insulators, various materials of composite insulators based on silicone, epoxy, or polyester materials for high voltage insulators, etc.).

The adhesion of pollutants may have very different values depending on the source of their production, for example: in the area of the conventional thermal power plants, the adhesion rate ranges from 0.1-2%, in agricultural areas with chemical treatment it is 2-3%, near steel, cement and magnesite enterprises it can reach tens of percent [6].

Different types of pollutants can have a substantially different type of sedimentation pattern on the surface of the insulators. Observations of the operating insulators and laboratory experiments confirmed that dry colloidal particles (e.g. fly ash) decompose relatively evenly on the top and the bottom surfaces of the insulator, due to the charging of particles in a strong electric field as well as by the triboelectric phenomenon. Similarly behaviour was observed with the layer of salts in laboratory tests in a fog chamber. In the case of heavier and more dimensional fractions of industrial pollutants (e.g. iron oxide, cement dust, etc.), it is possible to observe a different sedimentation pattern with a larger degree of uneven distribution. Substantially coarser layer is formed on the top surface of the insulator.

B. *Effect of the electric field on the polluted surface*

According to laws of physics, after applying high voltage on the insulator chain, the leakage current begins to flow on top of the contaminated surface of the insulator. However, the value of this current is ambiguous. The leakage current is determined by the immediate state of the surface conductivity, which depends on the various complicated and coexisting circumstances and conditions [6]:

- the impact of the complicated formation of the pollution layer,
- the method of humidifying this layer in various climatic and meteorological conditions,
- the shape and geometric dimensions of the insulator chain,
- the thermal effects of the leakage current itself lead to the heating of the polluting layer and thus the further change of its conductivity.

The last effect is the most discussed, because the increase of the temperature caused by flow of the leakage current has a twofold, even controversial effect on the change of the surface conductivity:

- the negative temperature coefficient of the resistance of the electrolytes, when by increasing the temperature of electrolytes, their electrical resistance decreases,
- on the other hand, at the same time, the increase of the temperature causes the decrease of the humidity and thus the increase of the resistance of the surface layer.

On the surface of the contaminated and sufficiently moistened insulator, the leakage current increases the temperature of the surface of the polluting layer. This higher temperature leads to the reduced resistance of the polluting layer and, as a result, the leakage current increases which leads to even higher temperature. Such rapid, uncontrolled increase of the leakage current and conductivity can only last until the boiling point of the surface electrolytes is reached. At this moment, intensive evaporation of the moisture takes place, thereby reducing the thickness of the electrolyte and at the same time decreasing the leakage current. Dry zones with the highest current density occur on the surface of the isolator. Nearly all high voltage is accumulated at the dry place. Next, the dry zone is overlapped - local skipping or a small arc is formed. A small arc overlaps the dry part of the insulator which is in series with the moistened portion. The arc may disappear when the phase current passes zero and the insulator returns to its original state. Forming of dry areas and re-moistening them may take a long time (in terms of hours).

If the intensity of precipitation increases, the dry surface may be moistened again and the process repeated. The emergence of short-term curves does not lead to a decrease in the electrical strength of the insulation. However, the long-term activity of the partial arcs can cause the disturbance of the insulation surface and the creation of carbonization tracks which leads to a reduction in the values of the breakdown voltages.

When the leakage current reaches a certain threshold, the partial arcs do not extinct but they expand to the opposite electrode. The flashover on the insulation surface occurs when the leakage current is sufficient to dry a certain part of the insulation surface and form a stable arc.

III. Experimental Setup

The experimental workplace shown in Fig. 1 was arranged for the measurement of the leakage currents flowing over the insulator surface. The tested insulators were placed in a fog chamber where the surface of the insulators was wetted.

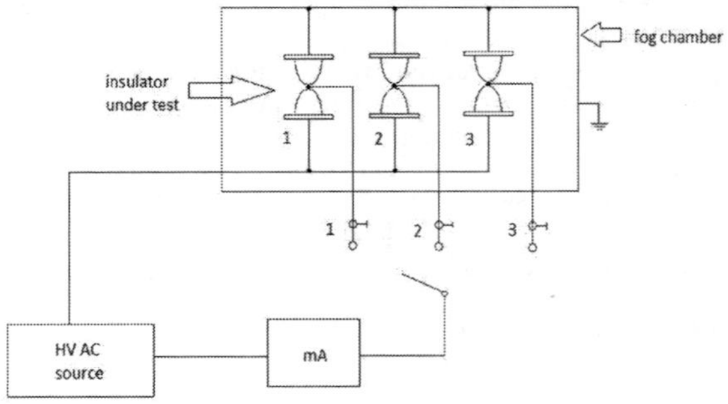

Figure 1. Diagram of the connection of the measuring devices

For the purpose of the experiment, three identical porcelain cap-and-pin insulators were chosen. The first insulator had a clean surface. The second had a medium dirty surface corresponding to the second degree of the pollution. The third as a very heavy dirty insulator, whose pollution corresponded to the fourth degree of the pollution. The insulators have been artificially polluted by the procedure in the IEC 815 with a salt brine prepared according to the relevant standard. Depending on how many grams of salt are added to the one liter of water a solution with different conductivity is obtained. This conductivity corresponds to a certain degree of contamination. For the second degree of contamination, 6g of salt was added to 1 liter of water. For the forth degree, 16g of salt was added to the water. A kaolin powder was added to the water solution in order to keep the salt brine on the insulator surface.

The prepared dry insulators were hung in a fog chamber in such way that the flow of the leakage current outside of the measuring devices was avoided. A voltage of 2 kV using an alternating high-voltage source was applied to the first insulator and immediately the current flowing over the insulator surface was read from the measuring device. Using the switch, the high voltage source was connected to the second and then to the third insulator. After finishing of the measurement of the dry insulators, water began to evaporate in the fog chamber, therefore wetting of the insulator surface began to change. Every two minutes, the value of the leakage current flowing over the insulator surface was recorded.

IV. Measuring Results

Surface conductivity σ according to the measured values of the leakage current can be calculated as:

$$\sigma = f\frac{I}{U} \tag{1}$$

where U is the applied voltage, I is the current flowing over the surface of the insulator, and f is the - "form-factor" dependent on the shape of the surface of the insulator. In order to correctly determine the value of the surface conductivity, the dimensions of the insulator must be taken into account. The way of the leak for the tested cap-and-pin insulator is shown in Fig.2. The indentations X and Y are assumed to be filled with a conducting material.

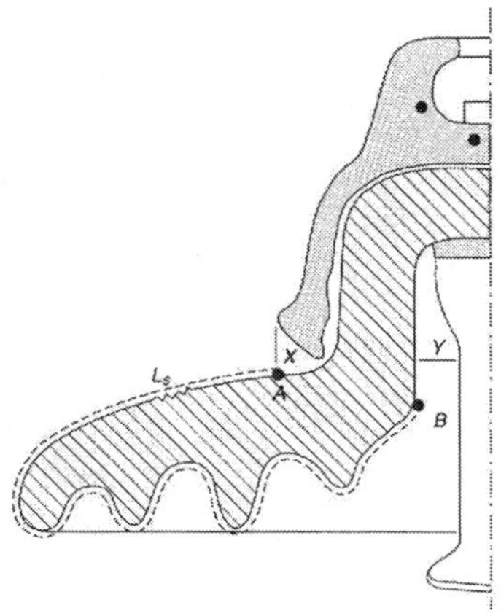

Figure 2. The leak path Ls for cap-and-pin insulator [7]

The form-factor includes the leakage path L_s, which is depends on the shape of the insulator. For the cap-and-pin insulator shown in Fig. 2 the form-factor f can be calculated as [7]:

$$f = \frac{1}{2\pi}\int_0^{L_s}\frac{dL_s}{r(L_s)} \tag{2}$$

where $r(L_s)$ is the radius corresponding to the path element dL_s.

The measured values of the leakage current and the corresponding surface conductivity of the insulators with different degrees of the pollution are shown in Table 1. The graphical dependences of the surface conductivity on the wetting time are shown in Fig. 3.

Table 1. Values of the leakage current and the surface conductivity of the cap-and-pin insulators

	clean insulator		insulator with medium pollution level		insulator with very heavy pollution level	
t [min]	I [mA]	σ.10-9 [S]	I [mA]	σ.10-9 [S]	I [mA]	σ.10-9 [S]
0	0.0025	0.488	0.0016	0.312	0.0035	0.6825
2	0.05	9.75	0.05	9.75	0.075	14.63
4	0.05	9.75	0.074	14.43	0.065	12.68
6	0.08	15.6	0.096	18.72	0.24	46.8
8	0.09	17.55	0.125	24.38	0.34	66.3
10	0.1	19.5	0.15	29.25	0.54	105.3
12	0.11	21.45	0.18	35.1	0.68	132.6
14	0.12	23.4	2.3	449	1.1	214.5
16	0.13	25.35	3.6	702	3.7	721.5
18	0.13	25.35	3.6	702	4.2	819
20	0.14	27.3	3.55	692	4.6	897
22	0.15	29.25	3.5	683	4.9	956
24	0.16	31.2	3.4	663	5.0	975
26	0.17	33.15	3.4	663	5.0	975
28	0.18	35.1	3.35	653	5.0	975
30	0.18	35.1	3.36	655	5.025	980
32	0.19	37.05	3.34	651	5.0	975
34	0.2	39	3.26	636	5.0	975

The measured values of the leakage currents of the dry insulators differ only minimally. The leakage current of the clean insulator was 0.0025mA, of the insulator with medium polluted surface was 0.0016mA, and of the insulator with very heavy polluted surface was 0.0035mA. After 12 minutes from the beginning of the wetting process, the leakage current increased sharply to 2.3mA and after 16 minutes it reached a maximum value of 3.6mA in the case of the medium polluted insulator. Subsequent humidification, lead to slight decrease of the leakage current. The leakage current of the very heavy polluted insulator increased sharply after 14 minutes from the beginning of humidification and the maximum value of 5mA was reached after 24 minutes.

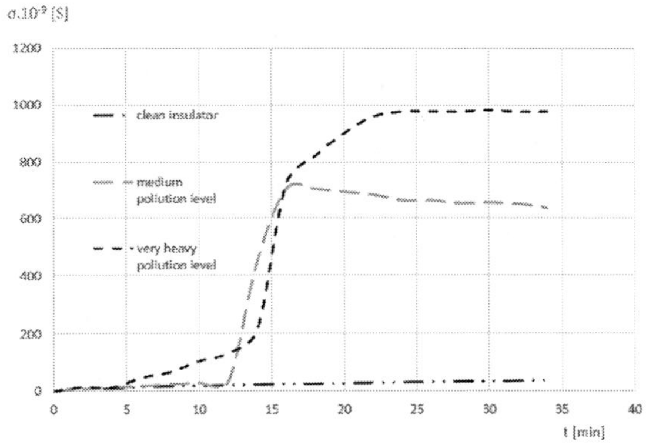

Figure 3. Dependence of surface conductivity on the time of wetting the contaminated layer of insulators

It can be seen from Fig. 3 that the surface conductivity of all insulators increases as the degree of moistening of the insulator surface increases. After half an hour from the start of wetting, insulator with the highest degree of pollution reached the highest conductivity, which 30 times exceeded the conductivity of the clean insulator.

V. Conclusions

The results of the experiment show that dry contamination has only marginal effect on the surface conductivity of the insulators and hence the value of the flashover voltage. With gradual wetting, the surface conductivity increases, whereby for clean insulator the increase of conductivity was very small:

from 0.488μS in dry state to 39μS after 34 minutes of humidification.

The conductivity of the medium polluted insulator reached a maximum of 702μS after 16minutes from the beginning of the wetting. Subsequently the conductivity began to decrease. After 34 minutes the conductivity dropped to the value 636μS. This decrease suggests that the surface of the insulator was self-washed. In the case of the very heavy polluted insulator, the maximum conductivity of 975μS was reached 24 minutes after the beginning of the wetting process and remained unchanged until the end of the experiment.

The high conductivity of the polluted insulators during their wetting reduces the value of the flashover voltage, which affects the reliability of the power supply. The slight wetting of the surface is especially dangerous during a mist or a mild rain. In the case of a heavy rain, the surface of the insulator can be self-cleaning and, therefore, the surface conductivity can be reduced.

Further research will focus on the measuring of the flashover voltage of the insulators with varying degrees of pollution during their wetting.

ACKNOWLEDGMENT

The authors would like to acknowledge Scientific Grant Agency of The Ministry of Slovak Republic and Slovak Academy of Science for funding experimental works in the frame of VEGA No. 1/0132/15, VEGA No. 1/0311/15 and VEGA No. 2/0141/16.

This work was supported by the Slovak Research and Development Agency under the contract No. APVV-15-0438.

REFERENCES

[1] E. Azordegan, B. Kordi. "Remote Assessment of High Voltage Insulators Using Wideband Electromagnetic Radiation Signature". *IEEE Transactions on Dielectrics and Electrical Insulation*, vol. 23, pp. 1467-1474, June 2016

[2] M.F. Palangar, M. Mirzaie. "Diagnosis of Porcelain and Glass Insulators Conditions Using Phase Angle Index". *IEEE Transactions on Dielectrics and Electrical Insulation*, vol. 23, pp. 1460-1466, June 2016

[3] A.H. El-Hag, L.H. Meyer, A. Nederian. "Experience With Salt – Fog and Inclined – Plane Tests for Aging Polymeric Insulators and Materials". *IEEE Electrical Insulation Magazine*, vol 26 No 2, pp. 42-50, March/April 2010

[4] S.S. Shunmugam, N. Vasudev, K.N. Ravi, K.A. Venkatesh. "Influence of Profile on the Pollution Performance of Cap-and-Pin Insulators – An Experimental Study". *IEEE Electrical Insulation Magazine,* vol. 32 No 6, pp. 20-27, November/December 2016

[5] R. Boudissa, S. Djafri, A. Haddad, R. Belaicha, R. Bearsch. "Effect of Insulator Shape on Surface Discharges and Flashover under Polluted Conditions". *IEEE Transactions on Dielectrics and Electrical Insulation*, vol. 12, No. 3, pp. 429-437, June 2005

[6] P. Csaji. *Určovanie stupňa znečistenia vonkajších izolátorov VVN*, kandidatská dizertáčna práca, Košice, máj 1986, p.171.

[7] E. Kuffel, W.S. Zaengl, J. Kuffel. *High Voltage Engineering, second edition*, published by Butterworth- Heinemann, 2000, ISBN 0 7506 3634 3, p. 539

On the Undesirable EMI of Large LED Information Board

Bystrík Dolník, Alexander Mészáros, Marek Pavlík, Ján Zbojovský, Michal Špes, Zsolt Čonka
Department of Electrical Power Engineering
Technical University of Kosice, Faculty of Electrical Engineering and Informatics
Košice, Slovakia 041 20
Email: bystrik.dolnik@tuke.sk

Abstract — **The quality of electrical equipment has become a critical concern in recent years. This concern has grown with the advent of positive industry trends to improve the proliferation of electronic equipment in households and commercial environments. However, the very technology that allows more efficient use of electrical energy with the switch-mode power conversion technology and implementation of IT technology, has also an influence that can negatively impact power quality and operation in the electromagnetic environment. Electrical equipment operating at high clock speeds may generate unacceptable levels of electromagnetic interference (EMI). The end result for electric and electronic products may be incorrect operation, failure or inability place on the market. In this article we will focus on reducing emissions of EMI generated from prototype of the LED information board designed for railways. We present mainly on EMI by radiated emissions in the frequency range from 30 MHz to 250 MHz for horizontal and vertical polarized electromagnetic wave. Various measures for reducing the EMI caused by galvanic coupling within the LED information board are compared. Finally, the EMI tests repeated in the anechoic chamber with electromagnetic compatibility receiving antenna pointed to the fact, that the largest source of EMI are generated by the LED display system receiving cards and power supply circuits for LED display units. Further measures to reduce EMI are discussed too.**

Keywords: insulation; pollution; surge arrester; leakage current; surface resistivity

I. Introduction

Failure-free operation of the electrical equipment and operation without producing disturbances are essential to product quality requirements. To meet these requirements a certain number of rules must be followed in the design and manufacture of the electrical equipment. The site and installation characteristics also play an important role in electromagnetic compatibility (EMC). The location and layout of power components, cable routing, shielding etc. are important right from the initial design phase [1], [2]. Although the equipment offers satisfactory EMC, a well designed installation can extend the compatibility safety margins. Only measurements requiring a high level of expertise and sophisticated equipment can produce valid results quantifying the EMC of equipment. Compliance with EMC standards provides the certainty that electric equipment will operate satisfactorily in surrounding electromagnetic environment [3].

EMC should be optimally in line with the general design of EMC. From the very beginning of the design, the designers should consider which sources of noise and sensitive circuits are used in their equipment. Consideration should also be given to cabling that takes into account recommendations for the length, cross-section, spacing and orientation used. Finally, EMC filters as an integral part of the overall EMC product design, as well as grounding concepts, shielding, and surge protection should be planned.

A. LED information board

The LED information boards are widely used for the displaying important information like pollution parameters, production figures, welcome messages, safety messages, weather conditions, results of sport events and in many other applications. The LED information boards tend to be equipped with the latest technologies, and comes with a clear display support, much low power consumption compared to traditional lighting, long serving life, reduced environmental impact, as well as suitability for indoor and outdoor use. Thanks to highly efficient and bright LEDs the LED information boards nowadays offers excellent legibility at great distances as well as close-up [4], [5].

With the increasing number of LED information boards, requirements for mechanical properties, weather resistance and EMC are also increasing. Most of the LED technology need to be paired with a switching drive circuit to provide constant current sourcing. These switching drive circuits operate at increasingly higher frequencies, and in order to maintain the highest efficiency, and highest LED longevity possible, they also need to maintain very high slew rates [6]. The steeper the voltage steps and the higher the switching frequency, the greater the high frequency content of generated disturbances. These disturbances can couple to other components in the power supply, producing noise sources radiating over a broad frequency spectrum, and contribute to the overall unwanted emissions from each individual lighting element [7].

B. EMI phenomena

Electromagnetic fields (signals) may be of natural origin or are generated by technical systems. One of the most famous phenomena is a lightning strike, the extreme electric current that usually flows from the cloud towards the Earth. This current causes large electromagnetic fields. Technical sources of electromagnetic signals are also important. LED information boards, as already mentioned, are one of these sources. In this category, three major sources of electromagnetic disturbance can be identified: power supplies, control signals, and processor operation. On the other hand, man deliberately generates signals (electromagnetic fields) to transmit them remotely, e.g. TV or broadcast, radar transmitters and others.

Any EM noise emission, natural or man-made, is potentially a disturbance to any other susceptible device in the environment. It may either put it out of action or, in many cases a worse problem, cause it to malfunction. If disruptive EM energy causes a malfunction of the electronics (device), it is considered to be interference, or so called electromagnetic interference (EMI). A key characteristic of EMI is its frequency. Currently, however, not all frequency ranges are completely regulated [8].

The first important frequency range is the power network frequency, which is 50 Hz or 60 Hz and their harmonics, generally considered up to the 40th (50th) order. The high frequency range starts above 9 kHz, also called the radio frequency (RF) range. EMC standards limit the frequency range to 400 GHz on the upper end. In the EMC the radiated range starts at 30 MHz and the upper limit of this range depends on the standard, generally about 1 GHz, for some products 2 GHz or 3 GHz.

C. *EMI mitigation*

There are several measures to mitigate EMI in electrical circuits. In particular, the elimination of undesirable couplings, shielding, grounding and overvoltage protection are included. Older analog technologies are not able to distinguish between desired and undesirable signals and are therefore more susceptible to EMI within the band. The latest technologies are more robust and use error correction technologies to minimize the impact of EMI. High frequency emission is mainly caused by harmonics of clock frequencies of digital circuits, or in special cases as partial discharges in high voltage electrical machines [9]. The transmission of electromagnetic energy from the source to the victim arises through the coupling paths: galvanic (electrical current), inductive (magnetic field), capacitive (electric field) and by radiation (electromagnetic field). An example of equivalent circuit of the EMI through coupling is depicted in Fig. 1.

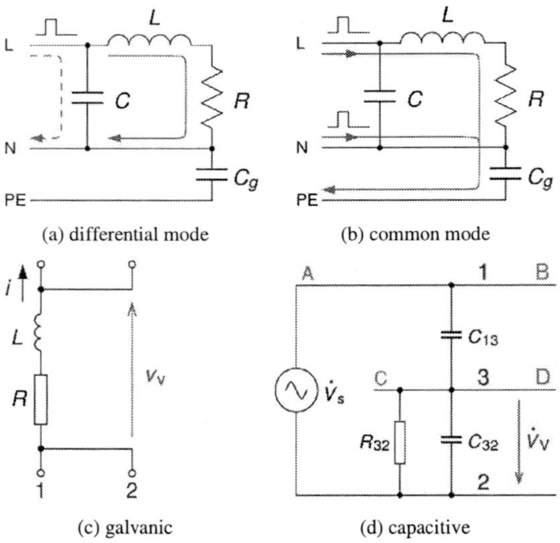

(a) differential mode (b) common mode

(c) galvanic (d) capacitive

Figure 1. The EMI in electrical circuit through couplings.

The EMI can occur between any two (or more) lines of the system. Such EMI is called differential mode or symmetrical mode. Differential mode is a result of parasitic components in a circuit. The EMI can also be conducted from any line in the system towards earth. This type of EMI is called common mode. The main difference is that common mode EMI travels in all lines in the same direction and then towards earth. Direct differential mode coupling of disturbances is normally prevented using twisted pair or coaxial cables. Common mode coupling results in common mode voltages and currents in a system. The coupling from common mode to differential mode can be reduced by symmetry of pair cables and by shielding.

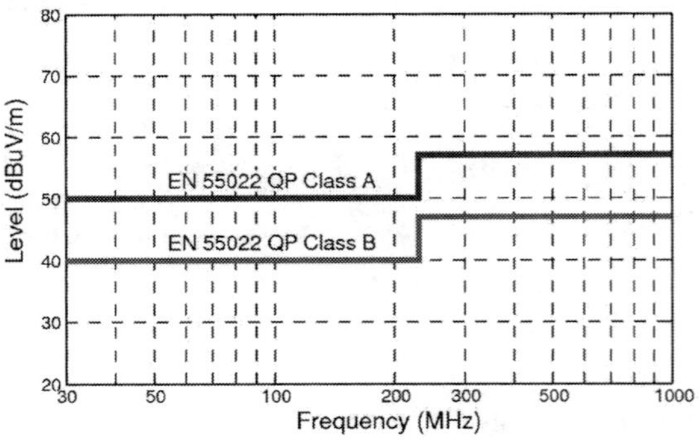

Figure 2. The disturbance field strength limits (quasi-peak) according to EN 55022 normalized to a distance of 3 m.

Common mode coupling can be reduced by increasing the distance of signal cables to power cables and by minimizing the area of earth loops. In cases where minimizing the area of common mode loops is not sufficient, lines may be provided with common mode chokes or ferrites to increase the impedance of the common mode circuit. In the Fig. 1 R and L is the resistance and the inductance of the wires respectively, C, C_{13} and C_{32} is the mutual capacity between conductors or systems, C_g is the parasitic capacity against ground (or grounding conductor), R_{32} is the high input resistance of the victim (e.g. in high voltage measuring circuits [10]), V_s is the voltage of the EMI source, v_v and V_v is the voltage of the victim. The interference voltage on the victim in the electrical circuit in Fig. 1c can be derived according to 2. Kirchhoff's law as:

$$v_v(t) = Ri(t) + L\mathrm{d}i(t)/\mathrm{d}t \tag{1}$$

and in the electrical circuit in Fig. 1d as:

$$v_v(t) = C_{13}R_{32}\,\mathrm{d}v_s(t)/\mathrm{d}t\,. \tag{2}$$

EMI reduction can be achieved by minimizing the right side in (1) and (2).

Shielding with RF seals or good conductive tapes is often the last choice at the expense of increased costs. To be effective at high frequencies, the shield of cables should be in contact with the metallic wall of an equipment enclosure. It is important that this connection be made with a good connector whose transfer impedance is adequately low. To achieve good EMC performance it is essential that the equipment is provided with a well designated grounding and bonding network [11], [12], [13].

D. *EMI limits*

Different EMC standards set recommended disturbance limits for conducted and radiated emissions. These disturbance limits are defined in dBμV for the conducted emission voltage and dBμV/m for the radiated emissions (field strength). The reference values for conducted emissions are 1 μV for 0 dBμV and for radiated emissions 1 μV/m for 0 dBμV/m. The disturbance limits are defined separately for residential, commercial, light-industrial environments and industrial environments.

Two disturbance limits exist for these two environments: Class A represents the industrial environment, and Class B defines the limits for residential environments. The class A and B limits of EN 55011 and EN 55022 have become the reference limits for most standards [14]. The disturbance field strength limits (quasi-peak) according to EN 55022 (information technology equipment), normalized to a distance of 3 m are shown in Fig. 2.

II. EXPERIMENT

In the experiment, measures to mitigate the couplings through AC mains wires and through the DC power supply of the LED modules to the LED information board have been investigated. The enclosure of the LED information board is made of galvanized steel, with protection rating IP65 and anti-reflective protective glass. The total weight of the LED information board is approximately 145 kg. In a folded state, the LED information board has the dimensions: 192 cm × 50 cm × 30 cm. The LED information board is powered from a single-phase AC low-voltage network that is converted by AC-to-DC converters with a maximum current draw of 20 A. A total of three AC-to-DC converters are installed in the LED information board. For receiving and sending data in several ways, an optical converter, router, and ADSL modem are installed. A mini PC is used for data processing and process control. The LED panels are controlled by a LED display system receiving card.

A. *Test procedure*

The measurements of the EMI by radiation (emissions) were carried out according to the standard EN 55022. The equipment under test (EUT) was put in the anechoic chamber 1.4 m high above the ground and connected to the AC mains through EMC filters. The distance between the EUT and the receiving EMC antenna was 3 m. At the beginning of the experiment, the initial state was tested, without EMI mitigation measures being used in order to obtain reference data.

The EUT was oriented towards the EMC antenna in the direction of the largest EMI emissions. The largest EMI emissions from EUT was experimentally detected for both, horizontal and vertical antenna polarization. After measuring the reference data, we have gradually used EMI mitigation measures, minimizing the galvanic coupling from the power circuits and from the DC power supply, as well as from the signal circuits used for controlling the LED panels, either alone or in combination.

B. *Evaluation*

As mentioned above, we measure at the beginning of the experiment the initial state. First, the EMI critical frequencies were detected. One or more EMI mitigation measures were then applied and the impact of the measure applied was compared. After testing the available measures to mitigate the EMI, we tested the EUT according to the limit shown in Fig. 2 for Class A. The ETS Lindgren 3142E model was used as a receiving EMI antenna connected by coaxial cable to the Agilent spectral analyser model N5183A.

III. RESULT AND DISCUSSION

The level of interference in the initial state without EMI mitigation measures is shown in Fig. 3. It has been found that the greatest EMI emissions is generated at frequencies of 37.5 MHz and 87.5 MHz, marked in the list of found EMI sources by the red letter F(ail). Other sources of EMI emissions were generated at frequencies of 62.5 MHz, 100 MHz, 112.5 MHz and 125 MHz. It is clear that listed frequencies are multiples of 12.5 MHz. In our experiments, we tested the EUT up to 1 GHz. We found, that the highest levels of EMI emissions are in the frequency range from 30 MHz to 250 MHz. It has also been demonstrated, as we assumed, that the EMI emissions in horizontal polarization is greater in comparison with EMI emissions in vertical polarization. To reduce the EMI emissions, the following measures were sequentially used:

- twisted power wires,
- chokes on wires (straight or twisted),
- suppression capacitors,
- impedance matching resistors,
- data wires on PCB with SMD LC filters,
- twisted data wires.

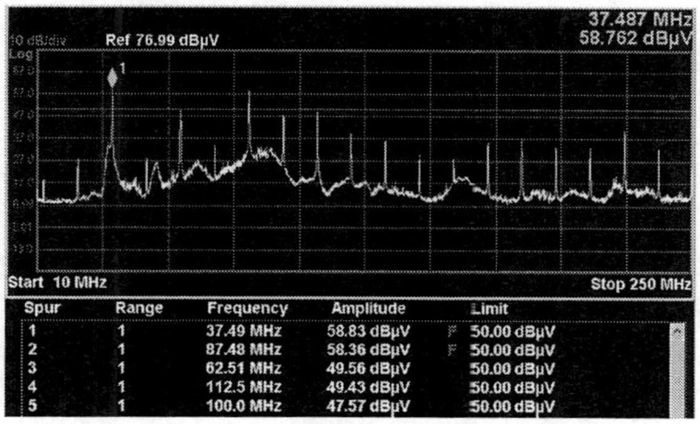

Figure 3. The initial state of the EMI measured on EUT.

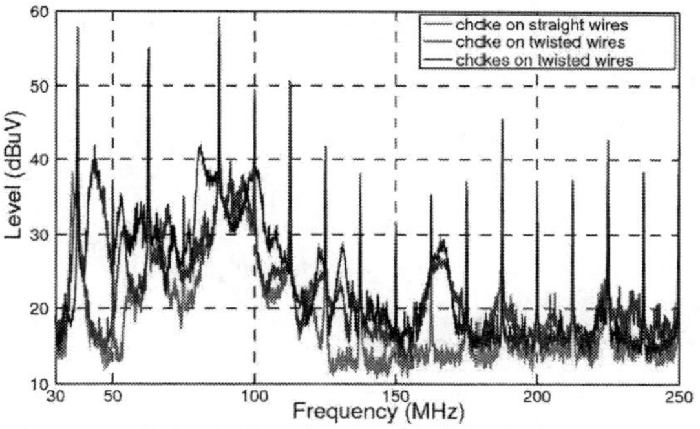

Figure 4. The level of interference with EMI mitigation measures.

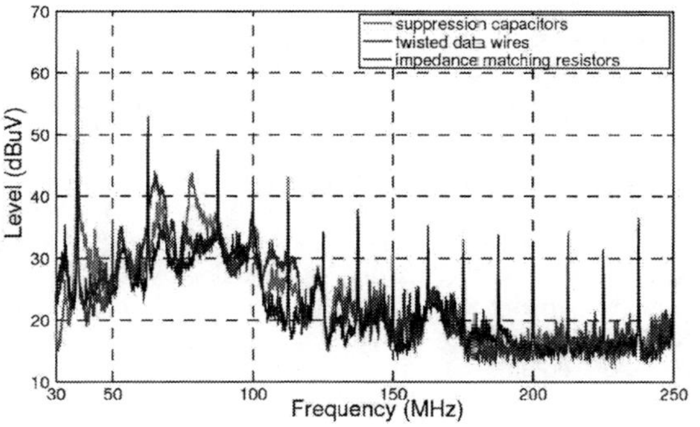

Figure 5. The level of interference with EMI mitigation measures.

In Fig. 4 the level of interference with three different EMI mitigation measures is shown. Using experimental testing, it was found that the most critical frequency range is from 30 MHz to 130 MHz. As can be seen from Fig. 4, each of the measures has different effects on the EMI level. It may be noted that the EMI decrease ranges from a few tenths of dB to a few units of dB. However, the mitigation effect is frequency dependent, and at certain frequencies, there is an increase in EMI. The best effect was when using a chokes at both ends of the twisted power wires.

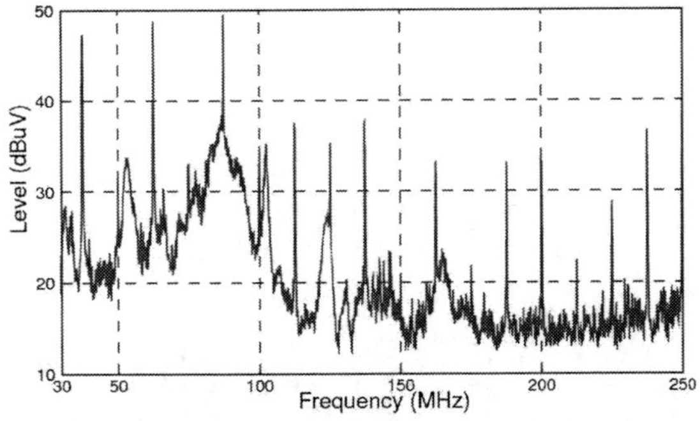

Figure 6. The level of interference with EMI mitigation measures.

The EMI emissions level of the EUT measured using three other measures is shown in Fig. 5. In this case, interference suppression capacitors at the input of the AC mains wires, twisted data wires, and better impedance matching of the data wires by resistors were used. By measuring, it was found that these three measures have a slightly better effect on reducing the EMI from the EUT compared with the previous case. The best effect was achieved by using impedance matching of the data wires.

The most optimal EMI radiation has been achieved by the simultaneous use of several EMI reduction measures. Under optimal EMI radiation, it should be understood a compromise between EMI levels (at multiples of 12.5 MHz) for the defined frequency range. An example of the measured EMI emissions levels for such case is shown in Fig. 6 where chokes on twisted pairs from AC-to-DC converters at both ends, impedance matching resistors and data wires on PCB with SMD LC filters are used. Other variants used in testing had only a slight effect, a few tenths of dB, at some multiples of the frequency 12.5 MHz.

IV. Conclusion

The widely used LED information boards equipped with the latest technologies have much low power consumption compared to traditional lighting, longer service life and are suitable for indoor and outdoor use. With the introduction of the latest IT technology into the LED information boards, the need for EMI testing of these devices also emerged. In the experiment the EMI emissions of LED information board for various measures to reduce the EMI emissions were observed.

It was found, that the highest levels of EMI emissions are in the frequency range from 30 MHz to 250 MHz, and that the most critical frequency range is from 30 MHz to 130 MHz. Several measures to reduce the EMI emissions were used. As a result, we found that each of the measures has different effects on the EMI level and the mitigation is frequency dependent. The most optimal EMI radiation has been achieved by the simultaneous use of several EMI reduction measures.

The best effect was achieved by using impedance matching of the data wires. An even more efficient reduction of EMI emissions is possible by optimally distributing of functional blocks in the LED information board and using shielding materials.

Acknowledgment

The author would like to thank to the Scientific Grant Agency of the Ministry of Education of the Slovak Republic and the Slovak Academy of Sciences under the project VEGA No. 1/0311/15, VEGA No. 2/0141/16. This work was supported by the Slovak Research and Development Agency under the contract No. APVV-15-0438.

References

[1] T. Williams. *EMC for Product Designers*. Ed. Newnes, 2017.

[2] D. Novák, J. Schneider, and D. Kocúr, "Static person detection and localization based on their respiratory motion using various antenna types," *Acta Electrotechnica et Informatica*, 2016, vol. 16, no. 3, pp. 54-59, doi: 10.15546/aeei-2016-0024.

[3] M. Tlig, J. B. H. Slama, and M. Belaid, "Conducted and radiated EMI evolution of power RF N-LDMOS after accelerated ageing tests," *Microelectronics Reliability*, 2013, vol. 53, no. 9-11, pp. 1793–1797, doi: 10.1016/j.microrel.2013.07.111.

[4] J.-H. Ahn, "Implementation of an LED tile controller for highquality image display," *Displays*, 2013, vol. 34, no. 1, pp. 17–26, doi: 10.1016/j.displa.2008.05.002.

[5] W. Huang, J.-M. Li, L.-M. Yang, Z.-L. Jin, Z.-G. Zhong, Y. Liu, Q.-Y. Chou, and F. Li, "Local dimming algorithm and color gamut calibration for RGB LED backlight LCD display," *Optics and Laser Technology*, 2011, vol. 43, no. 1, pp. 214–217, doi: 10.1016/j.optlastec.2010.06.016.

[6] C. Nelson and B. Essaff, Analog Circuit Design, B. Dobkin and J. Hamburger, Eds. Newnes, 2015, doi: 10.1016/B978-0-12-800001-4.00026-0.

[7] J.-H. Ahn, "LED brightness control for video display application," *Displays*, 2008, vol. 29, no. 5, pp. 506–511, doi:10.1016/j.displa.2008.05.002.

[8] Q. Dong and W. Ninghui, *Recent Developments in Applied Electrostatics*, 2004, S. Keping and Y. Gefei, Eds. Elsevier Science Ltd., doi: 10.1016/B978-008044584-7.50095-5.

[9] J. Kurimský, R. Cimbala, and I. Kolcunová, "Multi-scale decomposition for partial discharge analysis," *Przeglad Elektrotechniczny*, 2008, vol. 84, no. 9, pp. 169–173.

[10] J. Balogh, J. Petráš, and J. Džmura, "Measurement of polypropylene rope electrical strength," *Electric Power Engineering 2014*: Proceedings of the 15th International Scientific Conference, vol. 14. Brno, Czech Republic: Brno: University of Technology, May 2014.

[11] R. P. Clayton, *Introduction to Electromagnetic Compatibility*, 2nd ed. Wiley, 2006.

[12] W. H. Ott, *Electromagnetic Compatibility Engineering*, 1st ed. Wiley, Sep. 2009.

[13] W. P. Kodali, *Engineering Electromagnetic Compatibility: Principles, Measurements, Technologies, and Computer Models*, 2nd ed. Wiley, Jan. 2001.

[14] EN 55022: 2006, "Information technology equipment - Radio disturbance characteristics - Limits and methods of measurement." 2006, European Norm, 2006.

Measurement of Inrush Current Halogen Lamps

Ludvik Koval, Radek Martinek, Petr Bilik, Jan Vanus
Department of Cybernetics and Biomedical Engineering
VSB - Technical university of Ostrava,
Ostrava, Czech Republic
ludvik.koval@vsb.cz, radek.martinek@vsb.cz, petr.bilik@vsb.cz, jan.vanus@vsb.cz

Abstract — **Currently in the automotive industry, manufacturing industry and households still used as a light source halogen lamp. Compared with ordinary light bulb has halogen bulb a higher efficiency. Unlike other types light sources with higher efficiency as fluorescent-tube or LED, the halogen lamp has excellent properties of color rendition. The article discusses the problems of time measurement inrush current halogen lamps using LabVIEW. There are mentioned the reasons for the shorter life of the halogen bulbs and its possible extension. Part of this work is measurement and comparison the inrush current halogen bulbs in the circuit diagram with and without limiting inrush current.**

Keywords: Halogen Bulb, Bulb Measurements, Inrush Current, Limiting Inrush Current, LabView Measurements

I. Introduction

Light bulbs are simple and very cheap light sources, the discovery of which is traditionally attributed to T. Edison. Although today's effort to replace bulbs with more economical light sources with less power consumption and greater efficiency in converting electrical energy into light energy, they remain unsurpassed in several parameters. Despite the continuous improvement of alternative light sources, it is still true that bulb light is the most pleasant for the human eye and for color perception. Bulb light has the highest $R_a = 100$. The color rendering ratio of all artificial light sources since the radiated spectrum is continuous without visible spectral fluctuations and spectral selection in the visible area. Figure. 1 shows the difference between the color spectrum of tungsten bulbs and compact fluorescent lamps. In the spectrum of compact fluorescent lamps we see distinct blackouts of some colors, while the bulb spectrum is fluent throughout the visible area.

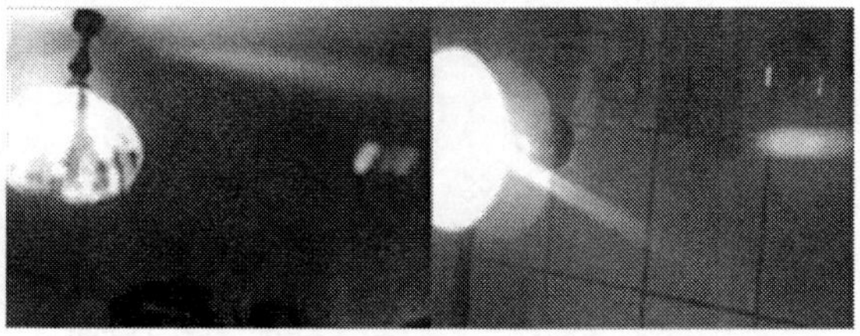

Figure 1 Comparison of the bulb spectrum (right) and fluorescent lamps [7]

On the other hand, however, the bulbs have considerable drawbacks - in particular, the low efficiency of converting electricity into light energy and a mean lifetime of only about 1000 hours (in the case of standard light bulbs), which is even diminished every switched on in real use. Their lifetime also greatly affects the operation at a voltage higher than the rated voltage. Decreasing the lifetime of the bulbs by the number of switching cycles is due to the fact that when the bulb is switched on, the current is several times higher than the rated current, thus changing the filament temperature rapidly with a high inrush current. The inrush current can be effectively reduced and thus extend the practical life of the bulb. The main goals of this work are to measure the inrush current and to design a circuit for its limitation that follow.

II. Principle of Action

The light bulb is a light source utilizing the physical principles of heating the metallic filament through the passage of electric current and the emission of radiation from the surface at high temperature. Emitted energy covers wavelengths over a range of overlapping spectrum of visible light, especially in the infrared range, and therefore we also perceive the bulb as a heat source. Stefan-Boltzmann's law says that the intensity of radiation increases with the fourth power of thermodynamic body temperature. With increasing body temperature, the amount of radiated energy is higher.

$$M = \sigma . T^4 \qquad (1)$$

M the radiation intensity ($W.m^{-2}$)
σ Stefan-Boltzmann constant $5.67 \cdot 10^{-8}$ ($Wm^{-2}K^{-4}$)
T Thermodynamic temperature (K)

Looking at the graph of the intensity of the black body radiation the wavelengths of radiation is in relation to the body temperature (Figure 2), it is possible to see the shift to shorter wavelengths with increasing temperature. The wavelength of the radiation with the maximum value λ_{max} is given by Wien's law to shift the wavelengths toward higher wavelengths at a lower temperature.

$$\lambda_{max} = \frac{2{,}898.10^{-3}}{T} \qquad (2)$$

λ_{max} (μm) the wavelength with the highest radiation intensity
T thermodynamic temperature (K)

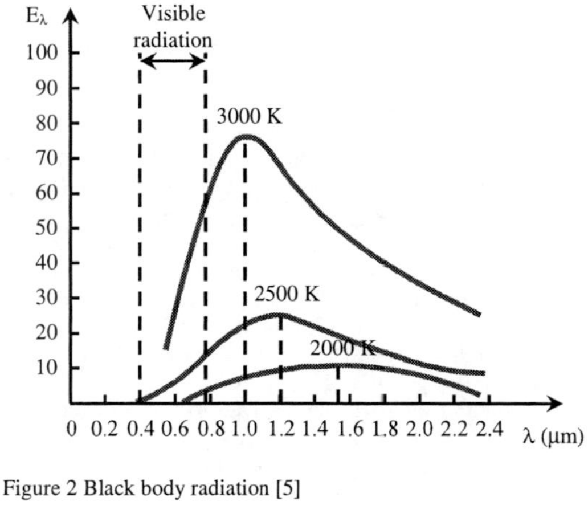

Figure 2 Black body radiation [5]

A. *Materials for the construction of bulbs*

The Filament and inlet holders are of molybdenum which, together with tungsten, have the most advantageous properties for glass traps, in particular the coefficient of thermal expansion. [1] For the filament lamp material, it is theoretically possible to use any conductor. In the case of the first bulbs, carbon filament was used, which, unlike the metal filaments, had a negative temperature coefficient of resistance. Various metals, such as platinum, tantalum, chromium, iridium, rhodium, ruthenium, osmium, molybdenum, have been tested and used. [2] It was necessary to find a material having a sufficiently high melting point to allow it to be heated to the highest temperature, while at the same time avoiding excessive sputtering or evaporation of the filament. In addition, the material must be easily machined to the shape of a coiled filament and must not be too expensive. For example platinum and rhodium are very precious metals, their price exceeds the price of gold. [3] The promising material was tantalum. It is well workable and its price is not high. Today, only tungsten is used as the filament material, which is even more advantageous for use as a filament lamp. The filament is wound with thin tungsten wire helically twisted into a spiral (Fig. 4). In this way, the winding of the filament makes it possible to achieve small bulb dimensions. The evaporation of tungsten filament can be reduced by doping pure tungsten with potassium, silicon or aluminum oxides, thus extending its life. [2]

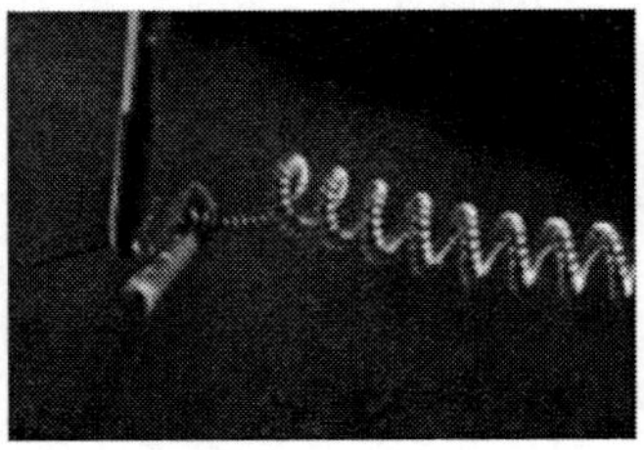

Figure 3 Detail of tungsten filament [6]

B. *Halogen bulbs*

Halogen bulbs represent an enhanced variant of the original light bulb, which they have many advantages over. Halogen bulbs, unlike ordinary bulbs, have an average life of 4,000 - 5,000 hours without tungsten deposition on the inner surface of the bulb and thus without the loss of light flux over their lifetime. Allow the filament to heat up to higher temperature.. As the luminous flux increases as the temperature rises, this makes it more efficient to convert electrical energy into visible light. The spectrum of a halogen bulb contains more short wavelengths and the light appears to be more white. The chromaticity temperature of the non-overheated halogen bulb is between 2900 and 3100 K. For special types of incandescent bulbs up to 3400 K. This is, however, the cost of a lower service life (only about 15 hours). [4] A common bulb has a chromaticity temperature of about 2500 K at a mean lamp life of 1000 hours.

C. *Disadvantages of halogen bulbs*

Halogen lamp works just like a normal bulb. Through the passage of the electric current there is a heating of the filament and the emission of photons. In addition, a tungsten-halogen cycle is applied which causes the vaporized tungsten atoms to return back to the filament to prevent its deposition on the inner surface of the flask and filament thinning. Tungsten-halogen cycle is called a phenomenon where the vaporized tungsten atoms combine near the inner surface of the flask at a temperature of less than 1700 K with halogen in a flask to the corresponding tungsten halide. Due to the distribution of the temperature field, this halide penetrates the filament, where it is again decomposed to halogen and tungsten due to the higher temperature. Tungsten is then deposited back on the filament. [4]

D. *Mechanical design of halogen bulbs*

Unlike a conventional bulb a halogen bulbs are substantially smaller and come in an always cylindrical or exceptionally spherical shapes. For conventional halogen bulbs with standard E14 and E27 Edison screw bases, the lamp is always a halogen bulb of standard dimensions enclosed in the outer shell of the bulb's classical dimensions. Because of the high temperatures and pressures, ordinary glass can´t be used to construct halogen bulbs. Therefore, special glass with a very high silica content (Vycor), pure silica (quartz glass) or hard glass is used for halogen bulb flasks, which further increases the working pressure and thus reduces evaporation of the tungsten atoms from the filament.

E. *Impact of the number of switching cycles on the life of the halogen bulb*

During the service life of a halogen bulb, it is usually re-ignited and switched off. The operating temperature of the filament is so high that the filament material is gradually recrystallized to increase its brittleness. The jumper connection of the rated voltage to the bulb is accompanied by a high impact current, rapid heating of the filament and its rapid thermal expansion. This recurrent dilation shock causes the interruption of the recrystallized filament to occur in the shortest time. [4] The inrush current is determined by the off-state (cold) filament bulb electrical resistance, the size of the connected electrical voltage and the internal resistance of the electrical power source. The halogen filament is made of tungsten. The dependence of the tensile strength of the tungsten filament on the temperature is linear, increasing with increasing temperature. The exact electrical resistance of the halogen filament in the off state is determined by measurement. Relatively accurately, we can determine it by calculating the resistance in operating state (rated voltage and rated current). The inrush current is then determined by the known cold resistance of the filament and the supply voltage.

III. LIMITING INRUSH CURRENT

The inrush current can be reduced in several ways. However, it is always a variant of preheating the filament of the bulb and thus bringing its temperature closer to the working temperature and thus increasing its electrical resistance. This is possible, especially with regard to the way of power supply, to realize in particular:

- By reducing the initial supply voltage
- Limiting the maximum supply current

The easiest method is to reduce the current by a light bulb. For low-power direct-current light bulbs, the scheme of Figure 4 is preferably used. It is assumed that the bulb's current will be limited to an acceptable value. Thus, the filament is gradually heated up to the operating temperature without any sudden changes in its length. The simple circuit of this design can be based on the principle of sequential opening of the MOSFET transistor. A possible alternative to this connection is to use the relay and its switching contact to short-circuit the resistor connected in

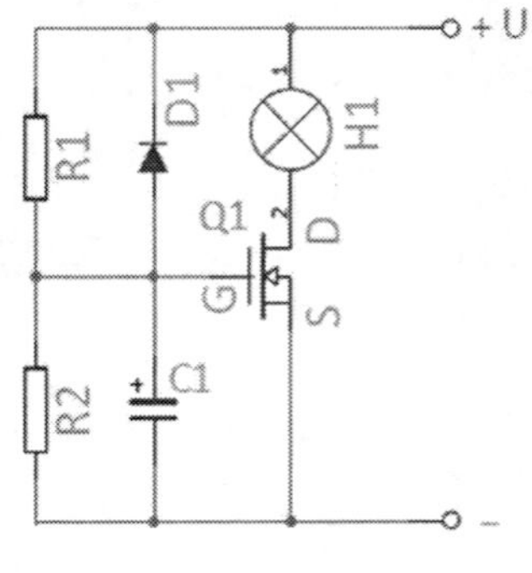

Figure 4 Simple limiter with reduced initial voltage

series with the bulb. This method is advantageous for high performance bulbs.

Pulse width modulation (PWM) can also be used to control the bulb current. A great advantage is the low loss. PWM is often used in dimmer constructions. It is also an advantageous way to control the light bulb's flow. We use so-called soft starters to realize the smooth running of the bulb. In the case of AC power, controlled rectifiers are used to limit the inrush current. Thyristors are used here as rectifying elements. The controlled rectifier allows to regulate the mean value of the rectified voltage. This also regulates the bulb current.

IV. Measurement of the Inrush current

The inrush current of the halogen bulb reaches high peak values and the transition time is very fast. It is necessary to take this aspect into consideration in order to capture it. In principle, the inrush current can be measured by means of current clamps and by means of a current shunt. Current clamps have the advantage of minimally affecting the current circuit and the galvanic separation of the measuring circuit. On the other hand, their accuracy is worse than in the case of the current shunt and also the dynamic properties are worse. The upper frequency band is usually limited by 10 to 100 kHz. Such a frequency range appears to be insufficient. Current clamps are usually designed as measuring transformers or magnetic circuits with a Hall probe or a combination of these. A more convenient method of measuring current when capturing fast processes is by using a resistance shunt. Generally, the shunt is a resistor with a very low resistance value. The current is measured as the output voltage of the shunt.

$$U_B = R_B . I \quad (V; \Omega, A) \tag{3}$$

U_B voltage of the shunt
I measured current
R_B resistance of the shunt

It is important that the resistance shunt have the least possible inductance. Therefore resistors wound with resistance wire are not used here, but bar resistors.

V. Capturing the Transition process

Due to the transient rate, it is suitable to use an oscilloscope. Digital oscilloscopes usually have the option of connecting to a personal computer or the possibility of directly storing measured data in an external memory. A disadvantage, on the contrary, is the low bit resolution of the converter. Most often, digital oscilloscopes are designed with an 8 bit resolution AD converter. To view the transition process, the 8-bit resolution is sufficient, but not from the point of view of the exact analysis.
Another option is to use a PC-based virtual oscilloscope with a higher resolution of A/D converter - 16-bit standard. The limiting factor is in this case the sampling rate. This is usually considerably lower than in the case of a digital oscilloscope. The connecting wires must have a sufficient cross section and, due to the expected fast increase in current, must have a minimum of their own inductance to avoid distortion. Interconnecting wires are preferably made from speaker Cu cables.

VI. Measurement of the transition process by a digital oscilloscope

In total, four measurements were performed. In the first measurement, a laboratory source was used as a DC voltage source. For other measurements, a 12 volt battery was used as a source. In the fourth measurement, the inductor was connected to the bulb. A bulb current and a bulb voltage were measured. The aim of the measurement was to get a better idea of the speed of the transition and the current peak value under different conditions.
The steady state current of the bulb filament at the operating temperature can be calculated from the nominal voltage and power ratings. For a 12 V/20 W lamp, the current will be I = 1.66 A. The filament resistance will be in the steady state R = 7.2 Ω. The temperature coefficient for tungsten is given α = 4.5 [10^{-3} K^{-1}]. The filament temperature under operating conditions is assumed $\Theta = 3000$ K. The temperature dependence of the filament resistance is given by the following formula:

$$R_\Theta = R_{20}(1 + \alpha\Delta\Theta) \ [\Omega; \ \Omega, 10^{-3}K^{-1}, K] \tag{4}$$

R_Θ resistance at working temperature
R_{20} resistance at a basic temperature of 20 °C
α temperature coefficient for tungsten
Θ operating temperature

The filament resistance at 20 °C will be approximately R_{20} = 0.55 Ω. The resistance ratio in the cold and warm state is approximately 13. The inrush current value also limits the internal resistance of the power supply. This was determined from measurements at different source loads. To determine its value, the relationship was used:

$$R_i = \frac{U_0 - U_L}{I_L} \ [\Omega; \ V, A] \tag{5}$$

R_i resistance of voltage source
U_L voltage at load
U_0 no load voltage
I_L current under load

The maximum current value will then be:

$$I_{max} = \frac{U_0}{R_i + R_S + R_B} \ [\mathrm{A}; \ V, \Omega] \tag{6}$$

I_{max} maximal inrush current
U_0 no load voltage
R_i resistance of voltage source
R_S resistance bulb in cold state
R_B resistance of the shunt

The calculation is based on a peak value of approximately 20A. In the case of measurement of the circuit with the laboratory source, the current limitation of the source occurred and the maximum current value was limited to about 16A. When measured with a inductor, the current was similarly limited to about 16A. The best results were when measured with the battery as a power source. The following figure shows the switching curves of a halogen bulb without a current limiter.

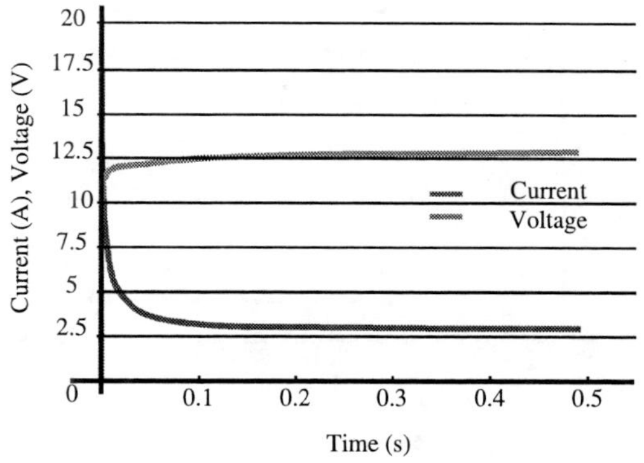

Figure 5 Measurement of the starting current of a halogen bulb without a limiter

VII. LIMITING CURRENT

From the measurements made, it is clear that all three methods of limiting current of a halogen bulb (resistor, inductor, limit switching or PWM) are usable. The most advantageous method of limiting the inrush current is the method of limiting the switching transistor.
The limiting resistor limitation method is loss-making. It limits the overall brightness of the bulb, or requires a resistor bridging after the filament is heated by relay contact.
The method of restricting the charging current by the inductor requires a demanding design of the inductor. The method of limiting the current through the switching transistor is more complicated, however, it is lossless and in the closed state almost does not affect the resulting brightness of the halogen bulb. The PWM limitation method has not been tested because the limitation by means of the switching transistor is simpler. And this method is efficient enough.
In the course of the test, the inrush current measurement for the different time constants of the switching transistor was performed. The higher the time constant, the greater the limitation of the inrush current, but also the longer switching cycle with the higher power load of the switching transistor. It was experimentally determined as the optimal value of the time constant $\tau = 0.474$ s where the inrush current is limited to twice the nominal current. In addition, it seems to be advantageous to connect the resistor to the circuit by preheating the bulb and then switching it on with a inrush current limiter. Delay in the enclosed charts captures the pre-heating bulb by this resistor. Subsequently, the resistor is bridged by the switching transistor. However, the circuit of the inrush current limitation by the filament lamp by the switching transistor is capable of operating efficiently even in the case of absence of a pre-heating resistor. The MOSFET BUZ11 transistor is used as the power element. In full open state, the power dissipation will not exceed 0.12 W.

VIII. MEASURING INSTRUMENT

For the purpose of measuring the inrush current, a measuring instrument has been designed and manufactured. The measuring device includes non-contact current switching circuits and an optional soft start circuit with an adjustable time constant. The wiring diagram of the measuring device is based on the principle of Figure 4. The power switch was implemented by a MOSFET power transistor.

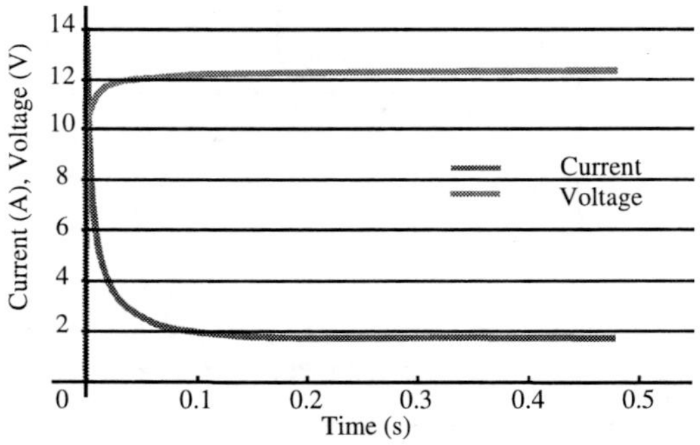

Figure 7 Measurement of the inrush current of a halogen bulb with a limiter, instantaneous switching

IX. SELECTING THE METER

Initial measurements have shown that the oscilloscope has a disadvantage to eight-bit depth of vertical resolution against computer-compatible measuring cards. On the contrary, the oscilloscope has an advantage in scanning speeds of up to 200 MS/s. Some oscilloscopes have a bit resolution greater than the basic 8 bits, but such a device was not available for measurement realization. Measuring cards have more advantageous properties at bit resolution, but this is at the expense of sampling rate.
The measurements were made using the NI 6221 card with a 16 bit resolution and sampling frequency of 250 kS/s. The measurement was two-channels, so the effective sampling frequency was maximum 125 kS/s. The scheme is shown in Figure 6.

The measurement application was programmed in LabVIEW. It consists of three main parts - data digitization,

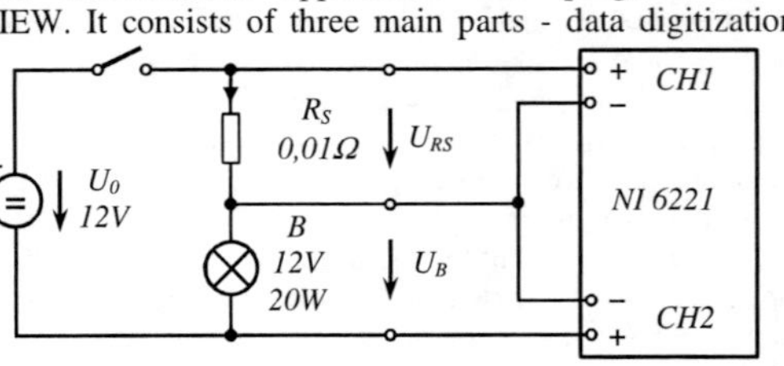

Figure 6 Measuring scheme

block analysis, and file-saving block. Applying both signals (current and voltage) continuously sampled and based on current trigger settings, the data were analyzed, displayed in a graph on the monitor and stored in a file. A 12V DC halogen bulb was used for the measurement. At a nominal voltage a current of 1.684 A was measured. The filament resistance during operation at rated voltage was determined to be 7.122 Ω. The next figure shows current and voltage curves with current limitation. The optimal time constant was set at $\tau = 0.474$ s.

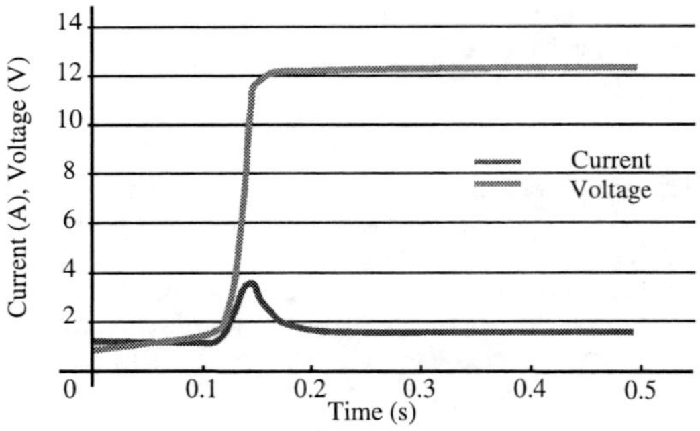

Figure 8 Measurement of the inrush current of a halogen bulb with a limiter, $\tau = 0.474$ s

CONCLUSION

In this work was studied the properties of halogen bulbs and the possibility of correct measurement of the time course of their inrush current. The timing of the inrush current was measured by means of a digital oscilloscope as well as a measuring card and a programmed measuring application in LabVIEW. In addition, a simple limiting current limiter was realized and comparative measurements for its various time constants were performed with this limiter.
The comparison of the devices used shows that a conventional digital oscilloscope has the disadvantage to measure the transition process of the inrush current of the bulb at a too small bit depth, having an adverse effect on the velocity of the

quantization error. On the other hand, the measuring card used was 16-bit with a guaranteed sampling rate of 250 kS/s.

When comparing the peak of the inrush current without the limiter and the limiter, it has been found that the slower the transition is, the more the filament is heated before the nominal current is reached. This results in less current overloading of the filament bulb. At the same time, the temperature of the filament increases more slowly and the dilation changes are slower and less destructive. The limiter can be designed so that the increase of the inrush current has a more appropriate characteristic. However, it would require a design with a limitation of the inrush current using a power transistor, or the use of a PWM limiter.

Acknowledgment

This work was supported by the Student Grant System of VSB-TU Ostrava [grant numbers SP2017/128].

References

[1] J. Drapala, M. Kursa, „Electrotechnical materials,": teaching text. First edition.Ostrava: VSB - TUO, 2012. ISBN 978-80-248-2570-0

[2] J. Habel. "Světelná technika a osvětlování", FCC Public, Praha, 1995

[3] Kitco, "Precious Metals". Internet: http://www.kitco.cz/rubriky/drahe-kovy [May 15, 2017]

[4] V. Dvořáček, "Light sources - Halogen bulbs", LIGHT 5/2008

[5] R. Goňo, V. Král, "Production and use of electric energy", teaching text, Ostrava, VŠB - TUO, 2012

[6] 21stoleti.cz, "Jsou úsporné žárovky nebezpečné? ". Internet: http://21stoleti.cz/2010/04/16/jsou-usporne-zarovky-nebezpecne Apr. 16, 2010 [May 28, 2017]

[7] Blog.cz, "Měření barvy světla zadarmo a správně". Internet: http://petr-kubac.blog.cz/1102/mereni-barvy-svetla-zadarmo-a-spravne Feb. 18, 2011 [May 25, 2017]

Comparison of Pseudorandom White-noise Generators Used as Signal Source for Wideband Analysis of Transformer Windings Impedance

Pawel Zydron, Maciej Kuniewski, Łukasz Fuśnik
Department of Electrical and Power Engineering
AGH University of Science and Technology
Krakow, Poland
pzydron@agh.edu.pl, maciej.kuniewski@agh.edu.pl, lfusnik@agh.edu.pl

Abstract — **One of the methods for diagnostic and assessment of the power transformers state is the measurement and analysis of broadband admittance/impedance characteristics of the windings. Such tests are realised by frequency response analysis (FRA) measurements, both in time and frequency domain. Usually Sweep Frequency Response Analysis (SFRA) method or Low Voltage Impulse (LVI) method with time-domain data transformation into frequency domain are used. However, other broadband signals may also be used for the identification of the broadband frequency characteristics. Theoretically, one of the signals that can be used to identify the transfer function parameters of tested circuits or systems is a *white noise* - random wideband signal. The report presents the results of the analysis of the signal properties of several pseudo-random white noise generators to evaluate their suitability in this type of study of transformers windings. The basic parameters of the voltage signals generated by them, important from the point of view of this diagnostic method, are compared. The influence of these parameters on the potential application of this type of stimulus for the study of transformer windings characteristics is analysed. The report presents and analyses the results of preliminary tests on the transformer windings, using low-voltage pseudo-random white noise signals.**

Keywords: : transformer windings; transfer function, diagnostics; frequency response analysis (FRA); white-noise

I. Transfer Function Analysis - Introduction

Diagnostic tests of power transformers windings have their very long history, starting from impulse methods with response waveform analysis in time domain [1, 2] to contemporary applied methods of admittance/impedance spectroscopy. Research and analysis of wideband frequency response characteristics of the transformer windings are used as a method for comparative assessment of their present condition [3-11]. This type of research is of great importance because power transformers are one of the most expensive components of the power systems, and their lifetime is up to 40+ or even 50+ years. The evaluation of the winding state together with the assessment of the state of their insulation is thus an important element of the commonly used *CBM* (*Condition Based Maintenance*) and *RCM* (*Reliability Centered Maintenance*). During analysis, transformer windings are considered as complex electrical networks, whose frequency-dependent electrical parameters reflect both their geometry and the material properties of the structural components. For diagnosing of defects, the winding equivalent schemes are analysed and comparisons of the measured characteristics with the reference characteristics obtained during previous measurements are made.

The *transfer function H(f)* (or *TF*), usually defined by the ratio of voltage and current spectra in different configurations [5, 11], is analysed and compared. Its evaluation is made by:

- acquisition of voltage and current waveforms in the time domain and determining of their spectra using Fourier transform (with DFT or FFT algorithms) for evaluation of the frequency response *H(f)* – used in *LVI* method;
- direct measurements in the frequency domain for determining winding characteristics *H(f)* by sequential spectrum scanning – used in *SFRA* method.

Typically, during measurements *the wideband admittance/ impedance* characteristics of the windings are determined and compared. As was mentioned, they are sensitive to changes in the geometrical dimensions of the windings, thereby detecting their deformation caused, for example, by the action of high Lorentz forces, arising in the windings during high overcurrent conditions. The causes of power transformer windings deformation can also be mechanical stresses resulting from accidents during the transport, leading to high acceleration values, stressing internal construction of the transformer.

Determination of transformer winding characteristics can be considered as a classic problem of determining an *unknown transfer function* of a system (Fig. 1). For power transformers it is usually an admittance characteristic *Y(f)*:

$$H(f) = Y(f) = I(f)/U(f) \tag{1}$$

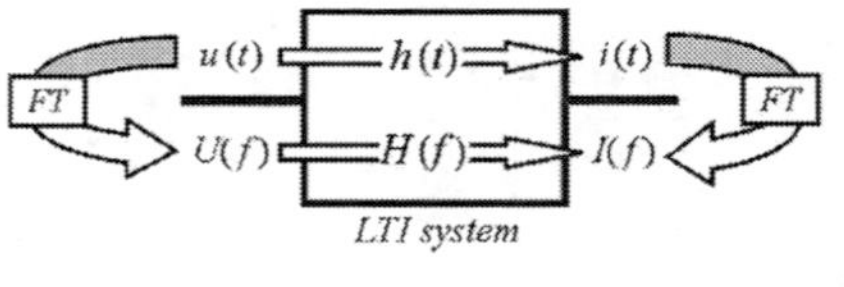

Figure 1. Impulse response *h*(*t*) and transfer function *H*(*f*) of an ivestigated *LTI* system (*FT* – Fourier Transform)

Evaluation of transfer function takes place by identifying the two frequency characteristics: e.g. the current *I*(*f*) – response signal, and the voltage *U*(*f*) – stimulating signal. Among the methods of tests used to determine the transfer function of the windings can distinguish a method using impulse excitation voltage with a broad spectrum - *LVI* (*Low Voltage Impulse*), and the *SFRA* (*Sweep Frequency Response*

Analysis) method, which tests the current response to a sequentially changed sinusoidal signal of increased or decreased frequency; in the range from f_{min} to f_{max}, with a constant or logarithmically-varying frequency step Δf.

In transfer function evaluation of transformer windings, it is assumed that the studied circuit is an object of the *LTI* (*Linear Time Invariant*) class. For this reason, usually only low-voltage excitations are used during the laboratory and off-line tests to prevent non-linear distortion in measurements. The results of on-line tests are also reported in which the stimulating voltages are the transition high level overvoltages generated during the switching operations in electric power network [6]. In recent years, new developments have been made and described on application of on-line transfer function assignment methods on in-service transformers [12-16]. Performed tests included, for example, the methods of injection of excitation signal on transformer windings, and verification of measurement sensitivity. Transfer function analysis is also used in the study of propagation of overvoltages in electrical power networks taking into account their transfer by transformer windings [17, 18].

In the study of transfer function calculated by digital acquisition and Fourier analysis of wideband electrical waveforms, excitation signals of various types can be used. The basic signal, used for many decades in the *LVI* method, is the voltage impulse. The paper reports results of the tests of several types of pseudorandom white-noise generators investigated for their use in measurements of transformer windings transfer function. Different wideband signal properties and requirements for data acquisition systems are also considered.

II. WIDEBAND SIGNALS FOR TF EVALUATION

In general, different types of voltage excitation can be used to determine *Y*(*f*), or Z(*f*), or *k*(*f*) characteristics (Fig. 2). In *SFRA* method analysed spectrum is "discrete" because continuous sine wave stimulus scan whole frequency range with selected frequency step Δf. When wideband voltage impulse or white-noise signal are used spectra of signals are continuous and their discretization is realised by sampling of signals with specified parameters. For proper selection of excitation signals, it is necessary to know their properties.

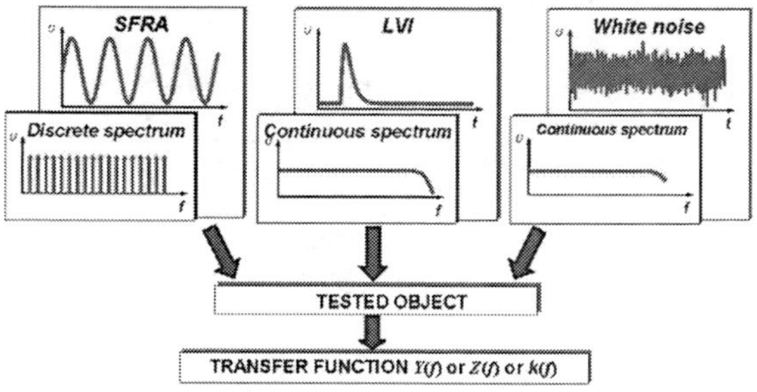

Figure 2. Selected methods of transfer function evaluation using different stimulating signals

A. *Low Voltage Impulse*

Low Voltage Impulse is a wideband, band-limited signal. For time domain description of full-wave lightning impulse waveform usually double-exponential curve is used:

$$u(t) = U_0 \cdot \left(e^{-\alpha \cdot t} - e^{-\beta \cdot t}\right) \quad (1)$$

where:

U_0 – constant voltage; α, β – attenuation coefficients, t – time.

By applying the integral Fourier transform, the spectrum of this signal can be determined [19]:

$$U(\omega) = \int_{-\infty}^{\infty} u(t) e^{-j\omega t} dt = \int_{0}^{\infty} U_0 \left(e^{-\alpha \cdot t} - e^{-\beta \cdot t}\right) \cdot e^{-j\omega t} dt = \\ = \frac{U_0(\beta - \alpha)}{(\alpha + j\omega)(\beta + j\omega)} = \frac{U_0(\beta - \alpha)}{\alpha\beta + j\omega(\alpha + \beta) - \omega^2} \quad (2)$$

This is the spectrum of a signal with limited band, having two characteristic frequencies associated with the values of attenuation coefficients of the first and second exponential component (Fig. 3).

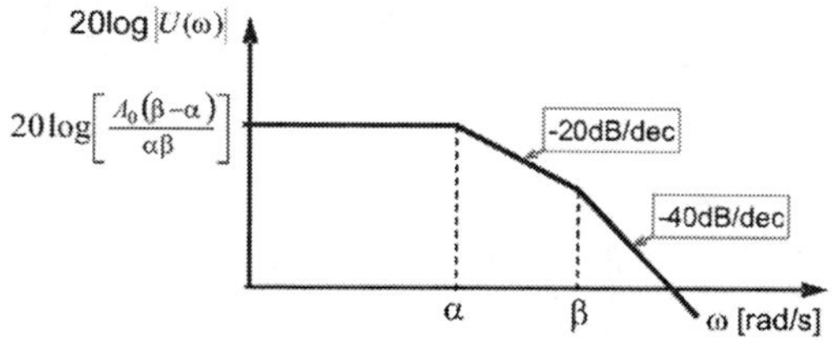

Figure 3. Spectrum of double-exponential impulse waveform

For the standard full lightning-impulse (defined in IEC 60060), the main part of its energy is accumulated in the range of relatively low frequencies (a dozen kHz). This range is considerably broadened for impulses chopped at front and is approximately inversely proportional to *time to chopping* T_C. Figure 5 illustrates the relationships between spectra for varying T_C time of impulses chopped at front and for full 1,2/50 μs impulse. Analysis of the spectra from Fig. 4 indicates that for the typical *TF* frequency ranges specified for windings of transformers (up to about 2 MHz) the excitation voltage impulse widths should be no greater than 0.5 μs. For wider impulses, frequency components closer to 2 MHz will have lower energy then lower frequency components.

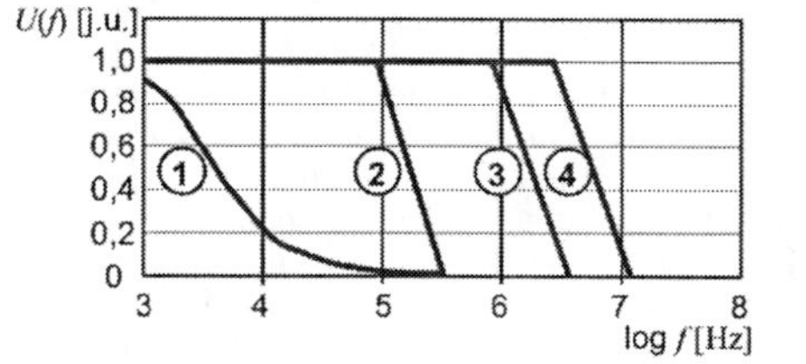

Figure 4. Approximated spectrum of selected voltage pulses [20]: 1) full lightning impulse 1,2/50 μs; chopped voltage impulses: 2) T_C=5 μs 3) T_C=0,5 μs 4) T_C=0,1 μs (T_C – time to chopping)

B. *White-noise*

White-noise is a special kind testing signal [21] having theoretically constant value of *power spectral density* (*PSD*) at all frequencies. This is random signal with zero mean value, and constant value of variance, whose autocorrelation function $R_x(\tau)$ is non-zero only for $\tau = 0$. These properties cause that white-noise is eliminated in measuring methods using time domain averaging [22]. When evaluating the transfer function, the averaging refers to successive approximations of this function obtained for the independent realization of the time sequences of pseudorandom white-noise. In practical situation pseudorandom white-noise signal with limited bandwidth and time duration are used. Considering properties of random white noise in time domain rounded relation between peak-to-peak voltage U_{pp} and effective voltage U_{rms} can be written [23]:

$$U_{pp} = (6 \div 8)\ U_{rms} \quad (3)$$

If pseudorandom white-noise signal is used to evaluate the transfer function of any system, it should have a constant PSD value in the frequency range for which this function is determined.

III. Measuring Instruments for SFRA Method

A relatively simple, but very precise measuring instrument/system for evaluation of the winding response as a function of frequency is the *Frequency Response Analyzer* (*FRA*). Basic design solutions of this type of instrument are based on a correlator circuit in which the selected output signal from tested object is correlated with two synchronous orthogonal $\sin\omega t$ and $\cos\omega t$ reference signals; the first refers to excitation voltage signal and the second is an additional reference signal used for proper complex signal analysis (Fig. 5). For longer correlation times, the value of the signal-to-noise ratio *SNR* is increased.

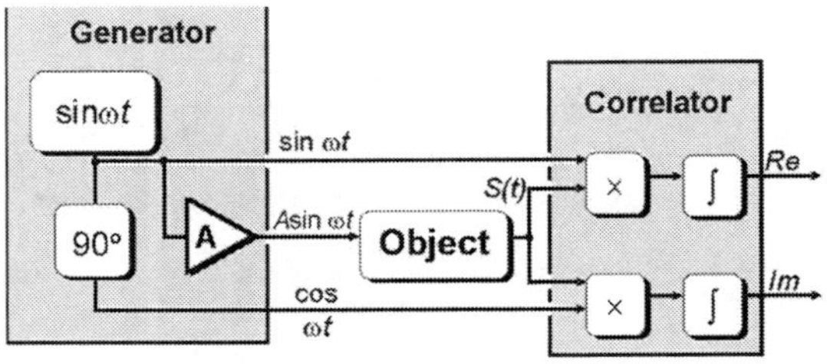

Figure 5. Block diagram of the Frequency Response Analyser

The *SFRA* method because its many advanteges: precision, high value of *SNR*, flexibility in freqency range setting, constant stimulus energy (U^2) at all frequencies, etc. is commonly used for transfer function evaluation in laboratory and off-line measurements.

IV. Transfer Function Evaluation by Acquisition of Signals in the Time Domain

Ideally, the input and output signals are not affected by the external random noise. In real situations, both the signal recorded at the input of the test system (e.g. voltage on winding) and its output (e.g. current in winding) are disturbed by the presence of uncorrelated random noise of unknown characteristic (Fig. 4). In this situation, the transfer function is estimated based on the sum of the individual signal and the accompanying noise [21].

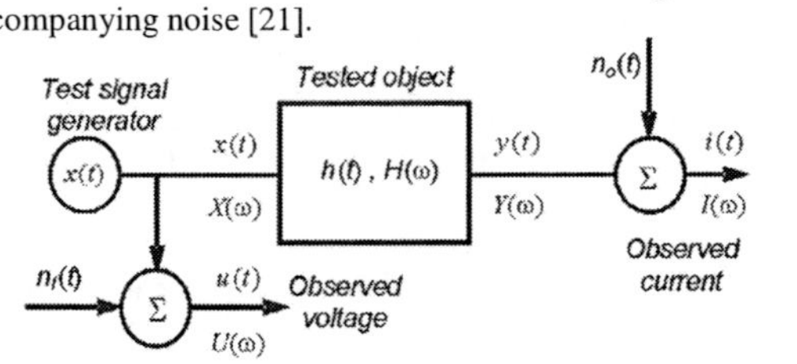

Figure 6. Estimation of impulse response $h(t)$ and transfer function $H(f)$ of an unknown *LTI* system with additive noises at input and output of tested system

Taking into account the relationship between the sampling frequency f_s, the number of samples N, the required spectral range (i.e. f_{max}) and the frequency analysis resolution Δf, they must be chosen in such a way that:

- sampling frequency f_s must satisfy Shannon's theorem and must also allow *TF* analysis within the required range - usually up to 2-3 MHz;
- the total recording time, depending on the number of samples N, must provide the required frequency resolution:

$$\Delta f = f_s/N \tag{4}$$

Knowing that the recorded waveform and spectrum of the signal always satisfy the Parseval's theorem (the energy of the spectrum is equal to the energy of the waveform), the frequency spectrum of the excitation signal should be chosen to fully utilize the available dynamic range of the excitation voltage. This means that the spectrum of the excitation signal should not exceed the range of TF analysis.

V. Results of Measurements

During laboratory test pseudorandom white-noise signal from three different generators were compared:

- Analogic model 2030,
- Tektronix model AFG 310,
- Tektronix model AFG 2021.

Measurements were performed with Tektronix 784D digital oscilloscope, in configuration with no-load on generator. Figures 7 to 10 presents frequency spectrum of measured white-noise. Calculations of Fourier spectrum were made for 10 and 100 ksamples. Results shows that analyzed generators have different upper band limit. For AFG 310 maximum stimulated frequency is approximately 10 MHz (Fig. 7), and for AFG 2021 is more than 50 MHz (Fig. 8). For Analogic 2030 generator maximum frequency band is related to chosen spectral line spacing; in measurements spectra for two values of line spacing (f_{ls}) were analyzed, 10 Hz and 1000 Hz, and for this values frequency limit is 300 kHz (Fig. 9) and 3 MHz (Fig. 10), respectively.

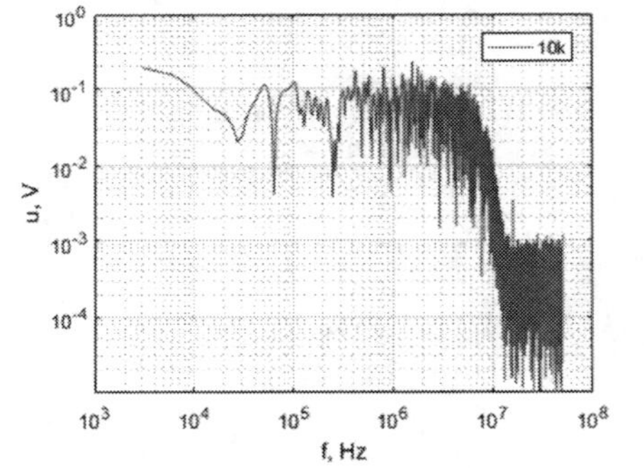

Figure 7. Frequency spectrum of output signal from Tektronix AFG310, calculated for 10k samples

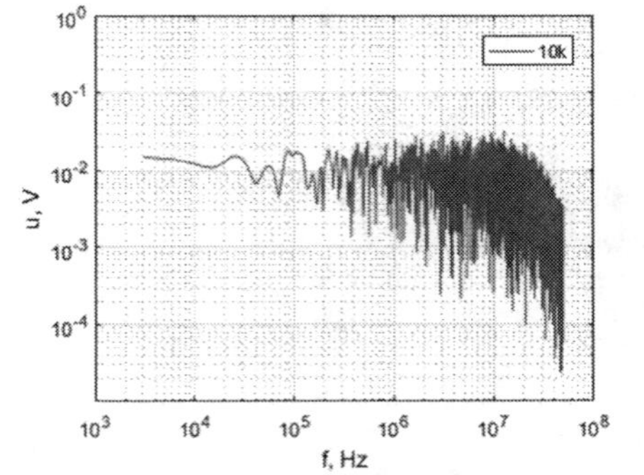

Figure 8. Frequency spectrum of output signal from Tektronix AFG2021, calculated for 10k samples

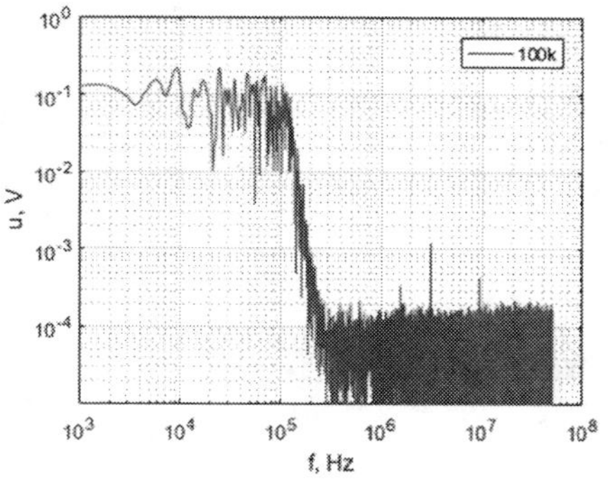

Figure 9. Frequency spectrum of output signal from Analogic 2030 (f_{ls} =10 Hz), calculated for 100k samples

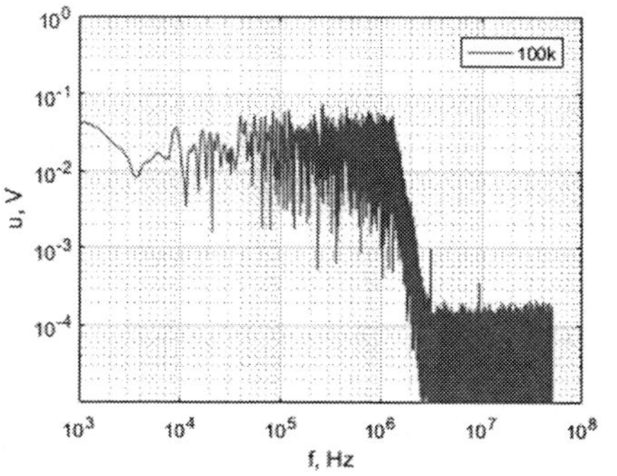

Figure 10. Frequency spectrum of output signal from Analogic 2030 (f_{ls} =1000 Hz), calculated for 100k samples

Second step in measurements was to compare frequency characteristics measured with white-noise stimulus to characteristics obtained with SFRA method. The SFRA stand consisted of generator Tektronix type AFG 310, a digital storage oscilloscope Tektronix type TDS 1000/2000 with analogue frequency band 100 MHz, sampling frequency 1 Gs/s, connected to the host computer by a GPIB-USB interface. The software based control, processing and analysis has been implemented in LabView™ (National Instruments) graphic programming environment. The sinusoidal voltage with amplitude of 20 V and stepwise variable frequency was applied to the terminals of the primary coil (HV). The frequency characteristics of the secondary coil voltage (LV) were recorded at discrete frequency steps for 200 measuring points in the frequency range from 10 kHz up to 1 MHz. The software was used to process and archive the measurement results and to determine the values of transformer wideband voltage ratio.

During laboratory tests, *TF* as a ratio of voltage at *Low Voltage* side to *High Voltage* side *k(f)* of tested 20 kVA distribution transformer were chosen. Results presented in figures Fig. 11 to Fig. 13 allow comparison of *TF* measured with SFRA sine wave stimulus and white-noise stimulus. During measurements of *TF* with white-noise averaging of frequency spectra was used to improve the signal to noise ratio. Results presented in Fig. 11 shows white-noise TF for single time data acquisition Characteristics in Fig. 12 and Fig. 13 were evaluated as averaged spectra of 10 measurements, for different length of data record 10 ksamples, and 50 ksamples, respectively. Averaging in frequency domain shows good influence on estimated *TF*. Transfer function obtained using white-noise stimulus shows similarity with the transfer function obtained with the SFRA method.

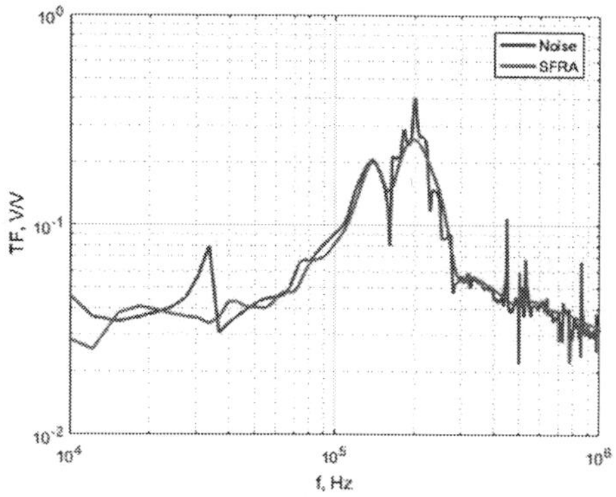

Figure 11. Comparison between measurement of transfer function obtained with SFRA method and with Noise method (Tektronix AFG310, 50 k samples, spectrum of single measurement)

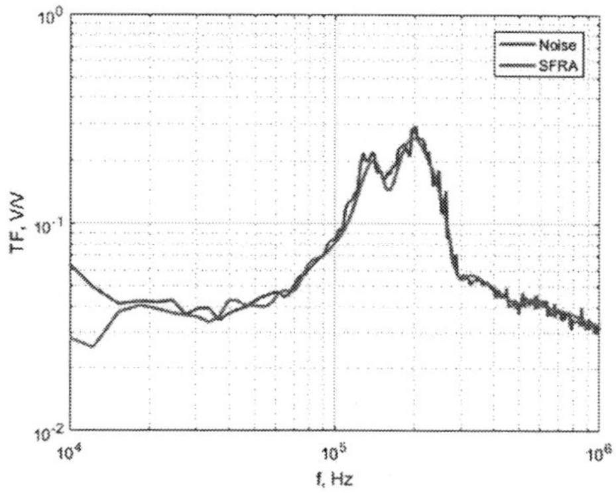

Figure 12. Comparison between measurement of transfer function obtained with SFRA method and with Noise method (Tektronix AFG310, 10 k samples, averaged spectrum from 10 measurements)

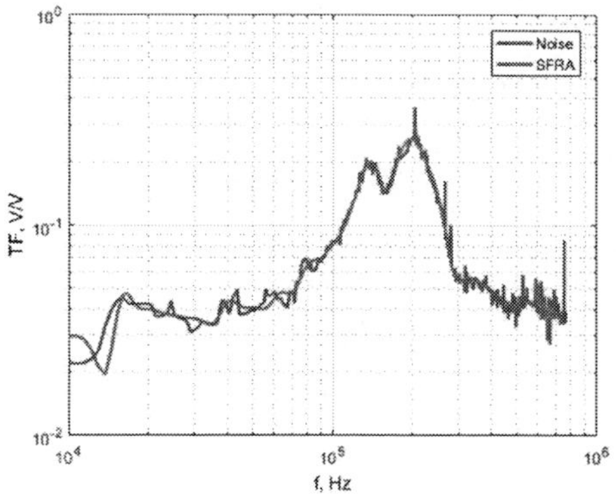

Figure 13. Comparison between measurement of transfer function obtained with SFRA method and with Noise method (Tektronix AFG310, 50 k samples, averaged spectrum from 10 measurements)

VI. Summary

Analysis of transfer function *TF* of power transformer windings is a powerful tool for determining their current state, based on reference data. For many years, work has been done on the routine use of these type of tests in on-line research, on transformers in operation. This paper describes the preliminary work on the estimation of the transfer function characteristic based on measurements using white-noise stimulus. Analysis of the presented results indicate that the obtained characteristics show similarity to the *TF*s recorded by SFRA method. Further work will be aimed at evaluating the quality of the obtained results based on the correlation analysis and the coherence function [21, 24].

ACKNOWLEDGMENT

The researches described in the paper were financed by the Polish Ministry of Science and Higher Education - grants for Faculty of Electrical Engineering, Automatics, Computer Science and Biomedical Engineering of AGH University of Science and Technology, Krakow, Poland.

REFERENCES

[1] J. H. Hagenguth, "Progress in impulse testing of transformers", Trans. of the American Institute of Electrical Engineers, vol. 63, pp 999-1005, 1944, doi: 10.1109/T-AIEE.1944.5058842.

[2] W. Lech and L. Tyminski, "Detecting transformer winding damage - the low voltage impulse method", Electr. Rev., vol. 179, 768–772, 1966.

[3] R. Malewski and B. Poulin, "Impulse testing of power transformers using the transfer function method", IEEE Trans. Power Deliv., vol. 3, pp. 476–489, 1988, doi: 10.1109/61.4283.

[4] T. Leibfried and K. Feser, "On-line monitoring of transformers by means of the transfer function method", Proc. 1994 IEEE ISEI, pp. 111-114, doi: 10.1109/ELINSL.1994.401455.

[5] J. Christian, K. Feser, U. Sunderman, and T. Leibfried, "Diagnostic of power transformers by using the transfer function method" ISH'99: 11th Int. Symp. on High Voltage Engineering, London, vol. 1, pp. 37–40, 1999, doi: 10.1049/cp:19990502.

[6] T. Leibfried and K. Feser, "Monitoring of power transformers using the transfer function method", IEEE Trans. Power Deliv., vol.: 14, pp. 1333-1341, 1999, doi: 10.1109/61.796226.

[7] E. P. Dick and C. C. Erven, "Transformer diagnostic testing by frequency response analysis", IEEE Trans. Power Apparatus and Systems, vol: PAS-97, pp. 2144–2153, 1978, doi: 10.1109/ TPAS.1978.354718.

[8] S. A. Ryder, "Diagnosing transformer faults using frequency response analysis", IEEE Electr. Insul. Mag., vol. 9, no. 2 , pp. 16-22, 2003, doi: 10.1109/MEI.2003.1192032.

[9] M. Florkowski and J. Furgał, "Detection of transformer winding deformations based on the transfer function – measurements and simulations", Meas. Sci. Technol., vol. 14, 2003, pp. 1986–1992, doi:10.1088/0957-0233/14/11/017.

[10] S. Tenbohlen and S. A. Ryder, "Making frequency response analysis measurements: A comparison of the Swept Frequency and Low Voltage Impulse methods", XIII Int. Symp. on High Voltage Engineering ISH, Netherlands, 2003.

[11] M. Florkowski and J. Furgał, "Transformer winding defects identification based on a high frequency method", Meas. Sci. Technol., vol. 18, pp. 2827-2835, 2007, doi:10.1088/0957-0233/18/9/012.

[12] T. De Rybel, A. Singh, J. A. Vandermaar, M. Wang, J. R. Marti, and K. D. Srivastava, "Apparatus for online power transformer winding monitoring using bushing tap injection", IEEE Trans. Power Deliv., vol. 24, pp. 996-1003, 2009, doi: 10.1109/TPWRD.2009.2022674.

[13] M. Bagheri, M. S. Naderi, and T. Blackburn, "Advanced transformer winding deformation diagnosis: moving from off-line to on-line", IEEE Trans. Dielectr.Electr. Insul., vol. 19, pp. 1860-1870, 2012, doi: 10.1109/TDEI.2012.6396941.

[14] M. Bagheri, M. S. Naderi, T. Blackburn, and T. Phung, "Frequency response analysis and short-circuit impedance measurement in detection of winding deformation within power transformers", IEEE Electr. Insul. Mag., vol. 29, no. 3, pp. 33-40, 2013, doi: 10.1109/MEI.2013.6507412.

[15] E. Gomez-Luna, G. A. Mayor, C. Gonzalez-Garcia, and J. P. Guerra, "Current status and future trends in Frequency-Response Analysis with a transformer in service", IEEE Trans. Power Deliv., vol. 28, no. 2, pp. 1024-1031, 2013, doi: 10.1109/ TPWRD.2012.2234141.

[16] Yong Liu et al., "A study of the sweep frequency impedance method and its application in the detection of internal winding short circuit faults in power transformers", IEEE Trans. Dielectr. Electr. Insul., vol. 22, pp. 2046-2056, 2015, doi: 10.1109/TDEI.2015.004977.

[17] M. Florkowski, J. Furgal, and M. Kuniewski, "Propagation of overvoltages in distribution transformers with silicon steel and amorphous cores", IET Gen. Trans. Distrib., vol. 9, pp. 2736-2742, Dec. 2015, doi: 10.1049/iet-gtd.2015.0395.

[18] M. Florkowski, J. Furgal, and M. Kuniewski, "Propagation of overvoltages transferred through distribution transformers in electric networks", IET Gen. Trans. Distrib., vol. 10, pp. 2531-2537, Jul. 2016, doi: 10.1049/iet-gtd.2015.1500.

[19] K. L. Kaiser, "Electromagnetic Compatibility Handbook", CRC Press, Boca Raton, 2005.

[20] P. Zydron, "Processing and analysis of the high voltage lightning impulses signals" (in Polish), Przeglad Elektrotechniczny (Electrical Review), vol. 86, no. 11B, pp. 304-307, 2010.

[21] J. S. Bendat and A. G. Piersol, "Random Data: Analysis and Measurement Procedures", 4th ed., Wiley, 2010.

[22] R. Wlodek, Z. Florkowski, P. Zydron, "Improvement of signal to noise ratio in partial discharge measurements", Proc. 4th ICPADM, vol. 2, pp. 590-593, 1994, doi: 10.1109/ICPADM. 1994.414079P.

[23] S.W. Smith, "The Scientist and Engineer's Guide to Digital Signal Processing", 2nd ed., California Technical Publishing, San Diego, 1999.

[24] R. Roth, "Effective measurements using digital signal analysis", IEEE Spectrum, vol. 8, no.4, pp. 62-70, 1971, doi: 10.1109/ MSPEC.1971.5218046.

A Fault Tree Cyber Security Analysis of Power System State Estimation Software

Irina Kolosok, Elena Korkina
Melentiev Energy Systems Institute
ESI SB RAS
Irkutsk, Russia
kolosok@isem.irk.ru, korkina@isem.sei.irk.ru

Anatolijs Mahnitko
Riga Technical University
RTU
Riga, Latvia
mahno@eef.rtu.lvl

Abstract— **The paper is concerned with the material devoted to the problem of resistance of the state estiamtion (SE) procedure to cyberattacks on the system for collection and processing of PMU measurements, i.e. wide area monitoring system (WAMS), and the SE procedure itself. Based on the analysis of potential cyberattacks, which reveals the most vulnerable WAMS components, we show the conditions for a failure of the SE procedure. The method of fault tree is proposed to describe the consequences of cyberattacks on the SE procedure. The fault tree method is applied to the software of power system state estimation in WAMS environment in real time. The upper level of the fault tree represents a group of system indices, i.e. the algorithms applied in the state estimation procedure, base scheme of the EPS and measurements. The middle level represents events that cause faults, and the lowest level is the types of actions on the fault causes. Currently, the results of the analysis of the received measurement data and results of the state estimation are used to assess the level of cyber-resistance of the state estimation software.**

Keywords: Power system state estimation; WAMS; PMU data; cyberattacks; fault tree.

I. Introduction

The IT specialists in many countries solve the problem of cyber security. Moreover, in some countries this problem is solved at the state level. Energy is one of the key production and social spheres. Therefore, cyber-security should be ensured at all facilities, to avoid cyberattack consequences in the form of wrong decisions on control systems of energy facilities.

An undoubted condition for the success of decisions on the control of electric power system is reliable information on state variables, which is obtained by processing measurement data by the state estimation (SE) methods. The main requirement for the SE procedure is high accuracy of the estimates to be obtained. Therefore, various algorithms for observability analysis of calculated scheme, methods for a priori and a posteriori input data validation and methods for robust SE [1-5 and others] are developed and applied to the SE procedure. The measurements involved in state estimation are SCADA measurements and/or PMU measurements [6,7]. The latter are preferable since in this case the state estimation problem is solved by linear algorithms without iterations, which provides time gain so important in an on-line mode. The PMU measurements, as well as SCADA measurements, need to be validated [8,9]. Especially the requirement for their validation is important when cyberattacks occur at the power industry facilities.

The paper presents a fault tree method to analyze cyber security of state estimation software in the event of cyberattacks.

II. Statement Of The State Estimation Problem

The state estimation procedure is aimed at obtaining the estimates of state variables of electric power system on the basis of telemetry data and/or synchronized phasor measurements. The state estimation procedure is based on the minimization of objective function [2,3]:

$$J(x) = [\bar{y} - y(\hat{x})]^T R_y^{-1} [\bar{y} - y(\hat{x})], \qquad (1)$$

where $x = \{U, \delta\}$ - state vector, $\hat{x}$ - estimates of state vector, $\bar{y} = \{U, P, Q, I\}$ - phasor measurements of voltage magnitudes U, currents I and powers P, Q, R_y - covariance matrix of measurement errors;

Methods and algorithms devised earlier for state estimation on the basis of telemetry data are adapted and updated because of the advent of next generation measurements, i.e. synchronized phasor measurements. Now, the procedure of state estimation apart from telemetry data can include synchronized phasor measurements $\bar{y} = \{U, \delta, I, \psi\}$ in the vector of measurements, which increases the redundancy of measurements. However, since telemetry data and synchronized phasor measurements have different accuracy, the heterogeneity of measurements increases, i.e. the conditioning of the Jacobian matrix, as the main component of the mathematical support of the state estimation, degrades. Therefore, solving the state estimation task on the basis of synchronized phasor measurements alone is more attractive [10].

When all measurements are considered in Cartesian coordinates, the model of measurements (2), where $x = \{U_{ai}, U_{ri}\}$, will become linear and the Jacobian matrix (3a) will be stationary. The estimates of state vector $\hat{x}$ can be obtained from:

$$\hat{x} = \left[H^T R_y^{-1} H \right]^{-1} H^T R_y^{-1} \bar{y}, \qquad (2)$$

without iterations, when the Jacobian matrix H is constant

III. Fault Tree

In 1962, the term "fault tree" first appeared to characterize a field-specific expert system in the military aviation, then in the nuclear energy, and later in the other sectors [11]. The fault tree represents a hierarchical structure expanding downwards.

- Root of the fault tree is always a considered object,
- System indices are characteristics of the object;
- Elements of system indices;
- Events that lead to a failure, the so called "leaves" (or technological problems) that go from the elements.
- Countermeasures to eliminate the problem.

The advantage of the fault tree over a more known tree of decisions [12] is a reduction in the total amount of possible problem solving variants to the number of problems that represent an obvious threat to the considered technical facility. Thus, the fault tree operates as a diagnostic method that suggests a real solution to a concrete problem situation.

The technology of fault tree analysis brought about the studies aimed at solving information problems, considering cyber threats. The cyberattack directed towards the information and communication system of an energy facility leads to a failure of measurement equipment; causes network traffic congestion, which results in either a delay in measurements or their failure to arrive at the computational procedures at all; corrupts measurements to cause wrong computations on failed input data.

The authors of [13] propose a fault tree structure to analyze information and communication system of wide area monitoring system (WAMS) of an energy facility (substation) in the event of observability loss. Structurally WAMS consists of the data measuring system that includes recorders of phasor measurements (PMU), phasor data concentrators (PDC) and data transfer channels (DTC). The method of observability analysis is used to calculate the estimates of state variables on the basis of available measurements. The observability analysis of an energy facility flow chart lies in the consideration of the amount and composition of measurements that come to the server or phasor data concentrator of the energy facility. Thus, in [13]:

- The root of fault tree is loss of substation observability;
- Level I is system indices (technical factors that lead to the observability loss: failure of data measuring system, failure of phasor data concentrator, failure in communication channels);
- Level II is a set of components (on the example of a failure in the data measuring system, i.e. current and voltage phasor measurements do not come);
- Level III (leaves) is a set of reasons (failures in the data measuring system: failure of an individual PMU, failures of instrument current and voltage transformers, and failures of the connections between the PMU and instrument transformers);
- Level IV is a set of measures on elimination of the failures.

Similarly to the fault tree in [14] we propose a tree of attacks for a systematic estimation of vulnerability of the internal network of the power system control center. The attack tree is used as a basis for numerous scenarios of testing the intrusions and studying the gaps in the cyber security system.

In [14] the authors calculate security of an energy facility, and in [13] the focus is made on the check of cyber security of local WAMS. In some of the studies [15-19] the authors describe the information problems that arose during state estimation in the case of cyberattacks. It is known that cyberattacks are most successful when aimed at unprotected components of an object. In terms of state estimation procedure these are: incomplete observability and poor quality of measurements and measuring equipment, downsides of algorithms, inhomogeneity of equivalent circuit components. In [20] the authors assess the observability preservation in the case of measurements failure or their poor quality. The research presented in [21] focuses on an analysis of observability reliability in the event of PMU failure. In [22] the authors demonstrate sensitivity of certain nodes of a scheme to external disturbances. All these cyberattacks are not mentioned directly in these studies, but what is a cyberattack if it is not a loss of measurements, failure of sensors, and presence of external disturbances? Bad data in measurements are detected by various methods [3,5,9,10,23,24,25], meanwhile their emergence can also be a result of a malicious action.

IV. A Probabilistic Approach To The Assessment Of State Estimation Procedure Cyber Security

Construction of a fault tree is based on the probabilistic approach: it is known with a certain probability $\boldsymbol{p}$ that the object operates without failures. Then we can conclude that with a probability $\boldsymbol{q=1-p}$ the object operation can fail. Here the certain factors that have some numerical form should coincide in time.

For example, for the failure of the local network of a facility [14], which represented malfunction of a backup control system of substation, it is necessary to damage communication servers (attack G_{14}) and/or the databases became accessible (attack G_{15}). The full scenario that leads to such a failure $V_1 = \prod_{i=1,14,15,16} G_i$, covers also "backdoors" in wireless communication (attack G_1), and damage of real-time servers (attack G_{16}). At any definite time instant the network security is expressed as $V_{sys} = \max\{V_j\}$, where V_j – the j-th set of intrusions characterized by different vulnerabilities of the network. After the attacks [14], the vulnerability indices are calculated considering preliminary assessment of cyber security (there was no attack/ there was an attack and measures were taken/ there was an attack and measures were not taken). Comparison of the obtained vulnerability indices, considering countermeasures, fosters the detection of the most unprotected places in the network.

Analysis of WAMS cyber security at a substation [13] the quantitative calculation of probabilistic measures demonstrates the most vulnerable sites in terms of cyber security of scheme observability, considering probabilities of failures of the measuring system Q_{meas}, phasor data concentrator Q_{PDC} and communications Q_{com} [14]:

$$Q_{sub\,i} = 1-(1-Q_{meas})(1-Q_{PDC})(1-Q_{com})) \quad (3)$$

The issue of state estimation cyber security is very serious, therefore, considering the experience of [15-25] we propose the statement of this problem as a tree of faults for the state estimation procedure.

V. Tree Of Faults For The Determination Of Cyber Security Of State Estimation Procedure

We will consider the software of electric power system state estimation procedure [25] as a failure-prone object. Let the software operate in the WAMS environment in real time.

The most important components, i.e. system indices, are: 1) current calculated scheme of an energy facility, 2) measurements and 3) algorithms applied to solve the state estimation problem. Each of the system indices can fail with some probability:

Failure of the measurement system, Q_{meas} : desynchronization of measurements; lack of measurements, which leads to the emergence of critical groups and critical measurements, and even to the loss of observability, and emergence of bad data.

Failure of the index "Calculated scheme", Q_{cs} , implies incorrect construction of a scheme down to its division into parts; erroneous values of the equivalent circuit parameters; corruption of base model which when overlapped with telesignals becomes a calculated scheme.

Failure of algorithms Q_{al} , is caused by highly inhomogeneous measurements and/or non-uniform equivalent circuit. It leads to non-convergence of the state estimation algorithm; a large number of the initial data errors, which the preprogramed bad data detection algorithm cannot cope with; insufficient amount of measurements, which results in the loss of total observability of the scheme, etc.

In [19] the authors show how many *physically vulnerable sites* there are in WAMS. Malicious attack on them will cause malfunction of the SE procedure, i.e. change in the functionality of PMU sensors; physical damage of communication channels, overload of network traffic (which prevents timely data transfer to the SE procedure); overwhelming the server with tasks, and impossibility of program startup; asynchronous telemetry and PMU measurements, etc. In terms of a fault tree, the above vulnerable sites should be located at the level of "leaves", i.e. events that cause the malfunction of SE procedure. Figure 1 presents a tree of faults for the power system state estimation software that operates in the WAMS environment in real time:

- Root of the fault tree represents malfunction of the SE procedure.
- Level I – system indices (incorrect measurements, distortion of a network design scheme, failure of algorithms);
- Level II – a set of elements (from desynchronization of measurements to incorrect completion of state estimation calculation);
- Level III – a set of reasons for the failures, including cyber-attacks: CA1 –hack of a network, CA2- false data injection, CA3- man-in-the-middle, CA4 –loss of synchronization in PMU measurements, CA5 – breach of database, CA6 – physical damage of channels, CA7- cyber-attack on software. The number of such faults, i.e. technological and organizational failures, as well as malicious intrusions, is great. Therefore, it is necessary to develop countermeasures to enhance the software fault tolerance.
- Level IV – countermeasures (c_number) to eliminate the faults:

c1- strict password policy to enter the system;

c21- generation of a response of scheme nodes to external disturbances (data injection);

c22- startup of the data validation program;

c3- monitoring of network processes;

c4- monitoring of a difference between the node voltage angles (loss of synchronization is seen in the WAMS graphical application on the screen of an operator);

c5- strict policy of database access rights;

c6- PMU testing;

c7- verification of instrument transformers;

c8- fiber optic cabling;

c9- channel integrity control;

c10- identification of equivalent network parameters (ENP) under weather conditions;

c11- antivirus and licensed software

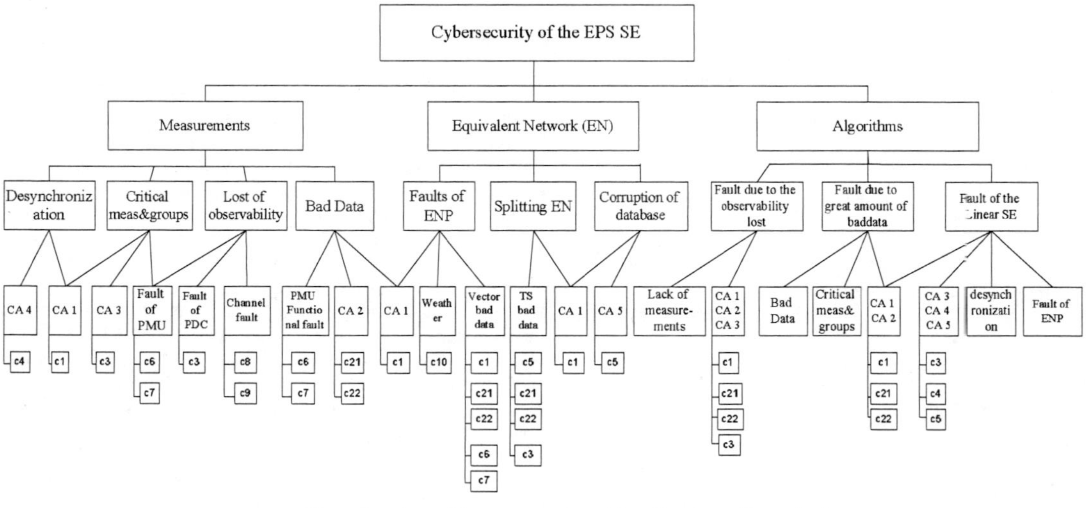

Figure 1. A tree of the power system state estimation software faults

The tree should be constructed highly accurately, since the available redundant objects make the calculations of the resultant probability of faults incorrect. Therefore, although Figure 1 is informative, it is necessary to remove recurring elements from it. The restructuring of the fault tree reveals that the algorithms become the most important system index of the **cyber-security** of the SE software and the calculated scheme and measurements shift to a lower significance level (remaining to be system indices).

Figure 2 presents the fault tree of the state estimation software after removal of recurring objects from it

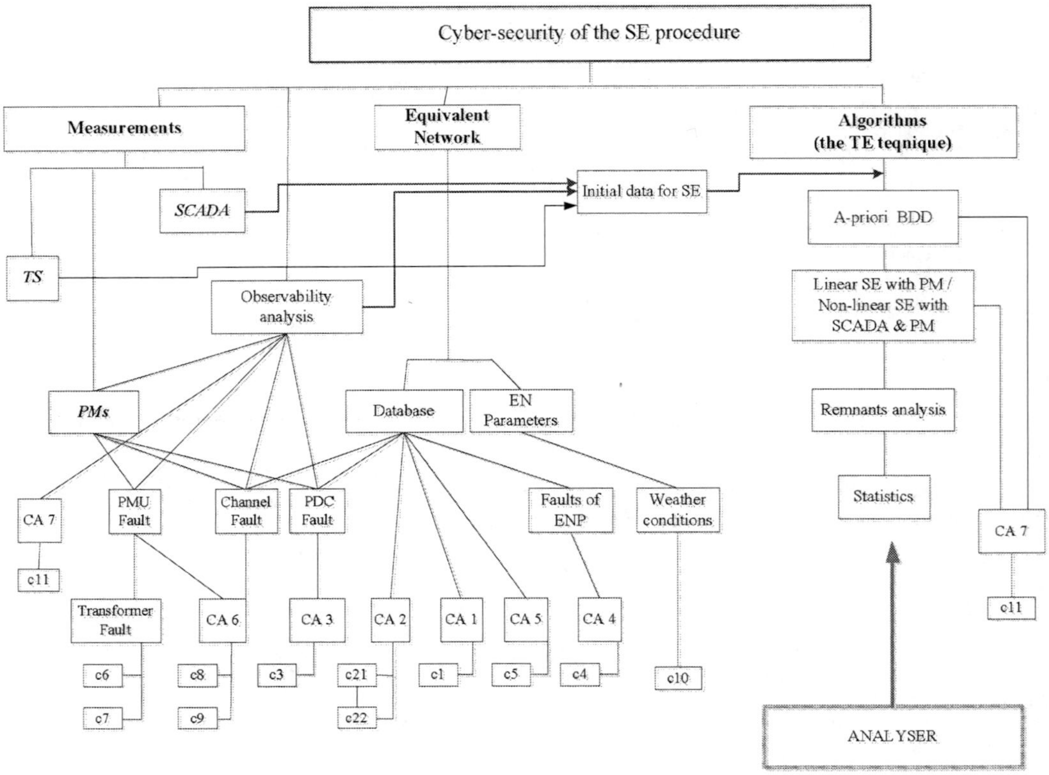

Figure 2. The final version of the fault tree

The technology of building such trees suggests 2 ways: analytical and computational. The analytical approach detects the most vulnerable links in the facility, whereas the computational one calculates a resultant probability value of the facility operability (in the range from 0 to 1) under cyber-attack.

VI. Case Study

Quantitatively the cyber-security of the SE software P_{SE} is the probability that this software will remain serviceable in the case of malicious intervention:

1. $P_{SE} = 1 - Q_{SE}$, where $Q_{SE} = 1 - \prod_i (1 - Q_{a\lg_i})$ - vulnerability (probability of faults) of the SE software, $Q_{a\lg_i}$ - fault probability of the programs.

2. Each system index is described by a group of system elements:

$$Q_{PMs} = 1 - (1 - Q_{PDC})(1 - Q_{channel})(1 - Q_{PMU})$$

$Q_{PDC} = 1 - (1 - Q_{CA3})$; $Q_{channel} = 1 - (1 - Q_{CA6})$;

$Q_{PMU} = 1 - (1 - Q_{Instr\_Tr-r})(1 - Q_{CA6})$;

$Q_{ENP} = 1 - (1 - Q_{weather\_cond})$;

$Q_{DataBase} = 1 - (1 - Q_{CA1})(1 - Q_{CA2})(1 - Q_{CA5})$;

$Q_{synchr} = 1 - (1 - Q_{CA4})$;

$Q_{observ} = 1 - (1 - Q_{PMU})(1 - Q_{PDC})(1 - Q_{channel})(1 - Q_{CA7})$

$Q_{apr\_BDD} = 1 - (1 - Q_{BDD\_initial})(1 - Q_{PMs})$;

$Q_{nonlin\_SE} = 1 - (1 - Q_{SE\_initial})(1 - Q_{apr\_BDD})(1 - Q_{CA7})$;

$Q_{lin\_SE} = 1 - (1 - Q_{SE\_initial})(1 - Q_{apr\_BDD})(1 - Q_{CA7})$;

The state estimation problem is solved by the test equation method [3]. The Observability analysis of synchronized phasor measurements is carried on first. If the synchronized phasor measurements are sufficient for the state estimation, then Validation of the initial data is performed a priori. In the absence of a great number of gross errors in synchronized phasor measurements the linear state estimation is used. Otherwise, the nonlinear state estimation with attraction of telemetry from SCADA is required, since it is more resistant to gross errors. Then the residuals (differences between the measurements and their estimates) are analyzed based on the results of any solution method. It is necessary to collect statistics on each performed run of the SE software – list of bad data, applied method (linear/nonlinear), availability of errant snapshots and emergency terminations. The SE software is run once every 5 seconds.

The **cyber-security** of the power system state estimation software (the program **ANALIZATOR**) can be calculated once every 5 minutes. During this time the server receives 15000 snapshots of WAMS (1 snapshot/20 mc).

The initial data for the program **ANALIZATOR**:

Q_{CA_i} - statistical index, initially equal to 0.

$Q_{BDD\_initial} = 0.001$, $Q_{nonlin\_initial} = 0.001$,

$Q_{lin\_initial} = 0.001$, Q_{PMU_i} for nodes of different connectivity are taken from [13] for 9 node IEEE scheme, $Q_{channel\,j}$ - from [26].

Calculation : for 5 min. of the software work 10 snapshots of synchronized phasor measurements received were errant, i.e.,

$$Q_{PDC} = \frac{10}{15000} = 0.00067 \text{ , then}$$

$$Q_{PMs} = 1-(1-0.00067)(1-0.001)(1-0.0879) = 0.0894 ;$$

$$Q_{observ} = 0.0894; \quad Q_{apr\_BDD} = 1-0.999*0.9105 = 0.0904,$$

$$Q_{lin\_SE} = 1-0.999*(1-0.0904)*(1-0) = 0.091;$$

$$Q_{SE} = 1-\prod_i (1-Q_{al_i}) = 1-(1-0.0894)(1-0.0904)(1-0.091) = 0.246$$

With such number of errant snapshots $P_{SE} = 0.754$.

Conclusion

The paper describes construction of the fault tree for the power system SE software in real time. The current SE software in the WAMS environment can experience cyberattacks. Therefore, its operation quality can be monitored by calculating the indices of failure-free operation of its components.

References

[1] A. Gamm, I. Golub, Observability of power systems. Moscow: Nauka, 1990. in Russian.

[2] A. Gamm, L. Gerasimov, I. Golub, Y. Grishin and I. Kolosok. *State Estimation in Electric Power Industry*. Moscow: Nauka, 1983. in Russian.

[3] A. Gamm, I. Kolosok, Bad data detection in measurements in electric power systems. Novosibirsk: Nauka, 2000. in Russian.

[4] A. Gamm, Y. Grishin, A. Glazunova, I. Kolosok, E. Korkina, "New EPS state estimation algorithms based on the technique of test equations and PMU measurements", Proc. "PowerTech'2007", pp 1670-1675. doi:10.1109/PCT.2007.4538566

[5] I. Kolosok, E. Korkina, A. Paltsev, "Bad Data Detection at Decomposition of State Estimation Problem", Proc. "PowerTech'2011", doi:10.1109/PTC.2011.6019459

[6] A. Phadke, "Synchronized Phasor Measurements. A Historical Overview", Proc. IEEE PES Transmission and Distribution Conf., 2002, №.1, pp.476-479.

[7] R. Nuqui, A. Phadke, "Hybrid Linear State Estimation Utilizing Synchronized Phasor Measurements", Proc. "PowerTech'2007", pp.1665-1669.

[8] Glazunova, I. Kolosok, E. Korkina, "Monitoring of EPS Operation by the State Estimation Methods", Proc. "Innovative Smart Grid Technologies" (ISGT-2011). doi:10.1109/ISGTEurope.2011.6162640

[9] M. Khokhlov, "Identifiability of errors in synchronized vector measurements" Modern approaches to ensuring the electric power system reliability (Sovremennye podkhody k obespecheniyu nadezhnosti elektroenergeticheskikh system). Syktyvkar: Komi RC UB RAS, 2014, pp.88-96.

[10] I. Kolosok, E. Korkina, E. Buchinsky, "Monitoring of EPS facilities using PMU-based linear state estimation", Proc. LMPS-2012. pp.245-253

[11] C. Ericson, "Fault tree analysis" http://www.eecs.ucf.edu/~hlugo/cop4331/ericson-ftatutorial.pdf (Accessed 10 January, 2017)

[12] J. Quinlan, "Simplifying decision trees", Intern. Journal of Man-Machine Studies – 1987, V.27 (3): 221. doi:10.1016/S0020-7373(87)80053-6

[13] O. Gomez, C. Portilla, M. Rios, "Reliability analysis of substation monitoring systems based on branch PMUs", IEEE Trans.on.Power Systems, vol.30, No.2, pp.962-969, 2015.

[14] C.Ten, C. Liu, M. Govindarasu, "Vulnerability assessment of cyber security for SCADA system using attack trees", IEEE Trans.on.Power Systems -2007.

[15] M.V. Khokhlov. "A matroid theory approach to constructing the sparse attacks on power system state estimation", Proc. Int. Conference on Problems of Critical Infrastructures, Irkutsk: ESI, pp.57-65, 2015, in Russian.

[16] I. Kolosok, L. Gurina, E. Korkina, "Reliability analysis of the state estimation by PMU measurements under cyberattacks on WAMS", Proc. Int. Scientific Workshop on Methodological Problems in Reliability Study of Large Energy Systems, pp. 231-237, 2014, in Russian.

[17] I. Kolosok, E. Korkina, L. Gurina, "Vulnerability analysis of the state estimation problem under cyberattacks on WAMS", Proc. Intern. Conf. on Problems of Critical Infrastructures. Irkutsk: ESI, pp. 73-83, 2015.

[18] I. Kolosok, L. Gurina, "Calculation of cyber security index in the problem of power system state estimation based on SCADA and WAMS measurements", Proc. of 9th Intern. Conf. on Critical Information Infrastructures Security, Limassol, Cyprus, ID 12.

[19] I. Kolosok, E. Korkina, "Role of the state estimation in ensuring cyber-physical security of electric power system", Proc. 88th Int. Scientific Workshop on Methodological Problems in Reliability Study of Large Energy Systems, 2014, Issue 67, Syktyvkar: Komi RC UB RAS, 2016, pp. 386-395, in Russian.

[20] I. Kolosok, E. Korkina, A. Mahnitko, "Supporting cyber-physical security of electric power system by the state estimation technique", Proc. RTUCON-2016.

[21] M. Khokhlov, I. Golub, "The unified approach for optimal PMU placement to ensure the reliability of power system observability", Proc. 86th Int. Scientific Workshop on Methodological Problems in Reliability Study of Large Energy Systems, 2014, pp. 591-601, 2015, in Russian.

[22] A. Gamm, I. Golub, "The problem of weak places in electric power systems", Proc. "PowerTech'1995", pp. 542-546.

[23] A.Glazunova, I. Kolosok, E. Korkina, "Study of test equations method's application for bad data detection in PMU measurements", Proc. PMAPS-2012, #106.

[24] I. Kolosok, E. Korkina, A. Mahnitko, "Detection of systematic errors in PMU measurements by the power system state estimation methods", Proc. RTUCON 2015, pp.289 – 292. doi: 10.1109/RTUCON.2015.7343131

[25] Y. Grishin, I. Kolosok, E. Korkina, L. Em, G.Ornov, N. Shelukhin, "New information technologies in state estimation software for analysis of EPS operating conditions", Proc. RNSPE-2001, pp. 271-275.

[26] F. Aminifar, M. Fotuhi-Firuzabad, V. Shahidehpour, A. Khodaei "Probabilistic multistage PMU placement in electric power systems", IEEE Trans. on power delivery, vol.26, N. 2, pp. 841-849, April, 2011.

Application of Thermodiagnostics and Numerical Models in Research of Energy Supply Units using Hydrogen Technologies

Petr Moldrik, Daniel Minarik
Centre ENET
VSB - Technical University of Ostrava
Ostrava, Czech Republic
petr.moldrik@vsb.cz, daniel.minarik@vsb.cz

Petr Kacor, Michal Ney
Department of Electrical Power Engineering
VSB - Technical University of Ostrava
Ostrava, Czech Republic
petr.kacor@vsb.cz, michal.ney.st@vsb.cz

Abstract — **This paper deals with utilisation of thermography for diagnostics of hydrogen fuel cells fitted with a proton exchange membrane, combined with photovoltaic panels to form a hybrid power system for storage of solar power into chemical energy in hydrogen with its subsequent conversion into electric power. Measurements taken using a thermal camera are used to identify distribution of the heat field over the stack of hydrogen fuel cells and any spots with insufficient cooling or even overheating of the stack. The cause of this phenomenon, certainly not desired for regular operation, is then disclosed using computer simulation employing a numerical model in the Ansys CFX program. The results obtained through measurement and PC simulations are processsed to suggest a suitable solution to enable a more effective and safer operation of the stack of hydrogen fuel cells subject to research within the hybrid power system.**

Keywords: Thermovision; fuel cell; hydrogen technology

I. Introduction

Hydrogen technologies currently find their use beyond the scope of mobile applications (vehicle driving gears), these can be actually found in stationary applications too, e.g. as uninterrupted power supply units for base transceiver stations (BTS). They are also starting to form assemblies with unstable renewable resources (solar and wind power plants), serving as the storage element to eliminate fluctuations of their power output. There are numerous hybrid power systems in the modern world, with on-grid or off-grid configurations. Our research unit at the VŠB - Technical University of Ostrava has also built one laboratory off-grid hybrid system using hydrogen gas as the storage fluid.

This system also includes hydrogen fuel cells suffering from losses (during conversion of chemical energy into electric power) demonstrated by the heat generated [1]. The heat dissi-pates from the cell through the cooling fins subject to cooling air flow. Inefficient heat dissipation brings the risk of bot excessive drying of electrolyte (the PEM polymer membrane in our case), as well as its degradation or even destruction due to heat. Inspections aimed at cooling of fuel cells in operation are therefore of vital importance. The very easy and prompt diagnostic methods include the use of thermographic display (thermal cameras). These are capable of providing the overall view of weak spots with excessive heating or even overheating.

II. Hydrogen Fuel Cells

The hydrogen fuel cells measured belong to the off-grid power system to ensure the uninterrupted supply of the defined load as well as storage of any surplus electric power generated by the photovoltaic (PV) panels for their subsequent utilisation. For a simplified diagram of this system, currently operated at the VŠB - Technical University of Ostrava, see the Fig. 1.

This system ensures storage of surplus electric power mainly by means of electrolysis of hydrogen which is stored in pressure vessels. Shortages of electric power generated by PV panels are covered by the hydrogen produced in the past and stored in hydrogen fuel cells used as backup supply units [2].

This system ensures storage of surplus electric power mainly by means of electrolysis of gaseous hydrogen, which is stored in pressure vessels. Shortages of electric power generated by PV panels are covered by the hydrogen produced in the past and stored in hydrogen fuel cells used as backup supply units [2].

Our research is focused on the stack of hydrogen fuel cells of PEM (Proton Exchange Membrane) type in this case. Together with the operation, measurement and control systems, this stack forms the production module NEXA illustrated by the photograph in the Fig. 2 and its essential technical characteristics are listed in the Tab. 1.

Table 1. The basic parameters of NEXA module of fuel cells

Rated Power	1200 W
Voltage	22 V to 50 V
Start-up Time	2 minutes
Noise	72 dBA
Waste Water	870 mL/hr
Dimensions	560 x 250 x 330 mm
Hydrogen Purity	99.99 %
Pressure	70 to 1720 kPa
Hydrogen Consumption	<18.5 L/min
Temperature	< 65 °C
Efficiency	50 %

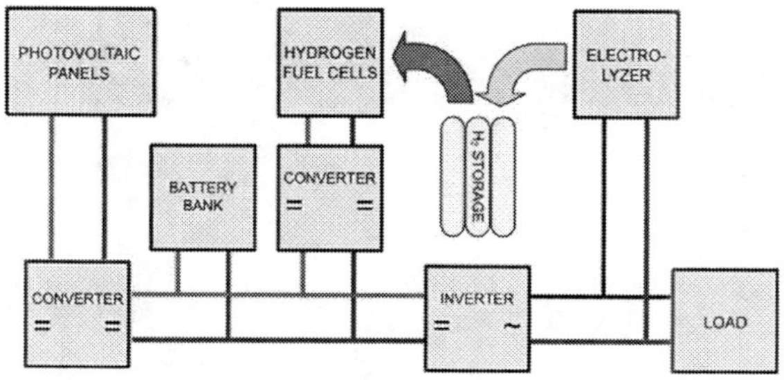

Figure 1. Block diagram of researched off-grid energetic system

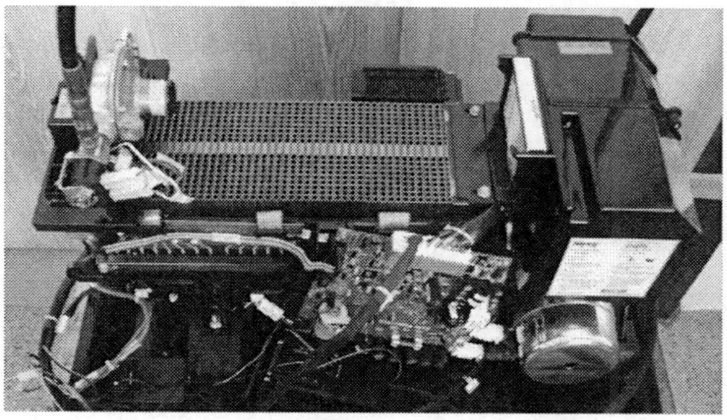

Figure 2. NEXA module with stack of PEM fuel cells

A. *PEM Fuel Cells*

The measured stack of PEM fuel cells NEXA is a device, which uses the electrochemical reaction to transform chemical energy held by the fuel (gaseous hydrogen), aided by the oxidizing agent, to electric power, water and heat. This transformation occurs within catalytic reactions on electrodes and it is mainly based on reversed principle of water electrolysis.

The physical structure of PEM fuel cell comprises of the MEA, plates with channels for reaction gases known as the FFP (Flow Field Plate) sometimes even called the BP (Bipolar Plates) and the Sealing and the terminal plates with electrodes (EP - End Plate). The electrodes, catalyst and membrane together form the membrane electrode assembly (MEA), where chemical reactions take place. The PEM is formed by a 50-200 µm thick rigid film (perfluorinated polymer of sulphonic acid) covered with a layer of Platinum catalyser. This solid electrolyte reduces corrosion and improves the life span of fuel cells [3]. Anodes and cathodes (paper with carbon fibres) are porous and allow for transmission of reaction gases, i.e. the gaseous hydrogen and the oxygen from ambient air. The flow field plates are made from graphite. They provide for minimising the contact of gas with the membrane assembly. The end plates with electrodes are conductive plates on steel basis. The arrangement of PEM fuel cell is shown on Fig. 3.

PEM fuel cell is provided with a simultaneous supply of the fuel gas (hydrogen in form of molecules H_2 on the anode side) and the oxidising agent (oxygen in form of molecules O_2 or air on the cathode side). The contact of H_2 hydrogen molecules with the platinum catalyser induces a reaction on the surface of proton membrane, whereas the hydrogen molecules decompose to individual atoms of H first, to be further degraded to H^+ protons and e^- electrons. The electrons are subject to the external electric load and they are received by the oxygen atoms on the cathode side to form the O_2^- ion, which has been formed upon degradation of the O_2 molecules using platinum catalyser. A polymer membrane is able to conduct protons H^+ from anode to cathode (see Fig. 3). PEM fuel cells provide for the chemical reaction listed below:

$$\text{Anode: } 2\,H_2 \Rightarrow 4\,H^+ + 4\,e^-$$

$$\text{Cathode: } O_2 + 4\,e^- + 4\,H^+ \Rightarrow 2\,H_2O$$

$$\text{Overall reaction: } 2\,H_2 + O_2 \Rightarrow 2\,H_2O$$

A single fuel cell produces about 1 V at open-circuit and about 0.6 V at full power output. The fuel is 99.99 % hydrogen and the hydrogen pressure is normally maintained at 0.3 bar [4]. PEM fuel cells are sensitive to the carbon monooxide content and sulphur compounds in fuel. Oxygen comes from the humidified ambient air under the pressure of 0.1 bar. A humidity exchanger transfers both fuel cell product water and heat from the wet cathode outlet to the dry incoming air. The excess product water is drawed on the system, as both liquid and vapor in the exhaust. The NEXA fuel cells module is equipped with a small compressor supplying excess oxidant air to the fuel cells, and the speed of the compressor can be adjusted to match the power demand from the fuel cell stack. The stack of Nexa module is air-cooled. The cooling fan draws air from the ambient surroundings in order to cool the stack and regulate the operating temperature.

III. THERMODIAGNOSTICS

The infrared thermography analyses the distribution of heat field across the surface of the body examined, utilising a contactless method and employing the so called "thermal camera". The thermal diagnostics of the stack of hydrogen fuel cells was conducted using the thermal camera FLIR E45 with a microbolometric detector.

Since the hydrogen fuel cells work as an electrochemical device converting chemical energy in fuel and oxidant into electric power directly, these are also a source of waste water and heat to be removed for the sake of safety of the whole system.

A. *Physical Theory of Heat Radiation*

Each body of mass with the temperature above the absolute zero (-273.16 °C), emits radiation in the infrared field. This radiation can be described using two theories, the corpuscular or wave theory of light, i.e. the radiation can be viewed from two different perspectives: as a flow of photons representing the basic energy quanta or as a radiation of electromagnetic waves.

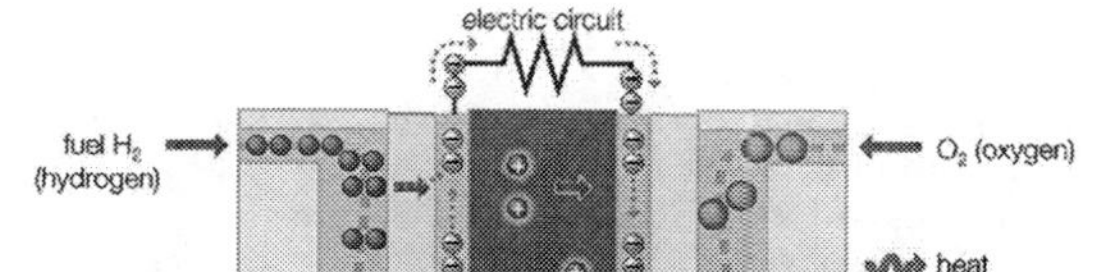

Figure 3. Principle and arrangement of hydrogen PEM fuel cell

When using the corpuscular theory, it is necessary to define the elementary energy quantum called the "photon". This can be pictured as the essential energy charge carrying the amount of energy depending on frequency. The energy value is then described by the Equation (1). The energy of one photon equals 1 eV, which translates into 1.6022 x 10^{-19} J [5].

$$e = h \cdot f \tag{1}$$

where h is a Planck´s constant (J/s) and f is a frequency (Hz).

As far as the wave radiation is concerned, it is characteristics for the frequency spectrum from audible waves up to the ionising radiation. The total radiant power at the temperature T can be then defined using the Equation (2).

$$E(T) = \int_0^\infty E_f(T, f)\ df \tag{2}$$

where E_f refers to the spectral radiant power dependent on characteristics of a general body and T represents the absolute temperature (K).

The infrared (heat) radiation is determined by the wavelength exceeding the value applicable to a visible light, i.e. from 750 nm to 1 mm. The band used for thermography is identified as LWIR (long wave infrared) with the wavelength of 8-15 μm.

B. Measurement of Fuel Cells with Thermal Camera

The thermal measurement conducted on the surface of stack of hydrogen fuel cells ran continuously for the interval of 2 hrs (7200 s), while the stack was actuated in the cold condition, supplying the power of 1200 W to the connected appliance for throughout the entire interval. This interval was also used for measurement of temperature inside the stack. The fuel cell stack temperature is measured at the cathode air exhaust [6]. The development of this temperature (T_{stack}) with the average surface temperature ($T_{surface}$) are shown in the graph in Fig. 4.

The thermograms obtained (one of these is illustrated in the Fig. 5) show evident temperature differences between individual parts of the surface on the stack of hydrogen fuel cells subject to measurement. That comprises 47 cells with identical output connected in series. The left end of this stack heats up significantly more (see the field "A" shown by Fig. 5), as it is closer to the entry of the cooling air supply channel. Following two hours of operation delivering the output of 1200 W, the average surface in this field equalled to 60.1 °C (333.3 K), whereas the value reached within the remaining part of the surface measured was 53.4 °C (326.6 K).

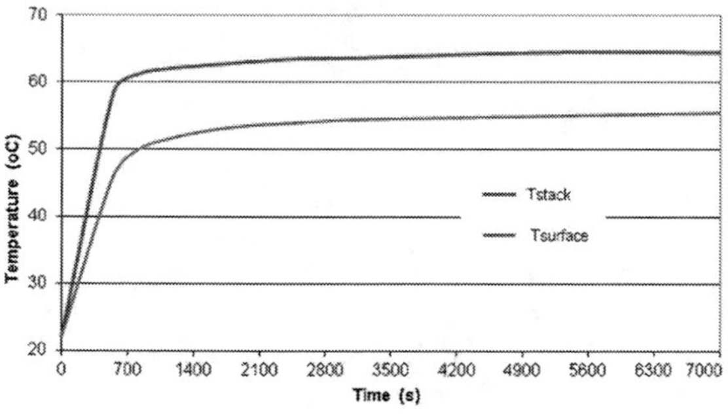

Figure 4. Time dependence of internal and surface temperatures of the stack of hydrogen fuel cells for the constant load of 1200 W.

Figure 5. Thermographic picture showing the surface of stack of hydrogen fuel cells after 2 hours of operation with the constant load of 1200 W

When the stack is in operation, its control system follows evaluation of the interior temperature to proceed with control of the active cooling system to prevent any temperature excess beyond 65 °C (338.2 K). The uneven heating, or even overheating, of the stack can then lead to gradual degradation of materials within the affected field. Any potential increase of refrigerating output will be logically resulting in a higher internal consumption of electric power.

IV. Computer Simulation

The aim of computer simulation was to identify the reason for occurrence of the undesired conditions described above during operation of the stack of hydrogen fuel cells subject to examination. The simulation was conducted using a numerical model and its analysis using the Ansys CFX program. The result obtained was to plot not only the temperature distribution inside the stack as a whole, yet also the distribution of field comprising velocities and pressures within the cooling air.

The simulation refers to a stable condition, when the stack supplies a constant output of 1200 W and the system generates waste heat exhausted through the cooling fins. The purpose of this analysis is to find the problematic spots with higher heating of the stack as well as to define its potential causes.

The calculation methodology is based on solving the basic equations (Navier-Stokes) that arise from the fundamental laws on conservation of mass, energy and mobility in case of gas flows [7].

A. 3D Model of the Stack of Hydrogen Fuel Cells

The 3D model of the stack of hydrogen fuel cells in our experiment was based on its actual dimensions and design. Each cell in this model was comprising a structure built with a polymer membrane enclosed tightly with bipolar electrically conductive plates with distinct vertical cooling fins on both sides. 47 cells connected in series form the stack. The mutually recumbent cells form a system of channels to guide the cooling air and exhaust the heat generated during the electrochemical reaction.

The supply of air into cooling channels is provided by means of a controlled radial fan via the plastic distributing channel below the stack. The 3D model with the distributing channel for cooling air has been drawn using the SolidWorks CAD software (see Fig. 6).

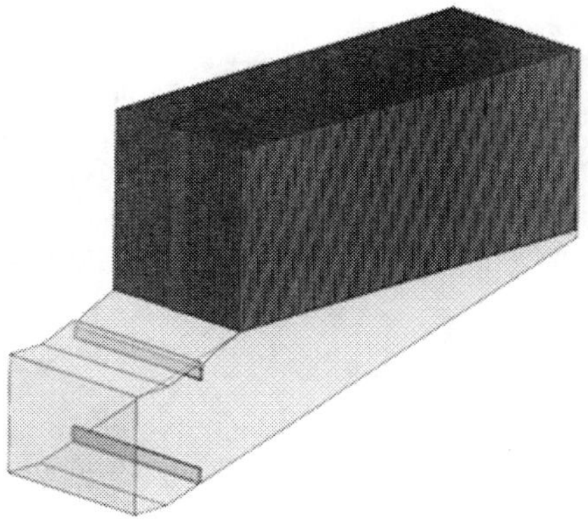

Figure 6. CAD model showing stack with the air distributing channel

B. *Input Data for CFD Analysis*

Correct configuration of the CFD (Computed Fluid Dynamics) model to simulate the stack required input of threshold limits and the particular model load. That concerned mainly applying the heat transfer coefficient α onto vertical surfaces of cooling fins and external surfaces of the stack. For the estimate of the heat transfer coefficient and its subsequent dependency on the ambient temperature were used the relations stated in [7].

The application of model load took into account the losses incurred by generation of heat within stack membranes and developing due to the electric current passing through the cooling fins. Application of load on the model is necessary to perform CFD analysis. The essential material parameters for CFD simulation in steady condition have been selected as follows: Thermal conductivity l (for fuel cell solids) and Density r, Thermal conductivity lAIR and Specific heat capacity depending on temperature rise cAIR (for ambient fluid) [8]. These parameters together with load values are shown in Tab. 2. In the next step we can apply the cooling fluid (air) velocity into the inlet of distribution channel.

Table 2. Material parameters and load values for CFD simulation

Object	PEM	Fins	Plates	Air
Heat Conductivity ($W \cdot m^{-1} \cdot K^{-1}$)	0.25	140	318	0.022
Density ($kg \cdot m^{-3}$)	2100	2300	19300	1.185
Specific heat Capacity ($J \cdot kg^{-1} \cdot K^{-1}$)	1100	837	129	1004
Loss (W)	1200	800	-	-
Volume (m^3)	0.000433	0.001087	-	-
Unit loss ($W \cdot m^{-3}$)	2730000	733000	-	-

C. *Assessment of CFD Analysis*

The calculated fuel cells heat model was analyzed by illustration of contours of temperature and velocity in selected cross-sections through the model volume. Fig. 7 shows the temperature distribution over the whole stack of fuel cells. As clearly indicated by this figure, the maximum temperature value can be found in the left section of the stack on fan side, where it reaches approx. 348 K. The temperature distribution is uneven. The coolant in bottom section is still cold and its temperature is approx. 308 K.

The air passing through cooling channels is heated gradually. That is associated with heat transfer from fins into the coolant; that has to cause a rise in temperature of the top fins. The temperature difference between the bottom and top sections of the stack ranges within the interval of approx. 15 K.

Fig. 8 shows volume contours of the velocity field for air used for cooling of the stack of fuel cell. Air enters the distribution channel in the front (from the left) at the velocity of approx. 10 $m \cdot s^{-1}$ with progressive ingress into the individual cooling channels between fins.

The inlet of distribution channel is fitted with cover plate to form a screen and narrow the coolant flow. There is a significant swirl occurring around these cover plates, resulting in basically zero airflow in the front (left) section of the stack. The cooling fluid flow is subject to strong distortion in the front (left) section of the stack and this coolant cannot enter the front (left) cooling fins effectively to drain the waste heat. That also explains the location of top temperature spot in the front (left) section.

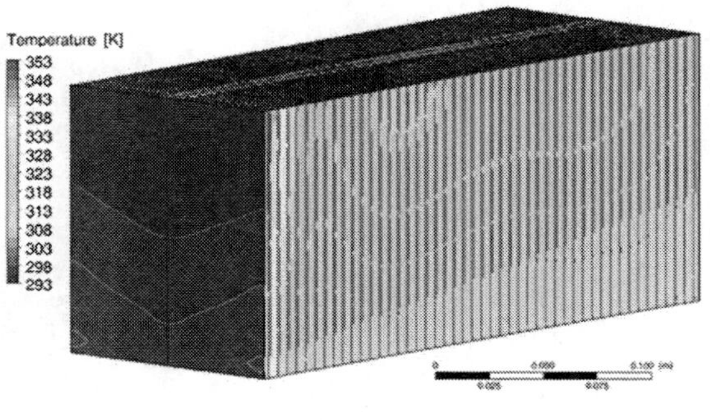

Figure 7. Temperature distribution over the hole stack of fuel cells (CFD model)

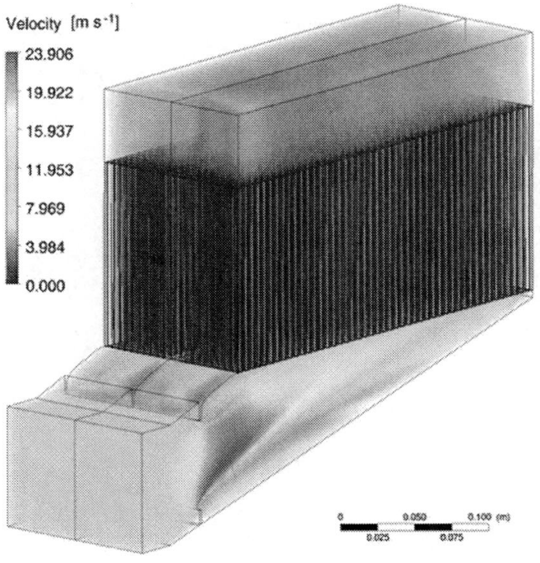

Figure 8. Course of velocity of the cooling fluid flow inside the distribution channel and stack of fuel cells

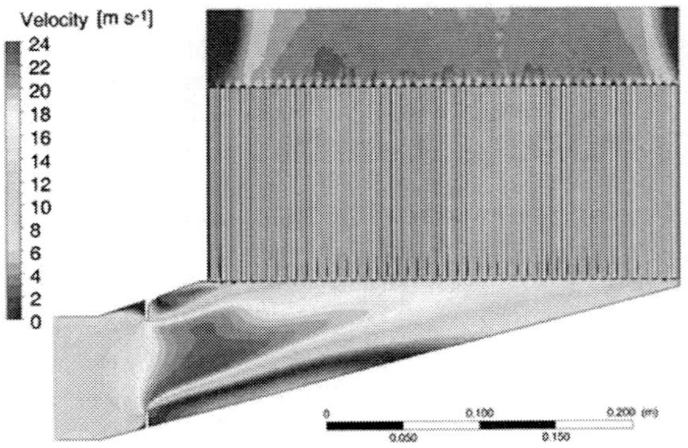

Figure 9. Course of velocity of the cooling fluid flow inside the distribution channel and stack of fuel cells at the cross-section

Fig. 9 shows the course of velocity on a longitudinal cross-section across the stack of fuel cells. Even this case indicates an evident negative activity of the screen that literally deflects the coolant flow from the front (left) section of the stack. This section is the most delicate part as far as the optimal cooling is concerned. There is an option to optimize the cooling installed, with further alterations and upgrades to such extent that would maximize the NEXA fuel cells module operating efficiency and to minimize overheating of front (left) section of the stack.

V. Conclusion

This paper has addressed the optional employment of thermal diagnostics in practice with regard to operation of PEM type hydrogen fuel cells integrated into a more complex hybrid energy system serving for storage of solar energy into hydrogen. We have also applied numerical methods to develop the numerical model in order to perform the control heat analysis focused on cooling of these cells that is basically the analysis of heat removal from this system. The computer simulation using the CFD model was used to determine the stable condition of an actual stack of hydrogen fuel cells subject to a contact load with the imperfect removal of the waste heat generated within, which is exhausted through the cooling fins. Our measurements and analyses helped identify the problematic spots, characteristics with an n increased heating of the stack, as well as the problematic locations inside the cooling air distributing channel causing its uneven distribution into channels within cooling fins, as the actual cause of such overheating. The established condition, considered as undesirable with respect to the stack operation, can be resolved by alteration of the inlet points of the plastic channel for distribution of cooling air, mainly in terms of modification to the cover plates located within, because these certainly form a screen reducing the cross section of air flow passage.

Acknowledgment

This research was partially supported by the SGS grant from VŠB - TU Ostrava (No. SP2017/54) and by the project LO1404: Sustainable development of ENET Centre.

References

[1] P. Srinivasan. *Fuel Cells From Fundamentals to Applications.* New York: Springer Science+Business Media, 2006, pp. 97-99.

[2] D. Minarik, B. Horak, P. Moldrik and Z. Slanina. "An Experimental Study of Laboratory Hybrid Power System with the Hydrogen Technologies." *Advances in Electrical and Electronic Engineering*, vol. 12, pp. 518-528, 2014.

[3] P. Moldrik and R. Chvalek. "Fuel cells reliability in energy storage system." in *Proc. of 12th International Scientific Conference on Electric Power Engineering (EPE)*, 2011, pp. 472-474.

[4] P.Moldrik and R. Chvalek. "PEM Fuel Cells - The Basic Characteristics." in *Proc. of 10th International Conference on Environment and Electrical Engineering*, 2011, pp. 8-11.

[5] M. Hrabcik and R. Gono. "Analysis of Issues in Long-term Thermovision Measurements in Distribution Networks." in *Proc. of 11th International Scientific Conference on Electric Power Engineering (EPE)*, 2010, pp. 721-725.

[6] J. Vaculik, Z. Hradilek, P. Moldrik and D. Minarik. "Utilisation of Thermograhy in the Field of PEM Fuel Cells." in *Proc. of 16th International Scientific Conference on Electric Power Engineering (EPE)*, 2015, pp. 488-492.

[7] P. Kacor, D. Minarik and P. Moldrik. "CFD Analysis of Temperature Rise of PEM Type Fuel Cell." in *Proc. of 8th International Scientific Symposium on Electrical Power Engineering*, 2015, pp. 180-183.

[8] W. A. N. Mohamed, R. Atan and A. A. Ismail. "Heat transfer simulation of a single channel air-cooled Polymer Electrolyte Membrane fuel cell stack with extended cooling surface." in *Proc. of 2010 International Conference on Science and Social Research (CSSR)*, 2010, pp. 91-96.

Assessment of the Normative Requirements for Ensuring the Safety of Powder Coating Booths in Terms of the Risk of Explosion

Richard Velicka
Department of Cybernetics and Biomedical Engineering
VSB – Technical University Ostrava
Ostrava, Czech Republic
richard.velicka@vsb.cz

Eva Velickova
Department of Occupational and Process Safety
VSB – Technical University Ostrava
Ostrava, Czech Republic
e.velickova@seznam.cz

Abstract—**During the process of electrostatic powder coating in coating booths, there is a probability of the risk of explosion. Compliance with the prescriptive requirements specified in the applicable legislation in order to minimize the risk associated with explosion risk. This article focuses on the analysis of the current safety level of coating booths in terms of the risk of explosion in relation to the lower explosive limit of the selected powder paints.**

***Keywords*—A powder coating booth, the normative requirement, explosive atmosphere, the lower explosive limit, fire-technical characteristics**

I. INTRODUCTION

With the continuous development of science and technology, newer and more modern technologies and materials emerge. These materials are being developed for the purpose of improved properties of the finished articles. Many materials may have different characteristics from those previously used, and during their processing, they may also affect the safety of the technological process. This paper focuses on the properties of powder paints used in coating booths in electrostatic application with regard to their explosiveness. These properties are expressed by fire-technical characteristics. The current situation in practice is that in many establishments, fire-technical characteristics are not available at all, or identical values are given for the full range of products, sometimes even in a relatively wide range. These values are taken from professional literature for "a kind of general" paint, or even based on measurement, but only of "some" of the produced paints. In any case, these values are not the same for the different paints, and especially for newly manufactured paints; they may have different values than those measured in the past and then presented.

II. MATERIALS AND METHODS

Fire-technical characteristics mean the properties of a substance expressed in quantifiable value or determined on the basis of measurable values of more partial properties or phenomena that describe the behavior of the substance in the process of combustion, or related to it, [1].

The basic fire-technical characteristics of powder materials include:

- Lower explosive limit (*LEL*)
- Ignition temperature of settled dust
- Ignition temperature of dust clouds
- The maximum explosion pressure
- The maximum rate of pressure rise (brisance)
- Explosion Constant
- Minimum ignition energy
- Susceptibility to spontaneous combustion

Furthermore, explosiveness of combustible dust is also influenced by its fineness, the concentration of combustible dust, the oxygen concentration in space, the pressure and temperature at the time of initiation, dust humidity, inert solid impurities, the state of movement of the mixture, the size and shape of the container, the arrangement of the containers, forming a hybrid mixture, and ventilation [2].

Powder coatings are a mixture of synthetic resins or a mixture polymers, pigments, fillers, and additives. Based on their physical and chemical properties, they belong to colloidal and dispersion systems and they are not diluted for application, not dissolved in any liquid. The most widely used powder coatings are epoxy, epoxy-polyester and polyester coatings.

The principle of the powder coating is application to the surface of the product, which is preheated, and subsequently, the paint is hardened in an oven. The process of electrostatic powder coating is based on the physical principle that oppositely charged particles attract each other, it means that an electric field is formed between the powder paint and coated object, and the particles easily adhere to the color of the grounded coated object. Powder coatings are conveyed from powder centers to spray guns. The powder particles are charged into the powder spray gun and applied in the form of the cloud on the grounded-coated object. Overspray that has not adhered is sucked, and according to the type of the separation system, it is either discharged to a separator or cyclone.

In a closed system, this overspray is sucked and removed through the suction pipe into the cyclone. In the cyclone, air is separated from the powder through a high rotation. The recycled powder is delivered back into the container in the powder center. The output air is purified from any residual powder in the terminal filter (Fig. 1). In the case of an open system, overspray is sucked into the separator, which consists of filter cartridges placed directly in the powder coating booth. The separated powder is gradually removed from the filter cartridges by compressed air and falls to the bottom of the filter chamber, where the powder paint is collected and continuously conveyed back into the powder container. Air from the space behind the filter cartridges is conveyed through the terminal filter to the production hall (Fig. 2).

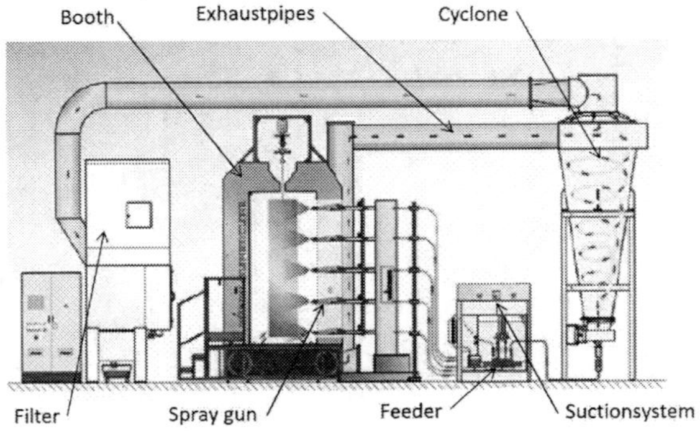

Figure 1. Circulation of powder paint inthe coating boothwith a closedsystem ofseparation [3]

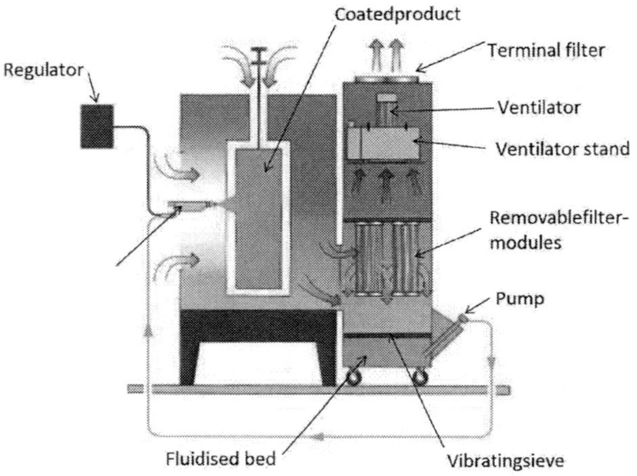

Figure 2. Circulation of powder paint in the coating boothwith an open system ofseparation [4]

Powder coating booths can be operated inside or outside the booth, with stands for automatic spraying equipment outside the booth, or with an operator and stands for automatic spraying equipment outside or inside the booth, [5], [6].

The explosion of dust-air mixtures results from the combination of the following conditions:

- the presence of flammable substances in concentration explosion limits,
- the presence of an oxidizing agent in sufficient quantity,
- the presence of an effective ignition source,
- combustible dust must be sufficiently dispersed in the mixture with the oxidant,
- combustible dust must be dispersed in an enclosed or partially enclosed space, [7].

According to Directive 1999/92/EC, which was implemented in the Czech Republic in Government Decree no. 406/2004 Coll. "On detailed requirements for ensuring safety and health at work in potentially explosive atmospheres" are assessing the risks of explosion. It is important to determine whether, under the circumstances, hazardous explosive atmosphere may occur, and whether this explosive atmosphere can be ignited. If a flammable substance is dispersed in the air and its concentration is within the explosion limits, it results in a hazardous explosive atmosphere. If it is not possible to completely eliminate its formation, measures must be taken to exclude the presence of effective ignition sources. In the event that a dangerous explosive atmosphere and effective ignition sources occur simultaneously, further measures to limit explosive effect must be taken in conjunction with organizational measures. Places with a hazardous explosive atmosphere are classified as hazardous areas. These hazardous areas are classified into zones according to the likelihood of hazardous explosive atmosphere occurrence:

- **zone 20** - the area where an explosive atmosphere formed by a cloud of dispersing combustible dust in air is present continuously, for long periods or frequently [5, 6, 8, 9, 10,11]
- **zone 21** -the area where an explosive atmosphere consisting of a cloud of raised combustible dust in air can occasionally occur in normal operation [5, 6, 8, 9, 10,11]
- **zone 22** - the area where forming explosive atmosphere consisting of a cloud of raised combustible dust in the air in normal operation is not likely, and if it occurs, it is present only for a short period [5, 6, 8, 9, 10,11]

For powder coating booths it applies that zone 20 is assigned the inner space of enclosed separating systems, and zone 22 the inner powder coating booths and open separating systems, as well as outer areas of the powder coating booths at a distance of 1 m from permanently open Decree 176/2008 Coll. "The technical requirements for machinery" it is stated that if the machinery fulfils the relevant provisions of the harmonized European standards which were referred to in the Official Journal of the European Union, or the harmonized Czech technical standards, or international technical standards in the member state of the European Union, which is related to the fundamental requirement, it is understood that a basic requirement to ensure the safety is met. [12, 13]

Risk of explosion in the powder coating booth may arise if a cloud of powder coating material produced by spraying equipment ignites due to raising of settled material or otherwise. An explosion is likely if effective initiator sources are present and the powder paint concentration in the air exceeds the ***Lower Explosive Limit*** (*LEL*).

In this case, an ignition source may be:

- electrostatic discharges
- electric sparks
- hot surfaces
- sparks generated by mechanical energy
- sources of thermal energy used in repair work
- flames transferred due to forced ventilation to the system for separating the powder

LEL may be approached or exceeded:

- if there is a malfunction of forced ventilation in the powder coating booth
- if the settled powder coating material is raised in the powder coating booths in the system for the separation of powder during automatic self-cleaning of the filter [5, 6, 14, 15, 16, 17]

Range of protective measures by the zone type is listed in the following table.holes. [5, 6]

To ensure safety in terms of the risk of explosion, a number of laws, government regulations and harmonized standards were issued in the Czech Republic, in which the individual requirements are specified in more detail. In the Government

TABLE I. RANGE OF PROTECTIVE MEASURES IN THE VARIOUS ZONES [18]

Zone	Sources of ignition must be reliably excluded
20	in normal operation (no fault) in case of expected faults in case of rare disorders
21	In normal operation (no fault) in case if expected faults
22	in normal operation (no fault)

In the above-mentioned zones, only equipment of the relevant category suitable for the particular zone may be used:

in zone 20, category 1 equipment,

in zone 21, category 1 or 2 equipment,

in zone 22, category 1, 2 or 3 [19, 20].

III. RESULTS

To determine the environment according to CSN 33 2000-1 ed. 2 and CSN EN 60079-0 ed. 4, it is important to know the value of the *LEL*. The lower explosion limit represents the lowest concentration of combustible dust in air, at which the mixture is explosive. [21, 22] Based on this quantity, protective measures against the danger of explosion at assessed facilities are implemented. If the performed test is positive, it is necessary to deal with explosion prevention for the particular facility.

Lower Explosive Limit (*LEL*) was set in the explosive autoclave VA-20L from the Swiss Company Kühner AG (see Fig. 3). It is an explosion proof hollow sphere having a volume of 20 l made of stainless steel. The water jacket serves to absorb heat from the explosion, so that the initial temperature T_i was 20°. In tests the dust is dispersed to the sphere from a pressurised container through the quick-opening valve and rebound nozzle jet. Quick-opening valve is opened pneumatically and closed by an auxiliary piston. The valve for the compressed air is activated electrically. The ignition sources placed in the centre of the sphere are two chemical lighters, each with energy of 1 kJ. The system for pressure measurement has two pressure sensors, recording and control equipment. Before dispersing dust, the sphere must be partially evacuated to a pressure of 0.4 bar, so that after the injection of dust, the pressure in the sphere (the initial pressure p_i) was 1,013 mbar. Dispersion overpressure p_z is 20 bars and the initiation delay t_v is 60 ms [23, 24, 25, 26, 21, 22]

Figure 3. Explosion autoclave VA-20L in laboratory *(author)*

The required amount of dust is put in the feeder, and the feeder is then pressurised to 20 bar. Before starting the test, must measure and record the temperature inside the chamber. When dispersing dust is initiated, the atmospheric pressure must be contained in the container. During initiation, the actual pressure (the initial pressure p_i) must be measured and recorded. The delay between the start of dispersing dust and the activation of the ignition source (delay in initiation t_v) must be (0.06 ± 0.01) s. The time course of pressure is recorded. From the curve of time versus pressure, explosion pressure p_{ex} is determined by calculating the arithmetic mean of the values measured with pressure sensors. Dust ignition occurred when the measured pressure (including the effects of chemical lighters) relative to the initial pressure:

$$p_i \geq 0.5 \text{ bar } [p_{ex} \geq (p_i + 0.5 \text{ bar})]. \quad (1)$$

The following equation for 20-l-apparatus defines the relationship between explosion pressure p_{ex} and the corrected pressure p_m due to the volume of the pressure vessel. Correction respects especially surface to volume ratio and a cooling vessel. For correction of the explosion overpressure at $p_{ex} > 5.5$ bar :

$$p_m = 0.775 \cdot p_{ex}^{1.15} \quad (2)$$

Another parameter is the size of the explosion pressure p_{ex}, which affects the way of correction, because at less than 5.5 bar to show the influence of the chemical igniters as pressure p_{ci}. For correction of the explosion overpressure at $p_{ex} < 5.5$ bar:

$$p_m = 5.5 \cdot (p_{ex} - p_{ci}) / (5.5 - p_{ci}) \text{ bar}, \quad (3)$$

where p_{ci} is pressure due to chemical igniters

When determining the lower explosive limit (*LEL*), we start with a concentration of 500 g/m^3 or other concentrations, at which the explosion occurs, and then the test is repeated to reduce the concentration in steps always to 50% of the previous concentration in a row: ....; 1,000; 750; 500; 250; 125; 60; ... g/m^3.

The highest concentrations of combustible dusts in which the explosion occurs in three consecutive tests, is taken as the lower explosive limit (*LEL*).[22]

To determine *LEL*, 7 samples of paints were selected, so that the most widely used paints, paints of different chemical composition, as well as paints with different grain size were represented. The test results are graphically presented in Fig. 4. – Fig. 10.

Paint sample number 1011810: *LEL* = 60 g/m^3

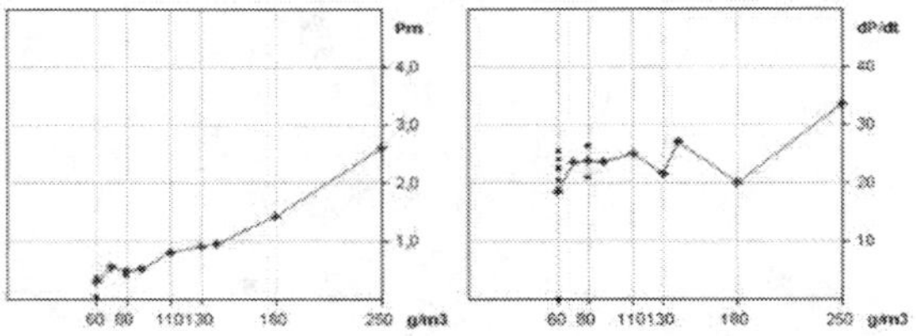

Figure 4. Graphical representation of data for determining the LEL of paint sample 1011810

Paint sample number 1012616: *LEL* = 40 g/m^3

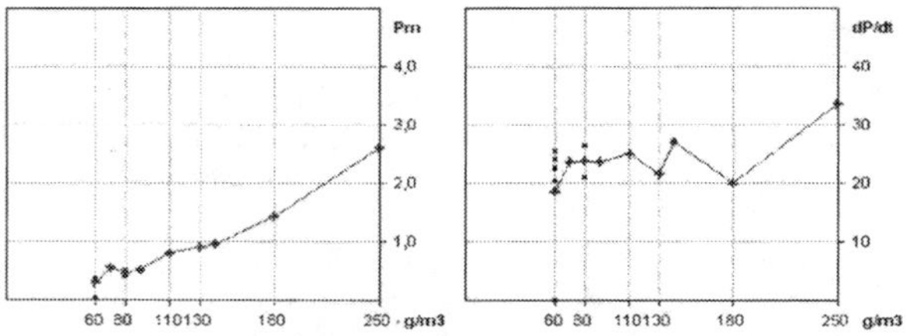

Figure 5. Graphical representation of data for determining the *LEL* of paint sample 1012616

Paint sample number 1012521: *LEL* = 20 g/m^3

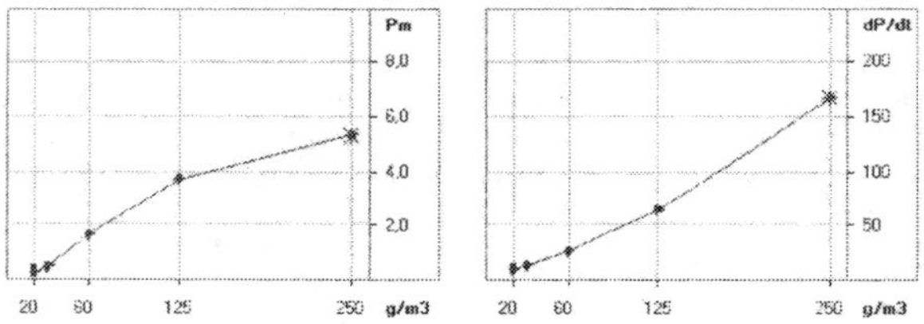

Figure 6. Graphical representation of data for determining the *LEL* of paint sample 1012521

Paint sample number 1013367: *LEL* = 30 g/m^3

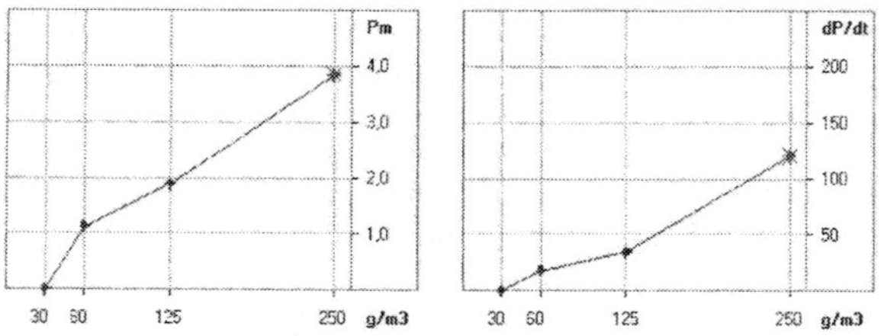

Figure 7. Graphical representation of data for determining the *LEL* of paint sample 1013367

Paint sample number 1028415: *LEL* = 30 g/m^3

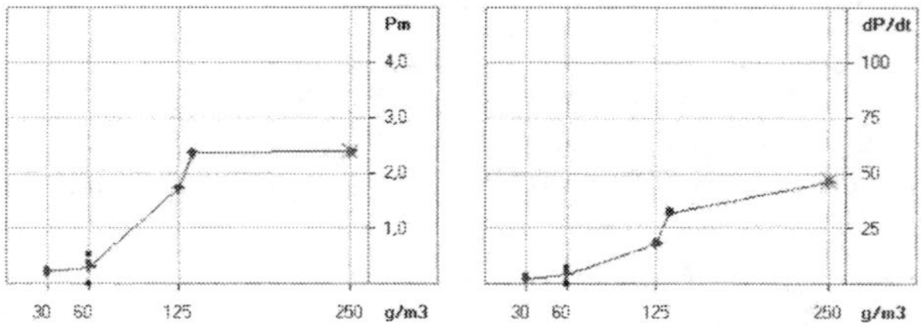

Figure 8. Graphical representation of data for determining the LEL of paint sample 1028415

Paint sample number 1027056: *LEL* = 10 g/m^3

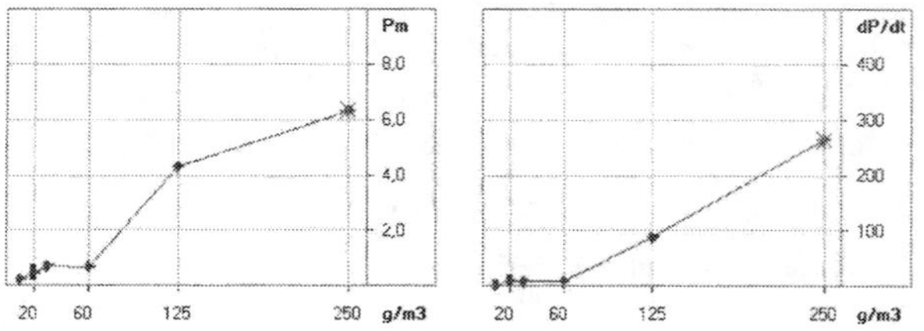

Figure 9. Graphical representation of data for determining the *LEL* of paint sample 1027056

Paint sample number 1027230: *LEL* = 80 g/m^3

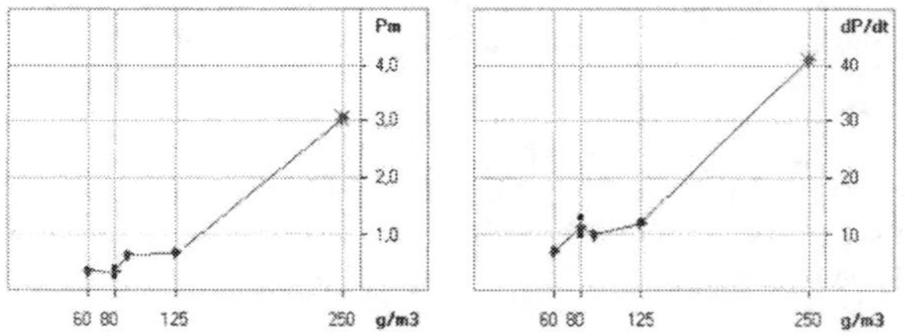

Figure 10. Graphical representation of data for determining the *LEL* of paint sample 1027230

Dependencies displayed on Figures 4. – 10. introduce the experimentally measured curve explosion selected samples of powder paints of different chemical composition and different granularity. They represent the dependence of pressure pm on the concentration and rate of explosion pressure rise d*p*/d*t* on the concentration of powder paints.

LEL measurement results, including average grain size of paint types are summarised in Table 2.

TABLE II. THE RESULTING VALUES *LEL* [27]

Paint designation	Paint type (Chemical composition)	Average grain size measured (μm)	*LEL* (g/m^3)
1011810	epoxy-polyester	32.480	60
1012616	epoxy-polyester	28.864	40
1012521	epoxy	28.468	20
1013367	polyester facade	38.086	30
1028415	epoxy-polyester	34.083	30
1027056	polyester facade	32.046	10
1027230	epoxy-polyester	41.647	80

These measurements were made to compare *LEL* values for paints used with the normative requirement for ensuring safety in terms of the risk of explosion. A normative requirement means the specific technical requirement contained in the Czech technical standard; if it is met, the requirement of the relevant provision of the decree is considered fulfilled. [1]

In the examined samples, the resulting lower explosive limit ranged from 10 to 80 g/m^3. According to ČSN EN 12981 + A1, the concentration of the powder paint must be is kept below the *LEL* by forced ventilation. In the powder coating booths, the systemof forced ventilation has to be ensured so that the average concentration of coating powder in the air does not exceed 50% of *LEL*. If there are no reliable data on the value of *LEL*, the average concentration must not exceed 10 g/m^3. For sample 1027056, which is a polyester paint, LEL is 10 g/m^3, so in this case, the *LEL*value of 5 g/m^3 should not be exceeded, which is the required value of 50% *LEL*. If the value is not known, it is 10 g/m^3, which is already the value at which the powder paint can explode, and it means that there is a real danger of explosion. This situation can easily occur because at most workplaces, *LEL* values are not available, as well as any other fire-technical characteristics.

For paint sample 1027056, this finding means that if fire-technical characteristics are not known, hazardous explosive atmosphere may be permanently present during the process of paint application, which would mean that according to the standard, the zone in the coating booth, is not zone 22, but 20 or 21.

ACKNOWLEDGMENT

This work was supported by the Student Grant System of VSB - TU Ostrava [grant number SP2017/128].

REFERENCES

[1] Decree of the Ministry of Interior no. 246/2001 Coll. establishing fire safety conditions and state fire surveillance (Fire Prevention Decree) of 29 06th 2001, (in Czech)

[2] S. Burian : Explosivity of combustible dust, 23.5.2008, [cit. 10.4.2014], Available at: <http://www.tlakinfo.cz/t.py?t=2&i=1585&h=12010.10.2012 >. (in Czech)

[3] Circulation of the powder paint during application,[cit. 10.3.2014], Available at: <http://www.wagner.cz/index.php?option=com_virtuemart&page=shop.browse&category_id=102&Itemid=103>,(in Czech)

[4] Painting booth with an open system separation [cit. 8.3.2014], Available at: <http://www.nordson.cz/NR/rdonlyres/2CC9DAC6-E91F-4092-99F6-E15CB4C07937/0/CARTRIDGE_Booths_English.pdf>.

[5] ČSN EN 12981+A1 Lakovny – Strikací kabiny pro nanaseni organickych praskových naterovych hmot – Bezpecnostni pozadavky, Praha, 2009 (in Czech)

[6] EN 12981+A1 Coating plants – Spray booths for application of organic powder rating material – Safety requirements

[7] I. Šimoníková, M. Kulich, Z. Pavelek, I. Bartlová, M. Váchová, J. Serafín, L. Kissiková, H. Věžníková, E. Veličková, A. Bernatík, S. Cáb. "Concept of the solution of explosion prevention in the conditions of industrial operations", In: *Koncepce reseni protivybuchove prevence v podminkach prumyslovych provozu*, SPBI Ostrava, 2012, 1st edition, ISBN 978-80-7385-120-0 (in Czech)

[8] CSN EN 1127-1 Vybusna prostredi – Prevence a ochrana proti vybuchu – Cast 1: Zakladni koncepce a metodika, Praha, 1997, (in Czech)

[9] EN 1127-1 Explosive atmospheres – Explosion prevention and protection –Part 1: Basic concepts and methodology

[10] CSN EN 50281-3 Elektricka zarizení pro prostory s horlavym prachem – Cast 3: Zarazování prostoru, kde jsou nebo mohou byt horlave prachy, Praha, 2003 (in Czech)

[11] EN 50281-3 Equipment for use in the presence of combustible dust – Part 3: Classification of areas where combustible dusts are or may be present, 2002

[12] Government Regulation 176/2008 Coll. of 21 April 2008 on technical requirements for machinery, (in Czech)

[13] Directive 2006/42/EC of the European Parliament and of the Council of 17 May 2006 on machinery, and amending Directive 95/16/EC (recast)

[14] CSN EN 50177 ed. 3 Stabilní elektrostaticke zarizeni pro nanaseni hořlavých praskovych naterovych hmot – Bezpecnostni pozadavky, Praha, 2010 (in Czech)

[15] EN 50177 ed. 3 Stationary electrostatic application equipment for ignitable coating powders – Safety requirements, 2009

[16] CSN EN 50050-2 Elektrostaticke rucni strikaci zarizeni – Bezpecnostni pozadavky – Cast 2: Rucni strikaci zarizeni pro horlave praskove naterové hmoty, Praha, 2010 (in Czech)

[17] EN 50050-2 Electrostatic hand-held spraying equipment – Safety requirements – Part 2: Hand-held spraying equipment for ignitable rating powder, 2009

[18] Ministry of Health of the Czech Republic: <https://osha.europa.eu/fop/czech-republic/cs/legislation/files/smernice1999_92_ec.pdf>.

[19] Directive 1999/92/EC of the Eeuropean Parliament and of the Council

[20] Narizeni vlady c. 406/2004 Sb. o blizsich pozadavcich na zajisteni bezpecnosti a ochrany zdravi pri praci v prostredí s nebezpecím vybuchu. (in Czech)

[21] CSN EN 14034-1+A1 Stanoveni vybuchovych charakteristik rozvireneho prachu – Cast 1: Stanoveni maximalniho vybuchoveho tlaku p_{max} rozvireneho prachu, Praha, 2011,(in Czech)

[22] EN 14034-1+A1 Determination of explosiv characteristics of dust clouds – Part 1: Determination of the maximum explosiv pressure p_{max} of dust clouds, 2011

[23] CSN EN 14034-2+A1 Stanoveni vybuchovych charakteristik rozvireneho prachu – Cast 2: Stanoveni maximalni rychlosti narustu vybuchoveho tlaku $(dp/dt)_{max}$ rozvireneho prachu, Praha, 2011, (in Czech)

[24] EN 14034-2+A1 Determination of explosiv characteristics of dust clouds –Part 2: Determination of the maximum rate of explosiv pressurerise $(dp/dt)_{max}$ of dust clouds, 2011

[25] CSN EN 14034-3+A1 Stanoveni vybuchovych charakteristik rozvireneho prachu – Cast 3: Stanoveni dolni meze vybusnosti LEL rozvířeného prachu, Praha, 2011 (in Czech)

[26] EN 14034-3+A1 Determination of explosiv characteristics of dust clouds – Part 3: Determination of the lower explosion limit LEL of dust clouds, 2011.

[27] J. Vanus, R. Martinek, P. Bilik, J. Koziorek, A. Dracka, "Smart home remote monitoring using PI system management tools". Paper presented at the Proceedings of the 8th International Scientific Symposium on Electrical Power Engineering, ELEKTROENERGETIKA 2015, 372-375.

[28] R. Martinek, J. Vanus, M. Kelnar, P. Bilik, "Control methods of active power filters using soft computing techniques". Paper presented at the Proceedings of the 8th International Scientific Symposium on Electrical Power Engineering, ELEKTROENERGETIKA 2015, 351-354.

[29] R. Martinek, J. Vanus, P. Bilik, "New strategies for application of recursive least square algorithm in active power filters". Paper presented at the Proceedings of the 8th International Scientific Symposium on Electrical Power Engineering, ELEKTROENERGETIKA 2015, 332-335.

[30] E. Velickova. *Explosiveness of powder coatings in paint booths.* VSB - Technical University of Ostrava, Faculty of Safety Engineering, Ostrava: Department of Occupational Safety and Processes, 2016. 91 pp. Dissertation thesis, (in Czech)

The Shielding Effectiveness Measuring of Brick Wall for Different Material Thickness

Marek Pavlík
Department of Electric Power Engineering
Technical University of Košice
Košice, Slovak republic
marek.pavlik@tuke.sk

Ján Zbojovský
Department of Electric Power Engineering
Technical University of Košice
Košice, Slovak republic
jan.zbojovsky@tuke.sk

Abstract — **This paper deals with measuring of shielding effectiveness of electromagnetic field and compare shielding effectiveness of different material thickness. The measurement was focused on measuring of shielding effectiveness in the range of high frequencies from 1 GHz to 9 GHz. The measured objects were a brick wall with a thickness 25 cm and 50 cm. Size of brick wall were 2x2 m. Measurement was preformed according IEEE 299-2006 Standard, Method for Measuring the Effectiveness of Electromagnetic Shielding Enclosures. Measurement was performed in the non-reflection chamber because exterior influence.**

Keywords: electromagnetic field; shielding; shielding effectiveness

I. Introduction

We can say that nowadays everyone is exposed to electric and magnetic fields generated from electricity transmission, telecommunication, broadcasting devices, aviation and from industrial area. The electromagnetic fields exposition has been continually increasing since the 20th century. Based on numerous scientific studies it is established, that exposition to electromagnetic field can affect the correct function of devices, and can cause harmful effects on biological organism. Therefore it is necessary to protect equipment's and biological organisms against the harmful effect. Electrical systems must be resistant against the activity of other systems and cannot adversely affect the normal operation of other systems and devices. Electromagnetic interference can cause severe problems, and this should be taken into account during the design of new equipment. Shielding is commonly used to protect against the harmful effect of electromagnetic field. It is therefore important to focus on the protection of human against exposure of electromagnetic radiation. People work and live in buildings that constitute the shielding against electromagnetic fields. The electromagnetic field penetrates through the wall of the building from the outside. Therefore, the experiment was focused on the brick wall [1][2][3].

The number of studies focused on the impact of mobile devices and Wireless sources to human health. However, problem is that also in the current studies, long-term mobile phone and Wireless internet users have had hardly more than 10 years of regular use of mobile phones. It still may be a relatively short latency period, particularly for slowly growing benign tumours. Among those long-term users, most were initially users of analogue mobile phone and thus, the number of long-term users of the digital technology is even smaller [3][4][5].

The trend of number of sources of electromagnetic radiation will increase. As ordinary persons we cannot perceive the electromagnetic radiation with our senses. This mean, human is so benevolent to potential threats. That is why it is a topic of increasing sources of electromagnetic radiation and its impact on human body for the general public so actual. Wide professional community is more and more focused on resources of electromagnetic fields, their impact and the associated concept of electromagnetic compatibility [4][5].

II. Shielding Effectiveness

Quality shielding materials are determined by three coefficients, shielding coefficient K_S, absorption coefficient A and a reflection coefficient R. The shielding effectiveness SE is closely related with coefficient K_S, R, and A. Shielding coefficient K_S is determined by the intensity of electric field strength E, possibly based on the intensity of the magnetic field H by the relation: [3][6]

$$K_S = \frac{E_2}{E_1}; K_S = \frac{H_2}{H_1} \text{ [-]} \tag{1}$$

where E_2 is intensity of electric field measured using the antenna placed in the prescribed configuration within the enclosure, E_1 is intensity of electric field measured using the antenna placed in the prescribed configuration in the absence of the enclosure, H_2 is intensity of magnetic field measured using the antenna placed in the prescribed configuration within the enclosure, H_1 is intensity of magnetic field measured using the antenna placed in the prescribed configuration in the absence of the enclosure [6][7][8].

Shielding effectiveness SE is calculated using the formula (2-4) if intensity of electric field and intensity of magnetic field are in basic units [6][9][10].

$$SE = 20\log\frac{1}{|K_S|} = 20\log\frac{|H_1|}{|H_2|} = 20\log\frac{|E_1|}{|E_2|} \text{ [dB]} \tag{2}$$

Formula varies for determining of the shielding effectiveness SE according to the frequency range. According to, the shielding effectiveness is determined by the relation:

$$SE = 20\log\frac{|H_1|}{|H_2|} = 20\log\frac{|V_1|}{|V_2|} \text{ [dB]} \tag{3}$$

For the frequency range from 50 Hz to 20 MHz and for the frequency range from 20 MHz to 300 MHz and also the same applies to the frequency range 300 MHz to 100 GHz:[11][12][13]

$$SE = 20\log\frac{|E_1|}{|E_2|} = 10\log\frac{P_1}{P_2}\ \text{[dB]} \tag{4}$$

where E_1 and H_1 are the intensity of electric field and magnetic field at any point in the space where there are no shielding materials respectively. E_2 and H_2 are the intensity of electric field and magnetic field where the shielding materials are in the same place. V_2 is voltage reading within the enclosure, V_1 is voltage reading in the absence of the enclosure and P_2 is power detected within the enclosure, P_1 is power detected in absence of the enclosure. Then *SE* of the material was calculated at specific frequency by using (4) [6][8][14].

If the energy field is less than 1/6 of a wavelength from a high impedance source, then the wave impedance is known as near field source; and thus a capacitive energy dominates the field during the near field effect, and this is because of the higher wave impedances, thus the electromagnetic loss tends to be greater. This is why it is possible to do more effective shielding from these impinging electric fields. A simpler way of looking at this effect is to understand that the culprit electric fields produce voltages onto the victim circuitry. For instance, if you suspect that a given analog interconnect is producing an EMI event, then try disconnecting the wiring from the circuit driving the line, and then short the signal pair together; any voltage differential should be shorted out and the input should quiet down, confirming that the electric field produced was the culprit [7][15][16][17].

If you are too close to a low impedance source, or a current source, then a near field energy source is produced; but what differs in this case, is that the inductive energy predominates. Reflection losses are much less effective here due to lower wave impedances and this effect continues as you drop in frequency; So, this is the main reason why shielding becomes less effective against low frequency magnetic fields, and why at this point in your design balanced circuits of twist pair wires are so important. Another way of looking at this effect is that magnetic fields produce currents over and onto their victim circuits. If you suspect that a given analog interconnect is magnetic field, then try disconnecting the wiring from the circuit driving the line and leaving the signal pair open; any current flowing should be stopped, this will confirm magnetic field coupling [10][11].

Shielding effectiveness can be calculated according to the relations (5-8) if the value of the transmitted signal is set in logarithmic unit.

$$SE = |E_{1\log}| - |E_{2\log}|\ \text{[dB]} \tag{5}$$

$$SE = |H_{1\log}| - |H_{2\log}|\ \text{[dB]} \tag{6}$$

$$SE = |V_{1\log}| - |V_{2\log}|\ \text{[dB]} \tag{7}$$

$$SE = P_{1\log} - P_{2\log}\ \text{[dB]} \tag{8}$$

Formula (5-8) is used according to the available measuring equipment.

When an electromagnetic wave traveling through space encounters a shield two things happen. First, much of the energy is reflected and then some of the energy that is not reflected is then absorbed by the shield; Only the residual energy emerges from the other side of the shield. These two effects of reflection and absorption are independent of each other, but when combined they produce the overall shields true effectiveness. A third factor called re-reflection occurs in very thin shields. This re-reflection occurs on the shield boundary on the far side of the shield material [3][8].

The main factors which determine the shielding effect are the capability of shielding materials (the electric and magnetic conductivity and the permeability), the thickness and the frequency of the incident wave. If we know all these factors, the materials shielding effects can be calculated by (9). If these factors are unknown, we can measure the intensity of electric field and magnetic field when there are shielding materials or not, and then *SE* could be calculated by (2). According to [8][9][10] shielding effectiveness is the sum of the reflection *R*, multiple reflections *B* and absorption *A* of electromagnetic field derived as: [3][5][10]

$$SE = A + R + B \tag{9}$$

$$SE = 15{,}4t\sqrt{f\mu\sigma} + 168{,}16 - 10\log\frac{\mu_R f}{\sigma_R} + 20\log\left(1 - e^{\frac{2t}{\delta}}\right)$$

where t is material thickness, σ is electrical conductivity of shielding material, σ_R is the relative conductivity, μ_R is the relative permeability of shielding material, μ is the magnetic permeability of shielding material, f is frequency, δ is depth of penetration. For the simplicity, it is possible to determine the shielding effectiveness *SE* also as (6) without the multiple reflections *B*.

$$SE = A + R \tag{10}$$

$$SE = 8{,}69\frac{t}{\sqrt{\frac{2}{\omega\mu\sigma}}} + 20\log\left(\frac{1}{4}\sqrt{\frac{\sigma}{\omega\mu_r\varepsilon_0}}\right)$$

where μ is permeability which included permeability of the shielding material, ε_0 is electric permittivity of vacuum. The expression (10) is simplified expression of (9). Both terms are correct except of other literatures. From equation (10) we can see, the relative permeability and relative conductivity affect shielding effectiveness *SE* [8][9].

III. Experiment

This workplace consists of an analog signal generator Agilent N5181A, spectrum analyzer Agilent N9038A MXE EMI, the receiving antenna and transmitting antenna horn type. Antennas were placed indoor of non-reflection chamber. Analog signal generator and spectrum analyzer were placed outdoor of non-reflection chamber. Workplace for purpose of measuring of shielding effectiveness of electromagnetic field is show on Fig.1. The workplace calibration was performed before the measurement. Measured object was placed at a distance of 0.3 m from the transmitting antenna.

The workplace calibration was performed before the measurement. The calibration of measuring workplace was based on radio communication equation (10), which is defined as follows:[1][10][15]

$$P_R = P_T - L_0 + G_T + G_R \tag{11}$$

where P_R is received power, P_T is transmit power, L_0 is loss of free space, G_T is gain of transmitting antenna and G_R is gain of receiving antenna. The loss of free space is defined as follows: [10][15]

$$L_0 = 20\log\frac{4\pi R}{\lambda} \tag{12}$$

where R is distance of antennas and λ is wavelength. The calibration was performed at a frequency of 1 GHz without shielding. The measurements were performed in two steps – with shielding and without shielding. And therefore, screening efficiency was calculated by the formula (10). The calibration was performed at a frequency of 1 GHz without shielding. The measured object was brick walls with thickness 25 cm and 50 cm – Fig. 2.

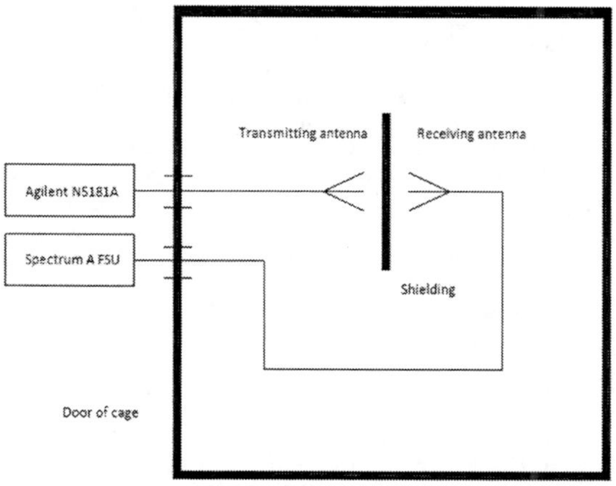

Figure 1. Workplace for purpose of measuring of shielding effectiveness of electromagnetic field

Figure 2. Measured object – brick with thinkness 25 cm a 50 cm

IV. Results

The resulting shielding effectiveness was calculated based on the formula (8). On the figure 3 is shown dependence of shielding effectiveness in the frequency range from 1 GHz to 9 GHz for brick wall with 25 cm and 50 cm thickness. The trend of shielding effectiveness is increasing for both brick wall. The values are higher for 50 cm brick wall in the whole frequency range. Shielding effectiveness has increasing character for both thicknesses.

The values of shielding effectiveness were in the range from 0.5 dB to 37 dB for 50 cm brick wall and the values of shielding effectiveness were in the range from 0.5 dB to 14.8 dB for 25 cm brick wall. The results show that the material thickness affects the shielding effectiveness. This fact is apparent from the relationship (10). However, the shielding effectiveness is not twice for twice material thickness. Measurement uncertainty was 0.65 dB. Measurement uncertainty affect shield effectiveness values, especially at lower frequencies in the measured range.

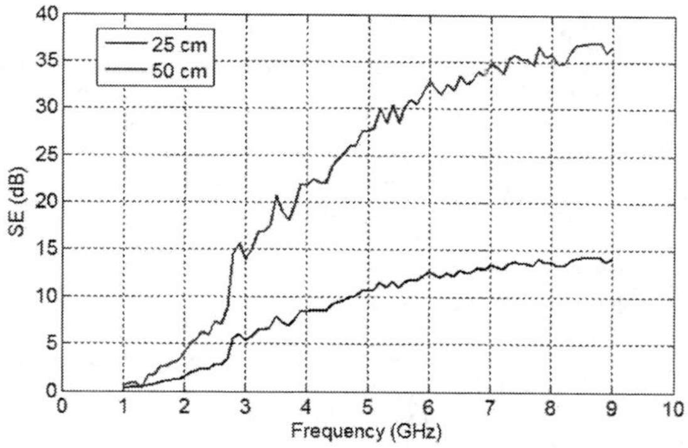

Figure 3. Shielding effectiveness for frequency range from 1 GHz to 9 GHz for brick wall with different thinkness

V. Conclusion

Material for electromagnetic shielding are mainly electrically conducting materials in the form of metals in bulk, porous, and coating forms. But this paper show on shielding ability of building material – brick wall.

This paper deals with measuring of shielding effectiveness of electromagnetic field and compare shielding effectiveness of different material thickness. The measurement was focused on measuring of shielding effectiveness in the range of high frequencies from 1 GHz to 9 GHz. The measured objects were a brick wall with a thickness 25 cm and 50 cm.

The increasing trend is mainly in the frequency range from 1 GHz to t' GHz. The trend is slower in the frequency range from 5 GHz to í GHz. The values of shielding effectiveness were in the range from 0.5 dB to 37 dB for 50 cm brick wall and the values of shielding effectiveness were in the range from 0.5 dB to 14.8 dB for 25 cm brick wall. The results show that the material thickness affects the shielding effectiveness. Twice of thickness means almost twice the shielding effectiveness.

Acknowledgment

The authors wish to acknowledge Scientific Grant Agency of The Ministry of Slovak Republic and Slovak Academy of Science for funding of experimental works in the frame of VEGA No.1/0132/15 grant Research of the penetrating of high frequency electromagnetic waves through ecological building materials.

This publication is the result of the Project implementation: Protection of the population of the Slovak Republic from the effects of electromagnetic fields, ITMS: 26220220145, Activity 2.1 supported by the Research & Development Operational Programme funded by the ERDF .

Európska únia

Agentúra Ministerstva školstva, vedy, výskumu a športu SR pre štrukturálne fondy EÚ

We support research activities in Slovakia / Project is cofinanced from EU funds.

References

[1] B. Dolník, "Electromagnetic compatibility. (Elektromagnetická kompatibilita)," TU of Košice, dec. 2013, monography. ISBN 978-80-8086-221-3.

[2] M. Kosterec, J. Kurimsky, M. Fol'ta, S. Bucko, and Z. Čonka, "Investigation of effects of non-ionizing electromagnetic fields interacting with biological systems," in Proceedings of the 8th International scientific symposium on electrical power engineering, 2015, pp. 564–567, ISBN 978-80-553-2187-5.

[3] L. Wang, Y. Gao, Y. Shen, F. Ruan,"Analysis of Shielding Effectiveness of Rectangular Cavity with Improved Aperture Model," International Symposium on Electromagnetic Compatibility, 2007, 23-26.10.2007, Qingdao, s.329 - 332, ISBN: 978-1-4244-1371-3.

[4] J. Besnoff, M. Chabalko, D. Ricketts, "A Frequency Selective Zero-permeability Metamaterial Shield for Reduction of Near-field Electromagnetic Energy," Antennas and Wireless Propagation Letters, IEEE, 2015, Volume: PP, Issue: 99, pp.1-4, ISSN : 1536-1225.

[5] C. Hao and L. Denghua, "Simplified Model of Shielding Effectiveness of a Cavity with Apertures on Different Sides," IEEE Transactions on Electromagnetic Compatibility, Vol.56, No.2, 2014, s.615-621, ISSN: 0018-9375.

[6] IEEE 299-2006 Standard, Method for Measuring the Effectiveness of Electromagnetic Shielding Enclosures, EMC Society, New York 2006, p.39.

[7] P. Liptai, M. Moravec, E. Lumnitzer, K. Lukáčová, "Impact analysis of the electromagnetic fields of transformer stations close to residential buildings," In: SGEM 2014, 14th international multidiscilinary scientific geoconference, Vol. 1, Issue 5 (2014), Albena, Bulgaria. p. 355-360., ISSN 1314-2704.

[8] Y. Guang, "Analysis of Electromagnetic Field on the Permeability Test of Concrete," Power and Energy Engineering Conference (APPEEC), Asia-Pacific, Shanghai, 27-29 march, 2012, pp. 1-2, ISSN:2157-4839.

[9] C. Hao, L. Denghua, "Simplified Model of Shielding Effectiveness of a Cavity with Apertures on Different Sides," IEEE Transactions on Electromagnetic Compatibility, Vol.56, No.2, 2014, s.615-621, ISSN: 0018-9375.

[10] L. Zhang, Y. Zhang, " Simulation Analysis for the Materials Shielding Effectiveness of EMP", Cross Strait Quad-Regional Radio Science and Wireless Technology Conference, July 26-30, pp. 32-35, 2011.

[11] P. Liptai, M. Moravec, M. Badida, "Research of possibilities of using the recycled materials based on rubber and textiles combined with vermiculite material in the area of noise reduction," Advanced Materials Research. Vol. 1001 (2014), p. 171-176. - ISSN 1022-6680.

[12] D. Medveď," Modeling of electromagnetic and thermal fields in induction heating," In: Elektroenergetika 2011 : proceedings of the 6th International Scientific Symposium on Electrical Power Engineering : 21. - 23.9.2011, Stará Lesná, Slovak Republic. - Košice : TU, 2011 S. 157-163. - ISBN 978-80-553-0724-4.

[13] V. Šlapák, K. Kyslan, M. Lacko, V., Fedák, F. Ďurovský, "Finite Control Set Model Predictive Speed Control of a DC Motor," (2016) Mathematical Problems in Engineering, 2016, art. no. 9571972, DOI: 10.1155/2016/9571972.

[14] J. Dziak, I. Tomčíková, "Models of ideal linear elements of DC and AC circuits and circuit simulation in MATLAB using Sparse Tableau analysis," 2016. In: SSIEE 2016 : proceeding of scientific and student's works in the field of Industrial Electrical Engineering : vol. 5. - Košice : TU, 2016 S. 139-141. - ISBN 978-80-553-2569-9.

[15] J. Zbojovský, P. Liptai, M. Moravec, "Modelling and calculating the shielding effectiveness of building materials," 2016. In: Technical sciences and technologies. Vol. 6, no. 4 (2016), p. 205-210. - ISSN 2411-5363.

[16] M. Kanálik, "Porovnanie medzinárodných noriem týkajúcich sa kvality elektriny," 2007. In: Elektroenergetika 2007. - Košice : Technical university, 2007 S. 235-237. - ISBN 9788080738440.

[17] M. Pavlík, I. Kolcunová, B. Dolník, J. Kurimský, A. Mészáros, D. Medveď, M. Kolcun, J. Zbojovský,"Measuring of Dependence of Shielding Effectiveness of Wet Materials on The Frequency of Electromagnetic Field in the High Frequency Range," Acta Electrotechnica et Informatica. Roč. 13, č. 3 (2013), s. 12-16. - ISSN 1335-8243.

Calculating the Shielding Effectiveness of Materials Based on the Simulation of Electromagnetic Field

Ján Zbojovský
Department of Electrical Power Engineering
Technical University of Košice
Košice, Slovak Republic
jan.zbojovsky@tuke.sk

Marek Pavlík
Department of Electrical Power Engineering
Technical University of Košice
Košice, Slovak Republic
marek.pavlik@tuke.sk

Abstract — **This paper deals with modelling the penetration of high frequency electromagnetic field through the specific materials. It is focused on the electromagnetic field distribution in frequency range from 1.5 to 5.0 GHz, penetrating through combination of these materials: polystyrene and brick, glass – vacuum – glass, and combinations with shielding material. Nowadays in domestic appliances and wide area exists a lot of electric and electronic devices, which are the sources of electromagnetic interference. To protect against harmful electromagnetic interference is in first case important design concept of the sources of electromagnetic interference, and some additional protection can be represented by construction works of rooms or building. For this reason, the article sets out to evaluate the shielding effectiveness of some specific materials and the ways how to improve the shielding effectiveness of these materials. Simulation of the penetration of electromagnetic field is created in the ANSYS HFSS.**

Keywords: shielding material, effectiveness, electromagnetic interference

I. INTRODUCTION

Nowadays everyone is exposed to electromagnetic radiation generated from telecommunication, broadcasting devices, electricity transmission, or from industrial area. Since the 20th century, the electromagnetic fields exposition has been continually increasing. Based on numerous scientific studies it is established, that exposition to electromagnetic field can affect the correct function of devices, and can cause harmful effects on biological organism. For example, one case from 1984. The fighter aircraft has crashed near Munich due to the interference of the electronic control system by electromagnetic waves. Because of this it is necessary to protect equipment's and biological organisms against the harmful effect. Electrical systems must be resistant against the activity of other systems and cannot adversely affect the normal operation of other systems and devices. Electromagnetic interference can cause severe problems, and this should be taken into account during the design of new equipment. Shielding is commonly used to protect against the harmful effect of electromagnetic field [1][2].

Modeling of electromagnetic fields around electrical equipment and in each part of these devices are now an important part of the development and analysis which are important to the quality of equipment with regard to electromagnetic compatibility. These analyzes are performed by computer programs. The simulation is performed in the program Ansys high frequency structural simulator. In this article is simulated and calculated shielding effectiveness of field penetration through combination of these materials: polystyrene and brick, glass – vacuum – glass, and combinations with conductive surface. Frequency range is from 1.5 GHz to 5 GHz and simulations are created in HFSS (High frequency structural simulator). In the conclusion is described the comparison of shielding effectiveness for common frequencies using in telecommunication area [3][4][5].

II. SHIELDING EFFECTIVENESS

Shielding effectiveness is important in valuation of shielding properties for the selected materials. Shielding effectiveness for our case is calculated according to the following relations if the value of the transmitted signal is set in logarithmic unit.

Shielding effectiveness can be expressed by:

$$SE=\left|E_{1\log}\right|-\left|E_{2\log}\right| \quad [\mathrm{dB}] \tag{1}$$

$$SE=\left|H_{1\log}\right|-\left|H_{2\log}\right| \quad [\mathrm{dB}] \tag{2}$$

$$SE=\left|V_{1\log}\right|-\left|V_{2\log}\right| \quad [\mathrm{dB}] \tag{3}$$

$$SE=P_{1\log}-P_{2\log} \quad [\mathrm{dB}] \tag{4}$$

Formula (1), (2), (3), (4) is used according to the available measuring equipment.

Where $E_{1\log}$, $H_{1\log}$, $V_{1\log}$, $P_{1\log}$ represents the magnitude of electric and magnetic field impinging on a shielding material (barrier) and $E_{2\log}$, $H_{2\log}$, $V_{2\log}$, $P_{2\log}$ represents the magnitude of electric and magnetic field at some specific point of shielded area. These equations are for case, if the value of the transmitted signal is set in logarithmic unit [5][6][7].

III. PREPARATION OF A MODEL FOR CALCULATING SHIELDING EFFECTIVENESS

The electromagnetic field simulations depending on frequency are realized in the program HFSS (High Frequency Structure Simulator). A model is represented by waveguide with input and output ports, in the center of which is located the shielding material. The input power is 1W and frequency range is set from 1.5 to 5 GHz, with step 0.01 GHz. Convergence was created to 5 steps. On the following picture the waveguide model is shown. The surrounding area is air. Simulations are created for these combination of materials: polystyrene + brick, glass + vacuum + glass, and combination with shielding material. On the following picture is shown the basic model of waveguide, and on the table 1. are described the parameters of selected materials [8][9][10].

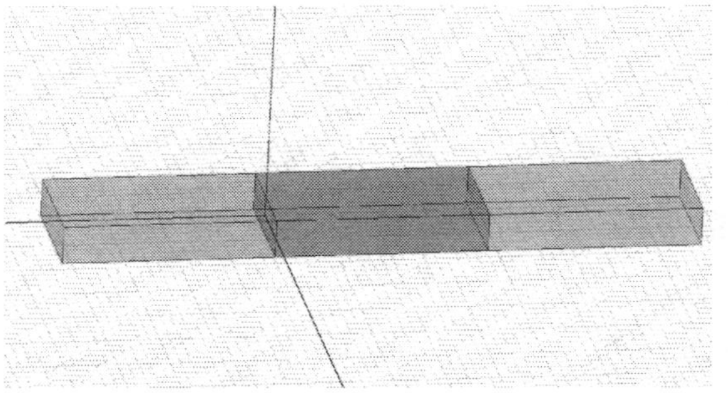

Figure 1. Model of waveguide

Table 1. Parameters of materials

Material	Relative permitivity	Relative permeability
Air	1.0006	1.0000004
Glass	5.5	1
Brick	4,01	1
Polystyrene	2,6	1

On the Figure 2 is shown propagation of electromagnetic wave through the combination "polystyrene + brick" with thickness 10cm

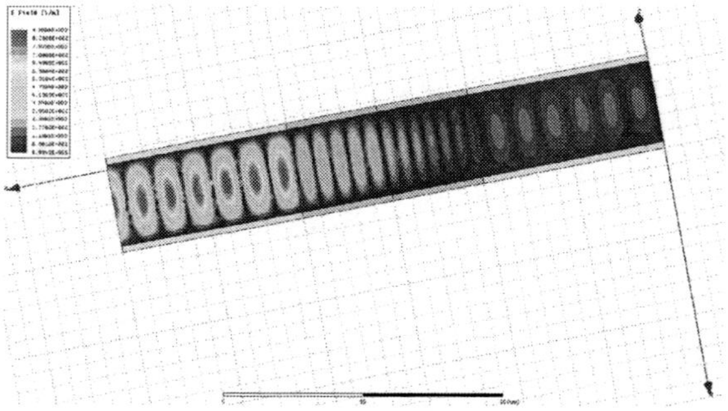

Figure 2. The electromagnetic wave propagation for combination "polystyrene + brick" with thickness 10cm

The average value of shielding effectiveness for this combination is 2.6806 dB. On the next picture is shown the S parameter (electromagnetic wave propagation).

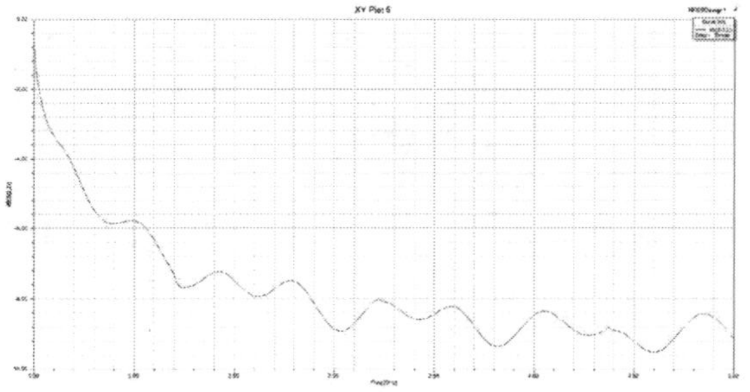

Figure 3. S parameter (electromagnetic wave propagation) for combination "polystyrene + brick

To achieve a better shielding effectiveness these materials was combined with shielding material as a conductive surface with resistance 50 ohm, shown on the Figure 4

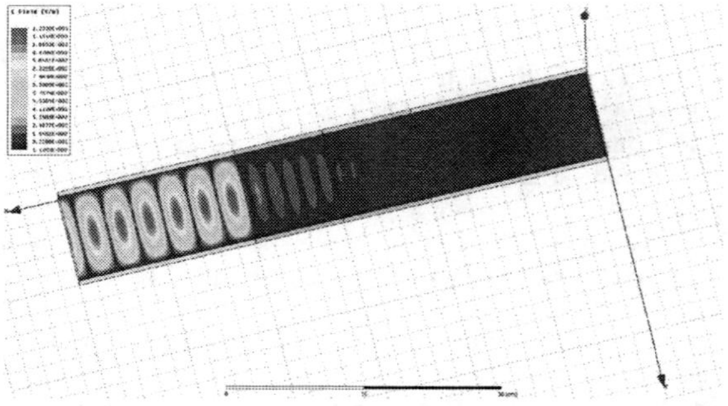

Figure 4. The electromagnetic wave propagation for combination "polystyrene + brick" with thickness 10cm and condictive surface with resistance 50 Ω.

The average value of shielding effectiveness in that case increased to 5,999 dB.

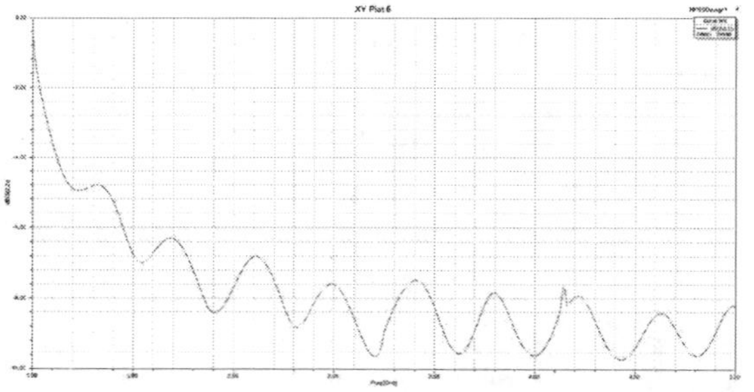

Figure 5. S parameter (electromagnetic wave propagation) for combination "polystyrene + brick and conductive surface

On the next example the combination glass 0,5 cm + vacuum 0,5 cm + glass 0,5 cm was used.

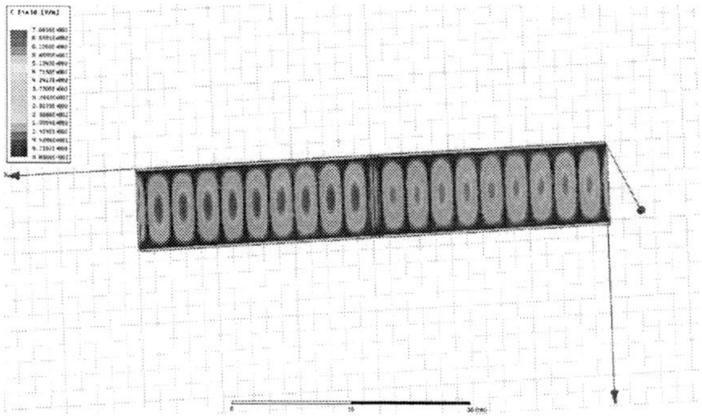

Figure 6. The electromagnetic wave propagation for glass 0,5 cm + vacuum 0,5 cm + glass 0,5 cm

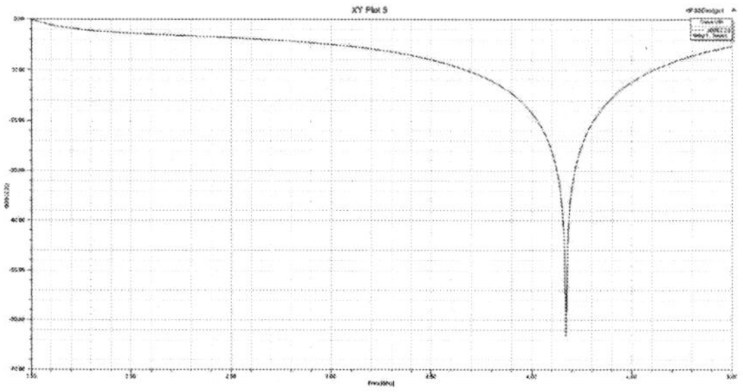

Figure 7. The S parameter for the combination glass 0,5 cm + vacuum 0,5 cm + glass 0,5 cm

The average value of shielding effectiveness for this combination is 0,0014 dB. Generally this value is very small. The next step was the combination of this materials with shielding material, in our case a layer with resistance 50 Ω.

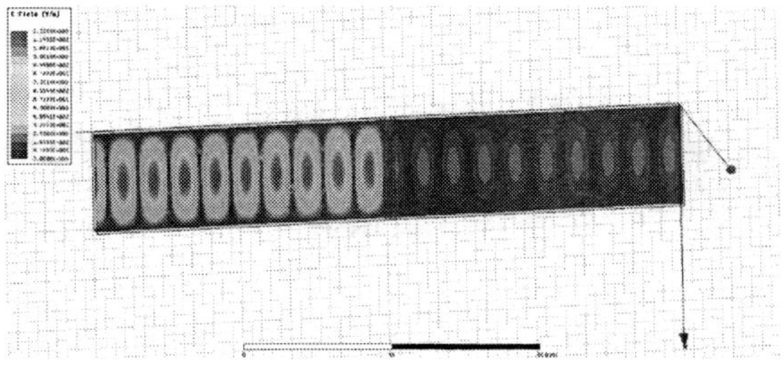

Figure 8. The electromagnetic wave propagation for glass 0,5 cm + vacuum 0,5 cm + glass 0,5 cm and conductive surface 50 Ω.

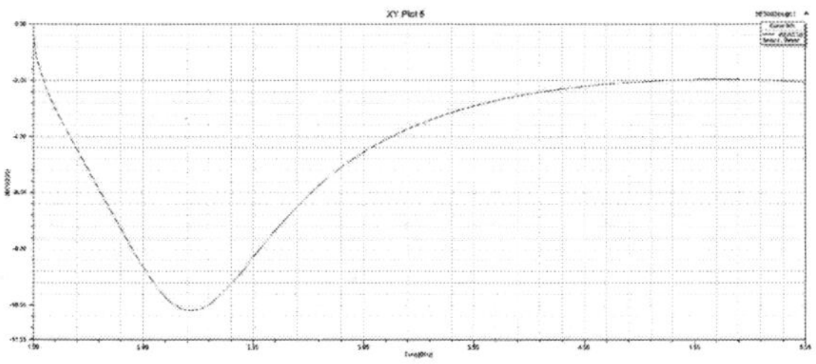

Figure 9. The S parameter for the combination glass 0,5 cm + vacuum 0,5 cm + glass 0,5 cm and conductive surface 50 Ω

Shielding effectiveness increased to 2,803 dB. The next step is comparison of shielding effectiveness for frequencies used in frequency range from 1.5 to 5 GHz shown in Table 2.

Table 2. Shielding effectiveness for selected frequencies

Frequency [GHz]	Material	Shielding effectiveness [dB]
1,8	"polystyrene + brick"	2,6678
	polystyrene + brick" and conductive surface	3,5686
	glass + vacuum + glass	0,0004
	glass + vacuum + glass and conductive surface 50 Ω.	4,2340
2,1	"polystyrene + brick"	0,7923
	polystyrene + brick" and conductive surface	5,2945
	glass + vacuum + glass	0,0002
	glass + vacuum + glass and conductive surface 50 Ω.	7,9789
2,4	"polystyrene + brick"	4,0746
	polystyrene + brick" and conductive surface	6,7256
	glass + vacuum + glass	0,0001
	glass + vacuum + glass and conductive surface 50 Ω.	7,6839
2,6	"polystyrene + brick"	4,7140
	polystyrene + brick" and conductive surface	4,9653
	glass + vacuum + glass	0,0001
	glass + vacuum + glass and conductive surface 50 Ω.	5,8526
5	"polystyrene + brick"	3,6277
	polystyrene + brick" and conductive surface	6,2327
	glass + vacuum + glass	0
	glass + vacuum + glass and conductive surface 50 Ω.	0,6479

IV. Conclusion

The main goal of this paper was to computing the shielding effectiveness of these materials: polystyrene and brick, glass – vacuum – glass, and combinations with conductive surface. This was achieved by simulation of electromagnetic field in computational software HFSS. Frequency range was from 1.5 to 5 GHz.

The average value of shielding effectiveness for combination polystyrene and brick is 2.6806 dB, for combination glass – vacuum – glass it was 0.0014 dB. Generally the building materials has the lower values of shielding effectiveness. This concern of the public to problematic of electromagnetic interference is enhanced by the increasing number of new sources of electromagnetic fields, particularly base stations for mobile operators and telecommunications equipment operators, which are placed directly in the towns and villages. People nowadays are finding solutions, how to decrease electromagnetic interference in our flats, houses, etc.... One of the way is combination of common materials, with specific shielding material and it was done in this article. The average value of shielding effectiveness for combination polystyrene – brick with conductive surface is 5.999 dB dB, for combination glass – vacuum – glass with conductive surface effectiveness increased to 2.803 dB

In the future the research will be focused on the possibilities how to increase shielding effectiveness of common building materials with combination of some specific shielding materials.

Acknowledgment

The authors wish to acknowledge Scientific Grant Agency of The Ministry of Slovak Republic and Slovak Academy of Science for funding of experimental works in the frame of VEGA No.1/0132/15 grant Research of the penetrating of high frequency electromagnetic waves through ecological building materials. This publication is the result of the Project implementation: Protection of the population of the Slovak Republic from the effects of electromagnetic fields, ITMS: 26220220145, Activity 2.1 supported by the Research & Development Operational Programme funded by the ERDF.

Európska únia
Európsky fond regionálneho rozvoja
Operačný program VÝSKUM a VÝVOJ
Agentúra Ministerstva školstva, vedy, výskumu a športu SR pre štrukturálne fondy EÚ

References

[1] B. Dolník, "Electromagnetic compatibility. (Elektromagnetická kompatibilita)," TU of Košice, dec. 2013, monography. ISBN 978-80-8086-221-3.

[2] Scientific Committee on Emerging and Newly Identified Health Risks SCENIHR, " Possible effects of Electromagnetic Fields (EMF) on Human Health", The SCENIHR adopted this opinion at the 16th plenary of 21 March 2007 after public consultation.

[3] P. Liptai, M. Moravec, E. Lumnitzer, K. Lukáčová, "Impact analysis of the electromagnetic fields of transformer stations close to residential buildings," In: SGEM 2014, vol. 1, pp. 355-360.

[4] M. Pavlík, I. Kolcunová, B. Dolník, J. Kurimský, A. Mészáros, D. Medveď, M. Kolcun and J. Zbojovský."Measuring of Dependence of Shielding Effectiveness of Wet Materials on The Frequency of Electromagnetic Field in the High Frequency Range," Acta Electrotechnica et Informatica. Roč. 13, č. 3 (2013), pp. 12-16. ISSN 1335-8243.

[5] L. Zhang, X. Hu, X. Lu, G. Zhu, Y. Zhang, "Simulation analysis for the materials shielding effectiveness of EMP," In: Cross Strait Quad- Regional Radio Science and Wireless Technology Conference (CSQRWC), 26-30.7.2011, Harbin, pp. 32-35, ISBN 978-1-4244-9792-8.

[6] R.A. Rosa, "Electromagnetic modeling, optimization and on-site verification of electromagnetic fields exposure from high power rf sources," In: Sevastopol. 17-21 Sept. 2007, 6th International Conference on Antenna Theory and Techniques, pp. 301-303, ISBN 978-1-4244-1584-7.

[7] L. Wang, Y. Gao, Y. Shen, F. Ruan, "Analysis of Shielding Effectiveness of Rectangular Cavity with Improved Aperture Model,"International Symposium on Electromagnetic Compatibility, 2007, 23-26.10.2007, Qingdao, pp. 329 - 332, ISBN 978-1-4244-1371-3.

[8] IEEE 299-2006 Standard, Method for Measuring the Effectiveness of Electromagnetic Shielding Enclosures, EMC Society, New York 2006, pp.39.

[9] Z. Čonka, P. Hocko, M. Novák, M. Kolcun and G. Morva. "Impact of renewable energy sources on stability of EWIS transmision system" In: 14th International Conference on Environment and Electrical Engineering : conference proceedings : Krakow, Poland, 10-12 May, 2014. - Danvers : IEEE, 2014 P. 75-79. - ISBN 978-1-4799-4661-7.

[10] M. Novák, R. Kravec, M. Kanálik, Z. Čonka and M. Kolcun. "UPFC influence to transient stability of power system" In: Elektro 2014 : 10th international conference : proceedings : Rajecké Teplice, May 19-20, 2014. - Žilina : ŽU, 2014 S. 343-346. - ISBN 978-1-4799-3720-2.

[11] V. Képeši, J. Labun, M. Češkovič, " Change of a monopole antenna's radiation due to finite dimensions of its ground plane", In: Acta Avionica. Roč. 17, 31 - č. 1 (2015), pp. 31-33. ISSN 1335-9479.

Corona Electrodes for Electrostatic Precipitators Supplied with Alternating Voltage

Jaroslav Džmura, Jaroslav Petráš
Department of Electrical Power Engineering
Faculty of Electrical Engineering and Informatics, Technical University of Košice
Košice, Slovakia
jaroslav.dzmura@tuke.sk, jaroslav.petras@tuke.sk

Abstract — **In this paper, various types of corona electrodes suitable for precipitators supplied with alternating voltage are described. For experiments spiral corona electrodes with different twist were designed. Thus the increase of corona electrode length is achieved at equal geometric dimensions of precipitator. Measurements of V-A characteristics were made together with separability measurement on the corona electrodes. Based on the results obtained from measurements it can be concluded that the proposed forms of corona electrodes are suitable for use in electrostatic precipitators supplied with alternating voltage.**

Keywords: corona discharge; electrical precipitator; corona electrode; V-A characteristic

I. INTRODUCTION

In fifties and seventies of the last century many results were published dealing with AC or combined electric field separation process (W. Paul a H. Steinwedel [1]). Theoretical basis was made for physical – mathematical analysis of electrodynamic processes for macroscopic charged particles in non-homogeneous AC electric field.

II. PARTICLE IN AC QUASI-HOMOGENEOUS FIELD

In AC electric high voltage field the process of charging at positive and negative voltage is described separately because due to the polarity effect the charge generation process is different. This is true also for electrically charged particle mobility [2, 3].

We assume that particles fall onto dielectric pad with electric permittivity ε_r and specific volume resistance ρ_V. The product of these values equals time needed for charge conduction from pad surface into earth ($\tau = \varepsilon_0 \cdot \varepsilon_r \cdot \rho_V$) [1]. If there is inorganic or organic particle sedimentation present, an additional layer is created on the pad with electric parameters which are different from those of the pad.

III. ELECTRIC CASCADE SEPARATOR SUPPLIED BY AC HIGH VOLTAGE

We designed a cascade electric separator supplied by AC high voltage (see Fig. 1) [4-6]. It consists of 4 sections, each of them having different linear corona electrodes and different geometrical layout [4, 6]. The collector electrode is made of insulation barrier on which solid particles are captured. Thin aluminum electrode is attached on the outer side of the tube and it creates the ground electrode. Inside of tube there is a fixed thin cooper wire serving as corona electrode. Sections of cascade can contain different number of corona electrodes and they can be placed in different distances from collecting electrode. The aim of corona electrode is to create strong inhomogeneous electric field. This field can be created by electrodes with small radius of bend. If corona electrode is positive new electrons are created. If corona electrode is negative the situation is analogical with the difference that avalanches emitted from negative point are moved to more homogeneous field, as a consequence of their mobility and ability to ionize decreases.

In electrical precipitator supplied by alternating voltage with metallic corona electrode insulating barrier as a collecting electrode is used. For maximal efficiency it is necessary to choose insulating materials with great value of resistance and permittivity. The choice of materials depends on another non electrical quantity e.g. temperature of flying gas, sufficient mechanical strength. Insulating barrier must not change its mechanical and dielectric properties under different temperature conditions.

The separability of the sections was tested by dust particles with specific resistivity spanning from Ω to MΩ values. The separability of the section was measured between 62% and 97.8% (Tab. 1).

Table 1. Average values of efficiency for each cascade section in tube-type precipitator supplied by alternating voltage

	Efficiency [%]			
	1st section	2nd section	3rd section	4th section
Powder with $\rho_c = 3\ \Omega.m$	83,16	70,72	73,94	62,02
Powder with $\rho_c = 4\ k\Omega.m$	96,82	81,72	83,7	72,12
Powder with $\rho_c = 4,8\ M\Omega.m$	93,45	93,56	97,82	95,14

IV. ELECTRIC SEPARATOR WITH SPIRAL CORONA ELECTRODE

To make the separability higher it is necessary to design other types of corona electrodes that could ensure higher amount of free charge carriers. One of the possibilities is to use longer corona electrodes for same separator dimension. This required electrode geometry change.

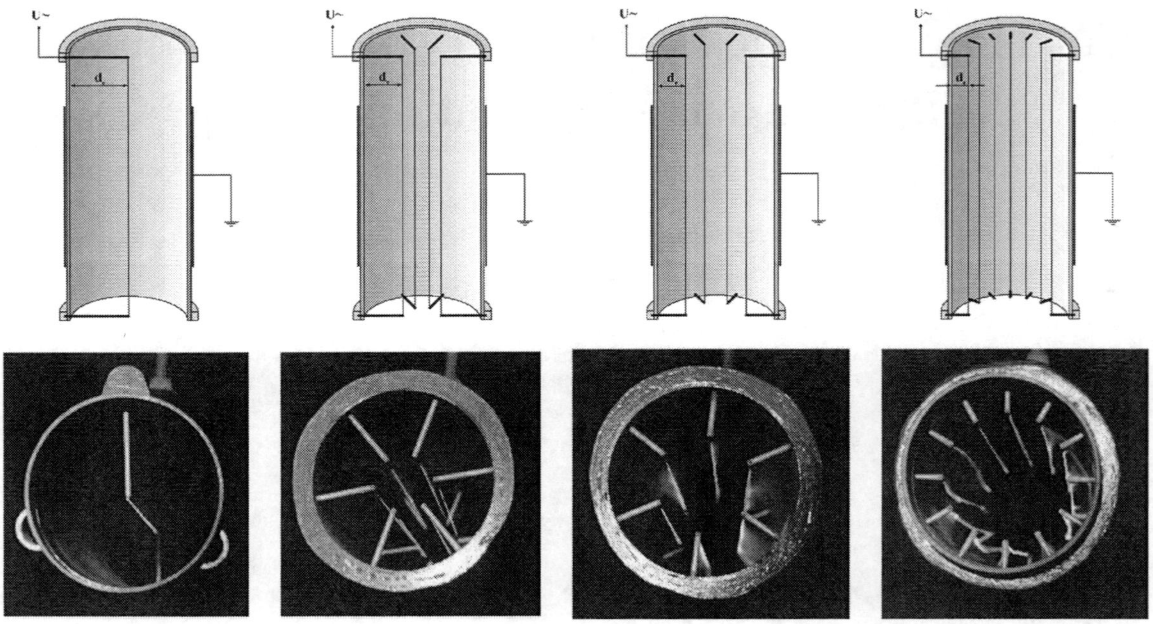

Section 1: 1 ionizing electrode, $d_e = 52{,}5\ \text{mm}$	Section 2: 6 ionizing electrodes, $d_e = 35\ \text{mm}$	Section 3: 6 ionizing electrodes, $d_e = 25\ \text{mm}$	Section 4: 12 ionizing electrodes, $d_e = 15\ \text{mm}$

Figure 1. Geometric arrangement and coronary discharge of individual sections of AC cascade tube separator

Unconventional dust separator designs included discharge activity optimization near electrode with extremely small radius of curvature. Coaxial wires were used in the shape of spiral with different twists (2.5; 4; 7 cm) (Fig. 2) which offered longer corona electrode in comparison to linear electrode (Fig. 3), supposing that the corona current used in discharge area for macroscopic particle charging is increased.

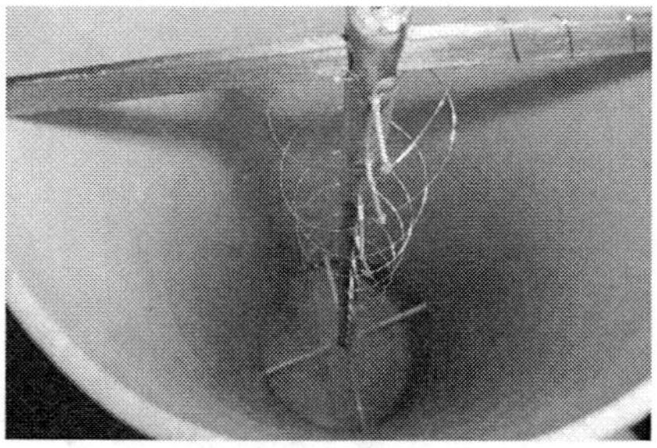

Figure 2. Spiral corona electrode

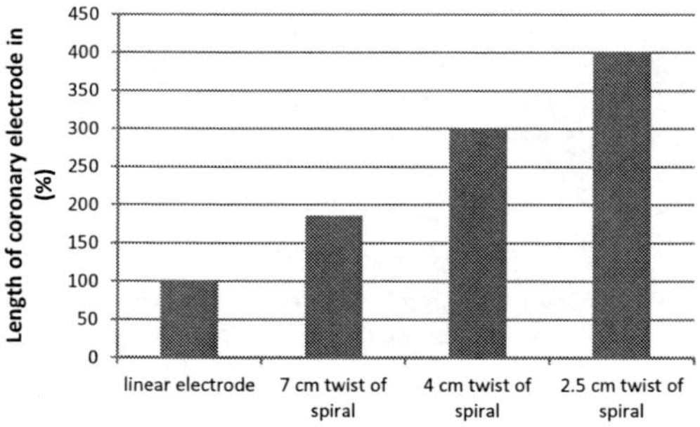

Figure 3. Percentage increase of lengths for different twists

When using plastic holders of electrode we encountered a problem with parasitic charge on the surface of the holders. It was partially eliminated by semiconductor layer on the surface of PVC pole connected to the electrode [7-8].

V. I-U CHARACTERISTIC OF CORONA ELECTRODES

Linear Cu wire or a spiral holder with spiral wire was placed into a valve with diameter 0.105 m made from PVC. Outer part of the valve was covered by metallic foil and earthed. AC voltage was connected to the electrode with the frequency of 50 Hz. The current was measured on the resistor placed in front of metallic foil earthing (Fig. 4).

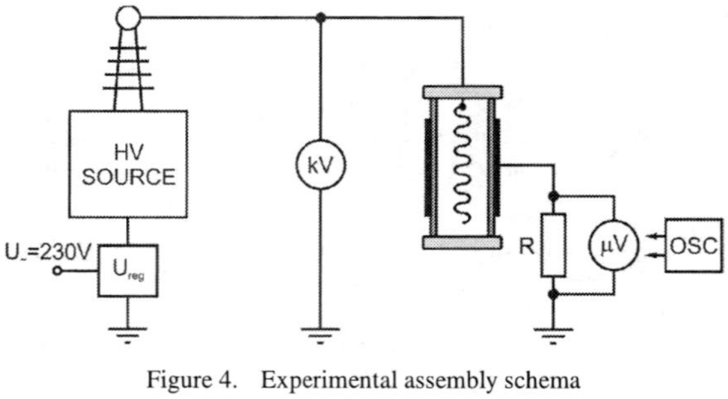

Figure 4. Experimental assembly schema

Fig. 4 shows I-U characteristics of separators with plastic electrode holders and a wire with painted microlayer.

As shown on Fig. 5, evenly graduated of electrode lengths by spiral with defined twist had different effect, because of local places if inhomogeneity of electric field, created by insulation layer damage on quasi random places. These places were created on distances where charge could occur without the influence of high electric field inhomogeneity (Fig. 6). For linear electrodes the distance was in average 11 mm and the lowest and highest values did not exceeded this average more than 2 mm. These characteristics were measured at voltages until 23 kV.

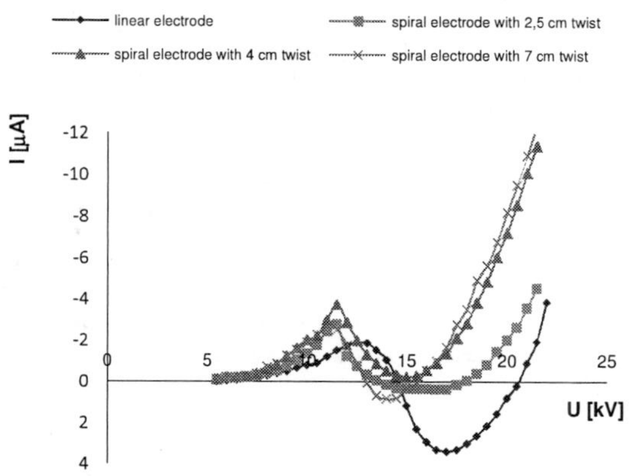

Figure 5. I-U characteristics of separators with plastic electrode holders and a wire with painted microlayer

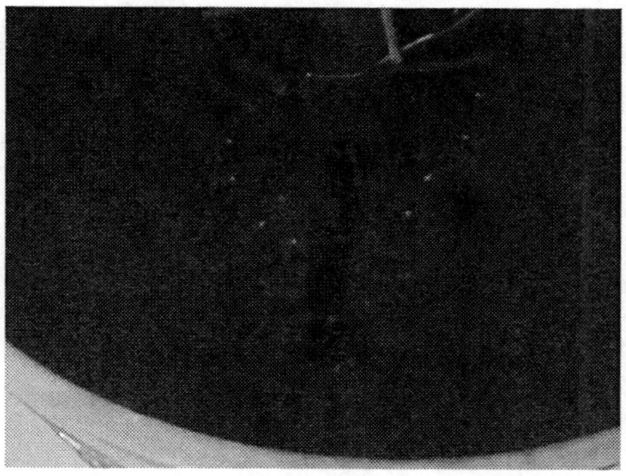

Figure 6. Corona electrode with an insulating micro-layer on its surface

In Fig. 7 there is the detail of the burnt place on the line corona electrode.

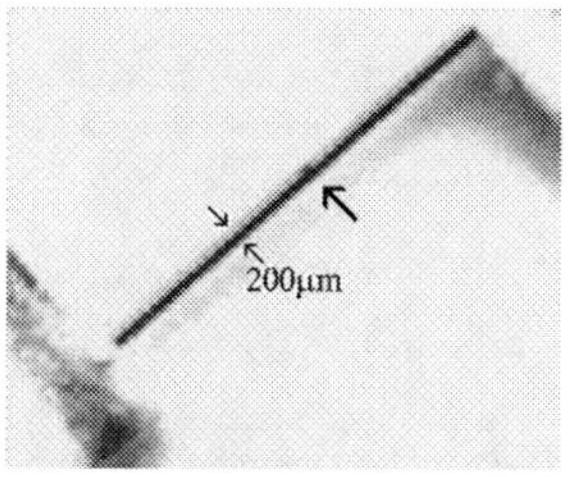

Figure 7. The detail of the microstrip on the line corona electrode

The influence of these places is significant and can be seen also on the graphs. Fig. 8 shows the dependences for linear electrodes.

It can be seen that the stress of electric field was repeated and therefore the places with corona were created repeatedly, the graphs slightly moved to negative current values.

We focused also on I/U characteristics to study corona discharges near:

a) pure Cu wire,

b) painted Cu wire with same radius.

As supposed in case a) I/U was analogous to the one during initial corona discharge as well as for the area of negative differential conductivity, after this area characteristics suddenly changed without charge commutation (Fig. 9), with continual increase of positive charge on the output.

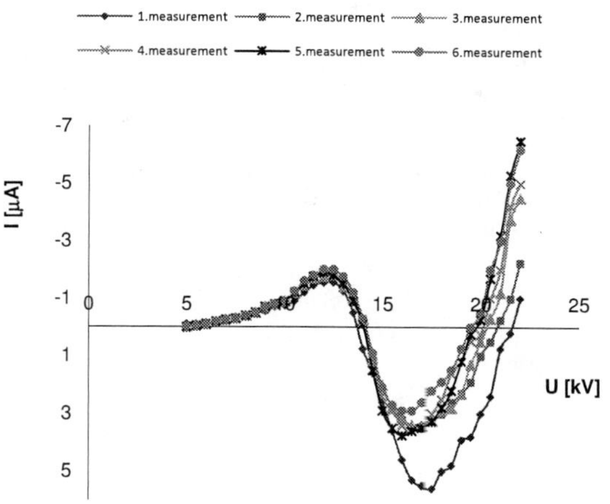

Figure 8. Continuous measurement on painted linear electrode

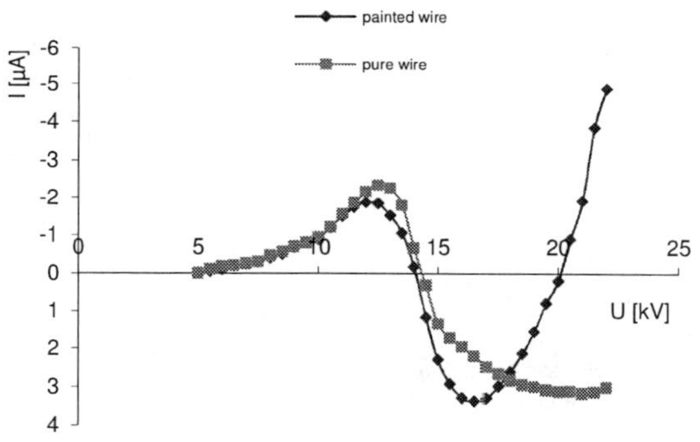

Figure 9. I/U characteristics of separators with linear electrodes with painted and clean surface

This can be explained by electron avalanches, which induced on the inner side negative charge and on the opposite side positive charge on the surface of insulation pole made of PVC. In case b), charge domains caused exponential increase of current because of barrier on insulated electrode.

VI. Experimental results

In our experiments we used following types of separators: linear separator with painted Cu wire, linear separator with clean Cu electrode, concentric spiral separator with 2.5cm twist with painted and clean wire, spiral separator with twist 4cm with painted and clean electrode, spiral separator with 7 cm twist with painted and clean wire.

We used dust from thermal power plant Vojany in Slovakia. The dust was gained from real electric chamber type separator placed in this power plant. The dust was collected from various parts of the separator and during various separator operation modes. We have chosen 3 samples for separation capability measurement. The specific resistivity of these samples is shown in Tab. 1.

Table 1. Dust parameters

Sample No.	Specific electric charge
1	3 Ω·m
2	13 Ω·m
3	800 Ω·m

The separation capability measurement was integrated into time-voltage measurements during separation. As expected, best results were achieved for dust No. 3 with specific resistivity 800 Ω·m (Tab. 2). It can be seen that beside specific conductivity, also other parameters, e.g. specific electric resistivity, relative permittivity and specific weight are important parameters for separation process. This fact can be seen also on I-t characteristics and it points out such parameters as particle shape, weight or size.

Table 2. Separation capability for different electrode types

	Sample No. 1	Sample No. 2	Sample No. 3
Linear painted electrode	77,689 %	75,422 %	97,641 %
Linear clean electrode	77,772 %	58,773 %	98,510 %
2.5 cm painted spiral	72,208 %	64,238 %	89,622 %
2.5 cm clean spiral	87,104 %	65,422 %	90,995 %
2.5 cm spiral, metallic holder	56,701 %	63,923 %	90,165 %
4 cm spiral painted wire	75,208 %	61,055 %	93,072 %
4 cm clean spiral	71,304 %	59,284 %	96,940 %
7 cm painted spiral	71,390 %	56,241 %	89,939 %
7 cm clean spiral	72,200 %	58,791 %	94,317 %

When observing the distribution of separated particles on coaxial separator perimeter, it can be seen that the particle copied the shape of spiral corona electrodes and some so called dead zones were observed. These zones were created by particle electric field interaction with same polarity. At higher voltages the electric field was contracted and the distribution of transported particles was changed. The fine grain particles of dust were concentrated in spiral axis on the sediment surface and coarse-grained were sedimented on marginal parts because of higher inertial force.

VII. Conclusions

According to numerous measurements and experiments the coaxial separator of dust with AC high voltage and with unconventional shape of corona electrode seems to be an alternative to conventional dust separators for dust with high specific resistivity.

There are some advantages like e.g. easy construction, light and comfortable manipulation with the possibility to exchange the separator.

Results achieved evoked also numerous new thoughts focusing on separator construction in order to exploit as much as possible the transport area of the unconventional separator. We suppose that more homogeneous distribution of macroscopic particles in volume can be achieved by smaller diameter of electrode spiral with constant twist. Thus the separation effect of strongly non-homogeneous electric field would be trimmed (particle distribution) and the so called dead zones would be minimized.

It is advisable to place the designed separator into parallel path of smoke gases for quick and non-conflict collection of dust particles from separator walls. This quick and non-conflict dust collection from separator walls could be achieved. It is not advisable to brush the particles from the walls of coaxial separator in this technological process. A better way is the electronic connection of electrode system to the generator of DC short pulses which can cause separation by opposite charge.

In production environment we cannot assume constant quality of combusted material, therefore cascade placement of designed separators is proposed.

Acknowledgment

The authors would like to acknowledge Scientific Grant Agency of The Ministry of Slovak Republic and Slovak Academy of Science for funding experimental works in the frame of VEGA No. 1/0311/15 and VEGA No. 2/0141/16.

This work was supported by the Slovak Research and Development Agency under the contract No. APVV-15-0438.

References

[1] K. Marton, "The study of unipolar corona discharge and its application in practice". SVŠT, Bratislava, 1989, 258 s. Doctoral Dissertation Thesis.

[2] K. Marton, "Wechselwirkung zwischen Gleich und Wechselspannungskorona bei der Aufladung von Pulvern", Beichte der X. VIuAt Elektrostatisches Beschichten, HfV Dresden, 1988

[3] K. Marton, "Auftrag anorganischer Pulver auf Keramikunterlagen im elektrischen 50Hz" - Hochspannungsfeld, Poster: Medzinárodné sypózium Elektrostatics 89, Budapesť, 1989

[4] J. Džmura, "Charging and transport of macroscopic particles in a strongly inhomogeneous electric field". Dissertation Thesis. Košice, 2002. 123 s.

[5] K. Marton, J. Balogh, J. Džmura, J. Petráš, "Elimination of dust particle sedimentation in industry environment". In: Journal of Electrostatics. Vol. 71, no. 3 (2013), p. 208-213. - ISSN 0304-3886

[6] J. Džmura, "Macroscopic particles in direct and alternating electric field". In: Przeglad Elektrotechniczny Konferencje. Vol. 5, no. 3 (2007), p. 78-80. - ISSN 1731-6103

[7] B. Dolník, "Contribution to analysis of daily diagram of supply voltage in low voltage network: working days versus non-working days". In: Electric Power Engineering (EPE). Ostrava: VŠB-TU, 2015 P. 373-376. ISBN 978-1-4673-6787-5

[8] J. Kurimský, R. Cimbala, I. Kolcunová, "Multi-scale decomposition for partial discharge analysis". In: Przeglad Elektrotechniczny. Vol. 84, no. 9 (2008), p. 169-173. - ISSN 0033-2097

C: Power Systems Relaying, Faults Analysis

The Effectiveness Increasing of Earth Fault Distance Protection Operation in High-Voltage Cable Line

Tatjana Lomane, Associate Professor
Riga Technical University,
Riga, Latvia
loman@eef.rtu.lv

Sergejs Rubcovs, Senior Lecturer
Riga Technical University,
Riga, Latvia
sergejs.rubcovs@ast.lv

Sergey Kovalenko, Leading Researcher
Riga Technical University,
Riga, Latvia
sergejs.kovalenko@rtu.lv

Kristina Berzina, Senior Lecturer
Riga Technical University,
Riga, Latvia
Kristina.Berzina@rtu.lv

Abstract — **In high voltage networks, the most common protection of transmission lines from all types of faults is distance protection. One of the main reasons for the widespread use of distance protection (DP) is the ability to provide near and far reservation of protection for adjacent objects. In the mixed cables and overhead lines networks there are problems of maintenance of the earth fault distance protection effective operation, because of differences in the characteristics of high-voltage cables and wires of overhead lines. This should be taken into account at the task of the setting the tripping parameters for the earth fault distance protection in the cable lines. Statistical analysis of the earth fault distance protection effective operation in cable and overhead lines showed that during the earth faults through the transitive resistance the classical type quadrilateral characteristics of operation does not provide the effective work of the DP in the cable lines. Based on probabilistic assessments of the earth faults programmable distance measuring units (PDMU) operation the new more efficient pentagonal tripping characteristic for the main DP zones for cable lines and combined tripping characteristic for complex network have been developed.**

Keywords: earth fault distance protection; high-voltage cable; tripping characteristic effectiveness

I. Introduction

Modern development of electrical networks is characterized by the use of high-voltage cable inserts and tendency of the cable lines construction (up to nominal voltage of 500 kV) instead of overhead line. This is the result of more stringent environmental policies, the rising cost of land, an increase the density of built-up territories or water obstacles presence in the transmission path. There are occur mixed cable and overhead lines of high-voltage network. In high voltage networks, the most common protection of power transmission lines from all types of faults the distance protection is used. One of the main reasons for the widespread use of distance protection (DP) is the ability to provide near and far reservation of adjacent objects protections. In cable and overhead lines mixed networks, there are problems in ensuring the effective operation of earth faults distance protection (DP) due to differences in the characteristics of high-voltage cables and wires of overhead lines [1]. Active and reactive impedance different ratios of positive and zero sequences for overhead and cable lines cause different currents distribution coefficients in the positive and zero sequence circuits in the line. As a result, the controlled by distance protection angles between the positive, negative and zero sequence currents vectors for overhead and cable lines during the earth faults are different. This must be taken into account when specifying the tripping conditions for starting elements of the earth fault DP in the cable lines. The statistical analysis of the earth fault DP operation in cable lines showed that in the earth fault modes through the transient resistance the quadrilateral tripping characteristic of the classical type [2-6] does not provide the DP effective operation. Based on probabilistic evaluations of PDMU operation in earth fault modes, a new pentagonal triggering characteristic has been developed, the use of which will increase the efficiency of its operation in the cable lines protection. In mixed networks it is possible to use the tripping characteristics of different shapes for the main and reserve protection zones.

II. Probabilistic Criteria for Assessing the Efficiency of Relay Protection

The PDMU can be considered to consist of three independent sequential algorithmic blocks:

1. Digital filtering (DF) of current and voltage input signals.

2. Measurement of the impedance vector to fault location.

3. Comparison of the measured impedance parameters with the tripping characteristic settings.

The individual blocks effectiveness operation can be evaluated on the criteria, which is dedicated to the many works [1, 2, 7 - 9]. The main criteria for the protective devices effectiveness operation in general are selectivity, speed and sensitivity. Satisfaction of these criteria is possible only with increasing the efficiency of individual units. In general, the PDMU functioning algorithm can be represented by a system of inequalities of the form [8]:

$$\chi[i(F,S), u(F,S), S, \Pi] > 0\,, \quad (1)$$

where χ is a set of operators specifying the arithmetic operations over samples;

$i(F,S), u(F,S)$ is a set of samples for controlled currents and voltages;

S is the vector of the power system regime parameters, which determine the current processes in the monitored object (the pre-emergency load current, the fault location and the value of the transient resistance at the fault point;

Π is a vector of the tripping settings (for example, the form of the tripping characteristic) and another setting parameters;

F is digital filters coefficients vector, etc.

The tasks of PDMU analyzing are reduced to the estimation of the χ operator for the given power system parameters Π and F in various modes, characterized by the multidimensional vector of the regime parameters S. Moreover, we can assume that for any time moment the values of these parameters are random. The set of samples $i(F,S), u(F,S)$ can be considered as a multidimensional system of random variables.

If it is assumed that the fulfilment of condition (1) determines the protection operation, and non-fulfilment (failure) of condition (1) determines protection non-operation, then the PDMU operation under the conditions of random changes of the regime parameter vector S can be estimated by probabilities coefficients [10]. These probabilistic estimates of the DP effectiveness operation can be obtained by the Monte Carlo method. The technique is based on the statistical modeling of the protected (controlled) object emergency mode and the PDMU operation in this modes with random values of the parameter vector S.

III. Earth Fault PDMU Algorithms of Operation

The PDMU operation is based on converting the input current and voltage into the impedance complex function according to a given algorithm. The classical earth fault PDMU algorithm is based on the positive sequence impedance calculation by the relay input signals of damaged phase voltage and the current function values. The positive sequence impedance of line distance between the protection installation point and the fault location is calculated by the classic algorithm [1-9]:

$$\dot{Z}1_m = \frac{\dot{U}_{ph}}{\dot{I}_{ph} + \dot{K}_N \cdot \dot{I}_E} \quad (2),$$

where $\dot{I}_{ph}$ is faulted phase current;

$\dot{I}_E$ is the earth current at the relay location;

$$\dot{I}_E = 3\dot{I}0_L\,;$$

$\dot{I}0_L$ is zero sequence currents at the relay location;

$\dot{K}_N$ is normal or classic compensation factor of voltage drops from zero sequence currents:

$$\dot{K}_N = \frac{\dot{Z}0_L - \dot{Z}1_L}{3 \cdot \dot{Z}1_L} \quad (3);$$

where $\dot{Z}1_L$, $\dot{Z}0_L$ are respectively positive and zero sequence impedances of the line per unit length;

The accuracy of the impedance calculation depends to a large extent on the compensation coefficient $\dot{K}_N$ setting value, the calculation principles of which can be different [1 - 2], but for the DP first and second zones are usually calculated by the expression (3). The resulting parameters of the calculated vector $\dot{Z}1_m$ are compared with the boundary conditions of the given region on the complex plane of the tripping characteristic, which is possible in the form of a circle or a polygon.

A more effective tripping characteristic is the characteristic of the quadrilateral form [2 - 6], since they provide an independent choice of the operating settings in the directions of the active and reactive resistances. This allows satisfy conflicting requirements for different axes on the complex impedance plane. In the standard software package of digital DP, there are 5 or 6 protection zones with their tripping areas and operating time settings [6, 7]. The operation time of protection is determined by the satisfaction of the complex impedance calculated vector to tripping conditions of the corresponding zone.

The earth fault PDMU have three complex tripping characteristics planes for estimating the positive sequence impedance $\dot{Z}1_m$ vector, which is calculated by the voltages and currents, that are applied to the relay, depending on the fault type, as indicated in Table 1

Table 1. Relay input of voltages un currents for Z1 calculation.

Fault type	Ur	Ir	Z1m
A0, AB0, AC0	$\dot{U}_A$	$\dot{I}_A + \dot{K}_N \dot{I}_0$	$\dot{U}_A / (\dot{I}_A + \dot{K}_N \dot{I}_0)$
B0, AB0, BC0	$\dot{U}_B$	$\dot{I}_B + \dot{K}_N \dot{I}_0$	$\dot{U}_B / (\dot{I}_B + \dot{K}_N \dot{I}_0)$
C0, AC0, BC0	$\dot{U}_C$	$\dot{I}_C + \dot{K}_N \dot{I}_0$	$\dot{U}_C / (\dot{I}_C + \dot{K}_N \dot{I}_0)$

IV. Analysis of Earth Faults Parameters for Cables and Overhead Lines

High voltage cable lines are characterized by significantly lower values of active and reactive resistances and significantly higher values of capacitive conductivity in comparison with overhead lines. In this case, the ratio of the parameters of the positive and zero sequence are fundamentally different. In Table 2 is shown that in overhead lines the ratio of the zero sequence reactive and active resistances to the positive sequence is close and in simple calculations can be taken equal to 2.5 - 3.5. Accordingly, the positive and zero sequences resistance angles are close, but the modules are different in 2.5 - 3.5 times. In cable lines, the zero sequence reactive impedances decreases by a factor of 2 - 5, and the values of the zero sequence active resistances increase by a factor of 10-15, so the angles of the positive and zero sequences resistance are significantly different, and the modules are close. Therefore, at earth faults the angles between different sequences current vectors (or voltages) in the overhead lines differ from these angles in cable lines.

Table 2. 330kV overhead lines and cables positive and zero sequences parameters

330 kV lines	X1, Ohms/km	Ratio X0/X1	R1, Ohms/km	Ratio R0/R1
Overhead	0.26-0.34	2.5-3.5	0.02-0.1	3-6
Cable	0.1-0.25	0.2-0.6	0.01-0.03	10-15

The DP starting units determine the damaged phases, the type of damage and the corresponding tripping characteristic by using the angles between the current vectors of the positive (I1), negative (I2) and zero (I0) sequences. As an example, it is shown the values of the controlled angles for individual fault types having a common axis of symmetry (special phase) in overhead lines [10] and cable lines in Table 3. The angles between the current vectors of the positive (I1), the negative (I2) and the zero (I0) sequences for all kinds of earth faults in the overhead lines have definite values that do not depend on the fault position on the overhead line [10].

In cable lines at earth faults, the values of the angles between the positive (I1), the negative (I2) and the zero (I0) sequences current vectors depending on the fault location on the line are varied. As a result of studies of the single-phase and two-phase earth faults in cable lines, the values of controlled angles between the vectors of the positive, negative, and zero sequence currents the species as $\arg\left[\dot{I}2_A/\dot{I}1_A\right]$, $\arg\left[\dot{I}0_A/\dot{I}1_A\right]$ and $\arg\left[\dot{I}2_A/\dot{I}0_A\right]$ depending at the fault location in the line are obtained.

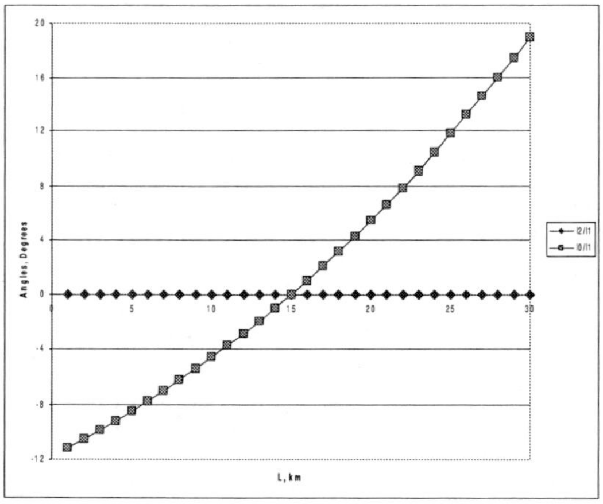

Figure 1. Angles $\arg\left[\dot{I}2_A/\dot{I}1_A\right]$ and $\arg\left[\dot{I}0_A/\dot{I}1_A\right]$ for single A phase to ground faults in cable.

These angles are shown in Fig. 1 for single-phase and in Fig. 2 for two-phase to ground. Moreover, for single-phase fault the angles values $\arg\left[\dot{I}0_A/\dot{I}1_A\right]$ in the cable lines change sign depending on the fault position on the line, and hence there is a point on the line where I0 and I1 current vectors coincide.

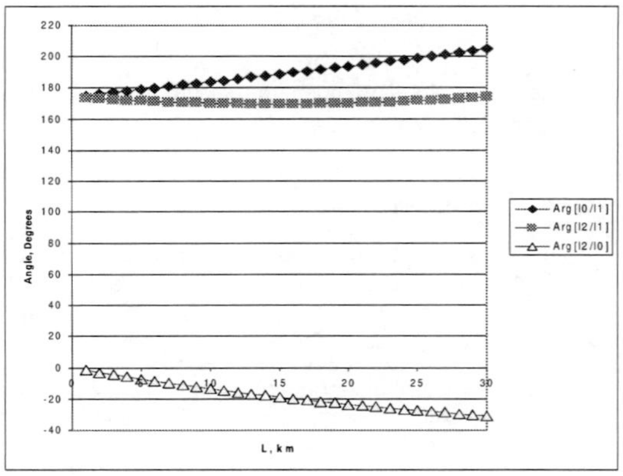

Figure 2. Angles $\arg\left[\dot{I}2_A/\dot{I}1_A\right]$, $\arg\left[\dot{I}0_A/\dot{I}1_A\right]$ and $\arg\left[\dot{I}2_A/\dot{I}0_A\right]$ for B and C phases to ground faults in cable.

As an example, Table 3 shows the ranges of angular variation between the vectors of the positive, negative and zero sequence currents or fault conditions for two fault types having a common symmetry axis (a special phase). This is the ground fault of phase A and two phases to ground B and C.

Table 3. Fault type and Fault conditions

Fault type	Fault conditions
A0, overhead line	$\arg\left[\dot{I}2_A/\dot{I}1_A\right]=0$ $\arg\left[\dot{I}2_A/\dot{I}0_A\right]=0$ $\arg\left[\dot{I}0_A/\dot{I}1_A\right]=0$
A0, cable	$\arg\left[\dot{I}2_A/\dot{I}1_A\right]=0$ $-\pi/6\le\arg\left[\dot{I}2_A/\dot{I}0_A\right]\le\pi/6$ $-\pi/6\le\arg\left[\dot{I}0_A/\dot{I}1_A\right]\le\pi/6$
BC0, overhead line	$\arg\left[\dot{I}2_A/\dot{I}1_A\right]=\pi$ $\arg\left[\dot{I}2_A/\dot{I}0_A\right]=0$ $\arg\left[\dot{I}0_A/\dot{I}1_A\right]=\pi$
BC0, cable	$\arg\left[\dot{I}2_A/\dot{I}1_A\right]=\pi$ $-\pi/6\le\arg\left[\dot{I}2_A/\dot{I}0_A\right]\le 0$ $\pi\le\arg\left[\dot{I}0_A/\dot{I}1_A\right]\le 7\pi/6$

V. Earth Fault PDMU Operation Analysis

Figure 3 shows the 330kV network diagram, where W is the high voltage transmission line with 30km long is considered as the DP controlled object. The earth fault DMU operation was analyzed in the line with two types of technical parameters of the transmission system:

- W - overhead line, with 2xACO-300 wire;
- W - cable line, with AHXCHHBMK-W 1x2500/90 190/345kV wire.

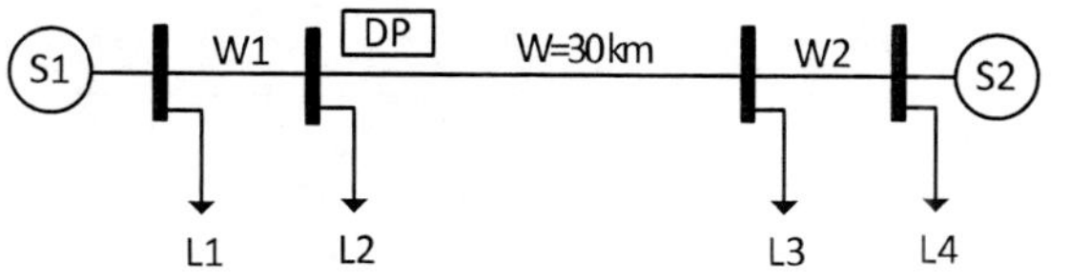

Figure 3. Electric power system test diagram.

It was simulated the earth faults along the entire length with an interval of 1 km on the protected line W. It is considered the possibility of the earth faults with a fault resistance R_{FN}, whose value varied from 0 to 5 Ohm.

In Table 4 are shown the positive, negative and zero sequence impedance of transmission line for earth faults calculation.

Table 4. Testing lines impedances

W	Positive, negative sequence impedance	Zero sequence impedance
Overhead line	$\dot{Z}1,2 = 9.72\,e^{j81.1^\circ}$	$\dot{Z}0 = 28.52\,e^{j75.4^\circ}$
Cable	$\dot{Z}1,2 = 5.11\,e^{j85.6^\circ}$	$\dot{Z}0 = 4.76\,e^{j22.2^\circ}$

Typically, the high-voltage power lines DP use tripping characteristics with five zones, of which 4 are directed per the line and one per the bus. In this study, a simplified characteristic with three zones was considered. This is sufficient for simulating the fault modes only within the protected line. The following limits of the operation zones were adopted. Zone 1 covers 85% of the line, zone 2 covers 120% of the line, Zone 3 covers 160% of the line.

It was analyzed the PDMU operation by the classical algorithm (2) with the using the classical compensation factor calculated by the parameters (3) for the overhead and cable lines respectively:

$$\dot{K}_{N,overheadLine} = 0.64e^{-j9} \text{ and } \dot{K}_{N,cable} = 0.34e^{-j125}. \quad (4)$$

In all the considered earth faults the positive sequence impedance measured by the PDMU (Z1m) are displayed on the complex plane of the tripping characteristic for the overhead line (Fig. 4) and cable line (Fig. 5). The tripping characteristics have the classical form with the limitation of the activation zones on top with the line of the parallel axis of the active resistance. Results of researches have shown, that in the presence of a transient resistance at the fault point in the overhead line, the angle measured by the protection of the resistance vector is less than the line angle due to the increase in the impedance vector coordinate along the active (R) axis more than along the reactive (X) axis. Therefore, to ensure the PDMU reliable operation in such modes, the tripping characteristic is expanded only along the active resistance axis (Fig.4.).

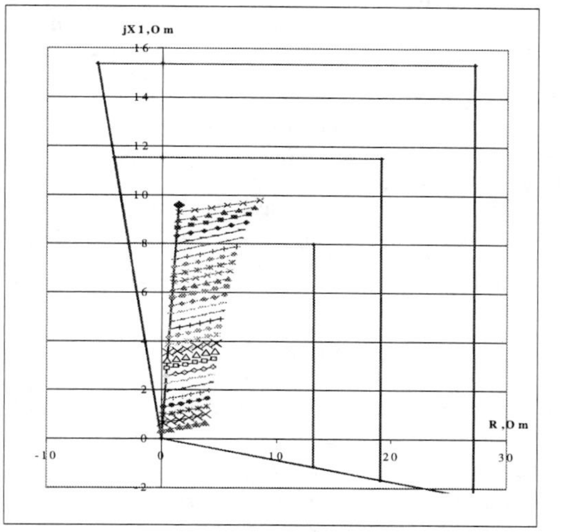

Figure 4. Classics tripping characteristic planes with estimations of DP operation in overhead line.

In the cable line, during earth faults with a transient resistance at the fault point, the coordinate of the impedance vector measured by the PDMU changes significantly along the active (R) axes and reactive (X) resistances. Therefore, the extension of the tripping characteristic only along the axis of the active resistance (Fig. 5) can not provide the PDMU effective operation in cable line even during earth faults with fault resistance in the first zone. In order to increase the PDMU efficiency in cable lines, it is necessary to change the type of the tripping characteristic parameters by increasing the slope of the top border line with the active resistance axis.

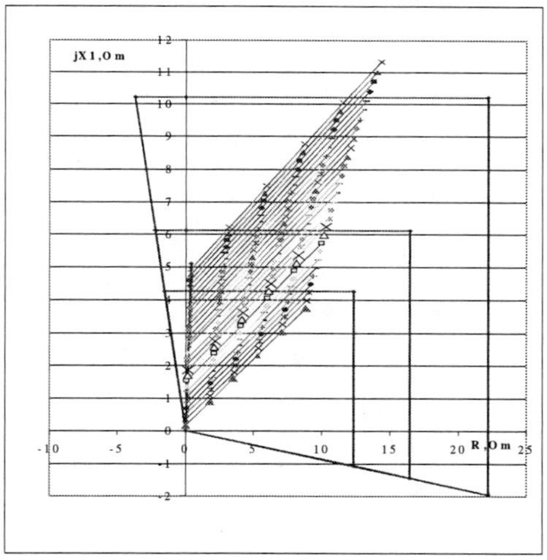

Figure 5. Classics tripping characteristic planes with estimations of DP operation in: A)- overhead line; B)- cable

VI. Improving the Efficienty of the Tripping Characteristics

To increase the PDMU efficiency a new tripping characteristic with the pentagonal form was developed. The new fifth top tripping characteristic is proposed to be introduced at the point of intersection of the line impedance vector with the horizontal border of the tripping characteristic. This will ensure the DP selective operation during the metallic faults at the end of the protected zone. In order to eliminate the failures of the protection operation in the earth faults modes with the transient resistance, it is proposed to increase the protection operation areas upwards from this point by increasing the angle of a slope of the top border with the active resistance axis. The optimal angle of a slope of the upper border for the tripping characteristic can be determined from the probabilistic criteria of the evaluation for the protection operation.

The statistical analysis of the PDMU operation is based on the Monte Carlo method. The criteria of the PDMU operation effectiveness estimation were adopted by the probabilistic coefficients:

Pexc_o is the probability coefficient of excessive operation when is required do not operate;

Pf_o is the probability coefficient of failure operation in the case of requirement to operate.

Fig.6. shows the probability coefficients Pexc_o and Pf_o dependences on angle of the tripping characteristic top border with the active resistance axis. The intersection point of Pexc_o(Angle) and Pf_o(Angle) dependences corresponds to the optimum value of the top angle. In this case there is minimum failures of tripping when is required to operate and minimal unnecessary tripping when is required do not operate. It has appeared, that for a considered cable line this angle is equal 20grad.

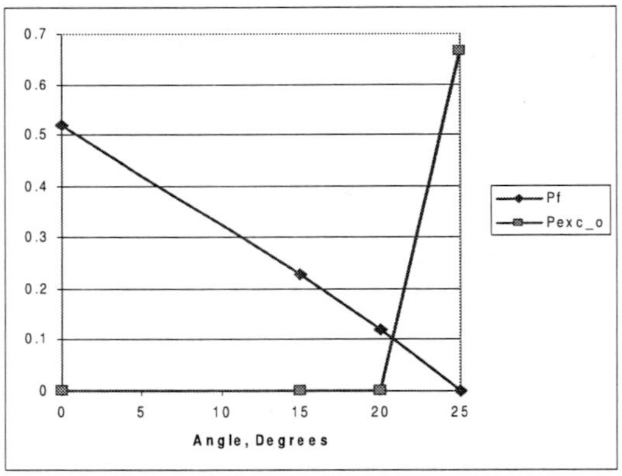

Figure 6. The probability coefficients of Pexc_o(Angle) and Pf_o(Angle) dependences of the DP operation effectiveness probabilistic assessments from the tripping characteristic top angle.

The optimum tripping characteristic for the cable line protection and the obtained estimates of the PDMU effectiveness operation at single phase to ground through the transition resistance at the faults point along the entire length of the cable line is shown in Fig.7. In this case, the vectors of calculated fault impedances at earth faults along the line does not go over the operation region of the second zone, i.e. the first zone earth faults with fault resistance will be disabled in accordance with the set time of first and second zone. In comparison with the classical tripping characteristic the new new fifth top tripping characteristic reduce the operation time of DP and it is an indicator of increasing the effectiveness of protection operation.

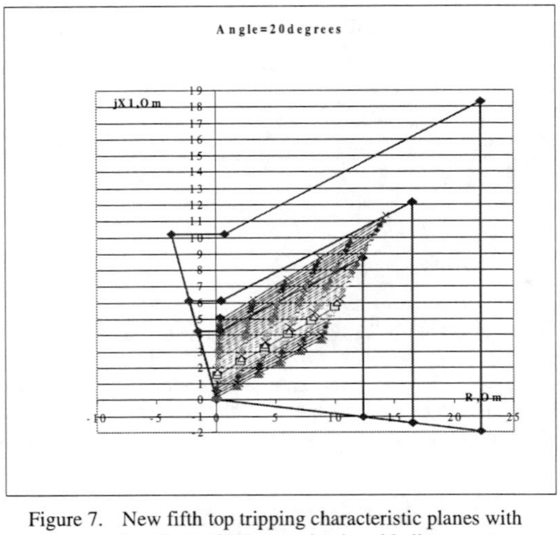

Figure 7. New fifth top tripping characteristic planes with estimations of DP operation in cable line.

On modern line digital protection terminals have the opportunity to install the parameters of the tripping characteristics for each protection zone separately. For effectiveness operation increasing the first and second PDMU zones must use a new pentagonal form of tripping characteristics, since in this case it is the minimum values of the probability coefficient of failure operation in case of requirement to operate (Pf_o) and the minimum values of the probability coefficient of excessive operation when is required do not operate Pexc_o.

The effectiveness of DP subsequent zones, providing near and far reservation in the direction of the protected line, is determined by the selectivity of the DP operation. Therefore, it is suggested, in the case of the coordination for cables protections operation, the pentagonal shape of tripping characteristics should be used for reservation zones, and when coordinating the protections of cable with overhead lines, it should be used the quadrilateral classical tripping characteristics for the reservation zones, as shown in Fig. 8.

VII. CONCLUTION

1. The analysis of the angles between current vectors of different sequences controlled by PDMU starting elements during the earth faults in cable lines showed the need to correct sets of taken for overhead lines possible ranges.

2.The received statistical estimates of the PDMU effectiveness operation in the overhead and cable lines showed that it was not effective to use the classical form quadrilateral characteristics for operation in the earth fault PDMU of high-voltage cable lines protections.

3. It is developed the new more effective pentagonal tripping characteristic for the cable lines PDMU based on probabilistic estimates of the DP operation.

4. It is suggested that in the case of the cable and overhead lines protections coordination must be used the complex tripping characteristics, where the first and second zones use the pentagonal tripping characteristics and a quadrilateral classic tripping characteristics should be used for the reservation zones.

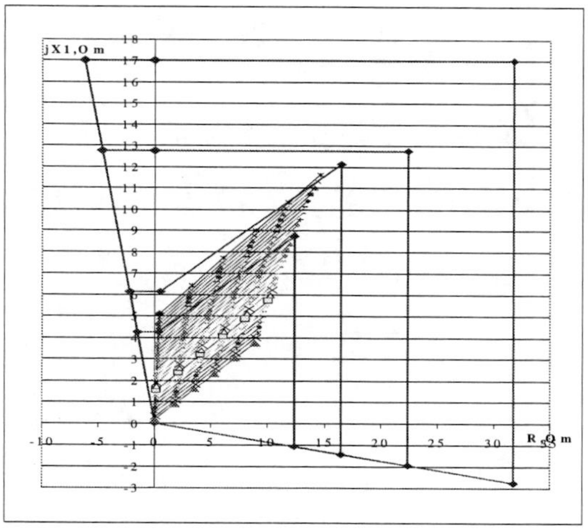

Figure 8. New tripping characteristic planes with estimations of DP operation in cable

REFERENCES

[1] T. Lomane, S. Rubcovs, "Estimation of Earth Fault Distance Protection for High Voltage Mixed Network,". Proceedings of 7th International Scientific Symposium on Electrical Power Engineering, EE2013, September 16-18, 2013, High Tatras - Stara Lesna, Slovak Republic, pp. 348-351.

[2] T. Lomane, V. Aditaja. "Ways Of Efficiency Increase For Distance Measuring Units", Proceedings of 8th International Scientific Symposium on Electrical Power Engineering, EE2015, September 16-18, 2015, High Tatras - Stara Lesna, Slovak Republic, pp. 348-351.

[3] SIPROTEC 7SA522 Distance Protection, Technical Manual, SIEMENS AG, 1999.

[4] Gerhard Ziegler , Numerical Distance Protection, Principles and Applications, SIEMENS, Germany, 2006.

[5] Application Guide on Protection of Complex Transmission Network Configurations, CIGRE, 1991.

[6] REL 511, Line distance protection terminal, User's Guide, ABB Relays AB, ABB Network Control & Protection, Sweden, 1995.

[7] Zima-Bockarjova, at al "Distance Protection Algorithm for Power Transmission Lines Based on Monte-Carlo Method". *IEEE Bucharest Power Tech Conference*, Rumanija, Bucharest, 28. Jun-2. Jul., 2009. IEEE, 2009, 1.-7.lpp.

[8] T. Loman, A. Voinovska. "The Effectiveness Estimation Of Digital Distance Relay Operation", 6th International Conference CPS'04, Štrbské Pleso, Slovak Republic, June 16-18, 2004 .

[9] T. Loman, A. Voinovska. "The Effectiveness Estimation Of Digital Distance Relay Operation", 6th International Conference CPS'04, Štrbské Pleso, Slovak Republic, June 16-18, 2004

[10] Fabrikant V. L., "Distace Protection", Russia, Moskow, High Scool, 1978, 215 pp. (Rus).

Location of Single-Phase Ground Fault by Positive, Negative and Zero Sequence of Overhead Power Lines in Ungrounded Power Grids of 6 - 10 - 35 kV Level Voltage

Alexsander Fedotov
Professor of department «Electrical power systems and grids»
Kazan State Power Engineering University
Kazan, Russia
fed.ai@mail.ru

Natalia Chernova
Associate professor of department «Electrical power systems and grids»
Kazan State Power Engineering University
Kazan, Russia
nv.chernova@list.ru

Georgii Vagapov – researcher of department «Methods of increase of power supply reliability and power quality in distributive electrical grids».
Kazan State Power Engineering University
Kazan, Russia
vagapov@list.ru

Abstract — **This paper reviews theoretical and practical points of the frequency spectrum characteristics of positive, negative and zero sequences of overhead power lines during the single-phase ground fault in ungrounded power grids of 6 - 10 - 35 kV level voltage. Evaluating the effectiveness is shown for each sequence mode settings for determine the fault location.**

Keywords: ungrounded distribution power grids; frequency spectrum characteristics; positive, negative and zero sequences; overhead power lines; single-phase ground fault.

I. Introduction

Ungrounded distribution power grids 6 - 10 - 35 kV level voltage power supply for tens of thousands of electricity consumers and .have for a total length of several tens of thousands of kilometers. Usually, overhead power lines of distribution power grids made with non-insulated wires. Self-supporting wires are not often install, because of their high cost. The single-phase ground faults are one of the most often types of faults in ungrounded distribution power grids 6 - 10 - 35 kV level voltage. Usually, single-phase ground faults starting with in the form of an intermittent arc with overvoltage dangerous. The main problem of detection of single-phase ground faults is low level of fault current in range of 1 -20 A with no depending from distance for fault place. Most of relay protection systems work by current of single-phase ground fault. The sensitivity of these types of systems are too low. Actually, problem of development of information characteristic of single-phase ground faults is open on the subject of distance between substation and fault place.

There are some papers [1, 2, and 3] shows relation between arcing fault and high harmonics in spectrum of voltage and current. There are some «resonance» harmonics caused by wave properties of power lines.

II. Mathematical model of power line

A. Mathematical model of electrical transmission power line

The arcing fault generate high harmonics in the place of fault, so it is possible to use high-harmonic voltage source for mathematical description, fig.1. There are main substation, overhead power line, voltage source connecting to phase A at an l_1 distance from the main substation and voltage source at fig.1.

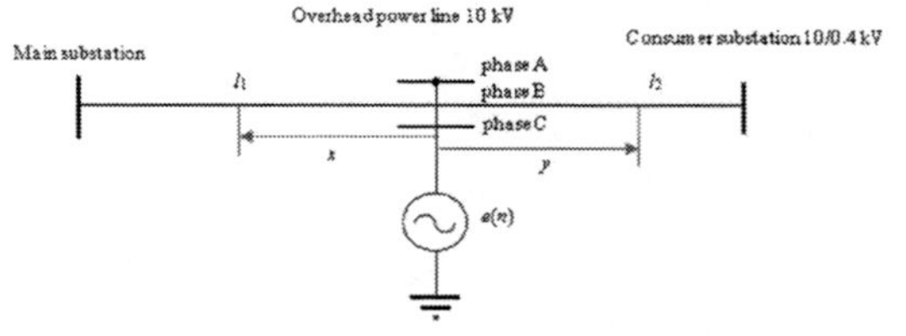

Figure 1. Schematic diagram of the power transmission model.

The start modeling conditions are following. The power line is without load and resistance of main substation equal zero. Equivalent grid shown on fig.2 in common view with resistance of positive, negative and zero sequences (PNZS) [4] for each power line section.

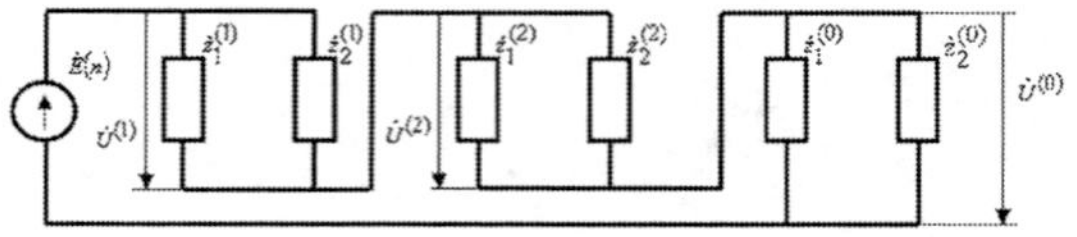

Figure 2. Substitution scheme of the power transmission line.

Superscripts are numbers of sequence of symmetric components and subscripts are numbers of section of overhead line on fig.2

Resistance of positive, negative and zero sequences have the following view

$$\dot{z}_1^{(1)} = \dot{z}_1^{(2)} = jz_c tg(n\beta l_1)\ \dot{z}_2^{(1)} = \dot{z}_2^{(2)} = -jz_c ctg(n\beta l_2).$$

$$\dot{z}_1^{(0)} = -j\dot{z}_c^{(0)} ctg\left(n\beta^{(0)} l_1\right)\ \dot{z}_2^{(0)} = -j\dot{z}_c^{(0)} ctg\left(n\beta^{(0)} l_2\right) \qquad (1)$$

Then it will be

$$\dot{z}_\Sigma^{(1)} = \dot{z}_\Sigma^{(2)} = \dot{z}_1^{(1)} // \dot{z}_2^{(1)} = jz_c \frac{\sin(n\beta l_1)\cos(n\beta l_2)}{\cos(n\beta l)}.$$

$$\dot{z}_\Sigma^{(0)} = \dot{z}_1^{(0)} // \dot{z}_2^{(0)} = -jz_c^{(0)} \frac{\cos\left(n\beta^{(0)} l_1\right)\cos\left(n\beta^{(0)} l_2\right)}{\sin\left(n\beta^{(0)} l\right)}. \qquad (2)$$

The resulting substitution scheme is shown in Fig. 3.

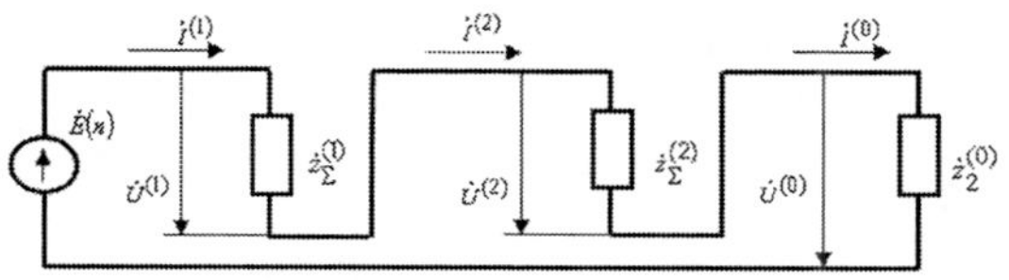

Figure 3. The resulting substitution scheme of the power transmission line.

Total current of n-th higher harmonic according fig.3

$$I = I^{(1)} = I^{(2)} = I^{(0)} = \frac{\dot{E}(n)}{\dot{z}_\Sigma^{(1)} + \dot{z}_\Sigma^{(2)} + \dot{z}_\Sigma^{(0)}} = \qquad (3)$$

$$= \frac{\dot{E}(n)}{j2z_c \frac{\sin(n\beta l_1)\cos(n\beta l_2)}{\cos(n\beta l)} \quad jz_c^{(0)} \frac{\cos\left(n\beta^{(0)} l_1\right)\cos\left(n\beta^{(0)} l_2\right)}{\sin\left(n\beta^{(0)} l\right)}}.$$

Equation (3) allows to highlight those informative signs of single-phase ground fault, which depend on the places of their measurement.

B. *The distribution of current and voltage along the power line in its start section*

Regularities of change of current and voltage depend from the distance x of fault place, fig.1. Currents and voltages for a section of an overhead line with distributed parameters have the following view

$$\begin{bmatrix} \dot{U}_x^{(1)} \\ \dot{I}_x^{(1)} \end{bmatrix} = \begin{bmatrix} \cos(n\beta x) & -jz_c \sin(n\beta x) \\ -j\frac{1}{z_c}\sin(n\beta x) & \cos(n\beta x) \end{bmatrix} \begin{bmatrix} \dot{U}^{(1)} \\ \dot{I}_1^{(1)} \end{bmatrix} \qquad (4)$$

According fig.2 and fig.3

$$\dot{I}_1^{(1)} = \dot{I}\frac{z_2^{(1)}}{z_1^{(1)} + z_2^{(1)}},\ \dot{U}_1^{(1)} = \dot{I}z_\Sigma^{(1)} = \dot{I}\frac{z_1^{(1)} z_2^{(1)}}{z_1^{(1)} + z_2^{(1)}} \qquad (5)$$

After mathematical transformations will be

$$\dot{U}_x^{(1)} = \cos(n\beta x)\dot{U}^{(1)} \quad jz_c \sin(n\beta x) I_1^{(1)} = = jz_c I \frac{\cos(n\beta l_2)}{\cos(n\beta l)} \sin[n\beta(l_1 \quad x)]$$

$$I_x^{(1)} = j\frac{1}{z_c}\sin(n\beta x)\dot{U}^{(1)} + \cos(n\beta x) I_1^{(1)} = = I\frac{\cos(n\beta l_2)}{\cos(n\beta l)}\cos[n\beta(l_1 \quad x)] \qquad (6)$$

For negative sequence equation $\dot{U}_x^{(2)} = \dot{U}_x^{(1)}$ and $I_x^{(2)} = I_x^{(1)}$ will be

$$\dot{U}_x^{(0)} = \cos\left(n\beta^{(0)} x\right)\dot{U}^{(0)} \quad jz_c^{(0)} \sin\left(n\beta^{(0)} x\right) I_1^{(0)} = = jz_c^{(0)} I \frac{\cos\left(n\beta^{(0)} l_2\right)}{\sin\left(n\beta^{(0)} l\right)}\cos\left[n\beta^{(0)}(l_1 \quad x)\right]$$

$$I_x^{(0)} = j\frac{1}{z_c^{(0)}}\sin\left(n\beta^{(0)} x\right)\dot{U}^{(0)} + \cos\left(n\beta^{(0)} x\right) I_1^{(0)} = = I\frac{\cos\left(n\beta^{(0)} l_2\right)}{\sin\left(n\beta^{(0)} l\right)}\sin\left[n\beta^{(0)}(l_1 \quad x)\right] \qquad .(7)$$

C. *The distribution of current and voltage along the power line in its end section*

Regularities of change of current and voltage depend from the distance y of fault place, fig.1. In this case, the initial values of the mode parameter are levels of current and voltage of the corresponding sequences and the final ones are current and voltage at the distance y from the connection point of higher harmonics source.

$$\begin{bmatrix} \dot{U}_y^{(1)} \\ \dot{I}_y^{(1)} \end{bmatrix} = \begin{bmatrix} \cos(n\beta y) & -jz_c \sin(n\beta x) \\ -j\frac{1}{z_c}\sin(n\beta y) & \cos(n\beta y) \end{bmatrix} \begin{bmatrix} \dot{U}^{(1)} \\ \dot{I}_2^{(1)} \end{bmatrix} \qquad (8)$$

The voltage equation in y point will be

$$\dot{U}_y^{(1)} = \cos(n\beta y)\dot{U}^{(1)} \quad jz_c \sin(n\beta y) I_1^{(1)} = = jz_c I \frac{\sin(n\beta l_1)}{\cos(n\beta l)}\cos[n\beta(l_2 \quad y)] \qquad (9)$$

The current equation in y point will be

$$I_y^{(1)} = j\frac{1}{z_c}\sin(n\beta y)\dot{U}^{(1)} + \cos(n\beta y) I_1^{(1)} = = I\frac{\sin(n\beta l_1)}{\cos(n\beta l)}\sin[n\beta(l_2 \quad y)] \qquad . (10)$$

Equations of voltage and current of zero sequence will be

$$\dot{U}_y^{(0)} = -jz_c^{(0)}\dot{I}\frac{\cos\left(n\beta^{(0)}l_1\right)}{\sin\left(n\beta^{(0)}l\right)}\cos\left[n\beta^{(0)}(l_2 - y)\right]$$

$$\dot{I}_y^{(0)} = \dot{I}\frac{\cos\left(n\beta^{(0)}l_1\right)}{\sin\left(n\beta^{(0)}l\right)}\sin\left[n\beta^{(0)}(l_2 - y)\right] . (11)$$

D. Assessment of efficiency of using parameters of a mode of each sequence for determination fault place.

Importantly pre-determine how exactly is measured mode parameter of the electrical grid for evaluate the effectiveness of the use of determine the place of fault. The problem is that current and voltage transformers are installed only in switchgears of main substitutions of 10 kV grids. Therefore, it is necessary to install special high-voltage sensors to obtain information about the parameters of the mode at other points of power line. Given their high cost, their using can be justified only in those cases, with no other way for getting information.

These "other methods" include the receive information on the side of 0.4 kV of the substations, connected to the power transmission line. Obviously, the current load of the power line on side 0.4 kV of 6 – 10 – 35 кV is not available for measurement. However, it is also obvious voltage measurement available for at any point of the overhead line, which connected the corresponding substation. Accordingly, further will be consider how exactly does this information be use to determine the location of the place of single-phase ground fault.

Delta - Star with grounded neutral or Star - Star with grounded neutral are typical windings of power transformers on consumer substations of power distribution grids of 6-10-35 kV level voltage. In all cases it means that the zero-sequence voltage not transform to 0.4 kV side. However, positive, negative sequences transform to 0.4 kV side and can contribute for solution of the problem of determining the location of fault. Use (4) voltage of positive and negative sequences change according sinusoidal law along the overhead power lines to place the transverse asymmetry.

$$U_x^{(1)} = U_x^{(2)} == z_c I\frac{\cos(n\beta l_2)}{\cos(n\beta l)}\sin[n\beta(l_1 \;\; x)] = . (12)$$
$$= A_1 \sin[n\beta(l_1 \;\; x)]$$

Use (11) voltage change according the sinusoidal law along the overhead power lines after the place of transverse asymmetry

$$U_y^{(1)} = U_y^{(2)} = z_c I\frac{\sin(n\beta l_1)}{\cos(n\beta l)}\cos[n\beta(l_2 \;\; y)] = . (13)$$
$$= B_1 \cos[n\beta(l_2 \;\; y)]$$

Graph 1 reproduces regularities of change of voltage of positive and negative sequences according (12) and (13). The amplitude of the trigonometric functions depends from the place of connection of the source of higher harmonics and level of current. In the general case equations (12) and (13) are valid for any harmonic, but preferably it is better to use resonance harmonics. Because they are clearly expressed in the signal spectrum and allow to tune up from higher harmonics caused by load and power lines. This circumstance makes it detection easier in the obtained spectrum with a transient resistance in fault place during the single-phase ground fault on overhead power line. In case of high-voltage sensors install on the overhead power line than it is possible to distinguish zero-sequence voltage.

Use (12) and (13) regularities of its change along the line can be described by the following expressions

$$U_x^{(0)} = z_c^{(0)} I\frac{\cos\left(n\beta^{(0)}l_2\right)}{\sin\left(n\beta^{(0)}l\right)}\cos\left[n\beta^{(0)}(l_1 \;\; x)\right] = . (14)$$
$$= A_0 \cos\left[n\beta^{(0)}(l_1 \;\; x)\right]$$

$$U_y^{(0)} = z_c^{(0)} I\frac{\cos\left(n\beta^{(0)}l_1\right)}{\sin\left(n\beta^{(0)}l\right)}\cos\left[n\beta^{(0)}(l_2 \;\; y)\right] = . (15)$$
$$= B_0 \cos\left[n\beta^{(0)}(l_2 \;\; y)\right]$$

Graph 2 reproduces equations (14) and (15). Conditionally points correspond places of installation of measuring sensors on substations on fig.4. Clearly, information can be obtained in this point only by the values of the voltages of each sequence. Measuring of positive and negative sequences can be possible only on 0.4 kV side of substations.

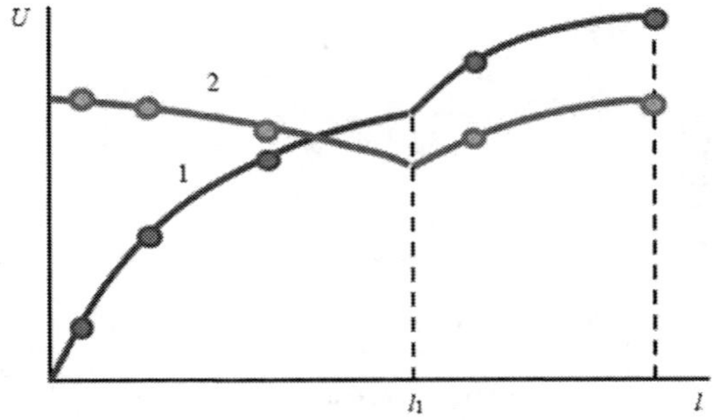

Figure 4. Changing of sequence voltage along power lines.

Graph 1 and 2 show the fracture at point of transverse asymmetry. This point identify by the laws of changing of the graphs, which are known, and boundary values of functions at ends of the line. This information is enough for determine the values of the coefficients A_1, B_1, A_0, B_0. Values of voltage in points of connection of substations to power lines allow to calculate the mentioned coefficients by the method of least squares and calculate the place of asymmetry.

Consider an example. Let be a feeder, fig. 5, with two connected consumer substations to power line. Fig. 6 show regularities of voltage changes along the power line of zero sequence in general. Fig. 7 show for positive (negative) sequence voltage.

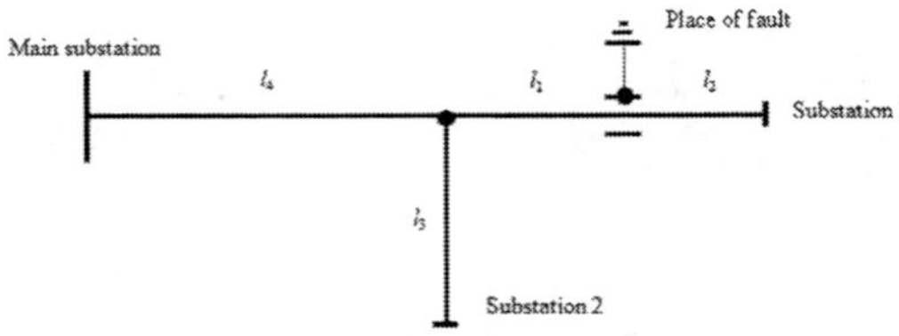

Figure 5. Schematic diagram of the feeder.

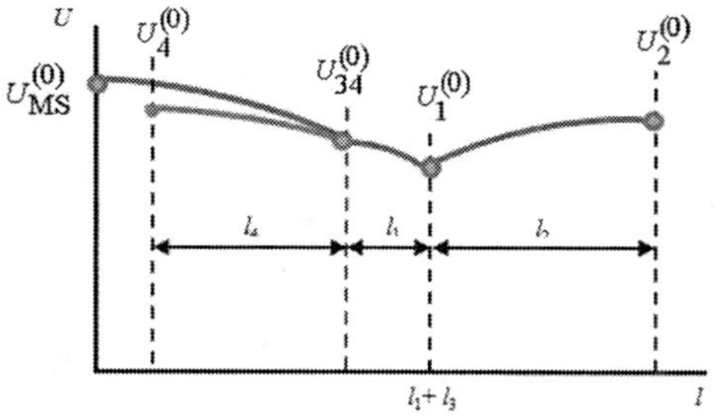

Figure 6. Graphs of voltage change of zero sequence along the power lines.

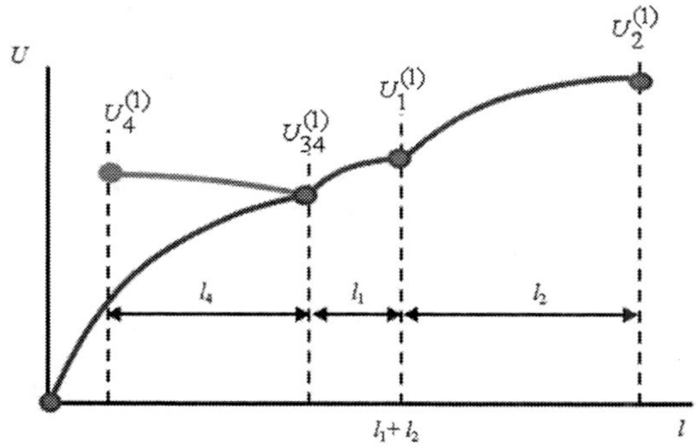

Figure 7. Graphs of voltage change of positive (negative) sequence along the power lines.

It can be case when the voltage of one sequences will be the same on substation 1 and 2, for example voltage of zero sequence fig. 6. In this case, checking voltage of positive (negative) sequences allows to uniquely identifying transverse asymmetry takes place on the line to substation 2. If measurements are made only on the low voltage side of substation, then there is no information about zero sequence of voltage. Nevertheless, it can be calculate the distance to the fault location by the mathematical model and uniquely identified fault line.

$$\frac{U_4^{(1)}}{U_2^{(1)}}=\frac{\cos(n\beta l_2)}{\cos(n\beta l_4)}\times\frac{1}{\cos(n\beta l_1)+\sin(n\beta l_1)\dfrac{\cos(n\beta l_{34})}{\cos(n\beta l_3)\sin(n\beta l_4)}} \quad (16)$$

Using value of currents of higher harmonics gives additional information about single-phase ground faults. However, this involves the installation of high-voltage sensors on the lines, because these currents will be small at the input of transformers. Inductive resistance of transformers are filters for higher harmonics.

Verification calculation on the simulation model of electricity transmission lines showed effectiveness of the proposed method for determining the location of single-phase ground fault by the voltages of the "resonant" harmonic measured along the line with connecting substations.

III. Results of field experiment

Field experiment was organize in existing electrical grids of 10 kV level voltage. The intermittent arc was reproduced in the "C" phase of overhead power line specially connected arrester [5]. Recording devices were placed at main substation with connecting to voltage transformer on 10 kV buses. Other recording devices were placed at distance 15 km from main substation near the place of fault with connecting to buses 0.4 kV buses. Results of the experiment presents on fig. 8 and 9.

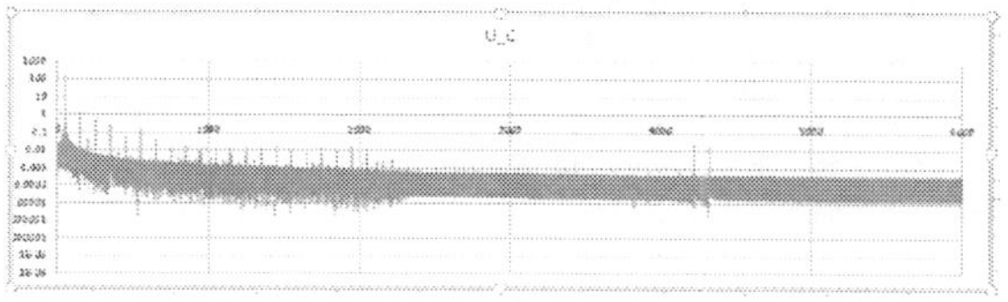

Figure 8. FFT-analysis of the voltage at the beginning of the feeder 10 kV during intermittent arcing.

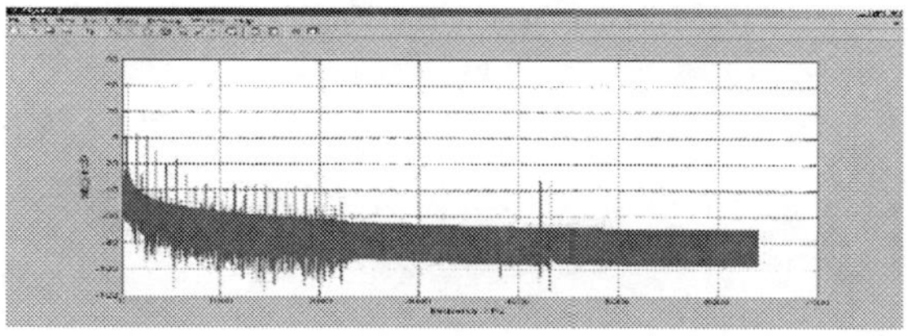

Figure 9. FFT-analysis of phase "C" voltage at 0.4 kV voltage.

"Resonant" harmonics observed both on the side 10 kV and 0.4 kV at points of measurement distant from each other for 15 km. Pre-formed electrical grid simulation showed the coincidence of the calculated "resonance" frequencies of the beginning of the spectrum with the experimental ones to within 10%. However, there is one limitation. Results of modeling in Matlab show that "resonant" harmonics observed at a resistance up to 100 ohm in the place of arcing fault.

IV. Conclusion.

The zero-sequence voltage is most convenient to use for the definition of the power transmission section where transverse asymmetry appeared. This is because the voltage of zero-sequence monotonically decrease from the beginning and end of the line to the place of fault. Moreover, it is enough to compare the voltage values at different points of power line to determine the line segment where there is fault.

It is necessary to install high-voltage sensors for determine the voltage of zero sequence which requires significant expenditure. Voltages of positive and negative sequences can be measured on the 0.4 kV side of consumer substations due to the use of many times cheaper equipment. However, the monotonous character of the change of voltage along the line requires the use of "resonant" harmonics of higher frequencies. Additional measurements serve to increase the accuracy of calculations, because transient resistance at the place of fault and resistance of ground influence on the results of the calculation of the fault location. Sensitivity increases in comparison with typical methods of relay protection.

References

[1] Zoran M. Radojevic; Joong-Rin Shin, "New Digital Algorithm for Adaptive Reclosing Based on the Calculation of the Faulted Phase Voltage Total Harmonic Distortion Factor," 2016 17th International Conference on Harmonics and Quality of Power (ICHQP), vol. 22, pp. 37–41, December 2016.

[2] Z.M. Radojevic ; V.V. Terzija ; M.B. Djuric, "Multipurpose overhead lines protection algorithm," IEE Proceedings - Generation, Transmission and Distribution, vol. 146, pp. 37–41, August 2002.

[3] N. Kagan; N. M. Matsuo, E. L. Ferrari; V. Vega-García; J. C. García-Arias ; T. P. Souza, "Computerized system for detection of high impedance faults in MV overhead distribution lines," IEE Proceedings - Generation, Transmission and Distribution, pp. 992–997, December 2016.

[4] C. Xiangyi, C. Yunping, Li Chunyan, T. Cheng, "Line measurement of positive, negative and zero sequence parameters of transmission lines with mutual inductance" 2007 International Power Engineering Conference, pp.1337 – 1342, December 2007.

[5] St. Kurt, A. Fedotov, G. Vagapov, N. Rozhentsova, R. Abdullazyanov, "Determination of breacks and single-phase ground faults in distribution electric grids by parameters of the mode on the side of 0.4 kV of substations", Industrial power engineering, pp. 34-40, April 2016.

Electrician Forms' Evaluation Using Machine Learning Methods

Peter Holcsik
Óbuda University
Research Group of Applied Disciplines
and Technologies in Energetics
Budapest, Hungary
peter.holcsik@elmu.hu

Miklós Tompa
Decision3 Hungary
Research Group of Applied Disciplines
and Technologies in Energetics
Budapest, Hungary
tompa.miklos@decision3.hu

Abstract — **Nowadays, the utility of an electrician must fill out a form for each work using their tablets. However, evaluation of these forms are limited, it is feasible only with manually interpretation of free text fields. The AD&TE Research Group's aim was to develop a new method for processing of the form field. For this, the group used random forest machine learning algorithms to analyze the forms filled in by the mechanics after power errors. The implemented method has over 90% accuracy in determining whether the form is filled after a certain type of fault (temporary short circuit). It has been proved, that in the investigated cases by using machine learning analysis workflows that require free text analysis can be radically simplified, and accelerated.**

Keywords: machine learning; random forest; smart switchboard; power companies; electricians record;

I. Introduction

Nowadays, the electricians of the ELMŰ-ÉMÁSZ Hungarian power company must fill in a form after each work on their tablets. Evaluating the data of these forms, the myriad of information you can be reached. However, today the evaluations of the free text fields are only limitedly feasible. The reason of it is that, the evaluator is required to open one by one the number of 100,000 forms annually, and interpret manually their contents. This manual process can be radically simplified by using machine learning methods.

Figure 1. The ELMŰ-ÉMÁSZ utility's electricians using their tablets[1]

The Research Group of Applied Disciplines and Technologies in Energetics (AD&TE) has analyzed the forms of 2014 and 2015 years. The analyzed forms had been filled in after network errors' corrections. On the question the group sought to answer whether the certain error was a so-called temporary short circuit. All this in order to fragment the effectiveness of the Smart switchboard.

II. The Smart switchboard Project [2]

The aim of the Smart switchboard (SSB) Project is to develop a new, innovative, low voltage (0,4kV) distribution unit of the distribution network which improves the network's quality.

A. The SSB concept

The SSB concept stands for a remote controlled LV switchboard which uses a circuit breaker for the dismantling of the short circuit current. The detection of the short circuit current is done by using detection equipment together with the corresponding measurement analysis system. It is suitable for remote switch-on (circuit breaker activation) which, if necessary, can be turned to clogging mode. It contains the possibility of visible interruption point and earthling as well.

The visible interruption point and the earthling functions are required for ensuring the life-, health and safety protection during the maintenance, reconstruction, etc. works. The remote monitoring functions could actively or passively monitor the current, the voltage and the performance of the LV system. The implementation of an automatic reclose, a so-called reclose function into the SSB is also possible.

B. The reclose function

When the reclose function is activated, the circuit breaker recloses after a previously specified time period following the defense action. If the defense continues to detect the short circuit, the circuit breaker opens again. Following this and after another pre-set time period, the automatics will close the contacts of the circuit breaker again. The automatics seek to switch back two times (two cycles). The final release is activated if the short circuit persists. Nowadays, the reclose technology is already in use for MV networks [3]. Its working principle and the pre-set time periods for MV networks are shown in Figure 3.

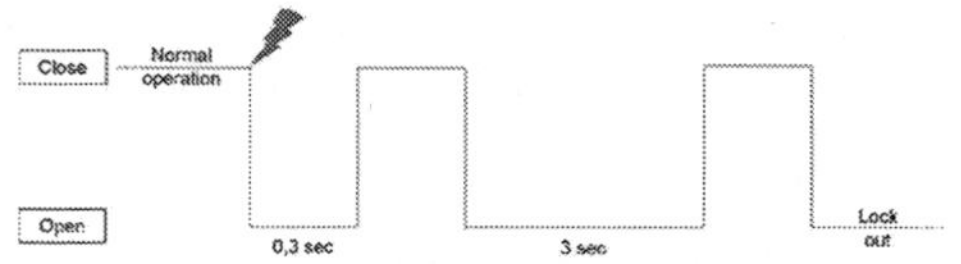

Figure 2. Working principle and pre-set time periods of the reclose function in case of MV networks

The implementation of the reclose function into the LV network will optimize the work of the electricians (they do not have to spend any more time going out to detect the faults in the network) and minimize the length of the LV power outages caused by short circuits. The earlier time consuming process (see Figure 4) is reduced to a few minutes (see Figure 3).

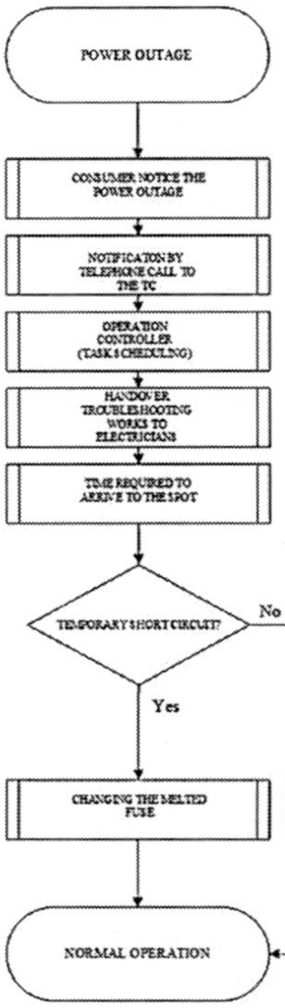

Figure 3. The LV power outage protection flow chart [2]

The AD&TE research team has analyzed the improvement of the quality of the LV networks following the implementation of the SSB's and the efficiency of these advanced tools.

C. Method used for analysing the efficiency of SSB's

When switchboards serving a large number of customers are upgraded to SSB's the related efficiency of these tools increases that is a small number of devices protect a large number of customers [2]. This means that the efficiency of the reclose function implementation will be higher for those switchboards of which melting down can impact a larger number of customers. The efficiency will decrease as the number of customers served by the upgraded switchboards is smaller.

The Research group of AD&TE recommends to the power supplier companies to begin the implementation of the SSB's with the switchboards connected to larger number of customers.

By carrying out the study of the efficiency the research group determined the number of costumers for which it is worthwhile upgrading the existing equipment. A cumulative efficiency function has been developed. This function shows the SAIDI improvement [4] which could be achieved when changing some of the fuses to SSB's.

The research group examined the data collected from the LV power outages during years 2014 and 2015.

The input data for the study was provided by ELMŰ Network ltd and ÉMÁSZ Network ltd.

D. The analysis of the worksheets

After the repair works has been completed, the electrician fills out the worksheet. This is done with the help of drop-down optional lists. Electrician will choose and complete the lists with information related to the cause of the error, whether one or more customers were affected, the type and category of the failed equipment (e.g. transformer station, switchboard, devices of the measuring station, etc.) and/or device (cable, overhead line, fuse, etc.).

In addition to the drop-down lists, free texts can be introduced as well. These could be notes and/or explanations related to the size of the replaced lining measured in ampere or the type of the repair which has been carried out (e.g. replacement of a circuit breaker, measuring equipment exchange, reconnection of a cable, etc.).

The information concerning the type of short circuits which caused the power outage (temporary or long term) is not recorded in the worksheets. Whether only a fuse was changed or a more complex repair work was carried out (e.g. cutting the branches of trees, reconnection of collapsed overhead lines, etc.) is not specified. Due to this reason, the database in its current form does not allow for a clear and rapid detection of the power outages caused by short circuits. In order to that, the worksheets must be analyzed one by one taking into account all the information which has been given in the free text notes and in the drop-down lists as well.

It can happen that the electrician replaced a fuse and the reason of the melt down of it was not a temporary short circuit but, for example, a cable of the network was cut by a branch of a tree or by an excavator working in the nearby.

During the two years period of the study a total number of 15634 pieces of worksheets were generated (after data cleaning and not including the faults of the public lighting and of the measuring stations). The examination of the worksheets one by one would have taken more than 521 working hours (a single empirical data based worksheet manual processing takes two minutes averagely.) Therefore the research group decided to apply the data mining technology [5].

III. Preparation for the Data Mining

To create data classification plate's model using the numerical and categorical variables stored data sheets were insufficient [2]. To improve the accuracy of free-text fields also had to been worked - like during manual processing. The text mining techniques allow the free text wrap and selection of the most characteristic words weighting. Though, the free text can be formed into a line of categorical variables which can be processed in machine learning algorithms.

A. Available Data

The research group's available data sheets contained the error events forms' data of the 2014-2015 years. Each record has been extracted from the database of ELMŰ-ÉMÁSZ Ltd. The panel consists of 15634 pcs of observation, or records all together, with 20 variables per observation. After the pre-clarification 10 variables has been omitted, because these variables could not describe the malfunctions squarely (e.g. region, settlement, group or not) or they could serve as unique identification (e.g. circuit's name). The variables left after the pre-clarification are shown in Table 1.

Table 1. The variables of the data sheet

Variable[1]	Description	Type
Hiba_Azon	Fault ID	Number
Hiba_ok	Faults' reason from drop-down list	Text
Hibahely_kat	Fault category from drop-down list	Text
Kiesett_EEGIS	Faulted network element ID (not noted in every case)	Text
Kiesés_kezdete	Time of outage detection and record at the call-center	Date and time
Kiesési_idő	Time between the reported fault and the repairing in minutes	Number
Helyszínen_tölt	Time between the electrician arrival and departure	Number
Munkair_megj	Dispatchers' notes, free-text	Text
Szerelö_megj	Electricians' notes, free-text	Text

[1] in original language

B. *The dataset of the model building*

Teaching the algorithm observations are needed. It means that, it is necessary to audit a certain amount of work sheets manually. The research group was able to examine six week data, within their 70 hours framework. During this period, they examined 2061 worksheets whether the error was caused by temporary short circuit or it was not. Nothing else was added to the original data sheet only with that the error could be averted with SSB: yes or no.

The resulting data set was used to train a model that has been designated the entire database errors caused temporary circuit SSB, with certain accuracy and error rate.

C. *Data transformation and data sanitization*

The data sheets contained multiple error types and places: incorrect data format, missing values, or content, from processing and modeling point of view.

For Example:

The left out _EEGIS data fields showed "NA" (missing data) values in many places, though they have been set to 0.

It can be stated on empirical basis, that the identification of an individual EEGIS (left out _EEGIS data field) is not relevant. The relevant is that any EEGIS identification has been recorded at all (based on characteristic initial hypothesis of the research group EEGIS ID application form in the case of temporary short circuit). Therefore, all different values from 0 have been modified to 1.

The *Helyszínen_tölt* data field's format has been modified. The time type variables have been changed to minutes. The reason is that the date hour: minute format processing for certain algorithms cannot be handled (possible fault exclusions)

IV. The Text Mining

The aim of the text mining phase is to make the free text fields analyzable with the help of machine learning tools. [6]

A. *Optimization of the characters and the content*

The Text Mining's (TM) first step was to convert the text into small letters and punctuation has been removed. The reason is that some algorithms cannot handle these characters, though that can distort measurements of the frequency of words, what is more they can appear as distinct variables (possible error exclusions)

Character lines with the same meaning have been replaced by the same strings. For example: „nki","betét" „biztosíték" have been exchanged to the word „bizti" which is the most commonly used expression for the previous words.

B. *Choosing relevant words*

Different words which appeared in the forms have been listed, and the frequency of their occurrence has been assigned. In other words, how many times a certain word appears in the technical forms all together. From all these words the general, (e.g. "no" or "street") and fragment words (e.g. linking words or affixes) which could not affect the final result, have been cleared.

From this cleared data set 65 most commonly occurred words have been chosen. (100 or more times match).

The most common words have been assigned to the data sheet as variables. Therefore, all observations of the nine variables (see Table 1) have received 65 further variables. The new variable field is TRUE if the observer technical *Szerelö_megj* field contains the certain word, and FALSE if it does not.

Before the construction of predictive models were needed to replace the names of the variables because as variable names (column names) containing accents, umlauts cannot be handled by the R h2o package, what the research group used.

The data sheet included *Kiesési_idö* and *Helyszínen_tölt* variables. These two variables are closely related, therefore only one of them could be taken into account. Time on site relates more to the types of the errors, so this variable was used in the analysis

V. Model Building

Constructing the model the research group used the n-fold random forest algorithm of R h2o package.

A. *The R software*

R is a free, open source, and free to use, professional and under constant development statistical software package, which is incredibly rich in developed procedures of math engines and work environments. [7] Such a math engine h2o.

B. *The h2o package*

The h2o is an open source math engine for big data that computes parallel distributed machine learning algorithms such as generalized linear models, gradient boosting machines, random forests, and neural networks (deep learning) within various cluster environments [8].

C. *The Random Forest*

Random Forest (RF) is a powerful classification tool. When given a set of data, RF generates a forest of classification trees, rather than a single classification tree. Each of these trees generates a classification for a given set of attributes. The classification from each h2o tree can be thought of as a vote; the most votes determine the classification. [11]

The Random Forest algorithm is one of the most commonly used algorithms. The random forest itself helps eliminate overfitting, for example, while generally accurate performance in the Gradient Boosting Machine (GBM) [x] algorithm is prone to this problem. Considering that a relative small sample was available, the overfitting problem is a major error possibility. Because of the number of elements, and data structures using the RF is also recommended, against the "fashionable" Neural Network (NN) algorithm. The NN algorithm would require a larger amount of data, therefore it would also require more calculating resources, and its performance is worse with heterogeneous variables what existed in our case.

D. *The N-Folds*

Over the n-folds cross validation of the data set is divided into n sections chosen randomly. One section will be the validation set, n-1 will be the training set. During the learning process of each sub-model validation will be set once, for a total of n times running down the learning process. The average of the n result will be the final model.

The procedure is to greatly reduce the chance of overfitting. Especially in case of small number of observations and in case of a high number of variables - that existed in this case.

VI. The Model

The parameters of the model:

- n-folds: 10
- number of trees: 200
- maximum depth: 5

Out of some variable parameter constructed model these parameters had the best accuracy. Mean accuracy: 89.65%, which means, that the model in correctly classifies the malfunctions on average 89.65% accuracy.

A. *The accuracy of the built model*

On the data sheet the n-fold random forest algorithm has been ran fifty times with the parameters above. The fifty model accuracy parameters draw the following pattern:

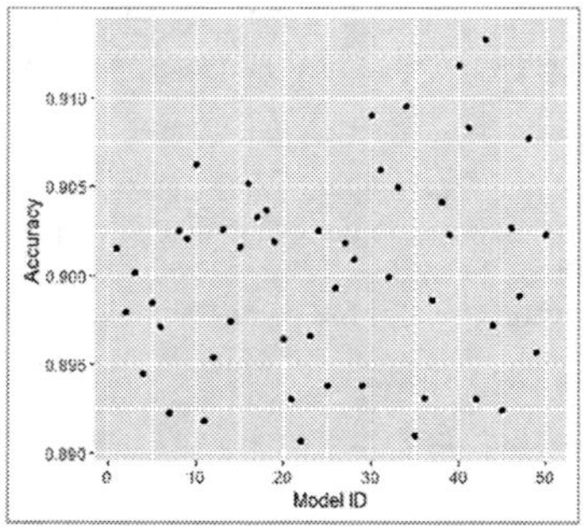

Figure 4. The deviation of the model's accuracy

The models' accuracy varied between 89% and 91.5%. The means and the medians of the values are almost the same, which means that the distribution is normal.

- Median: 0.9005302
- Mean: 0.9001264

Based on the above, the 50 model was ran provides accuracy of about 90%. The data deviation is 0.0056, which means that, assuming a perfect normal distribution with the accuracy of 95% probability is between 88.89% and 91.13% .

B. *The confusion matrix*

The confusion matrix shows, that after the test data was used in building the model - on data that did not participate in the building of the model (n-1) - how much erroneous classification arose.

In this case it meant, that, „a certain error could be handled using SSB?" in response to this question:

- The answer was yes, and yes was given in the observation (Yes/Yes) - TRUE
- The answer was yes, but no was given in the observation (Yes/No) - FALSE
- The answer was no and no was given in the observation (No/No) - TRUE
- The answer was no, but yes was given in the observation (No/Yes) - FALSE

The results of the confusion matrix are the following:

Table 2. Confusion matrix

	Yes	**No**	**Error**	**Rate**
Yes	222	6	0.026316	6 / 228
No	19	45	0.296875	19 / 64
SUM	**241**	**51**	**0.323191**	**25 / 292**

The lines show the errors' real classifications, while the columns represent the classifications predicted by the algorithm. 222 cases qualified temporary short-circuit cases in which they actually were. In 19 cases the algorithm gave "Yes" results in cases of in "No" marks.

The model classified 25 cases incorrectly in the observations out of 292 cases, which is 8.56% of error. This number refers to the given test's dataset, and illustrates the possibility of errors, but it is not a common characteristic of the model.

C. *ROC curve [9]*

The ROC curve generally represents the efficiency of the binary classifications. The y-axis represents the true-positive rate (TPR) on the x axis of the false-positive rate is shown (FPR).

$$TPR = \frac{TP}{(TP+FN)} = \frac{\sum True\ positive}{\sum Condition\ positive} \quad (1)$$

$$FPR = \frac{FP}{(FP+TN)} = \frac{\sum False\ positive}{\sum Condition\ negative} \quad (2)$$

where:
TP (true positive): Yes/Yes cases,
TN (true negative): No/No cases,
FP (false positive): Yes/No cases,
FN (false negative): No/Yes cases
(based on Chapter VI. /A)

The rate of TPR is the correctly predicted Yes results from all the actually Yes cases. This is also known as sensitivity or detection probability.

The rate of FPR is the incorrectly predicted Yes results compared to all No cases. – This is also called as probability of false alarm.

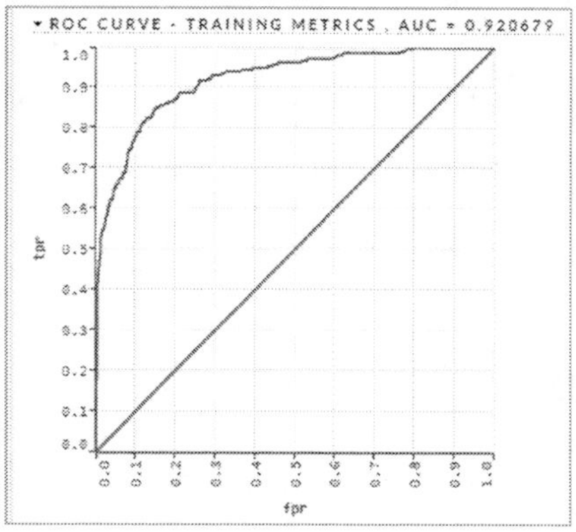

Figure 5. The test results' ROC curve

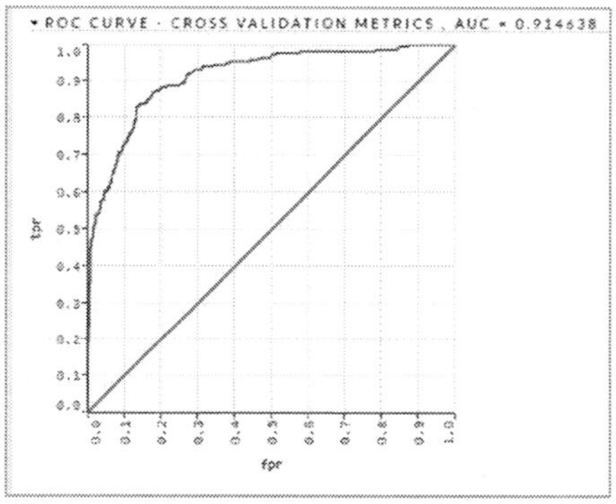

Figure 6. Learning datas' ROC curve

Thus, the ROC curve shows the change of FPR, depending on the TPR, the greater the area under the curve (AUC), the more accurate the model is. The AUC test data approaches the value of the ideal 100%: 92.07%. Thus, it can be stated that the model is accurate. The similarity of the learning's and test data's ROC curve shows that the received algorithm provides similar results to the search data used for testing, such as those used to form. The model is considered good in this respect as well.

D. *The most influential variables in classification*

According to the software's account the most influential facts on the algorithm and their types:

Table 3. The most influential variables in classification

Type	Variable[1]	Variable's description	Percentage
data field	Helyszinen_tolt	See on Table 1	37,93%
data field	Hibaok	See on Table 1	13,66%
data field	Hibahely	See on Table 1	10,86%
text	"szakad"	"tear"	4,34%
text	"oszlop"	"column"	3,61%
data field	Kieses_napja	See on Table 1	2,77%
text	"vezetek"	"line"	2,75%
text	"fa"	"tree"	2,44%
data field	Kieses_oraja	See on Table 1	2,43%
text	"csere"	"change"	2,34%
text	"bizti"	"fuse"	2,28%
text	"csatlakozo"	"connection cable"	1,55%
text	"kikapcsolva"	"switched off"	1,15%
text	"kabel"	"cable"	1,11%
text	"zarlat"	"short circuit"	1,00%
text	"levalasztva"	"disconnected"	0,92%
text	"szigetelo"	"insulator"	0,79%
text	"kosaras"	"aerial truck"	0,74%
text	"fogyaszto"	"consumer"	0,71%
text	"halozat"	"network"	0,68%
text	"volt"	"volage"	0,60%
text	"szukseges"	"necessary"	0,48%

[1] in original language

The data of Table 3 can be shown in one diagram according to the scaled percentages:

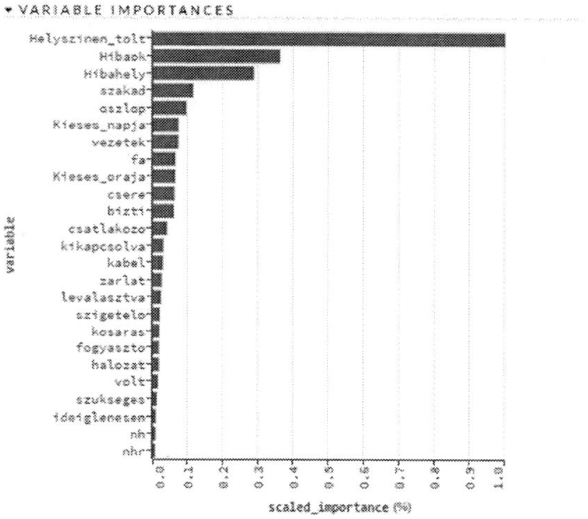

Figure 7. Scaled importance
The most influential factors according to Table 3.

Some interesting variables can be seen on Figure 7. There were some quite unexpected results for example "kosaras" ("aerial truck") or "szükséges" ("necessary") became the most influential variables in classification.

VII. THE IMPORTANCE OF THE MODEL

Using the model created by the research group, the whole error form database became automatically workable. The developed method made possible to decide with the probability of 90% that the technical forms were written about an error could be corrected with SSB. The prepared and classified databases retrieve individual error IDs, which can help determining the number of lost customers and the total downtime as well. Further processing of the data can help determine the optimal installation location of SSB and it can also predict future network quality improvement.

The further aim of the research group is to examine the contents of the technical forms expanded with regional and weather data using machine learning algorithms. [10]

Using the procedure described so far, technical forms can be used to answer a number of questions. The error cases are divided into several groups so-called clusters, depending on the analyzed problem, or the question asked. The analysis The more processed data are available, the easier to analyze them, but the use of text mining methods in this small data set significantly improved the algorithm works and has produced good results. So it can be said that if a similar yes-no questions' answers sought in the content of the forms the method has been described, can be practically applied regardless of the database size.

ACKNOWLEDGMENT

This research was supported by the ELMŰ Network Ltd. and the ÉMÁSZ Network Ltd. Special thanks for the support to József Fodor, Managing director of operations at ELMŰ DSO and ÉMÁSZ DSO, Tibor Fejes and Péter Bányai from the Network Documentary Department and the whole Services Coordination Department.

Special thanks for Tímea Kogyela for the English language support. Last but not least special thanks for our colleagues Judith Pálfi and Ferenc Novothny, PhD.

REFERENCES

[1] M. Kalász, L. Nagy, "Mobilized workforce management," MEE Conference, Mátraháza, 2013.09.12.

[2] J. Pálfi, M. Tompa, P. Holcsik, "Analysis of the efficiency of the recloser function of LV Smart Switchboards," ACTA Polytecnika Hungarica, in press.

[3] J. Gers, T. Holmes, "Protection of Electricity Distribution Networks," The Institution of Engineering and Technology, 2nd Edition, December 2004, ISBN-13: 978-0863415371

[4] "1366-2012 - IEEE Guide for Electric Power Distribution Reliability Indices" Revision of IEEE Std 1366-2003 (Revision of IEEE Std 1366-1998), May 31 2012., DOI:10.1109/IEEESTD.2012.6209381

[5] T. Portik, L Pokorádi, "The Summarized Weighted Mean of Maxima Defuzzification and Its Application at the End of the Risk Assessment Process" In: ACTA Polytechnica HUNGARICA Vol. 11, Issue 3, 2014, pp. 167-180, DOI: 10.12700/APH.11.03.2014.03.11, ISBN: 1785-8860

[6] S. Z. Mishu, S. M. Rafiuddin, "Performance analysis of supervised machine learning algorithms for text classification," Computer and Information Technology (ICCIT), 2016 19th International Conference, 18-20. December 2016, DOI: 10.1109/ICCITECHN.2016.7860233

[7] John Maindonald, W. John Braun "Data Analysis and Graphics Using R: An Example-Based Approach" Cambridge University Press; 3 edition, 2010, ISBN-13: 978-0521762939

[8] D. Cook, "Practical Machine Learning with H2O: Powerful, Scalable Techniques for Deep Learning and AI," O'Reilly Media, December 2016, ISBN-13: 978-1491964606

[9] L. Breiman, "Random Forests" Machine Learning, Springer Volume 45, Issue 1, pp 5–32, 2001, DOI: 10.1023/A:1010933404324

[10] R. Istók , "High Frequency Emissions of Electromagnetic and Electronic Fluorescent Lamps," 10th Jubilee IEEE International Symposium on Applied Computational Intelligence and Informatics (SACI 2015), Timisoara, Romania, 2015.05.21-2015.05.23., 2015. pp. 371-374. ISBN:978-1-4799-9910-1

Influence of Parallel Line on the Distance Protection Measurement

Jiří Bermann
Power Grids division
ABB s.r.o.
Prague, Czech Republic
jiri.bermann@cz.abb.com

Martin Procházka
Power Grids division
ABB s.r.o.
Trutnov, Czech Republic
martin.prochazka@cz.abb.com

Abstract — **The paper describes the error in line distance protection measurement due to the influence of a parallel line. It shows how this phenomenon can be compensated using the earth fault compensation factor. The paper introduces a simple method of compensation for sub-transmission network.**

Keywords: distance protection; parralel line; zero sequence conpensating factor

I. Introduction

The principle of a distance protection is based on the impedance calculation from voltage and current using Ohm's low. The term 'distance' means, that for all kind of shot circuits the measured impedance is proportional to the distance to fault.

With such a performance the fully selectivity between lines can be achieved. Additionally the sensitivity of a distance protection is much higher than simple overcurrent relays, because the fault impedance is evaluated even with current bellow the rated current.

For the phase to phase fault this condition is easily fulfilled by impedance calculation from phase to phase voltage and phase to phase current. For pour two-phase short circuit the necessary equations are shown in (1-3) bellow with phase voltages U_R, U_S, U_T and phase currents I_R, I_S, I_T.

$$Z_{RS} = \frac{U_R - U_S}{I_R - I_S} (\Omega / phase) \tag{1}$$

$$Z_{ST} = \frac{U_S - U_T}{I_S - I_T} (\Omega / phase) \tag{2}$$

$$Z_{TR} = \frac{U_T - U_R}{I_T - I_R} (\Omega / phase) \tag{3}$$

This equation is based on the Kirchhof's law applied to the loop from faulty phases. Because the impedances of both phases are equal, the resulting value gives directly phase impedance between the fault and installation point of the distance protection.

In another words, even we are evaluating a loop impedance, the result is half of the loop impedance i.e. the result gives impedance of just one phase. And such impedance is proportional to the distance to fault. The same is valid for 3-phase fault as well.

Distance measurement for single phase fault is also based on the Kirchhof's law, in this case it is applied to the loop from fault phase conductor and earth. Because the impedance of phase and earth are different, the earth compensation factor n is introduced, to achieve similar linear dependence between the fault and installation point of the distance protection. Additionally to the phase to phase measurement earth current I_E is introduced to the impedance calculation.

The earth compensating factor n can be calculated based on the positive and zero sequence line impedance. The ratio between zero and positive sequence impedance is called k_l in this paper.

$$Z_R = \frac{U_R - 0}{I_R - nI_E} (\Omega / phase) \tag{4}$$

$$Z_S = \frac{U_S - 0}{I_S - nI_E} (\Omega / phase) \tag{5}$$

$$Z_T = \frac{U_T - 0}{I_T - nI_E} (\Omega / phase) \tag{6}$$

$$n = \frac{k_l - 1}{3} = \frac{\frac{Z_0}{Z_1} - 1}{3} \tag{7}$$

If the earth compensating factor is correctly set based on the measured values, then the measurement is really proportional to the distance to fault as requested by the definition, and we can call the relay as distance protection.

$$Z_{ochr} = d.\ Z_{line} \tag{8}$$

where:

Z_{ochr} = impedance measured by the relay = positive sequence impedance to fault

Z_{line} = Total length of the line

d = relative distance to the fault. (Total length = 1)

More precisely, the measured impedance depends on the earth compensating factor and the measured ratio between phase current I and earth current I_E. But the linear proportion between fault location and measured impedance is still valid.

$$Z_{ochr} = \frac{d \cdot Z_{line}}{1 - n \cdot {I_E}/{I}} \tag{9}$$

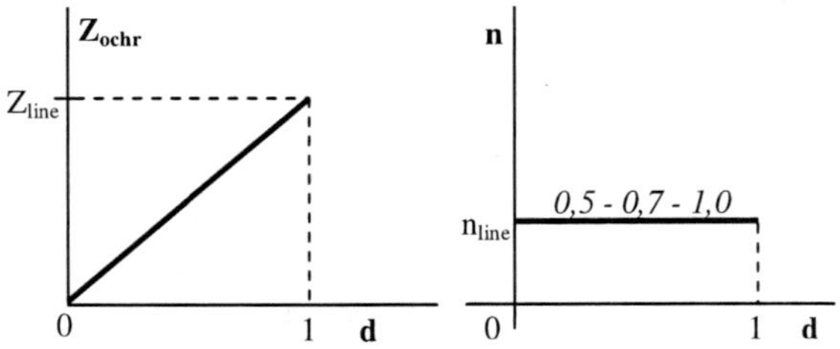

Figure 1. Distance protection measurement on a simple line

II. EARTH COMPENSATING FACTOR

The value of the earth compensating factor depends on the earth conductor (his presence and his material) and on the resistance of the landscape where the line is build (river valley or rocks). The typical values of n are:

- 0,5 = earth conductor 185mm2 AlFe,
- 0,66 = earth conductor 50mm2 Fe
- 1,0 = without earth conductor

Note:

Same distance relays, mostly machine protection relays, are not always using Ohm/phase measurement. In such a case different kinds of faults gives different measured distance.

III. NON-LINARITES

A simple homogenous line is very rare. Most common are double lines, where the measured impedance is no more proportional to the fault distance. Here we have to create dependences and method for correct relay setting.

A. *Galvanical influence*

Measurement is galvanicaly influenced by the parallel line when the fault is located on the following lines i.e. fault is not on the protected line. Distance protection creates a remote backup for such faults.

This influence is important during setting calculation of Zone 2 and Zone 3, i.e. zones, which are performing backup protection for the following line. The parallel line creates side infeed, which increases the measured impedance and the relay measures the fault as located more far away, then in reality. When the following line is also parallel line, it influences the measured impedance in opposite way, the parallel line causes decreasing of the measured impedance.

This phenomenon is quite complicated and we prefer discuss is more precisely in one of the next paper. This paper is concentrating on the second influence - the mutual zero sequence impedance between parallel lines.

B. *Mutual zero sequence impedance*

Mutual zero sequence impedance creates an inductive coupling between both parallel lines. This mutual impedance induces additional voltage drop to the parallel line and negatively influences the measurement of the distance protection. It sees the fault closer than the real location is. Because the mutual impedance is influencing just the zero sequence, this phenomenon has impact only to the phase to earth fault measurement.

To simplify the calculation we are considering just the reactance of the line and we are using just modulo of the earth compensating factor. So the results presented in this paper are not accrue, but give good overview about the result. Shure for the correct results the complex values has to be used. The line parameters are defined:

- X_l = Positive sequence reactance of the line
- $k_l.X_l$ = Zero sequence reactance of the line
- $X_{M0} = X_M$ = mutual zero sequence impedance of the parallel lines

The following sections are presenting just single phase faults of different locations on the line with the phase current I_R. The sequence circuit diagrams are not shown as well as the calculation of the final equation for the measured impedance called Z'_{ochr}. (Index ' means an impedance already compensated by the earth compensating factor.). We presume the fault located just on one of the parallel lines with the reactance as follows:

$d \,.\, X_l$ = reactance from beginning of the line to fault

$(1\text{-}d).X_l$ = reactance from fault to the end of the line

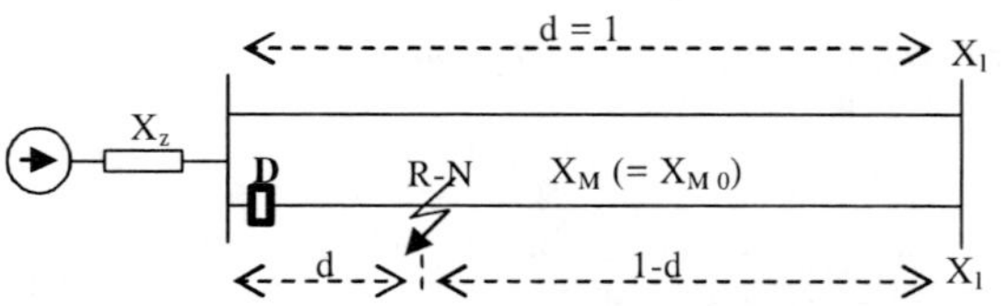

Figure 2. Two parallel lines with one side infeed from relay location

IV. TWO PARALLEL LINES WITH ONE SIDE INFEED, WITHOUT LOAD

Many authors discuss this phenomenon in such away, that they recommend using parallel line earth current compensation in the distance protection. In the reality the compensation is very rare used even the relay vendors are providing such functionality. Other papers explain the influence of the parallel line and they offers changing the security factor for distance zones setting. Mostly the security factor for zone 1 is decreased. Such approach prevents false relay operation and ensures selectivity. Then the distance principle is not fully followed, because the distance protection has different sensitivity for phase to phase fault and for single phase fault. The reach of the first zone is so much smaller than in case of single line.

In this paper we are proposing different way. The method we are describing remains the sensitivity of the first zone for phase to phase fault and for single phase fault. This is achieved by adjusting the earth fault compensation factor so that the reach of the first zone is more or less the same for all kind of faults.

The impedance measured by the distance protection and compensated by the earth compensating factor:

$$Z_R{}^{l}{}_{ochr} = dX_l \frac{2 + k_l}{3(1+n)} + \frac{d}{2-d} dX_M \frac{1}{3(1+n)} \tag{10}$$

The first part of the equation describes the impedance without influence of the mutual impedance, the second part shows the error created by parallel line. This is more clearly visible, if we insert the factor *n* equal to the earth compensating factor of a simple line as per (7).

Then the first part of the equation gives correct measurement $Z_{R'ochr} = d. X_l$

And the second part shows the parasite influence of the zero sequence mutual impedance X_M.

To achieve the correct $Z_{R'ochr}$ impedance measurement along the whole line for all d, the factor n would be non-linear depending on the distance to fault d as follows:

$$n = \frac{k_l - 1}{3} + \frac{d}{2-d}\frac{X_M}{X_l}\frac{1}{3} = \frac{k_l - 1}{3} + \frac{d}{2-d}\frac{k_M}{3} \quad (11)$$

So the first part is equal to the simple line earth compensating factor and the second part gives the error dependent on the distance to fault d.

In such a way we can define new factor k_M as ratio between zero sequence mutual reactance and positive sequence reactance. Similar as we have defined k_l in (7).

$$k_M = \frac{X_{M0}}{X_{l1}} = \frac{X_M}{X_l} \quad (12)$$

The dependency can be shown in the following graph for $k_l = 3$; $k_M = 2.4$. Here we can see that the necessary n is always higher than by a simple line and towards the end of the line the n is almost double of value. Unfortunately it is not possible implementing the inconstant value of n in the relay algorithm, so for the setting we have to use just constant value of n.

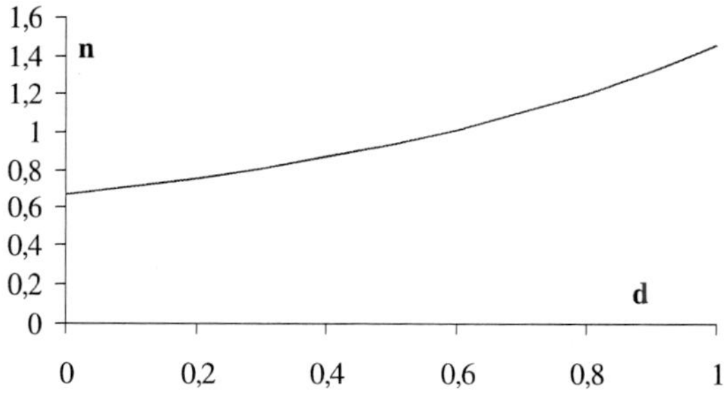

Figure 3. Necessery n needed for exact distance measurement

If set the n equal to the correct value of a simple line, the measured $Z_{R'ochr}$ impedance is reliable just for faults on the beginning of the line. For the faults far away on the line, the measured $Z_{R'ochr}$ impedance is always higher than the exact value.

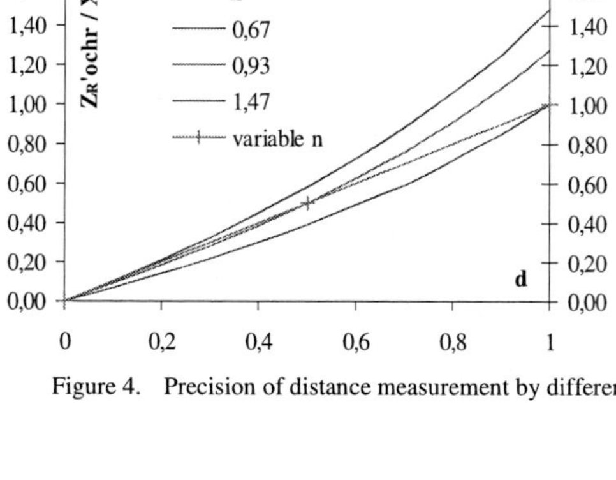

Figure 4. Precision of distance measurement by different n

If set n equal to the correct value in the middle of the line, the measured $Z_{R'ochr}$ impedance is reliable just at this point. Faults at the beginning of the line gives lower values, for the faults in the second half of the line, the measured $Z_{R'lochr}$ impedance is always higher than reliable value but no more than 150%.

As the best practice we can consider setting of n equal to the correct value at the end of the line where $d = 1$. Than all measured $Z_{R'ochr}$ impedances are lower, just faults at the end of the line are measured correctly. Additionally we achieve more sharply change in the impedance measurement between zone 1 and zone 2, which gives better performance of the distance protection relay.

V. Two parallel lines with one side infeed, without load with a transformer connected at the end

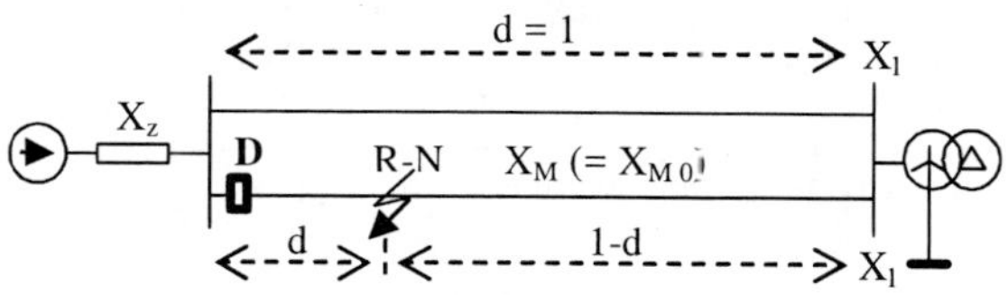

Figure 5. Two parallel lines with one side infeed and transformer

For this case the necessary n at the end of the line is equal like in the previous section, so we recommend using the same rule.

VI. Two parallel lines with infeed from both sites

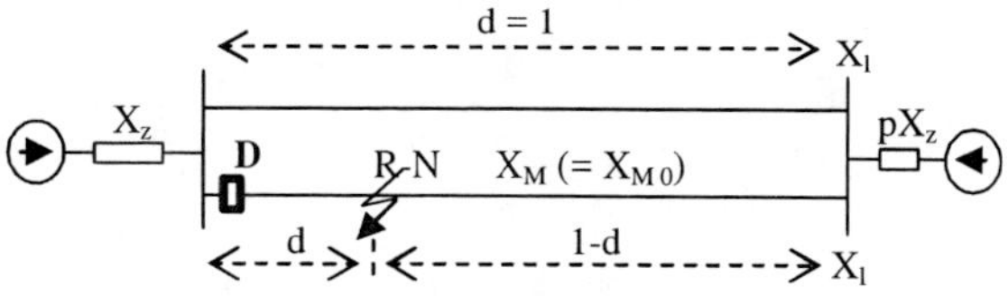

Figure 6. Two parallel lines with infeed from both sites with wight p

The ratio between short circuit impedances of both ends is defined as p and can be called as source weight factor. Similar as we have defined k_l in (7) we can define factors k_Z and as ratio between X_{Z0} and X_{Z1} where k_p is applied to the remote end source impedance. Factor l_z means ratio between X_Z and X_l.

The necessary earth fault compensation factor for correct distance protection measurement in each point d can be derived as follows:

$$n = \frac{k_l - 1}{3} + \frac{k_M}{3} \cdot \frac{k_p p - (1-d)(k_z + k_p p)}{k_p p + (1-d)\left[(k_z + k_p p) + \frac{1}{l_z}(k_l + k_M)\right]} \quad (13)$$

Again the resulting formula has two parts. The first part is equal to the factor valid for simple homogenous line and the second part of equation gives the error from the parallel line.

Figure 7 shows the resulting earth fault compensation factor for $k_l = 3$; $k_M = 2.4$; $k_Z = 0,3$; $k_P = 1$; $l_z = 0.3$

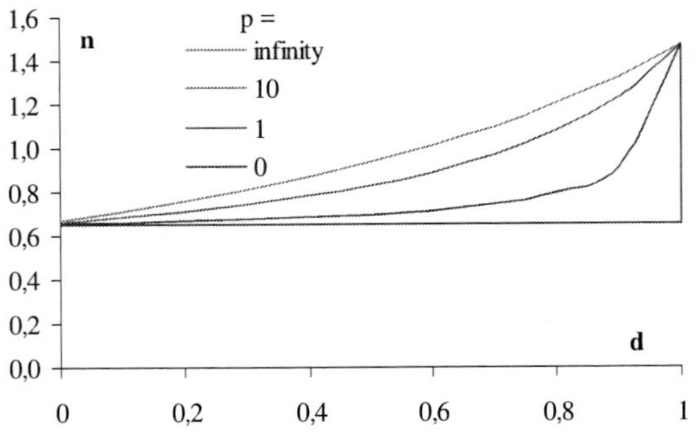

Figure 7. Necessery n needed for exact distance measurement

From the graph we can see, how the necessary n increase with the source weight factor p. The earth fault compensating factor is constant for case, where the double line is fed just from remote end. This is logical, as without the power source the currents in both lines are equal with opposite direction and the mutual zero sequence impedance takes no effect.

The earth fault compensating factor has so always between value valid for single homogenous line and case of double line with one infeed. As expected the n for remote busbar is constant independent from source weight factor. Therefore also here we recommend following the same rule for distance protection setting, i.e. setting the earth fault compensation factor according the fault situation at the end of the line.

VII. Fault Locator

The nonlinearity of distance measurement decreases the accuracy for fault locators as described. Without measuring the earth current of the parallel line the best accuracy can be achieved, if their compensation factor is based on the fault in the middle of the line (d = 0.5). In such a case the error at the beginning and at the end of the line has the same value with opposite sign.

When the measuring the earth current of the parallel line is introduced, the nonlinearity is reduced and the error of the fault locator is minimized.

VIII. Conclusion

The paper present a simple method for setting of the earth fault compensation factor under influence of the parallel line zero sequence mutual impedance X_M. Based on the short circuit calculation we can correct the set earth fault compensation factor as presented in the figure 7.

It is evident that the selectivity is not damaged when the presented method is followed. The practical experiences show, that such a simple method is successful.

References

[1] V. List a kol., Elektrotechnika- příručky, Elektrické sítě, SNTL, 1964

[2] J.Bermann, Transformer concept in CZ+SK, In: Referáty konference CK CIRED Tábor 2014

[3] J.Bermann, Distance prot. of generator, In: Referáty konference CK CIRED Tábor 2009

[4] J.Bermann, Protection of smaller transformers, In: Referáty konference CK CIRED Tábor 2004

[5] J.Bermann, Relationship between protections of transformers and substantions In: Referáty konference CK CIRED Tábor 2002

[6] J.Bermann, Cable inserted in HV line- protection and fault localisation In: Referáty konference CK CIRED Tábor 2016

[7] G. Ziegler, Digital distance protection, MCD Verl., 1999

[8] H. Ungrad, V. Narayan, Behaviour of distance relays under earth fault conditions on double-circuit lines, ABB Publication CH-ES 35-30.14 E

Use of the Nonlinearity of the Arc Resistance for the Designing Main and Backup Relay Protection

Vladimir Nagay
Dept. Electric power stations and electric power systems
Platov South-Russian State Polytechnic University (NPI)
Novocherkassk, Russian Federation
nvi53@mail.ru

Sergei Sarry
Dept. Electric power stations and electric power systems
Platov South-Russian State Polytechnic University (NPI)
Novocherkassk, Russian Federation
sv@sarry.ru

Ivan Nagay
Dept. Electric power stations and electric power systems
Platov South-Russian State Polytechnic University (NPI) – «NPP «RELDON» Ltd
Novocherkassk, Russian Federation
nagayiv@yandex.ru

Pavel Kireev
Dept. Electric power stations and electric power systems
Platov South-Russian State Polytechnic University (NPI)
Novocherkassk, Russian Federation
kireevps@yandex.ru

Alexander Ukraincev
Dept. Electric power stations and electric power systems
Platov South-Russian State Polytechnic University (NPI)
Novocherkassk, Russian Federation
xelandr@mail.ru

Abstract—**Often, faults in the electrical grids of modern power systems are accompanied by the appearance of an electric arc, which poses a significant danger for high-voltage equipment and maintenance personnel. The presence of a arc resistance at the fault location leads to a decrease in the short-circuit current and an increase in the resistance of the short-circuit and, accordingly, to a decrease in the sensitivity of the current and impedance protections. At the same time, the presence of a arc resistance, which has a non-linear character and its unstable parameters in time, makes it possible to use this as additional information signs when designing both main and backup protections. Taking into account the nonlinearity of the electric arc resistance in the algorithms of the protection function increases its technical perfection.**

Keywords: main and backup relay protection, short circuit, high-voltage electric arc, nonlinear arc resistance, harmonic voltage components. formatting

I. Introduction

Faults in the electrical networks of modern power systems often are accompanied by the appearance of an electric arc, which poses a significant danger for high-voltage equipment and maintenance personnel. The presence of an arc resistance at the fault location leads to a decrease in the short-circuit current and an increase in the resistance of the short-circuit loop and accordingly to decrease in the sensitivity of the current and distance protections. At the same time the presence of an arc resistance, which has a non-linear character and its unstable parameters in time, makes it possible to use this as additional information signs in the designing of both main and backup protections. Account of the nonlinearity of the electric arc resistance in the algorithms of the protection function increases its technical perfection. Authors are attempted to generalize their own experience in designing fast-acting and standby protections in view of the foregoing.

II. The Main Part

Brief characteristics of the electric arc. A high-voltage electric arc representing an independent arc discharge in a gas is characterized by a high current density and a low cathode potential drop [1-4]. When it's describing a cylindrically symmetric arc column, plasma is regarded as a continuous conductive medium, which parameters are depend only on the radial coordinate. In this case the temperature on the conductive surface of the column of the arc reaches 4000 K, and on the arc axis - several thousand degrees, respectively.

A qualitative assessment of the effectiveness of the use of the control of the parameters of information signs characterizing the fault in the presence of an arc resistance for the main and backup ones taking into account the function of local protection backup (transformers, bank of capacitors, reactors, switchgear) and distributed (overhead and cable lines) objects is in Table 1 that based on the researches of authors and works other scientists [1-7].

The successful solution of the problem of designing the relay protection of switchgear or overhead lines, transformers depends significantly on the choice of the principle of their operation and the optimal choice of information signs of such regime. The appearance of an electric arc is accompanied by a change in the parameters and characteristics of the electrical network, a significant increase in temperature, pressure, electrical conductivity, and thermal (light) emission at the arc fault location.

Table 1. Applicability of the use of monitoring the parameters of information signs characterizing the fault in the presence of an arc resistance

Information signs for faults accompanied by an electric arc	**Relay protection**			
	Main protection		*Backup protection*	
	Local object	***Distributed object***	***Local backup***	***Remote backup***
Group 1				
Phase currents	EU	EU	EU	EU
Phase and phase-to-phase voltages	EU	EU	EU	PU
Resistances	EU	EU	EU	EU
Ratios of phase currents	EU	EU	EU	PU
Symmetrical components	EU	EU	EU	EU
Harmonic components of currents and voltages	EU	PU	EU	PU
Orthogonal components of currents and voltages	EU	EU	EU	EU
Fault components of currents and voltages	EU	EU	EU	EU
Group 2	EU	EU	EU	EU
Luminous flux (thermal emission)	EU	UU	UU	UU
Pressure of the gaseous medium	EU	UU	UU	UU
Electromagnetic radiation	EU	UU	UU	UU
Electrical conductivity	EU	UU	UU	UU
Ambient temperature	EU	UU	UU	UU

The following notation is used in the table: EU - effective use, PU - possible to use, UU - unable to use.

These characteristics can underlie the methods of identifying the electric arc, which can be conditionally divided into two groups [6,7]. It is possible to classify methods based on the control of parameters and characteristics of an electrical network with an arc to the first group: the spectral composition of the current or voltage, the modules and the arguments of the phase and symmetrical components of the currents, voltages, unbalance levels of currents or voltages, resistances or volt-ampere characteristics of the short-circuited circuit. The disadvantage of these methods is that the controlled features can be observed not only with internal arc faults, but also with external faults, as well as in operating regimes of the network, which makes it difficult to recognize internal damage, i.e. it does not allow to perform protection with absolute selectivity.

The second group includes methods based on controlling the parameters of the electric arc and its attendant phenomena: temperature, pressure, electrical conductivity of the environment (the degree of ionization of gases), optical properties of the medium, electromagnetic radiation. The operation of the devices in this group is practically not affected by the operating regimes of the network.

Analysis of information signs characterizing the arc fault. A preliminary assessment of the ways to detect arc faults carried out above allows to determine the most perspective directions for designing both main and backup protection of local and distributed objects, depending on their design.

Monitoring of the electrical parameters of the circuit with the arc. The simplest technical solutions for protection against short-circuit accompanied by an open arc are the overcurrent or distance protection, which are currently the main ones for radial overhead lines and respectively backup for two-ended lines. Arc resistance is very important for the sensitivity of these protections. There are the dependences of the controlled parameters of the current, resistance, current argument (resistance) I_*, Z_*, φ_* by overcurrent and distance protection in Fig.1, built on the basis of the dependences given in [7]. It should be noted that the current decreases and, correspondingly, the resistance of the short-circuited circuit increases to 25%. Reduction of the argument of the fault current occurs even more significantly - up to 35%.

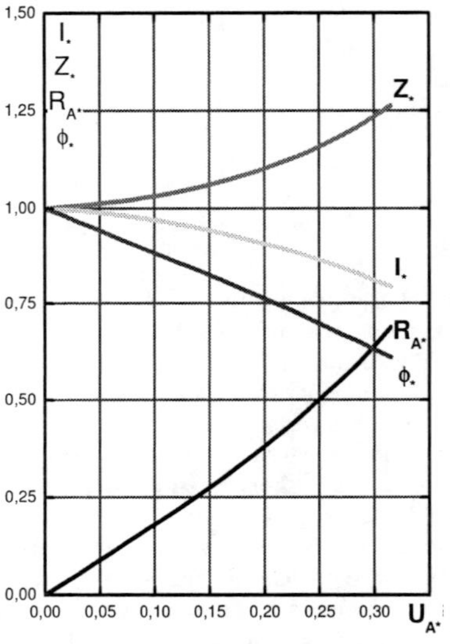

Figure 1. Dependences of the parameters of the short-circuit with the arc resistance on the voltage drop on its column

The magnitude of the current of the three-phase short-circuit with allowance for the arc resistance can be determined as (1):

$$I_* = U / \sqrt{3(x_T + x_S)^2 + (r_S + r_T + r_A)^2} \,. \quad (1)$$

where x_S, x_T, r_S, r_T – an inductive and active resistance of the system and transformer; $U_{*A} = U_A / U$ - voltage drop on the arc column related to the rated voltage U that burns on sides of the lower (medium) voltage of the transformer, and assuming that $r_A = U_{*A} / I_*$, (1) it is possible to convert and obtain the dependence of the arc resistance of the electric arc on the voltage drop on it, the parameters of the protected transformer and the supply network (2):

$$r_A = \frac{-6U_{A*}^2(r_S+r_T) - \sqrt{\left[6U_{A*}^2(r_S+r_T)\right]^2 - 12(3U_{A*}^2-1)\cdot U_{A*}^2 * \left[(x_S+x_T)^2+(r_S+r_T)^2\right]}}{2(3U_{A*}^2-1)}. \qquad (2)$$

Fig. 1 shows the dependences with (3-4):

$$x_S = 0 \; and \; r_S = 0 \,. \qquad (3)$$

$$r_A / z_T = f(U_{A*}), \quad \phi_* / \phi_T = f(U_{A*}), \quad z_* / z_T = f(U_{A*}), I_* / I_{DSC} = f(U_{\partial *}), \qquad (4)$$

With known parameters of the arc resistance r_A, it is not difficult to determine the modulus I_* (5) and the argument φ_* of the short-circuit current (6), the resistance of the short-circuit loop Z_* with allowance for the arc resistance (7):

$$I_* = U / \sqrt{3(x_T^2 + (r_T + r_A)^2)} \,. \qquad (5)$$

$$\phi_* = arctg(x_T /(r_T + r_A)) \,. \qquad (6)$$

$$z_* = \sqrt{x_T^2 + (r_T + r_A)^2} \,. \qquad (7)$$

where φ_T – argument of the transformer impedance Z_T, I_{DSC} – dead short-circuit current with r_A=0.

In this case, the absolute value of the arc resistance will be determined by the parameters of the protected transformer and the system, and the phase ratios of the short-circuit currents do not depend on the power of the transformer and are determined only by the voltage drop on the arc column, the change of the short-circuit current modulus does not exceed 20% of its value at the dead short-circuit, but the arguments of short-circuit currents are more susceptible to the influence of the arc resistance and are reduced by 20° -30°, what is shown in the graphs (Fig.1).

Due to the fact that the resistance of the arc is determined by the voltage drop and the short-circuit current, the parameters of the dead short circuit are taken as the base.

The nonlinearity of the volt-ampere characteristic of the arc and the instability of its resistance lead to a change in the shape of the current and especially in the shape of the network voltage [1-5] and to the associated increase in the level of high-frequency components, which was shown in Fig.2. The initial data for the defining dependencies (Fig. 2) were the results of mathematical simulation of arc burning with a length of up to 1 m in electrical switchgear of a 6 kV fed from a transformer with a power of up to 10 MVA and an amplitude of the induced current component of 5.5-6.0 kA. A method based on high-frequency components monitoring in voltage or current does not allow for protection with absolute selectivity and can only be used in conjunction with other control methods.

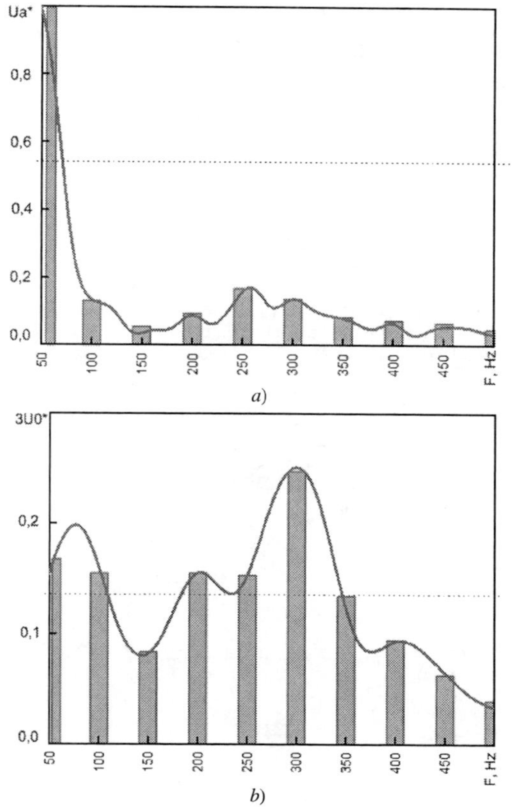

Figure 2. Amplitude-frequency characteristics of phase voltage $Ua*$ and zero sequence voltage $3U0*$ with a three-phase short-circuit in an electrical network with a voltage 6 kV when the arc column is touching with grounded metal structures

Devices that control the voltage asymmetry and distortion of their shape, as well as devices using current asymmetry do not have absolute selectivity, and their sensitivity depends both on the network parameters and on the inequality of the arc lengths. However, they can be used as starting or blocking units. The inequality of voltage drops on the columns of an arc of different phases leads to the appearance of negative sequence voltages and negative sequence currents (Fig. 3). In this case the voltage drops on the arc column is up to $0{,}25U_{nom}$ and the reduction of the positive sequence current does not exceed 5% of the current of the dead short circuit, and the negative sequence current does not exceed 8%, respectively. The presence of a negative sequence current in the specified range can be a sign of a three-phase fault through an electric arc.

An additional sign of a three-phase short-circuit through an electric arc in an electrical network with an isolated neutral is also the appearance of zero sequence voltage, which is often due to the touch of the arc column with the earthed metal structures, for example (Fig. 2), in the low voltage switchgears 6-10 kV.

Temperature control. The majority of the energy supplied to the arc column is converted to thermal energy [1-5]. The temperature change in the switchgear during arc burning is determined by the conditions for heat removal from the arc.

The calculated expressions for determining the currents of the positive and negative sequences are phase currents, determined taking into account the presence of the arc resistance (8)-(10):

$$\dot{I}_a = \frac{\dot{E}_{ab}(2Z_e + R_{Ab} + R_{Ac}) + \dot{E}_{bc}(Z_e + R_{Ab})}{(2Z_e + R_{Aa} + R_{Ab})(2Z_e + R_{Ab} + R_{Ac}) \quad (Z_e + R_{Ab})^2} \cdot (8)$$

$$\dot{I}_b = \frac{\dot{E}_{bc}(Z_e + R_{Aa}) \quad \dot{E}_{ab}(Z_e + R_{Ac})}{(2Z_e + R_{Aa} + R_{Ab})(2Z_e + R_{Ab} + R_{Ac}) \quad (Z_e + R_{Ab})^2} \cdot (9)$$

$$\dot{I}_c = -\frac{\dot{E}_{bc}(2\underline{Z}_e + R_{Aa} + R_{Ab}) + \dot{E}_{ab}(\underline{Z}_e + R_{Ab})}{(2\underline{Z}_e + R_{Aa} + R_{Ab})(2\underline{Z}_e + R_{Ab} + R_{Ac}) - (\underline{Z}_e + R_{Ab})^2} .(10)$$

where $R_{Aa}=U_{Aa}/I_{DSC}$, $R_{Ab}=U_{Ab}/I_{DSC}$, $R_{Ac}=U_{Ac}/I_{DSC}$ – resistances of the arc columns for phases A,B,C; I_{DSC} – dead short-circuit current with r_A=0; $\underline{Z}_e$– equivalent resistance of supply system; E_{ab}, E_{bc} – phase-to-phase power supply voltages; U_{Aa}, U_{Ab}, U_{Ac} – voltage drop on the arc column.

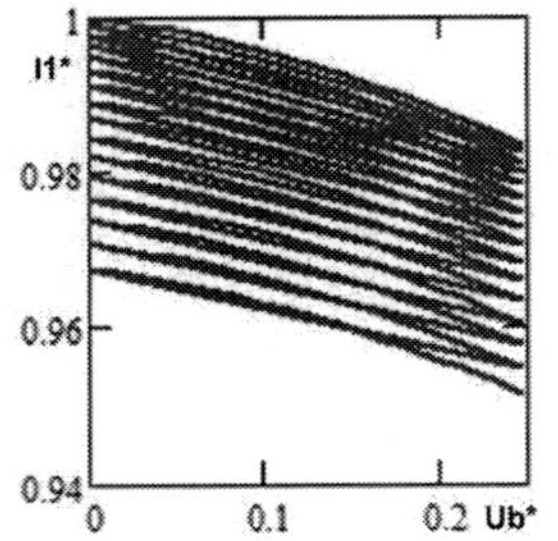

a)

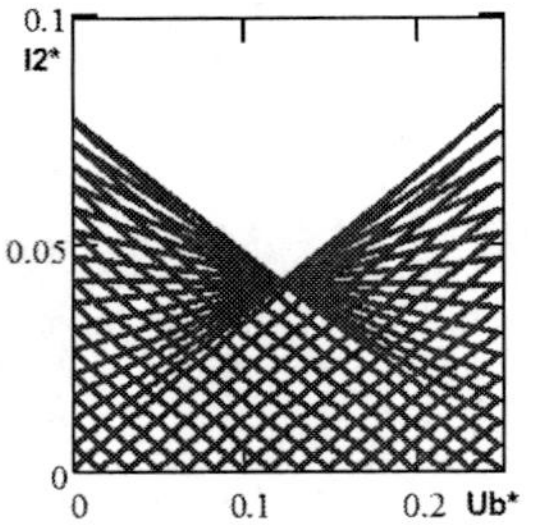

b)

Figure 3. Dependences of the positive sequence current (a) and the negative sequence current (b) on the voltage on the arc column of the middle phase *B* with fixed voltage on the arc columns at the extreme phases *A* and *C* in the a three-phase short circuit regime

When burning is in a closed volume, the heat transfer of arc energy to the environment varies exponentially with a time constant $\tau \approx 0{,}01$ c [9] $T=T_0+(T_1-T_0)e^{-t/\tau}$, T_1 where – value of the air temperature in the switchgear before the appearance of an arc fault; T_0 – arc temperature. The temperature for a sinusoidal current varies with respect to a certain average value T_{av} with pulsation of the doubled frequency of the network, which depends on the cooling conditions of the arc column [4,6]. However in the transition of two-phase short-circuits to three-phase short circuits, the second harmonic is significantly reduced.

Pressure control. The formation of an electric arc is associated with the release of a significant amount of energy, which leads in a closed space to an increase in pressure. The electrical installations of the cabinet structure, for example the switchgears, are generally not sealed, they have technological holes that limits the excess pressure in the switchgear compartments at the level of 0.02-0.04 MPa [1] instead of 300-500 MPa as is the case in sealed compartments [8]. It is shown in [9] that the pressure overincrease valves installed in the switchgear are steadily activated at currents exceeding 3.5 kA, which corresponds to a short-circuit with an arc at the lower voltage side of the 10 MVA transformer.

Control of the density of charged particles. The arc column is a highly ionized gas, the conductivity of which is large and approaches the conductivity of metals [1]. The opposite processes of ionization and deionization take place in the arc column. In the steady-state regime of arcing an equilibrium occurs between the newly formed and vanishing ionized particles. The degree of ionization of air around the arc decreases with distance from its column. This is due to the absorption of the energy released by the cold layers of air. Researches shows that it is possible to control the density of charged particles only at distances close to an arc in small-sized complete switchgears [7].

Control of thermal emission power. Due to the high temperature the electric arc is a powerful source of thermal emission. The line emission emitted by the electric arc carries information about practically all parameters of the arc column, such as the composition and concentration of neutral and charged particles, the strengths of the electric and magnetic fields, the energy of the particles and their temperature, etc.

It is known that the emission power of a single volume of an arc column P_i depends on the magnitude of the short circuit current I_{SC}, the value of which is affected by the arc length l_A, and, accordingly, the voltage drop on the electric arc column and the radius of the arc column, which also depends on the magnitude of the current. The relative value of the thermal (light) emission power of the electric arc column is proportional to the total power on the three arc columns and can be approximated $P_{r*}=(P_{an}+P_{bn}+P_{cn})/(P_{al}+P_{bl}+P_{cl})$, where P_{an}, P_{bn}, P_{cn} – power released on the arc column of phases *A*, *B*, *C* when the arc column is represented by a nonlinear resistance; P_{al},P_{bl},P_{cl} – power released on the arc column of phases *A*, *B*, *C* when the arc column is represented by a linear resistance.

Fig. 4 shows a time diagram of the relative power of emission when a column arc is represented by a nonlinear arc resistance. At the initial section (up to 0.01 sec) the aperiodic component of the current exerts a significant influence.

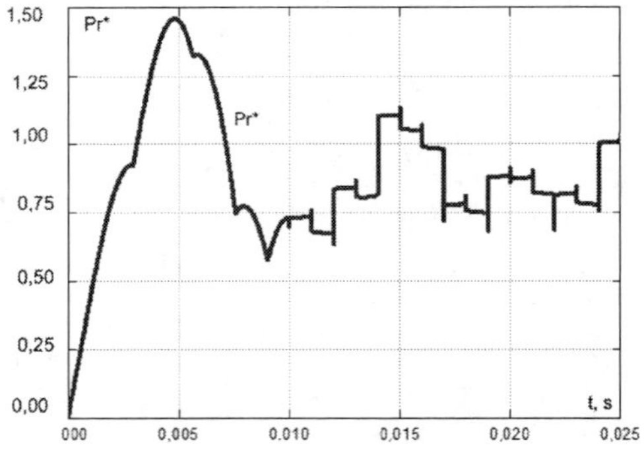

Figure 4. Time diagram of the relative power of emission when a column is represented by a nonlinear arc resistance

The emission power is approximately defined as the multiplication of the voltage drop on the arc by the current. The aperiodic component of the current exerts a significant influence on the initial section (up to 0.01 s), that is confirmed by the time diagrams of current and voltage on the arc column in one of the phases (phase A) (Fig. 5).

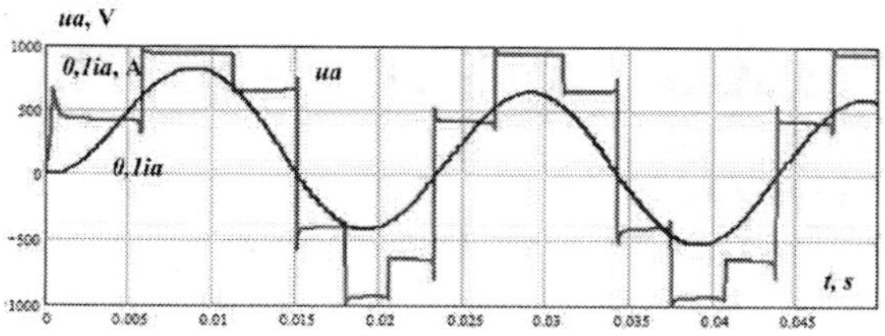

Figure 5. Voltage drop on the arc column and current phase *A*

In the following time it is the influence of the nonlinear nature of the arc resistance and its nonstationary of burning in time because of the formation of shunting loops and its possible inflating by gas flows. In this case it is also necessary to note the appearance of a changing harmonic composition in the power of the electric arc column emission, which is also an information sign of an arc short circuit.

III. Conclusions

The conducted researches allow to draw the following conclusions.

1. Account of the arc resistance in the form of nonlinear resistance is possible to expand the information base of relay protection and perform it with higher information and technical perfection.

2. The negative sequence current for voltage drops on the arc column is up to 25% of the rated voltage of the electrical installation does not exceed 10% of the positive sequence current for representing the arc resistance by a nonlinear characteristic.

3. The sensitivity of the main and backup protection is recommended to be checked taking into account the reduction of the short-circuit current, the increase in the resistance of the short-circuit and the voltage drop on the electric arc column. In this case, it is necessary to take into account the possible reduction of up to 25% of the short-circuit current and the increase in the resistance of the short-circuit loop to 25% when the voltage across the arc column drops to 30% of the rated voltage of the protected electrical equipment. The possible change in the arguments of the current and resistance can reach 30 °.

References

[1] O.B. Bron , L.K. Sushkov, "Plasma flows in the electric arc of switching devices", Energia, 1975, p. 212.

[2] G.V. Butkevitch, "Arc processes during switching circuits", Higher.sch., Moscow, 1967, p.196.

[3] A.M. Zalessky, "The electric arc of tripping", Gosenergoizdat, 1963, p. 267.

[4] I.K. Fedchenko, I.K. Sokolovsky, "Measurement of electric arc temperature", Kiev, Engineering,1966, p. 154.

[5] V.I. Nagay, "Classification of methods and analysis of information signs for detecting arc short circuits in electrical installations of the cabinet structure", University news. North-Caucasian Region. Technical Sciences Series, vol. 2, 2001, pp.50-54.

[6] S.V. Sarry, V.I. Nagay, "Investigation of the components of the light field in the complete switchgear with short circuit accompanied by an electric arc", Scientific and Technical Journal. Russian Electromechanics, vol. 1, 2000, pp. 92–95.

[7] V.I. Nagay, "Relay protection of the tapping substation in the electric networks", Moscow, Energoatomizdat, 2002, p.312.

[8] A.S. Kozlov, "About the causes and measures to prevent fires in power plants", Energetik, vol.4, 1998, p. 26.

[9] A.J. Zotov, "Protection of the busbar compartment of a 6-10 kV switchgear from an arc short-circuit with an arc-catching bus", Power engineering, vol.8, 1992, pp. 62-64

Earth Fault Detectors for Indicators in Medium Voltage Networks

Lubomir Marciniak
Institute of Electric Power Engineering
Czestochowa University of Technology
Czestochowa, Poland
lubmar@el.pcz.czest.pl

Abstract — **In the paper the effectiveness of earth fault protections within the detection of resistive short-circuits was estimated on the basis of the impulse increase measurements of the zero-sequence or phase currents. The influences of the network, compensation and forcing circuit parameters as well as the location of the measuring point were taken into account. It was demonstrated that the impulse-current protections are much more effective than the zero-sequence overcurrent ones.**

Keywords: earth faults; impuls-current protection; fault indicator; effectiveness

I. Introduction

For line section location and signaling faults in distributed medium voltage (MV) networks fault indicators (FIs) are more and more commonly used [1-6]. These are inexpensive devices intended for mass use, with simple protections or short circuit detectors. Supplementing indicators in communication equipment allows to realize the system of identification and location of short circuits [6-10] providing information directly from the damaged line section to the distribution management system. Based on this information it is possible to perform the switching in the network to isolate the damaged line section and to restore the power supply to the remainder. Such devices become the elements of smart grids [6-10]. Their use allows to shorten the duration of power interruptions several times and to reduce losses due to undelivered energy [11, 12].

For economic and technical reasons, both simple and effective earth fault protections should be used for fault indicators. High effectiveness of the earth faults detection in compensated MV networks with active current forcing automatics (ACFA) can be obtained using the impulse-current operating criteria [13]. These criteria are based on measurements of increments of the phase currents or earth fault current (equal to three times the zero-sequence current) occurring as a result of the forcing active earth current.

II. The Impulse-Current Criteria for Earth Fault Detection

Figure 1 illustrates the principle of impulse-current protections. These devices detect the earth faults by measuring the positive increase in earth or phase currents at their location point on the power line. The increases in currents are caused by the forcing active current during short-circuit by ACFA in a substation.

On a faulty line, when the short circuit occurs at any point F_1 after the measuring point P, the increase in an earth current is positive ($\Delta I_{Ep(1)} > 0$). On a healthy line, when the short circuit occurs at any point F_2 before measuring point P or on another line, there is a negative increase ($\Delta I_{Ep(2)} < 0$). The increase in the phase current in a damaged phase of a faulty line is positive and large ($\Delta I_{L1p(1)} > 0$). The increases in phase currents in healthy phases on a damaged line and in all phases on healthy lines are either negative ($\Delta I_{L1p(2)} < 0$) or small positive. These dependences allow for a selective detection of earth faults in compensated MV networks with ACFA.

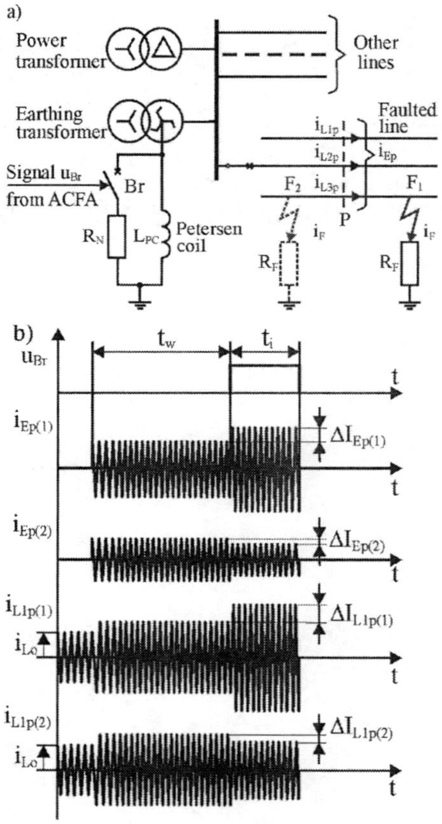

Figure 1. Simplified scheme of the compensated MV network (a) and illustration of impulse-current earth fault detection methods (b): L_{PC} – inductance of Petersen coil, R_N – forcing resistor, Br – circuit breaker for switching resistor R_N

Identification of a short circuit based on a zero-sequence impulse-current criterion comes down to detection the increase of the earth fault current ΔI_{Ep} that exceeds the setting value ΔI_s:

$$\Delta I_{Ep} = \left|\underline{I}_{Efp}\right| - \left|\underline{I}_{Ep}\right| \geq \Delta I_s \tag{1}$$

where: I_{Ep}, I_{Efp} – earth currents at the measuring point P before and after switching the forcing resistor.

The impulse of the current increment is verified in terms of its duration t_i:

$$t_{i\min} \leq t_i \leq t_{i\max} \tag{2}$$

where: $t_{i\,\min}$, $t_{i\,\max}$ - minimum and maximum duration of the impulse.

Time selection of the impulse is an additional condition for distinguishing an increase caused by the forcing active current from increases that arise at any point in the network when a short circuit appears. The lower interval of time selection must be less than the time delay of the power line protection at the electrical substation $t_{i\,\min} < t_{del}$, the upper range – less than the waiting time of the ACFA for autoextinction of the earth fault current $t_{i\,\max} < t_w$.

The setting value ΔI_s must be greater than the maximum increase of the current on the healthy line $\Delta I_{pa\,M}$ caused by the network asymmetry and can be estimated from the relationship [13]:

$$\Delta I_s \geq k_b \Delta I_{pa\,M} = k_s k_{as} \alpha I_{CS} e^{-\frac{k_{ap} k_{as} I_{CS}}{I_R + d_0 I_{CS}}} \tag{3}$$

where: I_{CS} – capacitive current in the network, $\alpha = I_{CL}/I_{CS}$ – ratio of the feeder capacitive current I_{CL} to the network capacitive current, I_R – rated current of the forcing resistor, d_0 – zero-sequence damping factor, $k_{as} = I_{as}/I_{CS}$ – asymmetry coefficient (the share of the asymmetry current I_{as} in the network capacitive current), $k_s = 1.5 \div 2$ – safety factor, k_{ap} – approximation coefficient dependent on damping factor (for $d_0 = 0.03 \div 0.05$ can be accepted $k_{ap} = 2$).

From equation (3) it follows that for the protection on the power line with the capacitive current $I_{CL} \leq 10$ A at any compensated network the setting value $\Delta I_s \leq 1$ A can be accepted.

Identification of a short circuit using the phase impulse-current criterion is based on the detection the increase in the current in a faulty phase ΔI_{Pp} exceeding the setting value ΔI_s:

$$\Delta I_{Pp} = \left|\underline{I}_{Pfp}\right| - \left|\underline{I}_{Pp}\right| \geq \Delta I_s \tag{4}$$

where: I_{Pp}, I_{Pfp} – phase currents at the measuring point P of a faulty phase before and after switching the forcing resistor.

The impulse increase in the phase current in a faulty phase is subjected to a time selection. It is an additional criterion for identification of an earth fault.

The current setting value for the phase impulse-current protection should be greater than the maximum increase of the phase current on a healthy line when external short circuit occurs [13]:

$$\Delta I_s \geq k_b \frac{I_{CL}}{3} \frac{\sqrt{I_{CS}\left(q_{PC}^{-1} + d_0\right) + I_R} - \sqrt{I_{CS}\left(q_{PC}^{-1} + d_0\right)}}{\sqrt{I_{CS}\left(q_{PC}^{-1} + d_0\right) + I_R} + \sqrt{I_{CS}\left(q_{PC}^{-1} + d_0\right)}} \tag{5}$$

where: q_{PC} – goodness of a compensation circuit.

When there is a large asymmetry current in the network $k_{as} = 0.03 \div 0.05$, the setting value should be estimated from the relationship:

$$\Delta I_s = I_{CL}/3 \tag{6}$$

III. The Effectiveness of the Impulse-Currrent Earth Fault Protections

The practical measure of the effectiveness of impulse-current protections, decisive about their usefulness in indicators, is the maximum value of the fault resistance R_{FM} at which the protection at a specific measuring point, with a preset threshold value ΔI_s resulting from its location in the network, is able to identify the earth fault inside the protected section of the line.

The maximum resistance R_{FM} can be estimated numerically from equations (1) and (4) comparing the increase in the current to the setting value. The phase and earth currents, necessary to calculate the resistance R_{FM}, can be determined from the parametric equations, which in the case of internal $L1$–E short-circuit take the forms:

$$\begin{cases} \underline{I}_{L1p} = \dfrac{I_{CS}\underline{b} - \underline{I}_{as}}{E + R_F I_{CS}\underline{b}}\left[E + \alpha(d_0 + j)I_{CL1}R_F\right] + \underline{I}_{Lo} \\ \underline{I}_{L1fp} = \dfrac{\left(I_{CS}\underline{b} + I_R - \underline{I}_{as}\right)\left(E + \alpha(d_0 + j)R_F I_{CL1}\right)}{E + R_F\left(I_{CS}\underline{b} + I_R\right)} + \underline{I}_{Lo} \end{cases} \tag{7}$$

$$\begin{cases} \underline{I}_{L2p} = \left(\underline{a}^2 - \dfrac{E + R_F\underline{I}_{as}}{E + R_F I_{CS}\underline{b}}\right)\alpha(d_0 + j)I_{CL2} + \underline{a}^2\underline{I}_{Lo} \\ \underline{I}_{L2fp} = \left(\underline{a}^2 - \dfrac{E + R_F\underline{I}_{as}}{E + R_F\left(I_{CS}\underline{b} + I_R\right)}\right)\alpha(d_0 + j)I_{CL2} + \underline{a}^2\underline{I}_{Lo} \end{cases} \tag{8}$$

$$\begin{cases} \underline{I}_{L3p} = \left(\underline{a} - \dfrac{E + R_F\underline{I}_{as}}{E + R_F I_{CS}\underline{b}}\right)\alpha(d_0 + j)I_{CL3} + \underline{a}\underline{I}_{Lo} \\ \underline{I}_{L3fp} = \left(\underline{a} - \dfrac{E + R_F\underline{I}_{as}}{E + R_F\left(I_{CS}\underline{b} + I_R\right)}\right)\alpha(d_0 + j)I_{CL3} + \underline{a}\underline{I}_{Lo} \end{cases} \tag{9}$$

$$\underline{I}_{Ep} = \sum_{i=1}^{3} k_i \underline{I}_{Lip}; \quad \underline{I}_{Efp} = \sum_{i=1}^{3} k_i \underline{I}_{Lifp}; \tag{10}$$

$$\underline{a} = e^{j2\pi/3}; \quad \underline{b} = \frac{1+\nu}{q_{PC}} + d_0 - j\nu; \quad \nu = \frac{I_{PC} - I_{CS}}{I_{CS}} \tag{11}$$

where: $E = U/\sqrt{3}$ – nominal phase voltage, U – rated line-to-line voltage, I_{Lo} – load current, I_{as} – asymmetry current, $I_{CL1} \div I_{CL3}$ – phase capacitive currents in the network, $k_i = K_i/K_n$ ($i = 1 \div 3$) – relative current transformation ratios of the zero-sequence filter (to the nominal current transformation ratio K_n), ν – compensation detuning, I_{PC} – Petersen coil current.

Equations (7)÷(10) have a general character. They take into account the essential parameters of the network, including the asymmetry current and error current caused by the current transformers (CT). With their help it is possible in a comprehensive way to estimate the effectiveness of the impulse-current protections.

The effectiveness of the zero-sequence impulse-current protection, calculated from these formulas, depending on the parameters of the network, compensation and forcing circuits is illustrated in Figures 2 to 5.

From the characteristics in Figure 2 it follows that the impulse protection can detect high-resistance earth fault in the well-compensated networks with capacitive current up to 150 A.

In the undercompensated networks with the detuning ratio $\nu \cong -(\alpha + d_0)$ it is possible to detect earth faults involving resistance $R_F > 2$ kΩ. This result is comparable to the effectiveness of admittance protections [14], currently considered as the best. This protection in the overcompensated networks can detect a short circuit involving resistance R_F to about 500 Ω. This result is worse than the one obtained for admittance protections, but better than that for zero sequence overcurrent ones, as shown below. From Figure 2 also shows that the impulse protection can be particularly effective in small cable networks, in which the location coefficient of measuring point reaches significant values even up to 0.5.

On the protection effectiveness a significant impact has the setting value of the current ΔI_s (Fig. 3). The strongest dependence R_{FM} from ΔI_s occurs in the undercompensated network with the detuning ratio $\nu \cong -(\alpha + d_0)$, i.e. when the maximum resistance is the highest. In most cases, the current setting can be chosen from the condition $\Delta I_s \leq 1$ A, which provides either a high or very high protection effectiveness.

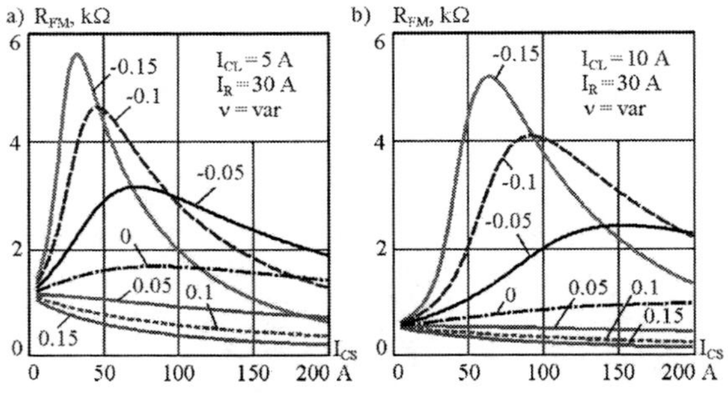

Figure 2. Dependences of the maximum fault resistance detected by the zero-sequence impulse-current protection with the threshold value $\Delta I_s = 1$ A as a function of the network capacitive current I_{CS} and various detuning coefficient ν. Line capacitive currents: a) $I_{CL} = 5$ A; b) $I_{CL} = 10$ A. Parameters: $U = 15$ kV, $d_0 = 0.05$, $q_{PC} = 80$

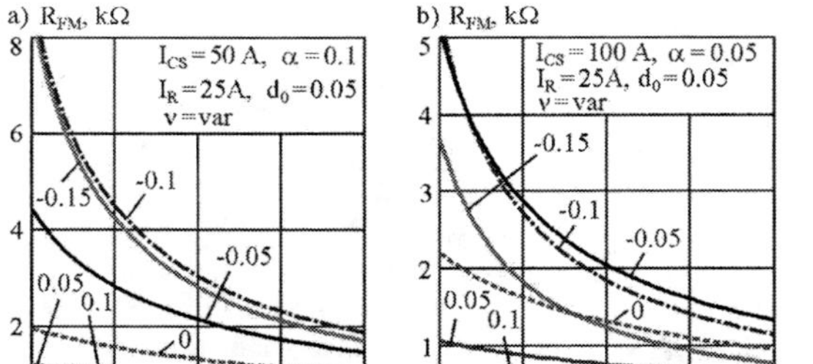

Figure 3. Dependences of the maximum fault resistance R_{FM} as a function of the current treshold value ΔI_s for the zero-seuence impulse-current protection: a) $I_{CS} = 50$ A, $\alpha = 0{,}1$; b) $I_{CS} = 100$ A, $\alpha = 0$. Parameters: $U = 15$ kV, $I_R = 25$ A, $d_0 = 0{,}05$, $q_{PC} = 80$

Improving the impulse protection effectiveness can be achieved by reducing the resistance in the forcing circuit, i.e. by increasing the forced current (Fig. 4a). A significant improvement occurs in the case of increasing current I_R to about 30 A. In the overcompensated network the leakage increase influences positively on the effectiveness of protection (Fig. 4b). In the undercompensated network the leakage increase may cause either an increase ($\nu << -\alpha$) or a decrease ($\nu \cong -\alpha$) of the protection effectiveness.

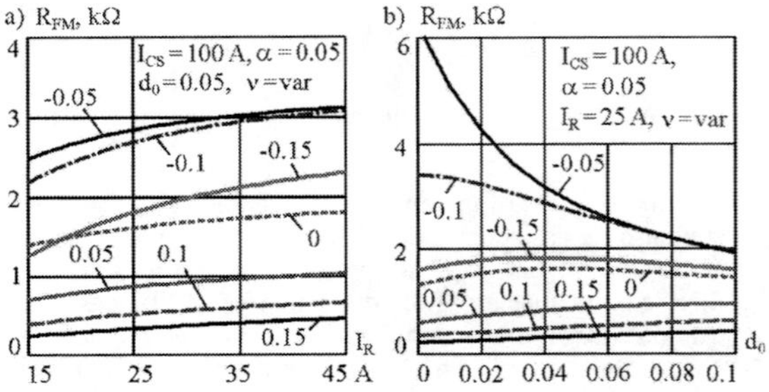

Figure 4. Dependences of the maximum fault resistance R_{FM} for the zero-sequence impulse protection with a threshold value $\Delta I_s = 1$ A as a function of the: a) forced current in the case $d_0 = 0.05$, b) damping coefficient d_0 in the case $I_R = 25$ A. Parameters: $U = 15$ kV, $I_{CS} = 100$ A, $\alpha = 0.05$, $q_{PC} = 80$

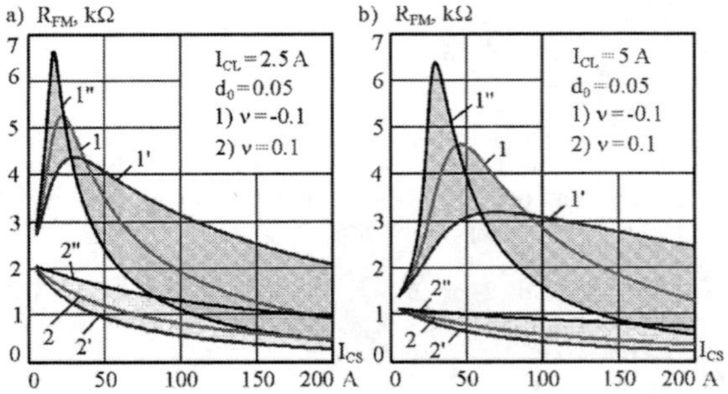

Figure 5. Dependences of the maximal fault resistance R_{FM} for the zero-sequence impulse protection with setting $\Delta I_s = 1$ A in asymmetrical network as a function of current I_{CS} in cases: a) $I_{CL} = 2.5$ A, b) $I_{CL} = 5$ A. Curves: 1 – $\nu = -0.1$, 2 – $\nu = 0.1$, 1', 2' – asymmetry current angle $\varphi_{as} = 57$ deg., 1", 2" – $\varphi_{as} = 237$ deg. Parameters: $U = 15$ kV, $I_R = 30$ A, $q_{PC} = 80$, $I_{as} = 3$ A

An important factor in the protection operation is the capacitive network asymmetry. As a result, during the single-phase short circuits an additional asymmetry current I_{as} occurs. As follows from Figure 5, the asymmetry may cause both a significant increase the fault resistance R_{FM}, detected by the protection, and its significant decrease. It depends on the ratio of the asymmetry current to the network capacitive current k_{as}, the phase angle of the asymmetry current φ_{as}, network parameters and the location of the measurement point α. A significant impact of the asymmetry on the impulse protection effectiveness occurs mainly in the undercompensated networks.

The effectiveness of the phase impulse-current protection illustrated in Figures 6 to 8.

Dependences on Figure 6 show that the phase impulse-current protection, like the zero-sequence impulse-current protection, gives the ability to detect high resistance earth faults in compensated networks, but its properties are slightly different. It shows the highest efficiency in the well-compensated networks with $\nu = -0.1 \div 0.1$. On non-loaded lines, it is equally effective both in overcompensated and undercompensated networks.

The effectiveness of the protection depends on the load current (Fig. 7) and asymmetry current (Fig. 8). The load current causes usually an increase R_{FM} (Fig. 7a). Only in cases where the detuning ratio and the angle of the load current φ_{Lo} are high and have opposite signs (Fig. 7.b), a slight decrease R_{FM} occurs.

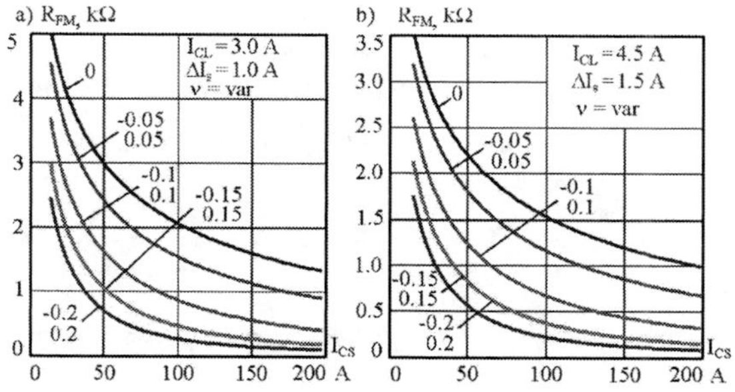

Figure 6. Dependencies of the maximum fault resistance R_{FM} for the phase impulse protection with a threshold value ΔI_s = 1 A (a) and ΔI_s = 1.5 A (b) as a function of the network capacitive current I_{CS} and various detuning coefficient s. Parameters: d_0 = 0,05, q_{PC} = 80; I_R = 15 A

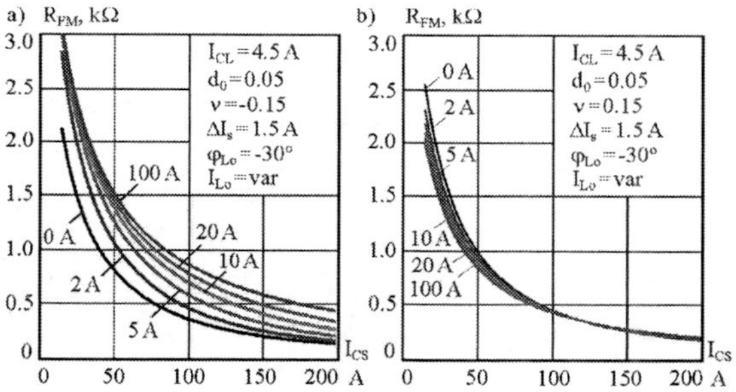

Figure 7. Impact of the load current I_{Lo} on maximum fault resistance for the phase impulse protection with a threshold value ΔI_s = 1.5 A as a function of the network capacitive current I_{CS}: a) detuning ratio ν = -0.15; b) ν = 0.15. Parameters: U = 15 kV, I_R = 15 A, q_{PC} = 80

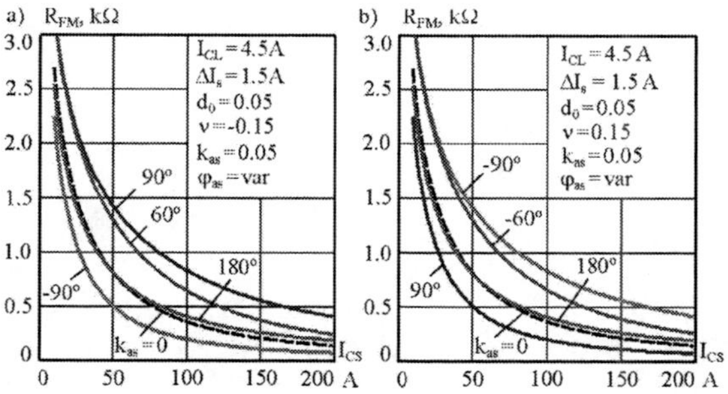

Figure 8. Impact of the asymmetry current and it angle on the maximum fault resistance for the phase impulse protection with a threshold value ΔI_s = 1.5 A as a function of the network capacitive current I_{CS}: a) ν = –0.15, k_{as} = 0.05; b) ν = 0.15, k_{as} = 0.05. Parameters: U = 15 kV, I_R = 15 A, q_{PC} = 80

The asymmetry current can affect both the improvement and the reduction of the protection effectiveness. In the undercompensated network, the highest increase R_{FM} occurs when the angle of the asymmetry current φ_{as} is close to 90 deg. (Fig. 8a) and the highest decrease R_{FM} when $\varphi_{as} \cong$ -90 deg. The impact of the asymmetry current on the protection effectiveness strongly depends on its share in the capacitive current of the network.

IV. Comparison of the Effectiveness of the Impulse-Current and Overcurrent Protections

In order to illustrate the advantages of impulse-current protections in relation to overcurrent ones a comparative analysis of their effectiveness was made. Figure 9 shows selected dependences of the maximum fault resistance R_{FM} detected by overcurrent protections and impulse-current ones as functions of a line capacitive current I_{CL} for several values of the detuning ratio ν from recommended range 0.05÷0.15, specified for the network capacitive current I_{CS} = 50 A (Fig. 9a) and I_{CS} = 100 A (Fig. 9b). The setting value of the overcurrent protection I_s was determined from the relationship:

$$I_s = k_s(I_{CL} + I_{err}) \tag{12}$$

where: I_{err} = 2 A – error current of the Ferranti filter converted to the primary winding, k_s = 2 – safety coefficient.

The setting value of the zero-sequence impulse-current protection was taken as 1 A, while setting values of the phase impulse-current protection was calculated from (5), assuming k_s = 2. In addition, to the calculations was assumed I_{Lo} = 50 A.

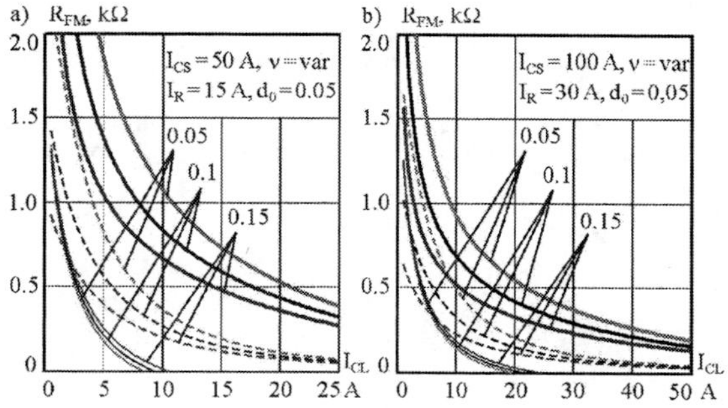

Figure 9. Comparison of the effectiveness of the overcurrent protection (thin solid lines), zero-sequence impulse-current (dashed lines) and phase impulse-current protection (bold solid lines) as a function of the capacitive current in the line I_{CL} for a few values of a detuning ratio in the network with: a) I_{CS} = 50 A; b) I_{CS} = 100 A

As can be seen from the Figure 9, the impulse-current protections are several times more effective than the zero-sequence overcurrent one. This particularly applies to the phase impulse-current protection. Overcurrent protections show good effectiveness only at the endpoints of the lines in which the capacitive currents are small. On the improvement of the overcurrent protection effectiveness positively influence the increase of the overcompensation detuning degree and the forced active current.

The phase impulse-current protection, unlike the overcurrent one, shows the highest effectiveness when the accurate compensation occurs, i.e. when $\nu \cong 0$. The zero-sequence impulse protection reaches the highest effectiveness, as it was shown earlier, in the case of undercompensation at $\nu = -(\alpha + d_0)$.

Impulse-current protections retain high effectiveness even in the case of location them at points with significant values of capacitive currents in which the overcurrent protection loses the ability to operate.

V. CONCLUSIONS

Single-phase short circuits inside of a compensated medium voltage network can be detected by measuring the impulse changes of the zero-sequence or phase currents.

Earth fault protections based on impulse-current criteria are distinguished by the simplicity of implementation and the high effectiveness in the detection of resistive short-circuits, especially in well-compensated networks. Their effectiveness is several times greater than the effectiveness of zero-sequence overcurrent protections and - in specific cases - is equal to the admittance protections, currently considered as the best.

Due to the simplicity of technical implementation, the impulse-current protections can be successfully used in fault indicators designed for installation in transformer substations or directly on the wires of overhead lines. Such devices, additionally equipped with transceiver systems, can be used in the central fault location systems and for control of reclosers and sectionalizers in smart grids.

REFERENCES

[1] H. Hänninen, "Single phase earth faults in high impedance grounded networks. Characteristics, indication and location", Technical Research Centre of Finland, Espoo 2001.

[2] L. Marciniak, "Indicators for short-circuit location in cable medium voltage networks", Automatyka Elektroenergetyczna, vol 2, 1998, pp. 4-7 (in Polish).

[3] M. Wądołowski, "Modern localization tools for short-circuits lokation in medium voltage networks - indicators of short-circuit current flow", Wiadomości Elektrotechniczne, vol. 9, 2004, pp. 54-57 (in Polish).

[4] E. Bjerkan, T. Venseth, "Locating earth-faults in compensated distribution networks by means of fault indicators", International Conference on Power Systems Transients (IPST'05), Montreal, 19-23 June 2005, paper No. IPST05 – 107.

[5] G. Verneau, Y. Chollot, P. Cumunel, "Auto-adaptive fault passage indicator with remote communication improves network availability", 21st International Conference on Electricity Distribution CIRED, Frankfurt, 6-9 June 2011, paper 0245.

[6] Fault Passage Indicators for overhead lines. Fault Passage Indicators for cable Networks, Nortroll product catalogs, Norway, [Online]. Available: www.nortroll.no., last see (2017, May)

[7] K. Cieślewicz, L. Marciniak, M. Kornatka, "Present capabilities and development trends of the central identification of short-circuits by CIZ system", Automatyka Elektroenergetyczna, par 1, vol. 3/1995, pp. 13-16, part 2, vol. 4/1995, pp. 21-24, part 3, vol. 1/1996, pp. 16-18 (in Polish).

[8] J. Byrczek, "Dispersion of protective automatics in MV network as a way to improve the quality of electricity", Wiadomości Elektrotechniczne, vol. 6, 2006, pp. 51-53 (in Polish).

[9] S.W. Luan, J.H. Teng, C. S. Chen, Y.H. Liu, "Development of an Intelligent Fault Indicator for Smart Grids", 2011 International Conference on Information and Electronics Engineering, IPCSIT vol.6 (2011) IACSIT Press, Singapore, pp. 23-27.

[10] Self-Healing Grid, Schnaider Electric Group broschure, [Online]. Available: www.schneider-electric.com., last see (2017, May)

[11] B. Grabarczyk, "Reducing SAIDI/SAIFI indicators using innovative restitution system Self Healing Grid", Wiadomości Elektrotechniczne, vol. 5, 2016, pp. 10-27 (in Polish).

[12] A. Grishkevich, Ł. Piątek, A. Burmutaew, "Frequency and outage duration in electric power systems", Przegląd Elektrotechniczny (Electrical Review), vol. 3/2009, pp. 220-222.

[13] L. Marciniak, Impulse-current and wavelet criteria of resistive and arc earth fault identification in the medium voltage networks. Wydawnictwo Politechniki Częstochowskiej, Monographs nr 273, 2013 (in Polish).

[14] J. Lorenc, „Admittance earth fault protections", Wydawnictwa Politechniki Poznańskiej, Poznań 2007 (in Polish).

Relay Protection Designing as a Solution to the Problem of Recognizing the Regimes of the Electrical Grid

Ivan Nagay
Dept. Electric power stations and electric power systems
Platov South-Russian State Polytechnic University (NPI) – «RELDON»
Novocherkassk, Russian Federation
nagayiv@mail.ru

Vladimir Nagay
Dept. Electric power stations and electric power systems
Platov South-Russian State Polytechnic University (NPI)
Novocherkassk, Russian Federation
nvi53@mail.ru

Pavel Kireev
Dept. Electric power stations and electric power systems
Platov South-Russian State Polytechnic University (NPI)
Novocherkassk, Russian Federation
kireevps@yandex.ru

Sergei.Sarry
Dept. Electric power stations and electric power systems
Platov South-Russian State Polytechnic University (NPI)
Novocherkassk, Russian Federation
sv@sarry.ru

Abstract—**This Electric grids 6-110 kV are one of the most widespread and extended in Russia and provides power to both industrial and municipal consumers. The detection of faults in electrical distribution grids with branching substations is a complex task, especially if there is an effect of the arc resistance, starting currents on the currents of fault regimes that are comparable in level with the currents of the load regimes. In this case the presence of a arc resistance at the fault location leads to a decrease in current and an increase in the resistance of the short-circuited circuit, i.e. to displacement of the fault regime area towards the load mode area. It is possible to cross these areas, especially if there is a power motor on adjacent substations in relation to the protected substation, which leads to a practical impossibility of recognizing fault regimes by traditional methods.**

Increasing use of microprocessor technology in the systems of relay protection and automation (RPA) of power systems allows to increase their technical perfection. The use of a programmable element base has a sense if not only well-proven algorithms of functioning are used, but also algorithms that increases the sensitivity, selectivity, speed and reliability of the relay protection devices. A purposeful definition of a strategy for recognizing normal, abnormal and fault regimes and making a decision based on the results of this recognition allows the designing of hardware and software relay protection complexes with better characteristics for the modern electric power systems.

In the paper are seen approaches to designing of relay protection of electric distribution grids with branching, intermediate substations and transformers of various power in view of pattern recognition task decision. The stages of designing a system for recognizing phase-to-phase short circuits, open-phase regimes and complicated faults against the background of loading and starting regimes of high-voltage electric motors. It are determined qualitative and quantitative criteria for recognizability of regimes. The technique of an estimation of the information signs characterizing regimes of operation of electric grids with use of linear and polynomial decision functions is offered.

Keywords: relay protection; information sign; remote backup; pattern recognition.

I. INTRODUCTION

Widely use of microprocessor technology in the systems of relay protection and automation (RPA) of power systems is allowed to increase significantly their information and technical perfection. Its application is effective if not only well-tested algorithms of functioning are used, but also new adaptive algorithms that use a much larger amount of information in comparison with RPA with direct algorithms of functioning and a limited amount of information. A purposeful definition of a strategy for recognizing normal, abnormal and fault regimes and making a decision based on the results of this recognition makes it possible to design hardware and software complexes of relay protection for modern electric power systems with better characteristics [1-7].

Electric networks 6-110 kV are among the most wided and provide power to both industrial and municipal consumers. Recognition of faults in distribution electrical networks in the presence of multi-ended overhead lines is a difficult task, especially when there is an effect of the arc resistance, starting currents on the currents of fault conditions comparable in level with the load regime currents [6]. In this case, the presence of the arc resistance at the fault location leads to a decrease in current and an increase in the resistance of the short-circuited circuit. Displacement of the fault regime area towards the load regime area. It is possible to cross these areas, especially in the presence of a power motor load [2-7]. The purpose of this paper is to investigate the possibility of using the mathematical apparatus of the theory of pattern recognition to improve relay protection [8-11] with the position of increasing sensitivity and selectivity.

II. Systems For Pattern Recognition For Relay Protection Purposes

The algorithms for the operation of relay protection as a system for recognition of fault regimes should ensure that the initial image x is correlated to one of the F_i classes (Fig. 1). The rules for assigning an image to one of the classes are implemented in the classification block. If the images correspond to the vectors - metric space elements, then the image can be assigned to the class, for example, by calculating the distance between the vector and the class [8-11], vectors of different classes. A signal should be formed about the belonging of the input image to a particular class or that the input image does not belong to any of the classes at the output of the classifier.

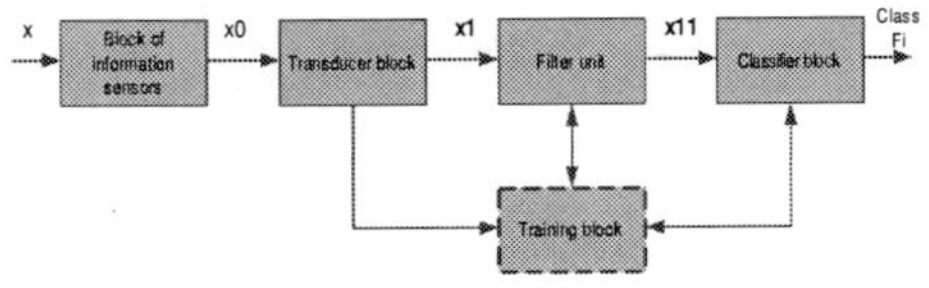

Figure 1. Recognition system of fault regimes

It is possible to addicting of block selection of the most informative signs of the image (in this case, a filter unit, for example, symmetrical, orthogonal or fault component signals) to the recognition system and included between the transducer block that converters analog signal $x1(t)$ in the vector form $\boldsymbol{x1}$ and the classifier block. Such inclusion of the filter unit block allows to reduce the dimensionality of vectors, the amount of information processed and, as a consequence, to increase the speed of RPA. The general recognition system can also have a training block that allows to formulate classification rules in one form or another.

The above justification allows us to formulate the following main tasks of detecting fault regimes of electric power systems based on the pattern recognition theory with respect to the designing of backup relay protection of electrical networks (Fig. 2):

1. Mathematical description of regimes. A convenient mathematical description for this problem can be a vector description of images (regimes). In this case, each vector x is associated with a vector $\boldsymbol{x} = (x, x1, x2, ..)$ of information signs xi of this image, which is an element of the vector space X [8-11].

2. The selection of the most informative signs is one of the main tasks ensuring finding the minimum number of signs most fully describing the images in this recognition system. The complete set of signs selected for recognition is considered to be the attribute alphabet [8,9], and the minimal set of characteristics (sufficient for solving the formulated recognition problem) is a vocabulary of signs. The effectiveness of the recognition system depends to a large extent on the correctness and optimality of the selection of the sign alphabet and the compilation of a vocabulary of signs. Their compilation is also determined by the ability to obtain practical information from the respective sensors, in this case from current and voltage measuring transformers.

3. Description of classes of recognizable images. This task is reduced to defining the class limits. Class limits can be explicitly specified at the stage of a mathematical description of the protected object, or the system must find them in the course of its work (in the presence of a training block).

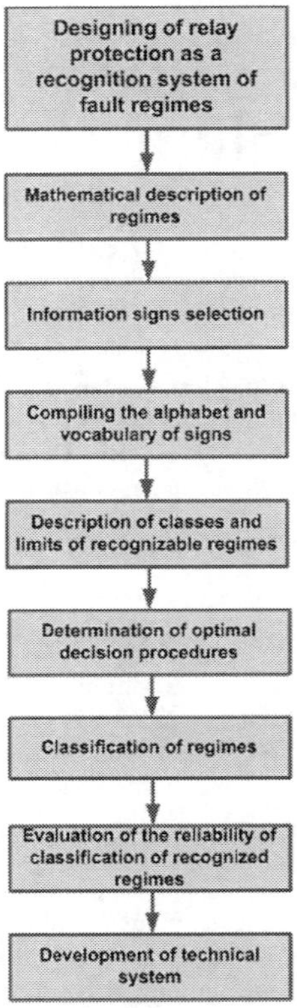

Figure 2. Stages of designing of fault regime recognition system

4. Determination of optimal deciding procedures (classification methods, algorithms for the operation of relay protection systems), i.e. methods of correlating the vector of attributes of an image to a certain class F_i. An example is the separation of load regimes, start-up and self-starting regimes of electric motors, magnetizing inrushes, short circuits and incomplete-phase regimes of the electrical network with using the control of symmetric, harmonic, orthogonal and fault components [6, 12-18].

5. Assessment of the reliability of the classification of images (regimes). This evaluation is necessary to determine the effectiveness of the developed system and is associated with the determination of possible and permissible losses associated with improper classification, for example, the impossibility of short-circuit selection for a particular transformer on branch substation if power of different transformers is equal or impossibility of detection of incomplete-phase regime in idle.

It is possible to single out several criteria for the classification of recognition systems: by the nature of information about the characteristics (deterministic, probabilistic, logical, structural, combined) and by the number of a priori information about the recognized objects. For relay protection and automation it is possible to use all the signs, except probabilistic, because in this case failure, excessive or their incorrect action is allowed, which is unacceptable for the requirements of reliability and selectivity.

In this case, three basic recognition systems are applicable: systems without learning (training) - the amount of a priori information is sufficient to determine the alphabet of signs, the formation of a vocabulary of signs and the definition of class limits; systems based on training with the teacher - the volume of a priori information is sufficient to select the alphabet of signs and formations of the sign vocabulary, but not enough to define limits and distance between classes; systems based on self-learning - the amount of a priori information is not enough even to form a vocabulary of signs.

Currently, relay systems are dominated by systems of the first type-a system without training.

The final stage of the solution of the problem under consideration is the development of a technical system.

III. Evaluation Of Information Signs Used In The Designing Of Relay Protection Of Electrical Networks

At the stage of the alphabet of signs, a certain subset of the attributes x is selected, which forms the sign space X. For example, in the set of images xi in the problem of recognizing the regimes of longitudinal-transverse asymmetry are interesting some subsets - the classes Fi (regimes: loading, starting (starting) of electric motors, magnetizing inrush, short circuits, multiple fault, incomplete phase regimes, etc.).

Classification with the decision functions. An important task of pattern recognition is the problem of class description [8,9]. If there is a set of classes F_i, then each image of x is described by some set of signs in the sign space by the vector $\boldsymbol{x}$ (vector $\boldsymbol{x11}$ in Fig. 1). The limits of the classes of recognizable images can be considered the limits of the areas X_i (i = 0,1, ..., m) and they can be determined, for example, using the solving function [8,9]. The separating surface must satisfy the conditions: the image x_i belongs to the class F_i and the distance L_i $(\boldsymbol{x})$> 0 between the classes F_i and F_j with $j \neq i$ exceeds the threshold value (setpoint). Solving functions of the form $d(\boldsymbol{x})=d(x_1,x_2,...x_n)=a_1x_1+...+a_2x_2+a_nx_n+a_{n+1}=d$ are linear decision functions (LDF), where $\boldsymbol{a}=(a_1,a_2,...a_n, a_{n+1})^T$ is a vector of LDF weighting factors, $\boldsymbol{x}=(x_1,..,x_n,1)^T$ is a vector of signs of the image (regime).

It is possible that not any two sets of vectors can be separated by a linear function. In this case, the classes are inseparable, and the recognition problem becomes linearly insoluble.

If the classes are not linearly separable, another problem can be posed: to find a LDF so that the error of incorrect classification is minimal. As a rule, such an approach in relay protection is unacceptable. Modification of the considered method for linear decision functions developing can be the use of several LDF with the verification of the location of a vector in the corresponding area, for example, in the logical AND or OR scheme, which is widely used in the designing of polygonal characteristics in the complex plane of resistance of the distance protection units and directional power relay units with a "narrow" angular characteristic. Also an example of the use of this approach (Fig. 3) is the selection of incomplete-phase regimes (the break of one phase wire, for example A, with possible ground fault on the branch substation side) on higher voltage side of a three-winding transformer with a simultaneous two-phase or three-phase fault on the lower voltage (LV) side (the windings are D-connected -Δ) or medium voltage (the windings are Y-connected) as a multiple fault. The use of symmetrical components of the currents of the negative I_{2*} and zero I_{0*} sequences does not allow selection of the regimes of phase-to-phase (two-phase) faults CAY, ABY on the medium voltage side and phase-to-phase (double-phase short circuit) faults BCΔ, ABΔ on the LV side with the decision functions LDF1 - LDF4 (Fig. 3, a). Expansion of the information base due to the use of the currents of undamaged phases I_{b*}, I_{c*} [13-15] and the introduction of LDF5-LDF10 provides a successful solution of this problem (Fig. 3,b). Nominal current of the protected transformer is taken as base current.

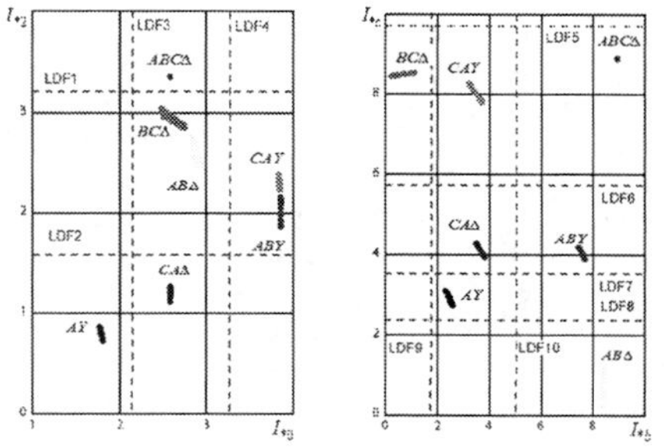

Figure 3. Separation of areas of open-phase regimes and multiple fault of a transformer on branch substation

Generalized Decision Functions (GDF). If classes can not be divided by LDF in the space C^n these classes must be embedded in the space C^m of bigger dimension $m> n$ by means of some φ: $C^n \rightarrow C^m$. This mapping should be such that the classes in C m can be linearly divided. It is possible to use generalized decision functions (GDF) for the rectifying mapping, for example, polynomial decision functions of the form $d(\boldsymbol{x})=d(x_1,x_2,...x_n)=x_1^{b1}+...+ x_2^{b2}+ x_n^{bn}+b_{n+1},b_1... b_{n+1}$ - degree of polynomial. An example of such recognition systems in relay protection is the characteristics of the impedance relay (distance organs) in the complex plane, representing a circle, ellipse, polygon.

IV. Classification With Distance Functions

This method involves the definition of a function that quantifies the measure of belonging of the image (regime) x to the class F_i, i.e some function L_i (x,F_i)=min $\|\boldsymbol{x}-\boldsymbol{y}\|$ $>k_rF_{ub}$, k_r is the reliability factor, F_{ub} is the possible unbalance due to the error of information sensors, converters, signal filters etc. It is necessary to choose a way of determining the measure of proximity between two images [8,9] between the images of the class or between two classes in order to determine the measure of the belonging of the image x to the class F_i. It is possible to define the measure of closeness $L(\boldsymbol{x},\boldsymbol{y})$ between the image vectors $\boldsymbol{x}$ and $\boldsymbol{y}$ of their signs due to the fact that each image of x is characterized by some vector of signs $\boldsymbol{x}$, then between the regimes (images) x and y.

Vectors of signs between which the distance is measured can have different dimensions, different ratio of magnitude, different priorities, which requires preliminary normalization and standardization of the meaning of the signs. It is necessary to take into account the sensitivity of a number of methods for decision function developing for standardization and normalization procedures, and it should be taken into account when choosing algorithms for recognition and comparison by this criterion in this case.

Methods for measuring distances between vectors of signs. If we consider images as elements of a metric space, then as the distance function we can use the metric of this space [8,9]. Following metrics can be used with regard to relay protection:

a) the Euclidean metric

$$L_2\,(\boldsymbol{x},\boldsymbol{y})=\|\boldsymbol{x}-\boldsymbol{y}\|_2=\sqrt{}(x_1-y_2)^2+...+(x_n-y_n)^2 \quad (1)$$

b) the uniform metric

$$L_\infty\,(\boldsymbol{x},\boldsymbol{y})=\|\boldsymbol{x}-\boldsymbol{y}\|_\infty=\max_{\leq i\leq n}\|x_i-y_i\| \quad (2)$$

Methods for determining the distance between the vector-image and the class.

After choosing a metric that measures the distance between images, you need to solve the problem of measuring the distance between the image and the class. To do this, you can use three methods. The first way is to determine the distance to the center of the class. This method is used if the class is described by one reference image - the center of the class, i.e. answers the "compactness principle". The distance between the image of x and the class F_i is defined as the distance between this image and the center of the class c_i. The second way is "the method of the nearest neighbor". In this method the distance is determined in accordance with the following expression $\| \boldsymbol{x}-\boldsymbol{y}_j \| = \min_{1 \leqslant i \leqslant n} \| \boldsymbol{x}-\boldsymbol{y}_j \|$ .This method is used when the error level of an incorrect classification is large and the errors in measurement and signal processing are within the permissible limits [8,9]. The main drawback of the "nearest neighbor method" is its sensitivity to the values of individual (possibly erroneous) data.

The third way is distance to the reference image determination. This method is used when the class is poorly described by one reference image - the center of the class. An example of a reference image may be the setting area of the overcurrent protection. In this case, the distance between the vector-image x and the class F_i is reduced to the calculation of the distance between this vector and the closest standard image of the class $L_i\,(x,F_i\ )=\min_k \| \boldsymbol{x}-\boldsymbol{c}_{ki} \|\ \ \| \boldsymbol{x}-\boldsymbol{y} \| > k_r F_{ub}$ where $\boldsymbol{c}_{ki}$ is the k-th reference vector-image of class F_i.

An example of an assessment of the recognizability of fault regimes (short circuits behind a 6.3 MVA transformer) and the starting regimes of high-voltage electric motors is Fig. 4, in which in the plane of active and reactive current components Ia_*, Ir_* and resistances Za_*, Zr_* are represents the areas of the regimes. The use of current increments ΔIa_*, ΔIr_* and resistances ΔZa_*, ΔZr_* allows to increase the distance between the areas of recognized regimes $L1$ and $L2$ and so increase the efficiency of recognition of fault regimes. It makes possible to develop the linear decision functions LDF1 and LDF2.

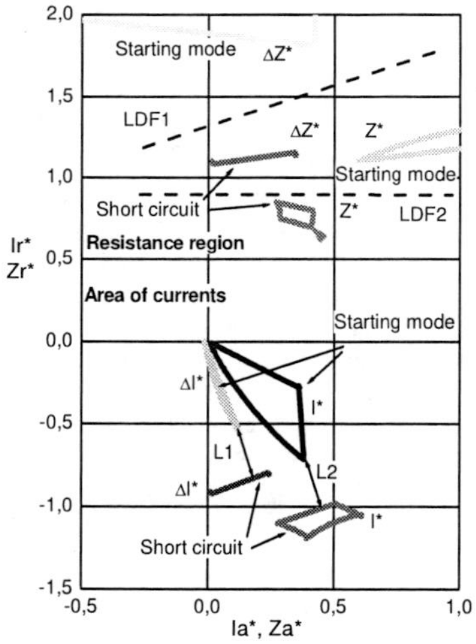

Figure 4. The recognizability of fault regimes behind a transformer 6.3 MVA in the presence of an electric load of comparable power at adjacent substations

The authors are implemented a project with backup protection of overhead lines with branch and intermediate substations in a 110 kV electric network that based on the methodology developed within the framework of research and development work. The main difference from traditional approaches is the use of the multiparameter principle of implementing algorithms for the operation of microprocessor relay protection systems. Examples of the application of the results of this article are the technical solutions of the KEDR-07M type protection system introduced in the distribution electric grids of the south of Russia and the APK RRZ system recommended for the installation in the electric system.

V. Conclusions

1. The proposed methodology for assessing information signs that are characterizing the regimes of electrical networks with use of linear and polynomial decision functions allows to provide a qualitative assessment of the recognizability of fault and permissible regimes.

2. Quantitative assessment of the recognition of fault and admissible regimes can be provided by measuring the distances between the vectors of signs (up to the "nearest neighbor"), by determining the distance to the center of the class, by determining the distance of the reference image.

3. An increase in the recognizability of fault regimes can be achieved by expanding the information base of relay protection.

4. These approaches to assess the recognizability of regimes of electrical networks and the formation of algorithms for the operation of relay protection have made it possible to implement a microprocessor multiparameter backup protection with increased sensitivity for overhead lines with branches (APK RRZ) in the research and development work.

References

[1] "The concept of the development of relay protection and automation of the power grid complex. Annex No. 1 to the protocol of the Management Board of JSC Rosseti 06.22.2015 №356pr," Releyshik, vol.3, pp. 28-43, 2016.

[2] D.S.Vasilyev, V.N. Tarasova ,"Perfection of long-range backup protection for power lines with tapping substations," Relay protection and automation, vol. 01 (22), pp. 26-31, 2016.

[3] R.M.Bajazitov "Experience in the operation of remote backup protection of dead-end lines in the Tatarstan grid company", Relay protection and automation, vol. 01 (22), pp. 50-58, 2016.

[4] S.Y. Petrov, "Problems of providing remote backup", Releyshik, vol.1, p.74, 2008.

[5] Y.Y.Lyamets, G.S. Nudelman, A.O. Pavlov, "Algorithmic models of electrical systems," Proceedings of the Academy of Electrotechnical Sciences of the Chuvash Republic. Chuvash university. Cheboksary, vol. 1-2, pp.10-21, 1999.

[6] V.I. Nagay, "Relay protection of the tapping substation in the electric networks", Moscow, Energoatomizdat, 2002, p.312.

[7] I.V. Nagay, "Load influence to the sensitivity of the current backup relay protection," Russian Electromechanics, vol. 1, pp.108-110, 2013.

[8] A.L.Gorelik, V.A. Skripkin, "Methods of recognition", Proc. allowance, Higher. HQ, Moscow, 1984.

[9] A.E. Lepsky, A.G. Bronevich, "Mathematical methods of image recognition", Lectures, Publishing house of the TTI SFU, Taganrog, 2009.

[10] A. Faure, "Perception et reconnaissance des formes",Transl. France, 1989, p. 272.

[11] J.T.Tou, R.C.Gonzales, "Pattern recognition principles", Moscow, Mir, 1978, p. 412

[12] D.S. Vasilyev, V.N.Kozlov, I.A. Rodionov, "Long-range backup protection of tapping substations in the open-phase mode", Relay protection and automation, vol. 3, pp. 62-64, 2011.

[13] I.V. Nagay, P.S. Kireev, G.N. Chmihalov, "Remote and local relay protection functions in the longitudinal-transverse asymmetry regimes", Russian Electromechanics, vol. 01, pp. 111-113, 2013.

[14] I.V.Nagay, V.I. Nagay, "Designing of multi-parameter backup protection of electric distribution grid 6-110 kV," Energetik, vol. 2, pp.18-21, 2013, in press.

[15] I.V. Nagai, "About perfection of protection from current networks open-phase modes", Global nuclear security, vol. 2. 2013, pp. 64-68.

[16] I.V.Nagay, P.S. Kireev, "Modeling of load regimes of branching substations", Russian Electromechanics, vol. 2, pp. 100-103, 2012.

[17] I.F. Maruda, "Relay protection of the overhead line 110-220 kV in the presence of the phase wire breaks", Power Stations, Moscow, vol.1, pp. 40-42, 2002.

[18] I.V.Nagay, V.I. Nagay, P.S. Kireev, S.V.Sarry, A.V. Ukraincev, "Recognition of Remote Short Circuits with the Transient Resistance of an Electric Arc", Proceedings of the 8th international scientific symposium on electrical power engineering Elektroenergetika 2015, Stara Lesna, SLOVAKIA, SEP 16-18, 2015, pp. 367-371.

Experimental Investigation of Electric Arc Characteristics

Pavel Kireev, Ivan Nagay, Vladimir Nagay
Dept. Electric power stations and electric power systems
Platov South-Russian State Polytechnic University (NPI)
Novocherkassk, Russian Federation
kireevps@yandex.ru, nagayiv@mail.ru, nvi53@mail.ru

Abstract — **The purpose of the investigation was creation of the mathematical tool for determining the parameters of the arc resistance at the fault location for the subsequent analysis of the operation of relay protection devices. The article presents the results of mathematical modeling and experimental studies of the arc resistance of an electric arc in the place of short circuit made at the Department of Electric power stations and electric power systems at the Platov SRSPU (NPI). A mathematical model of the electric arc column is developed on the base of the provisions of classical thermodynamics and physics of low-temperature plasma and takes into account a number of influencing factors: the nominal voltage and power of the source of electrical energy, the length of the arc, the material of the elements between which its existence is observed, composition and thermodynamic parameters of the environment. There is regression dependence of the electrical conductivity of the plasma of the low-voltage electric arc on the temperature at constant pressure, that was obtained by the authors experimentally. The results of calculating the combustion mode of the column of an low-voltage electric arc with a length of 18 mm. in the open air using this regression dependence are shown. The results of calculating the current-voltage characteristic of a column of an electric arc 0.6 m burning in an electric grid 6 kV between copper parts and parameters of a similar arc obtained experimentally are also given. The error in calculating the parameters of this high-voltage arc does not exceed 10%. The possibility of reducing the calculation error exist and is provided that the dependence of the electrical conductivity of the plasma of the high-voltage electric arc on the temperature and pressure is refined by additional experiments. A high degree of convergence of the experimental and calculated results confirms the possibility of using the results obtained in the analysis and development of new algorithms for the operation of high-speed relay protection devices.**

Keywords: relay protection, arc short circuit, volt-ampere characteristic.

I. INTRODUCTION

Arc resistance at the fault site is one of the factors affecting the parameters of the fault regime [1-5]. The electric arc is one of the kinds of an independent electric discharge, characterized by high current densities and a low cathode voltage drop in comparison with other types of independent discharges (dark and smoldering). The resistance of the electric arc has a pronounced nonlinear active character, which causes the appearance of higher harmonic voltage components in the network [1,6,9]. There are numerous arc math models [1,6,7], based on Cassie and Mayr works, some statistical and empirical data. Use of them provides calculation of typical arc fault characteristics and doesn't take to account the list of influencing factors (properties of power supply system, material type of damaged equipment, properties of surrounding air and so on), so provides low accuracy results.

The main difficulty in modeling the electric arc with nonlinear active resistance is the calculation of its volt-ampere characteristic (VAC), the form of which depends on a large number of factors. Authors proposed for these purposes a thermodynamic model of the electric arc column, developed on the basis of classical thermodynamics [10,11] and physics low-temperature plasmas [8,9,12-14] and taking into account a number of influencing factors: the nominal voltage and power of the source of electrical energy, the length of the arc, the material of the elements, between which are observed its existence, composition and thermodynamic parameters of the environment.

II. THE THEORETICAL PART

In the proposed thermodynamic model (Fig. 1) the electric arc in analogy with [6,8,12] is represented as a continuous horizontal plasma cylinder of radius r_A, length l_A and mass m_A. The arc burns between two electrodes, which, for simplicity, are taken to be the same and also in the form of cylinders of radius r_E, length l_E, and mass m_E. Combustion occurs in open ambient air at a temperature T_{Ai} and pressure p. The arc column has a temperature T_A same in its entire volume. The pressure inside the arc column is considered equal to the pressure of the surrounding air p (the Pinch effect is not taken into account).The electrodes are also in the ambient air and have a temperature T_E that is unique for each elementary volume of electrode material. The masses of the column of the electric arc and the electrodes are assumed to be constant (the processes of diffusion of matter and burnout of the electrodes are neglected). The length of the electric arc l_A in the combustion process is also considered constant.

The electrodes are supplied with voltage from a source of electrical energy consisting of an ideal internal source of sinusoidal EMF e_S and internal resistance r_S, the character of which is adopted as active for the purpose of simplifying the model. Electric current i flows through the arc and electrodes under the action of EMF.

The electric current leads to the release of energy in the column of the electric arc Q_{EA} and in the material of the electrodes Q_{EE}.

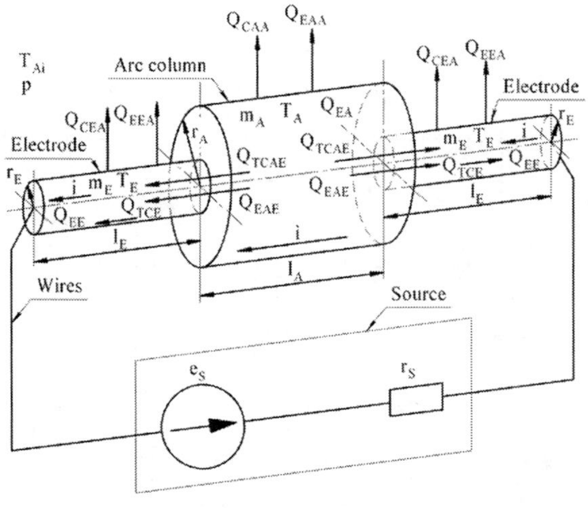

Figure 1. Thermodynamic model of electric arc column

Energy is removed from the arc column by emission Q_{EAA} and Q_{CAA} convection with atmospheric air, emission Q_{EAE} and thermal conductivity Q_{TCAE} with electrodes. Energy is diverted from the electrodes through emission Q_{EEA} and convection Q_{CEA} with atmospheric air. The transfer of energy inside the electrodes is due to the thermal conductivity Q_{TCE}. [15,17].

We assume that the ambient air is completely transparent ideal gas to simplify the model. Expression (1) is obtained by the expression of turning the corresponding expressions based on the Joule-Lenz law, Stefan-Boltzmann, Fourier, Newton-Richman for the electric arc column burning in the open air at a constant pressure p.

$$\frac{d(C_A(T_A,p)\cdot m_A\cdot T_A)}{dt} + p\cdot\frac{dV_A}{dt} = \frac{i^2\cdot l_A}{S_A\cdot\sigma_A(T_A,p)} - \alpha_A(T_A,p)\cdot S_{CAAi}\cdot(T_A-T_{Ai}) - \varepsilon_A(T_A,p)\cdot C_0\cdot S_{RAAi}\cdot\left(\frac{T_A}{100}\right)^4 - 2\cdot\left[\frac{1}{\frac{1}{\varepsilon_A(T_A,p)}+\frac{1}{\varepsilon_E}-1}\cdot C_0\cdot S_{RAE}\cdot\left(\left(\frac{T_A}{100}\right)^4-\left(\frac{T_E}{100}\right)^4\right) + \lambda_A(T_A,p)\cdot S_{HAE}\cdot\frac{dT_A}{dl_A}\right]; \quad (1)$$

where T_A – arc column temperature, T_{Ai} – ambient temperature, T_E – the temperature of the electrodes, t – time, $C_A(T_A,p)$ – the specific heat of the plasma arc, m_A – mass an arc discharge, V_A – arc column volume, S_A – cross-sectional area of the arc column, $\sigma_A(T_A,p)$ – specific conductivity of the plasma arc, $\alpha_A(T_A,p)$ – arc column heat transfer coefficient, S_{CAAi} – area of convective heat transfer between the arc and the air, $\varepsilon_A(T_A,p)$- the degree of blackness of the electric arc column, C_0 – emission coefficient blackbody, S_{RAAi} - area of radiation heat transfer between the arc and the air, ε_E – the degree of blackness of the electrode material, S_{RAE} - area of radiation heat transfer between the arc and the electrode, $\lambda_A(T_A,p)$ – specific thermal conductivity of the plasma arc, S_{HAE} - area of heat conduction between the arc and electrode.

Equation (2) is a similar expression describing the thermodynamic processes at the electrodes:

$$C_E\cdot m_E\cdot\frac{dT_E}{dt} = \frac{i^2\cdot l_E}{S_E\cdot\sigma_E} + \lambda_E\cdot S_{HAE}\cdot\frac{dT_E}{dl_E} + \frac{1}{\frac{1}{\varepsilon_A(T_A,p)}+\frac{1}{\varepsilon_E}-1}\cdot C_0\cdot S_{RAE}\cdot\left(\left(\frac{T_A}{100}\right)^4-\left(\frac{T_E}{100}\right)^4\right) - \alpha_E\cdot S_{CEAi}\cdot(T_E-T_{Ai}) - \varepsilon_E\cdot C_0\cdot S_{REAi}\cdot\left(\frac{T_E}{100}\right)^4 - \lambda_E\cdot S_{HEDE}\cdot\frac{dT_E}{dl_E}; \quad (2)$$

where C_E– the specific heat capacity of the electrode material, m_E – electrode weight, S_E – cross-sectional area of the electrode, σ_E – specific conductivity of the electrode material, λ_E – specific thermal conductivity of the electrode material, α_E - coefficient of heat transfer of the electrode, S_{CEAi} - area of convective heat transfer between the air electrode, S_{REAi} - area of radiation heat transfer between the air electrode, S_{HEDE} - area of heat conduction between the electrode and the elements of the damaged equipment.

Displacement arc column at different T_A can be determined from the equation of state of ideal gas if weight of arc discharge is constant.

$$V_A = \frac{m_A\cdot R\cdot T_A}{p\cdot M_A(T_A,p)}; \quad (3)$$

where R– universal gas constant, $M_A(T_A,p)$ – the molecular weight of the plasma arc column.

We obtain a system of equations describing the thermodynamic and electrical processes in the model by complementing the equation (1) - (3) with the current flow function, compiled on the base of Kirchhoff's and Ohm's laws.

$$i = \frac{e_S}{R_S + 2\cdot\frac{l_E}{S_E\cdot\sigma_E} + \frac{l_A}{S_A\cdot\sigma_A(T_A,p)}}; \quad (4)$$

A key role in this model verification are takes a nonlinear dependences of the thermodynamic and electrical parameters of the electric arc plasma on its temperature and pressure. The arc discharge plasma at the damage site consists of strongly ionized molecules of ambient air with a small (about 5-7%) admixture of electrode vapors [8], so the thermodynamic parameters of its plasma can be considered equal to the parameters of atmospheric air plasma under similar conditions. These parameters are given in special reference books [8,18-21, 23]. At that, vapors of electrodes are participate actively in the process of electric current, therefore, it is impossible to use the dependences of the specific electric conductivity of atmospheric air [8,22] when calculating this model. Authors carried out experimental researches [16] to determine the dependence of electrical conductivity of an electric arc plasma burning in the open air between carbon electrodes and temperature for its subsequent use in the proposed model.

III. RESULTS OF EXPERIMENTS

Direct current arc was used in the experimental researches to investigation by calculation method of arc temperature measuring. The researches were made using experimental stand (Fig. 2).

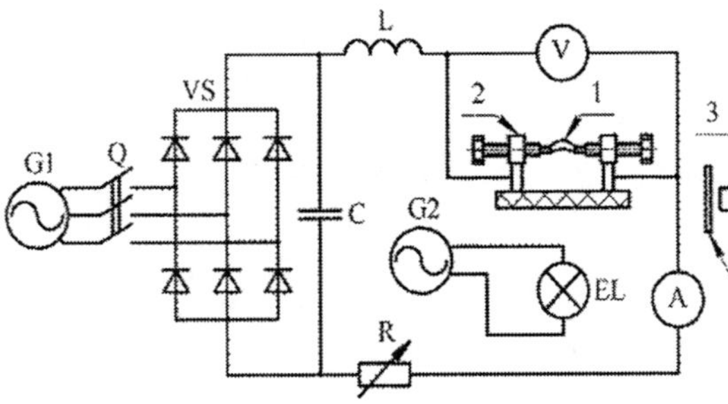

Figure 2. Scheme of experimental stand

The main element of the stand is device 2, which consist of two electrodes for electric arc creation. The distance between them can be changed for regulation of the arc column 1 length. The arc column current is regulated by variable resistor *R*. The stand allows to measure the current and voltage at the arc column using a voltmeter V and ampermeter A. Arc column dimensions can be measured by fixing them with camera 3 through welding glass 4 near with scaling element – filament of 100 watts lamp EL. Examples of obtained photos are given on the Fig. 3. The regression dependence is obtained during experimental researches and shown in Fig. 4.

The resulting regression dependence (Fig. 4) was tested in calculating the mode of a low-voltage electric arc burning with an arc length 18 mm on open air. The errors in the calculated parameters of the low-voltage electric arc in relation to the experimental data were as follows:

- amplitude of the current i flowing through the arc discharge: 1 A;
- instantaneous value of voltage on the electric arc column at the moment of its striking u_A: 11 V;
- instantaneous value of the voltage on the electric arc during its arcing and extinction u_A: 10 V.

Reduced calculated values of the voltage drop on the electric arc column and the increased current are due to the fact that the near-electrode voltage drop in the proposed thermodynamic model is not taken into account. The influence of this factor has a tangible effect on the parameters of short low-current arcs and becomes practically insignificant for long high-current arcs [8,9,12], whose calculation of the parameters was the goal of creating the model.

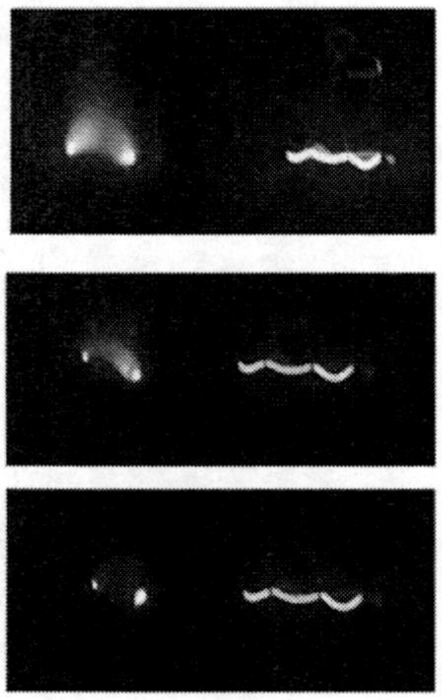

Figure 3. Examples of obtained photos of arc and lamp filament

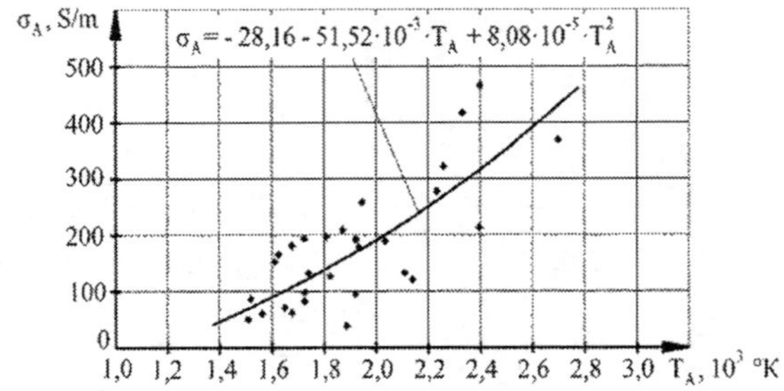

Figure 4. Dependence of electrical conductivity of a plasma of a carbon electric arc burning in the open air on its temperature

The histograms of the effective values of the harmonic components of the calculated and experimental voltage drops on the electric arc column with respect to the effective value of the fundamental harmonic voltage U_n/U_1 (n is the harmonic number) are shown in Fig. 5. The histograms of the experimental discharge are shown by solid lines, the calculated discharge are shown by dashed lines with shading.

The carried out experimental researches and modeling of processes in an electric arc have confirmed presence of odd harmonic components in an voltage on an arc of and it share decreases with growth of their order. The difference between the shares of the higher harmonics of the voltage of the experimental and calculated discharges does not exceed 7%.

Similar harmonic composition (Fig. 5), current and voltage parameters of the experimental and calculated arc discharges indicate the reliability of the proposed thermodynamic model (Fig. 1) and the truth of the regression curve of Fig. 4.

Authors made an approximate estimate of the electrical conductivity of the plasma of an fault arc burning between copper parts using the dependences of the electrical conductivity of plasma on atmospheric air [8] and the experimental data of arc interphase discharges in the switchgear. Calculation of the current-voltage characteristic of an arc column with length 0.6 m burning in a 6 kV network between copper parts with the use of this characteristic was carried out, which corresponds to the parameters of the experimental arc discharge (Fig. 6), organized by one of the authors of this article.

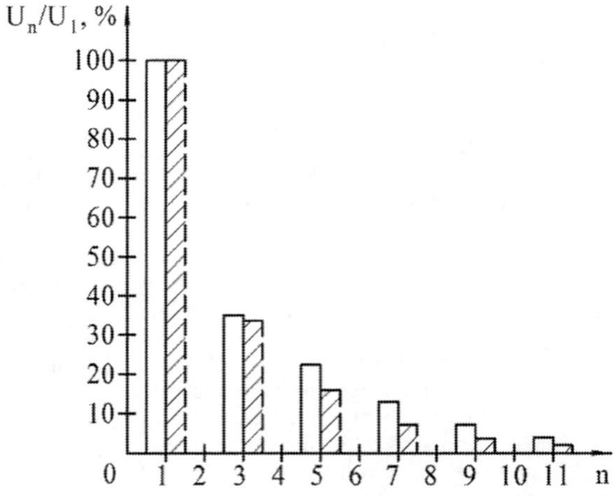

Figure 5. Histograms of the effective values of the harmonic voltage components with respect to the effective value of the fundamental harmonic voltage of the experimental and calculated arc discharges

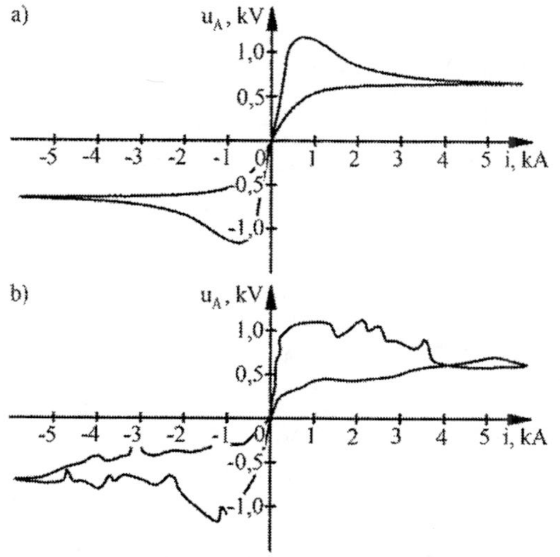

Figure 6. Calculated (a) and experimental (b) VAC of a 0.6 m arc burning in a 6 kV network between copper parts

The experimental arc discharge (Fig. 6, b) has an I_m current amplitude about of 6 kA, a steady-burning arc voltage u_{SBA} of about 600 V, an arc striking voltage u_{SA} of about 1100 V. It should be noted that there is a shortening and lengthening effect of the arc column by shunting its individual loops and the emergence of new ones, which introduces additional distortions and errors into the VAC under consideration.

The calculated arc discharge (Fig. 6, a) is characterized by the current amplitude I_m=5.8 kA (error 3,3%), the steady-burning arc voltage u_{SBA}=630 V (error 5,0%), arc striking voltage u_{SA}=1200 V (error 9,1 %). The errors in calculating the arc striking voltage of the arc discharge are due to the approximate nature of the dependence of the electrical conductivity of the electric arc column plasma on the temperature used in this calculation. The refinement of this dependence by the experimental method will make it possible to improve the accuracy of calculating the parameters of the arc resistance.

IV. Conclusion

A successful calculation of the high-voltage electric arc burning regime (similar shape of the VAC and parameters of stable-burning arc of the experimental and calculated arc discharges) attests to the physical reliability of the proposed thermodynamic model of the electric arc column and the possibility of its use in researching the features of the operation of the main and backup relay protection in electrical networks of 6-110 kV. The proposed model can be used in programs of electric networks regimes calculation, particularly in equivalent circuit blocks of furnaces and short circuits. Quantitative and qualitative voltage harmonic composition characteristics at the relay protection installation place for different types of faults, which obtained using the proposed model, can be used as information sign, used by the relay protection devices to recognize fault regimes and their types.

References

[1] V.I. Nagay, "Relay protection of the tapping substation in the electric networks," Energoatomizdat, Moscow, 2002.

[2] I.V. Nagay. "Remote backup in networks 6-110 kV. Problems and solutions", Electrical engineering news. Novosty elektrotehniky, vol. 6(66), 2010.

[3] B.N. Neklepaev, "Guidelines for Calculation of Short Circuit Currents and Selection of Electrical Equipment", Moscow, 2001, p.152 .

[4] N.V. Chernobrovov, V.A. Semenov, "Relay protection in power system. Tutorial for T. colleges", Energoatomizdat, Moscow,1998, p.800.

[5] M.A. Shabad, "Calculations of relay protection and automation of distribution networks", Federal State educational establishment, St.Petersburg, 2003, p.350 .

[6] V.V. Bourgsdorf, "Open high-power electric arcs", Electricity ,vol. 10, 1948, pp. 15 – 23.

[7] A.V. Lukonin, "Relay protection of closed electrical installations with a voltage 0.4-10 kV with recognition of damages, accompanied by an electric arc", The dissertation of a Cand.Tech.Sci. Novocherkassk, 2009, p. 200 .

[8] S.V. Dresvin, "Physics and equipment of the low-temperature plasma", Atomizdat, Moscow,1972, p 352.

[9] A. Engel, M. Steenbeck, "Physics and technology of electric discharges in gases. Volume Two. Properties of gas discharges. Technical applications", Moscow, Combining scientific and technical publishing house NKTP USSR, 1936, p. 380.

[10] V.P. Isachenko,"Heat transfer: Textbook for high schools", Energy, Moscow,1975, p.488.

[11] V.I. Lyashkov, "Theoretical basics of heat. Tutorial", Publisher Engineering 1, Moscow, 2005, p.260.

[12] V.L. Granovsky, "Electric current in the gas. Continuous current", The main edition of the physical and mathematical literature Publishing House "Science", Moscow,1971, p.490.

[13] L.S. Polak, "Essays on the physics and chemistry of low-temperature plasma", Science, Moscow, 1971, p.433.

[14] L. Spitzer, "Physics of fully ionized gases", M, Mir,1965, p.198

[15] P.S. Kireev, V.I. Nagay, I.V. Nagay, "Transient resistance model for evaluating the operation of relay protection", Scientific and Technical Journal. Russian Electromechanics, vol. 3, 2014. pp. 123–126.

[16] P.S. Kireev, I.V. Nagay, A.V. Ukraincev, A.V. Pilipenko, V.I. Nagay, "Experimental determination of the parameters of an electric arc plasma for the needs of relay protection", Cybernetics of energy systems. Proceedings of the XXXVII session seminar "Power supply", Novocherkassk, Oct., 2015, pp. 102–106.

[17] P.S. Kireev, "Experimental estimation of low voltage arc for relay protection purpose", Scientific and Technical Journal. Russian Electromechanics, vol. 1, 2016, pp. 78-81.

[18] N.B.Vargaftik, "Handbook of gases and liquids thermophysical properties", Fizmatgiz, Moscow, 1963, p.708.

[19] L.V. Gurvich, V.A. Weitz, V.A.Medvedev, "Thermodynamic properties of individual substances", Reference edition in 4 vols. vol.1, book 2, Moscow, Science, 1978, p. 328.

[20] L.M. Biberman, "Optical properties of hot air", Science, Moscow, 1970, p. 320.

[21] .V. Sychev, A.A. Wasserman, A.D .Kozlov, G.A. Spiridonov, V.A. Tsymarny, "Thermodynamic properties of air", Publishing Standards, Moscow, 1978, p.276 .

[22] I.A. Krinberg, "Electrical conductivity of air in the presence of an impurity", Journal of Applied Mechanics and Technical Physics, vol. 1, 1965, pp. 76–81.

[23] I.M. Chukhin, "Thermodynamic properties of air. Reference materials and guidelines for determination of air thermodynamic properties, taking into account the effect of temperature on their isobaric and isochoric heat capacity", Ivanovo, Ivanovo State Power University, 2001. 31p.

Analysis of the Functioning of the Earth Fault Protection With Active Influence on the Electrical Grid

Alexander Ukraincev
Dept. Electric power stations and electric power systems
Platov South-Russian State Polytechnic University (NPI)
Novocherkassk, Russian Federation
xelandr@mail.ru

Vladimir Nagay
Dept. Electric power stations and electric power systems
Platov South-Russian State Polytechnic University (NPI)
Novocherkassk, Russian Federation
nvi53@mail.ru

Ivan Nagay
Dept. Electric power stations and electric power systems
Platov South-Russian State Polytechnic University (NPI) – «NPP «RELDON» Ltd
Novocherkassk, Russian Federation
nagayiv@mail.ru

Sergei Sarry
Dept. Electric power stations and electric power systems
Platov South-Russian State Polytechnic University (NPI)
Novocherkassk, Russian Federation
sv@sarry.ru

Gennadij Chmihalov
Dept. Electric power stations and electric power systems
Platov South-Russian State Polytechnic University (NPI)
Novocherkassk, Russian Federation

Abstract— **Currently, in the distribution grids the main directional earth relay and non-directional earth fault protection current zero-sequence are used for the single-phase earth faults. At the same time in the last few years it has been possible to observe the processes of introduction of protection for the earth fault into the distribution grids based on the principles of imposing a signal of a non-grid frequency on the zero-sequence current. Such protections are most effective in view of the practical absence of the need to grading for the earth fault in the external grid with respect to the protected feeder. The most promising is the use as a superimposed signal of a special form or frequency other than an grid frequency, generated by special device, through a voltage transformer or arc suppression coils.**

The purpose of the analysis given in the article is to refine the options for designing such protections and then to propose the most effective scheme for their organization. A study was made of schemes of protection for earth fault based on systems with an impact on the electrical grid of various configurations and various complexity. The considered grids can be either with isolated neutral or with compensation of the capacitive current by means of a arc suppression coil, and should be equipped with voltage transformers with a secondary winding connected to an open-delta. When reliable protection from the earth fault installed in the schemes of electrical distribution networks, a unified system is formed selective protection of all connections from the earth fault, with the ability to trip damaged element. The presence of automatic transfer switch allows to carry out the uninterrupted operation of all consumers.

The article also focuses on the effectiveness of the protection for earth fault based on systems with active influence on the electrical network by the various signals. As an example, the method of superimposing an active component of a special form to a zero-sequence current by switching a low-resistance impedance in the voltage-transformer winding circuit is considered. Various options for designing such protections are presented: decentralized, distributed, individual and centralized by possible algorithms of operation with the use of absolute and / or relative measurement of the amplitude of the superimposed signal in the protected connections.

Keywords: relay protection, earth fault, zero sequence current, higher harmonic components.

I. Main Part

One of the possible options for the organization of protection against earth faults in 6-35 kV networks is the option in which the signal is superimposed on the protected network by acting on it through a voltage measuring transformer (winding connected to an open Delta) or through the signal winding of an earthing reactor. This is due to changes in the load parameters on the windings of the voltage transformer connected to an open Delta, then in the earth fault regime the current in the damaged feeder changes. Generation of a zero sequence current of a special form occurs in other words. The protection device responds to the additional components of the zero sequence current that occur when the voltage transformer control module operates. Research conducted on a specially developed network model has shown that this principle allows the selective detection of a damaged feeder with a high degree of reliability. The model is developed in the MATLAB software.

In this model, the control of the voltage transformer mode is realized with the help of the Pulsegenerator - signal generator blocks and the Breaker - key. The key is installed in the secondary winding of the voltage transformer connected in the open Delta circuit. The connection scheme of the secondary circuits of the voltage transformer is shown in Fig. 1.

The signal generator is switching the key, giving the logical signals of the specified duration and frequency to its logical input. If the signal from the generator is greater than zero, the key instantaneously closes and its resistance becomes 1 Ohm. When the signal from the generator becomes zero, the key is opened and its resistance increases to 100 ohms. Thus by adjusting the parameters of the signals supplied by the generator it is possible to adjust the frequency and shape of the additional zero sequence current to the fault site in case of a short circuit. It becomes possible to selectively determine the damaged feeder using special filtering algorithms designed to extract a signal of this shape and frequency.

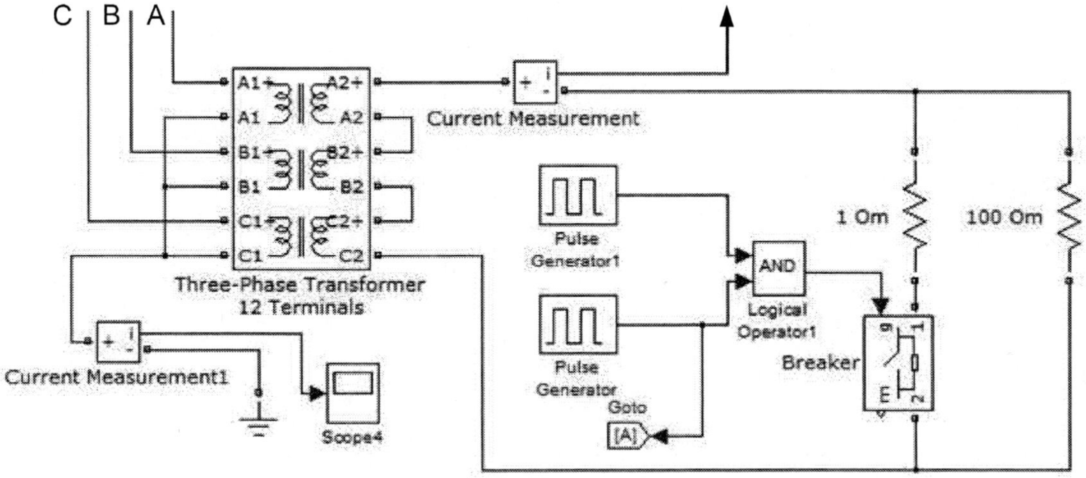

Fig. 1 Connection scheme of secondary circuits of a three-phase voltage transformer in the Matlab system

It is necessary to equip the protected network one centralized protection device and connect the zero sequence current transformers of all feeders to it or equip each feeder with an individual device with relative measurement of the value of the superimposed signal with connection of individual devices if it's need to monitoring the presence of a superimposed signal of a special form in the protected connections in earth fault regime. It is also possible regardless of the neutral point connection to use multi-functional protection devices that combine several principles. For example the absolute measurement of the zero sequence current (comparison with the setting) of the protected feeder (industrial frequency or its high-frequency harmonic components) and the principle of relative measurement (comparison of signals between protection devices protecting various feeders) of the zero sequence with a special shape. This protection must be highly sensitive regardless of the neutral point connection. Increasing the reliability of the operation of such protections for localizing various types of earth faults (metal circuit, shorting through the contact resistance, arc circuit, intermittent earth, intermittent (once in several periods of the industrial frequency resulting in a short-circuit) in 6-35 kV electrical distribution networks with neutral point connection can also be realized by using various methods of determining earth faults (in transient processes of the earth fault and in the steady-state regime of the fault) using the majority principle, using absolute and relative comparison of controlled signals (for example the spectral densities of earth fault currents with a certain frequency range).

The rated power of transformers installed in the system as a rule exceeds several times the nominal power of the generators. Since the magnetizing current of the transformers is non-sinusoidal, they can be considered as sources of higher harmonics. The phenomena associated with the higher harmonic components of currents and voltages can be due mainly to the nonlinear characteristics of the magnetization of the transformers. Transformers cause their fifth and seventh higher harmonics with significant amplitudes by their magnetizing current even with an average saturation of steel [1].

In Platov South-Russian State Polytechnic University (NPI) it is designing the microprocessor protection device (signaling) Renom-04T from the earth fault operating according to the principle of the relative measurement of the zero sequence current, having the ability to measure both the sum of the higher harmonics from the transformers installed in the network, and the value of the superimposed signal in the current of the protected feeder. In networks with capacitive current compensation even with resonance tuning of the earth reactor, the current of the earth fault at the fault location contains an uncompensated active component, higher harmonics and free components of the transient mode current. The fraction of higher harmonic components in the zero sequence current of the earth fault can reach 10% [2] and more of the total capacitive current of all feeders $I_{c\Sigma}$, depending on the features of the electrical network (the number and features of the sources of higher harmonics, their operation regimes, operating regimes of the network, etc.). In such networks it is possible to implement protection based on a relative comparison of the levels of higher harmonics of "natural origin" in the current of the earth fault of all feeders of the protected object. Level of higher harmonics for internal earth in the damaged feeder is always higher than in any of the undamaged feeder. This allows to selectively tripping feeder with earth fault using a non-linear inverse current-dependent characteristic. The family of operating characteristics used in the Renom-04T is shown in Fig. 2.

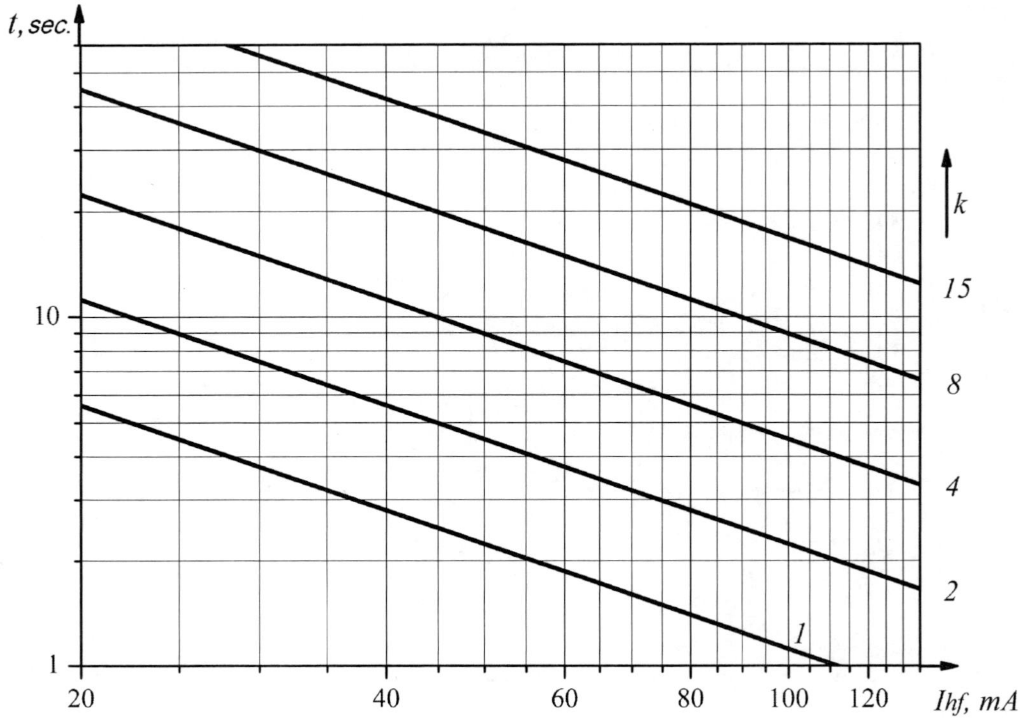

Fig.2. The characteristics of the Renome-04T

The dependence (Figure 2) of the protection tripping time on the magnitude of the higher harmonic components in the zero sequence current of the protected feeder is based on the following mathematical expression:

$$t = (\beta \cdot k)/(\alpha \cdot I_G) \quad (1)$$

where β is determined by AFC type of the high-frequency filter and for Renom-04T β = 0,112 [A•s] in the frequency range of the input current from 150 to 650 Hz, t– protection operation time, s; I_G – input earth fault current in the phase-to-ground fault regime, A; α – the relative value of the higher harmonic components at a frequency of 150 to 650 Hz in the input current; k – scale factor. The scale factor k can be set in the range from 1 to 15 in steps of 1 [3]. It determines from a family of characteristics a specific tripping characteristic in accordance with which the device operates with the required time delay, depending on the magnitude of the higher harmonic components in the zero sequence current.

One of the drawbacks of the protection devices against earth fault reacting to the higher harmonic components in the earth fault current is the lack of selectivity in the case of short-circuits on the bus of the complete switchgear and in the compartments of the high-voltage switch of the switchgear. It is also possible for them to be nonselective in the process of earth fault in adjacent sections of the radial distribution network.

The solution of the problem of selectivity of such protections installed at various substations can be the installation of a measuring zero sequence current transformer on the supply connection and the connection to it of one of the earth fault protection devices (KA10, KA30 and KA330 in Fig. 3). Below it is an example of a circuit diagram of an electrical network with an analysis of the functioning of the earth fault protection that monitors the magnitude of the higher harmonic components of natural origin in the currents of the protected connections at faults at various points.

Consider the earth fault at point S1 (Fig. 3). In this case on the substation 1 it is the higher zero sequence current and hence the higher level of the higher harmonic components (HHC), so it current will flow along the W1 line, it will be equal to the sum of the capacitive currents and HHC W2 and W3, while the earth fault protection device KA1 will operate. At the same time, the higher zero sequence current and the HHC will flow through the supply connection W1 (it will be equal to the sum of the capacitive currents and the HHC W11, W12, and W13) at the distribution point DP1, which will trigger the KA10 device. Thus, the operation of the protection device installed on the supply connection (KA10) indicates that the earth fault with respect to distribution point DP1 is external (located in the external network), and the protection KA1 tripping on Subst. 1 signals about detection of earth fault point on the protected W1 feeder. If it are earth faults at points S2 and S3 uses a similar algorithm.

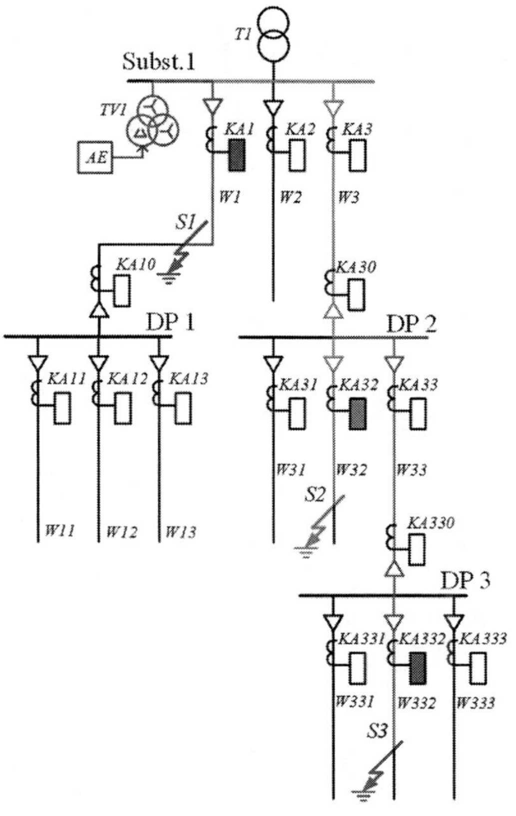

a)

	S1	S2	S3
F_{KA1}	1	0	0
F_{KA2}	0	0	0
F_{KA3}	0	1	1
F_{KA10}	1	1	1
F_{KA11}	0	0	0
F_{KA12}	0	0	0
F_{KA13}	0	0	0
F_{KA30}	1	0	0
F_{KA31}	0	0	0
F_{KA32}	0	1	0
F_{KA33}	0	0	1
F_{KA330}	1	1	0
F_{KA331}	0	0	0
F_{KA332}	0	0	1
F_{KA333}	0	0	0

b)

Fig. 3. The scheme for designing a earth fault protection system in the presence of distribution points DP: *a*) a network scheme with possible earth fault points, *b*) table of operations of the Renome-04T devices in the event of earth faults at points S1-S3.

Thus if the devices are triggered with a signal and the signal is transmitted from all devices to the duty personnel, it is possible to trace the direction of the protection from the buses to the line and unambiguously identify the damaged line.

In the case of equipping the network shown in Fig. 3,*a* by the unit for changing the load parameters on the windings of the voltage transformer connected to an open Delta (block AE on Figure 3) we can to increase the sensitivity and selection. So it is not necessary to install additional measuring zero sequence current transformer to ensure the selective action of the protection against earth fault. It is enough to agree the time of action of all protections on a step-by-step basis with the minimum time at the furthest away from the source of the superimposed protection signal.

II. Conclusions

If the earth fault protection system is implemented by one of the above methods, it becomes possible to organize a timely selective detection of the earth fault on the buses of the complete switchgear, which often turn into arc short circuits and then turn off the entire busbar section. When performing protection against earth fault is on a modern microprocessor-based basis, the implementation of algorithms for obtaining the sum of the currents of the protected connections as well as the action of protection taking into account the information received by the remote indication about the operation of protection devices on an adjacent substation is quite feasible and can be recommended for implementation in newly developed protections against earth fault.

The proposed method will make it possible to selectively trip the earth fault in the conditions of the constantly changing configuration of local electric grids (the length and number of lines).

References

[1] VM.S. Libkind, "Higher harmonics generated by transformers", Moscow, Publishing House of the USSR Academy of Sciences, 1959, p. 101.

[2] A.V. Ukraincev, G.N. Chmihalov, V.I. Nagay, S.V. Sarry, "Evaluation of information signs of protection against single-phase earth faults with relative measurement", Scientific and Technical Journal. Russian Electromechanics, vol. 4, 2011, pp. 129-132.

[3] A.V. Ukraincev, "Distributed earth fault protection system using the principle of relative measurement", Russian Electromechanics. Special Issue, 2010, pp. 71-72.

Distributed Simulation Models to Estimate Indices of Structural Reliability of Electric Power Systems

Andrey Grishkevich
Electrical Faculty
Czestochowa University of Technology
Czestochowa, Poland
a.grischkevich@el.pcz.czest.pl

Abstract—**The estimates of the contribution of failure states of an electric power system to the resulting indices of system reliability on the basis of the simulation model are dealt with. An algorithmic implementation of a distributed simulation model using the MPI library is proposed. The results of testing a distributed simulation model on a computing cluster are given. The laws of distributions different from exponential and the obtained distributions of duration of states are considered.**

***Keywords*: Monte Carlo methods; distributed simulation; MPI library; electric power system reliability; Markov's model**

I. Introduction

The calculation of the structural reliability of the power supply system [1] is based on the summation of the contribution of the various failure states of this system to the resulting reliability indices of the system. The contribution of states (for highly reliable systems, the simultaneous failures of one and two elements are usually considered) is calculated on the basis of models of the functioning of one and two elements of the system in terms of reliability.

Analytical methods for calculating reliability [2] use the Markov models of elements' functioning. The basis for these models is the assumption that the distributions of random variables are exponential. The distribution of some random variables in real systems, for example, the time of preventive repair, differs significantly from the exponential. At the same time, the maintenance discipline (methods of organizing and conducting repairs) has a significant influence on the distribution of a random variable, which varies very dynamically. Replacing the unknown distribution law by the exponential one leads to significant errors in results.

Accounting for uncertainties in the initial data has led to the development of methods for calculating the interval contribution of states to the resulting reliability indices [3]. However, taking into account the distribution law only by expanding the interval of initial data turns out to be too "rough".

The methods of mathematical modeling are very universal [4-6]. The construction of the simulation model is possible within the existing imitation environments, for example, in the Arena computer program [7]. However: 1) the graphical interface of the program, presenting significant opportunities for visualizing the process and ways of presenting the results of modeling, rather slows down the actual modeling process; 2) the specified package does not include tools for efficient parallel execution of modeling on the cluster. Achieving acceptable accuracy for the model of two elements requires extremely much time. Moreover, the description of the model in the Arena program, using its own way of specifying commands, does not give advantages to the description of the model in the high-level programming language, for example, C++. A computer program in the high-level programming language presents a wide range of opportunities for its modification and use of additional tools (program libraries).

The modeling algorithms are well developed, very simple and understandable. Such methods allow to take into account (to model) arbitrary laws of distribution of random variables. However, obtaining a satisfactory accuracy requires a large number of computational operations that are not achievable in an acceptable time on traditional computational systems with sequential organization of calculations. The time of computer modeling on a uniprocessor computer node with acceptable accuracy (a sufficiently large number of tests) is excessively long.

The distributed simulation models based on the universal parallel computing platform Star-P + MATLAB are ineffective [8]. The use of graphics accelerator is very effective [9], however, the cluster systems based on traditional processors are more accessible.

The modern software (MPI library [10]) allows connect the computing resources of several computers in a standard way, turning them into a system with parallel computing organization - a computing cluster, and changing the cluster size depending on the computing power requirements. Creating a simple cluster does not require the purchase of additional equipment, and is performed very quickly and simply on the basis of the existing computer network. If the processing power of the configured cluster is not enough, it is possible to use a more powerful cluster or a supercomputer. Currently, it is possible to access this technology via the Internet. The transformation of the program code of the modeling algorithm in a high-level language for working on a cluster requires the connection to an existing software project of a special library and the use of specific commands of this library. In this case, the program code is practically independent of the parameters of the cluster on which the calculations are performed. Thus, the computational efficiency is achieved, which ensures the practical applicability of the modeling methods.

The implementation of the simulation model in the C++ programming language using the MPI library allows for distributed modeling of the system behavior when using the capabilities of a computational cluster. The MPI library is a parallel programming library based on messaging. The calculation of the contribution of states to the resulting reliability indices of the system consists in the organization of an efficient exchange of messages (the results of intermediate calculations) between distributed processes of modeling.

II. Simulation Model of Functioning of One Element

Each element `E` of the electric power system (a transformer, a switch, ...) can be in one of four states (fig. 1) [2]. Suppose that `En (S2)` is the normal working state of the element, `Es (S4)` s the state between the failure of the element and the completion of the operative switching, `Er (S3)` is the emergency repair state of the element, `Em (S1)` – is the state of preventive maintenance (maintenance state) of an element; `tExy` – is the random transition time of the `E` element from the `Ex` state to the `Ey` state. The transitions between the states of one element in Fig. 1 are described by variables

```
double tInm, tInr, tIns, tIsr, tIrn;
```

in notation C++.

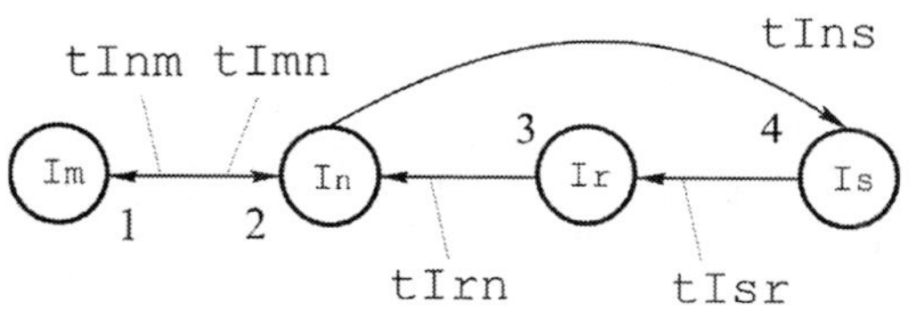

Figure 1. Diagram of transitions between the states for one element.

The total time `TEx (TSi)` spend in the `Ex (Si)` state is used as the statistics `S`; `NExy` – the total number of transitions from the `Ex` state to the `Ey` state. The collection of statistics for one element in Fig. 1 is described by variables

```
double TIm, TIn, TIr, TIs, NIns, NInm;
```

in notation C++.

The modeling process consists in generating a sequence of states of the system, starting from the initial `En` state. For each state, a random variable is generated that describes the output from this state, and the total residence time in the corresponding state `TSi`, `i=1, 2, ..., 4` is corrected for a random time in this state. The simulation process ends when `TIn > TIn_model`.

A fragment of the code of the simulation program for the transition between states for the model of one element in the C++ programming language is presented below

```
while(TIn<TIn_model)
{
if(tIns<=tInm)
   {
//In->Is->Ir->In
tInm = tInm - tIns;
StatSost(TIn, tIns);
tIns=SV_tIns();
//Is->Ir
StatSost(TIs, tIsr);
tIsr=SV_tIsr();//Ir->In
StatSost(TIr, tIrn);
tIrn=SV_tIrn();
StatSost(NIns, 1);
   }
else
   {
//In->Im->In
tIns = tIns - tInm;
StatSost(TIn, tInm);
tInm=SV_tInm();
//Im->In
StatSost(TIm, tImn);
tImn=SV_tImn();
StatSost(NInm, 1);
   }
}
```

where `SV_tExy()` is the procedure for generating a random variable `tExy`, `StatSost(S, p)` is the modification of the statistics `S` for the parameter `p`. As a procedure for collecting statistics, for example, the following procedure may be used

```
int StatSost(double &TSi, double tExy)
{
TSi+=tExy;
return 0;
}.
```

The total simulation time (model time)

$$TI = TIm + TIn + TIr + TIs$$

when the simulation process is sequentially organized, is described as the total time spent in the states `Si`, `i=1,2,...,4`.

The execution of some criterion (reaching the time `TIn` of the value of the `TIn_model`) in the state `In` leads to the end of the modeling process and the calculation of reliability indices. Accordingly, the variable

$$\overline{P}Si = TSi \; / \; TI$$

is the estimate of the probability of the system being in the state `Si` (an estimate of the state of contribution `Si` in the resulting reliability indices), `i=1, 2, ..., 4`, and

$$\overline{T}Inm0 = TIn \; / \; NInm,$$
$$\overline{T}Imn0 = TIm \; / \; NInm,$$
$$\overline{T}Ins0 = TIn \; / \; NIns,$$
$$\overline{T}Isr0 = TIs \; / \; NIns,$$
$$\overline{T}Irn0 = TIr \; / \; NIns,$$

the estimate of the average time of transition between states from the results of modeling.

III. Simulation Model of Functioning of Two Elements

The state of the system is determined by the state of each element of the system.

The state space and the transition diagram between states in the reliability simulation of functioning of two elements `I`, `K` are shown in Fig. 2 [2]. For convenience, the states are numbered, i.e. `IsKn=S4`. In total, 15 states are considered for the model of two elements.

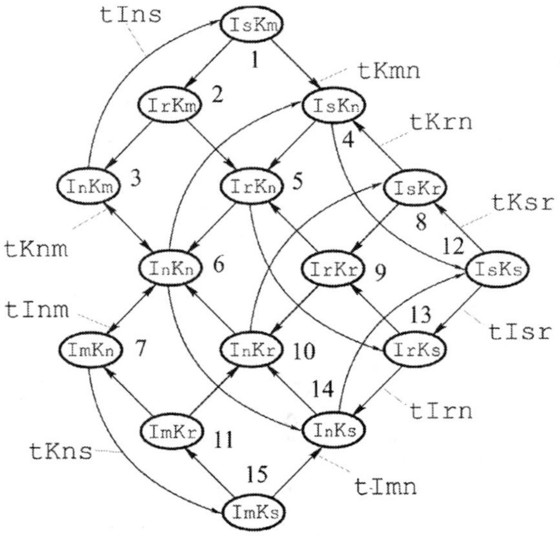

Figure 2. Diagram of transitions between the states for two elements.

For the state S4=IsKn the transitions to the state S5=IrKn through a random time interval tIsr or to the state S12=IsKs through a random time interval tKns are possible. The code fragment of the state transition simulation program for the two-element model in the C++ programming language as applied to the IsKn state is presented below

```
S4_IsKn:
if (tIsr-tIsr>=0&&tKns-tIsr>=0)
   {//IsKn->IrKn
tKns = tKns - tIsr;
StatSost(TIsKn, tIsr);
tIsr=SV_tIsr();
goto S5_IrKn;
   }
if (tIsr-tKns>=0&&tKns-tKns>=0)
   {//IsKn->IsKs
tIsr = tIsr - tKns;
StatSost(TIsKn, tKns);
tIsr=SV_tIsr();
goto S12_IsKs;
   }

S6_InKn:
if(TInKn>TInKn_model) end_of_modeling;.
```

IV. IMPLEMENTATION AND TESTING OF THE SIMULATION MODEL

Testing of the proposed simulation model was carried out under the assumption of the exponentiality of the laws of distribution of transitions between the states of the system in Fig. 1, 2. This made it possible to obtain exact analytical solutions, and on the basis of comparing the simulation results and the exact solution for the steady state probabilities, to choose the simulation interval.

The test numerical values of reliability of elements I, K of an electric power system [2]: TIns0=1/0.01 [rok] is the mean time of transition of the element I from the In state to the Is state; TInm0=1/2.2 [rok]; TIsr0=2/8760 [rok]; TIrn0=11.39/8760 [rok]; TImn0=7.96/8760 [rok]; TKns0=1/0.04 [rok]; TKnm0=1 [rok]; TKsr0=2/8760 [rok]; TKrn0=2.19/8760 [rok]; TKmn0=7/8760 [rok]. The presented numerical values during testing were interpreted as parameters of the exponential distribution of the corresponding random variable.

The pseudo-random numbers are generated based on a random number library (defined in the header file <random> C++11) [11]. The random number library provides a set of pseudorandom number generators (in particular, mt19937_64 - 64-bit Mersenne twister) that generate integer sequences with uniform distribution, and a set of random number distributions (in particular, exponential_distribution, lognormal_distribution), which transform uniform distributions of pseudo-random numbers into different statistical distributions. The process of generating the real values SV_tExy() in accordance with the exponential distribution is described by the code

```
#include <random>
...
std::mt19937_64 GEN_tExy(the number defining
the pseudo-random sequence);
std::exponential_distribution <double>
DIST_tExy(1/TExy0);
auto SV_tExy = std::bind(DIST_tExy,
GEN_tExy);
```

where 1/TExy0 is the parameter of the exponential distribution of the random variable tExy.

The installation of DeinoMPI 2.0 [10] allowed creating a 5-node cluster with 4-core processors Core(TM) i5-4570 CPU @ 3.20GHz on the local network (Fig. 3)

Based on the computer program [12], developed using the MPI library [10], 20 parallel processes are started at 5 compute nodes (the node includes 2 real and 2 virtual cores) of the network (cluster). That provides a maximum 20-fold acceleration of calculations.

In order to increase the thread safety in the uniform distribution generator, the process number is used to increase the entropy

```
#include <mpi.h>
...
int myid;
MPI_Comm_rank(MPI_COMM_WORLD, &myid);
std::mt19937_64 GEN_tExy(the number defining
the pseudo-random sequence + myid);.
```

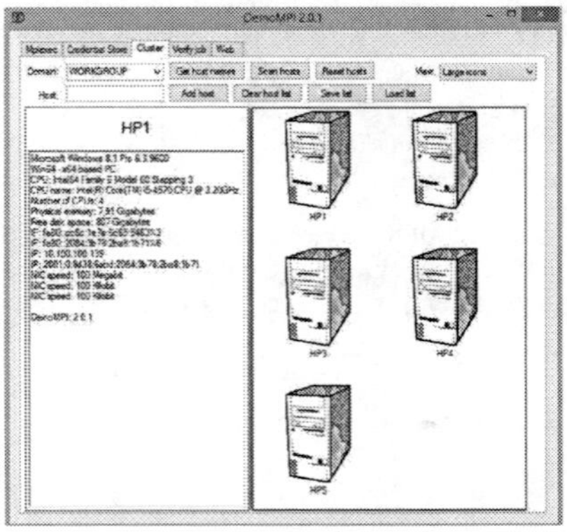

Figure 3. Computing cluster.

The greatest relative error reliability of states at the end of the simulation interval for the model of one and two elements is shown in Fig. 4, 5 respectively.

One elementary step of modeling approximately 9.07E8 (6.17E8) years for the model of one (two) elements is performed in 7.5 (9.5) seconds on the cluster of the above-mentioned configuration. The simulation of the 2.96E13 time interval (5.46E12) years required 64.5 (22.1) hours of operation of the used computational cluster.

As the value of the modeling interval increases, the relative error tends to decrease. The greatest relative error at the end of the simulation interval for the `IrKs` state of the two-element model is 0.05%. From the graph of Fig. 5 that for a simulation interval of more than 5E11 years the greatest relative error is less than 1%. This accuracy is achieved in 2 hours of operation of the computational cluster used, respectively. Simulation of one element on the specified interval of operation requires 1 hour of operation of the computational cluster used.

The calculation of the contribution of failure states to the resulting reliability indices by mathematical modeling is recommended to be carried out at the simulation interval `TIn_model` greater than 5E11 (E12) years. This allows to provide sufficient accuracy (1%) for an acceptable time (1-5 hours of operation of the cluster of the specified configuration).

The use of simulation methods in reliability calculations makes it possible to obtain a distribution of the duration of failures of states and to use distribution laws other than exponential ones.

V. Distribution of Duration of States

The numerical values of reliability indices of states (probability, average duration) obtained in the analytical calculations do not give complete information about the variability of the reliability indices. The distribution of reliability indices contains information about rare events (for example, long-term failures) that have very serious systemic consequences and is important for commercial customers with non-linear cost functions [13].

The change in the scalar statistics of `S` by the vector

```
#define INTERVAL 300
double S[INTERVAL];
```

and changing the procedure for collecting statistics

```
int StatSost(double *TSi, double tExy)
    {
if(tExy*8760.>=(double)INTERVAL)
TSi[INTERVAL-1]+=tExy;
else TSi[(int)(tExy*8760.)]+=tExy;
return 0;
    }
```

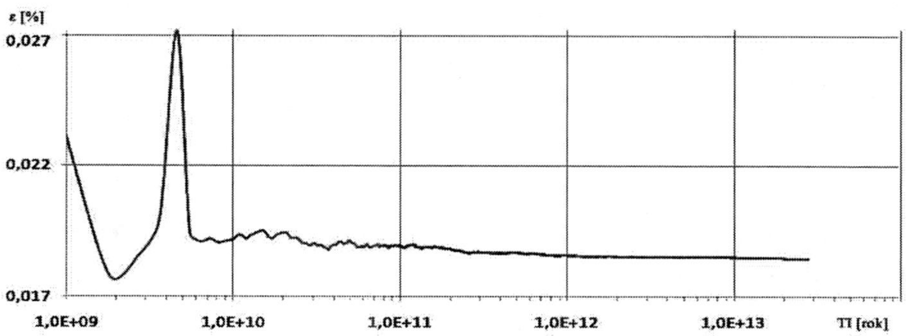

Figure 4. The greatest relative error reliability of states at the end of the modeling interval for model of one element.

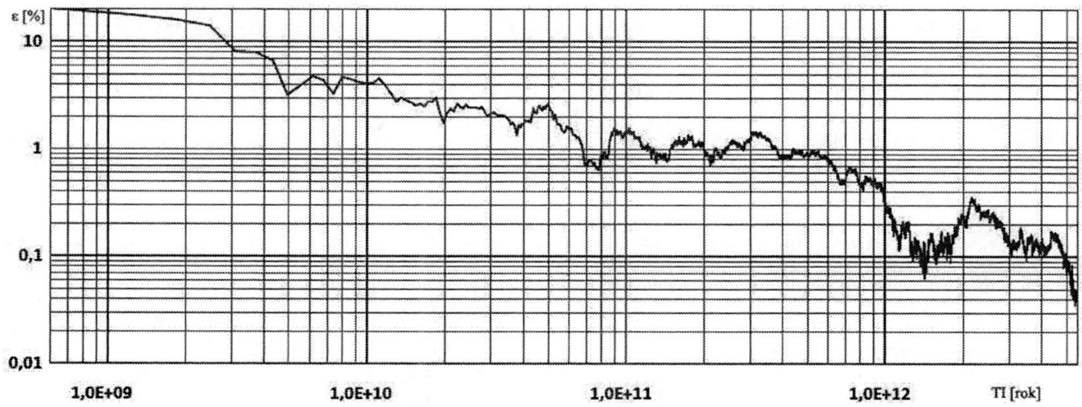

Figure 5. The greatest relative error reliability of states at the end of the modeling interval for model of two elements.

allows to investigate the intervals of the duration of the states [0; 1), [1; 2), ..., [299, 300) hours (the collection of statistics for other intervals requires a corresponding change in the algorithm). If `S[INTERVAL-1] >> S[INTERVAL-2]`, the variable `INTERVAL` must be increased.

Fig. 6 shows the simulation results of the duration of the cumulative state `ImKr+IrKr+IrKm` for the test exponential distribution of the reliability of the elements at `TIn_model=E12` (4.5 hours of operation of the cluster of the specified configuration). Such information is very useful, for example, when it is necessary to estimate the probability that the duration of a state exceeds a certain predetermined threshold value.

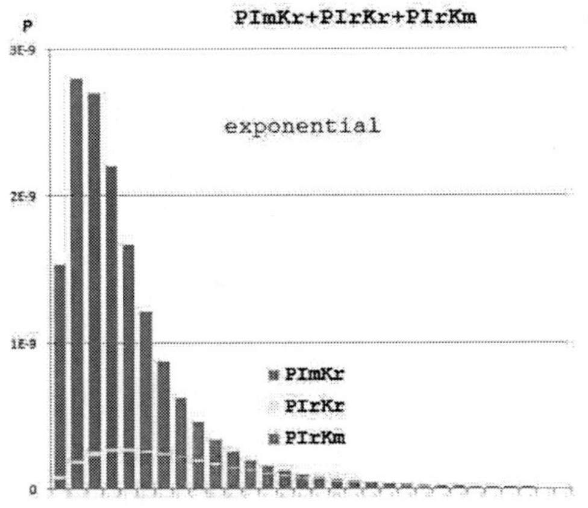

Figure 6. Distribution of duxration of states.

For the cumulative state `ImKr+IrKr+IrKm`, the average duration of one state is 1.95 hours, and for the `IrKm` state the largest single duration in the interval [69, 70) hours is fixed. The layout of Fig. 6 differs from the exponential distribution with a similar parameter. Analytic calculations of average values with subsequent interpretation of them as parameters of exponential distributions lead to errors.

VI. UNEXPONENT DISTRIBUTIONS

The simulation methods allow to perform reliability calculations for arbitrary distributions of random variables. This is an essential advantage of modeling methods, since the real distributions of transitions between system states differ from exponential ones [14].

For the description of emergency and preventive repairs of the elements in Fig. 1, 2, the logarithmic normal distribution law is chosen [11]

```
std::lognormal_distribution <double>
DIST_tExy (m, s);
```

with the `m`, `s` parameters. The parameters of the log-normal distribution are solutions of the non-linear system of equations

```
exp(m+(1/2)*s^2) = TExy0,
exp(s^2+2*m)*(exp(s^2)-1) = (TExy0/4)^2,
```

where the expressions on the left side represent the mathematical expectation and variance of the indicated distribution, respectively. For test numerical indicators ofreliability of the `I`, `K` elements of the electric power system [2]

```
DIST_tIsr(-8.41511631427871,
0.246220677069240);
DIST_tIrn(-6.67552771737957,
0.246220677069241);
DIST_tImn(-7.03383449498236,
0.246220677069240);
DIST_tKsr(-8.41511631427871,
0.246220677069240);
DIST_tKrn(-8.32436195101024,
0.246220677069240);
DIST_tKmn(-7.16235334578334,
0.246220677069246);.
```

The approach used made it possible to compare the simulation results for the exponential and logarithmically normal time distributions of maintenance. The results of simulation of the `PIm` state duration for a single element model at `TIn_model=E12` (2.7 and 5.7 hours of cluster operation for the exponential and log-normal maintenance time distributions, respectively) are shown in Fig. 7.

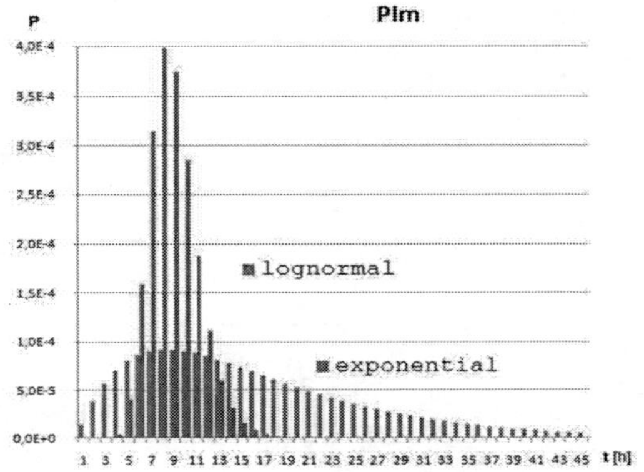

Figure 7. Probabilities of different durations of the `Pim` state.

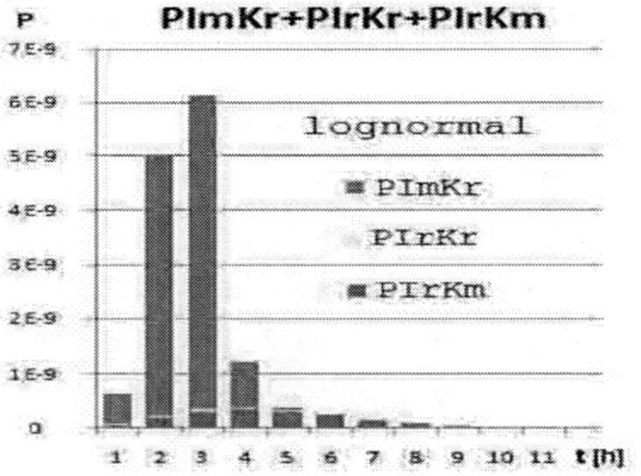

Figure 8. The greatest relative error of states at the end of the modeling interval for the model of 2 elements.

Fig. 8 shows the results of modeling the duration of the cumulative state `ImKr+IrKr+IrKm` for the test log-normal distribution of the emergency and preventive repair time of the elements at `TIn_model=E12` (8.7 hours of operation of the cluster of the specified configuration). For the cumulative state `ImKr+IrKr+IrKm`, the average duration of one state is 1.86 hours. In this case, for the state `IrKm`, the largest single duration in the interval [18; 19) hours is fixed.p

The law of log-normal distribution allows to obtain a more compressed interval of the simulated quantity, which more adequately describes the state of emergency and preventive repairs of the elements.

VII. Conclusion

A method for calculating the structural reliability of complex systems is proposed, in which the determination of the contribution of failure states to the resulting reliability indices is made on the basis of simulation models of functioning of the elements of the system. The results of testing the distributed simulation model are presented, which allow estimating the required parameters of a computational cluster for carrying out simulation calculations of the reliability of the electric power system with sufficient accuracy within the acceptable time.

The use of simulation models allows to take into account the laws of distribution of random variables, other than exponential ones. In particular, the ways of carrying out emergency and preventive repairs of the system elements. This ensures the adaptation of the method of calculating reliability for a specific power supply system. The distribution of the duration of states obtained on the basis of the simulation models makes it possible to calculate the reliability indices of the system on the basis of distributions characterizing the particular system being studied, rather than on the basis of the mean values under the assumption of exponential distribution.

Computer clusters have sufficient processing power to provide acceptable modeling accuracy in a reasonable amount of time. The wide availability and low price of cluster systems make it possible to practically use the methods of distributed simulation in calculating the structural reliability of complex electric power systems.

References

[1] A. Grishkevich and Ł. Piątek, "Combinatorial method of estimating the reliability of complex electric power systems," Proceedings of the IV International Scientific Symposium ELEKTROENERGETIKA 2007, Slovak Republic, Stara Lesna: Technical University of Kosice, pp. 664-674.

[2] A. Grishkevich, V. I. Hudym, A. M. Kruczinin, and A. Sawicki, Zagadnienia energetyczne wybranych współczesnych urządzeń i systemów elektrostalowniczych, Polska, Częstochowa: Wydawnictwo Politechniki Częstochowskiej, 2010.

[3] A. Grishkevich, and A. Burmutaev, "Modelowanie statystyczne oszacowań interwałowych wskaźników niezawodności strukturalnej układów elektrycznych," Przegląd Elektrotechniczny (Electrical Review), vol. R. 88(2012), nr. 8, pp. 77-79.

[4] T. Prauzner, "Interactive computer simulation as a response to contemporary problems of technical education," Society. Integration. Education, Proceedings of the International Scientific Conference, Volume II, Latvia: Rēzeknes Tehnoloģiju akadēmija, 2016, pp.579-588, doi: http://dx.doi.org/10.17770/sie2016vol2.1415.

[5] L. Borowik, and A. Sawicki, "Selected aspects of modeling the electric arc column in the gas at reduced pressure," High Temperature Materials and Processes, vol. 16(2012), no 4, pp. 275–296.

[6] L. Marciniak, "Modeling of Digital Earth Fault Protection Using PSCAD," Przegląd Elektrotechniczny (Electrical Review), vol. R. 85(2009), nr. 3, pp. 192-195.

[7] Arena Simulation Software. [online] Available: (https://www.arenasimulation.com/) last see (2017, May)

[8] H. Ge, and S. Asgarpoor, "Parallel Monte Carlo simulation for reliability and cost evaluation of equipment and systems," Electric Power Systems Research, 81 (2011), pp. 347–356.

[9] A. Grishkevich, "Modele symulacyjne do szacowania wskaźników niezawodności strukturalnej systemów elektroenergetycznych tworzone na bazie kart graficznych," Przegląd Elektrotechniczny (Electrical Review), vol. R. 93(2017), nr. 3, pp. 44-47, doi:10.15199/48.2017.03.11.

[10] DeinoMPI. [online] Available: (http://mpi.deino.net/), last see (2017, May)

[11] <random> [online] Available: (https://msdn.microsoft.com/en-us/library/bb982398.aspx), last see (2017, May)

[12] A. Grishkevich, "Rozproszone modele symulacyjne pozwalające oszacować wskaźniki niezawodności strukturalnej systemów elektroenergetycznych," Przegląd Elektrotechniczny (Electrical Review), vol. R. 91(2015), nr. 12, pp.106-109, doi:10.15199/48.2015.12.26.

[13] P. Wang, and L. Goel, "Power Distribution System Reliability Evaluation Using Both Analytical Reliability Network Equivalent Technique and Time-sequential Simulation Approach," in J. Faulin, A. A. Juan, S. Martorell, and J.-E. Ramírez-Márquez (Editors), Simulation Methods for Reliability and Availability of Complex Systems, London, Dordrecht, Heidelberg, New York: Springer, 2010, pp.145-173, doi:10.1007/978-1-84882-213-9.

[14] C. J. Zapata, L. P. Garcés, and O. Gómez, "Reliability Assessment of Energy Limited Systems Using Sequential Montecarlo Simulation," 2006 IEEE/PES Transmission and Distribution Conference and Exposition, Latin America: Venezuela, pp. 1-6, doi: 10.1109/TDCLA.2006.311443.

Danger of Re-Ingition of an Electric Arc in a Resonant Earthing Medium Voltage System

Stanislav Misak, Lukas Prokop
Enet Centre
VSB – Technical University of Ostrava
Ostrava, Czech Republic
{stanislav.misak, lukas.prokop}@ vsb.cz

Abstract —**There are medium voltage networks which have a small extent from the point of capacity current (around 20 Amps). These systems can be operated as isolated neutral systems, resistor earthing systems or resonance earthing systems in numerous other modifications for such an extent of capacity current. The option which is currently enforced is one based on the resonance earthing system. The aim of this paper is to point out the danger of an arc earthing fault which can arise in resonance earthing medium voltage systems, particularly for the concrete specific extent of a medium voltage system and the corresponding capacity current in the areas of (5-20) Amps. For this specific system extent and earthing method there can be multiple re-ignitions during a fault current half-period and this makes it impossible to detect such a fault using (conventional) digital relay protection which works on the principle of fault current individual parts evaluation. This can only occur in the case that a fault arises.**

Keywords: Single phase arcing earth fault; medium voltage distribution systems; resonance earthing system; isolated neutral systems

I. Introduction

The current trend of research and development in the area of electro power engineering is the development of so-called SMART GRIDS systems. From the point of extent, these are fairly small extent systems, unlike the conventional distribution systems, with a circuit order and consequently with the possibility of increasing sides supply. The basic parameter and characteristic which designers of such systems have to define, is apart from other things, the means of a neutral point earthing system. This means of system neutral point earthing also defines the protection conception and the corresponding means of fault or abnormal situations elimination which can arise in a concrete system. There are currently several basic earthing methods and the medium voltage (furthermore MV) systems can be based on these earthing methods generally divided into [2]: isolated neutral systems and impedance earthing systems. The impedance earthing systems are further divided into: (i) resistor earthing systems and (ii) resonance earthing systems. Each system method of neutral point earthing is specific and suitable for a definite system. We cannot say which one of earthing method is the most favorable for all types and the decision concerning system use suitability or changing a system neutral earthing method can be made on the condition that important factors are taken into consideration. The most important ones are: (i) fault current, (ii) fault overvoltage, (iii) overvoltage during earth fault interruption, (iv) fault elimination methods, (v) comparative investment or operating costs, and mainly (vi) safety factors. The first basic parameter (i) is the system extent, quantified by the capacity current value. Simply said, the capacity current value (hereinafter I_C) is decisive during the first selection of a concrete MV system earthing method. As mentioned above, in the case of SMART GRIDs systems it is about small extent energy systems with a IC of approximately (1-20) Amps. In classical MV systems this extent is equivalent to a power-plant operation, for example, of an own block consumption, eventually a mining system operation. All three basic earthing methods can be chosen. These kind of systems can be operated as isolated neutral systems or earthing systems. For an I_C extent of approximately 10 Amps the option of an isolated neutral point system or resonance earthing system is often used where the option with a resonance earthing system has recently been preferred [1, 2, 3]. This is valid for classic MV systems. The aim of this article is to point out possible dangerous overvoltage processes which can arise in the last mentioned earthing method if there is the most frequent fault asymmetry system, the so-called arc earth fault. This is an extremely dangerous fault which is for certain cases, despite f all the progress and modernization of protection conception elements, particularly difficult to detect, especially for the specific system extent and the value of the I_C in the new built MV systems of SMART GRIDs. We will specifically point out the dangerous accompanying effects of the arc earth fault, such as overvoltage, re-ignitions of arc discharge, etc. These accompanying effects will simultaneously be compared when arising in an isolated neutral system and resonance earthing system with the use of an analytical calculation and experimental measuring in an actual MV operation. The solution of this problem was initiated on the basis of a requirement formulated by a keeper of an actual MV system with the capacity extent of (5-20) Amps operated as a resonance earthing system. This system suffered frequent faults in the cable lines isolation system; there were specifically cable terminations puncturing. After an analysis of the cable terminations isolation system there was a detection of the isolation system destruction as a result of excessive electrical strength caused by overvoltage. Over the long-term time frame it analyzed the voltage time course and created a detailed record of the course at the moment of voltage values exceeding the determined limits. Several tens of determined limits were made exceeding records and the cable line isolation system of the 6-AYKCY type degraded by the arcing earth fault within the long - term measuring (about one month) was finally found. Despite the fact that the MV system was protected by a developed modern technique [3], the fault was not detected. After a detailed problem analysis of other MV system keepers with this I_C extent it was determined that this was not a rare case [4, 6]. The aim of the article is to clarify WHY the fault was not detected and suggest simple measures as to how to eliminate this type of fault.

II. An analysis of the system neural point earthing method

The following chapter clarifies the problem of the MV system of single phase earth faults. In the case of these faults it is the most frequent one and from the point of device dimensioning and protection system adjusting the most important asymmetry case caused by isolation resistance lowering in one phase of a system place. In an isolated neutral system or resonance earthing system this can be shown as earthing, while in resistor earthing networks this can be shown as a single phase short circuit [8]. The above-mentioned fault accompanying effects are overvoltage processes which negatively affect the isolation system condition of all the elements in the analyzed system. This is an extremely dangerous fault which despite all development in the protection area, is extremely poorly detected [2],[4],[5] for certain specific situations. The aim of the article is to specifically point out the problematic fault detection in the case of the single phase which means the fault with time resistance change. The next chapter initially the stabilized condition of the single phase earth fault and consequently analyses the temporary component of the fault and system recovery after the fault. Figure 1 schematically shows a general MV system where the earthing method is expressed by the general impedance $\hat{Z}_N$ in the MV system neutral. Phasors $\hat{U}_A$, $\hat{U}_B$ and $\hat{U}_C$ represent the electromotoric voltage power source $\hat{U}_{12}$, $\hat{U}_{23}$, $\hat{U}_{31}$ voltage on interphase capacities and corresponding current phasors. The nodes marked 1, 2, 3 represent the point for the asymmetrical transverse cell connection with the transverse asymmetry with voltage phasors against the earth in the asymmetry point ($\hat{U}_1$, $\hat{U}_2$, $\hat{U}_3$), or voltage on the circuit defining parallel connection of the capacity systems and conductivity. In order to solve ratios in a system through the symmetrical components method the following simplifying initial conditions can be accepted: there is a symmetrical source where voltage $\hat{U}$, interphase capacities are neglected, and the capacities of system individual phases against the earth are equal supposing symmetrical systems. During an earth fault, (see Figure 1), the symmetry impedances against the earth are destroyed. A fault current $\hat{I}_P$ begins to flow through the fault point with the resistance R_P, there will be a change of amplitudes size and I_C phase shifts in the individual phases and node voltage against the earth.

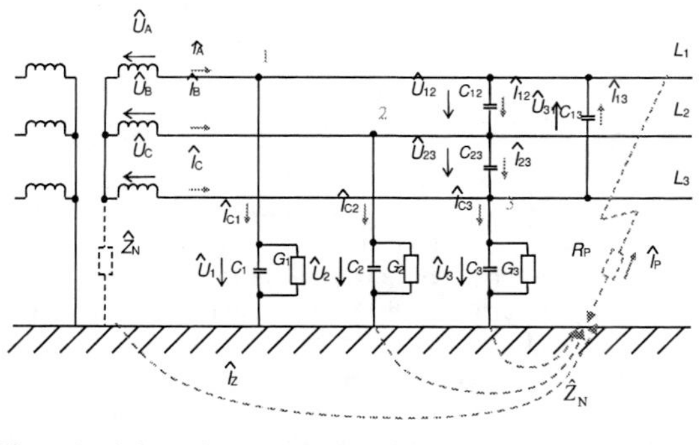

Figure 1. Schematic representation of medium voltage systems with general impedance in the moment of single phase earth fault origin.

For a mathematical description of voltage and current ratios we can work with the substitute The sequence impedance component. $\hat{Z}_{(1)}$ includes wiring and transformer sequent impedance components along with the impedance reverse component $\hat{Z}_{(2)}$ which includes the wiring and transformer reverse impedance components and the impedance zero component $\hat{Z}_{(0)}$. In the case of a detailed formulation it would by necessary to respect contributive external system impedance and generator impedance. These can be reglected. Figure 2 describes the system $\hat{Z}_{(0)}$, where $3\hat{Z}_N$ is general impedance in the system node, and $\hat{Z}_{(0)V}$ are zero transformer and wiring components in a set order. Impedance $\hat{Z}_C$ represents a total system capacity in the case of system leakage neglecting.

For the voltage at the asymmetry point against the earth we can write relations (hereafter r.) (1), (2), (3) according to the Figure 1:

$$\hat{U}_1 = \hat{U}_{(1)} + \hat{U}_{(2)} + \hat{U}_{(0)} = \frac{3R_P}{3R_P + 2\hat{Z}_{(1)} + \hat{Z}_{(0)}}\hat{U} \quad (1)$$

$$\hat{U}_2 = a^2\hat{U}_{(1)} + a\hat{U}_{(2)} + \hat{U}_{(0)} = \left(a^2 - \frac{\hat{Z}_{(0)} - \hat{Z}_{(1)}}{3R_P + 2\hat{Z}_{(1)} + \hat{Z}_{(0)}}\right)\hat{U} \quad (2)$$

$$\hat{U}_3 = a\hat{U}_{(1)} + a^2\hat{U}_{(2)} + \hat{U}_{(0)} = \left(a - \frac{\hat{Z}_{(0)} - \hat{Z}_{(1)}}{3R_P + 2\hat{Z}_{(1)} + \hat{Z}_{(0)}}\right)\hat{U} \quad (3)$$

In actuality the phase current is a fault current with a reversed flow direction. We can use the following ratio:

$$\hat{I}_P = -\hat{I}_A = -3\hat{I}_{(1)} = \frac{3\hat{U}}{2\hat{Z}_{(1)} + \hat{Z}_{(0)} + 3R_P} \quad (4)$$

The node voltage against the earth at the source point is identical to the zero component voltage by the scheme in (Figure 2) and can be written in the following form:

$$\hat{U}_0 = \frac{3\hat{Z}_N\hat{Z}_C}{\left(2\hat{Z}_{(1)} + \hat{Z}_{(0)} + 3R_P\right)\left(3\hat{Z}_N + \hat{Z}_C\right)}\hat{U} \quad (5)$$

The node voltage against the earth at the fault point has the following size:

$$\hat{U}_0' = \frac{\hat{Z}_{(0)} - \hat{Z}_{(1)}}{3R_P + 2\hat{Z}_{(1)} + \hat{Z}_{(0)}}\hat{U} \quad (6)$$

The r. (1) up to the r. (6) determines the voltage and current ratios in the system for the general system node. On the basis of these ratios we can describe the system parameters for any transformer node operation system. In accordance with the above-mentioned ratios, the only aspect in which particular operation methods differ is the size and character of the network node impedance which influences the zero component.

A. *An analysis of the single phase earth fault stabilized condition in isolated neutral systems*

The following: $\hat{Z}_N \rightarrow \infty$ is consequently valid for the scheme in (Figure 1) modified by the general impedance value in the system node in the case of an isolated neutral system operation [7]. For the system we can write the following for the $\hat{Z}_{(0)}$.

$$\hat{Z}_{(0)} = \hat{Z}_C + \hat{Z}_{(0)V} \quad (7)$$

If we consider the network extent with the IC of an ampere unit (of the order magnitude), the previous ratio can be simplified to:

$$\hat{Z}_{(0)i} \doteq \hat{Z}_C = \frac{1}{\frac{1}{R_C} + j\omega C} \doteq -\frac{j}{\omega C} \tag{8}$$

This means that the voltage ratios and current size do not depend on where the fault arises in the system. By completing the r. (8) into (4) for the fault zero resistance which means for the case of a metal single phase earth fault origin, we can obtain the current size in the earth fault point which represents the system I_C:

$$\hat{I}_{P(kap)} = j3\omega 3\,\hat{U} \tag{9}$$

The following formula is valid for the current moving through an against earth capacity:

$$\hat{I}^{[0]}{}_{C1} = \hat{Z}_C = \frac{\hat{U}^{[0]}{}_1}{-jX_C} \doteq j\frac{\hat{U}}{X_C} \tag{10}$$

The following is valid for the fault current:

$$\hat{I}_P = -\hat{I}_A = -3\hat{I}_{(1)} = -3j\frac{\hat{U}}{X_c} = -3\hat{I}^{[0]}{}_{C1} \tag{11}$$

The value of the I_C is one of the characteristic values of the system and the system node operation method is determined on the basis of this value. An isolated neutral system can be operated for a system configuration with a limited cables length. The entire capacity system current is flowing through the fault point. As the fault current has a capacity character, its extinguishing is difficult and interconnected with re-ignitions. Temporary processes which arise this way cause a network overvoltage. For the above-mentioned reasons the use of an isolated node is limited to small networks. The safety point of view, which is connected with the large step-voltages presence arising during earth currents passage, also needs to be considered. The I_C limit values and corresponding proposal of the MV system neutral point earthing method solve the regulations and norms, for example [6], of a MV system relevant keeper. According to [6] it is possible to operate, for instance, an isolated neutral system up until the 20 Amps of the I_C of the limit value where earth currents compensation is recommended from the 10 Amps value. Assuming an ideal metal earthing with $R_P = 0$, we can state that: (i) the fault phase I_C will increase 3 times, (ii) phases voltage to neutral without fault increases to the line-to-line value, (iii) node voltage increases to the line-to-ground value. From the point of frequency, this is a rare case assuming a metal single phase earth fault. Most single phase earth faults have the signs of a single phase. The fault resistance value has a significant effect on the character of the analysed quantities temporary processes (overvoltage and current impact component in the earth fault point) and vastly influences the voltage stationary condition in the asymmetry place and the current condition in the earth fault point.

B. *An analysis of the single phase earth fault in resonance earthing systems*

Compensation of capacity earth currents in MV networks must be carried out in order that only the so-called remaining current runs through the point of the random earth fault. The step-voltage in the fault surroundings can consequently be lowered and enable the affected wiring operation at the earth fault (the operation is limited by quenching coil allowed warming). The capacity currents compensation is created by obtaining a well-chosen inductance into the network node. This inductance is named according to the purpose "arc suppression" or "compensation" or by the theory author name "the Petersen coil". The isolated neutral network consequently changes into a network with an ineffectively or indirectly earthed node. For a resonance earthing system it is necessary to consider impedance in the transformer node to be an arc suppression coil impedance represented by the parallel connection of its L induction and loss resistance R_t. For the impedance zero component reads:

$$\hat{Z}_{(0)komp} \doteq \frac{1}{\frac{1}{R_C} + \frac{1}{3R_L} + j\left(\omega C - \frac{1}{3\omega L}\right)} \tag{12}$$

The fault current can be expressed by completing the r. (12) into the equation (4), see r. (13):

$$\hat{I}_P = \frac{3\hat{U}}{2\hat{Z}_{(1)} + \frac{1}{\frac{1}{R_C} + \frac{1}{3R_L} + j\left(\omega C - \frac{1}{3\omega\omega}\right)} + \hat{Z}_{(0)V} + 3R_P} \tag{13}$$

The r. (13) is derived for the case of the capacity asymmetry neglecting $1/R_P >> \omega C_{nes}$. The final fault current has an imaginary component and an uncompensated active component. The imaginary component is determined by the difference between the network I_C and the arc suppression coil inductive current. The uncompensated active component correspondents to the arc suppression coil losses and phases lead resistances against the earth.

$$\hat{I}_P = \hat{I}_R + j(\hat{I}_C - \hat{I}_L)\,\left|\hat{I}_P\right| = \min \Leftrightarrow \left|\hat{I}_C\right| = \left|\hat{I}_L\right| \tag{14}$$

This ratio describes the arc suppression coil basic function. The fault current has a minimal size for the resonance condition and is set by the network lead resistances and the coil losses. Another aspect which should be considered is the current of the non-compensated higher harmonious components whose size is not negligible in this case.

C. *The dynamic impacts of the single phase earth fault*

The previous chapter mathematically analysed the accompanying processes of the single phase earth fault for any system neutral earthing method under a stabilized condition. This was a stabilized condition analysis and indicates the condition after temporary processes end where these processes accompany the origin, course and termination of single phase earth faults. The majority of earth faults have a dynamic character. This is a result of fault resistance time change due to electrical arc burning [7] ,[9]. During the earth fault mentioned type, the isolation system can be stressed during one period of electrical arc burning. The voltage and current ratios change more often at the asymmetry point and the isolation system is not as stressed by the overvoltage processes amplitude but more by the potential dynamic change in the isolation structure. The following sub-chapter looks at the voltage and current ratios nature during a single phase earth fault.

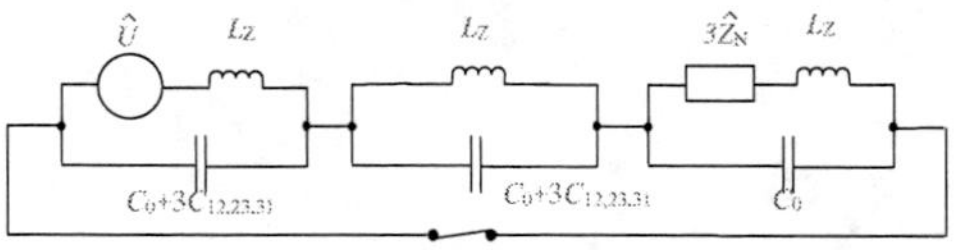

Figure 2. Substitute scheme with ascillation circuits for the overvoltage processes analysis.

The scheme for an analysis of overvoltage processes at the moment of the earth fault origin is indicated in (Figure 2), where the particular component impedances are component oscillation circuits. When an extreme case is considered, active resistances are neglected.

The sequent and reverse impedances are formed from the network short circuit inductance up to the earth fault point L_z and from the in-parallel connected network total capacity (consisting of the earth capacity C_0 and the triple mutual capacity among phases $C_{12,23,31}$). Zero impedance is formed from two parts. The first one is the inductance L_Z which is in the series along with the triple general impedance connected between the network node and the earth (e.g. the arc suppression coil induction) while the second one is the earth capacity C_0. The single phase earth fault origin is simulated by the switch connection. A temporary process consequently arises when the network phases symmetrical voltages before the fault transform into stabilized voltages, corresponding with the single phase earth fault described in the previous chapter. The mathematical formulation of the asymmetry point voltage condition and the fault current condition after the single phase arcing earth fault is possible on the basis of a circuit analysis, see Figure 2.The single phase arcing earth fault can generally be interrupted when either the 50 Hz frequency current component or the current temporary oscillation component moves through zero. Based on the assumed quenching mechanism the overvoltage origin theory differs. The following case presupposes an interruption in the 50 Hz stabilized zero current. The electrical turning off of the switch is substituted by a current injection with a reverse polarity than the single phase earth fault turned off current polarity. This injection is carried out between the switch disconnected contacts [11]. The circuit response is calculated and the arisen voltages (marked e, E) are superposed by voltages which existed in the circuit prior to the change, which means prior to the earth fault interruption. The voltages "before the change" are determined by the r. (1) up to (2). For the particular component impedances and the corresponding angle frequencies we write:

$$\hat{Z}_{(1)}(j\omega)=\hat{Z}_{(2)}(j\omega)=\frac{1}{C_0+3\cdot C_{12}}\cdot\frac{j\omega}{\omega_1^2-\omega}\qquad \omega_{(1)}^2=\frac{1}{L_Z\cdot(C_0+3\cdot C_{12})}\tag{15}$$

$$\hat{Z}_{(0)}(j\omega)=\frac{1}{C_0}\cdot\frac{j\omega}{\omega_0^2-\omega^2}\qquad \omega_{(0)}^2=\frac{1}{C_0\cdot(L_Z+3\cdot L_P)}\tag{16}$$

For the particular voltages at the asymmetry point prior to the earth fault interruption, the following scheme is valid in Figure 2.

$$\hat{U}_1=\frac{3\mathrm{R}_P}{3\mathrm{R}_P+2\left(\frac{1}{C_0+3\cdot C_{12}}\cdot\frac{j\omega}{\omega_1^2-\omega}\right)+\frac{1}{C_0}\cdot\frac{j\omega}{\omega_0^2-\omega^2}}\hat{U}$$

$$\hat{U}_2=\left(a^2-\frac{\hat{Z}_{(0)}-\hat{Z}_{(1)}}{3\mathrm{R}_P+2\left(\frac{1}{C_0+3\cdot C_{12}}\cdot\frac{j\omega}{\omega_1^2-\omega}\right)+\frac{1}{C_0}\cdot\frac{j\omega}{\omega_0^2-\omega^2}}\right)\hat{U}\tag{17}$$

$$\hat{U}_3=\left(a-\frac{\hat{Z}_{(0)}-\hat{Z}_{(1)}}{3\mathrm{R}_P+2\left(\frac{1}{C_0+3\cdot C_{12}}\cdot\frac{j\omega}{\omega_1^2-\omega}\right)+\frac{1}{C_0}\cdot\frac{j\omega}{\omega_0^2-\omega^2}}\right)\hat{U}$$

This formula $\omega_{(1)} >> \omega$, $\omega_{(0)} \approx \omega$ and thus $\hat{Z}_{(0)} >> \hat{Z}_{(1)}$ is valid in its own frequency for the resonance earthing systems, namely for the particular component circuits. When the switch in Figure 3 is turned on, the stabilized current moves through the circuit:

$$i_1(t)=\frac{U}{2\cdot Z_{(1)}(j\omega)+Z_{(0)}(j\omega)}\cdot\cos\omega t\tag{18}$$

where the influence of the electromotoric source immediate voltage is:

$$u(t)=U\cdot\cos\omega t$$

We can obtain an equation for the immediate current in the earth fault point, in a stabilized condition:

$$i_{(1)}(t)=\frac{U}{2\cdot L_Z\cdot\frac{\omega_1^2}{\omega_1^2-\omega^2}+(L_Z+3\cdot L_P)\cdot\frac{\omega_0^2}{\omega_0^2-\omega^2}}\cdot\cos\omega t\tag{19}$$

There are consequently defined two quantities: the asymmetry point voltage and the fault point current in the stabilized condition, the so-called "before the change", or before the single phase earth fault interruption and it is possible to derive formulas for the voltage and currents after the single phase earth fault interruption caused by the single phase arcing earth fault. The single phase earth fault interruption is simulated by a current injection with a reverse polarity between the switch contacts.

For the voltages after the single phase earth fault interruption is valid:

$$u_1(t)=\frac{3R_\mathrm{P}}{3R_\mathrm{P}+2\left(\frac{1}{C_0+3\cdot C_{12}}\cdot\frac{j\omega}{\omega_{(1)}^2-\omega}\right)+\frac{1}{\mathrm{C}_0}\cdot\frac{j\omega}{\omega_{(0)}^2-\omega^2}}\hat{U}+$$

$$+U\cdot\left(\cos\omega t-K\cos\omega_{(1)}t-(1-K)\cos\omega_{(0)}t\right)$$

(20)

$$u_2(t)=U\left[-\sqrt{3}\cos\left(\omega t+\frac{\pi}{6}\right)+(Q-P)\cos\omega t+P\cos\omega_{(1)}t-Q\cos\omega_{(0)}t\right]$$

$$u_3(t)=U\left[-\sqrt{3}\cos\left(\omega t-\frac{\pi}{6}\right)+(Q-P)\cos\omega t+P\cos\omega_{(1)}t-Q\cos\omega_{(0)}t\right]$$

Where

$$P=\frac{1}{N+2},\ Q=\frac{N}{N+2},\ Q-P\approx 1\quad N=\frac{L_Z+3L_P}{L_Z}\cdot\frac{\omega_{(0)}^2}{\omega_{(1)}^2}\cdot\frac{\omega_{(1)}^2-\omega^2}{\omega_{(0)}^2-\omega^2}$$

D. An analysis of overvoltage processes during an arcing earth fault in an actual MV operation.

The results from the experimental testing in an actual MV operation (power-plant block consumption) were used to verify mathematical formulations and a subsequent analysis of dangerous overvoltage processes for the system specific extent of the I_C (5-20) Amps. This MV testing system enabled two analyzed system earthing methods, namely the isolated neutral system and the resonance earthing system operation with a compensation possibility in the I_C (30-130)% extent. The aim of the MV testing system configuration design is a maximal approach to the MV system with a diagnosed fault of cable lines terminations. The MV testing system ran on a performance transformer of 110/6 kV, 40 MV·A in connection Dz1. The cable lines with a total charging current of 5,5 Amps and a I_C of 10 Amps is respected by the C_0 = 3,1 μF capacity with a length of approximately 3,6 km of the 6-AYKCY 240 mm^2 three-phase cable lines. The total serial component of the cable lines impedance for the length of 3,6 km has the value of (0,45 + j0,32) Ω. The impedance represents the general impedance which will actually be changed in the experiment by a selected method of system neutral point earthing.

The arcing earth fault were respected in the experiment by connecting one phase of the MV testing system to the earth using selected drive with an asynchronous engine via the sample of AXEKCY 75 mm^2 cable lines where the distance between the core and the cable lines grounding shielding is burnt. There was a simulated arcing earth fault, namely for a 5, 10, 15 and 20 Amps of I_C within the experimental measuring in a MV testing operation. The I_C extent of the MV testing system was altered when the outlets were gradually connected to the 6-AYKCY cable lines. Both methods of the system neutral point earthing were used for each value of the I_C. In the resonance earthing system, the level of the Petersen coil tuning was changed gradually by 20% of the inductive current. The experimental measuring conclusions can be the following:

1) In the case of an isolated testing system operation, it was a stabilized process for the entire extent of (5-20) Amps of I_C. As concerns the values of the non-fault phases voltage at the asymmetry point, this was also the case for the fault-phase voltage course and fault current value. The voltage ignition and the extinguishing peak on the arc discharge are almost identical and the electrical arc is naturally removed when moving through zero with the frequency of the power supply.

This is an electrical arc burning with a majority dominance of ionization processes above the re-combinational, in the area of the arc discharge. The electrical strength of the ionized space arc discharge increases according to the "E1" curve in Figure 3, with electrical puncturing and arc re-ignition. A detail of the arc voltage course in the case of an isolated testing system operation for the 10 Amps I_C is evident in Figure 4 which answers the length of a 5 mm arc discharge.

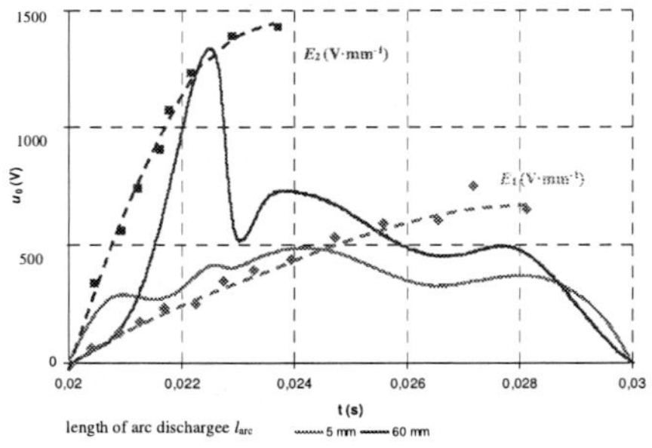

Figure 3. Arc voltage course in the area of electrical arc discharge for stable burning and burning stability limit of electrical arc.

The arch discharge detail for this case is presented in Figure 9. The sequence of the used pictures is at a time of 0 – 1, 2.5 and 5 ms.

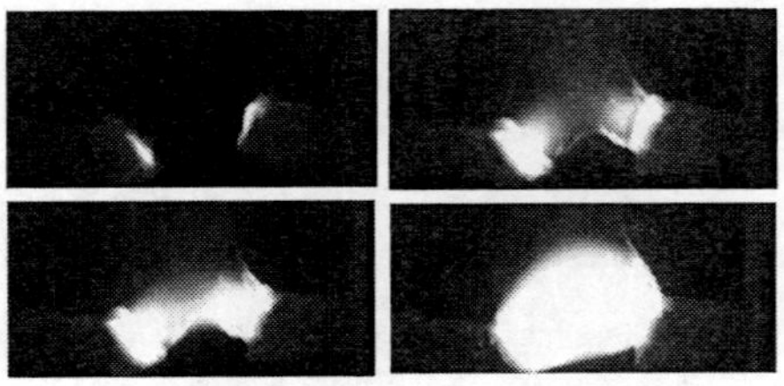

Figure 4. Photo detail of arc discharge – isolated neutral system (electrical arc length is 5 mm).

2) Regarding the voltage and current course, the situation is different in the case of the MV testing system operated as a resonance earthing system, namely a MV testing system with a (5-20) Amps I_C extent. An analysis of temporary processes is much more varied for this earthing method. For a different content of the I_C and the level of tuning it is also possible to obtain proof about the existence of the arc discharge for both basic zones (stable versus unstable zone). For an I_C extent of 15 Amps and more, the frequency of our own circuit zero component is close to the frequency of the power supply by (18). Simultaneously the coil loss-making resistance with the circuit resistance moderates the temporary process and voltage courses at the point of asymmetry and are similar to those in an isolated testing system operation. The stability of the electrical arc burning lowers if the system is overcompensated. This means that the current is overcompensated by the coil above 120% of the I_C. This never practically happen with respect to the extent of the current regulation by a coil maximally 130% I_C. The graphics presentation in Figure 5 indicates the experimental measuring results and the results of the zero component frequency analytical calculation.

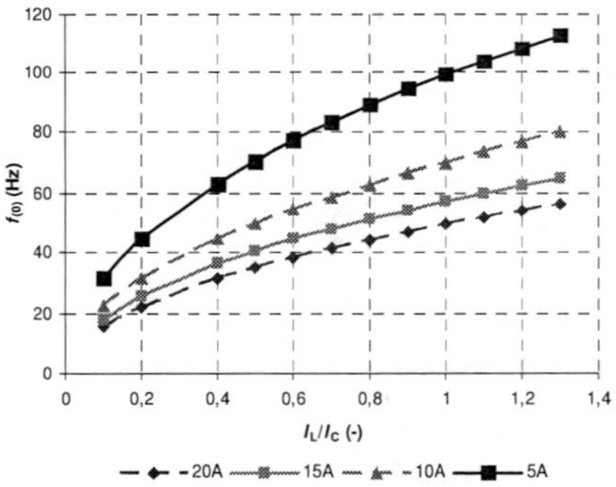

Figure 5. Zero component frequency for different methods of a MV testing system operation.

In this case it indicates the frequency of the temporary oscillation component double and during one half-period there are re-ignitions of electrical discharge which prevent the overvoltage process energy transformation into thermal energy due to arc discharge burning. Figure 8 shows the photo sequence of arc discharge for the time-track 0 – 1, 2.5 and 5 ms. Two maximum limits of the arc discharge current during one half-period are evident in the picture. This is the electrical arc limit condition in the area of the VA characteristics marked as the unease zone, namely the electrical arc stability limit. Here the ionization and recombination processes are balanced (curve "E2" in Figure 3). Due to a voltage temporary component with a double frequency, the value of the recovery voltage will be exceeded above the value of the dielectrical strength of the remaining plasma and the electrical arc will be re-ignitioned (length of a 60 mm arc discharge).

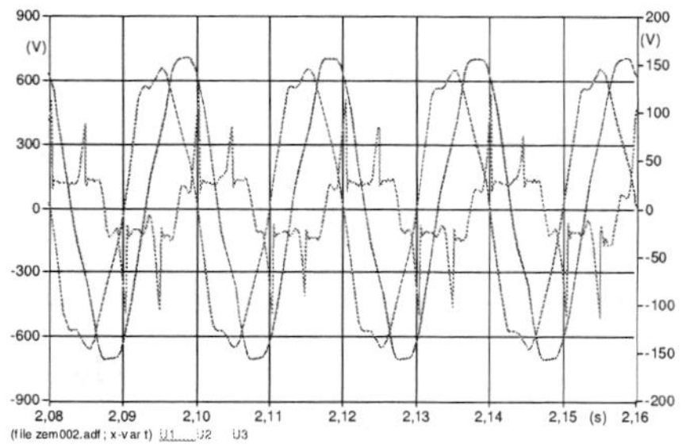

Figure 6. Courses of voltages (V) – case of DS indirectly grounded by inductor.

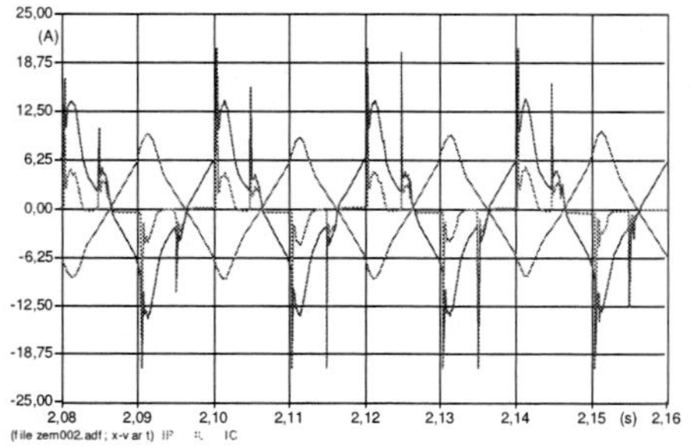

Figure 7. Course of currents. IP – fault current (A), IC – capacitive current (A), IL – current by inductor (A).

Fig. 6 and Fig. 7 represent the voltage and current courses at the point of asymmetry. A noticeable ignition and extinguishing peak for the arc voltage can be seen for the voltage course at the asymmetry point in Fig. 7 which can cause electrical arc re-ignition. There are more interesting courses of fault current, coil current and total system I_C, however. In the case of maximal capacity current compensation (I_L=I_C), only the first harmonic of the I_C is compensated by the Petersen coil.Due to this the fault current moves through the fault point and is limited by the lead admittance and the entry of the higher order harmonics. This radically complicates the situation as the ability of the circuit active elements to moderate temporary processes is negligible with respect to the small extent of the system. The higher order harmonics of the capacitors which respect the capacity against the earth and the cable lines interphase capacity, highlight the steepness of the current course in the analysed circuit for all current maximal limits, see Figure 8. This is an extremely dangerous situation as the detection of the dangerous arcing earth fault is impossible. This is due to the small energy value of the fault current with a strong amplitude but with such a great steepness and short duration time and the short timer of the digital relay protection impulse cells which detect the increase of fault current individual components which are just cleared. The currently used digital relay protections working on the principle of the fault current course evaluation are not able to detect this type of fault.

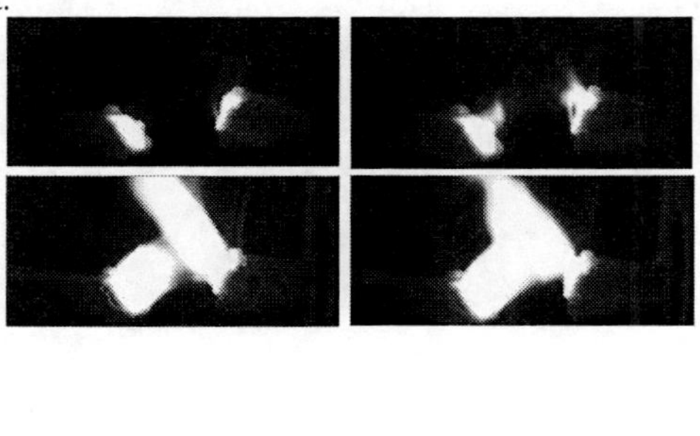

Figure 8. Photo detail of arc discharge – resonance earthing system (electrical arc length is 60 mm).

A current with so-called re-ignitions is injected over the long term to the fault point. All the devices isolation systems in a medium voltage system are strained by overvoltage processes which cause isolation system successive degradation. The consequence of this degradation can be the voltage and subsequent thermal puncturing with an interphase short circuit risk which is a fault with a electrical discharge high energy value. This fault brings the risk of direct danger to people or devices.

III. CONCLUSIONS

The aim of this paper was to point out the danger of single phase to arcing earth faults difficult detection, namely in resonance earthing systems for (5-20) Amps capacity current extent. The difficult detection of the above-mentioned fault type for the capacity current concrete extent is determined by the electrical arc re-ignitions over one half-period. This occurs for a concrete ratio of the inductive and capacity elements in the analysed system, for example, during MV system reconfiguration and corresponding capacity current decrease at the same value of the Petersen coil induction. For this case, the frequency of the temporary oscillation component of the fault current time course is double. The ionization and recombination processes in the electrical arc body are balanced and the recovered voltage will be exceeded above the value of the remaining plasma dielectric firmness with a double industrial frequency. This is an extremely dangerous condition which owing to the overvoltage processes tires the entire devices isolation system in a concrete MV system. One possibility for eliminating this undesirable condition is to place more Petersen compensation coils and the entire performance will correspond to the central coil original performance. This ensures a balanced ratio of the total capacity and MV system induction for a wide extent of system configuration options (or capacity current) and the system compensation amount. Where possible, another option is to operate these systems as isolated neutral ones with an added level drop detector as, for example, in MV mining networks. This issue is particularly relevant with regard to so-called SMART GRID systems development. These are energy systems with a galvanic isolation possibility where the isolation is from external energy systems, which can work autonomously. These systems with their extent are close to the values of (5-20) Amps and in the case of induction neutral point earthing, the above-mentioned problem of single phase to arcing earth fault difficult detection can arise.

ACKNOWLEDGMENT

This paper was supported by the following projects: Sustainable development of ENET Centre [LO1404]; Development of the ENET centre research infrastructure [CZ.1.05/2.1.00/19.0389]; Students Grant Competition SP2017/159] and Technology Agency of the Czech Republic. [TH01020426]

REFERENCES

[1] I. Zamora, A.J. Mazo´n, K.J. Sagastabeitia, J.J. Zamora: "*New method for detection of low current faults in electrical distribution systems*", IEEE Trans. Power Deliv., 2007, 22, (4), pp. 2072–2079

[2] Hanninen, S., Lehtonen, M., Pulkkinen, U.: "*A probabilistic method for detection and location of very high resistive earth faults*', Electr. Power System Research, Syst. Res., 2000, 54, (3), pp. 199–206

[3] T. Baldwin, F. Renovich, L. Saunders, "*Fault locating in ungrounded and high-resistance grounded systems*" IEEE Trans. Ind. Appl. 4, 1152–1159 (2001)

[4] Pazos, F.J., Amezua, A.: 'Electronic active earthing system for use in high-voltage distribution networks". European Patent 2128951A1, 2008

[5] K. J. Sagastabeitia; I. □amora; A. J. Mazón;□ Aginako;G. Buigues, "*Low-current fault detection in high impedance grounded distribution networks, using residual variations of asymmetries*" IET Generation, Transmission & Distribution (Volume: 6, Issue: 12, December 2012)

[6] K. J. Sagastabeitia;I. Zamora;A. J. Mazon;Z. Aginako;G. Buigues Phase Asymmetry "*A New Parameter for Detecting Single-Phase Earth Faults in Compensated MV Networks*" IEEE Transactions on Power Delivery (Volume: 26, Issue: 4, Oct. 2011)

[7] ČSN 33 3070 Suppression of earth-fault currents in high-voltage network

[8] ABB Company, [online] Available: www.abb.com, last see (2017, May)

[9] Liuming Jing, Dae-Hee Son, Sang-Hee Kang and Soon-Ryul Nam, "A Novel Protection Method for Single Line-to-Ground Faults in Ungrounded Low-Inertia Microgrids", Energies 2016, Vol.: 9, June 2016.

D: Power System Control, Liberalisation of Electricity Market

Frequency Stability in the Context of the European Network Codes

Karel Máslo
Transmission System Analysis Dpt.
ČEPS, a.s.
Prague, Czech Republic
maslo@ceps.cz

Abstract — **This paper deals dynamic models which is suitable for the frequency stability investigation of the present and future power system. Requirements on generators resulting from the new European network code are also presented.**

Keywords: Turbine dynamic mode, frequency stability, European network code, continental Europe (CE)

I. Introduction

European network codes take the form of Commission Regulation (EU) and they are published in the Official Journal of the European Union. Of the eight codes we focus on the generation, it is requirements for grid connection of generators [1]. This Regulation divides so called power generating modules into four categories. The category A starts from capacity of 800 W, and these modules must be able not only be connected to the network frequency in the range 47.5 to 51.5 Hz, and even reduce their power at over-frequency. The modules shall to cease active power output after a network operator instruction. For modules with capacity1 MW and more (type B) a synthetic inertia is mentioned. An active power frequency response capability (formerly called primary frequency control) is required for modules with a capacity above 50 MW (Type C). The respective system operator is entitled to nonsynchronous module C to require providing synthetic inertia during very fast frequency deviations. Performance of modules from B above with regard to the frequency stability and synthetic inertia shell be demonstrated be reports with the simulation results. Dynamic models can be applied to prove power-generating module capabilities and the power-generating facility owner shall produce and provide a validated simulation models.

Knowledge of dynamic models of individual sources (synchronous and non-synchronous) is therefore important and verified models will play a role in the future. The paper describes the various dynamic models of turbines and their governors, as defined in IEC standard [2] and used by transmission system operators of CE synchronous zone. These are models of steam, hydro and gas turbines used in synchronous modules. The used models can be unified into a few universal types. For this purpose the models are categorized according to the level of detail and the type of control. Then compatible alternatives are found in the MODES network simulator and its Siamese twin DMES, which is used as a simulation engine in the dispatcher training simulator DTS [3].

Requirements of the European network code for new non-synchronous sources in terms of frequency stability are described in the section III including so called synthetic inertia.

The paper is a follow-up to previous publications at the symposium [4]-[10].

II. Dynamics models of turbine

Contributions [11] - [13] showed detailed models of steam and hydro turbines. At that time there was no standard for these models, only the IEEE recommendations [14] and [15]. Now the situation is different because ENTSO-E has introduced the so-called Common Grid Model Exchange Standard (CGMES) to exchange information on dynamic models among network operators (see [16]) with the goal of creating a common dynamic model for CE synchronous zone. CGMES is based on a number of IEC standards. Dynamic models are covered by IEC 61970-302 [2]. The standard shows the block diagram of individual models of excitation and prime mover systems of synchronous generators wiring with a brief description of their parameters. This is a compilation of all possible models used in different programs and different transmission system operators. In this section, the often-used models are sorted and the related models are aligned in order to find equivalent models from the MODES library.

A. Simple steam turbine models in speed control

Three simple IEC models are depicted on the left side n Figure 1. In the right part is an equivalent scheme of a standard steam turbine model implemented in the MODES program

A varied designation of parameters of the same meaning (K≈1/bp, T3≈T1≈Tm) is seen in the IEC models., A universal schema corresponding to the standard steam turbine model STAN can be used for all three IEC models (with the appropriate choice of parameters as shown in the figure).

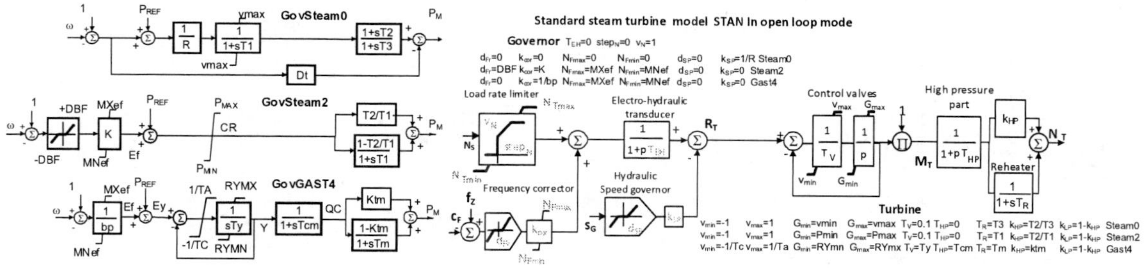

Figure 1. Simple steam turbine models: IEC standard (on the left modified according to [2]) and from the MODES library)

Although the GovGAST4 suggests the gas turbine usability, it has a typical structure of the steam turbine model. The mechanical power P_M consists of two parts that correspond to the passage of steam through the high pressure part of the turbine and the low pressure part (considering the delay in the reheater). In a gas turbine, these two delays usually do not occur.

It is noteworthy that IEC models are in the to speed control (there is a deviation ω-1 at the input of the model), whereas, the turbine usually operates in power control mode

B. Complex steam turbine models

While simple model turbines from the previous chapter modeled double body steam turbine, detailed models in this chapter allow to model separately high, medium and low pressure parts. The IEC models are shown in the left-hand section of Figure 2. The GovSteam1 even allows modeling of a two-shaft controlled cross-compound turbine. However, this solution is not much used in Europe.

The first three models are in the speed control mode, only the fourth GovSteamEU model allows to simulate the power control mode, which is a common control mode for synchronous interconnection and provides some ancillary services (primary and load frequency controls). This model also considers changes in admission pressure (using one time constant T_B) in a simplified way.

The right part of the figure shows a detailed model of the steam turbine (ST_A) implemented in the MODES simulator along with the universal controller model (for simplicity only the turbine part is shown. Variable N_S enables to connect unit to the remote load frequency control. The frequency correction enables to provide the primary frequency control.

Turbine model ST_A contains only the high and low pressure parts, but it is sufficient to simulate both the frequency and angle stability. In addition, it contains a detailed model of the control and intercept valves and allows detailed modeling of the fast valving.

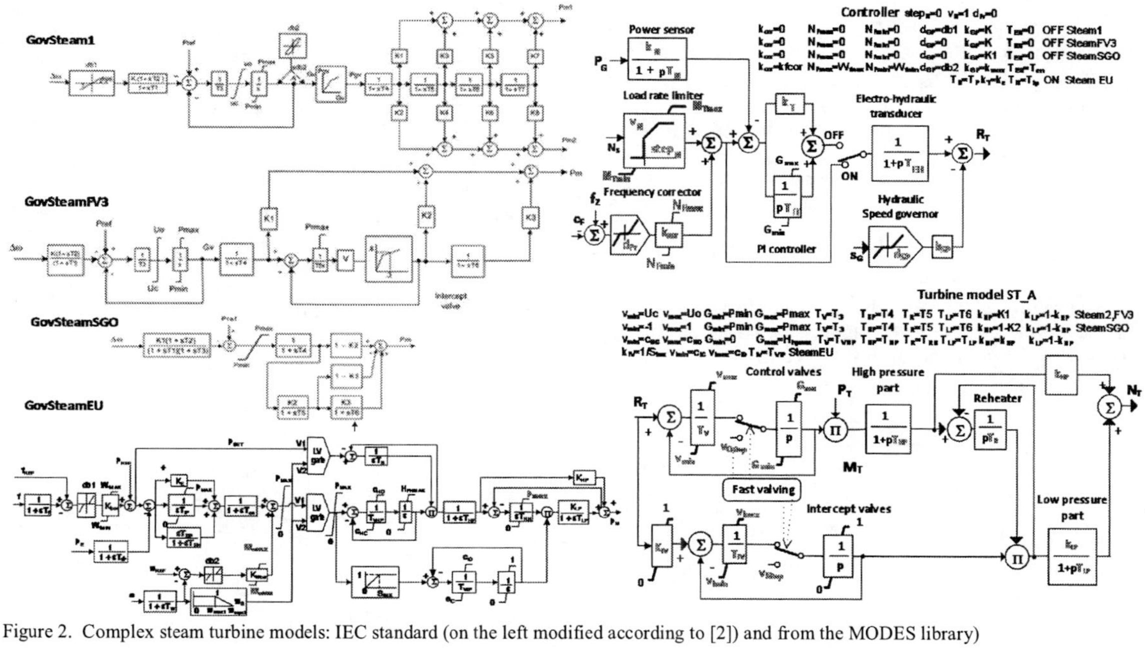

Figure 2. Complex steam turbine models: IEC standard (on the left modified according to [2]) and from the MODES library)

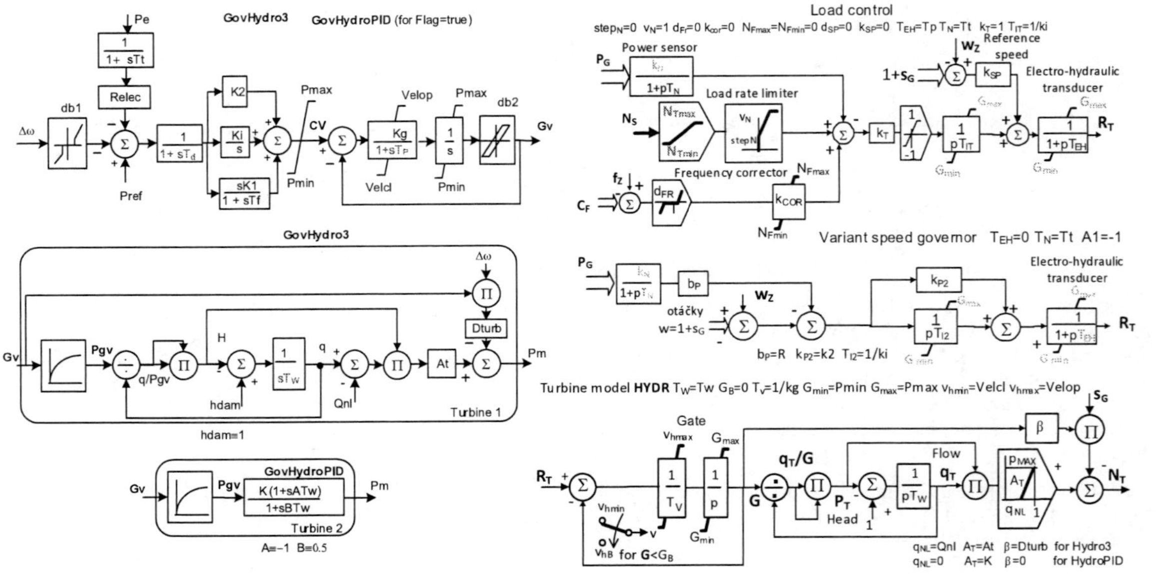

Figure 3. Hydro turbine models in power control mode: IEC standard (on the left modified according to [2]) and from the MODES library

C. Hydro turbine models in power control

Figure 3 shows IEC hydro turbine models in the left side. Both have more or less the same controller and differ in details of the turbine model. The first model is nonlinear with an inelastic water column (marked Turbine 1) and the second one is linear (Turbine 2). The structure of the controller is interesting. It has two controlled variables of speed ω and electric active power Pe. The controller enables to simulate three control modes:

1. With zero dead band db1, the controller regulates the power to Pe = (Pref-Δω)/R, where R is the speed droop and Pref is a basic gate opening. It is a **proportional speed control**.
2. With zero droop R it is **isochronous control** (it is suitable in separate operation to insulated load or in idle operation),
3. With db1≠0, the controller switches to the **power control** and for R = 1 it regulates the turbine output to Pref.

The controller shown in the figure at the top right is used for the **power control**. The variant speed controller is used **proportional and isochronous speed control**.

D. Hydro turbine models in speed control

Three IEC models of hydro turbines are shown in Figure 4. GovHydro2 and GovHydro4 have identical models of the speed governor and differ in detail of the turbine model. The GovHydro1 includes a simplified governor model (the dead band and hysteresis are not specified).

An equivalent model is shown in the right part of the figure. The time constant of the pilot servo motor Tp is much smaller compared to the time constant Tg, so this delay is neglected

The model HYDR, depicted in the right part of the figure, corresponds to the detailed IEC model (Turbine 1), with the exception of nonlinear dependence between gate position Gv and turbine power Pgv. This nonlinearity can be neglected for the angle and frequency stability calculations. The hydraulic head and nominal head is considered nominal (Hdam=1). Additionally, the model respects the reduced closing velocity of the gate in a small load (so called buffering) - more detailed description of the hydro turbine model is in [12].

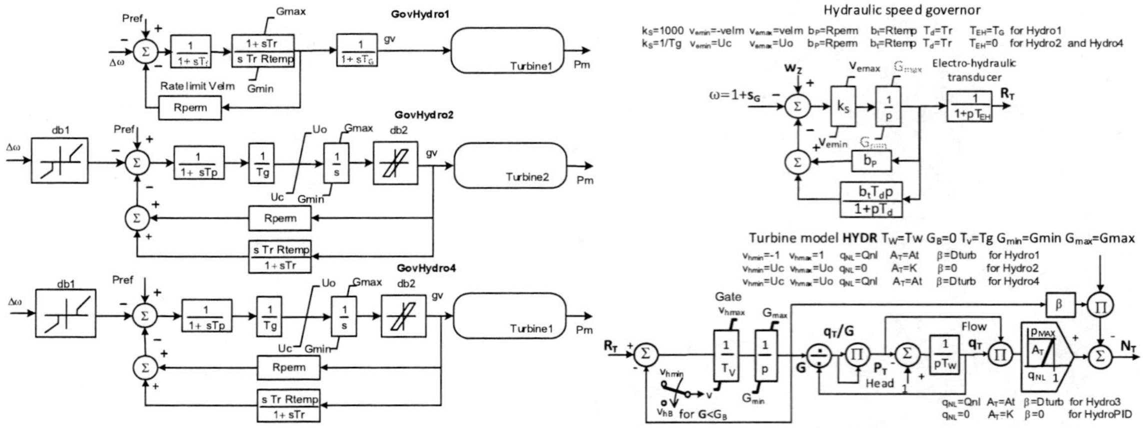

Figure 4. Hydro turbine models in speed control mode: IEC standard (on the left modified according to [2]) and from the MODES library

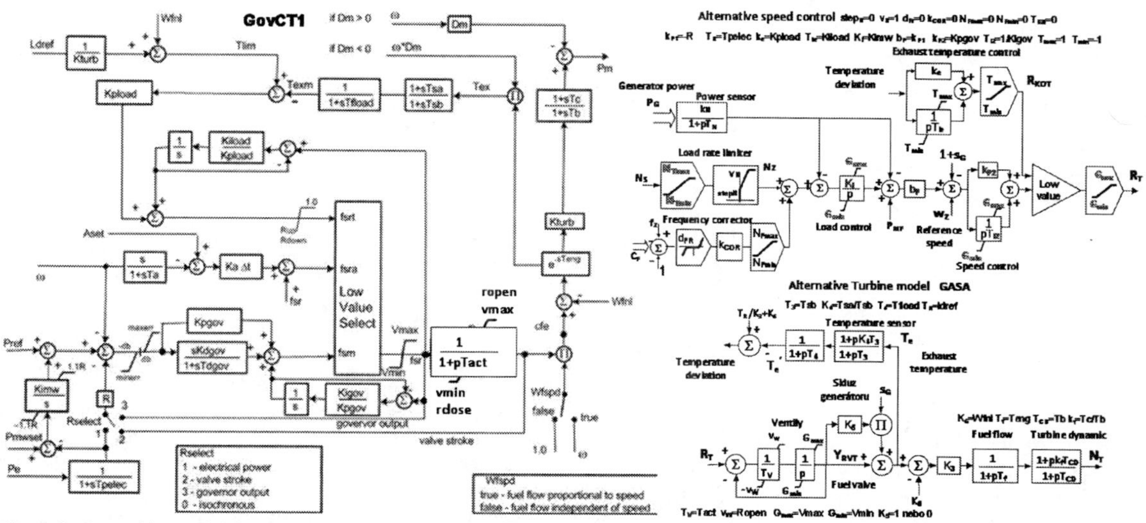

Figure 5. Gas turbine models in power control mode: IEC model GovCT1 (on the left modified according to [2]) and from the MODES library

E. *Gas turbine models in power control*

The IEC model in Figure 5. has three control loops for:

1. main governor,
2. acceleration control,
3. exhaust temperature control.

The turbine is controlled by the loop with the lowest value at the output. Smooth transition between power control and exhaust gases ensures integration elements (kiload and kigov). The main governor can work in several modes, depending on the Rselect switch. For values 2 and 3 it has a similar character to the speed governor of the hydro turbine (described in the chapter D). The **proportional and isochronous speed control** (described in the chapter C) are available for values 1 and 0. Additionally, the controller allows for remote control of power and changes the base opening Pref to adjust the power output to Pmwset. The turbine output may depend on the speed both on the fuel supply (depending on the Wfspd switch) and on the positive value Dm (this effect is mostly neglected). The model also includes the loop of the exhaust temperature Texm.

An equivalent model is depicted in the right part of the figure (again divided into part of the turbine and part of the controller). An alternative model of GASA turbine corresponds to the GovCT1 model with neglecting of the transport delay Teng (which does not occur in gas turbines) and the control effect Dm, which is not normally found in industrial heavy gas turbines (only for turbines derived from aero derivative turbines). An alternate speed controller allows both **proportional and isochronous speed control** as well as a change in the basic opening in the remote control. MODES contains a more detailed gas turbine model including residual heat that can be used in a steam turbine boiler of combined-gas cycle (see [8] for details).

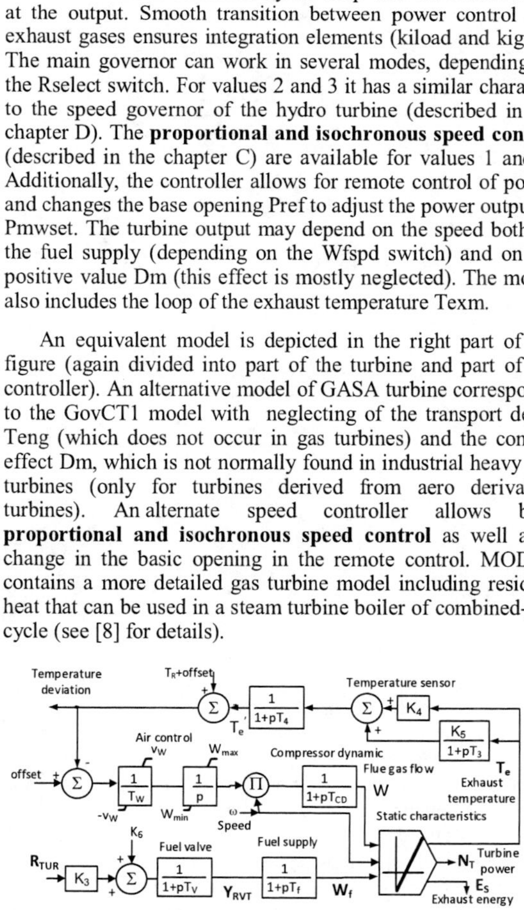

Figure 6. Detailed gas turbine model for combined gas cycle

III. Dynamic stability and Network codes

The new European Network Codes (EU Commission Regulation) will undoubtedly affect the dynamic stability of future European networks that are now undergoing major changes. These are the so-called connection codes (RfG-Requirement for Generators, DCC - Demand Connection Code and HVDC Code) and so-called operational codes (SO GL-Guideline for System Operation and NC ER - Emergency and Restoration Code). We will focus to the RfG and the requirements related to the power system frequency stability in this paper.

The RfG defines in detail the requirements for each type of source, both by size (already from 800 W for A-type), as well as by a prime mover technology (synchronous machines and non-synchronous production modules connected to the network via power electronics). However, the definition of certain parameters is left to the transmission system operator (TSO).

A limited frequency sensitive mode for over-frequency (LFSM-O) is required for all types of sources, starting with the smallest A type. The frequency threshold and speed droop shall be specified by the relevant TSO. Determining of the threshold frequency is logical for LFSM-O because the network frequency is maintained by the primary frequency control in the normal operation (with frequency deviations smaller than ±200 mHz). , The CE total primary reserve of 3000 MW (fully applied at frequency deviation ±200 mHz). Only when the deviation of the frequency is above +200 mHz, the primary control reserve is fully activated and there is a risk of a further increase in frequency, other sources (which in the normal operation do not respond to frequency deviations) should enter into play and begin to decrease power with the specified droop. Logically it is appropriate that this reduction was no delay and a sufficiently large, so we suggest the threshold 50.2 Hz and speed droop s=5% (standard value used for conventional sources).

Not only the threshold frequency and droop settings are important, but also the rate of response or dynamics plays a role. The RfG mentions it somewhat vaguely (the power-generating module shall be capable of activating a power frequency response with an initial delay that is as short as possible). The dynamics is more accurate in the report [17], which defines the time $t_{95\%}$, when the power change should be activated according to Figure 7.

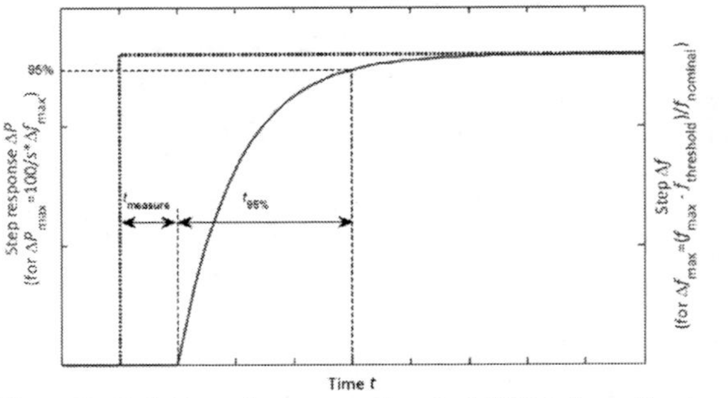

Figure 7. Definition of a response time for LFSM-O (according to [17])

Report [17] that $t_{95\%}$ should not exceed 1 s. It should also be noted that this time refers to non-synchronously connected sources. It is interesting that the report does not deal with the time of measurement $t_{measure}$. According to the manufacturer of the Czech frequency relays FREA, it is sufficient for several cycles of the rated frequency. This is confirmed by the IEEE standard [18], where the value of three cycles for modern electronic relays is given for secure frequency measurement. It means, that $t_{measure}$ corresponds to a maximum of 60 ms.

While observing these four parameters ($f_{threshold}$, s, $t_{95\%}$ and $t_{measure}$), it is assumed that by cooperation of conventional sources (providing a natural inertia) and non-synchronous resources (providing a fast power response) the frequency stability of the future system will be preserved and there will be no similar system disturbance, as a blackout in Southern Australia last year.

A. Blackout in South Australia on 28.9.2016l

The blackout in South Australia (SA) was greatly influenced by low power reserves and low system inertia. That is why we will mention it here. The description is based on the report [19].

Prior to the blackout, the SA area, with a load of 1895 MW, imported 612 MW by a double line 275 kV Heywood-South East. 883 MW of load covered wind turbines and 330 MW gas turbines. Extreme weather (strong winds, hailstorms and storms) caused disconnections of three lines (two-phase and one-phase short-circuits with unsuccessful reclosing) - see Figure 9.

Short circuits on the line caused a loss of generation 445 MW. Nine of the 13 wind farms were unable to withstand voltage drops ('voltage ride-through capability'). A high power transmission through Heywood's corridor activated automatic loss of synchronism protection mechanism that disconnected lines in the Heywood's corridor. The power deficit around 900 MW caused a rapid drop in frequency, as shown by the blue waveform in Figure 8.

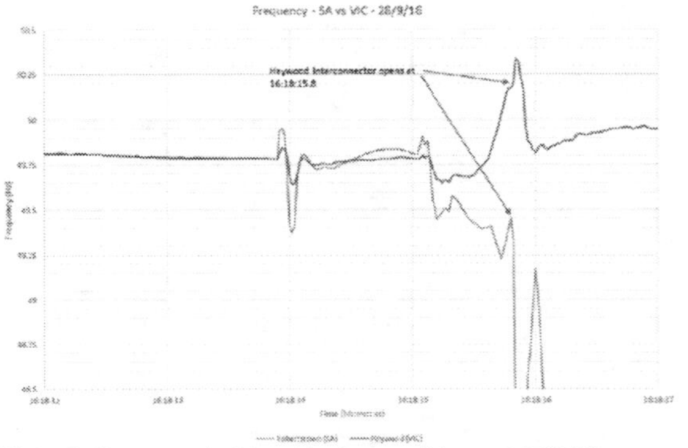

Figure 8. Frequency in SA and in the rest of the system ([19])

The rate of change of frequency has reached an extreme value of 6.25 Hz /s (caused by the large power deficiency and by a lack of inertia). According to report [19] the frequency collapsed so quickly that the under-frequency load shedding was unable to act and the SA region resulted in the blackout.

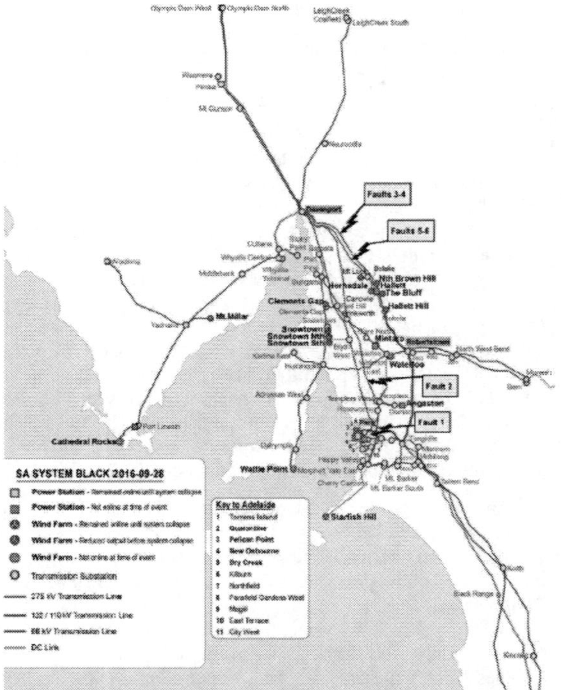

Figure 9. Scheme of SA transmission system with fault location ([19])

B. Synthetic inertia modelling

Although non-synchronous modules will be able to respond to frequency deviations, the role of inertia will remain important in the future, because it allows to compensate for power imbalance immediately. Conventional sources with synchronous generators cover the initial power imbalance from the kinetic energy of the rotated mass automatically without the need for any regulation. However, as conventional sources will be shut down as a consequence energy goals of the EU's (non-carbon policy) or states (Energiewende in Germany) and the share of non-synchronous sources will increase, synthetic inertia will be used. This capability is mentioned in RfG for non-synchronous modules of type C and D (TSO shall have the right to specify that these modules be capable of providing synthetic inertia including the operating principle of control systems and the associated performance parameters).

New control principles applied to on non-synchronous sources can be well tested on dynamic models. The original model of the photovoltaic power module with the LFSM-O function was presented already in 2013 at the previous symposium [5], with the threshold frequency and speed droop being programmed to 50.2 Hz and 5%, exactly what the ENTSO-E Working Group now recommends in [20].

The initial model of the wind power module was published in 2009 at the 5th symposium [6].

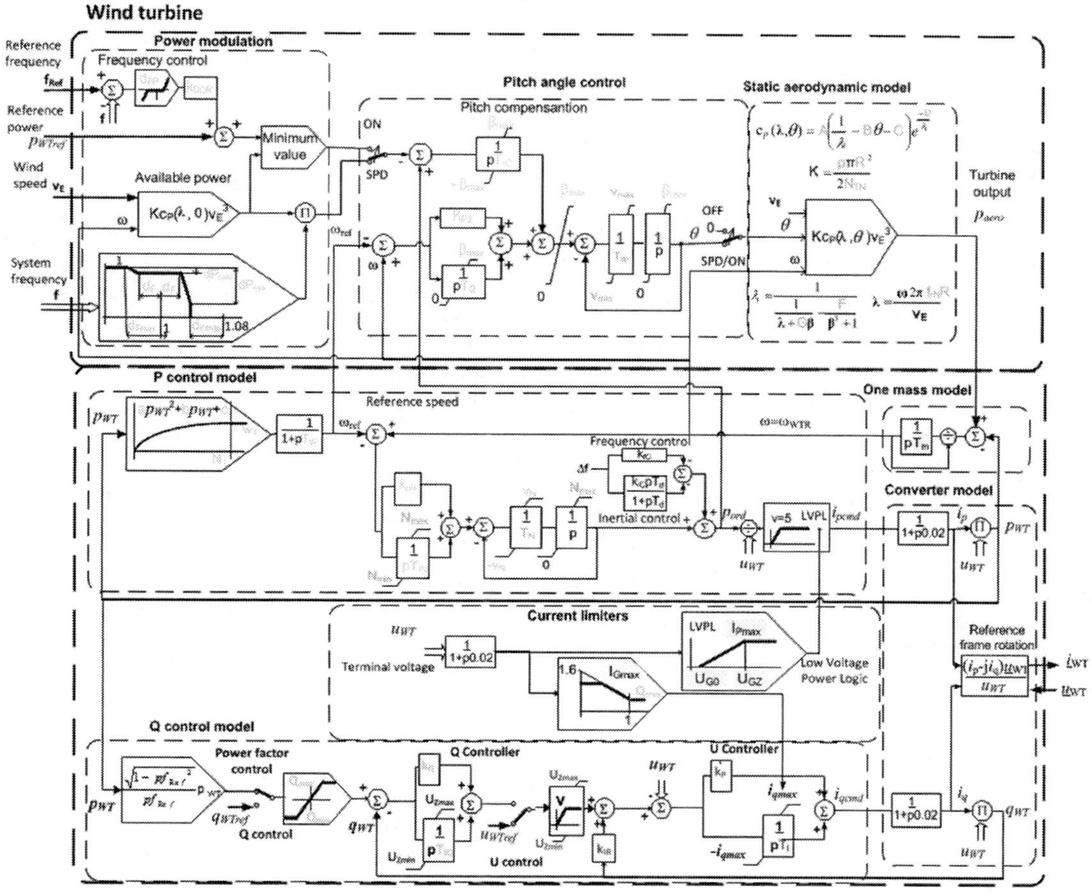

Figure 10. Wind power module with full power convertor and permanent magnet generator

This model was further enhanced by simulation of a full-power convertor equipped with control functions including synthetic inertia. Figure 10 shows a block diagram of the innovated model, as published in [21].

The wind turbine model (marked WIND model) responds to the frequency deviation either by the classical primary control (called frequency sensitive mode in RfG) or by the LFSM-O function and even by the LFSM-U function (when the turbine maintains a reserve dPmax to increase the output during the frequency decreasing). Control action is carried out by the turbine blades rotation ("Pitch Angle Control"). The converter control then adjusts the active power delivered to the network.

An inertial control loop is implemented in the inverter model, which, based on the frequency derivation, almost instantly changes of active power delivery and emulates the inertia of the synchronous machines.

IV. CONCLUSIONS

The new European network code (RfG) sets a number of requirements for power modules capability in terms of frequency stability. Many of these are new, in particular on the non-synchronous power modules. This is, for example, LFSM (the ability to change the power if the frequency is out of threshold) or the synthetic inertia.

A brief description of last year's blackout in South Australia demonstrates the importance of system inertia. If the system does not have enough natural inertia (kinetic energy accumulated in rotated inertia), it is necessary to ensure synthetic inertia on non-synchronous power modules, HVDCs and storage devices, where the power electronics allow fast change of the active power in response to the frequency derivation.

Frequency stability of the power system can be analyzed using dynamic model. That is why the paper contains a description of these models - not only with conventional sources (steam, hydro and gas turbines) but also non-synchronous sources (wind power plants). These models are implemented in the computational engine used in the MODES network simulator and the dispatcher training simulator DTS.

REFERENCES

[1] Commission regulation (EU) 2016/631: establishing a network code on requirements for grid connection of generators

[2] IEC 61970-302 Common Information Model (CIM) for Dynamics Specification

[3] K. Máslo, M. Kolcun "Simulation engine for dispatcher training and engineering network simulators", IFAC CIGRE/CIRED Workshop on Control of Transmission and Distribution Smart Grids (CTDSG'16), Prague 10/ 2016

[4] K. Máslo, "R&D projects and solution of power flows problems in the Central European region," in *Proc. 8th Int. Scientific Symp. on Electric Power Engineering (EE2015)*

[5] K. Máslo, "R&D projects and solution of power flows problems in the Central European region," in *Proc. 7th Int. Scientific Symp. on Electric Power Engineering (EE2013)*

[6] J. Rusnák, K. Máslo, P. Trubač, Modelování větrné elektrárny s dvojitě napájeným asynchronním generátorem, in *Proc. 5th Int. Scientific Symp. on Electric Power Engineering (EE2009)*

[7] K. Máslo, K. Witner, A. Kasembe, M. Škach, Z. Brettschneider: "Systémové stabilizátory – požadavky a jejich prověření", in *Proc. 4th Int. Scientific Symp. on Electric Power Engineering (EE2007)*

[8] K. Máslo, "Model a testování ostrovního provozu paroplynového cyklu", in *Proc. trd Int. Scientific Symp. on Electric Power Engineering (EE2005)*

[9] K. Máslo, M. Bíca, I. Petružela, "Zvýšení stability jaderné elektrárny Temelín pomocí rychlého řízení ventilů", in *Proc. 2nd Int. Scientific Symp. on Electric Power Engineering (EE2003)*

[10] J. Anděl, K. Máslo, "Využití modelu plynové turbíny při návrhu elektráren a tepláren", in *Proc. 1st Int. Scientific Symp. on Electric Power Engineering (EE2001)*

[11] K. Máslo, "Dynamický model ES pro dispečerský trenažér", in *Proc. 7th int. conference Control of power and heating systems*, Zlín 5/2006

[12] K. Máslo, K. Witner, A. Kasembe, L. Kočiš, M. Švancar, "Výpočet přechodných dějů při startu ze tmy z elektrárny Orlík", in *Proc. 8th int. conference* Electric Power Engineering, Dl. Stráně 6/2007

[13] K.Máslo, J.Feist, "Power system dynamics behavior, modeling and simulation of the primary control test", in *Proc. 2nd Int. workshop on Electric Power Control Centres*, Alghero, 1993

[14] IEEE Working Group Report: Dynamic Models for Fossil Fuelled Steam Units in Power System Studies; IEEE PAS No2; 1991

[15] IEEE Working Group Report: Hydraulic Turbine and Turbine Control Models for Systém Dynamic Studies; IEEE PAS No1;1992

[16] Common Grid Model Exchange Standard, Based on IEC Common Information Model, May 2014, https://www.entsoe.eu/Documents/CIM_documents/Grid_Model_CIM/140528_ENTSOE_CGMES_v2.4.14.pdf

[17] Frequency Stability Evaluation Criteria for the Synchronous Zone of Continental Europe, RG-CE SPD SG report, 3/2016, https://www.entsoe.eu/Documents/SOC%20documents/RGCE_SPD_frequency_stability_criteria_v10.pdf

[18] IEEE Std C37.117: Guide for the Application ofProtective Relays Used for Abnormal Frequency Load Shedding and Restoration, 2007

[19] Black system event in South Australia on 28 September 2016, AEMO report

[20] Specific Challenges on Frequency Stability for the Power System of Continental European, presentation on ENTSOE Public Workshop, Brussels 3/2017, https://www.entsoe.eu/news-events/events/Pages/Events/2017-03-09-cnc-freq.aspx?EventWorkshopId=283

[21] Operational validation of the grid reinforcements by 2050, Project e-Highway 2050 Deliverable D4.1., 2015. http://www.e-highway2050.eu/fileadmin/documents/Results/D4.1_Operational_validation_of_the_grid_reinforcements_by_2050.pdf

Information Technologies in Models for Assessing the Reliability of Electric Power Systems in Managing their Development

Iurii Chukreyev
Federal State Budgetary Scientific Institution - Institute of Socio-Economic and Energy Problems of the North in the Komi Scientific Center, Ural Branch of the RAS
Syktyvkar, Russia
chukreev@iespn.komisc.ru

Mikhail Chukreyev
Federal State Budgetary Scientific Institution - Institute of Socio-Economic and Energy Problems of the North in the Komi Scientific Center, Ural Branch of the RAS
Syktyvkar, Russia
mchukreyev@gmail.com

Abstract — **The determination of reliability criteria is an essential part of the work to establish the future power reserves in territorial areas of the United Power System of Russia. This task has been thoroughly studied. In this article, the main features of long term planning are discussed, and methods and models for the determination of reliability criteria are proposed.**

Keywords: UPS of Russia; reliability criteria; power system; adequacy parameter; parallel algorithm

I. INTRODUCTION

This template The development of the electrical power system is of major importance for the economic growth of a country. The United Power System (UPS) of Russia delivers energy to more than 95% of the consumers and covers almost the entire territory of the country. Nowadays, more than 680 power plants of different kind and size are connected to the network by 2 millions of kilometers of transmission lines, making the UPS a complex. The development of the UPS of Russia, as a united technological complex, needs to fulfill reliability and safety requirements at all time and in all territorial areas. The territorial areas are already indicated by the system operator (SO). In this paper, we are interested in the planning of the power system under reliability and safety constraints.

According to the Federal law on electrical power systems [1], the "SO- UPS" is responsibile for making the various development routes meet the safety and reliabiliyrequirements during power system planning. The determination of reliability criteria for the UPS of Russia is considered a cornerstone of the decision making process [2]. The above mentioned law and the Decree of the Russian Federation[1] have a strong impact on the technological rules of power system operation[2]. In this document, it is demonstrated that the determination of reliability criteria for long term planning should be performed for each year according to the UPS development program of the coming 5 to 7 years.

The determination of reliability criteria constitutes an essential task when establishing the power reserves necessary to keep the balance during a contingency. Fulfilling the long term planning requirements of the UPS of Russia requires development efforts in multiple directions:

- modeling of the UPS;
- building a database with information about the UPS network;
- developing methods for the determination of reliability criteria and their programming (taking into account the available data for variations in load, the occurrence of planned and emergency maintenance, the distribution of power reserves during long term outage of the power system).

The first of these assignments (modeling) is taken care of by "STC UPS". In this work, the main features of long term planning are discussed, and methods and models for the determination of reliability criteria are proposed.

In the rest of this paper, by the term "reliability" will be meant the ability of the UPS to satisfy the demand under given restrictions on the access to power resources considering planned and emergency outage of the power system components.

II. BRIEF ANALYSIS OF EXISTING MATHEMATICAL MODELS AND PROGRAMMING TOOLS FOR THE DETERMINATION OF RELIABILITY CRITERIA.

Problems related to the development of mathematical models based on a probabilistic analysis of the power system reliability are investigated in [4-12]. Solutions to these problems require the estimation of the demand in the different territorial areas, the random choice of the state of power generators and transmission lines (due to unplanned outages), as well as the analysis of the quality of power supply for the given system state. The development of mathematical models for the determination of reliability criteria is based on analytical methods or combinatorial and statistical models.

Models based on analytical methods rely on the probabilities of over- or under- production and of restricted transmission capacity between two neighboring areas. The analysis is performed sequentially from one node to another, thereby propagating across the entire network. The following models were developed using this method: АМОН/Д (ЭНИН) [4], АМОН/Н (КирНИОЭ) [5], TRELSS (EPRI), TPLAN

[1] Decree of the Russian Federation dated October 17, 2009 № 823 "On the schemes and programs of perspective development of electric power industry

[2] Approved at a joint meeting of the Scientific Council on the problems of reliability and safety of large-scale power systems and scientific-technical board of NP "STC UPS" 16.05.2011.

(Siemens) [10, 11]. On the one hand determination of reliability criteria of the power system by analytical methods significantly increases the computational efficiency; on the other hand it can be implemented only on a radial power system.

Models based on statistical modeling are more flexible and universal and have therefore found wide application for the determination of reliability criteria in complex power systems [6,7 8]. The determination of reliability criteria can be performed through the analysis of random events (models «Орион» [7, 8], Янтарь [6], ЭНИНа [4], MECORE [11], etc.) and of random processes (GE MARS, GridView, etc. [11]). The main drawback of these models lies in their poor computational efficiency. Nevertheless, the recent improvements in computational power of personal computers allow the determination of a wide range of reliability criteria, making it a method of choice for the analysis of complex power systems.

In practice, most of the programming tools used for the determination of reliability criteria are based on statistical modeling methods. In the following, issues with the development of technical guidelines and methods for the determination of reliability criteria in the long term planning of the UPS of Russia are discussed.

III. Technique for power system adequacy parameters estimation in the planning of the UPS of Russia.

Reservation is one of the basic ways to ensure power system reliability. It is aimed at partial compensation of all possible reasons of reliability degradation both during operation and for control of the power system development. Values of reserves needed depend on a set of factors causing decrease of the power system adequacy. Since it is not necessary and not expedient to reach absolute reliability, reservation in power system is an economic problem. The higher is the power system reliability, the more expensive will be the reliability increase.

In a pool of power systems, reservation of the generating capacity is obtained not only with the generating units of the territory under consideration but with the stand-by units of other territories due to the existence of the tie lines between the units. Therefore, power system reliability level depends on the distribution of standby capacity among individual territories and on spare transmission capability of the lines between them. In the market economy, the reservation level of an individual territory in the pool also depends on the regulations established for the market entities.

Thus, to ensure required power system adequacy, it is necessary to solve two interrelated problems. The first one is to determine adequacy parameters for a given set of generating units, power consumption and other power system characteristics mainly of random nature. The second problem is to determine values of required generating reserve margin of the territory power system and transfer reliability margin, which would ensure a preset adequacy level.

The process of determining reliability parameters in the power system pool, regardless of the interaction principles between the system entities, can be divided into two interrelated stages:

1. Define a set of system states and estimate probabilities of their operation caused by random states of the system elements with the time interval under consideration;

2. Determine consequences for the consumers caused by different system states in the form of generalized adequacy parameters (for the sake of brevity, this stage is called random state estimation).

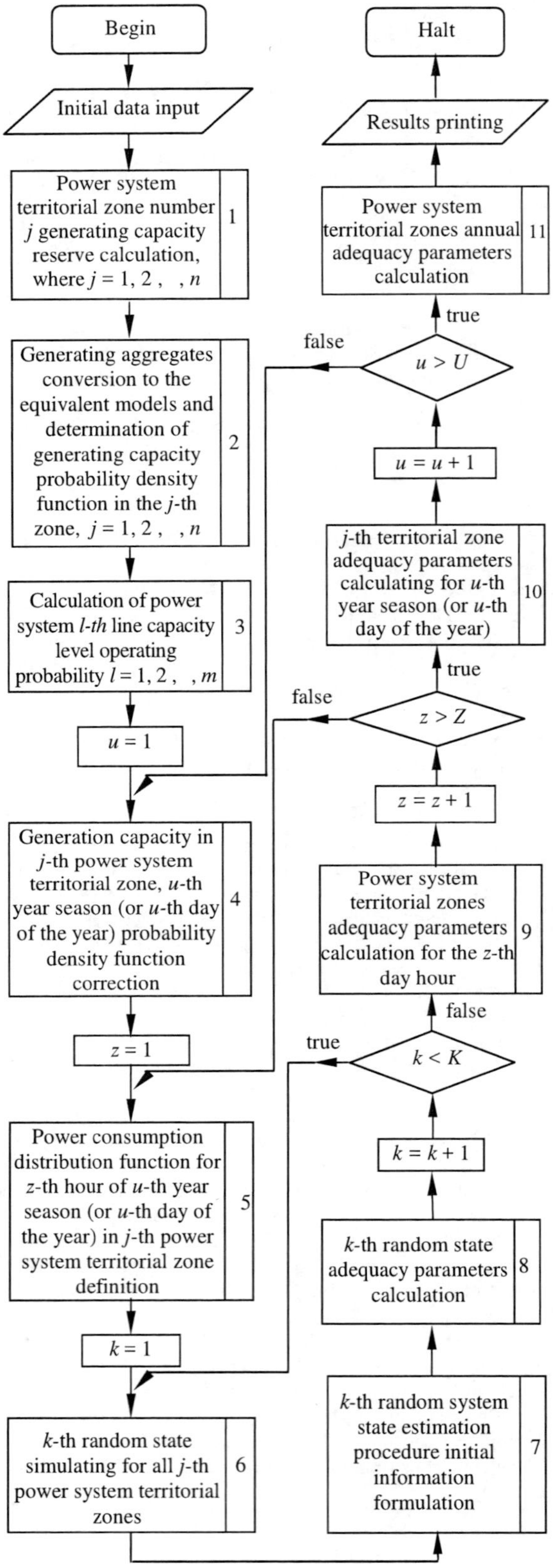

Figure 1. Outline flowchart for estimating reliability parameters of a power system pool

Approaches based on the combined application of the analytical methods [1,2,3] and methods of statistical modeling [4,5,6] are most widely used to determine power system adequacy parameters. In compliance with the above stages, an algorithm for power system pool adequacy estimation shall include models for defining system states and models for their evaluation (Fig. 1). Random states of the power system can be defined in two ways: with exhaustive search of all possible states of the generating capacity within the system individual territories and with the methods of statistical modeling.

System states to be considered in order to estimate adequacy parameters of the power system pool are very numerous. Therefore, the problem of evaluation of consequences for the consumers of power system territories, which is also called a problem of balancing power deficiency (BPD), shall be solved with effective methods. The resultant adequacy parameters for individual territories significantly depend on the problem statement and solution.

A flowchart in Fig. 1 shows that boxes from 2 to 6 are aimed at defining a set of power system states, boxes 7 and 8 evaluate consequences caused by these system states for the consumers, boxes 9, 10 and 11 determine annual parameters of the power system adequacy. To improve the proposed algorithm for determination of the adequacy parameters, it is necessary to develop methodical approaches, which would speed up the simulation process due to the use of computer hardware for parallel computations.

IV. Parallel algorithm of determining adequacy parameters

This section considers in more detail each possible strategy of making algorithm for determining reliability index (RI) fit different hardware architectures.

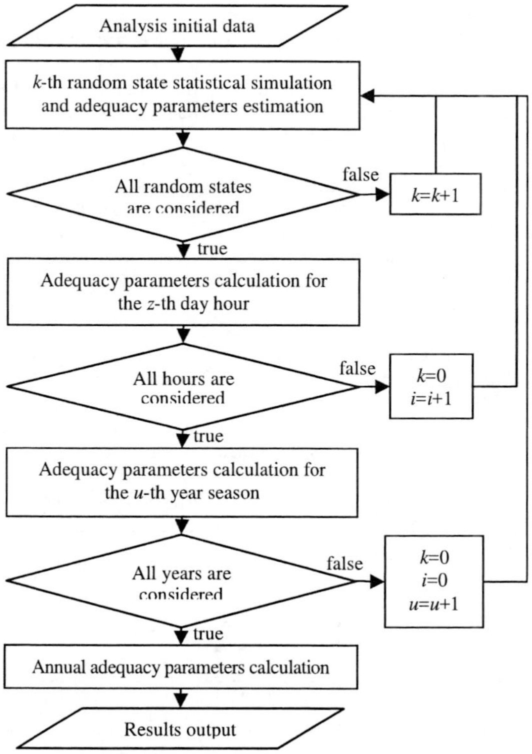

Figure 2. Algorithm for determining adequacy parameters with account of time-related aspect

As a first possible option, consider how an algorithm presented in Fig. 2 can be parallelized with respect to random states. A general view of the parallel algorithm in a simplified form is shown in Fig. 3. It is obvious from the flowchart that in the parallel algorithm the number of cycles decreases down to 2, and within them the set of random states $K=\{K_1, K_2, \ldots, K_k\}$ is simulated in parallel. An important aspect should be noted which plays a significant role during implementation of such paralleling of a computational process, i.e. dimension of the K set can make up to 10^3-10^5. The higher the K dimension, the more accurate results are obtained as the algorithm output. At low dimensions of the flowchart under analysis this value can be considerably reduced, however, for the practical purposes, the computations use an upper value of K dimension equal 10^4.

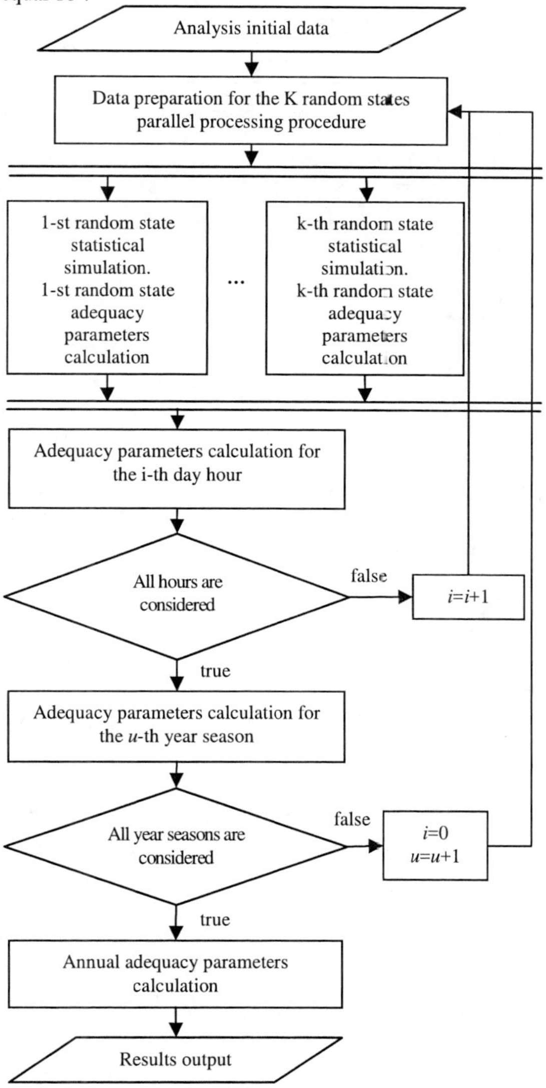

Figure 3. Paralleling a procedure of accumulating system random states in the adequacy parameters determining algorithm

As a second possible option to parallel the algorithm presented in Fig. 2, consider splitting a cycle with respect to the hours of the day. A flowchart of the simplified version of such algorithm is given in Fig. 4. Just like the algorithm in Fig. 3, the algorithm in Fig. 4 contains fewer cycles, parallel computations being done with respect to a set $I=\{I_1, I_2, \ldots, I_i\}$, which is a number of hours in a day rather than the set K of the system states. From the definition, $|I| = 24$. An important feature of this algorithm is that there is a cycle inside the parallel procedure. This feature, first, reduces the requirement to the available RAM size, since there is no need to preprocess data at the set K analysis; second, requires that the computer hardware architecture incorporates a more complex memory access system, an interim cache-memory, etc. In addition, significantly lower dimension of I compared to that of K shows that it will be impossible to get high parallelism of computations with the use of this algorithm.

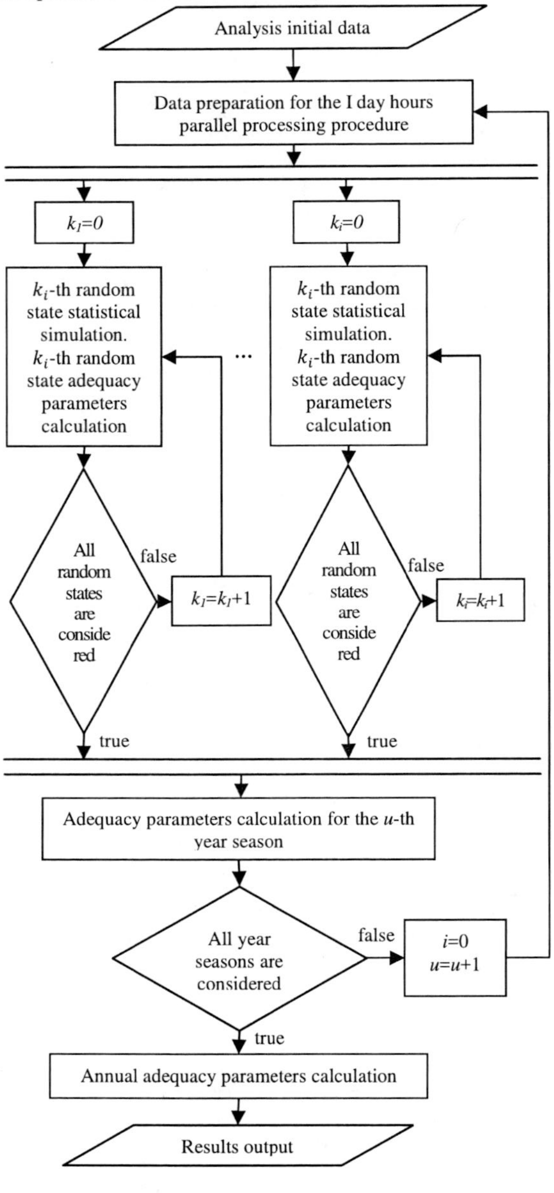

Figure 4. Paralleling a procedure of determining reliability indices at different hours of a day in the algorithm of determining adequacy parameters

Besides the presented two options of paralleling the algorithm given in Fig. 2, there is a third possible option, which considers the cycle for set of year seasons U={U_1, U_2, …, U_u}. From the standpoint of requirements to the computer hardware architecture, this option is similar to the one shown in Fig. 4 since $|U| = 1 \div 12$, therefore we will not consider it in detail.

Summarizing the above approaches to paralleling algorithm of determining adequacy parameters, it is possible to say the following: the first paralleling option better fits the compact GPU-class computers since currently such computers can run a significant number of parallel threads, have huge RAM and other features facilitating massively parallel computations; the second option (as the third one) better fits multi-core CPU with large-size cache-memory and modest, nowadays, parallelism of computations.

Consider the issues of practical implementation. As an example, consider implementation of the algorithm given in Fig.2 for the multi-core CPU. To estimate effectiveness of the algorithm paralleling, use as a criterion the speed-up of computations, which can be estimated according to the Amdahl's law in the following way:

$$\eta = \frac{T^S_{comp}}{T_{seq}+\frac{T_{par}}{p}+T_{comm}+T_{wait}} < \frac{T^S_{comp}}{T_{seq}+T_{comm}+T_{wait}} \quad (1)$$

where T^S_{comp} is total time of computations $=T_{seq} + T_{par}$, T_{seq} is run time of the sequential algorithm section, T_{par} is run time of the parallel algorithm section, T_{comm} is data communication time, T_{wait} is waiting time at synchronization, p is a number of processors. In the case of the problem under consideration T_{par} in percentage shall exceed T_{seq}, and T_{comm} and T_{wait} shall be insignificant if multi-core CPUs are used. For the modern multi-core CP, $p = 2 \div 12$ and in the near future will grow even higher. Due to this, it is possible to preliminary estimate η as equal to 1.7-10.5 (at $p = 2 \div 12$).

From the practical point of view, a process of adapting an algorithm to the target architecture consists in developing a code with the use of functions from different support libraries. Nowadays, the adaptation procedure, if applied to the multi-core CPU, is most effective in terms of time consumed, if it is based on the use of OpenMP library [9], embedded in a number of compilers, as well as the more universal library OpenCL [10]. These libraries are universal for different CP brands and do not depend on CP manufacturer and specifications. To say more, OpenCL is a new standard for the application development for heterogeneous systems. This standard is used for creating applications to be run on the system with CPUs, GPUs and expansion cards of different architectures. Due to this, there is no need to use different algorithms for the systems based on Intel, AMD and other platforms.

To validate the implementation of the algorithm, a number of test problems were computed for the test power system systems of various dimensions and configurations. Both OpenCL and OpenMP libraries were used as software for parallel computations. Results obtained with these two libraries are identical in many aspects.

Table 1. Coefficient of computation acceleration at different CP for considered test power system problems

CPU model	**Core frequency, GHz**	**Cores number**	**Threads number[3]**	**Random states number ($^{1000}/_{10000}$)**				
				Considered test problems				
				6	***10***	***21***	***51***	***81***
Core 2 Duo	3,0	2	2	1,82 1,78	1,85 1,85	1,96 1,96	1,9 1,89	1,92 1,95

[3] Thread is a process running at one processor. The number of processes done at the same time on one processor can be over one; their number depends on the process architecture.

Core i3 540	3,06	2	4	2,42 2,45	2,43 2,42	2,6 2,63	2,4 2,47	2,53 2,5
Core i5 760	2,8	4	4	3,27 3,48	3,07 3,44	3,25 3,50	3,46 3,39	3,43 3,42
Q9770	3,2	4	4	3,61 3,56	3,54 3,66	3,86 3,85	3,79 3,83	3,91 3,91
Core i56600	3,3	4	4	4,17 4,06	4,32 4,44	4,67 4,71	4,56 4.69	4.61 4,63
Xeon w3580	3,33	4	8	2,93 4,21	3,24 4,47	4,45 4,80	4,65 4,78	4,90 4,89
Core i7 6800K	3,4	6	12	7,87 7,91	8,01 8,08	8,06 8,11	8,22 8,2	8,33 8,35

Efficiency of paralleling is close to the theoretically estimated value and almost proportional to the number of simultaneously executed threads, and this makes paralleling the most promising direction for reducing time needed for computations, especially taking into account the current trend of increasing computational capabilities by quantitative growth of the procedures being executed at the same time.

V. Conclusions

The analysis of programming tools for the determination of reliabity criteria showed the similarities in the base method employed for power systems of different countries. Approaches are proposed for the development of programming tools for the determination of reliability criteria in the UPS of Russia. The system operator of the UPS is responsible for adapting the existing models for the determination of reliability criteria in system planning. For a practical implementation of the model and its application to actual power system planning, it is necessary to:

- create software for the determination of reliability criteria;
- refine the methods and programming tools for the determination of reliability criteria;
- develop the user interface for these programming tools.

References

[1] Federal'nyj zakon Rossijskoj Federacii ot 26.03.2003 g. № 35-03«Ob jelektrojenergetike».

[2] Je.P.Volkov, V.A. Barinov, "Metodicheskie principy obosnovanija razvitija jelektrojenergetiki v uslovijah ejo liberalizacii". // Izv. *RAN. Jenergetika*, 2006, – № 6.

[3] Ju.N.Rudenko, I.A. Ushakov, *Nadezhnost' sistem jenergetiki.* – M.: Nauka, 1986, p. 252.

[4] G.A. Volkov, *Optimizacija nadezhnosti jelektrojenergeticheskih sistem*, M.: Nauka, 1986. p.117.

[5] E.A Itkin, V.A Shadrin, "Postroenie modeli analiza nadezhnosti slozhnoj jelektricheskoj sistemy s ispol'zovaneim approksimirujushhih krivyh dlja opisanija sluchajnyh velichin, *Metodicheskie voprosy issledovanija nadezhnosti bol'shih sistem jenergetiki.* Irkutsk: SJeI SO AN SSSR, 1976. Vyp. 12. pp. 44-50.

[6] G.V.Kovalev, E.V. Sennova, M.B. Chel'cov i dr. / Pod red. N.I. Voropaja, "Nadezhnost' sistem jenergetiki: dostizhenija, problemy, perspektivy". Novosibirsk:,*Nauka. SO RAN*, 1999, pp. 434.

[7] Ju.Ja. Chukreev, "Modeli obespechenija nadezhnosti jelektrojenergeticheskih sistem". Syktyvkar, *Komi NC UrO RAN*, 1995, p. 176.

[8] Ju.Ja. Chukreev, M.Ju. Chukreev, "Obespechenie nadezhnosti pri upravlenii razvitiem jelektrojenergeticheskih sistem dlja uslovij reformirovanija jelektrojenergetiki", *Izv. RAN. Jenergetika*, 2008, № 4, pp. 39-48.

[9] OpenMP Reference Sheet for C/C++. (http://www.plutospin.com/files/OpenMP_reference.pdf) [May 29, 2017]

[10] NVIDIA OpenCL Best Practices Guide Version 2.3, NVIDIA Corporation, 2009

Problems of Rationing Reliability in Planning of Liberalized Electric Power Systems Development

Iurii Chukreyev, Mikhail Chukreyev
Federal State Budgetary Scientific Institution - Institute of Socio-Economic and Energy Problems of the North in the Komi Scientific Center, Ural Branch of the RAS
Syktyvkar, Russia
chukreev@iespn.komisc.ru

Michal Kolcun
Department of electric power engineering
Technical university in Košice
Košice, Slovakia
michal.kolcun@tuke.sk

Abstract — **The paper presents the methodology of assessment of adequacy of power grids for the circumstances of liberalization of power sector, gives the detailed comparative analysis of Russian and foreign indicators of adequacy and its criterion values, describes the issues of modelling the adequacy assessment of power grids and practical results of research.**

Keywords: UPS of Russia; reliability criteria; power system; adequacy parameter

I. Introduction

Ensuring adequacy in management of development of electric power systems (EPS) has always attracted attention. That is a requirement for strategic development of electric power engineering for the period of 5-20 years. The same is being done for the Unified energy system (UES) of Russia. A really strict hierarchical system comprising effective models was built in the 1970-ies of the XX century. They allowed solving the whole complex of problems in management of development of the UES power systems. Apart from solving optimization problems this system included problems of adequacy indicators assessment and justification of facilities [1].

The decisions about means of ensuring adequacy were made by the project Institute "Energosetproekt" and it was supervised by the Ministry of Energy of the Russian Federation. This Institute was in charge of standards development to ensure adequacy of the Russian EEC. They used to be the part of the normative technical documents (NTD) – guidelines for the energy systems design. In order to make the grounds for those standards some special software was developed. That approach allowed to create reserves of generating capacity in associations of EES for the short-term period of 5-7 years and for the long term period of 20 years.

The described arrangement of activities in justification of means to ensure reliability in our country is very close to that adopted in the Unified power system of North America. This work in North America is made by American Electric Reliability Corporation - NERC, they carry out annual analysis of the adequacy for the period of 2 - 10 years period of time. It also develops mandatory reliability standards, including adequacy. North American [2, 3] and Russian [4-6] school for ensuring adequacy of the EPS has a long traditional history. They were developed in parallel taking into account the specific characteristics of the electric power engineering of their home countries.

Electric power engineering has been reformed in Russia since the mid 1990-ies of the last century. It has led to reduced interest in ensuring adequacy in considering options for the UES development in Russia. Many factors have influenced that: rapid (one-third) reduction in electricity consumption, change of forms of ownership, deterioration of the quality management staff in electric power engineering, the closure of branch institutions etc. The Federal law "About electric power engineering" [7] described the need to solve this problem. Issues concerning the means of ensuring reliability became responsibility of the leading in power engineering sector organization of JSC "SO UES".

Since 2009 JSC "SO UES" jointly with JSC FGC UES annually has been implementing "Scheme and program of development of UES of Russia for a seven-year period" (hereinafter – SPD). In 2010 – 2012 JSC "so UES", attracting professionals in the field of reliability of other organizations, were developed regulating documents: the project of technological rules of work of the EPC (rules of functioning – PTF) and the new NTD. These documents clearly stated the purposes of the required calculations of adequacy indicators, i.e.:

- definition of the required power reserves and reserves of transmission capacity of electric networks for the period of 7-years (annually);
- forecast of development in the frame of the General scheme of power facilities location for the period of 15-years (once in 3 years).

II. Full power reserve and its components.

When planning development of the power energy sphere industry the balance of power for the EEC is always ready for the whole sector and for its components. Form of balance in relation to the NTD includes two positions (Fig. 1):

- the expenses side (demand): maximum power consumption, perspective statutory reserve generating capacity, export capacity in power system of foreign countries;
- incoming part (cover): rated power, power limitation to the maximum load, etc.

In the expenditure part of the balance normative perspective of reserve generating capacity is the most important point to ensure the reliability. In the NTD, it corresponds to the normative (estimated) reserve. Regulatory perspective, reserved of generating capacity is required to ensure reliable operation of the power system. It includes the repairing, compensatory, and strategic reserves. Such division is conditional, and these components are considered independent. That simplifies the solution of the problem of the justification of the normalized power reserve a lot.

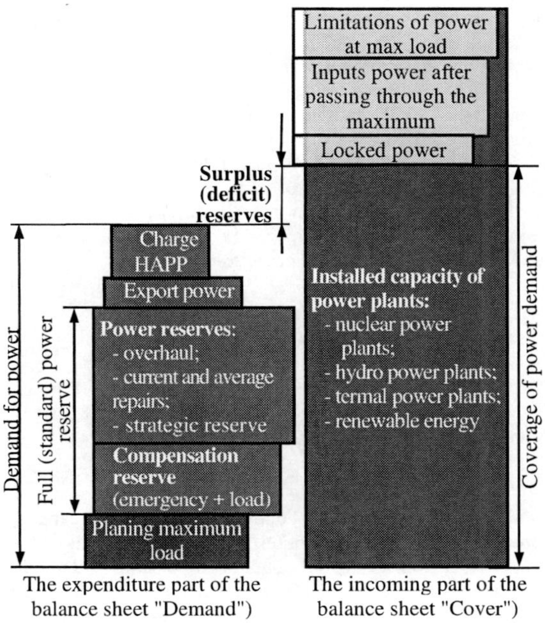

Fig. 1. The structure of the projected balance of power

Compensatory reserve is used to compensate: decrease of power generators at electric power stations in emergency situations, occasional excess load of electricity consumption. This might be caused by the forecast errors and temperature variations. Therefore, the compensation reserve is divided into emergency and loaded.

Repairing power reserve is made to compensate power generating equipment of power plants that is going to be maintained according to the schedule (average, current and capital). Taking the methodical approach in to consideration of mode of electric power consumption in a typical day of December peak load, which lasts during the year, is conventionally assumed that the target average and major repairs are carried out when the load is reduced during the summer. This allows not to consider them when assessing the performance adequacy of the EPS. In development of NTD of different years the repair reserve capacity was determined only by the normatives for current maintenance given in reference literature, for example, in [11].

In 2008, Russia has got a power market. These changes have led to the loss of the regulatory role of norms of duration and frequency of repairs. Similar things happened with the perspective five-years plans for repairs of the main equipment at power plants. As a result, in the current market conditions, scheduled maintenance during the autumn-winter period is defined by not the norms for carrying out repairs, but technical and financial capacity of energy companies. And for sure taking into account the given sensitive balance of the situation of energy system and the rules of the wholesale market of electric power (capacity).

Due to the listed reasons there is the need to update methodological approaches to determine perspective of maintenance component of the power reserve. It should be a compromise between pre-existing norms, statistical data on maintenance, balance between market relations and technical needs.

Introduction to balance of power strategic reserve is due to the conditions of the energy security of the state. In the latest edition of NTD its role was to ensure the balance of power due to unforeseen deviations of its components from the forecasted values. This is due to the inertia of the construction of energy facilities, as well as with volumes and rates of technical re-equipment of power plants.

Current development of economy, introduction of new energy technologies, reduction of terms of input of new and reconstruction of existing generating equipment and available for the recent 25 years, a significant backlog of excess capacity in UES of Russia questioned the need to maintain a strategic reserve of power in such volumes.

The regulatory definition of reserve generating capacity is difficult mainly due to the need to find compensatory reserve capacity. It depends on many random factors and events. Of course, all components of the regulatory reserve generating capacity affect the performance and adequacy. However, only the component of the compensatory reserve of capacity is possible and depends on random changes in load, generation, and transmission capacity of electric connections. Two other components that are planned are deterministic values. This is especially important when addressing the first of those listed tasks mentioned in the previous section, task adequacy.

Influence of random factors and events can be evaluated by certain indicators of reliability. Therefore, in accordance with [5], to the problems of adequacy of the EPS, including management and development of UES of Russia include only those whose solution is dealing with:

- analysis of failures due to accidental damages of the system elements (emergency component of compensatory reserve);
- taking into account irregular fluctuations and forecasting errors in load (load component of compensatory reserve).

Development of generating sources and electrical connections requires significant financial costs and has inertia. That should be reflected at the levels of redundancy taken in long-term development planning of energy systems. That will affect normative values of the received indicators of adequacy.

III. INDICATORS OF ADEQUACY.

Choosing the indicators characterizing the adequacy of the EPS, one should take into account that their number should be minimal and sufficient for making management decisions. The selected indicators of adequacy of EPS should respond to changes in the system which affect the decrease or increase of the system reliability. In domestic [4-6, 9, etc.] and foreign publications [2, 3], these guidelines best meet:

- the probability of deficit-free operation of territorial zones of the EPS ($p = 1 - J_{д}$), where $J_{д}$ is the integral probability of power shortage;
- the average number of days of power shortage (loss of the anticipated load – LOLE [2]);
- the average number of hours per year of power shortages (loss of load hours – LOLH).

The probabilistic indicator $J_{д}$ is widely used in domestic experience [4, 6, 8]. It is defined for all possible time intervals of load change by the following expression:

$$J_{\text{д}} = \sum_{i=1}^{T} Q_i \sum_{k=1}^{N} p_k (\overline{P}_{ik}^{\text{н}} - P_{ik}^{\text{н}}) = \sum_{i=1}^{T} \sum_{k=1}^{N} Q_i p_k (\overline{P}_{ik}^{\text{н}} - P_{ik}^{\text{н}}) \quad (1)$$

where Qi=1/T is the probability of deterministic steps of the load curve, usually equal 1/8760; T is number of stages of the graphics load (8760 hourly changes per year); $\overline{P}_{ik}^{\text{н}}$, $P_{ik}^{\text{н}}$ – respectively, required and provided available generating capacity and reserves capacity links load the k-th random state of the system; p_k – the probability of the k-th random state in which there is a shortage of capacity in the considered territorial area, i.e., when $\overline{P}_{ik}^{\text{н}} > P_{ik}^{\text{н}}$; N is the number of random states that are modeled on the i-th interval of the load change.

The indicators of adequacy in the form $J_{\text{з}}$ is carried out by simulation of random States generating capacity. They are called emergency exits equipment. Modeling is carried out for each hour of the diurnal variation of the load (24 steps) for

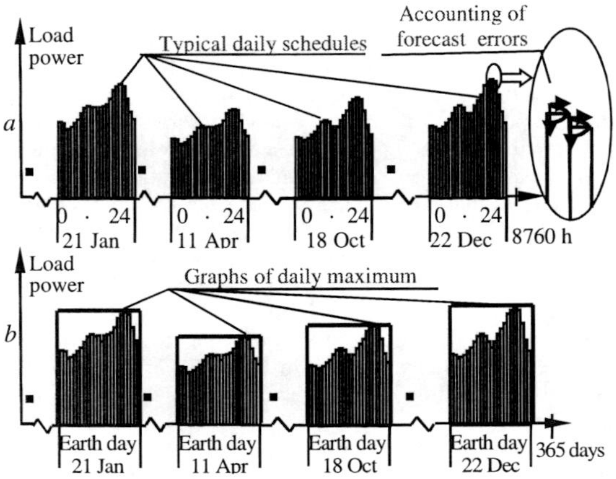

Fig. 2. Presentation mode energy consumption
a) daily schedule during the year per hour (365 days)
b) schedule of daily maximums of all days during the year

365 days of the year or 250 working days. When you define this probabilistic measure of adequacy of a multi-zone EPS, typically used in the presentation of the mode of electricity consumption, shown in Fig. 2, a.

In foreign experience to assess the adequacy two probabilistic measures of adequacy – LOLE and LOLH are mainly used. The definition of the indicator LOLE associated with simulation of random states generating capacity only for peak maximum values [2] (one step daily chart) for all 365 days of the year (Fig. 2, b). The definition of the indicator LOLH largely coincides with the definition of domestic rate $J_{\text{д}}$.

IV. Conditions for optimal reliability and criteria for the valuation indicators of the adequacy of the EPS

The optimal level of reliability of the EPS regardless of principles of management should meet the minimum functionality of given (discount) costs. Taking into account the compensation costs for damages by the unreliability of electricity supply:

$$З_{\Sigma}(\Pi) = З_R(\Pi) + З_L(\Pi) + M[У] \rightarrow \min, \quad (2)$$

where Π – indicators characterizing means of ensuring reliability of the power system (generating capacity reserves of the territorial zones and reserves bandwidth connections); $З_R(\Pi) = \sum_{j=1}^{n} з_{R_j}^{\text{уд}} R_j$ – the cost of the reserve generation capacity R_j all j power systems; $З_L(\Pi) = \sum_{l=1}^{m} з_{L_l}^{\text{уд}} P_l^L$ – costs increase in inventory capacity (P_l^L) l-x lines; $з_{R_j}^{\text{уд}}, з_{L_l}^{\text{уд}}$ – unit costs, respectively, for the creation of the reserve generating capacity in the j-th territorial area and to the strengthening of the capacity reserve (P_l^L) l-x lines; $M[У]$ – costs, compensatory damages consumers because of their long-term emergency limits (usually these costs are called the expected value of the damage to consumers); n or m is the number of territorial areas and of links connecting them in the model design scheme of the EPS.

In long-term development planning of the EPS is usually neglected the changes of generation. This allows us to consider all the components of the functional (2) continuous and convex. Then the optimal level of reliability of the EPS corresponds to the condition of equality to zero of the partial derivatives of total costs in the optimized variable of the system to the power reserves and reserves bandwidth connections. The complexity of solving such system of equations is mainly related to the difficulty of obtaining the partial derivatives from the quantitative indicator adequacy of mathematical expectations of damage to consumers from the undersupply of electric energy. In [6, 9] for the case of equality of the specific damages in the territorial zones, it is shown that the solution can be represented as:

$$\left.\begin{aligned} з_{R_j}^{\text{уд.}} - T_p у_0 J_{\text{д}_j}^{\text{опт.}} &= 0, \quad j = 1, 2, ..., n, \\ з_{L_l}^{\text{уд.}} - T_p у_0 J_{\text{п}_l}^{\text{опт.}} &= 0, \quad j = 1, 2, ..., m \end{aligned}\right\}. \quad (3)$$

The expression does not differ from the optimality conditions of concentrated EPS obtained in [8] for the concentrated EPS:

$$J_{\text{д}}^{\text{опт}} = з_R^{\text{уд}} / (y_0 T). \quad (4)$$

In our opinion in conditions of market relations in power industry optimization of reliability should be undertaken over a range of conditions: providing the required (normative) reliability, minimum costs for the creation of the reserve generating capacity in rural areas and strengthening of reserves bandwidth backbone links. In such case the functional (2) is [6. 9]:

$$З_{\Sigma}(\Pi) = З_R(\Pi) + З_L(\Pi) \rightarrow \min, \ \rho(\Pi) > \overline{\rho}, \quad (5)$$

where $\rho(\Pi)$, $\overline{\rho}$ – accordingly, the indicators of adequacy given the means to ensure reliability (П) and their normative significance (in guidelines 2003 and 2012 shall be equal to 0,996).

In developed countries the minimum value of reserve capacity from unplanned changes must also correspond to some pre-selected probability targets. Some of the normative values of LOLE and LOLH exist have at the expert level in North America and Western Europe have. They serve as indicators in decision-making. In the United States and Canada, the company NERC conducts annual review of adequacy for the period of 2 - 5 years or 6 - 10 years. The minimum amount of reserve power in these countries should correspond to the index of adequacy is not higher values of LOLE = 0,1 d./year. Common standards for some European countries [12]: France LOLH = 3 h/year, the UK and the Netherlands – LOLH = 4 h/year, Ireland – LOLH = 8 h/year, the Scandinavian countries – LOLP = 0,001.

V. Influence of various factors on decision-making while preparation of the NTD

In majority of all NTD, the problem of determination of magnitude of compensatory reserve capacity was carried out by OJSC "Energosetproekt Institute" with experts from Institute of Socio-Economic and Energy Problems of the North Komi scientific centre of the Ural branch of the Russian Academy of Sciences. Indicators of adequacy of the EPS depend mainly on the influence of the following conditions and factors, including probabilistic and unambiguous:

- model of the design scheme of UES of Russia, selected territorial areas and connections between them;
- available capacity of some territorial areas;
- structure of generating capacities;
- scheduled repairs of equipment;
- reduction of generation capacity in rural areas due to accidental damage of units of power plants;
- regular and irregular highs loads of territorial zones and graphs of their changes in the context of day and year;
- irregular fluctuations in the load and errors of forecasting of consumer demand;
- reserves bandwidth connections in normal and emergency modes between the selected in the model design scheme of UES of Russia territorial areas.

It is obvious that all of the above mentioned factors were taken into account when assessing the indicators of adequacy in the process of preparing the reference document for the design of energy systems. While the content of tasks in modern conditions more adequately reflect the actual processes. That should naturally influence the quality of decisions taken.

A. *Model of the designed scheme of UES of Russia*

Scheme with the number of territories was taken as a model of the design scheme of the EEC to justify a share of reserve capacity in NTD and it equals to the number of the United power system (UPS). A number of UPS presented two territorial areas. But the general design scheme does not exceed 10 territorial areas. In the conditions of formation of a share reserve of the 2012 edition of NTD a design scheme with 50 territorial areas was considered. It was developed by OJSC "Institute Energosetproekt" based on instructions of JSC "SO UES". Recently, experts of service of perspective development of JSC "SO UES" for evaluation of adequacy have used different scheme of UES of Russia, a number of territorial areas from 56 to 102. Obviously, a design scheme of UES of Russia affects the performance balance reliability and magnitude of compensatory reserve capacity. Therefore, in carrying out the work "SPD" it is required to carry out the correction values of the compensation reserve of the territorial areas based on the data in the form of fractions of a percent.

The points available capacities of separate territorial zones and structure of generating capacities in significant changes to the content of the task of assessment of indicators of adequacy are not observed.

In the following two factors (scheduled and emergency repairs of equipment) a positive trend in the work of the specialists of JSC "SO UES" should be noted. Information security of generating companies allows to carry out statistical processing of information about maintenance and to get average performance standards for scheduled and emergency load of maintenance. In this case, if earlier, such information was available for a wide range of specialists (guides), today it is only available for employees of JSC "SO UES". In our opinion there is a tendency to overestimate norms of planned maintenance and, conversely, lowering of the norms of emergency. This is due to the transfer of the generating companies of several emergency derived units in the category of planned maintenance and excess capacity. This allows to repair in the winter time periods at the moments of passing the maximum load.

The following two paragraphs relate to the modeling of the power consumption modes (load, irregular fluctuations and forecast errors). To evaluate indicators of adequacy in the justification of compensation of the power reserve used the load curve of the per hour territorial zone. It is typical for a winter working day. But it is necessary to consider its distribution only on the working days of the year (250 days). This is done conventionally. Foreign countries do that the same way. The difference is use of the maximum loads of all days of the year (365 values). As the maximum load in our country we use so-called "regular maximum" - the average load value during the hour of maximum consumption.

Irregular maximum load is determined by the values of the irregular fluctuations. They are usually subordinate to the normal distribution law. According to recent studies of specialists of JSC "SO UES" irregular fluctuations in the load to a large extent depend on weather conditions and are largely correlated between the territorial zones. It radically changes the previously used approach to modelling. Errors of forecast of the regular maximum load are also subordinate to the normal distribution law and depend on the planning horizon.

The last point refers to reserves bandwidth connections between local areas of the model design scheme of UES of Russia. The definition of these data is that there is always a certain margin of error. This can be explained by the fact that the model of the design scheme UES doesn't always reflect the existing processes. At the same time without creating a model design scheme of the UEC the task of performance assessment of adequacy for existing schemes is not possible.

VI. Conclusions

Currently, there is a need to develop new requirements to the content the problem of evaluating the adequacy for analysis of future options for development of the EPS. This wide range of works in the field of collecting and processing the retrospective information on the power consumption modes (especially the random component, caused by a temperature factor), the formation of planned and emergency repairs of the generating and network equipment, and other aspects.

Modern possibilities of information content of the problem of justification of means to ensure reliability, require updating of methodical approaches to the definition of all components of the regulatory reserve generating capacity. It is necessary to find a compromise between pre-existing standards for conducting emergency and scheduled repairs, the available statistical information reflecting the function of energy in the last decade, the balance between market relations and technical needs. From this standpoint, using domestic and foreign experience requires the development of a new guideline to assess adequacy and justification of the redundancy of perspective schemes of development of UES of Russia

References

[1] Je.P.Volkov, V.A. Barinov, "Upravlenie razvitiem i funkcionirovaniem jelektrojenergetiki v uslovijah formirovanija rynochnyh otnoshenij" Izv. RAN. *Jenergetika*, 2002. № 5. pp. 37–48.

[2] R. Billinton, *Power System Reliability Evaluation*. New York, London, Paris. Gordon and Breacbe Science Publishers, 1970. p. 299.

[3] R. Billinton, W. Li, *Reliability Assessment of Electric Power Systems Using Monte Carlo Methods*. New York and London, Plenum Press, 1994, p. 351.

[4] G.A. Volkov, "Optimizacija nadezhnosti jelektrojenergeticheskih sistem". M., *Nauka*, 1986, p. 117.

[5] Ju.N. Rudenko, I.A. Ushakov, "Nadezhnost' sistem jenergetiki". M.: Nauka, 1986. p. 252.

[6] Ju.Ja. Chukreev, "Modeli obespechenija nadezhnosti jelektrojenergeticheskih sistem". Syktyvkar: Komi NC UrO RAN, 1995. p. 176.

[7] Federal'nyj zakon Rossijskoj Federacii ot 26.03.2003 g. № pp. 35-03«Ob jelektrojenergetike».

[8] I.M. Markovich, *Rezhimy jenergeticheskih sistem.* M.: Jenergija, 1969. p. 351.

[9] Ju.Ja.Chukreev, M.Ju. Chukreev, "Modeli ocenki pokazatelej balansovoj nadezhnosti pri upravlenii razvitiem jelektrojenergeticheskih sistem". Syktyvkar: Komi NC UrO RAN, 2014. p. 207.

[10] *Metodicheskie ukazanija po proektirovaniju razvitija jenergosistem.* (Utverzhdeno Prikazom Minjenergo Rossii ot 30 ijunja 2003 g., № 281). – M.: Minjenergo RF, 2003.

[11] V.V.Ershevich, A.N.Zejliger, G.A.Illarionov i dr.; Pod red. S.S. Rokotjana i I.M.Shapiro. "Spravochnik po proektirovaniju jelektrojenergeticheskih sistem". M.: *Jenergoatomizdat*, 1985. p. 352.

[12] Ju.N. Kucherov, Ju.G Fedorov, "Razvitie normativnogo i metodicheskogo obespechenija nadezhnosti slozhnyh jenergosistem i jenergoob#edinenij v uslovijah liberalizovannoj jenergetiki" *Jelektro. Jelektrotehnika, jelektrojenergetika, jelektrotehnicheskaja promyshlennost'*, 2010, № 6. pp. 2-11.

Modeling the Set Mode of Complex Power Grid, Based on Infeed Coefficients

Dauren Akhmetbayev, Arman Akhmetbayev
Department Electric power supply S. Seyfullin KTU, Astana, Directorate of information systems of JSC Kaztelecom
Astana, Kazachstan
Axmetbaev46@mail.ru, Almaty.armik69@mail.ru

Azamat Suleymenov, Michal Kolcun
Department of electric power engineering
Technical university in Košice
Košice, Slovakia
suleymenovaz@gmail.com, michal.kolcun@tuke.sk

Abstract — **The paper states the methods for formation of node current balance equations of reversed form, based on matrix of infeed coefficients. The simplified method for calculation of given currents' distribution coefficients, based on any and all possible graph trees of complex power grid, is suggested. The algorithm for formation of node voltage equations of reversed form, based on matrix of infeed coefficients, is developed. The methodology, being developed, significantly reduces the scope of performed computing efforts and improves the efficiency in receipt of real solutions to equations of set modes.**

Keywords: Power grids, topology, graph theory, matrix of infeed coefficients.

I. Introduction

While analyzing the operation and control of electric energy system modes, there is the need in fulfillment of a great scope of works for calculation of set modes [1, 2, 3]. The equations of set modes for complex electric energy systems are non-linear and can be solved only by iterative methods. The known disadvantages of iterative methods are overcome by different, theoretically substantiated methods. Hence, there are a great number of commonly used programs for calculation of set modes [4, 5]. While choosing the most effective iterative methods, the special attention is paid to the provision and rapidity of convergence for iterative process.

The paper is devoted to solution of methodological problems in analysis and calculations of set modes for electrical power systems of higher dimension. The first papers in the sphere of set modes were based on use of bus admittance matrix as it is formed directly from the topological layouts, which turned out to be of little use for a complex power grid.

The researches, being performed [6, 7, 8], proved that the difficulties in receipt of real solutions at calculation of set modes may significantly be overcome, if we base on node equations of reversed form. The difficulties in formation of node equations of reversed form were to a certain degree overcome using the hybrid matrixes that allowed combining their advantages through presentation as the products of sparse matrixes applying different matrix factorization methods and iterative methods [9, 10, 11].

This paper improves the methodology for formation of node voltages of reversed form using the matrixes of infeed coefficients.

II. Topology of power grid and calculation of infeed coefficients

2.1 General provisions

The given current distribution coefficient of j-node in i-ranch is the ratio, defined by a complex number [12]:

$$- \qquad (1)$$

At, the expression (1) will have the following form:

$$(2)$$

It is enough to have all infeed and given currents coefficients to define the current distribution in layout [13]:

$$(3)$$

where J - given currents column matrix;
C- infeed coefficients rectangular matrix.

2.2 Topology and infeed coefficients in the grid

The topology of power grids allows establishing the analytical dependence between infeed coefficients and structure of linear graph in the grid. Such subgraphs as trees and 2-trees take the significant place in the topology of grids. A tree is a subgraph that does not contain any closed loops that unites all graph nodes. 2-tree is a subgraph that covers all nodes and consists of two unconnected parts without closed loops.

The numerators and denominators for given currents distribution coefficients in power grids are formed of values for trees and 2-trees of graphs [13, 14].

The paper [15] offers the new approach to definition of numerators for infeed coefficients in function of grid parameters without search and definition of value for 2-trees of graph in complex grid. The specific trees of oriented graph are formed on the basis of any and all possible trees in relation to that branch of layout where it is required to find a fraction of single current in the node under consideration. The elements for matrix of infeed coefficients are calculated pursuant to topological expression:

$$C_{ij} = \frac{\sum_{k=1}^{M} F_{kj}^{i}}{\sum_{k=1}^{N} F_k} \qquad (4)$$

Where $\sum_{k=1}^{N} F_k$ - total value for any and all possible trees of graph

$\sum_{k=1}^{M} F_{kj}^{i}$ - algebraic sum for specific trees of graph in i-branch.

The denominator for all infeed coefficients, included into the grid under study, is the same i.e. it is defined only once by totals value for possible trees of graph.

III. FORMING A MATRIX EQUATION OF SET MODE

3.1 Rearrangements of equation of state

The values for currents in branches of layout will remain to be unchanged, if we multiply and divide the right part of equation (3) in the left by branches resistance matrix, and write by the following way:

$$I = Z_g^{-1} \cdot Z_g \cdot C \cdot J \tag{5}$$

On the other part, the reversed form of node voltages equations looks like [16]:

$$U_\Delta = Z \cdot J, \tag{6}$$

which, taking into account the first Kirchhoff's law may be presented as it follows:

$$U_\Delta = Z \cdot M \cdot I = Z \cdot M \cdot Z_g^{-1} \cdot Z_g \cdot C \cdot J = C^t Z_g \cdot C \cdot J \tag{7}$$

The resultant equation (7) is the solution for node voltages equation and allows writing the identical equation as if follows:

$$Z = C^t \cdot Z_g \cdot C = Y^{-1} \tag{8}$$

3.2 Forming the reversed form of node voltages

The expression (7) allows directly binding matrix for given currents with matrix for voltages in nodes in relation to neutral. After substituting EMF in branches and cross branches with the corresponding given currents the voltages matrix in nodes in relation to neutral is defined by the following expression [17]:

$$U = U_0 + C^t \cdot Z_\pi \cdot C \cdot U_\pi^{-1} \cdot S \tag{9}$$

where

C - rectangular complex matrix of infeed coefficients;

Z_π - diagonal branches resistance matrix;

U_π - diagonal nodes associated voltages matrix

S - column matrix for associated powers of node loads and generators;

3.3 Calculation of set mode in 220kV real grid

The design layout for allocated part of grid for real electric energy system in Republic of Kazakhstan is shown on Figure 1. The branches resistances in equivalent circuit are shown on the figure and the full-rated powers of nodes are known.

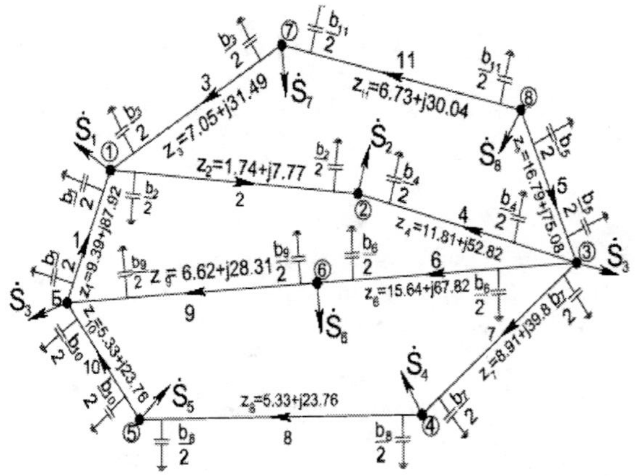

Figure 1. Design layout of power grid

The algorithms for formation of set mode are realized in MATLAB environment. Any and all possible trees of graph are defined pursuant to the theories of structural numbers [18], and the specific trees – pursuant to special developed algorithm for this purpose. The flow chart for realization of algorithms is shown on Figure 2.

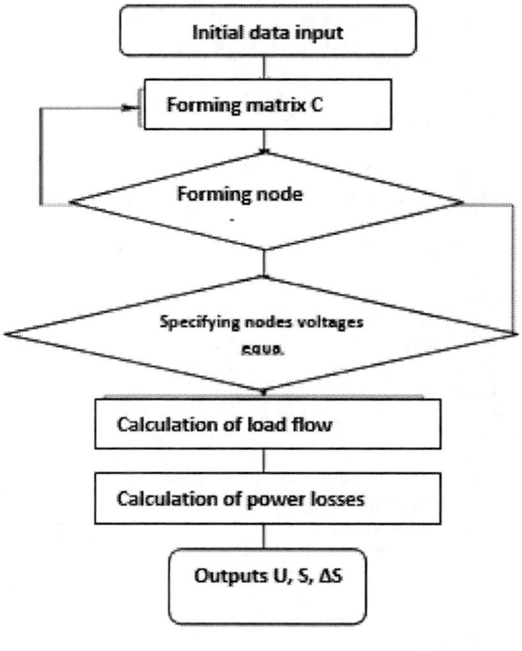

Figure 2. Flow chart for formation and calculation of set mode

Table 1. Calculated value of infeed coeficients

-0.0716 + 0.0043i	-0.0775 + 0.0048i	-0.4487 + 0.0259i	-0.3956 + 0.0229i
-0.0496 + 0.0029i	-0.0537 + 0.0033i	0.2222 + 0.0179i	0.1066 + 0.0159i
-0.0220 + 0.0013i	-0.0238 + 0.0015i	-0.6709 + 0.0079i	-0.5023 + 0.0070i
-0.0496 + 0.0029i	-0.0537 + 0.0033i	0.2222 + 0.0179i	0.1066 + 0.0159i
-0.0220 + 0.0013i	-0.0238 + 0.0015i	0.3291 + 0.0079i	0.4977 + 0.0070i
0.0954 + 0.0024i	-0.1914 + 0.0030i	0.2622 + 0.0133i	0.2874 + 0.0120i
-0.1669 + 0.0019i	0.1139 + 0.0018i	0.2892 + 0.0126i	0.3170 + 0.0109i
-0.1669 + 0.0019i	0.1139 + 0.0018i	0.2892 + 0.0126i	0.3170 + 0.0109i
0.0954 + 0.0024i	0.8086 + 0.0030i	0.2622 + 0.0133i	0.2874 + 0.0120i
0.8331 + 0.0019i	0.1139 + 0.0018i	0.2892 + 0.0126i	0.3170 + 0.0109i
-0.0220 + 0.0013i	-0.0238 + 0.0015i	0.3291 + 0.0079i	-0.5023 + 0.0070i

Table 2. Results from comparative calculation of node voltages

Topological method using matrix C		Checking calculation by RASTR program	
U, kV	δ, grade	U, kV	δ, grade
241.43	-4.8315	240,65	-4,79
241.53	-4.8433	240,78	-4,81
240.89	-4.3086	240,36	-4,28
237.38	-3.0902	237,1	-3,08
235.60	-1.8466	235,45	-1,84
236.27	-1.9230	236,12	-1,92
239.60	-6.2297	238,33	-6,17
239.12	-6.5359		-6,49

The allocated site of real power grid is in fact of low load that is confirmed by calculations. As it is seen from Table 2, the results, received by different methods, provide with good coincidence. The voltage modules, received by topological method, are assessed by changes to the higher side in comparison with results from "RASTR"-a. It should be mentioned that the results, received using ,matrix C, are close to real.

It is explained by the fact that the reactive powers in "RASTR" program, generated by the line in the calculation process, are taken into account by constant values and corrected due to voltage of set mode, whereas the reactive powers of line in the suggested method are specified in each iteration step. The distribution of reactive power along the line significantly influences on values of node voltages. The distribution of active power along the lines coincides as the values for phases of node voltages, received by suggested methods, are the same. Thus the suggested method may be used both within scientific and practical purposes.

IV. Conclusions

1. The improved method for calculation of given currents distribution coefficients was developed.
2. The method for calculation of specific trees in complex oriented graph was developed and implemented in MathLab environment for calculation of infeed coefficients.
3. The calculations of node voltages, made on the example of existing 220 kV grid, showed a good coincidence with the results from operational calculation in "RASTR" program.

References

[1] V.A. Venikov. "Modeling the large electric power supply systems," in *Electricity*, 1978, No.5

[2] V.S. Khachatrian. "Method and algorithm for calculation of set modes in large electricenergy system,s" In *Bulletin of AS of USSR. Energetics and transport*, 1973, No.4

[3] V.A. Venikov. "Calculation of mode for electric power systems due to the data of distance measured, basedon regularization method," In *Bulletin of AS of USSR, Energetics and transport*, 1976, No.2

[4] O.T. Geraskin. "Diakoptics and rarefaction in tasks for calculation of set modes in large electric power systems" In: *Bulletin of higher schools in the USSR. Energetics.* 1982, №1. pp. 7-12.

[5] S.K. Gurskyy. "Algorithmization of tasks to control the modes of complex systems in electrical energy industry," In: *Science and technology*, Minsk, 1977, pp. 367.

[6] K.A. Smirnov. "About convergence of iterative solutions for set mode in electrical power system at given powers of nodes," In: *Bulletin of AS of USSR. Energetics and transport.* 1975, №3. pp.120-129.

[7] Yu.V. Moroshkin, A. Yu. Pozdniakov, S G. Sokolov. "About convergence in methods for calculation of set modes at some linearization methods of node voltages equations," In: *Bulletin of AS of USSR. Energetics and transport.* 1978, №4. pp. 14-22.

[8] O.T. Geraskin. "Set modes in large electric power systems," In: *Bulletin of AS of USSR. Energetics and transport.* 1982, № 1. pp. 7-12.

[9] B. Z. Manusov. "Algorithms of Newton-Rafson method for solution of node equations in reversed form," In: *Bulletin of higher schools in the USSR. Energetics.* 1974, №9. pp. 3-7.

[10] Kh.F. Fazylov."Set mode of electric power systems and their optimization T" In: *Lightning*. 1999. p. 370.

[11] D. Medved, M. Kolcun, Z. Čonka, P. Hocko, Novák, M. "Simulation of faults in the power system and their elimination in the EMTP-ATP" *Proc. 7th International Scientific Symposium on Electrical Power Engineering,* (EE 2013), Stara Lesna, Slovakia, pp. 484-487

[12] N.A. Melnikov. "Application of distribution coefficients at calculations of complex equivalent networks for electrical circuits" In: *Electricity*, № 2, 1961. pp. 9-13.

[13] Z. Čonka. "Impact of renewable energy sources on stability of EWIS transmision system" *Proc. 14th International Conference on Environment and Electrical Engineering,* (IEEE 2014), Krakow, Poland, 10-12 May, 2014. Danvers, pp. 75-79, ISBN 978-1-4799-4661-7

[14] O.Z. Geraskin. "Methods for definition of 2-trees of graph in topology of power grids" In: *Bulletin of AS of USSR. Energetics and transport.* 1967.- №5, pp.106-112.

[15] M. Novák, M. Kolcun, P. Hocko, D. Medveď, R. Petrichenko, I. Zicmane. "Modelling of excitation system influence to transient stability of power system using PSLF," *Proc. 7th International Scientific Symposium on Electrical Power Engineering,* (EE 2013), Stara Lesna, Slovakia 2013, pp. 531-534, ISBN 978-80-553-1441-9

[16] O.T. Geraskin. "Applying the theories of signal graphs to calculation of infeed coefficients in power grids," In: *Bulletin of AS of USSR. Energetics and transport.* №5, 1965, pp. 33-39.

[17] D. Akhmetbayev. "Topological method of calculating current distribution indices," In: *World of Science. Praha:IADE*, 2010, pp. 12-16.

[18] N.A. Melnikov. " Matrix method for analysis of electrical circuits" In: *Energy*. 1972, pp. 232.

[19] D.S. Akhmetbayev. "Method for calculation of set modes in power grids, based on infeed coefficients" In: *Electricity*, № 11, 2010, pp. 23-27.

Supply Reliability of Industrial Distribution System

Jiri Drholec, Radomir Gono, Stanislav Rusek, Miroslava Gono, Tomas Novak
Department of Electrical Power Engineering
VSB - Technical University of Ostrava
Ostrava, Czech Republic
jiri.drholec.st@vsb.cz, radomir.gono@vsb.cz, stanislav.rusek@vsb.cz, tomas.novak1@vsb.cz

Abstract — **The purpose of the research was to determine the supply reliability which should lead to attempt to optimize the operation of the local distribution system in metallurgical plants to ensure high level of supply reliability. The paper deals with collecting data about failures and outages, evaluation of component reliability in metallurgy industry and its comparison with distribution network. Final results of calculated substation reliability supplying important consumer.**

Keywords: metallurgical plant; emission; reliability; distribution substation; backup; local distribution system

I. Introduction

Industries in the EU are subject to emission control at multiple levels. There is a legislation in place, further amended by integration of stricter permission policy within particular regions. The year 2016 brings application of the so called Best Available Technologies in the EU to enforce further reduction of the existing strict levels of permitted emission levels [1]. The agglomerate (ore sinter plant), which holds a vital importance for iron production, represents the dustiest part of the metallurgical plant. To ensure compliance with the required dust limits, metallurgical producers must invest not only into improvement of the environmental friendliness of their smelters yet especially into sustainable efficiency of dust elimination from combustion products. Construction of dust elimination facilities is very costly and it represents a significant financial burden to each metallurgical company. However, the dust pockets offer a brilliant solution to dust elimination needs, helping the metallurgical works reduce their emissions, since this filter is able to capture even the tiniest dust particles, benzo(a)pyrene and dioxins, emissions of polycyclic aromatic hydrocarbons can therefore drop by more than 90 %. The benefits of filters for the living environment are truly unquestionable.

Yet the subsequent operation of electrostatic filters is dependent on the high and especially reliable supply of electric power. Power blackouts impair proper functioning of the dust pocket, since it cannot deliver its full performance [2]. That has a major impact on the entire metallurgical plant operation, which needs to be reduced until full restoration of filter efficiency. The integrated permissions for metallurgy works shows certain level of direct links: if the dust elimination is not working, the production process must be limited or even suspended [3].

The reliability of power supply therefore remains the limiting factor for operation of not only iron production itself but of the whole metallurgical company [4]. This paper deals with operation of the local distribution system (LDS) situated within the premises of the company focused on production and processing of pig iron as well as the metallurgical and machine engineering production. The final part of this paper deals with reliability of power supply to the selected manufacturing plant (6 kV distribution substation).

II. Component Reliability

The most data on reliability of power supply within LDS can be obtained by monitoring and evaluation of its operation [5]. Efficient records of outages and failures with subsequent analyses of their consequences is based on procurement of the set of data for evaluation of various reliability indicators. However, application of the reliability theory requires gathering and processing of large amount of monitoring data. That is why collection of primary data is vital to ensure their further sorting and processing.

The primary data is collected by the head supervisor of control center maintaining permanent presence at this site. The records contain mainly this type of data:

- location and cause of failure,
- circumstances of the failure occurrence,
- course of the failure,
- further impact on operation of LDS,
- type of component affected,
- time of failure occurrence and extinction.

All events associated with operation of LDS are recorded by the dispatcher at the control center manually and it is not designed for filtering of any items or events. That is why software for dispatchers was developed at the VSB - Technical University of Ostrava to automate data collecting.

Precise enumeration of reliability indicators shall be also based on accurate information about the number of elements and their voltage levels. That is why these elements need to undergo the passportization process [6]. The local distribution system operated by the metallurgical company in question contains approximately 1400 cells and MV and HV power cabinets in 52 distribution substations.

Gathering and subsequent analysis of a large amount of data referring to operation of the metallurgical industrial LDS in the period of 2010 - 2014 allowed to design the five-year database to document the overall behavior of the system with respect to its operation, maintenance and failures, including time stamps for each event occurrence [7]. Calculation of certain maintenance indicators was performed using an approach different to the regular standard [8]. Using the actual completion of maintenance operations on equipment, as initially defined by the record at control center is more accurate than using values contained in the charts of internal regulations - the Revision Code to determine periodical performance of check-up and revision operations.

The volume of materials processed comprises 5 478 sheets in A4 format completed by the dispatcher. The number of events linked to operation of the metallurgy LDS directly to be processed was up to 10 000 over the five-year period. This was evidently not just a routine transcription of records into electronic form, yet rather a specialized procedure conducted by the expert familiar with the system wiring. He must be also able to make a decision on specific categorization of each event to filter anomalies that remain concealed to anyone not skilled in the field of electric power engineering.

Modest mathematical tool was employed for enumeration of the essential reliability indicators for elements [9] over the five-year period - Failure rate λ_p, Mean time to repair τ_p, Maintenance rate λ_u, Mean time to maintenance τ_u.

The importance of database of outages with the option to determine reliability indicators is also based on the fact that there are actually available certain databases showing failures within the distribution system networks, yet there is absolutely no database using industrial power network data. Unlike the reliability indicators for distribution networks are obviously different and cannot be subject to mutual comparisons.

Industrial grids differ by pattern, the method for connection into the grid or even geographical conditions, etc. Furthermore, power grids in metallurgy industry are also exposed to many adverse effects, e.g. presence of water, jolts, shaking, vibrations, heavier dust formation, high ambient temperatures or even occurrence of corrosive, chemical or other contaminating materials. That is why the component reliability results are very valuable for the existing voltage level of 6, 22 and 110 kV not subject to analysis yet.

Table 1 provides comparison of selected reliability indicators that have been enumerated from distribution networks by methodology CEZ22/80 for the period of 1975 - 1979, updated results from the period of 2000 - 2014 and from the five-year period of 2010 - 2014 with respect to the metallurgy industry LDS provided by the authors.

Table 1. Comparison of Reliability Indicators

Damaged equipment	Indicator	CEZ22/80	CEZ22/80 up.	LDS
Cable 6 kV	λ / year^{-1}	not included in the database	not included in the database	0.032
	τ / h			8.104
Cable 22 kV	λ / year^{-1}	14.5	4.661	0.022
	τ / h	215	5.710	11.417
Transformer 110 kV / MV	λ / year^{-1}	0.04	0.058	0.333
	τ / h	1 300	0.231	4.403
Transformer MV / MV	λ / year^{-1}	not included in the database	not included in the database	0.130
	τ / h			0.481
Transformer MV / LV	λ / year^{-1}	0.03	0.006	0.200
	τ / h	2 500	5.303	0.361
Power switch 6 kV	λ / year^{-1}	not included in the database	not included in the database	0.025
	τ / h			5.538
Power switch 22 kV	λ / year^{-1}	0.015	0.012	0.039
	τ / h	30	23.580	5.702

III. Substation Reliability

There are utilized obtained component indicators for reliability calculations of a selected distribution substation supplying important consumer in this chapter. Fig. 1 shows the electric power system diagram of supply to the industrial consumer (metallurgical works filters) using the distribution substation DS3 6 kV B. The aim was to determine reliability of supply provided to this substation.

The reliability level was determined using the modified method of reliability diagrams that assumed the following:

- design of a reliability diagram,
- assigning the relevant reliability values to individual electrical components,
- progressive simplification of the reliability diagram.

The uninterrupted supply of electric power to HV and MV distribution substations in metallurgical industrial grids is an essential prerequisite for faultless production process. Any power blackout must be removed and the power supply to the affected area restored as soon as possible. That is why parallel connection of elements is used as the cold backup. As far as this connection is concerned, the calculation will also need to include the switching time t_S.

The calculation itself was completed in several steps. It was necessary first explicitly define and analyze the possibilities of supply of the distribution substation. The next step involved design of the reliability diagram and its further simplification.

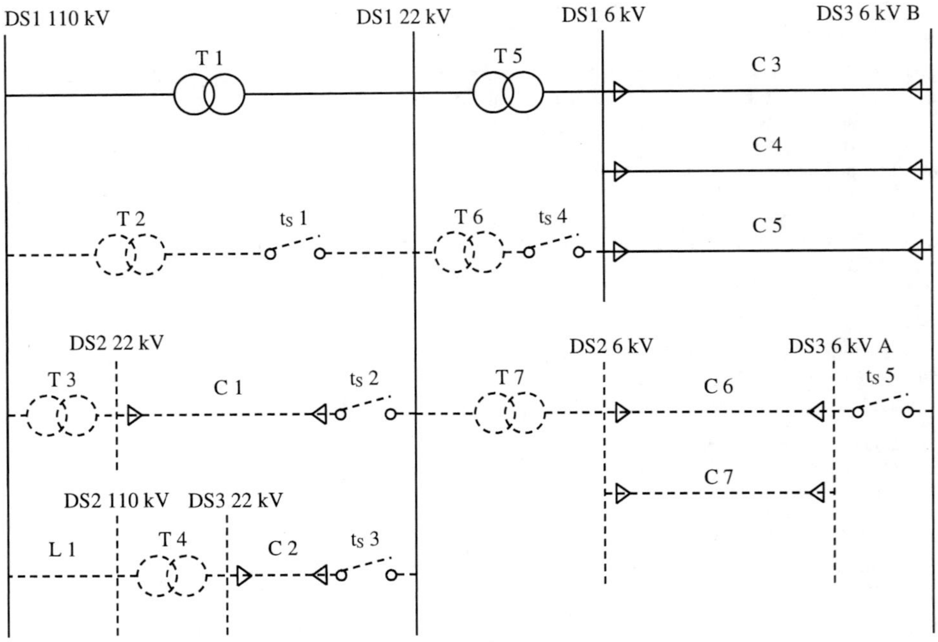

Figure 1. Electric power system diagram.

The results including several backup options obtained through calculations are summarized in Table 2. The table is split into three vertical parts. The first part defines probability of faultless operation of power supply to the distribution substation DS1 22 kV and its backup by three different means, the second one shows the probability of faultless operation from DS1 22 kV to the distribution substation DS3 6 kV B with one backup option. The third part shows the total probability of faultless operation from the superior system 110 kV distribution substation up to the distribution substation DS3 6 kV B. Such big differences among resulting faultless operation probabilities confirm that the backup systems are necessary in metallurgy industry.

Table 2. Overall Results of Faultless Operation Probability

DS1 110 kV → DS1 22 kV with the power supply transformer T 1 and cold backup:				DS1 22 kV → DS3 6 kV B with branch T 5, T 6, C 3, C 4, C 5 and cold backup:		Total DS1 110 kV → DS3 6 kV B
-----	T 2	T 3, K 1	L 1, T 4, C 2	-----	T 7, C 6, C 7	
0.98506849				0.99987806		0.98494837
					0.99993567	0.98500513
	0.99982116			0.99987806		0.99969924
					0.99993567	0.99975684
		0.99998002		0.99987806		0.99985808
					0.99993567	**0.99991569**
			0.99997985	0.99987806		0.99985791
					0.99993567	0.99991552

IV. Conclusions

The entire LDS of metallurgical plant is required to provide highly reliable service with a very low failure rate to avoid any failures in supply of electric power to end consumers, supported by the method of operation and backup employed.

Gathering of a large amount of data referring to operation of the metallurgical industrial LDS enables enumeration of the reliability indicators for significant elements within the local metallurgical distribution system and their comparison with reliability indicators of other distribution systems.

Calculation of certain reliability indicators was performed using an approach different to the regular standard. Those are values of the maintenance rate and the mean time to maintenance. Both values were calculated using the actual completion of maintenance operations on equipment recorded at control center. This approach is more accurate than using the Revision Code to determine periodical maintenance operations.

Rated voltage of the distribution substation selected for the reliability calculation was 6 kV. The supply of power to each distribution substation with any rated voltage is subject to the approved so-called modular connections kept during regular conditions of faultless operation. In case of maintenance request of any power component or its repair, the dispatcher at the control center decides to switch over to the alternate power supply to the particular distribution substation or the area affected. That means each component within the LDS is provided with a backup, this is an essential requirement for metallurgical grids.

Acknowledgment

This work was supported by the SGS grant from VSB - Technical University of Ostrava [grant number SP2017/54]; and the TUCENET [grant number LO1404].

References

[1] I. Klosok-Bazan, B. Gajdzik, J. Machnik-Słomka, W. Ocieczek, "Environmental aspects of innovation and new technology implementation in metallurgical industry", *Metalurgija 54* (2), pp. 433-436, 2015.

[2] L. Prokop, V. Král, Z. Hradílek, "Software OCENS - A tool for outage costs and energy not supply calculation". *Przeglad Elektrotechniczny*, 85 (3), pp. 227-230, 2009.

[3] J. Szkutnik, A. Gawlak, "Optimalization-based method of dividing the network development costs. Electrical Engineering" volume 93, number 3. Springer Electr Eng, 2011, 93: ISSN 0948-7921 doi: 10.1007/s00202-011-0199-1, pp 137-146

[4] M. Kašpírek, P. Krejčí, P. Santarius, K. Procházka, "Operation of Industry Forging Press and Impact on the Supply Grid". In *Sborník Elektroenergetika 2015*, Košice: TU Košice, 2015, pp. 46-49.

[5] M. Kornatka, "Analysis of unreliability of low voltage power networks some branches of Distribution System Operators", *Przeglad Elektrotechniczny* 2012, 88 (10B), pp. 303-306.

[6] R.E. Brown, *Electric power distribution reliability*, Marcel Dekker, New York, 2002, p. 365.

[7] M.T. Todinov, *Reliability and Risk Models: Setting Reliability Requirements*, Wiley, England, 2005, 322 p.

[8] R.E. Barlow, F. Proschan, *Statistical theory of reliability and life testing: probability models*, New York, USA: Holt, Rinehart and Winston, 1975.

[9] Z. Martínek, T. Klor, J. Holý, "Reliability of the electrical power system", *Proc. of the 2014 15th International Scientific Conference on Electric Power Engineering*, (EPE 2014), Brno, 2014, vol. 1, IEEE, pp. 81-84, doi: 10.1109/EPE.2014.6839427.

Optimal PI Fuzzy Controller for Nonlinear Systems

Daniela Perdukova
Dept. of Electrical Engineering and Mechatronics
Faculty of Electrical Eng. and Informatics TU Kosice
Kosice, Slovakia
daniela perdukova@tuke.sk

Pavol Fedor
Dept. of Electrical Engineering and Mechatronics
Faculty of Electrical Eng. and Informatics TU Kosice
Kosice, Slovakia
pavol.fedor@tuke.sk

Abstract — **The paper describes the procedure of designing a PI fuzzy controller for a nonlinear system that guarantees optimal control dynamics in terms of the selected criterion. The fuzzy controller design is based on finding the input signal time sequences that will lead to achieve the control target in an optimal way. On basis of this optimal input sequence it is possible to simply design a fuzzy controller for this system. Finding the optimal input vector involves repeated application of various input vectors to the controlled system and optimality criterion evaluation for each one of them. At this level the fuzzy controller design method enables the use of standard soft computing tools in the design process. Simulation results and experimental measurements have confirmed the correctness of the described method and its applicability to MIMO nonlinear dynamic systems in various fields of industry such printing and textile industry, chemical industry and also in the field of electric power engineering and power system control.**

Keywords: optimal control; fuzzy controller; nonliner system

I. Introduction

There are many processes in technological practice the analytical description of which is rather complicated. This can be due to their complexity, nonlinearity, transfer lags, complicated measurement of important parameters, etc. However, information on the performance of these processes can often be obtained experimentally (by suitably chosen measurements or by monitoring their responses to the control activities of the operator). In these situations, fuzzy systems can always be considered as an alternative for system modelling and control.

Applying fuzzy logic in fuzzy controller design is very often a suitable possibility of solving problem issues in control in various fields of industry because these controllers are an effective tool for achieving high quality properties of the controlled systems [1-5]. The disadvantage in this case is the unsystematic approach to their synthesis and a relatively demanding analysis of their stability. A fuzzy controller design is primarily based on the fuzzification of its range of inputs and the setting up of rules of its behaviour within this range. The behaviour of classic fuzzy controllers was designed on basis of linguistic rules obtained from experts. However, this knowledge is not always easily obtained, especially in cases of higher order nonlinear systems [6,7]. For this reason, special attention has been focused in recent years on the design of fuzzy control systems that are not based on the search for expert linguistic rules [8-11].

The paper is dealing with the design of nonlinear PI fuzzy controller for a two-axis mechatronic MIMO system. The proposed method only requires a suitable set of information on controlled system input/output relations. Fuzzy controller design is based on the idea that if there exists for a controlled system a satisfactory input signal time sequence that will achieve the control target in an optimal way, it is then possible to simply design a fuzzy controller for this system on basis of this sequence. The properties of the fuzzy controller design method have been verified by simulations in MATLAB programme and also by experimental laboratory measurements on a real nonlinear dual axis mechanical system.

II. A Dual Axis Mechanical System

In robotic applications often occur mechatronic multi-axis systems, where the axes influence each other and jointly in terms of control create a complex nonlinear MIMO system. The structure of dual axis mechanical system is shown in Fig. 1.

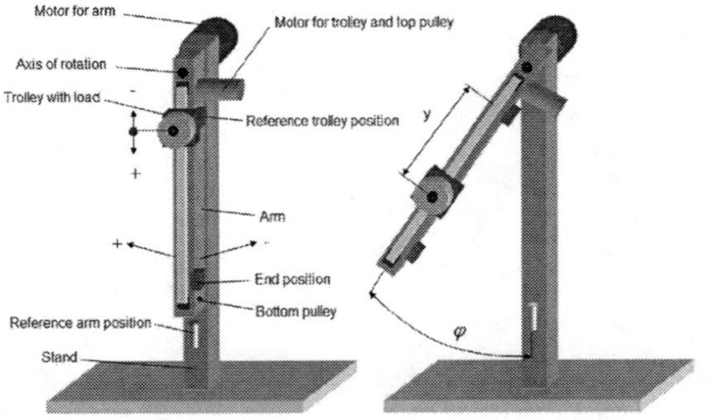

Figure 1. Structure of nonlinear dual axis mechanical system

The system consists of a swivel moving arm with an attached trolley with load that moves along the arm. Both arm and trolley movement are generated by dedicated drives with separately excited DC motors. Using Lagrange´s equations an analytical model of this system was derived and on basis of this a simulation model was constructed in the Matlab programme using the Simulink/SimMechanics tool [12]. It is a 4th order MIMO nonlinear mechanical system (NMS) with two inputs and four outputs ($y_1=x_1$, $y_2=x_2$, $y_3=x_3$, $y_4=x_4$) with structure as illustrated in Fig. 2, where two subsystems can be distinguished: the arm subsystem and the trolley subsystem. The physical denotation of the individual model quantities in Fig. 2 is as follows: u1–armature voltage–arm motor [V], u2–armature voltage–trolley motor [V], y1–arm position [degree], y2–arm movement speed [rad/s], y3–trolley position [m], y4–rotation speed of trolley motor shaft [rad/s].

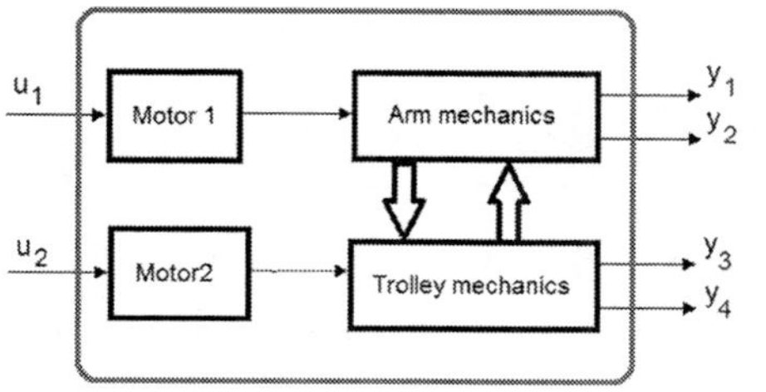

Figure 2. Nonlinear mechanical system as a MIMO system

The position of the arm is specified in the range from 0 to 180 (degree) in the positive direction and from 0 to -180 degree in the negative direction. The trolley position is from the interval <0, 460> mm, while the reference position for the trolley is the top position.

The goals of the system´s control are the following:

- Optimal dynamics during the movement of the arm in space in terms of the selected criterion
- Autonomous setting of the input quantities of the individual subsystems according to the desired values (unbonding of the subsystems)
- Robustness of control, especially with regard to the different moments of inertia of the load
- Compensation of the unmeasurable additive malfunctions.

III. OPTIMAL FUZZY CONTROLLER DESIGN METHOD

The whole algorithm of the nonlinear system fuzzy controller design can be summarized into the following two steps:

A. *Finding an optimal control sequence of input values for the controlled system.*

In this step, the NMS simulation model was used for identifying the optimal combination of discrete values of input voltages $du_{1opt}(t)$ and $du_{2opt}(t)$ for achieving the control goal.

When we are looking for the optimal input control signal we subject the simulation model of the controlled system to various input signal sequences which will secure the optimal transition of the controlled system into the desired state in accordance with the selected optimality criterion according Fig.3.

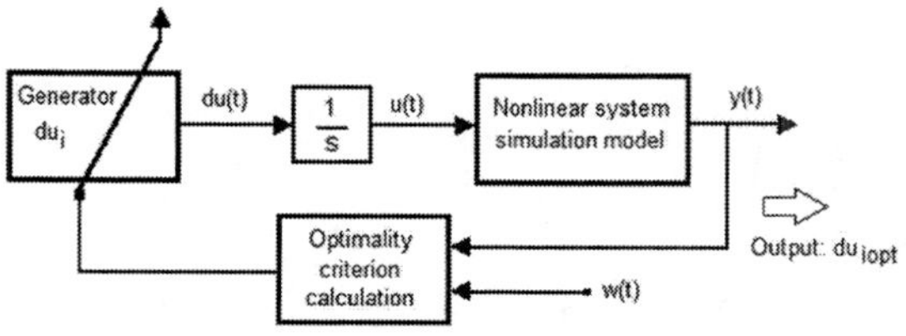

Figure 3. Finding the optimal input sequence du_{iop}

For the sake of simplicity we can chose the optimality criterion in the form of the integral of the quadratic deviation of system output *y* from desired value *w*:

$$J(e) = \int (w(t) - y(t))^2 dt \qquad (1)$$

Finding the optimal input vector involves repeated application of various input vectors to the controlled system fuzzy model and criterion (1) evaluation for each one of them. The sampling time of this sequence, i.e. the number of input vector samples has to be selected in terms of the Shanonn-Kotelnik theorem. However, if there is a simulation model of the controlled system available, the whole process of finding the optimal input vector can be fully algorithmed or it is also possible to apply methods exploiting genetic algorithms for the search [13,14]. The block diagram for identifying the optimal sequence of input values du_{1opt} (similar also applies to du_{2opt}) is shown in Fig. 4.

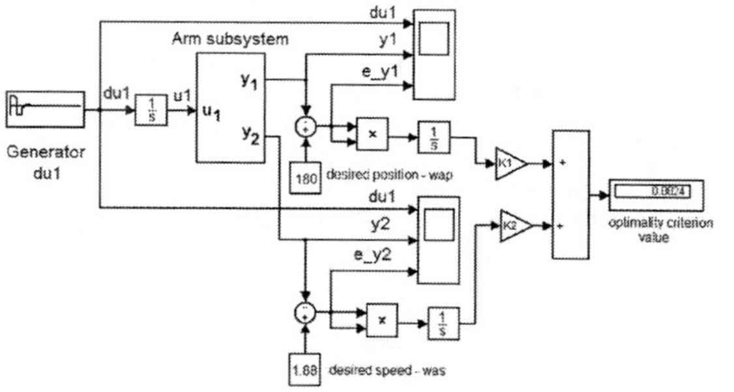

Figure 4. Block diagram for identifying optimal input vector du_{1opt}

For example, for the desired arm position value w_{ap}=180° and desired value of its speed w_{as}=1,88 s^{-1} (nominal value of motor speed) we found the optimal vector of the input sequence du_{1opt}=[24 24 20 -24 -24 -24 -6 0 3 2 1.5 1 1 0.5 0.1 0...] with sampling time T=0.5 *s*. The value of the criterion function for this identified input vector was du_{1opt} J=0.8824. Similarly, for the desired trolley position value w_{tp}=460 *mm* and the desired value of its speed w_{ts}=15.7 s^{-1} (nominal value of motor speed) we found the optimal vector of the input sequence du_{2opt}=[24 -9 0 -21 2.8 0 2 0.3 0.2 0.2 0.2 0.1 0.1 0.1 0 0 0 ...] with the same sampling time T=0.5 *s* and with value of criterion function J=0.8245. The sequences of optimal input vectors du_{1opt} and du_{2opt} were similarly identified for other desired input values of both subsystems, while the ranges for desired arm position and desired trolley position were divided into 10 levels.

B. *Design of fuzzy controller for the system´s fuzzy model.*

A PI-type discrete controller is generally described by the equation [15, 16]:

$$u_k = u_{k-1} + q_0 e_k + q_1 e_{k-1} \qquad (2)$$

where u_k, u_{k-1} are values of controller output in the relevant sampling steps, e_k, e_{k-1} are values of the control error, and q_0 ,q_1 are parameters of controller [17,18].

Its discrete characteristic can then be described by means of the following fuzzy rules:

$$\textbf{IF } e_k \textbf{ is ... AND } e_{k-1} \textbf{ is ... THEN } du_k \textit{ is ...} \qquad (3)$$

where the formulations e_k is ... , e_{k-1} is ... and du_k is ... represent the relevant fuzzy areas of the given variables.

According to equation (3), for the design of the concrete PI fuzzy controller we need to know the database of triplets [e_k , e_{k-1}, du_{1k}] which describe the relation between its input and output. We can obtain this database by applying the sequence of the optimal input vector du_{1opt} (or du_{2opt}, resp.) for all desired input values of both subsystems to the corresponding NMS simulation model inputs, as it is illustrated in Fig. 5.

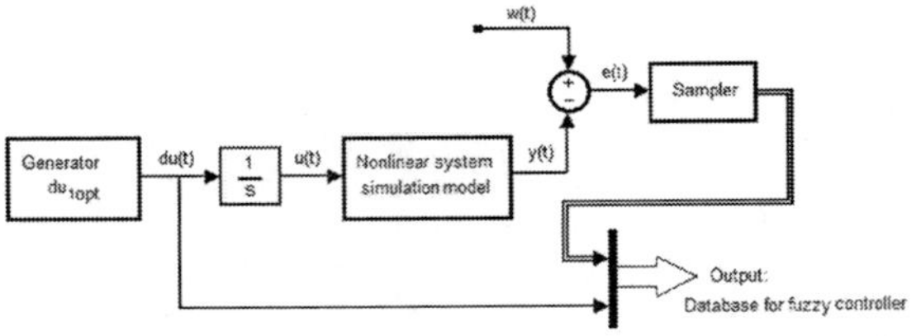

Figure 5. Identifying the database for fuzzy controller design

In this way we can obtain two databases (for the construction of fuzzy controllers for two subsystems: arm and trolley). The mentioned databases are then the foundation for the proposal of the fuzzy description of the controllers by means of the Anfis Editor GUI tool in the Matlab programme [19] where we search for FIS structures of fuzzy systems on basis of relations $[w_{apk}, e_k, e_{k-1}] \rightarrow du_{1k}$, or $[w_{tpk}, e_k, e_{k-1}] \rightarrow du_{2k}$ respectively. The method used for the fuzzy system generation was the subclustering method, having the following parameters: Range of influence=0.3, Squash factor=1.25, Accept ratio=0.3, Reject ratio=0. The proposed fuzzy controllers are represented by Sugeno type fuzzy systems with 10 rules for the subsystem of the arm and 9 rules for the subsystem of the trolley. Characteristic surfaces of the arm and trolley controller fuzzy systems are shown in Fig. 6.

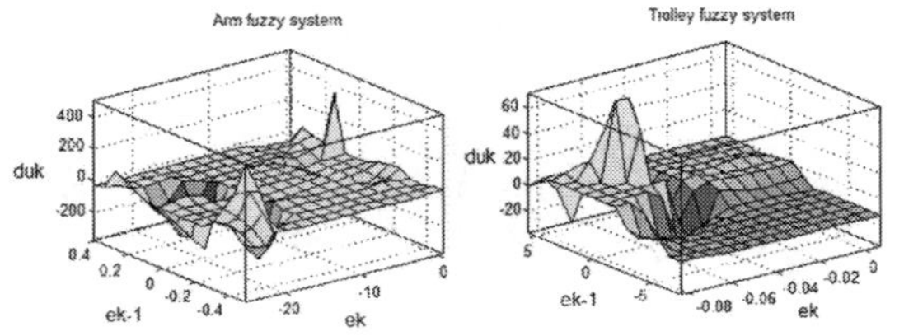

Figure 6. Characteristic surfaces of the arm and trolley fuzzy controller

IV. VERIFICATION OF THE PROPOSED FUZZY CONTROLLER AND DISCUSION

The proposed PI fuzzy controller has been verified by simulation and by experimental measurements on a real system physical model of which is in Fig.7.

Figure 7. Physical model of mechanical system

The block diagram showing the connection of the fuzzy controller for verification is presented in Fig. 8.

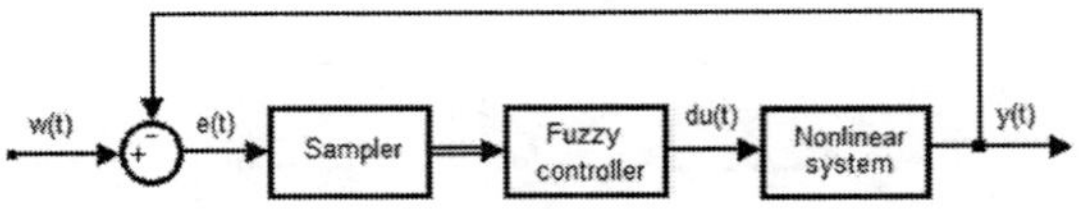

Figure 8. Fuzzy controller connection for verification

Figure 9 illustrates simulation results of the control of the arm and trolley on basis of which optimal input sequences were identified and fuzzy controller behaviour was proposed. It is clear that the proposed controller meets all the basic control targets stated in Chapter 2. The dynamics of setting the desired arm and trolley positions is without overshoot on the limit of aperiodicity, and there is practically no mutual affect between the settings of the arm and the trolley positions, whereby the requirement of autonomy for this MIMO system is fulfilled.

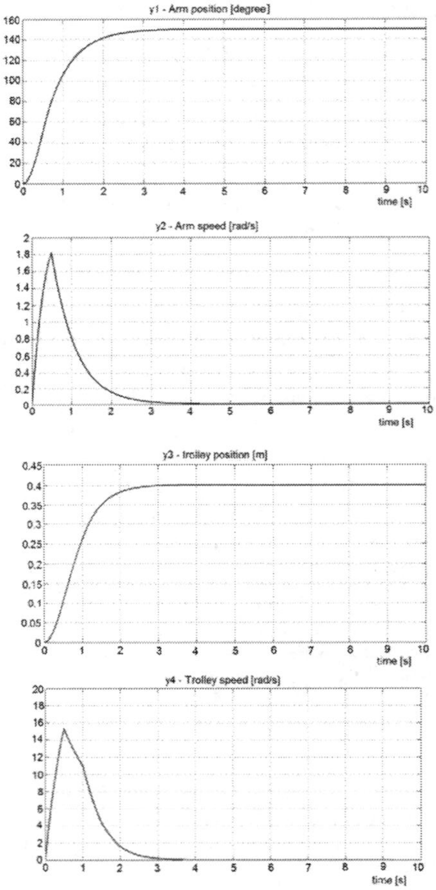

Figure 9. Optimal movement dynamics of NMS for individual axes

Results of experiments show that the proposed fuzzy controller, in spite of its simplicity and uncomplicated computer oriented design procedure, enables considerable improvement of the control circuit dynamic properties also in case of nonlinear MIMO controlled systems, where also the requirements for autonomy and robustness are met, as demonstrated in Fig. 10. The measured courses demonstrate the robustness of the proposed control, where the dynamics of control has been maintained also at the change of load weight in the ratio of 1:3. The robot reaches the desired final position within the time specified with very good accuracy with regard to the chosen PI character of the fuzzy controller.

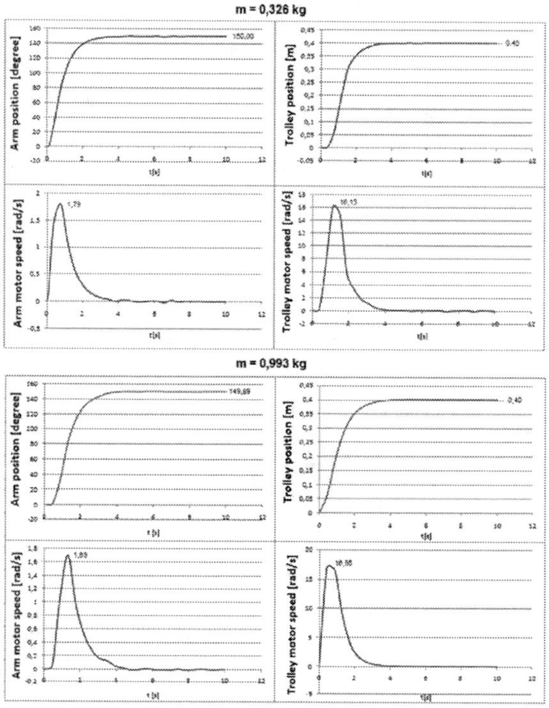

Figure 10. Experimentally measured dynamics of the proposed fuzzy controllers

The method of fuzzy controller construction is fully explained for one specific task, which is the transition of the system from one operating point (in this case the position of a trolley–load with weight m) to a second operating point. The whole construction process can be fully automated, e.g. in the MATLAB environment. During the automation process, suitable m-files were generated and standard tools for work with fuzzy systems were employed (Anfis Editor GUI). As this procedure can be fully algorithmized, it is in principle quite simple to use suitable computation means and design an optimal PI controller for all considered transitions within the selected operational workspace. This will increase the computational complexity, but the controller will be optimal for the whole operational workspace.

The presented method assumes optimization of a controller for a known value of $w(t)$ in equation (1). Although, for the sake of simplicity of presentation, responses for $w(t)$=const were given in the example, the presented method can similarly be applied to a prescribed known vector of the desired value $w(t)$ that varies in time, and in this way an optimal controller for its monitoring can be designed. In this sense the described method is also suitable for the tracking problem solution, similarly as it is stated in [20]. The difference lies in the approach to solving the problem of controlling an unknown nonlinear system. In the paper [20], an analytical approach and a standard feedback structure for tracking problem solution are chosen. The fuzzy model and fuzzy controller are a kind of complement to the analytical control structure and have to be described by the relevant mathematical membership functions, the controller structure has to be mathematically precisely defined and all state quantities of the controlled system have to be available. In the method presented in this paper a directly opposite (fuzzy) approach to the problem solution was chosen, based on the principle of *incompatibility* [8]. The fuzzy controller has a simple PI structure, and its parameters are determined by means of simulation of the found optimal vector for the concrete given task. The results obtained are reliable with "fuzzy certainty", i.e. with certainty that is sufficient for practical application in the majority of real systems.

V. CONCLUSION

The paper briefly describes the design procedure of a fuzzy controller that will establish optimal control dynamics in terms of the chosen optimisation criterion. The proposed procedure has been verified by experimental measurements on a concrete nonlinear mechatronic system. Results of experiments show that the controller, in spite of its simplicity and uncomplicated computer oriented design procedure, enables considerable improvement of the control circuit dynamic properties also in case of nonlinear MIMO controlled systems.

The method described enables us to choose various optimality criteria or their combinations according to the demands of the particular technology, which is a major advantage from the practical point of view. In practice, very often the desired trajectories of movement of the controlled system´s state vector are known in advance (e.g. work cycle in continuous material processing lines, in an automated robotic line, etc.), and therefore it is easily possible to optimize the controller directly for these trajectories. For example, the selection of a suitable energetic optimality criterion can bring about marked energy savings.

The main aim of the presented paper was to develop a fuzzy control design approach with as little system information as possible and to avoid the heuristic search for a rule-based fuzzy controller.

With this method no principal limitations for the investigated system´s nonlinearities are defined, and therefore there is good reason to assume that the presented method will find wide use in various sectors of industry, for example in steel industry, in printing, textile and chemical industries and also in electrical power engineering industry for power system control.

ACKNOWLEDGMENT

The authors wish to thank the project VEGA 1/0464/15 for its support.

REFERENCES

[1] S.A. Mousavi, M. Hashemipour, M. Sadeghi, J.S. Petrofsky, M.A. Prowse, "A fuzzy logic control system for the rotary dental instruments," *Iranian Journal of Science & Technology, Transaction B: Engineering*. 2010; Vol.34, No. B5, pp. 539-551.

[2] J. Amaral da Silva, G. Luiz de Oliveira Serra., "A new methodology for fuzzy model based robust control for nonlinear systems", *Proc. IEEE International Conference on Fuzzy Systems (FUZZ-IEEE)*, 2012, pp. 1-8, doi: 10.1109/FUZZ-IEEE.2012.6251329.

[3] H. Beirami, M.M. Zerafat,"Self-tuning of an interval type-2 fuzzy PID controller for a heat exchanger system", *Iranian Journal of Science & Technology Transactions of Mechanical Engineering*, 2015, Vol.39, No. M1, pp. 113-129.

[4] P. Fedor, D. Perduková., "A Simple Fuzzy Controller Structure", *Acta Electrotechnica et Informatica,* 2005; Vol. 5, No. 4, pp. 53-56.

[5] P. Girovský. "Fuzzy Control of Synchronous Motor with Permanent Magnet", *Acta Electrotechnica et Informatica.* 2016; Vol. 16, No. 4, pp. 17-20.

[6] S.E. Shafei, S. Sepasi, "Incorporating Sliding Mode and Fuzzy Controller with Bounded Torques for Set-Point Tracking of

Robot Manipulators", *Elektronika ir elektrotechnika*, 2010, No.8, pp. 3-8.

[7] B.V. Rami Redy, P. Sujatha, Y.V. Siva Reddy, "Fuzzy Logic and PSO Based Hybrid Technique formulation for optimal placement and sizing of interline power flow controller", *Acta Electrotechnica et Informatica*, 2015; Vol. 15, No. 1, pp. 50-60.

[8] P. Fedor, D. Perduková, "Model Based Fuzzy Control Applied to a Real Nonlinear Mechanical System", *Iranian Journal of Science and Technology, Transactions of Mechanical Engineering*, 2016, Vol. 40, No.2, pp. 113-124.

[9] P. Palacky, K. Baresova, M. Sobek, A. Havel, A., "The control system of electrical energy accumulation", *Proc. of 11th International Conference on ELEKTRO 2016*; Institute of Electrical and Electronics Engineers Inc. Press, May 2016, pp. 346-349.

[10] J. Vittek, P. Butko, B. Ftorek, P. Makyš, L. Gorel, "Energy Near-Optimal Control Strategies for Industrial and Traction Drives with A.C. Motors", *Mathematical Problems in Engineering*, 2017, 22 pages/ID 1857186. DOI: https://doi.org/10.1155/2017/1857186.

[11] B. Li, X. Fan, D. Zhang, G. Jiang, "Modeling and fuzzy sliding decoupling control of looper multivariable system", *Proc. of Chinese Control and Decision Conference (CCDC)*; 2016, pp. 2467-2472, doi: 10.1109/CCDC.2016.7531400.

[12] Š. Juško, *Nonlinear control of two axes mechanical system* Disertation work, 2012, pp.110.

[13] P. Brandstetter, M. Dobrovsky, "Speed Control of AC Drive with Induction Motor Using Genetic Algorithm," *Advances in Intelligent Systems and Computing*, Springer Press, Vol. 189, pp. 341-350, 2013.

[14] J. Žilková, P. Girovský, M. Batmend, „Modelling the technological part of a line by use of neural networks", *Advances in Intelligent Systems and Computing*, Springer Press, Vol. 239, pp. 349-358, 2014.

[15] P. Palacky, P. Brandstetter, P. Chlebis, V. Sladecek, P. Simonik, D. Slivka, "Control Algorithms of Propulsion Unit with Induction Motors for Electric Vehicle", *Advances in Electrical and Computer Engineering*, Vol. 14, No. 2, 2014, pp. 69-76, doi: 10.4316/AECE.2014.02012

[16] J. Dudrik, M. Pástor, M. Lacko, R. Žatkovič, "ZVZCS PWM DC-DC converter with controlled secondary rectifier for arc welding", *Acta Electrotechnica et Informatica*, Vol. 16, No. 2, 2016, pp. 33-41.

[17] I. Neborak, M. Kuchar, "Load torque impact on DC motor current control accuracy", *Proc. of 11th International Conference on ELEKTRO 2016*, Institute of Electrical and Electronics Engineers Inc. Press, pp. 341-345, May 2016.

[18] V. Šlapák, K. Kyslan, M. Lacko, V. Fedák, F. Ďurovský, "Finite control set model predictive speed control of a DC motor", *Mathematical Problems in Engineering*. Vol. 2016, no. ID 9571972, pp. 1-10, 2016

[19] A. Hüseyin, O. Faruk, "ANFIS & PID controller design and comparison for overhead cranes", *Indian Journal of Engineering & Materials Sciences*, Vol. 18, pp. 191-203, 2011.

[20] Chung-Shi Tseng, Bor-Sen Chen and Huey-Jian Uang, "Fuzzy tracking control design for nonlinear dynamic systems via TS fuzzy model", *IEEE Transactions on Fuzzy Systems*, Vol. 9, No. 3, pp. 381-392, 2001

Modeling of Electrical off-grid Network in the Simscape Power Systems

Dušan Medveď, Marián Klešč
Department of Electrical Power Engineering
Mäsiarska 74
042 00 Košice, Slovak Republic
dusan.medved@tuke.sk , Marian.Klesc@student.tuke.sk

Abstract — **This paper deals with the modeling and analysis of off-grid electrical networks. The library of Simscape Power Systems, part of the Simulink program, was used to model off-grid networks. The off-grid model includes electricity generation using diesel generators, photovoltaic panels and wind turbines, power consumption and basic electrical measurements. The impact of the different operating states on the modeled off-grid network has been analyzed, such as load coupling, load disconnection, load balancing, and changing of climatic conditions for photovoltaic panels and wind turbine.**

***Keywords:** off-grid network, Simscape Power Systems, diesel generator, photovoltaic panels, wind turbine*

I. Introduction

This paper presents the results of the simulation of the impact of different electricity sources on a small off-grid. Diesel generators, photovoltaic fields and wind turbines have been used as power sources. From the point of view of electricity consumption, the effect of disconnection or connection of a large load on the system and the effect of a dynamically changing load is described. Multiple circuits have been simulated to verify some of the network phenomena. The main monitored variables included network frequency, voltage at the point of consumption, and power produced by sources.

To simulate these phenomena, the Simscape Power Systems, which is an extension of Matab Simulink, was used. Based on the simulation analysis, a simple solution was developed to reduce the impact of transient phenomena. Since simulated transient phenomena of a short nature, i.e. they take a short time, the designed simulations simulate the time interval within 1000 seconds, which is about 17 minutes. A short time interval has also been chosen because the results that are written in matrices have some accuracy and can be processed with current common computing techniques.

As a basis for these considerations, papers from engineers involved in the IEEE association, the world's largest technological advancing organization, have been used.

II. Description of the Network Model in the Environment of Simscape Power Systems

A. Power sources

In the Simscape Power Systems, several electrical machines are implemented. Many of these electric machines can work in two states – as electricity generators or as motors, that is, as electric consumer appliances [1], [2].

A model of the synchronous machine with expressed poles was used. The synchronous generator is controlled by a hydraulic turbine combined with the PID control system and excited by the AC4A excitation system. The principal scheme of the G1 generator with a control and exciter system and generating output from the generator can be seen in Fig. 1.

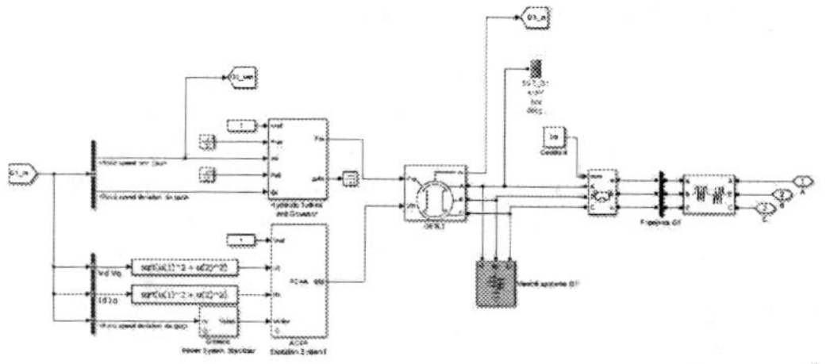

Figure 1. Principal scheme of the G1 generator connection

Output of the synchronous generator is a three-phase voltage at the terminals of the machine A, B and C and the measurement output marked with the letter m. The measurement output includes a vector with measured signals: stator currents, stator voltages, rotor angle deviation, rotor speed, electromagnetic torque, output active power *P*, output reactive power *Q*, and so on. These signals receive feedback from the generator that is input to the exciter winding input and the hydraulic turbine with the control. The label data of the simulated generator are shown in Tab. I.

Table 1. Data Of the Simulated Generator

		Reactances [pu]			
P_n	250 kW	x_d	2,84	x_q	2,44
V_n	415 V	$x_{d'}$	0,18	$x_{q'}$	0,36
f_n	50 Hz	$x_{d''}$	0,13	$x_{q''}$	0,09
Stator resistance R_s [pu]			0,0259375		
Coefficient of inertia *H* [s]			3,2		
Coefficient of friction *F* [pu]			0,01579		
Number of pole pairs *p* [–]			2		

In Fig. 2 is a model of a hydraulic turbine with PID control. This model has 5 inputs and 2 outputs. Inputs include reference speed, instantaneous mechanical speed, speed deviation, reference power and instantaneous power output. The output is the mechanical power P_m, which is also the input for the synchronous generator. In the mentioned model was set the reference speed $\omega_{ef} = 1$ pu, and the inputs of the immediate mechanical velocity ω_e and the velocity variation $d\omega$ were connected. This regulation ensures the regulation of the synchronous generator at the nominal frequency $f_n = 50$ Hz. Inputs of the reference mechanical power P_{ref} and instantaneous power P_{e0} are not connected. The circuit is set so that it does not take any feedback (or feedback from the gate output). It has been achieved that the turbine power was controlled only by rotor speed [7].

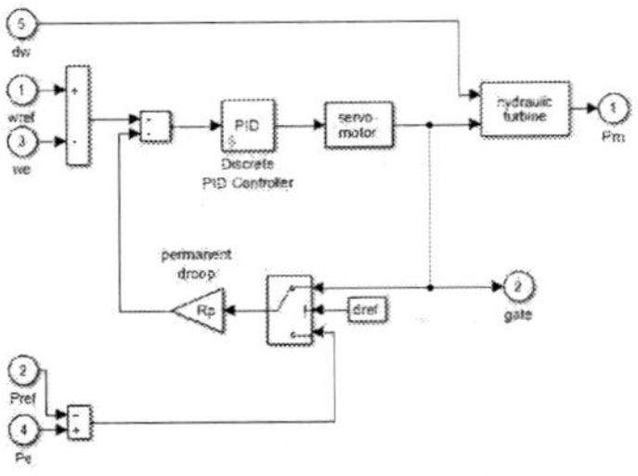

Figure 2. Hydraulic turbine diagram with PID control

As the excitation system, the AC4A excitation system was used, which is an excitation system with thyristor rectifier for AC alternator. The excitation system generates the excitation voltage that is the output of this model.

Generating power from the generator into the grid consists of a three-phase circuit breaker and a busbar measuring the terminal voltage and the supplied power to the grid. Behind these terminals there is a power line simulated by the resistance *R* and the inductance *L*. The mentioned generated power transmission into the output grid is shown in Fig. 1. Generator terminals are connected to the generator's own consumption, which consists of an active load of $P = 12.5$ kW. This load has been used because the synchronous generator simulated as a current source cannot be in series with the induction element of the three-phase line with which the generator is connected to the grid.

B. *Photovoltaic field*

Another source of electrical power that has been implemented in this paper is photovoltaic (PV) panels. The PV array is located in Simscape Power Systems block to simulate the photovoltaic panels. That's why the LG Electronics LG300N1-G3 photovoltaic panel has been selected. The label data of the modeled PV module are shown in Tab. II.

Table 2. Module Data for Photovoltaic Panel

Module data of PV array: LG Electronics LG300N1C-G3			
Max power [W]	305,226	Cell number in module (Ncell)	96
Open-circuit voltage V_{oc} [V]	64,2	Short-circuit current [A]	5,96
Voltage at max power V_{mp} [V]	54,7	Current at max power I_{mp} [A]	5,58
Temp. coefficient for V_{oc} [%/°C]	−0,273	Temp. coefficient for I_{sc} [%/°C]	0,062

In the PV array block setting, the PV panels were connected to a field containing 66 parallel strings that serve 5 modules in series. The PV array has 2 inputs. The first input is the solar radiation reported in watts per m^2 [W·m^{-2}], the second input is the temperature of the cell of the PV panel reported in Celsius degrees [°C]. The output of the PV array is a DC voltage and a measuring port *m* which contains the voltage and current of the PV cell, the diode current, the solar radiation and the temperature of the PV panel. Sunlight and solar panel temperature data were retrieved from the text file into the program. In Fig. 3 shows the I-V and P-V characteristics for the proposed PV array, under the solar radiation at three different intensities: 1000 W·m^{-2}, 500 W·m^{-2} and 100 W·m^{-2} when the PV panel temperature is constant $\vartheta = 25$ °C.

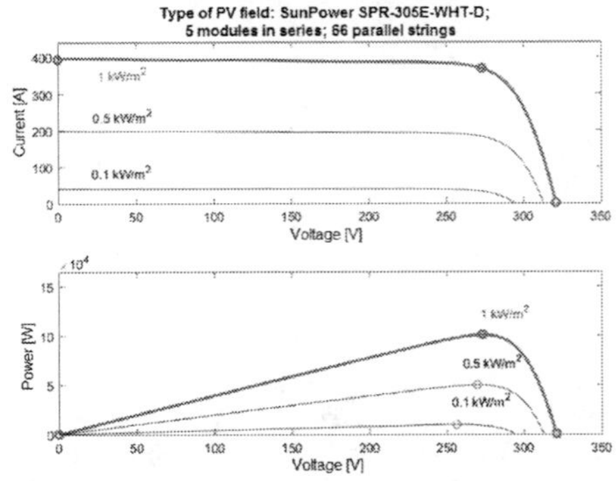

Figure 3. I-V and P-V characteristics for PV field at different solar radiation

Power output from the photovoltaic field is shown in Fig. 4. The DC current, which is generated by the PV panels, is controlled by a DC-DC amplifier, to which the MPPT controller operates. Subsequently, the current is changed to alternating current in a controlled three-phase VSC rectifier, including the inverter choke *RL*, the capacitive filter *C*, and the transformer TR1. The power is transferred out through a three-phase switch that is in the basic position "on". Block RL2 represents the power line by which is the PV field linked with the network. This model has been adapted and modified for the needs of this paper from MathWorks examples [8], [9].

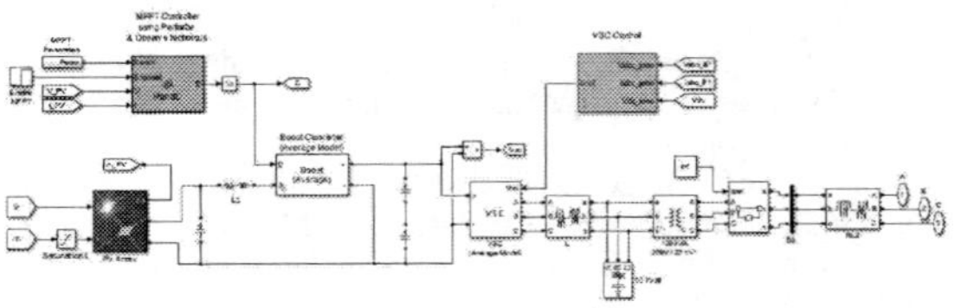

Figure 4. Power output from photovoltaic panels

C. *Wind turbine*

The wind turbine block with power-out transmission to the grid is considerably easier than a block of PV field. The wind turbine input is the *wind speed* reported in m·s^{-1} and a *Trip connector*. The wind speed for this model was retrieved from a text file. Trip connector serves to simulate the turbine protection system. Its input may be a logical zero or one. If at input is logical zero, the wind turbine is in operation and when at input is logical one, the turbine is disconnected. The wind turbine can have several protections. First of all, it is a wind turbine disconnection when there is slow/fast wind, but also overcurrent protection, undervoltage protection, overvoltage protection, or protection, acting in the unbalanced current or voltage.

The wind turbine output is a measuring port that contains the voltage and current at turbine terminals A, B and C, turbine output *P* and *Q*, turbine rotor speed, mechanical torque, and so on. The wind turbine used in this paper includes, in addition to the turbine, also an asynchronous engine that generates the electrical energy. In Fig 5 is shown the characteristic output of the designed turbine at different wind speeds.

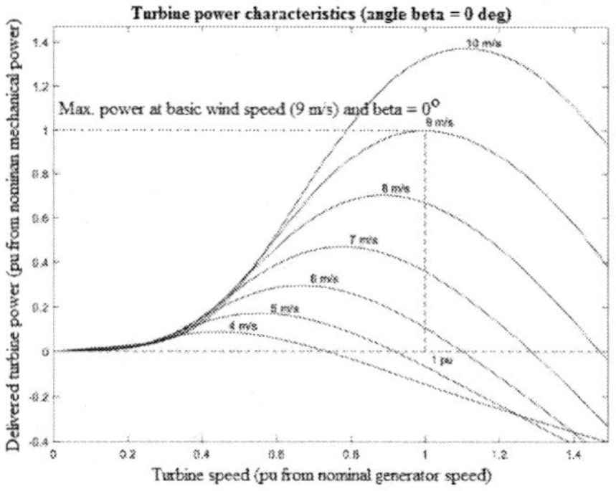

Figure 5. Characteristic turbine power at different wind speeds

Since during the simulation the disconnection of the wind turbine caused a mathematical error, the block "*Check static range*" was added to the scheme. This block stops the simulation if a wind speed is read at a speed that is not in the work range, and Matlab shows error message. The wind turbine operating range is in ranges from 4.5 $m \cdot s^{-1}$ to 12.5 $m \cdot s^{-1}$. The basic wind speed for the model turbine is 9 $m \cdot s^{-1}$.

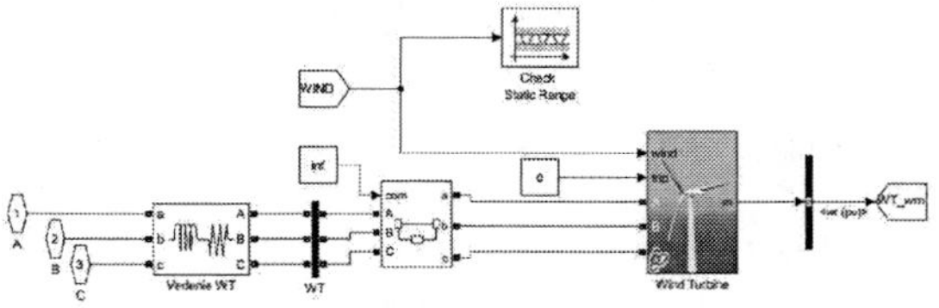

Figure 6. Wind turbine block with power delivery to the system

D. Loads

Simscape Power Systems offers several types of load. In this paper, three-phase serial *RLC* load and three-phase dynamic load were used. For both loads, the combined nominal voltage and nominal frequency of the network were entered.

For a static three-phase load, *PQ* power was input, which may be the same or specific for each load in all three phases. A static three-phase load contains also a voltage and current measurement that is optional [3], [4], [6].

For dynamic three-phase load, the *PQ* power was entered at the beginning of the simulation. The *PQ* power of a dynamic load can be controlled by an internal control that controls the amount of dynamic load based on the positive-sequence voltage component. If external control of power source is used, performance can be read from a file, and controlled by an external handling. The dynamic load contains also a measuring terminal *m*, the output of which is a vector with a positive-sequence voltage component, an active power *P* and a reactive power *Q* [5], [10].

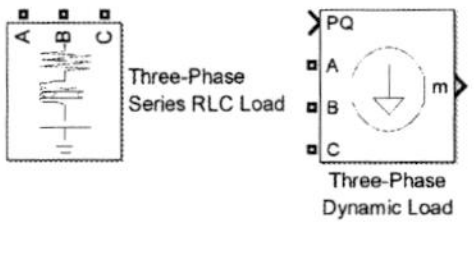

Figure 7. Loads in Simscape Power Systems

The loads were read using a Matlab script. In Fig 8, a proposed load block for the supply point A is shown. On the left, the load A1 and line A2 are shown, which are connected to the system via a three-phase circuit breaker and a power line simulated by the impedance R_a and L_a. Line A2 consists of a purely ohmical load, because the dynamic load line A1 cannot be connected in series with the inductive element of the three-phase line, which is the supply point connected to the system. On the right, the reading of block of line load A1 is displayed. If *init_const* = 1, the load, i.e. line A1 is set according to the vector from a text document. If *init_const* = 0, the load is set to the constant value, which is set in the text document for time $t = 0$. Current and voltage measurements were performed on bus-bar A. During the simulations, four consumption points A, B, C and D were considered. Each of these consumption points represents a part of the network. In some simulations, only static three-phase consumption points were used that were disconnected by a three-phase switch.

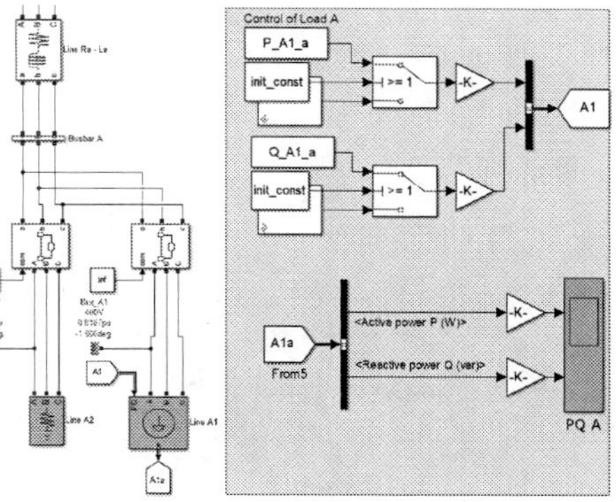

Figure 8. Designed load block with control

E. Measurement in Simscape Power Systems

In the simulations, electrical quantities were measured at selected locations in the network. Phase currents and voltages were measured using three-phase V-I measuring blocks, which were placed before loads and before the generators, resp. other sources. The measured output is the sinusoidal voltage/current depending on the time that has to be converted to the effective value (for comparison purposes). The scheme for measuring of the particular variables at the output of the G1 generator is shown in Fig. 9. The RMS current and voltage values for the L1 phase and the active and reactive power in the L1 phase were calculated from the measured currents and voltages.

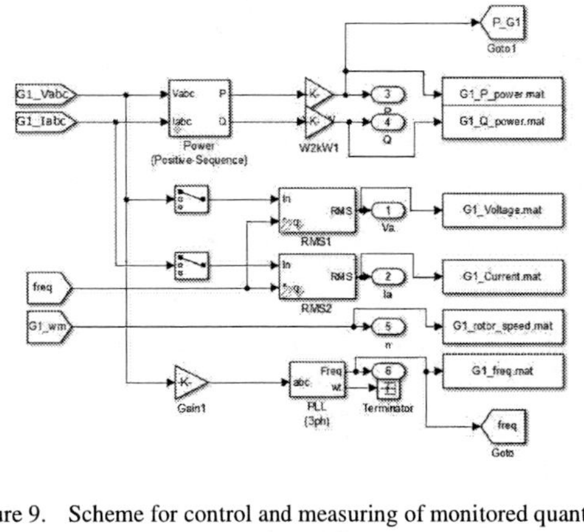

Figure 9. Scheme for control and measuring of monitored quantities

III. Mathematical Calculation of Off-Grid Network

A. *Model of a steady-state off-grid network*

In the off-grid steady-state model, the main aim was to point out that if no changes were made to the scheme and the correct initialization conditions were set, the network's frequency did not change and was 50 Hz. The phase voltage in phase L1 is equal to the portion of the line-to-line voltage and the square root of 3. If in a system were also considered losses on the line, the resulting voltage values were less than the expected 230 V. In the system were considered large losses on lines, so the phase voltages at the terminals were lower, namely: $U_{a\_A} = 220.5$ V, $U_{a\_B} = 221$ V, $U_{a\_C} = 223.1$ V, and $U_{a\_D} = 227$ V. The demand current depends on the size of the load being connected at the consumption point (load). The active and reactive *PQ* load was unchanged in the circuit.

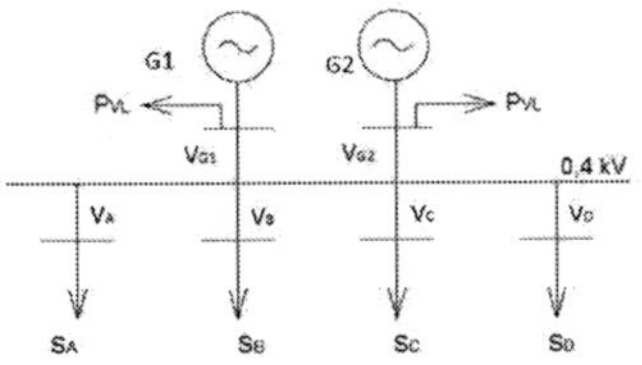

Figure 10. The steady-state model diagram

Table 3. Consumptions for Simulation of Steady-State Off-Grid Network

	U [V]	Demand power	Power factor cos *φ*	Overhead line *R* [Ω]	Overhead line *L* [μH]
G1, Self-consumption	400	12,5 kW	1	0,0025	11,27
G2, Self-consumption	400	12,5 kW	1	0,0025	11,27
Load A	400	100 kVA	0,95	0,0402	71,1
Load B	400	140 kVA	0,95	0,0268	47,4
Load C	400	180 kVA	0,95	0,0134	23,7
Load D	400	0	0,95	0,067	118,5

B. *Model of off-grid network with dynamic load*

In this part of the simulation there was modified model of loads. Instead of the loads modeled by the constant value, dynamic loads were used that were controlled by external input. Dynamic load operation is described in the previous chapter in part D. During these simulations, two generators with a nominal power of 250 kVA and four loads A, B, C and D were used in which the phase voltage and current in phase L1 and power value in L1 were measured.

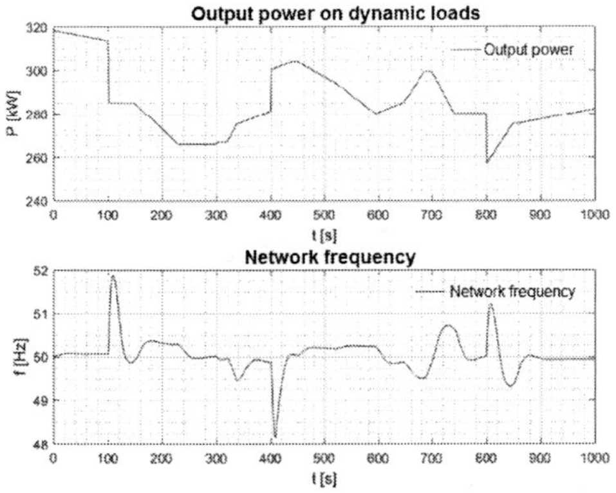

Figure 11. Network frequency response to output power

In Fig. 11, the abstraction performance is indicated by dynamic loads. Self-consumptions (2 × 12.5 kW) and parasitic loads to dynamic loads (3 × 9.5 kW + 4.75 kW), which are purely resistive, have to be added to the total output power. These parasitic loads are in the system because dynamic loads and synchronous generators cannot be in series with an inductive element of three-phase power lines. Those are described by the *RL* parameters listed in Tab. IV.

Table 4. Resistance and Inductance of Power Lines in Simulations with Dynamic Load

	R [Ω]	*L* [μH]
G1	0,0025	11,27
G2	0,0025	11,27
A	0,2010	355,50
B	0,0268	47,40
C	0,0134	23,70
D	0,0670	118,50
PV	0,0099	45,10
WT	0,0099	45,10

By a continual decreasing, respectively by increasing of the power consumption there has been observed, that the regulators of the synchronous generators respond to these changes, and there is a decrease, respectively increase in output power produced by synchronous generators, but the frequency is not regulated to the nominal value of $f_n = 50$ Hz. Thus, the frequency of the network will be short-lived at a different value near the nominal frequency due to the rate of decrease/increase of the consumed power. This can also be seen in Fig. 11, from 478 seconds to 595 seconds, the network's frequency was around 50.2 Hz. From 900 s to 1000 s the network frequency was stable at values between 49.93 and 49.95 Hz.

C. *Off-grid network model with photovoltaic field*

In the world, there is a tendency for off-grid networks that are powered by diesel generators to install renewable energy sources to reduce diesel consumption in diesel generators. In the mentioned model, photovoltaic panels with a total peak power of 100 kW were connected to the simulated off-grid network.

The Fig. 12 shows the solar radiation captured onto the photovoltaic field depending on the simulation time. At the beginning of the simulation, i.e. at time $t = 0$ s, the solar radiation captured on the PV field, is 857.4993 W·m^{-2}. At time $t = 276$ s, solar radiation decreases to 266.2146 W·m^{-2}, which is the first minimum of tested sunlight. At time $t = 700$ s, a second low of 242.5764 W·m^{-2} solar radiation occurs. Both decreasing of solar radiation were due to increased cloud cover. The photovoltaic cell temperature was around 27 °C. These data were measured in August 2013 as a part of a hybrid photovoltaic system operated by the Department of Electrical Power Engineering, FEI TU in Košice. These data represent actual data under the conditions for which the electricity from the PV field was produced.

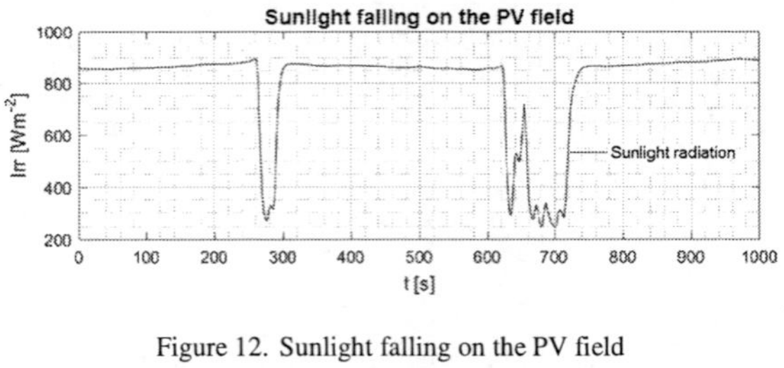

Figure 12. Sunlight falling on the PV field

B1. Off-grid network with PV-field and constant consumption

In order to best see the network's response to sudden drop in power output from the PV field, constant load values were set when simulating the photovoltaic field. In this case, the output produced by the diesel generators is adapted to the output produced by the PV field. The Fig 13 shows the PV field response to the test climatic conditions shown in Fig 11. At the same time, Fig 13 shows the response of generator no. 1. Generator no. 1 and 2 were connected in series, so generator 2 produces the same power as generator 1. So when the output of the PV field drops by 60 kW due to meteorological conditions, both generators must start to produce 30 kW more power.

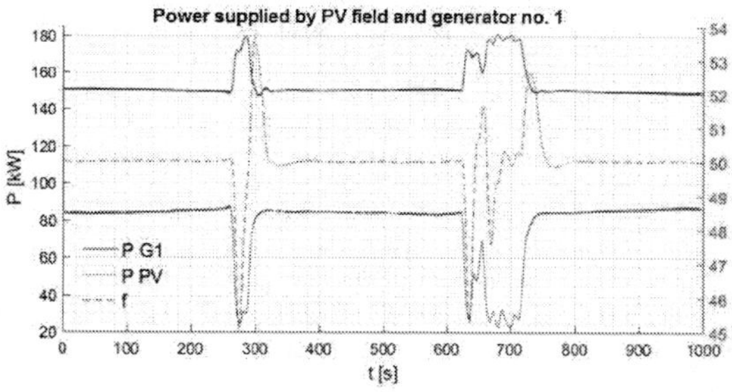

Figure 13. The frequency of the network and the power output of the PV array with the generator no. 1

This simulation shows that the PV field is capable of producing relatively high power, but due to the increased cloud, short-term power failures can occur in the PV field output. The off-grid network simulation was implemented without any added controllers that would run the network on the load side. If an off-grid network was operated in this way, it is likely that the quality and reliability of electricity would be limited. In the simulation, the network frequency dropped from 50 Hz to below 46 Hz. While diesel generators are capable of operating in a wide frequency range, a mentioned rapid drop in frequency would be at risk.

D. Off-grid model with a wind turbine that operates at a dynamic wind speed

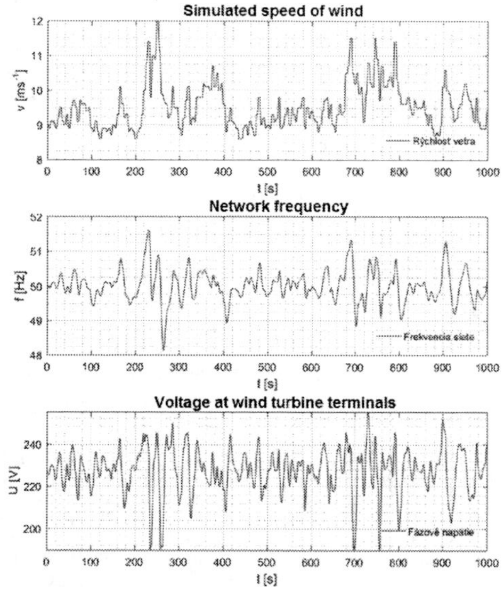

Figure 14. Simulated wind parameters

Wind simulation was used to simulate the wind-flow circuit as it is illustrated in Fig. 14. The wind loaded from the text file has a value of 9 m·s^{-1} at time $t = 0$, which is the nominal wind for the wind turbine used. Subsequently the wind varies around this value. Wind reaches a maximum value of 12 m·s^{-1}. The wind turbine operates with winds ranging from 4.5 m·s^{-1} to 12.5 m·s^{-1}.

Since the simulated wind turbine has no stabilizing mechanism, the supplied turbine power also varies around the nominal value. This was reflected negatively on network frequency and voltage. Since the simulated off-grid network is small in size, voltage fluctuations have been registered in all four A, B, C and D loads. Frequency of the grid and voltage at the wind turbine terminals are shown in Fig. 13. Referring to Fig 13 it can be seen that even with small wind changes, the frequency has risen above 51 Hz, or falls below 49 Hz. Voltage at wind turbine terminals is fluctuating. In case of a sudden change of wind, the voltage exceeds 250 V, respectively drops to 190 V. Since the voltage fluctuations are relatively strong, a digital flickermeter has been connected to load points A, B, C and D and to the wind turbine connection point.

C1. Flicker-effect measure in network with a wind turbine

In the previous section C there was a description of off-grid operation with a wind turbine with dynamic wind. In order to determine the flicker effect in the aforementioned network, a digital flickermeter was added at points A, B, C and D to find a short-term flicker rate that is calculated at simulation time of 5 to 605 s, representing a ten minute time period. In tab. V is the measured short-term rate of flicker and averaged percentiles. The smallest rate of flicker short-term perceptive was simulated with a constant wind velocity of 9 m·s^{-1} and a dynamic load. On the other side, the highest short-term flicker rate was simulated with dynamic wind speed and dynamic load. The limiting value for the short-term flicker was 1, for a long-term flicker value (for 2 hours) of 0.65. From the measured results it can be stated that the short-term flicker rate was in accordance with standard STN EN 50160.

The Fig 15 shows the measured instantaneous level of the flicker effect at the load point C for dynamic load simulation (Fig. 11) and the dynamic wind simulation (Fig. 14). From Fig. 15, it is apparent that the blink effect was occurred in the case of dynamic load simulation at a time when the load was connected or disconnected in the network. It was observed for example, at time $t_1 = 100$ s when a load of 30 kVA was disconnected at the load point B or at time $t_2 = 400$ s when a load of 20 kVA was connected at the load point C or at time $t_3 = 800$ s when the load of 20 kVA was disconnected from load point C. In the case when the dynamic wind acts on the wind turbine (see Fig. 14), the measured instantaneous level of flicker effect will appear as stochastic noise.

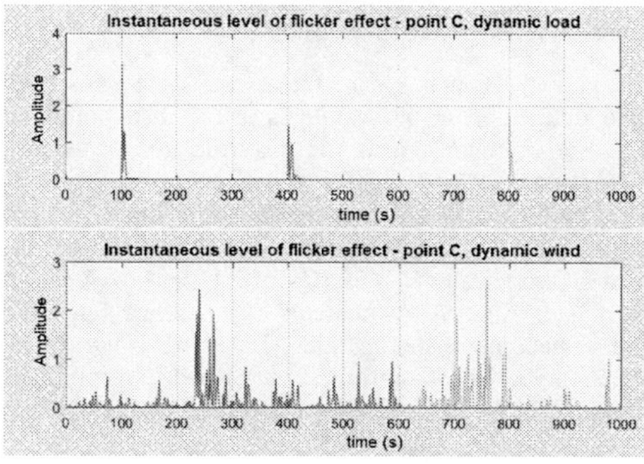

Figure 15. Measured instantaneous level of flicker effect

The Tab. V shows the short-term flicker rate response for the load points A, B, C and D and for the point on the wind turbine terminals. The particularity of these results is that in each simulated scheme, at the load point A, the highest degree of short-term flicker is measured. This is due to the fact that the point A is powered by a line whose resistance and reactance is much larger than the lines connecting the other points (see Tab. IV). The voltage at point A in these simulations was stabilized at $U = 189.6$ V (in real conditions, such a low voltage would be a problem for the operation of many devices).

Table 5. Flicker Effect in the Simulated Network

		$P_{0.1s}$	P_{1s}	P_{3s}	P_{10s}	P_{50s}	P_{st}
Simulation with WT, dynamic wind speed, dynamic load	Point A	7,94444	2,29646	1,09671	0,364422	0,043251	0,739981
	Point B	4,42942	1,38709	0,654338	0,225327	0,025813	0,565732
	Point C	4,12629	1,33554	0,640592	0,221686	0,025551	0,553068
	Point D	4,28825	1,38515	0,663718	0,22959	0,026457	0,563365
	Pt. WT	3,90262	1,27304	0,611244	0,212224	0,024508	0,539369
Simulation with WT, dynamic wind speed, constant load	Point A	4,76669	2,50239	1,21876	0,547187	0,068989	0,721009
	Point B	2,29082	1,2644	0,642569	0,286289	0,037393	0,5135
	Point C	2,18987	1,20936	0,615162	0,273998	0,035838	0,502251
	Point D	2,27869	1,25794	0,63941	0,284875	0,03721	0,512195
	Pt. WT	2,09109	1,15719	0,590026	0,262503	0,034409	0,491356
Simulation with WT, constant wind speed, dynamic load	Point A	4,96686	1,67448	0,38066	0,014602	0,011166	0,523317
	Point B	3,04421	0,966101	0,218407	0,00848	0,006955	0,404461
	Point C	2,77123	0,929307	0,212038	0,007767	0,00631	0,390405
	Point D	2,87724	0,959631	0,21867	0,008057	0,006548	0,397332
	Pt. WT	2,64433	0,879365	0,199992	0,007509	0,006025	0,380651

In order to reduce the influence of power line on the measured flicker effect, the simulations were repeated except that the line joining the load point A was simulated by resistance $R_c = 0.0134\ \Omega$ and inductance $L_c = 23.7\ \mu$H connecting the load point C. The results are given in Fig. 16.

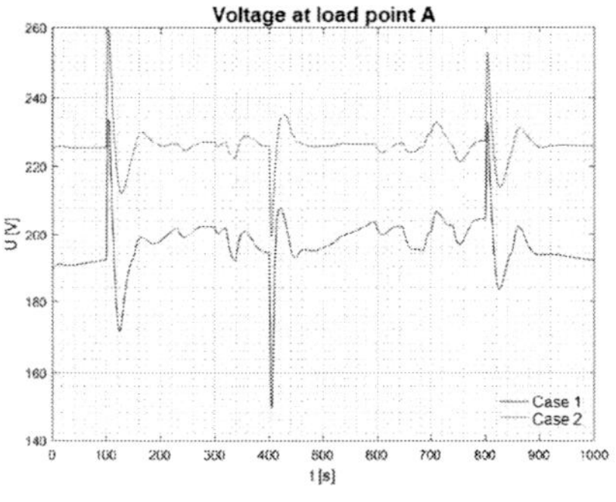

Figure 16. Voltage in simulated scheme for point A

In Fig. 16 is the voltage characteristics at the load point A in the case where was considered constant wind of 9 m·s^{-1} during the whole simulation and the dynamic load as described in section B. For the case 1 there was considered the original power line whose resistance was $R_a = 0.2010\ \Omega$ and inductance $L_a = 355.5\ \mu$H. In case 2, a point A was connected by a line with parameters $R_a = 0.0134\ \Omega$ and $L_a = 23.7\ \mu$H.

From Fig. 16, it is clear that in case 1 there is a greater voltage fluctuation at the terminals at the load point A as in case 2. For example, during the disconnection of the 20 kVA load from the load point C, there was observed (in case 1) at the load point A the short-term voltage drop from $U_f = 194.3$ V to $U_f = 149.5$ V, which is a drop of $\Delta U = 44.8$ V. In case 2, there was drop from $U_f = 225.7$ V to $U_f = 199.4$ V, which is drop of $\Delta U = 26.3$ V. As there is less voltage fluctuation in transient phenomena, the resulting flicker effect will be less. In test example 2, the value of the Short Term Perceptibility (P_{st}) of flicker effect in point A was $P_{st} = 0.335491$ (the original value, in case A was $P_{st} = 0.523317$).

IV. Conclusion

This paper presented the results of off-grid network simulations with consideration of renewable resources (PV source and wind turbine) and without considering RES. The simulated off-grid network consisted of two diesel generators with a nominal output of 250 kW and with loads A, B, C and D, representing 4 load points representing 4 off-grid sites. In the case of a load disconnection or connection, the generators are able to regulate the system so that the power output is equal to the power delivered. The regulation of diesel generators has ensured the control of the hydraulic turbine.

The problem of off-grid systems with a photovoltaic field is that the power output from the PV field cannot be regulated. The PV field produces electricity according to current climatic conditions. Therefore, in the case of rapid climate change, there is a rapid change in the output of the PV field. For example, with increased cloudiness, a sudden drop in the electricity produced from the PV field may occur. By adding a wind turbine into an off-grid, an increased flicker effect was observed. In addition to voltage fluctuations in the network, the network frequency also varies. Large frequency fluctuations can have a negative impact on diesel generators. Flicker effect occurs when disconnecting or connecting loads, resources, and off-grid networks. In both cases, it is necessary to consider how to remove the unfavorable phenomenon of blinking.

Acknowledgement

This work was supported by the Ministry of Education, Science, Research and Sport of the SR and the Slovak Academy of Sciences under the contract No. VEGA 1/0132/15.

References

[1] Z. Čonka, "Impact of renewable energy sources on stability of EWIS transmision system", In: *proc. 14th Int. Conf. on Env. and El. En.:* Krakow, Poland, 10-12 May, 2014. Danvers: IEEE, 2014 P. 75-79. ISBN 978-1-4799-4661-7.

[2] R. Cimbala, L. Kruželák, S. Bucko, J. Kurimský, M. Kosterec, "Influence of Electromagnetic Interference on Time-Domain Spectroscopy of Magnetic Nanofluids", In: *EPE 2016*. Danvers: IEEE, 2016, p. 279 - 282. ISBN 978-1-5090-0907-7.

[3] M. Pavlík, "Prediction of electricity price using RSI mechanism", In: *Scientific Letters of Academic Society of M. Baludansky*, Vol. 4, No. 6A (2016), p. 82-84. ISSN 1338-9432.

[4] J. Zbojovský, P. Liptai, M. Moravec, "Modelling and calculating the shielding effectiveness of building materials", In: *Technical sciences and technologies*. Vol. 6, no. 4 pp. 205-210. ISSN 2411-5363, 2016

[5] M. Kanálik, M. Pavlík, M. Kolcun, "The impact of multi-system overhead lines operation with different voltage levels to voltage unbalance", In: *Elektroenergetika 2015*. Košice: TU, pp. 73-76. ISBN 978-80-553-2187-5, 2015

[6] M. Kanálik, M. Kolcun, "Computation of harmonic flows in three-phase systems". In: *Acta Electrotechnica et Informatica*. Vol. 7, No. 3, pp. 41-45. ISSN 1335-8243, 2007

[7] MathWorks, "*Model hydraulic turbine and proportional-integral-derivative*", < https://www.mathworks.com/help/physmod/sps/powersys/ref/hydraulicturbineandgovernor.html >

[8] MathWorks, "*Three-Level Bridge*", < https://www.mathworks.com/help/physmod/sps/powersys/ref/threelevelbridge.html >.

[9] MathWorks, "*Average Model of a 100-kW Grid-Connected PV Array*",<https://www.mathworks.com/help/physmod/sps/examples/average-model-of-a-100-kw-grid-connected-pv-array.html >

[10] MathWorks, "*Implement three-phase dynamic load with active power and reactive power as function of voltage or controlled from external input*", < https://www.mathworks.com/help/physmod/sps/powersys/ref/threephasedynamicload.html >..

Perspectives of Cogeneration Plants and District Heating Sector – Assessment of the Impact of CO_2 Auctions and Other Legislative Changes

Jiří Vecka
Czech Technical University in Prague,
Fac. of Electrical Engineering, Dept. of Economics, Management and Humanities
Prague, Czech Republic
vecka.jiri@centrum.cz

Jaroslav Knápek
Czech Technical University in Prague
Fac. of Electrical Engineering, Dept. of Economics, Management and Humanities
Prague, Czech Republic
knapek@fel.cvut.cz

Jiří Vašíček
Czech Technical University in Prague
Fac. of Electrical Engineering, Dept. of Economics, Management and Humanities
Prague, Czech Republic
vasicek@fel.cvut.cz

Abstract — **The paper is describing major aspects of district heating technology and its position within energy sector in the context of the Czech Republic and EU as well and new emerging legislation with effects on district heating systems as such. Optimization model of differential NPV was chosen as optimal solution for comparing the development in business-as-usual situation and under the new requirements and each relevant legislation effect is transformed into risk factor influencing future heat prices. Model outcomes are formulated in the two legislation and three CO_2 price scenarios and confirm non-symmetrical effects on the heat market participants arising from the size of installations. There are various approaches how this distortion could be addressed. Among possible solutions indirect carbon tax based tool was identified as the most appropriate way how to remedy distortions on the heat market towards inclusion of CO_2 costs into price of fuels for installations outside EU ETS. This solution could be easily realized by using existing legislation tools and application of possible remedial tool on the heat market is described. Results of these analyses support the idea of necessity to introduce this type of tool as soon as possible in order to avoid undue competition distortions on the heat market in the Czech Republic.**

Keywords: district heating, combined heat and power, EU ETS, emission limits, CO_2 price

I. Introduction

Energy conversion for covering energy needs has always its environmental impacts, thus with consideration of economic perspective leading to seek for most efficient solutions. Taking into account the essential human needs for covering heat demand, the process heat, space-heating and hot water preparation are necessary for all sectors of the economy.

Relevant studies have confirmed the high efficiency of district heating compared to other heating options [1] especially by tapping the potential to utilize heat that would otherwise be of limited use or using combined heat and power technology [2].

The district heating sector could be defined as a branch of the state energy policy responsible for heat supply via heating networks. District heating (DH) covers heat needs of residential, amenity and industrial buildings and supplies heat for industrial processes as well.

DH sector is important part of EU energy sector, responsible for heat delivery to 60 million of EU inhabitants. Whole EU heat market for residential and service sector buildings based on IEA data has a volume about 11.5 EJ per year , share of district heating is 12% which equals to heat deliveries 1 370 PJ per year. DH sector also delivers heat to industry, 830 PJ per year. Industrial combined heat and power installations deliver another 790 PJ per year. Total heat delivery of EU DH sector is 2 990 PJ per year.

According to Ministry of Industry and Trade of the Czech Republic (MIT) the total annual heat consumption (445 PJ) is covered by the heat produced for district heating supplies by of approximately 150 PJ. Of that, approximately 110 PJ are centrally produced heat supplied by third parties via the district heating networks. The remaining 40 PJ of heat is consumption of auto-producers (eg. own consumption in technological processes within the facility which supplies heat outside the plant, supply of heat from the house boiler rooms within one building except heat sales to other objects, etc.). This heat is not considered as individual heating in the MIT statistics and forecasts of individual heat remains in the category of centrally produced heat. Totally 54 PJ of heat was delivered to households, which corresponds to approx. 1,6 million of households (around 38 % of households in the Czech Republic).

Case studies have confirmed significant potential for an increase in energy efficiency and considerable energy and emissions savings if all relevant actors are addressed and potential of district heating deployed [3].

II. Materials and Methods

A. Analysis of legislation effects

As a first step the effects affecting the heat market need to be thoroughly described. Legislation factors were identified as the most important in this respect because it is politically driven, independent on decisions of installations' operators and could severely influence conditions on the market. New environmental legislation requirements such as emission limits within the framework of Industrial Emissions Directive [4] and Medium Combustion Plant Directive [5] and EU ETS Directive [6] are main effects causing uncertainties on the heat market. Especially EU emission trading system (EU ETS) entails complexity of different issues influencing free allocation of emission rights (allowances) and thus influencing impacts of the CO_2 price on the installations within. Currently ongoing revision of EU ETS for 4th trading period could cause additional problematic impacts on certain heat market participants. There are also other non-legislative factors influencing heat market, mainly strategies and concepts focusing on district heating technology. Legislation requirements in relation to the installations' size are compared in the Table 1. Small scale emitters (below EU ETS thresholds) are usually in much favorable position compared to direct competitors on the heat market.

Table 1. Application of legislative tools based on the size of installation concerned

Legislative tool	Rated thermal input			
	Below 1 MW_{th}	*1-20 MW_{th}*	*20-50 MW_{th}*	*Above 50 MW_{th}*
Promotion of CHP	YES	YES	YES	YES
Emission limits	YES[a]	YES	YES	YES
Air Pollution fees	NO	NO[b]	YES	YES
Water Pollution fees	NO	NO[c]	YES	YES
CO_2 price	NO	NO	YES	YES
Energy taxation	YES	YES	YES	YES

[a] Below 0.3 MW_{th} only for new installations

[b] Air pollution fees are relevant only for certain fuels and energy production capable of exceed legislation limit of 50,000 CZK/year total sum of fees.

[c] Water pollution fees are relevant only for certain fuels and energy production capable of exceed legislation limit of 50,000 m^3/year total sum of water discharges.

Legislative tools such as air and water pollution fees are defined based on pollution discharged to environment and not related to the actual installation size. Thus application limit for these tools is defined based on expert estimation.

As is shown by the description of risk factors on the heat market, new environmental legislation is targeted mainly at larger unit and installations. All new legislation instruments are seeking for further limitation of polluting substances or greenhouse gasses.

B. Optimization modelling

In order to assess impact into heat price for modelled installation evaluation of incurred costs driven by effects on heat market needs to be done. The evaluation of future production from district heating plants is based on microeconomic approach to installation, which could be applied to various types of installations even across different sectors. The model is based on creation of future profit and loss statement in different scenarios. Main criterion is to keep constant profit margin and energy price increase is as variable. As district heating installations always compete with local heat sources, the crucial is identification of legislation induced risk factors, which is usually borne by district heating installations only. Other changes in expenditures – prices of fuels, labour costs etc. are applied to both categories of installations and could be disregarded in this respect.

Key aspect for general evaluation of investments is creation of excess financial resources (profit) for investors. This excess is defined as overall economic efficiency (Net Present Value) as follows:

$$NPV = DCF_L = \sum_{i=0}^{L} \frac{CF_i}{(1+r)^i} \qquad (1)$$

where:

L	lifetime of the project
DCF_L	discounted cash-flow during lifetime of the project [EUR]
CF_i	cash-flow in year i [EUR]
r	discount [-]

The final optimization criterion should enable selection of the optimal variant of system development from several variants of one strategy selected in preliminary optimization phase. One of the conditions for comparability of variants is that they cover the same time period. However the economic life of assessed objects can be different, usually exceed optimization period and will normally be completed in different years. This has to be reflected in the design of options of system development and allow that options are comparable.

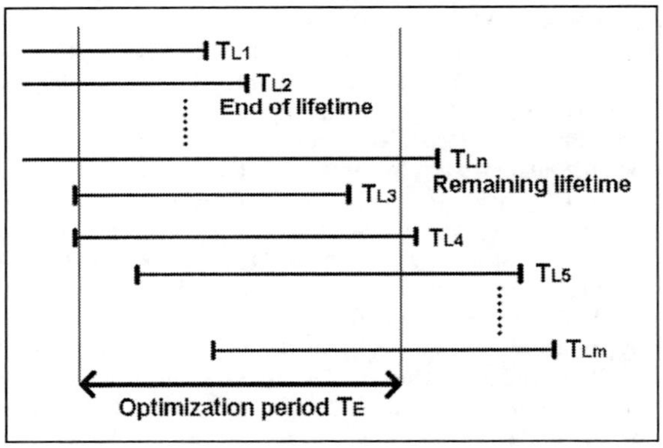

Figure 1. Time scheme development options for existing system

In case of evaluation period E < lifetime L:

$$DCF_L = \sum_{i=0}^{E} CF_i \cdot (1+r)^{-i} + CF_{TE} \sum_{i=E+1}^{L} (1+r)^{-i} \qquad (2)$$

In the decision making process project should be assessed by differential NPV, where the difference stems from the fact that without new legislation instruments plant would continue in business-as-usual operation. In case study example plant has to adjust its operation according to new requirements.

$$NPV_{Diff} = NPV_{New} - NPV_{BAU} \qquad (3)$$

where:

NPV_{New}	NPV in new circumstances [EUR]
NPV_{BAU}	NPV without new circumstances [EUR]

C. Economic model

Optimization model could be simplified to the economic one in order to properly assess future development of externalities for different actors on the heat market. Model should mirror impact of environmental instruments on economic situation of heat installation of all sizes and technology employed (e.g. combined heat and power technology). In case of district heating sector, assumption is that DH systems are already developed in the area where economical and technical opportunities were fulfilled. The system development is supposed to be carried out by operator of the installation and current connected DH system. At the installation level, there are frequently several types of units/devices in terms of fuel mix, commissioning, operation time etc., which has to be also taken into account.

Legislation-induced impacts could be defined as risk factors influencing heat price as follows:

$$I_{Total} = I_{CO2} + I_{ENV} + I_{PF} + I_{WF} + (I_{TAX} + I_{VAT}) \quad (4)$$

where:

I_{Total}	total impacts/risk factors influencing heat price
I_{CO2}	impact/risk factor caused by CO_2 price - emission trading scheme
I_{ENV}	impact/risk factor caused by new environmental performance levels
I_{PF}	impact/risk factor caused by pollution fees
I_{WF}	impact/risk factor caused by water fees
I_{TAX}	impact/risk factor caused by energy taxation
I_{VAT}	impact/risk factor caused by value added tax

Impact factors I_{CO2}, I_{ENV}, I_{PF}, I_{WF} are generally applicable. Impact factors I_{TAX} and I_{VAT} are driven by circumstances of heat delivery and are applicable only in certain scenarios.

General structure of the model could be summarized as input-output economic model as presented below.

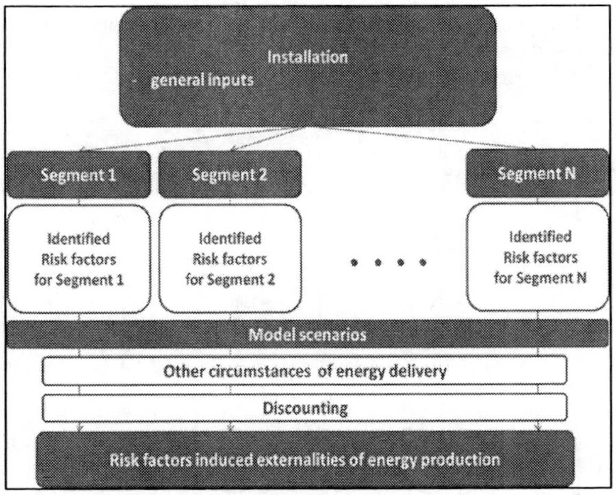

Figure 2. General structure of the model

Model uses installation general input to identify relevant model segment for calculation of induced externalities of energy production. There is a need for proper assessment of influencing risk factor for each reference case to correctly define segments and model scenarios.

This general structure was used to model the reference case of heat market of the Czech Republic, where structure inputs refer to installation parameters

- installation size,
- employed production technology,
- fuel input,
- energy parameters
  - efficiency of electricity and heat production,
  - water consumption,
- emission parameters,
- costs for reaching new emission parameters (based on legislation-induced requirements).

Model for reference case consist of 4 segments differentiated by size of respective installation: for installations below 1 MW thermal input, for installations from 1 to 20 MW thermal input, for installations from 20 to 50 MW thermal input and for installations above 50 MW thermal input. These segments were identified subject to the legislation analysis. Each of the segments reflects identified risk-factors in different way, based on relevant legislation provisions.

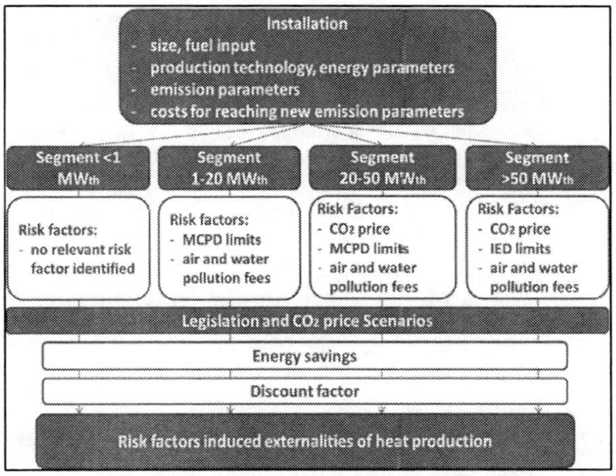

Figure 3. Structure of the model for reference case of the Czech Republic

There is a possibility to create different scenarios of legislation application and CO_2 price development based on various approaches. Model uses 2 legislation scenarios and 3 CO_2 price scenarios for reference case.

Model of reference case calculates 4 basic legislation-induced impacts:

- impact caused by CO_2 price - emission trading scheme [EUR]
- impact caused by new environmental performance levels [EUR]
- impact caused by pollution fees [EUR]
- impact caused by water fees [EUR]

These impacts are subsequently confronted with energy efficiency measures defined within State energy policy and reflecting of change in energy production. In order to reflect NPV calculation modelled impacts are also subject to discounting.

III. MODEL RESULTS

A. *Model summary – reference case*

The described economic model was used for calculating the implementation of the new EU and Czech environmental legislation in reference case of the situation in the Czech Republic and calculated legislation-induced influence on future heat prices could be summarized as follows:

- Model calculates the influence caused by CO_2 price, new emission levels, air pollution fees and water fees.
- Model can be applied to heat installations of whole range of installed thermal input and is applicable to combined heat and power generation and heat-only production as well. Model is applicable to solid (lignite/coal/biomass), liquid and gas fired installations.
- Certain inputs were set by using Association for district heating of the Czech Republic (ADH CR) internal database and expert estimation (e.g. efficiency of heat production, efficiency of electricity production etc.)
- A basic presumption is that heat and electricity production for the period 2015-2030 will be the same (or without significant changes) as average production during the period 2005-2008, which is the basic period for historical data according to relevant legislation.
- Model calculates the impacts on the energy produced (1 GJ of energy produced) for the whole Czech Republic based on fuels used in two basic legislation scenarios. The real impact on heat prices has to take into account the fuel mix used for heat generation in real CHP plant.
- Model uses uniform Discount factor for all model-segments to ensure comparability of outcomes.

B. *General inputs*

Table 2. Net calorific values and CO_2 emission factors

Fuel	Net Calorific Value	CO_2 Emission Factor[a]	
	[GJ/t] or [GJ/th.m³]	*[t CO_2/MWh of fuel calorific value]*	*[t CO_2/GJ of fuel calorific value]*
Lignite	14.5	0.360	0.100
Hard Coal	23	0.330	0.092
Fuel Oil	40.61	0.260	0.072
Natural Gas	34.08	0.200	0.056
Biomass	16.5/12[b]	0	0

[a] Values based on data from Ministry of Industry and Trade of the Czech Republic, Decree no. 480/2012 Coll. on the energy audits

[b] Values based on data from Ministry of Industry and Trade of the Czech Republic. Higher value for local installations below 1 MW_{th} firing biomass pellets.

Table 3. Heat demand development

Indicator		Heat demand[a]
Unit		*[%]*
Year	*2015*	100
	2020	96
	2025	91
	2030	85
	2035	84
	2040	82

[a] Values based on data from Update of state energy policy of the Czech Republic

Table 4. Water consumption

Installation size	Unit	Fuel			
		Lignite/ Hard Coal	*Liquid fuels*	*Natural Gas*	*Biomass*
Below 1 MW_{th}	[Th.m³/year]	400	200	100	200
1-20 MW_{th}		200	100	50	100
20-50 MW_{th}		40	20	10	20
Above 50 MW_{th}		0.4	-	0.1	0.2

Model uses several other more complex data sets - Energy production parameters, Emission parameters, Installation and Operational costs differentiated by respective model segments, which were developed based on the data from ADH CR internal database.

C. *Model outcomes*

Results of the model represent reference case of the situation in the Czech Republic concerning legislation-induced risk factors on the heat market participants. Given that heat is the main product of district heating plants, results is focused on the describing of externalities reflected in heat price. All the legislative aspects referred to hereinbefore will impact heat prices in the Czech Republic substantially.

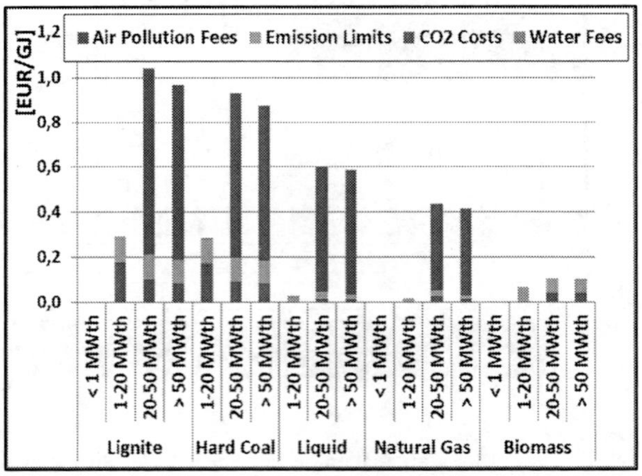

Figure 4. Comparison of risk factors in 2020, increase of costs per produced GJ of heat, optimization period 15 years, costs allocated to entire energy production, without VAT

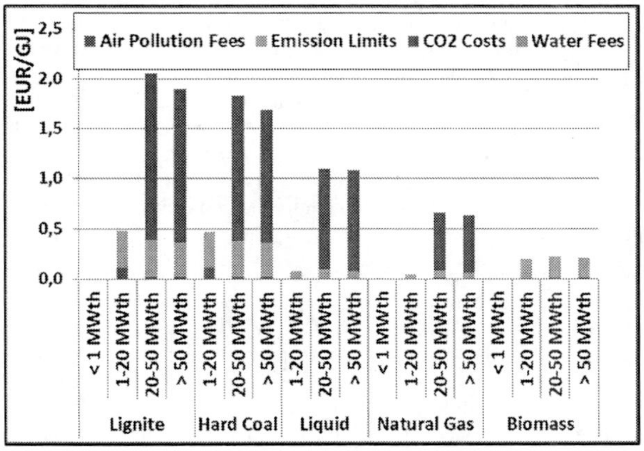

Figure 5. Comparison of risk factors in 2030, increase of costs per produced GJ of heat, optimization period 15 years, costs allocated to entire energy production, without VAT

Results of the analysis show among others:

- strong impact of CO_2 cost on heat price from installations under the EU ETS framework – this creates major detrimental effects on the heat market because of the fact that installation below the 20 MW limit are not subject of these cost. Cost of CO_2 are expected to be app. 80% of total price increase after the year 2020;
- new emission standards have the second highest potential impact, esp. for installation over 20 MW. Smaller installations have lower emission standards and, thus, increase of pollution fees has higher impact on their economy.

IV. Discussion

Results of the modelling clearly indicates that there is unequal treatment of polluters depending on the installed capacities which leads to the competitive distortions on the heat and power market with potential future escalation of this problem if no (legislative) remedial actions would be taken.

These remedial tools can be developed with the help of various instruments like the administrative (bans, limits, emission ceilings) or economic ones (taxes, charges, emissions trading), which offer polluters a certain degree of flexibility, and use market principles. The latter is used in several instances – esp. EU ETS and carbon taxation where relevant worldwide experience exists.

The emission trading subject to the research (for example [7]) and other administrative aspects in particular was found as inappropriate for addressing distortions effects of emerging environmental legislation. The EU ETS represents too complex system to accommodate huge number of new participants it is not realistically applicable to installations below 20 MW thermal input also from administrative and regulatory point of view.

On the contrary taking into account [8] carbon taxation could be seen as appropriate for addressing distortions effects of emerging environmental legislation. Indirect carbon taxation should not bring substantial increase in administrative costs thanks to the fact that existing legislative instruments could be used. Indirect carbon taxation as a tool for remedy of heat market distortions could be considered as realistically applicable to installations below 20 MW thermal input also from administrative and regulatory point of view.

A. Remedial tool application on the heat market

Following figures represent possible model situation on the reference case heat market with application of remedial tool – indirect carbon tax. Comparison is relevant in particular for fossil fuels based installations suffering from CO_2 costs, thus as major examples were chosen lignite and natural gas fired installations (the most common examples for reference case).

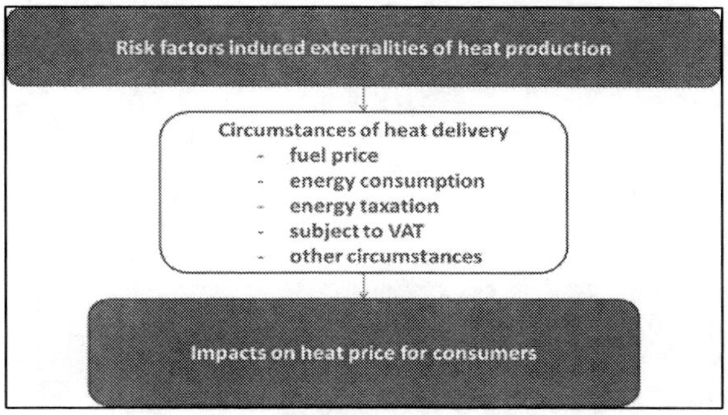

Figure 6. Transformation of risk factors into heat price for consumers

Model results are transformed into the change in heat price according to the Figure 6. Two basic types of consumers could be identified as:

- industrial consumers
  - VAT is not applicable – consumers are VAT payers (subject to VAT)
  - Ecology tax is not applicable – heat for industrial consumers is usually produced in CHP process which is excluded from energy taxation

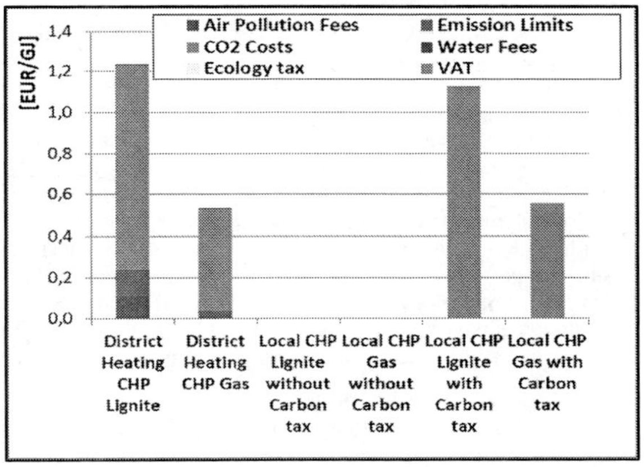

Figure 7. Comparison of externalities per produced GJ of heat for industrial consumers for District heating and Local boilers in year 2020 with inclusion of carbon tax tool

CO_2 costs dominate legislation induced factors and create major part of distortion effect on the heat market for reference case. New emission performance standards (emission limits) represent second most important risk factor. These are applicable only for district heating installation within defined assumptions for this reference case. Air pollution and water fees are having negligible effects on overall level of legislation induced costs. VAT and ecology tax are not applicable within defined assumptions for this reference case.

Columns 5 and 6 in the Figure 7 represents situation with inclusion of CO_2 tax based tool, which is allocating price of carbon also to installations below EU ETS limits. Comparison of columns 1 and 2 with columns 5 and 6 confirms the fact that

CO_2 tax based tool could remedy severe disproportions on the heat market for reference case towards uniform externalities for fossil fuels based heat market participants.

- residential consumers/households
  - VAT is applicable – consumers in households are not VAT payers (subject to VAT)
  - Ecology tax is applicable – heat for consumers in households connected to district heating is usually produced in CHP process (see assumptions for industrial consumers above) however energy taxation example is included for natural gas district heating installations, local heat solutions examples assumes heat-only installations

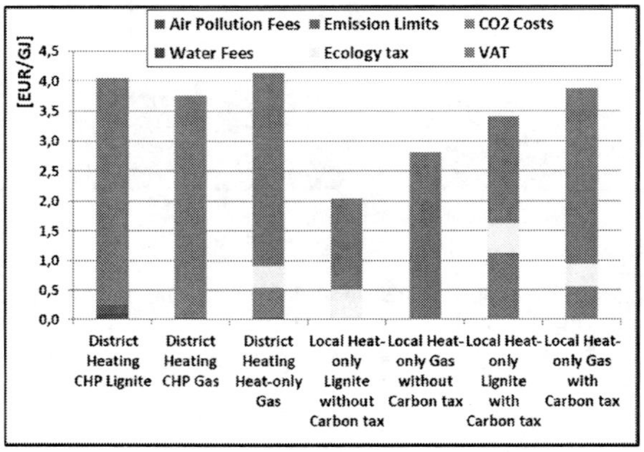

Figure 8. Comparison of externalities per produced GJ of heat for residential consumers for District heating and Local boilers in year 2020 with inclusion of carbon tax tool

In the case of heat market for households VAT dominates among legislation induced costs, this effect is derived from overall costs of heat production and higher legislation induced risk factors (such as CO_2 costs) also lead to higher VAT. CO_2 costs are representing second most important legislation induced externality and are becoming more important towards year 2030. Air pollution and water fees are having negligible effects on overall level of legislation induced costs. Ecology tax is applicable only in case of heat-only installations. In current situation heat-only natural gas fired installation in household are not subject of ecology tax. New emission performance standards (emission limits) have relevant effect only for district heating installation within defined assumptions for this reference case.

Major discrepancies in level of externalities caused by legislation are demonstrated by comparison of column 1 and 4 for lignite and 2, 3 and 4 for natural gas in Figure 8. Without remedial tools DH utilities will suffer from serious competition distortion and even reduced VAT rate is not efficient enough to remedy this deformation. Columns 6 and 7 represents situation with inclusion of CO_2 tax based tool, which is allocating price of carbon also to installations below EU ETS limits and thus balance out identified undue distortions on the heat market for households.

V. Conclusions

District heating sector and combined heat and power production effectively utilizes primary energy sources, reduce emissions related with covering of heat demand and can serve as the stabilization factor of heat prices both for households and industrial consumers. District heating sector is currently facing continuous implementation of the new legislation which will have the significant impact on its future economic competitiveness. Especially the revision of EU ETS scheme and CO_2 auctions will play the crucial role. Paper presents the analysis of current position of heating branch and identification of expected external factors influencing its economic competitiveness. The analysis is based on the effects of new legislation introduction on heat prices in mid-term period. Modelling approach is based on theory of differential NPV. Impact of following external factors is explicitly evaluated: new emission standards, air and water pollution fees and the revision of EU ETS scheme.

Analysis has been divided into the four segments – installations below 1 MW, installations in the range of 1-20 MW, 20-50 MW and over 50 MW. Results of the analysis show that the highest impacts of new legislation are onto installation in the range of 20-50 MW where EU ETS and emission standards have highest specific effects. This will negatively impact esp. on mid-size heating systems located in regional towns where new legislation impacts in full and, in contrary, there are only reduced effects of return from scale (as is in case of bigger installations).

Results of done analysis show the significant impacts of introduction of new EU legislation which could have potential significant impact on competitiveness of individual regional heating systems and heating branch as a whole. Presented analysis offers detailed comparison of different legislative tools which could lead to the weakening of overall economic and environmental effectiveness of proposed legislation. This could result, if there would not be adequate correction actions, in suboptimum alternative of heating branch development.

Acknowledgment

All the research described in this paper was conducted at the Czech Technical University in Prague, Fac. of Electrical Engineering, Dept. of Economics, Management and Humanities.

References

[1] A. Colmenar-Santos, E. Rosales-Asensio, D. Borge-Diez, J.-J. Blanes-Peiró, "District heating and cogeneration in the EU-28: Current situation, potential and proposed energy strategy for its generalisation," ELSEVIER, Renewable and Sustainable Energy Reviews 62 (2016), pp. 621–639, doi:10.1016/j.rser.2016.05.004

[2] B. Shi, L.-X. Yan, W. Wu, "Multi-objective optimization for combined heat and power economic dispatch with power transmission loss and emission reduction," ELSEVIER, Energy 56 (2013), pp. 135–143, doi:10.1016/j.energy.2013.04.066

[3] M. Gustafsson, M. Rönnelid, L. Trygg, B. Karlsson,"CO2 emission evaluation of energy conserving measures in buildings connected to a district heating system – Case study of a multi-dwelling building in Sweden," ELSEVIER, Energy 111 (2016), pp. 341–350, doi:10.1016/j.energy.2016.05.002

[4] Directive 2010/75/EU of the European Parliament and of the Council of 24 November 2010 on industrial emissions (integrated pollution prevention and control)

[5] Directive 2015/2193 of the European Parliament and of the Council of 25 November 2015 on the limitation of emissions of certain pollutants into the air from medium combustion plants

[6] Directive 2003/87/EC of the European Parliament and of the Council of 13 October 2003 establishing a scheme for greenhouse gas emission allowance trading within the Community including amendments

[7] Ch. Böhringer, A. Lange, "On the design of optimal grandfathering schemes for emission allowances," European Economic Review, 49 (8), pp. 2041–2055, doi: 10.1016/j.euroecorev.2004.06.006

[8] A. L. Bovenberg, R. A. Mooij, "Environmental tax reform and endogenous growth," ELSEVIER, Journal of Public Economics 63 (1997), pp. 207–237, doi:10.1016/S0047-2727(96)01596-4

Hoisting Electrical Drive Mathematical Model with Series Electromechanic Brake

Nikolay Djagarov
Nikola Vaptsarov Naval Academy
Electrical Department,
Varna, Bulgaria
jagarov@ieee.bg

Julia Djagarova
Technical University,
Electrical Department
Varna, Bulgaria
julija_d@abv.bg

***Abstract* — A mathematical model of an electric hoist`s asynchronous electrical drive with direct torque control is developed, using an electromechanical brake, the coil of which is in series connection with one of the phases. The mathematical model describes in detail the processes in the elements of the electric hoist, which allows the development of various methods for its control and the design of the frequency converter and the brake.**

Keywords* — *hoisting electrical drive; induction motor; electromechanical brake; electronic converter, direct torque control

I. Introduction

Electric hoists are an important part of lifting mechanisms and cranes. Frequent starts and stops are typical for them. Modern electric hoists usually use asynchronous motors with vector control or direct torque control. An important part of the electric hoist is the electromagnetic brake, which greatly determines the efficiency of its operation.

The electric hoist is used for lifting and lowering loads by means of a drum or a lift-wheel on which a rope is wound or a chain. In normal operation, the electric hoist must raise and lower the nominal load, stop and hold the load, limit the load-lowering speed in the presence or absence of electrical power, stop and hold the load in an emergency situation, withstand the temperature at the set-on / off frequency, the brake system must be adjusted if necessary.

The efficiency of the electric hoist operation highly depends on the used control method. In [1] two methods of efficiency optimization are analyzed: loss model control and search control. Search control technique is used in a steady state of drive and loss model during transient processes. As a result, power and energy losses are reduced.

The modeling of the processes in the electric hoist allows to examine the interaction processes between the different parts of the electric hoist, to choose the control method, to design the optimal lifting device.

Modern electric hoists perform complex tasks. They must control the speed of lifting and lowering the load, limit the starting current, hold the load stationary on command or when there is no power supply. To fulfill these tasks, a brake is essential to provide reliable and accurate braking to decelerate, stop or hold the load. Braking systems available for use on AC induction motors can be divided into two basic categories [2]: external braking; internal braking. External braking with electrical or mechanical devices mounted on the outside of the motor and includes such methods as: friction brake-spring-set, electrically released; magnetic brake-electrically set, spring-released; eddy current braking.

In [3] is proposed a method of breaking without additional elements in which the braking energy is absorbed by the kinetic energy of the drive at the applied second frequency, inducing motor losses. In [4] the various methods of stopping asynchronous drives are analyzed and is proposed to use combinations of two or more methods to accelerate the braking process. Different sequences of the methods used are analyzed, their characteristics and effectiveness are evaluated. In [5] a brake method is proposed that achieves a high motor slip mode between the rotary motor and the inverter output frequency. The engine current at braking is controlled by modified U / f samples. In [6], the particle swarm optimization method has been used to determine optimal supply voltage and motor stopping frequency, which minimizes braking losses. This method is particularly suitable for electric drives with frequent stops. In [7] is proposed a breaking method by discrete frequency control where the current inverter limits the currents. Discrete frequencies are subharmonic to grid frequency: 12.5Hz, 16.66Hz, 25Hz, achieving reverse frequency-torque ratio by which motor speed is controlled. Effective braking can be achieved in four steps using circuit electronic switching [8]. At first, a condenser is connected between two phases, providing single-phase self-excitation. After a certain drop in speed, a second condenser is connected to the same phases to support self-excitement and further reduction in speed. In the third stage the magnetic braking is performed by a short cut of the third phase. Finally, the engine is stopped by DC injection.

Electromechanical braking metod is similar to vehicle braking, but springs for lining the brake pads are used instead of hydraulic drive. A powerful coil is used to release the brake from the drum with a force opposite to spring`s one. When the winding is powered, the brake pads are released and the motor rotates. The winding receives power from the main contactor, thereby providing safety and fault protection against power failures in elevator and crane drives.

The creation of a mathematical model of an electric hoist with an asynchronous motor and an electromechanical brake is a complex task, as its component parts must be descrribed - motor, frequency converter based control system, electromechanical brake having different physical nature and different properties and characteristics.

In [9] is proposed simplified mathematical model of AC hoisting electric drive systems for determining the main quality parameters of the transient processes. By the quality parameters of the system transients is the control system is suggested.

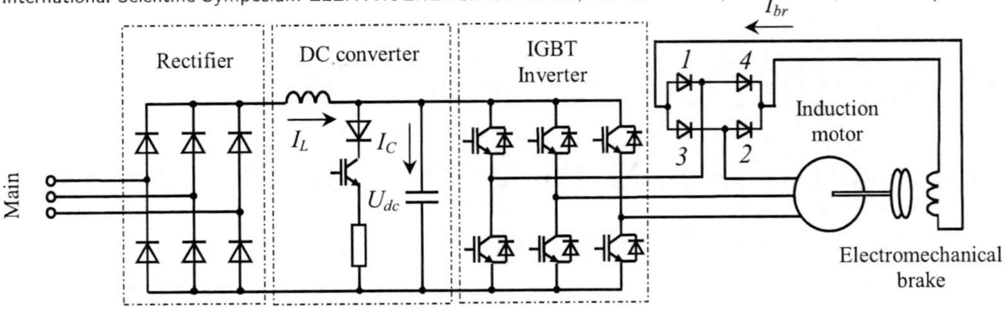

Fig.1. Scheme of investigated hoisting electrical drive

In this article is proposed a detailed mathematical model of an asynchronous electric hoist with direct torque control and an electromechanical brake in series connection with one of the phases.

II. Investigated Hoisting Electrical Drive

Figure 1 shows the circuit diagram of the investigated electric drive. The Frequency Converter consists of an uncontrolled rectifier, a DC converter, and an IGBT inverter. An electromagnetic brake is used, the coil of which is in series with the stator coil of the asynchronous motor.

The induction motor type is DM11-17-2; n=2850 rpm; U = 3x400 V; I = 11 A; cosφ= 0,83 with the following parameters: U_{DC}=600 V, R_s=0,01485 Ω, L_s=0,0003027 H, R_r=0,009295 Ω, L_r=0,0003027 H, L_m=0,01046 H, GD^2=3,1 kg.m^2, friction coefficient F=0,08 N.m.s, pairs of poles p=2.

III. Model of Investigated System

The modeling of the start and stop modes of the asynchronous drive provides characteristics acquisition of the transition conditions, which helps to select the method, control algorithm and the technical implementation of the controller performing the startup, operation and stopping of the electric hoist.

By using mathematical modeling, can be investigated the braking modes of an asynchronous drive by means of an electro-mechanic brake, the power of which is switched on consecutively in one of the phases.

Creating the mathematical model involves the following difficulties:

- The need for a precise model describing not only the electric hoist operation, but also the static converter processes, the asynchronous motor, the electromechanical-spring;
- The electric hoist consists of elements of different physical nature (electrical, mechanical, electronic), which determines the different time constants, which hamper the numerical study of the operation modes;
- Converter, engine and brake control and electric drive itself are complicated enough and requires great accuracy;
- Drive elements consist of continuous elements (motor, electromagnet) as well as discrete elements (static converters, rectifiers, DC regulator, inverter).

The mathematical model is designed using a non-iterative method for calculation of electromechanical processes in electrical energy systems [10,11,12]. The electric drive parts - the engine, the converters and the electromechanic brake are written in the canonical form of Cauchy in terms of their currents. With the help of the equations of the connections between them, the voltages in the connections of the investigated system are calculated. In this case, the derivatives of the currents are excluded and thus a non-iterative algorithm for calculation is obtained.

A. *Induction Motor Model*

In a phase coordinate system, the induction motor equations will have the following appearance:

$$
\begin{aligned}
u_a &= r_s.i_a + \frac{d}{dt}\psi_a; \\
u_b &= r_s.i_b + \frac{d}{dt}\psi_b; \\
u_c &= r_s.i_c + \frac{d}{dt}\psi_c; \\
0 &= r_r.i_{ra} + \frac{d}{dt}\psi_{ra}; \\
0 &= r_r.i_{rb} + \frac{d}{dt}\psi_{rb}; \\
0 &= r_r.i_{rc} + \frac{d}{dt}\psi_{rc}; \\
\frac{d}{dt}\omega_r &= \frac{1}{\tau_m}\left(T_M - T_{br}\right)
\end{aligned}
\qquad (1)
$$

where: s and r are indexes respectively for stator and rotor parameters and variables; T_M, T_{br} - engine electromagnetic moment and load resistance; l_s, l_r, m_s, m_r, m - coefficients of self and inter-inductance of stator and rotor; θ - angle between the phases of the stator and the rotor; $\theta_k = \int \omega_k dt$ - angle between the d-axis of the rotating coordinate system and the stator phase a; $\theta_r = \int \omega_r dt$ - angle between the d-axis of the rotating coordinate system and the rotor phase a;

$$\begin{bmatrix}\psi_a\\ \psi_b\\ \psi_c\end{bmatrix}=\begin{bmatrix}l_s & m_s & m_s\\ m_s & l_s & m_s\\ m_s & m_s & l_s\end{bmatrix}.\begin{bmatrix}i_a\\ i_b\\ i_c\end{bmatrix}+$$

$$+m.\begin{bmatrix}\cos\theta & \cos(\theta+120^\circ) & \cos(\theta-120^\circ)\\ \cos(\theta-120^\circ) & \cos\theta & \cos(\theta+120^\circ)\\ \cos(\theta+120^\circ) & \cos(\theta-120^\circ) & \cos\theta\end{bmatrix}.\begin{bmatrix}i_{r.a}\\ i_{r.b}\\ i_{r.c}\end{bmatrix}=$$

$$=\boldsymbol{\psi}_{abc}=\boldsymbol{M}_{ss}.\boldsymbol{I}_{abc}+\boldsymbol{M}_{sr}.\boldsymbol{I}_{r,abc}$$

;(2)

$$\begin{bmatrix}\psi_{r.a}\\ \psi_{r.b}\\ \psi_{r.c}\end{bmatrix}=m.\begin{bmatrix}\cos\theta & \cos(\theta+120^\circ) & \cos(\theta-120^\circ)\\ \cos(\theta-120^\circ) & \cos\theta & \cos(\theta+120^\circ)\\ \cos(\theta+120^\circ) & \cos(\theta-120^\circ) & \cos\theta\end{bmatrix}.\begin{bmatrix}i_a\\ i_b\\ i_c\end{bmatrix}$$

$$+\begin{bmatrix}l_r & m_r & m_r\\ m_r & l_r & m_r\\ m_r & m_r & l_r\end{bmatrix}.\begin{bmatrix}i_{r.a}\\ i_{r.b}\\ i_{r.c}\end{bmatrix}=\boldsymbol{\psi}_{r.abc}=\boldsymbol{M}_{rs}.\boldsymbol{I}_{abc}+\boldsymbol{M}_{rr}.\boldsymbol{I}_{r,abc}$$

(3)

$$\boldsymbol{P}=\frac{2}{3}.\begin{bmatrix}\cos\theta_k & \cos(\theta_k-120^\circ) & \cos(\theta_k+120^\circ)\\ -\sin\theta_k & -\sin(\theta_k-120^\circ) & -\sin(\theta_k+120^\circ)\\ 1/2 & 1/2 & 1/2\end{bmatrix};$$

(4)

$$\boldsymbol{P}_r=\frac{2}{3}.\begin{bmatrix}\cos(\theta_k-\theta_r) & \cos(\theta_k-\theta_r-120^\circ) & \cos(\theta_k-\theta_r+120^\circ)\\ -\sin(\theta_k-\theta_r) & -\sin(\theta_k-\theta_r-120^\circ) & -\sin(\theta_k-\theta_r+120^\circ)\\ 1/2 & 1/2 & 1/2\end{bmatrix}$$

(5)

After converting to a rotating coordinate system $d,q,0$ with the help of the matrix of Park`s direct transformation (4) and (5) and writing in the form of Cauchy, the following system of differential equations is obtained:

$$\frac{d}{dt}i_d=-a_{11}\cdot i_d+(\omega_k+a_{12}\cdot\omega_r)i_q-a_{13}\cdot i_{rd}+a_{14}\cdot\omega_r\cdot i_{rq}+b_{11}\cdot u_d;$$
$$\frac{d}{dt}i_q=-(\omega_k+a_{12}\cdot\omega_r)i_d-a_{11}\cdot i_q-a_{14}\cdot\omega_r\cdot i_{rd}+a_{13}\cdot i_{rq}+b_{11}\cdot u_q;$$
$$\frac{d}{dt}i_{rd}=a_{31}\cdot i_d-a_{32}\cdot\omega_r\cdot i_q-a_{33}\cdot i_{rd}+(\omega_k-a_{34}\cdot\omega_r)i_{rq}-b_{13}\cdot u_d;$$
$$\frac{d}{dt}i_{rq}=a_{32}\cdot\omega_r\cdot i_d+a_{31}\cdot i_q-(\omega_k-a_{34}\cdot\omega_r)i_{rd}-a_{33}\cdot i_{rq}-b_{13}\cdot u_q;$$

(6)

$$\frac{d}{dt}\omega_r=a_{41}\cdot T_M-a_{41}\cdot T_{br}. \quad (7)$$

where:

$T_M=x_{ad}\cdot(i_{rd}\cdot i_q-i_{rq}\cdot i_d)$ - the electromagnetic moment of IM; T_{br} - the torque of the motor shaft; τ_m - mechanical time constant of the mechanism and IM;

$$x_d{}'=x_s-\frac{x_{ad}^2}{x_r};\ b_{11}=\frac{1}{x_d{}'};\ b_{13}=\frac{x_{ad}}{x_r\cdot x_d{}'};\ a_{11}=r_s\cdot b_{11};$$
$$a_{12}=x_{ad}\cdot b_{13};\ a_{13}=r_r\cdot b_{13};a_{14}=x_{ad}\cdot b_{11};\ a_{31}=r_s\cdot b_{13};$$
$$a_{32}=x_s\cdot b_{13};\ a_{33}=r_r\cdot b_{33};\ a_{34}=x_s\cdot b_{11};\ a_{41}=\frac{1}{\tau_m};$$

(8)

r_s - stator resistance; x_s - stator reactance; x_{ad} - inter-inductance; r_r - rotor resistance; x_r - rotor reactance;

$$\boldsymbol{\psi}_{dq0}=\boldsymbol{P}.\boldsymbol{M}_{ss}.\boldsymbol{P}^{-1}.\boldsymbol{P}.\boldsymbol{I}_{abc}+\boldsymbol{P}.\boldsymbol{M}_{sr}.\boldsymbol{P}_r^{-1}.\boldsymbol{P}_r.\boldsymbol{I}_{r,abc}=$$
$$=\boldsymbol{L}_{dq0}.\boldsymbol{I}_{dq0}+\boldsymbol{M}_{dq0}.\boldsymbol{I}_{r.dq0};$$

(9)

$$\boldsymbol{\psi}_{r.dq0}=\boldsymbol{P}_r.\boldsymbol{M}_{rs}.\boldsymbol{P}^{-1}.\boldsymbol{P}.\boldsymbol{I}_{abc}+\boldsymbol{P}_r.\boldsymbol{M}_{rr}.\boldsymbol{P}_r^{-1}.\boldsymbol{P}_r.\boldsymbol{I}_{r,abc}=$$
$$=\boldsymbol{M}_{dq0}.\boldsymbol{I}_{dq0}+\boldsymbol{L}_{r.dq0}.\boldsymbol{I}_{r.dq0}$$

(10)

$$\boldsymbol{L}_{dq0}=\boldsymbol{P}.\boldsymbol{M}_{ss}.\boldsymbol{P}^{-1} \quad (11)$$

- matrix of self-inductances of the stator;

$$\boldsymbol{L}_{r.dq0}=\boldsymbol{P}_r.\boldsymbol{M}_{rr}.\boldsymbol{P}_r^{-1} \quad (12)$$

- matrix of self-inductances of the rotor;

$$\boldsymbol{P}.\boldsymbol{M}_{sr}.\boldsymbol{P}_r^{-1}=\boldsymbol{P}_r.\boldsymbol{M}_{rs}.\boldsymbol{P}^{-1}=\boldsymbol{M}_{dq0} \quad (13)$$

- matrix of inter-inductance between the stator and the rotor.

The induction motor model (6), (7) in matrix form:

$$\frac{d}{dt}\begin{bmatrix}\boldsymbol{I}_s\\ \boldsymbol{I}_r\end{bmatrix}_M=\begin{bmatrix}\boldsymbol{A}_{ss} & \boldsymbol{A}_{sr}\\ \boldsymbol{A}_{rs} & \boldsymbol{A}_{rr}\end{bmatrix}_M.\begin{bmatrix}\boldsymbol{I}_s\\ \boldsymbol{I}_r\end{bmatrix}_M+\begin{bmatrix}\boldsymbol{B}_{ss}\\ \boldsymbol{B}_{rs}\end{bmatrix}_M.\boldsymbol{U}_s=$$
$$=\begin{bmatrix}\boldsymbol{H}_s\\ \boldsymbol{H}_r\end{bmatrix}_M+\begin{bmatrix}\boldsymbol{B}_{ss}\\ \boldsymbol{B}_{rs}\end{bmatrix}_M.\boldsymbol{U}_s; \quad (14)$$
$$\frac{d}{dt}\omega_r=\frac{1}{\tau_m}(T_M-T_{br}),$$

where: s and r are indexes respectively for stator and rotor parameters; matrix elements and vectors $\boldsymbol{A}$ и $\boldsymbol{B}$ are functions of stator and rotor resistances and reactances and angle speeds of rotation of the coordinate system $d,q,0$ - ω_k and motor`s - ω_r; $\boldsymbol{U}_s$ - motor stator voltage vector.; T_{br} - resistive mechanical moment of the motor shaft;

B. *Frequency Converter Model*

The frequency converter consists of an uncontrolled rectifier, a DC converter for energy utilization returned via inverter and an inverter realized using IGBT transistors.

When designing mathematical models of converters, two groups of models are used. The first group includes models that simulate the non-linear characteristic of the switch by an ohmic-inductive bipolar that changes its parameters when changing its conductivity. The second group refers to models that, when switching switches, alter the topology of the converter circuits. An advantage of the first group of models is the simplicity of their design, but the disadvantages are the large dimension of the model and the change of the circuits time constants describing the converter when switching the switches. An advantage of the second group is the reduction of the order of the solved equation system (only the production chains are modeled), but the disadvantage is the more complex design of the converter model variable topology.

The mathematical model is designed using the second sub-process using the so-called switching functions [13,14].

Diode Rectifier Model

The electronic converter is a system with a variable topology, since switching switches on and off changes it. The so-called switching functions are used to design the switches model:

$$\frac{d}{dt}\begin{bmatrix} i_a\\ i_b\\ i_c\end{bmatrix} = \frac{1}{L}\cdot\begin{bmatrix}u_a\\ u_b\\ u_c\end{bmatrix} - \frac{R}{L}\cdot\begin{bmatrix}i_a\\ i_b\\ i_c\end{bmatrix} - \frac{U_{dc}}{L}\cdot\begin{bmatrix}(2S_a-S_b-S_c)/3\\ (2S_b-S_c-S_a)/3\\ (2S_c-S_a-S_b)/3\end{bmatrix} = \frac{d}{dt}\boldsymbol{I}_{abc} = \frac{1}{L}.\boldsymbol{U}_{abc} - \frac{R}{L}.\boldsymbol{I}_{abc} - \frac{U_{dc}}{L}.\boldsymbol{S} \qquad (15)$$

where: $\boldsymbol{U}_{abc}$, $\boldsymbol{I}_{abc}$ - stator voltage and current vectors; $\boldsymbol{S}$ - vector of the switching functions whose elements accept a value of 1 or 0 corresponding to the open or closed state of the valves; U_{dc} - rectifier output voltage; R, L - rectifier resistance and reactance.

After converting to a rotating coordinate system $d,q,0$, the equation system will have the following form:

$$\frac{d}{dt}\boldsymbol{I}_{dq} = \frac{1}{L}.\boldsymbol{U}_{dq} - \frac{R}{L}.\boldsymbol{I}_{dq} - \frac{U_R}{L}.\boldsymbol{P}.\boldsymbol{S} - \boldsymbol{W}.\boldsymbol{I}_{dq} \qquad (16)$$

where: $\boldsymbol{P}$ - matrix of Park`s direct transformation.

$$\boldsymbol{W} = \begin{bmatrix} 0 & -\omega_k\\ \omega_k & 0\end{bmatrix}.$$

Model of DC converter

The voltage of the condenser is calculated using the following equation:

$$\frac{d}{dt}U_{dc} = \frac{S_{dc}.I_C}{C}; \qquad (17)$$

where: S_{dc} is the converter switching function having two levels and controlling converter transistor`s condition – 1 for open and 0 for closed condition; $I_C = I_R - I_I$.

Mathematical Model of Inverter

Inverter output voltage equations:

$$\begin{bmatrix}u_a\\ u_b\\ u_c\end{bmatrix} = \frac{1}{3}\cdot\begin{bmatrix}2 & -1 & -1\\ -1 & 2 & -1\\ -1 & -1 & 2\end{bmatrix}\cdot\begin{bmatrix}S_{1a}.S_{2a}\\ S_{1b}.S_{2b}\\ S_{1c}.S_{2c}\end{bmatrix}.U_{dc}; \qquad (18)$$

where: S_{kj} - switching functions with values 1 or 0 in dependence of switches condition (open, closed); k - bridge arm (a,b,c); j - switch number in the bridge arm $(j=1,2)$.

Current on the DC side:

$$I_I = S_{1a}.S_{2a}.i_a + S_{1b}.S_{2b}.i_b + S_{1c}.S_{2c}.i_c. \qquad (19)$$

Output voltage of the inverter in the axes $d,q,0$:

$$\begin{bmatrix}u_d\\ u_q\\ u_0\end{bmatrix} = \frac{2}{3}\cdot\begin{bmatrix}\cos\theta_k & \cos(\theta_k-120^\circ) & \cos(\theta_k+120^\circ)\\ -\sin\theta_k & -\sin(\theta_k-120^\circ) & -\sin(\theta_k+120^\circ)\\ 1/2 & 1/2 & 1/2\end{bmatrix}\bullet \bullet\begin{bmatrix}u_a\\ u_b\\ u_c\end{bmatrix} = \boldsymbol{P}.\boldsymbol{U}_{abc} = \boldsymbol{U}_{dq0}. \qquad (20)$$

Mathematical Model of Electromagnetic Brake

One of the most used electromagnetic brakes uses the eddy currents to create a brake torque [15÷18]. Electromagnetic brakes [19,20], disc brakes [20], magneto-dust brakes [21], piezo-electric brakes [22] are also used.

In fig.2 is shown electromechanical brake scheme, used in the investigation. Armature and coil are used to generate magnetic field in air gap to move armature from maximum-air gap to zero-air gap state. In power off springs maintain maximum air gap by pushing the armature out. When the coil is supplied, electromagnetic flux generates magnetic force. When the resulting magnetic force exceeds total spring force then armature will start to move and air gap length will start to decrease. Lining provides friction for braking. Adjust nut is used to provide desired air gap length between housing and armature.

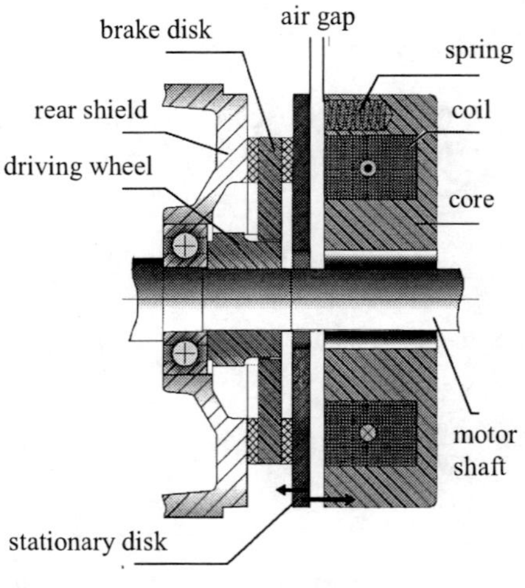

Fig.2. Electromechanical brake scheme

Electromagnets are complex electromechanical systems, which operation depends of the interaction of magnetic and electric fields and mechanical effects. Forces can be obtained from the electromagnetic field by numerical study using the following methods: the Maxwell stress tensor method; the virtual work method; the method of equivalent sources.

The calculation of the force acting on the piston can be done by any of the following methods. The effect of magnetic forces f_v within a given space V_Σ may be limited to a system of equivalent action on a surface Σ that limits the space under consideration. Then the resultant force will be [23]:

$$F = \int_{V_\Sigma} f_v dV = \int_\Sigma T_n dA \; ; \tag{21}$$

where: T_n - Maxwell acting tensor.

After the appropriate transformations for high magnetic inductivity, a simplified expression for the electromagnetic force can be obtained:

$$F = -\frac{1}{2}\mu_0 . \frac{N^2.I^2.A_0}{\delta^2} \; ; \tag{22}$$

where: μ_0 - gap magnetic inductivity; δ - electromagnet air gap; N - number of coil windings; I - coil current; A_0 - magnetic isolation.

It follows from (22) that: the electromagnetic force is direct proportional to the cross-section of the piston; the force is inversely proportional to the square of the length of the air gap, i.e. it must be minimal; there is a force of attraction between the two magnetic parts (stationary and movable), stationary mode is only possible at minimum value of the magnetic energy.

The moving masses of the mechanical system include: restraining the spring stopper; Anchor (plunger) with shaft; The parts of the spring. To simplify the model these parts are presented as a total mass m. The springs are preloaded during anchor movement. When supplied, the coil will generate a magnetic force F_{mag}^{-1}. The movement of the anchor will induce a damping force of friction $k_s.dz/dt$. Then the differential equation describing the mechanical sub-system will have the following form [24,25]:

$$m.\frac{d^2}{dt^2}z = -F_{mag}^{-1} - 2k_s.z - k_b.\frac{d}{dt}z - sign\left(\frac{d}{dt}z\right).F_f \tag{23}$$

where: z - anchor brake position; $2k_s.z$ - combined spring force; k_s - spring constant; k_b - damping coefficient; F_f - Coulomb`s friction.

The spring coil flux depends on the running current and the position of the brake plunger:

$$\frac{d}{dt}\psi_{cn} = l_{cn}.\frac{d}{dt}i_{cn} + i_{cn}.\frac{d}{dt}x.\frac{d}{dx}l_{cn} = l_{cn}.\frac{d}{dt}i_{cn} + v.i_{cn}.\frac{d}{dx}l_{cn} \; ; \tag{24}$$

where: l_{cn} - brake coil inductance; v - anchor displacement speed.

Then the voltage of the brake coil will be:

$$U_{cn} = r_{cn}.i_{cn} + l_{cn}.\frac{d}{dt}i_{cn} + v.i_{cn}.\frac{d}{dx}l_{cn} . \tag{25}$$

The magnetic force occurring between the brake coil and the magnetic plunger, the inductance l_{cn} and the flux ψ can be represented as a function of the current i and the displacement x. Calculation of the magnetic force can be obtained by the braking power:

$$W_{cn} = \int_{i(0)}^{i(t)} \psi(i,x)dt . \tag{26}$$

The inductance is calculated by:

$$l_{cn} = \frac{N.\phi}{i} = \frac{\psi}{i} . \tag{27}$$

The magnetic force can be calculated by the magnitude field strength:

$$F(i,x) = \frac{\partial}{\partial x}W_{cn} . \tag{28}$$

The working area of the brake is selected in the linear part of the magnetic system, but the model uses the whole feature, including the saturation.

When the brake is switched on and the displacement of the anchor is x_{on}, the anchor releases the relay, which releases the springs by pressing the brake disk. When the electric hoist is switched on, the magnetic force of the brake coil pulls on its anchor and releases the engine rotation.

Electromagnetic brake connection with electrical drive

The diode rectifier is a single-phase rectifier in series with phase a, and the electromagnetic brake coil is connected to the DC side. The rectifier sequentially in each half of the inverter output voltage supply the winding of the brake by two of the diodes in the opposite branch arms 1-2 and 3-4.

A way to connect the brake in series with phase a is to add the brake coil resistance to the first system equation (1) describing the motor. Then the total impedance of phase a will be:

$$r_a = r_s + r_{cn} \; ; \qquad l_a = l_s + l_{cn} \; ; \tag{29}$$

where: r_{cn}, l_{cn} - brake coil ohmic and inductive resistance.

Then it is necessary to re-calculate the coefficients a_{11}, a_{12}, a_{13}, a_{14} and b_{11} in the first equation of (8). The current through the winding of the brake will be calculated by motor current inverse Park transformation:

$$\begin{bmatrix} i_a \\ i_b \\ i_c \end{bmatrix} = \begin{bmatrix} cos\theta_k & -sin\theta_k & 1 \\ cos(\theta_k - 120°) & -sin(\theta_k - 120°) & 1 \\ cos(\theta_k + 120°) & -sin(\theta_k + 120°) & 1 \end{bmatrix} . \begin{bmatrix} i_d \\ i_q \\ i_0 \end{bmatrix} = \boldsymbol{I}_{abc} = \boldsymbol{P}^{-1}.\boldsymbol{I}_{dq0} \tag{30}$$

The current of the brake coil will be:

$$i_{cn} = |i_a| = |i_d.cos\theta_k - i_q.sin\theta_k| . \tag{31}$$

In (31) the absolute values indicate that the phase current is rectified from the brake rectifier.

The voltage of the brake coil:

$$u_{br} = r_{br}.i_{br} + l_{br}.\frac{d}{dt}i_{br}. \qquad (32)$$

At discrete time moments this voltage is calculated by finite differences between two steps of numerical integration. That is, at some point of time $t+1$ the brake voltage will be calculated using the ratio:

$$u_{br}^{t+1} = r_{br}\cdot i_{br}^{t+1} + l_{br}\cdot\left(i_{br}^{t+1} - i_{br}^{t}\right) \qquad (33)$$

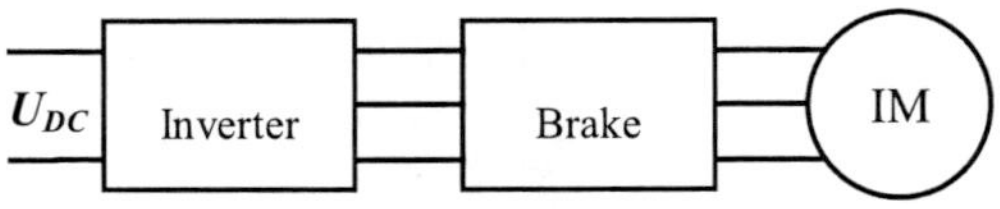

Fig.3. Block-scheme: inverter-brake-motor

Another way to model the processes in the brake is to present the brake as a three-phase resistance (fig. 3), for which the equation is:

$$\begin{bmatrix} u_a \\ u_b \\ u_c \end{bmatrix}_{inv} - \begin{bmatrix} u_a \\ u_b \\ u_c \end{bmatrix}_{M} = \begin{bmatrix} u_a \\ u_b \\ u_c \end{bmatrix}_{br} = \begin{bmatrix} r_{cn} & 0 & 0 \\ 0 & 0 & 0 \\ 0 & 0 & 0 \end{bmatrix} . \begin{bmatrix} i_a \\ i_b \\ i_c \end{bmatrix}_{br}$$

$$+ \begin{bmatrix} l_{br} & 0 & 0 \\ 0 & 0 & 0 \\ 0 & 0 & 0 \end{bmatrix} . \frac{d}{dt} \begin{bmatrix} i_a \\ i_b \\ i_c \end{bmatrix}_{br} = \boldsymbol{U}_{inv}^{abc} - \boldsymbol{U}_{M}^{abc} = \boldsymbol{U}_{br}^{abc} =$$

$$= \boldsymbol{R}_{br}^{abc}.\boldsymbol{I}_{br}^{abc} + \boldsymbol{L}_{br}^{abc}.\frac{d}{dt}\boldsymbol{I}_{br}^{abc}. \qquad (34)$$

After converting (34) to axis $d,q,0$ by Park transformation will obtain:

$$\boldsymbol{U}_{inv}^{dq0} - \boldsymbol{U}_{M}^{dq0} = \boldsymbol{U}_{br}^{dq0} = \boldsymbol{R}_{br}^{dq0}.\boldsymbol{I}_{br}^{dq0} + \boldsymbol{L}_{br}^{dq0}.\frac{d}{dt}\boldsymbol{I}_{br}^{dq0} + \boldsymbol{W}.\boldsymbol{L}_{br}^{dq0}.\boldsymbol{I}_{br}^{dq0} =$$

$$= \left(\boldsymbol{R}_{br}^{dq0} + \boldsymbol{W}.\boldsymbol{L}_{br}^{dq0}\right)\boldsymbol{L}_{br}^{dq0} + \boldsymbol{L}_{br}^{dq0}.\frac{d}{dt}\boldsymbol{I}_{br}^{dq0} \qquad (35)$$

where: $\boldsymbol{R}_{br}^{dq0} = \boldsymbol{P}.\boldsymbol{R}_{br}^{abc}.\boldsymbol{P}^{-1}$; $\boldsymbol{L}_{br}^{dq0} = \boldsymbol{P}.\boldsymbol{L}_{br}^{abc}.\boldsymbol{P}^{-1}$;

$$\boldsymbol{W} = \begin{bmatrix} 0 & -\omega_k & 0 \\ \omega_k & 0 & 0 \\ 0 & 0 & 0 \end{bmatrix}; \ \omega_k = \frac{d}{dt}\theta_k;$$

With the help of (34) the voltage vector supplied to the stator of the asynchronous motor $\boldsymbol{U}_{M}^{dq0}$ can be calculated by examining its model (6), (7). As can be seen from (26), for calculating this vector it is necessary to calculate the matrices $\boldsymbol{R}_{br}^{dq0}$ and $\boldsymbol{L}_{br}^{dq0}$, the calculation of which is obtained by multiplying the respective matrices containing the calculation of trigonometric functions $(sin\theta_k, cos\theta_k)$ of each step of numerical integration, which would greatly increase investigation time. For this reason, the first method for modeling the process in the brake coil is used.

As the induction resistance change during the movement of the anchor brake, it then follows a change in the coefficients of the equations (20), (25) and (8).

CONCLUSIONS

Reliable and efficient operation of electric hoists depends on the methods and control systems used. Modern electric hoists use frequency converters with DC power and vector control. An electromechanical brake is used to provide precise and reliable braking.

The article has developed a detailed mathematical model of an electric hoist using an asynchronous motor with an electro-mechanical brake. For the design of the mathematical model, a non-iterative method is used, allowing to make a model with minimal order describing in the best possible way the processes in the electric hoist components of different physical nature and described with different types of equations.

The designed mathematical model can be used to investigate electric hoist processes in normal and emergency modes of operation. The results obtained from the investigation can be used to select optimal operating modes and law and control system, and to design an electric hoist with the converter, brake and control system.

REFERENCES

[1] M. Rastogi, P.W. Hammond, “Dual-frequency braking in AC drives”, IEEE Transactions on Power Electronics, Vol.17, No.6, Nov 2002, pp.1032-1040.

[2] R. Shemanske, “Electronic motor braking”, IEEE Transactions on Industry Applications, Vol.IA-19, Apr 1983, pp.824-831

[3] P.L. Rongmei, S.L.Shimi, D.S.Chatterji, V.K.Sharmu, “Fast braking for induction motor”, International Journal of Engineering and Innovative Technology (IJEIT), Vol.1, No.6, June, 2012, pp.65-69.

[4] M.M. Swamy, T. Kume, Y. Yukihira, S. Fujii; M. Sawamura, “A novel stopping method for induction motors operating from variable frequency drives”, IEEE Transactions on Power Electronics, Vol.19, No.4, July, 2004, pp.1100-1107.

[5] M.M. Elkholy, M.A. Elhameed, “Braking of three phase induction motors by controlling applied voltage and frequency based on particle swarm optimization technique”, International Journal of Power Electronics and Drive System (IJPEDS) Vol.5, No.4, April 2015, pp.520~528.

[6] K. Rahane, R. Hardas, “Three phase induction motor braking using discrete frequency control”, .

[7] S.S. Murthy, G.J. Berg, C.S. Jha, A.K. Tandon, “A novel method of multistage dynamic braking of three-phase induction motors”, IEEE Transactions on Industry Applications, Vol.IA-20, No.2, March/April 1984, pp.328-334.

[8] A.S. Puttewar, N.U. Kakde, H.A. Fidvi, B. Nandeshwar, “Enhancement of Braking System in Automobile Using Electromagnetic Braking”, IOSR Journal of Mechanical and Civil Engineering (IOSR-JMCE), pp.54-59

[9] K. Khandakji, K. Zdrozis, “Determination of quality parameters of hoisting electric drive systems with 3-phase induction motors, Proc. 5th International Conference on Technology and Automation; Oct. 2005, pp.294-296.

[10] N.F. Djagarov, “Transient processes calculation in electrical systems with complex structure”, Electricity, No.1, 1990, Moscow, pp.9-16.

[11] N.F. Djagarov, Ship Electrical Power Systems, Lectures, Technical University of Varna, Bulgaria, 1997, 424 p.

[12] N.F Djagarov, “A method of transient electromechanical processes modeling in power systems, Proceedings of 2009

IEEE Bucharest PowerTech Conference, June 28th - July 2nd, Bucharest, Romania, art. No.5281949, 6 p.

[13] C.C.. Marouchos, The Switching Function: Analysis of power electronic circuits, The Institution of Engineering and Technology, 2006, 297 p.

[14] Djagarov N., Grozdev Zh., Bonev M., Djagarova J., Operation Study of Wind Permanent Magnet Synchronous Generator, Proceedings of 12th IEEE International Conference on Environment and Electrical Engineering EEEIC, Wroclaw, Poland, 5-8 May, 2013, pp.522-528.

[15] .M. Qian., P. Kachroo, "Modeling and control of electromagnetic brakes for enhanced braking capabilities for automated highway systems", IEEE Conference on Intelligent Transportation Systems, 1997, pp.391-396.

[16] D. Schieber, "Braking torque on rotating sheet in stationary magnetic field", IEE Proceedings, Vol.121, No.2, Feb 1974.

[17] J.H. Wouterse, "Critical torque of eddy current brake with widely separated soft iron poles", IEEE Proceedings, Vol.138, July 1991, pp.153-158.

[18] Q. Zhou, X. Guo, G. Tan, X. Shen, Y. Ye, Z. Wang, "Parameter analysis on torque stabilization for the eddy current brake: A developed model, simulation, and sensitive analysis", Hindawi Publishing Corporation Mathematical Problems in Engineering Vol. 2015, Article ID 436721, 10 pages.

[19] Y. Tenzer, B.L. Davies, F.R. Baena, "Four-state rotary joint control: results with a novel programmable brake", IEEE/ASME Transactions on Mechatronics, 2012, Vol.17, No.5, pp.915-923.

[20] H. Davis, W. Book, "Torque control of a redundantly actuated passive manipulator", Proceedings of the American Control Conference, 1997, Vol.2, pp.959–963.

[21] M. Russo, A. Tadros, "Controlling dissipative magnetic particle brakes in force reflective devices", American Society of Mechanical Engineers, Dynamic Systems and Control Division (Publication) DSC. 1992.

[22] M. Gogola, M. Goldfarb, "Design of a pzt-actuated proportional drum brake", IEEE/ASME Transactions on Mechatronics, 1999,.Vol.4, No.4, pp.409–413.

[23] D. Cazacu, C. Stanescu, "Performance analysis of a solenoidal electromagnet", Scientific Bulletin of the Electrical Engineering Faculty, Vol.11 No.2, pp.5-10, (16).

[24] Y. Wang, T. Megli, M. Haghgooie, K.S. Peterson, A.G. - Stefanopoulou, "Modeling and control of electromechanical valve actuator, 2002 Society of Automotive Engineers, SAE 2002-01-1106 Powertrain and Fluid Systems Conference, San Diego, CA, October 21-24, 2002.

[25] Y. Shiao, L.V. Dat, "Actuator control for a new hybrid electromagnetic valvetrain in spark ignition engines", Proceedings of IMech E Part D: Journal Automobile Engineering 227 (6) 2013, pp.789–799.

New Ways of Providing Ancillary Services in a Power System

Jiří Vašíček, Miroslav Vítek
Czech Technical University
Faculty of Electrical Engineering
Prague, Czech Republic
vasicek@ fel.cvut.cz, vitekm@ fel.cvut.cz

Abstract

This contribution focuses on new possibilities for balancing generation and load in an electrical power system.

The ever increasing proportion of intermittent power sources (wind and solar) in the power system changes the basic paradigm of power engineering economics whereby instead of adjusting generation output to the demand at the time, the focus is shifting towards adjusting consumption based on the output of intermittent generation. That is the goal of the so-called smart grids, where the consumer will be adapting to intermittent generation by responding to price signals.

In spite of or rather because of this, there will still be (at the moment, more and more) the need for power sources and appliances that will, at the dispatcher's request, quickly increase or decrease generation or instead of generators become consumers, i.e. their technical parameters enable them to maintain balance between generation and consumption and network frequency within set limits.

On the generation side, these are primarily gas turbines and closed cycle gas turbines, pumped storage hydro stations, heat plants with turbines equipped with heat accumulators or accumulating heat into a heat distribution network and networks of CHPs (combined heat and power) also with heat accumulators supplying heat into buildings in a wide area, but centrally controlled.

On the demand side, these are HDO systems (a form of ripple control in the Czech Republic) which control appliances that convert electricity into useful heat, which is easier to accumulate for water heating purposes, electrical boilers connected into heat networks and in the future also P2G systems using cheap electricity to break down water into oxygen and hydrogen. Being affiliated with current heat plants, they will utilize carbon from fossil or biofuel combustion for the methanisation of hydrogen and its injection into the natural gas network for the purposes of seasonal energy accumulation.

Nonetheless, for accumulation of alternate current over the course of a day to a week it is still the best to make use of the benefits of pumped storage hydro power stations which remain to be a proven form of electricity accumulation in the grid. The vast majority of them are designed and constructed in mountainous regions so that they can take advantage of the differences in elevation of the landscape. This appears to be at odds with nature protection interests. The paper concentrates on a location of an active lignite quarry in the northwest of the Czech Republic, where the interests of nature protection would not be affected.

From an economical perspective, in the case of a pumped storage hydro station, capital expenditure is key and therefore the paper also includes an assessment of the economic viability of the project, through the valuation of the effects resulting from the provision of ancillary services to the transmission system.

Keywords: ancillary services; system services; power balance; transmission power system; conventional power units; power to gas; pumped storage power plant; wind turbines; photovoltaic panels; accumulator of energy; flooded quarry; economic efficiency of investments; investment outlay; income from services sales; income from services provisions; reclamation after mining

I. ANCILLARY SERVICES (ANS) FOR THE TRANSMISSION NETWORK

In general, ancillary service is the ability of a power station block to either automatically or at the dispatchers request, regulate its output within a set timeframe, and thus help maintain the network frequency within prescribed limits.

Ancillary services can be divided into several categories, on the basis of the required response time to a rapid change of balance, the duration of the service and the actions of the transmission system operator (TSO), as per the table 1.

The growing capacity of intermittent power sources (primarily wind installations) is not securing sufficient volumes of electricity due to low utilisation levels (fewer than 2 000 hrs/year). This utilisation is decreasing further with the increase in installed capacity of wind power stations. This is caused primarily by the fact that at times of high wind speeds it is not possible to effectively utilise the electricity where it is generated and the transmission lines are overloaded. There is then no other option other than the curtailment of wind power plants (WPP). Curtailment is currently the only method of regulation of WPP. However, the necessity of the use of primary frequency controllers, similar to those at existing conventional plants (providing AnS), in new wind turbines is becoming apparent. The controller will be installed in order to gradually reduce output by setting the wind turbine blades in case of a sudden increase in network frequency [17]. This will further decrease the utilisation level of installed power of WPP.

Table 1. General classification of AnS services – operating reserves for frequency control [2]

type	primary	secondary	tertiary	time management
time	seconds	minutes	hours	days
from	0,01 seconds	cca 1 min.	5 min.	next day
to	1-2 min.	15 min.	steady state, aprox. 1 hr	reaching alignment of synchr. time with the standard CET
principle	solidarity	non-intervention	agreement	centrally controlled
who	all TSOs in proportion	TSO affected by the fault	affected TSO and neighb. TSOs based on the situation	all TSOs are sent revised frequency values
how	automatic frequency regulation of all spinning reserves (AFC)	automatic. through the control system of the affected TSO activating own reserves	dispatcher of the affected TSO activates cold reserves + arranges help with other TSOs	through the setting of a revised frequency for the AFC of all blocks of all TSOs
aim	to prevent a rapid drop in frequency and collapse of the network	to balance cross border flows based on the plan (electricity market trading)	dtto and return to steady state as per the plan	synchronise the sum of the number of cycles with time (max. deviation ± 20 seconds)
volume	3 GW proportionally split amongst all TSOs	as a minimum the nominal output of the biggest power station of each TSO	individually based on the current situation related to the transmission capabilities	sum of output of all transmission connected power stations

The only realistic possibility of **effectively** managing a further increase in output of intermittent power sources is the accumulation of electricity**.** Should the installed capacity of WPP exceed the annual maximum load, at times of high wind conditions there will not be any conventional power stations left to be curtailed. This growing effect of WPP is already noticeable i.e. that a short time lack of wind is not as much of an issue as strong wind is [3] [4].

Power sources that can at that time be used as sinks in order to reduce the excess output in the grid, are currently primarily pumped storage hydro stations (PSH). They are tried and tested sources with acceptable efficiency and capacity in relation to their installed capacity. Their task is to transfer electrical power in time from the times of excess to the times of lack of electrical output in the grid within its pumping cycle. This is given by the ratio of their installed power to output and is usually in the region of several hours. It would be needed to build a PSH with higher capacity, which could also operate on a weekly cycle. The combination of WPP and PSH is a partial solution to the efforts of securing electricity primarily from renewable sources. PSH is from this perspective ideal as it represents an emission free power source using only clean water.

An issue for further construction of PSH is the lack of suitable locations as these are mainly in mountainous regions which are often subject to nature protection designations. The construction of a PSH means a significant impact on nature primarily during of construction, the subsequent operation is generally without any negative ecological impacts.

It is therefore the construction process and the flooding of the area which raises the resistance of ecologists, local residents and other interested groups. The situation is similar to that of the necessary reinforcement/construction of transmission lines in order to enable the transport of electricity from windy regions to the windless ones. This social phenomenon is commonly referred to as the NIMBY effect (Not In My Back Yard). We therefore propose to locate a new PSH in an extracted lignite quarry, which will provide a high capacity power source and also minimize the NIMBY effect.

II. Proposal for a PSH in a flooded ČSA quarry following its closure

One of the first large quarries where extraction was curtailed and providing that territorial mining limits will not be lifted will cease its operation in 2024 [11], is the ČSA quarry run by Severní energetická, a.s. The anticipated volume of water in the residual pit after flooding is approx. 300 million m3 with maximum depth of 140 m and elevation of 180 m above sea level.

There are two main options for utilising the residual pit:

1) As a bottom reservoir only with the top reservoir constructed in nearby Ore mountains in order to increase the gradient
2) Direct damming of the quarry pit

The advantage of the first option is that due to a several times higher gradient (approx. 500 m) a smaller volume of water is required in order to achieve a certain level of accumulation capacity. This means that as the water level fluctuation due to the PSH operation would only be small (25 – 50 cm), the bottom reservoir could be used for recreational purposes. The disadvantage however is the ongoing NIMBY effect and the construction encroaching into a mountainous landscape. There is also the need to construct a longer pipeline connecting the top and bottom reservoir and higher pressure losses directly related to its length, resulting in higher CAPEX. Considering that the North Bohemian region already has a number of lakes suitable for recreation, we are leaning towards option 2 i.e. the damming of the quarry pit, which means a lower CAPEX, lower pipeline losses due to its length and higher accumulation potential with higher installed capacity, which has a positive effect on the economics of the project. The main advantage however is a smaller NIMBY effect as the land is already the property of the mining company [15]. The disadvantage is that the reservoir could not be, as already mentioned, used for recreational purposes due to high water level fluctuations in both reservoirs.

It would be worthwhile to form the top reservoir already during the mining process of the oldest part of the quarry pit where the overburden is at the moment (the internal spoil tip). It is necessary to strengthen the edges using stabilizer and energy gypsum as a building material suitable for reclamation purposes and produced in abundant quantities by coal-fired power plants. An important principle would be to achieve a maximum slope of the banks of 30 ° from the horizontal plane so that there is no threat of instability of the slopes. Raising the banks of the top reservoir compared to the surrounding terrain by about 10 m is also of consideration.

The bottom of the top reservoir could also be raised this way relative to the bed of the bottom reservoir, so that less water would be needed to fill them, however that would depend on the volume of overburden – whether there would be excess or lack of it.

Should there be a lack of overburden the bed of the bottom reservoir could be even lower however more water would be needed to fill it. This additional water would remain in the bottom reservoir and not be part of the pumping cycle. Nonetheless it would increase the volume of available water in the landscape at times of drought. Considering that reservoir beds as well as overburden usually consist of various types of clay, it would create an impermeable layer although only following a certain degree of water absorption.

In the centre of the quarry, it is important to achieve a certain thickness of reinforced reservoir bed with a stabilizer for the foundations of the concrete core of the dam, which will house the machine room. This concrete dam would be embedded in a sufficiently sized bulkhead that would support it at the same time. On the surface, this bulkhead would be strengthened with a strong surface layer of stabilizer, with similar pressure resistance to poor concrete.

Fig. 1 depicts the pit adjusted for the purposes of operating the PSH. The top reservoir is in place of the internal spoil tip, the bottom reservoir is in an area where active mining currently takes place. The advantage of this arrangement is a very short supply pipe to the turbines connecting the top and bottom reservoirs. This greatly reduces energy losses due to friction and the swirling of water in the pipework both during pumping as well as turbine operation. It can therefore be expected that the overall efficiency of the turnaround cycle would not be lower than 77 %.

The theoretical accumulation potential of a PSH is given by the useful volume of its reservoirs. The top reservoir has a volume of approximately 140 million m^3 with a middle gradient in the region of 100 m. The limiting factor is rather the useful volume of the bottom reservoir which is approximately half of the top one. At 90 % efficiency in the turbine mode this means that the maximum volume of energy that the PSH would be able to deliver in turbine mode during one turnaround/cycle would be:

$$W_t = h_s \cdot \varsigma \cdot g \cdot V \cdot \eta_c = 125 \cdot 1000\ 9{,}81 \cdot 70 \cdot 10^6 \cdot 0{,}9 \cdot 10^{-12} \cong$$
$$\cong 77 TJ \cong 21 GWh \quad (1)$$

where

- W_t energy provided by PSH in turbine mode in one cycle [GWh]
- h_s middle hydraulic gradient [m]
- ς water density [kg/m^3]
- g gravitational acceleration [m/s^2]
- V useful volume of bottom reservoir [m^3]
- η_t efficiency in turbine mode

That is roughly seven times more than our currently largest PSH Dlouhé Stráně! The volume of the top reservoir of the PSH would be slightly bigger than the volume of the top reservoir of PSH Dalešice (which is approximately 127 mil. m^3 [8][9]).

The energy consumed by pumping in order to fill the top reservoir from minimum to full with pumping efficiency of circa η_c = 87 % would roughly be:

$$W_c = \frac{h_s \cdot \varsigma \cdot g \cdot V}{\eta_c} \cong 99\ TJ \cong 27\ GWh \quad (2)$$

Figure 1. Design of the layout of the PSH in the ČSA lignite quarry [17]

This results in the overall turnaround/cycle efficiency of 21/27 ≅ 0,777.

Considering that, on average around 220 GWh (without self-consumption) of electricity is generated in the CZ per day, it is possible to say, that the new PSH would be able to accumulate one tenth of the daily output of the CZ [5] [8]!

III. Investment costs for the construction and running costs of PSH

Investment costs for the construction of PSH can be divided into two parts - construction and technological [9] [10]. The advantage of this proposal is the possibility of spreading the construction work over a longer period of time. The work on the top reservoir, in particular, will coincide with reclamation of the land, which must occur in order for the area to return to its "normal" function. Therefore it is possible to consider the expenditure related to this "slow" construction of the PSH as operating expenses for the reclamation of the quarry and not as investment costs for the PSH.

The shaping of the future PSH can take place whilst the quarry is still active. It would be advisable to plan the construction in such a way that the reclamation works are conducted in an intelligent way with the aim of discreetly creating the final work. The top reservoir could therefore be largely built whilst the quarry is still operational, when it is possible for the banks to be raised and strengthened, the foundations for the concrete core of the dam adjusted, the assembly of the technology planned, the corridor for the construction of the access routes and water supply (from the feeder connecting Ohře and Bílina) negotiated etc.

The investment costs of the construction part of the project would then only comprise the construction of the dam across the quarry pit and not the works done on the banks and the reservoir bed as these would be related to the reclamation of the land and would need to be done even in the case of a standard reclamation.

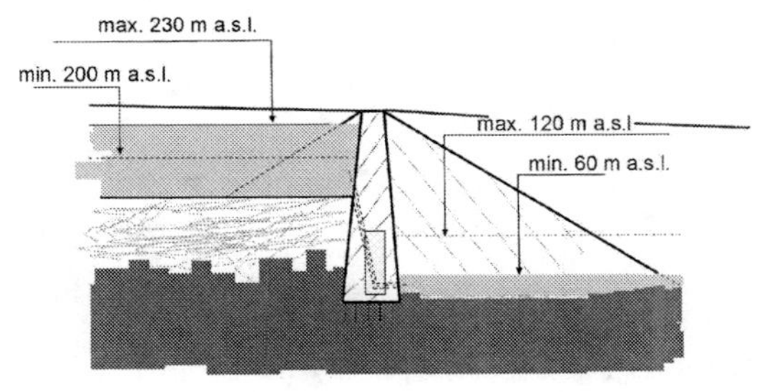

Figure 2. Detail of the dam

Considering that the equipment for the transport of soil for construction purposes would stay in place following the completion of the coal mining works, it is possible to estimate the total construction investment costs at a maximum of 10 billion CZK.

The technology expenditure will depend primarily on the installed capacity of the plant. At the assumed technology investment costs of 10K CZK / kW and 6 units (300 MW each) in the turbine (generator) mode, the total technology investment costs are 18 billion CZK. We assume with possibility of spin rotor little increasing about 10 % above the synchronous speed [13] [14] which causes the increase of power consumption in pump regime for quick exhausting unnecessary power from intermittent sources. The increase will be about 30 % over power in generation mode, that is 390 MW per unit, totally 2 340 MW.

It is therefore possible to estimate the total CAPEX at around 28 billion CZK, which compared to other mountain locations is a low value considering the much higher reservoir capacity. For example the most promising mountainous location for a PSH - Šumný důl - comes according to [12] to 27 billion at 880 MW installed capacity and 5.3 GWh output in one turnaround/cycle and is therefore comparatively more expensive. OPEX can be estimated at roughly 0.5% of CAPEX that is less than 150 million CZK.

IV. Economical viability of the project

Services that PSH offers can generally be divided into two groups:

1) static, which are of interest (value) for electricity traders
2) dynamic, which are of interest for real time power system dispatch control

Both types of services work towards minimising imbalances between planned and actual supply and demand.

A. *PSH static services revenues (electricity market trades)*

Efficiency of static services of a PSH, which at times of surplus output – i.e. cheap electricity works in a pumping mode as a consumer and at times of lack of output when electricity costs are high, switches to a turbine mode and supplies electricity as a power source. The difference in the electricity prices during those time periods ultimately determines the power plant's static services efficiency. The ratio of the sale and purchase price of electricity must, as a minimum, be equal to the inverse efficiency value of the turnaround/cycle, that is equal to the so-called coefficient of pumping.

$$k_c = \frac{1}{\eta_c} = \frac{1}{0{,}77} = 1{,}3 \qquad (3)$$

Of course to cover costs and make a profit, resp. the required return on capital investment, the required ratio is higher. Let assume that only half the storage capacity of the PSH, that is 13GWh for purchase and 10 GWh for sale (because 3 GWh are lost due to pumping) will be reserved for day ahead and intraday trading. Then should the e.g. average purchase price of electricity for pumping be 500 CZK and the price of the sold electricity 1000 CZK, the daily spread would equal to 3.5 million CZK. Should the power station work on those basis at least 285 days per year, the PSH would make 1 billion CZK per year through static services.

B. *PHS dynamic services revenues*

Revenues from static services would cover costs (operational cost + depreciation over 35 years), but would not bring sufficient return on capital investment (IRR only just under 1%). It is therefore necessary to also consider revenues from dynamic services, that is AnS for the transmission network. In order to achieve six percent[1] internal return on revenue IRR=6%, it is necessary to earn another billion CZK from dynamic services every year. If we consider half of the output regulation range of this PSH we come to respectable:

$$(2\ 340 + 1\ 800)/2 = 3\ 140/2 = 1\ 570 \text{ MW}$$

This is basically one eighth of the current annual maximum of the Czech network [5] [8]. It is therefore obvious that the significance of this PSH would go beyond the borders of the CZ and that the power station would be able to help address short term problems in the central European region. Of course, only if working on the assumption of increased cross-border transmission capacities. It could then help balance the network mainly at times of overproduction in German wind power plants in northern Europe [5].

Sufficiently high price for AnS MZ5 (minute reserve) supplied by this plant would be given by assuming 6 hrs of pumping (consumption) as a maximum and 6 hrs of electricity generation in turbine mode per day. This results in decreasing in market revenue from 1 billion to approximately 550 million CZK per year. The revenues from AnS make up for this loss several times over as we can assume min. 6 hrs per day MZ5+ at 900 MW, 6 hrs MZ5- at 1 120 MW and for the rest of the day i.e. 12 hrs MZ5+ at 1 800 MWh and MZ5- at 2 340 MW. In order to achieve IRR of at least 6 % it is necessary to earn 1.5 billion CZK a year through dynamic services. Assuming only stand by 300 days per year this equals to at least 5 million CZK per day. That means 81 CZK/MW per hour of stand-by[2].

Typical prices for this type of service are multiple times higher and therefore the provision of this service to TSOs could result in relatively high revenue and simple return on investment could be under 6 years. Of course the provision of AnS by this PSH would certainly push their price down over time, however, due to very low operational costs it would certainly be able to offer its operator/owner high cash flow and together with other PSHs provide certainty for dispatch control of the Czech power grid in the uncertain future full of intermittent power sources in Europe.

If the PSH was to reach annual revenues in the region of 1 – 2 billion CZK, at 35 years lifetime, the IRR would be between about 1 and 6 %. It is therefore possible to say, that the provision of AnS alone would be sufficient to secure this level of annual revenues. This would be attractive for the operator/owner of the PSH as it would mean income without the usual wear and tear of the equipment through the provision of static services.

V. Conclusions

Apart from the significance of this plant from an energy perspective it is also no less significant from a water management perspective. The bottom reservoir of the PSH would have available (retention) volume of 210 million m^3, which is a volume, which assuming sufficient inflow capacity can help deal with even a large scale flood. On the other hand at times of drought it can provide 140 million m3 of non-potable (service) water.

In conclusion it is only possible to list two examples from abroad. Firstly from Australia, where there is a plan to utilise a flooded gold mine in Queensland. 330 MW of PSH installed capacity at a cost of 280 million AUD [6].

In Arizona, there is also a plan to utilise a copper quarry following its closure and on the border of Nevada and California is a vast area affected by mining that is suitable for the construction of a PSH and solar power plants. PSH in the flooded quarries would be used to store surplus electricity generated in neighbouring solar plants [7].

Acknowledgment

The idea of this paper came from process of education process modernization in the Master's program in Economics and Management of Power Engineering on Faculty of Electrical Engineering Czech Technical University in Prague.

References

[1] T.D Thanh, D. Schulz, "Potential of new pumped-storage power plants in open cast mining structures and comparison of the storage capability with electro-mobility," Poster, Helmudt-Schmidt Universität, Hamburk 2009.

[2] Vítek M.: Ancillary Services in Power System and way of their providing. SGEM 2016.

[3] Agorameter (*data about operation Germany PN*). Internet: https://www.agora-energiewende.de, [May. 26, 2017].

[4] Fraunhofer ISE Energy Charts, data about electricity production and spot market prices I Germany. Internet: https://www.energy-charts.de/index.htm, [May. 26, 2017].

[5] Operation data ČEPS, a.s. Internet: https://www.ceps.cz/CZE/Data/Vsechna-data/Stranky/Default.aspx, [May. 26, 2017].

[6] G. Hering. "From mines to megawatts: The promise of 'conflict-free Big Solar' ". Internet: https://www.greenbiz.com/blog/2014/03/11/revitalizing-scarred-landscapes-clean-energy, [May. 26, 2017].

[7] S. Vorrath. "Genex plans pumped hydro storage plant, in old gold mine primed with solar". Internet: http://reneweconomy.com.au/2015/genex-plans-pumped-hydro-storage-plant-in-old-gold-mine-primed-with-solar-14500 Nov. 26. 2015 [May. 26, 2017].

[8] Annual and monthly reports on the CR's PS operation for the years 2008–2015. ERO, available from https://www.eru.cz/elektrina/statistika-a-sledovani-kvality/rocni-zpravy-o-provozu/...

[9] J. Budín, "Parametry a použití přečerpávacích vodních elektráren," Bakalářská práce, FEL ČVUT, Praha 2013.

[10] R.Portužák, "Přečerpávací vodní elektrárny," Prezentace, VŠB Ostrava, 2011.

[11] M. Říha, I. Dejmal, "Územní ekologické limity těžby v SHP," Společnost pro krajinu, Praha 2005.

[12] Synergion, a.s., "Přečerpávací vodní elektrárna Šumný důl," Základní informace o projektu, Praha 2008.

[13] T. Lorenz. *Regulace příkonu PVE v čerpadlovém provozu.* Dipl.práce, VÚT Brno, FEKT, 2008.

[14] V. Bartoš, J. Bendl, F. Žížek, "Dvojitě napájený asynchronní stroj pro přečerpávací vodní elektrárn," In: Elektroenergetika, Praha, 2000.

[15] "Těžba hnědého uhlí". http://www.sev-en.cz/cz/uhli/tezba.html [May. 26, 2017].

[16] Žilík J.: Využití zatopeného lomu pro PVE. Diploma thesis, ČVUT FEL, Praha 2016.

[17] K. Máslo, M. Vrba, P. Švejnar, L. Haňka, J. Veleba, M. Chladová, J. Vokál, B. Sadecký, V. Mach, Z. Brettschneider, Z. Hruška. *Řízení a stabilita ES.* ČEPS a.s., Praha 2013, ISBN 978-80-260-44671-1.

[1] Six percent IRR is during the current times of decreasing interest rates not a bad value of IRR.

[2] PSH could also offer other AnS, e.g. black start by supplying power necessary to start up coal fired plants connected to the Hradec substation (in the NW of CZ) following a collapse of the network.

Automated Parking System Supplied by Off-Grid Power System with an Intelligent Control System

Lukas Prokop, Tomas Vantuch, Stanislav Misak
ENET Centre
VSB – Technical university of Ostrava
Ostrava, Czech Republic
{lukas.prokop, tomas.vantuch, stanislav.misak}@vsb.cz

Abstract —**Automated car parking was built a few years ago at university campus at Technical university of Ostrava as a testbed for various innovative technology testing. Selected parking places will be equipped with fully automated charging station for e-vehicles. We introduce an innovative control system for energy management of off-grid power system, which supply the building and manage charging of e-vehicles. Control system is based on application of intelligent control tools and methods using neural networks and other similar algorithms.**

Keywords: Automated parking system; off-grid; e-vehicles;

I. INTRODUCTION

Essentially, automatic parking systems (hereinafter "APS") are parking houses with a computer-controlled automated process of the vehicle park-in/park-out by means of a sophisticated technological device. [1],[2] Except for the place intended for the vehicle handover/takeover, the object space is inaccessible to the driver. APS provides a progressive method of solution of covered parking in big town centres, at railway stations, housing estates, in administrative centres and wherever it is necessary to suggest an efficient way of short- and long-term parking of vehicles as a result of insufficient area for classic parking systems.[1],[2]

In principle, with regard to the manner of vehicle parking, two methods of vehicle movement in APS can be distinguished. The first e consists in the vehicle driving onto a palette on which it remains parked continually until subsequently leaving the APS. The second method makes use of parking carriages which locate the vehicle at the defined place and return to the home position, ready to locate another vehicle to another defined parking place. [1]

The particular advantage of APS is the high density of parked vehicles with regard to the built-up area of APS, which particularly predetermines APS for use in city centres with high building density and thus also lack of free space for the construction of classic parking sites. The high density of the built-up area also brings with it the high consumption of electric power.

Automatic parking systems enable more cars to be parked on a smaller area than traditional parking systems – such as parking houses, for the very reason that the parking process is provided automatically. In a way, therefore, this method prevents damage of cars in as much as there is, for instance, no risk of paintwork being scratched by other car users opening the doors. Furthermore, it is worth mentioning that the environment is not polluted unnecessarily as the car engines are not running during parking. Modern computer systems controlling these parking systems are extremely reliable, low-cost and safe.

At present, electric power is distributed in such localities by means of cable distribution networks, which, in most cases, nevertheless, were built decades ago and their transmission capacity, in many cases, is at the limit value for reliable and safe supply of electric power. Another negative factor is the ever-increasing consumption of electric power in urban agglomerations, which may have several causes, such as expansion of air-conditioning units or, for instance, the gradual development of electric cars with the continually rising input of charging stations. An overview of output parameters of charging stations is indicated in Table 1. [3 – 9]

Table 1. Comparison of nominal power of E-chargers.

Charging	**1-phase AC**		**3-phase AC**		**500 V DC**	**500 V DC + 400 V AC**
Nominal Power	3.7 kW	7.4 kW	11 kW	22 kW	50 kW	50 kW DC/ 43 kW AC
Nominal Current	16 A	32 A	16 A	32 A	100 A	100 A + 63 A

High-duty supercharger stations require transmission of large amounts of electric power. Powerful fast superchargers achieve the values of 50 kW, at the voltage level of 400 V; then the current can achieve the values of tens of amperes. [4]

The above-mentioned results in potential lack of electric power in city centres. To supply APS, therefore, it is suitable to employ a decentralized source of electric power, such as photovoltaic power plants, as they are used, for instance, for the supply of family houses. With regard to the stochastic characteristic of electric power production, it is suitable to supplement the supply system with a storage device capable of eliminating the stochastic character of electric power production partly and thus build up a hybrid supply system exploiting a renewable source of electric power; only in cases of shortage of power from the photovoltaic power plant, located mostly on the APS roof, is power supplied from the distribution network. [4, 6, 9]

This paper aims to present results of the pilot test operation of the APS supply system by means of utilization of a photovoltaic power plant and storage device. The first chapter contains a description of the automatic parking system. The second chapter presents the supply system with the photovoltaic power plant and storage system. The third chapter presents the assessment of the pilot test operation, including an analysis of consumption of the individual electric subsystems. The last chapter introduces the innovative control system.

II. Description of APS with off-grid supply system

The described automatic system of independent parking is a technological unit from the company, KOMA Industry s.r.o. This automated transport centre (ADC) is equipped with an automatic parking system in which there are 37 parking places for passenger vehicles located on four floors. The total capacity is 36 vehicles, 1 parking palette was removed from the parking system and the technology for off-grid operation of the APS building was installed in its place.

With its technological operation, the object ranks among the KOMA TOWER product series with the cars being stacked in square movements controlled by a highly sophisticated control system. In this way, the control system provides for real-time coordination and synergy of movements of several partial parking activities, resulting in a reduction and optimization of the parking time. The APS is unattended, which means that the driver parks the vehicle in and out on the basis of identification performed with a RFID magnetic card in the parking meter. The building is intended for parking vehicles of maximum vehicle dimensions of 1.9 m height, 2.2 m width and 5.4 m length, with the maximum weight of 3000 kg.

Figure 1. APS with Off-grid supply system.

The parking palettes on the APS ground floor are equipped with charging stations which enable charging of electric cars. This is an EVlink Wallbox EVH2S11P02K charging station with the maximum output of 22 kW.

The entire charging process consists in the electric car driver's driving into the area for vehicle handover, switching off the engine, applying the brake, engaging the gear, and connecting the electric car to the charging station. After the handover area has been left, the palette with the vehicle is located at the defined place, the charging station is simultaneously connected to the mains, upon which charging is started. During the stay at work, the electric car is being charged (usually for the period of 8 hours) until fully charged. At the end of the working hours, the owner gives the order to take out the vehicle by means of the RFID card, the palette with the charging station and the vehicle is disconnected from the network and transferred to the area of the vehicle handover. The driver disconnects the electric car from the charging station, gets in and drives out of the APS. The parking procedure of combustion engine vehicles is similar, without the vehicle being connected to the parking palette and charged.

The actual movement of the palette with the stationed vehicle is executed by means of skip conveyers (movement in the horizontal plane on the floor) and by means of an elevator located in one of the corners of the APS (movement between floors in the vertical plane).

By default, APS are supplied from the low-voltage distribution network; to ensure reliable operation, they are supplemented with a back-up source of electric power. In most cases, this back-up source is a diesel generator. With regard to the trends in the area of decentralized production [6, 7, 8], there is an upcoming possibility of using a hybrid supply system employing renewable sources of electric power and a storage system eliminating the stochastic character of supply from renewable sources. The APS presented in this paper is supplied from a system meeting these characteristics. A conceptual supply diagram is illustrated in Figure 2.

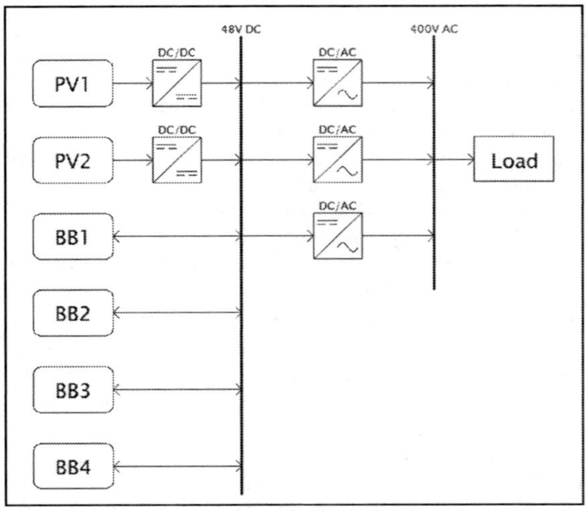

Figure 2. Topology of off-grid supply system.

Description of suuply system

A photovoltaic power plant is installed on the APS roof, with the total output of 10 kWp, divided into two strings, each with a peak output of 5 kWp. WINAICO WST-P6 polycrystalline panels are installed. The panel parameters are indicated in Table 2.

Table 2. Comparison of nominal power of E-chergers.

Electrical data (STC)		WST-250P6
Nominal power	P_{MAX}	250
Voltage at maximum power	V_{MP}	30,7
Current at maximum power	I_{MP}	8.15
Open circuit voltage	V_{OC}	37.5
Short circuit current	I_{SC}	8.53
Module efficiency	-	15.1

Subsequently, each string is connected to the Xantrex XW MPPT 80 600 solar charge controller. Settings of the solar charge controller are indicated in Table 3.

Table 3. Xantrex XW MPPT 80600 settings

Charger settings	
Charge Cycle	2 stage (no float)
Recharge voltage	53 V

Nominal Battery voltage	48 V
Low Battery Voltage Trigger Set	44 V
High Battery Voltage Trigger Set	56 V

The output of the solar charge controller is connected to a common DC bus bar of the nominal value of 48 V DC.

The 4 accu packs are connected to the same DC bus bar, each with the capacity of 360 A·h. To accumulate electric power, LiFePo4 cells of the LA3016 type are used. Each of the accu packs consists of 12 cells with the capacity of 30 A·h and is equipped with a balancer used to maintain the same voltage in every cell of the set. The entire accu pack is then equipped with a microprocessor-controlled active battery management system checking the voltage, current and temperature of every cell, and able to send a command to disconnect the accu pack as a result of exceeded limit values of voltage, current or temperature, damage, or on account of an external command. The above-mentioned conditions can also be used to disconnect the accu pack on the individual cells.

To the same DC bus bar, 3 Conext XW+ Inverter/Charger 6848 NA are connected, enabling the creation of a pure 3f AC sinusoidal system. Parameters of the Conext XW+ Inverter/Charger are indicated in Table 4.

Table 4. Parameters of the Conext XW+ Inverter/Charger

Inverter parameter	XW+ 6848 NA
Inverter AC output	
Output power (continuous) at 25°C	6800 W
Overload 30 min / 60 sec at 25°C	8500 W / 12000 W
Maximum output current 60 seconds (rms)	102 A (120 V); 52 A (240 V)
Output voltage	L-N: 120 V +/- 3%; L-L: 240 V +/- 3%
Input DC voltage range	42 to 60 V (48 V Nominal)
Maximum input DC current	180 A
Charger DC output	
Maximum output charge current	140 A
Output voltage range	40 - 64 V (48 V Nominal)

As there are two AC outputs on the Conext XW+, it is also possible, in addition to the back-up source in the form of, for instance, a diesel generator, to make use of direct connection to the distribution network used as a back-up form in case of lack of electric power produced by the photovoltaic system. Specific settings of the Conext XW+ are indicated in Table 5.

Table 5. Conext XW+ Inverter/Charger 6848 NA settings.

Conext XW+ Inverter/Charger 6848 NA	
Inverter settings	
Low Battery Cut Out	48 V
High Battery Cut Out	61 V
Chargers settings	
Battery Bank Capacity	4x360 A·h
Maximum Charge Current	140 A
Recharge Voltage	49,5 V
Custom Battery Settings	
Maximum Discharge Current	150 A
Grid Support	
Grid Support Voltage	50 V

It is, therefore, a hybrid form of connection, in which the fully off-grid mode is utilized for the period of sufficient production of the photovoltaic power plant (spring-to-autumn period in the given locality); in the event of insufficient production of the photovoltaic power plant, the distribution system is used to recharge the accumulation system (winter period in the given locality).

The above-mentioned hybrid off-grid system is used to supply the internal consumption of the APS. Parameters of the electric equipment of the APS are indicated in Table 6.

Table 6. Parameters of the devices

Device	**Number of Phases**	**Nominal current (A)**
Turntable	3+N	1.8
Traffic lights	1+N	1.08
Door	3+N	3.61
RFID card reader	1+N	2.71
EPS	1+N	2.71
Scales	1+N	0.54
Skip 1	1+N	16.2
Skip 2	1+N	16.2
Skip 3	1+N	16.2
Skip 4	1+N	16.2
Lift	3+N	41.5
Data boxes I-III	1+N	5.41
Lights	1+N	10.8

Consumption analysis

The pilot test operation of the hybrid system of APS supply was started in the spring months of 2016.

With regard to the APS operation, the primary item is its consumption in the so-called "stand-by" mode, where no vehicles are being parked. This consumption, evaluated on the basis of long-term measurement, achieves values within the range of 700 to 800 W. Subsequently, the energy intensity of an average parking cycle is characterized by the value of 1.5 kW·h for the entire parking cycle, depending on the position of the parking palette in the APS. Palettes located on the ground floor do not use the elevator, which is one of the appliances with the highest consumption of electric power; palettes located on the third floor, on the contrary, exploit the elevator to the maximum possible intensity. The consumption of electric power depends on the palette location and position. The graph in Figure 3 shows the course of active output drawn for a single cycle of vehicle park-in as well as for a single cycle of vehicle park-out. This concerns vehicle parking on a palette placed on the third floor of the APS; therefore, the elevator is utilized.

The graph in Figure 3 shows the course of the consumed active power. The first part of the graph represents the state when the APS is in the stand-by mode and the approx. active power of 760 W is drawn continually. This state is repeated in the time interval of 280 to 330 s as well as after the parking cycle is completed by parking out the vehicle at the approx.

time of 570 s. The actual parking cycle starts up at the approx. time of 30 s, when the parking meter has given a command upon application of the RFID card. At that time several devices started to consume electric power. First of all, the roll-up gate was closed;

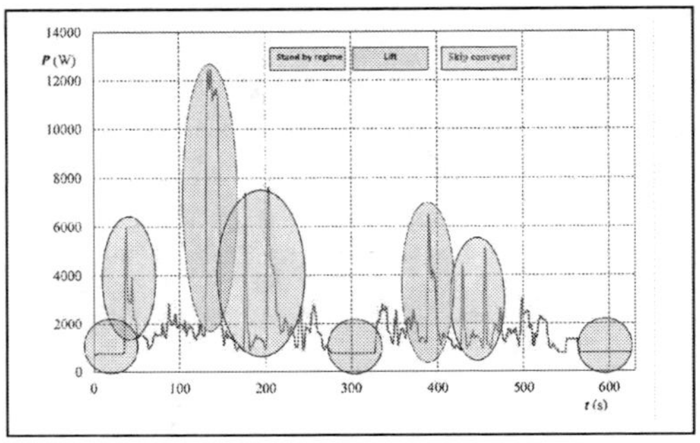

Figure 3. Course of active output during a single parking cycle (park-in and park-out)

simultaneously, the pertinent palette started to move and was shifted from the paring position onto the pertinent skip conveyor on the given floor. After the roll-up gate was shut, the turntable drive started to consume electric power, turning the parking palette to the position in which the parking palette could be transferred onto the skip conveyor (the first peak in the graph, approx. consumption of 6 kW) on the ground floor; in addition to being used for placing palettes on the ground floor, the conveyor is also used to transport palettes to/from the space of the palette elevator for palettes from higher floors of the APS. Subsequently, the palette was transferred by the skip conveyor to the elevator space and at the approx. time of 130 s, when the elevator started to consume electric power (the highest electric power consumption; the value of the consumed output exceeds 12 kW). After the elevator transferred the palette to the pertinent floor, the skip conveyor on the floor started to consume electric power (the third peak in the graph, approx. consumption of 7.3 kW); the skip conveyor transferred the palette with the vehicle before the target position. Subsequently, the palette with the vehicle was located to the pertinent parking position (the fourth peak in the graph, approx. consumption of 7.6 kW). At this moment, the palette with the vehicle is in its place already and the parking cycle is finishing; the skip conveyors are transferred to the home position. Thus the first part of the parking cycle is completed.

At the time from 280 s to 330 s, the APS is in stand-by mode again. At the approx. time of 330 s, a vehicle park-out command is given and the entire cycle is repeated in reverse order. The skip conveyor moves the palette out of the parking position and transfers it to the elevator area. The elevator transports the palette with the vehicle to the ground floor (approx. time of 390 s, consumption 6.5 kW). Then the ground-floor skip conveyor transports the palette to the area for the vehicle collection (approx. time of 430 s, approx. consumption of 4.3 kW) and the turntable swivels so that the vehicle is directed with its front part to the exit (approx. time of 460 s, approx. consumption of 5 kW). The driver leaves the APS handover area with the vehicle; subsequently, the parking cycle is completed automatically, when the roll-up gate is shut, the skip conveyors transferred to the home position and an empty palette is prepared to start up another parking cycle.

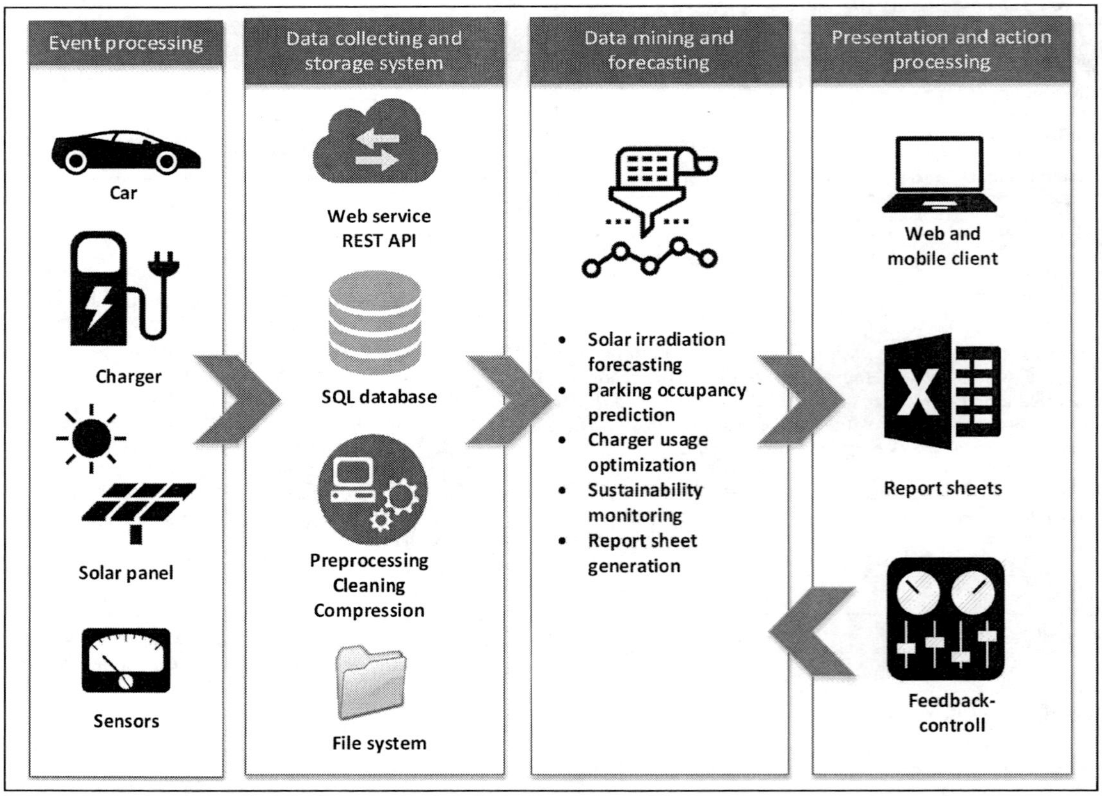

Figure 4. An intelligent control system scheme.

III. Intelligent Control System

Intelligent control system was developed and implemented to make APS more effective. The interaction of several stochastic entities such as dependency on weather conditions, parking lot occupancy and requirements for car recharging turns the problem into much more complex level. The issue of trafic and vehicle management driven by sensoric networks and intelligent algorithms is still much wider [10] and our solution should reflect this fact. The intelligent controlling software (ICS) should serve as a supportive tool for all of our issues combined and still should be able to be extended as well as to interact with other systems.

ICS consists of several layers of reposibilities, which are depicted in Figure 4 (diagram) and each of them is processing different important task. The sensoric network is the base of the event observing and data generating part. Each of the connected sensor is accesible thought the internet or it is able to call the API in defined time intervals to sent its state. The SQL database is filled by these unfiltered and unprocessed data. In an adjusted schedule, the data from database are loaded, filtered, statistically cleaned and compressed into the backup file to save space.

Further processing of the data is connected with statistical dataminig and machine learning to fulfill the needs of decision making and optimization processes. This decision making module creates several outputs for web and mobile client with possible autonomous or expert-driven feedback-controll of the entire system. The decision making module needs to be able to model and approximate several key aspects of the system, such as the future solar irradiation, future parking lot occupancy, optimization of vehicle charging, etc.

Conclucion

Automatic parking system presented in this paper is able to show a possible solution of two of most important problem in huge urban areas in large cities. The first problem is parking place in urbanized areas of world largest cities. The second problem is lack of electricity in extreme urbanized areas where electrification is a few tens of years old and it is not easy to extend and modernize cable lines in city center.

The advantage of presented solution is also intelligent control system which with high efficiency control APS and charging of electric vehicles with respect optimal energy utilization according to weather conditions.

The future work will be focused on optimization of intelligent control system. Also implementation of presented APS as a part future smart grids like "power balance device" which works like storage device which use capacity of e-vehicles to balance energy in grid.

Acknowledgment

This paper was supported by the following projects: Sustainable development of ENET Centre [LO1404]; Development of the ENET centre research infrastructure [CZ.1.05/2.1.00/19.0389]; Students Grant Competition SP2017/159] and Technology Agency of the Czech Republic. [TH01020426]

References

[1] M.Y.I. Idris, Y.Y. Leng, E.M. Tamil, N.M. Noor, Z. Razak. „Car park system: A review of smart parking system and its technology“ (2009) *Information Technology Journal*, 8 (2), pp. 101-113.

[2] A. Mathijssen, A. Johannes Pretorius. „Verified design of an automated parking garage“ (2007) *Lecture Notes in Computer Science (including subseries Lecture Notes in Artificial Intelligence and Lecture Notes in Bioinformatics)*, 4346 LNCS, pp. 165-180.

[3] M. Honarmand, A. Zakariazadeh, S. Jadid. "Self-scheduling of electric vehicles in an intelligent parking lot using stochastic optimization" (2015) *Journal of the Franklin Institute*, 352 (2), pp. 449-467.

[4] M. Honarmand, A. Zakariazadeh, S. Jadid. "Integrated scheduling of renewable generation and electric vehicles parking lot in a smart microgrid" (2014) *Energy Conversion and Management*, 86, pp. 745-755.

[5] U.C. Chukwu, S.M. Mahajan. "V2G parking lot with pv rooftop for capacity enhancement of a distribution system" (2014) *IEEE Transactions on Sustainable Energy*, 5 (1), art. no. 6588628, pp. 119-127. Cited 27 times.

[6] H. Lund, W. Kempton. "Integration of renewable energy into the transport and electricity sectors through V2G" (2008) *Energy Policy*, 36 (9), pp. 3578-3587.

[7] R. Loisel, G. Pasaoglu, C. Thiel. "Large-scale deployment of electric vehicles in Germany by 2030: An analysis of grid-to-vehicle and vehicle-to-grid concepts" (2014) *Energy Policy*, 65, pp. 432-443.

[8] L. Agarwal, W. Peng, L. Goel. "Probabilistic estimation of aggregated power capacity of EVs for vehicle-to-grid application" (2014) *International Conference on Probabilistic Methods Applied to Power Systems* (PMAPS), pp. 1-6.

[9] P. Kadurek, C. Loakimidis, P. Ferrão. *Electric vehicles and their impact to the electric grid in isolated systems*

[10] Md. Whaiduzzaman, M. Sookhak, A. Gani, R. Buyya. "A survey on vehicular cloud computing." *Journal of Network and Computer Applications* 40 (2014): 325-344.

Risk Assessment of Frequency Response and Reserve Requirement for a Power System with High Wind Power Penetration

Michael Negnevitsky
School of Engineering and ICT
University of Tasmania
Hobart, Australia
michael.negnevitsky@utas.edu.au

Hieu Nguyen
School of Engineering and ICT
University of Tasmania
Hobart, Australia

Marian Piekutowski
Hydro Tasmania
Hobart, Australia

Abstract — **Increasing role of wind power generation has created new challenges in power system planning and operation. Uncertainty of wind energy may lead to significant load-generation imbalances resulting in large frequency deviations, and hence increase system operation risk. Large wind penetration may also deteriorate the system primary frequency response due to its limited contribution to both system inertia and frequency control under the current practice of determining primary reserve. This paper proposes a risk-based approach to the adequacy assessment of primary frequency response. It can also assist in determining primary reserves requirements for a power system with significant penetration of wind power generation. A simplified dynamic model of frequency taking into account the system inertia and frequency control services is developed for fast evaluation of operational risks due to inadequate frequency response without performing dynamic simulations.**

***Keywords:** primary frequency response, reserve requirements, risk assessment, wind power generation.*

I. Introduction

Modern wind power generation (WPG) units, particularly electronic converter-based variable-speed wind turbines, have different mechanisms for regulating their output and reacting to changes associated with voltage or frequency disturbances, compared to conventional thermal and hydropower generating units. Firstly, the moving parts of a wind turbine are not synchronized with the system frequency. As a result, the kinetic energy stored in a wind turbine is independent of system frequency. In other words, the contribution of a wind turbine to the system inertia is limited. Secondly, since wind turbines are normally operated to extract maximum power from wind (while wind is an uncontrollable energy source), wind turbines power output cannot be increased in response to a frequency drop like the output from synchronous generators equipped with governors. Thus, the integration of wind power will reduce the average system inertia and governor ramping capability per unit of installed capacity. Especially, if this integration results in the decommissioning of conventional power plants, it will reduce the total system inertia and governor ramping capability significantly, leading to lower frequency nadirs and higher rate of change of frequency (RoCoF). This may cause unwanted load curtailment due to under-frequency load-shedding and trip of generators triggered by their RoCoF protection relays.

Operating reserve requirements are generally set using the deterministic criteria based on the size of the largest online infeed and may remain constant for all operating conditions during the year. However, recent studies [1]-[3] proposed that operating reserve requirements need to be set dynamically to cope better with the uncertain characteristics of a power system including wind generation variability and uncertainty. In [4], the author studied the impact of WPG on primary, secondary and tertiary reserves, and concluded that these reserve requirements increase proportionally to the installed WPG capacity. However, this conclusion does not take into account the cost and ability to provide additional reserves. Persaud *et al.* [5] concluded that the spinning reserve requirements are inversely proportional to the net demand (load minus WPG) forecasting accuracy. Therefore, when WPG is integrated to a power system, larger amounts of reserve would be required to maintain the same level of system security. However, this conclusion is based on the assumption that the reserve requirement is set based on a predefined security index, and hence the authors do not consider the balance that should be attained between costs of and benefits of spinning reserves for an efficient economic operation. Doherty and O'Malley [6] proposed to set spinning reserve requirements based on a predefined minimum system reliability level considering the installed wind capacity. However, setting a single level of reliability to be achieved at all periods of the optimization horizon results in suboptimal solutions as the cost and benefit of the reserve provision depends on several factors such as system demand, WPG output and units committed. In [7], the authors considered the wind forecast uncertainty and set the reserve requirements at 3.5 times of the standard deviation of the net demand error to capture more than 99.7% of system imbalances. However, this approach ignores the probability and extent of the system's contingencies and simply set larger amounts of spinning reserves as the total WPG production increases.

This paper proposes a novel approach to evaluating the adequacy of primary frequency response (PFR) in a power system with significant wind power generation. This approach is based on the risk assessment method presented in our previous work [8]. Firstly, for fast evaluation of PFR adequacy without performing dynamic simulations a mathematical model of frequency trajectory in frequency excursion events is developed taking into account system inertia and frequency control services. The evaluated risk of primary frequency response inadequacy is then included in a security-constrained economic dispatch to determine the primary reserve requirement. The effectiveness of the proposed approach is illustrated by its application to a test system under different scenarios.

II. RISK ASSESSMENT OF PRIMARY FREQUENCY RESPONSE

To evaluate the adequacy of primary frequency response (PFR), we use the risk assessment approach proposed in [8]. The PFR inadequacy risk is given by

$$Risk = \sum_j \sum_i P(C_i) \times P(S_j) \times Q(C_i, S_j) \quad (1)$$

where $P(C_i)$ is the probability of the *i*th contingency C_i, $P(S_j)$ is the probability of system operating condition S_j, and $Q(C_i, S_j)$ is the quantified consequence of the contingency C_i in the operating condition S_j.

To calculate the probability of random contingencies, it is assumed that the failure/repair cycle of a single unit is represented by a two-state Markov model and its transitions follow an exponential probability distribution. Thus, the probability of failure of a single component during a time period *T*, given that the repair process is neglected, can be expressed as [9]:

$$P(C_i) = 1 - e^{-\lambda_i T} \quad (2)$$

where λ_i is the failure rate of the given component.

In addition to random contingencies, power systems are subjected to unexpected variations in demand and power supply especially with the present of intermittent wind power generation. The load level at time *t*, l^t is the sum of the forecast load, l_f^t, and forecast error, ε_l^t. The load forecast error can be modelled as a zero-mean normally-distributed random variable with a standard deviation σ_l^t [6]:

$$\sigma_l^t = \frac{a}{100} l_f^t \quad (3)$$

where a is a constant representing the load forecast accuracy. Similarly, the wind generation level at time *t*, w^t, can be modelled as the sum of the forecast wind generation, w_f^t, and the forecast error, ε_w^t. As explained in [8], WPG forecast errors generally can be modelled as a normally-distributed random variable with zero mean and standard deviation σ_w^t which is given by:

$$\sigma_w^t = (\alpha + \beta \times w_f^t) W_I \quad (4)$$

where α and β are forecast parameters depending on the forecast horizon and the size of the region where wind farms are located, w_f is the forecast output of wind generation in per unit, and W_I is the total installed WPG capacity. To reduce computation time, instead of using the continuous normal probability distribution, the typical seven-interval discretization of the zero-mean continuous normal-distributed function is used. Therefore, at a given time *t* in the future, there are seven possible load levels and seven possible wind generation levels, and hence forty-nine possible operating conditions with the probability given by

$$P(S_{i,j}^t) = P(l_i^t) \times P(w_j^t) = P(\varepsilon_i) \times P(\varepsilon_j) \text{ for } i, j = 1, \ldots, 7 \quad (5)$$

The consequence of a contingency in an operating condition is quantified by the amount of expected load interruption cost (ELIC), which is the product of the unit interruption cost (UIC) or value of lost load (VoLL) in \$/MWh and the expected energy not supplied (EENS) in MWh. The EENS is estimated based on the amount of unwanted load curtailments such as under frequency load shedding (UFLS) due to frequency deviations using the mathematical models of frequency developed in the previous section. Detail explanation is provided in [8].

After evaluating the adequacy of PFR, the resulted risk given by (1) will then be used to determine the primary reserve requirement by co-optimizing with energy in a security-constrained economic dispatch in order to minimize the expected operation cost (EOC) given by

$$EOC = CoE + Risk + CoPR \quad (6)$$

where *CoE* is the cost of energy – the product of energy price (\$/MWh) and energy demand (MWh); and *CoPR* is the cost of primary reserve – the product of reserve price (\$/MWh) and reserve demand (MWh).

III. CASE STUDY

A case study is performed to evaluate the performance of the proposed risk-based assessment method. The test case system S1 is a small islanded power system with characteristics similar to the Tasmanian power system. The total installed generation capacity of the system is 2.9 GW with 2.2 GW of hydro, 400 MW of gas, and 300 MW of wind generation. The power demand of S1 is approximately in the range of 800-1700 MW [7]. It is connected to a larger power system S2 (maximum demand 9000 MW) via a single monopolar HVDC link with a capacity to transfer maximum 500 MW of power and minimum 50 MW of power in both direction. The HVDC link is also equipped with a frequency controller allowing it to transfer frequency control services between the two systems subjected to the margins between its power flow and transfer limits.

Figs 1, 2 and 2 show the system hourly pre-dispatch/forecast data for the next 24 hours sourced from [11]. Due to specific characteristics of the predominantly hydro-power system and its small size, the capability of providing fast frequency response in S1 is limited resulting in high cost of primary reserve, as can be seen in Fig. 3. In contrast, the cost of the primary reserve in S2 is much lower due to the dominance of thermal generation units that represent much cheaper sources of service in this system. In addition, the wind power generation level of S2 is negligible in comparison with the system size. The system inertia and capability of providing fast frequency responses are also strong. As a result, the reserve requirement for S2 is always set to the size of its largest online generator (550MW) and the PFR inadequacy risk of S2 is assumed to be negligible.

The study objective is to determine the interconnection flow (negative for import to and positive for export from S1) and the primary reserve requirement for S1 for each hour so that the operation cost of the two systems is minimized using the approach proposed in Section III. The results will then be compared with two other methods for primary reserve determination:

- Method A: $Reserve = P_{gen\_max}$, which is the size of the largest online generator.
- Method B: $Reserve = P_{gen\_max} + 3 \times \sigma_{netload}$ as proposed in [12] where $\sigma_{netload}$ is the standard deviation of net-load (demand minus wind generation) forecasting error given by

$$\sigma_{netload} = \sqrt{\sigma_l^2 + \sigma_w^2} \quad (7)$$

The contingency probabilities for S1 are calculated as explained in Section 2. The failure rates of system components are determined based on the reliability data of the Tasmanian power network provided by TasNetworks. It is assumed that the failure probability of the interconnection link is zero. Load forecast errors are given by (3) with $a = 1$. Wind generation forecast errors are given by (4) with $\alpha = 0.01$ and $\beta = 0.16$

based on the data of 1-hour WPG forecast in a region with a diameter of 360 kilometres [13]. Due to the small size of the test system it is assumed that the wind speeds at all wind farms in S1 are the same at any given moment.

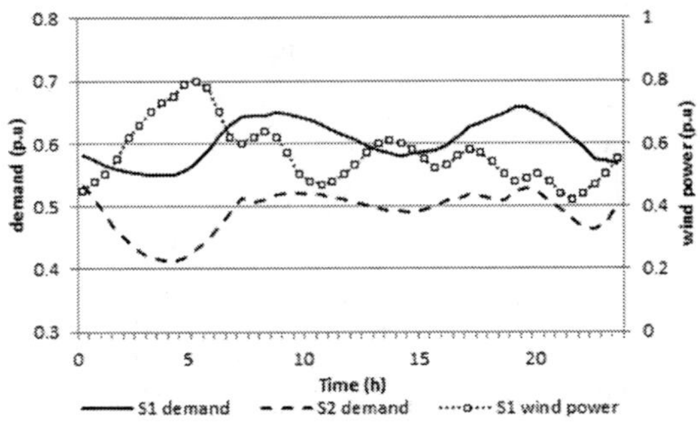

Fig. 1. Normalized forecast demand and wind power generation. The base values of S1 demand, S2 demand and S1 wind power are 1700 MW, 9000 MW and 300 MW, respectively.

The system nominal frequency is 50Hz. The primary frequency control is activated once the system frequency drops below 49.85 Hz, and it take 6 seconds to reach its maximum value. This corresponds to the fast rise frequency control ancillary service (R6 FCAS) in Australia [14]. There are three under-frequency relay settings for the period of the primary control shown in Table 1. These settings are similar to the automatic under-frequency load shedding configuration used in [8]. The load relief factor is $k = 1$. The unit interruption cost (UIC) is $12,500/MWh - the market price cap (MPC), which is often used in reliability studies in Australia [16].

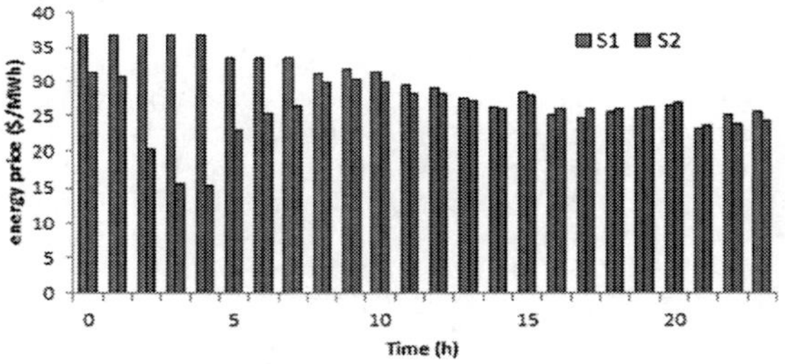

Fig. 2. Hourly forecast energy prices.

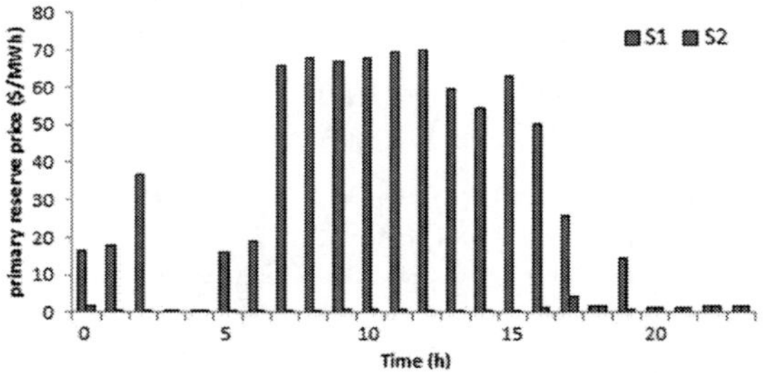

Fig. 3. Hourly forecast primary reserve prices.

Table 1. Under-Frequency Relay Settings

Relay	Load shedding block (Hz)	Frequency threshold 1 (Hz)	Time delay 1 (s)	Frequency threshold 2 (Hz)	Time delay 2 (s)
1	10	47.8	0.2	-	-
2	10	47.5	0.2	47.8	10
3	10	47.2	0.2	47.5	2

IV. Results and Discussion

Fig. 4 compares the hourly dispatch outcomes in terms of the expected operation cost (EOC) for the next 24 hours obtained using the proposed optimized approach against Method A and Method B. As can be seen, the optimized approach provides the least expensive solution for all 24 hours and can save up to 1.5% of EOC compared to the other two methods. This saving is rather modest due to a number of reasons. Firstly, as the size of S2 is much bigger than S1, its cost of energy (CoE) accounts for more than 80% of the total EOC. Secondly, the saving is achieved by optimizing the primary reserve requirement and PFR inadequacy risk for S1 only, whereas the primary reserve amount of S2 is fixed and its PFR inadequacy risk is negligible as explained in Section IV. If the wind power generation level in S2 is large enough, the same optimization approach can be applied to S2 and bigger saving can be achieved. Meanwhile, Method A, which sets the primary reserve amount equal to the largest online generator, results in the most expensive solution in 15 hours.

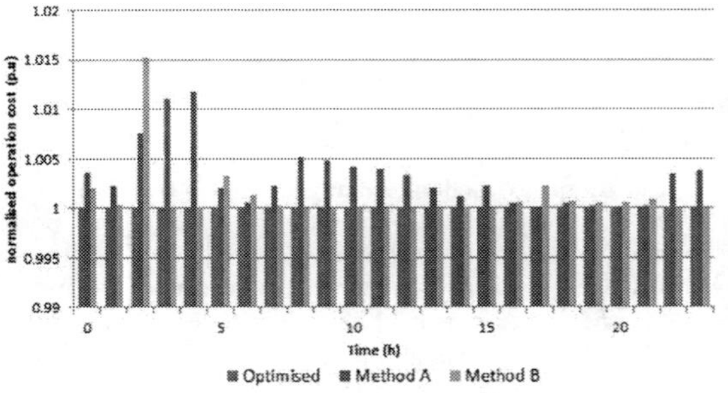

Fig. 4. Hourly expected operation costs obtained with the three approaches. These costs have been normalized based on the expected cost obtained with the proposed optimized approach.

V. Conclusions

The paper presented a risk-based approach to assess the adequacy of primary frequency response, and then determined the primary reserve requirement taking into account wind power generation and load forecast uncertainties. The amount of primary reserve provision is determined so that the total system expected operation cost (i.e., the sum of the actual operating cost and the risk of primary frequency response inadequacy in terms of load interruption cost) is minimized. A mathematical model of frequency responses is developed to dynamically evaluate the primary frequency response without running dynamic simulations. A case study is performed to demonstrate the performance of the proposed approach. The results show that the risk-based approach has potential to deliver lower cost dispatch solution compared with the traditional method setting the reserve requirement based on the largest generator infeed and a combination of loss of the largest generator and the net load uncertainty. The proposed approach is also able to assess the impact of the installed wind power capacity and the total system inertia on the expected cost of operation.

References

[1] F. Bouffard and F.D. Galiana, "Stochastic security for operations planning with significant wind power generation," IEEE Trans. Power Syst., vol. 23, no. 2, pp. 306-316, May 2008.

[2] M. A. Matos and R. J. Bessa, "Setting the operating reserve using probabilistic wind power forecasts," IEEE Trans. Power Syst., vol. 26, no. 2, pp. 594-603, May 2011.

[3] M. A. Ortega-Vaquez and D. S. Kirschen, "Estimating the spinning reserve requirements in systems with significant wind power generation penetration," IEEE Trans. Power Syst., vol. 24, no. 1, pp. 114-124, Feb. 2009.

[4] G. Dany, "Power reserve in interconnected systems with high wind power production," in Proc. IEEE Power Tech. Conf., Porto, Portugal, Sep. 2001.

[5] S. Persaud, B. Fox, and D. Flynn, "Effects of large scale wind power on total system variability and operation: Case study of Northern Ireland," Wind Engineering, vol. 27, pp. 3-20, 2003.

[6] R. Doherty and M. O'Malley, "A new approach to quantify reserve demand in systems with significant installed wind capacity," IEEE Trans. Power Syst., vol. 20, pp. 587-595, May 2005.

[7] M. Black and G. Strbac, "Value of bulk energy storage for managing wind power fluctuations," IEEE Trans. Energy Conversion, vol. 22, pp. 197-205, Mar. 2007.

[8] M. Negnevitsky, D. H. Nguyen, and M. Piekutowski, "Risk assessment for power system operation planning with high wind power penetration," IEEE Trans. Power Syst., to be published.

[9] R. Billinton and R. Allan, Reliability evaluation of power systems, 1996.

[10] TasNetworks, "Transend annual planning report 2013," Australia, 2013.

[11] Australian Energy Market Operator (AEMO) Data. Available: www.aemo.com.au/Electricity/Data.

[12] V. Silva, "Value of flexibility in systems with large wind penetration," Ph.D. dissertation, Imperial College London, University of London, London, UK, 2010.

[13] A. Fabbri, T. G. S. Roman, J. R. Abbad, and V. H. M. Quezada, "Assessment of the cost associated with wind generation prediction errors in a liberalized electricity market," IEEE Trans. Power Syst., vol. 20, no. 3, pp. 1440-1446, Aug. 2005.

[14] B. Blake, "ESOPP guide - FCAS constraint equations," Australian Energy Market Operator (AEMO), Dec. 2009.

[15] K. Mollah, M. Bahadornejad, and N. Nair, "Automatic under-frequency load shedding in New Zealand power system — A systematic review," in Proc. 21st AUPEC, Brisbane, QLD, Australia, Sept. 2011, pp. 1-7.

[16] AEMC Reliability Panel, "Final report: Reliability standard and reliability settings review," Australian Energy Market Commission, 2010.

Proposal for Defining of Some Optimal Parameters in Power Supply Systems of Industrial Branches in the Republic of Bulgaria

Valentin Gyurov
Department of Electric Power Supply and Equipment
Technical University of Varna
Varna, Republic of Bulgaria
e-mail: valentin.giurov@tu-varna.bg

Ilian Iliev
Department of Electrification of Mining Industry
University of mining and geology "St. Ivan Rilski"
Sofia, Republic of Bulgaria
e-mail: ilian.iliev@mgu.bg

Rumen Kirov
Department of Electric Power Supply and Equipment
Technical University of Varna
Varna, Republic of Bulgaria
e-mail: kvc_electroinvest@yahoo.com

Rosen Vasilev
Department of Electrical Fundamental and Electrical Measurements
Technical University of Varna
Varna, Republic of Bulgaria
e-mail: rsnvasilev@tu-varna.bg

Abstract— **The paper presents systematized and analyzed researches on energy indicators and power quality indicators for various industrial sites and branches in the Republic of Bulgaria. As a result of the study correlation dependencies and definition for optimal zones of variation of the studied indices were obtained. The study presents the specific energy consumption at different energy and voltage levels and defined areas of variation of parameters for achieving of maximum energy efficiency.**

Keywords: power and energy losses; electric power supply systems, energy efficiency

I. INTRODUCTION

The problems related to the improvement of the electric energy efficiency in industrial enterprises in the Republic of Bulgaria are particularly relevant. Medium-voltage networks in industrial power supply systems and urban electrification can make a significant contribution to achieving good performance in terms of energy efficiency, despite some features in their operation. The most significant features are presence of a large amount of elements (power lines, power transformers, switching and protection apparatus), lack of devices for regulating of the working regimes, incomplete unification of the schematic solutions, etc. These features cause some difficulties related to the larger number of calculations, analyzes, the ambiguity of the objectivity of the results obtained through determined approaches and reduction of possibilities for optimization. To solve these problems in energy efficiency studies, concepts related to the application of equivalence methods (e.g. equivalent resistances for power lines and power transformers) and solving of the assigned tasks in a summarized form can be applied. Such a generalized approach can be applied to the electricity supply system as a whole or to individual sub-schemes thereof. Based on this concept, a methodological approach is developed for experimental research in the industrial branch structure. With the use of this approach, results and recommendations are obtained to achieve electric energy efficiency.

II. EXPERIMENTAL RESEARCH

In this study, electric energy efficiency is assessed primarily by the "reactive load compensation" criterion by the coefficient of additional losses K_{AL} [1,3]

$$K_{AL} = \frac{|\Delta P_{PQ} - \Delta P_P|}{\Delta P_P} \cdot 100 = tg^2\varphi \tag{1}$$

Where: ΔP_{PQ} [kW] – active power loss, provoked by presence of active (P) and reactive (Q) loads; ΔP_P [kW] – active power loss, provoked by presence of active loads (P). The definitions of ΔP_{PQ} and ΔP_P are given in Formula 2 and Formula 3.

$$\Delta P_{PQ} = \frac{(P^2 + Q^2)\cdot R}{U^2} = \frac{P^2 \cdot R \cdot (1 + tg^2\varphi_{AVG})}{U^2} \tag{2}$$

$$\Delta P_P = \frac{P^2 \cdot R}{U^2} \tag{3}$$

The survey covers seven industrial sectors and the results are shown in Table 1. The presented results include data on average annual apparent power $S_{A,Y}$[MVA], active energy consumed per year W_Y,[10^3.MWh/year], average annual active power $P_{A,Y}$[MW], the average annual power factor $\cos\varphi_{A,Y}$ ($tg\varphi_{A,Y}$), required compensation power Q_{COMP}[MVAr] and investment costs for maintenance of reactive power compensation systems IC [BGN].

The use of reactive load compensation systems leads to an increasing of power factor and to a reduction of power losses in addition to the saving of funds from non-payment of sanctions for low power factor. The total annual economy EC [BGN] of funds for each industrial branch L is determined by Formula 4 and Formula 5

$$EC = EC_C + EC_S \tag{4}$$

$$EC_C = \delta P_{COMP} \cdot \tau \cdot C_E \tag{5}$$

Table 1. Energetic parameters for 7 industrial branches

Energetic parameters for 7 industrial branches								
Group		***1***	***2***	***3***	***4***	***5***	***6***	***7***
Branch		***Machine building***	***Chemical and mining***	***Building, construction and civil engineering***	***Electro technical industry***	***Paper, textile, clothing and footwear industry***	***Processing and manufacturing of metal, wood and plastic***	***Food industry***
$S_{A.Y}$	MVA	298	665	273	194	198	52	107
$\cos\varphi_{A.Y}$	-	0.874	0.862	0.86	0.876	0.859	0.839	0.843
$tg\varphi_{A.Y}$	-	0.556	0.588	0.553	0.55	0.596	0.648	0.638
$T_{A.Y}$	h	5558	8205	4322	3850	3674	3847	6385
W_Y	$.10^3$MWh/year	1448	4703	1009	654	625	168	576
$P_{A.Y}$	MW	260.4	573.2	243.8	169.9	170.1	43.6	90.2
Q_{COMP}	MVAr	50.25	128.97	46.322	31.77	39.63	12.43	24.81
IC	$.10^6$ BGN	4,271	10.962	3.937	2.7	3.369	1.057	2.109
W_{COMP}	$.10^3$MWh/year	31.42	105.34	22.7	14.85	14.125	3.88	13.248
W_{COMP}	%	2.17	2.24	2.25	2.27	2.26	2.31	2.3
EC_C	$.10^6$BGN	3.142	10.534	2.27	1.485	1.412	0.388	1.325
T_M	h	4950	7830	3680	3450	3220	3580	5970
τ	h	3356	7206	2120	1927	1743	2035	4554
EC_S	$.10^6$BGN	1.624	7.84	1.733	0.736	1.126	0.453	1.45
EC	$.10^6$BGN	4.766	17.934	3.943	2.22	2.539	0.841	2.775
$T_{PAYBACK}$	years	0.896	0.611	0.998	1.216	1.327	1.256	0.76

Where: EC_C [BGN] - economy of funds by reduction of active power consumption with reactive power compensation systems; EC_S [BGN] – economy of funds from non-payment of sanctions; δP_{COMP} [-] - difference between active power losses in maximum mode without and with reactive loads compensation system; $\tau=(0.124+T_M/10^3)^2.8760$ [h] – time period of maximum loss; T_M [h] – time period for use of maximum loss; 8760 [h] – total number of hours per year; C_E [BGN/kWh] – cost of electricity; $W_{COMP}=\delta_{COMP}.\tau$ [10^3.MWh /year] - annual economy of electricity with reactive power compensation.

The values of W_{COMP} are determined by two methodologies developed in [3] and [4]. In the first method, the additional losses of active power obtained after compensation of the reactive loads are determined by the Formula 1. In the second method, with the use of an ABB Power Plus-based system and a software algorithm in MatCAD platform, all active power losses are determined differentially by active (ΔP_P) and reactive (ΔP_Q) power, unbalanced power (ΔP_N), harmonics power (ΔP_D), and hidden power (ΔP_{S0}) in the neutral conductor. The values of W_{COMP} in absolute and relative units for different industry branches at different desired power factor $\cos\varphi_{DES}$ are presented in Table 2. The data show that the highest values of W_{COMP} can be achieved in the light industry sectors (branches 6 and 7) where the organization of the reactive load compensation is simplest and more efficient. The determination of the component EC_S is carried out in accordance with the applicable normative power factor tariff. The values of total annual economy EC are calculated by Formula 2 and presented in Table 1.

Table 2. Annual economy W_{COMP} for different industrial branches and desired power factor $\cos\varphi_{DES}$

Branch	**W_{COMP} - annual economy of electricity with reactive power compensation [10^3.MWh /year] and W_{COMP} [%]**			
	$\cos\varphi_{DES}$= 0.9	***$\cos\varphi_{DES}$= 0.92***	***$\cos\varphi_{DES}$= 0.94***	***$\cos\varphi_{DES}$= 0.96***
1	2.17	2.63	2.97	3.27
	31.42	38.08	43.01	47.35
2	2.24	2.59	3.08	3.52
	105.34	121.8	144.85	165.54
3	2.25	2.61	3.09	3.63
	22.7	26.33	31.178	36.33
4	2.27	2.69	3.15	3.72
	14.85	17.59	20.6	24.33
5	2.26	2.64	3.12	3.68
	14.12	16.44	19.6	23
6	2.31	2.75	3.21	3.79
	3.88	4.62	5.39	6.37
7	2.3	2.72	3.18	3.76
	13.25	15.67	13.32	21.66

Table 3. Energy efficiency assesments indicators with Present Value method

Parameters for 7 industrial branches								
Branch	***Indicator***	*Discount period t [years] for IR=6.5% and d=1.42%*						
		t=0	*t=0.25*	*t=0.5*	*t=0.75*	*t=1*	*t=2*	*t=3*
Machine building industry	***NDP***	-4.271	-3.976	-3.092	-1.623	0.428	5.062	9.63
	NPV	0.495	-3.084	-1.905	-0.734	0.428	0.36	0.297
	PI	-1	-0.931	-0.724	-0.38	0.1	1.185	2.255
Chemical and mining industry	***NDP***	-10.962	-9.851	-6.526	-0.998	6.72	24.155	41.347
	NPV	6.971	-6.494	-2.058	2.346	6.72	6.472	6.229
	PI	-1	-0.899	-0.595	-0.091	0.613	2.203	3.772
Building, construction and civil engineering	***NDP***	-3.937	-3.693	-2.962	-1.747	-0.049	3.784	7.563
	NPV	0.056	-2.955	-1.98	-1.011	-0.049	-0.104	-0.158
	PI	-1	-0.938	-0.752	-0.444	-0.013	0.961	1.921
Electro technical industry	***NDP***	-2.7	-2.563	-2.151	-1.466	-0.511	1.648	3.777
	NPV	-0.48	-2.147	-1.598	-1.052	-0.511	-0.541	-0.572
	PI	-1	-0.949	-0.797	-0.543	-0.189	0.61	1.399
Paper, textile, clothing and footwear industry	***NDP***	-3.368	-3.211	-2.74	-1.958	-0.865	1.603	4.037
	NPV	-2.83	-2.736	-2.108	-1.484	-0.865	-0.9	-0.935
	PI	-1	-0.953	-0.812	-0.581	-0.257	0.476	1.198
Processing and manufacturing of metal, wood and plastic	***NDP***	-1.056	-1.004	-0.848	-0.589	-0.227	0.59	1.396
	NPV	-0.216	-0.847	-0.639	-0.432	-0.227	-0.239	-0.251
	PI	-1	-0.951	-0.803	-0.556	-0.215	0.558	1.321
Food industry	***NDP***	-2.109	-1.937	-1.423	-0.567	0.627	3.325	5.985
	NPV	0.666	-1.417	-0.731	-0.049	0.627	0.589	0.055
	PI	-1	-0.918	-0.674	-0.269	0.297	1.577	2.838

The payback period $T_{PAYBACK}$ is defined with Formula 6 and shortest payback period was obtained for the chemical and mining industry – $T_{PAYBACK}$=0.611 years.

$$T_{PAYBACK} = \frac{IC}{EC} \tag{6}$$

III. Energy Efficiency Assessment with Present Value (PV) Method

The PV indicator is a basic feature of the "cost-benefit" method, which discounts investment cost, involved in building of the innovative project and evaluates the dynamics of the investment process. Net present value (NPV) is determined by calculating the costs (negative cash flows) and benefits (positive cash flows) for each period of an investment. The period is typically one year, but could be measured in quarter-years, half-years or months. NPV for determining of energy efficiency for each industrial branch for implementation of reactive power compensation system is defined by Formula 7 [7,8,9].

$$NPV = EC \cdot (1+d)^{-t} - IC \tag{7}$$

If the economic effect (total annual economy) EC is equal for each year (EC_1=EC_2=...=EC) then net discounted profit (NDP) can be calculated by Formula 8 [6,8]

$$NDP = EC \cdot \left(\frac{1-(1+d)^{-1}}{d} \right) - IC \tag{8}$$

The profit index PI, which shows how many NDP are generated by each currency investment IC, is determined with the use of Formula 9

$$PI = \frac{NDP}{IC} \tag{9}$$

The indicator for effective project is necessary to be PI>0. Real discount rate d [%], which is used for calculation of PI, NPV and NDP (Formulas 5, 6 and 7), depends on interest rate IR (calculated for 2016 by Bulgarian National Bank - IR=6.5%) and annual inflation rate b=5% [6].

$$d = \frac{IR-b}{1+b} \cdot 100 = 1.42 \tag{10}$$

Discounted payback period $T_{D,PAYBACK}$ which corresponds to NPV=0 is defined in Formula 11 [6].

$$T_{D,PAYBACK} = -\frac{\ln(1-T_{PAYBACK} \cdot IR)}{\ln(1+IR)} \tag{11}$$

In reference [9] are given tables for calculation of $T_{D,PAYBACK}$ in function of d[%] and annuity factor AF, which is calculated by Formula 12.

$$AF = \frac{EC}{IC} = \frac{1-(1+d)^{-t}}{d} \tag{12}$$

Where: t - duration of saving event by introducing innovation (e.g. energy savings, fuels, materials, etc.)

The values of NPV, PI and $T_{D,\ PAYBACK}$ for different industrial branches are calculated with formulas (7), (8), (9) and (11) and the results are shown in Table 3 and Figure 1.

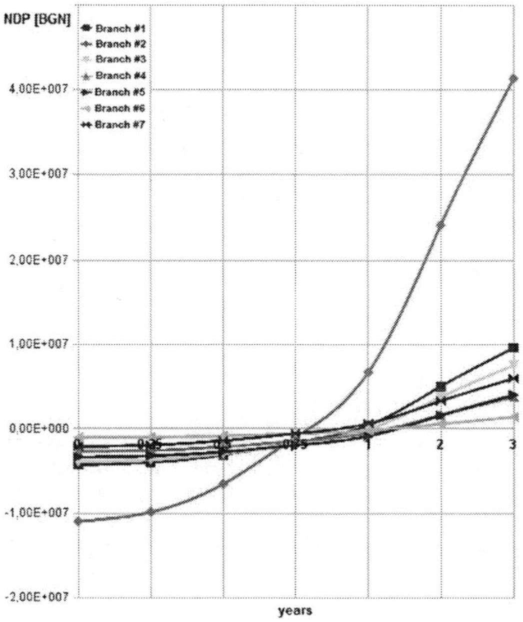

Figure 1. Net Discounted Profit – NDP [BGN] for different industrial branches

IV. Conclusion

The report presents an analysis of various technical and economic indicators characterizing the power supply systems in the branch structure of the Republic of Bulgaria. The study was performed according to the "reactive load compensation" criterion for a long period of time in reduced consumption mode. Various approaches for reduction of electricity consumption and power losses and for improvement of different regime parameters of power transformers are shown. The possibility for use of some technical and economic characteristics of the electricity supply systems as a means of assessing the electric energy efficiency is proved. The results of the study present an approach enabling to determine the effective modes of operation of electricity supply systems through the indicators NPV, NDP and PI.

References

[1] R. Kirov, V. Gyurov, V. Chikov and N. Makedonski "Electric Energy Efficiency in the Branches of Industrial Sector", Energy Forum 2011, Proceedings, pp. 270-275, 2011

[2] R. Kirov, N. Makedonski, V. Chikov and V. Gyurov "Techno-economic criteria for valuation of the electro-energy efficiency under compensation of reactive loads in branches of industry", 3th Scientific conference EF2011 Technical University of Sofia, Proc. vol 1., pp. 273-280, 2011

[3] R. Kirov, V. Gyurov, S. Stefanov, "Reactive power compensation – an effective apporach for economy of electric energy" Energy Forum 2006, Proceedings, 2006

[4] V. Gyurov, "Rationalization of reactive power compensation in unbalanced and non-linear contirions of electric power supply systems", PhD Thesys, Technical University of Varna, 2010

[5] G. Vagin, N. Golovin, E. Solncev and A. Lyamin "Methodic for techno-economic evalation of application of recources, energy efficiency tehcnologies and equipment in industry", Journal of Industrial Energetic, Moscow, vol.6, 2005

[6] V. Totmin, "Tasks in study on reactive power compensation systems" Journal of Industrial Energetic, Moscow, vol.8, 2006

[7] V. Gotman, "Criterions for assessment of reactive power compensation systems", Journal of Electricity, Moscow, vol.12, 2009

[8] N. Vasilev and B. Toshev, "Is the evalution of efficiency in street ligthing system is correct?", International conference "Unitech", Gabrovo, Proceedings, 2010

[9] "Methodological recommendations for assessment of efficiency of investment projects", Moscow, Economics, 2000

Optimization of Adaptive Filter Control Parameters for Shunt Active Power Filter

Radek Martinek, Petr Bilik, Jan Vanus, Ludvik Koval, Radana Kahankova
Department of Cybernetics and Biomedical Engineering
VSB - Technical university of Ostrava, Ostrava, Czech Republic
radek.martinek@vsb.cz, petr.bilik@vsb.cz, ludvik.koval@vsb.cz

Marcel Fajkus, Jan Nedoma
Department of Telecommunications
VSB - Technical university of Ostrava, Ostrava, Czech Republic
marcel.fajkus@vsb.cz, jan.nedoma@vsb.cz

Abstract — **This article is dedicated to the study of the influence of the optimal setting of the adaptive algorithm, Least Mean Squares (LMS) to manage Shunt Active Power Filter (SAPF) capable of reducing the Total Harmonic Distortion (THD) in the power system. Authors focused on the design, implementation, and verification of a novel method for optimization of the control parameters (such as Step Size and Filter Order) of adaptive filters. The main criterion for optimal parameter selection was THD. The results of these experiments indicate that the optimum setting of the Step Size for LMS algorithm has a major impact on the speed estimate compensating current, i.e. the ability of the SAPF to respond to dynamic changes in power system. The parameter such as the Filter Order has then essential impact on the value of the resulting THD, i.e. on the Power Quality. The original contribution of the paper is to define the working area of LMS algorithm, which is a compromise between quality and speed of the estimation the compensation current. The proposed system adaptively reduces distortion, dip a flicker. Real signals for the measurements were obtained on a sophisticated three-phase experimental workplace (programmable source HP 6834B, intelligent electronic device IED NI 9925).**

Keywords: Total Harmonic Distortion; Shunt Active Power Filter; Least Mean Squares; Step Size; Filter Order; Power Quality

I. Introduction

This paper deals with the issue of maintaining the quality of the supplied power with the increasing level of interference in distribution networks using Shunt Active Power Filters (SAPF). In general, Active Power Filters (APF) have been known to the public for several years, see [15], [16]. Nevertheless, the SAPF management area is experiencing a revolution in the last years [17], [18], [19] mainly thanks to the penetration of advanced signal processing techniques to the field of power engineering.

Within the publications [1], [2], [3], the group of authors focuses on the issue of control methods of SAPF using adaptive methods [4]. There are a number of approaches to adaptation, such as various soft computing techniques [2], Adaptive Linear Neuron, later Adaptive Linear Element - ADALINE [3], or Recursive Least Square Algorithm - RLS [1], [21], [22]. This article focuses primarily on the Least Mean Squares algorithm - LMS [8], which is most widespread in the area of adaptive control using SAPF, see [5], [6], and [7]. A major problem in the area that is currently not adequately addressed is the optimal setting of the control parameters of the LMS algorithm [14], i.e. filter order and step size. These parameters have a significant effect on the resulting Total Harmonic Distortion - THD [9], [10]. The authors draw on the experience from other application areas where adaptive systems are used such as fetal electrocardiography - ECG [8], adaptive channel equalization [11], [12], adaptive speech processing [13], etc.

The original contribution of this article is a study of the influence of control parameters (filter order and step size) of the LMS algorithm controlled SAPF on the total THD. The experiments demonstrated a significant influence of control parameters on the participation of filtration. For the LMS algorithm being investigated, the working area was deleted.

II. Shunt Active Power Filter

The function of parallel active filters is based on the subtraction of the basic 1st harmonic component of the current from the total nonharmonic process to the so-called residual curve. The controller then inverts this residual curve and the voltage inverter starts to produce the required voltages. The voltage produced is the difference between the voltage of the inverter and the mains voltage on the coupling filter and the compensation current is induced. This filter can modify the deformation voltage on the load, increase the power factor by compensating the reactive current component and also can symmetrize the current consumption. A reactive component is exchanged between the active filter and the load. All of these properties depend on the control algorithm and the performance of the control unit. The parallel active filter is connected to the grid via a passive LC low pass filter because the active filter itself is a harmonic source, [23], [24], [25].

III. Adaptive Filtration

The main idea of all adaptive filters is to continually alter parameters according to a predetermined algorithm, using both input information and auxiliary information coming mostly from the output of the entire system. Adaptive filters use so-called adaptive algorithms for iteratively changing the values of the pulse response vector so as to minimize the values of the error signal $e(n)$, which is given by the following equation:

$$e(n) = d(n) - y(n). \quad (1)$$

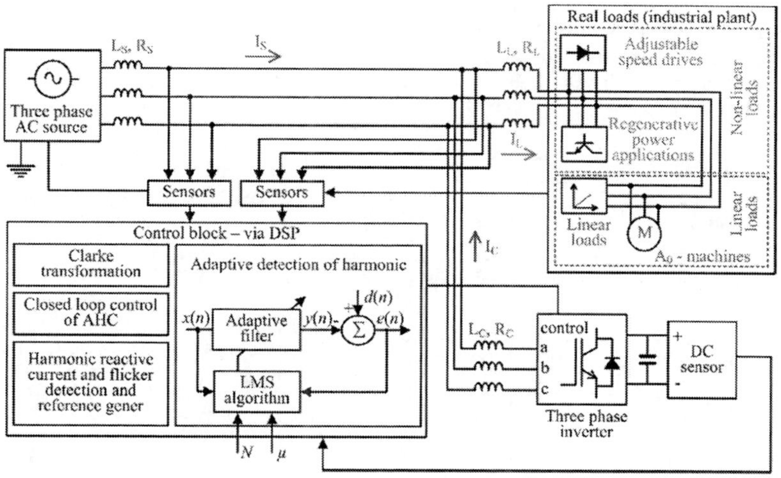

Figure 1. Simplified principal scheme of experiments.

where $d(n)$ is the desired signal and $y(n)$ indicates the existing output of the adaptive filter. The general diagram of the adaptive linear filter is shown in Fig. 1, [26], [27], [28].

In most cases, linear FIR filters with transversal structure are used, for the reasons of simple implementation and the modest mathematical apparatus.

The goal of adaptive algorithms is to achieve the best possible estimation of filter coefficients to get the output signal as close as possible to the input signal. In the process of finding a filter coefficient, optimization of the error function is usually solved and it can be stochastic or deterministic. The stochastic approach requires much more measurement to calculate statistics and is used by LMS and NLMS algorithms. The deterministic approach requires the calculation of the characteristics of a large number of samples and is used by RLS and QR-RLS algorithms [5].

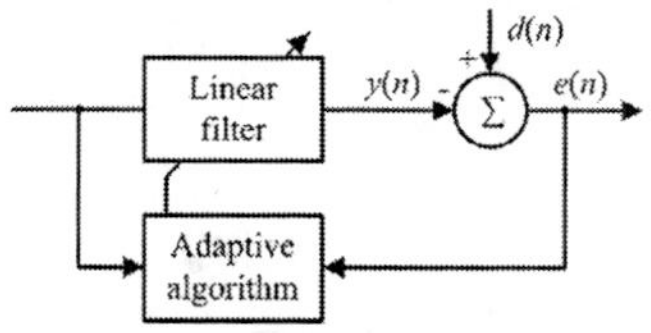

Figure 2. Generally scheme of linear adaptive filter

A. LMS algorithm

The LMS algorithm is one of the most widespread and most commonly adaptive algorithms used in current practice. The strength of the LMS algorithm lies mainly in mathematical simplicity and the wide range of applications of the LMS algorithm can be attributed to its simplicity and robustness. These algorithms are based on a gradient search algorithm, also called the highest gradient method. The dependence of the mean quadratic deviation of the output error signal of the adaptive FIR filter on the filter coefficients is a quadratic curve with one global minimum. The output signal is given by equations (2) and (3):

$$y(n) = \mathbf{w}(n)\mathbf{x}(n), \tag{2}$$

$$y(n) = \mathbf{w}^{\mathrm{T}}(n)\mathbf{x}(n). \tag{3}$$

The error signal calculated as the difference between the required and the output signal:

$$e(n) = d(n) - y(n). \tag{4}$$

Recursion of filter:

$$\mathbf{w}(n+1) = \mathbf{w}(n) + 2\mu e(n)\mathbf{x}(n), \tag{5}$$

where μ is the step size of the adaptive filter, $\mathbf{w}(n)$ is the filter coefficient vector, $\mathbf{x}(n)$ is the input filter vector. The filter step size is an essential parameter that has the ability to improve the adaptive filter convergence rate.

IV. Experimental Setup

Figure 2 illustrates a simplified schematic diagram of the experiments. This is an SAPF concept for a three-phase system where the LMS algorithm is used for DPS control. In general, SAPF can be considered as a supplementary power converter attached to the load (linear, nonlinear), a detailed description of the entire measurement concept, including the software and hardware used, can be found in [1], [2], and [3].

The actual experiments were carried out in the "simulation mode" (Three Phase Inverter was not used because the paper deals with the effect of the LMS algorithm parameters on the filtration efficiency) using real measured values on the experimental workstation with programmable source HP6834B [20].

The LMS adaptive algorithm is adaptable to the network current changes that are caused by the connection of primarily nonlinear loads (e.g. controlled drives, motor starters, electronic lighting, computers, UPS, and other electronic devices). The LMS algorithm is, therefore, able to drive the inverter so as to produce the same levels of harmonics as the nonlinear load but in the opposite phase. It is clear that at the connection point according to Fig. 3 the undesirable harmonic distortion is eliminated. In this way, the current drawn from the grid is filtered, and thus the voltage distortions caused by the load as well as the power factor of the grid are improved. More on the principles of SAPF management using adaptive systems can be found in [1], [2], and [3].

The experimental part deals with the influence of the filter order and the step size on the quality of filtration. The verification criterion is THD (%) value, see [1], [2], and [3]. The results of the experiments are summarized in Tab. 1. Experiments were performed for N = 10; 20; 30; 40; 50 and for μ = 0.01; 0.001; 0.0001; 0.00001.

TABLE I. RESULTS OF THE EXPERIMENTS

μ (-)	THD_L1 (%)	THD_L2 (%)	THD_L3 (%)	Conver. (s)
		N (-) = 10		
0,01	unstable	unstable	unstable	unstable
0,001	16,41	35,26	16,91	< 0.01
0,0001	7,39	6,65	6,79	0,02
0,00001	5,48	4,41	4,57	> 0.1
		N (-) = 20		
0,01	unstable	unstable	unstable	unstable
0,001	23,68	44,9	24,09	< 0.01
0,0001	8,15	7,07	7,81	0,015
0,00001	5,81	4,94	5,14	0,08
		N (-) = 30		
0,01	unstable	unstable	unstable	unstable
0,001	27,61	48,47	27,73	< 0.01
0,0001	8,79	7,57	8,17	0,01
0,00001	6,32	5,47	5,91	0,06
		N (-) = 40		
0,01	unstable	unstable	unstable	unstable
0,001	30,21	50,05	30,45	< 0.01
0,0001	9,32	7,81	8,66	< 0.01
0,00001	6,51	5,84	6,24	0,06
		N (-) = 50		
0,01	unstable	unstable	unstable	unstable
0,001	32,32	51,2	34,36	< 0.01
0,0001	9,54	8,11	9,27	< 0.01
0,00001	7,19	6,32	6,95	0,04

Figures 3 and 4 show selected results in time and frequency domain. The aim was to demonstrate the influence of N and μ on the quality of filtration and THD, respectively. The depicted waveforms are always shown without the filtration and the two extreme states and optimum values are highlighted. Fig. 3 shows the influence of magnitude μ and Fig. 4 shows the influence of size N.

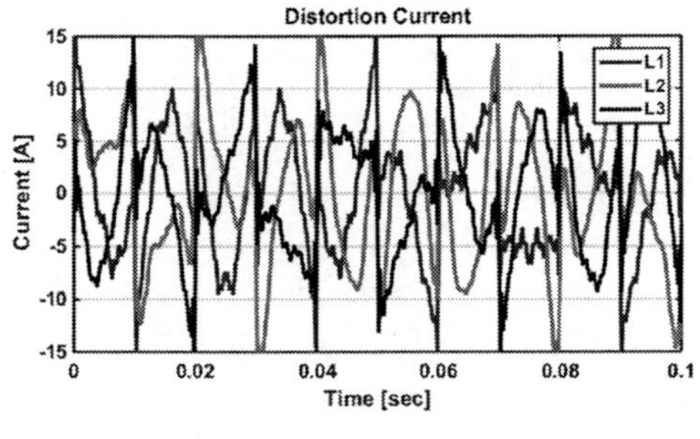

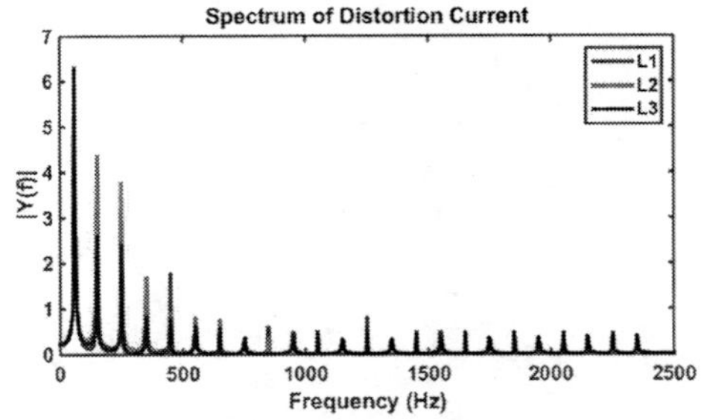

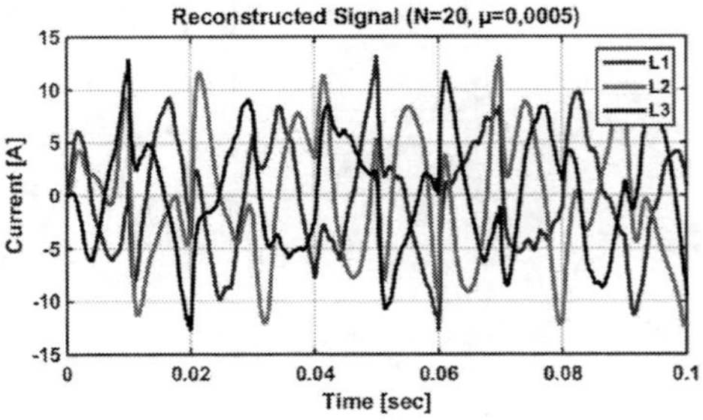

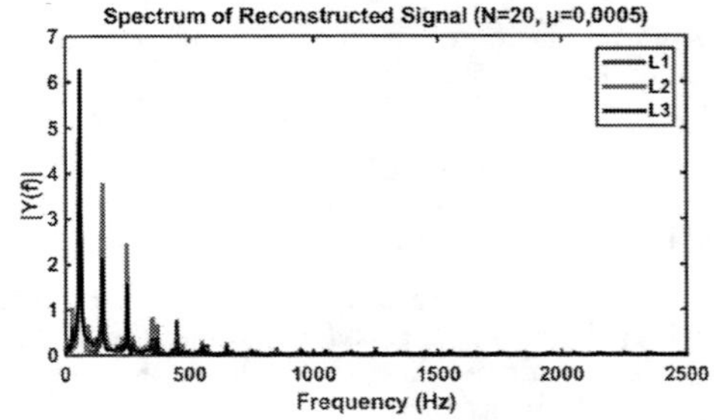

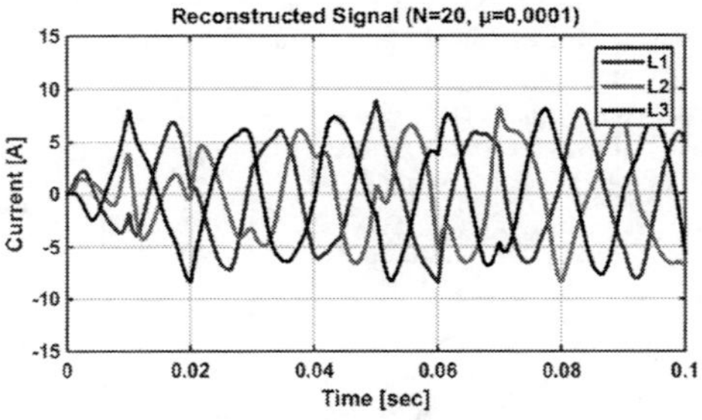

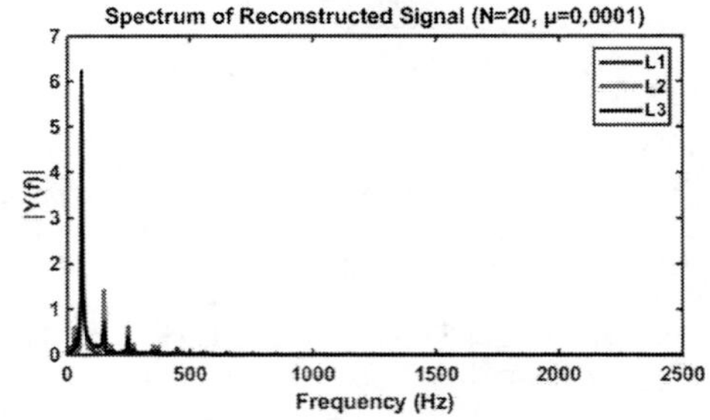

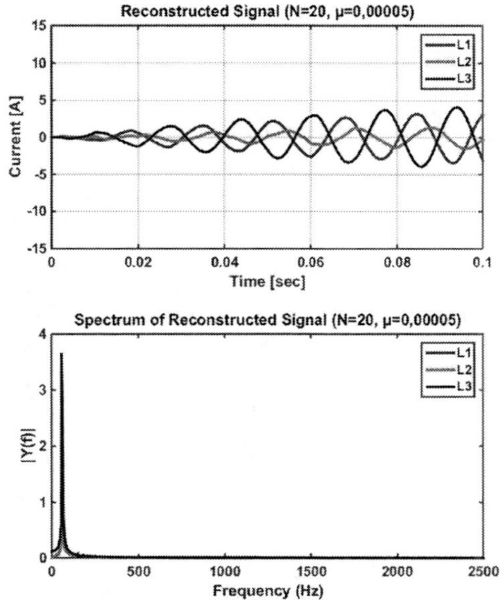

Figure 3. Effect of step size on filtration success.

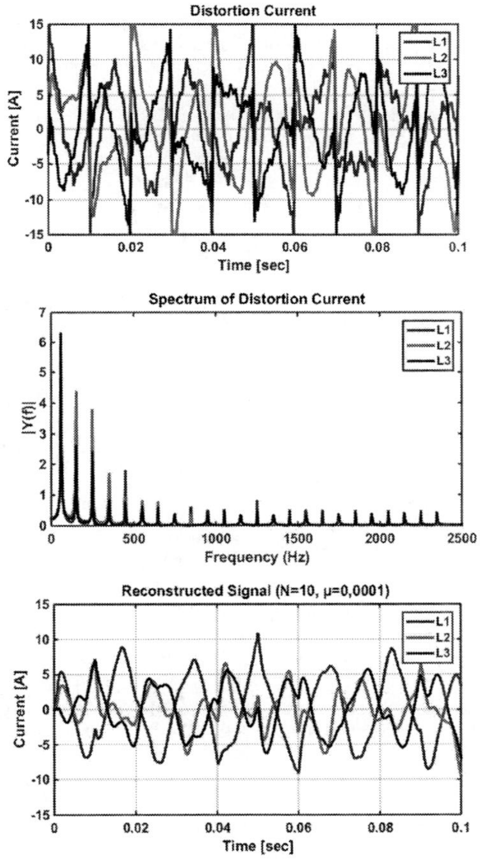

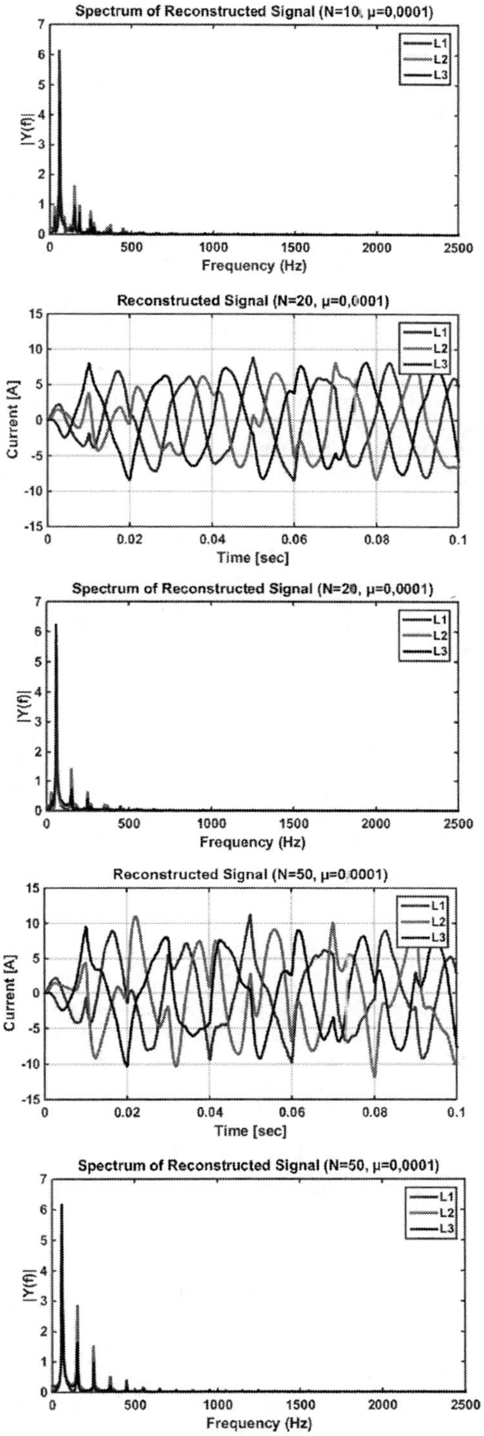

Figure 4. Effect of filter order on filtration success.

The step size has a major impact on the speed and stability of the LMS algorithm convergence, based on the experiments made:

a) If the selected μ value is too small, the time needed to find the optimal solution by LMS algorithm will be too long, see Fig. 3. From the point of view of the dynamic changes (load connection and disconnection) in the power grid, the demand is for the highest convergence rate.

b) If the selected μ value is too large, the adaptive filter will become unstable and the resulting output will exhibit deviations and thus the worse THD value, see Fig. 3.

Therefore, the appropriate choice μ is the question of a compromise between the rate of convergence and the quality of the filtration (THD). Experimental results have shown that for the LMS application of the SAPF control algorithm, the optimal value lies around the order of 0.0001.

Furthermore, it has been shown that the optimum value of the order of the filter is around the value of 20 since the higher order of the filter no longer improves the THD value, and hence the quality of the filtration, but on the contrary, the distortion is gradually increasing, see Fig. 4. It is also important to notice that the higher the filter order the higher are requirements for DSP computing performance. Small values of the filter order can lead to instability of the adaptive systems and to the significant deterioration in the value of THD.

CONCLUSION

This article was devoted to the use of the adaptive LMS algorithm for SAPF control. Within the realized experiments the influence of control parameters (i.e. step size, filter order) on the quality of filtering represented by the THD value was investigated. Experimental results confirmed the hypothesis that control parameters have a major effect on the efficiency of filtration and THD. Thus, it is obvious that parameter N and μ setting must be given a great deal of attention when designing an LMS-controlled SAPF. It turned out that a number of similar topical publications solve this problem only marginally and the area, therefore, offers a relatively wide space for further research and innovation

In general, SAPF-controlled adaptive LMS algorithm systems can very effectively compensate higher harmonic currents of non-linear loads, which are a major problem for modern power networks changing dynamically over time.

The subject of further research can be adaptive optimization of control parameters of the LMS algorithm for SAPF control. A similar concept has been successfully implemented in adaptive equalization of the transmission channel, see [14]. This approach would automatically set the N and μ parameters based on the currently measured THD.

ACKNOWLEDGMENT

This work was supported by the Student Grant System of VSB-TU Ostrava [grant number SP2017/128].

REFERENCES

[1] R. Martinek, J. Vanus, P. Bilik, "New strategies for application of recursive least square algorithm in active power filters," Proceedings of the 8th International Scientific Symposium on Electrical Power Engineering, ELEKTROENERGETIKA, pp. 332–335, 2015.

[2] R. Martinek, J. Vanus, M. Kelnar, P. Bilik, "Control methods of active power filters using soft computing techniques," Proceedings of the 8th International Scientific Symposium on Electrical Power Engineering, ELEKTROENERGETIKA 2015, pp. 351–354, 2015.

[3] R. Martinek, J. Vanus, P. Bilik, T. Stratil, J. Zidek, "An efficient control method of shunt active power filter using ADALINE," IFAC-PapersOnLine," vol. 49, pp. 352–357, 2016.

[4] S. Haykin, "Adaptive filter theory," Pearson Education India, 2008.

[5] M. Cirrincione, M. Pucci, G. Vitale, A. Miraoui, "Current harmonic compensation by a single-phase shunt active power filter controlled by adaptive neural filtering," IEEE Transactions on industrial electronics, vol.56, pp. 3128–3143, 2009.

[6] B. Singh, J. Solanki. "An implementation of an adaptive control algorithm for a three-phase shunt active filter," IEEE Transactions on Industrial Electronics, vol.56, pp. 2811–2820, 2009.

[7] M. Cirrincione, M. Pucci, G. Vitale. "A single-phase DG generation unit with shunt active power filter capability by adaptive neural filtering," IEEE Transactions on Industrial Electronics, vol.55, pp. 2093–2110, 2008.

[8] R. Martinek, R. Kahankova, H. Skutova, P. Koudelka, J. Zidek, J. Koziorek, "Adaptive signal processing techniques for extracting abdominal fetal electrocardiogram," 10th International Symposium on Communication Systems, Networks and Digital Signal Processing, 2016.

[9] M. Montero, M. I. Milans, E. R. Cadaval, F. B. Gonzalez, "Comparison of control strategies for shunt active power filters in three-phase four-wire systems," IEEE transactions on power electronics, vol. 22, pp. 229–236, 2007.

[10] R. Salem, N. Mendalek, K. Al-Haddad, "Experimental design of a nonlinear control technique for three-phase shunt active power filter," IEEE Transactions on Industrial Electronics, vol.57, pp. 3364–3375, 2010.

[11] R. Martinek, J. Vanus, P. Bilik, M. Al-Wohaishi, J. Zidek, H. Wen, "The implementation of equalization algorithms for real transmission channels," IEEE Instrumentation and Measurement Technology Conference, 2016.

[12] R. Martinek, J. Zidek, "The real implementation of NLMS channel equalizer into the system of software defined radio," Advances in Electrical and Electronic Engineering, vol.10, pp. 330–336, 2012.

[13] R. Martinek, M. Kelnar, J. Vanus, P. Bilik, J. Zidek, "A robust approach for acoustic noise suppression in speech using ANFIS," Journal of Electrical Engineering, vol.66, pp. 301–310. 2015.

[14] R. Martinek, J. Konecny, P. Koudelka, J. Zidek, H. Nazeran, "Adaptive optimization of control parameters for feed-forward software defined equalization," Wireless Personal Communications, vol.1, 2017.

[15] M. El-Habrouk, M. K. Darwish, P. Mehta. "Active power filters: A review," IEE Proceedings-Electric Power Applications vol.147, pp. 403–413, 2000.

[16] F. Z. Peng, "Application issues of active power filters." IEEE Industry applications magazine, vol. 4. pp. 21–30, 1998.

[17] T. Quoc-Nam, H. Lee, "An advanced current control strategy for three-phase shunt active power filters." IEEE Transactions on Industrial Electronics, vol. 60, pp. 5400–5410, 2013.

[18] S. Biricik, S. Redif, O.C. özerdem, S.K. Khadem, M. Basu, "Real-time control of shunt active power filter under distorted grid voltage and unbalanced load condition using self-tuning filter." IET Power Electronics, vol.7, pp. 1895–1905, 2014.

[19] P. Kanjiya, V. Khadkikar, H. H. Zeineldin, "Optimal control of shunt active power filter to meet IEEE Std. 519 current harmonic constraints under nonideal supply condition." IEEE Transactions on industrial electronics, vol. 62, pp. 724–734, 2015.

[20] "HP - Agilent 6834B, Agilent 6834B Specifications", Internet: http://www.keysight.com/en/pd-1000001649%3Aepsg%3Apro-pn-6834B/ac-power-source-4500-1500va-150-300v-30-15a?cc=CZ&lc=eng [May 15, 2017]

[21] M. Fajkus, J. Nedoma, R. Martinek, V. Vasinek, H. Nazeran and P. Siska," A Non-Invasive Multichannel Hybrid Fiber-Optic Sensor System for Vital Sign Monitoring," Sensors, vol. 17, 2017.

[22] R. Martinek, J. Nedoma, M. Fajkus, R. Kahankova, J. Konecny, P. Janku, S. Kepak, P. Bilik, H. Nazeran, "A phonocardiographic-based fiber-optic sensor and adaptive filtering system for noninvasive continuous fetal heart rate monitoring," Sensors (Switzerland), vol.17, 2017.

[23] V. Kus, J. Pikous, "The Effects of Connecting High Amounts of Nonlinear Appliances to the Supply Network". IEEE Conference: IEEE 5th International Conference on Power

Engineering, Energy and Electrical Drives Location: Riga, LATVIA Date: MAY 11-13, 2015, Pages: 201-206.

[24] V. Kus, T. Josefova, "The influence of the dead time duration and switching frequency on the input current distortion of voltage-source active rectifiers". International Conference on Applied Electronics (AE) Location: Pilsen, CZECH REPUBLIC Date: SEP 08-09, 2015, Pages: 131-134.

[25] V. Kus, T. Josefova, "New insights into the harmonic analysis of voltage-source active rectifier". 14th International Scientific Conference on Electric Power Engineering (EPE) : Kouty nad Desnou, CZECH REPUBLIC Date: MAY 28-30, 2013, Pages: 319-324.

[26] P. Garanayak, G. Panda, (2017). An adaptive linear neural network with least mean M-estimate weight updating rule employed for harmonics identification and power quality monitoring. Transactions of the Institute of Measurement and Control, 0142331217695402.

[27] H. Laib, A. E., Chaghi, P. Wira, (2017). A Neural and Fuzzy Logic Based Control Scheme for a Shunt Active Power Filter. In Recent Advances in Electrical Engineering and Control Applications (pp. 201-211). Springer International Publishing.

[28] C. Xu, K. Dai, X. Chen, L. Peng, Y. Zhang, Z. Dai, (2017). Parallel Resonance Detection and Selective Compensation Control for SAPF with Square-Wave Current Active Injection. IEEE Transactions on Industrial Electronics.

The Effect of the Forgetting Factor and Filter Length on the RLS Adaptive Control Algorithm in Shunt Active Power Filter

Radek Martinek, Petr Bilik, Jan Vanus, Ludvik Koval, Radana Kahankova
Department of Cybernetics and Biomedical Engineering
VSB - Technical university of Ostrava,
Ostrava, Czech Republic
radek.martinek@vsb.cz, petr.bilik@vsb.cz, ludvik.koval@vsb.cz

Marcel Fajkus, Jan Nedoma
Department of Telecommunications
VSB - Technical university of Ostrava,
Ostrava, Czech Republic
marcel.fajkus@vsb.cz, jan.nedoma@vsb.cz

Abstract — **This article deals with the use Recursive Least Squares (RLS) algorithm for the management of Shunt Active Power Filter (SAPF). When using the RLS algorithm the key parameter is a suitable choice of the Forgetting Factor. The selection of the forgetting factor has a significant influence on the performance of the SAPF, i.e. on the value of the resulting Signal to Noise Ratio (SNR). The value of the forgetting factor leads to a tradeoff between the stability and the tracking ability. In a system identification setting, both the filter length and a leakage phenomenon affect the selection of the forgetting factor. Within the performed experiments were defined and visualized the 3D loss function for the RLS algorithm which defines the working area based on the value of SNR. Experimental results show that the choice of the Forgetting Factor parameter has a significant impact on the quality and speed of the estimation the compensation current.**

Keywords: Recursive Least Squares (RLS); Forgetting Factor; Signal to Noise Ratio (SNR); Shunt Active Power Filter (SAPF)

I. Introduction

This article discusses the use of Recursive Least Squares (RLS) Algorithm [2] for control the Shunt Active Power Filter (SAPF), see [6], [7], [8]. This article builds on the author's publications [2], [3], [4] and extends them with new insights. The filter consists of a controlled current generator connected in parallel to the load [10], see Figure 1. It is used to compensate for undesirable higher harmonic currents by generating the same harmonic components of the opposite phase current and introducing them into the grid. The resulting current is then free of higher harmonics. In this way, the current drawn from the grid is filtered, and voltage distortion caused by the load and the power factor (fast reactive power compensation) are improved, [27], [28], [29].

Recent wide spread of power electronic equipment has caused an increase of the harmonic disturbances in the power systems. The nonlinear loads draw harmonic and reactive power components of current from ac mains, see [10]. For problems of harmonic pollution, passive and active power filters (APFs) are typical approaches that are used to improve the PF and to eliminate harmonics. In the past, passive LC filters are generally used to reduce these problems, but they have many demerits such as its being bulky and heavy, and its resonance, tuning problem, fixed compensation, noise, increased losses, etc. [11], [30], [31].

In this paper, we are presented SAPF for power quality improvement using RLS algorithm.

The main contribution of this article is the study of the influence of the control parameters (filter order and forgetting factor) of the RLS algorithm used to control the SAPF to the overall Signal to Noise Ratio (SNR) value. The experiments demonstrated a significant influence of control parameters on the filtration efficiency. For the RLS algorithm being investigated, the working area was deleted.

II. APPLICATION OF RLS ALGORITHM IN SAPF

Shunt active power filter is power electronic device that is used to compensate harmonics current by injecting equal but opposite harmonic compensating current with 180° phase shift with load current at the point of common coupling where the nonlinear load is connected, see [1], [24], and [25].

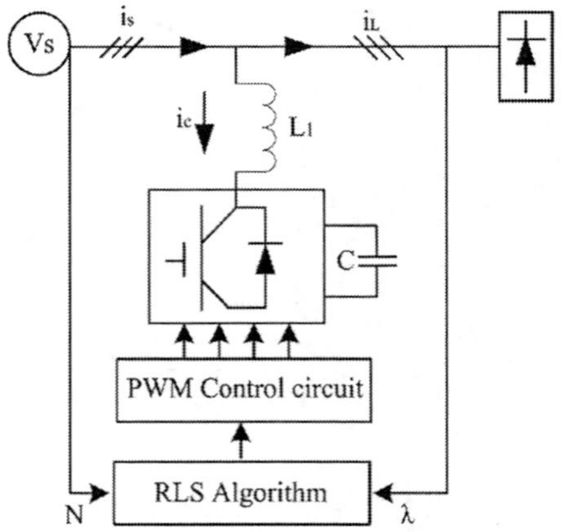

Figure 1. Control scheme: RLS adaptive control algorithm in SAPF.

A. *Reference signal generation for shunt active power filter using RLS algorithm*

In Figure 1 is shown generally principle application scheme of RLS algorithm in SAPF. Detailed description of the concept an efficient control method of shunt active power filter using adaptive filtering approach is described in [2], [3], and [4].

Parallel active filters in the distribution grid modify non-harmonic current from appliances connected to a common bus to one source. They are connected between consumers and the power source, which is the distribution transformer. The used power ratings are ranging from tens of kVA to several MVA. The distribution system may also be a power distribution within an industrial plant where rectifiers, cyclo-converters or other interfering devices are operating.

A parallel active filter can increase the power factor in such a grid by compensating the reactive current component and improve the flicker values (flashing of lighting devices due to higher harmonics). It is also used in conjunction with a static compensator where the static compensator regulates the power factor in rough jumps by switching the individual capacitor stages and the active filter performs the fine-tuning to the exact cos φ value. The static compensator can operate independently, or it can be controlled by an active filter, where the switching of the capacitor stages takes place via a thyristor.

B. *RLS algorithm*

The recursive variant of calculating the optimal adaptive filter coefficients according to the mean quadratic error is a set of adaptive filters based on a deterministic approach. A deterministic approach to filter design requires counting of multiple sample characteristics, this approach leads to the least squares recursive algorithm (RLS), which is the basic representative of the family of algorithms with recursive deterministic optimal adaptation. A detailed derivation of the RLS algorithm can be found [2].

The standard RLS algorithm preforms the following operations to update the coefficients of an adaptive filter.

1) Calculates the output signal y(n) of the adaptive filter.

2) Calculates the error signal e(n) by using the following equation: e(n) = d(n) – y(n).

3) Updates the filter coefficients by using the following equation:

$$\vec{w}(n+1) = \vec{w}(n) + e(n)\vec{K}(n). \tag{1}$$

where $\vec{w}(n)$ is the filter coefficients vector and $\vec{K}(n)$ is defined by the following equation:

$$\vec{K}(n) = \frac{P(n)\vec{u}(n)}{\lambda + \vec{u}^T(n)P(n)\vec{u}(n)}, \tag{2}$$

where λ is the forgetting factor and *P*(*n*) is the inverse correlation matrix of the input signal. Refer to the book Adaptive Filter Theory for more information about the inverse correlation matrix. *P*(n) has following initial value *P*(0):

$$P(0) = \begin{vmatrix} \delta^{-1} & 0 & 0 \\ 0 & \delta^{-1} & 0 \\ 0 & 0 & \delta^{-1} \end{vmatrix}. \tag{3}$$

Use the AFT Create FIR RLS VI to create an adaptive filter with the standard RLC algorithm.

C. *Discussion of choosing the coefficient value λ*

The coefficient λ is called the forgetting factor and takes values $\lambda \in (0,1]$. If $\lambda = 1$, then we are talking about the estimation without forgetting and thanks to the λ^{n-i} we can do weighing. The input values of each signal before $i = 1$ are considered zero; The significant values are the last n sets of values. Weighting is, therefore, preceding autocorrelation matrix or vector cross-correlations, and by adding a correction for the actual values of the autocorrelation matrix of the vector or cross-correlations. In a practical implementation, we usually consider from $\lambda = 0.98$ to $\lambda = 1$. For small values of the λ coefficient, there is a little contribution of previous samples to the covariance matrix. The filter is thus more sensitive to the current samples. In the case of $i = n$, i.e.$\lambda^{n-n} = \lambda^0 = 1$, the mathematical description of the RLS algorithm is reduced to the description of the LMS algorithm $\Lambda(n) = \lambda^{n-(i=n)} = 1$.

II. EXPERIMENTAL RESULTS

In this section, the loss functions for the tested adaptive system that uses the RLS algorithm will be introduced. Its advantage compared to the most commonly used LMS algorithm is greater accuracy, but it is redeemed by higher computational demands. A fundamental parameter of RLS algorithm is the forgetting factor λ (Forgetting Factor). It can reach values $0 < \lambda < 1$. In practice, the value of this coefficient is very close to 1. If $\lambda = 1$ it resembles the classical MSE concept (LMS) with increasing memory, and the estimate is performed without forgetting.

The effect of λ lies in the gradual forgetting of old data, with the highest weight being at the last measurement. Theoretically, for the convergence of parameters, it is necessary that $\lambda = 1$, but at the same time a requirement for the sensitivity of the algorithm to change the parameters (allowing for changes to be monitored) is required, which requires $\lambda < 1$.

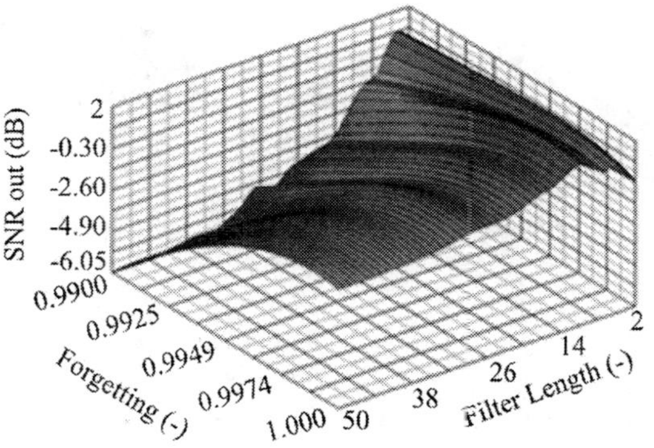

Figure 2. Graph of Loss Function $J(\lambda, N)$ with step $\Delta N = 1$, $\lambda \in [0.99, 1]$ with step $\Delta\lambda = 1 \times 10^{-3}$ for RLS algorithm based on SNR.

Figure 2 shows a 3D graph of the SNR dependence on parameters λ (forget factor) and *N* (filter order). It is clear from the graph that the RLS algorithm achieves better filtration results with increasing λ parameter (see also Figure 4), especially when $\lambda \rightarrow 1$, which corresponds to the theoretical assumptions that were discussed. Based on empirical experiences and experiments performed, it can be stated that: for waveforms that vary slowly over time (almost periodic signals) will be $\lambda = 1$. Conversely, in the case of the variable primary and reference signal (sudden changes in the grid, i.e. frequent load connections, disconnections, etc.) will be $\lambda < 1$.

This fact will again correspond to theoretical assumptions, because the RLS algorithm is able to monitor the changes of the NI-*f*EKG and *m*EKG variables by forgetting (i.e., $\lambda < 1$).

From the graph in Fig, 2 and Fig. 4, it is obvious that the value of the optimum filter order lies within the neighborhood of a local maximum ($N_{opt} = 12, \lambda_{opt} = 0.9987$). For higher filter order values, the quality of the filtration decreases gradually, for lower values, on the other hand, sharply.

Based on Fig. 3 it can be stated that it is unnecessary to set $N > 12$, because the higher order of the filter will no longer increase the SNR value, and hence the quality of the filtration. The RLS algorithm is computationally demanding. Therefore, when choosing and optimizing, an emphasis is placed on the choice of the lowest filter order *N*.

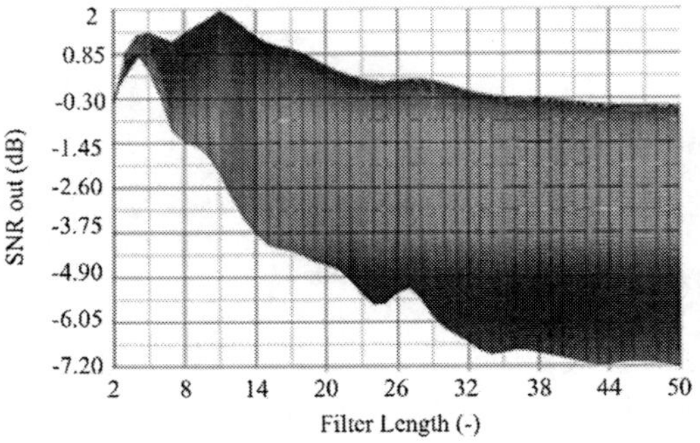

Figure 3. A detailed view of the working range of the RLS investigated filter order.

From Fig. 3 it is evident that for $N < 12$ decreases significantly. For this reason, it would be preferable to choose the range $N \in$ [global maximum, global maximum + 5] in step $\Delta N = 1$ when determining the optimal RLS algorithm setting for the given electrode combinations.

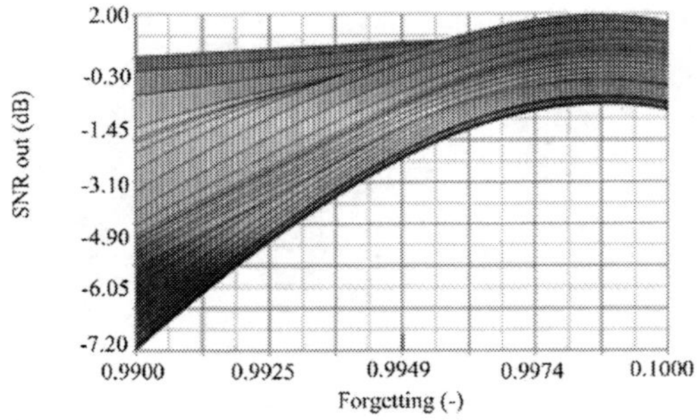

Figure 4. A detailed view of the working area of the forgotten factor of the RLS examined.

Table 1 summarizes the results of the RLS-based algorithm experiments based on the SNR. SNR values were determined for each phase L1, L2, and L3. Based on Fig. 3, the order of the filter was set to 14 for the experiments. Furthermore, the value of the forgetting factor was changed from 0.8 to 1.

λ (-)	SNR L1 (dB)	SNR L2 (dB)	SNR L3 (dB)
0.80000	unstable	unstable	unstable
0.90000	unstable	unstable	unstable
0.95000	0.0329	0.0044	1.4107
0.96000	0.0030	0.0044	1.4334
0.96500	1.3660	0.7367	1.4452
0.97000	1.5256	1.5826	1.4578
0.97500	1.4556	1.6714	1.4723
0.98000	1.4790	1.7969	1.4912
0.98500	1.5112	1.9824	1.5209
0.99000	1.5691	2.2817	1.5813
0.99500	1.7085	2.7507	1.7376
0.99990	1.9243	2.8443	1.9853
0.99995	1.9251	2.8392	1.9861
0.99999	1.9256	2.8350	1.9867
1.00000	1.9257	2.8339	1.9868

Figure 5 shows the results of selected experiments in the time and frequency domains. At first, the time course and the spectrum of the disturbed signal are shown. This is a simulated waveform where the emphasis was placed on transient phenomena, i.e. connection and disconnection of the nonlinear load (two different cases was simulated). Earlier experiments presented in [2] have shown that if we have a grid with no dynamic changes, i.e. frequent connection and disconnection of the load, it is advisable to choose the forget factor $\lambda = 1$. If frequent changes occur, the optimal value is $\lambda < 1$, see Fig. 4.

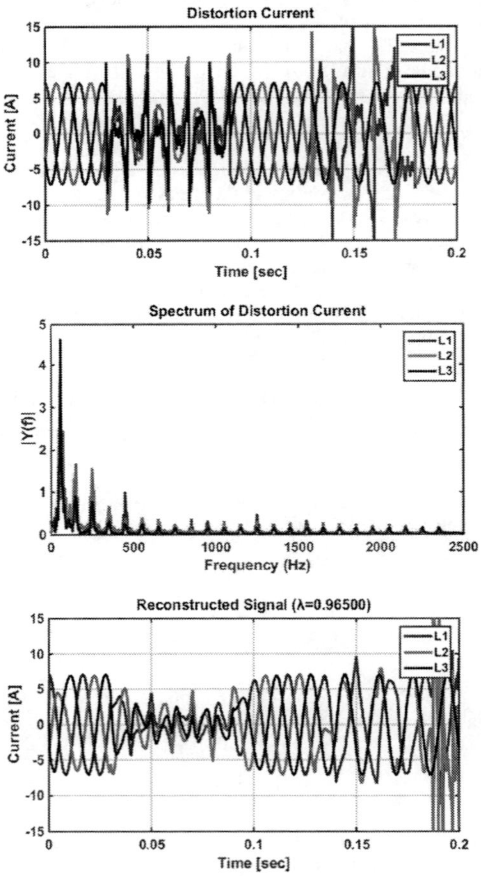

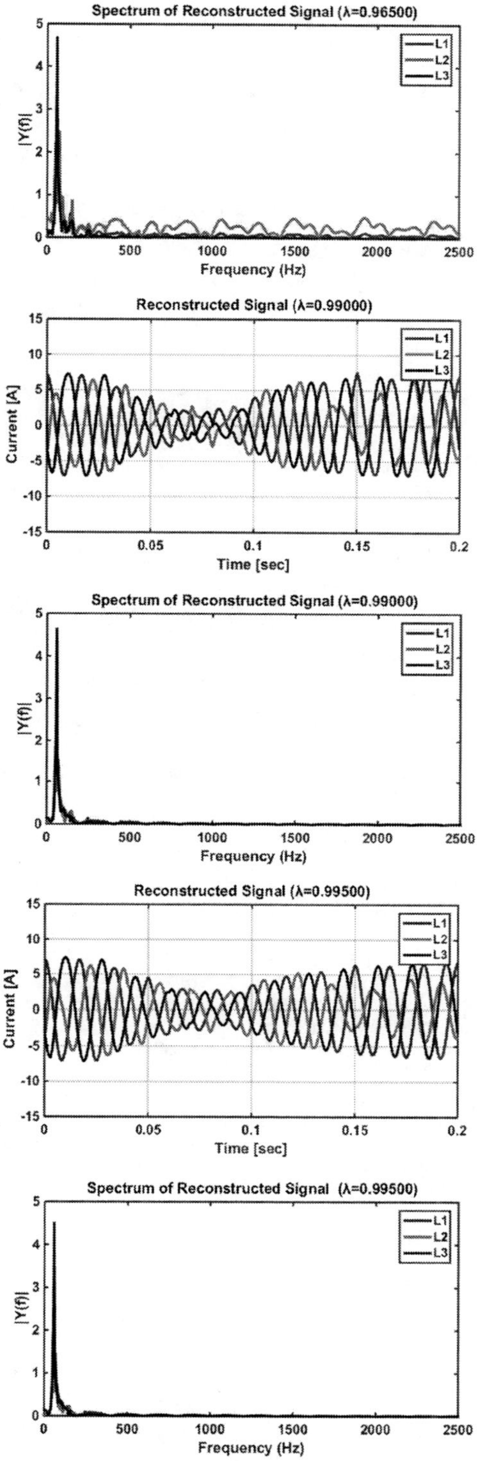

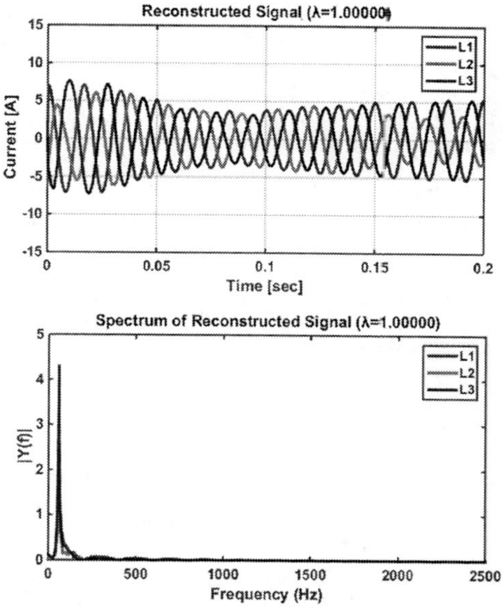

Figure 5. The effect of forgetting factor on the success of filtration.

Figure 5 also shows selected waveforms and spectra for selected λ values, so that the effect of this constant on the performance of the entire adaptive system is demonstrated.

III. Conclusion

This article introduces the SAPF management concept using the RLS adaptive algorithm. Experimental results confirmed the critical influence of filter parameters and forgetting factor on filtration efficiency. Experimental results were evaluated based on the SNR. The working area of the RLS algorithm was defined.

In general, the RLS algorithm achieves better results than the most widely used LMS algorithm [12], [13], and [14]. The great advantage of the RLS algorithm is the convergence rate. The disadvantage is the higher computational difficulty and the more difficult implementation, a more detailed description can be found in [15], [16], [17].

The new SAPF management methods are currently mentioned in a large number of publications as [18], [19], and [20]. This article has shown that the use of RLS algorithm may be one of the promising paths. However, it is absolutely essential to properly set the control parameters (filter order and forgetting factor). This is confirmed by the fact that the RLS algorithm is beginning to apply in other application areas which are based on the same principle, such as processing of fetal electrocardiography - ECG [23], adaptive channel equalization [21], adaptive speech processing [22], etc.

Acknowledgment

This work was supported by the Student Grant System of VSB-TU Ostrava [grant number SP2017/128].

References

[1] D.A. Tamboli, R. H. Chile. "Reference signal generation for shunt active power filter using adaptive filtering approach," International Conference on Industrial Instrumentation and Control, pp. 766–770, 2015.

[2] R. Martinek, J. Vanus, P. Bilik, "New strategies for application of recursive least square algorithm in active power filters," Proceedings of the 8th International Scientific Symposium on Electrical Power Engineering, ELEKTROENERGETIKA, pp. 332–335, 2015.

[3] R. Martinek, J. Vanus, M. Kelnar, P. Bilik, "Control methods of active power filters using soft computing techniques," Proceedings of the 8th International Scientific Symposium on Electrical Power Engineering, ELEKTROENERGETIKA 2015, pp. 351–354, 2015.

[4] R. Martinek, J. Vanus, P. Bilik, T. Stratil, J. Zidek, "An efficient control method of shunt active power filter using ADALINE," IFAC-PapersOnLine," vol. 49, pp. 352–357, 2016.

[5] S. Haykin, "Adaptive filter theory," Pearson Education India, 2008.

[6] M. Cirrincione, M. Pucci, G. Vitale, A. Miraoui, "Current harmonic compensation by a single-phase shunt active power filter controlled by adaptive neural filtering," IEEE Transactions on industrial electronics, vol.56, pp. 3128–3143, 2009.

[7] B. Singh, J. Solanki. "An implementation of an adaptive control algorithm for a three-phase shunt active filter," IEEE Transactions on Industrial Electronics, vol.56, pp. 2811–2820, 2009.

[8] M. Cirrincione, M. Pucci, G. Vitale. "A single-phase DG generation unit with shunt active power filter capability by adaptive neural filtering," IEEE Transactions on Industrial Electronics, vol.55, pp. 2093–2110, 2008.

[9] "HP Agilent 6834B, Agilent 6834B Specifications", Internet: http://www.keysight.com/en/pd-1000001649%3Aepsg%3Apro-pn-6834B/ac-power-source-4500-1500va-150-300v-30-15a?cc=CZ&lc=eng [May 15, 2017].

[10] K. Murat, E. Ozdemir, "An adaptive hysteresis band current controller for shunt active power filter, " Electric power systems research, vol. 73, pp. 113–119, 2005.

[11] K.K. Shyu, M.J. Yang, Y.M. Chen, Y.F. Lin, "Model reference adaptive control design for a shunt active-power-filter system." IEEE Transactions on Industrial Electronics, vol.55, pp. 97–106, 2008.

[12] A. Bhattacharya, Ch. Chakraborty, "A shunt active power filter with enhanced performance using ANN-based predictive and adaptive controllers." IEEE transactions on industrial electronics, vol.58, pp. 421–428, 2011.

[13] M. Cirrincione, M. Pucci, G. Vitale, "A single-phase DG generation unit with shunt active power filter capability by adaptive neural filtering," IEEE Transactions on Industrial Electronics, pp. 4373–4380, 2006.

[14] M. El-Habrouk, M. K. Darwish, P. Mehta. "Active power filters: A review," IEE Proceedings-Electric Power Applications vol.147, pp. 403–413, 2000.

[15] R. Martinek, R. Kahankova, H. Skutova, P. Koudelka, J. Zidek, J. Koziorek, "Adaptive signal processing techniques for extracting abdominal fetal electrocardiogram," 10th International Symposium on Communication Systems, Networks and Digital Signal Processing, 2016.

[16] R. Martinek, J. Vanus, M. Kelnar, P. Bilik, J. Zidek, "Application of recursive least square algorithm to adaptive channel equalization," XXI IMEKO World Congress "Measurement in Research and Industry, 2015.

[17] R. Martinek, J. Zidek, "The real implementation of NLMS channel equalizer into the system of software defined radio," Advances in Electrical and Electronic Engineering, vol.10, pp. 330–336, 2012.

[18] T. Quoc-Nam, H. Lee, "An advanced current control strategy for three-phase shunt active power filters." IEEE Transactions on Industrial Electronics, vol. 60, pp. 5400–5410, 2013.

[19] K. Antoniewicz, M. Jasinski, M.P. Kazmierkowski, M. Malinowski, "Model predictive control for three-level four-leg flying capacitor converter operating as shunt active power filter." IEEE Transactions on Industrial Electronics, vol.63, pp. 5255–5262, 2016.

[20] A. G. Exposito, A. J. Conejo, C. Canizares, "Electric Energy Systems: Analysis and Operation." CRC Press, ISBN 9780849373657, 2016.

[21] R. Martinek, J. Vanus, P. Bilik, M. Al-Wohaishi, J. Zidek, H. Wen, "The implementation of equalization algorithms for real transmission channels," IEEE Instrumentation and Measurement Technology Conference, 2016.

[22] R. Martinek, M. Kelnar, J. Vanus, P. Bilik, J. Zidek, "A robust approach for acoustic noise suppression in speech using ANFIS," Journal of Electrical Engineering, vol.66, pp. 301–310. 2015.

[23] R. Martinek, J. Nedoma, M. Fajkus, R. Kahankova, J. Konecny, P. Janku, S. Kepak, P. Bilik, H. Nazeran, "A phonocardiographic-based fiber-optic sensor and adaptive filtering system for noninvasive continuous fetal heart rate monitoring," Sensors (Switzerland), vol.17, 2017.

[24] M. Fajkus, J. Nedoma, R. Martinek, V. Vasinek, H. Nazeran and P. Siska, "A Non-Invasive Multichannel Hybrid Fiber-Optic Sensor System for Vital Sign Monitoring," Sensors, vol. 17, 2017.

[25] B. Singh, J. Solanki. "An implementation of an adaptive control algorithm for a three-phase shunt active filter," IEEE Transactions on Industrial Electronics, vol.56, pp. 2811–2820, 2009.

[26] R. Salem, N. Mendalek, K. Al-Haddad, "Experimental design of a nonlinear control technique for three-phase shunt active power filter," IEEE Transactions on Industrial Electronics, vol.57, pp. 3364–3375, 2010.

[27] V. Kus, J. Pikous, "The Effects of Connecting High Amounts of Nonlinear Appliances to the Supply Network". IEEE Conference: IEEE 5th International Conference on Power Engineering, Energy and Electrical Drives Location: Riga, LATVIA Date: MAY 11-13, 2015, Pages: 201-206.

[28] V. Kus, T. Josefova, "The influence of the dead time duration and switching frequency on the input current distortion of voltage-source active rectifiers". International Conference on Applied Electronics (AE) Location: Pilsen, CZECH REPUBLIC Date: SEP 08-09, 2015, Pages: 131-134.

[29] V. Kus, T. Josefova, "New insights into the harmonic analysis of voltage-source active rectifier". 14th International Scientific Conference on Electric Power Engineering (EPE) : Kouty nad Desnou, CZECH REPUBLIC Date: MAY 28-30, 2013, Pages: 319-324.

[30] L. Feng, W. Yong. "Modeling and Resonance Control of Modular Three Level Shunt Active Power Filter." IEEE Transactions on Industrial Electronics (2017).

[31] Y. Cao, al. "A Lyapunov Stability Theory-Based Control Strategy for Three-Level Shunt Active Power Filter." Energies 10.1 (2017): 112.

Use of the Software PI System Within the Concept of Smart Cities

Jan Vanus, Jiri Sykora, Radek Martinek, Petr Bilik, Ludvik Koval, Jan Zidek
Department of Cybernetics and Biomedical Engineering
VSB-TU, Faculty of Electrical Engineering and Computer Science
17. listopadu 15, 708 33 Ostrava-Poruba, Czech Republic

jan.vanus@vsb.cz, radek.martinek@vsb.cz, petr.bilik@vsb.cz, ludvik.koval@vsb.cz, jan.zidek@vsb.cz

Marcel Fajkus, Jan Nedoma
Department of Telecommunications
VSB-TU, Faculty of Electrical Engineering and Computer Science
17. listopadu 15, 708 33 Ostrava-Poruba, Czech Republic
marcel.fajkus@vsb.cz, jan.nedoma@vsb.cz,

Abstract — **The article focuses on the description of the possibility of using the PI System software for the monitoring and management of operating and technical functions in Intelligent Buildings (IB) and Smart Home (SH) with the aim of subsequent use for BMS Building Management System) in accordance with the requirements of the concept of Smart Cities (SC). This article describes the implementation of the PI System software for remote monitoring of HVAC control in the office building and monitoring HVAC, lights, and blinds control in the building of residential types. The original contribution of the article is the use of PI System software for large data processing in the real-life operation of buildings in order to efficiently use energy and reduce operational costs in buildings within the concept of SC. In the experimental part of this paper is described the indirect method of predicting CO_2 from the measured values of temperature and humidity sensors, using a gradient algorithm for error Backpropagation for the adaptation of a multilayer feed forward neural network. During the realized experiments the Levenberg-Marquardt method, the Bayesian Regulation method and the Regulation Scaled Conjugate gradient method for predicting the course of CO_2 concentration (ppm) from the measured indoor temperature *T* (°C) and relative humidity rH (%) and Outdoor temperature *T* (°C) were compared. The actual implementation was carried out in an experimental workplace (SH) using BACnet technology components to control heating, cooling, and ventilation. The obtained results were verified and classified using CA (Correlation Analysis) and the RMSE parameters (Root Mean Squared Error) and MAPE (Mean Absolute Percentage Error) were calculated. The information obtained can be used for the indirect monitoring of life activities of residents or for the optimization of technical operating functions. The indirect method for predicting CO_2 has the potential to reduce capital and operating costs by reducing the total number of sensors used in intelligent buildings. The experimental results confirm the suitability of the proposed method in real operation.**

Keywords: Smart Home, Intelligent Buildings, PI System, Interface, Smart Cities, Levenberg-Marquardt, the Bayesian Regulation method, Regulation Scaled Conjugate

Introduction

In the Moravian-Silesian Region, preparations for the implementation of Smart Region (SR) and Smart Cities Ostrava in cooperation with the Ministry of the Environment and the Ministry of Industry and Trade are in progress. One of the objectives of this project is to build a pilot region model in the Czech Republic in order to improve the quality of life and the environment of the Moravian-Silesian Region using the SC principle [1]. The SC concept is trying to make the most of modern information technologies and propose solutions for city-specific management in such a way that synergies between transport, logistics, safety, energy, building management, The difficulty and quality of life of citizens in the city. The individual criteria are different in each city and must be in line with strategic plans for the development of a specific city (development strategy, territorial plan, etc.). At the same time, the partial projects of a smart city must be in line with the development of the whole region or if it is in line with the concept of a SR. Chosen cross-cutting themes for SC and regions are as follows [2]:

- Sustainable urban mobility.
- Intelligent buildings and neighbour- hoods.
- Integrated infrastructures and processes in the energy sector.

There must be provide interoperability between the different technologies concepts within the SC in line with the Internet of Things (IoT), Smart Grids (SG), Smart Metering (SM), SH, Smart Home Care (SHC), or Green Buildings (GB) [3]. The authors deal with the use of PI System (PI) tools for monitoring, visualization and Big Data processing of measured quantities in the IB with possible use in other technological units within the SC concept. The article describes the implementation of the software PI System for remote monitoring of the HVAC and light control in office buildings and monitoring and control of HVAC, the lighting, and blinds in the building of residential types. In terms of the execution of the management of operation and technical features of individual structures and buildings located within the area of VSB TU Ostrava we can generally say that various types of technologies, standards, data protocols and bus systems for building automation (KNX, LonWorks, BACnet) have been used here [11]. PI System is used in applications for monitoring SG networks [4], in power plants [5], [6], [7] in industry [8], [9], [10] and for BMS in buildings [11] etc. For example in the Campus of Carnegie Mellon University (CMU) to support integrated ventilation (mechanical and natural), heating, cooling, lighting, and day lighting control strategies, OSIsoft's PI System is used as the real-time data platform in the laboratory. All systems installed in the facility, with more than 10 different BAS (Building Automation System) manufacturers (Siemens, Automated Logic, National Instrument, etc.) were integrated into a common platform using the PI Server [12].

IMPLEMENTATION OF PI SYSTEM IN OFFICE BUILDING

The block schema of the implementation of the PI System to monitor technical and operating functions (HVAC) at the new FEI building in the VSB TU Ostrava is shown below in the Figure 1. The acquired data measured by operating variables (HVAC) are stored in a database, from where they are stored on the PI server (PIAF) via the RDBMSPI interface. These data are then used in desktop and web applications on a remote computer, which communicates with the PI server. Applications that connect to PI, use PI SDK (Software Development Kit), PI API (Application Programming Integration) libraries or a combination of both. PI SDK and PI API are dynamically linked libraries, through which individual applications communicate with the PI server (Fig. 1). Libraries are installed on a computer as a part of the PI SDK installation process [11].

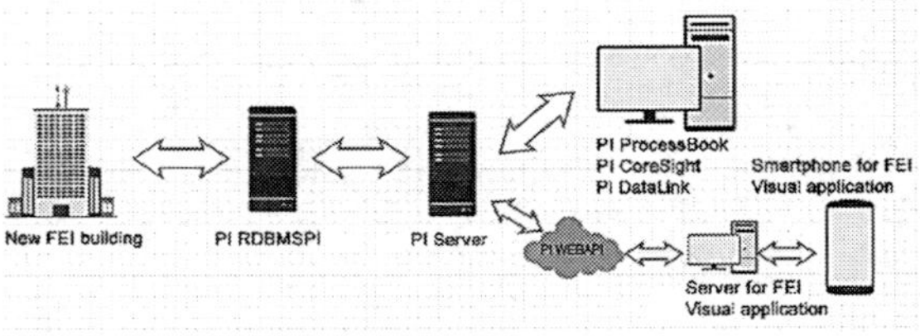

Figure 1. The block schema of the PI implementation at the new FEI building of TU Ostrava [11].

Figure 2. The administrative building (IB) visualization startup screen created in PI ProcessBook.

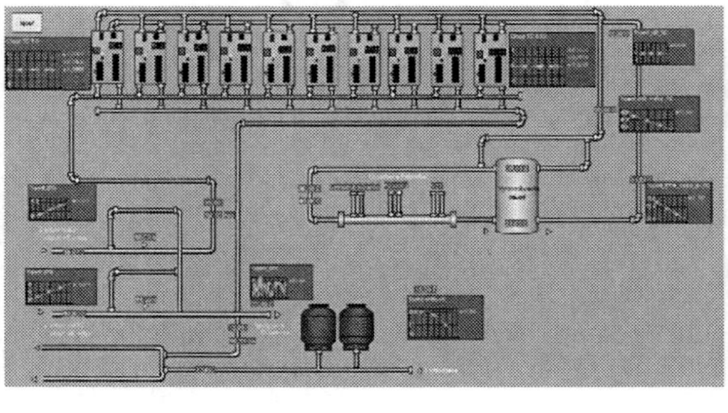

Figure 3. PI Processbook Display Heat pump system visualization.

IMPLEMENTATION OF PI SYSTEM IN SH

In SH at VŠB TU Ostrava, an OPC Server is used as a PI system data source, based on BACnet and KNX technologies and visualized using the Desigo Insight SW tool (Fig. 4).

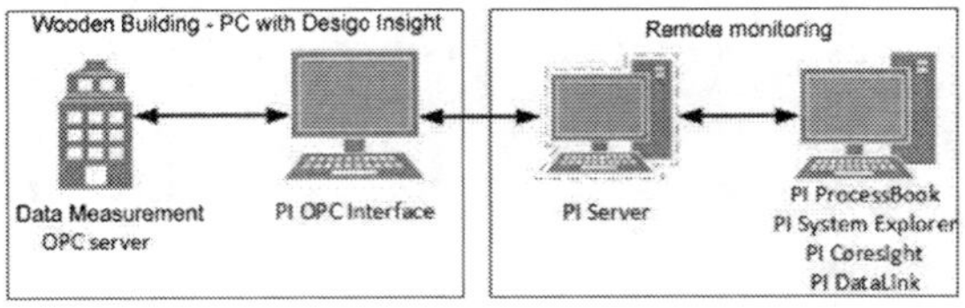

Figure 4. Block diagram illustrating data transfer from the OPC server at the SH (a wooden house) at VŠB TU Ostrava to the PI Server [13].

The OPC server communicates with the PI interface which forwards the data to the PI server. The PI server communicates with the client station that runs the individual applications.

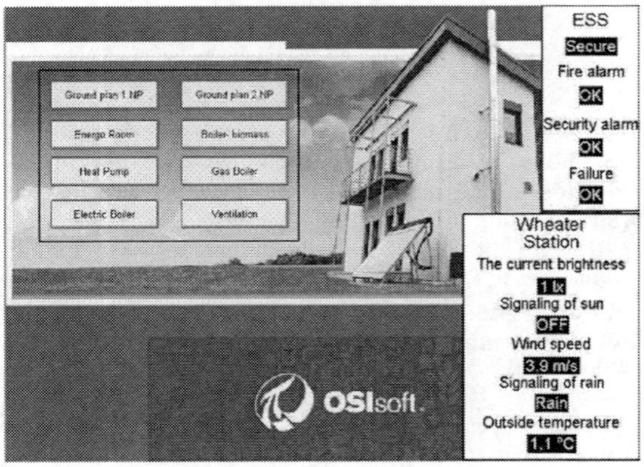

Figure 5. Introductory screen, SH visualization created in PI ProcessBook.

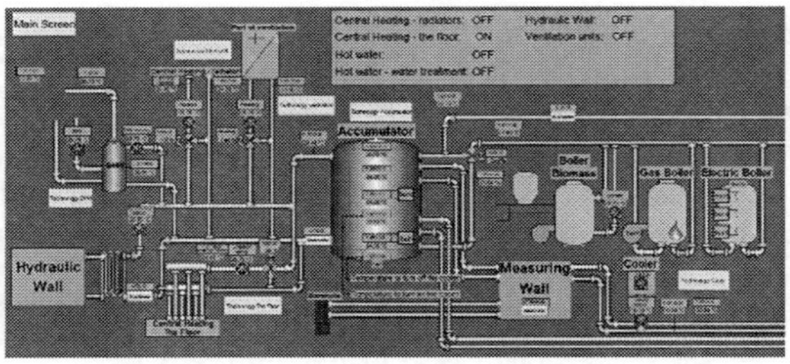

Figure 6. The Energo Room screen with a view of technology for heating and heat accumulation.

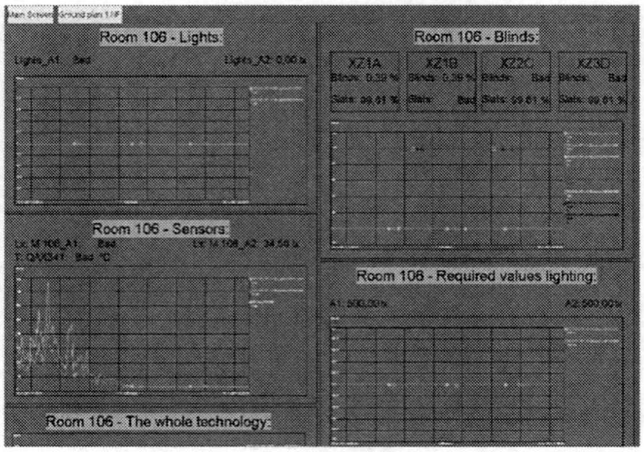

Figure 7. Monitoring of the state of the blinds, lighting and indoor environment control in the selected SH room.

The applications connected to the PI systems can be programmed using PI API, PI SDK or a combination of both. PI API (Application Programming Integration) and PI SDK (Software Development Kit) are dynamically connected libraries used by applications to communicate with the PI Server. The libraries are installed on a PC as a part of the PI SDK installation process. Generally speaking, PI SDK is used by any software with a graphic interface. Most interfaces are based on PI API, although some are based on a combination of these [13].

EXPERIMENTAL PART

The measured data are viewed using desktop or web applications of PI System. The desktop applications include:

- PI Data Link.
- PI Process Book (Fig. 2), (Fig. 3), Fig. 5), (Fig. 6), (Fig. 7).
- PI Data Access.

The web applications include:

- PI Web Parts.
- PI Active View.
- PI Coresight.

Information about the quality of internal environment in SH is provided by temperature sensors (indoor, outdoor) T_i, T_o (°C) and air humidity sensors rH (%). The technical parameters of the sensor are as follows:

- Temperature (T) measurement (range, accuracy), 0 to 50°C, + 1K,
- Relative humidity (rH) measurement (range, accuracy) 0 to 100%, + 5%,
- CO_2 measurement (range, accuracy), 0 to 2000 ppm, + 50 ppm.

More important is to know some facts about the concentrations of CO_2 (ppm). This information is crucial for the comfort and total habitability of separate rooms of IB. The cost of high-quality temperature and air humidity sensors is approximately in the range of Euro units. The cost of sensors for the measurement of CO_2 moves the price a rank higher to several tens of Euro. In the Czech Republic (CR), the emphasis is placed on the utilization of a good thermal insulation for large buildings, which can finally provide certain energy savings. This is realized for big office buildings, schools, hospitals, residential dwellings, family houses and blocks of flats. The internal environment of reconstructed insulated buildings is continuously getting worse with the increase in CO_2 concentrations and humidity rH. Many investors in the CR do not take this fact into account. To decrease the concentration of CO_2 in a room we can easily open a window and a door, or to use a forced ventilation as the part of a complex solution provided by HVAC technology (Heating, Ventilation and Air Conditioning)). Regarding the technology utilization in connection with the automatization of buildings, it is necessary to provide the measurement of CO_2 concentration before the implementation of HVAC controlled technology. To decrease the investment costs connected with the installation of missing sensors for the monitoring of CO_2 concentrations in big buildings or in huge residential dwellings, it is possible to use our proposed method for the prediction of CO_2 from the measurements of temperature *Ti*, *To* and air humidity rH, which are very often included in the standard equipment of IB.

To determine of the quality of internal environment in an intelligent building (IB) of each room was used the gradient algorithm of the back-propagation of error for the adaptation of the multilayer feed forward Neural network using prediction – the Levenberg-Marquardt (LM) method, the Bayesian Regulation (BR) method and the Scaled Conjugate Gradient (SCG) method. The CA and calculated RMSE parameters were used for the classification of prediction quality. The CA and calculated RMSE parameters and MAPE were used for the classification of prediction quality.

A. Methods used for processing of the measured data

Back propagation or the error back-propagation algorithm is a gradient algorithm for the adaptation of multi-layer feed forward networks. Organizational dynamics in the error back-propagation is equivalent to the multi-layer Perceptron. The network is formed by fixed acyclic topology with an input and output layer and one or two inner layers of hidden neurons. A universal solution for the number of neurons in the hidden layer so that the computing time is minimized and provides the most accurate result, has not yet been found [14]. There are three methods offered for an optimizing training method. The first two methods are among the exact methods where the analysis of validation data is used. This means that during the learning process of the network, the validation set is monitored. By the time the error rate starts to raise, the learning ceases. The first is the optimization function of SCG conjugate gradients, whose advantage is that it does not overburden the memory. Another function is LM, which is more time economical compared to SCG at the cost of a larger memory footprint (Tab. 1).

Table 1: Comparison of the quality of learning of the used NN methods (LM, BR, SCG).

	Number of neurons	Time *t* (hh:mm:ss)	RMSE training (ppm)	R Training
LM method	10	0:04:37	37.506	0.814
	100	0:16:28	26.030	0.915
	800	**0:10:00**	**18.551**	**0.953**
BR method	10	0:08:01	37.264	0.814
	100	0:36:05	36.377	0.882
	800	0:47:29	22.827	0.935
SCG method	10	0:00:57	43.706	0.703
	100	0:21:47	26 883	0.903
	800	0:29:42	36 711	0.835

The third algorithm is BR, which differs from the previous two in that it does not work with the validation set. It can be said about BR in general that it generalizes small or complex datasets or datasets containing noise rather well. The training ends with the minimizing of weights. In the next step the neural network itself is set up, specifically the number of neurons in the hidden layer [15]. The parameters of the neural network are selected sequentially with 10, 100 and 800 neurons in the hidden layer (Tab. 1, Tab. 2).

For the hidden layer, the hyperbolic tangent function was used and linear functions for the output layer. Table 1 lists the calculated parameters of RMSE, correlation coefficient **R** and learning time *t* (s) in order to compare the quality of learning of the individual methods - LM, BR and SCG. Table 1 shows that the results gradually improve with an increasing number of hidden neurons. Although this behavior does not represent a rule in the field of neural networks (it often happens that the setup of algorithms and data and the results are not unconditionally in direct proportion), NN (LM, BR, SGC) using measurement data exhibit behavior naturally expected for the task. The lowest values of MSE and at the same time the largest values of correlation coefficients **R** are shown in methods BR and LM. The disadvantage for BR is the relatively long time in learning NN. The SCG method shows the worst results.

Table 2 shows the calculated parameters RMSE, MAPE and coefficients **R** of CA in order to assess the predictive quality of the LM, BR and SCG methods for given number of neurons 10, 100 and 800 on individual predicted waveforms of CO_2 (ppm) concentration.

Table 2: Comparison and verification of predictive quality of the used methods (LM, BR, SCG).

	Number of neurons	**10**	**100**	**800**
LM method	RMSE (ppm)	37.68	26.588	**20.674**
	R training	0.791	0.885	**0.918**
	MAPE (%)	3.236	2.023	**1.607**
BR method	RMSE (ppm)	36.497	29.213	23.42
	R training	0.842	0.889	0.892
	MAPE (%)	3.509	2.252	1.634
SCG method	RMSE (ppm)	44.857	27.37	37.177
	R training	0.727	0.885	0.773
	MAPE (%)	3.991	2.074	3.857

The measure values of the CO_2 concentration waveform (ppm) were used as the reference waveform (number of samples 41760, time interval 1.6. 2015 – 30.6. 2015), that were compared with the predicted output waveforms of the CO_2 concentration calculated using the individual methods (LM, BR, SCG) from the measured values of T (°C) and rH (%), (Fig. 8).

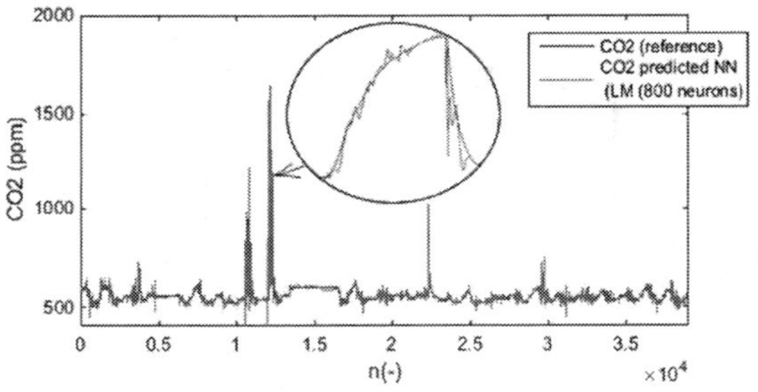

Figure 8. Waveforms of measured (reference, blue) and predicted CO_2 concentrations using NN (LM (800 neurons), red).

DISSCUSION

The LM method clearly demonstrated the best results in terms of correlation between the waveforms (Tab. 2), (Fig. 8), as well as in terms of comparison of the RMSE and MAPE parameters. Regression accuracy and thereby the estimation of the CO_2 content in the air is shown in Figure 8. The best achieved value of RMSE was 20.67 ppm (Tab. 2) measured with the LM method and 800 neurons. Other RMSE values were more than 23 ppm for the BR and LM methods (and SCG). However, the decreased time demand of the LM method is not negligible. In terms of relative accuracy/speed, the LM method appears to be the best. In practice, this will provide sufficient computing speed and usable accuracy. The blue line represents the actual (reference) output of the CO_2 concentration; the red line the predicted output of CO_2 concentration from BR. The measured values of CO_2 in the interior in the living area depend on the number of persons present, and also on the outdoor values of CO_2 especially during the winter months. The relative humidity in the interior is affected by the sources of moisture in the living area, the presence of people and measured values of relative outdoor humidity. The sensor set provides enough data to decide on air treatment in order to increase its quality. For automated prediction, the LM algorithm seems sufficient and can be applied in mechanisms for automatic control in IB, anticipating the habits of the occupants and the subsequent adequate automated response, or to create predictive models in order to reduce the number of installed physical sensors, thus reducing the IB end price.

CONCLUSION

In this article was described the implementation of the SW tool PI System for remote monitoring of the HVAC and Light control in office Intelligent buildings and monitoring of HVAC, lighting, and blinds control in the Smart Home - building of residential types with the aim of subsequent use for Building Management System in accordance with the requirements of the Smart Cities concept. In the second part of the article were compared the indirect methods (LM, BR and SCG) for prediction of CO_2 concentration from the measured values of temperature and humidity sensors using a gradient algorithm for error back-propagation for the adaptation of multilayer feed forward NN. Based on the long-term measurements and collected data of the abovementioned quantities, the LM and BR methods can be used to estimate the amount of CO_2 concentration in the air with sufficient accuracy in order to obtain information about the occupancy of each room (arrival time, time of departure, the number of people) in a Smart Home. The main contribution of the paper is the experimental verification of the proposed method in real operation. According to the authors' opinion, the use of CO_2 sensor in the operation of Intelligent Buildings is redundant, since the predicted value of CO_2 from the measured values of temperature and humidity correlates with the reference value of the measured CO_2 concentration. It is clear that before putting methods into technical practice it is necessary to calibrate the proposed method in real conditions. Implementation of the proposed indirect method of prediction of the CO_2 concentration has the potential to reduce capital and operating costs by reducing the total number of sensors used in Intelligent Buildings. The experimental results confirm the suitability of the proposed methods in real operation.

ACKNOWLEDGMENT

This work was supported by the Student Grant System of VSB-TU Ostrava [grant number SP2017/128] and "Promotion of Development of NEK activities in the field of energy saving energy concept" [grant number CZ.01.1.02/0.0/0.0/15_012/0002294].

REFERENCES

[1] R. Rožnovský. „Moravskoslezský kraj - cesta ke SMART regionu", Internet: http://www.france.cz/IMG/pdf/panel_2_-_7_-_m._roznovsky_smart_region.pdf?9290/9e760de2738466ace2d9490dce8cafe7d9fd3188., [Oct. 30, 2015],

[2] „Modrožlutá kniha Smart Písek", Internet: http://www.mupisek.cz/soubory/dokument24655.pdf., [Oct. 30, 2015],

[3] „Smart city – co to je, k čemu slouží a jak se vytváří", Internet: http://www.proelektrotechniky.cz/smart-city/Smart_City_bulletin1.pdf, [Oct 30, 2015],

[4] H. G. Aghamolki, Z. X. Miao, L. L. Fan, and Ieee, "A Hardware-in-the-Loop SCADA Testbed," in 2015 North American Power Symposium (North American Power Symposium, New York: Ieee, 2015.

[5] R. Berrios, F. Nunez, and A. Cipriano, "Fault tolerant measurement system based on Takagi-Sugeno fuzzy models for a gas turbine in a combined cycle power plant," (in English), Fuzzy Sets and Systems, Article vol. 174, no. 1, pp. 114-130, Jul 2011.

[6] A. Fras, T. Dang, and Ieee, Improving industrial application's performances with an Historian (2004 IEEE International Conference on Industrial Technology). New York: Ieee, 2004, pp. 718-721.

[7] M. McClintock, K. L. Cramblitt, and asme, Combined cycle plant performance monitoring (Proceedings of 2004 Asme Power). New York: Amer Soc Mechanical Engineers, 2004, pp. 291-294.

[8] A. Langford, D. Hunter, M. Australasian Inst, and Metallurgy, PIMS - The road to efficient, cost-effective plant operation (Eighth Mill Operators' Conference, Proceedings: EVOLUTION AND SURVIVAL IN THE MINERALS INDUSTRY, no. 4). Parkville Victoria: Australasian Inst Mining & Metallurgy, 2003, pp. 197-202.

[9] H. Liu, J. Guo, W. Yu, L. Zhu, Y. Liu, T. Xia, R. Sun, R.M. Gardner. "The Design and Implementation of the Enterprise Level Data Platform and Big Data Driven Applications and Analytics," (in English), 2016 Ieee/Pes Transmission and Distribution Conference and Exposition (T&D), Proceedings Paper p. 5, 2016.

[10] L. L. Simon and K. Hungerbuhler, "Industrial batch dryer data mining using intelligent pattern classifiers: Neural network, neuro-fuzzy and Takagi-Sugeno fuzzy models," (in English), Chemical Engineering Journal, Article vol. 157, no. 2-3, pp. 568-578, Mar 2010.

[11] J. Vanus, R. Martinek, P. Bilik, J. Zidek, and J. Machac, "Energy Management Strategies in Intelligent Office Building Using PI System," (in English), Proceedings of the 8th International Scientific Symposium on Electrical Power Engineering (Elektroenergetika 2015), Proceedings Paper pp. 428-431, 2015.

[12] "PI System for Monitoring Building Performance, Environmentally Sustainable Behavior, and Energy Conservation". Internet: http://www.osisoft.com/Presentations/PI-System-for-Monitoring-Building-Performance-Environmentally-Sustainable-Behavior--and-Energy-Conservation/ [May 3, 2017],

[13] J. Vanus, R. Martinek, P. Bilik, J. Zidek, and A. Dracka, " Smart Home Remote Monitoring Using PI System Management Tools," Proceedings of the 8th International Scientific Symposium on Electrical Power Engineering (Elektroenergetika 2015), Proceedings Paper pp. 384-387, 2015.

[14] S. Karsoliya, Approximating Number of Hidden layer neurons in Multiple Hidden Layer BPNN Architecture. International of Engineering Trends and Technology. 2012, 3(6) [cit. 2016-05-20]. ISSN 2231 – 5381, 2012.

[15] H. Gurecka, *Data classification using artificial neural networks*, Bachelor thesis, VUT Brno, 2016.

Design of an Application for the Monitoring and Visualization of Technological Processes with PI System in an Intelligent Building for Mobile Devices

Jan Vanus, Jaroslav Machac, Radek Martinek, Petr Bilik, Ludvik Koval, Jan Zidek
Department of Cybernetics and Biomedical Engineering
VSB-TU, Faculty of Electrical Engineering and Computer Science
17. listopadu 15, 708 33 Ostrava-Poruba, Czech Republic

jan.vanus@vsb.cz, radek.martinek@vsb.cz, petr.bilik@vsb.cz, ludvik.koval@vsb.cz, jan.zidek@vsb.cz

Marcel Fajkus, Jan Nedoma
Department of Telecommunications
VSB-TU, Faculty of Electrical Engineering and Computer Science
17. listopadu 15, 708 33 Ostrava-Poruba, Czech Republic
marcel.fajkus@vsb.cz, jan.nedoma@vsb.cz,

Abstract — **The article describes the design of a monitoring visualization application for mobile devices that monitors the processes in an administration building, namely in the new FEI building at VSB – TU Ostrava. The application is platform independent. The application has been developed using the Sencha Touch JavaScript framework and the PhoneGap wrapper. The application is optimized for Android devices. The application requires the execution of the application server, which communicates with the data server and with the developed application. The data of the created application is received from the data storage of the software tool PI System via the REST Web PI Client API. The application is used to visualize the operational and technical functions in an office building on an Android device, which allows the monitoring of available data in the FEI VSB building.**

Keywords: Sencha Touch, Apache Tomcat, Ajax, PhoneGap, SCADA, PI OSIsoft, JSON, Spring MVC, PI Web API

I. Introduction

The role played by intelligent or computerized systems in today's modern sophisticated buildings is essential. The monitoring and automatic control systems are important to achieve the required performance levels of the building. In terms of the development of the creation of web / mobile applications, the JavaScript application may be used for the visualization of technological processes in the administrative building. Javascript is in the forefront of the development of web browsers and hybrid applications and is becoming a very powerful language in which web applications are developed. More and more uses are being found for implementations of visualization and monitoring applications in everyday life. Albu used the MONNET visual surveillance system for tracking pedestrians over extended premises. The MONNET system is composed of intelligent nodes, which exchange information on the individually tracked pedestrians in an asynchronous manner [1]. The application can be used in Intelligent Buildings (Smart Home Care) supporting the independent living of older people or in homes for the elderly [2]. To visualize and monitor operating and technical functions in multiple Smart Home buildings or in Smart Home Care, more robust visualization and monitoring software tools can be utilized [3], which are also more efficient in larger office buildings. The possibilities of the development of visualizing and monitoring applications using wireless components are described in detail in articles. The monitoring of operating and technical data itself can be used to optimize the management of various operational and technical functions of buildings as lighting in order to reduce operating costs of the building, to set the optimal climate in different rooms of the building [4] with regard to the requirements of hygiene standards, or for classification [5], prediction [6] and the anticipation of building performance based on the measured and processed quantities [7]. This article describes the design of a monitoring visualization application for mobile devices running Android, which monitors technological processes in an intelligent building using the PI System software tools.

II. General Description of New FEI Building Automation

On the new FEI building at VSB – Technical University Ostrava the HVAC (Heating, Ventilation and Air Conditioning) system is controlled using the LonWorks bus system. Other implemented technological systems in the described building include:

- Control of lighting and outlets using KNX technology.
- Electronic Security System (ESS).
- Electronic fire alarm system (EPS).
- CCTV (Closed Circuit Television) circuit.
- Visualization of EPS, ESS.
- Visualization of HVAC.
- Visualization of the condition of lighting and outlets.
- Wireless system for controlling blinds.

In terms of the execution of the management of operation and technical features of individual structures and buildings located within the area of VSB TU Ostrava is possible generally say that various types of technologies, standards, data protocols and bus systems for building automation (KNX,

LonWorks) have been used here. At present, the need to implement a unifying tool (BMS - Building Management System) for processing and analysis of measured data from the implemented technology not only at the new FEI building at the Technical University of Ostrava, but also at other buildings of the whole area of the Technical University of Ostrava becomes obvious. A properly chosen BMS tool can serve to ensure monitoring of the condition of controlled devices, active alerting operators to the occurrence of atypical events – faults, exceeding threshold values with the monitored variables in the environment of the building (e.g. room temperature) or data archival for the retrospective analysis and subsequent optimization of the building operation in order to reduce operating costs and thus ensure energy savings.

III. PI SYSTEM DESCRIPTION

To unify the monitoring and control of various automation and electronic systems into a single BMS (Building Management System) environment it is necessary to use a software tool that allows you to consolidate several types of bus systems, standards and communication and data protocols into o single monitoring application. In our case, we took advantage of the opportunities offered by the PI System software tool (hereinafter referred to as PI) produced by OSIsoft. This is an industry standard designed for industrial enterprises and intelligent buildings for data reporting and real time events.

PI allows for easier retrieval, storage, distribution, analysis and visualization of data from various sources such as (database systems, SCADA systems, manual inputs, application software, actuators, sensors etc.). Individual PI programs or groups of PI programs can also exist separately. PI can be roughly divided into a data, server and client part. The client part of the PI (PI client) system is suitable for ensuring BMS in the described building. PI clients are PI software applications, with which end users work and which present the data in user-friendly environment with interesting graphics. These are desktop applications (PI DataLink, PI ProcessBook, PI Data Access) and web applications (PI WebParts, PI ActiveView, PI Coresight).

IV. PI SYSTEM IMPLEMENTATION

The block schema of the implementation of the PI System to monitor technical and operating functions (HVAC) at the new FEI building in the VSB Technical University Ostrava is shown below in the Figure 1.

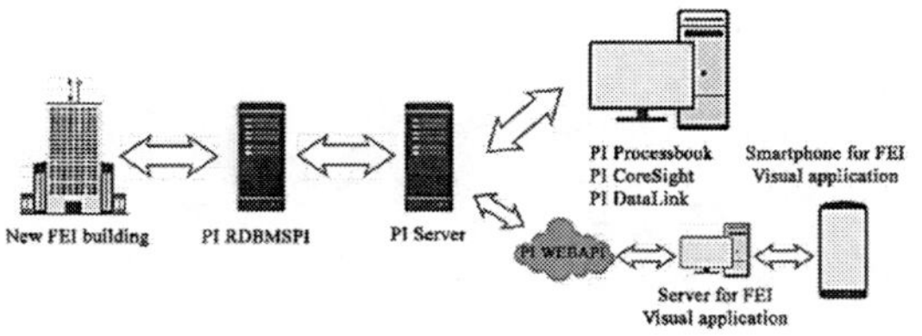

Figure 1. The block schema of the PI implementation at the new FEI building of VSB - TU Ostrava.

The acquired data measured by operating variables (HVAC) are stored in a database, from where they are stored on the PI server (PIAF) via the RDBMSPI interface. These data are then used in desktop and web applications on a remote computer, which communicates with the PI server. Applications that connect to PI, use PI SDK (Software Development Kit), PI API (Application Programming Integration) libraries or a combination of both. PI SDK and PI API are dynamically linked libraries, through which individual applications communicate with the PI server (Fig. 1). Libraries are installed on a computer as a part of the PI SDK installation process.

V. TECHNOLOGIES, TOOLS AND FRAMEWORKS USED FOR THE DESIGN OF THE APPLICATION FOR MOBILE DEVICES RUNNING ANDROID

A. *Server part - Spring MVC*

Spring MVC is a modular, open source J2EE application framework, also called a lightweight container. It is mainly used for creating web applications, but it can be used for other types of applications as well. The main advantage of Spring MVC is the simplification of the development of Java applications.

B. *Server part - Apache Tomcat*

Apache Tomcat is a web server and servlet container. It is being developed as an open source project. It is based on Java, Java servlets, Java Server Pages (JSP), and Enterprise JavaBeans (EJB). Tomcat excels in simplicity, transparency and small-footprint hardware. Apache Tomcat will be used as a container for the testing and development of custom applications.

C. *Server part - PI Web API*

PI Web API is a member of the PI system software tools. RESTfull is the PI system access layer that provides cross-platform programming interface for PI System. The RESTfull interface allows for the cross-platform development of web, desktop and mobile applications in various programming languages. PI Web API enables the retrieval and manipulation of data time-series from the PI data archive, and also loads and configures asset data and framework events from the PI AF server. It contains a set of REST services and APIs, which are being further developed. It is a web-based communication method. It is a commonly used tool for bilateral communication and communication with clients, because it allows companies to communicate between different IT infrastructures. In essence this creates a universal language for communication across IT infrastructures. The most commonly used models of web services are SOAP and REST. Figure 2 below describes a typical data stream of requests to the Web service, and responses to those requests. The Web service client in this case sends an HTTP GET request to the URL http://myserver/tasks for the Web server services. The URL indicates the source location which the client wants to access. Designations such as GET, POST or PUT indicate the action with which the client wants to query the server. When the server receives this request it processes it and sends a response back to the client as XML1 or JSON2. Web API has the following system requirements:

- Windows Server 2008 R2, 2012 or 2012R2 (2012 or newer),
- PI data archive 3.4.380 or newer (2012 or newer),
- PI Asset Server 2.1 or higher (2012 or newer).

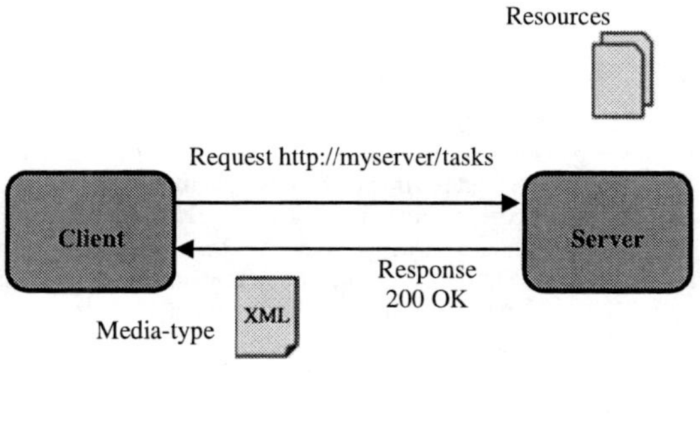

Figure 2. Client - server communication in PI Web API [8].

D. *Server part - Eclipse IDE*

Another important tool is the development environment. The officially supported one is the Eclipse IDE, currently the most popular Java development environment.

E. *Client part - Hybrid development*

Hybrid development combines both native and web approaches. The application consists of a native frame in which the web view and the web application are located. It is part of the application or it can be downloaded remotely via the Internet. The pros and cons of hybrid applications arise from the nature of the two approaches combined in the hybrid development. The advantage of hybrid applications against web ones is the possibility of publication in the official platform store. The user installs and runs them alongside the classic native applications. Another key advantage compared to traditional web applications is the ability to use frameworks which link the web application running in WebView with the system.

F. *Client part - Framework Sencha Touch*

The application programming is done using JavaScript and not the HTML5 markup. Although Sencha Touch takes full advantage of HTML5, the whole application code is placed in JavaScript files, and the HTML file contains only skeleton pages with the embedded application script. Sencha Touch has its own library named Ext. It is a robust library with various components. Sencha Touch makes writing application code a little harder than it is in ordinary JavaScript or e.g. jQuery framework, mainly due to its special syntax. In addition to writing code, however, the Sencha Touch framework is much more robust than e.g. the mentioned jQuery Mobile that contains more functionality; while a full-fledged Web application with all the trimmings can be created only with the help of this framework. Sencha Touch has so much functionality and settings that the programmer is not prevented from doing almost anything.

G. *Client part - PhoneGap*

One of the most successful so called "wrappers" is the PhoneGap. It is an application with inserted web code including HTML5, CSS3 and JavaScript, packaged for any operating system. This wrapper is often extended with the option of bridging some functionality that a website alone cannot handle. PhoneGap can bridge some of these. PhoneGap offers downloadable SDK on its website, through which application projects are created. Each project has its own special folder locations for native applications for different devices. Currently, PhoneGap version 4.2.0 supports the creation of applications for these operating systems: iOS, Android, Windows Phone, Bada, WebOS and BlackBerry.

VI. Design and Description of the Application and Server Architecture

A. *Design of the user interface*

Since the application is primarily used for mobile devices with touch screen, it is necessary to adapt the controls for this purpose. The Sencha Touch Framework offers controls and user interface components directly adapted for touch control. The interface layout was selected so that each part of the application has its own tab and these tabs are switched using menu buttons that swipe to the right, or using a button in the upper left corner (Fig. 3a). Fig. 3b) is the design of the visualization. It contains the view of real data and diagrams of the given room or system.

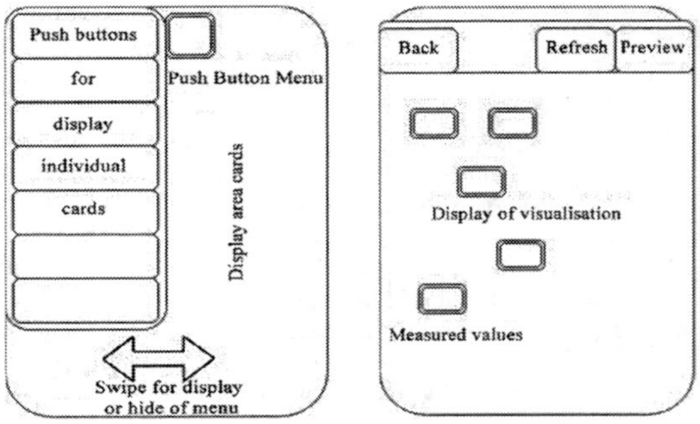

Figure 3. a) Main menu design, b) Visualization design.

Fig. 4 shows the chart design. These designs were discussed with the office building administrator.

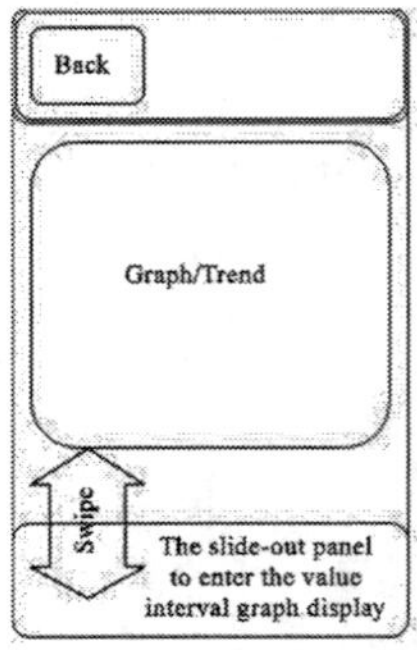

Figure 4. Chart design.

B. *Architecture, application communication, UseCase diagram*

The Sencha Touch architecture is based on the MVC model (Fig. 5). Model-View-Controller has its beginnings in the Smalltalk, where it is originally used to incorporate a traditional process: input, processing, output to the UI programs. It is equally suitable, however, for use in multi-tier Web applications.

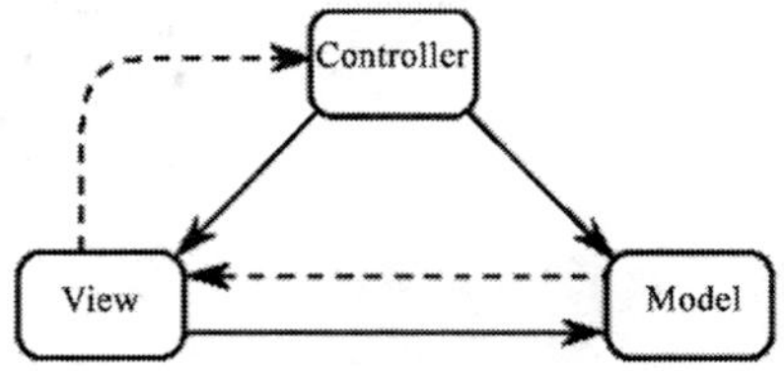

Figure 5. Model-View-Controller Template.

A description of the architecture and communication is shown in the following Fig. 6. The server component and the web client of the application are deployed on the application server Apache Tomcat, which is located on a PC within the VSB domain. The applications on mobile devices must be connected via WiFi to the VSB domain to function; without it they cannot receive any data necessary to create the visualization for the users. There is also the possibility of using the university VPN. A mobile device can also connect to the web client with a web browser after entering the proper URL, but they must also be located in the VSB domain or connected through the university VPN. With connection to the web client the initial startup is slow because the entire application must be

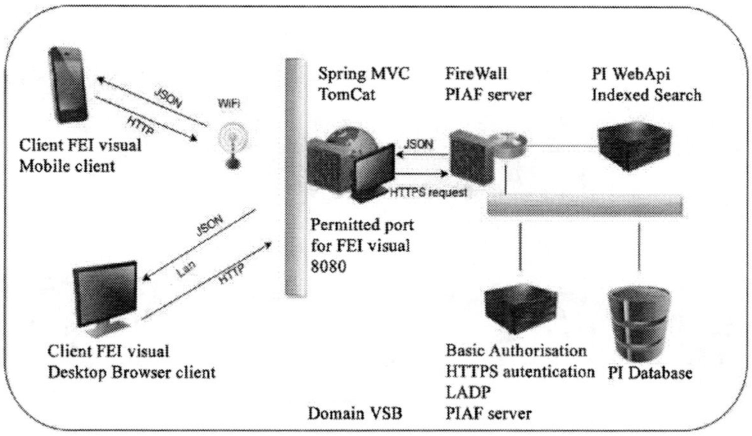

Figure 6. *Architecture and communication FeiVIsual.*

downloaded into the mobile or computer browser, and it is a major problem if the connection to the VPN is established using mobile data, where one usually tries to be economical. After connecting and starting the application, the HTTP GET request with the relevant parameters is sent to Tomcat that verifies the parameters and connects to the PI Web API using a certificate and HTTPS.

As parameters, the PI Web API receives the relevant attribute (value) web id and another parameter, which is the required interval or what historical data shall be returned, for example. -8 for historical data up to -8 hours. Web API processes the request and then sends the data as JSON to the client application. The client application receives the data. The model creates objects based on the received data which are then stored in the repository. From there the client application processes them and calculates the minimum and maximum as well as data for the waveform graph of various temperatures (Fig. 9b). Furthermore, they are processed for the visualization of equipment temperatures such as ventilation, engineering and visualization of the heat pumps.

C. Diagram of major use cases

This UseCase diagram (Fig. 7) shows the major use cases as data acquisition, visualization of the given space and the display of trends of attributes of individual spaces which are:

- VZT 1 auditorium EC3.
- VZT 2 auditorium EC2.
- VZT 3 auditorium EC1.
- VZT 4 depository.
- VZT 5 boardroom.
- Cooling plant.
- Heat pumps.

D. Deployment diagram

The deployment diagram (Fig. 8) shows the deployment of hardware and software resources and their major interprocesses.

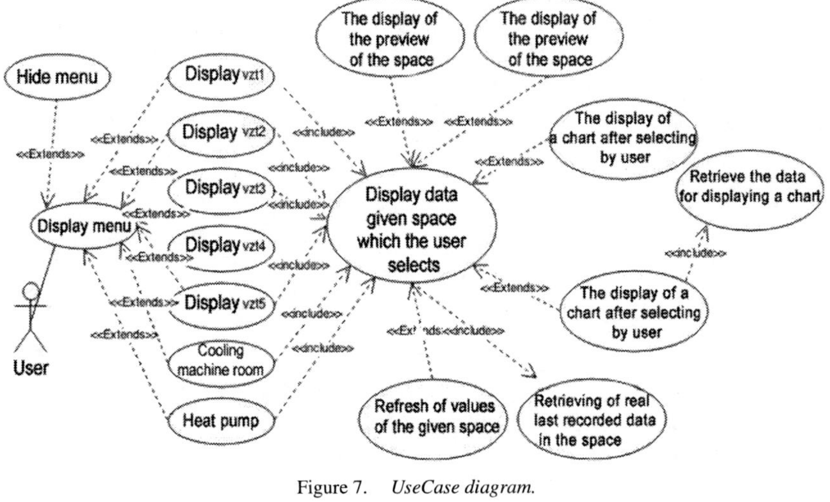

Figure 7. *UseCase diagram.*

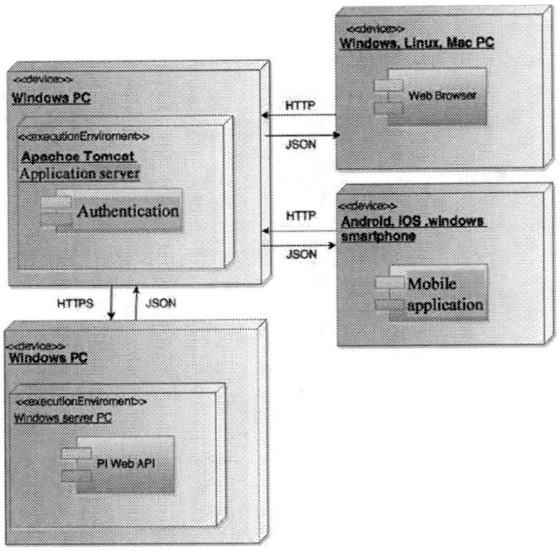

Figure 8. *Deployment diagram.*

Fig. 9 show the examples of applications displaying the cooling plant schematics in the intelligent building (Fig. 9a) and a graph of the measurement of the selected variable (temperature) within the selected interval (Fig. 9b).

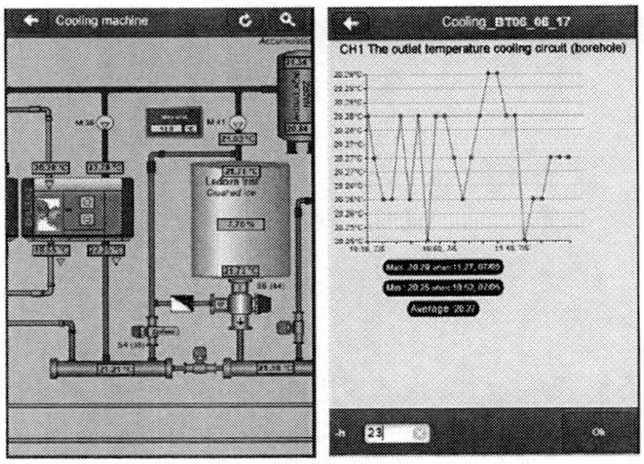

Figure 9. *a) Example of application - cooling plant Deployment diagram, b) Example of application – graph of measured value (Temperature) within the selected interval.*

CONCLUSION

The article describes the various tools and custom design of the user interface for the monitoring of technological processes in an intelligent building. The main feature of the development and implementation of a hybrid mobile application is that it is a platform independent application. The application is programmed using JavaScript Sencha Touch Framework 2. The application is deployed together with the server components on a PC, which is located in the VSB domain. The server part is the application server Apache Tomcat. The server part was implemented in Java using Java Spring MVC Framework. The application has been tested and used for the visualization and monitoring of data measured in the administrative building of the new FEI at the Technical University of Ostrava.

ACKNOWLEDGMENT

This work was supported by the Student Grant System of VSB-TU Ostrava [grant number SP2017/128] and “Promotion of Development of NEK activities in the field of energy saving energy concept” [grant number CZ.01.1.02/0.0/0.0/15_012/0002294].

REFERENCES

[1] A.B. Albu, D. Laurendeau, S. Comtois, D. Ouellet, P. Hebert, A. Zaccarin, M. Parizeau, R. Bergevin, X. Maldague, R. Drouin, S. Drouin, N. Martel-Brisson, F. Jean, H. Torresan, L. Gagnon, F. Laliberte. "MONNET: Monitoring pedestrians with a network of loosely-coupled cameras," in *18th International Conference on Pattern Recognition, Vol 4, Proceedings*, Y. Y. Tang, S. P. Wang, G. Lorette, D. S. Yeung, and H. Yan, Eds. (International Conference on Pattern Recognition, 2006, pp. 924-928.

[2] J. Vanus, M. Cerny, and J. Koziorek, "The proposal of the smart home care solution with KNX components," in *2015 38th International Conference on Telecommunications and Signal Processing, TSP 2015*, 2015.

[3] J. Vanus, R. Martinek, P. Bilik, J. Koziorek, and A. Dracka, "Smart home remote monitoring using PI System Management Tools," in *Proceedings of the 8th International Scientific Symposium on Electrical Power Engineering, ELEKTROENERGETIKA 2015*, 2015, pp. 372-375.

[4] J. Vanus, R. Martinek, P. Bilik, J. Zidek, and I. Skotnicova, "Evaluation of thermal comfort of the internal environment in smart home using objective and subjective factors," in *Proceedings - 2016 17th International Scientific Conference on Electric Power Engineering, EPE 2016*, 2016.

[5] J. Vanus and J. Koziorek, "Employment of Dynamic Time Warping method for monitoring states in Smart Home," in *2014 the 4th International Workshop on Computer Science and Engineering-Winter, WCSE 2014*, 2014.

[6] J. Vanus, R. Martinek, P. Bilik, J. Zidek, P. Dohnalek, and P. Gajdos, "New method for accurate prediction of CO2 in the Smart Home," in *Conference Record - IEEE Instrumentation and Measurement Technology Conference*, 2016, vol. 2016-July.

[7] S. Mahmoud, A. Lotfi, and C. Langensiepen, "Behavioural pattern identification and prediction in intelligent environments," *Applied Soft Computing*, vol. 13, no. 4, pp. 1813-1822, Apr 2013.

[8] OSIsoft, PI-Web-API_1.3.1.99_(2014_R2_Update_1)_User-Guide.pdf [May 15, 2017]

A Power Quality Parameters Setup for Power Quality Management Model as an Integrate Part of Active Demand Side Management

Jakub Kosmak, Stanislav Misak
Department of Electrical Power Engineering,
VSB – Technical University of Ostrava, Faculty of Electrical Engineering and Computer Science
Ostrava, Czech Republic
jakub.kosmak.st@vsb.cz, stanislav.misak@vsb.cz

Abstract*— This paper presents the determination of parameters for the power quality management system using Artificial Intelligence (probably multilayer neural networks with feedback learning algorithm). This predictive model will be used as a complementary tool for controlling and maintaining electrical power quality parameters within Off-Grid systems that are powered by renewable sources. Maintaining the power parameters within the limit in the Off-Grid system is very important and complicated because the stochastic nature of supply from renewable sources, which are used as dominant electric power sources in the system. Differences in short-circuit power can significantly affect the stability of the system and it may have a negative impact on the operation of sensitive appliances. We develop tools and methods to maintain energy quality parameters in Off-Grid systems within the limits by an intelligent approach based on artificial intelligence techniques. This paper presents the results of the most important and narrowest links of the parameters of the power quality to meteorological conditions, from which it can be deduced the prediction of the parameters of the power quality. The output data will be used for the Power Quality Management (PQM) model algorithm which will be implemented into the Active Demand Side Management system (ADSM) of the Off-Grid system.

Keywords—Power Quality Management; Active Demand Side Management; Artificial Intelligence, Neural Network

I. Introduction

Autonomous power systems (so called Off-Grid systems) are mainly supplied by renewable energy sources (RES), such as photovoltaic power plants (PV), wind turbines (WPPs), LPG generators, etc.

New issues related to the specific features of autonomous systems have been addressed for these systems. Particularly, these are: Demand Side Management technologies, which are intended to increase the efficiency of the power flow. Thanks to this tool, it is possible to optimize the operation in an autonomous system in order to cover the energy consumption as efficiently as possible. Problems or requirements that the Off-Grid system must meet:

- the prevailing use of RES,
- operation with a balanced power production and consumption,
- energy storage,
- Protection enabling bi – directional energy flow [1] and
- Active demand side management (ADSM). [2] - [4]

Another requirement is the power quality (PQ). PQ is specified by norms and standards. Some standards are unified for multi-countries, and some standards are intended for power system operators. [5] - **Chyba! Nenašiel sa žiaden zdroj odkazov.**

These standards are qualitative parameters of electrical quantities, including voltage, current, power and other derived quantities. For the correct and reliable operation of the equipment/appliances (also for their longest life), which are connected to the power system, it is very important to adhere to these PQ parameters, which has to be applied in power systems based on different voltage and power levels. [8] - [10] If PQ parameters are not within the allowed limits, the efficiency of electrical appliances may be reduced or even irreversible damage to sensitive appliances may be caused.

The main objective of this paper is to select power quality parameters based on their relationship to meteorological conditions. Furthermore, demonstrate the possibilities of predicting these PQ parameters in autonomous Off-Grid systems that will optimize the operation of the Off-Grid system with compliance with the power quality.

The analysis was performed at the supply voltage and power level, corresponding to the family house for best data relevance. For these analyzes, the power quality parameters were selected:

1) Short-Term Flicker severity - Pst (-)
2) Long-Term Flicker severity - Plt (-)
3) Total harmonic distortion of voltage - THDu (%)
4) Total harmonic distortion of current - THDi (%)
5) Power Frequency - f (Hz)
6) Voltage Variations – dV (V)

The power quality parameters [11] must be hold within the allowable range by providing predicted values for a time interval of at least 24 hours to optimize the balance of power production and consumption. Information about the power quality parameters is very important to ensure the proper operation of electrical equipment in the Off-Grid system.

It is necessary to ensure that information about PQ parameters exceeding the limits passed to the user in the form of warning messages. These reports inform that ADSM needs to make changes of the appliance plan or by modifying resources in the Off-Grid system to keep the quality parameters within limits. Each change in the Off-Grid system leads to a change in the ADSM control system. Because the Off-Grid system's control system is very complex, the best solution is to use Artificial Intelligence (AIT). This method will be able to map the stochastic behavior of consumers in relation to meteorological influences and time periods, including the relations between the productive nature of renewable sources and the weather.

It is also possible to infer relations between qualitative parameters of PQ and meteorological conditions, thus will lead to optimize ADSM. In the context of this paper, the relationship of PQ parameters to meteorological effects is analyzed by means of correlation analysis and data standardization. Henceforward these relations will be used to predict PQ parameters for ADSM. Long-time data collection has created a database of measured values in the Off-Grid system.

II. Physical Platform Description

The Experimental physical platform of the family house operated in Off-Grid mode, was developed in the premises of VSB - Technical University in Ostrava in 2011.

The Off-Grid experimental platform consists of a photovoltaic power plant and a rechargeable battery - used to store energy. PV is divided into 2 strings and the maximum installed power of each string is 2 kWp, i.e. 4 kWp. The first string uses polycrystalline solar cell technology and the panels are permanently installed on the roof of the building. The second PV string uses monocrystalline solar cell technology, and the panels are installed on a two-axis positioning device so called Tracker. The rechargeable battery is Ni-Cd type with an installed power of 750 A·h at 24 volts (DC bus bar).

In the Off-Grid, appliances have a significant impact on the parameters of installed power sources. Appliances in cooperation with power sources have an effect on PQ power parameters.

The appliances used in the Off-Grid experimental platform are the usual household appliances (such as refrigerator, microwave oven, kettle, heating system, air conditioning, etc.) including also office equipment (PCs, notebooks, LCDs, TVs, printers.

In 2012, the Off-Grid system was extended with a detailed monitoring system, ADSM, see detail in Figure 1. The system consists of all the specific requirements of the Off-Grid system:

- Autonomous operation independent of connection to outside grids.
- Well-balanced production and consumption of electric power.
- Energy storage system.
- Predominant use of RES.
- Ability to serve non-traditional loads.
- New type of network protection concept against short-circuit and for bi-directional power flows [1]
- Active Demand Side Management. [2] - [4]

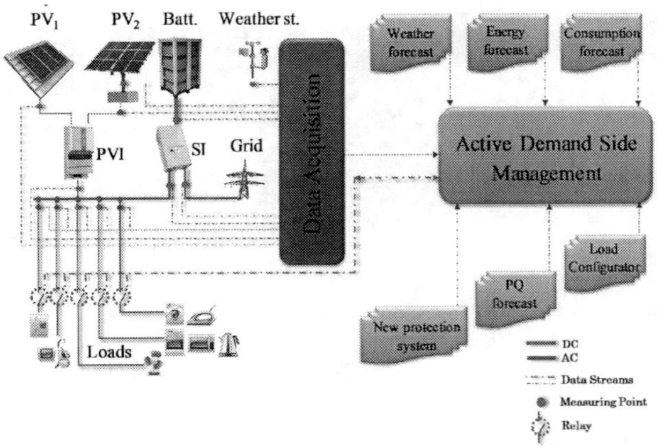

Figure 1. The block scheme of the Off-Grid system operated under ADSM.

Table 1. List of used Off-Grid inverters parameters

	Parameters	*Value*
PV Inverter	Maximal DC input voltage	600 V
	Rated DC input power	4,375 W
	Maximum DC input current / for each MPPT	32 A / 16 A
	Maximum AC output	4,600 W
	Maximum efficiency	97 %
	Output frequency range	45 – 55 Hz
	Total current harmonic distortion	< 3.5 %
Central Inverter	Nominal AC voltage	230 V
	Nominal AC Power	3,300 W
	DC Battery bank voltage	20 – 24 V
	Maximum charging battery current	140 A
	Maximum efficiency	94.5 %
	Continuous AC power output at 25°C for 30 / 5 / 1 min	4.2 / 4.6 / 5.0 kW
	Nominal output power frequency	50 Hz
	Output voltage harmonic distortion	< 3 %

The fundamental element of the physical platform is the central Off-Grid inverter. The Central Off-Grid Inverter is capable of generating bi-directional power flow through battery, inverter and charger, allowing stable operation of connected loads (appliances) and AC power generation. Basic parameters of the Central Inverter including individual PV inverters are summarized in 0

A fossil fuel generator may be used if the RES is out of power.

III. Correlation Analysis of Variables

This chapter is the most important chapter of the article. An analysis of the relationship between the variables is performed here. 95 parameters are stored in the database using a comprehensive monitoring system. Of course, only some of the parameters affect the PQ parameters. The variables used for correlation analysis [12] are determined to verify relationships and subsequently to discover possible hidden relationships [13] between input and output variables for the prediction model. The variables are listed in Table 2 below. The variables were selected in context to the hypothesis that there are relationships between the variable of electrical quantities and the weather conditions, which affect the PQ parameters. The exploration was divided into three areas according to the intensity of the sunshine - and therefore within cloudy, partly cloudy and clear days. These analyzes will be further examined.

Table 2. Input and Output variables of Forecasting model

Parameters	Type of variable	Description	Value
WD	Input	Wind Direction	°
WS	Input	Wind Speed	$m \cdot s^{-1}$
GSR	Input	Global Solar Radiance	$W \cdot m^{-2}$
Temp	Input	Air Temperature	°C
Hum	Input	Relative Humidity	%
AtmP	Input	Atmospheric Pressure	hPa
Pst	Output	Short Term Flicker Severity	-
Plt	Output	Long Term Flicker Severity	-
THDu	Output	Total Harmonic Voltage Distortion	%
THDi	Output	Total Harmonic Current Distortion	%
Freq	Output	Power Frequency	Hz
dV	Output	Voltage Variations	%

The graphical representations below show the correlation linkages between the power quantity and weather parameters during the different weather conditions.

For the comparison of the relations between the PQ parameters and the different weather conditions, as it was mentioned above, three types of weather were selected - cloudy, partly-cloudy and clear, which were divided into 6 days, i.e. 3 groups for two days.

Cloudy days:

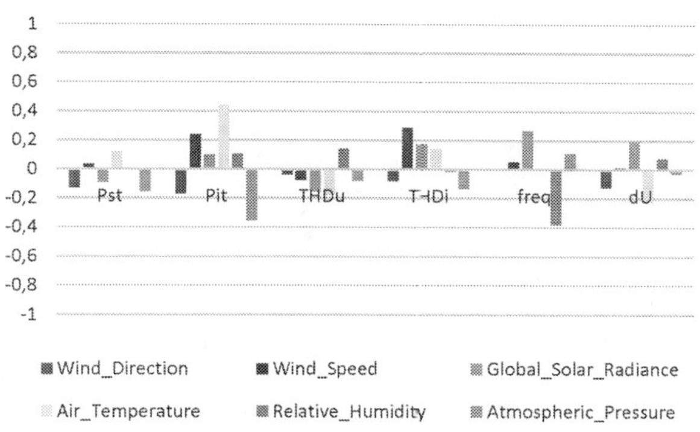

Figure 2. Cloudy weather graphic correlation date 13.4.2017

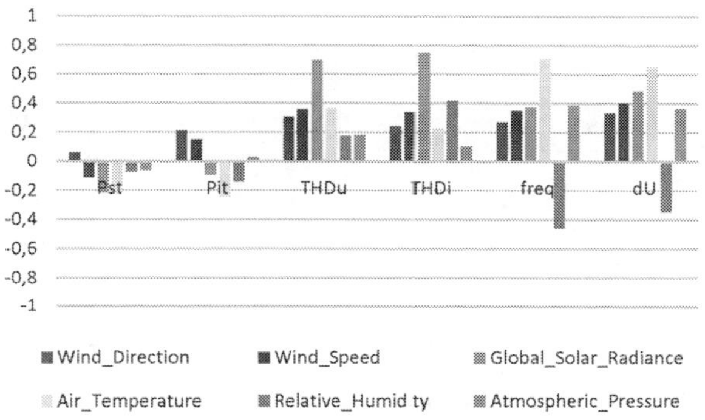

Figure 3. Cloudy weather graphic correlation date 14.5.2017

The first analysis is related to the cloudy days. It is possible to be stated that none of the PQ parameters has a dominant relationship to any weather parameter. As a result of renewable energy sources (in our case, PV), so almost no energy is supplied to the Off-Grid system. The battery bank supplies energy to the Off-Grid system through Central Inverter, which also checks its qualitative parameters. In both cases of cloudy days, we can find a uniform distribution of correlations of PQ parameters and weather variables.

One PQ quality parameter with a high correlation factor to weather variables is the Current Total Harmonic Distortion (THDi). The high correlation coefficient of THDi was found for Global Solar Radiation (GSR). GSR is low during cloudy days and it is therefore a characteristic parameter defining the weather trend and influencing consumer behavior in Off-Grid system.

Other parameters of PQ parameters include the long-term flicker severity (Plt) as well as the Voltage variations (dV). The high correlation coefficient of Plt on 13/4/2014 (see Figure 2.) and also coefficient of dV on 14/5/2014 (see Figure 3.), both were observed with the temperature which is the consequence of the weather trend and it influences the behavior of the consumer.

Partly cloudy days:

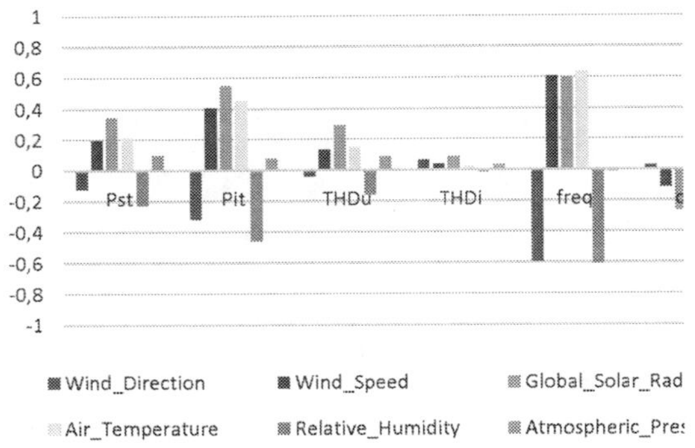

Figure 4. Partly cloudy weather graphic correlation from date 17.4.2014

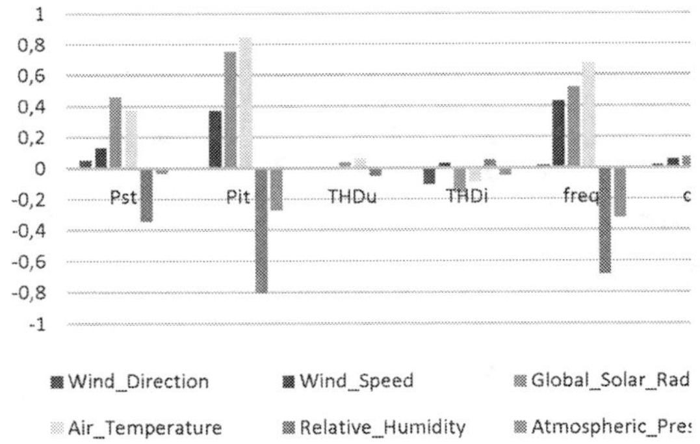

Figure 5. Partly cloudy weather graphic correlation from date 20.4.2014

For the second analysis, we chose the semi-cloudy weather, where the cloudiness was alternating more often with bright sunshine. This analysis demonstrates the dominant relationship between the power frequency and GSR parameters, as well as the long-term flicker and temperature respectively GSR. The graphical representation of the relation between frequency and GSR interprets the author's hypothesis on the relationships between the PQ parameters and the weather variables. The graphical dependences for partly cloudy days are visible in Figure 4. , Figure 5.

Frequency changes its value related to the load changes in the Off-Grid system and it has greater extent than in the electrical network. During the analysis, first the frequency ranged to 50 Hz. On the other hand, the Off-Grid works in the By-pass topology, which means the PV inverter supplies power to the system for the entire power consumption, but there is a time when power generation exceeds consumption. The Off-Grid system works as a lightened, and the frequency value reaches 55Hz. Even though this value is stable, the desired frequency value is not reached at 50 Hz. This process is called FSPC (Frequency Power Shift Control) [23], on which the Central Inverter operates. Furthermore, it is obvious from the analysis, that the analysis of the long-term flicker parameter (Plt) correlates with the temperature, or GSR as expected. As soon as the system changes the supply from the accumulator power bank (cloudy interval of the day) to a By-Pass topology, the value of long-term flicker is eliminated by optimized inverter control. If the accumulators are fully charged and part of the power is supplied by the PV inverter, the value of the long-term flicker rate is reduced to the desired value.

Clear weather days:

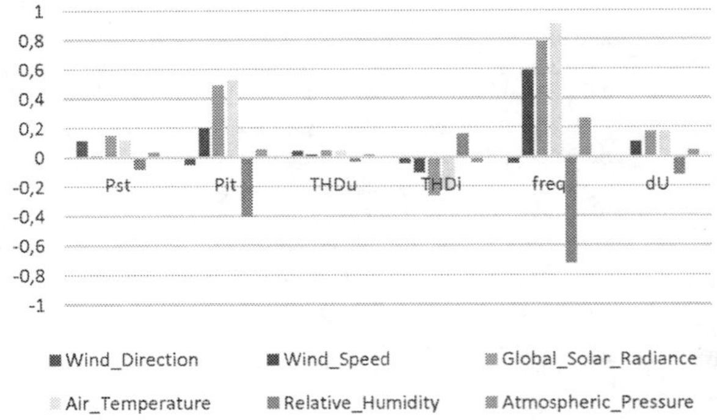

Figure 6. Clear weather graphic correlation from date 4.4.2014

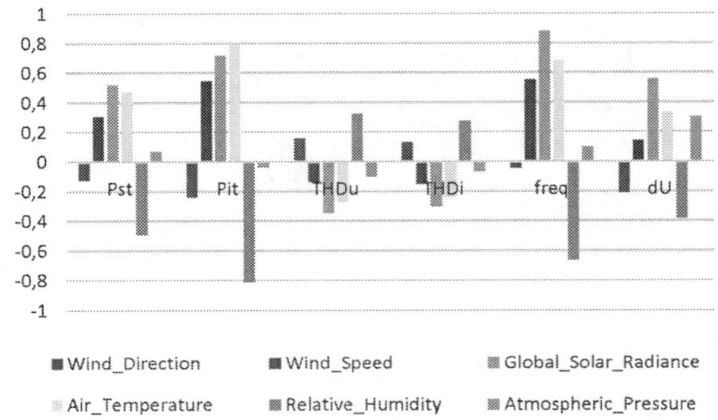

Figure 7. Clear weather graphic correlation from date 6.5.2014

For the third and last analyzed weather area, clear weather was selected when PV power supplies the most power to the Off-Grid system. Repeatedly here, in the case of a frequency parameter, this correlation depends on Off-Grid operating mode. Also, the dominant relationship between the power frequency and the GSR is possible to see from the Figure 6. and Figure 7. As mentioned before, the FSPC process exerts (Frequency Power Shift Control) [23]. For the Plt parameter, the value of the long-term flicker rate is also excluded by optimized inverter regulation control and, in the case of the PV inverter input while the accumulators are charged, the Plt value is reduced by the central inverter to the required value.

IV. Forecasting Model

The proposed prediction model in this paper will consist of two chained neural networks. Each of them processes different requirements and uses different input values. The pattern of the model is in Figure 8.

The first networks in this model will be Deep Believe Network (DBN) [14] based on the limited Restricted Boltzman Machines [15]. DBN met with success in case of time series predictions [16], [17] and also dimensionality reduction [18]. DBN will be designed to resolve the problem into a file of extracted functions. These functions will then be encoded in hidden network layers, giving a certain degree of abstraction to each layer. However, there is no direct approach that has been able to derive the correct number of layers or the number of neurons in the layers. DBN will be used to generate long-term predictions based on input variables whose values will be available in the future. This prediction will have a lower degree of accuracy due to the lower number of applied variables and their lower accuracy. The DBN method will therefore serve only as the next input variable of the second neuron network.

The second neural network is the Echo State Network (ESN), see [19]. It is a type of Recurrent Neural Network (RNN) with the principle of supervised learning and has been successfully used in many studies [20], [21]. ESN applies the actual values of all relevant variables whose accuracy is higher than the accuracy of DBN. This advantage and repeatability will serve to optimize DBN output. The resulting short-term prediction should be of the best possible quality as it will serve as an indicator of the state of the ADSM system that automatically responds to changes and adapt the Off-Grid system. The author's motivation is to create a suitable prediction model. The hypothetical adequacy of the model must be confirmed by testing the hypothesis whether a model based on a combination of selected networks can overcome the results of both independent networks.

Future Experiment will lead to determine if the Self-Organizing Migration Algorithm - SOMA - is appropriate solution for this task. The algorithm is based on social interactions within a group of individuals. Each individual is defined by the coordinate vectors. During initialization of the algorithm, all individuals are randomly assigned to the entire search space (all constants are selected at random). The algorithm runs through iterations, and during each of them, the position of all candidates is evaluated by the performance of the model with individual adjustments. The leader in the population determines for the following iterative development for non-leading elements, iterations are repeated in a given number of cycles, or until meeting the stop criteria (threshold error, etc.). The above is a major description of how the algorithm effectively achieves adaptation near the global extrema of the problem. Details of the algorithm are described in [22].

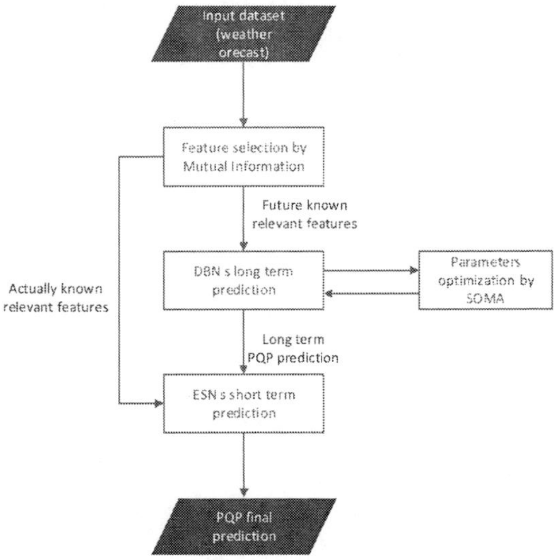

Figure 8. Diagram of the PQ forecasting model

V. CONCLUSION

In Off-Grid systems, the development of a control system is crucial to ensure that power flows are optimized and the power quality parameters are kept within the allowed limits. For this reason, our hypotheses of the relationship between power quality and weather variables have been demonstrated by correlation analysis. The results obtained from the analysis show the dominant relation of GSR parameter and Frequency in the Off-Grid system, and this parameter will be taken as the input variable for the frequency prediction. Furthermore, the analysis shows the relation of the GSR parameter and the long-term flicker severity (Plt). So this parameter was also selected as the input parameter for the long-term flicker severity predictor (Plt). For ability to predict power quality parameters the supplemental tool called Power Quality Management (PQM) containing artificial intelligence techniques will be implemented into the Active Demand Management System (ADSM) during the further researches.

The high-performance operating states of the Off-Grid system will be used for test methods related to the prediction of power quality parameters, which means operating states with the most demanding test conditions. The next step will be implementation of Power Quality Management (PQM) for ADSM. Therefore, interfacing the feedback with the control system for the production and consumption of electric energy with the possibility of eliminating the exceeding the power quality parameters limits.

ACKNOWLEDGMENT

This paper was supported by the following projects: Sustainable development of ENET Centre [LO1404]; Development of the ENET centre research infrastructure [CZ.1.05/2.1.00/19.0389]; Students Grant Competition SP2017/159] and Technology Agency of the Czech Republic. [TH01020426]

REFERENCES

[1] S. Mišák, J. Stuchlý, J. Vramba, T. Vantuch and D. Seidl, "A novel approach to adaptive active relay protection system in single phase AC coupling Off-Grid systems", *Electric Power Systems Research*, 2016, vol. 131, pp. 159-167, DOI: 10.1016/j.epsr.2015.10.015.

[2] Y. Chen, Z. Xu and J. Østergaard. "Islanding Control Architecture in future smart Grid with both demand and wind turbine control". *Electric Power Systems Research*. 2013, vol. 95, p. 214-224. DOI: 10.1016/j.epsr.2012.10.004.

[3] S. Mišák, J. Stuchlý, J. Platoš and P. Kromer, "A heuristic approach to Active Demand Side Management in Off-Grid systems operated in a Smart-Grid environment", *Energy and Buildings*, 2015, vol. 96, pp. 272-284, DOI: 10.1016/j.enbuild.2015.03.033.

[4] A. Urtasun, P. Sanchis, D. Barricarte, L. Marroyo. "Energy management strategy for a battery-diesel stand-alone system with distributed PV generation based on Grid frequency modulation". *Renewable Energy*. 2014, vol. 66, p. 325-336. DOI: 10.1016/j.renene.2013.12.020.

[5] EN 50160. Voltage Characteristics in Public Distribution Systems. European Union, 2011

[6] Series of standards IEC 61000. Electromagnetic compatibility. European Union: International Electrotechnical Commission, 2008. Internet: http://www.iec.ch/emc/basic_emc/basic_61000.htm [May 15, 2017]

[7] IEEE Recommended Practice for Monitoring Electric Power Quality," in *IEEE Std 1159-2009 (Revision of IEEE Std 1159-1995)* , vol., no., pp.c1-81, June 26 2009 doi: 10.1109/IEEESTD.2009.5154067

[8] W.G. Morsi, M.E. El-Hawary. "Power quality evaluation in smart Grids considering modern distortion in electric power

systems". *Electric Power Systems Research.* 2011, vol. 81, issue 5, p. 1117-1123. DOI: 10.1016/j.epsr.2010.12.013.

[9] A. Abrantes. "Overview of power quality aspects in wind generation". *2012 North American Power Symposium (NAPS).* IEEE, 2012, p. 1-6. DOI: 10.1109/NAPS.2012.6336402.

[10] M. Patsalides, A. Stavrou, V. Efthymiou, G.E. Georghiou. "Towards the establishment of maximum PV generation limits due to power quality constraints". *International Journal of Electrical Power.* 2012, vol. 42, issue 1, pp. 285-298. DOI: 10.1016/j.ijepes.2012.03.043.

[11] A. Baggini. *Handbook of power quality.* Chichester, England: John Wiley. ISBN 9780470754238.

[12] D. J. C. MacKay, *Information Theory, Inference & Learning Algorithms*, Cambridge University Press, New York, NY, USA, 2002.

[13] H. Almuallim, T. G. Dietterich, "Learning with many irrelevant features", *In Proceedings of the Ninth National Conference on Artificial Intelligence*, 1991, AAAI Press, pp. 547–552.

[14] G. E. Hinton, S. Osindero, Y.-W. Teh, A fast learning algorithm for deep belief nets, *Neural computation 18* (7), 2006, pp. 1527–1554.

[15] R. Salakhutdinov, A. Mnih, G. Hinton, "Restricted Boltzmann machines for collaborative filtering", in: *Proceedings of the 24th International Conference on Machine Learning,* (ICML '07), ACM, New York, NY, USA, 2007, pp. 791–798.

[16] T. Kuremoto, S. Kimura, K. Kobayashi, M., Obayashi, "Time series forecasting using a deep belief network with restricted boltzmann machines", Neurocomputing 137 2014, pp. 47 – 56, *advanced Intelligent Computing Theories and Methodologies Selected papers from the 2012 Eighth International Conference on Intelligent Computing* (ICIC 2012). doi:10.1016/j.neucom.2013.03.047.

[17] F. Shen, J. Chao, J. Zhao, "Forecasting exchange rate using deep belief networks and conjugate gradient method", Neurocomputing 167, 2015 pp. 243 – 253. doi:/10.1016/j.neucom.2015.04.071.

[18] G. E. Hinton, R. R. Salakhutdinov, "Reducing the dimensionality of data with neural networks", Science 313 (5786), 2006, pp. 504–507.

[19] H. Jaeger, The "echo state" approach to analysing and training recurrent neural networks-with an erratum note, Bonn, Germany: *German National Research Center for Information Technology GMD Technical Report 148*, 2001 34.

[20] F. M. Bianchi, S. Scardapane, A. Uncini, A. Rizzi, A. Sadeghian, "Prediction of telephone calls load using echo state network with exogenous variables", *Neural Networks 71*, 2015, pp. 204 – 213. doi: 10.1016/j.neunet.2015.08.010.

[21] C. Sheng, J. Zhao, Y. Liu, W. Wang, "Prediction for noisy nonlinear time series by echo state network based on dual estimation", *Neurocomputing* 82 (0), 2012, pp. 186 – 195. doi:10.1016/j.neucom.2011.11.021.

[22] I. Zelinka, "Soma—self-organizing migrating algorithm", in: *New optimization techniques in engineering*, Springer, 2004, pp. 167–217.

[23] Ecodirect, FSPC, 2014, Available from (http://www.ecodirect.com/SMA-Sunny-Island-4248U-4200-Watt-48-Volt-p/sma-sunny-island-4248u.htm) [last access 2014-11-18] [Online].

Development of Nuclear Power in Poland and COP 21

Iva Pavlova-Marciniak
Faculty of Electrical Engineering
Czestochowa University of Technology
Czestochowa, Poland
e-mail: iva@el.pcz.czest.pl

Abstract - **The paper will present the results of analysis of the possible development of modern nuclear energy in Poland, which is classified as an ecological source of energy. The use of such sources for electricity production, the demand for which is constantly growing, is one of the ways of meeting the provisions of the Paris Agreement, adopted at COP 21 in 2015. The agreement obliges states around the world to develop a low carbon economy so that the temperature rise on the Earth does not exceed 2^0C and preferably $1{,}5^0$C.**

Keywords: clean energy, nuclear reactor, biological protection, COP (Conference of Parties)

I. Introduction

Due to the rapid growth of atmospheric pollution in the second half of the past century, caused by industrial development in the world, in the mid-1990s industrialized countries joined international organizations and signed a number of agreements to reduce greenhouse gas emissions, including the Kyoto Protocol in December 1997. The protocol, however, began to take effect only after its ratification by 55 countries, including Poland, that is, from 16 February 2005. Nevertheless, it has never been ratified by the major polluters - China and the United States. The date of its validity was original until 2012, but the UN COP 14 (Poznan 2008) to COP 17 (Durban, South Africa, 2011) did not reach an agreement and no new document was signed.

II. Road to the New Climate Agreement

The greatest success achieved at the Poznan summit is an approved document called "Climate Package". The purpose of this document is to reduce CO_2 emissions by 20% by 2020 and to increase electricity from renewable sources to 20% *(3x20 Program)*. The deadline for the sale of all carbon dioxide emission allowances in auctions and trading on the stock markets, effective from 2013, was approved in the package. Such a decision would mean something big for Poland, as more than 80% of electricity produced is from conventional coal-fired power plants. As a result of the negotiations of the Polish government with the European Commision (EC) in December 2008, immediately after the end of COP 14, a positive decision was made by the energy sector to increase the number of permits for CO_2 emissions by 2020, the so-called transition period. Such a solution allow avoid a increase of electricity prices. On the other hand, it has been approved that by 2020 renewable energy in Poland must reach a level of at least 15% [1].

In March 2011, the European Commission adopted a *"Road Map for a Competitive Low Carbon Economy"*, covering a long-term reduction (CO_2) reduction plan of 80% by 2050.

At COP 17 (*Durban, South Africa, 2011*), not only the United States and China, but also developing countries, were against the signing of a new document. After long negotiations at this conference, the deadline for adoption of this important document was *postponed to 2015, and the Kyoto Protocol extended to 2020.*

At the UN COP 18 summit in Doha, Qatar in 2012, involving 17 000 participants (out of 194 countries), representatives of the International Atomic Energy Agency (IAEA) were invited. An expert on nuclear power, Alan Mc Donald presented report of agency *"Climate Change and Nuclear Power 2012"*, in which nuclear power was included in clean sources [2].

In November 2013, the next UN climate summit, COP 19, was attended by representatives from around 200 countries. The proposed reduction of CO_2 to 40% in 2030. Each country was tasked to develop and demonstrate how to meet these requirements at the key COP 21 summit in Paris in 2015, when the final agreement was postponed, following an expiration of the agreed agreement, replacing the expiring Kyoto Protocol. The earlier summit, the special COP 19 in Warsaw and the COP 20 in December 2014 in Lima (Peru), had the task of preparing the necessary documents [3].

On December 12, 2015, at the UN climate conference *COP 21* in Paris, a global climate agreement was signed with representatives from almost 200 countries, with the aim of halting global warming well below 2°C (1,5°C) [4]. The adopted Paris agreement creates a chance for sustainable development in the world and, unlike the Kyoto Protocol, seeks to reduce CO_2 and greenhouse gas emissions in two ways - by reducing emissions with new technologies and by sequestering greenhouse gases through forests. Paris Agreement adopted on 12 December 2015 The Paris Agreement is the first ever international agreement that obliges all countries in the world to work towards climate protection. The agreement was signed in New York by *Prime Minister Beata Szydlo on October 7, 2016*. Since Poland was one of the few EU Member States that ratified the agreement, it was possible to represent our country individually at COP 22.

The UN COP 22 was held in Marrakech, Morocco from 7 to 18 November 2016, the first of which was a meeting of the Paris Agreement Parties and a discussion on the implementation of the agreement was commenced. Poland actively participated in the negotiations because we wanted to spread the postulate to use forests to reduce carbon dioxide and other atmospheric greenhouse gases. At the High Level Segment (*HLS)* meeting on 15-18 November Poland was represented by prof. Jan Szyszko, Minister of the Environment. This was particularly important because the largest pollutants in Europe are in Poland, in Silesia and in Lesser Poland, according to data from the European Environment Agency *(EEA)* – Fig. 1 [5].

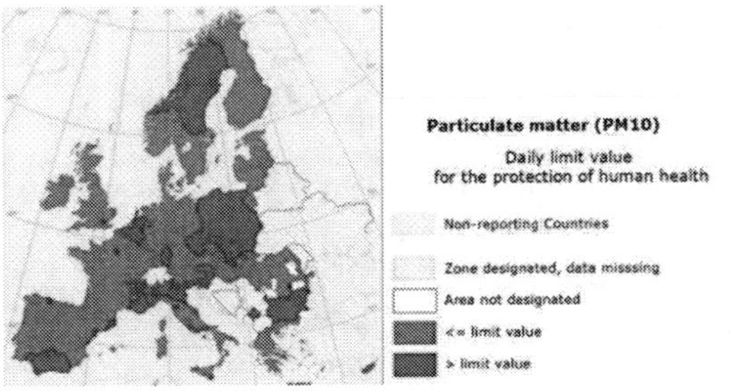

Figure 1. Areas of exceeded permissible daily concentration of suspended particulate matter PM10 (marked in red) [5]

Among the most polluted cities in Europe are as many as 9 in Poland, according to a report by the World Health Organization *(WHO),* as shown in Fig. 2 [6].

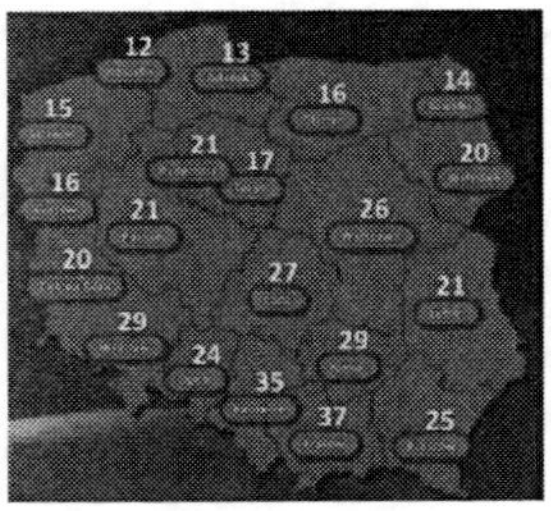

Figure 2. Air quality in Poland, amount of dust PM 2,5 $\mu g/m^3$ [6]

III. NUCLEAR POWER – NEW GENERATION

As already noted, at the UN COP 18 summit in Doha, Qatar in 2012, nuclear expert Alan McDonald presented a report of the International Atomic Energy Agency *(IAEA),* in which nuclear power was classified as pure sources. According to the installed plants in Europe, the leading place in the development of nuclear power is occupied by France, where the generated energy is about 85%. Air pollution is therefore the norm, and the demand for CO_2 emissions permits is negligible. Countries with nuclear power plants of the previous generation after the development of global nuclear reactors III and III+ generations resumed construction of new power.

The next-generation (III +) double-speed, water-pressure reactor, which was developed as a counterpart to the Pressurized Water Reactor *(PWR),* is the European Pressurized Reactor *(EPR)*. The project was originally developed and perfected by the French *company Areva* (*former Framatom*), Electrician de France *(EDF*) and German *Siemens AG*. The *EPR* reactor was used for the construction of the first two new generation power plants that were launched in 2005 - in Finland (*Olkiluoto)* and France (*Flamanville)*. The next two locations are in China and construction began in 2009 and 2010.

New types of safe and cheap reactors have been developed that have allowed to achieve very high plant availability ratios, up to *90%*. Modern nuclear technologies have been developed to minimize personnel radiation exposure and further minimize the transmission of radioactive substances into the environment. Technological solutions for the new generations of III and III + reactors have a very high level of safety, which ensures even in the event of a severe meltdown with the core of the reactor so that its location does not go beyond the power plant. The *"deep defense"* principle has been introduced in the nuclear power industry, according to which power plants should be designed with large security reserves. It is indeed the introduction of systems to prevent dangerous deviations from the nominal operating parameters and, if they occur, to prevent the development of breakdowns and safe shutdown of the power block. All of these principles are also related to the development and operation of a nuclear power plant at all levels and positions.

Ensuring radiation safety is still in the design stage. The fuel material, which holds more than *99.9%* of fission products, is tightly sealed in zirconia fuel jerseys and resistant to high temperatures of the order of one thousand Celsius degrees. The reactor tank, primary circuit and structural components are made of the finest quality steel – Fig. 3 [7].

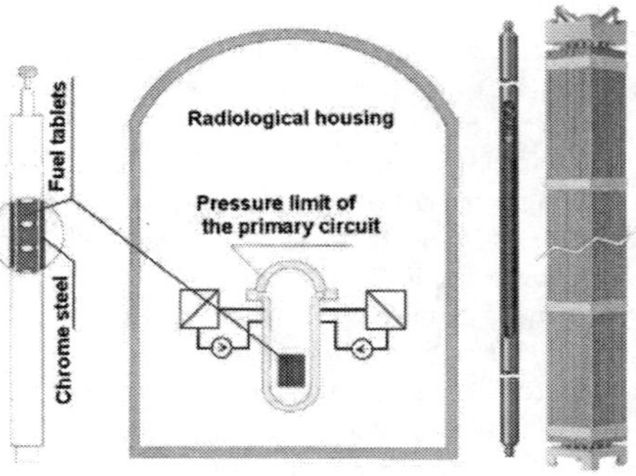

Figure 3. Radiation barrier system in a nuclear power plant [7]

Assembly work is controlled and carried out by the most qualified personnel, as well as technical control, all components are covered for the entire service life. The first circuit device includes a so-called safety enclosure that is resistant to pressure and temperature that can arise during and after a failure. All safety systems are designed so that if a system or component fails, the backup system is immediately switched on. All these systems provide modern nuclear power plants with resistance to damage and human errors.

The design and construction of modern reactors, including the *EPR*, and proper construction and control of power plants make the failure of the reactor core melting during operation is very unlikely, almost negligible. That is why it is very important: to have appropriate staff, to carry out training to maintain a high level of skills, and to regularly carry out control tests. Introduction of information systems to control the technological process with the use of modern solutions containing artificial intelligence additionally influences the high level of security. The design of nuclear power plants is always taken into account to ensure a high degree of seismic safety, even if tectonic movements are not specific to the location area. In nuclear power plants equipped with III and III+ generation reactors, the designed protection provides protection for personnel even in cases of severe melt failure.

Below is an illustration of the technological buildings location on the example of the *EPR r*eactor– Fig. 4 [7].

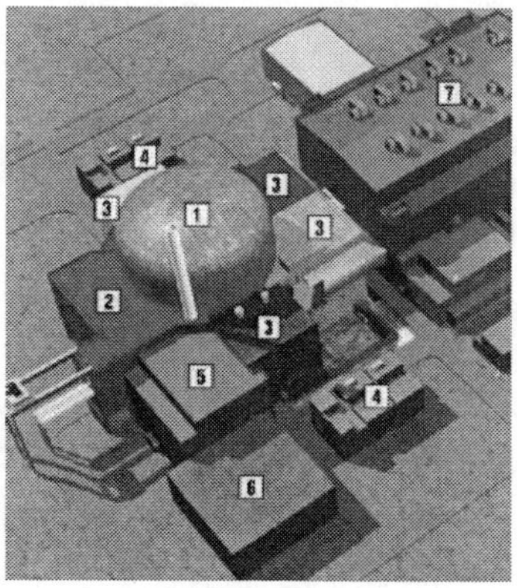

Figure 4. Distribution of technology buildings in EPR reactor: 1 – reactor building, 2 – fuel storage building, 3 – security system buildings, 4 – diesel generators, 5 – auxiliary systems, 6 – waste disposal, 7 – turbine plant [7]

Table 1 lists selected parameters of the power plant with modern, most commonly used new generation power reactors in the world [8].

Table1. Parameters of power units with III+ generation reactors [8]

Reactor type	Net Power	Circulation efficiency	Outlet pressure cooling	Steam pressure	Steam tempe-rature	Operating time
	[MWe]	*[%]*	*[MPa]*	*[MPa]*	*[°C]*	*[year]*
EPR	1600	36-37	15.5	7.7	293	60
AP1000	1117	32-35	15.41	5.76	272.9	60
VVER	1500/ 1600	33-35	16.2	7.8	329.7	50

The newly developed Russian technology of *VVER 1500/1600 (Pressurized Water Reactors)*, developed by *Rosatom*, meets the requirements of the state-of-the-art III+ generation power plant – Fig. 5 [9].

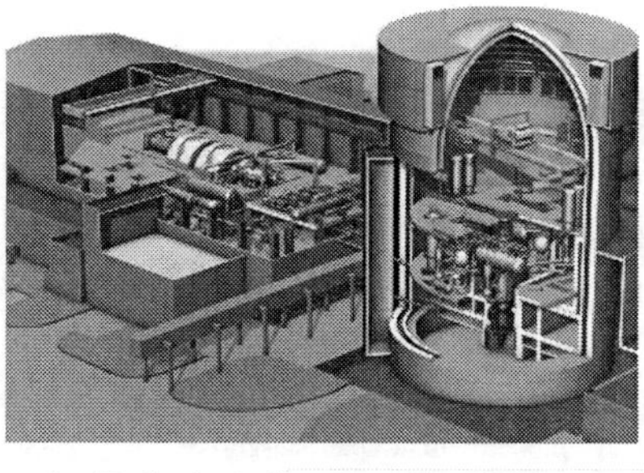

Figure 5. Technological system of the VVER reactor [9]

IV. NUCLEAR ENERGY IN EUROPE

EU Member States possessing or constructing nuclear power plants have an important section of electricity produced in their national energy balance. Finland, which has nuclear power plants with reactors of the previous generation *VVER*, is building new *EPR* reactors with *Areva*. However, the startup of this plant in Olkiluoto is delayed for financial and technical reasons. The concrete of the reactor well did not meet the safety requirements and some technological and safety systems were improved during the implementation.

On February 25, 2010, after the Russian Prime Minister approved the construction of a nuclear power plant in the Kaliningrad Oblast, a cornerstone was laid. The power plant will have two blocks of 1200 MW, or a total power of 2300 MW.

Countries that also have power reactors of the previous generation *VVER-440* and *VVER-1000*, such as Slovakia, have decided to build new nuclear power plants in Mechowiec. Two new blocks equipped with reactors of the next generation are being built. The next one is the construction of the Bohunice power unit.

In Bulgaria, after thirty years of operation, about 33% of the electricity produced comes from the Kozloduy nuclear power plant. At present there are two blocks equipped with Russian VVER-1000 reactors. After the modernization of the reactors and the purchase of TVS A12 (TBC A-12) fuel rods from the Russian TVEL, which can be used to extend the fuel live period from 12 to 18 months, to increase the depth of combustion of nuclear fuel, each reactor can reach a capacity of 1100 MW, or even 1200 MW [10]. Although their exploitation license has been extended by 20 years, it is already planned to build the next two blocks with modern III+ generation reactors – Fig. 6 [11].

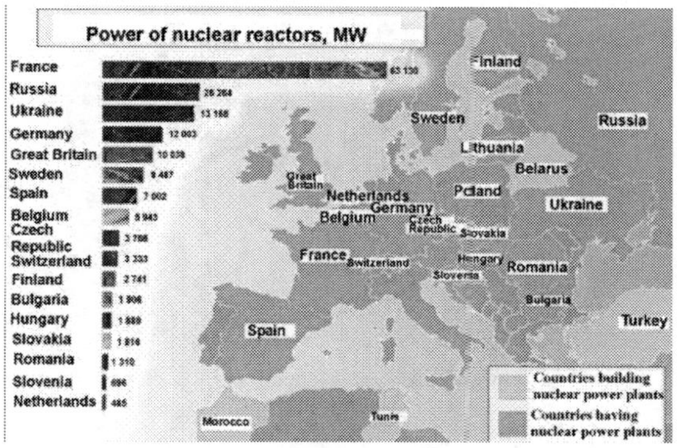

Figure 6. The installed power in nuclear power plants [11]

In September 2016, the European Commission completed its investigation into the Paks nuclear power plant project in Hungary and gave away green light on its implementation with Russia, after Budapest provided all the necessary explanations for the construction of two new reactors. The Hungarian-Russian agreement, signed in January 2014, obliges the Russia to provide a credit of up to 10 billion euros for a 21-year period with an average interest rate of 4.5% on the expansion of the Paks nuclear power plant to 2,4 GW.

In Romania about 18% of electricity demand is covered by nuclear power plants. It was expected that in 2015 the next two new generation reactors will be launched at the Cernavoda power plant, which will have in total four CANDU reactors.

V. NUCLEAR ENERGY IN POLAND

Poland's power industry is based primarily on coal, 80% of its electricity needs come from conventional power plants. Independently that renewable energy sources *(RES)* already occupy an important place in the energy balance, in order to meet the EU energy policy and the Paris Agreement, these solutions may be insufficient. The European Commission has already punished Bulgaria for crossing the permissible pollution standards in March 2017 and has also reminded Poland. The development of nuclear power can solve the problem of increasing electricity demand as well as the significant reduction of CO_2 and greenhouse gases.

The beginnings of the construction of nuclear power plants in Poland were still in the 1980's when the first location in Żarnowiec was to take place. Construction work was quite advanced, but the Chernobyl disaster in 1986 led to the detention of this construction. Since the Atomic Law in force then declared that if construction in a given area was abandoned for at least a year, it must be declared liquidation. The Council of Ministers adopted this Resolution on December 17, 1990, and the date of commencement of the liquidation was designated December 31, 1992.

Following the design and manufacture of next-generation reactors, the nuclear power generation renaissance in the world has followed. A number of talks and consultations were held in Poland, and on January 13, 2009 the Council of Ministers decided to begin work on the new Polish Nuclear Power Program. A number of amendments were made in the area of radiological protection, staff preparation, and a number of normative documents were created. The revised Atomic Law became effective on July 1, 2011, and in September 2014, its last amendments were announced in the Journal of Laws. Up until that point, however, location and technology were not selected – Fig. 7 [12].

Figure 7. Proposed locations of a nuclear power plant in Poland

The cost of building a 3 GW nuclear power plant (2x1500 MW) according to the current applicable prices was estimated at *PNL 50-60 billion*. The government estimates that three blocks of conventional coal can be built for this money. However, the trend is that CO_2 emission fees will increase, which means that they can reach *PLN 40-50 billion* over the course of the year. The final decision on the construction of a nuclear power plant in Poland may be halted in mid-2017, as the amount of energy required by 2030 is estimated at 10 GW. From the point of view of economic efficiency and energy security of the country unambiguous wins conventional blocks on coal, but taking into account the climatic requirements priority is nuclear power.

Final Conclusions

1. The Kyoto Protocol was in force until 2012, but due to a series of unsuccessful efforts at the UN COP 14 COP 20, the signing of the new document did not take place. In December 2015, at the UN climate conference COP 21 in Paris, with representatives from nearly 200 countries, a global climate agreement was reached, with the aim of halting global warming at well below 2°C. Current COP 21 countries would seek in the second half of the 21st century to offset greenhouse gas emissions and their removal, including their absorption through increased afforestation.

2. At the UN COP 18 summit in Doha, Qatar in 2012, nuclear expert Alan Mc Donald presented a report by the International Atomic Energy Agency *(IAEA),* in which nuclear power was classified as pure source.

3. New types of safe and cheap reactors have been developed that have allowed to achieve very high plant availability ratios of up to 90%. Technological solutions for the new generations of III and III+ reactors have a very high level of safety, which ensures even in the event of a core of the reactor meltdown, so that pollution does not go beyond the power plant.

4. In Poland after a series of talks on 13 January 2009, the Council of Ministers decided to begin work on the new Polish Nuclear Power Program. The revised Atomic Law became effective on July 1, 2011, and in September 2014, its last amendments were announced in the Journal of Laws.

5. For Poland from the point of view of economic efficiency and energy security of the country unambiguously wins the conventional blocks on coal, but taking into account the climatic requirements priority is nuclear power.

References

[1] I. Pavlova-Marciniak, "Implementation of climate policy assumptions in Poland", *8th International Scientific Symposium on Electrical Power Engineering*, 16 – 18 September 2015, Stara Lesna, Slovak Republic, pp. 455-458

[2] P. Pavlicek. "Climate Change and Nuclear Power: Interview with Alan McDonald". Internet: https://www.iaea.org/newscenter/multimedia/videos/climate-change-and-nuclear-power-interview-alan-mcdonald, Aug. 24, 2016 [Jun. 1, 20174]

[3] "COP20". Internet: http://www.cop20lima.org/about/cop20 [Jun. 1, 2017]

[4] "Porozumienie klimatyczne w Paryżu". Internet:: http://naukaoklimacie.pl/aktualnosci/porozumienie-klimatyczne-w-paryzu-124 [Jun. 1, 2017]

[5] "Raport EEA: jakość powietrza jednym z głównych problemów Polski". Internet: http://naukawpolsce.pap.pl/aktualnosci/news,404026,raport-eea-jakosc-powietrza-jednym-z-glownych-problemow-polski.html, Mar. 4, 2015, [Jun. 1, 2017]

[6] "Przerażający raport ws. jakości powietrza. Polskie miasta na czele listy". Internet: http://tvn24bis.pl/ze-swiata,75/raport-who-najbardziej-zanieczyszczone-miasta,643576.html, May 13, 2016 [Jun. 1, 2017]

[7] A. Strupczewski, "Porównanie reaktorów generacji III i III+, które mogłyby być oferowane Polsce", *Konferencja Naukowa Przyszłość energetyki jądrowej w Polsce*, Warszawa 2007, pp.95-120.

[8] Internet: http://www.atom.edu.pl/index.php/technologia/typy-reaktorow/pwr/epr.html [Jun. 1, 2017]

[9] "Nuclear Power in Russia". Internet: http://www.world-nuclear.org/information-library/country-profiles/countries-o-s/russia-nuclear-power.aspx Jun. 2017 [Jun. 19, 2017]

[10] "Русия доставя модернизирано ядрено гориво за шести блок на АЕЦ „Козлодуй"". Internet: http://epicenter.bg/article/Rusiya-dostavya-modernizirano-yadreno-gorivo-za-shesti-blok-na-AETs-Kozloduy-/102385/4/53 May 30, 2016 [Jun. 1, 2017]

[11] "Elektrownie jądrowe na świecie i wokół Polski". Internet: http://elektrownia-jadrowa.pl/elektrownie-jadrowe-na-swiecie-i-wokol-Polski-1.html [Jun. 1, 2017]

[12] Internet: http://www.nuclear.pl/ [Jun 1, 2017]

E: Works of PhD Students

Bad Data Detection by the Dynamic EPS State Estimation Methods

Anna Glazunova
Melentiev Energy Systems
Institute SB RAS
Irkutsk, Russia
glazunova@isem.irk.ru

Irina Kolosok
Melentiev Energy Systems
Institute SB RAS
Irkutsk, Russia
kolosok@isem.irk.ru

Evgenii Semshchikov
Melentiev Energy Systems
Institute SB RAS
Irkutsk, Russia
semshikove11@gmail.com

Abstract— **The dynamic state estimation plays an important role in solving the problem of error detection in measurements particularly in the context of low information redundancy and in the case of critical measurements. The paper presents a method for bad data detection which is based on the analysis of retrospective and forecasting information on the state variables.**

***Index Terms*-- bad data detection, dynamic state estimation, electric power system measurements, forecast**

I. Introduction

The operations of dispatching personnel at the on-line and emergency control of the power system states are supported by the hard- and software on the basis of the real-time calculation model of an electric power system (EPS).

The calculation model of the current EPS state is constructed using the state estimation (SE) method. The SE result is the calculation of the EPS steady state (the current state) based on the measurements of state variables and the data on the scheme topology state. The power system state is estimated by one snapshot of telemetry with the help of the static SE algorithms. Availability of several snapshots allows the application of the dynamic SE algorithms [1, 2, 3].

The SE result is correct, if there no gross errors in initial data. Such errors in initial information are the source of distortion of the calculated state obtained by SE, which can lead to wrong decisions at the EPS control and development of serious emergencies. Therefore the detection of gross errors (hereafter BDD – bad data detection) in measurements, suppression of their effect on the estimates of the EPS state variables is a most topical issue in the SE problem solution [4]. In real-time the preference is given to the algorithms of a priori BDD [4, 5 and others], which allow the bad data to be detected and eliminated before the calculation of estimates for state variables.

At present many BDD methods have been devised based on the analysis of:

- balance relations [4] ;
- innovations [7] – [10];
- estimation residues [8], [11];
- retrospective information on one and the same operating parameter [12];
- correlation among measurements [4], [12].

This work deals with the issues of applying the dynamic SE methods to test the validity of measurements in the areas of low information redundancy, which contain critical measurements and critical groups [6].

The second section presents information on the technique for forecasting of state variables with the help of dynamic state estimation (DSE). The third section describes the suggested BDD method. The fourth section presents the results. Conclusion is given at the close of the paper.

II. Forecasting of State Variables by the Dynamic State Estimation

Power system state estimation means the steady state calculation on the basis of measurements.

Online data processed using the state estimation methods are represented by the vector of measurements:

$$\bar{y} = (U_i, P_i, Q_i, P_{ij}, Q_{ij}, \delta_i), \tag{1}$$

where U_i – magnitudes of nodal voltages; P_i, Q_i – injections of active and reactive powers at nodes; P_{ij}, Q_{ij} – power flows in transformers and lines, δ_i – voltage phases. All the methods of EPS state estimation aim to achieve the minimum difference between the calculated estimates of state variables $\hat{y}$ and the measured values $\bar{y}$, provided the Ohm and Kirchhoff laws are observed.

Based on a priori knowledge of the system behavior, the state space representation for a discrete time-variant system is described by the following equations

$$x_{k+1} = F_k x_k + \xi_{Fk}, \quad (2)$$

$$\bar{y}_k = y(x_k) + \xi_{y_k} \quad (3)$$

where x_k – state vector at time instant k, F_k – transition matrix from state k to state $k+1$, $\bar{y}_k$ – measurements, ξ_{Fk} – model noise, ξ_{y_k} – measurement error, k – snapshot number.

The objective function in the dynamic state estimation looks as follows:

$$J(x) = (\bar{y} - y(x))^{T} R^{-1} (\bar{y} - y(x)) + (\tilde{x} - x)^{T} M^{-1} (\tilde{x} - x), \quad (4)$$

where $\tilde{x}$ – forecast of the measured state vector components, R – covariance matrix of the measurement error, M – covariance matrix of forecasting errors which is calculated during formation of the Kalman filter on the basis of equations (2), (3) and looks as follows:

$$M_{k+1} = F_k P_k F_k^{T} + W_k, \quad (5)$$

where W_k – covariance matrix of dynamics model noise, P_k – covariance matrix of the state vector component estimates.

In the objective function (4) the measured and forecasting values are represented by different terms.

The addend considers the process history. Criterion (4) is minimized by the state vector to determine the state vector component estimates and the dynamic SE problem is reduced to solving a system of nonlinear equations:

$$H_k^T R^{-1} (\bar{y}_k - y(x)) + H_k^T M_{k+1}^{-1} (\tilde{x}_k - x) = 0, \quad (6)$$

where H – Jacobian matrix.

Due to the nonlinear dependence $y(x)$ the problem is solved iteratively by Newton's method. The system is linearized at each iteration and a system of linear equations is solved

$$\Delta x_k^i = P_k^i (H_k^{T(i)} R^{-1} \Delta y + M_k^{-1} \Delta \tilde{x}), \quad (7)$$

where i – iteration number.

Forecasting of state vector components is considered as the dynamic state estimation, if there are no new measurements. The measurement covariances R in equation (7) are assumed to be equal to infinity. This assumption is treated as absence of measurements. Forecasts are calculated by the last estimated state vector on the basis of model (2). The accuracy of the state vector estimates is determined by the covariance matrix P_k .

The forecasts are calculated by the formula:

$$\tilde{x}_{k+1} = F_k \hat{x}_k + \Delta \tilde{x}_k = F_k \hat{x}_k + P_k M_{k+1}^{-1} (\tilde{x}_k - \bar{x}_k), \quad (8)$$

where $\hat{x}_k$ – state vector estimate.

The elements of matrix F are calculated in advance by the retrospective information. In [11] the transition matrix is calculated by using the exponential Holt smoothing. In this work matrix F is generated by searching, among the state vector measurements obtained for a certain time interval, for the distinct points, i.e. the snapshot numbers, with the change in state. For these snapshots the diagonal element of the transition matrix will differ from unity.

The forecasts of the unmeasurable state vector components are calculated by the formula:

$$\tilde{x}_{k+1} = \hat{x}_k. \quad (9)$$

III. DESCRIPTION OF THE DEVISED METHOD

The suggested method of bad data detection in measurements is based on the analysis of retrospective and forecasting information on state variables.

The measurements or estimates obtained at the previous snapshot are applied as retrospective information. The forecast value is calculated using DSE.

The retrospective information is tested in accordance with the conditions:

$$\left| \bar{y}_{i(k)} - \bar{y}_{i(k-1)} \right| \neq 0 \quad (10)$$

$$\left| \bar{y}_{i(k)} - \bar{y}_{i(k-1)} \right| < d_i \quad (11)$$

$$\left| \bar{y}_{i(k)} - \hat{y}_{i(k-1)} \right| < \hat{d}_i \quad (12)$$

Conditions (11), (12) are interchangeable. With the use of DSE preference is given to (12). In the case of change in the state the results of analysis of inequalities (11), (12) will be interpreted as an error in the measurement. This disadvantage is eliminated by forming an additional condition consisting of innovation that takes into account forecasting information and gives a correct answer in the case of change in the state:

$$\left|\bar{y}_{i(k)} - \tilde{y}_{i(k)}\right| < \tilde{d}_i . \tag{13}$$

In (10) – (13) $\bar{y}_{i(k)}$ – i-th measurement at snapshot k; $\tilde{y}_{i(k)}$ – forecast of the i-th measurement at snapshot k; $\bar{y}_{i(k-1)}$ – i-th measurement at snapshot k-1; $\hat{y}_{i(k-1)}$ – estimate of the i-th measurement at snapshot k–1; d_i, $\tilde{d}_i$, $\hat{d}_i$ – thresholds calculated by the formulas:

$$d_i = \gamma \cdot \sqrt{\sigma_i^2} \qquad \tilde{d}_i = \gamma \cdot \sqrt{\sigma_i^2 + \sigma_{N_i}{}^2},$$

$$\hat{d}_i = \gamma . \sqrt{\sigma_i^2 + Y_i},$$

where γ – quantile of distribution N(0,1), is determined by the specified probability of error of the first kind α (at $\gamma = 3$ the measurement with the probability 0.997 is accepted as erroneous at the threshold violation); σ_N^2 - diagonal element of matrix $\tilde{N}$, Y_i – diagonal element of matrix Y, where $Y = HPH^T$, $\tilde{N} = HMH^T$.

Conditions (10), (11), (12) and (13) can be processed in parallel. As a result of the analysis the three-digit error code – XYZ is formed, where the first element (X) indicates satisfaction – 1 or dissatisfaction – 0 of inequality (10), Y and Z are (11 or 12) and (13), respectively. The errors of the first and second kinds are possible by virtue of the erroneous results when using the BDD methods [13]:

1) The errors of the first kind appear in the situations, when as a result of using the method the true measurement was marked as erroneous and excluded from calculations. Such errors are not disastrous from the standpoint of the quality of SE results, but deteriorate observability or even lead to its loss.

2) The errors of the second kind occur as a result of incorrect work of the BDD methods, when the distorted measurements were not detected. Hence, the calculated state turns out to be far from the real one, and the higher the error value is, the worse is the result obtained.

The paper describes the studies performed to minimize the errors of the first and second kinds.

IV. Practical Application

The results of using the method were demonstrated on the example of searching for an error in the critical measurement U_1. (fig.1).

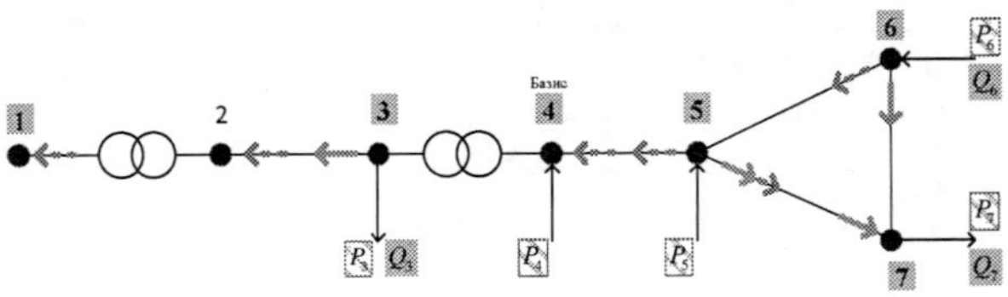

Figure 1. A test scheme (solid fill– reactiv parameters, liniar fill– activ parameters)

The archive with 145 snapshots was modeled in the application State+, where each snapshot was created on the basis of load curves at the consumption nodes and the random number generator simulating errors in measurements. The studied measurement U_1 was marked as erroneous at 9 snapshots comprising 6.2%. Table 1 presents the result obtained for measurement U_1 .

TABLE I. THE RESULT OF BDD METHOD

Numbers of snapshots	Type	Error code
10, 31, 53, 79, 107, 112, 113, 131, 132	Doubtful	101

Equation (12) using the retrospective information on the estimates instead of measurements was studied to reduce the errors of the first kind. In this case measurement U_1 proves to be doubtful at 2 snapshots (10, 79), which makes up 1.38%. The method ability not to miss errors is tested by modeling the gross errors of different multiplicity n :

$$\bar{y} = y_{ucm} + \xi_y + b_y , \tag{14}$$

where $b_y = n \cdot \sigma_y$.

The initial value of error multiplicity is accepted to be 4. The error is modeled at each 15th snapshot. In addition, if the measurement is marked as erroneous, its value is replaced by the forecast.

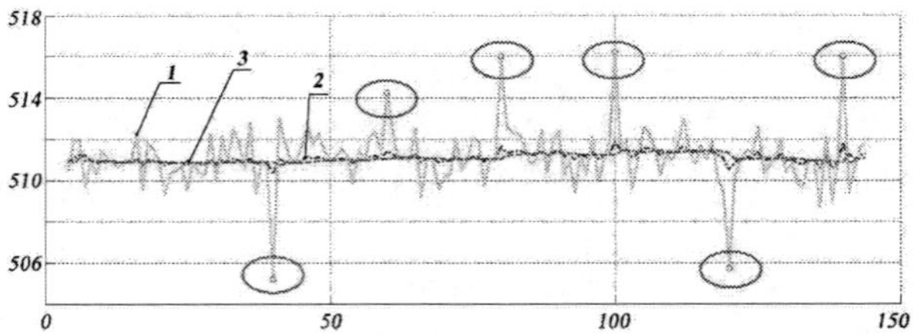

Figure 2. Measurements (1), estimates (2) and forecast (3) before replacing by the forescast

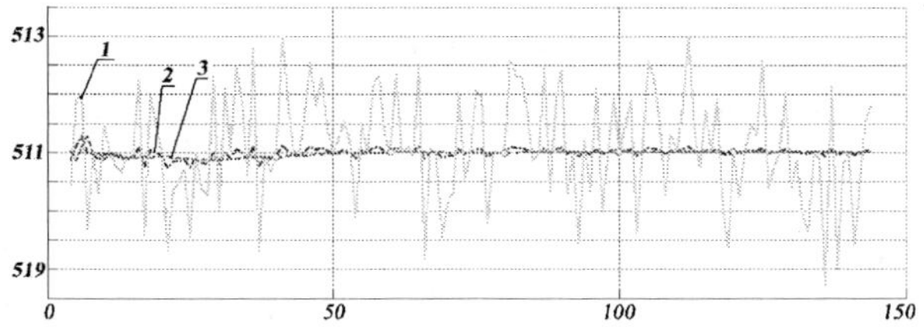

Figure 3. Measurements (1), estimates (2) and forecast (3) after replacing by the forescast

Figures 2 and 3 present the cases before and after replacing the erroneous measurement of voltage in the first node by the value of corresponding forecast. Errors were modeled every 20 snapshots using equation (14) with $n = 7$.

Figure 5 shows the results of using the error detection method. Multiplicity n is presented horizontally, the percentage of error detection – vertically.

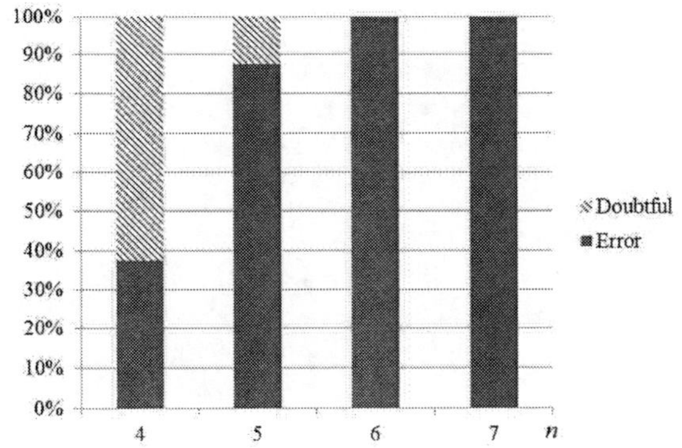

Figure 4. The result of using the method at different error multiplicity

The following example is illustrated on the basis of 3 node network with the real data obtained from SCADA. The case study (Figure 5) presents the situation when inequality (10) is violated. In this case the first variable in the error code equals to one, meaning that the same measurement has been sent twice at time t and $t+1$. From this it can concluded that the measuring device is not working properly or some unexpected error occurred during data transfer.

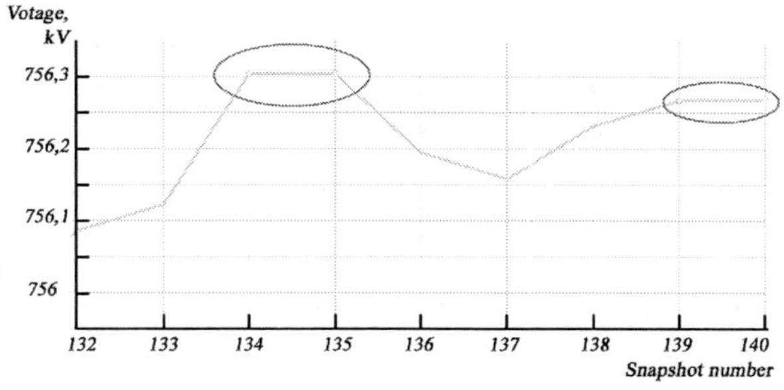

Figure 5. Erroneous measurements

Similar studies were carried out for the scheme consisting of 33 nodes.

V. Conclusion

The feasibility of the methods was tested in the simulation experiment on the schemes with 7and 33 nodes in different steady states.

The application of the suggested method allowed the decrease of the number of errors of the first kind. It becomes possible owing to the use of DSE to filter practically completely the errors in measurements of the state vector components.

The devised method makes it possible to detect errors of the first kind in the critical measurements, which is particularly important for observability provision.

The suggested method allows the gross errors to be detected in all measurements, including the critical ones. By the example of the schemes with 7 and 33nodes it was established that the method provides a correct solution in 100% of cases at the errors with the multiplicity above 6. With the multiplicity lower than 6 the errors are detected in 70% of cases.

The processes of the analysis of inequalities (10), (12), (13) are independent, therefore they can be organized in parallel, which ensures a higher speed of the method.

References

[1] A. Z. Gamm, Statistical methods for state estimation of electric power systems, M.: Nauka, 1976, p. 220 (in Russian)

[2] A. M. Glazunova, "Dynamic state estimation," in *the book* "Monitoring, control and protection of interconnected power systems". Springer-Verlag, Berlin, Hiedelberg, 2014, part 3.7. pp. 107-123

[3] M. B. Do Coutto Filho, J. C. Stacchini de Souza, R. S. Freund, "Forecasting-aided state estimation – Part 1: Implementation", *IEEE Trans. Power Syst.*, vol.24, P.1667, 2009.

[4] A. Z. Gamm, Yu. A. Grishin, A. M. Glazunova, I. N. Kolosok, E. S. Korkina, "New EPS state estimation algorithms based on the

technique of test equations and PMU measurements," in *Proc. 2007 International Conference PowerTech, Lausanne*, CDROM, # 256.

[5] K. Wilkosz, "Verification of the measurements of voltage magnitudes in electric power system," *Second Int. Symp. On Security Power System Operation*, Wroclaw, 1981, Paper E8, pp. 147-155.

[6] K. A. Clements, G.R. Krumpholz, P.W. Davis, "Power System State Estimation with Measurement Deficiency: An Observability/Measurement Placement Algorithm," *IEEE Trans. on Power Systems,* vol. PAS - 102, pp. 2012-2020, July 1983.

[7] M. B. Do Coutto Filho, J. C. Stacchini de Souza, R. S. Freund, "Forecasting-aided state estimation – Part 2: Implementation," *IEEE Trans. Power Syst.*, vol.24, pp. 1667-1677, 2009.

[8] Yun Yang, Wei Hu, Yong Min, "Projected unscented Kalman filter for dynamic state estimation and bad data detection in power system," in *Proc. 2014 12th IET International Conference on "Developments in Power System Protection",* pp. 1–6.

[9] M. Pignati, L. Zanni, S. Sarri, R. Cherkaoui, L–Y, Le Boudec, M. Paolone, "A pre-estimation filtering process of bad data for linear power systems state estimation using PMUs," in *Proc. 2014 18 th Power Systems Computation Conference, Wroclaw, Poland,* #94.

[10] H. Karimipour, V. Dinavahi, "Extended Kalman filer–based parallel dynamic state estimation." *IEEE transactions on smart grid*, vol. 6 NO.3, pp. 1539–1549, May 2015.

[11] M. Gol, A. Abur, "A modified chi–squares test for improved bad data detection," *in Proc. of International Conference "PowerTech'2015", Eindhoven, Holland,* June 2015, USB #448510

[12] V.G. Neuymin, A.S. Alexandrov, D.M. Maksimenko, "Verification of the model of scheme binding and identification of gross errors in telemetry data in the software "RastrWin3"," *in Proc. of the VI International Scientidic and Technical Conference "Power Energy through the Eyes of Youth", 9 – 13 November 2015.* pp. 526– 529.

[13] H. Cramer, Mathematical methods of statistics, *Moskow: Mir, 1975, p 575.*

New Database and Theoretical Model for Power Distribution Networks

Judith Pálfi
Óbuda University
Research Group of Applied Disciplines
and Technologies in Energetics
Budapest, Hungary
judith.palfi@kvk.uni-obuda.hu

Peter Holcsik
Óbuda University
Research Group of Applied Disciplines
and Technologies in Energetics
Budapest, Hungary
peter.holcsik@elmu.hu

Abstract — **ELMŰ-ÉMÁSZ network supplies electricity to 5 million customers in Hungary. The electrical network contains 2.4 million connection points represented by the power company in a GIS (Geographic Information System) based map. By cleaning and processing the data exported from the GIS map and using modern information technologies including Big data methods a new network model was created.**

The present paper presents the process by which this new database and network model was built. The significance of this model is based on the description and analysis of the essential characteristics of the power distribution networks.

The new model describes the complete electrical network including high-voltage substations, transmission networks, medium and low voltage distribution networks and the connection points of the consumers.

Keywords: electricity network model, GIS

I. Introduction

The structure of the electricity supply system is divided into three well defined parts: the high voltage (HV) power transmission network, the medium voltage (MV) distribution network and the low voltage (LV) distribution network. This structure is illustrated in the Fig. 1:

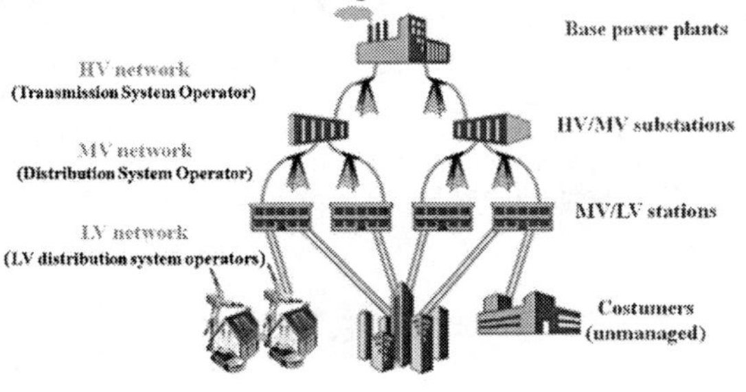

Figure 1. Structure of the electricity supply system

The Transmission System Operator (TSO) is on top of the structure. The MV distribution system operators are located below the TSO. At the bottom of the structure the operational control tasks are carried out by the LV distribution system operators. [1][2]

The application of advanced scientific methods and technologies is mandatory for ensuring an efficient functioning of the operation control system [3]. An example for such an applied method is the development of an automatic algorithm which can significantly decrease the time needed for identifying the faulty equipment.

The authors (founding members of the Research Group of Applied Disciplines and Technologies in Energetics - AD&TE at Óbuda University, Budapest) [4] describe the process of creation of a new network theory model built-up on a new data base. This process represents an important step in the development of the algorithm.

The paper presents the problems and the proposed solutions during the process of developing the database and the network theory model.

A. *Definition of the FLDa*

The location of the faulty equipment in the LV distribution network is currently identified manually, therefore, the time required for detecting the network fault location is considerable. The research group proposes a solution for the automatization of the searching process using an algorithm developed specifically for this purpose (Fault Location Determination algorithm – FLDa). The FLDa uses the data of sporadically installed smart sensors. As a result of the fast detection, the time of the fault localization on the LV distribution network can be decreased significantly. The algorithm sends to the operators of the LV network one line messages containing a unique identification number and the position of the defective device. The development and testing of the FLDa requires the creation of a new network database (NDB) and of a new transmission and distribution network model (TDNm).

B. *The content and usage of the NDB*

The NDB contains all the power network elements that are required and sufficient for the subsequent research tasks. These elements include HV cables and overhead lines, HV/MV substations, MV cables and overhead lines, live-line connectors, overhead switches, MV/MV stations, MV/LV stations, LV cables and overhead lines, connection cables, the connections between these elements and the consumers on the connection cables.

The NDB is used not only in the development of the FLDa but also in other research projects of the AD&TE group. These projects include among others the modelling of voltage problems caused by decentralized power production, investigation of the efficiency of implementing smart switchboards (SSB) in LV networks or the METEO project [4].

C. *The structure of the NDB*

The NDB is structured in data tables depending on the tasks to be solved. The following data tables were created:

1. nodal and connections data table of HV network
2. interconnection data table of HV and MV network (substations)
3. nodal and connections data table of MV network
4. interconnection data table of MV and LV network (stations)
5. nodal and connections data table of LV network
6. interconnection data table of LV network and consumers (connection cables).

D. Export and unification of the data

ELMŰ and ÉMÁSZ supported this research by allowing the research group to use the background database of their GIS based map (EÉGIS). "The EÉGIS system is in charge of the complete technical network registration of ELMŰ and ÉMÁSZ operated networks. EÉGIS includes the registration of the high, medium and low voltage electrical networks managed by ELMŰ and ÉMÁSZ, throughout the full service area. EÉGIS is a technical information system supported by Oracle 10/11g Spatial technology which is able to cooperate with the other IT systems of the company (SAP, work management, SCADA/DMS etc.)." [5]

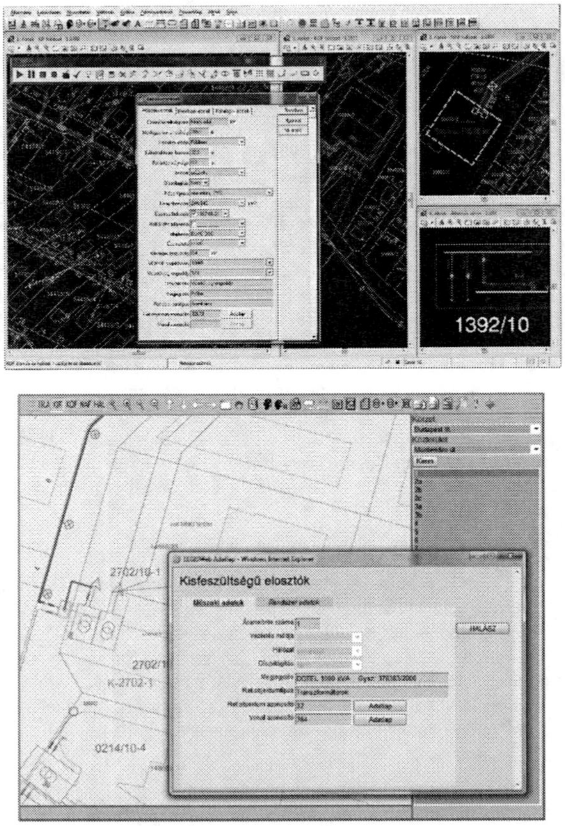

Figure 2. Screenshot from the EÉGIS system

Fig. 2 illustrates as an example a screenshot from the EÉGIS system about the LV switchboards.

The Network Documentary Department (NDD) of ELMŰ and ÉMÁSZ supplied the raw data for the research. The data was exported from the EÉGIS system by equipment type. The data tables were extracted in text files format (.txt) which could be handled both by the development software and by the database manager and analyser software. The data tables belonging to each other have been provided with linking fields.

During the processing and transformation of the data, the original identifier of each network element was saved in a masked form in order to ensure that the TDNm can be used in future for other research purposes, too.

II. THE NODAL AND CONNECTION DATA TABLE OF THE HV AND THE MV NETWORK

The HV and MV network was recreated by using the connection data table exported from the EÉGIS system. The system contained up to date information related to the nodes and connections of the networks.

The HV network is a looped network which „consists of multiple closed circles and its nodes are supplied from two or more sides." [6] The structure of the HV network is illustrated in Fig. 3.

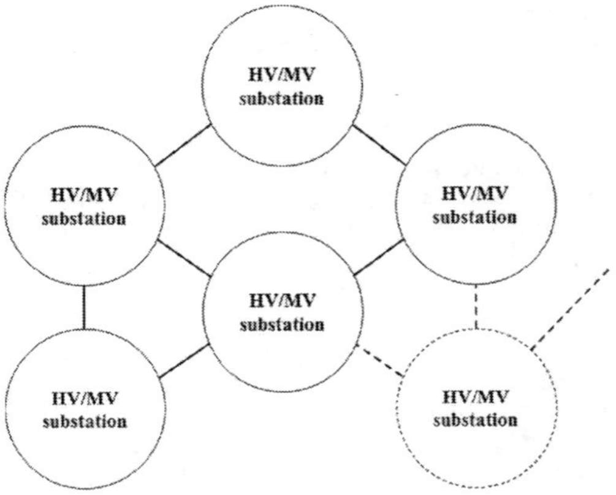

Figure 3. The looped structure of the HV network

In Fig. 3 the circles illustrate the HV/MV substations. The lines linking the circles represent the HV cables and the transmission systems which connect these substations.

Similarly to the HV network, the structure of the MV network is also looped, however the MV network is radially operated. This implies that the topology of the MV network is a simple tree structure. On a radial network "there is only one line connection between the power supply and the consumer point, the electricity supply can be done only from one side" [6]. The topology of the MV network is illustrated on Fig. 4:

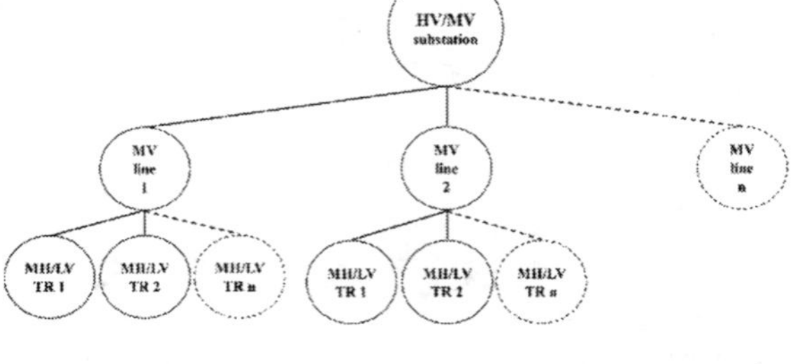

Figure 4. The topology of the MV network according to the TDNm

The circles in Fig. 4 represent the HV/MV substations and the MV/LV stations, while the lines between the circles illustrate the HV and MV cables and transmission systems.

By using the nodal and connection data table the one-level looped model of the HV network (Fig.3) and the two-level radial model of the MV network (Fig.4) could be created.

III. NODAL AND CONNECTION DATA TABLE OF THE LV NETWORK

The LV network, similar to the MV network, is operated radially and has a tree structure. However, its topology is variable in width and depth. Because of this, it was not possible to generate directly its nodal and connection data table using the tables extracted from the original data (see section II.).

In order to map the topology of the LV network, the network data was completed with GPS data. By this, the graphical representation of the individual network nodes appeared in one point. Fig. 5 shows an example for such an interconnection.

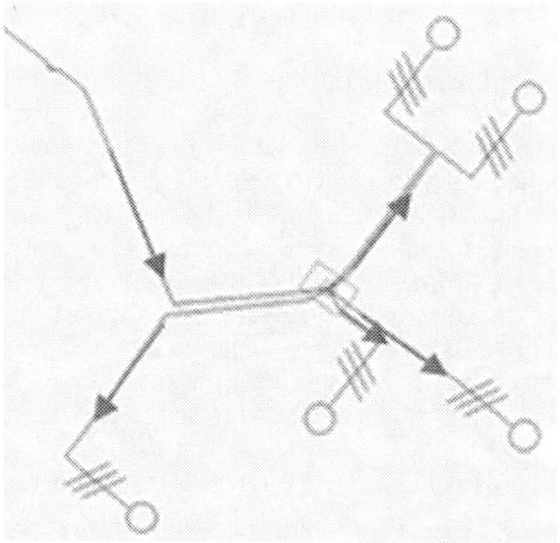

Figure 5. LV sample network showing the overlap of coordinates (screenshot from the EÉGIS system)

The green rectangle on Fig. 5 represents a switchboard connected to one LV input cable (blue arrow) and four output cables (green arrows). The network model built on the sample network presented in Fig. 5 can be seen in Fig. 6:

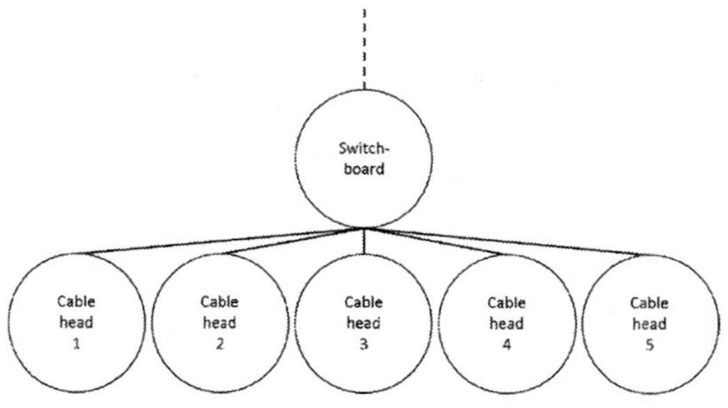

Figure 6. Sample network built on the network model

The nodal and connection tables of the LV network were built-up from this new representation of the LV network by Depth-first analysis [7][8] and by using the C++ programming language.

A. *Determination of the coordinates of swichboards*

In the exported data the location of the switchboards was defined by the coordinates of its two opposite corners. At the same time, the elements connected to the switchboards were linked to the centre of the graphical element. From the point of view of the network theory model the relevant points were the coordinates of the connection points, thus the centre coordinates of each switchboard had to be determined. This problem is illustrated in Fig. 7.:

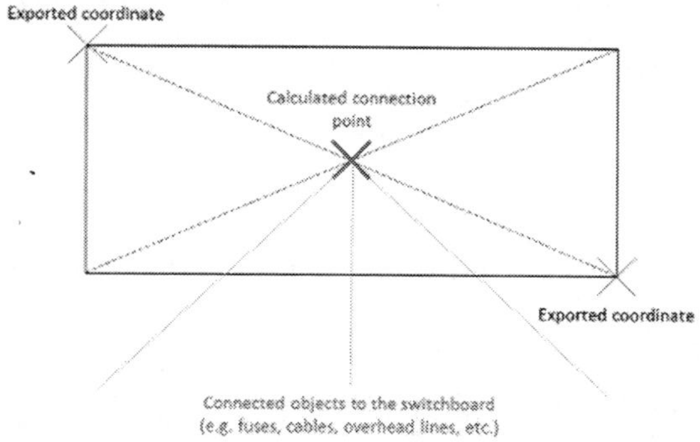

Figure 7. Calculated coordinates of the switchboard connection point

The centre of the switchboard shown on Fig. 7. is given by the intersection point of the diagonals. This intersection point was defined in the NDB as connection point in which fuses, cables, overhead lines, etc. connect.

B. *Analysis and validation*

The analysis and checking of the created data base was made by graphical visualisation. The newly generated map was compared to the original EÉGIS map. This analysis was carried out already during the creation of the data base and revealed that besides the originally defined element types, the implementation of further elements like fuses and switchboards was needed.

This process revealed also that within the LV network there are in certain cases discrepancies between the coordinates. In 5 % of the cases the end point of an element was not identical with the starting point of the next element. This was problematic as the Depth-frst analysis used for creating the LV network model was based on the assumption that the coordinates match. As a solution to the problem the authors introduced the notion of connection-space.

C. *The connection-space*

The connection-space was defined as a space in which by meeting certain conditions, two elements of the model get connected, even if originally there was no network element or connection between the two. Due to the fact that all circuits are ended by a consumer (see Fig. 1. household, office, factory), each point of the model with no consumer at the end had to be examined.

The data exported from the EÉGIS system contained the GPS coordinates and the network data of the elements. By this the exact definition of the connection-space around the blank ends became possible. This is shown in Fig. 8.

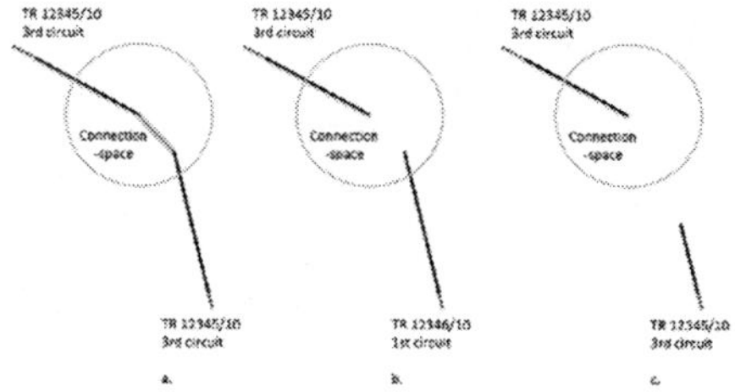

Figure 8. Representation of the connection-space

The software recognized in the connection-space the circuits belonging to each other and connected them with a fictive network element. This is shown by the green line in Fig.8a. If two circuit's end points were located in the same connection-space, however, based on further network information they do not belong together, they would not be connect by the software (Fig. 8b). Two independent circuit's end points are not connected together by the software when they are not located in the same connection-space (Fig. 8c).

IV. THE CONNECTION DATA TABLE OF THE CONSUMERS

The consumers can be supplied with electricity either from the LV, the MV or directly from the HV network. The connection point depends on the power needs of the consumer. The connection to different voltage levels requires different technologies. Residential or small business customers typically connect to LV networks. In order to do the connection to the network they only need a connection cable which is installed by the power company. Larger consumers, the so-called bus-consumers (for example hospitals, shopping malls, factories), need MV/LV stations. Industrial plants with very big energy consumption have to be equipped with own substations.

From the point of view of the TDNm this means that to given individual network elements consumers were assigned on every voltage level. By this it became visible how many pieces of consumers belong to the individual network elements. This number provided the extent of the lowest level in the TDNm graph.

ELMŰ and ÉMÁSZ supplied the data regarding the notification of the consumers about scheduled maintenance works causing the interruption of the power supply. According to the law [9][10], the power supply companies have to notify the consumers in all situations when scheduled maintenance works involve power outages. In order to be able to comply with this, the power supply companies have to maintain a database which contains the list of consumers who could be impacted by turning off given network elements. The databases with the number of consumers for each network elements are available.

V. INTERCONNECTION DATA TABLE OF DIFFERENT VOLTAGE LEVELS

In the current GIS system of the power suppliers the GPS data of the network elements are optimized for transparency. Therefore, the coordinates of individual network elements at the components linking different voltage levels are not identical in the graphical display. An example is shown in Figure 9.

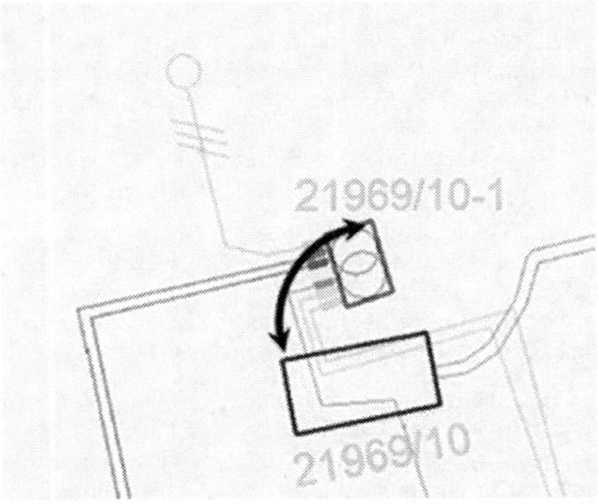

Figure 9. The shift of coordinates at different voltage levels (screenshot of the EÉGIS system)

The red rectangle on Fig. 9 represents the MV/LV station ID 21969/10 on LV distribution network view. The blue rectangle indicates the same MV/LV station on MV distribution network view. The upper left corner coordinates of the blue rectangle and the upper right corner coordinates of the red rectangle (the two corners indicated by the black arrow) should match. Because of this discrepancy, the connection data table of the different voltage levels could not be prepared by using the GPS coordinates data. In creating the model, the shifting between voltage levels was realized by using the connection data tables extracted from the EÉGIS system.

VI. THE NETWORK THEORY MODEL

The different data tables were linked together using the R software [11]. The result was the unified NDB. This was followed by the visualization of the TDNm.

The research team had multiple goals with the visualisation of the network: on one side, the correctness of the new data base (see III/B could be verified and validated [12]. On the other side, with the help of this model [13] [14], the authors were able to describe and document the properties of the network. [15]

The visualization of the TDNm is shown on Figure 10.

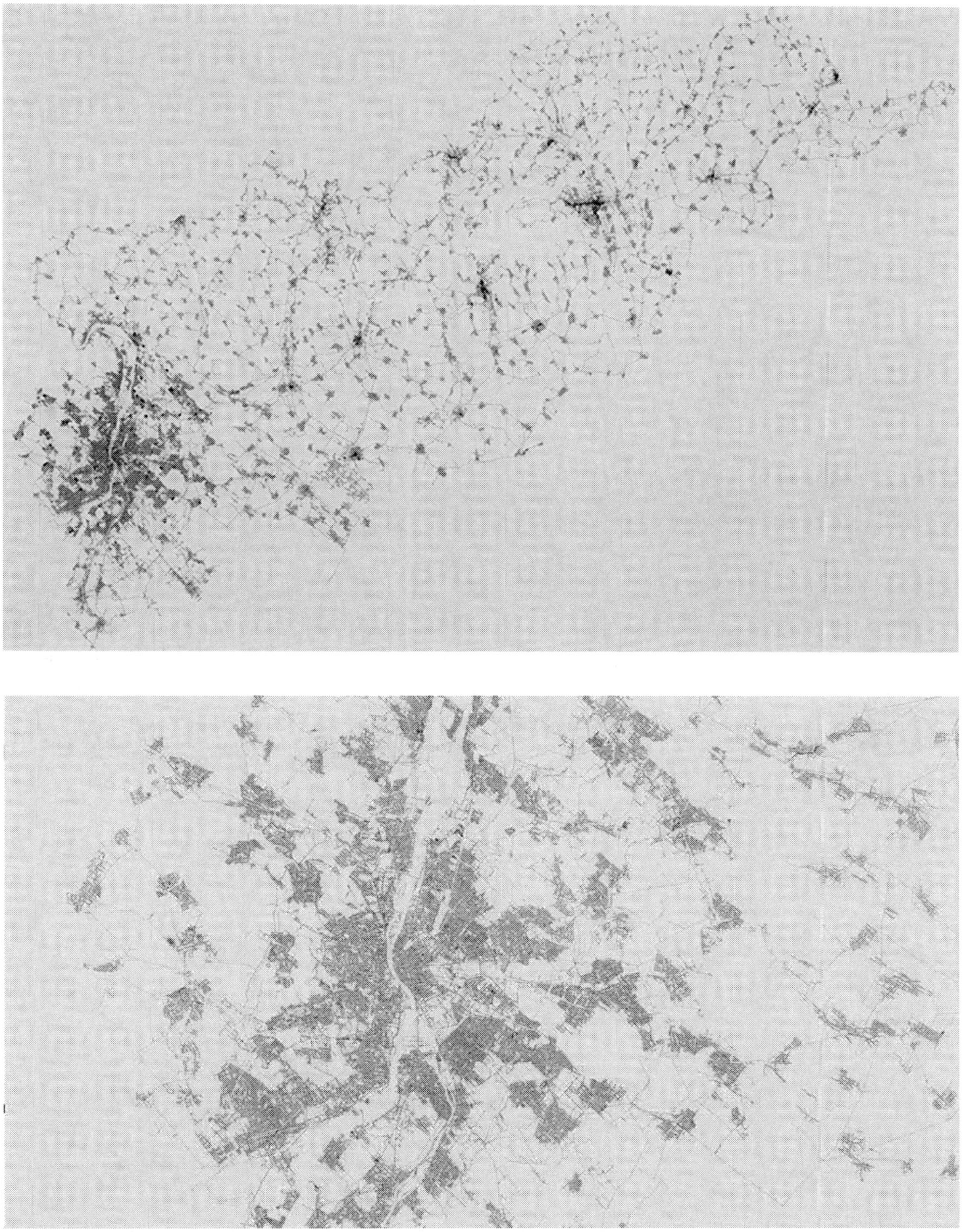

Figure 10. Transmission and Distribution network model - TDNm

Fig. 10 shows the result of the TDNm visualization which was created based on the unified NDB. The first picture is illustrating the TDNm showing the total area covered by the ELMŰ-ÉMÁSZ companies, while the second picture shows a zoom to the city of Budapest.

VII. Conclusion

The research project presented in this paper integrates different disciplines as the electrical systems, the graph theory and the IT. The research group aimed to create a database and a network theory model, which is suitable for analysing and verifying the hypothesises considered in the different research projects of the group.

Acknowledgment

The AD&TE research group would like to thank József Fodor, Managing Director Operations at ELMŰ DSO and ÉMÁSZ DSO and Péter Bakonyi, Head of Documentary Department for providing the required data to our research, and thus for supporting our investigations to a great extent. We would like to express our special thanks to Péter Soltész (AD&TE) and Miklós Tompa (AD&TE), Tibor Fejes (ELMŰ), Dr. Ferenc Novothny (Óbuda University), Dr. Péter Kádár (Óbuda University), Dr. Vajda István (Óbuda University), Dr. András Köpecsiri, Dr. Katalin Czöndör for their priceless work.

References

[1] T. Gonen, Electrical Power Transmission System Engineering: Analysis and Design, CRC Press, 3 edition, May 2014, ISBN: 978-1482232226

[2] R. Miller, Power System Operation, McGraw-Hill Education, 3 edition, January 1994, ISBN: 978-0070419773

[3] Z. Conka, M. Kolcun, G. Morva, "Utilizing of Phase Shift Transformer for increasing of Total Transfer Capacity", Acta Polytechnica Hungarica, 13(5):27-37, December 2016, DOI: 10.12700/APH.13.5.2016.5.2

[4] Offical home page of the Research Group of Applied Disciplines and Technologies in Energetics, short name: Research Group of AD&TE: http://adte.hu, last see: April 2017.

[5] Geometria ltd., The GIS system of the Elmű and Émász, product description, http://www.geometria.hu/?p=1357, last see: April 2017.

[6] J. Molnár, Electricity Operations, Technical Publisher, 1981, pp. 28, ISBN: 963-10-3815-7

[7] D. Jungnickel, Graphs, Networks and Algorithms, Springer, 4th ed. 2013 edition, November 2012, ISBN: 978-3642322778

[8] L. Pokorádi, J. Bera, Monte-carlo simulation of helicopter noise, Acta Polytechnica Hungarica Vol. 12, No. 2, 2015, 21–32, ISSN: 1785-8860, DOI: 10.12700/APH.12.2.2015.2.2

[9] Hungarian Electricity Act, LXXXVI 2007

[10] O. Fazekas, Operation and regulation of the electricity sector in Hungary, Complex Publish Ltd., 2010, ISBN 978-963-295-102-7

[11] H. Wickham, G. Grolemund, R for Data Science: Import, Tidy, Transform, Visualize, and Model Data, O'Reilly Media, 1 edition January 2017, ISBN: 978-1491910399

[12] P. Kádár, "Application of Optimization Techniques in the Power System Control", ACTA POLYTECHNICA HUNGARICA 10: 5, pp. 221-236, ISSN: 1785-8860, DOI: 10.12700/APH.10.05.2013.5.14

[13] J. Kopják, J. Kovács, "RTOS Program Models Used in Embedded Systems", Óbuda University e-Bulletin, Vol. 2, No. 1, pp. 161-171, 2011, ISSN 2062-2872

[14] D. Nagy, E. Rácz, A. Varga, H. Ruf, E. R. Neuchel, "Comparison of electric current -voltage characteristics and maximal power point values of differently and artificially aged solar panels", Proceedings of the11th IEEE International Symposium on Applied Computational Intelligence and Informatics SACI 2016. Budapest: IEEE, 2016. pp. 295-300.

[15] R. Istók , "High Frequency Emissions of Electromagnetic and Electronic Fluorescent Lamps," 10th Jubilee IEEE International Symposium on Applied Computational Intelligence and Informatics (SACI 2015), Timisoara, Romania, 2015.05.21-2015.05.23., 2015. pp. 371-374. ISBN:978-1-4799-9910-1

Comprehensive Mathematical Model of the Turbomachinery in the Czech Nuclear Power Plant Temelin

Verification of the Thermal power library model for the Dymola software by real measurement data from the power plant

Ales Hromadka
Department of Electrical Power Engineering and Ecology and Regional Innovation Centre for Electrical Engineering
University of West Bohemia
Pilsen, Czech Republic
aleshrom@kee.zcu.cz

Jana Jirickova
Department of Electrical Power Engineering and Ecology and Regional Innovation Centre for Electrical Engineering
University of West Bohemia
Pilsen, Czech Republic
jjiricko@kee.zcu.cz

Zbynek Martinek
Department of Electrical Power Engineering and Ecology
University of West Bohemia
Pilsen, Czech Republic
martinek@kee.zcu.cz

Martin Sirovy
Regional Innovation Centre for Electrical Engineering
University of West Bohemia
Pilsen, Czech Republic
sirovy@rice.zcu.cz

Abstract — **this paper concerns the verification of mathematical and simulation model of turbomachinery in the Czech nuclear power plant Temelin. Increasing of share of unpredictable renewable resources requires certain increase in the range of the regulation at thermal power plants. Hence, reliable mathematical and/or simulation models are needed to be able to simulate impact of increased power plant control range on the power plant technological units from the energy efficiency and lifetime point of view. The model has been composed in the simulation software Dymola. Verification and case study has been performed based on real measured data from the power plant.**

Keywords: thermodynamic model of turbomachinery, nuclear power plant, verification, case study, power generation, Dymola.

I. Introduction

This paper is focused on the modeling of the turbomachinery in lower power condition than the nominal one. It is especially useful when it comes to covering of the daily load diagram and power and frequency balance of the electrical network.

Current international environmental and political trends assume decreasing of carbon footprint in the atmosphere by decreasing share of the combustion of fossil fuels and increasing of renewable power sources. [1] However, this still requires a backup power capacity for grid control. One of whom will have to be also regulation from nuclear power plants which used to run typically only as baseload power sources. The motivation is further elaborated in the first part of the paper.

The second part of the paper is focused on composition of simulation model in simulation software Dymola from Dassault Systemes. By specific, the Thermal power library from Modelon has been used as a base library for this task. The structure of the state of the art simulation model is introduced including a future development of the model.

Next part of the paper is focused on mathematical modeling, employing Stodola's turbine model and its equations.

Further, the case study based on a real measurement from nuclear power plant Temelin has been elaborated. The first part of the case study describes the modeled technology. The next part describes the composed simulation model. The last part presents simulation results including simulation accuracy compared to real measurement.

In the conclusion, the results are summarized and the subject of further research is presented.

II. Motivation

A. *Increasing of share of renewable resources*

The usage of renewable resources in 2015 was 2.8% of global energy production – which is a new record in renewable resources production. [2] There is no doubt that increasing trend, which is shown in Figure 1. , will continue in several next years. Renewable resources are operated without pollution and carbon footprint. However, some of them are very unpredictable (dependence on wind, sun or flow) and it can be problem for regulation according to daily load diagram.

Therefore, with higher share of electrical energy from renewable resources, power control range of another sources must be increased. This is one of the core motivation for development of complex simulation models to be able to model energy efficiency effect of variable power control of the plant. Nuclear power plants have more complex regulation than classic thermal power plants.

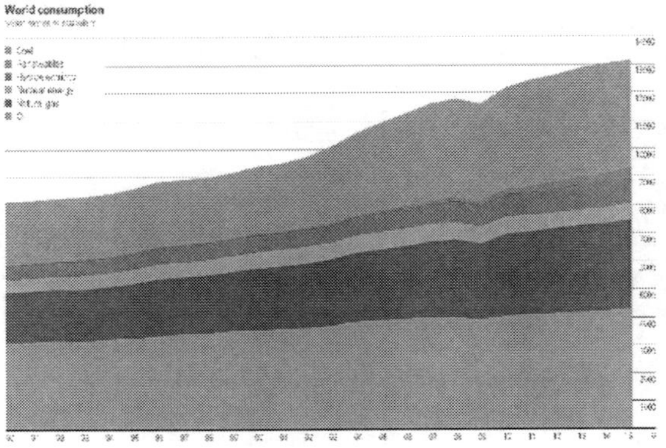

Figure 1. Coverage of World consuption of electrical energy from 1990 to 2015 [2]

It is caused by lower regulation ability of a nuclear reactor, which is not able to regulate as fast as boilers in classical thermal power plants. However, partial power control is possible at the cost of worsening energy efficiency and lifetime of some particular components. The aim of the developed simulation model is to quantify the efficiency decrease in lower power conditions.

B. Decreasing of share of combustion of fossil fuels

Classic thermal power plants produce majority electricity worldwide. However, current trend, which is focused on minimization pollution and carbon footprint in the atmosphere, classifies these power plants as unsustainable for future. Capacities of Nuclear power plants can possibly replace gradually outgoing capacities of classic thermal power plants.

From Figure 2. and Figure 3. , we can estimate future state of electrical power engineering in Czech Republic. Most probable scenario will look like following. Current thermal power plants will be in operation to the expiration of their potential. After that, the power plants must be replace by some cleaner technologies of electricity generation. It depends on too many aspects, what it will be. Currently, it seems that certain substitution for classic thermal power plants will be new nuclear technologies. From this perspective, the modeling of regulation of some current nuclear power plants is logical step for better imagination of options in future.

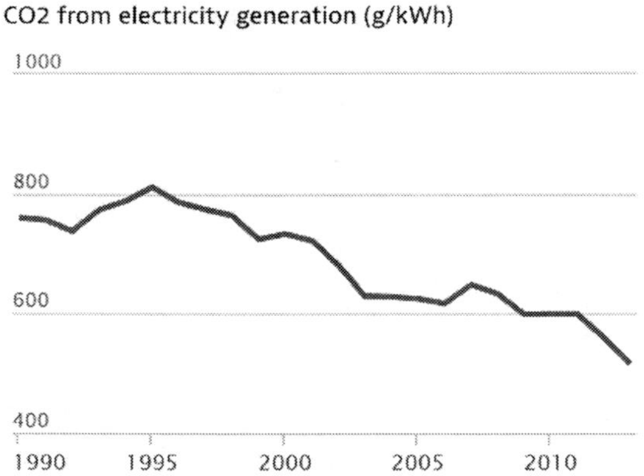

Figure 2. Progress of production of Carbon Dioxide-to-kWh ratios from 1990 to 2014 in Czech Republic [3]

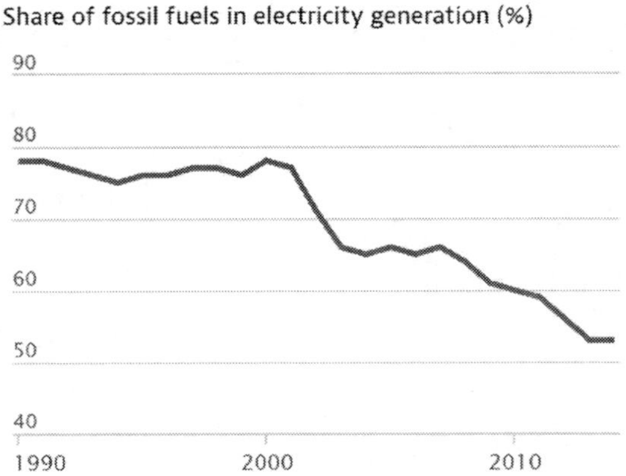

Figure 3. Progress of share of fossil fuels in electricity generation from 1990 to 2014 in Czech Republic [3]

C. Verification of the turbomachinery

The model of the turbomachinery has been developed by components from the library "thermal power" in the software Dymola. The Dymola is modeling software, which uses program language Modelica. Basic advantages of the Dymola are a multi-domain approach and usage of libraries with key components for creation of a model.

Developed model has been verified by real data from operation of Temelin. The verification has been performed by several states of the operation of Temelin mainly during the decline of performance. This paper is only focused on the verification of this model. However, long-term goal is to develop model of complete secondary circuit of the nuclear power plant Temelin.

III. Mathematical Modeling

The turbomachinery of both units in Temelin consists of high-pressure stage and three low-pressure stages. These low-pressure parts have been considered as same low-pressure turbines with same parameters. Therefore, these turbines could be modelled as one system, which contains these three turbines. High-pressure part has three partial arcs and lower-pressure parts have four partial arcs. Each of these arcs had to be modeled separately by the component for steam turbine.

A. Model of the Stodola's steam turbine

Most important component in the model has been steam turbine based on Stodola's ellipse law. One of the earliest attempts to assess the flow variation of a multistage turbine is credited to Stodola (1945). Main equation for the model of Stodola's turbine is given below: [4]

$$\frac{m\sqrt{T_{01}}}{p_{0I}} = k\sqrt{\left[1-\left(\frac{p_{02}}{p_{01}}\right)^2\right]}, \quad (1)$$

m means mass flow rate, T_{01} means inlet temperature, $\frac{p_{02}}{p_{01}}$ means a ratio between inlet pressure and outlet pressure and k is a typical constant for Stodola's equation, which is given by type of turbine. [4]

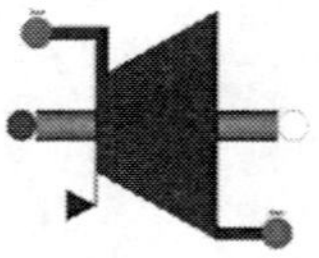

Figure 4. Graphical symbol of the steam turbine in Dymola

Overview of basic parameters for Stodola's turbine	
Parameter name	***Default***
Medium model	Water
Mechanical efficiency [%]	0.98
Nominal isentropic efficiency [%]	0.92
Efficiency degradation using Baumann's formula	false
Bauman coefficient ($\propto_B$)	0.8
Calculate flow coefficient	true
Flow area coefficient (Kt) [m^2]	0.003
Nominal inlet flowrate [kg/s]	variable
If true use T_nom else d_nom	true
Nominal inlet temperature [K]	variable
Nominal inlet specific enthalpy [J/kg]	variable
Nominal inlet density [kg/m^3]	variable
Nominal inlet pressure [Pa]	variable
Nominal outlet pressure [Pa]	variable

Table 1. Overview of crucial parameters for Stodola's turbine

The model of the Stodola's steam turbine respect the law of the ellipse, which works with an optional efficiency degradation using Baumann's formula. [5]

The inlet flowrate to turbine is controlled by using the partial arc. In this case, it is assumed that the flow rate is reduced by partial arc admission, not by throttling (i.e. no loss of thermodynamic efficiency occurs). To simulate throttling, insert a model of the valve assembly at the turbine inlet.

B. *Parametrization of the model*

Stodola's steam turbine have a several parameters, which are crucial to determine for its correct function. Parameters, which are variable, must be determined by external calculation. Two types of parameters are used in the model, i. e. nominal and initial. The parameters do not equal during dynamic modeling and it is necessary to determine both of them separately. It is crucial, to meet these assumptions for correct usage of the model: [5]

- Stodola's law (infinite number of stages).
- Constant isentropic efficiency with an optional efficiency degradation using Baumann's formula.
- No energy or mass storage.
- No shaft inertia.

Parameter "useNominalPoint" decides if the flow area coefficient is given as a parameter K_t or calculated from nominal values at an operating point. The flow area coefficient Kt, defined at design conditions: [5]

$$K_t = \frac{m * \sqrt{R*T}}{\sqrt{p_1^2 - p_2^2}}, \qquad (2)$$

K_t interpreted as effective turbine flow area. In our case, we used an option of calculation from nominal values.

Parameter of the isentropic efficiency equal to nominal η_{is} at no usage of Bauman's formula. However, if "use Bauman" is true, the efficiency is degraded if the fluid enters the two-phase region according to Bauman's formula: [5]

$$\eta_{is} = \eta_{is\_nom} * \left(1 - \propto_B * (1 - x)\right), \qquad (3)$$

, where $\propto_B$ an adjustable Bauman coefficient in range 0.5 to 2 and x is the inlet moisture of steam. [5]

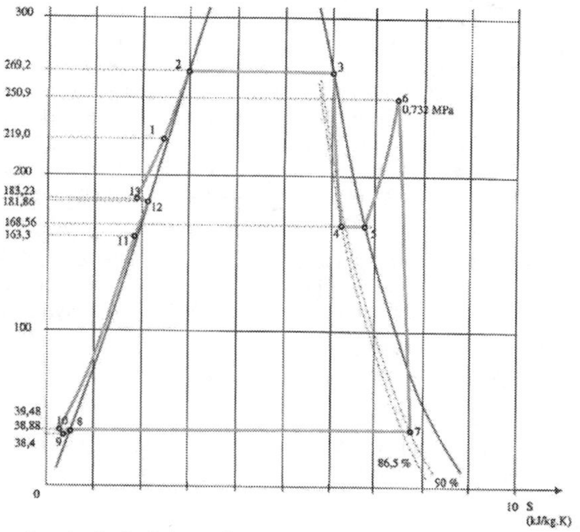

Figure 5. T-s diagram of the thermodynamic cycle in Temelin

IV. Case Study

A. *Description of modeled system*

Temelin is the biggest nuclear power plant in Czech Republic. Temelin has been in operation since 10th of June 2002 i.e. a trial operation of the first unit. In 18th of April 2003, the second unit started to work in trial operation. [6]

Firstly, Temelin was projected with overall nominal electrical performance 2000 MWe. However, later analysis have showed relatively high reserves of electrical performance in previous technology. Thanks to the reserves, the optimization of the technology in Temelin was performed. It meant that electrical performance at each of two units was raised from 1000 MWe to 1055 MWe. It was achieved by the change of the low-pressure stages of the turbomachinery. [6]

Later it was performed a reconstruction of the first unit, which has improved the electrical performance about 23 MWe. Currently the nominal electrical performance of the Temelin is 2133 MWe for each of two units. [6]

Currently, two units are in operation in the Temelin. Each of whom consist of primary, secondary and tertiary-cooling circuits. Basic parts of these circuits are described below.

The primary circuit contains: [6]

- Reactor VVER 1000 type V 320, which has the nominal heat performance 3120 MW_h. Nuclear fuel for the reactor is U 238 with percentage of enriched U 235 1.3-3.8%, which is placed into 163 fuel cassettes. Each of whom contain 312 fuel rods. Change of quarter of the fuel is performed once per year. The fuel rods remain in the reactor for mostly four years. However, their places are changed in the reactor every operating year. This operation is performed for optimization of their heat utilization.
- Cooling system of the primary circuit contains four cooling loops. This system has operation parameters: 15.7 MPa, 290 - 320°C and peak short-term

parameters: 17.6 MPa, 350°C. Flow of cooling light water in the reactor is 23.5 m^3/s.

- Set of four steam generators, which transfer heat from the primary circuit to the secondary circuit. The nominal inlet temperature in the primary circuit is 320°C and the nominal outlet temperature in the primary circuit is 278.5°C. Nominal inlet temperature in the secondary circuit is 220°C and nominal outlet temperature in the secondary circuit is 290°C. Nominal pressure in the secondary side of steam generator is 6.3 MPa. Each of whom create in nominal state approximately 408.33 kg steam/s.
- Set of four main feed pumps, which pump the cooling water back to the reactor. Each of whom have electrical consumption 5.1 MWe and hydraulic performance is 5.9 m^3/s.

The secondary circuit contains: [8]

- Main steam collector with valve assembly, which has a function to control of the thermodynamic parameters at the inlet to the high-pressure turbine.
- Turbomachinery which contains:
  - High-pressure part of the turbine (three partial arc).
  - Separator, which has a function preheater between the high-pressure and low-pressure stages.
  - Three lower-pressure parts of the turbine (three partial arc).
- System of the condensers.
- Low-pressure and high-pressure heaters for regeneration of the feed water.
- Feed tank.
- Set of secondary feed pumps.

Currently these units have a little bit difference in the nominal electrical performance. The first unit has 1078 MWe and second unit has 1055 MWe at ambient condition 18.5 °C. However, it depends on convenient ambient conditions. If the ambient temperature is lower than already mentioned 18.5 °C, the electrical Whole performance is higher. It is typical for winter season. On the other hand, at the higher ambient temperature means lower efficiency of the cycle and lower electrical performance.

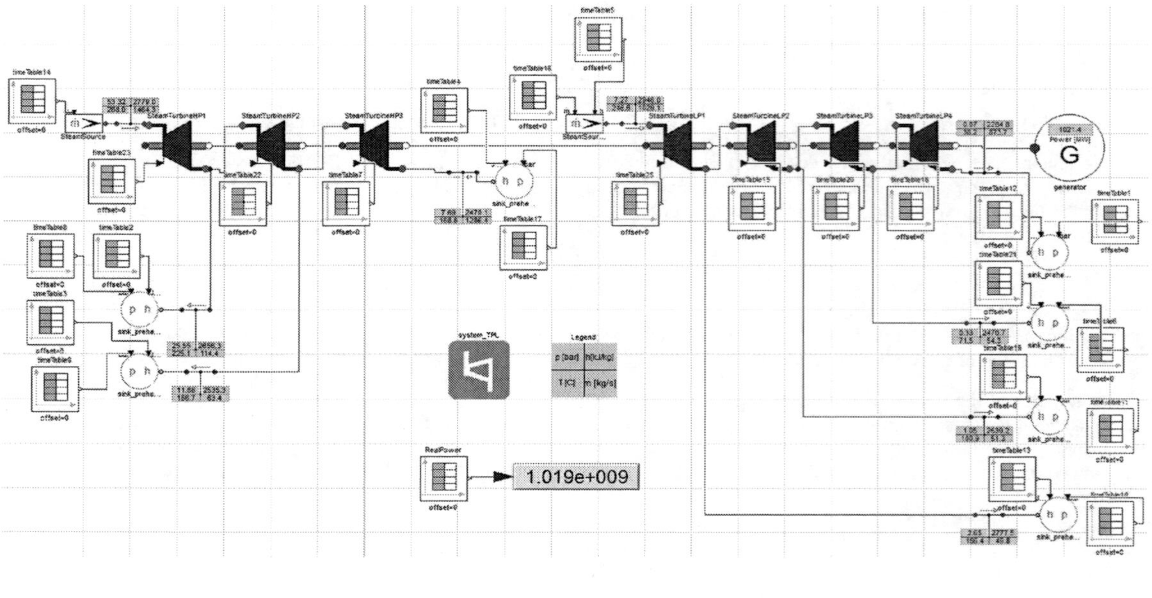

Figure 6. The thermodynamic model of the turbomachinery

The tertiary (cooling) circuit contains: [7]

- Cooling circuit feed pump.
- Pipelines and channels for cooling water.
- Cooling tower

B. *Solution of the model in Temelin*

This chapter is only focused on modeling of the secondary circuit especially turbomachinery in Temelin. Here is explained in more details, how the model in Temelin has been created:

- All turbine stages have certain inlet and certain outlet parameters.
  - The most important parameters are at the inlet of a turbine stage e.g. enthalpy, mass flow rate and pressure.
  - However, it is also necessary to know the outlet parameters for correct behavior determination of a turbine stage.
- The model algorithm has calculated inlet and outlet temperatures according to other thermodynamic parameters for the partial verification of the measurements.
- These thermodynamic inlet and outlet parameters, which have been necessary for the correct parametrization of our model, have been known thanks to real measurements in Temelin.
  - The measurements were performed during the operation in Temelin on 11st of May 2012 from 5 p.m. to 9 p.m. during drop of the reactor performance.
  - Twenty-five variable states have been obtained. Every ten minutes, one state has been obtained.
- The data from the measurements have been inserted into tables in the model.
  - As shown in Figure 6. , the tables determine all boundary conditions for our model.
- The thermodynamic efficiencies and all necessary coefficients for all turbine stages have been set at the

nominal state. The values for the nominal is shown in Table 2.

Nominal state of the turbomachinery in Temelin	
Parameter name	***Value***
Inlet mass flow rate HP stage [kg/s]	1643,1
Inlet pressure HP stage [bar]	54,72
Inlet enphalpy HP stage [kJ/kg]	2779
Outlet mass flow rate first partial arc HP [kg/s]	119,14
Outlet pressure first partial arc HP [bar]	26,22
Outlet mass flow rate second partial arc HP [kg/s]	65,633
Outlet pressure second partial arc HP [bar]	11,542
Outlet mass flow rate HP stage [kg/s]	1295,13
Outlet pressure HP stage [bar]	7,04
Outlet enphalpy HP stage [kJ/kg]	2467,7
Inlet mass flow rate LP stage [kg/s]	1035,8
Inlet pressure LP stage [bar]	6,97
Inlet enphalpy LP stage [kJ/kg]	2946
Outlet mass flow rate first partial arc LP [kg/s]	52,386
Outlet pressure first partial arc LP [bar]	2,802
Outlet mass flow rate second partial arc LP [kg/s]	50,871
Outlet pressure second partial arc LP [bar]	1,108
Outlet mass flow rate third partial arc LP [kg/s]	55,496
Outlet pressure third partial arc LP [bar]	0,352
Inlet mass flow rate LP stage [kg/s]	877,75
Inlet pressure LP stage [bar]	0,06
Inlet enphalpy LP stage [kJ/kg]	2322,1

Table 2. Nominal state of the turbomachinery in Temelin

Figure 7. Comparison of real with modelled performances

- The model configuration for all turbine stages have set at the nominal performance during twenty-five variable states.
- Our model should works with a relatively small error compared to the real measurements in the whole range of the reactor performance.
  - The correct setting of the Bauman coefficient ensures it.
  - The model contains 140 components, 1700 variables, 38 constants, 551 parameters, 1111 unknowns, 8 differentiated variables, 779 equations and from them 617 non trivial equations.

C. Achieved results of verification of the model

Verification of the model has been performed by aforementioned twenty-five variable thermodynamic states in Temelin. These measured parameters for the verification were collected each ten minutes continuously. The important values of the model are shown in **Chyba! Nenašiel sa žiaden zdroj odkazov.**.

Important values for the model	
Parameter name	***Value***
The state 1 electrical performance [MWe]	997,4
The state 25 electrical performance [MWe]	223,7
The lowest absolute error [-] (699,7 MWe)	1,07
The highest absolute error [-] (665,1 MWe)	7,07
The lowest relative error [%] (997,4 MWe)	0,139
The highest relative error [%] (264,3 MWe)	2,568
Ambient temperature [°C]	18,5
Ambient pressure [kPa]	101

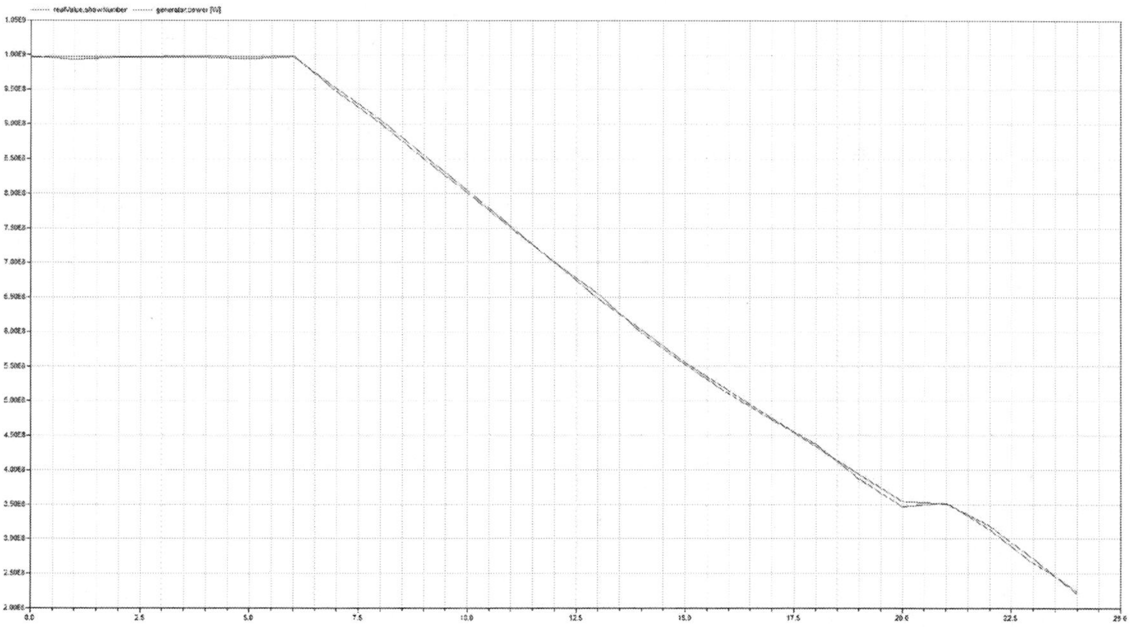

Table 3. Important values for the model

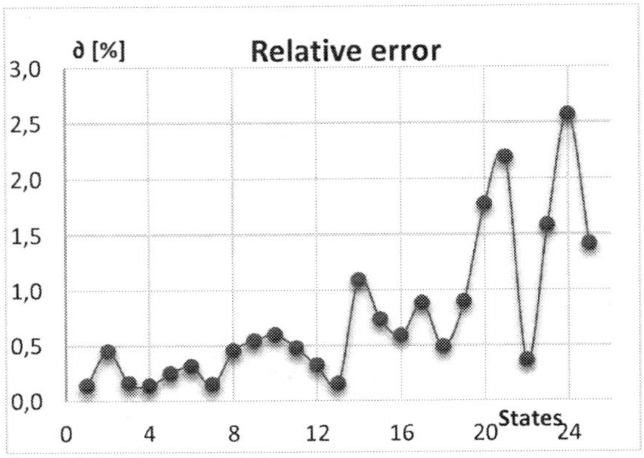

Figure 8. Expression of relative errors in states of the model

The biggest absolute error has been displayed during electrical performance 271, 27 MWe. It meant about 6,97 MWe lower performance than valid electrical performance i. e. relative error 2,568 %. Most of modeled values of the performance has located in the range of 1% relative error. As you can see in Figure 8.

V. Conclusion

This paper has been focused on the modeling of dynamic states of the turbomachinery in the nuclear power plant Temelin during lower electrical performance than the nominal. Firstly, motivation for usage of the model has been described by current trends. Further, the model of the turbomachinery has been introduced. At the end of the chapter have been mentioned information about our future aim with the model. Next part of the paper is focused on mathematical modeling in the software Dymola. Main part of this paper has been the case study.

The case study has been concerned a description of modeled technology, a solution of the model and achieved results of verification of the model.

Main conclusion of the paper can be summarized into following points:

- Future state of electrical power engineering in Czech Republic will be probably strongly depended on certain regulation in Czech nuclear power plants.
- Therefore, it is very important to model the whole thermodynamic cycle in Temelin and try to optimize them during operation with partial thermodynamic parameters.
- The model has been created according to the real measurements from Temelin.
- The results of the partial performances obtained in Temelin were used for the verification of the model in Dymola.
- All boundary conditions were given by the real measurements.
- The model has been parametrized at the nominal performance.
- Main goal of the paper was verification of the model during the partial load.

The contributions of the paper are:

- Graphical view of our thermodynamic model of the turbomachinery shown in Figure 6.
- Comparison of the performances determined from real generator measurement shown in Figure 7. in Temelin with modelled performances determined by our model.
- The comparison displayed in Figure 8. shows the accuracy in each state of the model.
- The comparison have proven that current model of the turbomachinery is relatively valid and most of modeled states has located in the range of 1% relative error. However, we have to develop the other components of the secondary circuit such as feed pumps, regenerative heaters, a steam generator, etc... All these parts must be interconnected. Finally, their accuracies have to be investigated for the whole cycle.
- If we acquire information about total accuracy of the, we can predict dynamic states, which can be required in future.

Acknowledgment

This paper is supported by the student research project SGS-2015-031, research project for advanced nuclear technologies TE01020455.

References

[1] Janicek, F.. Technologies for a sustainable future. N.p.: n.p., 2009. Print.

[2] "Statistical Review of World Energy." Bp.com. N.p., n.d. Web. 12 Apr. 2017.

[3] "Compare your country - Climate Change Mitigation Policies." By OECD. N.p., n.d. Web. 13 Apr. 2017.

[4] Dixon, S. L. Fluid mechanics, thermodynamics of turbomachinery. Amsterdam: *Elsevier-Butterworth-Heinemann*, 2006. Print.

[5] Cooke, D. H. "On Prediction of Off-Design Multistage Turbine Pressures by Stodola's Ellipse." Journal of Engineering for Gas Turbines and Power 107.3 (1985): 596. Web.

[6] Historie a současnost Elektrárny Temelín. Retrieved April 27, 2017, from https://www.cez.cz/cs/vyroba-elektriny/jaderna-energetika/jaderne-elektrarny-cez/ete/historie-a-soucasnost.html

[7] Technologie a zabezpečení. Retrieved April 27, 2017, from https://www.cez.cz/cs/vyroba-elektriny/jaderna-energetika/jaderne-elektrarny-cez/ete/technologie-a-zabezpeceni.html

[8] Hlavni technicke udaje. Retrieved April 27, 2017, from https://www.cez.cz/cs/vyroba-elektriny/jaderna-energetika/jaderne-elektrarny-cez/ete/technologie-a-zabezpeceni/2.html

Establishment of a New Reliability Model of the Hungarian Energy Transmission Network

Zsolt Mitrik
Óbuda University
Budapest, Hungary
mitrik.zsolt@kvk.uni-obuda.hu

Abstract — **The substations represent the HV system interconnection points, the reliability of them plays central role in the energy transmission. Accordingly, a smaller malfunction may cause significant disturbance in the energy transmission system. The weighing of the importance of the substations (the measure of the fallen out system in case of malfunction) is based on empirical data.**

The author –based on the theory that Prof. Dr. Nyepomjasij V.A. developed – researches the reliability that a typical substation connection schemes – used also in Hungary- provides. The importance of the article is that compares the reliability of the different substation connection schemes. Based on the received results, the author ranks the reliability on the different types providing assistance for the future selection of the substation type depending on the disposition within the energy transmission system.

***Keywords:** Reliabilty; Substation scheme; Energy transmission;*

I. Introduction

The main task of the transmission network is to transmission the electrical energy from the production unit to the consumers, in the quality according to the standard specifications. The secondary task is to connect the international networks and ensures the energy transmission between the concerned countries. Based on the above, it is clear that the fail safety has priority importance both in the national and international energy transmission point of view.

In Hungary, MAVIR Zrt. is liable to manage and supervise the energy transmission system. "By acting as licensee of transmission system operation, it is the responsibility of MAVIR to ensure the required permanent availability of the transmission network by operating the network and implementing remedial actions in case of contingencies. Complying with domestic and international requirements, MAVIR defines, plans and carries out the necessary development, renewal and maintenance activities and guarantees a non-discriminatory access for electricity market players in order to sustain operability of the system." [2] The substations and the connecting power lines constitute the energy transmission system.

The subject of the author's PhD research is the reliability of the energy transmission network. As part of this research, the aim of this article is the presentation of a new reliability calculation method on the high voltage transmission network.

II. The Energy Transmission System

A. The Hungairan energy transmission system

The author considers the MAVIR Zrt. operated network as one system and defines the substation in the network as interconnection points that are parts of the energy transmission network. At the same time, the author investigates the substation also alone that – however – constitutes sub-systems of the above described system; and therefore the author applies two-level decomposition.

Reviewing the energy transmission system, it can be determined that the system is divided into four voltage level:

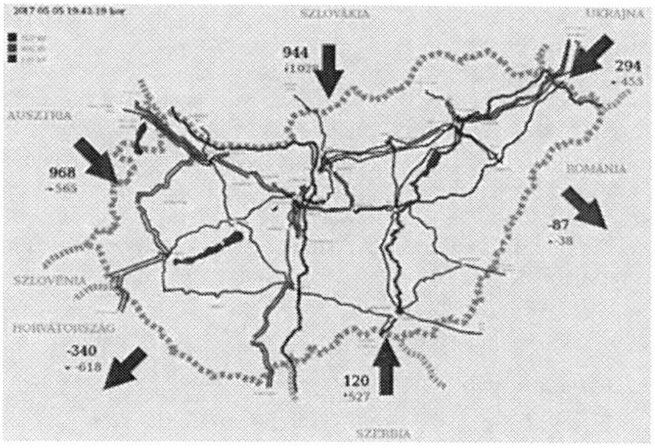

Figure 1. Hungarian energy transmission system [2]

Consumption	5252 MW
Production	3390 MW
Import-export saldo	
Schedule	1930 MW
Actual	1860 MW
Momentary deviation	-39 MW
Average deviation	MWh/h

Table 1. Hungarian energy transmission system momentary load value 2017.05.05 19:42 [2]

As Fig. 1 indicates the electric energy transmission is realized on 750 kV, 400 kV and 220 kV voltage levels. The distribution system operators (DSOs) connect to the transmission network at 120 kV voltage level. Above of all these, in special unique cases, high energy consuming plants and factory complexes may connect directly to the energy transmission network. These large consumers typically also connect to the 120 kV voltage level. Thirty substations constitute the interconnection points of the network.

B. *Substations*

The substation is defined as assembly of switchgears and transformers that constitutes power lines interconnections points according to IEC61936-1:2010. The task of the transformers is to convert the electric energy into different voltage levels, or from the research's point of view is such equipment that connects the different voltage levels.

C. *Typical substation layouts*

The typical layouts on 120 kV are the double busbar systems, on 220 kV are the double busbar, the breaker and half and "H" scheme, on 400 kV and on 750 kV only breaker and half can be found. Fig. 2 indicates these layouts:

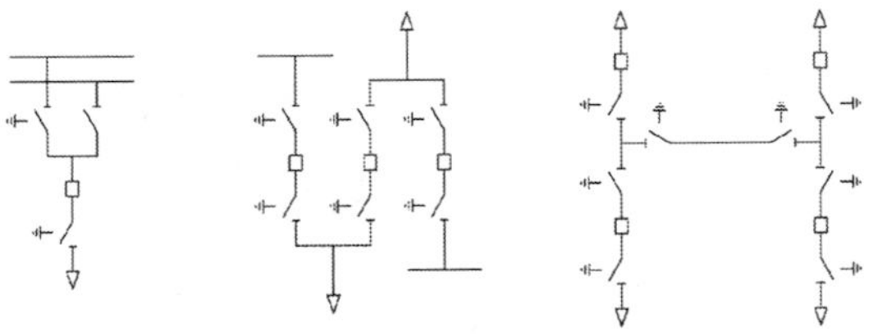

Figure 2. The most common substation layouts

The left scheme on Fig. 2 is the double busbar system, the middle one shows the breaker and half system and the right scheme shows the "H" scheme system. In case of a double busbar system the bay can be connected to any busbar through the busbar disconnectors. Hence, in case of one of the busbars failure, the energy flow is ensured even without solving the failure. In the breaker and half system the line also can be connected to a busbar in a way that even in case of the failure of one of the circuit breakers or busbar the energy flow is still ensured. In case of "H" scheme any transformer can be connected to any power line, the advantage is the simple structure and the little equipment need. Accordingly, except of some uniquely designed substation, the bays can be considered as typical, and this fact makes the calculations simplified.

III. Main equipments of substations

A. *Circuit breaker*

Fig. 3. displays the symbols of the circuit breaker.

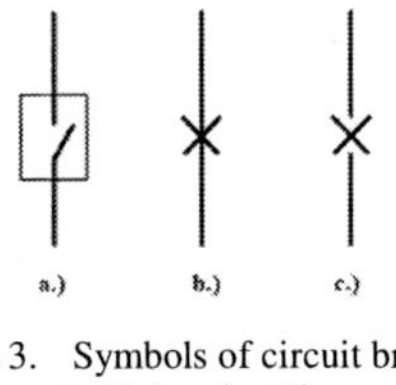

Figure 3. Symbols of circuit breaker
a.) the most common symbol b.) and c.) the same symbols in case of closed (b.)) and open (c.)) position

Single line schemes contain one of these symbols. Fig. 3a) is the most common symbol on the schemes. Fig. 3b) and 3c) are the same symbols indicating the closed (3b)) and the open (3c)) positions. The definition of the circuit breaker is the following: The circuit breaker is such a mechanic switching equipment that under normal circuit conditions is able to switch on, conduct and break the nominal current; in addition, if the circuit conditions differ from the normal condition (such as short circuit), is able to switch on, conduct up to determined time period and break the current. [3]

B. *Disconnector*

Fig. 4 displays the symbols of disconnectors.

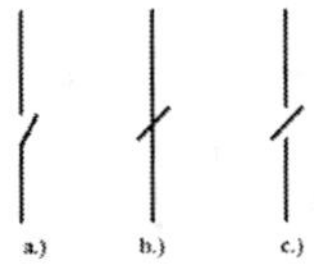

Figure 4. symbols of disconnectors a.) the most common symbol, b.) and c.) the same symbols in case of closed (b.)) and open (c.)) position

Single line schemes contain one of these symbols. The definition of the disconnector is the following: The task of the disconnector is to allow isolation of apparatus and transmission lines, for maintenance and the determination of the way of the current. The distance between the open contacts of the disconnector is always the safety isolation distance appropriate to the voltage level. The disconnector is unable to switch on and off the current, switching happens in a way to prevent the electrical arc ignition. [3] A typical substation layout is that disconnectors are installed on both sides of the circuit breaker. There is no disconnector built in the vicinity of the transformers due to the reason that typically there are circuit breaker both sides on the transformers and on the other side of the circuit breaker has disconnector.

C. *Transformer*

Fig. 5 displays the symbol of transformers.

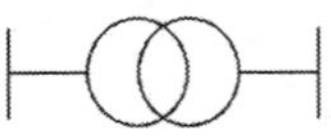

Figure 5. Symbol of transformer

Single line schemes contain this symbol. The definition of the transformer is the following: A transformer is electric equipment operating based on the stationary electro-magnetic induction which converts the current and the voltage of the alternating current.

D. *Current and voltage (potential) transformer*

Fig. 6. displays the symbols of the current and voltage transformer.

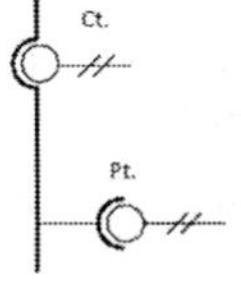

Figure 6. Symbols of current (Ct.) and voltage transformer (Pt.)

The current and voltage transformers (instrument transformer) provide information about the stationary and transient operation. The task in stationary operation condition is providing information about the voltage, current, active and reactive energy. [3]

IV. The new reliability model of energy transmission system

A. *Overview of substation layouts*

The substation layouts show the electric connections between the different bays and busbars. Based on the substation layouts supervised by MAVIR Zrt. the name of the substations, the number of bays by voltage level and types are indicated in the below table. The table shows the data according to 2013 status as this was the last year when MAVIR Zrt. published data about substations and power lines.

Substation name	120 kV		220 kV		400 kV		750 kV	
	Bay type	Pc.	Bay type	Pc.	Bay type	Pc.	Bay type	Pc.
Albertfalva	double busbar transformer branch	2	"H" switching with double disconnector	2				
Albertirsa					One and half CB system	7	One and half CB system	3
Békéscsaba	double busbar transformer branch	2			One and half CB system	5		
Bicske Dél	double busbar transformer branch	2			One and half CB system	4		
Debrecen	double busbar transformer branch	17	Power line branch	2				
Detk	double busbar	15	double busbar partly with auxiliary busbar	12				
Dunamenti	One and half CB system	19	One and half CB system	19				
Dunaújváros	double busbar	16	double busbar transformer branch	2				
Felsőzsolca	double busbar	14			One and half CB system	5		
Göd	double busbar partly with auxiliary busbar	18	Transformer branch	1	One and half CB system	7		
Gönyű					One and half CB system	3		
Győr	double busbar partly with auxiliary busbar	16	double busbar partly with auxiliary busbar	5	One and half CB system	7		
Hévíz	Split busbar	4			One and half CB system	6		
Kisvárda	double busbar	11	"H" switching with double disconnector					
Litér	double busbar with auxiliary busbar	13			One and half CB system	6		
Martonvásár			Transformer branch	2	One and half CB system	8		
Ócsa	Transformer branch	2	double busbar partly with auxiliary busbar	7				
Oroszlány	double busbar with auxiliary busbar	13	"H" switching with double disconnector	4				
Paks	double busbar with auxiliary busbar	16			One and half CB system	14		
Pécs	Transformer branch	2			One and half CB system	6		
Sajóivánka	double busbar with auxiliary busbar	11			Transformer branch	1		
Sajószöged	double busbar with auxiliary busbar	13	double busbar partly with auxiliary busbar	15	One and half CB system	6		
Sándorfalva	double busbar	10			One and half CB system	6		
Szeged	double busbar	12	Transformer branch	1				
Szolnok	double busbar	15	double busbar partly with auxiliary busbar	7	One and half CB system	6		
Szombathely	Transformer branch	2			One and half CB system	6		
Tiszaklök	double busbar	2	one busbar	4				
Toponár	Transformer branch	2			One and half CB system	4		
Zugló	Transformer branch	3	double busbar	10				

Table 2. Substation names, layouts and number of bays

The first column of Fig. 7 table indicates the name of the substation, the following 4 columns indicate the type of the bays by voltage level and the number of these.

B. Schematic diagram of energy transmission system

The substations in Figure 7 and the connecting power lines based on voltage level are displayed in the below figure.

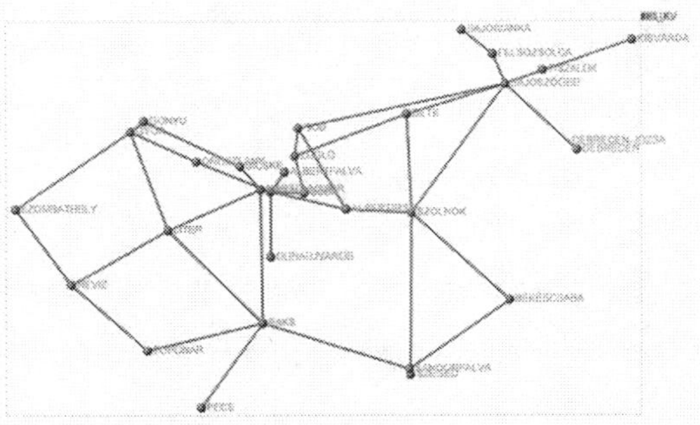

Figure 7. Schematic diagram of energy transmission system

The aggregation of the substations and power lines constituting the Hungarian energy transmission system is showed on Figure 8. This system can split up on voltage levels into different layers. The author designed the schematic diagram and the later new graph model by Creo 3.0 software. The schematic diagram of the energy transmission system provides a opportunity to consider the energy transmission system as a graph. The basis of the graph theory is that a network constitutes connection pointes (nodes) and edges. [4][5][6] Figure 9 shows the system of the nodes and edges.

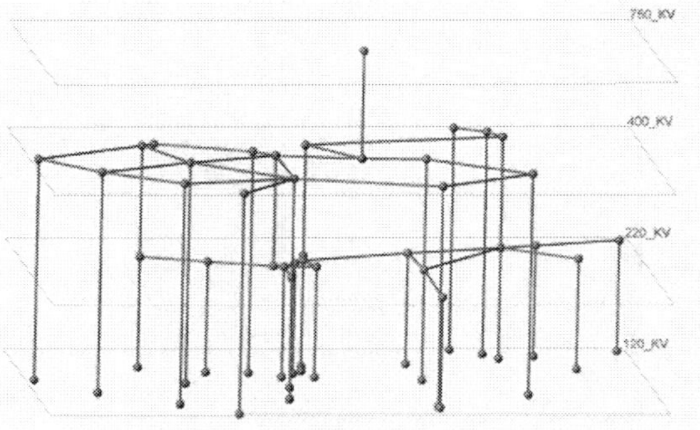

Figure 8. The new graph model of the energy transmission system

Applying the graph theory the author introduces that the nodes on the layers mean the substations (ball) and the edges connecting the nodes mean the power lines (horizontal lines). In addition, the connections between the layers such as the edges, the transformers (vertical lines) provide in this case.

V. Simplifications Applied During the Reseach

A. There are no outside connections of the investigated system.

There are no outside connections of the investigated network. The author considers it as an off-grid operation network. This can be done as the author researches not the reliability of the energy supply but the integrated items constituting the network. The aggregation of the installed items in the networks consists of primer and secondary equipment in the substation, the power lines and the transformers…

B. Normal operation in the investignated time moment

The author defines the normal operation as the bays are equally distributed to the busbars, the circuit breaker of the bus coupler and bus sectionalizer bays are in open position.

C. Data related to the installed switchgears

The author applied the published data of Anton Janssen [1] for the determination of the reliability of the installed switchgears. Naturally, the values of the equipments built into the Hungarian network may replace these data, and in case of refurbishing or replacement of the installed switchgears, one may receive real-time data and can update the database.

D. Valuation of failure

IEC 62271-1:2007 divides the high voltage equipment's failure into two groups based on the effect to the system.

Major failure (MaF): "... failure of a switchgear and control gear which causes the cessation of one or more of its fundamental functions. A major failure will result in an immediate change in the system operating conditions, e.g. the backup protective equipment will be required to remove the fault, or will result in mandatory removal from service within 30 minutes for unscheduled maintenance."

Minor failure (MiF): "... failure of equipment other than a major failure or any failure, even incomplete, of a constructional element or a sub-assembly which does not cause a major failure."

As MiFs do not cause the total loose of equipment, therefore the author excludes such failures from the research. The author notes that such failures decreases the time period between two maintenances.

VI. Reliability Flowchart

The reliability flowchart [7][8] displays the supply of the investigated point of the network, the connection order of the items in the scheme, the interactions on the operational reliability of the items to the investigated point of the network. Accordingly, the flowchart shows logical connections and not electrical connections. Any system can be split into the aggregation of serial and parallel elements. Serial elements are the aggregation of such equipments where the failure of one element makes the entire group inoperable. Parallel elements are the aggregation of such equipments where only all equipment's failure makes the entire group inoperable. For example, the reliability flowchart can be designed to one bay:

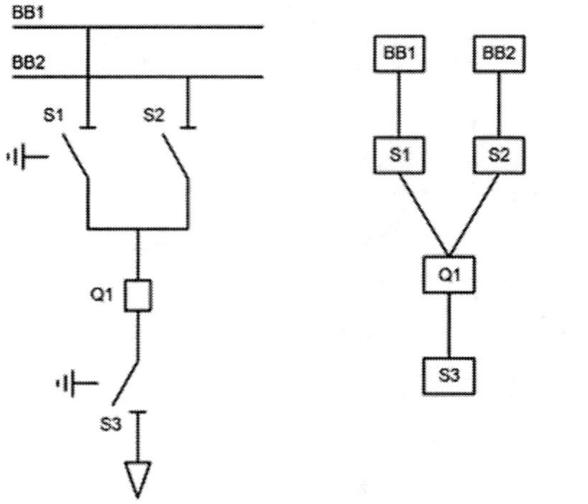

Figure 9. Reliability flowchart of a double busbar brach

VII. Conclusion

A new reliability model was necessary for the reliability investigation of the Hungarian energy transmission network. The author describes the creation steps of this new model in the article. Defining the model as a graph provides opportunity to handle the reliability problem as a mathematical problem [9][10] and as a result particular calculations can take place in the future. The author will describe the reliability calculation of the new model in his next article.

ACKNOWLEDGMENT

I wish to thank my colleagues, Prof. Dr. György Morva, Dr. Péter Kádár who gave me valuable advice and suggestions during the work.

References

[1] M. Runde, "Failure Frequencies for High-Voltage Circuit Breakers, Disconnectors, Earthing Switches, Instrument Transformers, and Gas-Insulated Switchgear," IEEE Transactions on Power Delivery, Vol. 28, Issue: 1, Jan. 2013, pp. 529 – 530, 2013, DOI: 10.1109/TPWRD.2012.2220638

[2] On-line chart of the actual electricity consumption and the actual supply of Hungary, MAVIR Ltd., avaible: https://www.mavir.hu/web/mavir/atvitel, unpublished, last see: April 2017

[3] K. Kiss, J. Kovács, F. Radvánszky, J. Tábori, A. Tombor, F. Weingart, „Automatisation and protection in the power systems," Hungarian Electrotechnican Publisher, Budapest, 1988, ISBN: 963-10-7554-0

[4] J. Pálfi, "Localization of faults in low voltage networks by the graph method," IEEE 11th International Symposium on Applied Computational Intelligence and Informatics (SACI), 12-14 May 2016, DOI: 10.1109/SACI.2016.7507410

[5] J. Pálfi, M. Tompa, P Holcsik, "Analysis of the Efficiency of the Recloser Function of LV Smart Switchboards," Acta Polytechnica Hungarica, Vol. 14, No. 2, 2017, DOI: 10.12700/APH.14.2.2017.2.7

[6] J. Pálfi, P. Holcsik, M. Tompa, „Network Science Tools applied to Low Voltage Networks," 17th IEEE International Symposium on Computational Intelligence and Informatics (CINTI 2016). Budapest, Hungary, 2016.11.17-2016.11.19., 2016. pp. 41-47. DOI: 10.1109/CINTI.2016.7846377

[7] P. Pausits, G. Szögi, M. Nallbani, I.J. Rudas, T. Haidegger, "Identification of Risks in the Application of Surgical Robotics," 19th IEEE International Conference on Intelligent Engineering Systems, Bratislava: INES 2015, pp. 147-151, ISBN:978-1-4673-7938-0

[8] T. Portik, L Pokorádi, "The Summarized Weighted Mean of Maxima Defuzzification and Its Application at the End of the Risk Assessment Process," in Acta Polytechnica Hungarica, Vol. 11, Issue 3, 2014, pp. 167- 180, DOI: 10.12700/APH.11.03.2014.03.11, ISBN: 1785-8860

[9] P. Kádár, "Application of Optimization Techniques in the Power System Control," Acta Polytechnica Hungarica 10: 5, pp. 221-236, ISSN: 1785-8860, DOI: 10.12700/APH.10.05.2013.5.14

[10] Z. Conka, M. Kolcun, G. Morva, „Utilizing of Phase Shift Transformer for increasing of Total Transmission Capacity," Acta Polytechnica Hungarica, 13(5):27-37, December 2016, DOI: 10.12700/APH.13.5.2016.5.2

GH_2 - Green Hydrogen
Hydrogen Energy in the Light of Sustainability

István Szén
Óbuda University, Kálmán Kandó Faculty of Electrical Engineering
Power System Department Budapest, Hungary
szen.istvan@kvk.uni-obuda.hu

Abstract — **Today the sustainable energy is a hot scientific topic but if we examine the indicators of the world's energy we will find that globally is still dominated the fossil fuels so it is totally unsustainable furthermore there are also considerable problems of social and economic levels that are related to energy supply. In the 21. century one of the most important questions are global warming, CO_2 emission and the security of energy supply.**

Renewable energy sources have an enormous potential, but their utilization is difficult due to their energy fluctuations and their stochastic - weather dependent - operation. We need a secondary energy sources that we can use as energy storage or we can use direct energy source. Of course, we consider it important to protect the environment and global access to energy sources. Maybe the hydrogen will be the future determinative energy source?

In my paper, I analyze the current state of the hydrogen energy, future prospects, and I will introduce our research model and our research plan.

Keywords: hydrogen, green hydrogen, sustainable energy, renewable energy sources, global warming, CO_2 emission, security of energy supply.

I. INTRODUCTION

Given the continued growth in the world's population from about 7,5 billion people today to over 9-10 billion by 2050 [1], plus the progressive industrialisation of developing nations, particularly in Asia but likewise in Africa and South America, the global demand for energy is expected to continue to grow in the coming decades as well by up to 50% until 2040, according to the International Energy Agency (IEA) [2]

At the same time, there is a growing global consensus that greenhouse gas emissions, which keep rising, need to be mitigated in order to prevent dangerous greenhouse gas induced climate change effects. Therefore, security of supply and climate change represent two major concerns about the future of the energy sector which give rise to the challenge of finding the best way to control in emissions while also providing the energy required to sustain economies.

The hydrogen is not only the strongest chemical fuel, it also serves as an energy carrier. The hydrogen is a globally accepted clean energy carrier as it is source-independent and has a high-energy content per mass compared to petroleum and other energy sources (Fig. 1). Due to these advantages, H_2 can be used also as energy sources for different appliances, such as hydrogen fuel cells where the reaction product is water and there is no CO_2 emission.

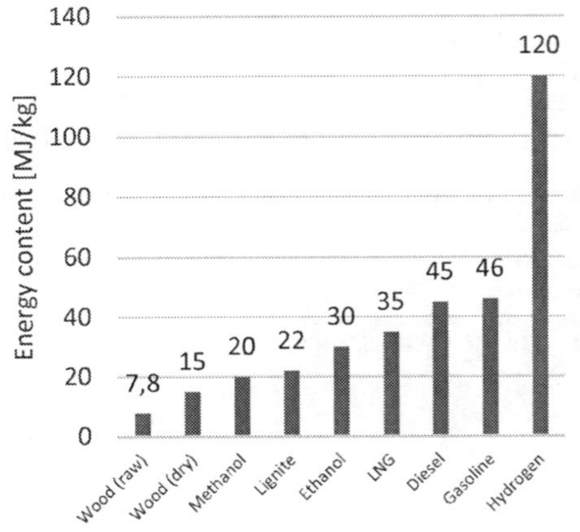

Figure 1.: Energy contetnts (MJ/kg) of different fuels

II. THE HYDROGEN AND THE HYDROGEN ECONOMY

A. *Short history of hydrogen*

In 1671, Robert Boyle discovered and described the reaction between iron filings and dilute acids, which was the first result at the field of production of hydrogen gas. [3] In 1766, Henry Cavendish was the first scientist who recognized hydrogen gas as a discrete substance by naming the gas from a metal-acid reaction "inflammable air". Later, after his further observations in 1781 the result was: the gas produces water during it was burning. In 1783, Antoine Lavoisier developed the element called Hydrogen [4]. In the experiment, he and Laplace reproduced Cavendish's finding, which was: water has been produced during hydrogen has been burned.

The hydrogen economy is a proposed system of delivering energy using hydrogen. The term hydrogen economy was coined by John Bockris during a talk he gave in 1970 at General Motors Technical Center. [5]

B. *Production of hydrogen*

Hydrogen is available on Earth only in compounds. In order to free hydrogen from these compounds needs energy. The hydrogen is considered as the "forever fuel", like electricity. Its secondary energy running mate can be produced from any primary energy fuel like coal, oil, natural gas, nuclear, all sorts of renewable energies, and from grid electricity. Natural gas reforming, coal gasification and water electrolysis are proven technologies for hydrogen production today and applied on an industrial scale. In Fig. 2 is presented the hydrogen production path.

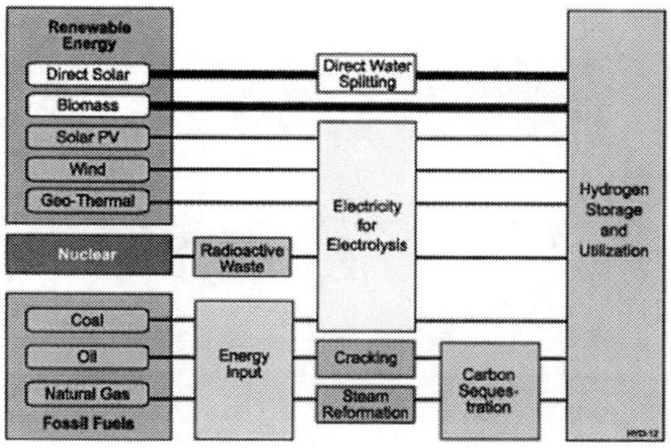

Figure 2.: Hydrogen production path [6]

Although hydrogen is a clean and accepted energy carrier but we need to look at the source of hydrogen.

Currently, hydrogen can be produced from natural gas by means of three different chemical processes: steam methane reforming (SMR), partial oxidation (POX) and auto-thermal reforming (ATR). Hydrogen can be produced from coal by gasification and the production of hydrogen from oil is achieved by partial oxidation.

In addition, the hydrogen can produce from biochemical process and some new technical method but their quantification is negligible at present.

Hydrogen and electricity are interchangeable via electrolyser and fuel cell; the one makes hydrogen from electricity, the other electricity from Hydrogen. Production of electrolytic Hydrogen is really necessary. The Hydrogen production adds an additional link to the energy conversion chain.

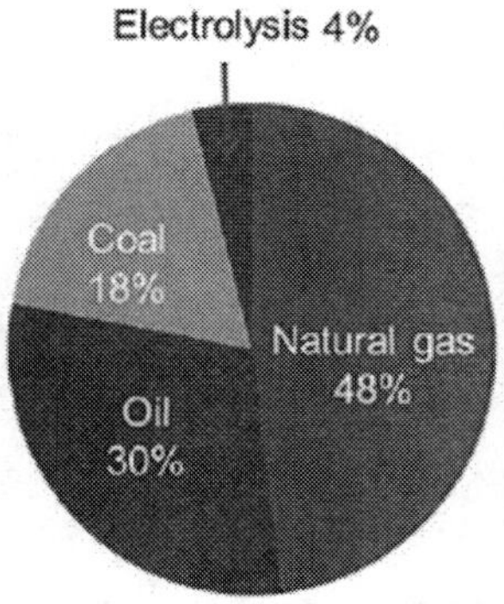

Figure 3.: World hydrogen production by fuels [6]

In Fig. 3 is presented the World hydrogen production. It can be seen that 96% of globally-produced hydrogen comes from fossil fuels. We make clean fuel from dirty fuels but at the same time CO_2 is still generated and the process is not sustainable. The CO_2 should be captured and in the industry has a ready-made technology with carbon capture and storage but the easily accessible fossil fuels are exhausted and difficult to access to the fossil fuels furthermore may raise some serious environmental concerns.

Certainly, due to its environmental and climatic cleanness, "solar-hydrogen" or "wind-hydrogen" – where hydrogen made from renewable energies: green hydrogen - is the ultimate goal.

Figure 4. World hydrogen production by fuels [6]

C. *Relationship between hydrogen, energy system and transportation.*

The currently used energy supply system is clearly centralized. Electricity, hydrogen, natural gas or oil are centrally produced in power stations, gas fields or oil refineries. The consequence of central production is the energy transportation that is causing significant losses.

Though hydrogen could well be transported pick-a-back in the operational natural gas grid avoiding an extra hydrogen transport system and decarbonizing process is much easier, when hydrogen from fossil fuels is generated in large scale systems rather than in distributed scale installations of much smaller capacities. Collecting the greenhouse gases from millions of distributed emitters are technically and commercially impossible.

On the other hand, we need to realize the growing presence of distributed generation and we also know that related to renewable energy sources there is a missing link. The development of energy storage is necessary. In addition, there is a need for a balancing control element for reducing energy fluctuation, because the solar radiation and the wind energy is a highly variable source of energy so the hydrogen could play in the energy system for the integration of renewable energies

The hydrogen is seen as one of the main alternative fuels for future road transport. The fuel cell vehicle capable of traveling 350-460 kilometer on a tank of hydrogen and refuellable in three-five minutes.

III. Our research plan – electrolysis and photovoltaic systems

We trust that the electrolysis can be a useful addition to decentralized renewable energy sources therefore in the initial phase of our research, we focus on electrolysis.

A. *Impact of ripples on the direct current (DC) used for electrolysis at hydrogen generation.*

Plenty of scientific papers talk about the electrolysis or electrolysis of clear water and generating hydrogen this way. Successful operation of electrolysis depends on many physical and technical factors. As it is well known, direct current (DC) needs for operating electrolysis. According to today´s widespread practice alternate current/direct current transformers (AC/DC transformers) and rectifiers can be used to get DC current. DC current generated by AC/DC transformers and power supplies is not an ideal electric source in the terms of operation of electrolyzing cell, because of its non-stable and continuously or frequently modifying cell voltage - i.e. ripples - output voltage came out from the mentioned devices. The ripples affects efficiency of the electrolyser cell power of the cell and the quantity of the generated gas.

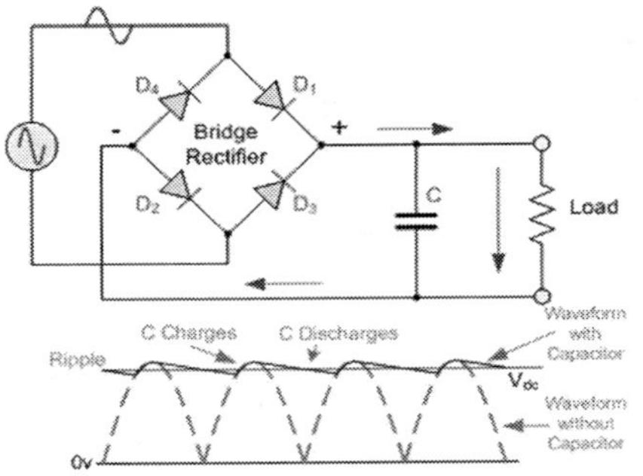

Figure 4.: Typical DC power supply – ripple voltage

In our research project, we would like to investigate the impact of the presence of the ripples, waveform distortions and higher harmonics of the voltage to the efficiency of the electrolysis process. Furthermore, we would like to investigate possible connections between the quality of the DC voltage generated by photovoltaic panels and the efficiency of the electrolysis.

We would like to realize our aims using three different research technologies: experimental, theoretical and computer modelling technique.

B. Optimizing of the output DC voltage of the photovoltaic devices in order to improve the efficiency of electrolysis process.

Direct current (DC) is needed for operating electrolysis. Shape of the DC current in function of time elapsed, or in other words, time dependence of the DC current plays important role at powerful electrolysis. In case of some DC current time shape the efficiency of the electrolysis process is low, but in case of some other cases the efficiency is high or higher. Our idea is to try to use DC current for electrolysis generated by photovoltaic systems.

A different idea to generate DC current is using photovoltaic panels as electric current generator. Solar panels are very good source for DC current generation, but the problematic point may be the time dependence or the time shape of the generated electric current.

Important goals of this part of the project are

- Characterizing by the photovoltaic system generated DC currents in function of time
- Trying to find the optimum shape of DC current for the electrolysis purpose
- Optimizing the efficiency of the electrolysis with the used DC current

In order to fulfil the goal written in the section here, we would like to develop new measurement technique or process in order to specify the time dependence of the DC voltage of the PV systems for electrolysis. How does the new measurement technique work? Does the new technique reliable? Can the new technique applicable for determination of stable time shape of the DC current? Another question may be that, how does the output DC current of the solar panels exactly depend on wavelengths of the incoming radiations (in case of natural and artificial lights as well)? Planned physical, technical and computer investigations could help at probing, testing, running and answering these questions.

Electrodes of the electrolyser in the industry are made from nickel-alloy, because these type of electrode materials are easily made and cheap.

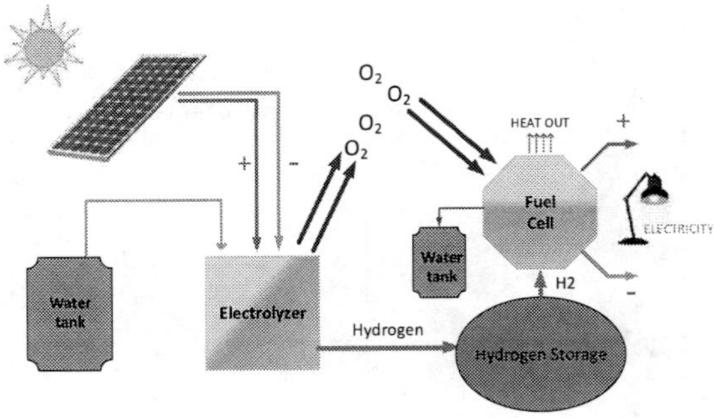

Figure 5.: The planned energy chain

C. Connections between the electrodes used for the electrolysis and the efficiency of the electrolyser

One of the main disadvantages of the clean nickel is that, the efficiency of the electrolysis is decreasing during the lifetime and uptime is running. In order to prevent the effect mentioned above, alloys are made for electrolysis purpose. Electrical conduction of alloys are more stable, so this way the efficiency of the electrolysis can be increased [8]. We would like to run experimental investigations with the use of many types of alloy electrodes made from different material properties (see as investigation of traditional and novel type electrodes).

Goal of the cell electrode research is to reach the optimum efficiency of the electrolyser. We would like to realize our aims using mainly experimental techniques, but as secondary technique we try to run theoretical technique and computer modelling as well.

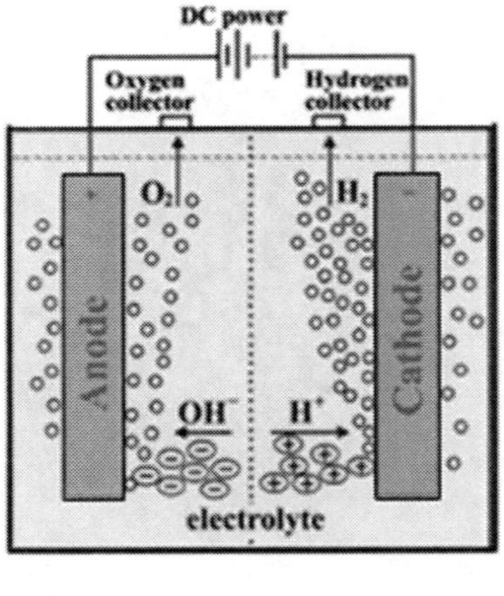

Figure 6.: Electrolyser cell

D. Connections between electrical conductivity, concentration and electrolysis voltage.

Electrical conductivity of the clear, distilled water is only 0.5 μS/cm at the temperature of 25 °C. In order to run effective and powerful operation of the electrolysis, increasing the electrical conductivity of the water should be done.

Experimentally used types of electrodes and concentration/composition of the used materials also affect the electrolyzing voltage of the electrolyser. In other words, types and concentration/composition of the used cell materials affect the efficiency of the electrolyser.

We would like to find correlations between the characteristics of the electrolytic solution, the cell material and the efficiency of the electrolyser. In order to find the best fitted results to this point, we plan to run experimental and theoretical modelling as well. Our research team plan to develop computer codes or series of routines based on MatLab environment to get answer the question of this point. Using computer modelling and experiments we might get successful results. We would like to realize our aims here using experimental methods.

E. Determining the string voltages of the photovoltaic systems optimized for electrolysis.

Using all the scientific sub-tasks detailed above and the results of scientific publications, we would like to determine the optimal string voltage of the photovoltaic systems used for electrolysing process. In order to determine the optimum string voltage of the photovoltaic system we will use experimental and theoretical processes.

F. Investigating of the operation of photovoltaic panels in function of the wavelength of the incoming light:

In order to answer these above-mentioned questions regarding to the DC voltage generated by solar panels, we would like to investigate photovoltaic cells by wavelength-selected irradiations in detail.

IV. Predicted possible results

The following results can be predicted:

- We could build the very first two-way hydrogen based energy storage system in Hungary.

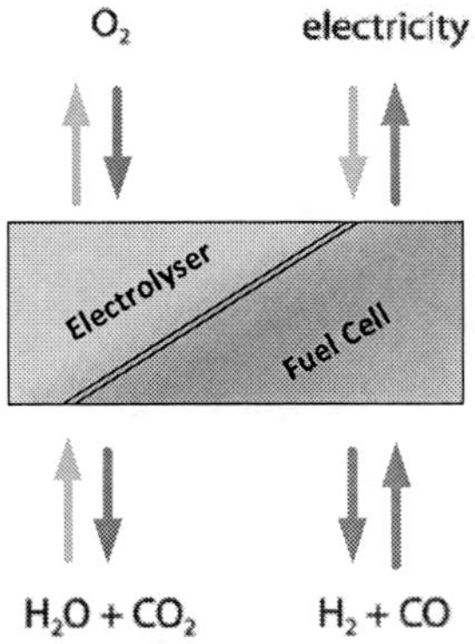

Figure 7.: Reversible electro-chemical cell

- Publication of the working physical model.
- Demonstration of carbon-dioxide-free "green" Hydrogen generation method.
- Making a good working physical and mathematical models, or in other words, laboratory experiments.
- Investigation of the energetic application of the non-used Hydrogen (which is industrial by-product in the real life)
- Investigation of the by-product oxygen came out from electrolysis as "secondary product".
- Testing from energetics point of view self-supporting water cleaning method.
- Investigation of reduction of the possible risks around environmental protection due to carbon-dioxide emission.
- As additional results: all developed experiments and computer modelling routines can be implemented to the engineering education (electrical engineering, energetics) at the university level.

V. Long-term plans

We plan to build a small scale integrated hydrogen system where we use only renewable energy sources and we use a two-way hydrogen based energy storage system. Of course, we will pay attention to the Sustainable Water and Sustainable Energy System.

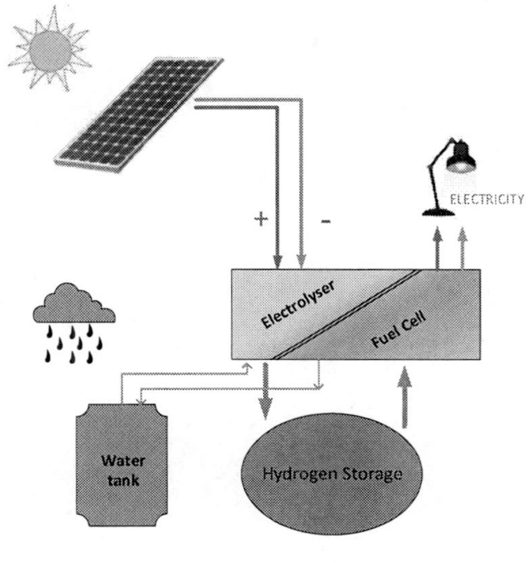

Figure 8.: Long-range, integrated solar-hydrogen system with sustainable water system

VI. Conclusion

At the global level, the industry of energy is currently fossil fuel based. We cannot change drastically this fact in the near future, but we must all focus on reducing greenhouse gas emissions. If there is a technical solution in our hands that can achieve this goal, then we have to do it.

We are responsible for technical development and the transfer of knowledge. We are responsible for the next generation's life.

I am confident that our research will help further sustainable development.

References

[1] United Nations, Department of Economic and Social Affairs, Population Division: "World Population Prospects: The 2015 Revision"

[2] International Energy Agency (IEA), "World energy outlook 2016.", Paris, OECD/IEA; 2016.

[3] Thomas Jefferson National Accelerator Facility - Office of Science Education, "The Element Hydrogen", Internet: http://education.jlab.org/itselemental/ele001.html, [May 9,2017].

[4] Stwertka, Albert: "A Guide to the Elements." Oxford University Press. pp. 16–21. ISBN 0-19-508083-1

[5] National Hydrogen Association; United States Department of Energy. "The History of Hydrogen". Internet: http://www.hydrogenassociation.org, [December 17, 2010].

[6] L.K. Shyuan: "Fuel Cell and Hydrogen Energy System" in *7th Asian School on Renewable Energy*, 2014, Fuel Cell Institute, University Kabangsaan Malaysia.

[7] R. Lan, J.T.S. Irvine, S. Tao, "Ammonia and related chemicals as potential indirect hydrogen storage materials", International Journal of Hydrogen Energy, Volume 37, Issue 2, 2012, Pages 1482-1494, ISSN 0360-3199, http://dx.doi.org/10.1016/j.ijhydene.2011.10.004.

[8] J.-H. Kim, J.-N. Lee, C.-Y. Yoo, K.-B. Lee, W.-M. Lee, "Low-cost and energy-efficient asymmetric nickel electrode for alkaline water electrolysis", International Journal of Hydrogen Energy, Volume 40, Issue 34, 2015, Pages 10720-10725, ISSN 0360-3199, http://dx.doi.org/10.1016/j.ijhydene.2015.07.025.

A New Method of Energy Calculation through Time Integration of a Vehicle Momentary Output using a Model in Matlab–Simulink

Stanislav Zajaczek, Jiří Cigánek, Maroš Ďurica
Department of Electrical Engineering
Faculty of Electrical Engineering and Computer Science
VŠB – TU Ostrava, Czech Republic
stanislav.zajaczek@vsb.cz, jiri.ciganek.fei@vsb.cz, maros.durica.st@vsb.cz

Abstract — **Energy calculations in the electric traction are of a cardinal importance for a preliminary economic evaluation of passenger and material transport, with a possibility of an optimal configuration of a traction vehicle loading for a given track. Another field of the use is development and production of new traction vehicles where the electrical energy consumption during the run can be reduced through a modification of the aerodynamic shape or a change of a drive type while maintaining the output. The aim of our research is an optimization of a calculation of tramway vehicle electrical energy consumption through a design of a model of a simplified connection of a vehicle and a feed array in the Matlab – Simulink environment. The entry data of the vehicle speed and traction power are arranged in matrices in Workspace in Matlab and used for gradual reading up of the model entry parameters in the Simulink environment. After starting the simulation, the data of the model feed array current and voltage is stored according to a chosen time step in a form of matrices and is used for a later graphic display of the energy calculation or for a verification of the model with corresponding measured values.**

Keywords: simulation; tramway; model; time; optimization

I. INTRODUCTION

In the future, the majority of large producers of transport vehicles in the European Union are planning to increase the production of vehicles with electric drives in order to meet legal emission limits. As a result, our research into the calculation of electrical energy consumption of a vehicle by a method of time integration of instantaneous power which is required for the overcoming of traction resistances in the individual driving modes has gained in importance. The model in the simulation performed by the Matlab – Simulink program includes many specific influences related to the tram journey dynamics as well as changes in the contact line parameters during a journey through a given track segment. A greater part of the input data is entered into the workspace; the data are arranged into matrices and used for a calculation of the main parameters for the simulation, as shown in the flowchart in Fig. 5. A tram load value, comprising both passengers and resistance components of a track, determines the required tractive power value on the wheel circumference. In order to make the calculation more accurate a model was created in the Matlab – Simulink software; its advantage is that it simulates a dynamic model behaviour which is affected by more influences at the same time. The simulation time can be shortened in an exact ratio to the real time and thus an evaluation and implementation of changes of the model parameters can be performed more quickly. A measurement on a unilaterally powered tram track in the Vřesinská – Zátiší segment of 9 km in length was performed in order to compare it with the calculated values of the electrical energy consumption during the tram journey which were obtained by using program applications. A tram track model used in the subsequent energy calculations was built in the Matlab – Simulink mathematical environment using GPS coordinates and corresponding data from geological websites.

II. TECHNICAL SPECIFICATION FOR TRAM AND TRACK DATA

Tram type T6A5, with registration number 1109, which currently provides transport for passengers on line No. 5, between the stations Vřesinská – Zátiší, was used for energy calculations. Tatra T6A5 is a unidirectional high-floor light rail vehicle designed for public transport in Europe and it is a sub-type of T6 standard trams produced in Tatra ČKD in Prague. The vehicle was derived from a bidirectional tram vehicle KT8D5. The body is mounted on two bogies, it has four axles and every axle is connected to one traction motor. The vehicle weighs 19.5 tons. The traction motors are direct-current four-pole series-wound electric motors TE 023, with power of 4×45 kW; they are controlled via pulse converters, with designation TV3 [2]. The maximum design speed is 65 km/h. A GPS satellite receiver in a mobile phone LG L40 with Android was used for recording the values of speed and position. The location of the GPS satellite receiver in the tram vehicle is shown in Fig. 1:

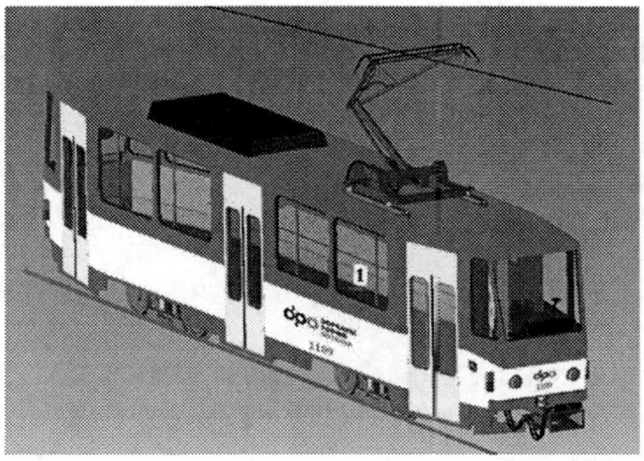

Figure 1. Gps receiver inside the tram, marked with number 1.

GPSLogger used the GPS signal of the Android phone to log the coordinates in a GPS or KML file at regular intervals. The application stores the recorded data on the MicroSD card in a file with the GPX extension; this file includes information about the latitude and longitude, speed, time and altitude. It also records the number of detected satellites and their spatial dispersion. The distances of the individual track sections were calculated by a command - distance on the surface of referential ellipsoid WGS84, this is a geodetic standard defined in 1984 which approximates the real shape of the Earth. It is related to the imperfect geometric shape of the Earth and the different distribution of its mass with regard to its centre. It is also used for calculating the distance in the GPS system. Track and time tachograms constitute especially important data for energy calculation by means of direct integration of vehicle power; it is shown in Fig. 2, which also shows the individual stops of the rail vehicle in relation to the length of the measured section and the journey time [3].

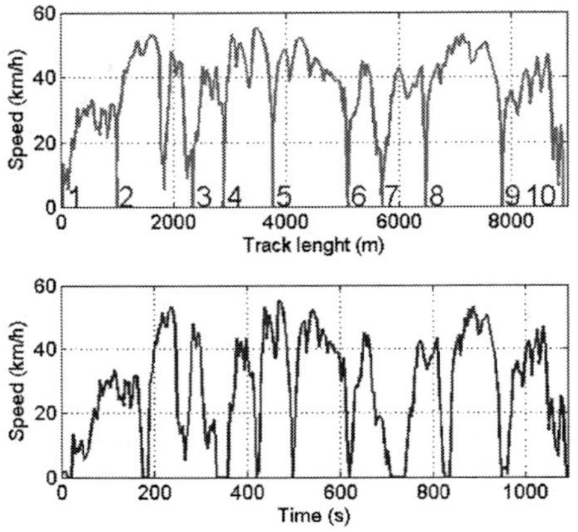

Figure 2. Track and time tachogram of a tram vehicle journey, the individual stops are shown in their sequence: 1 – Vřesinská, 2 – Poruba koupaliště, 3 – Vřesina, 4 – Nová Plzeň, 5 – Krásné Pole, 6 – U Obory, 7 – Dolní Lhota, 8 – Dolní Lhota osada, 9 – Horní Lhota, 10 – Zátiší.

Tram vehicle T6A5 has cushioned rubber wheels and it is therefore the Davis formula for calculating the specific rolling resistance which applies to it [1]:

$$p_0 = 3{,}65 + \frac{14{,}4}{M} + 0{,}045 \cdot V + \frac{44 \cdot s}{M \cdot n} \cdot V^2. \qquad (1)$$

where p_0 is specific rolling resistance for tram T6A5 (N/kN), M stands for weight per one axle (t), n is the number of axles, S stands for the frontal area of the vehicle (m^2) and V stands for the running speed (km/h).

GPS coordinates and corresponding data of altitudes taken from geological websites were used in the Matlab mathematical environment to construct a digital model of the tram track which was then used in energy calculations. The data were downloaded and saved directly in the Matlab work environment from the NASA server WorldWind WMS. The layer with the name mergedAsterElevations, which contains elevation data from three following map sources [4], was used for calculation:

- SRTM30 with Bathymetry 900m – an elevation map with accuracy of up to 900 m; the data are obtained by a radar located on a NASA satellite; it is regularly updated.
- ASTER 30m – an elevation map with accuracy of up to 30 m; the data are obtained by radar and by scanning the thermal emission in a visible and infrared bands. It is a joint project of the NASA, which provided the satellite, and the Japanese Ministry of Industry and Trade, which participated in the design of the equipment for scanning of the Earth's surface.
- USGS NED 10m – an elevation map with accuracy of up to 10 m; it was produced by filtering from the best available digital maps.

The layer mergedAsterElevations is defined on the NASA data server in the range of ±180° of longitude and ± 90° of latitude.

Figure 3. Digital model of the track with an elevation indicator; the red line shows the track. The tram journey started in the station Vřesinská, which is designated by number 1 in the Figure, and terminated in the station Zátiší, which is designated by number 2. The power supply of the track was provided by the substation Vřesina, which is designated by number 3.

The digital map of the vehicle track was used in the Matlab - Simulink mathematical environment to calculate the gradient of the tram track which is added up to the resistance from the curve in the given segment according to equation No. 2 and the result is then called reduced gradient; the progress is shown in Fig. 4.

$$p_{sred} = p_s + p_r = \pm s + pr \qquad (2)$$

where p_{sred} is reduced gradient of the tram track (N/kN), p_s is resistance from the track gradient (N/kN), p_r is resistance from the track curve (N/kN), s is the track gradient (‰).

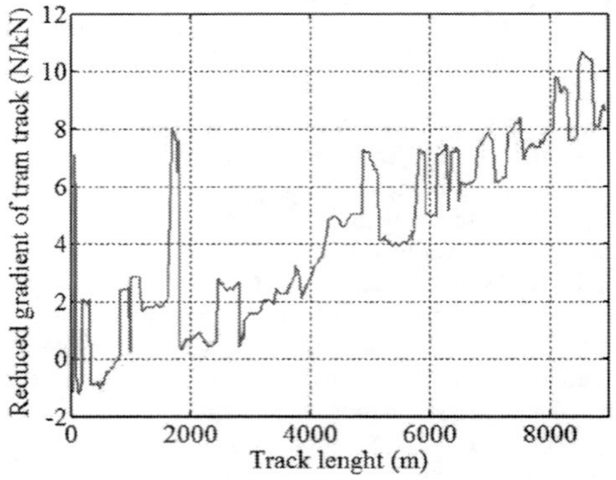

Figure 4. Reduced gradient of the tram track Vřesinská – Zátiší.

III. The Simulation of the T6A5 Type Tram Journey in the Vřesinská – Zátiší Track Segment

The calculation of the specific consumption of electric power by a rail vehicle can be done by the method of time integration of instantaneous power using the corresponding model in the Matlab - Simulink software. For this purpose Simulink can use the library Simscape Sim Power Systems, which includes elements commonly used in electric power engineering, such as network models, various electrical machines and model drive examples using different types of converters. The advantage is the possibility to load and save data during a simulation in the Matlab workspace which is called Workspace as shown in Fig. 5, which presents a flowchart of the simulation [5].

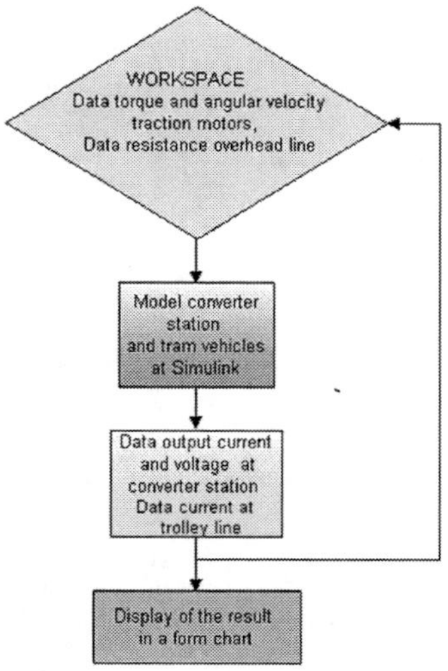

Figure 5. Flowchart simulation of a tram vehicle.

The model in Simulink is created by interconnection of discrete individual elements or composite parts with a data source from the workspace. The basis of the tram traction substation is a two-winding traction transformer 23 / 0.52 kV with a power output of 1.1 MVA, involving the winding Yd1. A 1- hour phase shift – 30° electrical degrees - is used for six-pulse rectification as shown in Fig. 6. Furthermore, the substation model uses measuring of substation voltage and current; the values are stored in the course of the simulation in the Workspace for later use. The model of the traction converter for the tram vehicle is shown in Fig. 7.

The basic components are:

- A GTO thyristor replaces the standard thyristor and trip circuit.
- A PI controller consists of a proportional and integration value of the controller, it is determined by the Ziegler - Nichols method and it is advantageous for the simulation since it is based on an experimental basis. The integral component is set to zero and the proportional component is adjusted so that the model will go beyond the stability limit and it will swing with undamped oscillations. By entering a corresponding critical value to the proportional element, the value of the integral component is calculated according to the oscillation period as stated in [6], [7]. Two PI controllers with a closed loop are used. The first one is a speed controller and in the simulation it substitutes for the activity of the driver and it is superior to the second controller of the traction motor current.
- The traction motors are substituted for by one motor with the corresponding parameters in the model. The traction motor, which in the model substitutes for the serial connection of bogie motor armatures, is connected with excitation in a series so that it will correspond to the real connection in the tram vehicle. After clicking on the motor icon, an editing window will open into which the parameters, such as the armature winding resistance, the excitation winding resistance, the armature winding inductance, the inertia, the viscous friction and the value of the initial speed, are entered.

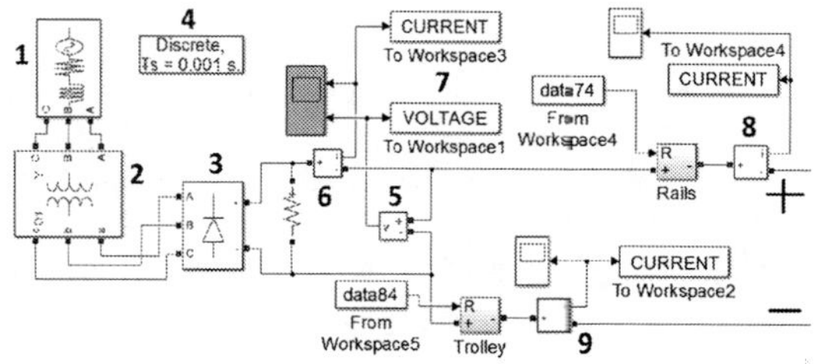

Figure 6. Traction substation connection, the individual parts are numbered according to their order: 1 – Source 23 kV, 2 – Traction transformer, 3 – Traction rectifier, 4 – Selection of simulation time - discrete time, 5 – Voltage measurement, 6 – Current measurement, 7 – Current and voltage data to Workspace, 8 – Measurement of current for rails, 9 – Measurement of current for traction contact lines.

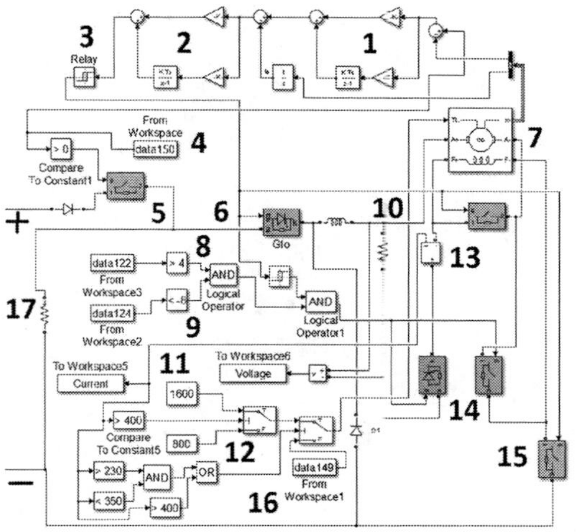

Figure 7. PWM converter with a series-wound DC motor, the individual parts are numbered according to their order: 1 – Speed controller, 2 – Current controller, 3 – Relay, switching hysteresis of the PWM converter, 4 – Reference data of the tram vehicle speed, reading from Workspace – (rad.s^{-1}), 5 – Main contactor, 6 – GTO thyristor, 7 – Series-wound DC motor, 8 – Lowest speed block during braking, 9 – Braking slowdown block, 10 – Braking resistance, 11 – Mechanical brake 1st stage, 12 – Mechanical brake 2nd stage, 13 – Brake current sensor, 14 – Brake mode switch, 15 – Run mode switch, 16 – Reference data of the corresponding torque on the shaft of the traction motor, reading from Workspace – (Nm), 17 – Auxiliary drives.

The traction motor simulation takes place in two quadrants, in the run mode and braking mode to resistance. Braking is active to the speed of 4 km/h – block No. 8 and in the slowdown higher than 8 km/h which will be ensured by block No. 9 in Fig. 7. In the simulation, block No. 11 and block No. 12 substitute for a two-stage mechanical brake, which is activated automatically in the case of brake current of 230 and 400 A. The switching components which are active in the run mode are in green colour and those which are active in the brake mode are in orange colour. The real time of the journey of the tram vehicle is too long for the simulation in Matlab - Simulink, therefore it was converted according to formula No. 3:

$$S_t = \frac{R_t}{10} \tag{3}$$

where S_t stands for simulation time (s) and R_t stands for real time (s).

IV. Conclusion

The contribution solved the problem of calculating the electrical energy consumption for selected tram vehicle T6A5 by means of a simulation in the Matlab – Simulink program. The model consists of a traction substation and a tram vehicle.

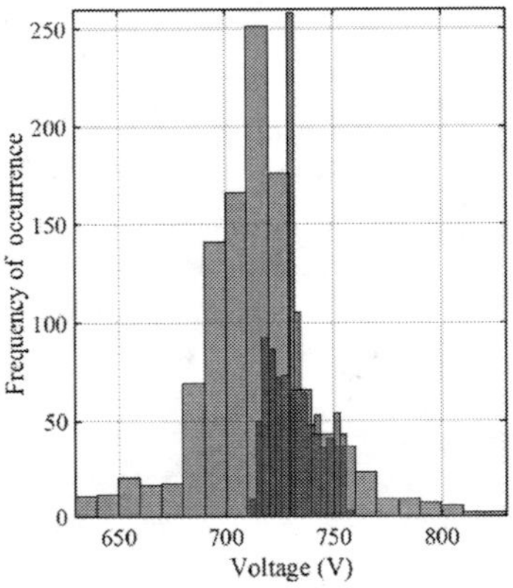

Figure 8. The histogram of the substation measured voltage is plotted in orange, the histogram of the substation simulated voltage is plotted in blue, the overlapping areas are plotted in brown.

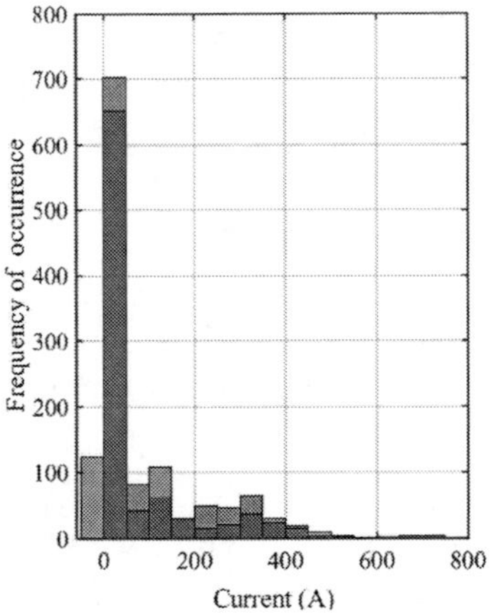

Figure 9. The histogram of the substation measured current is plotted in orange, the histogram of the substation simulated current is plotted in blue, the overlapping areas are plotted in brown.

In order to verify the model, measurement was carried out in the corresponding time at the substation Vřesina. Fig. 8 and 9 show the histograms of the measured and simulated substation output voltage and currents. We used the output values of voltage and current for the calculation of the electrical energy consumption by integrating the input of the vehicle in the given time. The result of the calculation of the electrical energy consumption by integration of the tram's instantaneous power by means of the simulation method is a 98% match with the measured values as can be seen in Fig. 10.

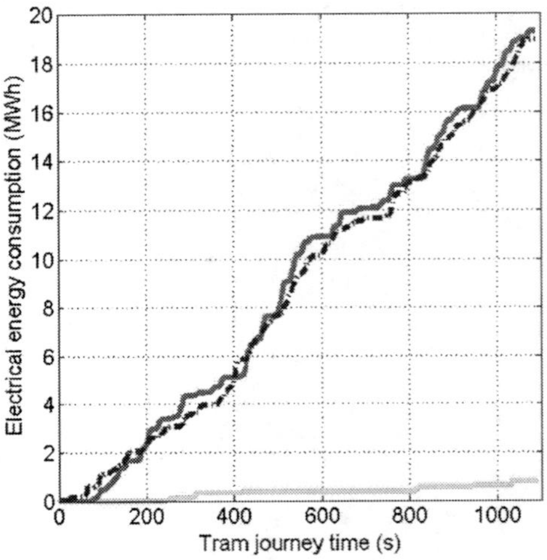

Figure 10. The measured consumption of electrical energy during the tram journey is shown in red, the simulated consumption is shown by a blue dashed line, the simulated consumption of electrical energy in the brake resistor is shown by a green line.

In our research, we wish to focus next on the improvement of the tram model in the Matlab – Simulink environment which will allow us to simulate more types of traction units and use the results for dimensioning traction contact lines.

References

[1] J. Palecek, "Electroenergetics in Transportation, study materials for lectures," VŠB – Technical University of Ostrava, pp. 1–36, Winter Semester 2010.

[2] J. Drábek, "Modern electric traction drives and their applications," Gdansk University of Technology, Poland, pp. 3–11, 2009, ISBN: 83-911669-6-1.

[3] G. Petrovski, "GPS, GLONASS, Galileo, and BeiDou for Mobile Devices," Cambridge University Press, United Kingdom, pp. 54–286, 2014, ISBN: 978-113-956-5455.

[4] M. Trauth, E. Sillmann, "MATLAB and Design Recipes for Earth Sciences," Springer – international publisher science and technology Heidelberg, Germany, pp. 134–146, 2013, ISBN: 978-3-642-32543-4.

[5] M. Perelmuter, "Electrotechnical Systems: Simulation with Simulink® and SimPower SystemsTM," CRC Press, USA, pp. 24–206, 2013, ISBN: 978-146-651- 4027.

[6] M. Bakshi, V. Bakshi,, "Elements Of Power Systems," Technical Publications Pune, India, pp. 204–346, 2009, ISBN: 978-81-8431-267-6.

[7] W. Liuping, Ch. Shan, Y. Dae, G. Lu, N. Ki, "PID and Predictive Control of Electrical Drives and Power Converters using Matlab/Simulink," Oxford University Press New Delhi, India, pp. 105–254, 2012, ISBN: 978-019-807-2447.

Determination of Current Probe Accuracy Depending on the Position of Conductor

Andrej Bolf, Pavol Belány, Filip Suško, Marek Roch, Juraj Altus, Michal Repák
Department of Power Electrical Systems
Faculty of Electrical Engineering, University of Žilina
Žilina, Slovak Republic
andrej.bolf@fel.uniza.sk

***Abstract*—This paper describes the current measurement by the indirect method. The value is measured for example with usage of current probe. This method is used in substations, where it is not possible to connect a measurement equipment to the circuit, because of the safety reasons. However, they can be used in circuits, which are not possible to be disconnected, because of the operation reasons. For this measurement were chosen three current probes of type ampFLEX for determination accuracy of these probes. From the results of this measurement, it is possible to make some recommendations for practice. Thank to well chosen measurement method, the error of measurement can be reduced to minimum. The graphical dependencies, which appear in this article, show the measurement errors of the current probes in the relation to the position of the conductor, from current value and from the harmonic order.**

Keywords—Rogowski coil; accuracy; measurement error; current probe

I. Introduction

Nowadays, with the boom of measurement technologies, there are changes in the way how to perform the measurement of the electrical quantities. The measurement of electrical quantities are closely related to the function of protections, which are used for example in substations. Therefore, accuracy of measurement is a very important factor. New technologies contributes to increasing the economy and accuracy of the measurements. The accuracy of the measurements is influenced by many factors, which even modern technologies can not completely eliminate and therefore, it is necessary to set certain methodical instructions for measurement [1].

One of the electrical quantities, which were affected by those changes, it is also the current measurement. Nowadays, direct and indirect methods are used for current measurements. The accuracy of the current measurement, which uses the direct method, depends mainly on the accuracy of the measurement equipment itself. The measurement, which is using the indirect method, is influenced by many factors. It consists of galvanically separated equipment from the circuit. For example, measurement by instrument transformers. This method leads to increased safety and it is mainly used in power engineering [2], [3]. The requirements for the instrument transformers differs according to application area. In power systems, modern ways of measuring currents and voltages for the security and monitoring are current sensors. These systems are the successors of classical current transformers. A large number of different shapes and sizes allows the usage of the current sensors in a variety of applications [4], [5]. Compact dimensions and low weight provide easy manipulation and installation. Current sensors are used in measuring and protection areas, where classical current transformers or direct methods can not be used. Large usage of current sensors are also in areas, with different protection of the machines, with large current flows [6].

The accuracy of the current sensors is large enough, but only when they are properly applied. In this case, that we are using these equipments for measurement of very small currents brings a mistake into the measurement results. Another of the factors, which are influencing the accuracy, is using unreliable current clips. It means that the location of the conductor is a wrong position. This is due to the fact, that current clips are used on large diameters of conductors and that the lead is not thin and therefore, not in the middle of the clip. Changing the position also changes the measurement error itself [1], [7].

II. Rogowski Coil Principle and Its Measurement Accuracy

Rogowski coil was named after Walter Rogowski. It is an electrical device for measuring the alternated current and high speed current pulses. From our point of view, the usage of the Rogowski coils is the best alternative to be used in medium voltage networks. They can be used instead of current transformer. Structure of Rogowski coil is shown in Fig. 1 [8], [9].

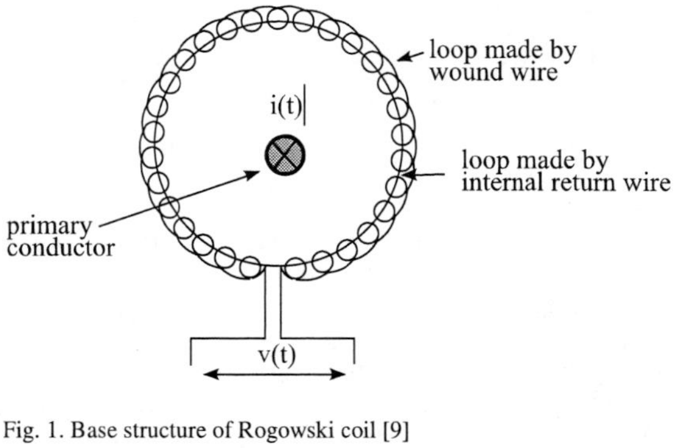

Fig. 1. Base structure of Rogowski coil [9]

The principle of functionality has been first published in 1992. It is uniformly wounded coil with non-magnetic core. It consists of a helical coil of wire with the lead from one end, returning through the center of the coil to the other end, so that both terminals are at the same end of the coil. The whole coil is then wrapped around the primary straight conductor, whose current is to be measured. The output voltage is proportional to the derivate of primary current and it is determined as [9]:

$$v(t) = M \cdot \frac{di(t)}{dt}, \tag{1}$$

where M is inductance and di(t)/dt is the derivation of the measured current over time.

As Rogowski coil does not consist of iron core, the saturation effect can not be present and the relationship between primary measured current. Secondary voltage has linear character, which is shown in Fig. 2. This picture shows comparison current transformer with Rogowski coil.

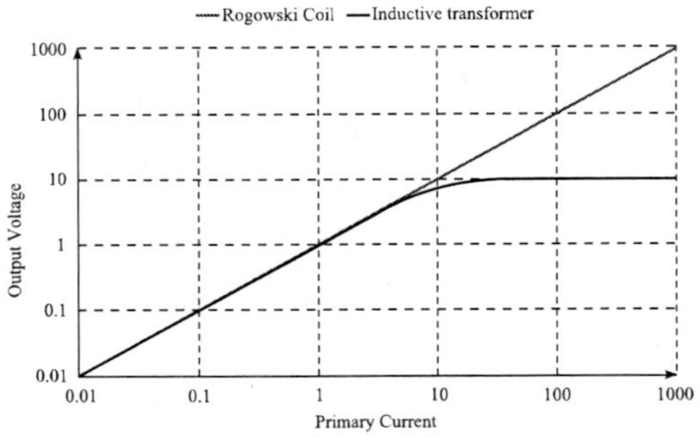

Fig. 2. Comparison of the Rogowski coil with Inductive transformer [9]

It is important to underline, that this comparison of the Rogowski coil with inductive transformer does not express accuracy changes, which comes to required dynamic range of the measurement. Inductive principle of the current measurement with ferromagnetic core reaches the saturation point at the certain current level. From this point, the accuracy of inductive current transformer is not guaranteed. Fig.3. shows the amplitude accuracy performance of Rogowski coil trough entire measurement range [8], [9].

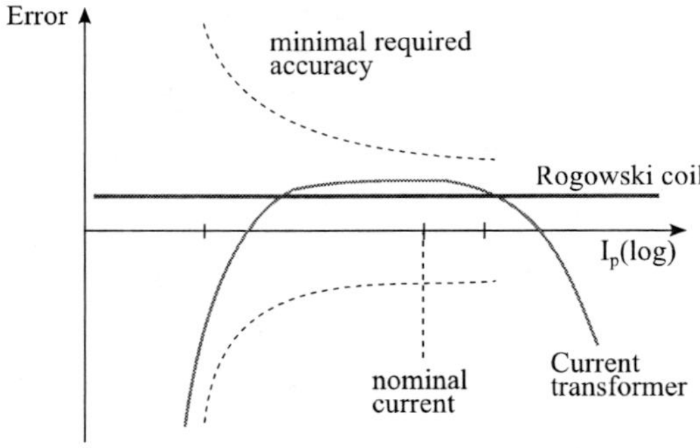

Fig. 3. Accuracy of Rogowski coil and current transformer from current size measurement

Amplitude accuracy of the Rogowski coil across entire dynamic range is constant. This is very important characteristic. Modern intelligent electronic devices provide possibility to set correction factors of the amplitude and phase measurements. Proper setting of the amplitude error correction in intelligent electronic devices brings error of the current measurement used by intelligent electronic devices close to 0 and this error is applicable across complete dynamic range [9].

III. Accuracy measurement of current probe

The accuracy measurement was realized with usage the current probe, type ampFLEX. For measurement were chosen three current probes. Each probe was tested in five positions face to primary conductor and measured with eight different values of the current. These positions of primary conductor are shown in Fig. 4. The current signal, which was needed for the measurement was supplied from the programmable source.

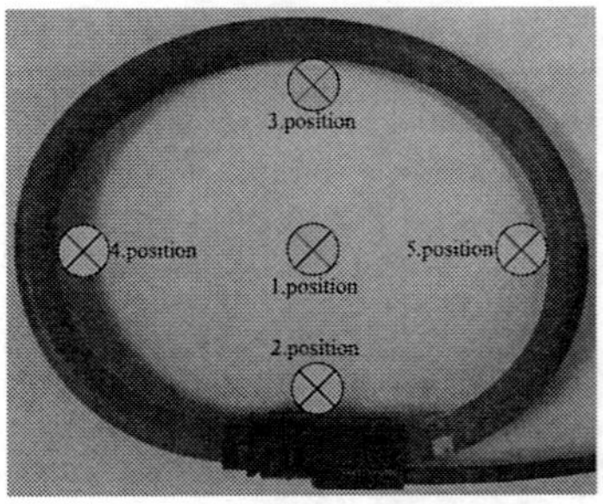

Fig. 4. Five positions of primary conductor into the current probe

In the next tables are shown the measured values of each current probe and their evaluated errors from the measurement. Table 1 to 3, presents the values of the first tested current probe. Table 4 to 6, presents the values of the second tested current probe and table 7 to 9, presents the values of the third tested current probe.

Table 1. Measured values of the ampFLEX current probe

L_1	$I_N = 1$ A					
I_0 [A]	Harmonic order	I_1 [A]	I_2 [A]	I_3 [A]	I_4 [A]	I_5 [A]
1,001	1.	1,002	0,990	1,007	1,010	1,007
1,001	3.	1,011	0,993	1,010	1,012	1,011
1,001	15.	1,008	0,998	1,012	1,014	1,012
1,001	25.	1,012	0,997	1,014	1,014	1,011

Absolute error of measurement				
ΔX_1	ΔX_2	ΔX_3	ΔX_4	ΔX_5
-0,001	0,011	-0,006	-0,009	-0,006
-0,010	0,008	-0,009	-0,011	-0,010
-0,007	0,003	-0,011	-0,013	-0,011
-0,011	0,004	-0,013	-0,013	-0,010
-0,016	-0,008	-0,019	-0,020	-0,022

Relative error of measurement				
ε_{r1} [%]	ε_{r2} [%]	ε_{r3} [%]	ε_{r4} [%]	ε_{r5} [%]
-0,100	1,059	-0,599	-0,899	-0,599
-0,999	0,799	-0,899	-1,099	-0,999
-0,699	0,310	-1,099	-1,299	-1,099
-1,099	0,410	-1,299	-1,299	-0,999
-1,598	-0,799	-1,898	-1,998	-2,198

Table 2. Measured values of the ampFLEX current probe

L_1	$I_N = 60$ A					
I_0 [A]	Harmonic order	I_1 [A]	I_2 [A]	I_3 [A]	I_4 [A]	I_5 [A]
60,000	1.	60,150	59,090	60,470	60,330	60,230
60,100	3.	60,130	59,210	60,610	60,380	60,350
60,120	15.	60,780	59,570	61,050	60,980	60,790
60,054	25.	62,090	60,880	62,350	62,330	61,160

Relative error of measurement				
ε_{r1} [%]	ε_{r2} [%]	ε_{r3} [%]	ε_{r4} [%]	ε_{r5} [%]
-0,250	1,517	-0,783	-0,550	-0,383
-0,050	1,481	-0,849	-0,466	-0,416
-1,098	0,915	-1,547	-1,430	-1,114
-3,390	-1,375	-3,823	-3,790	-1,842

Table 3. Measured values of the ampFLEX current probe

L1		I_N = 100 A				
I_0 [A]	Harmonic order	I_1 [A]	I_2 [A]	I_3 [A]	I_4 [A]	I_5 [A]
100,100	1.	100,200	98,610	100,700	100,500	100,300
95,882	3.	100,400	98,520	100,200	100,700	100,400
99,306	15.	101,700	99,980	102,200	102,100	101,800
101,230	25.	104,600	102,500	105,100	105,000	104,800

Relative error of measurement				
ε_{r1} [%]	ε_{r2} [%]	ε_{r3} [%]	ε_{r4} [%]	ε_{r5} [%]
-0,100	1,489	-0,599	-0,400	-0,200
-4,712	-2,751	-4,503	-5,025	-4,712
-2,411	-0,679	-2,914	-2,814	-2,511
-3,329	-1,255	-3,823	-3,724	-3,527

Table 4. Measured values of the ampFLEX current probe

L2		I_N = 1 A				
I_0 [A]	Harmonic order	I_1 [A]	I_2 [A]	I_3 [A]	I_4 [A]	I_5 [A]
1,008	1.	1,005	1,006	1,013	1,012	1,003
1,001	3.	1,005	1,006	1,013	1,012	1,004
1,001	15.	1,010	1,008	1,018	1,015	1,006
1,001	25.	1,013	1,005	1,022	1,021	1,009

Relative error of measurement				
ε_{r1} [%]	ε_{r2} [%]	ε_{r3} [%]	ε_{r4} [%]	ε_{r5} [%]
0,298	0,198	-0,496	-0,397	0,496
-0,400	-0,500	-1,199	-1,099	-0,300
-0,899	-0,699	-1,698	-1,399	-0,500
-1,199	-0,400	-2,098	-1,998	-0,799

Table 5. Measured values of the ampFLEX current probe

L2		I_N = 60 A				
I_0 [A]	Harmonic order	I_1 [A]	I_2 [A]	I_3 [A]	I_4 [A]	I_5 [A]
60,000	1.	60,610	60,260	61,050	60,840	60,280
60,100	3.	60,600	59,890	60,530	61,050	60,330
60,120	15.	61,250	60,680	61,710	61,590	60,950
60,054	25.	62,500	61,890	63,210	62,840	62,200

Relative error of measurement				
ε_{r1} [%]	ε_{r2} [%]	ε_{r3} [%]	ε_{r4} [%]	ε_{r5} [%]
-1,017	-0,433	-1,750	-1,400	-0,467
-0,832	0,349	-0,715	-1,581	-0,383
-1,880	-0,931	-2,645	-2,445	-1,381
-4,073	-3,057	-5,255	-4,639	-3,573

Table 6. Measured values of the ampFLEX current probe

L2		I_N = 100 A				
I_0 [A]	Harmonic order	I_1 [A]	I_2 [A]	I_3 [A]	I_4 [A]	I_5 [A]
100,100	1.	100,900	100,200	102,000	101,500	100,400
95,882	3.	101,000	99,980	101,800	101,500	100,600
99,306	15.	102,500	101,400	103,500	103,000	102,200
101,230	25.	105,400	104,300	106,300	105,600	104,800

Relative error of measurement				
ε_{r1} [%]	ε_{r2} [%]	ε_{r3} [%]	ε_{r4} [%]	ε_{r5} [%]
-0,799	-0,100	-1,898	-1,399	-0,300
-5,338	-4,274	-6,172	-5,859	-4,921
-3,216	-2,109	-4,223	-3,720	-2,914
-4,119	-3,033	-5,008	-4,317	-3,527

Table 7. Measured values of the ampFLEX current probe

L3		I_N = 1 A				
I_0 [A]	Harmonic order	I_1 [A]	I_2 [A]	I_3 [A]	I_4 [A]	I_5 [A]
1,001	1.	1,008	1,007	1,002	1,007	1,005
1,001	3.	1,005	1,004	1,003	1,005	1,007
1,001	15.	1,005	1,007	1,003	1,008	1,006
1,001	25.	1,011	1,012	1,006	1,011	1,011

Relative error of measurement				
ε_{r1} [%]	ε_{r2} [%]	ε_{r3} [%]	ε_{r4} [%]	ε_{r5} [%]
-0,689	-0,589	-0,090	-0,589	-0,390
-0,400	-0,300	-0,200	-0,400	-0,599
-0,400	-0,599	-0,200	-0,699	-0,500
-0,999	-1,099	-0,500	-0,999	-0,999

Table 8. Measured values of the ampFLEX current probe

L3		I_N = 60 A				
I_0 [A]	Harmonic order	I_1 [A]	I_2 [A]	I_3 [A]	I_4 [A]	I_5 [A]
60,000	1.	60,250	60,400	59,950	60,330	60,310
60,100	3.	60,310	59,850	59,970	60,330	60,270
60,120	15.	60,930	61,050	60,730	61,010	60,960
60,054	25.	62,170	61,860	61,880	62,220	62,310

Relative error of measurement				
ε_{r1} [%]	ε_{r2} [%]	ε_{r3} [%]	ε_{r4} [%]	ε_{r5} [%]
-0,417	-0,667	0,083	-0,550	-0,517
-0,349	0,416	0,216	-0,383	-0,283
-1,347	-1,547	-1,015	-1,480	-1,397
-3,523	-3,007	-3,041	-3,607	-3,757

Table 9. Measured values of the ampFLEX current probe

L3		I_N = 100 A				
I_0 [A]	Harmonic order	I_1 [A]	I_2 [A]	I_3 [A]	I_4 [A]	I_5 [A]
100,100	1.	100,400	99,700	99,870	100,500	100,600
95,882	3.	100,300	99,770	99,920	100,600	100,600
99,306	15.	102,000	101,300	101,300	101,900	101,900
101,230	25.	104,900	104,000	104,400	105,000	104,800

Relative error of measurement				
ε_{r1} [%]	ε_{r2} [%]	ε_{r3} [%]	ε_{r4} [%]	ε_{r5} [%]
-0,300	0,400	0,230	-0,400	-0,500
-4,608	-4,055	-4,211	-4,921	-4,921
-2,713	-2,008	-2,008	-2,713	-2,612
-3,625	-2,736	-3,133	-3,724	-3,527

However, when we performing any type of the measurement, we commit inaccuracies, so the measurement result differs from the actual value. From this reason, we have to determine an error of measurement. The difference between the actual value and the measured value is the absolute error of measurement ε, which is determined as:

$$\varepsilon = \Delta X = X_A - X_M, \qquad (2)$$

where X_A is the actual value and X_M is the measured value.

The error can be positive or negative. If the error is positive, we need to add it to the measured value, if we want to achieve the correct value. If the error is negative, we have to deduct it. It is determined as:

$$X = X_M \pm \varepsilon, \qquad (3)$$

where X_M is the measured value and ε is measurement error.

This formula means that the actual value of X is in the range of values from X_M - ε to X_M + ε.

If we want to express the error relative to the measured value, we can talk about relative error ε_r. This error is determined as:

$$\varepsilon_r = \frac{\Delta X}{X_A} = \frac{X_A - X_M}{X_A} \times 100\%, \qquad (4)$$

From the measured values were created graphic dependencies, which are shown on Figures 5 to 10.

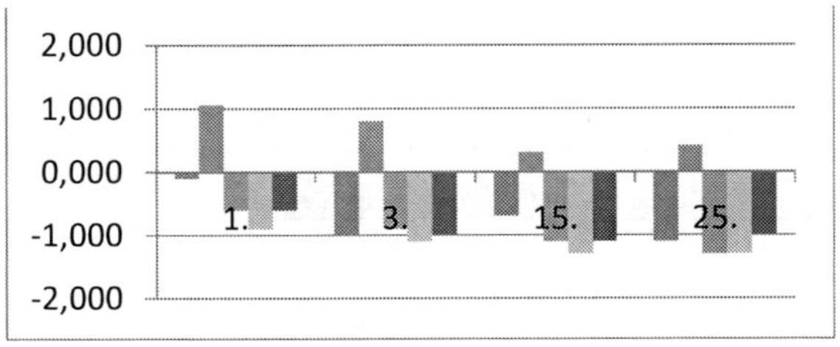

Fig. 5. Graphic dependence of relative measurement error from harmonic order for the current 1A

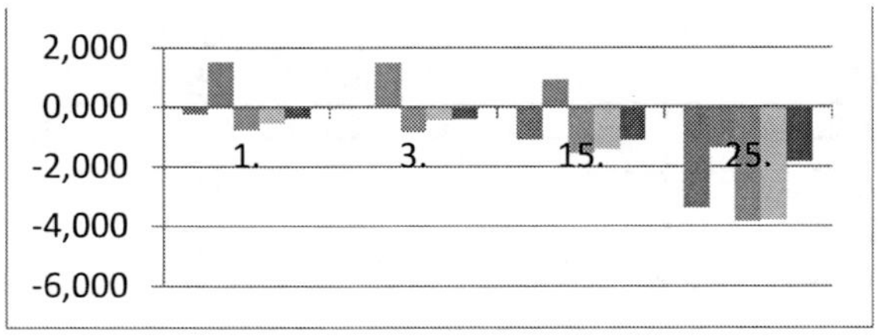

Fig. 6. Graphic dependence of relative measurement error from harmonic order for the current 60A

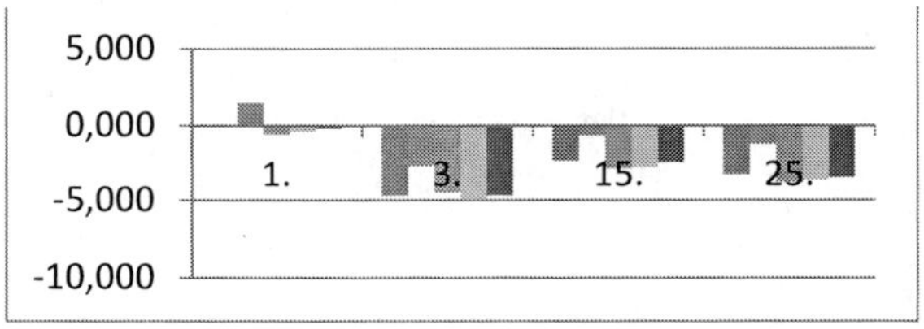

Fig. 7. Graphic dependence of relative measurement error from harmonic order for the current 100A

These Figures 5 to 7, presents how to changes the relative error of measurement from the harmonic order for the different range of currents. We can see, that with increasing the harmonic order or currents, also the relative error of measurement is increased respectively.

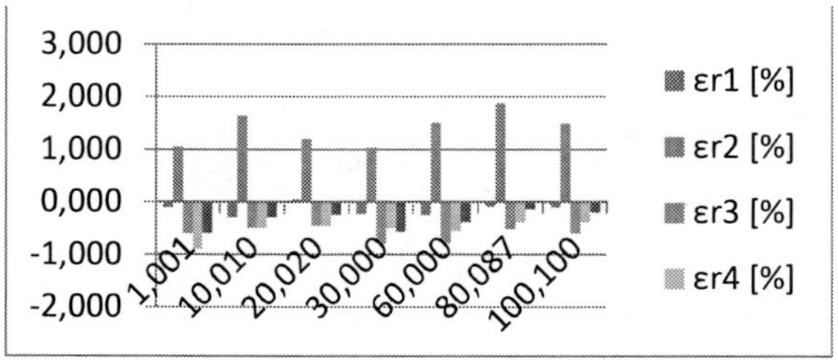

Fig. 8. Graphic dependence of relative measurement error from the current for the first harmonic order

Graphic dependencies are shown on Figures 8 to 10. They present how is the relative error of measurement from the current for the different harmonic order changed. The results is, that for the certain harmonic order and the different currents, the relative error of measurement changes. The higher the harmonic order is, the greater is the relative error of measurement.

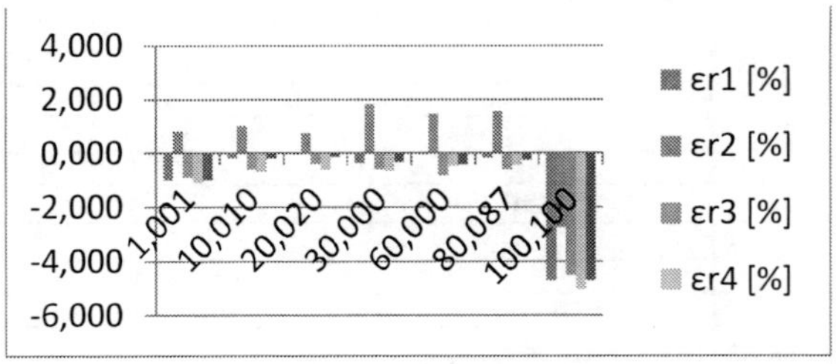

Fig. 9. Graphic dependence of relative measurement error from the current for the third harmonic order

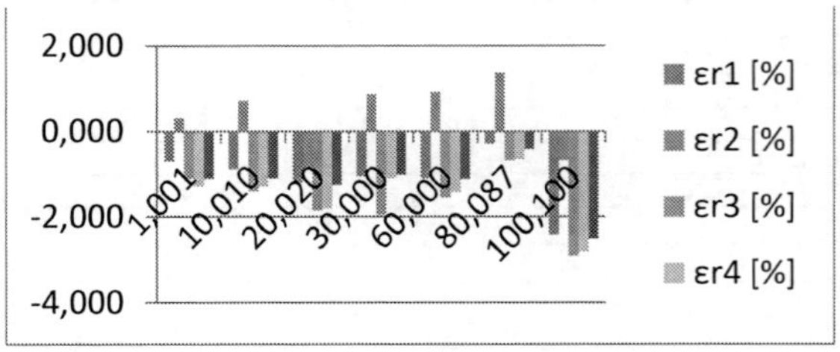

Fig. 10. Graphic dependence of relative measurement error from the current for the 15th harmonic order

IV. Conclusion

An indirect method was used to determine the accuracy of the current probe. Three current probes, with type of ampFLEX, were used for the measurement. It consists of placing the conductor in the field of the current probe. We found out, that the accuracy of current probe depends on the positions of conductor, the current value, the harmonic order, voltage inbalance and the type of probe. On the other hand it does not depend on the positions of another conductor near the probe, unless another conductor is not located in the probe field. From the measured values, which are shown in tables 1 to 9, several measurement errors, which occurred during the measurement were calculated. Graphical dependencies have been constructed from the calculated errors. These tables and graphs show, that if we measure higher current value, we commit the greater measurement error depending on the position of the measured conductor and also the harmonic order. The most accurate measurement was made, when the conductor was located at the center of current probe. This position is shown on Figure 4 on position number 1. Of course, such measurement cannot be carried out in practice, because the measured conductor is always located at the edge of the current probe and thus there is a certain inaccuracy. The most inaccurate measurements were, when the conductor was located at the junction of the current probe. This position is shown on Figure 4, in position number 2.

Acknowledgment

This work was supported by the Slovak Research and Development Agency under the contract No. APVV-15-0464.

References

[1] E. Oyarbide, C. Bernal and P. Molina-Gaudó, "New Current Measurement Procedure Using a Conventional Rogowski Transducer for the Analysis of Switching Transients in Transistors," in IEEE Transactions on Power Electronics, vol. 32, no. 4, pp. 2490-2492, April 2017.
doi: 10.1109/TPEL.2016.2624502

[2] R. Javora, M. Stefanka, P. Mahonen, T. Niemi and O. Rintamaki, "Protection in MV networks using electronic instrument transformers," CIRED 2009 - 20th International Conference and Exhibition on Electricity Distribution - Part 1, Prague, Czech Republic, 2009, pp. 1-4.

[3] J. P. Dupraz, A. Fanget, W. Grieshaber, G. F. Montillet, "Rogowski Coil: Exceptional Current Measurement Tool For Almost Any Application", IEEE General Meeting, Tampa, FL, June 24-28, 2007.S. Ghosh, "Electrical Machines," Pearson Education, India, 2005, pp. 243-246, ISBN 978-81-317-0509-4.

[4] P. Mähönen, V. Virtanen, T. Hakola, "The Rogowski coil and the voltage divider in power system protection and monitoring", Internet: https://library.e.abb.com/public/d9317de927099d73c2256c3e002bc491/CIGREreport.pdf, [May 16, 2017].

[5] L. A. Kojovic, M. T. Bishop, and T. R. Day, "Operational Performance of Relay Protection Systems Based on Low Power Current Sensors," in *Proceedings of the 36th Annual Western Protective Relay Conference*, Spokane, WA, 2009.

[6] W. Z. Fam, "A Combined Current and Voltage Sensor for Metering and Protection in High Voltage Power Systems", Proceedings of CCECE/CCGEI 1994, IEEE Catalog Number: 94TH8023, ISBN: 0-7803-2416-1.757.

[7] J. D. Ramboz, "Machinable Rogowski coil, design, and calibration," in IEEE Transactions on Instrumentation and Measurement, vol. 45, no. 2, pp. 511-515, Apr 1996.
doi: 10.1109/19.492777

[8] IEEE Guide for the Application of Rogowski Coils Used for Protective Relaying Purposes. IEEE Power Engineering Society. New York, NY, USA, ISBN: 978-0-7381-5714-6, 2008

[9] M. Štefanka, "Aplikace senzorů a digitalizace podle IEC 61850 v sítích a rozvodnách vysokého napětí" ("Application of Sensors and Digitalization Based on IEC 61850 in Medium Voltage Networks and Switchgears"), PhD. Thesis, Brno: Brno University of Technology, 2016

The Application Possibilities of a Multi-agent System in Distribution Grid Operation

Filip Suško, Andrej Bolf, Pavol Belány, Michal Repák, Juraj Altus
Department of Power Electrical Systems, Faculty of Electrical Engineering, University of Zilina,
Zilina, Slovakia
{filip.susko, andrej.bolf, pavol.belany, michal.repak, juraj.altus}@fel.uniza.sk

Abstract— **This paper presents control of distribution network using multi-agent system as well as network analysis. Multi-agent system itself consists of several agents with particular functions. The number of agents and their functions are closely related to the application area. Network analysis deals with the capabilities of network in different modes. The last chapter describes the proposed software. This can be used for real-time control and as well as simulation needs.**

Keywords: renewable energy; distribution grid; autonomous; multi-agent systems; structure; simulation; photovoltaic

I. Introduction

Nowadays, since trends in control and application of different types of renewable energy source are moving forward, it is necessary to ask what direction will follow the control system of distribution grids [1]. From the current development, it is clear that the impact of the utilization of modern grid components decreases the number of people needed to provide the necessary manipulations. Utilization of these components have opened up new possibilities in distribution grid control. However, it is necessary to establish requirements for the control system itself. As suitable solution seems to be the use of autonomous control. Autonomous control itself is closely connected with the field of robotics. Typical case is a robot that aims to find convenient path to the desired target with various external adverse effects [2].

The example of the autonomous system implementation is a multi-agent systems. Multi-agent systems and its application in distribution grids depend mainly on the chosen concept. The current trend of massive development of smart metering also contributes to utilization of multi-agent systems, as appropriate type of smart meter can provide highly relevant information required for operation of multi-agent system [3], [4]. It is also possible to control the demand using the smart meter, if it is equipped with switching relay or high-power transistor. The design of such a system is subjected to some requirements. It must be noted that the requirements are arising from certain historical control patterns and processes. It also pretty much depends on the topology of components and sources connected to the distribution grids. The number of agents and of their subordinate's services depends largely on the facts mentioned above. Therefore, it is necessary to make a model of the distribution grid using which it would be possible to determine the voltage drops under various loading conditions. Based on the results, it is possible to determine the ways of grid operation (on grid / off grid). Then, it would be possible to design very structure of multi-agent system with recommendations. However, structure itself must take into account the autonomy and the possibility of extending.

II. An Autonomous control and a Multi-agent systems

A. *An autonomous system*

Intelligence is a basic human ability to act individually and seek solutions to problems that it has not meet yet [5]. The origins of autonomous systems are associated with robotics. The simplest case is an example of a robot that is able to find the appropriate path to the specified target exposing to various external circumstances. Some examples of the ability to search for a suitable way can be seen in [6]. The first application designed for reverse analysis of events in the electrical power system were outlined in [7] and [8]. Later, there were also examples of the use of autonomous systems in the control of electrical power systems [9].

An autonomous system should have specific decision-making ability about themselves through the application of well-defined control procedures in the first place. It is able to adapt to changing conditions and creates new useful knowledge from the environment [3].

Such a system is usually made of mechanical structure, sensors and actuators, which are subjected to the control software. A typical example of such a system is self-moving pilotless vehicle designed to explore unknown terrain. In this case, we can talk about fully autonomous control. It is possible to utilize only a partial autonomous control, in the case of remote-controlled robotic vehicle, whose distance from the control center is so large that it causes delay during signals transmission. Remote object just gets the general commands (e. g. take samples) with such control and then they are translated to specific moves and acts in the autonomous control system of the vehicle itself [3].

Complex autonomous system consists of components, which are autonomous systems themselves, they just have a narrower scope of competence [3].

Autonomous systems can be divided into [3]:

- control systems,
- information systems.

The responsibility of the autonomous control system is focused on a specific process, for example movement of the vehicle. It also contains diagnostic functions that make possible to evaluate the behavior and inform of the real reachability of the prescribed target. Autonomous information system analyzes the behavior of another process or system and obtains comprehensive information about it [3].

The main advantage of control through autonomous system is long-term operation without manual intervention and adequate response even in an emergency and in rapidly changing conditions. Autonomous systems are of great importance especially in the control of complex processes and large-scale interconnected systems. The requirement of action despite incomplete access to all information has been associated with this.

Autonomous systems are systems operating with uncertain information, therefore they fall within the field of artificial intelligence [4]. Intelligent capabilities such as learning by experience, action planning and identification of faults that have been done by the system operator till now, are gradually becoming more inserted into autonomous systems. This fact then results in better prevention of disturbances, faster reaction times and overall simplification of system operators' work [3].

B. *Component of autonomous system and layers of autonomous component*

Component of the autonomous system, so called the autonomous component, should contain three layers - entity, coordination and implementation (Figure 1). There is a two-way flow of information between these layers. The diagnostic information are dominated in the way bottom-up and the command information mainly in the opposite direction, i.e. from top to bottom. These layers, especially their components, are mutually dependent and as individuals are useless [6]. Autonomous component operates independently on the basis of the entry requirements until these requirements are changed, even in the event of changes in working conditions. For example it can represent maintaining of the set power flow on the power line or load estimation. The output or the action of an autonomous component, which is in the system at a higher level, is used as an input, definition of the target for the lower-level component [3].

Each of the different levels of autonomous component contains a unit for sending and receiving information. It also includes a unit for logical reasoning, which derives new output values based on input data, all available information and instantaneous output values.

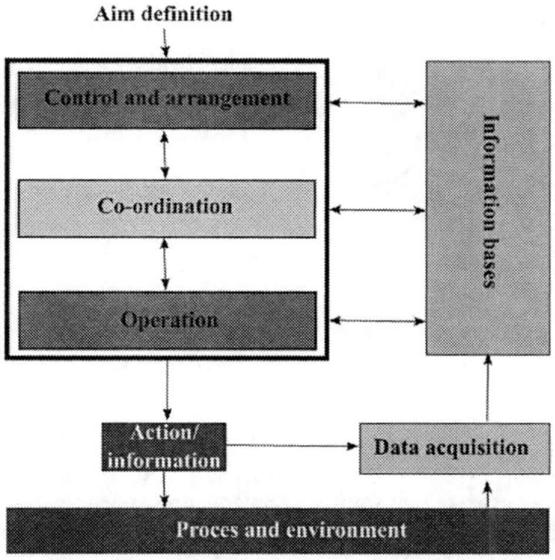

Figure 1. The structure of autonomous component

Each level of the autonomous components differs in level of abstraction, which increases from the executive level towards control and organizational level. Such structure separates each individual task and allows better concentration on the part of the problem corresponding to appropriate layer.

Hierarchical structure of the autonomous component layers can be compared to the translation of the source code in a higher programming language into low-level machine code that runs on the hardware architecture.

The basic source of information for all layers of an autonomous component is an information base that contains information of controlled system or process as well as general information from other autonomous components [3].

Term agent in the autonomous systems context refers to real or virtual object placed in an environment, in which an agent can act, perceive and partly imagine. This object is able to communicate with other agents and has an autonomous behavior that is based on his observation, knowledge and interaction with other agents [5]. Several authors have also reported other definitions, based on knowledge of problems resulting from the focus of their research. However, these definitions generally tend to conclude that the autonomous agent is placed in an environment, in which takes part in, perceives it with sensors and intervenes into it in order to influence the results to be recorded in the future [10]. There is a certain similarity between the concept of agent in artificial intelligence and of object in the field of software engineering. Despite their common features, the agent is more abstract and it also has some important features, in particular [11]:

- autonomy,
- existence in the particular environment,
- continuity of sensor's and actuator's links with the environment.

Agent is the basic building block of system that features intelligent behavior in the defined area.

The essential characteristic of an agent is that it operates independently that is why the concept of autonomous or intelligent agent is common. However, it does not mean that it works alone, but it has a dedicated responsibility. Intelligence in this case means that, for example, it is able to derived conclusions from known facts or acquires new knowledge by learning. Agent, in order to carry out their activities, must be able to evaluate the overall success of their actions. It is desirable to use an arrangement, where one agent only provides certain specific tasks and then there is a complex system of cooperating agents, which is called multi-agent systems (MAS). Multi-agent systems is relatively powerful tool, especially useful in areas, where the control strategy is hardly definable analytically or where exact analytical solution is possible, but it takes time and resources [5].

III. 22 KV DISTRIBUTION NETWORK

The basic characteristic of 22 kV distribution network is its structure. It is mostly a tree structure, but it can also have a lattice structure, mainly in urban cable networks. The distribution network includes many elements, for example power lines, transformers, switching elements, and more. Certain switching elements are permanently open in the fundamental connection. A change in their state occurs for example in the case of failure, when it is necessary to change network configuration to achieve restoration of power supply from another electrical substation. It means that an alternative connection is each connection, in which change in switching elements state took place compared to fundamental connection. This ability should be one of the fundamental capabilities of multi-agent system utilized for the control and operation of Smart Region. Multi-agent systems must also cope with changes in generation from renewable sources and consumption changes in this case.

A. *Renewable energy sources*

There are many renewable energy sources, which can be utilized in the distribution network. It is necessary to perform the analysis before their utilization, which consists of the network operation computing and determination of the impact of the renewable energy sources connection to the network. Unless the result of the analysis is positive, the source can be connected to the network. It should be noted that renewable energy sources are mostly distributed in the certain area and even if they are not placed side by side, they can work as one virtual power plant. It means that the sources apparently act as a single power plant with specified parameters. Renewable energy sources generation allocation is subjected to the availability of necessary input sources for generation. There are some predictive models used to calculate the achievable output, which level of accuracy depends on the time period of forecast.

Figure 2 shows the predicted values and the real output of wind power plant. These are the values for 24 hours interval.

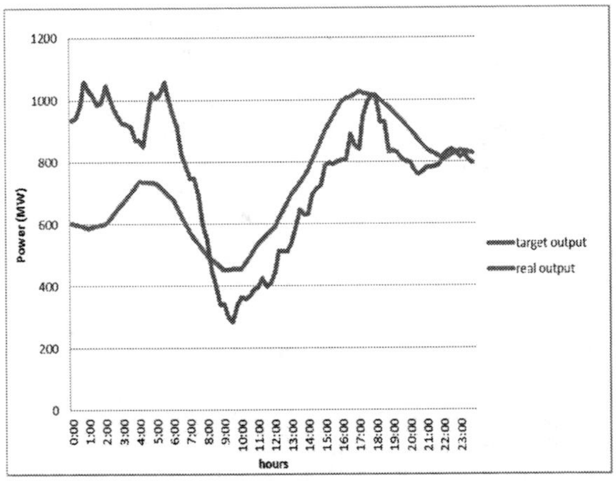

Figure 2. Prediction of wind power plant generation

Figure 3 presents predicted and real output of the photovoltaic power plant during one month.

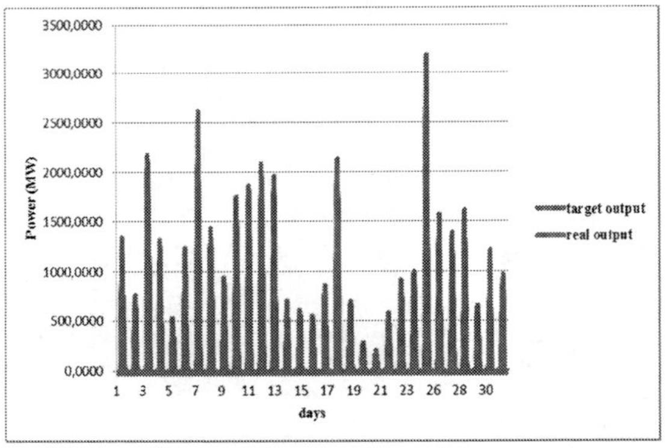

Figure 3. Prediction of photovoltaic power plant generation

B. Network operation computing

Computing of network operation is important in terms of behavior of the region or part of the network, voltage conditions and other important aspects. A given part of the network includes the generation from renewable energy sources, which are represented by the photovoltaic power plants with installed capacity 6.55 MW in this case. This output is considered to be the maximum achievable output under ideal weather condition with sufficient solar irradiance. If the network would operate in an island mode (off-grid), the sum of output power that can be supplied by renewable energy source is insufficient for case of maximum load. Another factor that adversely affects the ability of the network operation is the inability of this type renewable energy source generation at night and the fact that it is impossible to guarantee stable generation during day. It can be said that the generation from only one type of renewable energy source, in this case photovoltaic power plant, without back-up power supply is insufficient. Nowadays, there is a trend of using a battery with high power output, which can serve as a backup of renewable energy sources for a particular time and if the generation of renewable energy source is greater than the consumption, the battery can be charged.

The computation was carried out in several steps. The first step is to create loads and define their parameters. As discussed above, the actual loads have their maximum consumption defined, however they do not work at maximum, but with a smalller consumption (30% - 40%). Therefore, three scenarios of consumption are defined for computing:

- normal (designed) 100 %,
- winter load 75 %,
- summer load 35 %.

The names themselves are defined in the simulation environment. It means that, e.g. load connected to the bus N7 indicated in Figure 4 as L5, has consumption under normal mode 2,052 MW, in winter mode 1,539 MW and in summer mode 0,718 MW. Depending on the scenario, the simulation model also changes the voltages at particular nodes.

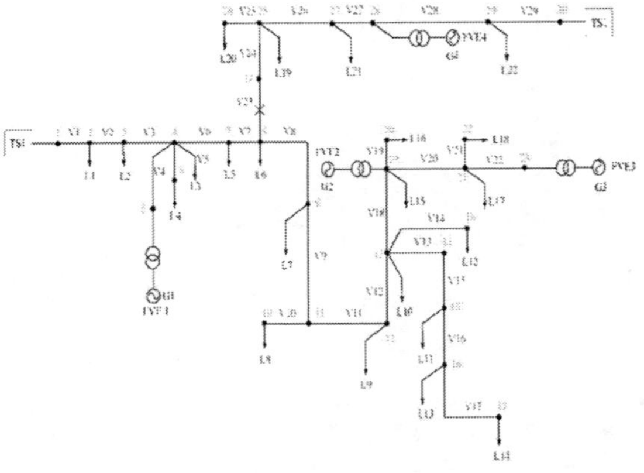

Figure 4. Computed distribution network

It can be seen in Figure 5 that the lowest voltage value was at the furthest point from the power supply, which is node N17 in this case, where the load is 0.42 MVA. This is due to the fact that the closest source is located 12.8 km away. N17 is the point with the highest voltage drop only in the case of 100 % or 75 % of consumption. When consumption is just 35% of the maximum value, the highest voltage drop is at the node N25. It should be noted that the smallest voltage changes are in case with actual load value of 75% of the nominal value. This condition represents fundamental connection and so this part of the network is fed from the upstream network of 110 kV. The high short-circuit power of upstream network was reflected positively on voltage drops. The actual voltage changes are not so high that it would not meet current standards. The voltage drops are also smaller due to the renewable energy sources generation. Note that the specified renewable energy sources used in computing do not respect the factors influencing the maximum power output, for example the case, when there is a decrease in the solar irradiance during the day. Thanks to that, it can be said that the behavior of the voltages in the network is correct just in case of maximum generation from renewable energy sources. However, this assumption is unrealistic with utilization of photovoltaic power plants only.

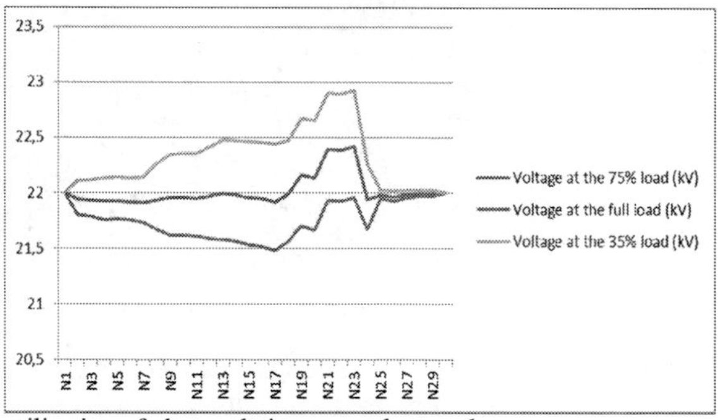

Figure 5. Voltage changes depending on the load in case of connection to a upstream network

The next step is simulation of network operation in the alternative connection, when the 22 kV network is disconnected from the upstream network and therefore actual consumption has to be covered only from local renewable sources. The network itself in off-grid mode is not able to operate at full load, because the balance of generation and consumption would not be met. The results are three characteristics shown in Figure 6. It can be said based on the Fig. 6 that the best operational condition of the network is at 35% of the nominal load value. In this state, the voltages in the nodes N1 - N18 are almost linear with the value of about 21.9 kV and the voltage in the nodes N19 - N23 is nearly equal to 22 kV in this case. The voltages start to increase from the node N24. It is caused by the connection of the power source in node N27. Generation of this source is sufficiently high relatively to the connected loads. It means that the network is capable of operation with voltage drops meeting the standard for a given voltage level with assumption of 35% of the nominal load. When analyzing two remaining characteristics at 100% load and 75% load, there are the highest voltage drops in nodes that have the highest voltage in previous situation. It is caused by the distance of these nodes from the source with highest power as well as by the fact that the source in the previous case has higher power compared to the loads and now it cannot cover consumption.

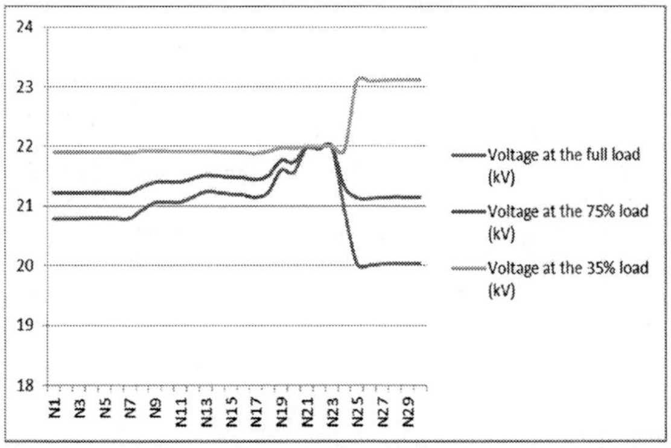

Figure 6. Voltage changes depending on the load during islanding

TABLE I. SIMULATION PARAMETERS

	Power	Power factor	Impedance [Ω]
Renewable energy FVE 1	0,95 MW	0,95	-
Renewable energy FVE 2	0,6MW	0,95	-
Renewable energy FVE 3	4 MW	0,95	-
Renewable energy FVE 4	1 MW	0,95	-
Superior system TS 1	180 MVA	0,114	0,306+2,671 i
Superior system TS 1	130 MVA	0,114	0,424+3,699 i

C. *Design of multi-agent system*

Multi-agent systems are the means of implementation of autonomous control. Before the structure designing, it is necessary to think about the concept of the system in terms of further applicability in networks with different structure. Multi-agent systems should be design for general use in networks with different generation from renewable sources and with different types of consumption.

The concept of multi-agent system shown in Figure 7 consists of 4 types of agents:

1. agent GRID,
2. agent LOAD,
3. agent SWITCH,
4. agent GENERATOR.

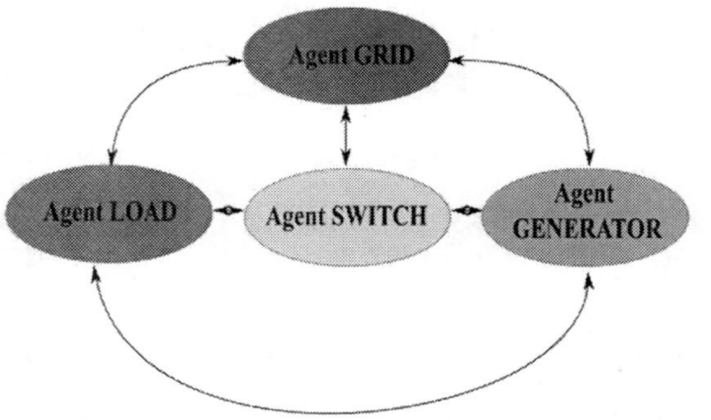

Figure 7. The structure of Multi-agent system

Multi-agent system seems to be composed only by these 4 given types of agents from the outside. However, each agent includes other subgroups. LOAD agent, for example, includes all types of loads that are connected to the network.

The very choice of agents is not the only step needed in the design. It is necessary to take into account the need for distribution of agents in that system. Based on that, it is possible to create a system where the agents are:

- centralized,

- decentralized.

Distribution of agents determines the time of response to the changes in the network as well as the actual requirements for computing equipment. The actual response time of such a system depends primarily on the selected type of agents, the number specific agent in such a system and on what kind of data is collected and transmitted between them. It is also important whether multiple agents communicate with supervisor agent or data are transmitted between all of them and each agent adds its own set of data to the previous one and moves it on. The last agent, which is nearest to the master agent, then forward the information to it. Rapid response to changes in the network is an essential aspect in the design because, for example, if power renewable energy source power transfer failure in the off-grid operation, the long reaction could results in synchronous machines' falling out of synchronism or high frequency changes etc. In the above mention problem, the fast reaction would have significant impact on network operation functionality because early disconnection of load or activation of additional generation would make this problem easily solvable.

D. Software for a managing multi-agent system

To control the distribution network using multi-agent system has been designed software to ensure agent-to-agent communication and communication between agents-to-sensors. The source code of software was written in C#.

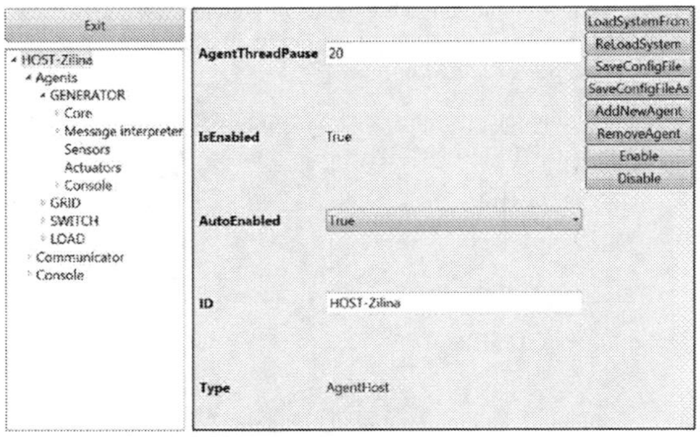

Figure 8. Software environment

The software environment is shown in Fig. 8. You can see the names of agents, on the left side of this image. The numbers and names follow from the previous design. Each agent contains some parts that may not be used by a particular agent. These parts then remain inactive. Because, this system must allow easy editing during running, it's equipped with a debugger. It makes the system adaptable for a variety of occasions. In the case of inter-agent communication, it's possible to monitor this and verify that the agent has communicated (for example to verify the available performance). At the core of specific agent, it is necessary to specify what agent will do. How it will communicate and whether the knowledge will be stored in the knowledge base.

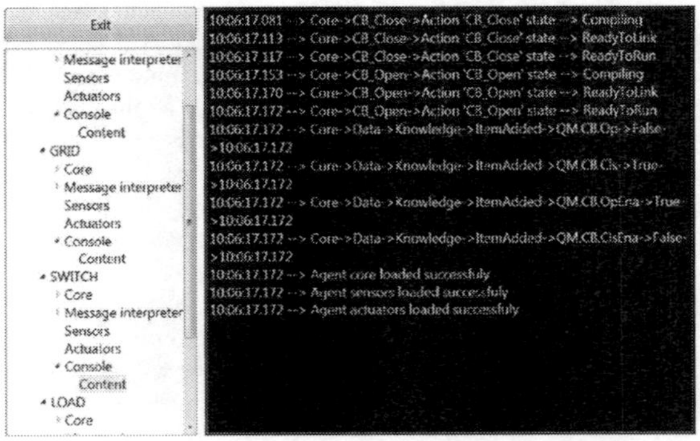

Figure 9. Console application for activity monitoring

The console application shown in Fig. 9 makes it possible to verify what the system is doing. The software includes a more consoles for clarity. Each agent has own console and one main console.

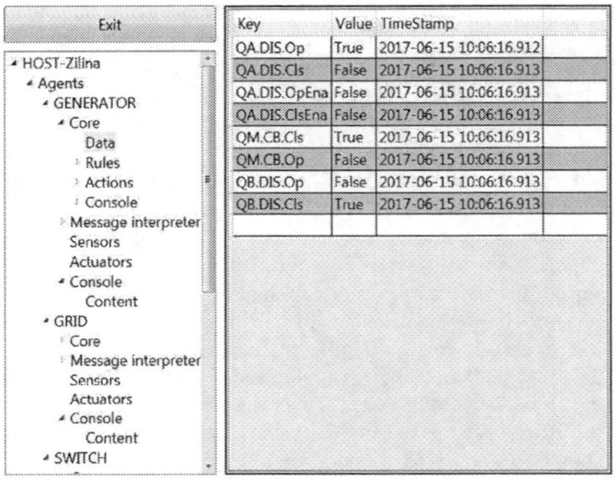

Figure 10. Data table

Figure 10 shows a table that is located at core of agent and displays and collects data that are accessed by other agents. The main difference between the console and this data table is that the data is retained for further use, but the console only displays the communications data etc. The main disadvantage of the data table is that it becomes unclear with a large number of data. So, it will be necessary to address this part more appropriately and modify the software.

One of the main advantages of this program is the possibility to use it for network operation as well as simulation purposes. The usage for the control purpose is conditioned by input of real-time values, because most of the data such as consumption information are not usable if they consist of an older value. If the software is used as a simulation tool, it is possible to use older data stored in the database. The simulation are suitable in the education process. Thanks to this, the software is universal but it is necessary to incorporate some deficiencies.

IV. Conclusion

Implementation of renewable energy sources with intermittent generation to the distribution system requires a deep study of their operation under varying operation conditions. This is very important in case that this kind of sources are considered as the main sources of Smart Region.

This paper described a study of the distribution network operation with renewable energy sources and utilization of autonomous control. Described software for multi-agent system management was tested only with simulation, and it was advisable to perform verification, e.g., oOn the- model distribution line. The results shows that this configuration of the network is capable of off-grid operation if certain requirements are fulfilled. One of these requirements is the utilization of various types of renewable energy sources. This is necessary in terms of backup. The autonomous control of such a network is applicable but it is needed to make certain legislative steps to allow load shedding, if necessary. In case of load shedding, it is necessary to define, which appliances can be turned off in their function and a possibility of damage point of view. Multi-agent system appears to be a suitable alternative for the control of the distribution network after resolving certain legislative and technical requirements.

Acknowledgment

This paper has been supported by the Educational grant agency (KEGA) Nr: 030ŽU-4/2014: The innovation of technology and education methods
oriented to area of intelligent control of power distribution networks (Smart Grids).

References

[1] A. M. T. Oo, G. M. Shafiullah, A. B. M. Shawkat Ali, P. Wolfs, M. T. Arif, *Renewable Energy Integration: Opportunities and Challenges*, London: Springer-Verlag , 2013.

[2] A. K. Marvasti, Y. Fu a S. DorMohammadi, "Optimal Operation of Active Distribution Grids: A System of Systems Framework," IEEE Transactions on Smart Grid, vol. 5, pp. 1228 - 1237, 2014.

[3] C. Rehtanz, *Autonomous Systems and Intelligent Agents in Power System Control and Operation*, Berlin: Springer Verlag, 2003.

[4] D. Fogel, T. Fukunda a L. Guan, "Scanning the Special Issue - Technology on Computatinal Intelligence," in. IEEE Proceedings Special Issue on Computational Intelligence, vol. 87 no. 9., pp. 1415-1421, September 1999.

[5] P. Návrat, M. Bieliková, Ľ. Beňušková, I. Kapustník, M. Unger, *Umelá inteligencia* (*Artificial intelligence*), Bratislava: STU, 2007, pp. 393.

[6] K. Passino, "Intelligent Control For Autonomous Systems," IEEE Spectrum, vol. 1, pp. 55-62, 1995.

[7] E. Abel, I. Laresgoiti, J. Perez, J. M. Corera, J. Echavarri, "Multiagent approach to analyze disturbances in electrical networks", in Proceedings of Expert System Application to Power System, 4th ESAP Conference, Melbourne, Australia, 1993. CRL Publishing Ltd.

[8] S. Talukdar, "Asynchronous Teams", in Proceedings of Expert System Application to Power System, 4th ESAP Conference, Melbourne, Australia, 1993. CRL Publishing Ltd.

[9] C. Rehtanz a C. Becker, "Autonomous Systems for Intelligent Coordinated Control of FACTS devices," in. Proceedings of Bulk Power Systems Dynamic and Control IV, Santorini, Greece, 1998.

[10] S. Franklin a A. Graesser, "The University of Memphis," Internet: http://www.msci.memphis.edu/~franklin/AgentProg.html, [Sept. 21, 2011].

[11] A. Kubík, *Inteligentní agenty* (*Intelligent Agents*), Brno: Computer Press, 2004, pp. 288.

Model of electric output from biogas station

Ladislav Novosad, Zdenek Hradilek
VSB – Technical University of Ostrava
Department of Electrical Power Engeneering
Ostrava – Poruba, Czech Republic
ladislav.novosad.st@vsb.cz, zdenek.hradilek@vsb.cz

Abstract —**The paper focuses primarily on verifying the functionality of the model, simulating the electrical output of biogas station. The main objective of the model itself is to create a simulation of electric power course which is supplied to 22kV distribution grid from biogas stations. Therefore, the article contains the basic principle of the draft BGS model and describes its main functional elements. It also describes the basic parameters of input data. The main objective of the model is to create a simulated course of electric power supply unlimited by time to 22kV distribution grid. Subsequently, a combination of this model output together with other models and simulations of renewable energy sources and 22kV systems is expected. The result of the cooperation of each model representing the selected elements of distribution grid, especially renewable sources, should then be a clarification of the possibility of connecting biogas stations to the 22 kV grids.**

Keywords: virtual BGS model, measurement database, biogas station, renewable source, co-generation, co-generation unit, simulation of electric power, distribution system

I. Introduction

The present article deals with the operation of biogas stations (BGS). As already mentioned, the aim of this article is to outline the possibility of creating a virtual model of the electric part of biogas station; then to verify the functionality of the model and use it to simulate the time dependence of the output power supplied to DG. The output of this model will subsequently be a course of supplied power to DG. Thus created course can be combined with other distribution system models, RES, simulating the operation of a comprehensive distribution grid part. Furthermore, the model will be used for automatic evaluation of measured data from individual biogas stations.

For the above reasons, the article will deal with the electrical part of BGS only. BGS power heat balance will not be considered in the model and we will not simulate the actual process of fermentation, or the operation of a non-electric BGS part.

II. Model Funcional Units

Designed virtual model consists of two main functional units, each of which has a separate output. The graphic part of BGS program was prepared using the Microsoft Visual Basic for Applications in MS Excel.

This research was supported by the SGS grant from VSB - Technical University of Ostrava (No. SP2017/54).

Model functional units:

- MEASURED DATA ANALYSER – evaluation of the measured data obtained by long-term measurements on BGS
- BGS ELECTRIC OUTPUT SIMULATOR

III. Measured Data Analyser

The aim of this part of the model is the automatic evaluation of a comprehensive set of measured data. A use of database of data obtained through long-term measurements from several BGSs. These measurements were processed in the context of long-term research into renewable sources at the Technical University – FEI. Biogas stations which were used for all three measurements were farm-type without connection to natural gas network and external container for storing biogas. The measurements were always carried out in the BGS main switchboard on the electric power output to the DG. The BGSs were connected to the DG via a 0.4/22kV transformer station. [1,2]

Data obtained through the aforementioned measurements must first be treated to the desired form (style). It is necessary to keep the exact form of the input data (sequence of quantities, spacing) for a proper function. At present, the model is able to retrieve data from three BGSs at the same time and analyse them. Measured data will be loaded into a simulator from a text file. Given the limited possibilities of MS Office Excel, it is possible to insert data with a maximum length of measurement of 30 days. Given that the total measurement times for each of the BGSs were different, the available data were shortened to a total period of 30 days of measurement. Also, the first and the last day of measurement were always omitted. The data were not complete for the whole 24 hours in this period (connecting and disconnecting the measuring instrument).

Individual displayed quantities can be selected both in groups and individually in the analyser program main menu. Then there is the option to choose a number of evaluated or compared BGSs (currently 3 data files maximum).

The outcome of the model evaluation will be both graphical information – diagrams and courses of selected quantities – and statistical information – the number of outages, decreases, total operating time, etc. which can be exported in the form of tables in a text file, as well as the juxtaposition of individual measurements at various BGSs. Examples of evaluation model outputs can be seen in the following examples (Figure 1).

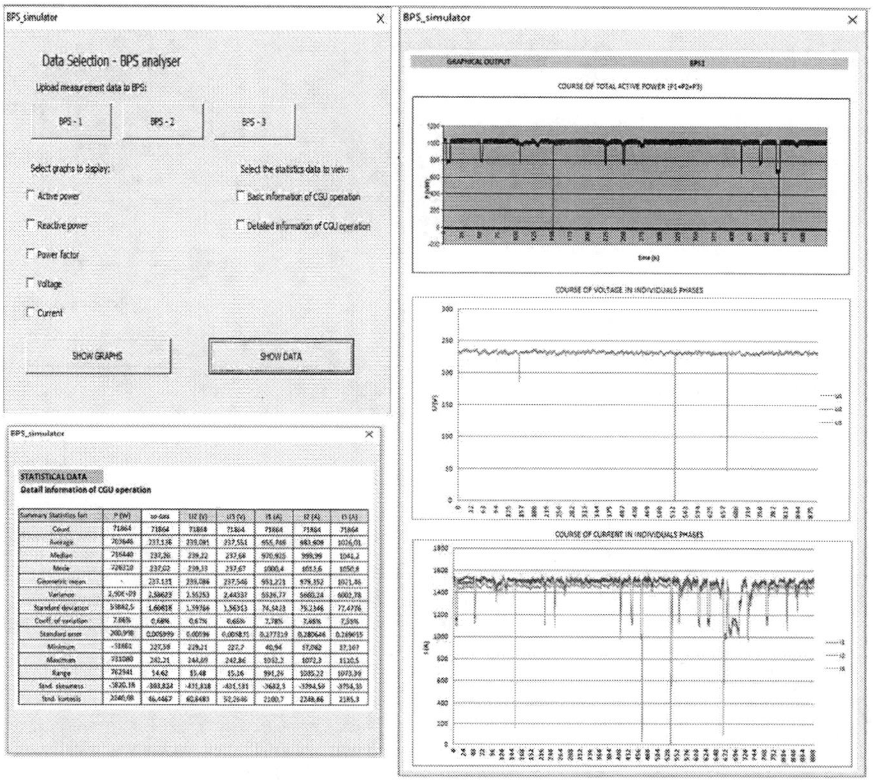

Figure 1. Example of graphical representation of measurement data in Excel and the main control panels of the program

The BGS simulator – measured data analyser is important also because of the functionality of other simulation blocks. Due to the fact that there is not the necessary database of BGS operation data, or the operation of co-generation units in biogas stations, several long-term measurements at BGS were carried out. These data will provide the basic statistics about the BGS operation. The number of outages or their frequency together with information about their time course will be primarily analysed. Subsequently, these data will be used for the purposes of electric output of BGS simulation. [4]

IV. BGS ELECTRIC OUTPUT SIMULATOR

However, the main purpose of the model will be, as already mentioned, to create a simulation of electric power supply course from the BGS.

The model itself is composed of the following basic parts (Figure 2):

1. Setting the basic BGS parameters;
2. Setting the time section of simulation
3. Measured data analyser – simulation data;
4. Measured data;
5. Calculation – a simulation of the model;
6. Output of the model – a simulation of active power course.

V. SETTING THE BASIC BGS PARAMETERS

This block is used to set the basic parameters of the model biogas station. In terms of electrical quantities, these parameters include the size (installed electrical power) and the number of co-generation units (CGU) located in the BGS, as well as their type, the consumption of biogas and other operational information (See Figure 3).

In the current state of development of the simulator, the following can be set:

- number of CGUs (1–5 pc)
- CGU powers (from about 100kW to 1000kW) – each CGU may have different powers
- gas tank size – or a time reserve of biogas for CGU operation

An important factor in the BGS operation is the supply of produced biogas – entered as a slack operation of all CGUs at 100% power in this case. The vast majority of CGUs in the Czech Republic are not equipped with gas tanks. Consequently, the co-generation units are able to operate when a failure of fermentation process occurs for a limited period only, in tens of minutes depending on the fermenter size. The fermenter size then affects the amount of gas stored. [2,3]

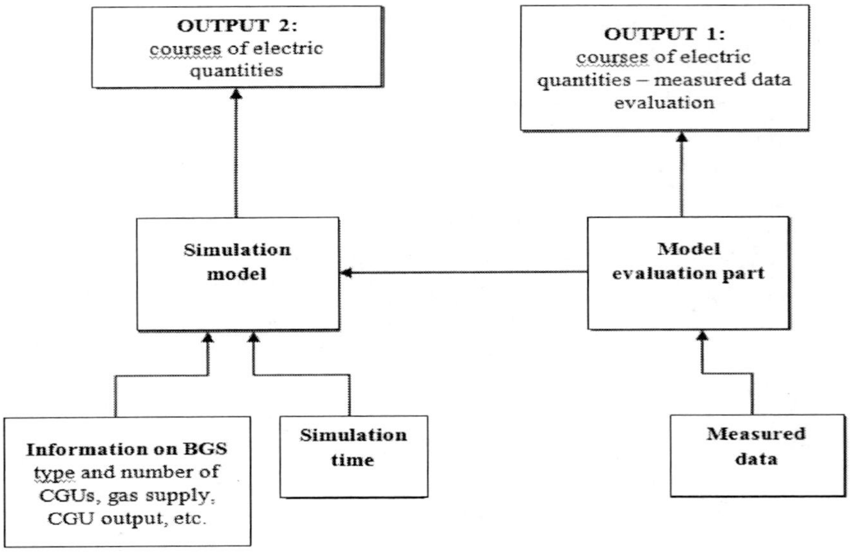

Figure 2. Block display of the main structural elements of the model

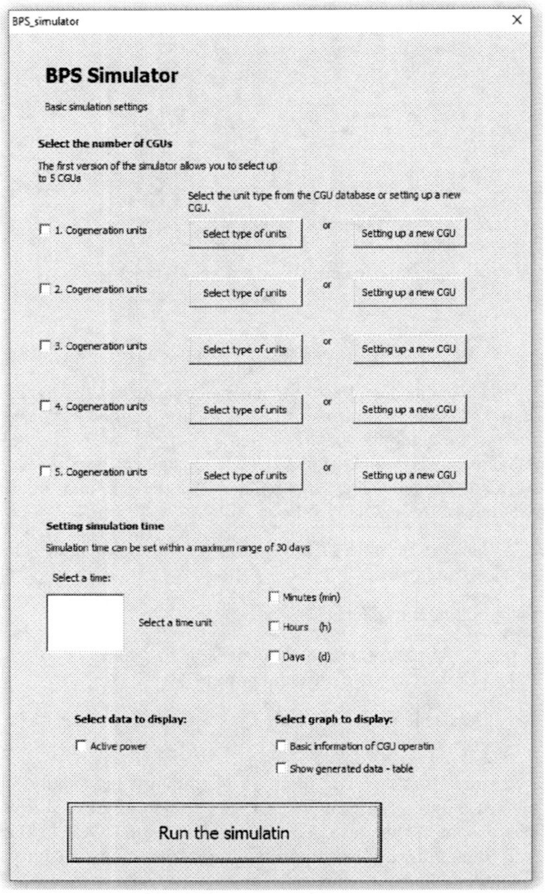

Figure 3. Main parameter settings panel for simulation

VI. Measured Data Analyser – Simulation Data, Measured Data

This section was described in the article above, including the importance of measured data for the simulation of the output power from the BGS.

VII. Calculation – A Simulation Of The Model

This is basically the most important part of the model. The simulation part of the model is designed for the actual calculation/generation of the electric power supply from the BGS simulated course. Basic input information which consist of data on reliability and outages are loaded to the simulation model from the measured data analyser output.

Given that the basic data on the reliability of the CGU operation are not available currently, the evaluation results of the three BGSs will be used for calculations. The calculation is made up of various mathematical operations based on probability and random generation of power outages. The result is then a set of data that can be characterized and displayed as virtual supplied power from BGS. These data can be stored separately in the form of text or graphical outputs.

The generated supplied power (see Figure 4) courses can be considered as a theoretical course of power supply from CGU and can be used for other simulations, e.g. in combination with electric power renewable sources. [2,5,6]

VIII. Conclusion

The article deals with a biogas station model. The model allows the creation of a theoretical simulated course of power supply from BGS to 22kV distribution grid. The created course can then be used in combination with other simulator outputs and prediction models of individual RES in DG. Cooperation of such models can then result in optimizing the operation of DG using the power available from the BGS. Above all, cooperation with PVP is assumed, thus providing additional options of stabilizing grids at 22kV level, possibly with a connection of BGS to DG in terms of possible support services.

A basic fact identified by both the long-term measurements and the output of the simulator is that BGSs with conventional combustion CGUs are reliable and stable sources of power supply.

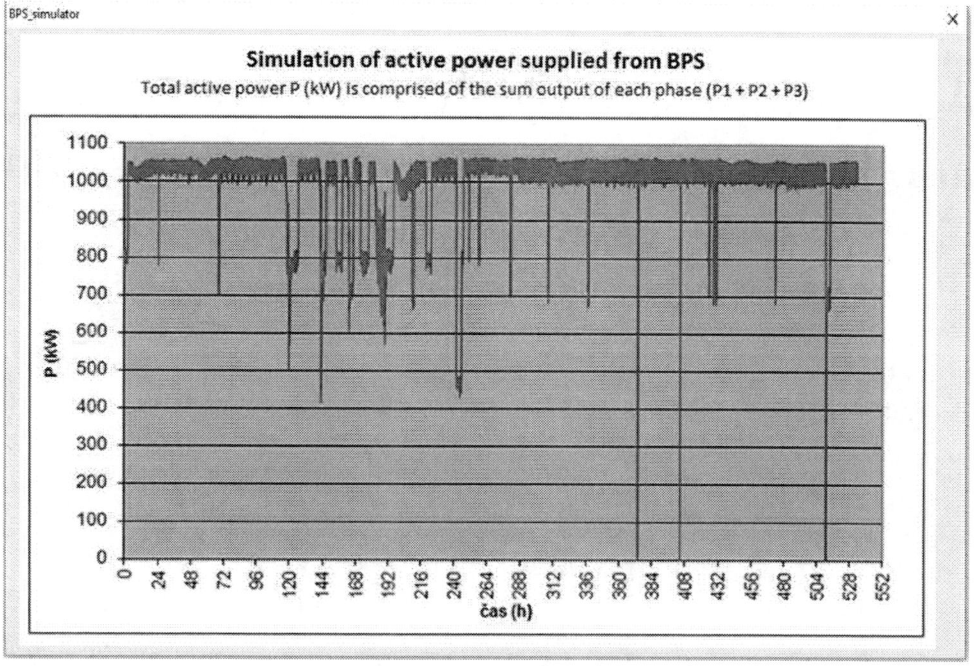

Figure 4. Graphical outputs of the simulation model

However, the above simulation process of the so-called general course of power supply from CGU is affected by several basic BGS disadvantages. These disadvantages include the dependence of supplied power on the CGU operation or on a sufficient amount of biogas. Due to the CGU high operational reliability (outages caused by CGU failure are very rare), the lack of biogas is one of the main reasons for DG power supply failures. This factor may be difficult to evaluate since it is a different factor among all the available measured BGSs as it depends on many factors. The most common causes of CGU failures include improper operation of BGS or CGU, particularly in terms of dosing biomass, and the very type of biomass as well as climatic conditions.

Acknowledgment

This research was supported by the SGS grant from VSB - Technical University of Ostrava (No. SP2017/54).

References

[1] L. Novosad, Z. Hradilek, "Power Analysis of Co-Generation Units at Biogas Station", Proceedings of the 16th International Scientific Conference on Electric Power Engineering (EPE) 2015 Dlouhé Stráně, 2015, ISBN 978-1-4673-6788-2

[2] L. Novosad, Z. Hradilek, "Analysis of Energy Balances Three Biogas Stations in Czech Republic", Proceedings of the 16th IEEE International Conference on Environment and Electrical Engineering, 2016

[3] M. Dummer, M. Litschmannova, "Statistika I.". Statistika I. VSB – Technical university of Ostrava, 1997, 80-7078-496-2

[4] Z. Hradílek, "Elektroenergetika průmyslových a distribučních zařízení," 1.vyd. VŠB – TU Ostrava. 2008. ISBN 978-80-7225-291-6.,p.315, 2008.

K.M.A. Islam, T. M. Khan, M. M. Rahman, M. S. Munna, "2015 3rd International Conference on Green Energy and Technology (ICGET)",2015, Pages: 1-6,

DOI: 10.1109/ICGET.2015.7315105

[5] M. Spacek, Z. Hradilek, P. Moldrik, "Cooperation between photovoltaic and biogas power plants within 22kV network", 2015 16th International Scientific Conference on Electric Power Engineering(EPE),2015,Pages: 341 -345

DOI: 10.1109/EPE.2015.7161098

Algorithm-based Analysis for the Charging Stations Impact Evaluation on the Low-voltage Distribution Networks

Larisa Grackova
Riga Technical University
Riga, Latvia
larisa.grackova@inbox.lv

Irina Oleinikova
Institute of Physical Energetics
Smart Grid Research Centre
Riga, Latvia
irina@edi.lv

Gaidis Klavs
Institute of Physical Energetics
Laboratory of Energy System
Analysis and Optimisation
Riga, Latvia
energy@edi.lv

Abstract— **The article deals with the principles of the analysis of the location for the charging stations and their impact on the distribution networks by using the algorithm consisting of four stages. During the first phase, quantitative and qualitative indicators and analysis were performed on the basis of mathematical statistics and probability theory. The second phase involved the model of optimizing and assessing the location of charging stations, the goal of which was to meet the necessary requirements for charging electric vehicles within the conventional parking spaces. To find the optimal connection points for charging stations, the optimal transport theory was used. The third phase is a decision making in accordance with the criteria in the normal and post-fault modes. During the fourth stage, the decision making parameters were tested on the base of the calculation process created before in the second block of this algorithm. MATLAB/Simulink environment (SimPowerSystems) was used to confirm the choice of places for connecting the charging stations. The simulation model is used to assess the deviations in voltage and power of the distribution network during the charging period of electric vehicles. The algorithm developed here to provide a basic framework for assessing the location of electrical vehicles charging infrastructure in obtaining a result for the existing and planned distribution networks. It is also assumed that new blocks will be introduced for future research.**

Keywords: Electric vehicle; distribution network; charging station

I. Introduction

The use of modern Battery electric vehicles (BEVs) and Plug-in hybrid vehicles (PEVs) is a potential option for reducing dependence on oil fuels and mitigating greenhouse gas (GHG) emissions in the transportation sector. However, the integration of charging (recharging) stations will not only increase energy demand, but can cause a weakening the reliability for the existing distribution networks. For this study, a survey of scientific papers on the operation of road electric transport, production and distribution of electricity to the consumer was made. The authors have an increasing interest in the following issues: battery technology for electric transport, types of road electric transport (passenger cars, trucks, buses and motorcycles), the behavior of electric transport users during the day, charging station regimes and their locations, payment charging principles for electricity consumption at charging stations, climatic zones and the geographical location of the state, the effect of charging stations on the quality of electricity (voltage gross variation, voltage harmonics, inter-harmonic voltage and voltage unbalance among others) and a local electrical network, etc.[1].

Particular attention was paid to the articles considering the methods and algorithms: the network slow and fast charging behaviour and the impact on the electricity system; a system for BEV charging monitoring and control of energy grids with distributed energy resources; monitoring and control for BEV, PHEV and vehicles to grid (V2G) techniques in an existing residential network; V2G technology to treats as new generators in the electricity generation portfolio and to address distributed energy resource generation oscillations; power management of a grid-connected DC Charging Station for BEVs using solar and wind power. All this information results in the conclusion that a wide range of models and algorithms exist for carrying out original studies.

In this area, the studies are conducted and are relevant for Latvia, since the relevant tools of state support adopted and being in the process of development are aimed at this. In Latvia, the development of a network of charging stations is in the initial stage at the present time. But, with the support of the state, 235 [1] are planned to be installed throughout the country until 2023 [2]. Within the framework of the public procurement[2] Road Traffic Safety Directorate, an analysis has been made of the placement of 150 fast charging stations on the main thoroughfares or close to them in accordance with the criteria for the number of inhabitants and their daily movements [3].

At the same time, according to statistical data, as of January 1, 2016, 227 passenger cars, 7 Light commercial vehicles (LCV), 3 motorcycles, 32 mopeds and 26 quads were

[1] http://www.e-transports.org/index.php/jaunumi/128-satiksmes-ministrija-latvija-plano-izbuvet-vel-235-elektrotransportlidzeklu-uzlades-stacijas
[2] http://etransports.lv/index.php/jaunumi/154-iepirkums-elektromobilu-atras-uzlades-staciju-uzstadisanas-vietu-izbuves-buvprojektu-izstrade-un-autoruzraudziba

registered, battery electric vehicles (BEVs) and plug-in hybrid electric vehicles (PHEVs). Popular passenger cars brands are: Volkswagen (VW e-up), Nissan Leaf, BMW i, and Tesla [4].

For comparison: on January 1, 2012 only 2 passenger cars PHEVs and 2 LCV BEVs were registered.

Thus, the aim proposed for the present research is to develop an algorithm for studying of a local distribution network in accordance with the needs of electric vehicles by applying of mathematical statistics methods and probability theory.

II. The tasks and architecture of algorithm

The main tasks are declared as follows:

- The behaviour of car owners at the parking places within 24 hours is reviewed and analyzed by cluster analysis.
- Technical indicators: charging types (standard, accelerated, rapid) and the most common models of road electric transport are accepted for calculations.
- Equalization of peak loads in the existing distribution network and increase of its reliability and efficiency, under the condition of additional load from charging stations.
- Choice of the location of the charging stations, taking into account their interaction with the energy company, including economic and environmental components
- Investigation of the distribution network in accordance with the needs of charging stations, as elements of an intelligent network (Smart Grid).

The algorithm consists of four steps. Each step has: input, calculation, analysis and decision of phases shown in Figure 1. It should be noted that in our case, the algorithm architecture is open for the involvement of possible changes when additional requirements are included, taking into account possible changes and requirements to all interested sides, including governmental and non-governmental organizations, the private sector and society in general.

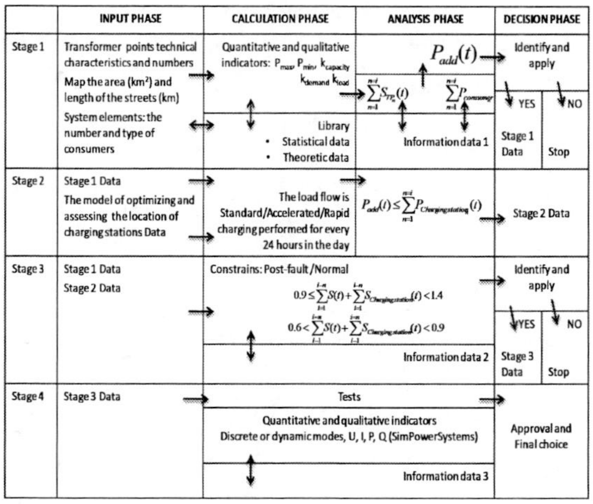

Fig.1. The architecture of algorithm

III. Methodology algorithm

In this section, the sequence of the research and the rationale for choosing the methods are described.

A. Stage 1

Within the Input phase the following was chosen: Scheme for the distribution network, technical specifications and numbers of the transformer substations, map of the area (km^2), length of the streets (km) and system elements (the number and type of objects). Input data is Library module containing a database (Excel document archive):

- technical specification of three-phase transformers 10/0.4 kV, (mode and nominal power $S_{n.tr}$)
- Quantitative indicators and load graphs: P - active power, kW; Q - reactive power, kVar; S - apparent power, kVA; coefficients of demand for lift installations of residential buildings of the k_d; estimated coefficients of reactive power of residential buildings; specific design electric loads of public buildings; daily load of consumers of the microdistrict; parameters of lighting installations of the transport and pedestrian system of streets and roads, etc.

The power consumption of the objects and their daily load schedules are estimated by means of two methods - statistical and calculating.

The statistical method is the most reliable method for the studying of the loads, since it is based on real measurements: indications of individual and general consumers' counters, as well as counters at network transformer points.

The calculation method is used to design the power supply system to select all its elements or in the absence of real measurements for individual objects. This method estimates the maximum loads on the basis of mathematical statistics and probability theory [5,6].

Library module contains eighteen profile objects. They are represented as follows: single family houses, multi-family houses (2-3-storey), multi-family houses (4-5-storey), multi-family houses (8-12-storey), garages (600-1000 auto), office (or small industries), supermarket, store and market, station (transports terminus), post-office, the local authorities and other institutions, petrol station, manufacture - 1 working shift, manufacture - shift work, catering establishment, school, primary school and kindergarten, health centre, street lighting.

Further, the transition to the Calculation phase is carried out, where the quantitative and qualitative indicators for the objects and for transformer substation (TS) according to the scheme for distribution network are determined.

For objects: maximum (peak), average (middle) and minimum (off-peak) demand and their daily time of use.

The load factor (f_{load}) shows the extent of use on the active power for the load curve in a specified time period and is determined by the expression (1).

$$f_{load} = \frac{P_1 t_1 + P_2 t_2 + ... + P_{24} t_{24}}{P_{max}(t_1 + t_2 + .. + t_{24})} = \frac{1}{P_{max} \times t} \int_0^{24} P(t) dt \qquad (1)$$

The irregularity factor ($f_{irreg.}$) is the ratio between the minimum loads by the average peak load in a specified time period (2).

$$f_{irreg.} = \frac{P_{\min}}{P_{\max}} \quad (2)$$

The total load and the schedule on the transformer substation from the consumer groups (3).

$$\sum_{i=1}^{i=n} S = \sqrt{\sum_{i=1}^{i=n} P^2 + \sum_{i=1}^{i=n} Q^2} \quad (3)$$

For transformer substations:

The substantiation of the choice of transformers (in our case, a two-transformer substation) is given to the load requirements according to:

$0.5 < k_{load}^{norm} < 0.9$ - the actual load factor of the transformer in the normal operating mode, which depends on the number of transformers in the substation and the load curve of the consumers' power supply (4);

$$k_{load}^{norm} = \sum_{i=1}^{i=n} S / S_{n.tr} \cdot n \quad (4)$$

$1 < k_{load}^{post-fauil} < 1.4$ - the load factor in the post-fault mode (5).

$$k_{load}^{post-fauil} = \sum_{i=1}^{i=n} S / S_{n.tr} \cdot (n/2) \quad (5)$$

where $\sum_{i=1}^{i=n} S$ - load from all consumers; $S_{n.tr}$ - transformer nominal power; n - the number of transformers.

Then, quantitative and qualitative indicators are generated in the report (in table and graphically) in information block 1.

The analysis phase examines the data from Information block 1, and the possible additional load is calculated for the temporary daily intervals by the expression (6).

$$S_{add}(t) = n \cdot S_{n.tr}(t) - \sum_{i=1}^{i=n} S(t) \cdot f_d \quad (6)$$

where

Diversity Factor (f_d) - the ratio, expressed as a numerical value or as a percentage, of the simultaneous maximum demand of a group of electrical appliances or consumers within a specified period, to the sum of their individual maximum demands within the same period, the value is always ≤ 1[3]. The factor k_s is applied to each group of loads and requires a detailed knowledge and investigation of the conditions. For this reason, $f_d = 0.9$ is the value for general application [7, 8].

The load increasing in peak, middle and off-peak demand periods and their daily time period of use are also quantitatively estimated.

[3]The World's Online Electrotechnical Vocabulary http://www.electropedia.org/

The Decision Phase. In the choice phase, we evaluate all the data obtained during the Analysis phase and choose one final decision - Yes or No. If Yes, then the transition to Stage 2 takes place. If No, then the algorithm is interrupted (Stop).

B. *Stage 2*

Input phase: Quantitative indicators Stage 1 (Decision Phase) and model of Electric Vehicles charging stations modes for low-voltage distribution network. The aim is to satisfy the necessary hourly (daily) requirements for charging electric transport under the conditions of traditional parking places. To find the optimal location of the charging stations, the optimal transport theory algorithm is performing. The model determines the optimal load demand EV for each hour. The algorithm is implemented as follows:

- The behaviour of car owners at the parking places within 24 hours is reviewed and analyzed by cluster analysis.
- Technical indicators: charging types (slow-medium-fast) and the most common models of road electric transport are accepted for the calculations.
- On the basis of the list of road transport, a park of possible electric transport is formed at geographic place taking into account the location of the transformer points. The dynamics of changes in harmful emissions from vehicles as the subject of the electric vehicles infrastructure development investigated using the COPERT IV program as well.
- Then, in accordance with two criteria - the optimal time of the battery charging (the degree of charging is necessary for the owner of vehicles) and the number of charging stations (charging types – slow, medium, fast) at the places of traditional parking - the load necessary for the charging process is estimated (power consumption).
- To provide the best solution during the process, the load is analyzed, several options are tested and then the final choice is presented.

Having received the final version of the location of the infrastructure for the electric transport fleet at the considered geographical location in the Calculation phase, the low, medium and high penetration demand for charges is estimated. Charging performed for every 24 hours per day is an additional consumer (7).

$$\sum_{i=1}^{i=n} S_{Chaging\ station}(t) = \sum_{i=1}^{i=n} S_{S\tan dart\ ch\arg ing}(t) + \\ + \sum_{i=1}^{i=n} S_{Accelerate\ ch\arg ing}(t) + \sum_{i=1}^{i=n} S_{Rapid\ ch\arg ing}(t) \quad (7)$$

where

$S_{standard\ charging}$: Slow chargers are typically rated at 3kW (1-phase, 13/16A), charging an EV on a slow charger usually takes 6-12 hours (depending on battery capacity).

$S_{Accelerated\ charging}$: Fast chargers are typically rated at 7-22kW (1- or 3-phase, 32A), charging an EV on a fast charger usually takes 1-4 hours (depending on battery capacity).

$S_{Rapid\ charging}$: DC/AC DC chargers are typically rated at 50kW and AC chargers are typically rated at 43kW (3-phase, 63A), charge points allow to charge 80% of a battery of electric car in less than an hour.

Subsequent actions are of the analysis phase there according to the expression, as well as the transition to the Decision phase (Stage 2 Data) where the quantitative indicators are illustrated (8).

$$S_{add}(t) \leq \sum_{i=1}^{i=n} S_{Charging\ station}(t) \tag{8}$$

C. Stage 3

During the third stage, according to the Electrical Codes and Standards, the technical characteristics of transformer substations are verified in Calculation phase and Analysis phase, which compares the total capacity of the objects with the permissible load of the transformers in the post-fault and normal modes.

The post-fault mode. In the case if one transformer is disconnected (at a two-transformer substation) during the repair or minimum loads, one transformer can be overloaded by 140% within 6 hours per day (no more than 5 days) (9).

$$0.9 \leq \sum_{i=1}^{i=n} S(t) + \sum_{i=1}^{i=n} S_{Charging\ station}(t) < 1.4 \tag{9}$$

The normal mode. At two-transformer substations it is customary to use the type of transformers with the equal (similar) technical characteristics that provide the following parameters: the maximum load of power supply objects; the daily schedule of loading of objects; the number of hours of maximum load according to the actual load schedule of the objects (10).

$$0.6 < \sum_{i=1}^{i=n} S(t) + \sum_{i=1}^{i=n} S_{Charging\ station}(t) < 0.9 \tag{10}$$

Input phase for Calculation phase and Analysis phase are the quantitative indicators Stage 1 Data and Stage 2 Data.

Formation of the calculated data is carried out in the form of a report in the information block 2. Analysis phase analyses the report and determines the fulfillment of the conditions (the normal and post-fault modes).

The final result of Stage 3 is Decision phase - Identify and apply - where the decision is Yes or No. If the conditions are met, go to Stage 4. If at least one of the conditions is not met, further studies are not performed, the algorithm is interrupted (Stop).

D. Stage 4

At the fourth stage, the input data is the decision made at Stage 3. In the Calculation phase, quantitative and qualitative indicators for each transformer substation are tested according to the following two conditions.

The first condition is compliance with reliability requirements of power supply to the local distribution network in accordance with the Latvian and European quality standards. To fulfil this condition the calculation process of Stage 1 of this algorithm (the parameters and coefficients of the local distribution network are calculated) should be used.

The second condition is a confirmation of the correct places connection at the charging stations with existing objects and assessment of the efficiency of their use for the power supply scheme under consideration [9].

For this condition, the distribution network is modelled and tested in discrete and dynamic modes, using Simulink Simscape Power Systems (SimPowerSystems). Simulation of the discrete and dynamic modes allows the evaluation of the operation of the existing distribution network (voltage, current, active, reactive and apparent power) in accordance with the electric vehicle demands of charging stations and as an element of the intelligent network (Smart Grid).

Information block 3 illustrates the data from Calculation phase used for the Analysis phase.

The final step is Approval and Final choice of an optimal scenario for a local distribution network in accordance with the needs of electric vehicles.

IV. Conclusions

In this study, the structure of the algorithm of the four stages for studying of a local distribution network in accordance with the needs of electric vehicles was considered and investigated.

At all the stages, the specific tasks are solved and realised. Each stage contains four phases (input, calculation, analysis and decision) with transitions and Information blocks.

The possible additional load for the time intervals was calculated during the first stage. Within the second stage, with the model of Electric Vehicles charging stations modes the maximum load on the local distribution network was estimated. This load cannot be higher than the additional load in Stage 1calculated before. The technical characteristics of transformer substations were verified in accordance with two criteria of Stage 3. The verification the final result was indicated. Stage 4 in this algorithm is necessary to test the local distribution network. The quantitative and qualitative indicators are calculated here; the models of them are developed with discrete and dynamic modes. The decision phase is the approval of the results and final selection of the optimal location of the charging stations within an optimal scenario.

To apply this algorithm an existing residential low-voltage system is proposed for its application. This framework can be adapted not only in the cities, but in the rural regions as well.

Identifying the locations of charging stations will help to identify the problems and reduce the cost of repair, reduce the potential damage associated with the accident in the distribution network.

Application of this algorithm will allow obtaining a result for the projected distribution networks, which are supposed to apply new technologies for road transport (Fuel Cell Electric - FCEV, Vehicle to Grid - V2G and Vehicle to Home - V2H, etc.).

However, there are some drawbacks in the algorithm that should be taken into account in future research. They are the following: the real overload abilities of the transformers, losses of power of the transformers, deviations in voltage and losses of cable lines, wide range of real readings of vehicles on charging modes and several others.

It should be noted that the accumulated experience of the countries where the network of charging stations already exists will be of great importance while creating an infrastructure for the charging stations in Latvia.

Acknowledgment

This paper has been supported by the Latvia National Research Programme 2014-2017 "LATENERGI".

REFERENCES

[1] H. Markiewicz, A. Klajn. *Power Quality Application Guide. Voltage Disturbances Standard EN 50160. Voltage Characteristics in Public Distribution Systems.*Wroclaw University of Technology, July 2004, p.16.

[2] Ministry of Environmental Protection and Regional Development of the Republic of Latvia. Open tender of Climate Change Financial Instrument. [Online]. Internet: http://www.varam.gov.lv/lat/darbibas_veidi/KPFI/projekti/?doc=17817, [June 8, 2017].

[3] A. Rubenis, D.Berjoza, A.Grislis, I.Francis. Latvijas nacionālā elektrotransportlīdzekļu uzlādes tīkla izveidošanas analīze. Rīga, 2016. 70 lpp. http://www.tf.llu.lv/getfile.php?id=5315.

[4] Road Traffic Safety Directorate. [Online]. "Reģistrēto transportlīdzekļu skaits | Transportlīdzekļi | Statistika | CSDD". Internet: https://www.csdd.lv/transportlidzekli/registreto-transportlidzeklu-skait, [June 8, 2017].

[5] V. Kholyanov, O. Kholyanova. *The power supply of non-industrial objects: Study guide.* Vladivostok. 2007. p. 121.

[6] A. Vanags. *Electrical networks and systems. Part I.* Riga. Petergailis, 1995. p. 438.

[7] Latvenergo, LEK energostandarti, [Online]. Internet: http://www.latvenergo.lv/lat/par_mums/normativie_dokumenti/lek_energostandarti/ [June 8, 2017].

[8] B. Brown, *PRM Design Guide*, Schneider Electric. [Online]. Internet: http://www.schneider-electric.us/en/download/document/0100SM1504/. [June 8, 2017].

[9] A. Lucas, F. Bonavitacola, E. Kotsakis, G. Fulli, "Grid harmonic impact of multiple electric vehicle fast charging", Electric Power Systems Research, Volume 127, 2015, pp. 13-21, ISSN 0378-7796, http://dx.doi.org/10.1016/j.epsr.2015.05.012.

Digital Overcurrent Relays in Industrial networks

Advanced fetures for optimization of tripping times

Marek Baňas, Marek Höger, Juraj Altus
Department of power electrical systems
University of Žilina
Žilina, Slovakia
marek.banas@fel.uniza.sk, marek.hoger@fel.uniza.sk, juraj.altus@fel.uniza.sk

Abstract— **This paper focuses on the possible use of additional features of the modern digital protective relays in medium voltage industrial distribution systems to reduce the tripping times and enhance the selectivity of the protection system. One of these additional features is the directional comparison. The paper describes the possible use of this function on a simplified real-world model of an industrial distribution system and compares the resulting tripping times with a conventional approach using only nondirectional and directional time-independent overcurrent protections. The reduction of tripping times is especially important in context of the continuous increase of short circuit power levels in the system. We can observe these phenomena in many industrial networks, due to upgrades in high voltage distribution networks powering the industrial networks, but also due to installation of new local electricity resources in the industrial networks. Reducing the tripping time reduces the risk of damage to the power equipment due to short circuit current, increases the reliability and lifetime of network equipment.**

Keywords*— *protective terminal, overcurrent relay, tripping time, selectivity, directional overcurrent relay, fault location, directional comparison

I. INTRODUCTION

Modern digital protection relays are complex devices providing not only wide spectrum of protective functions, but also bay control, measurement and automation functions, fault recording and data logging. Their software based logic offers flexibility incomparable with electromechanical or electronical protection relays. A single digital relay is capable to implement very complex protection schemes without the need of any additional specialized hardware (e.g. circuit breaker failure detection). Moreover, the relay is capable to adjust its settings automatically by switching different preconfigured setting groups, reflecting actual conditions in the network, e.g. changes in the network topology.

However, there is one more important aspect of modern digital protections significantly increasing their value compared to older technology – their ability of flexible communication. In previous generation of protection relays, main communication interface was a serial port and all most used communication protocols were optimized for use with serial communication in a strongly centralized communication architecture. This fact has significantly decreased the ability to share large amounts of data between devices, because of relative low communication speed (even when using some faster transport layers like metallic or optical Ethernet) and this architecture had also very limited potential for exchange of time critical messages (like tripping commands). Modern devices are using high speed metallic or optical Ethernet for communication and supports modern protocols like IEC 61850 for fast, flexible and efficient communication between devices. Flexible sharing of data between multiple protection relays enables to implement more robust protection schemes, improving the reliability, objectivity and selectivity of protection algorithms leading to possible decreasing of tripping times.

This ability of flexible communication between devices is nowadays implemented also in protective terminals for medium voltage networks. Peer-to-peer communication between protection relays enables efficient implementation of protection schemes with directional elements in networks supplied from two or more sides. Of course, the drawback of this solution is the necessity to build up new communications paths between the dispersed technology.

Another way how to improve the reliability of distribution system in urban or industrial environment is the change of overcurrent functions from time-independent to the time-dependent. However, to ensure the selectivity of tripping, this change requires a modification of the settings on most of the deployed protections relays in the given area.

Optimization of tripping times and control of selectivity is very important especially in the existing industrial networks. These industrial networks are usually supplied from the 110-kV distribution network and often incorporate some local generation for covering local demand. Power distribution equipment in these networks on the 110 kV and on the 6.3 kV voltage level is often more than 30 years old. Nowadays, the short circuit conditions in the 110 kV networks have changed and short circuit power levels are often significantly higher than it was expected during design time, so it is questionable, if the short circuit capability of the equipment is still appropriate.

Whereas the industrial factory distribution equipment is still the same for 30 years, it is doubtful whether at present or in the future in view of the resistance to the short-circuit currents is still appropriate. At the medium voltage level 6.3 kV is switchgear most often for economic reasons protected by only overcurrent protection. Therefore, this article is focuses mainly on the optimization of the tripping times with the use of advanced functions modern protection terminals.

This paper will discuss the possible improving of tripping times in industrial and urban high voltage networks using directional time depended overcurrent protections in combination with peer-to-peer communication between protection relays for remote tripping. Reducing of the tripping times will have a positive impact on the lifetime extension of the protected equipment. The second important fact is that shortening of tripping times reduces the probability of loss of dynamic stability network.

II. CHANGES OF SHORT CIRCUIT POWER LEVELS

This chapter describes the development of short circuit power in a selected 110 kV distribution substation and in an industrial medium voltage network (factory network) powered from this substation. Period of the investigation is from 10/2015 until 01/2016. Contributions from the distribution network (called NET) were obtained from operator of the distribution substation. Contributions from the industrial network were obtained from HV and MV network model, created in the Neplan program (according to IEC 60909). It should be noted that, contributions in the industrial network consist of, contributions from turbo generators (G) and asynchronous motors (ASM). For each calculation period, all changes in the distribution network (raising resources in nodal areas) and in the industrial network (raising resources and raising consumption) were taken into account. The single line diagram of the investigated network is shown in Fig 1.

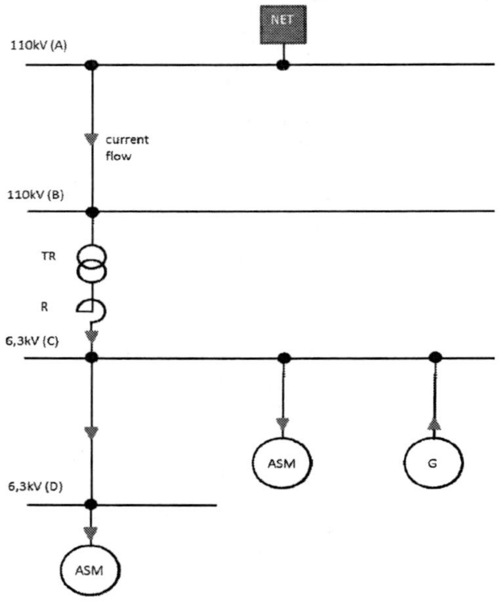

Fig. 1. Single line scheme of an industry distribution network

The short-circuit parameters of the components of an industrial distribution system are given and unchangeable for their lifetime. In our case, the parameters are taken from the real distribution system and located in table 1. Parameters are talking about short-circuit resistance of main bus bar switchgears.

TABLE I. PARAMETERS OF BUSBARS

Short circuit power on bus bars	S"k3max (MVA)
110 kV BB (B)	5995
6,3 kV BB (C)	750

Table 2. presents the contributions to overall short circuit power from all sources and motors for each investigation period. The contribution of short circuit power from the distribution network to the industrial network depends on the conditions and topology of nodal area providing power for the industrial network.

TABLE II. EXPECTED CHANGES OF SHORT CIRCUIT POWER

Fault location	Busbar B	Busbar B	Busbar B	Busbar B
Contibutions	S"k3max (MVA) from TG + ASM/SM + NET	S"k3max (MVA) from NET	S"k3max (MVA) from TG	S"k3max (MVA) from ASM/SM
Period				
1.10.2015	5810	5098	242	470
1.1.2016	6084	5100	483	500
1.11.2016	5854	4885	483	485
1.11.2017	5958	4990	481	487
1.1.2020	6289	5323	481	485
2.1.2025	6449	5482	481	485

From these results, it is evident, that there is gradual increase in short circuit power during the five-year period and most likely the increase will continue also after 01/2020.

III. ANALYSIS OF THE TRIPPING SCHEDULE OVERCURRENT PROTECTIONS

A. Tripping schedule for relays with definite time characteristic

Creating a tripping schedule, which is based on the time gradation, in the networks powered from two sides and with many sections, brings the problem of long tripping times. Most problematic are the sections which are fed during faults from both sides, for example the primary and secondary side of an medium and high voltage transformers and high voltage bus bars (C).

Tripping plan begins on the farthest place where is protection F2D with tripping time set to 0.4 s. This value represents a time of motor start up, when the inrush current of a motor falls under its nominal steady state current.

The recommended degree of selectivity for digital protection is 0.2 seconds, thus each next protection towards to the distribution network (NET) will operate with time of the previous protection plus degree of selectivity.

In case, when fault is fed only from one side (NET), protections will be operating selectively. In case, when fault is fed form two sides, for example between substations (A) and (B), there is a loss of selectivity on the side of the industrial network.

In Fig 2. are highlighted protections F1C (F2B, F11B), which can operate earlier than protection F1B. Pickup of these protections is function of their overcurrent value, which is set as threshold value.

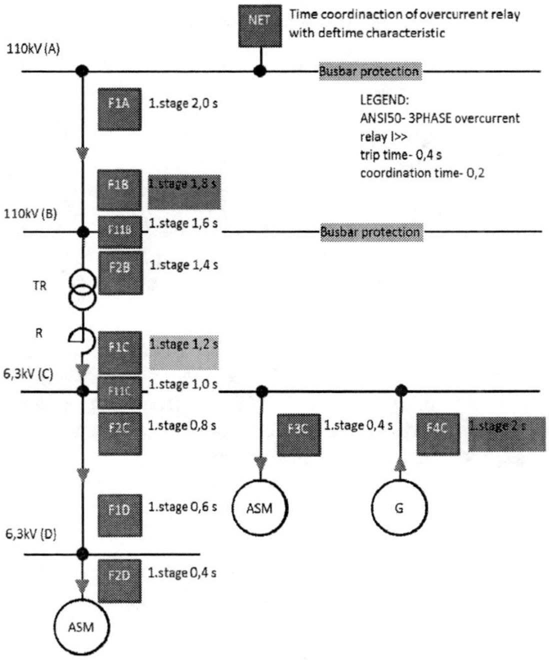

Fig. 2. Tripping schedule for relays with definite time characteristics

In this case, protection of turbo generator (G) must be set with greater tripping time than is tripping time of protection F1B to maintain his operation during fault between substations (A) and (B).

Switchgear equipment unnecessarily suffers by the high fault currents, because faults are switched-off in the long times and in addition also non-selective. From the previous chapter we know, that conditions in view of the stress from the short-circuit currents in the future will be even worse.

B. Tripping schedule for directional relay with definite time characteristic

In this scenario, a directional function was activated in protections F1A, F1B, F2B and F1C, with direction set to forward. Other protections stayed non-directional and their switch-off times are unchanged. The tripping times had been changed only for directional protections and for non-directional generator protection F4C. The new configuration of protections is in Fig. 3.

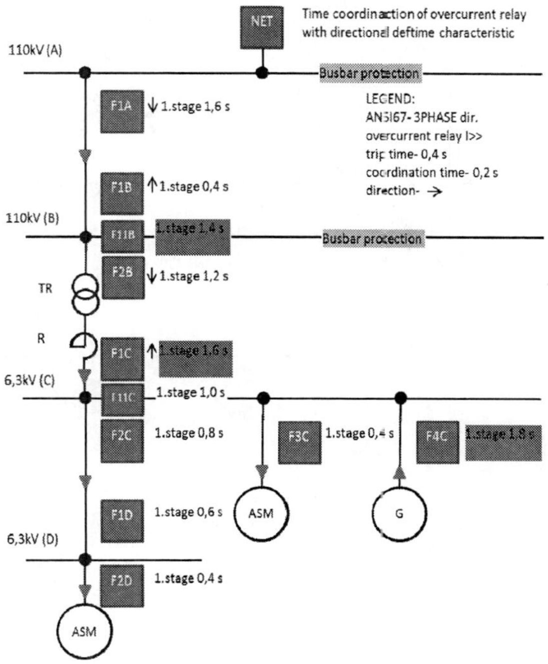

Fig. 3. Tripping schedule for directional relays with definite time characteristics

In case of a fault fed from the distribution network (NET) towards to farthest appliance is selectivity guaranteed. Busbar (A) respectively (B) are protected by their Busbar protection and opposite directional protection.

Busbar (C) does not include Busbar protection and is protected from the side (NET) with directional protection F2B and not closer protection F1C. This representing disadvantage for F2B e.g. setting of her threshold value. If protections between the resources are directional except protections F11B and F4C we can change their tripping times. In the principle, all times had been reduced except protection F1C. Its time was increased for reason of selectivity, for example if fault occur between relays F1A and F1B.

C. Tripping schedule for directional relay with definite time characteristic and direction comparison

Protections F1A, F1B, F2B and F1C are directional. Moreover, these protections were supplemented by function of directional comparison and second tripping stage.

The second stage for each protection is configured as described in the previous subchapter. The purpose of using this second stage is to create backup in case of loss of communication for directional comparison used by first stage protective functions. The configuration of protection system including directional comparison is shown in Fig 4.

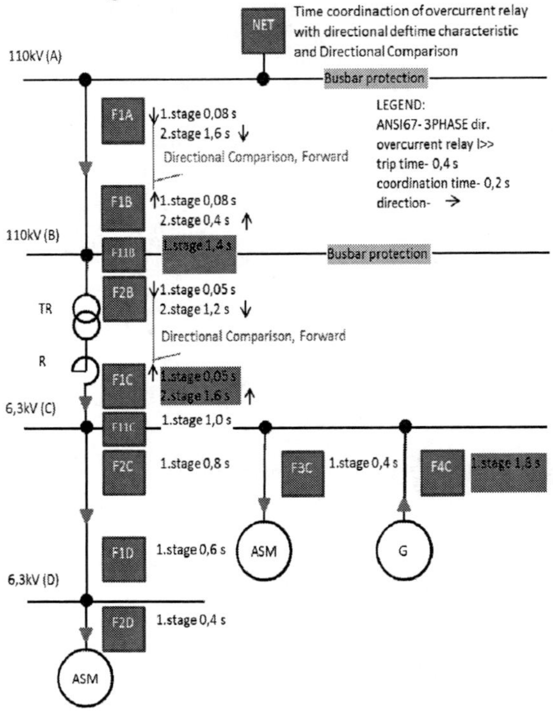

Fig. 4. Tripping schedule for relays with definite time characteristics

The directional comparison function was activated in the 1. stage of protections F1A - F1B and F2B - F1C. The role of the 1. stage of these protections is very fast tripping of high fault currents fed from both sides. Protections have been set with time delay 0.05 s respectively 0.08 s for safety reasons. In comparison with previous solutions, this new configuration enables to use significant shorter tripping times on the route between the both substations.

Similarly, to the preceding case, bus bars (A) and (B) are protected by the bus bar protection or 2. stage opposite directional protection. However, there is still one problematic place in the system. It's substation (C), where the faults are switched off in long times, 1.2 s from the side of distribution network (NET) and 1.8 s from the side of generators (G). Tripping time of the 2. stage of protections F1C and F4C remains the same.

Selectivity is achieved for tripping faults from both directions – from the 110-kV distribution substation (NET), but also from the side of the turbo generator (G).

Directional comparison principle

Directional comparison as a protective function is used for the selective isolation of a faulted line section (for example, subsections of a closed ring). Sections are isolated in very short time, so the stress of the equipment caused by high short circuits is significantly lower than in case of traditional protection schemes with the disadvantage of long tripping times.

The information about fault in forward direction is transferred between protection relays on both ends of the protected object (power line).

A directional definite time-overcurrent protection function is in operation in both devices in the forward direction (1st level). However, this function is not enabled in the idle state. The function is only activated when the information "fault in forward direction" is received from the opposite end. If the enabled function also defines the fault in the forward direction, the fault must be on the protected line section and the level trips immediately.

As this protection principle works with an enable procedure (and not with a blocking procedure), there is no need to delay the level.

Release principle doesn't work in these situations:

- Infeed at one end or weak infeed at one end: In this case, no release signal is generated.
- Failure of the communication route: In this case, the release signal is not transmitted.

Blocking principle is working in the idle state of the one relay or when communication channel is interrupted or when one of the ends has weak infeed. Under this circumstance, directional overcurrent stage needs a time delay of about 100 ms for his operations.

For our investigated system, the release principle is a better choice, because this principle needs only a minimum time delay for OC (overcurrent) stages and the protection system is functional even in case when communication is lost. In that case, the second OC stage creates backup for first stage with directional comparison. [1]

D. Tripping schedule for directional relay with definit time characteristic, directional comparison and tripping time optimization

Protection system is based on the previous subsection. The main difference is the reduction of degree of selectivity value from 0.2 s to 0.18 s. In theory, it would be possible to reduce the degree of selectivity to 0.15 s. The prerequisite for this reduction is the detailed analysis of response times of circuit breakers, tripping relays and protections, to prevent loss of selectivity.

The second change occurred in the adding of directional comparison between protections F1C - F4C which is set backward and ensures fast tripping of faults on the busbar (C).

In case of fault behind the protection F2C, it is necessary to create blocking logic for the "backward" stage of protections F1C and F4C. These functions are blocked by pickup signal from F2C relay. Tripping time of the second stage protections in relays F1C and F4C is delayed about 50 ms. This value is given by calculation time of algorithm in relay F2C, which is about 20 ms, with switching time of power relays which is about 8 ms and with communication delay which is about 1 ms.

This protection system is selective from both directions and is presented on figure 5. All directional protections with system of directional comparisons have created backup, in form of an additional directional tripping stage, for the case of communication loss.

Protections F1C and F4C are very important for the proper operation of the protection system, therefore also these protections have a backup function, in the form of additional non-directional stage. This non-directional stage provides protection in case, when the phase voltages have collapsed and the resulting voltage is too low for determination of fault direction.

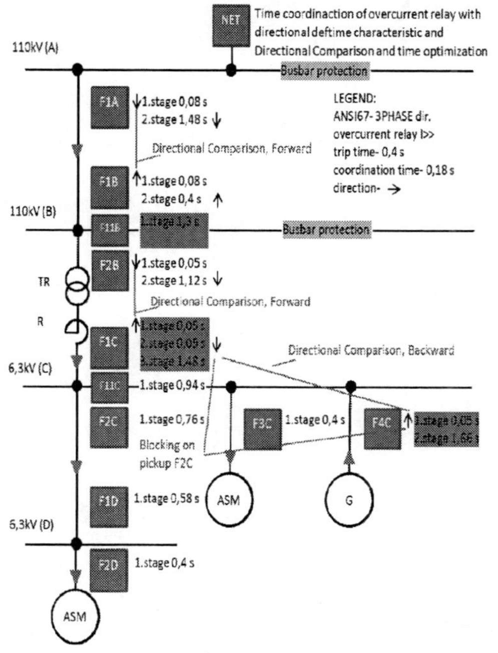

Fig. 5. Tripping schedule for relays with definite time characteristics

Protections F1C and F4C can trip high fault currents at busbars of switchgear (C) in short times. This system creates 100 % selective sections within the distribution system also provides backup in case of failure of a protection relay or it's circuit breaker.

The recommended communication interface between protection F1A and F1B resp. F2B and F1C is a direct optical connection through their optical interface. Between protection F1C and F4C, it is theoretically possible to use direct exchange of electrical signals via digital inputs and outputs of these protections. In cases of, environment with high electromagnetic interference, it is necessary to provide the exchange information between protections also over optical interface.

IV. Conclusion

The use of directional comparison in directional overcurrent protections causes considerable shortening of tripping times and creation of selective sections in the networks, which are powered from the two sides.

The protections operate selectively in both directions from the all sources. These improvements are most significant at medium voltage level, where switchgear is mainly protected with overcurrent protections. In the high voltage distribution system, overcurrent protection is used as backup protection to fast distance protection or differential protection. Protections systems using the directional comparisons principle are able to operate selective with very short tripping times, similar to the protection systems based on distance or differential protections.

Overcurrent protections (or feeder terminals with a protective functions) are, from an economic point of view, in comparison with fast distance protection approximately 30 % to 45 % cheaper. Directional comparison (similar to the use of differential protections) requires also optical communication network, but this form only ca. 0.5 % from prices of the overcurrent protection, which is a negligible.

If we compare tripping times from figure 2. with figure 5., we can clearly see the reduction of the tripping times for the most stressed sections of the distribution system during the faults.

<u>Comparison of tripping times on the section between (A) and (B):</u>

F1A (2 s) - F1B (1,8 s), Fig. 2.

F1A (0,08 s) - F1A (0,08 s), Fig. 5.

<u>Comparison of tripping times on the section between (B) and (C):</u>

F2B (1,4 s) - F1C (1,2 s), Fig. 2.

F2B (0,05 s) - F1C (0,05 s), Fig. 5.

<u>Comparison of tripping times on the section (C):</u>

F1C (1,2 s) - F4C (2,0 s), Fig. 2.

F1C (0,05 s) - F4C (0,05 s), Fig. 5.

To achieve the grade of selectivity less than 0.2 s in the tripping plan, it is necessary to make detailed analysis of reaction times of the tripping elements.

At very small grade of selectivity about 0.15 s, it may occur overlapping of tripping curves, so unwanted tripping of two consecutive protections can occur. This overlap can occur especially in systems with time-dependent overcurrent protection.

Different values of fault currents from the direction of the distribution substation (NET) and from the direction of the turbo generator (G) has a great impact on a setting currents threshold values and on the selectively of operation of non-directional time dependent or independent protections. These negative characteristics can be reduced by using directional overcurrent protections with a functions directional comparison.

Because industrial networks are in general very complex, it is important to validate the proposed protection system and its settings using computer simulation. The simulation should take into account also the accuracy and possible saturation of current transformers. In this way, we can verify the behavior of the protection system during different operating conditions and fault scenarios. All results presented in this paper were validated and obtained using Neplan simulation software.

REFERENCES

[1] M. Kezunovic, R. Jinfeng, S. Lotfifard, *Design, modelling and evaluation of protective relays for power systems*, Springer International Publishing, 2016.

[2] R. Patterson, E. Price, M. Sanders, "Directional comparison blocking system fundamentals," in 67th Annual Conference for Protective Relay Engineers, College Station, TX, 2014, pp. 50-71.
doi: 10.1109/CPRE.2014.6798994.

[3] D. Yuan, X. Dong, S. Chen, Z. Q. Bo, B. R. J. Caunce, A. Klimek, "An New Directional Comparison Scheme for Distribution Line Protection," in Proceedings of the 41st International Universities Power Engineering Conference, Newcastle-upon-Tyne, 2006, pp. 881-885.
doi: 10.1109/UPEC.2006.367607.

[4] A. G. Phadke, J. S. Thorp, *Computer Relaying for Power Systems*, John Wiley & Sons Ltd., 2009.

[5] M. Hraško, M. Höger, P. Hečko: "Modeling and identification of failures in distribution network", in Proceedings of the 5th International scientific symposium on Electric Power Engineering, ELEKTROENERGETIKA 2009, Stará Lesná, 2009, pp. 346 – 349.

The Impact of Power Quality on Smart Metering System

Pavol Belány, Filip Suško, Andrej Bolf, Michal Repák, Alena Otčenášová, Marek Roch
Department of Power Electrical Systems
Faculty of Electrical Engineering, University of Žilina
Žilina, Slovakia
pavol.belany@fel.uniza.sk, filip.susko@fel.uniza.sk, andrej.bolf@fel.uniza.sk,
michal.repak@fel.uniza.sk, alena.otcenasova@fel.uniza.sk, marek.roch@fel.uniza.sk

Abstract— **This article describes the effect of current harmonics on smart metering system measurement accuracy. The harmonics of current were minimal in the past. Due to higher utilization of converters in various applications, the impact of these harmonics on the metering system also increases. Because of the errors in measurement, the distribution companies can suffer income losses. This article contains measuring results that enable unambiguous determination of the harmonics impact. Graphical representation of the results is also shown to highlight the measuring errors as well as the differences in measured power according to the harmonic order.**

Keywords: current; smart meter; harmonic; non-harmonic; relative fault; electricity meter

I. Introduction

New trends in the electronics field start to significantly affect electrical power system. Adverse effects are caused by massive utilization of converters in various applications. It cannot be said that each type of the appliances has the same effect. The degree of influence is given by the design of appliance [1]. Currently, there are harmonics in the network and they have a significant impact [2].

Appliances with non-sinusoidal consumption have the most significant impact on the consumption measurement. Compact fluorescent lamps, drive converters are examples of these appliances. The error of analog disc type electricity meters is affected mostly by this non-sinusoidal consumption. This problem was partly solved by introduction of smart digital electric meters [1],[3]. These meters provide measurements that are less influenced by the non-sinusoidal consumption. Another advantage of modern smart meters is that they enable direct measuring of reactive power and evaluation of power quality according to STN EN 50160 [4].

Backward analysis of the data obtained according to the STN EN 50160 enables to determine source of declined power quality. Many connected devices are sensitive to this decline. Some of these devices even can stop operate because of that. The loss or change in functionality of these devices may cause incorrect evaluation of input data. This input error is distributed through the entire system and can causes a change to which the system was not prepared to. In the worst case scenario it can cause a malfunction, which can lead to an outage of a certain system and interruption of work [3].

II. The harmonics in distribution grid

Harmonics of current have been in the grid for many years. Original sources of harmonics were mercury rectifiers used in electric traction. Nowadays, appliances that generate harmonics are increasing in numbers. In the past, the harmonics affected only certain areas of communication and security technology. Today, the problem is becoming more general, because it affects customers, which are at the same time sources of the declined power quality. Harmonic frequencies are integer multiplies of the fundamental harmonic of the supply voltage. This means that a third harmonic with 150 Hz, fifth harmonic with 250 Hz, etc. are for fundamental harmonic with 50 Hz [1].

These components cause deformation of the supplied current and affect phase shift. This deformed current characteristic will have an impact on the metering devices accuracy.

A. Analysis of non-sinusoidal waveform

Each non-harmonic waveform can be decomposed to a sum of several sine waveforms. These waveforms have precisely given amplitude, frequency and phase. These parameters can be obtained by Fourier analysis. The basic advantage of using only sinusoidal waveforms is the ability to use conventional methods to calculate steady state [5].

If we analyze the grid in which the deformed waveforms of voltage and current occur, we perform the Fourier decomposition at first in order to obtain several waveforms. These waveforms will be purely sinusoidal with frequencies that are integer multiplies of the fundamental frequency. Another step is to use the conventional calculation methods for each harmonic independently [6]. The resulting system response is obtained by a suitable combination of individual results.

B. Adverse effects of harmonics

Harmonics cause adverse effects on devices. These effects can be divided to long-term and short-term.

Short-term effects are linked with malfunctions, erroneous functions or lower degree of quality during the device operation caused by the zero-crossing shift of voltage waveform. Large amplitudes of harmonics can cause erroneous operation of receivers of ripple control and protection devices [7].

Long-term effects are basically heat ones. Additional losses and overheating cause faster aging and can even cause damage to capacitors and rotating machines. They begin to have effect after 10 minutes of interference.

Adverse effects of harmonics of current waveform are caused primarily by the fact that effective value of non-harmonic current is higher than the value of fundamental harmonic, for which the devices are designed for [1], [5]. The most serious adverse effects in the power systems area are:

- erroneous operation of regulating devices,
- erroneous operation of ripple control and other grid signalization systems, protection systems and other means of control,
- additional losses in capacitors and rotating machines,
- additional noise of motors and other electric machines and devices,
- interference on data and telecommunication networks, both fixed and wireless lines,
- worse thermal mode of appliances,
- accelerating of aging of insulation, which leads to lower lifetime,
- reduced amount of supplied active power from generators and therefore their lower efficiency,
- creation of conditions favoring undesirable resonance phenomenon's in the grid,
- increase in registration and measuring errors,
- malfunction of batteries and capacitors due to overloading [5].

C. *Sources of harmonics*

Harmonics of currents are generated in small amounts by generators and transmission and distribution devices. However, high amounts of harmonics are generated by industrial and home appliances [7]. Usually there are only few dominant sources, which generate significant harmonic currents in the grid.

1) *Industrial appliances*

It's impossible to achieve a perfect sinusoidal waveform of voltage with generators. These problems occur because of technical deficits during manufacturing and can be almost eliminated by selecting suitable construction procedures. Another source is the asymmetrical operation, which causes generation of a third harmonic and other multiple of three harmonics.

Industrial devices, which can be a source of a significant level of harmonic distortion, include: regulated and non-regulated converters, welding units and arc and induction furnaces [8].

Power electronic devices can have a significant impact on the level of distortion in the electrical grid. The most widespread static converters are rectifiers, both controlled and uncontrolled. It is possible to calculate the shape and amplitude of the supplied electric current on the primary side of the rectifier's transformer for a specific rectifier layout. By the means of harmonic analysis of this waveform we can determine the amplitudes and harmonic spectrum of the current. On the AC side of the rectifier only certain harmonics, that follow, can appear:

$$h=k\cdot p\pm 1 \tag{1}$$

where h harmonic order,

p is the number of impulses of the rectifier,

k is a natural number [5].

These harmonics are called characteristic harmonics. The harmonics of different order can also appear due to asymmetries present in the rectifier, different control angles of the semiconductor parts and distorted supply voltage [1].

Harmonics amplitudes obey the amplitude law, which states that the amplitude should decrease with increasing harmonic order, as follows:

$$I_h = \frac{I_1}{h} \tag{2}$$

In practice, the rectifiers do not switch immediately and current waveforms do not have a precise rectangular shape. The harmonics amplitudes depend on the inductive voltage drop, which depends on the inductance of the circuit and the switching angle [5].

The size and direction of harmonic currents depends mostly on:

- rectifier circuit connections,
- winding connections of the rectifiers transformer,
- commutation angle γ,
- used control method and the size of the control angle α.

Arc furnaces can be considered as harmonics current sources with internal impedance consists of an inductance and a damping resistance [5].

2) *Loads in residential areas*

Loads in residential areas have lower rated power, but time they can become a source of harmonic distortion due to their large numbers and long utilization. Most significant contributors are semiconductor appliances (light dimmers, LED bulb converters and etc.). Computers are usually powered by switch mode power supplies. These supplies contain a rectifier and a smoothing capacitor [9]. This arrangement causes that the supplied current is in the form of short impulses. These impulses contain high amounts of harmonics [9].

III. SMART METER - LZQJ-XC

This three phase electric meter is used to measure active and reactive power in all 4 quadrants with frequency ranging from 50 to 60 Hz [10]. It has the following capabilities:

- two, three and four wire connection,
- Cl. 1, 0.5, 0.2 S accuracy class for active power,
- Cl. 2 accuracy class for reactive power,
- 32 tariff registers, maximum registers and 8 non-tariff registers,
- power profile with 32 channels,
- D0 optical interface,
- 7 impulse outputs,
- 2 independent series interfaces,
- wide measuring range, event log, measuring of instant values [10].

Accuracy class for measuring active power is 1 and for measuring reactive power is 2. Impulse output is implemented with the help of a LED interface. This interface has a constant 500/1000 imp/kWh. Calibration is done using the calibration – technical log. Real time clock is synchronized using the data interfaces. Six channels are available to the load management system. Data can be stored without power supply for a minimum of 10 years in the FLASH memory. Service is possible using a mechanical button, optical sensor to evoke a display or, after expansion with a Variomod module, using a TCP/IP protocol. The electric meters self-consumption per one phase is less than 1.3 VA and it is less than 0.01 VA in current mode. IP protection level is IP 51. Data interfaces are implemented using: a D0 optical data interface, RS-485 and RS-232. Maximum transfer rate is 9600 Baud. Other options are: measuring instant values, connection of optical head, installation check, cover manipulation check, electric meter terminals check, and influence of magnetic fields through the optical interface and from display during power outage. The electric meter has a defined constant of 500 Imp/kWh [10].

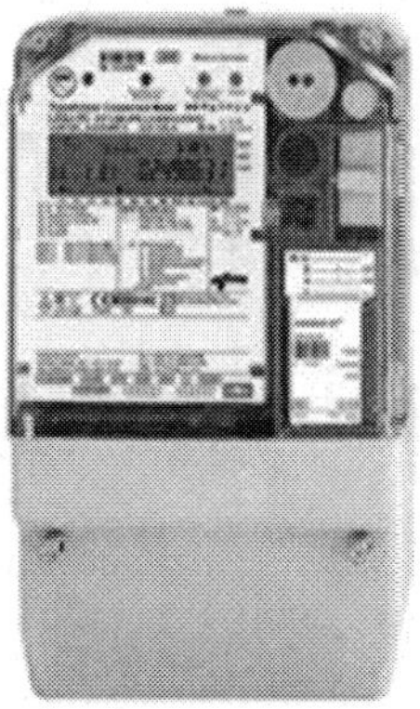

Fig. 1. Smart meter LZQJ-XC

IV. Reference standard 2330A

Measurement system, that is primary used to measure the accuracy of the electric meters, was used for measurement. As a source of electrical energy was used source from Applied Precision, whose basic parameters are given in Table 1 [11].

Table 1. Technical specification

General Parameters	
Basic Frequency Range	40 ..70 Hz
Test Voltage	30 .. 500 V
Voltage Ranges	continuous / autorange
Test Current	1 mA .. 120 A
Current Ranges	continuous / autorange
Basic Frequency	40 .. 70 Hz
Bandwidth	up to 4000 Hz
Power Factor Range	0 .. 1
Maximum Error	
Voltage	0.02 %
Current	0.02 %
Apparent Power	0.05 %
Active Power	0.05 %
Reactive Power	0.05 %
Frequency	0.005 Hz
Distortion	0.05 %

Reference standard RS 2330 is a three-phase accurate meter for measuring electrical energy. The reference standard meets all the requirements of the reference standard in three-phase testing and calibration systems. This standard can be set to actual or desired operation mode.

Reference standard RS 2330 is based on a precise 24-bit A / D conversion and digital signal processing technology that allows accurate evaluation of all variables. Variables that can be measured are as follows: voltage, current, phase, harmonic analysis and distortion.

The Reference Standard RS 2330 is equipped with three completely independent differential voltage input circuits, which allows the evaluation of signals on three independent channels [11].

V. Measurements

In order to verify the effect of harmonics on the accuracy of the Schrack LZQJ-XC, a measurement was carried out. We analyzed the effect of different combinations of harmonics on this specific type of a smart meter in this measurement. This particular electric meter has a certified accuracy of B (measuring accuracy class 1) for active power and class 2 for reactive power [10].

Each measurement was carried out as follows. The shape of the signals was predefined in the Elma software and these signals were used as input to the smart meter. After measurement, the errors were evaluated. The description of the shapes of the signal and the error evaluation is described in the following section.

Measurements were made with the following combinations of the fundamental harmonic and other harmonics. The levels of the harmonics were calculated according to the Eq. 2. The fundamental harmonic has level of 100%. The level of 100% was electrically expressed as 50 A. The harmonics had a decreasing tendency according to the described equation. Only the harmonics of the current was changed during the measurements. The voltage was set on level 230 V for each measurement. There is presented amplitudes dependence of harmonic order in Fig. 2.

The time of measurement was the same for every case. This time is not exactly determined because the measurement was set to count of impulses. For each measurement, 5000 impulses were measured. The 5000 impulses correspond to 10 kWh. The error calculation is performed by optical impulse output. The Elma control program counts the impulses and determines the measurement error.

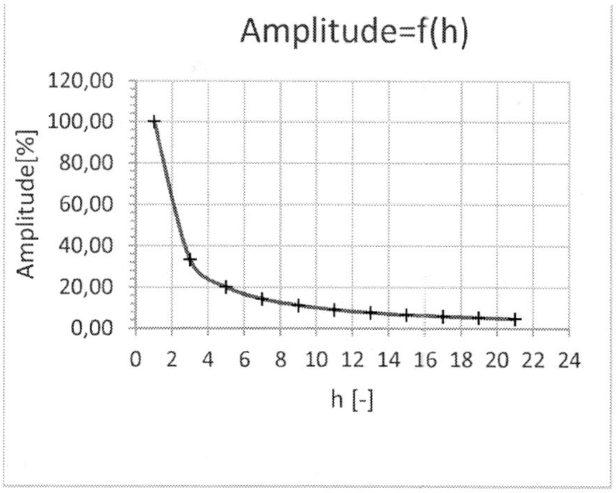

Fig. 2. The amplitudes dependence of harmonic order

These combinations were selected, because they are most common in the distribution network. Each combination is realized as a sum of the fundamental harmonic with gradual addition of harmonics up to a 21st harmonic. During the measurement, the boundary state was reached. After being overwhelmed, the smart meter Schrack LZQJ-XC was no longer able to measure the power consumption.

Table 2 shows results of the intelligent electric meter LZQJ-XC testing. This table is composed of the harmonics to which the electric meter was subjected.

Table 2. Resulting values of the smart electric meter testing

n[-]	Test	ε [%]
1	1	0,06
2	1+3	-1,1
3	1+3+5	-2,57
4	1+3+5+7	-4,47
5	1+3+5+7+9	-3,77
6	1+3+5+7+9+11	-4,94
7	1+3+5+7+9+11+13	-3,10
8	1+3+5+7+9+11+13+15	-2,77
9	1+3+5+7+9+11+13+15+17	-2,75
10	1+3+5+7+9+11+13+15+17+19	-2,61
11	1+3+5+7+9+11+13+15+17+19+21	-3,09

Fig. 3 shows the error function of the measurements. This function proves that the electric meter meets the accuracy requirement of its class only in one case – while being supplied by only the 50 Hz fundamental harmonic. With the addition of harmonics, the error has risen up to -4.94 %. This error was the maximum that was measured. During further addition of harmonics the error oscillated around -3 %.

These results show that from a given point the addition of more harmonics does not affect the measurement error. It has to be noted that for each of non-harmonic waveform that has been tested, the error has exceeded the limits of its accuracy class. Another interesting result is that the error is negative, meaning that the electric meter measured lower amount of electric energy than was actually supplied. This state is unwanted for the distribution companies, because they receive less revenue.

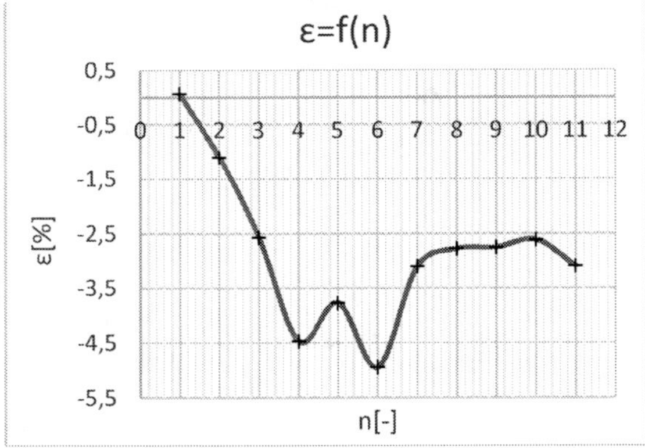

Fig. 3 The error function dependence of the measurement number

A following methodology was used to verify that the testing was done correctly. After each measurement the value from the electric meter was compared to the reference standard 2330 value. The results of these verification measurements are in table 3.

Active power is actually supplied from a reference standard 2330. This value was exported from the Elma control program. The active measured power is value that was read from register of a smart meter using an optical head and n is the number of the measurement.

It is obvious that the error values are not exactly the same. The electric meter is able to record power only in kWh using numbers with two decimal places. Based on the calculated error, we can conclude that the testing was correct because the calculated error is approximately equal to the error measured by the testing system.

Table 3. Comparison of actually supplied power and measured power

n[-]	Active power [W]	Active measured power [W]	ε [%]
1	10000	10010	0,10
2	10000	9870	-1,30
3	10000	9710	-2,90
4	10000	9550	-4,50
5	10000	9650	-3,50
6	10000	9520	-4,80
7	10000	9690	-3,10
8	10000	9750	-2,50
9	10000	9720	-2,80
10	10000	9750	-2,50
11	10000	9690	-3,10

Fig. 4 graphically shows the ratio between the real active power (red columns) with measured one (green columns) that was read-out from the electric meter. This graph takes values from the table 3.

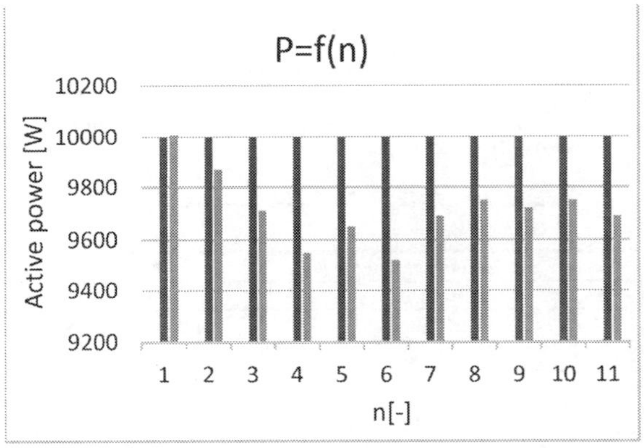

Fig. 4 Comparison of active power and active measured power

The mentioned states during the electric meter testing were chosen to verify if the selected electric meter is able to measure non-harmonic current and what is the error of this measurement.

Another measurement was to verify the accuracy on a real life example. Power quality measurement values from a residential housing area were obtained. These data include real values of harmonics up to the 50th harmonics. An error of -0.94% was obtained. This error is almost on the edge of the electric meter accuracy but it still achieves the accuracy requirements for the given non-harmonic waveform. The disadvantage is that the selected electric meter can also be used in industrial application, where the amount of harmonics can be much higher. If these values would be higher, the electric meter would not comply with its accuracy class.

Fig. 5 shows harmonic analysis of the signal and fig. 6 shows current curve used for this measurement. Because the dominant values of harmonics range only up to the 25th harmonics, the harmonics with higher orders were omitted for better readability.

It is obvious that odd harmonics are dominant in the grid. This confirms the Eq. 2 that describes harmonics developed in the network.

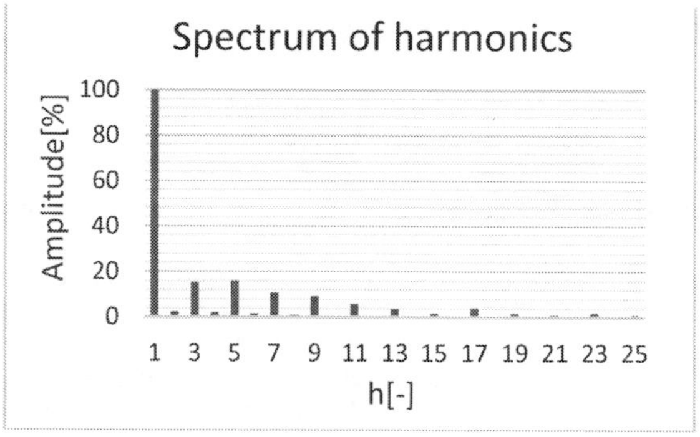

Fig. 5 Harmonic analysis

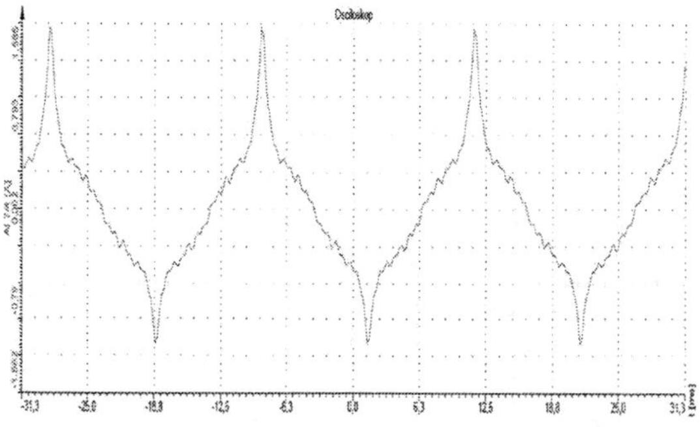

Fig. 6 Current curve

VI. Conclusions

Nowadays, much attention is given to the power quality. It is a one of the basic requirements that has to be fulfilled to secure reliable operation of individual systems. This article contains experimental results of an electric meter testing. The aim was to analyze if a smart electric meter can correctly work and measure within its prescribed accuracy tolerances while operating with declined power quality. We can conclude that the measurement accuracy is greatly affected by harmonics. Based on the experimental verification, we have proved that the electric meter is not able to measure a non-harmonic distorted signal, accurately. Many appliances, which decline power quality, are currently being used. This problematic will have to be accounted for in the future, because of the resulting effect of measuring systems.

An extension of systems that improve power quality will have to be secured in the future. Better control of appliances also needs to be carried out, so that these devices do not contribute to current deformation during their operation. There is a very thin line between meeting the accuracy requirements of the norms and exceeding the allowed error.

Acknowledgment

This work was supported by project ITMS: 26210120021, co-funded from EU sources and European Regional Development Fund.

References

[1] R.C. Dugan., M.F. McGranaghan, S. Santoso,,H.W. Beaty, *Electrical Power Systems Quality*, 2012, 3th edition, The McGraw-Hill Companies, Inc., NY, USA , ISBN:978-0-07-176155-0

[2] A. J. Berrisford, "New technology and power definitions make accurate revenue metering possible in the presence of harmonic distortion," in Electrical Power & Energy Conference (EPEC), 2009 IEEE, IEEE, doi: 10.1109/EPEC.2009.5420905.

[3] S. Makonin, F. Popowich, B. Gill, "The cognitive power meter: Looking beyond the smart meter," in 26th Annual IEEE Canadian Conference on Electrical and Computer Engineering (CCECE), 2013, IEEE, May 2013, doi: 10.1109/CCECE.2013.6567686

[4] B. Shahid, Z. Ahmed, A. Faroqi, R.M. Navid-ur-Rehman, "Implementation of smart system based on smart grid Smart Meter and smart appliances," in 2nd Iranian Conference on Smart Grids (ICSG), 2012, IEEE, MAY. 2012, ISBN: 978-1-4673-1399-5

[5] J. Altus, M. Novák, A. Otčenášová, M. Pokorný, *Elektrickomagnetická kompatibilita elektrizačných sústav*, EDIS Žilina, 2004, p. 328, ISBN 80-8070-227-6 ,

[6] I. Dronyuk; O. Fedevych; Z. Poplavska, "The generalized shift operator and non-harmonic signal analysis", in 14th International Conference The Experience of Designing and Application of CAD Systems in Microelectronics (CADSM),2017, pp-89-91, doi: 10.1109/CADSM.2017.7916092

[7] A. Y. Odintsov; Y. P. Salomatov; M. I. Sugak; K. V. Lemberg, "Formulation of the Hansen-Chu-McLean relations for the non-harmonic signals," in 2013 International Siberian Conference on Control and Communications (SIBCON), 2013, pp.1-4, doi: 10.1109/SIBCON.2013.6693597

[8] Y. Chen; R. Meng; L.Wang; Y. Yang; Y. Liu; X. Han, "An improved control strategy for three-phase four-wire active power filter based on vector-resonant control," in IEEE 11th Conference on Industrial Electronics and Applications (ICIEA), 2016, pp. 1440-1445, doi: 10.1109/ICIEA.2016.7603812

[9] N. A. Abidullah; A. R. Abdullah; N. H. Shamsudin; N. H. T. H. Ahmad; M. H. Jopri, "Real-time power quality signals monitoring system," in IEEE Student Conference on Research and Development, 2013, pp. 433-438, doi: 10.1109/SCOReD.2013.7002626

[10] EMH metering GmbH & Co. KG, "LZQJ-XC," [online] Internet: http://www.emh-metering.de/en/products/lzqj-xc1/, [May 14, 2017].

[11] Applied Precision, "Reference Standard RS 2x30," Product Catalog. [online], Internet: http://www.appliedp.com/download/catalog/rs2x30_pc_en.pdf/, [May 14, 2017].

Transient analysis of multi-circuit overhead transmission lines with two voltage levels

Tomáš Nazarčík, Zdeňka Benešová
Department of Theory of Electrical Engineering
University of West Bohemia
Pilsen, Czech Republic
nazarcik@kte.zcu.cz

Abstract—**In this paper a method for a common transient analysis for the multi-circuit overhead transmission line is presented. It takes into account the influences of the mutual couplings between the conductors. It is based on two-port approach, where the phase conductors and earth wires are modelled by inductive and capacitive coupled cascades of two-ports. For a considered network model, a system of ordinary differential equations (ODE) has been formulated and it has been solved in MATLAB using Runge-Kutta method of 4th order. It results in the time distribution of currents and voltages in all conductors. In this paper a system with four three-phase system of two voltage levels (EHV and HV) has been analysed. The calculations were performed for the typical transients that can occur on EHV/HV line - the single phase short circuit together with automatic reclose cycle.**

Keywords—multi-circuit line; transient analysis; mutual couplings; line switching; short circuit

I. Introduction

The growing demands of electric energy are connected with the requirement to higher power transmission. The inconveniences joined with finding new routes for the transmission lines made the transmission and distribution line operators build the multi-circuit overhead lines with four or six three-phase systems. This fact brings some problems that should be researched. The electromagnetic field distribution and phase optimization with respect to the safety standards has been analysed by several authors in [1],[2],and [3]. In the case of the quadruple transmission tower with four three-phase circuits, the line transposition seems very difficult. So that these multi-circuit lines are usually carried-out as non-transposed ones. The conductor's layout causes asymmetry of lines parameters and inductive and capacitive couplings between three-phase systems play an important role as well. During the standard operation these couplings cause phase current unbalance [4] that results in permanent current flowing through the earth wires. The transient analyses with regard to the mutual couplings was described in [5] using the frequency domain. In this paper the transient analyses has been researched in time domain. The typical transients that can occur on the transmission line (line switching, short circuit, automatic reclose cycle) were analysed. Based on obtained results it is possible to formulate some recommendations for the tower design and on setting of a protection system.

II. Model of the Transmission Line

A. Two-port model

The phase conductors of both voltage levels have been modelled with the help of cascades of gamma two-ports. As it is usual, the mentioned two-port involved parameters of the line (R, L, C and G). It is necessary to add the voltage and current source in each two-port. These sources are respecting the induced voltage via mutual inductive couplings and the induced current via mutual capacitive couplings. The final structure of the k-th two-port of the cascade of i-th phase conductor is depicted in Fig. 1.

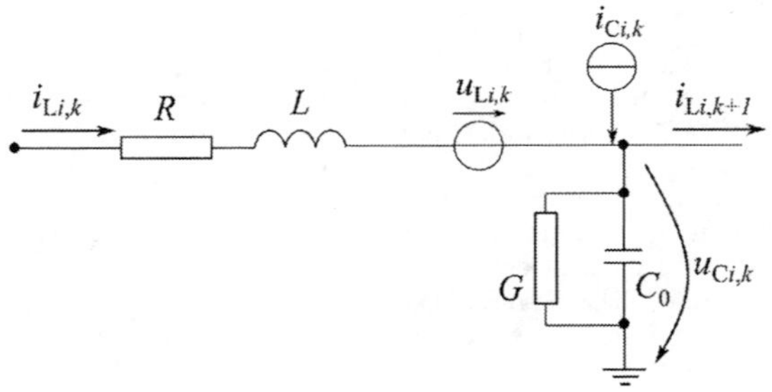

Figure 1. Basic gamma element of k-th two-port of i-th phase conductor

The value of the voltage source is given by the following formula:

$$u_{Li,k} = \sum_{j=1, j\neq i}^{n+m} L_{ij} \frac{di_{Lj,k}}{dt} \quad (1)$$

The L_{ij} is the mutual inductance between the i-th and j-th conductor. The $i_{Lj,k}$ is the current of the j-th conductor in k-th two-port. The $(n+m)$ is number of all conductors (phases+earth wires). The value of the current source can be expressed as:

$$i_{\mathrm{C}i,k} = \sum_{j=1,j\neq i}^{n+m} C_{ij}\left(\frac{du_{\mathrm{C}j,k}}{dt} - \frac{du_{\mathrm{C}i,k}}{dt}\right) \tag{2}$$

The C_{ij} is the mutual capacitance between the i-th and j-th conductor. The $u_{\mathrm{C}j,k}$ is the voltage of the j-th conductor in k-the two-port. For the evaluation of the mutual couplings it is necessary to know the geometric layout of the conductors. Using the algorithms described in [6] and [7] it is possible to obtain the matrices of inductances and capacitances. The earth wires were modelled as the two-ports without the capacitance and the conductance to the earth. The ground was respected by the resistance R_g and inductance L_g in series according to the Ruddenberg theory. The part of diagram

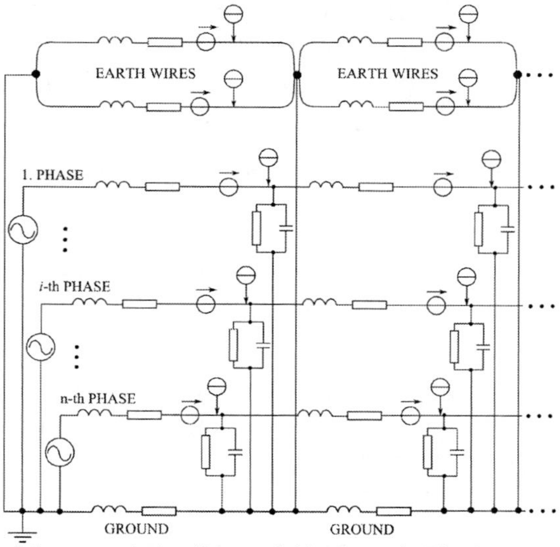

of the transmission line model is shown in Fig. 2.

Figure 2. Part of the two-port diagram

B. Mathematical model

The diagram can be described by differential equations, they must be written for the state variables (current of the inductance i_{L}, voltage on the capacitance u_{C}). In each two-port for the phase conductors there are 2 state variables, in each two-port for the earth wires or ground there is 1 state variable. Supposing a system with n phase conductors (of EHV and HV system together) and m earth wires we obtain $(n+m)$ cascades each of them consists of N two-ports. Considering this fact we obtain $N*(2*n+m+1)$ differential equations. Creating of the equations is demonstrated in following steps.

- Node between the k-th and $(k+1)$-th two-port of the i-th phase conductor

$$i_{\mathrm{L}i,k} + \sum_{j=1,j\neq i}^{n+m} C_{ij}\left(\frac{du_{\mathrm{C}j,k}}{dt} - \frac{du_{\mathrm{C}i,k}}{dt}\right) =$$

$$G_i u_{\mathrm{C}i,k} + C_{i0}\frac{du_{\mathrm{C}i,k}}{dt} + i_{\mathrm{L}i,k+1} \tag{3}$$

The equations for the nodes of grounding can by determined likewise (3).

- Loop phase conductor – ground of the i-th conductor in k-th two-port of the cascade

$$R_i i_{\mathrm{L}i,k} + L_i\frac{di_{\mathrm{L}i,k}}{dt} + \sum_{j=1,j\neq i}^{n+m} L_{ij}\frac{di_{\mathrm{L}j,k}}{dt} + u_{\mathrm{C}i,k}$$

$$-\left(R_g i_{\mathrm{L}g,k} + L_g\frac{di_{\mathrm{L}g,k}}{dt}\right) - u_{\mathrm{C}i,k-1} = 0 \tag{4}$$

In similar way we can write the equation for the loop earth wire – ground. The load is respected in the equation of last node in cascade or in a new loop, it depends on the type of load. For the solution of the system of ODE in MATLAB it is necessary to get the Jordan canonical form of the equations:

$$\frac{d}{dt}\boldsymbol{x} = \mathbf{A}\boldsymbol{x} + f \tag{5}$$

where $\boldsymbol{x}$ is a column vector of state variables, $\mathbf{A}$ is matrix of coefficients and f is a column vector of sources.

Equations that describe the transmission line contain more than one element with the derivative of the state variable in the equation. It requires sorting the equations on elements with derivative on the left side of equation, the rest elements on the right side:

$$\frac{d}{dt}\mathbf{A}_1\boldsymbol{x} = \mathbf{A}_2\boldsymbol{x} + f \tag{6}$$

where $\mathbf{A}_1$ is a matrix of coefficients of elements with derivative and $\mathbf{A}_2$ is a matrix of coefficients of elements without derivative. After that we can obtain the Jordan canonical form multiplying the equation (6) by the inverse matrix $\mathbf{A}_1$ from the left:

$$\frac{d}{dt}\boldsymbol{x} = \mathbf{A}_1{}^{-1}(\mathbf{A}_2\boldsymbol{x} + f) \tag{7}$$

The system of ODE in this form has been solved in MATLAB using the Runge-Kutta method of 4th order.

III. ILLUSTRATIVE EXAMPLES

We can find various types of basic phase arrangements of HV and EHV on the quadruple transmission tower. For the calculations the type Donau has been considered. It has both EHV circuits placed above the HV circuits that are in horizontal line – Fig. 3. The line length has been considered equal to 40 km. For the performed calculations it has been supposed the ideal phase configuration that was recommended on the base of results of electromagnetic field distribution published in [3]. This phase configuration provides the most significant reduction of the values of electric and magnetic field and it is depicted in Tab. I. For the calculations the circuit breaker has been modelled as an ideal switch (the influence of the arc has been neglected).

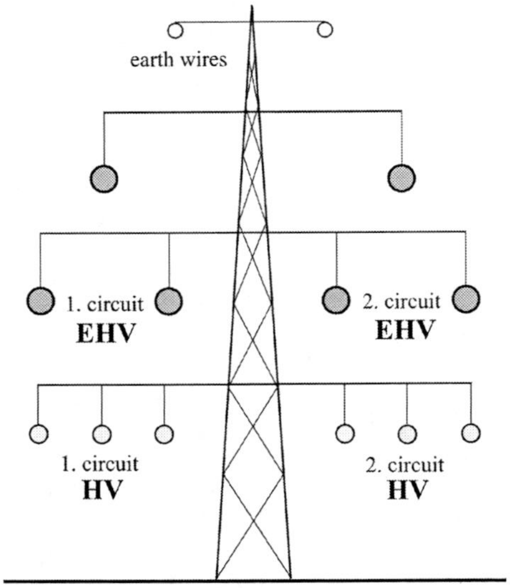

Figure 3. Design of the transmission tower Donau

TABLE I. CONSIDERED ASYMMETRICAL PHASE CONFIGURATION

1. circuit	2. circuit	1. circuit	2. circuit
EHV	**EHV**	B A C	B A C
HV	**HV**	c b a	c b a

A. EHV single phase short circuit with the automatic reclose cycle

Let's suppose following sequences of transients that should reflect the operation during the automatic reclose cycle (automatic RC):

- origin of the single phase short circuit - t_{SC}
- short circuit switching off - t_{OFF}
- phase switching on - t_{ON}
- end of the considered time interval - t_{END}

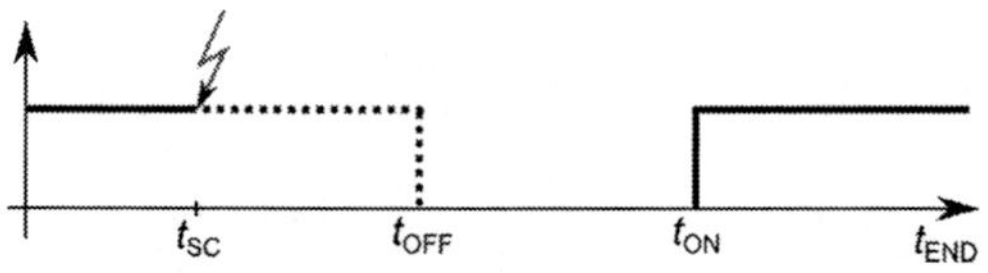

Figure 4. Switching diagram during the automatic RC

1) Phase C of the 1. EHV circuit

We considered the following specific times of the automatic RC → t_{SC} = 0.017 [s], t_{OFF} = 0.077 [s], t_{ON} = 0.107 [s], t_{END} = 0.14 [s]

In following figures selected EHV and HV phase currents and voltages during the mentioned time interval are shown.

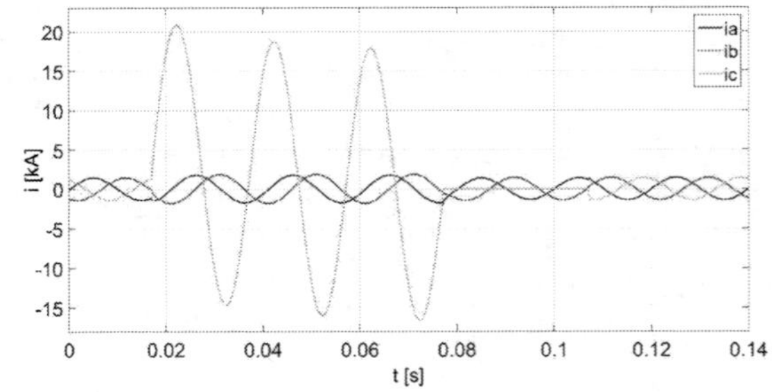

Figure 5. EHV phase currents of 1. circuit

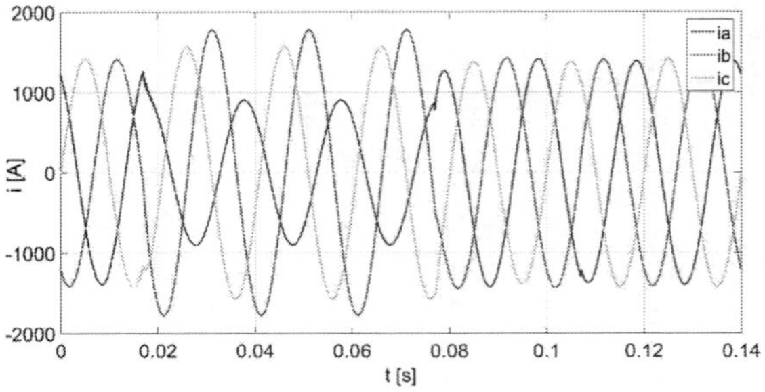

Figure 6. EHV phase currents of 2. circuit

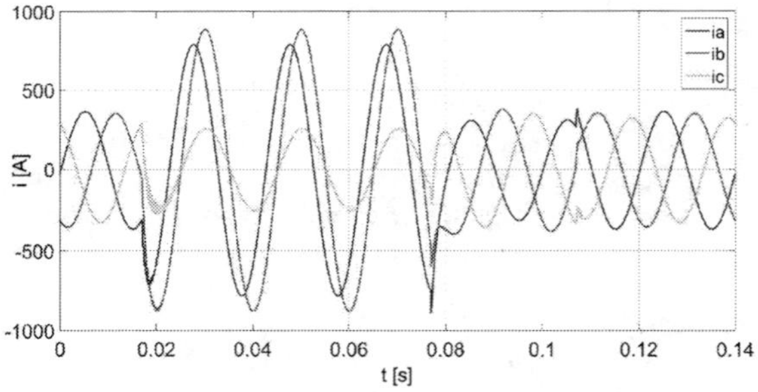

Figure 7. HV phase currents of 1. circuit

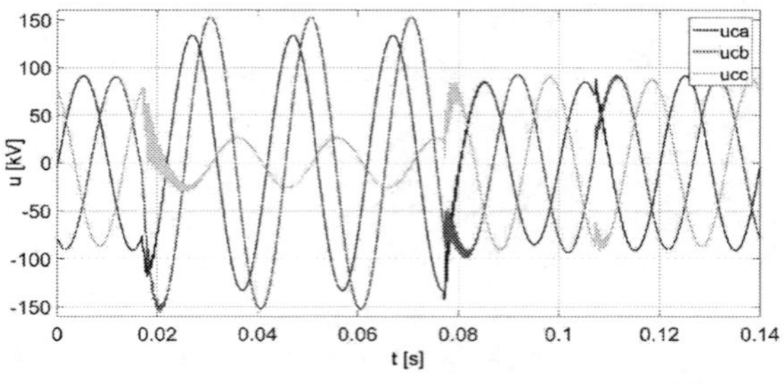

Figure 8. HV phase voltages of 1. circuit

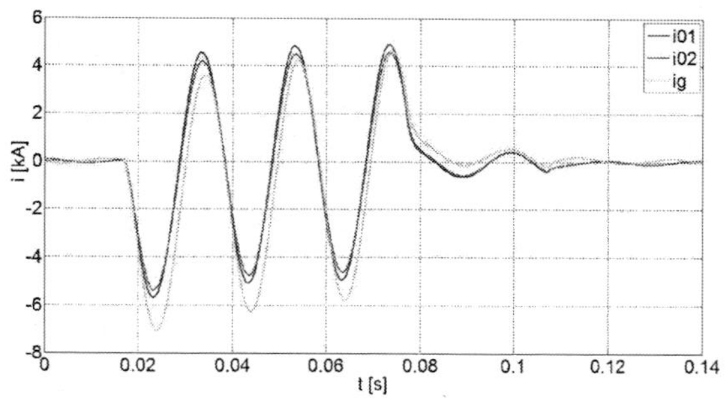

Figure 9. Currents in the earth wires and ground

2) *Phase B of the 1. EHV circuit*

We considered the following specific times of the automatic RC → t_{SC} = 0.017 [s], t_{OFF} = 0.080 [s], t_{ON} = 0.107 [s], t_{END} = 0.14 [s]

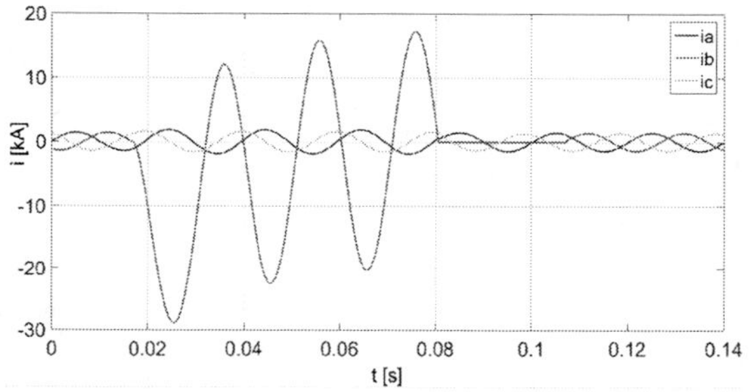

Figure 10. EHV phase currents of 1. circuit

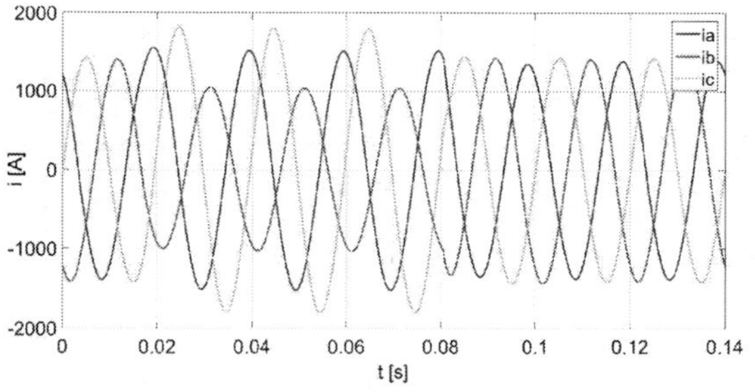

Figure 11. EHV phase currents of the 2. circuit

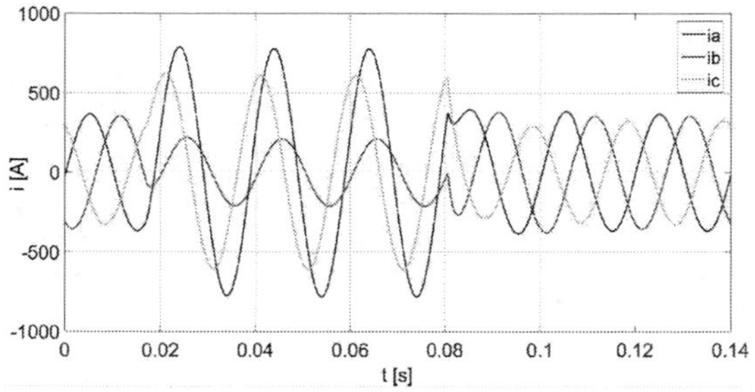

Figure 12. HV phase currents of the 1. circuit

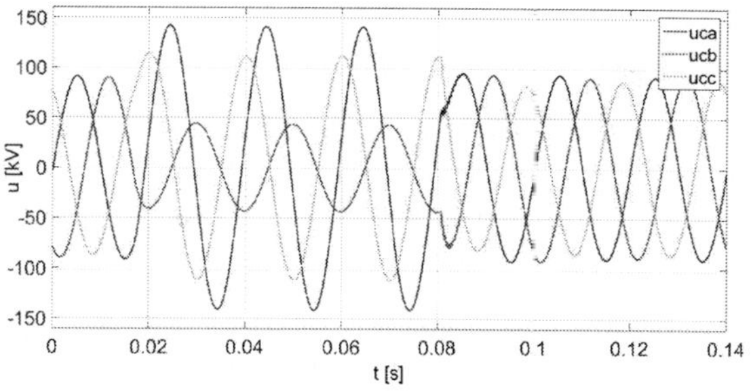

Figure 13. HV phase voltages of the 1. circuit

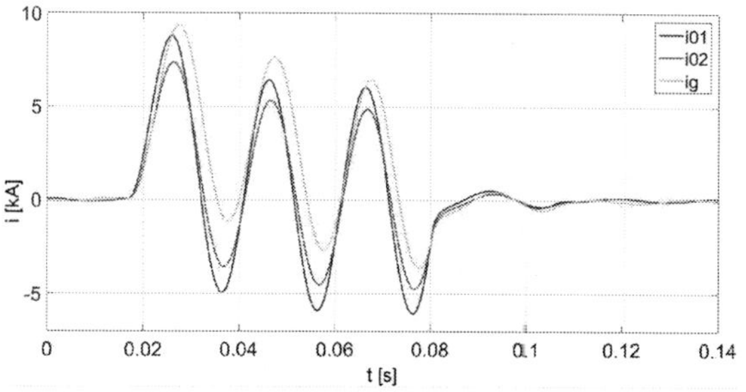

Figure 14. Currents in the earth wires and ground

Performed calculations of the automatic RC on the determined phases of the EHV circuit (subsection 1) and 2)) have shown some interesting results. Generally the lasting of the short circuit is connected with the considerable phase current asymmetry in the EHV circuits. The influence of the short circuit on the HV circuit is more significant. The lasting of the EHV short circuit is connected with huge phase current asymmetry and voltage unbalance of HV circuits. Comparing the results of the short circuit in phase *C* with phase *B* of the 1. EHV circuit, we can recognize that the impact of short circuit in the phase *C* on the HV conductors is bigger. The reason is the position of the conductor with phase *C* that is located near the axis of the tower close to the HV conductors. It determines the higher values of mutual inductances between the conductor with phase *C* and the all the HV conductors. Then the short circuit switching off is joined with the origin of the oscillating overvoltages on the HV circuits. The short circuit current is flowing through the both earth wires and the ground back. The division of this current between the earth wires and ground depends on the values of their impedances that is in the case of ground influenced by the ground conductivity. For the considered parameters in our calculations, more than 1/3 of the total value of short circuit current flows back through the ground, the rest part of short circuit current through the earth wires then. The division between two earth wires is not equal as well. In the case of the short circuit in the phase *B* of the 1. EHV circuit, this phenomenon is more significant. The bigger part of the short circuit current flows through the 1. earth wire due the smaller distance from the phase B that means the higher value of the mutual inductance and the induced voltage as well.

B. Evalution of the induced voltages

The current during the short circuit results in the higher value of induced voltage to the other conductors. The value of the induced voltage along the whole line is given by the time derivative of the short circuit current multiplied by the particular mutual inductance. The time derivative of the short circuit current in the case of phase B of the 1. EHV circuit is presented in Fig. 15.

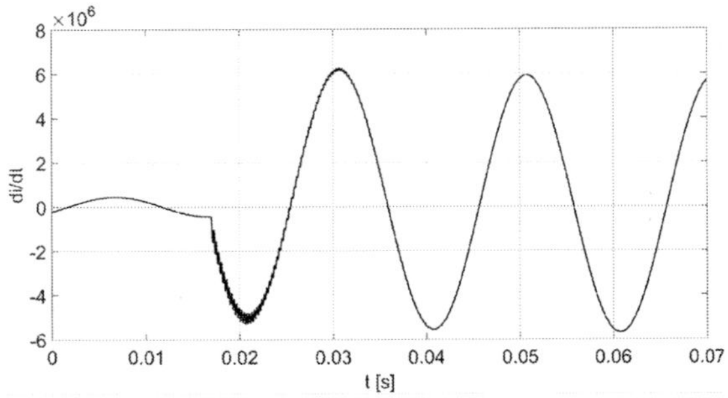

Figure 15. The time derivative of the short circuit current

The total induced voltage to the i-th conductor along the line is given by the sum:

$$u_{SCi} = L_{SCi}\left(\sum_{k=1}^{N} \frac{di_{SC,k}}{dt}\right)\frac{100}{U_{ni}} \ [\%] \tag{8}$$

TABLE II. INDUCED VOLTAGES DURING THE SHORT CIRCUIT IN PHASE C OF THE 1. EHV CIRCUIT

1. EHV circuit			2. EHV circuit		
42.92 %			35.38 %		
50.12 %			39.97 %		35.06 %
1. HV circuit			2. HV circuit		
162 %	170.4 %	173.9 %	138.2 %	131 %	125 %

Using the formula (8) we can evaluate the mutual influence between the conductor with short circuit and the remaining conductors. For the proper description of the mutual influence the relative values were used. The resulting induced voltages (maximal values) during the short circuit in phases C and B of the 1. EHV circuit are shown in Tab. II and III

TABLE III. INDUCED VOLTAGES DURING THE SHORT CIRCUIT IN PHASE B OF THE 1. EHV CIRCUIT

1. EHV circuit			2. EHV circuit		
			41.34 %		
50.13 %		50.13 %	41.34 %		37.51 %
1. HV circuit			2. HV circuit		
153.9 %	154.5 %	154.1 %	137.1 %	132.8 %	128.5 %

The relative values of the induced voltages between the phase with the short circuit and the remaining conductors have confirmed the assumption that the EHV short circuit originated on the conductor that is situated in smaller distance from the HV conductors has a more significant impact on those HV conductors. In this case the induced voltages via mutual inductive couplings reach higher values.

IV. CONLUSION

In this paper the algorithm for numerical transient analysis has been presented. It has been developed for multi-circuits transmission lines with two voltage levels. The algorithm can be used for any conductor's configuration. The geometric conductor's layout influences the line parameters including inductive and capacitive coupling that are evaluated in the first part of the code which was prepared in MATLAB. The multi-circuit line has been modelled by cascades of gamma two-ports which can be described with a system of differential equations. This approach allows solving various transients that can occur on the EHV/HV overhead transmission lines e.g. line switching, single phase short circuit, automatic reclose and etc. Their impact on the remaining circuits of the transmission line has been investigated. Performed calculations in this paper dealt with the single phase short circuit on the EHV and the automatic reclose cycle. The lasting of the short circuit is connected with the origin of voltage and current asymmetry that is on the HV circuits more significant. The short circuit switching off results in steep oscillating overvoltages. This knowledge is necessary to take into account in the process of planning of new lines and setting the protection systems. The mutual influence between the circuits during the faults should not lead to non-selective operation.

ACKNOWLEDGMENT

This paper has been supported by a project SGS-2015-035. Then the authors would like to thank the company ČEPS a.s. for providing data about the technical specifications of the transmission towers and conductors.

REFERENCES

[1] W. Liu, M. Pan, K. Liu, G. Xu, "Research on electromagnetic character of 500/220kV mixed-voltage quadruple-circuit transmission line," 2013 12th International Conference on Environment and Electrical Engineering, Wroclaw, 2013, pp. 227-231. doi: 10.1109/EEEIC.2013.654962 3

[2] G. Feng, Y. Wang, B. Zhang, "Study on Electromagnetic Environment of Multi-circuit Transmission Lines on Same Tower," 2008 Joint International Conference on Power System Technology and IEEE Power India Conference, New Delhi, 2008, pp. 1-5. doi: 10.1109/ICPST.2008.4745302

[3] T. Nazarčík., Z. Benešová. " Analysis of electromagnetic field in a vicinity of multi-circuit overhead lines with different voltage level", in Proceedings of the 19th International Symposium on High Voltage Engineering (ISH 2015). Plzeň: Západočeská univerzita, 2015. s. 1-6. ISBN: 978-80-261-0476-6

[4] A. Novitskiy, D. Westermann, "Interaction of multi-circuit overhead transmission lines of different voltages located on the same pylons," 2012 Electric Power Quality and Supply Reliability, Tartu, 2012, pp. 1-4. doi: 10.1109/PQ.2012.6256240

[5] B. Gustavsen, "Modal Domain-Based Modeling of Parallel Transmission Lines With Emphasis on Accurate Representation of Mutual Coupling Effects," in IEEE Transactions on Power Delivery, vol. 27, no. 4, pp. 2159-2167, Oct. 2012. doi: 10.1109/TPWRD.2012.2202923

[6] Z. Benešová, D. Mayer, "Algorithm for computation of inductances of three/phase overhead lines", in AMTEE ´03. Plzeň: Západočeská univerzita , 2003. pp. 1-6. ISBN: 80-7082-960-5

[7] Z. Benešová, L. Sroubova, "Capacitive Coupling in Double-Circuit Transmission Lines", Advances in Electrical and Electronic Engineering, pp. 157-160, Žilina 2004 [Online]. Internet: http://advances.utc.sk/index.php/AEEE/article/view/404. [May 16, 2017].

Integration of photovoltaic and wind power plants in low voltage networks using energy storages

Martin Vojtek, Michal Kolcun, Miroslav Mikita, Michal Špes
Department of Electric Power Engineering
Technical University of Košice,
Košice, Slovakia
martin.vojtek@tuke.sk, michal.kolcun@tuke.sk, miroslav.mikita@tuke.sk, michal.spes@tuke.sk

Abstract — **This paper deals with most common problems which occur when low voltage grids are highly penetrated with distributed generation based on renewable energy like solar or wind. Model of distribution grid in central Europe based on real data with high share of renewables is created and impacts are observed. Subsequently, simple strategy based on utilization of energy storage units to decrease or completely remove undesired effects is proposed and its behavior under real conditions is simulated. Results are discussed and evaluated, also some improvements are suggested.**

Keywords: energy storage; battery; renewable; photovoltaic; wind

I. Introduction

The electric power system is evolving constantly. At present, the changes are equally happening on the production side as well as on the side of consumption. Consumption increases its own efficiency; nevertheless it increases totally, mainly because of the increasing number of electric vehicles. Production is becoming more decentralized and production units are environmentally friendly, especially when based on renewable energy sources. With growing numbers of distributed generators, there is arising a need for their better integration.

Currently, the most used are photovoltaic power plants due to sufficient weather conditions, availability of this source and low operating costs. Their frequent installation in low-voltage networks deserves attention, especially in view of the fact that in the coming years there will be an even greater acceleration in the deployment of these resources at this voltage level [1]. The trend is to build photovoltaic power plants on the roofs of family houses or medium-sized sources suitable for powering of entire communities. According to current European Union policies which are being fulfilled successfully, the planned share is 20% of energy from renewables in final gross consumption by 2020 [2], [3], and at least 27% share by 2030 [4].

The biggest challenge related to these facts lies in the imbalance between production and consumption. The production of electricity by conventional production units is capable of adapting to the demand variations. In the case of photovoltaic and wind power plants, production depends on the availability of the primary source. For example, photovoltaic power plants produce electricity at a time when demand for electricity in low voltage networks with residential customers is low. In this case, demand peaks usually occur in the morning and in the evening, and production from photovoltaics has its peaks of production around the midday. The problem is most significant in the summer, as production is high and consumption low. During these periods in case of high photovoltaic penetration, surplus electricity occurs in low voltage grids, what cause electrical power backflow into superior system. Several studies and simulations show problems with voltage changes and keeping it within the prescribed limits in the case of a significant backflow of power [1], [5], [6]. In extreme cases, electrical equipment may be damaged due to increased voltage. From grid losses point of view of, it is most advantageous if the generated electricity is consumed directly at the place of production. These facts appear to be unfavorable mainly from the point of view of distribution system operators. The distribution system operators are obliged to maintain requirements on the quality and reliability of the electricity power supply and in order to this, it is necessary to increase the flexibility of grid operation [6], [7], [8]. In this context, accumulation of electricity area and the utilization of various technologies for accumulation of electricity have become more and more discussed. Energy storages are well known in other fields of electrical engineering such as electric vehicles or consumer electronics. In view of these trends, this area is dynamically developing, and nowadays there are a number of facilities and technologies (solid and flow batteries, pumped hydro power plants, hydrogen storages, flywheels ...) that offer a wide range of nominal power and capacity depending on the purpose of use [9], [10]. In addition, technical parameters such as energy density, number of cycles, ramp rates and other economic parameters are necessary to take into consideration.

This paper describes the example of energy storage utilization in low-voltage grid, which is highly penetrated with photovoltaics a wind power plants, in order to avoiding the production of unwanted effects such as backflow and subsequent excessive voltage fluctuation ,while also other beneficial services like reduction of peak demand are used. In next section, definition of main assumptions, which were needed to create the design of simple control strategy for community energy storage, is described. Proposed strategy is also described in detail. Section 3 deals with description of created model, which was used to verification of proposed strategy. Section 4 describes the input data required for the simulation derived from real measurements and comments simulation results. Finally, the simulation results and energy storage impact are discussed and evaluated.

II. Simple charge/discharge strategy

In order to verify a capability of energy storage devices to avoid undesired effects on low voltage grid and rest of electric power system, simple charge and discharge strategy was designed. Main task is to manage energy exchange between storage and community grid according to predefined rules and set of parameters. For design of this strategy, it was needed to define system states in every situation and therefore make a description of system behavior - its internal logic, which determines all of the circumstances and priorities when energy should be charged or discharged by storage device.

The basic assumptions about device functionality which were taken into consideration during the design of charge/discharge strategy had been defined as follows:

- Prevents back flow of energy to a higher voltage level and thus also utilize energy generated by photovoltaic and wind power plant in the same part of the distribution grid in which this source works
- Manage charge/discharge process according to current electricity pricing (low and high tariff)
- Stored energy can be used only during high tariff period
- Grid charging allowed only in low tariff period
- Allow to adjust the value of instantaneous power delivered from the superior high voltage grid during low and high tariff period
- Ensure minimal state of charge (SOC) value at the start of each day
- Enable to define allowed operating range of a storage (due to specific features of each technology)

Algorithm for control of community energy storage has been designed based on defined functionalities. The input values can be divided into two classes. Instantaneous values of considered variables which are continuously measured and loaded and there are also some charged or discharged by energy storage device.

Meaning of variables, which are used in proposed algorithm, is as follows:

- SOC – actual SOC (%)
- P_{LOAD} – active power consumed by load (kW)
- P_{PV} – active power generated by PV (kW)

Constant parameters, which must be chosen before the operation:

- SOC_{MAX} – maximal SOC (%)
- SOC_{MIN} – minimal SOC (%)
- SOC_{START} – required SOC at start of each day (%)
- LT – validity of low tariff (-)
- P_{HT} – desired active power delivered from the grid during high tariff (kW)
- P_{LT} – desired active power delivered from the grid during low tariff charging (kW)

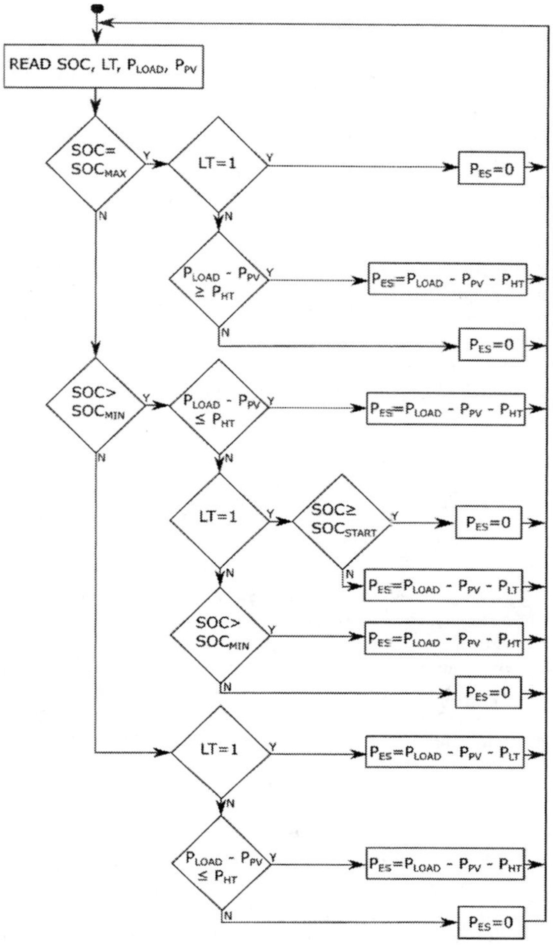

Figure 1. Proposed algorithm for control of community energy storage

III. Description of the model

It was created in Matlab Simulink environment employing Simpowersystems toolbox. Model consists of several parts coupled together to demonstrate and to verify strategy functionalities. Fundamental part of the system is community, which topology and parameters are inspired by real village situated in Eastern Slovakia. Other important parts of the proposed system are formed by photovoltaic and wind power plants and energy storage device. All of these parts are further described in next sub-sections and overview of the model can be seen in next figure.

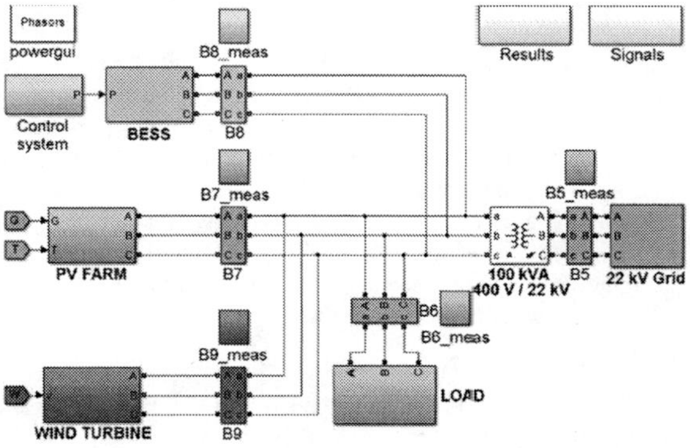

Figure 2. Model implementation in Matlab/Simulink environment

A. *Wind power plant*

In order to calculate a power output of wind turbine, model based on input wind speed was established. It was inspired by [11]. Real parameters of small scale wind turbine were used according to manufacturer's datasheet. Available power in wind is defined by (1).

$$P_{wind} = \frac{1}{2} * \rho * A * v^3 \tag{1}$$

In this case, Greek letter ρ is air density (1,255 kg/m³), A means blade swept area and v represents wind speed. Air density in normal conditions is changing in dependence on air pressure, but in this study this impact was neglected. Power of wind turbine delivered to the distribution grid of selected community is therefore calculated as follows:

$$P_{WP}(v) = \begin{cases} 0 & if \quad v < v_{cut-in} \\ P_{wind} * C_P & if \quad v_{cut-in} < v < v_{rated} \\ 15000 & if \quad v_{rated} < v < v_{cut-off} \\ 0 & if \quad v > v_{cut-off} \end{cases} \tag{2}$$

If wind speed is lower than starting speed (v_{cut-in}), turbine doesn´t produce any power. If wind speed is higher, but lower as rated wind speed (v_{rated}) at the same time, power delivered to the grid equals to available power in wind multiplied by turbine efficiency coefficient C_p. 15 000 is rated power of wind turbine in watts and this is a value of power which is delivered to the grid in case when wind speed is between rated speed and maximum wind speed ($v_{cut-off}$). If wind speed is higher than maximum speed, turbine is stopped to prevent its damage.

B. *Photovoltaic power plant*

It is made up of 8 parallel connected strings. Every string is composed by 20 modules connected into series. Photovoltaic module IBC Polysol 250 CS with nominal power of 250 Wp was used for single module modeling. It was modeled using one diode equivalent circuit including consideration of series and parallel resistance [12], [13].

Resulting photovoltaic current is given by (3). I_{PH} is photovoltaic current, I_S means saturation current, q is electrical charge, k is Boltzmann constant and T is operating temperature.

$$I = I_{PH} - I_S \left(e^{\left[\frac{q(V + I * R_S)}{kTA} \right]} - 1 \right) - \frac{(V + I * R_S)}{R_{SH}} \tag{3}$$

Since a producer does not give values of series (R_S) and parallel resistance (R_P), their values have been determined explicitly using the Lambert W-function and diode ideality factor (A) estimation [14]. All other values needed were found in datasheet provided by manufacturer. The input parameters for calculation of photovoltaic array current consist of irradiation, temperature and voltage. Voltage at which it is possible to use the maximum power was tracked by well-known perturb and observe technique. Output parameter is array power, which is injected into low voltage grid at $cos\varphi$=1. The efficiency of inverters used was set to 95%.

C. *Community load*

The load is modeled based on real data provided by the distribution system operator. It consists of 52 family houses, each of which was modeled separately and topology respects the actual status. Their peak load is between 0.35 and 3.78 kVA with power factor set to 0.95. Every customer has assigned one of three load diagram types which are defined in per units. Power consumed by each customer at any time is calculated by multiplying his peak load and corresponding value of typical load diagram (fig. 3). Subsequently, this value is fed to dynamic load block from Simulink library. It simulates the dynamically changing load and its power consumed depends on positive sequence voltage. The resulting load curve of all houses enters into control logic as P_{LOAD}. It is obtained by measurement at the interface of transformer and power line in which customers are connected and it represents the sum of active power consumed by all customers including losses.

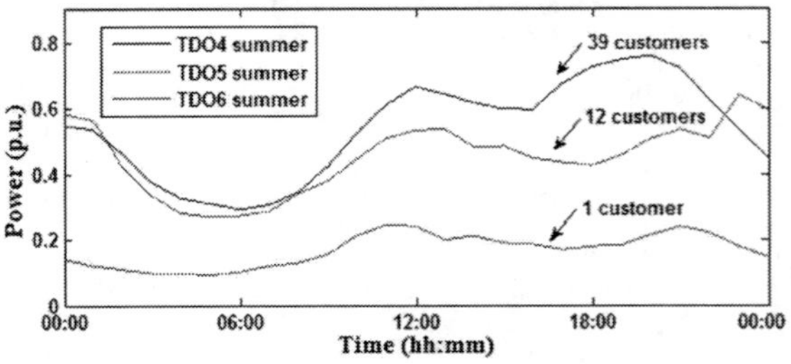

Figure 3. Typical daily load curves of simulated customers in summer

D. *Energy storage*

For calculating the energy and power balances of the system, any energy storage can be adequately described by its capacity, rated power and efficiency of electrical energy conversion into another form and back during charging and discharging.

This approach allows simulating different types of energy storage regardless of the technology used.

Besides these parameters, it is also necessary to know the state of charge of the storage device at any time, which was defined by (4). $SOC_{INITIAL}$ (%) indicates the initial state of charge as a percentage of total capacity, which must be defined before start of the simulation. E_{CH} (kWh) and E_{DISCH} (kWh) are total amounts of the energies stored during charging and discharged during the discharging, respectively. E_{FULL} (kWh) means total capacity of energy storage.

$$SOC(t) = SOC_{INIT} + \frac{100 * E_{CH}(t)}{E_{FULL}} - \frac{100 * E_{DISCH}(t)}{E_{FULL}} \tag{4}$$

Lithium ion battery pack was considered as energy storage in this model. It has total energy capacity of 750 kWh with 50 kW nominal power output. Efficiency of storage energy conversion was set to 95% for both charging and discharging what gives overall round trip efficiency about 90 %.

E. *Transformer*

Transformer model is represented by standard block from Simpowersystem library. Primary side is connected at the end of distribution feeder, secondary supplies low voltage grid. Parameters was set based on real values, its nominal voltage is 22/0.4 kV and apparent power of 100 kVA.

F. *Superior high voltage grid*

It consist of 110 kV grid equivalent, which is connected to distribution feeder through a 110/22 kV transformer. Distribution feeder is modeled as series of lines and loads, at the end of which the distribution transformer is connected.

IV. Simulation and Results

Based on created model, which allows observing the behavior of individual parts of the system and its power and energy balance, the simulations were performed in order to verify functionality of the proposed charge/discharge strategy and to demonstrate energy storage capabilities. Overall, two scenarios were simulated and both of them were simulated in time span of one week. Constant parameters, which must be set before the simulation, were the same in both cases. The difference between simulated scenarios lies only in climatic conditions and load demand data, which corresponds to different seasons of the year. First scenario works with data measured during third January week, therefore in winter. Second one simulates summer week, exactly first week of September.

Constant parameters (set-points) of control strategy were chosen the same for both scenarios and their values can be found in Tab I.

Table 1. Values of control strategy parameters

Parameter	Value
SOC_{MAX}	100 %
SOC_{MIN}	20 %
SOC_{START}	60 %
P_{HT}	10 kW
P_{LT}	60 kW
LT	22:00 – 06:00

Simulations were performed using Phasor simulation method with 50 Hz phasor frequency. This method counts only amplitude and phase of currents and voltages what improves simulation speed while sufficient accuracy is maintained. An Ode45 (Dormand-prince) solver with variable step was used.

A. *Winter scenario*

Input data for this scenario were measured from 19.1 to 27.1, 2016. It is a typical winter week. The temperature was measured by contact resistance sensor attached on the bottom of panel (Fig. 4) and irradiation in its plane (Fig. 5). Weather data were measured 3 meters above the ground at the roof of building (Fig. 6). Measurements were performed near studied locality by Department of electric power engineering of Technical University in Kosice.

Based on these values, model simulated power generated by renewable energy sources. These values are graphically represented in fig. 7 separately for power generated by photovoltaic and wind power plant.

Load curve is sum of consumed power by all of customers, and can be observed in the results of simulated scenario (Fig. 8). Sum of power generated by photovoltaic and wind power plant marked as renewables can be seen together with the load diagram. By comparing of these values it is immediately clear when the surplus energy is produced and without use of the storage facility, electricity would flow into the superior system. It can be seen that this situation is very rare, it did not occur during the simulated time range, and therefore utilization of surplus energy time-shifting in winter is unique.

Course of power charged or discharged by energy storage describes its reaction on events in the low-voltage according to the proposed strategy. It can be observed, that without the exception, all days the batteries are charged during the night and discharged during the day. Duration of low tariff charging depends on battery *SOC* when low tariff charging starts.

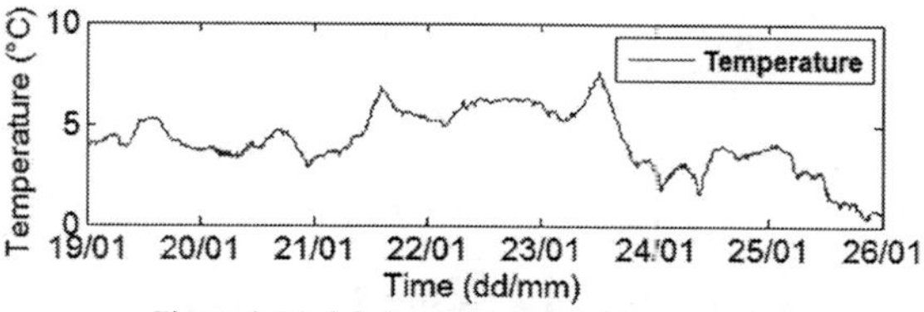

Figure 4. Module temperature in winter scenario

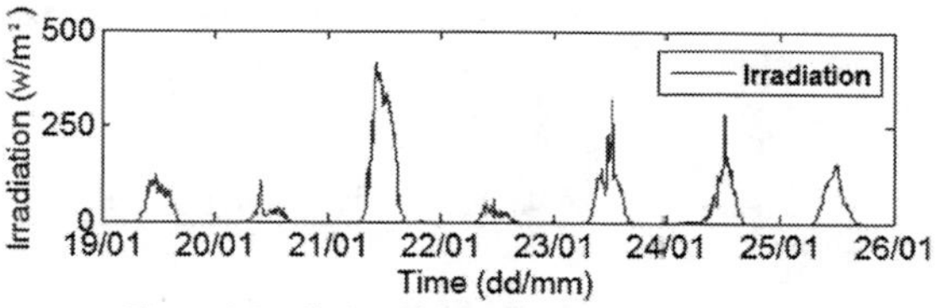

Figure 5. Irradiation of the module in winter scenario

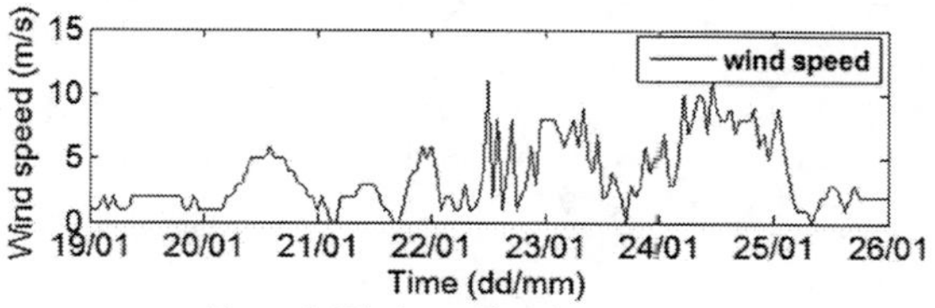

Figure 6. Wind speed in winter scenario

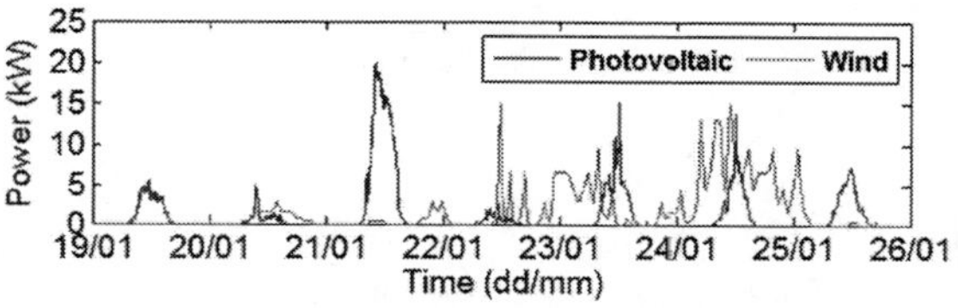

Figure 7. Power generated by photovoltaic and wind power plants

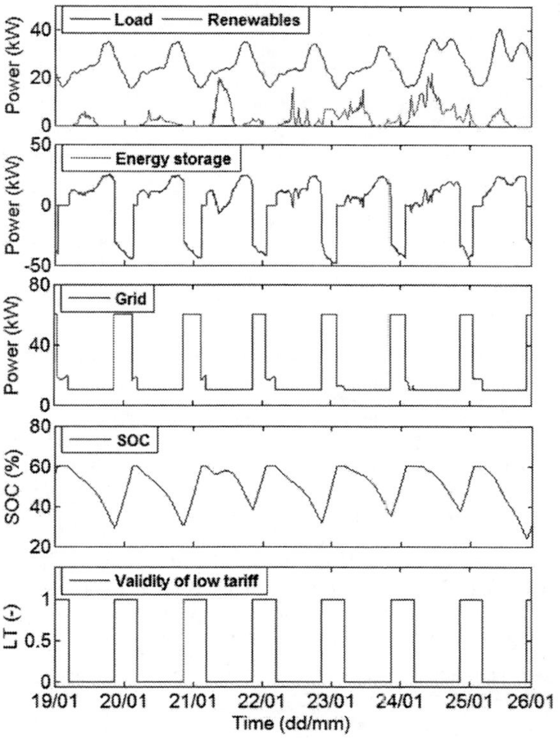

Figure 8. Winter scenario simulation results

Also power generation by wind turbine during the night helps energy storage to be charged more quickly to desired value. Thus, period of low tariff charging is shorter during these days, in which wind power generation occurred in low tariff, and consequently community drawn less power from superior grid.

This results show, that time-shifting of energy from low to high tariff prevails in winter. State of charge of the batteries was below 60% at the end of each day, what resulted into starting of low tariff charging to achieve this value. State of charge did not drop below 20%, which tells about sufficient sizing of energy storage and other system components.

The course of power delivered from the grid shows that the set values were not exceeded.

Based on energy balance of the system (Tab. 2), it can be seen that more than two thirds of the energy drawn from superior grid was delivered during low tariff period. Contribution of photovoltaic power plant during this season is almost negligible. Power generated by wind turbine is atypically also low for this season. The amount of energy delivered to the superior high voltage grid equals to zero, therefore power backflow did not occur. Storage energy balance is negative, but simulation ends at midnight and in real conditions energy storage would be charged till end of low tariff.

Table 2. Results of energy balance for winter scenario

Value	Energy [kWh]
Energy consumed by community	4268
Energy generated by photovoltaic power plant	217,3
Energy generated by wind turbine	285.4
Energy delivered from grid during HT	1167
Energy delivered from grid during LT	2573
Energy supplied to superior high voltage grid	0
Energy stored by energy storage	1450
Energy released by energy storage	1562
Storage energy balance	-112.1

B. *Summer scenario*

Input data for the simulation were measured from 1.9 to 7.9 2016 in the same manner as in the previous case (Fig. 9, 10 and11).

During the summer days can be observed much more power supplied to low-voltage grid by photovoltaic power plant while the community during this season is less loaded than in winter. Also wind power generation is higher and more evenly.

At the time of peak production of photovoltaic power plant, there is a significant surplus of electricity in the grid. This occurs around the noon when high tariff is valid. At this time, energy storage prevents the flow of electricity into superior grid by storing the excess energy in order to maintain required power delivered from the grid. Surplus energy stored is then delivered back even during the validity of high tariff. Surplus electricity time-shifting prevails in this case. It has a beneficial effect on the daily load leveling of the community, which can be seen in the course of power delivered from the grid.

The state of charge of the battery system is almost always above 60% when low tariff starts to be valid due to sufficient production of photovoltaic power plant during the day. On 6 September the state of charge has fallen below this value and the battery had to be charged from the grid.

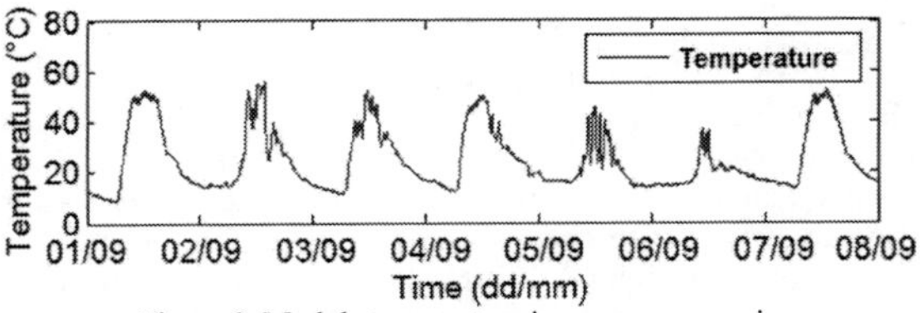

Figure 9. Module temperature in summer scenario

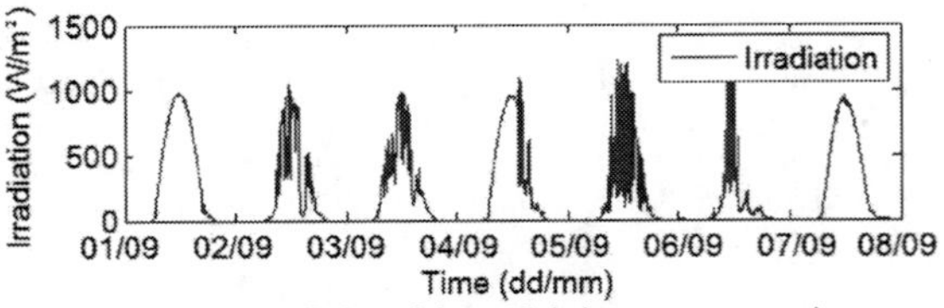

Figure 10. Irradiation of the module in summer scenario

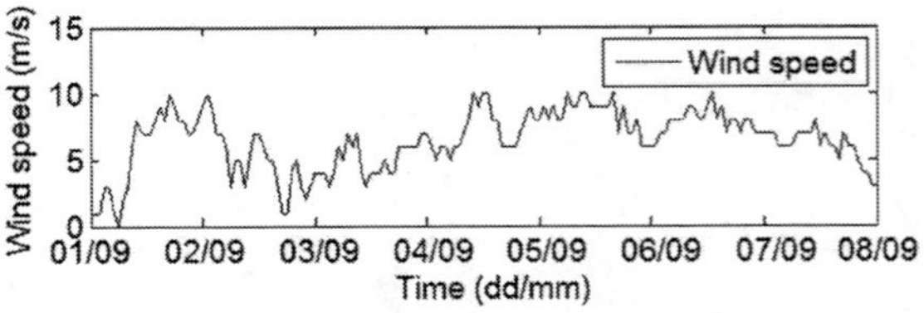

Figure 11. Wind speed in summer scenario

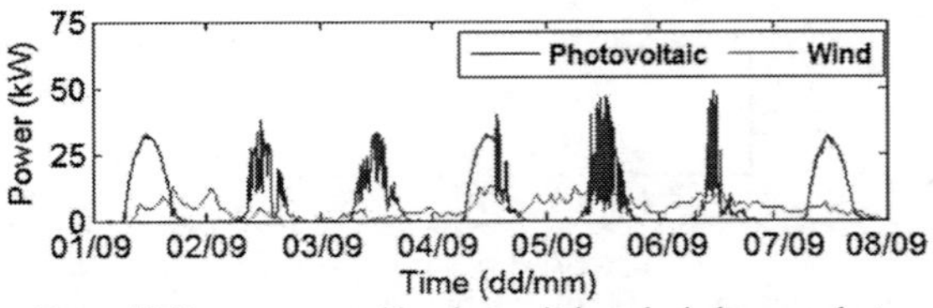

Figure 12. Power generated by photovoltaic and wind power plants

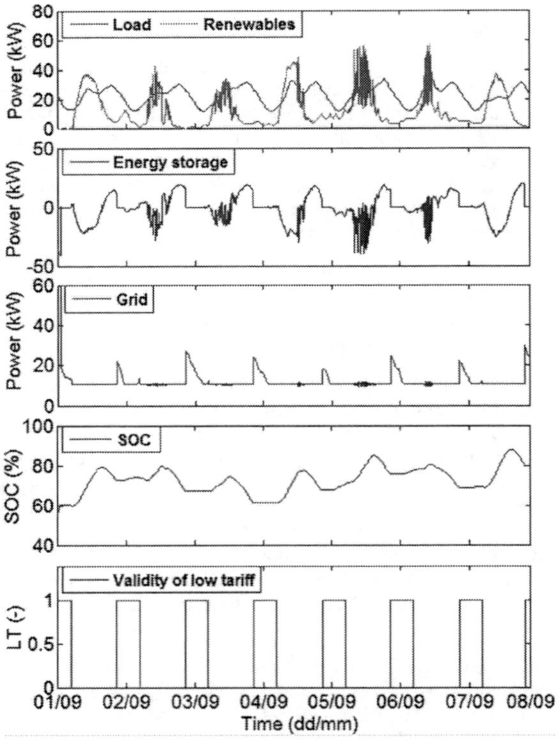

Figure 13. Simulation results for summer scenario

The energy balance shows that the amount of electricity supplied by photovoltaic is almost half of the total energy consumed. At the same time, energy supplied to the superior high voltage grid equals to zero, which means that the entire photovoltaic production is consumed in the place of production. More than half of the power delivered from the grid has been delivered during the low tariff. Energy balance of the storage unit is positive, and also the state of charge value is higher at the end of the week.

Table 3. Result of energy balance for summer scenario

Value	Energy [kWh]
Energy consumed by community	3762
Energy generated by photovoltaic power plant	1210
Energy generated by wind turbine	765,3
Energy delivered from grid during HT	1167
Energy delivered from grid during LT	944,2
Energy supplied to superior high voltage grid	0
Energy stored by energy storage	835,4
Energy released by energy storage	597,1
Storage energy balance	194,92

V. Conclusion

This paper describes the problems which are associated with high penetration of photovoltaic and wind power plants in low-voltage systems. Furthermore it explores the possibility of energy storage usage to address these issues and conducts demonstration of these devices capabilities. A strategy for the management of energy storage was proposed in order to prevent power back flow into the superior grid, energy consumption in place of production and exploitation of cheap energy taken in low tariff.

Simulation results show that the strategy works well for both tested seasons.

In winter, time shifting of energy from low to high tariffs prevails. There was not any amount of surplus electricity and all of the energy produced was consumed at the same time in the same place, consequently therefore renewable energy time shifting was not utilized This could have a positive economic impact. It can be also seen that the capacity of battery is used only in the range of 20-60%. Possible improvement of the system arises from this fact and it lies in the dynamic adjustment of values entering the control algorithm, for example, on the basis of prediction, which can reduce the battery capacity needed.

On the contrary, in summer time-shifting of excess energy generated from photovoltaic and wind power plant is the most frequently used, while energy storage simultaneously prevents the power back flow into the superior grid and thus avoids of energy losses arising by transfer of energy to the place of consumption and undesired voltage effects.

Power set points and predefined state of charge levels did not exceed the required values what proves adequate system design. This system including wind power in comparison to testing of control strategy in community only with photovoltaic power plants installed, provides generation more evenly and also during the night. This has positive impact on battery cycling.

Acknowledgment

This work was supported by Ministry of Education, Science, Research and Sport of the Slovak Republic and the Slovak Academy of Sciences under the contract No. VEGA 1/0132/15.

References

[1] K. M. M. Huq, M. E. Baran, S. Lukic and O. E. Nare, "An Energy Management System for a community energy storage system," IEEE Energy Conversion Congress and Exposition (ECCE), Raleigh, pp. 2759-2763, September 2012.

[2] P. Hocko, M. Novák, M. Kolcun and Z. Conka, "Influence of photovoltaic power plants on the power system to the prediction for year 2020," 2014 14th International Conference on Environment and Electrical Engineering, Krakow, 2014, pp. 116-121. doi: 10.1109/EEEIC.2014.683584.

[3] Eurostat – Statistics Explained. Official Eurostat webpage. http://ec.europa.eu/eurostat/statistics-explained/index.php/File:Share_of_renewables_in_gross_final_energy_consumption,_2014_and_2020_(%25)_YB16.png , (Accesed 2/2017).

[4] European comission – 2030 Energy strategy. Official webpage. https://ec.europa.eu/energy/en/topics/energy-strategy/2030-energy-strategy, (Accesed 2/2017).

[5] Z. Čonka et al., "Impact of renewable energy sources on stability of EWIS transmision system," 14th International Conference on Environment and Electrical Engineering, conference proceedings, Krakow, Poland, pp. 75-79, May, 2014. ISBN 978-1-4799-4661-7.

[6] J. Morávek, J. Drápela, V. Wasserbauer, P. Mastný, "Power Quality Issues Related to Power Flow Control in Systems with Renewable Energy Micro Sources," Proceedings of the 2016 17th International Scientific Conference on Electric Power Engineering (EPE), Praha, pp. 283-288, 2016, ISBN: 978-1-5090-0907- 7.

[7] D. Varajão, L. M. Miranda and R. E. Araújo, "Towards a new technological solution for community energy storage," 16th European Conference on Power Electronics and Applications, Lappeenranta, pp. 1-10, August 2014.

[8] D. Medveď, M. Kolcun, J. Petráš, R. Stolárik, Š. Vaško, "Use of Programmable Units in Photovoltaics," in Elektrotechnológie, Košice, pp. 6-10, 2015, ISBN 978-80-553-2139-4.

[9] W. Zhu, D. Garrett, J. Butkowski and Y. Wang, "Overview of distributive energy storage systems for residential communities, " IEEE Energytech, Cleveland, pp. 1-6, May 2012.

[10] L. Yao et al., "Challenges and progresses of energy storage technology and its application in power systems," J. Mod. Power Syst. Clean Energy, vol. 4, pp. 519–528, 2016.

[11] J. Wellmann, T. Morosuk, "Renewable Energy Supply and Demand for the City of El Gouna, Egypt," in Sustainability, vol.8, 314, March, 2016. doi: 10.3390/su8040314

[12] M. Kolcun, M. Vojtek, Z. Čonka, "The implementation of an ideal photovoltaic module in Matlab/Simulink using simpowersystems toolbox," 16th International Scientific Conference on Electric Power Engineering (EPE), Kouty nad Desnou, pp. 579-583, May, 2015.

[13] J. Morávek, P. Mastný, J. Pitron, "Modeling the Energy Systems in Matlab Simulink - Problematics of Energy System Control," Proceedings of the 16th International Scientific Conference on Electric Power Engineering (EPE), Ostrava, Czech Republic, May, 2015. s. 163-168. ISBN: 978-1-4673-6787-5.

[14] J. Cubas, S. Pindado, C. de Manuel, "Explicit Expressions for Solar Panel Equivalent Circuit Parameters Based on Analytical Formulation and the Lambert W-Function," in Energies, vol. 7, pp. 4098-4115, 2014.

Analysis of Thermal Field in Magnetic Fluid and Transformer Oil

Michal Kosterec, Juraj Kurimský
Department of Electric Power Engineering
Technical University of Košice, FEI
Košice, Slovakia
michal.kosterec@tuke.sk, juraj.kurimský@tuke.sk

Michal Rajňák, Milan Timko, Peter Kopčanský
Institute of Experimental Physics
Slovak Academy of Sciences
Košice, Slovakia
rajnak@saske.sk, timko@saske.sk, kopcan@saske.sk

Abstract —**Increasing interest in the use of magnetic fluids in power systems, especially in transformers, is nowadays registered. Magnetic fluid is used in cooling systems which have better heat transfer and dielectric properties as mineral transformer oil. It can be used in passive cooling system with limited dimensions because of its properties i.e. increased heat transfer coefficient, temperature dependent magnetization which can induce forced circulation without using any pumps. The circulation is caused by incident magnetic field and can be controlled by adjusting the magnetic field which leads to improving cooling capacity. This paper presents thermal distribution, fluid flow and cooling ability of mineral transformer oil and magnetic fluid based on mineral transformer oil. The concentration of nanoparticles Fe_3O_4 is 0.05 % volume of mineral transformer oil. The source of heat is copper conductor. Thermal distributions in mineral transformer oil and magnetic fluid are investigated and differences for both cases are discussed in the paper.**

Keywords: transformer oilt; magnetic fluid; temperature; thermal distribution

I. Introduction

Magnetic fluids are used in many important industrial applications for example as "silent" coils (noise reduction), sensors, sealing systems, various types of contacts, etc. Recently, increasing interest in their use in power equipment in electrical power engineering, particularly in transformers, has been registered with the use of magnetic fluid as a dielectric and a cooling medium. Magnetizable nanofluids, which are useful as an alternative to fluids used as cooling, are characterized by a lower concentration of magnetic nanoparticles [1].

Magnetic fluids are characterized by better heat transfer and dielectric properties than transformer oil and can be used to improve heat flux, thereby increasing the ability of active parts to withstand failures such as electromagnetic pulses. An external magnetic field may be used for creating the artificial circulation of magnetic fluids, and the magnetic forces inside the magnetic fluids can be controlled by an appropriate setting of the incident magnetic field [2],[3],[4].

More details, such as the local electric field spectrum, the flow and heat transfer, and the reactions of the magnetic fluid system to the change of the incident magnetic field, can sometimes be obtained using only numerical simulations [1],[3].

The insulating system of the power transformer consists of mineral oil and cellulose paper to withstand dielectric stresses. One of the functional requirements of mineral oil is not only as an insulator but also as a cooling medium. Although the current system has proven to be reliable and cost-effective, the low thermal conductivity of mineral oils used as coolant in transformers arises doubts about the ability to dissipate heat well. Magnetic fluids have proven to be a solution to this problem [5],[6].

Extended use of transformer oil as insulation and cooling in high-voltage equipment leads to extensive research work to increase its dielectric and thermal properties. An example of such work is the development of magnetic fluids. These fluids are made by adding magnetic nanoparticles to the transformer oil in order to increase the dielectric and thermal properties of this oil [7],[8].

Unique properties of magnetic fluid such as temperature-dependent magnetization, flexible properties and stability, make it an excellent medium for passive cooling systems. When the permanent magnet is placed on the hot side of the object, it will attract cold magnetic fluid to the hot side and thereby cause the hot magnetic fluid to be extruded. This is due to the fact that cold magnetic fluid having a temperature below Currie's temperature is characterized by higher magnetization than a hot magnetic fluid that has a temperature above Currie's temperature (becomes paramagnetic). Then the hot magnetic fluid is pushed towards the cold side of the contained and transfers the heat to the ambient. This phenomena is called thermomagnetic convection [9].

The main aim of this paper is to prove that the magnetic fluid is a cooling medium with better thermal properties than a transformer oil, and by using magnetic fluid the heat flux will increase and hence cooling efficiency will also increase.

II. Methods and Model

A. Methods

The method of magnetic induction was used for the analysis. The first approach to solving the magnetic induction equation is the derivation of the Ohm law and the Maxwell equations. The equation provides the relation between the fluid flow and the magnetic field.

The Ohm law, which defines the current density, is given:

$$\bar{J} = \sigma \bar{E} \qquad (1)$$

where: σ- electrical conductivity of material, $\bar{J}$- vector of current density, $\bar{E}$- electric field intensity

For the flow velocity of the fluid $\bar{U}$ in the magnetic field $\bar{B}$, Ohm's law is given:

$$\bar{J} = \sigma(\bar{E} + \bar{U} \times \bar{B}) \qquad (2)$$

From the Ohm law and the Maxwell equation we can derive the equation of induction as:

$$\frac{\partial \bar{B}}{\partial t} + (\bar{U} * \nabla)\bar{B} = \frac{1}{\mu\sigma}\nabla^2\bar{B} + (\bar{B} * \nabla)\bar{U} \qquad (3)$$

The current density $\bar{J}$ can be calculated using the Amperes law from the solved magnetic field $\bar{B}$ as:

$$\bar{J} = \frac{1}{\mu}\nabla \times \bar{B} \qquad (4)$$

In magneto-hydrodynamic the magnetic field $\bar{B}$ consist of the external field $\bar{B}_0$ and induced field $\bar{b}$ due to the movement of the fluid. Only induced field $\bar{b}$ needs to be solved.

The external field $\bar{B}_0$ from the Maxwell equations complies with the following equation:

$$\nabla^2\bar{B}_0 - \mu\sigma\frac{\partial \bar{B}_0}{\partial o} = 0 \qquad (5)$$

B. Model

For the simulation, a 3D model was created in AutoCAD and then loaded into the Ansys Workbench simulation program for further analysis. The model consisted of a conductor with a cross-section of 2.5 mm^2, 100 mm long, placed in a 20.82x15.42x100 mm container (Figure 1.). The loaded and meshed model in Ansys can be seen in Figure 2. The conductor was defined as a heat source of 1,255x10^7 W/m^3 (Figure 3.) which corresponds to the heat generated in the conductor by 16 A current flow (obtained from simulation in Electric module). The peripheral walls of the liquid had a defined heat flow of 100 W/m^2 and the ambient temperature was 20 °C. The analysis duration was 300 seconds.

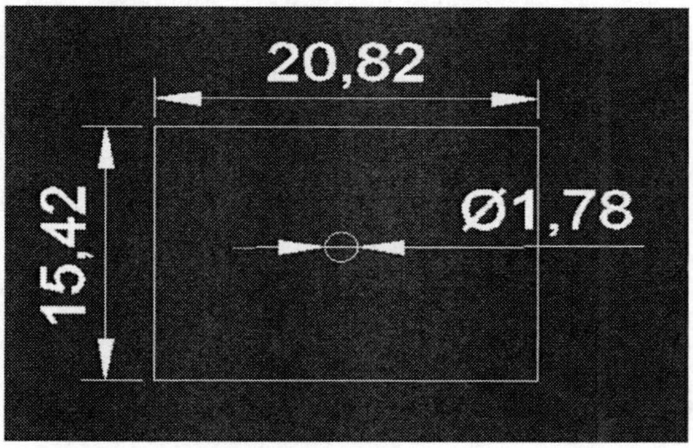

Figure 1. Dimensions of model

Figure 2. 3D mesh in Ansys

Three simulations were performed using this model. In the first simulation a transformer oil was used as a cooling medium. The gravity acted in the axis Z which is parallel to the copper conductor and the value was g= 9.81 m/s^2. Conductor was immersed in oil with temperature 20 °C. In the second simulation a magnetic fluid was used as a cooling medium. The gravity acted also in the Z axis. In this simulation copper conductor was defined as a source of magnetic field with B= 0.5 T and f= 50 Hz. In the third simulation a transformer oil was used as a cooling medium but in time 150 s a magnetic nanoparticles were added to the oil with concentration 0.05 % volume of transformer oil. Physical parameters of used material are listed in Table 1, Table 2 and Table 3.

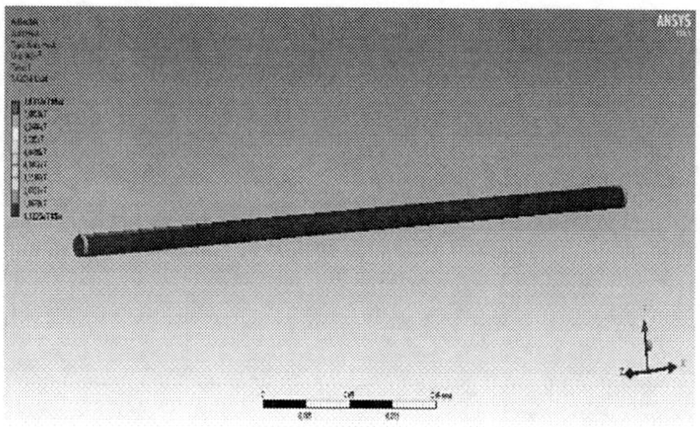

Figure 3. Solved joule heating in Electric module

Table 1 Copper conductor parametres

Density ρ	8978 kg/m^3
Specific heat C_p	381 J/(kgK)
Thermal conduction K_c	387,6 W/(mK)
Electric conductivity	5,8x10^7 1/Ωm
Magnetic permeability	1,257x10^{-6} H/m

Table 2 Transformer oil parametres

Density ρ	880 kg/m^3
Specific heat C_p	1860 J/(kg*K)
Thermal conductivity K_c	0,162 W/(m*K)
Viscosity	0,01936 kg/(m*s)
Thermal expansion coeff.	7,5x10^{-4} 1/K
Reference temperature	20 °C
Magnetic permeability	1,8472x10^{-6} H/m

Table 3 Nano particles parametres

Density ρ	5100 kg/m^3
Specific heat C_p	670 J/(kgK)
Magnetic permeability	1,8723x10^{-6} H/m
Concentration	0,05 % of volume

III. Results and Disscusion

At time 300 s, the conductor at the top reaches a surface temperature of 35.5 °C (308.65 K), while at the bottom it reaches a temperature of 27 ° C (300 K) (Figure 4.). The liquid is heated in the vicinity of the conductor and flows around the conductor upwards where high temperatures occur. There is an irregular distribution of temperature in both the liquid and the conductor. When the conductor is positioned perpendicular to the applied gravitational field (Figure 5.), its temperature is the same throughout its length, but it is no longer present when it is placed parallel to the gravitational field. This is due to the fact that the upper part of the conductor is cooled by a warm oil which has heated up from the conductor at the bottom of the container and has flowed upwards alongside the conductor where its temperature has increased. Natural circulation of transformer oil occurred as it can be seen by path lines in Figure 4. Fluid flow is greatest around the conductor and in the top and bottom of the liquid and it is approximately 1.24 mm/s (Figure 6.).

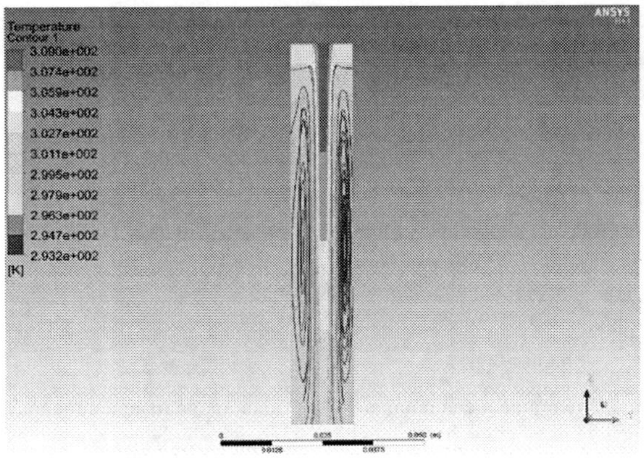

Figure 4. Temperature distribution and pathlines in oil at time 300 s

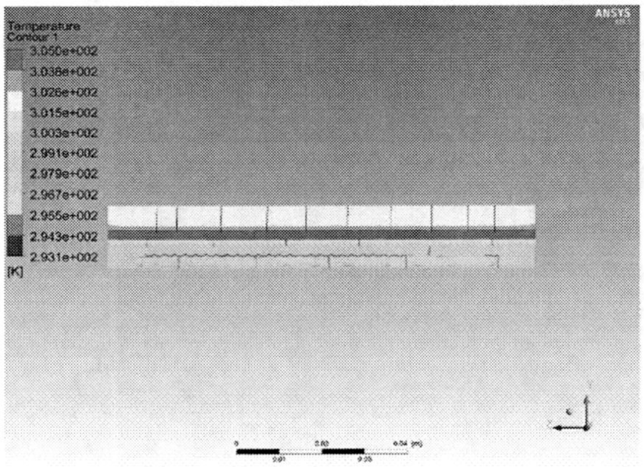

Figure 5. Temperature distribution and pathlines in oil at time 300 s with perpendiculary aplied gravitation

In the second simulation at time 300 s the conductor reaches a temperature of 28 °C (301 K), which is 6.5 °C lower than the highest conductor temperature when using transformer oil. The temperature is evenly distributed and local high temperature spots do not occur as it was when using oil (Figure 7). The conductor is only 1.5 °C warmer than the lowest temperature of magnetic fluid. Fluid flow is irregularly distributed throughout the volume of the fluid, due to the thermal flow of the liquid and the movement of the nanoparticles by the action of the magnetic field (Figure 8).

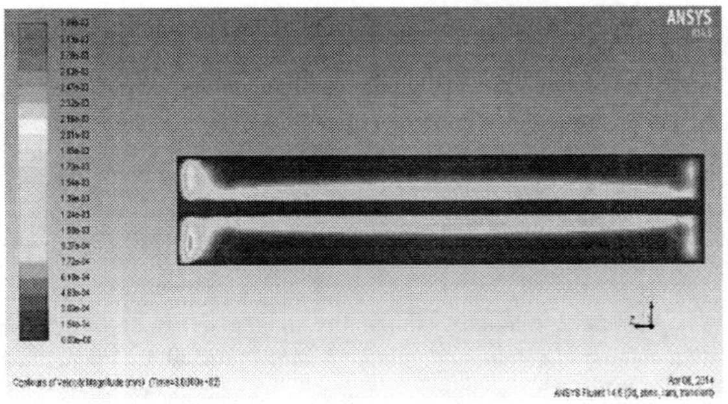

Figure 6. Contours of fluid flow velocity of oil at time 300 s, cross section at Z axis

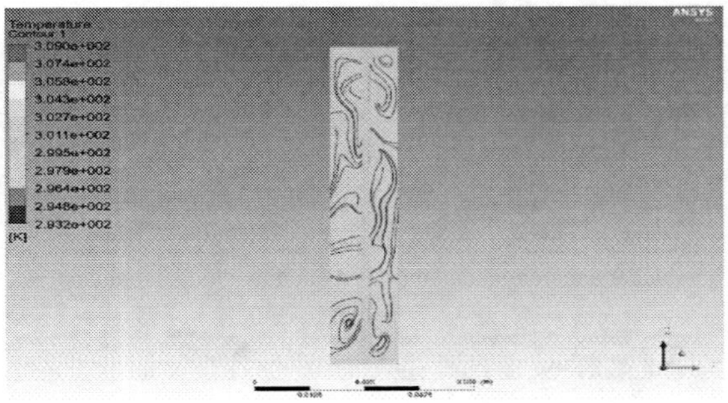

Figure 7. Temperature distribution and pathlines in magnetic fluid at time 300 s

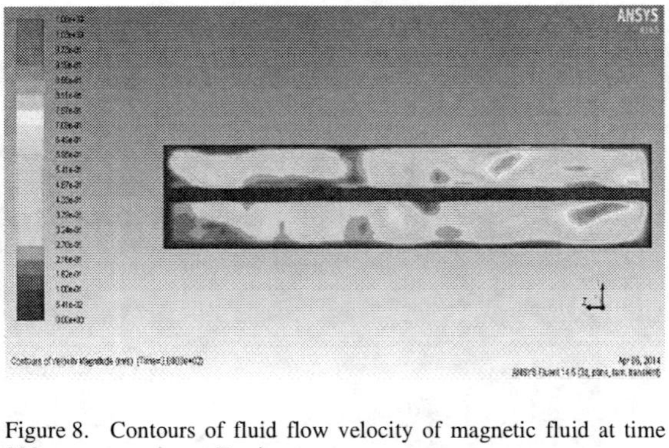

Figure 8. Contours of fluid flow velocity of magnetic fluid at time 300 s, cross section at Z axis

In the third simulation transformer oil was heated up with copper conductor and at time 150 s magnetic nanoparticles were added. Heating up from the conductor creates natural circulation in oil as it can be seen in Figure 9a at time 150 s. The oil in top of the container reached 28.35 °C (301.5 K) and the conductor has a temperature of 32.75 °C (305.9 K). On the bottom, the oil has a temperature of 21 °C (294.1 K). In 160 s observable changes began to show where the temperature distribution began to deform in the lower part of the container (Figure 9b). This is due to the fact that at the bottom of the container was cooler magnetic fluid than at the top of the container, and thus it was more attracted to the conductor. In Figure 8c can be seen how the cold magnetic fluid is attracted to the conductor at the bottom of the container. The movement of nanoparticles prevail over the natural circulation and the liquid began to mixing up and evenly distribution of temperature in volume occurred (Figure 9d and 9e).

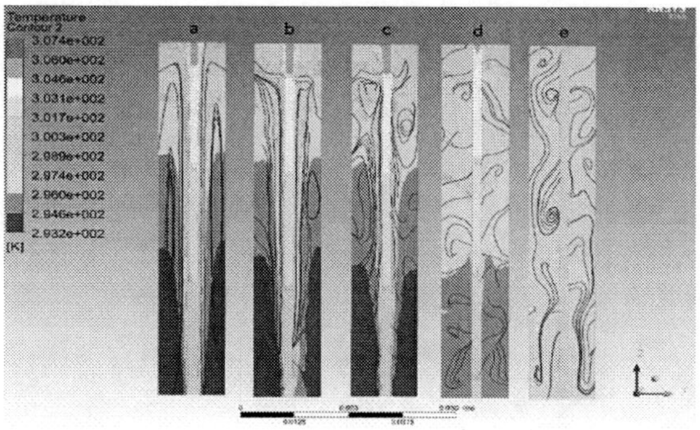

Figure 9. Temperature distribution and pathlines at time a: 150 s, b: 160 s, c: 180 s, d: 200 s, e: 270 s

IV. Conclusion

From the results of this simulation, it can be determined that magnetic fluid is a better cooling medium than transformer oil. The simulation revealed that the direction of gravity field does not affect the distribution of magnetic fluid temperature. The magnetic fluid evenly distributes the temperature in volume, unlike the oil. When using a magnetic fluid with the same volume as the oil, the conductor and the liquid reach a lower temperature. A smaller container with magnetic fluid may be used to achieve the same cooling effects as with oil cooling.

Acknowledgment

This work was supported by: Slovak Academy of Sciences and Ministry of Education in the framework of Projects VEGA No. 1/0311/15, and 2/0141/16; Ministry of Education Agency for structural funds of EU in frame of Projects No. 26220120055 and 313011D232. This work was supported by the Slovak Research and Development Agency under the contract No. APVV-15-0438.

References

[1] A. M. Morega, M. Morega, L. Pîslaru-Dănescu, V. Stoica, F. Nouraş and F. D. Stoian, "A novel, ferrofluid-cooled transformer. electromagnetic field and heat transfer by numerical simulation," 2010 12th International Conference on Optimization of Electrical and Electronic Equipment, Basov, 2010, pp. 140-146. DOI: 10.1109/OPTIM.2010.5510425.

[2] C. Tangthieng, B.A. Finlayson, J. Maulbetsch, T. Cader, "Heat transfer enhancement in ferrofluids subjected to steady magnetic fields", Journal of Magnetism and Magnetic Materials, Volume 201, Issue 1, 1999, pp. 252-255, ISSN 0304-8853, DOI: 10.1016/S0304-8853(99)00062-1.

[3] S.M. Snyder, T. Cader, B. A. Finlayson, "Finite element model of magnetoconvection of a ferrofluid", Journal of Magnetism and Magnetic Materials, Volume 262, Issue 2, 2003, pp- 269-279, ISSN 0304-8853, DOI:10.1016/S0304-8853(02)01502-0.

[4] Z. Čonka, M. Kolcun, M. Kosterec, J. Dudiak, M. Vojtek, M. Mikita, G. Morva, "Increasing Total Transfer Capacity (TTC) by PST in Central East Europe power system", in Proceedings of the 8th International Scientific Symposium on Electrical Power Engineering, ELEKTROENERGETIKA 2015, pp. 484-487, Technical University of Košice, 2015.

[5] S. Chaudhari, S. Patil, R. Zambare, S. Chakraborty: Exploration on Use of Ferrofluid in Power Transformers, IEEE 10th International Conference on the Properties and Applications of Dielectric Materials. Bangalore, India: 24-28 July 2012.

[6] M. Timko, K. Marton, L. Tomčo, J. Király, M. Molčan, M. Rajňák, P. Kopčanský, R. Cimbala, F. Stoian, S. Holotescu, A. Taculescu, "Magneto-dielectric properties of transformer oil based magnetic fluids in the frequency range up to 2 MHz", Magnetohydrodynamics. Vol. 48, no. 2 (2012), pp. 427-434. - ISSN 0024-998X.

[7] V. Segal, K. Raj, "An investigation of power transformer cooling with magnetic fluid", Indian J. Material Sci. Vol 5., December 1998, pp. 416-422.

[8] G.J. Hwang, M. Zahn, F. M. O'sullivan, R. Liu, "Electron Scavenging by conductive nanoparticles in oil insulated power transformers", Cambridge, 2009 Electrostatics Joint Conference-paper 1.1, 2009, pp. 1-13

[9] Z. Fang, R. O'handley, Y. Liu, M. Yang, "FEA simulation of passive ferrofluid cooling systems", in Proceedings of the COMSOL Conference 2010, Boston, pp. 1-4.

Resources adequacy – Analysis of renewable generation variability

Michaela Lachmanová
Dept. of Economics, Management and Humanities, Faculty of Electrical Engineering,
Czech Technical University,
Technická 2, 166 27 Praha, Czech Republic

Michaela.lachmanova@fel.cvut.cz

Abstract — **Integration of highly volatile intermittent sources bring new tasks to be solved. One of the most important terms related to renewables integration is generation adequacy, which is the ability of the power system to match the load at all times. Broadly conceived, generation adequacy requires appropriate portfolio of sources considering all technical and economic aspects of all types of energy sources. Analysis of short-term variability of renewable production is only the first step of setting this optimal portfolio.**

This paper deals with short-term analysis of renewable production, focusing on two approaches. First, capacity factor analysis, which provides ratio of produced energy and installed capacity. This analysis is provided in two ways – year capacity factor and hourly capacity factor analysis. The main analysis of paper lies in short-term variability of RES production analysis, which is provided by hourly gradient analysis. This paper introduces very first step of future research focusing on analysis of short-term variability of renewable production.

The main aim of this paper lies in discussion and comparison of German and Czech renewable variability and its impact on generation adequacy.

Keywords: Renewable production, capacity factor, gradient analysis, generation adequacy

I. Introduction

The enormous increase in the integration of renewables (RES) into the energy system with the system of subsidies (in particular the guaranteed purchase price, ie. Feed-in tariff) had serious consequences in the energy markets and the operation of the electrical system. The current way of supporting renewable energy sources and market mechanisms caused a surge supported sources, a significant change in the mix (conventional plants and operationally expensive resource are pushed out – called merit order effect) and last but not least, a significant increase in electricity prices for end users. Intermittent and highly volatile nature of the sources based on renewable energy sources additionally causes more and more problems with the management of the electricity system. Examples include unplanned overflows electricity through the Czech Republic due to production from wind farms in northern Germany to Austria and southern countries [1].

As we can see in last years, thanks to the European support of renewable sources, thanks to the merit order effect and the unpredictability of renewable energy production, renewable energy production begins to be preferred and also is preferred to be consumed first. Renewable production thus determines the character of the remaining sources covering the residual load. Residual load, as a random variable, describes the remaining electricity necessary for coverage. The residual load depends not only on renewable energy production, but also on the varying demand for electricity.

We can conclude, that to set a proper portfolio of sources ensuring generation adequacy, we need to know characteristics of sources covering residual load, because renewable production is satisfied and consumed first. Thus we need to know, what is the characteristics of demand for electricity and also characteristics of renewable production. This paper deals with first part of research, the analysis of short term variability of renewable production to be used to set the characteristics of residual load. Finally, the results are compared to results of German case study analysis, to set an impact of RES installation in Germany to Czech conditions.

II. Resources adequacy

This chapter deals with description of basics terms needed to better understanding of issue of renewable source production analysis. As we described previously, we need to describe characteristics of renewable production to ensure generation adequacy. These characteristics and also all terms are explained further.

At first we should summarize, what the term, generation adequacy, means. Which characteristics are used to fulfill this term (capacity credit and capacity factor). Also we should focus on statistic approach, used to evaluate numerically all these characteristics.

A. *Generation adequacy*

Due to this development and new European RES targets, it is necessary to find out proper solution how to integrate renewable sources to the grid with minimum negative impact. One of very important terms related to renewables integration is generation adequacy.

Generation adequacy is the ability of the generation in the power system to match the load on the power system at all times. Developing this definition more broadly in long-time period, it is also about finding the optimal structure of sources for electricity generation taking into consideration all economic, environmental and technical aspects of all types of sources. [5]

B. Capacity credit vs capacity factor

Capacity credit or capacity value of a generator is the amount of additional peak load that can be served by that generator. The capacity credit of RES is then calculated in order to measure the impact of uncertainty related to the contribution of renewable generation on resource adequacy. [6]

Concerning adequate generation capacity, it is important to introduce the concept usually designated as "capacity credit" of variable sources. It is often argued that the installation of 1 GW wind requires 1 GW of (flexible) back-up capacity, because, the reasoning goes, there are moments when there is no wind. This is not correct from a systems point of view. Indeed, because conventional power plants of the generation system also have a reliability that is lower than 100%, e.g. 90%, about 10% of the time, such plants are not always available when needed, especially in the event of unexpected outages caused by technical problems [5].

This paper calculates capacity credit of several renewable sources as a contribution to the production with some probability. The capacity credit of variable sources is defined as the equivalent of conventional generation that the variable production needs to back-up.

Capacity factor is the ratio of generator actual output over a period of time to its maximum nameplate generation capacity over the same period.

III. RENEWABLES PRODUCTION ANALYSIS

At first we should summarize the characteristics of renewable production in summary. This paper presents short-term variability analysis of RES production dataset from different sources as the first step of wide analysis of generation adequacy of electricity sources in the Czech Republic.

For this analysis quiet extended dataset was used. For results we chose dataset based on two photovoltaic plants with total installed capacity 2000 kW, production of 2 years. Dataset of four wind turbines with total installed capacity 10 000 kW was used, production of four years. Total portfolio of these sources called RES portfolio was then analyzed.

We can see plotted production of all sources on Fig.1–3.

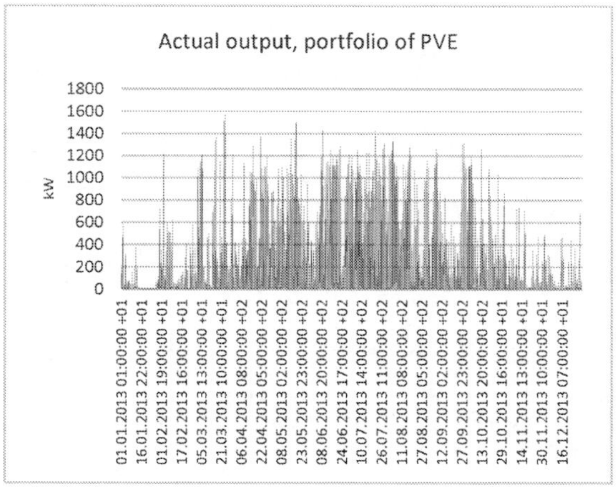

Fig. 1 Production of all analyzed PVE sources

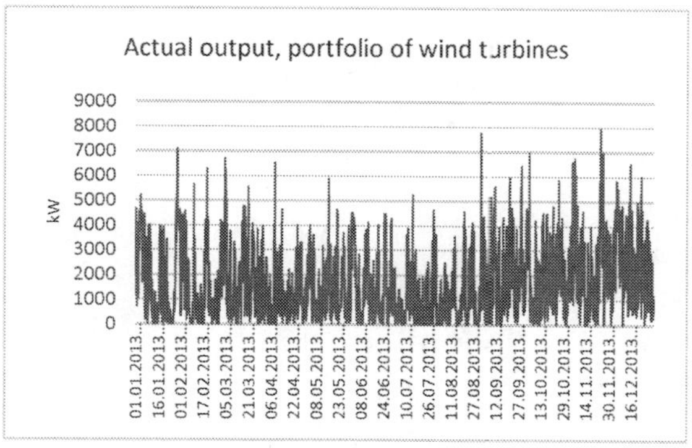

Fig. 2 Production of all analyzed wind turbines

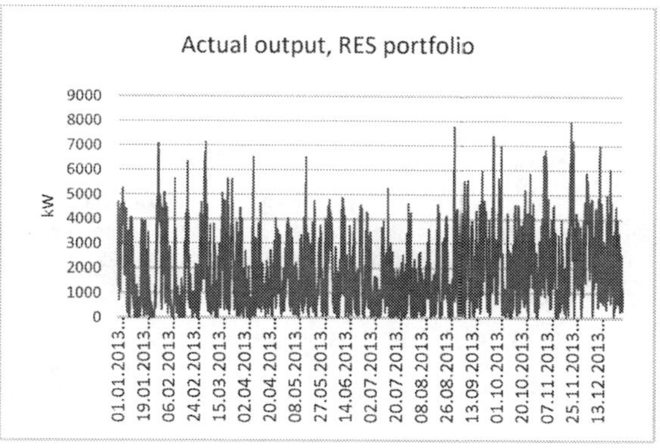

Fig. 3 Production of all RES sources

As we can see, variability of production seems very high, especially wind production highly varies.

For better understanding of using of wind and photovoltaic production, histograms of both production can be observed as follows.

At first we focused on one wind turbine and all descriptive characteristics and histogram of production can be observed as follows.

Tab. 1 Wind turbine 1, descriptive characteristics

Wind turbine 1	
Installed capacity (Pi)	2000 kW
Data year	2013
The height of the rotor axis	80 m
Arithmetic mean	488,2
Median	264,2
Standard deviation	550,17
Variance	302690,1
Coefficient of variation	1,12
Coefficient of variability	0,28
Time of use of source	2138,7

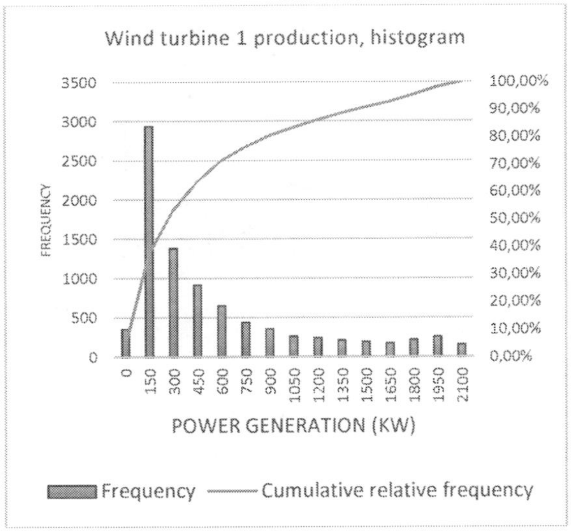

Fig. 4. Histogram of wind production

We can conclude that wind production is quite variable and highly left-skewed. That e.g. means, that 50% of measurements are below 300 kW, which is one seventh of installed capacity, or 80% is below one half of installed capacity.

Then we analyzed photovoltaic production. We divided our calculations to two cases – considering all data and only nonzero production, because 50% of time PVE does not produce.

Tab. 2 PVE 1, descriptive characteristics

PVE 1 installed capacity 927 kW		
	All data	Nonzero production
Arithmetic mean (Expected value)	89,63	196,69
Sample standard deviation	164,68	196,09
Median	0	120
Coefficient of variation (%)	183,7%	99,7%

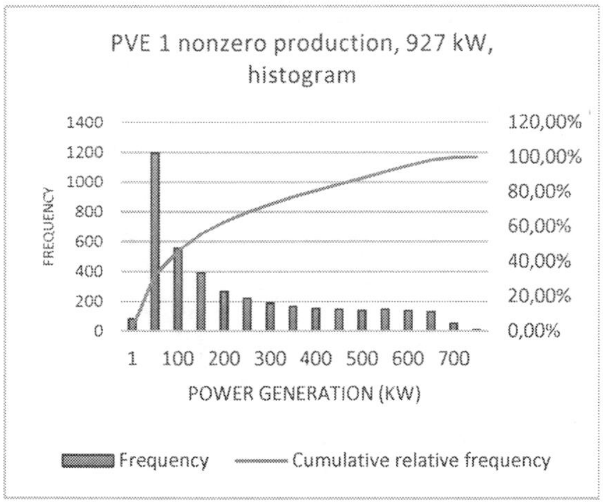

Fig. 5. Histogram of PVE 1, nonzero production

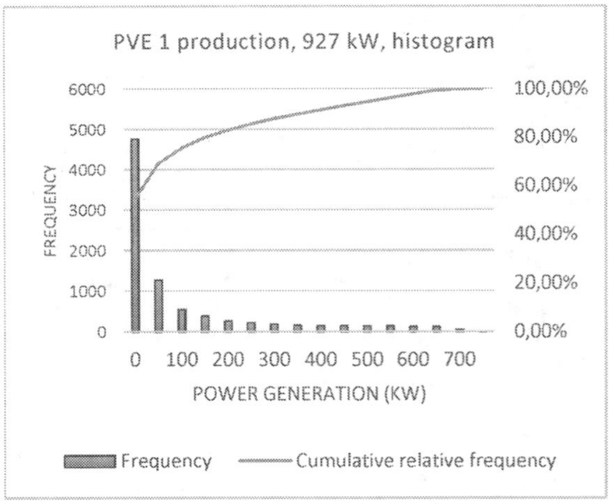

Fig. 6. Histogram of PVE 1 production

Basic statistical analysis of year hourly production can be seen in previous figures and tables. Fact, that almost 60% of production is zero, can be used for long-term grid operation planning, as well as fact, that 95% of hourly production is less than value 500 kW, which is slightly over half of installed capacity.

These results, verified on other sources show that renewable production is low, contribution to generation adequacy is also low and also the variability of production is very high. Further analysis focus on capacity factor and exact contribution to generation adequacy.

A. Capacity factor

Capacity factor can be expressed in two ways. The first approach is to express whole-year availability by one number. This number represents ratio of production per whole year. We will calculate these characteristics for photovoltaics panels and also for wind production as a basic characteristics of RES production [4].

Tab. 3 Basic characteristics of RES portfolio

	PVE sources (total)	**Wind turbines (total)**	**RES portfolio (total)**
Installed capacity (KW)	1946	10 050	11 996
Availability factor (h)	664	1398	1279
Capacity factor (%)	0,076	0,160	0,146

As we expected, capacity factor of both types of RES is very low, even in case of aggregation. For example, in average, capacity factor of photovoltaic production varies between 10-20% [3]. Our results shows that even large installation in the Czech Republic has lower capacity factor, thus the production is quite weak.

The more important research lies in hourly capacity factor analysis, which expresses ratio of an actual electrical energy output over a given period of time to the maximum possible electrical energy output over the same amount of time [2]. This number is highly related to the term availability factor, which is the amount of time that it is able to produce electricity over a certain period, divided by the amount of the time in the period.

We will set and calculate capacity factor of each type of source. Capacity factor expresses total availability of production in percentage.

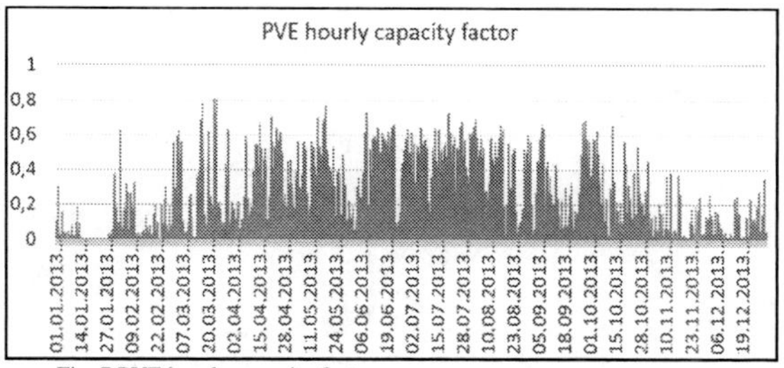

Fig. 7 PVE hourly capacity factor

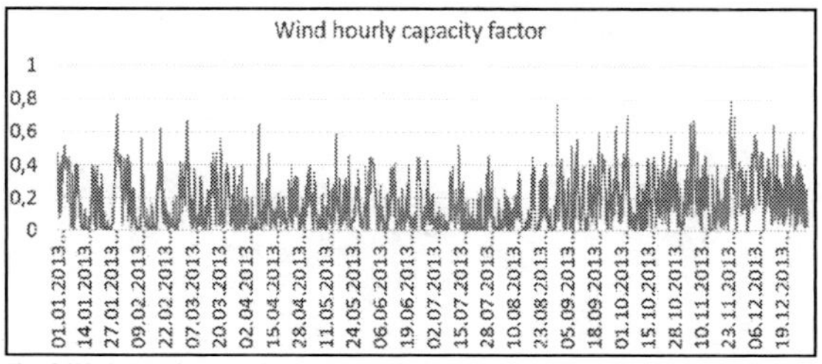

Fig. 8 Wind hourly capacity factor

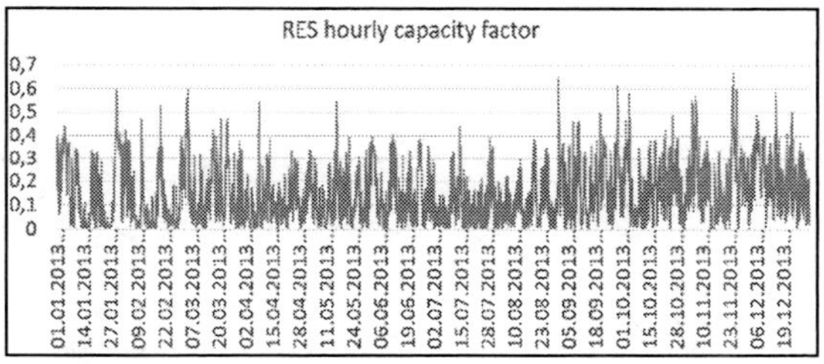

Fig. 9 Capacity factor of RES portfolio

B. Capacity credit (hourly gradients analysis)

The main idea of research lies in hourly changes analysis – so called gradients – which means that the main analysis lies in description of hourly changes of RES production. Hourly changes analysis is very important due to timeframe of short-term electricity markets and also due to timeframe of mechanisms of grid balancing.

So hourly gradients will be plotted and analyzed. We will focus especially on characteristics of maximum and minimum of gradients and also on probability of occurrence of extreme values. We should emphasize, that characteristics of short term variability is very important to grid balancing, backup sources dimensioning and also for short-term (intraday market) trading.

Hourly gradients are calculated as follows:

$$g_t = p_t - p_{t-1} \tag{1}$$

$p_t - production\ in\ time\ t$

And are plotted in following figures. All gradients are plotted considering whole year 2013.

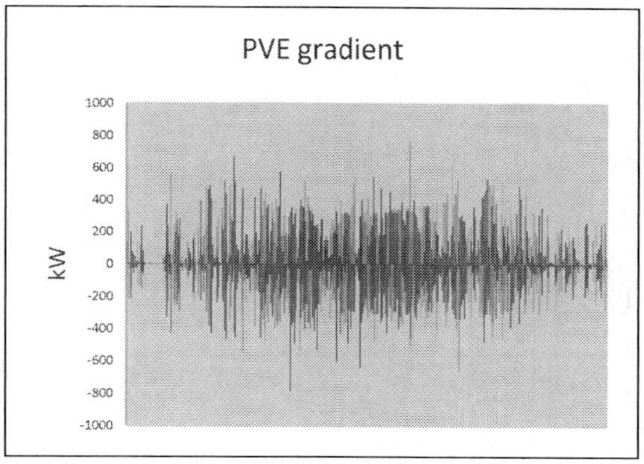

Fig. 10 PVE gradient, year 2013

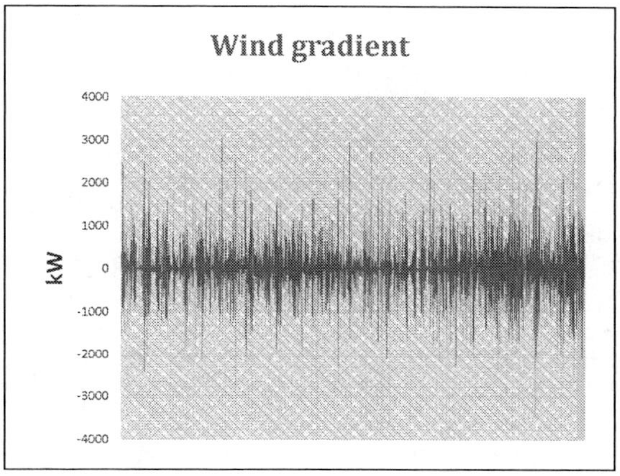

Fig. 11 Wind gradient, year 2013

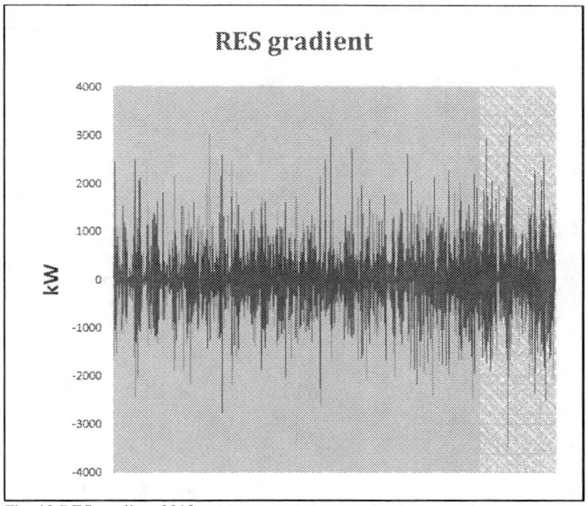

Fig. 12 RES gradient 2013

As we can see, the hourly changes are very significant and presented in following tables:

Tab. 4 Characteristics of hourly gradients

Hourly gradients			
Pinst	**1946 kW**	**10 050 kW**	**11 996 kW**
	PVE gradients	**Wind gradients**	**RES gradients**
MIN	-785	-3500,6	-3500,6
MAX	767	3282,97	3282,97
MEAN	0	-0,30	-0,30
MEDIAN	0	-1,14	-3,45
STDEV	109,5	502,07	515,07
95% bot	-214,62	-984,253	-1009,7
95% up	214,62	983,64	1009,9

The main results lies in follows: Minimum hourly gradient is in the case of PVE production -785 KW, which is approx. one third of installed capacity. Production of analyzed sources requires max third of installed capacity of backup sources. Previous results claim that we need more backup resources, but it is absolute production. If we only focus on variable production and inter-hour changes, it turns out that the amount of backup resources is lower. Peak analysis is analogous.

The second main result is presented by last two rows. It represents, that 95% of hourly gradients lies in interval set by <95% bot; 95% up>. That is very important information, because we see, that the hourly variability is not as high as expected.

The same results are shown in the following table, representing relative hourly changes related to installed capacity.

Tab. 5 Characteristics of relative hourly gradients

Relative hourly gradients			
	PVE gradients	**Wind gradients**	**RES gradients**
MIN	-0,403	-0,348	-0,292
MAX	0,394	0,327	0,274
MEAN	0,000	0,000	0,000
MEDIAN	0,000	0,000	0,000
STDEV	0,056	0,050	0,043
95% bot	-0,110	-0,098	-0,084
95% up	0,110	0,098	0,084

We can conclude, that renewables inter-hour changes in total vary between -30% and 27% of installed capacity and 95% of hourly gradients lies between -8.4% and 8.4%. That is not as bad as expected. The final conclusion can be represented by histograms of hourly gradients, representing distribution of hourly gradients. For better understanding histograms are given for relative values.

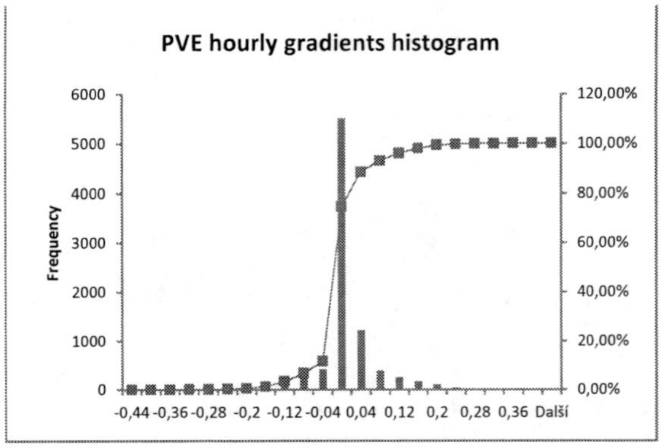

Fig. 13 PVE gradients histogram

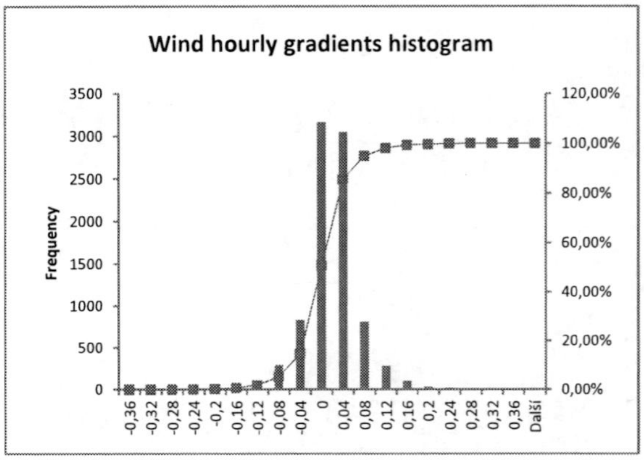

Fig. 14 Wind gradients histogram

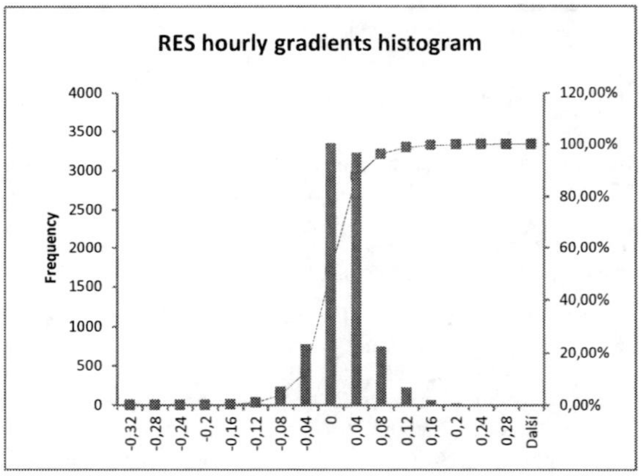

Fig. 15 RES gradients histogram

As we can see, in the case of renewables aggregated, almost 90% values does not exceed 0,04 value. That mean, that we can expect inter-hour cuts and occasional peaks. Also we can conclude, that almost 80% of values lies between -0,04 and 0,04 value, which means, that RES production in general mainly hourly vary in 4%. That is also not as high variability as expected.

IV. Comparison to Germany case study, Discussion

As we can compare with extended study presented in [6], capacity factor of renewables is 12% for wind generation and 9% for PVE production PVE production in Czech conditions has a lower capacity factor – only 7.6%. Author of this research explains this number as a consequence of high number of high-installed capacity sources in Germany. That means, that variability is much higher than in the Czech Republic. Also the capacity credit is much lower in Germany – the balancing of grid requires much more expensive and high-variable sources to ensure security supply.

But capacity factor of wind production is higher in the Czech conditions – 16%. It can be due to small installed capacity of wind production – the impact of high variability is not as negative as in Germany.

We should emphasize, that we also compared the hourly gradients characteristics, which is much interesting and useful. In German conditions, hourly gradients varies between **+88%-80%,** which means, that almost 90% of installed capacity should be still ready to cover hourly changes.

In Czech conditions the situation of gradients is better – gradient varies between **-29,2%-27,4%,** which requires only one third on installed capacity being ready to cover hourly changes. Even 95% of hourly gradients vary between **-8,4% and 8,4%**, which means that 95% hourly changes should not exceed 9%.

We chose for our research the comparison with Germany case, because German and Czech energy field is very interlinked. First, we talk about integrated electricity market, second we should mention the impact of the installation of RES in Germany on the Czech transmission system. We should realize that the variability of the German RES will always have a significant influence on the Czech energy industry. It can be assumed that the variability of the German resources is the variability that the resources in the Czech Republic are directed to and our results can be used as a model example for the dimensioning of the backup resources already.

V. Conclusion

The energy policy of the European Union has posed to the fore the issue of reducing dependence on imported energy and environmental issues, which resulted in, among other things, the requirement for the maximum possible utilization renewable resources. Although all Member States have taken steps to promote renewable energy resources, the result is not the same everywhere. This is mainly due to political, geopolitical and market reasons. Other obstacles are very high volatility of individual sources and consequences associated with it.

However, RES integration requires consideration of the all aspects of sources operation, especially for generation adequacy. Analysis of optimal national portfolio of sources requires analysis of short-term variability of renewables.

As results show, the inter-hour variability (hourly gradient) is not as high as expected. These results are only the first step, which is going to be verified on extended data set and also to be used to confrontation with model of electricity demand.

These all results will be further used to model of residual load and to describe all characteristics needed to set a backup sources and also to optimize all sources to ensure generation adequacy as described in first chapter. All these sources used in optimized portfolio should be evaluated from all points of view considering all technical and economic aspects of operation

Acknowledgment

Research described in the paper was supervised by prof. Ing. Jaroslav Knápek, CSc., Department of Economy, Management and Humanities, FEE CTU in Prague and supported by the Student Grant Competition under grant No SGS16/154/OHK5/2T/13.

References

[1] ČESKO V LÉTĚ ZAHLTILA NĚMECKÁ ELEKTŘINA (The Czech Republic was overloaded by German electricity). In:Www. zpravy.e15.cz [online]. Praha [cit. 2017-03-20]. Available at: http://zpravy.e15.cz/byznys/prumysl-a-energetika/cesko-v-lete-zahltila-nemecka-elektrina-1227966

[2] Capacity factor. NRC: Glossary [online]. [cit. 2017-03-21]. Available at: https://www.nrc.gov/reading-rm/basic-ref/glossary/capacity-factor-net.html

[3] Solar PV capacity factors in the US – the EIA data. Energy matters [online]. [cit. 2017-03-21]. Available at: http://euanmearns.com/solar-pv-capacity-factors-in-the-us-the-eia-data/

[4] Security of Supply and System Adequacy. Wind energy, the facts [online]. [cit. 2017-03-21]. Available at: https://www.wind-energy-the-facts.org/security-of-supply-and-system-adequacy.html

[5] Nuclear energy and renewables: system effects in low-carbon electricity systems. Paris: OECD, 2012. ISBN 9789264188518.

[6] MOGHANJOOGHI, Hamid Aghaie. *Model-based Analysis of Generation Resource Adequacy in Energy-only Markets*. Vienna, Austria, 2016. DISSERTATION. Faculty of Electrical Engineering and Information Technology, Vienna University of Technology. Supervisor: Prof. Dr. Reinhard Haas

Mesopic Vision Utilization During Public Illumination Realisation by Means of LED

Richard Baleja, Tomáš Novák
Department of Electrical Engineering
FEECS, VŠB-TU Ostrava
Ostrava, Czech Republic
bal0034@vsb.cz, tomas.novak1@vsb.cz

Karel Sokanský, Petr Bos
Department of Electrical Engineering
FEECS, VŠB-TU Ostrava
Ostrava, Czech Republic
karel.sokansky@vsb.cz, petr.bos@vsb.cz

Abstract — **This paper deals with the public lighting on roads. The lighting on the roads is a specific case of lighting systems, where an observer is located in an area with relatively low adaptation luminance. That means, that the human eye don´t assess the lighting parameters of road lighting according to photopic curve of human eye sensitivity as it does during the daylight, but it uses for this assessment a sensitivity curve, which is located in mesopic sensitivity of the human eye. The maximum sensitivity of the curve depends on the size of the adaptation luminance, in which the observer is located. To compare the differences between the values of photopic and mesopic light parameters were realized several measurements of roads lighting. For measurements were selected roads with different light levels and adaptation luminance background as well as different light sources which have different correlated color temperature. Light parameters on the measured roads were evaluated in terms of the luminance of the background for mesopic sensitivity. Within researching the changes in the sensitivity of the visual organ during the transition from photopic to mesopic vision, it is necessary to point out that this is a peripheral region of the visual organ. The central region of the eye (fovea) has to be excluded from changes researching, because in this region are located only photopic photoreceptors (cones). The article points out that lighting systems implemented with LED have significantly lower power consumption compared to lighting systems fitted with high-pressure sodium lamps. Since there are requests to further reduction of luminance (electric input) of the public lighting systems based on the usage of mesopic units, it is necessary to highlight the fact that the mesopic region prefers LED luminaires to high-pressure sodium lamps only in the peripheral region. Therefore, we cannot realize public lighting with LED for lower than regulatory requirements.**

Keywords: mesopic vision; photopic vision; luminous flux; correlated colour tempetarute; colour rendering index; spectrophotometer; illuminance meter; luminance camera;

I. Introduction

A design of public road networks is relatively complicated process by which is essential to respect several regulations and norms. It has come recently to the changes of two technical reports of the public road networks design. The technical report ČSN CEN/TR 13201-1 Public road networks lighting, part 1: Instructions for lighting category selection for road networks has come into force in the September 2016. This technical report has changed the procedure of lighting category selection for a certain road network. The changes make the previous norm different from the current one. The organization is being performed newly by means of point evaluation for the fixed road network. The second change is related to the norm ČSN EN 13201-2 Public road networks lighting, part 2: Demands. This norm has come into force in June 2016. Changes of required parameters for the individual categories of the lighting of road networks occurred in the framework of the norm.

Respective norms need to be abode, naturally, when designing the lighting systems of road networks. But it is convenient many times to respect also other scientific studies and papers which are concerned with issues of public lighting. It is necessary that a designed lighting system would provide requested figures of the norms for the fixed road network category. It is also essential to choose an appropriate type of lighting source. A couple of various lighting sources are recently accessible on the market. In particular, these are various types of discharge tubes and LED lighting sources. Such lighting sources are naturally indifferent to each other by different technical parameters. However, for the purposes of this article, it is sufficient if we pay our attention to one lighting spectrum and substitutive temperature of chromaticism.

Lighting systems are in most cases realised by means of discharge tubes (high-pressure sodium discharge tubes, mercury discharge tubes). This way of use of lighting systems is nowadays considered as very uneconomical. Old lighting systems are being reconstructed more and more and replaced by LED lighting sources. It is always necessary to choose correctly the colour of a lighting source (substitutive temperature of chromaticism) when reconstructing old lighting systems or designing a new system. The spectrum of lighting source relates closely to the substitutive temperature of chromaticism. Development of the spectrum of lighting source determines which wavelengths are for the fixed lighting source dominant and which are not dominant. Issues of blue light of LED sources is recently being examined. The blue light can inconveniently influence circadian rhythms of a human, so it is recommended to use in developed areas exclusively light sources with the warm colour of light. Cooler tones of light can be used in less developed and industrial areas; it is alternatively possible with these sources of light to realize lighting of motorized road networks with heavy traffic. The main focus of this paper is on the issue of mesopic vision which is related to the issue of the substitutive temperature of chromaticism. Examples of the measurement of the lighting parameters of various types of road networks are introduced in the article. There is also introduced their subsequent evaluation from which the evident difference of these parameters is apparent. For comparison were used measured and evaluated road networks which were lit by high-pressure sodium discharge tubes and by means of LED lighting sources. The spectrum of LED source is depicted in picture 1 and the spectrum of the sodium discharge tube is depicted in picture 2. These images are used for a better projection of the difference in the spectrums of these two lighting sources.

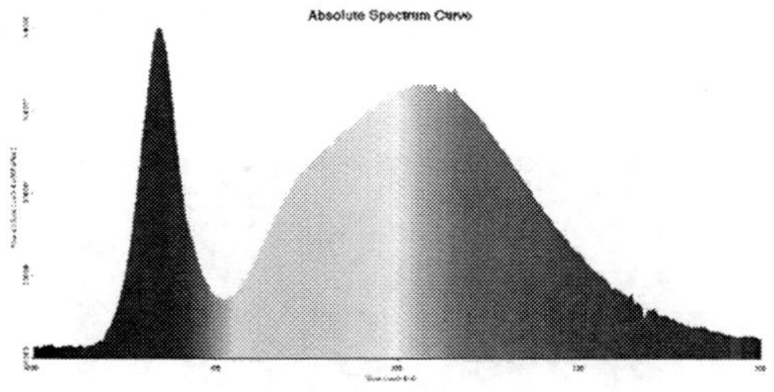

Figure 1. Absolute spectrum curve of the LED source

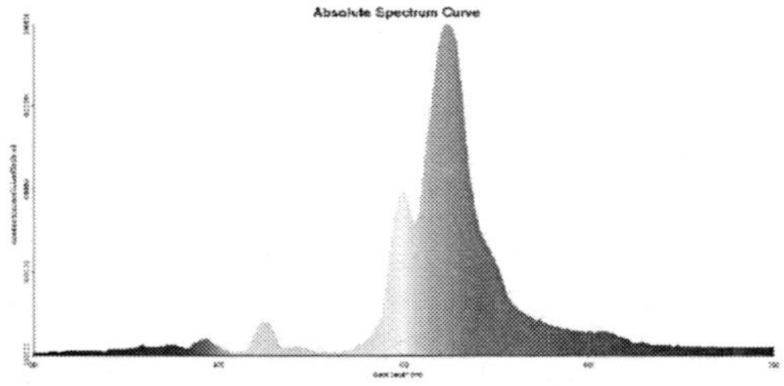

Figure 2. Absolute spectrum curve of high-pressure sodium lamps

II. Sensitivity of the human eyesight

The human eye is not equally sensitive to the radiance of different wavelengths. Development of this dependency, as well as limits of visibility, are also various among different people. The highest possible sensitivity of the human eye under sufficient lighting conditions is determined by the sensitivity of retinal cones and it oscillates during photopic vision around 555nm. The sensitivity of the human eye also depends on an environment in which is an observer located. There is a difference between an observer who is situated in an environment where is enough of light and an observer who is in dark environment. There are three vision categories of the human eyesight that we distinguish: photopic vision, scotopic vision, and mesopic vision.

The sensitivity of high brightness (photopic – day vision) is regulated in accordance to the photopic curve of the human eye sensitivity. The most of the devices which measure photometric quantities are calibrated in accordance with this curve. Light parameters are always stated for the photopic sensitivity of the human eyesight. These parameters are for instance light flow of the source or a lighting fixture.

A problem appears when an observer moves from a light environment to a dark environment. The curve of human eye sensitivity starts to move from the area of photopic sensitivity (day vision) across the myopic area to the area of scotopic sensitivity (night vision) which is characteristic of very low brightness (less than 0,005 cd/m2). The scotopic curve of the sensitivity of the human eyesight is practically not usable for appraisal of light parameters because the issue is that there are areas with very low brightness in which is practically zero visibility. The mesopic are is interesting for us from the functional perspective. The area is located between the photopic and scotopic curve of sensitivity. The mesopic area is formed by a collection of curves of sensitivity. In the contrary to this, photopic and scotopic areas are characterised only by one curve of sensitivity. One curve of the mesopic sensitivity exists for each one figure of brightness which means that a number of these curves is infinite. According to the sources found in the literature, the extent of the mesopic vision oscillates in the brightness range from 0,005 cd/m2 to 3 cd/m2. To be concerned with the issue of the mesopic vision has a purpose only in the cases when is an observer situated in an environment with brightness in the range that has been mentioned. So, these are mainly the area of the emergency lighting and the area of the lighting of road networks. If an observer is situated in the mentioned area and the sensitivity by any of the mesopic curve is being applied, the perception of the lighting flow of the source will be different than it would have been in the case of perception of the photopic curve of sensitivity. For instance, a perceived illumination on the road will be either bigger or smaller in the mesopic area in reliance on a type of lighting source. In practice, it is naturally essential to use such lighting sources so that perceived amount of light in the mesopic area would have been bigger as it is in the photopic area. If we compare two different light sources with the exact same light flow (in compliance with photopic curve) on the grounds of the chosen mesopic curve, and light flaw of sources which are evaluated through the mesopic curve, the sources will be different with reliance on the development of the radiant spectrum of each light source. The measurement to verify if this difference of light flaws is significant was carried out. The measurement was supposed to verify this issue. Bigger final light flaw results naturally in higher illuminance evaluated in the mesopic area.

III. Road networks lighting measurement

It was necessary to measure selected sections of road networks in order to verify the above-mentioned issue. It was essential to find areas with the level of photopic lighting approximately equal and where the lighting was realized by means of various lighting sources. Therefore was the selection of the suitable locations demanding. After a careful appraisal, a part of road networks on the Technologická Street in Ostrava – Poruba was selected for the consideration. The lighting is realized on this Street by means of high-pressure sodium discharge tubes with 70 W output and the substitutive temperature of chromaticism of 2000K. This road network continues to the K Myslivně Street where the lighting system is realized by means of LED lamps with 62 W output and the substitutive temperature of chromaticism of 4000K. The part of road network at the department store AVION in Ostrava, where the lighting system is realized by means of LED lamps with 75 W output and the substitutive temperature of chromaticism of 5000K, was also measured in order to compare the exact same LED sources which have the different substitutive temperature of chromaticism. It was not necessary to measure the lighting of road networks in accordance with ČSN EN 13201-4 road networks lighting – Part 4 Methods of measurement because our task was not discovering if the certain road network is correctly illuminated, but it was discovering of differences in the lighting of individual points are we able to achieve. The distribution of the measuring point on the particular road networks is evident in picture number 3. The positions of the measuring points were documented in order to verify the accuracy of the measuring device that was used. This device operated with two photosensitive sensors in order to measure lighting and radiated spectrum.

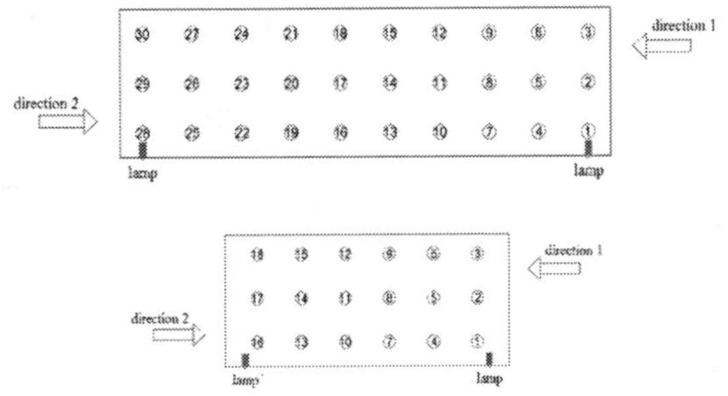

Figure 3. The position of the measured points on road

Lighting was measured by means of the spectrophotometer SPIC-200 by the Everfine company. It is possible to measure lighting with this device, radiated spectrum and a plenty of other lighting quantities at the same time. As it has been already mentioned above, this device uses for measurement two sensors. One sensor measures lighting and the other sensor measures radiated spectrum of a lighting source. Due to this fact, it was essential to compare these two sensors in order to examine if the lighting figure is the same at the both sensors. The first sensor output is lighting measured in lux, the output of the second sensor is the illuminated spectrum which is noted in the form of a chart. Figures from the measurement conducted by the second sensor were used for subsequent evaluation and calculation. Therefore, it was essential to finish counting of lighting in the individual points by the mathematic relation (1).

$$E_P = \sum(V_\gamma \cdot \theta_\gamma) \cdot 683 \text{ [lx]}$$

Where:

- E_P – is a photopic value of illuminance (lx)
- V_λ – is a photopic sensitivity of the human eye
- ϕ_λ – is a radiated output of the light source on a relevant wave length ($W/m^2/nm$)
- 683 – is a luminous constant for the photopic vision, (lm/W) – it characterizes maximum amount of luminous flux that we are able to get from one watt.

IV. Surrounding Brightness Measurement on the Chosen Road Networks

It was essential to carry out brightness analysis of the surroundings in which is located the road network that was being examined. This analysis was carried out in order to assign the corresponding mesopic curve of sensitivity. The brightness analysis was performed in both directions because it was a bidirectional road network with two different directions of coming drivers. Brightness analyser was placed in the height of 1,5 m above the roadway in the corresponding traffic lane in the distance of 60 meters from the beginning of the field on the road network being measured. Six brightness analyses of the surroundings were produced in total. Two directions were evaluated for each road network. The brightness analysis of the background that is surroundings was acquired by the brightness analyser CANON D450 from the LMK company. The measured results were subsequently processed in LMK LabSoft programme in which surfaces which need to be evaluated can be determined easily. Background brightness and surface brightness of a road network were evaluated at every brightness map. An example of brightness analysis, concretely for the system with LED lamps with the substitutive temperature of chromaticism of 4000K, is depicted in pictures 4 and 5.

Figure 4. A luminance analysis of the road with LED luminaires (possition 1)

Figure 5. A luminance analysis of the road with LED luminaires (possition 2)

Three figures of brightness were always calculated and recorded during the brightness analysis: maximal, minimal and average. The area of the measured part of the road network are visibly depicted in pictures 4 and 5. Width and lengths of the area of the road network is equal to the distance between two light points. Brightness analysis of the background (surroundings) which replaces adaptation brightness was evaluated for the whole area that was scanned by the brightness analyser. The measured and calculated figures of individual directions are situated in the tables number I and II. The same procedure of the analysis of the brightness parameters was used also with other two types of road networks.

Table 1. Luminancevalue for direction 1

Luminance cd/m2	Min	Max	Average
Background	0,068	21,051	0,422
Road	0,264	1,799	0,565

Table 2. Luminance value for direction 2

Luminance cd/m2	Min	Max	Average
Background	0,031	19,842	0,225
Road	0,192	1,243	0,571

V. Lighting Evaluation of Road Networks in Mesopic Area

For every brightness figure, there is one curve of the mesopic sensitivity present. Figures in charts for the photopic and the scotopic curves of sensitivity are accessible from the literature available. On the other hand, it is a little bit more complicated to obtain the mesopic curve of sensitivity for the stated brightness. The mesopic curves of sensitivity for the measured brightness of background were calculated to the end by the beforehand defined mesopic curves of sensitivity for the stated adaptation brightness which were evaluated in the photopic area.

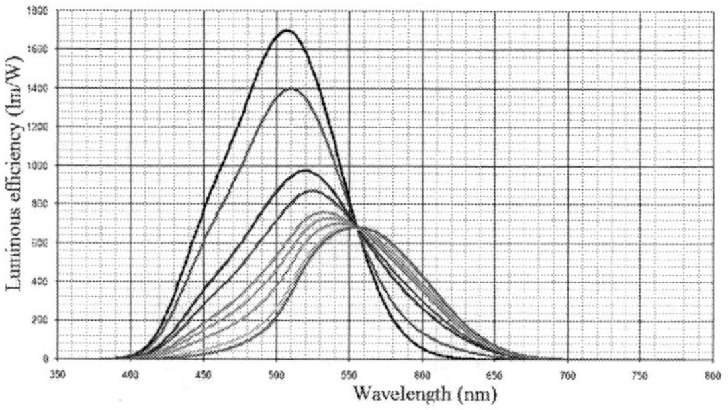

Figure 6. Course of photopic, scotopic and mesopic sensitivities of the human eye.

The visualisation of the sensitivity curves is depicted in picture number 6. There is also the curve for photopic vision (red curve) and the curve for scotopic vision (black curve) plotted. The curves which are placed between these two curves are the mesopic curves of sensitivity. The lower the adaptation brightness is, the closer the particular curve moves to the scotopic curve and vice versa. The invariable which defines which maximal amount of light flaw, which we are able to obtain the particular curve of sensitivity, has to be defined (except for the curve of sensitivity) for every adaptation brightness of the mesopic vision. This invariable entails figure of 683 lm/W of photopic vision. For the scotopic vision is this figure much bigger: 1700 lm/W. The figures of the mesopic curve of sensitivity oscillate in the range of these two figures in dependence on the measurement of the adaptation brightness.

LED lamp						
	Measered values		Calculated values			
			Photometric vision	Scotopic vision	Mesopic vision	
	E(lx)	S/P	E(lx)	E(lx)	E(lx)	E(lx)
1	14,61	1,65	14,61	24,03	16,34	16,93
2	16,63	1,65	16,23	26,74	18,16	18,82
3	15,85	1,64	15,86	26,04	17,72	18,37
4	11,89	1,58	11,90	18,75	13,15	13,59
5	14,19	1,62	14,19	22,93	15,79	16,35
6	14,67	1,63	14,68	23,88	16,37	16,95
7	8,03	1,49	8,03	11,99	8,76	9,01
8	9,38	1,48	9,36	13,86	10,19	10,47
9	9,00	1,46	9,01	13,17	9,77	10,04
10	5,06	1,15	5,11	5,85	5,24	5,29
11	6,63	1,19	6,66	7,96	6,90	6,98
12	7,19	1,24	7,24	8,96	7,55	7,66
13	5,14	1,03	8,03	11,99	8,76	9,01
14	6,17	1,09	5,37	5,51	5,39	5,40
15	7,59	1,09	7,86	8,56	7,99	8,03
16	5,64	1,03	5,96	6,15	5,99	6,00
17	7,01	1,10	7,18	7,91	7,31	7,36
18	8,66	1,12	8,80	9,86	8,99	9,06
19	8,64	1,18	8,80	9,86	8,99	9,06
20	9,56	1,23	9,70	11,97	10,12	10,26
21	11,74	1,23	11,82	14,52	12,32	12,49
22	11,53	1,21	11,63	14,11	12,08	12,24
23	13,39	1,31	13,46	17,64	14,23	14,49
24	15,29	1,33	15,35	20,44	16,28	16,60
25	14,82	1,36	14,88	20,18	15,85	16,19
26	17,23	1,41	17,28	24,31	18,56	19,01
27	18,11	1,40	18,14	25,39	19,47	19,93
28	15,77	1,32	15,83	20,90	16,76	17,08
29	18,22	1,41	18,26	25,65	19,61	20,08
30	18,05	1,41	18,09	25,44	19,43	19,90
Average	**11,52**	**1,33**	**11,64**	**16,15**	**12,47**	**12,75**

Figure 7. Measured and calculated values of illuminance on the road lighted by LED luminaires (4000K)

High-pressure sodium lamp						
	Measered values		Calculated values			
			Photometric vision	Scotopic vision	Mesopic vision	
	E(lx)	S/P	E(lx)	E(lx)	E(lx)	E(lx)
1	21,81	0,53	21,82	11,55	20,27	20,35
2	21,81	0,53	21,82	11,55	20,27	20,35
3	21,81	0,53	21,82	11,55	20,27	20,35
4	17,84	0,49	17,85	8,79	16,48	16,55
5	19,65	0,48	19,65	9,41	18,10	18,18
6	20,86	0,46	20,87	9,58	19,16	19,25
7	11,37	0,39	11,48	4,51	10,43	10,48
8	14,11	0,40	14,22	5,66	12,93	13,00
9	18,12	0,41	18,22	7,43	16,59	16,68
10	8,44	0,34	8,83	2,97	7,95	7,99
11	11,22	0,35	11,60	4,07	10,47	10,52
12	16,22	0,37	16,56	6,12	14,98	15,06
13	9,51	0,32	10,20	3,30	9,16	9,21
14	12,08	0,34	12,72	4,28	11,44	11,51
15	17,26	0,35	17,86	6,27	16,11	16,20
16	8,58	0,30	9,81	2,92	8,77	8,82
17	12,11	0,32	13,08	4,17	11,74	11,81
18	18,30	0,34	19,11	6,57	17,22	17,31
19	10,99	0,30	12,30	3,74	11,01	11,07
20	14,36	0,32	15,46	5,00	13,88	13,96
21	20,71	0,34	21,59	7,39	19,44	19,55
22	13,18	0,31	14,52	4,50	13,01	13,08
23	16,61	0,33	17,76	5,81	15,95	16,05
24	22,61	0,34	23,59	8,08	21,25	21,37
25	16,17	0,32	17,49	5,58	15,69	15,79
26	20,69	0,34	21,77	7,38	19,60	19,71
27	20,69	0,34	21,77	7,38	19,60	19,71
28	19,50	0,33	20,75	6,83	18,65	18,76
29	22,16	0,34	23,38	7,86	21,03	21,15
30	22,92	0,34	23,38	7,86	21,03	21,15
Average	**16,72**	**0,37**	**17,38**	**6,60**	**15,75**	**15,83**

Figure 8. Measured and calculated values of illuminance on the road lighted by high-pressure sodium lamps

LED lamp 5000K						
	Measered values		Calculated values			
			Photometric vision	Scotopic vision	Mesopic vision	
	E(lx)	S/P	E(lx)	E(lx)	E(lx)	E(lx)
1	22,31	1,86	22,33	41,61	28,36	26,27
2	21,33	1,87	21,34	39,89	27,15	25,13
3	20,07	1,85	20,09	37,17	25,43	23,58
4	18,18	1,82	18,19	33,04	22,84	21,23
5	19,78	1,83	19,79	36,19	24,93	23,15
6	18,71	1,81	18,73	33,91	23,48	21,83
7	15,07	1,72	15,09	25,95	18,49	17,31
8	15,82	1,74	15,84	27,48	19,48	18,22
9	14,31	1,72	14,34	24,62	17,56	16,44
10	13,00	1,67	13,06	21,79	15,79	14,84
11	14,00	1,67	14,05	23,39	16,97	15,96
12	13,72	1,66	13,77	22,88	16,62	15,63
13	14,06	1,66	14,14	23,48	17,06	16,05
14	13,63	1,67	13,71	22,90	16,59	15,59
15	12,47	1,65	12,58	20,79	15,15	14,26
16	13,23	1,65	13,34	22,02	16,06	15,11
17	14,67	1,67	14,76	24,64	17,85	16,78
18	15,60	1,67	15,67	26,13	18,94	17,81
Average	**16,11**	**1,73**	**16,16**	**28,22**	**19,93**	**18,62**

Figure 9. Measured and calculated values of illuminance on the road lighted by LED luminaires (5000K)

The results from the measurement of the brightness on the road networks are depicted in pictures 7 and 9. The first column represents the order of the measuring point on the grounds of the rough drawing in picture 3. Due to this, we are able to determine what is the difference between the measured and calculated figure of lighting at this point. The lighting figure is depicted in the next column (green column). This figure was measured by the SPIC-200 spectrophotometer by the sensor for lighting measurement. The blue column depicts S/P ratio. The S/P ratio is defined as the quotient of the radiant flaw which is evaluated by the photopic and the scotopic curve of sensitivity. The changes of the S/P ratio stated via changes of the spectral radiant of sources under various angles and eventual contributions by another lighting sources. The figures of lighting are depicted in the other columns. The figures were calculated to the end on the grounds of the measured spectrums of the measuring points. There are the figures of lighting evaluated via the photopic curve of sensitivity (day vision) on the grounds of the mathematical relation (1) are depicted. And there are as well the figures of lighting evaluated via the curve of scotopic sensitivity (night vision) on the grounds of the mathematical relation (2) and the figures of lighting which are evaluated via the calculated curves of the mesopic sensitivity on the grounds of the mathematical relation (3). There are depicted two figures of lighting for the mesopic area of vision due to the fact, as it was already mentioned, that adaptation radiance was at every measurement recorded in every direction from a driver's point of view. That is the reason why we must calculate lighting for the every mesopic curve separately in the final evaluation. To make the comparison simpler, there are always the average figures calculated in the lower line (orange line).

$$E_S = \sum(V_\gamma \cdot \theta_\gamma) \cdot 1700 \ [lx] \qquad (2)$$

Where:

- E_S – is a value of scotopic illuminance (lx)
- V'_γ – is scotopic sensitivity of the human eye
- ϕ_λ – is a radiated spectrum of the light source (W/m^2)
- 1700 – a light constant for scotopic vision, (lm/W) -it characterizes maximum amount of lumens that we are able to get from one watt.

$$E_M = \sum(V''_\gamma \cdot \theta_\gamma) \cdot X \ [lx] \qquad (3)$$

Where:

- E_M – is a value of mesopic illuminance (lx)
- V''_γ – is mesopic sensitivity of the human eye
- ϕ_λ – is radiated spectrum of the light source (W/m^2)
- X – a light constant for mesopic vision, (lm/W) - it characterizes maximum amount of lumens that we are able to get from one watt. The constant changes depending on the size of adaptation luminance of the background. It can pick up value from 683 lm/W to 1700 lm/W.

VI. Conclusion

It can be noted on the grounds of the comparison of the measured and calculated figures depicted in pictures 7 and 9 that the type of lighting source influences perception of lighting in the individual areas of vision. The high-pressure sodium discharge tubes have very low S/P ration in comparison with LED lamps (both colours). This is caused by the characteristics of the radiated spectrum (picture number 2) which dominantly radiates outside the area of the spectral sensitivity of the scotopic vision. This is the reason why the resulting scotopic lighting is significantly lower that the photopic lighting. The figures of LED in the mesopic areas of vision are bigger than in the photopic areas of the lighting. The situation is contrary in the case of high-pressure sodium discharge tubes. The mesopic vision is applied above all in the sphere of LED sources. It applies that the higher the substitutive temperature of the chromaticism or the lower the adaptation brightness of the background, the higher the increase of lighting in the mesopic area. The subsequent significant findings emerge from the measured differences between the photopic and mesopic evaluation of lighting.

- The increase of the sensitivity of the human eye appears in the area of the peripheral vision when LED is used with the higher temperature of chromaticism. This characteristic can manifest positively by the better perception of marginal areas around road networks, therefore by the better visibility of pedestrians who move (appear) outside a road network and they can enter a road network.
- The finding that the perception in the mesopic area can be significantly different from various types of perspective is highly important as well. If a driver comes to the area of high brightness the influence of the colour change decreases significantly in comparison to the situation where a driver comes to the area with low background brightness.

The paper points out to a possibility to verify figures of lighting on road networks in the mesopic area. It is essential to take into consideration also the other important facts which are related to the issue of the mesopic vision when designing lighting systems of public road networks.

Note 1: The above mentioned findings function only in the peripheral area, so it is not possible to lower normative demands for the lighting of road networks when LED are used.

Note 2: Significant increasing of the substitutive temperature of chromaticism at LED can lead to the negative perception of the colour of light by inhabitants. It is a psychological influence of the colour of light to the circadian rhythms of the human body. Therefore, it is recommended to use the maximal temperature o chromaticism ca. 4000K for installations of public lighting in inhabited areas.

Acknowledgment

This article was prepared with the support of the project "Remote Control of Public Lighting Luminaires via the Smart Technology Support" SP2017/97, by institution of VSB-TU Ostrava.

References

[1] J. L. Barbur, A. Stockman, "Photopic, mesopic and scotopic Vision and Changes in Visual Performace" [online] Internet: http://www.academia.edu/4491812/Photopic_Mesopic_and_Scotopic_Vision_and_Changes_in_Visual_Performance, [April 30, 2017].

[2] J. Habel, P. Žák, "Význam mezopického vidění pro prax," Internet: http://www.odbornecasopisy.cz/svetlo/casopis/tema/vyznam-mezopickeho-videni-pro-praxi--16105, [April 30, 2017]

[3] H. Hertenstein, M. Bach, N.J. Gross, F. Beisse, "Marked dissociation of photopic and mesopic contrast sensitivity even in normal observers," Graefe's Archive for Clinical ad Experimental Ophthalmology, 2016, vol. 254, pp. 373–384. ISSN 0721832X.

[4] J. Škoda, P. Baxant, M. Krbal, S. Sumec, T. Pavelka, "Photometry of LED sources," Przeglad Elektrotechniczny, 2013, vol. 89, no. 6, pp. 341–344. ISSN 00332097.

[5] J. Novotný, "Mobile Illuminance Measurement of Road Lighting," in Proceedings of the 21th International Conference LIGHT SVĚTLO 2015, pp. 129–131, ISBN 978-80-214-5244-2.

Assessment of Isolated Power System Frequency Defense Mechanisms

Vaclav Muzik, Zdenek Vostracky
Department of Electrical Power Engineering and Environmental Engineering
University of West Bohemia
Pilsen, Czech Republic
vmuzik@kee.zcu.cz, zdenekv@kee.zcu.cz

Abstract — **Under frequency load shedding (UFLS) is a mechanism that helps the TSO to preserve frequency as much close to 50 Hz during negative frequency changes greater than 1 Hz (resp. 1 - 1.9 Hz, 4 levels) by disconnection of 50% of the connected load in all.**

Unfortunately there are cases such as transition in to a stable island operation of a part of the electrical power system where the immediate disconnection of the connected load (via UFLS) during the primary regulation process might bring useless frequency collapse and black-start needs. This situation is presented on an example system where the generated electrical power in the area of electrical island is average but slightly lower than currently consumed electrical power in the moment of transition in to island operation. This state causes a transient process with frequency drop under the limit of first level of UFLS for few seconds resulting in the acceptable frequency range of 49 – 50 Hz.

Following paper describes generally situations that are threatened by the current settings of UFLS (given by the national grid code) based on real data from exemplary power system causing needless instability of the system and loss of operation primary thanks to sudden dangerous frequency jump over 50 Hz.

In the next part of the paper other approaches to isolated system frequency defense mechanisms are assessed (e.g. rate of change of frequency - RoCoF) in the view of safe transition in to the island operation. In the end examples of use in north parts of United Kingdom and Ireland are presented as possible approach to the question of frequency defense in states such as island operation of the part of the Czech power system. Influence of these mechanisms is modelled as well as the UFLS. (Abstract)

Keywords: under frequency load shedding, system stability, rate of change of frequency, island operation

I. Introduction (Heading 1)

With potential implementation of smart grid concept in the power grid the possibility of island operation of a part of the power system is a frequently discussed topic because of higher independence, self-sufficiency and also security. However the wide-public presented idea of small island operation in numbers of houses is quite different from the idea of isolated power grid with big working conventional power sources such as thermal power plants.

The ancillary service called "island operation" is being currently redefined on the European level because of specific needs of power systems such as Ireland or Scotland where island operation might be used as a tool to prevent the system from frequency collapses during lack of power or to prevent the system from spreading failure during surplus of active power caused for instance by strong winds generating power in wind farms (offshore windfarms – UK, North Germany, Netherlands).

In the countries without such wind farms the possibility of island operation is also discussed. For example the possible island operation in city of Pilsen, Czech Republic where the two installed power sources (in common 260.5 MW) are big enough to cover needs of the connected load through the whole year. Unfortunately current settings of the under frequency load shedding mechanism (UFLS) is not playing in the favor of reliable use of the ancillary service.

II. Under Frequency Load Shedding

A. *Currently used technology*

In the Czech Republic UFLS is defined in the grid code (revision 15, May 2015) as a tool preventing frequency collapse during frequency drops greater than -1 Hz via frequency relays installed in 110 kV and 22 kV substations (DSOs). In the UFLS is connected net 50% of the load. Triggering of the relays is divided in to four groups depending on the frequency drop. Following table describes levels of UFLS.

Table 1. UFLS levels in Czech Republic [1]

Level of UFLS	Level / Frequency			
	1./49	*2./48.7*	*3./48.4*	*4./48.1*
Amount of disconnected load	12	12	12	14

Disconnection is triggered by exceeding the frequency limit without any given delay (only respecting the amount of time to measure the frequency). The amount of connected load in to the UFLS is periodically controlled.

III. Island Operation

The need of island operation must be divided in to two categories – scheduled and sudden.

A. *Scheduled island opeation*

From previous studies and simulations is obvious that the best way to step in the island operation is a scheduled transition when the power sources have time to step in speed control regulation and increase the generation to approximately same amount as the current consumption. All this needs to be done under supervision of the DSO preparing the local network for this operation. In scheduled mode the availability of the power sources should be also solved in the view of primary sources – coal, oil, gas etc.

This scheduled mode might be used as a security precaution during near to collapse states – power overflows, weather conditions or cascade failure spreading. There is a strong need of communication between the DSO, TSO and the power plant operator. In this case the transition is possible in 10s of minutes. In this scenario the **transition is smooth** and the **risk of sudden frequency collapse** during the transition **is minimal.**

B. Sudden transition

Sudden transition might occur during frequency drops in the power grid caused by unexpected states such as short circuits, circuit breaker malfunctions, power generation drops etc. The difference in amount of active power in the moment of transition in to island operation determines the amount of frequency drop given by the following formula:

$$\frac{T_m \cdot S_{NG}}{f_n} \cdot \frac{df}{dt} = P_M - P_E = \Delta P \ [s; MVA; Hz; MW] \qquad (1)$$

Which describes the system via mechanical time constant T_m, nominal power of the generator set S_{NG}, system frequency f_n (50 Hz) and on the right side of the equation the difference between generated and consumed electrical power. For purpose of this paper, all generators are substituted by one equivalent power source represented by the nominal power (installed electrical power in the considered electrical island) and currently generated power P_M (mechanical power on the turbine). The current amount of connected electrical load is described by power P_E .

In the view of the power grid the difference between generated and consumed active power is maintained as much close to zero as possible in order to keep the frequency on 50 Hz. Unfortunately this is likely not true in the view of island operations because a part of the power grid (hypothetical islands) might be mostly importing or exporting throughout the year (parts of the grid with concentrated generation or consumption – capital city and so on). In case of major difference of active powers great frequency drops might happen during the transition and are representing a risk to generators, turbines and frequency dependent devices. In the scenario chosen for this paper the **transition is not smooth** and **represents a risk of frequency collapse of the system**.

IV. PROBLEM DESCRIPTION

The solved problem comes from the current definition of the UFLS which disconnects immediately the appropriate amount of load when the frequency limit is exceeded (see Table 1). Unfortunately this is a very unwanted intervention during the transition in to the island operation because of the possible stabilization of the frequency after fade out of the primary regulation state.

For the chosen system is picked city of Pilsen which represents the possibility of a small part of the power grid running islanded with approximately 200 thousand inhabitants. Following picture describes the model used for simulating the primary regulation state and the next table describes the chosen values for constants.

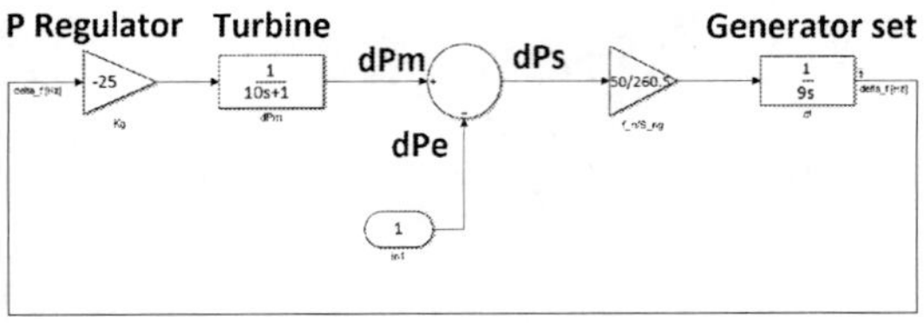

Figure 1. SIMULINK model of the descibred system

Table 2. Chosen values for the description of the system

Time constant T_M [s]	9
Time constant T_T [s]	10
Amplification of speed control K_G [MW/Hz]	25
Installed power S_{NG} [MW]	260.5

The influence of the self-regulation effect respecting the change of the consumed electrical of the frequency dependent devices is neglected for the simplicity of the described problem. The difference between active powers of consumption and generation is picked in each case individually.

Following figure shows an example of two transitions where one is made in a completely safe range of frequency (49-51 Hz) and the second one is exceeding the first level of UFLS but finally stabilizing above the level of 49 Hz after few seconds.

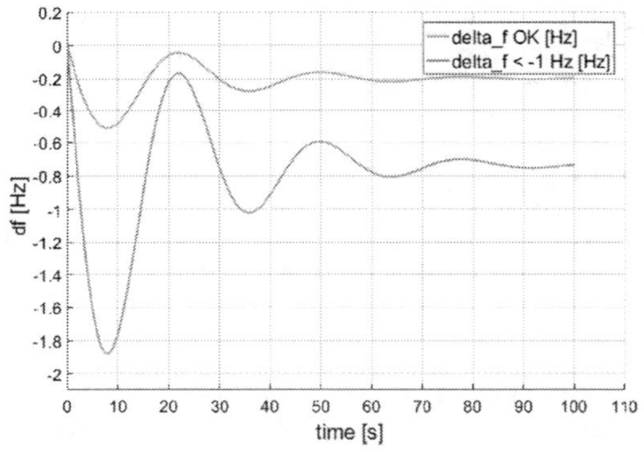

Figure 2. Primary regulation process

In the first case (green line, ΔP = 5MW) the frequency drops to 49.55 Hz and after approximately 50 seconds stabilizes on the level of 49.8 Hz. In the second case (blue line, ΔP = 18.4MW), the frequency drops to 48.18 Hz during the first 8 seconds and after approximately 80 seconds stabilizes on the level of 49.2 Hz. The control of the generation is neglected during the primary regulation process due to its small influence in the first deciding seconds. The trend of control is approximately 2 MW / min / block. Both of selected amount of ΔP are based on real service data.

In current settings of the UFLS there is a high probability after reaching 49 Hz level 12% of the load would be disconnected from the grid leading to certain discomfort of the consumer but mainly a frequency stability threat. The intervention of the UFLS is visible in the following figure.

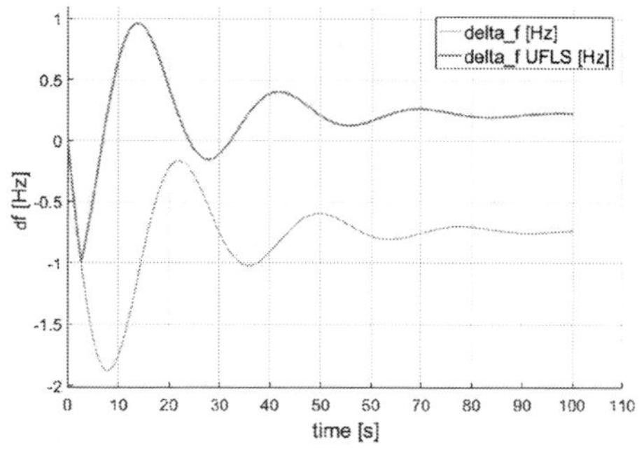

Figure 3. UFLS during primary regulation process

It is visible, that the frequency rapidly jumps above nominal frequency to approximately 51 Hz. It is clear that the current settings of the UFLS are very strict and definitely prevent the system from unstoppable frequency fall. But on the other hand during desired states such as island operation brings a problem during when generation is close to its installed maximum (260.5 MW) and is suddenly eased by 12% (up to 31.25 MW). Especially in situations when the active power difference is not a threat in the view of the steady state.

V. Rate of Change of Frequency

A possible approach to disconnection of the load during fast and rapid frequency changes might be installation of the Rate of Change of Frequency monitoring (RoCoF) which bring a certain delay to the decision of the shedding. RoCoF is well used in the United Kingdom for determining the disconnection of the wind farms from the rest of the grid. [2]

RoCoF based decision during load shedding is an available function in digital protection devices and it monitors the steepness of the change of the frequency $\frac{df}{ft}$ of the system. RoCoF condition is given in Hz/s and is different by the connected sources.

From the available reality based sources (UK and Ireland Grid Codes) the value of RoCoF is now set on the level of 0.5 Hz/s. To preserve the system stable, secure and more dynamic, the trend is to move the RoCoF condition to 1 Hz/s. This value seems to be unacceptable in global especially because of big conventional steam thermal power sources (above 150 MW per block). In case of wind energy, there is a trend to push the RoCoF condition to 2 Hz/s because of higher shock durability of the wind turbines. For example Denmark has a standard of 2.5 Hz/s RoCoF for newly connected units.

In the case of the island operation in the city of Pilsen with conventional power sources with less than 100 MW per block following conditions will be considered:

- RoCoF 0.2 Hz/s without delay
- RoCoF 0.5 Hz/s over 500ms
- RoCoF 1 Hz/s over 500ms

The RoCoF might be monitored many ways but for purpose of this paper following formula is used: [3]

$$\frac{df}{dt} = \frac{f_n - f_{n-3}}{T_{3cycles}} \quad [s; MVA; Hz; MW]$$

Where f_n is the currently measured frequency, f_{n-3} is the frequency measured 3 cycles earlier and $T_{3cycles}$ is the duration of the three measuring cycles. The measuring window is set to 60ms. [3]

VI. RoCoF Simulation

Following figure describes the upgraded simulation model in Simulink used for RoCoF based decision of the transition.

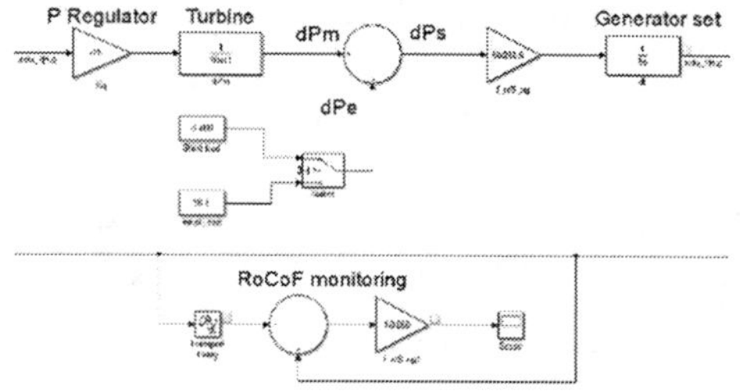

Figure 4. SIMULINK model of the descibred system with RoCoF monitoring

The following figure shows the RoCoF during the transition

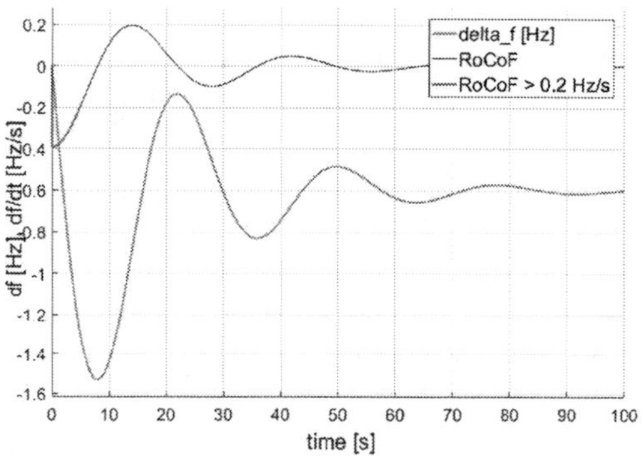

Figure 5. RoCoF during primary regulation process

It is visible that the RoCoF reaches only the first condition at 0.2 Hz/s and would be immediately disconnected. The standardly used condition of 0.5 Hz/s is not reached in this case.

Following figure shows the frequency changes when the RoCoF conditions 0.5 Hz/s is reached and also exceeded the 500ms time window.

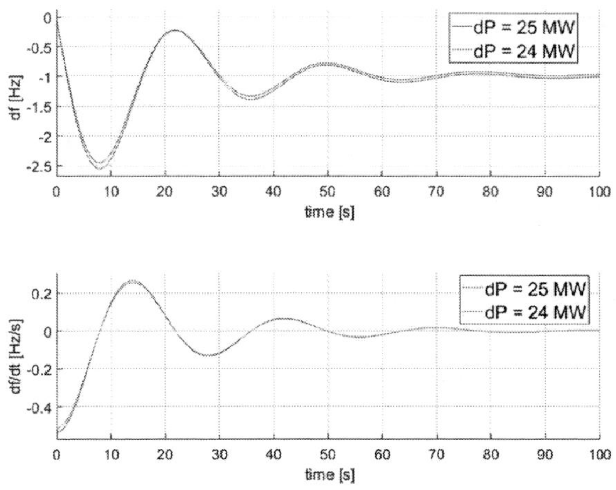

Figure 6. RoCoF reaching limits 0.5 Hz and 0.5 Hz for more than 500ms

Above displayed figure shows the state when the condition of RoCoF at 0.5 Hz/s is reached and the load needs to be disconnected at the active power difference equal to 25 MW. In the case of power difference equal to 24 MW is the RoCoF at 0.5 Hz/s reached but not for more than 500ms and the steady state is still not under the limit of 49 Hz.

VII. Conclusion

As seen in the previous chapters, use of RoCoF with UFLS might be a way to reach a stable transition during sudden island operations. Use of RoCoF extended the active power difference from 18.4 MW (Fig. 2) to approximately 24 MW (Fig. 6) respecting the limits given by current settings of UFLS (49 Hz).

The solution of this problem might be adding an island operation mode in to the frequency relays installed in 22 kV a 110 kV substations with a RoCoF monitoring to prevent from disconnecting the load during the first wave of the primary regulation process.

In conclusion this topic seems to be very important for the certification of island performance of a part of the power system. It is clear that during these situations the majority of the work is put on the DSO but on the other hand the island operation in a scale of a city (such as selected city of Pilsen) might be a very useful ancillary service for the TSO during emergency states of the grid and should be much more discussed and solved.

Acknowledgment

This paper was supported by the student research project SGS-2015-031

References

[1] PPA Energy, "Rate of Change of Frequency (ROCOF)-Review of TSO and Generator Submissions Final Report", PPA Energy, Guildford, Surrey, 2013.

[2] Commission for Energy Regulation, "Rate of Change of Frequency (RoCoF) Modification to the Grid Code", CER/14/081, Dublin, 2014.

[3] R. Bugdal, A. Dysko, G. M. Burt, J. R. McDonald, "Performance analysis of the ROCOF and Vector Shift methods using a dynamic protection modelling approach", in 15th International Conference on Power System Protection, 2006, pp. 139–144.

[4] A Dyśko, Dimitrios Tzelepis, Dr Campbell Booth, "Assessment of Risks Resulting from the Adjustment of ROCOF Based Loss of Mains Protection Settings", REF: ENA/LOM/TR/2015-001, University of Strathclyde, Glasgow, 2015.

[5] CEPS a.s., Rules for Transmission System Operation - Grid Code. Prague: CEPS a.s. (TSO), 2016, unpublished.

[6] J. Hajek, Prechodne jevy v elektrizacnich soustavach. Pilsen: Edicní stredisko VSSE. Czech Republic, 1983.

[7] P. Kundur, Power system stability and control. McGraw-Hill, Inc., 1993. ISBN: 978-0070359581.

[8] A. Chakrabart, S. Halder, Power System Analysis Operation and Control. PHI Learning Private Limited, 2012. ISBN: 978-8120340152.

[9] M. Young, M. Archer, "Appendix B: International Under-Frequency Events Resulting in System Splits" in System Operator Report: Automatic Under-Frequency Load Shedding Technical Report 2009, p. 27.

[10] V. Muzik, Z. Vostracky, M. Strelec and B. Rawn, "Control possibilities for island operation in city of Pilsen", in Proceedings of the 2016 17th International Scientific Conference on Electric Power Engineering (EPE). Prague: Czech Technical University in Prague, 2016. p. 47-52. ISBN: 978-1-5090-0907-7.

[11] K. Maslo, Rizeni a stabilita elektrizacni soustavy. Prague: Asociace energetickych manazeru, 2013. ISBN 978-80-260-4461-1.

Chalenges in Smart Grid

Michal Ivančák, Michal Kolcun, Martin Vojtek
Department of Electric Power Engineering
Technical University of Košice
Košice, Slovakia
michal.ivancak@tuke.sk, michal.kolcun@tuke.sk, martin.vojtek@tuke.sk

Abstract— **Smart Grid is a grid of tomorrow or the vision of future. It enables two-way communications and monitors & controls flow of energy, consumers demand, voltage, frequency and power factor on the regular basis. Other fact says that today's grids were designed to work in vertical structure from generation, transmission and distribution functions. On the other side, smart grids are going to operate in other way as horizontal system. In smart grid building is still very questions.**

Keywords: smart grid, renewable energy source, smart grid projects, smart grid building.

I. INTRODUCTION

Smart Grid is a grid of tomorrow or the vision of future. Smart grid enables two-way communications and monitors & controls flow of energy, consumers demand, voltage, frequency and power factor on the regular basis, for which smart meter plays an important role. It controls the use of different forms of energy that includes the renewable sources and energy efficient methodologies & technologies. It will also allow us to create more reliable, robust and secure electrical infrastructure. In this present environment, the smart grid technology is much needed due to the irregularities in the power demand and fuel cost. New threats and unanticipated conditions which are faced by large scale power delivery infrastructure in security, agility, and survivability is presented [1]. An important role of Distributed Energy Storage Systems (ESS) in combination with Power Electronics Devices (PED) is considered and will have a huge impact on future electrical supply systems and lead to many financial benefits. The challenges for a self-healing power grid in reliability, efficiency, security and resilience of energy infrastructure have been presented [3]. The transformation of Distribution systems has been shown that meet environmental targets, emphasis to support Plug-in Hybrid Electric Vehicles (PHEVs) as well as distributed generation and storage capabilities [4]. Smart grid functions that helps in interactions of agents like optimization to achieve adaptability, telecommunication, control self-healing, reliability and efficiency of power systems have been shown . An overview of the application of Wireless Sensor Networks (WSNs) for electric power systems with their merits and demerits have been presented . The various advantages of the smart grid have been presented and it is also discussed that how it will help in managing electricity demand growth, conserving energy, improving reliability and grid security, and also to reduce its carbon footprint. The concept and features of smart grid that involves the modernization of the present electric grid w.r.t generation, transmission and distribution in addition to the new challenges, and requirements have been introduced [8-9]. The critical issues on smart grid technologies primarily in terms of Information and Communication Technology (ICT) issues and opportunities have been discussed. The various smart grid concepts and different architectures implemented worldwide have been presented. The concepts of the smart grid, the role of computational intelligence in meeting their needs have been introduced. The communication requirements and the main technical challenges of the smart grid have been tackled. Some of the important issues and techniques for the power quality enhancement, its monitoring and analysis when deploying the smart grid have been introduced. For realizing the future-proof smart grid metering networks, security and privacy, some of the communications challenges have been presented [15]. The challenges and applications of communication technologies in smart grid have been discussed. The modeling and simulation of the Smart Grid in the Czech Republic have been presented and also elaborates the possible directions for the modeling of the smart grid and analysis research to reach the "quality by design" before the actual Smart Grid realization. An overview of the research on smart grid development has also been presented [18].

This paper presents a brief survey on smart grid with its pros and cons, changes and challenges in power sector. The paper has been divided into six sections. Section I introduces the role of smart grid in power system. Section I also sets the objective of the paper. Section II presents the key characteristics, its potential benefits and its role in environment. Section III presents the changes and different challenges in smart grid may be due to the use of latest technologies or evolutionary changes. Section IV presents some of the issues related with engineering economics i.e. cost analysis and its benefits. Section V presents some of the risks i.e. associated with smart grid and vision about smart grid. Finally, Section VI presents conclusions of the paper.

II. SMART GRID

Basically, the smart grid is the combination of information and communication technologies into the electric transmission and distribution networks as shown in Fig. 1.

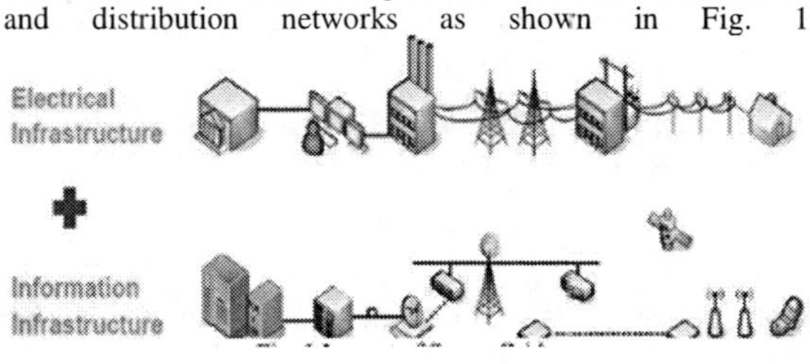

Figure 1. Smart grid

Smart grid uses two way communication technologies for delivering better electricity to the consumers. It is not only confined to the customers but also to the consumers. The various key drivers of the smart grid are:

- Increase in electricity demand
- Reliability
- Technologies advances
- Grid improvement
- Environment issues
- Efficiency

Smart grid gives the information regarding the consumption of energy in real time to all consumers. It helps the utilities in controlling the costs and improving their performances. To minimize the period of surpluses and outages, it enables the system so that customer can respond to the utility.

A. *Key characteristics of smart grid:*

For the conservation of energy and to satisfy the demand by the industrial and commercial consumers, smart grid has some very meaningful and important characteristics which are required by the utility. Fig. 2 shows the key features of the smart grid.

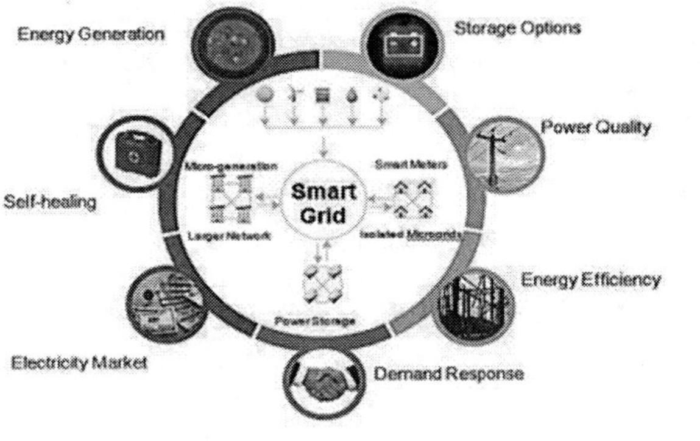

Figure 2. Key futures in Smart grid

As shown in Fig.2, the following functions are performed by smart grid:

1. *Incorporates & empowers the consumer:* it is the ability to incorporate the consumer equipment and behavior in grid designing and operation.
2. *Self-healing:* The grid itself automatically rapidly senses, detects, analyses, responds and then restores.
3. *Power quality needed by users:* The consistent good quality power to consumer and utility is provided by grid.
4. Use of latest energy storage devices like Super-conducting Magnetic Energy Storage (SMES), flywheels etc.
5. Fully enables and is supported by competitive electricity markets.
6. *Withstand attack:* The grid is resilient to physical/cyber-attacks.

The variety of resources are served by grid that includes demand response, photovoltaic, combined heat and power, wind and end-use efficiency.

B. *Potential benefits of the smart grid:*

Renewable energy resources in generation of power: Smart grid technology increases the use of renewable sources of energy like photovoltaic, wind, solar generation etc. and this contributes in the total power generation. These grids are known as micro grids

Optimization of the transmission and generation capacity: Smart grid technology also helps in reducing the losses in transmission system and increases the capacity of generation system.

Energy storage options: Smart grid technology uses the latest energy storage devices such as pumped-hydro storage, Super-conducting Magnetic Energy Storage, compressed air energy storage, super capacitors, ultra- capacitors, batteries, fly wheels that not only reduces the congestion, low utilization of power, volatility, security problems in power system but also helps in maintaining the stability and improves power quality as shown in Fig. 3.

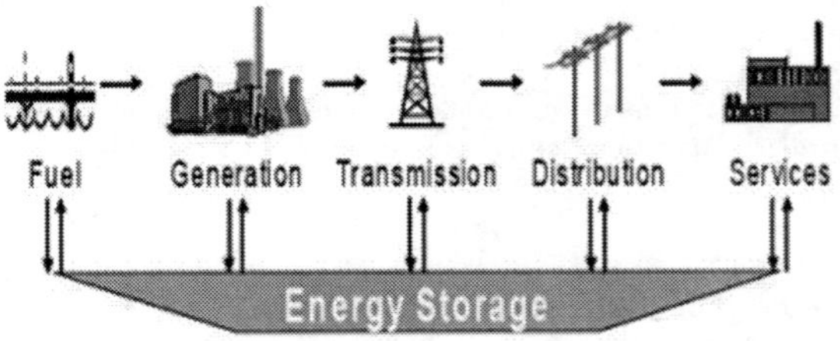

Figure 3. Cooperation in smart grid systems

Helps in improving a utilities power reliability, helps in management, operational performance, and the overall productivity.

By increasing the load and delivery efficiencies, reduction in carbon emission: By reducing the transmission congestion, blackouts and forced outages it is capable of increasing the efficiencies.

To manage the energy consumption, the consumers should be empowered and participate in several programs.

C. *Smart Grid and the Environment:*

Smart grid technologies are generally environmental friendly but besides the issues of primary energy supply, the major demerit to generate electricity with fossil fuels is that it emits various oxides and other pollutants when burnt. Today, the greatest economic and environmental challenges facing by humanity is greenhouse gases that contribute to the change in climate. Smart grids are also used to reduce air pollution and therefore playing an important role in global climatic change issue as maximum generation today results in emission of greenhouse gas.

Smart grids can bring about environmental improvements by:

- As from the latest survey, a smart grid could reduce transmission and distribution losses by 30%, so smart grid through better management of transmission and distribution network is more energy efficient.
- Manages peak load through demand response, thus avoiding spinning reserve.
- Saving of new infrastructure construction cost: It prevents us to build new power projects by managing the transmission and generation of the existing power supply.

Thus, conservation savings and energy efficiency will reduce our new infrastructure construction cost.

- To make consumers understand true cost of electricity by increasing transparency in electricity prices.

- Smart grid technology increases the contribution of renewable energy sources to reduce the CO2 emission. There has been fast development of co-generation and renewable energy, energy storage technologies, and increased interest in other distributed energy resources which is increasing day by day. As shown in fig. 4 the growth of renewable generation gets doubled in every three years for wind and solar power.

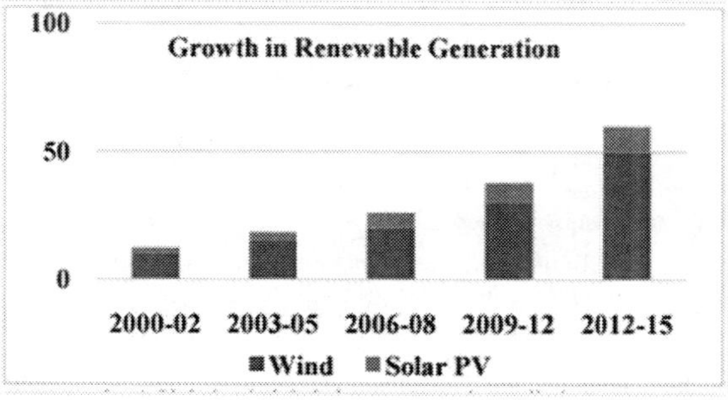

Figure 4. Global Wind & Solar PV Annual Installations

The oxides of carbon can be reduced by using different Smart Grid technologies by various methods. The use of energy and carbon impacts can be lowered and their effects can be quantified by using different mechanisms of smart grid.

These mechanisms are:

- The mechanisms in which both the residential and industrial sectors are incorporating smart grid enabled diagnostics and more electric vehicles are plugged in, is known as Direct Mechanism.

- The mechanisms in which smart grid capabilities are used to reduce the costs energy efficiency and renewable energy is known as Indirect mechanism.

III. Challenges and Changes in Smart Grid

A. *Changes due to Smart Grid:*

Smart grid brings a lot of challenges and changes in power system scenario in terms of new innovations and latest technologies as it includes the replacement of old existing technologies from the conventional grid and also the new storage devices. These vast changes in the power sector is very advantageous to the consumers as well as to the utilities.

a) Latest/New Technologies in Smart Grid:

Technology is one of the important constituents of Smart Grid which not only includes the hardware and the software but also the communication capabilities. It is still at very initial stage of development and are yet to be developed to a significant level although in some of the areas it is well developed.

Technologies which are needed to be implement in the Smart Grids are:

- Grid friendly Plug-in Hybrids
- Energy Storage
- Grid friendly loads
- Substation Automation
- Advanced distributed control
- Advanced Metering and Sensors
- Smart houses
- Communications
- Distribution Automation
- Demand Response
- Web Services and Grid Computing
- Advanced Conductors

Grid Friendly Appliances (GFA):

- GFA senses any change in the grid frequency and control the appliances to act as spinning reserve
- GFA is used only with switchable loads
- No communication is required

Energy Storage Devices:

- Energy storage has now became one of the important aspects of the smart grid and it allows the grid to:
- Shear off peak loads
- Damp system oscillations (transient stability support)
- Provide better integration of renewable systems
- Reduce the need for additional transmission assets
- Support voltage
- Increase the efficiency of existing power plant and transmission facilities
- Reduced energy cost
- Improve the reliability of electricity supply
- Reduce the investment required for new facilities

Advanced Metering and Sensors:

Smart meters plays an important role in implementing the smart grid technology, as it monitors the consumers demand, voltage, frequency and power factor on the regular basis. Generally, it involves the installation of an intelligent meter at residential customers which records these variables and sends them to Central Data Collection System through fast satellite communication link.

Goals:

- Load management & control capability
- Data matching system capacity to load requirements
- Tamper detection and theft identification
- Accurate load forecasting
- Immediate outage detection
- Enhanced distribution system optimization

Table 1. Comparing of metering systems

Metering System	Advance	Smart
Primary Function	Recording Interval	Recording Interval
Communication Capability	Network, two-way	Network, two-way into customer premise
Cost Range per Meter [excludes customer devices]	70€-150€	130€-250€
Home Area Network Gateway	Equipment piece or Separate system	Partially Integrated
Remotely Configurable Demand Limit Connect-Disconnect Service Switch	A separate piece of Equipment	Integrated
Support for Usage Displays	Remote Access Separate Service	Integrated Plus Separate Service
Rate Forms Supported	Flat, Tiered, Dynamic	Flat, Tiered, TOU, Dynamic
Support for Market Based Devices and Services	Open	"Gate Keeper" Potential
Data collection	Interval kWh	Interval kWh Customer device status
Obsolescence Ranking	Moderate from Low	Moderate to Uncertain

Metering system

b) Evolutionary changes in the network operations:

A better management of network flows is allowed by Smart grid technologies. Various network operation solutions includes:

- For a cross border interconnection capacity increasingly efficient allocation is done.
- By using FACTS devices, phase shifting transformers, HVDC devices etc., power flow is controlled.
- Real-time thermal monitoring exploitation for overhead lines and power cables.
- Including lower and high voltage level accurate analysis of the network.
- Coordination of several important operations across countries should be improved.
- Pre-contingency preventive actions should be activated after exceeding its pre-defined stability limits.
- Post-contingency corrective actions and defense schemes should be increased intelligently.

IV. Engineering Economic Issues

A. *Introduction:*

Smart grid enables two-way communication between the utility and the customers. The information is received to the electric power utilities from the consumer so as to provide better services and better access of data for making decision on the consumption of energy. Therefore, the economics of the project reflects profit to both the utility as well as to the consumers. Its benefit consists of:

- Less carbon emission
- Low operating and maintaining cost
- Low energy pricing
- Increased reliability and power quality
- Lower peak demand

B. *Cost benefits analysis:*

The cost-benefit analysis mainly has three main aims:

1. Across all smart grid a common cost-benefit methodology should be developed that can be applied.
2. An agreed upon methodology should be publish including all assumptions and different rules.
3. Changes and expansion can be easily accommodated in the methodology adopted.

C. *Cost analysis:*

According to the elements and the functions provided, the cost associated with smart grid will be categorized. The major cost items are:

- The design of the project and its study cost.
- Cost of a program and its management.
- Communication medium cost, installing and monitoring cost to enable two-way communication between the utility and the consumer.
- Installation of the smart meters cost.
- Costs associated with in-home devices and customer information systems
- Training and development cost of key staff.
- Number of transmission and distribution lines in the grid, and number of buildings to be covered will determine the size of the project.

D. *Benefits analysis:*

Few basic benefits of a smart grid are:

- Self-diagnosis and self-healing
- Reduced transmission congestion
- Reduced the outages and the blackouts
- Reduction in restoration time
- Shaving of the peak demand
- Increase the capacity of the system
- Reduce vulnerability
- Reduces problems related to power quality

E. *Economic Benefits:*

- Reduces the capacity cost
- Reduces the maintenance and operational cost
- From peak load reduction, cost saving should be there
- Commercial/industrial consumer cost should be reduced
- Security of the system should be improve

F. *Environmental related benefits:*

Smart grid reduces several emissions evolved from the oxides of carbon, nitrogen etc., at more economical way than many of the newest clean energy technologies.

The emissions can be reduced by the smart grid in the following four ways.

- By reducing the electrical losses
- By increasing the energy conservation
- Clean, renewable generation sources can be integrated and enabled.
- Distributed energy resources penetration can be increased.

G. *Calculating cost and benefits of the project:*

The approach for calculating the costs and benefits of the smart grid project has many steps:

- The elements that are deployed for the smart grids systems are to be identified.
- Assess each characteristics of the smart grid related to the project
- Identify the functions of each element of the project.
- Then, estimate the benefits
- Map each function onto a standardized set of benefit categories
- Define the baseline of the project and how it is estimated
- To calculate each type of benefit, identify and obtain the baseline and project data required.
- Then quantify, monetize and estimate the gains with the annualized costs
- Compare the cost benefits

V. Risks associated with smart grid

A. *Introduction:*

- *The Impact of Smart Grid projects complexity and scale–* Financial risk is created by the uncertainties that the actual benefits from the smart meter plan is less as compared to the stated projections.
- Depreciation of the technology–The continuous use of the same equipment for many decades in the electric industry that means once the equipment installed in any industry, it is expected to remain in service for its estimated useful life i.e. from 10 to 40 years.
- The effect of Systems Integration– The costs and benefits for implementation of partial or standalone smart grid is less attractive than the costs and benefits of a holistic program.
- Risk of stranded assets– This could involve equipment that was, at the time of installation, state-of-the art but before it reached to its useful life end it was eclipsed by newer technology costing much less.

B. *Challenges for Smart grid:*

- Lack of awareness: Consumers must be aware of the concepts of the smart grid, that what is the concept of smart grid? What are its benefits in terms of both economically as well as environmentally? Therefore:
  - Awareness among consumers should be made about their consumption of energy at home, offices etc.
  - The future prospects of Smart Grids must be clear.
  - The overall capabilities of Smart Grids should be focused firstly by the utilities then about the implementation and installation of smart meters.
- Knowledge & skills: Developing new skills in analytics, data management and decision support should be there before moving towards the smart grids and for this issue engineers and managers should be appointed and will need to be trained. For this, both time and money will be needed by the private as well as government organizations to support the various educational programs and this help in building managers and engineers for tomorrow. So, utilities have to think to bring such a change about how they can manage the transition in order to avoid over burdening of staff with change.
- Policy & regulation: Todays existing power network is good in the competition of generation and power supply but it does not promote the clean energy supply. So, by moving towards the smarter grid the policies must encourage incentives for investment and it should be match, to fulfill the needs of the consumers, utilities and suppliers to ensure that the societal goals are achieved at the lowest cost to the consumers.
- Cyber security & data privacy: Now a days the concerns on the invasion of security and privacy of the data has arises. Cyber-security and privacy of the data is one of the important issues. So, there should be proper management that can challenge the communication security and data management.
- Affordable capital: One of the major blockage in the implementation of Smart Grid is fund. So, the policy makers and regulators have to make such type of rules that can attract the private companies also. And the risk associated with the Smart Grid is more, but according to long run expectation the risk return profile will be more close to current scenario as new policy framework will be in place and optimal risk sharing across the chain will be there.

C. *Future vision about Smart Grid:*

The government around the world are planning to prepare the power sector for the introduction of the smart grid. Government organizations and the private sector companies have already started showing their interest towards smart grid technology and also, they are helping the market to understand the concept of the smart grid.

VI. Conclusions

Smart Grids are most comprehensive technology during recent years and it has been grown rapidly because of its benefits. It has many features and potential benefits in power sector. Its benefits related to grid operators, utilities environment has been explained. In this paper, different changes and challenges in smart grid due to the use of latest technologies or evolutionary changes, some of the issues related with engineering economics i.e. cost analysis, some of the risks i.e. associated with smart grid and India's vision about smart grid have been studied.

REFERENCES

[1] S.M. Amin, B.F. Wollenberg, "Toward a smart grid: power delivery for the 21st century," *IEEE Power and Energy Magazine,* Vol. 3, No. 5, pp. 34-41, October 2005.

[2] A. Mohd, D. Soest, E. Ortjohann, A. Schmelter, N. Hamsic, "Challenges in integrating distributed Energy storage systems into future smart grid," *IEEE International Symposium on Industrial Electronics*, Cambridge, pp. 1627-1632, July 2008.

[3] M. Amin, "Challenges in reliability, security, efficiency, and resilience of energy infrastructure: Toward smart self-healing electric power grid," *IEEE Power and Energy Society General Meeting - Conversion and Delivery of Electrical Energy*, Pittsburg, PA, pp. 1-5, July 2008.

[4] A. Ipakchi, F. Albuyeh, "Grid of the future," *IEEE Power and Energy Magazine,* Vol. 7, No. 2, pp. 52-62, April 2009.

[5] J. A. Momoh, "Smart grid design for efficient and flexible power networks operation and control," *IEEE Power Systems Conference and Exposition*, Seattle, WA, pp. 1-8, March 2009.

[6] V. C. Gungor, B. Lu, G. P. Hancke, "Opportunities and Challenges of Wireless Sensor Networks in Smart Grid," in *IEEE Transactions on Industrial Electronics*, vol. 57, no. 10, pp. 3557-3564, Oct. 2010. doi: 10.1109/TIE.2009.2039455.

[7] E. Santacana, G. Rackliffe, L. Tang, X. Feng, "Getting Smart," *IEEE Power and Energy Magazine,* Vol. 8, No. 2, pp. 41- 48, April 2010.

[8] H. Gharavi, R. Ghafurian, "Smart Grid: The Electric Energy System of the Future," in *Proceedings of the IEEE*, vol. 99, no. 6, pp. 917-921, June 2011. doi: 10.1109/JPROC.2011.2124210.

[9] K.M. Valsamma, "Smart Grid as a Desideratum in the Energy landscape: Keyaspects and Challenges," *IEEE International Conference on Engineering Education: Innovative Practices and Future Trends,* Kottayam, India, pp. 1-6, July 2011.

[10] V. C. Gungor, D. Sahin, T. Kocak, S. Ergut, C. Buccella, C. Cecati, G. P. Hancke, "Smart Grid Technologies: Communication Technologies and Standards," *IEEE Transactions on Industrial Informatics*, Vol. 7, No. 4, pp. 529-539, November 2011. doi: 10.1109/TII.2011.2166794

[11]M. Hashmi, S. Hanninen, K. Maki, "Survey of smart grid concepts, architectures, and technological demonstrations worldwide," *IEEE PES Conference on Innovative Smart Grid Technologies*, Medellin, pp. 1-7, October 2011.

[12]P.J. Werbos, "Computational Intelligence for the Smart Grid-History, Challenges, and Opportunities," *IEEE Computational Intelligence Magazine,* Vol. 6, No. 3, pp. 14-21, August 2011.

[13]F. Bouhafs, M. Mackay, M. Merabti, "Links to the Future: Communication Requirements and Challenges in the Smart Grid," *IEEE Power and Energy Magazine,* Vol. 10, No. 1, pp. 24-32, February 2012.

[14] G. W. Chang, S. Ranade, N. R. Watson, M. H. J. Bollen, "Monitoring issues and analysis techniques-smart grid aspect of power quality," *2012 IEEE 15th International Conference on Harmonics and Quality of Power*, Hong Kong, 2012, pp. 673-677. doi: 10.1109/ICHQP.2012.6381182

[15] Z. Fan, P. Kulkarni, S. Gormus, C. Efthymiou, G. Kalogridis, M. Sooriyabandara, Z. Zhu, S. Lambotharan, W. H. Chin, 'Smart Grid Communications: Overview of Research Challenges, Solutions, and Standardization Activities', *IEEE Communications Surveys Tutorials*, vol. 15, no. 1, pp. 21–38, 2013.

[16] R. Ma, H. H. Chen, Y. R. Huang, W. Meng, "Smart Grid Communication: Its Challenges and Opportunities," in *IEEE Transactions on Smart Grid*, vol. 4, no. 1, pp. 36-46, March 2013. doi: 10.1109/TSG.2012.2225851

[17] D. Gešvindr, B. Buhnova, J. Rosecky, "Overview of research challenges towards Smart Grid quality by design," *2014 Federated Conference on Computer Science and Information Systems*, Warsaw, 2014, pp. 1497-1504. doi: 10.15439/2014F318

[18] S. Bera, S. Misra and J. J. P. C. Rodrigues, "Cloud Computing Applications for Smart Grid: A Survey," in *IEEE Transactions on Parallel and Distributed Systems*, vol. 26, no. 5, pp. 1477-1494, May 1 2015. doi: 10.1109/TPDS.2014.2321378.

[19]M. Kornatka, "Analysis of unreliability of low voltage power networks some branches of Distribution System Operators" *Przegląd Elektrotechniczny* Volume 88 Issue 10B pp. 303-306, 2012.

[20]M. Kornatka, „Analysis of unreliability of medium voltage power networks chosen branches of distribution companies" *Rynek Energii Issue3* pp. 99-102, 2011.

Investigation of Photovoltaic Panel Losses using Computer Simulations

Andrea Varga
Doctoral School of Applied Informatics
Obuda University
Budapest, Hungary
varga.andrea@kvk.uni-obuda.hu

Peter Kadar
Power Systems Department
Kando Kalman Faculty of Electrical Engineering
Obuda University
Budapest, Hungary
kadar.peter@kvk.uni-obuda.hu

Ervin Racz
Power Systems Department
Kando Kalman Faculty of Electrical Engineering
Obuda University
Budapest, Hungary
racz.ervin@kvk.uni-obuda.hu

Abstract — **Nowadays, when we talk about energy generation, we must count with green energy as well. Photovoltaic panels play important role in this sector. Solar panels can be investigated in given time periods at the place where they are installed. Besides, pre-calculations or pre-designing of the solar systems have more and more important role. With the simulation of the photovoltaic panels a flexible environment can be created, in which any types of solar panels can be tested. This way not only the magnitude of the averaged energy generation can be given, but any effects of losses in the given system can also be tested. In the publication, a method how to find the losses of a real, working solar panel is introduced. Parts of the method are an experiment and a simulation technique connected to each other. Investigation of any effects of the losses in the use of PV panels is one of the most important points in the thesis of our Ph.D. student.**

Keywords: efficiency loss; solar panel; simulation; experiment; electric current-voltage characteristics

I. Introduction

Sun as energy source has one of the greatest potential in between the known renewable energy sources. This statement can be perceived by the following data: during 1.5 days to the Earth irradiated energy originating from the Sun is the same amount of energy which can be produced using all the available oils can be mined from the Earth [1]. Transforming radiation originating from the Sun to electric energy can go with direct and indirect ways. Photovoltaic cells, panels, modules or in other words photovoltaic devices can convert energy from the Sun to electric energy directly.

According to the analyzation of PV Market Alliance (PVMA) world energy/power market was 75 GW electric power in the year of 2016. This value was 50 GW electric power in the year 2015. Increase is 50% from the year of 2015 to 2016. (Other analysts estimated the value as 55-59 GW power.) Seeing these values it can be concluded that there was a very significant or serious increase in electric energy generation worldwide [2], [3].

Research emphasizes design of new generation of photovoltaic cells, which have better and better electric efficiency. Oher deal is to build solar panel parks with possible best electric efficiency. At design of these kinds of high efficiency solar panel parks many parameters must be taken into account, such as geographical location, tilting angle (angle of the incoming sunshine), clean connections, etc. In Figure 1. two different types of photovoltaic systems can be seen. Left side photo in Figure 1. shows a solar tracker system with six different types of solar panels, and right side photo introduces a household-size solar system modules. Both of them are installed on the roof of the Obuda University buildings.

Figure 1. Photovoltaic systems of Alternative Energy Sources Center at the roof top of Obuda University buildings. [photos made by the Authors]

In ideal case, we do not need to count with any electric losses, but in the real life, during the real operation of the solar panels the photovoltaic modules work with numerous electric losses i.e. metallization of the contact points and crystal defects. During ideal modeling of the photovoltaic panels, the above mentioned losses use to be neglected. But, in the case if we want to design a PV system with the possible best efficiency, the neglect mentioned above cannot be done anymore. In other words the losses must be taken into account.

In this paper we would like to show a method how to find the losses of a real, working solar panel. In order to describe the method an experiment made on a real PV panel, the results came out from the measurements, and a simulation technique connected to the experimental results will be presented. For this research work a solar panel with its nominal electric power of 20 W has been used. This solar panel can be purchased on the market. Type of it is MWG-20 and its most important data can be collected and read in Table 1.

Table 1. Data for MWG-20 solar panel

Parameters of the solar panel of MWG-20		
name of the parameter	***value***	***dimension***
Power	20	watts
Crystal structure	poly	-
Number of cells	36	pieces
Power tolerance	-0/+3	watts
Voltage at maximum power point	17.2	volts
Electric current at maximum power point	1.17	amps
Open circuit voltage	21.8	volts
Shortcut current	1.23	amps
Maximum system voltage	1000	volts
Size (h×w×d)	505×353×28	mm×mm×mm
Mass	2.3	kg
Warranty	5	years
Power warranty	25 (80%)	years

II. Electric Replacement Circuit and Equation of a Solar Cell

Electric replacement modeling circuit of a photovoltaic cell is shown in Figure 2. Using the replacement electric circuit of the solar cell, theoretical behavior of the photovoltaic device can be simulated.

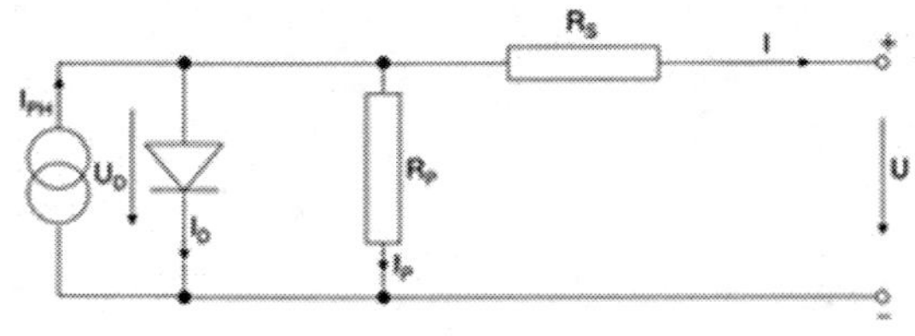

Figure 2. Electric replacement modeling circuit of a photovoltaic cell with usual notations. I_{PH}: photocurrent; I_D and U_D: electric current flows and voltage fells on the diode (dark current and dark voltage); R_P and I_P: parallel resistance and electric current flows on it; R_S shunt resistance; I: main current flows; U: output voltage. [drawing made by the Authors]

Source electric current of the current generator is proportional to the incoming photo power density (i.e. photo radiation intensity). Diode installed in parallel switching to the current generator unit is responsible for the simulation of the effects run at P-N junction. Serial and parallel resistances (R_S and R_P) hold any losses belong to the photovoltaic cell. Serial resistance (R_S) is the net ohmic resistance of the semiconductor domains, where if the current flowing through, then the voltage is decreasing. Using this serial resistance, transient resistances of the contacts (i.e. metallization of the contact points) can be considered. Electric current flows through the serial resistance causes losses in the electric power, and it warms up the device. Using the parallel resistance (R_P) the losses can be considered which decreases the flowing main current (current which can "take off" from the device). A part of the electron-hole pairs generated by the incoming radiation (light) can recombine inside the photovoltaic device. The PV cell can hold semiconductor domains, where there is no any P-N junction or diffusion potential is smaller. In case of some known technology used the density of the crystal defects can be found high. Electric currents generated at these places are leaking currents, and it causes losses in the PV system. Electric power generated on the parallel resistance will be transformed to heat.

Electric characteristics of a photovoltaic panel can be described analytically. Equation describes the characteristics of the solar sell is given here [4]:

$$I = I_{PH} - I_D\left(e^{\frac{U+I\cdot R_S}{U_T}} - 1\right) - \frac{U+I\cdot R_S}{R_P} \quad (1)$$

Where:

I: consumer current

I_{PH}: electric current can be generated by the incoming light photons.

I_D: dark current; the current flows through the P-N junction because of the U_D voltage (dark voltage).

U_T: thermic voltage

R_S, R_P: concentrated values of serial and parallel resistances, respectively.

Using equation (1) analytical simulations can be ran to describe the operation or the losses of a photovoltaic cell. In the next chapter a different simulations technique (non-analytical) will be introduced and described.

III. Effect of Losses on the Characteristics of the Photovoltaic Cells

Every photovoltaic cell has electric current-voltage characteristic plot, which must be measured by the manufacturer after manufacturing the solar cell module using International Electrotechnical Commission (IEC) standards. According to the given standard, intensity of the light irradiation, temperature of the photovoltaic panel, temperature of the environment neighboring the photovoltaic panel, incidence angle of the incoming radiation must be taken into account. The standard defines the light source itself, the reference device (a photovoltaic panel) and the whole process to measure the electric current-voltage characteristics.

In the case if there is no loss in the cell or there is loss in the cell the following plots (Figure 3.) can be given. Curve for an ideal solar cell (blue) and two curves for non-ideal solar cell (red, orange) can be compared to each other. Curve drawn by red ink belongs to smaller losses and curve painted by orange ink belongs to higher losses of the PV same panel in Figure 3. As Figure 3. shows it clearly, than curvature and slope of the non-ideal cell is very different than the same parameters of the ideal cell. If the serial resistance is nonzero ($R_S \neq 0$), then this loss does not give considerable influence on the open circuit voltage, because in this case the electric current does not flow through the serial resistance, but the electric current flows through the parallel resistance side.

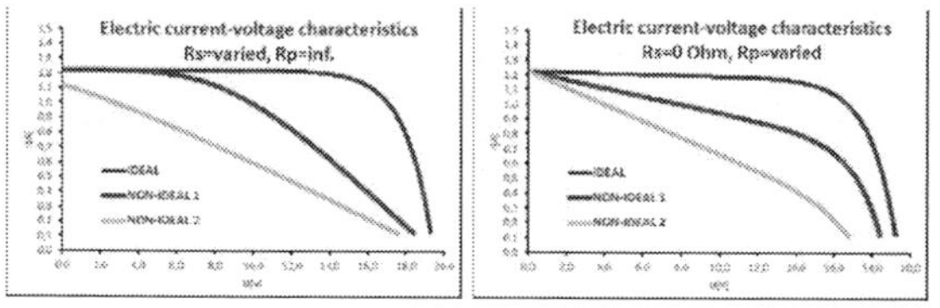

(3.a.) (3.b.)

Figure 3. Electric current–voltage characteristics for an ideal solar cell (blue curve) and a non-ideal solar cell (red and orange curve). Red curve belongs to smaller losses and orange curve belongs to stronger losses for the same PV cell. Figure 3.a. shows the case if R_S=varied and R_P=infinite. Figure 3.b. shows the case if R_S=0 Ohm and R_P=varied for the same photovoltaic cell.

In the case when the ohmic value of the parallel resistance is changing, the value of the flowing electric current is decreasing, because a deterministic part of the generated electric current is flowing through the parallel resistance. In case of very weak light irradiation this effect has strong influence on the generated power; in other words, the generated electric power is strongly decreasing in this case.

IV. MatLab Simulation

MatLab interactive software package and environment is very widespread on scientific and engineering fields. MatLab package and environment can be used for calculations, solving different types of equations and for modelling purposes as well. The basic software package is complemented with special tool boxes for different special purposes. Such toolboxes are for example the Control System Toolbox (CST) or the Simulink Toolbox. MatLab can run series of commands stored up in different data substances or files. CST is a collection of mathematical functions, which can be used for analyzation of finite dimension linear and time-variant systems. Simulink is a block-oriented language, it supports system modeling. Using these toolboxes hierarchical implementation of complex systems or simulation of system components can be made. In order to make our simulations done, the Simulink environment will be used by us. [5]

Using the tools build in the Simulink, the photovoltaic cell has been created. In other words, modeling circuit of the simulated photovoltaic panel has been programmed in Simulink environment. As first step of the programming work the block program using the pre-defined icons has been compiled. The collected block program with the necessary components is shown in Figure 6. Contacts of the photovoltaic cell has been closed by a variable resistance (see in Figure 6.).

As second step, communication and set-up window of the solar cell module in the programing environment has been opened. In order to open this window, double mouse click on Solar Cell icon in the block program has to be made. Figure 7. shows the communication and set-up window of the solar cell module in the programing environment. Choosing "by equivalent circuit parameters, 8 parameters" item, it opens the possibility to modify and set the values of the series and parallel resistance during the simulations. Parallel to set the values of the serial and parallel resistances other parameters was set according to the Standard Test Conditions (STC) such as 1000 W/m^2 irradiance, 25 °C as measurement temperature and device simulation temperature. After giving these parameters and values, the next step is to position sensors in the block program. In our case Current Sensor and Voltage Sensor are inserted in. Figure 8 displays the block program with the inserted sensors and their outputs in scope mode. Using this option, current-time, voltage-time functions and with the multiplication of the signals originated from the current and voltage sensors, the power-time function can also be plotted.

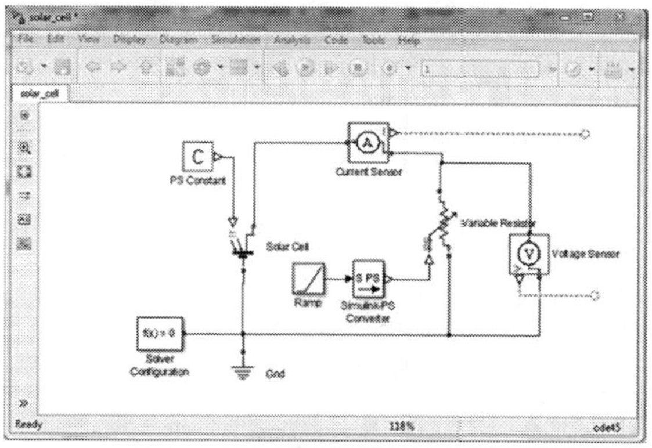

Figure 4. shows the programming window in the Simulink environment with the used components in the block program.

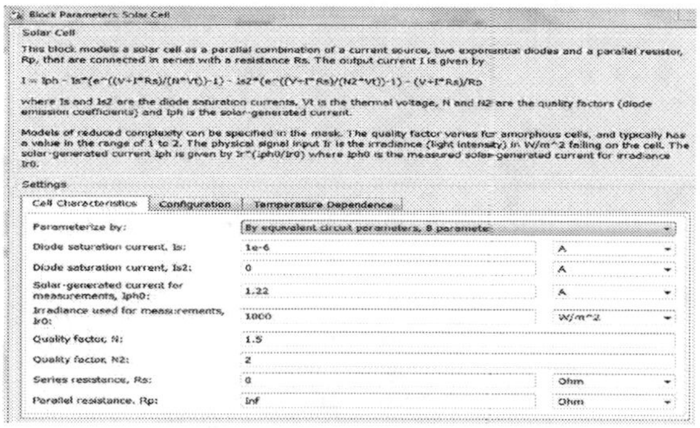

Figure 5. Communication and set-up window of Solar-Cell module. Using the screen the user can give the most important parameters of the solar cell. After giving the parameters run of the program can be started.

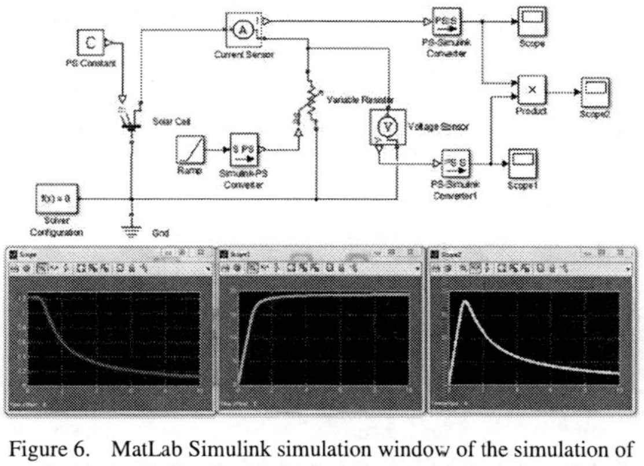

Figure 6. MatLab Simulink simulation window of the simulation of the photovoltaic cell with 4 subwindows: with the programing window with icons, with three other output windows Scope (electric current-time function), Scope1 (electric voltage-time function) and Scope2 (electric generated power-time function) recpectively.

V. EXPERIMENT

The small photovoltaic panel, which data and parameters are given in Table 1. is available in our laboratory. This panel was used for testing our method working. Nominal electric power of this photovoltaic panel is given as 20 W, and other parameters of it is listed in Table 1.. In this chapter the experiment with the given solar panel is described and introduced. Using and comparing the experimental data and measurement plots with the simulation results, the newly developed technique for losses finding for working solar panels can be presented.

As experiments, flash light irradiation of the investigated panel was designed. Using a Pasan-Flasher professional irradiation source with the accuracy class AAA+ different irradiation light intensities could be set starting from 1000 W/m^2 to 200 W/m^2. Marking AAA considers: homogeneity of the solar irradiation on the test field; homogeneity of the solar irradiation in the measuring time; conformity of the spectrum of the irradiated light with the AM 1.5 (air mass of 1.5) spectrum of the sunlight. The solar panel has been irradiated by these kinds of lights and during the irradiation electric current-voltage characteristics and generated electric power-voltage curves with the use of a very fast measuring electric chain unit have been measured. It is very important to note that, the experiment was made with 1000 W/m^2, 900 W/m^2, 800 W/m^2, 700 W/m^2, 400 W/m^2 and 200 W/m^2 light irradiation. The experimental arrangement with its most important parts is presented by Figure 7.

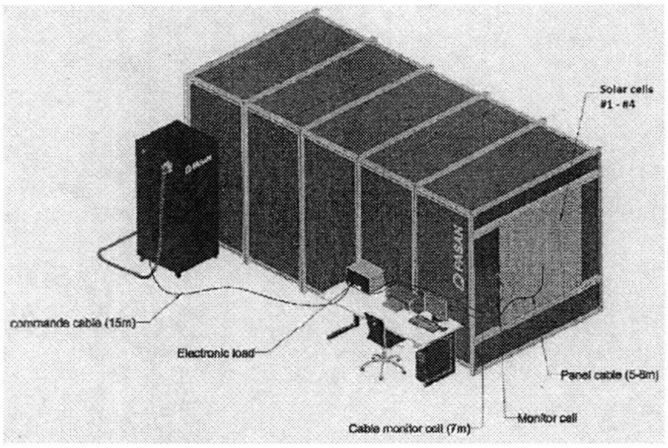

Figure 7. Schematics of the Pasan-Flasher system with its most important parts. Our solar panel was set up to the position "Solar cells #1-#4. as it is shown in the Figure."

In Figure 8. the measured electric current-voltage characteristics at different irradiations (detailed above) for the test solar panel is shown. Curves with different colors belong to different irradiations (see detailed color scheme in the Figure itself). Masurement plots show very nice electric current-voltage plots. It is clearly seen, that higher intensity irradiation (i.e. 1000 W/m^2 drawn by red ink) shows higher flowing electric currents and voltages, than lower intensity irradiation (such as for example light blue curve with 200 W/m^2 irradiation intensity). Complying with the standard test conditions (STC), the plot belongs to 1000 W/m^2 irradiation intensity will be investigated later. This plot will be compared to the MatLab simulation and use later on to introduce our losses finding technique. [6]

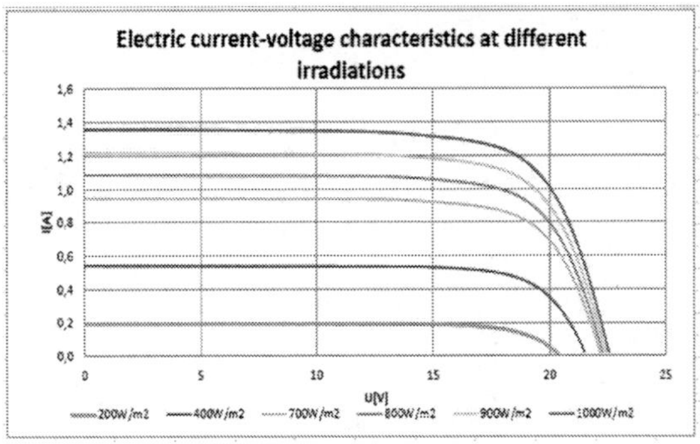

Figure 8. Experimental results for electric current-voltage characteristics. Color plots describe the characteristics for 1000 W/m^2, 900 W/m^2, 800 W/m^2, 700 W/m^2, 400 W/m^2 and 200 W/m^2 light irradiation, respectively.

Figure 9. shows the measured electric power-electric voltage characteristics at different light irradiations for our given solar panel. The exact light irradiation (see above) are also listed in the figure. Measurement plots in Figure 9 introduces very nicely running electric power-voltage curves for photovoltaics. It is clear, that higher generated power could be produced with irradiation of the highest light intensity (1000 W/m^2, red curve). In our case the maximum power value is approximately ~22.5 W, which is higher then the nominal value. Reason of the difference is the different shortcut current and open circuit voltage measured during the real experimental run.

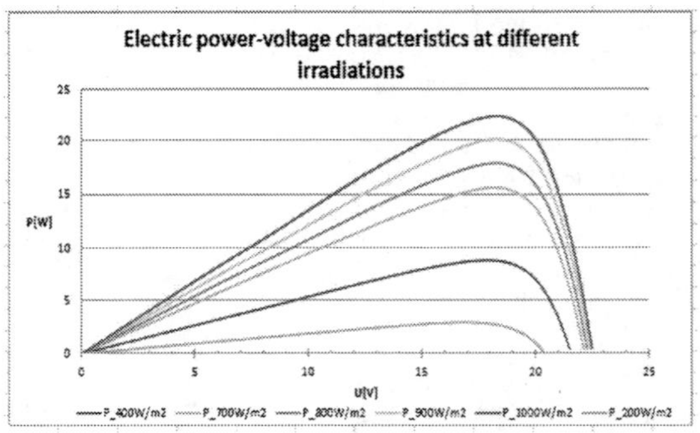

Figure 9. Experimental results for generating electric power-voltage plots. Color curves describe the plots for 1000 W/m^2, 900 W/m^2, 800 W/m^2, 700 W/m^2, 400 W/m^2 and 200 W/m^2 light irradiation, respectively.

VI. DETERMINATION OF THE LOSSES OF THE REAL SOLAR PANEL USING SIMULATIONS COMPARED TO EXPERIMENTAL RESULTS

In order to determine the losses of the solar panel the following technique has been developed. In the following part, the newly developed technique will be step by step introduced.

a) *Design of the ideal model*: Using and set the nominal data given by the data sheet of the test solar panel, or in other words, equivalently saying, the nominal data of our photovoltaic modul listed in

Table 1. and keeping the block parameters defined by the MatLab Simulink environment the introduced MatLab simulation has been programmed and running. It means, that the most important parameters were set as: number of solar cells in the whole panel is 36, diode saturation current is 10^{-6} Amps, irradiation is 1000 W/m^2, solar generated current for measurements is 1.23 Amps. After running the simulation the green curve shown by Figure 10 could be get. Figure 10 shows two plots: yellow plot is the measurement curve came out from Pasan-Flasher experiment using 1000 W/m^2 irradiation, and green plot is the curve came out after running the simulation.

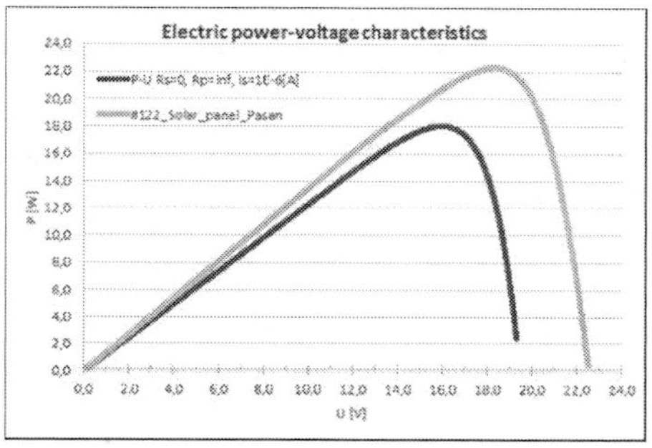

Figure 10. Plots came from simulation and experiments. Yellow plot is the curve came out from the Pasan-Flasher experiment at 1000 W/m^2 irradiation intensity. Green plot is the curve came out from the simulation run step a). Step a) means the design of the ideal model.

b) *Iteration of the ideal model*: In the case of step a) the open circuit voltage came out from the simulation was $U_{OC\text{-}sim}$=19.3 V. This value does not fit to the nominally given data of the panel, which was $U_{OC\text{-}nom}$=21.8 V. In order to set the value of $U_{OC\text{-}sim}$ equal to the $U_{OC\text{-}nom}$, the value of the saturation current at the block parameters of the programming window has been modified. After this iteration step, saturation current has been get as $I_S=1.8\cdot10^{-7}$ Amps.

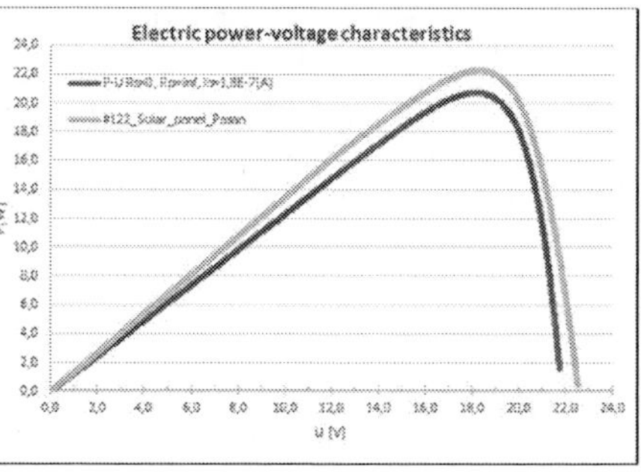

Figure 11. Plots came from simulation and experiments. Yellow plot is the curve came out from the Pasan-Flasher experiment at 1000 W/m^2 irradiation intensity. Green plot is the curve came out from the simulation run step b). Step b) means the iteration of the ideal model.

Comparing green plot to yellow plot in Figure 11. the following conclusion can be written: Actual solar panel "works better" (during the experiments) that it could be presumed based on nominal parameters (parameters listed in Table 1.). "Works better" means that the panel could give higher generated electric power from the measurements (P_{meas}=22.17 W) , than it was presumed from the nominal data. The simulated generating electric power (P_{sim_1}=20.75 W). This difference implies the modification of the short circuit current and open circuit voltage.

c) *Set up of the short circuit current and open circuit voltage to the actual value in the simulation*: Figure 12. introduces the difference between the experimental plot and the step b) simulated plot focusing on the short circuit current and open circuit voltage values. In step c) the simulation parameters (see dark blue values in Figure 12.) will be set to the measured values (see red values in Figure 12.).

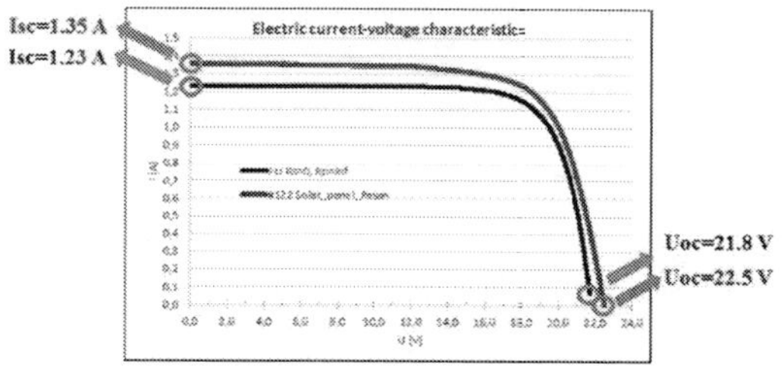

Figure 12. Plots came from simulation and experiments. Red plot is the electric current-voltage curve came out from the Pasan-Flasher experiment at 1000 W/m^2 irradiation intensity. Dark blue plot is the electric current-voltage curve came out from the simulation run step b). Step b) means the iteration of the ideal model.

d) *Simulation of the solar panel using the values measured by Pasan-Flasher system*: In this step the simulation has been running based on Pasan-Flasher measurement. This means that, data came from the experiments were used as initial data for the block parameters of the simulation. It is important to note here, that this step mentioned can be made after numerous iteration steps only. The result is shown in Figure 13. After this run the generating electric power came from the simulation became: P_{sim_2}=23.64 W. This value is higher than the measured, real number of the generated electric power.

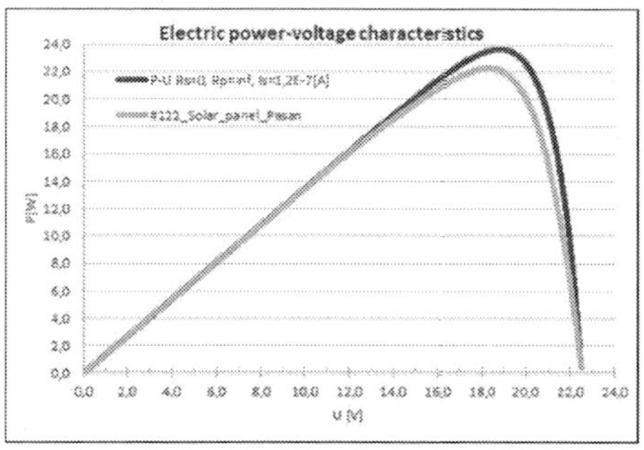

Figure 13. Plots came from simulation and experiments. Yellow plot is the curve came out from the Pasan-Flasher experiment at 1000 W/m^2 irradiation intensity. Green plot is the curve came out from the simulation run step d). Step

d) means the simulation of the solar panel using the values measured by Pasan-Flasher system.

e) *Giving electric replacement modeling circuit of a photovoltaic cell*: Result or in better word saying plots came from step d) does not fit to each other. The final goel is to fit the simulation plot to the measurement plot. Using the effects from the losses to the influence on electric power-voltage charecteristics, value of R_S (serial resistance) and value of R_P (parallel resistance) could be defined. In other words, this way the serial and parallel losses of the given solar panel could be described by concentrated electrical parameters. With the adjusted parameters measured and simulated plots in good fit could be received, which presented by Figure 14.

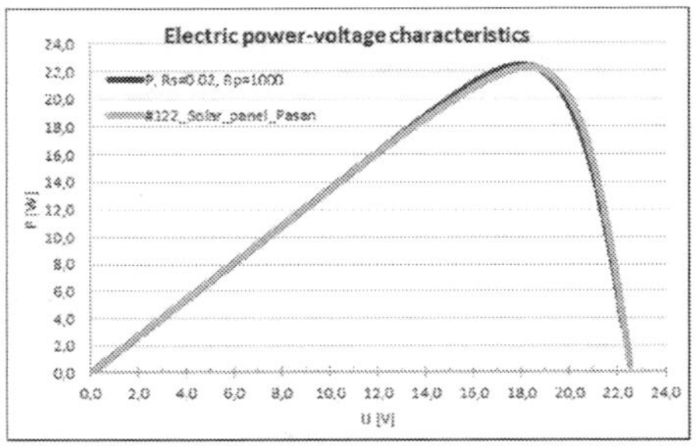

Figure 14. Plots came from simulation and experiments. Yellow plot is the curve came out from the Pasan-Flasher experiment at 1000 W/m^2 irradiation intensity. Green plot is the curve came out from the simulation run step e). Step e) means giving electric replacement modeling circuit of a photovoltaic cell.

Optimzed parameter values for the given solar cell are: serial resistance is R_S=0.02 Ohms; parallel resistance is: R_P=1000 Ohms; saturated current is: $I_S=1.2\cdot10^{-7}$ Amps. With the optimized parameter values the plots are fitting themselves, so the maximal power point values are identical, $P_{sim\_3}=P_{meas}$=22.17W.

VII. Conclusions and Summary

Goal of our scientific work was to determine serial and parallel resistances of a working, real solar panel in order to investigate losses of the panel. In order to fulfil the aim, a new technique has been developed. The technique is based on an experimental run and serial iterations of simulations. Electric current-voltage and electric power-voltage characteristics of the solar panel was recorded from an experiment using of a Pasan-Flasher irradiator system. Simulations made in MatLab environment based on the electric replacement modeling circuit of a photovoltaic panel. The main aim of the newly developed technique is to get the measured and simulated plots of electric power-voltage characteristics in total fit with numerous iteration steps mentioned before. After setting the fit, the losses or in other words the serial and parallel resistances of the photovoltaic panel can be read out from the simulations.

This technique can be used to develop a solar park with the maximum efficiency possible using given solar panels. With this method we are able to determine the panels with same losses. As it is well known, solar panels with same short circuit currents can be serial-connected to each other, but panels with same open circuit voltages can be parallel-connected to each other. Using this technique, optimal generating electric power can be get out from the system.

Acknowledgment

Authors are gratefully say thank you for the support of the Doctoral School of Applied Informatics and the Alternative Energy Sources Knowledge Center at the Obuda University. Authors say great thank you for Mr. László Nagy institute engineer at the Power System Department at the Obuda University, who helped our work with his great and useful remarks.

References

[1] 'Napenergia'. [Online]. Available: http://www.napenergia.hu/napelem.html. [Accessed: 10-Jun-2017].

[2] 'The PV Market Alliance'. [Online]. Available: http://www.pvmarketalliance.com/. [Accessed: 10-Jun-2017].

[3] 'Napelem piac és technológia - hírek, elemzések'. [Online]. Available: http://napelemek.blog.hu/. [Accessed: 29-Jun-2017].

[4] A. Nemcsics, "A napelem működése, fajtái és alkalmazása", BMF KVK, Budapest, 2005 (university book in Hungarian)

[5] H.-L. Tsai, C.-S. Tu, Y.-J. Su, 'Development of Generalized Photovoltaic Model Using MATLAB/SIMULINK', Lecture Notes in Engineering and Computer Science, vol. 2173, no. 1, pp. 846–851, Oct. 2008.

[6] R. Istók, P. Kádár, 'EMC measurement systems and noise suppression methods for led lamps and RF lighting controls', Pollack Periodica, vol. 11, no. 3, pp. 141–152, Nov. 2016. doi: 10.1556/606.2016.11.3.13

Assessment of the Breaking Characteristics of High Voltage Circuit Breakers with Respect to the Fault Circuit Interactions

Vladimir Vajnar, Zdenek Vostracky
Department of Electrical Power Engineering and Environmental Engineering
University of West Bohemia
Pilsen, Czech Republic
vajnar@kee.zcu.cz, zdenekv@kee.zcu.cz

Abstract — **The presented paper deals with the description, assessment and evaluation of the breaking characteristics of high voltage circuit breakers (HVCB). The paper aims at the relations to the internal and external breaking characteristics defined by the field parameters in the extinguishing chamber and the parameters of cleared fault circuit and its mutual interactions. The purpose of the paper is to introduce and to explain the tight relations between processes inside the circuit breaker and in the fault circuit during the breaking process and to determine the crucial aspects affecting the successful or unsuccessful breaking process. In the first part of the paper, the topic of breaking characteristics is introduced with respect to the breaker's internal and external processes, which contains the description of the key parameters of burning electric arc and mechanisms of its extinguishing. The second part of the paper deals with the time and space intervals of the breaking process and the major critical influences are introduced. The paper concludes with the detailed assessment of the specific HVCB's breaking characteristics, its limitary areas are discussed along with the critical arrangements of extinguishing chamber and the current and voltage values of the fault circuit. In the end of the paper, multiple directions for further studies are introduced with the emphasis on the state-of-the-art modeling software FEM and CFD tools and its various possibilities for the topic's progression.**

Keywords: circuit breaker; characteristics, breaking capability; interrupting time interval; breaker stressing

I. INTRODUCTION

In recent years power systems operate closer to their limits because of the growing demand and system consumers' requirements. Hand in hand with the growing complexity and continuous change in generation and consumption character it leads to the change of conditions of switching processes. In the field of switching apparatus (mainly circuit breakers) the breaking capability has been sufficient for most of the appliances thanks to adequate breaking ability due to the specific principle. The multiple efforts in recent decades nonetheless implied the possible savings of the extinguishing gas, which would mean lowering the design costs and requirements for circuit breaker's maintenance, etc. But with the continuous change of cleared parameters and breaking conditions of processes and fault circuits it leads to the fact, that a large number of switching is in the critical area of breaking characteristics and circuit breakers are experiencing hard situations. These can evolve to the complete failure of the circuit breaker and unsuccessful breaking process usually means the complete destruction of the breaker. Therefore it is important to aim the spotlight back at the correct assessment of theirs breaking characteristics with relation to the fault circuit and discuss the sufficiency of breaking capability with the identification of critical parameters both for circuit breaker and fault circuit [1, 2].

II. TIME INTERVALS OF BREAKING PROCESS

In regards of the breaking process of circuit breaker, we distinguish three different time intervals of burning arc within the breaking process: high current, interaction and dielectric.

A. High current interval

The high current interval takes the greatest part of time within the breaking process. It is the time area of burning electric arc. Contrary to the internal processes, the change rate of arc's parameters is very low and therefore we can assume the arc as a stationary. Based on the stationary characteristics of electric arc, the arc voltage u_a is constant and usually very low. The main parameter of arc within this interval is its diameter and heat content, which is proportionate with the arc current i_a (higher current leads to the larger diameter and higher heat contained in the arc). Thanks to differences between the time constants of arc and the harmonic current, we can assume the high current interval throughout the whole half-period except the *1 ms* area around the current zero. In high current interval, the initial conditions for interrupting are created.

B. Interaction interval

In interaction interval, there is a strong interaction between the arc and circuit in terms of energy flows in between. We can observe intensive cooling and the arc is has much smaller diameter compared to the nozzle diameter. In interaction interval, we require sufficient cooling for the successful corruption of the conductive arc channel. The key parameters in the interaction interval is the cleared current I_v and time derivatives of current and voltage - *di/dt* and *du/dt.*

C. Dielectric interval

Dielectric interval starts exactly in the moment when the current is successfully interrupted. In this moment, transient recovery voltage appears and its waveform is defined by the circuit parameters and the previous arc voltage too, which influent the initial rise of recovery voltage. The transient recovery voltage stresses the circuit breaker which must withstand. In the breaker, the area between contacts is recovering its dielectric features. During the recovering process, the dielectric strength has to be higher than local voltage stress in every spot in the area between contacts. Not fulfilling this condition leads to the *dielectric failure.*

The more detailed overview of each interval and description of dielectric recovery can be found in [1, 2, 4].

III. Characteristics of Breaking Capability

It can be understood from the previous chapter that the successful or unsuccessful breaking is dependent on all time intervals, in which each parameter affects the breaking in a different way. Regarding this, we divide the breaking characteristics into two groups, *internal* and *external*, which differs in the way of influencing the key parameters of dielectric failure: limitary rate of rise of recovery voltage s_{lim} and the breakdown voltage U_p. [1, 3, 5]

A. Internal characteristics

They define the breaking capability in dependence on design parameters of breaking chambers (i. e. gas pressure p, gas velocity, shape and diameter of a nozzle D_t, contact stroke z_k, etc.). They are deduced from the physical models of the arc. In terms of dielectric interval, they are defined for the s_{lim} as:

$$s_{\lim} = f\left(I_v, \frac{di}{dt}, p_1, D_t, z_k\right) \tag{1}$$

In dielectric interval, breaking characteristics as a set of breakdown voltages with dependence on input parameters are defined, as follows:

$$U_p = f(t, I_v, p, z_k) \tag{2}$$

B. External characteristics

External characteristics define the properties of the breaking chamber in different switched circuit arrangements. They are a set of limit parameters as a function of the circuit parameters. In interaction interval, they are defined for the s_{lim} as:

$$s_{\lim} = f\left(I_v, \frac{di}{dt}, t_d, Z_l\right) \tag{3}$$

In dielectric interval, they are defined for the breakdown voltage as:

$$U_p = f\left(I_v, \frac{du}{dt}\right) \tag{4}$$

C. Limitary rate of rise of recovery voltage

The formulas for the limitary rate of rise of recovery voltage s_{lim} are deduced from the interaction interval and generally it can have the following form (where C and D are the specific parameters of nozzle shape and diameter, and l_a is the arc length):

$$s_{\lim} = f\left(C, \frac{di}{dt}, v, \alpha, p, l_a\right) \tag{5}$$

$$s_{\lim} = C_{et} \cdot C_{tr} \cdot D_t \cdot p \cdot I_v^{-m} \tag{6}$$

$$I_{th\max} = C \cdot S_t \cdot p^s \cdot l_a^{-0,5} \tag{7}$$

We can also determine the congestion current I_{thmax}, when the arc's diameter is so large, it jams the nozzle and stops the gas flow. The congestion current is in equation (7), where C is the experimentally determined coefficient.

IV. Breaking Capability with Respect to the Breaking Chamber Parameters

With relation to the design of the breaking chamber, we distinguish two types of flows in the breaking chamber - single-sided and double-sided. Both are observable from the Figure 1.

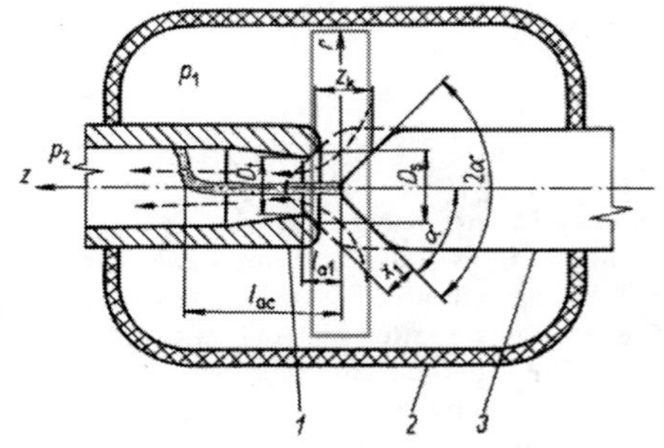

(a) single-sided flow

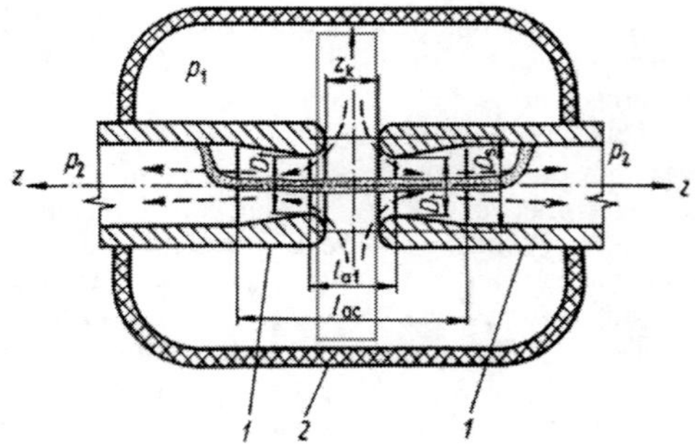

(b) double-sided flow

Figure 1. Schematic drawing of a breaking chamber

It's visible that double-sided flow can be assumed as two separate single-sided flows thanks to the axial symmetry with the r-axis. With relation to the equations (1) to (6) we can compare the s_{lim} of both designs with the following equation, where lower index *2* represents the double-sided flow:

$$s_{\lim 2} = 2 \cdot s_{\lim 1} \cdot \left(\frac{I_{v2}}{I_{v1}}\right)^{-m} \tag{8}$$

From the equation (8) it's possible to deduce ratio between break currents and with the proportional dependence between rate of rise of recovery voltage and breaking current $s = k{\cdot}I_v$ we get the following equations:

$$I_{v2} = 2 \cdot I_{v1} \cdot \left(\frac{s_{\lim 1}}{s_{\lim 2}}\right)^{1/m} \tag{9}$$

$$I_{v2} = 2^{\frac{1}{m+1}} \cdot I_{v1} \tag{10}$$

Based on the previous two equations we can compare the flow type and it's visible that s_{lim} of the double-sided flow is two times higher than single-sided. Regarding the breakdown voltage, the analogy is similar and therefore we can assume the ratio $U_{p2}/U_{p1} = 2$ thanks to the almost double length of the arc compared to the single-sided flow.

The previous values have been experimentally measured and tested and the description can be found in [1]. Modern computational tools provide perspective and dynamic way of the further studies. From the assessment it's visible that the double-sided flow means larger breaking capability. The main reason is the double drainage of hot ionized gas from the arc directly into exhausts (twice as much compared to the single-sided) which lowers their density in the space between contacts and the process of dielectric recovery is faster. Nonetheless, double-sided flow has a great requirements for the total amount of flowing gas, which enlarges the size of the equipment. For the appliances of regular distributional circuit breakers, both arrangements are used. For the breaking of the largest parameters (i. e. generator circuit breaker) double-sided arrangement is used. [7, 9]

V. Breaking Characteristics

Based on analyses of arc modeling, internal characteristics of the chamber and breaker's function in the grid, we obtain external limitary characteristics, which defines the properties and have a major significance for the optimal breaker choice.

A. *Characteristics in interaction inrerval*

Within the interaction interval, the polarity of cleared current changes and is the opposite from the polarity of recovery voltage, therefore the characteristics is related to the polarity of the current. The relation can be described with the following equation:

$$s_{\lim} = \frac{\frac{di}{dt}}{\left|\frac{di}{dt}\right|} \cdot k \cdot \left|\frac{di}{dt}\right|^{-m} \qquad (11)$$

In equation (11) the exponent m is deduced from theoretical models and usually is *1-3*. The dependence can be visualized in log-axes as shows the Fig. 2.

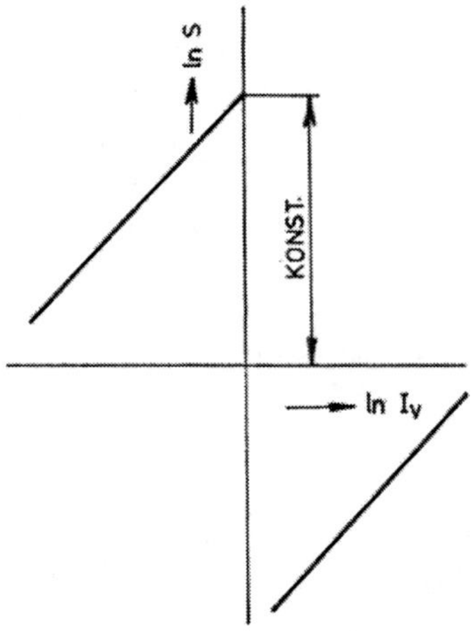

Figure 2. The limitary rate of rise of recovery voltage as a function of cleared current

B. *Characteristics in dielectric interval*

The description has been mentioned in the previous chapters. Similarly, within the dielectric interval, characteristic parameter U_p is dependent on the cleared current polarity. We can describe the characteristics with the following equation (12). We can observe similarities with the eqn. (11) therefore the illustration will be similar, which is visible from Fig. 3.

$$U_p = -\frac{\frac{di}{dt}}{\left|\frac{di}{dt}\right|} \cdot k \cdot I_v^{-\delta} \qquad (12)$$

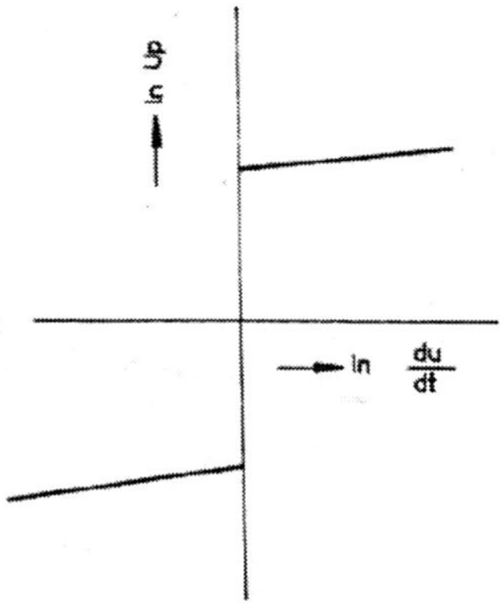

Figure 3. The relation between the breakdown voltage and initial rate of rise of recovery voltage

C. *Characteristics of breaking capability*

Based on the previous 2 subsections, we can assume the external characteristics as two functions $s_{lim} = f(I_v)$ and $U_p = f(I_v)$. Both of the mentioned characteristics can be joined together in the single resulting characteristics, which is in the Fig. 4.

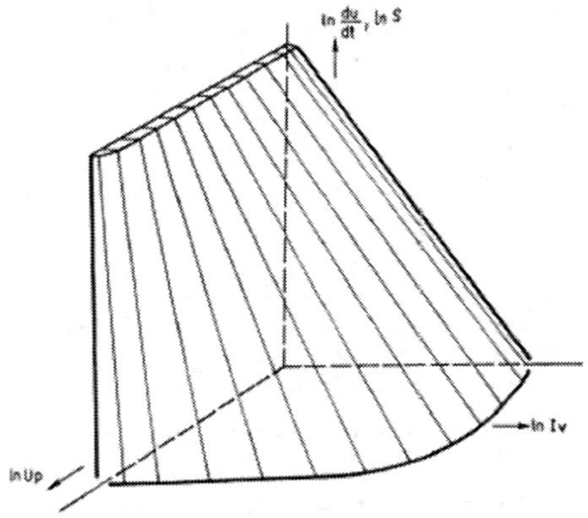

Figure 4. Characteristics of breaking capability in *log*-axes

The most important feature about breaking characteristics is their limitary area, which shows us the largest cleared parameters for the circuit breaker. The limitary areas are often induced from two critical scenarios for the circuit breaker -- terminal fault and short-line fault. Comparing the breaking characteristics with the cleared fault circuit parameters in relation with breaking time intervals, we obtain characteristics for the specific location in the grid defined for the maximal cleared current I_v, both for the interaction and dielectric interval. For the interaction interval, we get the critical current I_i as the intersection of initial RRRV on the cleared faulted line $s_L = f(I_v)$ and the characteristics $s_{lim} = f(I_v)$. On the other hand, for the dielectric interval, we get the critical cleared current I_U as the intersection of the peak value of recovery voltage U_C and characteristics $U_p = f(I_v)$. Both of these intersections are in the Figures 5 and 6.

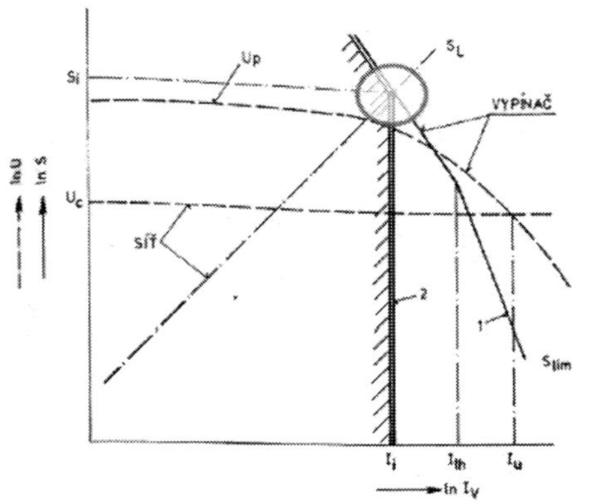

Figure 5. Limitary breaking characteristics for interaction interval

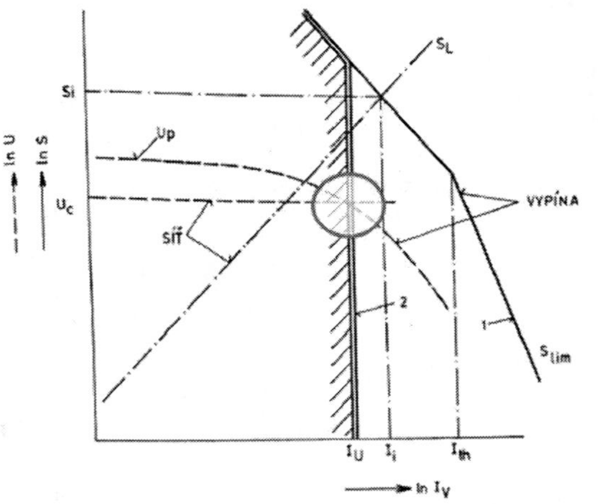

Figure 6. Limitary breaking characteristics for dielectric interval

The limitary cleared current is then picked as the lower one from I_i and I_U. The interaction interval criterion current I_i is often decisive for breakers with nominal voltages up to *150 kV* and nominal short-circuit currents above *25 kA*. The criterion current I_U is decisive for breakers above *150 kV* with small cleared currents.

Defining the critical values and parameters of breaking capability gives a way for slight variation of optional parameters of breaking chamber and nozzle or the possibility of connecting more chambers with lower parameters in the series. The assessment of the breaking characteristics and defining the limitary areas can provide a solid background for the optimal and economically viable usage of breaking chambers with sufficient reliability criteria for short-circuit currents and recovery voltage demands. Further assessment of optimal number of chambers can be found in [1, 3, 4, 6].

VI. New Influences on Breaking Characteristics

In recent decades, multiple reliability surveys have been done by the CIGRE association with important and diverse results (such as [8]). For a long time, the breaking characteristics description with short-circuit cleared current I_v and rate of rise of recovery voltage was sufficient. Nonetheless, a few cases of processes appeared in which the cleared current was much smaller and the circuit breaker failed due to the special time coherence between electrical and mechanical processes. This kind of randomness can be either clearance of the large asymmetrical currents as well as the breaking of small capacitive currents with failures within the dielectric interval before the maximal contacts' stroke. Both of these phenomena are very specific cases of switching and though their appearance is very rare, they can evolve into a complete failure of the breaker.

A. *Asymmetrical short-circuit currents*

The time delay of current zeros lead to the exhaustion of the flowing gas, and after the gas is completely exhausted, the breaking capability falls immediately to zero.

This phenomenon is the matter of recent efforts to save the SF6 gas as the greenhouse gas. Also, the total amount of installed power is growing and the breaking processes are getting harder and harder and breakings are in the critical area. The asymmetrical short-circuit currents are a great deal for the breaking characteristics. Authors have published several papers dealing with this topic in references mentioned at the bottom of this chapter.

The asymmetrical short-circuit current is assumed as the most dangerous case of switching according to the standards for insulated switching apparatus and HV circuit breakers (IEEE C37.04 and IEEE C37.20) because of its largest mechanical and thermal stresses on electrical and mechanical construction parts of circuit breakers and machine windings. The asymmetry is caused by the DC component which shifts the waveform above the horizontal zero axe. The DC component is exponentially decaying from its initial value with the time constant defined by the total resistance *R* and inductive reactance *X* of the fault circuit.

$$\tau = \frac{X_L}{2\pi f \cdot R} \tag{13}$$

The initial value of the DC component is proportionate to the value of current in the fault inception. The DC component is added to the fault waveform. The first effect is within the first peaks of the waveform and it must be secured that the peak current is smaller than a current in equation (7). The superposition of the fault current waveform with the DC component can lead to the thermal failure of the breaker due to the overload of the current I_{thmax}, shown in equation (7). The second effect is the delay between current zeros – longer time of burning arc and insufficient amount of flowing gas therefore can lead to the complete failure of the breaker.

It's recommended to involve the current asymmetry in the breaking characteristics either by adding an extra axis representing the DC component, or to find the closer relation between the flowing gas and time intervals between current zeros. The topic has been partially solved in [9, 10, 11].

B. Small capacitive currents

Breaking of the small capacitive current is specific for the processes of breaking of compensating shunt devices (such as capacitor batteries) or switching of the long unloaded overhead lines or large cable networks. Though the breaking process is not complete (as described in the Chapter II), the burning arc is quenched almost immediately, but the capacitive load is keeping the voltage at the specific value. It leads to the large dielectric stressing of the chamber – contacts are not at the maximal distance and the recovery process of the gas is not immediate, it can lead to restrike. Due to the remnants of the burning arc (hot and ionized non-homogenous gas) and fast transient recovery voltage on the terminals, restrike appears in the spot with highest values of electric field E and lowest values of gas' density ρ. [12]

This kind of measurement of electric and flow parameters inside of the breaker within the process would demand a tremendous amount of requirements for equipment and time and methodology investment, but nowadays' FEM and CFD simulation software provide a great tool for detailed analysis of physical phenomena within the chamber.

Based on the previous facts, several simulations have been done to obtain a rough overview of the dielectric stressing and flow field in the chamber. These models are shown in the Figures 7, 8 and 9.

The model has been created in ANSYS tool package. The distinct characters of various field properties imply the difficulties of the correct choice and the way of modelling. The ANSYS-Fluent has been chosen because of its strong mathematical equipment solving the high-pressure turbulent flows in the chamber using the finite volume method (FVM). This feature provides a great advantage of ANSYS-Fluent against other commercially successful tools.

The presented figures show the case of breaking of small capacitive current in the chamber of *72,5 kV* (nominal phase-to-phase voltage) circuit breaker. The initial condition of pressure inlet has been set to *500 kPa*, which is in accordance to the real pressure within the small cleared currents. Figures show the case with the distance *51 mm* between contacts, which is proportionate (with the contact separation velocity of *6 m/s*) to the time equivalent shortly after the arc is extinguished. Further description of the chamber and identification can't be provided as to secure the manufacturer's know-how.

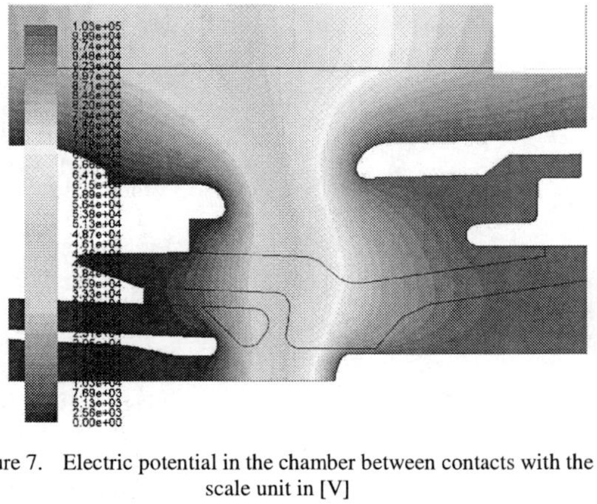

Figure 7. Electric potential in the chamber between contacts with the scale unit in [V]

We can observe that due to the geometry of contacts, the largest gradient of electric potential is at the contacts' surface. Also from the Figure 9 we can assume, that in the time interval before releasing the nozzle throat, the density between contacts is large, but after releasing the throat, the density decreases quickly, but with the large electric field, a restrike can appear in the moment the throat is released.

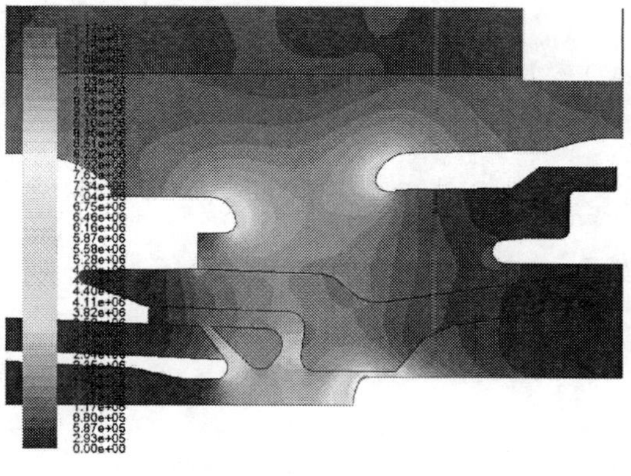

Figure 8. Electric field in the chamber with the scale unit in [V/m]

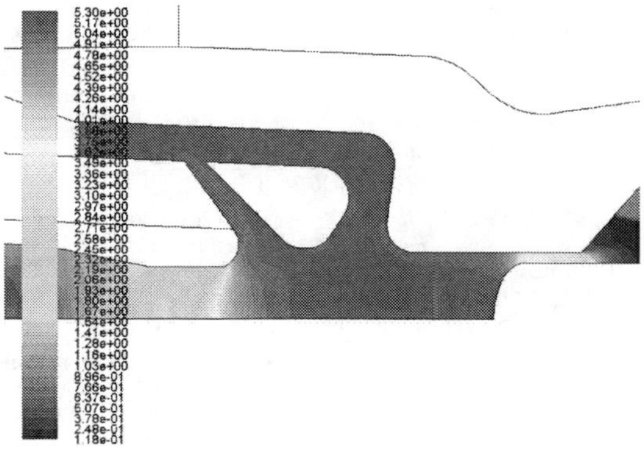

Figure 9. Gas density in the inlet channel and area between contacts with the scale unit in [kg/m^3]

It's apparent that simulation programs provide a great tool to effectively asset the physical fields and their relations during these specific cases. The closer look into the chamber's parameters during different scenarios is recommended so the more detailed description of the dielectric stress within the small capacitive current taken into account of breaking capabilities. From the evaluation of electric and flow field we can define any counter-measurement to suppress the dielectric failures to minimum and to include these specific cases in the characteristics of breaking capability.

VII. Conclusion

This paper introduced the topic of characteristics of breaking capability of high voltage circuit breakers, its definitions, constructions, key parameters, graphical interpretations and actual influences based on the state-of-the-art surveys and practical experience.

It's apparent that breaking characteristics (described in the Chapters III, IV and V) identify the critical applications of circuit breaker and its limiting parameters for the interaction and dielectric interval. The analysis of nominal short-circuit current and rate of rise of recovery voltage needs to be done to determine the successful or unsuccessful breaking. Breaking characteristics give an overview of the circuit breaker appliance with the relation to the internal parameters.

The last part of the paper introduced the modern and current affections of breakers' characteristics that distorts the proven methodology of the correct breaker choice and suggest the key parameters and phenomena to keep the topic of breaking characteristics actual and up-to-date. Though the topic of breaking characteristics is well known for several years, the possibility and efforts to join with the non-standard cases of switching is ingenious and in accordance to the actual needs of industry and service applications. Both parts, asymmetrical short-circuit currents and small capacitive currents are actively solved and analyzed by the authors.

ACKNOWLEDGEMENT

This paper has been financially supported by the project of University of West Bohemia in Pilsen with the grant number SGS-2015-031. The financial support is truly acknowledged.

REFERENCES

[1] Z. Vostracky, "Characteristics of switching abilities of circuit breakers and their uses in practise," Doctoral thesis (in Czech). Faculty of electrical engineering, Czech Technical University in Prague - Skoda Pilsen, 1986.

[2] K. Nakanishi, Ed., Switching phenomena in high-voltage circuit breakers. New York: M. Dekker, 1991. ISBN: 0-8247-8543-6

[3] L. Ramming, M. Aristizabal, "Cold Characteristic Development Test of a New SF6 High Voltage Circuit Breaker," 2006 IEEE/PES Transmission & Distribution Conference and Exposition: Latin America, Caracas, 2006, pp. 1-4.

doi: 10.1109/TDCLA.2006.311470

[4] V. Vajnar, Z. Vostracky, "Analysis of dielectric stress of SF6 high voltage circuit breaker", In: Proceedings of the 13th International Scientific Conference Energy-Ecology-Economy 2016. Bratislava: Slovak University of Technology in Bratislava, 2016, pp. 1-6

[5] L. van der Sluis, W. R. Rutgers, C. G. A. Koreman, "A physical arc model for the simulation of current zero behavior of high-voltage circuit breakers," in IEEE Transactions on Power Delivery, vol. 7, no. 2, pp. 1016-1022, Apr 1992.

doi: 10.1109/61.127112.

[6] S. Delie, D. Bešlija, M. Muratovic, M. H. Kim, M. Kapetanovic, H. Zildžo, "New approach to breakdown voltage estimation after interruption of capacitive currents," 2014 IEEE International Power Modulator and High Voltage Conference (IPMHVC), Santa Fe, NM, 2014, pp. 462-465.

doi: 10.1109/IPMHVC.2014.7287311.

[7] Z. Vostracky, V. Vajnar, and J. Sedlacek, "Partial discharges and their influence on service life of high voltage equipment", University of West Bohemia, Pilsen, 2015.

[8] CIGRE WG A3.06, „Final Report of the 2004 - 2007 International Enquiry on Reliability of High Voltage Equipment - Part 1: Summary and General Matters", Technical Brochure TB509, Paris, 2012.

[9] B. Kulicke, H. H. Schramm, "Clearance of Short-Circuits With Delayed Current Zeros in the Itaipu 550 kV-Substation," in IEEE Transactions on Power Apparatus and Systems, vol. PAS-99, no. 4, pp. 1406-1414, July 1980.

doi: 10.1109/TPAS.1980.319563.

[10] V. Vajnar, Z. Vostracky, J. Sedlacek, "Analysis of breaking capability within asymmetrical short circuits," 2015 16th International Scientific Conference on Electric Power Engineering (EPE), Kouty nad Desnou, 2015, pp. 657-661.

doi: 10.1109/EPE.2015.7161101.

[11] V. Vajnar, Z. Vostracky, "Reduced breaking capability of circuit breakers within operations with delayed current zeros," 2016 17th International Scientific Conference on Electric Power Engineering (EPE), Prague, 2016, pp. 1-5.

doi: 10.1109/EPE.2016.7521754.

[12] J. Sedlacek, "Stationary and dynamic dielectric stress of HV circuit breaker", Doctoral thesis (in Czech), Faculty of Electrical Engineering, University of West Bohemia, Pilsen, 1999.

Analysis of Dielectric Properties in High-Voltage Insulator by Dielectric Spectroscopy (DRS)

Roman Cimbala, Lukáš Kruželák, Samuel Bucko, Michal Kosterec
Department of Electrical Power Engineering
Technical University of Košice
Košice, Slovak Republic
roman.cimbala@tuke.sk, lukas.kruzelak@tuke.sk, samuel.bucko@tuke.sk, michal.kosterec@tuke.sk

Abstract—**The aim of this article is to investigate the dielectric properties of high-voltage insulator. Dielectric relaxation spectroscopy (DRS) was chosen as a method for this purpose and used on the plastic support insulator. Two measuring devices with different frequency range and different voltage range were used in this experiment, The first apparatus was with a frequency range from 0.1 mHz to 10 kHz, in depend on voltage value from 1V to 160 V. The other device had a frequency range from10 Hz to 2 MHz at a constant voltage of 2 V. We used these two devices to obtain a description of polarization processes in a broad spectrum from 0.1 mHz to 2 MHz at the voltage 2V.During the second experiment, a value of the voltage was increased only in one of the devices to explore polarization processes at increased voltage. This device can measure the surface resistance of the insulator, which has been also reviewed in this publication. Both, serial and parallel RC model were used for measurement in devices in order to compare the results of measurement.**

Keywords: arcing distance, creepage distance, form factor, dielectric losses.

I. Introduction

Electrical power engineering is one of the most important amongst the global industries, which represents a part of the economy of each country. It is characterized by great complexity and extensive system. The main goal of electrical power engineering is to produce and deliver electricity accessible to all customers. Great emphasis is therefore placed on reliable electricity supply while securing its lowest transmission costs of. For the distribution of electricity, we use overhead transmission line and cable transmission line. In the overhead transmission line, one of the most important components of the electro-insulator string is also called power transmission line insulator. The electrical insulator at the overhead transmission line performs two functions, namely a function of carrier, with attached conductor is and its second function is an electro-isolating function, where it separates the living and non-living part of the electrical line. Weather condition, environment, damage caused by birds and human activities have a significant influence on the outdoor insulation of transmission line [1]. The weather condition which makes the greatest threat to the safety of transmission lines´ operation contains mainly lightning strike, frost, strong wind, and other climate condition which can cause damage to its dielectric properties. Since the electrical insulator serves as a carrier for the overhead transmission line, it is constantly mechanically and electrically stressed [2]. This stress can worsen the dielectric properties of the electrical insulator. Therefore, in order to determine the isolating state of the outer insulators, it is necessary to choose a suitable diagnostic method which determines the state of the power transmission line insulators [3].

II. Outdoor Power Transmission Line Insulators

Electrical Insulators are used in all electrical devices, from a small capacitor to a huge generator. Technicians need to understand how to design a device with a proper insulator. Unfortunately, the damage of the insulator is the main reason why the electrical equipment stops working. In the context of power transmission, insulators describe the support structure for wires and other devices [4]. While knob and tube house wiring is transitioned into a flexible insulation, higher voltages found in HV power transmission remain the same, so there is no other economic way to insulate HV lines except by suspending the wires (using the surrounding air as the insulator) or burial. Frequently-used suspension insulators can be divided into porcelain insulators, glass insulators and composite insulators according to the materials they are made of. Different insulators are suitable for different areas, and they share the world market of insulators [5]. The composite insulators has been developed more rapidly than other two kinds in the recent 20 years for such reasons: light weight, high mechanical strength, strong anti-pollution flashover ability, and more economic properties with the raise of the high voltage level [6].

Electric insulators are mediate by two functions; the first function is that they mechanically fix the wires to the mast of the external power line. The other function of electrical-insulator is that it electrically separates electrical power conductors from the external power transmission mast. In order for electrical insulators to fulfill their purpose it is necessary to distinguish their particular parameters, which are:

- Arcing Distance - The shortest air-jump distance between the metal parts that have an operational voltage between them
- Creepage distance - the shortest distance on the surface of the insulator between its two conductive parts that have an operational voltage between them
- Form factor - quantifies the relationship between the distance of insulating roofs and the depth of insulator roof [7].

The form factor *f* can be mathematically expressed as:

$$f = \int_0^L \frac{dl}{2\pi R(l)} \tag{1}$$

Where L is the surface distance, $R\ (l)$ is the radius and dl is the elementary length.

During its operation, various impurities are spreaded throughe insulator´s surface. In the dry state, the impurities have a high resistance values and do not affect the discharging processes on the surface of the insulators. Moistening the dirty layer (rain, fog, dew) leads to a sharp drop in layer resistance due to the formation of weak electrolytes from water-soluble components of the contaminated substance [8].

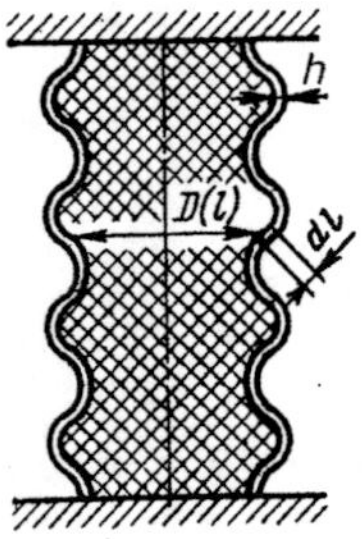

Figure 1. The surface of the insulator with a dirty layer of dirt [8].

Where h is the thickness of the dirty layer, $D\ (l)$ represents diameters of insulator length and dl is elementary length

On the surface of the insulator, a layer of impurities with thickness h has been formed with a specific volume resistance ρ_v. At the voltage U, the leakage current will flow through the insulator surface Iz. The value of leakage current is determined by the ohm law where Rz is the surface leakage resistance, which is determined as:

$$R_Z = \int_0^{Lz} \frac{\rho_v}{\pi h D(l)} dl \qquad (2)$$

Where l is distance from one electrode to the elementary length dl, Lz is the total length of the lead over the insulator surface and $D\ (l)$ is insulator diameter dependent on distance l

For simplicity, suppose h and ρ_v are not dependent on l. Then the relationship for Rz can be written:

$$R_Z = \frac{\rho_v L_Z}{\pi h D_e} \qquad (3)$$

Where D_e is equivalent to the diameter of the solid dielectric cylinder [9].

III. Measurement Capacity and Dissipation Factor

Isolation properties of high voltage devices can be imagined as a capacitor dielectric, whose electrodes are formed by a high voltage and a low voltage (grounded) part of the device. When an alternating voltage is applied on this capacitor, the current will flow through the dielectric. The ideal dielectric does not show any losses, and the dielectric charging current vector is naturally exclusively capacitive and displaced by 90 ° to the stress vector [10].

Since insulating materials used in electrical power plants are not ideal, the current flowing through the dielectric contains, in addition to the reactive component, an active component too, what is caused by the conductivity and polarization of the dielectric. A phasor diagram of a real capacitor with one polarization deck is shown in Fig.2

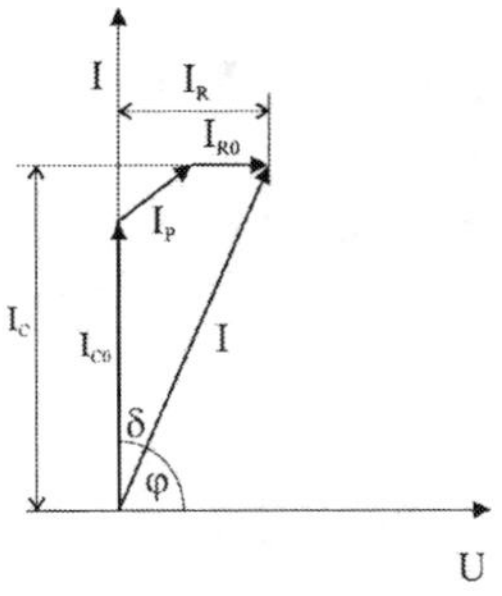

Figure 2. Phase chart of the real capacitor [11].

The dielectric loss factor (or dissipation factor) tgδ is defined as the tangent angle, which differentiates between the phase shift of the real dielectric current and the phase shift of the ideal dielectric current [11]. The value of the dielectric loss factor is related to dielectric losses and its value can be considered as a measurement of the quality of the whole insulation system. According to the phasor diagram (Fig. 2) we can state following:

$$tg\delta = \frac{I_R}{I_C} \qquad (4)$$

Dielectric losses make up the sum of conductivity losses, polarization losses and ionization losses. An increase in the values of dielectric loss usually points out the aging of insulation. The quantity tgδ depends on several factors. Depending on the frequency of the voltage measurement, we investigated a certain frequency domain that is affected by a group of elementary polarization processes [12]. It is necessary to determine which elementary polarization processes from the polarization spectrum affected the dielectric loss value. When measuring the voltage and frequency f_0. We know that the polarization process with settling time τ_0 will cause the greatest losses on the frequency f_0.

$$f_0 = \frac{1}{2\pi\tau_0} \qquad (4)$$

If the applied voltage rise or fall, the measured dielectric loss will decrease [13].

Frequency dependence of the loss factor is shown in Fig. 3. In the case of insulation with a relatively high value of conductivity, tgδ will decrease as a function of frequency (curve 1 Fig. 3). Isolation with one polarization deck has its maximum (curve 2 Fig.3). The dependence of $tg\delta = f(\omega)$ of the real dielectric represents the combination of the first two dependencies (curve 3 Fig. 3). In the case of complex dielectrics, there may be several polarization events with different settling times, so the dependence $tg\delta = f(\omega)$ can have several maximums [14].

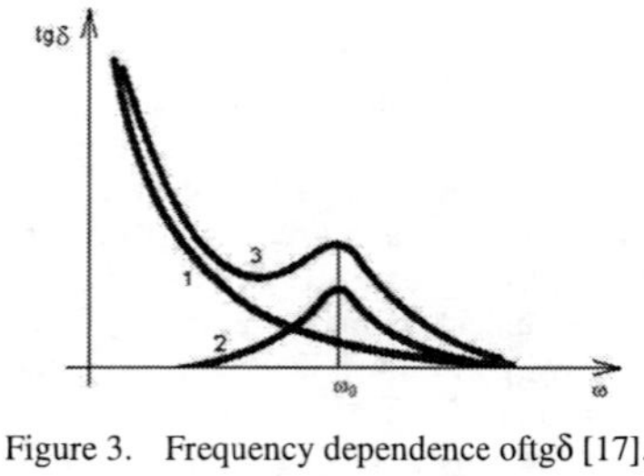

Figure 3. Frequency dependence oftgδ [17].

Then following quality of insulation is characterized by the lowest values of the dielectric loss factor. For quality of insulation, the values of the dielectric loss factor vary less, depending on the size of the test voltage [15],[16]. By increasing the intensity of the electric field acting on the insulator, the processes that affect the value of the dielectric loss factor will change [17]. This is especially noticeable if a completely different loss mechanism, caused by ionization processes, starts running above a certain voltage value. The factor of dielectric losses of insulation is strongly dependent on temperature [18]. Losses related to the electrical conductivity of the insulation are increase as the temperature increases. Since the values of the dielectric loss factor depend highly on temperature, it is necessary to perform the measurements under the same temperature conditions [19].

IV. Experimental Measurement

The loss factor and capacity were measured on the high-voltage supportive insulator. Two measuring instruments were selected for this experiment, each having a different frequency range. The main goal of using two devices was to measure the loss factor and capacity in a wide spectrum of frequencies. First used measuring instrument Megger IDAX 300 has a frequency range from 0.1 mHz to 10 kHz for a voltage range from 0 - 160 V. The second one was the Agilent E4980A Precision LCR meter with measuring range from 10 Hz to 2 MHz at a voltage of 2V. The measurements were made at room temperature (20 °C), we used clamps for the measurements were that were fastened around the insulator circumference and the distance between the clamps was 1 mm. By combining measured results from both instruments we have reached a visual overview of polarization processes in the frequency range from 0.1 mHz to 2 MHz. The measurements were made in both the serial and parallel model of plugging the sample device.

A. Graphical dependence of the loss factor

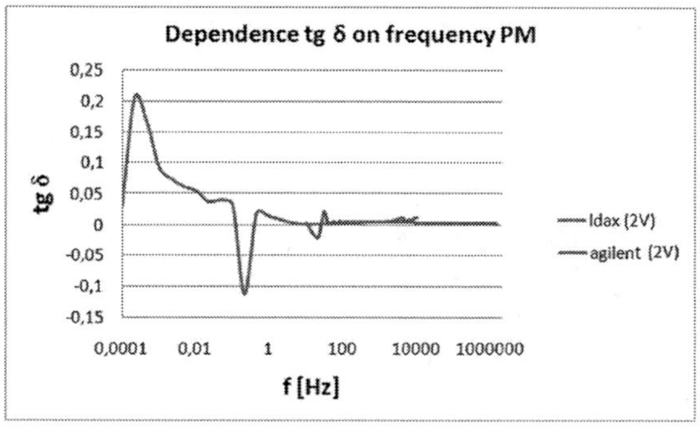

Figure 4. Dependence oftgδ on frequency, parallel model at the same voltage of the measuring instrument.

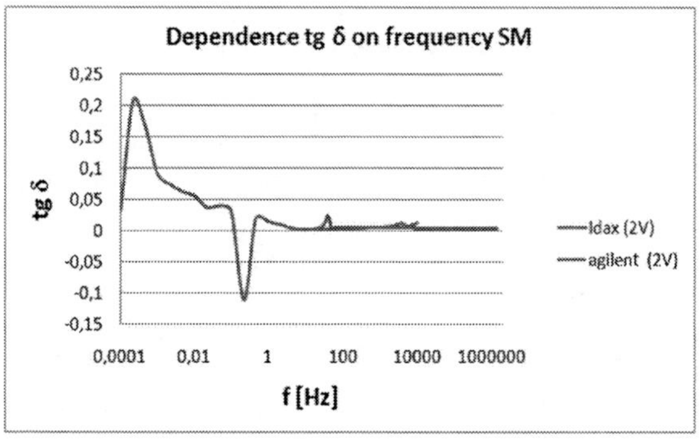

Figure 5. Dependence oftgδ on frequency, serial model at the same voltage of the measuring instrument.

When analyzing the graphical dependence of the coefficient of response on the frequency, it is possible to see that the elementary polarization process is observed at frequencies lower than 0.01 Hz when measuring a low voltage of 2 V. At such low frequencies and low voltage, the unidirectional component of the propulsion causes the polarization of the dielectric. When it reach a certain frequency, the dielectric losses are at the greatest level, and any further change of frequency causes a reduction in dielectric losses in the dielectric. In a further frequency increase there is an anomaly observed when the loss factor has declined sharply and returned to its original value again. This anomaly is probably caused due to the low value of voltage and depolarization processes resulting from the increase in frequency. Another growth of frequency lowers the value of the stratum factor. In the vicinity of the 50 Hz frequency, there is visible agitation on the Agilent device, which is caused due to the interference of the power supply from the power supply. By comparing the parallel and serialized models, it can be seen that the dielectric loss factor curves for both connections are identical.

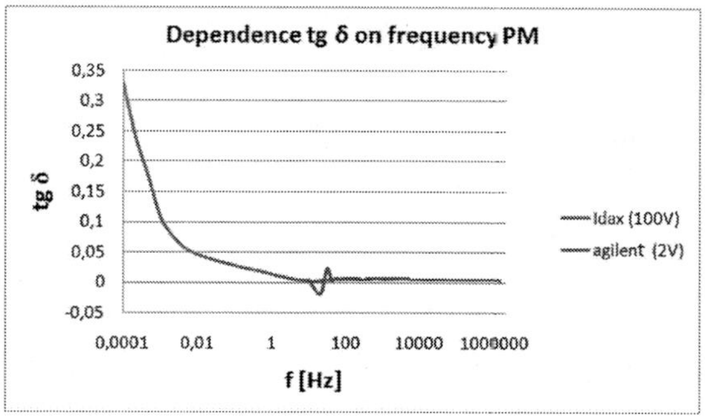

Figure 6. Dependenceof tgδ on frequency, parallel model at the different voltage of the measuring instrument.

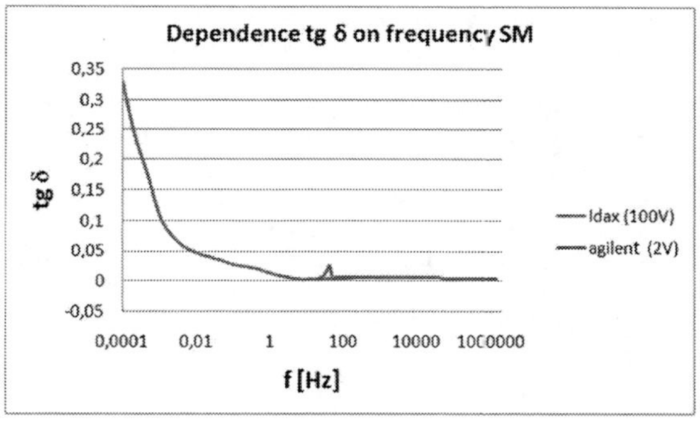

Figure 7. Dependence oftgδ on frequency, serial model at the different voltage of the measuring instrument.

When analyzing the loss factor from the frequency, it can be seen that the value of the loss factor decreases linearly with the rising f value of frequency when measuring the higher voltage. At a voltage of 100 V, the elementary polarization processes can be observed. Even with this measurement, the disturbance of the Agilent measuring instrument is due to interference from the power supply. Comparing the parallel and serial model of involvement, no changes in the measured values were observed.

B. Graphical dependence of the surface resistance

The Megger IDAX 300 can measure multiple parameters. One of these parameters is the resistance measurement of the explored sample. When measuring the support insulator, we used a serial and parallel model of connection to the measured sample. The measurement voltage was 2 V and 100 V for the frequency range from 0.1 mHz to 10 kHz. This frequency range was applicable only at a test voltage of 2 volts. When measured at a test voltage of 100 V, the instrument was able to measure only up to a frequency of 3 kHz which was caused by the small distance between the measuring terminals.

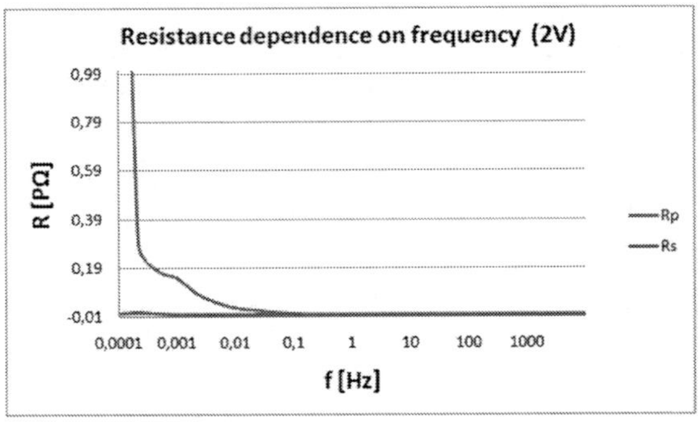

Figure 8. Graphical dependence of the surface resistivity from the frequency for parallel and serial models at a voltage of 2 V.

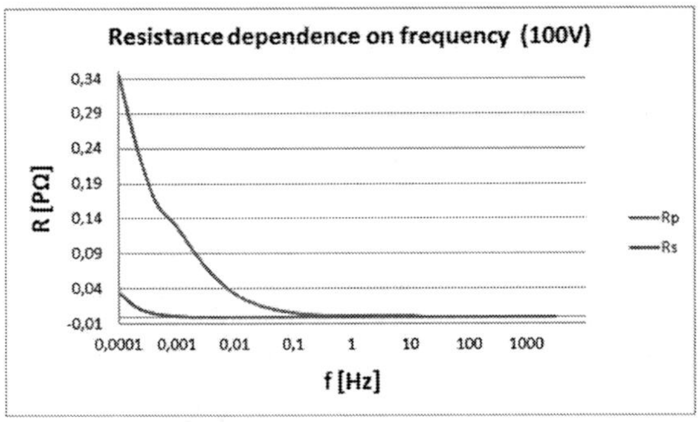

Figure 9. Graphical dependence of the surface resistivity from the frequency for parallel and serial models at a voltage of 100 V.

When analyzing the measured results it can be seen that using a 2 V measuring voltage, the value of resistance is infinitely large in the parallel model in the 100 mHz range. This phenomenon is caused by a small strain and low frequency to which the sample is apparently infinitely strongly resistent. With a parallel model, when increasing the frequency of the surface, the resistance decreases. The parallel compression resistance calculation device uses the formula:

$$R_P = \frac{1}{R_e\{\frac{1}{Z}\}} \tag{5}$$

Where $R_e$is the real impedance component. For calculating impedance Z, there is following relation:

$$Z = \frac{1}{2\pi f C} \tag{6}$$

Where C is the capacity of the measured object and f is the frequency. From this relationship, the surface resistance of the measured object decreases exponentially at frequencies lower than 1 Hz with increasing frequency. For the calculation of the surface resistance for the serial circuit of the apparatus, the relationship shall be as follows:

$$R_S = R_e\{Z\} \tag{7}$$

When comparing the surface resistance curve for serial and parallel model, it can be seen that low frequencies of less than 1 Hz, influence the value of surface resistance in parallel model more than in the serial one. This fact express the formula for a parallel model of measurement.. When examining surface resistance for a voltage of 100 V, at a frequency of less than 100 mHz, the instrument was able to measure the final value of the surface resistance in the parallel model. With a parallel model using 2 V, the final value failed to be measured. At higher tensions, the conductive current was sufficiently large, so that the instrument can measure it and the final resistance value of this device is reached. In the parallel model, a decreasing surface resistance curve at frequencies lower than 1 Hz can be observed.

V. Conclusion

This article deals with measuring the loss factor tgδ and the surface resistance of the insulator. Two devices with different wiring configurations were used for the measurement with respect to the measured sample. The used instruments used a different frequency range which we used for measurement in a wide bandwidth. During measurements, it was found out that when using a frequency of less than 1 Hz, the measured values were much greater than during measurements above 1 Hz. This phenomenon is caused by the fact that at such low frequencies we can approach the values of DC voltage, which causes the polarization of the dielectric in a one-dimensional field. This causes a higher conductivity of current that affects the value of the loss factor. Surface resistance measurement also involved the frequency and the used wiring model. Surface resistance values were stable at frequencies above 1 Hz. At lower values, the differences in surface resistance were observed by using a parallel and serial model.

Acknowledgments

This work was supported by Ministry of Education Agency for structural funds of EU in frame of projects No. 26220120055 and 26220220182, VEGA 1/0311/15 and 2/0141/16.

This work was supported by the Slovak Research and Development Agency under the contract No. APVV-15-0438.

Európska únia
Európsky fond regionálneho rozvoja
Operačný program VÝSKUM a VÝVOJ

References

[1] W. D. Shi, Z. C. Guan, L. M. Wang, J. Zhou, "A study on the selection of outdoor insulators used in high altitude areas," Annual Report Conference on Electrical Insulation and Dielectric Phenomena, 2003, pp. 391-393. doi: 10.1109/CEIDP.2003.1254875

[2] H. Saadati, P. Werle, E. Gockenbach, H. Borsi, J. M. Seifert, "Flashover performance of polluted composite insulators under AC and hybrid AC/DC field stress," 2016 IEEE Electrical Insulation Conference (EIC), Montreal, QC, 2016, pp. 170-173. doi: 10.1109/EIC.2016.7548687

[3] P. T. Dumont, "System engineering approach for diagnostics of electromechanical systems," 2015 Annual Reliability and Maintainability Symposium (RAMS), Palm Harbor, FL, 2015, pp. 1-7. doi: 10.1109/RAMS.2015.7105122

[4] Z. Guan, X. Wang, X. Bian, L. Wang, Z. Jia, "Analysis of causes of outdoor insulators damages on HV and UHV transmission lines in China," 2014 IEEE Electrical Insulation Conference (EIC), Philadelphia, PA, 2014, pp. 227-230. doi: 10.1109/EIC.2014.6869381

[5] C. Freitag, I. Mladenovic, C. Weindl, "Experiences of in field measured dissipation factor on MV PILC cables at 50 Hz," 22nd International Conference and Exhibition on Electricity

Distribution (CIRED 2013), Stockholm, 2013, pp. 1-4. doi: 10.1049/cp.2013.1145

[6] M. Kolcun, M. Kanálik, D. Medved', Z. Čonka, "Measuring of real value of short-circuit power in island operation condition," 2015 16th International Scientific Conference on Electric Power Engineering (EPE), Kouty nad Desnou, 2015, pp. 418-422. doi: 10.1109/EPE.2015.7161180

[7] J. Orsagova, P. Toman, "Evaluation of negative effect of distributed generation," 2008 Power Quality and Supply Reliability Conference, Parnu, 2008, pp. 131-135. doi: 10.1109/PQ.2008.4653749

[8] N. Hayakawa, K. Ishida, T. Matsuoka, H. Kojima, "Electrical Insulation Characteristics and Mechanisms of Low Dielectric Loss Materials for High-Voltage HTS Cables," in IEEE Transactions on Applied Superconductivity, vol. 27, no. 4, pp. 1-4, June 2017. doi: 10.1109/TASC.2016.2645796

[9] M. Abdus Salam, H. Ahmad, "Derivation of creepage distance in terms of ESDD and diameter of the contaminated insulator," in IEEE Power Engineering Review, vol. 20, no. 9, pp. 44-45, Sep 2000. doi: 10.1109/39.866870

[10] M. A. Salam et al., "Study of creepage distance of the contaminated insulator in correlation with salt deposit density," 2000 Annual Report Conference on Electrical Insulation and Dielectric Phenomena (Cat. No.00CH37132), Victoria, BC, 2000, pp. 215-217 vol.1. doi: 10.1109/CEIDP.2000.885265

[11] V. Kral, S. Rusek, R. Gono, "Processing of input databases of technical condition and importance and definition of order for maintenance of 110 kV / MV transformers," 2015 16th International Scientific Conference on Electric Power Engineering (EPE), Kouty nad Desnou, 2015, pp. 350-354. doi: 10.1109/EPE.2015.7161108

[12] S. Nuccio, R. Schifani, "AC power-loss measurements in solid dielectric materials at industrial frequency," in IEE Proceedings A - Physical Science, Measurement and Instrumentation, Management and Education - Reviews, vol. 128, no. 7, pp. 501-506, October 1981. doi: 10.1049/ip-a-1.1981.0076

[13] M. Pavlík, I. Kolcunová, B. Dolník, J. Kurimský, A. Mészáros, D. Medveď, M. Kolcun, J. Zbojovský, "Measuring of dependence of shielding effectiveness of wet materials on the frequency of electromagnetic field in the high frequency range", Acta Electrotechnica et Informatica, vol. 13, no. 3, pp. 12-16 (2013).

[14] E. Lumnitzer, P. Liptai, R. Drahoš, "Measurement and assessment of pulsed magnetic fields in the working environment," in Proceedings of the 8th International Scientific Symposium on Electrical Power Engineering, ELEKTROENERGETIKA 2015, 2015.

[15] J. Džmura, J. Petráš., J Balogh, "Modelling a Computer Simulation of Electrical Separation, 2016. In: EPE 2016 : proccedings of the 2016 17 th International Scientific Conference on Electric Power Engineering : May 16-18, 2016, Prague, 2016 pp.5377-542.

[16] M. Rajnak, J. Kurimsky, B. Dolnik, et al, "Dielectric-spectroscopy approach to ferrofluid nanoparticle clustering induced by an external electric field," Physical Review E - Statistical, Nonlinear, and Soft Matter Physics, vol. 90, no. 3, 2014.

[17] R. Janura, M. Gutten, D. Korenciak, M. Sebok, "Thermal processes in materials of oil transformers," 2016 Diagnostic of Electrical Machines and Insulating Systems in Electrical Engineering (DEMISEE), Papradno, 2016, pp. 81-84. ISBN 978-1-5090-1248-0

[18] M. Krbal, J. Škoda, S. Sumec, J. Štěpánek, "Laboratory high voltage capacitor with low PD level and low dissipation factor". 2016 17th International Scientific Conference on Electric Power Engineering (EPE), Prague, 2016, pp. 306-310. doi: 10.1109/EPE.2016.7521723

[19] R. Cimbala, J. Kurimský, I. Kolcunová, "Determination of thermal ageing influence on rotating machine insulation system using dielectric spectroscopy". Przegląd Elektrotechniczny. Vol. 87, no. 8 (2011), pp. 176-179. ISSN 0033-2097

Theoretical and Experimental Approach to the Issue of Limiting Short-circuit Current in the LV Circuits

Jozef Zuščak, František Janíček, Michal Váry, Ľudovít Hüttner
Institute of Power and Applied Electrical Engineering,
Faculty of Electrical Engineering and Information Technology, Slovak University of Technology in Bratislava,
Ilkovičova 3, 812 19 Bratislava, Slovak Republic
jozef.zuscak@stuba.sk, frantisek.janicek@stuba.sk, michal.vary@stuba.sk, ludovit.huttner@stuba.sk

Abstract — **Protection against short-circuit currents is an important factor in terms of safety, reliability and durability in electrical networks of all voltage levels. Well-designed protective device must be able to limit short-circuit current so that the Joule integral value does not exceed the standards set value. The paper analyzes experimentally measured arc voltage waveforms generated in arch chamber and its impact on the course of the current and the energy released in the quencher system. Experimental data obtained from switching short-circuit current in the LV circuits under different conditions of switching are compared with the results of theoretical simulation approach. Contribution shows the accuracy of calculation methods and circuit parameters which have an impact on limiting of short-circuit current, taking into account the proposals and verification activities of LV AC circuit breakers.**

Keywords: Short-circuit currents, Circuit breakers, Electrical safety, Power system protection

I. Introduction

The short - circuit current causes thermal and dynamic stresses in electrical circuits. Incorrectly designed circuit-breakers may damage the wiring or appliance. In our article we will deal with the limitation of the short-circuit current in AC electrical networks. We will describe the method of increasing the arc voltage resulting in the limitation of the short-circuit current and the interruption of the circuit before the current passes through the natural zero. On real samples, we will measure the current limitation and compare it with the theoretical calculation for the specified source parameters. The article should reflect the match of the theoretical results with the measurement. The results should help understand the current limitations in AC networks. By creating a script in Matlab, it is possible to simulate short circuits with different parameters of source and short circuit, thus limiting the number of measurements on circuit breaker.

II. Mathematic Description of the Problem

At the beginning, it was mentioned that the energy released during the passage of the short-circuit current displaces the device and the electrical wiring and cables connected to it. To limit this energy, it is necessary to reduce the value of Joule's integral [1-3].

$$JI = \int i^2 dt \tag{1}$$

It is clear from the equation that by limiting the current we will achieve a significant limitation of the released energy.

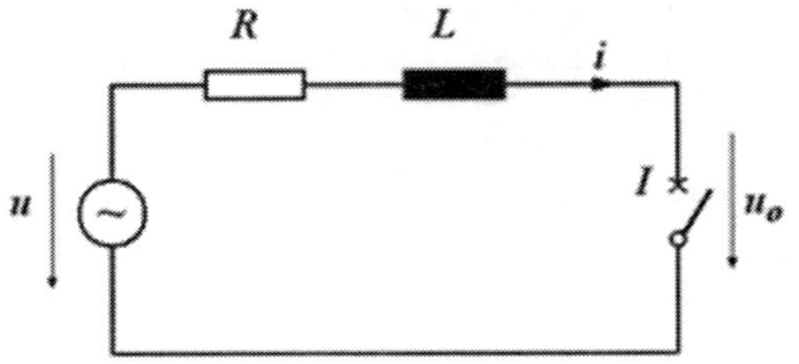

Fig.1 Equivalent circuit

Switching off the short-circuit current will be investigated on a simple circuit (Fig. 1). Circuit parameters are concentrated in RL elements. The internal impedance of the source is neglected in the calculation. An arc voltage is created on the circuit breaker. For simplicity, we assume that the switching device will create an arc voltage that will increase according to step function. For the circuit in Fig. (1) the 2nd Kirchhoff law applies.

$$u = U_o + Ri + L\frac{di}{dt} \tag{2}$$

After the adjustment, we get the differential non-homogeneous equation of 1st order:

$$\frac{di}{dt} = -\frac{R}{L}i + \frac{U_m.\sin(\omega t + \alpha) - U_o}{L} \tag{3}$$

After the short circuit the reaction time of switching device is t1, until the arc reaches its full value. At this time, the current will reach value [4]:

$$i(t_1)=\frac{U_m}{Z}[\sin(\omega t_1+\alpha-\varphi)-\sin(\alpha-\varphi)\exp(\frac{R}{L}t_1)] \quad (4)$$

In time t ϵ <0, t_1> the current value in the circuit is:

$$i(t)=\frac{U_m}{Z}[\sin(\omega t+\alpha-\varphi)-\sin(\alpha-\varphi)\exp(-\frac{R}{L}t)]=A \quad (5)$$

Which represents a steady and unidirectional (intermediate) component of the short-circuit current.

In time t > t_1 the arc voltage is constant (step rise).
We get the resulting relationship:

$$i(t)=A-\frac{U_o}{U_m\cos\varphi}(1-\exp(-\frac{R}{L}(t-t_1))] \quad (6)$$

U_m – amplitude of voltage source,
U_o – maximum value of arc voltage,
Z - circuit impedance
ω – angular frequency,
cosφ – power factor,
α – the moment of occurrence of the short circuit current with respect to the previous zero of the source voltage,
R, L – resistance and inductance of the circuit,
t_1 – time to reach full arc voltage

III. Creating the Simulation of Current Limitation at Step Voltage Rise

For the verification of the theoretical knowledge, a program for simulating short-circuit switching in the MATLAB program was designed. Theoretical relationships have been incorporated into the program. Graphical User Interface Fig. 2 was also created. The input parameters were the effective value of the short-circuit current, cos (φ), α, the effective value of the voltage U_{ef}, the maximum value of the arc voltage U_{o_max}, the total simulation time t, the arc voltage rise time t1 and the number of sampling steps [5].

The outputs of program are:

1. Time evolution of current and limited current
2. Time evolution of arc voltage
3. Joule integral of limited current

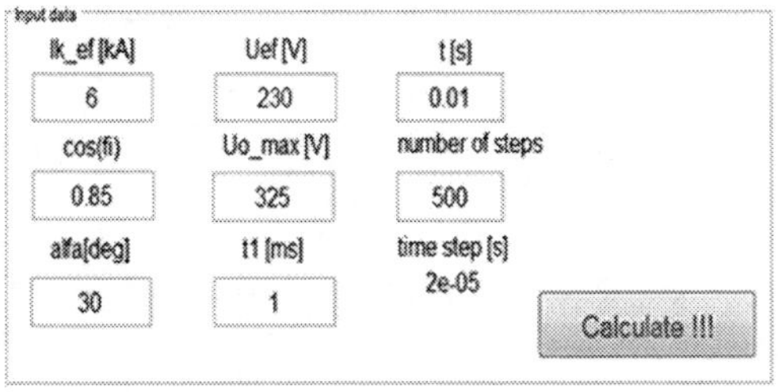

Fig.2 Input parameters

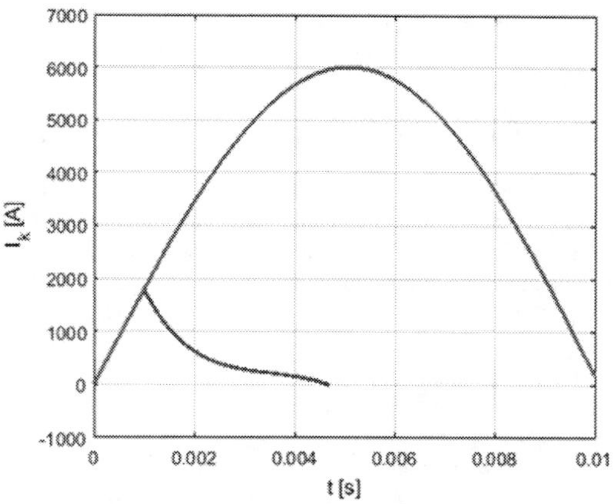

Fig.3 Current behavior

In Fig. 3, there is a zero current after 4 ms, if the arc current is not limited, so we would get a natural zero after 10 ms. Also, the maximum value of the limited current has reached only about 1.7 kA.

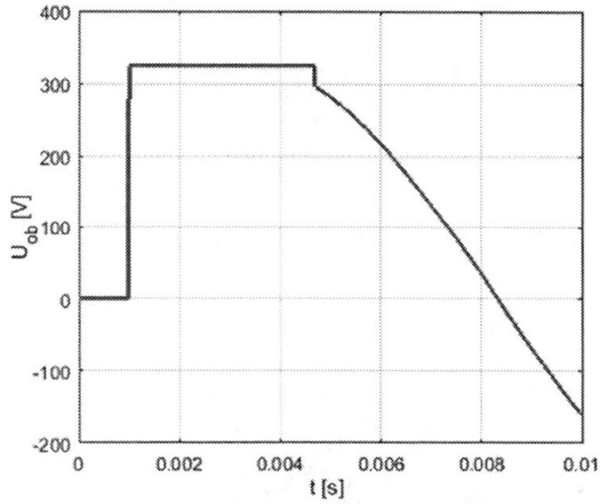

Fig.4 Arc voltage behavior

After the current limit in 4.4 ms, there regenerated voltage on the contacts of the protected device appears as shown in Fig. 4.

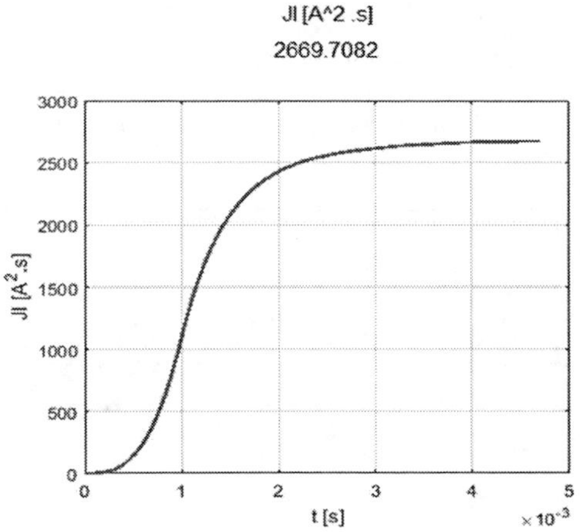

Fig.5 Joule integral (JI) behavior

Joule integral (JI) of the limited current has reached 2700 A^2s. The computational program can simulate various cases of short circuit.

PHYSICAL SOLUTION IN MINIATURE CIRCUIT BREAKERS

A miniature circuit breaker is a device that includes overcurrent and short-circuit triggers. The overcurrent trigger is formed by bimetallic, which deforms by thermal act of passing current and acts by force on the locking mechanism lock. But we will be interested in a short-circuit trigger. This is made up of a ferromagnetic core which is coupled with a non-magnetic hammer trip. The core is located inside the coil as shown in Fig. 6. When the short-circuit current passes through the coil, a magnetic field arises in the vicinity of the coil to form an opposite field of action in the core. The result is the force that acts on the core and gives it acceleration. The hammer coupled with the core activates the lock mechanism and the contacts are separated.

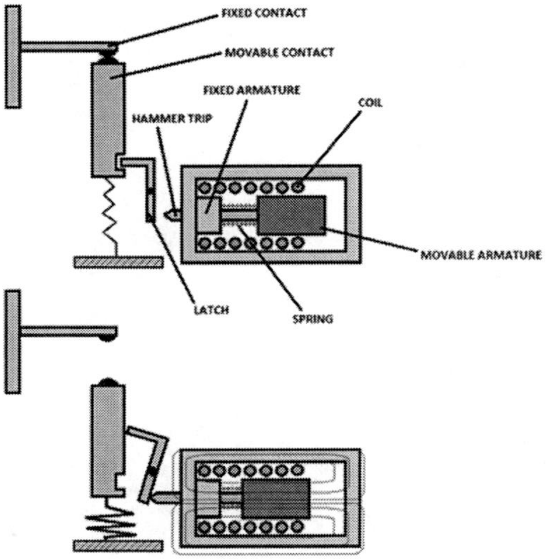

Fig. 6 Act of short-circuit trigger

It is necessary to realize that there is a delay in activation of the trigger, which is actually the reaction time that we mentioned in part 1. This reaction time depends on the size of the short-circuit current. In the past, a 3-D model of the short-circuit trigger was simulated using FEM to detect the reaction time at various multiples of the nominal current and various ferromagnetic core materials.

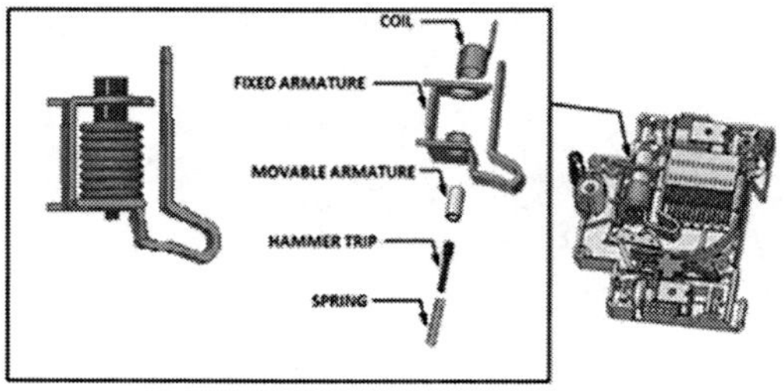

Fig. 7 Model of circuit breaker with detail of short-circuit trigger

For the circuit breaker the necessary parameters were found by measurements:

x_T = distance between armature and latch = 1,2 mm

F_t = tripping force = 2.18 N

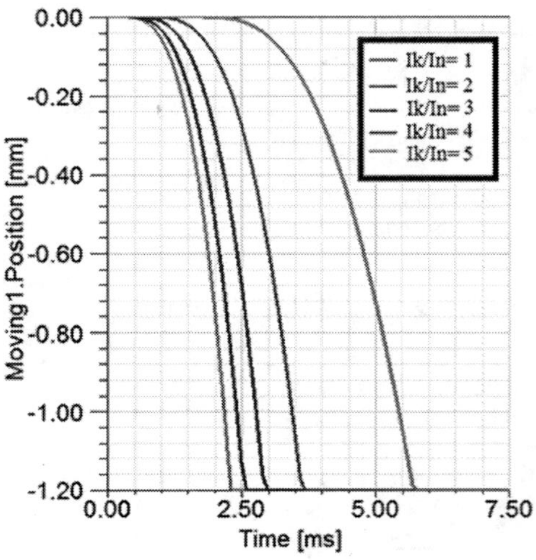

Fig. 8 Dependence of position of short-circuit trigger on time, parameter Ik/In .

It is clear from the figure that the reaction time decreases with increasing current.

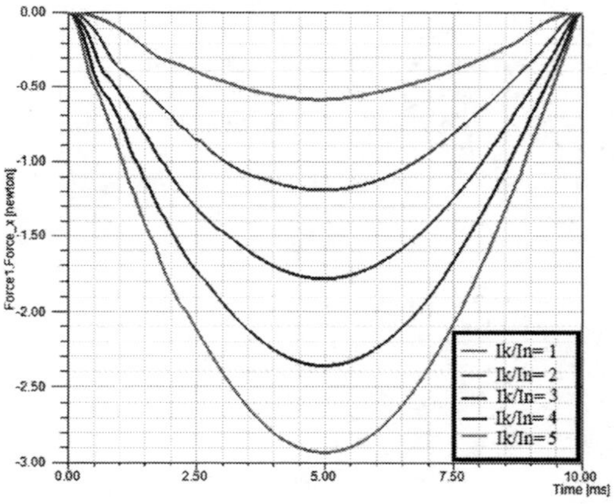

Fig. 9 Force acting on ferromagnetic core in dependence on time, parameter Ik/In

From Fig. 8 and Fig. 9, it is obvious that the pin is not sufficient to overcome the spring force but must have sufficient force to activate the locking mechanism lock.

IV. MEASUREMENTS ON VARIOUS SAMPLES OF MINIATURE CIRCUIT BREAKERS

The measurement was intended to demonstrate real arc behavior and short-circuit current limitations. In real circuit breakers it does not show a step rise in arc voltage. Moreover, this voltage does not remain constant even after reaching the maximum, but oscillates, what is caused by the movement of the arc in the deion chamber.

We have submitted to tests 3 circuit breakers from different manufacturers. The short-circuit current source was a capacitance battery circuit. The diagram of the measuring circuit is in Fig 10 [6].

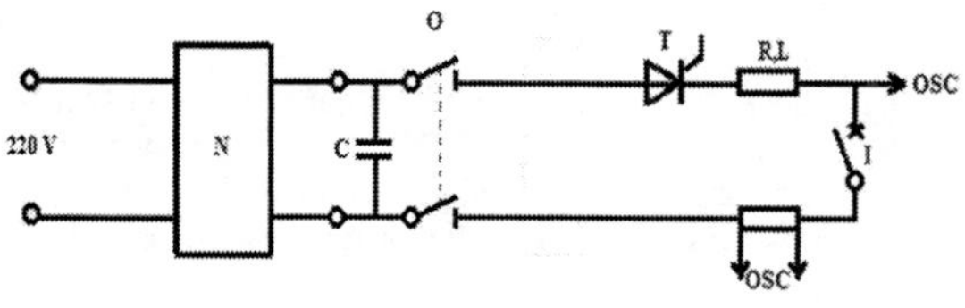

Fig.10 Test circuit with capacitance battery set-up

N= source / driver
C= capacitance battery system
O= disconnecting switch
T= thyristor
RL= resistance and inductance of the circuit
I = tested circuit breaker
OSC= oscilloscopic measurement of current and voltage

Using the N driver, which consist of an autotransformer and a rectifier, we controlled the charge voltage on the capacitor battery labeled C. The discharge of only one half of current wave is provided by the controlled semiconductor switch T in the wiring diagram. The wavelength and slope parameters are set by means of the RLC elements. Because we only charge one half of short-circuit current wave, we call this test non-destructive.

Tab. 1 Parameters of tested circuit breakers

Name	*Char.*	*In [A]*	*Un [V]*	*Poles*	*Breaking capacity [kA]*
Schrack C32/1,6kA	C	32	230/400	1	6
Schnaider Multi9 C60N	B	40	230/400	1	6
Schrack BR-C20/3	C	20	230/400	3	25

Results of measurements are shown on Fig. 11 - 19:

Schrack C32/1,6kA

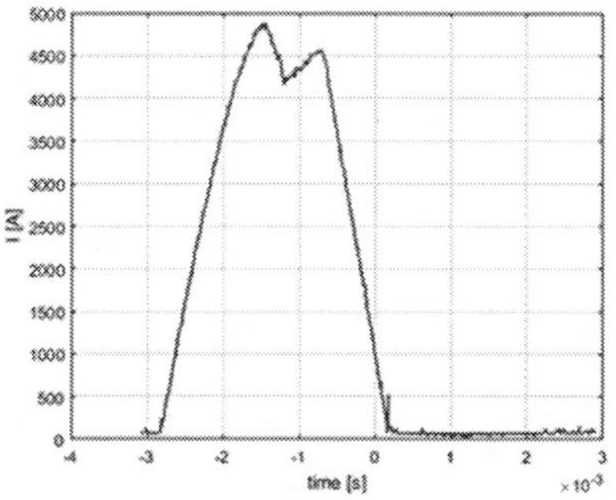

Fig.11 Current behavior

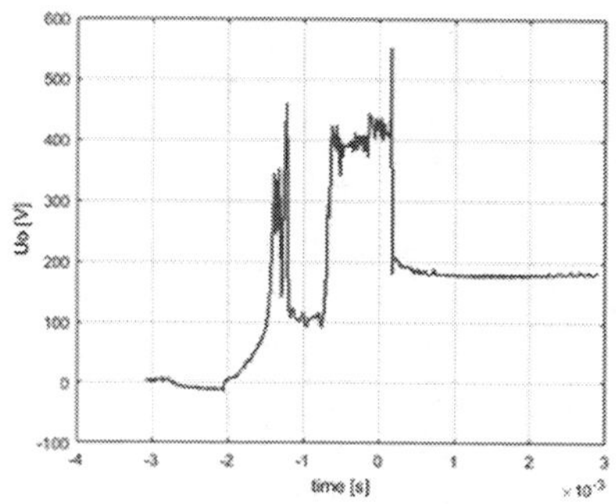

Fig.12 Arc voltage behavior

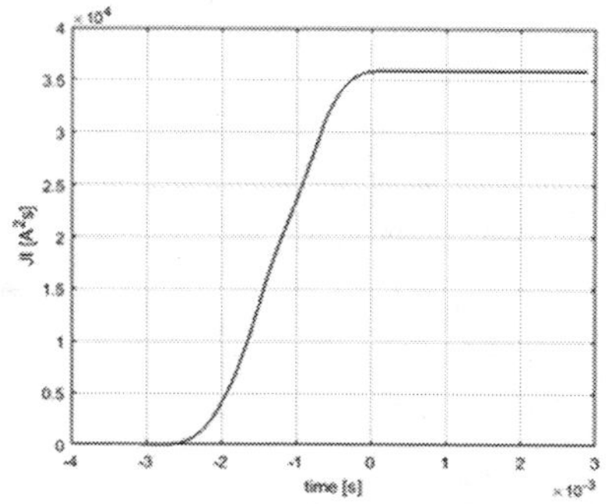

Fig.13 JI behavior

Schrack BR-C20/3

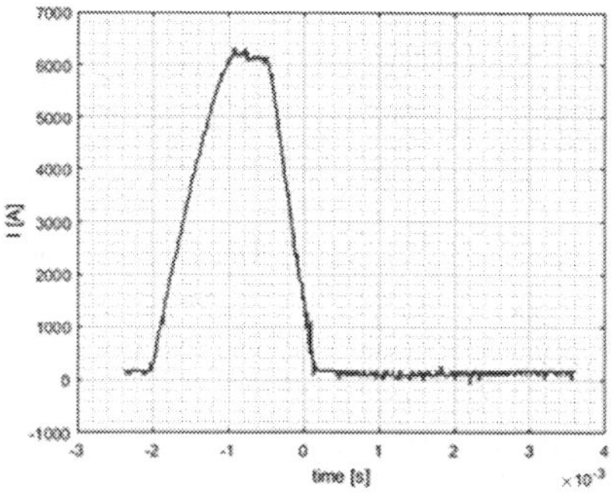

Fig.14 Current behavior

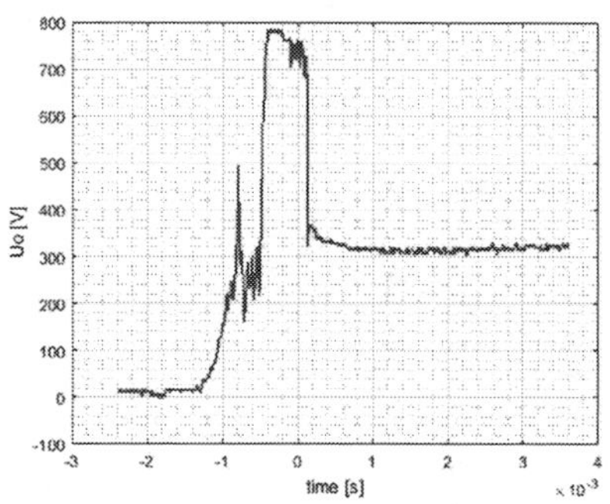

Fig.15 Arc voltage behavior

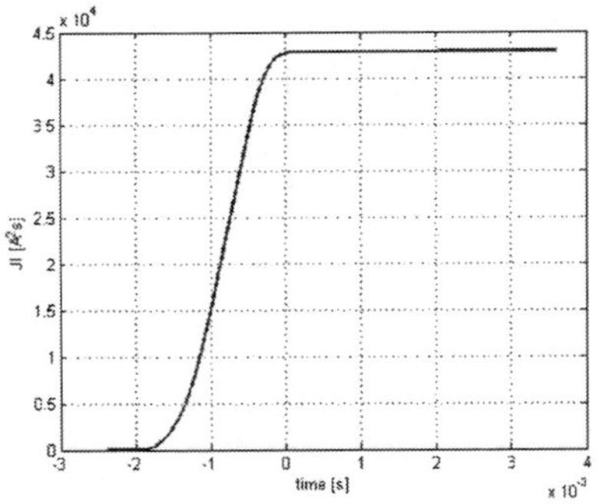

Fig.16 JI behavior

Schneider Multi9 C60N

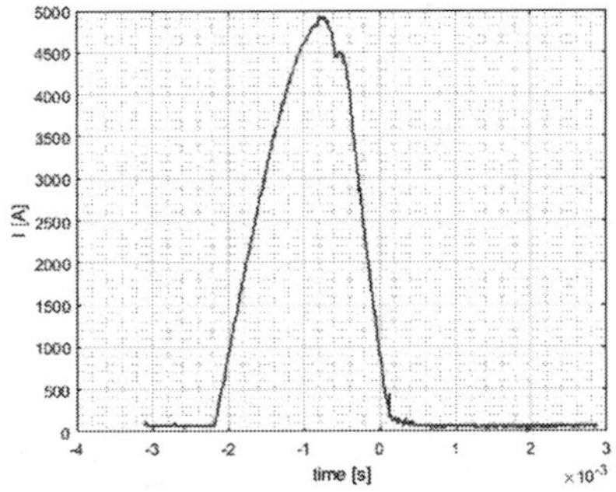

Fig.17 Current behavior

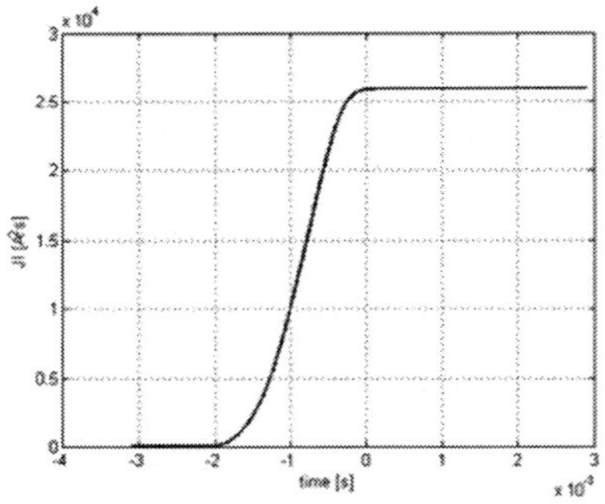

Fig.18 JI behavior

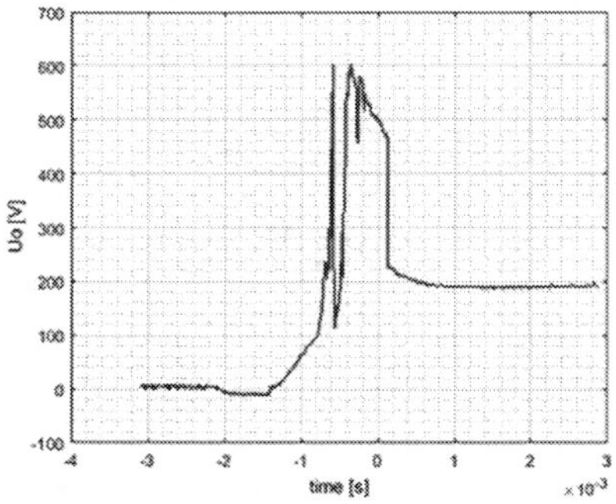

Fig.19 Arc voltage behavior

Tab. 2 Compare of maximal values of current, arc voltage and JI for tested circuit breakers

Name	*Imax [A]*	*Uomax [V]*	*JI*
Schrack C32/1,6kA	4800	550	36000
Schnaider Multi9 C60N	4900	600	26000
Schrack BR-C20/3	6200	780	43000

Fig. 12 shows the oscillations of the arc voltage and the step drop of the arc voltage from 520V to 100V due to the rise of the arc in the extinguishing chamber. This led to limitation of the current on Fig. 11, where the current was increased at the time when the arc entered the extinguishing chamber and the current was interrupted definitively.

V. Compare of Simulation and Measurements

In this chapter, we compare the outputs of our computational program with the measurement results. The input data for our program will be measured arc voltage data and the arc burning time. The output will be the comparison of Joule integral simulated current with the measured Joule integral. Also, the current limiting process is simulated and measured.

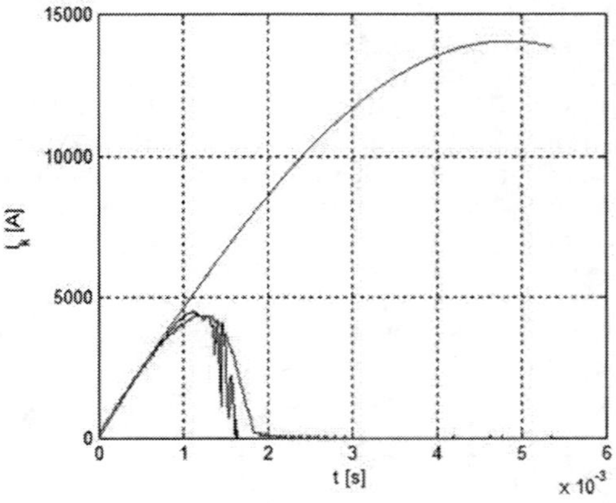

Fig.20 Behavior of current (green), measured limited current (red) and simulation (blue)

The simulated current limitation (blue) interrupts the circuit earlier. The current was interrupted after 1.6 milliseconds.

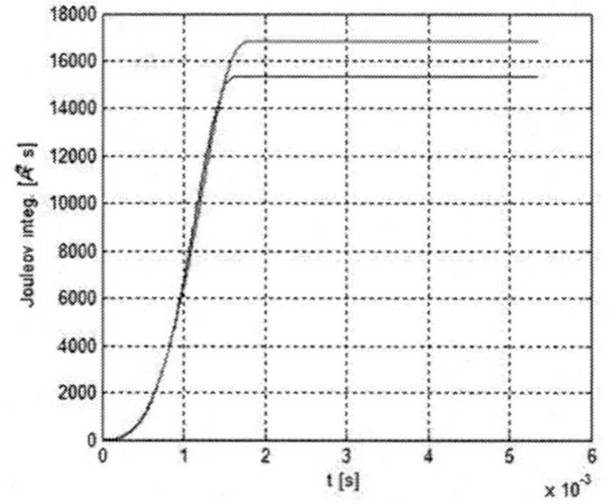

Fig.21 Behavior of JI measured (green) and simulated (blue)

The Joule integral of the simulated behavior has reached 2000 A^2s less value than the value measured. The simulation results are in good compliance to the measurement results.

VI. Conclusion

In this contribution we analyzed the effect of arc voltage on current limiting in LV electrical networks. We performed measurements on 3 samples of circuit breakers and watched the increase of the joule integral. From the measurements it was seen that the circuit breakers limit the short-circuit current by the increase of the arc voltage. We measured the phenomenon of oscillation and the movement of the arc in the chamber and thus the unstable nature of the arc voltage in the real circuit breakers. On the basis of the theory, we created a graphical user interface in Matlab to simulate a current limitation. The input parameter was the arc voltage from the measurement. The theoretical knowledge was verified by measurement and the result was in a relatively good match.

Acknowledgement

This work was supported by the Slovak Research and development Agency under the Contract No. APVV-15-0326.

References

[1] L. Hüttner, 'Solution of some electrical devices problems by simulation methods', Journal of Electrical Engineering, vol. 56, no. 3–4, pp. 89–94, 2005

[2] D. Mayer, B. Ulrich, Elektromagnetické aktuátory, 124 pages, BEN – technická literatúra, Praha, 2008

[3] HÜTTNER, Ľudovít - JURČACKO, Ľudovít - VALENT, Ferdinand. Problémy zhášania elektrického oblúka v elektrických prístrojoch a zariadeniach. In Elektrotechnický magazín. Roč. 16, č. 5-6 (2006), s.21-23. ISSN 1210-5422.

[4] D. Mayer D. Aplikovaný elektromagnetizmus. České Budějovice : Kopp nakladatelství, 2012, p. 531. ISBN: 978-80-7232-424-8.

[5] L. Hüttner L. Jurčacko F. Valent, "Skúmanie prerušenia prúdu v jednosmerných a fotovoltických rozvodoch", in ELOSYS. Elektrotechnika, informatika a telekomunikácie 2015 [CDROM]. Trenčín, Slovakia. 13. – 15. október 2015. 1. vyd. Bratislava : Nakladateľstvo STU v Bratislave, 2015, CD-ROM, pp. 92-96, ISBN 978-80-227-4437-9.

[6] J. Zuščak, F. Janíček, M. Váry, "Simulation of electromagnetic release used in miniature circuit breakers", . In ELITECH´15 [CDROM] : 17th Conference of doctoral students. Bratislava, Slovak Republic, May 25, 2015. 1. vyd. Bratislava : Nakladateľstvo STU, 2015, CD-ROM, p. 6. ISBN 978-80-227-4358-7.

[7] L. Hüttner, F. Valent, L. Jurčacko, "Short-circuit breaking capacity of miniature circuit breakers up to 125 A", in: XVIth symposium on physics of switching arc : Vol. 1 contributed papers. Vol. 2 invited papers : Brno, Czech Republic 5.-9. Sept. 2005. Brno : University of Technology, 2005. ISBN 80-214-2931-3. - S. 207-210

[8] L. Hüttner, M. Duč-Anci, "Temperature field during a short-time transient thermal process", in: XVIth symposium on physics of switching arc : Vol. 1 contributed papers. Vol. 2 invited papers : Brno, Czech Republic 5.-9. Sept. 2005. - Brno : University of Technology, 2005. ISBN 80-214-2931-3. - S. 97-100

[9] M. Rapšík, M. Smola, M. Boháč, M. Mucha, Základy energetiky. Slovenská technická univerzita, 2004. p 201., ISBN 80-227-2074-7

[10] S. Schreiter, A. Ehrhardt, M. Rock, "Apparatus for Switching Direct Currents in Photovoltaic Systems", XXth Symposium on Physicsof Switching Arc FSO 2013, 2th – 6th September 2013, Ski Hotel, Nove Mesto na Morave, Czech Republic, Invited Lectures and Contributed Papers, pp. 285 – 288.

[11] P. Zahlmann, J. Birkl, "Prüfung von Schaltgeräten für Photovoltaik-Generatoren", etz Elektrotechnik und Automation 8/2009, pp. 30-39.

The Impact of Power Line Parameters Asymmetry on Short-circuit Currents in the Comparison with the Standard IEC 60909

Marián Ivanič, Žaneta Eleschová, Jozef Bendík, Matej Cenký
Institute of Power and Applied Electrical Engineering
Faculty of Electrical Engineering and Information Technology, Slovak University of Technology
Bratislava, Slovak Republic
marian.ivanic@stuba.sk, zaneta.eleschova@stuba.sk, jozef.bendik@stuba.sk matej.cenky@stuba.sk

Abstract — **The paper analyses the differences arising between the values of the initial short-circuit currents on the parameter unsymmetrical and symmetrical power lines by the use of the relative deviations of short-circuit currents. The short-circuit currents on parameter symmetrical lines are solved according to the standard IEC 60909 which uses the symmetrical coordinate method. The impacts of ground resistivity and chosen towers of the 400 kV power lines are analysed on differences of short-circuit currents.**

Keywords: transmision lines, short-circuit currents, standard, error analysis

I. Introduction

The asymmetry of power system is neglected at solving short-circuit currents and the method of symmetrical components or the standard IEC 60909 is used for short-circuit calculation with the aim to determine what the maximal or minimal values of short-circuit current can arise when the individual types of short-circuits are considered. In this case the values of short-circuit currents of individual short-circuits will be the same without regard to which phase(s) cause short-circuit. But in the case, when the power system asymmetry is considered, the short-circuit current of individual short-circuits will not be the same, in addition the value of short-circuit current for certain short-circuits can be bigger than the value when the power system asymmetry is neglected. For example in the case of three-phase short-circuit and neglect of power system asymmetry the values of short-circuit currents in each phase will be the same, but with consideration of power system asymmetry the values of these currents will be different.

With regard to extensiveness of lines and the possibility to determine the parameter asymmetry of power line in comparison with other devices in power system, the values of short-circuit currents will be compared on 400 kV power lines with chosen tower types by relative deviations.

II. Calculation of AC Short-Circuit Currents

The standard IEC 60909 deals with the calculation of short-circuit currents, which by the use of symmetrical coordinate method neglects the parameter asymmetry of devices, also power lines. However the simplifications of short-circuit calculations according to this standard are applicable in the case of parameter asymmetry of short line.

The standard by the use of calculation determines two different short-circuit currents: The maximum short-circuit currents which determine the capacity or rating of electrical equipment. The minimum short-circuit currents which can be a basis, for example, for the selection of fuses, for the setting of protective devices and for checking the run-up of motors.

The time diagram of supposed short-circuit current is generally given by the sum of alternating component and direct component of short-circuit current. On the basis of the amplitude attenuation of alternating component two short-circuits are determined:

- far-from-generator short-circuits, the amplitude is constant during the short-circuit,
- near-to-generator short-circuits, the amplitude decays during the short-circuit.

The initial short circuit current I_k'' and peak current i_p are determined in most practical calculations. But the aim is to compare the short-circuit current values of individual fault types at consideration and neglect power system asymmetry, therefore initial short-circuit current will be only compared by the use of relative deviations.

- The type of short-circuit over its duration does not change.
- No load is assumed before short-circuit.
- Resistance of arc is not assumed.
- Shunt admittance of line is neglected.
- The ideal three phase symmetrical voltage source that complies with the conditions of equivalent voltage source is assumed at the beginning of line.
- The different types of short-circuit are assumed at the end of line.
- The short-circuit impedances of three phase alternating system are given by the matrix of phase impedances of line.

Then the relations amongst voltages and initial short-circuit currents can be written as follows

$$\begin{aligned}\bar{U}_a - \bar{U}'_a &= \bar{Z}_{aa}\bar{I}''_{ka} + \bar{Z}_{ab}\bar{I}''_{kb} + \bar{Z}_{ac}\bar{I}''_{kc}\\ \bar{U}_b - \bar{U}'_b &= \bar{Z}_{ba}\bar{I}''_{ka} + \bar{Z}_{bb}\bar{I}''_{kb} + \bar{Z}_{bc}\bar{I}''_{kc}\\ \bar{U}_c - \bar{U}'_c &= \bar{Z}_{ca}\bar{I}''_{ka} + \bar{Z}_{cb}\bar{I}''_{kb} + \bar{Z}_{cc}\bar{I}''_{kc}\end{aligned} \quad (1)$$

or in the matrix form

$$\begin{bmatrix}\bar{U}_a-\bar{U}'_a\\ \bar{U}_b-\bar{U}'_b\\ \bar{U}_c-\bar{U}'_c\end{bmatrix}=\begin{bmatrix}\bar{Z}_{aa}&\bar{Z}_{ab}&\bar{Z}_{ac}\\ \bar{Z}_{ba}&\bar{Z}_{bb}&\bar{Z}_{bc}\\ \bar{Z}_{ca}&\bar{Z}_{cb}&\bar{Z}_{cc}\end{bmatrix}\begin{bmatrix}\bar{I}''_{ka}\\ \bar{I}''_{kb}\\ \bar{I}''_{kc}\end{bmatrix} \quad (2)$$

List of parameters:

$\bar{U}_i$ Phase-to-ground voltages of assumed source
$\bar{U}'_i$ Phase-to-ground voltages in the place of fault
$\bar{I}''_{ki}$ Initial short-circuit currents
$\bar{Z}_{ij}$ Phase impedances of power line

The individual types of short-circuit will be distinguished by indices in the case of short-circuit current label.

Generally the self-impedances are different for the real power lines and the same is valid for mutual impedances so that the power line is parameter unsymmetrical.

The nominal phase-to-phase voltage of line U_n determines the phase-to-ground voltages of assumed three phase ideal symmetrical voltage source as follows

$$\bar{U}_a=\frac{cU_n}{\sqrt{3}};\ \bar{U}_b=\bar{a}^2\frac{cU_n}{\sqrt{3}};\ \bar{U}_c=\bar{a}\frac{cU_n}{\sqrt{3}};\bar{a}=1\angle 120° \quad (3)$$

where *c* is voltage factor.

Previous equations can be edited or simplified on the basis of conditions for certain type of short-circuit and then their solution gives the initial short-circuit currents.

The condition during the three-phase short-circuit with ground connection is as follows

$$\bar{U}'_a=\bar{U}'_b=\bar{U}'_c=0. \quad (4)$$

Then the initial short-circuit currents are given by relation

$$\begin{bmatrix}\bar{I}''_{k3Na}\\ \bar{I}''_{k3Nb}\\ \bar{I}''_{k3Nc}\end{bmatrix}=\begin{bmatrix}\bar{Z}_{aa}&\bar{Z}_{ab}&\bar{Z}_{ac}\\ \bar{Z}_{ba}&\bar{Z}_{bb}&\bar{Z}_{bc}\\ \bar{Z}_{ca}&\bar{Z}_{cb}&\bar{Z}_{cc}\end{bmatrix}^{-1}\begin{bmatrix}\bar{U}_a\\ \bar{U}_b\\ \bar{U}_c\end{bmatrix}. \quad (5)$$

During the three-phase short-circuit without ground connection the conditions are as follows

$$\bar{U}'_a=\bar{U}'_b=\bar{U}'_c=\bar{U},\\ \bar{I}''_{k3a}+\bar{I}''_{k3b}+\bar{I}''_{k3c}=0. \quad (6)$$

Then the initial short-circuit currents are given by relations

$$\begin{bmatrix}\bar{U}\\ \bar{I}''_{k3b}\\ \bar{I}''_{k3c}\end{bmatrix}=\begin{bmatrix}1&(\bar{Z}_{ab}-\bar{Z}_{aa})&(\bar{Z}_{ac}-\bar{Z}_{aa})\\ 1&(\bar{Z}_{bb}-\bar{Z}_{ba})&(\bar{Z}_{bc}-\bar{Z}_{ba})\\ 1&(\bar{Z}_{cb}-\bar{Z}_{ca})&(\bar{Z}_{cc}-\bar{Z}_{ca})\end{bmatrix}^{-1}\begin{bmatrix}\bar{U}_a\\ \bar{U}_b\\ \bar{U}_c\end{bmatrix} \quad (7)$$

and the current in the phase *a*

$$\bar{I}''_{k3a}=-(\bar{I}''_{k3b}+\bar{I}''_{k3c}). \quad (8)$$

The condition during the phase-to-ground short-circuit caused by the phase *a* are as follows

$$\bar{U}'_a=0;\ \bar{I}''_{k1b}=\bar{I}''_{k1c}=0 \quad (9)$$

Then the initial short-circuit currents in the phase *a* is given by relation

$$\bar{I}''_{k1a}=\frac{\bar{U}_a}{\bar{Z}_{aa}} \quad (10)$$

Analogical relations are valid for the initial short-circuit currents caused by other phases.

$$\bar{I}''_{k1b}=\frac{\bar{U}_b}{\bar{Z}_{bb}};\ \bar{I}''_{k1c}=\frac{\bar{U}_c}{\bar{Z}_{cc}} \quad (11)$$

The condition during the phase-to-phase short-circuit caused by the pair of phases *a, b* are as follows

$$\bar{U}'_a=\bar{U}'_b=\bar{U};\ \bar{I}''_{kc}=0;\ \bar{I}''_{k2b}=-\bar{I}''_{k2a} \quad (12)$$

The solution for the initial short-circuit currents caused by phases *a, b* is given by

$$\bar{I}''_{k2a}=\frac{\bar{U}_a-\bar{U}_b}{\bar{Z}_{aa}+\bar{Z}_{bb}-2\bar{Z}_{ab}}=-\bar{I}''_{k2b} \quad (13)$$

Similar relations are valid for other cases of phase-to-phase short-circuit.

The relations for the calculation of initial short-circuit currents of individual types of short-circuit are valid also for parameter symmetrical line when the self-impedances are equal and also mutual impedances. The only way how to get parameter symmetrical power line is the transposition of phase conductors – transposed line.

In the case of transposed lines the use of symmetrical components has sense due to simplification of relations which is applied in the standard IEC 60909. Then the relation for initial short-circuit currents of individual short-circuits are as follows:

- Three phase short-circuit:

$$I''_{k3}=I''_{k3N}=\frac{cU_n}{\sqrt{3}|\bar{Z}_1|} \quad (14)$$

- Phase to ground short-circuit:

$$I''_{k1}=\frac{c\sqrt{3}U_n}{|\bar{Z}_1+\bar{Z}_2+\bar{Z}_0|}=\frac{c\sqrt{3}U_n}{|2\bar{Z}_1+\bar{Z}_0|} \quad (15)$$

- Phase to phase short-circuit:

$$I''_{k2}=\frac{cU_n}{|\bar{Z}_1+\bar{Z}_2|}=\frac{cU_n}{|2\bar{Z}_1|} \quad (16)$$

List of parameters:

$\bar{Z}_1$ The positive sequence short-circuit impedance
$\bar{Z}_2$ The negative sequence short-circuit impedance
$\bar{Z}_0$ The zero sequence short-circuit impedance

In the case of transposed lines the currents of three-phase short-circuit with ground connection and without ground connections are the same.

Differences arising between short-circuit currents of untransposed and transposed power line will be compared by the relative deviations. The magnitudes of short-circuit currents of transposed lines are the same for individual short-circuits so the phases need not be distinguished. This means that the currents without label of phases are the currents of transposed line and then the relation for the relative deviation of initial short-circuit current is

$$\delta I''_{kSi} = \frac{I''_{kSi} - I''_{kS}}{I''_{kS}} 100 = \left(\frac{I''_{kSi}}{I''_{kS}} - 1\right) 100\ [\%] \tag{17}$$

The symbol S represents the type of short-circuits (possible values: *1*, *2*, *3*, *3N*) and i is the phase of power line (a, b or c).

The relative deviation of short-circuit currents do not depend on the length of the line when only the impedances of line are taken into account. This assumption is right when the capacity charging currents are negligible in comparison with the short-circuit currents.

The higher short-circuit currents of parameter asymmetrical power line are more important in the comparison of maximal short-circuit currents according to the standard from the view of rating of electrical equipment. So the maximal relative deviation of short-circuit currents will be analysed.

III. THE RELATIVE DEVIATIONS ON 400 KV POWER LINES

Individual short-circuit types are analysed on towers of one circuit three phase power line on voltage level 400 kV with the length 50 km. The voltage factor is c = 1.05. Analyse is done for different values of ground resistivity. The positions of conductors are in the following table. The towers Tubular, Cat I and III have the position of conductors in the vertices of the triangle. The towers Portal I, II and III have the horizontal position of conductors.

Table 1. The positions of conductors on the towers

Tower	Axis	a	b	c	g1	g2
Tubular	x [m]	-4.06	3.95	5.06	-2.50	2.50
	y [m]	22.00	25.50	18.00	33.30	33.30
Cat I	x [m]	-7.00	0.00	7.00	-4.00	4.00
	y [m]	18.50	26.80	18.50	30.50	30.50
Cat III	x [m]	-8.50	0.00	8.50	-4.00	4.00
	y [m]	18.50	26.80	18.50	30.50	30.50
Portal I	x [m]	-12.00	0.00	12.00	-6.00	6.00
	y [m]	18.00	18.00	18.00	29.80	29.80
Portal II	x [m]	-12.00	0.00	12.00	-6.00	6.00
	y [m]	18.80	18.80	18.80	31.80	31.80
Portal III	x [m]	-13.00	0.00	13.00	-7.00	7.00
	y [m]	18.80	18.80	18.80	31.80	31.80

a, b, c - phase conductors,
g1, g2 - ground wires,
x - horizontal distance of conductor from the middle of tower,
y - vertical distance of conductor above the ground.

The phase conductors consist of three bundle conductors per phase with bundle diameter 23.1 cm. Each conductor consists of ACSR (Aluminium Conductor Steel Reinforced) rope 434-AL1/56-ST1A. Ground wires consist of one conductor – ACSR rope 184-AL1/30-ST1A. Maximal sag of conductors is assumed 9 m. The required parameters of these ACSR ropes are in the following table.

Table 1. The ACSR rope parameters

Type of ACSR	ρ_{DC} [Ω/km]	d_l [cm]	ξ [-]
434-AL1/56-ST1A	0.0666	2.88	0.810
184-AL1/30-ST1A	0.1571	1.90	0.809

ρ_{DC} - The DC resistance of conductor
ξ - Rope coefficient
d_l - Outside diameter of rope

Individual phase impedances (self and mutual) were calculated according to Carson method.

The first fault type is analysed three-phase ground short-circuit. It is the simplest fault from the view of the calculation. This fault is the only symmetrical from the other ones. But in the case of parameter asymmetrical power system this three phase ground short-circuit is unsymmetrical as well.

Axis X is displayed in the logarithmic scale. The short-circuit currents are analysed also for the ideal ground $\rho = 0$, but for this case the logarithm of null does not exist. Due to this reason the value of ground resistivity $\rho = 0{,}1$ is used for the ideal ground in the figure(s).

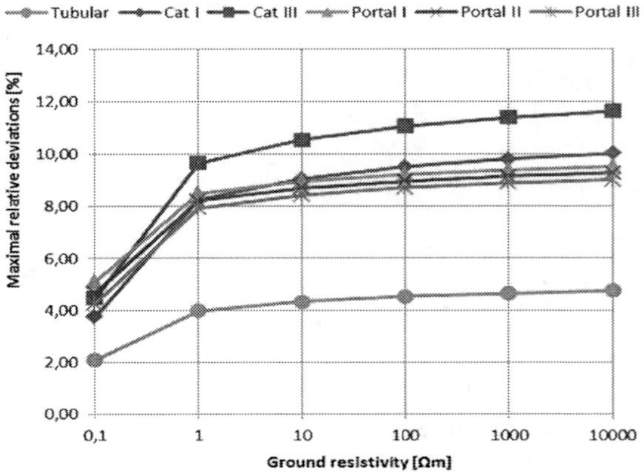

Figure 1. Three-phase ground short-circuit

The parameter asymmetry has significant impact on the differences of three-phase ground short-circuit currents between untransposed and transposed line. In addition the growth of ground resistivity increases the maximal relative deviations of short-circuit currents for all analysed towers. The smallest values of maximal relative deviations are in the case of ideal ground (ρ = 0), when the asymmetry of parameters and also currents are caused only by the conductors. Thus the real ground has negative impact on maximal relative deviations of three-phase ground short-circuit currents. The biggest values of maximal relative deviations are for the tower Cat III when the values are on the level about 11.5 %.

Three-phase short-circuit without ground connections is the next analysed fault. As this short-circuit has no connection with the ground, the short-circuit currents are closed only through the phases.

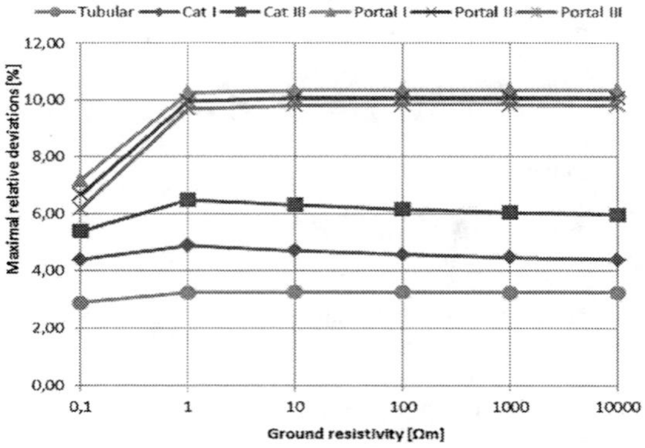

Figure 2. Three-phase short-circuit

Big differences of maximal relative deviations of three-phase short-circuit currents are between the ideal and real ground for towers Portal I, II and III. The next growth of ground resistivity has small impact on the differences of three-phase short-circuit currents between untransposed and transposed line. The highest values of maximal relative deviations are for the towers Portal I, II and III when the values are on the level about 10 %.

Phase to ground short-circuit is failure type which has the highest probability of occurrence in comparison with other short-circuit types.

Three phase short-circuit can arise only with one way and that is the connection of three phases. Opposite, one-phase short-circuit can arise with three ways and that is the connection of the first or second or third phase with ground.

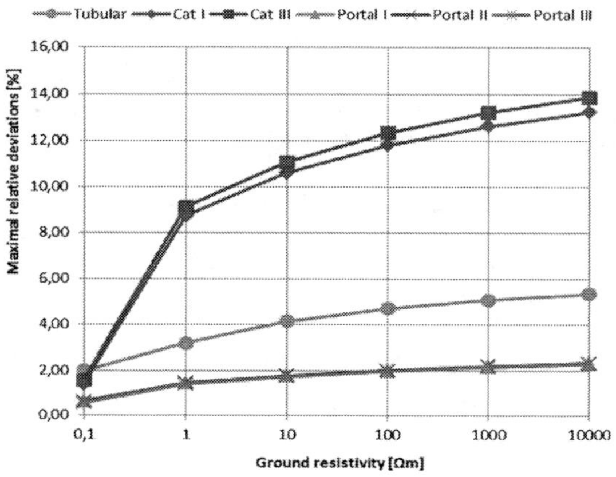

Figure 3. One-phase short-circuit

The growth of ground resistivity causes the growth of maximal relative deviations mainly for towers Cat I and III which can be up to 14 %. Although the growth of ground resistivity causes the growth ground impedances and then the initial phase-to-ground short-circuit currents are smaller.

The relative deviations are influenced by the ground wires too. In the case of towers Portal I, II and III the relative deviations will be the same and equal to zero due to two reasons. Phase-to-ground short-circuits depend only on the self-impedances and the heights of conductors of these towers are the same so the self-impedances must be the same.

The last analysed fault is two-phase short-circuit which can arise by three ways.

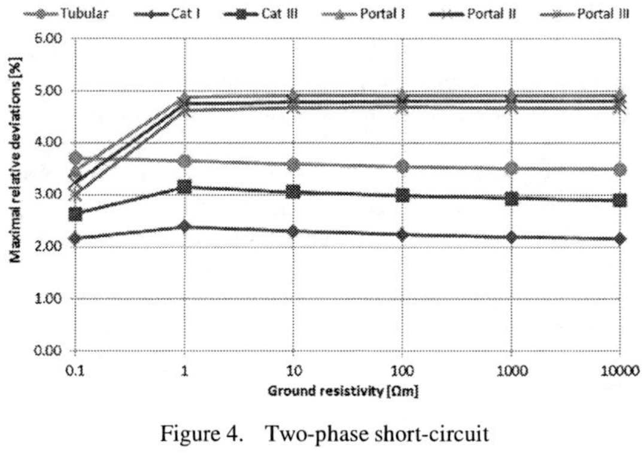

Figure 4. Two-phase short-circuit

The behaviour of the maximal relative deviations of two-phase short-circuit currents is similar to the maximal relative deviations of three-phase short-circuit currents. But in this case the maximal relative deviations are smaller. The highest values of maximal relative deviations are for the towers Portal I, II and III when the values are under the level 5 %.

IV. Conclusion

The initial short-circuit current values of parameter asymmetrical and symmetrical power line were compared by the use of the relative deviations of short-circuit currents. The maximal relative deviation of short-circuit currents occurs in the case of phase-to-ground short-circuit on the tower Cat III when the value of relative deviation can be up to 14 %.

The highest values of short-circuit currents are for the three-phase short-circuits with ground connection (Cat I, III) and without ground connection (Portal I, II, III). The relative deviations of short-circuit currents for these towers are on the level 10 %, but exception is the tower Cat III when the relative deviations are under 12 %.

The impact of real ground resistivity is almost negligible in the case of short-circuits without ground connection, but the maximal relative deviations of short-circuits with ground connections for towers Cat I and III was influenced more significantly. In the case of ideal ground the maximal relative deviations of short-circuits are smaller than for the real ground what can be seen mainly in the case of towers Cat I and III.

Acknowledgment

This publication is a result of implementation of the project: "Increase of Power Safety of the Slovak Republic" (ITMS: 26220220077) supported by the Research & Development Operational Programme funded by the ERDF.

References

[1] IEC International standard 60909-0, "Short-circuit currents in three-phase a.c. systems – Part 0: Calculation of currents", 2001.

[2] L. V. Strezoski and M. D. Prica, "Short-circuit analysis in large-scale distribution systems with high penetration of distributed generators," in IEEE/CAA Journal of Automatica Sinica, vol. 4, no. 2, pp. 243-251, April 2017. doi: 10.1109/JAS.2017.7510517.

[3] H. Ahmadi and M. Armstrong, "Transmission line impedance calculation using detailed line geometry and HEM soil resistivity measurements," 2016 IEEE Canadian Conference on Electrical and Computer Engineering (CCECE), Vancouver, BC, 2016, pp. 1-4. doi: 10.1109/CCECE.2016.7726808

[4] H. Dommel, "Electromagnetic Transients Program Reference Manual (EMTP Theory Book)", Bonneville Power Administration, Portland, 1986.

[5] A. J. Schwab, „Elektroenergiesysteme. Erzeugung, Transport, Übertragung und Verteilung elektrischer Energie", Berlin Heidelberg, Springer, 2012.

[6] P. Toman, J. Dvořák, D. Topolánek, J. Orságová, "Earth Fault Currents in MV Compensated Networks". In Proceedings of the 10th International Scientific Conference Electric Power Engineering 2010. Ostrava: VSB- TU OSTRAVA, 2009. s. 384-386. ISBN: 978-80-248-1947- 1.

Measurement of Transformer Harmonic No-load Currents Spectrum using Magnetic Field Spectral Analysis

Jozef Bendík, Matej Cenký, Peter Janiga, Mmarián Ivanič, Žaneta Eleschová
Faculty of Electrical Engineering and Information Technology STU
Institute of Power and Applied Electrical Engineering
Bratislava, Slovakia
jozef.bendik@stuba.sk, matej.cenky@stuba.sk, peter.janiga@stuba.sk, marian.ivanic@stuba.sk, zaneta.eleschova@stuba.

Abstract — **This paper presents a noninvasive and galvanic contactless method for the measurement of the harmonic current spectrum. The purpose of this research is to evaluate the simple and fast method of transformer diagnostics and software modeling. Advantages of using this method can be found in situations where contactless methods are preferred. Method is based on spectral analysis of superposed magnetic fields produced by transformer coils. The evaluation of this approach has been done by experimental measurement and by comparing the data of the current spectrum from each coil of the transformer and the magnetic field spectrum measured in one point of the observer in the vicinity of the power transformer in the same time. All measurements have been made on transformer 220/110 kV in no-load state. The results show a correlation between current harmonics and the magnetic field harmonic spectrum. However, the conditions such as geometry and measurement position are not negligible and must be chosen optimally.**

Keywords: magnetic flux density, power system security, power measurement

I. Introduction

It is well known fact that the voltage shape in most of electrical systems cannot be considered a sine wave. Distortion of voltage wave results in appearance of higher harmonics in frequency spectrum of network voltage (and sometimes non harmonics). The same applies for current wave absorbed by non-linear load. Harmonic disturbances are caused by non-linear systems. These systems generate voltages with frequencies different from the power frequency or absorb non-sinusoidal currents. The origin of harmonic disturbances are for example generators, transformers, tyristor controlled devices, power electronics ect. [1].

It is known that any cyclically repeating wave form can be represented as a sinusoidal wave at a fundamental frequency and sum of other sinusoidal waves, usually of smaller amplitudes than that of the fundamental, and of frequencies that are integral multiplies of the fundamental frequency. In general the additional sinusoids, or harmonics, each have a different phase relationship to the fundamental sinusoid at the origin of the time and tend to decrease in amplitude as the multiple of fundamental frequency, the harmonic order, increases. Using this approach a distorted wave equals fundamental wave plus harmonics, the distorted wave can be written as equation 1 [1] [2][6]. There are several well-known and easily applied mathematical methods that permit any finite, unvalued cyclically repetive wave shape to be analyzed into fundamental and harmonic components as defined below [1].

$$y = f(x) = a_0 + \sum_{n=1}^{\infty} b_n \sin(n\omega t + \phi_n) \tag{1}$$

Where:

- a_0 is a possible unidirectional component
- b_n is a peak value of n-th harmonic
- n is harmonics order
- φ_n is an phase angle of nth harmonics component when t = 0

This kind of representation of distorted wave lends itself to a quantities approach by application of mathematical techniques and offers several advantages:

- In a linear network each component of harmonic order can be considered separately. Investigation of problem is simplified [1].
- A meaningful comparison can be made between distorted voltage, current or electromagnetic field waveforms under different loads, system conditions or like in case of this article different measurement procedures [2]
- All the possible information about the distortion is given, that is, about the deviation of complex wave from a pure sinusoid waveform

A. General harmonics indicators

Not every time a knowledge of individual harmonic component is needed. Many times a single figure, indicator expressed by single value is required. The most common harmonic index indicator, which relates to the voltage and current waveform, is the THD (total harmonic distortion), which is defined as the root mean square (rms) of the harmonics expressed as a percentage of the fundamental component, equation 2 [3] [5][7][9]. Where I_n is the single frequency of rms. current at harmonic n, n is the maximum harmonic order to be considered and I_1 is the fundamental line to neutral rms. current. For majority of applications, it is sufficient enough to consider the harmonic range from the 2nd to the 50th.

$$THD = \frac{\sqrt{\sum_{n=2}^{N} I_n^2}}{I_1} \qquad (2)$$

Characterizing the current distortion level by THD can be misleading, in situations when the fundamental load current is low. A high THD value for input current may not be of significant concern if the load is light, since the magnitude of the harmonic current is low, though its relative distortion to the fundamental frequency is high [4] [6] [9]. To avoid such ambiguity a total demand distortion (TDD) factor is used, defined as:

$$TDD = \frac{\sqrt{\sum_{n=2}^{N} I_n^2}}{I_R} \qquad (3)$$

This factor is similar to THD except that the distortion is expressed as a percentage of some rated or maximum load current magnitude, rather than as a percentage of the fundamental current.

B. *Transformers no-load current analysys*

At no-load the primary winding of the transformer is practically balanced by the back electromagnetic field because the effect of winding resistance and leakage reactance is negligible at low currents. The impressed voltage V_1 for a sinusoidal supply is:

$$V_1 = -e_1 = -E_m \sin(\omega t) = N_1 \frac{d\varphi}{dt} \qquad (4)$$

Sinusoidal primary voltage produces a sinusoidal flux at no-load.

$$\varphi = -\int \frac{e_1}{N_1} dt = \frac{E_m}{N_1 \omega} \cos \omega t = \varphi_m \cos \omega t \qquad (5)$$

The primary current, however, will not be purely sinusoidal, because the flux is not linearly proportional to the magnetizing current. In an ideal core without hysteresis loss the flux φ and the magnetising current needed to produce it are related to each other by the magnetising curve of the steel used in the laminations, as shown in Figure 1(a). In Figure 1(b), where ϕ represents the sinusoidal flux needed to balance the primary voltage, the magnetising current is plotted against time for each value of ϕ and the resulting waveform is far from sinusoidal [6].

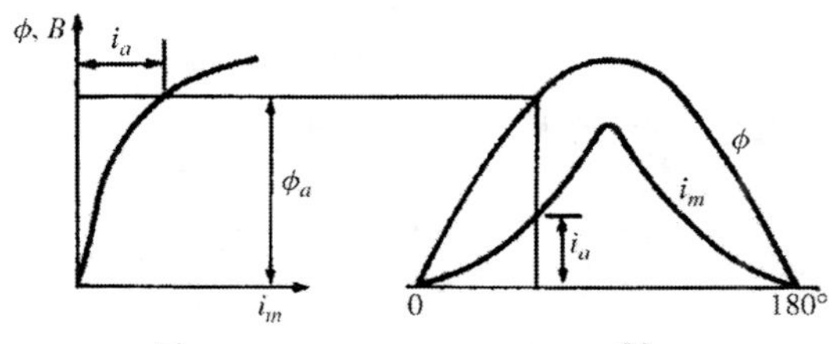

Figure 1. Transformer magnetisation (without hysteresis): (a) magnetisation curve; (b) flux and magnetisation current waveforms

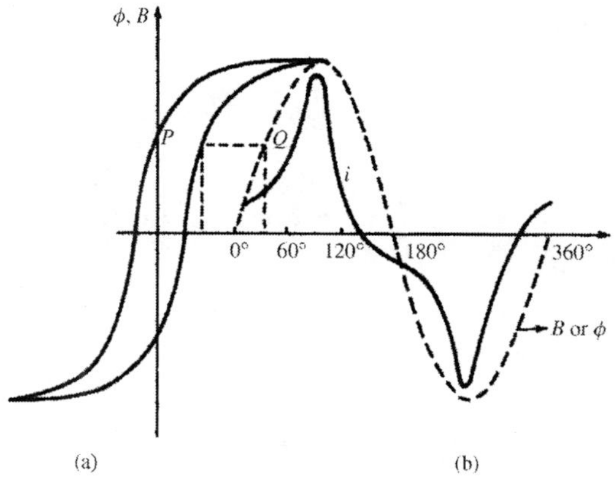

Figure 2. Transformer magnetisation (including hysteresis): (a) magnetisation curve; (b) flux and magnetisation current waveforms

C. *Analysys of time dependent data*

Next step is to implement harmonic analysis. Harmonic analysis is the process of calculating the magnitudes and phases of the fundamental and higher-order harmonics. The resulting series, known as the Fourier series, establishes a relationship between a time-domain function and that function in the frequency domain. When dealing with any kind of harmonic spectrum analysis, data for analysis is often available in the form of a sampled time function, represented by a time series of amplitudes, separated by fixed time intervals of limited duration. When dealing with such data, a modification of the Fourier transform, the discrete Fourier transform (DFT), is used. The implementation of the DFT by means of the so-called Fast Fourier transform (FFT) forms the basis of most modern spectral and harmonic analysis systems [2] [3] [6].

II. MAGNETIC FIELD REPRESENTATION OF NO-LOAD CURRENT

Fact that current passing thought any kind of conductor creates a magnetic field around creates the basic concept for of non-contact measurement and current harmonics spectrum analysis.

Magnetic flux density with frequency below 10 kHz created by any AC source can be described by phasors and vectors [8]. A vector is characterized by a magnitude and an angle in space or by three spacial components, equation 6.

$$\vec{B} = \left[B_x ; B_y ; B_z \right] \qquad (6)$$

Phasor on the other hand is a quantity with a sinusoidal time variation described by a magnitude and a phase angle, equation 7 for single frequency. Angle φ describes phase shift [4].

$$B_x(t) = B_x e^{j\varphi_x} e^{j\omega t} = B_x \cos(\omega t + \varphi_x) = B_x \angle \varphi_x \qquad (7)$$

Three orthogonal components of vector may be phasors with different magnitude and phase angles. This components are called phasor—vector, equation 8. In this article, a vector is indicated with an arrow and phasor-vectors with a hat over arrow.

$$\hat{\vec{B}} = \left[\hat{B}_x;\hat{B}_y;\hat{B}_z\right] = \left[B_x\angle\varphi_x;B_y\angle\varphi_y;B_z\angle\varphi_z\right] \quad (8)$$

To quantify magnetic flux density in single point of observer we use its the relationship between flux and current. This relations for single coil can be written by Biot-Sawart low as follows:

$$\hat{\vec{B}}_a = \vec{B}_a(t) = \frac{\mu_0}{4\pi}\int\frac{i_a(t)d\vec{l}_a \times \vec{r}_a}{\left|\vec{r}_a^{\,3}\right|} \quad (9)$$

Where i_a is phase current of first phase, $d\vec{l}_a$ is element of conductor and vector $\vec{r}_a$ points from conductor element to point of observer. By integrating along conductor path, in case of transformer, along its primary and secondary winding coil, and by summing the field from other two phases we get magnetic flux density in a single point of observer. For these type of calculations a finite element method is surely most suitable.

However most measurement device of magnetic flux density are not capable of displaying a phase-vector value of field for each individual frequency. Common type of evaluation the field is then made by rms value of the field, equation 10 [4] [8].

$$B_{rms} = \sqrt{B_x^{\,2} + B_y^{\,2} + B_z^{\,2}} \quad (10)$$

Magnetic flux density is a complex value, which can be described in various possibilities for each frequency. The total value of this is not only the current passing through the conductor, but also the winding, but also the geometry of the conductor and the position of measurement. It is important to have this fact on mind, because it can also be a source of misunderstanding and unappropriated error [4][8].

Finally, when measuring the magnetic flux density in locations like the power station, conductors that are not the parts of investigated devices also contribute to the final value of the field. It is good to keep the value of those fields as low as possible, or to measure them where their contribution is negligible.

III. Measurement and Comparison of Current and Magnetic Flux Density Frequency Spectrum

We assume that when magnetic flux density is measured properly and the distance between investigated device, and the measurement device is much higher then a distance between possible parasitic sources of magnetic flux and the measurement device, the magnetic flux density frequency spectrum will copy the trend of investigated current, in our case no-load power transformer 220/110 kV.

A. Current measurement

For current measurement a quality power quality analyzer BK-ELCOM ENA 500.22 has been used. The current was measured using measuring pliers placed on each phase of secondary winding of measurement transformer connected to the primary winding of power transformer. Parameters of analyzer for measurement were:

- sampling frequency of analyzer, 9600 samples per second
- 192 of samples per period
- DFT algorithm for spectral analysis
- 3 phase current measurement
- duration of analyzed time interval T = 0.2 s
- measurement of RMS values of current
- Ampflex measuring pliers

B. Magnetic flux density measurement

For magnetic flux measurement was used electromagnetic field EMC/EMI Spectrum Analyzer NF-5030. Device is capable to measure magnetic field in range frequency from 1Hz to 1MHz. Resolution of the device is 100pT to 20mT. The manufacturer states 3% accuracy. Parameters of the comparing measurement were:

- Frequency range 45 Hz – 2150 Hz
- 3- axis measurement
- RMS value of the field was evaluated
- position of measurement device was 1,8 m above ground
- measurement device was places 1 m from the transformer case, on the side of primary winding.

C. Data processing

It is important to underline the fact that current measurement gives us unique value only for each phase of the primary winding of the transformer. The value of the measured magnetic field is a superposition of all magnetic fields, which produces not only the current of the primary winding but also the secondary and third winding of the transformer. The Spectran measuring device was placed as close as possible to the primary winding coil of the power transformer.

Comparison of values from current measurement and magnetic flux measurement was made in the frequency domain, comparing the ratio to the first harmonic. However, for comparison we needed a single current value. The most optional and real combination of currents, which creates a magnetic flux density, based on phase shift and harmonic order. To achieve this, we used the following methodology.

Since the phase shift of every third harmonic is equal to the other two phases, we made a summation of phase currents for each harmonic order which is multiple of three.

$$\begin{aligned} i_{3\,phase1}(t) &= I_3\sin(3(\omega t+0^\circ)) = I_3\sin(3\omega t) \\ i_{3\,phase2}(t) &= I_3\sin(3(\omega t-120^\circ)) = I_3\sin(3\omega t) \\ i_{3\,phase3}(t) &= I_3\sin(3(\omega t+120^\circ)) = I_3\sin(3\omega t) \end{aligned} \quad (11)$$

This principle does not apply on orders different that multiple of three, equation 12 , 5th for example, so average of phase currents was calculated for comparison instead.

$$\begin{aligned} i_{5\,phase1}(t) &= I_5\sin(5(\omega t+0^\circ)) = I_5\sin(5\omega t) \\ i_{5\,phase2}(t) &= I_5\sin(5(\omega t-120^\circ)) = I_5\sin(5\omega t+120^\circ) \\ i_{5\,phase3}(t) &= I_5\sin(5(\omega t+120^\circ)) = I_5\sin(5\omega t-120^\circ) \end{aligned} \quad (12)$$

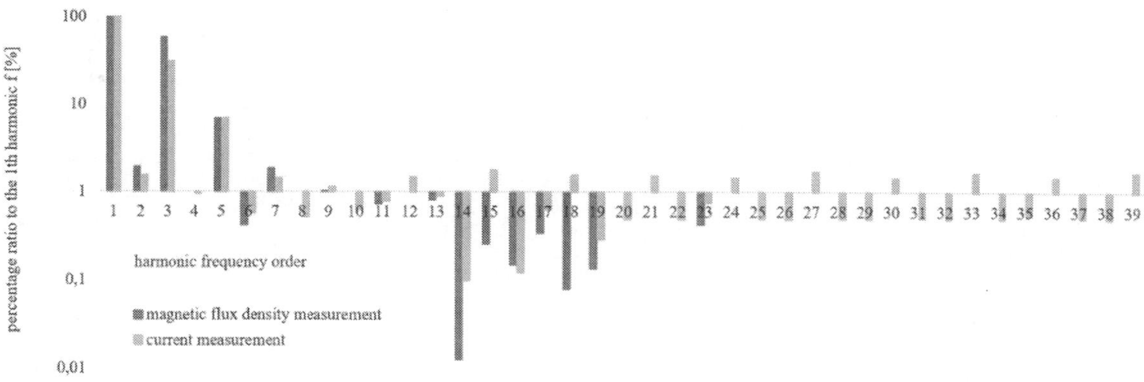

Figure 3. Comparison of frequensy spectrum of current and magnetic flux density measurement

Fig. 3 shows the results from the measurement and data post processing according to methodology above. Results are displayed in logarithmic scale. Magnetic flux density was measured only to 23th order because of measurement device time consumption. Table 1. shows the relative error of the measurement. Table 2. shows the comparison o THD values of both measurements.

Table 1. relative error of the measurement

harm. order	relative error	Flux measurement
1	0,00%	
2	23,65%	
3	88,92%	
4	100,00%	NO
5	0,27%	
6	26,61%	
7	30,02%	
8	100,00%	NO
9	11,73%	
10	100,00%	NO
11	7,25%	
12	100,00%	NO
13	9,43%	
14	87,34%	
15	85,58%	
16	21,59%	
17	46,02%	
18	95,01%	
19	53,55%	
20	100,00%	NO
21	100,00%	NO
22	100,00%	NO
23	43,66%	

Table 2. comparison of THD values

	THD
magnetic flux measurement	58,43%
current measurement	32,18%

Conclusion

This paper has shown the possibility of non-contact measurement for a power transformer. This principle can be applied to any power device. It has been shown that measuring magnetic flux density depends on the position, and surrounding power devices. Underestimating these facts can lead to a misinterpretation of measured data. In the shown experiment the magnetic flux density was measured only in one point of the observer and the current was measured only on the primary side of the power transformer. However, as shown, the trend for harmonics spectrum has a clear correlation. By multiplying simultaneous measurements of magnetic flux density, choosing the better position for measurement and knowing the geometry of conductors, it is possible to make these types of measurements even more accurate and suitable for diagnostic uses. We assume that this is not the most appropriate method for transformer diagnostics, but we assume that it can be useful for long period monitoring measurements.

Acknowledgement

This contribution is the result of the project implementation: National Centre for Research and Application of Renewable Energy Sources (ITMS: 26240120016), supported by the Research & Development Operational Programme funded by the ERDF.

References

[1] Working group 36-05",Harmonics, characteristic parameters, methods of study, estimates of existing values in the network" Cigre Study committee 36, 1981, Electra 077.

[2] Working group 05"Equipment producing harmonics and contidions governing their connnection to the mains power supply " Cigre Study committee 36, March 1989, Electra 124.

[3] Working group 05"Transformets and instruments for measuring harmonics " Cigre Study committee 36, May 1989, Electra 127.

[4] D. Mayer, J. Polák, "Metody řešení elektrických a magnetických polí", Praha : SNTL, 1983. pp.450 .

[5] EPRI. AC transmission line reference book - 200 kV and above - 3rd Ed, Chapter 7 pages ,442–560, 2005.

[6] P. A. Venikar, M. S. Ballal, B. S. Umre and H. M. Suryawanshi, "Transformer incipient inter-turn fault detection based on no-load current harmonic analysis," in 2015 IEEE Eindhoven PowerTech, Eindhoven, 2015, pp. 1-5. doi: 10.1109/PTC.2015.7232294

[7] S. A. Deokar and L. M. Waghmare, "Impact of power system harmonics on insulation failure of distribution transformer and its remedial measures," in 2011 3rd International Conference on Electronics Computer Technology, Kanyakumari, 2011, pp. 136-140. doi: 10.1109/ICECTECH.2011.5941817

[8] Mayer, D.: Aplikovaný elektromagnetizmus: úvod do makroskopické teorie elektromagnetického pole pro elektrotechnické inženýry. 2012. České Budějovice : Kopp, 2012. 531 s. ISBN - 978-80-7232-436-1

[9] C. H. Meng [1]C. H. Meng, C. W. Chang, C. C. Chiu, K. H. Chen, Y. H. Lin, T. Y. Tsai, C. C. Lee, "High efficiency and total harmonic distortion improvement by zero current prediction technique for transformer-free buck power factor corrector," 2014 IEEE Energy Conversion Congress and Exposition (ECCE), Pittsburgh, PA, 2014, pp. 1228-1231. doi: 10.1109/ECCE.2014.6953541

Reverse Interlocking Protection using IEC 61850 GOOSE

Mayada Daboul
Department of Electrical Power Engineering
Faculty of electrical engineering and communication
Brno, Czech Republic
xdabou00@stud.feec.vutbr.cz

Jaroslava Orsagova
Department of Electrical Power Engineering
Faculty of electrical engineering and communication
Brno, Czech Republic
orsagova@feec.vutbr.cz

Abstract — **Protection and Automation systems have been significantly developed as a result of advanced Intelligent Electronic Devices (IEDs) which are adapted to work with various communication interfaces. International standard IEC 61850 has facilitated high-speed peer-to-pee communications among IEDs by replacing analogue and binary signals by Ethernet messages so it reduces the cost of the system and improves the flexibility and the functionality of the system. This paper shows how to use the IEC 61850 within protection schemes which should be developed after assistance by communications system based on the new standard. The paper focuses on one type of protection schemes (Reverse Interlocking) that can be implemented using Generic Object-Oriented Substation Event (GOOSE) messages. A laboratory set-up including the software and hardware presented. Regarding to the terms of feasibility, speed, and dependability, laboratory results are discussed at the end.**

Keywords: IEC 61850; GOOSE; bus-bar protection.

I. INTRODUCTION

Substation automation systems (SASs) are defined as systems for managing, controlling, and protecting of power system. SASs based on creation a highly reliable communication system which responds quickly to real-time events with appropriate actions, thus maintaining uninterrupted power services to end users.

High reliable and available communication network is demand for functions of substation automation systems (SASs) such as control, measurement, disturbance recording and condition monitoring. Early substation automation systems were acquiring the information from utility and industrial substations through remote terminal units (RTUs). Conventional SAS were based on the hard wiring between microprocessor-based protective relays and serial peer-to-peer communications. This approach has its own disadvantages such as complicated substation wiring, using additional transducers, input and output contacts.

Modern power system has replaced the traditional microprocessors with the Intelligent Electronic Devices (IEDs) to provide data of protection, control, monitoring and recording at a much faster speed. IEDs provide qualitative and reliable information to system operators once these IEDs are networked together. Therefore, these IEDs need a quick and reliable communication system to realize substation automation functions [1]

IEC 61850 has defined data models and abstract communication services using standard communication protocols and hardware (e.g. Transmission Control Protocol (TCP), Manufacturing Message Specification (MMS) and Switched Ethernet) to be a future proof standard. (MMS) replaces traditional supervisory control and data acquisition (SCADA) communications.

IEC 61850-7-2 defined two time-critical services: the fast transmission of trip-type signals and sampled values (SVs) .These time-critical services enable exchanging the information between IEDs in order to provide rapid and reliable control and protection applications. SVs replaced the conventional current/ voltage transformer (CT/VT) signalling. IEC 61850 determined generic object oriented system event GOOSE as the service used for the speedy transmission of time critical information (status changes, blockings, and trips).

GOOSE messages reduce the various constraints of the traditionally hardwired system. Where GOOSE messages eliminates response time delay that was occurred by auxiliary relay on/off of the sending device and input filtering on the receiving device within hardwired system. Additionally, GOOSE enable to use a higher number of virtual inputs and outputs to protection and control devices without additional device or wiring costs [2].

GOOSE messaging methodology facilitates "real time" sharing of information between devices in the substation. "Real time" means that messages need to be sent and delivered rapidly, so GOOSE messaging uses a low-level Ethernet layer and then the packets are “tagged” to ensure them priority by network routers.

In comparison with traditional systems, enhanced performance and capacity introduce new and improved functionality, like faster reverse blocking scheme and coordinated arc protection [3], etc.

The paper focus on the integration of IEC 61850 GOOSE message communication between IEC 61850-enabled protection devices as defined in [4]. The paper presents how and for what purpose GOOSE can be used into substation design.

II. IEC 61850 GOOSE MESSAGE

GOOSE messaging is a peer-to-peer message service applied in strict timing requirements necessary of protection applications. GOOSE message exchange is based on the mechanism publisher/subscriber. All required information about the state of an IED events can be grouped as DataSet of GOOSE message to enhance the speed and reliability [5]. The data (status/ values) is sent by the publisher as an initial GOOSE message that is repeated to send with a long cycle time, even if the status have not changed in order to ensure that the subscribed IEDs will receive the GOOSE message in a specified time. When the GOOSE messages arrive to the receiving device, this device will receive information about the changing in the status and the time of the last status. Once the time of the last status changes, the receiving device will set local timers relating to a given event [6].

GOOSE messaging is a unidirectional multicast messaging service which does not wait for any acknowledgement whether the message has been received by the subscriber. But it is usually retransmitted with a minimum time interval to ensure the messaging service is reliable, even though no acknowledgement is ever received. Mechanism of the GOOSE continued retransmission ensures that at least one message will be received by its intended destination within the needed timespan. Multicasting nature of GOOSE communication enables simultaneously delivering the same event information to multiple IEDs in its peer group.

Performance Advantages of GOOSE Model

IEC 61850-8-1 specified a retransmission scheme to achieve a highest level of reliability of message delivery. Fig. 1 shows this mechanism of retransmission of GOOSE messages that can be related to two parameters:

- max time (*mt*): is defined during GOOSE configuration, it determines the interval of retransmission between consecutive GOOSE messages[7].
- time of transmission (*tot*): determines time of retransmission between messages after changing the value of one or more of data set elements.

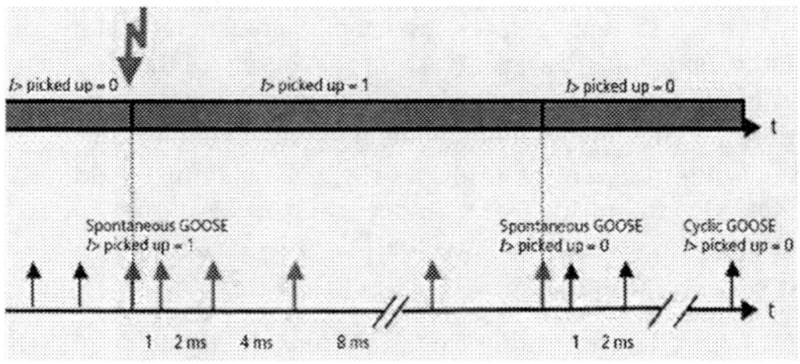

Figure 1 GOOSE telegram transmission

The initial time interval of *tot* is very short (4 milliseconds) and then slowly increase to reach *mt* till the dataset is again updated with refreshed data. For each message, publishers integrate a time to live (ttl) into the message to specify how long a GOOSE message is active before receiving the next message. *ttl* is calculated as multiples of the next *tot* to avoid alarms caused by the frequent and small Ethernet network delays.

On the subscriber side, subscriber constantly calculates time to wait (ttw) for each GOOSE message. The subscriber assumes that the message is lost or the communication channel is broken if ttw is greater than ttl.

As mentioned previously, for the initial message after a protection element in a data set changes state, the message is sent every 4 milliseconds. When a new data set event occurs, a new message is created, transmitted and repeated in the minimum time (Tmin). The retransmission time gradually increases until mt [5]. The stable retransmission time is shortened by occurrence the next new event.

The retransmission scheme is essential to perform transmission message from publisher to one/many subscribers and to ensure to the subscriber the health of communications channel.

GOOSE messages can be used for the following functionalities [8]:

- Breaker failure initiation.
- Reverse blocking (selectivity).
- Status for circuit breakers and disconnectors.
- Transfer tripping

III. Reverse Blocking Interlocking

Reverse interlocking provides a simple, clear and low-cost way of implementing substation bus protection in cooperation with protection devices. The interlocking principle is particularly used for short-circuit protection schemes.

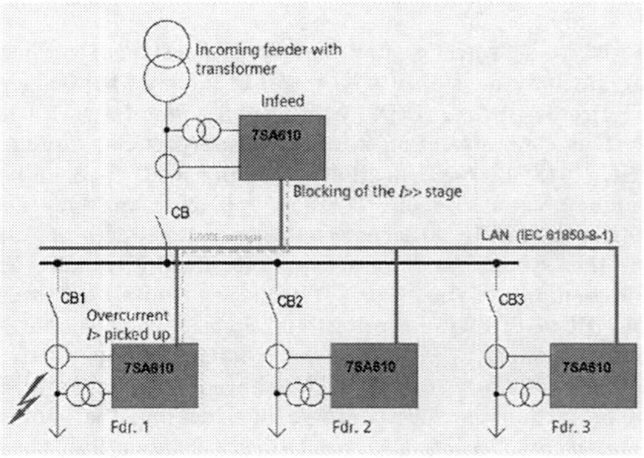

Figure 2 busbar protection using reverse interlocking

Fig. 2 presents a reverse blocking scheme where, the Incomer device uses definite time overcurrent element and the outgoing feeders use time overcurrent element. The Incomer relay selectively is allowed to trip or block depending on faults location as specified from outgoing relays. For a fault on an outgoing feeder, the protection relays of both the incoming and the faulty outgoing feeders start. On starting, the corresponding outgoing relay will send blocking signal to the Incomer relay after its time overcurrent element picks up in order to block the fast-acting stage (High-set current stage *I*>>) of the incoming feeder relay [9]. On the contrary, for a fault on the bus, the outgoing relays will not start and will not send a block signal to the incomer relay that will be allowed to trip the Circuit Breaker (CB) and clear the bus fault after a short time coordination delay.

In a conventional protection scheme, exchange of information between devices is realized by hardwiring. Where, the output contact terminals of sending IED are wired to inputs in the receiving IEDs [10]. Furthermore, the latency of the link is such that Protection Time OverCurrent (PTOC) of incomer feeder will trip the fault before it is aware that PTOC of outgoing feeder is already attempting to clear the fault. Thus, tripping CB of incomer feeder will cause power outage for all outgoing feeders even though the fault is confined to just one line.

Traditional interlocking scheme can be speeded up considerably using new the technology (e.g. GOOSE messages) that can offer an increased operational reliability and flexibility of the protection.

Blocking signals can be sent directly from relay-to-relay by transferring GOOSE messages between the interconnected IEDs without additional delay from auxiliary relays or input filters. Moreover, GOOSE messaging blocking signals can be sent to all the involved IEDs at the same time.

By applying GOOSE messaging into protection scheme, PTOC of outgoing device could send a GOOSE message to PTOC of incomer device to prevent trip and inform that outgoing protection is attempting to clear the fault. The advantage over the conventional scheme is that all outgoing feeders remain live, while faulty feeder must either clear the fault or lock out to isolate it.

applying GOOSE messaging afford a great advantage that is enhancing operational reliability significantly due to the inherent supervision of the GOOSE that manage the protection functionality priority in a particular application to obtain the appropriate protection.

IV. Laboratory Testing Setup

One of the common applications of GOOSE messages is implementing a reverse interlocking of bus protection scheme. Its most basic form is installation a directional overcurrent relay on the incomer. If the incomer relay detects a fault towards the bus, it will trip unless otherwise one of the feeder relays send a blocking signal to indicate that the fault is external to the bus.

The test procedure is verifying the programming for the GOOSE-based block tripping (I>> stage) of incomer feeder after pick-up (*I> stage*) of feeder1 in reverse interlocking of bus protection scheme.

A laboratory set-up (Fig.3) has been implemented using two protection relays, a communication device (Ethernet switch), DIGSI 4 software to configure IED and IEDScout software for evaluating the performance of GOOSE messages.

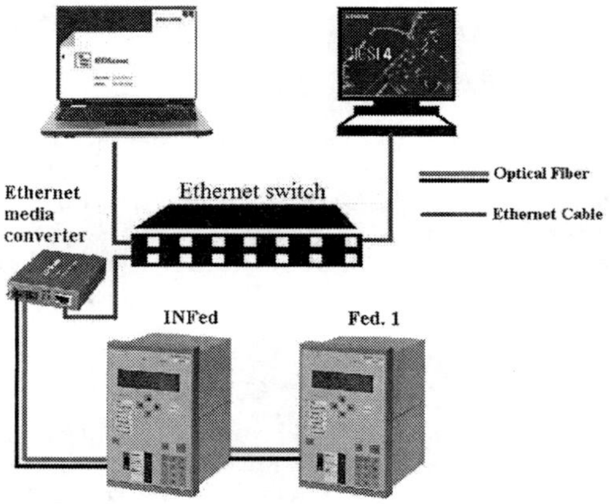

Figure 3 Laboratory Setup

The equipments and software are used for the laboratory setup are as follow:

1. IEC 61850 supporting Relays: two SIPROTEC4 7SA610 – Distance protection Relays, provide GOOSE message capability and Ethernet communication port [11].
2. DIGSI 4: Software used for configuration of SIPROTEC4 devices according to the protection scheme [12].
3. CFC (Continuous Function Chart): Software tool used to program the SIPROTEC devices for processing the incoming GOOSE messages.
4. IEC 61850 System Configurator: Software tool used for configuration of IEC 61850 stations, involved in the exchange of GOOSE messages.

For verifying IEC 61850 GOOSE between the relays after a fault on the outgoing feeder (Fed.1), the Fed.1should detect the fault by a Protection Time OverCurrent (PTOC) and create an indication about this fault. This information can be a part of the Data Set in the form of a GOOSE message that will send to the other relay (INFed.) that is configured to receive this GOOSE message and react on it by blocking trip. The short circuit should be cleared by a trip command (I> Trip / PTOC.Op) of Fed.1 after a parameterizable delay time has elapsed.

A. *Configuration Publishing GOOSE*

Fed.1was configured to create I> picked up (I> picked up / PTOC.str) indication after detecting short-circuit current. This indication is defined as a mandatory message in the IEC 61850 standard and it can be viewed under text (PROT/PTOC2/Str). This transmitted message automatically routed to the system interface S by DIGSI 4 and cannot be un-routed by the user.

B. *Configuration Subscribing GOOSE*

To subscribe the GOOSE messages from Fed.1, INFed was configured to receive the "External Signal" from the network system, and immediately to block tripping CB. The received message from system interface routed into the logic editor "CFC" in which logic blocks are used for mapping blocking of the I>> stage. In the GOOSE group, a further message is created to alarm if there is a disturbance in the GOOSE link as displayed in Fig.5.

C. *System Configuration*

Configuration System Tool used to set the IP addresses of the devices and network address, also to determine the publisher as a "source" of GOOSE and also the subscriber as "Destination" of GOOSE. Fig. 5 shows the devices with their signals. The data items AA1Q01KF1/PROT/PTOCC2/Str and AA1Q01KF2/CTRL/OCBlkGGIO1/SPCSO2 are parts of the DataSet that is exchanged in the form of the GOOSE message between IEDs.

By this tool, GOOSE messages is set to high priority to be repeated cyclically and spontaneously with high frequency: every 0.5 s cyclically, starting with 1 ms repetition on a spontaneous change to a message.

To exchange GOOSE message after configuration IEDs, there is need to upload this settings from DIGSI into the protection Devices by serial communication cable and also to connect them to a PC via Ethernet LAN to capture the GOOSE.

Both of these devices are connected to gather by multi-mode fibre-optic interface. The data traffic is controlled by special filters on the optical module to ensure that GOOSE will be only transmitted to the devices that expect information from a GOOSE. This reduces the processor load on the Ethernet module.

V. Measurement of Transmission GOOSE

IEC 61850-5 classifies application types depending on the speed of transmitted messages among IEDs. The standard also specifies the performance of each type of application, named as time duration of message transmission.

Table 1. Message Types Classes Group 1

Group1		
Protection and Control		
P1	**P2**	**P2**
for distribution bay	for transmission bay	for transmission bay with high performance synchronizing feature

For different communication requirements, there are two independent groups of performance classes. The first group is for protection and control, the second group is for metering and power quality applications [9] as is shown in Table I and Table II.

Table 2. Message Types Classes Group 2

Group 2		
Metering and Quality		
M_1	M_2	M_3
accuracy class 0.5	accuracy class 0.2	Quality metering up to the 40th harmonic.

IEC 61850-5 included "Transfer Time" to determine the maximum time allowed for the data exchange in the communication system. Transfer Time is crucial for any performance requirements. It has specified that transfer time should be less than 4ms for a Trip GOOSE command (class P2/P3) [13] and 20ms for a Block GOOSE command (class P2/P3) as shown in Table III

Table 3. Message Types and Performance Classes

Message type	Performance class	Application	GOOSE transmission time
1A	Fast Messages (Trip)	P_1	10 ms
		P_2/P_3	3 ms
1B	Fast Messages (Others)	P_1	100 ms
		P_2/P_3	20 ms
2	Medium Speed		100 ms
3	Low Speed		500 ms
4	Raw Data	P_1	10 ms
		P_2/P_3	3 ms
5	File Transfer		>1000 ms
6	Time Synchronization	T1 (time)	± 1 (Accuracy)
		T2 (time)	± 0.1 (Accuracy

In laboratory testing, GOOSE messages has captured using network analyser software as shown in Fig. 5; IEDScout tool from OMICRON was used to measure GOOSE transmission time among networked IEDs. For many measurements and records, the average value of the transmission time was 8 ms as is displayed in Fig.4.

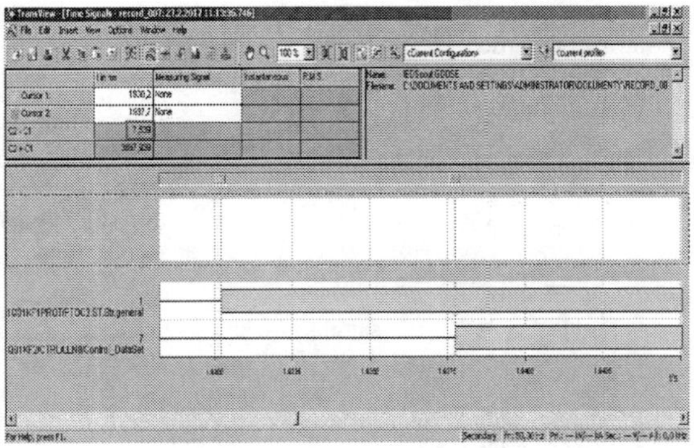

Figure 4 GOOSE transmission time

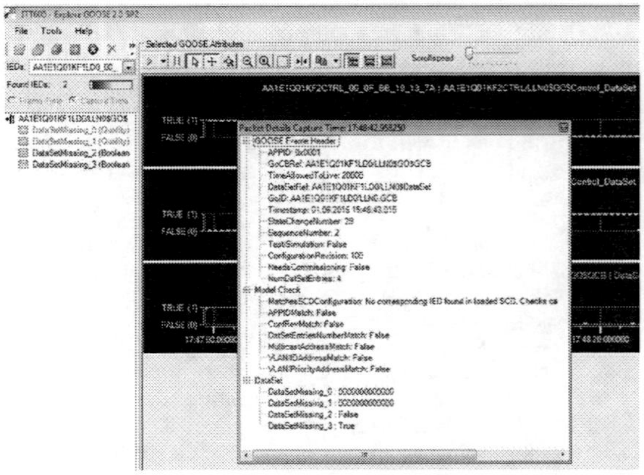

Figure 5 Captured GOOSE messages

ACKNOWLEDGMENT

This research paper has been carried out in the Centre for Research and Utilization of Renewable Energy (CVVOZE). Authors gratefully acknowledge financial support from the Ministry of Education, Youth and Sports of the Czech Republic under NPU I program (project No. LO1210) and from the Technology Agency of the Czech Republic (project No. TA04021490).

VI. CONCLUSIONS

It is proved that the GOOSE provides a very flexible, high-priority and reliable method for exchange the events substation between IEDs for protection and interlocking purpose. GOOSE message service model and the communication network features are deeply studied to find out its impact on the performance of GOOSE based protection devices.

Utilizing IEC 61850-GOOSE improves signalling performance and speed up more than traditional hardwired systems. Additionally, the use of IEC 61850 and GOOSE provides increased operating reliability of the protection through continuous supervision of the communication of the GOOSE messages.

This paper presented the laboratory setup and the method to evaluate the performance of GOOSE based revers interlocking scheme. The configuration of the IEC GOOSE and its performance testing are presented. The verification test of GOOSE performance is carried out to ensure the consistence of the GOOSE messages.

In the end, it should be noted that achieving the conformance does not necessarily mean achieving the interoperability between IEDs. In another words, different IEDs may conform to the standard but still might not be 100% interoperable to each other due to the difference of communication profiles, which defines the mandatory SUBSET of a standard with selected implemented function. Thus, Interoperability testing is very important to be the second step after the conformance testing. Interoperability testing of multivendor is the core of the current working that will present in near future.

REFERENCES

[1] International Electrotechnical Commission, "IEC 61850 Communication Networks and Systems in Substations", 2003.

[2] M. N. Noran, Z. Shukri, "Adaptive breaker failure protection scheme for double busbar substation using IEC 61850 GOOSE message communication," 2015 International Symposium on Technology Management and Emerging Technologies (ISTMET), Langkawai Island, 2015, pp. 207-212.
doi: 10.1109/ISTMET.2015.7359030.

[3] C. Cabrera, S. Chiu and N. K. C. Nair, "Implementation of Arc-Flash Protection using IEC 61850 Goose messaging," 2012 IEEE International Conference on Power System Technology (POWERCON), Auckland, 2012, pp. 1-6.
doi: 10.1109/PowerCon.2012.6401467.

[4] International Electrotechnical Commission, "IEC 61850-7-1: Communication Networks and Systems in Substations, Basic Communication Structure for Substation and Feeder Equipment-Compatible Logical Nodes Classes Introduction and Overview". 2003.

[5] T. Sidhu, M. Kanabar, P. Parikh, "Configuration and performance testing of IEC 61850 GOOSE," 2011 International Conference on Advanced Power System Automation and Protection, Beijing, 2011, pp. 1384-1389.
doi: 10.1109/APAP.2011.6180593.

[6] A. Apostolov, B. Vandiver, "IEC 61850 GOOSE applications to distribution protection schemes," 2011 64th Annual Conference for Protective Relay Engineers, College Station, TX, 2011, pp. 178-184. doi: 10.1109/CPRE.2011.6035618.

[7] J. Xu, C. W. Yang, G. Zhabelova, S. Berber, V. Vyatkin, "Towards implementation of IEC 61850 GOOSE messaging in IEC 61499 environment," 2013 11th IEEE International Conference on Industrial Informatics (INDIN), Bochum, 2013, pp. 464-470. doi: 10.1109/INDIN.2013.6622929.

[8] J. Mraz, A. Cowan, K. Gray, K. S. Shah, "Testing of IEC 61850-based protection and control schemes for a refinery electrical system," 2016 Petroleum and Chemical Industry Technical Conference (PCIC), Philadelphia, PA, 2016, pp. 1-9.
doi: 10.1109/PCICON.2016.7589210.

[9] C. H. Goff, A. H. Eltom, H. Saeed, "Fast bus protection using IEC 61850," 2015 IEEE Industry Applications Society Annual Meeting, Addison, TX, 2015, pp. 1-8.
doi: 10.1109/IAS.2015.7356855.

[10] A A. Apostolov, "Impact of IEC 61850 on the interoperability and reliability of protection schemes," 2013 IEEE Power & Energy Society General Meeting, Vancouver, BC, 2013, pp. 1-5.
doi: 10.1109/PESMG.2013.667305.

[11] Siemens Ltd, Reference Manual, "SIPROTECH 4".

[12] Siemens Ltd, Reference Manual, "DIGSI 4"

[13] M M. Daboul, J. Orsagova, T. Bajanek and V. Wasserbauer, "Testing protection relays based on IEC 61850 in Substation Automation Systems," 2015 16th International Scientific Conference on Electric Power Engineering (EPE), Kouty nad Desnou, 2015, pp. 335-340.
doi: 10.1109/EPE.2015.7161095.

[14] International Electrotechnical Commission, "International standard IEC 61850-5: Communication Networks and Systems in Substations, Communication requirements for functions and device models", 2003.

Effect Of Total Harmonics Distortion On Distance And Overcurrent Relays

Kinan Wannous
Brno University of Technology
Brno, Czech Republic
xwanno00@stud.feec.vutbr.cz

Petr Toman
Brno University of Technology
Brno, Czech Republic
toman@feec.vutbr.cz

Pedro Pozzobon
Federal University of Bahia
Salvador, Brazil
pedrohenrique_bp@hotmail.com

Abstract —**The main goal of this study is to determine how different magnitudes of harmonic distortion may influence the operation, more specifically, the tripping time of the most used digital protection relays. This paper describes the basic theory of protection function with harmonic content. Harmonic signal can cause problems in power system and protection relays function due to the voltage magnitude which is used as input in protection relay.**

Keywords: harmonic; protection relay; distance; overcurrent; CMC 256.

I. Introduction

As the usage of electronic and power electronic devices is rising worldwide, the level of harmonic distortion in the power grid rises as well, degrading the energy quality. Usually, harmonic distortion has many effects regarding the operation of many devices and machines. As relays are fundamental in order to maintain the safety and correct operation of our power grids, it is necessary to know how exactly harmonic distortion affects their operation. Protection relays are designed for 50 Hz; there were many studies focused on the wave distortion due to a presence of harmonics. Those studies found harmonics impact on electromechanical relays, some studies explained the harmonic impact on digital relays and the relay algorithms and they found that the algorithms are not completely resistant to all the harmonics. The harmonic impact can be noticed during the fault performance on digital relays. As is known the main difference between conventional relays and digital relays is in the way of input signal processing. In digital relays, the input signal is converted into the digital signal before begin processed. On the other hand, in conventional relays, the input signals (voltage-current) which come from current transformers and voltage ones are imposed to the electromagnetic winding.

II. Methodology of the Test

A. Total harmonic distortion (THD)

The harmonic distortion levels applied in the testing relays were taken from an E.ON Distribution study on harmonic levels in 30 different grids around Czech Republic; their measurements were taken in 2016 [1, 2]. The calculation of distortion as the following equation [3, 4, 5]:

$$I_{rms} = \sqrt{{I_{Fund,rms}}^2 + {I_{Harm,rms}}^2} \qquad (2)$$

Tripping time is compared with/without harmonics and calculated according to the following formula:

$$t_{(pu)} = \frac{t_{F+H}}{t_F} \qquad (3)$$

Three phase currents are used to test the relay and obtain the harmonic distortion; moreover, the configuration requires correct settings for each round of the test. The fundamental current value is 2 A, the time duration is 10 seconds, and the RMS of harmonic value is from 0% to approx. 4% of fundamental current; after that, tripping time is measured for each percentage of fundamental signal, an event is recorded and stored in the relay.

B. Signal generator and relay

The devices used to perform the tests were Omicron CMC 256 in order to generate electrical signals containing desired harmonic levels, and a digital relay. The exact setup can be seen in figure 1.

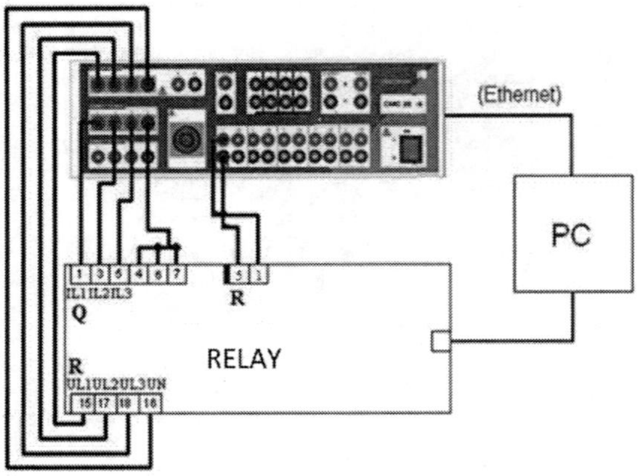

Figure 1. Connection Scheme

C. Overcurrent relay

Two different configurations were used on the overcurrent relay's tests: (i) Definite Time and (ii) Reverse Time.

D. Distance relay

Two different fault points were used in the tests and we choose single phase with ground (slg), the first point was taken in the zone 1, and the second fault point was located near to the border from zone 1 to zone 2. The characteristic zones of the distance relay and chosen points can be seen in figure 2. The figure 2 shows the relay characteristics. As is known most digital relays use quadrilateral characteristics due to the many advantages which those characteristics can provide. The configuration and settings of relay should be done in the first place, after that, exporting these settings to CMC 256 is important to evaluate the relay functions. CMC 256 provides different platforms to test the relay with its zones setting, transmission line parameters and time delay. Moreover, advanced distance test displays the amplitude and phase of the tree phase currents and voltages determined by the fault type and fault location.

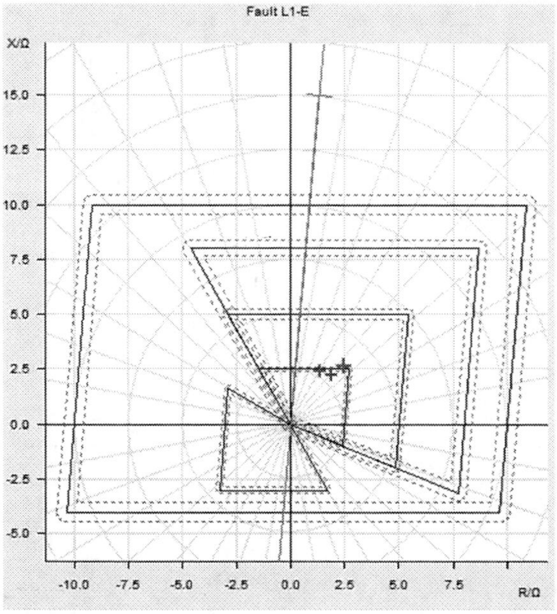

Figure 2. Characteristic zones of distance relay and fault points

III. Faults and Abnormal Phenomena in Power System

Power plant produces the electric energy and then the transmission lines transmit power over long distance and distribute it to the customers

The protection devices can distinguish the ordinary operation from the uncommon ones that can damage or harm the system's components. Normal operation of the power system can be differentiated of the following:

- The nominal frequency equal 50 Hz.
- The currents and voltages waves should be symmetrical.
- No loss of system stability and supply.
- Voltage level is close to the nominal value.
- Total harmonic distortion (THD) within allowed limits.

Types of electrical protection relays:

Every protection device can be characterized by the following issues:

- Increasing of the currents beyond permissible limits.
- Increasing of harmonic contents.
- Increasing of the negative and zero sequence components.
- Changes of the voltage level.

The relay can take the decision tripping/warning according to the fault impact; also it can mention the fault type and zone.

The first step was to test the overcurrent relay when inputted with different high harmonic distortions and there could be observed the behavior of the tripping time. The next step was to input the average harmonic distortions of Czech's power grid to see if there would be any major change on the base tripping time of the overcurrent relay. The next step was the test the influence of harmonic distortion on the distance relay.

A. *Overcurrnet relay*

Overcurrent relay is used the fault current as a basis for selectivity. Since overcurrent measures and monitors the current and when the current magnitude exceeds the permissible limits which will be determined according to the protected item, after that, the relay should send tripping order to isolate the part where that fault happened and protect the power system [10].

B. *Differential relay*

Differential relay is comparing between two phases of input and output currents for power system and it is used to protect the transformers.

C. *Impedance relay*

This relay measures the impedance of the line which is protected. Impedance protection is usually protecting the transmission lines of the power system. This relay can stop the fault in the beginning and protect the line from any damage occur in case of short circuit fault. The relay operates once the measured impedance of the transmission line drops under the set relay impedance.

The protection devices react the events with tripping, warning signal depending on the event itself which include abnormal operation.

IV. Standards in Order to Determine Harmonic Limit Values

The IEC (International Electrotechnical Commission) is an organization for standardization comprising all national electrotechnical committees.

There are three European standards to define the limitation of harmonic currents injected to the power system. They define limits of harmonic component of the input current which may be produced by the equipment:

- IEC 61000-3-2:2014 Electromagnetic compatibility (EMC) - Part 3-2: Limits - Limits for harmonic current emissions (equipment input current ≤ 16 A per phase). This standard is applicable to the electrical equipment which required an input current up to 16 Ampere.
- IEC 61000-4-7: Testing and measurement techniques – General guide on harmonics and interharmonics measurements and instrumentation, for power supply systems and equipment connected thereto. New requirement lunched for measuring harmonic should get 10 cycles of the fundamental current. Furthermore no gap and no overlapping between successive measuring windows are allowed.
- IEEE 519: the standard provides limitations for harmonic distortion (limitation of the harmonic voltage distortion provided by the distributor to the customers, limitation of harmonic current distortion superimposed to the system by a customer).

Individual Voltage Distortion limits for the bus voltage level 69 kV and below should be not exceed 3% and total voltage distortion THD should be not exceed 5%. Moreover, according to IEEE 519 there is more conditions with regard to the effects of harmonics on protective relays (electromechanical, static and digital relays).

It is important to bear in mind that the impact of the power quality on protection relays which can cause incorrect trips and relay maloperation, meanwhile, incorrect tripping time occurrence depends on the power frequency variations and harmonic distortion [12].

V. FUNCTIONAL DEVELOPMENT OF DIGITAL PROTECTION RELAYS

The applications of the protection relays have many steps regarding the processing the input signals of the relay. We can summarize these steps with the following [7,11]:

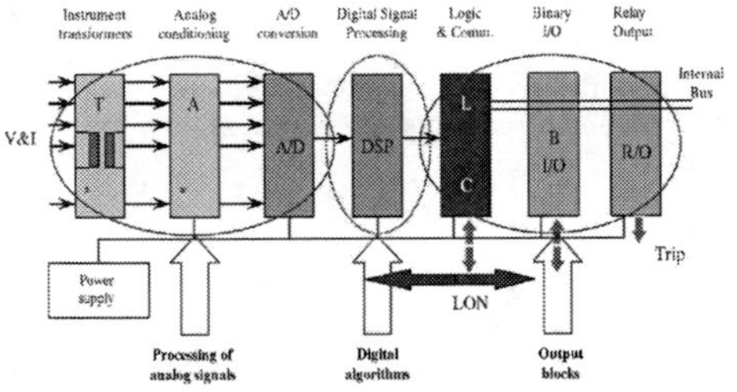

Figure 3. Digital relay structure

- Processor
- Analog input system
- Digital output system
- Analog-digital converter

VI. EVALUATION OF V & I MEASUREMENST AND THD

Thirty representative LV distribution grids with connected PV plants were chosen and these grids were measured in [2].

The previous study evaluated the harmonic level for thirty LV grids and it found the level of (3rd, 5th and 7th) harmonics on voltage are less than 2% from the measured values. Nevertheless, the study found greater effect with the 15th voltage harmonic. Regarding the total harmonic distortion (THD) there is no impact. In this paper, we took the harmonic ratios from the previous study and used them as an input signal for digital protection. The aim is to find and analyze the harmonic effect on digital protection functions. The input signal is the sum of the current or voltage signal with the harmonics. Figure 4 shows the total harmonic distortion which evaluated in E.ON Distribution Company in Czech Republic [2].

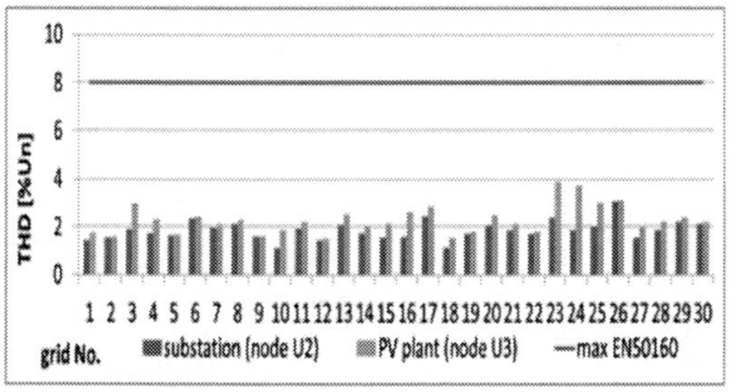

Figure 4. Evaluation of total harmonic distortion THD [1]

A. *Overcurrent relay*

- Definite Time Configuration: Absolutely no changes on tripping time were noted when applying harmonic distortion to the overcurrent relay in this configuration.
- Reverse Time Configuration: The tests performed on reverse time configuration have led to figure 5. From the values shown on the figure, it can be concluded that for THD values lower than 5%, the effects of harmonics on tripping time of overcurrent relay (on reverse time configuration) are minimal, as the difference from the lowest tripping time to the highest is less than 0.06 seconds as shown in figure 5 and table1 (* t_{avg} is the tripping time average for repeating the test 5th time).

Table 1: Data from overcurrent relay's tests with trip time.

Grid No	*Irms = 5 A*	*Harmonics (%)*				*Trip Time (s)*
	If (A)	*I3*	*I5*	*I7*	*I15*	t_{avg} *of 5 tests**
0	5	0.00%	0.00%	0.00%	0.00%	13.529
1	4.999038	0.80%	1.60%	0.80%	0.10%	13.566
2	4.999338	0.20%	0.80%	1.40%	0.10%	13.540
3	4.997206	1.90%	2.20%	1.60%	0.42%	13.570
4	4.998545	1.00%	1.80%	1.20%	0.38%	13.526
5	4.999028	0.60%	1.70%	0.80%	0.02%	13.541
6	4.997786	0.40%	2.20%	1.90%	0.50%	13.584
7	4.998478	0.70%	1.90%	1.40%	0.18%	13.570
8	4.998251	0.70%	1.80%	1.80%	0.17%	13.540
9	4.998959	0.40%	1.60%	1.20%	0.08%	13.533
10	4.998844	1.00%	1.60%	1.00%	0.26%	13.552
11	4.998368	0.80%	1.80%	1.60%	0.30%	13.545
12	4.998828	0.40%	1.40%	1.60%	0.10%	13.549
13	4.997979	1.40%	1.40%	2.00%	0.41%	13.532
14	4.998381	0.80%	1.80%	1.60%	0.20%	13.544
15	4.998766	1.00%	1.80%	0.80%	0.24%	13.576
16	4.998029	1.60%	1.80%	1.40%	0.36%	13.554
17	4.99742	1.00%	2.60%	1.60%	0.08%	13.563
18	4.999291	0.60%	1.00%	1.20%	0.19%	13.550
19	4.999021	0.60%	0.80%	1.70%	0.16%	13.543
20	4.998087	1.00%	1.80%	1.80%	0.42%	13.587
21	4.998788	0.60%	1.70%	1.20%	0.40%	13.562
22	4.998897	0.40%	1.00%	1.80%	0.12%	13.552
23	4.99581	2.50%	2.40%	2.10%	0.60%	13.566
24	4.996663	2.40%	2.00%	1.80%	0.60%	13.568
25	4.997055	1.60%	2.40%	1.80%	0.48%	13.571
26	4.997144	0.40%	1.00%	3.20%	0.18%	13.563
27	4.998682	1.00%	1.00%	1.80%	0.18%	13.528
28	4.998235	1.00%	2.00%	1.40%	0.32%	13.537
29	4.998377	0.60%	0.60%	2.40%	0.12%	13.563
30	4.998606	0.40%	1.00%	2.10%	0.08%	13.5732

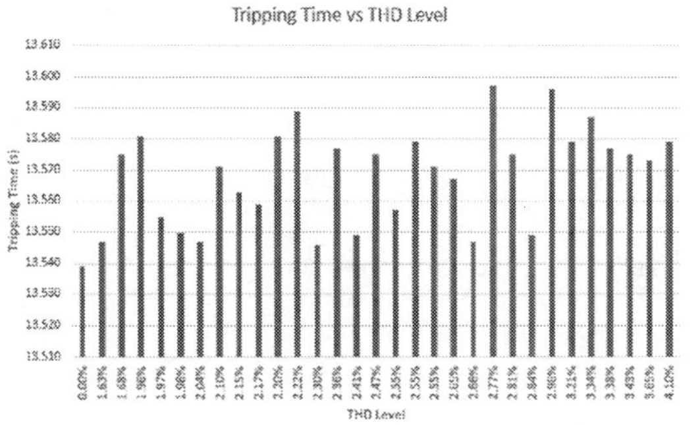

Figure 5. Overcurrent relay: tripping time (seconds) vs THD level

B. *Distance relay*

Table 2: Data from distance relay's tests with fault zone limit.

THD Level	Tripping Time (ms)					
	Test 1	*Test 2*	*Test 3*	*Test 4*	*Test 5*	*Average*
0.00%	0	20	30	20	1	14.2
1.63%	20	21	20	50	20	26.2
1.68%	1005	1000	26	1000	1000	806.2
1.96%	1000	170	1000	1000	15	637.0
1.97%	30	1000	1005	25	1005	613.0
1.98%	1005	1000	1005	80	1000	818.0
2.04%	1000	130	180	30	1005	469.0
2.10%	291	70	20	1000	1000	476.2
2.15%	1000	1000	70	1000	1000	814.0
2.17%	1000	1000	50	15	15	416.0
2.20%	1000	1000	1000	40	210	650.0
2.22%	20	1001	20	1000	1000	608.2
2.30%	1000	160	1000	1005	60	645.0
2.36%	80	20	1000	30	1000	426.0
2.41%	100	1000	1000	1000	1000	820.0
2.47%	60	40	20	1000	1005	425.0
2.55%	1006	1000	160	1000	1000	833.2
2.56%	1001	1000	20	1005	1000	805.2
2.65%	1000	80	160	70	1000	462.0
2.66%	1000	1000	1000	1000	1000	1000.0
2.77%	1000	1000	90	1005	1000	819.0
2.81%	1000	1000	20	1000	1000	804.0
2.84%	1000	30	1000	1001	1000	806.2
2.98%	1000	1000	40	15	1000	611.0
3.21%	1000	46	40	1000	1000	617.2
3.34%	1000	1000	26	1000	1000	805.2
3.38%	50	1000	70	20	30	234.0
3.43%	1000	1000	20	280	45	469.0
3.66%	70	1000	1000	1000	1005	815.0
4.10%	1000	1000	1000	1000	50	810.0

- Fault located near the border from zone 1 to zone 2: The tests were performed with zone 1 tripping time set to 0 ms and zone 2 tripping time set to 1000 ms. From this test's data, on table 2, it is possible to see how harmonic distortion can affect the distance relay's accuracy to sense where the fault took place.
- Behavior of the distance protection with harmonic distortion: During harmonic distortion, a portion of the current is missing or effect, so that principally *(Z=U/I)* an impedance which is too large, is measured. In provided measurements on distance protection, the zone reach is reduced. This acceptable for close in faults, because the distance to the zone limit is large [6, 7].
- For faults close to the zone limit, an under reach is not permitted, the relay will trip in the second zone with time delay.

Faults sensed by distance relay as zone 1 are highlighted in green and faults sensed as zone 2 are highlighted in red.

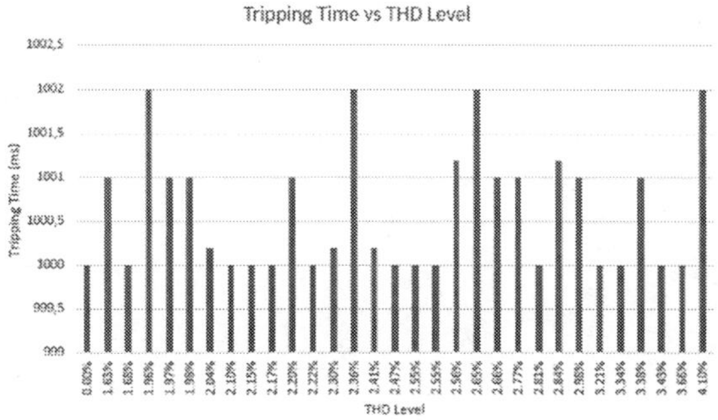

Figure 6. Distance Relay: Tripping Time (seconds) vs THD Level

CONCLUSIONS

Digital relays are limited in a way that they can only respond to changes in magnitude of the fundamental current or voltage. Regarding overcurrent relays, low level harmonic distortion may not affect their operation. Though, concerning distance relays, whilst the relay's ability to find faults away from zone's limit may still be reliable, when it comes to faults located near the limit of the zone, there is a chance the distance relay can be misguided regarding the location of the fault.

ACKNOWLEDGMENT

This research work has been carried out in the Centre for Research and Utilization of Renewable Energy (CVVOZE). Authors gratefully acknowledge financial support from the Ministry of Education, Youth and Sports of the Czech Republic under NPU I programme (project No. LO1210).

REFERENCES

[1] M. Kaspirek, L.Mikulas, D. Mezera, K. Prochazka, P. Santarius, P. Krejci, "Analysis Of Voltage Quality Parameters In MV Distribution Grid," in 24th International Conference on Electricity Distribution CIRED, Glasgow, June 2017. Paper. 0148.

[2] M. Kaspirek, L.Mikulas, D. Mezera, "Analysis Of Voltage Quality Parameters In LV Distribution Grids With Connected Des," in 24th International Conference on Electricity Distribution CIRED, Glasgow, June 2017. Paper. 0111.

[3] K. Wannous, P. Toman, "The effects of harmonics on overcurrent relays," in 17th International Scientific Conference on Electric Power Engineering (EPE), Prague, 2016, pp. 1-6. doi: 10.1109/EPE.2016.7521794.

[4] Z. Muller, J. Svec, M. Cernan, J.Kyncl, "The Use of Regression Methods for Measurement and Diagnostics in Electrical Power Engineering", in Proceedings of the 13th International Scientific

Conference EPE 2012. pp. 413-417, 2012, ISBN 978-80-214-4514-7.

[5] L. Al-Musawi, N. Al-Mutawaly and N. Schofield, "Shunt active power filtering for smart appliances," in IEEE 28th Canadian Conference on Electrical and Computer Engineering (CCECE), Halifax, NS, 2015, pp. 466-470.
doi: 10.1109/CCECE.2015.7129321.

[6] G. Ziegler, "Numerical distance protection: principles and applications", John Wiley & Sons, 2011.

[7] W. Rebizant, J. Szafran, A. Wiszniewski, "Digital signal processing in power system protection and control", Springer Science & Business Media, 2011.

[8] M. Weeks, "Digital signal processing using MATLAB & wavelets", Jones and Bartlett Publishers, Inc., USA, 2010.

[9] S.J. Orfanidis, "Introduction to signal processing", Rutgers University, 2010; pp.795.

[10] J. R. S. S. Kumara, A. Atputharajah, J. B. Ekanayake, F. J. Mumford, "Over current protection coordination of distribution networks with fault current limiters," in IEEE Power Engineering Society General Meeting, Montreal, Que., 2006, pp. 8. doi: 10.1109/PES.2006.1709251

[11] Jirickova, J "IEC61850: Implementation in new Generation of Electrical Protection", Proceedings of The 4th International Scientific Symposium on electric Power Engineering – Elektroenergetika 2007, pp 471-473, 2007.

[12] I. Zamora, A. J. Mazón, V. Valverde, E. Torres, and A. Dysko, 'Power quality and digital protection relays', in International Conference on Renewable Energies and Power Quality (ICREPQ'04). Barcelona, 2004.

Electrical Characteristics of Incompletely Unrolled Three-phase Power Cable on Wooden Drum

Matej Cenký
Institute of Electrical Engineering
Slovak University of Technology
Bratislava, Slovakia
matej.cenky@stuba.sk

Marián Ivanič
Institute of Electrical Engineering
Slovak University of Technology
Bratislava, Slovakia
marian.ivanic@stuba.sk

Jozef Bendík
Institute of Electrical Engineering
Slovak University of Technology
Bratislava, Slovakia
jozef.bendik@stuba.sk

Žaneta Eleschová
Institute of Electrical Engineering
Slovak University of Technology
Bratislava, Slovakia
zaneta.eleschova@stuba.sk

Abstract —**The article presents a theoretical computational approach for electrical parameters of a general 3-phase power cable. The power cable used in this article is described, as well as the wooden drum on which is wired before installation. Using the theoretical background, the electrical parameters of the cable are computed and compared to the reference measured values from the manufacturer. The error is evaluated and discussed. After the initial evaluation, the second scenario is considered, with cable incompletely developed and partially still wrapped on the wooden drum. Solution for this problem is achieved through simulation in Ansys Maxwell software. Model of the power cable is described and the results are discussed in comparison to the perfectly developed power cable. All the results are also compared to the accordingly selected overhead line conductor to confirm the expected parametrical differences.**

Keywords: power cables, electrical parameters, wooden drum, simulation

I. Introduction

All engineering aspects are taken into consideration to establish optimal solution or design. In the field of power engineering, the choice is often limited only to overhead power lines and power cables. Both of the offered possibilities have its own positive and negative sides. For economic reasons, overhead lines are mostly used for transmission of the electricity in the rural areas, where environmental and safety requirements are met. On the other hand, power cables are used mostly in more populated areas. It is so because of more reasons like: no visual smog, practically no safety distances, many different types of cables for different applications and so on. Whatever the requirements are, the same basic parts for all power cables are:

- Low resistance conductor(s)
- Insulation of the conductor(s)

Also with the higher voltage, the complexity of the cable construction rises. Looking even at the basic construction it is clear, that the power cables will be definitely more expensive alternative for long distance or bulk transmission in comparison to the overhead lines. The overhead line itself is using only bare conductors, typically consisted of Aluminium (as conductive part) and Steel (as reinforcing part).

Important parametrical differences between the power cables and overhead lines are:

- Much lower cable inductance
- Much higher cable capacitance
- Much higher cable charging current

These general differences are expected to be seen as a results, when comparing electrical parameters computed for overhead line conductors and according three-phase power cable later in this article.

II. Evolution of Cable Electrical Characteristics Computation

Electromagnetic theory of transmission lines has been developed quite early by Maxwell himself, Rayleigh, Heaviside, Thomson and other famous scientists. Later, in 1934, Schelkunoff worked out a study on coaxial structures [1]. He derived original Maxwell equations for the impedances and admittances which link voltages and currents in such systems.

The main assumptions were:

- Homogeneous system
- Ideal isotropic materials – constant conductivity, permittivity and permeability
- Longitudal currents in the insulating materials can be neglected

Pollaczek then in 1943 studied the electromagnetic field propagation in the ground [2], filling the gap for a modelling of a buried single-core cable. Schlkunoff's model was limited to surrounding infinite medium. From now on, modelling of a three-phase systems as computed assuming the conductors (cables) are parallel and the superstition principle can be used on their electrical field.

All the above mentioned formulas are not easy to handle, because of the plenty of Bessel functions included. This have changed Wedepohl and Wilcox [3] in 1973. They approximated Bessel's functions and the final formulas are valid up to 100 kHz. Formulas are originally defined for a buried 3-phase link witch singe core cables. With three-core cables, impedances and admittances can be derived the same way from Maxwell equations, but with additional assumptions.

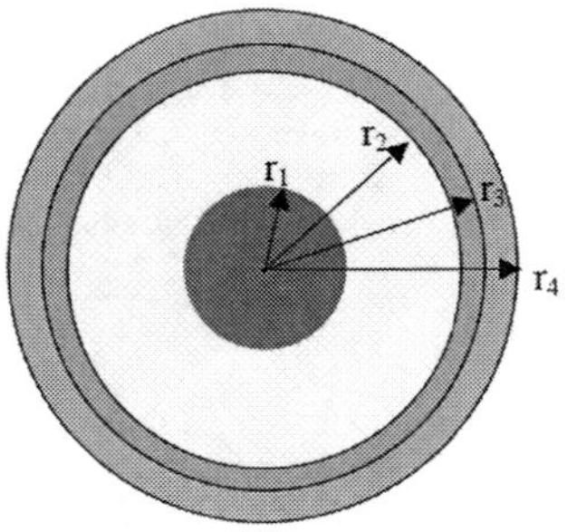

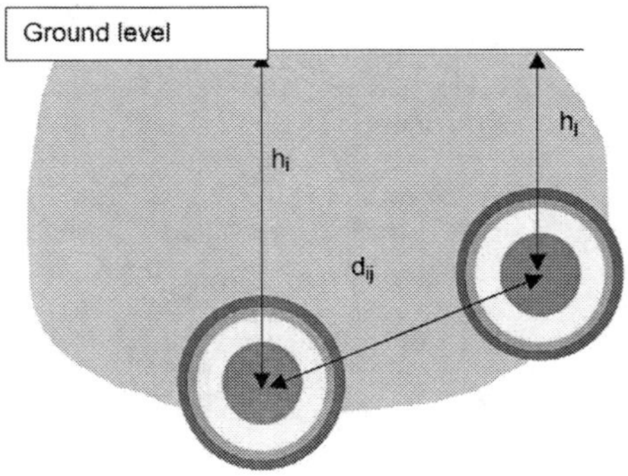

Figure 1. Wilcox's and Wedepohl's model of a conductor (upper part) and a system of conductors in ground (lower part)

This model based on the hyperbolical approximations of Bessel functions is still very accurate, with maximum error at 5% value (with frequency in 100 ~ 200 Hz range) dependent also on material used.

A. *Single core cables*

Formulas for impedance of a single core cables are too complex to be written in this paper. They can be found in the original paper by Wedepohl and Wilcox [3].

The computation of admittance is achieved through a well-known formula:

$$Y = \frac{j.\omega.2.\pi.\varepsilon}{\ln\left(\frac{r_2}{r_1}\right)} \tag{1}$$

$$\varepsilon = \varepsilon_i + \frac{g_i}{j.\omega} \tag{2}$$

Where:

- ε_i – permittivity of the insulation
- g_i – conductivity of the insulation
- r_1 – inner radius of the insulation
- r_2 – outer radius of the insulation

B. *Three core cables*

The simplification in this case lies in the approximation of the steel armor. A steel pipe is used as model, so it is the same modelling as the modelling of the pipe-type cables. This solution is not completely correct, but is assumed as an acceptable simplification.

The general theory was summarized Ametani [4] and its assumptions are:

- constant magnetic permeability of steel
- armor thickness is greater than its skin depth
- cores diameter are small compared to inner diameter of the armor
- eccentric positions of the cores does not affect their internal impedances and admittances

The impedances are achieved in a similar manner as in the single core cables, but with different assumptions when deriving the formulae. Again, for the extensive content, they are not here explicitly defined, but can be found in academic literature [5].

On the other hand, generally accurate formulae for the capacitances between the cores and the armor of the three core cables does not exist. These capacitances are mostly derived from potential coefficients, because the capacitance matrix $[\boldsymbol{C}]$ is obtained by inverting the matrix of potential coefficients $[\lambda]$.

Firstly, we have to compute the prerequisite potential coefficients on diagonal and also on non-diagonal parts of potential coefficients matrix:

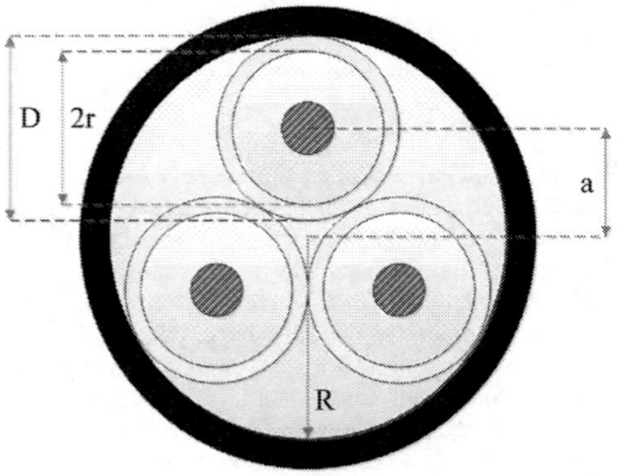

Figure 2. Components in the three core cable model for capacitance computations

$$\lambda = \frac{1}{2\pi\varepsilon}\ln\left[\left(\frac{R}{r}\right)\cdot\left(1-\left(\frac{a}{R}\right)^2\right)\right] \quad (3)$$

$$\mu = \frac{1}{2\pi\varepsilon}\ln\left(\frac{\sqrt{R^4+a^4+a^2R^2}}{aR\sqrt{3}}\right) \quad (4)$$

$$C_k = -C_m = \frac{\mu}{(\lambda-\mu)(\lambda+2\mu)} \quad (5)$$

$$C_p = C_0 + 2C_k = \frac{\lambda+\mu}{(\lambda-\mu)(\lambda+2\mu)} \quad (6)$$

Then arrange to the capacitance and potential coefficients matrix.

$$[C] = \begin{bmatrix} C_p & C_m & C_m \\ C_m & C_p & C_m \\ C_m & C_m & C_p \end{bmatrix} \quad (7)$$

$$[\lambda] = \begin{bmatrix} \lambda & \mu & \mu \\ \mu & \lambda & \mu \\ \mu & \mu & \lambda \end{bmatrix} \quad (8)$$

And finally invert the potential coefficient matrix.

$$[C] = [\lambda]^{-1} \quad (9)$$

The quicker approach is proposed by Simon [6]. The key variables in his expression are shown in Fig. 3 and those are:

- t – thickness of sheath insulation [mm]
- T – thickness of conductor insulation [mm]
- d – diameter of conductor [mm]
- ε_r – relative permittivity [-]

$$C = \frac{0.03\varepsilon_r}{\log_{10}\left[\left(0.52\left(\frac{t}{T}\right)^2 - 1.7\left(\frac{t}{T}\right) + 3.84\right)\left(\frac{T+t}{d}\right)+1\right]} \quad (10)$$

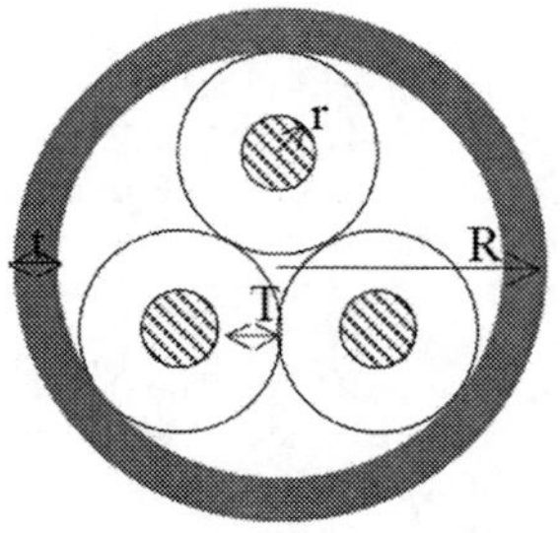

Figure 3. A three core cable model used in Simon's simplified expression

III. Positive Sequence Impedance for Power Frequency Concerns

Thanks to low frequency used in power engineering industry (50 or 60 Hz), huge simplifications may be achieved, which leads to much simpler formulae for impedance than above mentioned standard methods.

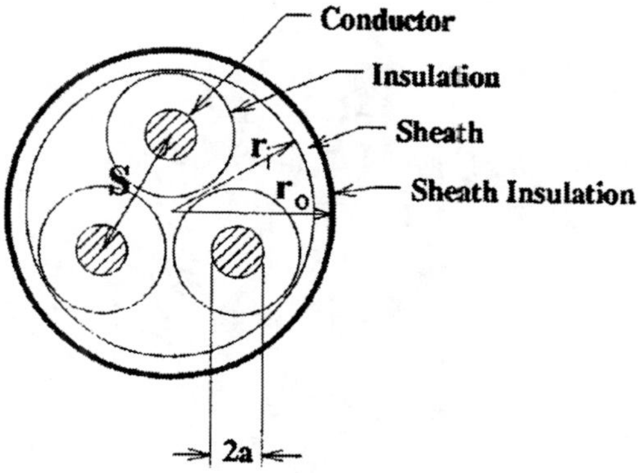

Figure 4. A three core cable model for positive sequence impedance calculations

The positive sequence resistance R_1 of the conductors is defined by the materials and the cross-section – therefore it will not be computed and only the manufacturer's measurement will be taken into account. Also with small difference, the DC resistance can be taken as positive sequence resistance because of the low frequency.

The positive sequence reactance formulae in terms of the Companion document to NZCCPTS [7] are as follows:

$$X_1 = 2.893x10^{-3} f.\log_{10}\left(\frac{S}{GMR}\right) \quad (11)$$

$$GMR = \xi r_{con} \quad (12)$$

Where:

- f – frequency [Hz]
- S – separation of conductors in cable [m]
- GMR – Geometric Mean Radius
- ξ – parameter of the conductor dependent on its construction (0.7788 for solid round conductor) [m]
- r_{con} – radius of the conductor [m]

Therefore the positive sequence impedance of a three core cable is defined by following expression:

$$\overline{Z}_1 = R_1 + jX_1 \quad (13)$$

IV. Investigated Cable

As the investigated cable was chosen the three core copper cable with parameters in Table 1.

Table 1. Parameters of the Selected Cable

FXCEL 12/20(24)kV 3x16/10 cable		
Parameter	***Value***	***Unit***
r_o	43	[mm]
r_i	40.4	[mm]
S	17	[mm]
ξ	0.726	[-]
r_{con}	3	[mm]
R_{con}	1.15	[Ω/km]
ε_{ins}	2.8	[-]

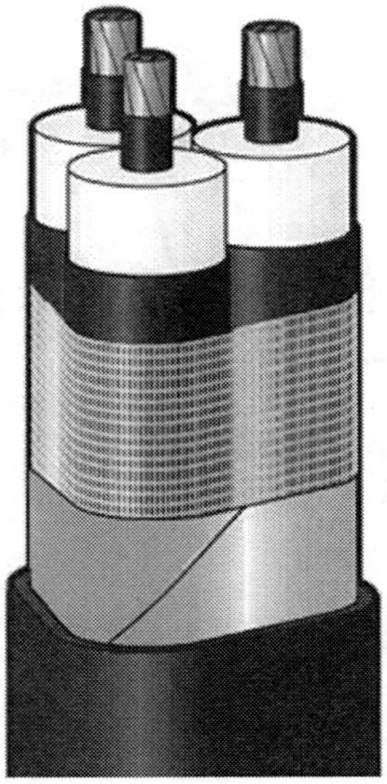

Figure 5. Construction of FXCEL cable used in the simulations and computations in this paper

V. POSITIVE SEQUENCE IMPEDANCE RESULTS FOR PERFECTLY DEVELOPED CONDUCTOR

The computed and measured values of the electrical parameters of the three core cable are in Table 2. We can observe, that the only parameters, which can be compared to the measured values is the capacitance and inductance of the cable. The difference between the measured and computed capacitance values using Simon's expression is only 6.36%, which is highly dependent on the permittivity of the insulation. Usually the manufacturer doesn't provide such detailed specification, so the computed values are probably always only an approximation of the real capacitance value. The difference in the inductance is even higher (14.38%), which tells us, that the approximation of the method was probably too high, and more complex calculations are needed.

Table 2. Computed and measured electrical parameters of a three core selected cable

FXCEL 12/20(24)kV 3x16/10 cable				
Parameter	***Computed***	***Measured***	***Unit***	***Difference [%]***
R_1	-	1.150	Ω/km	-
L_1	0.411	0.480	mH/km	**14.38**
C_1	0.117	0.110	μF/km	**6.36**

VI. MODELLING OF THE INCOMPLETELY DEVELOPED CABLE

There are specific installations of cables, in which they are not completely developed and are still partially wrapped on the drum (mostly temporary installations or other academic reasons). Obviously, all the above mentioned formulas are assuming, that the power cable is buried, but more importantly – completely developed in its length and therefore is perfectly straight. The results will not be correct, if this assumption would be used on this case. In the scientific literature is no mentioning of such case of computation, so simulation using the finite element method was only suitable solution. As the simulation tool was chosen the Ansys Maxwell software.

A. *Defining the cable drum*

Cable drums are generally divided into wooden or steel drums. In every cable's technical specification should be defined the size and type of the transportation drum. Selected copper cable in this paper have prescribed a K20 type wooden drum. Important dimension of the mentioned wooden drum with cable are:

- Radius of the cable curve – 681.36 [mm]
- Maximum number of helixes – 17 [-]
- Cable overhang – 2x700 = 1400 [mm]
- Length of 1 helix – 5681.11 [mm]
- Length of 2 helixes – 9962.22 [mm]
- Length of 3 helixes – 14243.33 [mm]
- Length of 4 helixes – 18524.44 [mm]

Based on the simulaton restrictions, the final model is similar to following Figure:

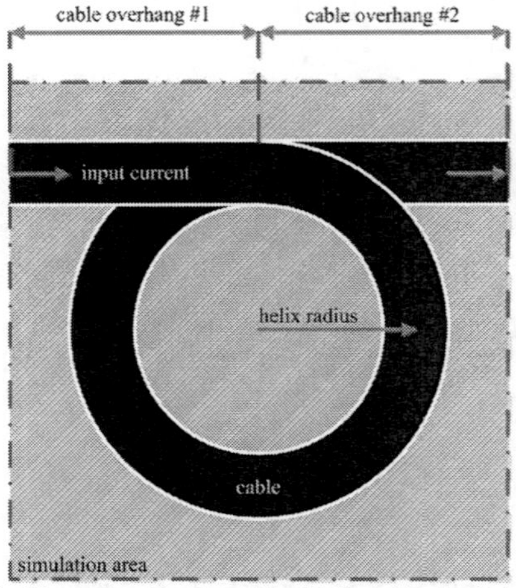

Figure 6. Illustrative side view of final helix cable simulation model used in Ansys Maxwell software

B. *Defining the simulation procedure*

The division of the problem was necessary while simulation software generally requires a high-end hardware or server installation. Dividing the problem into separate parts enlightened the computing demands from the computer and made the simulations quicker and even possible. The simulations progress was as follows:

1. Straight cable (2D simulation)
2. Cable with one helix on drum (3D simulation)
3. Cable with two helixes on drum (3D simulation)
4. Cable with three helixes on drum (3D simulation)
5. Cable with four helixes on drum (3D simulation)

Due to the low computing power, it was not possible to achieve simulation results with more than four cable helixes. An approximation using the obtained values from simulation was made, to determine the difference between the impedances of the straight and wrapped cable. The capacitance was not taken into account, because of the shielding of the cable, which is during operational state effectively grounded.

Figure 7. An illustration of simulation model in Ansys Maxwell software – the detailed view on conductor (upper) and complete wooden drum wrapping (lower)

VII. Sequence Impedance Results for Partially Wrapped Conductor

The ratio between the straight and wrapped conductor was made on the base of the simulation results, which was as follows:

Table 3. Ratio of positive sequence electrical parameters from straight and helix-type simulation results

FXCEL 12/20(24)kV 3x16/10 cable					
parameter	***1h***	***2h***	***3h***	***4h***	***Average***
$R_{helix}/R_{straight}$ [-]	1.04	1.03	1.03	1.03	**1.03**
$L_{helix}/L_{straight}$ [-]	1.07	1.05	1.06	1.05	**1.06**

It is clear, that the resistance of the conductor is affected minimally (~3% higher than straight conductor), as well as the positive sequence inductance (~6% higher than straight conductor).

Table 4. Ratio of zero sequence electrical parameters from straight and helix-type simulation results

FXCEL 12/20(24)kV 3x16/10 cable					
parameter	***1h***	***2h***	***3h***	***4h***	***Average***
$R_{helix}/R_{straight}$ [-]	1.04	1.03	1.03	1.03	**1.03**
$L_{helix}/L_{straight}$ [-]	5.58	5.97	6.19	6.35	**f(x)**

The results in the case of the zero sequence inductance are more complicated. While the resistance keeps solid still at the 3% difference, the inductance is more complex in this case. The average value cannot be achieved, instead the function of the inductance ratio is graphically shown in Figure 8 and numerically evaluated in Equation 14.

$$\frac{L_{0h}}{L_0} = 0.5525\ln(helixNumber) + 5.5855 \tag{14}$$

So it is obvious, that the parameters of the three-core three-phase cable when unrolled on drum are radically different in the matter of the zero sequence inductance. Therefore it is important to carefully measure the electrical parameters of such installations. Particularly the zero sequence inductance if desired.

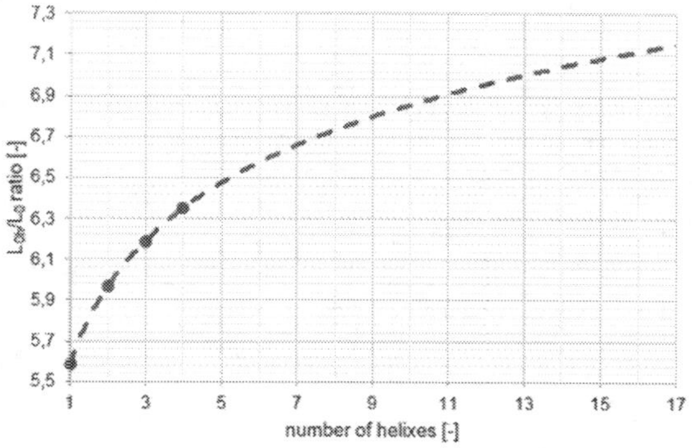

Figure 8. Function describing the dependency of the ratio of the zero sequence inductance from the number of helixes of the cable on the wooden drum

VIII. Comparison to the Overhead Line Electrical Parameters

An according ACSR conductor was chosen and general 22kV tower was used to compute the electrical parameters of the overhead line as an alternative to this three core cable. Every conductor on the tower is positioned 8m above terrain and the conductors are separated with 1.44m distance. The ACSR conductor is of type 70/11-1 with these parameters:

- Outside diameter – d – 11.7 [mm]
- T/D ratio – 0.32 [-]
- DC Resistance – $R_{DC} = R_1$ – 0.431 [Ω/km]

For quicker computations, Matlab SimPowerSystems library was used for this procedure. Obtained electrical parameters in positive sequence are as follows:

Table 5. Computed electrical parameters of the ACSR overhead line

Parameter	***Computed***	***Unit***
R_1	0.431	Ω.km
L_1	1.187	mH/km
C_1	9.786	nF/km

The parametrical differences between the cable and overhead lines are apparent. They are shown in the table below:

Table 6. Parametrical Comparison between the three-core cable and overhead line

Parameter	***OHL***	***Cable***	***Unit***	***OHL - Cable [unit]***
R_1	0.431	1.150	Ω/km	**-0.72**
L_1	1.187	0.480	mH/km	**0.70**
C_1	9.786	110	nF/km	**-100**

The assumptions from the beginning of this paper was therefore confirmed.

Conclusion

This paper was aimed on the electrical characteristics of the power cables and the theoretical background how to compute them. The single core as well as three core cables were point of interest, and for a selected three-core cables also its electrical parameters evaluated. They were obtained by the simplest computational method for positive sequence impedance, which lead to the higher error in inductance of a three core cable when compared to the manufacturer's specifications. On the other hand, the capacitance was very close to the specifications.

In the second part, the wrapped conductor on the wooden drum was modeled in the simulation software Ansys Maxwell. The ratio of the wrapped and straight three-core cable was achieved and discussed. It was shown, that the helixes around the drum will increase the zero sequence inductance more than 6 times in comparison to the straight conductor and can be described by a function with natural logarithm. The same effect of helixes can be basically neglected in inspected positive sequence parameters as well as in zero sequence resistance, where the difference was from 3 to 5% from the straight conductor reference values.

The last part of this article was aimed on the parametrical comparison of the three-core cable and accordingly selected ACSR power line. The assumptions from the beginning of the article were confirmed. The greatest difference occurred in capacitance (~10 nF/km for overhead line and ~110 nF/km for power cable), which has also influence on the charging current of the line/cable. It is therefore very important to know the electrical characteristics of the not only newly installed systems to safely and effectively operate the power grid.

Acknowledgement

This publication is a result of implementation of the project: "Increase of Power Safety of the Slovak Republic" (ITMS: 26220220077) supported by the Research & Development Operational Programme funded by the ERDF.

References

[1] S. A. Schelkunoff, "The electromagnetic theory of coaxial transmission lines and cylindrical shields," in The Bell System Technical Journal, vol. 13, no. 4, pp. 532-579, Oct. 1934. doi: 10.1002/j.1538-7305.1934.tb00679.x

[2] F. Pollaczek, "Uber das feld einer unendlich langen wechsel stromdurchflossenen einfachtleitung", Elektrische NachrichtenTechnik, Vol. 3, No. 9, 1926.

[3] L. M. Wedepohl and D. J. Wilcox, "Transient analysis of underground power-transmission systems. System-model and wave-propagation characteristics," in Electrical Engineers, Proceedings of the Institution of, vol. 120, no. 2, pp. 253-260, February 1973. doi: 10.1049/piee.1973.

[4] A. Ametani, "A General Formulation of Impedance and Admittance of Cables," in IEEE Transactions on Power Apparatus and Systems, vol. PAS-99, no. 3, pp. 902-910, May 1980. doi: 10.1109/TPAS.1980.319718.

[5] WG B1.30, "Cable System Electrical Characteristics", CIGRE Technical Brochure, ISBN: 978-2-85873-225-8, April 2013.

[6] J. R. Lucas, High Voltage Cables [chapter], in book "High Voltage Engineering", University of Moratuwa, Sri Lanka, 2001, p.204.

[7] A. K. Parsotam, Fundamentals of Calculation of Earth Potential Rise in the Underground Power Distribution Cable Network. New Zealand Committee for the Co-ordination of Power and Telecommunication Systems, 1997.

Measuring Current Distribution of Phase Conductor and Current Intesity Induced in Ground Wire on the 400 kV Transmission Line

Practical Experiences

Tamás Bagi
Óbuda University, Department of Safety Science
ELINOR Engineering Ltd.
Budapest, Hungary
bagi.tamas@elinor.hu

Abstract — **Due to global warming-up impacts of climate changes could be observed in the National Hungarian Electricity Network. Extreme weather conditions (glaciations, extreme squall) are more common and their impacts are more significant that is why they need to be taken into consideration as well. It is also confirmed by interruptions in the transmission and distribution network in 2013 and 2014. Consequently The Hungarian Transmission System Operator Company Ltd. (hereunder MAVIR Ltd.) requires an emergency plan to find solutions to avoid and prevent similar equipment failures in future. (havaria)**

Keywords: deiceing; transmission line; ground wire; measure

I. Introduction

As during their everyday lives people may experience, the general climate has changed dramatically since the 1950s-1960s. A significant part of Hungary's electrical grid network was built during the second part of the 1900s, based on the economic and technical requirements of that era. This of course means that the relevant engineering specifications and regulations were a lot different since they had been drafted for the weather conditions experienced back then.

Moreover, it is also apparent that the guiding principle of past constructions was that each of the power lines were created to achieve an average lifespan of 50-70 years, and therefore they are nearing the end of their planned operation time. Due to this fact and the extreme weather conditions of March 2013 and December 2014, many key power lines were critically damaged or rendered unserviceable.

II. Presentation Incidents

A. *14 March 2013*

By analyzing the significant damage caused by the extreme weather conditions it can be stated that the 400 kV transmission line between Munkács and Sajószöged, and the 220 kV lines between Munkács - Tiszalök and Kisvárda – Sajószöged on the power grid were damaged considerably. (20 towers were collapsed, 89 cross-arms were damaged and the base of 14 other columns was also subjected to the adverse weather effects.)

Figure 1. Fallen Steel Towers

Figure 2. Damaged Steel Towers

B. 01-December-2014

Following 2013 and another harsh winter, the power grid and relay network of Hungary had once again fallen into severe disrepair. In 2014, Pest County's 220 kV transmission line between Ócsa and Zugló and the 400 kV line between Albertirsa – Göd I-II became significantly damaged, with each of them having a section falling onto the nearby electric power distribution network and the M31 highway, rendering it unfit for traffic for a while.

Figure 3. Damaged Steel Structures on the M31 Highway

Figure 4. Damaged Steel Towers on The M31 Highway

III. CONCLUSION

Due to climate change, extreme weather conditions are increasingly endangering safe operation of the electric power system. In recent years, considerable damage and emergencies were caused due to the glaciations (freezing rain) damage to transmission lines (high voltage transmission line poles collapsed, wire disruptions, overload for transmission line's covering ice, unbalancing tension between two adjacent spans, transmission line conductor galloping etc.). These not only threaten the security of electricity supply in the domestic supply but it may also mean the loss of international connections. One of the methods by which the damage can be reduced or avoided by the defrosting the ice of phase or ground wire.

These issues shed light on the weaknesses of the current network and inspired me to choose this as the topic of my research.

IV. POSSIBLE SOLUTIONS

After investigating certain international networks, we noted that ice formation and/or extreme weather-related disrepairs only occur in a few regions, and within each of them there are different strategies and techniques to avoid subsequent malfunctions:

- Canada,
- Russia,
- China,
- and Carpathian Basin.

As a conclusion, the applied solutions can be classified into two groups. One of these uses Direct Current while the other one uses Alternating Current during deicing procedures.

A. DC deicing technical features in comparison with AC method

- No need of considering line's stray inductance and capacitance influence, don't need reactive power compensation for the transmission line;
- Making electric current through line being much evener, heating high efficiency;
- Carrying electric current stably, regardless of the line length;
- Requiring converter device.

B. DC Solutions

In march 2016, the representatives of NR Electric Co., Ltd. visited Hungary as part of an endeavor to establish an Eastern-European division (center) within Hungary. They have also heard about Hungary's ice-related problems and wanted to demonstrate effective ways of coping with this issue in China, using their own products.

Since China boasts a vast and expansive HVDC network, their use of a Direct Current deicing method is and understandable strategy.

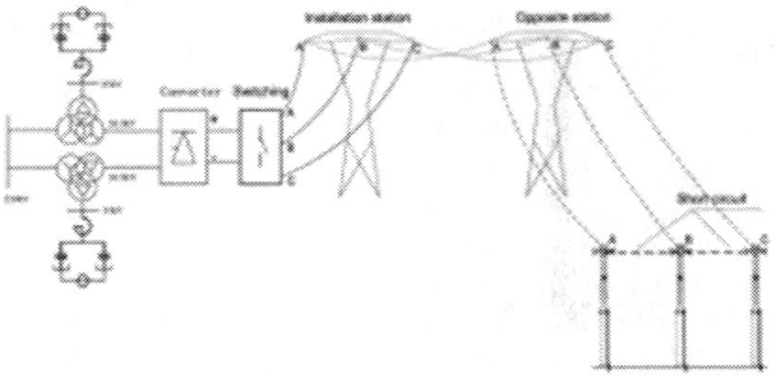

Figure 5. Typical topology for DC for Deicing of one transmission line [1]

They provided three examples, with each of them being a possible deicing solution that they use:

- DC de-icer with SVC (Static Var Compensator) [1]

Figure 6. Substaion scheme

- Movable DC de-icer – Cabinet Scheme [1]

Figure 7. Cabinet scheme [1]

Figure 8. Cabinet scheme

- Movable DC de-icer – Container Scheme [1]

Figure 9. Container scheme [1]

Figure 10. Container scheme [1]

Table 1. Properties of the de-icer system's [1]

	Rated direct current (A)	*Rated direct voltage (kV)*	*Rated Capacity (MW)*	*SVC Capacity (MVar)*
Substation	4500 A	+/- 22,5 kV	200 MW	-60 / +180 MVar
Container Scheme	2000 A	12,5 kV	25 MW	-
Cabinet Scheme	1000 A	12 kV	12 MW	-

Sadly, using Direct Current deicing on the Hungarian transmission and distribution networks is not possible due to a variety of reasons. The first and most significant factor that prohibits the use of this method is the fact that we simply lack the Direct Current intake with a performance that is high enough to enable such procedures. Another important obstacle in the way of introducing this technique is the structure of the network itself, since the Hungarian distribution and transmission systems are effectively grounded. Due to these facts, the only viable solution would be the use of Alternating Currents, but even this would require special preparations.

C. AC Solution

Deicing using Alternating Current would not be a trivial task either. It would be easier to perform in the case of line wires since the electricity flowing through the line conductors of power lines could be improved through an increased load and/or capacity regrouping, but this method would not be viable in the case of earth wires since – under normal operating circumstances – they possess no electrical currents and are grounded as well.

Under normal operating circumstances, the loop currents generated by the electromagnetic field of the currents flowing within the line conductors would conflict with each other, making this form of solution inadequate for the deicing of guard conductors.

Figure 11. Conceptual scheme [2]

In order to have the line wires conduct zero-sequence current, a special state of operation would have to be established during which zero-sequence current flows through the conductors. According to previous estimations, 50-60% of this zero-sequence current would have to be induced to render it adequate for deicing.

In cooperation with MAVIR Ltd., the practical tests of the assumptions listed above was made possible by performing measurements during realistic circumstances. The planned measures were divided into two parts.

First, we wanted to discover how the currents flowing through the line and specific earth wires is divided. The second phase was to check whether the current adequate for deicing becomes induced within the earth lines. The location designated by MAVIR for taking the measurements was the 400 kV THE – Sajószöged power line which is currently out of operation.

1. Measuring current division

Initial data:

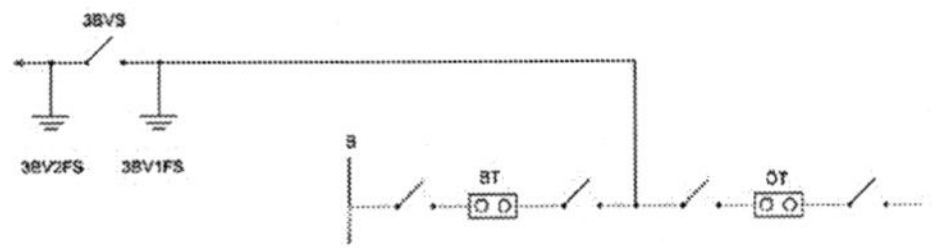

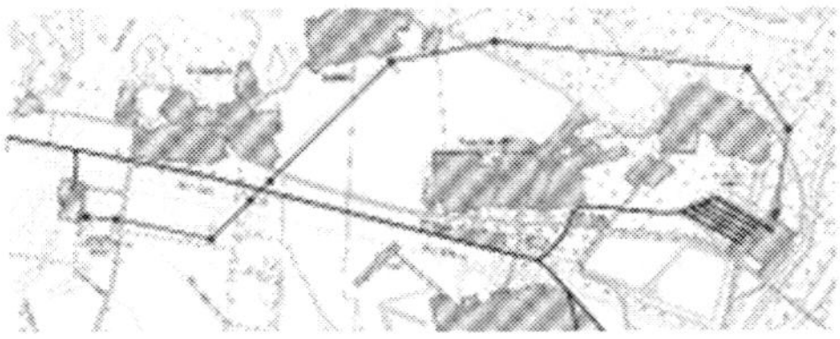

Figure 12. Overall layout [2],[3]

- Length of the power line: 11072m
- Type of the live conductor: 3x2x500/65 mm^2 ACSR
- Type of the guard conductor: 2x95/55 mm^2 ACSR
- Type of supporting devices (OT): Generator

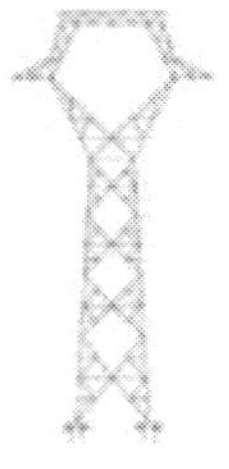

Figure 13. Generator Steel Tower [3]

- Type of anchor poles (OSF, OVSF): Ipoly

Figure 14. Ipoly Steel Tower [3]

Measurement Process

During the preparations for the measurements, the employees of MAVIR Ltd. had grounded the line conductors of pole no. 2 to its body and supplied the required power supply (0,4 kV) from the auxiliary circuit of transformer no. 2 to the Sajószöged 400/220/120 kV station's 3B field and informed us about the status of the field, the power line and the disconnecting switches. These devices had been checked personally as well.

For the safe, secure and successful procedure we formed two groups. The first group was stationed at the substation and provided the required circuits, the supply voltage and kept informing the second group (which was taking measurements on the power line) about the supplied current.

Figure 15. 3 Phase 0,4kV power supply [2]

The second group responsible for the power lines determined three locations for taking the measurements, based on previous discussions. The measurements were taken on the first part (towers 2 and 3), the middle (towers 11 and 12) and towards the end (towers 21 and 22).

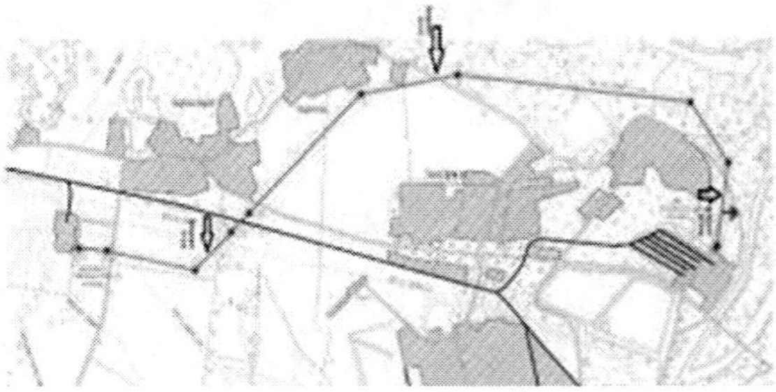

Figure 16. Measurement sites [3]

Measurement Results:

Table 2.– Measuring location 1. (between 2. and 3. towers) [4]

	I_{L1a}	I_{L1b}	I_{L2a}	I_{L2b}	I_{L3a}	I_{L4b}
Measured Value (A)	25,4	40,6	39,2	32,3	36,8	27,6
Measured Value (%)	37,9%	60,6%	54,4%	44,9%	56,6%	42,5%
Supplied Value (A)	67		72		65	

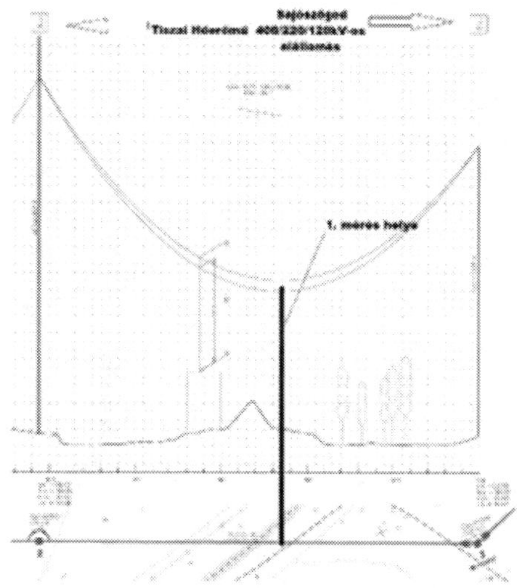

Figure 17. Measuring location 1.[3]

Table 3.– Measuring location 2. (between 11. and 12. towers) [4]

	I_{L1a}	I_{L1b}	I_{L2a}	I_{L2b}	I_{L3a}	I_{L4b}
Measured Value (A)	32,0	32,5	35,4	34,9	32,0	31,4
Measured Value (%)	47,8%	48,5%	49,2%	48,5%	49,2%	48,3%
Supplied Value (A)	67		72		65	

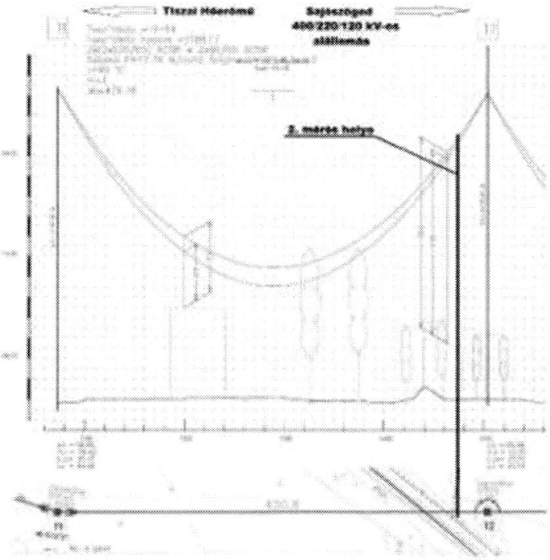

Figure 18. Measuring location 2. [3]

Table 4.– Measuring location 3. (between 21. and 22. towers) [4]

	I_{L1a}	I_{L1b}	I_{L2a}	I_{L2b}	I_{L3a}	I_{L4b}
Measured Value (A)	32,6	32,3	35,2	35,2	31,9	31,4
Measured Value (%)	48,7%	48,2%	48,9%	48,9%	49,1%	48,3%
Supplied Value (A)	67		72		65	

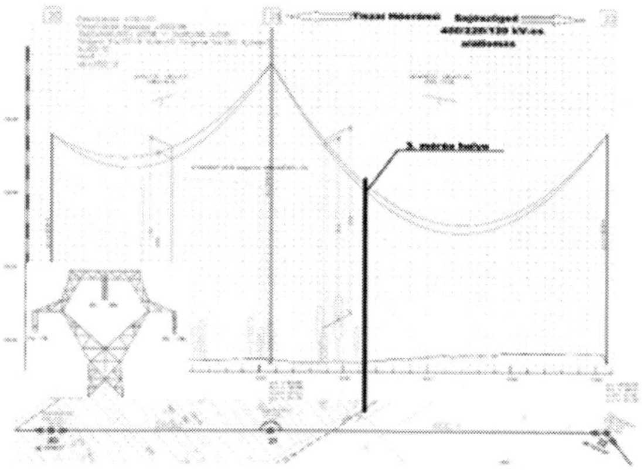

Figure 19. Measuring location 3. [3]

The line conductor was measured using a type MW 3500 clamp meter with a measuring limit of 2000 A and a car with a lifting platform supplied by MAVIR.

Figure 20. Measuring process

Based on the measurements, it can be noted that the currents flowing through the line conductors spread among the bundles evenly.

2. Measuring induced currents:

Initial data:

The initial status, the measured power line, the place of grounding and the measurement locations were same as during the previous measurements. The only difference was the method of supply.

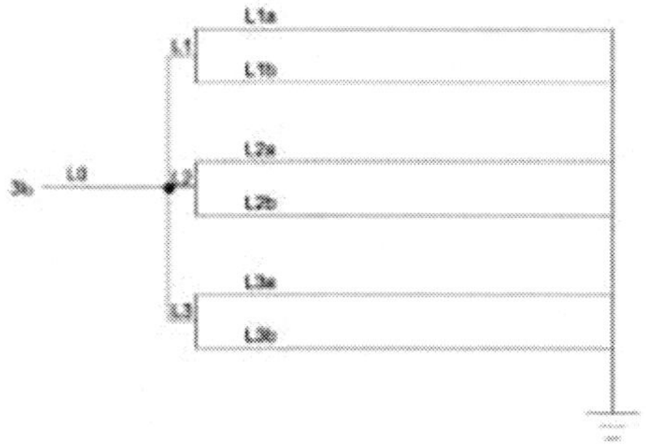

Figure 21. Zero Sequence 1 Phase 0,4kV power supply

Measurement Results:

Table 5.– Measuring location 1. (between 2. and 3. towers) [4]

	I_{VV1}	I_{VV2}
Measured Value (I [A])	46,8	6,2
Measured Value (I_0 [%])	159,7	21,1
Supplied Value ($3I_0$ [A])	88	

Table 6.– Measuring location 2. (between 11. and 12. towers) [4]

	I_{VV1}	I_{VV2}
Measured Value (I [A])	19,9	16,6
Measured Value (I_0 [%])	67,9	56,6
Supplied Value ($3I_0$ [A])	88	

Table 7.– Measuring location 3. (between 21. and 22. towers) [4]

	I_{VV1}	I_{VV2}
Measured Value (I [A])	15,1	13,9
Measured Value (I_0 [%])	51,5	47,4
Supplied Value ($3I_0$ [A])	88	

The measurements on the guard conductors were taken with the Mw 3500 clamp meter and the platform car provided by MAVIR Ltd.

The measurements validated our previous assumptions, namely that the amount of the current induced through a $3I_0$ (zero-sequence) supply within the guard conductor was around 50-60% of the supplied I_0. The only significant difference was around pole no. 2, which was grounded.

Summary:

Based on the first measurement, we can declare that the current division within the bundles of the line conductor was approximately symmetrical, with negligible differences. The greatest discrepancy was noted in the case of the 67A L_1 line, where instead of 33,5 A – 33,5 A, one of the bundles exhibited 40,6 A, while the other 25,37 A. The resulting discrepancy was ~8,1A (~12%).

The second measurement proved our previous assumptions that a 3I0 zero-sequence supply will cause current to flow through the guard conductors of the power line with an estimated amount of 50-60% when compared to the I_0 current.

The discrepancy between towers 2-3 was due to the fact that that the line conductors of pole 2. were grounded into the steel assembly of the pole itself.

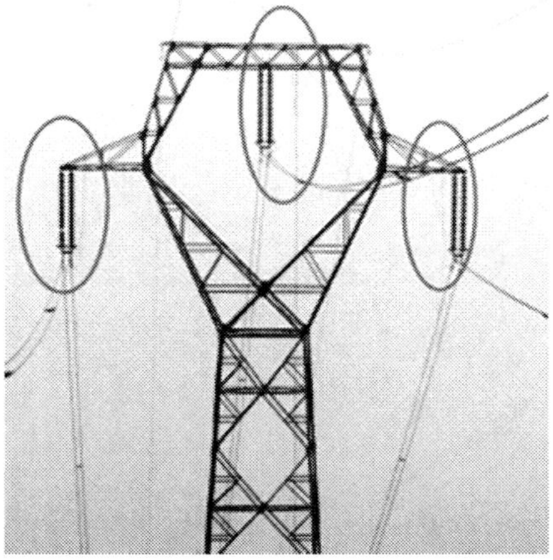

Figure 22. The Grounded Tower

The other two measurement locations were as expected with the lowest value being 47%, the highest one being 68%.

ACKNOWLEDGMENT

I would like to thank NR. Electric Ltd., Óbuda University and MAVIR Ltd. for allowing of measurements and preparations.

REFERENCES

[1] "NR Electric Co., Ltd." [Online]. Available: http://www.nrec.com/en/. [Accessed: 3-May-2017].

[2] T. Bagi, Gy. Morva, A jegesedés hatása a villamosenergia-rendszer biztonságos üzemvitélere, Elektrotechnika, Volume 109. 2016/3, pp.- 5-7, ISSN:0367-0708.

[3] MAHALIA database: THE – Sajószöged 400 kV, unpublished.

[4] T. Bagi, A. Bálint, I. Horváth, "Measurement Report–2016.01.28". Óbuda University, 2016, unpublished.

Energetically Passive Fiber-optic Sensor System for the Security Windows of the Buildings

Marcel Fajkus, Jan Nedoma,
Stanislav Kepak, Vladimir Vasinek
Department of Telecommunications
VSB - Technical university of Ostrava,
Ostrava, Czech Republic
marcel.fajkus@vsb.cz, jan.nedoma@vsb.cz,
stanislav.kepak@vsb.cz, vladimir.vasinek@vsb.cz

Radek Martinek, Petr Bilik,
Jan Vanus
Department of Cybernetics and Biomedical Engineering
VSB - Technical university of Ostrava,
Ostrava, Czech Republic
radek.martinek@vsb.cz, petr.bilik@vsb.cz,
jan.vanus@vsb.cz

Abstract —**An important element within any buildings, halls and warehouses is protection against intrusions and theft of expensive materials and funds stored in these areas. Currently, there are the many solutions for detecting unwanted intrusion. This article provides a simple solution based on the performed real measurements of the security the perimeter, which is based on fiber Bragg grating that is additionally implemented into structure of the plastic windows. In the case of attempts to break into the protected object, the system records the vibration response in the form of deformation which is transmitted to the Bragg sensor. The article presents two approaches to evaluating the data: the power measurement approach (it uses the point chirped FBG sensor with possible use temperature range from -30 °C up to 60 °C) and spectral measurement approach (it can monitor up to 80 FBG sensors simultaneously). The advantages of the fiber-optic technology against the conventional solutions are obvious: presented solution is not detectable by conventional methods (EMI-free) and it is energetically passive.**

Keywords: fiber-optic sensor; perimeter; security; energetically passive sensors; fiber Bragg sensor (FBG)

I. Introduction

The perimeter protection is an outdoor protection of the areas or the special applications, such as technical, electronic or electronic-mechanical outdoor security systems. It is an important part of guarding large complexes of buildings and areas such as power grids, airports, prisons, industrial buildings, waterworks, etc. The purpose of perimeter guarding is to capture the potential intruders by technical means promptly, that is, at the moment when they do not commit criminal activity in guarded premises. The basic requirement for elements of perimeter protection is the independence of function on the climatic conditions. At present, we use, for example, passive infrared detectors [1], infrared barriers [2], microwave barriers [3], vibration-sensitive microphone cables and slit cables [4], deformation sensors [5], pressure underground hoses [6], motion video detectors, and radars [7]. A very interesting alternative seems fiber-optic systems, which are used in many other applications such as vital sign monitoring, automotive industry etc. [8-13]. These systems for security of objects and infrastructure use several advantages of fiber-optic sensors respectively fiber-optic cables. These are their dielectric nature, i.e., the inability to easily find where they are installed, and thus easier to disable them from the security system. A very important advantage is the fact that they are not influenced by electromagnetic interferences (EMI), which occurs e.g. in the energy industry. Other advantages of fiber-optic systems include their long life, almost maintenance-free operation, easy installation, small dimensions and great reach (Units up to tens of km, depending on the cable attenuation and power of the radiation source). Very important is also deposited fibers due to the sensitivity of the system [14-18].

II. Methods

A. *Fiber Bragg grating*

Fiber Bragg grating is formed by periodic change of refractive index in core of optical fiber. This structure reflects one wavelength, called Bragg wavelength, and another wavelengths are transmitted with relative small attenuation. Bragg wavelength is given by equation (1):

$$\lambda_B = 2n_{\text{eff}}\Lambda, \tag{1}$$

where n_{eff} is effective refractive index and Λ is period of changes of refractive index in core of optical fiber. Structure of Bragg grating is shown in Fig. 1.

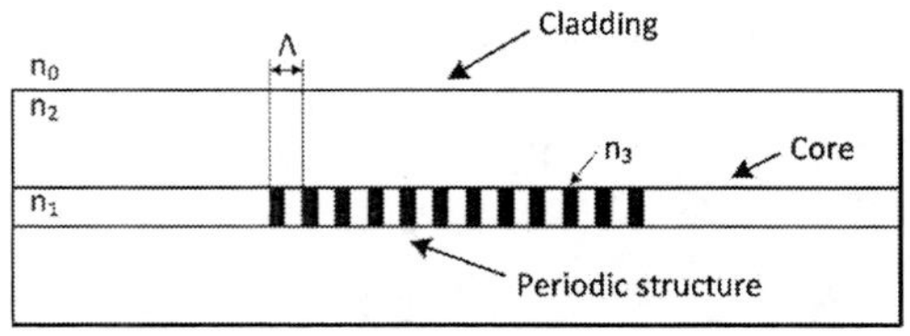

Figure 1. Structure of uniform fiber Bragg grating.

Core refractive index n_1 is modulated with depth $\delta n = (n_3 - n_1)$, with typical value from 10^{-5} to 10^{-3}. The value of refractive index modulation depends on the properties of the photosensitive optical fiber and the pre-processing of the fiber before the production of the FBG (Flame Brushing or hydrogenation). By using these techniques, modulated refractive index as 10^{-2} and higher has been obtained. [19-20].

Fiber Bragg grating belongs to group of single point optical fiber sensors. Individual sensors can be coupled to a sensory array using a time or wavelength division multiplexing method [21-22]. In sensory applications, we use a thermal and deformation dependencies of structure on external influence, which is given by equation (2):

$$\frac{\Delta\lambda}{\lambda_B} = (1 - p_e)\Delta\varepsilon + (\alpha_\Lambda + \alpha_n)\Delta T, \quad (2)$$

where p_e is photo-elastic coefficient, $\Delta\varepsilon$ is applied deformation, α_Λ is thermal expansion coefficient, α_Λ is thermal-optic coefficient and ΔT is temperature change. The temperature and deformation coefficient is dependent on the spectral position of the Bragg wavelength. Therefore, in literature is reported the normalized deformation coefficient at constant temperature by equation (3):

$$\frac{1}{\lambda_B}\frac{\Delta\lambda_B}{\Delta\varepsilon} = 0.78 \times 10^{-6} \ \mu\varepsilon^{-1}, \quad (3)$$

where $\Delta\lambda_B$ is Bragg wavelength change, $\Delta\varepsilon$ is deformation change. Then, normalized thermal coefficient at constant deformation is given by equation (4):

$$\frac{1}{\lambda_B}\frac{\Delta\lambda_B}{\Delta T} = 6.678 \times 10^{-6} \ {}^{\circ}C^{-1}. \quad (4)$$

where ΔT is temperature change. [23]

B. *Interferometric sensor*

Interferometers are widely used in science and industry for the measurement of small displacements or refractive index changes etc. These instruments that are based on interference of two light beams and their phase change, which is given by equation (5)

$$\Phi = 2\pi L \frac{n}{\lambda}, \quad (5)$$

where n is the refractive index, L is the length of the path and λ is wavelength of light. When we use the coherent light source, the phenomenon (light interference) can occur. An interference minimum is a place where two waves with the opposite phase are joined, and it is given by:

$$\Delta s = (2k + 1)\frac{\lambda}{2} \quad (6)$$

and an interference maximum is a place where two waves with the same phase are joined, and it is given as following:

$$\Delta s = 2k\frac{\lambda}{2}, \quad (7)$$

where k is the size of the wave vector and Δs is the path difference. We can express the output intensity of the interferometer by:

$$I = \frac{I_0 \cdot \alpha}{2}(1 + \cos\Delta\phi), \quad (8)$$

where α expresses the optical losses of the interferometer [24].

III. Experimental Setup

The prototype of a plastic window was used for the experiment. The Mach-Zehnder interferometer and fiber Bragg grating sensor were implemented within the window structure. The Mach-zehnder interferometer was inserted into the plastic frame around the entire transparent surface, The FBG sensor was placed on a transparent surface on the right side of the prototype, see Fig. 2. To measure vibrations on the window using the Bragg grating, a method of the power measurements were used (not spectral evaluation). Narrow-spectral light from the tunable laser was tuned to the central portion of the descending edge of the broadband chirped Bragg grating. The light power reflected from the Bragg grating, which is dependent on the spectral shift of the FBG was captured by a photodetector (PD), digitized by the A/D converter (National Instruments Card) and processed by an application implemented in LabVIEW. In the interferometric measurement, a narrow-spectrum DFB (Distributed Feedback Laser) was used. The signal from the interferometer was again detected by the photodetector and digitized and processed by the NI card and application written in LabVIEW, see Fig. 2.

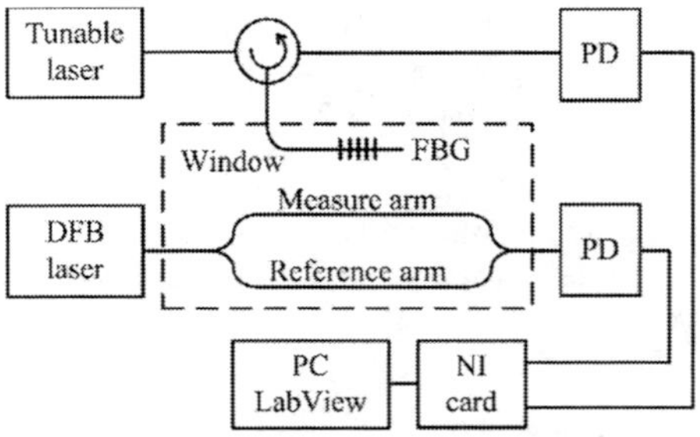

Figure 2. Measurement scheme with FBG a interferometric sensor.

IV. Results

The sensitivity of the interferometric and FBG sensor was analyzed by the vibration response to the transparent part of the prototype window. The response was caused by constant pressure parameters. This was accomplished by triggering a test item from a constant height on the transparent part of the window. The constant height was secured by the Thorlabs holders. Figure 3 shows defined places (9 points in total).

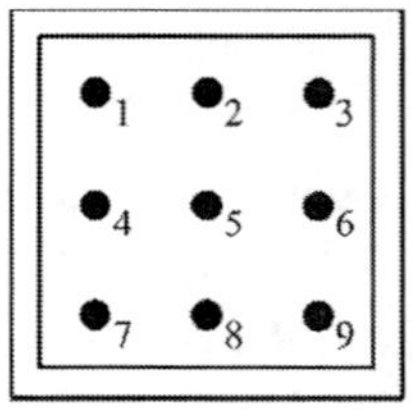

Figure 3. The position of the defined 9 points.

For each impact site, 50 impacts of the test subject were recorded. The real configuration of measurement scheme is shown in Fig. 4.

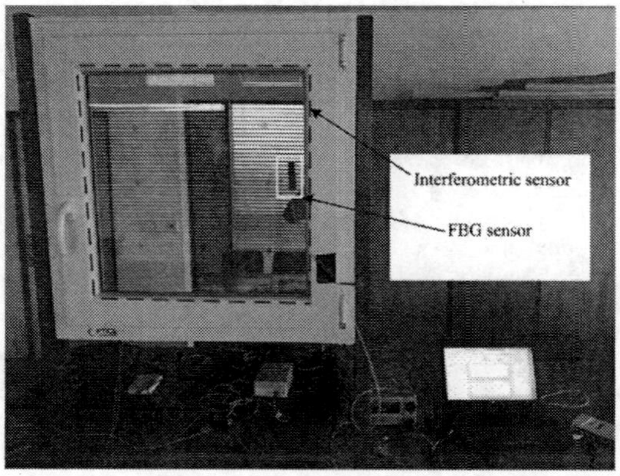

Figure 4. The real configuration of measurement scheme.

For better clarity, the data analyzed from both sensor types is displayed with 5 impacts, see Fig. 5. The graphs show the dependence of the amplitude response of the test subject on the time.

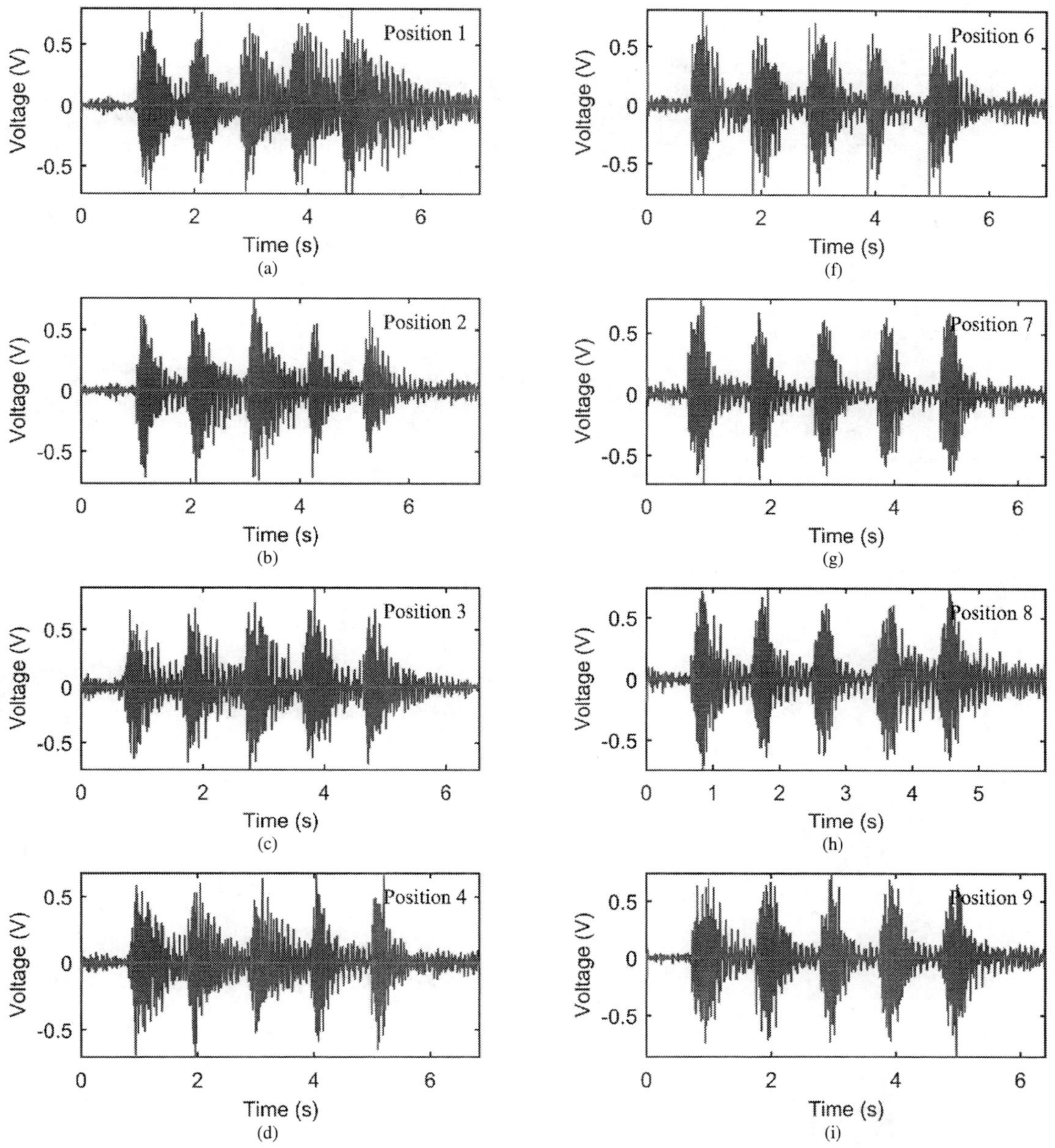

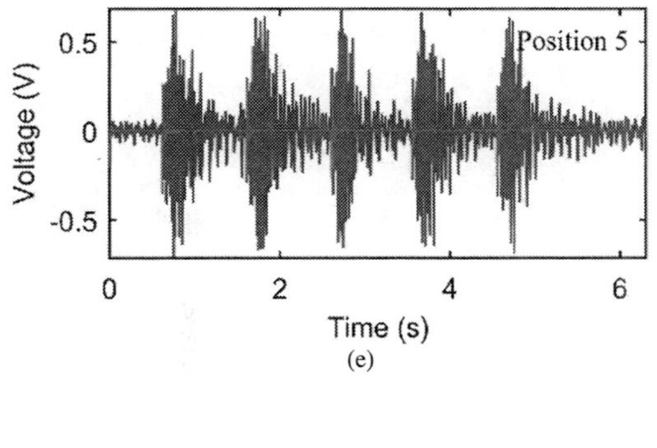

Figure 5. Results of measurements with interferometric (blue line) and FBG (red line) sensor.

The results show that in most cases the interferometric sensor achieves up to 4 times higher sensitivity to the vibration response (mean value about 1.511 V) than FBG sensor (mean value 0.456 V). Signal to noise ratio at the interferometer is balanced and reaches mean value about 22.982 dB. While the signal-to-noise ratio of the FBG is in the range from 20.730 dB to 39.660 dB. Due to the location of the interferometric sensor around the entire part of the transparent window area, this sensor has better parameters in terms of complex sensitivity comparison. For achieving a complex coverage of the transparent window surface by using FBG, it would be necessary to use a minimum of 4 FBG sensor (each on one side).

TABLE I. SENSOR SENSITIVITIES IMPLEMENTED ON SECURITY WINDOW FOR VARIOU PLACES

Position	FBG		Interferometer	
	Peak-Peak (V)	SNR (dB)	Peak-Peak (V)	SNR (dB)
1	0.207	23.050	1.521	21.633
2	0.343	23.067	1.524	26.004
3	0.187	23.493	1.600	20.730
4	0.307	22.217	1.381	25.045
5	0.579	22.326	1.399	30.562
6	1.651	23.381	1.580	39.660
7	0.209	22.983	1.509	21.710
8	0.342	22.800	1.478	25.980
9	0.280	23.518	1.605	24.243

Figure 6 shows the responses from the FBG sensor (Fig. 6a) and the interferometric sensor (Fig. 6b) in the 3D view. It is clear from the figure that the sensitivity of the FBG sensor is strongly dependent on the response distance from the location of the FBG sensor. In the case of the interferometric sensor, the sensitivity is more balanced, which corresponds to the location of the measuring arm around the entire area of the transparent window surface. Also, we can see that the higher sensitivity is achieved at the edges of the window and smaller at the center of the transparent part of the window prototype.

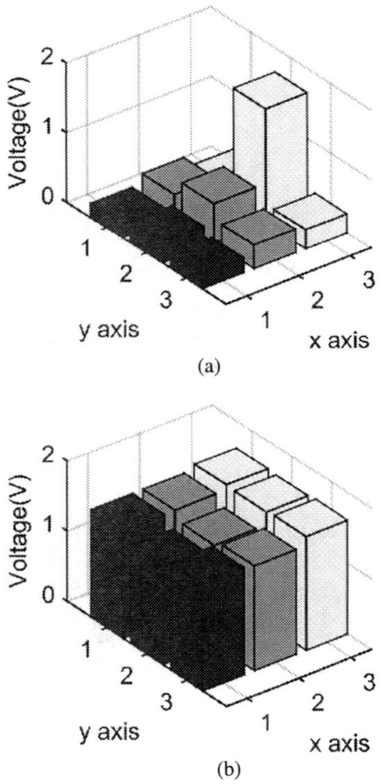

Figure 6. Windows sensitivities in 3D graph; (a) for FBG sensor; (b) for interferometric sensor.

CONSLUSION

This paper deals with the analysis of the use of fiber-optic technology in perimeter applications, namely window protection. The proposed idea offers against to the conventional solutions two are obvious advantages: presented solution is not detectable by conventional methods (EMI-free) and it is energetically passive. The article presents a statistical analysis and comparison of the two types of fiber-optic sensor (interferometric and fiber Bragg grating) which could be additionally implemented into the structure of the secured window.

ACKNOWLEDGMENT

This article was supported by the project of the Technology Agency of the Czech Republic TA04021263 and by Ministry of Education of the Czech Republic within the projects Nos. SP2017/128 and SP2017/79. The research has been partially supported by the Ministry of Education, Youth and Sports of the Czech Republic through the grant project no. CZ.1.07/2.3.00/20.0217 within the frame of the operation programme Education for competitiveness financed by the European Structural Funds and from the state budget of the Czech Republic. This article was also supported by the Ministry of the Interior of the Czech Republic within the projects Nos. VI20152020008 and VI2VS/444. The research has been partially supported by the COST action MP1401. This article was also supported by the Ministry of Industry and Trade of the Czech Republic within the project No. FV 10396.

REFERENCES

[1] J.Q. Ni, S. Liu, J.S. Radcliffe, C. Vonderohe, "Evaluation and characterisation of Passive Infrared Detectors to monitor pig activities in an environmental research building," Biosystems Engineering, vol. 158, pp. 86–94, 2017.

[2] F. M. Raimondi, M. Trapanese, P. Martorana, V. Franzitta, V. Di Dio, "A security and surveillance application of the innovative Monitoring Underwater Buoy Systems (MUnBuS) on the Protected Marine Area (AMP) of capo gallo (PA-IT)," OCEANS 2015 - MTS/IEEE Washington, Washington, DC, 2015, pp. 1-4. doi: 10.23919/OCEANS.2015.7404560.

[3] M. Caris, S. Stanko, W. Johannes, S. Sieger, N. Pohl, "Detection and tracking of micro aerial vehicles with millimeter wave radar," EuMW 2016 - Conference Proceedings 46th European Microwave Conference, pp. 1553–1555, 2017.

[4] T. Wang, J. Cao, J. Wang, X. Lai, Z. Lin, "Excavation devices classification using enhanced acoustics by MVDR beamforming with a cross microphone array," International Conference on Digital Signal Processing, DSP, pp. 388–392, 2017.

[5] Y.K. Fuh, Z.M. Huang, B.S. Wang, S.C. Li, "Self-Powered Active Sensor with Concentric Topography of Piezoelectric Fibers," Nanoscale Res Lett. 2017; 12: 44. doi: 10.1186/s11671-016-1786-x.

[6] Zhang, G., Xie, C., Lu, S., Xie, S. "Development of underground sprinkler irrigation device based on steel ball driving," Nongye Gongcheng Xuebao/Transactions of the Chinese Society of Agricultural Engineering, vol. 32, pp. 102–106, 2016.

[7] A. P. Prathosh, P. Praveena, L. K. Mestha, S. Bharadwaj, "Estimation of Respiratory Pattern From Video Using Selective Ensemble Aggregation," in IEEE Transactions on Signal Processing, vol. 65, no. 11, pp. 2902-2916, June1, 1 2017. doi: 10.1109/TSP.2017.2664048.

[8] J. Nedoma, M. Fajkus, L. Bednarek, V. Vasinek, "Non-destructive fiber-optic sensor system for the measurement of speed in road traffic," Advances in Electrical and Electronic Engineering, vol. 14, pp. 602–608, 2016.

[9] R. Martinek, J. Nedoma, M. Fajkus, R. Kahankova, J. Konecny, P. Janku, S. Kepak, P. Bilik, H. Nazeran, "A phonocardiographic-based fiber-optic sensor and adaptive filtering system for noninvasive continuous fetal heart rate monitoring," Sensors (Basel). 2017 Apr 18;17(4). pii: E890. doi: 10.3390/s17040890.

[10] J. Nedoma, M. Fajkus, P. Siska, R. Martinek, V. Vasinek, "Non-invasive fiber optic probe encapsulated into PolyDiMethylSiloxane for measuring respiratory and heart rate of the human body," Advances in Electrical and Electronic Engineering, vol.15, pp. 93–100, 2017.

[11] M. Fajkus, J. Nedoma, P. Siska, V. Vasinek, "FBG sensor of breathing encapsulated into polydimethylsiloxane," in Proc. SPIE 9994, Optical Materials and Biomaterials in Security and Defence Systems Technology XIII, 99940N (October 24, 2016); doi:10.1117/12.2241663.

[12] M. Fajkus, J. Nedoma, R. Martinek, V. Vasinek, H. Nazeran, P. Siska," A Non-Invasive Multichannel Hybrid Fiber-Optic Sensor System for Vital Sign Monitoring," Sensors (Basel). 2017 Jan; 17(1): 111. doi: 10.3390/s17010111.

[13] M. Fajkus, J. Nedoma, P. Siska, L. Bednarek, S. Zabka, V. Vasinek, "Perimeter system based on a combination of a Mach-Zehnder interferometer and the bragg gratings," Advances in Electrical and Electronic Engineering, vol. 14, pp. 318–324, 2016.

[14] J. Nedoma, M. Fajkus, V. Vasinek, "Influence of PDMS encapsulation on the sensitivity and frequency range of fiber-optic interferometer," Proc. SPIE 9994, Optical Materials and Biomaterials in Security and Defence Systems Technology XIII, 99940P (24 October 2016); doi: 10.1117/12.2243170.

[15] J. Nedoma, M. Fajkus, L. Bednarek, J. Frnda, J. Zavadil, V. Vasinek," Encapsulation of FBG sensor into the PDMS and its effect on spectral and temperature characteristics," Adv. Electr. Electron. Eng., vol. 14, 2016. doi: 10.15598/aeee.v14i4.1786

[16] M. Fajkus, J. Nedoma, L. Bednarek, J. Frnda, V. Vasinek, "Analysis of the applicability of singlemode optical fibers for measurement of deformation with distributed systems BOTDR," Advances in Electrical and Electronic Engineering, vol.14, pp. 453–459, 2016. doi: 10.15598/aeee.v14i4.1785

[17] M. Fajkus, J. Nedoma, S. Kepak, J. Jaros, J. Cubik, O. Zboril, M. Novak, V. Vasinek," Effect of the geometric deformations on the Brillouin scattering in the standard single-mode optical fiber," Proc. SPIE 9889, Optical Modelling and Design IV, 98892E (April 27, 2016); doi:10.1117/12.2239550.

[18] J. Cubik, S. Kepak, M. Fajkus, O. Zboril, J. Nedoma, A. Davidson, V. Vasinek, "Fixing methods for the use of optical fibers in interferometric arrangements," Proc. SPIE 10142, 20th Slovak-Czech-Polish Optical Conference on Wave and Quantum Aspects of Contemporary Optics, 101421A (December 23, 2016); doi:10.1117/12.2257060.

[19] P.J. Lemaire, R.M. Atkins, V. Mizrahi, W.A. Reed, "High pressure H2 loading as a technique for achieving ultrahigh UV photosensitivity and thermal sensitivity in GeO2 doped optical fibers," Electron. Lett., vol. 29 pp. 1191–1193, 1993.

[20] F. Bilodeau, B. Malo, J. Albert, D.C. Johnson, K.O. Hill, Y. Hibino, M. Abe M. Kawachi, "Photosensitization of optical fiber and silica-on-silicon/silica waveguides," Opt. Lett., vol. 18 pp. 953–955, 1993.

[21] M. Fajkus, J. Nedoma, S. Kepak, L. Rapant, R. Martinek, L. Bednarek, M. Novak and V. Vasinek, "Mathematical model of optimized design of multi-point sensoric measurement with Bragg gratings using wavelength divison multiplex," In Proc. SPIE 9889, Optical Modelling and Design IV, 98892F (April 27, 2016); doi:10.1117/12.2239551.

[22] M. Fajkus, I. Navruz, S. Kepak, A. Davidson, P. Siska, J. Cubik, and V. Vasinek, "Capacity of wavelength and time division multiplexing for quasi-distributed measurement using fiber bragg gratings," Advances in Electrical and Electronic Engineering, vol. 13, 2015, DOI: 10.15598/aeee.v13i5.1508.

[23] A. D. Kersey, M. A. Davis, H.J. Patrick, M. LeBlanc, K. P. Koo, C. G. Askins, M. A. Putnam and E. J. Friebele, "Fiber grating sensors," J. Light. Technol., vol. 15, pp. 1442–1463, 1997.

[24] E.P. Goodwin, J.C. Wyant, "Field guide to interferometric optical testing," Bellingham, ISBN 9780819465108, 2006.

Analysis Usability of the Energetically Passive Sensors for the Load Monitoring of Production Press Machines

Marcel Fajkus, Jan Nedoma,
Lukas Bednarek, Vladimir Vasinek
Department of Telecommunications
VSB - Technical university of Ostrava
Ostrava, Czech Republic
marcel.fajkus@vsb.cz, jan.nedoma@vsb.cz,
lukas.bednarek@vsb.cz, vladimir.vasinek@vsb.cz

Radek Martinek, Petr Bilik,
Jan Vanus
Department of Cybernetics and Biomedical Engineering
VSB - Technical university of Ostrava
Ostrava, Czech Republic
radek.martinek@vsb.cz, petr.bilik@vsb.cz,
jan.vanus@vsb.cz

Abstract — **This article focuses on the analysis usability of the energetically passive fiber-optic sensors for the monitoring operating condition of the pressing machines. Nowadays the conventional approaches of the load monitoring condition of the pressing machines are based on using the strain gauges and piezoelectric sensors. These sensors are sensitive to electromagnetic interference, subject to corrosion and action of moisture and require the electrical power. The alternative way how to monitor operation condition of the press machines could be using the fiber-optic sensors that those failures have. Within the real performed experiments has been verified and analyzed fiber Bragg grating (FBG) sensors and classic foil strain gauges (as a reference sensor). The measurement was performed on the hydraulic press ALPHA 4 from the company FORM + TEST Seidner, during a load of press in the range from 0 up to 2000 kN with a step of 500 kN.**

Keywords: fiber Bragg grating, deformation, tonnage, press machines, monitoring load of production press machines

I. Introduction

Metal stamping presses are widely used in current manufacturing processes. These machines make everything from deep-drawn automobile body panels to the change in our pockets. The press belongs to the category of forming machines which are designed for the realization of the shaping process. Forming is a process in which the workpieces are plastically deformed by tools. The objective is achieving the desired shape of the element (forging, compact) without taking splinters. In the field of the forging industry, we use different types of presses which can be classified according to their technological use for forging, extrusion, squirting, punching, straightening, tensile, workshop or mounting, etc. At the pressing process, large forces of press effect on the machined material, which is accompanied by the deformation of the individual parts of the press machine. For this reason, it is necessary to ensure that the nominal load of the press is not exceeded. The construction of the press must be rigid, and the precisely defined position between the tool and the compact must be maintained under all circumstances. For mentioned reasons, machines monitoring is carried out already for several decades. The primary objective of machines monitoring is obtaining information about their current load or tonnage. Appropriate tonnage choice can also ensure better results during the forming process, higher product quality, and increasing machine life.

Forming machines are also divided according to the type of construction. The most common constructions of the presses are C-frame (Fig. 1a) and straight-side (Fig. 1b). The press of C-frame type uses one or two piston rolls, straight-side press uses four columns at the corners. Knowledge of load distribution on individual columns makes possible to prevent failures and to increasing machine life.

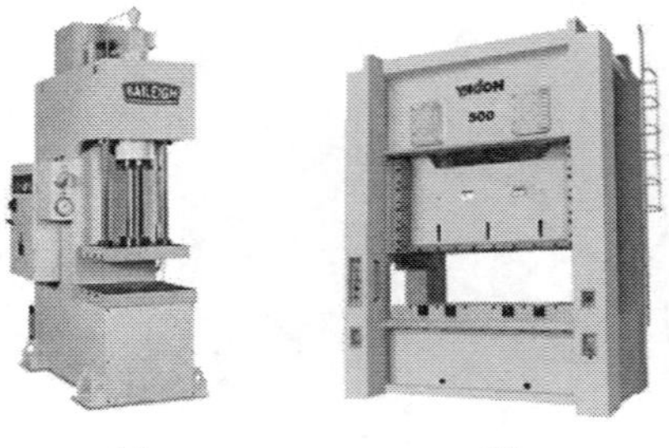

(a) (b)

Figure 1. Basic construction of presses (a) C-frame press; (b) straight-side press.

II. Load Monitoring of Production Machines

A. *Working metal properties*

The properties of the metal formed by the stamping press have a great impact on the load the press is subjected to. The shear strength of the working metal has a large effect on the press loading. The shear strength is determined mostly by the toughness and hardness of the metal. Soft metals like aluminum take a little force to punch and form. Hard metals such as steel will require a higher press loading.

The relationship between the metal material properties and the press loading force may be calculated using the following equation:

$$F = S \times t \times L, \quad (1)$$

where F is press load force, S is the shear strength of the formed material, t is the thickness of the material and L is length of cut.

B. *Background of tonnage signal*

Monitoring the load distribution on individual press parts provides good feedback for die designers. Load distribution information is also useful for machines maintenance because uneven loads cause a shorter life of the press. [1]

The authors of the article [2] carried out load distribution monitoring of the four-column press when one sensor was located on each column as shown in Fig. 2a. An example of tonnage signals recorded from four strain gauges is shown in Fig. 2b. In the figure, the x axis is the crank angle of the main shaft of the press and the y axis is the tonnage measured.

Fig. 2b shows tonnage signals from four channels of tonnage sensors. There are two errors, first is parallelism error and second is die off-center on the bed of the press. Parallelism occurs when both dies (upper and lower) are not parallel. In this case, the two dies will contact at one corner first. This will be reflected as an earlier increase of one tonnage signal than the others. Die off-center error occurs when a die is not positioned at the center of the bed of a press due to die setup error or material feeding problems, which is resulting in an unbalanced load of individual columns.

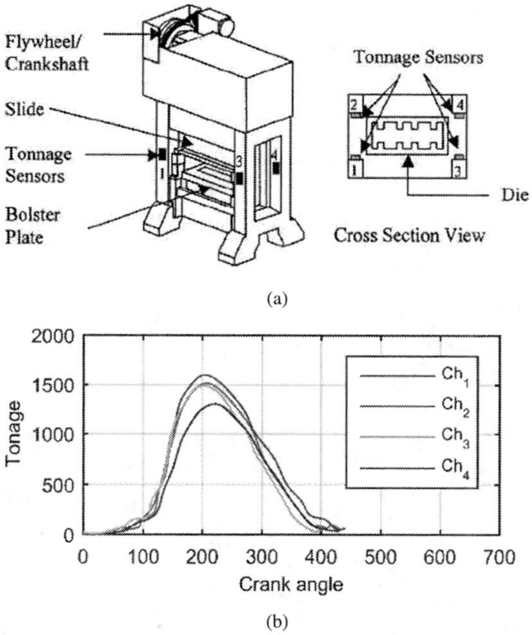

Figure 2. Monitoring the load distribution of the straight-side press; (a) location of sensors; (b) tonnage signals from four channels of tonnage sensors.

In addition to monitoring the load, reverse load monitoring is performed. This reverse load occurs when puncturing or dividing the material. When strokes or ramming steel, a huge amount of energy is released, which cause that the ram and the upper part of the die are bounced. These vibrations reduce the life of the embossing press.

III. Methods

The original load monitoring systems were complicated and susceptible to electromagnetic interference. Today's systems are simpler to implement, use, and are more reliable. These systems allow setting both the lower and upper load limits, and then the stamping process is interrupted if limits are exceeded.

A. *Conventional sensors for load monitoring*

The oldest load monitoring concepts come out from the measurement of the pressure of the liquid acting on the rolls. These systems are calibrated so that the fluid pressure is recalculated to the mass using the roll area. Measurement accuracy is affected by the friction inside the rolls and the precision of the pressure gauges. Later, resistive and semiconductor strain gauges (Fig. 3) started used.

Figure 3. A uni-axial strain gauge.

Strain gauge load sensing systems are the standard method of load sensing in solid mechanical members. Output signal of strain gauge signal is proportional to the affecting deformation. The nominal resistance of most gages is 350 ohms. As the gage is stretched or compressed, the resistance increases or decreases by a slight predictable amount due to the change in the length and cross-sectional area of the gage foil tracks. The amount of resistance change per length movement is predictable and termed gage factor. Gage factor (GF) is defined following equation (2):

$$GF = (\Delta R/R)/(\Delta L/L) \tag{2}$$

where ΔR is strained resistance in ohms, R is unstrained resistance in ohms, ΔL is strained gage length and L is unstrained gage length. [3]

The condition is the location of the strain gauge to a place which is linearly deformed with the load size. Suitable places are the columns or frame of the press. Another type of sensors is piezoelectric sensor whose output voltage is dependent on the affecting deformation. The advantage of these sensors against strain gauges is that they do not require any electrical power supply. However, the output signal is an electrical character still. The location of the deformation sensors must be in places where the linear dependence of the deformation on the force is manifested, there are large deformation effects, and mechanical noise is petite. Presently, load monitoring systems from IMCO company are on the market. IMCO systems are not universal. They are designed for presses based on their type, not tonnage. They allow permanent load monitoring, and they may end the pressing process in case of overloading. Another feature is the ability to set the maximum load for certain kinds of pressing. These systems are based on piezoelectric sensors.

B. *Fiber Bragg grating sensor for load monitoring*

Optical fibers are primarily designed for data transmission in telecommunication and data networks. The telecommunications are focused on minimizing the affecting transmission of light in the optical fiber by external influences. Sensory applications are looking for ways to increase these external influences. According to the spatial distribution of the measurand (the quantity to be measured), fiber optic sensors can be classified as point sensor (the measurement is carried out at a single point in space), quasi-distributed sensor (the measuring value is determined by several fixed, discrete points along a single fibre optical cable) and distributed sensor (the parameter of interest is measured with a certain spatial resolution at any point along a single optical cable).

Fiber Bragg grating sensor belongs to the group of point sensors. Fiber Bragg grating is formed by a periodic change of refractive index in core optical fiber (Fig. 4). This structure reflected small spectral part (called Bragg wavelength) back and other wavelengths are transmitted by this structure without attenuation. [4]

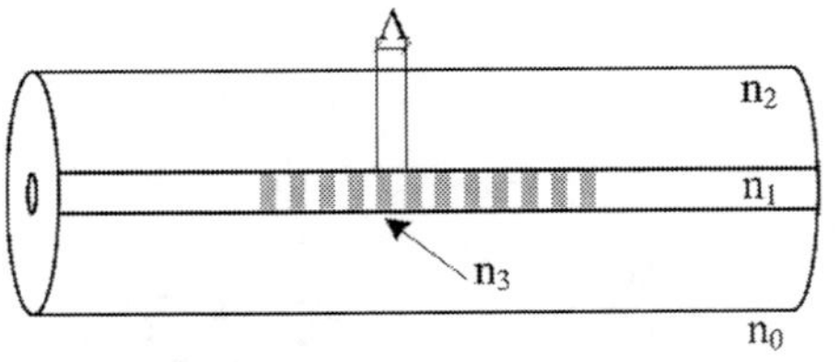

Figure 4. Structure of Fiber Bragg grating.

The spectral location of Bragg wavelength is given by equation (3)

$$\lambda_B = 2n_{eff}\Lambda \quad (3)$$

where n_{eff} is the effective refraction index of Bragg structure in optical fiber, and Λ is the period of changes of refractive index.

The spectral location of Bragg wavelength is influenced by external influences. The shift of the Bragg wavelength occurs by causing an external deformation or temperature change.

This approach is used in sensor applications [5-14]. Bragg's sensors are used in many areas such as civil engineering, rail transport, energy, biomedical applications and mechanical engineering. Number of quantities such as deformation, pressure, force, vibration, displacement, acceleration, temperature, electrical current, humidity is possible to measure using these sensors. Increasing the required sensitivity on the desired measured quantity is done by a suitable manner of mechanical encapsulation.

Deformation or temperature acting on the FBG results in Bragg wavelength change. The normalized FBG strain response at constant temperature is given by (4):

$$\frac{1}{\lambda_B}\frac{\Delta\lambda_B}{\Delta\varepsilon} = 0.78 \times 10^{-6} \quad \mu\varepsilon^{-1} \quad (4)$$

and the normalized FBG temperature response at constant strain is given by (5):

$$\frac{1}{\lambda_B}\frac{\Delta\lambda_B}{\Delta T} = 0.678 \times 10^{-6} \quad {}^{\circ}C^{-1} \quad (5)$$

where λ_B is the Bragg wavelength, $\Delta\lambda_B$ is the shift of the Bragg wavelength, $\Delta\varepsilon$ is the change of deformation and ΔT is change of temperature.

One of the advantages of FBGs is its intrinsic multiplexing capability. In fact, hundreds of fiber Bragg gratings can be written on a single optical fiber, which can be as close as millimeters or separated by a few kilometers. The wavelength-division-multiplexing (WDM) technique [15-16] can be easily employed to multiplex and interrogate FBG sensor array. The advantage of this approach with several tens sensors on one optical fiber is that only one optical fiber can serve to collect information from all sensors

As described in the above in this chapters, existing load monitoring solutions of production machines are implemented by conventional piezoelectric sensors and foil strain gauges. These sensors are small and inexpensive, and they can measure deformations on a large range. The disadvantages of these sensors are their sensitivity to electromagnetic interference, the need for electrical power or a small range of operating temperatures. Fiber-optic Bragg's sensors reach minuscule dimensions (their cross-section is 250 μm). These sensors are very solid in tensile, immune to electromagnetic interference, do not require electrical power, corrosion and moisture resistant, and allow measurements of long distances. Therefore, fiber-optic Bragg's sensors are a very convenient alternative with higher added features than conventional piezoelectric sensors and strain gauges.

IV. Experimental Setup

Bragg's sensors can be applied to a measured structure (piston roll, machine frame, etc.) by gluing method. For the case of acceleration of implementation, the FBG sensor can be stuck onto a metallic shim and welded by a spot welder to the desired location then.

The sensitivity of the FBG sensors to the deformation is very high and reaches values around 1.2 μm/μstrain. The temperature range of the fiber-optic sensors is from -40 °C to 150 °C. For higher temperatures, ceramic encapsulation of the sensor increases the temperature range up to 1100 °C [17].

In cooperation with the Faculty of Civil Engineering at VŠB-TU Ostrava, the principle of monitoring load of press (ALPHA 4 from FORM + TEST Seidner) was verified using a foil strain gauge sensor and a fiber-optic FBG sensor (Fig. 5). Fig. 6 shows implemented FBG sensor on one of the four columns using a two-component polymer adhesive.

Figure 5. FBG sensor glued on column of press for load monitoring of machine.

Figure 6. Fiber Bragg sensor implementation of one column of press machine.

In the press, an iron cube was burdened in the range of forces of 0 to 2000 kN in 500 kN steps (Fig. 7a). Fig. 7b shows the resulting measured deformations using the foil strain gauge sensor and the FBG sensor.

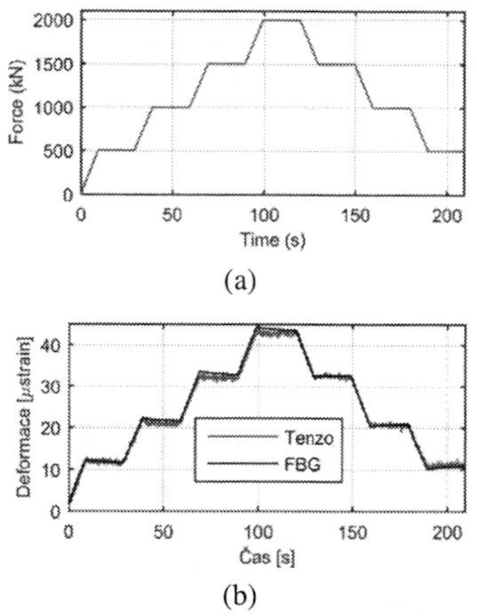

Figure 7. Press load (ALPHA 4); (a) adjusted load; (b) measured load using foil strain gauge and Bragg's sensor (FBG).

The dependence between the deformation of one press column and the pressure produced by the hydraulic press was determined to be 0.0235 μstrain/kN. The presented results show that Bragg sensors can be used for monitoring the press load and distribution of tonnage between the columns.

Conclusion

The manufacturing processes, in which the product is processed by the high forces of the hydraulic presses, require the monitoring the load of the individual parts of the forming and production machines. This monitoring is critical not only because of the resulting quality of the products but also because of the lifespan of the machines. Nowadays, piezoelectric sensors and strain gauges which show many inappropriate features are used for monitoring purposes. The measurements verified their deployment to monitoring the load of the press. The results show that FBG sensors create an alternative for load monitoring of manufacturing machines.

Acknowledgements

This article was supported by the project of the Technology Agency of the Czech Republic TA04021263. This work was partly supported by the Student Grant System of VSB-TU Ostrava [grant numbers SP2017/128, SP2017/79]. The research has been partially supported by the Ministry of Education, Youth and Sports of the Czech Republic through the grant project no. CZ.1.07/2.3.00/20.0217 within the frame of the operation programme Education for competitiveness financed by the European Structural Funds and from the state budget of the Czech Republic. This article was also supported by the Ministry of the Interior of the Czech Republic within the projects Nos. VI20152020008 and VI2VS/444. The research has been partially supported by the COST action MP1401. This article was also supported by the Ministry of Industry and Trade of the Czech Republic within the project No. FV 10396.

References

[1] D. Falcone, "Equipping Your Press With the Right Tonnage Monitor," Stamping Journal, November 2002.

[2] J. Kim, Q. Huang, J. Shi and T. S. Chang, "Online multichannel forging tonnage monitoring and fault pattern discrimination using principal curve," J. Manuf. Sci. Eng. Trans. ASME, vol. 128, 2006.

[3] D. A. Smith, "Basic principles of press force monitors," Smith & Associates, March 2005.

[4] A. D. Kersey, M. A. Davis, H.J. Patrick, M. LeBlanc, K. P. Koo, C. G. Askins, M. A. Putnam and E. J. Friebele, "Fiber grating sensors," J. Light. Technol., vol. 15, pp. 1442–1463, 1997.

[5] J. Nedoma, M. Fajkus, L. Bednarek, J. Frnda, J. Zavadil, V. Vasinek," Encapsulation of FBG sensor into the PDMS and its effect on spectral and temperature characteristics," Adv. Electr. Electron. Eng., vol. 14, 2016. doi: 10.15598/aeee.v14i4.1786.

[6] M. Fajkus, J. Nedoma, P. Siska, L. Bednarek, S. Zabka, V. Vasinek, "Perimeter system based on a combination of a Mach-Zehnder interferometer and the bragg gratings," Advances in Electrical and Electronic Engineering, vol. 14, pp. 318–324, 2016.

[7] J. Nedoma, M. Fajkus, L. Bednarek, V. Vasinek, "Non-destructive fiber-optic sensor system for the measurement of speed in road traffic," Advances in Electrical and Electronic Engineering, vol. 14, pp. 602–608, 2016.

[8] M. Fajkus, J. Nedoma, P. Siska, V. Vasinek, "FBG sensor of breathing encapsulated into polydimethylsiloxane," in Proc. SPIE 9994, Optical Materials and Biomaterials in Security and Defence Systems Technology XIII, 99940N (October 24, 2016); doi:10.1117/12.2241663.

[9] M. Fajkus, J. Nedoma, R. Martinek, V. Vasinek, H. Nazeran, P. Siska," A Non-Invasive Multichannel Hybrid Fiber-Optic Sensor System for Vital Sign Monitoring," Sensors (Basel). 2017 Jan; 17(1): 111. doi: 10.3390/s17010111.

[10] R. Martinek, J. Nedoma, M. Fajkus, R. Kahankova, J. Konecny, P. Janku, S. Kepak, P. Bilik, H. Nazeran, "A phonocardiographic-based fiber-optic sensor and adaptive filtering system for noninvasive continuous fetal heart rate monitoring," Sensors (Basel). 2017 Apr 18;17(4). pii: E890. doi: 10.3390/s17040890.

[11] R.P. Chen, Y.W. Wang, X.W. Ye, X.C. Bian, X.P. Dong, "Tensile force of geogrids embedded in pile-supported reinforced embankment: A full-scale experimental study", Geotextiles and Geomembranes, Volume 44, Issue 2, 2016, Pages 157-169, ISSN 0266-1144, doi: 10.1016/j.geotexmem.2015.08.001..

[12] J. Nedoma, M. Fajkus, P. Siska, R. Martinek, V. Vasinek, "Non-invasive fiber optic probe encapsulated into PolyDiMethylSiloxane for measuring respiratory and heart rate of the human body," Advances in Electrical and Electronic Engineering, vol.15, pp. 93–100, 2017.

[13] T. T. Sun, S. X. Pei, J. H. Li and L. L. Zhao, "A transverse strain sensor based on tilted fiber Bragg grating," Guangdianzi Jiguang/Journal Optoelectron. Laser, vol. 26, 2015.

[14] M. F. Domingues, C. Tavares, C. Leitao, A. Frizera-Neto, N. Alberto, C. Marques, A. Radwan, J. Rodriguez, O. Postolache, E. Rocon, P. Andre, and P. Antunes, "Insole optical fiber Bragg grating sensors network for dynamic vertical force monitoring," J. Biomed. Opt. 22(9), 091507 (Feb 28, 2017). doi:10.1117/1.JBO.22.9.091507.

[15] M. Fajkus, I. Navruz, S. Kepak, A. Davidson, P. Siska, J. Cubik, and V. Vasinek, "Capacity of wavelength and time division multiplexing for quasi-distributed measurement using fiber bragg gratings," Adv. Electr. Electron. Eng., vol. 13, 2015.

[16] M. Fajkus, J. Nedoma, S. Kepak, L. Rapant, R. Martinek, L. Bednarek, M. Novak and V. Vasinek, "Mathematical model of optimized design of multi-point sensoric measurement with Bragg gratings using wavelength divison multiplex," In Proc. SPIE. 9889, Optical Modelling and Design IV, 98892F. (April 27, 2016) doi: 10.1117/12.2239551.

[17] D. Barrera, V. Finazzi, J. Villatoro, S. Sales and V. Pruneri, "Packaged optical sensors based on regenerated fiber bragg gratings for high temperature applications," IEEE Sens. J., vol 12, pp. 107–112, 2012.

Feasibility Study for Demand Response in Commercial Buildings

Anna Mutule, Artjom Obushev, Ervins Grebesh, Aleksandrs Lvovs [1-4]
[1-4] Smart Grid Research Centre
Institute of Physical Energetics
Riga, Latvia
amutule@edi.lv [1], a.obushev@gmail.com [2], ervins.grebess@gmail.com [3], aleksandrs.lvovs@inbox.lv [4]

***Abstract*—This paper gives an overview on the potential of demand side management of the existing commercial building located in Latvia. It aims to achieve the rational use of energy and cost reduction based on price responsive demand and demonstrate first steps of implementation of commercial building automation system with automated energy consumption scheduling units for load flexibility and controllable portion estimation. To activate demand response potential and enable independent aggregator participation in the Latvian scale, relationships between suppliers, BRPs, TSO and DSO are shown within general electricity market structure, taking into account the current status and promising areas of DR in the Baltic region.**

Keywords*— *Commercial building, demand response, demand side management

I. INTRODUCTION

As a result of energy policy developments aimed to increase the share of energy production from renewable, the need for balancing production and consumption in electricity grids has increased. Balancing energy may either come from centralized energy production side, which is still dominating the balancing energy market, or from the energy demand side where demand response (DR) or demand side management (DSM) resources can stem from either distributed generators or switchable loads in various end-use sectors. Since the importance of DR is not just the profitability but also the reliability of the grid, it is very likely that DR will be mandatory for certain types of electricity customers. In such conditions, the requirements of DR will establish a unique relationship between the customers and the suppliers of electricity where the customers will play a functional part in the reliable operation of the grid. This will require the integration of the loads into the control structure of the grid.

Commercial buildings are good candidates for implementing demand response because they usually incorporate building management systems which are able to control and monitor various electrical devices, from lighting to Heating, Ventilation and Air Conditioning (HVAC), security to power management [1]. On the other hand, DR in commercial buildings comes with a monetary cost because all the loads in an office building serve the needs of humans that produce work therefore there are limitations on how much load can be shed. In order to carry out DR effectively and minimize disruption to workers, a special system is required. To develop such a system, the relationship between work and amount of energy consumed needs to be established. This is a difficult problem because it involves multidisciplinary elements.

The relevance of practical implementation of DR is growing, and every new project is unique - with its own vision and approach based on local area system and market operation. The concept of DR was developed in the 1990s [2]. In [3], two types of DR commonly used nowadays are classified: price-based programs and incentive-based programs. An approach to data analysis with low data rate and the developed model for load forecasting are described in [4]. A model with added thermal impact of outside heating is considered in [5]. Paper [6] focuses on various methods of frequency regulation DR, papers [7,8] are concerned with HVAC potential in commercial buildings. From papers [5-8], it can be concluded that the main focus area of DR control load potential is HVAC systems that can reach 70% of total load.

The paper [9] presents demand side management platform for optimal energy management strategies, and discusses topicality of smart metering and control including deployment of the price responsive demand. The tests described were performed in Smart Grid Research Centre research facility Demand Side Management-Smart Home Lab with the aim to achieve the rational use of energy and demonstrate first steps of implementation of home automation system with automated energy consumption scheduling units for load flexibility and controllable portion estimation. The current solution makes it possible to monitor aspects of building performance without costly investments in wiring and helps to identify where improvements in energy efficiency can be done (using different energy management strategies based on particle swarm optimization [10], genetic algorithms [11], linear optimization techniques [12], Monte Carlo simulations [13], Markov decision process [14], dynamic programming [15], Multiagent control [16], with new equipment, distributed generation sources, storage technologies etc.) and then see and analyze the effects of implementation. Smart monitoring and control implementation enable competitive development of demand reduction and demand responsiveness by the investments in advanced infrastructure like that.

In recent years, various methodologies have been applied to investigate the potential application of demand response in different countries, customers and environments. As the literature in this field is rapidly expanding in breadth and depth, an extensive review is outside the scope of this study. However, a brief description of a limited number of distinct related publications is provided as a hint of the advances related to the specific area of interest.

II. THE STATUS OF DR AND DSM IN THE BALTIC REGION

Despite DR development activities in neighboring Northern countries, as well as in Poland [17], in the Baltic States the usage of DR and DSM is very limited nowadays, and there is no specific legal framework for flexibility services and roes of aggregators. This could be explained by historical background – Baltic power systems are synchronously connected with Russian/Belarusian power systems that provide primary regulation reserves (FCR) to the Baltic power systems and relatively small amount of production from wind/solar power plants.

Taking into account future plans for desynchronization from Russian/ Belarusian power systems, frequency regulation and balancing issues will ask for finding additional flexibility in the Baltic power systems, and this could be achieved by involvement of DR and DSM.

In Estonia, the potential of DR and DSM, as well as exploration of flexibility possibilities have been evaluated in several studies, e.g. [18,19]. The evaluated Estonian potential of DR varies significantly in range from 51MW - 180MW in the nearest future (year 2020) and potentially could reach 260MW in more distant future (year 2030). In [18] industry, public and commercial sectors have been analysed, as the most perspective DR providers. Industrial DR potential has been evaluated on the basis of interviews, analysis and measurements. The study of public and commercial sectors (e.g. office buildings and cold storage/ventilation) was based on the studies and analysis of similar constructions' consumption. Another study on Estonian DR potential [19], uses Pöyry's power market model BID3 to model optimized usage of DR. The total capacity of DR in aforementioned study has been guided by Estonian transmission system operator Elering AS. Prerequisite for active usage of DR/DSM provided possibilities is full scale roll-out of smart meters (legal requirement of full-scale roll-out of smart metering system by the end of 2016 is in place in Estonia). Latvia and Lithuania are one step behind from Estonia - full-scale analyses of DR /DSM potential evaluation have not been performed previously. However, similarly like in Estonia, from legal framework side there are no limitations for flexibility services introduction in Latvia and Lithuania. It shall also be noted that in Latvia full-scale roll-out of smart metering system begun in year 2014 with the aim to install smart meters to all customers with consumption over 2500kWh/year till year 2017, and till year 2023 replace all power meters with smart meters.

Therefore, in order to identify promising areas of DR/DSM application in all Baltic States, as well as to foster more active introduction of aggregators and flexibility technologies, additional studies in Latvia and Lithuania are necessary. Broader scope of studies will also allow researchers to identify similarities and differences in DR/DSM application profile in each country.

III. THE OBJECT OF RESEARCH

A. Theoretical part

The building where Institute of Physical Energetics is located, was chosen as a place for pilot project of DR realization. DR realization for this commercial building has some limitations, due to object age. The observed object was built more than 60 years ago. While most of the papers present their studies on new buildings with controllable electrical heating [20], the building under consideration is connected to the central heating system, which does not affect the electric load and is not controllable. Moreover, since the thermal insulation works have not been done yet, the observed object is not energy efficient. Based on the above-mentioned, the pilot project provides many opportunities for DR profitable realization. In Table (1), information about the pilot project object is given.

TABLE I. OBJECT INFORMATION

Parameter	Value or description
Year of construction	1956
Area, square meters	~5000
Floors	3+basement+meeting hall
Amount of workers	~500
Peak consumption/month	~55 MWh
Load type	Office and manufactory
Peak load	160 kW

Figure 1 shows electricity consumption of building during the year. The parallel axis shows outside daily-average temperature in Celsius degrees, and daylight duration in hours and change during the year.

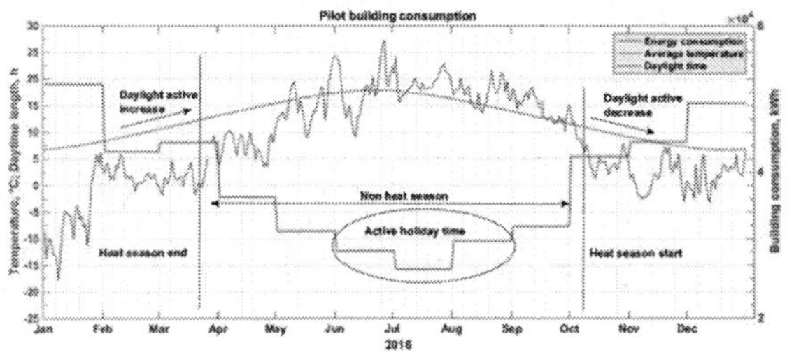

Fig. 1. Consumption, outside temperature and daylight change.

As can be seen from Figure 1, energy consumption of the building is decreasing during the summer period. Activity fall can be explained by the summer vacations and the longest daylight hours, when lightning is not used so frequently. Starting from the autumn season, energy consumption is increasing gradually.

Geographical location of the chosen object is hold in Riga, Latvia. Latvian climate is separated into four specific seasons. Therefore, when the outside temperature falls below zero the heating season starts. From Figure 1 it can be observed, that heat season in Latvia potentially starts very early. As it was mentioned previously, there is no possibility to control the obtained heat amount and coupled with low building energy efficiency, in the observed building. Electrical heaters are massively used for second order heat sources, which increase energy consumption amount proportionally when outside temperature is decreasing. For Latvian latitude, the daylight time is slightly short in winter period. The peak month when daylight time is increasing very fast is March, while in September it is rapidly decreasing. In this case, second evident event of consumption increase is lightning. In mid autumn, winter and mid spring seasons lights are used during all working daytime, which results in consumption increase.

The main task during the first part of pilot project is to identify possible amount of controllable load for DR, to share this load on categories and reveal how flexible they are.

There are numerous types of loads in office buildings but not all of them are suitable for controlling within a DR scenario. Some loads might not have facilities to be controlled automatically, whereas others would not be worth controlling for DR because of the little contribution they make to overall power consumption. For this reason, the loads that are suitable for controlling during a DR scenario are identified.

Figure 2 shows the possible concept of energy consumption parts. Parts of energy consumption was defined before in [9].

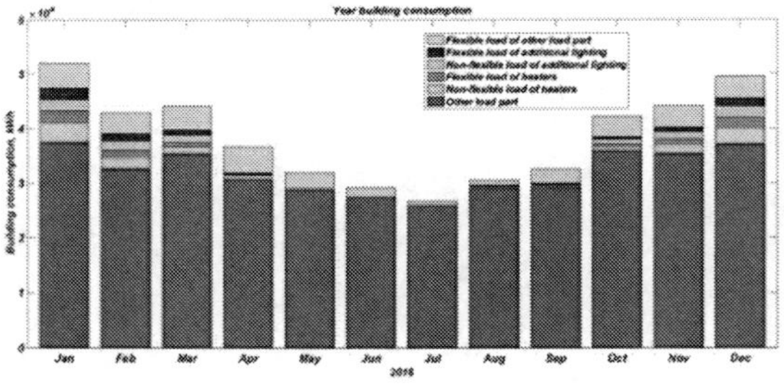

Fig. 2. Building consumption theoretical segregation.

From the year consumption data, geographical and climate aspects consumption can be allocated on 3 controllable and 3 uncontrollable loads. Electrical heaters and lighting that creates additional load starting from autumn period can be both controllable and uncontrollable load. The percentile distribution of their part depends on workers comfort level [21] from thermal part of view; daytime and room location from lightning part of view. The load which was not classified at this moment can reach significant part of potentially controllable load. This load will be classified after practical realization of the first stage of pilot project.

Commonly, the papers related to DR show that the main controllable load goes from heating, and/or from HVAC in total. The observed pilot project building does not have modern HVAC systems and cannot realize potential from ventilation or air conditioning systems.

B. *Realization of Stage one*

It was decided to use the top-down strategy for the practical realisation of the first part of the pilot project - monitoring equipment installation. Thereby, the building was divided into blocks in order to implement the primary task - object overview. It allowed revealing load behavior and energy consumption of each such block. In the next step, attention was paid to most interesting blocks, which had potential for DR. Top-down strategy implementation helped to solve several tasks:

Primary:

- Observed object load verification;
- Blocks behavior;
- Load consumers for DR is simplified and will be faster.

Secondary:

- Object energy efficiency validation [22];
- Electrical and thermal wiring verification.

For the practical realization of the first stage a decision was made to connect monitoring equipment to the main cable lines. Main lines are going from feeding transformer of 10/0.4 kV lower voltage. A schematic plan of connection is shown in Figure 3.

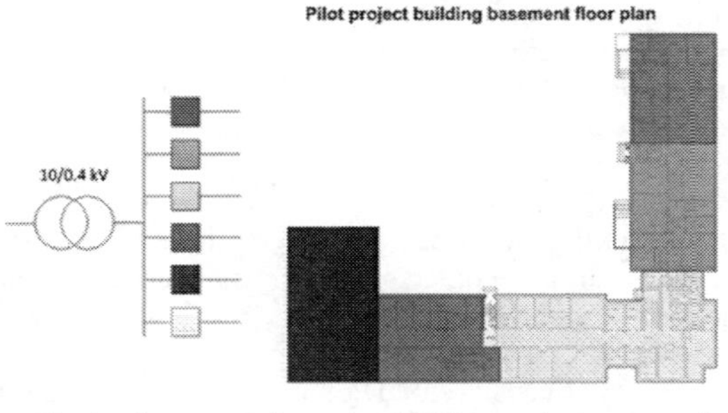

Fig. 3. Concept of pilot project building monitoring connection

The implementation of further steps is outlined in Chapter VI related to future work.

IV. Market Models

Demand side response is a key feature of the European electricity system for achieving its ambitious energy policy objectives. The trend to strengthen consumer flexibility in Electricity market environment is conditioned by the wider expansion of Smart Grid Technologies including Intelligent Metering Systems.

Prospectively, bulk installations of Smart Meters alongside with development of more flexible appliances should lead to the widespread public acceptance and deployment of price responsive demand. ENTSO-E contributes to the on-going discussion on the integration of DSR in all electricity markets, which is currently progressing at a fast pace thanks to the works of the Smart Grid Task Force EG3 led by the European Commission and the involvement of stakeholders [23]. Enabling independent aggregation is important for the healthy growth of market competition. In today's European energy markets, aggregators are faced with many stakeholders to whom a contractual relationship needs to be established before the flexibility potential of any single participating site (whether it be an industrial, commercial, institutional or residential customer) can be commercialized into the markets. An aggregator can only succeed when their customers succeed and benefit from Demand Response. To enable independent aggregators to enter the market in a safe and scalable manner, it is critical that the role and responsibilities of these new entrants are clarified. In particular, it is important that the relationships between suppliers, Balance Responsible Parties (BRPs), and independent aggregators are clear, fair, and allow for fair competition. Figure 6 presents author's point of view of general electricity market design in future to enable independent aggregator participation in the Latvian scale. Currently, electricity market design in the Latvian scale not includes independent aggregator participation. It should also be mentioned that overall electricity market design is more complex with different relations between actors and their roles and responsibilities: consumers/prosumers (small, medium, big) connected at different voltage levels (LV,MV,HV), Producers, Energy retailers, TSO's, DSO's etc.

In the scope of this study, detailed consumer/prosumer tariffs are used: for working days (Fig. 4) and weekends (Fig. 5). Tariffs consist of four main parts: fixed part – DSO three time zone tariff (this part of tariff depends on consumer size and connection point; it is fixed for working days and weekend) [24], State Feed-in-Tariff part (RES and Cogeneration power plant subsidy schemes) [25] and VAT, variable part includes Nord pool spot Latvian zone day-ahead prices [26]. In some cases additional fixed part is set, defining trading expenses; however, this part is much smaller than previously defined ones.

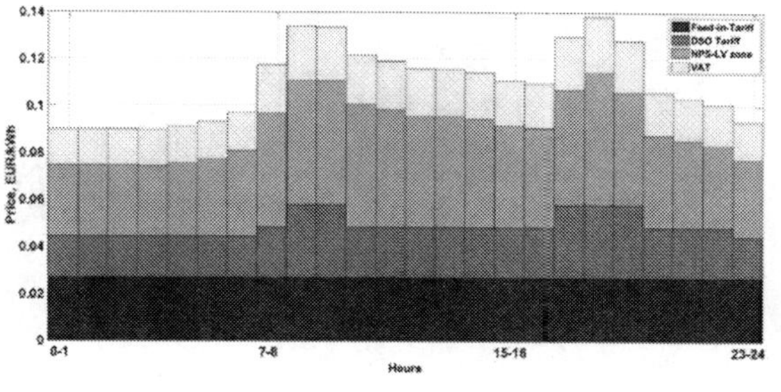

Fig. 4. Power consumer tariff at Working days (07.03.17)

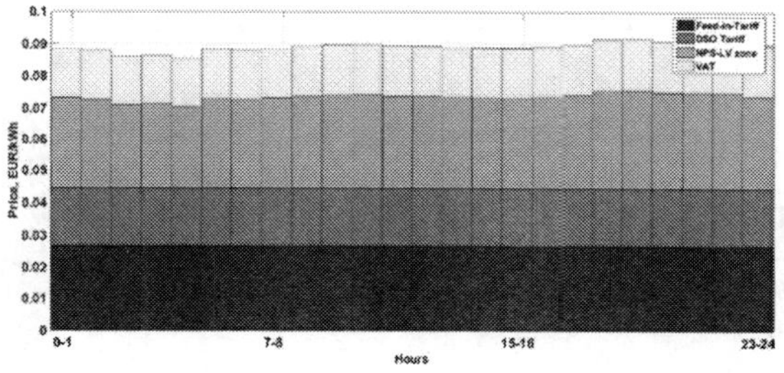

Fig. 5. Power consumer tariff at Weekends (12.03.17)

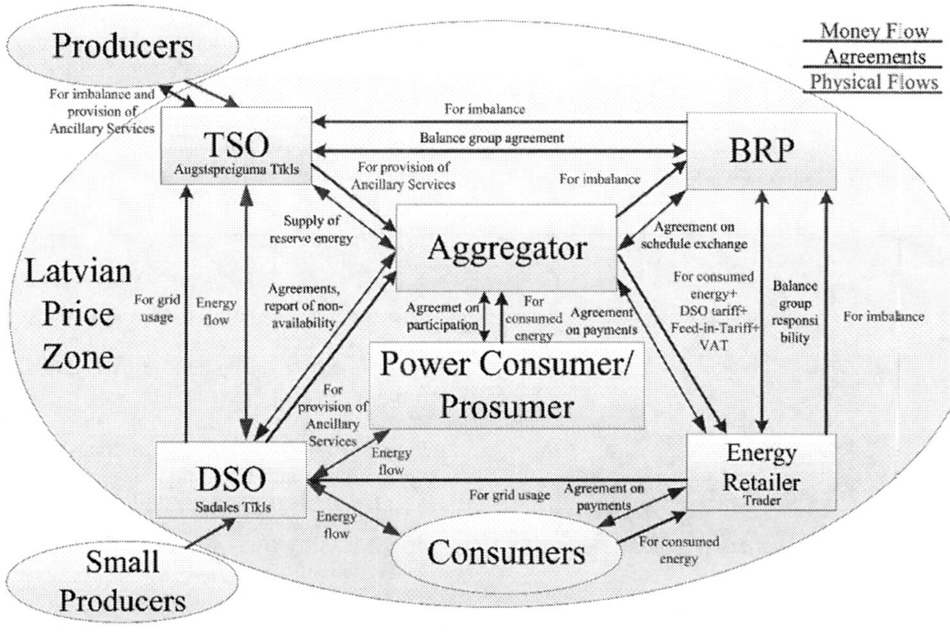

Fig. 6. General electricity market design structure

From commercial building consumption profiles depicted in Figure 7 it can be noticed that significant DR activity can be realized during working days due to higher concentration of flexible loads and dynamic prices. At weekend, prices are unified without high price peaks and low concentration of flexible loads. The concentration of flexibility is estimated approximately based on different appliance availability in observed commercial building.

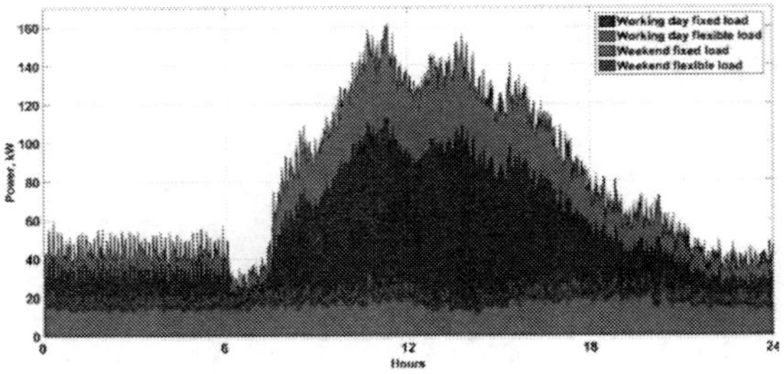

Fig. 7. Commercial building consumption profiles (07/12.03.17)

V. Optimal Energy Strategy

Energy management strategy (EMS) is widely recognized as one of instruments for the optimal operation of the distribution grid and optimization of the management of Distributed Energy Resources from a system perspective, and assessment of market design in terms of suitability for integration of renewable energy. The EMS solution provided in this paper is based on [9] platform as instrument for total energy consumption costs reduction, considering energy saving and a comfortable work/lifestyle. EMS is based on the load management systems and allows switching on/off appliances when particular circumstances occur in order to maintain the net power exchange with the LV network within a desired range. In the current study, load profile is changed by load management system according to saving functional condition: shedding loads when energy price exceeds a fixed threshold and moving that amount of energy to next hours with lower prices. Figures 8 and 9 show load shifting based on price based DR program, where low priority loads are shifted from hours with high energy prices.

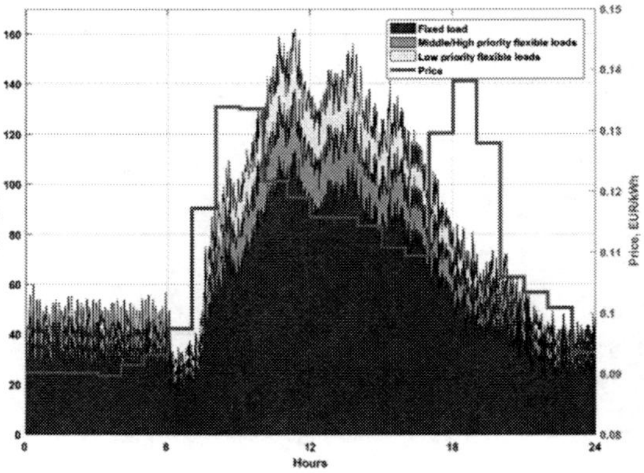

Fig. 8. Consumption without load shifting

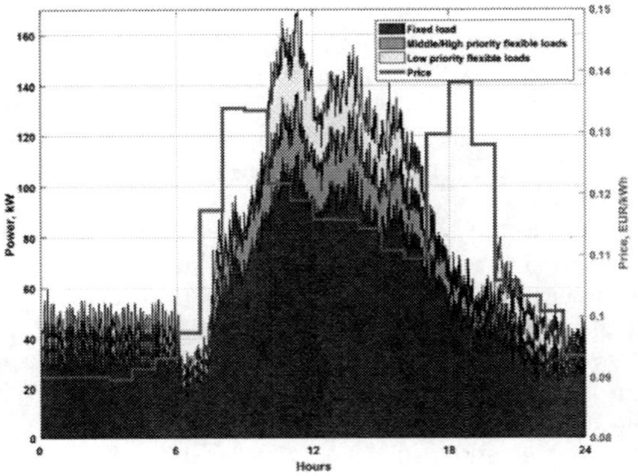

Fig. 9. Consumption with load shifting using price based DR program

Total shifted low priority flexible load was 51 kWh, the cost without shift was 214.32 EUR, the cost with shift - 213.24 EUR. These results represent a condition when all low priority flexible load energy is shifted to lower price hours. However, such a condition is not always fulfilled in real life, due to the reason that such kind of load is not needed anymore in other time periods. In such way it's possible to decrease costs more significantly, convincing personal not to use appliance during high price hours, but shift their behavior and needs to other hours, not influencing personal work and comfort level. In case when NPS_LV zone price difference is higher, cost reduction will be more significant.

Currently, load management system is organized so as to shift loads forward to hours with prices lower than the defined threshold. It is also possible to organize other kind of functional conditions: Comfort - when power exceeds contractual limit and Emergency - shedding of loads according to their priority as a result of an emergency signal received from the distributor (Critical Peak Price or by Voltage and Frequency set points). In this way implemented incentive-based scheme in synergy with price-based scheme makes it possible to receive extra cost reduction providing middle and high priority flexible load activation for incentive-based scheme.

VI. Conclusion and Future Work

DR strategy planning from the early phase of its implementation can reduce the risk of missing effective DR strategies or programming unwanted strategies. Performing building simulations will reduce costs and efforts to manually demonstrate the effect of DR strategies. Analysis of the combination of demand, energy, comfort and economics is important to achieve the best DR strategy for any building.

The feasibility study shows that buildings should not have fixed control schemes for DR. This is because buildings have different energy requirements throughout the year and the proportion of power that is curtailed from lighting or HVAC changes with outside temperatures. For this reason, an automatic control system needs to be smart enough to adjust the mixture of contribution from these two categories of loads.

The results of this study supplement the studies performed in Estonia. However, additional studies would be necessary to cover all possible DR/DSR application areas so as to achieve results comparable in the Baltic States context, which in its turn will promote coordinated adjustments in legislation for creation of liquid and transparent market of flexibility services.

ACKNOWLEDGMENT

The research has been supported by Smart Grids Plus ERA-Net project CLOUDGRID No.77547 (Transnational CLOUD for Interconnection of Demonstration Facilities for Smart GRID Lab Research & Development).

REFERENCES

[1] N. Motegi, M.A. Piette, D.S. Watchon, S. Kiliccote and P. Xu, Introduction to Commercial Building Control Strategies and Techniques for Demand Response, Demand Response Research Center, Lawrence Berkeley National Laboratory, 2007. [Accessed: Mar. 10, 2017].

[2] EPRI, "Principles and Practice of Demand Side Management," [Online]. Available: http://www.epri.com/Pages/Default.aspx

[3] T. Woolf, E. Malone, L. Schwartz and J.Shenot, "A Framework for Evaluating the Cost-Effectiveness of Demand Response," [Online]. Available: http://emp.lbl.gov/sites/all/files/napdr-cost-effectiveness.pdf. [Accessed: Mar. 10, 2017].

[4] J. L. Mathieu, P. N. Price, S. Kiliccote and M. A. Piette, "Quantifying changes in building electricity use, with application to demand response," IEEE Transaction on Smart Grid, vol. 2, No. 3, pp. 507-519.

[5] W. Mai and C. Y. Chung, "Model Predictive Control based on thermal dynamic building model in the demand-side management," IEEE Power and Energy Society General Meeting (PESGM), 2016, pp. 1-5.

[6] I. Beil, I. Hiskens and S. Backaus, "Frequency Regulation from Commercial Building HVAC Demand Response," Proceedings of the IEEE, 2016, pp. 745-757.

[7] G. Goddard, J. Klosem and S. Backhaus, "Model Development and Identification for Fast Demand Response in Commercial HVAC Systems," IEEE Transaction of Smart Grid, 2014, pp. 2084-2092.

[8] H. Hao, C. D. Corbin, K. Kalsi and R. G. Pratt, "Transactive Control of Commercial Buildings for Demand Response," IEEE Transactions on Power Systems, 2017, pp. 774-783.

[9] A. Obushevs, I. Oleinikova and A.Mutule. Demand side management platform for HAN flexibility estimation with agent control // 13th International Conference on the European Energy Market "EEM16", Porto, Portugal, 6 - 9 June 2016, DOI: 10.1109/EEM.2016.7521185.

[10] M. A. A. Pedrasa, T. D. Spooner and I. F. MacGill, "Coordinated scheduling of residential distributed energy resources to optimize smart home energy services," IEEE Trans. Smart Grid, vol. 1, no. 2, Sep. 2010, pp. 134–143.

[11] Z. Zhao, W. C. Lee, Y. Shin and K.-B. Song, "An optimal power scheduling method for demand response in home energy management system," IEEE Trans. Smart Grid, vol. 4, no. 3, Sep. 2013, pp. 1391–1400.

[12] M. Pipattanasomporn, M. Kuzlu and S. Rahman, "An algorithm for intelligent home energy management and demand response analysis," IEEE Trans. Smart Grid, vol. 3, no. 4, Dec. 2012, pp. 2166–2173.

[13] Z. Chen, L. Wu and Y. Fu, "Real-time price based demand response management for residential appliances via stochastic optimization and robust optimization," IEEE Trans. Smart Grid, vol. 3, no. 4, Dec. 2012, pp. 1822–1831.

[14] T. T. Kim and H. V. Poor, "Scheduling power consumption with price uncertainty," IEEE Trans. Smart Grid, vol. 2, no. 3, Sep. 2011, pp. 519–527.

[15] C. Vivekananthan, Y. Mishra and F. Li, "Real-time price based home energy management scheduler," IEEE Trans. Power Syst. , vol. PP, no. 99, Sep. 2014, pp. 1-11.

[16] J. K. Kok, C. J. Warmer and I. G. Kamphuis, ''Powermatcher: Multiagent control in the electricity infrastructure,'' in *Proc. ACM 4th Int. Joint Conf.Autonomous Agents Multiagent Syst*, Jul. 2005, pp. 75-82.

[17] Smart Energy Demand Coalition (SEDC). "Mapping Demand Response in Europe Today", 2015. [Online]. Available: http://www.smartenergydemand.eu/wp-content/uploads/2015/09/Mapping-Demand-Response-in-Europe-Today-2015.pdf. [Accessed: Mar. 10, 2017].

[18] I. Drovtar, A. Rosin and J. Kilter. "Demand Side Management in Small Power Systems – The Estonian Case Study." ELEKTRONIKA IR ELEKTROTECHNIKA, ISSN 1392-1215, vol. 22, no. 3, 2016.

[19] S. Carter, R. Desai, J. Forsman, M. Martin, O. Pearce, B. Steel and M. Vestli. *Demand-side response as a source for flexibility.* Tallinn: Pöyry, 2015.

[20] M. Ali, A. Safdarian and M. Lehtonen, "Demand response potential of residential HVAC loads considering users preferences," ISGT Europe, 2014, pp. 1-5.

[21] L.A. Hurtado, P. H. Nguyen and W. L. Kling, "Agent-based control for building energy management in the smart grid framework," IEEE PES Innovative Smart Grid Technologies Europe, 2014, pp. 1-6.

[22] S. Puradbhat and R. Banrjee, "Estimating demand side management impacts on buildings in smart grid," ISGT Asia, 2014, pp. 1-5.

[23] ENTSO-E Working Group Market Design & RES, "Market Design for Demand Side Response," [Online]. Available: https://www.entsoe.eu. [Accessed: Mar. 10, 2017].

[24] AS "Sadales tikls" tariffs [Online]. Available: https://www.sadalestikls.lv/eng/sistemas_pakalpojuma_tarifi/notice_on_the_tariffs_proposal/. [Accessed: Mar. 10, 2017].

[25] Feed-in-Tariff [Online]. Available: https://www.sprk.gov.lv/jaunums/obligata-iepirkuma-komponensu-oik-kopeja-vertiba-no-2017-gada-1-aprila-nemainas. [Accessed: Mar. 10, 2017].

[26] Latvian area Elspot prices [Online]. Available: http://www.nordpoolspot.com/Market-data1/Elspot/Area-Prices/LV/Hourly/?view=table. [Accessed: Mar. 10, 2017].

Effect of Average Environmental Conditions in Slovakia at the Current Conductor Capacity of Overhead Transmission Line

Michal Špes
Department of Electric Power Engineering
Faculty of Electrical Engineering and Informatics, TUKE
Košice, Slovakia
Michal.Spes@tuke.sk

Michal Kosterec
Department of Electric Power Engineering
Faculty of Electrical Engineering and Informatics, TUKE
Košice, Slovakia
Michal.Kosterec@tuke.sk

Ľubomír Beňa
Department of Electric Power Engineering
Faculty of Electrical Engineering and Informatics, TUKE
Košice, Slovakia
Lubomir.Bena@tuke.sk

Michal Márton
Department of Electronics and Multimedia
Telecommunications, Faculty of Electrical Engineering
and Informatics, TUKE
Košice, Slovakia
Michal.Marton@tuke.sk

Martin Vojtek
Department of Electric Power Engineering
Faculty of Electrical Engineering and Informatics, TUKE
Košice, Slovakia
Martin.Vo
jtek@tuke.sk

Zdeněk Müller
Department of Electrical Power Engineering, Faculty of
Electrical Engineering, CVUT
Prague, Czech Republic
zdenek.muller@fel.cvut.cz

Abstract — **Current capacity of conductors is defined as the maximum amount of current that can flow without damaging the conductor. It depends primarily on the electrical and mechanical properties of the conductor, their ability to spread the heat generated and the ambient conditions. The proposal of powerlines as each electricity device are subject to the valid standards. Standard EN 50341 defines the environmental conditions for calculating the maximum current capacity. In general, the environmental conditions are not constant over time. Depend on the climatic conditions of the country, area, that are different climatic nature. This paper deals with the calculation of the maximum current capacity of conductor of overhead transmission line based on ambient climatic conditions according to the equations defined in the IEEE Std 738-1993. The article describes average weather conditions in Slovakia and its impact on the determination of the maximum current capacity. Article compares the maximum current capacity given by ambient conditions referred to a standard and current capacity determined by average weather conditions.**

Keywords: current capacity; ambient conditions; overhead lines; power lines; EN 50341; ACSR Conductor

I. INTRODUCTION

Power system consists of electric power components, which are used for generation, transformation and electricity transmission. For transmission of electricity, we distinguish the transmission and distribution power lines. Transmission lines are represented as ultra-high voltage system, which provides transmission of power from sources to the three local areas of distribution system. Cross-border power lines belongs to the transmission system of power lines too [1] [2].

Current capacity of all power lines depends on their construction and types of used the conductor. Manufacturer of the conductors specified maximum allowable temperature for the operation of the power line. The maximum allowable temperature is related the current value that can flow through the conductor and leads to the maximum operating temperature [3] [4] [5].

The maximum current value that flow through the conductors when the maximum permitted allowable temperature is referred as capacity or ampacity of power lines [3] [4] [5].

Capacity of the power lines depends on the electrical and mechanical features, thermal features of the conductor insulation and ambient conditions [3] [4] [5].

Every construction of electricity facilities is subject to the standard. This standard describes methods for construction these facilities and to determination of maximum allowable conditions, that are non-destructive for their operation.

During the preparation of the project for construction of power lines, it is necessary to determine the capacity of power lines. In order to determine the capacity of power lines, there are standards EN 50341 where are defined ambient conditions, that have main influence on the value of maximal current.

II. Climatic conditions of Slovak Republic

Climatic conditions or the climate are understood as a long-term weather conditions with all the peculiarities that may occur at a given location or area. In determining the climate and weather conditions, the geographical location of the area must be based [6].

The Slovak Republic belongs to the northern temperate climatic zone with regular alternation of the seasons with an even distribution of rainfall during the year. The climate is influenced by the prevailing western airflow that brings wet ocean air of moderate zones. It moderates temperature amplitude of day and year and brings atmospheric precipitation. Continental air of moderate zones brings warm, sunny and less humid summers and cold winters with low rainfall [6].

A. *Solar radiation*

Solar radiation consists of direct and diffuse radiation incident on the surface. This is most affected by the duration of sunshine and cloud cover. The annual average values of solar radiation in lowlands are in the range 1200 to 1300 kWh.m^{-2}. In the highest position, it is the 1100-1200 kWh.m^{-2}. The middle mountain positions, the average annual solar radiation is 1050-1100 kWh.m^{-2} [6].

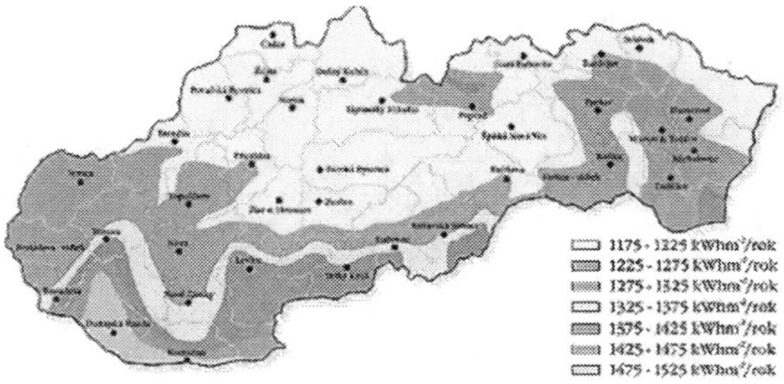

Figure 1 The amount of incident solar radiation in Slovakia during the year [6]

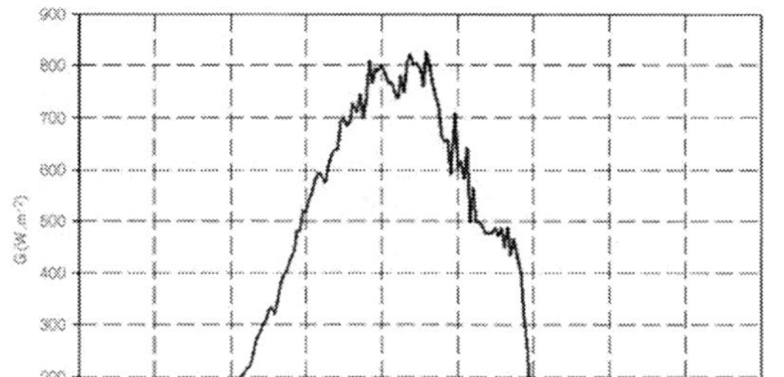

Figure 2 The course of solar radiation intensity during the day [7]

B. *Air temperature*

The air temperature is the major climatic factor, as well as the main factors affecting the actual current load of conductor [6].

In a long-term measurement of air temperature in the region of Slovakia warmest zone is Danubian lowland with average air temperature in January -1 to -2 °C, in the month of July 18 to 21 °C. The annual average of this area is in the range of 9-11 °C. The Eastern plains, the average temperature is slightly lower. The basins and valleys of is average annual air temperature 6-8 °C. In the upstream basins below 6°C. For the altitude of 1000 meters above sea level average value reaches the interval 4-5°C, at a height of 2000 meters above sea level around -1°C [6].

The average temperature during the months of June to August is shown in Figure 3 and during the months of December to February is shown in Figure 4.

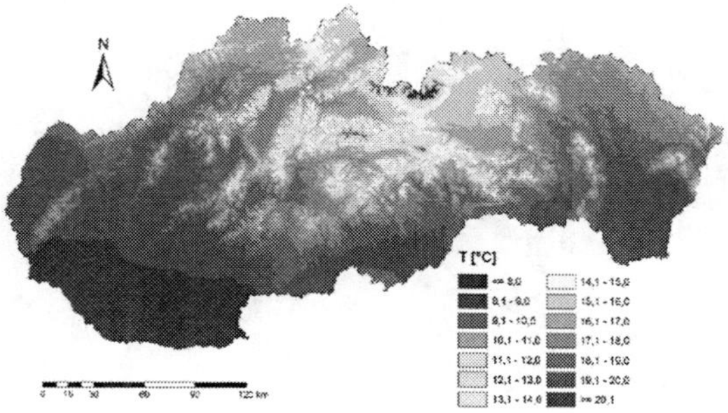

Figure 3 Average air temperature from June to August [6]

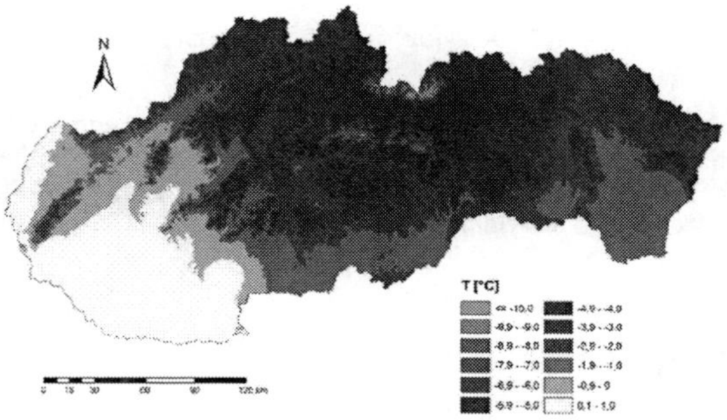

Figure 4 Average temperature from December to February [6]

C. *Wind conditions*

Wind conditions of Slovakia are considerable complicated. The main impact has the variability of weather during the year. In the lowland of western Slovakia, the annual average speed is in the range 3-4 m/s. In Eastern Slovakia 2-3 m/s. In the basin, wind speed is influenced by openness or closeness to the flow. In the more open basins, the wind speed is in the range 2-3 m/s. In closed basins, the average wind speed is of 1-2 m/s. In the mountains, depending on the altitude, the average annual speed is in the interval 4-8 m/s [6]

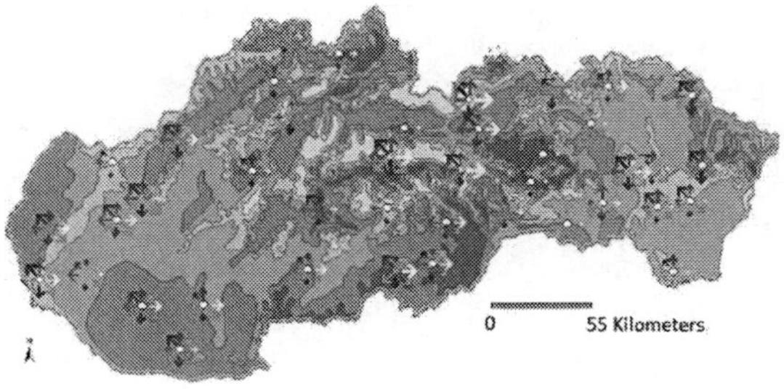

Figure 5 Map of wind areas of Slovakia [6]

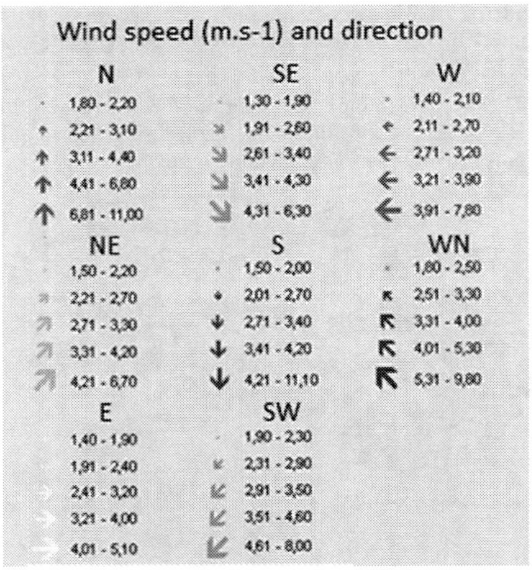

Figure 6 Wind speed and direction of wind [6]

III. DEFINING RESEARCH PROBLEMS

1) Current capacity

Current capacity of the power lines means a current value that can be transmitted through the power line without destroying conductors. This state is defined by equation, which depends on determinants that represents the produced and consumed heat (1) [8] [9].

$$P_Z + P_S = P_k + P_r \tag{1}$$

where:

P_Z is heating the conductor due to current flow (W/m),

P_S is heating the conductor due to solar radiation (W/m),

P_k is cooling the conductor due to airflow (W/m),

P_r is cooling the conductor due to natural radiation (W/m).

By substitution equation above, we obtained a following expression for determining current capacity [8] [9]:

$$I = \sqrt{\frac{P_r + P_k - P_S}{R_{ac}}} \tag{2}$$

where:

R_{ac} is AC resistance of conductor (Ω/m).

For the design of power lines is currently applicable standard EN 50341 where are defined ambient conditions for calculation of maximum allowable current value. This standard recommended maximum operating temperature 70°C of conductor [10] [11].

Ambient conditions for calculation of maximum allowable current value of conductor according to the standard are:

- The ambient temperature is 35 ° C,
- Wind speed is 0.5 m / s at 45 ° angle of impact,
- Global temperature of sunlight is 1000 W/m^2,
- Absorption coefficient is 0.5,
- Emissivity coefficient of 0.5 [11].

It is necessary to notice that conditions given by a standard defined the worst case of ambient conditions, which is rarely found. On the base of this hypothesis, we compare current value given by the ambient conditions of Slovakia and current value determined by a calculation given by the ambient conditions accordance a standard EN 50341.

2) Construction of ultra high voltage power lines

In practical terms, for the line of 400 kV voltage level are used trunked conductors where one phase consists of three conductors each technically and electrically connected at a distance, thereby enhanced radius of the conductor of one phase [16].

As conductors of transmission lines are used aluminum cables with steel core. Their advantage is greater mechanical strength, that allows cable to be used for long distance. Among their other advantages include greater flexibility, more uniform structure. Material error may degrade the entire wire, but using ACSR ropes, breaking of one wire will not damage the entire conductor [16].

The conductor may also include an optical fiber to provide device communication. [13][15].

Parameters of the examined conductor are in the table below.

Table 1 Parameters of the examined conductor [16]

Parameters	ACSR 680/73
Rope diameter (mm)	35,8
Rope cross-section (mm^2)	761,69
Nominal weight (kg.km^{-1})	2556,12
Specific gravity (MN.m^{-3})	0,03291
The maximum permissible stresses (MPa)	93,072
Elastic modulus (MPa)	69447
The coefficient of thermal expansion (1/°C)	19,46
Rated DC resistance (Ω/km)	0,0433

A. *Impact of solar radiation to the current capacity of conductor*

Total solar radiation incident on the general plane is the sum of direct radiation, diffuse radiation and reflected radiation. With sufficient precision, it is possible to neglect diffuse and reflected radiation as they affect 2% - 4% [17] [18] [19].

Heat gain due to solar radiation we can obtain from following equation [17] [18] [19]:

$$P_S = \varepsilon_a \cdot D \cdot I_S \tag{3}$$

where

ε_a is radiation absorption coefficient (-),

D is diameter of the conductor (m)

I_s is intensity of solar radiation (W/m^2)

As is apparent from equation above most influence determines argument *Is*, which is variable in the time. Other arguments are of constant character. Argument *D* is significant for a design of new power lines, where can by choose the type of used conductor [17] [18] [19].

For purpose of this article we examine contribution of the solar radiation to the current capacity in Slovakia ambient conditions. Input data of solar intensity radiation are from the Figure 2.

Results of the calculation for ACSR conductor 680/73 with comparing current capacity given by a calculation accordance a standard EN 50341 are shown in figure bellow (Figure 7).

As is shown in the Figure 7, current capacity given by a calculation accordance to the ambient conditions intended by a standard is 632,02 A for the one conductor in three beam connections (Red line).

Current capacity of ACSR conductor 680/73 for average value of solar radiation intensity during the day in Slovakia are in range from 0 to 820 W/m^2 and it does not achieve a significant change of current value. The resulting current capacity of intensity of solar radiation 0 W/m^2 is 759,917 A that is 20,24% higher than the capacities given by the standard. On the other side, for intensity of solar radiation 820 W/m^2 is current capacity 656,77 A, what is decrease in current capacity for ACSR 680/73 only about 3,92%.

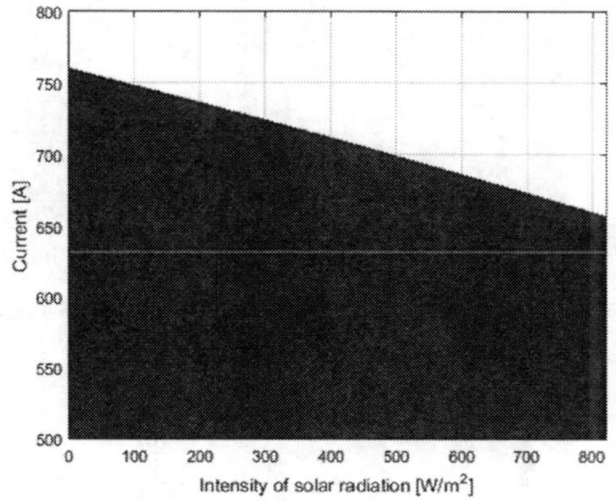

Figure 7 Comparing resulting current capacity based on the average intensity of solar radiation

B. Impact of air flow to the current capacity of conductor

Airflow cooling is a physically complex process that depends on the nature of the hydrodynamic and boundary temperature layer, the shape and size of which is affected by the velocity and the flow direction. which are formed near the heat of the exchange surface and the material quantities of the fluid [17] [18] [19].

The magnitude of the thermal loss of the conductor due to airflow is defined by equitation: [17] [18] [19]:

$$P_k = 0{,}0119 \cdot \left(\frac{D \cdot \rho \cdot w}{\nu} \right)^{0,6} \cdot \lambda \cdot K_{uhol} \cdot (\Theta_v - \Theta_0) \qquad (4)$$

$$K_{uhol} = 1{,}194 - \cos(\varphi) + 0{,}194 \cdot \cos(2\varphi) + +0{,}368 \cdot \cos(2\varphi) \qquad (5)$$

where

ρ is air density (kg/m^3),

D is diameter of conductor (m),

Θ_v is conductor temperature (K),

Θ_O is ambient temperature (K),

λ is thermal conductivity of the fluid (W/(m.K)),

w is velocity of the flowing fluid (m/s),

ν is kinematic viscosity (m^2/s),

f is the angle between the wind direction and the conductor's axis (°).

From equation above (4) it follows that most influence on the magnitude of the thermal loss of the conductor has air velocity w. A significant parameter is also ambient temperature, which decreases the value of the thermal losses.

For purpose of our research are another parameters constant. A magnitude Kuhol represents coefficient that represents the influence of wind on the value of thermal loss.

To determine impact of air velocity to the maximum allowable current capacity of conductor we use map of wind areas of Slovakia (Figure 5).

Map of wind areas has defined for Slovakia territory points with specific air velocity and direction of air flow.

To determine current capacity of conductor ACSR 680/73 we create an interval of average values of air velocity from 1,30 to 11,10 m/s.

Results from the calculation for ACSR conductor 680/73 with comparing current capacity given by a calculation accordance a standard EN 50341 are on the Figure 8.

As is shown on the figure, change of air velocity has influence on the current value. For air velocity in the range 1,30 m/s to 6,0 m/s we experienced an exponential growth of the current capacity. In the range 6,0 m/s to 11,10 m/s there was a linear growth of the current capacity.

A difference between the current ampacity given by calculation standard EN 50341 and the current capacity of conductor ACSR 680/73 obtained by a calculation for average values of air velocity were as follows:

- For air velocity 1,3 m/s is difference between calculation standard and average value 31,90%. Current value in this point is equal 833,40 A.

  Current value given by a calculation accordance a standard is 632,02 A (Red line).

- For an exponential growth of current capacity of conductor for air velocity 6,0 m/s there was an increase towards a value 631,85 A 106,66%.
- For air velocity 11,10 m/s is growth in current capacity equal to 147,39% of the capacity given by an ambient conditions contained in the standard.

Based on these results, we can say that a minimal change of air velocity leads to increase a current capacity of conductor.

The most pronounced increase is in the initial values in the range of average air velocity in Slovakia weather conditions.

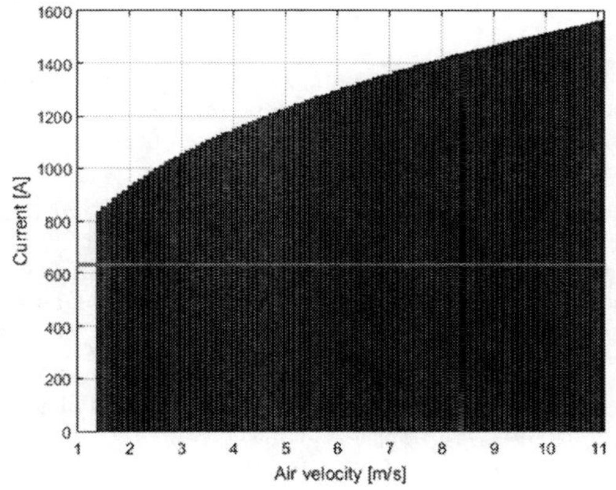

Figure 8 Comparing resulting current capacity based on the average air velocity

C. Impact of natural radiation to the current capacity of conductor

The radiation represents a mechanism of heat transfer, which consists of the emission and absorption of the electromagnetic radiation. An object with a non-zero temperature emits electromagnetic radiation according to Planck's law [17] [18] [19].

The total amount of energy emitted from the surface of the object increases with surface temperature. Depending on the temperature of the body surface, the emission spectrum changes. By the increase of the temperature there is a change of the spectrum to shorter wavelengths. Each object in addition to its own radiation captures the photons radiated by nearby objects. The resulting energy balance of the process is given by the difference of radiated and received energy. As the amount of radiated energy increases with temperature, the result radiation is the transfer of energy from warmer units to cooler [17] [18] [19].

The magnitude of the thermal loss of the conductor due to natural radiation is then equal to [17] [18] [19]:

$$P_r = \varepsilon_S \cdot S \cdot \sigma \cdot (\Theta_V^4 - \Theta_0^4) \quad (6)$$

where:

ε_S is a coefficient of emissivity of radiation (-),

S is cross section of conductor (mm$^{2)}$,

σ is Stefan-Boltzmann constant (W/(m^2.K^4)),

Θ_v is conductor temperature (K),

Θ_O is ambient temperature (K).

To determine impact of natural radiation to the maximum allowable current capacity of conductor we average temperature for summer season (Figure 3) and winter season (Figure 4).

To determine the current capacity of conductor we choose a two range for examining. First is temperature range for summer season. It is from 8°C to 20°C. Second studying temperature range for winter season is from -10°C to 0°C.

Results from the calculation are on the following figures (Figure 9, Figure 10).

As figures show (Figure 9, Figure 10), when ambient temperature changes occur equally to change current capacity for conductor ACSR 680/73. This change current capacity occurs as a direct result of changes in the coefficient of heat loss by natural radiation (P_r), wherein in increasing the ambient temperature leads to inversely proportional heat rejection to the environment.

In terms of size of current capacity in the summer season, there is a decrease current value from 902,30 A to 795,50 A. For the winter season, there was a decrease of current capacity from 1038,08 A to 965,50A.

As we see from the Figure 9, for ambient temperature 8°C is current capacity value higher against the current capacity given by a standard about 42,78%. Due to an increase of ambient temperature to 20°C occur a change in difference between current values in this point of temperature and current value given by a standard to 25,90%.

Magnitude of ambient temperature has most influence on the value of current capacity (equation 6). This theoretical knowledge has been confirmed by a Figure 10.

At a lower temperature occurred an increase of current capacity. For average ambient temperature in Slovakia conditions -20°C was current value equal to 1038,09 A what is higher compared to the current value given by a calculation with ambient conditions accordance a standard about 64,29%.

In the second point of examining range 0°C is current capacity equal to 965,50 A, what is about 52,80% higher than current value given by a calculation with ambient conditions accordance a standard.

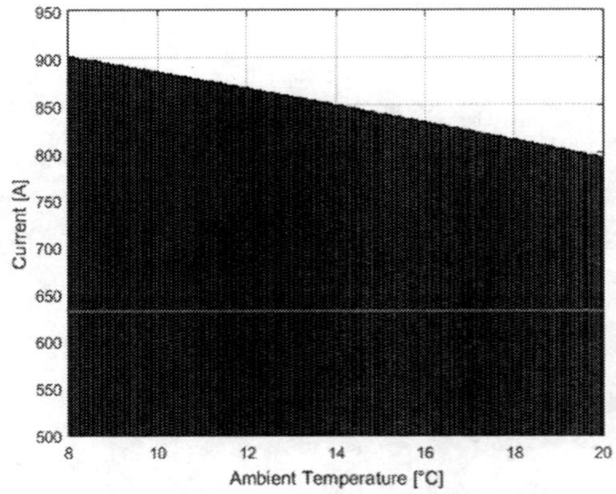

Figure 9 Comparing resulting current capacity based on the average temperature in summer season

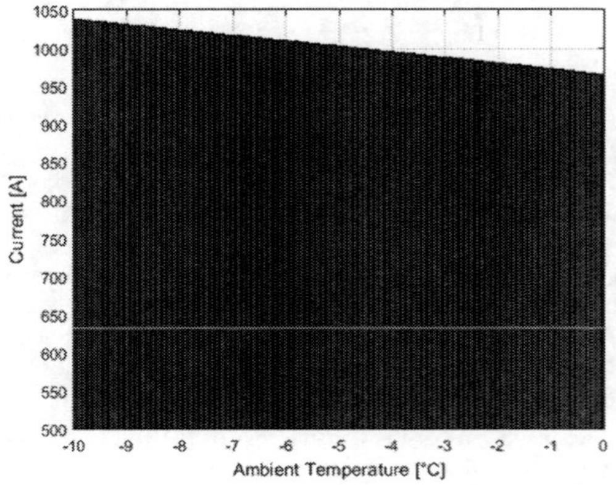

Figure 10 Comparing resulting current capacity based on the average temperature in winter season

Conclusion

Power lines are the most important part of power systems. Their role is not substitutable and their construction determines the size of the current capacity. As every construction of electricity facilities is subject to the standard, also construction of power lines and design is subject of standard. Current capacity of the conductors depends also on ambient conditions, which are described in standard EN 50341. These ambient conditions are given for the worse case that can occur. As is know these ambient conditions or their combination are very rare in nature.

This article described method to determining the current capacity for ACSR conductor 680/73 accordance average ambient conditions in Slovakia.

Results show that ambient conditions have most influence on the actual value of current capacity. If we can accurately determine the ambient conditions in real time, we can determine the current capacity under these terms and adapt operation of power system or power lines.

Acknowledgment

This work was supported by the Ministry of Education, Science, Research and Sport of the Slovak Republic and the Slovak Academy of Sciences under the contract No. VEGA 1/0132/15.

References

[1] M. Kolcun, V. Griger: "Controlling the operation of the power system" (Riadenie prevádzky elektrizačnej sústavy), Mercury - Smékal, Košice, 2003, 288 pages, ISBN 80-89061-57-5.

[2] Working Group B2.12, "Guide for selection of weather parameters for bare overhead conductor rating," CIGRE, 2006

[3] A. Bracale, "Probabilistic index for increasing hourly transmission line rating," in Int. Journal of Emerging Electric Power Systems. 2007, pp. 119.

[4] T. Košický, Ľ. Beňa, M. Kolcun: "Analysis of utilisation energy storage systems for frequency regulation", Acta Electrotechnica et Informatica. Vol. 14, Nr. 3 (2014), p. 36-42. - ISSN 1335-8243

[5] R. Jakubčak, Ľ. Beňa, M. Kmec: "Possibilities of Using Facts Devices In Power System", Acta Electrotechnica et Informatica. Vol. 13, No. 3 (2013), p. 8-11. - ISSN 1335-8243

[6] Slovenský hydrometeorologický ústav,"Klimatické podmienky na Slovensku" < http://www.shmu.sk/sk/?page=1064>

[7] A. Klenovcanova, T. Brestovic: "Possibilities of utilization of photovoltaic cells for electricity production in Kosice area " (Možnosti využitia fotovoltickych člankov na výrobu elektrickej energie v oblasti Košíc). Acta Mechanica Slovaca 2007, ELFA, s.r.o., 11, 4-D, 511-516, ISSN: 1335-2393

[8] M. Špes: "Powerline ampacity system" Proc. SCYR 2016. - Košice : TU, 2016 pages 202-205. - ISBN 978-80-553-2566-8

[9] M. Špes, Ľ. Beňa, M. Mikita, M. Márton, H. Wachta: "Possibilities of increasing transmission capacity overhead lines", Acta Electrotechnica et Informatica. Vol. 16, No. 3 (2016), p. 20-25. - ISSN 1335-8243

[10] L. Varga, P. Leščinsky, Ľ. Beňa: "Mechanical calculation of ovearhead power line conductor under combined load", Acta Electrotechnica et Informatica. Vol. 1, no. 1 (2001), p. 28-31. - ISSN 1335-8243

[11] Z. Čonka, M. Kolcun, G. Morva: "Utilizing of Phase Shift Transformer for increasing of Total Transfer Capacity" In: Acta Polytechnica Hungarica Vol. 13, no. 5(2016), p. 27-37 2016 – ISSN 1785-8860

[12] EN 50341-1 ED.2 (333300): Elektrická venkovní vedení s napětím nad AC 1 kV - Část 1: Obecné požadavky - Společné specifikace

[13] Z. Čonka, M. Kolcun, "Increasing Net Transfer Capacity (NTC) by UPFC in Central East Europe power system". In: Electric Power Engineering (EPE). - Ostrava : VŠB-TU, 2015 P. 132-136. - ISBN 978-1-4673-6787-5

[14] D. Solus, L. Ovseník, J. Turán, "Optical correlator in vertical traffic signs inventory system". Paper presented at the 2015 IEEE 13th International Scientific Conference on Informatics, INFORMATICS 2015 - Proceedings, 247-251. doi:10.1109/Informatics.2015.7377841

[15] D. Solus, L. Ovseník, J. Turán, "Signal processing - object detection methods with usage of optical correlator". Paper presented at the 2016 26th International Conference Radioelektronika, RADIOELEKTRONIKA 2016, 315-318. doi:10.1109/RADIOELEK.2016.7477434

[16] Š. Fecko, J. Žiaran,L. Varga: "Power lines – Overhead power lines" (Elektrické siete - Vonkajšie silové vedenia), SVŠT Bratislava, 1990.

[17] CIGRE Task Force B2.12.3, "Sag-tension Calculation Methods for Overhead Lines." In Technical Brochure 324, June 2007.

[18] CIGRE Working Group B2.12, "Alternating Current (AC) Resistance of Helically Stranded Conductors." In Technical Brochure 345, April 2008.

[19] IEEE Power Engineering Society, "IEEE standard for calculating the current temperature of bare overhead conductors," 2007

Issues of Mercury Hg Emissions in the Czech Republic in Accordance With New Legislation Valid from 2021

Zuzana Vávrová, Zbyszek Szeliga, Pavel Dvořák, Ondřej Palička, Radim Fojtů
Centrum ENET
VSB – Technical University of Ostrava
Czech Republic
zuzana.vavrova@vsb.cz, zbyszek.szeliga@vsb.cz, pavel.dvorak1@vsb.cz, ondrej.palicka@vsb.cz

***Abstract*—This article discusses about new mercury emission limits that will be valid from 2021. Nowadays, problematics of mercury emission and the current value of mercury emissions from large combustion plants in the Czech Republic are unknown. Therefore, here are the forms of mercury, their formation during the coal combustion and the measures for mercury reduction. There is also a methodology for measurement of mercury emissions.**

***Keywords*—mercury emission, legislation, coal combustion, power plant**

I. Introduction

The year 2021 will bring into force new legislation relating pollutant emissions. New emissions limits will be more strict and that is obvious because worldwide trend is to reduce emissions.

What is new is monitoring and controlling mercury emissions that are emitted from large combustion plants (with heat input higher than 50 MWt). For large combustion plants in the Czech Republic, issues of mercury emissions are new and in a relatively short time it will be necessary to find out the amount of mercury emissions that are emitted into the air by individual large combustion plants. Based on this information, in case the new emissions limits are not met, it will be necessary to apply the available technologies/measures that achieve these new mercury emissions limits. [1].

II. Mercury Emissions for Existing and New Energy Sources Valid from 2021

New mercury emissions limits will be valid from 2021. The values set by new legislation are listed in the table I. Limits are set for large combustion plants and are divided by heat input from 50 MWt to 300 MWt. Further division is based on combusted coal (lignite or black coal) and commissioning. Mercury emissions will be reported as yearly average [ηg/Nm3], for sources with heat input to 300 MWt. And for sources with heat input higher than 300 MWt mercury emissions will be reported as an average of few samples [ηg/Nm3] obtained during one year. [1].

III. Mercury Emissions from Coal Combustion Processes

Mercury is one of the most dangerous elements for human health and environment. Mercury is potent neurotoxin, which accumulates in the human body. Mercury then causes damage to the nervous system, kidneys damage, damage of respiratory system, and others. [2,3].

TABLE I. Hg EMISSIONS LIMITS VALUES VALID FROM 2021 [1]

Yearly average or average of samples obtained during one year Hg [ηg/Nm3]				
Heat input [MWt]	**New plant**		**Existing plant**	
	Hg (coal)	***Hg (lignite)***	***Hg (coal)***	***Hg (lignite)***
< 300	< 1 - 3	< 1 - 5	< 1 - 9	< 1- 10
≥ 300	< 1 - 2	< 1 - 4	< 1 - 4	< 1- 7

Human activity, especially mining and coal combustion, has increased the amount of mercury in the environment.

Anthropogenic sources of mercury emissions account approx. 30% of the total amount of mercury entering the atmosphere each year. The updated inventory of emissions to air confirms coal combustion as a continuing major source of mercury emissions. [2].

Coal does not contain high concentrations of mercury, but the combination of the large volume of combusted coal and the fact that a significant amount of the mercury present in coal is emitted to the atmosphere yield large overall emissions from this sector.

Mercury is present in coal in different trace amounts. Therefore, during the combustion process, mercury passes in the form of vapor into flue gas.

Mercury in flue gas is in three primary forms, each of these forms of mercury has its specific properties. Knowledge of these properties is important for other possibilities of reduction and capture. The amount of emitted mercury by the large combustion plants depends on the type of combusted coal and on the emission control devices that is used. [1,2,3].

- Gaseous oxidized mercury Hg^{2+}
- Gaseous elemental mercury Hg^0
- Mercury bound to particles Hg_p

Gaseous oxidized mercury Hg^{2+} is water soluble. Therefore, the oxidized form of mercury Hg^{2+} in the atmosphere is removed by rainfall. Oxidized form of mercury presented in flue gas can be removed by wet flue gas desulphurization systems. It all results in that the oxidized form of mercury Hg^{2+} represents only 3-5% of the total mercury in the atmosphere. [1, 2, 3].

Gaseous elemental mercury stays in the atmosphere long enough, an average of 0.5 - 2 years, for it to be transported around the world. In order to remove the elemental form of mercury, it is necessary to oxidize it first. For that reason elemental form of mercury Hg^0 is major problem in terms of flue gas resulting from the coal combustion. [1, 2, 3].

Mercury bound to particles Hg_p is adsorbed onto surfaces –slag, fly ash. It can be easily capture by dust control devices.

IV. Mercury Distribution During Ccoal Combustion

Mercury is presented in the coal in trace amounts, most often from 0,02 to 0,25 ppm. The amount of mercury varies depending on the type of coal (black, lignite) and the origin of the coal. [3]

In table II. you can see mercury values in mg / kg_{dry} and chlorine content % $_{dry}$ from individual lignite samples. This lignite is used in the Czech Republic. Knowledge of the fuel base is important.

TABLE II. Mercury Content in Lignite Used for Combustion in the Czech Republic [5]

Lignite sample from quarry	*Hg mg/kgdry*	*Cl %dry*
Bílina	0,22	0,0074
Nástup Tušimice	0,203	0,0088
Jiří	0,482	0,0095
Vršany	0,262	0,0061
Mibrag	0,29	0,0162

The table III. shows the analysis of black coal sample. This black coal is burned in only large combustion plant in the Czech Republic.

Table II. and table III. show the difference between lignite and black coal in terms of mercury content and chlorine content, which is important for the oxidation of the elemental form of mercury. Mercury content in black coal used for combustion in the Czech Republic [5]

TABLE III. Mercury Content in Black Coal Used for Combustion in the Czech Republic [5]

Black coal sample	*Hg mg/kg_{dry}*	*Cl $\%_{dry}$*
	0,516	0,19

During coal combustion, mercury easily evaporates and passes in the gaseous form to the flue gas. The flue gas then continues through the heat exchange surfaces of the boiler. The flue gas is cooled and oxidation of elemental mercury stars to oxidized with the presence of halide, especially $HgCl_2$, Hg_2Cl_2, $HgBr_2$ or HgO. [3, 4].

The content of chlorine Cl and other halogens is important for the oxidation of elemental form of mercury Hg^0, which affects the final ratio between Hg^0 and Hg^{2+}.

Fig. 1 shows the percent composition of flue gas, measured on selected real large combustion plants. Black coal and lignite was burned during the tests. Measurements took place at the boiler outlet.

Flue gas produced by the combustion of black coal has a higher proportion of the oxidized form of mercury because black coal contains more chlorine. [4, 5]

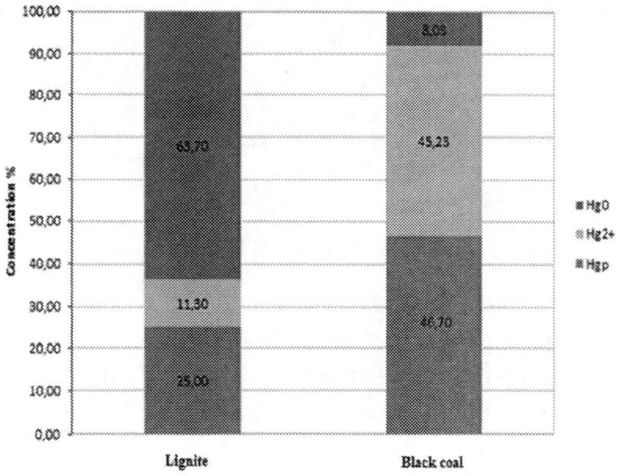

Fig. 1 Comparing the measurement results of individual forms of mercury in the flue gas at a selected combustion plans for black coal and for lignite [5,6]

During further cooling, near the air heater (LUVO), elemental form of mercury is adsorbed on the surface of solid particles (fly ash). Formation of mercury bound to particles Hg_p. This form of mercury is captured mainly in dust control devices.

Mercury in the flue gas continues through the flue gas desulfurization unit, where the acidic components of the flue gas are generally neutralized. If this is a wet flue gas desulfurization unit, then a considerable part of the oxidized form of mercury Hg^{2+} is captured.

Only the elemental form of mercury Hg^0 is emitted to the atmosphere. Elemental form of mercury is major problem in terms of capturing and reducing.

The whole process of formation of mercury emissions and its forms during coal combustion is shown in Fig. 2. [5]

V. Techniques to Prevent and / or Reduce Mercury Emissions

From the point of view of further developments, it will be necessary to consider measures that will lead to the reduction of mercury emissions.

One possible solution for mercury reduction is to limit the combustion of existing fuel or replace it with fuel with less mercury content or co-firing more fuels. This is unrealistic in most cases. For this reason, it is important to use other technologies and their combinations to achieve the required emission limits for mercury. [7]

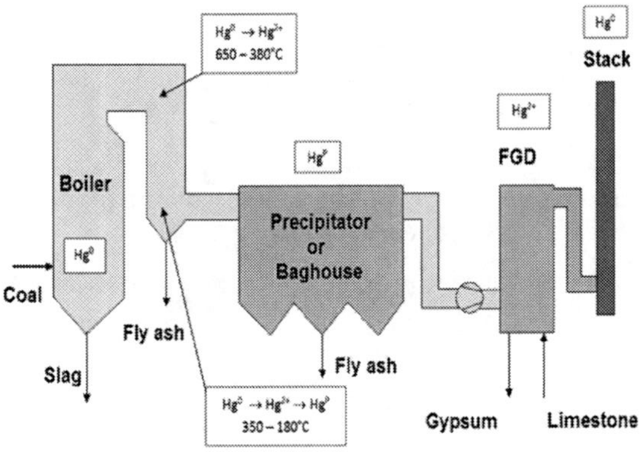

Fig. 2 Formation of mercury emissions during coal combustion [5]

A. Dust control devices

The electrostatic precipitator is one of the most widespread devices used in combustion devices to separate solid particles. However, it can only separate Hg_P.

Another technology for removing solid pollutants is a fabric filter which, with respect to the removal of mercury emissions, has a higher efficiency than an electrostatic precipitator. [7]

B. Desulphurization systems

Desulfurization devices operate on the principle of neutralizing acidic components of flue gas with alkaline components of the sorbent. In the case of wet flue gas desulfurization, oxidized form of mercury is largely removed from flue gas, because this form of mercury is water soluble. However, the elemental form of mercury passes through the desulfurization unit. Therefore, the use of selective catalytic reduction (SCR) and wet flue gas desulfurization methods are considered. During the measurement of mercury emission concentrations on the desulphurization device, it was found that mercury was re-emitted. That means, that at the output of the unit there was an increase in the elemental form of mercury. [7]

C. Selective catalytic reduction SCR

SCR technology is capable of oxidizing the elemental form of mercury. This fact allows us to subsequently removed mercury in the desulphurization reactor. However, it is important that flue gas contains sufficient amount chlorine (halogens) to oxidize mercury in the catalyst. SCR is now used to remove nitrogen oxides from flue gas. Ammonia/urea is injected into the flue gas stream, and nitrogen oxides then react to form of elemental nitrogen. However, this reaction takes place on the catalyst faster, because the catalyst binds primarily ammonia particles. The efficiency of the SCR depends primarily on the properties of the catalyst (active area), the residence time of the flue gas on the catalyst and the stoichiometry.

D. Sorbent injection

Another way to reduce mercury emissions is to use sorbents. Here mercury adsorbs sorbents. An important parameter is the flue gas temperature, the sorbent properties (specific surface). In practice, we can use calcium-based sorbents, activated charcoal sorbents, fly ash sorbents with high content of carbon, etc. In view of the higher efficiency of capture of mercury emissions, it is advisable to use a sorbent combination with a fabric filter, there is a stronger interaction between flue gas and sorbent.[3,7]

VI. Measuring Methods for Mecury in Flue Gas Generated from Coal Combustion

Mercury in flue gas is presented in several forms, which complicates its determination. There are two methodologies for determining the concentration of mercury in the flue gas. The first is a manual method for determining mercury emissions, and the second one is a continuous method for determining mercury emissions.

A. Ontario hydro method

This is a manual method for measuring mercury concentrations. By this method, we are able to determine individual forms of mercury. The principle of this method is the isokinetic sampling of the flue gas stream. The sample is taken thought heated probe. The flue gas sample goes further through series of absorbers that contain various chemicals that capture individual forms of mercury. Samples are then sent to the laboratory for analysis. The Ontario-hydro method determines very accurately mercury concentrations from the collected sample. Schematic diagram of the measuring device is shown in Fig. No 3. [8].

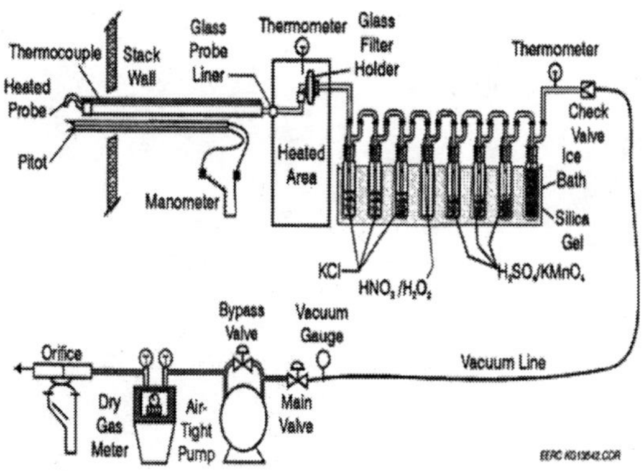

Fig. 3 Schematic diagram of mercury sampling train [8]

In the case of continuous measurement devices, mercury emissions can be measured inaccurately. The size of measuring inaccuracy depends on the composition of the flue gas. Especially it is influenced by content of sulfur oxides.

VII. Measurement on real large combustion plant

To determine the content of each mercury form, it must be during the measurement the samples of fuel, samples of slag, samples of ash, and samples of gypsum slurry from wet flue gas desulphurization taken. These samples are used for the determination of mercury bound to the fine particles Hg_p. [5, 6]

On the basis of the Ontario hydro method is determined the content of gaseous forms of mercury, namely the elemental forms of mercury and the oxidized form of mercury. Using isokinetic sampling, the samples are taken from stream of flue gas before and after the electrical precipitator and on the output of flue gas desulphurization technology.

Based on the results of measurement on a real combustion plan burning lignite there is assembled balance. This balance is shown in the Fig. No.4. There is shown values of individual forms of mercury emissions Hg^0, Hg^{2+}, Hg_p and Hg^{all}.

The measurement shows the efficiency of individual devices in terms of reduction of individual forms of mercury Hg^0. Most of mercury bound to particles is captured in electrostatic precipitator (ESP).The wet flue gas desulphurization device (wet FGD) is supposed to reduce the oxidized form of mercury Hg^{2+}, and there is also an unwanted re-emission of the elemental form of mercury. [5, 6]

Total mercury emissions emitted by the combustion plant Hg^{all} are measured behind the flue gas desulfurization device. This value will be monitored from 2021.

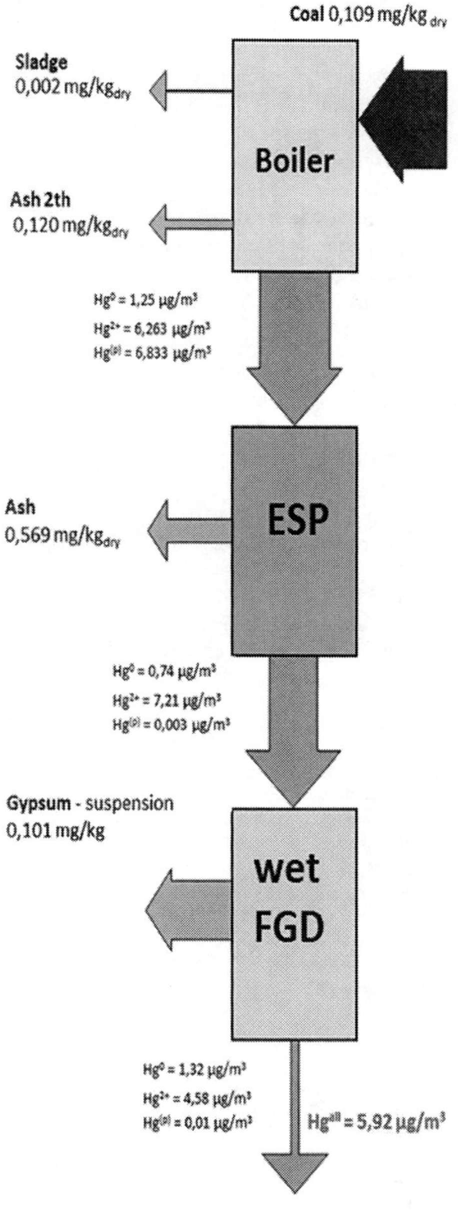

Fig. 4 Mercury balance – measurement on a real large combustion plant [5,6]

VIII. Conclusions

In 2021, a new legislation will come into force which not only reduces existing emission limits but requires monitoring of mercury emissions for large combustion plants with a thermal input of more than 50 MWt.

However, in the Czech Republic, the mercury emissions values of individual combustion plants are not known. This means that, in the first instance, it will be necessary to measure the amount of mercury emitted for each large combustion plant. Based on these measuring, it will be necessary to take measures to fulfil the emission limits. Everything must be done in a relatively short time, both the measurement and investment associated with the installation of new technologies.

As far as the measurement itself is concerned, it will be necessary to develop a sufficiently reliable and accurate measurement method for the continuous measurement of mercury emissions.

Acknowledgment

This research was conducted within the framework of the project TUCENET Sustainable Development of Centre ENET [LO1404] and by the SGS grant from VSB – Technical University of Ostrava [Grant No. SP2016/95, SP2017/54 and SP2017/127].

References

[1] P. Frolka, "BREF Large combustion plants," XVII forum for information exchange about BAT: MPO, 24.11.2016, 2016.

[2] UNEP. Global Mercury Assesment [online]. 2013 [cit. 2014-10-10]. ISBN DTI/1636/GE.

[3] M. Stach, Z. Klika, L. Bartoňová. „Distribution and removing of mercury during the solid fuel combustion," In. Proceedings of scientific papers at the VŠB - Technical University of Ostrava: Mining and Geological Series. Ostrava: College of Mining, 2005, s. 27-42. ISSN 0474-8467

[4] K. Krotla, K. Orehovsky. "Use of catalytic reduction of flue gas to reduce mercury emissions," 2013

[5] L. Pilař, Z. Szeliga, R. Zbieg. „Monitoring conversions of Hg^0/Hg^{2+} in flue gas from energy sources", upublished.

[6] Technical reports – Accredited workplace for diagnostics and operation of power generation equipment center ENET.

[7] Best Available Techniques (BAT) Reference Document for Large Combustion Plants: Industrial Emission Directive 2010/75/EU. Final Draft. JOINT RESEARCH CENTRE Institute for Prospective Technological Studies Sustainable Production and Consumption Unit European IPPC Bureau, 2016.

[8] Standard Test Method for Elemental, Oxidized, Particle – Bound and Total Mercury in Flue Gas Generated from Coal – Fired Stattionary Sources (Ontario Hydro Method), ASTM D6784-02 (2008)

Studies on Interactions in Multi-Infeed Systems

Mateusz Polewaczyk, Sylwester Robak
Electrical Power Engineering Institute
Warsaw University of Technology
Warsaw, Poland
mateusz.polewaczyk@ien.pw.edu.pl, sylwester.robak@ien.pw.edu.pl

Abstract — **In a Multi-Infeed HVDC systems HVDC links may be situated close to each other in the electric power grid. This fact causes some problems during operation of adjacent converter stations such as: voltage interactions, commutation failures, transient overvoltages or control interactions. Faults in power system may also impact adjacent HVDC links. In this paper the vulnerability of HVDC links in a Multi-Infeed system to AC system disturbances is examined. The complexity of interactions in Multi-Infeed systems indicate that location of a new HVDC link in a close electrical proximity to other HVDC link should be preceded by detailed studies with the use of proper indices. This paper presents interactions in Multi-Infeed systems, indicators used in the analyses and simulation results of such systems.**

Keywords: interactions in HVDC systems, HVDC transmission systems, Multi-Infeed HVDC, Dual-Infeed HVDC,

I. Introduction

The properties of High Voltage Direct Current (HVDC) transmission systems are considered advantageous in comparison with conventional High Voltage Alternating Current (HVAC) transmission systems. Current development of power electronics devices leads to the increase of the number of HVDC systems. In the world the two main technologies are used. The first one is thyristor-based Line Commutated Converters (LCC) and it is the most popular and conventional one. The second one is Voltage Source Converters (VSC) and it is a new, emerging and transistor-based technology.

Today it is observed that HVDC links play a major role in bulk power transmission systems. HVDC transmission systems are one of the basic components of the future Smart Grids. The versatile properties of HVDC systems make them applicable in both land and offshore power engineering [1]. A further development of HVDC technology will allow us to integrate large scale renewables and in the future may let us connect independent HVDC links into a common HVDC grid. Nowadays HVDC systems are becoming more common in modern power systems. This, in turn, makes converter stations of HVDC links be situated in a close electrical proximity in a power system. The system in which one HVDC system affects the operation of another HVDC system through a common AC system where AC impedance between the two HVDC systems is small is referred to as a Multi-Infeed HVDC system [2]. Thus, HVDC links in a Multi-Infeed system cannot be considered as independent links and should be considered as a whole group. The degree of interaction depends mainly on the distance between the converter stations and the short-circuit power level at the connection nodes [3].

Operation of HVDC links in a Multi-Infeed system causes some problems connected with a proper operation of HVDC links. Typical interactions that occur in the Multi-Infeed system include: commutation failures, overvoltage, impact on harmonic and voltage performance.

II. Multi-Infeed HVDC Systems

A Multi-Infeed system is created when two or more HVDC links are located in a close electrical proximity. In this way converter stations (belonging to different HVDC links) may interact with each other causing additional difficulties connected with proper operation of HVDC links [4].

Fig. 1. shows the schematic representation of Multi-Infeed HVDC system. Impedances Z_{12} and Z_n correspond to AC line length between HVDC links.

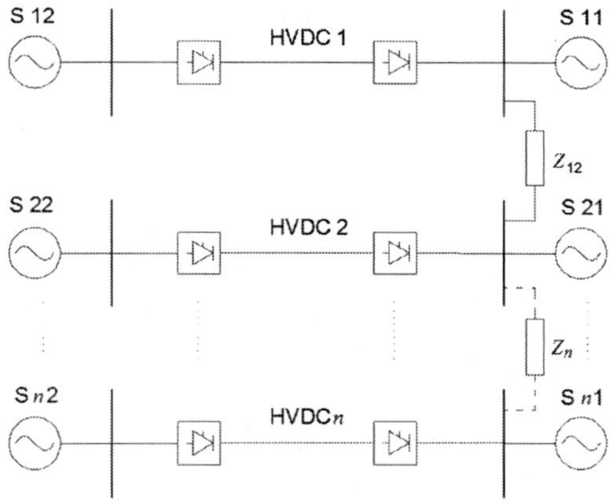

Figure 1. Schematic diagram of the Multi-Infeed HVDC system

A. Dual-Infeed system

A Dual-Infeed HVDC system (DI-HVDC) is a variant of a Multi-Infeed system where there are only two HVDC links situated in a close electrical proximity. Converter stations of different HVDC links may be connected to one AC node or nodes being in a close electrical proximity [3].

There are two common types of a Dual-Infeed HVDC system depending on a converter station operation mode [4]:

- IIMI (Inverter-Inverter Multi-Infeed) – where in a close electrical proximity there are two converter stations operating as an inverter.
- IRMI (Inverter-Rectifier Multi-Infeed) – where one converter station operates as an inverter and another as a rectifier.

B. *HMIDC (Hybrid Multi-Infeed HVDC)*

Hybrid Multi-Infeed HVDC systems are structures in modern power systems where some of the adjacent links are based on LCC and some of them are based on VSC.

III. Interaction types in Multi-Infeed HVDC System

In Multi-Infeed systems the interactions between converter stations are a complex issue. This is due to nonlinear characteristic of AC/DC operation. The determination of level of interactions in Multi-Infeed systems is important from the point of view of stable and safe operation of the power system. Generally speaking, with an increasing electrical distance between converter stations, interactions decrease.

There are some interaction types which can be observed during Multi-Infeed system operation. The following sections include short descriptions of each type.

A. *Voltage Interactions*

In order to assess a voltage interaction between HVDC links, the voltage change at one of nodes and the voltage change in an adjacent node should be determined. The voltage interaction strongly depends on an electrical distance between converter stations.

B. *Commutation Failure*

Commutation Failure (CF) is a phenomenon occurring in LCC HVDC converters typically during an inverter operation mode. The risk of commutation failure increases when there is too low short-circuit power level at AC nodes where HVDC converters are connected. CF event occurs when the time available for commutation is exceeded.

CF usually appears when [6]:

- voltage in AC grid decreases, for example during AC system faults,
- phase angle of the commutation voltage suddenly changes,
- incorrect operation of firing control occurs.

Decrease of the voltage, increase of the current in HVDC link and also additional stresses in thyristor valves (overcurrent) are the effects of the commutation failure phenomenon.

A continuously appearing commutation failure may lead to blocking of a HVDC converter and consequently even cause power system instability.

A commutation failure in one of the converter stations may cause a commutation failure in the converter located in a close electrical proximity. A concurrent commutation failure (CCF) appears when both converters are affected by these events.

C. *Fault Time Recovery*

Operation of adjacent HVDC link may increase the time required to return to proper operation of several HVDC links located in a close electrical proximity.

D. *Harmonic Performance*

LCC converters are harmonic source in both – AC and DC side of HVDC link. The harmonic effect can propagate over a long distance within AC grid and can be reinforced easily by resonance phenomenon.

A harmonic interaction is a complex issue which requires the detailed analysis. Especially strong interaction occurs when adjacent converter stations work as inverters. When there is more than one harmonic source it is possible that they will interact with each other.

E. *Transient Overvoltage (TOV)*

In the case of HVDC links based on LCC converters, there is a considerable demand for reactive power consumption of the converter stations. The demand is met by compensation equipment, AC side filters and shunt capacitors. Reactive power compensation by the means of shunt capacitors is one of the basic methods of improving power network operating conditions [7]. Reactive power consumption of the LCC converter is about half of the rated power of a HVDC link. Thus, sudden interruption of power transmission in the HVDC link may lead to surges in AC grid. Even a worse case occurs where several HVDC links are located in electrical proximity. Simultaneous power interruption by such localized HVDC links results in a very high overvoltage that exceeds the overvoltage that could occur during a single HVDC operation.

F. *Control Interaction*

Utilization of fast control systems in adjacent HVDC converters may cause an adverse effect in proper operation of HVDC link. Control interactions occur especially when short-circuit power level at AC connected node is decreased. Strong interactions between control systems appear when advanced functions such as stability improvement or VDCOL (Voltage Dependent Current Order Limiter) systems are used. Conventional controllers do not cause adverse effect in adjacent control systems. Dynamic behavior of the control system depends on the mode of operation of the converter [4].

IV. Mitigation strategies

Some mitigation strategies should be used in order to decrease the level of interactions between converter stations in Multi-Infeed system. The following countermeasures can decrease the probability of occurrence of commutation failures (both local and concurrent) [4]:

- increasing the electrical distance between HVDC links,
- increasing gamma (extinction angle) value during inverter operation,
- using smoothing reactor with higher inductance,
- application of VSC or CCC (Capacitor Commutated Converter) topology converter rather than LCC.

The harmonic interaction can be alleviated by [4]:

- increasing the electrical distance between HVDC links,
- taking into account design of a new filter and the existing HVDC system's filter located in a close electrical proximity,
- applying VSC topology converter rather than LCC.

The problem of control interaction may be alleviated by:

- coordination of the control systems,
- optimization of the control functions,
- introduction of several levels of control.

V. Interaction indices for Multi-Infeed systems

In order to evaluate the interaction level, the following factors are introduced [4]:

- Multi-Infeed Interaction Factor (*MIIF*) – the index used for quantify the voltage interaction between converters of Multi-Infeed system,
- Commutation Failure Immunity Index (*CFII*) – the index used to evaluate susceptibility to commutation failure of LCC converter,
- Effective Short Circuit Ratio (*ESCR*), also used in a Multi Infeed context as so-called Multi Infeed Effective Short Circuit Ratio (*MIESCR*).

A. Multi-Infeed Interaction Factor (MIIF)

MIIF index was introduced in order to evaluate the voltage interaction between any two converter AC buses voltage. *MIIF* is an empirical index obtained by simulation results. It is defined as a ratio of approximately 1% voltage change at one of the converter AC bus (induced by artificial connection of shunt reactive element) and observed percentage voltage change in an adjacent node with another converter.

Assuming that nodes *i*-th and *j*-th are adjacent and different HVDC converters are connected to them, the *MIIF* index is calculated according to an equation below:

$$MIIF_{ji} = \frac{\Delta V_j}{\Delta V_i} \qquad (1)$$

where: ΔV_i – is the induced approximately 1% voltage change at AC bus *i*-th, ΔV_j – is the observed voltage change at AC bus *j*-th.

According to the definition of *MIIF*, if two AC buses are infinitely far apart then *MIIF* value equals 0. If converter AC buses are connected to the same AC bus then *MIIF* is 1.

B. Commutation Failure Immunity Index (CFII)

The *CFII* index defined by (2) is used in the analysis of the susceptibility of the LCC converter to commutation failure. This indicator is determined by simulation of a 3-phase to ground fault through an inductive element with the lowest inductance never results in CF.

$$CFII = \frac{V_{ac}^2}{\omega \cdot L_{min} \cdot P_{DC}} \cdot 100\% \qquad (2)$$

where: V_{ac} – the AC voltage of the connected grid, ω – the angular frequency, L_{min} – the lowest inductance never results in CF, P_{DC} – the nominal power of the HVDC link.

In general it was proved that immunity against commutation failure is the lowest during inductive faults, hence the decision to use this type of fault in the analysis [8]. A higher value of the *CFII* means that the converter is more resistant to occurrence of the commutation failure (with higher value of *CFII* there is necessary short-circuit with higher power to appear CF phenomenon).

C. System Strength Indices

Generally, in power system operation the short circuit power (also referred to as short circuit capacity) becomes an important and useful factor [9]. In many cases very simple criteria based on short-circuit power can be used to assess the power system features [10].

One of the most commonly used index in the analysis of system strength with HVDC link is Effective Short-Circuit Ratio (*ESCR*), calculated according to the following equation [11]:

$$ESCR = \frac{\text{SCC} - Q_c + S_l}{P_{DC}} \qquad (3)$$

where: SCC – the short-circuit capacity available on the converter bus, Q_c – the reactive power from filters and capacitor banks, S_l – the synchronous condenser capacity, P_{DC} – the nominal power of the HVDC link

It is also possible to calculate *ESCR* index in the Multi-Infeed context when HVDC links are connected to a common AC bus or adjacent AC buses. A so-called Multi-Infeed Effective Short-Circuit Ratio takes into consideration the influence of other links being in a close electrical proximity. It can be calculated according to an equation below [4]:

$$MIESCR = \frac{\text{SCC}_i - Q_{ci}}{P_{DC_i} + \sum_j (MIIF_{j,i} \times P_{DC_j})} \qquad (4)$$

where: SCC_i – the short-circuit capacity available on the converter bus, Q_{ci} – the reactive power from filters and capacitor banks, P_{DC_i} – the nominal power of the HVDC link, $MIIF_{j,i}$ – Multi-Infeed Interaction Factor

VI. Test System Description

In order to analyze different factors which can affect interactions in Multi-Infeed system, the computer simulations using PSCAD program were performed. Simulation studies have been performed on the test system shown in Fig. 2. The impedance is marked as *Z* in the diagram corresponds to AC line length *l* between node 1 and 2 of HVDC links.

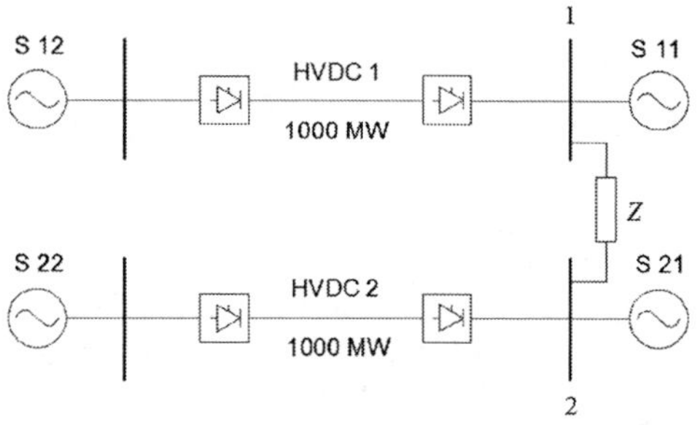

Figure 2. Schematic representation of Multi-Infeed test system

The test model is a Multi-Infeed HVDC system in a Dual-Infeed variant. HVDC links are based on Cigre HVDC Benchmark model [12] with use of LCC converters. Rated parameters of monopolar HVDC links are 1000 MW and 500 kV. A control system of rectifier maintains constant current (CC) and of inverter maintains constant extinction angle (CEA). Two types of Multi-Infeed system: IRMI and IIMI systems were examined. A node number 1 in the test system is the node where an inverter is connected (for IIMI systems) or rectifier (for IRMI systems) and a node with number 2 is the node where an inverter is connected for all studied cases. Hence, ESCR1 corresponds to short circuit power level of the node 1, whereas ESCR2 corresponds to short-circuit power level of the node 2. The influence of: length of AC transmission line connecting HVDC substations, short-circuit power level at node 1 and node 2 and inductance of smoothing reactor was studied.

VII. Simulation Results

A. Voltage interactions

The influence of short-circuit power level and AC line length on *MIIF* factors were verified. Fig. 3 presents the $MIIF_{12}$ values as a function of short-circuit power level expressed in *ESCR* values for IIMI and IRMI systems.

In the test model AC terminal with number 1 was a bus with the rectifier station (IRMI system) or the inverter station (IIMI system). The terminal with number 2 (node in a close electrical proximity) in all studied cases was connected to inverter station (both IRMI and IIMI system).

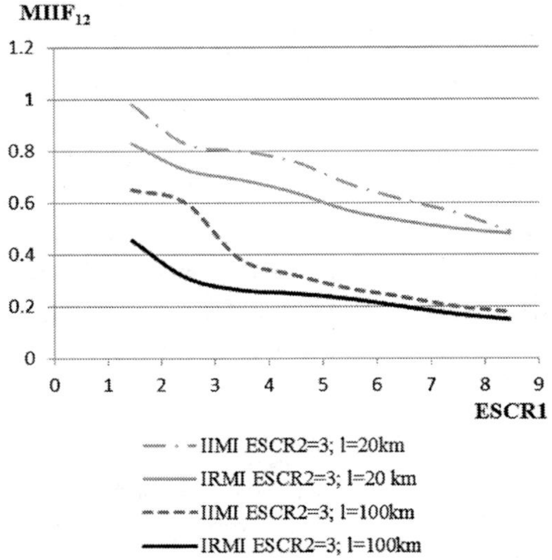

Figure 3. $MIIF_{12}$ in IRMI and IIMI system with different line length between adjacent links as function of ESCR at node 1

The analysis of Fig. 3 leads to the following observations:

- voltage interaction on the rectifier station is smaller (IRMI system) than on inverter station (IIMI system) in the corresponding variants,
- a much higher voltage interaction is between converter station being in a closer electrical proximity (smaller length of AC line).

Smaller values of $MIIF_{12}$ for IRMI systems are connected with the reactive power demand of a rectifier station and the operation characteristics. During a voltage dip at AC rectifier node reactive power demand decreases because of maintaining constant current in HVDC link (it is associated with the reduction of firing angle). This operation leads to improving power factor at a rectifier node. In turn, an inverter operation during voltage dip leads to decreasing power factor.

Fig. 4 shows the $MIIF_{21}$ values as a function of short-circuit power level expressed in *ESCR* values for both IIMI and IRMI systems.

Simulation results shown in Fig. 4 leads to very important conclusions:

- the smaller short circuit capacity in an adjacent node, the smaller voltage interactions,

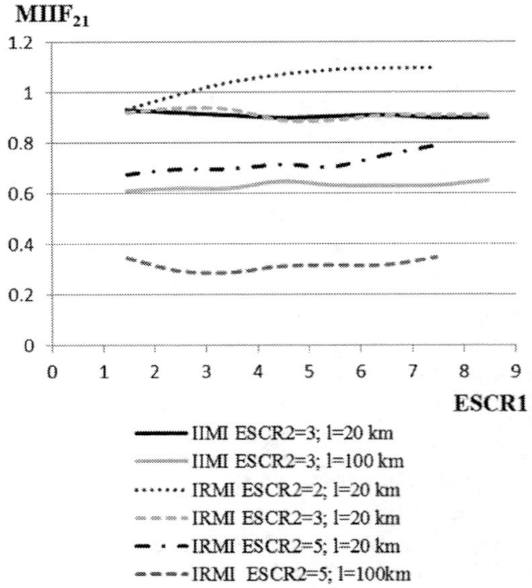

Figure 4. $MIIF_{21}$ in IRMI and IIMI system with different line length between adjacent links as function of ESCR at node 1

- for the very small value of *ESCR* (small short circuit capacity) at the grid node in IRMI variants there is very high voltage interaction (*MIIF* values are higher than 1). It means that voltage change at adjacent node is higher than at the studied grid node,
- the considerable influence on value of $MIIF_{21}$ has AC line length between converter stations. It can therefore be stated that *MIIF* can be used to estimate the electrical proximity between AC nodes with HVDC converters.

The obtained results of $MIIF_{21}$ indicate that interaction on inverter station is almost not dependent on the converter station operating mode (similar values for IRMI and IIMI systems).

The values of $MIIF_{21}$ are similar for both IIMI and IRMI systems. Thus, voltage interactions on adjacent inverter station's node are almost independent from the mode of operation of converter located in a close electrical proximity.

From the analysis of Fig. 3. and 4. it can be concluded that voltage interactions in IRMI systems are lower in comparison to interactions in IIMI systems. Moreover, *MIIF* index can be used to estimate electrical distance between HVDC converters. It is an empiric index which it is difficult to calculate from analytical equations because of non-linear properties of AC/DC operations during disturbances in power system. Moreover, *MIIF* has an asymmetrical feature – generally speaking the value of $MIIF_{12}$ is not the same as $MIIF_{21}$.

B. Commuation failure phenomenon

Fig. 5. shows that *CFII* values depend on the inductance of the smoothing reactor in HVDC link. The selection of a proper value of inductance should be determined by detailed analysis taking into account *CFII* index. Fig. 5 shows that as the L_{DC} inductance increases, the CFII index value increases at the same time. It means that it is possible to decrease the susceptibility of HVDC converter by the use of the reactor with higher inductance.

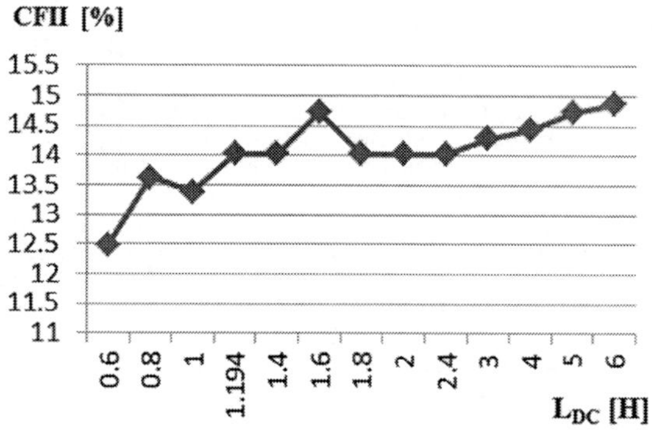

Figure 5. *CFII* [%] as a function of inductance of smoothing reactor [H]

As it is shown in Fig. 6. there is a strong dependence of electrical distance between HVDC links and short circuit power levels on *CFII* values. Both local (*LCFII*) and concurrent commutation failure immunity index (*CCFII*) were examined. *CCFII* reached big values because there was necessary fault with enough power to cause commutation failure in both local and adjacent converter stations.

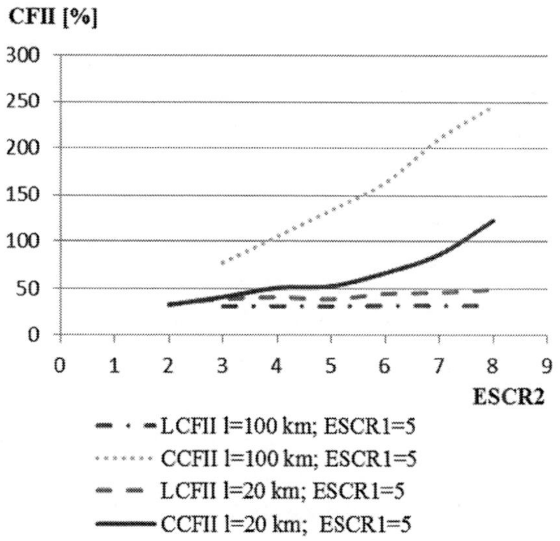

Figure 6. *CFII* [%] as a function of short-circuit power level of the inverter AC node – IIMI system

C. *Fault time recovery in Multi-Infeed system*

In order to examine the influence of short circuit power levels on the fault time recovery, three phase to ground 100 ms short circuits at AC node with inverter of HVDC 1 link were simulated. The distance between converter stations equaled 100 km. The dependences of fault time recovery on *ESCR* values of adjacent node are shown in Fig. 7.

Recovery time after disturbances is determined as the time needed to recover 90% of the nominal power of the HVDC link after fault clearance.

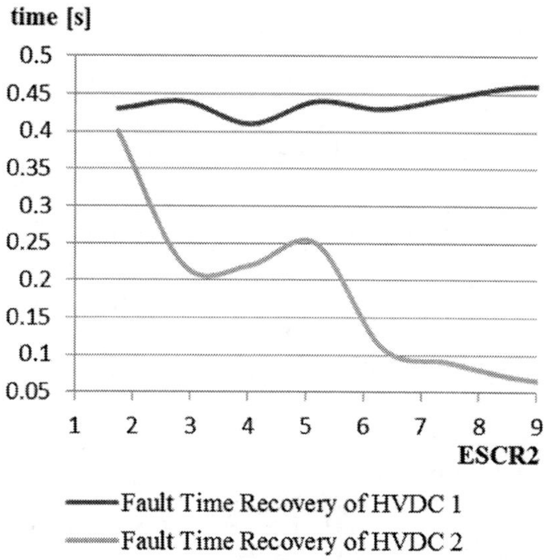

Figure 7. Fault time recovery of HVDC links terminated in close electrical proximity as a function of ESCR 2

Fault time recovery of adjacent link after short-circuits at connected AC node of inverter 1 depends strongly on short-circuit power level at node of inverter 2. The higher ESCR value at adjacent node, the smaller time necessary to recovery transmission power of adjacent HVDC link.

VIII. Conclusions

Interactions in Multi-Infeed systems are a complex issue. HVDC links forming part of Multi-Infeed system cannot be treated as independent links. The decision of a location of the HVDC link in a close electrical proximity to others HVDC links should be preceded by detailed technical studies. Such a proceeding may allow us to avoid or alleviate adverse effect of operations of adjacent HVDC links.

Simulation results indicate that the level of interactions between adjacent HVDC converter stations depends on a number of factors such as: short-circuit power at connected node, converter station operation mode, electrical distance between HVDC links or inductance of smoothing reactor.

The results of simulations indicate that voltage interactions are higher between two inverter stations (IIMI system) in a comparison with one inverter and one rectifier station (IRMI system).

The commutation failure phenomenon is different in Multi-Infeed systems compared to single HVDC link. This kind of disturbance may appear in adjacent links. The susceptibility of HVDC system to a commutation failure should be evaluated by using *CFII*.

Considerable progress in the development of power electronics devices will gradually increase the available nominal parameters, reliability and expansion of HVDC applications in modern power systems. Therefore a further increase of HVDC links number is expected.

In order to reduce negative effects of interactions occurring in Multi-Infeed systems, it is necessary to intensify research work leading to elaborate methods for alleviating an impact of adjacent HVDC links.

ACKNOWLEDGMENT

The publication was supported by the statutory funds of Faculty of Electrical Engineering of Warsaw University of Technology.

REFERENCES

[1] P. J. Piotrowski, S. Robak, M. M. Polewaczyk, R. Raczkowski, "Offshore Substation workers' exposure to harmful factors – actions minimizing risk of hazards," Medycyna pracy, vol. 67, pp. 51-72, Jan. 2016, doi:10.13075/mp.5893.00320

[2] H.P. Lips, "Aspects of Multiple Infeed of HVDC Inverter Stations into a Common AC System," IEEE Transactions on Power Apparatus Systems, vol. 92, pp. 775-779, Apr. 1973.

[3] M. Polewaczyk, S. Robak, "HVDC transmission systems in modern power systems," Przeglad Elektrotechniczny, pp. 1-10, Jul. 2016, doi:10.15199/48.2016.07.01

[4] B. Davis, A. Williamson, Gole A., Ek B., Long B., Burton B., "Systems with multiple DC Infeed," CIGRE Working Group B4.41, TB364, Paris 2008

[5] X. Chen , A. M. Gole, C. Guo, "Multi-infeed HVDC system with inverter and rectifier infeeds," 10th IET International Conference on AC and DC Power Transmission (ACDC 2012), pp. 1-6, doi:10.1049/cp.2012.1998

[6] C. Bayliss , B. Hardy, "Transmission and Distribution Electrical Engineering," 4th Edition, Newnes 2011

[7] J. Szczeglow, W. Wysocki, S. Robak, "The influence of control strategy choice on effectiveness of reactive power compensation in distribution network," Przeglad Elektrotechniczny, vol. 9, pp. 158-161, Sept. 2012

[8] E. Rahimi "Commutation Failure in Single and Multi-Infeed systems" The 8th IEE International Conference on AC and DC Power Transmission, 2006. ACDC 2006 doi:10.1049/cp:20060037

[9] J. Machowski, P. Kacejko, S Robak, P. Miller, "Simplified angle and voltage stability criteria for power system planning based on the short-circuit power," INTERNATIONAL TRANSACTIONS ON ELECTRICAL ENERGY SYSTEMS, vol. 25, pp. 3096-3108, Nov. 2015, doi: 10.1002/etep.2024

[10] S. Robak, K. Gryszpanowicz, "Rotor angle small signal stability assessment in transmission network expansion planning," ELECTRIC POWER SYSTEMS RESEARCH, vol. 128, pp. 144-150 , Nov. 2015, doi: 10.1016/j.epsr.2015.07.003

[11] Y. Changyou , B. Guangquan, S. Huadong, "The FrameWork of Voltage Stability in Multi-infeed AC/DC Power System," Advanced Materials Research 2012, No 354-355, pp. 1137-1143 doi: 10.4028/www.scientific.net/AMR.354-355.1137

[12] M. Szechtman, T. Wess, C. V. Thio: First Benchmark Model for HVDC control studies, CIGRE -WG 14.02., ELECTRA, No 135, 1991, pp. 54-73

Technical University of Kosice
Department of Electrical Power Engineering
Letna 9
042 00 Kosice
Slovak Republic

CURRAN ASSOCIATES INC.
proceedings.com

ISBN 978-1-5108-4653-1